HELPING TEACHERS
AND STUDENTS
SUCCEED TOGETHER

WILEY

Principles of
ANATOMY &
PHYSIOLOGY

13th Edition

Gerard J. Tortora

Bergen Community College

Bryan Derrickson

Valencia Community College

John Wiley & Sons, Inc.

WILEY

Vice President & Publisher	Kaye Pace
Executive Editor	Bonnie Roesch
Executive Marketing Manager	Clay Stone
Developmental Editor	Mary Berry
Senior Media Editor	Linda Muriello
Project Editor	Lorraina Raccuia
Program Assistant	Lauren Morris
Production Manager	Dorothy Sinclair
Production Editor	Sandra Dumas
Senior Illustration Editor	Anna Melhorn
Illustration Coordinator	Claudia Volano
Senior Designer	Madelyn Lesure
Text Designer	Brian Salisbury
Photo Department Manager	Hilary Newman
Production Management Services	Ingrao Associates

Page layout was completed by Laura Ierardi, LCI Design.

Photo Credits
Front cover photo credits (bottom left to right): (muscle) ©Mark Nielsen; (neuron cell body) ©Dr. Don Fawcett/Visuals Unlimited/Getty Images, Inc.; (ear cross-section) ©MedicalRF.com/Getty Images, Inc.; (red blood cells) ©Dr. Don Fawcett/Visuals Unlimited/Getty Images, Inc.; (spinal cord) ©Ron Boardman/Stone/Getty Images, Inc.
Back cover photo credits: (top left) ©Dr. Don Fawcett/Visuals Unlimited/Getty Images, Inc.; (top right) ©Ron Boardman/Stone/Getty Images, Inc.; (center left) ©Mark Nielsen; (center right) ©Dr. Don Fawcett/Visuals Unlimited/Getty Images, Inc.; (bottom) ©MedicalRF.com/Getty Images, Inc.

This book was typeset in 10.5/12.5 Times at Aptara®, Inc. and printed and bound by Quad Graphics. The cover was printed by Quad Graphics.

Founded in 1807, John Wiley & Sons, Inc., has been a valued source of knowledge and understanding for more than 200 years, helping people around the world meet their needs and fulfill their aspirations. Our company is built on a foundation of principles that include responsibility to the communities we serve and where we live and work. In 2008, we launched a Corporate Citizenship Initiative, a global effort to address the environmental, social, economic, and ethical challenges we face in our business. Among the issues we are addressing are carbon impact, paper specifications and procurement, ethical conduct within our business and among our vendors, and community and charitable support. For more information, please visit our website: www.wiley.com/go/citizenship.

The paper in this book was manufactured by a mill whose forest management programs include sustained yield harvesting of its timberlands. Sustained yield harvesting principles ensure that the number of trees cut each year does not exceed the amount of new growth.

This book is printed on acid-free paper.

ISBN 13 978-0470-56510-0
ISBN 13 978-0470-91777-0

Printed in the United States of America.

10 9 8 7 6 5 4 3 2 1

HELPING TEACHERS AND STUDENTS SUCCEED TOGETHER

An anatomy and physiology course can be the gateway to a gratifying career in a whole host of health-related professions. It can also be an incredible challenge. Through years of collaboration with students and instructors alike, we have come to intimately understand not only the material but also the evolving dynamics of teaching and learning A&P. So with every new edition, it's our goal to find new ways to help instructors teach more easily and effectively and students to learn in a way that sticks.

We believe we bring together experience and innovation like no one else, offering a unique solution for A&P designed to help instructors and students succeed together. From constantly evolving animations and visualizations to design based on optimal learning to lessons firmly grounded in learning outcomes, everything is designed with the goal of helping instructors like you teach in a way that inspires confidence and resilience in students and better learning outcomes.

The thirteenth edition of **Principles of Anatomy and Physiology,** integrated with **WileyPLUS,** builds students' confidence; it takes the guesswork out of studying by providing students with a clear roadmap (one that tells them what to do, how to do it, and if they did it right). Students will take more initiative, so instructors can have greater impact.

Principles of Anatomy and Physiology 13e continues to offer a balanced presentation of content under the umbrella of our primary and unifying theme of homeostasis, supported by relevant discussions of disruptions to homeostasis. In addition, years of student feedback have convinced us that readers learn anatomy and physiology more readily when they remain mindful of the relationship between structure and function. As a writing team—an anatomist and a physiologist—our very different specializations offer practical advantages in fine-tuning the balance between anatomy and physiology.

On the following pages students will discover the tips and tools needed to make the most of their study time using the integrated text and media. Instructors will gain an overview of the changes to this edition and of the resources available to create dynamic classroom experiences as well as build meaningful assessment opportunities. Both students and instructors will be interested in the outstanding resources available to seamlessly link laboratory activity with lecture presentation and study time.

The challenges of learning anatomy and physiology can be complex and time-consuming. This textbook and *WileyPLUS for Anatomy and Physiology* have been carefully designed to maximize your study time by simplifying the choices you make in deciding what to study and how to study it, and in assessing your understanding of the content.

Anatomy and Physiology Is a Visual Science

Studying the figures in this book is as important as reading the narrative. The tools described here will help you understand the concepts being presented in any figure and ensure that you get the most out of the visuals.

❶ **LEGEND.** Read this first. It explains what the figure is about.

❷ **KEY CONCEPT STATEMENT.** Indicated by a "key" icon, this reveals a basic idea portrayed in the figure.

❸ **ORIENTATION DIAGRAM.** Added to many figures, this small diagram helps you understand the perspective from which you are viewing a particular piece of anatomical art.

❹ **FIGURE QUESTION.** Found at the bottom of each figure and accompanied by a "question mark" icon, this serves as a self-check to help you understand the material as you go along.

❺ **FUNCTIONS BOX.** Included with selected figures, these provide a brief summary of the functions of the anatomical structure or system depicted.

Figure 24.11 External and internal anatomy of the stomach.

The four regions of the stomach are the cardia, fundus, body, and pyloric part.

(a) Anterior view of regions of stomach

FUNCTIONS OF THE STOMACH

1. Mixes saliva, food, and gastric juice to form chyme.
2. Serves as reservoir for food before release into small intestine.
3. Secretes gastric juice, which contains HCl (kills bacteria and denatures protein), pepsin (begins the digestion of proteins), intrinsic factor (aids absorption of vitamin B_{12}), and gastric lipase (aids digestion of triglycerides).
4. Secretes gastrin into blood.

(b) Anterior view of internal anatomy

After a very large meal, does your stomach still have rugae?

Figure 23.17 Changes in partial pressures of oxygen and carbon dioxide (in mmHg) during external and internal respiration.

Gases diffuse from areas of higher partial pressure to areas of lower partial pressure.

(a) External respiration: pulmonary gas exchange

(b) Internal respiration: systemic gas exchange

What causes oxygen to enter pulmonary capillaries from alveoli and to enter tissue cells from systemic capillaries?

❻ **MP3 DOWNLOADS.** In each chapter you will find that several illustrations are marked with an icon that looks like an iPod. This indicates that an audio file that narrates and discusses the important elements of that particular illustration is available. You can access these downloads on the student companion website or within *WileyPLUS*.

Studying physiology requires an understanding of the sequence of processes. Correlation of sequential processes in text and art is achieved through the use of special numbered lists in the narrative that correspond to numbered segments in the accompanying figure. This approach is used extensively throughout the book to lend clarity to the flow of complex processes.

Physiology of Hearing

The following events are involved in hearing (Figure 17.22):

1. The auricle directs sound waves into the external auditory canal.

2. When sound waves strike the tympanic membrane, the alternating waves of high and low pressure in the air cause the tympanic membrane to vibrate back and forth. The tympanic membrane vibrates slowly in response to low-frequency (low-pitched) sounds and rapidly in response to high-frequency (high-pitched) sounds.

3. The central area of the tympanic membrane connects to the malleus, which vibrates along with the tympanic membrane. This vibration is transmitted from the malleus to the incus and then to the stapes.

4. As the stapes moves back and forth, its oval-shaped footplate, which is attached via a ligament to the circumference of the oval window, vibrates in the oval window. The vibrations at the oval window are about 20 times more vigorous than the tympanic membrane because the auditory ossicles efficiently transmit small vibrations spread over a large surface area (the tympanic membrane) into larger vibrations at a smaller surface (the oval window).

5. The movement of the stapes at the oval window sets up fluid pressure waves in the perilymph of the cochlea. As the oval window bulges inward, it pushes on the perilymph of the scala vestibuli.

6. Pressure waves are transmitted from the scala vestibuli to the scala tympani and eventually to the round window, causing it to bulge outward into the middle ear. (See 9 in the figure.)

7. The pressure waves travel through the perilymph of the scala vestibuli, then the vestibular membrane, and then move into the endolymph inside the cochlear duct.

8. The pressure waves in the endolymph cause the basilar membrane to vibrate, which moves the hair cells of the spiral organ against the tectorial membrane. This leads to bending of the stereocilia and ultimately to the generation of nerve impulses in first-order neurons in cochlear nerve fibers.

9. Sound waves of various frequencies cause certain regions of the basilar membrane to vibrate more intensely than other regions. Each segment of the basilar membrane is "tuned" for

Figure 17.22 Events in the stimulation of auditory receptors in the right ear. The numbers correspond to the events listed in the text. The cochlea has been uncoiled to more easily visualize the transmission of sound waves and their distortion of the vestibular and basilar membranes of the cochlear duct.

🔓 Hair cells of the spiral organ (organ of Corti) convert a mechanical vibration (stimulus) into an electrical signal (receptor potential).

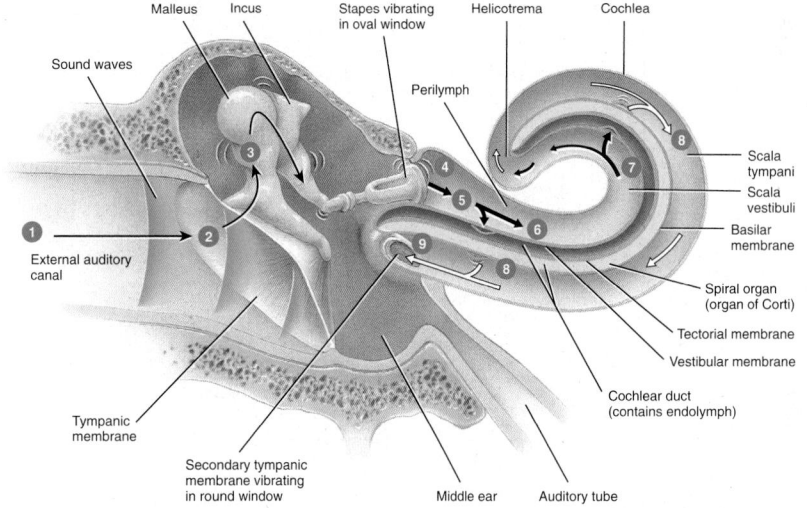

❓ Which part of the basilar membrane vibrates most vigorously in response to high-frequency (high-pitched) sounds?

WILEY PLUS

There are many visual resources within *WileyPLUS,* in addition to the art from your text. These visual aids can help you master the topic you are studying. Examples closely integrated with the reading material include *animations, cadaver video clips,* and *Real Anatomy Views. Anatomy Drill and Practice* lets you test your knowledge of structures with simple-to-use drag-and-drop labeling exercises or fill-in-the-blank labeling. You can drill and practice on these activities using illustrations from the text, cadaver photographs, histology micrographs, or lab models.

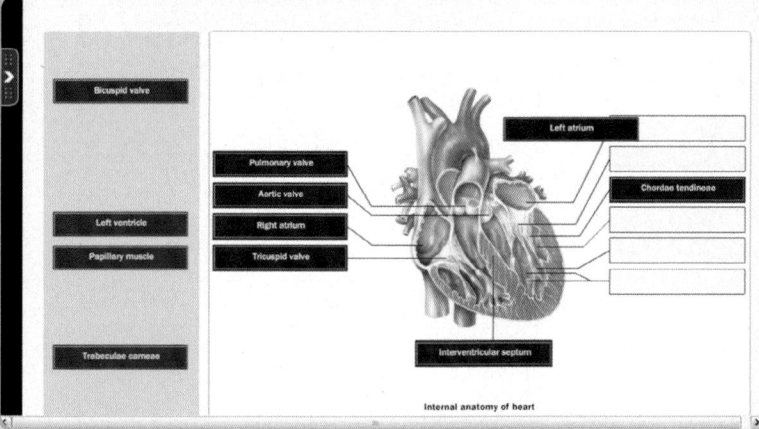

Exhibits Organize Complex Anatomy into Manageable Modules

Many topics in this text have been organized into **Exhibits** that bring together all of the information and elements that you need to learn the complex terminology, anatomy, and the relevance of the anatomy into a simple-to-navigate content module. You will find Exhibits for tissues, bones, joints, skeletal muscles, nerves, and blood vessels. Most exhibits include the following:

❶ Objective to focus your study.

❷ Overview narrative of the structure(s).

❸ Table summarizing key features of the structure(s).

❹ Illustrations and photographs.

❺ Checkpoint Question to assess your understanding.

❻ Clinical Connection to provide relevance for learning the details.

EXHIBIT 11.B

Muscles of the Head That Move the Eyeballs (Extrinsic Eye Muscles) and Upper Eyelids *(Figure 11.5)*

● OBJECTIVE

• Describe the origin, insertion, action, and innervation of the extrinsic eye muscles that move the eyeballs and upper eyelids.

Muscles that move the eyeballs are called **extrinsic eye muscles** because they originate outside the eyeballs (in the orbit) and insert on the outer surface of the sclera ("white of the eye") (Figure 11.5). The extrinsic eye muscles are some of the fastest contracting and most precisely controlled skeletal muscles in the body.

Three pairs of extrinsic eye muscles control movements of the eyeballs: (1) superior and inferior recti, (2) lateral and medial recti, and (3) superior and inferior obliques. The four recti muscles (superior, inferior, lateral, and medial) arise from a tendinous ring in the orbit and insert into the sclera of the eye. As their names imply, the **superior** and **inferior recti** move the eyeballs superiorly and inferiorly; the **lateral** and **medial recti** move the eyeballs laterally and medially, respectively.

The actions of the oblique muscles cannot be deduced from their names. The **superior oblique** muscle originates posteriorly near the tendinous ring, then passes anteriorly superior to the medial rectus muscle, and ends in a round tendon. The tendon extends through a pulleylike loop of fibrocartilaginous tissue called the *trochlea* (= pulley) on the anterior and medial part of the roof of the orbit. Finally, the tendon turns and inserts on the posterolateral aspect of the eyeball. Accordingly, the superior oblique muscle moves the eyeballs inferiorly and laterally. The **inferior oblique** muscle originates on the maxilla at the anteromedial aspect of the floor of the orbit. It then passes posteriorly and laterally and inserts

on the posterolateral aspect of the eyeball. Because of this arrangement, the inferior oblique muscle moves the eyeballs superiorly and laterally.

Unlike the recti and oblique muscles, the **levator palpebrae superioris** does not move the eyeballs, since its tendon passes/inserts into the upper eyelid. Rather, it raises the upper/opens the eyes. It is therefore an antagonist to the/which closes the eyes.

✚ CLINICAL CONNECTION | Strabismus

Strabismus (stra-BIZ-mus; *strabismos* = squin/tion in which the two eyeballs are not properly alig/hereditary or it can be due to birth injuries, poor att/muscles, problems with the brain's control center, or l/Strabismus can be constant or intermittent. In strab/sends an image to a different area of the brain and b/usually ignores the messages sent by one of the eyes,/becomes weaker; hence "lazy eye," or *amblyopia*, de/*strabismus* results when a lesion in the oculomotor/to move the eyeball medially and inferiorly. A lesion/(VI) nerve results in *internal strabismus*, a condition i/ball moves medially when at rest and cannot move la/
Treatment options for strabismus depend on the/problem and include surgery, visual therapy (retraining t/center), and orthoptics (eye muscle training to straight/

MUSCLE	ORIGIN	INSERTION	ACTION	INN
Superior rectus (*rectus* = fascicles parallel to midline)	Common tendinous ring (attached to orbit around optic foramen).	Superior and central part of eyeballs.	Moves eyeballs superiorly (elevation) and medially (adduction), and rotates them medially.	Ocul
Inferior rectus	Same as above.	Inferior and central part of eyeballs.	Moves eyeballs inferiorly (depression) and medially (adduction), and rotates them medially.	Ocul
Lateral rectus	Same as above.	Lateral side of eyeballs.	Moves eyeballs laterally (abduction).	Abdu
Medial rectus	Same as above.	Medial side of eyeballs.	Moves eyeballs medially (adduction).	Ocul
Superior oblique (*oblique* = fascicles diagonal to midline)	Sphenoid bone, superior and medial to common tendinous ring in orbit.	Eyeball between superior and lateral recti. Muscle inserts into superior and lateral surfaces of eyeball via tendon that passes through trochlea.	Moves eyeballs inferiorly (depression) and laterally (abduction), and rotates them medially.	Troch
Inferior oblique	Maxilla in floor of orbit.	Eyeballs between inferior and lateral recti.	Moves eyeballs superiorly (elevation) and laterally (abduction), and rotates them laterally.	Ocul
Levator palpebrae superioris (le-VĀ-tor PAL-pe-brē soo-per'-ē-OR-is; *palpebrae* = eyelids)	Roof of orbit (lesser wing of sphenoid bone).	Skin and tarsal plate of upper eyelids. (opens eyes).	Elevates upper eyelids	Ocul

EXHIBIT 11.B

EXHIBIT 11.B

Muscles of the Head That Move the Eyeballs (Extrinsic Eye Muscles) and Upper Eyelids *(Figure 11.5)* CONTINUED

RELATING MUSCLES TO MOVEMENTS

Arrange the muscles in this exhibit according to their actions on the eyeballs: (1) elevation, (2) depression, (3) abduction, (4) adduction, (5) medial rotation, and (6) lateral rotation. The same muscle may be mentioned more than once.

✔ CHECKPOINT

Which muscles that move the eyeballs contract and relax as you look to your left without moving your head?

Figure 11.5 **Muscles of the head that move the eyeballs (extrinsic eye muscles) and upper eyelid.**

❻ The extrinsic muscles of the eyeball are among the fastest contracting and most precisely controlled skeletal muscles in the body.

Trochlea
SUPERIOR OBLIQUE
LEVATOR PALPEBRAE SUPERIORIS (cut)
SUPERIOR RECTUS
MEDIAL RECTUS
Optic (II) nerve
Common tendinous ring
LATERAL RECTUS
Sphenoid bone
INFERIOR RECTUS
INFERIOR OBLIQUE

Frontal bone
Eyeball
Cornea
Maxilla

(a) Right lateral view of right eyeball

INFERIOR OBLIQUE   SUPERIOR RECTUS
Trochlea
LATERAL RECTUS   MEDIAL RECTUS
SUPERIOR OBLIQUE   INFERIOR RECTUS

(b) Movements of right eyeball in response to contraction of extrinsic muscles

Frontal bone (cut)
SUPERIOR RECTUS
MEDIAL RECTUS
LATERAL RECTUS
INFERIOR RECTUS
Zygomatic bone (cut)

SUPERIOR OBLIQUE
LEVATOR PALPEBRAE SUPERIORIS
INFERIOR OBLIQUE

(c) Right lateral view of right eyeball

❓ How does the inferior oblique muscle move the eyeball superiorly and laterally?

Clinical Discussions Make Your Study Relevant

The relevance of the anatomy and physiology that you are studying is best understood when you make the connection between normal structure and function and what happens when the body doesn't work the way it should. Throughout the chapters of the text you will find **Clinical Connections** that introduce you to interesting clinical perspectives related to the text discussion. In addition, at the end of each body system chapter you will find the **Disorders: Homeostatic Imbalances** section, which includes concise discussions of major diseases and disorders. These provide answers to many of your questions about medical problems. The **Medical Terminology** section that follows includes selected terms dealing with both normal and pathological conditions.

CLINICAL CONNECTION | *Arthroplasty*

Joints that have been severely damaged by diseases such as arthritis, or by injury, may be replaced surgically with artificial joints in a procedure referred to as **arthroplasty** (AR-thrō-plas'-tē; *arthr-*= joint; *plasty*=plastic repair of). Although most joints in the body can be repaired by arthroplasty, the ones most commonly replaced are the hips, knees, and shoulders. About 400,000 hip replacements and 300,000 knee replacements are performed annually in the United States. During the procedure, the ends of the damaged bones are removed and metal, ceramic, or plastic components are fixed in place. The goals of arthroplasty are to relieve pain and increase range of motion.

Partial hip replacements involve only the femur. **Total hip replacements** involve both the acetabulum and head of the femur (Figures A–C). The damaged portions of the acetabulum and the head of the femur are replaced by prefabricated prostheses (artificial devices). The acetabulum is shaped to accept the new socket, the head of the femur is removed, and the center of the femur is shaped to fit the femoral component. The acetabular component consists of a plastic such as polyethylene, and the femoral component is composed of a metal such as cobalt-chrome, titanium alloys, or stainless steel. These materials are designed to withstand a high degree of stress and to prevent a response by the immune system. Once the appropriate acetabular and femoral components are selected, they are attached to the healthy portion of bone with acrylic cement, which forms an interlocking mechanical bond.

Knee replacements are actually a resurfacing of cartilage and, like hip replacements, may be partial or total. In a **partial knee replacement (PKR)**, also called a **unicompartmental knee replacement**, only one side of the knee joint is replaced. Once the damaged cartilage is removed from the distal end of the femur, the femur is reshaped and a metal femoral component is cemented in place. Then the damaged cartilage from the proximal end of the tibia is removed, along with the meniscus. The tibia is reshaped and fitted with a plastic tibial component that is cemented into place. If the posterior surface of the patella is badly damaged, the patella is replaced with a plastic patellar component.

(A) Preparation for total hip replacement

(B) Components of an artificial hip joint prior to implantation

(C) Radiograph of an artificial hip joint

WILEY PLUS

WileyPLUS offers you opportunities for even further Clinical Connections with animated and interactive case studies that relate specifically to one body system. Look for these under additional chapter resources as an interesting and engaging break from traditional study routines.

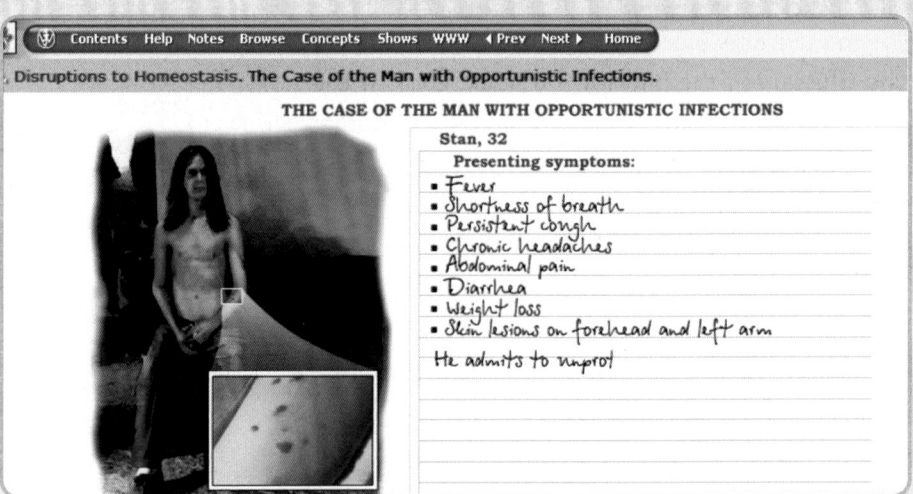

Contents   Help   Notes   Browse   Concepts   Shows   WWW   ◀ Prev   Next ▶   Home

Disruptions to Homeostasis. The Case of the Man with Opportunistic Infections.

THE CASE OF THE MAN WITH OPPORTUNISTIC INFECTIONS

Stan, 32

Presenting symptoms:
- Fever
- Shortness of breath
- Persistent cough
- Chronic headaches
- Abdominal pain
- Diarrhea
- Weight loss
- Skin lesions on forehead and left arm

He admits to unprot

Chapter Resources Help You Focus and Review

Your book has a variety of special features that will make your time studying anatomy and physiology a more rewarding experience. These have been developed based on feedback from students—like you—who have used previous editions of the text. Their effectiveness is even further enhanced within *WileyPLUS for Anatomy and Physiology*.

Chapter Introductions set the stage for the content to come. Each chapter starts with a succinct overview of the particular system's role in maintaining homeostasis in your body, followed by an introduction to the chapter content. This opening page concludes with a question that always begins with "Did you ever wonder...?" These questions will capture your interest and encourage you to find the answer in the chapter material to come.

Objectives at the start of each section help you focus on what is important as you read. All of the content within *WileyPLUS* is tagged to these specific learning objectives so that you can organize your study or review what is still not clear in simple, more meaningful ways.

Checkpoint Questions at the end of each section help you assess if you have absorbed what you have read. Take time to review these questions or answer them within the Practice section of each *WileyPLUS* concept module,

where your answers will automatically be graded to let you know where you stand.

Mnemonics are a memory aid that can be particularly helpful when learning specific anatomical features. Mnemonics are included throughout the text—some displayed in figures, tables, or Exhibits, and some included within the text discussion. We encourage you not only to use the mnemonics provided, but also to create your own to help you learn the multitude of terms involved in your study of human anatomy.

Chapter Review and Resource Summary is a helpful table at the end of chapters that offers you a concise summary of the important concepts from the chapter and links each section to the media resources available in *WileyPLUS for Anatomy and Physiology*.

Self-Quiz Questions give you an opportunity to evaluate your understanding of the chapter as a whole. Within *WileyPLUS*, use **Progress Check** to quiz yourself on individual or multiple chapters in preparation for exams or quizzes.

Critical Thinking Questions are word problems that allow you to apply the concepts you have studied in the chapter to specific situations.

Mastering the Language of Anatomy and Physiology

Throughout the text we have included **Pronunciations** and, sometimes, **Word Roots** for many terms that may be new to you. These appear in parentheses immediately following the new words. The pronunciations are repeated in the Glossary at the back of the book. Look at the words carefully and say them out loud several times. Learning to pronounce a new word will help you remember it and make it a useful part of your medical vocabulary. Take a few minutes to read the Pronunciation Key, found at the beginning of the Glossary at the end of this text, so it will be familiar as you encounter new words.

To provide more assistance in learning the language of anatomy, a full **Glossary** of terms with phonetic pronunciations appears at the end of the book. The basic building blocks of medical terminology—**Combining Forms, Word**

Roots, Prefixes, and Suffixes—are listed inside the back cover, accompanied by **Eponyms,** traditional terms that include reference to a person's name, along with the current terminology.

WileyPLUS houses help for you in building your new language skills as well. The **Audio Glossary,** which is always available to you, lets you hear all these new, unfamiliar terms pronounced. Throughout the e-text, these terms can be clicked on and heard pronounced as you read. In addition, you can use the helpful **Mastering Vocabulary** program, which creates electronic flashcards for you of the key terms within each chapter for practice, as well as take a self-quiz specifically on the terms introduced in each chapter.

NOTES TO INSTRUCTORS

As active teachers of the course, we recognize both the rewards and challenges in providing a strong foundation for understanding the complexities of the human body. We believe that teaching goes beyond just sharing information. *How* we share information makes all the difference—especially, if as we do, you have an increasingly diverse population of students with varying learning abilities. As we revised this text we focused on those areas that we knew we could enhance to provide greater impact in terms of better learning outcomes. Feedback from many of you, as well as from the students we interact with in our own classrooms, guided us in ensuring that the revisions to the text, along with the powerful new *WileyPLUS for Anatomy and Physiology*, support the needs and challenges you face day to day in your own classrooms.

We focused on several key areas for revision: enhancing the all-important visuals, both drawings and photographs; increasing the use of Exhibits that provide a focused and functional organization of detailed content; adding some new and revising many of the tables to increase their effectiveness; updating and adding clinical material that helps students relate what they are learning to their desired career goals and the world around them; and making narrative changes aimed at increasing student engagement with—and comprehension of—the material. For a detailed list of revisions for each chapter please visit our website at www.wiley.com/college/sc/aandp and click on the text cover.

The Art of Anatomy and Physiology

Illustrations throughout the text have been refined. The color palette for the skulls in Chapter 7, and for the brain and spinal cord throughout the text, has been adjusted for greater impact. Illustrations in each chapter have been revised and updated to provide greater clarity and more saturated colors. Particular emphasis was placed on revised drawings of joints, muscles, and blood vessels.

Cadaver photographs are included throughout the text to help students relate the content to real-life images. These are often paired with diagrams to help make the connections. Most of the meticulous dissections and outstanding photography come from Mark Nielsen's lab at the University of Utah.

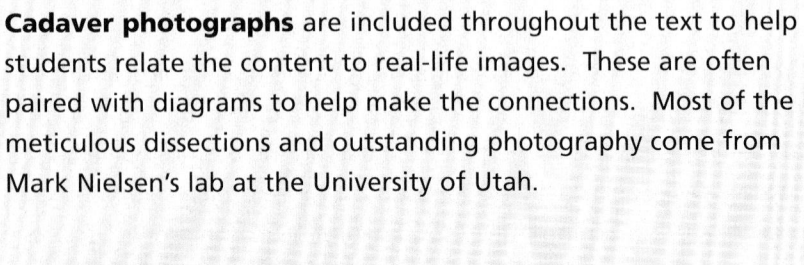

Most tissue **Photomicrographs** have been replaced with exceptionally clear photomicrographs with high-magnification blowups.

LM 240x

LM 630x

LM 400x

LM 630x

Exhibits and Tables

The use of the pedagogically designed **Exhibits** has been expanded to include the axial and appendicular skeletons, as well as cranial nerves, providing students with simplified presentations of complex content.

New **Tables,** including Skin Glands, Common Bone Fractures, Summary of the Levels of Organization within a Skeletal Muscle, and Summary of the Respiratory System, have been added, in addition to refinement of many of the existing tables with either new illustrations or rewritten text.

EXHIBIT 14.H Vagus (X) Nerve *(Figure 14.24)*

OBJECTIVE
• Identify the origin of the vagus nerve in the brain, the foramen through which it exits the skull, and its function.

The **vagus (X) nerve** (VĀ-gus = vagrant or wandering) is a mixed cranial nerve that is distributed from the head and neck into the thorax and abdomen. In the neck, it lies medial and posterior to the internal jugular vein and common carotid artery.

Sensory axons in the vagus nerve arise from the skin of the external ear for touch, pain, and thermal sensations; a few taste buds in the epiglottis and pharynx; and proprioceptors in muscles of the neck and throat. Also, sensory axons come from baroreceptors in the carotid sinus and chemoreceptors in the carotid and aortic bodies. The majority of sensory neurons come from visceral sensory receptors in most organs of the thoracic and abdominal cavities that convey sensations (such as hunger, fullness, and discomfort) from these organs. The sensory neu-

rons have cell bodies in the **superior** and **inferior ganglia** and then pass through the jugular foramen to end in the medulla and pons.

The branchial motor neurons, which run briefly with the accessory nerve, arise from nuclei in the medulla oblongata and supply muscles of the pharynx, larynx, and soft palate that are used in swallowing, vocalization, and coughing. Historically these motor neurons have been called the cranial accessory nerve, but these fibers actually belong to the vagus (X) nerve.

Axons of autonomic motor neurons in the vagus nerve originate in nuclei of the medulla and supply the lungs, heart, glands of the g... testinal (GI) tract, and smooth muscle of the respiratory passa... esophagus, stomach, gallbladder, small intestine, and most of t... intestine (see Figure 15.3). Autonomic motor axons initiate smoo... cle contractions in the gastrointestinal tract to aid motility and s... secretion by digestive glands; activate smooth muscle to constri... ratory passageways; and decrease heart rate.

CHECKPOINT
On what basis is the vagus nerve named?

Figure 14.24 Vagus (X) nerve.

The vagus nerve is widely distributed in the head, neck, thorax, and abdomen.

Carotid sinus
Carotid body
Aortic bodies
GLOSSOPHARYNGEAL (IX) NERVE
Anterior
Medulla oblongata
Posterior
Inferior surface of brain
Inferior ganglion
Superior ganglion
VAGUS (X) NERVE
Pancreas
Small intestine
Larynx
Heart
Lungs
Liver and gallbladder
Stomach
Pancreas (behind stomach)
Colon
Vagus (X) nerve

CLINICAL CONNEC...
Vagal Paralysis, Dysphagia, and Tachycardia

Injury to the vagus (X) nerve c... conditions such as trauma or l... causes **vagal paralysis,** or int... tions of sensations from many o... in the thoracic and abdomina... ties; **dysphagia** (dis-FĀ-gē-a), or... culty in swallowing; and **tachy...** (tak'-i-KAR-dē-a), or increased ... rate. •

? Where is the vagus nerve located in the neck region?

TABLE 6.1

Some Common Fractures

FRACTURE	DESCRIPTION	ILLUSTRATION	RADIOGRAPH
Open (Compound)	The broken ends of the bone protrude through the skin. Conversely, a **closed (simple) fracture** does not break the skin.	Humerus	Radius / Ulna
Comminuted (KOM-i-noo-ted; *com-* = together; *-minuted* = crumbled)	The bone is splintered, crushed, or broken into pieces at the site of impact, and smaller bone fragments lie between the two main fragments.	Humerus	
Greenstick	A partial fracture in which one side of the bone is broken and the other side bends; similar to the way a green twig breaks on one side while the other side stays whole, but bends; occurs only in children, whose bones are not fully ossified and contain more organic material than inorganic material.	Ulna / Radius / Wrist bones	
Impacted	One end of the fractured bone is forcefully driven into the interior of the other.	Humerus	

Clinical Connections

Your students are fascinated by the **Clinical Connections** to the normal anatomy and physiology that they are learning. You'll find that the text is liberally peppered with engaging discussions of a wide variety of clinical scenarios from disease coverage to tests and procedures. As always, we have updated all of the Clinical Connections and Disorders: Homeostatic Imbalances sections to reflect the most current information. We have added several new Clinical Connections, such as a feature on fibromyalgia, to the text. A complete reference list of the Clinical Connections within each chapter follows the Table of Contents.

WileyPLUS *and You*

WileyPLUS for Anatomy and Physiology is an innovative, research-based online environment designed for both effective teaching and learning. Utilizing *WileyPLUS* in your course provides your students with an accessible, affordable, and active learning platform and gives you tools and resources to efficiently build presentations for a dynamic classroom experience and to create and manage effective assessment strategies. The underlying principles of **design, engagement,** and **measurable outcomes** provide the foundation for this powerful, new release of *WileyPLUS*.

DESIGN

- New research-based design helps students manage their time better and develop better study skills.
- Course Calendars help track assignments for both students and teachers.
- New Course Plan makes it easier to assign reading, activities, and assessment. Simple drag-and-drop tools make it easy to assign the course plan as-is or in any way that best reflects your course syllabus.

The new design makes it easy for students to know *what* it is they need to do, boosting their confidence and preparing them for greater engagement in class and lab.

ENGAGEMENT

- Complete online version of the textbook allows for seamless integration of all content.
- Relevant student study tools and learning resources ensure positive learning outcomes.
- Immediate feedback boosts confidence and helps students see a return on investment for each study session.
- Precreated activities encourage learning outside of the classroom.
- Course materials, including PowerPoint stacks with animations and Wiley's Visual Library for Anatomy and Physiology, help you personalize lessons and optimize your time.

Concept mastery in this discipline is directly related to students keeping up with the work and not falling behind. The new Concept Modules, Animations and Activities, Self Study, and Progress Checks in *WileyPLUS* will ensure that students know how to study effectively so they will remain engaged and stay on task.

MEASURABLE OUTCOMES

- Progress Check enables students to hone in on areas of weakness for increased success.
- Self-assessment and remediation for all Learning Objectives let students know exactly how their efforts have paid off.
- Instant reports monitor trends in class performance, use of course materials, and student progress toward learning objectives.

With new detailed reporting capabilities, students will know that they are doing it right. With increased confidence, motivation is sustained so students stay on task—success will follow.

Please contact your Wiley representative for details about these and other resources or visit our website at www.wiley.com/college/sc/aandp and click on the text cover to explore the assets more fully.

RESOURCES FOR INTEGRATING LABORATORY EXPERIENCES

Laboratory Manual for Anatomy and Physiology, 4e

Connie Allen and Valerie Harper

Newly revised, **Laboratory Manual for Anatomy and Physiology** with **WileyPLUS 5.0** engages your students in active learning and focuses on the most important concepts in A&P. Exercises reflect the multiple ways in which students learn and provide guidance for anatomical exploration and application of critical thinking to analyzing physiological processes. A concise narrative, self-contained exercises that include a wide variety of activities and question types, and two types of lab reports for each exercise keep students focused on the task at hand. Depending on your needs, a *Cat Dissection Manual* or *Fetal Pig Dissection Manual* accompanies the main text. Rich media within *WileyPLUS* further enhance the student experience and include dissection videos, animations, and illustrated drill and practice exercises with illustrations, micrographs, cadaver photos, and popular lab models. Each lab text comes with access to *PowerPhys 2.0*.

Atlas of Human Anatomy, 1e

Mark Nielsen and Shawn Miller

This new atlas filled with outstanding photographs of meticulously executed dissections of the human body has been developed to be a strong teaching and learning solution, not just a catalog of photographs. Organized around body systems, each chapter includes a narrative overview of the body system and is then followed with detailed photographs that accurately and realistically represent the anatomical structures. Histology is included. *Atlas of Human Anatomy* will work well in your laboratories, as a study companion to your textbook, and as a print companion to the *Real Anatomy* DVD.

Photographic Atlas of the Human Body, 2e

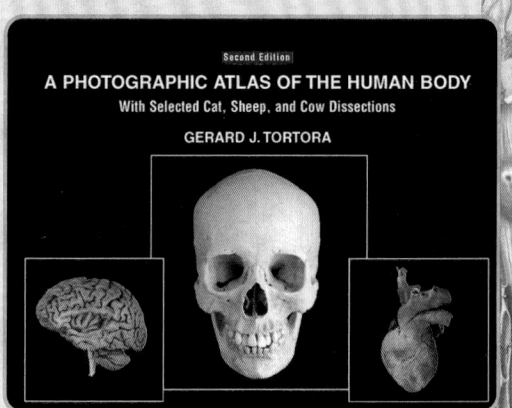

Gerard J. Tortora

Like the new atlas from Nielsen and Miller, this popular atlas is also systemic in its approach to the photographic review of the human body. In addition to the excellent cadaver photographs and micrographs, this atlas also contains selected cat and sheep heart dissections. The high-quality imagery can be used in the classroom, in the laboratory, or for study and review.

RESOURCES FOR INTEGRATING LABORATORY EXPERIENCES

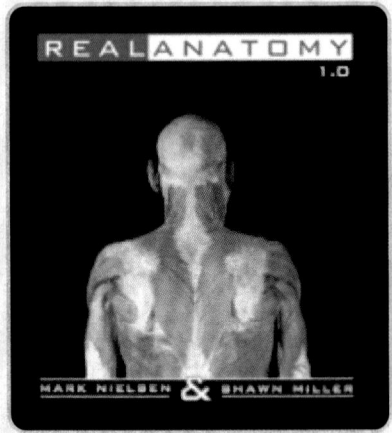

Real Anatomy

Mark Nielsen and Shawn Miller

Real Anatomy is 3-D imaging software that allows you to dissect through multiple layers of a three-dimensional real human body to study and learn the anatomical structures of all body systems.

- Dissect through up to 40 layers of the body and discover the relationships of the structures to the whole.

- Rotate the body, as well as major organs, to view the image from multiple perspectives.

- Use a built-in zoom feature to get a closer look at detail.

- A unique approach to highlighting and labeling structures does not obscure the real anatomy in view.

RESOURCES FOR INTEGRATING LABORATORY EXPERIENCES

- Related images provide multiple views of structures being studied.

- View histology micrographs at varied levels of magnification with the virtual microscope.

- Snapshots of any image can be saved for use in PowerPoints, quizzes, or handouts.

- Audio pronunciation of all labeled structures is readily available.

Virtual Dissection—100% Real

RESOURCES FOR INTEGRATING LABORATORY EXPERIENCES

PowerPhys 2.0

Connie Allen, Valerie Harper, Thomas Lancraft, and Yuri Ivlev

PowerPhys 2.0 provides a simulated laboratory experience for students, giving them the opportunity to review their knowledge of core physiological concepts, predict outcomes of an experiment, collect data, analyze it, and report on their findings. This revised edition features a new activity on Homeostatic Imbalance of Thyroid Function and revised lab report questions throughout. An easy-to-use and intuitive interface guides students through the experiments from basic review to laboratory reports. All experiments contain randomly generated data, allowing students to experiment multiple times but still arrive at the same conclusions. A perfect addition to distant learning or hybrid courses, *PowerPhys 2.0* is a stand-alone web-based program and is fully integrated with Allen and Harper's laboratory manual.

Interactions: Exploring the Functions of the Human Body 3.0

Thomas Lancraft and Frances Frierson

Interactions 3.0 is the most complete program of interactive animations and activities available for anatomy and physiology. A series of modules encompassing all body systems focuses on a review of anatomy, the examination of physiological processes using animations and interactive exercises, and clinical correlations to enhance student understanding. At the heart of **Interactions** is a focus on core principles—*homeostasis, communication, energy flow, fluid flow,* and *boundaries*—that underscore the key relationships between structure and function as well as interrelationships between systems. It is the reinforcement of these fundamental organizing principles that sets this series apart from others. **Interactions** is available on DVD, web-based, or fully integrated within *WileyPLUS*.

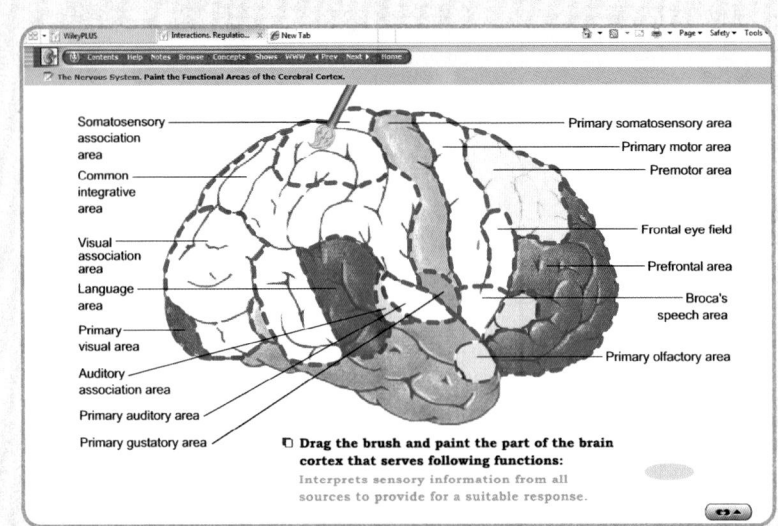

ACKNOWLEDGMENTS

We wish to especially thank several academic colleagues for their helpful contributions to this edition. Creating and implementing the integration of this text with *WileyPLUS for Anatomy and Physiology* was only possible because of the expertise and fine work of the following group of people. We are so very grateful to you:

Sarah Bales
Moraine Valley Community College

Celina Bellanceau
University of South Florida

Curtis DeFriez
Weber State University

Alan Erickson
South Dakota State University

Gibril Fadika
Hampton University

Pamela Fouche
Walters State Community College

Sophia Garcia
Tarrant County College–Trinity River

Clare Hays
Metropolitan State College of Denver

Jason Hunt
Brigham Young University–Idaho

Judy Learn
North Seattle Community College

Jerri K. Lindsey
Tarrant County College

Todd Miller
Hunter College

Erin Morrey
Georgia Perimeter College

Gus Pita
Hunter College

Susan Puglisi
Norwalk Community College

Saeed Rahmanian
Roane State Community College

Lori A. Smith
American River College

Randall Tracy
Worcester State College

Jay Zimmer
South Florida Community College

We are also very grateful to our colleagues who have reviewed the manuscript or participated in focus groups and offered numerous suggestions for improvement:

Charles J. Biggers
University of Memphis

Gladys Bolding
Georgia Perimeter College

Lois Borek
Georgia State University

Betsy Brantley
Lansing Community College

Arthur R. Buckley
University of Cincinnati

Alex Cheroske
Mesa Community College

Robert Comegys
Old Dominion University

Curtis DeFriez
Weber State University

William Dunscombe
Union County College

Heather Dy
Long Beach Community College

Christine Ross Earls
Fairfield University

Angela Edwards
University of South Carolina Allendale

Sharon Ellerton
Queensborough Community College

David Evans
Pennsylvania College of Technology

Gibril Fadika
Hampton University

Sandy Garrett
Texas Woman's University

Michael Harman
Lone Star College

Jane Horlings
Saddleback College

Barbara Hunnicutt
Seminole State College

Jason Hunt
Brigham Young University–Idaho

Alexander T. Imholtz
Prince George's Community College

Amy E. Jetton
Middle Tennessee State University

Becky Keck
Clemson University

Marc LaBella
Ocean County College

Ellen Lathrop-Davis
Community College of Baltimore County

Billy Bob Long
Del Mar College

Wayne M. Mason
Western Kentucky University

Karen McLellan
Indiana Purdue University Fort Wayne

Marie McMahon
Miramar College

Erin Morrey
Georgia Perimeter College

Maria Oehler
Florida State College at Jacksonville

Betsy Ott
Tyler Junior College

Gilbert Pitts
Austin Peay State University

Saeed Rahmanian
Roane State Community College

Terrence J. Ravine
University of South Alabama

Philip D. Reynolds
Troy University

John Roufaiel
SUNY Rockland Community College

Kelly Sexton
North Lake College

Colleen Sinclair
Towson University

Lori A. Smith
American River College

Nora Stevens
Portland Community College

Leo B. Stouder
Broward College

Dennis Strete
McLennan Community College

Peter Susan
Trident Technical College

Jared Taglialatela
Clayton State College

Bonnie J. Tarricone
Ivy Tech Community College

Heather Walker
Clemson University

Janice Webster
Ivy Tech Community College

Delores Wenzel
University of Georgia

Matthew A. Williamson
Georgia Southern University

Finally, our hats are off to everyone at Wiley. We enjoy collaborating with this enthusiastic, dedicated, and talented team of publishing professionals. Our thanks to the entire team: Bonnie Roesch, Executive Editor; Mary Berry and Karen Trost, Developmental Editors; Lorraina Raccuia, Project Editor; Lauren Morris, Program Assistant; Suzanne Ingrao, Outside Production Editor; Hilary Newman, Photo Manager; Claudia Volano, Illustration Coordinator; Anna Melhorn, Senior Illustration Editor; Madelyn Lesure, Senior Designer; Laura Ierardi, LCI Design; Linda Muriello, Senior Media Editor; and Clay Stone, Executive Marketing Manager.

Gerard J. Tortora
Department of Science and Health, S229
Bergen Community College
400 Paramus Road
Paramus, NJ 07652

Bryan Derrickson
Science Department
Valencia Community College
1800 S. Kirkman Rd.
Orlando, FL 32811

ABOUT THE AUTHORS

Courtesy of Heidi Chung.

Gerard J. Tortora is Professor of Biology and former Biology Coordinator at Bergen Community College in Paramus, New Jersey, where he teaches human anatomy and physiology as well as microbiology. He received his bachelor's degree in biology from Fairleigh Dickinson University and his master's degree in science education from Montclair State College. He is a member of many professional organizations, including the Human Anatomy and Physiology Society (HAPS), American Society of Microbiology (ASM), American Association for the Advancement of Science (AAAS), National Education Association (NEA), and Metropolitan Association of College and University Biologists (MACUB).

Above all, Jerry is devoted to his students and their aspirations. In recognition of this commitment, Jerry was the recipient of MACUB's 1992 President's Memorial Award. In 1996, he received a National Institute for Staff and Organizational Development (NISOD) excellence award from the University of Texas and was selected to represent Bergen Community College in a campaign to increase awareness of the contributions of community colleges to higher education.

Jerry is the author of several best-selling science textbooks and laboratory manuals, a calling that often requires an additional 40 hours per week beyond his teaching responsibilities. Nevertheless, he still makes time for four or five weekly aerobic workouts that include biking and running. He also enjoys attending college basketball and professional hockey games and performances at the Metropolitan Opera House.

To my mother, Angelina M. Tortora.
(August 20, 1913–August 14, 2010).
Her love, guidance, faith, support, and example continue to be the cornerstone of my personal and professional life. **G.J.T.**

Bryan Derrickson is Professor of Biology at Valencia Community College in Orlando, Florida, where he teaches human anatomy and physiology as well as general biology and human sexuality. He received his bachelor's degree in biology from Morehouse College and his doctorate in cell biology from Duke University. Bryan's study at Duke was in the Physiology Division within the Department of Cell Biology, so while his degree is in cell biology, his training focused on physiology. At Valencia, he frequently serves on faculty hiring committees. He has served as a member of the Faculty Senate, which is the governing body of the college, and as a member of the Teaching and Learning Academy, which sets the standards for the acquisition of tenure by faculty members. Nationally, he is a member of

the Human Anatomy and Physiology Society (HAPS) and the National Association of Biology Teachers (NABT).

Bryan has always wanted to teach. Inspired by several biology professors while in college, he decided to pursue physiology with an eye to teaching at the college level. He is completely dedicated to the success of his students. He particularly enjoys the challenges of his diverse student population, in terms of their age, ethnicity, and academic ability, and he finds being able to reach all of them, despite their differences, a rewarding experience. His students continually recognize Bryan's efforts and care by nominating him for a campus award known as the "Valencia Professor Who Makes Valencia a Better Place to Start." Bryan has received this award three times.

To my family: Rosalind, Hurley, Cherie, and Robb.
Your support and motivation have been invaluable. **B.H.D.**

BRIEF CONTENTS

CONTENTS

5 THE INTEGUMENTARY SYSTEM 153

6 THE SKELETAL SYSTEM: BONE TISSUE 182

7 THE SKELETAL SYSTEM: THE AXIAL SKELETON 208

CLINICAL CONNECTIONS AND DISORDERS

1 AN INTRODUCTION TO THE HUMAN BODY

THE HUMAN BODY AND HOMEOSTASIS *Humans have many ways to maintain homeostasis, the state of relative stability of the body's internal environment. Disruptions to homeostasis often set in motion corrective cycles, called feedback systems, that help restore the conditions needed for health and life.*

Our fascinating journey through the human body begins with an overview of the meanings of anatomy and physiology, followed by a discussion of the organization of the human body and the properties that it shares with all living things. Next, you will discover how the body regulates its own internal environment; this unceasing process, called homeostasis, is a major theme in every chapter of this book. Finally, we introduce the basic vocabulary that will help you speak about the body in a way that is understood by scientists and health-care professionals alike.

? *Did you ever wonder why an autopsy is performed?*

1.1 ANATOMY AND PHYSIOLOGY DEFINED

◉ **OBJECTIVE**

• Define anatomy and physiology, and name several subspecialties of these sciences.

Two branches of science—anatomy and physiology—provide the foundation for understanding the body's parts and functions. **Anatomy** (a-NAT-ō-mē; *ana-* = up; *-tomy* = process of cutting) is the science of body *structures* and the relationships among them. It was first studied by **dissection** (dis-SEK-shun; *dis-* = apart; *-section* = act of cutting), the careful cutting apart of body structures to study their relationships. Today, a variety of imaging techniques (see Table 1.3) also contribute to the advancement of anatomical knowledge. Whereas anatomy deals with structures of the body, **physiology** (fiz′-ē-OL-ō-jē; *physio-* = nature; *-logy* = study of) is the science of body *functions*—how the body parts work. Table 1.1 describes several subspecialties of anatomy and physiology.

Because structure and function are so closely related, you will learn about the human body by studying its anatomy and physiology together. The structure of a part of the body often reflects its functions. For example, the bones of the skull join tightly to form a rigid case that protects the brain. The bones of the fingers are more loosely joined to allow a variety of movements. The walls of the air sacs in the lungs are very thin, permitting rapid movement of inhaled oxygen into the blood. The lining of the urinary bladder is much thicker to prevent the escape of urine into the pelvic cavity, yet its construction allows for considerable stretching.

✔ **CHECKPOINT**

1. What body function might a respiratory therapist strive to improve? What structures are involved?
2. Give your own example of how the structure of a part of the body is related to its function.

1.2 LEVELS OF STRUCTURAL ORGANIZATION AND BODY SYSTEMS

◉ **OBJECTIVES**

• Describe the body's six levels of structural organization.
• List the 11 systems of the human body, representative organs present in each, and their general functions.

The levels of organization of a language—letters, words, sentences, paragraphs, and so on—can be compared to the levels of

---

TABLE 1.1

Selected Subspecialties of Anatomy and Physiology

SUBSPECIALTIES OF ANATOMY	STUDY OF	SUBSPECIALTIES OF PHYSIOLOGY	STUDY OF
Embryology (em′-brē-OL-ō-jē; *embry-* = embryo; *-logy* = study of)	The first eight weeks of development after fertilization of a human egg.	**Neurophysiology** (NOOR-ō-fiz-ē-ol′-ō-jē; *neuro-* = nerve)	Functional properties of nerve cells.
Developmental biology	The complete development of an individual from fertilization to death.	**Endocrinology** (en′-dō-kri-NOL-ō-jē; *endo-* = within; *-crin* = secretion)	Hormones (chemical regulators in the blood) and how they control body functions.
Cell biology	Cellular structure and functions.	**Cardiovascular physiology** (kar-dē-ō-VAS-kū-lar; *cardi-* = heart; *-vascular* = blood vessels)	Functions of the heart and blood vessels.
Histology (hiss′-TOL-ō-jē; *hist-* = tissue)	Microscopic structure of tissues.		
Gross anatomy	Structures that can be examined without a microscope.	**Immunology** (im′-ū-NOL-ō-jē; *immun-* = not susceptible)	The body's defenses against disease-causing agents.
Systemic anatomy	Structure of specific systems of the body such as the nervous or respiratory systems.	**Respiratory physiology** (RES-pir-a-to′-rē; *respira-* = to breathe)	Functions of the air passageways and lungs.
Regional anatomy	Specific regions of the body such as the head or chest.	**Renal physiology** (RĒ-nal; *ren-* = kidney)	Functions of the kidneys.
Surface anatomy	Surface markings of the body to understand internal anatomy through visualization and palpation (gentle touch).	**Exercise physiology**	Changes in cell and organ functions due to muscular activity.
Radiographic anatomy (rā′-dē-ō-GRAF-ik; *radio-* = ray; *-graphic* = to write)	Body structures that can be visualized with x-rays.	**Pathophysiology** (PATH-ō-fiz-ē-ol′-ō-jē)	Functional changes associated with disease and aging.
Pathological anatomy (path′-ō-LOJ-i-kal; *path-* = disease)	Structural changes (gross to microscopic) associated with disease.		

organization of the human body. Your exploration of the human body will extend from atoms and molecules to the whole person. From the smallest to the largest, six levels of organization will help you to understand anatomy and physiology: the chemical, cellular, tissue, organ, system, and organismal levels of organization (Figure 1.1).

1 **Chemical level.** This very basic level can be compared to the *letters of the alphabet* and includes **atoms,** the smallest

units of matter that participate in chemical reactions, and **molecules,** two or more atoms joined together. Certain atoms, such as carbon (C), hydrogen (H), oxygen (O), nitrogen (N), phosphorus (P), calcium (Ca), and sulfur (S), are essential for maintaining life. Two familiar molecules found in the body are deoxyribonucleic acid (DNA), the genetic material passed from one generation to the next, and glucose, commonly known as blood sugar. Chapters 2 and 25 focus on the chemical level of organization.

Figure 1.1 **Levels of structural organization in the human body.**

The levels of structural organization are chemical, cellular, tissue, organ, system, and organismal.

1 CHEMICAL LEVEL

Atoms (C, H, O, N, P)

Molecule (DNA)

2 CELLULAR LEVEL

Smooth muscle cell

3 TISSUE LEVEL

Smooth muscle tissue

Epithelial and connective tissues

4 ORGAN LEVEL

Smooth muscle tissue layers

Epithelial tissue

Stomach

Stomach

5 SYSTEM LEVEL

Salivary glands
Mouth
Pharynx (throat)
Esophagus
Liver
Stomach
Gallbladder
Pancreas (behind stomach)
Large intestine
Small intestine

Digestive system

6 ORGANISMAL LEVEL

? **Which level of structural organization is composed of two or more different types of tissues that work together to perform a specific function?**

2 **Cellular level.** Molecules combine to form **cells,** the basic structural and functional units of an organism that are composed of chemicals. Just as *words* are the smallest elements of language that make sense, cells are the smallest living units in the human body. Among the many kinds of cells in your body are muscle cells, nerve cells, and epithelial cells. Figure 1.1 shows a smooth muscle cell, one of the three types of muscle cells in the body. The cellular level of organization is the focus of Chapter 3.

3 **Tissue level.** **Tissues** are groups of cells and the materials surrounding them that work together to perform a particular function, similar to the way words are put together to form *sentences.* There are just four basic types of tissues in your body: epithelial tissue, connective tissue, muscular tissue, and nervous tissue. *Epithelial tissue* covers body surfaces, lines hollow organs and cavities, and forms glands. *Connective tissue* connects, supports, and protects body organs while distributing blood vessels to other tissues. *Muscular tissue* contracts to make body parts move and generates heat. *Nervous tissue* carries information from one part of the body to another through nerve impulses. Chapter 4 describes the tissue level of organization in greater detail. Shown in Figure 1.1 is smooth muscle tissue, which consists of tightly packed smooth muscle cells.

4 **Organ level.** At the organ level different types of tissues are joined together. Similar to the relationship between sentences and *paragraphs,* **organs** are structures that are composed of two or more different types of tissues; they have specific functions and usually have recognizable shapes. Examples of organs are the stomach, skin, bones, heart, liver, lungs, and brain. Figure 1.1 shows how several tissues make up the stomach. The stomach's outer covering is a layer of epithelial tissue and connective tissue that reduces friction when the stomach moves and rubs against other organs. Underneath are three layers of a type of muscular tissue called *smooth muscle tissue,* which contracts to churn and mix food and then push it into the next digestive organ, the small intestine. The innermost lining is an *epithelial tissue layer* that produces fluid and chemicals responsible for digestion in the stomach.

5 **System level.** A **system** (or *chapter* in our language analogy) consists of related organs (paragraphs) with a common function. An example of the system level, also called the *organsystem level,* is the digestive system, which breaks down and absorbs food. Its organs include the mouth, salivary glands, pharynx (throat), esophagus (food tube), stomach, small intestine, large intestine, liver, gallbladder, and pancreas. Sometimes an organ is part of more than one system. The pancreas, for example, is part of both the digestive system and the hormone-producing endocrine system.

6 **Organismal level.** An **organism** (OR-ga-nizm), any living individual, can be compared to a *book* in our analogy. All the parts of the human body functioning together constitute the total organism.

In the chapters that follow, you will study the anatomy and physiology of the body systems. Table 1.2 lists the components and introduces the functions of these systems. You will also discover that all body systems influence one another. As you study

---

TABLE 1.2

The Eleven Systems of the Human Body

INTEGUMENTARY SYSTEM (CHAPTER 5)

Components: **Skin** and associated structures, such as **hair, fingernails** and **toenails, sweat glands,** and **oil glands.**

Functions: Protects body; helps regulate body temperature; eliminates some wastes; helps make vitamin D; detects sensations such as touch, pain, warmth, and cold.

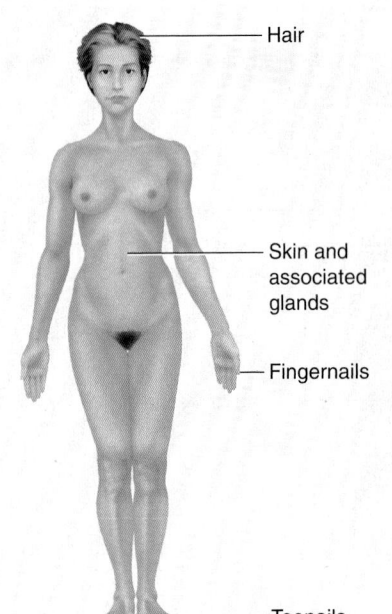

— Hair

— Skin and associated glands

— Fingernails

— Toenails

SKELETAL SYSTEM (CHAPTERS 6–9)

Components: **Bones** and **joints** of the body and their associated **cartilages.**

Functions: Supports and protects body; provides surface area for muscle attachments; aids body movements; houses cells that produce blood cells; stores minerals and lipids (fats).

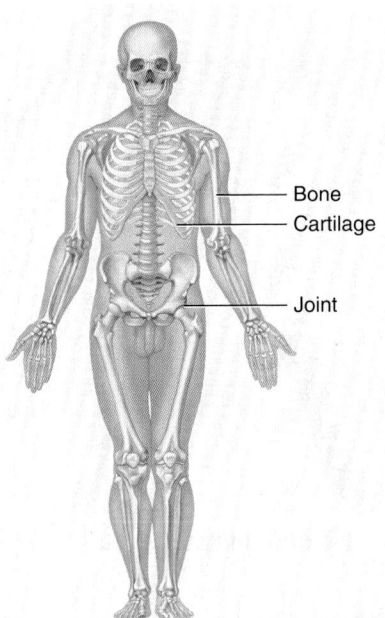

— Bone

— Cartilage

— Joint

each of the body systems in more detail, you will discover how they work together to maintain health, provide protection from disease, and allow for reproduction of the human species.

Health-care professionals and students of anatomy and physiology commonly use several noninvasive diagnostic techniques to assess certain aspects of body structure and function. A **noninvasive diagnostic technique** is one that does not involve insertion of an instrument or device through the skin or a body opening. In **inspection,** the examiner observes the body for any changes that deviate from normal. For example, a physician may examine the mouth cavity for evidence of disease. Following inspection, one or more additional techniques may be employed. In **palpation** (pal-PĀ-shun; *palp-* = gently touching) the examiner feels body surfaces with the hands. An example is palpating the abdomen to detect enlarged or tender internal organs or abnormal masses. In **auscultation** (aws-kul-TĀ-shun; *auscult-* = listening) the examiner listens to body sounds to evaluate the functioning of certain organs, often using a stethoscope to amplify the sounds. An example is auscultation of the lungs during breathing to check for crackling sounds associated with abnormal fluid accumulation. In **percussion** (pur-KUSH-un; *percus-* = beat through) the examiner taps on the body surface with the fingertips and listens to the resulting echo. For example, percussion may reveal the abnormal presence of fluid in the lungs or air in the intestines. It may also provide information about the size, consistency, and position of an underlying structure. An understanding of anatomy is important for the effective application of most of these diagnostic techniques. •

✔ CHECKPOINT

3. Define the following terms: atom, molecule, cell, tissue, organ, system, and organism.
4. At what levels of organization would an exercise physiologist study the human body? *(Hint: Refer to Table 1.1.)*
5. Referring to Table 1.2, which body systems help eliminate wastes?

1.3 CHARACTERISTICS OF THE LIVING HUMAN ORGANISM

◉ OBJECTIVES
• Define the important life processes of the human body.
• Define homeostasis and explain its relationship to interstitial fluid.

Basic Life Processes

Certain processes distinguish organisms, or living things, from nonliving things. Following are the six most important life processes of the human body:

1. **Metabolism** (me-TAB-ō-lizm) is the sum of all the chemical processes that occur in the body. One phase of metabolism is **catabolism** (ka-TAB-ō-lizm; *catabol-* = throwing down; *-ism* = a condition), the breakdown of complex chemical substances into simpler components. The other phase of

---

MUSCULAR SYSTEM (CHAPTERS 10, 11)

Components: Specifically, **skeletal muscle** tissue—muscle usually attached to bones (other muscle tissues include smooth and cardiac).

Functions: Participates in body movements, such as walking; maintains posture; produces heat.

Skeletal muscle

Tendon

NERVOUS SYSTEM (CHAPTERS 12–17)

Components: **Brain, spinal cord, nerves,** and special sense organs, such as **eyes** and **ears.**

Functions: Generates action potentials (nerve impulses) to regulate body activities; detects changes in body's internal and external environments, interprets changes, and responds by causing muscular contractions or glandular secretions.

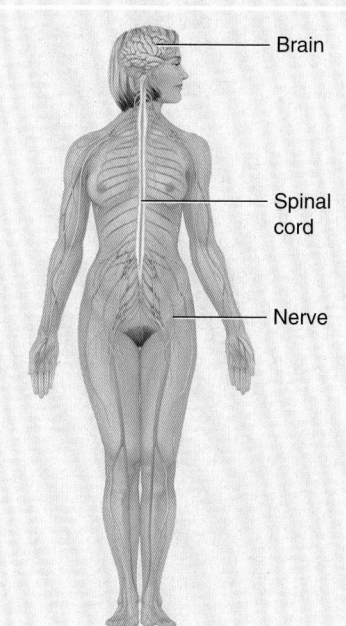

Brain

Spinal cord

Nerve

TABLE 1.2 CONTINUES

TABLE 1.2 CONTINUED

The Eleven Systems of the Human Body

ENDOCRINE SYSTEM (CHAPTER 18)

Components: Hormone-producing glands (**pineal gland, hypothalamus, pituitary gland, thymus, thyroid gland, parathyroid glands, adrenal glands, pancreas, ovaries,** and **testes**) and hormone-producing cells in several other organs.

Functions: Regulates body activities by releasing hormones (chemical messengers transported in blood from endocrine gland or tissue to target organ).

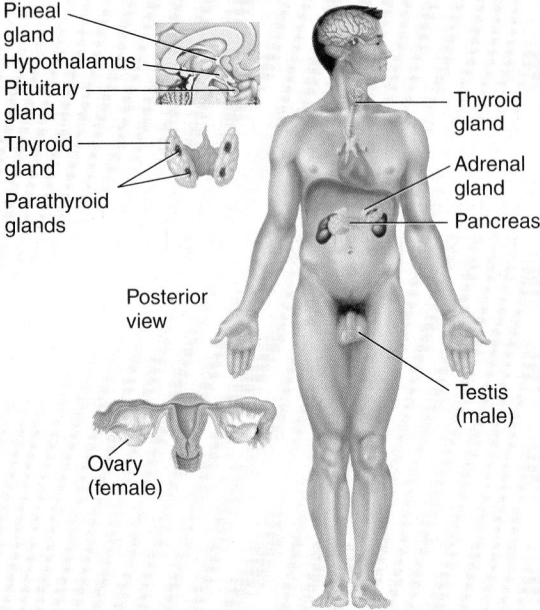

CARDIOVASCULAR SYSTEM (CHAPTERS 19–21)

Components: **Blood, heart,** and **blood vessels**.

Functions: Heart pumps blood through blood vessels; blood carries oxygen and nutrients to cells and carbon dioxide and wastes away from cells and helps regulate acid–base balance, temperature, and water content of body fluids; blood components help defend against disease and repair damaged blood vessels.

LYMPHATIC SYSTEM AND IMMUNITY (CHAPTER 22)

Components: **Lymphatic fluid** and **vessels; spleen, thymus, lymph nodes,** and **tonsils;** cells that carry out immune responses (**B cells, T cells,** and others).

Functions: Returns proteins and fluid to blood; carries lipids from gastrointestinal tract to blood; contains sites of maturation and proliferation of B cells and T cells that protect against disease-causing microbes.

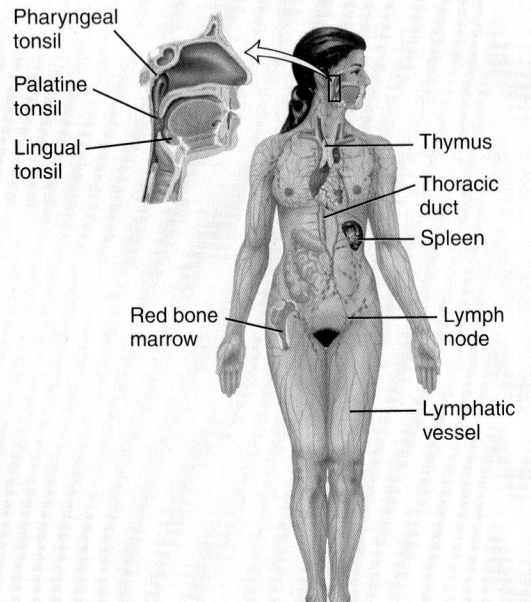

RESPIRATORY SYSTEM (CHAPTER 23)

Components: **Lungs** and air passageways such as the **pharynx (throat), larynx (voice box), trachea (windpipe),** and **bronchial tubes** leading into and out of lungs.

Functions: Transfers oxygen from inhaled air to blood and carbon dioxide from blood to exhaled air; helps regulate acid–base balance of body fluids; air flowing out of lungs through vocal cords produces sounds.

DIGESTIVE SYSTEM (CHAPTER 24)

Components: Organs of gastrointestinal tract, a long tube that includes the **mouth, pharynx** (throat), **esophagus** (food tube), **stomach, small** and **large intestines,** and **anus;** also includes accessory organs that assist in digestive processes, such as **salivary glands, liver, gallbladder,** and **pancreas.**

Functions: Achieves physical and chemical breakdown of food; absorbs nutrients; eliminates solid wastes.

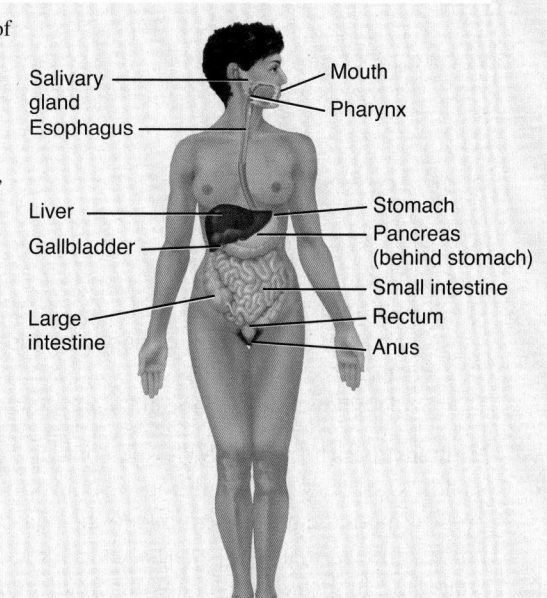

Salivary gland
Mouth
Pharynx
Esophagus
Liver
Stomach
Gallbladder
Pancreas (behind stomach)
Small intestine
Large intestine
Rectum
Anus

URINARY SYSTEM (CHAPTER 26)

Components: **Kidneys, ureters, urinary bladder,** and **urethra.**

Functions: Produces, stores, and eliminates urine; eliminates wastes and regulates volume and chemical composition of blood; helps maintain the acid–base balance of body fluids; maintains body's mineral balance; helps regulate production of red blood cells.

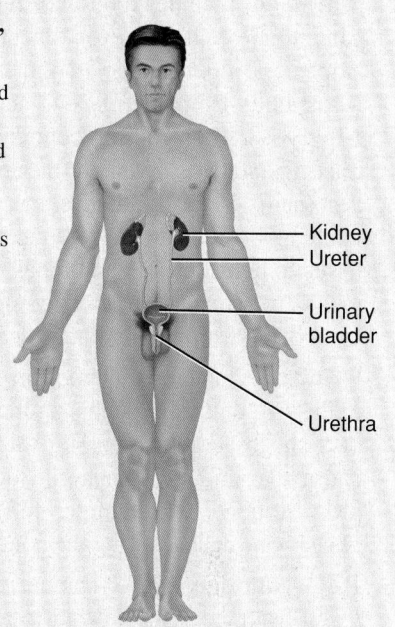

Kidney
Ureter
Urinary bladder
Urethra

REPRODUCTIVE SYSTEMS (CHAPTER 28)

Components: **Gonads** (**testes** in males and **ovaries** in females) and associated organs (**uterine tubes, uterus, vagina,** and **mammary glands** in females and **epididymides, ductus deferens, seminal vesicles, prostate,** and **penis** in males).

Functions: Gonads produce gametes (sperm or oocytes) that unite to form a new organism; gonads also release hormones that regulate reproduction and other body processes; associated organs transport and store gametes; mammary glands produce milk.

Mammary gland

Ovary
Uterine (fallopian) tube
Uterus
Vagina

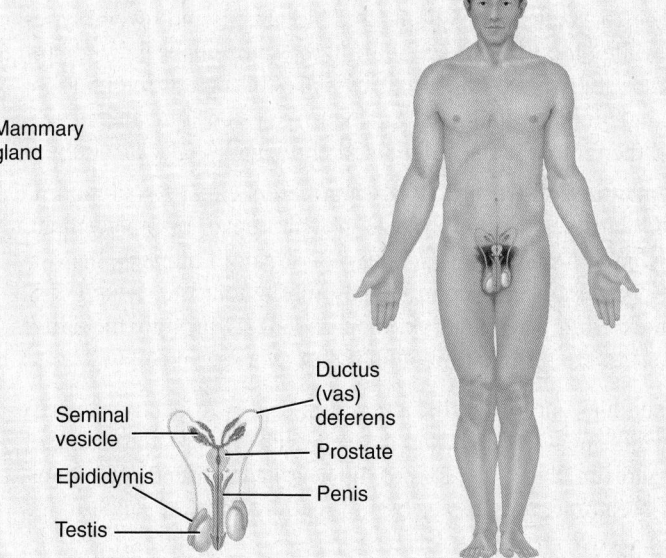

Ductus (vas) deferens
Seminal vesicle
Prostate
Epididymis
Penis
Testis

metabolism is **anabolism** (a-NAB-ō-lizm; *anabol-* = a raising up), the building up of complex chemical substances from smaller, simpler components. For example, digestive processes catabolize (split) proteins in food into amino acids. These amino acids are then used to anabolize (build) new proteins that make up body structures such as muscles and bones.

2. **Responsiveness** is the body's ability to detect and respond to changes. For example, an increase in body temperature during a fever represents a change in the internal environment (within the body), and turning your head toward the sound of squealing brakes is a response to a change in the external environment (outside the body) to prepare the body for a potential threat. Different

cells in the body respond to environmental changes in characteristic ways. Nerve cells respond by generating electrical signals known as nerve impulses (action potentials). Muscle cells respond by contracting, which generates force to move body parts.

3. **Movement** includes motion of the whole body, individual organs, single cells, and even tiny structures inside cells. For example, the coordinated action of leg muscles moves your whole body from one place to another when you walk or run. After you eat a meal that contains fats, your gallbladder contracts and releases bile into the gastrointestinal tract to help digest them. When a body tissue is damaged or infected, certain white blood cells move from the bloodstream into the affected tissue to help clean up and repair the area. Inside the cell, various parts, such as secretory vesicles (see Figure 3.20), move from one position to another to carry out their functions.

4. **Growth** is an increase in body size that results from an increase in the size of existing cells, an increase in the number of cells, or both. In addition, a tissue sometimes increases in size because the amount of material between cells increases. In a growing bone, for example, mineral deposits accumulate between bone cells, causing the bone to grow in length and width.

5. **Differentiation** (dif′-er-en-shē-Ā-shun) is the development of a cell from an unspecialized to a specialized state. Such precursor cells, which can divide and give rise to cells that undergo differentiation, are known as **stem cells.** As you will see later in the text, each type of cell in the body has a specialized structure and function that differs from that of its precursor (ancestor) cells. For example, red blood cells and several types of white blood cells all arise from the same unspecialized precursor cells in red bone marrow. Also through differentiation, a single fertilized human egg (ovum) develops into an embryo, and then into a fetus, an infant, a child, and finally an adult.

6. **Reproduction** (rē-prō-DUK-shun) refers either to (1) the formation of new cells for tissue growth, repair, or replacement, or (2) the production of a new individual. In humans, the former process occurs continuously throughout life, which continues from one generation to the next through the latter process, the fertilization of an ovum by a sperm cell.

When any one of the life processes ceases to occur properly, the result is death of cells and tissues, which may lead to death of the organism. Clinically, loss of the heartbeat, absence of spontaneous breathing, and loss of brain functions indicate death in the human body.

CLINICAL CONNECTION | *Autopsy*

An **autopsy** (AW-top-sē = seeing with one's own eyes) or *necropsy* is a postmortem (after death) examination of the body and dissection of its internal organs to confirm or determine the cause of death. An autopsy can uncover the existence of diseases not detected during life, determine the extent of injuries, and explain how those injuries may have contributed to a person's death. It also may provide more information about a disease, assist in the accumulation of statistical data, and educate health-care students. Moreover, an autopsy can reveal conditions that may affect offspring or siblings (such as congenital heart defects). Sometimes an autopsy is legally required, such as during a criminal investigation. It may also be useful in resolving disputes between beneficiaries and insurance companies about the cause of death. •

✔**CHECKPOINT**

6. List the six most important life processes in the human body.

1.4 HOMEOSTASIS

 OBJECTIVES

- Define homeostasis.
- Describe the components of a feedback system.
- Contrast the operation of negative and positive feedback systems.
- Explain how homeostatic imbalances are related to disorders.

Homeostasis (hō′-mē-ō-STĀ-sis; *homeo-* = sameness; *-stasis* = standing still) is the condition of equilibrium (balance) in the body's internal environment due to the constant interaction of the body's many regulatory processes. Homeostasis is a dynamic condition. In response to changing conditions, the body's equilibrium can shift among points in a narrow range that is compatible with maintaining life. For example, the level of glucose in blood normally stays between 70 and 110 milligrams of glucose per 100 milliliters of blood.* Each structure, from the cellular level to the system level, contributes in some way to keeping the internal environment of the body within normal limits.

Homeostasis and Body Fluids

An important aspect of homeostasis is maintaining the volume and composition of **body fluids,** dilute, watery solutions containing dissolved chemicals that are found inside cells as well as surrounding them. The fluid within cells is **intracellular fluid** (*intra-* = inside), abbreviated **ICF.** The fluid outside body cells is **extracellular fluid** (*extra-* = outside), abbreviated **ECF.** The ECF that fills the narrow spaces between cells of tissues is known as **interstitial fluid** (in′-ter-STISH-al; *inter-* = between). As you progress with your studies, you will learn that the ECF differs depending on where it occurs in the body: ECF within blood vessels is termed **blood plasma,** within lymphatic vessels it is called **lymph,** in and around the brain and spinal cord it is known as **cerebrospinal fluid,** in joints it is referred to as **synovial fluid,** and the ECF of the eyes is called **aqueous humor** and **vitreous body.**

The proper functioning of body cells depends on precise regulation of the composition of the interstitial fluid surrounding them. Because of this, interstitial fluid is often called the body's *internal environment.* The composition of interstitial fluid changes as substances move back and forth between it and blood plasma. Such exchange of materials occurs across the thin walls of the smallest blood vessels in the body, the *blood capillaries.* This movement in both directions across capillary walls provides needed materials,

---

*Appendix A describes metric measurements.

such as glucose, oxygen, ions, and so on, to tissue cells. It also removes wastes, such as carbon dioxide, from interstitial fluid.

Control of Homeostasis

Homeostasis in the human body is continually being disturbed. Some disruptions come from the external environment in the form of physical insults such as the intense heat of a hot summer day or a lack of enough oxygen for that two-mile run. Other disruptions originate in the internal environment, such as a blood glucose level that falls too low when you skip breakfast. Homeostatic imbalances may also occur due to psychological stresses in our social environment—the demands of work and school, for example. In most cases the disruption of homeostasis is mild and temporary, and the responses of body cells quickly restore balance in the internal environment. However, in some cases the disruption of homeostasis may be intense and prolonged, as in poisoning, overexposure to temperature extremes, severe infection, or major surgery.

Fortunately, the body has many regulating systems that can usually bring the internal environment back into balance. Most often, the nervous system and the endocrine system, working together or independently, provide the needed corrective measures. The nervous system regulates homeostasis by sending electrical signals known as *nerve impulses (action potentials)* to organs that can counteract changes from the balanced state. The endocrine system includes many glands that secrete messenger molecules called *hormones* into the blood. Nerve impulses typically cause rapid changes, but hormones usually work more slowly. Both means of regulation, however, work toward the same end, usually through negative feedback systems.

Feedback Systems

The body can regulate its internal environment through many feedback systems. A **feedback system** or *feedback loop* is a cycle of events in which the status of a body condition is monitored, evaluated, changed, remonitored, reevaluated, and so on. Each monitored variable, such as body temperature, blood pressure, or blood glucose level, is termed a *controlled condition.* Any disruption that changes a controlled condition is called a *stimulus.* A feedback system includes three basic components: a receptor, a control center, and an effector (Figure 1.2).

1. A **receptor** is a body structure that monitors changes in a controlled condition and sends input to a control center. This pathway is called an *afferent pathway* (AF-er-ent; *af-* = toward; *-ferrent* = carried), since the information flows *toward* the control center. Typically, the *input* is in the form of nerve impulses or chemical signals. For example, certain nerve endings in the skin sense temperature and can detect changes, such as a dramatic drop in temperature.

2. A **control center** in the body, for example, the brain, sets the range of values within which a controlled condition should be maintained (set point), evaluates the input it receives from receptors, and generates output commands when they are needed. *Output* from the control center typically occurs as

nerve impulses, or hormones or other chemical signals. This pathway is called an *efferent pathway* (EF-er-ent; *ef-* = away from), since the information flows *away from* the control center. In our skin temperature example, the brain acts as the control center, receiving nerve impulses from the skin receptors and generating nerve impulses as output.

3. An **effector** (e-FEK-tor) is a body structure that receives output from the control center and produces a *response* or effect that changes the controlled condition. Nearly every organ or tissue in the body can behave as an effector. When your body temperature drops sharply, your brain (control center) sends nerve impulses

Figure 1.2 Operation of a feedback system. The dashed return arrow symbolizes negative feedback.

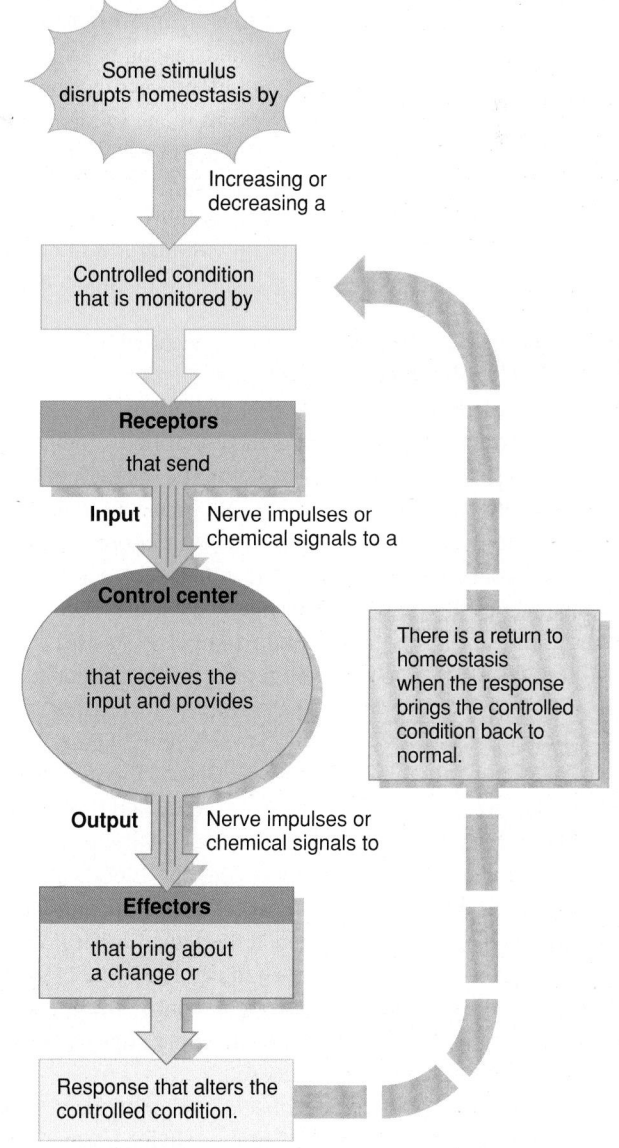

The three basic components of a feedback system are the receptor, control center, and effector.

What is the main difference between negative and positive feedback systems?

(output) to your skeletal muscles (effectors). The result is shivering, which generates heat and raises your body temperature.

A group of receptors and effectors communicating with their control center forms a feedback system that can regulate a controlled condition in the body's internal environment. In a feedback system, the response of the system "feeds back" information to change the controlled condition in some way, either negating it (negative feedback) or enhancing it (positive feedback).

NEGATIVE FEEDBACK SYSTEMS A **negative feedback system** *reverses* a change in a controlled condition. Consider the regulation of blood pressure. Blood pressure (BP) is the force exerted by blood as it presses against the walls of blood vessels. When the heart beats faster or harder, BP increases. If some internal or external stimulus causes blood pressure (controlled condition) to rise, the following sequence of events occurs (Figure 1.3). *Baroreceptors* (the receptors), pressure-sensitive nerve cells located in the walls of certain blood vessels, detect the higher pressure. The baroreceptors send nerve impulses (input) to the brain (control center), which interprets the impulses and responds by sending nerve impulses (output) to the heart and blood vessels (the effectors). Heart rate decreases and blood vessels dilate (widen), which cause BP to decrease (response). This sequence of events quickly returns the controlled condition—blood pressure—to normal, and homeostasis is restored. Notice that the activity of the effector causes BP to drop, a result that negates the original stimulus (an increase in BP). This is why it is called a negative feedback system.

POSITIVE FEEDBACK SYSTEMS Unlike a negative feedback system, a **positive feedback system** tends to *strengthen* or *reinforce* a change in one of the body's controlled conditions. In a positive feedback system, the response affects the controlled condition differently than in a negative feedback system. The control center still provides commands to an effector, but this time the effector produces a physiological response that adds to or *reinforces* the initial change in the controlled condition. The action of a positive feedback system continues until it is interrupted by some mechanism.

Normal childbirth provides a good example of a positive feedback system (Figure 1.4). The first contractions of labor (stimulus) push part of the fetus into the cervix, the lowest part of the uterus, which opens into the vagina. Stretch-sensitive nerve cells (receptors) monitor the amount of stretching of the cervix (controlled condition). As stretching increases, they send more nerve impulses (input) to the brain (control center), which in turn releases the hormone oxytocin (output) into the blood. Oxytocin causes muscles in the wall of the uterus (effector) to contract even more forcefully. The contractions push the fetus farther down the uterus, which stretches the cervix even more. The cycle of stretching, hormone release, and ever-stronger contractions is interrupted only by the birth of the baby. Then, stretching of the cervix ceases and oxytocin is no longer released.

Another example of positive feedback is what happens to your body when you lose a great deal of blood. Under normal conditions, the heart pumps blood under sufficient pressure to body cells to provide them with oxygen and nutrients to maintain homeostasis. Upon severe blood loss, blood pressure drops and blood cells

Figure 1.3 **Homeostatic regulation of blood pressure by a negative feedback system.** Note that the response is fed back into the system, and the system continues to lower blood pressure until there is a return to normal blood pressure (homeostasis).

If the response reverses the stimulus, a system is operating by negative feedback.

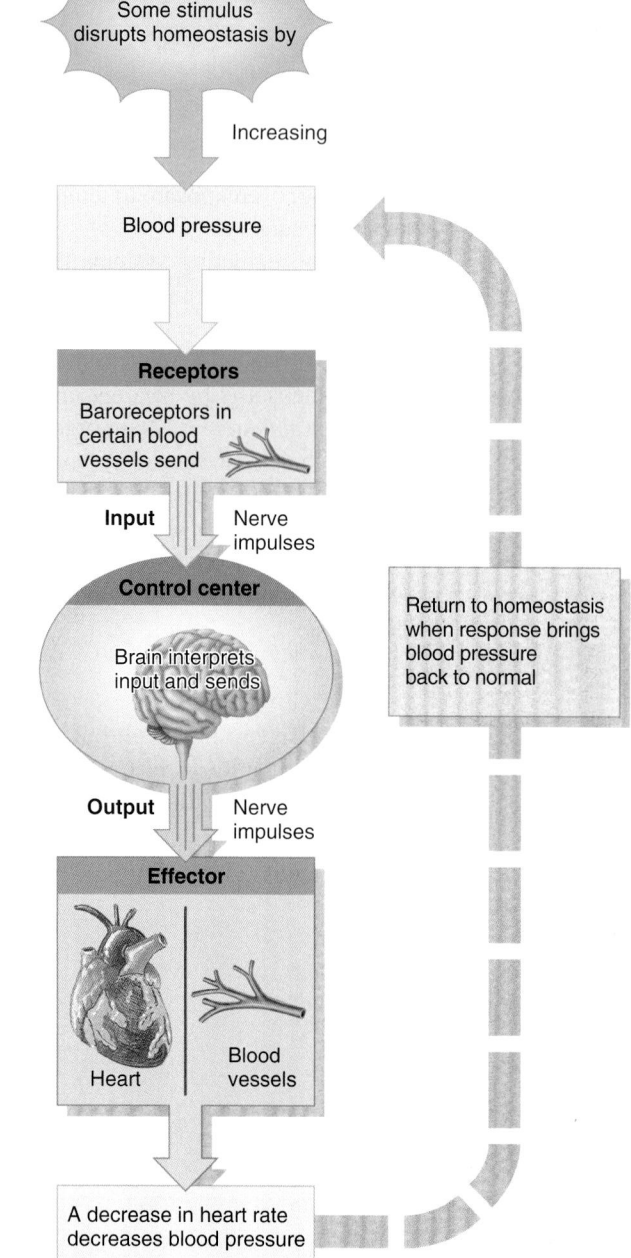

What would happen to heart rate if some stimulus caused blood pressure to decrease? Would this occur by way of positive or negative feedback?

Figure 1.4 Positive feedback control of labor contractions during birth of a baby. The solid return arrow symbolizes positive feedback.

🔑 If the response enhances or intensifies the stimulus, a system is operating by positive feedback.

Contractions of wall of uterus force baby's head or body into the cervix, thus

Increasing

Stretching of cervix

Receptors
Stretch-sensitive nerve cells in cervix send

Input　Nerve impulses

Control center
Brain interprets input and releases

Positive feedback: Increased stretching of cervix causes release of more oxytocin, which results in more stretching of the cervix

Output　Oxytocin

Effectors
Muscles in wall of uterus contract more forcefully

Baby's body stretches cervix more

Interruption of cycle: Birth of baby decreases stretching of cervix, thus breaking the positive feedback cycle

❓ **Why do positive feedback systems that are part of a normal physiological response include some mechanism that terminates the system?**

(including heart cells) receive less oxygen and function less efficiently. If the blood loss continues, heart cells become weaker, the pumping action of the heart decreases further, and blood pressure continues to fall. This is an example of a positive feedback cycle that has serious consequences and may even lead to death if there is no medical intervention. As you will see in Chapter 19, blood clotting is also an example of a positive feedback system.

These examples suggest some important differences between positive and negative feedback systems. Because a positive feedback system continually reinforces a change in a controlled condition, some event outside the system must shut it off. If the action of a positive feedback system is not stopped, it can "run away" and may even produce life-threatening conditions in the body. The action of a negative feedback system, by contrast, slows and then stops as the controlled condition returns to its normal state. Usually, positive feedback systems reinforce conditions that do not happen very often, and negative feedback systems regulate conditions in the body that remain fairly stable over long periods.

Homeostatic Imbalances

You've seen homeostasis defined as a condition in which the body's internal environment remains relatively stable. The body's ability to maintain homeostasis gives it tremendous healing power and a remarkable resistance to abuse. The physiological processes responsible for maintaining homeostasis are in large part also responsible for your good health.

For most people, lifelong good health is not something that happens effortlessly. The many factors in this balance called health include the following:

- The environment and your own behavior.
- Your genetic makeup.
- The air you breathe, the food you eat, and even the thoughts you think.

The way you live your life can either support or interfere with your body's ability to maintain homeostasis and recover from the inevitable stresses life throws your way.

Many diseases are the result of years of poor health behavior that interferes with the body's natural drive to maintain homeostasis. An obvious example is smoking-related illness. Smoking tobacco exposes sensitive lung tissue to a multitude of chemicals that cause cancer and damage the lung's ability to repair itself. Because diseases such as emphysema and lung cancer are difficult to treat and are very rarely cured, it is much wiser to quit smoking—or never start—than to hope a doctor can "fix" you once you are diagnosed with a lung disease. Developing a lifestyle that works with, rather than against, your body's homeostatic processes helps you maximize your personal potential for optimal health and well-being.

As long as all of the body's controlled conditions remain within certain narrow limits, body cells function efficiently, homeostasis is maintained, and the body stays healthy. Should one or more components of the body lose their ability to contribute to homeostasis, however, the normal balance among all of the body's processes may be disturbed. If the homeostatic imbalance is moderate, a disorder or disease may occur; if it is severe, death may result.

A **disorder** is any abnormality of structure or function. **Disease** is a more specific term for an illness characterized by a recognizable set of signs and symptoms. A *local disease* affects one part or a limited region of the body (for example, a sinus infection); a *systemic disease* affects either the entire body or several parts of it (for example, influenza). Diseases alter body structures and functions in characteristic ways. A person with a disease may experience **symptoms,** *subjective* changes in body functions that are not apparent to an observer. Examples of symptoms are headache, nausea, and anxiety. *Objective* changes that a clinician can observe and measure are called **signs.** Signs of disease can be either anatomical, such as swelling or a rash, or physiological, such as fever, high blood pressure, or paralysis.

The science that deals with why, when, and where diseases occur and how they are transmitted among individuals in a community is known as **epidemiology** (ep′-i-dē-mē-OL-ō-jē; *epi-* = upon; *-demi* = people). **Pharmacology** (far′-ma-KOL-ō-jē; *pharmac-* = drug) is the science that deals with the effects and uses of drugs in the treatment of disease.

CLINICAL CONNECTION | *Diagnosis of Disease*

Diagnosis (di′-ag-NŌ-sis; *dia-* = through; *-gnosis* = knowledge) is the science and skill of distinguishing one disorder or disease from another. The patient's symptoms and signs, his or her medical history, a physical exam, and laboratory tests provide the basis for making a diagnosis. Taking a *medical history* consists of collecting information about events that might be related to a patient's illness. These include the chief complaint (primary reason for seeking medical attention), history of present illness, past medical problems, family medical problems, social history, and review of symptoms. A *physical examination* is an orderly evaluation of the body and its functions. This process includes the noninvasive techniques of inspection, palpation, auscultation, and percussion that you learned about earlier in the chapter, along with measurement of vital signs (temperature, pulse, respiratory rate, and blood pressure), and sometimes laboratory tests. •

✔CHECKPOINT

7. Describe the locations of intracellular fluid, extracellular fluid, interstitial fluid, and blood plasma.
8. Why is interstitial fluid called the internal environment of the body?
9. What types of disturbances can act as stimuli that initiate a feedback system?
10. Define receptor, control center, and effector.
11. What is the difference between symptoms and signs of a disease? Give examples of each.

1.5 BASIC ANATOMICAL TERMINOLOGY

◉ OBJECTIVES

• Describe the anatomical position.
• Relate the anatomical names and the corresponding common names for various regions of the human body.

• Define the anatomical planes, anatomical sections, and directional terms used to describe the human body.
• Outline the major body cavities, the organs they contain, and their associated linings.

Scientists and health-care professionals use a common language of special terms when referring to body structures and their functions. The language of anatomy they use has precisely defined meanings that allow us to communicate clearly and precisely. For example, is it correct to say, "The wrist is above the fingers"? This might be true if your upper limbs (described shortly) are at your sides. But if you hold your hands up above your head, your fingers would be above your wrists. To prevent this kind of confusion, anatomists use a standard anatomical position and a special vocabulary for relating body parts to one another.

Body Positions

Descriptions of any region or part of the human body assume that it is in a standard position of reference called the **anatomical position** (an′-a-TOM-i-kal). In the anatomical position, the subject stands erect facing the observer, with the head level and the eyes facing directly forward. The feet are flat on the floor and directed forward, and the upper limbs are at the sides with the palms turned forward (Figure 1.5). In the anatomical position, the body is upright. Two terms describe a reclining body. If the body is lying face down, it is in the **prone** position. If the body is lying face up, it is in the **supine** position.

Regional Names

The human body is divided into several major regions that can be identified externally. The principal regions are the head, neck, trunk, upper limbs, and lower limbs (Figure 1.5). The **head** consists of the skull and face. The *skull* encloses and protects the brain; the *face* is the front portion of the head that includes the eyes, nose, mouth, forehead, cheeks, and chin. The **neck** supports the head and attaches it to the trunk. The **trunk** consists of the chest, abdomen, and pelvis. Each **upper limb** attaches to the trunk and consists of the shoulder, armpit, arm (portion of the limb from the shoulder to the elbow), forearm (portion of the limb from the elbow to the wrist), wrist, and hand. Each **lower limb** also attaches to the trunk and consists of the buttock, thigh (portion of the limb from the buttock to the knee), leg (portion of the limb from the knee to the ankle), ankle, and foot. The *groin* is the area on the front surface of the body marked by a crease on each side, where the trunk attaches to the thighs.

Figure 1.5 shows the anatomical and common names of major parts of the body. The anatomical term in the form of an adjective for each part appears first and is followed in parentheses by the corresponding common name. For example, if you receive a tetanus shot in your *gluteal region,* it is an injection in your *buttock.* Because the anatomical term for a body part usually is based on a Greek or Latin word, it may look different

Figure 1.5 The anatomical position. The anatomical names and corresponding common names (in parentheses) are indicated for specific body regions. For example, the cephalic region is the head.

In the anatomical position, the subject stands erect facing the observer with the head level and the eyes facing forward. The feet are flat on the floor and directed forward, and the upper limbs are at the sides with the palms facing forward.

CEPHALIC (HEAD)
- Cranial (skull)
- Facial (face)

CERVICAL (NECK)

Frontal (forehead)
Temporal (temple)
Orbital or ocular (eye)
Otic (ear)
Buccal (cheek)
Nasal (nose)
Oral (mouth)
Mental (chin)
Sternal (breastbone)

Axillary (armpit)
Brachial (arm)
Antecubital (Front of elbow)
Antebrachial (forearm)
Carpal (wrist)
Palmar or volar (palm)
Digital or phalangeal (fingers)

Thoracic (chest)
Mammary (breast)
Abdominal (abdomen)
Umbilical (navel)
Coxal (hip)
Inguinal (groin)

TRUNK

Pelvic (pelvis)

Pollex (thumb)
Manual (hand)
Pubic (pubis)

Femoral (thigh)
Patellar (anterior surface of knee)
Crural (leg)

Tarsal (ankle)
Pedal (foot)
Digital or phalangeal (toes)

Dorsum (top of foot)
Hallux (great toe)

(a) Anterior view

Occipital (base of skull)
Acromial (shoulder)
Scapular (shoulder blade)
Vertebral (spinal column)
Olecranal or cubital (back of elbow)
Sacral (between hips)
Coccygeal (tailbone)
Gluteal (buttock)
Perineal (region between anus and external genitals)
Popliteal (hollow behind knee)
Sural (calf)
Plantar (sole)

Lumbar (loin)

CEPHALIC (HEAD)
CERVICAL (NECK)
UPPER LIMB
Dorsum (back of hand)
LOWER LIMB
Calcaneal (heel)

(b) Posterior view

? **What is the usefulness of defining one standard anatomical position?**

from the common name for the same part or area. For example, the Latin word for *axilla* (ak-SIL-a) is the common name armpit. Thus, the axillary nerve is one of the nerves passing within the armpit. You will learn more about the Greek and Latin word roots of anatomical and physiological terms as you read this book.

Directional Terms

To locate various body structures, anatomists use specific **directional terms,** words that describe the position of one body part relative to another. Several directional terms are grouped in pairs that have opposite meanings, such as anterior (front) and posterior (back). Exhibit 1.A and Figure 1.6 present the main directional terms.

EXHIBIT 1.A Directional Terms *(Figure 1.6)*

OBJECTIVE

- Define each directional term used to describe the human body.

OVERVIEW

Most of the directional terms used to describe the relationship of one part of the body to another can be grouped into pairs that have opposite meanings. For example, **superior** means toward the upper part of the body, and **inferior** means toward the lower part of the body. It is important to understand that directional terms have relative meanings; they make sense only when used to describe the position of one structure relative to another. For example, your knee is superior to your ankle, even though both are located in the inferior half of the body. Study the directional terms below and the example of how each is used. As you read the examples, look at Figure 1.6 to see the location of each structure.

CHECKPOINT

Which directional terms can be used to specify the relationships between (1) the elbow and the shoulder, (2) the left and right shoulders, (3) the sternum and the humerus, and (4) the heart and the diaphragm?

DIRECTIONAL TERM	DEFINITION	EXAMPLE OF USE
Superior (soo'-PĒR-ē-or) (**cephalic** or **cranial**)	Toward the head, or the upper part of a structure.	The heart is superior to the liver.
Inferior (in'-FĒR-ē-or) (**caudal**)	Away from the head, or the lower part of a structure.	The stomach is inferior to the lungs.
Anterior (an-TĒR-ē-or) (**ventral**)*	Nearer to or at the front of the body.	The sternum (breastbone) is anterior to the heart.
Posterior (pos-TĒR-ē-or) (**dorsal**)	Nearer to or at the back of the body.	The esophagus (food tube) is posterior to the trachea (windpipe).
Medial (MĒ-dē-al)	Nearer to the midline.†	The ulna is medial to the radius.
Lateral (LAT-er-al)	Farther from the midline.	The lungs are lateral to the heart.
Intermediate (in'-ter-MĒ-dē-at)	Between two structures.	The transverse colon is intermediate between the ascending and descending colons.
Ipsilateral (ip-si-LAT-er-al)	On the same side of the body as another structure.	The gallbladder and ascending colon are ipsilateral.
Contralateral (CON-tra-lat-er-al)	On the opposite side of the body from another structure.	The ascending and descending colons are contralateral.
Proximal (PROK-si-mal)	Nearer to the attachment of a limb to the trunk; nearer to the origination of a structure.	The humerus (arm bone) is proximal to the radius.
Distal (DIS-tal)	Farther from the attachment of a limb to the trunk; farther from the origination of a structure.	The phalanges (finger bones) are distal to the carpals (wrist bones).
Superficial (soo'-per-FISH-al) (**external**)	Toward or on the surface of the body.	The ribs are superficial to the lungs.
Deep (**internal**)	Away from the surface of the body.	The ribs are deep to the skin of the chest and back.

*Note that the terms *anterior* and *ventral* mean the same thing in humans. However, in four-legged animals *ventral* refers to the belly side and is therefore *inferior*. Similarly, the terms *posterior* and *dorsal* mean the same thing in humans, but in four-legged animals *dorsal* refers to the back side and is therefore *superior*.

†Recall that the **midline** is an imaginary vertical line that divides the body into equal right and left sides.

Figure 1.6 Directional terms.

Directional terms precisely locate various parts of the body relative to one another.

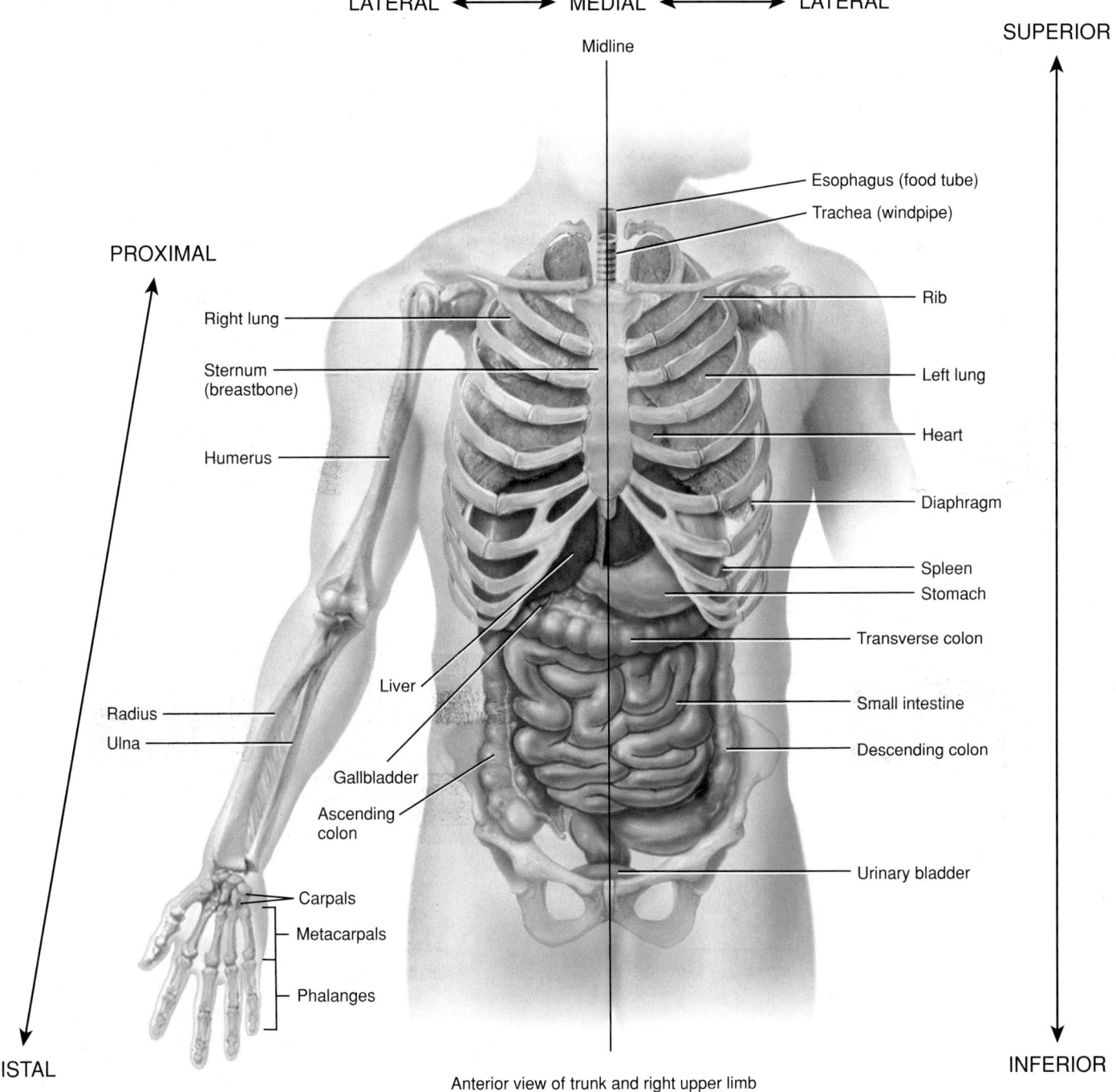

Anterior view of trunk and right upper limb

Is the radius proximal to the humerus? Is the esophagus anterior to the trachea? Are the ribs superficial to the lungs? Is the urinary bladder medial to the ascending colon? Is the sternum lateral to the descending colon?

EXHIBIT 1.A **15**

Planes and Sections

You will also study parts of the body relative to **planes,** imaginary flat surfaces that pass through the body parts (Figure 1.7). A **sagittal plane** (SAJ-i-tal; *sagitt-* = arrow) is a vertical plane that divides the body or an organ into right and left sides. More specifically, when such a plane passes through the midline of the body or an organ and divides it into *equal* right and left sides, it is called a **midsagittal plane** or a **median plane.** The **midline** is an imaginary vertical line that divides the body into equal left and right sides. If the sagittal plane does not pass through the midline but instead divides the body or an organ into *unequal* right and left sides, it is called a **parasagittal plane** (*para-* = near). A **frontal** or **coronal plane** (kō-RŌ-nal; *corona* = crown) divides the body or an organ into anterior (front) and posterior (back) portions. A **trans-** verse plane divides the body or an organ into superior (upper) and inferior (lower) portions. Other names for a transverse plane are a **cross-sectional** or **horizontal plane.** Sagittal, frontal, and transverse planes are all at right angles to one another. An **oblique plane** (ō-BLĒK), by contrast, passes through the body or an organ at an oblique angle (any angle other than a 90-degree angle).

Figure 1.8 Planes and sections through different parts of the brain. The diagrams (left) show the planes, and the photographs (right) show the resulting sections. Note: The arrows in the diagrams indicate the direction from which each section is viewed. This aid is used throughout the book to indicate viewing perspectives.

Planes divide the body in various ways to produce sections.

(a)

(b)

(c)

Figure 1.7 Planes through the human body.

Frontal, transverse, sagittal, and oblique planes divide the body in specific ways.

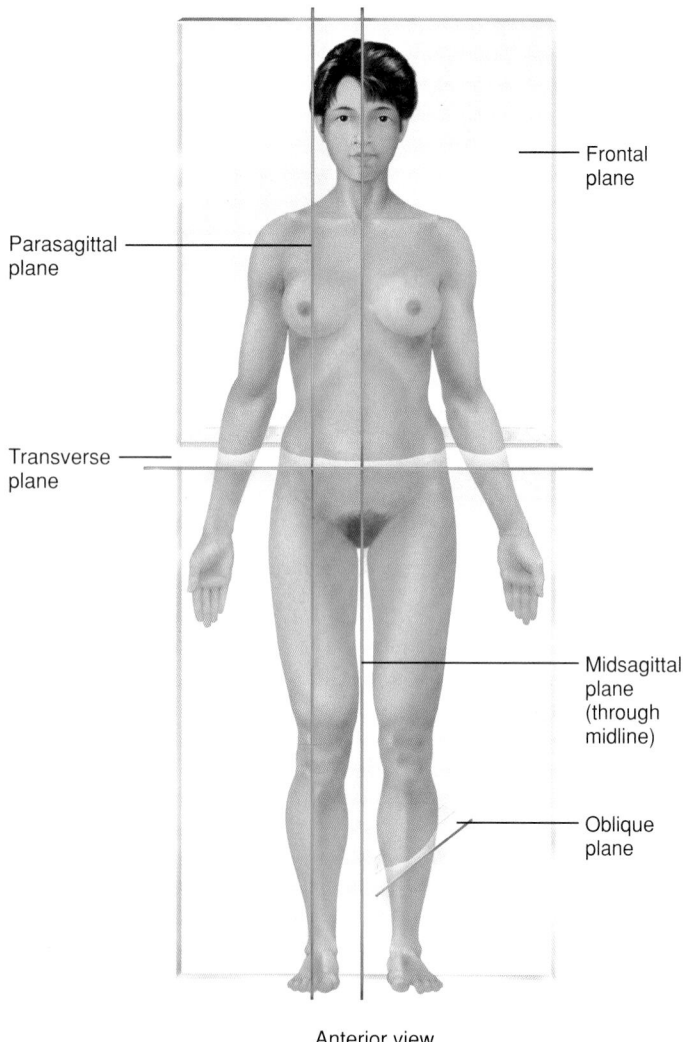

Anterior view

? Which plane divides the heart into anterior and posterior portions?

? Which plane divides the brain into unequal right and left portions?

When you study a body region, you often view it in section. A **section** is a cut of the body or one of its organs made along one of the planes just described. It is important to know the plane of the section so you can understand the anatomical relationship of one part to another. Figure 1.8a–c indicates how three different sections—*transverse, frontal,* and *midsagittal*—provide different views of the brain.

Body Cavities

Body cavities are spaces within the body that help protect, separate, and support internal organs. Bones, muscles, ligaments, and other structures separate the various body cavities from one another. Here we discuss several body cavities (Figure 1.9).

The cranial bones form a hollow space of the head called the **cranial cavity** (KRĀ-nē-al), which contains the brain. The bones of the vertebral column (backbone) form the **vertebral (spinal)**

canal (VER-te-bral), which contains the spinal cord. The cranial cavity and vertebral canal are continuous with one another. Three layers of protective tissue, the **meninges** (me-NIN-jēz), and a shock-absorbing fluid surround the brain and spinal cord.

The major body cavities of the trunk are the thoracic and abdominopelvic cavities. The **thoracic cavity** (thor-AS-ik; *thorac-* = chest) or chest cavity (Figure 1.10) is formed by the ribs, the muscles of the chest, the sternum (breastbone), and the thoracic portion of the vertebral column. Within the thoracic cavity are the **pericardial cavity** (per'-i-KAR-dē-al; *peri-* = around; *-cardial* = heart), a fluid-filled space that surrounds the heart, and two fluid-filled spaces called **pleural cavities** (PLOOR-al; *pleur-* = rib or side), one around each lung. The central part of the thoracic cavity is an anatomical region called the **mediastinum** (mē'-dē-as-TĪ-num; *media-* = middle; *-stinum* = partition). It is between the lungs, extending from the sternum to the vertebral column and from the first rib to the diaphragm (Figure 1.10a, b). The mediastinum

Figure 1.9 Body cavities. The dashed line in (a) indicates the border between the abdominal and pelvic cavities.

The major cavities of the trunk are the thoracic and abdominopelvic cavities.

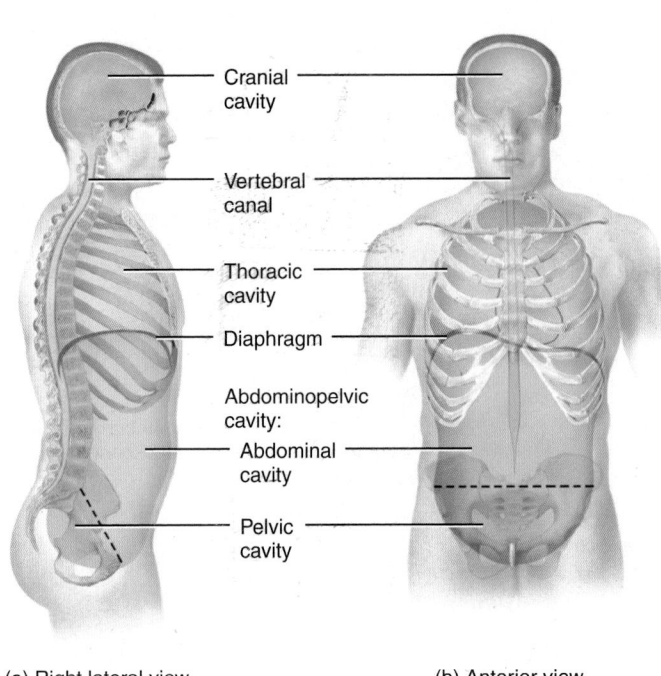

(a) Right lateral view   (b) Anterior view

CAVITY	COMMENTS
Cranial cavity	Formed by cranial bones and contains brain.
Vertebral canal	Formed by vertebral column and contains spinal cord and the beginnings of spinal nerves.
Thoracic cavity*	Chest cavity; contains pleural and pericardial cavities and mediastinum.
Pleural cavity	Each surrounds a lung; the serous membrane of each pleural cavity is the pleura.
Pericardial cavity	Surrounds the heart; the serous membrane of the pericardial cavity is the pericardium.
Mediastinum	Central portion of thoracic cavity between the lungs; extends from sternum to vertebral column and from first rib to diaphragm; contains heart, thymus, esophagus, trachea, and several large blood vessels.
Abdominopelvic cavity	Subdivided into abdominal and pelvic cavities.
Abdominal cavity	Contains stomach, spleen, liver, gallbladder, small intestine, and most of large intestine; the serous membrane of the abdominal cavity is the peritoneum.
Pelvic cavity	Contains urinary bladder, portions of large intestine, and internal organs of reproduction.

* See Figure 1.10 for details of the thoracic cavity.

? In which cavities are the following organs located: urinary bladder, stomach, heart, small intestine, lungs, internal female reproductive organs, thymus, spleen, liver? Use the following symbols for your responses: T = thoracic cavity, A = abdominal cavity, or P = pelvic cavity.

Figure 1.10 The thoracic cavity. The dashed lines indicate the borders of the mediastinum. Note: When transverse sections are viewed inferiorly (from below), the anterior aspect of the body appears on top and the left side of the body appears on the right side of the illustration.

🔑 The thoracic cavity contains three smaller cavities and the mediastinum.

Right pleural cavity

Parietal pleura

Visceral pleura

Diaphragm

Mediastinum

Pericardial cavity

Parietal pericardium

Visceral pericardium

Left pleural cavity

(a) Anterior view of thoracic cavity

Transverse plane

View

ANTERIOR

Muscle

Heart

PERICARDIAL CAVITY

Right lung

Aorta

RIGHT PLEURAL CAVITY

Rib

Sternum (breastbone)

Thymus

Left lung

Esophagus (food tube)

Vertebral column (backbone)

LEFT PLEURAL CAVITY

POSTERIOR

(b) Inferior view of transverse section of thoracic cavity

❓ **What is the name of the cavity that surrounds the heart? Which cavities surround the lungs?**

contains all thoracic organs except the lungs themselves. Among the structures in the mediastinum are the heart, esophagus, trachea, thymus, and several large blood vessels that enter and exit the heart. The **diaphragm** (DĪ-a-fram = partition or wall) is a dome-shaped muscle that separates the thoracic cavity from the abdominopelvic cavity.

The **abdominopelvic cavity** (ab-dom′-i-nō-PEL-vik; see Figure 1.9) extends from the diaphragm to the groin and is encircled

Figure 1.11 The abdominopelvic cavity. The dashed lower line shows the approximate boundary between the abdominal and pelvic cavities.

 The abdominopelvic cavity extends from the diaphragm to the groin.

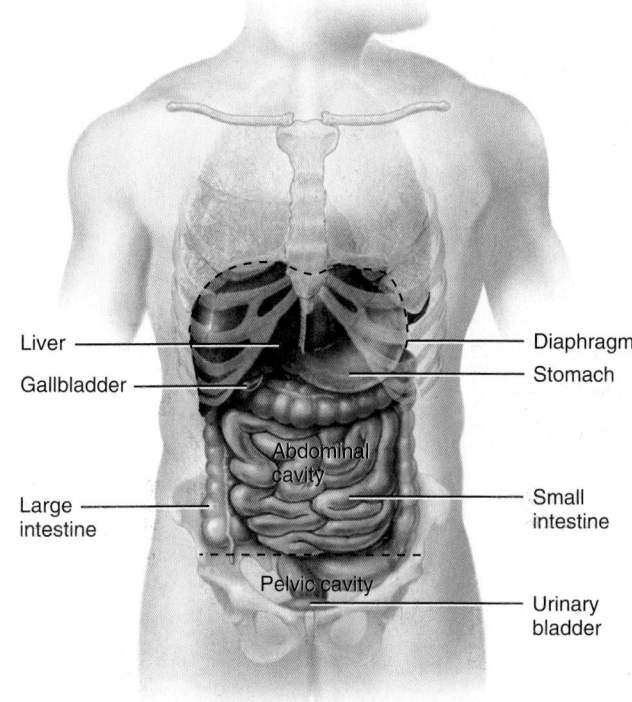

Liver

Gallbladder

Abdominal cavity

Large intestine

Pelvic cavity

Diaphragm

Stomach

Small intestine

Urinary bladder

Anterior view

? To which body systems do the organs shown here within the abdominal and pelvic cavities belong? (*Hint:* Refer to Table 1.2.)

by the abdominal muscular wall and the bones and muscles of the pelvis. As the name suggests, the abdominopelvic cavity is divided into two portions, even though no wall separates them (Figure 1.11). The superior portion, the **abdominal cavity** (ab-DOM-i-nal; *abdomin-* = belly), contains the stomach, spleen, liver, gallbladder, small intestine, and most of the large intestine. The inferior portion, the **pelvic cavity** (PEL-vik; *pelv-* = basin), contains the urinary bladder, portions of the large intestine, and internal organs of the reproductive system. Organs inside the thoracic and abdominopelvic cavities are called **viscera** (VIS-er-a).

Thoracic and Abdominal Cavity Membranes

A **membrane** is a thin, pliable tissue that covers, lines, partitions, or connects structures. One example is a slippery, double-layered membrane called a **serous membrane** (SĒR-us), which covers the viscera within the thoracic and abdominal cavities and also lines the walls of the thorax and abdomen. The parts of a serous membrane are (1) the

parietal layer (pa-RĪ-e-tal), a thin epithelium that lines the walls of the cavities, and (2) the *visceral layer* (VIS-er-al), a thin epithelium that covers and adheres to the viscera within the cavities. A small amount of lubricating fluid (serous fluid) between the two layers reduces friction, allowing the viscera to slide somewhat during movements, such as when the lungs inflate and deflate during breathing.

The serous membrane of the pleural cavities is called the **pleura** (PLOO-ra). The *visceral pleura* clings to the surface of the lungs, and the anterior part of the *parietal pleura* lines the chest wall, covering the superior surface of the diaphragm (see Figure 1.10a). In between is the *pleural cavity*, filled with a small amount of lubricating fluid (see Figure 1.10). The serous membrane of the pericardial cavity is the **pericardium** (per′-i-KAR-dē-um). The *visceral pericardium* covers the surface of the heart; the *parietal pericardium* lines the chest wall. Between them is the *pericardial cavity*, filled with a small amount of lubricating fluid (see Figure 1.10). The **peritoneum** (per-i-tō-NĒ-um) is the serous membrane of the abdominal cavity. The *visceral peritoneum* covers the abdominal viscera, and the *parietal peritoneum* lines the abdominal wall, covering the inferior surface of the diaphragm. Between them is the *peritoneal cavity*, which contains a small amount of lubricating fluid. Most abdominal organs are located in the peritoneal cavity. Some are located between the parietal peritoneum and the posterior abdominal wall. Such organs are said to be *retroperitoneal* (re′-trō-per-i-tō-NĒ-al; *retro-* = behind). The kidneys, adrenal glands, pancreas, duodenum of the small intestine, ascending and descending colons of the large intestine, and portions of the abdominal aorta and inferior vena cava are retroperitoneal.

In addition to the major body cavities just described, you will also learn about other body cavities in later chapters. These include the *oral (mouth) cavity*, which contains the tongue and teeth (see Figure 24.5); the *nasal cavity* in the nose (see Figure 23.1); the *orbital cavities (orbits)*, which contain the eyeballs (see Figure 7.3); the *middle ear cavities (middle ears)*, which contain small bones (see Figure 17.4); and the *synovial cavities*, which are found in freely movable joints and contain synovial fluid (see Figure 9.3).

A summary of the major body cavities and their membranes is presented in the table included in Figure 1.9.

Abdominopelvic Regions and Quadrants

To describe the location of the many abdominal and pelvic organs more easily, anatomists and clinicians use two methods of dividing the abdominopelvic cavity into smaller areas. In the first method, two horizontal and two vertical lines, aligned like a tic-tac-toe grid, partition this cavity into nine **abdominopelvic regions** (Figure 1.12a,b). The top horizontal line, the *subcostal line* (*sub-* = under; *costal* = rib), is drawn just inferior to the rib cage, across the inferior portion of the stomach; the bottom horizontal line, the *transtubercular line* (trans-too-BER-kū-lar), is drawn just inferior to the tops of the hip bones. Two vertical lines, the left and right *midclavicular lines* (mid-kla-VIK-ū-lar), are drawn through the midpoints of the clavicles (collar bones), just medial to the nipples. The four lines divide the abdominopelvic cavity into a larger middle section and smaller left and right sections.

Figure 1.12 Regions and quadrants of the abdominopelvic cavity.

The nine-region designation is used for anatomical studies; the quadrant designation is used to locate the site of pain, tumors, or some other abnormality.

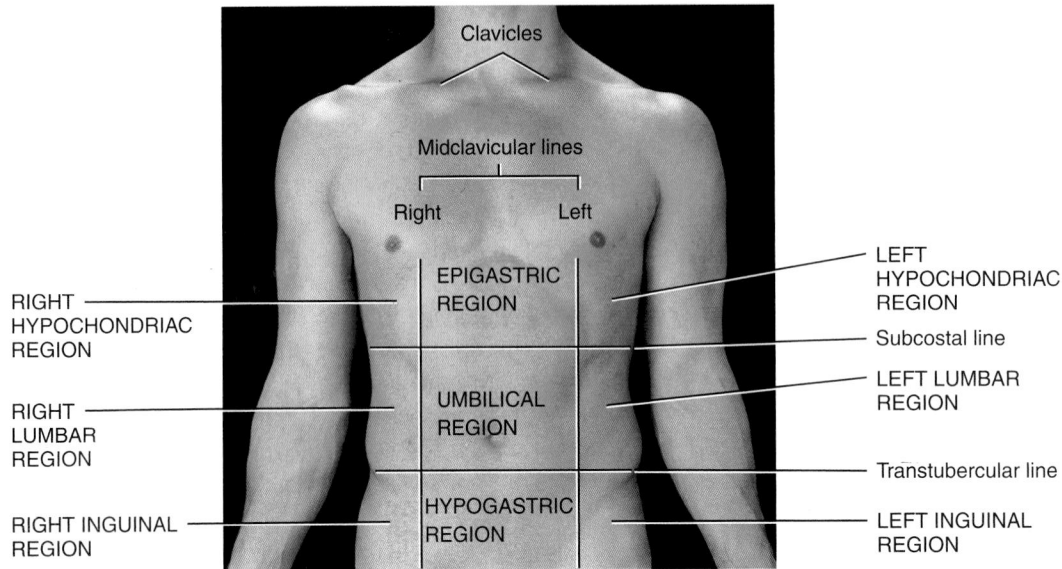

(a) Anterior view showing abdominopelvic regions

(b) Anterior view showing location of abdominopelvic regions

(c) Anterior view showing location of abdominopelvic quadrants

In which abdominopelvic region is each of the following found: most of the liver, ascending colon, urinary bladder, and most of the small intestine? In which abdominopelvic quadrant would pain from appendicitis (inflammation of the appendix) be felt?

The names of the nine abdominopelvic regions are **right hypochondriac** (hī′-pō-KON-drē-ak), **epigastric** (ep-i-GAS-trik), **left hypochondriac, right lumbar, umbilical** (um-BIL-i-kal) **left lumbar, right inguinal (iliac)** (IN-gwi-nal), **hypogastric (pubic),** and **left inguinal (iliac).**

The second method is simpler and divides the abdominopelvic cavity into **quadrants** (KWOD-rantz; *quad-* = one-fourth), as shown in Figure 1.12c. In this method, a midsagittal line (the *median line*) and a transverse line (the *transumbilical line*) are passed through the **umbilicus** (um-BI-li-kus; *umbilic-* = navel) or *belly button.* The names of the abdominopelvic quadrants are **right upper quadrant (RUQ), left upper quadrant (LUQ), right lower quadrant (RLQ),** and **left lower quadrant (LLQ).** The nine-region division is more widely used for anatomical studies, and quadrants are more commonly used by clinicians for describing the site of abdominopelvic pain, a tumor, or another abnormality.

✔ **CHECKPOINT**

12. Locate each region shown in Figure 1.5 on your own body, and then identify it by its anatomical name and the corresponding common name.
13. What structures separate the various body cavities from one another?
14. Locate the nine abdominopelvic regions and the four abdominopelvic quadrants on yourself, and list some of the organs found in each.

1.6 MEDICAL IMAGING

◉ **OBJECTIVE**

• Describe the principles and importance of medical imaging procedures in the evaluation of organ functions and the diagnosis of disease.

Medical imaging refers to techniques and procedures used to create images of the human body. Various types of medical imaging allow visualization of structures inside our bodies and are increasingly helpful for precise diagnosis of a wide range of anatomical and physiological disorders. The grandparent of all medical imaging techniques is conventional radiography (x-rays), in medical use since the late 1940s. The newer imaging technologies not only contribute to diagnosis of disease, but they also are advancing our understanding of normal anatomy and physiology. Table 1.3 describes some commonly used medical imaging techniques. Other imaging methods, such as cardiac catheterization, will be discussed in later chapters.

✔ **CHECKPOINT**

15. Which forms of medical imaging would be used to show a blockage in an artery of the heart?
16. Of the medical imaging techniques outlined in Table 1.3, which one best reveals the physiology of a structure?
17. Which medical imaging technique would you use to determine whether a bone was broken?

TABLE 1.3

Common Medical Imaging Procedures

RADIOGRAPHY

Procedure: A single barrage of x-rays passes through the body, producing an image of interior structures on x-ray–sensitive film. The resulting two-dimensional image is a *radiograph* (RĀ-dē-ō-graf′), commonly called an x-ray.

Comments: Relatively inexpensive, quick, and simple to perform; usually provides sufficient information for diagnosis. X-rays do not easily pass through dense structures, so bones appear white. Hollow structures, such as the lungs, appear black. Structures of intermediate density, such as skin, fat, and muscle, appear as varying shades of gray. At low doses, x-rays are useful for examining soft tissues such as the breast (**mammography**) and for determining bone density (**bone densitometry**).

Radiograph of the thorax in anterior view

— Left clavicle

— Rib

— Left lung

— Heart

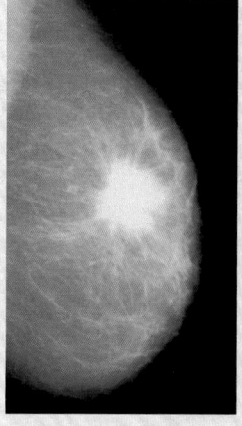

Mammogram of a female breast showing a cancerous tumor (white mass with uneven border)

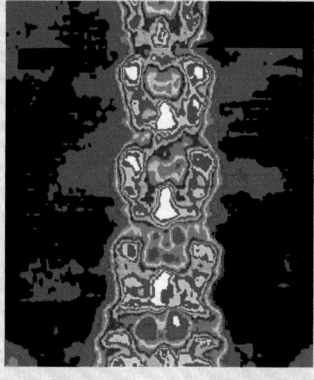

Bone densitometry scan of the lumbar spine in anterior view

TABLE 1.3 CONTINUES

TABLE 1.3 CONTINUED

Common Medical Imaging Procedures

RADIOGRAPHY (CONTINUED)

It is necessary to use a substance called a contrast medium to make hollow or fluid-filled structures visible (appear white) in radiographs. X-rays make structures that contain contrast media appear white. The medium may be introduced by injection, orally, or rectally, depending on the structure to be imaged. Contrast x-rays are used to image blood vessels (**angiography**), the urinary system (**intravenous urography**), and the gastrointestinal tract (**barium contrast x-ray**).

Angiogram of an adult human heart showing a blockage in a coronary artery (arrow)

Intravenous urogram showing a kidney stone (arrow) in the right kidney

Barium contrast x-ray showing a cancer of the ascending colon (arrow)

MAGNETIC RESONANCE IMAGING (MRI)

Procedure: The body is exposed to a high-energy magnetic field, which causes protons (small positive particles within atoms, such as hydrogen) in body fluids and tissues to arrange themselves in relation to the field. Then a pulse of radio waves "reads" these ion patterns, and a color-coded image is assembled on a video monitor. The result is a two- or three-dimensional blueprint of cellular chemistry.

Comments: Relatively safe, but cannot be used on patients with metal in their bodies. Shows fine details for soft tissues but not for bones. Most useful for differentiating between normal and abnormal tissues. Used to detect tumors and artery-clogging fatty plaques; reveal brain abnormalities; measure blood flow; and detect a variety of musculoskeletal, liver, and kidney disorders.

Magnetic resonance image of the brain in sagittal section

DIFFUSION TENSOR IMAGING (DTI)

Procedure: **Diffusion tensor imaging (DTI)** is a variation of MRI that tracks the movement of water molecules along the length of axons, long processes of nerve cells that form the white matter in the brain. The white matter is organized into bundles of axons called tracts that connect one part of the brain with another. DTI provides images of the brain's white matter.

Comments: Used in the diagnosis of addictions, epilepsy, brain tumors, traumatic brain injury, stroke, multiple sclerosis, and neurodegenerative diseases.

Diffusion tensor image of transverse section of the brain showing white matter (various colors)

COMPUTED TOMOGRAPHY (CT)
[formerly called computerized axial tomography (CAT) scanning]

Procedure: In this form of computer-assisted radiography, an x-ray beam traces an arc at multiple angles around a section of the body. The resulting transverse section of the body, called a *CT scan,* is shown on a video monitor.

Comments: Visualizes soft tissues and organs with much more detail than conventional radiographs. Differing tissue densities show up as various shades of gray. Multiple scans can be assembled to build three-dimensional views of structures (described next). Whole-body CT scanning typically targets the torso and appears to provide the most benefit in screening for lung cancers, coronary artery disease, and kidney cancers.

ANTERIOR

Heart
Aorta
Backbone
Left rib
Left scapula

POSTERIOR

Computed tomography scan of the thorax in inferior view

CORONARY (CARDIAC) COMPUTED TOMOGRAPHY ANGIOGRAPHY (CCTA) SCAN

Procedure: In this form of computer-assisted radiography, an iodine-containing contrast medium is injected into a vein and a beta-blocker is given to decrease heart rate. Then, numerous x-ray beams trace an arc around the heart and a scanner detects the x-ray beams and transmits them to a computer, which transforms the information into a three-dimensional image of the coronary blood vessels on a monitor. The image produced is called a *CCTA scan* and can be generated in less than 20 seconds.

Comments: Used primarily to determine if there are any coronary artery blockages (for example, atherosclerotic plaque or calcium) that may require an intervention such as angioplasty or stent. The CCTA scan can be rotated, enlarged, and moved at any angle. The procedure can take thousands of images of the heart within the time of a single heartbeat, so it provides a great amount of detail about the heart's structure and function.

Normal left coronary artery

Blocked right coronary artery

CCTA scan of coronary arteries

ULTRASOUND SCANNING

Procedure: High-frequency sound waves produced by a handheld wand reflect off body tissues and are detected by the same instrument. The image, which may be still or moving, is called a *sonogram* (SON-ō-gram) and is shown on a video monitor.

Comments: Safe, noninvasive, painless, and uses no dyes. Most commonly used to visualize the fetus during pregnancy. Also used to observe the size, location, and actions of organs and blood flow through blood vessels (**Doppler ultrasound**).

Forehead

Eye

Hand

Sonogram of a fetus (Courtesy of Andrew Joseph Tortora and Damaris Soler)

TABLE 1.3 CONTINUES

TABLE 1.3 CONTINUED

Common Medical Imaging Procedures

POSITRON EMISSION TOMOGRAPHY (PET)

Procedure: A substance that emits positrons (positively charged particles) is injected into the body, where it is taken up by tissues. The collision of positrons with negatively charged electrons in body tissues produces gamma rays (similar to x-rays) that are detected by gamma cameras positioned around the subject. A computer receives signals from the gamma cameras and constructs a *PET scan* image, displayed in color on a video monitor. The PET scan shows where the injected substance is being used in the body. In the PET scan image shown here, the black and blue colors indicate minimal activity; the red, orange, yellow, and white colors indicate areas of increasingly greater activity.

Comments: Used to study the physiology of body structures, such as metabolism in the brain or heart.

ANTERIOR

POSTERIOR

Positron emission tomography scan of a transverse section of the brain (circled area at upper left indicates where a stroke has occurred)

RADIONUCLIDE SCANNING

Procedure: A *radionuclide* (radioactive substance) is introduced intravenously into the body and carried by the blood to the tissue to be imaged. Gamma rays emitted by the radionuclide are detected by a gamma camera outside the subject, and the data are fed into a computer. The computer constructs a *radionuclide image* and displays it in color on a video monitor. Areas of intense color take up a lot of the radionuclide and represent high tissue activity; areas of less intense color take up smaller amounts of the radionuclide and represent low tissue activity. **Single-photon-emission computerized tomography (SPECT) scanning** is a specialized type of radionuclide scanning that is especially useful for studying the brain, heart, lungs, and liver.

Comments: Used to study activity of a tissue or organ, such as searching for malignant tumors in body tissue or scars that may interfere with heart muscle activity.

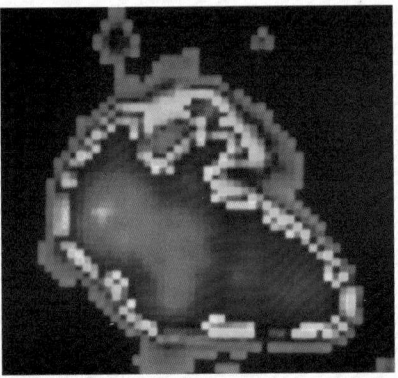

Radionuclide (nuclear) scan of a normal human heart

Single-photon-emission computerized tomography (SPECT) scan of a transverse section of the brain (green area at lower left indicates a migraine attack)

ENDOSCOPY

Procedure: Endoscopy involves the visual examination of the inside of body organs or cavities using a lighted instrument with lenses called an *endoscope*. The image is viewed through an eyepiece on the endoscope or projected onto a monitor.

Comments: Examples include *colonoscopy* (used to examine the interior of the colon, which is part of the large intestine), *laparoscopy* (used to examine the organs within the abdominopelvic cavity), and *arthroscopy* (used to examine the interior of a joint, usually the knee).

Interior view of the colon as shown by colonoscopy

CHAPTER REVIEW AND RESOURCE SUMMARY

Review

Resource

1.1 Anatomy and Physiology Defined

1. Anatomy is the science of body structures and the relationships among structures; physiology is the science of body functions.
2. Dissection is the careful cutting apart of body structures to study their relationships.
3. Some subspecialties of anatomy are embryology, developmental biology, cell biology, histology, gross anatomy, systemic anatomy, regional anatomy, surface anatomy, radiographic anatomy, and pathological anatomy (see Table 1.1).
4. Some subspecialties of physiology are neurophysiology, endocrinology, cardiovascular physiology, immunology, respiratory physiology, renal physiology, exercise physiology, and pathophysiology (see Table 1.1).

1.2 Levels of Structural Organization and Body Systems

1. The human body consists of six levels of structural organization: chemical, cellular, tissue, organ, system, and organismal.
2. Cells are the basic structural and functional living units of an organism and the smallest living units in the human body.
3. Tissues are groups of cells and the materials surrounding them that work together to perform a particular function.
4. Organs are composed of two or more different types of tissues; they have specific functions and usually have recognizable shapes.
5. Systems consist of related organs that have a common function.
6. An organism is any living individual.
7. Table 1.2 introduces the 11 systems of the human organism: the integumentary, skeletal, muscular, nervous, endocrine, cardiovascular, lymphatic, respiratory, digestive, urinary, and reproductive systems.

1.3 Characteristics of the Living Human Organism

1. All organisms carry on certain processes that distinguish them from nonliving things.
2. Among the life processes in humans are metabolism, responsiveness, movement, growth, differentiation, and reproduction.

1.4 Homeostasis

1. Homeostasis is a condition of equilibrium in the body's internal environment produced by the interplay of all the body's regulatory processes.
2. Body fluids are dilute, watery solutions. Intracellular fluid (ICF) is inside cells, and extracellular fluid (ECF) is outside cells. Plasma is the ECF within blood vessels. Interstitial fluid is the ECF that fills spaces between tissue cells; because it surrounds all body cells, interstitial fluid is called the body's internal environment.
3. Disruptions of homeostasis come from external and internal stimuli and psychological stresses. When disruption of homeostasis is mild and temporary, responses of body cells quickly restore balance in the internal environment. If disruption is extreme, regulation of homeostasis may fail.
4. Most often, the nervous and endocrine systems acting together or separately regulate homeostasis. The nervous system detects body changes and sends nerve impulses to counteract changes in controlled conditions. The endocrine system regulates homeostasis by secreting hormones.
5. Feedback systems include three components: (1) Receptors monitor changes in a controlled condition and send input to a control center (afferent pathway). (2) The control center sets the value (set point) at which a controlled condition should be maintained, evaluates the input it receives from receptors (efferent pathway), and generates output commands when they are needed. (3) Effectors receive output from the control center and produce a response (effect) that alters the controlled condition.
6. If a response reverses the original stimulus, the system is operating by negative feedback. If a response enhances the original stimulus, the system is operating by positive feedback.
7. One example of negative feedback is the regulation of blood pressure. If a stimulus causes blood pressure (controlled condition) to rise, baroreceptors (pressure-sensitive nerve cells, the receptors) in blood vessels send impulses (input) to the brain (control center). The brain sends impulses (output) to the heart (effector). As a result, heart rate decreases (response) and blood pressure decreases to normal (restoration of homeostasis).

Anatomy Overview - The Integumentary System
Anatomy Overview - The Skeletal System
Anatomy Overview - The Muscular System
Anatomy Overview - The Nervous System
Anatomy Overview - The Endocrine System
Anatomy Overview - The Cardiovascular System
Anatomy Overview - The Lymphatic and Immune Systems
Anatomy Overview - The Respiratory System
Anatomy Overview - The Digestive System
Anatomy Overview - The Urinary System
Anatomy Overview - The Reproductive Systems
Exercise - Concentrate on Systemic Functions
Exercise - Find the System Outsiders

Animation - Communication, Regulation, and Homeostasis

Animation - Communication, Regulation, and Homeostasis
Animation - Homeostatic Relationships
Animation - Negative Feedback Control of Blood Pressure
Animation - Negative Feedback Control of Temperature
Animation - Positive Feedback Control of Labor
Exercise - System Contributions to Body Homeostasis
Concepts and Connections - Negative Feedback Loop
Concepts and Connections - Regulation and Communication

Review

8. One example of positive feedback occurs during the birth of a baby. When labor begins, the cervix of the uterus is stretched (stimulus), and stretch-sensitive nerve cells in the cervix (receptors) send nerve impulses (input) to the brain (control center). The brain responds by releasing oxytocin (output), which stimulates the uterus (effector) to contract more forcefully (response). Movement of the fetus further stretches the cervix, more oxytocin is released, and even more forceful contractions occur. The cycle is broken with the birth of the baby.

9. Disruptions of homeostasis—homeostatic imbalances—can lead to disorders, diseases, and even death. A disorder is a general term for any abnormality of structure or function. A disease is an illness with a definite set of signs and symptoms.

10. Symptoms are subjective changes in body functions that are not apparent to an observer; signs are objective changes that can be observed and measured.

1.5 Basic Anatomical Terminology

Anatomy Overview - Serous Membrane
Figure 1.5 - The Anatomical Position
Figure 1.6 - Directional Terms
Figure 1.9 - Body Cavities
Figure 1.10 - The Thoracic Cavity

1. Descriptions of any region of the body assume the body is in the anatomical position, in which the subject stands erect facing the observer, with the head level and the eyes facing directly forward. The feet are flat on the floor and directed forward, and the upper limbs are at the sides, with the palms turned forward. A body lying facedown is prone; a body lying faceup is supine.

2. Regional names are terms given to specific regions of the body. The principal regions are the head, neck, trunk, upper limbs, and lower limbs. Within the regions, specific body parts have anatomical names and corresponding common names. Examples are thoracic (chest), nasal (nose), and carpal (wrist).

3. Directional terms indicate the relationship of one part of the body to another. Exhibit 1.A summarizes commonly used directional terms.

4. Planes are imaginary flat surfaces that are used to divide the body or organs to visualize interior structures. A midsagittal plane divides the body or an organ into *equal* right and left sides. A parasagittal plane divides the body or an organ into *unequal* right and left sides. A frontal plane divides the body or an organ into anterior and posterior portions. A transverse plane divides the body or an organ into superior and inferior portions. An oblique plane passes through the body or an organ at an oblique angle.

5. Sections are cuts of the body or its organs made along a plane. They are named according to the plane along which the cut is made and include transverse, frontal, and sagittal sections.

6. Figure 1.9 summarizes body cavities and their membranes. Body cavities are spaces in the body that help protect, separate, and support internal organs. The cranial cavity contains the brain, and the vertebral canal contains the spinal cord. The meninges are protective tissues that line the cranial cavity and vertebral canal. The diaphragm separates the thoracic cavity from the abdominopelvic cavity. Viscera are organs within the thoracic and abdominopelvic cavities. A serous membrane lines the wall of the cavity and adheres to the viscera.

7. The thoracic cavity is subdivided into three smaller cavities: a pericardial cavity, which contains the heart, and two pleural cavities, each of which contains a lung. The central part of the thoracic cavity is an anatomical region called the mediastinum. It is located between the pleural cavities, extending from the sternum to the vertebral column and from the first rib to the diaphragm. It contains all thoracic viscera except the lungs.

8. The abdominopelvic cavity is divided into a superior abdominal and an inferior pelvic cavity. Viscera of the abdominal cavity include the stomach, spleen, liver, gallbladder, small intestine, and most of the large intestine. Viscera of the pelvic cavity include the urinary bladder, portions of the large intestine, and internal organs of the reproductive system.

9. Serous membranes line the walls of the thoracic and abdominal cavities and cover the organs within them. They include the pleura, associated with the lungs; the pericardium, associated with the heart; and the peritoneum, associated with the abdominal cavity.

10. To describe the location of organs more easily, the abdominopelvic cavity is divided into nine regions: right hypochondriac, epigastric, left hypochondriac, right lumbar, umbilical, left lumbar, right inguinal (iliac), hypogastric (pubic), and left inguinal (iliac). To locate the site of an abdominopelvic abnormality in clinical studies, the abdominopelvic cavity is divided into quadrants: right upper quadrant (RUQ), left upper quadrant (LUQ), right lower quadrant (RLQ), and left lower quadrant (LLQ).

1.6 Medical Imaging

1. Medical imaging refers to techniques and procedures used to create images of the human body. They allow visualization of internal structures to diagnose abnormal anatomy and deviations from normal physiology.

2. Table 1.3 summarizes and illustrates several medical imaging techniques.

SELF-QUIZ QUESTIONS

Fill in the blanks in the following statements.

1. A(n) _____ is a group of similar cells and their surrounding materials performing specific functions.

2. The sum of all of the body's chemical processes is _____. It consists of two parts: the phase that builds up new substances is _____, and the phase that breaks down substances is _____.

3. The fluid located within cells is the _____, whereas the fluid located outside of the cells is _____.

Indicate whether the following statements are true or false.

4. In a positive feedback system, the response enhances or intensifies the original stimulus.

5. A person lying face down would be in the supine position. *Prone*

6. The highest level of structural organization is the system level.

Choose the one best answer to the following questions.

7. A plane that separates the body into unequal right and left sides is a
   (a) transverse plane   (b) frontal plane
   (c) midsagittal plane   (d) coronal plane
   (e) parasagittal plane

8. Midway through a 5-mile workout, a runner begins to sweat profusely. The sweat glands producing the sweat would be considered which part of a feedback loop?
   (a) controlled condition (b) receptors   (c) stimulus
   (d) effectors   (e) control center

9. An unspecialized stem cell becomes a brain cell during fetal development. This is an example of
   (a) differentiation   (b) growth   (c) organization
   (d) responsiveness   (e) homeostasis.

10. A radiography technician needs to x-ray a growth on the urinary bladder. To accomplish this, the camera must be positioned on the _____ region.
   (a) left inguinal   (b) epigastric   (c) hypogastric
   (d) right inguinal   (e) umbilical

11. Which of the following would not be associated with the thoracic cavity? (1) pericardium, (2) mediastinum, (3) peritoneum, (4) pleura,
   (a) 2 and 3   (b) 2   (c) 3
   (d) 1 and 4   (e) 3 and 4

12. Match the following common names and anatomical descriptive adjectives:
   ___ (a) axillary   (1) skull
   ___ (b) inguinal   (2) eye
   ___ (c) cervical   (3) cheek
   ___ (d) cranial   (4) armpit
   ___ (e) oral   (5) arm
   ___ (f) brachial   (6) groin
   ___ (g) orbital   (7) buttock
   ___ (h) gluteal   (8) neck
   ___ (i) buccal   (9) mouth
   ___ (j) coxal   (10) hip

13. Choose the term that best fits the blank in each statement. Some answers may be used more than once.
   ___ (a) Your eyes are _____ to your chin.   (1) superior
   ___ (b) Your skin is _____ to your heart.   (2) inferior
   ___ (c) Your right shoulder is _____ and _____ from your umbilicus (belly button).   (3) anterior
                                                                       (4) posterior
   ___ (d) In the anatomical position, your thumb is _____.   (5) medial
   ___ (e) Your buttocks are _____.   (6) lateral
   ___ (f) Your right foot and right hand are _____.   (7) intermediate
                                                                       (8) ipsilateral
   ___ (g) Your knee is _____ between your thigh and toes.   (9) contralateral
                                                                       (10) proximal
   ___ (h) Your lungs are _____ to your spinal column.   (11) distal
   ___ (i) Your breastbone is _____ to your chin.   (12) superficial
                                                                       (13) deep
   ___ (j) Your calf is _____ to your heel.

14. Match the following cavities to their definitions:
   ___ (a) a fluid-filled space that surrounds the heart   (1) cranial cavity
   ___ (b) the cavity that contains the brain   (2) vertebral canal
   ___ (c) a cavity formed by the ribs, muscles of the chest, sternum, and part of the vertebral column   (3) thoracic cavity
                                                                       (4) pericardial cavity
                                                                       (5) pleural cavity
                                                                       (6) abdominal cavity
                                                                       (7) pelvic cavity
   ___ (d) a cavity that contains the stomach, spleen, liver, gallbladder, small intestine, and most of the large intestine
   ___ (e) fluid-filled space that surrounds a lung
   ___ (f) the cavity that contains the urinary bladder, part of the large intestine, and the organs of the reproductive system
   ___ (g) the canal that contains the spinal cord

15. Match the following systems with their functions:

_____ (a) nervous system
_____ (b) endocrine system
_____ (c) urinary system
_____ (d) cardiovascular system
_____ (e) muscular system
_____ (f) respiratory system
_____ (g) digestive system
_____ (h) skeletal system
_____ (i) integumentary system
_____ (j) lymphatic system and immunity
_____ (k) reproductive system

(1) regulates body activities through hormones (chemicals) transported in the blood to various target organs of the body
(2) produces gametes; releases hormones from gonads
(3) protects against disease; returns fluids to blood
(4) protects body by forming a barrier to the outside environment; helps regulate body temperature
(5) transports oxygen and nutrients to cells; protects against disease; carries wastes away from cells
(6) regulates body activities through action potentials (nerve impulses); receives sensory information; interprets and responds to the information
(7) carries out the physical and chemical breakdown of food and absorption of nutrients
(8) transfers oxygen and carbon dioxide between air and blood
(9) supports and protects the body; provides internal framework; provides a place for muscle attachment
(10) powers movements of the body and stabilizes body position
(11) eliminates wastes; regulates the volume and chemical composition of blood

CRITICAL THINKING QUESTIONS

1. You are studying for your first anatomy and physiology exam and want to know which areas of your brain are working hardest as you study. Your classmate suggests that you could have a computed tomography (CT) scan done to assess your brain activity. Would this be the best way to determine brain activity levels?

2. There is much interest in using stem cells to help in the treatment of diseases such as type I diabetes, which is due to a malfunction of some of the normal cells in the pancreas. What would make stem cells useful in disease treatment?

3. On her first anatomy and physiology exam, Heather defined homeostasis as "the condition in which the body approaches room temperature and stays there." Do you agree with Heather's definition?

? ANSWERS TO FIGURE QUESTIONS

1.1 Organs are composed of two or more different types of tissues that work together to perform a specific function.

1.2 The difference between negative and positive feedback systems is that in negative feedback systems, the response reverses the original stimulus, but in positive feedback systems, the response enhances the original stimulus.

1.3 When something causes blood pressure to decrease, then heart rate increases due to operation of this negative feedback system.

1.4 Because positive feedback systems continually intensify or reinforce the original stimulus, some mechanism is needed to end the response.

1.5 Having one standard anatomical position allows directional terms to be clearly defined so that any body part can be described in relation to any other part.

1.6 No, the radius is distal to the humerus. No, the esophagus is posterior to the trachea. Yes, the ribs are superficial to the lungs. Yes, the urinary bladder is medial to the ascending colon. No, the sternum is medial to the descending colon.

1.7 The frontal plane divides the heart into anterior and posterior portions.

1.8 The parasagittal plane (not shown in the figure) divides the brain into unequal right and left portions.

1.9 Urinary bladder = P, stomach = A, heart = T, small intestine = A, lungs = T, internal female reproductive organs = P, thymus = T, spleen = A, liver = A.

1.10 The pericardial cavity surrounds the heart, and the pleural cavities surround the lungs.

1.11 The illustrated abdominal cavity organs all belong to the digestive system (liver, gallbladder, stomach, small intestine, and most of the large intestine). Illustrated pelvic cavity organs belong to the urinary system (the urinary bladder) and the digestive system (part of the large intestine).

1.12 The liver is mostly in the epigastric region; the ascending colon is in the right lumbar region; the urinary bladder is in the hypogastric region; most of the small intestine is in the umbilical region. The pain associated with appendicitis would be felt in the right lower quadrant (RLQ).

2
THE CHEMICAL LEVEL OF ORGANIZATION

CHEMISTRY AND HOMEOSTASIS *Maintaining the proper assortment and quantity of thousands of different chemicals in your body, and monitoring the interactions of these chemicals with one another, are two important aspects of homeostasis.*

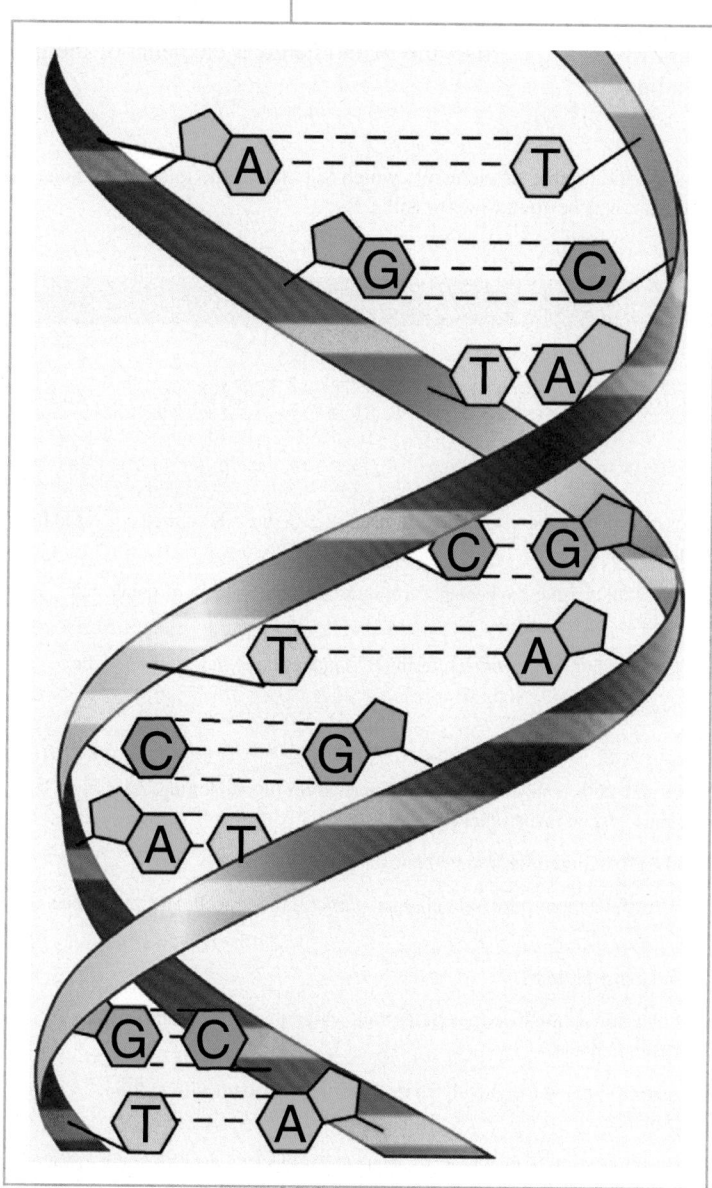

You learned in Chapter 1 that the chemical level of organization, the lowest level of structural organization, consists of atoms and molecules. These letters of the anatomical alphabet ultimately combine to form body organs and systems of astonishing size and complexity. In this chapter, we consider how atoms bond together to form molecules, and how atoms and molecules release or store energy in processes known as chemical reactions. You will also learn about the vital importance of water, which accounts for nearly two-thirds of your body weight, in chemical reactions and the maintenance of homeostasis. Finally, we present several groups of molecules whose unique properties contribute to assembly of your body's structures and to powering the processes that enable you to live.

Chemistry (KEM-is-trē) is the science of the structure and interactions of matter. All living and nonliving things consist of **matter,** which is anything that occupies space and has **mass.** Mass is the amount of matter in any object, which does not change. *Weight,* the force of gravity acting on matter, does change. When objects are farther from Earth, the pull of gravity is weaker; this is why the weight of an astronaut is close to zero in outer space.

? *Did you ever wonder how fatty acids relate to health and disease?*

2.1 HOW MATTER IS ORGANIZED

● OBJECTIVES

• Identify the main chemical elements of the human body.
• Describe the structures of atoms, ions, molecules, free radicals, and compounds.

Chemical Elements

Matter exists in three states: solid, liquid, and gas. *Solids,* such as bones and teeth, are compact and have a definite shape and volume. *Liquids,* such as blood plasma, have a definite volume and assume the shape of their container. *Gases,* like oxygen and carbon dioxide, have neither a definite shape nor volume. *All* forms of matter—both living and nonliving—are made up of a limited number of building blocks called **chemical elements.** Each element is a substance that cannot be split into a simpler substance by ordinary chemical means. Scientists now recognize 117 elements. Of these, 92 occur naturally on Earth. The rest have been produced from the natural elements using particle accelerators or nuclear reactors. Each element is designated by a **chemical sym-bol,** one or two letters of the element's name in English, Latin, or another language. Examples of chemical symbols are H for hydrogen, C for carbon, O for oxygen, N for nitrogen, Ca for calcium, and Na for sodium (*natrium* = sodium).*

Twenty-six different chemical elements normally are present in your body. Just four elements, called the **major elements,** constitute about 96% of the body's mass: oxygen, carbon, hydrogen, and nitrogen. Eight others, the **lesser elements,** contribute about 3.6% to the body's mass: calcium, phosphorus (P), potassium (K), sulfur (S), sodium, chlorine (Cl), magnesium (Mg), and iron (Fe). An additional 14 elements—the **trace elements**—are present in tiny amounts. Together, they account for the remaining about 0.4% of the body's mass. Several trace elements have important functions in the body. For example, iodine is needed to make thyroid hormones. The functions of some trace elements are unknown. Table 2.1 lists the main chemical elements of the human body.

*The periodic table of elements, which lists all of the known chemical elements, can be found in Appendix B.

TABLE 2.1		
Main Chemical Elements in the Body		
CHEMICAL ELEMENT (SYMBOL)	**% OF TOTAL BODY MASS**	**SIGNIFICANCE**
MAJOR ELEMENTS	(about 96)	
Oxygen (O)	65.0	Part of water and many organic (carbon-containing) molecules; used to generate ATP, a molecule used by cells to temporarily store chemical energy.
Carbon (C)	18.5	Forms backbone chains and rings of all organic molecules: carbohydrates, lipids (fats), proteins, and nucleic acids (DNA and RNA).
Hydrogen (H)	9.5	Constituent of water and most organic molecules; ionized form (H^+) makes body fluids more acidic.
Nitrogen (N)	3.2	Component of all proteins and nucleic acids.
LESSER ELEMENTS	(about 3.6)	
Calcium (Ca)	1.5	Contributes to hardness of bones and teeth; ionized form (Ca^{2+}) needed for blood clotting, release of some hormones, contraction of muscle, and many other processes.
Phosphorus (P)	1.0	Component of nucleic acids and ATP; required for normal bone and tooth structure.
Potassium (K)	0.35	Ionized form (K^+) is the most plentiful cation (positively charged particle) in intracellular fluid; needed to generate action potentials.
Sulfur (S)	0.25	Component of some vitamins and many proteins.
Sodium (Na)	0.2	Ionized form (Na^+) is the most plentiful cation in extracellular fluid; essential for maintaining water balance; needed to generate action potentials.
Chlorine (Cl)	0.2	Ionized form (Cl^-) is the most plentiful anion (negatively charged particle) in extracellular fluid; essential for maintaining water balance.
Magnesium (Mg)	0.1	Ionized form (Mg^{2+}) needed for action of many enzymes, molecules that increase the rate of chemical reactions in organisms.
Iron (Fe)	0.005	Ionized forms (Fe^{2+} and Fe^{3+}) are part of hemoglobin (oxygen-carrying protein in red blood cells) and some enzymes.
TRACE ELEMENTS	(about 0.4)	Aluminum (Al), boron (B), chromium (Cr), cobalt (Co), copper (Cu), fluorine (F), iodine (I), manganese (Mn), molybdenum (Mo), selenium (Se), silicon (Si), tin (Sn), vanadium (V), and zinc (Zn).

Structure of Atoms

Each element is made up of **atoms,** the smallest units of matter that retain the properties and characteristics of the element. Atoms are extremely small. Two hundred thousand of the largest atoms would fit on the period at the end of this sentence. Hydrogen atoms, the smallest atoms, have a diameter less than 0.1 nanometer (0.1×10^{-9} m = 0.0000000001 m), and the largest atoms are only five times larger.

Dozens of different **subatomic particles** compose individual atoms. However, only three types of subatomic particles are important for understanding the chemical reactions in the human body: protons, neutrons, and electrons (Figure 2.1). The dense central core of an atom is its **nucleus.** Within the nucleus are positively charged **protons (p^+)** and uncharged (neutral) **neutrons (n^0).** The tiny, negatively charged **electrons (e^-)** move about in a large space surrounding the nucleus. They do not follow a fixed path or orbit but instead form a negatively charged "cloud" that envelops the nucleus (Figure 2.1a).

Even though their exact positions cannot be predicted, specific groups of electrons are most likely to move about within certain regions around the nucleus. These regions, called **electron shells,** are depicted as simple circles around the nucleus. Because each electron shell can hold a specific number of electrons, the electron

Figure 2.1 Two representations of the structure of an atom. Electrons move about the nucleus, which contains neutrons and protons. (a) In the electron cloud model of an atom, the shading represents the chance of finding an electron in regions outside the nucleus. (b) In the electron shell model, filled circles represent individual electrons, which are grouped into concentric circles according to the shells they occupy. Both models depict a carbon atom, with six protons, six neutrons, and six electrons.

 An atom is the smallest unit of matter that retains the properties and characteristics of its element.

- Protons (p^+) ⎤
- Neutrons (n^0) ⎦ Nucleus
- Electrons (e^-)

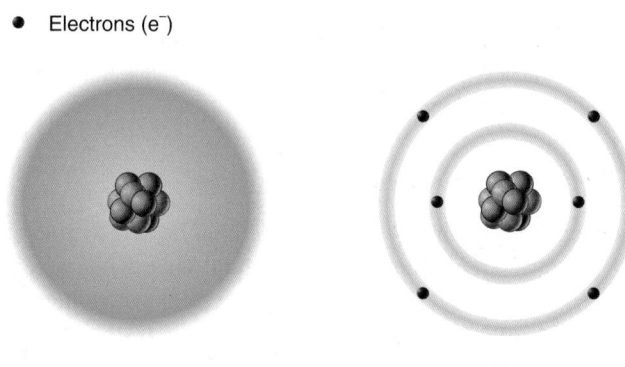

(a) Electron cloud model   (b) Electron shell model

❓ **How are the electrons of carbon distributed between the first and second electron shells?**

shell model best conveys this aspect of atomic structure (Figure 2.1b). The first electron shell (nearest the nucleus) never holds more than 2 electrons. The second shell holds a maximum of 8 electrons, and the third can hold up to 18 electrons. The electron shells fill with electrons in a specific order, beginning with the first shell. For example, notice in Figure 2.2 that sodium (Na), which has 11 electrons total, contains 2 electrons in the first shell, 8 electrons in the second shell, and 1 electron in the third shell. The most massive element present in the human body is iodine, which has a total of 53 electrons: 2 in the first shell, 8 in the second shell, 18 in the third shell, 18 in the fourth shell, and 7 in the fifth shell.

The number of electrons in an atom of an element always equals the number of protons. Because each electron and proton carries one charge, the negatively charged electrons and the positively charged protons balance each other. Thus, each atom is electrically neutral; its total charge is zero.

Atomic Number and Mass Number

The *number of protons* in the nucleus of an atom is an atom's **atomic number.** Figure 2.2 shows that atoms of different elements have different atomic numbers because they have different numbers of protons. For example, oxygen has an atomic number of 8 because its nucleus has 8 protons, and sodium has an atomic number of 11 because its nucleus has 11 protons.

The **mass number** of an atom is the sum of its protons and neutrons. Because sodium has 11 protons and 12 neutrons, its mass number is 23 (Figure 2.2). Although all atoms of one element have the same number of protons, they may have different numbers of neutrons and thus different mass numbers. **Isotopes** are atoms of an element that have different numbers of neutrons and therefore different mass numbers. In a sample of oxygen, for example, most atoms have 8 neutrons, and a few have 9 or 10, but all have 8 protons and 8 electrons. Most isotopes are stable, which means that their nuclear structure does not change over time. The stable isotopes of oxygen are designated ^{16}O, ^{17}O, and ^{18}O (or O-16, O-17, and O-18). As you may already have determined, the numbers indicate the mass number of each isotope. As you will discover shortly, the number of electrons of an atom determines its chemical properties. Although the isotopes of an element have different numbers of neutrons, they have identical chemical properties because they have the same number of electrons.

Certain isotopes called **radioactive isotopes** are unstable; their nuclei decay (spontaneously change) into a stable configuration. Examples are H-3, C-14, O-15, and O-19. As they decay, these atoms emit radiation—either subatomic particles or packets of energy—and in the process often transform into a different element. For example, the radioactive isotope of carbon, C-14, decays to N-14. The decay of a radioisotope may be as fast as a fraction of a second or as slow as millions of years. The **half-life** of an isotope is the time required for half of the radioactive atoms in a sample of that isotope to decay into a more stable form. The half-life of C-14, which is used to determine the age of organic samples, is about 5730 years; the half-life of I-131, an important clinical tool, is 8 days.

Figure 2.2 Atomic structures of several stable atoms.

The atoms of different elements have different atomic numbers because they have different numbers of protons.

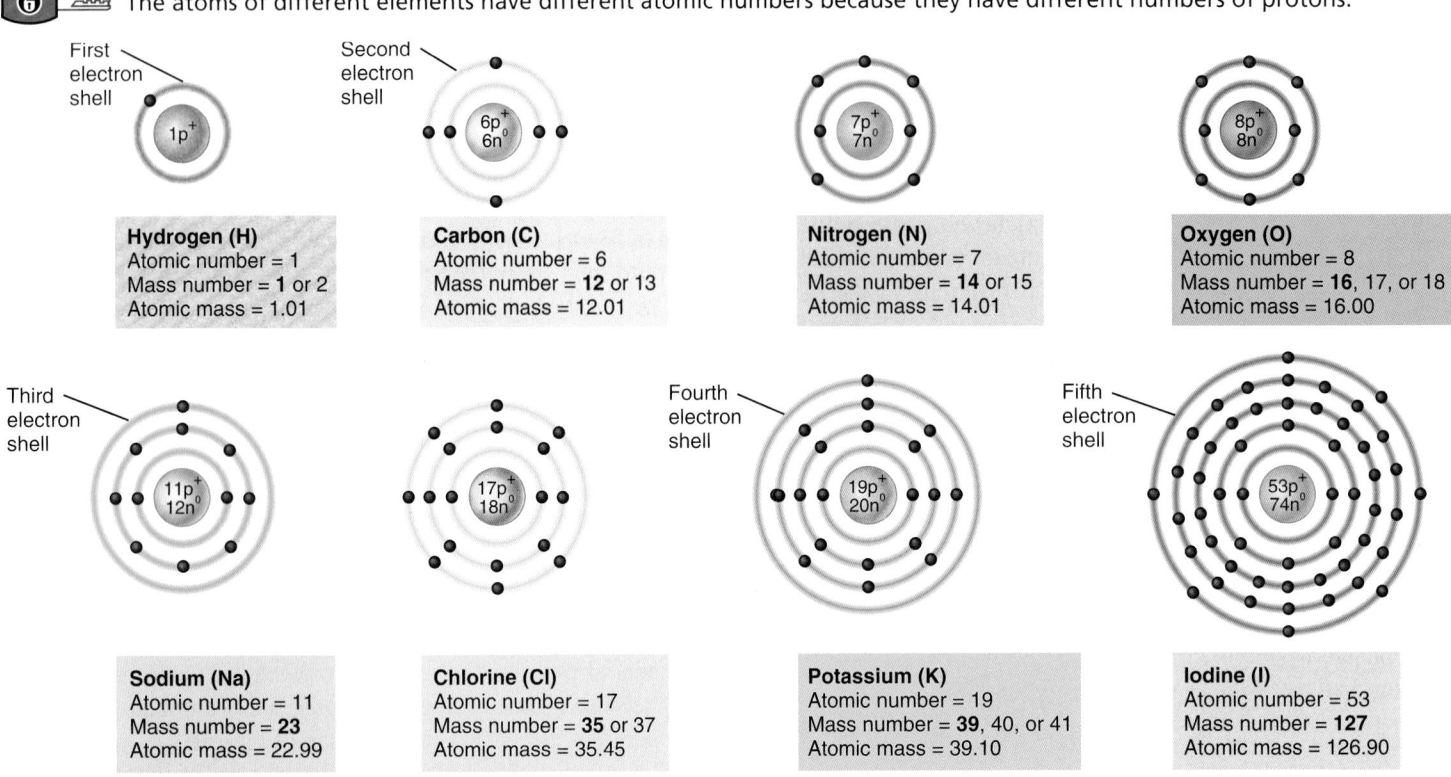

Hydrogen (H)
Atomic number = 1
Mass number = **1** or 2
Atomic mass = 1.01

Carbon (C)
Atomic number = 6
Mass number = **12** or 13
Atomic mass = 12.01

Nitrogen (N)
Atomic number = 7
Mass number = **14** or 15
Atomic mass = 14.01

Oxygen (O)
Atomic number = 8
Mass number = **16**, 17, or 18
Atomic mass = 16.00

Sodium (Na)
Atomic number = 11
Mass number = **23**
Atomic mass = 22.99

Chlorine (Cl)
Atomic number = 17
Mass number = **35** or 37
Atomic mass = 35.45

Potassium (K)
Atomic number = 19
Mass number = **39**, 40, or 41
Atomic mass = 39.10

Iodine (I)
Atomic number = 53
Mass number = **127**
Atomic mass = 126.90

Atomic number = number of protons in an atom
Mass number = number of protons and neutrons in an atom (boldface indicates most common isotope)
Atomic mass = average mass of all stable atoms of a given element in daltons

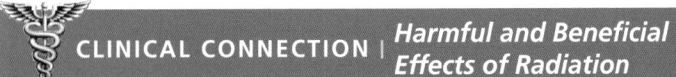 **Which four of these elements are present most abundantly in living organisms?**

Atomic Mass

The standard unit for measuring the mass of atoms and their subatomic particles is a **dalton,** also known as an *atomic mass unit (amu)*. A neutron has a mass of 1.008 daltons, and a proton has a mass of 1.007 daltons. The mass of an electron, at 0.0005 dalton, is almost 2000 times smaller than the mass of a neutron or proton. The **atomic mass** (also called the *atomic weight*) of an element is the average mass of all its naturally occurring isotopes. Typically, the atomic mass of an element is close to the mass number of its most abundant isotope.

Ions, Molecules, and Compounds

As we discussed, atoms of the same element have the same number of protons. The atoms of each element have a characteristic way of losing, gaining, or sharing their electrons when interacting with other atoms to achieve stability. The way that electrons behave enables atoms in the body to exist in electrically charged forms called ions, or to join with each other into complex combinations called molecules. If an atom either *gives up* or *gains* electrons, it becomes an ion. An **ion** is an atom that has a positive or negative charge because it has unequal numbers of protons and electrons. *Ionization* is the process of giving up or gaining electrons. An ion of an atom is symbolized by writing its chemical symbol followed by the number of its positive (+) or negative (−) charges. Thus, Ca^{2+} stands for a calcium ion that has two positive charges because it has lost two electrons.

When two or more atoms *share* electrons, the resulting combination is called a **molecule** (MOL-e-kūl). A *molecular formula* indicates the elements and the number of atoms of each element that make up a molecule. A molecule may consist of two atoms of the same kind, such as an oxygen molecule (Figure 2.3a). The molecular formula for a molecule of oxygen is O_2. The subscript 2 indicates that the molecule contains two atoms of oxygen. Two or more different kinds of atoms may also form a molecule, as in a water molecule (H_2O). In H_2O one atom of oxygen shares electrons with two atoms of hydrogen.

A **compound** is a substance that contains atoms of two or more different elements. Most of the atoms in the body are joined into compounds. Water (H_2O) and sodium chloride (NaCl), common table salt, are compounds. However, a molecule of oxygen (O_2) is not a compound because it consists of atoms of only one element.

A **free radical** is an atom or group of atoms with an unpaired electron in the outermost shell. A common example is superoxide, which is formed by the addition of an electron to an oxygen molecule (Figure 2.3b). Having an unpaired electron makes a free radical unstable, highly reactive, and destructive to nearby molecules. Free radicals become stable by either giving up their unpaired electron to, or taking on an electron from, another molecule. In so doing, free radicals may break apart important body molecules.

CLINICAL CONNECTION | *Free Radicals and Antioxidants*

There are several sources of free radicals, including exposure to ultraviolet radiation in sunlight, exposure to x-rays, and some reactions that occur during normal metabolic processes. Certain harmful substances, such as carbon tetrachloride (a solvent used in dry cleaning), also give rise to free radicals when they participate in metabolic reactions in the body. Among the many disorders, diseases, and conditions linked to oxygen-derived free radicals are cancer, atherosclerosis, Alzheimer disease, emphysema, diabetes mellitus, cataracts, macular degeneration, rheumatoid arthritis, and deterioration associated with aging. Consuming more **antioxidants**—substances that inactivate oxygen-derived free radicals—is thought to slow the pace of damage caused by free radicals. Important dietary antioxidants include selenium, zinc, beta-carotene, and vitamins C and E. Red, blue, or purple fruits and vegetables contain high levels of antioxidants. •

✔ **CHECKPOINT**

1. List the names and chemical symbols of the 12 most abundant chemical elements in the human body.
2. What are the atomic number, mass number, and atomic mass of carbon? How are they related?
3. Define isotopes and free radicals.

Figure 2.3 **Atomic structures of an oxygen molecule and a superoxide free radical.**

 A free radical has an unpaired electron in its outermost electron shell.

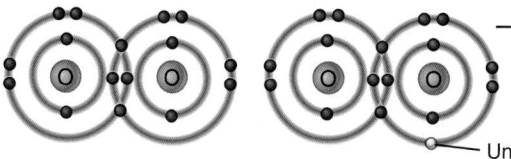

— Unpaired electron

(a) Oxygen molecule (O_2) (b) Superoxide free radical (O_2^-)

? **What substances in the body can inactivate oxygen-derived free radicals?**

2.2 CHEMICAL BONDS

◉ **OBJECTIVES**

- Describe how valence electrons form chemical bonds.
- Distinguish among ionic, covalent, and hydrogen bonds.

The forces that hold together the atoms of a molecule or a compound are **chemical bonds.** The likelihood that an atom will form a chemical bond with another atom depends on the number of electrons in its outermost shell, also called the **valence shell.** An atom with a valence shell holding eight electrons is *chemically stable,* which means it is unlikely to form chemical bonds with other atoms. Neon, for example, has eight electrons in its valence shell, and for this reason it does not bond easily with other atoms. The valence shell of hydrogen and helium is the first electron shell, which holds a maximum of two electrons. Because helium has two valence electrons, it too is stable and seldom bonds with other atoms. Hydrogen, on the other hand, has only one valence electron (see Figure 2.2), so it binds readily with other atoms.

The atoms of most biologically important elements do not have eight electrons in their valence shells. Under the right conditions, two or more atoms can interact in ways that produce a chemically stable arrangement of eight valence electrons for each atom. This chemical principle, called the **octet rule** (*octet* = set of eight), helps explain why atoms interact in predictable ways. One atom is more likely to interact with another atom if doing so will leave both with eight valence electrons. For this to happen, an atom either empties its partially filled valence shell, fills it with donated electrons, or shares electrons with other atoms. The way that valence electrons are distributed determines what kind of chemical bond results. We will consider three types of chemical bonds: ionic bonds, covalent bonds, and hydrogen bonds.

Ionic Bonds

As you have already learned, when atoms lose or gain one or more valence electrons, ions are formed. Positively and negatively charged ions are attracted to one another—opposites attract. The force of attraction that holds together ions with opposite charges is an **ionic bond.** Consider sodium and chlorine atoms, the components of common table salt. Sodium has one valence electron (Figure 2.4a). If sodium *loses* this electron, it is left with the eight electrons in its second shell, which becomes the valence shell. As a result, however, the total number of protons (11) exceeds the number of electrons (10). Thus, the sodium atom has become a **cation** (KAT-ī-on), or positively charged ion. A sodium ion has a charge of 1+ and is written Na^+. By contrast, chlorine has seven valence electrons (Figure 2.4b). If chlorine *gains* an electron from a neighboring atom, it will have a complete octet in its third electron shell. After gaining an electron, the total number of electrons (18) exceeds the number of protons (17), and the chlorine atom has become an **anion** (AN-ī-on), a negatively charged ion. The ionic form of chlorine is called a *chloride* ion. It has a charge of 1− and is written Cl^-. When an atom of sodium donates its sole valence electron to an atom of chlorine, the resulting positive and negative charges pull both ions tightly together, forming an ionic bond (Figure 2.4c). The resulting compound is sodium chloride, written NaCl.

In general, ionic compounds exist as solids, with an orderly, repeating arrangement of the ions, as in a crystal of NaCl (Figure 2.4d). A crystal of NaCl may be large or small—the total number of ions can vary—but the ratio of Na^+ to Cl^- is always 1:1. In the body, ionic bonds are found mainly in teeth and bones,

where they give great strength to these important structural tissues. An ionic compound that breaks apart into positive and negative ions in solution is called an **electrolyte** (e-LEK-trō-līt). Most ions in the body are dissolved in body fluids as electrolytes, so named because their solutions can conduct an electric current. (In Chapter 27 we will discuss the chemistry and importance of electrolytes.) Table 2.2 lists the names and symbols of common ions in the body.

TABLE 2.2

Common Ions in the Body

CATIONS		ANIONS	
NAME	SYMBOL	NAME	SYMBOL
Hydrogen ion	H^+	Fluoride ion	F^-
Sodium ion	Na^+	Chloride ion	Cl^-
Potassium ion	K^+	Iodide ion	I^-
Ammonium ion	NH_4^+	Hydroxide ion	OH^-
Magnesium ion	Mg^{2+}	Bicarbonate ion	HCO_3^-
Calcium ion	Ca^{2+}	Oxide ion	O^{2-}
Iron(II) ion	Fe^{2+}	Sulfate ion	SO_4^{2-}
Iron(III) ion	Fe^{3+}	Phosphate ion	PO_4^{3-}

Figure 2.4 Ions and ionic bond formation. (a) A sodium atom can have a complete octet of electrons in its outermost shell by losing one electron. (b) A chlorine atom can have a complete octet by gaining one electron. (c) An ionic bond may form between oppositely charged ions. (d) In a crystal of NaCl, each Na^+ is surrounded by six Cl^-. In (a), (b), and (c), the electron that is lost or accepted is colored red.

An ionic bond is the force of attraction that holds together oppositely charged ions.

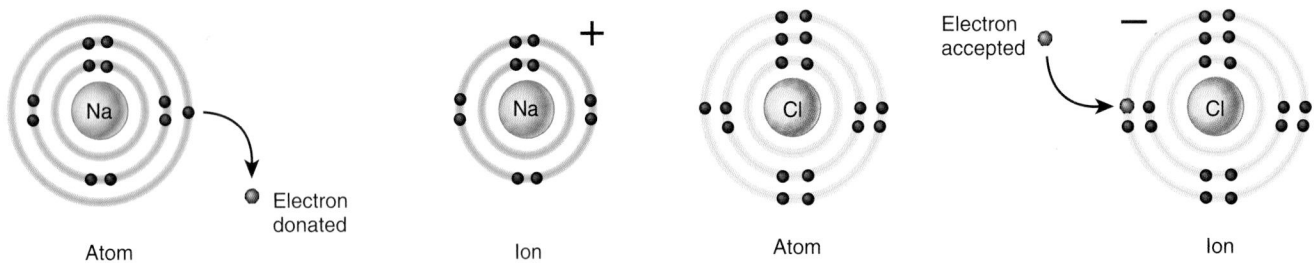

(a) Sodium: 1 valence electron

(b) Chlorine: 7 valence electrons

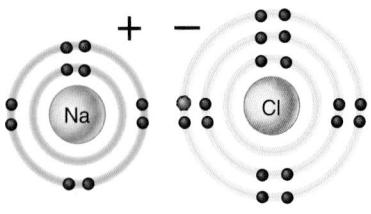

(c) Ionic bond in sodium chloride (NaCl)

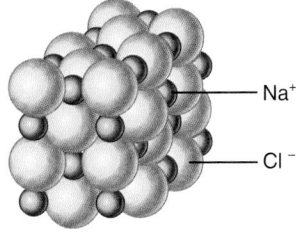

(d) Packing of ions in a crystal of sodium chloride

? **What are cations and anions?**

Covalent Bonds

When a **covalent bond** forms, two or more atoms *share* electrons rather than gaining or losing them. Atoms form a covalently bonded molecule by sharing one, two, or three pairs of valence electrons. The larger the number of electron pairs shared between two atoms, the stronger the covalent bond. Covalent bonds may form between atoms of the same element or between atoms of different elements. They are the most common chemical bonds in the body, and the compounds that result from them form most of the body's structures.

A **single covalent bond** results when two atoms share one electron pair. For example, a molecule of hydrogen forms when two hydrogen atoms share their single valence electrons (Figure 2.5a), which allows both atoms to have a full valence shell at

Figure 2.5 Covalent bond formation. The red electrons are shared equally in (a)–(d) and unequally in (e). In writing the structural formula of a covalently bonded molecule, each straight line between the chemical symbols for two atoms denotes a pair of shared electrons. In molecular formulas, the number of atoms in each molecule is noted by subscripts.

🔑 In a covalent bond, two atoms share one, two, or three pairs of valence electrons.

What is the principal difference between an ionic bond and a covalent bond?

least part of the time. A **double covalent bond** results when two atoms share two pairs of electrons, as happens in an oxygen molecule (Figure 2.5b). A **triple covalent bond** occurs when two atoms share three pairs of electrons, as in a molecule of nitrogen (Figure 2.5c). Notice in the *structural formulas* for covalently bonded molecules in Figure 2.5 that the number of lines between the chemical symbols for two atoms indicates whether the bond is a single (—), double (=), or triple (≡) covalent bond.

The same principles of covalent bonding that apply to atoms of the same element also apply to covalent bonds between atoms of different elements. The gas methane (CH_4) contains covalent bonds formed between the atoms of two different elements, one carbon and four hydrogens (Figure 2.5d). The valence shell of the carbon atom can hold eight electrons but has only four of its own. The single electron shell of a hydrogen atom can hold two electrons, but each hydrogen atom has only one of its own. A methane molecule contains four separate single covalent bonds. Each hydrogen atom shares one pair of electrons with the carbon atom.

In some covalent bonds, two atoms share the electrons equally—one atom does not attract the shared electrons more strongly than the other atom. This type of bond is a **nonpolar covalent bond.** The bonds between two identical atoms are always nonpolar covalent bonds (Figure 2.5a–c). The bonds between carbon and hydrogen atoms are also nonpolar, such as the four C—H bonds in a methane molecule (Figure 2.5d).

In a **polar covalent bond,** the sharing of electrons between two atoms is unequal—the nucleus of one atom attracts the shared electrons more strongly than the nucleus of the other atom. When polar covalent bonds form, the resulting molecule has a partial negative charge near the atom that attracts electrons more strongly. This atom has greater **electronegativity,** the power to attract electrons to itself. At least one other atom in the molecule then will have a partial positive charge. The partial charges are indicated by a lowercase Greek delta with a minus or plus sign: δ^- or δ^+. A very important example of a polar covalent bond in living systems is the bond between oxygen and hydrogen in a molecule of water (Figure 2.5e); in this molecule, the nucleus of the oxygen atom attracts the electrons more strongly than do the nuclei of the hydrogen atoms, so the oxygen atom is said to have greater electronegativity. Later in the chapter, we will see how polar covalent bonds allow water to dissolve many molecules that are important to life. Bonds between nitrogen and hydrogen and those between oxygen and carbon are also polar bonds.

Hydrogen Bonds

The polar covalent bonds that form between hydrogen atoms and other atoms can give rise to a third type of chemical bond, a hydrogen bond (Figure 2.6). A **hydrogen bond** forms when a hydrogen atom with a partial positive charge (δ^+) attracts the partial negative charge (δ^-) of neighboring electronegative atoms, most often larger oxygen or nitrogen atoms. Thus, hydrogen bonds result from attraction of oppositely charged parts of molecules

rather than from sharing of electrons as in covalent bonds, or the loss or gain of electrons as in ionic bonds. Hydrogen bonds are weak compared to ionic and covalent bonds. Thus, they cannot bind atoms into molecules. However, hydrogen bonds do establish important links between molecules or between different parts of a large molecule, such as a protein or nucleic acid (both discussed later in this chapter).

The hydrogen bonds that link neighboring water molecules give water considerable *cohesion,* the tendency of like particles to stay together. The cohesion of water molecules creates a very high **surface tension,** a measure of the difficulty of stretching or breaking the surface of a liquid. At the boundary between water and air, water's surface tension is very high because the water molecules are much more attracted to one another than they are attracted to molecules in the air. This is readily seen when a spider walks on water or a leaf floats on water. The influence of water's surface tension on the body can be seen in the way it increases the work required for breathing. A thin film of watery fluid coats the air sacs of the lungs. So, each inhalation must have enough force to overcome the opposing effect of surface tension as the air sacs stretch and enlarge when taking in air.

Even though single hydrogen bonds are weak, very large molecules may contain thousands of these bonds. Acting collectively, hydrogen bonds provide considerable strength and stability and help determine the three-dimensional shape of large molecules. As you will see later in this chapter, a large molecule's shape determines how it functions.

Figure 2.6 Hydrogen bonding among water molecules. Each water molecule forms hydrogen bonds (indicated by dotted lines) with three to four neighboring water molecules.

Hydrogen bonds occur because hydrogen atoms in one water molecule are attracted to the partial negative charge of the oxygen atom in another water molecule.

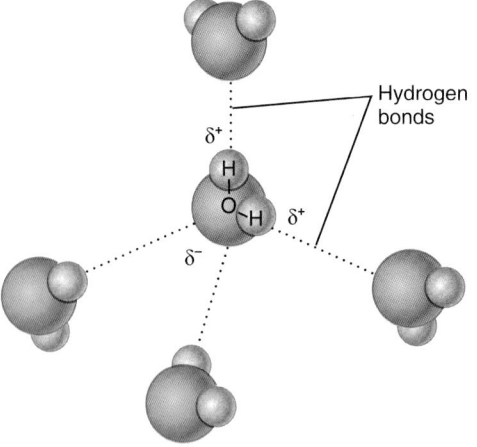

? **Why would you expect ammonia (NH_3) to form hydrogen bonds with water molecules?**

✔CHECKPOINT

4. Which electron shell is the valence shell of an atom, and what is its significance?
5. Compare the properties of ionic, covalent, and hydrogen bonds.
6. What information is conveyed when you write the molecular or structural formula for a molecule?

2.3 CHEMICAL REACTIONS

◉ OBJECTIVES

• Define a chemical reaction.
• Describe the various forms of energy.
• Compare exergonic and endergonic chemical reactions.
• Describe the role of activation energy and catalysts in chemical reactions.
• Describe synthesis, decomposition, exchange, and reversible reactions.

A **chemical reaction** occurs when new bonds form or old bonds break between atoms. Chemical reactions are the foundation of all life processes, and as we have seen, the interactions of valence electrons are the basis of all chemical reactions. Consider how hydrogen and oxygen molecules react to form water molecules (Figure 2.7). The starting substances—two H_2 and one O_2—are known as the **reactants.** The ending substances—two molecules of H_2O—are the **products.** The arrow in the figure indicates the direction in which the reaction proceeds. In a chemical reaction, the total mass of the reactants equals the total mass of the products. Thus, the number of atoms of each element is the same before and after the reaction. However, because the atoms are rearranged, the reactants and products have different chemical properties. Through thousands of different chemical reactions, body structures are built and body functions are carried out. The

Figure 2.7 The chemical reaction between two hydrogen molecules (H_2) and one oxygen molecule (O_2) to form two molecules of water (H_2O). Note that the reaction occurs by breaking old bonds and making new bonds.

 The number of atoms of each element is the same before and after a chemical reaction.

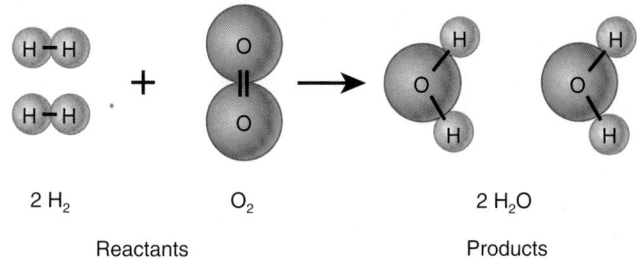

Reactants   Products

❓ **Why does this reaction require two molecules of H_2?**

term **metabolism** refers to all the chemical reactions occurring in the body.

Forms of Energy and Chemical Reactions

Each chemical reaction involves energy changes. **Energy** (*en-* = in; *-ergy* = work) is the capacity to do work. Two principal forms of energy are **potential energy,** energy stored by matter due to its position, and **kinetic energy,** the energy associated with matter in motion. For example, the energy stored in water behind a dam or in a person poised to jump down some steps is potential energy. When the gates of the dam are opened or the person jumps, potential energy is converted into kinetic energy. **Chemical energy** is a form of potential energy that is stored in the bonds of compounds and molecules. The total amount of energy present at the beginning and end of a chemical reaction is the same. Although energy can be neither created nor destroyed, it may be converted from one form to another. This principle is known as the **law of conservation of energy.** For example, some of the chemical energy in the foods we eat is eventually converted into various forms of kinetic energy, such as mechanical energy used to walk and talk. Conversion of energy from one form to another generally releases heat, some of which is used to maintain normal body temperature.

Energy Transfer in Chemical Reactions

Chemical bonds represent stored chemical energy, and chemical reactions occur when new bonds are formed or old bonds are broken between atoms. The *overall reaction* may either release energy or absorb energy. **Exergonic reactions** (*ex-* = out) release more energy than they absorb. By contrast, **endergonic reactions** (*end-* = within) absorb more energy than they release.

A key feature of the body's metabolism is the coupling of exergonic reactions and endergonic reactions. Energy released from an exergonic reaction often is used to drive an endergonic one. In general, exergonic reactions occur as nutrients, such as glucose, are broken down. Some of the energy released may be trapped in the covalent bonds of adenosine triphosphate (ATP), which we describe more fully later in this chapter. If a molecule of glucose is completely broken down, the chemical energy in its bonds can be used to produce as many as 38 molecules of ATP. The energy transferred to the ATP molecules is then used to drive endergonic reactions needed to build body structures, such as muscles and bones. The energy in ATP is also used to do the mechanical work involved in the contraction of muscle or the movement of substances into or out of cells.

Activation Energy

Because particles of matter such as atoms, ions, and molecules have kinetic energy, they are continuously moving and colliding with one another. A sufficiently forceful collision can disrupt the movement of valence electrons, causing an existing chemical bond to break or a new one to form. The collision energy needed to break the chemical bonds of the reactants is called the

activation energy of the reaction (Figure 2.8). This initial energy "investment" is needed to start a reaction. The reactants must absorb enough energy for their chemical bonds to become unstable and their valence electrons to form new combinations. Then, as new bonds form, energy is released to the surroundings.

Both the concentration of particles and the temperature influence the chance that a collision will occur and cause a chemical reaction.

- *Concentration.* The more particles of matter present in a confined space, the greater the chance that they will collide (think of people crowding into a subway car at rush hour). The concentration of particles increases when more are added to a given space or when the pressure on the space increases, which forces the particles closer together so that they collide more often.

- *Temperature.* As temperature rises, particles of matter move about more rapidly. Thus, the higher the temperature of matter, the more forcefully particles will collide, and the greater the chance that a collision will produce a reaction.

Catalysts

As we have seen, chemical reactions occur when chemical bonds break or form after atoms, ions, or molecules collide with one another. Body temperature and the concentrations of molecules in body fluids, however, are far too low for most chemical reactions to occur rapidly enough to maintain life. Raising the temperature

Figure 2.9 Comparison of energy needed for a chemical reaction to proceed with a catalyst (blue curve) and without a catalyst (red curve).

Catalysts speed up chemical reactions by lowering the activation energy.

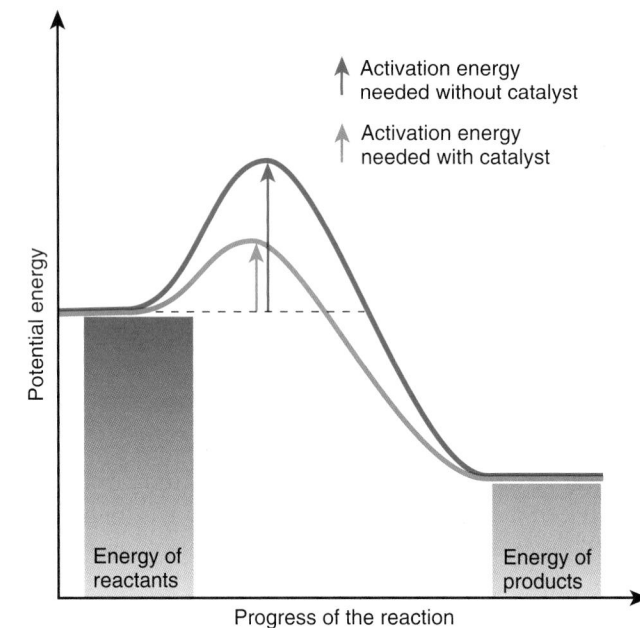

? Does a catalyst change the potential energies of the products and reactants?

and the number of reacting particles of matter in the body could increase the frequency of collisions and thus increase the rate of chemical reactions, but doing so could also damage or kill the body's cells.

Substances called catalysts solve this problem. **Catalysts** are chemical compounds that speed up chemical reactions by lowering the activation energy needed for a reaction to occur (Figure 2.9). The most important catalysts in the body are enzymes, which we will discuss later in this chapter.

A catalyst does not alter the difference in potential energy between the reactants and the products. Rather, it lowers the amount of energy needed to start the reaction.

For chemical reactions to occur, some particles of matter—especially large molecules—not only must collide with sufficient force, but they must hit one another at precise spots. A catalyst helps to properly orient the colliding particles. Thus, they interact at the spots that make the reaction happen. Although the action of a catalyst helps to speed up a chemical reaction, the catalyst itself is unchanged at the end of the reaction. A single catalyst molecule can assist one chemical reaction after another.

Types of Chemical Reactions

After a chemical reaction takes place, the atoms of the reactants are rearranged to yield products with new chemical properties.

Figure 2.8 Activation energy.

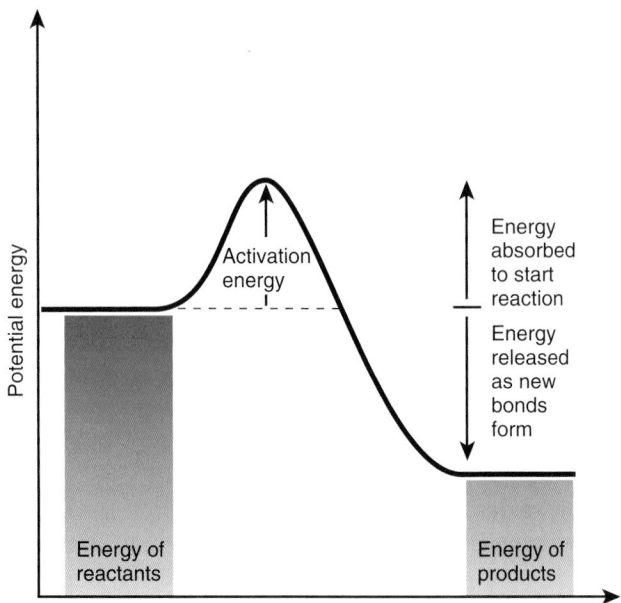

Activation energy is the energy needed to break chemical bonds in the reactant molecules so a reaction can start.

? Why is the reaction illustrated here exergonic?

In this section we will look at the types of chemical reactions common to all living cells. Once you have learned them, you will be able to understand the chemical reactions so important to the operation of the human body that are discussed throughout the book.

Synthesis Reactions—Anabolism

When two or more atoms, ions, or molecules combine to form new and larger molecules, the processes are called **synthesis reactions.** The word *synthesis* means "to put together." A synthesis reaction can be expressed as follows:

$$A + B \xrightarrow{\text{Combine to form}} AB$$

Atom, ion, or molecule A   Atom, ion, or molecule B   New molecule AB

One example of a synthesis reaction is the reaction between two hydrogen molecules and one oxygen molecule to form two molecules of water (see Figure 2.7). Another example of a synthesis reaction is the formation of ammonia from nitrogen and hydrogen:

$$N_2 + 3H_2 \xrightarrow{\text{Combine to form}} 2NH_3$$

One nitrogen molecule   Three hydrogen molecules   Two ammonia molecules

All the synthesis reactions that occur in your body are collectively referred to as **anabolism** (a-NAB-ō-lizm). Overall, anabolic reactions are usually endergonic because they absorb more energy than they release. Combining simple molecules like amino acids (discussed shortly) to form large molecules such as proteins is an example of anabolism.

Decomposition Reactions—Catabolism

Decomposition reactions split up large molecules into smaller atoms, ions, or molecules. A decomposition reaction is expressed as follows:

$$AB \xrightarrow{\text{Breaks down into}} A + B$$

Molecule AB   Atom, ion, or molecule A   Atom, ion, or molecule B

The decomposition reactions that occur in your body are collectively referred to as **catabolism** (ka-TAB-ō-lizm). Overall, catabolic reactions are usually exergonic because they release more energy than they absorb. For instance, the series of reactions that break down glucose to pyruvic acid, with the net production of two molecules of ATP, are important catabolic reactions in the body. These reactions will be discussed in Chapter 25.

Exchange Reactions

Many reactions in the body are **exchange reactions;** they consist of both synthesis and decomposition reactions. One type of exchange reaction works like this:

$$AB + CD \longrightarrow AD + BC$$

The bonds between A and B and between C and D break (decomposition), and new bonds then form (synthesis) between A and D and between B and C. An example of an exchange reaction is

$$HCl + NaHCO_3 \longrightarrow H_2CO_3 + NaCl$$

Hydrochloric acid   Sodium bicarbonate   Carbonic acid   Sodium chloride

Notice that the ions in both compounds have "switched partners": The hydrogen ion (H^+) from HCl has combined with the bicarbonate ion (HCO_3^-) from $NaHCO_3$, and the sodium ion (Na^+) from $NaHCO_3$ has combined with the chloride ion (Cl^-) from HCl.

Reversible Reactions

Some chemical reactions proceed in only one direction, from reactants to products, as previously indicated by the single arrows. Other chemical reactions may be reversible. In a **reversible reaction,** the products can revert to the original reactants. A reversible reaction is indicated by two half-arrows pointing in opposite directions:

$$AB \underset{\text{Combines to form}}{\overset{\text{Breaks down into}}{\rightleftharpoons}} A + B$$

Some reactions are reversible only under special conditions:

$$AB \underset{\text{Heat}}{\overset{\text{Water}}{\rightleftharpoons}} A + B$$

In that case, whatever is written above or below the arrows indicates the condition needed for the reaction to occur. In these reactions, AB breaks down into A and B only when water is added, and A and B react to produce AB only when heat is applied. Many reversible reactions in the body require catalysts called enzymes. Often, different enzymes guide the reactions in opposite directions.

Oxidation–Reduction Reactions

You will learn in Chapter 25 that chemical reactions called oxidation–reduction reactions are essential to life, since they are the reactions that break down food molecules to produce energy. These reactions are concerned with the transfer of electrons between atoms and molecules. **Oxidation** refers to the loss of electrons, and in the process the oxidized substance releases energy. **Reduction** refers to the gain of electrons, and in the process the reduced substance gains energy. **Oxidation–reduction reactions** are always parallel; when one substance is oxidized, another is reduced at the same time. When a food molecule, such as glucose, is oxidized, the energy produced is used by a cell to carry out its various functions.

✔CHECKPOINT

7. What is the relationship between reactants and products in a chemical reaction?
8. Compare potential energy and kinetic energy.
9. How do catalysts affect activation energy?
10. How are anabolism and catabolism related to synthesis and decomposition reactions, respectively?
11. Why are oxidation–reduction reactions important?

2.4 INORGANIC COMPOUNDS AND SOLUTIONS

◉ OBJECTIVES

- Describe the properties of water and those of inorganic acids, bases, and salts.
- Distinguish among solutions, colloids, and suspensions.
- Define pH and explain the role of buffer systems in homeostasis.

Most of the chemicals in your body exist in the form of compounds. Biologists and chemists divide these compounds into two principal classes: inorganic compounds and organic compounds. **Inorganic compounds** usually lack carbon and are structurally simple. Their molecules also have only a few atoms and cannot be used by cells to perform complicated biological functions. They include water and many salts, acids, and bases. Inorganic compounds may have either ionic or covalent bonds. Water makes up 55–60% of a lean adult's total body mass; all other inorganic compounds combined add 1–2%. Examples of inorganic compounds that contain carbon are carbon dioxide (CO_2), bicarbonate ion (HCO_3^-), and carbonic acid (H_2CO_3). **Organic compounds** always contain carbon, usually contain hydrogen, and always have covalent bonds. Most are large molecules and many are made up of long chains of carbon atoms. Organic compounds make up the remaining 38–43% of the human body.

Water

Water is the most important and abundant inorganic compound in all living systems. Although you might be able to survive for weeks without food, without water you would die in a matter of days. Nearly all the body's chemical reactions occur in a watery medium. Water has many properties that make it such an indispensable compound for life. We have already mentioned the most important property of water, its polarity—the uneven sharing of valence electrons that confers a partial negative charge near the one oxygen atom and two partial positive charges near the two hydrogen atoms in a water molecule (see Figure 2.5e). This property alone makes water an excellent solvent for other ionic or polar substances, gives water molecules cohesion (the tendency to stick together), and allows water to resist temperature changes.

Water as a Solvent

In medieval times people searched in vain for a "universal solvent," a substance that would dissolve all other materials. They found nothing that worked as well as water. Although it is the most versatile solvent known, water is not the universal solvent sought by medieval alchemists. If it were, no container could hold it because it would dissolve all potential containers! What exactly is a solvent? In a **solution,** a substance called the **solvent** dissolves another substance called the **solute.** Usually there is more solvent than solute in a solution. For example, your sweat is a di-

lute solution of water (the solvent) plus small amounts of salts (the solutes).

The versatility of water as a solvent for ionized or polar substances is due to its polar covalent bonds and its bent shape, which allows each water molecule to interact with several neighboring ions or molecules. Solutes that are charged or contain polar covalent bonds are **hydrophilic** (*hydro-* = water; *-philic* = loving), which means they dissolve easily in water. Common examples of hydrophilic solutes are sugar and salt. Molecules that contain mainly nonpolar covalent bonds, by contrast, are **hydrophobic** (*-phobic* = fearing). They are not very water soluble. Examples of hydrophobic compounds include animal fats and vegetable oils.

To understand the dissolving power of water, consider what happens when a crystal of a salt such as sodium chloride (NaCl) is placed in water (Figure 2.10). The electronegative oxygen

Figure 2.10 How polar water molecules dissolve salts and polar substances. When a crystal of sodium chloride is placed in water, the slightly negative oxygen end (red) of water molecules is attracted to the positive sodium ions (Na^+), and the slightly positive hydrogen portions (gray) of water molecules are attracted to the negative chloride ions (Cl^-). In addition to dissolving sodium chloride, water also causes it to dissociate, or separate into charged particles, which is discussed shortly.

 Water is a versatile solvent because its polar covalent bonds, in which electrons are shared unequally, create positive and negative regions.

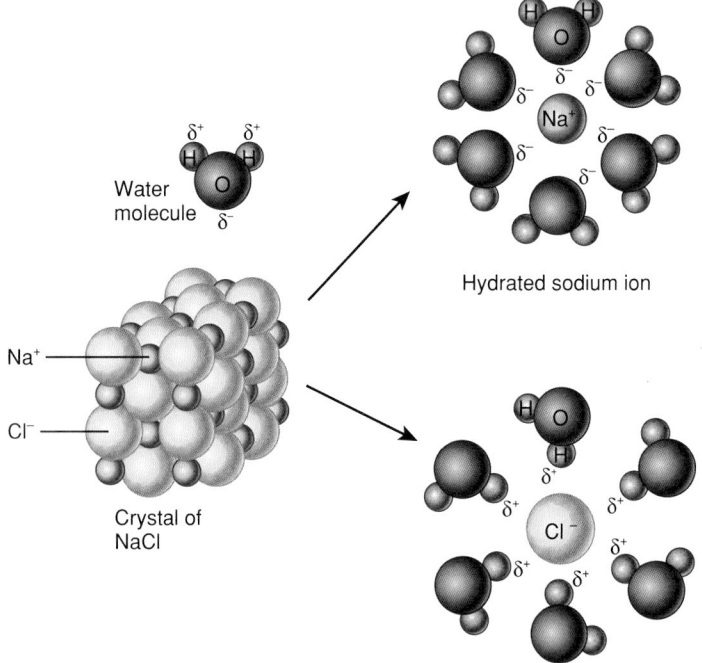

Hydrated sodium ion

Hydrated chloride ion

? Table sugar (sucrose) easily dissolves in water but is not an electrolyte. Is it likely that all the covalent bonds between atoms in table sugar are nonpolar bonds? Why or why not?

atom in water molecules attracts the sodium ions (Na^+), and the electropositive hydrogen atoms in water molecules attract the chloride ions (Cl^-). Soon, water molecules surround and separate Na^+ and Cl^- ions from each other at the surface of the crystal, breaking the ionic bonds that held NaCl together. The water molecules surrounding the ions also lessen the chance that Na^+ and Cl^- will come together and re-form an ionic bond.

The ability of water to form solutions is essential to health and survival. Because water can dissolve so many different substances, it is an ideal medium for metabolic reactions. Water enables dissolved reactants to collide and form products. Water also dissolves waste products, which allows them to be flushed out of the body in the urine.

Water in Chemical Reactions

Water serves as the medium for most chemical reactions in the body and participates as a reactant or product in certain reactions. During digestion, for example, decomposition reactions break down large nutrient molecules into smaller molecules by the addition of water molecules. This type of reaction is called **hydrolysis** (hī-DROL-i-sis; *-lysis* = to loosen or break apart). Hydrolysis reactions enable dietary nutrients to be absorbed into the body. By contrast, when two smaller molecules join to form a larger molecule in a **dehydration synthesis reaction** (*de-* = from, down, or out; *hydra-* = water), a water molecule is one of the products formed. As you will see later in the chapter, such reactions occur during synthesis of proteins and other large molecules (for example, see Figure 2.21).

Thermal Properties of Water

In comparison to most substances, water can absorb or release a relatively large amount of heat with only a modest change in its own temperature. For this reason, water is said to have a high *heat capacity*. The reason for this property is the large number of hydrogen bonds in water. As water absorbs heat energy, some of the energy is used to break hydrogen bonds. Less energy is then left over to increase the motion of water molecules, which would increase the water's temperature. The high heat capacity of water is the reason it is used in automobile radiators; it cools the engine by absorbing heat without its own temperature rising to an unacceptably high level. The large amount of water in the body has a similar effect: It lessens the impact of environmental temperature changes, helping to maintain the homeostasis of body temperature.

Water also requires a large amount of heat to change from a liquid to a gas. Its *heat of vaporization* is high. As water evaporates from the surface of the skin, it removes a large quantity of heat, providing an important cooling mechanism.

Water as a Lubricant

Water is a major component of mucus and other lubricating fluids throughout the body. Lubrication is especially necessary in the chest (pleural and pericardial cavities) and abdomen (peritoneal cavity), where internal organs touch and slide over one another. It is also needed at joints, where bones, ligaments, and tendons rub against one another. Inside the gastrointestinal tract, mucus and other watery secretions moisten foods, which aids their smooth passage through the digestive system.

Solutions, Colloids, and Suspensions

A **mixture** is a combination of elements or compounds that are physically blended together but not bound by chemical bonds. For example, the air you are breathing is a mixture of gases that includes nitrogen, oxygen, argon, and carbon dioxide. Three common liquid mixtures are solutions, colloids, and suspensions.

Once mixed together, solutes in a solution remain evenly dispersed among the solvent molecules. Because the solute particles in a solution are very small, a solution looks clear and transparent.

A **colloid** differs from a solution mainly because of the size of its particles. The solute particles in a colloid are large enough to scatter light, just as water droplets in fog scatter light from a car's headlight beams. For this reason, colloids usually appear translucent or opaque. Milk is an example of a liquid that is both a colloid and a solution: The large milk proteins make it a colloid, whereas calcium salts, milk sugar (lactose), ions, and other small particles are in solution.

The solutes in both solutions and colloids do not settle out and accumulate on the bottom of the container. In a **suspension,** by contrast, the suspended material may mix with the liquid or suspending medium for some time, but eventually it will settle out. Blood is an example of a suspension. When freshly drawn from the body, blood has an even, reddish color. After blood sits for a while in a test tube, red blood cells settle out of the suspension and drift to the bottom of the tube (see Figure 19.1a). The upper layer, the liquid portion of blood, appears pale yellow and is called blood plasma. Blood plasma is both a solution of ions and other small solutes and a colloid due to the presence of larger plasma proteins.

The **concentration** of a solution may be expressed in several ways. One common way is by a mass per volume **percentage,** which gives the relative mass of a solute found in a given volume of solution. For example, you may have seen the following on the label of a bottle of wine: "Alcohol 14.1% by volume." Another way expresses concentration in units of **moles per liter (mol/L),** which relate to the total number of molecules in a given volume of solution. A **mole** is the amount of any substance that has a mass in grams equal to the sum of the atomic masses of all its atoms. For example, 1 mole of the element chlorine (atomic mass = 35.45) is 35.45 grams and 1 mole of the salt sodium chloride (NaCl) is 58.44 grams (22.99 for Na + 35.45 for Cl). Just as a dozen always means 12 of something, a mole of anything has the same number of particles: 6.023×10^{23}. This huge number is called *Avogadro's number.* Thus, measurements of substances that are stated in moles tell us about the numbers of atoms, ions, or molecules present. This is important when chemical reactions are occurring because each reaction requires a set number of

TABLE 2.3

Percentage and Molarity

DEFINITION	EXAMPLE
Percentage (mass per volume) Number of grams of a substance per 100 milliliters (mL) of solution	To make a 10% NaCl solution, take 10 g of NaCl and add enough water to make a total of 100 mL of solution.
Molarity: 5 moles (mol) per liter A 1 molar (1 M) solution = 1 mole of a solute in 1 liter of solution	To make a 1 molar (1 M) solution of NaCl, dissolve 1 mole of NaCl (58.44 g) in enough water to make a total of 1 liter of solution.

atoms of specific elements. Table 2.3 describes these ways of expressing concentration.

Inorganic Acids, Bases, and Salts

When inorganic acids, bases, or salts dissolve in water, they **dissociate** (dis'-sō-sē-ĀT); that is, they separate into ions and become surrounded by water molecules. An **acid** (Figure 2.11a) is a substance that dissociates into one or more **hydrogen ions (H$^+$)** and one or more anions. Because H$^+$ is a single proton with one positive charge, an acid is also referred to as a **proton donor.** A **base,** by contrast (Figure 2.11b), removes H$^+$ from a solution and is therefore a **proton acceptor.** Many bases dissociate into one or more **hydroxide ions (OH$^-$)** and one or more cations.

A **salt,** when dissolved in water, dissociates into cations and anions, neither of which is H$^+$ or OH$^-$ (Figure 2.11c). In the body, salts such as potassium chloride are electrolytes that are im-

Figure 2.11 Dissociation of inorganic acids, bases, and salts.

 Dissociation is the separation of inorganic acids, bases, and salts into ions in a solution.

(a) Acid (b) Base (c) Salt

? The compound CaCO$_3$ (calcium carbonate) dissociates into a calcium ion (Ca^{2+}) and a carbonate ion (CO$_3^{2-}$). Is it an acid, a base, or a salt? What about H$_2$SO$_4$, which dissociates into two H$^+$ and one SO$_4^{2-}$?

portant for carrying electrical currents (ions flowing from one place to another), especially in nerve and muscular tissues. The ions of salts also provide many essential chemical elements in intracellular and extracellular fluids such as blood, lymph, and the interstitial fluid of tissues.

Acids and bases react with one another to form salts. For example, the reaction of hydrochloric acid (HCl) and potassium hydroxide (KOH), a base, produces the salt potassium chloride (KCl) and water (H$_2$O). This exchange reaction can be written as follows:

$$HCl + KOH \longrightarrow H^+ + Cl^- + K^+ + OH^- \longrightarrow KCl + H_2O$$
Acid Base Dissociated ions Salt Water

Acid–Base Balance: The Concept of pH

To ensure homeostasis, intracellular and extracellular fluids must contain almost balanced quantities of acids and bases. The more hydrogen ions (H$^+$) dissolved in a solution, the more acidic the solution; the more hydroxide ions (OH$^-$), the more basic (alkaline) the solution. The chemical reactions that take place in the body are very sensitive to even small changes in the acidity or alkalinity of the body fluids in which they occur. Any departure from the narrow limits of normal H$^+$ and OH$^-$ concentrations greatly disrupts body functions.

A solution's acidity or alkalinity is expressed on the **pH scale,** which extends from 0 to 14 (Figure 2.12). This scale is based on the concentration of H$^+$ in moles per liter. A pH of 7 means that a solution contains one ten-millionth (0.0000001) of a mole of hydrogen ions per liter. The number 0.0000001 is written as 1×10^{-7} in scientific notation, which indicates that the number is 1 with the decimal point moved seven places to the left. To convert this value to pH, the negative exponent (–7) is changed to a positive number (7). A solution with a H$^+$ concentration of 0.0001 (10^{-4}) moles per liter has a pH of 4; a solution with a H$^+$ concentration of 0.000000001 (10^{-9}) moles per liter has a pH of 9; and so on. It is important to realize that a change of one whole number on the pH scale represents a *tenfold* change in the number of H$^+$. A pH of 6 denotes 10 times more H$^+$ than a pH of 7, and a pH of 8 indicates 10 times fewer H$^+$ than a pH of 7 and 100 times fewer H$^+$ than a pH of 6.

The midpoint of the pH scale is 7, where the concentrations of H$^+$ and OH$^-$ are equal. A substance with a pH of 7, such as pure water, is neutral. A solution that has more H$^+$ than OH$^-$ is an **acidic solution** and has a pH below 7. A solution that has more OH$^-$ than H$^+$ is a **basic (alkaline) solution** and has a pH above 7.

Maintaining pH: Buffer Systems

Although the pH of body fluids may differ, as we have discussed, the normal limits for each fluid are quite narrow. Table 2.4 shows the pH values for certain body fluids along with those of some common substances outside the body. Homeostatic mechanisms maintain the pH of blood between 7.35 and 7.45, which is slightly more basic than pure water. You will learn in Chapter 27 that if the

Figure 2.12 The pH scale. A pH below 7 indicates an acidic solution—more H^+ than OH^-. A pH above 7 indicates a basic (alkaline) solution; that is, there are more OH^- than H^+.

 The lower the numerical value of the pH, the more acidic is the solution because the H^+ concentration becomes progressively greater. The higher the pH, the more basic the solution.

At pH 7 (neutrality), the concentrations of H^+ and OH^- are equal (10^{-7} mol/liter). **What are the concentrations of H^+ and OH^- at pH 6? Which pH is more acidic, 6.82 or 6.91? Which pH is closer to neutral, 8.41 or 5.59?**

TABLE 2.4

pH Values of Selected Substances

SUBSTANCE*	pH VALUE
• Gastric juice (found in the stomach)	1.2–3.0
Lemon juice	2.3
Vinegar	3.0
Carbonated soft drink	3.0–3.5
Orange juice	3.5
• Vaginal fluid	3.5–4.5
Tomato juice	4.2
Coffee	5.0
• Urine	4.6–8.0
• Saliva	6.35–6.85
Milk	6.8
Distilled (pure) water	7.0
• Blood	7.35–7.45
• Semen (fluid containing sperm)	7.20–7.60
• Cerebrospinal fluid (fluid associated with nervous system)	7.4
• Pancreatic juice (digestive juice of the pancreas)	7.1–8.2
• Bile (liver secretion that aids fat digestion)	7.6–8.6
Milk of magnesia	10.5
Lye (sodium hydroxide)	14.0

*Bullets (•) denote substances in the human body.

pH of blood falls below 7.35, a condition called *acidosis* occurs, and if the pH rises above 7.45, it results in a condition called *alkalosis;* both conditions can seriously compromise homeostasis. Saliva is slightly acidic, and semen is slightly basic. Because the kidneys help remove excess acid from the body, urine can be quite acidic.

Even though strong acids and bases are continually taken into and formed by the body, the pH of fluids inside and outside cells remains almost constant. One important reason is the presence of **buffer systems,** which function to convert strong acids or bases into weak acids or bases. Strong acids (or bases) ionize easily and contribute many H^+ (or OH^-) to a solution. Therefore, they can change pH drastically, which can disrupt the body's metabolism. Weak acids (or bases) do not ionize as much and contribute fewer H^+ (or OH^-). Hence, they have less effect on the pH. The chemical compounds that can convert strong acids or bases into weak ones are called **buffers.** They do so by removing or adding protons (H^+).

One important buffer system in the body is the **carbonic acid–bicarbonate buffer system.** Carbonic acid (H_2CO_3) can act as a weak acid, and the bicarbonate ion (HCO_3^-) can act as a weak base. Hence, this buffer system can compensate for either an excess or a shortage of H^+. For example, if there is an excess of H^+ (an acidic condition), HCO_3^- can function as a weak base and remove the excess H^+, as follows:

$$H^+ \quad + \quad HCO_3^- \quad \longrightarrow \quad H_2CO_3$$

Hydrogen ion   Bicarbonate ion (weak base)   Carbonic acid

If there is a shortage of H^+ (an alkaline condition), by contrast, H_2CO_3 can function as a weak acid and provide needed H^+ as follows:

$$H_2CO_3 \longrightarrow H^+ + HCO_3^-$$

Carbonic acid (weak acid)　　Hydrogen ion　　Bicarbonate ion

Chapter 27 describes buffers and their roles in maintaining acid–base balance in more detail.

✔ CHECKPOINT

12. How do inorganic compounds differ from organic compounds?
13. Describe two ways to express the concentration of a solution.
14. What functions does water perform in the body?
15. How do bicarbonate ions prevent buildup of excess H^+?

2.5 ORGANIC COMPOUNDS

◉ OBJECTIVES

• Describe the functional groups of organic molecules.
• Identify the building blocks and functions of carbohydrates, lipids, and proteins.
• Describe the structure and functions of deoxyribonucleic acid (DNA), ribonucleic acid (RNA), and adenosine triphosphate (ATP).

Many organic molecules are relatively large and have unique characteristics that allow them to carry out complex functions. Important categories of organic compounds include carbohydrates, lipids, proteins, nucleic acids, and adenosine triphosphate (ATP).

Carbon and Its Functional Groups

Carbon has several properties that make it particularly useful to living organisms. For one thing, it can form bonds with one to thousands of other carbon atoms to produce large molecules that can have many different shapes. Due to this property of carbon, the body can build many different organic compounds, each of which has a unique structure and function. Moreover, the large size of most carbon-containing molecules and the fact that some do not dissolve easily in water make them useful materials for building body structures.

Organic compounds are usually held together by covalent bonds. Carbon has four electrons in its outermost (valence) shell. It can bond covalently with a variety of atoms, including other carbon atoms, to form rings and straight or branched chains. Other elements that most often bond with carbon in organic compounds are hydrogen, oxygen, and nitrogen. Sulfur and phosphorus are also present in organic compounds. The other elements listed in Table 2.1 are present in a smaller number of organic compounds.

The chain of carbon atoms in an organic molecule is called the **carbon skeleton.** Many of the carbons are bonded to hydrogen

atoms, yielding a **hydrocarbon.** Also attached to the carbon skeleton are distinctive **functional groups,** other atoms or molecules bound to the hydrocarbon skeleton. Each type of functional group has a specific arrangement of atoms that confers characteristic chemical properties on the organic molecule attached to it. Table 2.5 lists the most common functional groups of organic molecules and describes some of their properties. Because organic molecules often are big, there are shorthand methods for representing

TABLE 2.5

Major Functional Groups of Organic Molecules

NAME AND STRUCTURAL FORMULA*	OCCURRENCE AND SIGNIFICANCE
Hydroxyl R—O—H	*Alcohols* contain an —OH group, which is polar and hydrophilic due to its electronegative O atom. Molecules with many —OH groups dissolve easily in water.
Sulfhydryl R—S—H	*Thiols* have an —SH group, which is polar and hydrophilic due to its electronegative S atom. Certain amino acids (for example, cysteine) contain —SH groups, which help stabilize the shape of proteins.
Carbonyl R—C(=O)—R or R—C(=O)—H	*Ketones* contain a carbonyl group within the carbon skeleton. The carbonyl group is polar and hydrophilic due to its electronegative O atom. *Aldehydes* have a carbonyl group at the end of the carbon skeleton.
Carboxyl R—C(=O)—OH or R—C(=O)—O⁻	*Carboxylic acids* contain a carboxyl group at the end of the carbon skeleton. All amino acids have a —COOH group at one end. The negatively charged form predominates at the pH of body cells and is hydrophilic.
Ester R—C(=O)—O—R	*Esters* predominate in dietary fats and oils and also occur in our body as triglycerides. Aspirin is an ester of salicylic acid, a pain-relieving molecule found in the bark of the willow tree.
Phosphate R—O—P(=O)(O⁻)—O⁻	*Phosphates* contain a phosphate group ($—PO_4^{2-}$), which is very hydrophilic due to the dual negative charges. An important example is adenosine triphosphate (ATP), which transfers chemical energy between organic molecules during chemical reactions.
Amino R—N(H)(H) or R—N⁺(H)(H)(H)	*Amines* have an —NH2 group, which can act as a base and pick up a hydrogen ion, giving the amino group a positive charge. At the pH of body fluids, most amino groups have a charge of 1+. All amino acids have an amino group at one end.

*R = variable group.

Figure 2.13 Alternative ways to write the structural formula for glucose.

 In standard shorthand, carbon atoms are understood to be at locations where two bond lines intersect, and single hydrogen atoms are not indicated.

| All atoms written out | Standard shorthand |

How many hydroxyl groups does a molecule of glucose have? How many carbon atoms are part of glucose's carbon skeleton?

their structural formulas. Figure 2.13 shows two ways to indicate the structure of the sugar glucose, a molecule with a ring-shaped carbon skeleton that has several hydroxyl groups attached.

Small organic molecules can combine into very large molecules that are called **macromolecules** (*macro-* = large). Macromolecules are usually **polymers** (*poly-* = many; *-mers* = parts). A polymer is a large molecule formed by the covalent bonding of many identical or similar small building-block molecules called **monomers** (*mono-* = one). Usually, the reaction that joins two monomers is a dehydration synthesis. In this type of reaction, a hydrogen atom is removed from one monomer and a hydroxyl group is removed from the other to form a molecule of water (see Figure 2.15a). Macromolecules such as carbohydrates, lipids, proteins, and nucleic acids are assembled in cells via dehydration synthesis reactions.

Molecules that have the same molecular formula but different structures are called **isomers** (Ī-so-merz; *iso-* = equal or the same). For example, the molecular formulas for the sugars glucose and fructose are both $C_6H_{12}O_6$. The individual atoms, however, are positioned differently along the carbon skeleton (see Figure 2.15a), giving the sugars different chemical properties.

Carbohydrates

Carbohydrates include sugars, glycogen, starches, and cellulose. Even though they are a large and diverse group of organic compounds and have several functions, carbohydrates represent only 2–3% of your total body mass. In humans and animals, carbohydrates function mainly as a source of chemical energy for generating ATP needed to drive metabolic reactions. Only a few carbohydrates are used for building structural units. One example is deoxyribose, a type of sugar that is a building block of deoxyribonucleic acid (DNA), the molecule that carries inherited genetic information.

Carbon, hydrogen, and oxygen are the elements found in carbohydrates. The ratio of hydrogen to oxygen atoms is usually

TABLE 2.6

Major Carbohydrate Groups

TYPE OF CARBOHYDRATE	EXAMPLES
Monosaccharides (simple sugars that contain from 3 to 7 carbon atoms)	Glucose (the main blood sugar). Fructose (found in fruits). Galactose (in milk sugar). Deoxyribose (in DNA). Ribose (in RNA).
Disaccharides (simple sugars formed from the combination of two monosaccharides by dehydration synthesis)	Sucrose (table sugar) = glucose + fructose. Lactose (milk sugar) = glucose + galactose. Maltose = glucose + glucose.
Polysaccharides (from tens to hundreds of monosaccharides joined by dehydration synthesis)	Glycogen (stored form of carbohydrates in animals). Starch (stored form of carbohydrates in plants and main carbohydrates in food). Cellulose (part of cell walls in plants that cannot be digested by humans but aids movement of food through intestines).

2:1, the same as in water. Although there are exceptions, carbohydrates generally contain one water molecule for each carbon atom. This is the reason they are called carbohydrates, which means "watered carbon." The three major groups of carbohydrates, based on their sizes, are monosaccharides, disaccharides, and polysaccharides (Table 2.6).

Monosaccharides and Disaccharides: The Simple Sugars
Monosaccharides and disaccharides are known as **simple sugars.** The monomers of carbohydrates, **monosaccharides** (mon′-ō-SAK-a-rīds; *sacchar-* = sugar), contain from three to seven carbon atoms. They are designated by names ending in "-ose" with a prefix that indicates the number of carbon atoms. For example, monosaccharides with three carbons are called *trioses* (*tri-* = three). There are also *tetroses* (four-carbon sugars), *pentoses* (five-carbon sugars), *hexoses* (six-carbon sugars), and *heptoses* (seven-carbon sugars). Examples of pentoses and hexoses are illustrated in Figure 2.14. Cells throughout the body break down the hexose glucose to produce ATP.

A **disaccharide** (dī-SAK-a-rīd; *di-* = two) is a molecule formed from the combination of two monosaccharides by dehydration synthesis (Figure 2.15). For example, molecules of the monosaccharides glucose and fructose combine to form a molecule of the disaccharide sucrose (table sugar), as shown in Figure 2.15a. Glucose and fructose are isomers. As you learned earlier in the chapter, isomers have the same molecular formula, but the relative positions of the oxygen and carbon atoms are different, causing the compounds to have different chemical properties. Notice that the formula for sucrose is $C_{12}H_{22}O_{11}$, not $C_{12}H_{24}O_{12}$, because a molecule of water is removed as the two monosaccharides are joined.

reactions. Unlike simple sugars, polysaccharides usually are insoluble in water and do not taste sweet. The main polysaccharide in the human body is **glycogen,** which is made entirely of glucose monomers linked to one another in branching chains (Figure 2.16). A limited amount of carbohydrates is stored as glycogen in the liver and skeletal muscles. **Starches** are polysaccharides formed from glucose by plants. They are found in foods such as pasta and potatoes and are the major carbohydrates in the diet. Like disaccharides, polysaccharides such as glycogen and starches can be broken down into monosaccharides through hydrolysis reactions. For example, when the blood glucose level falls, liver cells break down glycogen into glucose and release it into the blood, making it available to body cells, which break it down to synthesize ATP. **Cellulose** is a polysaccharide formed from glucose by plants that cannot be digested by humans but does provide bulk to help eliminate feces.

Lipids

A second important group of organic compounds is **lipids** (*lip-* = fat). Lipids make up 18–25% of body mass in lean adults. Like carbohydrates, lipids contain carbon, hydrogen, and oxygen. Unlike carbohydrates, they do not have a 2:1 ratio of hydrogen to oxygen. The proportion of electronegative oxygen atoms in lipids is usually smaller than in carbohydrates, so there are fewer polar covalent bonds. As a result, most lipids are insoluble in polar solvents such as water; they are *hydrophobic*. Because they are hydrophobic, only the smallest lipids (some fatty acids) can dissolve in watery blood plasma. To become more soluble in blood plasma, other lipid molecules join with hydrophilic protein molecules. The resulting lipid/protein complexes are termed **lipoproteins.** Lipoproteins are soluble because the proteins are on the outside and the lipids are on the inside.

The diverse lipid family includes fatty acids, triglycerides (fats and oils), phospholipids (lipids that contain phosphorus), steroids (lipids that contain rings of carbon atoms), eicosanoids (20-carbon lipids), and a variety of other lipids, including fat-soluble vitamins (vitamins A, D, E, and K) and lipoproteins. Table 2.7 introduces the various types of lipids and highlights their roles in the human body.

Figure 2.16 Part of a glycogen molecule, the main polysaccharide in the human body.

🔑 Glycogen is made up of glucose monomers and is the stored form of carbohydrate in the human body.

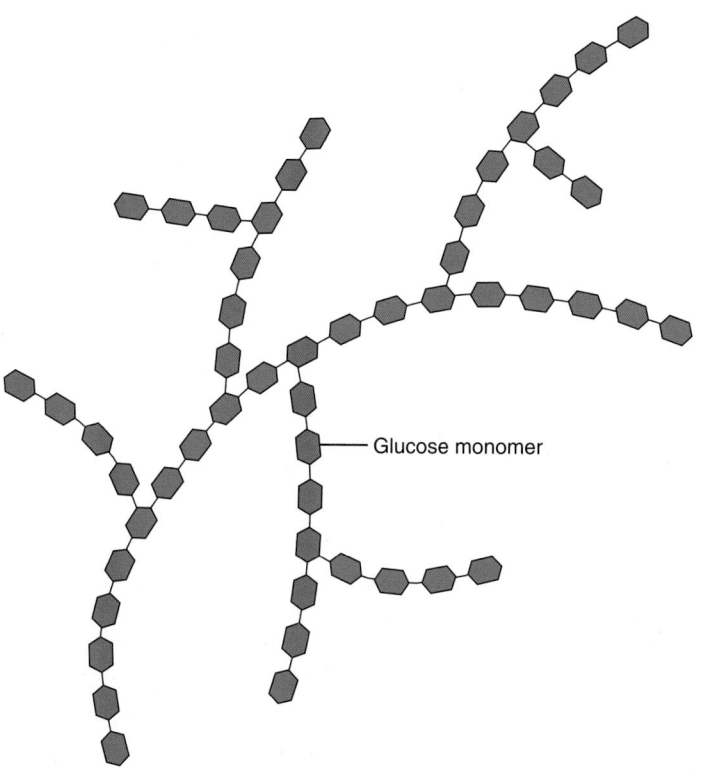

Glucose monomer

❓ **Which body cells store glycogen?**

TABLE 2.7

Types of Lipids in the Body

TYPE OF LIPID	FUNCTIONS
Fatty acids	Used to synthesize triglycerides and phospholipids or catabolized to generate adenosine triphosphate (ATP).
Triglycerides (fats and oils)	Protection, insulation, energy storage.
Phospholipids	Major lipid component of cell membranes.
Steroids	
Cholesterol	Minor component of all animal cell membranes; precursor of bile salts, vitamin D, and steroid hormones.
Bile salts	Needed for digestion and absorption of dietary lipids.
Vitamin D	Helps regulate calcium level in body; needed for bone growth and repair.
Adrenocortical hormones	Help regulate metabolism, resistance to stress, and salt and water balance.
Sex hormones	Stimulate reproductive functions and sexual characteristics.
Eicosanoids (prostaglandins and leukotrienes)	Have diverse effects on modifying responses to hormones, blood clotting, inflammation, immunity, stomach acid secretion, airway diameter, lipid breakdown, and smooth muscle contraction.
Other lipids	
Carotenes	Needed for synthesis of vitamin A (used to make visual pigments in eye); function as antioxidants.
Vitamin E	Promotes wound healing, prevents tissue scarring, contributes to normal structure and function of nervous system, and functions as antioxidant.
Vitamin K	Required for synthesis of blood-clotting proteins.
Lipoproteins	Transport lipids in blood, carry triglycerides and cholesterol to tissues, and remove excess cholesterol from blood.

Fatty Acids

Among the simplest lipids are the **fatty acids,** which are used to synthesize triglycerides and phospholipids. Fatty acids can also be catabolized to generate adenosine triphosphate (ATP). A fatty acid consists of a carboxyl group and a hydrocarbon chain (Figure 2.17a). Fatty acids can be either saturated or unsaturated. A

saturated fatty acid contains only *single covalent bonds* between the carbon atoms of the hydrocarbon chain. Because they lack double bonds, each carbon atom of the hydrocarbon chain is *saturated with hydrogen atoms* (see, for example, palmitic acid in Figure 2.17a). An **unsaturated fatty acid** contains one or more *double covalent bonds* between the carbon atoms of the hydrocarbon chain.

Figure 2.17 Fatty acid structure and triglyceride synthesis. Shown in (a) are the structures of a saturated fatty acid and an unsaturated fatty acid. Each time a glycerol and a fatty acid are joined in dehydration synthesis (b), a molecule of water is removed. An ester linkage joins the glycerol to each of the three molecules of fatty acids, which vary in length and in number and location of double bonds between carbon atoms (C=C). Shown here (c) is a triglyceride molecule that contains two saturated fatty acids and a monounsaturated fatty acid. The kink (bend) in the oleic acid occurs at the double bond.

One glycerol and three fatty acids are the building blocks of triglycerides.

(a) Structures of saturated and unsaturated fatty acids

(b) Dehydration synthesis involving glycerol and a fatty acid

(c) Triglyceride (fat) molecule

Does the oxygen in the water molecule removed during dehydration synthesis come from the glycerol or from a fatty acid?

Thus, the fatty acid is not completely saturated with hydrogen atoms (see, for example, oleic acid in Figure 2.17a). The unsaturated fatty acid has a kink (bend) at the site of the double bond. If the fatty acid has just one double bond in the hydrocarbon chain, it is *monounsaturated* and it has just one kink. If a fatty acid has more than one double bond in the hydrocarbon chain, it is *polyunsaturated* and it contains more than one kink.

Triglycerides

The most plentiful lipids in your body and in your diet are the **triglycerides** (trī-GLI-ser-īds; *tri-* = three), also known as **triacylglycerols.** A triglyceride consists of two types of building blocks, a single glycerol molecule and three fatty acid molecules. A three-carbon **glycerol** molecule forms the backbone of a triglyceride (Figure 2.17b, c). Three fatty acids are attached by dehydration synthesis reactions, one to each carbon of the glycerol backbone. The chemical bond formed where each water molecule is removed is an *ester linkage* (see Table 2.5). The reverse reaction, hydrolysis, breaks down a single molecule of a triglyceride into three fatty acids and glycerol.

Triglycerides can be either solids or liquids at room temperature. A **fat** is a triglyceride that is a solid at room temperature. The fatty acids of a fat are mostly saturated. Because these saturated fatty acids lack double bonds in their hydrocarbon chains, they can closely pack together and solidify at room temperature. A fat that mainly consists of saturated fatty acids is called a **saturated fat.** Although saturated fats occur mostly in meats (especially red meats) and nonskim dairy products (whole milk, cheese, and butter), they are also found in a few plant products, such as cocoa butter, palm oil, and coconut oil. Diets that contain large amounts of saturated fats are associated with disorders such as heart disease and colorectal cancer.

An **oil** is a triglyceride that is a liquid at room temperature. The fatty acids of an oil are mostly unsaturated. Recall that unsaturated fatty acids contain one or more double bonds in their hydrocarbon chains. The kinks at the sites of the double bonds prevent the unsaturated fatty acids of an oil from closely packing together and solidifying. The fatty acids of an oil can be either monounsaturated or polyunsaturated. **Monounsaturated fats** contain triglycerides that mostly consist of monounsaturated fatty acids. Olive oil, peanut oil, canola oil, most nuts, and avocados are rich in triglycerides with monounsaturated fatty acids. **Polyunsaturated fats** contain triglycerides that mostly consist of polyunsaturated fatty acids. Corn oil, safflower oil, sunflower oil, soybean oil, and fatty fish (salmon, tuna, and mackerel) contain a high percentage of polyunsaturated fatty acids. Both monounsaturated and polyunsaturated fats are believed to decrease the risk of heart disease.

Triglycerides are the body's most highly concentrated form of chemical energy. Triglycerides provide more than twice as much energy per gram as do carbohydrates and proteins. Our capacity to store triglycerides in adipose (fat) tissue is unlimited for all practical purposes. Excess dietary carbohydrates, proteins, fats, and oils all have the same fate: They are deposited in adipose tissue as triglycerides.

As its name implies, a group of fatty acids called **essential fatty acids (EFAs)** is essential to human health. However, they cannot be made by the human body and must be obtained from foods or supplements. Among the more important EFAs are *omega-3 fatty acids, omega-6 fatty acids,* and cis-*fatty acids.*

Omega-3 and omega-6 fatty acids are polyunsaturated fatty acids that are believed to work together to promote health. They may have a protective effect against heart disease and stroke by lowering total cholesterol, raising HDL (high-density lipoproteins or "good cholesterol") and lowering LDL (low-density lipoproteins or "bad cholesterol"). In addition, omega-3 and omega-6 fatty acids decrease bone loss by increasing calcium utilization by the body; reduce symptoms of arthritis due to inflammation; promote wound healing; improve certain skin disorders (psoriasis, eczema, and acne); and improve mental functions. Primary sources of omega-3 fatty acids include flaxseed, fatty fish, oils that have large amounts of polyunsaturated fatty acids, fish oils, and walnuts. Primary sources of omega-6 fatty acids include most processed foods (cereals, breads, white rice), eggs, baked goods, oils with large amounts of polyunsaturated fatty acids, and meats (especially organ meats, such as liver).

Note in Figure 2.17a that the hydrogen atoms on either side of the double bond in oleic acid are on the same side of the unsaturated fatty acid. Such an unsaturated fatty acid is called a *cis*-fatty acid. *Cis*-fatty acids are nutritionally beneficial unsaturated fatty acids that are used by the body to produce hormonelike regulators and cell membranes. However, when *cis*-fatty acids are heated, pressurized, and combined with a catalyst (usually nickel) in a process called *hydrogenation,* they are changed to unhealthy *trans*-fatty acids. In *trans*-fatty acids the hydrogen atoms are on opposite sides of the double bond of an unsaturated fatty acid. Hydrogenation is used by manufacturers to make vegetable oils solid at room temperature and less likely to turn rancid. Hydrogenated or *trans*-fatty acids are common in commercially baked goods (crackers, cakes, and cookies), salty snack foods, some margarines, and fried foods (donuts and french fries). When oil is used for frying and if the oil is reused (like in fast food french fry machines), *cis*-fatty acids are converted to *trans*-fatty acids. If a product label contains the words hydrogenated or partially hydrogenated, then the product contains *trans*-fatty acids. Among the adverse effects of *trans*-fatty acids are an increase in total cholesterol, a decrease in HDL, an increase in LDL, and an increase in triglycerides. These effects, which can increase the risk of heart disease and other cardiovascular diseases, are similar to those caused by saturated fats. •

Phospholipids

Like triglycerides, **phospholipids** have a glycerol backbone and two fatty acid chains attached to the first two carbons. In the third position, however, a phosphate group (PO_4^{3-}) links a small charged group that usually contains nitrogen (N) to the backbone (Figure 2.18). This portion of the molecule (the "head") is polar and can form hydrogen bonds with water molecules. The two fatty acids (the "tails"), by contrast, are nonpolar and can interact only with other lipids. Molecules that have both polar and nonpolar parts are said to be **amphipathic** (am-fē-PATH-ik;

Figure 2.18 Phospholipids. (a) In the synthesis of phospholipids, two fatty acids attach to the first two carbons of the glycerol backbone. A phosphate group links a small charged group to the third carbon in glycerol. In (b), the circle represents the polar head region, and the two wavy lines represent the two nonpolar tails. Double bonds in the fatty acid hydrocarbon chain often form kinks in the tail.

🔑 Phospholipids are amphipathic molecules, having both polar and nonpolar regions.

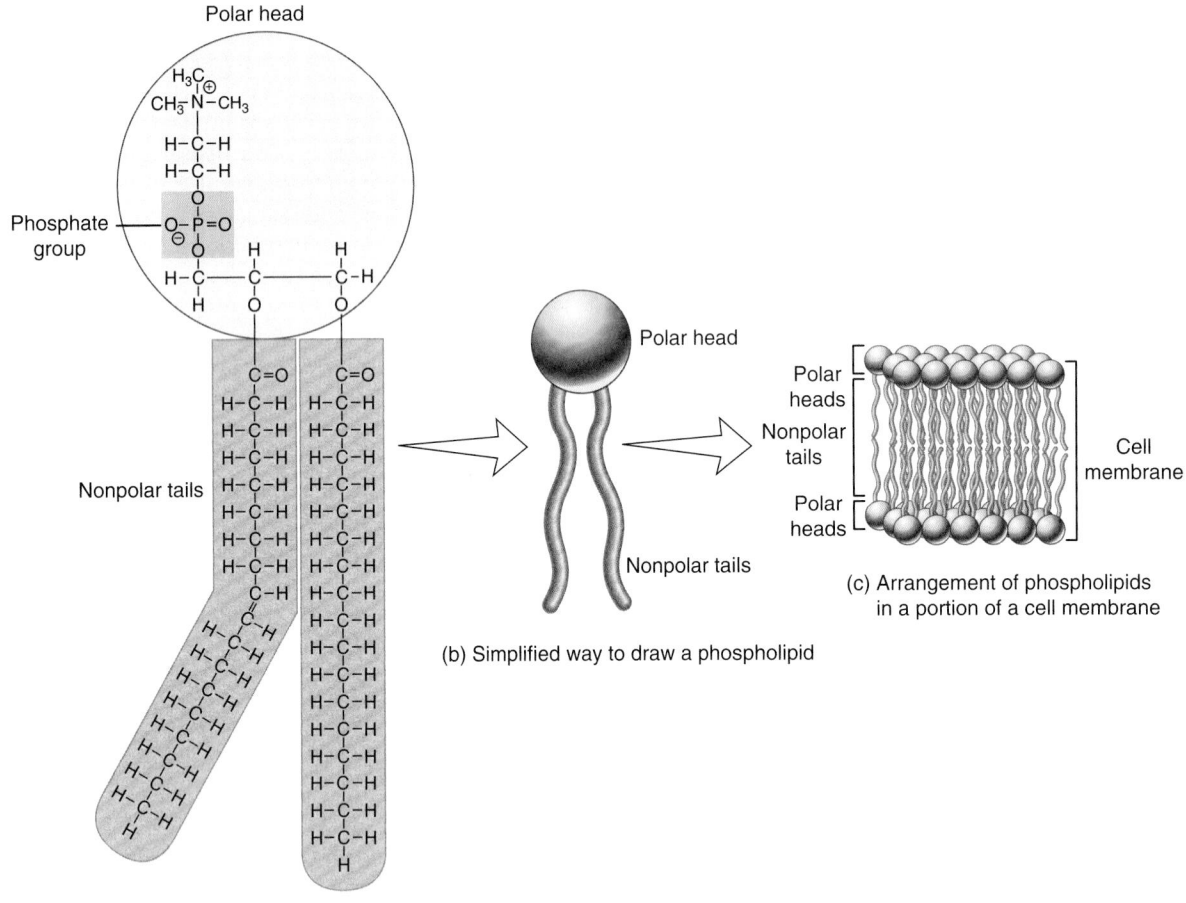

(a) Chemical structure of a phospholipid

(b) Simplified way to draw a phospholipid

(c) Arrangement of phospholipids in a portion of a cell membrane

❓ **Which portion of a phospholipid is hydrophilic, and which portion is hydrophobic?**

amphi- = on both sides; *-pathic* = feeling). Amphipathic phospholipids line up tail-to-tail in a double row to make up much of the membrane that surrounds each cell (Figure 2.18c).

Steroids

The structure of **steroids** differs considerably from that of the triglycerides. Steroids have four rings of carbon atoms (colored gold in Figure 2.19). Body cells synthesize other steroids from cholesterol (Figure 2.19a), which has a large nonpolar region consisting of the four rings and a hydrocarbon tail. In the body, the commonly encountered steroids, such as cholesterol, estrogens, testosterone, cortisol, bile salts, and vitamin D, are known as **sterols** because they also have at least one hydroxyl (alcohol) group (—OH). The polar hydroxyl groups make sterols weakly amphipathic. Cholesterol is needed for cell membrane structure; estrogens and testosterone are required for regulating sexual functions; cortisol is necessary for maintaining normal blood sugar levels; bile salts are needed for lipid

digestion and absorption; and vitamin D is related to bone growth. In Chapter 10, we will discuss the use of anabolic steroids by athletes to increase muscle size, strength, and endurance.

Other Lipids

Eicosanoids (ī-KŌ-sa-noyds; *eicosan-* = twenty) are lipids derived from a 20-carbon fatty acid called arachidonic acid. The two principal subclasses of eicosanoids are the **prostaglandins** (pros′-ta-GLAN-dins) and the **leukotrienes** (loo′-kō-TRĪ-ēnz). Prostaglandins have a wide variety of functions. They modify responses to hormones, contribute to the inflammatory response (Chapter 22), prevent stomach ulcers, dilate (enlarge) airways to the lungs, regulate body temperature, and influence formation of blood clots, to name just a few. Leukotrienes participate in allergic and inflammatory responses.

Other lipids also include fat-soluble vitamins such as beta-carotenes (the yellow-orange pigments in egg yolk, carrots, and

Figure 2.19 Steroids. All steroids have four rings of carbon atoms. The individual rings are designated by the letters A, B, C, and D.

 Cholesterol, which is synthesized in the liver, is the starting material for synthesis of other steroids in the body.

(a) Cholesterol

(b) Estradiol (an estrogen or female sex hormone)

(c) Testosterone (a male sex hormone)

(d) Cortisol

How is the structure of estradiol different from that of testosterone?

tomatoes that are converted to vitamin A); vitamins D, E, and K; and lipoproteins.

✔ CHECKPOINT

16. How are carbohydrates classified?
17. How are dehydration synthesis and hydrolysis reactions related?
18. What is the importance to the body of triglycerides, phospholipids, steroids, lipoproteins, and eicosanoids?
19. Distinguish among saturated, monounsaturated, and polyunsaturated fats.

Proteins

Proteins are large molecules that contain carbon, hydrogen, oxygen, and nitrogen. Some proteins also contain sulfur. A normal, lean adult body is 12–18% protein. Much more complex in structure than carbohydrates or lipids, proteins have many roles in the body and are largely responsible for the structure of body tissues. Enzymes are proteins that speed up most biochemical reactions. Other proteins work as "motors" to drive muscle contraction.

Antibodies are proteins that defend against invading microbes. Some hormones that regulate homeostasis also are proteins. Table 2.8 describes several important functions of proteins.

Amino Acids and Polypeptides

The monomers of proteins are **amino acids** (a-MĒ-nō). Each of the 20 different amino acids has a hydrogen (H) atom and three important functional groups attached to a central carbon atom (Figure 2.20a): (1) an amino group ($-NH_2$), (2) an acidic carboxyl group ($-COOH$), and (3) a side chain (R group). At the normal pH of body fluids, both the amino group and the carboxyl group are ionized (Figure 2.20b). The different side chains give each amino acid its distinctive chemical identity (Figure 2.20c).

A protein is synthesized in stepwise fashion—one amino acid is joined to a second, a third is then added to the first two, and so on. The covalent bond joining each pair of amino acids is a **peptide bond.** It always forms between the carbon of the carboxyl group ($-COOH$) of one amino acid and the nitrogen of the amino group ($-NH_2$) of another. As the peptide bond is formed, a molecule of water is removed (Figure 2.21), making this a

TABLE 2.8	
Functions of Proteins	
TYPE OF PROTEIN	**FUNCTIONS**
Structural	Form structural framework of various parts of body.
	Examples: collagen in bone and other connective tissues; keratin in skin, hair, and fingernails.
Regulatory	Function as hormones that regulate various physiological processes; control growth and development; as neurotransmitters, mediate responses of nervous system.
	Examples: the hormone insulin (regulates blood glucose level); the neurotransmitter known as substance P (mediates sensation of pain in nervous system).
Contractile	Allow shortening of muscle cells, which produces movement.
	Examples: myosin; actin.
Immunological	Aid responses that protect body against foreign substances and invading pathogens.
	Examples: antibodies; interleukins.
Transport	Carry vital substances throughout body.
	Example: hemoglobin (transports most oxygen and some carbon dioxide in blood).
Catalytic	Act as enzymes that regulate biochemical reactions.
	Examples: salivary amylase; sucrase; ATPase.

Figure 2.20 Amino acids. (a) In keeping with their name, amino acids have an amino group (shaded blue) and a carboxyl (acid) group (shaded red). The side chain (R group) is different in each amino acid. (b) At pH close to 7, both the amino group and the carboxyl group are ionized. (c) Glycine is the simplest amino acid; the side chain is a single H atom. Cysteine is one of two amino acids that contain sulfur (S). The side chain in tyrosine contains a six-carbon ring. Lysine has a second amino group at the end of its side chain.

Body proteins contain 20 different amino acids, each of which has a unique side chain.

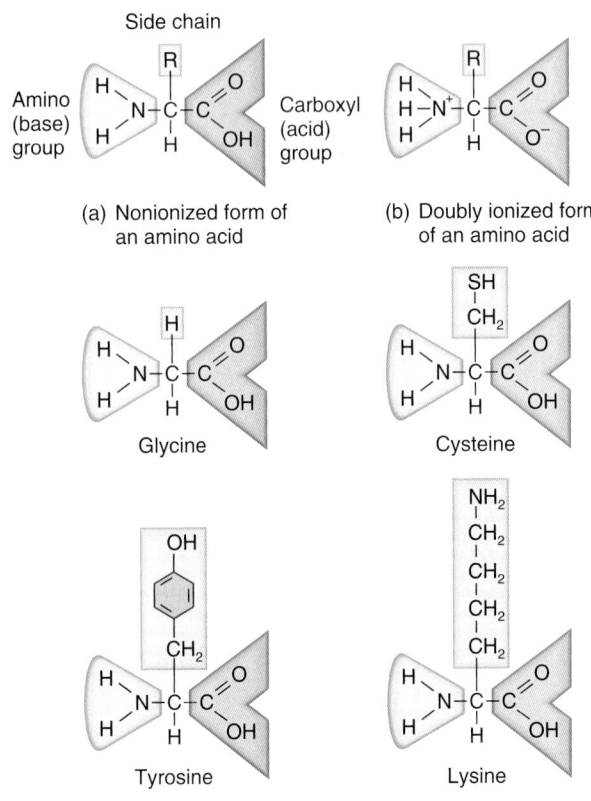

(c) Representative amino acids

In an amino acid, what is the minimum number of carbon atoms? Of nitrogen atoms?

dehydration synthesis reaction. Breaking a peptide bond, as occurs during digestion of dietary proteins, is a hydrolysis reaction (Figure 2.21).

When two amino acids combine, a **dipeptide** results. Adding another amino acid to a dipeptide produces a **tripeptide.** Further additions of amino acids result in the formation of a chainlike **peptide** (4–9 amino acids) or **polypeptide** (10–2000 or more amino acids). Small proteins may consist of a single polypeptide chain with as few as 50 amino acids. Larger proteins have hundreds or thousands of amino acids and may consist of two or more polypeptide chains folded together.

Because each variation in the number or sequence of amino acids can produce a different protein, a great variety of proteins is possible. The situation is similar to using an alphabet of 20 letters to form words. Each different amino acid is like a letter, and their various combinations give rise to a seemingly endless diversity of words (peptides, polypeptides, and proteins).

Levels of Structural Organization in Proteins

Proteins exhibit four levels of structural organization. The **primary structure** is the unique sequence of amino acids that are linked by covalent peptide bonds to form a polypeptide chain (Figure 2.22a). A protein's primary structure is genetically determined, and any changes in a protein's amino acid sequence can have serious consequences for body cells. In **sickle-cell disease,** for example, a nonpolar amino acid (valine) replaces a polar amino acid (glutamate) through two mutations in the oxygen-carrying protein hemoglobin. This change of amino acids diminishes hemoglobin's water solubility. As a result, the altered hemoglobin tends to form crystals inside red blood cells, producing deformed, sickle-shaped cells that cannot properly squeeze through narrow blood vessels. The symptoms and treatment of sickle-cell disease are discussed in Disorders: Homeostatic Imbalances in Chapter 19.

The **secondary structure** of a protein is the repeated twisting or folding of neighboring amino acids in the polypeptide chain (Figure 2.22b). Two common secondary structures are *alpha helixes* (clockwise spirals) and *beta pleated sheets.* The secondary structure of a protein is stabilized by hydrogen bonds, which form at regular intervals along the polypeptide backbone.

Figure 2.21 Formation of a peptide bond between two amino acids during dehydration synthesis. In this example, glycine is joined to alanine, forming a dipeptide (read from left to right). Breaking a peptide bond occurs via hydrolysis (read from right to left).

Amino acids are the monomers used to build proteins.

What type of reaction takes place during catabolism of proteins?

Figure 2.22 **Levels of structural organization in proteins.** (a) The primary structure is the sequence of amino acids in the polypeptide. (b) Common secondary structures include alpha helixes and beta pleated sheets. For simplicity, the amino acid side groups are not shown here. (c) The tertiary structure is the overall folding pattern that produces a distinctive, three-dimensional shape. (d) The quaternary structure in a protein is the arrangement of two or more polypeptide chains relative to one another.

The unique shape of each protein permits it to carry out specific functions.

Amino acids

Peptide bond

Hydrogen bond

Polypeptide chain

(a) Primary structure
(amino acid sequence)

Alpha helix

(b) Secondary structure
(twisting and folding of neighboring amino acids, stabilized by hydrogen bonds)

Beta pleated sheet

(c) Tertiary structure
(three-dimensional shape of polypeptide chain)

(d) Quaternary structure
(arrangement of two or more polypeptide chains)

? **Do all proteins have a quaternary structure?**

The **tertiary structure** (TUR-shē-er′-ē) refers to the three-dimensional shape of a polypeptide chain. Each protein has a unique tertiary structure that determines how it will function. The tertiary folding pattern may allow amino acids at opposite ends of the chain to be close neighbors (Figure 2.22c). Several types of bonds can contribute to a protein's tertiary structure. The strongest but least common bonds, S—S covalent bonds called *disulfide bridges,* form between the sulfhydryl groups of two monomers of the amino acid cysteine. Many weak bonds—hydrogen bonds, ionic bonds, and hydrophobic interactions—also help determine the folding pattern. Some parts of a polypeptide are attracted to water (hydrophilic), and other parts are repelled by it (hydrophobic). Because most proteins in our body exist in watery surroundings, the folding process places most amino acids with hydrophobic side chains in the central core, away from the protein's surface. Often, helper molecules known as *chaperones* aid the folding process.

In those proteins that contain more than one polypeptide chain (not all of them do), the arrangement of the individual polypeptide chains relative to one another is the **quaternary structure** (KWA-ter-ner′-ē; Figure 2.22d). The bonds that hold polypeptide chains together are similar to those that maintain the tertiary structure.

Proteins vary tremendously in structure. Different proteins have different architectures and different three-dimensional shapes. This variation in structure and shape is directly related to their diverse functions. In practically every case, the function of a protein depends on its ability to recognize and bind to some other molecule. Thus, a hormone binds to a specific protein on a cell in order to alter its function, and an antibody protein binds to a foreign substance (antigen) that has invaded the body. A protein's unique shape permits it to interact with other molecules to carry out a specific function.

On the basis of overall shape, proteins are classified as fibrous or globular. **Fibrous proteins** are insoluble in water and their polypeptide chains form long strands that are parallel to each other. Fibrous proteins have many structural functions. Examples include *collagen* (strengthens bones, ligaments, and tendons), *elastin* (provides stretch in skin, blood vessels, and lung tissue), *keratin* (forms structure of hair and nails and waterproofs the skin), *dystrophin* (reinforces parts of muscle cells), *fibrin* (forms blood clots), and *actin* and *myosin* (are involved in contraction of muscle cells, division in all cells, and transport of substances within cells). **Globular proteins** are more or less soluble in water and their polypeptide chains are spherical (globular) in shape. Globular proteins have metabolic functions. Examples include *enzymes,* which function as catalysts; *antibodies* and *complement proteins,* which help protect us against disease; *hemoglobin,* which transports oxygen; *lipoproteins,* which transport lipids and cholesterol; *albumins,* which help regulate blood pH; *membrane proteins,* which transport substances into and out of cells; and some *hormones* such as *insulin,* which helps regulate blood sugar level.

Homeostatic mechanisms maintain the temperature and chemical composition of body fluids, which allow body proteins to keep their proper three-dimensional shapes. If a protein encounters an altered environment, it may unravel and lose its characteristic shape (secondary, tertiary, and quaternary structure). This process is called **denaturation.** Denatured proteins are no longer functional. Although in some cases denaturation can be reversed, a frying egg is a common example of permanent denaturation. In a raw egg the soluble egg-white protein (albumin) is a clear, viscous fluid. When heat is applied to the egg, the protein denatures, becomes insoluble, and turns white.

Enzymes

In living cells, most catalysts are protein molecules called **enzymes** (EN-zīms). Some enzymes consist of two parts—a protein portion, called the **apoenzyme** (ā′-pō-EN-zīm), and a nonprotein portion, called a **cofactor.** The cofactor may be a metal ion (such as iron, magnesium, zinc, or calcium) or an organic molecule called a *coenzyme.* Coenzymes often are derived from vitamins. The names of enzymes usually end in the suffix *-ase.* All enzymes can be grouped according to the types of chemical reactions they catalyze. For example, *oxidases* add oxygen, *kinases* add phosphate, *dehydrogenases* remove hydrogen, *ATPases* split ATP, *anhydrases* remove water, *proteases* break down proteins, and *lipases* break down triglycerides.

Enzymes catalyze specific reactions. They do so with great efficiency and with many built-in controls. Three important properties of enzymes are as follows:

1. *Enzymes are highly specific.* Each particular enzyme binds only to specific **substrates**—the reactant molecules on which the enzyme acts. Of the more than 1000 known enzymes in your body, each has a characteristic three-dimensional shape with a specific surface configuration, which allows it to recognize and bind to certain substrates. In some cases, the part of the enzyme that catalyzes the reaction, called the **active site,** is thought to fit the substrate like a key fits in a lock. In other cases the active site changes its shape to fit snugly around the substrate once the substrate enters the active site. This change in shape is known as an *induced fit.*

   Not only is an enzyme matched to a particular substrate; it also catalyzes a specific reaction. From among the large number of diverse molecules in a cell, an enzyme must recognize the correct substrate and then take it apart or merge it with another substrate to form one or more specific products.

2. *Enzymes are very efficient.* Under optimal conditions; enzymes can catalyze reactions at rates that are from 100 million to 10 billion times more rapid than those of similar reactions occurring without enzymes. The number of substrate molecules that a single enzyme molecule can convert to product molecules in one second is generally between 1 and 10,000 and can be as high as 600,000.

3. *Enzymes are subject to a variety of cellular controls.* Their rate of synthesis and their concentration at any given time are under the control of a cell's genes. Substances within the cell may either enhance or inhibit the activity of a given enzyme. Many enzymes have both active and inactive forms in

cells. The rate at which the inactive form becomes active or vice versa is determined by the chemical environment inside the cell.

Enzymes lower the activation energy of a chemical reaction by decreasing the "randomness" of the collisions between molecules. They also help bring the substrates together in the proper orientation so that the reaction can occur. Figure 2.23 depicts how an enzyme works:

1 The substrates make contact with the active site on the surface of the enzyme molecule, forming a temporary intermediate compound called the **enzyme–substrate complex.** In this reaction the two substrate molecules are sucrose (a disaccharide) and water.

2 The substrate molecules are transformed by the rearrangement of existing atoms, the breakdown of the substrate molecule, or the combination of several substrate molecules into the products of the reaction. Here the products are two monosaccharides: glucose and fructose.

3 After the reaction is completed and the reaction products move away from the enzyme, the unchanged enzyme is free to attach to other substrate molecules.

Sometimes a single enzyme may catalyze a reversible reaction in either direction, depending on the relative amounts of the substrates and products. For example, the enzyme *carbonic anhydrase* catalyzes the following reversible reaction:

$$CO_2 + H_2O \underset{}{\overset{\text{Carbonic anhydrase}}{\rightleftharpoons}} H_2CO_3$$

Carbon dioxide Water Carbonic acid

During exercise, when more CO_2 is produced and released into the blood, the reaction flows to the right, increasing the amount of carbonic acid in the blood. Then, as you exhale CO_2, its level in the blood falls and the reaction flows to the left, converting carbonic acid to CO_2 and H_2O.

Figure 2.23 How an enzyme works.

An enzyme speeds up a chemical reaction without being altered or consumed.

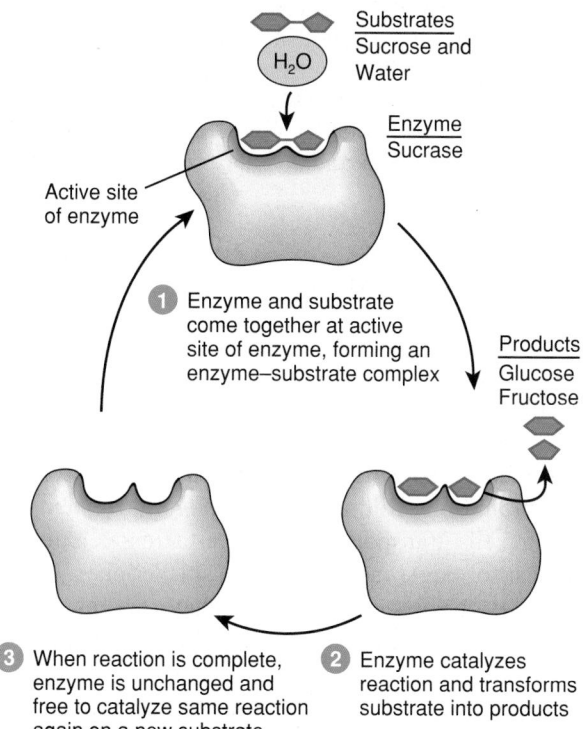

Substrates
Sucrose and
Water

H_2O

Enzyme
Sucrase

Active site
of enzyme

1 Enzyme and substrate come together at active site of enzyme, forming an enzyme–substrate complex

Products
Glucose
Fructose

3 When reaction is complete, enzyme is unchanged and free to catalyze same reaction again on a new substrate

2 Enzyme catalyzes reaction and transforms substrate into products

(a) Mechanism of enzyme action

Substrate
(glucose)

Enzyme
(hexokinase)

Substrate

Enzyme

(b) Molecular model of enzyme and substrate uncombined (left) and enzyme–substrate complex (right)

? **Why is it that sucrase cannot catalyze the formation of sucrose from glucose and fructose?**

Nucleic Acids: Deoxyribonucleic Acid (DNA) and Ribonucleic Acid (RNA)

Nucleic acids (noo-KLĒ-ic), so named because they were first discovered in the nuclei of cells, are huge organic molecules that contain carbon, hydrogen, oxygen, nitrogen, and phosphorus. Nucleic acids are of two varieties. The first, **deoxyribonucleic acid (DNA)** (dē-ok′-sē-rī-bō-nū-KLĒ-ik), forms the inherited genetic material inside each human cell. In humans, each **gene** (JĒN) is a segment of a DNA molecule. Our genes determine the traits we inherit, and by controlling protein synthesis they regulate most of the activities that take place in body cells throughout our lives. When a cell divides, its hereditary information passes on to the next generation of cells. **Ribonucleic acid (RNA),** the second type of nucleic acid, relays instructions from the genes to guide each cell's synthesis of proteins from amino acids.

A nucleic acid is a chain of repeating monomers called **nucleotides.** Each nucleotide of DNA consists of three parts (Figure 2.24a):

1. *Nitrogenous base.* DNA contains four different nitrogenous bases, which contain atoms of C, H, O, and N. In DNA the four **nitrogenous bases** are adenine (A), thymine (T), cytosine (C), and guanine (G). Adenine and guanine are larger, double-ring bases called **purines** (PŪR-ēnz); thymine and cytosine are smaller, single-ring bases called **pyrimidines** (pī-RIM-i-dēnz). The nucleotides are named according to the base that is present. For instance, a nucleotide containing thymine is called a thymine nucleotide, one containing adenine is called an adenine nucleotide, and so on.

2. *Pentose sugar.* A five-carbon sugar called **deoxyribose** attaches to each base in DNA.

3. *Phosphate group.* Phosphate groups (PO_4^{3-}) alternate with pentose sugars to form the "backbone" of a DNA strand; the bases project inward from the backbone chain (Figure 2.24b).

In 1953, F.H.C. Crick of Great Britain and J.D. Watson, a young American scientist, published a brief paper describing how these three components might be arranged in DNA. Their insights into data gathered by others led them to construct a model so elegant and simple that the scientific world immediately knew it was correct! In the Watson–Crick **double helix** model, DNA resembles a spiral ladder (Figure 2.24b). Two strands of alternating phosphate groups and deoxyribose sugars form the uprights of the ladder. Paired bases, held together by hydrogen bonds, form the rungs. Because adenine always pairs with thymine, and cytosine always pairs with guanine, if you know the sequence of bases in one strand of DNA, you can predict the sequence on the complementary (second) strand. Each time DNA is copied, as when living cells divide to increase their number, the two strands unwind. Each strand serves as the template or mold on which to construct a new second strand. Any change that occurs in the base sequence of a DNA strand is called a *mutation.* Some mutations can result in the death of a cell, cause cancer, or produce genetic defects in future generations.

RNA, the second variety of nucleic acid, differs from DNA in several respects. In humans, RNA is single-stranded. The sugar in the RNA nucleotide is the pentose **ribose,** and RNA contains the pyrimidine base uracil (U) instead of thymine. Cells contain three different kinds of RNA: messenger RNA, ribosomal RNA, and transfer RNA. Each has a specific role to perform in carrying out the instructions coded in DNA (see Figure 3.29).

CLINICAL CONNECTION | DNA Fingerprinting

A technique called **DNA fingerprinting** is used in research and in courts of law to ascertain whether a person's DNA matches the DNA obtained from samples or pieces of legal evidence such as blood stains or hairs. In each person, certain DNA segments contain base sequences that are repeated several times. Both the number of repeat copies in one region and the number of regions subject to repeat are different from one person to another. DNA fingerprinting can be done with minute quantities of DNA—for example, from a single strand of hair, a drop of semen, or a spot of blood. It also can be used to identify a crime victim or a child's biological parents and even to determine whether two people have a common ancestor. •

Adenosine Triphosphate

Adenosine triphosphate (a-DEN-ō-sēn trī-FOS-fāt) or **ATP** is the "energy currency" of living systems (Figure 2.25). ATP transfers the energy liberated in exergonic catabolic reactions to power cellular activities that require energy (endergonic reactions). Among these cellular activities are muscular contractions, movement of chromosomes during cell division, movement of structures within cells, transport of substances across cell membranes, and synthesis of larger molecules from smaller ones. As its name implies, ATP consists of three phosphate groups attached to adenosine, a unit composed of adenine and the five-carbon sugar ribose.

When a water molecule is added to ATP, the third phosphate group (PO_4^{3-}), symbolized by Ⓟ in the following discussion, is removed, and the overall reaction liberates energy. The enzyme that catalyzes the hydrolysis of ATP is called *ATPase.* Removal of the third phosphate group produces a molecule called **adenosine diphosphate (ADP)** in the following reaction:

$$\text{ATP} + \text{H}_2\text{O} \xrightarrow{\textit{ATPase}} \text{ADP} + \text{Ⓟ} + \text{E}$$

Adenosine triphosphate Water Adenosine diphosphate Phosphate group Energy

As noted previously, the energy supplied by the catabolism of ATP into ADP is constantly being used by the cell. As the supply of ATP at any given time is limited, a mechanism exists to replenish it: The enzyme *ATP synthase* catalyzes the addition of a phosphate group to ADP in the following reaction:

$$\text{ADP} + \text{Ⓟ} + \text{E} \xrightarrow{\textit{ATP synthase}} \text{ATP} + \text{H}_2\text{O}$$

Adenosine diphosphate Phosphate group Energy Adenosine triphosphate Water

Where does the cell get the energy required to produce ATP? The energy needed to attach a phosphate group to ADP is supplied

Figure 2.24 DNA molecule. (a) A nucleotide consists of a base, a pentose sugar, and a phosphate group. (b) The paired bases project toward the center of the double helix. The structure is stabilized by hydrogen bonds (dotted lines) between each base pair. There are two hydrogen bonds between adenine and thymine and three between cytosine and guanine.

Nucleotides are the monomers of nucleic acids.

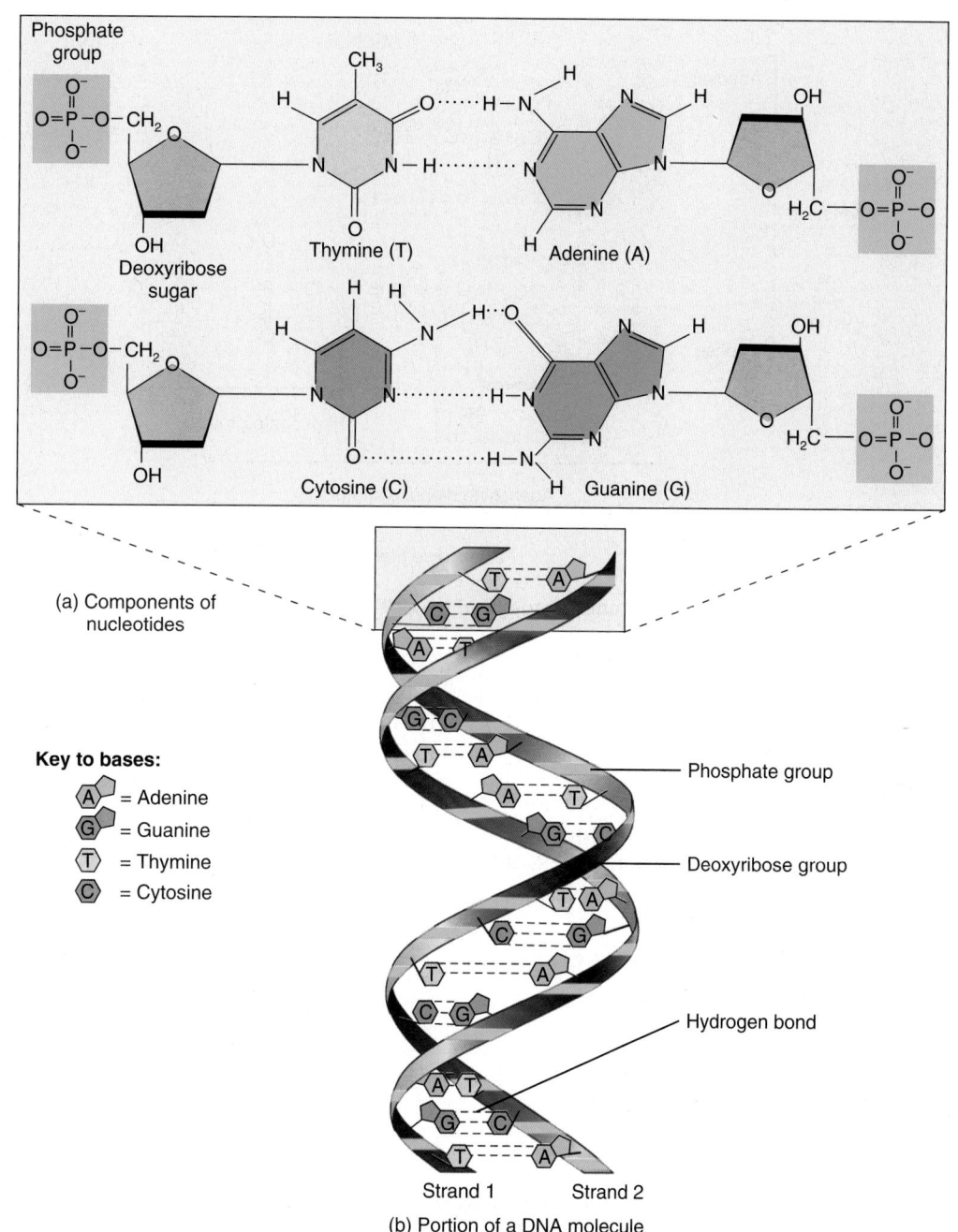

(a) Components of nucleotides

Key to bases:

(A) = Adenine
(G) = Guanine
(T) = Thymine
(C) = Cytosine

Phosphate group

Deoxyribose group

Hydrogen bond

Strand 1   Strand 2

(b) Portion of a DNA molecule

? **Which bases always pair with each other?**

Figure 2.25 Structures of ATP and ADP. "Squiggles" (~) indicate the two phosphate bonds that can be used to transfer energy. Energy transfer typically involves hydrolysis of the last phosphate bond of ATP.

ATP transfers chemical energy to power cellular activities.

? **What are some cellular activities that depend on energy supplied by ATP?**

mainly by the catabolism of glucose in a process called cellular respiration. Cellular respiration has two phases, anaerobic and aerobic:

1. **Anaerobic phase.** In a series of reactions that do not require oxygen, glucose is partially broken down by a series of catabolic reactions into pyruvic acid. Each glucose molecule that is converted into a pyruvic acid molecule yields two molecules of ATP.

2. **Aerobic phase.** In the presence of oxygen, glucose is completely broken down into carbon dioxide and water. These reactions generate heat and 36 or 38 ATP molecules.

Chapters 10 and 25 cover the details of cellular respiration.

In Chapter 1, you learned that the human body comprises various levels of organization; this chapter has just showed you the alphabet of atoms and molecules that is the basis for the language of the body. Now that you have an understanding of the chemistry of the human body, you are ready to form words; in Chapter 3 you will see how atoms and molecules are organized to form structures of cells and perform the activities of cells that contribute to homeostasis.

✔ CHECKPOINT

20. Define a protein. What is a peptide bond?
21. Outline the levels of structural organization in proteins.
22. Distinguish between fibrous and globular proteins in terms of structure and function.
23. How do DNA and RNA differ?
24. In the reaction catalyzed by ATP synthase, what are the substrates and products? Is this an exergonic or endergonic reaction?

CHAPTER REVIEW AND RESOURCE SUMMARY

Review

2.1 How Matter Is Organized

1. All forms of matter are composed of chemical elements.
2. Oxygen, carbon, hydrogen, and nitrogen make up about 96% of body mass.
3. Each element is made up of small units called atoms. Atoms consist of a nucleus, which contains protons and neutrons, plus electrons that move about the nucleus in regions called electron shells.
4. The number of protons (the atomic number) distinguishes the atoms of one element from those of another element.
5. The mass number of an atom is the sum of its protons and neutrons.

Resource

Anatomy Overview - Common Biomolecules
Animation - Atomic Structure and the Basis of Bonds

Review   **Resource**

6. Different atoms of an element that have the same number of protons but different numbers of neutrons are called isotopes. Radioactive isotopes are unstable and decay.

7. The atomic mass of an element is the average mass of all naturally occurring isotopes of that element.

8. An atom that *gives up* or *gains* electrons becomes an ion—an atom that has a positive or negative charge because it has unequal numbers of protons and electrons. Positively charged ions are cations; negatively charged ions are anions.

9. If two atoms share electrons, a molecule is formed. Compounds contain atoms of two or more elements.

10. A free radical is an atom or group of atoms with an unpaired electron in its outermost shell. A common example is superoxide, an anion which is formed by the addition of an electron to an oxygen molecule.

2.2 Chemical Bonds

1. Forces of attraction called chemical bonds hold atoms together. These bonds result from gaining, losing, or sharing electrons in the valence shell.

2. Most atoms become stable when they have an octet of eight electrons in their valence (outermost) electron shell.

3. When the force of attraction between ions of opposite charge holds them together, an ionic bond has formed.

4. In a covalent bond, atoms share pairs of valence electrons. Covalent bonds may be single, double, or triple and either nonpolar or polar.

5. An atom of hydrogen that forms a polar covalent bond with an oxygen atom or a nitrogen atom may also form a weaker bond, called a hydrogen bond, with an electronegative atom. The polar covalent bond causes the hydrogen atom to have a partial positive charge (δ^+) that attracts the partial negative charge (δ^-) of neighboring electronegative atoms, often oxygen or nitrogen.

Animation - Chemical Bonding
Figure 2.6 - Hydrogen
    Bonding among
    Water Molecules
Exercise - Bond
    Boulevard
Concepts and Connections - Chemical
    Bonds

2.3 Chemical Reactions

1. When atoms combine with or break apart from other atoms, a chemical reaction occurs. The starting substances are the reactants, and the ending ones are the products.

2. Energy, the capacity to do work, is of two principal kinds: potential (stored) energy and kinetic energy (energy of motion).

3. Endergonic reactions require energy; exergonic reactions release energy. ATP couples endergonic and exergonic reactions.

4. The initial energy investment needed to start a reaction is the activation energy. Reactions are more likely when the concentrations and the temperatures of the reacting particles are higher.

5. Catalysts accelerate chemical reactions by lowering the activation energy. Most catalysts in living organisms are protein molecules called enzymes.

6. Synthesis reactions involve the combination of reactants to produce larger molecules. The reactions are anabolic and usually endergonic.

7. In decomposition reactions, a substance is broken down into smaller molecules. The reactions are catabolic and usually exergonic.

8. Exchange reactions involve the replacement of one atom or atoms by another atom or atoms.

9. In reversible reactions, end products can revert to the original reactants.

Animation - Types of Reactions and
    Equilibrium
Exercise - Reaction Race

2.4 Inorganic Compounds and Solutions

1. Inorganic compounds usually are small and usually lack carbon. Organic substances always contain carbon, usually contain hydrogen, and always have covalent bonds.

2. Water is the most abundant substance in the body. It is an excellent solvent and suspending medium, participates in hydrolysis and dehydration synthesis reactions, and serves as a lubricant. Because of its many hydrogen bonds, water molecules are cohesive, which causes a high surface tension. Water also has a high capacity for absorbing heat and a high heat of vaporization.

3. Inorganic acids, bases, and salts dissociate into ions in water. An acid ionizes into hydrogen ions (H^+) and anions and is a proton donor; many bases ionize into cations and hydroxide ions (OH^-), and all are proton acceptors. A salt ionizes into neither H^+ nor OH^-.

4. Mixtures are combinations of elements or compounds that are physically blended together but are not bound by chemical bonds. Solutions, colloids, and suspensions are mixtures with different properties.

Anatomy Overview - Common
    Biomolecules: Water
Anatomy Overview - Common
    Biomolecules: Blood Gases
Anatomy Overview - Common
    Biomolecules: Electrolytes
Animation - Acids and Bases
Animation - Water and Fluid Flow
Exercise - Destination: Acid–Base
    Balance

Review

5. Two ways to express the concentration of a solution are percentage (mass per volume), expressed in grams per 100 mL of a solution, and moles per liter. A mole (abbreviated mol) is the amount in grams of any substance that has a mass equal to the combined atomic mass of all its atoms.
6. The pH of body fluids must remain fairly constant for the body to maintain homeostasis. On the pH scale, 7 represents neutrality. Values below 7 indicate acidic solutions, and values above 7 indicate alkaline solutions. Normal blood pH is 7.35–7.45.
7. Buffer systems remove or add protons (H^+) to help maintain pH homeostasis.
8. One important buffer system is the carbonic acid–bicarbonate buffer system. The bicarbonate ion (HCO_3^-) acts as a weak base and removes excess H^+, and carbonic acid (H_2CO_3) acts as a weak acid and adds H^+.

2.5 Organic Compounds

1. Carbon, with its four valence electrons, bonds covalently with other carbon atoms to form large molecules of many different shapes. Attached to the carbon skeletons of organic molecules are functional groups that confer distinctive chemical properties.
2. Small organic molecules are joined together to form larger molecules by dehydration synthesis reactions in which a molecule of water is removed. In the reverse process, called hydrolysis, large molecules are broken down into smaller ones by the addition of water.
3. Carbohydrates provide most of the chemical energy needed to generate ATP. They may be monosaccharides, disaccharides, or polysaccharides.
4. Lipids are a diverse group of compounds that include fatty acids, triglycerides (fats and oils), phospholipids, steroids, and eicosanoids. Triglycerides protect, insulate, provide energy, and are stored. Phospholipids are important cell membrane components. Steroids are important in cell membrane structure, regulating sexual functions, maintaining normal blood sugar level, aiding lipid digestion and absorption, and helping bone growth. Eicosanoids (prostaglandins and leukotrienes) modify hormone responses, contribute to inflammation, dilate airways, and regulate body temperature.
5. Proteins are constructed from amino acids. They give structure to the body, regulate processes, provide protection, help muscles contract, transport substances, and serve as enzymes. Levels of structural organization among proteins include primary, secondary, tertiary, and (sometimes) quaternary. Variations in protein structure and shape are related to their diverse functions.
6. Deoxyribonucleic acid (DNA) and ribonucleic acid (RNA) are nucleic acids consisting of nitrogenous bases, five-carbon (pentose) sugars, and phosphate groups. DNA is a double helix and is the primary chemical in genes. RNA takes part in protein synthesis.
7. Adenosine triphosphate (ATP) is the principal energy-transferring molecule in living systems. When it transfers energy to an endergonic reaction, it is decomposed to adenosine diphosphate (ADP) and a phosphate group. ATP is synthesized from ADP and a phosphate group using the energy supplied by various decomposition reactions, particularly those of glucose.

Anatomy Overview - Carbohydrates
Anatomy Overview - Lipids
Anatomy Overview - Proteins
Anatomy Overview - Nucleic Acids
Animation: Enzyme Functions and ATP Parts 1 & 2
Figure 2.21 - Formation of a Peptide Bond between Two Amino Acids
Figure 2.23 - How an Enzyme Works
Exercise - Enzyme Anticipation
Concepts and Connections - Organic Molecules

SELF-QUIZ QUESTIONS

Fill in the blanks in the following statements.

1. An atom with a mass number of 18 that contains 10 neutrons would have an atomic number of _____.
2. Matter exists in three forms: _____, _____, and _____.
3. The building blocks of carbohydrates are the monomers _____ while the building blocks of proteins are the monomers _____.

Indicate whether the following statements are true or false.

4. The elements that compose most of the body's mass are carbon, hydrogen, oxygen, and nitrogen.
5. Ionic bonds are created when atoms share electrons in the valence shell.
6. Human blood has a normal pH between 7.35 and 7.45 and is considered slightly alkaline.

Choose the one best answer to the following questions.

7. Which of the following would be considered a compound? (1) $C_6H_{12}O_6$, (2) O_2, (3) Fe, (4) H_2, (5) CH_4.
   (a) all are compounds   (b) 1, 2, 4, and 5   (c) 1 and 5
   (d) 2 and 4   (e) 3
8. The monosaccharides glucose and fructose combine to form the disaccharide sucrose by a process known as
   (a) dehydration synthesis   (b) hydrolysis
   (c) decomposition   (d) hydrogen bonding
   (e) ionization.

9. Which of the following is *not* a function of proteins?
   (a) provide structural framework
   (b) bring about contraction
   (c) transport materials throughout the body
   (d) store energy
   (e) regulate many physiological processes

10. Which of the following organic compounds are classified as lipids?
    (1) polysaccharides, (2) triglycerides, (3) steroids, (4) enzymes, (5) eicosanoids.
    (a) 1, 2, and 4   (b) 2, 3, and 5   (c) 2 and 5
    (d) 2, 3, 4, and 5   (e) 2 and 3

11. A compound dissociates in water and forms a cation other than H^+ and an anion other than OH^-. This substance most likely is a(n)
    (a) acid   (b) base   (c) enzyme
    (d) buffer   (e) salt.

12. Which of the following statements regarding ATP are *true*? (1) ATP is the energy currency for the cell. (2) The energy supplied by the hydrolysis of ATP is constantly being used by cells. (3) Energy is required to produce ATP. (4) The production of ATP involves both aerobic and anaerobic phases. (5) The process of producing energy in the form of ATP is termed the law of conservation of energy.
    (a) 1, 2, 3, and 4   (b) 1, 2, 3, and 5   (c) 2, 4, and 5
    (d) 1, 2, and 4   (e) 3, 4, and 5

13. During the course of analyzing an unknown chemical, a chemist determines that the chemical is composed of carbon, hydrogen, and oxygen in the proportion of 1 carbon to 2 hydrogens to 1 oxygen. The chemical is probably
    (a) an amino acid   (b) DNA
    (c) a triglyceride   (d) a protein
    (e) a monosaccharide.

14. Match the following reactions with the term that describes them:
    ____ (a) $H_2 + Cl_2 \longrightarrow 2HCl$ (1) synthesis
    ____ (b) $3\,NaOH + H_3PO_4 \longrightarrow$ reaction
             $Na_3PO_4 + 3\,H_2O$ (2) exchange
    ____ (c) $CaCO_3 + CO_2 + H_2O \longrightarrow$ reaction
             $Ca(HCO_3)_2$ (3) decomposition
    ____ (d) $NH_3 + H_2O \rightleftharpoons NH_4^+ + OH^-$ reaction
    ____ (e) $C_{12}H_{22}O_{11} + H_2O \longrightarrow$ (4) reversible
             $C_6H_{12}O_6 + C_6H_{12}O_6$ reaction

15. Match the following:
    ____ (a) an abundant polar covalent molecule that serves as a solvent, has a high heat capacity, creates a high surface tension, and serves as a lubricant
    ____ (b) a substance that dissociates into one or more hydrogen ions and one or more anions
    ____ (c) a substance that dissociates into cations and anions, neither of which is a hydrogen ion or a hydroxide ion
    ____ (d) a proton acceptor
    ____ (e) a measure of hydrogen ion concentration
    ____ (f) a chemical compound that can convert strong acids and bases into weak ones
    ____ (g) a catalyst for chemical reactions that is specific, efficient, and under cellular control
    ____ (h) a single-stranded compound that contains a five-carbon sugar, and the bases adenine, cytosine, guanine, and uracil
    ____ (i) a compound that functions to temporarily store and then transfer energy liberated in exergonic reactions to cellular activities that require energy
    ____ (j) a double-stranded compound that contains a five-carbon sugar, the bases adenine, thymine, cytosine, and guanine, and the body's genetic material
    ____ (k) a charged atom
    ____ (l) a charged atom with an unpaired electron in its outermost shell

    (1) acid
    (2) free radical
    (3) base
    (4) buffer
    (5) enzyme
    (6) ion
    (7) pH
    (8) salt
    (9) RNA
    (10) ATP
    (11) water
    (12) DNA

CRITICAL THINKING QUESTIONS

1. Your best friend has decided to begin frying his breakfast eggs in margarine instead of butter because he has heard that eating butter is bad for his heart. Has he made a wise choice? Are there other alternatives?

2. A 4-month-old baby is admitted to the hospital with a temperature of 102°F (38.9°C). Why is it critical to treat the fever as quickly as possible?

3. During chemistry lab, Maria places sucrose (table sugar) in a glass beaker, adds water, and stirs. As the table sugar disappears, she loudly proclaims that she has chemically broken down the sucrose into fructose and glucose. Is Maria's chemical analysis correct?

? ANSWERS TO FIGURE QUESTIONS

2.1 In carbon, the first shell contains two electrons and the second shell contains four electrons.

2.2 The four most plentiful elements in living organisms are oxygen, carbon, hydrogen, and nitrogen.

2.3 Antioxidants such as selenium, zinc, beta-carotene, vitamin C, and vitamin E can inactivate free radicals derived from oxygen.

2.4 A cation is a positively charged ion; an anion is a negatively charged ion.

2.5 An ionic bond involves the *loss* and *gain* of electrons; a covalent bond involves the *sharing* of pairs of electrons.

2.6 The N atom in ammonia is electronegative. Because it attracts electrons more strongly than do the H atoms, the nitrogen end of ammonia acquires a slight negative charge, allowing H atoms in water molecules (or in other ammonia molecules) to form hydrogen bonds with it. Likewise, O atoms in water molecules can form hydrogen bonds with H atoms in ammonia molecules.

2.7 The number of hydrogen atoms in the reactants must equal the number in the products—in this case, four hydrogen atoms total. Put another way, two molecules of H_2 are needed to react with each molecule of O_2 so that the number of H atoms and O atoms in the reactants is the same as the number of H atoms and O atoms in the products.

2.8 This reaction is exergonic because the reactants have more potential energy than the products.

2.9 No. A catalyst does not change the potential energies of the products and reactants; it only lowers the activation energy needed to get the reaction going.

2.10 No. Because sugar easily dissolves in a polar solvent (water), you can correctly predict that it has several polar covalent bonds.

2.11 $CaCO_3$ is a salt, and H_2SO_4 is an acid.

2.12 At pH = 6, $[H^+] = 10^{-6}$ mol/liter and $[OH^-] = 10^{-8}$ mol/liter. A pH of 6.82 is more acidic than a pH of 6.91. Both pH = 8.41 and pH = 5.59 are 1.41 pH units from neutral (pH = 7).

2.13 Glucose has five —OH groups and six carbon atoms.

2.14 Hexoses are six-carbon sugars; examples include glucose, fructose, and galactose.

2.15 There are 6 carbons in fructose and 12 in sucrose.

2.16 Cells in the liver and in skeletal muscle store glycogen.

2.17 The oxygen in the water molecule comes from a fatty acid.

2.18 The polar head is hydrophilic, and the nonpolar tails are hydrophobic.

2.19 The only differences between estradiol and testosterone are the number of double bonds and the types of functional groups attached to ring A.

2.20 An amino acid has a minimum of two carbon atoms and one nitrogen atom.

2.21 Hydrolysis occurs during catabolism of proteins.

2.22 No. Proteins consisting of a single polypeptide chain do not have a quaternary structure.

2.23 Sucrase has specificity for the sucrose molecule and thus would not "recognize" glucose and fructose.

2.24 In DNA, thymine always pairs with adenine, and cytosine always pairs with guanine.

2.25 Cellular activities that depend on energy supplied by ATP include muscular contractions, movement of chromosomes, transport of substances across cell membranes, and synthesis (anabolic) reactions.

3 THE CELLULAR LEVEL OF ORGANIZATION

CELLS AND HOMEOSTASIS *Cells carry out a multitude of functions that help each system contribute to the homeostasis of the entire body. At the same time, all cells share key structures and functions that support their intense activity.*

In the previous chapter you learned about the atoms and molecules that compose the alphabet of the language of the human body. These are combined into about 200 different types of words called **cells**—living structural and functional units enclosed by a membrane. All cells arise from existing cells by the process of **cell division,** in which one cell divides into two identical cells. Different types of cells fulfill unique roles that support homeostasis and contribute to the many functional capabilities of the human organism. **Cell biology** or *cytology* is the study of cellular structure and function. As you study the various parts of a cell and their relationships to one another, you will learn that cell structure and function are intimately related. In this chapter, you will learn that cells carry out a dazzling array of chemical reactions to create and maintain life processes—in part, by isolating specific types of chemical reactions within specialized cellular structures.

Did you ever wonder why cancer is so difficult to treat?

3.1 PARTS OF A CELL

◉ **OBJECTIVE**

• Name and describe the three main parts of a cell.

Figure 3.1 provides an overview of the typical structures found in body cells. Most cells have many of the structures shown in this diagram. For ease of study, we divide the cell into three main parts: plasma membrane, cytoplasm, and nucleus.

1. The **plasma membrane** forms the cell's flexible outer surface, separating the cell's *internal environment* (everything inside the cell) from the *external environment* (everything outside the cell). It is a selective barrier that regulates the flow of materials into and out of a cell. This selectivity helps establish and maintain the appropriate environment for normal cellular activities. The plasma membrane also plays a key role in communication among cells and between cells and their external environment.

2. The **cytoplasm** (SĪ-tō-plasm; *-plasm* = formed or molded) consists of all the cellular contents between the plasma membrane and the nucleus. This compartment has two components: cytosol and organelles. **Cytosol** (SĪ-tō-sol), the fluid portion of cytoplasm, contains water, dissolved solutes, and suspended particles. Within the cytosol are several different types of **organelles** (or-gan-ELZ = little organs). Each type of organelle has a characteristic shape and specific functions. Examples include the cytoskeleton, ribosomes, endoplasmic reticulum, Golgi complex, lysosomes, peroxisomes, and mitochondria.

3. The **nucleus** (NOO-klē-us = nut kernel) is a large organelle that houses most of a cell's DNA. Within the nucleus, each **chromosome** (KRŌ-mō-sōm; *chromo-* = colored), a single molecule of DNA associated with several proteins, contains thousands of hereditary units called **genes** that control most aspects of cellular structure and function.

✔ **CHECKPOINT**

1. List the three main parts of a cell and explain their functions.

Figure 3.1 **Typical structures found in body cells.**

The cell is the basic living, structural, and functional unit of the body.

Flagellum   Cilium

Cytoskeleton:
  Microtubule
  Microfilament
  Intermediate filament
Microvilli
Centrosome:
  Pericentriolar material
  Centrioles
PLASMA MEMBRANE
Secretory vesicle
Lysosome
Smooth endoplasmic reticulum (ER)
Peroxisome
Mitochondrion
Microtubule

Proteasome   Free ribosomes

NUCLEUS:
  Chromatin
  Nuclear pore
  Nuclear envelope
  Nucleolus
Glycogen granules
CYTOPLASM (cytosol plus organelles except the nucleus)
Rough endoplasmic reticulum (ER)
Membrane-bound ribosome
Golgi complex

Microfilament

Sectional view

❓ **What are the three principal parts of a cell?**

3.2 THE PLASMA MEMBRANE

⊙ OBJECTIVES

- Distinguish between cytoplasm and cytosol.
- Explain the concept of selective permeability.
- Define the electrochemical gradient and describe its components.

The **plasma membrane,** a flexible yet sturdy barrier that surrounds and contains the cytoplasm of a cell, is best described by using a structural model called the **fluid mosaic model.** According to this model, the molecular arrangement of the plasma membrane resembles a continually moving sea of fluid lipids that contains a mosaic of many different proteins (Figure 3.2). Some proteins float freely like icebergs in the lipid sea, whereas others are anchored at specific locations like islands. The membrane lipids allow passage of several types of lipid-soluble molecules but act as a barrier to the entry or exit of charged or polar substances. Some of the proteins in the plasma membrane allow movement of polar molecules and ions into and out of the cell. Other proteins can act as signal receptors or as molecules that link the plasma membrane to intracellular or extracellular proteins.

Structure of the Plasma Membrane

The Lipid Bilayer

The basic structural framework of the plasma membrane is the **lipid bilayer,** two back-to-back layers made up of three types of lipid molecules—phospholipids, cholesterol, and glycolipids (Figure 3.2). About 75% of the membrane lipids are **phospholipids,** lipids that contain phosphorus. Present in smaller amounts are **cholesterol** (about 20%), a steroid with an attached —OH (hydroxyl) group, and various **glycolipids** (about 5%), lipids with attached carbohydrate groups.

The bilayer arrangement occurs because the lipids are **amphipathic** (am-fē-PATH-ik) molecules, which means that they have both polar and nonpolar parts. In phospholipids (see Figure 2.18), the polar part is the phosphate-containing "head," which is *hydrophilic* (*hydro-* = water; *-philic* = loving). The nonpolar parts are the two long fatty acid "tails," which are *hydrophobic* (*-phobic* = fearing) hydrocarbon chains. Because "like seeks like," the phospholipid molecules orient themselves in the bilayer with their hydrophilic heads facing outward. In this way, the heads face a watery fluid on either side—cytosol on the inside and extracellular fluid on the outside. The hydrophobic fatty acid tails in each half

Figure 3.2 **The fluid mosaic arrangement of lipids and proteins in the plasma membrane.**

 Membranes are fluid structures because the lipids and many of the proteins are free to rotate and move sideways in their own half of the bilayer.

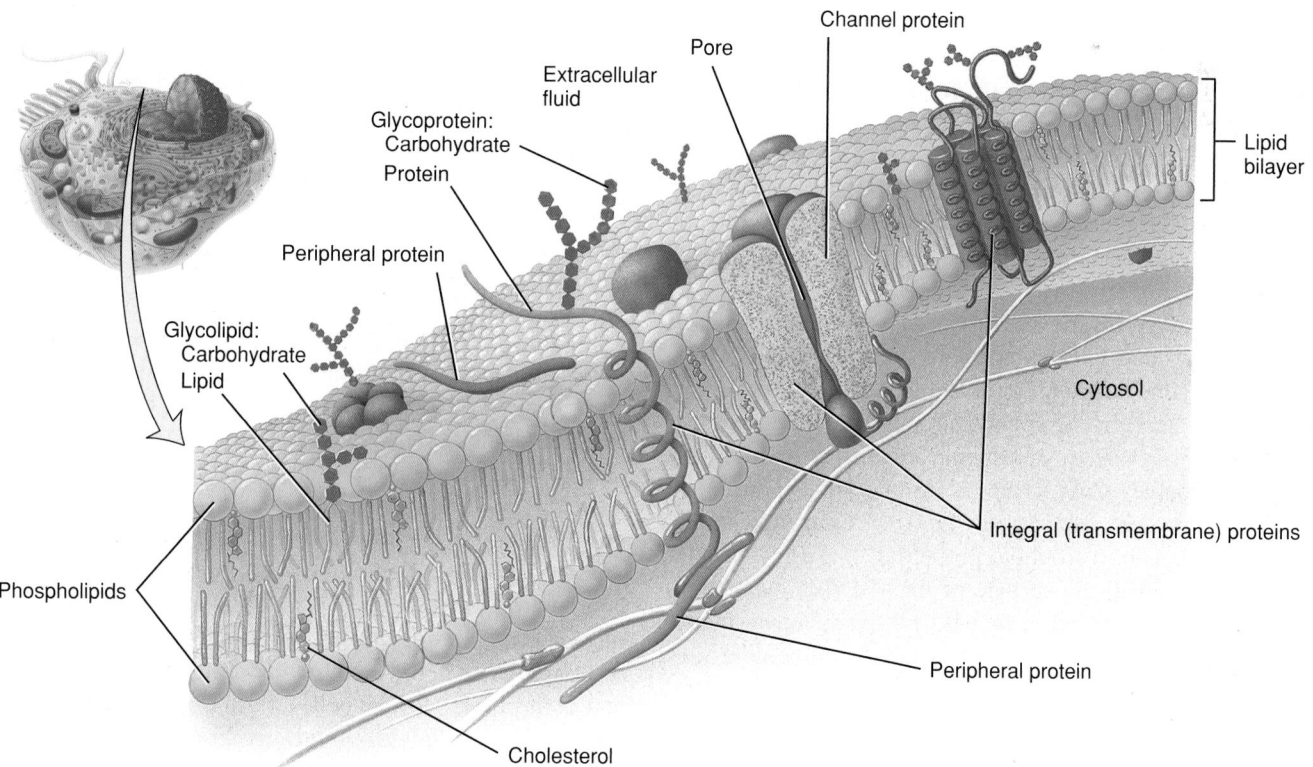

? **What is the glycocalyx?**

of the bilayer point toward one another, forming a nonpolar, hydrophobic region in the membrane's interior.

Cholesterol molecules are weakly amphipathic (see Figure 2.19a) and are interspersed among the other lipids in both layers of the membrane. The tiny —OH group is the only polar region of cholesterol, and it forms hydrogen bonds with the polar heads of phospholipids and glycolipids. The stiff steroid rings and hydrocarbon tail of cholesterol are nonpolar; they fit among the fatty acid tails of the phospholipids and glycolipids. The carbohydrate groups of glycolipids form a polar "head"; their fatty acid "tails" are nonpolar. Glycolipids appear only in the membrane layer that faces the extracellular fluid, which is one reason the two sides of the bilayer are asymmetric, or different.

Arrangement of Membrane Proteins

Membrane proteins are classified as integral or peripheral according to whether they are firmly embedded in the membrane (Figure 3.2). **Integral proteins** extend into or through the lipid bilayer among the fatty acid tails and are firmly embedded in it. Most integral proteins are **transmembrane proteins,** which means that they span the entire lipid bilayer and protrude into both the cytosol and extracellular fluid. A few integral proteins are tightly attached to one side of the bilayer by covalent bonding to fatty acids. Like membrane lipids, integral membrane proteins are amphipathic. Their hydrophilic regions protrude into either the watery extracllular fluid or the cytosol, and their hydrophobic regions extend among the fatty acid tails.

As their name implies, **peripheral proteins** (pe-RIF-er-al) are not as firmly embedded in the membrane. They are attached to the polar heads of membrane lipids or to integral proteins at the inner or outer surface of the membrane.

Many integral proteins are **glycoproteins,** proteins with carbohydrate groups attached to the ends that protrude into the extracellular fluid. The carbohydrates are *oligosaccharides* (*oligo-* = few; *-saccharides* = sugars), chains of 2 to 60 monosaccharides that may be straight or branched. The carbohydrate portions of glycolipids and glycoproteins form an extensive sugary coat called the **glycocalyx** (glī-kō-KĀL-iks). The pattern of carbohydrates in the glycocalyx varies from one cell to another. Therefore, the glycocalyx acts like a molecular "signature" that enables cells to recognize one another. For example, a white blood cell's ability to detect a "foreign" glycocalyx is one basis of the immune response that helps us destroy invading organisms. In addition, the glycocalyx enables cells to adhere to one another in some tissues and protects cells from being digested by enzymes in the extracellular fluid. The hydrophilic properties of the glycocalyx attract a film of fluid to the surface of many cells. This action makes red blood cells slippery as they flow through narrow blood vessels and protects cells that line the airways and the gastrointestinal tract from drying out.

Functions of Membrane Proteins

Generally, the types of lipids in cellular membranes vary only slightly. In contrast, the membranes of different cells and various intracellular organelles have remarkably different assortments of proteins that determine many of the membrane's functions (Figure 3.3).

- Some integral proteins form **ion channels,** *pores* or holes that specific ions, such as potassium ions (K^+), can flow through to get into or out of the cell. Most ion channels are *selective;* they allow only a single type of ion to pass through.

- Other integral proteins act as **carriers,** selectively moving a polar substance or ion from one side of the membrane to the other. Carriers are also known as *transporters.*

- Integral proteins called **receptors** serve as cellular recognition sites. Each type of receptor recognizes and binds a specific type of molecule. For instance, insulin receptors bind the hormone insulin. A specific molecule that binds to a receptor is called a **ligand** (LĪ-gand; *liga* = tied) of that receptor.

- Some integral proteins are **enzymes** that catalyze specific chemical reactions at the inside or outside surface of the cell.

- Integral proteins may also serve as **linkers** that anchor proteins in the plasma membranes of neighboring cells to one another or to protein filaments inside and outside the cell. Peripheral proteins also serve as enzymes and linkers.

- Membrane glycoproteins and glycolipids often serve as **cell-identity markers.** They may enable a cell to (1) recognize other cells of the same kind during tissue formation or (2) recognize and respond to potentially dangerous foreign cells. The ABO blood type markers are one example of cell-identity markers. When you receive a blood transfusion, the blood type must be compatible with your own, or red blood cells may clump together.

In addition, peripheral proteins help support the plasma membrane, anchor integral proteins, and participate in mechanical activities such as moving materials and organelles within cells, changing cell shape in dividing and muscle cells, and attaching cells to one another.

Membrane Fluidity

Membranes are fluid structures; that is, most of the membrane lipids and many of the membrane proteins easily rotate and move sideways in their own half of the bilayer. Neighboring lipid molecules exchange places about 10 million times per second and may wander completely around a cell in only a few minutes! Membrane fluidity depends both on the number of double bonds in the fatty acid tails of the lipids that make up the bilayer, and on the amount of cholesterol present. Each double bond puts a "kink" in the fatty acid tail (see Figure 2.18), which increases membrane fluidity by preventing lipid molecules from packing tightly in the membrane. Membrane fluidity is an excellent compromise for the cell; a rigid membrane would lack mobility, and a completely fluid membrane would lack the structural organization and mechanical support required by the cell. Membrane fluidity allows interactions to occur within the plasma membrane, such as the assembly of membrane proteins. It also

Figure 3.3 Functions of membrane proteins.

Membrane proteins largely reflect the functions a cell can perform.

Extracellular fluid	Plasma membrane	Cytosol

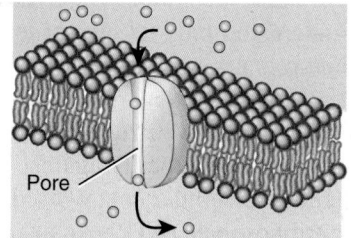

Ion channel (integral)
Allows specific ion (○) to move through water-filled pore. Most plasma membranes include specific channels for several common ions.

Pore

Carrier (integral)
Transports specific substances (○) across membrane by changing shape. For example, amino acids, needed to synthesize new proteins, enter body cells via carriers. Carrier proteins are also known as *transporters*.

Ligand

Receptor (integral)
Recognizes specific ligand (▽) and alters cell's function in some way. For example, antidiuretic hormone binds to receptors in the kidneys and changes the water permeability of certain plasma membranes.

Substrate
Products

Enzyme (integral and peripheral)
Catalyzes reaction inside or outside cell (depending on which direction the active site faces). For example, lactase protruding from epithelial cells lining your small intestine splits the disaccharide lactose in the milk you drink.

Linker (integral and peripheral)
Anchors filaments inside and outside the plasma membrane, providing structural stability and shape for the cell. May also participate in movement of the cell or link two cells together.

MHC protein

Cell identity marker (glycoprotein)
Distinguishes your cells from anyone else's (unless you are an identical twin). An important class of such markers are the major histocompatibility (MHC) proteins.

? **When stimulating a cell, the hormone insulin first binds to a protein in the plasma membrane. This action best represents which membrane protein function?**

enables the movement of the membrane components responsible for cellular processes such as cell movement, growth, division, and secretion, and the formation of cellular junctions. Fluidity allows the lipid bilayer to self-seal if torn or punctured. When a needle is pushed through a plasma membrane and pulled out, the puncture site seals spontaneously, and the cell does not burst. This property of the lipid bilayer allows a procedure called intracytoplasmic sperm injection to help infertile couples conceive a child; scientists can fertilize an oocyte by injecting a sperm cell through a tiny syringe. It also permits removal and replacement of a cell's nucleus in cloning experiments, such as the one that created Dolly, the famous cloned sheep.

Despite the great mobility of membrane lipids and proteins in their own half of the bilayer, they seldom flip-flop from one half of the bilayer to the other, because it is difficult for hydrophilic parts of membrane molecules to pass through the hydrophobic core of the membrane. This difficulty contributes to the asymmetry of the membrane bilayer.

Because of the way it forms hydrogen bonds with neighboring phospholipid and glycolipid heads and fills the space between bent fatty acid tails, cholesterol makes the lipid bilayer stronger but less fluid at normal body temperature. At low temperatures, cholesterol has the opposite effect—it increases membrane fluidity.

Membrane Permeability

The term *permeable* means that a structure permits the passage of substances through it, while *impermeable* means that a structure does not permit the passage of substances through it. The permeability of the plasma membrane to different substances varies. Plasma membranes permit some substances to pass more readily than others. This property of membranes is termed **selective permeability** (per′-mē-a-BIL-i-tē).

The lipid bilayer portion of the membrane is permeable to nonpolar, uncharged molecules, such as oxygen, carbon dioxide, and steroids, but is impermeable to ions and large, uncharged polar molecules such as glucose. It is also *slightly* permeable to small, uncharged polar molecules such as water and urea, a waste product from the breakdown of amino acids. The slight permeability to water and urea is an unexpected property since they are polar molecules. These two small molecules are thought to pass through the lipid bilayer in the following way. As the fatty acid tails of membrane phospholipids and glycolipids randomly move about, small gaps briefly appear in the hydrophobic environment of the membrane's interior. Water and urea molecules are small enough to move from one gap to another until they have crossed the membrane.

Transmembrane proteins that act as channels and carriers increase the plasma membrane's permeability to a variety of ions and uncharged polar molecules that, unlike water and urea molecules, cannot cross the lipid bilayer unassisted. Channels and carriers are very selective. Each one helps a specific molecule or ion to cross the membrane. Macromolecules, such as proteins, are so large that they are unable to pass across the plasma membrane except by endocytosis and exocytosis (discussed later in this chapter).

Gradients across the Plasma Membrane

The selective permeability of the plasma membrane allows a living cell to maintain different concentrations of certain substances on either side of the plasma membrane. A **concentration gradient** is a difference in the concentration of a chemical from one place to another, such as from the inside to the outside of the plasma membrane. Many ions and molecules are more concentrated in either the cytosol or the extracellular fluid. For instance, oxygen molecules and sodium ions (Na^+) are more concentrated in the extracellular fluid than in the cytosol; the opposite is true of carbon dioxide molecules and potassium ions (K^+).

The plasma membrane also creates a difference in the distribution of positively and negatively charged ions between the two sides of the plasma membrane. Typically, the inner surface of the plasma membrane is more negatively charged and the outer surface is more positively charged. A difference in electrical charges between two regions constitutes an **electrical gradient.** Because it occurs across the plasma membrane, this charge difference is termed the **membrane potential.**

As you will see shortly, the concentration gradient and electrical gradient are important because they help move substances across the plasma membrane. In many cases a substance will move across a plasma membrane *down its concentration gradient.* That is to say, a substance will move "downhill," from where it is more concentrated to where it is less concentrated, to reach equilibrium. Similarly, a positively charged substance will tend to move toward a negatively charged area, and a negatively charged substance will tend to move toward a positively charged area. The combined influence of the concentration gradient and the electrical gradient on movement of a particular ion is referred to as its **electrochemical gradient.**

✔ CHECKPOINT

2. How do hydrophobic and hydrophilic regions govern the arrangement of membrane lipids in a bilayer?
3. What substances can and cannot diffuse through the lipid bilayer?
4. "The proteins present in a plasma membrane determine the functions that a membrane can perform." Is this statement true or false? Explain your answer.
5. How does cholesterol affect membrane fluidity?
6. Why are membranes said to have selective permeability?
7. What factors contribute to an electrochemical gradient?

3.3 TRANSPORT ACROSS THE PLASMA MEMBRANE

◉ OBJECTIVE

• Describe the processes that transport substances across the plasma membrane.

Transport of materials across the plasma membrane is essential to the life of a cell. Certain substances must move into the cell to support metabolic reactions. Other substances that have been pro-

duced by the cell for export or as cellular waste products must move out of the cell.

Substances generally move across cellular membranes via transport processes that can be classified as passive or active, depending on whether they require cellular energy. In **passive processes,** a substance moves down its concentration or electrical gradient to cross the membrane using only its own kinetic energy (energy of motion). Kinetic energy is intrinsic to the particles that are moving. There is no input of energy from the cell. An example is simple diffusion. In **active processes,** cellular energy is used to drive the substance "uphill" against its concentration or electrical gradient. The cellular energy used is usually in the form of adenosine triphosphate (ATP). An example is active transport. Another way that some substances may enter and leave cells is an active process in which tiny, spherical membrane sacs referred to as **vesicles** are used. Examples include endocytosis, in which vesicles detach from the plasma membrane while bringing materials into a cell, and exocytosis, the merging of vesicles with the plasma membrane to release materials from the cell.

Passive Processes

The Principle of Diffusion

Learning why materials diffuse across membranes requires an understanding of how diffusion occurs in a solution. **Diffusion** (di-FŪ-zhun; *diffus-* = spreading) is a passive process in which the random mixing of particles in a solution occurs because of the particles' kinetic energy. Both the *solutes,* the dissolved substances, and the *solvent,* the liquid that does the dissolving, undergo diffusion. If a particular solute is present in high concentration in one area of a solution and in low concentration in another area, solute molecules will diffuse toward the area of lower concentration—they move *down their concentration gradient.* After some time, the particles become evenly distributed throughout the solution and the solution is said to be at equilibrium. The particles continue to move about randomly due to their kinetic energy, but their concentrations do not change.

For example, when you place a crystal of dye in a water-filled container (Figure 3.4), the color is most intense in the area closest to the dye because its concentration is higher there. At increasing distances, the color is lighter and lighter because the dye concentration is lower. Some time later, the solution of water and dye will have a uniform color, because the dye molecules and water molecules have diffused down their concentration gradients until they are evenly mixed in solution—they are at equilibrium (ē-kwi-LIB-rē-um).

In this simple example, no membrane was involved. Substances may also diffuse through a membrane, if the membrane is permeable to them. Several factors influence the diffusion rate of substances across plasma membranes:

• **Steepness of the concentration gradient.** The greater the difference in concentration between the two sides of the membrane, the higher is the rate of diffusion. When charged particles are diffusing, the steepness of the electrochemical gradient determines the diffusion rate across the membrane.

- *Temperature.* The higher the temperature, the faster the rate of diffusion. All of the body's diffusion processes occur more rapidly in a person with a fever.
- *Mass of the diffusing substance.* The larger the mass of the diffusing particle, the slower its diffusion rate. Smaller molecules diffuse more rapidly than larger ones.
- *Surface area.* The larger the membrane surface area available for diffusion, the faster is the diffusion rate. For example, the air sacs of the lungs have a large surface area available for diffusion of oxygen from the air into the blood. Some lung diseases, such as emphysema, reduce the surface area. This slows the rate of oxygen diffusion and makes breathing more difficult.
- *Diffusion distance.* The greater the distance over which diffusion must occur, the longer it takes. Diffusion across a plasma membrane takes only a fraction of a second because the membrane is so thin. In pneumonia, fluid collects in the lungs; the additional fluid increases the diffusion distance because oxygen must move through both the built-up fluid and the membrane to reach the bloodstream.

Now that you have a basic understanding of the nature of diffusion, we will consider three types of diffusion: simple diffusion, facilitated diffusion, and osmosis.

Simple Diffusion

Simple diffusion is a passive process in which substances move freely through the lipid bilayer of the plasma membranes of cells without the help of membrane transport proteins (Figure 3.5). Nonpolar, hydrophobic molecules move across the lipid bilayer through the process of simple diffusion. Such molecules include oxygen, carbon dioxide, and nitrogen gases; fatty acids; steroids; and fat-soluble vitamins (A, D, E, and K). Small, uncharged polar

Figure 3.4 Principle of diffusion. At the beginning of our experiment, a crystal of dye placed in a cylinder of water dissolves (a) and then diffuses from the region of higher dye concentration to regions of lower dye concentration (b). At equilibrium (c), the dye concentration is uniform throughout, although random movement continues.

🔑 In diffusion, a substance moves down its concentration gradient.

Beginning   Intermediate   Equilibrium
(a)   (b)   (c)

❓ **How would having a fever affect body processes that involve diffusion?**

Figure 3.5 Simple diffusion, channel-mediated facilitated diffusion, and carrier-mediated facilitated diffusion.

🔑 In simple diffusion, a substance moves across the lipid bilayer of the plasma membrane without the help of membrane transport proteins. In facilitated diffusion, a substance moves across the lipid bilayer aided by a channel protein or a carrier protein.

Extracellular fluid

Concentration gradient

Cytosol

Simple diffusion

Channel-mediated facilitated diffusion

Carrier-mediated facilitated diffusion

❓ **What types of molecules move across the lipid bilayer of the plasma membrane via simple diffusion?**

molecules such as water, urea, and small alcohols also pass through the lipid bilayer by simple diffusion. Simple diffusion through the lipid bilayer is important in the movement of oxygen and carbon dioxide between blood and body cells, and between blood and air within the lungs during breathing. It also is the route for absorption of some nutrients and excretion of some wastes by body cells.

Facilitated Diffusion

Solutes that are too polar or highly charged to move through the lipid bilayer by simple diffusion can cross the plasma membrane by a passive process called **facilitated diffusion.** In this process, an integral membrane protein assists a specific substance across the membrane. The integral membrane protein can be either a membrane channel or a carrier.

CHANNEL-MEDIATED FACILITATED DIFFUSION In **channel-mediated facilitated diffusion,** a solute moves down its concentration gradient across the lipid bilayer through a membrane channel (Figure 3.5). Most membrane channels are *ion channels,* integral transmembrane proteins that allow passage of small, inorganic ions that are too hydrophilic to penetrate the nonpolar interior of the lipid bilayer. Each ion can diffuse across the membrane only at certain sites. In typical plasma membranes, the most numerous ion channels are selective for K^+ (potassium ions) or Cl^- (chloride ions); fewer channels are available for Na^+ (sodium ions) or Ca^{2+} (calcium ions). Diffusion of ions through channels is generally slower than free diffusion through the lipid bilayer because channels occupy a smaller fraction of the membrane's total surface area than lipids. Still, facilitated diffusion through channels is a very fast process: More than a million potassium ions can flow through a K^+ channel in one second!

A channel is said to be *gated* when part of the channel protein acts as a "plug" or "gate," changing shape in one way to open the pore and in another way to close it (Figure 3.6). Some gated channels randomly alternate between the open and closed positions; others are regulated by chemical or electrical changes inside and outside the cell. When the gates of a channel are open, ions diffuse into or out of cells, down their electrochemical gradients. The plasma membranes of different types of cells may have different numbers of ion channels and thus display different permeabilities to various ions.

CARRIER-MEDIATED FACILITATED DIFFUSION In **carrier-mediated facilitated diffusion,** a *carrier* (also called a *transporter*) moves a solute down its concentration gradient across the plasma membrane (see Figure 3.5). Since this is a passive process, no cellular energy is required. The solute binds to a specific carrier on one side of the membrane and is released on the other side after the carrier undergoes a change in shape. The solute binds more

often to the carrier on the side of the membrane with a higher concentration of solute. Once the concentration is the same on both sides of the membrane, solute molecules bind to the carrier on the cytosolic side and move out to the extracellular fluid as rapidly as they bind to the carrier on the extracellular side and move into the cytosol. The rate of carrier-mediated facilitated diffusion (how quickly it occurs) is determined by the steepness of the concentration gradient across the membrane.

The number of carriers available in a plasma membrane places an upper limit, called the *transport maximum,* on the rate at which facilitated diffusion can occur. Once all of the carriers are occupied, the transport maximum is reached, and a further increase in the concentration gradient does not increase the rate of facilitated diffusion. Thus, much like a completely saturated sponge can absorb no more water, the process of carrier-mediated facilitated diffusion exhibits *saturation.*

Figure 3.6 Channel-mediated facilitated diffusion of potassium ions (K^+) through a gated K^+ channel. A gated channel is one in which a portion of the channel protein acts as a gate to open or close the channel's pore to the passage of ions.

 Channels are integral membrane proteins that allow specific, small, inorganic ions to pass across the membrane by facilitated diffusion.

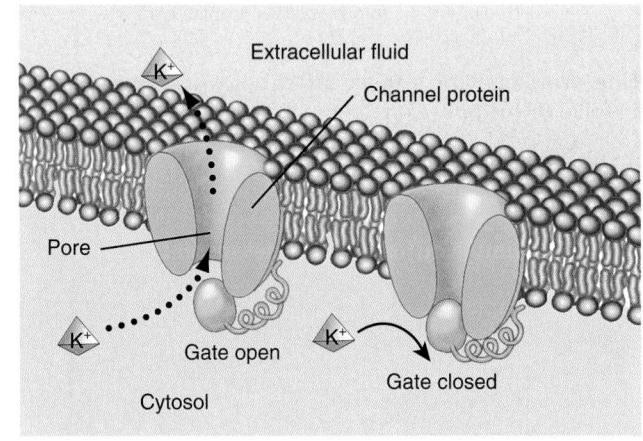

Details of the K⁺ channel

Is the concentration of K^+ in body cells higher in the cytosol or in the extracellular fluid?

Substances that move across the plasma membrane by carrier-mediated facilitated diffusion include glucose, fructose, galactose, and some vitamins. Glucose, the body's preferred energy source for making ATP, enters many body cells by carrier-mediated facilitated diffusion as follows (Figure 3.7):

1. Glucose binds to a specific type of carrier protein called the *glucose transporter* (GluT) on the outside surface of the membrane.

2. As the transporter undergoes a change in shape, glucose passes through the membrane.

3. The transporter releases glucose on the other side of the membrane.

The selective permeability of the plasma membrane is often regulated to achieve homeostasis. For instance, the hormone in-sulin, via the action of the insulin receptor, promotes the insertion of many copies of glucose transporters into the plasma membranes of certain cells. Thus, the effect of insulin is to elevate the transport maximum for facilitated diffusion of glucose into cells. With more glucose transporters available, body cells can pick up glucose from the blood more rapidly. An inability to produce or utilize insulin is called diabetes mellitus (Chapter 18).

Osmosis

Osmosis (oz-MŌ-sis) is a type of diffusion in which there is net movement of a solvent through a selectively permeable membrane. Like the other types of diffusion, osmosis is a passive process. In living systems, the solvent is water, which moves by osmosis across plasma membranes from an area of *higher water concentration* to an area of *lower water concentration*. Another way to understand this idea is to consider the solute concentration: In osmosis, water moves through a selectively permeable membrane from an area of *lower solute concentration* to an area of *higher solute concentration*. During osmosis, water molecules pass through a plasma membrane in two ways: (1) by moving between neighboring phospholipid molecules in the lipid bilayer via simple diffusion, as previously described, and (2) by moving through **aquaporins** (ak-wa-POR-ins; *aqua-* = water), integral membrane proteins that function as water channels.

Osmosis occurs only when a membrane is permeable to water but is not permeable to certain solutes. A simple experiment can demonstrate osmosis. Consider a U-shaped tube in which a selectively permeable membrane separates the left and right arms of the tube. A volume of pure water is poured into the left arm, and the same volume of a solution containing a solute that cannot pass through the membrane is poured into the right arm (Figure 3.8a). Because the *water* concentration is higher on the left and lower on the right, net movement of water molecules—osmosis—occurs from left to right, so that the water is moving down its concentration gradient. At the same time, the membrane prevents diffusion of the solute from the right arm into the left arm. As a result, the volume of water in the left arm decreases, and the volume of solution in the right arm increases (Figure 3.8b).

You might think that osmosis would continue until no water remained on the left side, but this is *not* what happens. In this experiment, the higher the column of solution in the right arm becomes, the more pressure it exerts on its side of the membrane. Pressure exerted in this way by a liquid, known as **hydrostatic pressure,** forces water molecules to move back into the left arm. Equilibrium is reached when just as many water molecules move from right to left due to the hydrostatic pressure as move from left to right due to osmosis (Figure 3.8b).

To further complicate matters, the solution with the impermeable solute also exerts a force, called the **osmotic pressure.** The osmotic pressure of a solution is proportional to the concentration of the solute particles that cannot cross the membrane—the higher the solute concentration, the higher the solution's osmotic

Figure 3.7 Carrier-mediated facilitated diffusion of glucose across a plasma membrane. The carrier protein binds to glucose in the extracellular fluid and releases it into the cytosol.

Carriers are integral membrane proteins that undergo changes in shape in order to move substances across the membrane by facilitated diffusion.

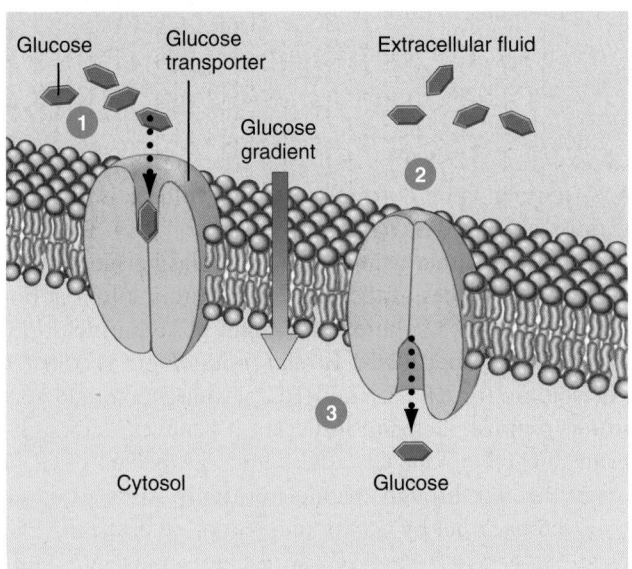

Does insulin alter glucose transport by facilitated diffusion?

Figure 3.8 Principle of osmosis. Water molecules move through the selectively permeable membrane; the solute molecules in the right arm cannot pass through the membrane. (a) As the experiment starts, water molecules move from the left arm into the right arm, down the water concentration gradient. (b) After some time, the volume of water in the left arm has decreased and the volume of solution in the right arm has increased. At equilibrium, there is no net osmosis: Hydrostatic pressure forces just as many water molecules to move from right to left as osmosis forces water molecules to move from left to right. (c) If pressure is applied to the solution in the right arm, the starting conditions can be restored. This pressure, which stops osmosis, is equal to the osmotic pressure.

Osmosis is the movement of water molecules through a selectively permeable membrane.

(a) Starting conditions   (b) Equilibrium   (c) Restoring starting conditions

Will the fluid level in the right arm rise until the water concentrations are the same in both arms?

pressure. Consider what would happen if a piston were used to apply more pressure to the fluid in the right arm of the tube in Figure 3.8. With enough pressure, the volume of fluid in each arm could be restored to the starting volume, and the concentration of solute in the right arm would be the same as it was at the beginning of the experiment (Figure 3.8c). The amount of pressure needed to restore the starting condition equals the osmotic pressure. So, in our experiment osmotic pressure is the pressure needed to stop the movement of water from the left tube into the right tube. Notice that the osmotic pressure of a solution does not produce the movement of water during osmosis. Rather it is the pressure that would *prevent* such water movement.

Normally, the osmotic pressure of the cytosol is the same as the osmotic pressure of the interstitial fluid outside cells. Because the osmotic pressure on both sides of the plasma membrane (which is selectively permeable) is the same, cell volume remains relatively constant. When body cells are placed in a solution having a different osmotic pressure than cytosol, however, the shape and volume of the cells change. As water moves by osmosis into or out of the cells, their volume increases or decreases. A solution's **tonicity** (tō-NIS-i-tē; *tonic* = tension) is a measure of the solution's ability to change the volume of cells by altering their water content.

Any solution in which a cell—for example, a red blood cell (RBC)—maintains its normal shape and volume is an **isotonic solution** (ī′-sō-TON-ik; *iso-* = same) (Figure 3.9). The concentrations of solutes that cannot cross the plasma membrane are the same on both sides of the membrane in this solution. For instance, a 0.9% NaCl solution (0.9 gram of sodium chloride in 100 mL of solution), called a *normal (physiological) saline solution,* is isotonic for RBCs. The RBC plasma membrane permits the water to move back and forth, but it behaves as though it is impermeable to Na^+ and Cl^-, the solutes. (Any Na^+ or Cl^- ions that enter the cell through channels or transporters are immediately moved back out by active transport or other means.) When RBCs are bathed in 0.9% NaCl, water molecules enter and exit at the same rate, allowing the RBCs to keep their normal shape and volume.

A different situation results if RBCs are placed in a **hypotonic solution** (hī′-pō-TON-ik; *hypo-* = less than), a solution that has a *lower* concentration of solutes than the cytosol inside the RBCs (Figure 3.9). In this case, water molecules enter the cells faster than they leave, causing the RBCs to swell and eventually to burst. The rupture of RBCs in this manner is called **hemolysis** (hē-MOL-i-sis; *hemo-* = blood; *-lysis* = to loosen or split apart); the rupture of other types of cells due to placement in a hypotonic solution is referred to simply as **lysis.** Pure water is very hypotonic and causes rapid hemolysis.

A **hypertonic solution** (hī′-per-TON-ik; *hyper-* = greater than) has a *higher* concentration of solutes than does the cytosol inside RBCs (Figure 3.9). One example of a hypertonic solution is a 2% NaCl solution. In such a solution, water molecules move out of the cells faster than they enter, causing the cells to shrink. Such shrinkage of cells is called **crenation** (kre-NĀ-shun).

CLINICAL CONNECTION | *Medical Uses of Isotonic, Hypertonic, and Hypotonic Solutions*

RBCs and other body cells may be damaged or destroyed if exposed to hypertonic or hypotonic solutions. For this reason, most **intravenous (IV) solutions,** liquids infused into the blood of a vein, are isotonic. Examples are isotonic saline (0.9% NaCl) and D5W, which stands for dextrose 5% in water. Sometimes infusion of a hypertonic solution such as mannitol is useful to treat patients who have *cerebral edema,* excess interstitial fluid in the brain. Infusion of such a solution relieves fluid overload by causing osmosis of water from interstitial fluid into the blood. The kidneys then excrete the excess water from the blood into the urine. Hypotonic solutions, given either orally or through an IV, can be used to treat people who are dehydrated. The water in the hypotonic solution moves from the blood into interstitial fluid and then into body cells to rehydrate them. Water and most sports drinks that you consume to "rehydrate" after a workout are hypotonic relative to your body cells. •

✔**CHECKPOINT**

8. What factors can increase the rate of diffusion?
9. How does simple diffusion compare with facilitated diffusion?
10. What is osmotic pressure?
11. Distinguish among isotonic, hypotonic, and hypertonic solutions.

Active Processes

Active Transport

Some polar or charged solutes that must enter or leave body cells cannot cross the plasma membrane through any form of passive transport because they would need to move "uphill," *against* their concentration gradients. Such solutes may be able to cross the membrane by a process called **active transport.** Active transport is considered an active process because energy is required for carrier proteins to move solutes across the membrane against a concentration gradient. Two sources of cellular energy can be used to drive active transport: (1) Energy obtained from hydrolysis of adenosine triphosphate (ATP) is the source in *primary active transport;* (2) energy stored in an ionic concentration gradient is the source in *secondary active transport.* Like carrier-mediated facilitated diffusion, active transport processes exhibit a transport maximum and saturation. Solutes actively transported across the plasma membrane include several ions, such as Na^+, K^+, H^+, Ca^{2+}, I^- (iodide ions), and Cl^-; amino acids; and monosaccha-

Figure 3.9 Tonicity and its effects on red blood cells (RBCs). The arrows indicate the direction and degree of water movement into and out of the cells. One example of an isotonic solution for RBCs is 0.9% NaCl.

Cells placed in an isotonic solution maintain their shape because there is no net water movement into or out of the cells.

Isotonic solution   Hypotonic solution   Hypertonic solution

(a) Illustrations showing direction of water movement

Normal RBC shape   RBC undergoes hemolysis   RBC undergoes crenation

SEM

(b) Scanning electron micrographs (all 15,000x)

 Will a 2% solution of NaCl cause hemolysis or crenation of RBCs? Why?

rides. (Note that some of these substances also cross the membrane via facilitated diffusion when the proper channel proteins or carriers are present.)

PRIMARY ACTIVE TRANSPORT In **primary active transport,** energy derived from hydrolysis of ATP changes the shape of a carrier protein, which "pumps" a substance across a plasma membrane against its concentration gradient. Indeed, carrier proteins that mediate primary active transport are often called **pumps.** A typical body cell expends about 40% of the ATP it generates on primary active transport. Chemicals that turn off ATP production—for example, the poison cyanide—are lethal because they shut down active transport in cells throughout the body.

The most prevalent primary active transport mechanism expels sodium ions (Na^+) from cells and brings potassium ions (K^+) in. Because of the specific ions it moves, this carrier is called the **sodium-potassium pump.** Because a part of the sodium-potassium pump acts as an *ATPase,* an enzyme that hydrolyzes ATP, another name for this pump is **Na^+/K^+ ATPase.**

All cells have thousands of sodium-potassium pumps in their plasma membranes. These sodium-potassium pumps maintain a low concentration of Na^+ in the cytosol by pumping these ions into the extracellular fluid against the Na^+ concentration gradient. At the same time, the pumps move K^+ into cells against the K^+ concentration gradient. Because K^+ and Na^+ slowly leak back across the plasma membrane down their electrochemical gradients—through passive transport or secondary active transport—the sodium-potassium pumps must work nonstop to maintain a low concentration of Na^+ and a high concentration of K^+ in the cytosol.

Figure 3.10 depicts the operation of the sodium-potassium pump:

❶ Three Na^+ in the cytosol bind to the pump protein.

❷ Binding of Na^+ triggers the hydrolysis of ATP into ADP, a reaction that also attaches a phosphate group ⓟ to the pump protein. This chemical reaction changes the shape of the pump protein, expelling the three Na^+ into the extracellular fluid. Now the shape of the pump protein favors binding of two K^+ in the extracellular fluid to the pump protein.

❸ The binding of K^+ triggers release of the phosphate group from the pump protein. This reaction again causes the shape of the pump protein to change.

❹ As the pump protein reverts to its original shape, it releases K^+ into the cytosol. At this point, the pump is again ready to bind three Na^+, and the cycle repeats.

The different concentrations of Na^+ and K^+ in cytosol and extracellular fluid are crucial for maintaining normal cell volume and for the ability of some cells to generate electrical signals such as action potentials. Recall that the tonicity of a solution is proportional to the concentration of its solute particles that cannot penetrate the membrane. Because sodium ions that diffuse into a cell or enter through secondary active transport are immediately pumped out, it is as if they never entered. In effect, sodium ions behave as if they cannot penetrate the membrane. Thus, sodium ions are an important contributor to the tonicity of the extracellular fluid. A similar condition holds for K^+ in the cytosol. By helping to maintain normal tonicity on each side of the plasma membrane, the sodium-potassium pumps ensure that cells neither shrink nor swell due to the movement of water by osmosis out of or into cells.

SECONDARY ACTIVE TRANSPORT In **secondary active transport,** the energy stored in a Na^+ or H^+ concentration gradient is used to drive other substances across the membrane against their own concentration gradients. Because a Na^+ or H^+ gradient is established by primary active transport, secondary active transport *indirectly* uses energy obtained from the hydrolysis of ATP.

The sodium-potassium pump maintains a steep concentration gradient of Na^+ across the plasma membrane. As a result, the sodium ions have stored or potential energy, just like water behind a dam. Accordingly, if there is a route for Na^+ to leak back in, some of the stored energy can be converted to kinetic energy (energy of motion) and used to transport other substances *against their concentration gradients.* In essence, secondary active transport proteins harness the energy in the Na^+ concentration gradient by providing routes for Na^+ to leak into cells. In secondary active transport, a carrier protein simultaneously binds to Na^+ and another substance and then changes its shape so that both substances cross the membrane at the same time. If these transporters move two substances in the same direction they are called **symporters** (sim-PORT-ers; *sym-* = same); **antiporters** (an′-tē-PORT-ers), in contrast, move two substances in opposite directions across the membrane (*anti-* = against).

Figure 3.10 **The sodium-potassium pump (Na^+/K^+ATPase) expels sodium ions (Na^+) and brings potassium ions (K^+) into the cell.**

🔑 Sodium-potassium pumps maintain a low intracellular concentration of sodium ions.

❓ **What is the role of ATP in the operation of this pump?**

Plasma membranes contain several antiporters and symporters that are powered by the Na^+ gradient (Figure 3.11). For instance, the concentration of calcium ions (Ca^{2+}) is low in the cytosol because Na^+/Ca^{2+} antiporters eject calcium ions. Likewise, Na^+/H^+ antiporters help regulate the cytosol's pH (H^+ concentration) by expelling excess H^+. By contrast, dietary glucose and amino acids are absorbed into cells that line the small intestine by Na^+/glucose and Na^+/amino acid symporters (Figure 3.11b). In each case, sodium ions are moving down their concentration gradient while the other solutes move "uphill," against their concentration gradients. Keep in mind that all these symporters and antiporters can do their job because the sodium–potassium pumps maintain a low concentration of Na^+ in the cytosol.

CLINICAL CONNECTION | *Digitalis Increases Ca²⁺ in Heart Muscle Cells*

Digitalis often is given to patients with *heart failure,* a condition of weakened pumping action by the heart. Digitalis exerts its effect by slowing the action of the sodium–potassium pumps, which lets more Na^+ accumulate inside heart muscle cells. The result is a decreased Na^+ concentration gradient across the plasma membrane, which causes the Na^+/Ca^{2+} antiporters to slow down. As a result, more Ca^{2+} remains inside heart muscle cells. The slight increase in the level of Ca^{2+} in the cytosol of heart muscle cells increases the force of their contractions and thus strengthens the force of the heartbeat. •

Transport in Vesicles

A **vesicle** (VES-i-kul = little blister or bladder), as noted earlier, is a small, spherical sac. As you will learn later in this chapter, a variety of substances are transported in vesicles from one structure to another within cells. Vesicles also import materials from and release materials into extracellular fluid. During **endocytosis** (en′-dō-sī-TŌ-sis; *endo-* = within), materials move into a cell in a vesicle formed from the plasma membrane. In **exocytosis** (ek′-sō-sī-TŌ-sis; *exo-* = out), materials move out of a cell by the fusion with the plasma membrane of vesicles formed inside the cell. Both endocytosis and exocytosis require energy supplied by ATP. Thus, transport in vesicles is an active process.

ENDOCYTOSIS Here we consider three types of endocytosis: receptor-mediated endocytosis, phagocytosis, and bulk-phase endocytosis. **Receptor-mediated endocytosis** is a highly selective type of endocytosis by which cells take up specific ligands. (Recall that ligands are molecules that bind to specific receptors.) A vesicle forms after a receptor protein in the plasma membrane recognizes and binds to a particular particle in the extracellular fluid. For instance, cells take up cholesterol-containing low-density lipoproteins (LDLs), transferrin (an iron-transporting protein in the blood), some vitamins, antibodies, and certain hormones by receptor-mediated endocytosis. Receptor-mediated endocytosis of LDLs (and other ligands) occurs as follows (Figure 3.12):

1 *Binding.* On the extracellular side of the plasma membrane, an LDL particle that contains cholesterol binds to a specific receptor in the plasma membrane to form a receptor–LDL complex. The receptors are integral membrane proteins that are concentrated in regions of the plasma membrane called *clathrin-coated pits.* Here, a protein called *clathrin* attaches to the membrane on its cytoplasmic

Figure 3.11 Secondary active transport mechanisms. (a) Antiporters carry two substances across the membrane in opposite directions. (b) Symporters carry two substances across the membrane in the same direction.

 Secondary active transport mechanisms use the energy stored in an ionic concentration gradient (here, for Na^+). Because primary active transport pumps that hydrolyze ATP maintain the gradient, secondary active transport mechanisms consume ATP indirectly.

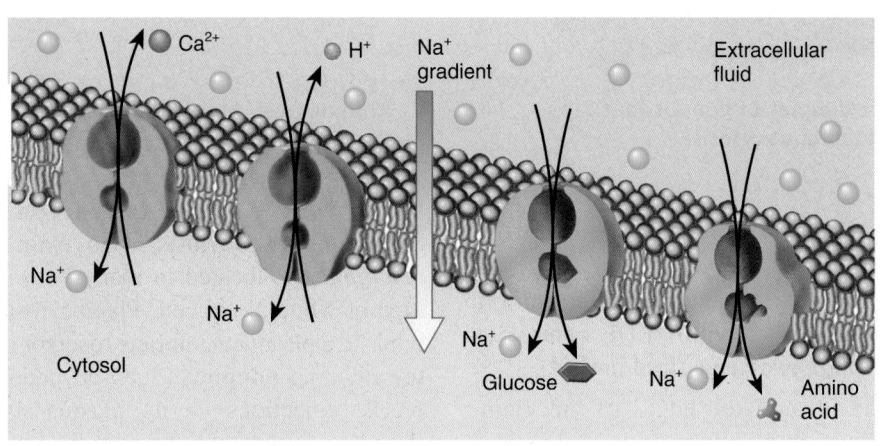

(a) Antiporters   (b) Symporters

? **What is the main difference between primary and secondary active transport mechanisms?**

Figure 3.12 Receptor-mediated endocytosis of a low-density lipoprotein (LDL) particle.

 Receptor-mediated endocytosis imports materials that are needed by cells.

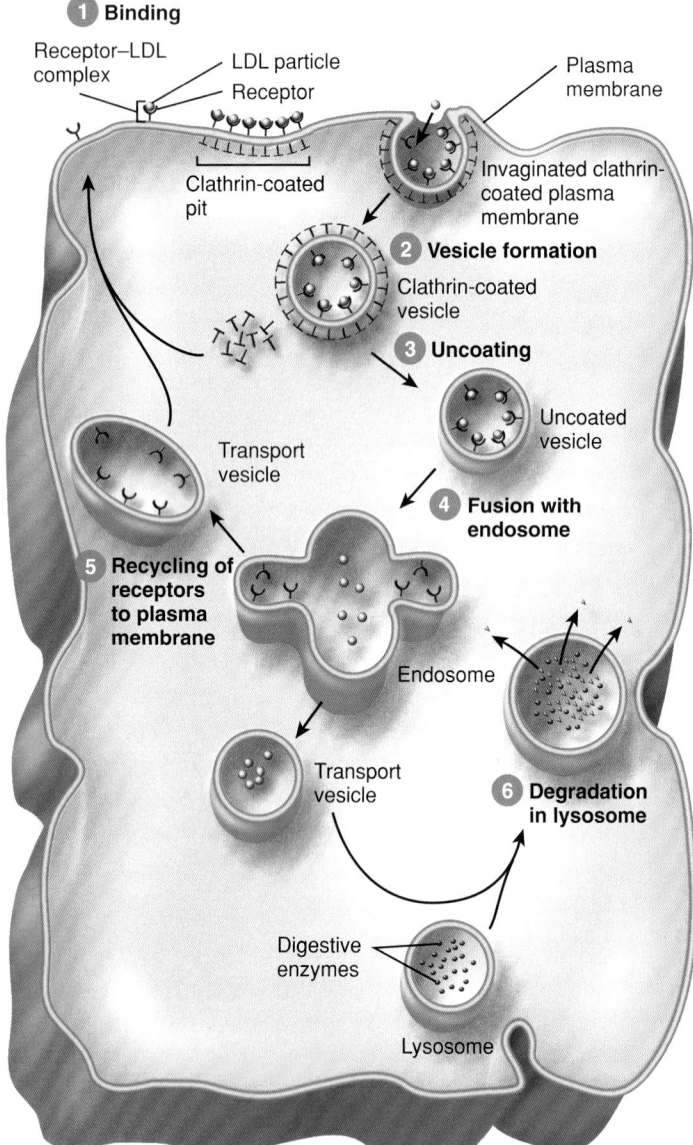

? **What are several other examples of ligands that can undergo receptor-mediated endocytosis?**

side. Many clathrin molecules come together, forming a basketlike structure around the receptor–LDL complexes that causes the membrane to invaginate (fold inward).

2 *Vesicle formation.* The invaginated edges of the membrane around the clathrin-coated pit fuse, and a small piece of the membrane pinches off. The resulting vesicle, known as a *clathrin-coated vesicle,* contains the receptor–LDL complexes.

3 *Uncoating.* Almost immediately after it is formed, the clathrin-coated vesicle loses its clathrin coat to become an *uncoated vesicle.* Clathrin molecules either return to the inner surface of the plasma membrane or help form coats on other vesicles inside the cell.

4 *Fusion with endosome.* The uncoated vesicle quickly fuses with a vesicle known as an *endosome.* Within an endosome, the LDL particles separate from their receptors.

5 *Recycling of receptors to plasma membrane.* Most of the receptors accumulate in elongated protrusions of the endosome (the arms of the cross-shaped vesicle at the center of the figure). These pinch off, forming transport vesicles that return the receptors to the plasma membrane. An LDL receptor is returned to the plasma membrane about 10 minutes after it enters a cell.

6 *Degradation in lysosomes.* Other transport vesicles, which contain the LDL particles, bud off the endosome and soon fuse with a *lysosome.* Lysosomes contain many digestive enzymes. Certain enzymes break down the large protein and lipid molecules of the LDL particle into amino acids, fatty acids, and cholesterol. These smaller molecules then leave the lysosome. The cell uses cholesterol for rebuilding its membranes and for synthesis of steroids, such as estrogen. Fatty acids and amino acids can be used for ATP production or to build other molecules needed by the cell.

CLINICAL CONNECTION | *Viruses and Receptor-Mediated Endocytosis*

Although receptor-mediated endocytosis normally imports needed materials, some viruses are able to use this mechanism to enter and infect body cells. For example, the human immunodeficiency virus (HIV), which causes acquired immunodeficiency syndrome (AIDS), can attach to a receptor called CD4. This receptor is present in the plasma membrane of white blood cells called helper T cells. After binding to CD4, HIV enters the helper T cell via receptor-mediated endocytosis. •

Phagocytosis (fag′-ō-sī-TŌ-sis; *phago-* = to eat) is a form of endocytosis in which the cell engulfs large solid particles, such as worn-out cells, whole bacteria, or viruses (Figure 3.13). Only a few body cells, termed **phagocytes** (FAG-ō-sīts), are able to carry out phagocytosis. Two main types of phagocytes are *macrophages,* located in many body tissues, and *neutrophils,* a type of white blood cell. Phagocytosis begins when the particle binds to a plasma membrane receptor on the phagocyte, causing it to extend **pseudopods** (SOO-dō-pods; *pseudo-* = false; *-pods* = feet), projections of its plasma membrane and cytoplasm. Pseudopods surround the particle outside the cell, and the membranes fuse to form a vesicle called a *phagosome,* which enters the cytoplasm. The phagosome fuses with one or more lysosomes, and lysosomal enzymes break down the ingested material.

Figure 3.13 Phagocytosis. Pseudopods surround a particle, and the membranes fuse to form a phagosome.

 Phagocytosis is a vital defense mechanism that helps protect the body from disease.

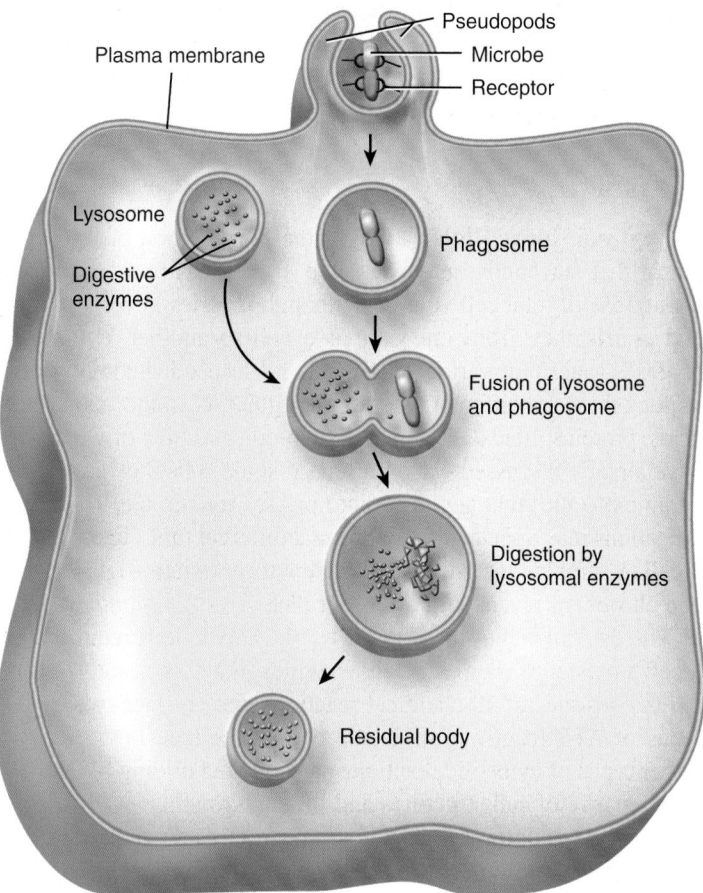

(a) Diagram of the process

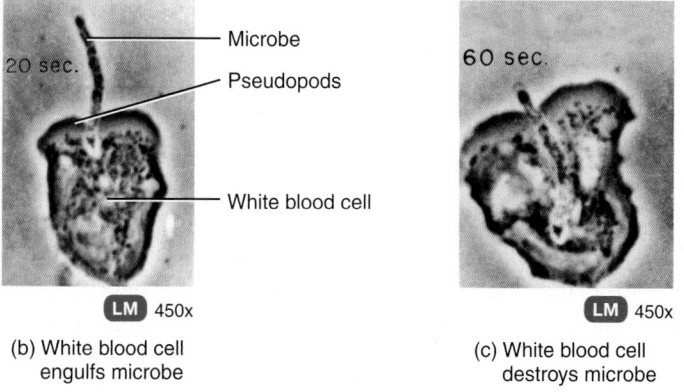

(b) White blood cell engulfs microbe

(c) White blood cell destroys microbe

? **What triggers pseudopod formation?**

In most cases, any undigested materials in the phagosome remain indefinitely in a vesicle called a *residual body*. The residual bodies are then either secreted by the cell via exocytosis or they remain stored in the cell as lipofuscin granules.

The process of phagocytosis is a vital defense mechanism that helps protect the body from disease. Through phagocytosis, macrophages dispose of invading microbes and billions of aged, worn-out red blood cells every day; neutrophils also help rid the body of invading microbes. **Pus** is a mixture of dead neutrophils, macrophages, and tissue cells and fluid in an infected wound. •

Most body cells carry out **bulk-phase endocytosis,** also called **pinocytosis** (pi-nō-sī-TŌ-sis; *pino-* = to drink), a form of endocytosis in which tiny droplets of extracellular fluid are taken up (Figure 3.14). No receptor proteins are involved; all solutes dissolved in the extracellular fluid are brought into the cell. During bulk-phase endocytosis, the plasma membrane folds inward and forms a vesicle containing a droplet of extracellular fluid. The vesicle detaches or "pinches off" from the plasma membrane and

Figure 3.14 Bulk-phase endocytosis. The plasma membrane folds inward, forming a vesicle.

 Most body cells carry out bulk-phase endocytosis, the nonselective uptake of tiny droplets of extracellular fluid.

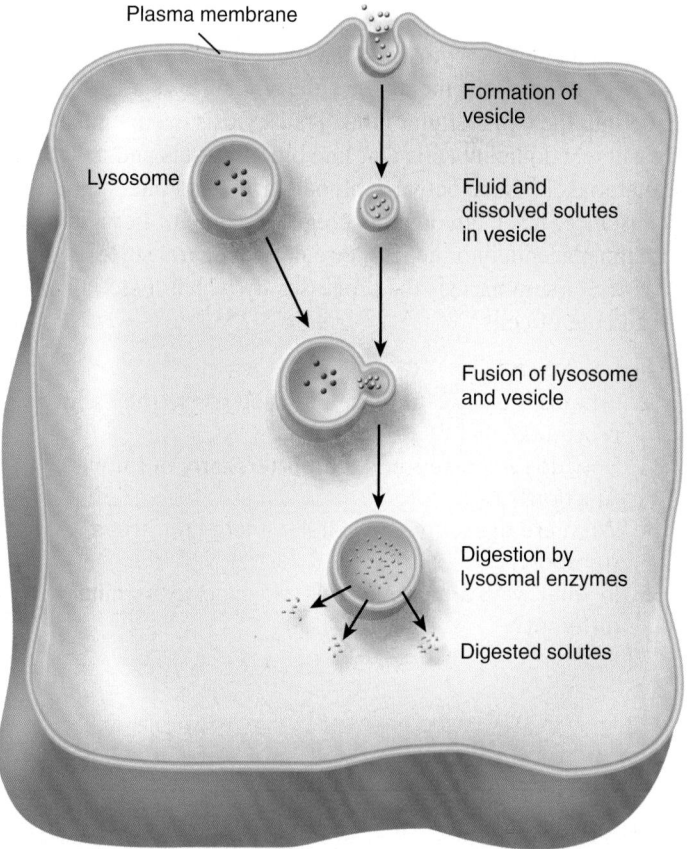

? **How do receptor-mediated endocytosis and phagocytosis differ from bulk-phase endocytosis?**

enters the cytosol. Within the cell, the vesicle fuses with a lysosome, where enzymes degrade the engulfed solutes. The resulting smaller molecules, such as amino acids and fatty acids, leave the lysosome to be used elsewhere in the cell. Bulk-phase endocytosis occurs in most cells, especially absorptive cells in the intestines and kidneys.

EXOCYTOSIS In contrast with endocytosis, which brings materials into a cell, **exocytosis** releases materials from a cell. All cells carry out exocytosis, but it is especially important in two types of cells: (1) secretory cells that liberate digestive enzymes, hormones, mucus, or other secretions and (2) nerve cells that release substances called *neurotransmitters* (see Figure 12.23). In some cases, wastes are also released by exocytosis. During exocytosis, membrane-enclosed vesicles called *secretory vesicles* form inside the cell, fuse with the plasma membrane, and release their contents into the extracellular fluid.

Segments of the plasma membrane lost through endocytosis are recovered or recycled by exocytosis. The balance between endocytosis and exocytosis keeps the surface area of a cell's plasma membrane relatively constant. Membrane exchange is quite extensive in certain cells. In your pancreas, for example, the cells that secrete digestive enzymes can recycle an amount of plasma membrane equal to the cell's entire surface area in 90 minutes.

TRANSCYTOSIS Transport in vesicles may also be used to successively move a substance into, across, and out of a cell. In this active process, called **transcytosis** (tranz′-sī-TŌ-sis)**,** vesicles undergo endocytosis on one side of a cell, move across the cell, and then undergo exocytosis on the opposite side. As the vesicles fuse with the plasma membrane, the vesicular contents are released into the extracellular fluid. Transcytosis occurs most often across the endothelial cells that line blood vessels and is a means for materials to move between blood plasma and interstitial fluid. For instance, when a woman is pregnant, some of her antibodies cross the placenta into the fetal circulation via transcytosis.

Table 3.1 summarizes the processes by which materials move into and out of cells.

✔ **CHECKPOINT**

12. What is the key difference between passive and active processes?
13. How do symporters and antiporters carry out their functions?
14. What are the sources of cellular energy for active transport?
15. In what ways are endocytosis and exocytosis similar and different?

3.4 CYTOPLASM

◉ **OBJECTIVE**

• Describe the structure and function of cytoplasm, cytosol, and organelles.

Cytoplasm consists of all the cellular contents between the plasma membrane and the nucleus, and has two components: (1) the cytosol and (2) organelles, tiny structures that perform different functions in the cell.

Cytosol

The **cytosol (intracellular fluid)** is the fluid portion of the cytoplasm that surrounds organelles (see Figure 3.1) and constitutes about 55% of total cell volume. Although it varies in composition and consistency from one part of a cell to another, cytosol is 75–90% water plus various dissolved and suspended components. Among these are different types of ions, glucose, amino acids, fatty acids, proteins, lipids, ATP, and waste products, some of which we have already discussed. Also present in some cells are various organic molecules that aggregate into masses for storage. These aggregations may appear and disappear at different times in the life of a cell. Examples include *lipid droplets* that contain triglycerides, and clusters of glycogen molecules called *glycogen granules*.

The cytosol is the site of many chemical reactions required for a cell's existence. For example, enzymes in cytosol catalyze *glycolysis,* a series of 10 chemical reactions that produce two molecules of ATP from one molecule of glucose (see Figure 25.4). Other types of cytosolic reactions provide the building blocks for maintenance of cell structures and for cell growth.

The **cytoskeleton** is a network of protein filaments that extends throughout the cytosol (see Figure 3.1). Three types of filamentous proteins contribute to the cytoskeleton's structure, as well as the structure of other organelles. In the order of their increasing diameter, these structures are microfilaments, intermediate filaments, and microtubules.

MICROFILAMENTS Microfilaments (mī-krō-FIL-a-ments) are the thinnest elements of the cytoskeleton. They are composed of the proteins *actin* and *myosin* and are most prevalent at the edge of a cell (Figure 3.15a). Microfilaments have two general functions: They help generate movement and provide mechanical support. With respect to movement, microfilaments are involved in muscle contraction, cell division, and cell locomotion, such as occurs during the migration of embryonic cells during development, the invasion of tissues by white blood cells to fight infection, or the migration of skin cells during wound healing.

Microfilaments provide much of the mechanical support that is responsible for the basic strength and shapes of cells. They anchor the cytoskeleton to integral proteins in the plasma membrane. Microfilaments also provide mechanical support for cell extensions called **microvilli** (mī-krō-VIL-ī; *micro-* = small; *-villi* = tufts of hair; singular is *microvillus*), nonmotile, microscopic fingerlike

TABLE 3.1

Transport of Materials into and out of Cells

TRANSPORT PROCESS	DESCRIPTION	SUBSTANCES TRANSPORTED
PASSIVE PROCESSES	Movement of substances down a concentration gradient until equilibrium is reached; do not require cellular energy in the form of ATP.	
Diffusion	Movement of molecules or ions down a concentration gradient due to their kinetic energy until they reach equilibrium.	
Simple diffusion	Passive movement of a substance down its concentration gradient through the lipid bilayer of the plasma membrane without the help of membrane transport proteins.	Nonpolar, hydrophobic solutes: oxygen, carbon dioxide, and nitrogen gases; fatty acids; steroids; and fat-soluble vitamins. Polar molecules such as water, urea, and small alcohols.
Facilitated diffusion	Passive movement of a substance down its concentration gradient through the lipid bilayer by transmembrane proteins that function as channels or carriers.	Polar or charged solutes: glucose; fructose; galactose; some vitamins; and ions such as K^+, Cl^-, Na^+, and Ca^{2+}.
Osmosis	Passive movement of water molecules across a selectively permeable membrane from an area of higher to lower water concentration until equilibrium is reached.	Solvent: water in living systems.
ACTIVE PROCESSES	Movement of substances against a concentration gradient; requires cellular energy in the form of ATP.	
Active Transport	Active process in which a cell expends energy to move a substance across the membrane against its concentration gradient by transmembrane proteins that function as carriers.	Polar or charged solutes.
Primary active transport	Active process in which a substance moves across the membrane against its concentration gradient by pumps (carriers) that use energy supplied by hydrolysis of ATP.	Na^+, K^+, Ca^{2+}, H^+, I^-, Cl^-, and other ions.
Secondary active transport	Coupled active transport of two substances across the membrane using energy supplied by a Na^+ or H^+ concentration gradient maintained by primary active transport pumps. Antiporters move Na^+ (or H^+) and another substance in opposite directions across the membrane; symporters move Na^+ (or H^+) and another substance in the same direction across the membrane.	Antiport: Ca^{2+}, H^+ out of cells. Symport: glucose, amino acids into cells.
Transport in Vesicles	Active process in which substances move into or out of cells in vesicles that bud from plasma membrane; requires energy supplied by ATP.	
Endocytosis	Movement of substances into a cell in vesicles.	
Receptor-mediated endocytosis	Ligand–receptor complexes trigger infolding of a clathrin-coated pit that forms a vesicle containing ligands.	Ligands: transferrin, low-density lipoproteins (LDLs), some vitamins, certain hormones, and antibodies.
Phagocytosis	"Cell eating"; movement of a solid particle into a cell after pseudopods engulf it to form a phagosome.	Bacteria, viruses, and aged or dead cells.
Bulk-phase endocytosis	"Cell drinking"; movement of extracellular fluid into a cell by infolding of plasma membrane to form a vesicle.	Solutes in extracellular fluid.
Exocytosis	Movement of substances out of a cell in secretory vesicles that fuse with the plasma membrane and release their contents into the extracellular fluid.	Neurotransmitters, hormones, and digestive enzymes.
Transcytosis	Movement of a substance through a cell as a result of endocytosis on one side and exocytosis on the opposite side.	Substances, such as antibodies, across endothelial cells. This is a common route for substances to pass between blood plasma and interstitial fluid.

Figure 3.15 Cytoskeleton.

The cytoskeleton is a network of three types of protein filaments—microfilaments, intermediate filaments, and microtubules—that extend throughout the cytoplasm.

Microvillus

MICROFILAMENTS

Nucleus

(a) Microfilament

Microfilaments

Nucleus

IFM 1500x

INTERMEDIATE FILAMENTS

(b) Intermediate filament

Nucleus

Intermediate filaments (green)

IFM 800x

MICROTUBULES

Centrosome

(c) Microtubule

Microtubules (green)

Nucleus

IFM 500x

FUNCTIONS OF THE CYTOSKELETON

1. Serves as a scaffold that helps to determine a cell's shape and to organize the cellular contents.
2. Aids movement of organelles within the cell, of chromosomes during cell division, and of whole cells such as phagocytes.

? Which cytoskeletal component helps form the structure of centrioles, cilia, and flagella?

projections of the plasma membrane. Within each microvillus is a core of parallel microfilaments that supports it. Because they greatly increase the surface area of the cell, microvilli are abundant on cells involved in absorption, such as the epithelial cells that line the small intestine.

INTERMEDIATE FILAMENTS As their name suggests, **intermediate filaments** are thicker than microfilaments but thinner than microtubules (Figure 3.15b). Several different proteins can compose intermediate filaments, which are exceptionally strong. They are found in parts of cells subject to mechanical stress; they help sta-

bilize the position of organelles such as the nucleus and help attach cells to one another.

MICROTUBULES Microtubules (mī-krō-TOO-būls′) are the largest of the cytoskeletal components and are long, unbranched hollow tubes composed mainly of the protein *tubulin*. The assembly of microtubules begins in an organelle called the centrosome (discussed shortly). The microtubules grow outward from the centrosome toward the periphery of the cell (Figure 3.15c). Microtubules help determine cell shape. They also function in the movement of organelles such as secretory vesicles, of chromosomes during cell division, and of specialized cell projections, such as cilia and flagella.

Organelles

As noted earlier, **organelles** are specialized structures within the cell that have characteristic shapes, and they perform specific functions in cellular growth, maintenance, and reproduction. Despite the many chemical reactions going on in a cell at any given time, there is little interference among reactions because they are confined to different organelles. Each type of organelle has its own set of enzymes that carry out specific reactions, and serves as a functional compartment for specific biochemical processes. The numbers and types of organelles vary in different cells, depending on the cell's function. Although they have different functions, organelles often cooperate to maintain homeostasis. Even though the nucleus is a large organelle, it is discussed in a separate section because of its special importance in directing the life of a cell.

Centrosome

The **centrosome** (SEN-trō-sōm), located near the nucleus, consists of two components: a pair of centrioles and pericentriolar material (Figure 3.16a). The two **centrioles** (SEN-trē-ōls) are cylindrical structures, each composed of nine clusters of three microtubules (triplets) arranged in a circular pattern (Figure 3.16b). The long axis of one centriole is at a right angle to the long axis of the other (Figure 3.16c). Surrounding the centrioles is **pericentriolar material** (per′-ē-sen′-trē-Ō-lar), which contains hundreds of ring-shaped complexes composed of the protein *tubulin*. These tubulin complexes are the organizing centers for growth of the mitotic spindle, which plays a critical role in cell division, and for microtubule formation in nondividing cells. During cell division, centrosomes replicate so that succeeding generations of cells have the capacity for cell division.

Cilia and Flagella

Microtubules are the dominant components of cilia and flagella, which are motile projections of the cell surface. **Cilia** (SIL-ē-a = eyelashes; singular is *cilium*) are numerous, short, hairlike projections that extend from the surface of the cell (see Figures 3.1 and 3.17b). Each cilium contains a core of 20 microtubules surrounded by plasma membrane (Figure 3.17a). The microtubules are arranged such that one pair in the center is surrounded by nine clusters of two fused microtubules (doublets). Each cilium is anchored to a *basal body* just below the surface of the plasma membrane. A basal body is similar in structure to a centriole and functions in initiating the assembly of cilia and flagella.

A cilium displays an oarlike pattern of beating; it is relatively stiff during the power stroke (oar digging into the water), but more flexible during the recovery stroke (oar moving above the water preparing for a new stroke) (Figure 3.17d). The coordinated movement of many cilia on the surface of a cell causes the steady movement of fluid along the cell's surface. Many cells of the respiratory tract, for example, have hundreds of cilia that help sweep foreign particles

Figure 3.16 Centrosome.

 Located near the nucleus, the centrosome consists of a pair of centrioles and pericentriolar material.

FUNCTIONS OF CENTROSOMES
1. The pericentriolar material of the centrosome contains tubulins that build microtubules in nondividing cells.
2. The pericentriolar material of the centrosome forms the mitotic spindle during cell division.

— Pericentriolar material
— Centrioles
— Microtubules (triplets)

(a) Details of a centrosome

(b) Arrangement of microtubules in centrosome

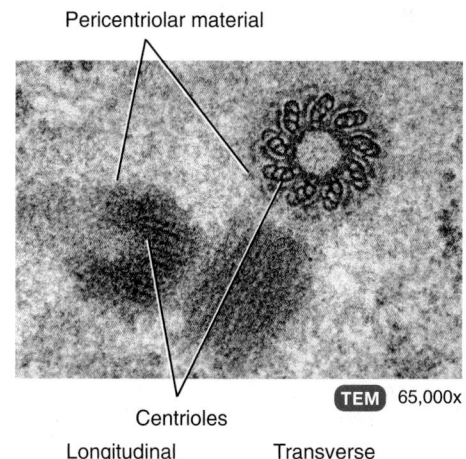

Pericentriolar material

TEM 65,000x

Centrioles

Longitudinal sections   Transverse section

(c) Centrioles

❓ **If you observed that a cell did not have a centrosome, what could you predict about its capacity for cell division?**

trapped in mucus away from the lungs. In cystic fibrosis, the extremely thick mucous secretions that are produced interfere with ciliary action and the normal functions of the respiratory tract.

Figure 3.17 **Cilia and flagella.**

A cilium contains a core of microtubules with one pair in the center surrounded by nine clusters of doublet microtubules.

FUNCTIONS OF CILIA AND FLAGELLA

1. Cilia move fluids along a cell's surface.
2. A flagellum moves an entire cell.

Cilium or flagellum

Doublet microtubules

Central pair of microtubules

Plasma membrane

Basal body

(a) Arrangement of microtubules in a cilium or flagellum

SEM 3000x

(b) Cilia lining the trachea

Cilia

SEM 4000x

(c) Flagellum of a sperm cell

Flagellum

Movement of liquid

Cilium

Cell surface

Power stroke
Recovery stroke

(d) Ciliary movement

Movement of cell

Cell surface

(e) Flagellar movement

? **What is the functional difference between cilia and flagella?**

CLINICAL CONNECTION | *Cilia and Smoking*

The movement of cilia is also paralyzed by nicotine in cigarette smoke. For this reason, smokers cough often to remove foreign particles from their airways. Cells that line the uterine (fallopian) tubes also have cilia that sweep oocytes (egg cells) toward the uterus, and females who smoke have an increased risk of ectopic (outside the uterus) pregnancy. •

Flagella (fla-JEL-a = whip; singular is *flagellum*) are similar in structure to cilia but are typically much longer. Flagella usually move an entire cell. A flagellum generates forward motion along its axis by rapidly wiggling in a wavelike pattern (Figure 3.17e). The only example of a flagellum in the human body is a sperm cell's tail, which propels the sperm toward the oocyte in the uterine tube (Figure 3.17c).

Ribosomes

Ribosomes (RĪ-bō-sōms; *-somes* = bodies) are the sites of protein synthesis. The name of these tiny structures reflects their high content of one type of ribonucleic acid, **ribosomal RNA (rRNA),**

but each one also includes more than 50 proteins. Structurally, a ribosome consists of two subunits, one about half the size of the other (Figure 3.18). The large and small subunits are made separately in the nucleolus, a spherical body inside the nucleus. Once produced, the large and small subunits exit the nucleus separately, then come together in the cytoplasm.

Some ribosomes are attached to the outer surface of the nuclear membrane and to an extensively folded membrane called the endoplasmic reticulum. These ribosomes synthesize proteins destined for specific organelles, for insertion in the plasma membrane, or for export from the cell. Other ribosomes are "free" or unattached to other cytoplasmic structures. Free ribosomes synthesize proteins used in the cytosol. Ribosomes are also located within mitochondria, where they synthesize mitochondrial proteins.

Endoplasmic Reticulum

The **endoplasmic reticulum** (en'-dō-PLAS-mik re-TIK-ū-lum; *-plasmic* = cytoplasm; *reticulum* = network) or **ER** is a network of membranes in the form of flattened sacs or tubules (Figure 3.19). The ER extends from the nuclear envelope (membrane around the

Figure 3.18 Ribosomes.

Ribosomes are the sites of protein synthesis.

Large subunit   Small subunit   Complete functional ribosome

Details of ribosomal subunits

FUNCTIONS OF RIBOSOMES

1. Ribosomes associated with endoplasmic reticulum synthesize proteins destined for insertion in the plasma membrane or secretion from the cell.
2. Free ribosomes synthesize proteins used in the cytosol.

Where are subunits of ribosomes synthesized and assembled?

Figure 3.19 Endoplasmic reticulum.

The endoplasmic reticulum is a network of membrane-enclosed sacs or tubules that extend throughout the cytoplasm and connect to the nuclear envelope.

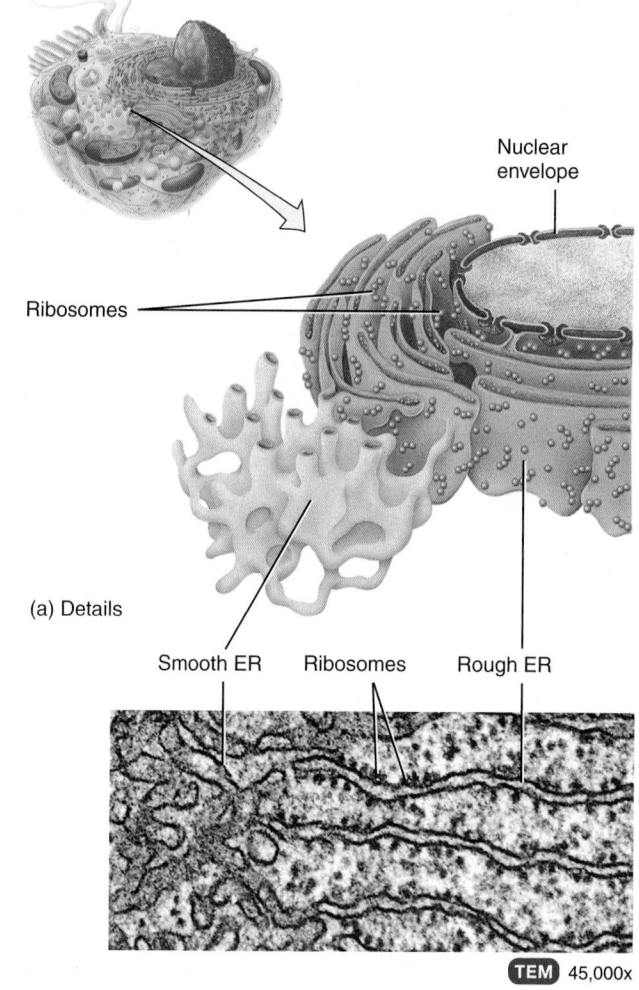

Nuclear envelope

Ribosomes

(a) Details

Smooth ER   Ribosomes   Rough ER

TEM 45,000x

(b) Transverse section

FUNCTIONS OF ENDOPLASMIC RETICULUM

1. Rough ER synthesizes glycoproteins and phospholipids that are transferred into cellular organelles, inserted into the plasma membrane, or secreted during exocytosis.
2. Smooth ER synthesizes fatty acids and steroids, such as estrogens and testosterone; inactivates or detoxifies drugs and other potentially harmful substances; removes the phosphate group from glucose-6-phosphate; and stores and releases calcium ions that trigger contraction in muscle cells.

What are the structural and functional differences between rough and smooth ER?

nucleus), to which it is connected, throughout the cytoplasm. The ER is so extensive that it constitutes more than half of the membranous surfaces within the cytoplasm of most cells.

Cells contain two distinct forms of ER, which differ in structure and function. **Rough ER** is continuous with the nuclear membrane and usually is folded into a series of flattened sacs. The outer surface of rough ER is studded with ribosomes, the sites of protein synthesis. Proteins synthesized by ribosomes attached to rough ER enter spaces within the ER for processing and sorting. In some cases, enzymes attach the proteins to carbohydrates to form glycoproteins. In other cases, enzymes attach the proteins to phospholipids, also synthesized by rough ER. These molecules (glycoproteins and phospholipids) may be incorporated into the membranes of organelles, inserted into the plasma membrane, or secreted via exocytosis. Thus rough ER produces secretory proteins, membrane proteins, and many organellar proteins.

Smooth ER extends from the rough ER to form a network of membrane tubules (Figure 3.19). Unlike rough ER, smooth ER does not have ribosomes on the outer surfaces of its membrane. However, smooth ER contains unique enzymes that make it functionally more diverse than rough ER. Because it lacks ribosomes,

smooth ER does not synthesize proteins, but it does synthesize fatty acids and steroids, such as estrogens and testosterone. In liver cells, enzymes of the smooth ER help release glucose into the bloodstream and inactivate or detoxify lipid-soluble drugs or potentially harmful substances, such as alcohol, pesticides, and *carcinogens* (cancer-causing agents). In liver, kidney, and intestinal cells a smooth ER enzyme removes the phosphate group from glucose-6-phosphate, which allows the "free" glucose to enter the bloodstream. In muscle cells, the calcium ions (Ca^{2+}) that trigger contraction are released from the sarcoplasmic reticulum, a form of smooth ER.

CLINICAL CONNECTION | Smooth ER and Drug Tolerance

One of the functions of smooth ER, as noted earlier, is to detoxify certain drugs. Individuals who repeatedly take such drugs, such as the sedative phenobarbital, develop changes in the smooth ER in their liver cells. Prolonged administration of phenobarbital results in increased tolerance to the drug; the same dose no longer produces the same degree of sedation. With repeated exposure to the drug, the amount of smooth ER and its enzymes increases to protect the cell from its toxic effects. As the amount of smooth ER increases, higher and higher dosages of the drug are needed to achieve the original effect. This could result in an increased possibility of overdose and increased drug dependence. •

Golgi Complex

Most of the proteins synthesized by ribosomes attached to rough ER are ultimately transported to other regions of the cell. The first step in the transport pathway is through an organelle called the **Golgi complex** (GOL-jē). It consists of 3 to 20 **cisternae** (sis-TER-nē = cavities; singular is *cisterna*), small, flattened membranous sacs with bulging edges that resemble a stack of pita bread (Figure 3.20). The cisternae are often curved, giving the Golgi complex a cuplike shape. Most cells have several Golgi complexes, and Golgi complexes are more extensive in cells that secrete proteins, a clue to the organelle's role in the cell.

The cisternae at the opposite ends of a Golgi complex differ from each other in size, shape, and enzymatic activity. The convex **entry** or *cis* **face** is a cisterna that faces the rough ER. The concave **exit** or *trans* **face** is a cisterna that faces the plasma membrane. Sacs between the entry and exit faces are called **medial cisternae.** Transport vesicles (described shortly) from the ER merge to form the entry face. From the entry face, the cisternae are thought to mature, in turn becoming medial and then exit cisternae.

Figure 3.20 Golgi complex.

The opposite faces of a Golgi complex differ in size, shape, content, and enzymatic activities.

FUNCTIONS OF GOLGI COMPLEX
1. Modifies, sorts, packages, and transports proteins received from the rough ER.
2. Forms secretory vesicles that discharge processed proteins via exocytosis into extracellular fluid; forms membrane vesicles that ferry new molecules to the plasma membrane; forms transport vesicles that carry molecules to other organelles, such as lysosomes.

(a) Details
(b) Transverse section
TEM 65,000x

How do the entry and exit faces differ in function?

Different enzymes in the entry, medial, and exit cisternae of the Golgi complex permit each of these areas to modify, sort, and package proteins into vesicles for transport to different destinations. The entry face receives and modifies proteins produced by the rough ER. The medial cisternae add carbohydrates to proteins to form glycoproteins and lipids to proteins to form lipoproteins. The exit face modifies the molecules further and then sorts and packages them for transport to their destinations.

Proteins arriving at, passing through, and exiting the Golgi complex do so through maturation of the cisternae and exchanges that occur via transfer vesicles (Figure 3.21):

① Proteins synthesized by ribosomes on the rough ER are surrounded by a piece of the ER membrane, which eventually buds from the membrane surface to form transport vesicles.

② Transport vesicles move toward the entry face of the Golgi complex.

③ Fusion of several transport vesicles creates the entry face of the Golgi complex and releases proteins into its lumen (space).

④ The proteins move from the entry face into one or more medial cisternae. Enzymes in the medial cisternae modify the proteins to form glycoproteins, glycolipids, and lipoproteins. **Transfer vesicles** that bud from the edges of the cisternae move specific enzymes back toward the entry face and move some partially modified proteins toward the exit face.

⑤ The products of the medial cisternae move into the lumen of the exit face.

⑥ Within the exit face cisterna, the products are further modified and are sorted and packaged.

⑦ Some of the processed proteins leave the exit face and are stored in **secretory vesicles.** These vesicles deliver the proteins to the plasma membrane, where they are discharged by exocytosis into the extracellular fluid. For example, certain pancreatic cells release the hormone insulin in this way.

⑧ Other processed proteins leave the exit face in **membrane vesicles** that deliver their contents to the plasma membrane for incorporation into the membrane. In doing so, the Golgi complex adds new segments of plasma membrane as existing segments are lost and modifies the number and distribution of membrane molecules.

⑨ Finally, some processed proteins leave the exit face in transport vesicles that will carry the proteins to another cellular destination. For instance, transport vesicles carry digestive enzymes to lysosomes; the structure and functions of these important organelles are discussed next.

Figure 3.21 **Processing and packaging of proteins by the Golgi complex.**

All proteins exported from the cell are processed in the Golgi complex.

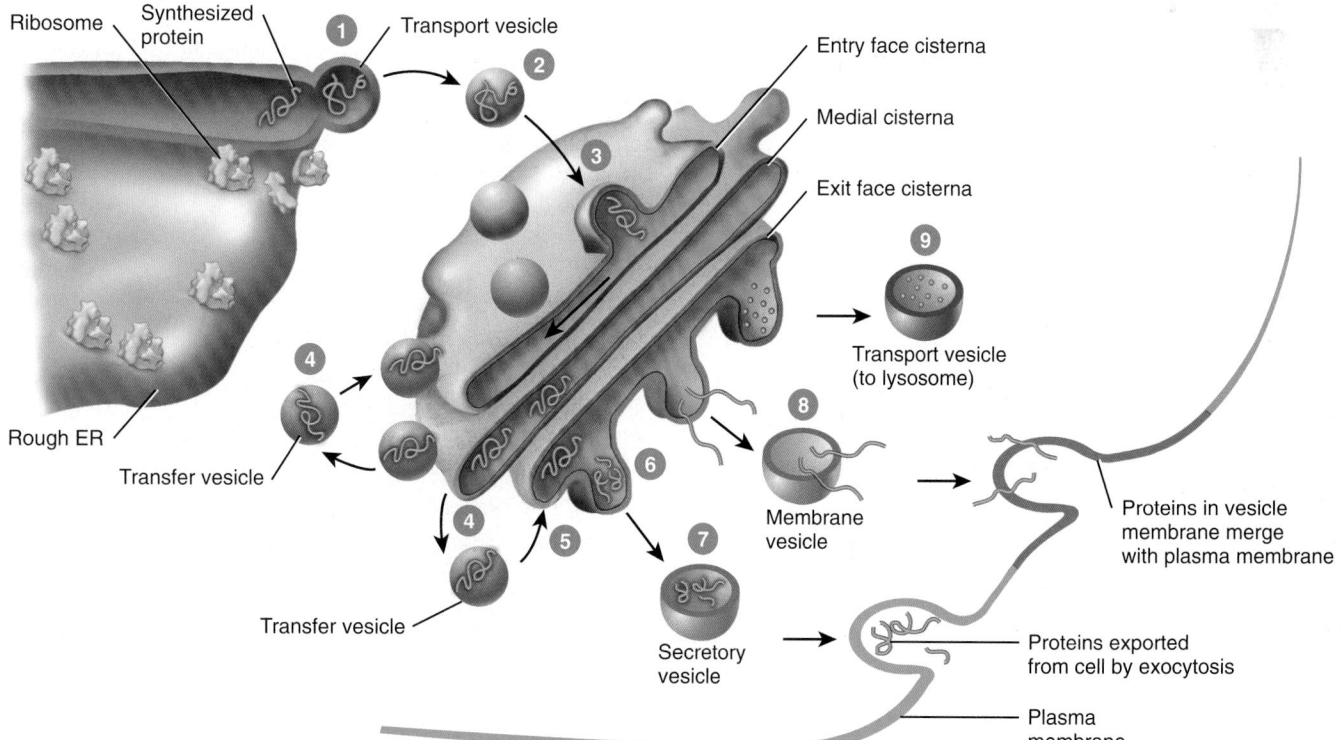

What are the three general destinations for proteins that leave the Golgi complex?

Lysosomes

Lysosomes (LĪ-sō-sōms; *lyso-* = dissolving; *-somes* = bodies) are membrane-enclosed vesicles that form from the Golgi complex (Figure 3.22). They can contain as many as 60 kinds of powerful digestive and hydrolytic enzymes that can break down a wide variety of molecules once lysosomes fuse with vesicles formed during endocytosis. Because lysosomal enzymes work best at an acidic pH, the lysosomal membrane includes active transport pumps that import hydrogen ions (H^+). Thus, the lysosomal interior has a pH of 5, which is 100 times more acidic than the pH of the cytosol (pH 7). The lysosomal membrane also includes transporters that move the final products of digestion, such as glucose, fatty acids, and amino acids, into the cytosol.

Figure 3.22 Lysosomes.

Lysosomes contain several types of powerful digestive enzymes.

FUNCTIONS OF LYSOSOMES

1. Digest substances that enter a cell via endocytosis and transport final products of digestion into cytosol.
2. Carry out autophagy, the digestion of worn-out organelles.
3. Implement autolysis, the digestion of an entire cell.
4. Accomplish extracellular digestion.

Digestive enzymes

(a) Lysosome

Lysosomes

TEM 16,000x

(b) Several lysosomes

? **What is the name of the process by which worn-out organelles are digested by lysosomes?**

Lysosomal enzymes also help recycle worn-out cell structures. A lysosome can engulf another organelle, digest it, and return the digested components to the cytosol for reuse. In this way, old organelles are continually replaced. The process by which entire worn-out organelles are digested is called **autophagy** (aw-TOF-a-jē; *auto-* = self; *-phagy* = eating). In autophagy, the organelle to be digested is enclosed by a membrane derived from the ER to create a vesicle called an **autophagosome** (aw-tō-FĀ-gō-sōm); the vesicle then fuses with a lysosome. In this way, a human liver cell, for example, recycles about half of its cytoplasmic contents every week. Autophagy is also involved in cellular differentiation, control of growth, tissue remodeling, adaptation to adverse environments, and cell defense. Lysosomal enzymes may also destroy the entire cell that contains them, a process known as **autolysis** (aw-TOL-i-sis). Autolysis occurs in some pathological conditions and also is responsible for the tissue deterioration that occurs immediately after death.

As we just discussed, most lysosomal enzymes act within a cell. However, some operate in extracellular digestion. One example occurs during fertilization. The head of a sperm cell releases lysosomal enzymes that aid its penetration of the oocyte by dissolving its protective coating in a process called the acrosomal reaction (see Section 29.1).

CLINICAL CONNECTION | *Tay-Sachs Disease*

Some disorders are caused by faulty or absent lysosomal enzymes. For instance, **Tay-Sachs disease** (TĀ-SAKS), which most often affects children of Ashkenazi (eastern European Jewish) descent, is an inherited condition characterized by the absence of a single lysosomal enzyme called Hex A. This enzyme normally breaks down a membrane glycolipid called ganglioside G_{M2} that is especially prevalent in nerve cells. As the excess ganglioside G_{M2} accumulates, the nerve cells function less efficiently. Children with Tay-Sachs disease typically experience seizures and muscle rigidity. They gradually become blind, demented, and uncoordinated and usually die before the age of 5. Tests can now reveal whether an adult is a carrier of the defective gene. •

Peroxisomes

Another group of organelles similar in structure to lysosomes, but smaller, are the **peroxisomes** (pe-ROKS-i-sōms; *peroxi-* = peroxide; *-somes* = bodies; see Figure 3.1). Peroxisomes, also called *microbodies,* contain several *oxidases,* enzymes that can oxidize (remove hydrogen atoms from) various organic substances. For instance, amino acids and fatty acids are oxidized in peroxisomes as part of normal metabolism. In addition, enzymes in peroxisomes oxidize toxic substances, such as alcohol. Thus, peroxisomes are very abundant in the liver, where detoxification of alcohol and other damaging substances occurs. A by-product of the oxidation reactions is hydrogen peroxide (H_2O_2), a potentially toxic compound, and associated free radicals such as superoxide. However, peroxisomes also contain the enzyme *catalase,* which decomposes H_2O_2. Because production and degradation of H_2O_2 occur within the same organelle, peroxisomes protect other parts of the cell from the toxic effects of H_2O_2. Peroxisomes also contain enzymes that destroy superoxide. Without peroxisomes, by-products of metabolism could accumulate inside a cell and result in cellular death. Peroxisomes can self-replicate. New peroxisomes may form from preexisting ones by enlarging and dividing. They may also form by a process in which components accumulate at a given site in the cell and then assemble into a peroxisome.

Proteasomes

As you have just learned, lysosomes degrade proteins delivered to them in vesicles. Cytosolic proteins also require disposal at certain times in the life of a cell. Continuous destruction of unneeded, damaged, or faulty proteins is the function of tiny barrel-shaped structures consisting of four stacked rings of proteins around a central core called **proteasomes** (PRŌ-tē-a-sōms = protein bodies). For example, proteins that are part of metabolic pathways need to be degraded after they have accomplished their function. Such protein destruction plays a part in negative feedback by halting a pathway once the appropriate response has been achieved. A typical body cell contains many thousands of proteasomes, in both the cytosol and the nucleus. Discovered only recently because they are far too small to discern under the light microscope and do not show up well in electron micrographs, proteasomes were so named because they contain myriad *proteases,* enzymes that cut proteins into small peptides. Once the enzymes of a proteasome have chopped up a protein into smaller chunks, other enzymes then break down the peptides into amino acids, which can be recycled into new proteins.

CLINICAL CONNECTION | *Proteasomes and Disease*

Some diseases could result from failure of proteasomes to degrade abnormal proteins. For example, clumps of misfolded proteins accumulate in brain cells of people with Parkinson disease and Alzheimer disease. Discovering why the proteasomes fail to clear these abnormal proteins is a goal of ongoing research. •

Mitochondria

Because they generate most of the ATP through aerobic (oxygen-requiring) respiration, **mitochondria** (mī-tō-KON-drē-a; *mito-* = thread; *-chondria* = granules; singular is *mitochondrion*) are referred to as the "powerhouses" of the cell. A cell may have as few as a hundred or as many as several thousand mitochondria, depending on its activity. Active cells, such as those found in the muscles, liver, and kidneys, which use ATP at a high rate, have a large number of mitochondria. For example, regular exercise can lead to an increase in the number of mitochondria in muscle cells, and this allows muscle cells to function more efficiently. Mitochondria are usually located within the cell where oxygen enters the cell or where the ATP is used, for example, among the contractile proteins in muscle cells.

A mitochondrion consists of an **outer mitochondrial membrane** and an **inner mitochondrial membrane** with a small fluid-filled

space between them (Figure 3.23). Both membranes are similar in structure to the plasma membrane. The inner mitochondrial membrane contains a series of folds called **mitochondrial cristae** (KRIS-tē = ridges). The central fluid-filled cavity of a mitochondrion, enclosed by the inner mitochondrial membrane, is the **mitochondrial matrix.** The elaborate folds of the cristae provide an enormous surface area for the chemical reactions that are part of the aerobic phase of *cellular respiration,* the reactions that produce most of a cell's ATP (see Chapter 25). The enzymes that catalyze these reactions are located on the cristae and in the matrix of the mitochondria.

Mitochondria also play an important and early role in **apoptosis** (ap'-ōp-TŌ-sis or ap-ō-TŌ-sis = a falling off), the orderly, genetically programmed death of a cell. In response to stimuli such as large numbers of destructive free radicals, DNA damage, growth factor deprivation, or lack of oxygen and nutrients, certain chemicals are released from mitochondria following the formation of a pore in the outer mitochondrial membrane. One of the chemicals released into the cytosol of the cell is cytochrome *c,* which while inside the mitochondria is involved in aerobic cellular respiration. In the cytosol, however, cytochrome *c* and other substances initiate a cascade of activation of protein-digesting enzymes that bring about apoptosis.

Like peroxisomes, mitochondria self-replicate, a process that occurs during times of increased cellular energy demand or before cell division. Synthesis of some of the proteins needed for mitochondrial functions occurs on the ribosomes that are present in the mitochondrial matrix. Mitochondria even have their own DNA, in the form of multiple copies of a circular DNA molecule that contains 37 genes. These mitochondrial genes control the synthesis of 2 ribosomal RNAs, 22 transfer RNAs, and 13 proteins that build mitochondrial components.

Although the nucleus of each somatic cell contains genes from both your mother and your father, mitochondrial genes are inherited only from your mother. This is due to the fact that all mitochondria in a cell are descendants of those that were present in the oocyte (egg) during the fertilization process. The head of a sperm (the part that penetrates and fertilizes an oocyte) normally lacks most organelles, such as mitochondria, ribosomes, endoplasmic reticulum, and the Golgi complex, and any sperm mitochondria that do enter the oocyte are soon destroyed. Since all mitochondrial genes are inherited from the maternal parent, mitochondrial DNA can be used to trace maternal lineage (in other words, to determine whether two or more individuals are related through their mother's side of the family).

✔CHECKPOINT

16. What are some of the chemicals present in cytosol?
17. What is the function of cytosol?
18. Define an organelle.
19. Which organelles are surrounded by a membrane and which are not?
20. Which organelles contribute to synthesizing protein hormones and packaging them into secretory vesicles?
21. What happens on the cristae and in the matrix of mitochondria?

Figure 3.23 **Mitochondria.**

Within mitochondria, chemical reactions of aerobic cellular respiration generate ATP.

FUNCTIONS OF MITOCHONDRIA

1. Generate ATP through reactions of aerobic cellular respiration.
2. Play an important early role in apoptosis.

(a) Details

(b) Transverse section

TEM 80,000x

? **How do the mitochondrial cristae contribute to its ATP-producing function?**

3.5 NUCLEUS

◉ **OBJECTIVE**

• Describe the structure and function of the nucleus.

The **nucleus** is a spherical or oval-shaped structure that usually is the most prominent feature of a cell (Figure 3.24). Most cells have a single nucleus, although some, such as mature red blood cells, have none. In contrast, skeletal muscle cells and a few other types of cells have multiple nuclei. A double membrane called the **nuclear envelope** separates the nucleus from the cytoplasm. Both layers of the nuclear envelope are lipid bilayers similar to the plasma membrane. The outer membrane of the nuclear envelope is continuous with rough ER and resembles it in structure. Many openings called **nuclear pores** extend through the nuclear envelope. Each nuclear pore consists of a circular arrangement of proteins surrounding a large central opening that is about 10 times wider than the pore of a channel protein in the plasma membrane.

Nuclear pores control the movement of substances between the nucleus and the cytoplasm. Small molecules and ions move through the pores passively by diffusion. Most large molecules, such as RNAs and proteins, cannot pass through the nuclear pores by diffusion. Instead, their passage involves an active transport process in which the molecules are recognized and selectively transported through the nuclear pore into or out of the nucleus. For example, proteins needed for nuclear functions move from the cytosol into the nucleus; newly formed RNA molecules move from the nucleus into the cytosol in this manner.

Inside the nucleus are one or more spherical bodies called **nucleoli** (noo′-KLĒ-ō-li; singular is *nucleolus*) that function in producing ribosomes. Each nucleolus is simply a cluster of protein, DNA, and RNA; it is not enclosed by a membrane. Nucleoli are the sites of synthesis of rRNA and assembly of rRNA and proteins into ribosomal subunits. Nucleoli are quite prominent in cells that synthesize large amounts of protein, such as muscle and liver cells. Nucleoli disperse and disappear during cell division and reorganize once new cells are formed.

Within the nucleus are most of the cell's hereditary units, called **genes,** which control cellular structure and direct cellular activities. Genes are arranged along **chromosomes** (KRŌ-mō-sōms;

Figure 3.24 Nucleus.

The nucleus contains most of the cell's genes, which are located on chromosomes.

Chromatin
Nuclear envelope
Nucleolus
Nuclear pore
Polyribosome
Rough endoplasmic reticulum

FUNCTIONS OF NUCLEUS

1. Controls cellular structure.
2. Directs cellular activities.
3. Produces ribosomes in nucleoli.

Nuclear envelope
Nuclear pore

(a) Details of the nucleus

(b) Details of the nuclear envelope

Chromatin
Nuclear envelope
Nucleolus
Nuclear pore

about 10,000x **TEM**

(c) Transverse section of the nucleus

? **What is chromatin?**

chromo- = colored). Human somatic (body) cells have 46 chromosomes, 23 inherited from each parent. Each chromosome is a long molecule of DNA that is coiled together with several proteins (Figure 3.25). This complex of DNA, proteins, and some RNA is called **chromatin** (KRŌ-ma-tin). The total genetic information carried in a cell or an organism is its **genome** (JĒ-nōm).

In cells that are not dividing, the chromatin appears as a diffuse, granular mass. Electron micrographs reveal that chromatin has a beads-on-a-string structure. Each bead is a **nucleosome** (NOO-klē-ō-sōm) that consists of double-stranded DNA wrapped twice around a core of eight proteins called **histones,** which help

organize the coiling and folding of DNA. The string between the beads is called **linker DNA,** which holds adjacent nucleosomes together. In cells that are not dividing, another histone promotes coiling of nucleosomes into a larger-diameter **chromatin fiber,** which then folds into large loops. Just before cell division takes place, however, the DNA replicates (duplicates) and the loops condense even more, forming a pair of **chromatids** (KRŌ-matids). As you will see shortly, during cell division a pair of chromatids constitutes a chromosome.

The main parts of a cell, their structure, and their functions are summarized in Table 3.2.

Figure 3.25 **Packing of DNA into a chromosome in a dividing cell.** When packing is complete, two identical DNA molecules and their histones form a pair of chromatids, which are held together by a centromere.

A chromosome is a highly coiled and folded DNA molecule that is combined with protein molecules.

(a) Illustration

SEM 6050x

(b)

? **What are the components of a nucleosome?**

TABLE 3.2

Cell Parts and Their Functions

PART	DESCRIPTION	FUNCTIONS
PLASMA MEMBRANE	Fluid mosaic lipid bilayer (phospholipids, cholesterol, and glycolipids) studded with proteins; surrounds cytoplasm.	Protects cellular contents; makes contact with other cells; contains channels, transporters, receptors, enzymes, cell-identity markers, and linker proteins; mediates entry and exit of substances.
CYTOPLASM	Cellular contents between plasma membrane and nucleus—cytosol and organelles.	Site of all intracellular activities except those occurring in the nucleus.
Cytosol	Composed of water, solutes, suspended particles, lipid droplets, and glycogen granules. Within the cytoplasm is the cytoskeleton, a network of microfilaments, intermediate filaments, and microtubules.	Fluid in which many of cell's metabolic reactions occur. Maintains shape and general organization of cellular contents; responsible for cellular movements.
Organelles	Specialized structures with characteristic shapes.	Each organelle has specific functions.
Centrosome	Pair of centrioles plus pericentriolar material.	The pericentriolar material contains tubulins, which are used for growth of the mitotic spindle and microtubule formation.
Cilia and flagella	Motile cell surface projections that contain 20 microtubules and a basal body.	Cilia: move fluids over cell's surface; flagella: move entire cell.
Ribosome	Composed of two subunits containing ribosomal RNA and proteins; may be free in cytosol or attached to rough ER.	Protein synthesis.
Endoplasmic reticulum (ER)	Membranous network of flattened sacs or tubules. Rough ER is covered by ribosomes and is attached to the nuclear envelope; smooth ER lacks ribosomes.	Rough ER: synthesizes glycoproteins and phospholipids that are transferred to cellular organelles, inserted into plasma membrane, or secreted during exocytosis; smooth ER: synthesizes fatty acids and steroids, inactivates or detoxifies drugs, removes phosphate group from glucose-6-phosphate, and stores and releases calcium ions in muscle cells.
Golgi complex	Consists of 3–20 flattened membranous sacs called cisternae; structurally and functionally divided into entry (*cis*) face, medial cisternae, and exit (*trans*) face.	Entry (*cis*) face accepts proteins from rough ER; medial cisternae form glycoproteins, glycolipids, and lipoproteins; exit (*trans*) face modifies molecules further, then sorts and packages them for transport to their destinations.
Lysosome	Vesicle formed from Golgi complex; contains digestive enzymes.	Fuses with and digests contents of endosomes, pinocytic vesicles, and phagosomes and transports final products of digestion into cytosol; digests worn-out organelles (autophagy), entire cells (autolysis), and extracellular materials.
Peroxisome	Vesicle containing oxidases (oxidative enzymes) and catalase (decomposes hydrogen peroxide); new peroxisomes bud from preexisting ones.	Oxidizes amino acids and fatty acids; detoxifies harmful substances, such as hydrogen peroxide and associated free radicals.
Proteasome	Tiny barrel-shaped structure that contains proteases (proteolytic enzymes).	Degrades unneeded, damaged, or faulty proteins by cutting them into small peptides.
Mitochondrion	Consists of an outer and an inner mitochondrial membrane, cristae, and matrix; new mitochondria form from preexisting ones.	Site of aerobic cellular respiration reactions that produce most of a cell's ATP. Plays an important early role in apoptosis.
NUCLEUS	Consists of a nuclear envelope with pores, nucleoli, and chromosomes, which exist as a tangled mass of chromatin in interphase cells.	Nuclear pores control the movement of substances between the nucleus and cytoplasm, nucleoli produce ribosomes, and chromosomes consist of genes that control cellular structure and direct cellular functions.

Flagellum — Cilium

Proteasome
Intermediate filament
Centrosome

Lysosome
Smooth ER

Peroxisome

Microtubule

NUCLEUS
CYTOPLASM
PLASMA MEMBRANE
Ribosome on rough ER
Golgi complex
Mitochondrion

Microfilament

In the last decade of the twentieth century, the genomes of humans, mice, fruit flies, and more than 50 microbes were sequenced. As a result, research in the field of **genomics** (jē-NŌM-iks), the study of the relationships between the genome and the biological functions of an organism, has flourished. The Human Genome Project began in June 1990 as an effort to sequence all of the nearly 3.2 billion nucleotides of our genome, and was completed in April 2003. More than 99.9% of the nucleotide bases are identical in everyone. Less than 0.1% of our DNA (1 in each 1000 bases) accounts for inherited differences among humans. Surprisingly, at least half of the human genome consists of repeated sequences that do not code for proteins, so-called junk DNA. The average gene consists of 3000 nucleotides, but sizes vary greatly. The largest known human gene, with 2.4 million nucleotides, codes for the protein dystrophin. Scientists now know that the total number of genes in the human genome is about 30,000, far fewer than the 100,000 previously predicted to exist. Information regarding the human genome and how it is affected by the environment seeks to identify and discover the functions of the specific genes that play a role in genetic diseases. Genomic medicine also aims to design new drugs and to provide screening tests to enable physicians to provide more effective counseling and treatment for disorders with significant genetic components, such as hypertension (high blood pressure), obesity, diabetes, and cancer. •

✔ CHECKPOINT

22. How do large particles enter and exit the nucleus?
23. Where are ribosomes produced?
24. How is DNA packed in the nucleus?

3.6 PROTEIN SYNTHESIS

◉ OBJECTIVE

• Describe the sequence of events in protein synthesis.

Although cells synthesize many chemicals to maintain homeostasis, much of the cellular machinery is devoted to synthesizing large numbers of diverse proteins. The proteins in turn determine the physical and chemical characteristics of cells and, therefore, of the organisms formed from them. Some proteins help assemble cellular structures such as the plasma membrane, the cytoskeleton, and other organelles. Others serve as hormones, antibodies, and contractile elements in muscular tissue. Still others act as enzymes, regulating the rates of the numerous chemical reactions that occur in cells, or transporters, carrying various materials in the blood. Just as genome means all of the genes in an organism, **proteome** (PRŌ-te-ōm) refers to all of an organism's proteins.

In the process called **gene expression,** a gene's DNA is used as a template for synthesis of a specific protein. First, in a process aptly named *transcription,* the information encoded in a specific region of DNA is *transcribed* (copied) to produce a specific molecule of RNA (ribonucleic acid). In a second process, referred to as translation, the RNA attaches to a ribosome, where the information contained in RNA is *translated* into a corresponding sequence of amino acids to form a new protein molecule (Figure 3.26).

Figure 3.26 Overview of gene expression. Synthesis of a specific protein requires transcription of a gene's DNA into RNA and translation of RNA into a corresponding sequence of amino acids.

🔑 Transcription occurs in the nucleus; translation occurs in the cytoplasm.

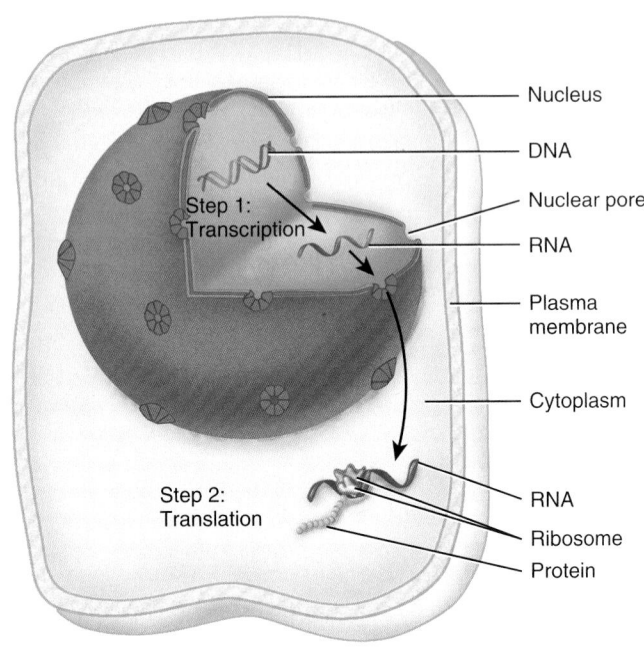

? Why are proteins important in the life of a cell?

DNA and RNA store genetic information as sets of three nucleotides. A sequence of three such nucleotides in DNA is called a **base triplet.** Each DNA base triplet is transcribed as a complementary sequence of three nucleotides, called a **codon.** A given codon specifies a particular amino acid. The **genetic code** is the set of rules that relate the base triplet sequence of DNA to the corresponding codons of RNA and the amino acids they specify.

Transcription

During **transcription,** which occurs in the nucleus, the genetic information represented by the sequence of base triplets in DNA serves as a template for copying the information into a complementary sequence of codons. Three types of RNA are made from the DNA template:

1. **Messenger RNA (mRNA)** directs the synthesis of a protein.

2. **Ribosomal RNA (rRNA)** joins with ribosomal proteins to make ribosomes.

3. **Transfer RNA (tRNA)** binds to an amino acid and holds it in place on a ribosome until it is incorporated into a protein during translation. One end of the tRNA carries a specific amino acid, and the opposite end consists of a triplet of nucleotides

called an **anticodon.** By pairing between complementary bases, the tRNA anticodon attaches to the mRNA codon. Each of the more than 20 different types of tRNA binds to only one of the 20 different amino acids.

The enzyme **RNA polymerase** (po-LIM-er-ās) catalyzes transcription of DNA. However, the enzyme must be instructed where to start the transcription process and where to end it. Only one of the two DNA strands serves as a template for RNA synthesis. The segment of DNA where transcription begins, a special nucleotide sequence called a **promoter,** is located near the beginning of a gene (Figure 3.27a). This is where RNA polymerase attaches to the DNA. During transcription, bases pair in a complementary manner: The bases cytosine (C), guanine (G), and thymine (T) in the DNA template pair with guanine, cytosine, and adenine (A), respectively, in the RNA strand (Figure 3.27b). However, adenine in the DNA template pairs with uracil (U), not thymine, in RNA:

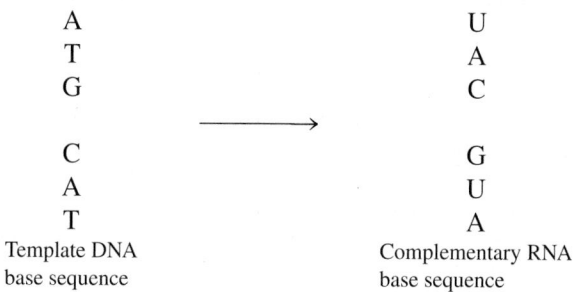

Template DNA base sequence		Complementary RNA base sequence
A		U
T		A
G	→	C
C		G
A		U
T		A

Transcription of the DNA strand ends at another special nucleotide sequence called a **terminator,** which specifies the end of the gene (Figure 3.27a). When RNA polymerase reaches the terminator, the enzyme detaches from the transcribed RNA molecule and the DNA strand.

Not all parts of a gene actually code for parts of a protein. Regions within a gene called **introns** *do not* code for parts of proteins. They are located between regions called **exons** that *do* code for segments of a protein. Immediately after transcription, the transcript includes information from both introns and exons and is called **pre-mRNA.** The introns are removed from pre-mRNA by **small nuclear ribonucleoproteins** (snRNPs, pronounced "snurps"; Figure 3.27b). The snRNPs are enzymes that cut out the introns and splice together the exons. The resulting product is a functional mRNA molecule that passes through a pore in the nuclear envelope to reach the cytoplasm, where translation takes place.

Although the human genome contains around 30,000 genes, there are probably 500,000 to 1 million human proteins. How can so many proteins be coded for by so few genes? Part of the answer lies in **alternative splicing** of mRNA, a process in which the pre-mRNA transcribed from a gene is spliced in different ways to produce several different mRNAs. The different mRNAs are then translated into different proteins. In this way, one gene may code for 10 or more different proteins. In addition, chemical modifications are made to proteins after translation, for example, as proteins pass through the Golgi complex. Such chemical alterations can produce two or more different proteins from a single translation.

Figure 3.27 Transcription. DNA transcription begins at a promoter and ends at a terminator.

During transcription, the genetic information in DNA is copied to RNA.

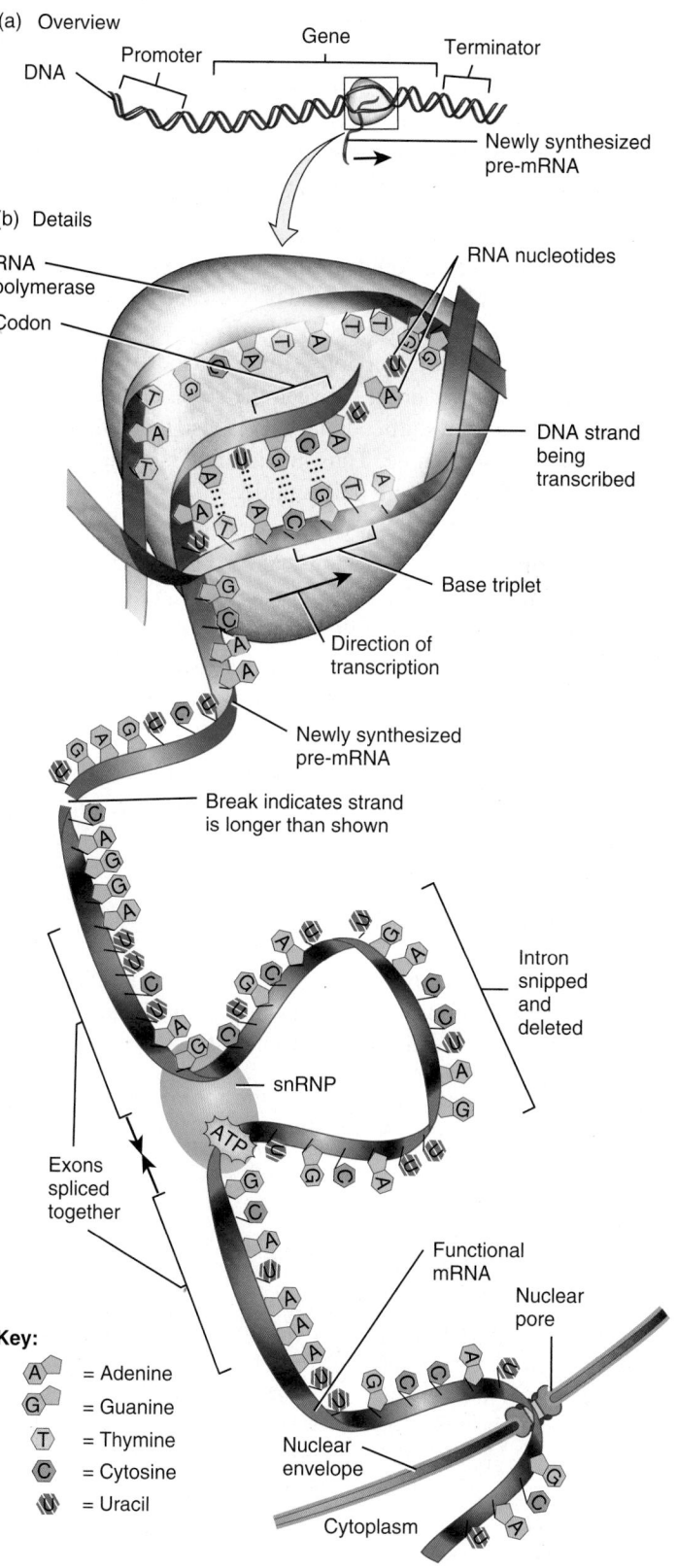

(a) Overview

DNA — Promoter — Gene — Terminator
Newly synthesized pre-mRNA

(b) Details

RNA polymerase
Codon
RNA nucleotides
DNA strand being transcribed
Base triplet
Direction of transcription
Newly synthesized pre-mRNA
Break indicates strand is longer than shown
Intron snipped and deleted
snRNP
Exons spliced together
Functional mRNA
Nuclear pore
Nuclear envelope
Cytoplasm

Key:

A	= Adenine
G	= Guanine
T	= Thymine
C	= Cytosine
U	= Uracil

If the DNA template had the base sequence AGCT, what would be the mRNA base sequence, and what enzyme would catalyze DNA transcription?

Translation

In the process of **translation,** the nucleotide sequence in an mRNA molecule specifies the amino acid sequence of a protein. Ribosomes in the cytoplasm carry out translation. The small subunit of a ribosome has a *binding site* for mRNA; the large subunit has two binding sites for tRNA molecules, a *P site* and an *A site* (Figure 3.28). The first tRNA molecule bearing its specific amino acid attaches to mRNA at the P site. The A site holds the next tRNA molecule bearing its amino acid. Translation occurs in the following way (Figure 3.29):

1 An mRNA molecule binds to the small ribosomal subunit at the mRNA binding site. A special tRNA, called *initiator tRNA,* binds to the start codon (AUG) on mRNA, where translation begins. The tRNA anticodon (UAC) attaches to the mRNA codon (AUG) by pairing between the complementary bases. Besides being the start codon, AUG is also the codon for the amino acid methionine. Thus, methionine is always the first amino acid in a growing polypeptide.

2 Next, the large ribosomal subunit attaches to the small ribosomal subunit–mRNA complex, creating a functional ribosome. The initiator tRNA, with its amino acid (methionine), fits into the P site of the ribosome.

3 The anticodon of another tRNA with its attached amino acid pairs with the second mRNA codon at the A site of the ribosome.

4 A component of the large ribosomal subunit catalyzes the formation of a peptide bond between methionine, which separates from its tRNA at the P site, and the amino acid carried by the tRNA at the A site.

5 After peptide bond formation, the tRNA at the P site detaches from the ribosome, and the ribosome shifts the mRNA strand by one codon. The tRNA in the A site bearing the two-peptide protein shifts into the P site, allowing another tRNA with its amino acid to bind to a newly exposed codon at the A site. Steps **3** through **5** occur repeatedly, and the protein lengthens progressively.

6 Protein synthesis ends when the ribosome reaches a stop codon, which causes the completed protein to detach from the final tRNA. When the tRNA vacates the ribosome, the ribosome splits into its large and small subunits.

Protein synthesis progresses at a rate of about 15 peptide bonds per second. As the ribosome moves along the mRNA and before it completes synthesis of the whole protein, another ribosome may attach behind it and begin translation of the same mRNA strand. Several ribosomes attached to the same mRNA constitute a **polyribosome.** The simultaneous movement of several ribosomes along the same mRNA molecule permits the translation of one mRNA into several identical proteins at the same time.

CLINICAL CONNECTION | *Recombinant DNA*

Scientists have developed techniques for inserting genes from other organisms into a variety of host cells. Manipulating the cell in this way can cause the host organism to produce proteins it normally does not synthesize. Organisms so altered are called **recombinants** (rē-KOM-bi-nants), and their DNA—a combination of DNA from different sources—is called **recombinant DNA.** When recombinant DNA functions properly, the host will synthesize the protein specified by the new gene it has acquired. The technology that has arisen from the manipulation of genetic material is referred to as **genetic engineering.**

The practical applications of recombinant DNA technology are enormous. Strains of recombinant bacteria now produce large quantities of many important therapeutic substances, including *human growth hormone (hGH),* required for normal growth and metabolism; *insulin,* a hormone that helps regulate blood glucose level and is used by diabetics; *interferon (IFN),* an antiviral (and possibly anticancer) substance; and *erythropoietin (EPO),* a hormone that stimulates production of red blood cells. •

✔ CHECKPOINT

25. What is meant by the term gene expression?
26. What is the difference between transcription and translation?

3.7 CELL DIVISION

◉ **OBJECTIVES**
• Discuss the stages, events, and significance of somatic and reproductive cell division.
• Describe the signals that induce somatic cell division.

Most cells of the human body undergo **cell division,** the process by which cells reproduce themselves. The two types of cell division—somatic cell division and reproductive cell division—accomplish different goals for the organism.

Figure 3.28 Translation. During translation, an mRNA molecule binds to a ribosome. Then, the mRNA nucleotide sequence specifies the amino acid sequence of a protein.

🔑 Ribosomes have a binding site for mRNA and a P site and an A site for attachment of tRNAs.

(a) Components of a ribosome and their relationship to mRNA and protein during translation

(b) Interior view of tRNA binding sites

? What roles do the P and A sites serve?

Figure 3.29 Protein elongation and termination of protein synthesis during translation.

During protein synthesis the small and large ribosomal subunits join to form a functional ribosome. When the process is complete, they separate.

1 Initiator tRNA attaches to a start codon.

2 Large and small ribosomal subunits join to form a functional ribosome and initiator tRNA fits into position on the ribosome.

3 Anticodon of incoming tRNA pairs with next mRNA codon beside initiator tRNA.

4 Amino acid on initiator tRNA forms a peptide bond with amino acid beside it.

5 tRNA leaves the ribosome; ribosome shifts by one codon; tRNA binds to newly exposed codon; steps **3** – **5** repeat.

6 Protein synthesis stops when the ribosome reaches stop codon on mRNA.

Key:
- A = Adenine
- G = Guanine
- T = Thymine
- C = Cytosine
- U = Uracil

Summary of movement of ribosome along mRNA

? What is the function of a stop codon?

A **somatic cell** (sō-MAT-ik; *soma* = body) is any cell of the body other than a germ cell. A **germ cell** is a gamete (sperm or oocyte) or any precursor cell destined to become a gamete. In **somatic cell division,** a cell undergoes a nuclear division called **mitosis** (mī-TŌ-sis; *mitos* = thread) and a cytoplasmic division called **cytokinesis** (sī-tō-ki-NĒ-sis; *cyto-* = cell; *-kinesis* = movement) to produce two genetically identical cells, each with the same number and kind of chromosomes as the original cell. Somatic cell division replaces dead or injured cells and adds new ones during tissue growth.

Reproductive cell division is the mechanism that produces gametes, the cells needed to form the next generation of sexually reproducing organisms. This process consists of a special two-step division called **meiosis,** in which the number of chromosomes in the nucleus is reduced by half.

Somatic Cell Division

The **cell cycle** is an orderly sequence of events in which a somatic cell duplicates its contents and divides in two. Some cells divide more than others. Human cells, such as those in the brain, stomach, and kidneys, contain 23 pairs of chromosomes, for a total of 46. One member of each pair is inherited from each parent. The two chromosomes that make up each pair are called **homologous chromosomes** (hō-MOL-ō-gus; *homo-* = same) or **homologs;** they contain similar genes arranged in the same (or almost the same) order. When examined under a light microscope, homologous chromosomes generally look very similar. The exception to this rule is one pair of chromosomes called the **sex chromosomes,** designated X and Y. In females the homologous pair of sex chromosomes consists of two large X chromosomes; in males the pair consists of an X and a much smaller Y chromosome. Because somatic cells contain two sets of chromosomes, they are called **diploid cells** (DIP-loyd; *dipl-* = double; *-oid* = form), symbolized **2n.**

When a cell reproduces, it must replicate (duplicate) all its chromosomes to pass its genes to the next generation of cells. The cell cycle consists of two major periods: interphase, when a cell is not dividing, and the mitotic (M) phase, when a cell is dividing (Figure 3.30).

Interphase

During **interphase** (IN-ter-fāz) the cell replicates its DNA through a process that will be described shortly. It also produces additional organelles and cytosolic components in anticipation of cell division. Interphase is a state of high metabolic activity; it is during this time that the cell does most of its growing. Interphase consists of three phases: G_1, S, and G_2 (Figure 3.30). The S stands for *synthesis* of DNA. Because the G phases are periods when there is no activity related to DNA duplication, they are thought of as *gaps* or interruptions in DNA duplication.

The **G_1 phase** is the interval between the mitotic phase and the S phase. During G_1, the cell is metabolically active; it replicates most of its organelles and cytosolic components but not its DNA. Replication of centrosomes also begins in the G_1 phase. Virtually all the cellular activities described in this chapter happen during

Figure 3.30 The cell cycle. Not illustrated is cytokinesis (division of the cytoplasm), which usually occurs during late anaphase of the mitotic phase.

 In a complete cell cycle, a starting cell duplicates its contents and divides into two identical cells.

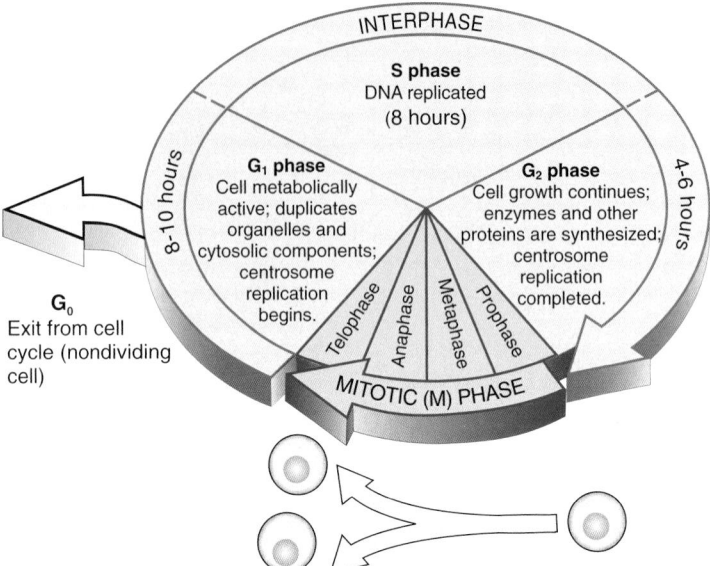

? During which phase of the cell cycle does DNA replication occur?

G_1. For a cell with a total cell cycle time of 24 hours, G_1 lasts 8 to 10 hours. However, the duration of this phase is quite variable. It is very short in many embryonic cells or cancer cells. Cells that remain in G_1 for a very long time, perhaps destined never to divide again, are said to be in the **G_0 phase.** Most nerve cells are in the G_0 phase. Once a cell enters the S phase, however, it is committed to go through the rest of the cell cycle.

The **S phase,** the interval between G_1 and G_2, lasts about 8 hours. During the S phase, DNA replication occurs. As a result of DNA replication, the two identical cells formed during cell division later in the cell cycle will have the same genetic material. The **G_2 phase** is the interval between the S phase and the mitotic phase. It lasts 4 to 6 hours. During G_2, cell growth continues, enzymes and other proteins are synthesized in preparation for cell division, and replication of centrosomes is completed. When DNA replicates during the S phase, its helical structure partially uncoils, and the two strands separate at the points where hydrogen bonds connect base pairs (Figure 3.31). Each exposed base of the old DNA strand then pairs with the complementary base of a newly synthesized nucleotide. A new DNA strand takes shape as chemical bonds form between neighboring nucleotides. The uncoiling and complementary base pairing continues until each of the two original DNA strands is joined with a newly formed complementary DNA strand. The original DNA molecule has become two identical DNA molecules.

Figure 3.31 Replication of DNA. The two strands of the double helix separate by breaking the hydrogen bonds (shown as dotted lines) between nucleotides. New, complementary nucleotides attach at the proper sites, and a new strand of DNA is synthesized alongside each of the original strands. Arrows indicate hydrogen bonds forming again between pairs of bases.

🔑 Replication doubles the amount of DNA.

Key:
- Ⓐ = Adenine
- Ⓖ = Guanine
- Ⓣ = Thymine
- Ⓒ = Cytosine

Hydrogen bond
Phosphate group
Deoxyribose sugar

Old strand   New strand   New strand   Old strand

❓ **Why must DNA replication occur before cytokinesis in somatic cell division?**

A microscopic view of a cell during interphase shows a clearly defined nuclear envelope, a nucleolus, and a tangled mass of chromatin (Figure 3.32a). Once a cell completes its activities during the G_1, S, and G_2 phases of interphase, the mitotic phase begins.

Mitotic Phase

The **mitotic (M) phase** of the cell cycle, which results in the formation of two identical cells, consists of a nuclear division (mitosis) and a cytoplasmic division (cytokinesis) to form two identical cells. The events that occur during mitosis and cytokinesis are plainly visible under a microscope because chromatin condenses into discrete chromosomes.

NUCLEAR DIVISION: MITOSIS **Mitosis,** as noted earlier, is the distribution of two sets of chromosomes into two separate nuclei. The process results in the *exact* partitioning of genetic information. For convenience, biologists divide the process into four stages: prophase, metaphase, anaphase, and telophase. However, mitosis is a continuous process; one stage merges seamlessly into the next.

1. **Prophase** (PRŌ-fāz). During early prophase, the chromatin fibers condense and shorten into chromosomes that are visible under the light microscope (Figure 3.32b). The condensation process may prevent entangling of the long DNA strands as they move during mitosis. Because longitudinal DNA replication took place during the S phase of interphase, each prophase chromosome consists of a pair of identical strands called *chromatids*. A constricted region called a **centromere** (SEN-trō-mēr) holds the chromatid pair together. At the outside of each centromere is a protein complex known as the **kinetochore** (ki-NET-ō-kor). Later in prophase, tubulins in the pericentriolar material of the centrosomes start to form the **mitotic spindle,** a football-shaped assembly of microtubules that attach to the kinetochore (Figure 3.32b). As the microtubules lengthen, they push the centrosomes to the poles (ends) of the cell so that the spindle extends from pole to pole. The mitotic spindle is responsible for the separation of chromatids to opposite poles of the cell. Then, the nucleolus disappears and the nuclear envelope breaks down.

2. **Metaphase** (MET-a-fāz). During metaphase, the microtubules of the mitotic spindle align the centromeres of the chromatid pairs at the exact center of the mitotic spindle (Figure 3.32c). This midpoint region is called the **metaphase plate.**

3. **Anaphase** (AN-a-fāz). During anaphase, the centromeres split, separating the two members of each chromatid pair, which move toward opposite poles of the cell (Figure 3.32d). Once separated, the chromatids are termed *chromosomes.* As the chromosomes are pulled by the microtubules of the mitotic spindle during anaphase, they appear V-shaped because the centromeres lead the way, dragging the trailing arms of the chromosomes toward the pole.

4. **Telophase** (TEL-ō-fāz). The final stage of mitosis, telophase, begins after chromosomal movement stops (Figure 3.32e). The identical sets of chromosomes, now at opposite poles of the cell, uncoil and revert to the threadlike chromatin form. A nuclear envelope forms around each chromatin mass, nucleoli reappear in the identical nuclei, and the mitotic spindle breaks up.

CYTOPLASMIC DIVISION: CYTOKINESIS As noted earlier, division of a cell's cytoplasm and organelles into two identical cells is called **cytokinesis.** This process usually begins in late anaphase with the formation of a **cleavage furrow,** a slight indentation of the plasma membrane, and is completed after telophase. The

Figure 3.32 Cell division: mitosis and cytokinesis. Begin the sequence at ❶ at the top of the figure and read clockwise to complete the process.

🔑 In somatic cell division, a single starting cell divides to produce two identical diploid cells.

(a) INTERPHASE

LM all at 700x

Centrosome:
Centrioles
Pericentriolar material
Nucleolus
Nuclear envelope
Chromatin
Plasma membrane
Cytosol

Kinetochore
Mitotic spindle (microtubules)
Fragments of nuclear envelope

Centromere
Chromosome (two chromatids joined at centromere)

Early   Late
(b) PROPHASE

Metaphase plate

(c) METAPHASE

(f) IDENTICAL CELLS IN INTERPHASE

Cleavage furrow

(e) TELOPHASE

Cleavage furrow

Chromosome

Late   Early
(d) ANAPHASE

❓ **When does cytokinesis begin?**

cleavage furrow usually appears midway between the centrosomes and extends around the periphery of the cell (Figure 3.32d, e). Actin microfilaments that lie just inside the plasma membrane form a *contractile ring* that pulls the plasma membrane progressively inward. The ring constricts the center of the cell, like tightening a belt around the waist, and ultimately pinches it in two. Because the plane of the cleavage furrow is always perpendicular to the mitotic spindle, the two sets of chromosomes end up in separate cells. When cytokinesis is complete, interphase begins (Figure 3.32f).

The sequence of events can be summarized as

$$G_1 \longrightarrow S \text{ phase} \longrightarrow G_2 \text{ phase} \longrightarrow \text{mitosis} \longrightarrow \text{cytokinesis}$$

Table 3.3 summarizes the events of the cell cycle in somatic cells.

CLINICAL CONNECTION | *Mitotic Spindle and Cancer*

One of the distinguishing features of cancer cells is uncontrolled division, which results in the formation of a mass of cells called a *neoplasm* or *tumor*. One of the ways to treat cancer is by *chemotherapy*, the use of anticancer drugs. Some of these drugs stop cell division by inhibiting the formation of the mitotic spindle. Unfortunately, these types of anticancer drugs also kill all types of rapidly dividing cells in the body, causing side effects such as nausea, diarrhea, hair loss, fatigue, and decreased resistance to disease. •

Control of Cell Destiny

A cell has three possible destinies: (1) to remain alive and functioning without dividing, (2) to grow and divide, or (3) to die. Homeostasis is maintained when there is a balance between cell proliferation and cell death. The signals that tell a cell when to exist in the G_0 phase, when to divide, and when to die have been the subjects of intense and fruitful research in recent years.

Within a cell, there are enzymes called **cyclin-dependent protein kinases (Cdk's)** that can transfer a phosphate group from ATP to a protein to activate the protein; other enzymes can remove the phosphate group from the protein to deactivate it. The activation and deactivation of Cdk's at the appropriate time is crucial in the initiation and regulation of DNA replication, mitosis, and cytokinesis.

Switching the Cdk's on and off is the responsibility of cellular proteins called **cyclins** (SĪK-lins), so named because their levels rise and fall during the cell cycle. The joining of a specific cyclin and Cdk molecule triggers various events that control cell division.

The activation of specific cyclin–Cdk complexes is responsible for progression of a cell from G_1 to S to G_2 to mitosis in a specific order. If any step in the sequence is delayed, all subsequent steps are delayed in order to maintain the normal sequence. The levels of cyclins in the cell are very important in determining the timing and sequence of events in cell division. For example, the level of the cyclin that helps drive a cell from G_2 to mitosis rises throughout the G_1, S, and G_2 phases and into mitosis. The high level triggers mitosis, but toward the end of mitosis, the level declines rapidly and mitosis ends. Destruction of this cyclin, as well as others in the cell, is by proteasomes.

Cellular death is also regulated. Throughout the lifetime of an organism, certain cells undergo apoptosis, an orderly, genetically programmed death (see the discussion of mitochondria in Section 3.4). In apoptosis, a triggering agent from either outside or inside the cell causes "cell-suicide" genes to produce enzymes that damage the cell in several ways, including disruption of its cytoskeleton and nucleus. As a result, the cell shrinks and pulls away from neighboring cells. Although the plasma membrane

TABLE 3.3

Events of the Somatic Cell Cycle

PHASE	ACTIVITY
Interphase	Period between cell divisions; chromosomes not visible under light microscope.
G_1 phase	Metabolically active cell duplicates most of its organelles and cytosolic components; replication of chromosomes begins. (Cells that remain in the G_1 phase for a very long time, and possibly never divide again, are said to be in the G_0 phase.)
S phase	Replication of DNA and centrosomes.
G_2 phase	Cell growth, enzyme and protein synthesis continue; replication of centrosomes complete.
Mitotic phase	Parent cell produces identical cells with identical chromosomes; chromosomes visible under light microscope.
Mitosis	Nuclear division; distribution of two sets of chromosomes into separate nuclei.
Prophase	Chromatin fibers condense into paired chromatids; nucleolus and nuclear envelope disappear; each centrosome moves to an opposite pole of the cell.
Metaphase	Centromeres of chromatid pairs line up at metaphase plate.
Anaphase	Centromeres split; identical sets of chromosomes move to opposite poles of cell.
Telophase	Nuclear envelopes and nucleoli reappear; chromosomes resume chromatin form; mitotic spindle disappears.
Cytokinesis	Cytoplasmic division; contractile ring forms cleavage furrow around center of cell, dividing cytoplasm into separate and equal portions.

remains intact, the DNA within the nucleus fragments and the cytoplasm shrinks. Nearby phagocytes then ingest the dying cell via a complex process that involves a receptor protein in the plasma membrane of the phagocyte that binds to a lipid in the plasma membrane of the dying cell. Apoptosis removes unneeded cells during fetal development, such as the webbing between digits. It continues to occur after birth to regulate the number of cells in a tissue and eliminate potentially dangerous cells such as cancer cells.

Apoptosis is a normal type of cell death; in contrast, **necrosis** (ne-KRŌ-sis = death) is a pathological type of cell death that results from tissue injury. In necrosis, many adjacent cells swell, burst, and spill their cytoplasm into the interstitial fluid. The cellular debris usually stimulates an inflammatory response by the immune system, a process that does not occur in apoptosis.

CLINICAL CONNECTION | *Tumor-Suppressor Genes*

Abnormalities in genes that regulate the cell cycle or apoptosis are associated with many diseases. For example, damage to genes called **tumor-suppressor genes,** which produce proteins that normally inhibit cell division, causes some types of cancer. Loss or alteration of a tumor-suppressor gene called *p53* on chromosome 17 is the most common genetic change leading to a wide variety of tumors, including breast and colon cancers. The normal *p53* protein arrests cells in the G_1 phase, which prevents cell division. Normal *p53* protein also assists in repair of damaged DNA and induces apoptosis in the cells where DNA repair was not successful. For this reason, the *p53* gene is nicknamed "the guardian angel of the genome." •

Reproductive Cell Division

In the process called sexual reproduction, each new organism is the result of the union of two different gametes (fertilization), one produced by each parent. If gametes had the same number of chromosomes as somatic cells, the number of chromosomes would double at fertilization. **Meiosis** (mī-Ō-sis; *mei-* = lessening; *-osis* = condition of), the reproductive cell division that occurs in the gonads (ovaries and testes), produces gametes in which the number of chromosomes is reduced by half. As a result, gametes contain a single set of 23 chromosomes and thus are **haploid (*n*) cells** (HAP-loyd; *hapl-* = single). Fertilization restores the diploid number of chromosomes.

Meiosis

Unlike mitosis, which is complete after a single round, meiosis occurs in two successive stages: **meiosis I** and **meiosis II.** During the interphase that precedes meiosis I, the chromosomes of the diploid cell start to replicate. As a result of replication, each chromosome consists of two sister (genetically identical) chromatids, which are attached at their centromeres. This replication of chromosomes is similar to the one that precedes mitosis in somatic cell division.

MEIOSIS I   Meiosis I, which begins once chromosomal replication is complete, consists of four phases: prophase I, metaphase I,

anaphase I, and telophase I (Figure 3.33a). Prophase I is an extended phase in which the chromosomes shorten and thicken, the nuclear envelope and nucleoli disappear, and the mitotic spindle forms. Two events that are not seen in mitotic prophase occur during prophase I of meiosis (Figure 3.33b). First, the two sister chromatids of each pair of homologous chromosomes pair off, an event called **synapsis** (sin-AP-sis). The resulting four chromatids form a structure called a **tetrad.** Second, parts of the chromatids of two homologous chromosomes may be exchanged with one another. Such an exchange between parts of nonsister (genetically different) chromatids is called **crossing-over.** This process, among others, permits an exchange of genes between chromatids of homologous chromosomes. Due to crossing-over, the resulting cells are genetically unlike each other and genetically unlike the starting cell that produced them. Crossing-over results in *genetic recombination*—that is, the formation of new combinations of genes—and accounts for part of the great genetic variation among humans and other organisms that form gametes via meiosis.

In metaphase I, the tetrads formed by the homologous pairs of chromosomes line up along the metaphase plate of the cell, with homologous chromosomes side by side (Figure 3.33a). During anaphase I, the members of each homologous pair of chromosomes separate as they are pulled to opposite poles of the cell by the microtubules attached to the centromeres. The paired chromatids, held by a centromere, remain together. (Recall that during mitotic anaphase, the centromeres split and the sister chromatids separate.) Telophase I and cytokinesis of meiosis are similar to telophase and cytokinesis of mitosis. The net effect of meiosis I is that each resulting cell contains the haploid number of chromosomes because it contains only one member of each pair of the homologous chromosomes present in the starting cell.

MEIOSIS II   The second stage of meiosis, meiosis II, also consists of four phases: prophase II, metaphase II, anaphase II, and telophase II (Figure 3.33a). These phases are similar to those that occur during mitosis; the centromeres split, and the sister chromatids separate and move toward opposite poles of the cell.

In summary, meiosis I begins with a diploid starting cell and ends with two cells, each with the haploid number of chromosomes. During meiosis II, each of the two haploid cells formed during meiosis I divides; the net result is four haploid gametes that are genetically different from the original diploid starting cell.

Figure 3.34 compares the events of meiosis and mitosis.

✔CHECKPOINT

27. Distinguish between somatic and reproductive cell division and explain the importance of each.
28. What is the significance of interphase?
29. Outline the major events of each stage of the mitotic phase of the cell cycle.
30. How are apoptosis and necrosis similar? How do they differ?
31. How are haploid and diploid cells different?
32. What are homologous chromosomes?

Figure 3.33 Meiosis, reproductive cell division. Details of each of the stages are discussed in the text.

🔒 In reproductive cell division, a single diploid starting cell undergoes meiosis I and meiosis II to produce four haploid gametes that are genetically different from the starting cell that produced them.

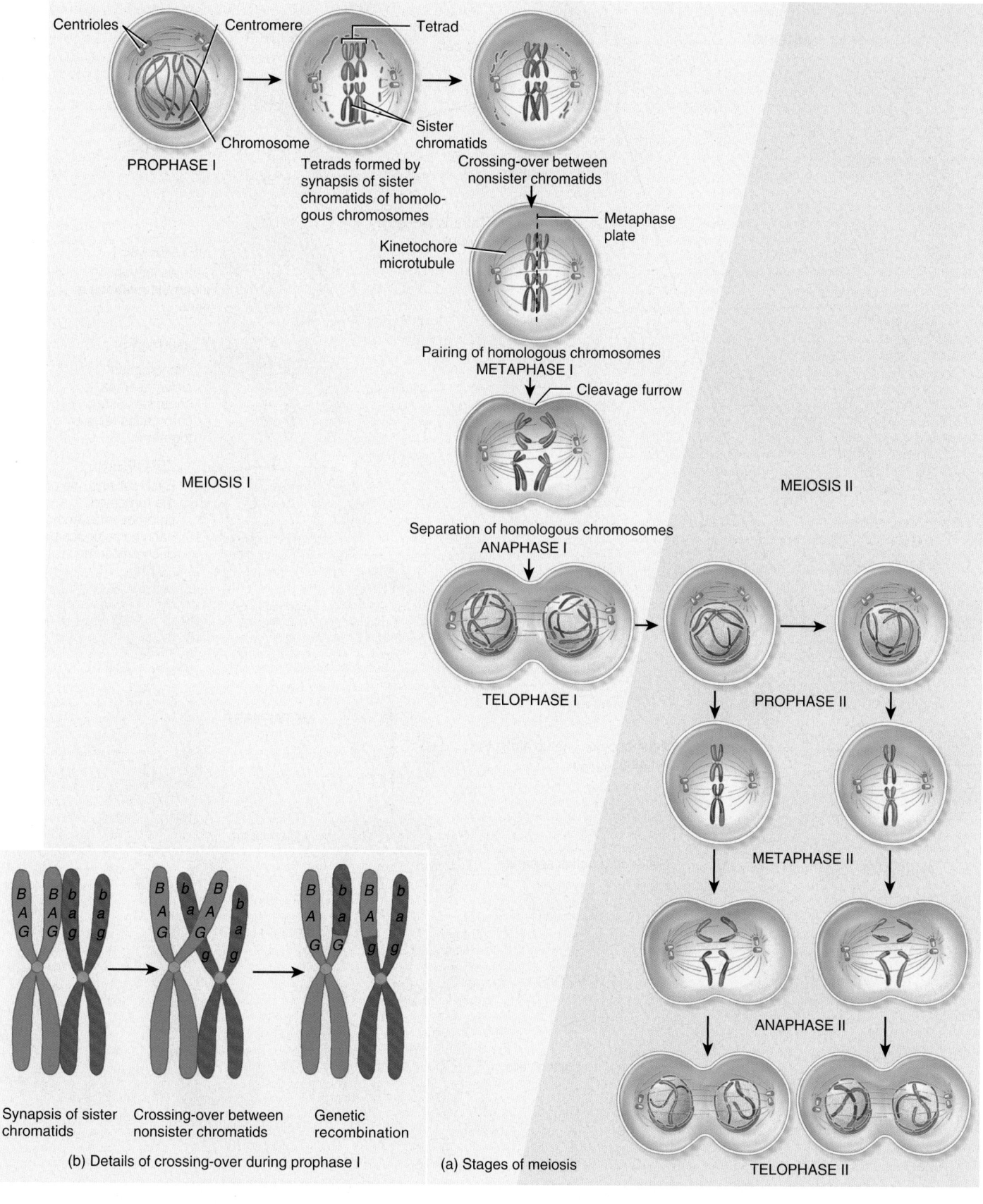

Centrioles
Centromere
Tetrad
Chromosome
Sister chromatids
PROPHASE I

Tetrads formed by synapsis of sister chromatids of homologous chromosomes

Crossing-over between nonsister chromatids

Kinetochore microtubule
Metaphase plate

Pairing of homologous chromosomes
METAPHASE I

MEIOSIS I

Cleavage furrow

Separation of homologous chromosomes
ANAPHASE I

MEIOSIS II

TELOPHASE I
PROPHASE II

METAPHASE II

ANAPHASE II

TELOPHASE II

Synapsis of sister chromatids
Crossing-over between nonsister chromatids
Genetic recombination

(b) Details of crossing-over during prophase I

(a) Stages of meiosis

❓ **How does crossing-over affect the genetic content of the haploid gametes?**

Figure 3.34 Comparison between mitosis (left) and meiosis (right) in which the starting cell has two pairs of homologous chromosomes.

🔑 ⚷ The phases of meiosis II and mitosis are similar.

MITOSIS

MEIOSIS

Starting cell

2n
Chromosomes
already replicated

Crossing-over

PROPHASE I

Tetrads formed by
synapsis

METAPHASE I
Tetrads line up
along the metaphase
plate

ANAPHASE I
Homologous
chromosomes
separate (sister
chromatids remain
together)

TELOPHASE I
Each cell has one of
the replicated
chromosomes from
each homologous pair
of chromosomes (n)

PROPHASE

PROPHASE
II

METAPHASE   Chromosomes line up along the
metaphase plate

METAPHASE
II

ANAPHASE   Sister chromatids separate

ANAPHASE
II

TELOPHASE   Cytokinesis

TELOPHASE
II

Resulting cells

INTERPHASE

2n   2n   n   n   n   n

Somatic cells with diploid number
of chromosomes (not replicated)

Gametes with haploid number of
chromosomes (not replicated)

❓ How does anaphase I of meiosis differ from anaphase of mitosis?

3.8 CELLULAR DIVERSITY

◉ **OBJECTIVE**

- Describe how cells differ in size and shape.

The body of an average human adult is composed of nearly 100 trillion cells. All of these cells can be classified into about 200 different cell types. Cells vary considerably in size. The sizes of cells are measured in units called *micrometers* (mī-KROM-i-ters). One micrometer (μm) is equal to 1 one-millionth of a meter, or 10^{-6} m (1/25,000 of an inch). High-powered microscopes are needed to see the smallest cells of the body. The largest cell, a single oocyte, has a diameter of about 140 μm and is barely visible to the un-aided eye. A red blood cell has a diameter of 8 μm. To better visualize this, an average hair from the top of your head is approximately 100 μm in diameter.

The shapes of cells also vary considerably (Figure 3.35). They may be round, oval, flat, cube-shaped, column-shaped, elongated, star-shaped, cylindrical, or disc-shaped. A cell's shape is related to its function in the body. For example, a sperm cell has a long whiplike tail (flagellum) that it uses for locomotion. The disc shape of a red blood cell gives it a large surface area that enhances its ability to pass oxygen to other cells. The long, spindle shape of a relaxed smooth muscle cell shortens as it contracts. This change in shape allows groups of smooth muscle cells to narrow or widen the passage for blood flowing through blood vessels. In this way, they regulate blood flow through various tissues. Recall that some cells contain microvilli, which greatly increase their surface area. Microvilli are common in the epithelial cells that line the small intestine, where the large surface area speeds the absorption of digested food. Nerve cells have long extensions that permit them to conduct nerve impulses over great distances. As you will see in the following chapters, cellular diversity also permits organization of cells into more complex tissues and organs.

✔**CHECKPOINT**

33. How is cell shape related to function? Give several of your own examples.

Figure 3.35 Diverse shapes and sizes of human cells. The relative difference in size between the smallest and largest cells is actually much greater than shown here.

🔑 The nearly 100 trillion cells in an average adult can be classified into about 200 different cell types.

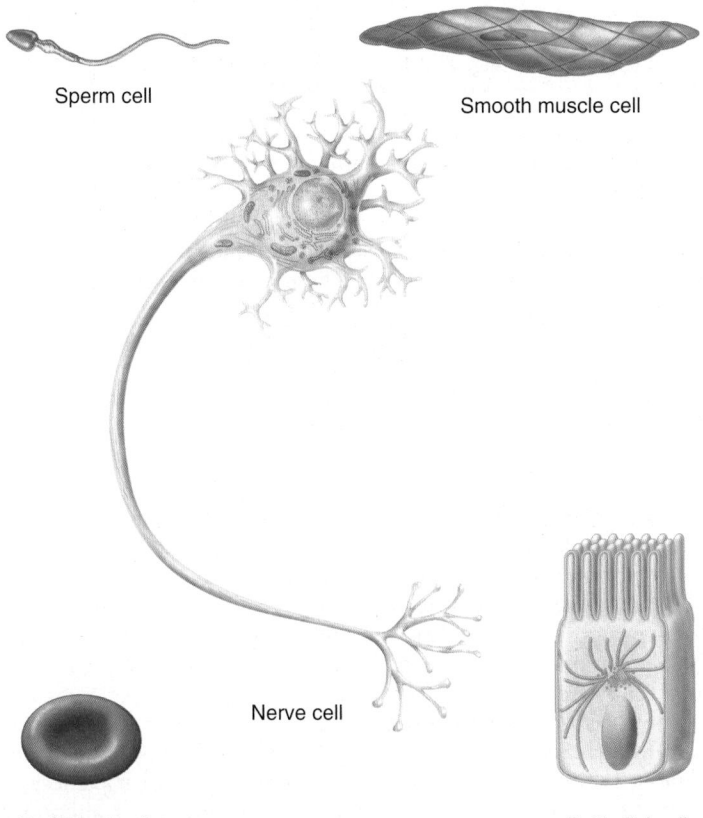

Sperm cell

Smooth muscle cell

Nerve cell

Red blood cell

Epithelial cell

❓ Why are sperm the only body cells that need to have a flagellum?

3.9 AGING AND CELLS

◉ **OBJECTIVE**

- Describe the cellular changes that occur with aging.

Aging is a normal process accompanied by a progressive alteration of the body's homeostatic adaptive responses. It produces observable changes in structure and function and increases vulnerability to environmental stress and disease. The specialized branch of medicine that deals with the medical problems and care of elderly persons is **geriatrics** (jer'-ē-AT-riks; *ger-* = old age; *-iatrics* = medicine). **Gerontology** (jer'-on-TOL-ō-jē) is the scientific study of the process and problems associated with aging.

Although many millions of new cells normally are produced each minute, several kinds of cells in the body—skeletal muscle cells and nerve cells—do not divide because they are arrested permanently in the G_0 phase (see the discussion of interphase earlier in the chapter). Experiments have shown that many other cell types have only a limited capability to divide. Normal cells grown outside the body divide only a certain number of times and then stop. These observations suggest that cessation of mitosis is a normal, genetically programmed event. According to this view, "aging genes" are part of the genetic blueprint at birth. These genes have an important function in normal cells, but their activities slow over time. They bring about aging by slowing down or halting processes vital to life.

Another aspect of aging involves **telomeres** (TĒ-lō-mērz), specific DNA sequences found only at the tips of each chromosome. These pieces of DNA protect the tips of chromosomes from erosion and from sticking to one another. However, in most normal body cells each cycle of cell division shortens the telomeres. Eventually, after many cycles of cell division, the telomeres can be completely gone and even some of the functional chromosomal material may be lost. These observations suggest that erosion of DNA from the tips of our chromosomes contributes greatly to aging and death of cells. It has recently been learned that individuals who experience high levels of stress have significantly shorter telomere length.

Glucose, the most abundant sugar in the body, plays a role in the aging process. It is haphazardly added to proteins inside and outside cells, forming irreversible cross-links between adjacent protein molecules. With advancing age, more cross-links form, which contributes to the stiffening and loss of elasticity that occur in aging tissues.

CLINICAL CONNECTION | *Free Radicals*

Free radicals produce oxidative damage in lipids, proteins, or nucleic acids by "stealing" an electron to accompany their unpaired electrons. Some effects are wrinkled skin, stiff joints, and hardened arteries. Normal metabolism—for example, aerobic cellular respiration in mitochondria—produces some free radicals. Others are present in air pollution, radiation, and certain foods we eat. Naturally occurring enzymes in peroxisomes and in the cytosol normally dispose of free radicals. Certain dietary substances, such as vitamin E, vitamin C, beta-carotene, zinc, and selenium, are referred to as *antioxidants* because they inhibit the formation of free radicals. •

Some theories of aging explain the process at the cellular level, while others concentrate on regulatory mechanisms operating within the entire organism. For example, the immune system may start to attack the body's own cells. This *autoimmune response*

might be caused by changes in cell-identity markers at the surface of cells that cause antibodies to attach to and mark the cell for destruction. As changes in the proteins on the plasma membrane of cells increase, the autoimmune response intensifies, producing the well-known signs of aging. In the chapters that follow, we will discuss the effects of aging on each body system in sections similar to this one.

CLINICAL CONNECTION | *Progeria and Werner Syndrome*

Progeria (prō-JER-ē-a) is a disease characterized by normal development in the first year of life followed by rapid aging. It is caused by a genetic defect in which telomeres are considerably shorter than normal. Symptoms include dry and wrinkled skin, total baldness, and birdlike facial features. Death usually occurs around age 13.

Werner syndrome (VER-ner) is a rare, inherited disease that causes a rapid acceleration of aging, usually while the person is only in his or her twenties. It is characterized by wrinkling of the skin, graying of the hair and baldness, cataracts, muscular atrophy, and a tendency to develop diabetes mellitus, cancer, and cardiovascular disease. Most afflicted individuals die before age 50. Recently, the gene that causes Werner syndrome has been identified. Researchers hope to use the information to gain insight into the mechanisms of aging, as well as to help those suffering from the disorder. •

✔ CHECKPOINT

34. What is one reason that some tissues become stiffer as they age?

DISORDERS: HOMEOSTATIC IMBALANCES

Most chapters in the text are followed by concise discussions of major diseases and disorders that illustrate departures from normal homeostasis. These discussions provide answers to many questions that you might ask about medical problems.

Cancer

Cancer is a group of diseases characterized by uncontrolled or abnormal cell division. When cells in a part of the body divide without control, the excess tissue that develops is called a **tumor** or **neoplasm** (NĒ-ō-plazm; *neo-* = new). The study of tumors is called **oncology** (on-KOL-ō-jē; *onco-* = swelling or mass). Tumors may be cancerous and often fatal, or they may be harmless. A cancerous neoplasm is called a **malignant tumor** (ma-LIG-nant) or **malignancy.** One property of most malignant tumors is their ability to undergo **metastasis** (me-TAS-ta-sis), the spread of cancerous cells to other parts of the body. A **benign tumor** (be-NĪN) is a neoplasm that does not metastasize. An example is a wart. Most benign tumors may be removed surgically if they interfere with normal body function or become disfiguring. Some benign tumors can be inoperable and perhaps fatal.

Types of Cancer

The name of a cancer is derived from the type of tissue in which it develops. Most human cancers are **carcinomas** (kar-si-NŌ-maz; *carcin* = cancer; *-omas* = tumors), malignant tumors that arise from epithelial cells. **Melanomas** (mel-a-NŌ-maz; *melan-* = black), for example, are cancerous growths of melanocytes, skin epithelial cells that produce the pigment melanin. **Sarcoma** (sar-KŌ-ma; *sarc-* = flesh) is a general term for any cancer arising from muscle cells or connective tissues. For example, **osteogenic sarcoma** (*osteo-* = bone; *-genic* = origin), the most frequent type of childhood cancer, destroys normal bone tissue. **Leukemia** (loo-KĒ-mē-a; *leuk-* = white; *-emia* = blood) is a cancer of blood-forming organs characterized by rapid growth of abnormal leukocytes (white blood cells). **Lymphoma** (lim-FŌ-ma) is a malignant disease of lymphatic tissue—for example, of lymph nodes.

SEM

Lung cancer cell dividing

Growth and Spread of Cancer

Cells of malignant tumors duplicate rapidly and continuously. As malignant cells invade surrounding tissues, they often trigger **angiogenesis** (an'-jē-ō-JEN-e-sis), the growth of new networks of blood vessels. Proteins that stimulate angiogenesis in tumors are called **tumor angiogenesis factors (TAFs).** The formation of new blood vessels can occur either by overproduction of TAFs or by the lack of naturally occurring angiogenesis inhibitors. As the cancer grows, it begins to compete with normal tissues for space and nutrients. Eventually, the normal tissue decreases in size and dies. Some malignant cells may detach from the initial (primary) tumor and invade a body cavity or enter the blood or lymph, then circulate to and invade other body tissues, establishing secondary tumors. Malignant cells resist the antitumor defenses of the body. The pain associated with cancer develops when the tumor presses on nerves or blocks a passageway in an organ so that secretions build up pressure, or as a result of dying tissue or organs.

Causes of Cancer

Several factors may trigger a normal cell to lose control and become cancerous. One cause is environmental agents: substances in the air we breathe, the water we drink, and the food we eat. A chemical agent or radiation that produces cancer is called a **carcinogen** (car-SIN-ō-jen). Carcinogens induce **mutations** (mū-TĀ-shuns), permanent changes in the DNA base sequence of a gene. The World Health Organization estimates that carcinogens are associated with 60–90% of all human cancers. Examples of carcinogens are hydrocarbons found in cigarette tar, radon gas from the earth, and ultraviolet (UV) radiation in sunlight.

Intensive research efforts are now directed toward studying cancer-causing genes, or **oncogenes** (ON-kō-jēnz). When inappropriately activated, these genes have the ability to transform a normal cell into a cancerous cell. Most oncogenes derive from normal genes called **proto-oncogenes** that regulate growth and development. The proto-oncogene undergoes some change that causes it (1) to be expressed inappropriately, (2) to make its products in excessive amounts, or (3) to make its products at the wrong time. Some oncogenes cause excessive production of growth factors, chemicals that stimulate cell growth. Others may trigger changes in a cell-surface receptor, causing it to send signals as though it were being activated by a growth factor. As a result, the growth pattern of the cell becomes abnormal.

Proto-oncogenes in every cell carry out normal cellular functions until a malignant change occurs. It appears that some proto-oncogenes are activated to oncogenes by mutations in which the DNA of the proto-oncogene is altered. Other proto-oncogenes are activated by a rearrangement of the chromosomes so that segments of DNA are exchanged. Rearrangement activates proto-oncogenes by placing them near genes that enhance their activity.

Some cancers have a viral origin. Viruses are tiny packages of nucleic acids, either RNA or DNA, that can reproduce only while inside the cells they infect. Some viruses, termed **oncogenic viruses,** cause cancer by stimulating abnormal proliferation of cells. For instance, the *human papillomavirus (HPV)* causes virtually all cervical cancers in women. The virus produces a protein that causes proteasomes to destroy p53, a protein that normally suppresses unregulated cell division. In the absence of this suppressor protein, cells proliferate without control.

Recent studies suggest that certain cancers may be linked to a cell having abnormal numbers of chromosomes. As a result, the cell could potentially have extra copies of oncogenes or too few copies of tumor-suppressor genes, which in either case could lead to uncontrolled cell proliferation. There is also some evidence suggesting that cancer may be caused by normal stem cells that develop into cancerous stem cells capable of forming malignant tumors.

Later in the book, we will discuss the process of inflammation, which is a defensive response to tissue damage. It appears that inflammation contributes to various steps in the development of cancer. Some evidence suggests that chronic inflammation stimulates the proliferation of mutated cells and enhances their survival, promotes angiogenesis, and contributes to invasion and metastasis of cancer cells. There is a clear relationship between certain chronic inflammatory conditions and the transformation of inflamed tissue into a malignant tissue. For example, chronic gastritis (inflammation of the stomach lining) and peptic ulcers may be a causative factor in 60–90% of stomach cancers. Chronic hepatitis (inflammation of the liver) and cirrhosis of the liver are believed to be responsible for about 80% of liver cancers. Colorectal cancer is ten times more likely to occur in patients with chronic inflammatory diseases of the colon, such as ulcerative colitis and Crohn's disease. And the relationship between asbestosis and silicosis, two chronic lung inflammatory conditions, and lung cancer has long been recognized. Chronic inflammation is also an underlying contributor to rheumatoid arthritis, Alzheimer disease, depression, schizophrenia, cardiovascular disease, and diabetes.

Carcinogenesis: A Multistep Process

Carcinogenesis (kar'-si-nō-JEN-e-sis) is a multistep process of cancer development in which as many as 10 distinct mutations may have to accumulate in a cell before it becomes cancerous. The progression of genetic changes leading to cancer is best understood for colon (colorectal) cancer. Such cancers, as well as lung and breast cancer, take years or decades to develop. In colon cancer, the tumor begins as an area of increased cell proliferation that results from one mutation. This growth then progresses to abnormal, but noncancerous, growths called adenomas. After two or three additional mutations, a mutation of the tumor-suppressor gene *p53* occurs and a carcinoma develops. The fact that so many mutations are needed for a cancer to develop indicates that cell growth is normally controlled with many sets of checks and balances. Thus, it is not surprising that a compromised immune system contributes significantly to carcinogenesis.

Treatment of Cancer

Many cancers are removed surgically. However, cancer that is widely distributed throughout the body or exists in organs with essential functions, such as the brain, which might be greatly harmed by surgery, may be treated with chemotherapy and radiation therapy instead. Sometimes surgery, chemotherapy, and radiation therapy are used in combination. Chemotherapy involves administering drugs that cause death of cancerous cells. Radiation therapy breaks chromosomes, thus blocking cell division. Because cancerous cells divide rapidly, they are more vulnerable to the destructive effects of chemotherapy and radiation therapy than are normal cells. Unfortunately for the patients, hair follicle cells, red bone marrow cells, and cells lining the gastrointestinal tract also are rapidly dividing. Hence, the side effects of chemotherapy and radiation therapy include hair loss due to death of hair follicle cells, vomiting and nausea due to death of cells lining the stomach and intestines, and susceptibility to infection due to slowed production of white blood cells in red bone marrow.

Treating cancer is difficult because it is not a single disease and because the cells in a single tumor population rarely behave all in the same way. Although most cancers are thought to derive from a single abnormal cell, by the time a tumor reaches a clinically detectable size, it may contain a diverse population of abnormal cells. For example, some cancerous cells metastasize readily, and others do not. Some are sensitive to chemotherapy drugs and some are drug-resistant. Because of differences in drug resistance, a single chemotherapeutic agent may destroy susceptible cells but permit resistant cells to proliferate.

Another potential treatment for cancer that is currently under development is *virotherapy,* the use of viruses to kill cancer cells. The viruses employed in this strategy are designed so that they specifically target cancer cells without affecting the healthy cells of the body. For example, proteins (such as antibodies) that specifically bind to receptors found only in cancer cells are attached to viruses. Once inside the body, the viruses bind to cancer cells and then infect them. The cancer cells are eventually killed once the viruses cause cellular lysis.

Researchers are also investigating the role of *metastasis regulatory genes* that control the ability of cancer cells to undergo metastasis. Scientists hope to develop therapeutic drugs that can manipulate these genes and, therefore, block metastasis of cancer cells.

MEDICAL TERMINOLOGY

Most chapters in this text are followed by a glossary of key medical terms that include both normal and pathological conditions. You should familiarize yourself with these terms because they will play an essential role in your medical vocabulary.

Some of these conditions, as well as ones discussed in the text, are referred to as local or systemic. A *local disease* is one that affects one part or a limited area of the body. A *systemic disease* affects the entire body or several parts.

Anaplasia (an'-a-PLĀ-zē-a; *an-* = not; *-plasia* = to shape) The loss of tissue differentiation and function that is characteristic of most malignancies.

Atrophy (AT-rō-fē; *a-* = without; *-trophy* = nourishment) A decrease in the size of cells, with a subsequent decrease in the size of the affected tissue or organ; wasting away.

Dysplasia (dis-PLĀ-zē-a; *dys-* = abnormal) Alteration in the size, shape, and organization of cells due to chronic irritation or inflammation; may progress to neoplasia (tumor formation, usually malignant) or revert to normal if the irritation is removed.

Hyperplasia (hī-per-PLĀ-zē-a; *hyper-* = over) Increase in the number of cells of a tissue due to an increase in the frequency of cell division.

Hypertrophy (hī-PER-trō-fē) Increase in the size of cells without cell division.

Metaplasia (met'-a-PLĀ-zē-a; *meta-* = change) The transformation of one type of cell into another.

Progeny (PROJ-e-nē; *pro-* = forward; *-geny* = production) Offspring or descendants.

Proteomics (prō'-tē-Ō-miks; *proteo-* = protein) The study of the proteome (all of an organism's proteins) in order to identify all the proteins produced; it involves determining the three-dimensional structure of proteins so that drugs can be designed to alter protein activity to help in the treatment and diagnosis of disease.

Tumor marker A substance introduced into circulation by tumor cells that indicates the presence of a tumor, as well as the specific type. Tumor markers may be used to screen, diagnose, make a prognosis, evaluate a response to treatment, and monitor for recurrence of cancer.

CHAPTER REVIEW AND RESOURCE SUMMARY

WILEY PLUS

Review

Resource

Introduction

1. A cell is the basic, living, structural and functional unit of the body.
2. Cell biology is the scientific study of cellular structure and function.

3.1 Parts of a Cell

1. Figure 3.1 provides an overview of the typical structures in body cells.
2. The principal parts of a cell are the plasma membrane; the cytoplasm, the cellular contents between the plasma membrane and nucleus; and the nucleus.

Anatomy Overview - Cell Structure and Function
Figure 3.1 - Structures in Body Cells
Concepts and Connections - Human Cell

3.2 The Plasma Membrane

1. The plasma membrane, which surrounds and contains the cytoplasm of a cell, is composed of proteins and lipids.
2. According to the fluid mosaic model, the membrane is a mosaic of proteins floating like icebergs in a lipid bilayer sea.
3. The lipid bilayer consists of two back-to-back layers of phospholipids, cholesterol, and glycolipids. The bilayer arrangement occurs because the lipids are amphipathic, having both polar and nonpolar parts.
4. Integral proteins extend into or through the lipid bilayer; peripheral proteins associate with membrane lipids or integral proteins at the inner or outer surface of the membrane.

Anatomy Overview - Plasma Membrane Structure
Figure 3.2 - The Fluid Mosaic Arrangement of the Plasma Membrane
Exercise - Paint a Cell Membrane
Concepts and Connections - Membrane Functions

Review   **Resource**

5. Many integral proteins are glycoproteins, with sugar groups attached to the ends that face the extra-cellular fluid. Together with glycolipids, the glycoproteins form a glycocalyx on the extracellular surface of cells.

6. Membrane proteins have a variety of functions. Integral proteins are channels and carriers that help specific solutes cross the membrane; receptors that serve as cellular recognition sites; enzymes that catalyze specific chemical reactions; and linkers that anchor proteins in the plasma membranes to protein filaments inside and outside the cell. Peripheral proteins serve as enzymes and linkers; support the plasma membrane; anchor integral proteins; and participate in mechanical activities. Membrane glycoproteins function as cell-identity markers.

7. Membrane fluidity is greater when there are more double bonds in the fatty acid tails of the lipids that make up the bilayer. Cholesterol makes the lipid bilayer stronger but less fluid at normal body temperature. Its fluidity allows interactions to occur within the plasma membrane, enables the movement of membrane components, and permits the lipid bilayer to self-seal when torn or punctured.

8. The membrane's selective permeability permits some substances to pass more readily than others. The lipid bilayer is permeable to most nonpolar, uncharged molecules. It is impermeable to ions and charged or polar molecules other than water and urea. Channels and carriers increase the plasma membrane's permeability to small- and medium-sized polar and charged substances, including ions, that cannot cross the lipid bilayer.

9. The selective permeability of the plasma membrane supports the existence of concentration gradients, differences in the concentrations of chemicals between one side of the membrane and the other.

3.3 Transport across the Plasma Membrane

1. In passive processes, a substance moves down its concentration gradient across the membrane using its own kinetic energy of motion. In active processes, cellular energy is used to drive the substance "uphill" against its concentration gradient.

2. In diffusion, molecules or ions move from an area of higher concentration to an area of lower concentration until an equilibrium is reached. The rate of diffusion across a plasma membrane is affected by the steepness of the concentration gradient, temperature, mass of the diffusing substance, surface area available for diffusion, and the distance over which diffusion must occur.

3. Nonpolar, hydrophobic molecules such as oxygen, carbon dioxide, nitrogen, steroids, and fat-soluble vitamins (A, E, D, and K) plus small, polar, uncharged molecules such as water, urea, and small alcohols diffuse through the lipid bilayer of the plasma membrane via simple diffusion.

4. In channel-mediated facilitated diffusion, a solute moves down its concentration gradient across the lipid bilayer through a membrane channel. Examples include ion channels that allow specific ions such as K^+, Cl^-, Na^+, or Ca^{2+} (which are too hydrophilic to penetrate the membrane's nonpolar interior) to move across the plasma membrane. In carrier-mediated facilitated diffusion, a solute such as glucose binds to a specific carrier protein on one side of the membrane and is released on the other side after the carrier undergoes a change in shape.

5. Osmosis is a type of diffusion in which there is net movement of water through a selectively permeable membrane from an area of higher water concentration to an area of lower water concentration. In an isotonic solution, red blood cells maintain their normal shape; in a hypotonic solution, they undergo hemolysis; in a hypertonic solution, they undergo crenation.

6. Substances can cross the membrane against their concentration gradient by active transport. Actively transported substances include ions such as Na^+, K^+, H^+, Ca^{2+}, I^-, and Cl^-; amino acids; and monosaccharides. Two sources of energy drive active transport: Energy obtained from hydrolysis of ATP is the source in primary active transport, and energy stored in a Na^+ or H^+ concentration gradient is the source in secondary active transport. The most prevalent primary active transport pump is the sodium-potassium pump, also known as Na^+/K^+ ATPase. Secondary active transport mechanisms include both symporters and antiporters that are powered by either a Na^+ or H^+ concentration gradient. Symporters move two substances in the same direction across the membrane; antiporters move two substances in opposite directions.

7. In endocytosis, tiny vesicles detach from the plasma membrane to move materials across the membrane into a cell; in exocytosis, vesicles merge with the plasma membrane to move materials out of a cell. Receptor-mediated endocytosis is the selective uptake of large molecules and particles (ligands) that bind to specific receptors in membrane areas called clathrin-coated pits. In bulk-phase endocytosis (pinocytosis), the ingestion of extracellular fluid, a vesicle surrounds the fluid to take it into the cell.

8. Phagocytosis is the ingestion of solid particles. Some white blood cells destroy microbes that enter the body in this way.

Animation - Membrane Functions
Animation - Transport across the
  Plasma Membrane
Figure 3.5 - Types of
  Diffusion
Figure 3.9 - Tonicity and
  Its Effects on Red
  Blood Cells
Figure 3.12 - Receptor-Mediated
  Endocytosis of a Low-Density
  Lipoprotein Particle
Exercise - Membrane Function Matchup
Exercise - Create a Transport Condition
Exercise - Osmosis—Move It!
Concepts and Connections - Membrane
  Transport Processes

Review

9. In transcytosis, vesicles undergo endocytosis on one side of a cell, move across the cell, and undergo exocytosis on the opposite side.

3.4 Cytoplasm

1. Cytoplasm—all the cellular contents within the plasma membrane except for the nucleus—consists of cytosol and organelles. Cytosol is the fluid portion of cytoplasm, containing water, ions, glucose, amino acids, fatty acids, proteins, lipids, ATP, and waste products. It is the site of many chemical reactions required for a cell's existence.

2. Components of the cytoskeleton, a network of several kinds of protein filaments that extend throughout the cytoplasm, include microfilaments, intermediate filaments, and microtubules. The cytoskeleton provides a structural framework for the cell and is responsible for cell movements.

3. Organelles are specialized structures with characteristic shapes that have specific functions. The centrosome consists of pericentriolar material and a pair of centrioles. The pericentriolar material organizes microtubules in nondividing cells and the mitotic spindle in dividing cells.

4. Cilia and flagella, motile projections of the cell surface, are formed by basal bodies. Cilia move fluid along the cell surface; flagella move an entire cell.

5. Ribosomes consist of two subunits made in the nucleus that are composed of ribosomal RNA and ribosomal proteins. They serve as sites of protein synthesis.

6. Endoplasmic reticulum (ER) is a network of membranes that form flattened sacs or tubules; it extends from the nuclear envelope throughout the cytoplasm. Rough ER is studded with ribosomes that synthesize proteins; the proteins then enter the space within the ER for processing and sorting. Rough ER produces secretory proteins, membrane proteins, and organelle proteins; forms glycoproteins; synthesizes phospholipids; and attaches proteins to phospholipids. Smooth ER lacks ribosomes. It synthesizes fatty acids and steroids; inactivates or detoxifies drugs and other potentially harmful substances; removes phosphate from glucose-6-phosphate; and releases calcium ions that trigger contraction in muscle cells.

7. The Golgi complex consists of flattened sacs called cisternae. The entry, medial, and exit regions of the Golgi complex contain different enzymes that permit each to modify, sort, and package proteins for transport in secretory vesicles, membrane vesicles, or transport vesicles to different cellular destinations.

8. Lysosomes are membrane-enclosed vesicles that contain digestive enzymes. Endosomes, phagosomes, and pinocytic vesicles deliver materials to lysosomes for degradation. Lysosomes function in digestion of worn-out organelles (autophagy), digestion of a host cell (autolysis), and extracellular digestion.

9. Peroxisomes contain oxidases that oxidize amino acids, fatty acids, and toxic substances; the hydrogen peroxide produced in the process is destroyed by catalase. The proteases contained in proteasomes, another kind of organelle, continually degrade unneeded, damaged, or faulty proteins by cutting them into small peptides.

10. Mitochondria consist of a smooth outer membrane, an inner membrane containing mitochondrial cristae, and a fluid-filled cavity called the mitochondrial matrix. These so-called powerhouses of the cell produce most of a cell's ATP and can play an important early role in apoptosis.

Anatomy Overview - Cell Structure and Function
Figure 3.21 - Processing and Packaging of Proteins by the Golgi Complex
Exercise - Concentrate on Cellular Functions
Exercise - Target Practice
Exercise - Protein - Producing Processes
Exercise - Synthesize and Transport a Protein
Concepts and Connections - Human Cell
Concepts and Connections - Membranous Organelles

3.5 Nucleus

1. The nucleus consists of a double nuclear envelope; nuclear pores, which control the movement of substances between the nucleus and cytoplasm; nucleoli, which produce ribosomes; and genes arranged on chromosomes, which control cellular structure and direct cellular activities.

2. Human somatic cells have 46 chromosomes, 23 inherited from each parent. The total genetic information carried in a cell or an organism is its genome.

3.6 Protein Synthesis

1. Cells make proteins by transcribing and translating the genetic information contained in DNA.

2. The genetic code is the set of rules that relates the base triplet sequences of DNA to the corresponding codons of RNA and the amino acids they specify.

3. In transcription, the genetic information in the sequence of base triplets in DNA serves as a template for copying the information into a complementary sequence of codons in messenger RNA. Transcription begins on DNA in a region called a promoter. Regions of DNA that code for protein synthesis are called exons; those that do not are called introns.

Review

Resource

4. Newly synthesized pre-mRNA is modified before leaving the nucleus.

5. In the process of translation, the nucleotide sequence of mRNA specifies the amino acid sequence of a protein. The mRNA binds to a ribosome, specific amino acids attach to tRNA, and anticodons of tRNA bind to codons of mRNA, bringing specific amino acids into position on a growing polypeptide. Translation begins at the start codon and ends at the stop codon.

3.7 Cell Division

1. Cell division, the process by which cells reproduce themselves, consists of nuclear division (mitosis or meiosis) and cytoplasmic division (cytokinesis). Cell division that replaces cells or adds new ones is called somatic cell division and involves mitosis and cytokinesis. Cell division that results in the production of gametes (sperm and ova) is called reproductive cell division and consists of meiosis and cytokinesis.

2. The cell cycle, an orderly sequence of events in which a somatic cell duplicates its contents and divides in two, consists of interphase and a mitotic phase. Human somatic cells contain 23 pairs of homologous chromosomes and are thus diploid ($2n$). Before the mitotic phase, the DNA molecules, or chromosomes, replicate themselves so that identical sets of chromosomes can be passed on to the next generation of cells.

3. A cell between divisions that is carrying on every life process except division is said to be in interphase, which consists of three phases: G_1, S, and G_2. During the G_1 phase, the cell replicates its organelles and cytosolic components, and centrosome replication begins; during the S phase, DNA replication occurs; during the G_2 phase, enzymes and other proteins are synthesized and centrosome replication is completed.

4. Mitosis is the splitting of the chromosomes and the distribution of two identical sets of chromosomes into separate and equal nuclei; it consists of prophase, metaphase, anaphase, and telophase.

5. In cytokinesis, which usually begins in late anaphase and ends once mitosis is complete, a cleavage furrow forms at the cell's metaphase plate and progresses inward, pinching in through the cell to form two separate portions of cytoplasm.

6. A cell can either remain alive and functioning without dividing, grow and divide, or die. The control of cell division depends on specific cyclin-dependent protein kinases and cyclins.

7. Apoptosis is normal, programmed cell death. It first occurs during embryological development and continues throughout the lifetime of an organism.

8. Certain genes regulate both cell division and apoptosis. Abnormalities in these genes are associated with a wide variety of diseases and disorders.

9. In sexual reproduction, each new organism is the result of the union of two different gametes, one from each parent. Gametes contain a single set of chromosomes (23) and thus are haploid (n).

10. Meiosis is the process that produces haploid gametes; it consists of two successive nuclear divisions, called meiosis I and meiosis II. During meiosis I, homologous chromosomes undergo synapsis (pairing) and crossing-over; the net result is two haploid cells that are genetically unlike each other and unlike the starting diploid parent cell that produced them. During meiosis II, two haploid cells divide to form four haploid cells.

Animation - The Cell Cycle and Division Processes
Figure 3.34 - Comparison between Mitosis and Meiosis
Exercise - Mitosis Matchup
Exercise - Meiosis Matchup
Exercise - Cell Cycle Quiz

3.8 Cellular Diversity

1. The almost 200 different types of cells in the body vary considerably in size and shape.

2. The sizes of cells are measured in micrometers. One micrometer (μm) equals 10^{-6} m (1/25,000 of an inch). Cells in the body range from 8 μm to 140 μm in size.

3. A cell's shape is related to its function.

Anatomy Overview - Cell Structure and Function
Concepts and Connections - Human Cell

3.9 Aging and Cells

1. Aging is a normal process accompanied by progressive alteration of the body's homeostatic adaptive responses.

2. Many theories of aging have been proposed, including genetically programmed cessation of cell division, buildup of free radicals, and an intensified autoimmune response.

Fill in the blanks in the following statements.

1. The three major parts of the cell are the _____, _____, and _____.

2. Cell death that is genetically programmed is known as _____, while cell death which is due to tissue injury is known as _____.

3. _____ are special DNA sequences located at the ends of chromosomes and whose erosion contributes to cellular aging and death.

4. The mRNA base sequence that is complementary to the DNA base sequence ATC would be _____.

Indicate whether the following statements are true or false.

5. A small membrane surface area will increase the rate of diffusion across the cell membrane.

6. The cells created during meiosis are genetically different from the original cell.

7. An important and abundant active mechanism that helps maintain cellular tonicity is the Na^+/K^+ ATPase pump.

Choose the one best answer to the following questions.

8. If the concentration of solutes in the ECF and ICFs are equal, the cell is in a(n) _____ solution.
   (a) hypertonic   (b) hydrophobic   (c) saturated
   (d) hypotonic   (e) isotonic

9. Which membrane protein is *incorrectly* matched with its function?
   (a) receptor: allows recognition of specific molecules
   (b) ion channel: allows passage of specific ions through the membrane
   (c) carrier: allows cells to recognize each other and foreign cells
   (d) linker: allows binding of one cell to another and provides stability and shape to a cell
   (e) enzyme: catalyzes cellular reactions

10. Place the following steps in protein synthesis in the correct order.
    (a) binding of anticodons of tRNA to codons of mRNA
    (b) modification of newly synthesized pre-mRNA by snRNPs before leaving the nucleus and entering the cytoplasm
    (c) attachment of RNA polymerase at promoter
    (d) binding of mRNA to a ribosome's small subunit
    (e) amino acids joined by peptide bonds
    (f) joining of large and small ribosomal subunits to create a functional ribosome
    (g) transcription of a segment of DNA onto mRNA
    (h) detachment of protein from ribosome when ribosome reaches stop codon on mRNA
    (i) detachment of RNA polymerase after reaching terminator
    (j) attachment of specific amino acids to tRNA
    (k) binding of initiator tRNA to start codon on mRNA

11. Which of the following organelles function primarily in decomposition reactions? (1) ribosomes, (2) proteasomes, (3) lysosomes, (4) centrosomes, (5) peroxisomes
    (a) 2, 3, and 5   (b) 3 and 5   (c) 2, 4, and 5
    (d) 1 and 4   (e) 2 and 5

12. Which of the following statements regarding the nucleus are true? (1) Nucleoli within the nucleus are the sites of ribosome synthesis. (2) The nucleus contains the cell's hereditary units. (3) The nuclear membrane is a solid, impermeable membrane. (4) Protein synthesis occurs within the nucleus. (5) In nondividing cells, DNA is found in the nucleus in the form of chromatin.
    (a) 1, 2, and 3   (b) 1, 2, and 4   (c) 1, 2, and 5
    (d) 2, 4, and 5   (e) 2, 3, and 4

13. Match the following:
    ____ (a) mitosis
    ____ (b) meiosis
    ____ (c) prophase
    ____ (d) metaphase
    ____ (e) anaphase
    ____ (f) telophase
    ____ (g) cytokinesis
    ____ (h) interphase

    (1) cytoplasmic division
    (2) somatic cell division resulting in the formation of two identical cells
    (3) reproductive cell division that reduces the number of chromosomes by half
    (4) stage of cell division when replication of DNA occurs
    (5) stage when chromatin fibers condense and shorten to form chromosomes
    (6) stage when centromeres split and chromatids move to opposite poles of the cell
    (7) stage when centromeres of chromatid pairs line up at the center of the mitotic spindle
    (8) stage when chromosomes uncoil and revert to chromatin

14. Match the following:

_____ (a) cytoskeleton
_____ (b) centrosome
_____ (c) ribosomes
_____ (d) rough ER
_____ (e) smooth ER
_____ (f) Golgi complex
_____ (g) lysosomes
_____ (h) peroxisomes
_____ (i) mitochondria
_____ (j) cilia
_____ (k) flagellum
_____ (l) proteasomes
_____ (m) vesicles

(1) membrane-enclosed vesicles formed in the Golgi complex that contain strong hydrolytic and digestive enzymes
(2) network of protein filaments that extend throughout the cytoplasm, providing cellular shape, organization, and movement
(3) sites of protein synthesis
(4) contain enzymes that break apart unneeded, damaged, or faulty proteins into small peptides
(5) site where secretory proteins and membrane molecules are synthesized
(6) membrane-enclosed vesicles that contain enzymes that oxidize various organic substances
(7) short microtubular structures extending from the plasma membrane and involved in movement of materials along the cell's surface
(8) modifies, sorts, packages, and transports molecules synthesized in the rough ER
(9) an organizing center for growth of the mitotic spindle
(10) function in ATP generation
(11) functions in synthesizing fatty acids and steroids, helping liver cells release glucose into the bloodstream, and detoxification
(12) membrane-bound sacs that transport, transfer, or secrete proteins
(13) long microtubular structure extending from the plasma membrane and involved in movement of a cell

15. Match the following:

_____ (a) diffusion
_____ (b) osmosis
_____ (c) facilitated diffusion
_____ (d) primary active transport
_____ (e) secondary active transport
_____ (f) vesicular transport
_____ (g) phagocytosis
_____ (h) pinocytosis
_____ (i) exocytosis
_____ (j) receptor-mediated endocytosis
_____ (k) transcytosis

(1) passive transport in which a solute binds to a specific carrier on one side of the membrane and is released on the other side
(2) movement of materials out of the cell by fusing of secretory vesicles with the plasma membrane
(3) the random mixing of particles in a solution due to the kinetic energy of the particles; substances move from high to low concentrations until equilibrium is reached
(4) transport of substances either into or out of the cell by means of small, spherical membranous sacs formed by budding off from existing membranes
(5) uses energy derived from hydrolysis of ATP to change the shape of a carrier protein, which "pumps" a substance across a cellular membrane against its concentration gradient
(6) vesicular movement involving endocytosis on one side of a cell and subsequent exocytosis on the opposite side of the cell
(7) type of endocytosis that involves the nonselective uptake of tiny droplets of extracellular fluid
(8) type of endocytosis in which large solid particles are taken in
(9) movement of water from an area of higher to an area of lower water concentration through a selectively permeable membrane
(10) process that allows a cell to take specific ligands from the ECF by forming vesicles
(11) indirectly uses energy obtained from the breakdown of ATP; involves symporters and antiporters

CRITICAL THINKING QUESTIONS

1. Mucin is a protein present in saliva and other secretions. When mixed with water, it becomes the slippery substance known as mucus. Trace the route taken by mucin through the cell, from its synthesis to its secretion, listing all the organelles and processes involved.

2. Sam does not consume alcohol, while his brother Sebastian regularly drinks large quantities of alcohol. If we could examine the liver cells of each of these brothers, would we see a difference in smooth ER and peroxisomes? Explain.

3. Marathon runners can become dehydrated due to the extreme physical activity. What types of fluids should they consume in order to rehydrate their cells?

? ANSWERS TO FIGURE QUESTIONS

3.1 The three main parts of a cell are the plasma membrane, cytoplasm, and nucleus.

3.2 The glycocalyx is the sugary coat on the extracellular surface of the plasma membrane. It is composed of the carbohydrate portions of membrane glycolipids and glycoproteins.

3.3 The membrane protein that binds to insulin acts as a receptor.

3.4 Because fever involves an increase in body temperature, the rates of all diffusion processes would increase.

3.5 Nonpolar, hydrophobic molecules (oxygen, carbon dioxide, and nitrogen gases; fatty acids; steroids; and fat-soluble vitamins) plus small, uncharged polar molecules (water, urea, and small alcohols) move across the lipid bilayer of the plasma membrane through the process of simple diffusion.

3.6 The concentration of K^+ is higher in the cytosol of body cells than in extracellular fluids.

3.7 Yes. Insulin promotes insertion of glucose transporter (GluT) in the plasma membrane, which increases cellular glucose uptake by carrier-mediated facilitated diffusion.

3.8 No. The water concentrations can never be the same in the two arms because the left arm contains pure water and the right arm contains a solution that is less than 100% water.

3.9 A 2% solution of NaCl will cause crenation of RBCs because it is hypertonic.

3.10 ATP adds a phosphate group to the pump protein, which changes the pump's three-dimensional shape. ATP transfers energy to power the pump.

3.11 In secondary active transport, hydrolysis of ATP is used indirectly to drive the activity of symporter or antiporter proteins; this reaction directly powers the pump protein in primary active transport.

3.12 Transferrin, vitamins, and hormones are other examples of ligands that can undergo receptor-mediated endocytosis.

3.13 The binding of particles to a plasma membrane receptor triggers pseudopod formation.

3.14 Receptor-mediated endocytosis and phagocytosis involve receptor proteins; bulk-phase endocytosis does not.

3.15 Microtubules help to form centrioles, cilia, and flagella.

3.16 A cell without a centrosome probably would not be able to undergo cell division.

3.17 Cilia move fluids across cell surfaces; flagella move an entire cell.

3.18 Large and small ribosomal subunits are synthesized separately in the nucleolus in the nucleus and then assembled in the cytoplasm.

3.19 Rough ER has attached ribosomes; smooth ER does not. Rough ER synthesizes proteins that will be exported from the cell; smooth ER is associated with lipid synthesis and other metabolic reactions.

3.20 The entry face receives and modifies proteins from rough ER; the exit face modifies, sorts, and packages molecules for transport to other destinations.

3.21 Some proteins are secreted from the cell by exocytosis, some are incorporated into the plasma membrane, and some occupy storage vesicles that become lysosomes.

3.22 Digestion of worn-out organelles by lysosomes is called autophagy.

3.23 Mitochondrial cristae increase the surface area available for chemical reactions and contain some of the enzymes needed for ATP production.

3.24 Chromatin is a complex of DNA, proteins, and some RNA.

3.25 A nucleosome is a double-stranded molecule of DNA wrapped twice around a core of eight histones (proteins).

3.26 Proteins determine the physical and chemical characteristics of cells.

3.27 The DNA base sequence AGCT would be transcribed into the mRNA base sequence UCGA by RNA polymerase.

3.28 The P site holds the tRNA attached to the growing polypeptide. The A site holds the tRNA carrying the next amino acid to be added to the growing polypeptide.

3.29 When a ribosome encounters a stop codon at the A site, it releases the completed protein from the final tRNA.

3.30 DNA replicates during the S phase of interphase of the cell cycle.

3.31 DNA replication must occur before cytokinesis so that each of the new cells will have a complete genome.

3.32 Cytokinesis usually starts in late anaphase.

3.33 The result of crossing-over is that four haploid gametes are genetically unlike each other and genetically unlike the starting cell that produced them.

3.34 During anaphase I of meiosis, the paired chromatids are held together by a centromere and do not separate. During anaphase of mitosis, the paired chromatids separate and the centromeres split.

3.35 Sperm, which use the flagella for locomotion, are the only body cells required to move considerable distances.

4 THE TISSUE LEVEL OF ORGANIZATION

TISSUES AND HOMEOSTASIS *The four basic types of tissues in the human body contribute to homeostasis by providing diverse functions including protection, support, communication among cells, and resistance to disease, to name just a few.*

As you learned in Chapter 3, a cell is a complex collection of compartments, each of which carries out a host of biochemical reactions that make life possible. However, a cell seldom functions as an isolated unit in the body. Instead, cells usually work together in groups called tissues. A **tissue** is a group of cells that usually have a common origin in an embryo and function together to carry out specialized activities. The structure and properties of a specific tissue are influenced by factors such as the nature of the extracellular material that surrounds the tissue cells and the connections between the cells that compose the tissue. Tissues may be hard, semisolid, or even liquid in their consistency, a range exemplified by bone, fat, and blood. In addition, tissues vary tremendously with respect to the kinds of cells present, how the cells are arranged, and the types of fibers present, if any.

Histology (hiss′-TOL-ō-jē; *histo-* = tissue; *-logy* = study of) is the science that deals with the study of tissues. A **pathologist** (pa-THOL-ō-jist; *patho-* = disease) is a physician who specializes in laboratory studies of cells and tissues to help other physicians make accurate diagnoses. One of the principal functions of a pathologist is to examine tissues for any changes that might indicate disease.

Did you ever wonder whether the complications of liposuction outweigh the benefits?

4.1 TYPES OF TISSUES

⦿ OBJECTIVE

• Name the four basic types of tissues that make up the human body, and state the characteristics of each.

Body tissues can be classified into four basic types according to their structure and function (Figure 4.1):

1. **Epithelial tissues** cover body surfaces and line hollow organs, body cavities, and ducts; they also forms glands. This tissue allows the body to interact with both its internal and external environments.

2. **Connective tissues** protect and support the body and its organs. Various types of connective tissues bind organs together, store energy reserves as fat, and help provide the body with immunity to disease-causing organisms.

3. **Muscular tissues** are composed of cells specialized for contraction and generation of force. In the process, muscular tissues generate heat that warms the body.

4. **Nervous tissue** detects changes in a variety of conditions inside and outside the body and responds by generating electrical signals called nerve action potentials (nerve impulses) that activate muscular contractions and glandular secretions.

Epithelial tissues and most types of connective tissues, except cartilage, bone, and blood, are more general in nature and have a wide distribution in the body. These tissues are components of most body organs and have a wide range of structure and function. We will look at epithelial tissues and connective tissues in some detail in this chapter. The general features of bone tissue and blood will be introduced here, but their detailed discussion is presented in Chapters 6 and 19, respectively. Similarly, the structure and function of muscular tissues and nervous tissue are introduced here and examined in detail in Chapters 10 and 12, respectively.

Normally, most cells within a tissue remain anchored to other cells or structures. Only a few cells, such as phagocytes, move freely through the body, searching for invaders to destroy. However, many cells migrate extensively during the growth and development process before birth.

⚕ **CLINICAL CONNECTION | Biopsy**

A **biopsy** (BĪ-op-sē; *bio-* = life; *-opsy* = to view) is the removal of a sample of living tissue for microscopic examination. This procedure is used to help diagnose many disorders, especially cancer, and to discover the cause of unexplained infections and inflammations. Both normal and potentially diseased tissues are removed for purposes of comparison. Once the tissue samples are removed, either surgically or through a needle and syringe, they may be preserved, stained to highlight special properties, or cut into thin sections for microscopic observation. Sometimes a biopsy is conducted while a patient is anesthetized during surgery to help a physician determine the most appropriate treatment. For example, if a biopsy of thyroid tissue reveals malignant cells, the surgeon can proceed immediately with the most appropriate procedure. •

✔ CHECKPOINT

1. Define a tissue.
2. What are the four basic types of human tissues?

4.2 CELL JUNCTIONS

⦿ OBJECTIVE

• Describe the structure and functions of the five main types of cell junctions.

Before looking more specifically at the types of tissues, we will first examine how cells are held together to form tissues. Most epithelial cells and some muscle and nerve cells are tightly joined into functional units. **Cell junctions** are contact points between the plasma membranes of tissue cells. Here we consider the five most important types of cell junctions: tight junctions, adherens junctions, desmosomes, hemidesmosomes, and gap junctions (Figure 4.2).

Figure 4.1 Types of tissues.

🔓 ▭ Each of the four types of tissues has different cells that vary in shape, structure, function, and distribution.

(a) Epithelial tissue   (b) Connective tissue   (c) Muscular tissue   (d) Nervous tissue

❓ **What are some key differences in function among the four tissue types?**

Figure 4.2 Cell junctions.

Most epithelial cells and some muscle and nerve cells contain cell junctions.

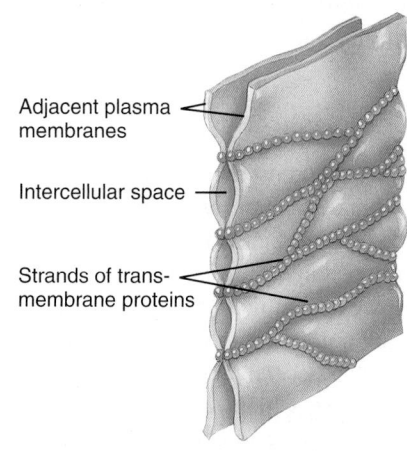

(a) Tight junctions

(e) Gap junction

(b) Adherens junction

(d) Hemidesmosome

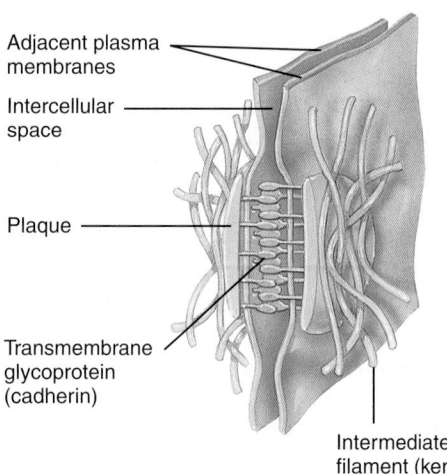

(c) Desmosome

? **Which type of cell junction functions in communication between adjacent cells?**

Tight Junctions

Tight junctions consist of weblike strands of transmembrane proteins that fuse together the outer surfaces of adjacent plasma membranes to seal off passageways between adjacent cells (Figure 4.2a). Cells of epithelial tissues that line the stomach, intestines, and urinary bladder have many tight junctions. They inhibit the passage of substances between cells and prevent the contents of these organs from leaking into the blood or surrounding tissues.

Adherens Junctions

Adherens junctions (ad-HER-ens) contain *plaque* (PLAK), a dense layer of proteins on the inside of the plasma membrane that attaches both to membrane proteins and to microfilaments of the cytoskeleton (Figure 4.2b). Transmembrane glycoproteins called **cadherins** join the cells. Each cadherin inserts into the plaque from the opposite side of the plasma membrane, partially crosses the intercellular space (the space between the cells), and connects to cadherins of an adjacent cell. In epithelial cells, adherens junctions often form extensive zones called **adhesion belts** because they encircle the cell similar to the way a belt encircles your waist. Adherens junctions help epithelial surfaces resist separation during various contractile activities, as when food moves through the intestines.

Desmosomes

Like adherens junctions, **desmosomes** (DEZ-mō-sōms; *desmo-* = band) contain plaque and have transmembrane glycoproteins (cadherins) that extend into the intercellular space between adjacent cell membranes and attach cells to one another (Figure 4.2c). However, unlike adherens junctions, the plaque of desmosomes does not attach to microfilaments. Instead, a desmosome plaque attaches to elements of the cytoskeleton known as intermediate filaments, which consist of the protein keratin. The intermediate filaments extend from desmosomes on one side of the cell across the cytosol to desmosomes on the opposite side of the cell. This structural arrangement contributes to the stability of the cells and tissue. These spot-weld-like junctions are common among the cells that make up the epidermis (the outermost layer of the skin) and among cardiac muscle cells in the heart. Desmosomes prevent epidermal cells from separating under tension and cardiac muscle cells from pulling apart during contraction.

Hemidesmosomes

Hemidesmosomes (*hemi-* = half) resemble desmosomes, but they do not link adjacent cells. The name arises from the fact that they look like half of a desmosome (Figure 4.2d). However, the transmembrane glycoproteins in hemidesmosomes are **integrins** rather than cadherins. On the inside of the plasma membrane, integrins attach to intermediate filaments made of the protein keratin. On the outside of the plasma membrane, the integrins attach to the protein *laminin,* which is present in the basement membrane (discussed shortly). Thus, hemidesmosomes anchor cells not to each other but to the basement membrane.

Gap Junctions

At **gap junctions,** membrane proteins called **connexins** form tiny fluid-filled tunnels called *connexons* that connect neighboring cells (Figure 4.2e). The plasma membranes of gap junctions are not fused together as in tight junctions but are separated by a very narrow intercellular gap (space). Through the connexons, ions and small molecules can diffuse from the cytosol of one cell to another, but the passage of large molecules such as vital intracellular proteins is prevented. The transfer of nutrients, and perhaps wastes, takes place through gap junctions in avascular tissues such as the lens and cornea of the eye. Gap junctions allow the cells in a tissue to communicate with one another. In a developing embryo, some of the chemical and electrical signals that regulate growth and cell differentiation travel via gap junctions. Gap junctions also enable nerve or muscle impulses to spread rapidly among cells, a process that is crucial for the normal operation of some parts of the nervous system and for the contraction of muscle in the heart, gastrointestinal tract, and uterus.

✔ CHECKPOINT

3. Which type of cell junction prevents the contents of organs from leaking into surrounding tissues?
4. Which types of cell junctions are found in epithelial tissues?

4.3 COMPARISON BETWEEN EPITHELIAL AND CONNECTIVE TISSUES

◉ OBJECTIVE

• State the main differences between epithelial and connective tissues.

Before examining epithelial tissues and connective tissues in more detail, let's compare these two widely distributed tissues (Figure 4.3). Major structural differences between an epithelial tissue and a connective tissue are immediately obvious under a light microscope. The first obvious difference is the number of cells in relation to the extracellular matrix (the substance between cells). In an epithelial tissue many cells are tightly packed together with little or no extracellular matrix, whereas in a connective tissue a large amount of extracellular material separates cells that are usually widely scattered. The second obvious difference is that an epithelial tissue has no blood vessels, whereas most connective tissues have significant networks of blood vessels. Another key difference is that epithelial tissues almost always form surface layers and are not covered by another tissue. An exception is the epithelial lining of blood vessels where blood constantly passes over the epithelium. While these key structural distinctions account for some of the major functional differences between these tissue types, they also lead to a common bond. Because epithelial tissues lack blood vessels and form surfaces, they are always found immediately adjacent to blood-vessel-rich connective tissues, which enable them to make the exchanges with blood necessary for the delivery of oxygen and nutrients and the removal of wastes that are critical processes for their survival and function.

✔ CHECKPOINT

5. Why are epithelial and connective tissues found adjacent to each other?

4.4 EPITHELIAL TISSUES

◉ OBJECTIVES

• Describe the general features of epithelial tissues.
• List the location, structure, and function of each different type of epithelial tissues.

An **epithelial tissue** (ep-i-THĒ-lē-al) or **epithelium** (plural is *epithelia*) consists of cells arranged in continuous sheets, in either

Figure 4.3 Comparison between epithelial tissues and connective tissues.

 The ratio of cells to extracellular matrix is a major difference between epithelial tissues and connective tissues.

(a) Epithelial tissue with many cells tightly packed together and little to no extracellular matrix.

(b) Connective tissue with a few scattered cells surrounded by large amounts of extracellular matrix.

? **What relationship between epithelial tissues and connective tissues is important for the survival and function of epithelial tissues?**

single or multiple layers. Because the cells are closely packed and are held tightly together by many cell junctions, there is little intercellular space between adjacent plasma membranes. Epithelial tissues form coverings and linings throughout the body. They are rarely covered by another tissue, so they always have a free surface. Epithelial tissues have three major functions. They serve as (1) selective barriers that limit or aid the transfer of substances into and out of the body; (2) secretory surfaces that release products produced by the cells onto their free surfaces; and (3) protective surfaces that resist the abrasive influences of the environment.

The various surfaces of epithelial cells often differ in structure and have specialized functions. The **apical (free) surface** of an epithelial cell faces the body surface, a body cavity, the lumen (interior space) of an internal organ, or a tubular duct that receives cell secretions (Figure 4.4). Apical surfaces may contain cilia or microvilli. The **lateral surfaces** of an epithelial cell, which face the adjacent cells on either side, may contain tight junctions, adherens junctions, desmosomes, and/or gap junctions. The **basal surface** of an epithelial cell is opposite the apical surface. The basal surfaces of the deepest layer of epithelial cells adhere to extracellular materials such as the basement membrane. Hemidesmosomes in the basal surfaces of the deepest layer of epithelial cells anchor the epithelium to the basement membrane (described next). In discussing epithelia with multiple layers, the term *apical layer* refers to the most superficial layer of cells, and the *basal layer* is the deepest layer of cells.

Figure 4.4 Surfaces of epithelial cells and the structure and location of the basement membrane.

 The basement membrane is found between epithelial tissues and connective tissues.

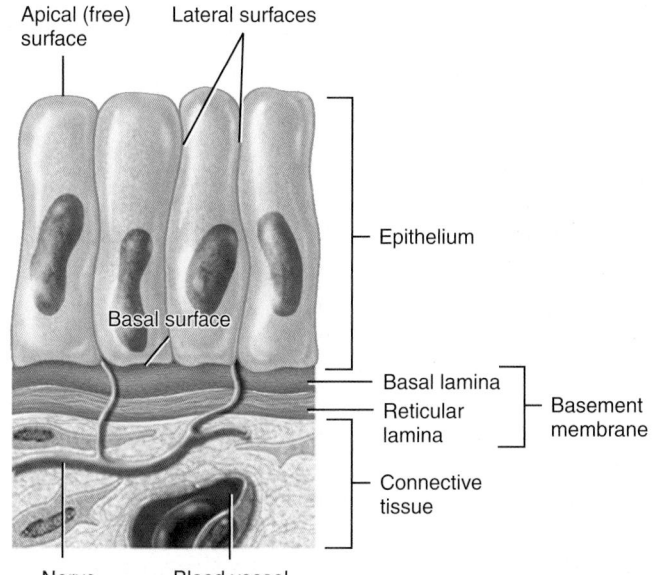

Apical (free) surface
Lateral surfaces
Basal surface
Epithelium
Basal lamina
Reticular lamina
Basement membrane
Connective tissue
Nerve
Blood vessel

? **What are the functions of the basement membrane?**

The **basement membrane** is a thin extracellular layer that commonly consists of two layers, the basal lamina and reticular lamina. The *basal lamina* (*lamina* = thin layer) is closer to—and secreted by—the epithelial cells. It contains proteins such as laminin and collagen (described shortly), as well as glycoproteins and proteoglycans (also described shortly). As you have already learned, the laminin molecules in the basal lamina adhere to integrins in hemidesmosomes and thus attach epithelial cells to the basement membrane (see Figure 4.2d). The *reticular lamina* is closer to the underlying connective tissue and contains proteins such as collagen produced by connective tissue cells called *fibroblasts* (see Figure 4.8). In addition to attaching to and supporting the overlying epithelial tissue, basement membranes have other functions. They form a surface along which epithelial cells migrate during growth or wound healing, restrict passage of larger molecules between epithelium and connective tissue, and participate in filtration of blood in the kidneys.

CLINICAL CONNECTION | Basement Membranes and Disease

Under certain conditions, basement membranes become markedly thickened, due to increased production of collagen and laminin. In untreated cases of diabetes mellitus, the basement membrane of small blood vessels (capillaries) thickens, especially in the eyes and kidneys. Because of this the blood vessels cannot function properly and blindness and kidney failure may result. •

Epithelial tissues have their own nerve supply, but as mentioned previously are **avascular** (*a-* = without; *-vascular* = vessel), relying on the blood vessels of the adjacent connective tissue to bring nutrients and remove wastes. Exchange of substances between epithelial tissues and connective tissues occurs by diffusion.

Because epithelial tissues form boundaries between the body's organs, or between the body and the external environment, they are repeatedly subjected to physical stress and injury. A high rate of cell division allows epithelial tissues to constantly renew and repair themselves by sloughing off dead or injured cells and replacing them with new ones. Epithelial tissues play many different roles in the body; the most important are protection, filtration, secretion, absorption, and excretion. In addition, epithelial tissues combine with nervous tissue to form special organs for smell, hearing, vision, and touch.

Epithelial tissues may be divided into two types. (1) **Covering and lining epithelium** forms the *outer* covering of the skin and some internal organs. It also forms the *inner* lining of blood vessels, ducts, and body cavities, and the interior of the respiratory, digestive, urinary, and reproductive systems. (2) **Glandular epithelium** makes up the secreting portion of glands such as the thyroid gland, adrenal glands, and sweat glands.

Classification of Epithelial Tissues

Types of covering and lining epithelial tissues are classified according to two characteristics: the arrangement of cells into layers and the shapes of the cells (Figure 4.5).

Figure 4.5 Cell shapes and arrangement of layers for covering and lining epithelium.

Cell shapes and arrangement of layers are the bases for classifying covering and lining epithelium.

Which cell shape is best adapted for the rapid movement of substances from one cell to another?

1. ***Arrangement of cells in layers*** (Figure 4.5). The cells are arranged in one or more layers depending on function:

   a. *Simple epithelium* is a single layer of cells that functions in diffusion, osmosis, filtration, secretion, or absorption. **Secretion** is the production and release of substances such as mucus, sweat, or enzymes. **Absorption** is the intake of fluids or other substances such as digested food from the intestinal tract.

   b. *Pseudostratified epithelium* (*pseudo-* = false) appears to have multiple layers of cells because the cell nuclei lie at different levels and not all cells reach the apical surface, but it is actually a simple epithelium because all its cells rest on the basement membrane. Cells that do extend to the apical surface may contain cilia; others (goblet cells) secrete mucus.

   c. *Stratified epithelium* (*stratum* = layer) consists of two or more layers of cells that protect underlying tissues in locations where there is considerable wear and tear.

2. ***Cell shapes*** (Figure 4.5). Epithelial cells vary in shape depending on their function:

   a. *Squamous* cells (SKWĀ-mus = flat) are thin, which allows for the rapid passage of substances through them.

   b. *Cuboidal* cells are as tall as they are wide and are shaped like cubes or hexagons. They may have microvilli at their apical surface and function in either secretion or absorption.

   c. *Columnar* cells are much taller than they are wide, like columns, and protect underlying tissues. Their apical surfaces may have cilia or microvilli, and they often are specialized for secretion and absorption.

   d. *Transitional* cells change shape, from squamous to cuboidal and back, as organs such as the urinary bladder stretch (distend) to a larger size and then collapse to a smaller size.

When we combine the two characteristics (arrangements of layers and cell shapes), we come up with the following types of epithelial tissues:

I. Simple epithelium
   A. Simple squamous epithelium
   B. Simple cuboidal epithelium
   C. Simple columnar epithelium (nonciliated and ciliated)
   D. Pseudostratified columnar epithelium (nonciliated and ciliated)

II. Stratified epithelium
   A. Stratified squamous epithelium (keratinized, when surface cells are dead and become hardened, and nonkeratinized, when surface cells remain alive)*
   B. Stratified cuboidal epithelium*
   C. Stratified columnar epithelium*
   D. Transitional epithelium

We will now examine the important features of covering and lining epithelium.

Covering and Lining Epithelium

As noted earlier, covering and lining epithelium forms the outer covering of the skin and some internal organs. It also forms the inner lining of blood vessels, ducts, and body cavities, and the interior of the respiratory, digestive, urinary, and reproductive systems. Table 4.1 describes covering and lining epithelium in more detail. The discussion of each type consists of a photomicrograph, a corresponding diagram, and an inset that identifies a major location of the tissue in the body. Descriptions, locations, and functions of the tissues accompany each illustration.

CLINICAL CONNECTION | *Papanicolaou Test*

A **Papanicolaou test** (pa-pa-NI-kō-lō), also called a **Pap test** or **Pap smear,** involves collection and microscopic examination of epithelial cells that have been scraped off the apical layer of a tissue. A very common type of Pap test involves examining the cells from the nonkeratinized stratified squamous epithelium of the vagina and cervix (inferior portion) of the uterus. This type of Pap test is performed mainly to detect early changes in the cells of the female reproductive system that may indicate a precancerous condition or cancer. In performing a Pap smear, the cells are scraped from the tissue and then smeared on a microscope slide. The slides are then sent to a laboratory for analysis. Pap tests should be started within three years of the onset of sexual activity, or age 21, whichever comes first. Annual screening is recommended for females ages 21–30 and every 2–3 years for females age 30 or older following three consecutive negative Pap tests. •

*This classification is based on the shape of the cells at the apical surface.

TABLE 4.1

Epithelial Tissues: Covering and Lining Epithelium

A. SIMPLE SQUAMOUS EPITHELIUM

Description	Single layer of flat cells that resembles a tiled floor when viewed from apical surface; centrally located nucleus that is flattened and oval or spherical in shape.
Location	Most commonly (1) lines the cardiovascular and lymphatic system (heart, blood vessels, lymphatic vessel linings), where it is known as **endothelium** (en′-dō-THĒ-lē-um; *endo-* = within; *-thelium* = covering), and (2) forms the epithelial layer of serous membranes (peritoneum, pleura, pericardium), where it is called **mesothelium** (mez′-ō-THĒ-lē-um; *meso-* = middle). Also found in air sacs of lungs, glomerular (Bowman's) capsule of kidneys, inner surface of tympanic membrane (eardrum).
Function	Present at sites of filtration (such as blood filtration in kidneys) or diffusion (such as diffusion of oxygen into blood vessels of lungs) and at site of secretion in serous membranes. Not found in body areas subject to mechanical stress (wear and tear).

Surface view of simple squamous epithelium
of mesothelial lining of peritoneum

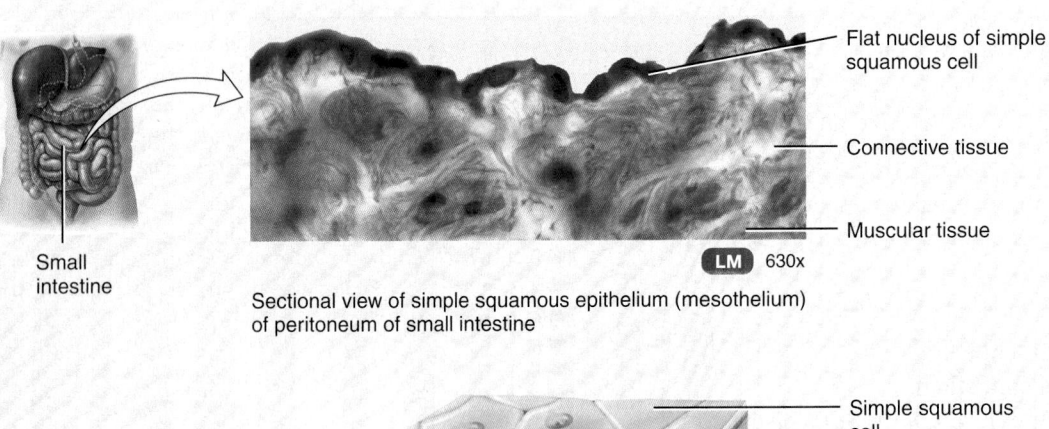

Sectional view of simple squamous epithelium (mesothelium)
of peritoneum of small intestine

Simple squamous epithelium

B. SIMPLE CUBOIDAL EPITHELIUM

Description Single layer of cube-shaped cells; round, centrally located nucleus. Cuboidal cell shape is obvious when tissue is sectioned and viewed from the side. (Note: Strictly cuboidal cells could not form small tubes; these cuboidal cells are more pie-shaped but still nearly as high as they are wide at the base.)

Location Covers surface of ovary; lines anterior surface of capsule of lens of the eye; forms pigmented epithelium at posterior surface of retina of the eye; lines kidney tubules and smaller ducts of many glands; makes up secreting portion of some glands, such as thyroid gland and ducts of some glands such as pancreas.

Function Secretion and absorption.

Sectional view of simple cuboidal epithelium of urinary tubules

Simple cuboidal epithelium

C. NONCILIATED SIMPLE COLUMNAR EPITHELIUM

Description Single layer of nonciliated columnlike cells with oval nuclei near base of cells; contains (1) columnar epithelial cells with microvilli at apical surface and (2) goblet cells. **Microvilli,** fingerlike cytoplasmic projections, increase surface area of plasma membrane (see Figure 3.1), thus increasing cell's rate of absorption. **Goblet cells** are modified columnar epithelial cells that secrete mucus, a slightly sticky fluid, at their apical surfaces. Before release, mucus accumulates in upper portion of cell, causing it to bulge and making the whole cell resemble a goblet or wine glass.

Location Lines gastrointestinal tract (from stomach to anus), ducts of many glands, and gallbladder.

Function Secretion and absorption; larger columnar cells contain more organelles and thus are capable of higher level of secretion and absorption than are cuboidal cells. Secreted mucus lubricates linings of digestive, respiratory, and reproductive tracts, and most of urinary tract; helps prevent destruction of stomach lining by acidic gastric juice secreted by stomach.

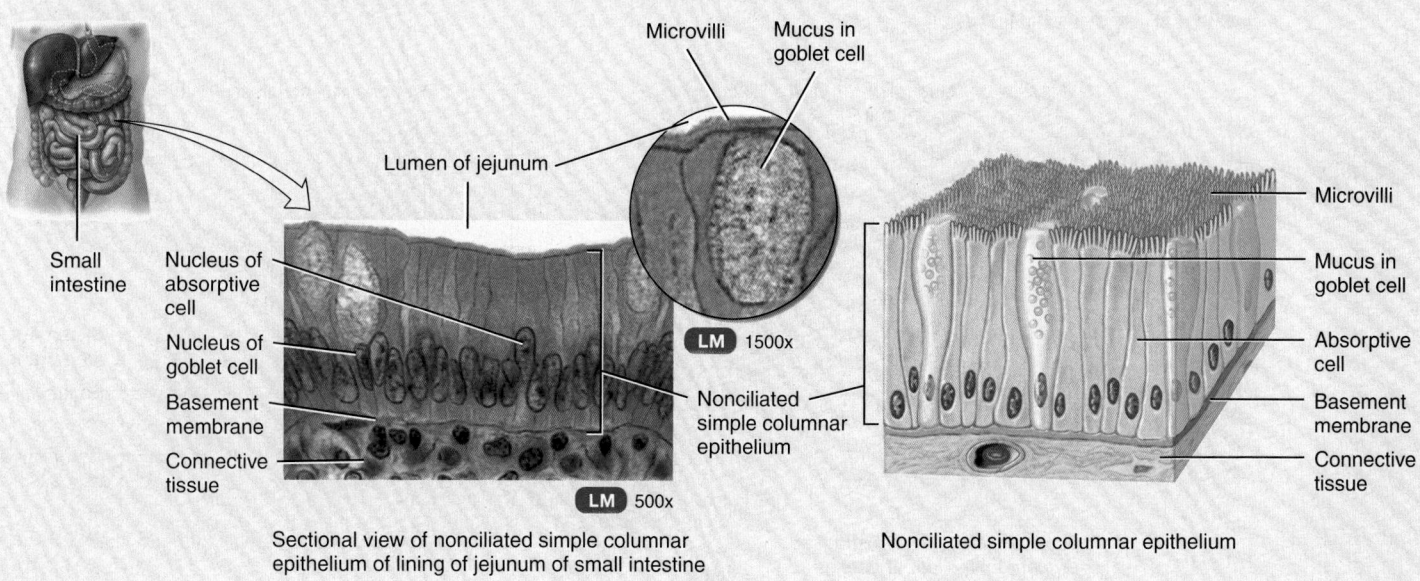

Sectional view of nonciliated simple columnar epithelium of lining of jejunum of small intestine

Nonciliated simple columnar epithelium

TABLE 4.1 CONTINUES

TABLE 4.1 CONTINUED

Epithelial Tissues: Covering and Lining Epithelium

D. CILIATED SIMPLE COLUMNAR EPITHELIUM

Description   Single layer of ciliated columnlike cells with oval nuclei near base of cells. Goblet cells are usually interspersed among ciliated columnar epithelia.

Location   Lines some bronchioles (small tubes) of respiratory tract, uterine (fallopian) tubes, uterus, some paranasal sinuses, central canal of spinal cord, and ventricles of brain.

Function   Cilia beat in unison, moving mucus and foreign particles toward throat, where they can be coughed up and swallowed or spit out. Coughing and sneezing speed up movement of cilia and mucus. Cilia also help move oocytes expelled from ovaries through uterine (fallopian) tubes into uterus.

Sectional view of ciliated simple columnar epithelium of uterine tube

Ciliated simple columnar epithelium

E. PSEUDOSTRATIFIED COLUMNAR EPITHELIUM

Description   Appears to have several layers because cell nuclei are at various levels. All cells are attached to basement membrane in a single layer, but some cells do not extend to apical surface. When viewed from side, these features give false impression of a multilayered tissue (thus the name pseudostratified; *pseudo* = false).

Pseudostratified ciliated columnar epithelium contains cells that extend to surface and secrete mucus (goblet cells) or bear cilia.

Pseudostratified nonciliated columnar epithelium contains cells without cilia and lacks goblet cells.

Location   Ciliated variety lines airways of most of upper respiratory tract; nonciliated variety lines larger ducts of many glands, epididymis, and part of male urethra.

Function   Ciliated variety secretes mucus that traps foreign particles, and cilia sweep away mucus for elimination from body; nonciliated variety functions in absorption and protection.

Sectional view of pseudostratified columnar epithelium of trachea

Pseudostratified ciliated columnar epithelium

F. STRATIFIED SQUAMOUS EPITHELIUM

Description   Two or more layers of cells; cells in apical layer and several layers deep to it are squamous; cells in deeper layers vary from cuboidal to columnar. As basal cells divide, daughter cells arising from cell divisions push upward toward apical layer. As they move toward surface and away from blood supply in underlying connective tissue, they become dehydrated and less metabolically active. Tough proteins predominate as cytoplasm is reduced, and cells become tough, hard structures that eventually die. At apical layer, after dead cells lose cell junctions they are sloughed off, but they are replaced continuously as new cells emerge from basal cells.

Keratinized stratified squamous epithelium develops tough layer of keratin in apical layer of cells and several layers deep to it (see Figure 5.3). (**Keratin** is a tough, fibrous intracellular protein that helps protect skin and underlying tissues from heat, microbes, and chemicals.) Relative amount of keratin increases in cells as they move away from nutritive blood supply and organelles die.

Nonkeratinized stratified squamous epithelium does not contain large amounts of keratin in apical layer and several layers deep and is constantly moistened by mucus from salivary and mucous glands; organelles are not replaced.

Location   Keratinized variety forms superficial layer of skin; nonkeratinized variety lines wet surfaces (lining of mouth, esophagus, part of epiglottis, part of pharynx, and vagina) and covers tongue.

Function   Protection against abrasion, water loss, ultraviolet radiation, and foreign invasion. Both types form first line of defense against microbes.

Sectional view of nonkeratinized stratified squamous epithelium of lining of vagina

Nonkeratinized stratified squamous epithelium

Sectional view of keratinized stratified squamous epithelium of epidermis

TABLE 4.1 CONTINUES

TABLE 4.1 CONTINUED

Epithelial Tissues: Covering and Lining Epithelium

G. STRATIFIED CUBOIDAL EPITHELIUM

Description	Two or more layers of cells; cells in apical layer are cube-shaped; fairly rare type.
Location	Ducts of adult sweat glands and esophageal glands, part of male urethra.
Function	Protection; limited secretion and absorption.

Sectional view of stratified cuboidal epithelium
of the duct of an esophageal gland

Stratified cuboidal epithelium

H. STRATIFIED COLUMNAR EPITHELIUM

Description	Basal layers usually consist of shortened, irregularly shaped cells; only apical layer has columnar cells; uncommon.
Location	Lines part of urethra; large excretory ducts of some glands, such as esophageal glands; small areas in anal mucous membrane; part of conjunctiva of eye.
Function	Protection and secretion.

Sectional view of stratified columnar epithelium
of lining of pharynx

Stratified columnar epithelium

I. TRANSITIONAL EPITHELIUM

Description Variable appearance (transitional). In relaxed or unstretched state, looks like stratified cuboidal epithelium, except apical layer cells tend to be large and rounded. As tissue is stretched, cells become flatter, giving the appearance of stratified squamous epithelium. Multiple layers and elasticity make it ideal for lining hollow structures (urinary bladder) subject to expansion from within.

Location Lines urinary bladder and portions of ureters and urethra.

Function Allows urinary organs to stretch and maintain protective lining while holding variable amounts of fluid without rupturing.

Urinary bladder

Lumen of urinary bladder

Lumen of urinary bladder

Rounded surface cell in relaxed state

Nucleus of transitional cell

LM 630x

Transitional epithelium

Apical surface

Basement membrane

Connective tissue

Connective tissue

Relaxed transitional epithelium

LM 400x

Sectional view of transitional epithelium of urinary bladder in relaxed (empty) state

Lumen of urinary bladder

Lumen of urinary bladder

Flattened surface cell in filled state

LM 1000x

Transitional epithelium

Connective tissue

LM 630x

Sectional view of transitional epithelium of urinary bladder in filled state

Glandular Epithelium

The function of glandular epithelium is secretion, which is accomplished by glandular cells that often lie in clusters deep to the covering and lining epithelium. A **gland** may consist of a single cell or a group of cells that secrete substances into ducts (tubes), onto a surface, or into the blood. All glands of the body are classified as either endocrine or exocrine.

The secretions of **endocrine glands** (EN-dō-krin; *endo-* = inside; *-crine* = secretion; Table 4.2), called hormones, enter the

---

TABLE 4.2

Epithelial Tissues: Glandular Epithelium

A. ENDOCRINE GLANDS

Description	Secretions (*hormones*) enter interstitial fluid and diffuse directly into bloodstream without flowing through a duct. Endocrine glands will be described in detail in Chapter 18.
Location	Examples include pituitary gland at base of brain, pineal gland in brain, thyroid and parathyroid glands near larynx (voice box), adrenal glands superior to kidneys, pancreas near stomach, ovaries in pelvic cavity, testes in scrotum, thymus in thoracic cavity.
Function	Hormones regulate many metabolic and physiological activities to maintain homeostasis.

Sectional view of endocrine gland (thyroid gland)

LM 630x

Endocrine gland (thyroid gland)

B. EXOCRINE GLANDS

Description	Secretory products released into ducts that empty onto surface of a covering and lining epithelium, such as skin surface or lumen of hollow organ.
Location	Sweat, oil, and earwax glands of skin; digestive glands such as salivary glands (secrete into mouth cavity) and pancreas (secretes into small intestine).
Function	Produce substances such as sweat to help lower body temperature, oil, earwax, saliva, or digestive enzymes.

Sectional view of the secretory portion of an exocrine gland (eccrine sweat gland)

LM 400x

Exocrine gland (eccrine sweat gland)

interstitial fluid and then diffuse directly into the bloodstream without flowing through a duct. Endocrine glands will be described in detail in Chapter 18. Endocrine secretions have far-reaching effects because they are distributed throughout the body by the bloodstream.

Exocrine glands (EKS-ō-krin; *exo-* = outside; Table 4.2) secrete their products into ducts that empty onto the surface of a covering and lining epithelium such as the skin surface or the lumen of a hollow organ. The secretions of exocrine glands have limited effects and some of them would be harmful if they entered the bloodstream. As you will learn later in the text, some glands of the body, such as the pancreas, ovaries, and testes, are mixed glands that contain both endocrine and exocrine tissue.

Structural Classification of Exocrine Glands

Exocrine glands are classified as unicellular or multicellular. As the name implies, **unicellular glands** are single-celled glands. Goblet cells are important unicellular exocrine glands that secrete mucus directly onto the apical surface of a lining epithelium. Most exocrine glands are **multicellular glands,** composed of

many cells that form a distinctive microscopic structure or macroscopic organ. Examples include sudoriferous (sweat), sebaceous (oil), and salivary glands.

Multicellular glands are categorized according to two criteria: (1) whether their ducts are branched or unbranched and (2) the shape of the secretory portions of the gland (Figure 4.6). If the duct of the gland does not branch, it is a **simple gland.** If the duct branches, it is a **compound gland.** Glands with tubular secretory parts are **tubular glands;** those with rounded secretory portions are **acinar glands** (AS-i-nar; *acin-* = berry), also called *alveolar glands.* **Tubuloacinar glands** have both tubular and more rounded secretory parts.

Combinations of these features are the criteria for the following structural classification scheme for multicellular exocrine glands:

I. Simple glands
  A. **Simple tubular.** Tubular secretory part is straight and attaches to a single unbranched duct. Example: glands in the large intestine.

Figure 4.6 Multicellular exocrine glands. Pink represents the secretory portion; lavender represents the duct.

 Structural classification of multicellular exocrine glands is based on the branching pattern of the duct and the shape of the secreting portion.

Duct

Secretory portion

Simple tubular   Simple branched tubular   Simple coiled tubular   Simple acinar   Simple branched acinar

Compound tubular   Compound acinar   Compound tubuloacinar

? **How do simple multicellular exocrine glands differ from compound ones?**

B. **Simple branched tubular.** Tubular secretory part is branched and attaches to a single unbranched duct. Example: gastric glands.

C. **Simple coiled tubular.** Tubular secretory part is coiled and attaches to a single unbranched duct. Example: sweat glands.

D. **Simple acinar.** Secretory portion is rounded and attaches to a single unbranched duct. Example: glands of the penile urethra.

E. **Simple branched acinar.** Rounded secretory part is branched and attaches to a single unbranched duct. Example: sebaceous glands.

II. Compound glands

A. **Compound tubular.** Secretory portion is tubular and attaches to a branched duct. Example: bulbourethral (Cowper's) glands.

B. **Compound acinar.** Secretory portion is rounded and attaches to a branched duct. Example: mammary glands.

C. **Compound tubuloacinar.** Secretory portion is both tubular and rounded and attaches to a branched duct. Example: acinar glands of the pancreas.

Functional Classification of Exocrine Glands

The functional classification of exocrine glands is based on how their secretions are released. Each of these secretory processes begins with the endoplasmic reticulum and Golgi complex working together to form intracellular secretory vesicles that contain the secretory product. Secretions of **merocrine glands** (MER-ō-krin; *mero-* = a part) are synthesized on ribosomes attached to rough ER; processed, sorted, and packaged by the Golgi complex; and released from the cell in secretory vesicles via exocytosis (Figure 4.7a). Most exocrine glands of the body are merocrine glands. Examples include the salivary glands and pancreas. **Apocrine glands** (AP-ō-krin; *apo-* = from) accumulate their secretory product at the apical surface of the secreting cell. Then, that portion of the cell pinches off by exocytosis from the rest of the cell to release the secretion (Figure 4.7b). The remaining part of the cell repairs itself and repeats the process. Electron microscopy has recently confirmed that this is the mechanism of secretion of milk fats in the mammary glands. Recent evidence reveals that the sweat glands of the skin, named apocrine sweat glands after this mode of secretion, actually undergo merocrine secretion. The cells of **holocrine glands** (HŌ-lō-krin; *holo-* = entire) accumulate a secretory product in their cytosol. As the secretory cell matures, it ruptures and becomes the secretory product (Figure 4.7c). Because the cell ruptures in this mode of secretion, the secretion contains large amounts of lipids from the plasma membrane and intracellular membranes. The sloughed off cell is replaced by a new cell. One example of a holocrine gland is a sebaceous gland of the skin.

✔CHECKPOINT

6. Describe the various layering arrangements and cell shapes of epithelium.
7. What characteristics are common to all epithelial tissues?

8. How is the structure of the following epithelial tissues related to their functions: simple squamous, simple cuboidal, simple columnar (nonciliated and ciliated), pseudostratified columnar (ciliated and nonciliated), stratified squamous (keratinized and nonkeratinized), stratified cuboidal, stratified columnar, and transitional?
9. Where are endothelium and mesothelium located?
10. What is the difference between endocrine glands and exocrine glands? Name and give examples of the three functional classes of exocrine glands based on how their secretions are released.

4.5 CONNECTIVE TISSUES

◉ OBJECTIVES

• Describe the general features of connective tissues.
• Describe the structure, location, and function of the various types of connective tissues.

Connective tissues are one of the most abundant and widely distributed tissues in the body. In their various forms, connective tissues have a variety of functions. They bind together, support, and strengthen other body tissues; protect and insulate internal organs; compartmentalize structures such as skeletal muscles; serve as the major transport system within the body (blood, a fluid connective tissue); are the primary locations of stored energy reserves (adipose, or fat, tissue); and are the main source of immune responses.

General Features of Connective Tissues

Connective tissues consist of two basic elements: extracellular matrix and cells. A connective tissue's **extracellular matrix** (MĀ-triks) is the material located between its widely spaced cells. The extracellular matrix consists of *protein fibers* and *ground substance,* the material between the cells and the fibers. The extracellular fibers are secreted by the connective tissue cells and account for many of the functional properties of the tissue in addition to controlling the surrounding watery environment via specific proteoglycan molecules (described shortly). The structure of the extracellular matrix determines much of the tissue's qualities. For instance, in cartilage, the extracellular matrix is firm but pliable. The extracellular matrix of bone, by contrast, is hard and inflexible.

Recall that, in contrast to epithelial tissues, connective tissues do not usually occur on body surfaces. Also, unlike epithelial tissues, connective tissues usually are highly vascular; that is, they have a rich blood supply. Exceptions include cartilage, which is avascular, and tendons, with a scanty blood supply. Except for cartilage, connective tissues, like epithelial tissues, are supplied with nerves.

Connective Tissue Cells

Embryonic cells called mesenchymal cells give rise to the cells of connective tissues. Each major type of connective tissue contains

Figure 4.7 Functional classification of multicellular exocrine glands.

The functional classification of exocrine glands is based on whether a secretion is a product of a cell or consists of an entire or a partial glandular cell.

(a) Merocrine secretion

(b) Apocrine secretion

(c) Holocrine secretion

? **What class of glands are sebaceous (oil) glands? Salivary glands?**

an immature class of cells with a name ending in -*blast,* which means "to bud or sprout." These immature cells are called *fibroblasts* in loose and dense connective tissue (described shortly), *chondroblasts* in cartilage, and *osteoblasts* in bone. Blast cells retain the capacity for cell division and secrete the extracellular matrix that is characteristic of the tissue. In cartilage and bone, once the extracellular matrix is produced, the immature cells differentiate into mature cells with names ending in -*cyte,* namely, chondrocytes and osteocytes. Mature cells have reduced capacities for cell division and extracellular matrix formation and are mostly involved in monitoring and maintaining the extracellular matrix.

The types of cells in connective tissues vary according to the type of tissue and include the following (Figure 4.8):

1. **Fibroblasts** (FĪ-brō-blasts; *fibro-* = fibers) are large, flat cells with branching processes. They are present in all the general connective tissues, and usually are the most numerous. Fibroblasts migrate through the connective tissues, secreting the fibers and certain components of the ground substance of the extracellular matrix.

Figure 4.8 Representative cells and fibers present in connective tissues.

Fibroblasts are usually the most numerous connective tissue cells.

Macrophage

Collagen
fiber

Elastic
fiber

Ground
substance

Blood vessel

Adipocyte

Reticular
fiber

Fibroblast

Eosinophil

Mast cell

Neutrophil

Plasma cell

? What is the function of fibroblasts?

2. **Macrophages** (MAK-rō-fā-jez; *macro-* = large; *-phages* = eaters) develop from *monocytes,* a type of white blood cell. Macrophages have an irregular shape with short branching projections and are capable of engulfing bacteria and cellular debris by phagocytosis. *Fixed macrophages* reside in a particular tissue; examples include alveolar macrophages in the lungs or splenic macrophages in the spleen. *Wandering macrophages* have the ability to move throughout the tissue and gather at sites of infection or inflammation to carry on phagocytosis.

3. **Plasma cells** are small cells that develop from a type of white blood cell called a *B lymphocyte.* Plasma cells secrete antibodies, proteins that attack or neutralize foreign substances in the body. Thus, plasma cells are an important part of the body's immune response. Although they are found in many places in the body, most plasma cells reside in connective tissues, especially in the gastrointestinal and respiratory tracts. They are also abundant in the salivary glands, lymph nodes, spleen, and red bone marrow.

4. **Mast cells** are abundant alongside the blood vessels that supply connective tissue. They produce histamine, a chemical that dilates small blood vessels as part of the inflammatory response, the body's reaction to injury or infection. In addition, researchers have recently discovered that mast cells can bind to, ingest, and kill bacteria.

5. **Adipocytes** (AD-i-pō-sīts), also called fat cells or *adipose* cells, are connective tissue cells that store triglycerides (fats). They are found deep to the skin and around organs such as the heart and kidneys.

6. **White blood cells** are not found in significant numbers in normal connective tissues. However, in response to certain conditions they migrate from blood into connective tissues. For example, *neutrophils* gather at sites of infection, and *eosinophils* migrate to sites of parasitic invasions and allergic responses.

Connective Tissue Extracellular Matrix

Each type of connective tissue has unique properties, based on the specific extracellular materials between the cells. The extracellular matrix consists of two major components: (1) the ground substance and (2) the fibers.

Ground Substance

As noted earlier, the **ground substance** is the component of a connective tissue between the cells and fibers. The ground substance may be fluid, semifluid, gelatinous, or calcified. It supports cells, binds them together, stores water, and provides a medium for exchange of substances between the blood and cells. It plays an active role in how tissues develop, migrate, proliferate, and change shape, and in how they carry out their metabolic functions.

Ground substance contains water and an assortment of large organic molecules, many of which are complex combinations of polysaccharides and proteins. The polysaccharides include hyaluronic acid, chondroitin sulfate, dermatan sulfate, and keratan sulfate. Collectively, they are referred to as **glycosaminoglycans** (glī-kōs-a-mē′-nō-GLĪ-kans) or **GAGs.** Except for hyaluronic acid, the GAGs are associated with proteins called **proteoglycans** (prō-tē-ō-GLĪ-kans). The proteoglycans form a core protein and the GAGs project from the protein like the bristles of a brush. One of the most important properties of GAGs is that they trap water, making the ground substance more jellylike.

Hyaluronic acid (hī′-a-loo-RON-ik) is a viscous, slippery substance that binds cells together, lubricates joints, and helps maintain the shape of the eyeballs. White blood cells, sperm cells, and some bacteria produce *hyaluronidase,* an enzyme that breaks apart hyaluronic acid, thus causing the ground substance of connective tissue to become more liquid. The ability to produce hyaluronidase helps white blood cells move more easily through connective tissues to reach sites of infection and aids penetration of an oocyte by a sperm cell during fertilization. It also accounts for the rapid spread of bacteria through connective tissues. **Chondroitin sulfate** (kon-DROY-tin) provides support and adhesiveness in cartilage, bone, skin, and blood vessels. The skin, tendons, blood vessels, and heart valves contain **dermatan sulfate;** bone, cartilage, and the cornea of the eye contain **keratan sulfate.** Also present in the ground substance are **adhesion proteins,** which are responsible for linking components of the ground substance to one another and to the surfaces of cells. The main adhesion protein of connective tissues is **fibronectin,** which binds to both collagen fibers (discussed shortly) and ground substance, linking them together. Fibronectin also attaches cells to the ground substance.

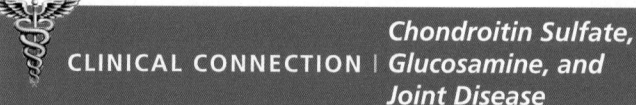

CLINICAL CONNECTION | Chondroitin Sulfate, Glucosamine, and Joint Disease

In recent years, **chondroitin sulfate** and **glucosamine** (a proteoglycan) have been used as nutritional supplements either alone or in combination to promote and maintain the structure and function of joint cartilage, to provide pain relief from osteoarthritis, and to reduce joint inflammation. Although these supplements have benefited some individuals with moderate to severe osteoarthritis, the benefit is minimal in lesser cases. More research is needed to determine how they act and why they help some people and not others. •

Fibers

Three types of **fibers** are embedded in the extracellular matrix between the cells: collagen fibers, elastic fibers, and reticular fibers (Figure 4.8). They function to strengthen and support connective tissues.

Collagen fibers (KOL-a-jen; *colla* = glue) are very strong and resist pulling forces (tension), but they are not stiff, which allows tissue flexibility. The properties of different types of collagen fibers vary from tissue to tissue. For example, the collagen fibers found in cartilage and bone form different associations with surrounding molecules. As a result of these associations, the collagen fibers in cartilage are surrounded by more water molecules than those in bone, which gives cartilage a more cushioning effect. Collagen fibers often occur in parallel bundles (see Table 4.5, dense regular connective tissue). The bundle arrangement adds great tensile strength to the tissue. Chemically, collagen fibers consist of the protein *collagen,* which is the most abundant protein in your body, representing about 25% of the total. Collagen fibers are found in most types of connective tissues, especially bone, cartilage, tendons (which attach muscle to bone), and ligaments (which attach bone to bone).

Elastic fibers, which are smaller in diameter than collagen fibers, branch and join together to form a fibrous network within a connective tissue. An elastic fiber consists of molecules of the protein *elastin* surrounded by a glycoprotein named *fibrillin,* which adds strength and stability. Because of their unique molecular structure, elastic fibers are strong but can be stretched up to 150% of their relaxed length without breaking. Equally important, elastic fibers have the ability to return to their original shape after being stretched, a property called *elasticity.* Elastic fibers are plentiful in skin, blood vessel walls, and lung tissue.

Reticular fibers (*reticul-* = net), consisting of *collagen* arranged in fine bundles with a coating of glycoprotein, provide support in the walls of blood vessels and form a network around the cells in some tissues, such as areolar connective tissue (a-RĒ-ō-lar; *areol-* = small space), adipose tissue, nerve fibers, and smooth muscle tissue. Produced by fibroblasts, reticular fibers are much thinner than collagen fibers and form branching networks. Like collagen fibers, reticular fibers provide support and strength. Reticular fibers are plentiful in reticular connective tissue, which forms the **stroma** (STRŌ-ma = bed or covering) or supporting framework of many soft organs, such as the spleen and lymph nodes. These fibers also help form the basement membrane.

CLINICAL CONNECTION | Marfan Syndrome

Marfan syndrome (MAR-fan) is an inherited disorder caused by a defective fibrillin gene. The result is abnormal development of elastic fibers. Tissues rich in elastic fibers are malformed or weakened. Structures affected most seriously are the covering layer of bones (periosteum), the ligament that suspends the lens of the eye, and the walls of the large arteries. People with Marfan syndrome tend to be tall and have disproportionately long arms, legs, fingers, and toes. A common symptom is blurred vision caused by displacement of the lens of the eye. The most life-threatening complication of Marfan syndrome is weakening of the aorta (the main artery that emerges from the heart), which can suddenly burst. •

Classification of Connective Tissues

Because of the diversity of cells and extracellular matrix and the differences in their relative proportions, the classification of connective tissues is not always clear-cut and several classifications exist. We offer the following classification scheme:

I. Embryonic connective tissues
   A. Mesenchyme
   B. Mucous connective tissue
II. Mature connective tissues
   A. Loose connective tissues
      1. Areolar connective tissue
      2. Adipose tissue
      3. Reticular connective tissue
   B. Dense connective tissues
      1. Dense regular connective tissue
      2. Dense irregular connective tissue
      3. Elastic connective tissue

C. Cartilage
   1. Hyaline cartilage
   2. Fibrocartilage
   3. Elastic cartilage
D. Bone tissue
E. Liquid connective tissue
   1. Blood tissue
   2. Lymph

Embryonic Connective Tissues

Note that our classification scheme has two major subclasses of connective tissues: embryonic and mature. **Embryonic connective tissues** are present primarily in the *embryo,* the developing human from fertilization through the first two months of pregnancy, and in the *fetus,* the developing human from the third month of pregnancy to birth (Table 4.3).

TABLE 4.3

Embryonic Connective Tissues

A. MESENCHYME

Description	Irregularly shaped mesenchymal cells embedded in semifluid ground substance that contains delicate reticular fibers.
Location	Almost exclusively under skin and along developing bones of embryo; some in adult connective tissue, especially along blood vessels.
Function	Forms almost all other types of connective tissue.

LM 1000x

Ground substance — Reticular fiber
— Nucleus of mesenchymal cell

Embryo

Ground substance
Nucleus of mesenchymal cell
Reticular fiber

LM 300x

Sectional view of mesenchyme of a developing embryo

Mesenchyme

B. MUCOUS CONNECTIVE TISSUE

Description	Widely scattered fibroblasts embedded in viscous, jellylike ground substance that contains fine collagen fibers.
Location	Umbilical cord of fetus.
Function	Support.

Epithelial surface cell of umbilical cord
Ground substance
Nucleus of fibroblast
Collagen fiber

Fetus

Umbilical cord

Epithelial surface cell of umbilical cord

LM 1000x

Collagen fiber
Nucleus of fibroblast
Ground substance

LM 200x

Sectional view of mucous connective tissue of the umbilical cord

Mucous connective tissue

Mature Connective Tissues

The second major subclass of connective tissues, **mature connective tissues,** are present in the newborn. Their cells arise primarily from mesenchyme. In the next section we explore the numerous types of mature connective tissues. The five types of mature connective tissues are (1) loose connective tissues, (2) dense connective tissues, (3) cartilage, (4) bone tissue, and (5) liquid connective tissue (blood tissue and lymph). We now examine each in detail.

Loose Connective Tissues

The fibers of **loose connective tissues** are *loosely* arranged between cells. The types of loose connective tissues are areolar connective tissue, adipose tissue, and reticular connective tissue (Table 4.4).

CLINICAL CONNECTION | *Liposuction*

A surgical procedure called **liposuction** (LIP-ō-suk′-shun; *lip-* = fat) or **suction lipectomy** (*-ectomy* = to cut out) involves suctioning out small amounts of adipose tissue from various areas of the body. After an incision is made in the skin, the fat is removed through a stainless steel tube, called a cannula, with the assistance of a powerful vacuum pressure unit that suctions out the fat. The technique can be used as a body-contouring procedure in regions such as the thighs, buttocks, arms, breasts, and abdomen, and to transfer fat to another area of the body. Postsurgical complications that may develop include fat that may enter blood vessels broken during the procedure and obstruct blood flow, infection, loss of feeling in the area, fluid depletion, injury to internal structures, and severe postoperative pain. •

Dense Connective Tissues

Dense connective tissues contain more fibers, which are thicker and more *densely* packed, but have considerably fewer cells than loose connective tissues. There are three types: dense regular connective tissue, dense irregular connective tissue, and elastic connective tissue (Table 4.5).

Cartilage

Cartilage (KAR-ti-lij) consists of a dense network of collagen fibers and elastic fibers firmly embedded in chondroitin sulfate, a gel-like component of the ground substance. Cartilage can endure considerably more stress than loose and dense connective tissues. The strength of cartilage is due to its collagen fibers, and its *resilience* (ability to assume its original shape after deformation) is due to chondroitin sulfate.

Like other connective tissues, cartilage has few cells and large quantities of extracellular matrix. It differs from other connective tissues, however, in not having nerves or blood vessels in its extracellular matrix. Interestingly, cartilage does not have a blood supply because it secretes an *antiangiogenesis* factor (an′-tē-an′-jē-ō-JEN-e-sis; *anti-* = against; *angio-* = vessel; *-genesis* = production), a substance that prevents blood vessel growth. Because of this property, antiangiogenesis factor is being studied as a possible cancer treatment. If cancer cells can be stopped from promoting new blood vessel growth, their rapid rate of cell division and expansion can be slowed or even halted.

TABLE 4.4

Mature Connective Tissues: Loose Connective Tissues

A. AREOLAR CONNECTIVE TISSUE

Description	One of the most widely distributed connective tissues; consists of fibers (collagen, elastic, reticular) arranged randomly and several kinds of cells (fibroblasts, macrophages, plasma cells, adipocytes, mast cells, and a few white blood cells) embedded in semifluid ground substance (hyaluronic acid, chondroitin sulfate, dermatan sulfate, and keratan sulfate).
Location	In and around nearly every body structure (thus, called "packing material" of the body): in subcutaneous layer deep to skin; papillary (superficial) region of dermis of skin; lamina propria of mucous membranes; around blood vessels, nerves, and body organs.
Function	Strength, elasticity, support.

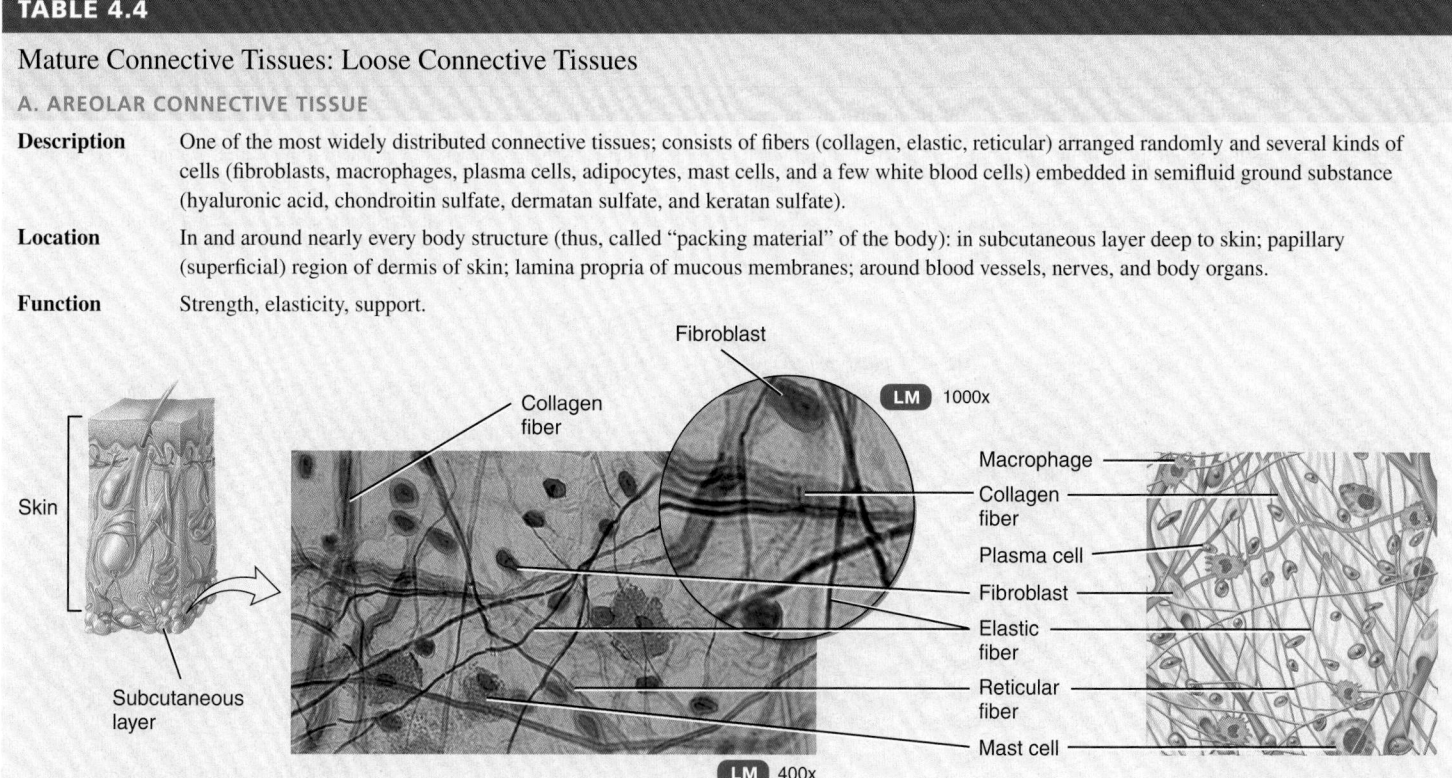

Sectional view of subcutaneous areolar connective tissue

Areolar connective tissue

TABLE 4.4 CONTINUES

TABLE 4.4 CONTINUED

Mature Connective Tissues: Loose Connective Tissues

B. ADIPOSE TISSUE

Description	Has cells derived from fibroblasts (called **adipocytes**) that are specialized for storage of triglycerides (fats) as a large, centrally located droplet. Cell fills up with a single, large triglyceride droplet, and cytoplasm and nucleus are pushed to periphery of cell. With weight gain, amount of adipose tissue increases and new blood vessels form. Thus, an obese person has many more blood vessels than does a lean person, a situation that can cause high blood pressure, since the heart has to work harder. Most adipose tissue in adults is *white adipose tissue* (just described). *Brown adipose tissue* (BAT) is darker due to very rich blood supply and numerous pigmented mitochondria that participate in aerobic cellular respiration. BAT is widespread in the fetus and infant; adults have only small amounts.
Location	Wherever areolar connective tissue is located: subcutaneous layer deep to skin, around heart and kidneys, yellow bone marrow, padding around joints and behind eyeball in eye socket.
Function	Reduces heat loss through skin; serves as an energy reserve; supports and protects organs. In newborns, BAT generates heat to maintain proper body temperature.

Sectional view of adipose tissue showing adipocytes of white fat and details of an adipocyte

Adipose tissue

C. RETICULAR CONNECTIVE TISSUE

Description	Fine interlacing network of reticular fibers (thin form of collagen fiber) and reticular cells.
Location	Stroma (supporting framework) of liver, spleen, lymph nodes; red bone marrow; reticular lamina of basement membrane; around blood vessels and muscles.
Function	Forms stroma of organs; binds smooth muscle tissue cells; filters and removes worn-out blood cells in spleen and microbes in lymph nodes.

Sectional view of reticular connective tissue of a lymph node

Reticular connective tissue

TABLE 4.5

Mature Connective Tissues: Dense Connective Tissues

D. DENSE REGULAR CONNECTIVE TISSUE

Description Shiny white extracellular matrix; mainly collagen fibers *regularly* arranged in bundles with fibroblasts in rows between bundles. Collagen fibers are not living (are protein structures secreted by fibroblasts), so damaged tendons and ligaments heal very slowly.

Location Forms tendons (attach muscle to bone), most ligaments (attach bone to bone), and aponeuroses (sheetlike tendons that attach muscle to muscle or muscle to bone).

Function Provides strong attachment between various structures. Tissue structure withstands pulling (tension) along long axis of fibers.

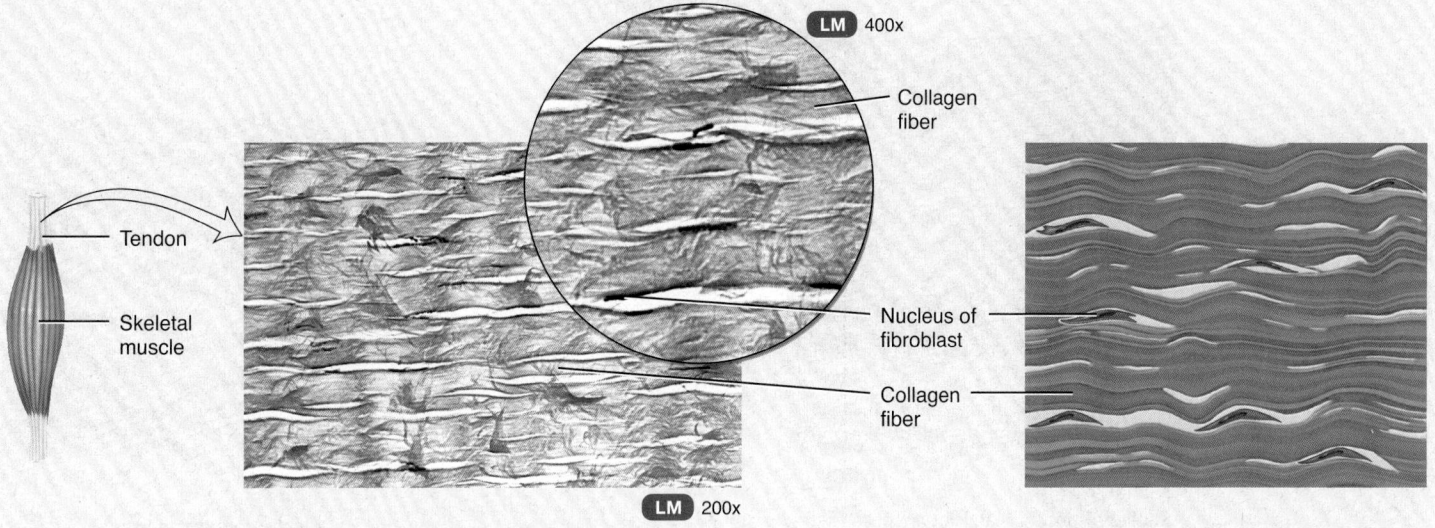

Sectional view of dense regular connective tissue of a tendon

Dense regular connective tissue

E. DENSE IRREGULAR CONNECTIVE TISSUE

Description Collagen fibers; usually *irregularly* arranged with a few fibroblasts.

Location Often occurs in sheets, such as fasciae (tissue beneath skin and around muscles and other organs), reticular (deeper) region of dermis of skin, fibrous pericardium of heart, periosteum of bone, perichondrium of cartilage, joint capsules, membrane capsules around various organs (kidneys, liver, testes, lymph nodes); also in heart valves.

Function Provides tensile (pulling) strength in many directions.

Sectional view of dense irregular connective tissue of reticular region of dermis

Dense irregular connective tissue

TABLE 4.5 CONTINUES

TABLE 4.5 CONTINUED

Mature Connective Tissues: Dense Connective Tissues

F. ELASTIC CONNECTIVE TISSUE

Description   Predominantly elastic fibers with fibroblasts between fibers; unstained tissue is yellowish.

Location   Lung tissue, walls of elastic arteries, trachea, bronchial tubes, true vocal cords, suspensory ligaments of penis, some ligaments between vertebrae.

Function   Allows stretching of various organs; is strong and can recoil to original shape after being stretched. Elasticity is important to normal functioning of lung tissue (recoils in exhaling) and elastic arteries (recoil between heartbeats to help maintain blood flow).

Sectional view of elastic connective tissue of aorta

Elastic connective tissue

The cells of mature cartilage, called **chondrocytes** (KON-drō-sīts; *chondro-* = cartilage), occur singly or in groups within spaces called **lacunae** (la-KOO-nē = little lakes; singular is *lacuna,* pronounced la-KOO-na) in the extracellular matrix. A covering of dense irregular connective tissue called the **perichondrium** (per′-i-KON-drē-um; *peri-* = around) surrounds the surface of most cartilage and contains blood vessels and nerves and is the source of new cartilage cells. Since cartilage has no blood supply, it heals poorly following an injury.

The cells and collagen-embedded extracellular matrix of cartilage form a strong, firm material that resists tension (stretching), compression (squeezing), and shear (pushing in opposite directions). The chondroitin sulfate in the extracellular matrix is largely responsible for cartilage's resilience. Because of these properties, cartilage plays an important role as a support tissue in the body. It is also a precursor to bone, forming almost the entire embryonic skeleton. Though bone gradually replaces cartilage during further development, cartilage persists after birth as the growth plates within bone that allow bones to increase in length during the growing years. Cartilage also persists throughout life as the lubricated articular surfaces of most joints.

There are three types of cartilage: hyaline cartilage, fibrocartilage, and elastic cartilage (Table 4.6).

Repair and Growth of Cartilage

Metabolically, cartilage is a relatively inactive tissue that grows slowly. When injured or inflamed, cartilage repair proceeds slowly, in large part because cartilage is avascular. Substances needed for repair and blood cells that participate in tissue repair must diffuse or migrate into the cartilage. The growth of cartilage follows two basic patterns: interstitial growth and appositional growth.

In **interstitial growth** (in′-ter-STISH-al), there is growth from within the tissue. When cartilage grows by interstitial growth, the cartilage increases rapidly in size due to the division of existing chondrocytes and the continuous deposition of increasing amounts of extracellular matrix by the chondrocytes. As the chondrocytes synthesize new matrix, they are pushed away from each other. These events cause the cartilage to expand from within like bread rising, which is the reason for the term *inter*stitial. This growth pattern occurs while the cartilage is young and pliable, during childhood and adolescence.

In **appositional growth** (ap-ō-ZISH-un-al), there is growth at the outer surface of the tissue. When cartilage grows by appositional growth, cells in the inner cellular layer of the perichondrium differentiate into chondroblasts. As differentiation continues, the chondroblasts surround themselves with extracellular matrix and become chondrocytes. As a result, matrix accumulates

TABLE 4.6

Mature Connective Tissues: Cartilage

A. HYALINE CARTILAGE

Description	Hyaline (*hyalinos* = glassy) cartilage contains a resilient gel as ground substance and appears in the body as a bluish-white, shiny substance (can stain pink or purple when prepared for microscopic examination). Fine collagen fibers are not visible with ordinary staining techniques; prominent chondrocytes are found in lacunae surrounded by perichondrium (exceptions: articular cartilage in joints and cartilage of epiphyseal plates, where bones lengthen during growth).
Location	Most abundant cartilage in body; at ends of long bones, anterior ends of ribs, nose, parts of larynx, trachea, bronchi, bronchial tubes, embryonic and fetal skeleton.
Function	Provides smooth surfaces for movement at joints, flexibility, and support; weakest type of cartilage.

Skeleton

Fetus

Perichondrium

LM 400x

Lacuna containing chondrocyte

Nucleus of chondrocyte

Ground substance

LM 200x

Sectional view of hyaline cartilage of a developing fetal bone

Hyaline cartilage

B. FIBROCARTILAGE

Description	Chondrocytes scattered among clearly visible thick bundles of collagen fibers within extracellular matrix; lacks perichondrium.
Location	Pubic symphysis (where hip bones join anteriorly), intervertebral discs, menisci (cartilage pads) of knee, portions of tendons that insert into cartilage.
Function	Support and joining structures together. Strength and rigidity make it the strongest type of cartilage.

Nucleus of chondrocyte

LM 400x

Lacuna containing chondrocyte

Vertebra

Intervertebral disc

Collagen fibers in ground substance

Nucleus of chondrocyte

Lacuna containing chondrocyte

LM 200x

Sectional view of fibrocartilage of intervertebral disc

Fibrocartilage

TABLE 4.6 CONTINUES

TABLE 4.6 CONTINUED

Mature Connective Tissues: Cartilage

C. ELASTIC CARTILAGE

Description	Chondrocytes in threadlike network of elastic fibers within extracellular matrix; perichondrium present.
Location	Lid on top of larynx (epiglottis), part of external ear (auricle), auditory (eustachian) tubes.
Function	Provides strength and elasticity; maintains shape of certain structures.

Sectional view of elastic cartilage of auricle of ear

Elastic cartilage

beneath the perichondrium on the outer surface of the cartilage, causing it to grow in width. Appositional growth starts later than interstitial growth and continues through adolescence.

Bone Tissue

Cartilage, joints, and bones make up the skeletal system. The skeletal system supports soft tissues, protects delicate structures, and works with skeletal muscles to generate movement. Bones store calcium and phosphorus; house red bone marrow, which produces blood cells; and contain yellow bone marrow, a storage site for triglycerides. Bones are organs composed of several different connective tissues, including **bone** or **osseous tissue** (OS-ē-us), the periosteum, red and yellow bone marrow, and the endosteum (a membrane that lines a space within bone that stores yellow bone marrow). Bone tissue is classified as either compact or spongy, depending on how its extracellular matrix and cells are organized.

The basic unit of **compact bone** is an **osteon** or **haversian system** (Table 4.7). Each osteon has four parts:

1. The **lamellae** (la-MEL-lē = little plates; singular is *lamella*) are concentric rings of extracellular matrix that consist of mineral salts (mostly calcium and phosphates), which give bone its hardness and compressive strength, and collagen fibers, which give bone its tensile strength. The lamellae are responsible for the compact nature of this type of bone tissue.

2. **Lacunae** (la-KOO-nē; singular is *lacuna*, pronounced la-KOO-na) are small spaces between lamellae that contain mature bone cells called **osteocytes.**

3. Projecting from the lacunae are **canaliculi** (kan-a-LIK-ū-lī = little canals), networks of minute canals containing the processes of osteocytes. Canaliculi provide routes for nutrients to reach osteocytes and for wastes to leave them.

4. A **central (haversian) canal** contains blood vessels and nerves.

Spongy bone lacks osteons. Rather, it consists of columns of bone called **trabeculae** (tra-BEK-ū-lē = little beams), which contain lamellae, osteocytes, lacunae, and canaliculi. Spaces between trabeculae are filled with red bone marrow. Chapter 6 presents bone tissue histology in more detail.

CLINICAL CONNECTION | Tissue Engineering

Tissue engineering is a technology that combines synthetic material with cells and has allowed scientists to grow new tissues in the laboratory to replace damaged tissues in the body. Tissue engineers have already developed laboratory-grown versions of skin and cartilage using scaffolding beds of biodegradable synthetic materials or collagen as substrates that permit body cells to be cultured. As the cells divide and assemble, the scaffolding degrades; the new, permanent tissue is then implanted in the patient. Other structures currently under development include bones, tendons, heart valves, bone marrow, and intestines. Work is also under way to develop insulin-producing cells for diabetics, dopamine-producing cells for Parkinson disease patients, and even entire livers and kidneys. •

TABLE 4.7

Mature Connective Tissues: Bone Tissue

Description	Compact bone tissue consists of osteons (haversian systems) that contain lamellae, lacunae, osteocytes, canaliculi, and central (haversian) canals. By contrast, spongy bone tissue (see Figure 6.3a, b) consists of thin columns called trabeculae; spaces between trabeculae are filled with red bone marrow.
Location	Both compact and spongy bone tissue make up the various parts of bones of the body.
Function	Support, protection, storage; houses blood-forming tissue; serves as levers that act with muscle tissue to enable movement.

Femur

Canaliculi
Central (haversian) canal
Lacuna
Lamellae

LM 400x

Sectional view of several osteons (haversian systems) of femur (thigh bone)

LM 100x

Calcified extracellular matrix
Osteocyte
Canaliculi
Lacuna

Details of an osteocyte

Liquid Connective Tissue

BLOOD TISSUE **Blood tissue** (or simply **blood**) is a connective tissue with a liquid extracellular matrix and formed elements. The extracellular matrix is called blood plasma. The **blood plasma** is a pale yellow fluid that consists mostly of water with a wide variety of dissolved substances—nutrients, wastes, enzymes, plasma proteins, hormones, respiratory gases, and ions. Suspended in the blood plasma are **formed elements**—red blood cells (erythrocytes), white blood cells (leukocytes), and platelets (thrombocytes) (Table 4.8). **Red blood cells** transport oxygen to body cells and remove some carbon dioxide from them. **White blood cells** are involved in phagocytosis, immunity, and allergic reactions. **Platelets** (PLĀT-lets) participate in blood clotting. The details of blood are considered in Chapter 19.

LYMPH **Lymph** is the extracellular fluid that flows in lymphatic vessels. It is a connective tissue that consists of several types of cells in a clear liquid extracellular matrix that is similar to blood plasma but with much less protein. The composition of lymph varies from one part of the body to another. For example, lymph leaving lymph nodes includes many lymphocytes, a type of white blood cell, in contrast to lymph from the small intestine, which has a high content of newly absorbed dietary lipids. The details of lymph are considered in Chapter 22.

✔ CHECKPOINT

11. In what ways do connective tissues differ from epithelia?
12. What are the features of the cells, ground substance, and fibers that make up connective tissues?
13. How are connective tissues classified? List the various types.
14. Describe how the structures of the following connective tissues are related to their functions: areolar connective tissue, adipose tissue, reticular connective tissue, dense regular connective tissue, dense irregular connective tissue, elastic connective tissue, hyaline cartilage, fibrocartilage, elastic cartilage, bone tissue, blood tissue, and lymph.
15. What is the difference between interstitial and appositional growth of cartilage?

4.6 MEMBRANES

⊙ **OBJECTIVES**
• Define a membrane.
• Describe the classification of membranes.

Membranes are flat sheets of pliable tissue that cover or line a part of the body. The majority of membranes consist of an epithelial layer and an underlying connective tissue layer and are called **epithelial membranes.** The principal epithelial membranes of the

TABLE 4.8

Mature Connective Tissues: Blood

Description	Blood plasma and formed elements: red blood cells (erythrocytes), white blood cells (leukocytes), platelets (thrombocytes).
Location	Within blood vessels (arteries, arterioles, capillaries, venules, veins), within chambers of heart.
Function	Red blood cells: transport oxygen and some carbon dioxide; white blood cells: carry on phagocytosis and mediate allergic reactions and immune system responses; platelets: essential for blood clotting.

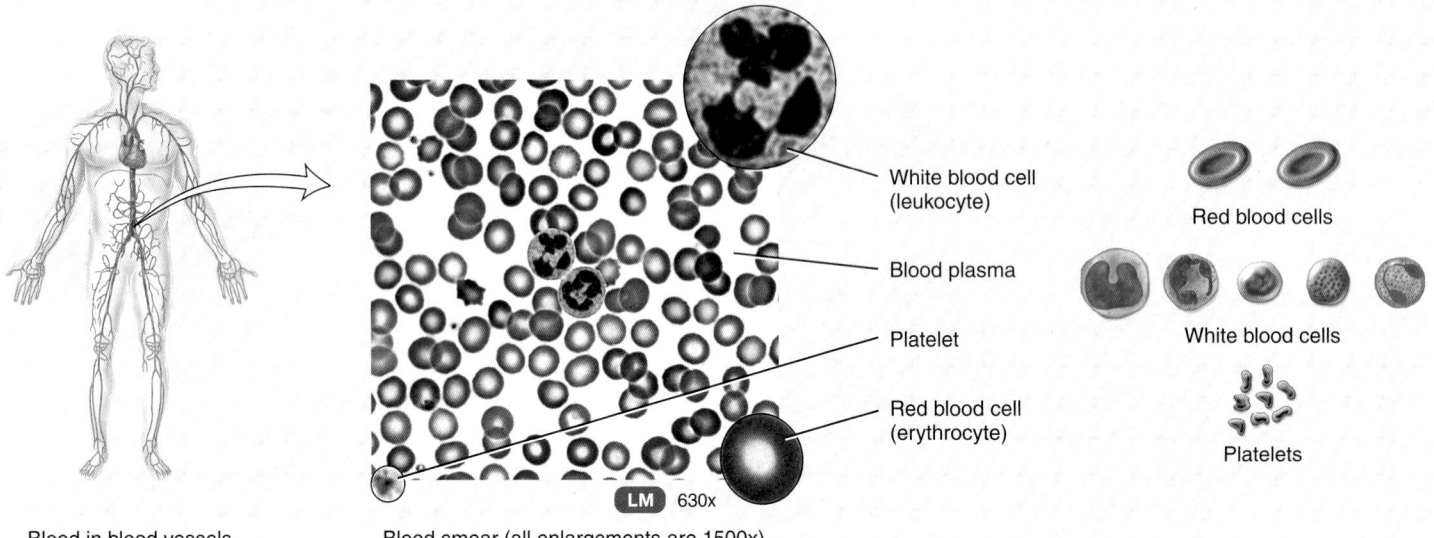

White blood cell (leukocyte)

Blood plasma

Platelet

Red blood cell (erythrocyte)

Red blood cells

White blood cells

Platelets

LM 630x

Blood in blood vessels

Blood smear (all enlargements are 1500x)

body are mucous membranes, serous membranes, and the cutaneous membrane, or skin. Another type of membrane, a synovial membrane, lines joints and contains connective tissue but no epithelium.

Epithelial Membranes

Mucous Membranes

A **mucous membrane** or **mucosa** (mū-KŌ-sa) lines a body cavity that opens directly to the exterior. Mucous membranes line the entire digestive, respiratory, and reproductive tracts, and much of the urinary tract. They consist of a lining layer of epithelium and an underlying layer of connective tissue (Figure 4.9a).

The epithelial layer of a mucous membrane is an important feature of the body's defense mechanisms because it is a barrier that microbes and other pathogens have difficulty penetrating. Usually, tight junctions connect the cells, so materials cannot leak in between them. Goblet cells and other cells of the epithelial layer of a mucous membrane secrete mucus, and this slippery fluid prevents the cavities from drying out. It also traps particles in the respiratory passageways and lubricates food as it moves through the gastrointestinal tract. In addition, the epithelial layer secretes some of the enzymes needed for digestion and is the site of food and fluid absorption in the gastrointestinal tract. The epithelia of mucous membranes vary greatly in different parts of the body. For example, the mucous membrane of the small intestine is nonciliated simple columnar epithelium, and the large airways to the lungs consist of pseudostratified ciliated columnar epithelium (see Table 4.1).

The connective tissue layer of a mucous membrane is areolar connective tissue and is called the **lamina propria** (LAM-i-na PRŌ-prē-a; *propria* = one's own), so named because it belongs to (is owned by) the mucous membrane. The lamina propria supports the epithelium, binds it to the underlying structures, allows some flexibility of the membrane, and affords some protection for underlying structures. It also holds blood vessels in place and is the vascular source for the overlying epithelium. Oxygen and nutrients diffuse from the lamina propria to the covering epithelium; carbon dioxide and wastes diffuse in the opposite direction.

Serous Membranes

A **serous membrane** (SĒR-us = watery) or **serosa** lines a body cavity that does not open directly to the exterior (thoracic or abdominal cavities), and it covers the organs that are within the cavity. Serous membranes consist of areolar connective tissue covered by mesothelium (simple squamous epithelium) (Figure 4.9b). You will recall from Chapter 1 that serous membranes have two layers: The layer attached to and lining the cavity wall is called the **parietal layer** (pa-RĪ-e-tal; *pariet-* = wall); the layer that covers and adheres to the organs within the cavity is the **visceral layer** (*viscer-* = body organ) (see Figure 1.10a). The mesothelium of a serous membrane secretes **serous fluid,** a watery lubricant that allows organs to glide easily over one another or to slide against the walls of cavities.

Recall from Chapter 1 that the serous membrane lining the thoracic cavity and covering the lungs is the **pleura.** The serous membrane lining the heart cavity and covering the heart is the

Figure 4.9 Membranes.

A membrane is a flat sheet of pliable tissues that covers or lines a part of the body.

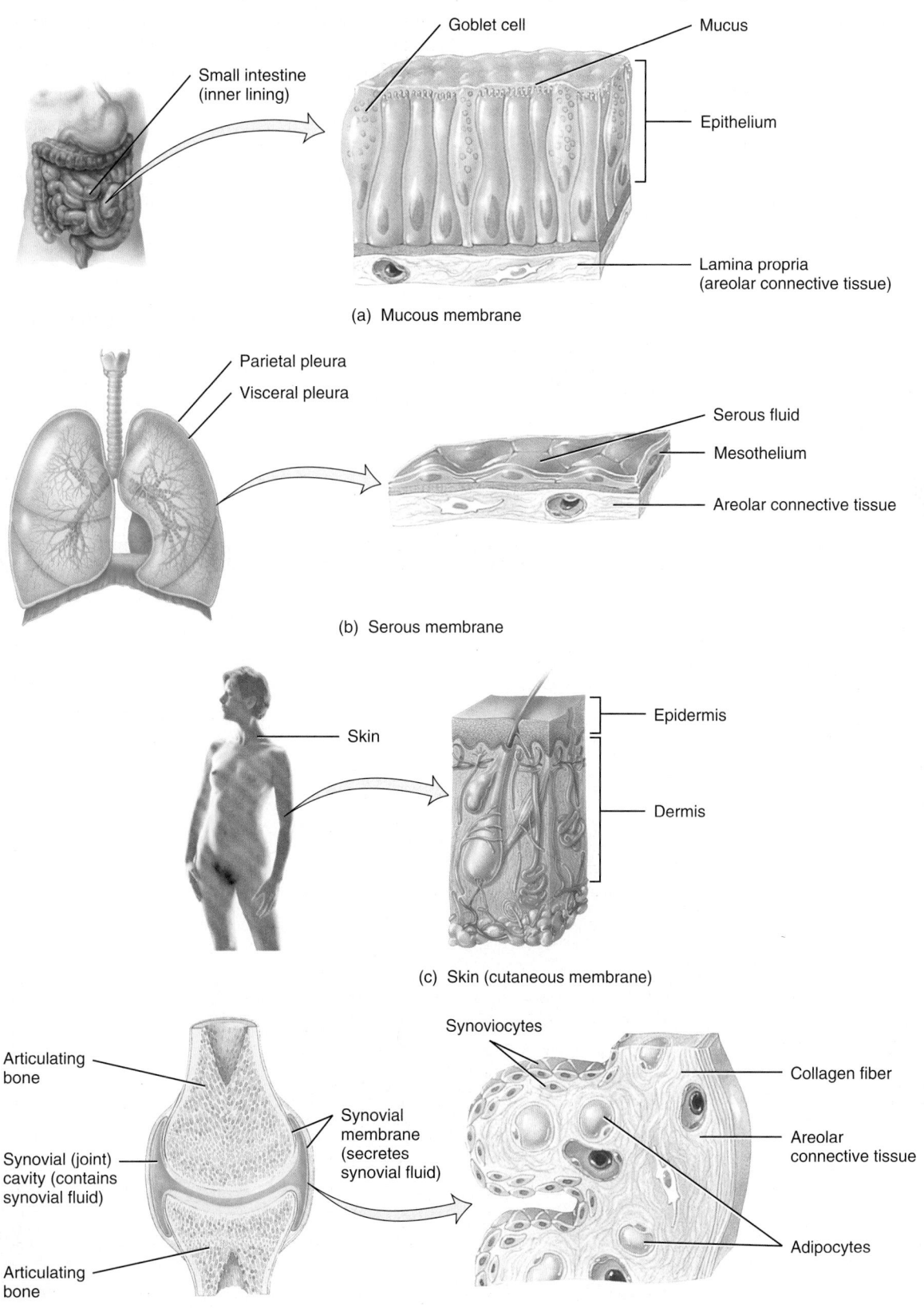

(a) Mucous membrane

(b) Serous membrane

(c) Skin (cutaneous membrane)

(d) Synovial membrane

? **What is an epithelial membrane?**

pericardium. The serous membrane lining the abdominal cavity and covering the abdominal organs is the **peritoneum.**

Cutaneous Membrane

The **cutaneous membrane** or **skin** covers the entire surface of the body and consists of a superficial portion called the *epidermis* and a deeper portion called the *dermis* (Figure 4.9c). The epidermis consists of keratinized stratified squamous epithelium, which protects underlying tissues. The dermis consists of dense irregular connective tissue and areolar connective tissue. Details of the cutaneous membrane are presented in Chapter 5.

Synovial Membranes

Synovial membranes (si-NŌ-vē-al; *syn-* = together, referring here to a place where bones come together; *-ova* = egg, because of their resemblance to the slimy egg white of an uncooked egg) line the cavities of freely movable joints (joint cavities). Like serous membranes, synovial membranes line structures that do not open to the exterior. Unlike mucous, serous, and cutaneous membranes, they lack an epithelium and are therefore not epithelial membranes. Synovial membranes are composed of a discontinuous layer of cells called **synoviocytes** (si-NŌ-vē-ō-sīts), which are closer to the synovial cavity (space between the bones), and a layer of connective tissue (areolar and adipose) deep to the synoviocytes (Figure 4.9d). Synoviocytes secrete some of the components of synovial fluid. **Synovial fluid** lubricates and nourishes the cartilage covering the bones at movable joints and contains macrophages that remove microbes and debris from the joint cavity.

✔ **CHECKPOINT**

16. Define the following kinds of membranes: mucous, serous, cutaneous, and synovial. How do they differ from one another?
17. Where is each type of membrane located in the body? What are their functions?

TABLE 4.9

Muscular Tissues

A. SKELETAL MUSCLE TISSUE

Description	Long, cylindrical, striated fibers (*striations* are alternating light and dark bands within fibers that are visible under a light microscope). Skeletal muscle fibers vary greatly in length, from a few centimeters in short muscles to 30–40 cm (about 12–16 in.) in longest muscles. A muscle fiber is a roughly cylindrical, multinucleated cell with nuclei at periphery. Skeletal muscle is considered *voluntary* because it can be made to contract or relax by conscious control.
Location	Usually attached to bones by tendons.
Function	Motion, posture, heat production, protection.

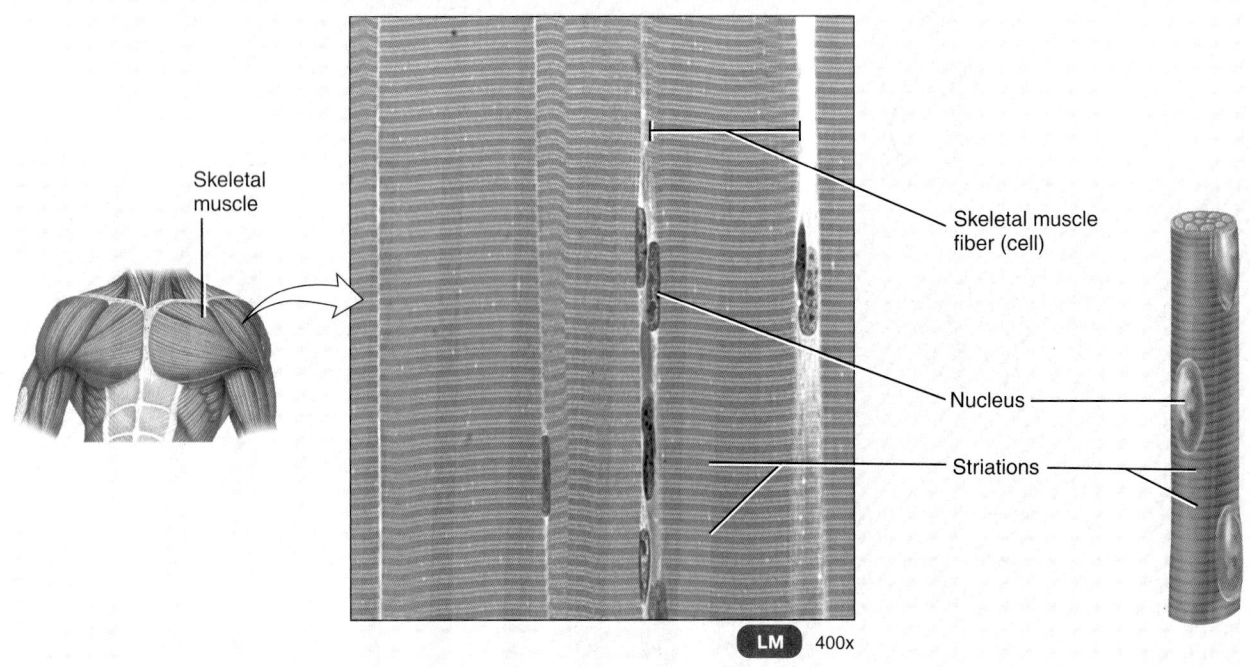

Skeletal muscle
Skeletal muscle fiber (cell)
Nucleus
Striations

LM 400x

Longitudinal section of skeletal muscle tissue

Skeletal muscle fiber

4.7 MUSCULAR TISSUES

◉ OBJECTIVES

• Describe the general features of muscular tissues.
• Contrast the structure, location, and mode of control of skeletal, cardiac, and smooth muscle tissue.

Muscular tissues consist of elongated cells called *muscle fibers* or *myocytes* that can use ATP to generate force. As a result, muscular tissues produce body movements, maintain posture, and generate heat. They also provides protection. Based on their location and certain structural and functional features, muscular tissues are classified into three types: skeletal, cardiac, and smooth (Table 4.9).

Chapter 10 provides a more detailed discussion of muscular tissue.

✔CHECKPOINT

18. Which types of muscular tissues are striated? Which are smooth?
19. Which types of muscular tissues have gap junctions?

4.8 NERVOUS TISSUE

◉ OBJECTIVE

• Describe the structural features and functions of nervous tissue.

Despite the awesome complexity of the nervous system, it consists of only two principal types of cells: neurons and neuroglia. **Neurons** (NOO-rons; *neuro-* = nerve), or nerve cells, are sensitive to various stimuli. They convert stimuli into electrical signals called **nerve action potentials (nerve impulses)** and conduct these action potentials to other neurons, to muscle tissue, or to glands. Most neurons consist of three basic parts: a cell body and two kinds of cell processes—dendrites and axons (Table 4.10). The **cell body** contains the nucleus and other organelles. **Dendrites** (*dendr-* = tree) are tapering, highly branched, and usually short cell processes (extensions). They are the major receiving or input portion of a neuron. The **axon** (*axo-* = axis) of a neuron is a single, thin, cylindrical process that may be very long. It is the

B. CARDIAC MUSCLE TISSUE

Description Branched, striated fibers with usually only one centrally located nucleus (occasionally two). Attach end to end by transverse thickenings of plasma membrane called *intercalated discs* (in-TER-ka-lāt-ed; *intercalate* = to insert between), which contain desmosomes and gap junctions. Desmosomes strengthen tissue and hold fibers together during vigorous contractions. Gap junctions provide route for quick conduction of electrical signals (muscle action potentials) throughout heart. *Involuntary* (not conscious) control.

Location Heart wall.

Function Pumps blood to all parts of body.

LM 500x

Longitudinal section of cardiac muscle tissue

Cardiac muscle fibers

TABLE 4.9 CONTINUES

TABLE 4.9 CONTINUED

Muscular Tissues

C. SMOOTH MUSCLE TISSUE

Description   Fibers usually *involuntary,* nonstriated (lack striations, hence the term *smooth*). Smooth muscle fiber is a small spindle-shaped cell thickest in middle, tapering at each end, and containing a single, centrally located nucleus. Gap junctions connect many individual fibers in some smooth muscle tissues (for example, in wall of intestines). Can produce powerful contractions as many muscle fibers contract in unison. Where gap junctions are absent, such as iris of eye, smooth muscle fibers contract individually, like skeletal muscle fibers.

Location   Iris of eyes; walls of hollow internal structures such as blood vessels, airways to lungs, stomach, intestines, gallbladder, urinary bladder, and uterus.

Function   Motion (constriction of blood vessels and airways, propulsion of foods through gastrointestinal tract, contraction of urinary bladder and gallbladder).

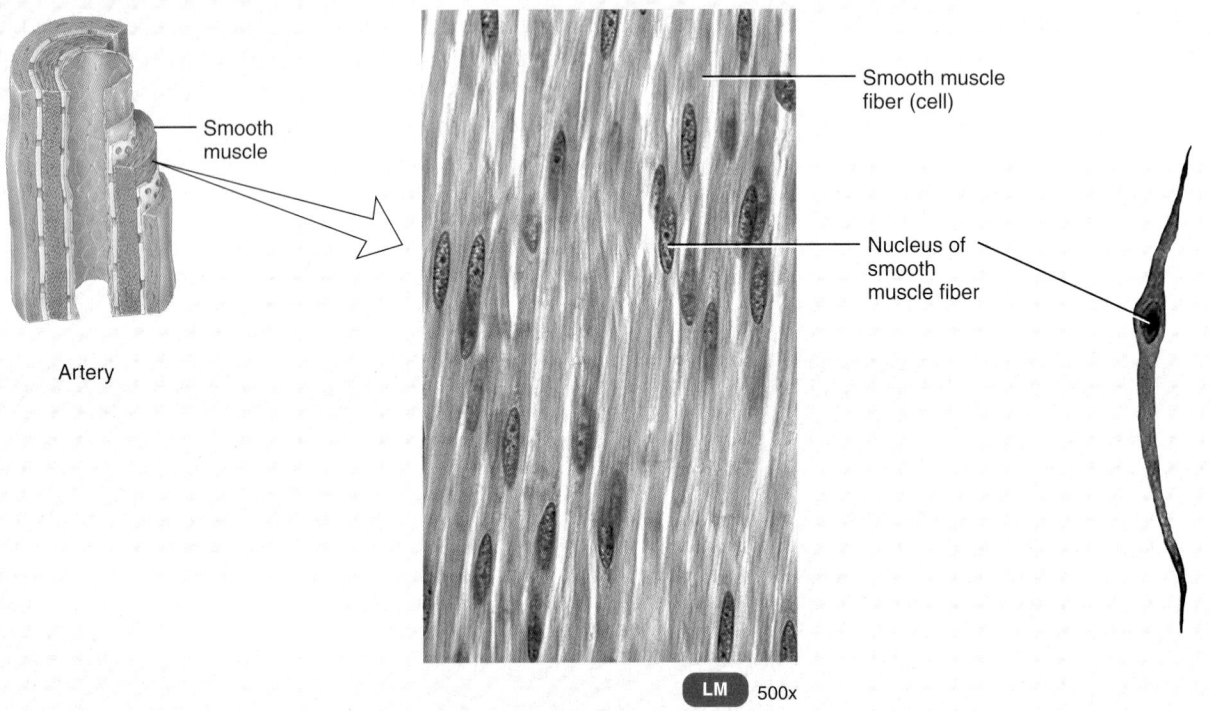

Artery

Smooth muscle fiber (cell)

Nucleus of smooth muscle fiber

LM 500x

Longitudinal section of smooth muscle tissue

Smooth muscle fiber

output portion of a neuron, conducting nerve impulses toward another neuron or to some other tissue.

Even though **neuroglia** (noo-RŌG-lē-a; *-glia* = glue) do not generate or conduct nerve impulses, these cells do have many important supportive functions. The detailed structure and function of neurons and neuroglia are considered in Chapter 12.

✔CHECKPOINT

20. What are the functions of the dendrites, cell body, and axon of a neuron?

4.9 EXCITABLE CELLS

◉ OBJECTIVE

• Explain the concept of electrical excitability.

Neurons and muscle fibers are considered **excitable cells** because they exhibit **electrical excitability,** the ability to respond to certain stimuli by producing electrical signals such as *action potentials.* Action potentials can propagate (travel) along the plasma membrane of a neuron or muscle fiber due to the presence of specific voltage-gated ion channels. When an action potential forms in a neuron, the neuron releases chemicals called *neurotransmitters,* which allow neurons to communicate with other neurons, muscle fibers, or glands. When an action potential occurs in a muscle fiber, the muscle fiber contracts, resulting in activities such as the movement of the limbs, the propulsion of food through the small intestine, and the movement of blood out of the

TABLE 4.10	
Nervous Tissue	
Description	(1) Neurons (nerve cells), which consist of cell body and processes extending from cell body (one to multiple dendrites and a single axon); and (2) neuroglia, which do not generate or conduct nerve impulses but have other important supporting functions.
Location	Nervous system.
Function	Exhibits sensitivity to various types of stimuli; converts stimuli into nerve impulses (action potentials); conducts nerve impulses to other neurons, muscle fibers, or glands.

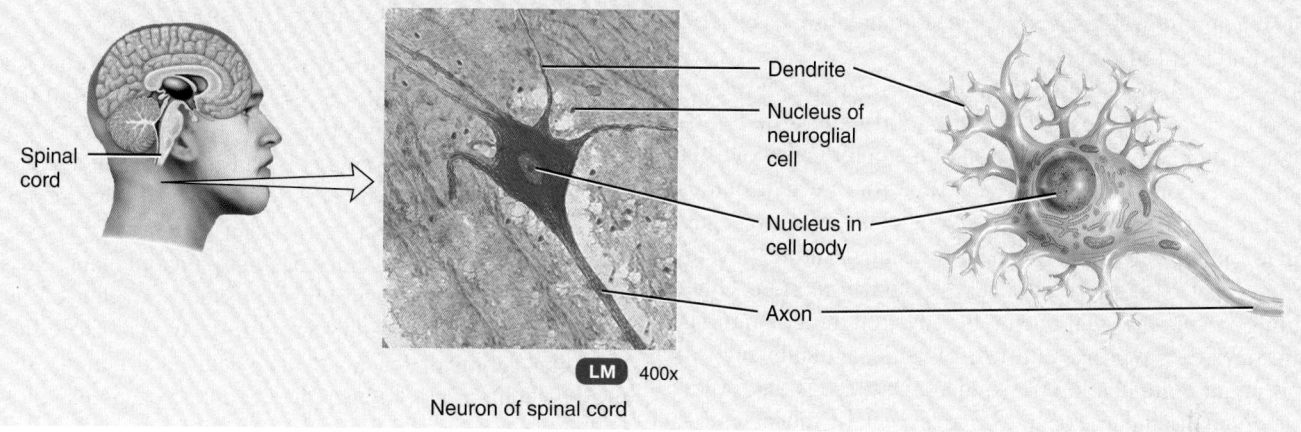

LM 400x

Neuron of spinal cord

heart and into the blood vessels of the body. The muscle action potential and the nerve action potential are discussed in detail in Chapters 10 and 12, respectively.

CHECKPOINT

21. Why is electrical excitability important to neurons and muscle fibers?

4.10 TISSUE REPAIR: RESTORING HOMEOSTASIS

OBJECTIVE

• Describe the role of tissue repair in restoring homeostasis.

Tissue repair is the replacement of worn-out, damaged, or dead cells. New cells originate by cell division from the **stroma,** the supporting connective tissue, or from the **parenchyma** (pa-REN-ki-ma), cells that constitute the functioning part of the tissue or organ. In adults, each of the four basic tissue types (epithelial, connective, muscle, and nervous) has a different capacity for replenishing parenchymal cells lost by damage, disease, or other processes.

Epithelial cells, which endure considerable wear and tear (and even injury) in some locations, have a continuous capacity for renewal. In some cases, immature, undifferentiated cells called **stem cells** divide to replace lost or damaged cells. For example, stem cells reside in protected locations in the epithelia of the skin and gastrointestinal tract to replenish cells sloughed from the apical layer, and stem cells in red bone marrow continually provide new red and white blood cells and platelets. In other cases, mature, differentiated cells can undergo cell division; examples include hepatocytes (liver cells) and endothelial cells in blood vessels.

Some connective tissues also have a continuous capacity for renewal. One example is bone, which has an ample blood supply. Connective tissues such as cartilage can replenish cells much less readily, in part because of a smaller blood supply.

Muscular tissues have a relatively poor capacity for renewal of lost cells. Even though skeletal muscle tissue contains stem cells called *satellite cells,* they do not divide rapidly enough to replace extensively damaged muscle fibers. Cardiac muscle tissue lacks satellite cells, and existing cardiac muscle fibers do not undergo mitosis to form new cells. Recent evidence suggests that stem cells do migrate into the heart from the blood. There, they can differentiate and replace a limited number of cardiac muscle fibers and endothelial cells in heart blood vessels. Smooth muscle fibers can proliferate to some extent, but they do so much more slowly than the cells of epithelial or connective tissues.

Nervous tissue has the poorest capacity for renewal. Although experiments have revealed the presence of some stem cells in the brain, they normally do not undergo mitosis to replace damaged neurons. Discovering why this is so is a major goal of researchers who seek ways to repair nervous tissue damaged by injury or disease.

The restoration of an injured tissue or organ to normal structure and function depends entirely on whether parenchymal cells are active in the repair process. If parenchymal cells accomplish the repair, **tissue regeneration** is possible, and a near-perfect reconstruction of the injured tissue may occur. However, if fibroblasts of the stroma are active in the repair, the replacement tissue will be a new connective tissue. The fibroblasts synthesize collagen and other extracellular matrix materials that aggregate to form scar tissue, a process known as **fibrosis.** Because scar tissue is not specialized to perform the functions of the parenchymal tissue, the original function of the tissue or organ is impaired.

When tissue damage is extensive, as in large, open wounds, both the connective tissue stroma and the parenchymal cells are active in repair; fibroblasts divide rapidly, and new collagen fibers are manufactured to provide structural strength. Blood capillaries also sprout new buds to supply the healing tissue with the materials it needs. All these processes create an actively growing connective tissue called **granulation tissue** (gran-ū-LĀ-shun). This new tissue forms across a wound or surgical incision to provide a framework (stroma) that supports the epithelial cells that migrate into the open area and fill it. The newly formed granulation tissue also secretes a fluid that kills bacteria.

At times, a small but significant number of patients develop a complication of surgery called **wound dehiscence** (dē-HISS-ens), the partial or complete separation of the outer layers of a sutured incision. A common cause is surgical error in which sutures or staples are placed too far apart, too close to the incision edges, or under too much pressure. It can also occur if sutures are removed too early or if there is a deep wound infection. Other contributing factors are age, chemotherapy, coughing, straining, vomiting, obesity, smoking, and use of anticoagulants such as aspirin. A major complication of wound dehiscence is the protrusion of an organ through the open wound, especially the intestines. This can lead to peritonitis (inflammation of the peritoneum) and septic shock (shock that results from bacterial toxins due to vasodilation).

Three factors affect tissue repair: nutrition, blood circulation, and age. Nutrition is vital because the healing process places a great demand on the body's store of nutrients. Adequate protein in the diet is important because most of the structural components of a tissue are proteins. Several vitamins also play a direct role in wound healing and tissue repair. For example, vitamin C directly affects the normal production and maintenance of matrix materials, especially collagen, and strengthens and promotes the formation of new blood vessels. In a person with vitamin C deficiency, even superficial wounds fail to heal, and the walls of the blood vessels become fragile and are easily ruptured.

Proper blood circulation is essential to transport oxygen, nutrients, antibodies, and many defensive cells to the injured site. The blood also plays an important role in the removal of tissue fluid, bacteria, foreign bodies, and debris, elements that would otherwise interfere with healing. The third factor in tissue repair, age, is the topic of the next section.

CLINICAL CONNECTION | Adhesions

Scar tissue can form **adhesions** (ad-HĒ-zhuns; *adhaero* = to stick to), abnormal joining of tissues. Adhesions commonly form in the abdomen around a site of previous inflammation such as an inflamed appendix, and they can develop after surgery. Although adhesions do not always cause problems, they can decrease tissue flexibility, cause obstruction (such as in the intestine), and make a subsequent operation, such as a cesarean section (c-section), more difficult. In rare cases adhesions can result in infertility. An *adhesiotomy*, the surgical release of adhesions, may be required. •

✔ CHECKPOINT

22. How are stromal and parenchymal repair of a tissue different?
23. What is the importance of granulation tissue?

4.11 AGING AND TISSUES

◉ OBJECTIVE

• Describe the effects of aging on tissues.

In later chapters, the effects of aging on specific body systems will be considered. With respect to tissues, epithelial tissues get progressively thinner and connective tissues become more fragile with aging. This is evidenced by an increased incidence of skin and mucous membrane disorders, wrinkles, more susceptibility to bruises, increased loss of bone density, higher rates of bone fractures, and increased episodes of joint pain and disorders. There is also an effect of aging on muscle tissue as evidenced by loss of skeletal muscle mass and strength, decline in the efficiency of pumping action of the heart, and decreased activity of smooth muscle–containing organs, for example, organs of the gastrointestinal tract.

Generally, tissues heal faster and leave less obvious scars in the young than in the aged. In fact, surgery performed on fetuses leaves no scars at all. The younger body is generally in a better nutritional state, its tissues have a better blood supply, and its cells have a higher metabolic rate. Thus, its cells can synthesize needed materials and divide more quickly. The extracellular components of tissues also change with age. Glucose, the most abundant sugar in the body, plays a role in the aging process. As the body ages, glucose is haphazardly added to proteins inside and outside cells, forming irreversible cross-links between adjacent protein molecules. With advancing age, more cross-links form, which contributes to the stiffening and loss of elasticity that occur in aging tissues. Collagen fibers, responsible for the strength of tendons, increase in number and change in quality with aging. Changes in the collagen of arterial walls affect the flexibility of arteries as much as the fatty deposits associated with atherosclerosis (see Coronary Artery Disease in Disorders: Homeostatic Imbalances section of Chapter 20). Elastin, another extracellular component, is responsible for the elasticity of blood vessels and skin. It thickens, fragments, and acquires a greater affinity for calcium with age—changes that may also be associated with the development of atherosclerosis.

✔ CHECKPOINT

24. What common changes occur in epithelial and connective tissues with aging?

DISORDERS: HOMEOSTATIC IMBALANCES

Disorders of epithelial tissues are mainly specific to individual organs, such as peptic ulcer disease (PUD), which erodes the epithelial lining of the stomach or small intestine. For this reason, epithelial disorders are described along with their relevant body systems throughout the text. The most prevalent disorders of connective tissues are **autoimmune diseases**—diseases in which antibodies produced by the immune system fail to distinguish what is foreign from what is self and attack the body's own tissues. One of the most common autoimmune disorders is rheumatoid arthritis, which attacks the synovial membranes of joints. Because connective tissues are one of the most abundant and widely distributed of the four main types of tissues, disorders related to them often affect multiple body systems. Common disorders of muscular tissues and nervous tissue are described at the ends of Chapters 10 and 12, respectively.

Sjögren's Syndrome

Sjögren's syndrome (SHO-grenz) is a common autoimmune disorder in which your own immune system attacks the lacrimal (tear) glands and salivary glands, especially, and causes inflammation and destruction of these exocrine glands. Signs include dryness of the eyes, mouth, nose, ears, skin, and vagina, and salivary gland enlargement. Systemic effects include fatigue, arthritis, difficulty in swallowing, pancreatitis (inflammation of the pancreas), pleuritis (inflammation of the pleurae of the lungs), and muscle and joint pain. The disorder affects females more than males by a ratio of 9 to 1. About 20% of older adults experience some signs of Sjögren's. Treatment is supportive and includes using artificial tears to moisten the eyes, sipping fluids, chewing sugarless gum, using a saliva substitute to moisten the mouth, and applying moisturizing creams for the skin. If symptoms or complications are severe, medications may be used. These include cyclosporine eyedrops, pilocarpine to increase saliva production, immunosuppressants, nonsteroidal anti-inflammatory drugs, and corticosteroids.

Systemic Lupus Erythematosus

Systemic lupus erythematosus (er-i-thē-ma-TŌ-sus), **SLE**, or simply lupus, is a chronic inflammatory disease of connective tissue occurring mostly in nonwhite women during their childbearing years. It is an autoimmune disease that can cause tissue damage in every body system. The disease, which can range from a mild condition in most patients to a rapidly fatal disease, is marked by periods of exacerbation and remission. The prevalence of SLE is about 1 in 2000, with females more likely to be afflicted than males by a ratio of 8 or 9 to 1.

Although the cause of SLE is unknown, genetic, environmental, and hormonal factors all have been implicated. The genetic component is suggested by studies of twins and family history. Environmental factors include viruses, bacteria, chemicals, drugs, exposure to excessive sunlight, and emotional stress. Sex hormones, such as estrogens, may also trigger SLE.

Signs and symptoms of SLE include painful joints, low-grade fever, fatigue, mouth ulcers, weight loss, enlarged lymph nodes and spleen, sensitivity to sunlight, rapid loss of large amounts of scalp hair, and anorexia. A distinguishing feature of lupus is an eruption across the bridge of the nose and cheeks called a "butterfly rash." Other skin lesions may occur, including blistering and ulceration. The erosive nature of some SLE skin lesions was thought to resemble the damage inflicted by the bite of a wolf—thus, the name *lupus* (= wolf). The most serious complications of the disease involve inflammation of the kidneys, liver, spleen, lungs, heart, brain, and gastrointestinal tract. Because there is no cure for SLE, treatment is supportive, including anti-inflammatory drugs such as aspirin, and immunosuppressive drugs.

MEDICAL TERMINOLOGY

Atrophy (AT-rō-fē; *a-* = without; *-trophy* = nourishment) A decrease in the size of cells, with a subsequent decrease in the size of the affected tissue or organ.

Hypertrophy (hī-PER-trō-fē; *hyper-* = above or excessive) Increase in the size of a tissue because its cells enlarge without undergoing cell division.

Tissue rejection An immune response of the body directed at foreign proteins in a transplanted tissue or organ; immunosuppressive drugs, such as cyclosporine, have largely overcome tissue rejection in heart-, kidney-, and liver-transplant patients.

Tissue transplantation The replacement of a diseased or injured tissue or organ. The most successful transplants involve use of a person's own tissues or those from an identical twin.

Xenotransplantation (zen'-ō-trans-plan-TĀ-shun; *xeno-* = strange, foreign) The replacement of a diseased or injured tissue or organ with cells or tissues from an animal. Porcine (from pigs) and bovine (from cows) heart valves are used for some heart-valve replacement surgeries.

CHAPTER REVIEW AND RESOURCE SUMMARY

Review

4.1 Types of Tissues

1. A tissue is a group of cells, usually with a similar embryological origin, that is specialized for a particular function.
2. The various tissues of the body are classified into four basic types: epithelial, connective, muscular, and nervous tissue.

Resource

Anatomy Overview - Epithelial Tissues
Anatomy Overview - Connective Tissues
Anatomy Overview - Muscle Tissue
Anatomy Overview - Nervous Tissue
Concepts and Connections - Human Tissue

Review

4.2 Cell Junctions

1. Cell junctions are points of contact between adjacent plasma membranes.
2. Tight junctions form fluid-tight seals between cells; adherens junctions, desmosomes, and hemidesmosomes anchor cells to one another or to the basement membrane; and gap junctions permit electrical and chemical signals to pass between cells.

Animation - Intracellular Junctions
Figure 4.2 - Cell Junctions
Exercise - Getting It
   Together

4.3 Comparison between Epithelial and Connective Tissues

1. Epithelial tissues have many cells tightly packed together and are avascular.
2. Connective tissues have relatively few cells with lots of extracellular material.

Anatomy Overview - Epithelial Tissues
Anatomy Overview - Connective Tissues

4.4 Epithelial Tissues

1. The subtypes of epithelial tissues include covering and lining epithelia and glandular epithelia.
2. Epithelial tissues consist mostly of cells with little extracellular material between adjacent plasma membranes. The apical, lateral, and basal surfaces of epithelial cells are modified in various ways to carry out specific functions. Epithelium is arranged in sheets and attached to a basement membrane. Although epithelial tissues are avascular, they have a nerve supply. The high rate of cell division gives epithelial tissues a high capacity for renewal.
3. Epithelial layers can be simple (one layer), pseudostratified (appears to be several layers but is actually a single layer), or stratified (several layers). The cell shapes may be squamous (flat), cuboidal (cubelike), columnar (rectangular), or transitional (variable). The subtypes of epithelia include covering and lining epithelia and glandular epithelia.
4. Simple squamous epithelium, which consists of a single layer of flat cells (Table 4.1), is found in parts of the body where filtration or diffusion is a priority process. One type, called endothelium, lines the heart and blood vessels. Another type, called mesothelium, forms the serous membranes that line the thoracic and abdominopelvic cavities and cover the organs within them.
5. Simple cuboidal epithelium consists of a single layer of cube-shaped cells that function in secretion and absorption (Table 4.1). It is found covering the ovaries, in the kidneys and eyes, and lining some glandular ducts.
6. Nonciliated simple columnar epithelium, a single layer of nonciliated rectangular cells (Table 4.1), lines most of the gastrointestinal tract and contains specialized cells that perform absorption and secrete mucus. Ciliated simple columnar epithelium, which consists of a single layer of ciliated rectangular cells (Table 4.1), is found in a few portions of the upper respiratory tract, where it moves foreign particles trapped in mucus out of the respiratory tract. A ciliated variety of pseudostratified columnar epithelium (Table 4.1) contains goblet cells and lines most of the upper respiratory tract; a nonciliated variety has no goblet cells and lines ducts of many glands, the epididymis, and part of the male urethra. The ciliated variety moves mucus in the respiratory tract. The nonciliated variety functions in absorption and protection.
7. Stratified epithelium consists of several layers of cells: Cells of the apical layer of stratified squamous epithelium and several layers deep to it are flat (Table 4.1); a nonkeratinized variety lines the mouth, and a keratinized variety forms the epidermis, the most superficial layer of the skin. Cells at the apical layer of stratified cuboidal epithelium are cube-shaped (Table 4.1); found in adult sweat glands and in a portion of the male urethra, stratified cuboidal epithelium protects and provides limited secretion and absorption. Cells of the apical layer of stratified columnar epithelium have a columnar shape (Table 4.1); this type of stratified epithelium is found in a portion of the male urethra and in large excretory ducts of some glands, and functions in protection and secretion.
8. Transitional epithelium consists of several layers of cells whose appearance varies with the degree of stretching (Table 4.1). It lines the urinary bladder.
9. A gland is a single cell or a group of epithelial cells adapted for secretion. There are two types of glands: endocrine and exocrine. Endocrine glands secrete hormones into interstitial fluid and then into the blood (Table 4.2). Exocrine glands (mucous, sweat, oil, and digestive glands) secrete into ducts or directly onto a free surface (Table 4.2).
10. The structural classification of exocrine glands includes unicellular and multicellular glands. The functional classification of exocrine glands includes merocrine, apocrine, and holocrine glands.

Anatomy Overview - Epithelial Tissues
Anatomy Overview - Epithelial
   Membranes
Figure 4.5 - Cell Shapes
   and Arrangement of
   Layers for Covering
   and Lining Epithelium

Review

4.5 Connective Tissues

1. Connective tissues, one of the most abundant body tissues, consist of relatively few cells and an abundant extracellular matrix of ground substance and protein fibers. They do not usually occur on free surfaces, they have a nerve supply (except for cartilage), and they are highly vascular (except for cartilage, tendons, and ligaments).
2. Cells in connective tissues are derived primarily from mesenchymal cells. Cell types include fibroblasts (secrete extracellular matrix), macrophages (perform phagocytosis), plasma cells (secrete antibodies), mast cells (produce histamine), adipocytes (store fat), and white blood cells (migrate from blood in response to infections).
3. The ground substance and fibers make up the extracellular matrix. The ground substance supports and binds cells together, provides a medium for the exchange of materials, stores water, and actively influences cell functions. Substances found in the ground substance include water and polysaccharides such as hyaluronic acid, chondroitin sulfate, dermatan sulfate, and keratan sulfate (glycosaminoglycans). Also present are proteoglycans and adhesion proteins.
4. The fibers in the extracellular matrix provide strength and support and are of three types: (a) Collagen fibers (composed of collagen) are found in large amounts in bone, tendons, and ligaments. (b) Elastic fibers (composed of elastin, fibrillin, and other glycoproteins) are found in skin, blood vessel walls, and lungs. (c) Reticular fibers (composed of collagen and glycoprotein) are found around fat cells, nerve fibers, and skeletal and smooth muscle cells.
5. The two major subclasses of connective tissues are embryonic connective tissues (found in the embryo and fetus) and mature connective tissues (present in the newborn). The embryonic connective tissues (summarized in Table 4.3) are mesenchyme, which forms almost all other connective tissues, and mucous connective tissue, found in the umbilical cord of the fetus, where it gives support. Mature connective tissues differentiate from mesenchyme and are subdivided into several types: loose or dense connective tissue, cartilage, bone tissue, and liquid connective tissue.
6. Loose connective tissue includes areolar connective tissue, adipose tissue, and reticular connective tissue. Areolar connective tissue consists of the three types of fibers (collagen, elastic, and reticular), several types of cells, and a semifluid ground substance (Table 4.4); it is found in the subcutaneous layer, in mucous membranes, and around blood vessels, nerves, and body organs. Adipose tissue consists of adipocytes, which store triglycerides (Table 4.4); it is found in the subcutaneous layer, around organs, and in yellow bone marrow. Brown adipose tissue (BAT) generates heat. Reticular connective tissue consists of reticular fibers and reticular cells and is found in the liver, spleen, and lymph nodes (Table 4.4).
7. Dense connective tissue includes dense regular connective tissue, dense irregular connective tissue, and elastic connective tissue. Dense regular connective tissue consists of parallel bundles of collagen fibers and fibroblasts (Table 4.5); it forms tendons, most ligaments, and aponeuroses. Dense irregular connective tissue usually consists of randomly arranged collagen fibers and a few fibroblasts (Table 4.5); it is found in fasciae, the dermis of skin, and membrane capsules around organs. Elastic connective tissue consists of branching elastic fibers and fibroblasts (Table 4.5) and is found in the walls of large arteries, lungs, trachea, and bronchial tubes.
8. Cartilage contains chondrocytes and has a rubbery extracellular matrix (chondroitin sulfate) containing collagen and elastic fibers. Hyaline cartilage, which consists of a gel-like ground substance and appears bluish white in the body, is found in the embryonic skeleton, at the ends of bones, in the nose, and in respiratory structures (Table 4.6); it is flexible, allows movement, provides support, and is usually surrounded by a perichondrium. Fibrocartilage is found in the pubic symphysis, intervertebral discs, and menisci (cartilage pads) of the knee joint (Table 4.6); it contains chondrocytes scattered among clearly visible bundles of collagen fibers. Elastic cartilage maintains the shape of organs such as the epiglottis of the larynx, auditory (eustachian) tubes, and external ear (Table 4.6); its chondrocytes are located within a threadlike network of elastic fibers, and it has a perichondrium.
9. Bone or osseous tissue consists of an extracellular matrix of mineral salts and collagen fibers that contribute to the hardness of bone, and osteocytes that are located in lacunae (Table 4.7). It supports and protects the body, provides a surface area for muscle attachment, helps the body move, stores minerals, and houses blood-forming tissue.
10. There are two types of liquid connective tissues: blood and lymph. Blood tissue consists of blood plasma and formed elements—red blood cells, white blood cells, and platelets (Table 4.8); its cells transport oxygen and carbon dioxide, carry on phagocytosis, participate in allergic reactions, provide immunity, and bring about blood clotting. Lymph, the extracellular fluid that flows in lymphatic vessels, is a clear fluid similar to blood plasma but with less protein.

Anatomy Overview - Connective Tissues
Figure 4.7 - Functional Classification of Multicellular Exocrine Glands
Figure 4.8 - Representative Cells and Fibers in Connective Tissues

| Review | Resource |

4.6 Membranes

1. An epithelial membrane consists of an epithelial layer overlying a connective tissue layer. Types include mucous, serous, and cutaneous membranes.
2. Mucous membranes line cavities that open to the exterior, such as the gastrointestinal tract.
3. Serous membranes line closed cavities (pleura, pericardium, peritoneum) and cover the organs in the cavities. These membranes consist of parietal and visceral layers.
4. The cutaneous membrane is the skin. It covers the entire body and consists of a superficial epidermis (epithelium) and a deep dermis (connective tissue).
5. Synovial membranes line joint cavities and consist of areolar connective tissue; they do not have an epithelial layer.

Anatomy Overview - Epithelial Membranes

4.7 Muscular Tissues

1. Muscular tissues consist of cells called muscle fibers or myocytes that are specialized for contraction. They provide motion, maintenance of posture, heat production, and protection.
2. Skeletal muscle tissue is attached to bones and is striated and voluntary (Table 4.9).
3. The action of cardiac muscle tissue, which forms most of the heart wall and is striated, is involuntary (Table 4.9).
4. Smooth muscle tissue is found in the walls of hollow internal structures (blood vessels and viscera) and is nonstriated and involuntary (Table 4.9).

Anatomy Overview - Muscle Tissue

4.8 Nervous Tissue

1. The nervous system is composed of neurons (nerve cells) and neuroglia (protective and supporting cells) (Table 4.10).
2. Neurons respond to stimuli by converting the stimuli into electrical signals called nerve action potentials (nerve impulses), and conducting nerve impulses to other cells.
3. Most neurons consist of a cell body and two types of processes, dendrites and axons.

Anatomy Overview - Nervous Tissue
Exercise - Concentrate on Tissue Function
Exercise - Name That Tissue
Exercise - Tissue Identification

4.9 Excitable Cells

1. Electrical excitability is the ability to respond to certain stimuli by producing electrical signals such as action potentials.
2. Because neurons and muscle fibers exhibit electrical excitability, they are considered excitable cells.

Anatomy Overview - Nervous Tissue

4.10 Tissue Repair: Restoring Homeostasis

1. Tissue repair is the replacement of worn-out, damaged, or dead cells by healthy ones.
2. Stem cells may divide to replace lost or damaged cells.
3. If the injury is superficial, tissue repair involves parenchymal regeneration; if damage is extensive, granulation tissue is involved.
4. Good nutrition and blood circulation are vital to tissue repair.

4.11 Aging and Tissues

1. Tissues heal faster and leave less obvious scars in the young than in the aged; surgery performed on fetuses leaves no scars.
2. The extracellular components of tissues, such as collagen and elastic fibers, also change with age.

SELF-QUIZ QUESTIONS

Fill in the blanks in the following statements.

1. The four types of tissues are _____ , _____ , _____ , and _____ .

2. Epithelial tissue tends to be classified according to two criteria: _____ and _____ .

Indicate whether the following statements are true or false.

3. Epithelial tissue cells have an apical surface at the top and are attached to a basement membrane at the bottom.

4. Connective tissue fibers that are arranged in bundles and lend strength and flexibility to a tissue are collagen fibers.

Choose the one best answer to the following questions.

5. Which of the following muscular tissues can be voluntarily controlled? (1) cardiac, (2), smooth, (3) skeletal.
   (a) 1, 2, and 3   (b) 2   (c) 1
   (d) 1 and 3   (e) 3

6. Which of the following tissues is avascular?
   (a) cardiac muscle   (b) stratified squamous epithelial
   (c) compact bone   (d) skeletal muscle
   (e) adipose

7. If the lining of an organ produces and releases mucus, which of the following cells would likely be found in the tissue lining the organ?
   (a) goblet cells   (b) mast cells   (c) macrophages
   (d) osteoblasts   (e) fibroblasts

8. Why does damaged cartilage heal slowly?
   (a) Damaged cartilage undergoes fibrosis, which interferes with the movement of materials needed for repair.
   (b) Cartilage does not contain fibroblasts, which are needed to produce the fibers in cartilage tissue.
   (c) Cartilage is avascular, so materials needed for repair must diffuse from surrounding tissue.
   (d) Chondrocytes cannot be replaced once they are damaged.
   (e) Chondrocytes undergo mitosis slowly, which delays healing.

9. Which of the following is *true* concerning serous membranes?
   (a) A serous membrane lines a body part that opens directly to the body's exterior.
   (b) The parietal portion of a serous membrane attaches to the organ.
   (c) The visceral portion of a serous membrane attaches to a body cavity wall.
   (d) The serous membrane covering the heart is known as the peritoneum.
   (e) The serous membrane covering the lungs is known as the pleura.

10. The type of exocrine gland that forms its secretory product and simply releases it from the cell by exocytosis is the
    (a) apocrine gland   (b) merocrine gland (c) holocrine gland
    (d) endocrine gland   (e) tubular gland.

11. Tissue changes that occur with aging can be due to (1) cross-links forming between glucose and proteins, (2) a decrease in the amount of collagen fibers, (3) a decreased blood supply, (4) improper nutrition, (5) a higher cellular metabolic rate.
    (a) 1, 2, 3, 4, and 5   (b) 1, 2, 3, and 4   (c) 1 and 4
    (d) 1, 3, and 4   (e) 1, 2, and 3

12. What type of cell junction would be required for cells to communicate with one another?
    (a) adherens junction   (b) desmosome   (c) gap junction
    (d) tight junction   (e) hemidesmosome

13. Match the following epithelial tissues to their descriptions:
    ____(a) contains a single layer of flat cells; found in the body where filtration (kidney) or diffusion (lungs) are priority processes
    ____(b) found in the superficial part of skin; provides protection from heat, microbes, and chemicals
    ____(c) contains cube-shaped cells functioning in secretion and absorption
    ____(d) lines the lower respiratory tract and uterine tubes; wavelike motion of cilia propels materials through the lumen
    ____(e) contains cells with microvilli and goblet cells; found in linings of the digestive, reproductive, and urinary tracts
    ____(f) found in the urinary bladder; contains cells that can change shape (stretch or relax)
    ____(g) contains cells that are all attached to the basement membrane, although some do not reach the surface; those cells that do extend to the surface secrete mucus or contain cilia
    ____(h) a fairly rare type of epithelium that has a mainly protective function

    (1) pseudostratified ciliated columnar epithelium
    (2) ciliated simple columnar epithelium
    (3) transitional epithelium
    (4) simple squamous epithelium
    (5) simple cuboidal epithelium
    (6) nonciliated simple columnar epithelium
    (7) stratified cuboidal epithelium
    (8) keratinized stratified squamous epithelium

14. For each of the following items, indicate the tissue type with which they are associated. Use **E** for epithelial tissues, **C** for connective tissues, **M** for muscle tissues, and **N** for nervous tissue.
    ____(a) bind, support
    ____(b) contains elongated cells that generate force
    ____(c) neuroglia
    ____(d) avascular
    ____(e) may contain fibroblasts
    ____(f) tightly packed cells
    ____(g) intercalated discs
    ____(h) goblet cells
    ____(i) contain extracellular matrix
    ____(j) striated
    ____(k) generate action potentials
    ____(l) cilia
    ____(m) ground substance
    ____(n) apical surface
    ____(o) excitable

15. Match the following connective tissues to their descriptions:

____(a) the tissue from which all other connective tissues eventually arise

____(b) connective tissue with a clear, liquid matrix that flows in lymphatic vessels

____(c) connective tissue consisting of several kinds of cells, containing all three fiber types randomly arranged, and found in the subcutaneous layer deep to the skin

____(d) a loose connective tissue specialized for triglyceride storage

____(e) tissue that contains reticular fibers and reticular cells and forms the stroma of certain organs such as the spleen

____(f) tissue with irregularly arranged collagen fibers found in the dermis of the skin

____(g) tissue found in the lungs that is strong and can recoil back to its original shape after being stretched

____(h) tissue that affords flexibility at joints and reduces joint friction

____(i) tissue that provides strength and rigidity and is the strongest of the three types of cartilage

____(j) bundles of collagen arranged in parallel patterns; compose tendons and ligaments

____(k) tissue that forms the internal framework of the body and works with skeletal muscle to generate movement

____(l) tissue that contains a network of elastic fibers, providing strength, elasticity, and maintenance of shape; located in the external ear

____(m) connective tissue with formed elements suspended in a liquid matrix called plasma

(1) blood
(2) fibrocartilage
(3) mesenchyme
(4) dense regular connective tissue
(5) lymph
(6) hyaline cartilage
(7) dense irregular connective tissue
(8) areolar connective tissue
(9) reticular connective tissue
(10) bone (osseous tissue)
(11) elastic connective tissue
(12) elastic cartilage
(13) adipose tissue

CRITICAL THINKING QUESTIONS

1. Imagine that you live 50 years in the future, and that you can custom-design a human to suit the environment. Your assignment is to customize the human's tissues so that the individual can survive on a large planet with gravity, a cold, dry climate, and a thin atmosphere. What adaptations would you incorporate into the structure and/or amount of tissues, and why?

2. You are entering a "Cutest Baby Contest" and have asked your colleagues to help you choose the most adorable picture of yourself as a baby. One of your colleagues rudely points out that you were quite chubby as an infant. You, however, are not offended. Explain to your colleague the benefit of that "baby fat."

3. You've been on a "bread-and-water" diet for three weeks and have noticed that a cut on your shin won't heal and bleeds easily. Why?

? ANSWERS TO FIGURE QUESTIONS

4.1 Epithelial tissues cover the body, line various structures, and form glands. Connective tissues protect, support, bind organs together, store energy, and help provide immunity. Muscular tissue contracts and generates force and heat. Nervous tissue detects changes in the environment and generates nerve impulses that activate muscular contraction and glandular secretion.

4.2 Gap junctions allow cellular communication via passage of electrical and chemical signals between adjacent cells.

4.3 Since epithelial tissues are avascular, they depend on the blood vessels in connective tissues for oxygen, nutrients, and waste disposal.

4.4 The basement membrane provides physical support for the epithelium and plays a part in growth and wound healing,
restriction of molecule movement between tissues, and blood filtration in the kidneys.

4.5 Because the cells are so thin, substances move most rapidly through squamous cells.

4.6 Simple multicellular exocrine glands have a nonbranched duct; compound multicellular exocrine glands have a branched duct.

4.7 Sebaceous (oil) glands are holocrine glands, and salivary glands are merocrine glands.

4.8 Fibroblasts secrete the fibers and ground substance of the extracellular matrix.

4.9 An epithelial membrane is a membrane that consists of an epithelial layer and an underlying layer of connective tissue.

5 THE INTEGUMENTARY SYSTEM

THE INTEGUMENTARY SYSTEM AND HOMEOSTASIS *The integumentary system contributes to homeostasis by protecting the body and helping regulate body temperature. It also allows you to sense pleasurable, painful, and other stimuli in your external environment.*

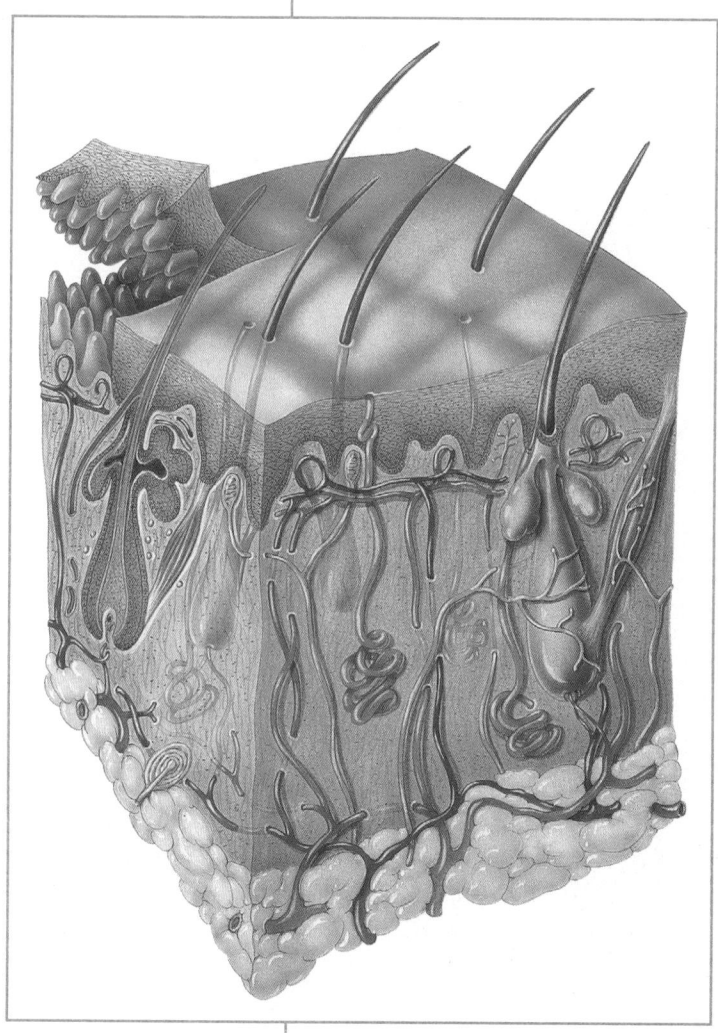

Recall from Chapter 1 that a system consists of a group of organs working together to perform specific activities. The **integumentary system** (in-teg-ū-MEN-tar-ē; *in-* = inward; *-tegere* = to cover) is composed of the skin, hair, oil and sweat glands, nails, and sensory receptors. The integumentary system helps maintain a constant body temperature, protects the body, and provides sensory information about the surrounding environment. Of all the body's organs, none is more easily inspected or more exposed to infection, disease, and injury than the skin. Although its location makes it vulnerable to damage from trauma, sunlight, microbes, and pollutants in the environment, the skin's protective features ward off such damage. Because of its visibility, skin reflects our emotions (frowning, blushing) and some aspects of normal physiology (such as sweating). Changes in skin color may also indicate homeostatic imbalances in the body. For example, the bluish skin color associated with hypoxia (oxygen deficiency at the tissue level) is one sign of heart failure as well as other disorders. Abnormal skin eruptions or rashes such as chickenpox, cold sores, or measles may reveal systemic infections or diseases of internal organs, while other conditions, such as warts, age spots, or pimples, may involve the skin alone. So important is the skin to self-image that many people spend a great deal of time and money to restore it to a more normal or youthful appearance. **Dermatology** (der′-ma-TOL-ō-jē; *dermato-* = skin; *-logy* = study of) is the medical specialty that deals with the diagnosis and treatment of integumentary system disorders.

? *Did you ever wonder why it is so difficult to save the life of someone with extensive third-degree burns?*

5.1 STRUCTURE OF THE SKIN

◉ **OBJECTIVES**

- Describe the layers of the epidermis and the cells that compose them.
- Compare the composition of the papillary and reticular regions of the dermis.
- Explain the basis for different skin colors.

The **skin,** also known as the **cutaneous membrane** (kū-TĀ-nē-us), covers the external surface of the body and is the largest organ of the body in both surface area and weight. In adults, the skin covers an area of about 2 square meters (22 square feet) and weighs 4.5–5 kg (10–11 lb), about 7% of total body weight. It ranges in thickness from 0.5 mm (0.02 in.) on the eyelids to 4.0 mm (0.16 in.) on the heels. Over most of the body it is 1–2 mm (0.04–0.08 in.) thick. The skin consists of two main parts (Figure 5.1). The superficial,

Figure 5.1 **Components of the integumentary system.** The skin consists of a superficial, thin epidermis and a deep, thicker dermis. Deep to the skin is the subcutaneous layer, which attaches the dermis to underlying fascia.

🔑 The integumentary system includes the skin, hair, oil and sweat glands, nails, and sensory receptors.

(a) Sectional view of skin and subcutaneous layer

thinner portion, which is composed of *epithelial tissue,* is the **epidermis** (ep'-i-DERM-is; *epi-* = above). The deeper, thicker *connective tissue* portion is the **dermis.** While the epidermis is avascular, the dermis is vascular. For this reason, if you cut the epidermis there is no bleeding, but if you cut the dermis there is bleeding.

Deep to the dermis, but not part of the skin, is the **subcutaneous (subQ) layer.** Also called the **hypodermis** (*hypo-* = below), this layer consists of areolar and adipose tissues. Fibers that extend from the dermis anchor the skin to the subcutaneous layer, which in turn attaches to underlying fascia, the connective tissue

around muscles and bones. The subcutaneous layer serves as a storage depot for fat and contains large blood vessels that supply the skin. This region (and sometimes the dermis) also contains nerve endings called **pacinian (lamellated) corpuscles** (pa-SIN-ē-an) that are sensitive to pressure (Figure 5.1).

Epidermis

The **epidermis** is composed of keratinized stratified squamous epithelium. It contains four principal types of cells: keratinocytes, melanocytes, Langerhans cells, and Merkel cells

(b) Sectional view of skin

(c) Epidermal ridges and sweat pores

(d) Sectional view of dermal papillae and epidermal ridges

FUNCTIONS OF THE INTEGUMENTARY SYSTEM

1. Regulates body temperature.
2. Stores blood.
3. Protects body from external environment.
4. Detects cutaneous sensations.
5. Excretes and absorbs substances.
6. Synthesizes vitamin D.

? **What types of tissues make up the epidermis and the dermis?**

(Figure 5.2). About 90% of epidermal cells are **keratinocytes** (ker-a-TIN-ō-sīts′; *keratino-* = hornlike; *-cytes* = cells), which are arranged in four or five layers and produce the protein **keratin** (KER-a-tin) (Figure 5.2a). Recall from Chapter 4 that keratin is a tough, fibrous protein that helps protect the skin and underlying tissues from abrasions, heat, microbes, and chemicals. Keratinocytes also produce lamellar granules, which release a water-repellent sealant that decreases water entry and loss and inhibits the entry of foreign materials.

About 8% of the epidermal cells are **melanocytes** (MEL-a-nō-sīts′; *melano-* = black), which develop from the ectoderm of a developing embryo and produce the pigment melanin (Figure 5.2b). Their long, slender projections extend between the keratinocytes and transfer melanin granules to them. **Melanin** (MEL-a-nin) is a yellow-red or brown-black pigment that contributes to skin color and absorbs damaging ultraviolet (UV) light. Once inside keratinocytes, the melanin granules cluster to form a protective veil over the nu-

cleus, on the side toward the skin surface. In this way, they shield the nuclear DNA from damage by UV light. Although their melanin granules effectively protect keratinocytes, melanocytes themselves are particularly susceptible to damage by UV light.

Langerhans cells (LANG-er-hans), also called *epidermal dendritic cells,* arise from red bone marrow and migrate to the epidermis (Figure 5.2c), where they constitute a small fraction of the epidermal cells. They participate in immune responses mounted against microbes that invade the skin, and are easily damaged by UV light. Their role in the immune response is to help other cells of the immune system recognize an invading microbe and destroy it.

Merkel cells (MER-kel) are the least numerous of the epidermal cells. They are located in the deepest layer of the epidermis, where they contact the flattened process of a sensory neuron (nerve cell), a structure called a **Merkel (tactile) disc** (Figure 5.2d). Merkel cells and their associated Merkel discs detect touch sensations.

Several distinct layers of keratinocytes in various stages of development form the epidermis (Figure 5.3). In most regions of the body the epidermis has four strata or layers—stratum basale, stratum spinosum, stratum granulosum, and a thin stratum corneum. This is called **thin skin.** Where exposure to friction is greatest, such as in the fingertips, palms, and soles, the epidermis has five layers—stratum basale, stratum spinosum, stratum granulosum, stratum lucidum, and a thick stratum corneum. This is called **thick skin.** The details of thin and thick skin are discussed later in the chapter.

Stratum Basale

The deepest layer of the epidermis is the **stratum basale** (ba-SA-lē; *basal* = base), composed of a single row of cuboidal or columnar keratinocytes. Some cells in this layer are *stem cells* that undergo cell division to continually produce new keratinocytes. The nuclei of keratinocytes in the stratum basale are large, and their cytoplasm contains many ribosomes, a small Golgi complex, a few mitochondria, and some rough endoplasmic reticulum. The cytoskeleton within keratinocytes of the stratum basale includes scattered intermediate filaments, called *keratin intermediate filaments (tonofilaments).* The keratin intermediate filaments form the tough protein keratin in its more superficial epidermal layers. Keratin protects the deeper layers from injury. Keratin intermediate filaments attach to desmosomes, which bind cells of the stratum basale to each other and to the cells of the adjacent stratum spinosum, and to hemidesmosomes, which bind the keratinocytes to the basement membrane positioned between the epidermis and the dermis. Melanocytes and Merkel cells with their associated Merkel discs are scattered among the keratinocytes of the basal layer. The stratum basale is also known as the **stratum germinativum** (jer′-mi-na-TĒ-vum; *germ-* = sprout) to indicate its role in forming new cells.

Figure 5.2 Cells in the epidermis. Besides keratinocytes, the epidermis contains melanocytes, which produce the pigment melanin; Langerhans cells, which participate in immune responses; and Merkel cells, which function in the sensation of touch.

Most of the epidermis consists of keratinocytes, which produce the protein keratin (protects underlying tissues) and lamellar granules (contain a waterproof sealant).

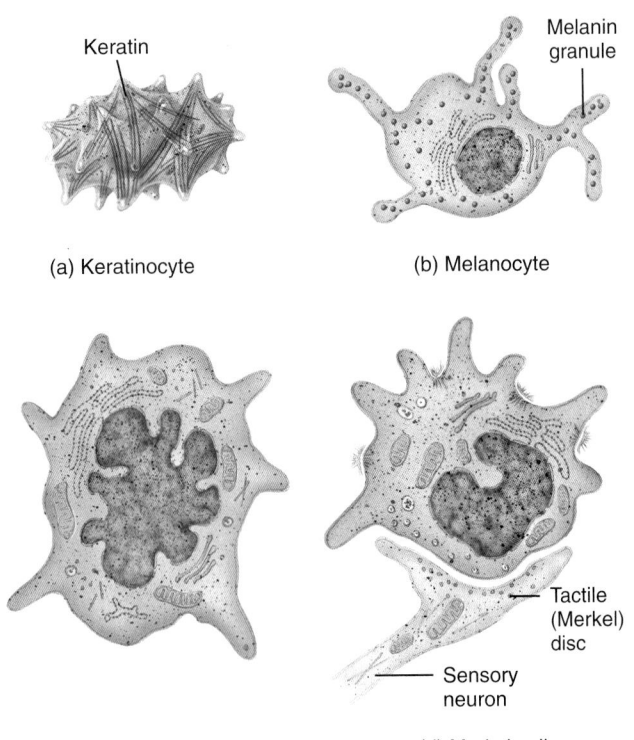

(a) Keratinocyte

(b) Melanocyte

(c) Langerhans cell

(d) Merkel cell

? **What is the function of melanin?**

New skin cannot regenerate if an injury destroys a large area of the stratum basale and its stem cells. Skin wounds of this magnitude require skin grafts in order to heal. A **skin graft** is the transfer of a

Figure 5.3 Layers of the epidermis.

The epidermis consists of keratinized stratified squamous epithelium.

(a) Four principal cell types in epidermis

(b) Photomicrograph of a portion of the skin

 Which epidermal layer includes stem cells that continually undergo cell division?

patch of healthy skin taken from a donor site to cover a wound. A skin graft is performed to protect against fluid loss and infection, to promote tissue healing, to reduce scar formation, to prevent loss of function, and for cosmetic reasons. To avoid tissue rejection, the transplanted skin is usually taken from the same individual (*autograft*) or an identical twin (*isograft*). If skin damage is so extensive that an autograft would cause harm, a self-donation procedure called *autologous skin transplantation* (aw-TOL-ō-gus) may be used. In this procedure, performed most often for severely burned patients, small amounts of an individual's epidermis are removed, and the keratinocytes are cultured in the laboratory to produce thin sheets of skin. The new skin is transplanted back to the patient so that it covers the burn wound and generates a permanent skin. Also available as skin grafts for wound coverage are products (Apligraft and Transite) grown in the laboratory from the foreskins of circumcised infants. •

Stratum Spinosum

Superficial to the stratum basale is the **stratum spinosum** (spi-NŌ-sum; *spinos-* = thornlike). This stratum mainly consists of numerous keratinocytes arranged in 8–10 layers. Cells in the more superficial layers become somewhat flattened. The keratinocytes in the stratum spinosum, which are produced by the stem cells in

the basal layer, have the same organelles as cells of the stratum basale and some retain their ability to divide. The keratinocytes of this layer produce coarser bundles of keratin in intermediate filaments than those of the basal layer. Although they are rounded and larger in living tissue, cells of the stratum spinosum shrink and pull apart when prepared for microscopic examination so that they appear to be covered with thornlike spines (thus, the name) (Figure 5.3a). At each spinelike projection, bundles of keratin intermediate filaments insert into desmosomes, which tightly join the cells to one another. This arrangement provides both strength and flexibility to the skin. Langerhans cells and projections of melanocytes are also present in the stratum spinosum.

Stratum Granulosum

At about the middle of the epidermis, the **stratum granulosum** (gran-ū-LŌ-sum; *granulos-* = little grains) consists of three to five layers of flattened keratinocytes that are undergoing apoptosis. (Recall from Chapter 3 that *apoptosis* is an orderly, genetically programmed cell death in which the nucleus fragments before the cells die.) The nuclei and other organelles of these cells begin to degenerate as they move farther from their source of nutrition (the dermal blood vessels). Even though keratin intermediate filaments are no longer being produced by these cells, they become more apparent because the organelles in the cells are regressing. A distinctive feature of cells in this layer is the presence of darkly staining granules of a protein called **keratohyalin** (ker'-a-tō-HĪ-a-lin), which assembles keratin intermediate filaments into keratin. Also present in the keratinocytes are membrane-enclosed **lamellar granules** (la-MEL-ar), which fuse with the plasma membrane and release a lipid-rich secretion. This secretion is deposited in the spaces between cells of the stratum granulosum, stratum lucidum, and stratum corneum. The lipid-rich secretion acts as a water-repellent sealant, retarding loss and entry of water and entry of foreign materials. As their nuclei break down during apoptosis, the keratinocytes of the stratum granulosum can no longer carry on vital metabolic reactions, and they die. Thus, the stratum granulosum marks the transition between the deeper, metabolically active strata and the dead cells of the more superficial strata.

Stratum Lucidum

The **stratum lucidum** (LOO-si-dum; *lucid-* = clear) is present only in the thick skin of areas such as the fingertips, palms, and soles. It consists of four to six layers of flattened clear, dead keratinocytes that contain large amounts of keratin and thickened plasma membranes. This probably provides an additional level of toughness in this region of thick skin.

Stratum Corneum

The **stratum corneum** (KOR-nē-um; *corne-* = horn or horny) consists on average of 25 to 30 layers of flattened dead keratinocytes, but can range in thickness from a few cells in thin skin to 50 or more cell layers in thick skin. The cells are extremely thin, flat, plasma membrane–enclosed packages of keratin that no longer contain a nucleus or any internal organelles. They are the final product of the differentiation process of the keratinocytes. The cells within each layer overlap one another like the scales on the skin of a snake. Neighboring layers of cells also form strong connections with one another. The plasma membranes of adjacent cells are arranged in complex, wavy folds that fit together like pieces of a jigsaw puzzle to hold the layers together. In this outer stratum of the epidermis, cells are continuously shed and replaced by cells from the deeper strata. Its multiple layers of dead cells help the stratum corneum to protect deeper layers from injury and microbial invasion. Constant exposure of skin to friction stimulates increased cell production and keratin production that results in the formation of a **callus,** an abnormal thickening of the stratum corneum.

Keratinization and Growth of the Epidermis

Newly formed cells in the stratum basale are slowly pushed to the surface. As the cells move from one epidermal layer to the next, they accumulate more and more keratin, a process called **keratinization** (ker'-a-tin-i-ZĀ-shun). Then they undergo apoptosis. Eventually the keratinized cells slough off and are replaced by underlying cells that in turn become keratinized. The whole process by which cells form in the stratum basale, rise to the surface, become keratinized, and slough off takes about four to six weeks in an average epidermis of 0.1 mm (0.004 in.) thickness. Nutrients and oxygen diffuse to the avascular epidermis from blood vessels in the dermis. The epidermal cells of the stratum basale are closest to these blood vessels and receive most of the nutrients and oxygen. These cells are the most active metabolically and continuously undergo cell division to produce new keratinocytes. As the new keratinocytes are pushed farther from the blood supply by continuing cell division, the epidermal strata above the basale receive fewer nutrients and the cells become less active and eventually die. The rate of cell division in the stratum basale increases when the outer layers of the epidermis are stripped away, as occurs in abrasions and burns. The mechanisms that regulate this remarkable growth are not well understood, but hormonelike proteins such as **epidermal growth factor (EGF)** play a role. An excessive amount of keratinized cells shed from the skin of the scalp is called **dandruff.**

Table 5.1 summarizes the distinctive features of the epidermal strata.

CLINICAL CONNECTION | *Psoriasis*

Psoriasis (sō-RĪ-a-sis) is a common and chronic skin disorder in which keratinocytes divide and move more quickly than normal from the stratum basale to the stratum corneum. They are shed prematurely in as little as 7 to 10 days. The immature keratinocytes make an abnormal keratin, which forms flaky, silvery scales at the skin surface, most often on the knees, elbows, and scalp (dandruff). Effective treatments—various topical ointments and ultraviolet phototherapy—suppress cell division, decrease the rate of cell growth, or inhibit keratinization. •

TABLE 5.1

Summary of Epidermal Strata (see Figure 5.3)

STRATUM	DESCRIPTION
Basale	Deepest layer, composed of single row of cuboidal or columnar keratinocytes that contain scattered keratin intermediate filaments (tonofilaments); stem cells undergo cell division to produce new keratinocytes; melanocytes and Merkel cells associated with Merkel discs are scattered among keratinocytes.
Spinosum	Eight to ten rows of many-sided keratinocytes with bundles of keratin intermediate filaments; contains armlike processes of melanocytes and Langerhans cells.
Granulosum	Three to five rows of flattened keratinocytes, in which organelles are beginning to degenerate; cells contain the protein keratohyalin (converts keratin intermediate filaments into keratin) and lamellar granules (release lipid-rich, water-repellent secretion).
Lucidum	Present only in skin of fingertips, palms, and soles; consists of four to six rows of clear, flat, dead keratinocytes with large amounts of keratin.
Corneum	Twenty-five to thirty rows of dead, flat keratinocytes that contain mostly keratin.

Dermis

The second, deeper part of the skin, the **dermis,** is composed of dense irregular connective tissue containing collagen and elastic fibers. This woven network of fibers has great tensile strength (resists pulling or stretching forces). The dermis also has the ability to stretch and recoil easily. It is much thicker than the epidermis, and this thickness varies from region to region in the body, reaching its greatest thickness on the palms and soles. Leather, which we use for belts, shoes, baseball gloves, and basketballs, is the dried and treated dermis of other animals. The few cells present in the dermis include predominantly fibroblasts, with some macrophages, and a few adipocytes near its boundary with the subcutaneous layer. Blood vessels, nerves, glands, and hair follicles (epithelial invaginations of the epidermis) are embedded in the dermal layer. The dermis is essential to the survival of the epidermis, and these adjacent layers form many important structural and functional relations. Based on its tissue structure, the dermis can be divided into a thin superficial papillary region and a thick deeper reticular region.

The **papillary region** makes up about one-fifth of the thickness of the total layer (see Figure 5.1). It consists of thin collagen and fine elastic fibers. Its surface area is greatly increased by **dermal papillae** (pa-PIL-ē = nipples), small, nipple-shaped structures that project into the undersurface of the epidermis. All dermal papillae contain **capillary loops** (blood vessels). Some also contain tactile receptors called **Meissner corpuscles** (MĪS-ner) or *corpuscles of touch,* nerve endings that are sensitive to touch. Still other dermal papillae also contain **free nerve endings,** dendrites that lack any apparent structural specialization. Different free nerve endings initiate signals that give rise to sensations of warmth, coolness, pain, tickling, and itching.

The **reticular region** (*reticul-* = netlike), which is attached to the subcutaneous layer, consists of bundles of thick collagen fibers, scattered fibroblasts, and various wandering cells (such as macrophages). Some adipose cells can be present in the deepest part of the layer, along with some coarse elastic fibers (see Figure 5.1). The collagen fibers in the reticular region are arranged in a netlike manner and have a more regular arrangement than those in the papillary region. The more regular orientation of the large collagen fibers helps the skin resist stretching. Blood vessels, nerves, hair follicles, sebaceous (oil) glands, and sudoriferous (sweat) glands occupy the spaces between fibers.

The combination of collagen and elastic fibers in the reticular region provides the skin with strength, **extensibility** (ek-sten′-si-BIL-i-tē), the ability to stretch, and **elasticity** (e-las-TISS-i-tē), the ability to return to original shape after stretching. The extensibility of skin can be readily seen around joints and in pregnancy and obesity.

CLINICAL CONNECTION | *Stretch Marks*

Because of the collagenous, vascular structure of the dermis, **striae** (STRĪ-ē = streaks) or **stretch marks,** a form of internal scarring, can result from the internal damage to this layer that occurs when the skin is stretched too much. When the skin is overstretched, the lateral bonding between adjacent collagen fibers is disrupted and small dermal blood vessels rupture. This is why stretch marks initially appear as reddish streaks at these sites. Later, after scar tissue (which is poorly vascularized) forms at these sites of dermal breakdown, the stretch marks appear as silvery white streaks. Stretch marks often occur in the abdominal skin during pregnancy, on the skin of weight-lifters where the skin is stretched by a rapid increase in muscle mass, and in the stretched skin accompanying gross obesity. •

The surfaces of the palms, fingers, soles, and toes have a series of ridges and grooves. They appear either as straight lines or as a pattern of loops and whorls, as on the tips of the digits. These **epidermal ridges** are produced during the third month of fetal development as downward projections of the epidermis into the dermis between the dermal papillae of the papillary region (see Figure 5.1). The epidermal ridges create a strong bond between the epidermis and dermis in a region of high mechanical stress. The epidermal ridges also increase the surface area of the epidermis and thus increase the grip of the hand or foot by increasing friction. Finally, the epidermal ridges greatly increase surface area, which increases the number of Meissner corpuscles and thus increases tactile sensitivity. Because the ducts of sweat glands open on the tops of the epidermal ridges as sweat pores, the sweat and ridges form **fingerprints** (or **footprints**) on touching a smooth object. The epidermal ridge pattern is in part genetically determined and is unique for each individual. Even identical twins have different patterns. Normally, the ridge pattern does not change during life, except to enlarge, and thus can serve as the basis for identification. The study of the pattern of epidermal ridges is called **dermatoglyphics** (der′-ma-tō-GLIF-iks; *glyphe* = carved work).

In addition to forming epidermal ridges, the complex papillary surface of the dermis has other functional properties. The dermal papillae greatly increase the surface contact between the dermis and epidermis. This increased dermal contact surface, with its extensive network of small blood vessels, serves as an important source of nutrition for the overlying epidermis. Molecules diffuse from the small blood capillaries in the dermal papillae to the cells of the stratum basale, allowing the basal epithelial stem cells to divide and the keratinocytes to grow and develop. As keratinocytes push toward the surface and away from the dermal blood source, they are no longer able to obtain the nutrition they require, leading to the eventual breakdown of their organelles.

The dermal papillae fit together with the complementary epidermal ridge to form an extremely strong junction between the two layers. This jigsaw puzzle–like connection strengthens the skin against shearing forces (forces that laterally shift in relation to each other) that attempt to separate the epidermis from the dermis.

In certain regions of the body, collagen fibers within the reticular region tend to orient more in one direction than another because of natural tension experienced by these regions of the skin resulting from bony projections, orientation of muscles, and movements of joints. **Tension lines (lines of cleavage)** in the skin indicate the predominant direction of underlying collagen fibers. Knowledge of tension lines is especially important to plastic surgeons. For example, a surgical incision running parallel to the collagen fibers will heal with only a fine scar. A surgical incision made across the rows of fibers disrupts the collagen, and the wound tends to gape open and heal in a broad, thick scar. •

Table 5.2 summarizes the structural features of the papillary and reticular regions of the dermis.

The Structural Basis of Skin Color

Melanin, hemoglobin, and carotene are three pigments that impart a wide variety of colors to skin. The amount of **melanin** causes the skin's color to vary from pale yellow to reddish-brown to black. The difference between the two forms of melanin, *pheomelanin* (fē-ō-MEL-a-nin) (yellow to red) and *eumelanin* (ū-MEL-a-nin) (brown to black), is most apparent in the hair. Melanocytes, the melanin-producing cells, are most plentiful in the epidermis of the penis, nipples of the breasts, area just around the nipples (areolae), face, and limbs. They are also present in mucous membranes. Because the *number* of melanocytes is about the same in all people, differences in skin color are due mainly to the *amount of pigment* the melanocytes produce and transfer to keratinocytes. In some people who are genetically predisposed, melanin accumulates in patches called *freckles*. Freckles typically are reddish or brown and tend to be more visible in the summer than the winter. As a person ages, *age (liver) spots* may develop. These flat blemishes have nothing to do with the liver. They look like freckles and range in color from light brown to black. Like freckles, age spots are accumulations of melanin. Age spots are darker than freckles and build up over time due to exposure to sunlight. Age spots do not fade away during the winter months and are more common in adults over forty. A round, flat, or raised area that represents a benign localized overgrowth of melanocytes and usually develops in childhood or adolescence is called a **nevus** (NĒ-vus), or a **mole.**

Melanocytes synthesize melanin from the amino acid *tyrosine* in the presence of an enzyme called *tyrosinase*. Synthesis occurs in an organelle called a **melanosome** (MEL-an-ō-sōm). Exposure to ultraviolet (UV) light increases the enzymatic activity within melanosomes and thus increases melanin production. Both the amount and darkness of melanin increase on UV exposure, which gives the skin a tanned appearance and helps protect the body against further UV radiation. Melanin absorbs UV radiation, prevents damage to DNA in epidermal cells, and neutralizes free radicals that form in the skin following damage by UV radiation. Thus, within limits, melanin serves a protective function. In response to DNA damage, melanin production increases. As you will see later, exposing the skin to a *small* amount of UV light is actually necessary for the skin to begin the process of vitamin D synthesis. However, repeatedly exposing the skin to a *large* amount of UV light may cause skin cancer. A tan is lost when the melanin-containing keratinocytes are shed from the stratum corneum.

Albinism (AL-bin-izm; *albin-* = white) is the inherited inability of an individual to produce melanin. Most **albinos** (al-BĪ-nōs), people affected by albinism, have melanocytes that are unable to synthesize tyrosinase. Melanin is missing from their hair, eyes, and skin. This results in problems with vision and a tendency of the skin to burn easily on overexposure to sunlight.

In another condition, called **vitiligo** (vit-i-LĪ-gō), the partial or complete loss of melanocytes from patches of skin produces irregular white spots. The loss of melanocytes may be related to an immune system malfunction in which antibodies attack the melanocytes. •

Dark-skinned individuals have large amounts of melanin in the epidermis, so their skin color ranges from yellow to reddish-brown to black. Light-skinned individuals have little melanin in the epidermis. Thus, the epidermis appears translucent, and skin color ranges from pink to red depending on the oxygen content of the blood moving through capillaries in the dermis. The red color is due to **hemoglobin** (hē-mō-GLŌ-bin), the oxygen-carrying pigment in red blood cells.

Carotene (KAR-ō-tēn; *carot* = carrot) is a yellow-orange pigment that gives egg yolks and carrots their color. This precursor of vitamin A, which is used to synthesize pigments needed for vision, is stored in the stratum corneum and fatty areas of the dermis and subcutaneous layer in response to excessive dietary intake. In fact, so much carotene may be deposited in the skin after eating large amounts of carotene-rich foods that the skin actually turns orange, which is especially apparent in light-skinned individuals. Decreasing carotene intake eliminates the problem.

TABLE 5.2

Summary of Papillary and Reticular Regions of the Dermis (see Figure 5.1b)

REGION	DESCRIPTION
Papillary	Superficial portion of dermis (about one-fifth); consists of dense irregular connective tissue with thin collagen and fine elastic fibers; contains dermal ridges that house capillaries, Meissner corpuscles, and free nerve endings.
Reticular	Deeper portion of dermis (about four-fifths); consists of dense irregular connective tissue with bundles of thick collagen and some coarse elastic fibers. Spaces between fibers contain some adipose cells, hair follicles, nerves, sebaceous glands, and sudoriferous glands.

The color of skin and mucous membranes can provide clues for diagnosing certain conditions. When blood is not picking up an adequate amount of oxygen from the lungs, as in someone who has stopped breathing, the mucous membranes, nail beds, and skin appear bluish or **cyanotic** (sī-a-NOT-ik; *cyan-* = blue). **Jaundice** (JON-dis; *jaund-* = yellow) is due to a buildup of the yellow pigment bilirubin in the skin. This condition gives a yellowish appearance to the skin and the whites of the eyes, and usually indicates liver disease. **Erythema** (er-e-THĒ-ma; *eryth-* = red), redness of the skin, is caused by engorgement of capillaries in the dermis with blood due to skin injury, exposure to heat, infection, inflammation, or allergic reactions. **Pallor** (PAL-or), or paleness of the skin, may occur in conditions such as shock and anemia. All skin color changes are observed most readily in people with lighter-colored skin and may be more difficult to discern in people with darker skin. However, examination of the nail beds and gums can provide some information about circulation in individuals with darker skin. •

Tattooing and Body Piercing

Tattooing is a permanent coloration of the skin in which a foreign pigment is deposited with a needle into the dermis. It is believed that the practice originated in ancient Egypt between 4000 and 2000 B.C. Today, tattooing is performed in one form or another by nearly all peoples of the world, and it is estimated that about one in five U.S. college students has one or more tattoos. They are created by injecting ink with a needle that punctures the epidermis, moves between 50 and 3000 times per minute, and deposits the ink in the dermis. Since the dermis is stable (unlike the epidermis, which is shed about every four to six weeks), tattoos are permanent. However, they can fade over time due to exposure to sunlight, improper healing, picking scabs, and flushing away of ink particles by the lymphatic system. Tattoos can be removed by lasers, which use concentrated beams of light. In the procedure, which requires a series of treatments, the tattoo inks and pigments selectively absorb the high-intensity laser light without destroying normal surrounding skin tissue. The laser causes the tattoo to dissolve into small ink particles that are eventually removed by the immune system. Laser removal of tattoos involves a considerable investment in time and money and can be quite painful.

Body piercing, the insertion of jewelry through an artificial opening, is also an ancient practice employed by Egyptian pharaohs and Roman soldiers, and a current tradition among many Americans. Today it is estimated that about one in three U.S. college students has had a body piercing. For most piercing locations, the piercer cleans the skin with an antiseptic, retracts the skin with forceps, and pushes a needle through the skin. Then the jewelry is connected to the needle and pushed through the skin. Total healing can take up to a year. Among the sites that are pierced are the ears, nose, eyebrows, lips, tongue, nipples, navel, and genitals. Potential complications of body piercing are infections, allergic reactions, and anatomical damage (such as nerve damage or cartilage deformation). In addition, body piercing jewelry may interfere with certain medical procedures such as masks used for resuscitation, airway management procedures, urinary catheterization, radiographs, and delivery of a baby.

✔ **CHECKPOINT**

1. What structures are included in the integumentary system?
2. How does the process of keratinization occur?
3. What are the structural and functional differences between the epidermis and dermis?
4. How are epidermal ridges formed?
5. What are the three pigments in the skin and how do they contribute to skin color?
6. What is a tattoo? What are some potential problems associated with body piercing?

5.2 ACCESSORY STRUCTURES OF THE SKIN

◉ **OBJECTIVE**

• Contrast the structure, distribution, and functions of hair, skin glands, and nails.

Accessory structures of the skin—hair, skin glands, and nails—develop from the embryonic epidermis. They have a host of important functions. For example, hair and nails protect the body, and sweat glands help regulate body temperature.

Hair

Hairs, or *pili* (PĪ-lī), are present on most skin surfaces except the palms, palmar surfaces of the fingers, the soles, and plantar surfaces of the feet. In adults, hair usually is most heavily distributed across the scalp, in the eyebrows, in the axillae (armpits), and around the external genitalia. Genetic and hormonal influences largely determine the thickness and the pattern of hair distribution.

Although the protection it offers is limited, hair on the head guards the scalp from injury and the sun's rays. It also decreases heat loss from the scalp. Eyebrows and eyelashes protect the eyes from foreign particles, similar to the way hair in the nostrils and in the external ear canal defends those structures. Touch receptors (hair root plexuses) associated with hair follicles are activated whenever a hair is moved even slightly. Thus, hairs also function in sensing light touch.

Anatomy of a Hair

Each hair is composed of columns of dead, keratinized epidermal cells bonded together by extracellular proteins. The **shaft** is the superficial portion of the hair, which projects above the surface of the skin (Figure 5.4a). The **root** is the portion of the hair deep to the shaft that penetrates into the dermis, and sometimes into the subcutaneous layer. The shaft and root of the hair both consist of three concentric layers of cells: medulla, cortex, and cuticle of the hair (Figure 5.4c, d). The inner *medulla*, which may be lacking in thinner hair, is composed of two or three rows of irregularly shaped cells that contain large amounts of pigment granules in

Figure 5.4 Hair.

Hairs are growths of epidermis composed of dead, keratinized epidermal cells.

(b) Several hair shafts showing the shinglelike cuticle cells

SEM 70x

(a) Hair and surrounding structures

(c) Frontal and transverse sections of hair root

(d) Transverse section of hair root

? **Why does it hurt when you pluck a hair out but not when you have a haircut?**

dark hair, small amounts of pigment granules in gray hair, and a lack of pigment granules and the presence of air bubbles in white hair. The middle *cortex* forms the major part of the shaft and consists of elongated cells. The *cuticle of the hair,* the outermost layer, consists of a single layer of thin, flat cells that are the most heavily keratinized. Cuticle cells on the shaft are arranged like shingles on the side of a house, with their free edges pointing toward the end of the hair (Figure 5.4b).

Surrounding the root of the hair is the **hair follicle** (FOL-i-kul), which is made up of an external root sheath and an internal root sheath, together referred to as an **epithelial root sheath** (Figure 5.4c, d). The *external root sheath* is a downward continuation of the epidermis. The *internal root sheath* is produced by the matrix (described shortly) and forms a cellular tubular sheath of epithelium between the external root sheath and the hair. The dense dermis surrounding the hair follicle is called the **dermal root sheath.**

The base of each hair follicle and its surrounding dermal root sheath is an onion-shaped structure, the **bulb** (Figure 5.4c). This structure houses a nipple-shaped indentation, the **papilla of the hair,** which contains areolar connective tissue and many blood vessels that nourish the growing hair follicle. The bulb also contains a germinal layer of cells called the **hair matrix.** The hair matrix cells arise from the stratum basale, the site of cell division. Hence, hair matrix cells are responsible for the growth of existing hairs, and they produce new hairs when old hairs are shed. This replacement process occurs within the same follicle. Hair matrix cells also give rise to the cells of the internal root sheath.

CLINICAL CONNECTION | Hair Removal

A substance that removes hair is called a **depilatory** (de-PIL-a-tō-rē). It dissolves the protein in the hair shaft, turning it into a gelatinous mass that can be wiped away. Because the hair root is not affected, regrowth of the hair occurs. In **electrolysis,** an electric current is used to destroy the hair matrix so the hair cannot regrow. **Laser treatments** may also be used to remove hair. •

Sebaceous (oil) glands (discussed shortly) and a bundle of smooth muscle cells are also associated with hairs (Figure 5.4a). The smooth muscle is the **arrector pili** (a-REK-tor PĪ-lī; *arrect-* = to raise). It extends from the superficial dermis of the skin to the dermal root sheath around the side of the hair follicle. In its normal position, hair emerges at an angle to the surface of the skin. Under physiological or emotional stress, such as cold or fright, autonomic nerve endings stimulate the arrector pili muscles to contract, which pulls the hair shafts perpendicular to the skin surface. This action causes "goose bumps" or "gooseflesh" because the skin around the shaft forms slight elevations.

Surrounding each hair follicle are dendrites of neurons that form a **hair root plexus** (PLEK-sus), which is sensitive to touch (Figure 5.4a). The hair root plexuses generate nerve impulses if their hair shafts are moved.

Hair Growth

Each hair follicle goes through a growth cycle, which consists of a growth stage, a regression stage, and a resting stage. During the **growth stage,** cells of the hair matrix divide. As new cells from the hair matrix are added to the base of the hair root, existing cells of the hair root are pushed upward and the hair grows longer. While the cells of the hair are being pushed upward, they become keratinized and die. Following the growth stage is the **regression stage,** when the cells of the hair matrix stop dividing, the hair follicle atrophies (shrinks), and the hair stops growing. After the regression stage, the hair follicle enters a **resting stage.** Following the resting stage, a new growth cycle begins. The old hair root falls out or is pushed out of the hair follicle, and a new hair begins to grow in its place. Scalp hair is in the growth stage for 2 to 6 years, the regression stage for 2 to 3 weeks, and the resting stage for about 3 months. At any time, about 85% of scalp hairs are in the growth stage. Visible hair is dead, but until the hair is pushed out of its follicle by a new hair, portions of its root within the scalp are alive.

Normal hair loss in the adult scalp is about 70–100 hairs per day. Both the rate of growth and the replacement cycle may be altered by illness, radiation therapy, chemotherapy, age, genetics, gender, and severe emotional stress. Rapid weight-loss diets that severely restrict calories or protein increase hair loss. The rate of shedding also increases for three to four months after childbirth. **Alopecia** (al′-ō-PĒ-shē-a), the partial or complete lack of hair, may result from genetic factors, aging, endocrine disorders, chemotherapy, or skin disease.

CLINICAL CONNECTION | Chemotherapy and Hair Loss

Chemotherapy is the treatment of disease, usually cancer, by means of chemical substances or drugs. Chemotherapeutic agents interrupt the life cycle of rapidly dividing cancer cells. Unfortunately, the drugs also affect other rapidly dividing cells in the body, such as the hair matrix cells of a hair. It is for this reason that individuals undergoing chemotherapy experience hair loss. Since about 15% of the hair matrix cells of scalp hairs are in the resting stage, these cells are not affected by chemotherapy. Once chemotherapy is stopped, the hair matrix cells replace lost hair follicles and hair growth resumes. •

Types of Hairs

Hair follicles develop at about twelve weeks after fertilization. Usually by the fifth month of development, the follicles produce very fine, nonpigmented, downy hairs called **lanugo** (la-NOO-gō = wool or down) that cover the body of the fetus. Prior to birth, the lanugo of the eyebrows, eyelashes, and scalp are shed and replaced by long, coarse, heavily pigmented hairs called **terminal hairs.** The lanugo of the rest of the body are replaced by **vellus hairs** (VEL-us = fleece), commonly called "peach fuzz," which are short, fine, pale hairs that are barely visible to the naked eye. During childhood, vellus hairs cover most of the body except for the hairs of the eyebrows, eyelashes, and scalp, which are terminal hairs. In response to hormones (androgens) secreted at puberty, terminal hairs replace vellus hairs in the axillae (armpits) and pubic regions of boys and girls and they replace vellus hairs on the face, limbs, and chests of boys, which leads to the formation of a mustache, a beard, hairy arms and legs, and a hairy chest. During adulthood, about 95% of body hair on males is terminal hair and 5% is vellus hair; on females, about 35% of body hair is terminal hair and 65% is vellus hair.

Hair Color

The color of hair is due primarily to the amount and type of melanin in its keratinized cells. Melanin is synthesized by melanocytes scattered in the matrix of the bulb and passes into cells of the cortex and medulla of the hair (Figure 5.4c). Dark-colored hair contains mostly eumelanin (brown to black); blond and red hair contain variants of pheomelanin (yellow to red). Hair becomes gray because of a progressive decline in melanin production; gray hair contains only a few melanin granules. White hair results from the lack of melanin and the accumulation of air bubbles in the shaft.

Hair coloring is a process that adds or removes pigment. Temporary hair dyes coat the surface of a hair shaft and usually wash out within 2 or 3 shampoos. Semipermanent dyes penetrate the hair shaft moderately and do fade and wash out of hair after about 5 to 10 shampoos. Permanent hair dyes penetrate deeply into the hair shaft and don't wash out but are eventually lost as the hair grows out.

CLINICAL CONNECTION | *Hair and Hormones*

At puberty, when the testes begin secreting significant quantities of androgens (masculinizing sex hormones), males develop the typical male pattern of hair growth throughout the body, including a beard and a hairy chest. In females at puberty, the ovaries and the adrenal glands produce small quantities of androgens, which promote hair growth throughout the body including the axillae and pubic region. Occasionally, a tumor of the adrenal glands, testes, or ovaries produces an excessive amount of androgens. The result in females or prepubertal males is **hirsutism** (HER-soo-tizm; *hirsut-* = shaggy), excessive body hair or body hair in areas that usually are not hairy.

Surprisingly, androgens also must be present for occurrence of the most common form of baldness, **androgenic alopecia** (an'-drō-JEN-ik al'-ō-PĒ-shē-a) or **male-pattern baldness.** In genetically predisposed adults, androgens inhibit hair growth. In men, hair loss usually begins with a receding hairline followed by hair loss in the temples and crown. Women are more likely to have thinning of hair on top of the head. The first drug approved for enhancing scalp hair growth was minoxidil (Rogaine). It causes vasodilation (widening of blood vessels), thus increasing circulation; direct stimulation of hair follicle cells to pass into growth stage follicles; and inhibition of androgens. In about a third of the people who try it, minoxidil improves hair growth, causing scalp follicles to enlarge and lengthening the growth cycle. For many, however, the hair growth is meager. Minoxidil does not help people who already are bald. •

Skin Glands

Recall from Chapter 4 that glands are epithelial cells that secrete a substance. Several kinds of exocrine glands are associated with the skin: sebaceous (oil) glands, sudoriferous (sweat) glands, and ceruminous glands. Mammary glands, which are specialized sudoriferous glands that secrete milk, are discussed in Chapter 28 along with the female reproductive system.

Sebaceous Glands

Sebaceous glands (se-BĀ-shus; *sebace-* = greasy) or **oil glands** are simple, branched acinar (rounded) glands. With few exceptions, they are connected to hair follicles (see Figures 5.1 and 5.4a). The secreting portion of a sebaceous gland lies in the dermis and usually opens into the neck of a hair follicle. In some locations, such as the lips, glans penis, labia minora, and tarsal glands of the eyelids, sebaceous glands open directly onto the surface of the skin. Absent in the palms and soles, sebaceous glands are small in most areas of the trunk and limbs, but large in the skin of the breasts, face, neck, and superior chest.

Sebaceous glands secrete an oily substance called **sebum** (SĒ-bum), a mixture of triglycerides, cholesterol, proteins, and inorganic salts. Sebum coats the surface of hairs and helps keep them from drying and becoming brittle. Sebum also prevents excessive evaporation of water from the skin, keeps the skin soft and pliable, and inhibits the growth of some (but not all) bacteria.

CLINICAL CONNECTION | *Acne*

During childhood, sebaceous glands are relatively small and inactive. At puberty, androgens from the testes, ovaries, and adrenal glands stimulate sebaceous glands to grow in size and increase their production of sebum. **Acne** is an inflammation of sebaceous glands that usually begins at puberty, when the sebaceous glands are stimulated by androgens. Acne occurs predominantly in sebaceous follicles that have been colonized by bacteria, some of which thrive in the lipid-rich sebum. The infection may cause a cyst or sac of connective tissue cells to form, which can destroy and displace epidermal cells. This condition, called **cystic acne,** can permanently scar the epidermis. Treatment consists of gently washing the affected areas once or twice daily with a mild soap, topical antibiotics (such as clindamycin and erythromycin), topical drugs such as benzoyl peroxide or tretinoin, and oral antibiotics (such as tetracycline, minocycline, erythromycin, and isotretinoin). Contrary to popular belief, foods such as chocolate or fried foods do not cause or worsen acne. •

Sudoriferous Glands

There are three million to four million **sweat glands,** or **sudoriferous glands** (soo'-dor-IF-er-us; *sudor-* = sweat; *-ferous* = bearing) in the body. The cells of these glands release sweat, or perspiration, into hair follicles or onto the skin surface through pores. Sweat glands are divided into two main types, eccrine and apocrine, based on their structure and type of secretion.

Eccrine sweat glands (EK-rin; *eccrine* = secreting outwardly) are simple, coiled tubular glands that are much more common than apocrine sweat glands (see Figures 5.1 and 5.4a). They are distributed throughout the skin of most regions of the body, especially in the skin of the forehead, palms, and soles. Eccrine sweat glands are not present, however, in the margins of the lips, nail beds of the fingers and toes, glans penis, glans clitoris, labia minora, or eardrums. The secretory portion of eccrine sweat glands is located mostly in the deep dermis (sometimes in the upper subcutaneous layer). The excretory duct projects through the dermis and epidermis and ends as a pore at the surface of the epidermis (see Figure 5.1).

The sweat produced by eccrine sweat glands (about 600 mL per day) consists of water, ions (mostly Na^+ and Cl^-), urea, uric acid, ammonia, amino acids, glucose, and lactic acid. The main function of eccrine sweat glands is to help regulate body temperature through evaporation. As sweat evaporates, large quantities of heat energy leave the body surface. The homeostatic regulation of body temperature is known as **thermoregulation.** This role of eccrine sweat glands in helping the body to achieve thermoregulation is known as **thermoregulatory sweating.** During thermoregulatory sweating, sweat first forms on the forehead and scalp and then extends to the rest of the body, forming last on the palms and soles. Sweat that evaporates from the skin before it is perceived as moisture is termed **insensible perspiration** (*in-* = not). Sweat that is excreted in larger amounts and is seen as moisture on the skin is called **sensible perspiration.**

The sweat produced by eccrine sweat glands also plays a small role in eliminating wastes such as urea, uric acid, and ammonia from the body. However, the kidneys play more of a role in the excretion of these waste products from the body than eccrine sweat glands.

Eccrine sweat glands also release sweat in response to an emotional stress such as fear or embarrassment. This type of sweating is referred to as **emotional sweating** or a *cold sweat*. In contrast to thermoregulatory sweating, emotional sweating first occurs on the palms, soles, and axillae and then spreads to other areas of the body. As you will soon learn, apocrine sweat glands are also active during emotional sweating.

Apocrine sweat glands (AP-ō-krin; *apo-* = separated from) are also simple, coiled tubular glands (see Figures 5.1 and 5.4a). They are found mainly in the skin of the axilla (armpit), groin, areolae (pigmented areas around the nipples) of the breasts, and bearded regions of the face in adult males. These glands were once thought to release their secretions in an apocrine manner (see text coverage in Chapter 4 and Figure 4.7b)—by pinching off a portion of the cell. We now know, however, that their secretion is via exocytosis, which is characteristic of eccrine glands (see Figure 5.4a). Nevertheless, the term *apocrine* is still used. The secretory portion of these sweat glands is located in the lower dermis or upper subcutaneous layer, and the excretory duct opens into hair follicles (see Figure 5.1).

Compared to eccrine sweat, apocrine sweat appears milky or yellowish in color. Apocrine sweat contains the same components as eccrine sweat plus lipids and proteins. Sweat secreted from apocrine sweat glands is odorless. However, when apocrine sweat interacts with bacteria on the surface of the skin, the bacteria metabolize its components, causing apocrine sweat to have a musky odor that is often referred to as body odor. Eccrine sweat glands start to function soon after birth, but apocrine sweat glands do not begin to function until puberty.

Apocrine sweat glands, along with eccrine sweat glands, are active during emotional sweating. In addition, apocrine sweat glands secrete sweat during sexual activities. In contrast to eccrine sweat glands, apocrine sweat glands are not active during thermoregulatory sweating and, therefore, do not play a role in thermoregulation.

Ceruminous Glands

Modified sweat glands in the external ear, called **ceruminous glands** (se-RŪ-mi-nus; *cer-* = wax), produce a waxy lubricating secretion. The secretory portions of ceruminous glands lie in the subcutaneous layer, deep to sebaceous glands. Their excretory ducts open either directly onto the surface of the external auditory canal (ear canal) or into ducts of sebaceous glands. The combined secretion of the ceruminous and sebaceous glands is a yellowish material called **cerumen** (se-ROO-men), or earwax. Cerumen, together with hairs in the external auditory canal, provides a sticky barrier that impedes the entrance of foreign bodies and insects. Cerumen also waterproofs the canal and prevents bacteria and fungi from entering cells.

Table 5.3 presents a summary of skin glands.

⚕ CLINICAL CONNECTION | *Impacted Cerumen*

Some people produce an abnormally large amount of cerumen in the external auditory canal. If it accumulates until it becomes impacted (firmly wedged), sound waves may be prevented from reaching the eardrum. Treatments for **impacted cerumen** include periodic ear irrigation with enzymes to dissolve the wax and removal of wax with a blunt instrument by trained medical personnel. The use of cotton-tipped swabs or sharp objects is not recommended for this purpose because they may push the cerumen further into the external auditory canal and damage the eardrum. •

Nails

Nails are plates of tightly packed, hard, dead, keratinized epidermal cells that form a clear, solid covering over the dorsal surfaces of the distal portions of the digits. Each nail consists of a nail body, a free edge, and a nail root (Figure 5.5). The **nail body (plate)** is the visible portion of the nail. It is comparable to the stratum corneum of the general epidermis, with the exception that its flattened, keratinized cells fill with a harder type of keratin and the cells are not shed. Below the nail body is a region of epithelium and a deeper layer of dermis. Most of the nail body appears pink because of blood flowing through the capillaries in the underlying dermis. The **free edge** is the part of the nail body that may extend past the distal end of the digit. The free edge is white because there are no underlying capillaries. The **nail root** is the portion of the nail that is buried in a fold of skin. The whitish, crescent-shaped area of the proximal end of the nail body is called the **lunula** (LOO-noo-la = little moon). It appears whitish because the vascular tissue underneath does not show through due to a thickened region of epithelium in the area. Beneath the free edge is a thickened region of stratum corneum called the **hyponychium** (hī′-pō-NIK-ē-um; *hypo-* = below; *-onych* = nail), or **nail bed,** which secures the nail to the fingertip. The **eponychium** (ep′-ō-NIK-ē-um; *ep-* = above) or **cuticle** is a narrow band of epidermis that extends from and adheres to the margin (lateral border) of the nail wall. It occupies the proximal border of the nail and consists of stratum corneum.

The proximal portion of the epithelium deep to the nail root is the **nail matrix.** The superficial nail matrix cells divide mitotically

TABLE 5.3

Summary of Skin Glands (see Figures 5.1 and 5.4a)

FEATURE	SEBACEOUS (OIL) GLANDS	ECCRINE SWEAT GLANDS	APOCRINE SWEAT GLANDS	CERUMINOUS GLANDS
Distribution	Largely in lips, glans penis, labia minora, and tarsal glands; small in trunk and limbs; absent in palms and soles.	Throughout skin of most regions of body, especially skin of forehead, palms, and soles.	Skin of axillae, groin, areolae, bearded regions of face, clitoris, and labia minora.	External auditory canal.
Location of secretory portion	Dermis.	Mostly in deep dermis (sometimes in upper subcutaneous layer).	Mostly in deep dermis and upper subcutaneous layer.	Subcutaneous layer.
Termination of excretory duct	Mostly connected to hair follicles.	Surface of epidermis.	Hair follicle.	Surface of external auditory canal or into ducts of sebaceous glands.
Secretion	Sebum (mixture of triglycerides, cholesterol, proteins, and inorganic salts).	Perspiration, which consists of water, ions (Na^+, Cl^-), urea, uric acid, ammonia, amino acids, glucose, and lactic acid.	Perspiration, which consists of same components as eccrine sweat glands plus lipids and proteins.	Cerumen, a waxy material.
Functions	Prevent hairs from drying out, prevent water loss from skin, keep skin soft, inhibit growth of some bacteria.	Regulation of body temperature, waste removal, stimulated during emotional stress.	Stimulated during emotional stress and sexual excitement.	Impede entrance of foreign bodies and insects into external ear canal, waterproof canal, prevent microbes from entering cells.
Onset of function	Relatively inactive during childhood; activated during puberty.	Soon after birth.	Puberty.	Soon after birth.

Figure 5.5 Nails. Shown is a fingernail.

Nail cells arise by transformation of superficial cells of the nail matrix.

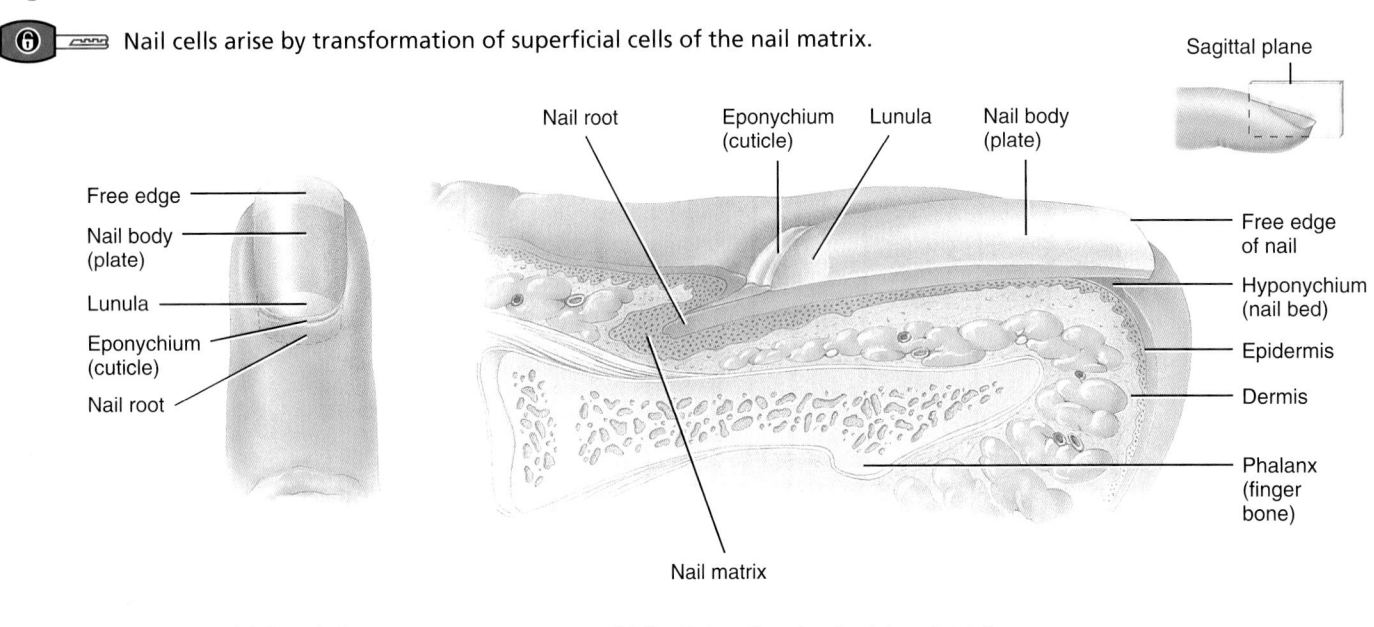

(a) Dorsal view   (b) Sagittal section showing internal detail

Why are nails so hard?

to produce new nail cells. Nail growth occurs by the transformation of superficial cells of the matrix into new nail cells. The growth rate of nails is determined by the rate of mitosis in matrix cells, which is influenced by factors such as a person's age, health, and nutritional status. Nail growth also varies according to the season, the time of day, and environmental temperature. The average growth in the length of fingernails is about 1 mm (0.04 in.) per week. The growth rate is somewhat slower in toenails. The longer the digit the faster the nail grows.

Nails have a variety of functions:

1. They protect the distal end of the digits.

2. They provide support and counterpressure to the palmar surface of the fingers to enhance touch perception and manipulation.

3. They allow us to grasp and manipulate small objects, and they can be used to scratch and groom the body in various ways.

✔ CHECKPOINT
7. Describe the structure of a hair. What causes "goose bumps"?
8. Contrast the locations and functions of sebaceous (oil) glands, sudoriferous (sweat) glands, and ceruminous glands.
9. Describe the parts of a nail.

5.3 TYPES OF SKIN

◉ OBJECTIVE
• Compare structural and functional differences in thin and thick skin.

Although the skin over the entire body is similar in structure, there are quite a few local variations related to thickness of the epidermis, strength, flexibility, degree of keratinization, distribution and type of hair, density and types of glands, pigmentation, vascularity (blood supply), and innervation (nerve supply). Two major types of skin are recognized on the basis of certain structural and functional properties: **thin (hairy) skin** and **thick (hairless) skin.** The greatest contributor to epidermal thickness is the increased number of layers in the stratum corneum. This arises in response to the greater mechanical stress in regions of thick skin. Table 5.4 presents a comparison of the features of thin and thick skin.

✔ CHECKPOINT
10. What criteria are used to distinguish thin and thick skin?

5.4 FUNCTIONS OF THE SKIN

◉ OBJECTIVE
• Describe how the skin contributes to the regulation of body temperature, storage of blood, protection, sensation, excretion and absorption, and synthesis of vitamin D.

Now that you have a basic understanding of the structure of the skin, you can better appreciate its many functions, which were introduced at the beginning of this chapter. The numerous functions of the integumentary system (mainly the skin) include thermoregulation, storage of blood, protection, cutaneous sensations, excretion and absorption, and synthesis of vitamin D.

Thermoregulation

Recall that **thermoregulation** is the homeostatic regulation of body temperature. The skin contributes to thermoregulation in two ways: by liberating sweat at its surface and by adjusting the flow of blood in the dermis. In response to high environmental temperature or heat produced by exercise, sweat production from eccrine

TABLE 5.4		
Comparison of Thin and Thick Skin		
FEATURE	**THIN SKIN**	**THICK SKIN**
Distribution	All parts of body except areas such as palms, palmar surface of digits, and soles.	Areas such as palms, palmar surface of digits, and soles.
Epidermal thickness	0.10–0.15 mm (0.004–0.006 in.).	0.6–4.5 mm (0.024–0.18 in.), due mostly to a thicker stratum corneum.
Epidermal strata	Stratum lucidum essentially lacking; thinner strata spinosum and corneum.	Stratum lucidum present; thicker strata spinosum and corneum.
Epidermal ridges	Lacking due to poorly developed, fewer, and less-well-organized dermal papillae.	Present due to well-developed and more numerous dermal papillae organized in parallel rows.
Hair follicles and arrector pili muscles	Present.	Absent.
Sebaceous glands	Present.	Absent.
Sudoriferous glands	Fewer.	More numerous.
Sensory receptors	Sparser.	Denser.

sweat glands increases; the evaporation of sweat from the skin surface helps lower body temperature. In addition, blood vessels in the dermis of the skin dilate (become wider); consequently, more blood flows through the dermis, which increases the amount of heat loss from the body (see Figure 25.19). In response to low environmental temperature, production of sweat from eccrine sweat glands is decreased, which helps conserve heat. Also, the blood vessels in the dermis of the skin constrict (become narrow), which decreases blood flow through the skin and reduces heat loss from the body. And, skeletal muscle contractions generate body heat.

Blood Reservoir

The dermis houses an extensive network of blood vessels that carry 8–10% of the total blood flow in a resting adult. For this reason, the skin acts as a **blood reservoir.**

Protection

The skin provides **protection** to the body in various ways. Keratin protects underlying tissues from microbes, abrasion, heat, and chemicals, and the tightly interlocked keratinocytes resist invasion by microbes. Lipids released by lamellar granules inhibit evaporation of water from the skin surface, thus guarding against dehydration; they also retard entry of water across the skin surface during showers and swims. The oily sebum from the sebaceous glands keeps skin and hairs from drying out and contains *bactericidal chemicals* (substances that kill bacteria). The acidic pH of perspiration retards the growth of some microbes. The pigment melanin helps shield against the damaging effects of ultraviolet light. Two types of cells carry out protective functions that are immunological in nature. Epidermal Langerhans cells alert the immune system to the presence of potentially harmful microbial invaders by recognizing and processing them, and macrophages in the dermis phagocytize bacteria and viruses that manage to bypass the Langerhans cells of the epidermis.

Cutaneous Sensations

Cutaneous sensations are sensations that arise in the skin, including tactile sensations—touch, pressure, vibration, and tickling—as well as thermal sensations such as warmth and coolness. Another cutaneous sensation, pain, usually is an indication of impending or actual tissue damage. There is a wide variety of nerve endings and receptors distributed throughout the skin, including the tactile discs of the epidermis, the corpuscles of touch in the dermis, and hair root plexuses around each hair follicle. Chapter 16 provides more details on the topic of cutaneous sensations.

Excretion and Absorption

The skin normally has a small role in **excretion,** the elimination of substances from the body, and **absorption,** the passage of materials from the external environment into body cells. Despite the almost waterproof nature of the stratum corneum, about 400 mL

of water evaporates through it daily. A sedentary person loses an additional 200 mL per day as sweat; a physically active person loses much more. Besides removing water and heat from the body, sweat also is the vehicle for excretion of small amounts of salts, carbon dioxide, and two organic molecules that result from the breakdown of proteins—ammonia and urea.

The absorption of water-soluble substances through the skin is negligible, but certain lipid-soluble materials do penetrate the skin. These include fat-soluble vitamins (A, D, E, and K), certain drugs, and the gases oxygen and carbon dioxide. Toxic materials that can be absorbed through the skin include organic solvents such as acetone (in some nail polish removers) and carbon tetrachloride (dry-cleaning fluid); salts of heavy metals such as lead, mercury, and arsenic; and the substances in poison ivy and poison oak. Since topical (applied to the skin) steroids, such as cortisone, are lipid-soluble, they move easily into the papillary region of the dermis. Here, they exert their anti-inflammatory properties by inhibiting histamine production by mast cells (recall that histamine contributes to inflammation). Certain drugs that are absorbed by the skin may be administered by applying adhesive patches to the skin.

CLINICAL CONNECTION | *Transdermal Drug Administration*

Most drugs are either absorbed into the body through the digestive system or injected into subcutaneous tissue or muscle. An alternative route, **transdermal (transcutaneous) drug administration,** enables a drug contained within an adhesive skin patch to pass across the epidermis and into the blood vessels of the dermis. The drug is released continuously at a controlled rate over a period of one to several days. This method of administration is especially useful for drugs that are quickly eliminated from the body because such drugs, if taken in other forms, would have to be taken quite frequently. Because the major barrier to penetration of most drugs is the stratum corneum, transdermal absorption is most rapid in regions of the skin where this layer is thin, such as the scrotum, face, and scalp. A growing number of drugs are available for transdermal administration, including nitroglycerin, for prevention of angina pectoris (chest pain associated with heart disease); scopolamine, for motion sickness; estradiol, used for estrogen-replacement therapy during menopause; ethinyl estradiol and norelgestromin in contraceptive patches; nicotine, used to help people stop smoking; and fentanyl, used to relieve severe pain in cancer patients.   •

Synthesis of Vitamin D

Synthesis of vitamin D requires activation of a precursor molecule in the skin by ultraviolet (UV) rays in sunlight. Enzymes in the liver and kidneys then modify the activated molecule, finally producing *calcitriol,* the most active form of vitamin D. Calcitriol is a hormone that aids in the absorption of calcium from foods in the gastrointestinal tract into the blood. Only a small amount of exposure to UV light (about 10 to 15 minutes at least twice a week) is required for vitamin D synthesis. People who avoid sun exposure and individuals who live in colder, northern climates may require vitamin D supplements to avoid vitamin D deficiency. Most cells of the immune system have vitamin D receptors, and the cells activate vitamin D in response to an infection, especially a respiratory infection, such as influenza. Vitamin D is believed to enhance phagocytic activity, increase the production of antimicrobial substances in phagocytes, regulate immune functions, and help reduce inflammation.

✔ CHECKPOINT

11. In what two ways does the skin help regulate body temperature?
12. How does the skin serve as a protective barrier?
13. What sensations arise from stimulation of neurons in the skin?
14. What types of molecules can penetrate the stratum corneum?

5.5 MAINTAINING HOMEOSTASIS: SKIN WOUND HEALING

◉ OBJECTIVE

• Explain how epidermal wounds and deep wounds heal.

Skin damage sets in motion a sequence of events that repairs the skin to its normal (or near-normal) structure and function. Two kinds of wound-healing processes can occur, depending on the depth of the injury. Epidermal wound healing occurs following wounds that affect only the epidermis; deep wound healing occurs following wounds that penetrate the dermis.

Epidermal Wound Healing

Even though the central portion of an epidermal wound may extend to the dermis, the edges of the wound usually involve only slight damage to superficial epidermal cells. Common types of epidermal wounds include abrasions, in which a portion of skin has been scraped away, and minor burns.

In response to an epidermal injury, basal cells of the epidermis surrounding the wound break contact with the basement membrane. The cells then enlarge and migrate across the wound (Figure 5.6a). The cells appear to migrate as a sheet until advancing cells from opposite sides of the wound meet. When epidermal cells encounter one another, they stop migrating due to a cellular response called **contact inhibition.** Migration of the epidermal cells stops completely when each is finally in contact with other epidermal cells on all sides.

As the basal epidermal cells migrate, a hormone called *epidermal growth factor* stimulates basal stem cells to divide and replace the ones that have moved into the wound. The relocated basal epidermal cells divide to build new strata, thus thickening the new epidermis (Figure 5.6b).

Deep Wound Healing

Deep wound healing occurs when an injury extends to the dermis and subcutaneous layer. Because multiple tissue layers must be repaired, the healing process is more complex than in epidermal wound healing. In addition, because scar tissue is formed, the healed tissue loses some of its normal function. Deep wound healing occurs in four phases: an inflammatory phase, a migratory phase, a proliferative phase, and a maturation phase.

During the **inflammatory phase,** a blood clot forms in the wound and loosely unites the wound edges (Figure 5.6c). As its name implies, this phase of deep wound healing involves **inflammation,** a vascular and cellular response that helps eliminate microbes, foreign material, and dying tissue in preparation for repair. The vasodilation and increased permeability of blood vessels associated with inflammation enhance delivery of helpful cells. These include phagocytic white blood cells called neutrophils; monocytes, which develop into macrophages that

Figure 5.6 Skin wound healing.

🔑 ⌦ In an epidermal wound, the injury is restricted to the epidermis; in a deep wound, the injury extends deep into the dermis.

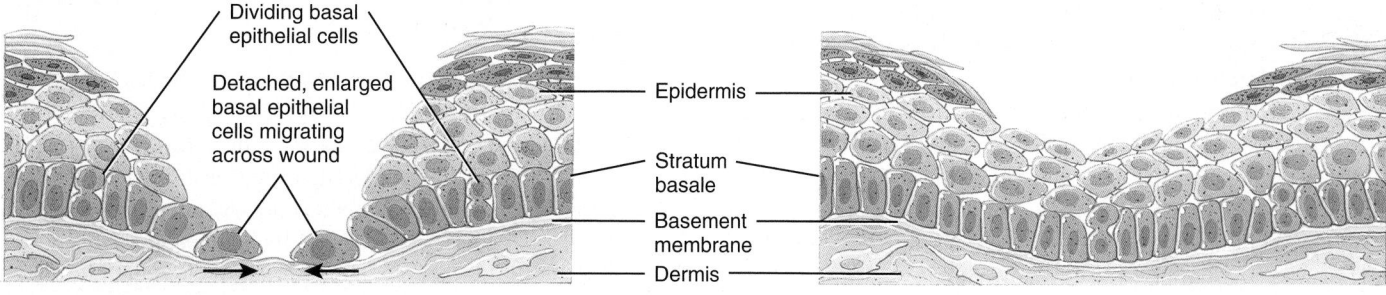

(a) Division of basal epithelial cells and migration across wound

(b) Thickening of epidermis

Epidermal wound healing

FIGURE 5.6 CONTINUES ▶

■ **FIGURE 5.6 CONTINUED** ▶

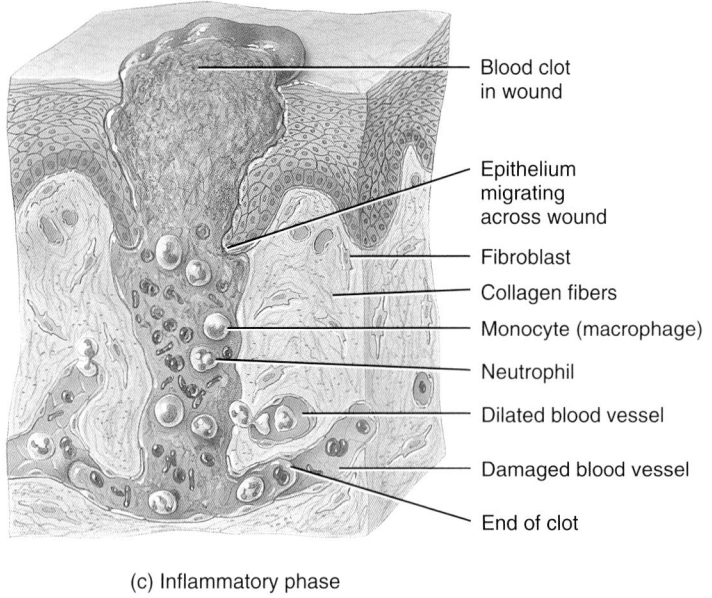

Blood clot in wound
Epithelium migrating across wound
Fibroblast
Collagen fibers
Monocyte (macrophage)
Neutrophil
Dilated blood vessel
Damaged blood vessel
End of clot

(c) Inflammatory phase

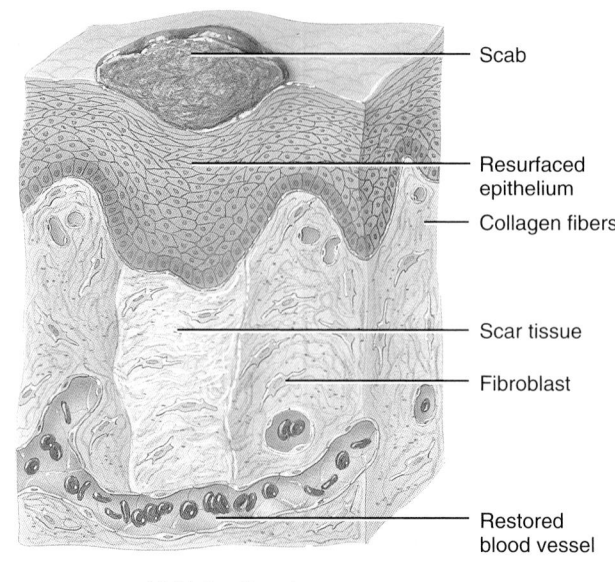

Scab
Resurfaced epithelium
Collagen fibers
Scar tissue
Fibroblast
Restored blood vessel

(d) Maturation phase

Deep wound healing

? **Would you expect an epidermal wound to bleed? Why or why not?**

phagocytize microbes; and mesenchymal cells, which develop into fibroblasts.

The three phases that follow do the work of repairing the wound. In the **migratory phase,** the clot becomes a scab, and epithelial cells migrate beneath the scab to bridge the wound. Fibroblasts migrate along fibrin threads and begin synthesizing scar tissue (collagen fibers and glycoproteins), and damaged blood vessels begin to regrow. During this phase, the tissue filling the wound is called **granulation tissue.** The **proliferative phase** is characterized by extensive growth of epithelial cells beneath the scab, deposition by fibroblasts of collagen fibers in random patterns, and continued growth of blood vessels. Finally, during the **maturation phase,** the scab sloughs off once the epidermis has been restored to normal thickness. Collagen fibers become more organized, fibroblasts decrease in number, and blood vessels are restored to normal (Figure 5.6d).

The process of scar tissue formation is called **fibrosis.** Sometimes, so much scar tissue is formed during deep wound healing that a raised scar—one that is elevated above the normal epidermal surface—results. If such a scar remains within the boundaries of the original wound, it is a **hypertrophic scar.** If it extends beyond the boundaries into normal surrounding tissues, it is a **keloid scar,** also called a **cheloid scar.** Scar tissue differs from normal skin in that its collagen fibers are more densely arranged, it has decreased elasticity, it has fewer blood vessels, and it may or may not contain the same number of hairs, skin glands, or sensory structures as undamaged skin. Because of the

arrangement of collagen fibers and the scarcity of blood vessels, scars usually are lighter in color than normal skin.

✔**CHECKPOINT**
15. Why doesn't epidermal wound healing result in scar formation?

5.6 DEVELOPMENT OF THE INTEGUMENTARY SYSTEM

◉ **OBJECTIVE**
• Describe the development of the epidermis, its accessory structures, and the dermis.

The *epidermis* is derived from the **ectoderm,** which covers the surface of the embryo. Initially, at about the fourth week after fertilization, the epidermis consists of only a single layer of ectodermal cells (Figure 5.7a). At the beginning of the seventh week the single layer, called the **basal layer,** divides and forms a superficial protected layer of flattened cells called the **periderm** (Figure 5.7b). The peridermal cells are continuously sloughed off, and by the fifth month of development secretions from sebaceous glands mix with them and hairs to form a fatty substance called **vernix caseosa** (VER-niks KĀ-sē-ō-sa; *vernix* = varnish; *caseosa* = cheese). This substance covers and protects the skin of the fetus from the constant exposure to the amniotic fluid in which it is bathed. In addition, the

Figure 5.7 Development of the integumentary system.

The epidermis develops from ectoderm, and the dermis develops from mesoderm.

(a) Fourth week
- Ectoderm
- Mesenchyme

(b) Seventh week
- Periderm
- Basal layer

(c) Eleven weeks
- Intermediate layer
- Epidermal ridge
- Basal layer
- Dermal papilla
- Melanoblast
- Developing collagen and elastic fibers

(d) Twelve weeks
- Basal layer
- Bud of developing sudoriferous gland
- Hair bud

(e) Fourteen weeks
- Developing sudoriferous gland
- Developing sebaceous gland
- Hair bulb

(f) Sixteen weeks
- Developing sebaceous gland
- Hair shaft
- Papilla of the hair

(g) Eighteen weeks
- Hair shaft
- Sweat pore
- Duct of sudoriferous gland
- Arrector pili muscle
- Epithelial root sheath
- Dermal root sheath
- Secretory portion of sudoriferous gland
- Sebaceous gland
- Bulb
- Papilla of the hair
- Blood vessels

(h) At birth
- Stratum corneum
- Stratum lucidum
- Stratum granulosum
- Stratum spinosum
- Stratum basale
- Melanocyte
- Epidermis
- Dermis

What is the composition of vernix caseosa?

171

vernix caseosa facilitates the birth of the fetus because of its slippery nature and protects the skin from being damaged by the nails.

By about 11 weeks (Figure 5.7c), the basal layer forms an intermediate layer of cells. Proliferation of the basal cells eventually forms all layers of the epidermis, which are present at birth (Figure 5.7h). *Epidermal ridges* form along with the epidermal layers. By about the eleventh week, cells from the ectoderm migrate into the dermis and differentiate into *melanoblasts*. These cells soon enter the epidermis and differentiate into *melanocytes*. Later in the first trimester of pregnancy, *Langerhans cells,* which arise from red bone marrow, invade the epidermis. *Merkel cells* appear in the epidermis in the fourth to sixth months; their origin is unknown.

The *dermis* arises from **mesoderm** located deep to the surface ectoderm. The mesoderm gives rise to a loosely organized embryonic connective tissue called **mesenchyme** (MEZ-en-kīm; see Figure 5.7a). By 11 weeks, the mesenchymal cells differentiate into fibroblasts and begin to form collagen and elastic fibers. As the epidermal ridges form, parts of the superficial dermis project into the epidermis and develop into the *dermal papillae,* which contain capillary loops, corpuscles of touch, and free nerve endings (Figure 5.7c).

Hair follicles develop at about 12 weeks as downgrowths of the basal layer of the epidermis into the deeper dermis. The downgrowths are called **hair buds** (Figure 5.7d). As the hair buds penetrate deeper into the dermis, their distal ends become club-shaped and are called **hair bulbs** (Figure 5.7e). Invaginations of the hair bulbs, called **papillae of the hair,** fill with mesoderm in which blood vessels and nerve endings develop (Figure 5.7f). Cells in the center of a hair bulb develop into the *matrix,* which forms the *hair,* and the peripheral cells of the hair bulb form the *epithelial root sheath;* mesenchyme in the surrounding dermis develops into the *dermal root sheath* and *arrector pili muscle* (Figure 5.7g). By the fifth month, the hair follicles produce lanugo (delicate fetal hair; see Types of Hairs earlier in the chapter). It is produced first on the head and then on other parts of the body, and is usually shed prior to birth.

Most *sebaceous (oil) glands* develop as outgrowths from the sides of hair follicles at about four months and remain connected to the follicles (Figure 5.7e). Most *sudoriferous (sweat) glands* are derived from downgrowths **(buds)** of the stratum basale of the epidermis into the dermis (Figure 5.7d). As the buds penetrate into the dermis, the proximal portion forms the duct of the sweat gland and the distal portion coils and forms the secretory portion of the gland (Figure 5.7g). Sweat glands appear at about five months on the palms and soles and a little later in other regions.

Nails are developed at about 10 weeks. Initially they consist of a thick layer of epithelium called the **primary nail field.** The nail itself is keratinized epithelium and grows distally from its base. It is not until the ninth month that the nails actually reach the tips of the digits.

✔**CHECKPOINT**

16. Which structures develop as downgrowths of the stratum basale?

5.7 AGING AND THE INTEGUMENTARY SYSTEM

◉ **OBJECTIVE**
• Describe the effects of aging on the integumentary system.

The pronounced effects of skin aging do not become noticeable until people reach their late forties. Most of the age-related changes occur in the proteins in the dermis. Collagen fibers in the dermis begin to decrease in number, stiffen, break apart, and disorganize into a shapeless, matted tangle. Elastic fibers lose some of their elasticity, thicken into clumps, and fray, an effect that is greatly accelerated in the skin of smokers. Fibroblasts, which produce both collagen and elastic fibers, decrease in number. As a result, the skin forms the characteristic crevices and furrows known as *wrinkles.*

With further aging, Langerhans cells dwindle in number and macrophages become less-efficient phagocytes, thus decreasing the skin's immune responsiveness. Moreover, decreased size of sebaceous glands leads to dry and broken skin that is more susceptible to infection. Production of sweat diminishes, which probably contributes to the increased incidence of heat stroke in the elderly. There is a decrease in the number of functioning melanocytes, resulting in gray hair and atypical skin pigmentation. Hair loss increases with aging as hair follicles stop producing hairs. About 25% of males begin to show signs of hair loss by age 30 and about two-thirds have significant hair loss by age 60. Both males and females develop pattern baldness. An increase in the size of some melanocytes produces pigmented blotching (age spots). Walls of blood vessels in the dermis become thicker and less permeable, and subcutaneous adipose tissue is lost. Aged skin (especially the dermis) is thinner than young skin, and the migration of cells from the basal layer to the epidermal surface slows considerably. With the onset of old age, skin heals poorly and becomes more susceptible to pathological conditions such as skin cancer and pressure sores. **Rosacea** (ro-ZĀ-shē-a = rosy) is a skin condition that affects mostly light-skinned adults between the ages of 30 and 60. It is characterized by redness, tiny pimples, and noticeable blood vessels, usually in the central area of the face.

Growth of nails and hair slows during the second and third decades of life. The nails also may become more brittle with age, often due to dehydration or repeated use of cuticle remover or nail polish.

Several cosmetic anti-aging treatments are available to diminish the effects of aging or sun-damaged skin. These include the following:

1. **Topical products** that bleach the skin to tone down blotches and blemishes (hydroquinone) or decrease fine wrinkles and roughness (retinoic acid).
2. **Microdermabrasion** (mī-krō-DER-ma-brā'-zhun; *mikros-* = small; *-derm-* = skin; *-abrasio* = to wear away), the use of tiny crystals under pressure to remove and vacuum the skin's surface cells to improve skin texture and reduce blemishes.

3. **Chemical peel,** the application of a mild acid (such as gly-colic acid) to the skin to remove surface cells to improve skin texture and reduce blemishes.

4. **Laser resurfacing,** the use of a laser to clear up blood vessels near the skin surface, even out blotches and blemishes, and decrease fine wrinkles.

5. **Dermal fillers,** injections of collagen from cows, hyaluronic acid, calcium hydroxylapatite (Radiesse®), or polylactic acid (Sculptra®) that plumps up the skin to smooth out wrinkles and fill in furrows, such as those around the nose and mouth and between the eyebrows.

6. **Fat transplantation,** in which fat from one part of the body is injected into another location such as around the eyes.

7. **Botulinum toxin** or **Botox**®, a diluted version of a toxin that is injected into the skin to paralyze skeletal muscles that cause the skin to wrinkle.

8. **Radio frequency nonsurgical facelift,** the use of radio frequency emissions to tighten the deeper layers of the skin of the jowls, neck, and sagging eyebrows and eyelids.

9. **Facelift, browlift,** or **necklift,** invasive surgery in which loose skin and fat are removed surgically and the underlying connective tissue and muscle are tightened.

CLINICAL CONNECTION | *Sun Damage, Sunscreens, and Sunblocks*

Although basking in the warmth of the sun may feel good, it is not a healthy practice. There are two forms of ultraviolet radiation that affect the health of the skin. Longer-wavelength ultraviolet A (UVA) rays make up nearly 95% of the ultraviolet radiation that reaches the earth. UVA rays are not absorbed by the ozone layer. They penetrate the furthest into the skin, where they are absorbed by melanocytes and thus are involved in sun tanning. UVA rays also depress the immune system. Shorter-wavelength ultraviolet B (UVB) rays are partially absorbed by the ozone layer and do not penetrate the skin as deeply as UVA rays. UVB rays cause sunburn and are responsible for most of the tissue damage (production of oxygen free radicals which disrupt collagen and elastic fibers) that results in wrinkling and aging of the skin and cataract formation. Both UVA and UVB rays are thought to cause skin cancer. Long-term overexposure to sunlight results in dilated blood vessels, age spots, freckles, and changes in skin texture.

Exposure to ultraviolet radiation (either natural sunlight or the artificial light of a tanning booth) may also produce **photosensitivity,** a heightened reaction of the skin after consumption of certain medications or contact with certain substances. Photosensitivity is characterized by redness, itching, blistering, peeling, hives, and even shock. Among the medications or substances that may cause a photosensitivity reaction are certain antibiotics (tetracycline), nonsteroidal anti-inflammatory drugs (ibuprofen or naproxen), certain herbal supplements (St. John's wort), some birth control pills, some high blood pressure medications, some antihistamines, and certain artificial sweeteners, perfumes, aftershaves, lotions, detergents, and medicated cosmetics.

Self-tanning lotions (sunless tanners), topically applied substances, contain a color additive (dihydroxyacetone) that produces a tanned appearance by interacting with proteins in the skin.

Sunscreens are topically applied preparations that contain various chemical agents (such as benzophenone or one of its derivatives) that absorb UVB rays but let most of the UVA rays pass through.

Sunblocks are topically applied preparations that contain substances such as zinc oxide that reflect and scatter both UVB and UVA rays.

Both sunscreens and sunblocks are graded according to a *sun protection factor (SPF)* rating, which measures the level of protection they supposedly provide against UV rays. The higher the rating, presumably the greater the degree of protection. As a precautionary measure, individuals who plan to spend a significant amount of time in the sun should use a sunscreen or a sunblock with an SPF of 15 or higher. Although sunscreens protect against sunburn, there is considerable debate as to whether they actually protect against skin cancer. In fact, some studies suggest that sunscreens increase the incidence of skin cancer because of the false sense of security they provide. •

✔ CHECKPOINT

17. What factors contribute to the susceptibility of aging skin to infection?

• • •

To appreciate the many ways that skin contributes to homeostasis of other body systems, examine *Focus on Homeostasis: The Integumentary System.* This feature is the first of 11, found at the end of selected chapters, that explain how the body system under consideration contributes to the homeostasis of all other body systems. Next, in Chapter 6, we will explore how bone tissue is formed and how bones are assembled into the skeletal system, which, like the skin, protects many of our internal organs.

BODY SYSTEM	CONTRIBUTION OF THE INTEGUMENTARY SYSTEM
For all body systems	Skin and hair provide barriers that protect all internal organs from damaging agents in external environment; sweat glands and skin blood vessels regulate body temperature, needed for proper functioning of other body systems.
Skeletal system	Skin helps activate vitamin D, needed for proper absorption of dietary calcium and phosphorus to build and maintain bones.
Muscular system 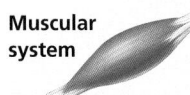	Skin helps provide calcium ions, needed for muscle contraction.
Nervous system	Nerve endings in skin and subcutaneous tissue provide input to brain for touch, pressure, thermal, and pain sensations.
Endocrine system	Keratinocytes in skin help activate vitamin D to calcitriol, a hormone that aids absorption of dietary calcium and phosphorus.
Cardiovascular system 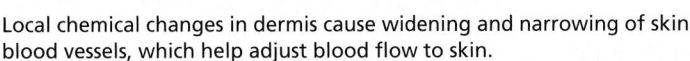	Local chemical changes in dermis cause widening and narrowing of skin blood vessels, which help adjust blood flow to skin.
Lymphatic system and immunity	Skin is "first line of defense" in immunity, providing mechanical barriers and chemical secretions that discourage penetration and growth of microbes; Langerhans cells in epidermis participate in immune responses by recognizing and processing foreign antigens; macrophages in dermis phagocytize microbes that penetrate skin surface.
Respiratory system	Hairs in nose filter dust particles from inhaled air; stimulation of pain nerve endings in skin may alter breathing rate.
Digestive system	Skin helps activate vitamin D to the hormone calcitriol, which promotes absorption of dietary calcium and phosphorus in small intestine.
Urinary system	Kidney cells receive partially activated vitamin D hormone from skin and convert it to calcitriol; some waste products are excreted from body in sweat, contributing to excretion by urinary system.
Reproductive systems	Nerve endings in skin and subcutaneous tissue respond to erotic stimuli, thereby contributing to sexual pleasure; suckling of a baby stimulates nerve endings in skin, leading to milk ejection; mammary glands (modified sweat glands) produce milk; skin stretches during pregnancy as fetus enlarges.

THE INTEGUMENTARY SYSTEM

DISORDERS: HOMEOSTATIC IMBALANCES

Skin Cancer

Excessive exposure to ultraviolet radiation from the sun or tanning beds has caused virtually all of the one million cases of **skin cancer** diagnosed annually in the United States. One-half of all cancers in the United States are skin cancers. There are three common forms of skin cancer. **Basal cell carcinomas** account for about 78% of all skin cancers. The tumors arise from cells in the stratum basale of the epidermis and rarely metastasize. **Squamous cell carcinomas,** which account for about 20% of all skin cancers, arise from the stratum spinosum of the epidermis, and they have a variable tendency to metastasize. Basal and squamous cell carcinomas are together known as *nonmelanoma skin cancer.*

Malignant melanomas arise from melanocytes and account for about 2% of all skin cancers. The estimated lifetime risk of developing melanoma is now 1 in 75, double the risk only 20 years ago. In part, this increase is due to depletion of the ozone layer, which absorbs some UV light high in the atmosphere. But the main reason for the increase is that more people are spending more time in the sun and in tanning beds. Malignant melanomas metastasize rapidly and can kill a person within months of diagnosis.

The key to successful treatment of malignant melanoma is early detection. The early warning signs of malignant melanoma are identified by the acronym ABCDE (Figure 5.8). *A* is for *asymmetry;* malignant melanomas tend to lack symmetry. This means that they have irregular shapes, such as two very different looking halves. *B* is for *border;* malignant melanomas have irregular—notched, indented, scalloped, or indistinct—borders. *C* is for *color;* malignant melanomas have uneven coloration and may contain several colors. *D* is for *diameter;* ordinary moles typically are smaller than 6 mm (0.25 in.), about the size of a pencil eraser. *E* is for *evolving;* malignant melanomas change in size, shape, and color. Once a malignant melanoma has the characteristics of A, B, and C, it is usually larger than 6 mm.

Among the risk factors for skin cancer are the following:

1. **Skin type.** Individuals with light-colored skin who never tan but always burn are at high risk.

2. **Sun exposure.** People who live in areas with many days of sunlight per year and at high altitudes (where ultraviolet light is more intense) have a higher risk of developing skin cancer. Likewise, people who engage in outdoor occupations and those who have suffered three or more severe sunburns have a higher risk.

3. **Family history.** Skin cancer rates are higher in some families than in others.

4. **Age.** Older people are more prone to skin cancer owing to longer total exposure to sunlight.

5. **Immunological status.** Immunosuppressed individuals have a higher incidence of skin cancer.

Burns

A **burn** is tissue damage caused by excessive heat, electricity, radioactivity, or corrosive chemicals that denature (break down) proteins in the skin. Burns destroy some of the skin's important contributions to homeostasis—protection against microbial invasion and dehydration, and thermoregulation.

Burns are graded according to their severity. A *first-degree burn* involves only the epidermis (Figure 5.9a). It is characterized by mild pain and erythema (redness) but no blisters. Skin functions remain intact. Immediate flushing with cold water may lessen the pain and damage caused by a first-degree burn. Generally, healing of a first-degree burn will occur in 3 to 6 days and may be accompanied by flaking or peeling. One example of a first-degree burn is mild sunburn.

A *second-degree burn* destroys the epidermis and part of the dermis (Figure 5.9b). Some skin functions are lost. In a second-degree burn, redness, blister formation, edema, and pain result. In a blister the epidermis separates from the dermis due to the accumulation of tissue fluid between them. Associated structures, such as hair follicles, sebaceous glands, and sweat glands, usually are not injured. If there is no infection, second-degree burns heal without skin grafting in about 3 to 4 weeks, but scarring may result. First- and second-degree burns are collectively referred to as *partial-thickness burns.*

A *third-degree burn* or *full-thickness burn* destroys the epidermis, dermis, and subcutaneous layer (Figure 5.9c). Most skin functions are lost. Such burns vary in appearance from marble-white to mahogany colored to charred, dry wounds. There is marked edema, and the burned region is numb because sensory nerve endings have been destroyed. Regeneration occurs slowly, and much granulation tissue forms before being covered by epithelium. Skin grafting may be required to promote healing and to minimize scarring.

The injury to the skin tissues directly in contact with the damaging agent is the *local effect* of a burn. Generally, however, the *systemic effects* of a major burn are a greater threat to life. The systemic effects of a burn may include (1) a large loss of water, plasma, and plasma proteins, which causes shock; (2) bacterial infection; (3) reduced circulation of blood; (4) decreased production of urine; and (5) diminished immune responses.

The seriousness of a burn is determined by its depth and extent of area involved, as well as the person's age and general health. According to the American Burn Association's classification of burn injury, a major burn includes third-degree burns over 10% of body surface area; or second-degree burns over 25% of body surface area; or any third-degree burns on the face, hands, feet, or *perineum* (per-i-NĒ-um, which includes the anal and urogenital regions). When the burn

Figure 5.8 Comparison of a normal nevus (mole) and a malignant melanoma.

 Excessive exposure to ultraviolet radiation from the sun or tanning beds accounts for almost all cases of skin cancer.

(a) Normal nevus (mole)   (b) Malignant melanoma

 Which is the most common type of skin cancer?

Figure 5.9 Burns.

🔑 ⎓⎓ A burn is tissue damage caused by agents that destroy the proteins in the skin.

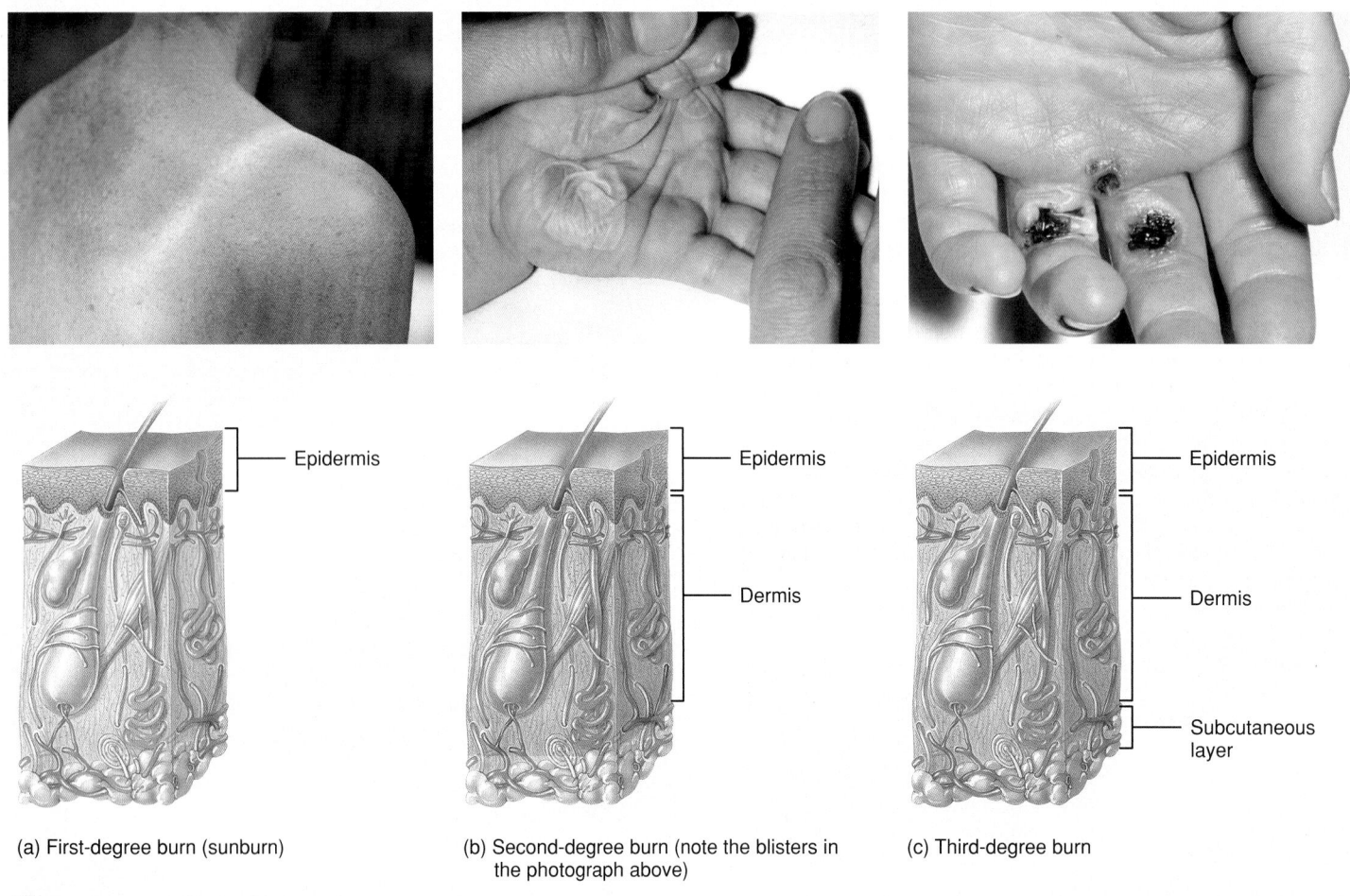

(a) First-degree burn (sunburn)

(b) Second-degree burn (note the blisters in the photograph above)

(c) Third-degree burn

❓ **What factors determine the seriousness of a burn?**

area exceeds 70%, more than half the victims die. A quick means for estimating the surface area affected by a burn in an adult is the **rule of nines** (Figure 5.10):

1. Count 9% if both the anterior and posterior surfaces of the head and neck are affected.

2. Count 9% for both the anterior and posterior surfaces of each upper limb (total of 18% for both upper limbs).

3. Count four times nine, or 36%, for both the anterior and posterior surfaces of the trunk, including the buttocks.

4. Count 9% for the anterior and 9% for the posterior surfaces of each lower limb as far up as the buttocks (total of 36% for both lower limbs).

5. Count 1% for the perineum.

In severely burned patients with full-thickness or deep partial-thickness burns where there is not sufficient autograft, a tissue engineered product called Integra® Dermal Regeneration Template (DRT) is available. It is designed to promote organized regeneration of the dermis while providing a protective barrier against fluid loss

and microbes. Integra® DRT consists of two layers, just like human skin. The bottom layer, called the matrix layer, is composed of bovine (cow) collagen and the carbohydrate glycosaminoglycan (GAG). It mimics the dermis, functions as an extracellular layer, and induces the body's own dermal cells to migrate into the area and regenerate a new dermis. The outer layer, called the silicone layer, consists of a thin layer of silicone that mimics the epidermis. Its role is to close the wound, control fluid loss, and serve as a protective barrier. Once the dermis has regenerated sufficiently (about three weeks), the silicone layer is removed and a thin sheet of the patient's own epidermal cells is applied.

Many people who have been burned in fires also inhale smoke. If the smoke is unusually hot or dense or if inhalation is prolonged, serious problems can develop. The hot smoke can damage the trachea (windpipe), causing its lining to swell. As the swelling narrows the trachea, airflow into the lungs is obstructed. Further, small airways inside the lungs can also narrow, producing wheezing or shortness of breath. A person who has inhaled smoke is given oxygen through a face mask, and a tube may be inserted into the trachea to assist breathing.

Figure 5.10 Rule-of-nines method for determining the extent of a burn. The percentages are the approximate proportions of the body surface area.

 The rule of nines is a quick way for estimating the surface area affected by a burn in an adult.

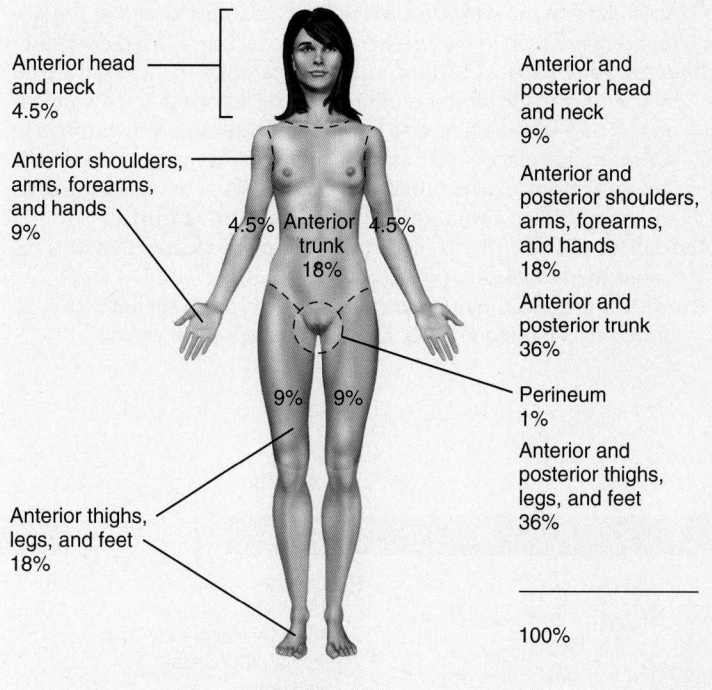

Anterior head and neck 4.5%

Anterior shoulders, arms, forearms, and hands 9%

4.5% Anterior trunk 18% 4.5%

9% 9%

Anterior thighs, legs, and feet 18%

Anterior and posterior head and neck 9%

Anterior and posterior shoulders, arms, forearms, and hands 18%

Anterior and posterior trunk 36%

Perineum 1%

Anterior and posterior thighs, legs, and feet 36%

100%

Anterior view

❓ What percentage of the body would be burned if only the anterior trunk and anterior left upper limb were involved?

Pressure Ulcers

Pressure ulcers, also known as *decubitus ulcers* (dē-KŪ-bi-tus) or *bedsores,* are caused by a constant deficiency of blood flow to tissues (Figure 5.11). Typically the affected tissue overlies a bony projection that has been subjected to prolonged pressure against an object such as a bed, cast, or splint. If the pressure is relieved in a few hours, redness occurs but no lasting tissue damage results. Blistering of the affected area may indicate superficial damage; a reddish-blue discoloration may indicate deep tissue damage. Prolonged pressure causes tissue ulceration. Small breaks in the epidermis become infected, and the sensitive subcutaneous layer and deeper tissues are damaged. Eventually, the tissue dies. Pressure ulcers occur most often in bedridden patients. With proper care, pressure ulcers are preventable, but they can develop very quickly in patients who are very old or very ill.

Figure 5.11 Pressure ulcer.

 A pressure ulcer is a shedding of epithelium caused by a constant deficiency of blood flow to tissues.

Pressure ulcer on heel

❓ What parts of the body are usually affected by pressure ulcers?

MEDICAL TERMINOLOGY

Abrasion (a-BRĀ-shun; *ab-* = away; *-rasion* = scraped) An area where skin has been scraped away.

Blister A collection of serous fluid within the epidermis or between the epidermis and dermis, due to short-term but severe friction. The term **bulla** (BUL-a) refers to a large blister.

Callus (KAL-lus = hard skin) An area of hardened and thickened skin that is usually seen in palms and soles and is due to persistent pressure and friction.

Cold sore A lesion, usually in an oral mucous membrane, caused by Type 1 herpes simplex virus (HSV) transmitted by oral or respiratory routes. The virus remains dormant until triggered by factors such as ultraviolet light, hormonal changes, and emotional stress. Also called a **fever blister.**

Comedo (KOM-ē-dō = to eat up) A collection of sebaceous material and dead cells in the hair follicle and excretory duct of the sebaceous (oil) gland. Usually found over the face, chest, and back, and more commonly during adolescence. Also called a **blackhead.**

Contact dermatitis (der-ma-TĪ-tis; *dermat-* = skin; *-itis* = inflammation of) Inflammation of the skin characterized by redness, itching, and swelling and caused by exposure of the skin to chemicals that bring about an allergic reaction, such as poison ivy toxin.

Contusion (kon-TOO-shun; *contundere* = to bruise) Condition in which tissue deep to the skin is damaged, but the epidermis is not broken.

Corn A painful conical thickening of the stratum corneum of the epidermis found principally over toe joints and between the toes, often caused by friction or pressure. Corns may be hard or soft, depending on their location. Hard corns are usually found over toe joints, and soft corns are usually found between the fourth and fifth toes.

Cyst (SIST = sac containing fluid) A sac with a distinct connective tissue wall, containing a fluid or other material.

Eczema (EK-ze-ma; *ekzeo-* = to boil over) An inflammation of the skin characterized by patches of red, blistering, dry, extremely itchy skin. It occurs mostly in skin creases in the wrists, backs of the knees, and fronts of the elbows. It typically begins in infancy and many children outgrow the condition. The cause is unknown but is linked to genetics and allergies.

Frostbite Local destruction of skin and subcutaneous tissue on exposed surfaces as a result of extreme cold. In mild cases, the skin is blue and swollen and there is slight pain. In severe cases there is considerable swelling, some bleeding, no pain, and blistering. If untreated, gangrene may develop. Frostbite is treated by rapid rewarming.

Hemangioma (hē-man′-jē-Ō-ma; *hem-* = blood; *-angi-* = blood vessel; *-oma* = tumor) Localized benign tumor of the skin and subcutaneous layer that results from an abnormal increase in blood vessels. One type is a **portwine stain,** a flat, pink, red, or purple lesion present at birth, usually at the nape of the neck.

Hives Reddened elevated patches of skin that are often itchy. Most commonly caused by infections, physical trauma, medications, emotional stress, food additives, and certain food allergies. Also called **urticaria** (ūr-ti-KAR-ē-a).

Keloid (KĒ-loid; *kelis* = tumor) An elevated, irregular darkened area of excess scar tissue caused by collagen formation during healing. It extends beyond the original injury and is tender and frequently painful. It occurs in the dermis and underlying subcutaneous tissue, usually after trauma, surgery, a burn, or severe acne; more common in people of African descent.

Keratosis (ker'-a-TŌ-sis; *kera-* = horn) Formation of a hardened growth of epidermal tissue, such as *solar keratosis,* a premalignant lesion of the sun-exposed skin of the face and hands.

Laceration (las-er-Ā-shun; *lacer-* = torn) An irregular tear of the skin.

Lice Contagious arthropods that include two basic forms. **Head lice** are tiny, jumping arthropods that suck blood from the scalp. They lay eggs, called nits, and their saliva causes itching that may lead to complications. **Pubic lice** are tiny arthropods that do not jump; they look like miniature crabs.

Papule (PAP-ūl; *papula* = pimple) A small, round skin elevation less than 1 cm in diameter. One example is a pimple.

Pruritus (proo-RĪ-tus; *pruri-* = to itch) Itching, one of the most common dermatological disorders. It may be caused by skin disorders (infections), systemic disorders (cancer, kidney failure), psychogenic factors (emotional stress), or allergic reactions.

Psoriasis (sō-RĪ-a-sis) A common and chronic skin disorder in which keratinocytes divide and move more quickly than normal from the stratum basale to the stratum corneum. As a result, the surface cells never get a chance to cycle into the later keratinizing stages. The surface cells are shed immaturely and on the scalp are called dandruff.

Tinea corporis (TIN-ē-a KOR-po-ris) A fungal infection characterized by scaling, itching, and sometimes painful lesions that may appear on any part of the body, also known as **ringworm.** Fungi thrive in warm, moist places such as skin folds of the groin, where it is known as **tinea cruris** (KROO-ris) **(jock itch)** or between the toes, where it is called **tinea pedis** (PE-dis) **(athlete's foot).**

Topical In reference to a medication, applied to the skin surface rather than ingested or injected.

Wart Mass produced by uncontrolled growth of epithelial skin cells; caused by a papillomavirus. Most warts are noncancerous.

CHAPTER REVIEW AND RESOURCE SUMMARY

Review

Resource

5.1 Structure of the Skin

1. The integumentary system consists of the skin, hair, oil and sweat glands, nails, and sensory receptors.
2. The skin is the largest organ of the body in surface area and weight. The principal parts of the skin are the epidermis (superficial) and dermis (deep).
3. The subcutaneous layer (hypodermis) is deep to the dermis and not part of the skin. It anchors the dermis to underlying tissues and organs, and it contains pacinian (lamellated) corpuscles.
4. The types of cells in the epidermis are keratinocytes, melanocytes, Langerhans cells, and Merkel cells.
5. The epidermal layers, from deep to superficial, are the stratum basale, stratum spinosum, stratum granulosum, stratum lucidum (in thick skin only), and stratum corneum (see Table 5.1). Stem cells in the stratum basale undergo continuous cell division, producing keratinocytes for the other layers.
6. The dermis is composed of dense irregular connective tissue containing collagen and elastic fibers. It is divided into papillary and reticular regions. The papillary region contains thin collagen and fine elastic fibers, dermal papillae, and Meissner corpuscles. The reticular region contains bundles of thick collagen and some coarse elastic fibers, fibroblasts and macrophages, adipose tissue, hair follicles, nerves, sebaceous (oil) glands, and sudoriferous (sweat) glands. (See Table 5.2.)
7. Epidermal ridges provide the basis for fingerprints and footprints.
8. The color of skin is due to melanin, carotene, and hemoglobin.
9. In tattooing, a pigment is deposited with a needle in the dermis. Body piercing is the insertion of jewelry through an artificial opening.

Anatomy Overview - The Skin
Figure 5.2 - Cells in the Epidermis
Figure 5.3 - Layers of the Epidermis

5.2 Accessory Structures of the Skin

1. Accessory structures of the skin—hair, skin glands, and nails—develop from the embryonic epidermis.
2. A hair consists of a shaft, most of which is superficial to the surface, a root that penetrates the dermis and sometimes the subcutaneous layer, and a hair follicle.
3. Associated with each hair follicle is a sebaceous (oil) gland, an arrector pili muscle, and a hair root plexus.
4. New hairs develop from division of hair matrix cells in the bulb; hair replacement and growth occur in a cyclical pattern consisting of growth, regression, and resting stages.
5. Hairs offer a limited amount of protection—from the sun, heat loss, and entry of foreign particles into the eyes, nose, and ears. They also function in sensing light touch.

Anatomy Overview - Hair
Anatomy Overview - Nails
Figure 5.4 - Hair

Review   **Resource**

6. Lanugo of the fetus is shed before birth. Most body hair on males is terminal (coarse, pigmented); most body hair on females is vellus (fine).
7. Sebaceous (oil) glands are usually connected to hair follicles; they are absent from the palms and soles. Sebaceous glands produce sebum, which moistens hairs and waterproofs the skin. Clogged sebaceous glands may produce acne.
8. There are two types of sudoriferous (sweat) glands: eccrine and apocrine. Eccrine sweat glands have an extensive distribution; their ducts terminate at pores at the surface of the epidermis. Eccrine sweat glands are involved in thermoregulation and waste removal and are stimulated during emotional stress. Apocrine sweat glands are limited to the skin of the axillae, groin, and areolae; their ducts open into hair follicles. Apocrine sweat glands are stimulated during emotional stress and sexual excitement. (See Table 5.3.)
9. Ceruminous glands are modified sudoriferous glands that secrete cerumen. They are found in the external auditory canal (ear canal).
10. Nails are hard, dead keratinized epidermal cells over the dorsal surfaces of the distal portions of the digits. The principal parts of a nail are the nail body, free edge, nail root, lunula, eponychium, and nail matrix. Cell division of the nail matrix cells produces new nails.

5.3   Types of Skin

1. Thin skin covers all parts of the body except for the palms, palmar surfaces of the digits, and the soles.
2. Thick skin covers the palms, palmar surfaces of the digits, and soles. (See Table 5.4.)

5.4   Functions of the Skin

Anatomy Overview - The Integument and Disease Resistance
Animation - Nonspecific Disease Resistance

1. Skin functions include body temperature regulation, blood storage, protection, sensation, excretion and absorption, and synthesis of vitamin D.
2. The skin participates in thermoregulation by liberating sweat at its surface and by adjusting the flow of blood in the dermis.
3. The skin provides physical, chemical, and biological barriers that help protect the body.
4. Cutaneous sensations include tactile sensations, thermal sensations, and pain.

5.5   Maintaining Homeostasis: Skin Wound Healing

Figure 5.6 - Skin Wound Healing

1. In an epidermal wound, the central portion of the wound usually extends down to the dermis; the wound edges involve only superficial damage to the epidermal cells.
2. Epidermal wounds are repaired by enlargement and migration of basal cells, contact inhibition, and division of migrating and stationary basal cells.
3. During the inflammatory phase of deep wound healing, a blood clot unites the wound edges, epithelial cells migrate across the wound, vasodilation and increased permeability of blood vessels enhance delivery of phagocytes, and mesenchymal cells develop into fibroblasts.
4. During the migratory phase, fibroblasts migrate along fibrin threads and begin synthesizing collagen fibers and glycoproteins.
5. During the proliferative phase, epithelial cells grow extensively.
6. During the maturation phase, the scab sloughs off, the epidermis is restored to normal thickness, collagen fibers become more organized, fibroblasts begin to disappear, and blood vessels are restored to normal.

5.6   Development of the Integumentary System

1. The epidermis develops from the embryonic ectoderm, and the accessory structures of the skin (hair, nails, and skin glands) are epidermal derivatives.
2. The dermis is derived from mesodermal cells.

5.7   Aging and the Integumentary System

1. Most effects of aging begin to occur when people reach their late forties.
2. Among the effects of aging are wrinkling, loss of subcutaneous adipose tissue, atrophy of sebaceous glands, and decrease in the number of melanocytes and Langerhans cells.

Fill in the blanks in the following statements.

1. The epidermal layer that is found in thick skin but not in thin skin is the _____.

2. The most common sweat glands that release a watery secretion are _____ sweat glands; modified sweat glands in the ear are _____ glands; sweat glands located in the axillae, groin, areolae, and beards of males and that release a slightly viscous, lipid-rich secretion are _____ sweat glands.

Indicate whether the following statements are true or false.

3. An individual with a dark skin color has more melanocytes than a fair-skinned person.

4. In order to permanently prevent growth of an unwanted hair, you must destroy the hair matrix.

Choose the one best answer to the following questions.

5. The layer of the epidermis that contains stem cells undergoing mitosis is the
   (a) stratum corneum.   (b) stratum lucidum.
   (c) stratum basale.   (d) stratum spinosum.
   (e) stratum granulosum.

6. The substance that helps promote mitosis in epidermal skin cells is
   (a) keratohyalin.   (b) melanin.   (c) carotene.
   (d) collagen.   (e) epidermal growth factor.

7. Which of the following is *not* a function *of* skin?
   (a) calcium production   (b) vitamin D synthesis
   (c) protection   (d) excretion of wastes
   (e) temperature regulation

8. To expose underlying tissues in the bottom of the foot, a foot surgeon must first cut through the skin. Place the following layers in the order that the scalpel would cut. (1) stratum lucidum, (2) stratum corneum, (3) stratum basale, (4) stratum granulosum, (5) stratum spinosum.
   (a) 3, 5, 4, 1, 2   (b) 2, 1, 5, 4, 3   (c) 2, 1, 4, 5, 3
   (d) 1, 3, 5, 4, 2   (e) 3, 4, 5, 1, 2

9. Aging of the skin can result in
   (a) an increase in collagen and elastic fibers.
   (b) a decrease in the activity of sebaceous glands.
   (c) a thickening of the skin.
   (d) an increased blood flow to the skin.
   (e) an increase in toenail growth.

10. Which of the following is *not* true?
   (a) Albinism is an inherited inability of melanocytes to produce melanin.
   (b) Striae occurs when the dermis is overstretched to the point of tearing.
   (c) In order to prevent excessive scarring, surgeons should cut parallel to the lines of cleavage.
   (d) The papillary layer of the dermis is responsible for fingerprints.
   (e) Much of the body's fat is located in the dermis of the skin.

11. A patient is brought into the emergency room suffering from a burn. The patient does not feel any pain at the burn site. Using a gentle pull on a hair, the examining physician can remove entire hair follicles from the patient's arm. This patient is suffering from what type of burn?
   (a) third degree   (b) second degree   (c) first degree
   (d) partial thickness   (e) localized

12. Which of the following statements are *true?* (1) Nails are composed of tightly packed, hard, keratinized cells of the epidermis that form a clear, solid covering over the dorsal surface of the terminal end of digits. (2) The free edge of the nail is white due to the absence of capillaries. (3) Nails help us grasp and manipulate small objects. (4) Nails protect the ends of digits from trauma. (5) Nail color is due to a combination of melanin and carotene.
   (a) 1, 2, and 3   (b) 1, 3, and 4   (c) 1, 2, 3, and 4
   (d) 2, 3, and 4   (e) 1, 3, and 5

13. Match the following:
   _____ (a) produce the protein that helps protect the skin and underlying tissues from light, heat, microbes, and many chemicals
   _____ (b) produce a pigment that contributes to skin color and absorbs ultraviolet light
   _____ (c) cells that arise from red bone marrow, migrate to the epidermis, and participate in immune responses
   _____ (d) cells thought to function in the sensation of touch
   _____ (e) located in the dermis, they function in the sensations of warmth, coolness, pain, itching, and tickling
   _____ (f) smooth muscles associated with the hair follicles; when contracted, they pull the hair shafts perpendicular to the skin's surface
   _____ (g) an abnormal thickening of the epidermis
   _____ (h) release a lipid-rich secretion that functions as a water-repellent sealant in the stratum granulosum
   _____ (i) pressure-sensitive cells found mostly in the subcutaneous layer
   _____ (j) a fatty substance that covers and protects the skin of the fetus from the constant exposure to amniotic fluid
   _____ (k) associated with hair follicles, these secrete an oily substance that helps prevent hair from becoming brittle, prevents evaporation of water from the skin's surface, and inhibits the growth of certain bacteria

   (1) Merkel cells
   (2) callus
   (3) keratinocytes
   (4) Langerhans cells
   (5) melanocytes
   (6) free nerve endings
   (7) sebaceous glands
   (8) lamellar granules
   (9) pacinian (lamellated) corpuscles
   (10) vernix caseosa
   (11) arrector pili

14. Match the following:
    ___ (a) deep region of the dermis composed primarily of dense irregular connective tissue
    ___ (b) composed of keratinized stratified squamous epithelial tissue
    ___ (c) not considered part of the skin, it contains areolar and adipose tissues and blood vessels; attaches skin to underlying tissues and organs
    ___ (d) superficial region of the dermis; composed of areolar connective tissue

    (1) subcutaneous layer (hypodermis)
    (2) papillary region
    (3) reticular region
    (4) epidermis

15. Match the following and place the phases of deep wound healing in the correct order:
    ___ (a) epithelial cells migrate under scab to bridge the wound; formation of granulation tissue
    ___ (b) sloughing of scab; reorganization of collagen fibers; blood vessels return to normal
    ___ (c) vasodilation and increased permeability of blood vessels to deliver cells involved in phagocytosis; clot formation
    ___ (d) extensive growth of epithelial cells beneath scab; random deposition of collagen fibers; continued growth of blood vessels

    (1) proliferative phase
    (2) inflammatory phase
    (3) maturation phase
    (4) migratory phase

CRITICAL THINKING QUESTIONS

1. The amount of dust that collects in a house with an assortment of dogs, cats, and people is truly amazing. A lot of these dust particles had a previous "life" as part of the home's living occupants. Where did the dust originate on the human body?

2. Josie reassures her mother that the tattoo she received at the tattoo parlor will eventually disappear. She knows this because she has learned in biology class that skin cells are shed every four weeks. Is Josie correct?

3. Six months ago, Chef Eduardo sliced through the end of his right thumbnail. Although the surrounding nail grows normally, this part of his nail remains split and doesn't seem to want to "heal." What has happened to cause this?

? ANSWERS TO FIGURE QUESTIONS

5.1 The epidermis is composed of epithelial tissue; the dermis is made up of connective tissue.

5.2 Melanin protects DNA of keratinocytes from the damaging effects of UV light.

5.3 The stratum basale is the layer of the epidermis with stem cells that continually undergo cell division.

5.4 Plucking a hair stimulates hair root plexuses in the dermis, some of which are sensitive to pain. Because the cells of a hair shaft are already dead and the hair shaft lacks nerves, cutting hair is not painful.

5.5 Nails are hard because they are composed of tightly packed, hard, keratinized epidermal cells.

5.6 Since the epidermis is avascular, an epidermal wound would not produce any bleeding.

5.7 Vernix caseosa consists of secretions from sebaceous glands, sloughed off peridermal cells, and hairs.

5.8 Basal cell carcinoma is the most common type of skin cancer.

5.9 The seriousness of a burn is determined by the depth and extent of the area involved, the individual's age, and general health.

5.10 About 22.5% of the body would be involved (4.5% [anterior arm] + 18% [anterior trunk]).

5.11 Pressure ulcers typically develop in tissues that overlie bony projections subjected to pressure, such as the shoulders, hips, buttocks, heels, and ankles.

6 THE SKELETAL SYSTEM: BONE TISSUE

BONE TISSUE AND HOMEOSTASIS *Bone tissue is continuously growing, remodeling, and repairing itself. It contributes to homeostasis of the body by providing support and protection, producing blood cells, and storing minerals and triglycerides.*

A bone is composed of several different tissues working together: bone or osseous tissue, cartilage, dense connective tissues, epithelium, adipose tissue, and nervous tissue. For this reason, each individual bone in your body is considered an organ. Bone tissue, a complex and dynamic living tissue, continually engages in a process called *remodeling*—the construction of new bone tissue and breaking down of old bone tissue. The entire framework of bones and their cartilages, along with ligaments and tendons, constitutes the **skeletal system.** In this chapter we will survey the various components of bones to help you understand how bones form and age, and how exercise affects their density and strength. The study of bone structure and the treatment of bone disorders is called **osteology** (os-tē-OL-ō-jē; *osteo-* = bone; *-logy* = study of).

Did you ever wonder why more females than males are affected by osteoporosis?

6.1 FUNCTIONS OF BONE AND THE SKELETAL SYSTEM

◉ **OBJECTIVE**

• Describe the six main functions of the skeletal system.

Bone tissue makes up about 18% of the weight of the human body. The skeletal system performs several basic functions:

1. *Support.* The skeleton serves as the structural framework for the body by supporting soft tissues and providing attachment points for the tendons of most skeletal muscles.

2. *Protection.* The skeleton protects the most important internal organs from injury. For example, cranial bones protect the brain, vertebrae (backbones) protect the spinal cord, and the rib cage protects the heart and lungs.

3. *Assistance in movement.* Most skeletal muscles attach to bones; when they contract, they pull on bones to produce movement. This function is discussed in detail in Chapter 10.

4. *Mineral homeostasis (storage and release).* Bone tissue stores several minerals, especially calcium and phosphorus, which contribute to the strength of bone. Bone tissue stores about 99% of the body's calcium. On demand, bone releases minerals into the blood to maintain critical mineral balances (homeostasis) and to distribute the minerals to other parts of the body.

5. *Blood cell production.* Within certain bones, a connective tissue called **red bone marrow** produces red blood cells, white blood cells, and platelets, a process called **hemopoiesis** (hēm-ō-poy-Ē-sis; *hemo-* = blood; *-poiesis* = making). Red bone marrow consists of developing blood cells, adipocytes, fibroblasts, and macrophages within a network of reticular fibers. It is present in developing bones of the fetus and in some adult bones, such as the hip (pelvic) bones, ribs, sternum (breastbone), vertebrae (backbones), skull, and ends of the bones of the humerus (arm bone) and femur (thigh bone). In a newborn, all bone marrow is red and is involved in hemopoiesis. With increasing age, much of the bone marrow changes from red to yellow. Blood cell production is considered in detail in Section 19.2.

6. *Triglyceride storage.* **Yellow bone marrow** consists mainly of adipose cells, which store triglycerides. The stored triglycerides are a potential chemical energy reserve.

✔**CHECKPOINT**

1. How does the skeletal system function in support, protection, movement, and storage of minerals?
2. Describe the role of bones in blood cell production.
3. Which bones contain red bone marrow?
4 How do red bone marrow and yellow bone marrow differ in composition and function?

6.2 STRUCTURE OF BONE

◉ **OBJECTIVE**

• Describe the structure and functions of each part of a long bone.

We will now examine the structure of bone at the macroscopic level. Macroscopic bone structure may be analyzed by considering the parts of a long bone, such as the humerus (the arm bone) shown in Figure 6.1a. A *long bone* is one that has greater length than width. A typical long bone consists of the following parts:

1. The **diaphysis** (dī-AF-i-sis = growing between) is the bone's shaft or body—the long, cylindrical, main portion of the bone.

2. The **epiphyses** (e-PIF-i-sēz = growing over; singular is *epiphysis*) are the proximal and distal ends of the bone.

3. The **metaphyses** (me-TAF-i-sēz; *meta-* = between; singular is *metaphysis*) are the regions between the diaphysis and the epiphyses. In a growing bone, each metaphysis contains an *epiphyseal (growth) plate* (ep'-i-FIZ-ē-al), a layer of hyaline cartilage that allows the diaphysis of the bone to grow in length (described later in the chapter). When a bone ceases to grow in length at about ages 18–21, the cartilage in the epiphyseal plate is replaced by bone; the resulting bony structure is known as the *epiphyseal line.*

4. The **articular cartilage** is a thin layer of hyaline cartilage covering the part of the epiphysis where the bone forms an articulation (joint) with another bone. Articular cartilage reduces friction and absorbs shock at freely movable joints. Because articular cartilage lacks a perichondrium and lacks blood vessels, repair of damage is limited.

5. The **periosteum** (per-ē-OS-tē-um; *peri-* = around) is a tough connective tissue sheath and its associated blood supply that surrounds the bone surface wherever it is not covered by articular cartilage. It is composed of an *outer fibrous layer* of dense irregular connective tissue and an *inner osteogenic layer* that consists of cells. Some of the cells enable bone to grow in thickness, but not in length. The periosteum also protects the bone, assists in fracture repair, helps nourish bone tissue, and serves as an attachment point for ligaments and tendons. The periosteum is attached to the underlying bone by **perforating (Sharpey's) fibers,** thick bundles of collagen that extend from the periosteum into the bone extracellular matrix.

6. The **medullary cavity** (MED-ū-lar-ē; *medulla-* = marrow, pith), or **marrow cavity,** is a hollow, cylindrical space within the diaphysis that contains fatty yellow bone marrow and numerous blood vessels in adults. This cavity minimizes the weight of the bone by reducing the dense bony material where it is least needed. The long bones' tubular design provides maximum strength with minimum weight.

Figure 6.1 Parts of a long bone. The spongy bone tissue of the epiphyses and metaphyses contains red bone marrow, and the medullary cavity of the diaphysis contains yellow bone marrow (in adults).

A long bone is covered by articular cartilage at the articular surfaces of its proximal and distal epiphyses and by periosteum around all other parts of the bone.

(a) Partially sectioned humerus (arm bone)

(b) Partially sectioned humerus

FUNCTIONS OF BONE TISSUE

1. Supports soft tissue and provides attachment for skeletal muscles.
2. Protects internal organs.
3. Assists in movement, along with skeletal muscles.
4. Stores and releases minerals.
5. Contains red bone marrow, which produces blood cells.
6. Contains yellow bone marrow, which stores triglycerides (fats).

? **What is the functional significance of the periosteum?**

7. The **endosteum** (end-OS-tē-um; *endo-* = within) is a thin membrane that lines the medullary cavity. It contains a single layer of bone-forming cells and a small amount of connective tissue.

✔ CHECKPOINT

5. Diagram the parts of a long bone, and list the functions of each part.

6.3 HISTOLOGY OF BONE TISSUE

◉ OBJECTIVES

• Explain why bone tissue is classified as a connective tissue.
• Describe the cellular composition of bone tissue and the functions of each type of cell.
• Compare the structural and functional differences between compact and spongy bone tissue.

We will now examine the structure of bone at the microscopic level. Like other connective tissues, **bone,** or **osseous tissue** (OS-ē-us), contains an abundant extracellular matrix that surrounds widely separated cells. The extracellular matrix is about 15% water, 30% collagen fibers, and 55% crystallized mineral salts. The most abundant mineral salt is calcium phosphate $[Ca_3(PO_4)_2]$. It combines with another mineral salt, calcium hydroxide $[Ca(OH)_2]$, to form crystals of **hydroxyapatite** [Ca$_{10}$(PO$_4$)$_6$(OH)$_2$] (hī-drok-sē-AP-a-tīt). As the crystals form, they combine with still other mineral salts, such as calcium carbonate (CaCO$_3$), and ions such as magnesium, fluoride, potassium, and sulfate. As these mineral salts are deposited in the framework formed by the collagen fibers of the extracellular matrix, they crystallize and the tissue hardens. This process, called **calcification** (kal′-si-fi-KĀ-shun), is initiated by bone-building cells called osteoblasts (described shortly).

It was once thought that calcification simply occurred when enough mineral salts were present to form crystals. We now know that the process requires the presence of collagen fibers. Mineral salts first begin to crystallize in the microscopic spaces between collagen fibers. After the spaces are filled, mineral crystals accumulate around the collagen fibers. The combination of crystallized salts and collagen fibers is responsible for the characteristics of bone.

Although a bone's *hardness* depends on the crystallized inorganic mineral salts, a bone's *flexibility* depends on its collagen fibers. Like reinforcing metal rods in concrete, collagen fibers and other organic molecules provide *tensile strength,* resistance to being stretched or torn apart. Soaking a bone in an acidic solution, such as vinegar, dissolves its mineral salts, causing the bone to become rubbery and flexible. As you will see shortly, when the need for particular minerals arises or as part of bone formation or breakdown, bone cells called osteoclasts secrete enzymes and

acids that break down both the mineral salts and the collagen fibers of the extracellular matrix of bone.

Four types of cells are present in bone tissue: osteogenic cells, osteoblasts, osteocytes, and osteoclasts (Figure 6.2).

1. **Osteogenic cells** (os′-tē-ō-JEN-ik; *-genic* = producing) are unspecialized bone stem cells derived from mesenchyme, the tissue from which almost all connective tissues are formed. They are the only bone cells to undergo cell division; the resulting cells develop into osteoblasts. Osteogenic cells are found along the inner portion of the periosteum, in the endosteum, and in the canals within bone that contain blood vessels.

2. **Osteoblasts** (OS-tē-ō-blasts′; *-blasts* = buds or sprouts) are bone-building cells. They synthesize and secrete collagen fibers and other organic components needed to build the extracellular matrix of bone tissue, and they initiate calcification (described shortly). As osteoblasts surround themselves with extracellular matrix, they become trapped in their secretions and become osteocytes. (Note: The ending *-blast* in the name of a bone cell or any other connective tissue cell means that the cell secretes extracellular matrix.)

3. **Osteocytes** (OS-tē-ō-sīts′; *-cytes* = cells), mature bone cells, are the main cells in bone tissue and maintain its daily metabolism, such as the exchange of nutrients and wastes with the blood. Like osteoblasts, osteocytes do not undergo cell division.

Figure 6.2 **Types of cells in bone tissue.**

Osteogenic cells undergo cell division and develop into osteoblasts, which secrete bone extracellular matrix.

From bone cell lineage

From white blood cell lineage

Ruffled border

Osteogenic cell (develops into an osteoblast)

Osteoblast (forms bone extracellular matrix)

Osteocyte (maintains bone tissue)

Osteoclast (functions in resorption, the breakdown of bone extracellular matrix)

SEM 1160x

SEM 1160x
Dr. Richard Kessel and Randy Kardon/Tissues & Organs/Visuals Unlimited

SEM 5626x
Dr. Richard Kessel and Randy Kardon/Tissues & Organs/Visuals Unlimited

? **Why is bone resorption important?**

(Note: The ending *-cyte* in the name of a bone cell or any other tissue cell means that the cell maintains the tissue.)

4. **Osteoclasts** (OS-tē-ō-klasts′; *-clast* = break) are huge cells derived from the fusion of as many as 50 monocytes (a type of white blood cell) and are concentrated in the endosteum. On the side of the cell that faces the bone surface, the osteoclast's plasma membrane is deeply folded into a *ruffled border*. Here the cell releases powerful lysosomal enzymes and acids that digest the protein and mineral components of the underlying extracellular bone matrix. This breakdown of bone extracellular matrix, termed **resorption** (rē-SORP-shun), is part of the normal development, maintenance, and repair of bone. (Note: The ending *-clast* means that the cell breaks down extracellular matrix.) As you will see later, in response to certain hormones, osteoclasts help regulate blood calcium level (see Section 6.7). They are also target cells for drug therapy used to treat osteoporosis (see Disorders: Homeostatic Imbalances at the end of this chapter).

A mnemonic that will help you remember the difference between the function of osteoblasts and osteoclasts is as follows: osteo**B**lasts **B**uild bone, while osteo**C**lasts **C**arve out bone.

Bone is not completely solid but has many small spaces between its cells and extracellular matrix components. Some spaces serve as channels for blood vessels that supply bone cells with nutrients. Other spaces act as storage areas for red bone marrow. Depending on the size and distribution of the spaces, the regions of a bone may be categorized as compact or spongy (see Figure 6.1). Overall, about 80% of the skeleton is compact bone and 20% is spongy bone.

Compact Bone Tissue

Compact bone tissue contains few spaces (Figure 6.3a) and is the strongest form of bone tissue. It is found beneath the periosteum of all bones and makes up the bulk of the diaphyses of long bones. Compact bone tissue provides protection and support and resists the stresses produced by weight and movement.

Compact bone tissue is composed of repeating structural units called **osteons,** or **haversian systems** (ha-VER-shan). Each osteon consists of concentric lamellae arranged around a **central (haversian) canal.** Resembling the growth rings of a tree, the **concentric lamellae** (LA-mel-ē) are circular plates of mineralized extracellular matrix of increasing diameter, surrounding a small network of blood vessels, lymphatics, and nerves located in the central canal (Figure 6.3a). These tube-like units of bone generally form a series of parallel cylinders that, in long bones, tend to run parallel to the long axis of the bone. Between the concentric lamellae are small spaces called **lacunae** (la-KOO-nē = little lakes; singular is *lacuna*), which contain osteocytes. Radiating in all directions from the lacunae are tiny **canaliculi** (kan-a-LIK-ū-lī = small channels), which are filled with extracellular fluid. Inside the canaliculi are slender fingerlike processes of osteocytes (see inset at right of Figure 6.3a). Neighboring osteocytes communicate via gap junctions (see Section 4.2). The canaliculi connect lacunae with one another and with the central canals, forming an intricate, miniature system of interconnected canals throughout

the bone. This system provides many routes for nutrients and oxygen to reach the osteocytes and for the removal of wastes.

Osteons in compact bone tissue are aligned in the same direction and are parallel to the length of the diaphysis. As a result, the shaft of a long bone resists bending or fracturing even when considerable force is applied from either end. Compact bone tissue tends to be thickest in those parts of a bone where stresses are applied in relatively few directions. The lines of stress in a bone are not static. They change as a person learns to walk and in response to repeated strenuous physical activity, such as weight training. The lines of stress in a bone also can change because of fractures or physical deformity. Thus, the organization of osteons is not static but changes over time in response to the physical demands placed on the skeleton.

The areas between neighboring osteons contain lamellae called **interstitial lamellae** (in′-ter-STISH-al), which also have lacunae with osteocytes and canaliculi. Interstitial lamellae are fragments of older osteons that have been partially destroyed during bone rebuilding or growth.

Blood vessels, lymphatic vessels, and nerves from the periosteum penetrate the compact bone through transverse **perforating canals** or *Volkmann's canals* (FOLK-mans). The vessels and nerves of the perforating canals connect with those of the medullary cavity, periosteum, and central canals.

Arranged around the entire outer and inner circumference of the shaft of a long bone are lamellae called **circumferential lamellae** (ser′-kum-fer-EN-shē-al). They develop during initial bone formation. The circumferential lamellae directly deep to the periosteum are called *outer circumferential lamellae*. They are connected to the periosteum by **perforating** (*Sharpey's*) **fibers.** The circumferential lamellae that line the medullary cavity are called *inner circumferential lamellae* (Figure 6.3a).

Spongy Bone Tissue

In contrast to compact bone tissue, **spongy bone tissue,** also referred to as *trabecular* or *cancellous bone tissue,* does not contain osteons (Figure 6.3b, c). Spongy bone tissue is always located in the *interior* of a bone, protected by a covering of compact bone. It consists of lamellae that are arranged in an irregular pattern of thin columns called **trabeculae** (tra-BEK-ū-lē = little beams; singular is *trabecula*). Between the trabeculae are spaces that are visible to the unaided eye. These macroscopic spaces are filled with red bone marrow in bones that produce blood cells, and yellow bone marrow (adipose tissue) in other bones. Both types of bone marrow contain numerous small blood vessels that provide nourishment to the osteocytes. Each trabecula consists of concentric lamellae, osteocytes that lie in lacunae, and canaliculi that radiate outward from the lacunae.

Spongy bone tissue makes up most of the interior bone tissue of short, flat, sesamoid, and irregularly shaped bones. In long bones it forms the core of the epiphyses beneath the paper-thin layer of compact bone, and forms a variable narrow rim bordering the medullary cavity of the diaphysis. Spongy bone is always covered by a layer of compact bone for protection.

At first glance, the trabeculae of spongy bone tissue may appear to be less organized than the osteons of compact bone tissue.

Figure 6.3 **Histology of compact and spongy bone.** (a) Sections through the diaphysis of a long bone, from the surrounding periosteum on the right, to compact bone in the middle, to spongy bone and the medullary cavity on the left. The inset at the upper right shows an osteocyte in a lacuna. (b, c) Details of spongy bone. See Table 4.7 for a photomicrograph of compact bone tissue and Figure 6.11a for a scanning electron micrograph of spongy bone tissue.

Bone tissue is organized in concentric lamellae around a central (haversian) canal in compact bone and in irregularly arranged lamellae in the trabeculae in spongy bone.

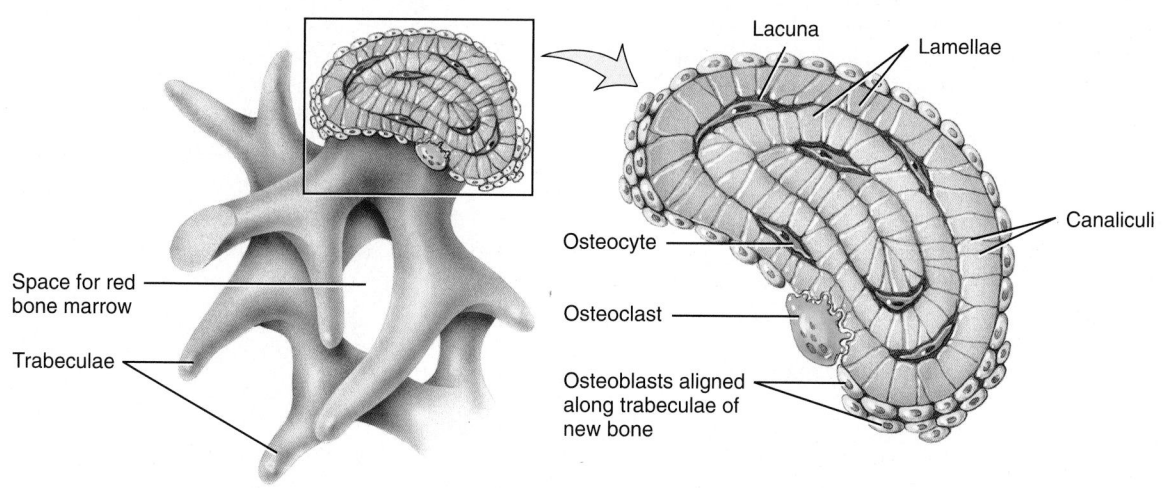

(a) Osteons (haversian systems) in compact bone and trabeculae in spongy bone

(b) Enlarged aspect of spongy bone trabeculae

(c) Details of a section of a trabecula

? **As people age, some central (haversian) canals may become blocked. What effect would this have on the surrounding osteocytes?**

187

However, they are precisely oriented along lines of stress, a characteristic that helps bones resist stresses and transfer force without breaking. Spongy bone tissue tends to be located where bones are not heavily stressed or where stresses are applied from many directions. The trabeculae do not achieve their final arrangement until locomotion is completely learned. In fact, the arrangement can even be altered as lines of stress change due to a poorly healed fracture or a deformity.

Spongy bone tissue is different from compact bone tissue in two respects. First, spongy bone tissue is light, which reduces the overall weight of a bone. This reduction in weight allows the bone to move more readily when pulled by a skeletal muscle. Second, the trabeculae of spongy bone tissue support and protect the red bone marrow. Spongy bone in the hip bones, ribs, sternum (breastbone), vertebrae, and the proximal ends of the humerus and femur is the only site where red bone marrow is stored and, thus, the site where hemopoiesis (blood cell production) occurs in adults.

CLINICAL CONNECTION | *Bone Scan*

A **bone scan** is a diagnostic procedure that takes advantage of the fact that bone is living tissue. A small amount of a radioactive tracer compound that is readily absorbed by bone is injected intravenously. The degree of uptake of the tracer is related to the amount of blood flow to the bone. A scanning device (gamma camera) measures the radiation emitted from the bones, and the information is translated into a photograph that can be read like an x-ray on a monitor. Normal bone tissue is identified by a consistent gray color throughout because of its uniform uptake of the radioactive tracer. Darker or lighter areas may indicate bone abnormalities. Darker areas called "hot spots" are areas of increased metabolism that absorb more of the radioactive tracer due to increased blood flow. Hot spots may indicate bone cancer, abnormal healing of fractures, or abnormal bone growth. Lighter areas called "cold spots" are areas of decreased metabolism that absorb less of the radioactive tracer due to decreased blood flow. Cold spots may indicate problems such as degenerative bone disease, decalcified bone, fractures, bone infections, Paget's disease, and rheumatoid arthritis. A bone scan detects abnormalities 3 to 6 months sooner than standard x-ray procedures and exposes the patient to less radiation. A bone scan is the standard test for bone density screening, particularly important in screening for osteoporosis in females. •

✔CHECKPOINT

6. Why is bone considered a connective tissue?
7. What factors contribute to the hardness and tensile strength of bone?
8. List the four types of cells in bone tissue and their functions.
9. What is the composition of the extracellular matrix of bone tissue?
10. How are compact and spongy bone tissues different in microscopic appearance, location, and function?
11. What is a bone scan and how is it used clinically?

6.4 BLOOD AND NERVE SUPPLY OF BONE

◉ OBJECTIVE

• Describe the blood and nerve supply of bone.

Bone is richly supplied with blood. Blood vessels, which are especially abundant in portions of bone containing red bone marrow, pass into bones from the periosteum. We will consider the blood supply of a long bone such as the mature tibia (shin bone) shown in Figure 6.4.

Periosteal arteries (per-ē-OS-tē-al), small arteries accompanied by nerves, enter the diaphysis through many perforating (Volkmann's) canals and supply the periosteum and outer part of the compact bone (see Figure 6.3a). Near the center of the diaphysis, a large **nutrient artery** passes through a hole in compact bone called the **nutrient foramen.** On entering the medullary cavity, the nutrient artery divides into proximal and distal branches that course toward each end of the bone. These branches supply both the inner part of compact bone tissue of the diaphysis and the

Figure 6.4 Blood supply of a mature long bone.

Bone is richly supplied with blood vessels.

Partially sectioned tibia (shin bone)

❓ **Where do periosteal arteries enter bone tissue?**

spongy bone tissue and red bone marrow as far as the epiphyseal plates (or lines). Some bones, like the tibia, have only one nutrient artery; others, like the femur (thigh bone), have several. The ends of long bones are supplied by the metaphyseal and epiphyseal arteries, which arise from arteries that supply the associated joint. The **metaphyseal arteries** (met-a-FIZ-ē-al) enter the metaphyses of a long bone and, together with the nutrient artery, supply the red bone marrow and bone tissue of the metaphyses. The **epiphyseal arteries** (ep′-i-FIZ-ē-al) enter the epiphyses of a long bone and supply the red bone marrow and bone tissue of the epiphyses.

Veins that carry blood away from long bones are evident in three places: (1) One or two **nutrient veins** accompany the nutrient artery and exit through the diaphysis; (2) numerous **epiphyseal veins** and **metaphyseal veins** accompany their respective arteries and exit through the epiphyses; and (3) many small **periosteal veins** accompany their respective arteries and exit through the periosteum.

Nerves accompany the blood vessels that supply bones. The periosteum is rich in sensory nerves, some of which carry pain sensations. These nerves are especially sensitive to tearing or tension, which explains the severe pain resulting from a fracture or a bone tumor. For the same reason there is some pain associated with a bone marrow needle biopsy. In this procedure, a needle is inserted into the middle of the bone to withdraw a sample of red bone marrow to examine it for conditions such as leukemias, metastatic neoplasms, lymphoma, Hodgkin's disease, and aplastic anemia. As the needle penetrates the periosteum, pain is felt. Once it passes through, there is little pain.

✔CHECKPOINT

12. Explain the location and roles of the nutrient arteries, nutrient foramina, epiphyseal arteries, and periosteal arteries.
13. Which part of a bone contains sensory nerves associated with pain?
14. Describe one situation in which these sensory neurons are important.
15. How is a bone marrow needle biopsy performed? What conditions are diagnosed through this procedure?

6.5 BONE FORMATION

◉ OBJECTIVES
• Describe the steps of intramembranous and endochondral ossification.
• Explain how bone grows in length and thickness.
• Describe the process involved in bone remodeling.

The process by which bone forms is called **ossification** (os′-i-fi-KĀ-shun; *ossi-* = bone; *-fication* = making) or **osteogenesis** (os′-tē-ō-JEN-e-sis). Bone formation occurs in four principal situations: (1) the initial formation of bones in an embryo and fetus, (2) the growth of bones during infancy, childhood, and adolescence until their adult sizes are reached, (3) the remodeling of bone (replacement of old bone by new bone tissue throughout life), and (4) the repair of fractures (breaks in bones) throughout life.

Initial Bone Formation in an Embryo and Fetus

We will first consider the initial formation of bone in an embryo and fetus. The embryonic "skeleton," initially composed of mesenchyme in the general shape of bones, is the site where cartilage formation and ossification occur during the sixth week of embryonic development. Bone formation follows one of two patterns.

The two methods of bone formation, which both involve the replacement of a preexisting connective tissue with bone, do not lead to differences in the structure of mature bones, but are simply different methods of bone development. In the first type of ossification, called **intramembranous ossification** (in′-tra-MEM-bra-nus; *intra-* = within; *-membran-* = membrane), bone forms directly within mesenchyme, which is arranged in sheetlike layers that resemble membranes. In the second type, **endochondral ossification** (en′-dō-KON-dral; *endo-* = within; *-chondral* = cartilage), bone forms within hyaline cartilage that develops from mesenchyme.

Intramembranous Ossification

Intramembranous ossification is the simpler of the two methods of bone formation. The flat bones of the skull, most of the facial bones, mandible (lower jawbone), and the medial part of the clavicle (collar bone) are formed in this way. Also, the "soft spots" that help the fetal skull pass through the birth canal later harden as they undergo intramembranous ossification, which occurs as follows (Figure 6.5):

❶ *Development of the ossification center.* At the site where the bone will develop, specific chemical messages cause the mesenchymal cells to cluster together and differentiate, first into osteogenic cells and then into osteoblasts. The site of such a cluster is called an **ossification center.** Osteoblasts secrete the organic extracellular matrix of bone until they are surrounded by it.

Figure 6.5 Intramembranous ossification. Refer to this figure as you read the corresponding numbered paragraphs in the text. Illustrations **1** and **2** show a smaller field of vision at higher magnification than illustrations **3** and **4**.

🔑 ▭ Intramembranous ossification involves the formation of bone within mesenchyme arranged in sheetlike layers that resemble membranes.

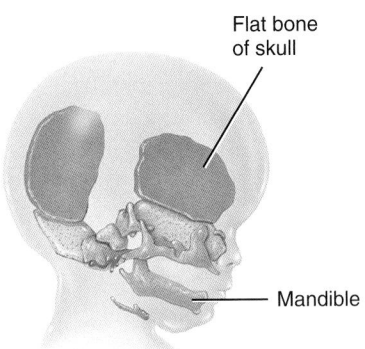

Flat bone of skull

Mandible

Blood capillary
Ossification center
Mesenchymal cell
Osteoblast
Collagen fiber

1 Development of ossification center: osteoblasts secrete organic extracellular matrix.

Osteocyte in lacuna
Canaliculus
Osteoblast
Newly calcified bone matrix

2 Calcification: calcium and other mineral salts are deposited and extracellular matrix calcifies (hardens).

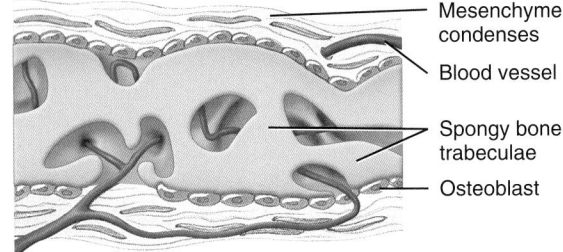

Mesenchyme condenses
Blood vessel
Spongy bone trabeculae
Osteoblast

3 Formation of trabeculae: extracellular matrix develops into trabeculae that fuse to form spongy bone.

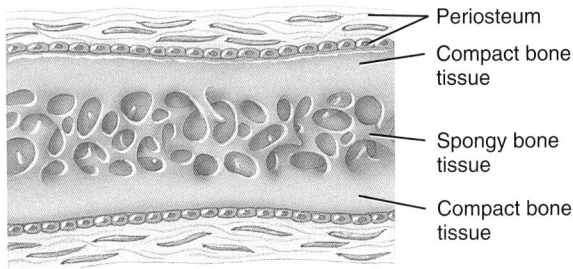

Periosteum
Compact bone tissue
Spongy bone tissue
Compact bone tissue

4 Development of the periosteum: mesenchyme at the periphery of the bone develops into the periosteum.

2 ***Calcification.*** Next, the secretion of extracellular matrix stops, and the cells, now called osteocytes, lie in lacunae and extend their narrow cytoplasmic processes into canaliculi that radiate in all directions. Within a few days, calcium and other mineral salts are deposited and the extracellular matrix hardens or calcifies (calcification).

3 ***Formation of trabeculae.*** As the bone extracellular matrix forms, it develops into trabeculae that fuse with one another to form spongy bone around the network of blood vessels in the tissue. Connective tissue that is associated with the blood vessels in the trabeculae differentiates into red bone marrow.

4 ***Development of the periosteum.*** In conjunction with the formation of trabeculae, the mesenchyme condenses at the periphery of the bone and develops into the periosteum. Eventually, a thin layer of compact bone replaces the surface layers of the spongy bone, but spongy bone remains in the center. Much of the newly formed bone is remodeled (destroyed and reformed) as the bone is transformed into its adult size and shape.

Endochondral Ossification

The replacement of cartilage by bone is called endochondral ossification. Although most bones of the body are formed in this

? **Which bones of the body develop by intramembranous ossification?**

way, the process is best observed in a long bone. It proceeds as follows (Figure 6.6):

1 ***Development of the cartilage model.*** At the site where the bone is going to form, specific chemical messages cause the mesenchymal cells to crowd together in the general shape of the future bone, and then develop into chondroblasts. The chondroblasts secrete cartilage extracellular matrix, producing a **cartilage model** consisting of hyaline cartilage. A covering called the **perichondrium** (per′-i-KON-drē-um) develops around the cartilage model.

Figure 6.6 Endochondral ossification.

During endochondral ossification, bone gradually replaces a cartilage model.

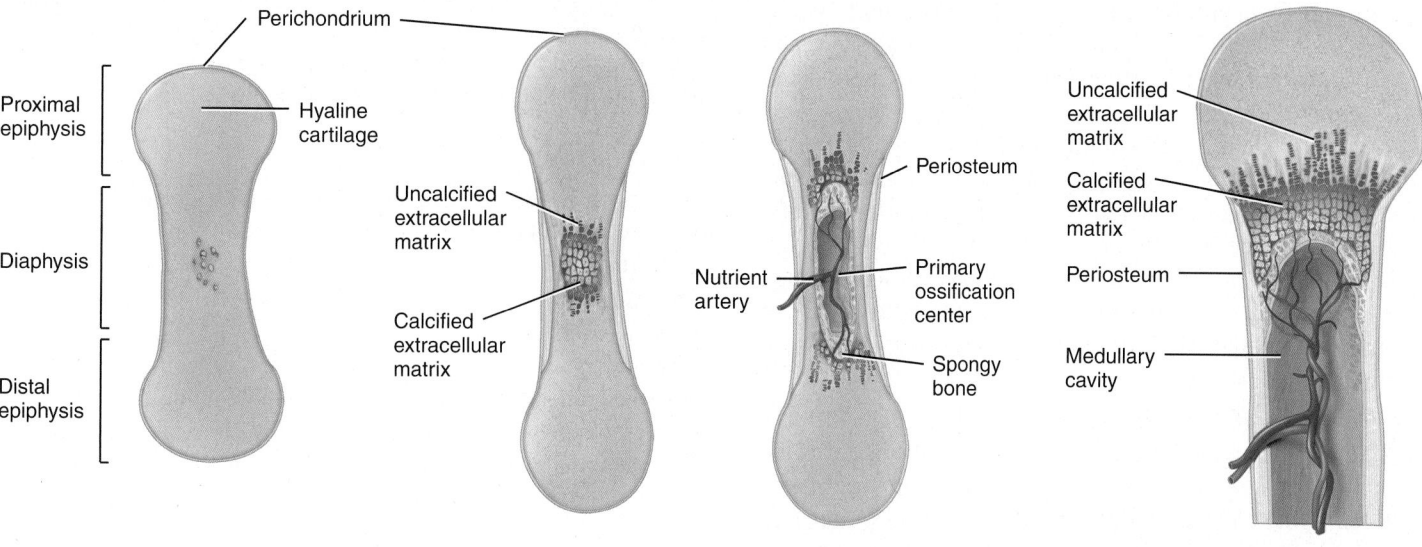

1. Development of cartilage model: mesenchymal cells develop into chondroblasts, which form the cartilage model.

2. Growth of cartilage model: growth occurs by cell division of chondrocytes.

3. Development of primary ossification center: in this region of the diaphysis, bone tissue replaces most of the cartilage.

4. Development of the medullary (marrow) cavity: bone breakdown by osteoclasts forms the medullary cavity.

5. Development of secondary ossification centers: these occur in the epiphyses of the bone.

6. Formation of articular cartilage and epiphyseal plate: both structures consist of hyaline cartilage.

(a) Sequence of events

(b) Fifteen-week fetus. The red areas represent bones that are forming (calcified). Dark gray areas represent cartilage (uncalcified).

? **Where in the cartilage model do secondary ossification centers develop during endochondral ossification?**

2. *Growth of the cartilage model.* Once chondroblasts become deeply buried in the cartilage extracellular matrix, they are called chondrocytes. The cartilage model grows in length by continual cell division of chondrocytes, accompanied by fur-ther secretion of the cartilage extracellular matrix. This type of cartilaginous growth, called **interstitial (endogenous) growth** (growth from within), results in an increase in length. In contrast, growth of the cartilage in thickness is due mainly

to the deposition of extracellular matrix material on the cartilage surface of the model by new chondroblasts that develop from the perichondrium. This process is called **appositional (exogenous) growth** (a-pō-ZISH-o-nal), meaning growth at the outer surface. Interstitial growth and appositional growth of cartilage are described in more detail in Section 4.5.

As the cartilage model continues to grow, chondrocytes in its mid-region hypertrophy (increase in size) and the surrounding cartilage extracellular matrix begins to calcify. Other chondrocytes within the calcifying cartilage die because nutrients can no longer diffuse quickly enough through the extracellular matrix. As these chondrocytes die, the spaces left behind by dead chondrocytes merge into small cavities called lacunae.

③ *Development of the primary ossification center.* Primary ossification proceeds *inward* from the external surface of the bone. A nutrient artery penetrates the perichondrium and the calcifying cartilage model through a nutrient foramen in the midregion of the cartilage model, stimulating osteogenic cells in the perichondrium to differentiate into osteoblasts. Once the perichondrium starts to form bone, it is known as the **periosteum.** Near the middle of the model, periosteal capillaries grow into the disintegrating calcified cartilage, inducing growth of a **primary ossification center,** a region where bone tissue will replace most of the cartilage. Osteoblasts then begin to deposit bone extracellular matrix over the remnants of calcified cartilage, forming spongy bone trabeculae. Primary ossification spreads from this central location toward both ends of the cartilage model.

④ *Development of the medullary (marrow) cavity.* As the primary ossification center grows toward the ends of the bone, osteoclasts break down some of the newly formed spongy bone trabeculae. This activity leaves a cavity, the medullary (marrow) cavity, in the diaphysis (shaft). Eventually, most of the wall of the diaphysis is replaced by compact bone.

⑤ *Development of the secondary ossification centers.* When branches of the epiphyseal artery enter the epiphyses, **secondary ossification centers** develop, usually around the time of birth. Bone formation is similar to what occurs in primary ossification centers. However, in the secondary ossification centers spongy bone remains in the interior of the epiphyses (no medullary cavities are formed here). In contrast to primary ossification, secondary ossification proceeds *outward* from the center of the epiphysis toward the outer surface of the bone.

⑥ *Formation of articular cartilage and the epiphyseal (growth) plate.* The hyaline cartilage that covers the epiphyses becomes the articular cartilage. Prior to adulthood, hyaline cartilage remains between the diaphysis and epiphysis as the epiphyseal (growth) plate, the region responsible for the lengthwise growth of long bones that you will learn about next.

Bone Growth during Infancy, Childhood, and Adolescence

During infancy, childhood, and adolescence, bones throughout the body grow in thickness by appositional growth, and long bones lengthen by the addition of bone material on the diaphyseal side of the epiphyseal plate by interstitial growth.

Growth in Length

The growth in length of long bones involves the following two major events: (1) interstitial growth of cartilage on the epiphyseal side of the epiphyseal plate and (2) replacement of cartilage on the diaphyseal side of the epiphyseal plate with bone by endochondral ossification.

To understand how a bone grows in length, you need to know some of the details of the structure of the epiphyseal plate. The **epiphyseal (growth) plate** (ep-i-FIZ-ē-al) is a layer of hyaline cartilage in the metaphysis of a growing bone that consists of four zones (Figure 6.7b):

1. *Zone of resting cartilage.* This layer is nearest the epiphysis and consists of small, scattered chondrocytes. The term "resting" is used because the cells do not function in bone growth. Rather, they anchor the epiphyseal plate to the epiphysis of the bone.

2. *Zone of proliferating cartilage.* Slightly larger chondrocytes in this zone are arranged like stacks of coins. These chondrocytes undergo interstitial growth as they divide and secrete extracellular matrix. The chondrocytes in this zone divide to replace those that die at the diaphyseal side of the epiphyseal plate.

3. *Zone of hypertrophic cartilage* (hī-per-TRŌ-fik). This layer consists of large, maturing chondrocytes arranged in columns.

4. *Zone of calcified cartilage.* The final zone of the epiphyseal plate is only a few cells thick and consists mostly of chondrocytes that are dead because the extracellular matrix around them has calcified. Osteoclasts dissolve the calcified cartilage, and osteoblasts and capillaries from the diaphysis invade the area. The osteoblasts lay down bone extracellular matrix, replacing the calcified cartilage by the process of endochondral ossification. Recall that endochondral ossification is the replacement of cartilage with bone. As a result, the zone of calcified cartilage becomes the "new diaphysis" that is firmly cemented to the rest of the diaphysis of the bone.

The activity of the epiphyseal plate is the only way that the diaphysis can increase in length. As a bone grows, chondrocytes proliferate on the epiphyseal side of the plate. New chondrocytes replace older ones, which are destroyed by calcification. Thus, the cartilage is replaced by bone on the diaphyseal side of the plate. In this way the thickness of the epiphyseal plate remains relatively constant, but the bone on the diaphyseal side increases

(a) Radiograph showing the epiphyseal plate of the femur of a 3-year-old

— Femur

— EPIPHYSEAL PLATE

— Tibia

Diaphyseal side

— Developing bone of diaphysis

— Zone of calcified cartilage

— Zone of hypertrophic cartilage

— Zone of proliferating cartilage

— Zone of resting cartilage

Epiphyseal side **LM** 400x

(b) Histology of the epiphyseal plate

Figure 6.7 **Epiphyseal (growth) plate.** The epiphyseal (growth) plate appears as a dark band between whiter calcified areas in the radiograph (x-ray) shown in part (a).

🔑 The epiphyseal (growth) plate allows the diaphysis of a bone to increase in length.

in length (Figure 6.7c). If a bone fracture damages the epiphyseal plate, the fractured bone may be shorter than normal once adult stature is reached. This is because damage to cartilage, which is avascular, accelerates closure of the epiphyseal plate due to the cessation of cartilage cell division, thus inhibiting lengthwise growth of the bone.

When adolescence comes to an end (at about age 18 in females and age 21 in males), the epiphyseal plates close; that is, the epiphyseal cartilage cells stop dividing and bone replaces all remaining cartilage. The epiphyseal plate fades, leaving a bony structure called the **epiphyseal line.** With the appearance of the epiphyseal line, bone growth in length stops completely.

Closure of the epiphyseal plate is a gradual process and the degree to which it occurs is useful in determining bone age, predicting adult height, and establishing age at death from skeletal remains, especially in infants, children, and adolescents. For example, an open epiphyseal plate indicates a younger person, while a partially closed epiphyseal plate or a completely closed one indicates an older person. It should also be kept in mind that closure of the epiphyseal plate, on average, takes place 1–2 years earlier in females.

(c) Lengthwise growth of bone at epiphyseal plate

❓ **How does the epiphyseal (growth) plate account for the lengthwise growth of the diaphysis?**

Growth in Thickness

Like cartilage, bone can grow in thickness (diameter) only by appositional growth (Figure 6.8a):

1 At the bone surface, periosteal cells differentiate into osteoblasts, which secrete the collagen fibers and other organic molecules that form bone extracellular matrix. The osteoblasts become surrounded by extracellular matrix and develop into osteocytes. This process forms bone ridges on either side of a periosteal blood vessel. The ridges slowly enlarge and create a groove for the periosteal blood vessel.

2 Eventually, the ridges fold together and fuse, and the groove becomes a tunnel that encloses the blood vessel. The former periosteum now becomes the endosteum that lines the tunnel.

3 Osteoblasts in the endosteum deposit bone extracellular matrix, forming new concentric lamellae. The formation of additional concentric lamellae proceeds inward toward the periosteal blood vessel. In this way, the tunnel fills in, and a new osteon is created.

4 As an osteon is forming, osteoblasts under the periosteum deposit new circumferential lamellae, further increasing the thickness of the bone. As additional periosteal blood vessels become enclosed as in step **1**, the growth process continues.

Recall that as new bone tissue is being deposited on the outer surface of bone, the bone tissue lining the medullary cavity is destroyed by osteoclasts in the endosteum. In this way, the medullary cavity enlarges as the bone increases in thickness (Figure 6.8b).

Remodeling of Bone

Like skin, bone forms before birth but continually renews itself thereafter. **Bone remodeling** is the ongoing replacement of old bone tissue by new bone tissue. It involves **bone resorption,** the removal of minerals and collagen fibers from bone by osteoclasts, and **bone deposition,** the addition of minerals and collagen fibers to bone by osteoblasts. Thus, bone resorption results in the destruction of bone extracellular matrix, while bone deposition results in the formation of bone extracellular matrix. At any given time, about 5% of the total bone mass in the body is being remodeled. The renewal rate for compact bone tissue is about 4% per year, and for spongy bone tissue it is about 20% per year. Remodeling also takes place at different rates in different regions of the body. The distal portion of the femur is replaced about every four months. By contrast, bone in certain areas of the shaft of the femur will not be replaced completely during an individual's life. Even after bones have reached their adult shapes and sizes, old bone is continually destroyed and new bone is formed in its place. Remodeling also removes injured bone, replacing it with new bone tissue. Remodeling may

be triggered by factors such as exercise, sedentary lifestyle, and changes in diet.

Remodeling has several other benefits. Since the strength of bone is related to the degree to which it is stressed, if newly formed bone is subjected to heavy loads, it will grow thicker and therefore be stronger than the old bone. Also, the shape of a bone can be altered for proper support based on the stress patterns experienced during the remodeling process. Finally, new bone is more resistant to fracture than old bone.

CLINICAL CONNECTION | *Remodeling and Orthodontics*

Orthodontics (or-thō-DON-tiks) is the branch of dentistry concerned with the prevention and correction of poorly aligned teeth. The movement of teeth by braces places a stress on the bone that forms the sockets that anchor the teeth. In response to this artificial stress, osteoclasts and osteoblasts remodel the sockets so that the teeth align properly. •

During the process of bone resorption, an osteoclast attaches tightly to the bone surface at the endosteum or periosteum and forms a leakproof seal at the edges of its ruffled border (see Figure 6.2). Then it releases protein-digesting lysosomal enzymes and several acids into the sealed pocket. The enzymes digest collagen fibers and other organic substances while the acids dissolve the bone minerals. Working together, several osteoclasts carve out a small tunnel in the old bone. The degraded bone proteins and extracellular matrix minerals, mainly calcium and phosphorus, enter an osteoclast by endocytosis, cross the cell in vesicles, and undergo exocytosis on the side opposite the ruffled border. Now in the interstitial fluid, the products of bone resorption diffuse into nearby blood capillaries. Once a small area of bone has been resorbed, osteoclasts depart and osteoblasts move in to rebuild the bone in that area.

CLINICAL CONNECTION | *Paget's Disease*

A delicate balance exists between the actions of osteoclasts and osteoblasts. Should too much new tissue be formed, the bones become abnormally thick and heavy. If too much mineral material is deposited in the bone, the surplus may form thick bumps, called *spurs,* on the bone that interfere with movement at joints. Excessive loss of calcium or tissue weakens the bones, and they may break, as occurs in osteoporosis, or they may become too flexible, as in rickets and osteomalacia. In **Paget's disease,** there is an excessive proliferation of osteoclasts so that bone resorption occurs faster than bone deposition. In response, osteoblasts attempt to compensate, but the new bone is weaker because it has a higher proportion of spongy to compact bone, mineralization is decreased, and the newly synthesized extracellular matrix contains abnormal proteins. The newly formed bone, especially that of the pelvis, limbs, lower vertebrae, and skull, becomes enlarged, hard, and brittle and fractures easily. •

Figure 6.8 Bone growth in thickness.

As new bone is deposited on the outer surface of bone by osteoblasts, the bone tissue lining the medullary cavity is destroyed by osteoclasts in the endosteum.

Periosteal ridges
Periosteum
Periosteal blood vessel
Perforating canal
Groove

1 Ridges in periosteum create groove for periosteal blood vessel.

Endosteum
Tunnel

2 Periosteal ridges fuse, forming an endosteum-lined tunnel.

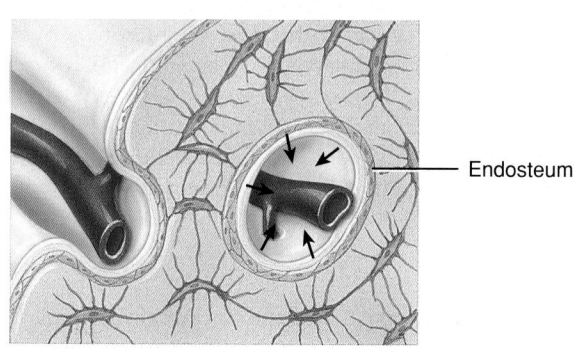

Endosteum

3 Osteoblasts in endosteum build new concentric lamellae inward toward center of tunnel, forming a new osteon.

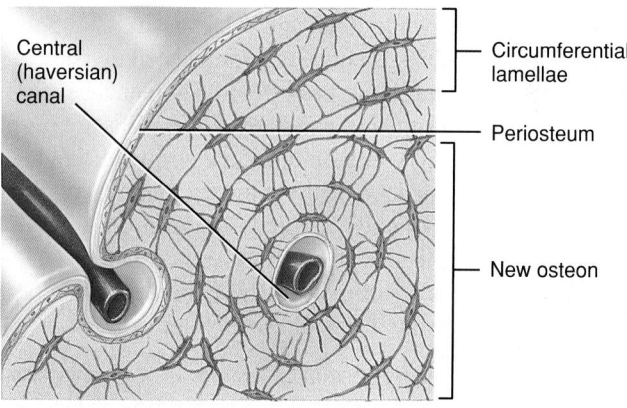

Central (haversian) canal
Circumferential lamellae
Periosteum
New osteon

4 Bone grows outward as osteoblasts in periosteum build new circumferential lamellae. Osteon formation repeats as new periosteal ridges fold over blood vessels.

(a) Microscopic details

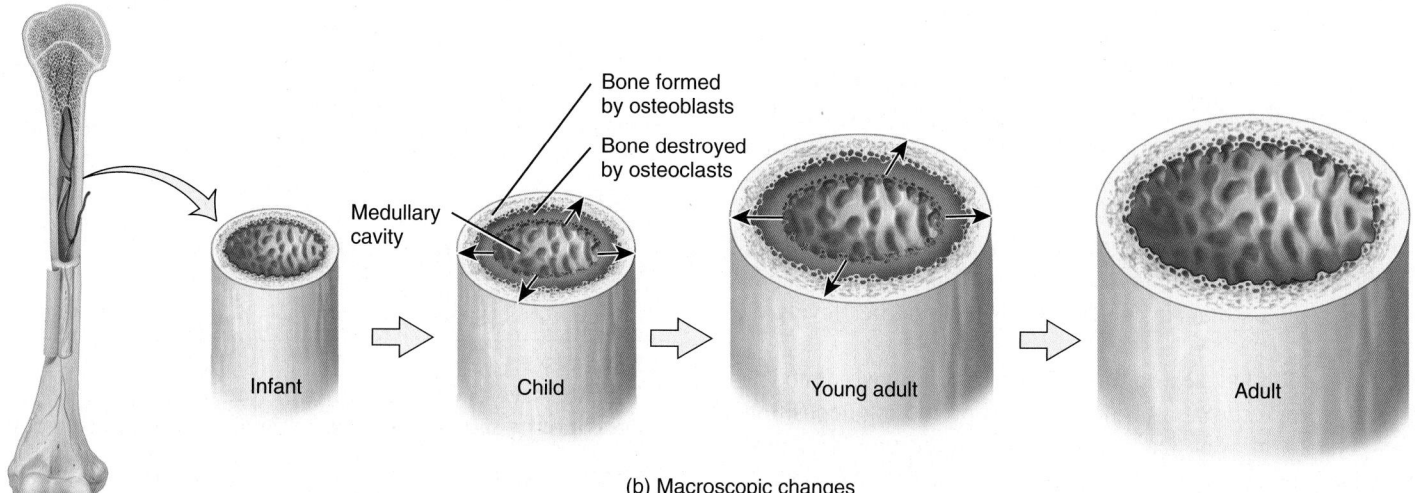

Bone formed by osteoblasts
Bone destroyed by osteoclasts
Medullary cavity

Infant   Child   Young adult   Adult

(b) Macroscopic changes

? **How does the medullary cavity enlarge during growth in thickness?**

Factors Affecting Bone Growth and Bone Remodeling

Normal bone metabolism—growth in the young and bone remodeling in the adult—depends on several factors. These include adequate dietary intake of minerals and vitamins, as well as sufficient levels of several hormones.

1. *Minerals.* Large amounts of calcium and phosphorus are needed while bones are growing, as are smaller amounts of magnesium, fluoride, and manganese. These minerals are also necessary during bone remodeling.

2. *Vitamins.* Vitamin A stimulates activity of osteoblasts. Vitamin C is needed for synthesis of collagen, the main bone protein. As you will soon learn, vitamin D helps build bone by increasing the absorption of calcium from foods in the gastrointestinal tract into the blood. Vitamins K and B_{12} are also needed for synthesis of bone proteins.

3. *Hormones.* During childhood, the hormones most important to bone growth are the insulinlike growth factors (IGFs), which are produced by the liver and bone tissue (see Section 18.6). IGFs stimulate osteoblasts, promote cell division at the epiphyseal plate and in the periosteum, and enhance synthesis of the proteins needed to build new bone. IGFs are produced in response to the secretion of human growth hormone (hGH) from the anterior lobe of the pituitary gland (see Section 18.6). Thyroid hormones (T_3 and T_4) from the thyroid gland also promote bone growth by stimulating osteoblasts. In addition, the hormone insulin from the pancreas promotes bone growth by increasing the synthesis of bone proteins.

At puberty, the secretion of hormones known as sex hormones causes a dramatic effect on bone growth. The **sex hormones** include estrogens (produced by the ovaries) and androgens such as testosterone (produced by the testes). Although females have much higher levels of estrogens and males have higher levels of androgens, females also have low levels of androgens, and males have low levels of estrogens. The adrenal glands of both sexes produce androgens, and other tissues, such as adipose tissue, can convert androgens to estrogens. These hormones are responsible for increased osteoblast activity and synthesis of bone extracellular matrix and the sudden "growth spurt" that occurs during the teenage years. Estrogens also promote changes in the skeleton that are typical of females, such as widening of the pelvis. Ultimately sex hormones, especially estrogens in both sexes, shut down growth at epiphyseal (growth) plates, causing elongation of the bones to cease. Lengthwise growth of bones typically ends earlier in females than in males due to their higher levels of estrogens.

During adulthood, sex hormones contribute to bone remodeling by slowing resorption of old bone and promoting deposition of new bone. One way that estrogens slow resorption is by promoting apoptosis (programmed death) of osteoclasts. As you will see shortly, parathyroid hormone, calcitriol (the active form of vitamin D), and calcitonin are other hormones that can affect bone remodeling.

Moderate weight-bearing exercises maintain sufficient strain on bones to increase and maintain their density.

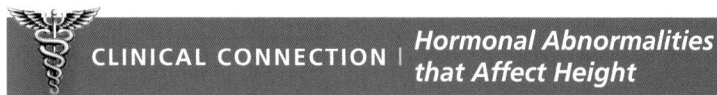

Excessive or deficient secretion of hormones that normally control bone growth can cause a person to be abnormally tall or short. Oversecretion of hGH during childhood produces **giantism,** in which a person becomes much taller and heavier than normal. Undersecretion of hGH produces **pituitary dwarfism,** in which a person has short stature. (The usual adult height of a *dwarf* is under 4 feet 10 inches.) Although the head, trunk, and limbs of a pituitary dwarf are smaller than normal, they are proportionate. The condition can be treated medically with hGH until epiphyseal plate closure. Oversecretion of hGH during adulthood is called **acromegaly** (ak′-rō-MEG-a-lē). Although hGH cannot produce further lengthening of the long bones because the epiphyseal (growth) plates are already closed, the bones of the hands, feet, and jaws thicken and other tissues enlarge. In addition, the eyelids, lips, tongue, and nose enlarge, and the skin thickens and develops furrows, especially on the forehead and soles. **Achondroplasia** (ā-kon-dro-PLĀ-zē-a; *a-* = without; *-chondro-* = cartilage; *-plasia* = to mold) is an inherited condition in which the conversion of cartilage to bone is abnormal. It results in the most common type of dwarfism, called **achondroplastic dwarfism.** These individuals are typically about 4 feet tall as adults. They have an average-size trunk, short limbs, and a slightly enlarged head with a prominent forehead and flattened nose at the bridge. The condition is essentially untreatable, although some individuals opt for limb-lengthening surgery. •

✔CHECKPOINT

16. What are the major events of intramembranous ossification and endochondral ossification, and how are they different?
17. Describe the zones of the epiphyseal (growth) plate and their functions, and the significance of the epiphyseal line.
18. Explain how bone growth in length differs from bone growth in thickness.
19. How could the metaphyseal area of a bone help determine the age of a skeleton?
20. Define remodeling, and describe the roles of osteoblasts and osteoclasts in the process.
21. What factors affect bone growth and bone remodeling?

6.6 FRACTURE AND REPAIR OF BONE

◉ OBJECTIVES

- Describe several common types of fractures.
- Describe the sequence of events involved in fracture repair.

A **fracture** (FRAK-choor) is any break in a bone. Fractures are named according to their severity, the shape or position of the fracture line, or even the physician who first described them. Some of the common types of fractures are shown in Table 6.1.

TABLE 6.1

Some Common Fractures

FRACTURE	DESCRIPTION	ILLUSTRATION	RADIOGRAPH
Open (Compound)	The broken ends of the bone protrude through the skin. Conversely, a **closed (simple) fracture** does not break the skin.		Humerus Radius Ulna
Comminuted (KOM-i-noo-ted; *com-* = together; *-minuted* = crumbled)	The bone is splintered, crushed, or broken into pieces at the site of impact, and smaller bone fragments lie between the two main fragments.		Humerus
Greenstick	A partial fracture in which one side of the bone is broken and the other side bends; similar to the way a green twig breaks on one side while the other side stays whole, but bends; occurs only in children, whose bones are not fully ossified and contain more organic material than inorganic material.		Ulna Radius Wrist bones
Impacted	One end of the fractured bone is forcefully driven into the interior of the other.		Humerus

TABLE 6.1 CONTINUES

In some cases, a bone may fracture without visibly breaking. A **stress fracture** is a series of microscopic fissures in bone that forms without any evidence of injury to other tissues. In healthy adults, stress fractures result from repeated, strenuous activities such as running, jumping, or aerobic dancing. Stress fractures are quite painful and also result from disease processes that disrupt normal bone calcification, such as osteoporosis (discussed in Disorders: Homeostatic Imbalances at the end of this chapter). About 25% of stress fractures involve the tibia. Although standard x-ray images often fail to reveal the presence of stress fractures, they show up clearly in a bone scan.

TABLE 6.1 CONTINUED

Some Common Fractures

FRACTURE	DESCRIPTION	ILLUSTRATION	RADIOGRAPH
Pott	Fracture of the distal end of the lateral leg bone (fibula), with serious injury of the distal tibial articulation.		
Colles' (KOL-ēz)	Fracture of the distal end of the lateral forearm bone (radius) in which the distal fragment is displaced posteriorly.		

CLINICAL CONNECTION | *Treatments for Fractures*

Treatments for fractures vary according to age, type of fracture, and the bone involved. The ultimate goals of fracture treatment are realignment of the bone fragments, immobilization to maintain realignment, and restoration of function. For bones to unite properly, the fractured ends must be brought into alignment. This process, called **reduction,** is commonly referred to as setting a fracture. In **closed reduction,** the fractured ends of a bone are brought into alignment by manual manipulation, and the skin remains intact. In **open reduction,** the fractured ends of a bone are brought into alignment by a surgical procedure using internal fixation devices such as screws, plates, pins, rods, and wires. Following reduction, a fractured bone may be kept immobilized by a cast, sling, splint, elastic bandage, external fixation device, or a combination of these devices. •

The repair of a bone fracture involves the following steps (Figure 6.9):

1. ***Formation of fracture hematoma.*** Blood vessels crossing the fracture line are broken. As blood leaks from the torn ends of the vessels, a mass of blood (usually clotted) forms around the site of the fracture. This mass of blood, called a **fracture hematoma** (hē′-ma-TŌ-ma; *hemat-* = blood; *-oma* = tumor), usually forms 6 to 8 hours after the injury. Because the circulation of blood stops at the site where the fracture hematoma forms, nearby bone cells die. Swelling and inflammation occur in response to dead bone cells, producing additional cellular debris. Phagocytes (neutrophils and macrophages) and osteoclasts begin to remove the dead or damaged tissue in and around the fracture hematoma. This stage may last up to several weeks.

2. ***Fibrocartilaginous callus formation.*** Fibroblasts from the periosteum invade the fracture site and produce collagen fibers. In addition, cells from the periosteum develop into chondroblasts and begin to produce fibrocartilage in this region. These events lead to the development of a **fibrocartilaginous (soft) callus** (fī-brō-kar-ti-LAJ-i-nus), a mass of repair tissue consisting of collagen fibers and cartilage that bridges the broken ends of the bone. Formation of the fibrocartilaginous callus takes about 3 weeks.

3. ***Bony callus formation.*** In areas closer to well-vascularized healthy bone tissue, osteogenic cells develop into osteoblasts, which begin to produce spongy bone trabeculae. The trabeculae join living and dead portions of the original bone fragments. In time, the fibrocartilage is converted to spongy bone, and the callus is then referred to as a **bony (hard) callus.** The bony callus lasts about 3 to 4 months.

4. ***Bone remodeling.*** The final phase of fracture repair is **bone remodeling** of the callus. Dead portions of the original fragments of broken bone are gradually resorbed by osteoclasts. Compact bone replaces spongy bone around the periphery of the fracture. Sometimes, the repair process is so thorough that the fracture line is undetectable, even in a radiograph (x-ray).

However, a thickened area on the surface of the bone remains as evidence of a healed fracture.

Although bone has a generous blood supply, healing sometimes takes months. The calcium and phosphorus needed to strengthen and harden new bone are deposited only gradually, and bone cells generally grow and reproduce slowly. The temporary disruption in their blood supply also helps explain the slowness of healing of severely fractured bones.

✔CHECKPOINT

22. List the types of fractures and outline the four steps involved in fracture repair.

Figure 6.9 **Steps in repair of a bone fracture.**

🔒 ▱ Bone heals more rapidly than cartilage because its blood supply is more plentiful.

Osteon
Periosteum
Compact bone
Spongy bone
Fracture hematoma

Blood vessel

Phagocyte
Fracture hematoma
Red blood cell
Bone fragment
Osteocyte

1 Formation of fracture hematoma

Fibroblast
Phagocyte
Fibrocartilaginous callus
Osteoblast
Collagen fiber
Chondroblast
Cartilage

2 Fibrocartilaginous callus formation

Bony callus
Osteoblast
Spongy bone
Osteocyte

3 Bony callus formation

New compact bone
Osteoclast

4 Bone remodeling

❓ **Why does it sometimes take months for a fracture to heal?**

6.7 BONE'S ROLE IN CALCIUM HOMEOSTASIS

● OBJECTIVES

• Describe the importance of calcium in the body.
• Explain how blood calcium level is regulated.

Bone is the body's major calcium reservoir, storing 99% of total body calcium. One way to maintain the level of calcium in the blood is to control the rates of calcium resorption from bone into blood and of calcium deposition from blood into bone. Both nerve and muscle cells depend on a stable level of calcium ions (Ca^{2+}) in extracellular fluid to function properly. Blood clotting also requires Ca^{2+}. Also, many enzymes require Ca^{2+} as a cofactor (an additional substance needed for an enzymatic reaction to occur). For this reason, the blood plasma level of Ca^{2+} is very closely regulated between 9 and 11 mg/100 mL. Even small changes in Ca^{2+} concentration outside this range may prove fatal—the heart may stop (cardiac arrest) if the concentration goes too high, or breathing may cease (respiratory arrest) if the level falls too low. The role of bone in calcium homeostasis is to help "buffer" the blood Ca^{2+} level, releasing Ca^{2+} into blood plasma (using osteoclasts) when the level decreases, and absorbing Ca^{2+} (using osteoblasts) when the level rises.

Ca^{2+} exchange is regulated by hormones, the most important of which is **parathyroid hormone (PTH)** secreted by the parathyroid glands (see Figure 18.13). This hormone increases blood Ca^{2+} level. PTH secretion operates via a negative feedback system (Figure 6.10). If some stimulus causes the blood Ca^{2+} level to decrease, parathyroid gland cells (receptors) detect this change and increase their production of a molecule known as cyclic adenosine monophosphate (cyclic AMP). The gene for PTH within the nucleus of a parathyroid gland cell (the control center) detects the intracellular increase in cyclic AMP (the input). As a result, PTH synthesis speeds up, and more PTH (the output) is released into the blood. The presence of higher levels of PTH increases the number and activity of osteoclasts (effectors), which step up the pace of bone resorption. The resulting release of Ca^{2+} from bone into blood returns the blood Ca^{2+} level to normal.

PTH also acts on the kidneys (effectors) to decrease loss of Ca^{2+} in the urine, so more is retained in the blood. And PTH stimulates formation of **calcitriol** (the active form of vitamin D), a hormone that promotes absorption of calcium from foods in the gastrointestinal tract into the blood. Both of these actions also help elevate blood Ca^{2+} level.

Another hormone works to decrease blood Ca^{2+} level. When blood Ca^{2+} rises above normal, *parafollicular cells* in the thyroid gland secrete **calcitonin (CT)** (kal-si-TŌ-nin). CT inhibits activity of osteoclasts, speeds blood Ca^{2+} uptake by bone, and accelerates Ca^{2+} deposition into bones. The net result is that CT promotes bone formation and decreases blood Ca^{2+} level. Despite these effects, the role of CT in normal calcium homeostasis is uncertain because it can be completely absent without causing symptoms. Nevertheless, calcitonin harvested from salmon (Miacalcin®) is an effective drug for treating osteoporosis because it slows bone resorption.

Figure 18.14 summarizes the roles of parathyroid hormone, calcitriol, and calcitonin in regulation of blood Ca^{2+} level.

Figure 6.10 Negative feedback system for the regulation of blood calcium (Ca^{2+}) concentration. PTH = parathyroid hormone.

Release of calcium from bone matrix and retention of calcium by the kidneys are the two main ways that blood calcium level can be increased.

Some stimulus disrupts homeostasis by

Decreasing

Blood calcium (Ca^{2+}) level

Receptors
Parathyroid gland cells detect lowered Ca^{2+} concentration

Input — Increased production of cyclic AMP

Control center
PTH gene "turned on"

Output — Increased release of PTH

Effectors
Osteoclasts increase bone resorption

Kidneys retain Ca^{2+} in blood, excrete phosphate in urine, and produce calcitriol

Return to homeostasis when response brings blood Ca^{2+} level back to normal

Increase in blood Ca^{2+} level

? **What body functions depend on proper levels of Ca^{2+}?**

✔ CHECKPOINT

23. How do hormones act on bone to regulate calcium homeostasis?

6.8 EXERCISE AND BONE TISSUE

◉ **OBJECTIVE**

- Describe how exercise and mechanical stress affect bone tissue.

Within limits, bone tissue has the ability to alter its strength in response to changes in mechanical stress. When placed under stress, bone tissue becomes stronger through increased deposition of mineral salts and production of collagen fibers by osteoblasts. Without mechanical stress, bone does not remodel normally because bone resorption occurs more quickly than bone formation. Research has shown that high-impact intermittent strains more strongly influence bone deposition as compared with lower-impact constant strains. Therefore, running and jumping stimulate bone growth more dramatically than walking.

The main mechanical stresses on bone are those that result from the pull of skeletal muscles and the pull of gravity. If a person is bedridden or has a fractured bone in a cast, the strength of the unstressed bones diminishes because of the loss of bone minerals and decreased numbers of collagen fibers. Astronauts subjected to the microgravity of space also lose bone mass. In any of these cases, bone loss can be dramatic—as much as 1% per week. In contrast, the bones of athletes, which are repetitively and highly stressed, become notably thicker and stronger than those of astronauts or nonathletes. Weight-bearing activities, such as walking or moderate weight lifting, help build and retain bone mass. Adolescents and young adults should engage in regular weight-bearing exercise prior to the closure of the epiphyseal plates to help build total mass prior to its inevitable reduction with aging. However, people of all ages can and should strengthen their bones by engaging in weight-bearing exercise.

✔ **CHECKPOINT**

24. How do mechanical stresses strengthen bone tissue?
25. Would children raised in space ever be able to return to Earth?
26. Why is it important to engage in weight-bearing exercises before the epiphyseal plates close?

6.9 AGING AND BONE TISSUE

◉ **OBJECTIVE**

- Describe the effects of aging on bone tissue.

From birth through adolescence, more bone tissue is produced than is lost during bone remodeling. In young adults the rates of bone deposition and resorption are about the same. As the level of sex hormones diminishes during middle age, especially in women after menopause, a decrease in bone mass occurs because bone resorption by osteoclasts outpaces bone deposition by osteoblasts. In old age, loss of bone through resorption occurs more rapidly than bone gain. Because women's bones generally are smaller and less massive than men's bones to begin with, loss of bone mass in old age typically has a greater adverse effect in females. These factors contribute to the higher incidence of osteoporosis in females.

There are two principal effects of aging on bone tissue: loss of bone mass and brittleness. Loss of bone mass results from **demineralization** (dē-min′-er-al-i-ZĀ-shun), the loss of calcium and other minerals from bone extracellular matrix. This loss usually begins after age 30 in females, accelerates greatly around age 45 as levels of estrogens decrease, and continues until as much as 30% of the calcium in bones is lost by age 70. Once bone loss begins in females, about 8% of bone mass is lost every 10 years. In males, calcium loss typically does not begin until after age 60, and about 3% of bone mass is lost every 10 years. The loss of calcium from bones is one of the problems in osteoporosis (described shortly).

The second principal effect of aging on the skeletal system, brittleness, results from a decreased rate of protein synthesis. Recall that the organic part of bone extracellular matrix, mainly collagen fibers, gives bone its tensile strength. The loss of tensile strength causes the bones to become very brittle and susceptible to fracture. In some elderly people, collagen fiber synthesis slows, in part due to diminished production of human growth hormone. In addition to increasing the susceptibility to fractures, loss of bone mass also leads to deformity, pain, loss of height, and loss of teeth.

Table 6.2 summarizes the factors that influence bone metabolism.

✔ **CHECKPOINT**

27. What is demineralization, and how does it affect the functioning of bone?
28. What changes occur in the organic part of bone extracellular matrix with aging?

TABLE 6.2

Summary of Factors That Affect Bone Growth

FACTOR	COMMENT
MINERALS	
Calcium and phosphorus	Make bone extracellular matrix hard.
Magnesium	Helps form bone extracellular matrix.
Fluoride	Helps strengthen bone extracellular matrix.
Manganese	Activates enzymes involved in synthesis of bone extracellular matrix.
VITAMINS	
Vitamin A	Needed for the activity of osteoblasts during remodeling of bone; deficiency stunts bone growth; toxic in high doses.
Vitamin C	Needed for synthesis of collagen, the main bone protein; deficiency leads to decreased collagen production, which slows down bone growth and delays repair of broken bones.
Vitamin D	Active form (calcitriol) is produced by the kidneys; helps build bone by increasing absorption of calcium from gastrointestinal tract into blood; deficiency causes faulty calcification and slows down bone growth; may reduce the risk of osteoporosis but is toxic if taken in high doses.
Vitamins K and B$_{12}$	Needed for synthesis of bone proteins; deficiency leads to abnormal protein production in bone extracellular matrix and decreased bone density.
HORMONES	
Human growth hormone (hGH)	Secreted by the anterior lobe of the pituitary gland; promotes general growth of all body tissues, including bone, mainly by stimulating production of insulinlike growth factors.
Insulinlike growth factors (IGFs)	Secreted by the liver, bones, and other tissues upon stimulation by human growth hormone; promotes normal bone growth by stimulating osteoblasts and by increasing the synthesis of proteins needed to build new bone.
Thyroid hormones (thyroxine and triiodothyronine)	Secreted by thyroid gland; promote normal bone growth by stimulating osteoblasts.
Insulin	Secreted by the pancreas; promotes normal bone growth by increasing the synthesis of bone proteins.
Sex hormones (estrogens and testosterone)	Secreted by the ovaries in women (estrogens) and by the testes in men (testosterone); stimulate osteoblasts and promote the sudden "growth spurt" that occurs during the teenage years; shut down growth at the epiphyseal plates around age 18–21, causing lengthwise growth of bone to end; contribute to bone remodeling during adulthood by slowing bone resorption by osteoclasts and promoting bone deposition by osteoblasts.
Parathyroid hormone (PTH)	Secreted by the parathyroid glands; promotes bone resorption by osteoclasts; enhances recovery of calcium ions from urine; promotes formation of the active form of vitamin D (calcitriol).
Calcitonin (CT)	Secreted by the thyroid gland; inhibits bone resorption by osteoclasts.
EXERCISE	
	Weight-bearing activities stimulate osteoblasts and, consequently, help build thicker, stronger bones and retard loss of bone mass that occurs as people age.
AGING	
	As the level of sex hormones diminishes during middle age to older adulthood, especially in women after menopause, bone resorption by osteoclasts outpaces bone deposition by osteoblasts, which leads to a decrease in bone mass and an increased risk of osteoporosis.

DISORDERS: HOMEOSTATIC IMBALANCES

Osteoporosis

Osteoporosis (os'-tē-ō-pō-RŌ-sis; *por* = passageway; *-osis* = condition), literally a condition of porous bones, affects 10 million people a year in the United States (Figure 6.11). In addition, 18 million people have low bone mass (*osteopenia*), which puts them at risk for osteoporosis. The basic problem is that bone resorption (breakdown) outpaces bone deposition (formation). In large part this is due to depletion of calcium from the body—more calcium is lost in urine, feces, and sweat than is absorbed from the diet. Bone mass becomes so depleted that bones fracture, often spontaneously, under the mechanical stresses of everyday living. For example, a hip fracture might result from simply sitting down too quickly. In the United States, osteoporosis causes more than one and a half million fractures a year, mainly in the hips, wrists, and vertebrae. Osteoporosis afflicts the entire skeletal system. In addition to fractures, osteoporosis causes shrinkage of vertebrae, height loss, hunched backs, and bone pain.

Osteoporosis primarily affects middle-aged and elderly people, 80 percent of them women. Older women suffer from osteoporosis more often than men for two reasons: (1) Women's bones are less massive than men's bones, and (2) production of estrogens in women declines dramatically at menopause, while production of the main androgen, testosterone, in older men wanes gradually and only slightly. Estrogens and testosterone stimulate osteoblast activity and synthesis

Figure 6.11 Comparison of spongy bone tissue from (a) a normal young adult and (b) a person with osteoporosis. Notice the weakened trabeculae in (b). Compact bone tissue is similarly affected by osteoporosis.

 In osteoporosis, bone resorption outpaces bone formation, so bone mass decreases.

(a) Normal bone   (b) Osteoporotic bone

? If you wanted to develop a drug to lessen the effects of osteoporosis, would you look for a chemical that inhibits the activity of osteoblasts or that of osteoclasts?

of bone matrix. Besides gender, risk factors for developing osteoporosis include a family history of the disease, European or Asian ancestry, thin or small body build, an inactive lifestyle, cigarette smoking, a diet low in calcium and vitamin D, more than two alcoholic drinks a day, and the use of certain medications.

Osteoporosis is diagnosed by taking a family history and undergoing a *bone mineral density* (BMD) test. Performed like x-rays, BMD tests measure bone density. They can also be used to confirm a diagnosis of osteoporosis, determine the rate of bone loss, and monitor the effects of treatment.

Treatment options for osteoporosis are varied. With regard to nutrition, a diet high in calcium is important to reduce the risk of fractures. Vitamin D is necessary for the body to utilize calcium. In terms of exercise, regular performance of weight-bearing exercises has been shown to maintain and build bone mass. These exercises include walking, jogging, hiking, climbing stairs, playing tennis, and dancing. Resistance exercises, such as weight lifting, also build bone strength and muscle mass.

Medications used to treat osteoporosis are generally of two types: (1) **antireabsorptive drugs** slow down the progression of bone loss and (2) **bone-building drugs** promote increasing bone mass. Among the antireabsorptive drugs are (1) *bisphosphonates,* which inhibit osteoclasts (Fosamax®, Actonel®, Boniva®, and calcitonin); (2) *selective estrogen receptor modulators*, which mimic the effects of estrogens without unwanted side effects (Raloxifene®, Evista®); and (3) estrogen replacement therapy (ERT), which replaces estrogens lost during and after menopause (Premarin®), and hormone replacement therapy (HRT), which replaces estrogens and progesterone lost during and after menopause (Prempro®). ERT helps maintain and increase bone mass after menopause. Women on ERT have a slightly increased risk of stroke and blood clots. HRT also helps maintain and increase bone mass. Women on HRT have increased risks of heart disease, breast cancer, stroke, blood clots, and dementia.

Among the bone-building drugs is parathyroid hormone (PTH), which stimulates osteoblasts to produce new bone (Forteo®). Others are under development.

Rickets and Osteomalacia

Rickets and **osteomalacia** (os'-tē-ō-ma-LĀ-shē-a; *malacia* = softness) are two forms of the same disease that result from inadequate calcification of the extracellular bone matrix, usually caused by a vitamin D deficiency. Rickets is a disease of children in which the growing bones become "soft" or rubbery and are easily deformed. Because new bone formed at the epiphyseal (growth) plates fails to ossify, bowed legs and deformities of the skull, rib cage, and pelvis are common. Osteomalacia is the adult counterpart of rickets, sometimes called *adult rickets*. New bone formed during remodeling fails to calcify, and the person experiences varying degrees of pain and tenderness in bones, especially the hip and legs. Bone fractures also result from minor trauma. Prevention and treatment for rickets and osteomalacia consists of the administration of adequate vitamin D and exposure to moderate amounts of sunlight.

MEDICAL TERMINOLOGY

Osteoarthritis (os'-tē-ō-ar-THRĪ-tis; *arthr* = joint) The degeneration of articular cartilage such that the bony ends touch; the resulting friction of bone against bone worsens the condition. Usually associated with the elderly.

Osteomyelitis (os'-tē-ō-mī-e-LĪ-tis) An infection of bone characterized by high fever, sweating, chills, pain, nausea, pus formation, edema, and warmth over the affected bone and rigid overlying muscles. It is often caused by bacteria, usually *Staphylococcus* *aureus.* The bacteria may reach the bone from outside the body (through open fractures, penetrating wounds, or orthopedic surgical procedures); from other sites of infection in the body (abscessed teeth, burn infections, urinary tract infections, or upper respiratory infections) via the blood; and from adjacent soft tissue infections (as occurs in diabetes mellitus).

Osteopenia (os'-tē-ō-PĒ-nē-a; *penia* = poverty) Reduced bone mass due to a decrease in the rate of bone synthesis to a level too low to compensate for normal bone resorption; any decrease in bone mass below normal. An example is osteoporosis.

Osteosarcoma (os'-tē-ō-sar-KŌ-ma; *sarcoma* = connective tissue tumor) Bone cancer that primarily affects osteoblasts and occurs most often in teenagers during their growth spurt; the most common sites are the metaphyses of the thigh bone (femur), shin bone (tibia), and arm bone (humerus). Metastases occur most often in lungs; treatment consists of multidrug chemotherapy and removal of the malignant growth, or amputation of the limb.

CHAPTER REVIEW AND RESOURCE SUMMARY

Review	Resource

Introduction

1. A bone is made up of several different tissues: bone or osseous tissue, cartilage, dense connective tissue, epithelium, adipose tissue, and nervous tissue.
2. The entire framework of bones and their cartilages constitutes the skeletal system.

6.1 Functions of Bone and the Skeletal System

1. The skeletal system functions in support, protection, movement, mineral homeostasis, blood cell production, and triglyceride storage.

Anatomy Overview - Bone Structure and Tissues

6.2 Structure of Bone

1. Parts of a typical long bone are the diaphysis (shaft), proximal and distal epiphyses (ends), metaphyses, articular cartilage, periosteum, medullary (marrow) cavity, and endosteum.

Exercise - Growing Long Bone

6.3 Histology of Bone Tissue

1. Bone tissue consists of widely separated cells surrounded by large amounts of extracellular matrix.
2. The four principal types of cells in bone tissue are osteogenic cells, osteoblasts (bone-building cells), osteocytes (maintain daily activity of bone), and osteoclasts (bone-destroying cells).
3. The extracellular matrix of bone contains abundant mineral salts (mostly hydroxyapatite) and collagen fibers.
4. Compact bone tissue consists of osteons (haversian systems) with little space between them.
5. Compact bone tissue lies over spongy bone tissue in the epiphyses and makes up most of the bone tissue of the diaphysis. Functionally, compact bone tissue is the strongest form of bone and protects, supports, and resists stress.
6. Spongy bone tissue does not contain osteons. It consists of trabeculae surrounding many red bone marrow–filled spaces.
7. Spongy bone tissue forms most of the structure of short, flat, and irregular bones, and the interior of the epiphyses in long bones. Functionally, spongy bone tissue trabeculae offer resistance along lines of stress, support and protect red bone marrow, and make bones lighter for easier movement.

Anatomy Overview - Compact Bone
Anatomy Overview - Spongy Bone
Figure 6.3 - Histology of Compact and Spongy Bone
Exercise - Concentrate on Bone Tissue

6.4 Blood and Nerve Supply of Bone

1. Long bones are supplied by periosteal, nutrient, metaphyseal, and epiphyseal arteries; veins accompany the arteries.
2. Nerves accompany blood vessels in bone; the periosteum is rich in sensory neurons.

6.5 Bone Formation

1. The process by which bone forms, called ossification, occurs in four principal situations: (1) the initial formation of bones in an embryo and fetus; (2) the growth of bones during infancy, childhood, and adolescence until their adult sizes are reached; (3) the remodeling of bone (replacement of old bone by new bone tissue throughout life); and (4) the repair of fractures (breaks in bones) throughout life.
2. Bone development begins during the sixth or seventh week of embryonic development. The two types of ossification, intramembranous and endochondral, involve the replacement of a preexisting connective tissue with bone. Intramembranous ossification refers to bone formation directly within mesenchyme arranged in sheetlike layers that resemble membranes. Endochondral ossification refers to

Animation - Bone Formation
Animation - Bone Elongation and Bone Widening
Animation - Bone Remodeling
Animation - Regulation of Blood Calcium
Animation - Bone Dynamics and Tissue
Animation - Regulation of Bone Growth

Review

Resource

bone formation within hyaline cartilage that develops from mesenchyme. The primary ossification center of a long bone is in the diaphysis. Cartilage degenerates, leaving cavities that merge to form the medullary cavity. Osteoblasts lay down bone. Next, ossification occurs in the epiphyses, where bone replaces cartilage, except for the epiphyseal (growth) plate.

3. The epiphyseal plate consists of four zones: zone of resting cartilage, zone of proliferating cartilage, zone of hypertrophic cartilage, and zone of calcified cartilage. Because of the cell division in the epiphyseal (growth) plate, the diaphysis of a bone increases in length.

4. Bone grows in thickness or diameter due to the addition of new bone tissue by periosteal osteoblasts around the outer surface of the bone (appositional growth).

5. Bone remodeling is an ongoing process in which osteoclasts carve out small tunnels in old bone tissue and then osteoblasts rebuild it.

6. In bone resorption, osteoclasts release enzymes and acids that degrade collagen fibers and dissolve mineral salts.

7. Dietary minerals (especially calcium and phosphorus) and vitamins (A, C, D, K, and B_{12}) are needed for bone growth and maintenance. Insulinlike growth factors (IGFs), human growth hormone, thyroid hormones, and insulin stimulate bone growth.

8. Sex hormones slow resorption of old bone and promote new bone deposition.

Figure 6.5 - Intramembranous Ossification
Figure 6.6 - Enchondral Ossification
Figure 6.7 - Epiphyseal (Growth) Plate
Exercise - Observe the Ossification
Exercise - Bone Growth Sequencing
Exercise - Concentrate on Bone Tissue
Exercise - Regulation Sequences

6.6 Fracture and Repair of Bone

1. A fracture is any break in a bone. Types of fractures include closed (simple), open (compound), comminuted, greenstick, impacted, stress, Pott's, and Colles'.

2. Fracture repair involves formation of a fracture hematoma, a fibrocartilaginous callus and a bony callus, and bone remodeling.

Animation - Bone Remodeling
Figure 6.9 - Steps in Repair of a Bone Fracture

6.7 Bone's Role in Calcium Homeostasis

1. Bone is the major reservoir for calcium in the body.

2. Parathyroid hormone (PTH) secreted by the parathyroid glands increases blood Ca^{2+} level. Calcitonin (CT) from the thyroid gland has the potential to decrease blood Ca^{2+} level. Vitamin D enhances absorption of calcium and phosphate and thus raises the blood levels of these substances.

Animation - Regulation of Blood Calcium
Concepts and Connections - Blood Calcium Regulation
Exercise - Calcium Homeostasis
Exercise - Regulation Sequencing

6.8 Exercise and Bone Tissue

1. Mechanical stress increases bone strength by increasing deposition of mineral salts and production of collagen fibers.

2. Removal of mechanical stress weakens bone through demineralization and collagen fiber reduction.

Animation - Response to Stress in Adult Bones

6.9 Aging and Bone Tissue

1. The principal effect of aging is demineralization, a loss of calcium from bones, which is due to reduced osteoblast activity.

2. Another effect is decreased production of extracellular matrix proteins (mostly collagen fibers), which makes bones more brittle and thus more susceptible to fracture.

SELF-QUIZ QUESTIONS

Fill in the blanks in the following statements.

1. Bone growth in length is called _____ growth, and bone growth in diameter (thickness) is called _____ growth.

2. The crystallized inorganic mineral salts in bone contribute to bone's _____, while the collagen fibers and other organic molecules provide bone with _____.

Indicate whether the following statements are true or false.

3. Bone resorption involves increased activity of osteoclasts.

4. The formation of bone from cartilage is known as endochondral ossification.

5. The growth of bone is controlled primarily by hormones.

Choose the one best answer to the following questions.

6. Place in order the steps involved in intramembranous ossification. (1) Bony matrices fuse to form trabeculae. (2) Clusters of osteoblasts form a center of ossification that secretes the organic extracellular matrix. (3) Spongy bone is replaced with compact bone on the bone's surface. (4) Periosteum develops on the bone's periphery. (5) The extracellular matrix hardens by deposition of calcium and mineral salts.
(a) 2, 4, 5, 1, 3　　(b) 4, 3, 5, 1, 2　　(c) 1, 2, 5, 4, 3
(d) 2, 5, 1, 4, 3　　(e) 5, 1, 3, 4, 2

7. Place in order the steps involved in endochondral ossification. (1) Nutrient artery invades the perichondrium. (2) Osteoclasts create a marrow cavity. (3) Chondrocytes enlarge and calcify. (4) Secondary ossification centers appear at epiphyses. (5) Osteoblasts become active in the primary ossification center.
   (a) 3, 1, 5, 2, 4   (b) 3, 1, 5, 4, 2   (c) 1, 3, 5, 2, 4
   (d) 1, 2, 3, 5, 4   (e) 2, 5, 4, 3, 1

8. Spongy bone differs from compact bone because spongy bone
   (a) is composed of numerous osteons (haversian systems).
   (b) is found primarily in the diaphyses of long bones, and compact bone is found primarily in the epiphyses of long bones.
   (c) contains osteons all aligned in the same direction along lines of stress.
   (d) does not contain osteocytes contained in lacunae.
   (e) is composed of trabeculae that are oriented along lines of stress.

9. A primary effect of weight-bearing exercise on bones is to
   (a) provide oxygen for bone development.
   (b) increase the demineralization of bone.
   (c) maintain and increase bone mass.
   (d) stimulate the release of sex hormones for bone growth.
   (e) utilize the stored triglycerides from the yellow bone marrow.

10. Place in order the steps involved in the repair of a bone fracture. (1) Osteoblast production of trabeculae and bony callus formation; (2) formation of a hematoma at the site of fracture; (3) resorption of remaining bone fragments and remodeling of bone; (4) migration of fibroblasts to the fracture site; (5) bridging of broken ends of bones by a fibrocartilaginous callus.
    (a) 2, 4, 5, 1, 3   (b) 2, 5, 4, 1, 3   (c) 1, 2, 5, 4, 3
    (d) 2, 5, 1, 3, 4   (e) 5, 2, 4, 1, 3

11. Match the following:
    ____ (a) column-like layer of maturing chondrocytes
    ____ (b) layer of small, scattered chondrocytes anchoring the epiphyseal (growth) plate to the bone
    ____ (c) layer of actively dividing chondrocytes
    ____ (d) region of dead chondrocytes

    (1) zone of hypertrophic cartilage
    (2) zone of calcified cartilage
    (3) zone of proliferating cartilage
    (4) zone of resting cartilage

12. Match the following:
    ____ (a) decreases blood calcium levels by accelerating calcium deposition in bones and inhibiting osteoclasts
    ____ (b) required for collagen synthesis
    ____ (c) during childhood, it promotes growth at epiphyseal plate; production stimulated by human growth hormone
    ____ (d) involved in bone growth by increasing osteoblast activity; causes long bones to stop growing in length
    ____ (e) required for protein synthesis
    ____ (f) active form of vitamin D; raises blood calcium levels by increasing absorption of calcium from digestive tract
    ____ (g) raises blood calcium levels by increasing bone resorption

    (1) PTH
    (2) CT
    (3) calcitriol
    (4) insulinlike growth factors
    (5) sex hormones
    (6) vitamin C
    (7) vitamin K

13. Match the following:
    ____ (a) space within the shaft of the bone that contains yellow bone marrow
    ____ (b) triglyceride storage tissue
    ____ (c) hemopoietic tissue
    ____ (d) thin layer of hyaline cartilage covering the ends of bones where they form a joint
    ____ (e) proximal and distal ends of bones
    ____ (f) the long, cylindrical main portion of the bone; the shaft
    ____ (g) in a growing bone, the region that contains the epiphyseal (growth) plate
    ____ (h) the tough covering that surrounds the bone surface wherever cartilage is not present
    ____ (i) a layer of hyaline cartilage in the area between the shaft and end of a growing bone
    ____ (j) membrane lining the medullary cavity
    ____ (k) a remnant of the active epiphyseal (growth) plate; a sign that the bone has stopped growing in length
    ____ (l) bundles of collagen fibers that attach periosteum to bone

    (1) articular cartilage
    (2) endosteum
    (3) medullary cavity
    (4) diaphysis
    (5) epiphyses
    (6) metaphysis
    (7) periosteum
    (8) red bone marrow
    (9) yellow bone marrow
    (10) perforating (Sharpey's) fibers
    (11) epiphyseal line
    (12) epiphyseal (growth) plate

14. Match the following:
    ____ (a) a broken bone in which one end of the fractured bone is driven into the other end
    ____ (b) a condition of porous bones characterized by decreased bone mass and increased susceptibility to fractures
    ____ (c) splintered bone, with smaller fragments lying between main fragments
    ____ (d) a broken bone that does not break through the skin
    ____ (e) a partial break in a bone in which one side of the bone is broken and the other side bends
    ____ (f) a broken bone that protrudes through the skin
    ____ (g) microscopic bone breaks resulting from inability to withstand repeated stressful impact
    ____ (h) a degeneration of articular cartilage allowing the bony ends to touch; worsens due to friction between the bones
    ____ (i) condition characterized by failure of new bone formed by remodeling to calcify in adults
    ____ (j) an infection of bone

    (1) closed (simple) fracture
    (2) open (compound) fracture
    (3) impacted fracture
    (4) greenstick fracture
    (5) stress fracture
    (6) comminuted fracture
    (7) osteoporosis
    (8) osteomalacia
    (9) osteoarthritis
    (10) osteomyelitis

15. Match the following:
_____ (a) small spaces between lamellae that contain osteocytes
_____ (b) tiny canals that penetrate compact bone; carry blood vessels, lymphatic vessels, and nerves from the periosteum
_____ (c) areas between osteons; fragments of old osteons
_____ (d) cells that secrete the components required to build bone
_____ (e) microscopic unit of compact bone tissue
_____ (f) interconnected, tiny canals filled with extracellular fluid; connect lacunae to each other and to the central canal
_____ (g) canals that extend longitudinally through the bone and carry blood vessels and nerves to the central canal
_____ (h) large cells derived from numerous monocytes; involved in resorption by release of lysosomal enzymes and acids
_____ (i) irregular lattice of thin columns of bone found in spongy bone tissue
_____ (j) rings of hard calcified matrix found just beneath the periosteum and lining the medullary cavity
_____ (k) mature cells that maintain the daily metabolism of bone
_____ (l) an opening in the shaft of the bone allowing an artery to pass into the bone
_____ (m) unspecialized stem cells derived from mesenchyme

(1) osteogenic cells
(2) osteocytes
(3) osteon (haversian system)
(4) perforating (Volkmann's) canals
(5) circumferential lamellae
(6) osteoblasts
(7) trabeculae
(8) interstitial lamellae
(9) canaliculi
(10) osteoclasts
(11) nutrient foramen
(12) lacunae
(13) haversian (central) canals

CRITICAL THINKING QUESTIONS

1. Taryn is a high school senior who is undergoing a strenuous running regimen for several hours a day in order to qualify for her state high school track meet. Lately she has experienced intense pain in her right leg that is hindering her workouts. Her physician performs an examination of her right leg. The doctor doesn't notice any outward evidence of injury; he then orders a bone scan. What does her doctor suspect the problem is?

2. While playing basketball, nine-year-old Marcus fell and broke his left arm. The arm was placed in a cast and appeared to heal normally. As an adult, Marcus was puzzled because it seemed that his right arm is longer than his left arm. He measured both arms and he was correct—his right arm is longer! How would you explain to Marcus what happened?

3. Astronauts in space exercise as part of their daily routine, yet they still have problems with bone weakness after prolonged stays in space. Why does this happen?

? ANSWERS TO FIGURE QUESTIONS

6.1 The periosteum is essential for growth in bone thickness, bone repair, and bone nutrition. It also serves as a point of attachment for ligaments and tendons.

6.2 Bone resorption is necessary for the development, maintenance, and repair of bone.

6.3 The central (haversian) canals are the main blood supply to the osteocytes of an osteon (haversian system), so their blockage would lead to death of the osteocytes.

6.4 Periosteal arteries enter bone tissue through perforating (Volkmann's) canals.

6.5 Flat bones of the skull, most facial bones, the mandible (lower jawbone), and the medial part of the clavicle develop by intramembranous ossification.

6.6 Secondary ossification centers develop in the regions of the cartilage model that will give rise to the epiphyses.

6.7 The lengthwise growth of the diaphysis is caused by cell divisions in the zone of proliferating cartilage and replacement of the zone of calcified cartilage with bone (new diaphysis).

6.8 The medullary cavity enlarges by activity of the osteoclasts in the endosteum.

6.9 Healing of bone fractures can take months because calcium and phosphorus deposition is a slow process, and bone cells generally grow and reproduce slowly.

6.10 Heartbeat, respiration, nerve cell functioning, enzyme functioning, and blood clotting all depend on proper levels of calcium.

6.11 A drug that inhibits the activity of osteoclasts might lessen the effects of osteoporosis, since osteoclasts are responsible for bone resorption.

7 THE SKELETAL SYSTEM: THE AXIAL SKELETON

THE AXIAL SKELETON AND HOMEOSTASIS *The bones of the axial skeleton contribute to homeostasis by protecting many of the body's organs such as the brain, spinal cord, heart, and lungs. They are also important in support and calcium storage and release.*

Without bones, you could not survive. You would be unable to perform movements such as walking or grasping, and the slightest blow to your head or chest could damage your brain or heart. Because the skeletal system forms the framework of the body, a familiarity with the names, shapes, and positions of individual bones will help you locate and name many other anatomical features. For example, the radial artery, the site where the pulse is usually taken, is named for its closeness to the radius, the lateral bone of the forearm. The ulnar nerve is named for its proximity to the ulna, the medial bone of the forearm. The frontal lobe of the brain lies deep to the frontal (forehead) bone. The tibialis anterior muscle lies along the anterior surface of the tibia (shin bone). Parts of certain bones also serve to locate structures within the skull and to outline the lungs, heart, and abdominal and pelvic organs.

Movements such as throwing a ball, biking, and walking require interactions between bones and muscles. To understand how muscles produce different movements, from high-fives to three-point shots, you will need to learn where the muscles attach on individual bones and what types of joints are involved. Together, the bones, muscles, and joints form an integrated system called the **musculoskeletal system.** The branch of medical science concerned with the prevention or correction of disorders of the musculoskeletal system is called **orthopedics** (or'-thō-PĒ-diks; *ortho-* = correct; *-pedi* = child).

 Did you ever wonder what causes people to become measurably shorter as they age?

208

7.1 DIVISIONS OF THE SKELETAL SYSTEM

● OBJECTIVE

- Describe how the skeleton is organized into axial and appendicular divisions.

The adult human skeleton consists of 206 named bones, most of which are paired, with one member of each pair on the right and left sides of the body. The skeletons of infants and children have more than 206 bones because some of their bones (sacrum and coccyx) fuse later in life. Examples are the hip bones and some bones (sacrum and coccyx) of the vertebral column (backbone).

Bones of the adult skeleton are grouped into two principal divisions: the **axial skeleton** and the **appendicular skeleton** (*appendic-* = to hang onto). Table 7.1 presents the 80 bones of the axial skeleton and the 126 bones of the appendicular skeleton. Figure 7.1 shows how both divisions join to form the complete skeleton (the bones of the axial skeleton are shown in blue). You can remember the names of the divisions if you think of the axial skeleton as consisting of the bones that lie around the longitudinal *axis* of the human body, an imaginary vertical line that runs through the body's center of gravity from the head to the space between the feet: skull bones, auditory ossicles (ear bones), hyoid bone (see Figure 7.5), ribs, sternum (breastbone), and bones of the vertebral column. The appendicular skeleton consists of the bones of the **upper** and **lower** limbs (extremities or ap-

pendages), plus the bones forming the **girdles** that connect the limbs to the axial skeleton. Functionally, the auditory ossicles in the middle ear, which vibrate in response to sound waves that strike the eardrum, are not part of either the axial or appendicular skeleton, but they are grouped with the axial skeleton for convenience (see Chapter 17).

We will organize our study of the skeletal system around the two divisions of the skeleton, with emphasis on how the many bones of the body are interrelated. In this chapter we focus on the axial skeleton, looking first at the skull and then at the bones of the vertebral column and the chest. In Chapter 8 we explore the appendicular skeleton, examining in turn the bones of the pectoral (shoulder) girdle and upper limbs, and then the pelvic (hip) girdle and the lower limbs. Before we examine the axial skeleton, we direct your attention to some general characteristics of bones.

✔ CHECKPOINT

1. On what basis is the skeleton grouped into the axial and appendicular divisions?

7.2 TYPES OF BONES

● OBJECTIVE

- Classify bones based on their shape or location.

Almost all bones of the body can be classified into five main types based on shape: long, short, flat, irregular, and sesamoid

TABLE 7.1

The Bones of the Adult Skeletal System

DIVISION OF THE SKELETON	STRUCTURE	NUMBER OF BONES	DIVISION OF THE SKELETON	STRUCTURE	NUMBER OF BONES
Axial skeleton	**Skull**		Appendicular skeleton	**Pectoral (shoulder) girdles**	
	Cranium	8		Clavicle	2
	Face	14		Scapula	2
	Hyoid	1		**Upper limbs**	
	Auditory ossicles	6		Humerus	2
	Vertebral column	26		Ulna	2
	Thorax			Radius	2
	Sternum	1		Carpals	16
	Ribs	24		Metacarpals	10
	Number of bones = 80			Phalanges	28
				Pelvic (hip) girdle	
				Hip, pelvic, or coxal bone	2
				Lower limbs	
				Femur	2
				Patella	2
				Fibula	2
				Tibia	2
				Tarsals	14
				Metatarsals	10
				Phalanges	28
				Number of bones = 126	
				Total bones in an adult skeleton = 206	

Figure 7.1 Divisions of the skeletal system. The axial skeleton is indicated in blue. (Note the position of the hyoid bone in Figure 7.5.)

The adult human skeleton consists of 206 bones grouped into two divisions: the axial skeleton and the appendicular skeleton.

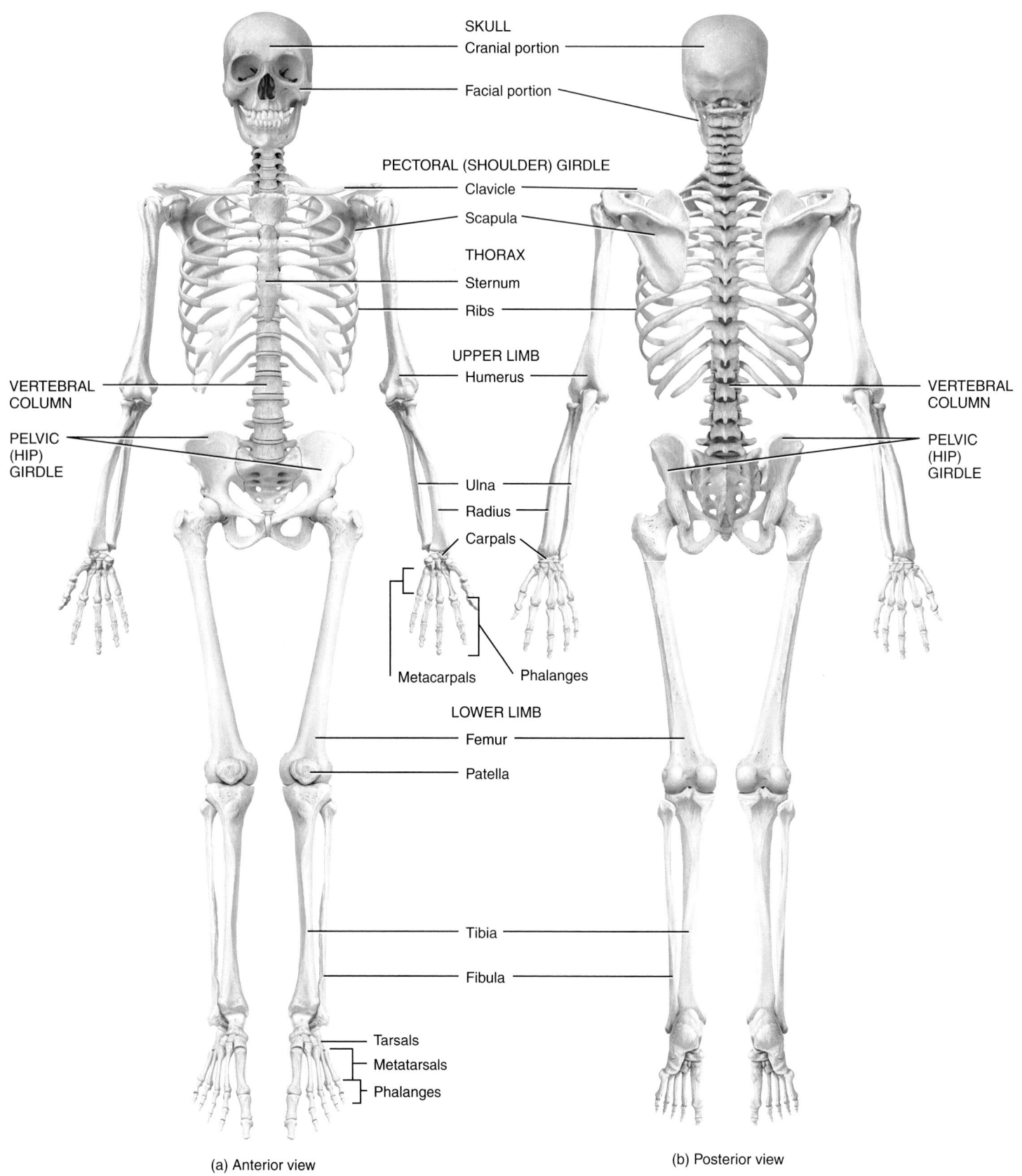

SKULL
Cranial portion
Facial portion

PECTORAL (SHOULDER) GIRDLE
Clavicle
Scapula

THORAX
Sternum
Ribs

UPPER LIMB
Humerus

VERTEBRAL COLUMN

PELVIC (HIP) GIRDLE

VERTEBRAL COLUMN

PELVIC (HIP) GIRDLE

Ulna
Radius
Carpals

Metacarpals   Phalanges

LOWER LIMB
Femur
Patella

Tibia
Fibula

Tarsals
Metatarsals
Phalanges

(a) Anterior view

(b) Posterior view

Which of the following structures are part of the axial skeleton, and which are part of the appendicular skeleton? Skull, clavicle, vertebral column, shoulder girdle, humerus, pelvic girdle, and femur.

(Figure 7.2). As you learned in Chapter 6, **long bones** have greater length than width, consist of a shaft and a variable number of extremities or epiphyses (ends), and are slightly curved for strength. A curved bone absorbs the stress of the body's weight at several different points, so that it is evenly distributed. If bones were straight, the weight of the body would be unevenly distributed, and the bone would fracture more easily. Long bones consist mostly of *compact bone tissue* in their diaphyses but have considerable amounts of *spongy bone tissue* in their epiphyses. Long bones vary tremendously in size and include those in the femur (thigh bone), tibia and fibula (leg bones), humerus (arm bone), ulna and radius (forearm bones), and phalanges (finger and toe bones).

Short bones are somewhat cube-shaped and are nearly equal in length and width. They consist of spongy bone tissue except at the surface, which has a thin layer of compact bone tissue. Examples of short bones are most carpal (wrist) bones and most tarsal (ankle) bones.

Flat bones are generally thin and composed of two nearly parallel plates of compact bone tissue enclosing a layer of spongy bone tissue. Flat bones afford considerable protection and provide extensive areas for muscle attachment. Flat bones include the cra-

nial bones, which protect the brain; the sternum (breastbone) and ribs, which protect organs in the thorax; and the scapulae (shoulder blades).

Irregular bones have complex shapes and cannot be grouped into any of the previous categories. They vary in the amount of spongy and compact bone present. Such bones include the vertebrae (backbones), hip bones, certain facial bones, and the calcaneus.

Sesamoid bones (SES-a-moyd = shaped like a sesame seed) develop in certain tendons where there is considerable friction, tension, and physical stress, such as the palms and soles. They may vary in number from person to person, are not always completely ossified, and typically measure only a few millimeters in diameter. Notable exceptions are the two patellae (kneecaps), large sesamoid bones located in the quadriceps femoris tendon (see Figure 11.20a) that are normally present in everyone. Functionally, sesamoid bones protect tendons from excessive wear and tear, and they often change the direction of pull of a tendon, which improves the mechanical advantage at a joint.

An additional type of bone is classified by location rather than shape. **Sutural bones** (SOO-chur-al; *sutur-* = seam) are small bones located in sutures (joints) between certain cranial bones (see Figure 7.6). Their number varies greatly from person to person.

Recall from Chapter 6 that in adults, red bone marrow is restricted to flat bones such as the ribs, sternum (breastbone), and skull; irregular bones such as vertebrae (backbones) and hip bones; long bones such as the proximal epiphyses of the femur (thigh bone) and humerus (arm bone); and some short bones.

✔ CHECKPOINT
2. Give examples of long, short, flat, and irregular bones.

7.3 BONE SURFACE MARKINGS

◉ OBJECTIVE
• Describe the principal surface markings on bones and the functions of each.

Bones have characteristic **surface markings,** structural features adapted for specific functions. Most are not present at birth but develop in response to certain forces and are most prominent in the adult skeleton. In response to tension on a bone surface from tendons, ligaments, aponeuroses, and fasciae, new bone is deposited, resulting in raised or roughened areas. Conversely, compression on a bone surface results in a depression.

There are two major types of surface markings: (1) *depressions and openings,* which allow the passage of soft tissues (such as blood vessels, nerves, ligaments, and tendons) or form joints, and (2) *processes,* projections or outgrowths that either help form joints or serve as attachment points for connective tissue (such as ligaments and tendons). Table 7.2 describes the various surface markings and provides examples of each.

✔ CHECKPOINT
3. What are surface markings? What are their general functions?

Figure 7.2 Types of bones based on shape. The bones are not drawn to scale.

🔓 The shapes of bones largely determine their functions.

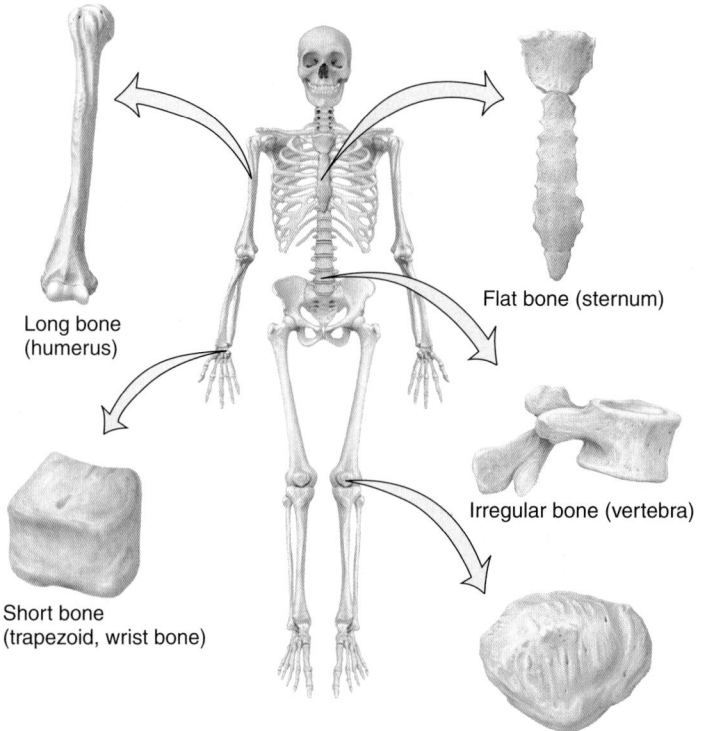

Long bone (humerus)

Short bone (trapezoid, wrist bone)

Flat bone (sternum)

Irregular bone (vertebra)

Sesamoid bone (patella)

❓ **Which type of bone primarily provides protection and a large surface area for muscle attachment?**

TABLE 7.2

Bone Surface Markings

MARKING	DESCRIPTION	EXAMPLE
DEPRESSIONS AND OPENINGS: SITES ALLOWING THE PASSAGE OF SOFT TISSUE (NERVES, BLOOD VESSELS, LIGAMENTS, TENDONS) OR FORMATION OF JOINTS		
Fissure (FISH-ur)	Narrow slit between adjacent parts of bones through which blood vessels or nerves pass.	Superior orbital fissure of sphenoid bone (Figure 7.12).
Foramen (fō-RĀ-men = hole; plural is *foramina*)	Opening through which blood vessels, nerves, or ligaments pass.	Optic foramen of sphenoid bone (Figure 7.12).
Fossa (FOS-a = trench; plural is *fossae*, FOS-ē)	Shallow depression.	Coronoid fossa of humerus (Figure 8.5a).
Sulcus (SUL-kus = groove; plural is *sulci*, SUL-sī)	Furrow along bone surface that accommodates blood vessel, nerve, or tendon.	Intertubercular sulcus of humerus (Figure 8.5a).
Meatus (mē-Ā-tus = passageway; plural is *meati*, mē-Ā-tī)	Tubelike opening.	External auditory meatus of temporal bone (Figure 7.4b).
PROCESSES: PROJECTIONS OR OUTGROWTHS ON BONE THAT FORM JOINTS OR ATTACHMENT POINTS FOR CONNECTIVE TISSUE, SUCH AS LIGAMENTS AND TENDONS		
Processes that form joints		
Condyle (KON-dīl; *condylus* = knuckle)	Large, round protuberance with a smooth articular surface at end of bone.	Lateral condyle of femur (Figure 8.13a).
Facet (FAS-et or fa-SET)	Smooth, flat, slightly concave or convex articular surface.	Superior articular facet of vertebra (Figure 7.18d).
Head	Usually rounded articular projection supported on neck (constricted portion) of bone.	Head of femur (Figure 8.13a).
Processes that form attachment points for connective tissue		
Crest	Prominent ridge or elongated projection.	Iliac crest of hip bone (Figure 8.10b).
Epicondyle (*epi-* = above)	Typically roughened projection above condyle.	Medial epicondyle of femur (Figure 8.13a).
Line (*linea*)	Long, narrow ridge or border (less prominent than crest).	Linea aspera of femur (Figure 8.13b).
Spinous process	Sharp, slender projection.	Spinous process of vertebra (Figure 7.17).
Trochanter (trō-KAN-ter)	Very large projection.	Greater trochanter of femur (Figure 8.13b).
Tubercle (TOO-ber-kul; *tuber-* = knob)	Variably sized rounded projection.	Greater tubercle of humerus (Figure 8.5a).
Tuberosity	Variably sized projection that has a rough, bumpy surface.	Ischial tuberosity of hip bone (Figure 8.10b).

7.4 SKULL

◉ OBJECTIVES

• Name the cranial and facial bones and indicate whether they are paired or single.
• Describe the following special features of the skull: sutures, paranasal sinuses, and fontanels.

The **skull** is the bony framework of the head. It contains 22 bones (not counting the bones of the middle ears) and rests on the superior end of the vertebral column (backbone). The bones of the skull are grouped into two categories: cranial bones and facial bones. The **cranial bones** (*crani-* = brain case) form the cranial cavity, which encloses and protects the brain. The eight cranial bones are the frontal bone, two parietal bones, two temporal bones, the occipital bone, the sphenoid bone, and the ethmoid bone. Fourteen **facial bones** form the face: two nasal bones, two maxillae (or maxillas), two zygomatic bones, the mandible, two lacrimal bones, two palatine bones, two inferior nasal conchae, and the vomer. Exhibits 7.A through 7.G illustrate the bones of the skull from different views.

EXHIBIT 7.A   Cranial Bones—Frontal Bone *(Figure 7.3)*

● OBJECTIVE

• Identify the location and surface features of the frontal bone.

The **frontal bone** forms the forehead (the anterior part of the cranium), the roofs of the *orbits* (eye sockets), and most of the anterior part of the cranial floor (Figure 7.3). Soon after birth, the left and right sides of the frontal bone are united by the *metopic suture,* which usually disappears between the ages of six and eight.

Note the *frontal squama,* a scalelike plate of bone that forms the forehead in the anterior view of the skull shown in Figure 7.3a. It gradually slopes inferiorly from the coronal suture, on the top of the skull, then angles abruptly and becomes almost vertical above the orbits. At the superior border of the orbits, the frontal bone thickens, forming the *supraorbital margin* (*supra-* = above; *-orbi* = circle). From this margin, the frontal bone extends posteriorly to form the roof of the orbit, which is part of the floor of the cranial cavity. Within the supraorbital margin, slightly

Figure 7.3 Anterior view of the skull.

 The skull consists of cranial bones and facial bones.

FRONTAL BONE

Frontal squama

PARIETAL BONE

Supraorbital notch

Squamous suture

SPHENOID BONE

Orbit

ETHMOID BONE

PALATINE BONE

LACRIMAL BONE

Zygomaticofacial foramen

ZYGOMATIC BONE

Perpendicular plate of ethmoid bone

INFERIOR NASAL CONCHA

VOMER

Mental foramen

Coronal suture

Supraorbital foramen

Supraorbital margin

Optic foramen (canal)

Superior orbital fissure

TEMPORAL BONE

NASAL BONE

Inferior orbital fissure

Middle nasal concha

Infraorbital foramen

MAXILLA

Alveolar process of maxilla

Alveolar process of mandible

MANDIBLE

(a) Anterior view

EXHIBIT 7.A CONTINUES ▶

EXHIBIT 7.A   **213**

EXHIBIT 7.A   Cranial Bones—Frontal Bone   *(Figure 7.3)*   CONTINUED

medial to its midpoint, is a hole called the *supraorbital foramen.* Sometimes the foramen is incomplete and is called the *supraorbital notch.* As you read about each foramen associated with a cranial bone, refer to Table 7.3 to note which structures pass through it. The *frontal sinuses* lie deep to the frontal squama. Sinuses, or more technically paranasal sinuses, are mucous membrane–lined cavities within certain skull bones that will be discussed later.

✔ CHECKPOINT

What structures pass through the supraorbital foramen?

CLINICAL CONNECTION | *Black Eye*

A **black eye** is a bruising around the eye, commonly due to an injury to the face, rather than an eye injury. In response to trauma, blood and other fluids accumulate in the space around the eye, causing the swelling and dark discoloration. One cause might be a blow to the sharp ridge just superior to the supraorbital margin that fractures the frontal bone, resulting in bleeding. Another is a blow to the nose. Certain surgical procedures (face lift, eyelid surgery, jaw surgery, or nasal surgery) can also result in black eyes.   •

⬤ FIGURE 7.3 CONTINUED ▶

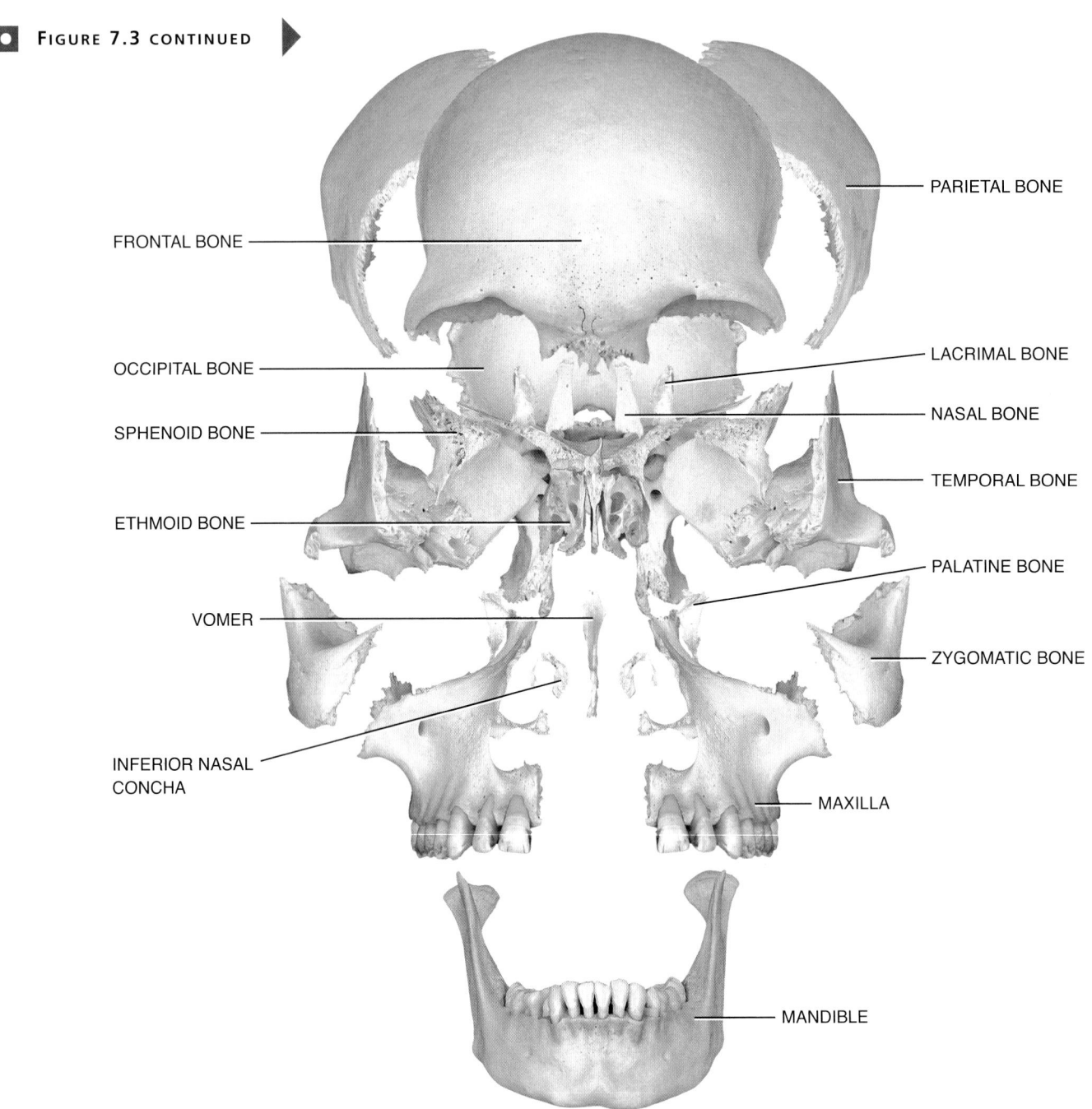

(b) Anterior view of disarticulated skull

❓ **Which of the bones shown here are cranial bones?**

EXHIBIT 7.B Cranial Bones–Parietal Bones *(Figure 7.4)*

● **OBJECTIVE**

• Identify the location and surface features of the parietal bones.

The two **parietal bones** (pa-RĪ-e-tal; *pariet-* = wall) form the greater portion of the sides and roof of the cranial cavity (Figure 7.4). The internal surfaces of the parietal bones contain many protrusions and depressions that accommodate the blood vessels supplying the dura mater, the superficial connective tissue (meninx) covering of the brain.

✔**CHECKPOINT**

How do the parietal bones relate to the cranial cavity?

Figure 7.4 Superior and right lateral views of the skull.

 The zygomatic arch is formed by the zygomatic process of the temporal bone and the temporal process of the zygomatic bone.

View

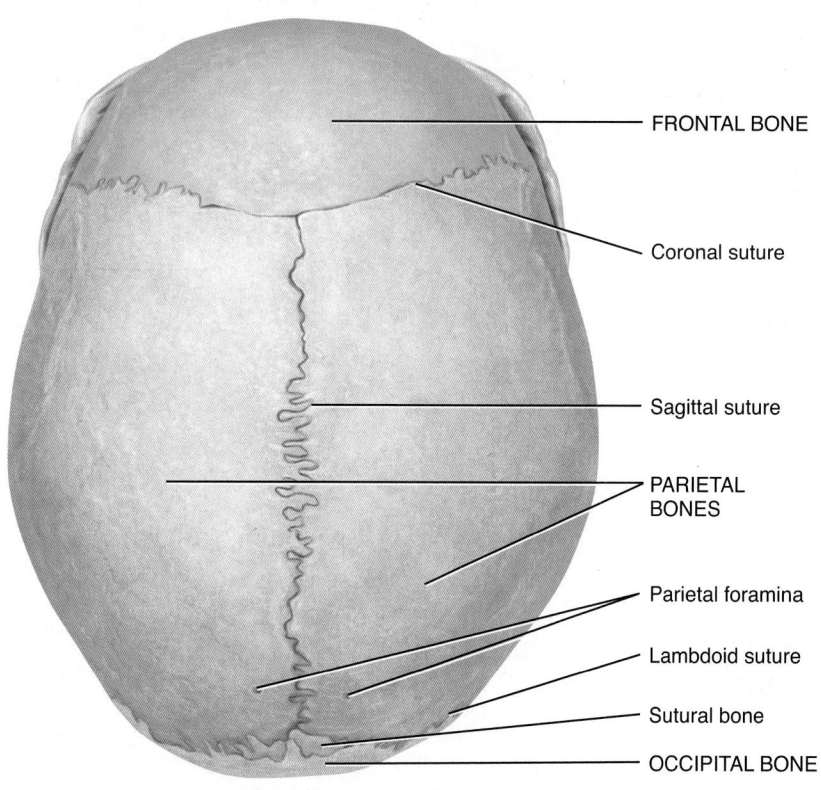

FRONTAL BONE

Coronal suture

Sagittal suture

PARIETAL BONES

Parietal foramina

Lambdoid suture

Sutural bone

OCCIPITAL BONE

(a) Superior view

EXHIBIT 7.B CONTINUES ▶

EXHIBIT 7.B **215**

EXHIBIT 7.B | Cranial Bones–Parietal Bones *(Figure 7.4)* **CONTINUED**

FIGURE 7.4 CONTINUED ▶

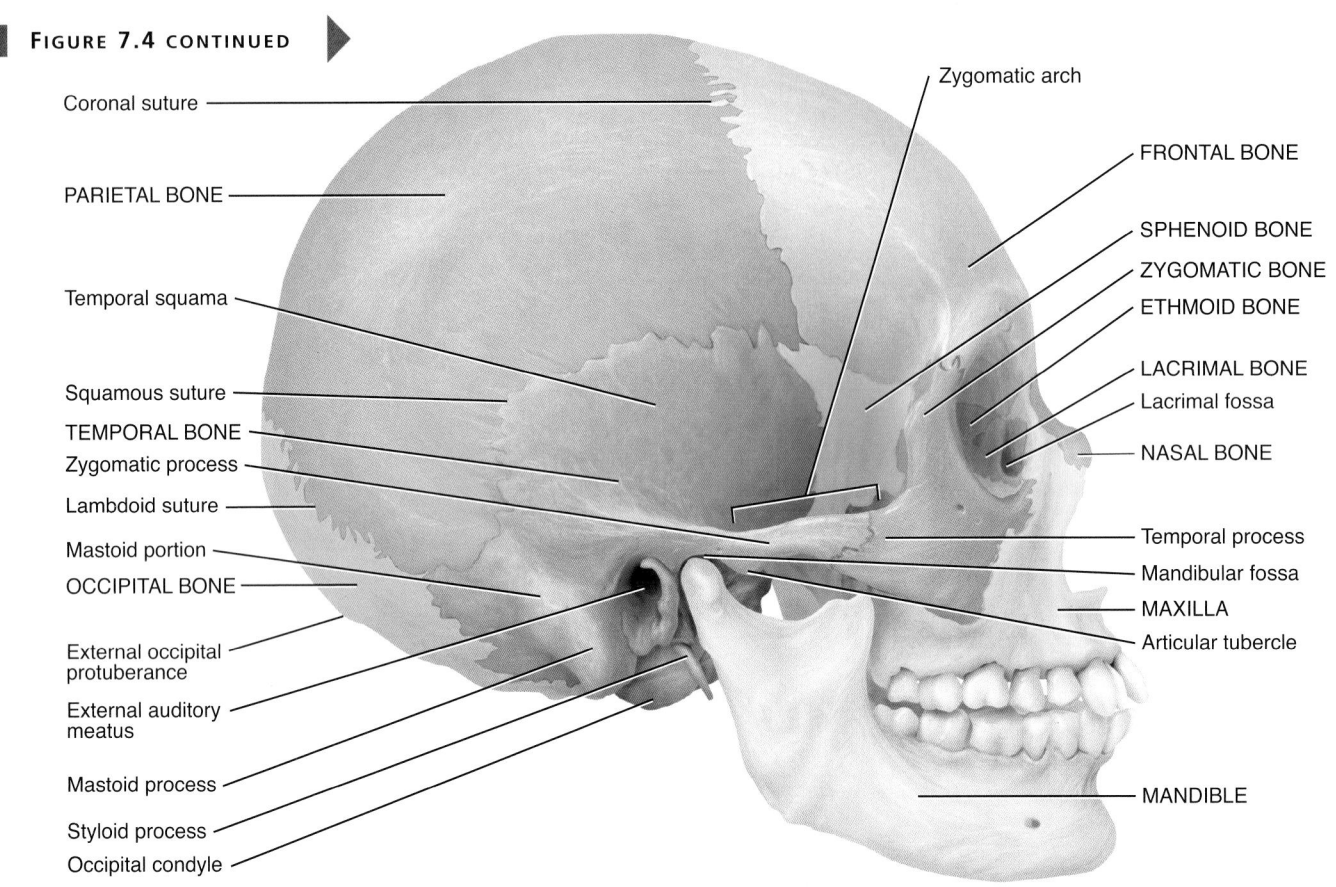

Coronal suture

PARIETAL BONE

Temporal squama

Squamous suture

TEMPORAL BONE

Zygomatic process

Lambdoid suture

Mastoid portion

OCCIPITAL BONE

External occipital protuberance

External auditory meatus

Mastoid process

Styloid process

Occipital condyle

Zygomatic arch

FRONTAL BONE

SPHENOID BONE

ZYGOMATIC BONE

ETHMOID BONE

LACRIMAL BONE

Lacrimal fossa

NASAL BONE

Temporal process

Mandibular fossa

MAXILLA

Articular tubercle

MANDIBLE

(b) Right lateral view

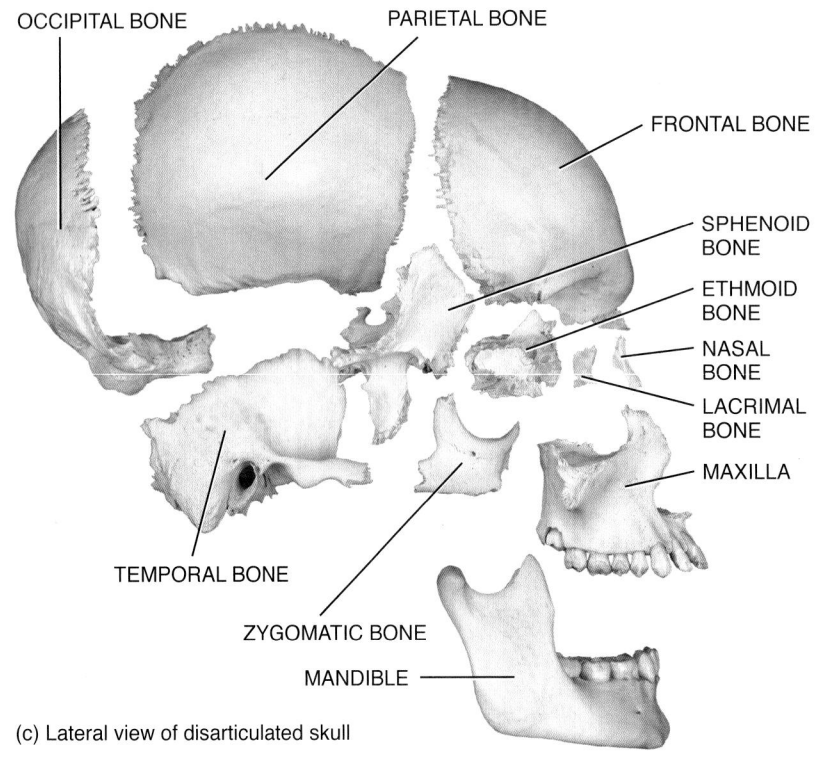

OCCIPITAL BONE

PARIETAL BONE

FRONTAL BONE

SPHENOID BONE

ETHMOID BONE

NASAL BONE

LACRIMAL BONE

MAXILLA

TEMPORAL BONE

ZYGOMATIC BONE

MANDIBLE

(c) Lateral view of disarticulated skull

? **What major bones are joined by (1) the squamous suture, (2) the lambdoid suture, and (3) the coronal suture?**

EXHIBIT 7.C   Cranial Bones—Temporal Bones *(Figure 7.5)*

■ **OBJECTIVE**

• Identify the location and surface features of the temporal bones.

The paired **temporal bones** (*tempor-* = temple) form the inferior lateral aspects of the cranium and part of the cranial floor. (The terms *temporal* and *temple* are derived from the Latin word *tempus,* meaning "time," in reference to the graying of hair in the temple area, a sign of time's passing.) In Figure 7.3a, note the *temporal squama* (= scale), the thin, flat part of the temporal bone that forms the anterior and superior part of the *temple* (the region of the cranium around the ear). Projecting from the inferior portion of the temporal squama is the *zygomatic process,* which articulates (forms a joint) with the temporal process of the zygomatic (cheek) bone. Together, the zygomatic process of the temporal bone and the temporal process of the zygomatic bone form the *zygomatic arch.*

A socket called the *mandibular fossa* is located on the inferior posterior surface of the zygomatic process of each temporal bone. Anterior to the mandibular fossa is a rounded elevation, the *articular tubercle* (Figure 7.4b). The mandibular fossa and articular tubercle articulate with the mandible (lower jawbone) to form the *temporomandibular joint (TMJ).*

The *mastoid portion* (*mastoid* = breast-shaped; Figure 7.4b) of the temporal bone is located posterior and inferior to the *external auditory meatus* (*meatus* = passageway), or ear canal, which directs sound waves into the ear. In an adult, this portion of the bone contains several *mastoid air cells* that communicate with the hollow space of the middle ear.

These tiny air-filled compartments are separated from the brain by thin bony partitions. Middle ear infections that go untreated can spread into the mastoid air cells, causing a painful inflammation called **mastoiditis** (mas′-toy-DĪ-tis).

The *mastoid process* is a rounded projection of the mastoid portion of the temporal bone posterior and inferior to the external auditory meatus. It is the point of attachment for several neck muscles. The *internal auditory meatus* (Figure 7.5) is the opening through which the facial (VII) nerve and vestibulocochlear (VIII) nerve pass. The *styloid process* (*styl-* = stake or pole) projects inferiorly from the inferior surface of the temporal bone and serves as a point of attachment for muscles and ligaments of the tongue and neck (see Figure 7.4b). Between the styloid process and the mastoid process is the *stylomastoid foramen,* through which the facial (VII) nerve and stylomastoid artery pass (see Figure 7.7).

At the floor of the cranial cavity (see Figure 7.8a) is the *petrous portion* (*petrous* = rock) of the temporal bone. This triangular part, located at the base of the skull between the sphenoid and occipital bones, houses the internal ear and the middle ear, structures involved in hearing and equilibrium (balance). It also contains the *carotid foramen,* through which the carotid artery passes (see Figure 7.7). Posterior to the carotid foramen and anterior to the occipital bone is the *jugular foramen,* a passageway for the jugular vein.

✔ **CHECKPOINT**

What structures form the zygomatic arch?

Figure 7.5 Medial view of sagittal section of the skull. Although the hyoid bone is not part of the skull, it is included here for reference.

🔒 The cranial bones are the frontal, parietal, temporal, occipital, sphenoid, and ethmoid bones. The facial bones are the nasal bones, maxillae, zygomatic bones, lacrimal bones, palatine bones, inferior nasal conchae, mandible, and vomer.

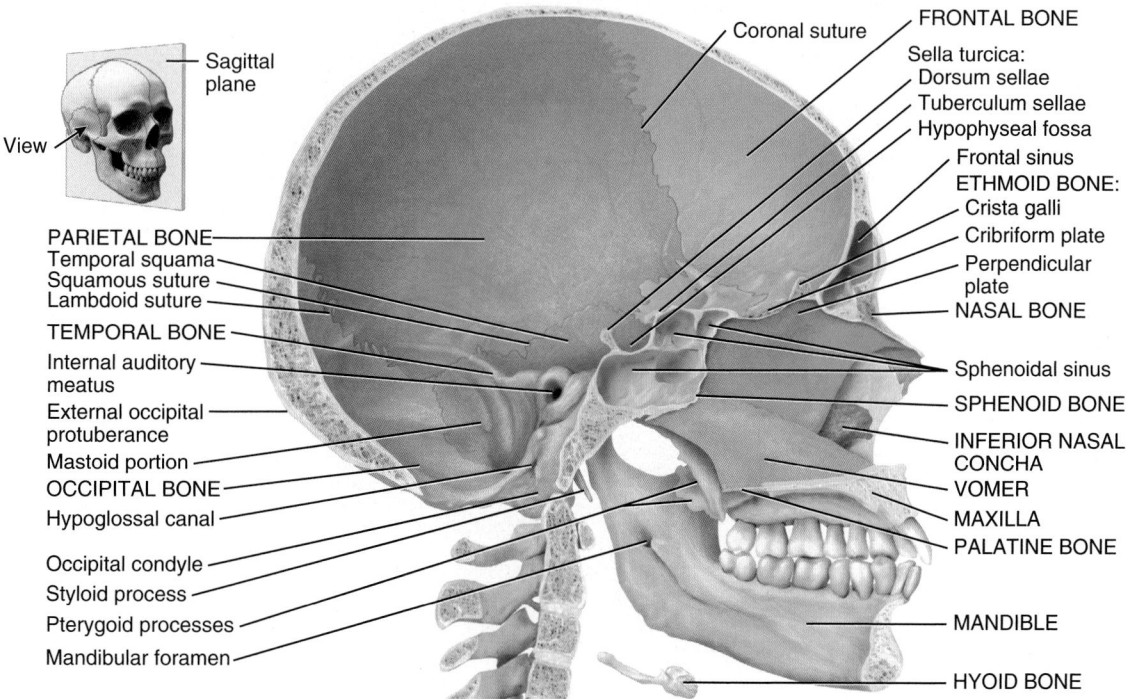

Medial view of sagittal section

? **With which bones does the temporal bone articulate?**

EXHIBIT 7.C **217**

EXHIBIT 7.D Cranial Bones—Occipital Bone *(Figures 7.6 and 7.7)*

■ **OBJECTIVE**

• Identify the location and surface features of the occipital bone.

The **occipital bone** (ok-SIP-i-tal; *occipit-* = back of head) forms the posterior part and most of the base of the cranium (Figure 7.6; also see Figure 7.4). Also view the occipital bone and surrounding structures in the inferior view of the skull in Figure 7.7. The *foramen magnum* (= large hole) is in the inferior part of the bone. The medulla oblongata (inferior part of the brain) connects with the spinal cord within this foramen, and the vertebral and spinal arteries also pass through it along with the accessory (XI) nerve. The *occipital condyles,* oval processes with convex surfaces on either side of the foramen magnum (Figure 7.7), articulate with depressions on the first cervical vertebra (atlas) to form the *atlanto-occipital joint,* which allows you to nod your head "yes." Superior to each occipital condyle on the inferior surface of the skull is the *hypoglossal canal (hypo-* = under; *-glossal* = tongue). (See Figure 7.5.)

The *external occipital protuberance* is the most prominent midline projection on the posterior surface of the bone just above the foramen magnum. You may be able to feel this structure as a bump on the back

Figure 7.6 Posterior view of the skull. The sutures are exaggerated for emphasis.

The occipital bone forms most of the posterior and inferior portions of the cranium.

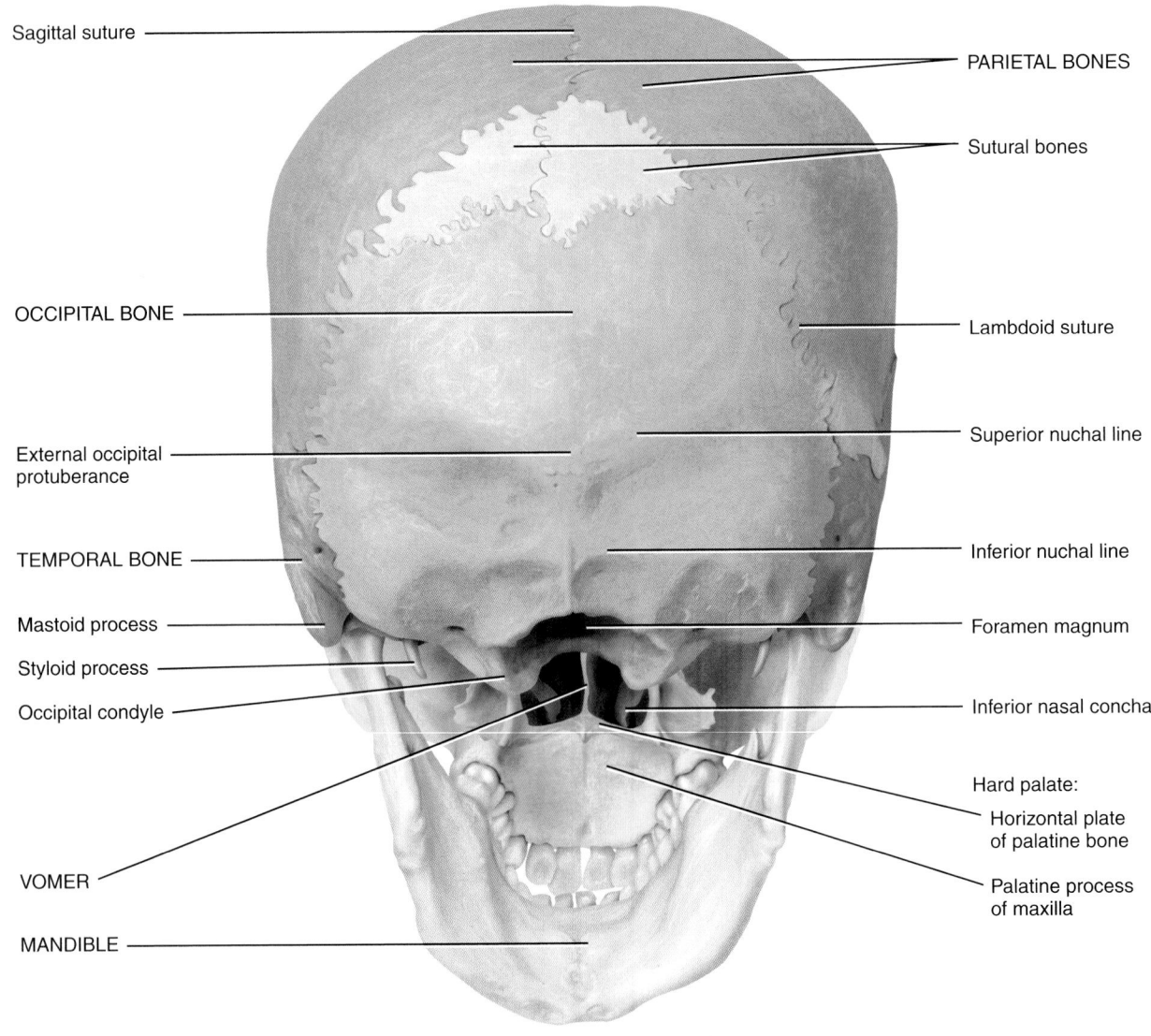

Posteroinferior view

❓ **Which bones form the posterior, lateral portion of the cranium?**

of your head, just above your neck. (See Figure 7.4b.) A large fibrous, elastic ligament, the *ligamentum nuchae* (*nucha-* = nape of neck), extends from the external occipital protuberance to the seventh cervical vertebra to help support the head. Extending laterally from the protuberance are two curved ridges, the *superior nuchal lines,* and below these are two *inferior nuchal lines,* which are areas of muscle attachment (Figure 7.7).

✔ CHECKPOINT
What structures pass through the hypoglossal canal?

Figure 7.7 **Inferior view of the skull.** The mandible (lower jawbone) has been removed.

🔑 The occipital condyles of the occipital bone articulate with the first cervical vertebra to form the atlanto-occipital joint.

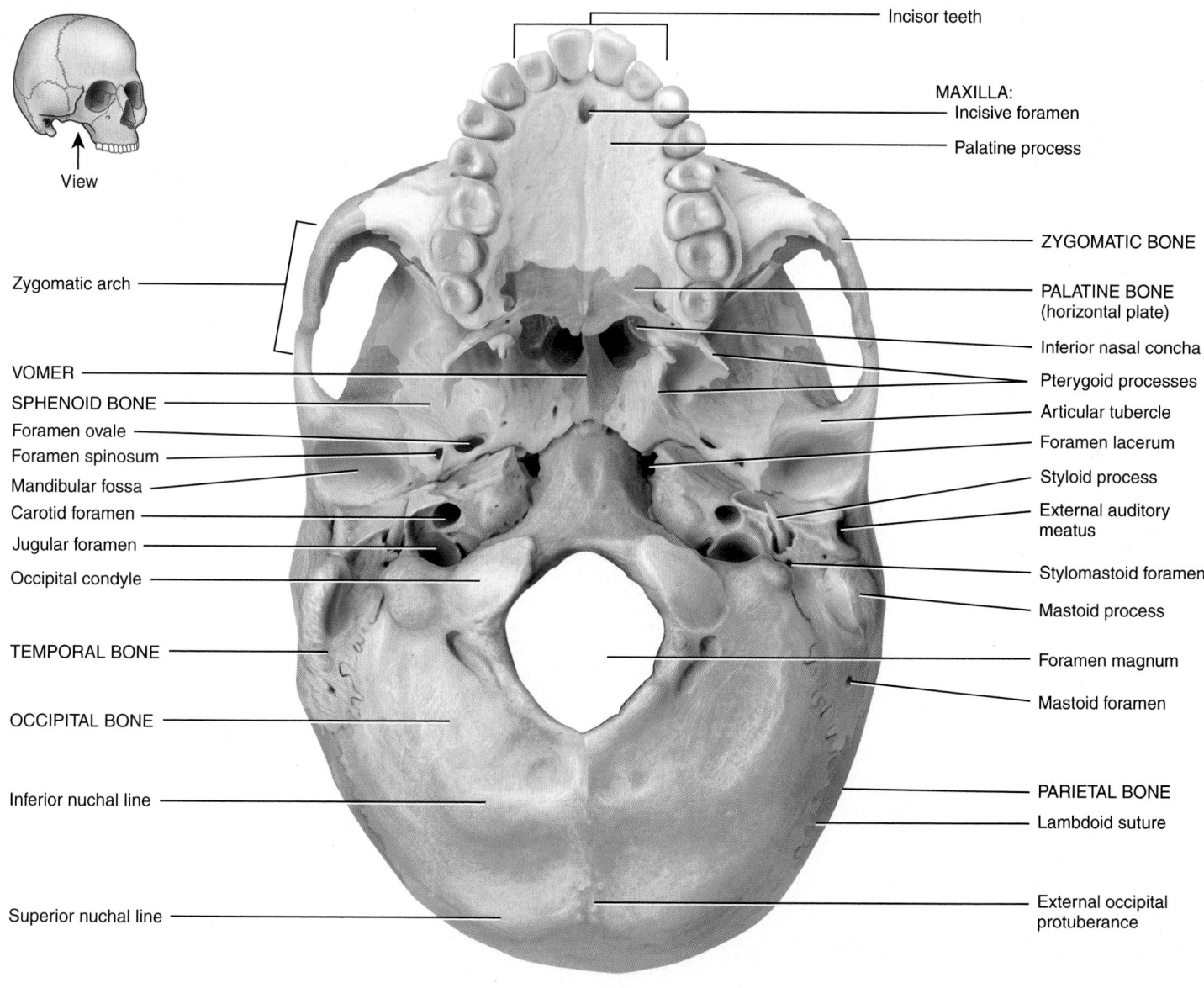

View

Incisor teeth

MAXILLA:
Incisive foramen
Palatine process

ZYGOMATIC BONE

PALATINE BONE
(horizontal plate)

Inferior nasal concha

Pterygoid processes

Articular tubercle

Foramen lacerum

Styloid process

External auditory meatus

Stylomastoid foramen

Mastoid process

Foramen magnum

Mastoid foramen

PARIETAL BONE

Lambdoid suture

External occipital protuberance

Zygomatic arch

VOMER
SPHENOID BONE
Foramen ovale
Foramen spinosum
Mandibular fossa
Carotid foramen
Jugular foramen
Occipital condyle

TEMPORAL BONE

OCCIPITAL BONE

Inferior nuchal line

Superior nuchal line

Inferior view

❓ **What organs of the nervous system join together within the foramen magnum?**

EXHIBIT 7.D **219**

EXHIBIT 7.E Cranial Bones—Sphenoid Bone *(Figure 7.8)*

● **OBJECTIVE**

- Identify the location and surface features of the sphenoid bone.

The **sphenoid bone** (SFĒ-noyd = wedge-shaped) lies at the middle part of the base of the skull (Figures 7.7 and 7.8). This bone is the keystone of the cranial floor because it articulates with all the other cranial bones, holding them together. View the floor of the cranium superiorly (Figure 7.8a) and note the sphenoid articulations. The sphenoid bone joins anteriorly with the frontal and ethmoid bones, laterally with the temporal bones, and posteriorly with the occipital bone. The sphenoid lies posterior and slightly superior to the nasal cavity and forms part of the floor, side walls, and rear wall of the orbit (see Figure 7.12).

The shape of the sphenoid resembles a butterfly with outstretched wings (Figure 7.8b). The *body* of the sphenoid is the hollowed cube-like medial portion between the ethmoid and occipital bones. The space inside the body is the *sphenoidal sinus,* which drains into the nasal cavity (see Figure 7.13). The *sella turcica* (SEL-a TUR-si-ka; *sella* = saddle; *turcica* = Turkish) is a bony saddle-shaped structure on the superior surface of the body of the sphenoid (Figure 7.8a). The anterior part of the sella turcica, which forms the horn of the saddle, is a ridge called the *tuberculum sellae.* The seat of the saddle is a depression, the *hypophyseal fossa* (hī-pō-FIZ-ē-al), which contains the pituitary gland. The posterior part of the sella turcica, which forms the back of the saddle, is another ridge called the *dorsum sellae.*

Figure 7.8 **Sphenoid bone.**

🔑 ⊂▭ The sphenoid bone is called the keystone of the cranial floor because it articulates with all other cranial bones, holding them together.

View

Transverse plane

FRONTAL BONE

ETHMOID BONE:
  Crista galli
  Olfactory foramina
  Cribriform plate

SPHENOID BONE:
  Lesser wing
  Tuberculum sellae ⎤
  Hypophyseal fossa ⎬ Sella turcica
  Dorsum sellae ⎦

Coronal suture

Superior orbital fissure

Foramen rotundum

Foramen ovale

Foramen spinosum

Foramen lacerum

Squamous suture

TEMPORAL BONE:
  Petrous portion
  Internal auditory meatus

Hypoglossal canal

Foramen magnum

Jugular foramen

PARIETAL BONE

Lambdoid suture

OCCIPITAL BONE

(a) Superior view of sphenoid bone in floor of cranium

The *greater wings* of the sphenoid project laterally from the body and form the anterolateral floor of the cranium. The greater wings also form part of the lateral wall of the skull just anterior to the temporal bone and can be viewed externally. The *lesser wings,* which are smaller, form a ridge of bone anterior and superior to the greater wings. They form part of the floor of the cranium and the posterior part of the orbit of the eye.

Between the body and lesser wing just anterior to the sella turcica is the *optic foramen* or *canal* (*optic* = eye), through which the optic (II) nerve and ophthalmic artery pass into the orbit. Lateral to the body between the greater and lesser wings is a triangular slit called the *superior orbital fissure.* This fissure may also be seen in the anterior view of the orbit in Figure 7.12. Blood vessels and cranial nerves pass through this fissure.

The *pterygoid processes* (TER-i-goyd = winglike) project inferiorly from the points where the body and greater wings of the sphenoid bone

unite; they form the lateral posterior region of the nasal cavity (see Figures 7.7 and 7.8b). Some of the muscles that move the mandible attach to the pterygoid processes. At the base of the lateral pterygoid process in the greater wing is the *foramen ovale* (= oval hole). The *foramen lacerum* (= lacerated), covered in part by a layer of fibrocartilage in living subjects, is bounded anteriorly by the sphenoid bone and medially by the sphenoid and occipital bones. It transmits a branch of the ascending pharyngeal artery. Another foramen associated with the sphenoid bone is the *foramen rotundum* (= round hole) located at the junction of the anterior and medial parts of the sphenoid bone. The maxillary branch of the trigeminal (V) nerve passes through the foramen rotundum.

✔ CHECKPOINT

Why is the sphenoid bone called the keystone of the cranial floor?

Frontal plane

View

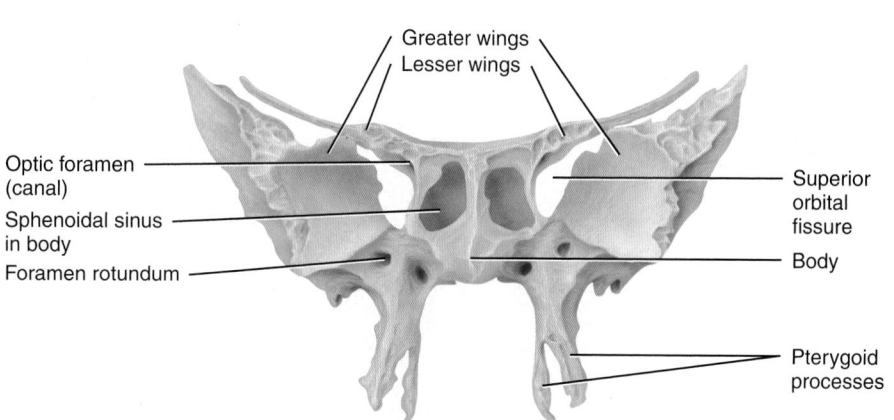

Greater wings
Lesser wings

Optic foramen (canal)

Sphenoidal sinus in body

Foramen rotundum

Superior orbital fissure

Body

Pterygoid processes

(b) Anterior view of sphenoid bone

? **Name the bones that articulate with the sphenoid bone, starting at the crista galli of the ethmoid bone and going in a clockwise direction.**

EXHIBIT 7.E   **221**

EXHIBIT 7.F   Cranial Bones—Ethmoid Bone   *(Figure 7.9)*

● **OBJECTIVE**

• Identify the location and surface features of the ethmoid bone.

The **ethmoid bone** (ETH-moyd = like a sieve) is a delicate bone located in the anterior part of the cranial floor medial to the orbits and is sponge-like in appearance (Figure 7.9). It is anterior to the sphenoid and posterior to the nasal bones. The ethmoid bone forms (1) part of the anterior portion of the cranial floor; (2) the medial wall of the orbits; (3) the superior portion of the **nasal septum,** a partition that divides the nasal cavity into right and left sides; and (4) most of the superior sidewalls of the nasal cavity. The ethmoid bone is a major superior supporting structure of the nasal cavity and forms an extensive surface area in the nasal cavity.

The *cribriform plate* (*cribri-* = sieve) of the ethmoid bone lies in the anterior floor of the cranium and forms the roof of the nasal cavity. The cribriform plate contains the *olfactory foramina* (*olfact-* = to smell) through which the olfactory nerves pass. Projecting superiorly from the cribriform plate is a triangular process called the *crista galli* (*crista* = crest; *galli* = cock), which serves as a point of attachment for the falx cerebri, the membrane that separates the two sides of the brain. Projecting inferiorly from the cribriform plate is the *perpendicular plate*, which forms the superior portion of the nasal septum (see Figure 7.11).

The *lateral masses* of the ethmoid bone compose most of the wall between the nasal cavity and the orbits. They contain 3 to 18 air spaces called ethmoidal cells. The ethmoidal cells together form the *ethmoidal sinuses* (see Figure 7.13). The lateral masses contain two thin, scroll-

Figure 7.9 Ethmoid bone.

The ethmoid bone forms part of the anterior portion of the cranial floor, the medial wall of the orbits, the superior portions of the nasal septum, and most of the side walls of the nasal cavity.

Sagittal plane

View

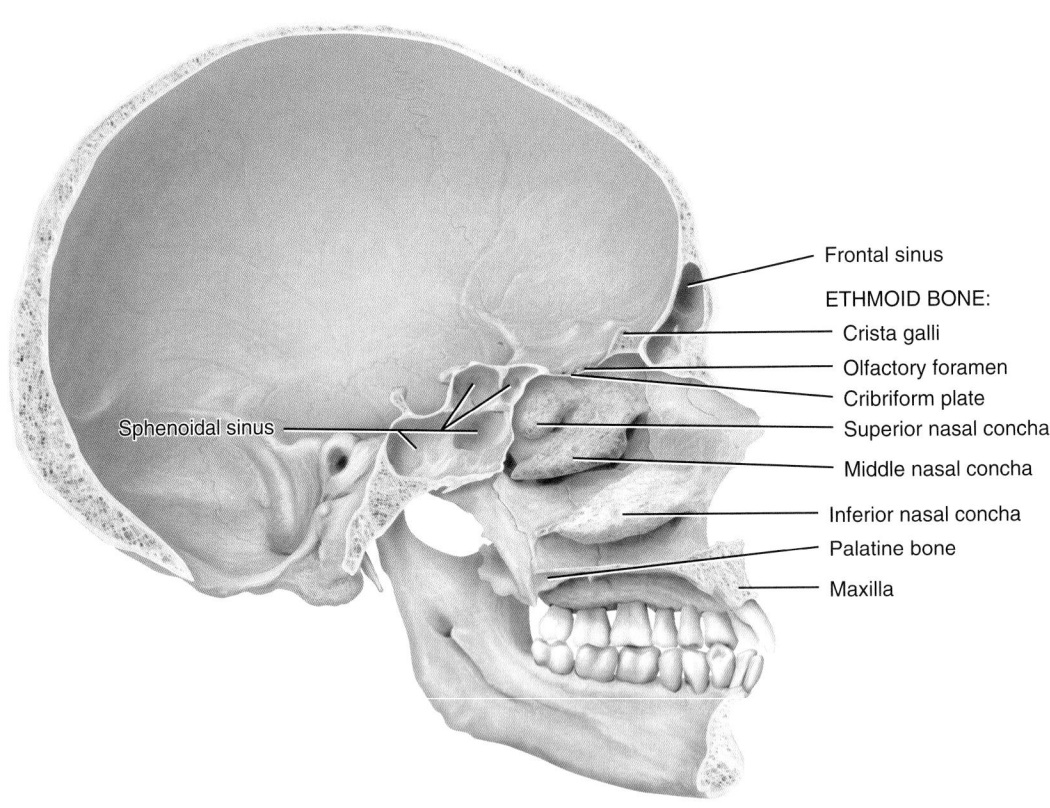

Frontal sinus

ETHMOID BONE:
Crista galli
Olfactory foramen
Cribriform plate
Superior nasal concha
Middle nasal concha
Inferior nasal concha
Palatine bone
Maxilla

Sphenoidal sinus

(a) Medial view of sagittal section

shaped projections lateral to the nasal septum. These are called the *superior nasal concha* (KON-ka = shell) or *turbinate* and the *middle nasal concha (turbinate)*. The plural form is *conchae* (KON-kē). A third pair of conchae, the inferior nasal conchae, are separate bones (discussed shortly). The conchae greatly increase the vascular and mucous membrane surface area in the nasal cavity, which warms and moistens (humidifies) inhaled air before it passes into the lungs. The conchae also cause inhaled air to swirl, and the result is that many inhaled particles become trapped in the mucus that lines the nasal cavity. This action of the conchae helps cleanse inhaled air before it passes into the rest of the respiratory passageways. The superior nasal conchae are near the olfactory foramina of the cribriform plate where the sensory receptors for olfaction (smell) terminate in the mucous membrane of the superior nasal conchae. Thus, they increase the surface area for the sense of smell.

✔CHECKPOINT

The ethmoid bone forms which other cranial structures?

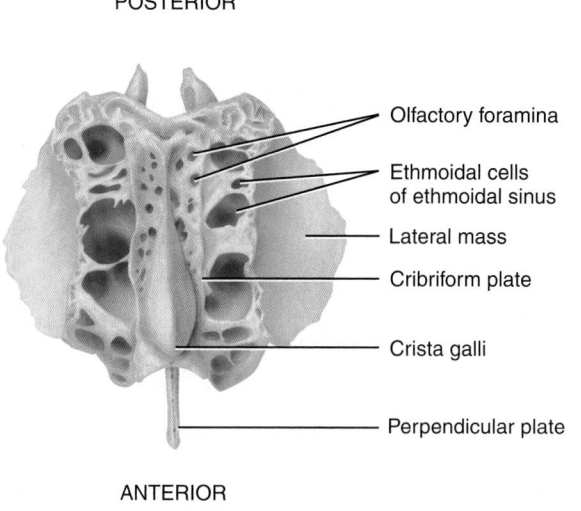

POSTERIOR

- Olfactory foramina
- Ethmoidal cells of ethmoidal sinus
- Lateral mass
- Cribriform plate
- Crista galli
- Perpendicular plate

ANTERIOR

(b) Superior view

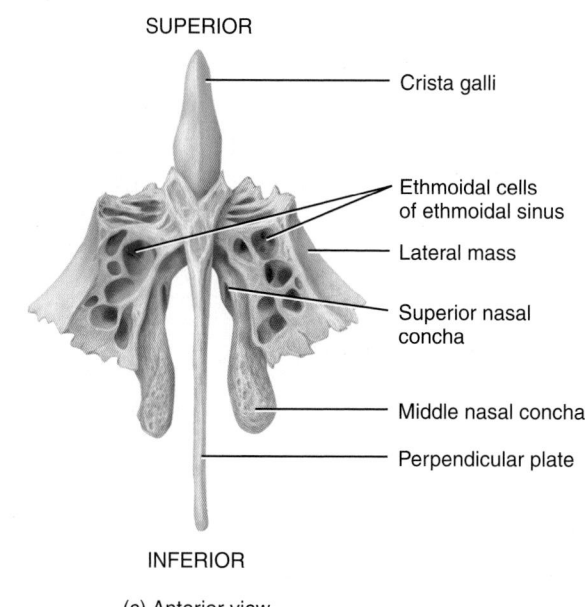

SUPERIOR

- Crista galli
- Ethmoidal cells of ethmoidal sinus
- Lateral mass
- Superior nasal concha
- Middle nasal concha
- Perpendicular plate

INFERIOR

(c) Anterior view

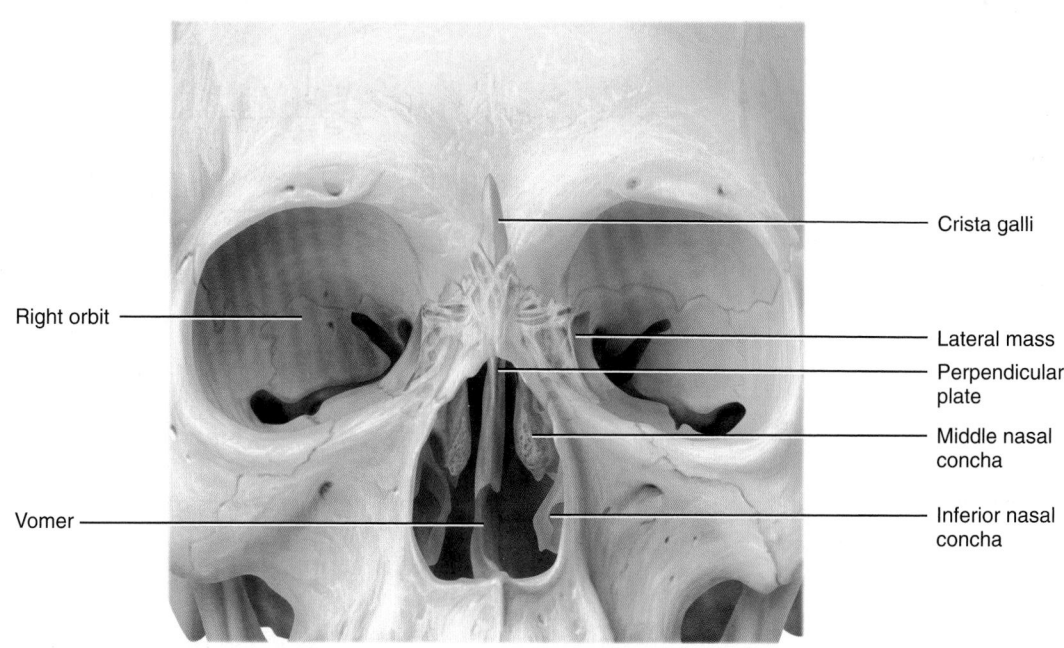

- Crista galli
- Right orbit
- Lateral mass
- Perpendicular plate
- Middle nasal concha
- Vomer
- Inferior nasal concha

(d) Anterior view of position of ethmoid bone in skull
(projected to the surface)

Exhibit 7.F continues ▶

EXHIBIT 7.F **223**

EXHIBIT 7.F   Cranial Bones—Ethmoid Bone *(Figure 7.9)*   CONTINUED

● **FIGURE 7.9 CONTINUED** ▶

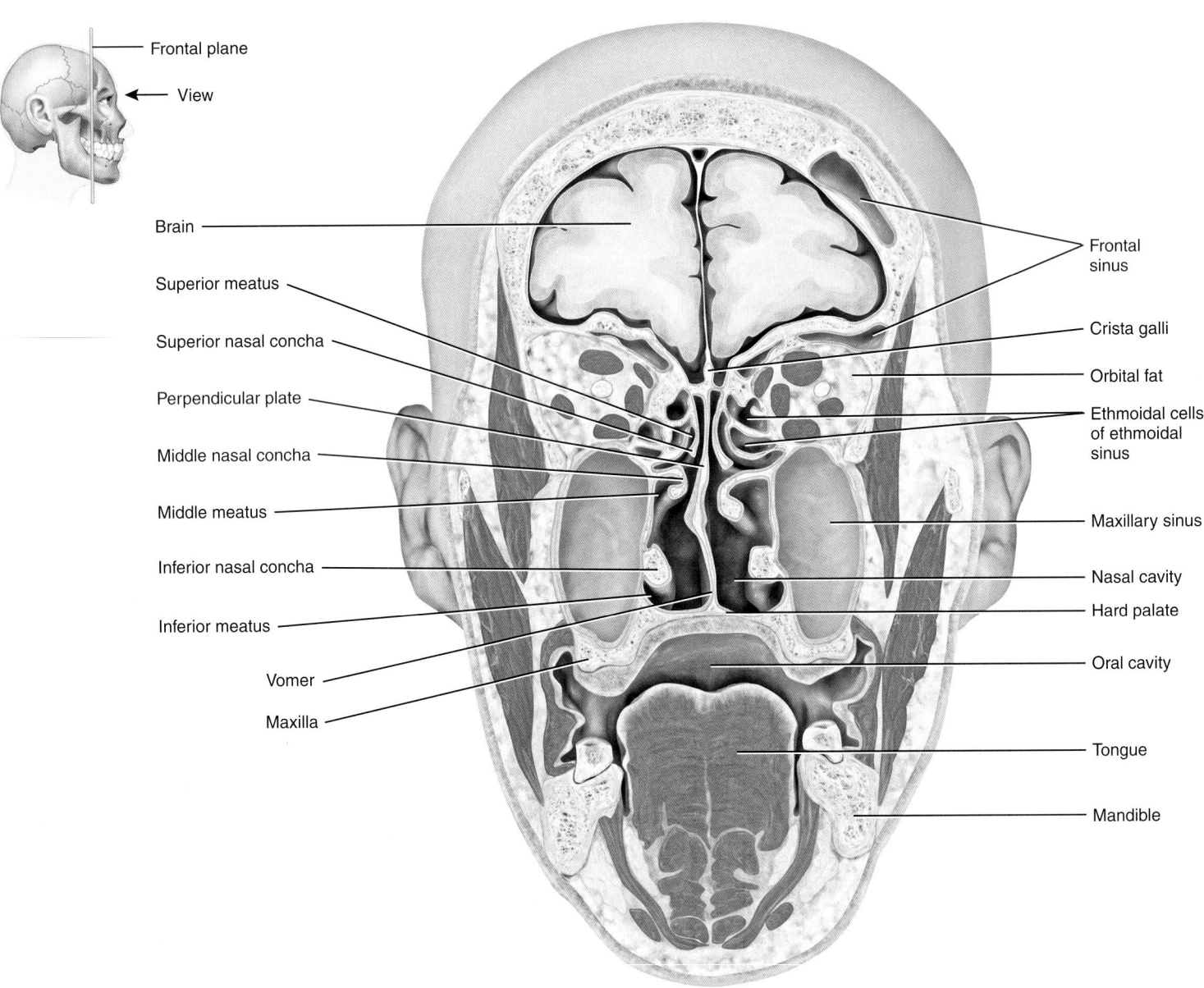

Frontal plane
View

Brain
Superior meatus
Superior nasal concha
Perpendicular plate
Middle nasal concha
Middle meatus
Inferior nasal concha
Inferior meatus
Vomer
Maxilla

Frontal sinus
Crista galli
Orbital fat
Ethmoidal cells of ethmoidal sinus
Maxillary sinus
Nasal cavity
Hard palate
Oral cavity
Tongue
Mandible

(e) Frontal section through the head

? **What part of the ethmoid bone forms the superior part of the nasal septum? The medial walls of the orbits?**

EXHIBIT 7.G

Facial Bones—Nasal, Lacrimal, Palatine, Inferior Nasal Conchae, Vomer, Maxillae, Zygomatic, and Mandible *(Figure 7.10)*

▣ OBJECTIVE

- Identify the location and surface features of the following bones: nasal, lacrimal, palatine, inferior nasal conchae, vomer, maxillae, zygomatic, and mandible.

The shape of the face changes dramatically during the first two years after birth. The brain and cranial bones expand, the first set of teeth form and erupt (emerge), and the paranasal sinuses increase in size. Growth of the face ceases at about 16 years of age. The 14 facial bones include two nasal bones, two maxillae (or maxillas), two zygomatic bones, the mandible, two lacrimal bones, two palatine bones, two inferior nasal conchae, and the vomer.

Nasal Bones

The paired **nasal bones** are small, flattened, rectangular-shaped bones that form the bridge of the nose (see Figure 7.3a). These small bones protect the upper entry to the nasal cavity and provide attachment for a couple of thin muscles of facial expression. For those of you who wear glasses, they are the bones that form the resting place for the bridge of the glasses. The major structural portion of the nose consists of cartilage.

Lacrimal Bones

The paired **lacrimal bones** (LAK-ri-mal; *lacrim-* = teardrops) are thin and roughly resemble a fingernail in size and shape (see Figures 7.3, 7.4b, and 7.12). These bones, the smallest bones of the face, are posterior and lateral to the nasal bones and form a part of the medial wall of each orbit. The lacrimal bones each contain a *lacrimal fossa,* a vertical tunnel formed with the maxilla, that houses the lacrimal sac, a structure that gathers tears and passes them into the nasal cavity (see Figure 7.12).

Palatine Bones

The two L-shaped **palatine bones** (PAL-a-tīn) form the posterior portion of the hard palate, part of the floor and lateral wall of the nasal cavity, and a small portion of the floors of the orbits (see Figures 7.7 and 7.12). The posterior portion of the hard palate is formed by the *horizontal plates* of the palatine bones (see Figures 7.6 and 7.7).

Inferior Nasal Conchae

The two **inferior nasal conchae,** which are inferior to the middle nasal conchae of the ethmoid bone, are separate bones, not part of the ethmoid bone (see Figures 7.3 and 7.9a). These scroll-like bones form a part of the inferior lateral wall of the nasal cavity and project into the nasal cavity. All three pairs of nasal conchae (superior, middle, and inferior) increase the surface area of the nasal cavity and help swirl and filter air before it passes into the lungs. However, only the superior nasal conchae of the ethmoid bone are involved in the sense of smell.

Vomer

The **vomer** (VŌ-mer = plowshare) is a roughly triangular bone on the floor of the nasal cavity that articulates superiorly with the perpendicular plate of the ethmoid bone and sphenoid bone and inferiorly with both the maxillae and palatine bones along the midline (see Figures 7.3a, 7.7, and 7.11). It forms the inferior portion of the bony nasal septum, the partition that divides the nasal cavity into right and left sides.

⚕ CLINICAL CONNECTION | *Cleft Palate and Cleft Lip*

Usually the palatine processes of the maxillary bones unite during weeks 10 to 12 of embryonic development. Failure to do so can result in one type of **cleft palate.** The condition may also involve incomplete fusion of the horizontal plates of the palatine bones (see Figure 7.7). Another form of this condition, called **cleft lip,** involves a split in the upper lip. Cleft lip and cleft palate often occur together. Depending on the extent and position of the cleft, speech and swallowing may be affected. In addition, children with cleft palate tend to have many ear infections, which can lead to hearing loss. Facial and oral surgeons recommend closure of cleft lip during the first few weeks following birth, and surgical results are excellent. Repair of cleft palate typically is completed between 12 and 18 months of age, ideally before the child begins to talk. Because the palate is important for pronouncing consonants, speech therapy may be required, and orthodontic therapy may be needed to align the teeth. Recent research strongly suggests that supplementation with folic acid (one of the B vitamins) during early pregnancy decreases the incidence of cleft palate and cleft lip. The mechanism behind this is not yet understood. •

Maxillae

The paired **maxillae** (mak-SIL-ē = jawbones; singular is *maxilla*) unite to form the upper jawbone. They articulate with every bone of the face except the mandible (lower jawbone) (see Figures 7.3a, 7.4b, and 7.7). The maxillae form part of the floors of the orbits, part of the lateral walls and floor of the nasal cavity, and most of the hard palate. The **hard palate** is the bony roof of the mouth, and is formed by the palatine processes of the maxillae and horizontal plates of the palatine bones. The hard palate separates the nasal cavity from the oral cavity.

Each maxilla contains a large *maxillary sinus* that empties into the nasal cavity (see Figure 7.13). The *alveolar process* (al-VĒ-ō-lar; *alveol-* = small cavity) of the maxilla is a ridgelike arch that contains the *alveoli* (sockets) for the maxillary (upper) teeth. The *palatine process* is a horizontal projection of the maxilla that forms the anterior three-quarters of the hard palate. The union and fusion of the maxillary bones normally is completed before birth. If this fusion fails, this condition is referred to as a cleft palate.

The *infraorbital foramen* (*infra-* = below; *-orbital* = orbit; see Figure 7.3a), an opening in the maxilla inferior to the orbit, allows passage of the infraorbital blood vessels and nerve, a branch of the maxillary division of the trigeminal (V) nerve. Another prominent foramen in the maxilla is the *incisive foramen* (= incisor teeth) just posterior to the incisor teeth (see Figure 7.7). It transmits branches of the greater palatine blood vessels and nasopalatine nerve. A final structure associated with the maxilla and sphenoid bone is the *inferior orbital fissure,* located between the greater wing of the sphenoid and the maxilla (see Figure 7.12).

Zygomatic Bones

The two **zygomatic bones** (*zygo-* = yokelike), commonly called cheekbones, form the prominences of the cheeks and part of the lateral wall and floor of each orbit (see Figure 7.12). They articulate with the frontal, maxilla, sphenoid, and temporal bones.

EXHIBIT 7.G CONTINUES

EXHIBIT 7.G **225**

EXHIBIT 7.G

Facial Bones—Nasal, Lacrimal, Palatine, Inferior Nasal Conchae, Vomer, Maxillae, Zygomatic, and Mandible *(Figure 7.10)* CONTINUED

The *temporal process* of the zygomatic bone projects posteriorly and articulates with the zygomatic process of the temporal bone to form the *zygomatic arch* (see Figure 7.4b).

Mandible

The **mandible** (*mand-* = to chew), or lower jawbone, is the largest, strongest facial bone (Figure 7.10). It is the only movable skull bone (other than the auditory ossicles, the small bones of the ear). In the lateral view, you can see that the mandible consists of a curved, horizontal portion, the *body,* and two perpendicular portions, the *rami* (RĀ-mī = branches; singular is *ramus*). The *angle* of the mandible is the area where each *ramus* meets the body. Each ramus has a posterior *condylar process* (KON-di-lar) that articulates with the mandibular fossa and articular tubercle of the temporal bone (see Figure 7.4b) to form the **temporomandibular joint (TMJ),** and an anterior *coronoid process* (KOR-ō-noyd) to which the temporalis muscle attaches. The depression between the coronoid and condylar processes is called the *mandibular notch.* The *alveolar process* is the ridgelike arch containing the *alveoli* (sockets) for the mandibular (lower) teeth.

The *mental foramen* (*ment-* = chin) is approximately inferior to the second premolar tooth. It is near this foramen that dentists reach the mental nerve when injecting anesthetics. Another foramen associated with the mandible is the *mandibular foramen* on the medial surface of each ramus, another site often used by dentists to inject anesthetics. The mandibular foramen is the beginning of the *mandibular canal,* which runs obliquely in the ramus and anteriorly to the body. Through the canal pass the inferior alveolar nerves and blood vessels, which are distributed to the mandibular teeth.

CLINICAL CONNECTION | *Temporomandibular Joint Syndrome*

One problem associated with the temporomandibular joint is **temporomandibular joint (TMJ) syndrome.** It is characterized by dull pain around the ear, tenderness of the jaw muscles, a clicking or popping noise when opening or closing the mouth, limited or abnormal opening of the mouth, headache, tooth sensitivity, and abnormal wearing of the teeth. TMJ syndrome can be caused by improperly aligned teeth, grinding or clenching the teeth, trauma to the head and neck, or arthritis. Treatments include application of moist heat or ice, limiting the diet to soft foods, administration of pain relievers such as aspirin, muscle retraining, use of a splint or bite plate to reduce clenching and teeth grinding (especially when worn at night), adjustment or reshaping of the teeth (orthodontic treatment), and surgery. •

✔CHECKPOINT

Which bones form the hard palate? Which bones form the nasal septum?

Figure 7.10 Mandible.

🔑 The mandible is the largest and strongest facial bone.

Right lateral view

❓ **What is the distinctive functional feature of the mandible among all the skull bones?**

General Features and Functions

Besides forming the large cranial cavity, the skull also forms several smaller cavities, including the nasal cavity and orbits (eye sockets), which open to the exterior. Certain skull bones also contain cavities called paranasal sinuses that are lined with mucous membranes and open into the nasal cavity. Also within the skull are small middle ear cavities in the temporal bones that house the structures that are involved in hearing and equilibrium (balance).

Other than the auditory ossicles (tiny bones involved in hearing), which are located within the temporal bones, the mandible is the only movable bone of the skull. Joints called sutures attach most of the skull bones together and are especially noticeable on the outer surface of the skull.

The skull has many surface markings, such as foramina (rounded passageways) and fissures (slit-like openings) through which blood vessels and nerves pass. You will learn the names of important skull bone surface markings as we describe each bone.

In addition to protecting the brain, the cranial bones stabilize the positions of the brain, blood vessels, lymphatic vessels, and nerves through the attachment of their inner surfaces to meninges (membranes). The outer surfaces of cranial bones provide large areas of attachment for muscles that move various parts of the head. The bones also provide attachment for some muscles that produce facial expressions such as the frown of concentration you wear when studying this book. The facial bones form the framework of the face and provide support for the entrances to the di-

gestive and respiratory systems. Together, the cranial and facial bones protect and support the delicate special sense organs for vision, taste, smell, hearing, and equilibrium (balance).

Nasal Septum

The nasal cavity is a space inside the skull that is divided into right and left sides by a vertical partition called the **nasal septum,** which consists of bone and cartilage. The three components of the nasal septum are the vomer, septal cartilage, and the perpendicular plate of the ethmoid bone (Figure 7.11). The anterior border of the vomer articulates with the septal cartilage, which is hyaline cartilage, to form the anterior portion of the septum. The superior border of the vomer articulates with the perpendicular plate of the ethmoid bone to form the remainder of the nasal septum. The term "broken nose," in most cases, refers to damage to the septal cartilage rather than the nasal bones themselves.

CLINICAL CONNECTION | *Deviated Nasal Septum*

A **deviated nasal septum** is one that does not run along the midline of the nasal cavity. It deviates (bends) to one side. A blow to the nose can easily damage, or break, this delicate septum of bone and displace and damage the cartilage. Often, when a broken nasal septum heals, the bones and cartilage deviate to one side or the other. This deviated septum can block airflow into the constricted side of the nose, making it difficult to breathe through that half of the nasal cavity. The deviation usually occurs at the junction of the

Figure 7.11 Nasal septum.

The structures that form the nasal septum are the perpendicular plate of the ethmoid bone, the vomer, and septal cartilage.

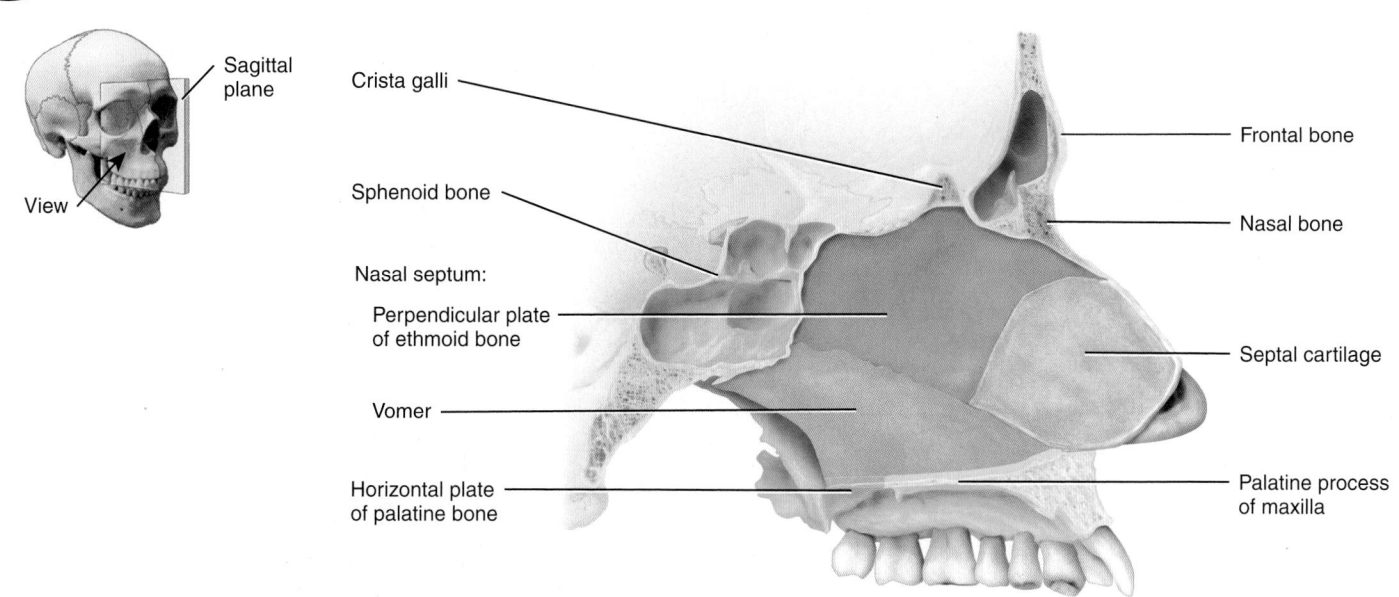

Sagittal section

? **What is the function of the nasal septum?**

vomer bone with the septal cartilage. Septal deviations may also oc-cur due to developmental abnormality. If the deviation is severe, it may block the nasal passageway entirely. Even a partial blockage may lead to infection. If inflammation occurs, it may cause nasal conges-tion, blockage of the paranasal sinus openings, chronic sinusitis, headache, and nosebleeds. The condition usually can be corrected or improved surgically. •

Orbits

Seven bones of the skull join to form each **orbit** (eye socket) or *orbital cavity,* which contains the eyeball and associated struc-tures (Figure 7.12). The three cranial bones of the orbit are the frontal, sphenoid, and ethmoid; the four facial bones are the pala-tine, zygomatic, lacrimal, and maxilla. Each pyramid-shaped or-bit has four regions that converge posteriorly:

1. Parts of the frontal and sphenoid bones comprise the *roof* of the orbit.
2. Parts of the zygomatic and sphenoid bones form the *lateral wall* of the orbit.
3. Parts of the maxilla, zygomatic, and palatine bones make up the *floor* of the orbit.

4. Parts of the maxilla, lacrimal, ethmoid, and sphenoid bones form the *medial wall* of the orbit.

Associated with each orbit are five openings:

1. The *optic foramen (canal)* is at the junction of the roof and medial wall.
2. The *superior orbital fissure* is at the superior lateral angle of the apex.
3. The *inferior orbital fissure* is at the junction of the lateral wall and floor.
4. The *supraorbital foramen* is on the medial side of the supraor-bital margin of the frontal bone.
5. The *lacrimal fossa* is in the lacrimal bone.

Foramina

We mentioned most of the **foramina** (openings for blood vessels, nerves, or ligaments) of the skull in the descriptions of the cranial and facial bones that they penetrate. As preparation for studying other systems of the body, especially the nervous and cardiovas-cular systems, these foramina and the structures passing through them are listed in Table 7.3. For your convenience and for future reference, the foramina are listed alphabetically.

Figure 7.12 **Details of the orbit (eye socket).**

The orbit is a pyramid-shaped structure that contains the eyeball and associated structures.

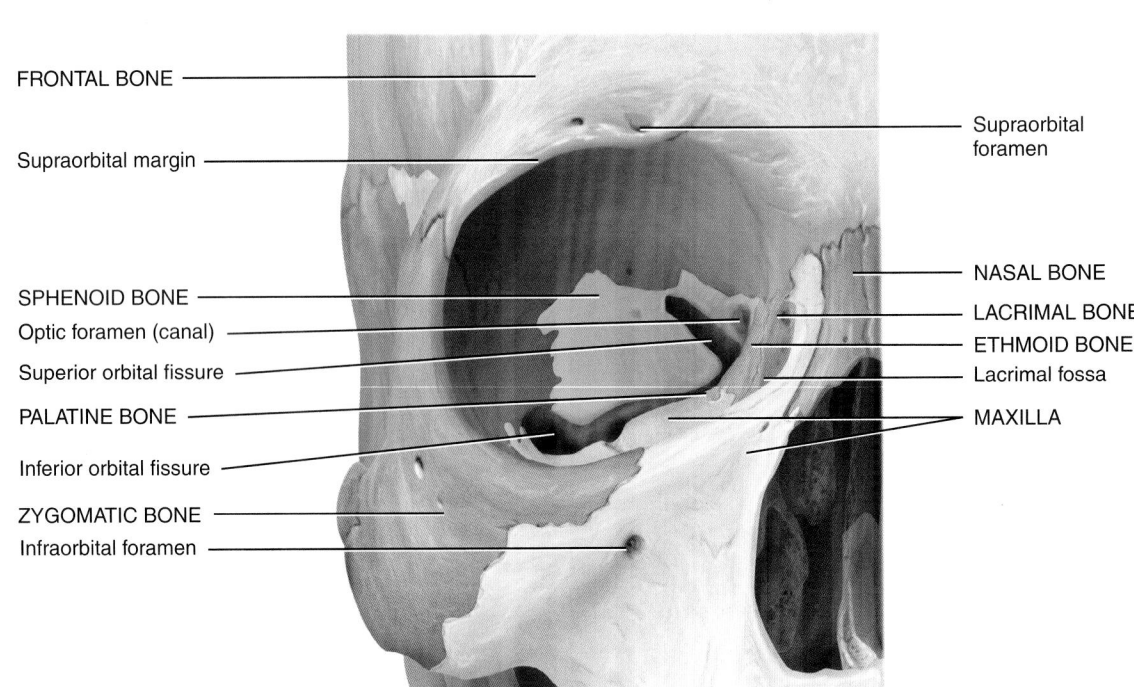

Anterior view showing the bones of the right orbit

? **Which seven bones form the orbit?**

TABLE 7.3

Principal Foramina of the Skull

FORAMEN	LOCATION	STRUCTURES PASSING THROUGH*
Carotid (relating to carotid artery in neck)	Petrous portion of temporal bone (Figure 7.7).	Internal carotid artery, sympathetic nerves for eyes.
Hypoglossal (*hypo-* = under; *-glossus* = tongue)	Superior to base of occipital condyles (Figure 7.8a).	Cranial nerve XII (hypoglossal), branch of ascending pharyngeal artery.
Infraorbital (*infra-* = below)	Inferior to orbit in maxilla (Figure 7.12).	Infraorbital nerve and blood vessels, branch of maxillary division of cranial nerve V (trigeminal).
Jugular (*jugul-* = throat)	Posterior to carotid canal between petrous portion of temporal bone and occipital bone (Figure 7.8a).	Internal jugular vein; cranial nerves IX (glossopharyngeal), X (vagus), XI (accessory).
Lacerum (*lacerum* = lacerated)	Bounded anteriorly by sphenoid bone, posteriorly by petrous portion of temporal bone, medially by sphenoid and occipital bones (Figure 7.8a).	Branch of ascending pharyngeal artery.
Magnum (= large)	Occipital bone (Figure 7.7).	Medulla oblongata and its membranes (meninges), cranial nerve XI (accessory), vertebral and spinal arteries.
Mandibular (*mand-* = to chew)	Medial surface of ramus of mandible (Figure 7.10).	Inferior alveolar nerve and blood vessels.
Mastoid (= breast-shaped)	Posterior border of mastoid process of temporal bone (Figure 7.7).	Emissary vein to transverse sinus, branch of occipital artery to dura mater.
Mental (*ment-* = chin)	Inferior to second premolar tooth in mandible (Figure 7.10).	Mental nerve and vessels.
Olfactory (*olfact-* = to smell)	Cribriform plate of ethmoid bone (Figure 7.8a).	Cranial nerve I (olfactory).
Optic (= eye)	Between superior and inferior portions of small wing of sphenoid bone (Figure 7.12).	Cranial nerve II (optic), ophthalmic artery.
Ovale (= oval)	Greater wing of sphenoid bone (Figure 7.8a).	Mandibular branch of cranial nerve V (trigeminal).
Rotundum (= round)	Junction of anterior and medial parts of sphenoid bone (Figure 7.8a, b).	Maxillary branch of cranial nerve V (trigeminal).
Stylomastoid (*stylo-* = stake or pole)	Between styloid and mastoid processes of temporal bone (Figure 7.7).	Cranial nerve VII (facial), stylomastoid artery.
Supraorbital (*supra-* = above)	Supraorbital margin of orbit in frontal bone (Figure 7.12).	Supraorbital nerve and artery.

*The cranial nerves listed here are described in Table 14.4.

Unique Features of the Skull

The skull exhibits several unique features not seen in other bones of the body. These include sutures, paranasal sinuses, and fontanels.

Sutures

A **suture** (SOO-chur = seam) is an immovable joint in most cases in an adult skull that holds most skull bones together. Sutures in the skulls of infants and children, however, often are movable and function as important growth centers in the developing skull. The names of many sutures reflect the bones they unite. For example, the frontozygomatic suture is between the frontal bone and the zygomatic bone. Similarly, the sphenoparietal suture is between the

sphenoid bone and the parietal bone. In other cases, however, the names of sutures are not so obvious. Of the many sutures found in the skull, we will identify only four prominent ones:

1. The **coronal suture** (KO-rō-nal; *coron-* = relating to the frontal or coronal plane) unites the frontal bone and both parietal bones (see Figure 7.4b).

2. The **sagittal suture** (SAJ-i-tal; *sagitt-* = arrow) unites the two parietal bones on the superior midline of the skull (see Figure 7.4a). The sagittal suture is so named because in the infant, before the bones of the skull are firmly united, the suture and the fontanels (soft spots) associated with it resemble an arrow.

3. The **lambdoid suture** (LAM-doyd) unites the two parietal bones to the occipital bone. This suture is so named because of its resemblance to the capital Greek letter lambda (Λ), as can be seen in Figure 7.6 (with the help of a little imagination). Sutural bones may occur within the sagittal and lambdoid sutures.

4. The two **squamous sutures** (SKWĀ-mus; *squam-* = flat, like the flat overlapping scales of a snake) unite the parietal and temporal bones on the lateral aspects of the skull (see Figure 7.4b).

Paranasal Sinuses

The **paranasal sinuses** (par'-a-NĀ-zal SĪ-nus-ez; *para-* = beside) are cavities within certain cranial and facial bones near the nasal cavity. They are most evident in a sagittal section of the skull (Figure 7.13). The paranasal sinuses are lined with mucous membranes that are continuous with the lining of the nasal cavity. Secretions produced by the mucous membranes of the paranasal sinuses drain into the lateral wall of the nasal cavity. Paranasal

Figure 7.13 Paranasal sinuses projected to the surface.

Paranasal sinuses are mucous membrane–lined spaces in the frontal, sphenoid, ethmoid, and maxillary bones that connect to the nasal cavity.

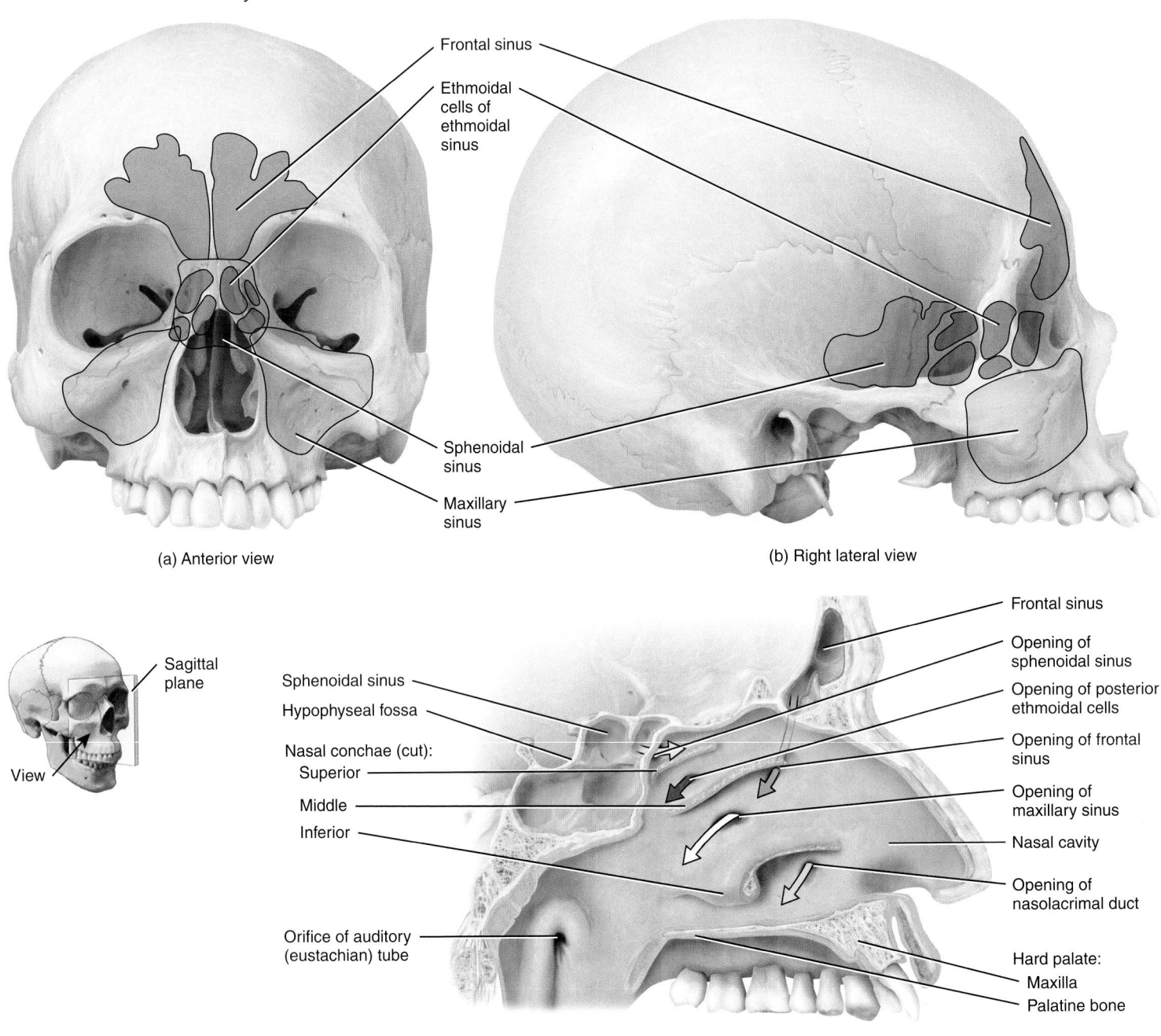

(a) Anterior view

(b) Right lateral view

(c) Sagittal section

? What are the functions of the paranasal sinuses?

sinuses are quite small or absent at birth, but increase in size during two periods of facial enlargement—during the eruption of the teeth and at the onset of puberty. They arise as outgrowths of the nasal mucosa that project into the surrounding bones. Skull bones containing the paranasal sinuses are the frontal, sphenoid, ethmoid, and maxillae. The paranasal sinuses allow the skull to increase in size without a change in the mass (weight) of the bone. The paranasal sinuses increase the surface area of the nasal mucosa, thus increasing the production of mucus to help moisten and cleanse inhaled air. In addition, the paranasal sinuses serve as resonating (echo) chambers within the skull that intensify and prolong sounds, thereby enhancing the quality of the voice. The influence of the paranasal sinuses on your voice becomes obvious when you have a cold; the passageways through which sound travels into and out of the paranasal sinuses become blocked by excess mucus production, changing the quality of your voice.

CLINICAL CONNECTION | *Sinusitis*

Sinusitis (sīn-ū-SĪ-tis) is an inflammation of the mucous membrane of one or more paranasal sinuses. It may be caused by a microbial infection (virus, bacterium, or fungus), allergic reactions, nasal polyps, or a severely deviated nasal septum. If the inflammation or an obstruction blocks the drainage of mucus into the nasal cavity, fluid pressure builds up in the paranasal sinuses, and a sinus headache may develop. Other symptoms may include nasal congestion, inability to smell, fever, and cough. Treatment options include decongestant sprays or drops, oral decongestants, nasal corticosteroids, antibiotics, analgesics to relieve pain, warm compresses, and surgery. •

Fontanels

The skull of a developing embryo consists of cartilage and mesenchyme arranged in thin plates around the developing brain. Gradually, ossification occurs, and bone slowly replaces most of the cartilage and mesenchyme. At birth, bone ossification is incomplete, and the mesenchyme-filled spaces become dense connective tissue regions between incompletely developed cranial bones called **fontanels** (fon-ta-NELZ = little fountains), commonly called "soft spots" (Figure 7.14). Fontanels are areas where unossified mesenchyme develops into the dense connective tissues of the skull. As bone formation continues after birth, the fontanels are eventually replaced with bone by intramembranous ossification, and the thin collagenous connective tissue junctions that remain between neighboring bones become the sutures. Functionally, the fontanels serve as spacers for the growth of neighboring skull bones and provide some flexibility to the fetal skull, allowing the skull to change shape as it passes through the birth canal and later permitting rapid growth of the brain during infancy. Although an infant may have many fontanels at birth, the form and location of six are fairly constant:

- The unpaired **anterior fontanel,** the largest fontanel, is located at the midline among the two parietal bones and the frontal bone, and is roughly diamond-shaped. It usually closes 18 to 24 months after birth.

- The unpaired **posterior fontanel** is located at the midline among the two parietal bones and the occipital bone. Because it is much smaller than the anterior fontanel, it generally closes about 2 months after birth.

Figure 7.14 Fontanels at birth.

Fontanels are mesenchyme-filled spaces between cranial bones that are present at birth.

Right lateral view

? **Which fontanel is bordered by four different skull bones?**

• The paired **anterolateral fontanels,** located laterally among the frontal, parietal, temporal, and sphenoid bones, are small and irregular in shape. Normally, they close about 3 months after birth.

• The paired **posterolateral fontanels,** located laterally among the parietal, occipital, and temporal bones, are irregularly shaped. They begin to close 1 to 2 months after birth, but closure is generally not complete until 12 months.

The amount of closure in fontanels helps a physician gauge the degree of brain development. In addition, the anterior fontanel serves as a landmark for withdrawal of blood for analysis from the superior sagittal sinus (a large midline vein within the covering tissues that surround the brain). (See Figure 21.24.)

✔ CHECKPOINT
4. Describe the cavities within the skull and the nasal septum.
5. Which foramina and fissures are associated with the orbit?
6. What structures make up the nasal septum?
7. Define the following: foramen, suture, paranasal sinus, and fontanel.

7.5 HYOID BONE

◉ OBJECTIVE
• Describe the relationship of the hyoid bone to the skull.

The single **hyoid bone** (= U-shaped) is a unique component of the axial skeleton because it does not articulate with any other bone. Rather, it is suspended from the styloid processes of the temporal bones by ligaments and muscles. Located in the anterior neck between the mandible and larynx (Figure 7.15a), the hyoid bone supports the tongue, providing attachment sites for some tongue muscles and for muscles of the neck and pharynx. The hyoid bone consists of a horizontal *body* and paired projections called the *lesser horns* and the *greater horns* (Figure 7.15b, c). Muscles and ligaments attach to the body and these paired projections.

The hyoid bone and the cartilages of the larynx and trachea are often fractured during strangulation. As a result, they are carefully examined at autopsy when manual strangulation is a suspected cause of death.

✔ CHECKPOINT
8. What are the functions of the hyoid bone?

Figure 7.15 Hyoid bone.

 The hyoid bone supports the tongue, providing attachment sites for muscles of the tongue, neck, and pharynx.

(a) Position of hyoid

(b) Anterior view

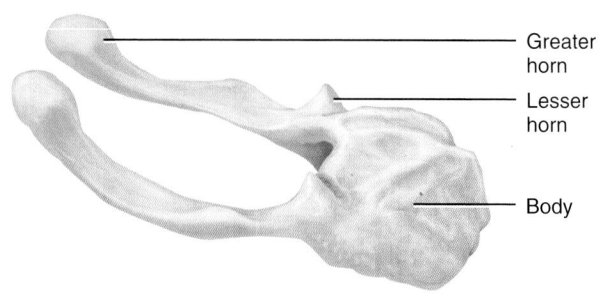

(c) Right lateral view

? **In what way is the hyoid bone different from all the other bones of the axial skeleton?**

7.6 VERTEBRAL COLUMN

◉ OBJECTIVES

- Identify the regions and normal curves of the vertebral column.
- Describe the structural and functional features of the bones in various regions of the vertebral column.

The **vertebral column,** also called the *spine, backbone,* or *spinal column,* makes up about two-fifths of your total height and is composed of a series of bones called **vertebrae** (VER-te-brē; singular is *vertebra*). The vertebral column, the sternum, and the ribs form the skeleton of the trunk of the body. The vertebral column consists of bone and connective tissue; the spinal cord that it surrounds and protects consists of nervous and connective tissues. At about 71 cm (28 in.) in an average adult male and about 61 cm (24 in.) in an average adult female, the vertebral column functions as a strong, flexible rod with elements that can move forward, backward, and sideways, and rotate. In addition to enclosing and protecting the spinal cord, it supports the head and serves as a point of attachment for the ribs, pelvic girdle, and muscles of the back and upper limbs.

The total number of vertebrae during early development is 33. As a child grows, several vertebrae in the sacral and coccygeal regions fuse. As a result, the adult vertebral column typically contains 26 vertebrae (Figure 7.16a). These are distributed as follows:

- 7 **cervical vertebrae** (*cervic-* = neck) are in the neck region.
- 12 **thoracic vertebrae** (*thorax* = chest) are posterior to the thoracic cavity.

Figure 7.16 Vertebral column. The numbers in parentheses in part (a) indicate the number of vertebrae in each region. In part (d), the relative size of the disc has been enlarged for emphasis.

🔑 The adult vertebral column typically contains 26 vertebrae.

POSTERIOR   ANTERIOR

Cervical vertebrae (7)
Thoracic vertebrae (12)
Lumbar vertebrae (5)
Intervertebral disc
Sacrum (1)
Coccyx (1)

Cervical curve (formed by 7 cervical vertebrae)
Thoracic curve (formed by 12 thoracic vertebrae)
Intervertebral disc
Lumbar curve (formed by 5 lumbar vertebrae)
Intervertebral foramen
Sacrum
Sacral curve (formed by 5 fused sacral vertebrae)
Coccyx

(a) Anterior view showing regions of the vertebral column

(b) Right lateral view showing four normal curves

FIGURE 7.16 CONTINUES ▶

FIGURE 7.16 CONTINUED

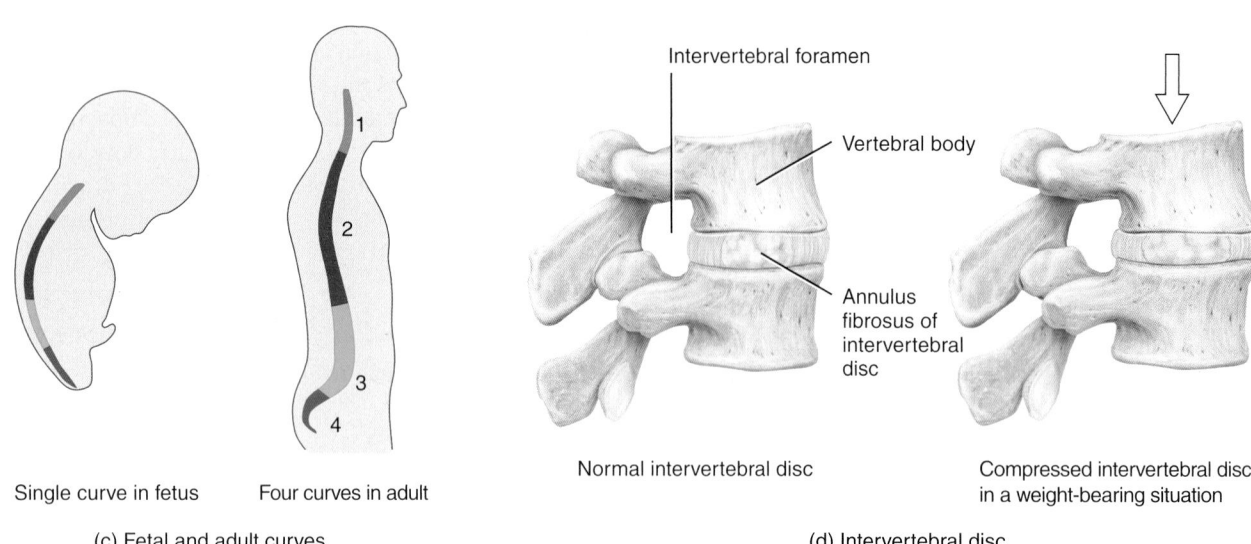

Single curve in fetus   Four curves in adult

(c) Fetal and adult curves

Normal intervertebral disc

Compressed intervertebral disc in a weight-bearing situation

(d) Intervertebral disc

Which curves of the adult vertebral column are concave (relative to the anterior side of the body)?

- 5 **lumbar vertebrae** (*lumb-* = loin) support the lower back.

- 1 **sacrum** (SĀ-krum = sacred bone) consists of five fused **sacral vertebrae.**

- 1 **coccyx** (KOK-siks = cuckoo, because the shape resembles the bill of a cuckoo bird) usually consists of four fused **coccygeal vertebrae** (kok-SIJ-ē-al).

The cervical, thoracic, and lumbar vertebrae are movable, but the sacrum and coccyx are not. We will discuss each of these regions in detail shortly.

Normal Curves of the Vertebral Column

When viewed from the anterior or posterior, a normal adult vertebral column appears straight. But when viewed from the side, it shows four slight bends called **normal curves** (Figure 7.16b). Relative to the front of the body, the *cervical* and *lumbar curves* are convex (bulging out); the *thoracic* and *sacral curves* are concave (cupping in). The curves of the vertebral column increase its strength, help maintain balance in the upright position, absorb shocks during walking, and help protect the vertebrae from fracture.

The fetus has a single anteriorly concave curve throughout the length of the entire vertebral column (Figure 7.16c). At about the third month after birth, when an infant begins to hold its head erect, the anteriorly convex cervical curve develops. Later, when the child sits up, stands, and walks, the anteriorly convex lumbar curve develops. The thoracic and sacral curves are called *primary curves* because they retain the original curvature of the embryonic vertebral column. The cervical and lumbar curves are known as *secondary curves* because they begin to form later, several months after birth. All curves are fully developed by age 10. However, secondary curves may be progressively lost in old age.

Various conditions may exaggerate the normal curves of the vertebral column, or the column may acquire a lateral bend, resulting in **abnormal curves** of the vertebral column. Three such abnormal curves—kyphosis, lordosis, and scoliosis—are described in Disorders: Homeostatic Imbalances at the end of this chapter.

Intervertebral Discs

Intervertebral discs (in'-ter-VER-te-bral; *inter-* = between) are found between the bodies of adjacent vertebrae from the second cervical vertebra to the sacrum (Figure 7.16d) and account for about 25% of the height of the vertebral column. Each disc has an outer fibrous ring consisting of fibrocartilage called the *annulus fibrosus* (*annulus* = ringlike) and an inner soft, pulpy, highly elastic substance called the *nucleus pulposus* (*pulposus* = pulplike). The discs form strong joints, permit various movements of the vertebral column, and absorb vertical shock. Under compression, they flatten and broaden.

During the course of a day the discs compress and lose water from their cartilage so that we are a bit shorter at night. While we are sleeping there is less compression and rehydration occurs, so that we are taller when we awaken in the morning. With age, the nucleus pulposus hardens and becomes less elastic. Decrease in vertebral height with age results from bone loss in the vertebral bodies and not a decrease in thickness of the intervertebral discs.

Since intervertebral discs are avascular, the annulus fibrosus and nucleus pulposus rely on blood vessels from the bodies of vertebrae to obtain oxygen and nutrients and remove wastes.

Certain stretching exercises, such as yoga, decompress discs and increase general blood circulation, both of which speed up the uptake of oxygen and nutrients by discs and the removal of wastes.

Parts of a Typical Vertebra

Vertebrae in different regions of the spinal column vary in size, shape, and detail, but they are similar enough that we can discuss the structures (and the functions) of a typical vertebra (Figure 7.17). Vertebrae typically consist of a vertebral body, a vertebral arch, and several processes.

Vertebral Body

The **vertebral body,** the thick, disc-shaped anterior portion, is the weight-bearing part of a vertebra. Its superior and inferior surfaces are roughened for the attachment of cartilaginous intervertebral discs. The anterior and lateral surfaces contain nutrient foramina, openings through which blood vessels deliver nutrients and oxygen and remove carbon dioxide and wastes from bone tissue.

Vertebral Arch

Two short, thick processes, the *pedicles* (PED-i-kuls = little feet), project posteriorly from the vertebral body and then unite with the flat *laminae* (LAM-i-nē = thin layers) to form the **vertebral arch.** The vertebral arch extends posteriorly from the body of the vertebra; together, the vertebral body and the vertebral arch surround the spinal cord by forming the *vertebral foramen.* The vertebral foramen contains the spinal cord, adipose tissue, areolar connective tissue, and blood vessels. Collectively, the vertebral foramina of all vertebrae form the **vertebral (spinal) canal.** The pedicles exhibit superior and inferior indentations called *vertebral notches.* When the vertebral notches are stacked on top of one another, they form an opening between adjoining vertebrae on both sides of the column. Each opening, called an *intervertebral foramen,* permits the passage of a single spinal nerve carrying information to and from the spinal cord.

Processes

Seven **processes** arise from the vertebral arch. At the point where a lamina and pedicle join, a *transverse process* extends laterally

Figure 7.17 Structure of a typical vertebra, as illustrated by a thoracic vertebra. In part b, only one spinal nerve has been included, and it has been extended beyond the intervertebral foramen for clarity.

🔑 A vertebra consists of a vertebral body, a vertebral arch, and several processes.

(a) Superior view

(b) Right posterolateral view of articulated vertebrae

? **What are the functions of the vertebral and intervertebral foramina?**

on each side. A single *spinous process (spine)* projects posteriorly from the junction of the laminae. These three processes serve as points of attachment for muscles. The remaining four processes form joints with other vertebrae above or below. The two *superior articular processes* of a vertebra articulate (form joints) with the two inferior articular processes of the vertebra immediately above them. In turn, the two *inferior articular processes* of that vertebra articulate with the two superior articular processes of the vertebra immediately below them, and so on. The articulating surfaces of the articular processes, which are referred to as *facets* (FAS-ets or fa-SETS = little faces), are covered with hyaline cartilage. The articulations formed between the vertebral bodies and articular facets of successive vertebrae are termed *intervertebral joints.*

Regions of the Vertebral Column

Exhibits 7.H through 7.K present the five regions of the vertebral column, beginning superiorly and moving inferiorly. The regions are the cervical, thoracic, lumbar, sacral, and coccygeal. Note that vertebrae in each region are numbered in sequence, from superior to inferior. When you actually view the bones of the vertebral column, you will notice that the transition from one region to the next is not abrupt but gradual, a feature that helps the vertebrae fit together.

Age-Related Changes in the Vertebral Column

With advancing age the vertebral column undergoes changes that are characteristic of the skeletal system in general. These changes include reduction in the mass and density of the bone along with a reduction in the collagen-to-mineral content within the bone, changes that make the bones more brittle and susceptible to damage. The articular surfaces, those surfaces where neighboring bones move against one another, lose their covering cartilage as they age; in their place rough bony growths form that lead to arthritic conditions. In the vertebral column, bony growths around the intervertebral discs, called *osteophytes,* can lead to a narrowing (stenosis) of the vertebral canal. This narrowing can lead to compression of spinal nerves and the spinal cord, which can manifest as pain and decreased muscle function in the back and lower limbs.

✔ **CHECKPOINT**

9. What are the functions of the vertebral column?
10. Describe the four curves of the vertebral column.
11. What are the three main parts of a typical vertebra?
12. What are the principal distinguishing characteristics of the bones of the various regions of the vertebral column?

7.7 THORAX

◉ **OBJECTIVE**

• Identify the bones of the thorax and their functions.

The term **thorax** refers to the entire chest region. The skeletal part of the thorax, the **thoracic cage,** is a bony enclosure formed by the sternum, ribs and their costal cartilages, and the bodies of the thoracic vertebrae (see Exhibits 7.L and 7.M). The costal cartilages attach the ribs to the sternum. The thoracic cage is narrower at its superior end and broader at its inferior end and is flattened from front to back. It encloses and protects the organs in the thoracic and superior abdominal cavities, provides support for the bones of the upper limbs, and, as you will see in Chapter 23, plays a role in breathing.

✔ **CHECKPOINT**

13. What bones form the skeleton of the thorax?
14. What are the functions of the bones of the thorax?
15. How are ribs classified?

EXHIBIT 7.H   Vertebral Regions—Cervical Vertebrae *(Figure 7.18)*

◉ OBJECTIVE
• Identify the location and surface features of the cervical vertebrae.

The bodies of the **cervical vertebrae** (C1–C7) are smaller than all other vertebrae except those that form the coccyx (Figure 7.18a). Their vertebral arches, however, are larger. All cervical vertebrae have three foramina: one vertebral foramen and two transverse foramina (Figure 7.18c). The *vertebral foramina* of cervical vertebrae are the largest in the spinal column because they house the cervical enlargement of the spinal cord. Each cervical transverse process contains a *transverse foramen* through which the vertebral artery and its accompanying vein and nerve fibers pass. The spinous processes of C2 through C6 are often *bifid*—that is, they branch into two small projections at the tips (Figure 7.18a, c).

The first two cervical vertebrae differ considerably from the others. The **atlas** (C1), named after the mythological Atlas who supported the world on his shoulders, is the first cervical vertebra inferior to the skull (Figure 7.18a, b). The atlas is a ring of bone with *anterior* and *posterior arches* and large *lateral masses*. It lacks a body and a spinous process. The superior surfaces of the lateral masses, called *superior articular facets*, are concave. They articulate with the occipital condyles of the occipital bone to form the paired *atlanto-occipital joints*. These articulations permit you to move your head to signify "yes." The inferior surfaces of the lateral masses, the *inferior articular facets,* articulate with the second cervical vertebra. The transverse processes and transverse foramina of the atlas are quite large.

The second cervical vertebra (C2), the **axis** (see Figure 7.18a, d, e), does have a vertebral body. A peglike process called the *dens* (= tooth) or *odontoid process* projects superiorly through the anterior portion of the vertebral foramen of the atlas. The dens makes a pivot on which the atlas and head rotate. This arrangement permits side-to-side movement of the head, as when you move your head to signify "no." The articulation formed between the anterior arch of the atlas and dens of the axis, and between their articular facets, is called the *atlanto-axial joint.* In some instances of trauma, the dens of the axis may be driven into the medulla oblongata of the brain. This type of injury is the usual cause of death from whiplash injuries.

The third through sixth cervical vertebrae (C3–C6), represented by the vertebra in Figure 7.18c, correspond to the structural pattern of the

Figure 7.18 Cervical vertebrae.

🔑 The cervical vertebrae are found in the neck region.

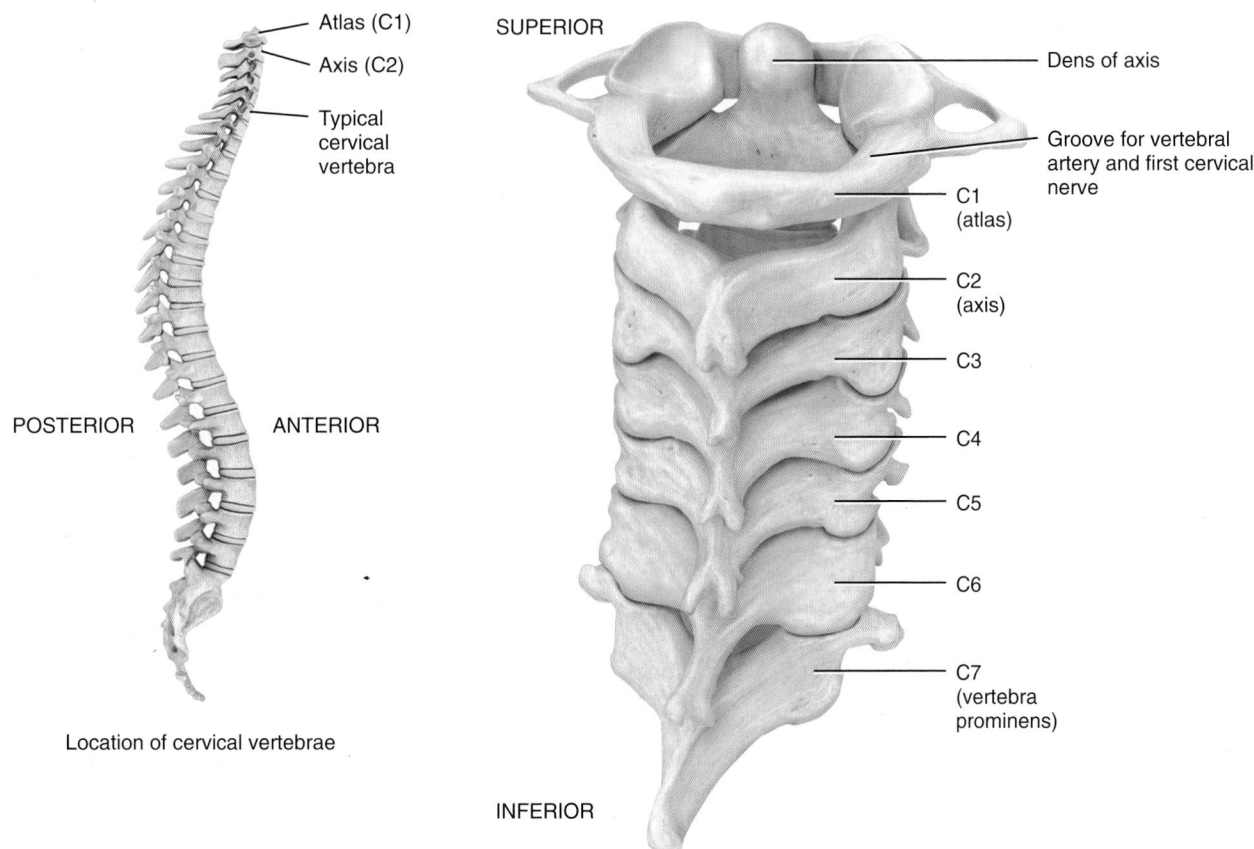

Atlas (C1)
Axis (C2)
Typical cervical vertebra

POSTERIOR   ANTERIOR

Location of cervical vertebrae

SUPERIOR

Dens of axis
Groove for vertebral artery and first cervical nerve
C1 (atlas)
C2 (axis)
C3
C4
C5
C6
C7 (vertebra prominens)

INFERIOR

(a) Posterior view of articulated cervical vertebrae

EXHIBIT 7.H CONTINUES ▶

EXHIBIT 7.H **237**

EXHIBIT 7.H   Vertebral Regions—Cervical Vertebrae *(Figure 7.18)*   CONTINUED

typical cervical vertebra previously described. The seventh cervical vertebra (C7), called the *vertebra prominens,* is somewhat different (see Figure 7.18a). It has a large, nonbifid spinous process that may be seen and felt at the base of the neck, but otherwise is typical.

✔CHECKPOINT
How do the atlas and axis differ from the other cervical vertebrae?

⦿ FIGURE 7.18 CONTINUED ▶

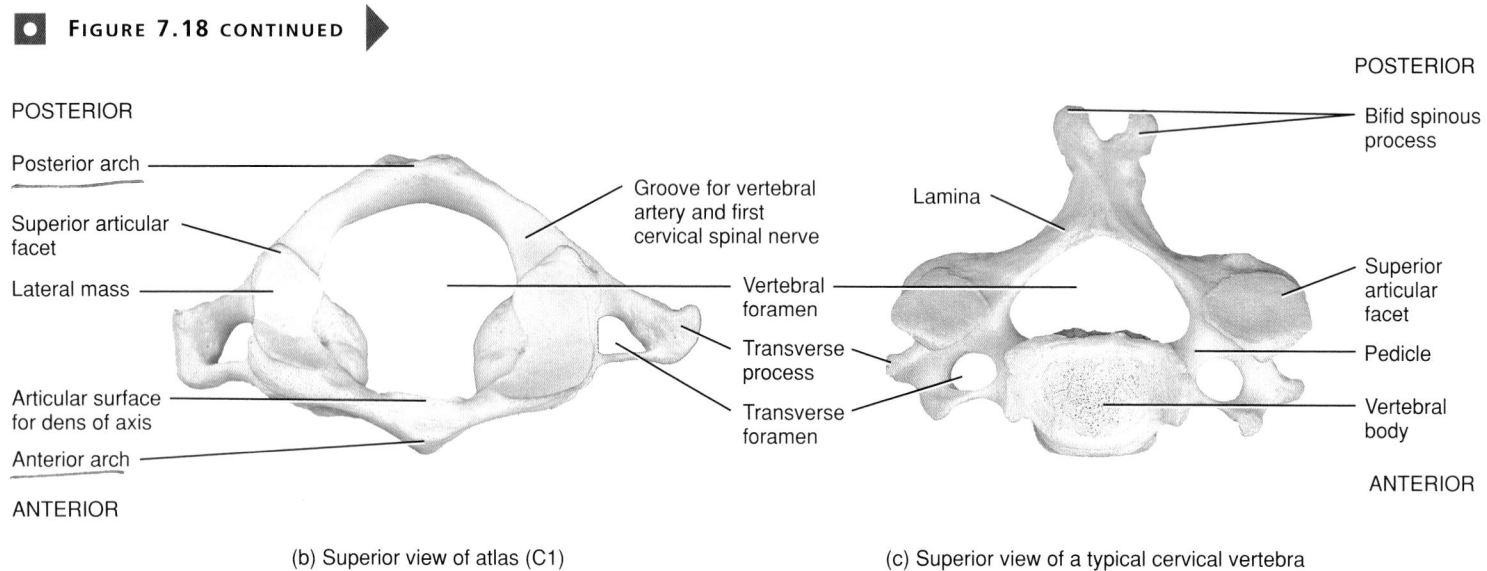

(b) Superior view of atlas (C1)

(c) Superior view of a typical cervical vertebra

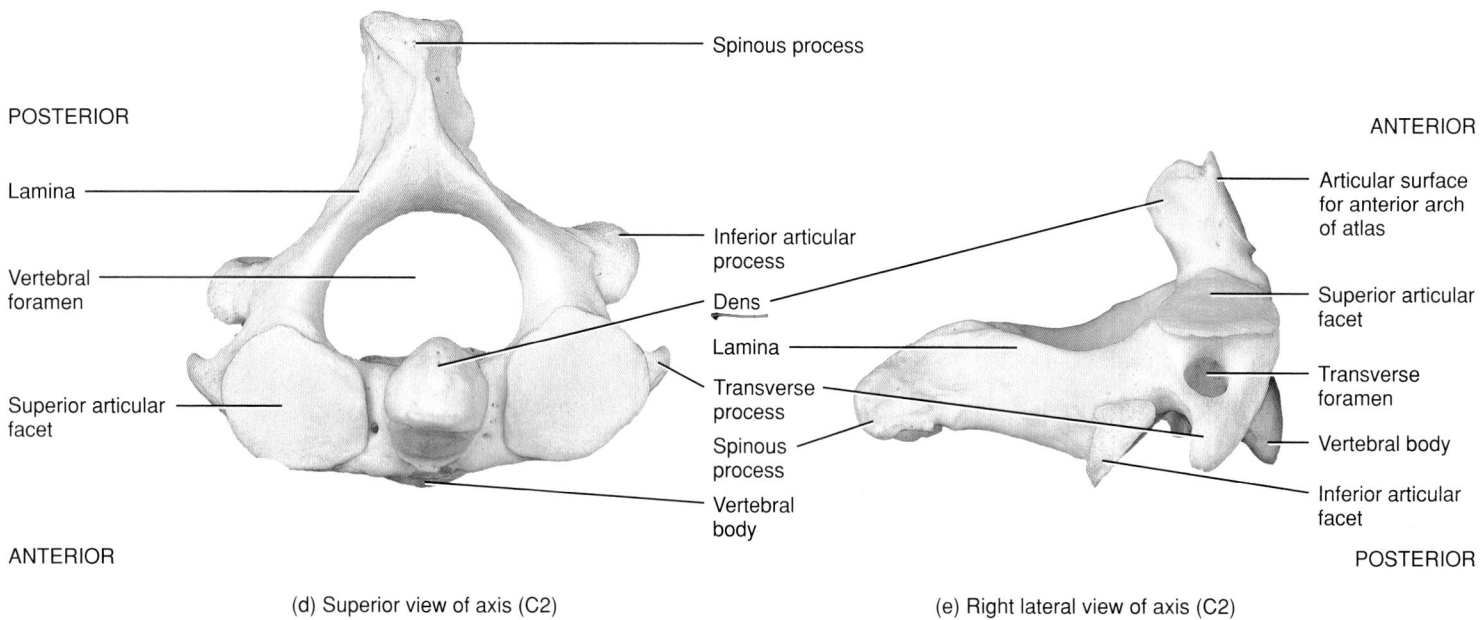

(d) Superior view of axis (C2)

(e) Right lateral view of axis (C2)

❓ Which joint permits you to move your head to signify "no"? Which bones are involved?

EXHIBIT 7.1 Vertebral Regions—Thoracic Vertebrae *(Figure 7.19)*

● **OBJECTIVE**

• Identify the location and surface features of the thoracic vertebrae.

Thoracic vertebrae (T1–T12; Figure 7.19) are considerably larger and stronger than cervical vertebrae. In addition, the spinous processes on T1 through T10 are long, laterally flattened, and directed inferiorly. In contrast, the spinous processes on T11 and T12 are shorter, broader, and directed more posteriorly. Compared to cervical vertebrae, thoracic vertebrae also have longer and larger transverse processes. They are easily identified by their *costal facets* (*cost-* = rib), which are articular surfaces for the ribs.

The feature of the thoracic vertebrae that distinguishes them from other vertebrae is that they articulate with the ribs. Except for T11 and T12, the transverse processes of thoracic vertebrae have costal facets that articulate with the *tubercles* of the ribs. Additionally, the vertebral bodies of thoracic vertebrae have articular surfaces that form articulations with the *heads* of the ribs (see Figure 7.23). The articular surfaces on the vertebral bodies are called either facets or demifacets. A *facet* is formed when the head of a rib articulates with the body of one vertebra. A *demifacet* is formed when the head of a rib articulates with two adjacent vertebral bodies. As you can see in Figure 7.19, on each side of the vertebral body T1 has a superior facet for the first rib and an inferior demifacet for the second rib. On each side of the vertebral body of T2–T8, there is a

Figure 7.19 Thoracic vertebrae.

The thoracic vertebrae are found in the chest region and articulate with the ribs.

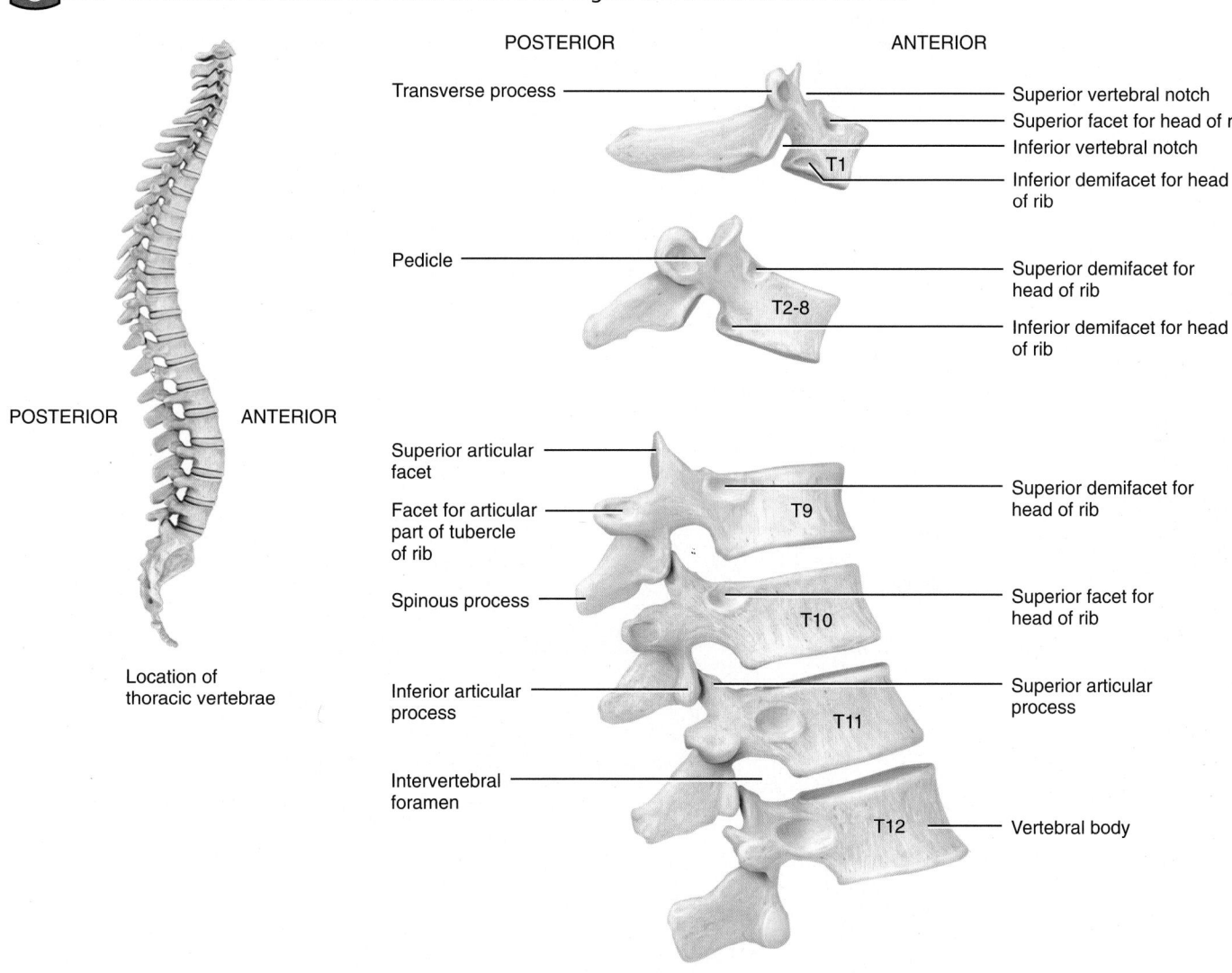

(a) Right lateral view of several articulated thoracic vertebrae

EXHIBIT 7.1 CONTINUES ▶

EXHIBIT 7.1 **239**

EXHIBIT 7.I Vertebral Regions—Thoracic Vertebrae *(Figure 7.19)* CONTINUED

superior demifacet and an inferior demifacet as ribs two through nine articulate with two vertebrae, and T10–T12 have a facet on each side of the vertebral body for ribs 10–12. These articulations between the thoracic vertebrae and ribs, called *vertebrocostal joints,* are distinguishing features of thoracic vertebrae. Movements of the thoracic region are limited by the attachment of the ribs to the sternum.

✔CHECKPOINT

Describe several distinguishing features of thoracic vertebrae.

● **FIGURE 7.19** CONTINUED ▶

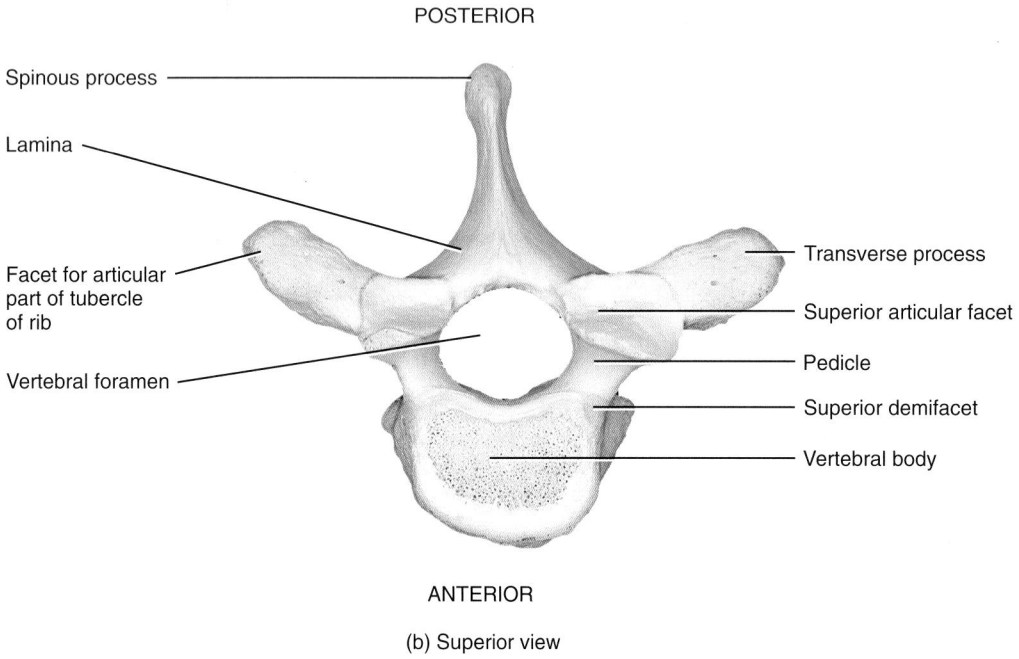

POSTERIOR

Spinous process

Lamina

Facet for articular part of tubercle of rib

Vertebral foramen

Transverse process

Superior articular facet

Pedicle

Superior demifacet

Vertebral body

ANTERIOR

(b) Superior view

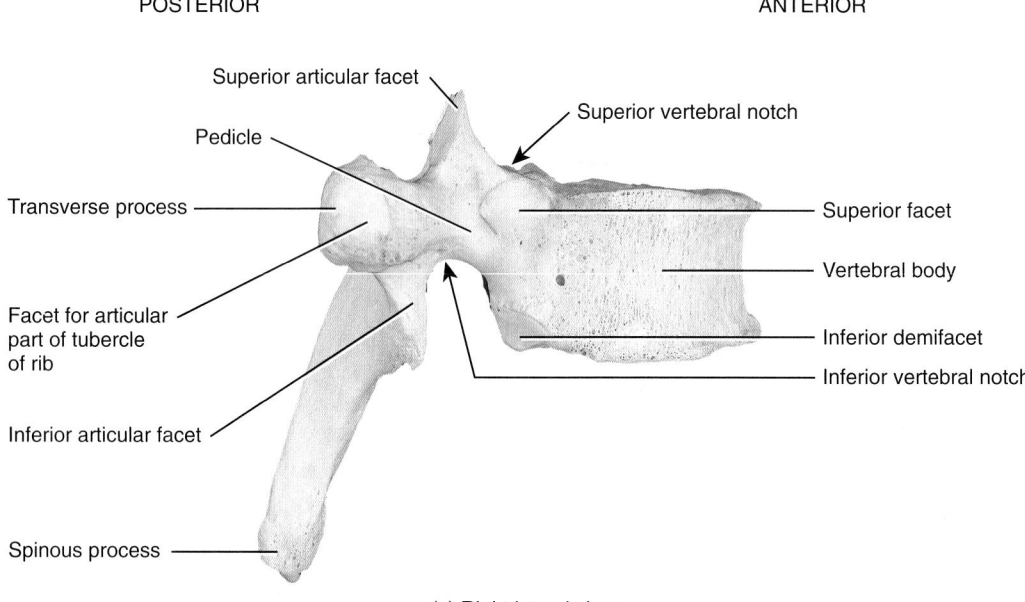

POSTERIOR

ANTERIOR

Superior articular facet

Superior vertebral notch

Pedicle

Transverse process

Superior facet

Vertebral body

Facet for articular part of tubercle of rib

Inferior demifacet

Inferior vertebral notch

Inferior articular facet

Spinous process

(c) Right lateral view

? **Which parts of the thoracic vertebrae articulate with the ribs?**

⦿ **OBJECTIVE**

• Identify the location and surface features of the lumbar vertebrae.

The **lumbar vertebrae** (L1–L5) are the largest and strongest of the unfused bones in the vertebral column (Figure 7.20) because the amount of body weight supported by the vertebrae increases toward the inferior end of the backbone. Their various projections are short and thick. The superior articular processes are directed medially instead of superiorly, and the inferior articular processes are directed laterally instead of inferiorly. The spinous processes are quadrilateral in shape, are thick and broad, and project nearly straight posteriorly. The spinous processes are well adapted for the attachment of the large back muscles.

Figure 7.20 Lumbar vertebrae.

🔑 ⫘ Lumbar vertebrae are found in the lower back.

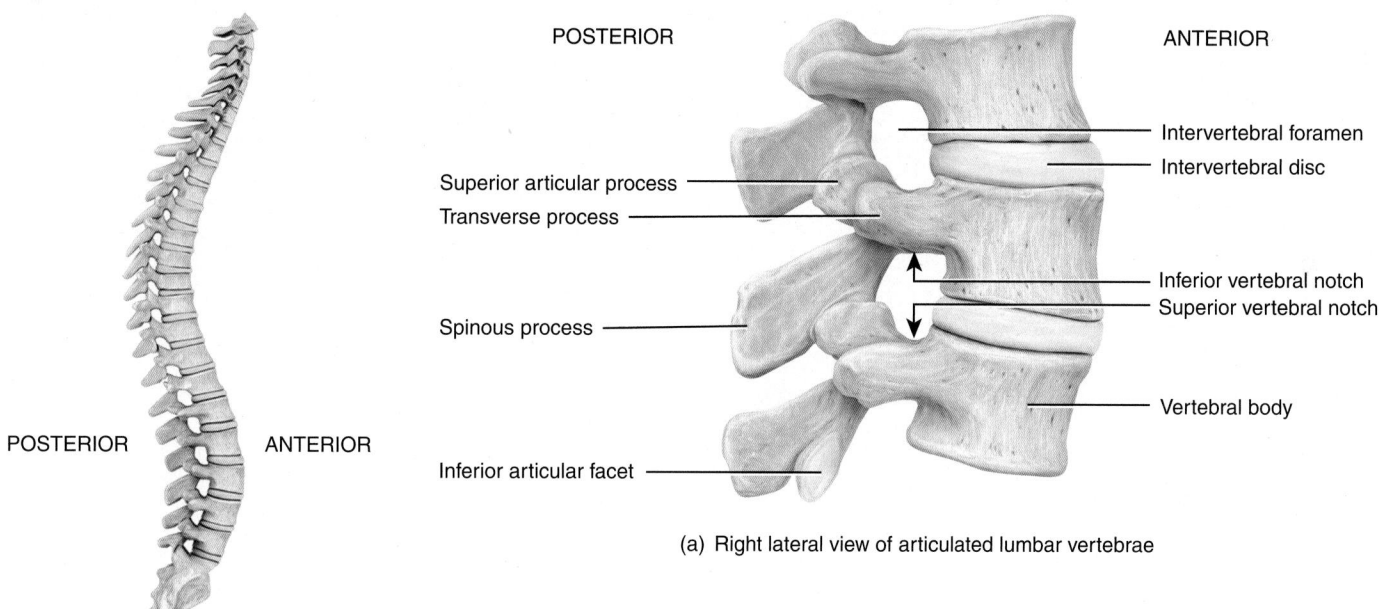

(a) Right lateral view of articulated lumbar vertebrae

Location of lumbar vertebrae

(b) Superior view

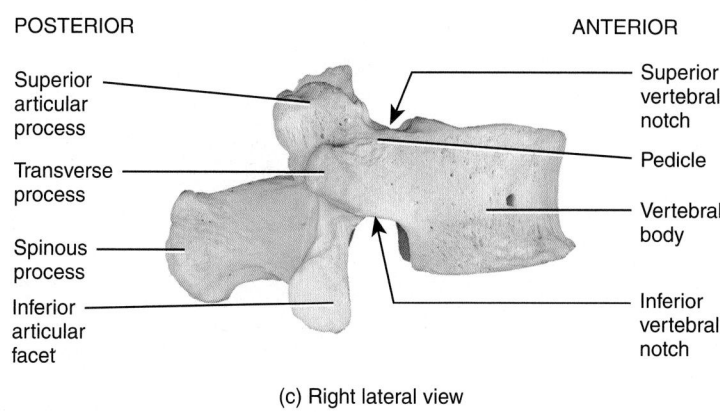

(c) Right lateral view

❓ **Why are the lumbar vertebrae the largest and strongest in the vertebral column?**

EXHIBIT 7.J CONTINUES

EXHIBIT 7.J **241**

A summary of the major structural differences among cervical, thoracic, and lumbar vertebrae is presented in Table 7.4.

✔**CHECKPOINT**

What are the distinguishing features of the lumbar vertebrae?

TABLE 7.4

Comparison of Major Structural Features of Cervical, Thoracic, and Lumbar Vertebrae

CHARACTERISTIC	CERVICAL	THORACIC	LUMBAR
Overall structure			
Size	Small.	Larger.	Largest.
Foramina	One vertebral and two transverse.	One vertebral.	One vertebral.
Spinous processes	Slender, often bifid (C2–C6).	Long, fairly thick (most project inferiorly).	Short, blunt (project posteriorly rather than inferiorly).
Transverse processes	Small.	Fairly large.	Large and blunt.
Articular facets for ribs	Absent.	Present.	Absent.
Direction of articular facets			
Superior	Posterosuperior.	Posterolateral.	Medial.
Inferior	Anteroinferior.	Anteromedial.	Lateral.
Size of intervertebral discs	Thick relative to size of vertebral bodies.	Thin relative to size of vertebral bodies.	Thickest.

EXHIBIT 7.K

Vertebral Regions—Sacral and Coccygeal Vertebrae *(Figure 7.21)*

OBJECTIVE

- Identify the location and surface features of the sacral and coccygeal vertebrae.

Sacrum

The **sacrum** (SĀ-krum) is a triangular bone formed by the union of five sacral vertebrae (S1–S5) (Figure 7.21a). The sacral vertebrae begin to fuse in individuals between 16 and 18 years of age, a process usually completed by age 30. Positioned at the posterior portion of the pelvic cavity medial to the two hip bones, the sacrum serves as a strong foundation for the pelvic girdle. The female sacrum is shorter, wider, and more curved between S2 and S3 than the male sacrum (see Table 8.1).

The concave anterior side of the sacrum faces the pelvic cavity. It is smooth and contains four *transverse lines (ridges)* that mark the joining of the sacral vertebral bodies (Figure 7.21a). At the ends of these lines are four pairs of *anterior sacral foramina*. The lateral portion of the su-

perior surface of the sacrum contains a smooth surface called the *sacral ala* (ĀL-a = wing; plural is *alae*, ĀL-ē), which is formed by the fused transverse processes of the first sacral vertebra (S1).

The convex, posterior surface of the sacrum contains a *median sacral crest*, the fused spinous processes of the upper sacral vertebrae; a *lateral sacral crest*, the fused transverse processes of the sacral vertebrae; and four pairs of *posterior sacral foramina* (Figure 7.21b). These foramina connect with anterior sacral foramina to allow passage of nerves and blood vessels. The *sacral canal* is a continuation of the vertebral cavity. The laminae of the fifth sacral vertebra, and sometimes the fourth, fail to meet. This leaves an inferior entrance to the vertebral canal called the *sacral hiatus* (hī-Ā-tus = opening). On either side of the sacral hiatus is a *sacral cornu* (KOR-noo; *cornu* = horn; plural is *cornua*, KOR-noo-a), the inferior articular processes of the fifth sacral vertebra. They are connected by ligaments to the coccyx.

The narrow inferior portion of the sacrum is known as the *apex*. The broad superior portion of the sacrum is called the *base*. The anteriorly

Figure 7.21 Sacrum and coccyx.

🔑 The sacrum is formed by the union of five sacral vertebrae, and the coccyx is formed by the union of usually four coccygeal vertebrae.

(a) Anterior view

(b) Posterior view

? How many foramina pierce the sacrum, and what is their function?

EXHIBIT 7.K CONTINUES ▶

EXHIBIT 7.K **243**

projecting border of the base, called the *sacral promontory* (PROM-on-tō-rē), is one of the points used for measurements of the pelvis. On both lateral surfaces the sacrum has a large ear-shaped *auricular surface* that articulates with the ilium of each hip bone to form the *sacroiliac joint* (see Figure 8.9). Posterior to the auricular surface is a roughened surface, the *sacral tuberosity,* which contains depressions for the attachment of ligaments. The sacral tuberosity unites with the hip bones to form the sacroiliac joints. The *superior articular processes* of the sacrum articulate with the inferior articular processes of the fifth lumbar vertebra, and the base of the sacrum articulates with the body of the fifth lumbar vertebra to form the *lumbosacral joint.*

CLINICAL CONNECTION | *Caudal Anesthesia*

Anesthetic agents that act on the sacral and coccygeal nerves are sometimes injected through the sacral hiatus, a procedure called **caudal anesthesia.** The procedure is used most often to relieve pain during labor and to provide anesthesia to the perineal area. Because the sacral hiatus is between the sacral cornua, the cornua are important bony landmarks for locating the hiatus. Anesthetic agents may also be injected through the posterior sacral foramina. Since the injection site is inferior to the lowest portion of the spinal cord, there is little danger of damaging the cord. •

Coccyx

The **coccyx,** like the sacrum, is triangular in shape. It is formed by the fusion of usually four coccygeal vertebrae, indicated in Figure 7.21 as Co1–Co4. The coccygeal vertebrae fuse somewhat later than the sacral vertebrae, between the ages of 20 and 30. The dorsal surface of the body of the coccyx contains two long *coccygeal cornua* that are connected by ligaments to the sacral cornua. The coccygeal cornua are the pedicles and superior articular processes of the first coccygeal vertebra. They are on the lateral surfaces of the coccyx, formed by a series of *transverse processes;* the first pair are the largest. The coccyx articulates superiorly with the apex of the sacrum. In females, the coccyx points inferiorly to allow the passage of a baby during birth; in males, it points anteriorly (see Table 8.1).

✔CHECKPOINT

How many vertebrae fuse to form the sacrum and coccyx?

EXHIBIT 7.L   Thoracic Bones—Sternum *(Figure 7.22)*

● OBJECTIVE

- Identify the location and surface features of the sternum.

Sternum

The **sternum,** or breastbone, is a flat, narrow bone located in the center of the anterior thoracic wall that measures about 15 cm (6 in.) in length and consists of three parts (Figure 7.22). The superior part is the **manubrium** (ma-NOO-brē-um = handlelike); the middle and largest part is the **body;** and the inferior, smallest part is the **xiphoid process** (ZĪ-foyd = sword-shaped). The segments of the sternum typically fuse by age 25, and the points of fusion are marked by transverse ridges.

The junction of the manubrium and body forms the *sternal angle.* The manubrium has a depression on its superior surface, the *suprasternal notch.* Lateral to the suprasternal notch are *clavicular notches* that articulate with the medial ends of the clavicles to form the *sternoclavicular*

joints. The manubrium also articulates with the costal cartilages of the first and second ribs. The body of the sternum articulates directly or indirectly with the costal cartilages of the second through tenth ribs. The xiphoid process consists of hyaline cartilage during infancy and childhood and does not completely ossify until about age 40. No ribs are attached to it, but the xiphoid process provides attachment for some abdominal muscles. Incorrect positioning of the hands of a rescuer during cardiopulmonary resuscitation (CPR) may fracture the xiphoid process, driving it into internal organs. During thoracic surgery, the sternum may be split along the midline and the halves spread apart to allow surgeons access to structures in the thoracic cavity such as the thymus, heart, and great vessels of the heart. After surgery, the halves of the sternum are held together with wire sutures.

✔ CHECKPOINT

What is the clinical significance of the xiphoid process?

Figure 7.22 Skeleton of the thorax.

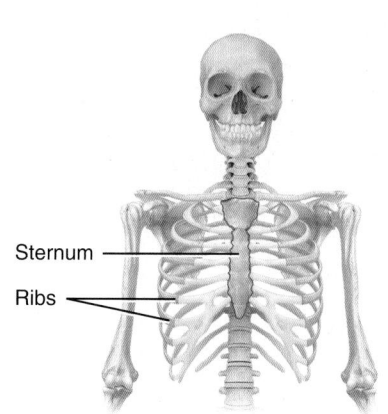

🔑 ⚏ The bones of the thorax enclose and protect organs in the thoracic cavity and in the superior abdominal cavity.

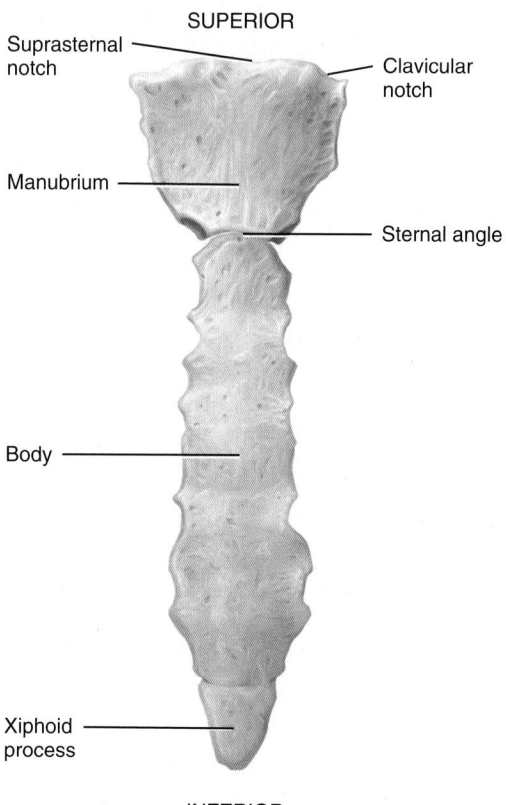

(a) Anterior view of sternum

EXHIBIT 7.L CONTINUES ▶

EXHIBIT 7.L **245**

EXHIBIT 7.L   Thoracic Bones—Sternum   *(Figure 7.22)*   CONTINUED

⬛ **FIGURE 7.22 CONTINUED** ▶

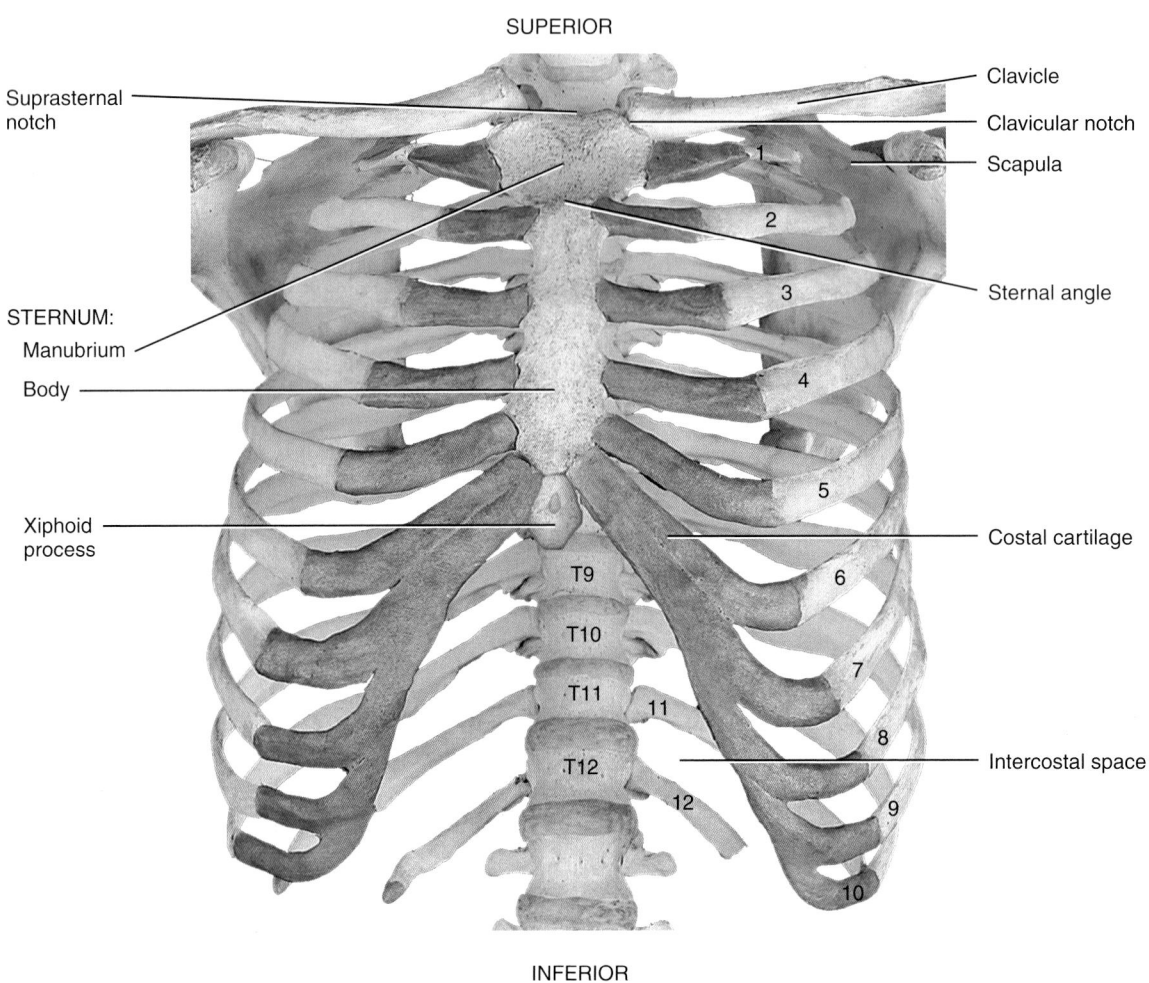

SUPERIOR

Suprasternal notch

Clavicle

Clavicular notch

Scapula

1

2

3

Sternal angle

STERNUM:

Manubrium

Body

4

5

Xiphoid process

T9

T10

T11

T12

6

7

8

9

10

11

12

Costal cartilage

Intercostal space

INFERIOR

(b) Anterior view of skeleton of thorax

❓ **With which ribs does the body of the sternum articulate?**

EXHIBIT 7.M   Thoracic Bones—Ribs   *(Figure 7.23)*

⊙ **OBJECTIVE**

• Identify the location and surface features of the ribs.

Twelve pairs of **ribs,** numbered 1–12 from superior to inferior, give structural support to the sides of the thoracic cavity (Figure 7.22b). The ribs increase in length from the first through seventh, and then decrease in length to the twelfth rib. Each rib articulates posteriorly with its corresponding thoracic vertebra.

The first through seventh pairs of ribs have a direct anterior attachment to the sternum by a strip of hyaline cartilage called *costal cartilage* (*cost-* = rib). The costal cartilages contribute to the elasticity of the thoracic cage and prevent various blows to the chest from fracturing the sternum and/or ribs. The ribs that have costal cartilages and attach directly to the sternum are called *true (vertebrosternal) ribs.* The articulations formed between the true ribs and the sternum are called *sternocostal joints.* The remaining five pairs of ribs are termed *false ribs* because their costal cartilages either attach indirectly to the sternum or

do not attach to the sternum at all. The cartilages of the eighth, ninth, and tenth pairs of ribs attach to one another and then to the cartilages of the seventh pair of ribs. These false ribs are called *vertebrochondral ribs.* The eleventh and twelfth pairs of ribs are false ribs designated as *floating (vertebral) ribs* because the costal cartilages at their anterior ends do not attach to the sternum at all. These ribs attach only posteriorly to the thoracic vertebrae. Inflammation of one or more costal cartilages, called *costochondritis,* is characterized by local tenderness and pain in the anterior chest wall that may radiate. The symptoms mimic the chest pain associated with a heart attack (angina pectoris).

Figure 7.23a shows the parts of a typical (third through ninth) rib. The *head* is a projection at the posterior end of the rib that contains a pair of articular *facets* (superior and inferior). The facet of the head may fit either into a facet on the body of a single vertebra or into the demifacets of two adjoining vertebrae to form a *vertebrocostal joint.* The *neck* is a constricted portion of a rib just lateral to the head. A knoblike structure on the posterior surface, where the neck joins the body, is called a *tubercle*

Figure 7.23 **The structure of ribs.**

🔑 Each rib articulates posteriorly with its corresponding thoracic vertebra.

(a) Posterior view of left rib

(b) Posterior view of left ribs articulated with thoracic vertebrae and sternum

EXHIBIT 7.M CONTINUES

EXHIBIT 7.M   **247**

EXHIBIT 7.M Thoracic Bones—Ribs *(Figure 7.23)* CONTINUED

(TOO-ber-kul). The *nonarticular part* of the tubercle attaches to the transverse process of a vertebra by a ligament (lateral costotransverse ligament). The *articular part* of the tubercle articulates with the facet of a transverse process of a vertebra (Figure 7.23c) to form vertebrocostal joints. The *body (shaft)* is the main part of the rib. A short distance beyond the tubercle, an abrupt change in the curvature of the shaft occurs. This point is called the *costal angle.* The inner surface of the rib has a *costal groove* that protects the intercostal blood vessels and a small nerve.

Spaces between ribs, called *intercostal spaces,* are occupied by intercostal muscles, blood vessels, and nerves. Surgical access to the lungs or other structures in the thoracic cavity is commonly obtained through an intercostal space. Special rib retractors are used to create a wide separation between ribs. The costal cartilages are sufficiently elastic in younger individuals to permit considerable bending without breaking.

In summary, the posterior portion of the rib connects to a thoracic vertebra by its head and the articular part of a tubercle. The facet of the head fits into either a facet on the body of one vertebra (T1 only) or into the demifacets of two adjoining vertebrae. The articular part of the tubercle articulates with the facet of the transverse process of the vertebra.

CLINICAL CONNECTION | *Rib Fractures, Dislocations, and Separations*

Rib fractures are the most common chest injuries. They usually result from direct blows, most often from impact with a steering wheel, falls, or crushing injuries to the chest. Ribs tend to break at the point where the greatest force is applied, but they may also break at their weakest point—the site of greatest curvature, just anterior to the costal angle. The middle ribs are the most commonly fractured. In some cases, fractured ribs may puncture the heart, great vessels of the heart, lungs, trachea, bronchi, esophagus, spleen, liver, and kidneys. Rib fractures are usually quite painful. Rib fractures are no longer bound with bandages because of the pneumonia that would result from lack of proper lung ventilation.

Dislocated ribs, which are common in body contact sports, involve displacement of a costal cartilage from the sternum, with resulting pain, especially during deep inhalations.

Separated ribs involve displacement of a rib and its costal cartilage; as a result, a rib may move superiorly, overriding the rib above and causing severe pain. •

✔ CHECKPOINT
How are ribs classified?

● **FIGURE 7.23 CONTINUED** ▶

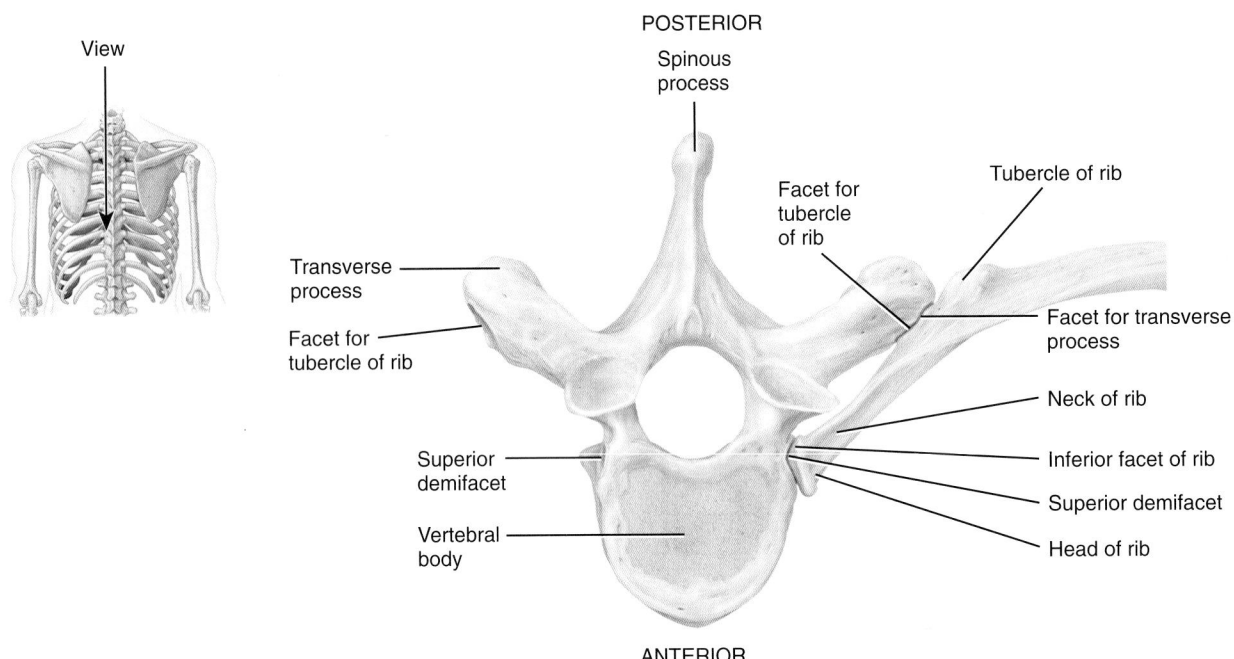

(c) Superior view of left rib articulated with thoracic vertebra

❓ **How does a rib articulate with a thoracic vertebra?**

Herniated (Slipped) Disc

In their function as shock absorbers, intervertebral discs are constantly being compressed. If the anterior and posterior ligaments of the discs become injured or weakened, the pressure developed in the nucleus pulposus may be great enough to rupture the surrounding fibrocartilage (annulus fibrosus). If this occurs, the nucleus pulposus may herniate (protrude) posteriorly or into one of the adjacent vertebral bodies (Figure 7.24). This condition is called a **herniated (slipped) disc.** Because the lumbar region bears much of the weight of the body, and is the region of the most flexing and bending, herniated discs most often occur in the lumbar area.

Frequently, the nucleus pulposus slips posteriorly toward the spinal cord and spinal nerves. This movement exerts pressure on the spinal nerves, causing local weakness and acute pain. If the roots of the sciatic nerve, which passes from the spinal cord to the foot, are compressed, the pain radiates down the posterior thigh, through the calf, and occasionally into the foot. If pressure is exerted on the spinal cord itself, some of its neurons may be destroyed. Treatment options include bed rest, medications for pain, physical therapy and exercises, and *percutaneous endoscopic discectomy* (removal of disc material using a laser). A person with a herniated disc may also undergo a laminectomy, a procedure in which parts of the laminae of the vertebra and intervertebral disc are removed to relieve pressure on the nerves.

Abnormal Curves of the Vertebral Column

Various conditions may exaggerate the normal curves of the vertebral column, or the column may acquire a lateral bend, resulting in **abnormal curves** of the vertebral column.

Scoliosis (skō-lē-Ō-sis; *scolio-* = crooked), the most common of the abnormal curves, is a lateral bending of the vertebral column, usually in the thoracic region (Figure 7.25a). It may result from congenitally (present at birth) malformed vertebrae, chronic sciatica (pain in the lower back and lower limb), paralysis of muscles on one side of the vertebral column, poor posture, or one leg being shorter than the other.

Signs of scoliosis include uneven shoulders and waist, one shoulder blade more prominent than the other, one hip higher than the other, and leaning to one side. In severe scoliosis (a curve greater than 70 degrees), breathing is more difficult and the pumping action of the heart is less efficient. Chronic back pain and arthritis of the vertebral column may also develop. Treatment options include wearing a back brace, physical therapy, chiropractic care, and surgery (fusion of vertebrae and insertion of metal rods, hooks, and wires to reinforce the surgery).

Figure 7.24 Herniated (slipped) disc.

Most often the nucleus pulposus herniates posteriorly.

POSTERIOR

- Spinous process of vertebra
- Spinal cord
- Spinal nerve
- Herniated disc
- Nucleus pulposus
- Annulus fibrosus

ANTERIOR   Superior view

? **Why do most herniated discs occur in the lumbar region?**

Figure 7.25 Abnormal curves of the vertebral column.

An abnormal curve is the result of the exaggeration of a normal curve.

(a) Scoliosis   (b) Kyphosis   (c) Lordosis

 Which abnormal curve is common in women with advanced osteoporosis?

Kyphosis (kī-FŌ-sis; *kyphoss-* = hump; *-osis* = condition) is an increase in the thoracic curve of the vertebral column that produces a "hunchback" look (Figure 7.25b). In tuberculosis of the spine, vertebral bodies may partially collapse, causing an acute angular bending of the vertebral column. In the elderly, degeneration of the intervertebral discs leads to kyphosis. Kyphosis may also be caused by rickets and poor posture. It is also common in females with advanced osteoporosis.

Lordosis (lor-DŌ-sis; *lord-* = bent backward), sometimes called *hollow back,* is an increase in the lumbar curve of the vertebral column (Figure 7.25c). It may result from increased weight of the abdomen as in pregnancy or extreme obesity, poor posture, rickets, osteoporosis, or tuberculosis of the spine.

Spina Bifida

Spina bifida (SPĪ-na BIF-i-da) is a congenital defect of the vertebral column in which laminae of L5 and/or S1 fail to develop normally and unite at the midline. The least serious form is called *spina bifida occulta.* It occurs in L5 or S1 and produces no symptoms. The only evidence of its presence is a small dimple with a tuft of hair in the overlying skin. Several types of spina bifida involve protrusion of meninges (membranes) and/or spinal cord through the defect in the laminae and are collectively termed *spina bifida cystica* because of the presence of a cystlike sac protruding from the backbone (Figure 7.26). If the sac contains the meninges from the spinal cord and cerebrospinal fluid, the condition is called *spina bifida with meningocele* (me-NING-gō-sēl). If the spinal cord and/or its nerve roots are in the sac, the condition is called *spina bifida with meningomyelocele* (me-ning-gō-MĪ-ē-lō-sēl). The larger the cyst and the number of neural structures it contains, the more serious the neurological problems. In severe cases, there may be partial or complete paralysis, partial or complete loss of urinary bladder and bowel control, and the absence of reflexes. An increased risk of spina bifida is associated with low levels of a B vitamin called folic acid during pregnancy. Spina bifida may be diagnosed prenatally by a test of the mother's blood for a substance produced by the fetus called alpha-fetoprotein, by sonography, or by amniocentesis (withdrawal of amniotic fluid for analysis).

Figure 7.26 Spina bifida with meningomyelocele.

 Spina bifida is caused by a failure of laminae to unite at the midline.

 Deficiency of which B vitamin is linked to spina bifida?

Fractures of the Vertebral Column

Fractures of the vertebral column often involve C1, C2, C4–T7, and T12–L2. Cervical or lumbar fractures usually result from a flexion–compression type of injury such as might be sustained in landing on the feet or buttocks after a fall or having a weight fall on the shoulders. Cervical vertebrae may be fractured or dislodged by a fall on the head with acute flexion of the neck, as might happen on diving into shallow water or being thrown from a horse. Spinal cord or spinal nerve damage may occur as a result of fractures of the vertebral column if the fractures compromise the foramina.

MEDICAL TERMINOLOGY

Chiropractic (kī-rō-PRAK-tik; *cheir-* = hand; *-praktikos* = efficient) A holistic health-care discipline that focuses on nerves, muscles, and bones. A **chiropractor** is a health-care professional who is concerned with the diagnoses, treatment, and prevention of mechanical disorders of the musculoskeletal system and the effects of these disorders on the nervous system and health in general. Treatment involves using the hands to apply specific force to adjust joints of the body (manual adjustment), especially the vertebral column. Chiropractors may also use massage, heat therapy, ultrasound, electric stimulation, and acupuncture. Chiropractors often provide information about diet, exercise, changes in lifestyle, and stress management. Chiropractors do not prescribe drugs or perform surgery.

Craniostenosis (krā-nē-ō-sten-Ō-sis; *cranio-* = skull; *-stenosis* = narrowing) Premature closure of one or more cranial sutures during the first 18 to 20 months of life, resulting in a distorted skull. Premature closure of the sagittal suture produces a long narrow skull; premature closure of the coronal suture results in a broad skull. Premature closure of all sutures restricts brain growth and development; surgery is necessary to prevent brain damage.

Craniotomy (krā-nē-OT-ō-mē; *cranio-* = skull; *-tome* = cutting) Surgical procedure in which part of the cranium is removed. It may be performed to remove a blood clot, a brain tumor, or a sample of brain tissue for biopsy.

Laminectomy (lam'-i-NEK-tō-mē; *lamina-* = layer) Surgical procedure to remove a vertebral lamina. It may be performed to access the vertebral cavity and relieve the symptoms of a herniated disc.

Lumbar spine stenosis (*sten-* = narrowed) Narrowing of the spinal canal in the lumbar part of the vertebral column, due to hypertrophy of surrounding bone or soft tissues. It may be caused by arthritic changes in the intervertebral discs and is a common cause of back and leg pain.

Spinal fusion (FŪ-zhun) Surgical procedure in which two or more vertebrae of the vertebral column are stabilized with a bone graft or synthetic device. It may be performed to treat a fracture of a vertebra or following removal of a herniated disc.

Whiplash injury Injury to the neck region due to severe hyperextension (backward tilting) of the head followed by severe hyperflexion (forward tilting) of the head, usually associated with a rear-end automobile collision. Symptoms are related to stretching and tearing of ligaments and muscles, vertebral fractures, and herniated vertebral discs.

CHAPTER REVIEW AND RESOURCE SUMMARY

Review

Resource

Introduction

1. Bones protect soft body parts and make movement possible; they also serve as landmarks for locating parts of other body systems.
2. The musculoskeletal system is composed of the bones, joints, and muscles working together.

7.1 Divisions of the Skeletal System (see Table 7.1)

1. The axial skeleton consists of bones arranged along the longitudinal axis. The parts of the axial skeleton are the skull, auditory ossicles (ear bones), hyoid bone, vertebral column, sternum, and ribs.
2. The appendicular skeleton consists of the bones of the girdles and the upper and lower limbs (extremities). The parts of the appendicular skeleton are the pectoral (shoulder) girdles, bones of the upper limbs, pelvic (hip) girdles, and bones of the lower limbs.

Anatomy Overview - The Axial Skeleton

7.2 Types of Bones

1. On the basis of shape, bones are classified as long, short, flat, irregular, or sesamoid. Sesamoid bones develop in tendons or ligaments.
2. Sutural bones are found within the sutures of some cranial bones.

7.3 Bone Surface Markings

1. Surface markings are structural features visible on the surfaces of bones.
2. Each marking—whether a depression, an opening, or a process—is structured for a specific function, such as joint formation, muscle attachment, or passage of nerves and blood vessels (see Table 7.2).

7.4 Skull

1. The 22 bones of the skull include cranial bones and facial bones.
2. The eight cranial bones are the frontal, parietal (2), temporal (2), occipital, sphenoid, and ethmoid.
3. The 14 facial bones are the nasal (2), maxillae (2), zygomatic (2), lacrimal (2), palatine (2), inferior nasal conchae (2), vomer, and mandible.
4. The nasal septum consists of the vomer, perpendicular plate of the ethmoid, and septal cartilage. The nasal septum divides the nasal cavity into left and right sides.
5. Seven skull bones form each of the orbits (eye sockets).
6. The foramina of the skull bones provide passages for nerves and blood vessels (see Table 7.3).
7. Sutures are immovable joints in adults that connect most bones of the skull. Examples are the coronal, sagittal, lambdoid, and squamous sutures.
8. Paranasal sinuses are cavities in bones of the skull that are connected to the nasal cavity. The frontal, sphenoid, and ethmoid bones and the maxillae contain paranasal sinuses.
9. Fontanels are mesenchyme-filled spaces between the cranial bones of fetuses and infants. The major fontanels are the anterior, posterior, anterolaterals (2), and posterolaterals (2). After birth, the fontanels fill in with bone and become sutures.

Anatomy Overview - Facial Bones
Anatomy Overview - Fibrous Joints and Sutures
Figure 7.8 - Sphenoid Bone
Figure 7.12 - Details of the Orbit

7.5 Hyoid Bone

1. The hyoid bone is a U-shaped bone that does not articulate with any other bone.
2. It supports the tongue and provides attachment for some tongue muscles and for some muscles of the pharynx and neck.

7.6 Vertebral Column

1. The vertebral column, sternum, and ribs constitute the skeleton of the body's trunk.
2. The 26 bones of the adult vertebral column are the cervical vertebrae (7), the thoracic vertebrae (12), the lumbar vertebrae (5), the sacrum (5 fused vertebrae), and the coccyx (usually 4 fused vertebrae).
3. The adult vertebral column contains four normal curves (cervical, thoracic, lumbar, and sacral) that provide strength, support, and balance.
4. Each vertebra usually consists of a vertebral body, vertebral arch, and seven processes. Vertebrae in the different regions of the column vary in size, shape, and detail.

Anatomy Overview - The Vertebral Column
Figure 7.16 - Vertebral Column
Table 7.4 - Comparison of Major Structural Features of Cervical, Thoracic, and Lumbar Vertebrae
Figure 7.21 - Sacrum and Coccyx

Review

7.7 Thorax

1. The thoracic skeleton consists of the sternum, ribs, costal cartilages, and thoracic vertebrae.
2. The thoracic cage protects vital organs in the chest area and upper abdomen.

Resource

Anatomy Overview - The Thorax
Figure 7.22 - Skeleton of the Thorax
Exercise - Bones, Bones, Bones

SELF-QUIZ QUESTIONS

Fill in the blanks in the following statements.

1. Membrane-filled spaces between cranial bones that enable the fetal skull to modify its size and shape for passage through the birth canal are called _____.

2. The hypophyseal fossa of the sella turcica of the sphenoid bone contains the _____.

3. The regions of the vertebral column that consist of fused vertebrae are the _____ and the _____.

Indicate whether the following statements are true or false.

4. The atlanto-occipital joints allow you to rotate the head, as in signifying "no."

5. Ribs that are not attached to the sternum are known as the true ribs.

Choose the one best answer to the following questions.

6. In which of the following bones are paranasal sinuses *not* found?
   (a) frontal bone   (b) sphenoid bone   (c) lacrimal bones
   (d) ethmoid bone   (e) maxillae

7. Which of the following pairs are mismatched?
   (a) mandible: only movable bone in the skull
   (b) hyoid: bone that does not articulate with any other bone
   (c) sacrum: supports lower back
   (d) thoracic vertebrae: articulate with thoracic ribs posteriorly
   (e) inferior nasal conchae: classified as facial bones

8. Which of the following bones are *not* paired?
   (a) vomer   (b) palatine   (c) lacrimal
   (d) maxilla   (e) nasal

9. The suture located between a parietal and temporal bone is the
   (a) lambdoid.   (b) sagittal.   (c) coronal.
   (d) anterolateral. (e) squamous.

10. The primary vertebral curves that appear during fetal development are the (1) cervical curve, (2) thoracic curve, (3) lumbar curve, (4) coccyx curve, (5) sacral curve.
    (a) 2 and 3   (b) 1 and 2   (c) 2 and 4
    (d) 2 and 5   (e) 1 and 3

11. Which of the following are functions of the cranial bones? (1) protection of the brain; (2) attachment of muscles that move the head; (3) protection of the special sense organs; (4) attachment to the meninges; (5) attachment of muscles that produce facial expressions.
    (a) 1, 2, and 5   (b) 1, 2, 4, and 5   (c) 2 and 5
    (d) 1, 2, 3, and 5   (e) 1, 2, 3, 4, and 5

12. Match the following:
    ____ (a) prominent ridge or elongated projection
    ____ (b) tubelike opening
    ____ (c) large round protuberance at the end of a bone
    ____ (d) smooth, flat articular surface
    ____ (e) sharp, slender projection
    ____ (f) opening for passage of blood vessels, nerves, or ligaments
    ____ (g) large, rounded, rough projection
    ____ (h) shallow depression
    ____ (i) narrow slit between adjacent parts of bones for passage of blood vessels or nerves

    (1) foramen
    (2) tuberosity
    (3) spinous process
    (4) crest
    (5) facet
    (6) fissure
    (7) condyle
    (8) fossa
    (9) meatus

13. Match the following:
    ____ (a) supraorbital foramen
    ____ (b) temporomandibular joint
    ____ (c) external auditory meatus
    ____ (d) foramen magnum
    ____ (e) optic foramen
    ____ (f) cribriform plate
    ____ (g) palatine process
    ____ (h) ramus, body, and condylar process
    ____ (i) transverse foramina, bifid spinous processes
    ____ (j) dens
    ____ (k) promontory
    ____ (l) costal cartilages
    ____ (m) xiphoid process

    (1) temporal bone
    (2) sphenoid bone
    (3) cervical vertebrae
    (4) ethmoid bone
    (5) articulation of mandibular fossa and articular tubercle of the temporal bone to the mandible
    (6) occipital bone
    (7) frontal bone
    (8) maxillae
    (9) mandible
    (10) axis
    (11) sacrum
    (12) sternum
    (13) ribs

14. Match the following (the same answer may be used more than once):

_____ (a) bones that have greater length than width and consist of a shaft and a variable number of extremities

_____ (b) cube-shaped bones that are nearly equal in length and width

_____ (c) bones that develop in certain tendons where there is considerable friction, tension, and physical stress

_____ (d) small bones located within joints between certain cranial bones

_____ (e) thin bones composed of two nearly parallel plates of compact bone enclosing a layer of spongy bone

_____ (f) bones with complex shapes, including the vertebrae and some facial bones

_____ (g) patella is an example

_____ (h) bones that provide considerable protection and extensive areas for muscle attachment

_____ (i) include femur, tibia, fibula, humerus, ulna, and radius

_____ (j) include cranial bones, sternum, and ribs

_____ (k) include almost all of the carpal (wrist) and tarsal (ankle) bones

(1) irregular bones
(2) long bones
(3) short bones
(4) flat bones
(5) sesamoid bones
(6) sutural bones

15. Match the following:

_____ (a) forms the forehead

_____ (b) form the inferior lateral aspects of the cranium and part of the cranial floor; contain zygomatic process and mastoid process

_____ (c) forms part of the anterior portion of the cranial floor, medial wall of the orbits, superior portions of nasal septum, most of the side walls of the nasal cavity; is a major supporting structure of the nasal cavity

_____ (d) form the prominence of the cheek and part of the lateral wall and floor of each orbit

_____ (e) the largest, strongest facial bone; is the only movable skull bone

_____ (f) a roughly triangular bone on the floor of the nasal cavity; one of the components of the nasal septum

_____ (g) form greater portion of the sides and roof of the cranial cavity

_____ (h) forms the posterior part and most of the base of the cranium; contains the foramen magnum

_____ (i) called the keystone of the cranial floor; contains the sella turcica, optic foramen, and pterygoid processes

_____ (j) form the bridge of the nose

_____ (k) the smallest bones of the face; contain a vertical groove that houses a structure that gathers tears and passes them into the nasal cavity

_____ (l) does not articulate with any other bone

_____ (m) unite to form the upper jawbone and articulate with every bone of the face except the lower jawbone

_____ (n) form the posterior part of the hard palate, part of the floor and lateral wall of the nasal cavity, and a small portion of the floors of the orbits

_____ (o) scroll-like bones that form a part of the lateral walls of the nasal cavity; functions in the turbulent circulation and filtration of air

(1) temporal bones
(2) parietal bones
(3) frontal bone
(4) occipital bone
(5) sphenoid bone
(6) ethmoid bone
(7) nasal bones
(8) maxillae
(9) zygomatic bones
(10) lacrimal bones
(11) palatine bones
(12) vomer
(13) mandible
(14) inferior nasal conchae
(15) hyoid bone

CRITICAL THINKING QUESTIONS

1. Jimmy is in a car accident. He can't open his mouth and has been told that he suffers from the following: black eye, broken nose, broken cheek, broken upper jaw, damaged eye socket, and punctured lung. Describe *exactly* what structures have been affected by his car accident.

2. Bubba is a tug-of-war expert. He practices day and night by pulling on a rope attached to an 800-lb anchor. What kinds of changes would you expect that he develops in his bone structure?

3. A new mother brings her newborn infant home and has been told by her well-meaning friend not to wash the baby's hair for several months because the water and soap could "get through that soft area in the top of the head and cause brain damage." Explain to her why this is not true.

? ANSWERS TO FIGURE QUESTIONS

7.1 The skull and vertebral column are part of the axial skeleton. The clavicle, shoulder girdle, humerus, pelvic girdle, and femur are part of the appendicular skeleton.

7.2 Flat bones protect underlying organs and provide a large surface area for muscle attachment.

7.3 The frontal, parietal, sphenoid, ethmoid, temporal, and occipital bones are all cranial bones.

7.4 The parietal and temporal bones are joined by the squamous suture, the parietal and occipital bones are joined by the lambdoid suture, and the parietal and frontal bones are joined by the coronal suture.

7.5 The temporal bone articulates with the mandible and the parietal, sphenoid, zygomatic, and occipital bones.

7.6 The parietal bones form the posterior, lateral portion of the cranium.

7.7 The medulla oblongata of the brain connects with the spinal cord in the foramen magnum.

7.8 From the crista galli of the ethmoid bone, the sphenoid articulates with the frontal, parietal, temporal, occipital, temporal, parietal, and frontal bones, ending again at the crista galli of the ethmoid bone.

7.9 The perpendicular plate of the ethmoid bone forms the superior part of the nasal septum, and the lateral masses compose most of the medial walls of the orbits.

7.10 The mandible is the only movable skull bone, other than the auditory ossicles.

7.11 The nasal septum divides the nasal cavity into right and left sides.

7.12 Bones forming the orbit are the frontal, sphenoid, zygomatic, maxilla, lacrimal, ethmoid, and palatine.

7.13 The paranasal sinuses produce mucus and serve as resonating chambers for vocalizations.

7.14 The paired anterolateral fontanels are bordered by four different skull bones: the frontal, parietal, temporal, and sphenoid bones.

7.15 The hyoid bone does not articulate with any other bone.

7.16 The thoracic and sacral curves of the vertebral column are concave relative to the anterior of the body.

7.17 The vertebral foramina enclose the spinal cord; the intervertebral foramina provide spaces through which spinal nerves exit the vertebral column.

7.18 The atlas moving on the axis at the atlanto-axial point permits movement of the head to signify "no."

7.19 The facets and demifacets on the vertebral bodies of the thoracic vertebrae articulate with the heads of the ribs, and the facets on the transverse processes of these vertebrae articulate with the tubercles of the ribs.

7.20 The lumbar vertebrae are the largest and strongest in the body because the amount of weight supported by vertebrae increases toward the inferior end of the vertebral column.

7.21 There are four pairs of sacral foramina, for a total of eight. Each anterior sacral foramen joins a posterior sacral foramen at the intervertebral foramen. Nerves and blood vessels pass through these tunnels in the bone.

7.22 The body of the sternum articulates directly or indirectly with ribs 2–10.

7.23 The facet on the head of a rib fits into a facet on the body of a vertebra, and the articular part of the tubercle of a rib articulates with the facet of the transverse process of a vertebra.

7.24 Most herniated discs occur in the lumbar region because it bears most of the body weight and most flexing and bending occur there.

7.25 Kyphosis is common in individuals with advanced osteoporosis.

7.26 Deficiency of folic acid is associated with spina bifida.

8 THE SKELETAL SYSTEM: THE APPENDICULAR SKELETON

THE APPENDICULAR SKELETON AND HOMEOSTASIS *The bones of the appendicular skeleton contribute to homeostasis by providing attachment points and leverage for muscles, which aids body movements; by providing support and protection of internal organs, such as the reproductive organs; and by storing and releasing calcium.*

As noted in Chapter 7, the two main divisions of the skeletal system are the axial skeleton and the appendicular skeleton. As you learned in that chapter, the general function of the axial skeleton is the protection of internal organs; the primary function of the appendicular skeleton, the focus of this chapter, is movement. The appendicular skeleton includes the bones that make up the upper and lower limbs as well as the bones of the two girdles that attach the limbs to the axial skeleton. The bones of the appendicular skeleton are connected with one another and with skeletal muscles, permitting you to do things such as walk, write, use a computer, dance, swim, and play a musical instrument.

? *Did you ever wonder what causes runner's knee?*

8.1 PECTORAL (SHOULDER) GIRDLE

● OBJECTIVE

• Identify the bones of the pectoral (shoulder) girdle, their functions, and their principal markings.

The human body has two **pectoral** (PEK-tō-ral) or **shoulder girdles** that attach the bones of the upper limbs to the axial skeleton (Figure 8.1). Each of the two pectoral girdles consists of a clavicle and a scapula. The *clavicle* is the anterior bone and articulates with the manubrium of the sternum at the *sternoclavicular joint*. The scapula articulates with the clavicle at the *acromioclavicular joint* and with the humerus at the *glenohumeral (shoulder) joint*. The pectoral girdles do not articulate with the vertebral column and are held in position and stabilized by a group of large muscles that extend from the vertebral column and ribs to the scapula.

The bones of the pectoral girdle are discussed in Exhibits 8.A and 8.B.

✔CHECKPOINT

1. What is the function of the pectoral girdle?

Figure 8.1 Right pectoral (shoulder) girdle.

🔓 The clavicle is the anterior bone of the pectoral girdle, and the scapula is the posterior bone.

(a) Anterior view of pectoral girdle

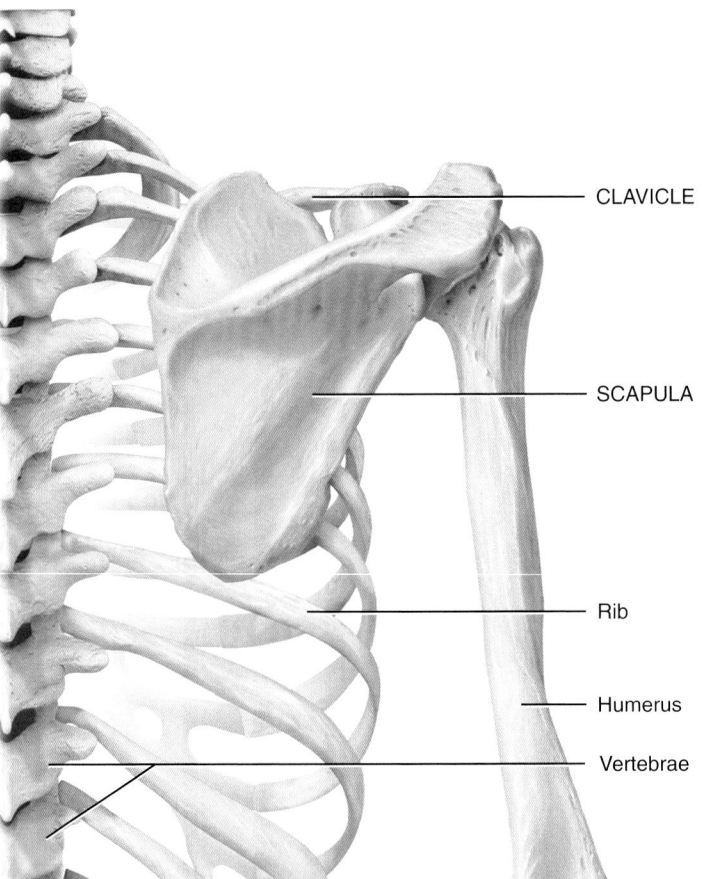

(b) Posterior view of pectoral girdle

❓ **What is the function of the pectoral girdle?**

EXHIBIT 8.A Pectoral Girdle—Clavicle *(Figure 8.2)*

OBJECTIVE

• Describe the location and surface features of the clavicle.

Each slender, **S**-shaped **clavicle** (KLAV-i-kul = key), or *collarbone,* lies horizontally across the anterior part of the thorax superior to the first rib (Figure 8.2). It is subcutaneous (under the skin) and easily palpable along its length. The bone is **S**-shaped because the medial half is convex anteriorly (curves toward you when viewed in the anatomical position), and the lateral half is concave anteriorly (curves away from you). It is rougher and more curved in males.

The medial end, called the *sternal end,* is rounded and articulates with the manubrium of the sternum to form the *sternoclavicular joint.* The broad, flat, lateral end, the *acromial end* (a-KRŌ-mē-al), articulates with the acromion of the scapula to form the *acromioclavicular joint* (see Figure 8.1). The *conoid tubercle* (KŌ-noyd = conelike) on the inferior surface of the lateral end of the bone is a point of attachment for the conoid ligament, which attaches the clavicle and scapula. As its name implies, the *impression for the costoclavicular ligament* on the inferior surface of the sternal end is a point of attachment for the costoclavicular ligament (Figure 8.2b), which attaches the clavicle and first rib.

CLINICAL CONNECTION | *Fractured Clavicle*

The clavicle transmits mechanical force from the upper limb to the trunk. If the force transmitted to the clavicle is excessive, as when you fall on your outstretched arm, a **fractured clavicle** may result. A fractured clavicle may also result from a blow to the superior part of the anterior thorax, for example, as a result of an impact following an automobile accident. The clavicle is one of the most frequently broken bones in the body. Because the junction of the two curves of the clavicle is its weakest point, the clavicular midregion is the most frequent fracture site. Even in the absence of fracture, compression of the clavicle as a result of automobile accidents involving the use of shoulder harness seatbelts often causes damage to the brachial plexus (the network of nerves that enter the upper limb), which lies between the clavicle and the second rib. A fractured clavicle is usually treated with a figure-of-8 sling to keep the arm from moving outward. •

✔CHECKPOINT

Which joints are formed by the articulation of the clavicle with other bones? Which areas of the clavicle are involved in each joint?

Figure 8.2 Right clavicle.

The clavicle articulates medially with the manubrium of the sternum and laterally with the acromion of the scapula.

(a) Superior view

(b) Inferior view

? Which part of the clavicle is its weakest point?

EXHIBIT 8.A **257**

EXHIBIT 8.B Pectoral Girdle—Scapula *(Figure 8.3)*

■ OBJECTIVE
• Describe the location and surface features of the scapula.

Each **scapula** (SCAP-ū-la; plural is *scapulae*), or *shoulder blade,* is a large, triangular, flat bone situated in the superior part of the posterior thorax between the levels of the second and seventh ribs (Figure 8.3).

A prominent ridge called the *spine* runs diagonally across the posterior surface of the scapula. The lateral end of the spine projects as a flattened, expanded process called the *acromion* (a-KRŌ-mē-on; *acrom-* = topmost;

-omos = shoulder), easily felt as the high point of the shoulder. Tailors measure the length of the upper limb from the acromion. As noted earlier, the acromion articulates with the acromial end of the clavicle to form the *acromioclavicular joint.* Inferior to the acromion is a shallow depression, the *glenoid cavity,* that accepts the head of the humerus (arm bone) to form the *glenohumeral (shoulder) joint* (see Figure 8.1).

The thin edge of the scapula closer to the vertebral column is called the *medial (vertebral) border.* The thick edge of the scapula closer to the arm is called the *lateral (axillary) border.* The medial and lateral borders

Figure 8.3 **Right scapula (shoulder blade).**

🔑 The glenoid cavity of the scapula articulates with the head of the humerus to form the glenohumeral (shoulder) joint.

Scapula

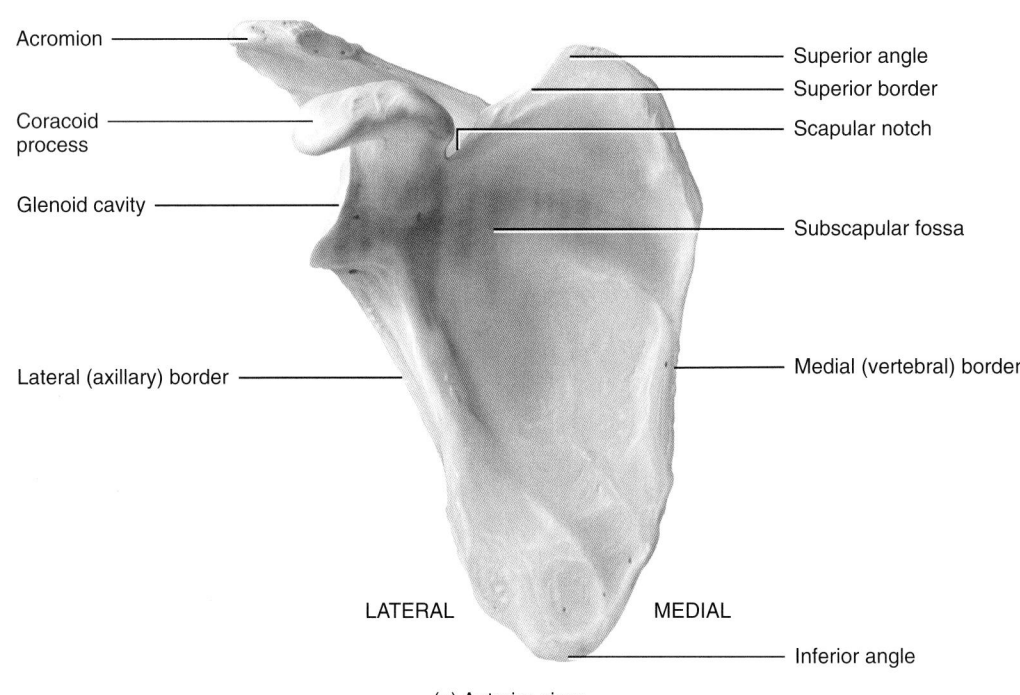

Acromion

Coracoid process

Glenoid cavity

Lateral (axillary) border

Superior angle

Superior border

Scapular notch

Subscapular fossa

Medial (vertebral) border

LATERAL   MEDIAL

Inferior angle

(a) Anterior view

join at the *inferior angle.* The superior edge of the scapula, called the *superior border,* joins the medial border at the *superior angle.* The *scapular notch* is a prominent indentation along the superior border through which the suprascapular nerve passes.

At the lateral end of the superior border of the scapula is a projection of the anterior surface called the *coracoid process* (KOR-a-koyd = like a crow's beak), to which the tendons of muscles (pectoralis minor, coracobrachialis, and biceps brachii) and ligaments (coracoacromial, conoid, and trapezoid) attach. Superior and inferior to the spine on the posterior surface of the scapula are two fossae: The *supraspinous fossa* (sū-pra-SPĪ-nus) is a surface of attachment for the supraspinatus muscle of the shoulder, and the *infraspinous fossa* (in-fra-SPĪ-nus) serves as a surface of attachment for the infraspinatus muscle of the shoulder. On the anterior surface of the scapula is a slightly hollowed-out area called the *subscapular fossa,* a surface of attachment for the subscapularis muscle.

✔ CHECKPOINT

Which joints are formed by the scapula with other bones? What are the names of the parts of the scapula that form each joint?

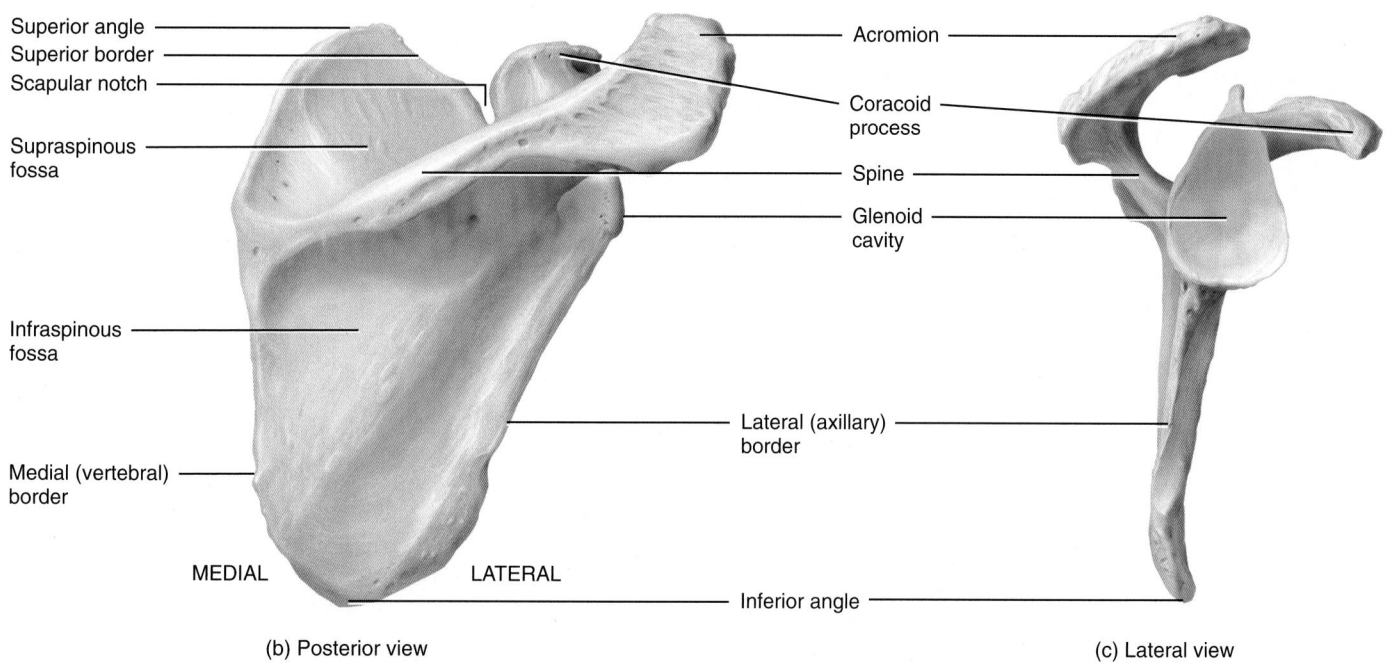

(b) Posterior view

(c) Lateral view

? **Which part of the scapula forms the high point of the shoulder?**

EXHIBIT 8.B **259**

8.2 UPPER LIMB (EXTREMITY)

⊙ **OBJECTIVE**

• Identify the bones of the upper limb and their principal markings.

Each **upper limb (upper extremity)** has 30 bones in three locations—(1) the humerus in the arm; (2) the ulna and radius in the forearm; and (3) the 8 carpals in the carpus (wrist), the 5 metacarpals in the metacarpus (palm), and the 14 phalanges (bones of the digits) in the hand (Figure 8.4).

The bones of the upper limb are discussed in Exhibits 8.C through 8.E.

✔**CHECKPOINT**

2. Name the bones that form the upper limb, from proximal to distal.

Figure 8.4 Right upper limb.

🔒 ⚊ Each upper limb includes a humerus, an ulna, a radius, carpals, metacarpals, and phalanges.

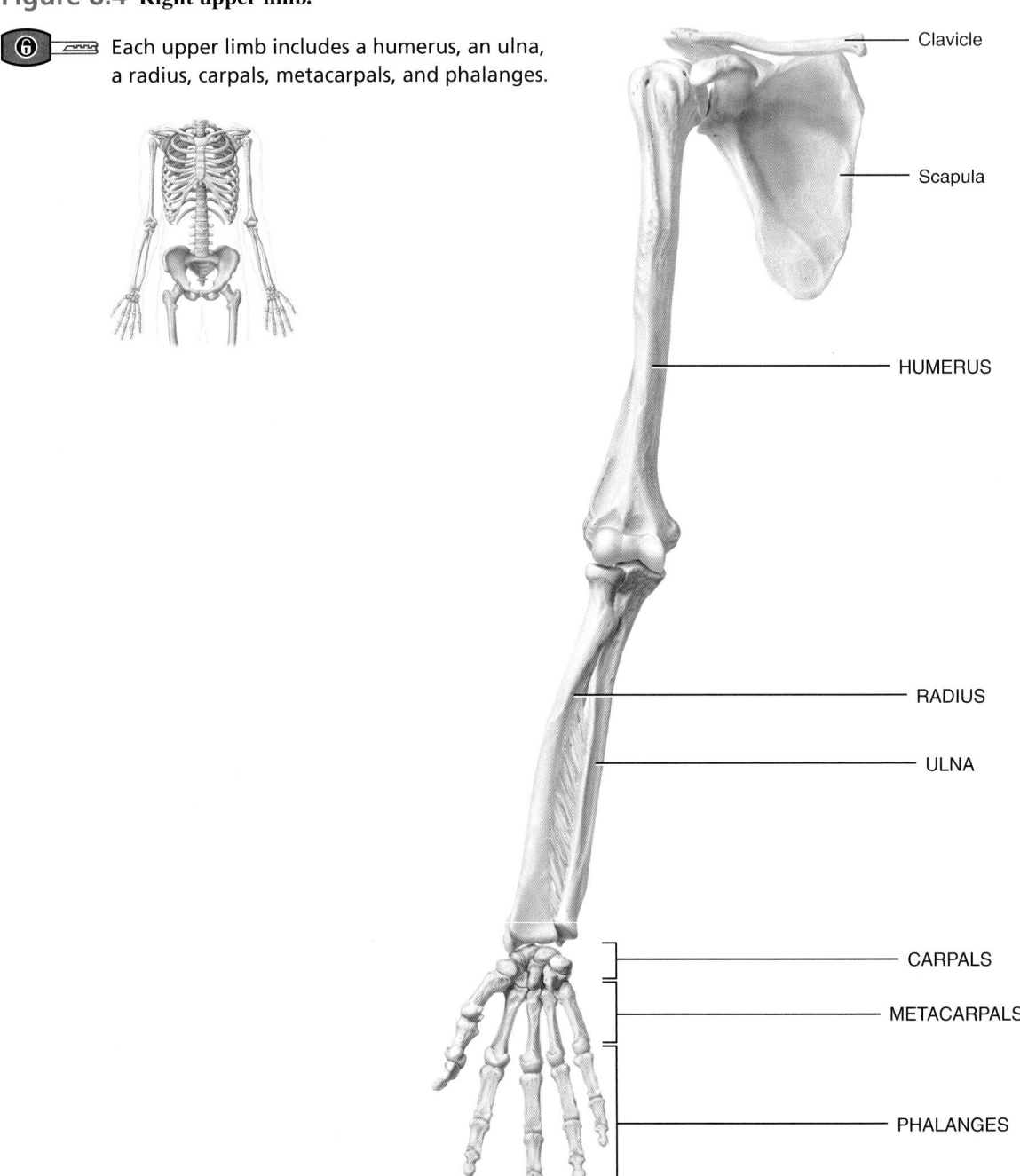

Clavicle

Scapula

HUMERUS

RADIUS

ULNA

CARPALS

METACARPALS

PHALANGES

Anterior view of free upper limb

❓ **How many bones make up each upper limb?**

EXHIBIT 8.C Skeleton of the Arm—Humerus *(Figure 8.5)*

⊙ **OBJECTIVE**

• Identify the location and surface landmarks of the humerus.

The **humerus** (HŪ-mer-us), or arm bone, is the longest and largest bone of the upper limb (Figure 8.5). It articulates proximally with the scapula and distally with two bones, the ulna and the radius, to form the elbow joint.

The proximal end of the humerus features a rounded *head* that articulates with the glenoid cavity of the scapula to form the *glenohumeral (shoulder) joint.* Distal to the head is the *anatomical neck,* which is visible as an oblique groove. It is the former site of the epiphyseal (growth) plate in an adult humerus. The *greater tubercle* is a lateral projection distal to the anatomical neck. It is the most laterally palpable bony landmark of the shoulder region and is immediately inferior to the palpable

Figure 8.5 **Right humerus in relation to the scapula, ulna, and radius.**

🔑 ⊏══ The humerus is the longest and largest bone of the upper limb.

(a) Anterior view   (b) Posterior view

❓ **Which parts of the humerus articulate with the radius at the elbow? With the ulna at the elbow?**

EXHIBIT 8.C CONTINUES

EXHIBIT 8.C **261**

acromion of the scapula mentioned earlier. The *lesser tubercle* projects anteriorly. Between the two tubercles there is a groove named the *intertubercular sulcus.* The *surgical neck* is a constriction in the humerus just distal to the tubercles, where the head tapers to the shaft; it is so named because fractures often occur here.

The *body (shaft)* of the humerus is roughly cylindrical at its proximal end, but it gradually becomes triangular until it is flattened and broad at its distal end. Laterally, at the middle portion of the shaft, there is a roughened, V-shaped area called the *deltoid tuberosity.* This area serves as a point of attachment for the tendons of the deltoid muscle. On the posterior surface of the humerus is the *radial groove,* which runs along the deltoid tuberosity and contains the radial nerve.

Several prominent features are evident at the distal end of the humerus. The *capitulum* (ka-PIT-ū-lum; *capit-* = head) is a rounded knob on the lateral aspect of the bone that articulates with the head of the radius. The *radial fossa* is an anterior depression above the capitulum that articulates with the head of the radius when the forearm is flexed (bent). The *trochlea* (TROK-lē-a = pulley), located medial to the capitulum, is a spool-shaped surface that articulates with the trochlear notch of the ulna.

The *coronoid fossa* (KOR-ō-noyd = crown-shaped) is an anterior depression that receives the coronoid process of the ulna when the forearm is flexed. The *olecranon fossa* (ō-LEK-ra-non = elbow) is a large posterior depression that receives the olecranon of the ulna when the forearm is extended (straightened). The *medial epicondyle* and *lateral epicondyle* are rough projections on either side of the distal end of the humerus to which the tendons of most muscles of the forearm are attached. The ulnar nerve may be palpated by rolling a finger over the skin surface above the posterior surface of the medial epicondyle. This nerve is the one that makes you feel a very severe pain when you hit your elbow, which for some reason is commonly referred to as the funnybone, even though this event is anything but funny.

✔ CHECKPOINT

Distinguish between the anatomical neck and the surgical neck of the humerus. Name the proximal and distal points formed by the humerus, and indicate which parts of the bones are involved.

Skeleton of the Forearm—Ulna and Radius

EXHIBIT 8.D

(Figures 8.6 and 8.7)

◉ OBJECTIVE

• Identify the location and surface landmarks of the ulna and radius.

The **ulna** is located on the medial aspect (the little-finger side) of the forearm and is longer than the radius (Figure 8.6). You may find it convenient to use an aid called a *mnemonic device* (nē-MON-ik = memory) to learn new or unfamiliar information. One such mnemonic to help you remember the location of the ulna in relation to the hand is "p.u." (the **p**inky is on the **u**lna side).

At the proximal end of the ulna (Figure 8.6b) is the *olecranon,* which forms the prominence of the elbow. With the olecranon, an anterior projection called the *coronoid process* (Figure 8.6a) articulates with the trochlea of the humerus. The *trochlear notch* is a large curved area between the olecranon and coronoid process that forms part of the elbow joint (see Figure 8.7b). Lateral and inferior to the trochlear notch is a depression, the *radial notch,* which articulates with the head of the radius. Just inferior to the coronoid process is the *ulnar tuberosity,* to which the biceps brachii muscle attaches. The distal end of the ulna consists of a *head* that is separated from the wrist by a disc of fibrocartilage. A *styloid process* is located on the posterior side of the ulna's distal end. It provides attachment for the ulnar collateral ligament to the wrist.

The **radius** is the smaller bone of the forearm and is located on the lateral aspect (thumb side) of the forearm (Figure 8.6a). In contrast to the ulna, the radius is narrow at its proximal end and widens at its distal end.

The proximal end of the radius has a disc-shaped *head* that articulates with the capitulum of the humerus and the radial notch of the ulna. Inferior to the head is the constricted *neck.* A roughened area inferior to the neck on the anteromedial side, called the *radial tuberosity,* is a point of

attachment for the tendons of the biceps brachii muscle. The shaft of the radius widens distally to form a *styloid process* on the lateral side, which can be felt proximal to the thumb. The styloid process provides attachment for the brachioradialis muscle and for attachment of the radial collateral ligament to the wrist. Fracture of the distal end of the radius is the most common fracture in adults older than 50 years.

The ulna and radius articulate with the humerus at the *elbow joint.* The articulation occurs in two places (Figure 8.7a, b): where the head of the radius articulates with the capitulum of the humerus, and where the trochlear notch of the ulna articulates with the trochlea of the humerus.

The ulna and the radius connect with one another at three sites. First, a broad, flat, fibrous connective tissue called the *interosseous membrane* (in-ter-OS-ē-us; *inter-* = between, *-osse* = bone) joins the shafts of the two bones (see Figure 8.6). This membrane also provides a site of attachment for some tendons of deep skeletal muscles of the forearm. The ulna and radius articulate directly at their proximal and distal ends (Figure 8.7b, c). Proximally, the head of the radius articulates with the ulna's radial notch. This articulation is the *proximal radioulnar joint.* Distally, the head of the ulna articulates with the *ulnar notch* of the radius. This articulation is the *distal radioulnar joint.* Finally, the distal end of the radius articulates with three bones of the wrist—the lunate, the scaphoid, and the triquetrum—to form the *radiocarpal (wrist) joint.*

✔ CHECKPOINT

How many joints are formed between the ulna and radius, what are their names, and what parts of the bones are involved?

Figure 8.6 **Right ulna and radius in relation to the humerus and carpals.**

In the forearm, the longer ulna is on the medial side, and the shorter radius is on the lateral side.

Radius

Ulna

Humerus

Radial fossa

Coronoid fossa

Capitulum

Trochlea

HEAD OF RADIUS

CORONOID PROCESS

NECK OF RADIUS

ULNAR TUBEROSITY

RADIAL TUBEROSITY

Olecranon fossa

OLECRANON

HEAD OF RADIUS

NECK OF RADIUS

RADIUS

ULNA

Interosseous membrane

HEAD OF ULNA

STYLOID PROCESS OF ULNA

STYLOID PROCESS OF RADIUS

Carpals

STYLOID PROCESS OF RADIUS

LATERAL

MEDIAL

LATERAL

(a) Anterior view

(b) Posterior view

? **What part of the ulna is called the "elbow"?**

EXHIBIT 8.D CONTINUES ▶

EXHIBIT 8.D **263**

Figure 8.7 Articulations formed by the ulna and radius. (a) Elbow joint. (b) Joint surfaces at proximal end of the ulna. (c) Joint surfaces at distal ends of radius and ulna.

The elbow joint is formed by two articulations: (1) the trochlear notch of the ulna with the trochlea of the humerus and (2) the head of the radius with the capitulum of the humerus.

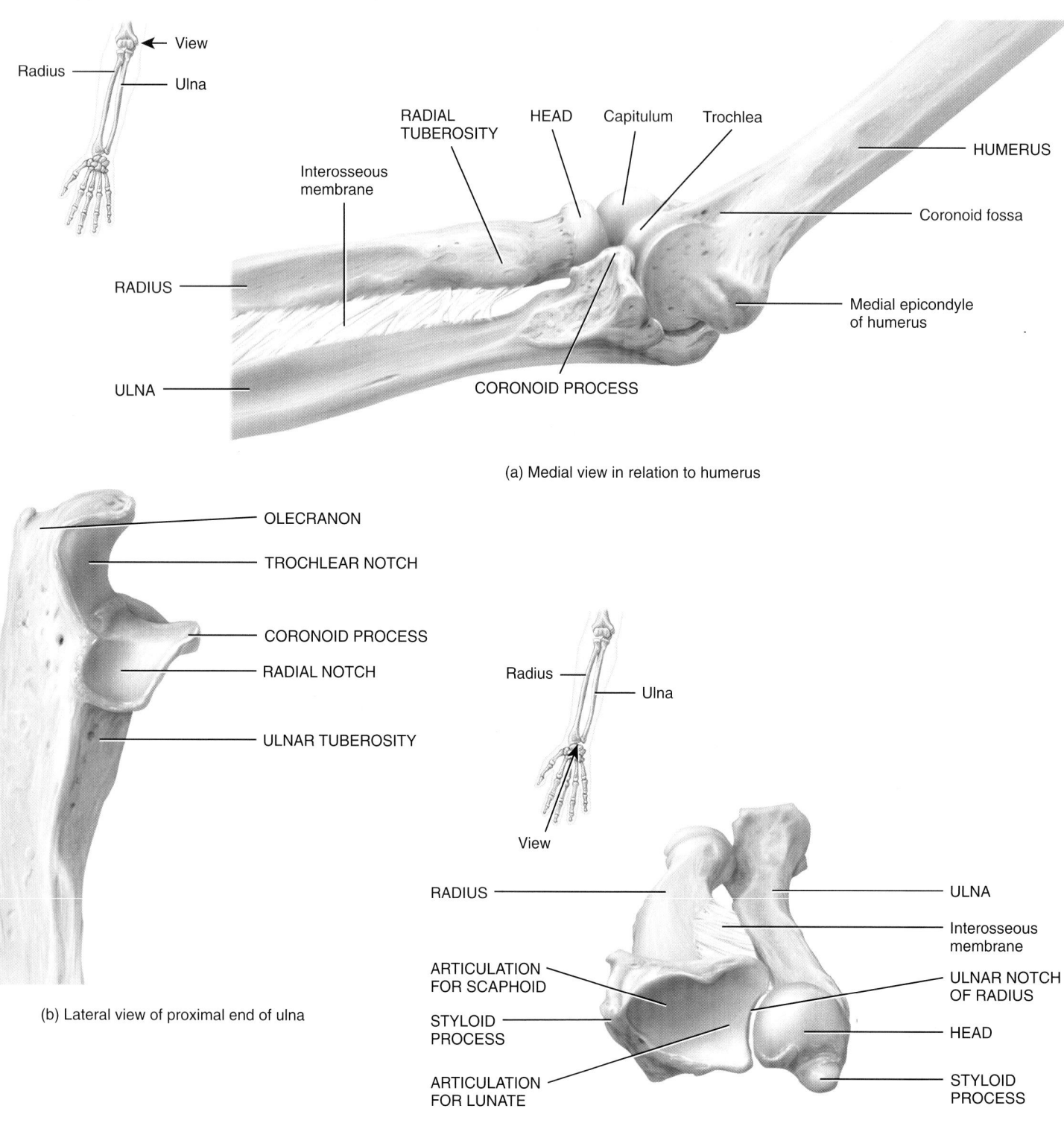

(a) Medial view in relation to humerus

(b) Lateral view of proximal end of ulna

(c) Inferior view of distal ends of radius and ulna

How many points of attachment are there between the radius and ulna?

OBJECTIVE

• Identify the location and surface landmarks of the bones of the hand.

Carpals

The **carpus** (*wrist*) is the proximal region of the hand and consists of eight small bones, the **carpals,** joined to one another by ligaments (Figure 8.8). Articulations among carpal bones are called *intercarpal joints.* The carpals are arranged in two transverse rows of four bones each.

Figure 8.8 **Right wrist and hand in relation to the ulna and radius.**

The skeleton of the hand consists of the proximal carpals, the intermediate metacarpals, and the distal phalanges.

Carpals
Metacarpals
Phalanges

Radius
Ulna

CARPALS:
Lunate
Pisiform
Triquetrum
Hamate

CARPALS:
Scaphoid
Trapezium
Capitate
Trapezoid
Sesamoid bones

CARPALS:
Scaphoid
Capitate
Trapezium
Trapezoid
Base
Shaft
Head

I
II
III
IV
V

Thumb (pollex)
Index finger
Middle finger
Ring finger
Little finger

METACARPALS
Base
Shaft
Head

PHALANGES
Proximal
Middle
Distal

(a) Anterior view

(b) Posterior view

MNEMONIC for carpal bones*:	
Stop **L**etting **T**hose **P**eople **T**ouch **T**he **C**adaver's **H**and.	
Scaphoid **L**unate **T**riquetrum **P**isiform	**T**rapezium **T**rapezoid **C**apitate **H**amate
Proximal row	Distal row
Lateral ————→ Medial	Lateral ————→ Medial

* *Edward Tanner, University of Alabama, School of Medicine.*

Which is the most frequently fractured wrist bone?

EXHIBIT 8.E CONTINUES

EXHIBIT 8.E **265**

Their names reflect their shapes. The carpals in the proximal row, from lateral to medial, are the

- **scaphoid** (SKAF-oyd = boatlike)
- **lunate** (LOO-nāt = moon-shaped)
- **triquetrum** (trī-KWĒ-trum = three-cornered)
- **pisiform** (PĪS-i-form = pea-shaped).

The proximal row of carpals articulates with the distal ends of the ulna and radius to form the *wrist joint*. The carpals in the distal row, from lateral to medial, are the

- **trapezium** (tra-PĒ-zē-um = four-sided figure with no two sides parallel)
- **trapezoid** (TRAP-e-zoyd = four-sided figure with two sides parallel)
- **capitate** (KAP-i-tāt = head-shaped)
- **hamate** (HAM-āt = hooked).

The capitate is the largest carpal bone; its rounded projection, the head, articulates with the lunate. The hamate is named for a large hook-shaped projection on its anterior surface. In about 70% of carpal fractures, only the scaphoid is broken. This is because the force of a fall on an outstretched hand is transmitted from the capitate through the scaphoid to the radius.

The anterior concave space formed by the pisiform and hamate (on the ulnar side), and the scaphoid and trapezium (on the radial side), with the rooflike covering of the *flexor retinaculum* (strong fibrous bands of connective tissue) is the **carpal tunnel.** The long flexor tendons of the digits and thumb and the median nerve pass through the carpal tunnel. Narrowing of the carpal tunnel, due to such factors as inflammation, may give rise to a condition called *carpal tunnel syndrome* (described in Clinical Connection: Carpal Tunnel Syndrome in Chapter 11).

There is a useful mnemonic for learning the names of the carpal bones provided in Figure 8.8. The first letter of the carpal bones from lateral to medial (proximal row, then distal row) corresponds to the first letter of each word in the mnemonic.

Metacarpals

The **metacarpus** (*meta-* = beyond), or *palm,* is the intermediate region of the hand and consists of five bones called **metacarpals.**

Each metacarpal bone consists of a proximal *base,* an intermediate *shaft,* and a distal *head* (Figure 8.8b). The metacarpal bones are numbered I to V (or 1–5), starting with the thumb, from lateral to medial. The bases articulate with the distal row of carpal bones to form the *carpometacarpal joints.* The heads articulate with the proximal phalanges to form the *metacarpophalangeal joints.* The heads of the metacarpals, commonly called "knuckles," are readily visible in a clenched fist.

A **boxer's fracture** is a fracture of the fifth metacarpal, usually near the head of the bone. It frequently occurs after a person punches another person or an object, such as a wall. It is characterized by pain, swelling, and tenderness. There may also be a bump on the side of the hand. Treatment is either by casting or surgery, and the fracture usually heals in about 6 weeks. •

Phalanges

The **phalanges** (fa-LAN-jēz; *phalan-* = a battle line), or bones of the *digits,* make up the distal part of the hand. There are 14 phalanges in the five digits of each hand and, like the metacarpals, the digits are numbered I to V (or 1–5), beginning with the thumb, from lateral to medial. A single bone of a digit is referred to as a *phalanx* (FĀ-lanks).

Each phalanx consists of a proximal *base,* an intermediate *shaft,* and a distal *head.* The thumb (*pollex*) has two phalanges, and there are three phalanges in each of the other four digits. In order from the thumb, these other four digits are commonly referred to as the index finger, middle finger, ring finger, and little finger. The first row of phalanges, the *proximal row,* articulates with the metacarpal bones and second row of phalanges. The second row of phalanges, the *middle row,* articulates with the proximal row and the third row, called the *distal row.* The thumb has no middle phalanx. Joints between phalanges are called *interphalangeal joints.*

✔ **CHECKPOINT**

Which is more distal, the base or the head of the carpals?
With which bones do the proximal phalanges articulate?

8.3 PELVIC (HIP) GIRDLE

◉ OBJECTIVE

• Identify the bones of the pelvic girdle and their principal markings.

The **pelvic (hip) girdle** consists of the two **hip bones,** also called **coxal** (KOK-sal; *cox-* = hip) or **pelvic bones** or **os coxa** (Figure 8.9). The hip bones unite anteriorly at a joint called the **pubic symphysis** (PŪ-bik SIM-fi-sis). They unite posteriorly with the sacrum at the *sacroiliac joints.* The complete ring composed of the hip bones, pubic symphysis, and sacrum forms a deep, basin-like structure called the **bony pelvis** (*pelv-* = basin). The plural is *pelves* (PEL-vēz) or *pelvises.* Functionally, the bony pelvis provides a strong and stable support for the vertebral column and pelvic and lower abdominal organs. The pelvic girdle of the bony pelvis also connects the bones of the lower limbs to the axial skeleton.

Each of the two hip bones of a newborn consists of three bones separated by cartilage: a superior *ilium,* an inferior and anterior *pubis,* and an inferior and posterior *ischium.* By age 23, the three separate bones fuse together (Figure 8.10a). Although the hip bones function as single bones, anatomists commonly discuss each hip bone as three separate bones.

The bones of the pelvic girdle are discussed in Exhibit 8.F.

✔CHECKPOINT

3. Describe the distinguishing characteristics of the individual bones of the pelvic girdle.
4. Which bones form the acetabulum? What is its function?

Figure 8.9 Bony pelvis. Shown here is the female bony pelvis.

🔑 The hip bones unite anteriorly at the public symphysis and posteriorly at the sacrum to form the bony pelvis.

Pelvic (hip) girdle

Hip bone

Sacrum

Coccyx

Pubic symphysis

Sacroiliac joint

Sacral promontory

Pelvic brim

Acetabulum

Obturator foramen

Anterosuperior view of pelvic girdle

❓ **What are the functions of the bony pelvis?**

Figure 8.10 **Right hip bone.** The lines of fusion of the ilium, ischium, and pubis depicted in part (a) are not always visible in an adult.

The acetabulum is the socket for the head of the femur, where the three parts of the hip bone converge and ossify.

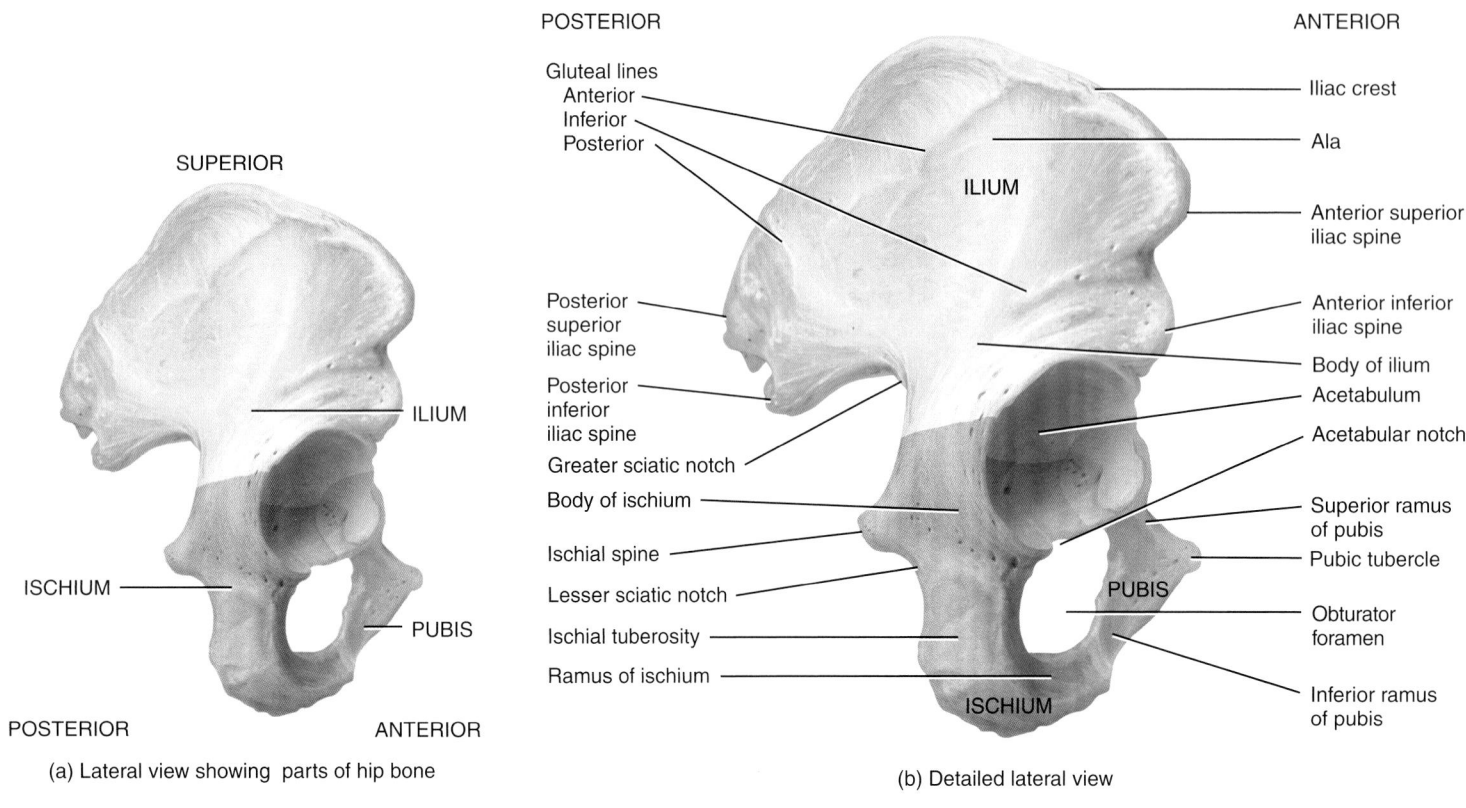

(a) Lateral view showing parts of hip bone

(b) Detailed lateral view

(c) Detailed medial view

? **Which part of the hip bone articulates with the femur? With the sacrum?**

EXHIBIT 8.F Bones of the Pelvic Girdle

• Identify the locations and surface features of the three components of the hip bone.

Ilium

The **ilium** (IL-ē-um = flank), the largest of the three components of the hip bone (Figure 8.10b), is composed of a superior *ala* (= wing) and an inferior *body*. The body is one of the components of the *acetabulum,* the socket for the head of the femur. The superior border of the ilium, the *iliac crest,* ends anteriorly in a blunt *anterior superior iliac spine.* Bruising of the anterior superior iliac spine and associated soft tissues, such as occurs in body contact sports, is called a **hip pointer.** Below this spine is the *anterior inferior iliac spine.* Posteriorly, the iliac crest ends in a sharp *posterior superior iliac spine.* Below this spine is the *posterior inferior iliac spine.* The spines serve as points of attachment for the tendons of the muscles of the trunk, hip, and thighs. Below the posterior inferior iliac spine is the *greater sciatic notch* (sī-AT-ik), through which the sciatic nerve, the longest nerve in the body, passes.

The medial surface of the ilium contains the *iliac fossa,* a concavity where the tendon of the iliacus muscle attaches. Posterior to this fossa are the *iliac tuberosity,* a point of attachment for the sacroiliac ligament, and the *auricular surface* (*auric-* = ear-shaped), which articulates with the sacrum to form the *sacroiliac joint* (see Figure 8.9). Projecting anteriorly and inferiorly from the auricular surface is a ridge called the *arcuate line* (AR-kū-āt; *arc-* = bow).

The other conspicuous markings of the ilium are three arched lines on its lateral surface called the *posterior gluteal line* (*glut-* = buttock), the *anterior gluteal line,* and the *inferior gluteal line.* The gluteal muscles attach to the ilium between these lines.

Ischium

The **ischium** (IS-kē-um = hip), the inferior, posterior portion of the hip bone (Figure 8.10b, c), comprises a superior *body* and an inferior *ramus* (*ram-* = branch; plural is *rami*). The ramus is the portion of the ischium that fuses with the pubis. Features of the ischium include the prominent *ischial spine,* a *lesser sciatic notch* below the spine, and a rough and thickened *ischial tuberosity.* Because this prominent tuberosity is just deep to the skin, it commonly begins hurting after a relatively short time when you sit on a hard surface. Together, the ramus and the pubis surround the *obturator foramen* (OB-too-rā-tōr; *obtur-* = closed up), the largest foramen in the skeleton. The foramen is so named because, even though blood vessels and nerves pass through it, it is nearly completely closed by the fibrous *obturator membrane.*

Pubis

The **pubis** (PŪ-bis; plural is *pubes*), meaning pubic bone, is the anterior and inferior part of the hip bone (Figure 8.10b, c). A *superior ramus,* an *inferior ramus,* and a *body* between the rami make up the pubis. The anterior, superior border of the body is the *pubic crest,* and at its lateral end is a projection called the *pubic tubercle.* This tubercle is the beginning of a raised line, the *pectineal line* (pek-TIN-ē-al), which extends superiorly and laterally along the superior ramus to merge with the arcuate line of the ilium. These lines, as you will see shortly, are important landmarks for distinguishing the superior (false) and inferior (true) portions of the bony pelvis.

The *pubic symphysis* is the joint between the two pubes of the hip bones (see Figure 8.9). It consists of a disc of fibrocartilage. Inferior to this joint, the inferior rami of the two pubic bones converge to form the *pubic arch.* In the later stages of pregnancy, the hormone relaxin (produced by the ovaries and placenta) increases the flexibility of the pubic symphysis to ease delivery of the baby. Weakening of the joint, together with an already altered center of gravity due to an enlarged uterus, also changes the mother's gait during pregnancy.

The *acetabulum* (as-e-TAB-ū-lum = vinegar cup) is a deep fossa formed by the ilium, ischium, and pubis. It functions as the socket that accepts the rounded head of the femur. Together, the acetabulum and the femoral head form the *hip (coxal) joint.* On the inferior side of the acetabulum is a deep indentation, the *acetabular notch,* that forms a foramen through which blood vessels and nerves pass and serves as a point of attachment for ligaments of the femur (for example, the ligament of the head of the femur).

✔ **CHECKPOINT**

Why is the obturator foramen so named? Which joints are formed by the union of the hip bones with other bones?

EXHIBIT 8.F **269**

8.4 FALSE AND TRUE PELVES

● **OBJECTIVES**

- Distinguish between the false and true pelves.
- Explain why the false and true pelves are important clinically.

The bony pelvis is divided into superior and inferior portions by a boundary called the *pelvic brim* or *inlet* (Figure 8.11a). You can trace the pelvic brim by following the landmarks around parts of the hip bones to form the outline of an oblique plane. Beginning posteriorly at the *sacral promontory* of the sacrum, trace laterally and inferiorly along the *arcuate lines* of the ilium. Continue

Figure 8.11 **True and false pelves.** Shown here is the female pelvis. For simplicity, in part (a) the landmarks of the pelvic brim are shown only on the left side of the body, and the outline of the pelvic brim is shown only on the right side. The entire pelvic brim is shown in Table 8.1.

🔑 ▭ The true and false pelves are separated by the pelvic brim.

Hip bone

Sacrum

Pelvic brim

Pelvic brim landmarks:
Sacral promontory

Arcuate line

Pectineal line

Pubic crest

Pubic symphysis

(a) Anterior view of pelvic girdle

Midsagittal plane

POSTERIOR

ANTERIOR

Sacral canal

Sacrum

TRUE PELVIS

Coccyx

Plane of pelvic outlet

Sacral promontory

FALSE PELVIS

Plane of pelvic brim

Pelvic axis

Pubic symphysis

(b) Midsagittal section indicating locations of true and false pelves inferiorly along the *pectineal lines* of the pubis. Finally, trace anteriorly along the *pubic crest* to the superior portion of the *pubic symphysis.* Together, these points form an oblique plane that is higher in the back than in the front. The circumference of this plane is the pelvic brim.

The portion of the bony pelvis superior to the pelvic brim is referred to as the **false (greater) pelvis** (Figure 8.11b). It is bordered by the lumbar vertebrae posteriorly, the upper portions of the hip bones laterally, and the abdominal wall anteriorly. The space enclosed by the false pelvis is part of the lower abdomen; it contains the superior portion of the urinary bladder (when it is full) and the lower intestines in both genders and the uterus, ovaries, and uterine tubes of the female.

The portion of the bony pelvis inferior to the pelvic brim is the **true (lesser) pelvis** (Figure 8.11b). It has an inlet, an outlet, and a cavity. It is bounded by the sacrum and coccyx posteriorly, inferior portions of the ilium and ischium laterally, and the pubic bones anteriorly. The true pelvis surrounds the pelvic cavity, which was described in Chapter 1 (see Figure 1.9). The true pelvis contains the rectum and urinary bladder in both genders, the vagina and cervix of the uterus in females, and the prostate in males. The superior opening of the true pelvis, bordered by the pelvic brim, is called the *pelvic inlet;* the inferior opening of the true pelvis is the *pelvic outlet.* The *pelvic axis* is an imaginary line that

curves through the true pelvis from the central point of the plane of the pelvic inlet to the central point of the plane of the pelvic outlet. During childbirth the pelvic axis is the route taken by the baby's head as it descends through the pelvis.

CLINICAL CONNECTION | *Pelvimetry*

Pelvimetry is the measurement of the size of the inlet and outlet of the birth canal, which may be done by ultrasonography or physical examination. Measurement of the pelvic cavity in pregnant females is important because the fetus must pass through the narrower opening of the pelvis at birth. A cesarean section is usually planned if it is determined that the pelvic cavity is too small to permit passage of the baby. •

✔CHECKPOINT
5. Why are the false and true pelves important clinically?

8.5 COMPARISON OF FEMALE AND MALE PELVES

◉ OBJECTIVE
• Compare the principal differences between female and male pelves.

Generally, the bones of males are larger and heavier and possess larger surface markings than those of females of comparable age and physical stature. Sex-related differences in the features of bones are readily apparent when comparing the adult female and male pelves. Most of the structural differences in the pelves are adaptations to the requirements of pregnancy and childbirth. The female's pelvis is wider and shallower than the male's. Consequently, there is more space in the true pelvis of the female, especially in the pelvic inlet and pelvic outlet, to accommodate the passage of the infant's head at birth. Other significant structural differences between the pelves of females and males are listed and illustrated in Table 8.1.

✔CHECKPOINT
6. How is the female pelvis adapted for pregnancy and childbirth?
7. Using Table 8.1 as a guide; select the three ways that are easiest for you to distinguish a female from a male pelvis.

(c) Anterior view of false pelvis (pink)

(d) Anterior view of true pelvis (blue)

 What is the significance of the pelvic axis?

TABLE 8.1

Comparison of Female and Male Pelves

POINT OF COMPARISON	FEMALE	MALE
General structure	Light and thin.	Heavy and thick.
False (greater) pelvis	Shallow.	Deep.
Pelvic brim (inlet)	Wide and more oval.	Narrow and heart-shaped.
Acetabulum	Small and faces anteriorly.	Large and faces laterally.
Obturator foramen	Oval.	Round.
Pubic arch	Greater than 90° angle.	Less than 90° angle.

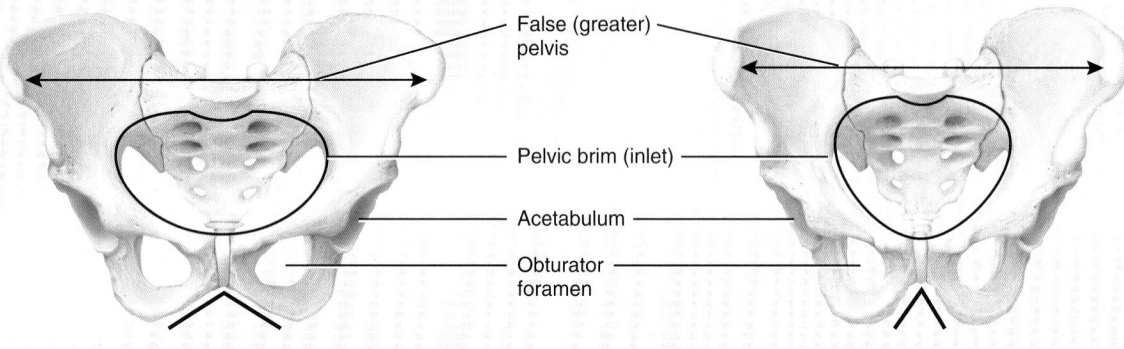

Anterior views

POINT OF COMPARISON	FEMALE	MALE
Iliac crest	Less curved.	More curved.
Ilium	Less vertical.	More vertical.
Greater sciatic notch	Wide (almost 90°).	Narrow (about 70°; inverted V).
Coccyx	More movable and more curved anteriorly.	Less movable and less curved anteriorly.
Sacrum	Shorter, wider (see anterior views), and less curved anteriorly.	Longer, narrower (see anterior views), and more curved anteriorly.

Right lateral views

POINT OF COMPARISON	FEMALE	MALE
Pelvic outlet	Wider.	Narrower.
Ischial tuberosity	Shorter, farther apart, and more medially projecting.	Longer, closer together, and more laterally projecting.

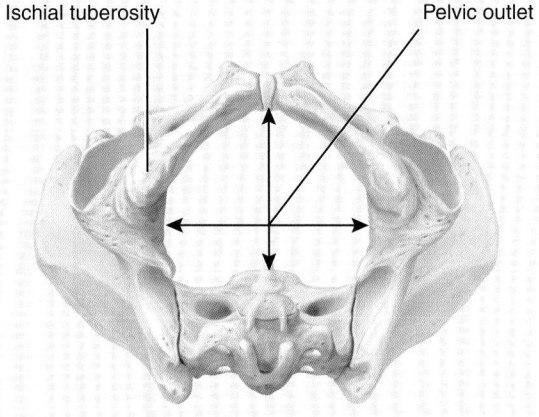

Inferior views

8.6 LOWER LIMB (EXTREMITY)

 OBJECTIVE

- Identify the bones of the lower limb and their principal markings.

Each **lower limb (lower extremity)** has 30 bones in four locations—(1) the femur in the thigh; (2) the patella (kneecap); (3) the tibia and fibula in the leg; and (4) the 7 tarsals in the tarsus (ankle), the 5 metatarsals in the metatarsus, and the 14 phalanges (bones of the digits) in the foot (Figure 8.12).

The bones of the lower limb are discussed in Exhibits 8.G through 8.I.

✔**CHECKPOINT**

8. Name the bones that form the lower limb, from proximal to distal.
9. What are the names and functions of the arches of the foot?

Figure 8.12 Right lower limb.

🔒 ◁▭ Each lower limb includes a femur, a patella (kneecap), a tibia, a fibula, tarsals (ankle bones), metatarsals, and phalanges (bones of the digits).

- Hip bone
- Sacrum
- FEMUR
- PATELLA
- TIBIA
- FIBULA
- TARSALS
- METATARSALS
- PHALANGES

Anterior view of free lower limb

❓ **How many bones make up each lower limb?**

EXHIBIT 8.G

Skeleton of the Thigh—Femur and Patella
(Figures 8.13 and 8.14)

◉ **OBJECTIVE**

• Identify the location and surface features of the femur and patella.

Femur

The **femur,** or thigh bone, is the longest, heaviest, and strongest bone in the body (Figure 8.13). Its proximal end articulates with the acetabulum of the hip bone. Its distal end articulates with the tibia and patella. The

Figure 8.13 Right femur in relation to the hip bone, patella, tibia, and fibula.

🔑 The acetabulum of the hip bone and head of the femur articulate to form the hip joint.

Femur

Hip bone

GREATER TROCHANTER

HEAD

NECK

GREATER TROCHANTER

INTERTROCHANTERIC LINE   CREST

GLUTEAL TUBEROSITY

LESSER TROCHANTER

FEMUR

LINEA ASPERA

BODY (SHAFT)

ADDUCTOR TUBERCLE

LATERAL EPICONDYLE

MEDIAL EPICONDYLE

LATERAL EPICONDYLE

LATERAL CONDYLE

MEDIAL CONDYLE

INTERCONDYLAR FOSSA

Patella

LATERAL CONDYLE

Fibula

Fibula

Tibia

(a) Anterior view

(b) Posterior view

body (shaft) of the femur angles medially and, as a result, the knee joints are closer to the midline than the hip joints. This angle of the femoral shaft *(angle of convergence)* is greater in females because the female pelvis is broader.

The proximal end of the femur consists of a rounded *head* that articulates with the acetabulum of the hip bone to form the *hip (coxal) joint*. The head contains a small central depression (pit) called the *fovea capitis* (FŌ-vē-a CAP-i-tis; *fovea* = pit, *capitis* = of the head). The ligament of the head of the femur connects the fovea capitis of the femur to the acetabulum of the hip bone. The *neck* of the femur is a constricted region distal to the head. A "broken hip" is more often associated with a fracture in the neck of the femur than with fractures of the hip bones. The *greater trochanter* (trō-KAN-ter) and *lesser trochanter* are projections from the junction of the neck and shaft that serve as points of attachment for the tendons of some of the thigh and buttock muscles. The greater trochanter is the prominence felt and seen anterior to the hollow on the side of the hip. It is a landmark commonly used to locate the site for intramuscular injections into the lateral surface of the thigh. The lesser trochanter is inferior and medial to the greater trochanter. Between the anterior surfaces of the trochanters is a narrow *intertrochanteric line* (Figure 8.13a). A ridge called the *intertrochanteric crest* appears between the posterior surfaces of the trochanters (Figure 8.13b).

Inferior to the intertrochanteric crest on the posterior surface of the body of the femur is a vertical ridge called the *gluteal tuberosity*. It blends into another vertical ridge called the *linea aspera* (LIN-ē-a AS-per-a; *asper* = rough). Both ridges serve as attachment points for the tendons of several thigh muscles.

The expanded distal end of the femur includes the *medial condyle* (= knuckle) and the *lateral condyle*. These articulate with the medial and lateral condyles of the tibia. Superior to the condyles are the *medial epicondyle* and the *lateral epicondyle*, to which ligaments of the knee joint attach. A depressed area between the condyles on the posterior surface is called the *intercondylar fossa* (in-ter-KON-di-lar). The *patellar surface* is located between the condyles on the anterior surface. Just superior to the medial epicondyle is the *adductor tubercle*, a roughened projection that is a site of attachment for the adductor magnus muscle.

Femur —

View

HEAD

FOVEA CAPITIS

GREATER TROCHANTER

NECK

INTERTROCHANTERIC CREST

LESSER TROCHANTER

(c) Medial view of proximal end of femur

? **Why is the angle of convergence of the femurs greater in females than males?**

EXHIBIT 8.G CONTINUES ▶

EXHIBIT 8.G **275**

Patella

The **patella** (= little dish), or kneecap, is a small, triangular bone located anterior to the knee joint (Figure 8.14). The broad proximal end of this sesamoid bone, which develops in the tendon of the quadriceps femoris muscle, is called the *base;* the pointed distal end is referred to as the *apex.* The posterior surface contains two *articular facets,* one for the medial condyle of the femur and another for the lateral condyle of the femur. The patellar ligament attaches the patella to the tibial tuberosity. The *patellofemoral joint,* between the posterior surface of the patella and the patellar surface of the femur, is the intermediate component of the *tibiofemoral (knee) joint.* The patella increases the leverage of the tendon of the quadriceps femoris muscle, maintains the position of the tendon when the knee is bent (flexed), and protects the knee joint.

CLINICAL CONNECTION | *Patellofemoral Stress Syndrome*

Patellofemoral stress syndrome (runner's knee) is one of the most common problems runners experience. During normal flexion and extension of the knee, the patella tracks (glides) superiorly and inferiorly in the groove between the femoral condyles. In patellofemoral stress syndrome, normal tracking does not occur; instead, the patella tracks laterally as well as superiorly and inferiorly, and the increased pressure on the joint causes aching or tenderness around or under the patella. The pain typically occurs after a person has been sitting for a while, especially after exercise. It is worsened by squatting or walking down stairs. One cause of runner's knee is constantly walking, running, or jogging on the same side of the road. Other predisposing factors include running on hills, running long distances, and an anatomical deformity called genu valgum, or knock-knee (see Medical Terminology at the end of the chapter). •

✔ **CHECKPOINT**

What is the clinical importance of the greater trochanter?
Which joints are formed by the femur with other bones?

Figure 8.14 Right patella.

The patella articulates with the lateral and medial condyles of the femur.

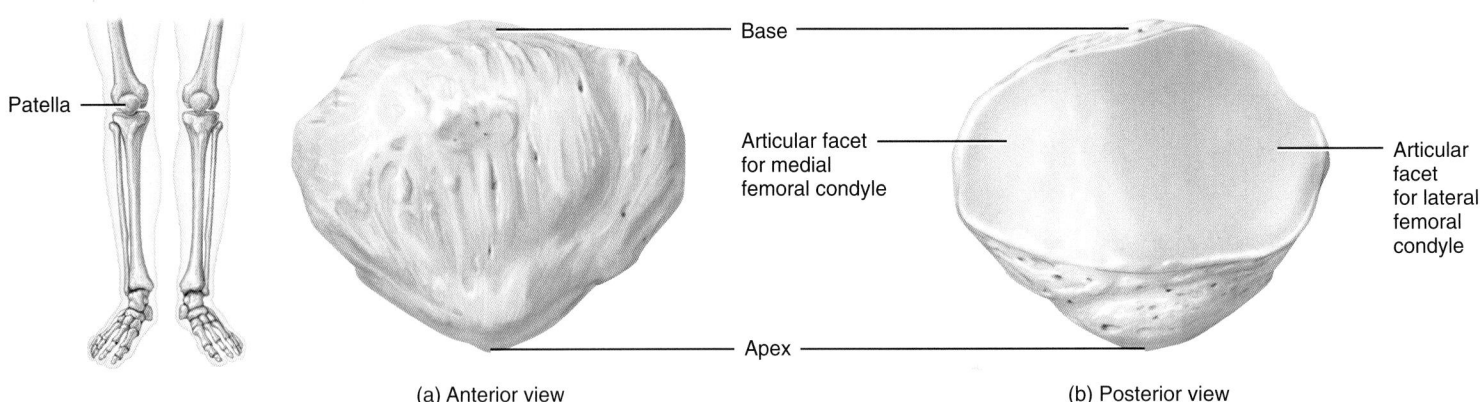

(a) Anterior view   (b) Posterior view

Base

Articular facet for medial femoral condyle

Articular facet for lateral femoral condyle

Apex

Patella

? **The patella is classified as which type of bone? Why?**

EXHIBIT 8.H Skeleton of the Leg—Tibia and Fibula *(Figure 8.15)*

◉ OBJECTIVE

- Identify the location and surface features of the tibia and fibula.

Tibia

The **tibia,** or shin bone, is the larger, medial, weight-bearing bone of the leg (Figure 8.15). The term *tibia* means flute, because the tibial bones of birds were used in ancient times to make musical instruments. The tibia

Figure 8.15 Right tibia and fibula in relation to the femur, patella, and talus.

The tibia articulates with the femur and fibula proximally, and with the fibula and talus distally.

Tibia

Fibula

Femur

Patella

INTERCONDYLAR EMINENCE

LATERAL CONDYLE

MEDIAL CONDYLE

LATERAL CONDYLE

HEAD

HEAD

TIBIAL TUBEROSITY

FIBULA

TIBIA

Interosseous membrane

FIBULA

ANTERIOR BORDER (CREST)

MEDIAL MALLEOLUS

LATERAL MALLEOLUS

Talus

LATERAL MALLEOLUS

Calcaneus

MNEMONIC for location of tibia and fibula:
The fibu**LA** is **LA**teral.

(a) Anterior view

(b) Posterior view

EXHIBIT 8.H CONTINUES

EXHIBIT 8.H **277**

EXHIBIT 8.H Skeleton of the Leg—Tibia and Fibula *(Figure 8.15)* CONTINUED

FIGURE 8.15 CONTINUED

POSTERIOR   ANTERIOR

Tibia

View

FIBULAR NOTCH

MEDIAL MALLEOLUS

(c) Lateral view of distal end of tibia

? Which leg bone bears the weight of the body?

articulates at its proximal end with the femur and fibula, and at its distal end with the fibula and the talus bone of the ankle. The tibia and fibula, like the ulna and radius, are connected by an interosseous membrane.

The proximal end of the tibia is expanded into a *lateral condyle* and a *medial condyle.* These articulate with the condyles of the femur to form the lateral and medial *tibiofemoral (knee) joints.* The inferior surface of the lateral condyle articulates with the head of the fibula. The slightly concave condyles are separated by an upward projection called the *intercondylar eminence* (Figure 8.15b). The *tibial tuberosity* on the anterior surface is a point of attachment for the patellar ligament. Inferior to and continuous with the tibial tuberosity is a sharp ridge that can be felt below the skin, known as the *anterior border (crest)* or *shin.*

The medial surface of the distal end of the tibia forms the *medial malleolus* (mal-LĒ-ō-lus = hammer). This structure articulates with the talus of the ankle and forms the prominence that can be felt on the medial surface of the ankle. The *fibular notch* (Figure 8.15c) articulates with the distal end of the fibula to form the *distal tibiofibular joint.* Of all the long bones of the body, the tibia is the most frequently fractured and is also the most frequent site of an open (compound) fracture.

Fibula

The **fibula** is parallel and lateral to the tibia, but it is considerably smaller. (See Figure 8.15 for a mnemonic describing the relative positions of the tibia and fibula.) Unlike the tibia, the fibula does not articulate with the femur, but it does help stabilize the ankle joint.

The *head* of the fibula, the proximal end, articulates with the inferior surface of the lateral condyle of the tibia below the level of the knee joint to form the *proximal tibiofibular joint.* The distal end is more arrowhead-shaped and has a projection called the *lateral malleolus* that articulates with the talus of the ankle. This forms the prominence on the lateral surface of the ankle. As noted previously, the fibula also articulates with the tibia at the fibular notch to form the distal tibiofibular joint.

CLINICAL CONNECTION | *Bone Grafting*

Bone grafting generally consists of taking a piece of bone, along with its periosteum and nutrient artery, from one part of the body to replace missing bone in another part of the body. The transplanted bone restores the blood supply to the transplanted site, and healing occurs as in a fracture. The fibula is a common source of bone for grafting because even after a piece of the fibula has been removed, walking, running, and jumping can be normal. Recall that the tibia is the weight-bearing bone of the leg. •

✔ CHECKPOINT

Which structures form the medial and lateral prominences of the ankle? Which joints are formed by the tibia and fibula with other bones?

EXHIBIT 8.1 Skeleton of the Foot *(Figures 8.16 and 8.17)*

OBJECTIVE

- Identify the location and surface features of the bones of the foot.

Tarsals, Metatarsals, and Phalanges

The **tarsus** (ankle) is the proximal region of the foot and consists of seven **tarsal bones** (Figure 8.16). They include the **talus** (TĀ-lus = ankle bone) and **calcaneus** (kal-KĀ-nē-us = heel), located in the posterior part of the foot. The calcaneus is the largest and strongest tarsal bone.

The anterior tarsal bones are the **navicular** (na-VIK-ū-lar = like a little boat), three **cuneiform bones** (KYOO-nē-i-form = wedge-shaped) called the **third (lateral), second (intermediate),** and **first (medial) cuneiforms,** and the **cuboid** (KŪ-boyd = cube-shaped). (A mnemonic to help you remember the names of the tarsal bones is included in Figure 8.16.) Joints between tarsal bones are called *intertarsal joints.* The talus, the most superior tarsal bone, is the only bone of the foot that articulates with the fibula and tibia. It articulates on one side with the medial malleolus of the tibia and on the other side with the lateral malleolus of the fibula. These articulations form the *talocrural (ankle) joint.*

Figure 8.16 Right foot.

🔑 The skeleton of the foot consists of the proximal tarsals, the intermediate metatarsals, and the distal phalanges.

Superior view

Inferior view

LATERAL   POSTERIOR   MEDIAL   POSTERIOR   LATERAL

TARSALS:
Calcaneus

Tarsals
Metatarsals
Phalanges

Cuboid

Base

Shaft

Head

TARSALS:
Calcaneus

TARSALS:
Talus
Navicular
Third (lateral) cuneiform
Second (intermediate) cuneiform
First (medial) cuneiform

TARSALS:
Calcaneus

Cuboid

METATARSALS:

V IV III II I   I II III IV V

Sesamoid bones

PHALANGES:
Proximal

Middle

Distal

Great toe (hallux)

(a) Superior view   (b) Inferior view

MNEMONIC for tarsals:

Tall	Centers	Never	Take	Shots	From	Corners.
Talus	Calcaneus	Navicular	Third cuneiform	Second cuneiform	First cuneiform	Cuboid

❓ **Which tarsal bone articulates with the tibia and fibula?**

EXHIBIT 8.1 CONTINUES ▶

EXHIBIT 8.1   **279**

EXHIBIT 8.1   Skeleton of the Foot *(Figures 8.16 and 8.17)*   CONTINUED

During walking, the talus transmits about half the weight of the body to the calcaneus. The remainder is transmitted to the other tarsal bones.

The **metatarsus,** the intermediate region of the foot, consists of five **metatarsal bones** numbered I to V (or 1–5) from the medial to lateral position (Figure 8.16). Like the metacarpals of the palm of the hand, each metatarsal consists of a proximal *base,* an intermediate *shaft,* and a distal *head.* The metatarsals articulate proximally with the first, second, and third cuneiform bones and with the cuboid to form the *tarsometatarsal joints.* Distally, they articulate with the proximal row of phalanges to form the *metatarsophalangeal joints.* The first metatarsal is thicker than the others because it bears more weight.

CLINICAL CONNECTION | *Fractures of the Metatarsals*

Fractures of the metatarsals occur when a heavy object falls on the foot or when a heavy object rolls over the foot. Such fractures are also common among dancers, especially ballet dancers. If a ballet dancer is on the tip of her toes and loses her balance, the full body weight is placed on the metatarsals, causing one or more of them to fracture.   •

The **phalanges** comprise the distal component of the foot and resemble those of the hand both in number and arrangement. The toes are numbered I to V (or 1–5) beginning with the great toe, from medial to lateral. Each *phalanx* (singular) consists of a proximal *base,* an intermediate *shaft,* and a distal *head.* The great or big toe (*hallux;* HAL-eks) has two large, heavy phalanges called proximal and distal phalanges. The other four toes each have three phalanges—proximal, middle, and distal. Joints between phalanges of the foot, like those of the hand, are called *interphalangeal joints.*

Arches of the Foot

The bones of the foot are arranged in two **arches** that are held in position by ligaments and tendons (Figure 8.17). The arches enable the foot to support the weight of the body, provide an ideal distribution of body weight over the soft and hard tissues of the foot, and provide leverage while walking. The arches are not rigid; they yield as weight is applied and spring back when the weight is lifted, thus storing energy for the next step and helping to absorb shocks. Usually, the arches are fully developed by age 12 or 13.

The **longitudinal arch** has two parts, both of which consist of tarsal and metatarsal bones arranged to form an arch from the anterior to the posterior part of the foot. The *medial part* of the longitudinal arch, which originates at the calcaneus, rises to the talus and descends through the navicular, the three cuneiforms, and the heads of the three medial metatarsals. The *lateral part* of the longitudinal arch also begins at the calcaneus. It rises at the cuboid and descends to the heads of the two lateral metatarsals. The medial portion of the longitudinal arch is so high that the medial portion of the foot between the ball and heel does not touch the ground when you walk on a hard surface.

The **transverse arch** is found between the medial and lateral aspects of the foot and is formed by the navicular, three cuneiforms, and the bases of the five metatarsals.

As noted earlier, one function of the arches is to distribute body weight over the soft and hard tissues of the body. Normally, the ball of the foot carries about 40% of the weight and the heel carries about 60%. The ball of the foot is the padded portion of the sole superficial to the heads of the metatarsals. When a person wears high-heeled shoes, however, the distribution of weight changes so that the ball of the foot may carry up to 80% and the heel 20%. As a result, the fat pads at the ball of the foot are damaged, joint pain develops, and structural changes in bones may occur.

CLINICAL CONNECTION | *Flatfoot and Clawfoot*

The bones composing the arches of the foot are held in position by ligaments and tendons. If these ligaments and tendons are weakened, the height of the medial longitudinal arch may decrease or "fall." The result is **flatfoot,** the causes of which include excessive weight, postural abnormalities, weakened supporting tissues, and genetic predisposition. Fallen arches may lead to inflammation of the fascia of the sole (plantar fasciitis), Achilles tendinitis, shin splints, stress fractures, bunions, and calluses. A custom-designed arch support often is prescribed to treat flatfoot.

Clawfoot is a condition in which the medial longitudinal arch is abnormally elevated. It is often caused by muscle deformities, such as may occur in diabetics whose neurological lesions lead to atrophy of muscles of the foot.   •

✔CHECKPOINT

Which tarsal bone articulates with both the tibia and fibula?
What are the names and functions of the arches of the foot?

Figure 8.17 Arches of the right foot.

Arches help the foot support and distribute the weight of the body and provide leverage during walking.

Lateral view of arches

? **What structural feature of the arches allows them to absorb shocks?**

EXHIBIT 8.I **281**

8.7 DEVELOPMENT OF THE SKELETAL SYSTEM

◉ OBJECTIVE

• Describe the development of the skeletal system.

Most skeletal tissue arises from *mesenchymal cells,* connective tissue cells derived from **mesoderm.** However, much of the skeleton of the skull arises from **ectoderm.** The mesenchymal cells condense and form models of bones in areas where the bones themselves will ultimately form. In some cases, the bones form directly within the mesenchyme (intramembranous ossification; see Figure 6.5). In other cases, the bones form within hyaline cartilage that develops from mesenchyme (endochondral ossification; see Figure 6.6).

The *skull* begins development during the fourth week after fertilization. It develops from mesenchyme around the developing brain and consists of two major portions: **neurocranium** (mesodermal in origin), which forms the bones of the skull, and **viscerocranium** (ectodermal in origin), which forms the bones of the face (Figure 8.18a). The neurocranium is divided into two parts:

1. The **cartilaginous neurocranium** consists of hyaline cartilage developed from mesenchyme at the base of the developing skull. It later undergoes endochondral ossification to form the *bones at the base of the skull.*

2. The **membranous neurocranium** consists of mesenchyme and later undergoes intramembranous ossification to form the *flat bones that make up the roof and sides of the skull.*

During fetal life and infancy the flat bones are separated by membrane-filled spaces called fontanels (see Figure 7.14).

The viscerocranium, like the neurocranium, is divided into two parts:

1. The **cartilaginous viscerocranium** is derived from the cartilage of the first two pharyngeal (branchial) arches (see Figure 29.13). Endochondral ossification of these cartilages forms the *ear bones* and *hyoid bone.*

2. The **membranous viscerocranium** is derived from mesenchyme in the first pharyngeal arch and, following intramembranous ossification, forms the facial bones.

Vertebrae and *ribs* are derived from portions of cube-shaped masses of mesoderm called somites (see Figure 10.17). Mesenchymal cells from these regions surround the notochord (see Figure 10.17) at about 4 weeks after fertilization. The **notochord** is a solid cylinder of mesodermal cells that induces (stimulates) the mesenchymal cells to form the *vertebral bodies, costal (rib) centers,* and *vertebral arch centers.* Between the vertebral bodies, the notochord induces mesenchymal cells to form the *nucleus pulposus* of an intervertebral disc and surrounding mesenchymal cells form the *annulus fibrosus* of an intervertebral disc. As development continues, other parts of a vertebra form and the *vertebral arch* surrounds the spinal cord (failure of the vertebral arch to develop properly results in a condition called spina bifida; see Disorders: Homeostatic Imbalances in Chapter 7). In the thoracic region, processes from the vertebrae develop into the *ribs.* The *sternum* develops from mesoderm in the anterior body wall.

Figure 8.18 Development of the skeletal system. Bones that develop from the cartilaginous neurocranium are indicated in light blue; from the cartilaginous viscerocranium in dark blue; from the membranous neurocranium in dark red; and from the membranous viscerocranium in light red.

 After the limb buds develop, endochondral ossification of the limb bones begins by the end of the eighth embryonic week.

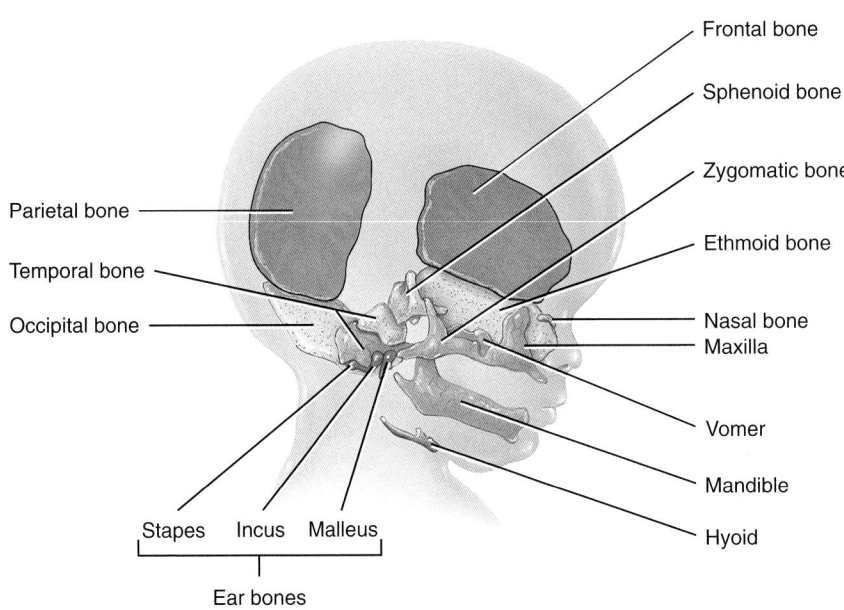

(a) Development of the skull

The *skeleton of the limbs* is derived from mesoderm. During the middle of the fourth week after fertilization, the upper limbs appear as small elevations at the sides of the trunk called **upper limb buds** (Figure 8.18b). About 2 days later, the **lower limb buds** appear. The limb buds consist of **mesenchyme** covered by **ectoderm.** At this point, a mesenchymal skeleton exists in the limbs; some of the masses of mesoderm surrounding the developing bones will become the skeletal muscles of the limbs.

By the sixth week, the limb buds develop a constriction around the middle portion. The constriction produces flattened distal segments of the upper buds called **hand plates** and distal segments of the lower buds called **foot plates** (Figure 8.18c). These plates represent the beginnings of the hands and feet, respectively. At this stage of limb development, a cartilaginous skeleton formed from mesenchyme is present. By the seventh week (Figure 8.18d), the *arm, forearm,* and *hand* are evident in the upper limb bud, and the *thigh, leg,* and *foot* appear in the

lower limb bud. By the eighth week (Figure 8.18e), as the shoulder, elbow, and wrist areas become apparent, the upper limb bud is appropriately called the upper limb, and the free lower limb bud is now the free lower limb.

Endochondral ossification of the limb bones begins by the end of the eighth week after fertilization. By the twelfth week, primary ossification centers are present in most of the limb bones. Most secondary ossification centers appear after birth.

✔CHECKPOINT

10. When and how do the limbs develop?

• • •

To appreciate the skeletal system's contributions to homeostasis of other body systems, examine *Focus on Homeostasis: The Skeletal System.* Next, in Chapter 9, we will see how joints both hold the skeleton together and permit it to participate in movements.

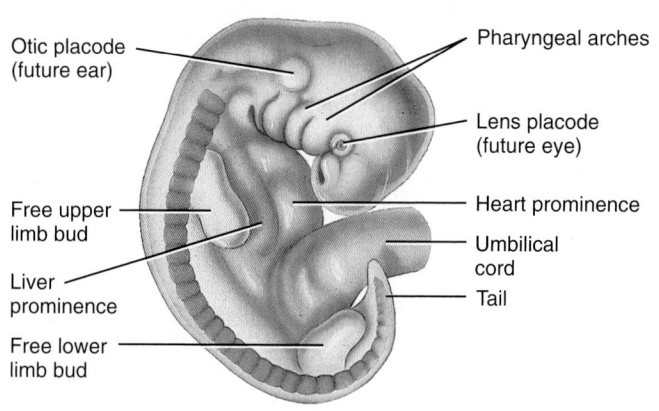

(b) Four-week embryo showing development of free limb buds

(c) Six-week embryo showing development of hand and foot plates

(d) Seven-week embryo showing development of arm, forearm, and hand in free upper limb bud and thigh, leg, and foot in free lower limb bud

(e) Eight-week embryo in which free limb buds have developed into free upper and lower limbs

? **Which of the three basic embryonic tissues—ectoderm, mesoderm, and endoderm—gives rise to most of the skeletal system?**

For all body systems Bones provide support and protection for internal organs; bones store and release calcium, which is needed for proper functioning of most body tissues.

Integumentary system Bones provide strong support for overlying muscles and skin.

Muscular system Bones provide attachment points for muscles and leverage for muscles to bring about body movements; contraction of skeletal muscle requires calcium ions.

Nervous system Skull and vertebrae protect brain and spinal cord; normal blood level of calcium is needed for normal functioning of neurons and neuroglia.

Endocrine system Bones store and release calcium, needed during exocytosis of hormone-filled vesicles and for normal actions of many hormones.

Cardiovascular system Red bone marrow carries out hemopoiesis (blood cell formation); rhythmic beating of the heart requires calcium ions.

Lymphatic system and immunity Red bone marrow produces lymphocytes, white blood cells that are involved in immune responses.

Respiratory system Axial skeleton of thorax protects lungs; rib movements assist in breathing; some muscles used for breathing attach to bones via tendons.

Digestive system Teeth masticate (chew) food; rib cage protects esophagus, stomach, and liver; pelvis protects portions of the intestines.

Urinary system Ribs partially protect kidneys; pelvis protects urinary bladder and urethra.

Reproductive systems Pelvis protects ovaries, uterine (fallopian) tubes, and uterus in females and part of ductus (vas) deferens and accessory glands in males; bones are an important source of calcium needed for milk synthesis during lactation.

THE SKELETAL SYSTEM

DISORDERS: HOMEOSTATIC IMBALANCES

Hip Fracture

Although any region of the hip girdle may fracture, the term **hip fracture** most commonly applies to a break in the bones associated with the hip joint—the head, neck, or trochanteric regions of the femur, or the bones that form the acetabulum. In the United States, 300,000 to 500,000 people sustain hip fractures each year. The incidence of hip fractures is increasing, due in part to longer life spans. Decreases in bone mass due to osteoporosis (which occurs more often in females), along with an increased tendency to fall, predispose elderly people to hip fractures.

Hip fractures often require surgical treatment, the goal of which is to repair and stabilize the fracture, increase mobility, and de-crease pain. Sometimes the repair is accomplished by using surgical pins, screws, nails, and plates to secure the head of the femur. In severe hip fractures, the femoral head or the acetabulum of the hip bone may be replaced by prostheses (artificial devices). The procedure of replacing either the femoral head or the acetabulum is *hemiarthroplasty* (hem-ē-AR-thrō-plas-tē; *hemi-* = one-half; *-arthro-* = joint; *-plasty* = molding). Replacement of both the femoral head and acetabulum is *total hip arthroplasty*. The acetabular prosthesis is made of plastic, and the femoral prosthesis is metal; both are designed to withstand a high degree of stress. The prostheses are attached to healthy portions of bone with acrylic cement and screws (see Figure 9.16).

MEDICAL TERMINOLOGY

Clubfoot or talipes equinovarus (TA-li-pēz ĒK-wīn-ō-VA-rus; *-pes* = foot; *equino-* = horse) An inherited deformity in which the foot is twisted inferiorly and medially, and the angle of the arch is increased; occurs in 1 of every 1000 births. Treatment consists of manipulating the arch to a normal curvature by casts or adhesive tape, usually soon after birth. Corrective shoes or surgery may also be required.

Genu valgum (JĒ-noo VAL-gum; *genu* = knee; *valgum* = bent outward) A deformity in which the knees are abnormally close together and the space between the ankles is increased due to a lateral angulation of the tibia in relation to the femur. Also called **knock-knee.**

Genu varum (JĒ-noo VAR-um; *varum* = bent toward the midline) A deformity in which the knees are abnormally separated and there is a medial angulation of the tibia in relation to the femur and the lower limbs are bowed laterally. Also called **bowleg.**

Hallux valgus (HAL-uks VAL-gus; *hallux* = great toe) Angulation of the great toe away from the midline of the body, typically caused by wearing tightly fitting shoes. When the great toe angles toward the next toe, there is a bony protrusion at the base of the great toe. Also called a **bunion.**

CHAPTER REVIEW AND RESOURCE SUMMARY

Review

Resource

8.1 Pectoral (Shoulder) Girdle

1. Each of the body's two pectoral (shoulder) girdles consists of a clavicle and scapula.
2. Each pectoral girdle attaches an upper limb to the axial skeleton (see Exhibits 8.A and 8.B).

Anatomy Overview - The Appendicular Skeleton: The Pectoral Girdle
Figure 8.1 - Right Pectoral (Shoulder) Girdle

8.2 Upper Limb (Extremity)

1. Each of the two upper limbs (extremities) contains 30 bones.
2. The bones of each upper limb include the humerus, ulna, radius, carpals, metacarpals, and phalanges (see Exhibits 8.C through 8.E).

Anatomy Overview - The Upper Limb
Figure 8.5 - Right Humerus

8.3 Pelvic (Hip) Girdle

1. The pelvic (hip) girdle consists of two hip bones.
2. Each hip bone consists of three parts: the ilium, pubis, and ischium (see Exhibit 8.F).
3. The hip bones, sacrum, and pubic symphysis form the bony pelvis. It supports the vertebral column and pelvic viscera and attaches the free lower limbs to the axial skeleton.

Anatomy Overview - The Pelvic Girdle

8.4 False and True Pelves

1. The false pelvis is separated from the true pelvis by the pelvic brim.
2. The true pelvis surrounds the pelvic cavity and houses the rectum and urinary bladder in both genders, the vagina and cervix of the uterus in females, and the prostate in males.
3. The false pelvis is the lower portion of the abdomen that is situated superior to the pelvic brim. It contains the superior portion of the urinary bladder (when full) and the lower intestines in both genders and the uterus, uterine tubes, and ovaries in the female.

Anatomy Overview - The Pelvic Girdle

Review

8.5 Comparison of Female and Male Pelves

Anatomy Overview - The Pelvic Girdle

1. Bones of the male skeleton are generally larger and heavier than bones of the female skeleton. They also have more prominent markings for muscle attachments.
2. The female pelvis is adapted for pregnancy and childbirth. Sex-related differences in pelvic structure are listed and illustrated in Table 8.1.

8.6 Lower Limb (Extremity)

Anatomy Overview - The Lower Limb
Figure 8.13 - Right Femur

1. Each of the two lower limbs (extremities) contains 30 bones.
2. The bones of each lower limb include the femur, the patella, the tibia, the fibula, the tarsals, the metatarsals, and the phalanges (see Exhibits 8.G through 8.I).
3. The bones of the foot are arranged in two arches, the longitudinal arch and the transverse arch, to provide support and leverage.

8.7 Development of the Skeletal System

1. Most bones form from mesoderm by intramembranous or endochondral ossification; much of the skeleton of the skull arises from ectoderm.
2. Bones of the limbs develop from limb buds, which consist of mesoderm and ectoderm.

SELF-QUIZ QUESTIONS

Fill in the blanks in the following statements.

1. The bones that make up the palm are the _____.
2. List the three bones that fuse to form a hip (coxal) bone: _____, _____, _____.
3. The portion of the bony pelvis that is inferior to the pelvic brim is the _____ pelvis; the portion that is superior to the pelvic brim is the _____ pelvis.

Indicate whether the following statements are true or false.

4. The largest carpal bone is the lunate.
5. The anterior joint formed by the two coxal (hip) bones is the pubic symphysis.

Choose the one best answer to the following questions.

6. Which of the following statements are *true?* (1) The pectoral girdle consists of the scapula, the clavicle, and the sternum. (2) Although the joints of the pectoral girdle are not very stable, they allow free movement in many directions. (3) The anterior component of the pectoral girdle is the scapula. (4) The pectoral girdle articulates directly with the vertebral column. (5) The posterior component of the pectoral girdle is the sternum.
(a) 1, 2, and 3 (b) 2 only (c) 4 only
(d) 2, 3, and 5 (e) 3, 4, and 5

7. Which of the following are *true* concerning the elbow joint? (1) When the forearm is extended, the olecranon fossa receives the olecranon. (2) When the forearm is flexed, the radial fossa receives the coronoid process. (3) The head of the radius articulates with the capitulum. (4) The trochlea articulates with the trochlear notch. (5) The head of the ulna articulates with the ulnar notch of the radius.
(a) 1, 2, 3, 4, and 5 (b) 1, 3, and 4 (c) 1, 3, 4, and 5
(d) 1, 2, 3, and 4 (e) 2, 3, and 4

8. Which of the following is the most superior of the tarsals and articulates with the distal end of the tibia?
(a) calcaneus (b) navicular (c) cuboid
(d) cuneiform (e) talus

9. Which is (are) *not true* concerning the scapula? (1) The lateral border is also known as the axillary border. (2) The scapular notch accommodates the head of the humerus. (3) The scapula is also known as the collarbone. (4) The acromion process articulates with the clavicle. (5) The coracoid process is utilized for muscle attachment.
(a) 1, 2, and 3 (b) 3 only (c) 2 and 3
(d) 3 and 4 (e) 2, 3, and 5

10. Which of the following is *false?*
(a) A decrease in the height of the medial longitudinal arch creates a condition known as clawfoot.
(b) The transverse arch is formed by the navicular, cuneiforms, and bases of the five metatarsals.
(c) The longitudinal arch has medial and lateral parts, both of which originate at the calcaneus.
(d) Arches help to absorb shocks.
(e) Arches enable the foot to support the body's weight.

11. Which of the following is (are) involved in the knee joint? Choose the one best answer. (1) fibular notch of the tibia (2) lateral condyle of tibia (3) head of fibula (4) greater trochanter of femur (5) medial condyle of femur.
(a) 1 and 2 (b) 2 and 5
(c) 1, 2, 3, and 5 (d) 3 and 4
(e) 4 only

12. The greater sciatic notch is located on the
(a) ilium (b) ischium (c) femur
(d) pubis (e) sacrum

13. Match the following:

_____ (a) a large, triangular, flat bone found in the posterior part of the thorax

_____ (b) an S-shaped bone lying horizontally in the superior and anterior part of the thorax

_____ (c) articulates proximally with the scapula and distally with the radius and ulna

_____ (d) located on the medial aspect of the forearm

_____ (e) located on the lateral aspect of the forearm

_____ (f) the longest, heaviest, and strongest bone of the body

_____ (g) the larger, medial bone of the leg

_____ (h) the smaller, lateral bone of the leg

_____ (i) heel bone

_____ (j) sesamoid bone that articulates with the femur and tibia

(1) calcaneus
(2) scapula
(3) patella
(4) radius
(5) femur
(6) clavicle
(7) ulna
(8) tibia
(9) humerus
(10) fibula

14. Match the following:

_____ (a) largest and strongest tarsal bone

_____ (b) most medial bone in the distal row of carpals; has a hook-shaped projection on anterior surface

_____ (c) most medial, pea-shaped bone located in the proximal row of carpals

_____ (d) articulate with metatarsals I–III and cuboid

_____ (e) located in the proximal row of carpals; its name means "moon-shaped"

_____ (f) most lateral bone in the distal row of carpals

_____ (g) largest carpal bone

_____ (h) generally classified as proximal, middle, and distal

_____ (i) most lateral bone in the proximal row of carpals

_____ (j) articulates with the tibia and fibula

_____ (k) located in the proximal row of carpals; its name indicates that it is "three-cornered"

_____ (l) lateral bone that articulates with the calcaneus and metatarsals; IV–V

_____ (m) articulates with metacarpal II

_____ (n) boat-shaped bone that articulates with the talus

(1) cuboid
(2) triquetrum
(3) calcaneus
(4) pisiform
(5) capitate
(6) phalanges
(7) trapezoid
(8) hamate
(9) lunate
(10) scaphoid
(11) cuneiforms
(12) navicular
(13) trapezium
(14) talus

15. Match the following (some answers will be used more than once):

_____ (a) olecranon
_____ (b) olecranon fossa
_____ (c) trochlea
_____ (d) greater trochanter
_____ (e) medial malleolus
_____ (f) acromial end
_____ (g) capitulum
_____ (h) acromion
_____ (i) radial tuberosity
_____ (j) acetabulum
_____ (k) lateral malleolus
_____ (l) glenoid cavity
_____ (m) coronoid process
_____ (n) linea aspera
_____ (o) anterior border
_____ (p) anterior superior iliac spine
_____ (q) fovea capitis
_____ (r) greater tubercle
_____ (s) trochlear notch
_____ (t) obturator foramen
_____ (u) styloid process

(1) clavicle
(2) scapula
(3) humerus
(4) ulna
(5) radius
(6) femur
(7) tibia
(8) fibula
(9) hip bone

CRITICAL THINKING QUESTIONS

1. Mr. Smith's dog Rover dug up a complete set of human bones in the woods near his house. After examining the scene, the local police collected the bones and transported them to the coroner's office for identification. Later, Mr. Smith read in the newspaper that the bones belonged to an elderly female. How was this determined?

2. A proud dad holds his 5-month-old baby girl upright on her feet while supporting her under her arms. He states that she can never be a dancer because her feet are too flat. Is this true? Why or why not?

3. The local newspaper reported that Farmer White caught his hand in a piece of machinery last Tuesday. He lost the lateral two fingers of his left hand. His daughter, who is taking high school science, reports that Farmer White has three remaining phalanges. Is she correct, or does she need a refresher course in anatomy? Support your answer.

❓ ANSWERS TO FIGURE QUESTIONS

8.1 The pectoral girdles attach the upper limbs to the axial skeleton.

8.2 The weakest part of the clavicle is its midregion at the junction of the two curves.

8.3 The acromion of the scapula forms the high point of the shoulder.

8.4 Each upper limb has 30 bones.

8.5 The radius articulates at the elbow with the capitulum and radial fossa of the humerus. The ulna articulates at the elbow with the trochlea, coronoid fossa, and olecranon fossa of the humerus.

8.6 The olecranon is the "elbow" part of the ulna.

8.7 The radius and ulna form the proximal and distal radioulnar joints. Their shafts are also connected by the interosseous membrane.

8.8 The scaphoid is the most frequently fractured wrist bone.

8.9 The bony pelvis attaches the lower limbs to the axial skeleton and supports the backbone and pelvic viscera.

8.10 The femur articulates with the acetabulum of the hip bone; the sacrum articulates with the auricular surface of the hip bone.

8.11 The pelvic axis is the course taken by a baby's head as it descends through the pelvis during childbirth.

8.12 Each lower limb has 30 bones.

8.13 The angle of convergence of the femurs is greater in females than males because the female pelvis is broader.

8.14 The patella is classified as a sesamoid bone because it develops in a tendon (the tendon of the quadriceps femoris muscle of the thigh).

8.15 The tibia is the weight-bearing bone of the leg.

8.16 The talus is the only tarsal bone that articulates with the tibia and the fibula.

8.17 Because the arches are not rigid, they yield when weight is applied and spring back when weight is lifted, allowing them to absorb the shock of walking.

8.18 Most of the skeletal system arises from embryonic mesoderm.

9 JOINTS

JOINTS AND HOMEOSTASIS   The joints of the skeletal system contribute to homeostasis by holding bones together in ways that allow for movement and flexibility.

Bones are too rigid to bend without being damaged. Fortunately, flexible connective tissues form joints that hold bones together while still permitting, in most cases, some degree of movement. A **joint,** also called an **articulation** (ar-tik-ū-LĀ-shun) or **arthrosis** (ar-THRŌ-sis), is a point of contact between two bones, between bone and cartilage, or between bone and teeth. When we say one bone *articulates* with another bone, we mean that the bones form a joint. You can appreciate the importance of joints if you have ever had a cast over your knee joint, which makes walking difficult, or a splint on your finger, which limits your ability to manipulate small objects. The scientific study of joints is termed **arthrology** (ar-THROL-ō-jē; *arthr-* = joint; *-logy* = study of). The study of motion of the human body is called **kinesiology** (ki-nē-sē′-OL-ō-jē; *kinesi* = movement).

? *Did you ever wonder why pitchers so often require rotator cuff surgery?*

9.1 JOINT CLASSIFICATIONS

⦿ **OBJECTIVE**
- Describe the structural and functional classifications of joints.

Joints are classified structurally, based on their anatomical characteristics, and functionally, based on the type of movement they permit.

The structural classification of joints is based on two criteria: (1) the presence or absence of a space between the articulating bones, called a synovial cavity, and (2) the type of connective tissue that binds the bones together. Structurally, joints are classified as one of the following types:

- **Fibrous joints** (FĪ-brus): There is no synovial cavity, and the bones are held together by dense irregular connective tissue that is rich in collagen fibers.
- **Cartilaginous joints** (kar′-ti-LAJ-i-nus): There is no synovial cavity, and the bones are held together by cartilage.
- **Synovial joints** (si-NŌ-vē-al): The bones forming the joint have a synovial cavity and are united by the dense irregular connective tissue of an articular capsule, and often by accessory ligaments.

The functional classification of joints relates to the degree of movement they permit. Functionally, joints are classified as one of the following types:

- **Synarthrosis** (sin′-ar-THRŌ-sis; *syn-* = together): An immovable joint. The plural is *synarthroses.*

- **Amphiarthrosis** (am′-fē-ar-THRŌ-sis; *amphi-* = on both sides): A slightly movable joint. The plural is *amphiarthroses.*
- **Diarthrosis** (dī-ar-THRŌ-sis = movable joint): A freely movable joint. The plural is *diarthroses.* All diarthroses are synovial joints. They have a variety of shapes and permit several different types of movements.

The following sections present the joints of the body according to their structural classifications. As we examine the structure of each type of joint, we will also outline its functions.

✔ **CHECKPOINT**
1. On what basis are joints classified?

9.2 FIBROUS JOINTS

⦿ **OBJECTIVE**
- Describe the structure and functions of the three types of fibrous joints.

As previously noted, **fibrous joints** lack a synovial cavity, and the articulating bones are held very closely together by dense irregular connective tissue. Fibrous joints permit little or no movement. The three types of fibrous joints are sutures, syndesmoses, and interosseous membranes.

Sutures

A **suture** (SOO-chur; *sutur* = seam) is a fibrous joint composed of a thin layer of dense irregular connective tissue; sutures occur only between bones of the skull. An example is the coronal suture

Figure 9.1 Fibrous joints.

 At a fibrous joint the bones are held together by dense irregular connective tissue.

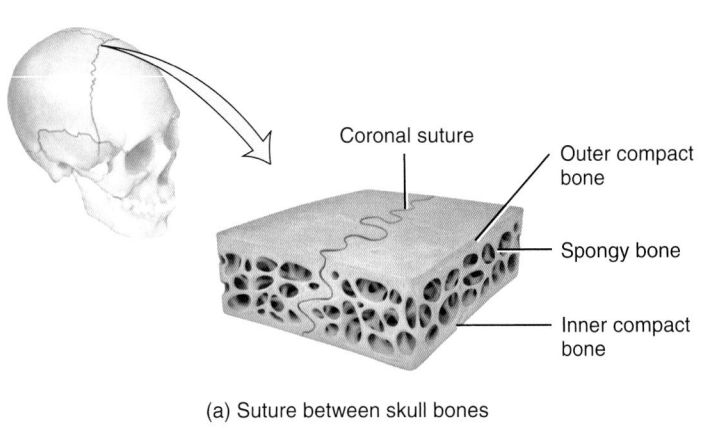

(a) Suture between skull bones

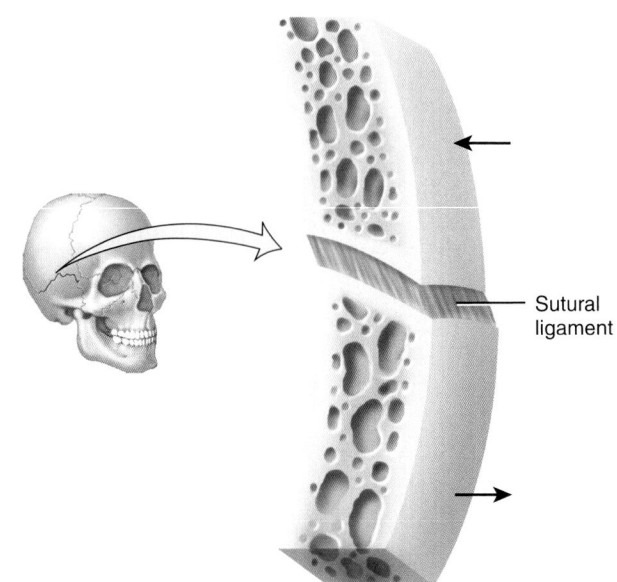

(b) Slight movement at suture

between the parietal and frontal bones (Figure 9.1a). The irregular, interlocking edges of sutures give them added strength and decrease their chance of fracturing. Sutures are joints that form as the numerous bones of the skull come in contact during development. They are immovable or slightly movable. In older individuals, sutures are immovable (synarthroses), but in infants and children they are slightly movable (amphiarthroses) (Figure 9.1b). Sutures play important roles in shock absorption in the skull.

Some sutures, although present during growth of the skull, are replaced by bone in the adult. Such a suture is called a **synostosis** (sin'-os-TŌ-sis; *os-* = bone), or bony joint—a joint in which there is a complete fusion of two separate bones into one. For example, the frontal bone grows in halves that join together across a suture line. Usually they are completely fused by age 6 and the suture becomes obscure. If the suture persists beyond age 6, it is called a **frontal** or **metopic suture** (me-TŌ-pik; *metopon* = forehead). A synostosis is classified as a synarthrosis because it is immovable.

Syndesmoses

A **syndesmosis** (sin'-dez-MŌ-sis; *syndesmo-* = band or ligament) is a fibrous joint in which there is a greater distance between the articulating surfaces and more dense irregular connective tissue than in a suture. The dense irregular connective tissue is typically arranged as a bundle (ligament), allowing the joint to permit limited movement. One example of a syndesmosis is the distal tibiofibular joint, where the anterior tibiofibular ligament connects the tibia and fibula (Figure 9.1c, left). It permits slight movement (amphiarthrosis). Another example of a syndesmosis is called a **gomphosis** (gom-FŌ-sis; *gompbo-* = bolt or nail) or *dentoalveolar joint,* in which a cone-shaped peg fits into a socket. The only examples of gomphoses in the human body are the articulations between the roots of the teeth and their sockets (alveoli) in the maxillae and mandible (Figure 9.1c, right). The dense irregular connective tissue between a tooth and its socket is the thin periodontal ligament (membrane). A healthy gomphosis permits no movement (synarthrosis). Inflammation and degeneration of the gums, periodontal ligament, and bone is called *periodontal disease.*

Interosseous Membranes

The final category of fibrous joint is the **interosseous membrane** (in'-ter-OS-ē-us), which is a substantial sheet of dense irregular connective tissue that binds neighboring long bones and permits slight movement (amphiarthrosis). There are two principal interosseous membrane joints in the human body. One occurs between the radius and ulna in the forearm (see Figure 8.6) and the other occurs between the tibia and fibula in the leg (Figure 9.1d).

✔CHECKPOINT

2. Which fibrous joints are synarthroses? Which are amphiarthroses?

Syndesmosis between tibia and fibula at distal tibiofibular joint

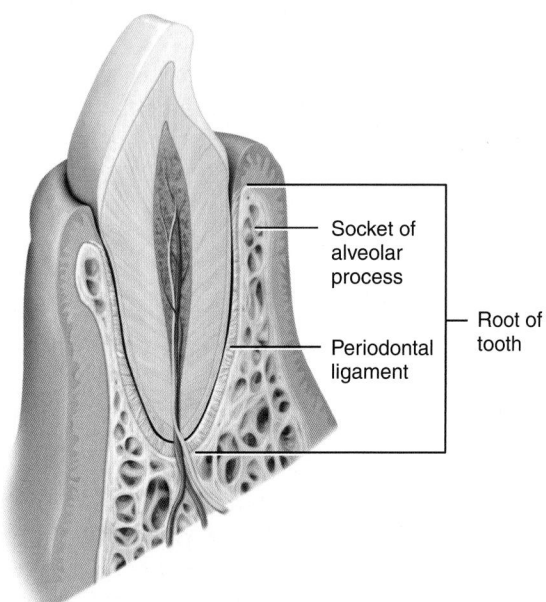

Syndesmosis between tooth and socket of alveolar process (gomphosis)

Socket of alveolar process

Root of tooth

Periodontal ligament

(c) Syndesmosis

Fibula

Interosseous membrane

Tibia

(d) Interosseous membrane between diaphyses of tibia and fibula

 Functionally, why are sutures classified as synarthroses, and syndesmoses as amphiarthroses?

9.3 CARTILAGINOUS JOINTS

◉ **OBJECTIVE**

- Describe the structure and functions of the two types of cartilaginous joints.

Like a fibrous joint, a **cartilaginous joint** (kar′-ti-LAJ-i-nus) lacks a synovial cavity and allows little or no movement. Here the articulating bones are tightly connected by either hyaline cartilage or fibrocartilage (see Table 4.6). The two types of cartilaginous joints are synchondroses and symphyses.

Synchondroses

A **synchondrosis** (sin′-kon-DRŌ-sis; *chondro-* = cartilage) is a cartilaginous joint in which the connecting material is hyaline cartilage. An example of a synchondrosis is the epiphyseal (growth) plate that connects the epiphysis and diaphysis of a growing bone (Figure 9.2a). A photomicrograph of the epiphyseal plate is shown in Figure 6.7b. Functionally, a synchondro-

sis is immovable (*synarthrosis*). When bone elongation ceases, bone replaces the hyaline cartilage, and the synchondrosis becomes a synostosis, a bony joint. Another example of a synchondrosis is the joint between the first rib and the manubrium of the sternum, which also ossifies during adult life and becomes an immovable (synarthrosis) synostosis, or bony joint (see Figure 7.22b).

In an x-ray of a young person's skeleton, the synchondroses are easily seen as thin dark areas between the white-appearing bone tissue (see Figure 6.7a). This is how a medical professional can view an x-ray and determine that a person still has some degree of growth remaining. Breaks in a bone that extend into the epiphyseal plate and damage the cartilage of the synchondrosis can affect further growth of the bone, leading to abbreviated development of the length of the bone that leads to a bone of shortened length.

Symphyses

A **symphysis** (SIM-fi-sis = growing together) is a cartilaginous joint in which the ends of the articulating bones are covered with hyaline cartilage, but a broad, flat disc of fibrocartilage connects the bones. All symphyses (plural) occur in the midline of the body. The pubic symphysis between the anterior surfaces of the hip bones is one example of a symphysis (Figure 9.2b). This type of joint is also found at the junction of the manubrium and body of the sternum (see Figure 7.22) and at the intervertebral joints between the bodies of vertebrae (see Figure 7.20a). A portion of the intervertebral disc is composed of fibrocartilage. A symphysis is a slightly movable joint (amphiarthrosis).

✔ **CHECKPOINT**

3. Which cartilaginous joints are synarthroses? Which are amphiarthroses?

Figure 9.2 Cartilaginous joints.

🔑 ⚎ At a cartilaginous joint the bones are held together by cartilage.

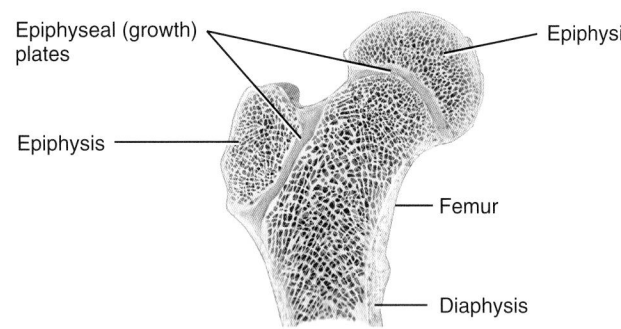

Epiphyseal (growth) plates

Epiphysis

Epiphysis

Femur

Diaphysis

(a) Synchondrosis

Hip bones

Pubic symphysis

(b) Symphysis

❓ **What is the structural difference between a synchondrosis and a symphysis?**

9.4 SYNOVIAL JOINTS

◉ **OBJECTIVES**

- Describe the structure of synovial joints.
- Describe the structure and function of bursae and tendon sheaths.

Structure of Synovial Joints

Synovial joints (si-NŌ-vē-al) have certain characteristics that distinguish them from other joints. The unique characteristic of a synovial joint is the presence of a space called a **synovial (joint) cavity** between the articulating bones (Figure 9.3). Because the synovial cavity allows considerable movement at a joint, all synovial joints are classified functionally as freely movable (diarthroses). The bones at a synovial joint are covered by a layer of hyaline cartilage called **articular cartilage.** The cartilage covers the articulating surface of the bones with a smooth, slippery

Figure 9.3 **Structure of a typical synovial joint.** Note the two layers of the articular capsule—the fibrous membrane and the synovial membrane. Synovial fluid lubricates the synovial cavity, which is located between the synovial membrane and the articular cartilage.

🔑 The distinguishing feature of a synovial joint is the synovial cavity between the articulating bones.

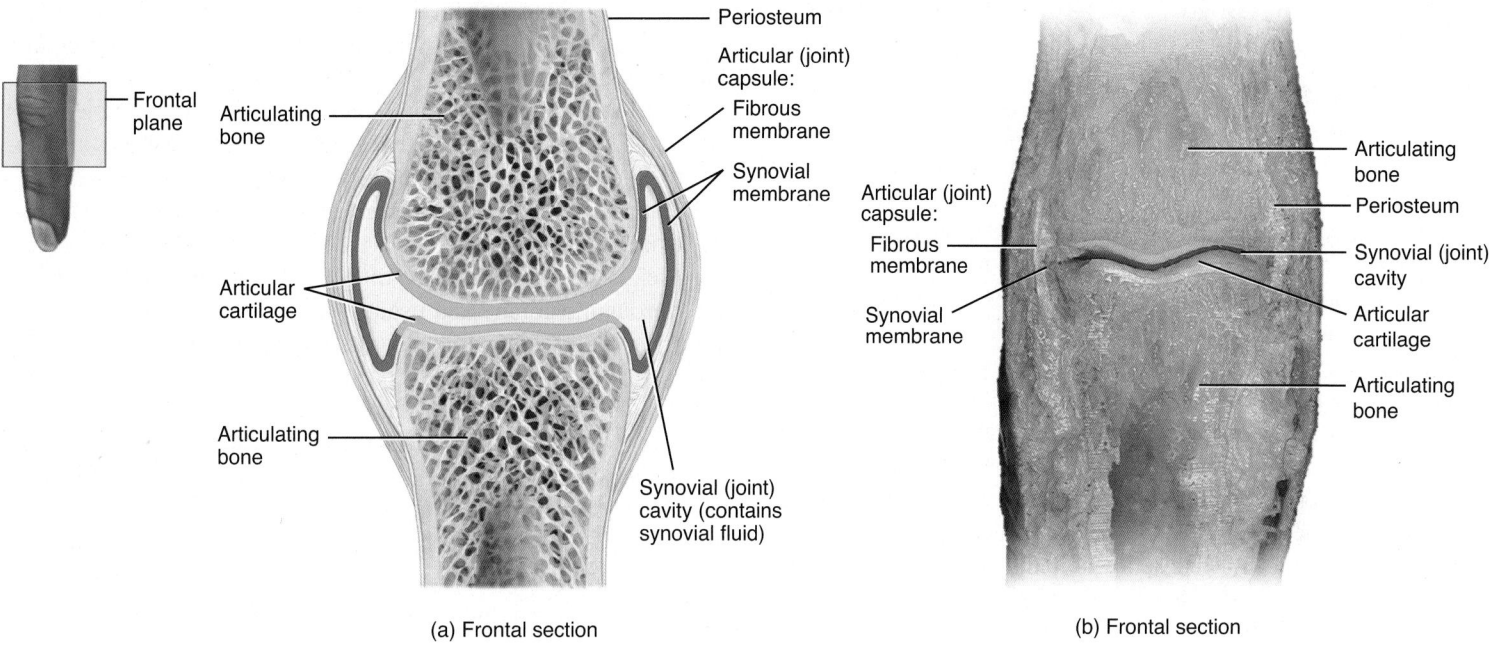

(a) Frontal section

(b) Frontal section

❓ **What is the functional classification of synovial joints?**

surface but does not bind them together. Articular cartilage reduces friction between bones in the joint during movement and helps to absorb shock.

CLINICAL CONNECTION | *Autologous Chondrocyte Implantation*

When there is damage to articular cartilage in the knee joint, especially involving the femur, there is an alternative to partial or total knee replacement (see Section 9.10) called **autologous chondrocyte implantation** (aw-TOL-o-gus), or **ACI**. Candidates for ACI have cartilage damage due to acute or repetitive trauma, not arthritis. In the procedure, healthy chondrocytes (cartilage cells) are taken from an area of the femoral condyle that is not weight-bearing and sent to a laboratory, where they are cultured for 4 to 5 weeks to generate between 5 million and 10 million cells. When the cultured cells are available, the implantation takes place. The damaged area is prepared by removing dead cartilage from the defect, which is covered by a piece of periosteum, usually taken from the tibia. Then the cultured chondrocytes are injected under the periosteum, where they will grow and mature over time. The patient can put the full weight of the body on the knee in about 10 to 12 weeks. •

Articular Capsule

A sleevelike **articular (joint) capsule** surrounds a synovial joint, encloses the synovial cavity, and unites the articulating bones. The articular capsule is composed of two layers, an outer fibrous membrane and an inner synovial membrane (Figure 9.3a). The **fibrous membrane** usually consists of dense irregular connective tissue (mostly collagen fibers) that attaches to the periosteum of the articulating bones. In fact, the fibrous membrane is literally a thickened continuation of the periosteum between the bones. The flexibility of the fibrous membrane permits considerable movement at a joint, while its great tensile strength (resistance to stretching) helps prevent the bones from dislocating. The fibers of some fibrous membranes are arranged as parallel bundles of dense regular connective tissue that are highly adapted for resisting strains. The strength of these fiber bundles, called **ligaments** (*liga-* = bound or tied), is one of the principal mechanical factors that hold bones close together in a synovial joint. Ligaments are often designated by individual names. The inner layer of the articular capsule, the **synovial membrane,** is composed of areolar connective tissue with elastic fibers. At many synovial joints the synovial membrane includes accumulations of adipose tissue, called **articular fat pads.** An example is the infrapatellar fat pad in the knee (see Figure 9.15c).

A **"double-jointed"** person does not really have extra joints. Individuals who are double-jointed have greater flexibility in their articular capsules and ligaments; the resulting increase in range of motion allows them to entertain fellow partygoers with activities such as touching their thumbs to their wrists and putting their ankles or elbows behind their necks. Unfortunately, such flexible joints are less structurally stable and are more easily dislocated.

Synovial Fluid

The synovial membrane secretes **synovial fluid** (*ov-* = egg), a viscous, clear or pale yellow fluid named for its similarity in appearance and consistency to uncooked egg white. Synovial fluid consists of hyaluronic acid secreted by fibroblastlike cells in the synovial membrane and interstitial fluid filtered from blood plasma. It forms a thin film over the surfaces within the articular capsule. Its functions include reducing friction by lubricating the joint, absorbing shocks, and supplying oxygen and nutrients to and removing carbon dioxide and metabolic wastes from the chondrocytes within articular cartilage. (Recall that cartilage is an avascular tissue, so it does not have blood vessels to perform the latter function.) Synovial fluid also contains phagocytic cells that remove microbes and the debris that results from normal wear and tear in the joint. When a synovial joint is immobile for a time, the fluid becomes quite viscous (gel-like), but as joint movement increases, the fluid becomes less viscous. One of the benefits of warming up before exercise is that it stimulates the production and secretion of synovial fluid; more fluid means less stress on the joints during exercise.

We are all familiar with the cracking sounds heard as certain joints move, or the popping sounds that arise when a person pulls on his fingers to crack his knuckles. According to one theory, when the synovial cavity expands, the pressure inside the synovial cavity decreases, creating a partial vacuum. The suction draws carbon dioxide and oxygen out of blood vessels in the synovial membrane, forming bubbles in the fluid. When the fingers are flexed (bent) the volume of the cavity decreases and the pressure increases; this bursts the bubbles and creates cracking or popping sounds as the gases are driven back into solution.

As a result of various injuries or diseases, there may be an excessive buildup of synovial fluid in a joint cavity, resulting in pain and decreased mobility. In order to relieve the pressure and ease the pain, **aspiration of synovial fluid** (as'-pi-RĀ-shun) may be necessary. In this procedure, a needle is inserted into the joint cavity and the fluid is withdrawn into a syringe; the fluid is then analyzed for diagnostic purposes. For example, the fluid may contain bacteria, which confirms a diagnosis of infection, or it may contain urea crystals, which confirms a diagnosis of gout. It is also possible to inject medication into a joint cavity; anti-inflammatory drugs such as cortisol are also used to decrease joint swelling caused by retention of fluid from the blood that entered the joint. •

Accessory Ligaments, Articular Discs, and Labra

Many synovial joints also contain **accessory ligaments** called extracapsular ligaments and intracapsular ligaments (see Figure 9.15d). *Extracapsular ligaments* lie outside the articular capsule. Examples are the fibular and tibial collateral ligaments of the knee joint. *Intracapsular ligaments* occur within the articular capsule but are excluded from the synovial cavity by folds of the synovial membrane. Examples are the anterior and posterior cruciate ligaments of the knee joint.

Inside some synovial joints, such as the knee, crescent-shaped pads of fibrocartilage lie between the articular surfaces of the bones and are attached to the fibrous capsule. These pads are called **articular discs** or **menisci** (me-NIS-sī or me-NIS-kī; singular is *meniscus*). Figure 9.15d depicts the lateral and medial menisci in the knee joint. The discs bind strongly to the inside of the fibrous membrane and usually subdivide the synovial cavity into two spaces, allowing separate movements to occur in each space. As you will see later, separate movements also occur in the respective compartments of the temporomandibular joint (TMJ) (see Exhibit 9.A). The functions of the menisci are not completely understood but are known to include the following: (1) shock absorption; (2) a better fit between articulating bony surfaces; (3) providing adaptable surfaces for combined movements; (4) weight distribution over a greater contact surface; and (5) distribution of synovial lubricant across the articular surfaces of the joint.

A **labrum** (LĀ-brum; the plural is *labra*), prominent in the ball-and-socket joints of the shoulder and hip (see Figures 9.12c, d; 9.14c), is the fibrocartilaginous lip that extends from the edge of the joint socket. The labrum helps deepen the joint socket and increases the area of contact between the socket over the ball-like surface of the head of the humerus or femur.

The tearing of menisci in the knee, commonly called **torn cartilage,** occurs often among athletes. Such damaged cartilage will begin to wear and may cause arthritis to develop unless the damaged cartilage is treated surgically. Years ago, if a patient had torn cartilage, the entire meniscus was removed by a procedure called a *meniscectomy* (men'-i-SEK-tō-mē). The problem was that over time the articular cartilage was worn away more quickly. Currently, surgeons perform a partial meniscectomy, in which only the torn segment of the meniscus is removed. Surgical repair of the torn cartilage may be assisted by **arthroscopy** (ar-THROS-kō-pē; *-scopy* = observation). This minimally invasive procedure involves examination of the interior of a joint, usually the knee, with an arthroscope, a lighted, pencil-thin fiber optic camera used for visualizing the nature and extent of damage. Arthroscopy is also used to monitor the progression of disease and the effects of therapy. The insertion of surgical instruments through other incisions also enables a physician to remove torn cartilage and repair damaged cruciate ligaments in the knee; obtain tissue samples for analysis; and perform surgery on other joints, such as the shoulder, elbow, ankle, and wrist. •

Nerve and Blood Supply

The nerves that supply a joint are the same as those that supply the skeletal muscles that move the joint. Synovial joints contain many nerve endings that are distributed to the articular capsule and associated ligaments. Some of the nerve endings convey information about pain from the joint to the spinal cord and brain for processing. Other nerve endings respond to the degree of movement and stretch at a joint, such as when a physician strikes the tendon below your kneecap to test your reflexes. The spinal cord and brain respond by sending impulses through different nerves to the muscles to adjust body movements.

Although many of the components of synovial joints are avascular, arteries in the vicinity send out numerous branches that penetrate the ligaments and articular capsule to deliver oxygen and nutrients. Veins remove carbon dioxide and wastes from the joints. The arterial branches from several different arteries typically merge around a joint before penetrating the articular capsule. The chondrocytes in the articular cartilage of a synovial joint receive oxygen and nutrients from synovial fluid derived from blood; all other joint tissues are supplied directly by capillaries. Carbon dioxide and wastes pass from chondrocytes of articular cartilage into synovial fluid and then into veins; carbon dioxide and wastes from all other joint structures pass directly into veins.

CLINICAL CONNECTION | Sprain and Strain

A **sprain** is the forcible wrenching or twisting of a joint that stretches or tears its ligaments but does not dislocate the bones. It occurs when the ligaments are stressed beyond their normal capacity. Severe sprains may be so painful that the joint cannot be moved. There is considerable swelling, which results from chemicals released by the damaged cells and hemorrhage of ruptured blood vessels. The lateral ankle joint is most often sprained; the wrist is another area that is frequently sprained. A **strain** is a stretched or partially torn muscle or muscle and tendon. It often occurs when a muscle contracts suddenly and powerfully—such as the leg muscles of sprinters when they spring from the blocks.

Initially sprains should be treated with **PRICE:** protection, rest, ice, compression, and elevation. PRICE therapy may be used on muscle strains, joint inflammation, suspected fractures, and bruises. The five components of PRICE therapy are

- **Protection** means protecting the injury from further damage; for example, stop the activity and use padding and protection, and use splints or a sling, or crutches, if necessary.

- **Rest** the injured area to avoid further damage to the tissues. Stop the activity immediately. Avoid exercise or other activities that cause pain or swelling to the injured area. Rest is needed for repair. Exercising before an injury has healed may increase the probability of re-injury.

- **Ice** the injured area as soon as possible. Applying ice slows blood flow to the area, reduces swelling, and relieves pain. Ice works effectively when applied for 20 minutes, off for 40 minutes, back on for 20 minutes, and so on.

- **Compression** by wrap or bandage helps to reduce swelling. Care must be taken to compress the injured area but not to block blood flow.

- **Elevation** of the injured area above the level of the heart, when possible, will reduce potential swelling. •

Bursae and Tendon Sheaths

The various movements of the body create friction between moving parts. Saclike structures called **bursae** (BER-sē = purses; singular is *bursa*) are strategically situated to alleviate friction in some joints, such as the shoulder and knee joints (see Figures 9.12 and 9.15c). Bursae are not strictly part of synovial joints, but they do resemble joint capsules because their walls consist of an outer fibrous membrane of thin, dense connective tissue lined by a synovial membrane. They are filled with a small amount of fluid that is similar to synovial fluid. Bursae can be located between the skin and bones, tendons and bones, muscles and bones, or ligaments and bones. The fluid-filled bursal sacs cushion the movement of these body parts against one another.

CLINICAL CONNECTION | Bursitis

An acute or chronic inflammation of a bursa, called **bursitis** (bur-SĪ-tis), is usually caused by irritation from repeated, excessive exertion of a joint. The condition may also be caused by trauma, by an acute or chronic infection (including syphilis and tuberculosis), or by rheumatoid arthritis (described in Disorders: Homeostatic Imbalances at the end of this chapter). Symptoms include pain, swelling, tenderness, and limited movement. Treatment may include oral anti-inflammatory agents and injections of cortisol-like steroids. •

Structures called tendon sheaths also reduce friction at joints. **Tendon (synovial) sheaths** are tubelike bursae that wrap around certain tendons that experience considerable friction as they pass through tunnels formed by connective tissue and bone. The inner layer of a tendon sheath, the *visceral layer*, is attached to the surface of the tendon. The outer layer, known as the *parietal layer*, is attached to bone (see Figure 11.18a). Between the layers is a cavity that contains a film of synovial fluid. A tendon sheath protects all sides of a tendon from friction as the tendon slides back and forth. Tendon sheaths are found where tendons pass through synovial cavities, such as the tendon of the biceps brachii muscle at the shoulder joint (see Figure 9.12c). Tendon sheaths are also found at the wrist and ankle, where many tendons come together in a confined space (see Figure 11.18a), and in the fingers and toes, where there is a great deal of movement (see Figure 11.18).

CLINICAL CONNECTION | Tenosynovitis

Tenosynovitis (ten'-o-sīn-ō-VĪ-tis) is an inflammation of the tendons, tendon sheaths, and synovial membranes surrounding certain joints. The tendons most often affected are at the wrists, shoulders, elbows (resulting in *tennis elbow*), finger joints (resulting in *trigger finger*), ankles, and feet. The affected sheaths sometimes become visibly swollen because of fluid accumulation. Tenderness and pain are frequently associated with movement of the body part. The condition often follows trauma, strain, or excessive exercise. Tenosynovitis of the dorsum of the foot may be caused by tying shoelaces too tightly. Gymnasts are prone to developing the condition as a result of chronic, repetitive, and maximum hyperextension at the wrists. Other repetitive movements involving activities such as typing, haircutting, carpentry, and assembly line work can also result in tenosynovitis. •

✔**CHECKPOINT**

4. How does the structure of synovial joints classify them as diarthroses?
5. What are the functions of articular cartilage, synovial fluid, and articular discs?
6. What types of sensations are perceived at joints, and from what sources do joints receive nourishment?
7. In what ways are bursae similar to joint capsules? How do they differ?

9.5 TYPES OF MOVEMENTS AT SYNOVIAL JOINTS

◉ **OBJECTIVE**

• Describe the types of movements that can occur at synovial joints.

Anatomists, physical therapists, and kinesiologists (professionals who study the science of human movement and look for ways to improve the efficiency and performance of the human body at work, in sports, and in daily activities) use specific terminology to designate the movements that can occur at synovial joints. These precise terms may indicate the form of motion, the direction of movement, or the relationship of one body part to another during movement. Movements at synovial joints are grouped into four main categories: (1) gliding, (2) angular movements, (3) rotation, and (4) special movements, which occur only at certain joints.

Gliding

Gliding is a simple movement in which relatively flat bone surfaces move back-and-forth and from side-to-side with respect to one another (Figure 9.4). There is no significant alteration of the angle between the bones. Gliding movements are limited in range due to the structure of the articular capsule and associated ligaments and bones; however, these sliding movements can also be combined with rotation. The intercarpal and intertarsal joints are examples of articulations where gliding movements occur.

Angular Movements

In **angular movements,** there is an increase or a decrease in the angle between articulating bones. The major angular movements are flexion, extension, lateral flexion, hyperextension, abduction, ad-

duction, and circumduction. These movements are discussed with respect to the body in the anatomical position (see Figure 1.5).

Flexion, Extension, Lateral Flexion, and Hyperextension

Flexion and extension are opposite movements. In **flexion** (FLEK-shun; *flex-* = to bend) there is a decrease in the angle between articulating bones; in **extension** (eks-TEN-shun; *exten-* = to stretch out) there is an increase in the angle between articulating bones, often to restore a part of the body to the anatomical position after it has been flexed (Figure 9.5). Both movements usually occur along the sagittal plane. All of the following are examples of flexion (as you have probably already guessed, extension is simply the reverse of these movements):

• Bending the head toward the chest at the atlanto-occipital joint between the atlas (the first vertebra) and the occipital bone of the skull, and at the cervical intervertebral joints between the cervical vertebrae (Figure 9.5a)

• Bending the trunk forward at the intervertebral joints

• Moving the humerus forward at the shoulder joint, as in swinging the arms forward while walking (Figure 9.5b)

• Moving the forearm toward the arm at the elbow joint between the humerus, ulna, and radius (Figure 9.5c)

• Moving the palm toward the forearm at the wrist or radiocarpal joint between the radius and carpals, as in the upward movement when doing wrist curls (Figure 9.5d)

• Bending the digits of the hand or feet at the interphalangeal joints between phalanges

• Moving the femur forward at the hip joint between the femur and hip bone, as in walking (Figure 9.5e)

Figure 9.4 Gliding movements at synovial joints.

🔒 ⟳ Gliding movements consist of side-to-side and back-and-forth motions.

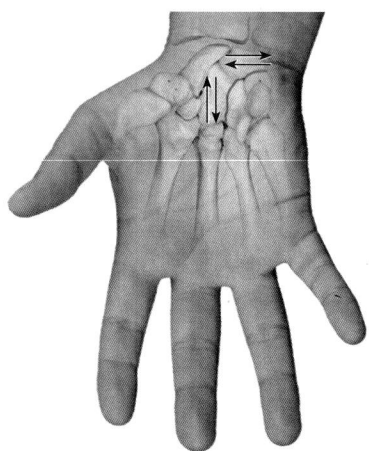

Gliding between intercarpals (arrows)

❓ **What are two examples of joints that permit gliding movements?**

Figure 9.5 Angular movements at synovial joints—flexion, extension, hyperextension, and lateral flexion.

In angular movements, there is an increase or decrease in the angle between articulating bones.

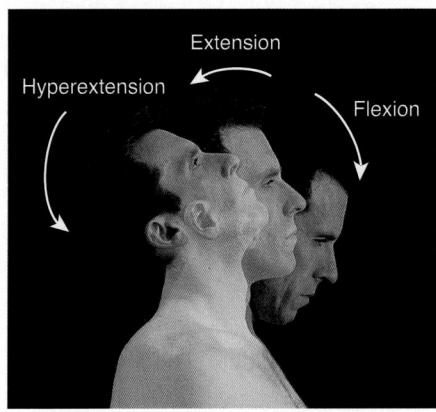

(a) Atlanto-occipital and cervical intervertebral joints

(b) Shoulder joint

(c) Elbow joint

(d) Wrist joint

(e) Hip joint

(f) Knee joint

(g) Intervertebral joints

What are two examples of flexion that do not occur along the sagittal plane?

- Moving the heel toward the buttock at the tibiofemoral joint between the tibia, femur, and patella, as occurs when bending the knee (Figure 9.5f)

Although flexion and extension usually occur along the sagittal plane, there are a few exceptions. For example, flexion of the thumb involves movement of the thumb medially across the palm at the carpometacarpal joint between the trapezium and metacarpal of the thumb, as when you touch your thumb to the opposite side of your palm (see Figure 11.18g). Another example is movement of the trunk sideways to the right or left at the waist. This movement, which occurs along the frontal plane and involves the intervertebral joints, is called **lateral flexion** (Figure 9.5g).

Continuation of extension beyond the anatomical position is called **hyperextension** (hī-per-ek-STEN-shun; *hyper-* = beyond or excessive). Examples of hyperextension include:

- Bending the head backward at the atlanto-occipital and cervical intervertebral joints (Figure 9.5a)
- Bending the trunk backward at the intervertebral joints
- Moving the humerus backward at the shoulder joint, as in swinging the arms backward while walking (Figure 9.5b)

- Moving the palm backward at the wrist joint (Figure 9.5d)
- Moving the femur backward at the hip joint, as in walking (Figure 9.5e)

Hyperextension of hinge joints, such as the elbow, interphalangeal, and knee joints, is usually prevented by the arrangement of ligaments and the anatomical alignment of the bones.

Abduction, Adduction, and Circumduction

Abduction (ab-DUK-shun; *ab-* = away; *-duct* = to lead) is the movement of a bone away from the midline; **adduction** (ad-DUK-shun; *ad-* = toward) is the movement of a bone toward the midline. Both movements usually occur along the frontal plane. Examples of abduction include moving the humerus laterally at the shoulder joint, moving the palm laterally at the wrist joint, and moving the femur laterally at the hip joint (Figure 9.6a–c). The movement that returns each of these body parts to the anatomical position is adduction (Figure 9.6a–c).

The midline of the body is *not* used as a point of reference for abduction and adduction of the digits. In abduction of the fingers (but not the thumb), an imaginary line is drawn through the longitudinal axis of the middle (longest) finger, and the fingers move

Figure 9.6 Angular movements at synovial joints—abduction and adduction.

🔓 Abduction and adduction usually occur along the frontal plane.

(a) Shoulder joint

(b) Wrist joint

(c) Hip joint

(d) Metacarpophalangeal joints of the fingers (not the thumb)

❓ **In what way is considering adduction as "adding your limb to your trunk" an effective learning device?**

Figure 9.7 Angular movements at synovial joints—circumduction.

🔑 ▭ Circumduction is the movement of the distal end of a body part in a circle.

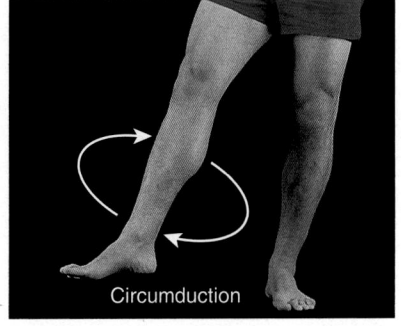

(a) Shoulder joint

(b) Hip joint

❓ **Which movements in continuous sequence produce circumduction?**

away (spread out) from the middle finger (Figure 9.6d). In abduction of the thumb, the thumb moves away from the palm in the sagittal plane (see Figure 11.18g). Abduction of the toes is relative to an imaginary line drawn through the second toe. Adduction of the fingers and toes returns them to the anatomical position. Adduction of the thumb moves the thumb toward the palm in the sagittal plane (see Figure 11.18g).

Circumduction (ser-kum-DUK-shun; *circ-* = circle) is movement of the distal end of a body part in a circle (Figure 9.7). Circumduction is not an isolated movement by itself but rather a continuous sequence of flexion, abduction, extension, adduction, and rotation of the joint (or in the opposite order). It does not occur along a separate axis or plane of movement. Examples of circumduction are moving the humerus in a circle at the shoulder joint (Figure 9.7a), moving the hand in a circle at the wrist joint, moving the thumb in a circle at the carpometacarpal joint, moving the fingers in a circle at the metacarpophalangeal joints (between the metacarpals and phalanges), and moving the femur in a circle at

the hip joint (Figure 9.7b). Both the shoulder and hip joints permit circumduction. Flexion, abduction, extension, and adduction are more limited in the hip joints than in the shoulder joints due to the tension on certain ligaments and muscles and the depth of the acetabulum in the hip joint (see Exhibits 9.B and 9.D).

Rotation

In **rotation** (rō-TĀ-shun; *rota-* = revolve), a bone revolves around its own longitudinal axis. One example is turning the head from side to side at the atlanto-axial joint (between the atlas and axis), as when you shake your head "no" (Figure 9.8a). Another is turning the trunk from side to side at the intervertebral joints while keeping the hips and lower limbs in the anatomical position. In the limbs, rotation is defined relative to the midline, and specific qualifying terms are used. If the anterior surface of a bone of the limb is turned toward the midline, the movement is called *medial (internal) rotation*. You can medially rotate the humerus at the

Figure 9.8 Rotation at synovial joints.

🔑 ▭ In rotation, a bone revolves around its own longitudinal axis.

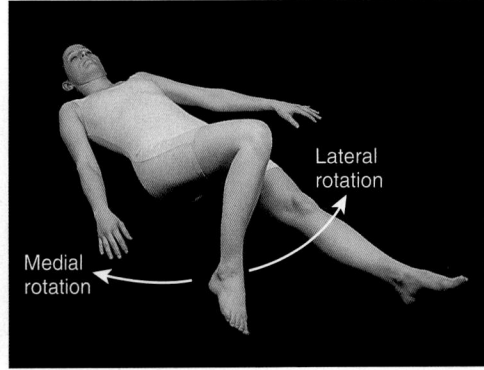

(a) Atlanto-axial joint

(b) Shoulder joint

(c) Hip joint

 How do medial and lateral rotation differ?

shoulder joint as follows: Starting in the anatomical position, flex your elbow and then move your palm across the chest (Figure 9.8b). You can medially rotate the femur at the hip joint as follows: Lie on your back, bend your knee, and then move your leg and foot laterally from the midline. Although you are moving your leg and foot laterally, the femur is rotating medially (Figure 9.8c). Medial rotation of the leg at the knee joint can be produced by sitting on a chair, bending your knee, raising your lower limb off the floor, and turning your toes medially. If the anterior surface of the bone of a limb is turned away from the midline, the movement is called *lateral (external) rotation* (see Figure 9.8b, c).

Special Movements

Special movements occur only at certain joints. They include elevation, depression, protraction, retraction, inversion, eversion, dorsiflexion, plantar flexion, supination, pronation, and opposition (Figure 9.9):

- **Elevation** (el-e-VĀ-shun = to lift up) is a superior movement of a part of the body, such as closing the mouth at the temporomandibular joint (between the mandible and temporal bone) to elevate the mandible (Figure 9.9a) or shrugging the shoulders at the acromioclavicular joint to elevate the scapula and clavicle. Its opposing movement is depression. Other bones that may be elevated (or depressed) include the hyoid and ribs.

- **Depression** (de-PRESH-un = to press down) is an inferior movement of a part of the body, such as opening the mouth to depress the mandible (Figure 9.9b) or returning shrugged shoulders to the anatomical position to depress the scapula and clavicle.

- **Protraction** (prō-TRAK-shun = to draw forth) is a movement of a part of the body anteriorly in the transverse plane. Its

Figure 9.9 Special movements at synovial joints.

Special movements occur only at certain synovial joints.

(a)   Temporomandibular joint   (b)   Elevation / Depression

(c)   Temporomandibular joint   (d)   Protraction / Retraction

(e)   Intertarsal joints   (f)   Inversion / Eversion

(g) Ankle joint   Dorsiflexion / Plantar flexion

(h) Radioulnar joint   Palm posterior / Palm anterior / Pronation / Supination

(i) Carpometacarpal joint   Opposition

❓ **What movement of the shoulder girdle occurs when you bring your arms forward until the elbows touch?**

opposing movement is retraction. You can protract your mandible at the temporomandibular joint by thrusting it outward (Figure 9.9c) or protract your clavicles at the acromioclavicular and sternoclavicular joints by crossing your arms.

- **Retraction** (rē-TRAK-shun = to draw back) is a movement of a protracted part of the body back to the anatomical position (Figure 9.9d).

- **Inversion** (in-VER-zhun = to turn inward) is movement of the sole medially at the intertarsal joints (between the tarsals) (Figure 9.9e). Its opposing movement is eversion. Physical therapists also refer to inversion of the feet as *supination.*

- **Eversion** (ē-VER-zhun = to turn outward) is a movement of the sole laterally at the intertarsal joints (Figure 9.9f). Physical therapists also refer to eversion of the feet as *pronation.*

- **Dorsiflexion** (dor-si-FLEK-shun) refers to bending of the foot at the ankle or talocrural joint (between the tibia, fibula, and talus) in the direction of the dorsum (superior surface) (Figure 9.9g). Dorsiflexion occurs when you stand on your heels. Its opposing movement is plantar flexion.

- **Plantar flexion** (PLAN-tar) involves bending of the foot at the ankle joint in the direction of the plantar or inferior surface (see Figure 9.9g), as when you elevate your body by standing on your toes.

- **Supination** (soo-pi-NĀ-shun) is a movement of the forearm at the proximal and distal radioulnar joints in which the palm is turned anteriorly (Figure 9.9h). This position of the palms is one of the defining features of the anatomical position. Its opposing movement is pronation.

- **Pronation** (prō-NĀ-shun) is a movement of the forearm at the proximal and distal radioulnar joints in which the distal end of the radius crosses over the distal end of the ulna and the palm is turned posteriorly (Figure 9.9h).

- **Opposition** (op-ō-ZISH-un) is the movement of the thumb at the carpometacarpal joint (between the trapezium and metacarpal of the thumb) in which the thumb moves across the palm to touch the tips of the fingers on the same hand (Figure 9.9i). These "opposable thumbs" allow the distinctive digital movement that gives humans and other primates the ability to grasp and manipulate objects very precisely.

A summary of the movements that occur at synovial joints is presented in Table 9.1.

✔ **CHECKPOINT**

8. What are the four major categories of movements that occur at synovial joints?
9. On yourself or with a partner, demonstrate each movement listed in Table 9.1.

TABLE 9.1

Summary of Movements at Synovial Joints

MOVEMENT	DESCRIPTION	MOVEMENT	DESCRIPTION
Gliding	Movement of relatively flat bone surfaces back and forth and side to side over one another; little change in angle between bones.	**Rotation**	Movement of bone around longitudinal axis; in limbs, may be medial (toward midline) or lateral (away from midline).
Angular	Increase or decrease in angle between bones.	**Special**	Occurs at specific joints.
Flexion	Decrease in angle between articulating bones, usually in sagittal plane.	Elevation	Superior movement of body part.
Lateral flexion	Movement of trunk in frontal plane.	Depression	Inferior movement of body part.
Extension	Increase in angle between articulating bones, usually in sagittal plane.	Protraction	Anterior movement of body part in transverse plane.
Hyperextension	Extension beyond anatomical position.	Retraction	Posterior movement of body part in transverse plane.
Abduction	Movement of bone away from midline, usually in frontal plane.	Inversion	Medial movement of sole.
		Eversion	Lateral movement of sole.
Adduction	Movement of bone toward midline, usually in frontal plane.	Dorsiflexion	Bending foot in direction of dorsum (superior surface).
Circumduction	Flexion, abduction, extension, adduction, and rotation in succession (or in the opposite order); distal end of body part moves in circle.	Plantar flexion	Bending foot in direction of plantar surface (sole).
		Supination	Movement of forearm that turns palm anteriorly.
		Pronation	Movement of forearm that turns palm posteriorly.
		Opposition	Movement of thumb across palm to touch fingertips on same hand.

9.6 TYPES OF SYNOVIAL JOINTS

⦿ OBJECTIVE

• Describe the six subtypes of synovial joints.

Although all synovial joints share many characteristics in common, the shapes of the articulating surfaces vary; thus, many types of movements are possible. Synovial joints are divided into six categories based on type of movement: plane, hinge, pivot, condyloid, saddle, and ball-and-socket.

Plane Joints

The articulating surfaces of bones in a **plane joint** (PLĀN), also called a *planar joint* (PLĀ-nar), are flat or slightly curved (Figure 9.10a). Plane joints primarily permit back-and-forth and side-to-side movements between the flat surfaces of bones, but they may also rotate against one another. Many plane joints are *biaxial,* meaning that they permit movement in two axes. An *axis* is a straight line around which a bone rotates (revolves) or slides. If plane joints rotate in addition to sliding, then they are *triaxial (multiaxial),* permitting movement in three axes. Some examples of plane joints are the intercarpal joints (between carpal bones at the wrist), intertarsal joints (between tarsal bones at the ankle), sternoclavicular joints (between the manubrium of the sternum and the clavicle), acromioclavicular joints (between the acromion of the scapula and the clavicle), sternocostal joints (between the sternum and ends of the costal cartilages at the tips of the second through seventh pairs of ribs), and vertebrocostal joints (between the heads and tubercles of ribs and transverse processes of thoracic vertebrae).

Hinge Joints

In a **hinge joint,** or *ginglymus joint* (JIN-gli-mus), the convex surface of one bone fits into the concave surface of another bone (Figure 9.10b). As the name implies, hinge joints produce an angular, opening-and-closing motion like that of a hinged door. In most joint movements, one bone remains in a fixed position while the other moves around an axis. Hinge joints are *uniaxial (monaxial)* because they typically allow motion around a single axis. Hinge joints permit only flexion and extension. Examples of hinge joints are the knee (actually a modified hinge joint, which will be described later), elbow, ankle, and interphalangeal joints (between the phalanges of the fingers and toes).

Pivot Joints

In a **pivot joint,** or *trochoid* (TRŌ-koyd) *joint,* the rounded or pointed surface of one bone articulates with a ring formed partly by another bone and partly by a ligament (Figure 9.10c).

A pivot joint is *uniaxial* because it allows rotation only around its own longitudinal axis. Examples of pivot joints are the atlanto-axial joint, in which the atlas rotates around the axis and permits the head to turn from side to side as when you shake your head "no" (see Figure 9.8a), and the radioulnar joints that enable the palms to turn anteriorly and posteriorly as the head of the radius pivots around its long axis in the radial notch of the ulna (see Figure 9.9h).

Condyloid Joints

In a **condyloid joint** (KON-di-loyd; *condyl-* = knuckle) or *ellipsoidal joint,* the convex oval-shaped projection of one bone fits into the oval-shaped depression of another bone (Figure 9.10d). A condyloid joint is *biaxial* because the movement it permits is around two axes (flexion–extension and abduction–adduction), plus limited circumduction (remember that circumduction is not an isolated movement). Examples of condyloid joints are the radiocarpal (wrist) and metacarpophalangeal joints (between the metacarpals and proximal phalanges) of the second through fifth digits.

Saddle Joints

In a **saddle joint** or *sellar joint* (SEL-ar), the articular surface of one bone is saddle-shaped, and the articular surface of the other bone fits into the "saddle" as a sitting rider would sit (Figure 9.10e). The movements at a saddle joint are the same as those at a condyloid joint: *biaxial* (flexion–extension and abduction–adduction) plus limited circumduction. An example of a saddle joint is the carpometacarpal joint between the trapezium of the carpus and metacarpal of the thumb.

Ball-and-Socket Joints

A **ball-and-socket joint** or *spheroid* (SFĒ-royd) *joint* consists of the ball-like surface of one bone fitting into a cuplike depression of another bone (Figure 9.10f). Such joints are *triaxial (multiaxial),* permitting movements around three axes (flexion–extension, abduction–adduction, and rotation). Examples of ball-and-socket joints are the shoulder and hip joints. At the shoulder joint, the head of the humerus fits into the glenoid cavity of the scapula. At the hip joint, the head of the femur fits into the acetabulum of the hip (coxal) bone.

Table 9.2 summarizes the structural and functional categories of joints.

✔ CHECKPOINT

10. Which types of joints are uniaxial, biaxial, and triaxial?

Figure 9.10 Types of synovial joints. For each type, a drawing of the actual joint and a simplified diagram are shown.

🔒 ⚼ Synovial joints are classified into six principal types based on the shapes of the articulating bone surfaces.

(a) Plane joint between navicular and second and third cuneiforms of tarsus in foot

(b) Hinge joint between trochlea of humerus and trochlear notch of ulna at the elbow

(c) Pivot joint between head of radius and radial notch of ulna

(d) Condyloid joint between radius and scaphoid and lunate bones of carpus (wrist)

(e) Saddle joint between trapezium of carpus (wrist) and metacarpal of thumb

(f) Ball-and-socket joint between head of femur and acetabulum of hip bone

 Which of the joints shown here are biaxial?

TABLE 9.2

Summary of Structural and Functional Classifications of Joints

STRUCTURAL CLASSIFICATION	DESCRIPTION	FUNCTIONAL CLASSIFICATION	EXAMPLE
FIBROUS No Synovial Cavity; Articulating Bones Held Together by Fibrous Connective Tissue.			
Suture	Articulating bones united by thin layer of dense irregular connective tissue, found between skull bones; with age, some sutures replaced by synostosis (separate cranial bones fuse into single bone).	Synarthrosis (immovable) and amphiarthrosis (slightly movable).	Coronal suture.
Syndesmosis	Articulating bones united by more dense irregular connective tissue, usually a ligament.	Amphiarthrosis (slightly movable).	Distal tibiofibular joint.
Interosseous membrane	Articulating bones united by substantial sheet of dense irregular connective tissue.	Amphiarthrosis (slightly movable).	Between tibia and fibula.
CARTILAGINOUS No Synovial Cavity; Articulating Bones United by Hyaline Cartilage or Fibrocartilage.			
Synchondrosis	Connecting material: hyaline cartilage; becomes synostosis when bone elongation ceases.	Synarthrosis (immovable).	Epiphyseal plate between diaphysis and epiphysis of long bone.
Symphysis	Connecting material: broad, flat disc of fibrocartilage.	Amphiarthrosis (slightly movable).	Pubic symphysis and intervertebral joints.
SYNOVIAL Characterized by Synovial Cavity, Articular Cartilage, and Articular (Joint) Capsule; May Contain Accessory Ligaments, Articular Discs, and Bursae.			
Plane	Articulated surfaces flat or slightly curved.	Many biaxial diarthroses (freely movable): back-and-forth and side-to-side movements. Some triaxial diarthrosis: back-and-forth, side-to-side, rotation.	Intercarpal, intertarsal, sternocostal (between sternum and second to seventh pairs of ribs), and vertebrocostal joints.
Hinge	Convex surface fits into concave surface.	Uniaxial diarthrosis: flexion–extension.	Knee (modified hinge), elbow, ankle, and interphalangeal joints.
Pivot	Rounded or pointed surface fits into ring formed partly by bone and partly by ligament.	Uniaxial diarthrosis: rotation.	Atlanto-axial and radioulnar joints.
Condyloid	Oval-shaped projection fits into oval-shaped depression.	Biaxial diarthrosis: flexion–extension, abduction–adduction.	Radiocarpal and metacarpophalangeal joints.
Saddle	Articular surface of one bone is saddle-shaped; articular surface of other bone "sits" in saddle.	Biaxial diarthrosis: flexion–extension, abduction–adduction.	Carpometacarpal joint between trapezium and metacarpal of thumb.
Ball-and-socket	Ball-like surface fits into cuplike depression.	Triaxial diarthrosis: flexion–extension, abduction–adduction, rotation.	Shoulder and hip joints.

9.7 FACTORS AFFECTING CONTACT AND RANGE OF MOTION AT SYNOVIAL JOINTS

◉ OBJECTIVE

• Describe six factors that influence the type of movement and range of motion possible at a synovial joint.

The articular surfaces of synovial joints contact one another and determine the type and possible range of motion. **Range of motion (ROM)** refers to the range, measured in degrees of a circle, through which the bones of a joint can be moved. The following factors contribute to keeping the articular surfaces in contact and affect range of motion:

1. *Structure or shape of the articulating bones.* The structure or shape of the articulating bones determines how closely they can fit together. The articular surfaces of some bones have a complementary relationship. This spatial relationship is very obvious at the hip joint, where the head of the femur articulates with the acetabulum of the hip bone. An interlocking fit allows rotational movement.

2. *Strength and tension (tautness) of the joint ligaments.* The different components of a fibrous capsule are tense or taut only when the joint is in certain positions. Tense ligaments not only restrict the range of motion but also direct the movement of the articulating bones with respect to each other. In the knee joint, for example, the anterior cruciate ligament is taut and the posterior cruciate ligament is loose when the knee is straightened, and the reverse occurs when the knee is bent. In the hip joint, certain ligaments become taut when standing and more firmly attach the head of the femur to the acetabulum of the hip bone.

3. *Arrangement and tension of the muscles.* Muscle tension reinforces the restraint placed on a joint by its ligaments, and thus restricts movement. A good example of the effect of muscle tension on a joint is seen at the hip joint. When the thigh is flexed with the knee extended, the flexion of the hip joint is restricted by the tension of the hamstring muscles on the posterior surface of the thigh, so most of us can't raise a straightened leg more than a 90-degree angle from the floor. But if the knee is also flexed, the tension on the hamstring muscles is lessened, and the thigh can be raised farther, allowing you to raise your thigh to touch your chest.

4. *Contact of soft parts.* The point at which one body surface contacts another may limit mobility. For example, if you bend your arm at the elbow, it can move no farther after the anterior surface of the forearm meets with and presses against the biceps brachii muscle of the arm. Joint movement may also be restricted by the presence of adipose tissue.

5. *Hormones.* Joint flexibility may also be affected by hormones. For example, relaxin, a hormone produced by the placenta and ovaries, increases the flexibility of the fibrocartilage of the pubic symphysis and loosens the ligaments between the sacrum, hip bone, and coccyx toward the end of pregnancy. These changes permit expansion of the pelvic outlet, which assists in delivery of the baby.

6. *Disuse.* Movement at a joint may be restricted if a joint has not been used for an extended period. For example, if an elbow joint is immobilized by a cast, range of motion at the joint may be limited for a time after the cast is removed. Disuse may also result in decreased amounts of synovial fluid, diminished flexibility of ligaments and tendons, and *muscular atrophy,* a reduction in size or wasting of a muscle.

✔ CHECKPOINT

11. How do the strength and tension of ligaments determine range of motion?

9.8 SELECTED JOINTS OF THE BODY

◉ OBJECTIVE

• Identify the major joints of the body by location, classification, and movements.

In Chapters 7 and 8, we discussed the major bones and their markings. In this chapter we have examined how joints are classified according to their structure and function, and we have introduced the movements that occur at joints. Table 9.3 (selected joints of the axial skeleton) and Table 9.4 (selected joints of the appendicular skeleton) will help you integrate the information you have learned in all three chapters. These tables list some of the major joints of the body according to their articular components (the bones that enter into their formation), their structural and functional classification, and the type(s) of movement that occur(s) at each joint.

Next we examine in detail several selected joints of the body in a series of exhibits. Each exhibit considers a specific synovial joint and contains (1) a definition—a description of the type of joint and the bones that form the joint; (2) the anatomical components—a description of the major connecting ligaments, articular disc (if present), articular capsule, and other distinguishing features of the joint; and (3) the joint's possible movements. Each exhibit also refers you to a figure that illustrates the joint. The joints described are the temporomandibular joint (TMJ), shoulder (humeroscapular or glenohumeral) joint, elbow joint, hip (coxal) joint, and knee (tibiofemoral) joint. Because these joints are described in Exhibits 9.A–9.E (Figures 9.11–9.15), they are not included in Tables 9.3 and 9.4.

✔ CHECKPOINT

12. Using Tables 9.3 and 9.4 as a guide, identify only the cartilaginous joints.

TABLE 9.3

Selected Joints of the Axial Skeleton

JOINT	ARTICULAR COMPONENTS	CLASSIFICATION	MOVEMENTS
Suture	Between skull bones.	*Structural:* fibrous. *Functional:* amphiarthrosis and synarthrosis.	None.
Atlanto-occipital	Between superior articular facets of atlas and occipital condyles of occipital bone.	*Structural:* synovial (condyloid). *Functional:* diarthrosis.	Flexion and extension of head; slight lateral flexion of head to either side.
Atlanto-axial	(1) Between dens of axis and anterior arch of atlas; (2) between lateral masses of atlas and axis.	*Structural:* synovial (pivot) between dens and anterior arch; synovial (planar) between lateral masses. *Functional:* diarthrosis.	Rotation of head.
Intervertebral	(1) Between vertebral bodies; (2) between vertebral arches.	*Structural:* cartilaginous (symphysis) between vertebral bodies; synovial (planar) between vertebral arches. *Functional:* amphiarthrosis between vertebral bodies; diarthrosis between vertebral arches.	Flexion, extension, lateral flexion, and rotation of vertebral column.
Vertebrocostal	(1) Between facets of heads of ribs and facets of bodies of adjacent thoracic vertebrae and intervertebral discs between them; (2) between articular part of tubercles of ribs and facets of transverse processes of thoracic vertebrae.	*Structural:* synovial (planar). *Functional:* diarthrosis.	Slight gliding.
Sternocostal	Between sternum and first seven pairs of ribs.	*Structural:* cartilaginous (synchondrosis) between sternum and first pair of ribs; synovial (plane) between sternum and second through seventh pairs of ribs. *Functional:* synarthrosis between sternum and first pair of ribs; diarthrosis between sternum and second through seventh pairs of ribs.	None between sternum and first pair of ribs; slight gliding between sternum and second through seventh pairs of ribs.
Lumbosacral	(1) Between body of fifth lumbar vertebra and base of sacrum; (2) between inferior articular facets of fifth lumbar vertebra and superior articular facets of first vertebra of sacrum.	*Structural:* cartilaginous (symphysis) between body and base; synovial (planar) between articular facets. *Functional:* amphiarthrosis between body and base; diarthrosis between articular facets.	Flexion, extension, lateral flexion, and rotation of vertebral column.

TABLE 9.4

Selected Joints of the Appendicular Skeleton

JOINT	ARTICULAR COMPONENTS	CLASSIFICATION	MOVEMENTS
Sternoclavicular	Between sternal end of clavicle, manubrium of sternum, and first costal cartilage.	*Structural:* synovial (plane, pivot). *Functional:* diarthrosis.	Gliding, with limited movements in nearly every direction.
Acromioclavicular	Between acromion of scapula and acromial end of clavicle.	*Structural:* synovial (plane). *Functional:* diarthrosis.	Gliding and rotation of scapula on clavicle.
Radioulnar	Proximal radioulnar joint between head of radius and radial notch of ulna; distal radioulnar joint between ulnar notch of radius and head of ulna.	*Structural:* synovial (pivot). *Functional:* diarthrosis.	Rotation of forearm.
Wrist (radiocarpal)	Between distal end of radius and scaphoid, lunate, and triquetrum of carpus.	*Structural:* synovial (condyloid). *Functional:* diarthrosis.	Flexion, extension, abduction, adduction, circumduction, and slight hyperextension of wrist.
Intercarpal	Between proximal row of carpal bones, distal row of carpal bones, and between both rows of carpal bones (midcarpal joints).	*Structural:* synovial (plane), except for hamate, scaphoid, and lunate (midcarpal) joint, which is synovial (saddle). *Functional:* diarthrosis.	Gliding plus flexion, extension, abduction, adduction, and slight rotation at midcarpal joints.
Carpometacarpal	Carpometacarpal joint of thumb between trapezium of carpus and first metacarpal; carpometacarpal joints of remaining digits formed between carpus and second through fifth metacarpals.	*Structural:* synovial (saddle) at thumb; synovial (plane) at remaining digits. *Functional:* diarthrosis.	Flexion, extension, abduction, adduction, and circumduction at thumb; gliding at remaining digits.
Metacarpophalangeal and metatarsophalangeal	Between heads of metacarpals (or metatarsals) and bases of proximal phalanges.	*Structural:* synovial (condyloid). *Functional:* diarthrosis.	Flexion, extension, abduction, adduction, and circumduction of phalanges.
Interphalangeal	Between heads of phalanges and bases of more distal phalanges.	*Structural:* synovial (hinge). *Functional:* diarthrosis.	Flexion and extension of phalanges.
Sacroiliac	Between auricular surfaces of sacrum and ilia of hip bones.	*Structural:* synovial (plane). *Functional:* diarthrosis.	Slight gliding (even more so during pregnancy).
Pubic symphysis	Between anterior surfaces of hip bones.	*Structural:* cartilaginous (symphysis). *Functional:* amphiarthrosis.	Slight movements (even more so during pregnancy).
Tibiofibular	Proximal tibiofibular joint between lateral condyle of tibia and head of fibula; distal tibiofibular joint between distal end of fibula and fibular notch of tibia.	*Structural:* synovial (plane) at proximal joint; fibrous (syndesmosis) at distal joint. *Functional:* diarthrosis at proximal joint; amphiarthrosis at distal joint.	Slight gliding at proximal joint; slight rotation of fibula during dorsiflexion of foot.
Ankle (talocrural)	(1) Between distal end of tibia and its medial malleolus and talus; (2) between lateral malleolus of fibula and talus.	*Structural:* synovial (hinge). *Functional:* diarthrosis.	Dorsiflexion and plantar flexion of foot.
Intertarsal	Subtalar joint between talus and calcaneus of tarsus; talocalcaneonavicular joint between talus and calcaneus and navicular of tarsus; calcaneocuboid joint between calcaneus and cuboid of tarsus.	*Structural:* synovial (plane) at subtalar and calcaneocuboid joints; synovial (saddle) at talocalcaneonavicular joint. *Functional:* diarthrosis.	Inversion and eversion of foot.
Tarsometatarsal	Between three cuneiforms of tarsus and bases of five metatarsal bones.	*Structural:* synovial (plane). *Functional:* diarthrosis.	Slight gliding.

EXHIBIT 9.A   Temporomandibular Joint *(Figure 9.11)*

OBJECTIVE

- Describe the anatomical components of the temporomandibular joint and explain the movements that can occur at this joint.

Definition

The **temporomandibular joint (TMJ)** (tem′-po-rō-man-DIB-ū-lar) is a combined hinge and planer joint formed by the condylar process of the mandible and the mandibular fossa and articular tubercle of the temporal bone. The temporomandibular joint is the only freely movable joint between skull bones (with the exception of the ear ossicles); all other skull joints are sutures and therefore immovable or slightly movable.

Anatomical Components

1. *Articular disc (meniscus).* Fibrocartilage disc that separates the synovial cavity into superior and inferior compartments, each with a synovial membrane (Figure 9.11c).

2. *Articular capsule.* Thin, fairly loose envelope around the circumference of the joint (Figure 9.11a, b).

3. *Lateral ligament.* Two short bands on the lateral surface of the articular capsule that extend inferiorly and posteriorly from the inferior border and tubercle of the zygomatic process of the temporal bone to the lateral and posterior aspect of the neck of the mandible. The lateral ligament is covered by the parotid gland and helps strengthen the joint laterally and prevent displacement of the mandible (Figure 9.11a).

4. *Sphenomandibular ligament* (sfē-nō-man-DIB-ū-lar). Thin band that extends inferiorly and anteriorly from the spine of the sphenoid bone to the ramus of the mandible (Figure 9.11b). It does not contribute significantly to the strength of the joint.

5. *Stylomandibular ligament* (stī-lō-man-DIB-ū-lar). Thickened band of deep cervical fascia that extends from the styloid process of the temporal bone to the inferior and posterior border of the ramus of the mandible. This ligament separates the parotid gland from the submandibular gland and limits movement of the mandible at the TMJ (Figure 9.11a, b).

Movements

In the temporomandibular joint, only the mandible moves because the temporal bone is firmly anchored to other bones of the skull by sutures. Accordingly, the mandible may function in depression (jaw opening) and elevation (jaw closing), which occurs in the inferior compartment, and protraction, retraction, lateral displacement, and slight rotation, which occur in the superior compartment (see Figure 9.9a–d).

CLINICAL CONNECTION | Dislocated Mandible

A **dislocation** (dis′-lō-KĀ-shun; *dis-* = apart) or *luxation* (luks-Ā-shun; *luxatio* = dislocation) is the displacement of a bone from a joint with tearing of ligaments, tendons, and articular capsules. A **dislocated mandible** can occur in several ways. *Anterior displacements* are the most common and occur when the condylar processes of the mandible pass anterior to the articular tubercles. Common causes are extreme mouth opening, as in yawning or taking a large bite, dental procedures, or general anesthesia. *Posterior displacement* can be caused by a direct blow to the chin. *Superior displacements* are typically caused by a direct blow to a partially opened mouth. *Lateral dislocations* are usually associated with mandibular fractures. •

✔ CHECKPOINT

What distinguishes the temporomandibular joint from the other joints of the skull?

Figure 9.11 **Right temporomandibular joint (TMJ).**

The TMJ is the only movable joint between skull bones.

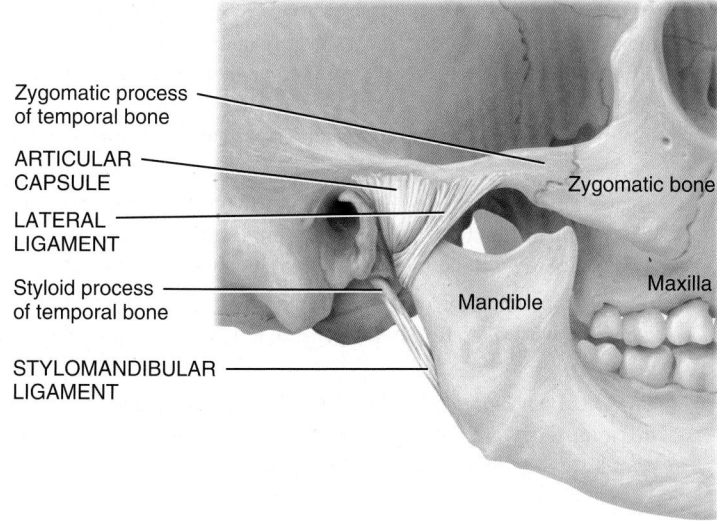

Zygomatic process of temporal bone

ARTICULAR CAPSULE

LATERAL LIGAMENT

Styloid process of temporal bone

STYLOMANDIBULAR LIGAMENT

Zygomatic bone

Maxilla

Mandible

(a) Right lateral view

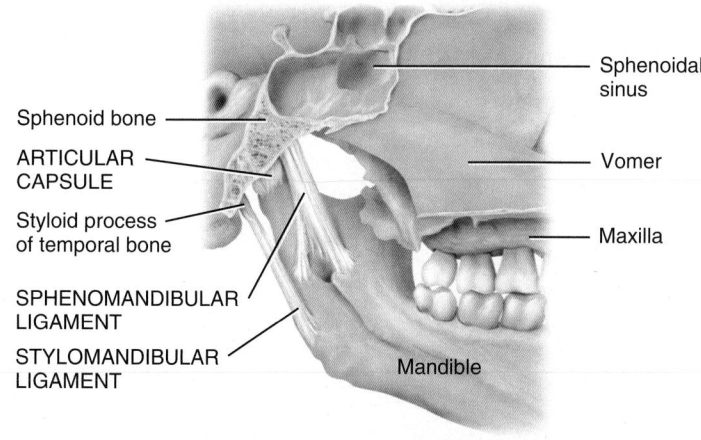

Sphenoid bone

ARTICULAR CAPSULE

Styloid process of temporal bone

SPHENOMANDIBULAR LIGAMENT

STYLOMANDIBULAR LIGAMENT

Sphenoidal sinus

Vomer

Maxilla

Mandible

(b) Left medial view

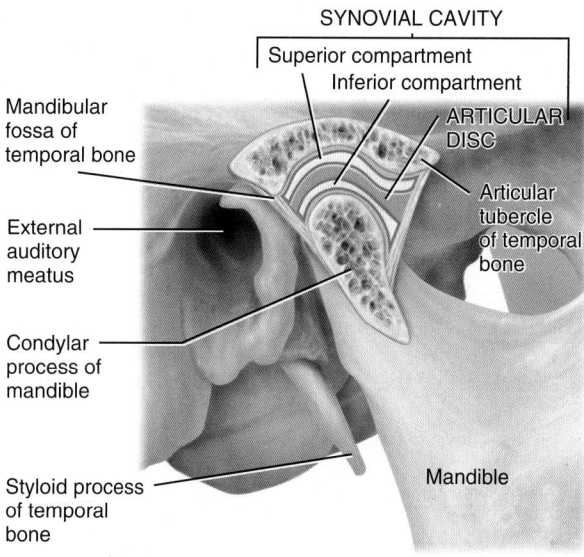

SYNOVIAL CAVITY

Superior compartment

Inferior compartment

Mandibular fossa of temporal bone

External auditory meatus

Condylar process of mandible

Styloid process of temporal bone

ARTICULAR DISC

Articular tubercle of temporal bone

Mandible

(c) Sagittal section viewed from right

? **Which ligament prevents displacement of the mandible?**

EXHIBIT 9.A **309**

EXHIBIT 9.B   Shoulder Joint   *(Figure 9.12)*

● OBJECTIVE

- Describe the anatomical components of the shoulder joint and the movements that can occur at this joint.

Definition

The **shoulder joint** is a ball-and-socket joint formed by the head of the humerus and the glenoid cavity of the scapula. It also is referred to as the *humeroscapular* or *glenohumeral joint.*

Anatomical Components

1. *Articular capsule.* Thin, loose sac that completely envelops the joint and extends from the glenoid cavity to the anatomical neck of the humerus. The inferior part of the capsule is its weakest area (Figure 9.12).

2. *Coracohumeral ligament* (kor′-a-kō-Ū-mer-al). Strong, broad ligament that strengthens the superior part of the articular capsule and extends from the coracoid process of the scapula to the greater tuber-

cle of the humerus (Figure 9.12a, b). The ligament strengthens the superior part of the articular capsule and reinforces the anterior aspect of the articular capsule.

3. *Glenohumeral ligaments* (glē-no-Ū-mer-al). Three thickenings of the articular capsule over the anterior surface of the joint that extend from the glenoid cavity to the lesser tubercle and anatomical neck of the humerus. These ligaments are often indistinct or absent and provide only minimal strength (Figure 9.12a, b). They play a role in joint stabilization when the humerus approaches or exceeds its limits of motion.

4. *Transverse humeral ligament.* Narrow sheet extending from the greater tubercle to the lesser tubercle of the humerus (Figure 9.12a). The ligament functions as a retinaculum (retaining band of connective tissue) to hold the long head of the biceps brachii muscle.

5. *Glenoid labrum.* Narrow rim of fibrocartilage around the edge of the glenoid cavity that slightly deepens and enlarges the glenoid cavity (Figure 9.12b, c).

6. *Bursae.* Four *bursae* (see Section 9.4) are associated with the shoulder joint. They are the *subscapular bursa* (Figure 9.12a), *subdeltoid bursa, subacromial bursa* (Figure 9.12a–c), and *subcoracoid bursa.*

Figure 9.12 **Right shoulder (humeroscapular or glenohumeral) joint.**

🔒 ⚏ Most of the stability of the shoulder joint results from the arrangement of the rotator cuff muscles.

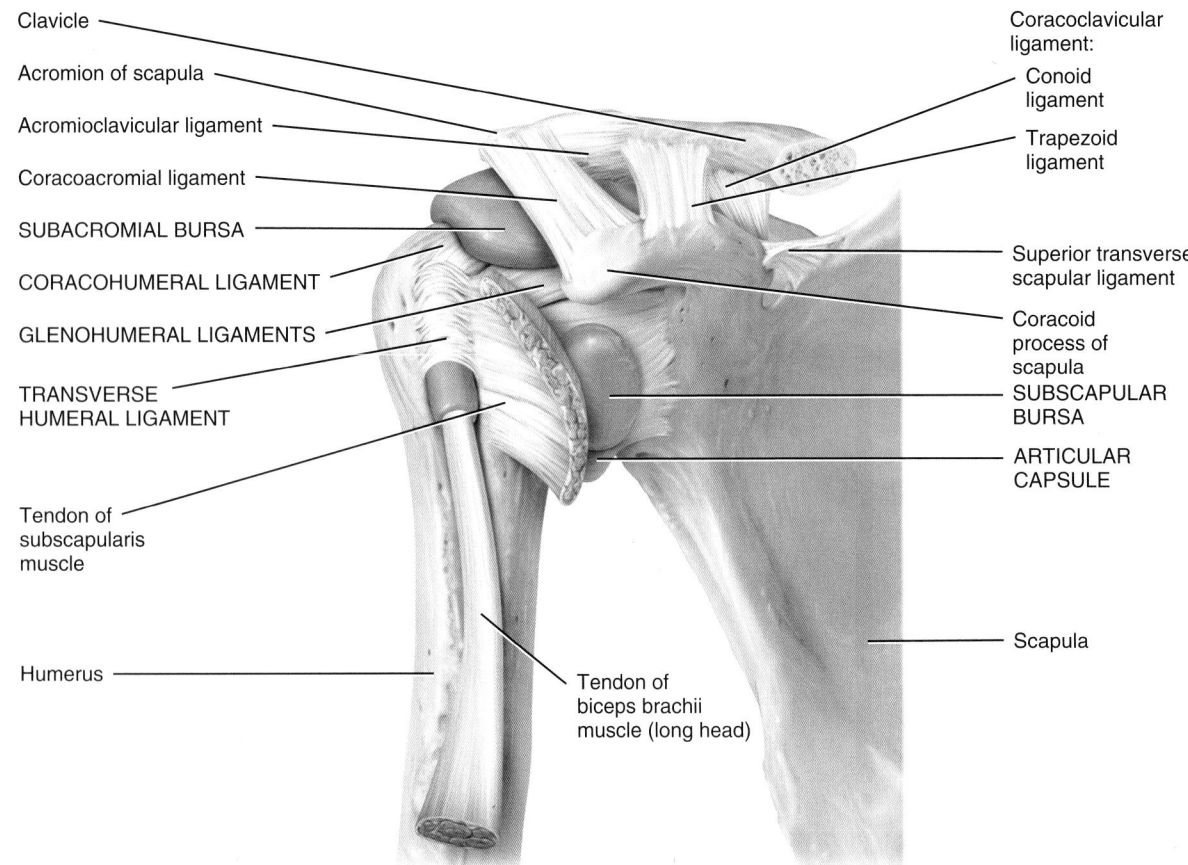

(a) Anterior view

Movements

The shoulder joint allows flexion, extension, hyperextension, abduction, adduction, medial rotation, lateral rotation, and circumduction of the arm (see Figures 9.5–9.8). It has more freedom of movement than any other joint of the body. This freedom results from the looseness of the articular capsule and the shallowness of the glenoid cavity in relation to the large size of the head of the humerus.

Although the ligaments of the shoulder joint strengthen it to some extent, most of the strength results from the muscles that surround the joint,

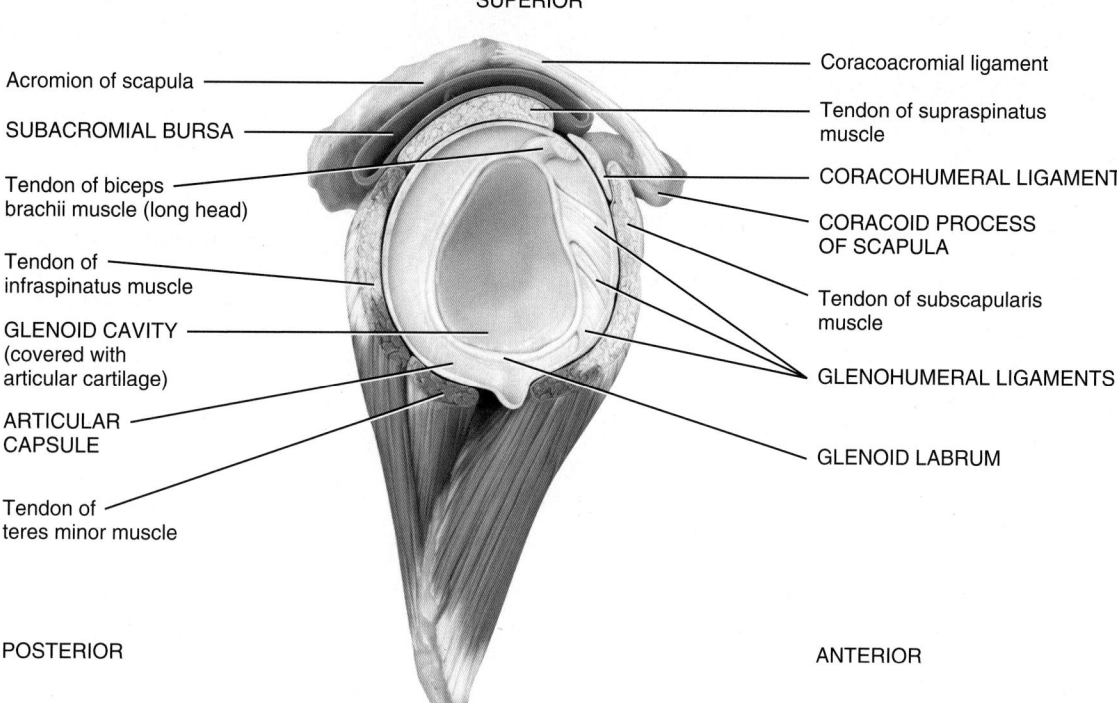

View

SUPERIOR

Acromion of scapula

SUBACROMIAL BURSA

Tendon of biceps brachii muscle (long head)

Tendon of infraspinatus muscle

GLENOID CAVITY (covered with articular cartilage)

ARTICULAR CAPSULE

Tendon of teres minor muscle

POSTERIOR

Coracoacromial ligament

Tendon of supraspinatus muscle

CORACOHUMERAL LIGAMENT

CORACOID PROCESS OF SCAPULA

Tendon of subscapularis muscle

GLENOHUMERAL LIGAMENTS

GLENOID LABRUM

ANTERIOR

(b) Lateral view (opened)

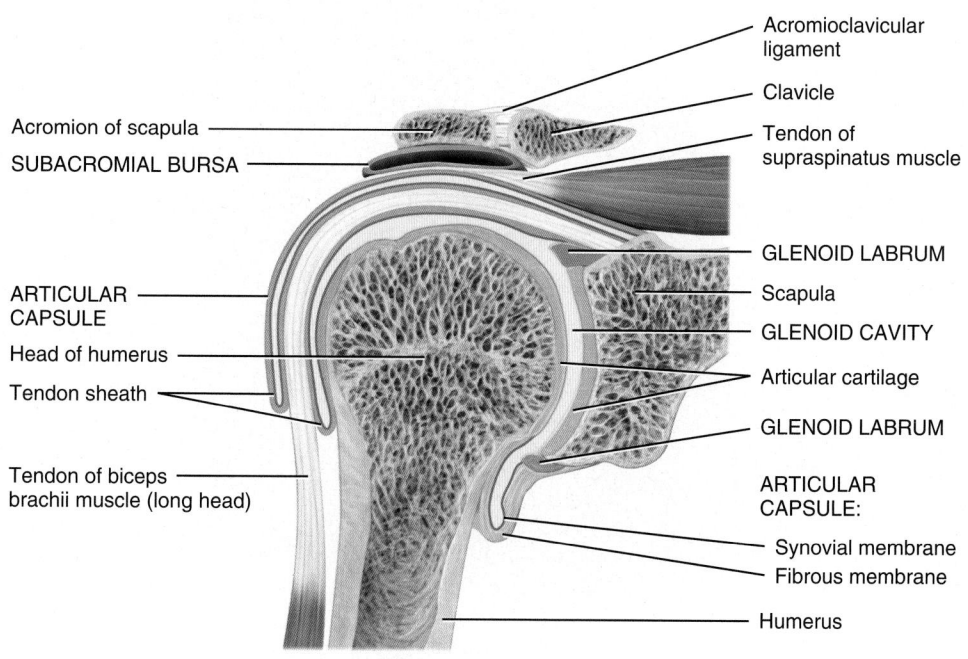

Frontal plane

Acromion of scapula

SUBACROMIAL BURSA

ARTICULAR CAPSULE

Head of humerus

Tendon sheath

Tendon of biceps brachii muscle (long head)

Acromioclavicular ligament

Clavicle

Tendon of supraspinatus muscle

GLENOID LABRUM

Scapula

GLENOID CAVITY

Articular cartilage

GLENOID LABRUM

ARTICULAR CAPSULE:

Synovial membrane

Fibrous membrane

Humerus

(c) Frontal section

EXHIBIT 9.B CONTINUES ▶

EXHIBIT 9.B **311**

EXHIBIT 9.B   Shoulder Joint *(Figure 9.12)*   CONTINUED

especially the *rotator cuff muscles*. These muscles (supraspinatus, infraspinatus, teres minor, and subscapularis) join the scapula to the humerus (see also Figure 11.15). The tendons of the rotator cuff muscles encircle the joint (except for the inferior portion) and intimately surround the articular capsule. The rotator cuff muscles work as a group to hold the head of the humerus in the glenoid cavity.

CLINICAL CONNECTION | *Rotator Cuff Injury, Dislocated and Separated Shoulder, and Torn Glenoid Labrum*

Rotator cuff injury is a strain or tear in the rotator cuff muscles and is a common injury among baseball pitchers, volleyball players, racket sports players, swimmers, and violinists, due to shoulder movements that involve vigorous circumduction. It also occurs as a result of wear and tear, aging, trauma, poor posture, improper lifting, and repetitive motions in certain jobs, such as placing items on a shelf above your head. Most often, there is tearing of the supraspinatus muscle tendon of the rotator cuff. This tendon is especially predisposed to wear and tear because of its location between the head of the humerus and acromion of the scapula, which compresses the tendon during shoulder movements. Poor posture and poor body mechanics also increase compression of the supraspinatus muscle tendon.

The joint most commonly dislocated in adults is the shoulder joint because its socket is quite shallow and the bones are held together by supporting muscles. Usually in a **dislocated shoulder,** the head of the humerus becomes displaced inferiorly, where the articular capsule is least protected. Dislocations of the mandible, elbow, fingers, knee, or hip are less common. Dislocations are treated with rest, ice, pain relievers, manual manipulation, or surgery followed by use of a sling and physical therapy.

A **separated shoulder** actually refers to an injury not to the shoulder joint but to the acromioclavicular joint, a joint formed by the acromion of the scapula and the acromial end of the clavicle. This condition is usually the result of forceful trauma to the joint, as when the shoulder strikes the ground in a fall. Treatment options are similar to those for treating a dislocated shoulder, although surgery is rarely needed.

In a **torn glenoid labrum,** the fibrocartilaginous labrum may tear away from the glenoid cavity. This causes the joint to catch or feel like it's slipping out of place. The shoulder may indeed become dislocated as a result. A torn labrum is reattached to the glenoid surgically with anchors and sutures. The repaired joint is more stable.   •

✔**CHECKPOINT**

Which tendons at the shoulder joint of a baseball pitcher are most likely to be torn due to excessive circumduction?

FIGURE 9.12 CONTINUED ▶

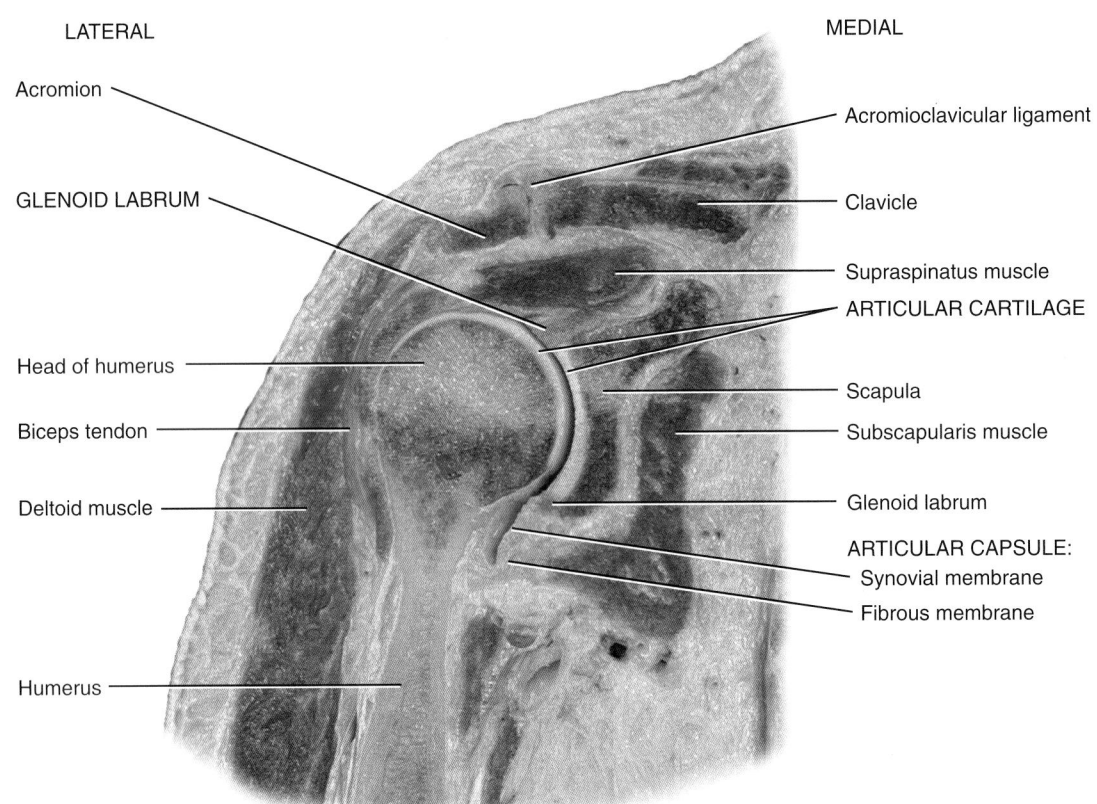

(d) Frontal section

? **Why does the shoulder joint have more freedom of movement than any other joint of the body?**

EXHIBIT 9.C   Elbow Joint   *(Figure 9.13)*

◉ OBJECTIVE

• Describe the anatomical components of the elbow joint and the movements that can occur at this joint.

Definition

The **elbow joint** is a hinge joint formed by the trochlea and capitulum of the humerus, the trochlear notch of the ulna, and the head of the radius.

Anatomical Components

1. *Articular capsule.* The anterior part of the articular capsule covers the anterior part of the elbow joint, from the radial and coronoid fossae of the humerus to the coronoid process of the ulna and the anular ligament of the radius. The posterior part extends from the capitulum, olecranon fossa, and lateral epicondyle of the humerus to the anular ligament of the radius, the olecranon of the ulna, and the ulna posterior to the radial notch (Figure 9.13a, b).

2. *Ulnar collateral ligament.* Thick, triangular ligament that extends from the medial epicondyle of the humerus to the coronoid process and olecranon of the ulna (Figure 9.13a). Part of this ligament deepens the socket for the trochlea of the humerus.

3. *Radial collateral ligament.* Strong, triangular ligament that extends from the lateral epicondyle of the humerus to the anular ligament of the radius and the radial notch of the ulna (Figure 9.13b).

4. *Anular ligament of the radius.* Strong band that encircles the head of the radius. It holds the head of the radius in the radial notch of the ulna (Figure 9.13a, b).

Movements

The elbow joint allows flexion and extension of the forearm (see Figure 9.5c).

⚕ CLINICAL CONNECTION | *Tennis Elbow, Little-League Elbow, and Dislocation of the Radial Head*

Tennis elbow most commonly refers to pain at or near the lateral epicondyle of the humerus, usually caused by an improperly executed backhand. The extensor muscles strain or sprain, resulting in pain. **Little-League elbow,** inflammation of the medial epicondyle, typically develops as a result of a heavy pitching schedule and/or a schedule that involves throwing curve balls, especially among youngsters. In this disorder, the elbow joint may enlarge, fragment, or separate.

A **dislocation of the radial head** is the most common upper limb dislocation in children. In this injury, the head of the radius slides past or ruptures the radial anular ligament, a ligament that forms a collar around the head of the radius at the proximal radioulnar joint. Dislocation is most apt to occur when a strong pull is applied to the forearm while it is extended and supinated, for instance, while swinging a child around with outstretched arms. •

✔ CHECKPOINT

At the elbow joint, which ligaments connect (a) the humerus and the ulna, and (b) the humerus and the radius?

Figure 9.13 Right elbow joint.

🔒 ⚏ The elbow joint is formed by parts of three bones: humerus, ulna, and radius.

(a) Medial aspect

(b) Lateral aspect

❓ **Which movements are possible at a hinge joint?**

EXHIBIT 9.C   **313**

EXHIBIT 9.D

Hip Joint (*Figure 9.14*)

OBJECTIVE
• Describe the anatomical components of the hip joint and the movements that can occur at this joint.

Definition

The **hip joint** *(coxal joint)* is a ball-and-socket joint formed by the head of the femur and the acetabulum of the hip bone.

Anatomical Components

1. ***Articular capsule.*** Very dense and strong capsule that extends from the rim of the acetabulum to the neck of the femur (Figure 9.14c). With its accessory ligaments, this is one of the strongest structures of the body. The articular capsule consists of circular and longitudinal fibers. The circular fibers, called the zona orbicularis, form a collar around the neck of the femur. Accessory ligaments known as the

Figure 9.14 Right hip (coxal) joint.

The articular capsule of the hip joint is one of the strongest structures in the body.

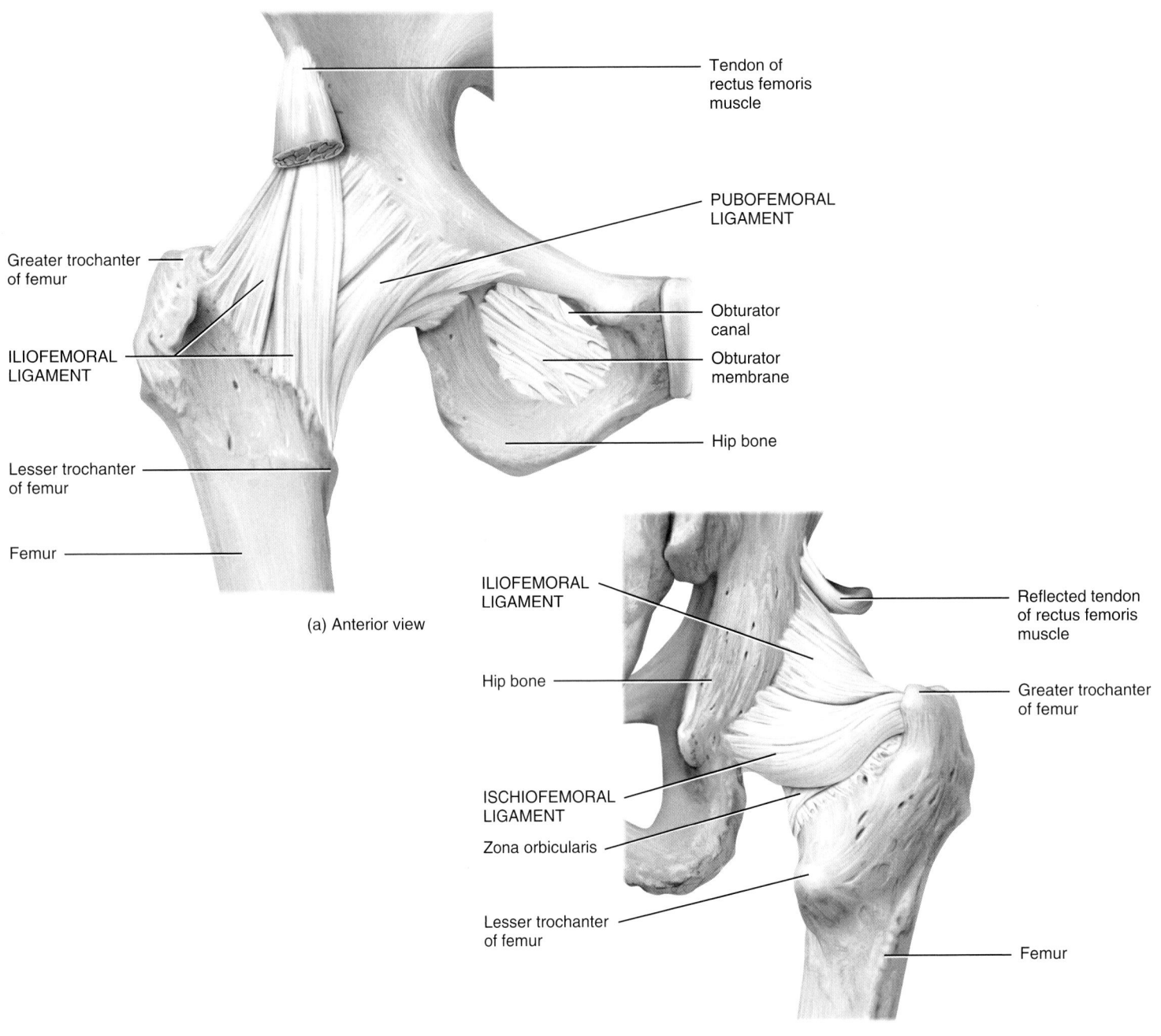

(a) Anterior view

(b) Posterior view

iliofemoral ligament, pubofemoral ligament, and *ischiofemoral liga-ment* reinforce the longitudinal fibers of the articular capsule.

2. **Iliofemoral ligament** (il'-ē-ō-FEM-ō-ral). Thickened portion of the articular capsule that extends from the anterior inferior iliac spine of the hip bone to the intertrochanteric line of the femur (Figure 9.14a, b). This ligament is said to be the body's strongest and prevents hyperextension of the femur at the hip joint during standing.

3. **Pubofemoral ligament** (pū'-bō-FEM-ō-ral). Thickened portion of the articular capsule that extends from the pubic part of the rim of the acetabulum to the neck of the femur (Figure 9.14a). This ligament prevents overabduction of the femur at the hip joint and strengthens the articular capsule.

4. **Ischiofemoral ligament** (is'-kē-ō-FEM-ō-ral). Thickened portion of the articular capsule that extends from the ischial wall of the acetabulum to the neck of the femur (Figure 9.14b). This ligament slackens during adduction, tenses during abduction, and strengthens the articular capsule.

5. **Ligament of the head of the femur.** Flat, triangular band (primarily a synovial fold) that extends from the fossa of the acetabulum to the fovea capitis of the head of the femur (Figure 9.14c). The ligament usually contains a small artery that supplies the head of the femur.

6. **Acetabular labrum** (as-e-TAB-ū-lar LĀ-brum). Fibrocartilage rim attached to the margin of the acetabulum that enhances the depth of the acetabulum. As a result, dislocation of the femur is rare (Figure 9.14c).

7. **Transverse ligament of the acetabulum.** Strong ligament that crosses over the acetabular notch. It supports part of the acetabular labrum and is connected with the ligament of the head of the femur and the articular capsule (Figure 9.14c).

Movements

The hip joint allows flexion, extension, abduction, adduction, circumduction, medial rotation, and lateral rotation of the thigh (see Figures 9.5–9.8). The extreme stability of the hip joint is related to the very strong articular capsule and its accessory ligaments, the manner in which the femur fits into the acetabulum, and the muscles surrounding the joint. Although the shoulder and hip joints are both ball-and-socket joints, the hip joints do not have as wide a range of motion. Flexion is limited by the anterior surface of the thigh coming into contact with the anterior abdominal wall when the knee is flexed and by tension of the hamstring muscles when the knee is extended. Extension is limited by tension of the iliofemoral, pubofemoral, and ischiofemoral ligaments. Abduction is limited by the tension of the pubofemoral ligament, and adduction is limited by contact with the opposite limb and tension in the ligament of the head of the femur. Medial rotation is limited by the tension in the ischiofemoral ligament, and lateral rotation is limited by tension in the iliofemoral and pubofemoral ligaments.

✔CHECKPOINT
What factors limit the degree of flexion and abduction at the hip joint?

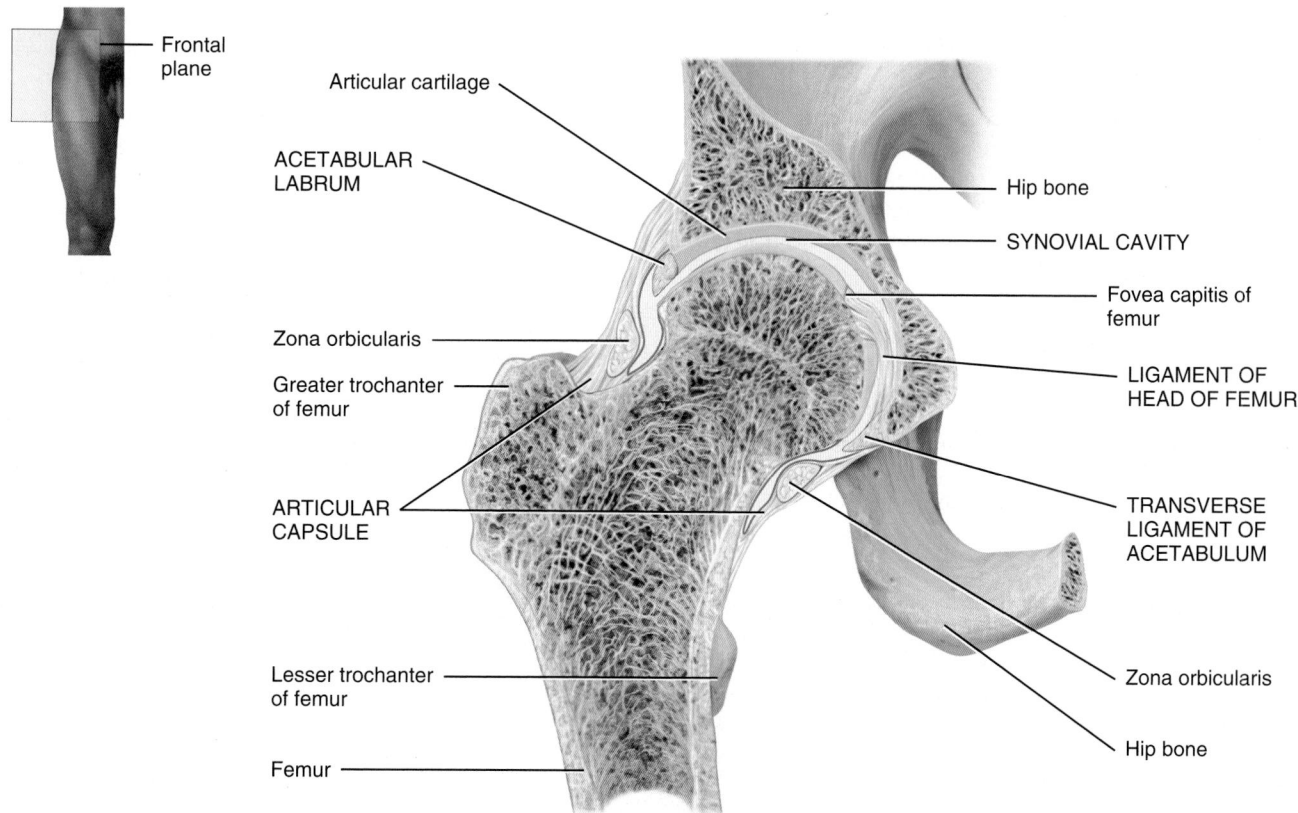

(c) Frontal section

? **Which ligaments limit the degree of extension that is possible at the hip joint?**

EXHIBIT 9.D **315**

EXHIBIT 9.E   Knee Joint   *(Figure 9.15)*

○ **OBJECTIVE**

- Describe the main anatomical components of the knee joint and explain the movements that can occur at this joint.

Definition

The **knee joint** *(tibiofemoral joint)* is the largest and most complex joint of the body (Figure 9.15). It is a modified hinge joint (because its primary movement is a uniaxial hinge movement) that consists of three joints within a single synovial cavity:

1. Laterally is a *tibiofemoral joint,* between the lateral condyle of the femur, lateral meniscus, and lateral condyle of the tibia, which is the weight-bearing bone of the leg.

2. Medially is another *tibiofemoral joint,* between the medial condyle of the femur, medial meniscus, and medial condyle of the tibia.

3. An intermediate *patellofemoral joint* is between the patella and the patellar surface of the femur.

Anatomical Components

1. *Articular capsule.* No complete, independent capsule unites the bones of the knee joint. The ligamentous sheath surrounding the joint consists mostly of muscle tendons or their expansions (Figure 9.15e–g). There are, however, some capsular fibers connecting the articulating bones.

2. *Medial and lateral patellar retinacula* (ret′-i-NAK-ū-la). Fused tendons of insertion of the quadriceps femoris muscle and the fascia lata (deep fascia of thigh) that strengthen the anterior surface of the joint (Figure 9.15e).

3. *Patellar ligament.* Continuation of the common tendon of insertion of the quadriceps femoris muscle that extends from the patella to the tibial tuberosity. This ligament also strengthens the anterior surface of the joint. The posterior surface of the ligament is separated from the synovial membrane of the joint by an infrapatellar fat pad (Figure 9.15c–e).

4. *Oblique popliteal ligament* (pop-LIT-ē-al). Broad, flat ligament that extends from the intercondylar fossa and lateral condyle of the femur to the head and medial condyle of the tibia (Figure 9.15f, h). The ligament strengthens the posterior surface of the joint.

5. *Arcuate popliteal ligament.* Extends from the lateral condyle of the femur to the styloid process of the head of the fibula. It strengthens the lower lateral part of the posterior surface of the joint (Figure 9.15f).

6. *Tibial collateral ligament.* Broad, flat ligament on the medial surface of the joint that extends from the medial condyle of the femur to the medial condyle of the tibia (Figure 9.15a, e–h). Tendons of the sartorius, gracilis, and semitendinosus muscles, all of which strengthen the medial aspect of the joint, cross the ligament. The tibial collateral ligament is firmly attached to the medial meniscus.

7. *Fibular collateral ligament.* Strong, rounded ligament on the lateral surface of the joint that extends from the lateral condyle of the femur to the lateral side of the head of the fibula (Figure 9.15a, e–h). It strengthens the lateral aspect of the joint. The ligament is covered by the tendon of the biceps femoris muscle. The tendon of the popliteal muscle is deep to the ligament.

8. *Intracapsular ligaments* (in′-tra-KAP-sū-lar). Ligaments within the capsule that connect the tibia and femur. The anterior and posterior **cruciate ligaments** (KROO-shē-āt = like a cross) are named based on their origins relative to the intercondylar area of the tibia. From their origins, they cross on their way to their destinations on the femur.

   a. *Anterior cruciate ligament (ACL).* Extends posteriorly and laterally from a point anterior to the intercondylar area of the tibia to the posterior part of the medial surface of the lateral condyle of the femur (Figure 9.15a, b, h). The ACL limits hyperextension of the knee (which normally does not occur at this joint) and prevents the anterior sliding of the tibia on the femur. This ligament is stretched or torn in about 70% of all serious knee injuries.

      ACL injuries are much more common in females than males, perhaps as much as 3 to 6 times. The reasons are unclear but may be related to less space between the femoral condyle in females so that the space for ACL movement is limited; the wider pelvis of females that creates a greater angle between the femur and tibia and increases the risk for an ACL tear; female hormones that allow for greater flexibility of ligaments, muscles, and tendons but which do not permit them to absorb the stresses put on them and thus the stresses are transferred to the ACL; and females having less muscle strength so that they rely less on muscles and more on the ACL to hold the knee in place.

   b. *Posterior cruciate ligament (PCL).* Extends anteriorly and medially from a depression on the posterior intercondylar area of the tibia and lateral meniscus to the anterior part of the lateral surface of the medial condyle of the femur (Figure 9.15a, b, h). The PCL prevents the posterior sliding of the tibia (and anterior sliding of the femur) when the knee is flexed. This is very important when walking down stairs or a steep incline.

9. *Articular discs (menisci).* Two fibrocartilage discs between the tibial and femoral condyles help compensate for the irregular shapes of the bones and circulate synovial fluid.

   a. *Medial meniscus.* Semicircular piece of fibrocartilage (C-shaped). Its anterior end is attached to the anterior intercondylar fossa of the tibia, anterior to the anterior cruciate ligament. Its posterior end is attached to the posterior intercondylar fossa of the tibia between the attachments of the posterior cruciate ligament and lateral meniscus (Figure 9.15a, b, d, h).

   b. *Lateral meniscus.* Nearly circular piece of fibrocartilage (approaches an incomplete O in shape) (Figure 9.15a, b, d, h). Its anterior end is attached anteriorly to the intercondylar eminence of the tibia, and laterally and posteriorly to the anterior cruciate ligament. Its posterior end is attached posteriorly to the intercondylar eminence of the tibia, and anteriorly to the posterior end of the medial meniscus. The anterior surfaces of the medial and lateral menisci are connected to each other by the *transverse ligament of the knee* (Figure 9.15a) and to the margins of the head of the tibia by the *coronary ligaments* (not illustrated).

10. The more important bursae of the knee include the following:

    a. *Prepatellar bursa* between the patella and skin (Figure 9.15c, d).

    b. *Infrapatellar bursa* between superior part of tibia and patellar ligament (Figure 9.15c–e).

    c. *Suprapatellar bursa* between inferior part of femur and deep surface of quadriceps femoris muscle (Figure 9.15c–e).

Figure 9.15 Right knee (tibiofemoral) joint.

The knee joint is the largest and most complex joint in the body.

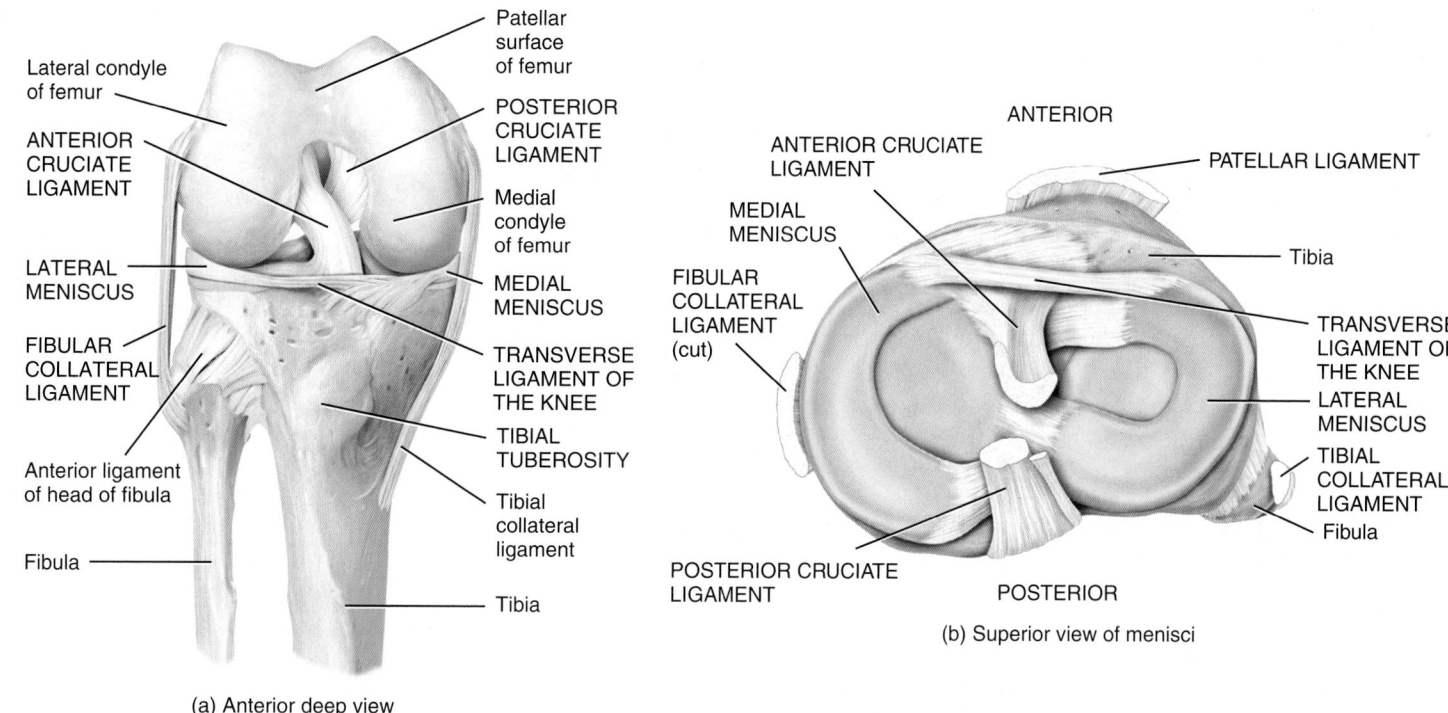

(a) Anterior deep view

Labels for (a):
- Lateral condyle of femur
- ANTERIOR CRUCIATE LIGAMENT
- LATERAL MENISCUS
- FIBULAR COLLATERAL LIGAMENT
- Anterior ligament of head of fibula
- Fibula
- Patellar surface of femur
- POSTERIOR CRUCIATE LIGAMENT
- Medial condyle of femur
- MEDIAL MENISCUS
- TRANSVERSE LIGAMENT OF THE KNEE
- TIBIAL TUBEROSITY
- Tibial collateral ligament
- Tibia

(b) Superior view of menisci

Labels for (b):
- ANTERIOR
- ANTERIOR CRUCIATE LIGAMENT
- MEDIAL MENISCUS
- FIBULAR COLLATERAL LIGAMENT (cut)
- POSTERIOR CRUCIATE LIGAMENT
- POSTERIOR
- PATELLAR LIGAMENT
- Tibia
- TRANSVERSE LIGAMENT OF THE KNEE
- LATERAL MENISCUS
- TIBIAL COLLATERAL LIGAMENT
- Fibula

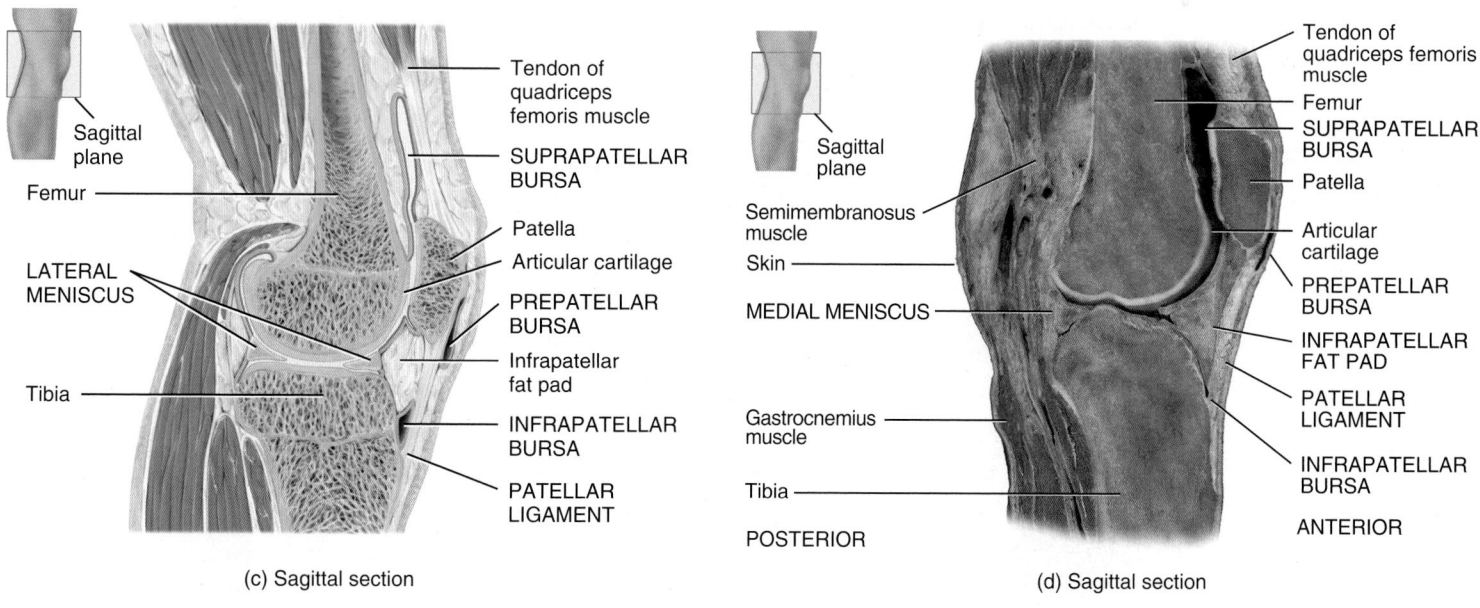

(c) Sagittal section

Labels for (c):
- Sagittal plane
- Femur
- LATERAL MENISCUS
- Tibia
- Tendon of quadriceps femoris muscle
- SUPRAPATELLAR BURSA
- Patella
- Articular cartilage
- PREPATELLAR BURSA
- Infrapatellar fat pad
- INFRAPATELLAR BURSA
- PATELLAR LIGAMENT

(d) Sagittal section

Labels for (d):
- Sagittal plane
- Semimembranosus muscle
- Skin
- MEDIAL MENISCUS
- Gastrocnemius muscle
- Tibia
- POSTERIOR
- Tendon of quadriceps femoris muscle
- Femur
- SUPRAPATELLAR BURSA
- Patella
- Articular cartilage
- PREPATELLAR BURSA
- INFRAPATELLAR FAT PAD
- PATELLAR LIGAMENT
- INFRAPATELLAR BURSA
- ANTERIOR

EXHIBIT 9.E CONTINUES ▶

EXHIBIT 9.E **317**

EXHIBIT 9.E   Knee Joint *(Figure 9.15)*   CONTINUED

Movements

The knee joint allows flexion, extension, slight medial rotation, and lateral rotation of the leg in the flexed position (see Figures 9.5f and 9.8c).

⬤ FIGURE 9.15 CONTINUED ▶

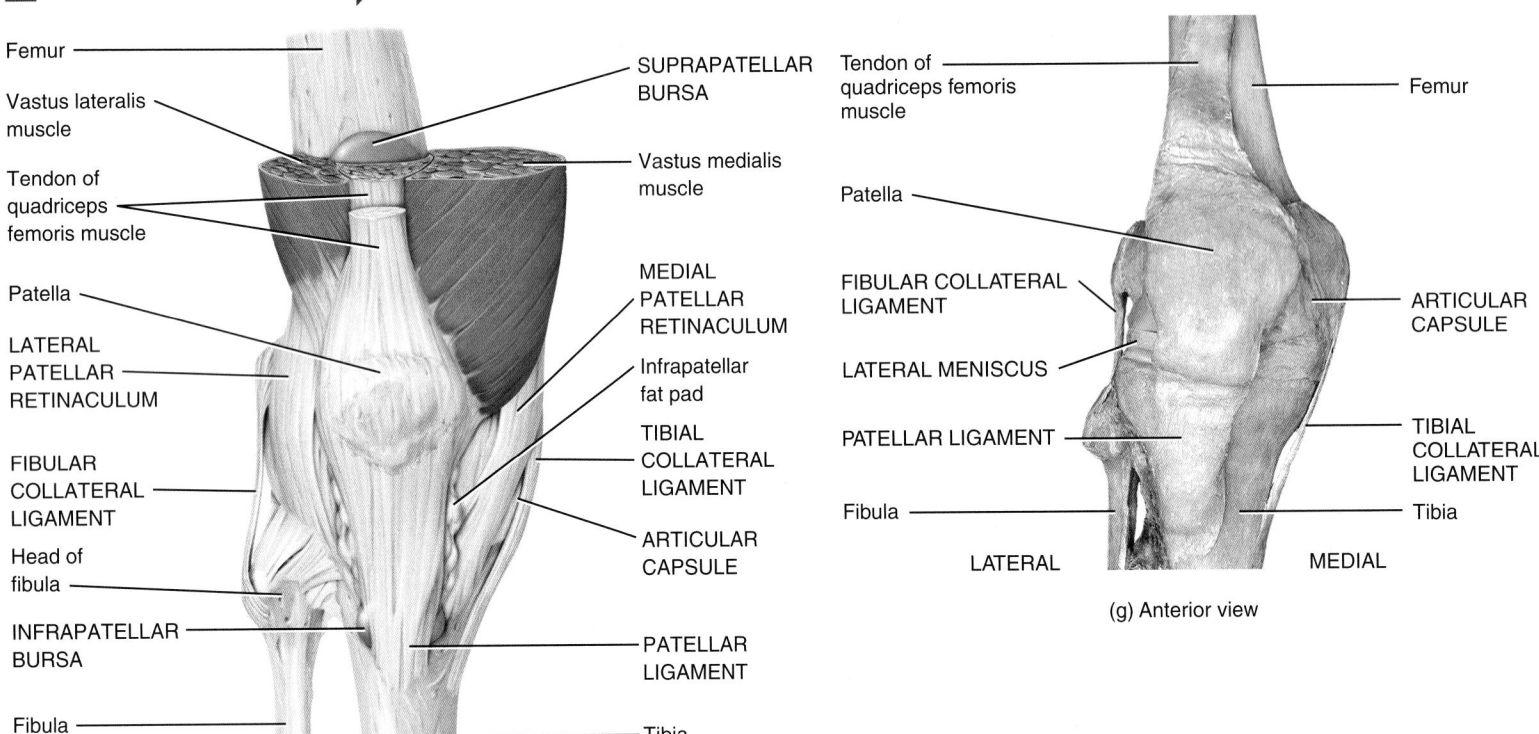

Femur

Vastus lateralis muscle

Tendon of quadriceps femoris muscle

Patella

LATERAL PATELLAR RETINACULUM

FIBULAR COLLATERAL LIGAMENT

Head of fibula

INFRAPATELLAR BURSA

Fibula

SUPRAPATELLAR BURSA

Vastus medialis muscle

MEDIAL PATELLAR RETINACULUM

Infrapatellar fat pad

TIBIAL COLLATERAL LIGAMENT

ARTICULAR CAPSULE

PATELLAR LIGAMENT

Tibia

(e) Anterior superficial view

Tendon of quadriceps femoris muscle

Patella

FIBULAR COLLATERAL LIGAMENT

LATERAL MENISCUS

PATELLAR LIGAMENT

Fibula

Femur

ARTICULAR CAPSULE

TIBIAL COLLATERAL LIGAMENT

Tibia

LATERAL

MEDIAL

(g) Anterior view

Femur

Tendon of adductor magnus muscle

Medial head of gastrocnemius muscle

TIBIAL COLLATERAL LIGAMENT

Popliteus muscle

Semimembranosus tendon

Tibia

ARTICULAR CAPSULE

Lateral head of gastrocnemius muscle

OBLIQUE POPLITEAL LIGAMENT

ARCUATE POPLITEAL LIGAMENT

FIBULAR COLLATERAL LIGAMENT

Posterior ligament of head of fibula

Fibula

(f) Posterior deep view

Articular cartilage

POSTERIOR CRUCIATE LIGAMENT (PCL)

TIBIAL COLLATERAL LIGAMENT

MEDIAL MENISCUS

OBLIQUE POPLITEAL LIGAMENT (cut)

Tibia

Femur

ANTERIOR CRUCIATE LIGAMENT (ACL)

FIBULAR COLLATERAL LIGAMENT

LATERAL MENISCUS

Posterior ligament of head of fibula

Fibula

MEDIAL

LATERAL

(h) Posterior view

CLINICAL CONNECTION | *Knee Injuries*

The knee joint is the joint most vulnerable to damage because it is a mobile, weight-bearing joint and its stability depends almost entirely on its associated ligaments and muscles. Further, there is no complementary fit between the surfaces of the articulating bones. Following are several kinds of **knee injuries.** A **swollen knee** may occur immediately or hours after an injury. The initial swelling is due to escape of blood from damaged blood vessels adjacent to areas of injury, including rupture of the anterior cruciate ligament, damage to synovial membranes, torn menisci, fractures, or collateral ligament sprains. Delayed swelling is due to excessive production of synovial fluid, a condition commonly referred to as "water on the knee."

The firm attachment of the tibial collateral ligament to the medial meniscus is clinically significant because tearing of the ligament typically also results in tearing of the meniscus. Such an injury may occur in sports such as football and rugby when the knee receives a blow from the lateral side while the foot is fixed on the ground. The force of the blow may also tear the anterior cruciate ligament, which is also connected to the medial meniscus. The term **"unhappy triad"** is applied to a knee injury that involves damage to the three components of the knee at the same time: the tibial collateral ligament, medial meniscus, and anterior cruciate ligament.

A **dislocated knee** refers to the displacement of the tibia relative to the femur. The most common type is dislocation anteriorly, resulting from hyperextension of the knee. A frequent consequence of a dislocated knee is damage to the popliteal artery.

If no surgery is required, treatment of knee injuries involves PRICE (protection, rest, ice, compression, and elevation) with some strengthening exercises and perhaps physical therapy. •

✔CHECKPOINT

What are the opposing functions of the anterior and posterior cruciate ligaments?

(i) Medial view

(j) Lateral view

? **What movement occurs at the knee joint when the quadriceps femoris (anterior thigh) muscles contract?**

EXHIBIT 9.E **319**

9.9 AGING AND JOINTS

⊙ **OBJECTIVE**

- Explain the effects of aging on joints.

Aging usually results in decreased production of synovial fluid in joints. In addition, the articular cartilage becomes thinner with age, and ligaments shorten and lose some of their flexibility. The effects of aging on joints are influenced by genetic factors and by wear and tear, and vary considerably from one person to another. Although degenerative changes in joints may begin as early as age 20, most changes do not occur until much later. By age 80, almost everyone develops some type of degeneration in the knees, elbows, hips, and shoulders. It is also common for elderly individuals to develop degenerative changes in the vertebral column, resulting in a hunched-over posture and pressure on nerve roots. One type of arthritis, called osteoarthritis (see Disorders: Homeostatic Imbalances at the end of this chapter), is at least partially age-related. Nearly everyone over age 70 has evidence of some osteoarthritic changes. Stretching and aerobic exercises that attempt to maintain full range of motion are helpful in minimizing the effects of aging. They help to maintain the effective functioning of ligaments, tendons, muscles, synovial fluid, and articular cartilage.

✔ **CHECKPOINT**

13. Which joints show evidence of degeneration in nearly all individuals as aging progresses?

9.10 ARTHROPLASTY

⊙ **OBJECTIVE**

- Explain the procedures involved in arthroplasty, and describe how a total hip replacement is performed.

Joints that have been severely damaged by diseases such as arthritis, or by injury, may be replaced surgically with artificial joints in a procedure referred to as **arthroplasty** (AR-thrō-plas′-tē; *arthr-* = joint; *-plasty* = plastic repair of). Although most joints in the body can be repaired by arthroplasty, the ones most commonly replaced are the hips, knees, and shoulders. About 400,000 hip replacements and about 300,000 knee replacements are performed annually in the United States. During the procedure, the ends of the damaged bones are removed and metal, ceramic, or plastic components are fixed in place. The goals of arthroplasty are to relieve pain and increase range of motion.

Hip Replacements

Partial hip replacements involve only the femur. **Total hip replacements** involve both the acetabulum and head of the femur (Figure 9.16a–c). The damaged portions of the acetabulum and the head of the femur are replaced by prefabricated prostheses (artificial devices). The acetabulum is shaped to accept the new socket, the head of the femur is removed, and the center of the femur is shaped to fit the femoral component. The acetabular component consists of a plastic such as polyethylene, and the femoral component is composed of a metal such as cobalt-chrome, titanium alloys, or stainless steel. These materials are designed to withstand a high degree of stress and to prevent a response by the immune system. Once the appropriate acetabular and femoral components are selected, they are attached to the healthy portion of bone with acrylic cement, which forms an interlocking mechanical bond.

Knee Replacements

Knee replacements are actually a resurfacing of cartilage and, like hip replacements, may be partial or total. In a **total knee replacement,** the damaged cartilage is removed from the distal end of the femur, the proximal end of the tibia, and the back surface of the patella (if the back surface of the patella is not badly damaged, it may be left intact) (Figure 9.16d–f). The femur is reshaped and fitted with a metal femoral component and cemented in place. The tibia is reshaped and fitted with a plastic tibial component that is cemented in place. If the back surface of the patella is badly damaged, it is replaced with a plastic patellar implant.

In a **partial knee replacement,** also called a **unicompartmental knee replacement,** only one side of the knee joint is replaced. Once the damaged cartilage is removed from the distal end of the femur, the femur is reshaped and a metal femoral component is cemented in place. Then the damaged cartilage from the proximal end of the tibia is removed, along with the meniscus. The tibia is reshaped and fitted with a plastic tibial component that is cemented into place. If the back surface of the patella is badly damaged, it is replaced with a plastic patellar component.

Researchers are continually seeking to improve the strength of the cement and devise ways to stimulate bone growth around the implanted area. Potential complications of arthroplasty include infection, blood clots, loosening or dislocation of the replacement components, and nerve injury.

With increasing sensitivity of metal detectors at airports and other public areas, it is possible that metal point replacements may activate metal detectors.

✔ **CHECKPOINT**

14. Which joints of the body most commonly undergo arthroplasty?

Figure 9.16 Total hip and knee replacement.

 In a total hip replacement, damaged portions of the acetabulum and the head of the femur are replaced by prostheses.

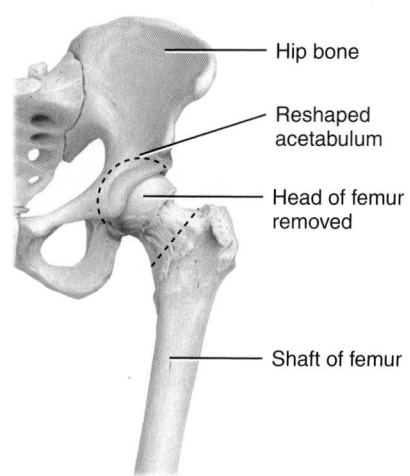

Hip bone

Reshaped acetabulum

Head of femur removed

Shaft of femur

(a) Preparation for total hip replacement

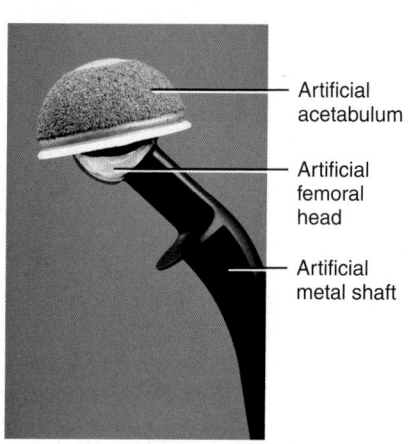

Artificial acetabulum

Artificial femoral head

Artificial metal shaft

(b) Components of an artificial hip joint prior to implantation

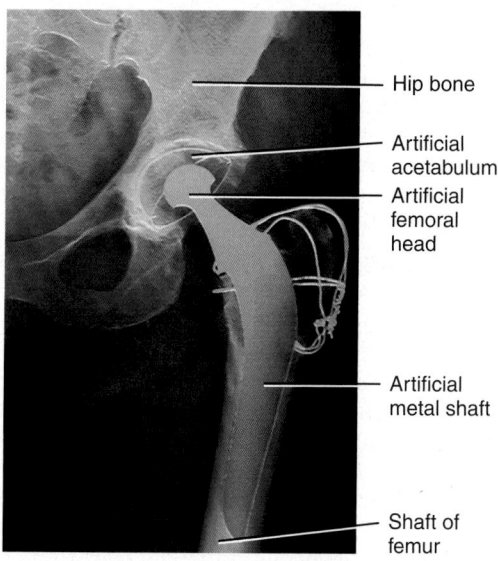

Hip bone

Artificial acetabulum

Artificial femoral head

Artificial metal shaft

Shaft of femur

(c) Radiograph of an artificial hip joint

Femoral surfaces removed

Tibial surfaces removed

Patellar surface removed

(d) Preparation for total knee replacement

Plastic spacer

Femoral component   Tibial component   Patellar component

Femoral component in place

Tibial component in place

(e) Components of artificial knee joint prior to implantation (left) and implanted (right)

Femur

Femoral component

Tibial component

Tibia

Fibula

(f) Radiograph of a total knee replacement

? **What is the purpose of arthroplasty?**

DISORDERS: HOMEOSTATIC IMBALANCES

Rheumatism and Arthritis

Rheumatism (ROO-ma-tizm) is any painful disorder of the supporting structures of the body—bones, ligaments, tendons, or muscles—that is not caused by infection or injury. **Arthritis** is a form of rheumatism in which the joints are swollen, stiff, and painful. It afflicts about 45 million people in the United States and is the leading cause of physical disability among adults over age 65.

Osteoarthritis

Osteoarthritis (OA) (os′-tē-ō-ar-THRĪ-tis) is a degenerative joint disease in which joint cartilage is gradually lost. It results from a combination of aging, obesity, irritation of the joints, muscle weakness, and wear and abrasion. Commonly known as "wear-and-tear" arthritis, osteoarthritis is the most common type of arthritis.

Osteoarthritis is a progressive disorder of synovial joints, particularly weight-bearing joints. Articular cartilage deteriorates and new bone forms in the subchondral areas and at the margins of the joint. The cartilage slowly degenerates, and as the bone ends become exposed, spurs (small bumps) of new osseous tissue are deposited on them in a misguided effort by the body to protect against increased friction. These spurs decrease the space of the joint cavity and restrict joint movement. Unlike rheumatoid arthritis (described next), osteoarthritis affects mainly the articular cartilage, although the synovial membrane often becomes inflamed late in the disease. Two major distinctions between osteoarthritis and rheumatoid arthritis are that osteoarthritis first afflicts the larger joints (knees, hips) and is due to wear and tear, whereas rheumatoid arthritis first strikes smaller joints and is an active attack on the cartilage. Osteoarthritis is the most common reason for hip- and knee-replacement surgery.

Rheumatoid Arthritis

Rheumatoid arthritis (RA) is an autoimmune disease in which the immune system of the body attacks its own tissues—in this case, its own cartilage and joint linings. RA is characterized by inflammation of the joint, which causes swelling, pain, and loss of function. Usually, this form of arthritis occurs bilaterally: If one wrist is affected, the other is also likely to be affected, although they are often not affected to the same degree.

The primary symptom of RA is inflammation of the synovial membrane. If untreated, the membrane thickens, and synovial fluid accumulates. The resulting pressure causes pain and tenderness. The membrane then produces an abnormal granulation tissue, called pannus, that adheres to the surface of the articular cartilage and sometimes erodes the cartilage completely. When the cartilage is destroyed, fibrous tissue joins the exposed bone ends. The fibrous tissue ossifies and fuses the joint so that it becomes immovable—the ultimate crippling effect of RA. Growth of the granulation tissue causes the distortion of the fingers that characterizes hands of RA sufferers.

Gouty Arthritis

Uric acid (a substance that gives urine its name) is a waste product produced during the metabolism of nucleic acid (DNA and RNA) subunits. A person who suffers from **gout** (GOWT) either produces excessive amounts of uric acid or is not able to excrete as much as normal. The result is a buildup of uric acid in the blood. This excess acid then reacts with sodium to form a salt called sodium urate. Crystals of this salt accumulate in soft tissues such as the kidneys and in the cartilage of the ears and joints.

In **gouty arthritis,** sodium urate crystals are deposited in the soft tissues of the joints. Gout most often affects the joints of the feet, especially at the base of the big toe. The crystals irritate and erode the cartilage, causing inflammation, swelling, and acute pain. Eventually, the crystals destroy all joint tissues. If the disorder is untreated, the ends of the articulating bones fuse, and the joint becomes immovable. Treatment consists of pain relief (ibuprofen, naproxen, colchicine, and cortisone) followed by administration of allopurinol to keep uric acid levels low so that crystals do not form.

Lyme Disease

A spiral-shaped bacterium called *Borrelia burgdorferi* causes **Lyme disease,** named for the town of Lyme, Connecticut, where it was first reported in 1975. The bacteria are transmitted to humans mainly by deer ticks (*Ixodes dammini*). These ticks are so small that their bites often go unnoticed. Within a few weeks of the tick bite, a rash may appear at the site. Although the rash often resembles a bull's-eye target, there are many variations, and some people never develop a rash. Other symptoms include joint stiffness, fever and chills, headache, stiff neck, nausea, and low back pain. In advanced stages of the disease, arthritis is the main complication. It usually afflicts the larger joints such as the knee, ankle, hip, elbow, or wrist. Antibiotics are generally effective against Lyme disease, especially if they are given promptly. However, some symptoms may linger for years.

Ankylosing Spondylitis

Ankylosing spondylitis (ang′-ki-LŌ-sing spon′-di-LĪ-tis; *ankyle* = stiff; *spondyl* = vertebra) is an inflammatory disease of unknown origin that affects joints between vertebrae (intervertebral) and between the sacrum and hip bone (sacroiliac joint). The disease, which is more common in males, sets in between ages 20 and 40. It is characterized by pain and stiffness in the hips and lower back that progress upward along the backbone. Inflammation can lead to *ankylosis* (severe or complete loss of movement at a joint) and *kyphosis* (hunchback). Treatment consists of anti-inflammatory drugs, heat, massage, and supervised exercise.

MEDICAL TERMINOLOGY

Arthralgia (ar-THRAL-jē-a; *arthr-* = joint; *-algia* = pain) Pain in a joint.

Bursectomy (bur-SEK-tō-mē; *-ectomy* = removal of) Removal of a bursa.

Chondritis (kon-DRĪ-tis; *chondr-* = cartilage) Inflammation of cartilage.

Subluxation (sub-luks-Ā-shun) A partial or incomplete dislocation.

Synovitis (sin′-ō-VĪ-tis) Inflammation of a synovial membrane in a joint.

CHAPTER REVIEW AND RESOURCE SUMMARY

Review

Resource

Introduction

1. A joint (articulation or arthrosis) is a point of contact between two bones, between bone and cartilage, or between bone and teeth.
2. A joint's structure may permit no movement, slight movement, or free movement.

Anatomy Overview - Joints

9.1 Joint Classifications

1. Structural classification is based on the presence or absence of a synovial cavity and the type of connective tissue. Structurally, joints are classified as fibrous, cartilaginous, or synovial.
2. Functional classification of joints is based on the degree of movement permitted. Joints may be synarthroses (immovable), amphiarthroses (slightly movable), or diarthroses (freely movable).

Anatomy Overview - Joints

9.2 Fibrous Joints

1. The bones of fibrous joints are held together by dense irregular connective tissue.
2. These joints include immovable or slightly movable sutures (found between skull bones), immovable to slightly movable syndesmoses (such as roots of teeth in the sockets in the mandible and maxilla and the distal tibiofibular joint), and slightly movable interosseous membranes (found between the radius and ulna in the forearm and the tibia and fibula in the leg).

Anatomy Overview - Fibrous Joints

9.3 Cartilaginous Joints

1. The bones of cartilaginous joints are held together by cartilage.
2. These joints include immovable synchondroses united by hyaline cartilage (epiphyseal plates between diaphyses and epiphyses) and slightly movable symphyses united by fibrocartilage (pubic symphysis).

Anatomy Overview - Cartilaginous Joints

9.4 Synovial Joints

1. Synovial joints contain a space between bones called the synovial cavity. All synovial joints are diarthroses.
2. Other characteristics of synovial joints are the presence of articular cartilage and an articular capsule, made up of a fibrous membrane and a synovial membrane.
3. The synovial membrane secretes synovial fluid, which forms a thin, viscous film over the surfaces within the articular capsule.
4. Many synovial joints also contain accessory ligaments (extracapsular and intracapsular) and articular discs (menisci).
5. Synovial joints contain an extensive nerve and blood supply. The nerves convey information about pain, joint movements, and the degree of stretch at a joint. Blood vessels penetrate the articular capsule and ligaments.
6. Bursae are saclike structures, similar in structure to joint capsules, that alleviate friction in joints such as the shoulder and knee joints.
7. Tendon sheaths are tubelike bursae that wrap around tendons where there is considerable friction.

Anatomy Overview - Synovial Joints
Anatomy Overview - Structural Subtypes of Synovial Joints
Figure 9.3 - Structure of a Typical Synovial Joint
Exercise - Judge That Joint

9.5 Types of Movements at Synovial Joints

1. In a gliding movement, the nearly flat surfaces of bones move back and forth and side to side.
2. In angular movements, a change in the angle between bones occurs. Examples are flexion–extension, lateral flexion, hyperextension, and abduction–adduction. Circumduction refers to the movement of the distal end of a body part in a circle and involves a continuous sequence of flexion, abduction, extension, adduction, and rotation of the joint (or in the opposite direction).
3. In rotation, a bone moves around its own longitudinal axis.
4. Special movements occur at specific synovial joints. Examples are elevation–depression, protraction–retraction, inversion–eversion, dorsiflexion–plantar flexion, supination–pronation, and opposition.
5. Table 9.1 summarizes the various types of movements at synovial joints.

Anatomy Overview - Movements Produced at Synovial Joints
Figure 9.5 - Angular Movements at Synovial Joints
Figure 9.6 - Angular Movements at Synovial Joints— Abduction and Adduction
Figure 9.7 - Angular Movements at Synovial Joints—Circumduction
Figure 9.8 - Rotation at Synovial Joints
Figure 9.9 - Special Movements at Synovial Joints
Exercise - Manage Movements

9.6 Types of Synovial Joints

1. Types of synovial joints are plane, hinge, pivot, condyloid, saddle, and ball-and-socket.
2. In a plane joint the articulating surfaces are flat, and the bones primarily glide back and forth and side to side (many are biaxial); they may also permit rotation (triaxial); examples are joints between carpals and tarsals.

Concepts and Connections - Joint Classifications

Review

3. In a hinge joint, the convex surface of one bone fits into the concave surface of another, and the motion is angular around one axis (uniaxial); examples are the elbow, knee (a modified hinge joint), and ankle joints.

4. In a pivot joint, a round or pointed surface of one bone fits into a ring formed by another bone and a ligament, and movement is rotational (uniaxial); examples are the atlanto-axial and radioulnar joints.

5. In a condyloid joint, an oval projection of one bone fits into an oval cavity of another, and motion is angular around two axes (biaxial); examples include the wrist joint and metacarpophalangeal joints of the second through fifth digits.

6. In a saddle joint, the articular surface of one bone is shaped like a saddle and the other bone fits into the saddle like a sitting rider; movement is biaxial. An example is the carpometacarpal joint between the trapezium and the metacarpal of the thumb.

7. In a ball-and-socket joint, the ball-shaped surface of one bone fits into the cuplike depression of another; motion is around three axes (triaxial). Examples include the shoulder and hip joints.

8. Table 9.2 summarizes the structural and functional categories of joints.

9.7 Factors Affecting Contact and Range of Motion at Synovial Joints

1. The ways that articular surfaces of synovial joints contact one another determine the type of movement that is possible.

2. Factors that contribute to keeping the surfaces in contact and affect range of motion are structure or shape of the articulating bones, strength and tension of the joint ligaments, arrangement and tension of the muscles, apposition of soft parts, hormones, and disuse.

9.8 Selected Joints of the Body

1. A summary of selected joints of the body, including articular components, structural and functional classifications, and movements, is presented in Tables 9.3 and 9.4.

2. The temporomandibular joint (TMJ) is between the condyle of the mandible and mandibular fossa and articular tubercle of the temporal bone (Exhibit 9.1).

3. The shoulder (humeroscapular or glenohumeral) joint is between the head of the humerus and glenoid cavity of the scapula (Exhibit 9.2).

4. The elbow joint is between the trochlea of the humerus, the trochlear notch of the ulna, and the head of the radius (Exhibit 9.3).

5. The hip (coxal) joint is between the head of the femur and acetabulum of the hip bone (Exhibit 9.4).

6. The knee (tibiofemoral) joint is between the patella and patellar surface of the femur; the lateral condyle of the femur, the lateral meniscus, and the lateral condyle of the tibia; and the medial condyle of the femur, the medial meniscus, and the medial condyle of the tibia (Exhibit 9.5).

9.9 Aging and Joints

1. With aging, a decrease in synovial fluid, thinning of articular cartilage, and decreased flexibility of ligaments occur.

2. Most individuals experience some degeneration in the knees, elbows, hips, and shoulders due to the aging process.

9.10 Arthroplasty

1. Arthroplasty refers to the surgical replacement of joints.

2. The most commonly replaced joints are the hips, knees, and shoulders.

Resource

Anatomy Overview - Specific Synovial Joints
Anatomy Overview - Specific Synovial Diarthrotic Joints
Figure 9.11 - Right Temporomandibular Joint
Figure 9.15 - Right Knee Joint

Homeostatic Imbalances - The Case of the Painful Hands

SELF-QUIZ QUESTIONS

Fill in the blanks in the following statements.

1. A point of contact between two bones, between bone and cartilage, or between bone and teeth is called a(n) _____.

2. The surgical procedure in which a severely damaged joint is replaced with an artificial joint is known as _____.

Indicate whether the following statements are true or false.

3. Menisci are fluid-filled sacs located outside of the joint cavity to ease friction between bones and softer tissue.

4. Shrugging your shoulders involves flexion and extension.

5. Synovial fluid becomes more viscous (thicker) as movement at the joint increases.

Choose the one best answer to the following questions.

6. Which of the following are structural classifications of joints? (1) amphiarthrosis, (2) cartilaginous, (3) synovial, (4) synarthrosis, (5) fibrous.
   (a) 1, 2, 3, 4, and 5   (b) 2 and 5   (c) 1 and 4
   (d) 1, 2, 4, and 5   (e) 2, 3, and 5

7. Which of the following joints could be classified functionally as synarthroses? (1) syndesmosis, (2) symphysis, (3) synovial, (4) gomphosis, (5) suture.
   (a) 1 and 2   (b) 3 and 5   (c) 1, 2, and 3
   (d) 4 and 5   (e) 5 only

8. The most common degenerative joint disease in the elderly, often caused by wear-and-tear, is
   (a) rheumatoid arthritis.   (b) osteoarthritis.   (c) rheumatism.
   (d) gouty arthritis.   (e) ankylosing spondylitis.

9. Chewing your food involves (1) flexion, (2) extension, (3) hyperextension, (4) elevation, (5) depression.
   (a) 1 and 2   (b) 1 and 3   (c) 4 and 5
   (d) 3 and 5   (e) 1 and 4

10. Synovial fluid functions to (1) absorb shocks at joints, (2) lubricate joints, (3) form a blood clot in a joint injury, (4) supply oxygen and nutrients to chondrocytes, (5) provide phagocytes to remove debris from joints.
    (a) 1, 2, 4, and 5   (b) 1, 2, 3, 4, and 5   (c) 1, 2, and 4
    (d) 3 and 4   (e) 2, 4, and 5

11. Which of the following statements are *true* concerning a synovial joint? (1) The bones at a synovial joint are covered by a mucous membrane. (2) The articular capsule surrounds a synovial joint, encloses the synovial cavity, and unites the articulating bones. (3) The fibrous membrane of the articular capsule permits considerable movement at a joint. (4) The tensile strength of the fibrous membrane helps prevent bones from disarticulating. (5) All synovial joints contain a fibrous membrane.
    (a) 1, 2, 3, and 4   (b) 2, 3, 4, and 5   (c) 2, 3, and 4
    (d) 1, 2, and 3   (e) 2, 4, and 5

12. Which of the following keep the articular surfaces of synovial joints in contact and affect range of motion? (1) structure or shape of the articulating bones, (2) strength and tension of the joint ligaments, (3) arrangement and tension of muscles, (4) lack of use, (5) contact of soft parts.
    (a) 1, 2, 3, and 5   (b) 2, 3, 4, and 5   (c) 1, 3, 4, and 5
    (d) 1, 3, and 5   (e) 1, 2, 3, 4, and 5

13. Match the following:
    ____ (a) a fibrous joint that unites the bones of the skull; a synarthrosis
    ____ (b) a fibrous joint between the tibia and fibula; an amphiarthrosis
    ____ (c) the articulation between bone and teeth
    ____ (d) the epiphyseal (growth) plate
    ____ (e) joint between the two pubic bones
    ____ (f) joint with a cavity between the bones; diarthrosis
    ____ (g) a bony joint

    (1) synostosis
    (2) synchondrosis
    (3) syndesmosis
    (4) synovial
    (5) suture
    (6) symphysis
    (7) gomphosis

14. Match the following:
    ____ (a) rounded or pointed surface of one bone articulates with a ring formed by another bone and a ligament; allows rotation around its own axis
    ____ (b) articulating bone surfaces are flat or slightly curved; permit gliding movement
    ____ (c) convex, oval projection of one bone fits into oval depression of another bone; permits movement in two axes
    ____ (d) convex surface of one bone articulates with concave surface of another bone; permits flexion and extension
    ____ (e) ball-shaped surface of one bone articulates with cuplike depression of another bone; permits largest degree of movement in three axes
    ____ (f) modified condyloid joint where articulating bones resemble a rider sitting in a saddle

    (1) hinge joint
    (2) saddle joint
    (3) ball-and-socket joint
    (4) plane joint
    (5) condyloid joint
    (6) pivot joint

15. Match the following:

____ (a) upward movement of a body part

____ (b) downward movement of a body part

____ (c) movement of bone toward midline

____ (d) movement in which relatively flat bone surfaces move back-and-forth and side-to-side with respect to one another

____ (e) movement of a body part anteriorly in the transverse plane

____ (f) decrease in angle between bones

____ (g) movement of an anteriorly projected body part back to the anatomical position

____ (h) movement of the sole medially

____ (i) movement of the sole laterally

____ (j) movement of bone away from midline

____ (k) action that occurs when you stand on your heels

____ (l) action that occurs when you stand on your toes

____ (m) movement of the forearm to turn the palm anteriorly

____ (n) movement of the forearm to turn the palm posteriorly

____ (o) movement of thumb across the palm to touch the tips of the fingers of the same hand

____ (p) increase in angle between bones

____ (q) movement of distal end of a part of the body in a circle

____ (r) bone revolves around its own longitudinal axis

(1) pronation
(2) plantar flexion
(3) eversion
(4) abduction
(5) rotation
(6) retraction
(7) opposition
(8) elevation
(9) flexion
(10) adduction
(11) depression
(12) inversion
(13) gliding
(14) extension
(15) protraction
(16) dorsiflexion
(17) circumduction
(18) supination

CRITICAL THINKING QUESTIONS

1. Katie loves pretending that she's a human cannonball. As she jumps off the diving board, she assumes the proper position before she pounds into the water: head and thighs tucked against her chest; back rounded; arms pressed against her sides while her forearms, crossed in front of her shins, hold her legs tightly folded against her chest. Use the proper anatomical terms to describe the position of Katie's back, head, and free limbs.

2. During football practice, Jeremiah was tackled and twisted his lower leg. There was a sharp pain, followed immediately by swelling of the knee joint. The pain and swelling worsened throughout the remainder of the afternoon until Jeremiah could barely walk. The coach told Jeremiah to see a doctor who might want to "drain the water off his knee." What was the coach referring to and what specifically do you think happened to Jeremiah's knee joint to cause these symptoms?

3. After lunch, during a particularly long and dull class video, Antonio became sleepy and yawned. To his dismay, he was then unable to close his mouth. Explain what happened and what should be done to correct this problem.

? ANSWERS TO FIGURE QUESTIONS

9.1 Functionally, sutures are classified as synarthroses because they are immovable; syndesmoses are classified as amphiarthroses because they are slightly movable.

9.2 The structural difference between a synchondrosis and a symphysis is the type of cartilage that holds the joint together: hyaline cartilage in a synchondrosis and fibrocartilage in a symphysis.

9.3 Functionally, synovial joints are diarthroses, freely movable joints.

9.4 Gliding movements occur at intercarpal joints and at intertarsal joints.

9.5 Two examples of flexion that do not occur along the sagittal plane are flexion of the thumb and lateral flexion of the trunk.

9.6 When you adduct your arm or leg, you bring it closer to the midline of the body, thus "adding" it to the trunk.

9.7 Circumduction involves flexion, abduction, extension, adduction, and rotation in a continuous sequence (or in the opposite order).

9.8 The anterior surface of a bone or limb rotates toward the midline in medial rotation, and away from the midline in lateral rotation.

9.9 Bringing your arms forward until the elbows touch is an example of protraction.

9.10 Many plane, saddle, and condyloid joints are biaxial joints.

9.11 The lateral ligament prevents displacement of the mandible.

9.12 The shoulder joint is the most freely movable joint in the body because of the looseness of its articular capsule and the shallowness of the glenoid cavity in relation to the size of the head of the humerus.

9.13 A hinge joint permits flexion and extension.

9.14 Tension in three ligaments—iliofemoral, pubofemoral, and ischiofemoral—limits the degree of extension at the hip joint.

9.15 Contraction of the quadriceps femoris muscle causes extension at the knee joint.

9.16 The purpose of arthroplasty is to relieve joint pain and permit greater range of motion.

10 MUSCULAR TISSUE

MUSCULAR TISSUE AND HOMEOSTASIS *Muscular tissue contributes to homeostasis by producing body movements, moving substances through the body, and producing heat to maintain normal body temperature.*

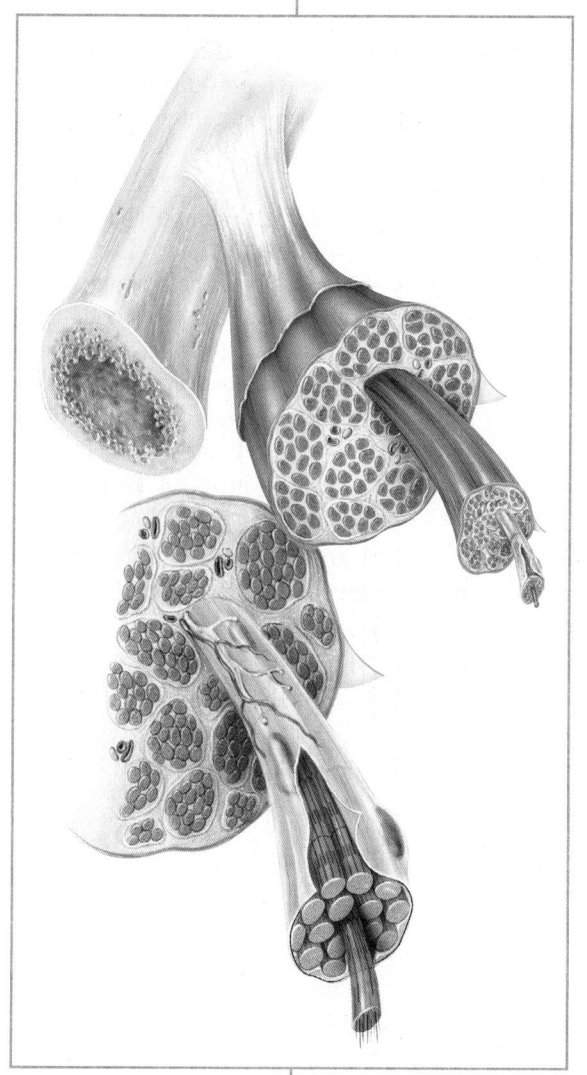

Although bones provide leverage and form the framework of the body, they cannot move body parts by themselves. Motion results from the alternating contraction and relaxation of muscles, which make up 40–50% of total adult body weight (depending on the percentage of body fat, gender, and exercise regimen). Your muscular strength reflects the primary function of muscle—the transformation of chemical energy into mechanical energy to generate force, perform work, and produce movement. In addition, muscle tissues stabilize body position, regulate organ volume, generate heat, and propel fluids and food matter through various body systems. The scientific study of muscles is known as **myology** (mī-OL-ō-jē; *myo-* = muscle; *-logy* = study of).

? *Did you ever wonder what causes rigor mortis?*

10.1 OVERVIEW OF MUSCULAR TISSUE

• Explain the structural differences between the three types of muscular tissue.
• Compare the functions and special properties of the three types of muscular tissue.

Types of Muscular Tissue

The three types of muscular tissue—skeletal, cardiac, and smooth—were introduced in Chapter 4 (see Table 4.9). Although the different types of muscular tissue share some properties, they differ from one another in their microscopic anatomy, location, and how they are controlled by the nervous and endocrine systems.

Skeletal muscle tissue is so named because most skeletal muscles move bones of the skeleton. (A few skeletal muscles attach to and move the skin or other skeletal muscles.) Skeletal muscle tissue is *striated:* Alternating light and dark protein bands (*striations*) are seen when the tissue is examined with a microscope (see Table 4.9). Skeletal muscle tissue works mainly in a *voluntary* manner. Its activity can be consciously controlled by neurons (nerve cells) that are part of the somatic (voluntary) division of the nervous system. (Figure 12.10 depicts the divisions of the nervous system.) Most skeletal muscles also are controlled subconsciously to some extent. For example, your diaphragm continues to alternately contract and relax without conscious control so that you don't stop breathing. Also, you do not need to consciously think about contracting the skeletal muscles that maintain your posture or stabilize body positions.

Only the heart contains **cardiac muscle tissue,** which forms most of the heart wall. Cardiac muscle is also *striated,* but its action is *involuntary.* The alternating contraction and relaxation of the heart is not consciously controlled. Rather, the heart beats because it has a pacemaker that initiates each contraction. This built-in rhythm is termed **autorhythmicity** (aw′-tō-rith-MISS-i-tē). Several hormones and neurotransmitters can adjust heart rate by speeding or slowing the pacemaker.

Smooth muscle tissue is located in the walls of hollow internal structures, such as blood vessels, airways, and most organs in the abdominopelvic cavity. It is also found in the skin, attached to hair follicles. Under a microscope, this tissue lacks the striations of skeletal and cardiac muscle tissue. For this reason, it looks *nonstriated,* which is why it is referred to as *smooth.* The action of smooth muscle is usually *involuntary,* and some smooth muscle tissue, such as the muscles that propel food through your gastrointestinal tract, has autorhythmicity. Both cardiac muscle and smooth muscle are regulated by neurons that are part of the autonomic (involuntary) division of the nervous system and by hormones released by endocrine glands.

Functions of Muscular Tissue

Through sustained contraction or alternating contraction and relaxation, muscular tissue has four key functions: producing body movements, stabilizing body positions, storing and moving substances within the body, and generating heat.

1. ***Producing body movements.*** Movements of the whole body such as walking and running, and localized movements such as grasping a pencil, keyboarding, or nodding the head as a result of muscular contractions, rely on the integrated functioning of skeletal muscles, bones, and joints.

2. ***Stabilizing body positions.*** Skeletal muscle contractions stabilize joints and help maintain body positions, such as standing or sitting. Postural muscles contract continuously when you are awake; for example, sustained contractions of your neck muscles hold your head upright when you are listening intently to your anatomy and physiology lecture.

3. ***Storing and moving substances within the body.*** Storage is accomplished by sustained contractions of ringlike bands of smooth muscle called *sphincters,* which prevent outflow of the contents of a hollow organ. Temporary storage of food in the stomach or urine in the urinary bladder is possible because smooth muscle sphincters close off the outlets of these organs. Cardiac muscle contractions of the heart pump blood through the blood vessels of the body. Contraction and relaxation of smooth muscle in the walls of blood vessels help adjust blood vessel diameter and thus regulate the rate of blood flow. Smooth muscle contractions also move food and substances such as bile and enzymes through the gastrointestinal tract, push gametes (sperm and oocytes) through the passageways of the reproductive systems, and propel urine through the urinary system. Skeletal muscle contractions promote the flow of lymph and aid the return of blood in veins to the heart.

4. ***Generating heat.*** As muscular tissue contracts, it produces heat, a process known as **thermogenesis** (ther′-mō-JEN-e-sis). Much of the heat generated by muscle is used to maintain normal body temperature. Involuntary contractions of skeletal muscles, known as *shivering,* can increase the rate of heat production.

Properties of Muscular Tissue

Muscular tissue has four special properties that enable it to function and contribute to homeostasis:

1. **Electrical excitability** (ek-sīt′-a-BIL-i-tē), a property of both muscle and nerve cells that was introduced in Chapter 4, is the ability to respond to certain stimuli by producing electrical signals called **action potentials (impulses).** Action potentials in muscles are referred to as *muscle action potentials;* those in nerve cells are called *nerve action potentials.* Chapter 12 provides more detail about how action potentials arise (see Section 12.3). For muscle cells, two main types of stimuli trigger action potentials. One is autorhythmic *electrical signals* arising in the muscular tissue itself, as in the heart's pacemaker. The other is *chemical stimuli,* such as neurotransmitters released by neurons, hormones distributed by the blood, or even local changes in pH.

2. **Contractility** (kon′-trak-TIL-i-tē) is the ability of muscular tissue to contract forcefully when stimulated by an action potential. When a skeletal muscle contracts, it generates tension (force of contraction) while pulling on its attachment points. In some muscle contractions, the muscle develops tension (force of contraction) but does not shorten. An example is holding this book in an outstretched hand. In other muscle contractions, the tension generated is great enough to overcome the load (resistance) of the object being moved so the muscle shortens and movement occurs. An example is lifting a book off a table.

3. **Extensibility** (ek-sten′-si-BIL-i-tē) is the ability of muscular tissue to stretch, within limits, without being damaged. The connective tissue within the muscle limits the range of extensibility and keeps it within the contractile range of the muscle cells. Normally, smooth muscle is subject to the greatest amount of stretching. For example, each time your stomach fills with food, the smooth muscle in the wall is stretched. Cardiac muscle also is stretched each time the heart fills with blood.

4. **Elasticity** (e-las-TIS-i-tē) is the ability of muscular tissue to return to its original length and shape after contraction or extension.

Skeletal muscle is the focus of much of this chapter. Cardiac muscle and smooth muscle are described briefly here. Cardiac muscle is discussed in more detail in Chapter 20 (the heart), and smooth muscle is included in Chapter 15 (the autonomic nervous system), as well as in discussions of the various organs containing smooth muscle.

✔ CHECKPOINT

1. What features distinguish the three types of muscular tissue?
2. List the general functions of muscular tissue.
3. Describe the four properties of muscular tissue.

10.2 SKELETAL MUSCLE TISSUE

◉ OBJECTIVES

- Explain the importance of connective tissue components, blood vessels, and nerves to skeletal muscles.
- Describe the microscopic anatomy of a skeletal muscle fiber.
- Distinguish thick filaments from thin filaments.
- Describe the functions of skeletal muscle proteins.

Each of your skeletal muscles is a separate organ composed of hundreds to thousands of cells, which are called **muscle fibers** because of their elongated shapes. Thus, *muscle cell* and *muscle fiber* are two terms for the same structure. Skeletal muscle also contains connective tissues surrounding muscle fibers and whole muscles, and blood vessels and nerves (Figure 10.1). To understand how contraction of skeletal muscle can generate tension, you must first understand its gross and microscopic anatomy.

Connective Tissue Components

Connective tissue surrounds and protects muscular tissue. The **subcutaneous layer** or **hypodermis,** which separates muscle from skin (see Figure 11.21), is composed of areolar connective tissue and adipose tissue. It provides a pathway for nerves, blood vessels, and lymphatic vessels to enter and exit muscles. The adipose tissue of the subcutaneous layer stores most of the body's triglycerides, serves as an insulating layer that reduces heat loss, and protects muscles from physical trauma. **Fascia** (FASH-ē-a = bandage) is a dense sheet or broad band of irregular connective tissue that lines the body wall and limbs and supports and surrounds muscles and other organs of the body. As you will see, fascia holds muscles with similar functions together (see Figure 11.21). Fascia allows free movement of muscles; carries nerves, blood vessels, and lymphatic vessels; and fills spaces between muscles.

Three layers of connective tissue extend from the fascia to protect and strengthen skeletal muscle (Figure 10.1):

- The outermost layer of dense, irregular connective tissue, encircling the entire muscle, is the **epimysium** (ep-i-MĪZ-ē-um; *epi-* = upon).
- **Perimysium** (per-i-MĪZ-ē-um; *peri-* = around) is also a layer of dense, irregular connective tissue, but it surrounds groups of 10 to 100 or more muscle fibers, separating them into bundles called **fascicles** (FAS-i-kuls = little bundles). Many fascicles are large enough to be seen with the naked eye. They give a cut of meat its characteristic "grain"; if you tear a piece of meat, it rips apart along the fascicles.
- **Endomysium** (en′-dō-MĪZ-ē-um; *endo-* = within) penetrates the interior of each fascicle and separates individual muscle fibers from one another. The endomysium is mostly reticular fibers.

The epimysium, perimysium, and endomysium all are continuous with the connective tissue that attaches skeletal muscle to other structures, such as bone or another muscle. For example, all three connective tissue layers may extend beyond the muscle fibers to form a ropelike **tendon** that attaches a muscle to the periosteum of a bone. An example is the *calcaneal (Achilles) tendon* of the gastrocnemius (calf) muscle, which attaches the muscle to the calcaneus (heel bone) (shown in Figure 11.22c). When the connective tissue elements extend as a broad, flat sheet, it is called an **aponeurosis** (ap-ō-noo-RŌ-sis; *apo-* = from; *-neur-* = a sinew). An example is the *epicranial aponeurosis* on top of the skull between the frontal and occipital bellies of the occipitofrontalis muscle (shown in Figure 11.4a, c).

⚕ **CLINICAL CONNECTION | *Fibromyalgia***

Fibromyalgia (fī-brō-mī-AL-jē-a; *algia* = painful condition) is a chronic, painful, nonarticular rheumatic disorder that affects the fibrous connective tissue components of muscles, tendons, and ligaments. A striking sign is pain that results from gentle pressure at specific "tender points." Even without pressure, there is pain, tenderness, and stiffness of muscles, tendons, and surrounding soft tissues. Besides muscle pain, those with fibromyalgia report severe fatigue, poor sleep, headaches, depression, irritable bowel syndrome, and inability to carry out their daily activities. There is no specific identifiable cause. Treatment consists of stress reduction, regular exercise, application of heat, gentle massage, physical therapy, medication for pain, and a low-dose antidepressant to help improve sleep. •

Figure 10.1 Organization of skeletal muscle and its connective tissue coverings.

A skeletal muscle consists of individual muscle fibers (cells) bundled into fascicles and surrounded by three connective tissue layers that are extensions of the fascia.

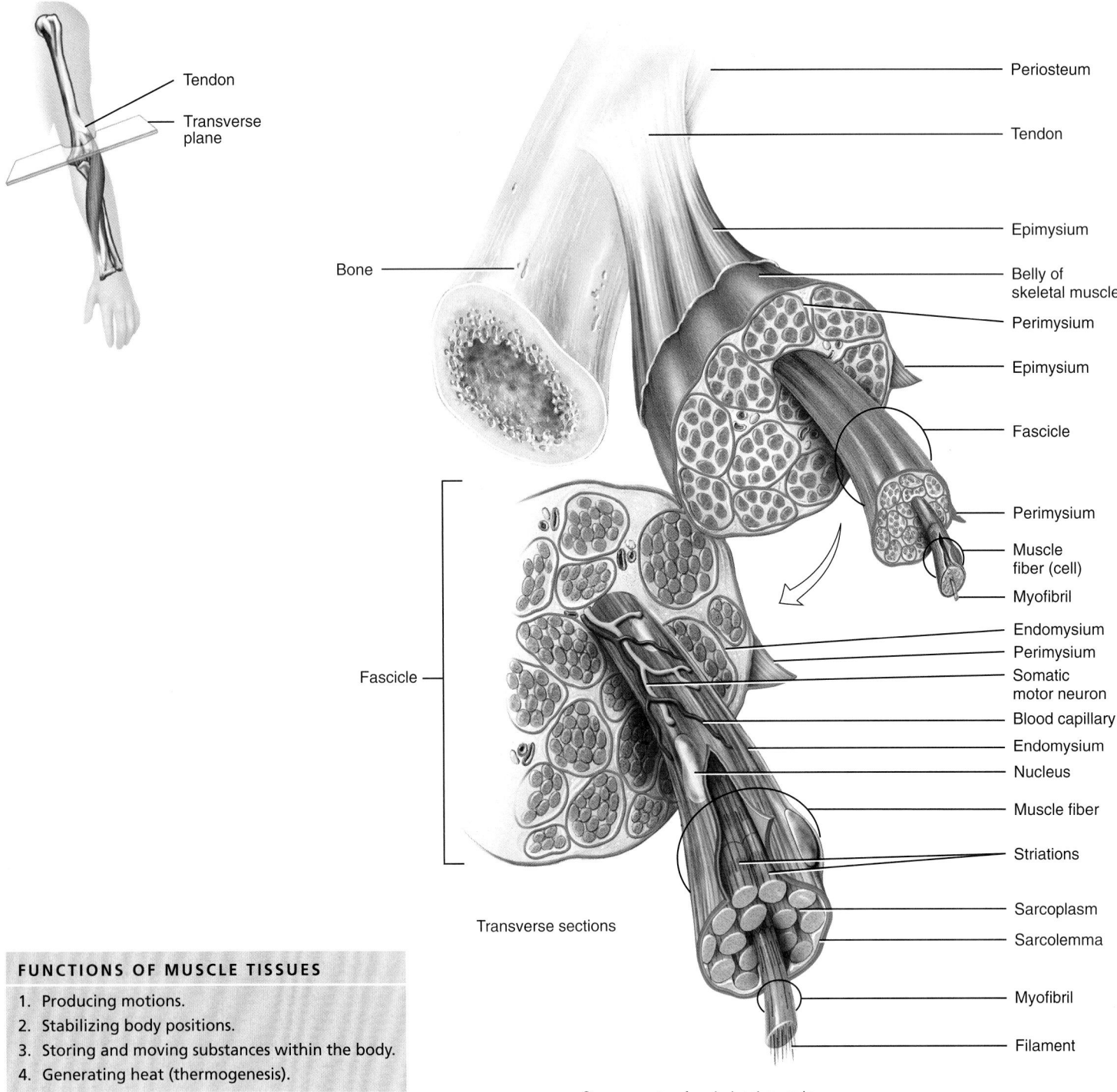

Tendon

Transverse plane

Bone

Fascicle

Transverse sections

Periosteum

Tendon

Epimysium

Belly of skeletal muscle

Perimysium

Epimysium

Fascicle

Perimysium

Muscle fiber (cell)

Myofibril

Endomysium

Perimysium

Somatic motor neuron

Blood capillary

Endomysium

Nucleus

Muscle fiber

Striations

Sarcoplasm

Sarcolemma

Myofibril

Filament

Components of a skeletal muscle

FUNCTIONS OF MUSCLE TISSUES

1. Producing motions.
2. Stabilizing body positions.
3. Storing and moving substances within the body.
4. Generating heat (thermogenesis).

Which connective tissue coat surrounds groups of muscle fibers, separating them into fascicles?

Nerve and Blood Supply

Skeletal muscles are well supplied with nerves and blood vessels. Generally, an artery and one or two veins accompany each nerve that penetrates a skeletal muscle. The neurons that stimulate skeletal muscle to contract are *somatic motor neurons*. Each somatic motor neuron has a threadlike axon that extends from the brain or spinal cord to a group of skeletal muscle fibers (see Figure 10.9d). The axon of a somatic motor neuron typically branches many times, each branch extending to a different skeletal muscle fiber.

Microscopic blood vessels called capillaries are plentiful in muscular tissue; each muscle fiber is in close contact with one or more capillaries (see Figure 10.9d). The blood capillaries bring in oxygen and nutrients and remove heat and the waste products of muscle metabolism. Especially during contraction, a muscle fiber synthesizes and uses considerable ATP (adenosine triphosphate). These reactions, which you will learn more about later on, require oxygen, glucose, fatty acids, and other substances that are delivered to the muscle fiber in the blood.

Microscopic Anatomy of a Skeletal Muscle Fiber

The most important components of a skeletal muscle are the muscle fibers themselves. The diameter of a mature skeletal muscle fiber ranges from 10 to 100 μm.* The typical length of a mature skeletal muscle fiber is about 10 cm (4 in.), although some are as long as 30 cm (12 in.). Because each skeletal muscle fiber arises during embryonic development from the fusion of a hundred or more small mesodermal cells called *myoblasts* (MĪ-ō-blasts) (Figure 10.2a), each mature skeletal muscle fiber has a hundred or more nuclei. Once fusion has occurred, the muscle fiber loses its ability to undergo cell division. Thus, the number of skeletal muscle fibers is set before you are born, and most of these cells last a lifetime.

Sarcolemma, Transverse Tubules, and Sarcoplasm

The multiple nuclei of a skeletal muscle fiber are located just beneath the **sarcolemma** (sar'-kō-LEM-ma; *sarc-* = flesh; *-lemma* = sheath), the plasma membrane of a muscle cell (Figure 10.2b, c). Thousands of tiny invaginations of the sarcolemma, called **transverse (T) tubules,** tunnel in from the surface toward the center of each muscle fiber. Because T tubules are open to the outside of the fiber, they are filled with interstitial fluid. Muscle action potentials travel along the sarcolemma and through the T tubules, quickly spreading throughout the muscle fiber. This arrangement ensures that an action potential excites all parts of the muscle fiber at essentially the same instant.

Within the sarcolemma is the **sarcoplasm** (SAR-kō-plazm), the cytoplasm of a muscle fiber. Sarcoplasm includes a substantial amount of glycogen, which is a large molecule composed of many glucose molecules (see Figure 2.16). Glycogen can be used for synthesis of ATP. In addition, the sarcoplasm contains a red-colored protein called **myoglobin** (mī-ō-GLŌB-in). This protein, found only in muscle, binds oxygen molecules that diffuse into muscle fibers from interstitial fluid. Myoglobin releases

*One micrometer (μm) is 10^{-6} meter (1/25,000 in.).

oxygen when it is needed by the mitochondria for ATP production. The mitochondria lie in rows throughout the muscle fiber, strategically close to the contractile muscle proteins that use ATP during contraction so that ATP can be produced quickly as needed (Figure 10.2c).

Myofibrils and Sarcoplasmic Reticulum

At high magnification, the sarcoplasm appears stuffed with little threads. These small structures are the **myofibrils** (mī-ō-FĪ-brils; *myo-* = muscle; *-fibrilla* = little fiber), the contractile organelles of skeletal muscle (Figure 10.2c). Myofibrils are about 2 μm in diameter and extend the entire length of a muscle fiber. Their prominent striations make the entire skeletal muscle fiber appear striped (striated).

A fluid-filled system of membranous sacs called the **sarcoplasmic reticulum** (sar'-kō-PLAZ-mik re-TIK-ū-lum) or **SR** encircles each myofibril (Figure 10.2c). This elaborate system is similar to smooth endoplasmic reticulum in nonmuscular cells. Dilated end sacs of the sarcoplasmic reticulum called **terminal cisterns** (SIS-terns = reservoirs) butt against the T tubule from

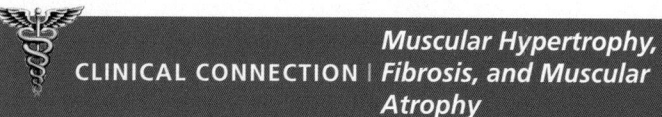

CLINICAL CONNECTION | *Muscular Hypertrophy, Fibrosis, and Muscular Atrophy*

The muscle growth that occurs after birth occurs by enlargement of existing muscle fibers, called **muscular hypertrophy** (hī-PER-trō-fē; *hyper-* = above or excessive; *-trophy* = nourishment), rather than by **muscular hyperplasia** (hī-per-PLĀ-zē-a; *-plasis* = molding), an increase in the number of fibers. Muscular hypertrophy is due to increased production of myofibrils, mitochondria, sarcoplasmic reticulum, and other organelles. It results from very forceful, repetitive muscular activity, such as strength training. Because hypertrophied muscles contain more myofibrils, they are capable of more forceful contractions. During childhood, human growth hormone and other hormones stimulate an increase in the size of skeletal muscle fibers. The hormone testosterone (from the testes in males and in small amounts from other tissues such as the ovaries in females) promotes further enlargement of muscle fibers.

A few myoblasts do persist in mature skeletal muscle as *satellite cells* (Figure 10.2a, b). Satellite cells retain the capacity to fuse with one another or with damaged muscle fibers to regenerate functional muscle fibers. However, the number of new skeletal muscle fibers that can be formed by satellite cells is not enough to compensate for significant skeletal muscle damage or degeneration. In such cases, skeletal muscle tissue undergoes **fibrosis,** the replacement of muscle fibers by fibrous scar tissue. Thus, regeneration of skeletal muscle tissue is limited.

Muscular atrophy (AT-rō-fē; *a-* = without, *-trophy* = nourishment) is a wasting away of muscles. Individual muscle fibers decrease in size as a result of progressive loss of myofibrils. Muscular atrophy that occurs because muscles are not used is termed *disuse atrophy.* Bedridden individuals and people with casts experience disuse atrophy because the flow of nerve impulses (nerve action potentials) to inactive skeletal muscle is greatly reduced, but the condition is reversible. If instead the nerve supply to a muscle is disrupted or cut, the muscle undergoes *denervation atrophy.* Over a period of 6 months to 2 years, the muscle shrinks to about one-fourth its original size, and the muscle fibers are irreversibly replaced by fibrous connective tissue. •

Figure 10.2 Microscopic organization of skeletal muscle. (a) During embryonic development, many myoblasts fuse to form one skeletal muscle fiber. Once fusion has occurred, a skeletal muscle fiber loses the ability to undergo cell division, but satellite cells retain this ability. (b–d) The sarcolemma of the fiber encloses sarcoplasm and myofibrils, which are striated. Sarcoplasmic reticulum wraps around each myofibril. Thousands of transverse tubules, filled with interstitial fluid, invaginate from the sarcolemma toward the center of the muscle fiber. A triad is a transverse tubule and the two terminal cisterns of the sarcoplasmic reticulum on either side of it. A photomicrograph of skeletal muscle tissue is shown in Table 4.9.

The contractile elements of muscle fibers, the myofibrils, contain overlapping thick and thin filaments.

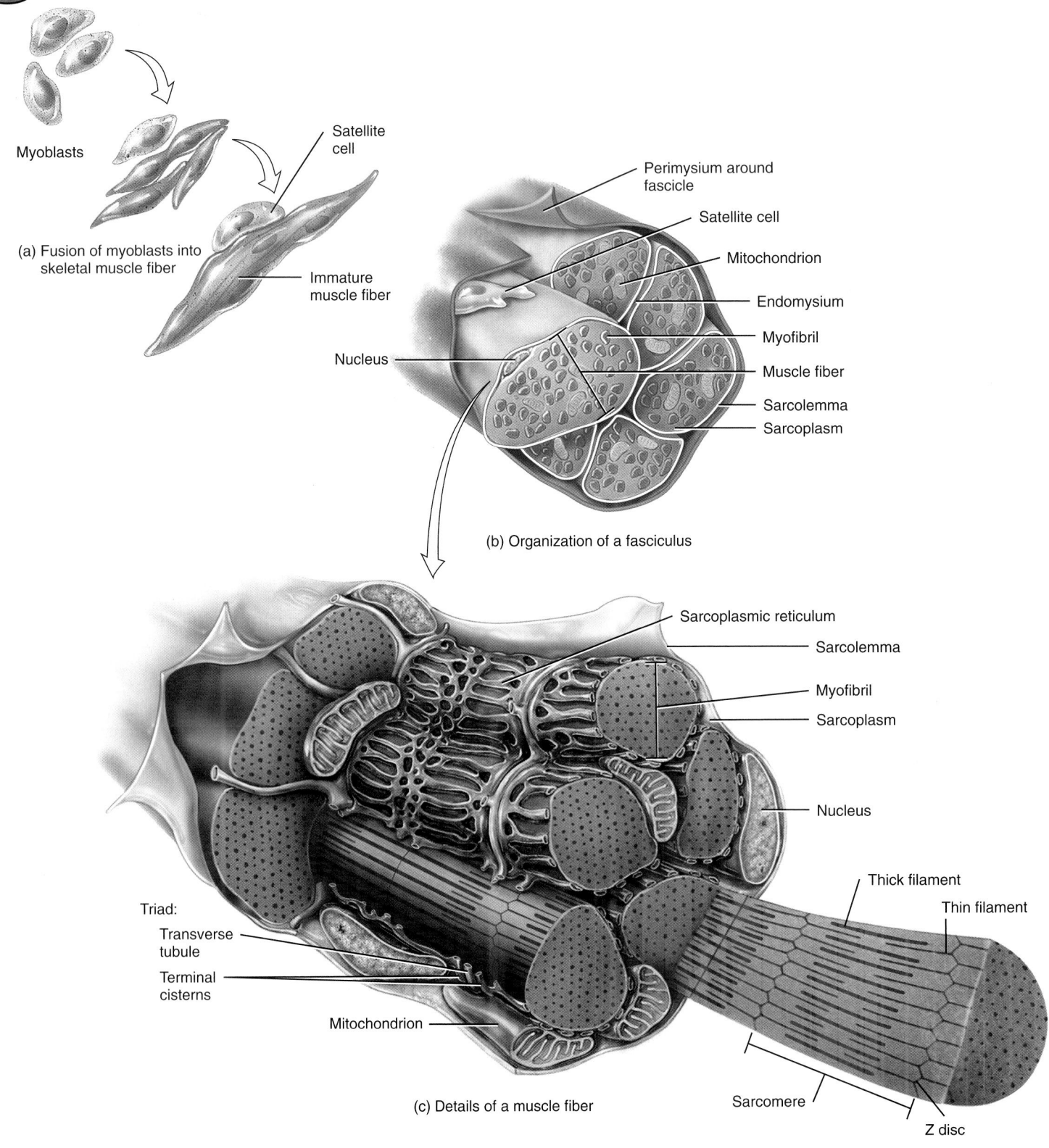

Myoblasts

Satellite cell

(a) Fusion of myoblasts into skeletal muscle fiber

Immature muscle fiber

Perimysium around fascicle

Satellite cell

Mitochondrion

Endomysium

Myofibril

Muscle fiber

Nucleus

Sarcolemma

Sarcoplasm

(b) Organization of a fasciculus

Sarcoplasmic reticulum

Sarcolemma

Myofibril

Sarcoplasm

Nucleus

Thick filament

Thin filament

Triad:

Transverse tubule

Terminal cisterns

Mitochondrion

Sarcomere

Z disc

(c) Details of a muscle fiber

Sarcolemma

Sarcoplasmic
reticulum (SR)

Transverse
tubule

Terminal cistern
of SR

Sarcoplasm

Nucleus

Membrane
protein

Z disc   Thick   Thin   Z disc
          filament   filament

Dystrophin

Sarcomere

Myofibril

Mitochondrion   Myoglobin   Glycogen granules

Key:

○ = Ca²⁺

⊗ = Ca²⁺ active
    transport pumps

< = Ca²⁺ release
    channels

(d) Simplistic representation of a muscle fiber

❓ **Which structure shown here releases calcium ions to trigger muscle contraction?**

both sides. A transverse tubule and the two terminal cisterns on either side of it form a **triad** (*tri-* = three). In a relaxed muscle fiber, the sarcoplasmic reticulum stores calcium ions (Ca^{2+}). Release of Ca^{2+} from the terminal cisterns of the sarcoplasmic reticulum triggers muscle contraction.

Filaments and the Sarcomere

Within myofibrils are smaller protein structures called **filaments** or **myofilaments** (Figure 10.2c). *Thin filaments* are 8 nm in diameter and 1–2 μm long[†] and composed mostly of the protein actin, while *thick filaments* are 16 nm in diameter and 1–2 μm long and composed mostly of the protein myosin. Both thin and thick filaments are directly involved in the contractile process. Overall, there are two thin filaments for every thick filament in the regions of filament overlap. The filaments inside a myofibril do not extend the entire length of a muscle fiber. Instead, they are arranged in compartments called **sarcomeres** (SAR-kō-mērs; *-mere* = part), which are the basic functional units of a myofibril (Figure 10.3a).

[†]One nanometer (nm) is 10^{-9} meter (0.001 μm); one micrometer (μm) = 1/25,000 of an inch.

Narrow, plate-shaped regions of dense protein material called **Z discs** separate one sarcomere from the next. Thus, a sarcomere extends from one Z disc to the next Z disc.

The extent of overlap of the thick and thin filaments depends on whether the muscle is contracted, relaxed, or stretched. The pattern of their overlap, consisting of a variety of zones and bands (Figure 10.3b), creates the striations that can be seen both in single myofibrils and in whole muscle fibers. The darker middle part of the sarcomere is the **A band,** which extends the entire length of the thick filaments (Figure 10.3b). Toward each end of the A band is a *zone of overlap,* where the thick and thin filaments lie side by side. The **I band** is a lighter, less dense area that contains the rest of the thin filaments but no thick filaments (Figure 10.3b), and a Z disc passes through the center of each I band. A narrow **H zone** in the center of each A band contains thick but not thin filaments. A mnemonic that will help you to remember the composition of the I and H bands is as follows: the letter I is thin (contains thin filaments), while the letter H is thick (contains thick filaments). Supporting proteins that hold the thick filaments together at the center of the H zone form the **M line,** so named because it is at the *middle* of the sarcomere. Table 10.1 summarizes the components of the sarcomere.

Figure 10.3 The arrangement of filaments within a sarcomere. A sarcomere extends from one Z disc to the next.

Myofibrils contain two types of filaments: thick filaments and thin filaments.

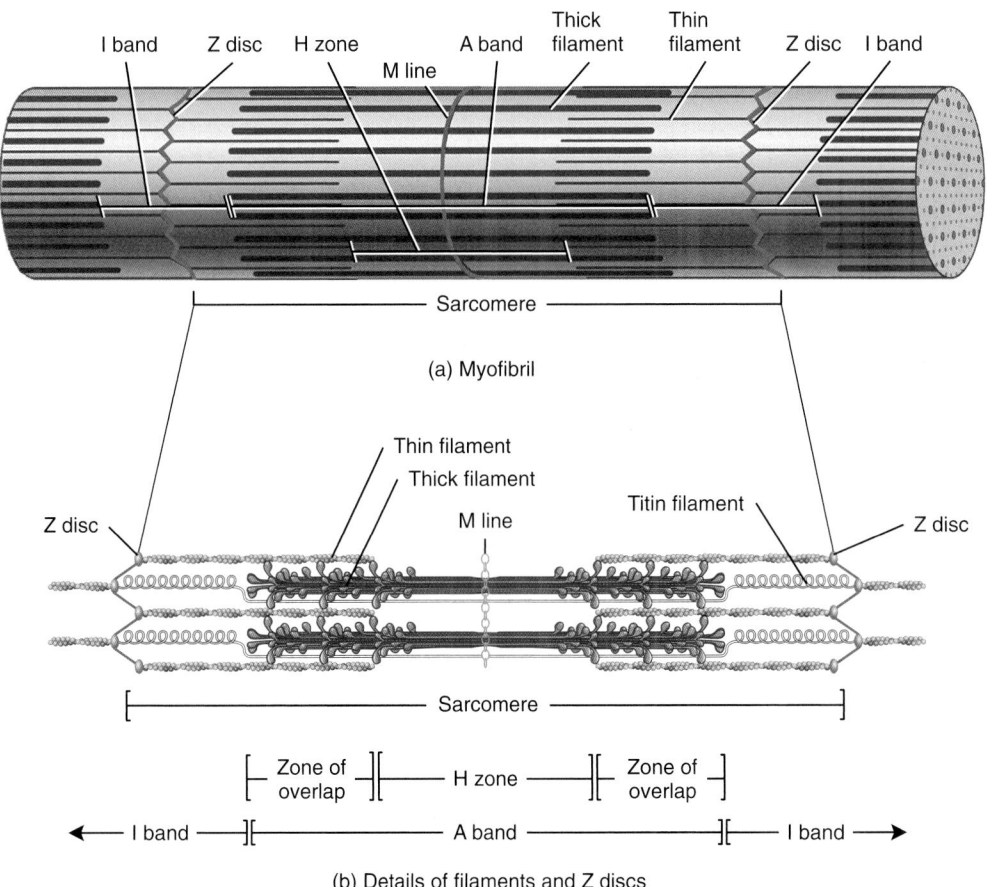

(a) Myofibril

(b) Details of filaments and Z discs

Which of the following is the smallest: muscle fiber, thick filament, or myofibril? Which is largest?

TABLE 10.1

Components of a Sarcomere

COMPONENT	DESCRIPTION
Z discs	Narrow, plate-shaped regions of dense material that separate one sarcomere from the next.
A band	Dark, middle part of sarcomere that extends entire length of thick filaments and includes those parts of thin filaments that overlap thick filaments.
I band	Lighter, less dense area of sarcomere that contains remainder of thin filaments but no thick filaments. A Z disc passes through center of each I band.
H zone	Narrow region in center of each A band that contains thick filaments but no thin filaments.
M line	Region in center of H zone that contains proteins that hold thick filaments together at center of sarcomere.

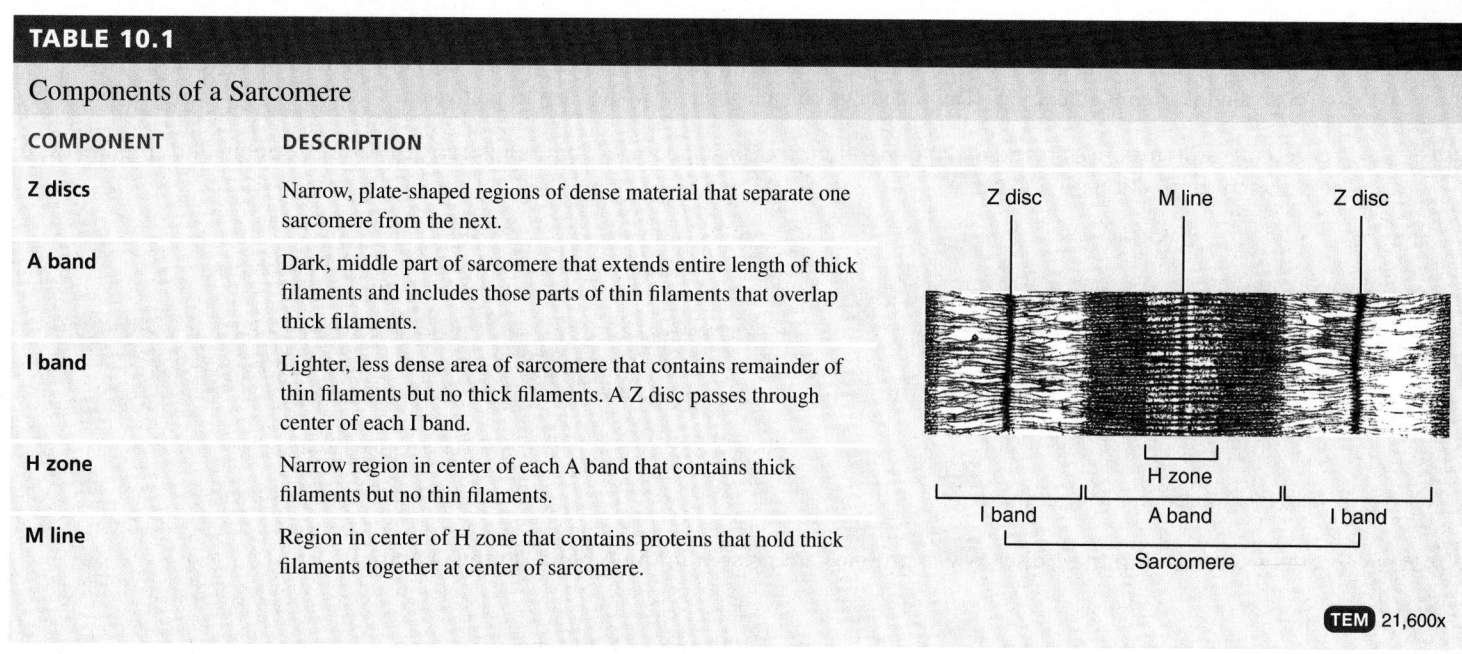

TEM 21,600x

CLINICAL CONNECTION | *Exercise-Induced Muscle Damage*

Comparison of electron micrographs of muscle tissue taken from athletes before and after intense exercise reveals considerable **exercise-induced muscle damage,** including torn sarcolemmas in some muscle fibers, damaged myofibrils, and disrupted Z discs. Microscopic muscle damage after exercise also is indicated by increases in blood levels of proteins, such as myoglobin and the enzyme creatine kinase, that are normally confined within muscle fibers. From 12 to 48 hours after a period of strenuous exercise, skeletal muscles often become sore. Such **delayed onset muscle soreness (DOMS)** is accompanied by stiffness, tenderness, and swelling. Although the causes of DOMS are not completely understood, microscopic muscle damage appears to be a major factor. In response to exercise-induced muscle damage, muscle fibers undergo repair: New regions of sarcolemma are formed to replace torn sarcolemmas, and more muscle proteins (including those of the myofibrils) are synthesized in the sarcoplasm of the muscle fibers. •

Muscle Proteins

Myofibrils are built from three kinds of proteins: (1) contractile proteins, which generate force during contraction; (2) regulatory proteins, which help switch the contraction process on and off; and (3) structural proteins, which keep the thick and thin filaments in the proper alignment, give the myofibril elasticity and extensibility, and link the myofibrils to the sarcolemma and extracellular matrix.

The two *contractile proteins* in muscle are myosin and actin, components of thick and thin filaments, respectively. **Myosin** (MĪ-ō-sin) is the main component of thick filaments and functions as a motor protein in all three types of muscle tissue. *Motor proteins* pull various cellular structures to achieve movement by converting the chemical energy in ATP to the mechanical energy of motion, that is, the production of force. In skeletal muscle, about 300 molecules of myosin form a single thick filament. Each myosin molecule

is shaped like two golf clubs twisted together (Figure 10.4a). The *myosin tail* (twisted golf club handles) points toward the M line in the center of the sarcomere. Tails of neighboring myosin molecules lie parallel to one another, forming the shaft of the thick filament. The two projections of each myosin molecule (golf club heads) are called *myosin heads*. The heads project outward from the shaft in a spiraling fashion, each extending toward one of the six thin filaments that surround each thick filament.

Thin filaments are anchored to Z discs (see Figure 10.3b). Their main component is the protein **actin** (AK-tin). Individual actin molecules join to form an actin filament that is twisted into a helix (Figure 10.4b). On each actin molecule is a *myosin-binding site*, where a myosin head can attach.

Smaller amounts of two *regulatory proteins*—**tropomyosin** (trō-pō-MĪ-ō-sin) and **troponin** (TRŌ-pō-nin)—are also part of the thin filament. In relaxed muscle, myosin is blocked from binding to actin because strands of tropomyosin cover the myosin-binding sites on actin. The tropomyosin strands in turn are held in place by troponin molecules. You will soon learn that when calcium ions (Ca^{2+}) bind to troponin, it undergoes a change in shape; this change moves tropomyosin away from myosin-binding sites on actin and muscle contraction subsequently begins as myosin binds to actin.

Besides contractile and regulatory proteins, muscle contains about a dozen *structural proteins,* which contribute to the alignment, stability, elasticity, and extensibility of myofibrils. Several key structural proteins are titin, α-actinin, myomesin, nebulin, and dystrophin. **Titin** (*titan* = gigantic) is the third most plentiful protein in skeletal muscle (after actin and myosin). This molecule's name reflects its huge size. With a molecular mass of about 3 million daltons, titin is 50 times larger than an average-sized protein. Each titin molecule spans half a sarcomere, from a Z disc to an M line (see Figure 10.3b), a distance of 1 to 1.2 μm in relaxed muscle. Each titin molecule connects a Z disc to the M line of the sarcomere, thereby

Figure 10.4 Structure of thick and thin filaments. (a) A thick filament contains about 300 myosin molecules, one of which is shown enlarged. The myosin tails form the shaft of the thick filament, and the myosin heads project outward toward the surrounding thin filaments. (b) Thin filaments contain actin, troponin, and tropomyosin.

Contractile proteins (myosin and actin) generate force during contraction; regulatory proteins (troponin and tropomyosin) help switch contraction on and off.

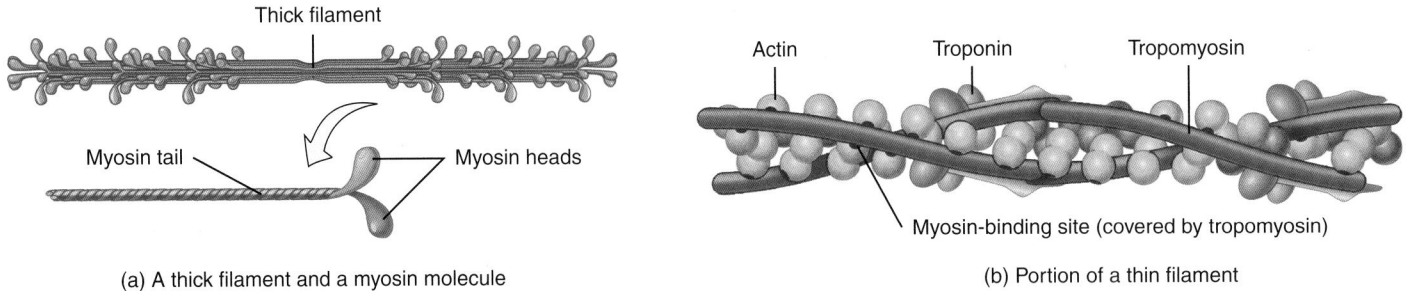

(a) A thick filament and a myosin molecule

(b) Portion of a thin filament

? **Which proteins connect into the Z disc? Which proteins are present in the A band? In the I band?**

helping stabilize the position of the thick filament. The part of the titin molecule that extends from the Z disc is very elastic. Because it can stretch to at least four times its resting length and then spring back unharmed, titin accounts for much of the elasticity and extensibility of myofibrils. Titin probably helps the sarcomere return to its resting length after a muscle has contracted or been stretched, may help prevent overextension of sarcomeres, and maintains the central location of the A bands.

The dense material of the Z discs contains molecules of **α-actinin,** which bind to actin molecules of the thin filament and to titin. Molecules of the protein **myomesin** (mī-ō-MĒ-sin) form the M line. The M line proteins bind to titin and connect adjacent thick filaments to one another. Myosin holds the thick filaments in alignment at the M line. **Nebulin** (NEB-ū-lin) is a long, nonelastic protein wrapped around the entire length of each thin filament. It helps anchor the thin filaments to the Z discs and regulates the

TABLE 10.2

Summary of Skeletal Muscle Fiber Proteins

TYPE OF PROTEIN	DESCRIPTION
Contractile proteins	Proteins that generate force during muscle contractions.
Myosin	Contractile protein that makes up thick filament; molecule consists of a tail and two myosin heads, which bind to myosin-binding sites on actin molecules of thin filament during muscle contraction.
Actin	Contractile protein that is the main component of thin filament; each actin molecule has a myosin-binding site where myosin head of thick filament binds during muscle contraction.
Regulatory proteins	Proteins that help switch muscle contraction process on and off.
Tropomyosin	Regulatory protein that is a component of thin filament; when skeletal muscle fiber is relaxed, tropomyosin covers myosin-binding sites on actin molecules, thereby preventing myosin from binding to actin.
Troponin	Regulatory protein that is a component of thin filament; when calcium ions (Ca^{2+}) bind to troponin, it changes shape; this conformational change moves tropomyosin away from myosin-binding sites on actin molecules, and muscle contraction subsequently begins as myosin binds to actin.
Structural proteins	Proteins that keep thick and thin filaments of myofibrils in proper alignment, give myofibrils elasticity and extensibility, and link myofibrils to sarcolemma and extracellular matrix.
Titin	Structural protein that connects Z disc to M line of sarcomere, thereby helping to stabilize thick filament position; can stretch and then spring back unharmed, and thus accounts for much of the elasticity and extensibility of myofibrils.
α-Actinin	Structural protein of Z discs that attaches to actin molecules of thin filaments and to titin molecules.
Myomesin	Structural protein that forms M line of sarcomere; binds to titin molecules and connects adjacent thick filaments to one another.
Nebulin	Structural protein that wraps around entire length of each thin filament; helps anchor thin filaments to Z discs and regulates length of thin filaments during development.
Dystrophin	Structural protein that links thin filaments of sarcomere to integral membrane proteins in sarcolemma, which are attached in turn to proteins in connective tissue matrix that surrounds muscle fibers; is thought to help reinforce sarcolemma and help transmit tension generated by sarcomeres to tendons.

length of thin filaments during development. **Dystrophin** (dis-TRŌ-fin) links thin filaments of the sarcomere to integral membrane proteins of the sarcolemma, which are attached in turn to proteins in the connective tissue extracellular matrix that surrounds muscle fibers (see Figure 10.2d). Dystrophin and its associated proteins are thought to reinforce the sarcolemma and help

transmit the tension generated by the sarcomeres to the tendons. The relationship of dystrophin to muscular dystrophy is discussed in Disorders: Homeostatic Imbalances at the end of the chapter.

Table 10.2 summarizes the different types of skeletal muscle fiber proteins, and Table 10.3 summarizes the levels of organization within a skeletal muscle.

TABLE 10.3

Levels of Organization within a Skeletal Muscle

LEVEL	DESCRIPTION
Skeletal muscle	Organ made up of fascicles that contain muscle fibers (cells), blood vessels, and nerves; wrapped in epimysium.
Fascicle	Bundle of muscle fibers wrapped in perimysium.
Muscle fiber (cell)	Long cylindrical cell covered by endomysium and sarcolemma; contains sarcoplasm, myofibrils, many peripherally located nuclei, mitochondria, transverse tubules, sarcoplasmic reticulum, and terminal cisterns. The fiber has a striated appearance.
Myofibril	Threadlike contractile elements within sarcoplasm of muscle fiber that extend entire length of fiber; composed of filaments.
Filaments (myofilaments)	Contractile proteins within myofibrils that are of two types: thick filaments composed of myosin and thin filaments composed of actin, tropomyosin, and troponin; sliding of thin filaments past thick filaments produces muscle shortening.

✔CHECKPOINT

✔CHECKPOINT

4. What types of fascia cover skeletal muscles?
5. Why is a rich blood supply important for muscle contraction?
6. How are the structures of thin and thick filaments different?

10.3 CONTRACTION AND RELAXATION OF SKELETAL MUSCLE FIBERS

◉ OBJECTIVES

• Outline the steps involved in the sliding filament mechanism of muscle contraction.
• Describe how muscle action potentials arise at the neuromuscular junction.

When scientists examined the first electron micrographs of skeletal muscle in the mid-1950s, they were surprised to see that the lengths of the thick and thin filaments were the same in both relaxed and contracted muscle. It had been thought that muscle contraction must be a folding process, somewhat like closing an accordion. Instead, researchers discovered that skeletal muscle shortens during contraction because the thick and thin filaments slide past one another. The model describing this process is known as the **sliding filament mechanism.**

The Sliding Filament Mechanism

Muscle contraction occurs because myosin heads attach to and "walk" along the thin filaments at both ends of a sarcomere, progressively pulling the thin filaments toward the M line (Figure 10.5). As a result, the thin filaments slide inward and meet at the center of a sarcomere. They may even move so far inward that their ends overlap (Figure 10.5c). As the thin filaments slide inward, the Z discs come closer together, and the sarcomere shortens. However, the lengths of the individual thick and thin filaments do not change. Shortening of the sarcomeres causes shortening of the whole muscle fiber, which in turn leads to shortening of the entire muscle.

The Contraction Cycle

At the onset of contraction, the sarcoplasmic reticulum releases calcium ions (Ca^{2+}) into the sarcoplasm. There, they bind to troponin. Troponin then moves tropomyosin away from the myosin-binding sites on actin. Once the binding sites are "free," the **contraction cycle**—the repeating sequence of events that

Figure 10.5 Sliding filament mechanism of muscle contraction, as it occurs in two adjacent sarcomeres.

🔓 ⬭⬭⬭ During muscle contractions, thin filaments move toward the M line of each sarcomere.

(a) Relaxed muscle

(b) Partially contracted muscle

(c) Maximally contracted muscle

❓ What happens to the I band and H zone as muscle contracts? Do the lengths of the thick and thin filaments change?

causes the filaments to slide—begins. The contraction cycle consists of four steps (Figure 10.6):

❶ ATP hydrolysis. The myosin head includes an ATP-binding site and an ATPase, an enzyme that hydrolyzes ATP into ADP (adenosine diphosphate) and a phosphate group. This hydrolysis reaction reorients and energizes the myosin head. Notice that the products of ATP hydrolysis—ADP and a phosphate group—are still attached to the myosin head.

❷ Attachment of myosin to actin to form cross-bridges. The energized myosin head attaches to the myosin-binding site on actin and releases the previously hydrolyzed phosphate group. When the myosin heads attach to actin during contraction, they are referred to as **cross-bridges.**

❸ Power stroke. After the cross-bridges form, the power stroke occurs. During the power stroke, the site on the cross-bridge where ADP is still bound opens. As a result, the cross-bridge rotates and releases the ADP. The cross-bridge generates force as it rotates toward the center of the sarcomere, sliding the thin filament past the thick filament toward the M line.

❹ Detachment of myosin from actin. At the end of the power stroke, the cross-bridge remains firmly attached to actin until

it binds another molecule of ATP. As ATP binds to the ATP-binding site on the myosin head, the myosin head detaches from actin.

The contraction cycle repeats as the myosin ATPase hydrolyzes the newly bound molecule of ATP, and continues as long as ATP is available and the Ca^{2+} level near the thin filament is sufficiently high. The cross-bridges keep rotating back and forth with each power stroke, pulling the thin filaments toward the M line. Each of the 600 cross-bridges in one thick filament attaches and detaches about five times per second. At any one instant, some of the myosin heads are attached to actin, forming cross-bridges and generating force, and other myosin heads are detached from actin, getting ready to bind again.

As the contraction cycle continues, movement of cross-bridges applies the force that draws the Z discs toward each other, and the sarcomere shortens. During a maximal muscle contraction, the distance between two Z discs can decrease to half the resting length. The Z discs in turn pull on neighboring sarcomeres, and the whole muscle fiber shortens. Some of the components of a muscle are elastic: They stretch slightly before they transfer the tension generated by the sliding filaments. The elastic components include titin molecules, connective tissue around the muscle fibers (endomysium, perimysium, and epimysium), and tendons

Figure 10.6 The contraction cycle. Sarcomeres exert force and shorten through repeated cycles during which the myosin heads attach to actin (cross-bridges), rotate, and detach.

🔑 During the power stroke of contraction, cross-bridges rotate and move the thin filaments past the thick filaments toward the center of the sarcomere.

? **What would happen if ATP suddenly were not available after the sarcomere had started to shorten?**

that attach muscle to bone. As the cells of a skeletal muscle start to shorten, they first pull on their connective tissue coverings and tendons. The coverings and tendons stretch and then become taut, and the tension passed through the tendons pulls on the bones to which they are attached. The result is movement of a part of the body. You will soon learn, however, that the contraction cycle does not always result in shortening of the muscle fibers and the whole muscle. In some contractions, the cross-bridges rotate and generate tension, but the thin filaments cannot slide inward because the tension they generate is not large enough to move the load on the muscle (such as trying to lift a whole box of books with one hand).

Excitation–Contraction Coupling

An increase in Ca^{2+} concentration in the sarcoplasm starts muscle contraction, and a decrease stops it. When a muscle fiber is relaxed, the concentration of Ca^{2+} in its sarcoplasm is very low, only about 0.1 micromole per liter (0.1 μmol/L). However, a

huge amount of Ca^{2+} is stored inside the sarcoplasmic reticulum (Figure 10.7a). As a muscle action potential propagates along the sarcolemma and into the T tubules, it causes **Ca^{2+} release channels** in the SR membrane to open (Figure 10.7b). When these channels open, Ca^{2+} flows out of the SR into the sarcoplasm around the thick and thin filaments. As a result, the Ca^{2+} concentration in the sarcoplasm rises tenfold or more. The released calcium ions combine with troponin, causing it to change shape. This conformational change moves tropomyosin away from the myosin-binding sites on actin. Once these binding sites are free, myosin heads bind to them to form cross-bridges, and the contraction cycle begins. The events just described are referred to collectively as **excitation–contraction coupling,** as they are the steps that connect excitation (a muscle action potential propagating along the sarcolemma and into the T tubules) to contraction (sliding of the filaments).

The sarcoplasmic reticulum membrane also contains **Ca^{2+} active transport pumps** that use ATP to move Ca^{2+} constantly

Figure 10.7 The role of Ca^{2+} in the regulation of contraction by troponin and tropomyosin. (a) During relaxation, the level of Ca^{2+} in the sarcoplasm is low, only 0.1 μM (0.0001 mM), because calcium ions are pumped into the sarcoplasmic reticulum by Ca^{2+} active transport pumps. (b) A muscle action potential propagating along a transverse tubule opens Ca^{2+} release channels in the sarcoplasmic reticulum, calcium ions flow into the cytosol, and contraction begins.

An increase in the Ca^{2+} level in the sarcoplasm starts the sliding of thin filaments. When the level of Ca^{2+} in the sarcoplasm declines, sliding stops.

Troponin holds tropomyosin in position to block myosin-binding sites on actin.

(a) Relaxation

Key:
○ = Ca^{2+}
⊗ = Ca^{2+} active transport pumps
< = Ca^{2+} release channels

Ca^{2+} binds to troponin, which changes the shape of the troponin–tropomyosin complex and uncovers the myosin-binding sites on actin.

(b) Contraction

? **What are three functions of ATP in muscle contraction?**

from the sarcoplasm into the SR (Figure 10.7). While muscle action potentials continue to propagate through the T tubules, the Ca^{2+} release channels are open. Calcium ions flow into the sarcoplasm more rapidly than they are transported back by the pumps. After the last action potential has propagated throughout the T tubules, the Ca^{2+} release channels close. As the pumps move Ca^{2+} back into the SR, the concentration of calcium ions in the sarcoplasm quickly decreases. Inside the SR, molecules of a calcium-binding protein, appropriately called **calsequestrin** (kal′-se-KWES-trin), bind to the Ca^{2+}, enabling even more Ca^{2+} to be sequestered or stored within the SR. As a result, the concentration of Ca^{2+} is 10,000 times higher in the SR than in the cytosol in a relaxed muscle fiber. As the Ca^{2+} level in the cytosol drops, tropomyosin covers the myosin-binding sites, and the muscle fiber relaxes.

CLINICAL CONNECTION | *Rigor Mortis*

After death, cellular membranes become leaky. Calcium ions leak out of the sarcoplasmic reticulum into the sarcoplasm and allow myosin heads to bind to actin. ATP synthesis ceases shortly after breathing stops, however, so the cross-bridges cannot detach from actin. The resulting condition, in which muscles are in a state of rigidity (cannot contract or stretch), is called **rigor mortis** (rigidity of death). Rigor mortis begins 3–4 hours after death and lasts about 24 hours; then it disappears as proteolytic enzymes from lysosomes digest the cross-bridges. •

Length–Tension Relationship

Figure 10.8 shows the **length–tension relationship** for skeletal muscle, which indicates how the forcefulness of muscle contraction depends on the length of the sarcomeres within a muscle *before contraction begins*. At a sarcomere length of about 2.0–2.4 μm (which is very close to the resting length in most muscles), the zone of overlap in each sarcomere is optimal, and the muscle fiber can develop maximum tension. Notice in Figure 10.8 that maximum tension (100%) occurs when the zone of overlap between a thick and thin filament extends from the edge of the H zone to one end of a thick filament.

As the sarcomeres of a muscle fiber are stretched to a longer length, the zone of overlap shortens, and fewer myosin heads can make contact with thin filaments. Therefore, the tension the fiber can produce decreases. When a skeletal muscle fiber is stretched to 170% of its optimal length, there is no overlap between the thick and thin filaments. Because none of the myosin heads can bind to thin filaments, the muscle fiber cannot contract, and tension is zero. As sarcomere lengths become increasingly shorter than the optimum, the tension that can develop again decreases. This is because thick filaments crumple as they are compressed by the Z discs, resulting in fewer myosin heads making contact with thin filaments. Normally, resting muscle fiber length is held very close to the optimum by firm attachments of skeletal muscle to bones (via their tendons) and to other inelastic tissues.

Figure 10.8 **Length–tension relationship in a skeletal muscle fiber.** Maximum tension during contraction occurs when the resting sarcomere length is 2.0–2.4 μm.

 A muscle fiber develops its greatest tension when there is an optimal zone of overlap between thick and thin filaments.

? Why is tension maximal at a sarcomere length of 2.2 μm?

The Neuromuscular Junction

As noted earlier in the chapter, the neurons that stimulate skeletal muscle fibers to contract are called **somatic motor neurons.** Each somatic motor neuron has a threadlike axon that extends from the brain or spinal cord to a group of skeletal muscle fibers. A muscle fiber contracts in response to one or more action potentials propagating along its sarcolemma and through its system of T tubules. Muscle action potentials arise at the **neuromuscular junction (NMJ)** (noo-rō-MUS-kū-lar), the synapse between a somatic motor neuron and a skeletal muscle fiber (Figure 10.9a). A **synapse** is a region where communication occurs between two neurons, or between a neuron and a target cell—in this case, between a somatic motor neuron and a muscle fiber. At most synapses a small gap, called the **synaptic cleft,** separates the two cells. Because the cells do not physically touch, the action potential cannot "jump the gap" from one cell to another. Instead, the first cell communicates with the second by releasing a chemical called a **neurotransmitter.**

At the NMJ, the end of the motor neuron, called the **axon terminal,** divides into a cluster of **synaptic end bulbs** (Figure 10.9a, b), the neural part of the NMJ. Suspended in the cytosol within each synaptic end bulb are hundreds of membrane-enclosed sacs called **synaptic vesicles.** Inside each synaptic vesicle are thousands of molecules of **acetylcholine** (as′-ē-til-KŌ-lēn), abbreviated **ACh,** the neurotransmitter released at the NMJ.

The region of the sarcolemma opposite the synaptic end bulbs, called the **motor end plate** (Figure 10.9b, c), is the muscle fiber part of the NMJ. Within each motor end plate are 30 million to 40 million **acetylcholine receptors,** integral transmembrane proteins to which ACh specifically binds. These receptors are abundant in **junctional folds,** deep grooves in the motor end plate that provide a large surface area for ACh. As you will see, the ACh receptors are ligand-gated ion channels. A neuromuscular junction

thus includes all the synaptic end bulbs on one side of the synaptic cleft, plus the motor end plate of the muscle fiber on the other side.

A nerve impulse (nerve action potential) elicits a muscle action potential in the following way (Figure 10.9c):

1 ***Release of acetylcholine.*** Arrival of the nerve impulse at the synaptic end bulbs stimulates voltage-gated channels to open. Because calcium ions are more concentrated in the extracellular fluid, Ca^{2+} flows inward through the open

Figure 10.9 Structure of the neuromuscular junction (NMJ), the synapse between a somatic motor neuron and a skeletal muscle fiber.

Synaptic end bulbs at the tips of axon terminals contain synaptic vesicles filled with acetylcholine (ACh).

(a) Neuromuscular junction

(b) Enlarged view of the neuromuscular junction

(c) Binding of acetylcholine to ACh receptors in the motor end plate

Blood capillary

Axon terminal

Axon collateral (branch)

Synaptic end bulbs

Somatic motor neuron

Axon collateral (branch)

Skeletal muscle fiber

Synaptic end bulbs

SEM 1650x

(d) Neuromuscular junction

? What part of the sarcolemma contains acetylcholine receptors?

channels. The entering Ca^{2+} in turn stimulates the synaptic vesicles to undergo exocytosis. During exocytosis, the synaptic vesicles fuse with the motor neuron's plasma membrane, liberating ACh into the synaptic cleft. The ACh then diffuses across the synaptic cleft between the motor neuron and the motor end plate.

2 *Activation of ACh receptors.* Binding of two molecules of ACh to the receptor on the motor end plate opens an ion channel in the ACh receptor. Once the channel is open, small cations, most importantly Na^+, can flow across the membrane.

3 *Production of muscle action potential.* The inflow of Na^+ (down its electrochemical gradient) makes the inside of the muscle fiber more positively charged. This change in the membrane potential triggers a muscle action potential. Each nerve impulse normally elicits one muscle action potential. The muscle action potential then propagates along the sarcolemma into the system of T tubules. This causes the sarcoplasmic reticulum to release its stored Ca^{2+} into the sarcoplasm and the muscle fiber subsequently contracts.

4 *Termination of ACh activity.* The effect of ACh binding lasts only briefly because ACh is rapidly broken down by an enzyme called **acetylcholinesterase,** or **AChE** (as′-ē-til-kō′-lin-ES-ter-ās). This enzyme is attached to collagen fibers in the extracellular matrix of the synaptic cleft. AChE breaks down ACh into acetyl and choline, products that cannot activate the ACh receptor.

If another nerve impulse releases more acetylcholine, steps **2** and **3** repeat. When action potentials in the motor neuron cease, ACh is no longer released, and AChE rapidly breaks down the ACh already present in the synaptic cleft. This ends the production of muscle action potentials, the Ca^{2+} moves from the sarcoplasm of the muscle fiber back into the sarcoplasmic reticulum, and the Ca^{2+} release channels in the sarcoplasmic reticulum membrane close.

The NMJ usually is near the midpoint of a skeletal muscle fiber. Muscle action potentials that arise at the NMJ propagate toward both ends of the fiber. This arrangement permits nearly simultaneous activation (and thus contraction) of all parts of the muscle fiber.

Figure 10.10 summarizes the events that occur during contraction and relaxation of a skeletal muscle fiber.

Several plant products and drugs selectively block certain events at the NMJ. *Botulinum toxin* (bot-u-LĪN-um), produced by the bacterium *Clostridium botulinum,* blocks exocytosis of synaptic vesicles at the NMJ. As a result, ACh is not released, and muscle contraction does not occur. The bacteria proliferate in improperly canned foods, and their toxin is one of the most lethal chemicals known. A tiny amount can cause death by paralyzing skeletal muscles. Breathing stops due to paralysis of respiratory muscles, including the diaphragm. Yet it is also the first bacterial toxin to be used as a medicine (Botox®). Injections of Botox into the affected muscles can help patients who have strabismus (crossed eyes), blepharospasm (uncontrollable blinking), or spasms of the vocal cords that interfere with speech. It is also used to alleviate chronic back pain due to muscle spasms in the lumbar region and as a cosmetic treatment to relax muscles that cause facial wrinkles.

The plant derivative *curare,* a poison used by South American Indians on arrows and blowgun darts, causes muscle paralysis by binding to and blocking ACh receptors. In the presence of curare, the ion channels do not open. Curare-like drugs are often used during surgery to relax skeletal muscles.

A family of chemicals called *anticholinesterase agents* have the property of slowing the enzymatic activity of acetylcholinesterase, thus slowing removal of ACh from the synaptic cleft. At low doses, these agents can strengthen weak muscle contractions. One example is neostigmine, which is used to treat patients with myasthenia gravis (see Disorders: Homeostatic Imbalances at the end of this chapter). Neostigmine is also used as an antidote for curare poisoning and to terminate the effects of curare-like drugs after surgery.

Figure 10.10 Summary of the events of contraction and relaxation in a skeletal muscle fiber.

Acetylcholine released at the neuromuscular junction triggers a muscle action potential, which leads to muscle contraction.

1 Nerve impulse arrives at axon terminal of motor neuron and triggers release of acetylcholine (ACh).

Nerve impulse
Ca²⁺

ACh receptor

Synaptic vesicle filled with ACh

2 ACh diffuses across synaptic cleft, binds to its receptors in the motor end plate, and triggers a muscle action potential (AP).

3 Acetylcholinesterase in synaptic cleft destroys ACh so another muscle action potential does not arise unless more ACh is released from motor neuron.

Muscle action potential

Transverse tubule

4 Muscle AP traveling along transverse tubule opens Ca²⁺ release channels in the sarcoplasmic reticulum (SR) membrane, which allows calcium ions to flood into the sarcoplasm.

Ca²⁺

SR

9 Muscle relaxes.

8 Troponin–tropomyosin complex slides back into position where it blocks the myosin binding sites on actin.

5 Ca²⁺ binds to troponin on the thin filament, exposing the binding sites for myosin.

Elevated Ca²⁺

Ca²⁺ active transport pumps

7 Ca²⁺ release channels in SR close and Ca²⁺ active transport pumps use ATP to restore low level of Ca²⁺ in sarcoplasm.

6 Contraction: power strokes use ATP; myosin heads bind to actin, swivel, and release; thin filaments are pulled toward center of sarcomere.

? **Which numbered steps in this figure are part of excitation–contraction coupling?**

Electromyography (e-lek′-trō-mī-OG-ra-fē; *electro-* = electricity; *-myo-* = muscle; *-graph* = to write) or **EMG** is a test that measures the electrical activity (muscle action potentials) in resting and contracting muscles. Normally, resting muscle produces no electrical activity; a slight contraction produces some electrical activity; and a more forceful contraction produces increased electrical activity. In the procedure, a ground electrode is placed over the muscle to be tested to eliminate background electrical activity. Then, a fine needle attached by wires to a recording instrument is inserted into the muscle. The electrical activity of the muscle is displayed as waves on an oscilloscope and heard through a loudspeaker.

EMG helps to determine if muscle weakness or paralysis is due to a malfunction of the muscle itself or the nerves supplying the muscle. EMG is also used to diagnose certain muscle disorders, such as muscular dystrophy, and to understand which muscles function during complex movements. •

✔ CHECKPOINT

7. What roles do contractile, regulatory, and structural proteins play in muscle contraction and relaxation?
8. How do calcium ions and ATP contribute to muscle contraction and relaxation?
9. How does sarcomere length influence the maximum tension that is possible during muscle contraction?
10. How is the motor end plate different from other parts of the sarcolemma?

10.4 MUSCLE METABOLISM

◉ OBJECTIVES

• Describe the reactions by which muscle fibers produce ATP.
• Distinguish between anaerobic and aerobic cellular respiration.
• Describe the factors that contribute to muscle fatigue.

Production of ATP in Muscle Fibers

Unlike most cells of the body, skeletal muscle fibers often switch between a low level of activity, when they are relaxed and using only a modest amount of ATP, and a high level of activity, when they are contracting and using ATP at a rapid pace. A huge amount of ATP is needed to power the contraction cycle, to pump Ca^{2+} into the sarcoplasmic reticulum, and for other metabolic reactions involved in muscle contraction. However, the ATP present inside muscle fibers is enough to power contraction for only a few seconds. If muscle contractions continue past that time, the muscle fibers must make more ATP. Muscle fibers have three ways to produce ATP: (1) from creatine phosphate, (2) by anaerobic cellular respiration, and (3) by aerobic cellular respiration (Figure 10.11). The use of creatine phosphate for ATP production is unique to muscle fibers, but all body cells make ATP by the reactions of anaerobic and aerobic cellular respiration. We consider the events of cellular respiration briefly here and then in detail in Chapter 25.

Creatine Phosphate

While muscle fibers are relaxed, they produce more ATP than they need for resting metabolism. Most of the excess ATP is used to synthesize **creatine phosphate** (KRĒ-a-tēn), an energy-rich molecule that is found in muscle fibers (Figure 10.11a). The enzyme creatine kinase (CK) catalyzes the transfer of one of the high-energy phosphate groups from ATP to creatine, forming creatine phosphate and ADP. **Creatine** is a small, amino acid–like molecule that is synthesized in the liver, kidneys, and pancreas and then transported to muscle fibers. Creatine phosphate is three to six times more plentiful than ATP in the sarcoplasm of a relaxed muscle fiber. When contraction begins and the ADP level starts to rise, CK catalyzes the transfer of a high-energy phosphate group from creatine phosphate back to ADP. This direct phosphorylation reaction quickly generates new ATP molecules. Since the formation of ATP from creatine phosphate occurs very rapidly, creatine phosphate is the first source of energy when muscle contraction begins. The other energy-generating mechanisms in a muscle fiber (the pathways of anaerobic and aerobic cellular respiration) take a relatively longer period of time to produce ATP compared to creatine phosphate. Together, stores of creatine phosphate and ATP provide enough energy for muscles to contract maximally for about 15 seconds.

Creatine is both synthesized in the body and derived from foods such as milk, red meat, and some fish. Adults need to synthesize and ingest a total of about 2 grams of creatine daily to make up for the urinary loss of creatinine, the breakdown product of creatine. Some studies have demonstrated improved performance from **creatine supplementation** during explosive movements, such as sprinting. Other studies, however, have failed to find a performance-enhancing effect of creatine supplementation. Moreover, ingesting extra creatine decreases the body's own synthesis of creatine, and it is not known whether natural synthesis recovers after long-term creatine supplementation. In addition, creatine supplementation can cause dehydration and may cause kidney dysfunction. Further research is needed to determine both the long-term safety and the value of creatine supplementation. •

Anaerobic Cellular Respiration

Anaerobic cellular respiration (an′-ar-Ō-bik) is a series of ATP-producing reactions that do not require oxygen. When muscle activity continues and the supply of creatine phosphate within the muscle fibers is depleted, glucose is catabolized to generate ATP. Glucose easily passes from the blood into contracting muscle fibers via facilitated diffusion, and it is also produced by the breakdown of glycogen within muscle fibers (Figure 10.11b). Then, a series of reactions known as *glycolysis* quickly breaks down each glucose molecule into two molecules of pyruvic acid. Glycolysis occurs in the cytosol and produces a net gain of two molecules of ATP.

Figure 10.11 Production of ATP for muscle contraction. (a) Creatine phosphate, formed from ATP while the muscle is relaxed, transfers a high-energy phosphate group to ADP, forming ATP, during muscle contraction. (b) Breakdown of muscle glycogen into glucose and production of pyruvic acid from glucose via glycolysis produce both ATP and lactic acid. Because no oxygen is needed, this is an anaerobic pathway. (c) Within mitochondria, pyruvic acid, fatty acids, and amino acids are used to produce ATP via aerobic cellular respiration, an oxygen-requiring set of reactions.

During a long-term event such as a marathon race, most ATP is produced aerobically.

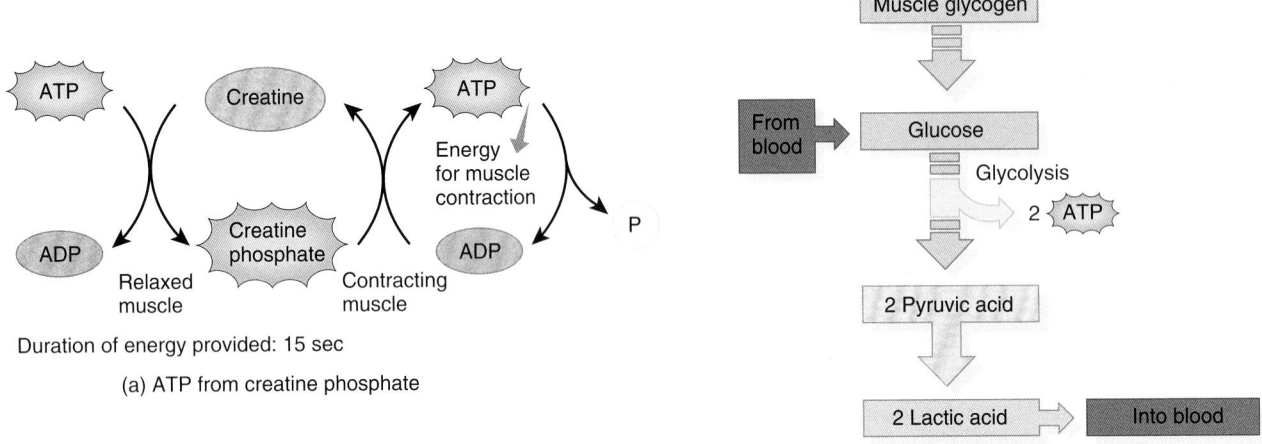

Duration of energy provided: 15 sec

(a) ATP from creatine phosphate

Duration of energy provided: 30–40 sec

(b) ATP from anaerobic glycolysis

Duration of energy provided: Minutes to hours

(c) ATP from aerobic cellular respiration

? **Where inside a skeletal muscle fiber are the events shown here occurring?**

Ordinarily, the pyruvic acid formed by glycolysis enters the mitochondria, where it undergoes a series of oxygen-requiring reactions called aerobic cellular respiration (described next) that produce a large amount of ATP. However, during periods of heavy exercise, not enough oxygen is available to skeletal muscle fibers. When this occurs, anaerobic reactions convert most of the pyruvic acid to lactic acid in the cytosol. About 80% of the lactic acid produced in this way diffuses out of the skeletal muscle fibers into the blood. Liver cells can convert some of the lactic acid back to glucose. In addition to providing new glucose molecules, this conversion reduces acidity of the blood. Anaerobic cellular respiration can provide enough energy for about 30 to 40 seconds of maximal muscle activity.

Aerobic Cellular Respiration

During periods of rest or light to moderate exercise, a sufficient amount of oxygen is available to skeletal muscle fibers. In such cases, ATP used for muscular activity is produced from a series of oxygen-requiring reactions called **aerobic cellular respiration** (ār-Ō-bik). During this process, pyruvic acid enters the mitochondria, where it is completely oxidized in reactions that generate ATP, carbon dioxide, water, and heat (Figure 10.11c). Although aerobic cellular respiration is slower than glycolysis, it yields much more ATP. One molecule of glucose yields about 36 molecules of ATP via aerobic cellular respiration, but yields only 2 molecules of ATP via anaerobic glycolysis.

Muscle tissue has two sources of oxygen: (1) oxygen that diffuses into muscle fibers from the blood and (2) oxygen released by myoglobin within muscle fibers. Both myoglobin (found only in muscle cells) and hemoglobin (found only in red blood cells) are oxygen-binding proteins. They bind oxygen when it is plentiful and release oxygen when it is scarce.

Aerobic cellular respiration supplies enough ATP for prolonged activity provided sufficient oxygen and nutrients are available. These nutrients include the pyruvic acid obtained from the glycolysis of glucose, fatty acids from the breakdown of triglycerides in adipose cells, and amino acids from the breakdown of proteins. In activities that last more than 10 minutes, the aerobic system provides more than 90% of the needed ATP. At the end of an endurance event such as a marathon race, nearly 100% of the ATP is being produced by aerobic cellular respiration.

Muscle Fatigue

The inability of a muscle to maintain force of contraction after prolonged activity is called **muscle fatigue** (fa-TĒG). Fatigue results mainly from changes within muscle fibers. Even before actual muscle fatigue occurs, a person may have feelings of tiredness and the desire to cease activity; this response, called *central fatigue,* is caused by changes in the central nervous system (brain and spinal cord). Although its exact mechanism is unknown, it may be a protective mechanism to stop a person from exercising before muscles become damaged. As you will see, certain types of skeletal muscle fibers fatigue more quickly than others.

Although the precise mechanisms that cause muscle fatigue are still not clear, several factors are thought to contribute. One is inadequate release of calcium ions from the SR, resulting in a decline of Ca^{2+} concentration in the sarcoplasm. Depletion of creatine phosphate also is associated with fatigue, but surprisingly, the ATP levels in fatigued muscle often are not much lower than those in resting muscle. Other factors that contribute to muscle fatigue include insufficient oxygen, depletion of glycogen and other nutrients, buildup of lactic acid and ADP, and failure of action potentials in the motor neuron to release enough acetylcholine.

Oxygen Consumption after Exercise

During prolonged periods of muscle contraction, increases in breathing rate and blood flow enhance oxygen delivery to muscle tissue. After muscle contraction has stopped, heavy breathing continues for a while, and oxygen consumption remains above the resting level. Depending on the intensity of the exercise, the recovery period may be just a few minutes, or it may last as long as several hours. The term **oxygen debt** refers to the added oxygen, over and above the resting oxygen consumption, that is taken into the body after exercise. This extra oxygen is used to "pay back" or restore metabolic conditions to the resting level in three ways: (1) to convert lactic acid back into glycogen stores in the liver, (2) to resynthesize creatine phosphate and ATP in muscle fibers, and (3) to replace the oxygen removed from myoglobin.

The metabolic changes that occur *during exercise* can account for only some of the extra oxygen used *after exercise.* Only a

small amount of glycogen resynthesis occurs from lactic acid. Instead, most glycogen is made much later from dietary carbohydrates. Much of the lactic acid that remains after exercise is converted back to pyruvic acid and used for ATP production via aerobic cellular respiration in the heart, liver, kidneys, and skeletal muscle. Oxygen use after exercise also is boosted by ongoing changes. First, the elevated body temperature after strenuous exercise increases the rate of chemical reactions throughout the body. Faster reactions use ATP more rapidly, and more oxygen is needed to produce the ATP. Second, the heart and the muscles used in breathing are still working harder than they were at rest, and thus they consume more ATP. Third, tissue repair processes are occurring at an increased pace. For these reasons, **recovery oxygen uptake** is a better term than oxygen debt for the elevated use of oxygen after exercise.

✔CHECKPOINT

11. Which ATP-producing reactions are aerobic and which are anaerobic?
12. Which sources provide ATP during a marathon race?
13. What factors contribute to muscle fatigue?
14. Why is the term recovery oxygen uptake more accurate than oxygen debt?

10.5 CONTROL OF MUSCLE TENSION

⊙ OBJECTIVES

- Describe the structure and function of a motor unit, and define motor unit recruitment.
- Explain the phases of a twitch contraction.
- Describe how frequency of stimulation affects muscle tension, and how muscle tone is produced.
- Distinguish between isotonic and isometric contractions.

A single nerve impulse in a somatic motor neuron elicits a single muscle action potential in all the skeletal muscle fibers with which it forms synapses. Action potentials always have the same size in a given neuron or muscle fiber. In contrast, the force of muscle fiber contraction does vary; a muscle fiber is capable of producing a much greater force than the one that results from a single action potential. The total force or tension that a single muscle fiber can produce depends mainly on the rate at which nerve impulses arrive at the neuromuscular junction. The number of impulses per second is the *frequency of stimulation.* Maximum tension is also affected by the amount of stretch before contraction (see Figure 10.8) and by nutrient and oxygen availability. The total tension a whole muscle can produce depends on the number of muscle fibers that are contracting in unison.

Motor Units

Even though each skeletal muscle fiber has only a single neuromuscular junction, the axon of a somatic motor neuron branches out and forms neuromuscular junctions with many different muscle fibers. A **motor unit** consists of a somatic motor neuron plus all the skeletal muscle fibers it stimulates (Figure 10.12). A single somatic motor neuron makes contact with an average of 150 skeletal muscle fibers, and all of the muscle fibers in one motor unit contract in unison. Typically, the muscle fibers of a motor unit are dispersed throughout a muscle rather than clustered together.

Whole muscles that control precise movements consist of many small motor units. For instance, muscles of the larynx (voice box) that control voice production have as few as two or three muscle fibers per motor unit, and muscles controlling eye movements may have 10 to 20 muscle fibers per motor unit. In contrast, skeletal muscles responsible for large-scale and powerful movements, such as the biceps brachii muscle in the arm and the gastrocnemius muscle in the calf of the leg, have as many as 2000 to 3000 muscle fibers in some motor units. Because all the muscle fibers of a motor unit contract and relax together, the total strength of a contraction depends, in part, on the size of the motor units and the number that are activated at a given time.

Twitch Contraction

A **twitch contraction** is the brief contraction of all the muscle fibers in a motor unit in response to a single action potential in its motor neuron. In the laboratory, a twitch can be produced by direct electrical stimulation of a motor neuron or its muscle fibers. The record of a muscle contraction, called a **myogram** (MĪ-ō-gram), is shown in Figure 10.13. Twitches of skeletal muscle fibers last anywhere from 20 to 200 msec. This is very long compared to the brief 1–2 msec[‡] that a muscle action potential lasts.

Note that a brief delay occurs between application of the stimulus (time zero on the graph) and the beginning of contraction.

[‡]One millisecond (msec) is 10^{-3} second (0.001 sec).

The delay, which lasts about 2 msec, is termed the **latent period.** During the latent period, the muscle action potential sweeps over the sarcolemma and calcium ions are released from the sarcoplasmic reticulum. The second phase, the **contraction period,** lasts 10–100 msec. During this time, Ca^{2+} binds to troponin, myosin-binding sites on actin are exposed, and cross-bridges form. Peak tension develops in the muscle fiber. During the third phase, the **relaxation period,** also lasting 10–100 msec, Ca^{2+} is actively transported back into the sarcoplasmic reticulum, myosin-binding sites are covered by tropomyosin, myosin heads detach from actin, and tension in the muscle fiber decreases. The actual duration of these periods depends on the type of skeletal muscle fiber. Some fibers, such as the fast-twitch fibers that move the eyes (described shortly), have contraction periods as brief as 10 msec and equally brief relaxation periods. Others, such as the slow-twitch fibers that move the legs, have contraction and relaxation periods of about 100 msec each.

If two stimuli are applied, one immediately after the other, the muscle will respond to the first stimulus but not to the second. When a muscle fiber receives enough stimulation to contract, it temporarily loses its excitability and cannot respond for a time. The period of lost excitability, called the **refractory period** (rē-FRAK-tō-rē), is a characteristic of all muscle and nerve cells. The duration of the refractory period varies with the muscle involved. Skeletal muscle has a short refractory period of about 5 msec; cardiac muscle has a longer refractory period of about 300 msec.

Frequency of Stimulation

When a second stimulus occurs after the refractory period of the first stimulus is over, but before the skeletal muscle fiber has relaxed, the second contraction will actually be stronger than the first (Figure 10.14b). This phenomenon, in which stimuli arriving at different times cause larger contractions, is

Figure 10.12 Motor units. Two somatic motor neurons (one purple and one green) are shown, each supplying the muscle fibers of its motor unit.

🔑 A motor unit consists of a somatic motor neuron plus all the muscle fibers it stimulates.

❓ **What is the effect of the size of a motor unit on its strength of contraction? (Assume that each muscle fiber can generate about the same amount of tension.)**

Figure 10.13 Myogram of a twitch contraction. The arrow indicates the time at which the stimulus occurred.

🔑 A myogram is a record of a muscle contraction.

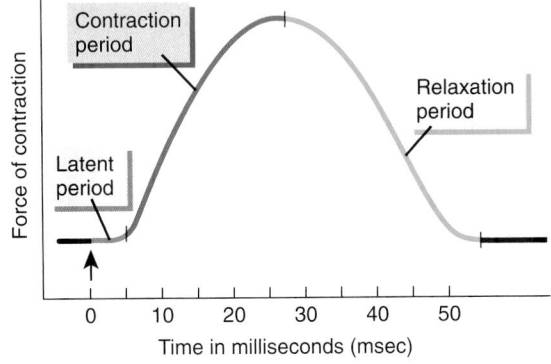

❓ **What events occur during the latent period?**

Figure 10.14 **Myograms showing the effects of different frequencies of stimulation.** (a) Single twitch. (b) When a second stimulus occurs before the muscle fiber has relaxed, the second contraction is stronger than the first, a phenomenon called wave summation. (The solid black line indicates the force of contraction expected in a single twitch.) (c) Unfused tetanus produces a jagged curve due to partial relaxation of the muscle fiber between stimuli. (d) In fused tetanus, which occurs when there are 80–100 stimuli per second, the myogram line, like the contraction force, is steady and sustained.

 Due to wave summation, the tension produced during a sustained contraction is greater than that produced by a single twitch.

(a) Single twitch   (b) Wave summation   (c) Unfused tetanus   (d) Fused tetanus

? Would the peak force of the second contraction in part b be larger or smaller if the second stimulus were applied a few milliseconds later?

called **wave summation.** When a skeletal muscle fiber is stimulated at a rate of 20 to 30 times per second, it can only partially relax between stimuli. The result is a sustained but wavering contraction called **unfused (incomplete) tetanus** (*tetan-* = rigid, tense; Figure 10.14c). When a skeletal muscle fiber is stimulated at a higher rate of 80 to 100 times per second, it does not relax at all. The result is **fused (complete) tetanus,** a sustained contraction in which individual twitches cannot be detected (Figure 10.14d).

Wave summation and both kinds of tetanus occur when additional Ca^{2+} is released from the sarcoplasmic reticulum by subsequent stimuli while the levels of Ca^{2+} in the sarcoplasm are still elevated from the first stimulus. Because of the buildup in the Ca^{2+} level, the peak tension generated during fused tetanus is 5 to 10 times larger than the peak tension produced during a single twitch. Even so, smooth, sustained voluntary muscle contractions are achieved mainly by out-of-synchrony unfused tetanus in different motor units.

The stretch of elastic components, such as tendons and connective tissues around muscle fibers, also affects wave summation. During wave summation, elastic components are not given much time to spring back between contractions, and thus remain taut. While in this state, the elastic components do not require very much stretching before the beginning of the next

muscular contraction. The combination of the tautness of the elastic components and the partially contracted state of the filaments enables the force of another contraction to be greater than the one before.

Motor Unit Recruitment

The process in which the number of active motor units increases is called **motor unit recruitment.** Typically, the different motor units of an entire muscle are not stimulated to contract in unison. While some motor units are contracting, others are relaxed. This pattern of motor unit activity delays muscle fatigue and allows contraction of a whole muscle to be sustained for long periods. The weakest motor units are recruited first, with progressively stronger motor units added if the task requires more force.

Recruitment is one factor responsible for producing smooth movements rather than a series of jerks. As mentioned, the number of muscle fibers innervated by one motor neuron varies greatly. Precise movements are brought about by small changes in muscle contraction. Therefore, the small muscles that produce precise movements are made up of small motor units. For this reason, when a motor unit is recruited or turned off, only slight changes occur in muscle tension. By contrast, large motor units are active when a large amount of tension is needed and precision is less important.

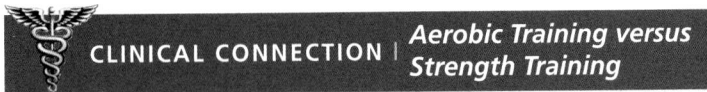

Regular, repeated activities such as jogging or aerobic dancing increase the supply of oxygen-rich blood available to skeletal muscles for aerobic cellular respiration. By contrast, activities such as weight lifting rely more on anaerobic production of ATP through glycolysis. Such anaerobic activities stimulate synthesis of muscle proteins and result, over time, in increased muscle size (muscle hypertrophy). Athletes who engage in anaerobic training should have a diet that includes an adequate amount of proteins. This protein intake will allow the body to synthesize muscle proteins and to increase muscle mass. As a result, aerobic training builds endurance for prolonged activities; in contrast, anaerobic training builds muscle strength for short-term feats. **Interval training** is a workout regimen that incorporates both types of training—for example, alternating sprints with jogging. •

Muscle Tone

Even at rest, a skeletal muscle exhibits **muscle tone** (*tonos* = tension), a small amount of tautness or tension in the muscle due to weak, involuntary contractions of its motor units. Recall that skeletal muscle contracts only after it is activated by acetylcholine released by nerve impulses in its motor neurons. Hence, muscle tone is established by neurons in the brain and spinal cord that excite the muscle's motor neurons. When the motor neurons serving a skeletal muscle are damaged or cut, the muscle becomes **flaccid** (FLAK-sid or FLAS-sid = flabby), a state of limpness in which muscle tone is lost. To sustain muscle tone, small groups of motor units are alternatively active and inactive in a constantly shifting pattern. Muscle tone keeps skeletal muscles firm, but it does not result in a force strong enough to produce movement. For example, when you are awake, the muscles in the back of the neck are in normal tonic contraction; they keep the head upright and prevent it from slumping forward on the chest. Muscle tone also is important in smooth muscle tissues, such as those found in the gastrointestinal tract, where the walls of the digestive organs maintain a steady pressure on their contents. The tone of smooth muscle fibers in the walls of blood vessels plays a crucial role in maintaining blood pressure.

Hypotonia (hī′-pō-TŌ-nē-a; *hypo-* = below) refers to decreased or lost muscle tone. Such muscles are said to be flaccid. Flaccid muscles are loose and appear flattened rather than rounded. Certain disorders of the nervous system and disruptions in the balance of electrolytes (especially sodium, calcium, and, to a lesser extent, magnesium) may result in **flaccid paralysis** (pa-RAL-i-sis), which is characterized by loss of muscle tone, loss or reduction of tendon reflexes, and atrophy (wasting away) and degeneration of muscles.

Hypertonia (hī′-per-TŌ-nē-a; *hyper-* = above) refers to increased muscle tone and is expressed in two ways: spasticity or rigidity. **Spasticity** (spas-TIS-i-tē) is characterized by increased muscle tone (stiffness) associated with an increase in tendon reflexes and pathological reflexes (such as the Babinski sign, in which the great toe extends with or without fanning of the other toes in response to stroking the outer margin of the sole). Certain disorders of the nervous system and electrolyte disturbances such as those previously noted may result in **spastic paralysis**, partial paralysis in which the muscles exhibit spasticity. **Rigidity** refers to increased muscle tone in which reflexes are not affected, as occurs in tetanus. Tetanus is a disease caused by a bacterium, *Clostridium tetani,* that enters the body thorough exposed wounds. It leads to muscle stiffness and spasms that can make breathing difficult and can become life threatening as a result. The bacteria produce a toxin that interferes with the nerves controlling the muscles. The first signs are typically spasms and stiffness in the muscles of the face and jaws. •

Isotonic and Isometric Contractions

Muscle contractions may be either isotonic or isometric. In an **isotonic contraction** (ī′-sō-TON-ik; *iso-* = equal; *-tonic* = tension), the *tension* (force of contraction) developed in the muscle remains almost constant while the muscle changes its length. Isotonic contractions are used for body movements and for moving objects. The two types of isotonic contractions are concentric and eccentric. In a **concentric isotonic contraction** (kon-SEN-trik), if the tension generated is great enough to overcome the resistance of the object to be moved, the muscle shortens and pulls on another structure, such as a tendon, to produce movement and to reduce the angle at a joint. Picking a book up off a table involves concentric isotonic contractions of the biceps brachii muscle in the arm (Figure 10.15a). By contrast, as you lower the book to place it back on the table, the previously shortened biceps lengthens in a controlled manner while it continues to contract. When the length of a muscle increases during a contraction, the contraction is an **eccentric isotonic contraction** (ek-SEN-trik) (Figure 10.15b). During an eccentric contraction, the tension exerted by the myosin cross-bridges resists movement of a load (the book, in this case) and slows the lengthening process. For reasons that are not well understood, repeated eccentric isotonic contractions (for example, walking downhill) produce more muscle damage and more delayed-onset muscle soreness than do concentric isotonic contractions.

In an **isometric contraction** (ī′-sō-MET-rik; *metro* = measure or length), the tension generated is not enough to exceed the resistance of the object to be moved, and the muscle does not change its length. An example would be holding a book steady using an outstretched arm (Figure 10.15c). These contractions are important for maintaining posture and for supporting objects in a fixed position. Although isometric contractions do not result in body movement, energy is still expended. The book pulls the arm downward, stretching the shoulder and arm muscles. The isometric contraction of the shoulder and arm muscles counteracts the stretch. Isometric contractions are important because they stabilize some joints as others are moved. Most activities include both isotonic and isometric contractions.

Figure 10.15 Comparison between isotonic (concentric and eccentric) and isometric contractions. (a and b) Isotonic contractions of the biceps brachii muscle in the arm. (c) Isometric contraction of shoulder and arm muscles.

 In an isotonic contraction, tension remains constant as muscle length decreases or increases; in an isometric contraction, tension increases greatly without a change in muscle length.

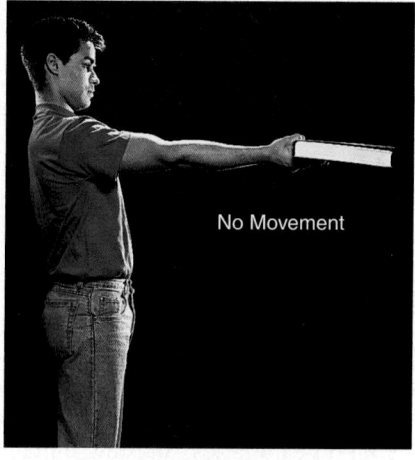

(a) Concentric contraction while picking up a book

(b) Eccentric contraction while lowering a book

(c) Isometric contraction while holding a book steady

? **What type of contraction occurs in your neck muscles while you are walking?**

✔ CHECKPOINT

15. How are the sizes of motor units related to the degree of muscular control they allow?
16. What is motor unit recruitment?
17. Why is muscle tone important?
18. Define each of the following terms: concentric isotonic contraction, eccentric isotonic contraction, and isometric contraction.
19. Demonstrate an isotonic contraction. How does it feel? What do you think causes the physical discomfort you are experiencing?

10.6 TYPES OF SKELETAL MUSCLE FIBERS

 OBJECTIVE

• Compare the structure and function of the three types of skeletal muscle fibers.

Skeletal muscle fibers are not all alike in composition and function. For example, muscle fibers vary in their content of myoglobin, the red-colored protein that binds oxygen in muscle fibers. Skeletal muscle fibers that have a high myoglobin content are termed *red muscle fibers* and appear darker (the dark meat in chicken legs and thighs); those that have a low content of myoglobin are called *white muscle fibers* and appear lighter (the white meat in chicken breasts). Red muscle fibers also contain more mitochondria and are supplied by more blood capillaries.

Skeletal muscle fibers also contract and relax at different speeds, and vary in which metabolic reactions they use to gener-

ate ATP and in how quickly they fatigue. For example, a fiber is categorized as either slow or fast depending on how rapidly the ATPase in its myosin heads hydrolyzes ATP. Based on all these structural and functional characteristics, skeletal muscle fibers are classified into three main types: (1) slow oxidative fibers, (2) fast oxidative–glycolytic fibers, and (3) fast glycolytic fibers.

Slow Oxidative Fibers

Slow oxidative (SO) fibers are smallest in diameter and thus are the least powerful type of muscle fibers. They appear dark red because they contain large amounts of myoglobin and many blood capillaries. Because they have many large mitochondria, SO fibers generate ATP mainly by aerobic cellular respiration, which is why they are called oxidative fibers. These fibers are said to be "slow" because the ATPase in the myosin heads hydrolyzes ATP relatively slowly and the contraction cycle proceeds at a slower pace than in "fast" fibers. As a result, SO fibers have a slow speed of contraction. Their twitch contractions last from 100 to 200 msec, and they take longer to reach peak tension. However, slow fibers are very resistant to fatigue and are capable of prolonged, sustained contractions for many hours. These slow-twitch, fatigue-resistant fibers are adapted for maintaining posture and for aerobic, endurance-type activities such as running a marathon.

Fast Oxidative–Glycolytic Fibers

Fast oxidative–glycolytic (FOG) fibers are intermediate in diameter between the other two types of fibers. Like slow oxidative fibers, they contain large amounts of myoglobin and many blood capillaries. Thus, they also have a dark red appearance. FOG fibers can generate considerable ATP by aerobic cellular respiration, which gives them a moderately high resistance to fatigue. Because their intracellular glycogen level is high, they also generate ATP by anaerobic glycolysis. FOG fibers are "fast" because the ATPase in their myosin heads hydrolyzes ATP three to five times faster than the myosin ATPase in SO fibers, which makes their speed of contraction faster. Thus, twitches of FOG fibers reach peak tension more quickly than those of SO fibers but are briefer in duration—less than 100 msec. FOG fibers contribute to activities such as walking and sprinting.

Fast Glycolytic Fibers

Fast glycolytic (FG) fibers are largest in diameter and contain the most myofibrils. Hence, they can generate the most powerful contractions. FG fibers have low myoglobin content, relatively few blood capillaries, and few mitochondria, and appear white in color. They contain large amounts of glycogen and generate ATP mainly by glycolysis. Due to their large size and their ability to hydrolyze ATP rapidly, FG fibers contract strongly and quickly. These fast-twitch fibers are adapted for intense anaerobic movements of short duration, such as weight lifting or throwing a ball, but they fatigue quickly. Strength training programs that engage a person in activities requiring great strength for short times increase the size, strength, and glycogen content of fast glycolytic fibers. The FG fibers of a weight lifter may be 50% larger than those of a sedentary person or an endurance athlete because of increased synthesis of muscle proteins. The overall result is muscle enlargement due to hypertrophy of the FG fibers.

Distribution and Recruitment of Different Types of Fibers

Most skeletal muscles are a mixture of all three types of skeletal muscle fibers; about half the fibers in a typical skeletal muscle are SO fibers. However, the proportions vary somewhat, depending on the action of the muscle, the person's training regimen, and genetic factors. For example, the continually active postural muscles of the neck, back, and legs have a high proportion of SO fibers. Muscles of the shoulders and arms, in contrast, are not constantly active but are used briefly now and then to produce large amounts of tension, such as in lifting and throwing. These muscles have a high proportion of FG fibers. Leg muscles, which not only support the body but are also used for walking and running, have large numbers of both SO and FOG fibers.

Within a particular motor unit, all of the skeletal muscle fibers are of the same type. The different motor units in a muscle are recruited in a specific order, depending on need. For example, if weak contractions suffice to perform a task, only SO motor units are activated. If more force is needed, the motor units of FOG fibers are also recruited. Finally, if maximal force is required, motor units of FG fibers are also called into action with the other two

types. Activation of various motor units is controlled by the brain and spinal cord.

Table 10.4 summarizes the characteristics of the three types of skeletal muscle fibers.

✔ **CHECKPOINT**

20. Why are some skeletal muscle fibers classified as "fast" and others are said to be "slow"?
21. In what order are the various types of skeletal muscle fibers recruited when you sprint to make it to the bus stop?

10.7 EXERCISE AND SKELETAL MUSCLE TISSUE

◉ **OBJECTIVE**

• Describe the effects of exercise on different types of skeletal muscle fibers.

The relative ratio of fast glycolytic (FG) and slow oxidative (SO) fibers in each muscle is genetically determined and helps account for individual differences in physical performance. For example, people with a higher proportion of FG fibers (see Table 10.4) often excel in activities that require periods of intense activity, such as weight lifting or sprinting. People with higher percentages of SO fibers are better at activities that require endurance, such as long-distance running.

Although the total number of skeletal muscle fibers usually does not increase with exercise, the characteristics of those present can change to some extent. Various types of exercises can induce changes in the fibers in a skeletal muscle. Endurance-type (aerobic) exercises, such as running or swimming, cause a gradual transformation of some FG fibers into fast oxidative–glycolytic (FOG) fibers. The transformed muscle fibers show slight increases in diameter, number of mitochondria, blood supply, and strength. Endurance exercises also result in cardiovascular and respiratory changes that cause skeletal muscles to receive better supplies of oxygen and nutrients but do not increase muscle mass. By contrast, exercises that require great strength for short periods produce an increase in the size and strength of FG fibers. The increase in size is due to increased synthesis of thick and thin filaments. The overall result is muscle enlargement (hypertrophy), as evidenced by the bulging muscles of body builders.

A certain degree of elasticity is an important attribute of skeletal muscles and their connective tissue attachments. Greater elasticity contributes to a greater degree of flexibility, increasing the range of motion of a joint. When a relaxed muscle is physically stretched, its ability to lengthen is limited by connective tissue structures, such as fasciae. Regular stretching gradually lengthens these structures, but the process occurs very slowly. To see an improvement in flexibility, stretching exercises must be performed regularly—daily, if possible—for many weeks.

Effective Stretching

Stretching cold muscles does not increase flexibility and may cause injury. Tissues stretch best when slow, gentle force is

TABLE 10.4			
Characteristics of the Three Types of Skeletal Muscle Fibers			

Slow oxidative fiber
Fast glycolytic fiber
Fast oxidative–glycolytic fiber

LM 440x

Transverse section of three types of skeletal muscle fibers

	SLOW OXIDATIVE (SO) FIBERS	FAST OXIDATIVE–GLYCOLYTIC (FOG) FIBERS	FAST GLYCOLYTIC (FG) FIBERS
STRUCTURAL CHARACTERISTIC			
Fiber diameter	Smallest.	Intermediate.	Largest.
Myoglobin content	Large amount.	Large amount.	Small amount.
Mitochondria	Many.	Many.	Few.
Capillaries	Many.	Many.	Few.
Color	Red.	Red-pink.	White (pale).
FUNCTIONAL CHARACTERISTIC			
Capacity for generating ATP and method used	High, by aerobic cellular respiration.	Intermediate, by both aerobic cellular respiration and anaerobic cellular respiration (glycolysis).	Low, by anaerobic cellular respiration (glycolysis).
Rate of ATP hydrolysis by myosin ATPase	Slow.	Fast.	Fast.
Contraction velocity	Slow.	Fast.	Fast.
Fatigue resistance	High.	Intermediate.	Low.
Creatine kinase	Lowest amount.	Intermediate amount.	Highest amount.
Glycogen stores	Low.	Intermediate.	High.
Order of recruitment	First.	Second.	Third.
Location where fibers are abundant	Postural muscles such as those of neck.	Lower limb muscles.	Upper limb muscles.
Primary functions of fibers	Maintaining posture and aerobic endurance activities.	Walking, sprinting.	Rapid, intense movements of short duration.

applied at elevated tissue temperatures. An external source of heat, such as hot packs or ultrasound, may be used, but 10 or more minutes of muscular contraction is also a good way to raise muscle temperature. Exercise heats muscle more deeply and thoroughly than external measures. That's where the term "warm-up" comes from. Many people stretch before they engage in exercise, but it's important to warm up (for example, walking, jogging, easy swimming, or easy aerobics) *before* stretching to avoid injury.

Strength Training

Strength training refers to the process of exercising with progressively heavier resistance for the purpose of strengthening the musculoskeletal system. This activity results not only in stronger muscles, but in many other health benefits as well. Strength training also helps to increase bone strength by increasing the deposition of bone minerals in young adults and helping to prevent, or at least slow, their loss in later life. By increasing muscle mass, strength training raises resting metabolic rate, the amount of energy expended at rest, so a person can eat more food without gaining weight. Strength training helps to prevent back injury and other injuries from participation in sports and other physical activities. Psychological benefits include reductions in feelings of stress and fatigue. As repeated training builds exercise tolerance, it takes increasingly longer before lactic acid is produced in the muscle, resulting in a reduced probability of muscle spasms.

The use of **anabolic** (to build up proteins) **steroids,** also called **"roids,"** by athletes has received widespread attention. These steroid hormones, similar to testosterone, are taken to increase muscle size by increasing the synthesis of proteins in muscle and thus increasing strength during athletic contests. However, the large doses needed to produce an effect have damaging, sometimes even devastating side effects, including liver cancer, kidney damage, increased risk of heart disease, stunted growth, wide mood swings, increased acne, and increased irritability and aggression. Additionally, females who take anabolic steroids may experience atrophy of the breasts and uterus, menstrual irregularities, sterility, facial hair growth, and deepening of the voice. Males may experience diminished testosterone secretion, atrophy of the testes, sterility, and baldness. •

✔**CHECKPOINT**

22. On a cellular level, what causes muscle hypertrophy?

10.8 CARDIAC MUSCLE TISSUE

◉ **OBJECTIVE**

• Describe the main structural and functional characteristics of cardiac muscle tissue.

The principal tissue in the heart wall is **cardiac muscle tissue** (described in more detail in Chapter 20 and illustrated in Figure 20.9). Between the layers of **cardiac muscle fibers,** the contractile cells of the heart, are sheets of connective tissue that contain blood vessels, nerves, and the conduction system of the heart. Cardiac muscle fibers have the same arrangement of actin and myosin and the same bands, zones, and Z discs as skeletal muscle fibers. However, *intercalated discs* (in-TER-ka-lāt-ed; *intercal-* = to insert between) are unique to cardiac muscle fibers. These microscopic structures are irregular transverse thickenings of the sarcolemma that connect the ends of cardiac muscle fibers to one another. The discs contain *desmosomes,* which hold the fibers together, and *gap junctions,* which allow muscle action potentials to spread from one cardiac muscle fiber to another (see Figure 4.2e). Cardiac muscle tissue has an endomysium and perimysium, but lacks an epimysium.

In response to a single action potential, cardiac muscle tissue remains contracted 10 to 15 times longer than skeletal muscle tissue (see Figure 20.11). The long contraction is due to prolonged delivery of Ca^{2+} into the sarcoplasm. In cardiac muscle fibers, Ca^{2+} enters the sarcoplasm both from the sarcoplasmic reticulum (as in skeletal muscle fibers) and from the interstitial fluid that bathes the fibers. Because the channels that allow inflow of Ca^{2+} from interstitial fluid stay open for a relatively long time, a cardiac muscle contraction lasts much longer than a skeletal muscle twitch.

We have seen that skeletal muscle tissue contracts only when stimulated by acetylcholine released by a nerve impulse in a motor neuron. In contrast, cardiac muscle tissue contracts when stimulated by its own autorhythmic muscle fibers. Under normal resting conditions, cardiac muscle tissue contracts and relaxes about 75 times a minute. This continuous, rhythmic activity is a major physiological difference between cardiac and skeletal muscle tissue. The mitochondria in cardiac muscle fibers are larger and more numerous than in skeletal muscle fibers. This structural feature correctly suggests that cardiac muscle depends largely on aerobic cellular respiration to generate ATP, and thus requires a constant supply of oxygen. Cardiac muscle fibers can also use lactic acid produced by skeletal muscle fibers to make ATP, a benefit during exercise. Like skeletal muscle, cardiac muscle fibers can undergo hypertrophy in response to an increased workload. This is called a *physiological enlarged heart* and it is why many athletes have enlarged hearts. By contrast, a *pathological enlarged heart* is related to significant heart disease.

✔**CHECKPOINT**

23. What are the similarities among and differences between skeletal and cardiac muscle?

10.9 SMOOTH MUSCLE TISSUE

◉ **OBJECTIVE**

• Describe the main structural and functional characteristics of smooth muscle tissue.

Like cardiac muscle tissue, **smooth muscle tissue** is usually activated involuntarily. Of the two types of smooth muscle tissue, the more common type is **visceral (single-unit) smooth muscle tissue** (Figure 10.16a). It is found in the skin and in tubular arrangements that form part of the walls of small arteries and veins and of hollow organs such as the stomach, intestines, uterus, and urinary bladder. Like cardiac muscle, visceral smooth muscle is autorhythmic. The fibers connect to one another by gap junctions, forming a network through which muscle action potentials can spread. When a neurotransmitter, hormone, or autorhythmic signal stimulates one fiber, the muscle action potential is transmitted to neighboring fibers, which then contract in unison, as a single unit.

The second type of smooth muscle tissue, **multiunit smooth muscle tissue** (Figure 10.16b), consists of individual fibers, each with its own motor neuron terminals and with few gap junctions between neighboring fibers. Stimulation of one visceral muscle fiber causes contraction of many adjacent fibers, but stimulation of one multiunit fiber causes contraction of that fiber only. Multiunit smooth muscle tissue is found in the walls of large arteries, in airways to the lungs, in the arrector pili muscles that attach to hair follicles, in the muscles of the iris that adjust pupil diameter, and in the ciliary body that adjusts focus of the lens in the eye.

Figure 10.16 **Smooth muscle tissue.** (a) One autonomic motor neuron synapses with several visceral smooth muscle fibers, and action potentials spread to neighboring fibers through gap junctions. (b) Three autonomic motor neurons synapse with individual multiunit smooth muscle fibers; stimulation of one multiunit fiber causes contraction of that fiber only. (c) Relaxed and contracted smooth muscle fiber. A photomicrograph of smooth muscle tissue is shown in Table 4.9.

 Visceral smooth muscle fibers connect to one another by gap junctions and contract as a single unit. Multiunit smooth muscle fibers lack gap junctions and contract independently.

(a) Visceral (single-unit) smooth muscle tissue

(b) Multiunit smooth muscle tissue

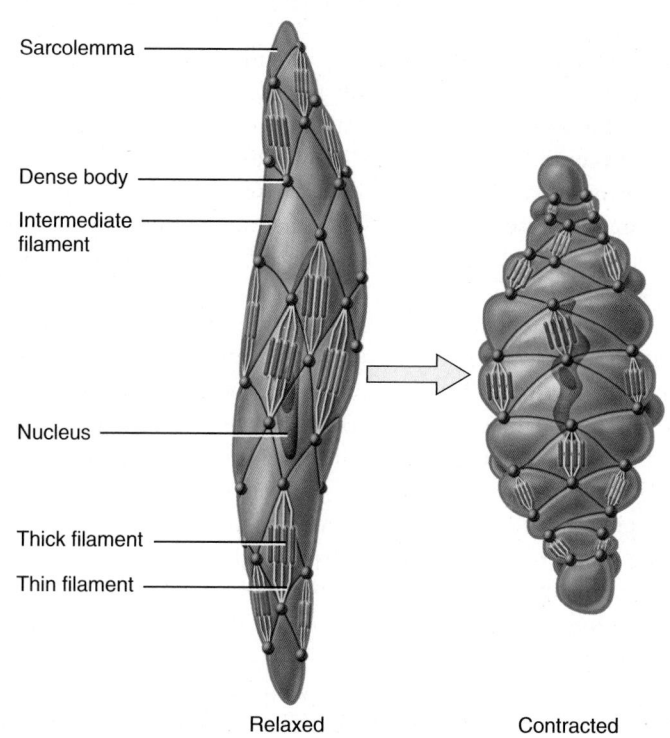

(c) Microscopic anatomy of a relaxed and contracted smooth muscle fiber

? Which type of smooth muscle is more like cardiac muscle than skeletal muscle, with respect to both its structure and function?

Microscopic Anatomy of Smooth Muscle

A single relaxed smooth muscle fiber is 30–200 μm long. It is thickest in the middle (3–8 μm) and tapers at each end (Figure 10.16c). Within each fiber is a single, oval, centrally located nucleus. The sarcoplasm of smooth muscle fibers contains both *thick filaments* and *thin filaments,* in ratios between 1:10 and 1:15, but they are not arranged in orderly sarcomeres as in striated muscle. Smooth muscle fibers also contain *intermediate filaments.* Because the various filaments have no regular pattern of overlap, smooth muscle fibers do not exhibit striations (see Table 4.9), causing a smooth appearance. Smooth muscle fibers also lack transverse tubules and have only a small amount of sarcoplasmic reticulum for storage of Ca^{2+}. Although there are no transverse tubules in smooth muscle tissue, there are small pouchlike invaginations of the plasma membrane called **caveolae** (kav'-ē-Ō-lē; *cavus* = space) that contain extracellular Ca^{2+} that can be used for muscular contraction.

In smooth muscle fibers, the thin filaments attach to structures called **dense bodies,** which are functionally similar to Z discs in striated muscle fibers. Some dense bodies are dispersed throughout the sarcoplasm; others are attached to the sarcolemma. Bundles of intermediate filaments also attach to dense bodies and stretch from one dense body to another (Figure 10.16c). During contraction, the sliding filament mechanism involving thick and thin filaments generates tension that is transmitted to intermediate filaments. These in turn pull on the dense bodies attached to the sarcolemma, causing a lengthwise shortening of the muscle fiber. As a smooth muscle fiber contracts, it rotates as a corkscrew turns. The fiber twists in a helix as it contracts, and rotates in the opposite direction as it relaxes.

Physiology of Smooth Muscle

Although the principles of contraction are similar, smooth muscle tissue exhibits some important physiological differences from cardiac and skeletal muscle tissue. Contraction in a smooth muscle fiber starts more slowly and lasts much longer than skeletal muscle fiber contraction. Another difference is that smooth muscle can both shorten and stretch to a greater extent than the other muscle types.

An increase in the concentration of Ca^{2+} in the cytosol of a smooth muscle fiber initiates contraction, just as in striated muscle. Sarcoplasmic reticulum (the reservoir for Ca^{2+} in striated muscle) is found in small amounts in smooth muscle. Calcium ions flow into smooth muscle cytosol from both the interstitial fluid and sarcoplasmic reticulum. Because there are no transverse tubules in smooth muscle fibers (there are caveolae instead), it takes longer for Ca^{2+} to reach the filaments in the center of the fiber and trigger the contractile process. This accounts, in part, for the slow onset of contraction of smooth muscle.

Several mechanisms regulate contraction and relaxation of smooth muscle cells. In one such mechanism, a regulatory protein called **calmodulin** (cal-MOD-ū-lin) binds to Ca^{2+} in the cytosol. (Recall that troponin takes this role in striated muscle fibers.) After binding to Ca^{2+}, calmodulin activates an enzyme called *myosin light chain kinase*. This enzyme uses ATP to add a phosphate group to a portion of the myosin head. Once the phosphate group is attached, the myosin head can bind to actin, and contraction can occur. Because myosin light chain kinase works rather slowly, it contributes to the slowness of smooth muscle contraction.

Not only do calcium ions enter smooth muscle fibers slowly, they also move slowly out of the muscle fiber, which delays relaxation. The prolonged presence of Ca^{2+} in the cytosol provides for **smooth muscle tone,** a state of continued partial contraction. Smooth muscle tissue can thus sustain long-term tone, which is important in the gastrointestinal tract, where the walls maintain a steady pressure on the contents of the tract, and in the walls of blood vessels called *arterioles,* which maintain a steady pressure on blood.

Most smooth muscle fibers contract or relax in response to action potentials from the autonomic nervous system. In addition, many smooth muscle fibers contract or relax in response to stretching, hormones, or local factors such as changes in pH, oxygen and carbon dioxide levels, temperature, and ion concentrations. For example, the hormone epinephrine, released by the adrenal medulla, causes relaxation of smooth muscle in the airways and in some blood vessel walls (those that have so-called β_2 receptors; see Table 15.2).

Unlike striated muscle fibers, smooth muscle fibers can stretch considerably and still maintain their contractile function. When smooth muscle fibers are stretched, they initially contract, developing increased tension. Within a minute or so, the tension decreases. This phenomenon, which is called the **stress–relaxation response,** allows smooth muscle to undergo great changes in length while retaining the ability to contract effectively. Thus, even though smooth muscle in the walls of blood vessels and hollow organs such as the stomach, intestines, and urinary bladder can stretch, the pressure on the contents within them changes very little. After the organ empties, the

smooth muscle in the wall rebounds, and the wall retains its firmness.

✔ CHECKPOINT

24. What are the differences between visceral and multiunit smooth muscle?
25. How are skeletal and smooth muscle similar? How do they differ?

10.10 REGENERATION OF MUSCULAR TISSUE

◉ OBJECTIVE
• Explain how muscle fibers regenerate.

Because mature skeletal muscle fibers have lost the ability to undergo cell division, growth of skeletal muscle after birth is due mainly to **hypertrophy** (hī-PER-trō-fē), the enlargement of existing cells, rather than to **hyperplasia** (hī-per-PLĀ-zē-a), an increase in the number of fibers. Satellite cells divide slowly and fuse with existing fibers to assist both in muscle growth and in repair of damaged fibers. Thus, skeletal muscle tissue can regenerate only to a limited extent.

Until recently it was believed that damaged cardiac muscle fibers could not be replaced and that healing took place exclusively by fibrosis, the formation of scar tissue. New research described in Chapter 20 indicates that, under certain circumstances, cardiac muscle tissue can regenerate. In addition, cardiac muscle fibers can undergo hypertrophy in response to increased workload. Hence, many athletes have enlarged hearts.

Smooth muscle tissue, like skeletal and cardiac muscle tissue, can undergo hypertrophy. In addition, certain smooth muscle fibers, such as those in the uterus, retain their capacity for division and thus can grow by hyperplasia. Also, new smooth muscle fibers can arise from cells called *pericytes,* stem cells found in association with blood capillaries and small veins. Smooth muscle fibers can also proliferate in certain pathological conditions, such as occur in the development of atherosclerosis (see Disorders: Homeostatic Imbalances in Chapter 20). Compared with the other two types of muscle tissue, smooth muscle tissue has considerably greater powers of regeneration. Such powers are still limited when compared with other tissues, such as epithelium.

Table 10.5 summarizes the major characteristics of the three types of muscular tissue.

✔ CHECKPOINT

26. Which type of muscular tissue has the highest capacity for regeneration?

10.11 DEVELOPMENT OF MUSCLE

◉ OBJECTIVE
• Describe the development of muscles.

Except for muscles such as those of the iris of the eyes and the arrector pili muscles attached to hairs, all muscles of the body are derived from **mesoderm.** As the mesoderm develops, part of it

TABLE 10.5

Summary of the Major Features of the Three Types of Muscular Tissue

CHARACTERISTIC	SKELETAL MUSCLE	CARDIAC MUSCLE	SMOOTH MUSCLE
Microscopic appearance and features	Long cylindrical fiber with many peripherally located nuclei; unbranched; striated.	Branched cylindrical fiber with one centrally located nucleus; intercalated discs join neighboring fibers; striated.	Fiber thickest in middle, tapered at each end, and with one centrally positioned nucleus; not striated.
Location	Most commonly attached by tendons to bones.	Heart.	Walls of hollow viscera, airways, blood vessels, iris and ciliary body of eye, arrector pili muscles of hair follicles.
Fiber diameter	Very large (10–100 μm).	Large (10–20 μm).	Small (3–8 μm).
Connective tissue components	Endomysium, perimysium, and epimysium.	Endomysium and perimysium.	Endomysium.
Fiber length	Very large (100 μm–30 cm = 12 in.).	Large (50–100 μm).	Intermediate (30–200 μm).
Contractile proteins organized into sarcomeres	Yes.	Yes.	No.
Sarcoplasmic reticulum	Abundant.	Some.	Very little.
Transverse tubules present	Yes, aligned with each A–I band junction.	Yes, aligned with each Z disc.	No.
Junctions between fibers	None.	Intercalated discs contain gap junctions and desmosomes.	Gap junctions in visceral smooth muscle; none in multiunit smooth muscle.
Autorhythmicity	No.	Yes.	Yes, in visceral smooth muscle.
Source of Ca^{2+} for contraction	Sarcoplasmic reticulum.	Sarcoplasmic reticulum and interstitial fluid.	Sarcoplasmic reticulum and interstitial fluid.
Regulator proteins for contraction	Troponin and tropomyosin.	Troponin and tropomyosin.	Calmodulin and myosin light chain kinase.
Speed of contraction	Fast.	Moderate.	Slow.
Nervous control	Voluntary (somatic nervous system).	Involuntary (autonomic nervous system).	Involuntary (autonomic nervous system).
Contraction regulation	Acetylcholine released by somatic motor neurons.	Acetylcholine and norepinephrine released by autonomic motor neurons; several hormones.	Acetylcholine and norepinephrine released by autonomic motor neurons; several hormones; local chemical changes; stretching.
Capacity for regeneration	Limited, via satellite cells.	Limited, under certain conditions.	Considerable (compared with other muscle tissues, but limited compared with epithelium), via pericytes.

becomes arranged in dense columns on either side of the developing nervous system. These columns of mesoderm undergo segmentation into a series of cube-shaped structures called **somites** (SŌ-mīts) (Figure 10.17a). The first pair of somites appears on the 20th day of embryonic development. Eventually, 42 to 44 pairs of somites are formed by the end of the fifth week.

The number of somites can be correlated to the approximate age of the embryo.

The cells of a somite differentiate into three regions: (1) a **myotome** (MĪ-ō-tōm), which, as the name suggests, forms the skeletal muscles of the head, neck, and limbs; (2) a **dermatome** (DER-ma-tōm), which forms the connective tissues, including the

Figure 10.17 Location and structure of somites, key structures in the development of the muscular system.

Most muscles are derived from mesoderm.

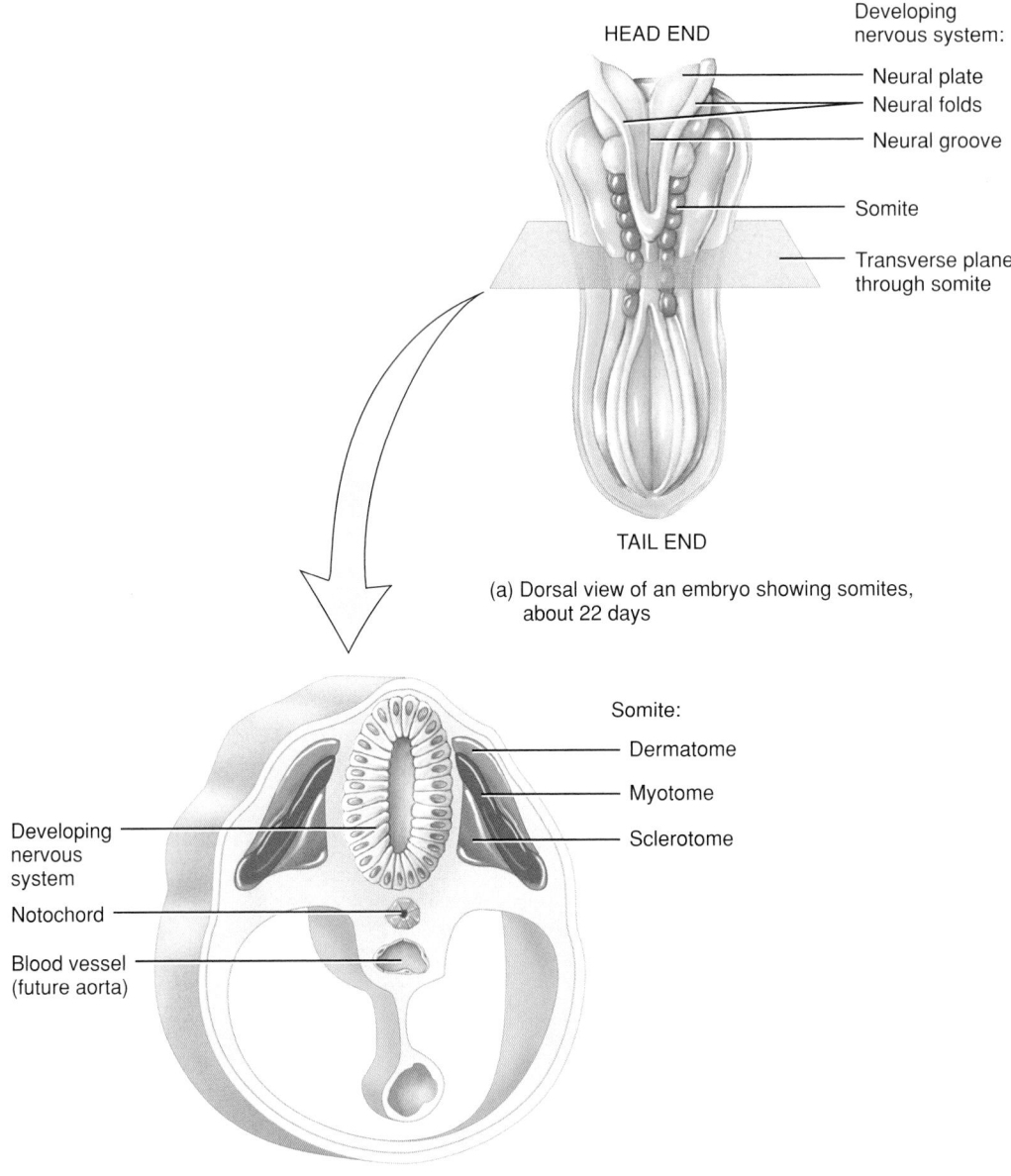

(a) Dorsal view of an embryo showing somites, about 22 days

(b) Transverse section through a somite showing its subdivisions

? Which part of a somite differentiates into skeletal muscle?

dermis of the skin; and (3) a **sclerotome** (SKLE-rō-tōm), which gives rise to the vertebrae (Figure 10.17b, c).

Cardiac muscle develops from **mesodermal cells** that migrate to and envelop the developing heart while it is still in the form of endocardial heart tubes (see Figure 20.19).

Smooth muscle develops from **mesodermal cells** that migrate to and envelop the developing gastrointestinal tract and viscera.

✔CHECKPOINT

27. Which structures develop from myotomes, dermatomes, and sclerotomes?

10.12 AGING AND MUSCULAR TISSUE

◉ OBJECTIVE

• Explain the effects of aging on skeletal muscle.

Between the ages of 30 and 50, humans undergo a slow, progressive loss of skeletal muscle mass that is replaced largely by fibrous connective tissue and adipose tissue. An estimated 10% of muscle mass is lost during these years. In part, this decline may be due to decreased levels of physical activity. Accompanying the loss of muscle mass is a decrease in maximal strength, a slowing of muscle reflexes, and a loss of flexibility. With aging, the relative number of slow oxidative (SO) fibers appears to increase. This could be due to either the atrophy of the other fiber types or their conversion into slow oxidative fibers. Another 40% of muscle is typically lost between the ages of 50 and 80. Loss of muscle strength is usually not perceived by persons until they reach the age of 60 to 65. At that point it is most common for muscles of the lower limbs to weaken before those of the upper limbs. Thus the independence of the elderly may be affected when it becomes difficult to climb stairs or get up from a seated position.

Assuming that there is not a chronic medical condition for which exercise is contraindicated, exercise has been shown to be effective at any age. Aerobic activities and strength training programs are effective in older people and can slow or even reverse the age-associated decline in muscular performance.

✔CHECKPOINT

28. Why does muscle strength decrease with aging?
29. Why do you think a healthy 30-year-old can lift a 25-lb load much more comfortably than an 80-year-old?

DISORDERS: HOMEOSTATIC IMBALANCES

Abnormalities of skeletal muscle function may be due to disease or damage of any of the components of a motor unit: somatic motor neurons, neuromuscular junctions, or muscle fibers. The term **neuromuscular disease** encompasses problems at all three sites; the term **myopathy** (mī-OP-a-thē; -*pathy* = disease) signifies a disease or disorder of the skeletal muscle tissue itself.

Myasthenia Gravis

Myasthenia gravis (mī-as-THĒ-nē-a GRAV-is; *mys-* = muscle; *-aisthesis* = sensation) is an autoimmune disease that causes chronic, progressive damage of the neuromuscular junction. The immune system inappropriately produces antibodies that bind to and block some ACh receptors, thereby decreasing the number of functional ACh receptors at the motor end plates of skeletal muscles (see Figure 10.9). Because 75% of patients with myasthenia gravis have hyperplasia or tumors of the thymus, it is thought that thymic abnormalities cause the disorder. As the disease progresses, more ACh receptors are lost. Thus, muscles become increasingly weaker, fatigue more easily, and may eventually cease to function.

Myasthenia gravis occurs in about 1 in 10,000 people and is more common in women, typically ages 20 to 40 at onset; men usually are ages 50 to 60 at onset. The muscles of the face and neck are most often affected. Initial symptoms include weakness of the eye muscles, which may produce double vision, and weakness of the throat muscles that may produce difficulty in swallowing. Later, the person has difficulty chewing and talking. Eventually the muscles of the limbs may become involved. Death may result from paralysis of the respiratory muscles, but often the disorder does not progress to that stage.

Anticholinesterase drugs such as pyridostigmine (Mestinon) or neostigmine, the first line of treatment, act as inhibitors of acetylcholinesterase, the enzyme that breaks down ACh. Thus, the inhibitors raise the level of ACh that is available to bind with still-functional receptors. More recently, steroid drugs such as prednisone have been used with success to reduce antibody levels. Another treatment is plasmapheresis, a procedure that removes the antibodies from the blood. Often, surgical removal of the thymus (thymectomy) is helpful.

Muscular Dystrophy

The term **muscular dystrophy** (DIS-trō-fē′; *dys-* = difficult; *-trophy* = nourishment) refers to a group of inherited muscle-destroying diseases that cause progressive degeneration of skeletal muscle fibers. The most common form of muscular dystrophy is *Duchenne muscular dystrophy* (doo-SHĀN) or *DMD*. Because the mutated gene is on the X chromosome, and because males have only one X chromosome, DMD strikes boys almost exclusively. (Sex-linked inheritance is described in Chapter 29.) Worldwide, about 1 in every 3500 male babies—about 21,000 in all—are born with DMD each year. The disorder usually becomes apparent between the ages of 2 and 5, when parents notice the child falls often and has difficulty running, jumping, and hopping. By age 12 most boys with DMD are unable to walk. Respiratory or cardiac failure usually causes death by age 20.

In DMD, the gene that codes for the protein dystrophin is mutated, so little or no dystrophin is present in the sarcolemma. Without the reinforcing effect of dystrophin, the sarcolemma tears easily during muscle contraction, causing muscle fibers to rupture and die. The dystrophin gene was discovered in 1987, and by 1990 the first attempts were made to treat DMD patients with gene therapy. The muscles of three boys with DMD were injected with myoblasts bearing functional dystrophin genes, but only a few muscle fibers gained the ability to produce dystrophin. Similar clinical trials with additional patients have also failed. An alternative approach to the problem is to find a way to induce muscle fibers to produce the protein *utrophin,* which is similar to dystrophin. Experiments with dystrophin-deficient mice suggest this approach may work.

Abnormal Contractions of Skeletal Muscle

One kind of abnormal muscular contraction is a **spasm,** a sudden involuntary contraction of a single muscle in a large group of muscles. A painful spasmodic contraction is known as a **cramp.** Cramps may be caused by inadequate blood flow to muscles, overuse of a muscle, dehydration, injury, holding a position for prolonged periods, and low blood levels of electrolytes, such as potassium. A **tic** is a spasmodic twitching made involuntarily by muscles that are ordinarily under voluntary control. Twitching of the eyelid and facial muscles are examples of tics. A **tremor** is a rhythmic, involuntary, purposeless contraction that produces a quivering or shaking movement. A **fasciculation** (fa-sik-ū-LĀ-shun) is an involuntary, brief twitch of an entire motor unit that is visible under the skin; it occurs irregularly and is not associated with movement of the affected muscle. Fasciculations may be seen in multiple sclerosis (see Disorders: Homeostatic Imbalances in Chapter 12) or in amyotrophic lateral sclerosis (Lou Gehrig's disease, see Clinical Connection: Amyotrophic Lateral Sclerosis in Chapter 16). A **fibrillation** (fi-bri-LĀ-shun) is a spontaneous contraction of a single muscle fiber that is not visible under the skin but can be recorded by electromyography. Fibrillations may signal destruction of motor neurons.

MEDICAL TERMINOLOGY

Myalgia (mī-AL-jē-a; *-algia* = painful condition) Pain in or associated with muscles.

Myoma (mī-Ō-ma; *-oma* = tumor) A tumor consisting of muscle tissue.

Myomalacia (mī'-ō-ma-LĀ-shē-a; *-malacia* = soft) Pathological softening of muscle tissue.

Myositis (mī'-ō-SĪ-tis; *-itis* = inflammation of) Inflammation of muscle fibers (cells).

Myotonia (mī'-ō-TŌ-nē-a; *-tonia* = tension) Increased muscular excitability and contractility, with decreased power of relaxation; tonic spasm of the muscle.

Volkmann's contracture (FŌLK-manz kon-TRAK-chur; *contra-* = against) Permanent shortening (contracture) of a muscle due to replacement of destroyed muscle fibers by fibrous connective tissue, which lacks extensibility. Typically occurs in forearm flexor muscles. Destruction of muscle fibers may occur from interference with circulation caused by a tight bandage, a piece of elastic, or a cast.

CHAPTER REVIEW AND RESOURCE SUMMARY

Review	Resource

Introduction

1. Motion results from alternating contraction and relaxation of muscles, which constitute 40–50% of total body weight.
2. The prime function of muscle is changing chemical energy into mechanical energy to perform work.

10.1 Overview of Muscular Tissue

1. The three types of muscular tissue are skeletal, cardiac, and smooth. Skeletal muscle tissue is primarily attached to bones; it is striated and voluntary. Cardiac muscle tissue forms the wall of the heart; it is striated and involuntary. Smooth muscle tissue is located primarily in internal organs; it is non-striated (smooth) and involuntary.
2. Through contraction and relaxation, muscular tissue performs four important functions: producing body movements; stabilizing body positions; moving substances within the body and regulating organ volume; and producing heat.
3. Four special properties of muscular tissues are (1) electrical excitability, the property of responding to stimuli by producing action potentials; (2) contractility, the ability to generate tension to do work; (3) extensibility, the ability to be extended (stretched); and (4) elasticity, the ability to return to original shape after contraction or extension.

Anatomy Overview - Muscular Tissue

10.2 Skeletal Muscle Tissue

1. The subcutaneous layer separates skin from muscles, provides a pathway for blood vessels and nerves to enter and exit muscles, and protects muscles from physical trauma. Fascia lines the body wall and limbs that surround and support muscles, allows free movement of muscles, carries nerves and blood vessels, and fills space between muscles.
2. Tendons and aponeuroses are extensions of connective tissue beyond muscle fibers that attach the muscle to bone or to other muscle. A tendon is generally ropelike in shape; an aponeurosis is wide and flat.
3. Skeletal muscles are well supplied with nerves and blood vessels. Generally, an artery and one or two veins accompany each nerve that penetrates a skeletal muscle.
4. Somatic motor neurons provide the nerve impulses that stimulate skeletal muscle to contract.
5. Blood capillaries bring in oxygen and nutrients and remove heat and waste products of muscle metabolism.
6. The major cells of skeletal muscle tissue are termed skeletal muscle fibers. Each muscle fiber has 100 or more nuclei because it arises from the fusion of many myoblasts. Satellite cells are myoblasts that persist after birth. The sarcolemma is a muscle fiber's plasma membrane; it surrounds the sarcoplasm. Transverse tubules are invaginations of the sarcolemma.
7. Each muscle fiber (cell) contains hundreds of myofibrils, the contractile elements of skeletal muscle. Sarcoplasmic reticulum (SR) surrounds each myofibril. Within a myofibril are thin and thick filaments, arranged in compartments called sarcomeres.

Anatomy Overview - Skeletal Muscle
Anatomy Overview - Cross-section of Skeletal Muscle
Animation - Muscle Cell Structures

Review	Resource

8. The overlapping of thick and thin filaments produces striations. Darker A bands alternate with lighter I bands. Table 10.1 summarizes the components of the sarcomere.

9. Myofibrils are composed of three types of proteins: contractile, regulatory, and structural. The contractile proteins are myosin (thick filament) and actin (thin filament). Regulatory proteins are tropomyosin and troponin, both of which are part of the thin filament. Structural proteins include titin (links Z disc to M line and stabilizes thick filament), myomesin (forms M line), nebulin (anchors thin filaments to Z discs and regulates length of thin filaments during development), and dystrophin (links thin filaments to sarcolemma). Table 10.2 summarizes the different types of skeletal muscle fiber proteins. Table 10.3 summarizes the levels of organization within a skeletal muscle.

10. Projecting myosin heads contain actin-binding and ATP-binding sites and are the motor proteins that power muscle contraction.

10.3 Contraction and Relaxation of Skeletal Muscle Fibers

1. Muscle contraction occurs because cross-bridges attach to and "walk" along the thin filaments at both ends of a sarcomere, progressively pulling the thin filaments toward the center of a sarcomere. As the thin filaments slide inward, the Z discs come closer together, and the sarcomere shortens.

2. The contraction cycle is the repeating sequence of events that causes sliding of the filaments: (1) Myosin ATPase hydrolyzes ATP and becomes energized; (2) the myosin head attaches to actin, forming a cross-bridge; (3) the cross-bridge generates force as it rotates toward the center of the sarcomere (power stroke); and (4) binding of ATP to the myosin head detaches it from actin. The myosin head again hydrolyzes the ATP, returns to its original position, and binds to a new site on actin as the cycle continues.

3. An increase in Ca^{2+} concentration in the cytosol starts filament sliding; a decrease turns off the sliding process.

4. The muscle action potential propagating into the T tubule system causes opening of Ca^{2+} release channels in the SR membrane. Calcium ions diffuse from the SR into the sarcoplasm and combine with troponin. This binding causes tropomyosin to move away from the myosin-binding sites on actin.

5. Ca^{2+} active transport pumps continually remove Ca^{2+} from the sarcoplasm into the SR. When the concentration of calcium ions in the sarcoplasm decreases, tropomyosin slides back over and blocks the myosin-binding sites, and the muscle fiber relaxes.

6. A muscle fiber develops its greatest tension when there is an optimal zone of overlap between thick and thin filaments. This dependency is the length–tension relationship.

7. The neuromuscular junction (NMJ) is the synapse between a somatic motor neuron and a skeletal muscle fiber. The NMJ includes the axon terminals and synaptic end bulbs of a motor neuron, plus the adjacent motor end plate of the muscle fiber sarcolemma.

8. When a nerve impulse reaches the synaptic end bulbs of a somatic motor neuron, it triggers exocytosis of the synaptic vesicles, which releases acetylcholine (ACh). ACh diffuses across the synaptic cleft and binds to ACh receptors, initiating a muscle action potential. Acetylcholinesterase then quickly breaks down ACh into its component parts.

Resource for section 10.3:
Animation - Contraction and Movement
Animation - Neuromuscular Junctions
Animation - Contraction of a Sarcomere
Animation - Control of Muscle Tension
Figure 10.5 - Sliding Filament Mechanism of Muscle Contraction
Figure 10.6 - The Contraction Cycle
Figure 10.10 - Summary of Events of Contraction and Relaxation
Exercise - Contraction Connections
Exercise - Fueling Contraction and Recovery
Exercise - Muscle Car
Concepts and Connections - Events at the Neuromuscular Junction
Concepts and Connections - Excitation of Skeletal Muscle
Concepts and Connections - Skeletal Muscle Contraction Cycle

10.4 Muscle Metabolism

1. Muscle fibers have three sources for ATP production: creatine, anaerobic cellular respiration, and aerobic cellular respiration.

2. Creatine kinase catalyzes the transfer of a high-energy phosphate group from creatine phosphate to ADP to form new ATP molecules. Together, creatine phosphate and ATP provide enough energy for muscles to contract maximally for about 15 seconds.

3. Glucose is converted to pyruvic acid in the reactions of glycolysis, which yield two ATPs without using oxygen. Such anaerobic cellular respiration can provide enough energy for 30–40 seconds of maximal muscle activity.

4. Muscular activity that occurs over a prolonged period of time depends on aerobic cellular respiration, mitochondrial reactions that require oxygen to produce ATP.

5. The inability of a muscle to contract forcefully after prolonged activity is muscle fatigue.

6. Elevated oxygen use after exercise is called recovery oxygen uptake.

Resource for section 10.4:
Animation - Muscle Metabolism
Concepts and Connections - Muscle Metabolism
Figure 10.11 - Production of ATP for Muscle Contraction

10.5 Control of Muscle Tension

1. A motor neuron and the muscle fibers it stimulates form a motor unit. A single motor unit may contain as few as 2 or as many as 3000 muscle fibers.

Resource for section 10.5:
Animation - Control of Muscle Tension
Exercise - Increase Muscle Tension

Review	**Resource**

2. Recruitment is the process of increasing the number of active motor units.

3. A twitch contraction is a brief contraction of all the muscle fibers in a motor unit in response to a single action potential.

4. A record of a contraction is called a myogram. It consists of a latent period, a contraction period, and a relaxation period.

5. Wave summation is the increased strength of a contraction that occurs when a second stimulus arrives before the muscle fiber has completely relaxed after a previous stimulus.

6. Repeated stimuli can produce unfused (incomplete) tetanus, a sustained muscle contraction with partial relaxation between stimuli. More rapidly repeating stimuli produce fused (complete) tetanus, a sustained contraction without partial relaxation between stimuli.

7. Continuous involuntary activation of a small number of motor units produces muscle tone, which is essential for maintaining posture.

8. In a concentric isotonic contraction, the muscle shortens to produce movement and to reduce the angle at a joint. During an eccentric isotonic contraction, the muscle lengthens.

9. Isometric contractions, in which tension is generated without muscle changing its length, are important because they stabilize some joints as others are moved.

10.6 Types of Skeletal Muscle Fibers

Anatomy Overview - Skeletal Muscle
Animation - Muscle Cell Structures

1. On the basis of their structure and function, skeletal muscle fibers are classified as slow oxidative (SO), fast oxidative–glycolytic (FOG), and fast glycolytic (FG) fibers.

2. Most skeletal muscles contain a mixture of all three fiber types. Their proportions vary with the typical action of the muscle.

3. The motor units of a muscle are recruited in the following order: first SO fibers, then FOG fibers, and finally FG fibers.

4. Table 10.4 summarizes the three types of skeletal muscle fibers.

10.7 Exercise and Skeletal Muscle Tissue

1. Various types of exercises can induce changes in the fibers in a skeletal muscle. Endurance-type (aerobic) exercises cause a gradual transformation of some fast glycolytic (FG) fibers into fast oxidative–glycolytic (FOG) fibers.

2. Exercises that require great strength for short periods produce an increase in the size and strength of fast glycolytic (FG) fibers. The increase in size is due to increased synthesis of thick and thin filaments.

10.8 Cardiac Muscle Tissue

Anatomy Overview - Cardiac Muscle
Anatomy Overview - Muscle Tissue

1. Cardiac muscle is found only in the heart. Cardiac muscle fibers have the same arrangement of actin and myosin and the same bands, zones, and Z discs as skeletal muscle fibers. The fibers connect to one another through intercalated discs, which contain both desmosomes and gap junctions.

2. Cardiac muscle tissue remains contracted 10 to 15 times longer than skeletal muscle tissue due to prolonged delivery of Ca^{2+} into the sarcoplasm.

3. Cardiac muscle tissue contracts when stimulated by its own autorhythmic fibers. Due to its continuous, rhythmic activity, cardiac muscle depends greatly on aerobic cellular respiration to generate ATP.

10.9 Smooth Muscle Tissue

Anatomy Overview - Smooth Muscle
Tissue
Anatomy Overview - Muscle Tissue
Figure 10.16 - Smooth
Muscle Tissue

1. Smooth muscle is nonstriated and involuntary.

2. Smooth muscle fibers contain intermediate filaments and dense bodies; the function of dense bodies is similar to that of the Z discs in striated muscle.

3. Visceral (single-unit) smooth muscle is found in the walls of hollow viscera and of small blood vessels. Many fibers form a network that contracts in unison.

4. Multiunit smooth muscle is found in large blood vessels, large airways to the lungs, arrector pili muscles, and the eye, where it adjusts pupil diameter and lens focus. The fibers operate independently rather than in unison.

5. The duration of contraction and relaxation of smooth muscle is longer than in skeletal muscle since it takes longer for Ca^{2+} to reach the filaments.

6. Smooth muscle fibers contract in response to nerve impulses, hormones, and local factors.

7. Smooth muscle fibers can stretch considerably and still maintain their contractile function.

Review	Resource

10.10 Regeneration of Muscular Tissue

1. Skeletal muscle fibers cannot divide and have limited powers of regeneration; cardiac muscle fibers can regenerate under limited circumstances; and smooth muscle fibers have the best capacity for division and regeneration.
2. Table 10.5 summarizes the major characteristics of the three types of muscular tissue.

10.11 Development of Muscle

1. With few exceptions, muscles develop from mesoderm.
2. Skeletal muscles of the head and limbs develop from general mesoderm. Other skeletal muscles develop from the mesoderm of somites.

10.12 Aging and Muscular Tissue

1. With aging, there is a slow, progressive loss of skeletal muscle mass, which is replaced by fibrous connective tissue and fat.
2. Aging also results in a decrease in muscle strength, slower muscle reflexes, and loss of flexibility.

SELF-QUIZ QUESTIONS

Fill in the blanks in the following statements.

1. A single somatic motor neuron and all of the muscle fibers it stimulates is known as a _____.

2. The wasting away of muscle due to lack of use is known as _____ while the replacement of skeletal muscle fibers with scar tissue is known as _____.

3. The synaptic end bulbs of somatic motor neurons contain synaptic vesicles filled with the neurotransmitter _____.

Indicate whether the following statements are True *or* False.

4. The ability of muscle cells to respond to stimuli and produce electrical signals is known as excitability.

5. The sequence of events resulting in skeletal muscle contraction are
   (a) generation of a nerve impulse;
   (b) release of the neurotransmitter acetylcholine;
   (c) generation of a muscle action potential;
   (d) release of calcium ions from the sarcoplasmic reticulum;
   (e) calcium ion binding to troponin;
   (f) power stroke with actin and myosin binding and release.

Choose the one best answer to the following questions.

6. In muscle physiology, the latent period refers to
   (a) the period of lost excitability that occurs when two stimuli are applied immediately one after the other.
   (b) the brief contraction of a motor unit.
   (c) the period of elevated oxygen use after exercise.
   (d) an inability of a muscle to contract forcefully after prolonged activity.
   (e) a brief delay that occurs between application of a stimulus and the beginning of contraction.

7. Which of the following muscle proteins and their descriptions are mismatched?
   (a) titin: regulatory protein that holds troponin in place
   (b) myosin: contractile motor protein

   (c) tropomyosin: regulatory protein that blocks myosin-binding sites
   (d) actin: contractile protein that contains myosin-binding sites
   (e) calsequestrin: calcium-binding protein

8. During muscle contraction all of the following occur *except*
   (a) cross-bridges are formed when the energized myosin head attaches to actin's myosin-binding site.
   (b) ATP undergoes hydrolysis.
   (c) the thick filaments slide inward toward the M line.
   (d) calcium concentration in the cytosol increases.
   (e) the Z discs are drawn toward each other.

9. Which of the following is *not* true concerning sarcomeres (before contraction begins) and muscle fiber length–tension relationships?
   (a) If sarcomeres are stretched, the tension in the fiber decreases.
   (b) If a muscle cell is stretched so that there is no overlap of the filaments, no tension is generated.
   (c) Extremely compressed sarcomeres result in less muscle tension.
   (d) Maximum tension occurs when the zone of overlap between a thick and thin filament extends from the edge of the H zone to one end of a thick filament.
   (e) If sarcomeres shorten, the tension in them increases.

10. Which of the following are sources of ATP for muscle contraction? (1) creatine phosphate, (2) glycolysis, (3) anaerobic cellular respiration, (4) aerobic cellular respiration, (5) acetylcholine.
    (a) 1, 2, and 3   (b) 2, 3, and 4   (c) 2, 3, and 5
    (d) 1, 2, 3, and 4   (e) 2, 3, 4, and 5

11. What would happen if ATP were suddenly unavailable after the sarcomere had begun to shorten?
    (a) Nothing. The contraction would proceed normally.
    (b) The myosin heads would be unable to detach from actin.
    (c) Troponin would bind with the myosin heads.
    (d) Actin and myosin filaments would separate completely and be unable to recombine.
    (e) The myosin heads would detach completely from actin.

12. Match the following:

_____ (a) a sheath of areolar connective tissue that wraps around individual skeletal muscle fibers

_____ (b) dense irregular connective tissue that separates a muscle into groups of individual muscle fibers

_____ (c) bundles of muscle fibers

_____ (d) the outermost connective tissue layer that encircles an entire skeletal muscle

_____ (e) dense irregular connective tissue that lines the body wall and limbs and holds functional muscle units together

_____ (f) a cord of dense regular connective tissue that attaches muscle to the periosteum of bone

_____ (g) muscle cell

_____ (h) areolar and adipose connective tissue that separates muscle from skin

_____ (i) connective tissue elements extended as a broad, flat layer

_____ (j) a two-layer tube of fibrous connective tissue enclosing certain tendons

(1) aponeurosis
(2) fascia
(3) subcutaneous layer
(4) tendon
(5) endomysium
(6) perimysium
(7) epimysium
(8) tendon (synovial) sheath
(9) fascicles
(10) muscle fiber

13. Match the following:

_____ (a) synapse between a motor neuron and a muscle fiber

_____ (b) invaginations of the sarcolemma from the surface toward the center of the muscle fiber

_____ (c) myoblasts that persist in mature skeletal muscle

_____ (d) plasma membrane of a muscle fiber

_____ (e) oxygen-binding protein found only in muscle fibers

_____ (f) Ca^{2+}-storing tubular system similar to smooth endoplasmic reticulum

_____ (g) contracting unit of a skeletal muscle fiber

_____ (h) middle area in the sarcomere where thick and thin filaments are found

_____ (i) area in the sarcomere where only thin filaments are present but thick filaments are not

_____ (j) separates the sarcomeres from each other

_____ (k) area of only thick filaments

_____ (l) cytoplasm of a muscle fiber

_____ (m) composed of supporting proteins holding thick filaments together at the H zone

(1) A band
(2) I band
(3) Z disc
(4) H zone
(5) M line
(6) sarcomere
(7) neuromuscular junction
(8) myoglobin
(9) satellite cells
(10) transverse tubules
(11) sarcoplasmic reticulum
(12) sarcolemma
(13) sarcoplasm

14. Match the following (some questions will have more than one answer):

_____ (a) has fibers joined by intercalated discs

_____ (b) thick and thin filaments are not arranged as orderly sarcomeres

_____ (c) uses satellite cells to repair damaged muscle fibers

_____ (d) striated

_____ (e) contraction begins slowly but lasts for long periods

_____ (f) has an extended contraction due to prolonged calcium delivery from both the sarcoplasmic reticulum and the interstitial fluid

_____ (g) does not exhibit autorhythmicity

_____ (h) uses pericytes to repair damaged muscle fibers

_____ (i) uses troponin as a regulatory protein

_____ (j) can be classified as single-unit or multiunit

_____ (k) can be autorhythmic

_____ (l) uses calmodulin as a regulatory protein

(1) skeletal muscle
(2) cardiac muscle
(3) smooth muscle

15. Match the following:

_____ (a) the smooth muscle action that allows the fibers to maintain their contractile function even when stretched

_____ (b) a brief contraction of all the muscle fibers in a motor unit of a muscle in response to a single action potential in its motor neuron

_____ (c) sustained contraction of a muscle, with no relaxation between stimuli

_____ (d) larger contractions resulting from stimuli arriving at different times

_____ (e) process of increasing the number of activated motor units

_____ (f) contraction in which the muscle shortens

_____ (g) inability of a muscle to maintain its strength of contraction or tension during prolonged activity

_____ (h) sustained, but wavering contraction with partial relaxation between stimuli

_____ (i) produced by the continual involuntary activation of a small number of skeletal muscle motor units; results in firmness in skeletal muscle

_____ (j) contraction in which muscle tension is generated without shortening of the muscle

_____ (k) amount of oxygen needed to restore the body's metabolic conditions back to resting levels after exercise

_____ (l) contraction in which a muscle lengthens

(1) muscle fatigue
(2) twitch contraction
(3) wave summation
(4) fused (complete) tetanus
(5) concentric isotonic contraction
(6) motor unit recruitment
(7) muscle tone
(8) eccentric isotonic contraction
(9) isometric contraction
(10) stress–relaxation response
(11) recovery oxygen uptake
(12) unfused (incomplete) tetanus

CRITICAL THINKING QUESTIONS

1. Weight lifter Jamal has been practicing many hours a day, and his muscles have gotten noticeably bigger. He tells you that his muscle cells are "multiplying like crazy and making him get stronger and stronger." Do you believe his explanation? Why or why not?

2. Chicken breasts are composed of "white meat" while chicken legs are composed of "dark meat." The breasts and legs of migrating ducks are dark meat. The breasts of both chickens and ducks are used in flying. How can you explain the differences in the color of the meat (muscles)? How are they adapted for their particular functions?

3. Polio is a disease caused by a virus that can attack the somatic motor neurons in the central nervous system. Individuals who suffer from polio can develop muscle weakness and atrophy. In a certain percentage of cases, the individuals may die due to respiratory paralysis. Relate your knowledge of how muscle fibers function to the symptoms exhibited by infected individuals.

? ANSWERS TO FIGURE QUESTIONS

10.1 Perimysium bundles groups of muscle fibers into fascicles.

10.2 The sarcoplasmic reticulum releases calcium ions to trigger muscle contraction.

10.3 The following are arranged from smallest to largest: thick filament, myofibril, muscle fiber.

10.4 Actin and titin anchor into the Z disc. A bands contain myosin, actin, troponin, tropomyosin, and titin; I bands contain actin, troponin, tropomyosin, and titin.

10.5 The I bands and H zones disappear during muscle contraction; the lengths of the thin and thick filaments do not change.

10.6 If ATP were not available, the cross-bridges would not be able to detach from actin. The muscles would remain in a state of rigidity, as occurs in rigor mortis.

10.7 Three functions of ATP in muscle contraction are the following: (1) Its hydrolysis by an ATPase activates the myosin head so it can bind to actin and rotate; (2) its binding to myosin causes detachment from actin after the power stroke; and (3) it powers the pumps that transport Ca^{2+} from the cytosol back into the sarcoplasmic reticulum.

10.8 A sarcomere length of 2.2 μm gives a generous zone of overlap between the parts of the thick filaments that have myosin heads and the thin filaments without the overlap's being so extensive that sarcomere shortening is limited.

10.9 The part of the sarcolemma that contains acetylcholine receptors is the motor end plate.

10.10 Steps 4 through 6 are part of excitation–contraction coupling (muscle action potential through binding of myosin heads to actin).

10.11 Glycolysis, exchange of phosphate between creatine phosphate and ADP, and glycogen breakdown occur in the cytosol. Oxidation of pyruvic acid, amino acids, and fatty acids (aerobic cellular respiration) occurs in mitochondria.

10.12 Motor units having many muscle fibers are capable of more forceful contractions than those having only a few fibers.

10.13 During the latent period, the muscle action potential sweeps over the sarcolemma and calcium ions are released from the sarcoplasmic reticulum.

10.14 If the second stimulus were applied a little later, the second contraction would be smaller than the one illustrated in part b.

10.15 Holding your head upright without movement involves mainly isometric contractions.

10.16 Visceral smooth muscle is more like cardiac muscle; both contain gap junctions, which allow action potentials to spread from each cell to its neighbors.

10.17 The myotome of a somite differentiates into skeletal muscle.

11 | THE MUSCULAR SYSTEM

THE MUSCULAR SYSTEM AND HOMEOSTASIS *The muscular system and muscular tissue of your body contribute to homeostasis by stabilizing body position, producing movements, regulating organ volume, moving substances within the body, and producing heat.*

Together, the voluntarily controlled muscles of your body compose the **muscular system.** Almost all of the 700 individual muscles that make up the muscular system, such as the biceps brachii muscle, include both skeletal muscle tissue and connective tissue. The function of most muscles is to produce movements of body parts. A few muscles function mainly to stabilize bones so that other skeletal muscles can execute a movement more effectively. This chapter presents many of the major skeletal muscles in the body, most of which are found on both the right and left sides. We will identify the attachment sites and innervation (the nerve or nerves that stimulate contraction) of each muscle described. Developing a working knowledge of these key aspects of skeletal muscle anatomy will enable you to understand how normal movements occur. This knowledge is especially crucial for professionals, such as those in the allied health and physical rehabilitation fields, who work with patients whose normal patterns of movement and physical mobility have been disrupted by physical trauma, surgery, or muscular paralysis.

? *Did you ever wonder how carpal tunnel syndrome occurs?*

11.1 HOW SKELETAL MUSCLES PRODUCE MOVEMENTS

● OBJECTIVES

- Describe the relationship between bones and skeletal muscles in producing body movements.
- Define lever and fulcrum, and compare the three types of levers based on location of the fulcrum, effort, and load.
- Identify the types of fascicle arrangements in a skeletal muscle, and relate the arrangements to strength of contraction and range of motion.
- Explain how the prime mover, antagonist, synergist, and fixator in a muscle group work together to produce movement.

Muscle Attachment Sites: Origin and Insertion

Those skeletal muscles that produce movements do so by exerting force on tendons, which in turn pull on bones or other structures (such as skin). Most muscles cross at least one joint and are usually attached to articulating bones that form the joint (Figure 11.1a).

When a skeletal muscle contracts, it pulls one of the articulating bones toward the other. The two articulating bones usually do not move equally in response to contraction. One bone remains stationary or near its original position, either because other muscles stabilize that bone by contracting and pulling it in the opposite direction or because its structure makes it less movable. Ordinarily, the attachment of a muscle's tendon to the stationary bone is called the **origin** (OR-i-jin); the attachment of the muscle's other tendon to the movable bone is called the **insertion** (in-SER-shun). A good analogy is a spring on a door. In this example, the part of the spring attached to the frame is the origin; the part attached to the door represents the insertion. A useful rule of thumb is that the origin is usually proximal and the insertion distal; the insertion is usually pulled toward the origin. The fleshy portion of the muscle between the tendons is called the **belly (body),** the coiled middle portion of the spring in our example. The **actions** of

Figure 11.1 **Relationship of skeletal muscles to bones.** Muscles are attached to bones by tendons at their origin and insertion. Skeletal muscles produce movements by pulling on bones. Bones serve as levers, and joints act as fulcrums for the levers. Here the lever–fulcrum principle is illustrated by the movement of the forearm. Note where the load (resistance) and effort are applied in (b).

In the limbs, the origin of a muscle is usually proximal and the insertion is usually distal.

(a) Origin and insertion of a skeletal muscle

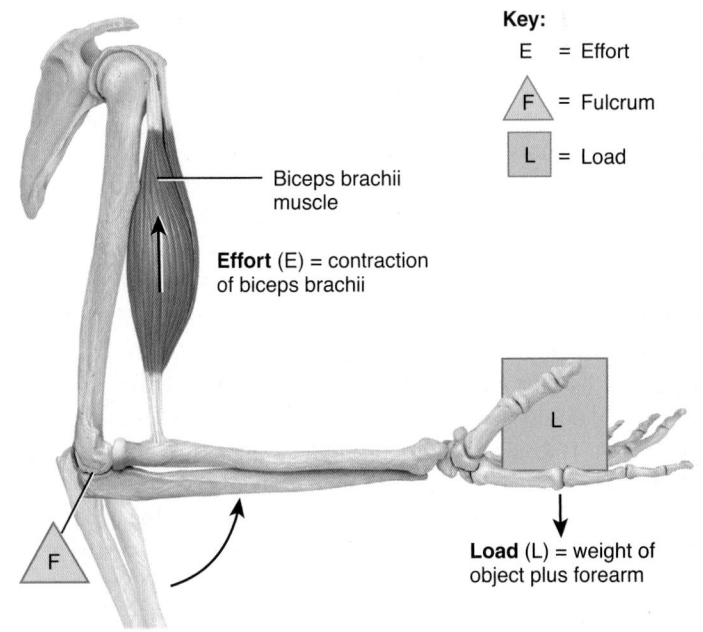

Key:

E = Effort

F = Fulcrum

L = Load

Biceps brachii muscle

Effort (E) = contraction of biceps brachii

Load (L) = weight of object plus forearm

Fulcrum (F) = elbow joint

(b) Movement of the forearm lifting a weight

 Where is the belly of the muscle that extends the forearm located?

a muscle are the main movements that occur when the muscle contracts. In our spring example, this would be the closing of the door. Certain muscles are also capable of **reverse muscle action (RMA).** This means that during specific movements of the body the actions are reversed and therefore the positions of the origin and insertion of a specific muscle are switched.

Muscles that move a body part often do not cover the moving part. Figure 11.1b shows that although one of the functions of the biceps brachii muscle is to move the forearm, the belly of the muscle lies over the humerus, not over the forearm. You will also see that muscles that cross two joints, such as the rectus femoris and sartorius of the thigh, have more complex actions than muscles that cross only one joint.

Lever Systems and Leverage

In producing movement, bones act as levers, and joints function as the fulcrums of these levers. A **lever** is a rigid structure that can move around a fixed point called a **fulcrum,** symbolized by ⛰. A lever is acted on at two different points by two different forces: the **effort** (E), which causes movement, and the **load** L or **resistance,** which opposes movement. The effort is the force exerted by muscular contraction; the load is typically the weight of the body part that is moved or some resistance that the moving body part is trying to overcome (such as the weight of a book you might be picking up). Motion occurs when the effort applied to the bone at the insertion exceeds the load. Consider the biceps brachii flexing the forearm at the elbow as an object is lifted (Figure 11.1b). When the forearm is raised, the elbow is the fulcrum. The weight of the forearm plus the weight of the object in the hand is the load. The force of contraction of the biceps brachii pulling the forearm up is the effort.

The relative distance between the fulcrum and load and the point at which the effort is applied determine whether a given lever operates at a mechanical advantage or a mechanical disadvantage. For example, if the load is closer to the fulcrum and the effort farther from the fulcrum, then only a relatively small effort is required to move a large load over a small distance. This is called a **mechanical advantage.** If, instead, the load is farther from the fulcrum and the effort is applied closer to the fulcrum, then a relatively large effort is required to move a small load (but at greater speed). This is called a **mechanical disadvantage.** Compare chewing something hard (the load) with your front teeth and the teeth in the back of your mouth. It is much easier to crush the hard food item with the back teeth because they are closer to the fulcrum (the jaw or temporomandibular joint) than are the front teeth. Here is one more example you can try. Straighten out a paper clip. Now get a pair of scissors and try to cut the paper clip with the tip of the scissors (mechanical disadvantage) versus near the pivot point of the scissors (mechanical advantage).

Levers are categorized into three types according to the positions of the fulcrum, the effort, and the load:

1. The fulcrum is between the effort and the load in **first-class levers** (Figure 11.2a). (Think E*F*L.) Scissors and seesaws are examples of first-class levers. A first-class lever can produce either a mechanical advantage or mechanical disadvantage depending on whether the effort or the load is closer to the fulcrum. (Think of an adult and a child on a seesaw.) As we've seen in the preceding examples, if the effort (child) is farther from the fulcrum than the load (adult), a heavy load can be moved, but not very far or fast. If the effort is closer to the fulcrum than the load, only a lighter load can be moved, but it moves far and fast. There are few first-class levers in the body. One example is the lever formed by the head resting on the vertebral column (Figure 11.2a). When the head is raised, the contraction of the posterior neck muscles provides the effort E, the joint between the atlas and the occipital bone (atlanto-occipital joint) forms the fulcrum ⛰, and the weight of the anterior portion of the skull is the load L.

2. The load is between the fulcrum and the effort in **second-class levers** (Figure 11.2b). (Think E*L*F.) Second-class levers operate like a wheelbarrow. They always produce a mechanical advantage because the load is always closer to the fulcrum than the effort. This arrangement sacrifices speed and range of motion for force; this type of lever produces the most force. This class of lever is uncommon in the human body. An example is standing up on your toes. The fulcrum ⛰ is the ball of the foot. The load is the weight of the body. The effort (E) is the contraction of the muscles of the calf, which raise the heel off the ground.

3. The effort is between the fulcrum and the load in **third-class levers** (Figure 11.2c). (Think F*E*L.) These levers operate like a pair of forceps and are the most common levers in the body. Third-class levers always produce a mechanical disadvantage because the effort is always closer to the fulcrum than the load. In the body, this arrangement favors speed and range of motion over force. The elbow joint, the biceps brachii muscle, and the bones of the arm and forearm are one example of a third-class lever (Figure 11.2c). As we have seen, in flexing the forearm at the elbow, the elbow joint is the fulcrum ⛰, the contraction of the biceps brachii muscle provides the effort E, and the weight of the hand and forearm is the load L.

Effects of Fascicle Arrangement

Recall from Chapter 10 that the skeletal muscle fibers (cells) within a muscle are arranged in bundles known as **fascicles** (FAS-i-kuls). Within a fascicle, all muscle fibers are parallel to one another. The fascicles, however, may form one of five patterns with respect to the tendons: parallel, fusiform (spindle-shaped, narrow toward the ends and wide in the middle), circular, triangular, or pennate (shaped like a feather) (Table 11.1).

Fascicular arrangement affects a muscle's power and range of motion. As a muscle fiber contracts, it shortens to about 70% of its resting length. The longer the fibers in a muscle, the greater the range of motion it can produce. However, the power of a muscle depends not on length but on its total cross-sectional area, because a short fiber can contract as forcefully as a long one. So the more fibers per unit of cross-sectional area a muscle has, the more power it can produce. Fascicular arrangement often represents a compromise

Figure 11.2 Lever structure and types of levers.

Levers are divided into three types based on the placement of the fulcrum, effort, and load (resistance).

Key:

E = Effort

F = Fulcrum

L = Load

(a) First-class lever

(b) Second-class lever

(c) Third-class lever

? **Which type of lever produces the most force?**

between power and range of motion. Pennate muscles, for instance, have a large number of short-fibered fascicles distributed over their tendons, giving them greater power but a smaller range of motion. In contrast, parallel muscles have comparatively fewer fascicles, but they have long fibers that extend the length of the muscle, so they have a greater range of motion but less power.

Coordination among Muscles

Movements often are the result of several skeletal muscles acting as a group. Most skeletal muscles are arranged in opposing (antagonistic) pairs at joints—that is, flexors–extensors, abductors–adductors, and so on. Within opposing pairs, one muscle, called the **prime mover** or **agonist** (= leader), contracts to cause an action while the other muscle, the **antagonist** (*anti-* = against), stretches and yields to the effects of the prime mover. In the process of flexing the forearm at the elbow, for instance, the biceps brachii is the prime mover, and the triceps brachii is the antagonist (see Figure 11.1a). The antagonist and prime mover are usually located on opposite sides of the bone or joint, as is the case in this example.

With an opposing pair of muscles, the roles of the prime mover and antagonist can switch for different movements. For example, while extending the forearm at the elbow against resistance (i.e., lowering the load shown in Figure 11.2c), the triceps brachii becomes the prime mover, and the biceps brachii is the antagonist. If a prime mover and its antagonist contract at the same time with equal force, there will be no movement.

Sometimes a prime mover crosses other joints before it reaches the joint at which its primary action occurs. The biceps brachii, for

TABLE 11.1

Arrangement of Fascicles

PARALLEL

Fascicles parallel to longitudinal axis of muscle; terminate at either end in flat tendons.

Example: Sternohyoid muscle (see Figure 11.8a)

FUSIFORM

Fascicles nearly parallel to longitudinal axis of muscle; terminate in flat tendons; muscle tapers toward tendons, where diameter is less than at belly.

Example: Digastric muscle (see Figure 11.8a)

CIRCULAR

Fascicles in concentric circular arrangements form sphincter muscles that enclose an orifice (opening).

Example: Orbicularis oculi muscle (see Figure 11.4a)

TRIANGULAR

Fascicles spread over broad area converge at thick central tendon; gives muscle a triangular appearance.

Example: Pectoralis major muscle (see Figure 11.3a)

PENNATE

Short fascicles in relation to total muscle length; tendon extends nearly entire length of muscle.

UNIPENNATE

Fascicles arranged on only one side of tendon.

Example: Extensor digitorum longus muscle (see Figure 11.22b)

BIPENNATE

Fascicles arranged on both sides of centrally positioned tendons.

Example: Rectus femoris muscle (see Figure 11.20a)

MULTIPENNATE

Fascicles attach obliquely from many directions to several tendons.

Example: Deltoid muscle (see Figure 11.10a)

---

example, spans both the shoulder and elbow joints, with primary action on the forearm. To prevent unwanted movements at intermediate joints or to otherwise aid the movement of the prime mover, muscles called **synergists** (SIN-er-jists; *syn-* = together; *-ergon* = work) contract and stabilize the intermediate joints. As an example, muscles that flex the fingers (prime movers) cross the intercarpal and radiocarpal joints (intermediate joints). If movement at these intermediate joints were unrestrained, you would not be able to flex your fingers without flexing the wrist at the same time. Synergistic contraction of the wrist extensor muscles stabilizes the wrist joint and prevents unwanted movement, while the flexor muscles of the fingers contract to bring about the primary action, efficient flexion of the fingers. Synergists are usually located close to the prime mover.

Some muscles in a group also act as **fixators,** stabilizing the origin of the prime mover so that the prime mover can act more efficiently. Fixators steady the proximal end of a limb while movements occur at the distal end. For example, the scapula is a freely movable bone that serves as the origin for several muscles that move the arm. When the arm muscles contract, the scapula must be held steady. In abduction of the arm, the deltoid muscle serves as the prime mover, and fixators (pectoralis minor, trapezius, subclavius, serratus anterior muscles, and others) hold the scapula firmly against the back of the chest (see Figure 11.14a, b). The insertion of the deltoid muscle pulls on the humerus to abduct the arm. Under different conditions—that is, for different movements—and at different times, many muscles may act as prime movers, antagonists, synergists, or fixators.

In the limbs, a **compartment** is a group of skeletal muscles, their associated blood vessels, and associated nerves, all of which have a common function. In the upper limbs, for example, flexor compartment muscles are anterior, and extensor compartment muscles are posterior.

The overall goal of **stretching** is to achieve normal range of motion of joints and mobility of soft tissues surrounding the joints. For most individuals, the best stretching routine involves *static stretching,* that is, slow sustained stretching that holds a muscle in a lengthened position. The muscles should be stretched to the point of slight discomfort (not pain) and held for about 30 seconds. Stretching should be done after warming up to increase the range of motion most effectively.

1. *Improved physical performance.* A flexible joint has the ability to move through a greater range of motion, which improves performance.

2. *Decreased risk of injury.* Stretching decreases resistance in various soft tissues so there is less likelihood of exceeding maximum tissue extensibility during an activity (i.e., injuring the soft tissues).

3. *Reduced muscle soreness.* Stretching can reduce some of the muscle soreness that results after exercise.

4. *Improved posture.* Poor posture results from improper position of various parts of the body and the effects of gravity over a number of years. Stretching can help realign soft tissues to improve and maintain good posture. •

✔ CHECKPOINT

1. Using the terms origin, insertion, and belly, describe how skeletal muscles produce body movements by pulling on bones.
2. List the three types of levers, and give an example of a first-, second-, and third-class lever found in the body.
3. Describe the various arrangements of fascicles.
4. Define the roles of the prime mover (agonist), antagonist, synergist, and fixator in producing various movements of the free upper limb.
5. What is a muscle compartment?

11.2 HOW SKELETAL MUSCLES ARE NAMED

◉ OBJECTIVE

• Explain seven features used in naming skeletal muscles.

The names of most of the skeletal muscles contain combinations of the word roots of their distinctive features. This works two ways. You can learn the names of muscles by remembering the terms that refer to muscle features, such as the pattern of the muscle's fascicles; the size, shape, action, number of origins, and location of the muscle; and the sites of origin and insertion of the muscle. Knowing the names of a muscle will then give you clues about its features. Study Table 11.2 to become familiar with the terms used in muscle names.

✔ CHECKPOINT

6. Select 10 muscles in Figure 11.3 and identify the features on which their names are based. (*Hint*: Use the prefix, suffix, and root of each muscle's name as a guide.)

11.3 PRINCIPAL SKELETAL MUSCLES

Exhibits 11.A–11.T will assist you in learning the names of the principal skeletal muscles in various regions of the body. The muscles in the exhibits are divided into groups according to the part of the body on which they act. As you study groups of muscles in the exhibits, refer to Figure 11.3 to see how each group is related to the others.

The exhibits contain the following elements:

• *Objective.* This statement describes what you should learn from the exhibit.

• *Overview.* These paragraphs provide a general introduction to the muscles under consideration and emphasize how the muscles are organized within various regions. The discussion also highlights any distinguishing features of the muscles.

• *Muscle names.* The word roots indicate how the muscles are named. As noted previously, once you have mastered the naming of the muscles, you can more easily understand their actions.

• *Origins, insertions, and actions.* You are also given the origin, insertion, and actions of each muscle.

• *Relating muscles to movements.* These exercises will help you organize the muscles in the body region under consideration according to the actions they produce.

• *Innervation.* This section lists the nerve or nerves that cause contraction of each muscle. In general, cranial nerves, which arise from the lower parts of the brain, serve muscles in the head region. Spinal nerves, which arise from the spinal cord within the vertebral column, innervate muscles in the rest of the body. Cranial nerves are designated by both a name and a Roman numeral: the facial (VII) nerve, for example. Spinal nerves are numbered in groups according to the part of the spinal cord from which they arise: C = cervical (neck region), T = thoracic (chest region), L = lumbar (lower-back region), and S = sacral (buttocks region). An example is T1, the first thoracic spinal nerve.

• *Questions.* These knowledge checkpoints relate specifically to information in each exhibit, and take the form of review, critical thinking, and/or application questions.

• *Clinical Connections.* Selected exhibits include clinical applications, which explore the clinical, professional, or everyday relevance of a particular muscle or its function through descriptions of disorders or clinical procedures.

• *Figures.* The figures in the exhibits may present superficial and deep, anterior and posterior, or medial and lateral views to show each muscle's position as clearly as possible. The muscle names in all capital letters are specifically referred to in the table part of the exhibit.

TABLE 11.2

Characteristics Used to Name Muscles

NAME	MEANING	EXAMPLE	FIGURE
DIRECTION: ORIENTATION OF MUSCLE FASCICLES RELATIVE TO THE BODY'S MIDLINE			
Rectus	Parallel to midline	Rectus abdominis	11.10c
Transverse	Perpendicular to midline	Transversus abdominis	11.10c
Oblique	Diagonal to midline	External oblique	11.10a
SIZE: RELATIVE SIZE OF THE MUSCLE			
Maximus	Largest	Gluteus maximus	11.20c
Minimus	Smallest	Gluteus minimus	11.20d
Longus	Long	Adductor longus	11.20a
Brevis	Short	Adductor brevis	11.20b
Latissimus	Widest	Latissimus dorsi	11.15b
Longissimus	Longest	Longissimus capitis	11.19a
Magnus	Large	Adductor magnus	11.20b
Major	Larger	Pectoralis major	11.10a
Minor	Smaller	Pectoralis minor	11.14a
Vastus	Huge	Vastus lateralis	11.20a
SHAPE: RELATIVE SHAPE OF THE MUSCLE			
Deltoid	Triangular	Deltoid	11.15b
Trapezius	Trapezoid	Trapezius	11.3b
Serratus	Saw-toothed	Serratus anterior	11.14b
Rhomboid	Diamond-shaped	Rhomboid major	11.15d
Orbicularis	Circular	Orbicularis oculi	11.4a
Pectinate	Comblike	Pectineus	11.20a
Piriformis	Pear-shaped	Piriformis	11.20d
Platys	Flat	Platysma	11.4c
Quadratus	Square, four-sided	Quadratus femoris	11.20d
Gracilis	Slender	Gracilis	11.20a
ACTION: PRINCIPAL ACTION OF THE MUSCLE			
Flexor	Decreases joint angle	Flexor carpi radialis	11.17a
Extensor	Increases joint angle	Extensor carpi ulnaris	11.17d
Abductor	Moves bone away from midline	Abductor pollicis longus	11.17e
Adductor	Moves bone closer to midline	Adductor longus	11.20a
Levator	Raises or elevates body part	Levator scapulae	11.14a
Depressor	Lowers or depresses body part	Depressor labii inferioris	11.4b
Supinator	Turns palm anteriorly	Supinator	11.17c
Pronator	Turns palm posteriorly	Pronator teres	11.17a
Sphincter	Decreases size of an opening	External anal sphincter	11.12
Tensor	Makes body part rigid	Tensor fasciae latae	11.20a
Rotator	Rotates bone around longitudinal axis	Rotatore	11.19b
NUMBER OF ORIGINS: NUMBER OF TENDONS OF ORIGIN			
Biceps	Two origins	Biceps brachii	11.16a
Triceps	Three origins	Triceps brachii	11.16b
Quadriceps	Four origins	Quadriceps femoris	11.20a
LOCATION: STRUCTURE NEAR WHICH A MUSCLE IS FOUND			
Example: Temporalis, muscle near temporal bone.			11.4c
ORIGIN AND INSERTION: SITES WHERE MUSCLE ORIGINATES AND INSERTS			
Example: Sternocleidomastoid, originating on sternum and clavicle and inserting on mastoid process of temporal bone.			11.3a

Figure 11.3 Principal superficial skeletal muscles.

Most movements require several skeletal muscles acting in groups rather than individually.

Occipitofrontalis (frontal belly)

Nasalis

Orbicularis oris

Depressor anguli oris

Platysma

Sternohyoid

Serratus anterior

Rectus abdominis

External oblique

Brachioradialis

Tensor fasciae latae

Iliacus

Psoas major

Pectineus

Adductor longus

Sartorius

Gracilis

Vastus lateralis

Rectus femoris

Vastus medialis

Tendon of quadriceps femoris

Patella

Gastrocnemius

Soleus

Epicranial aponeurosis

Temporalis

Orbicularis oculi

Masseter

Sternocleidomastoid

Trapezius

Deltoid

Pectoralis major

Triceps brachii

Biceps brachii

Brachialis

Pronator teres

Brachioradialis

Flexor carpi radialis

Palmaris longus

Flexor digitorum superficialis

Flexor carpi ulnaris

Thenar muscles

Hypothenar muscles

Iliotibial tract

Patellar ligament

Tibialis anterior

Fibularis longus

Tibia

(a) Anterior view

FIGURE 11.3 CONTINUES ▶

FIGURE 11.3 CONTINUED

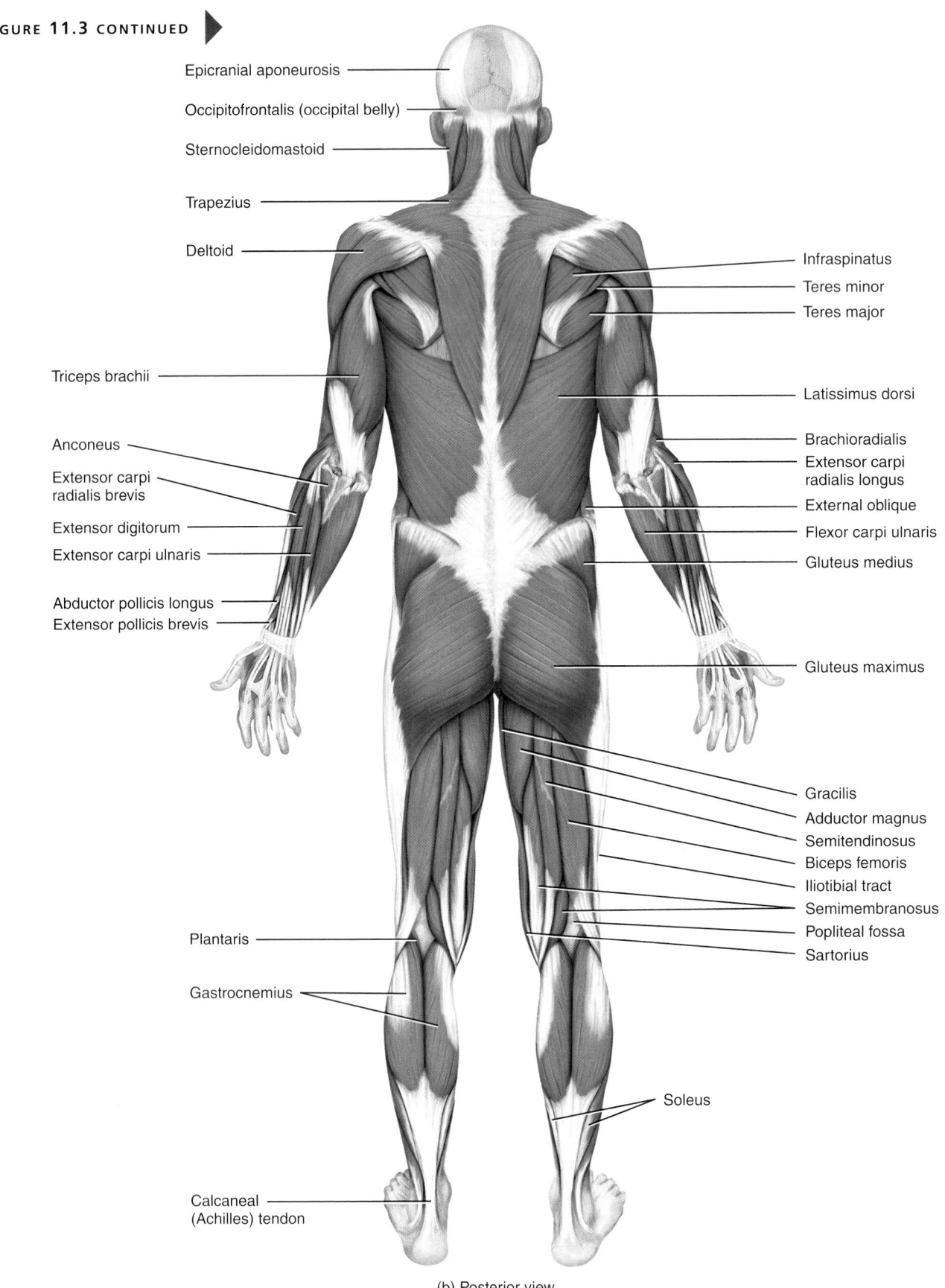

Epicranial aponeurosis

Occipitofrontalis (occipital belly)

Sternocleidomastoid

Trapezius

Deltoid

Triceps brachii

Anconeus

Extensor carpi
radialis brevis

Extensor digitorum

Extensor carpi ulnaris

Abductor pollicis longus

Extensor pollicis brevis

Plantaris

Gastrocnemius

Calcaneal
(Achilles) tendon

Infraspinatus

Teres minor

Teres major

Latissimus dorsi

Brachioradialis

Extensor carpi
radialis longus

External oblique

Flexor carpi ulnaris

Gluteus medius

Gluteus maximus

Gracilis

Adductor magnus

Semitendinosus

Biceps femoris

Iliotibial tract

Semimembranosus

Popliteal fossa

Sartorius

Soleus

(b) Posterior view

❓ **Give an example of a muscle named for each of the following characteristics: direction of fibers, shape, action, size, origin and insertion, location, and number of tendons of origin.**

EXHIBIT 11.A

Muscles of the Head That Produce Facial Expressions *(Figure 11.4)*

OBJECTIVE

• Describe the origin, insertion, action, and innervation of the muscles of facial expression.

The muscles of facial expression, which provide us with the ability to express a wide variety of emotions, lie within the subcutaneous layer (Figure 11.4). They usually originate in the fascia or bones of the skull and insert into the skin. Because of their insertions, the muscles of facial expression move the skin rather than a joint when they contract.

Among the noteworthy muscles in this group are those surrounding the orifices (openings) of the head such as the eyes, nose, and mouth. These muscles function as *sphincters* (SFINGK-ters), which close the orifices, and *dilators* (DĪ-lā-tors), which dilate or open the orifices. For example, the **orbicularis oculi** muscle closes the eye, and the levator palpebrae superioris muscle opens it (discussed in Exhibit 11.B). The **occipitofrontalis** is an unusual muscle in this group because it is made up of two parts: an anterior part called the **frontal belly (frontalis),** which is superficial to the frontal bone, and a posterior part called the **occipital belly (occipitalis),** which is superficial to the occipital bone. The two muscular portions are held together by a strong **aponeurosis** (sheetlike tendon), the **epicranial aponeurosis** (ep-i-KRĀ-nē-al ap′-ō-noo-RŌ-sis), also called the **galea aponeurotica** (GĀ-lē-a ap′-ō-noo′-RŌ-ti-ka), that covers the superior and lateral surfaces of the skull. The **buccinator** muscle forms the major muscular portion of the cheek. The duct of the parotid gland (a salivary gland) passes through the buccinator muscle to reach the oral cavity. The buccinator muscle is so named because it compresses the cheeks (*bucc-* = cheek) during blowing—for example, when a musician plays a brass instrument such as a trumpet. It functions in whistling, blowing, and sucking and assists in chewing.

CLINICAL CONNECTION | *Bell's Palsy*

Bell's palsy, also known as **facial paralysis,** is a unilateral paralysis of the muscles of facial expression. It is due to damage or disease of the facial (VII) nerve. Possible causes include inflammation of the facial nerve due to an ear infection, ear surgery that damages the facial nerve, or infection by the herpes simplex virus. The paralysis causes the entire side of the face to droop in severe cases. The person cannot wrinkle the forehead, close the eye, or pucker the lips on the affected side. Drooling and difficulty in swallowing also occur. Eighty percent of patients recover completely within a few weeks to a few months. For other, paralysis is permanent. The symptoms of Bell's palsy mimic those of a stroke. •

RELATING MUSCLES TO MOVEMENTS

Arrange the muscles in this exhibit into two groups: (1) those that act on the mouth and (2) those that act on the eyes.

✔CHECKPOINT

Why do the muscles of facial expression move the skin rather than a joint?

EXHIBIT 11.A CONTINUES ▶

EXHIBIT 11.A **375**

EXHIBIT 11.A

Muscles of the Head That Produce Facial Expressions (*Figure 11.4*) CONTINUED

MUSCLE	ORIGIN	INSERTION	ACTION	INNERVATION
SCALP MUSCLES				
Occipitofrontalis (ok-sip'-i-tō-frun-TĀ-lis)				
Frontal belly	Epicranial aponeurosis.	Skin superior to supraorbital margin.	Draws scalp anteriorly, raises eyebrows, and wrinkles skin of forehead horizontally as in look of surprise.	Facial (VII) nerve.
Occipital belly (*occipit-* = back of the head)	Occipital bone and mastoid process of temporal bone.	Epicranial aponeurosis.	Draws scalp posteriorly.	Facial (VII) nerve.
MOUTH MUSCLES				
Orbicularis oris (or-bi'-kū-LAR-is OR-is; *orb-* = circular; *oris* = of the mouth)	Muscle fibers surrounding opening of mouth.	Skin at corner of mouth.	Closes and protrudes lips, as in kissing; compresses lips against teeth; and shapes lips during speech.	Facial (VII) nerve.
Zygomaticus major (zī-gō-MA-ti-kus; *zygomatic* = cheek bone; *rnajor* = greater)	Zygomatic bone.	Skin at angle of mouth and orbicularis oris.	Draws angle of mouth superiorly and laterally, as in smiling.	Facial (VII) nerve.
Zygomaticus minor (*minor* = lesser)	Zygomatic bone.	Upper lip.	Raises (elevates) upper lip, exposing maxillary (upper) teeth.	Facial (VII) nerve.
Levator labii superioris (le-VĀ-tor LĀ-bē-ī soo-per'-ē-OR-is; *levator* = raises or elevates; *labii* = lip; *superioris* = upper)	Superior to infraorbital foramen of maxilla.	Skin at angle of mouth and orbicularis oris.	Raises upper lip.	Facial (VII) nerve.
Depressor labii inferioris (de-PRE-sor LĀ-bē-ī; *depressor* = depresses or lowers; *inferioris* = lower)	Mandible.	Skin of lower lip.	Depresses (lowers) lower lip.	Facial (VII) nerve.
Depressor anguli oris (ANG-ū-lī; *angul* = angle or corner; *oris* = mouth)	Mandible.	Angle of mouth.	Draws angle of mouth laterally and inferiorly, as in opening mouth.	Facial (VII) nerve.
Levator anguli oris	Inferior to infraorbital foramen.	Skin of lower lip and orbicularis oris.	Draws angle of mouth laterally and superiorly.	Facial (VII) nerve.
Buccinator (BUK-si-nā'-tor; *bucc-* = cheek)	Alveolar processes of maxilla and mandible and pterygomandibular raphe (fibrous band extending from pterygoid process of sphenoid bone to mandible).	Orbicularis oris.	Presses cheeks against teeth and lips, as in whistling, blowing, and sucking; draws corner of mouth laterally; and assists in mastication (chewing) by keeping food between the teeth (and not between teeth and cheeks).	Facial (VII) nerve.
Risorius (ri-ZOR-ē-us; *risor* = laughter)	Fascia over parotid (salivary) gland.	Skin at angle of mouth.	Draws angle of mouth laterally, as in grimacing.	Facial (VII) nerve.
Mentalis (men-TĀ-lis; *mental* = chin)	Mandible.	Skin of chin.	Elevates and protrudes lower lip and pulls skin of chin up, as in pouting.	Facial (VII) nerve.
NECK MUSCLES				
Platysma (pla-TIZ-ma; *platys* = flat, broad)	Fascia over deltoid and pectoralis major muscles.	Mandible, blends with muscles around angle of mouth, and skin of lower face.	Draws outer part of lower lip inferiorly and posteriorly as in pouting; depresses mandible.	Facial (VII) nerve.

MUSCLE	ORIGIN	INSERTION	ACTION	INNERVATION
ORBIT AND EYEBROW MUSCLES				
Orbicularis oculi (or-bi′-kū-LAR-is OK-ū-lī; *oculi* = eye)	Medial wall of orbit.	Circular path around orbit.	Closes eye.	Facial (VII) nerve.
Corrugator supercilii (KOR-a-gā′-tor soo-per-SIL-ē-ī; *corrugat* = wrinkle; *supercilii* = eyebrow)	Medial end of superciliary arch of frontal bone.	Skin of eyebrow.	Draws eyebrow inferiorly and wrinkles skin of forehead vertically as in frowning.	Facial (VII) nerve.

Figure 11.4 Muscles of the head that produce facial expressions.

When they contract, muscles of facial expression move the skin rather than a joint.

(a) Anterior superficial view   (b) Anterior deep view

EXHIBIT 11.A CONTINUES ▶

EXHIBIT 11.A **377**

● FIGURE **11.4** CONTINUED ▶

Epicranial aponeurosis

TEMPORALIS

OCCIPITOFRONTALIS
(OCCIPITAL BELLY)

Zygomatic arch

Posterior auricular

Mandible

MASSETER

Splenius capitis

Sternocleidomastoid

Splenius cervicis

Trapezius

Levator scapulae

Middle
scalene

OCCIPITO-
FRONTALIS
(FRONTAL
BELLY)

ORBICULARIS
OCULI

ZYGOMATICUS
MINOR

NASALIS

LEVATOR LABII
SUPERIORIS

ZYGOMATICUS
MAJOR

LEVATOR
ANGULI ORIS

BUCCINATOR

ORBICULARIS
ORIS

RISORIUS

DEPRESSOR
LABII INFERIORIS

MENTALIS

DEPRESSOR
ANGULI ORIS

PLATYSMA

(c) Right lateral superficial view

❓ **Which muscles of facial expression cause frowning, smiling, pouting, and squinting?**

EXHIBIT 11.B

Muscles of the Head That Move the Eyeballs (Extrinsic Eye Muscles) and Upper Eyelids *(Figure 11.5)*

◉ OBJECTIVE

• Describe the origin, insertion, action, and innervation of the extrinsic eye muscles that move the eyeballs and upper eyelids.

Muscles that move the eyeballs are called **extrinsic eye muscles** because they originate outside the eyeballs (in the orbit) and insert on the outer surface of the sclera ("white of the eye") (Figure 11.5). The extrinsic eye muscles are some of the fastest contracting and most precisely controlled skeletal muscles in the body.

Three pairs of extrinsic eye muscles control movements of the eyeballs: (1) superior and inferior recti, (2) lateral and medial recti, and (3) superior and inferior obliques. The four recti muscles (superior, inferior, lateral, and medial) arise from a tendinous ring in the orbit and insert into the sclera of the eye. As their names imply, the **superior** and **inferior recti** move the eyeballs superiorly and inferiorly; the **lateral** and **medial recti** move the eyeballs laterally and medially, respectively.

The actions of the oblique muscles cannot be deduced from their names. The **superior oblique** muscle originates posteriorly near the tendinous ring, then passes anteriorly superior to the medial rectus muscle, and ends in a round tendon. The tendon extends through a pulleylike loop of fibrocartilaginous tissue called the *trochlea* (= pulley) on the anterior and medial part of the roof of the orbit. Finally, the tendon turns and inserts on the posterolateral aspect of the eyeball. Accordingly, the superior oblique muscle moves the eyeballs inferiorly and laterally. The **inferior oblique** muscle originates on the maxilla at the anteromedial aspect of the floor of the orbit. It then passes posteriorly and laterally and inserts on the posterolateral aspect of the eyeball. Because of this arrangement, the inferior oblique muscle moves the eyeballs superiorly and laterally.

Unlike the recti and oblique muscles, the **levator palpebrae superioris** does not move the eyeballs, since its tendon passes the eyeball and inserts into the upper eyelid. Rather, it raises the upper eyelids, that is, opens the eyes. It is therefore an antagonist to the orbicularis oculi, which closes the eyes.

⚕ CLINICAL CONNECTION | *Strabismus*

Strabismus (stra-BIZ-mus; *strabismos* = squinting) is a condition in which the two eyeballs are not properly aligned. This can be hereditary or it can be due to birth injuries, poor attachments of the muscles, problems with the brain's control center, or localized disease. Strabismus can be constant or intermittent. In strabismus, each eye sends an image to a different area of the brain and because the brain usually ignores the messages sent by one of the eyes, the ignored eye becomes weaker; hence "lazy eye," or *amblyopia*, develops. *External strabismus* results when a lesion in the oculomotor (III) nerve causes the eyeball to move laterally when at rest, and results in an inability to move the eyeball medially and inferiorly. A lesion in the abducens (VI) nerve results in *internal strabismus*, a condition in which the eyeball moves medially when at rest and cannot move laterally.

Treatment options for strabismus depend on the specific type of problem and include surgery, visual therapy (retraining the brain's control center), and orthoptics (eye muscle training to straighten the eyes). •

MUSCLE	ORIGIN	INSERTION	ACTION	INNERVATION
Superior rectus (*rectus* = fascicles parallel to midline)	Common tendinous ring (attached to orbit around optic foramen).	Superior and central part of eyeballs.	Moves eyeballs superiorly (elevation) and medially (adduction), and rotates them medially.	Oculomotor (III) nerve.
Inferior rectus	Same as above.	Inferior and central part of eyeballs.	Moves eyeballs inferiorly (depression) and medially (adduction), and rotates them medially.	Oculomotor (III) nerve.
Lateral rectus	Same as above.	Lateral side of eyeballs.	Moves eyeballs laterally (abduction).	Abducens (VI) nerve.
Medial rectus	Same as above.	Medial side of eyeballs.	Moves eyeballs medially (adduction).	Oculomotor (III) nerve.
Superior oblique (*oblique* = fascicles diagonal to midline)	Sphenoid bone, superior and medial to common tendinous ring in orbit.	Eyeball between superior and lateral recti. Muscle inserts into superior and lateral surfaces of eyeball via tendon that passes through trochlea.	Moves eyeballs inferiorly (depression) and laterally (abduction), and rotates them medially.	Trochlear (IV) nerve.
Inferior oblique	Maxilla in floor of orbit.	Eyeballs between inferior and lateral recti.	Moves eyeballs superiorly (elevation) and laterally (abduction), and rotates them laterally.	Oculomotor (III) nerve.
Levator palpebrae superioris (le-VĀ-tor PAL-pe-brē soo-per'-ē-OR-is; *palpebrae* = eyelids)	Roof of orbit (lesser wing of sphenoid bone).	Skin and tarsal plate of upper eyelids. (opens eyes).	Elevates upper eyelids	Oculomotor (III) nerve.

EXHIBIT 11.B CONTINUES ▶

EXHIBIT 11.B **379**

RELATING MUSCLES TO MOVEMENTS

Arrange the muscles in this exhibit according to their actions on the eyeballs: (1) elevation, (2) depression, (3) abduction, (4) adduction, (5) medial rotation, and (6) lateral rotation. The same muscle may be mentioned more than once.

✔ CHECKPOINT

Which muscles that move the eyeballs contract and relax as you look to your left without moving your head?

Figure 11.5 **Muscles of the head that move the eyeballs (extrinsic eye muscles) and upper eyelid.**

The extrinsic muscles of the eyeball are among the fastest contracting and most precisely controlled skeletal muscles in the body.

(a) Right lateral view of right eyeball

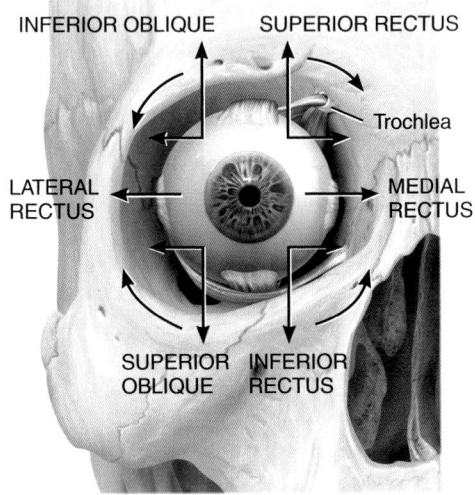

(b) Movements of right eyeball in response to contraction of extrinsic muscles

(c) Right lateral view of right eyeball

❓ How does the inferior oblique muscle move the eyeball superiorly and laterally?

◉ OBJECTIVE

- Describe the origin, insertion, action, and innervation of the muscles that move the mandible and assist in mastication and speech.

The muscles that move the mandible (lower jawbone) at the temporomandibular joint (TMJ) are known as the **muscles of mastication (chewing)** (Figure 11.6). Of the four pairs of muscles involved in mastication, three are powerful closers of the jaw and account for the strength of the bite: **masseter, temporalis,** and **medial pterygoid.** Of these, the masseter is the strongest muscle of mastication. The **medial** and **lateral pterygoid** muscles assist in mastication by moving the mandible from side to side to help grind food. Additionally, the lateral pterygoid muscles protract (protrude) the mandible. The masseter muscle has been removed in Figure 11.6 to illustrate the deeper pterygoid muscles; the masseter can be seen in Figure 11.4c. Note the enormous bulk of the temporalis and masseter muscles compared to the smaller mass of the two pterygoid muscles.

Note: A mnemonic for muscles of mastication is **T**eeny **M**ice **M**ake **P**etite **L**ittle **P**rints = **T**emporalis, **M**asseter, **M**edial **P**terygoid, and **L**ateral **P**terygoid.

⚕ CLINICAL CONNECTION | *Gravity and the Mandible*

As just noted, three of the four muscles of mastication close the mandible and only the lateral pterygoid opens the mouth. The force of **gravity on the mandible** offsets this imbalance. When the masseter, temporalis, and medial pterygoid muscles relax, the mandible drops. Now you know why the mouth of many persons, particularly the elderly, is open while the person is asleep in a chair. In contrast, astronauts in zero gravity must work hard to open their mouths. •

RELATING MUSCLES TO MOVEMENTS

Arrange the muscles in this exhibit according to their actions on the mandible: (1) elevation, (2) depression, (3) retraction, (4) protraction, and (5) side-to-side movement. The same muscle may be mentioned more than once.

✔ CHECKPOINT

What would happen if you lost tone in the masseter and temporalis muscles?

MUSCLE	ORIGIN	INSERTION	ACTION	INNERVATION
Masseter (MA-se-ter = chewer) (see Figure 11.4c)	Maxilla and zygomatic arch.	Angle and ramus of mandible.	Elevates mandible, as in closing mouth.	Mandibular division of trigeminal (V) nerve.
Temporalis (tem′-pō-RĀ-lis; *tempor-* = time or temples)	Temporal bone.	Coronoid process and ramus of mandible.	Elevates and retracts mandible.	Mandibular division of trigeminal (V) nerve.
Medial pterygoid (TER-i-goyd; *medial* = closer to midline; *pterygoid* = winglike)	Medial surface of lateral portion of pterygoid process of sphenoid bone; maxilla.	Angle and ramus of mandible.	Elevates and protracts (protrudes) mandible and moves mandible from side to side.	Mandibular division of trigeminal (V) nerve.
Lateral pterygoid (TER-i-goyd; *lateral* = farther from midline)	Greater wing and lateral surface of lateral portion of pterygoid process of sphenoid bone.	Condyle of mandible; temporomandibular joint (TMJ).	Protracts mandible, depresses mandible as in opening mouth, and moves mandible from side to side.	Mandibular division of trigeminal (V) nerve.

EXHIBIT 11.C CONTINUES ▶

EXHIBIT 11.C **381**

Figure 11.6 Muscles that move the mandible (lower jawbone) and assist in mastication (chewing) and speech.

The muscles that move the mandible are also known as muscles of mastication.

Right lateral superficial view

? **Which is the strongest muscle of mastication?**

EXHIBIT 11.D

Muscles of the Head That Move the Tongue and Assist in Mastication and Speech *(Figure 11.7)*

◉ OBJECTIVE

- Describe the origin, insertion, action, and innervation of the muscles that move the tongue and assist in mastication and speech.

The tongue is a highly mobile structure that is vital to digestive functions such as *mastication* (chewing), detection of taste, and *deglutition* (swallowing). It is also important in speech. The tongue's mobility is greatly aided by its attachment to the mandible, styloid process of the temporal bone, and hyoid bone.

The tongue is divided into lateral halves by a median fibrous septum. The septum extends throughout the length of the tongue. Inferiorly, the septum attaches to the hyoid bone. Muscles of the tongue are of two principal types: extrinsic and intrinsic. **Extrinsic tongue muscles** originate outside the tongue and insert into it (Figure 11.7). They move the entire tongue in various directions, such as anteriorly, posteriorly, and laterally. **Intrinsic tongue muscles** originate and insert within the tongue. These muscles alter the shape of the tongue rather than moving the entire tongue. The extrinsic and intrinsic muscles of the tongue insert into both lateral halves of the tongue.

When you study the extrinsic tongue muscles, you will notice that all of their names end in *glossus*, meaning tongue. You will also notice that the actions of the muscles are obvious, considering the positions of the mandible, styloid process, hyoid bone, and soft palate, which serve as origins for these muscles. For example, the **genioglossus** (origin: the mandible) pulls the tongue downward and forward, the **styloglossus** (origin: the styloid process) pulls the tongue upward and backward, the **hyoglossus** (origin: the hyoid bone) pulls the tongue downward and flattens it, and the **palatoglossus** (origin: the soft palate) raises the back portion of the tongue.

MUSCLE	ORIGIN	INSERTION	ACTION	INNERVATION
Genioglossus (jē'-nē-ō-GLOS-us; *genio-* = chin; *-glossus* = tongue)	Mandible.	Undersurface of tongue and hyoid bone.	Depresses tongue and thrusts it anteriorly (protraction).	Hypoglossal (XII) nerve.
Styloglossus (stī'-lō-GLOS-us; *stylo* = stake or pole; styloid process of temporal bone)	Styloid process of temporal bone.	Side and undersurface of tongue.	Elevates tongue and draws it posteriorly (retraction).	Hypoglossal (XII) nerve.
Hyoglossus (hī'-ō-GLOS-us; *hyo* = U-shaped)	Greater horn and body of hyoid bone.	Side of tongue.	Depresses tongue and draws down its sides.	Hypoglossal (XII) nerve.
Palatoglossus (pal'-a-tō-GLOS-us; *palato* = roof of mouth or palate)	Anterior surface of soft palate.	Side of tongue.	Elevates posterior portion of tongue and draws soft palate down on tongue.	Pharyngeal plexus, which contains axons from the vagus (X) nerve.

EXHIBIT 11.D CONTINUES ▶

EXHIBIT 11.D **383**

CLINICAL CONNECTION | *Intubation during Anesthesia*

When general anesthesia is administered during surgery, a total relaxation of the muscles results. Once the various types of drugs for anesthesia have been given (especially the paralytic agents), the patient's airway must be protected and the lungs ventilated because the muscles involved with respiration are among those paralyzed. Paralysis of the genioglossus muscle causes the tongue to fall posteriorly, which may obstruct the airway to the lungs. To avoid this, the mandible is either manually thrust forward and held in place (known as the "sniffing position"), or a tube is inserted from the lips through the laryngopharynx (inferior portion of the throat) into the trachea **(endotracheal intubation).** People can also be intubated nasally (through the nose). •

RELATING MUSCLES TO MOVEMENTS

Arrange the muscles in this exhibit according to the following actions on the tongue: (1) depression, (2) elevation, (3) protraction, and (4) retraction. The same muscle may be mentioned more than once.

✔ **CHECKPOINT**

When your physician says, "Open your mouth, stick out your tongue, and say ahh," to examine the inside of your mouth for possible signs of infection, which muscles do you contract?

Figure 11.7 Muscles of the head that move the tongue and assist in mastication (chewing) and speech—extrinsic tongue muscles.

The extrinsic and intrinsic muscles of the tongue are arranged in both lateral halves of the tongue.

Superior constrictor
Styloid process of temporal bone
Mastoid process of temporal bone
Digastric (posterior belly cut)
Middle constrictor
Stylohyoid
Stylopharyngeus
HYOGLOSSUS
Hyoid bone
Inferior constrictor
Thyroid cartilage of larynx

STYLOGLOSSUS
PALATOGLOSSUS
Hard palate (cut)
Tongue
GENIOGLOSSUS
Mandible (cut)
GENIOHYOID
Mylohyoid
Intermediate tendon of digastric (cut)
Fibrous loop for intermediate tendon of digastric
Thyrohyoid membrane (connects hyoid bone to larynx)

Right side deep view

❓ **What are the functions of the tongue?**

EXHIBIT 11.E

Muscles of the Anterior Neck That Assist in Deglutition and Speech *(Figure 11.8)*

◉ OBJECTIVE

- Describe the origin, insertion, action, and innervation of the muscles of the anterior neck that assist in deglutition and speech.

Two groups of muscles are associated with the anterior aspect of the neck: (1) the **suprahyoid muscles,** so called because they are located superior to the hyoid bone, and (2) the **infrahyoid muscles,** named for their position inferior to the hyoid bone (Figure 11.8). Both groups of muscles stabilize the hyoid bone, allowing it to serve as a firm base on which the tongue can move.

As a group, the suprahyoid muscles elevate the hyoid bone, floor of the oral cavity, and tongue during deglutition (swallowing). As its name suggests, the **digastric** muscle has two bellies, anterior and posterior, united by an intermediate tendon that is held in position by a fibrous loop. This muscle elevates the hyoid bone and larynx (voice box) during swallowing and speech. In a *reverse muscle action (RMA),* when the hyoid is stabilized, the digastric depresses the mandible and is therefore synergistic to the lateral pterygoid in the opening of the mouth. The **stylohyoid** muscle elevates and draws the hyoid bone posteriorly, thus elongating the floor of the oral cavity during swallowing. The **mylohyoid** muscle elevates the hyoid bone and helps press the tongue against the roof of the oral cavity during swallowing to move food from the oral cavity into the throat. The **geniohyoid** muscle (see Figure 11.7) elevates and draws the hyoid bone anteriorly to shorten the floor of the oral cavity and to widen the throat to receive food that is being swallowed. It also depresses the mandible.

The infrahyoid muscles are sometimes called "strap" muscles because of their ribbonlike appearance. Most of the infrahyoid muscles

MUSCLE	ORIGIN	INSERTION	ACTION	INNERVATION
SUPRAHYOID MUSCLES				
Digastric (dī′-GAS-trik; *di-* = two; *-gastr-* = belly)	Anterior belly from inner side of inferior border of mandible; posterior belly from temporal bone.	Body of hyoid bone via an intermediate tendon.	Elevates hyoid bone. RMA: Depresses mandible, as in opening mouth.	Anterior belly: mandibular division of trigeminal (V) nerve. Posterior belly: facial (VII) nerve.
Stylohyoid (stī′-lō-HĪ-oyd; *stylo-* = stake or pole, styloid process of temporal bone; *-hyo-* = U-shaped, pertaining to hyoid bone)	Styloid process of temporal bone.	Body of hyoid bone.	Elevates hyoid bone and draws it posteriorly.	Facial (VII) nerve.
Mylohyoid (mī′-lō-HĪ-oyd; *mylo* = mill)	Inner surface of mandible.	Body of hyoid bone.	Elevates hyoid bone and floor of mouth and depresses mandible.	Mandibular division of trigeminal (V) nerve.
Geniohyoid (jē′-nē-ō-HĪ-oyd; *genio* = chin) (see Figure 11.7)	Inner surface of mandible.	Body of hyoid bone.	Elevates hyoid bone, draws hyoid bone and tongue anteriorly. Depresses mandible.	First cervical spinal nerve (C1).
INFRAHYOID MUSCLES				
Omohyoid (ō-mō-HĪ-oyd; *omo* = relationship to shoulder)	Superior border of scapula and superior transverse ligament.	Body of hyoid bone.	Depresses hyoid bone.	Branches of spinal nerves C1–C3.
Sternohyoid (ster′-nō-HĪ-oyd; *sterno* = sternum)	Medial end of clavicle and manubrium of sternum.	Body of hyoid bone.	Depresses hyoid bone.	Branches of spinal nerves C1–C3.
Sternothyroid (ster′-nō-THĪ-royd; *thyro* = thyroid gland)	Manubrium of sternum.	Thyroid cartilage of larynx.	Depresses thyroid cartilage of larynx.	Branches of spinal nerves C1–C3.
Thyrohyoid (thī′-rō-HĪ-oyd)	Thyroid cartilage of larynx.	Greater horn of hyoid bone.	Depresses hyoid bone. RMA: Elevates thyroid cartilage.	Branches of spinal nerves C1–C2 and descending hypoglossal (XII) nerve.

EXHIBIT 11.E CONTINUES ▶

EXHIBIT 11.E **385**

depress the hyoid bone and some move the larynx during swallowing and speech. The **omohyoid** muscle, like the digastric muscle, is composed of two bellies connected by an intermediate tendon. In this case, however, the two bellies are referred to as *superior* and *inferior*, rather than anterior and posterior. Together, the omohyoid, **sternohyoid,** and **thyrohyoid** muscles depress the hyoid bone. In addition, the **sternothyroid** muscle depresses the thyroid cartilage (Adam's apple) of the larynx to produce low sounds; the RMA of the thyrohyoid muscle elevates the thyroid cartilage to produce high sounds.

CLINICAL CONNECTION | *Dysphagia*

Dysphagia (dis-FĀ-jē-a; *dys-* = abnormal; *-phagia* = to eat) is a clinical term for difficulty in swallowing. Some individuals are unable to swallow while others have difficulty swallowing liquids, foods, or saliva. Causes include nervous system disorders that weaken or damage muscles of deglutition (stroke, Parkinson disease, cerebral palsy); infections; cancer of the head, neck, or esophagus; and injuries to the head, neck, or chest.   •

Figure 11.8 Muscles of the anterior neck that assist in deglutition (swallowing) and speech.

The suprahyoid muscles elevate the hyoid bone, the floor of the oral cavity, and the tongue during swallowing.

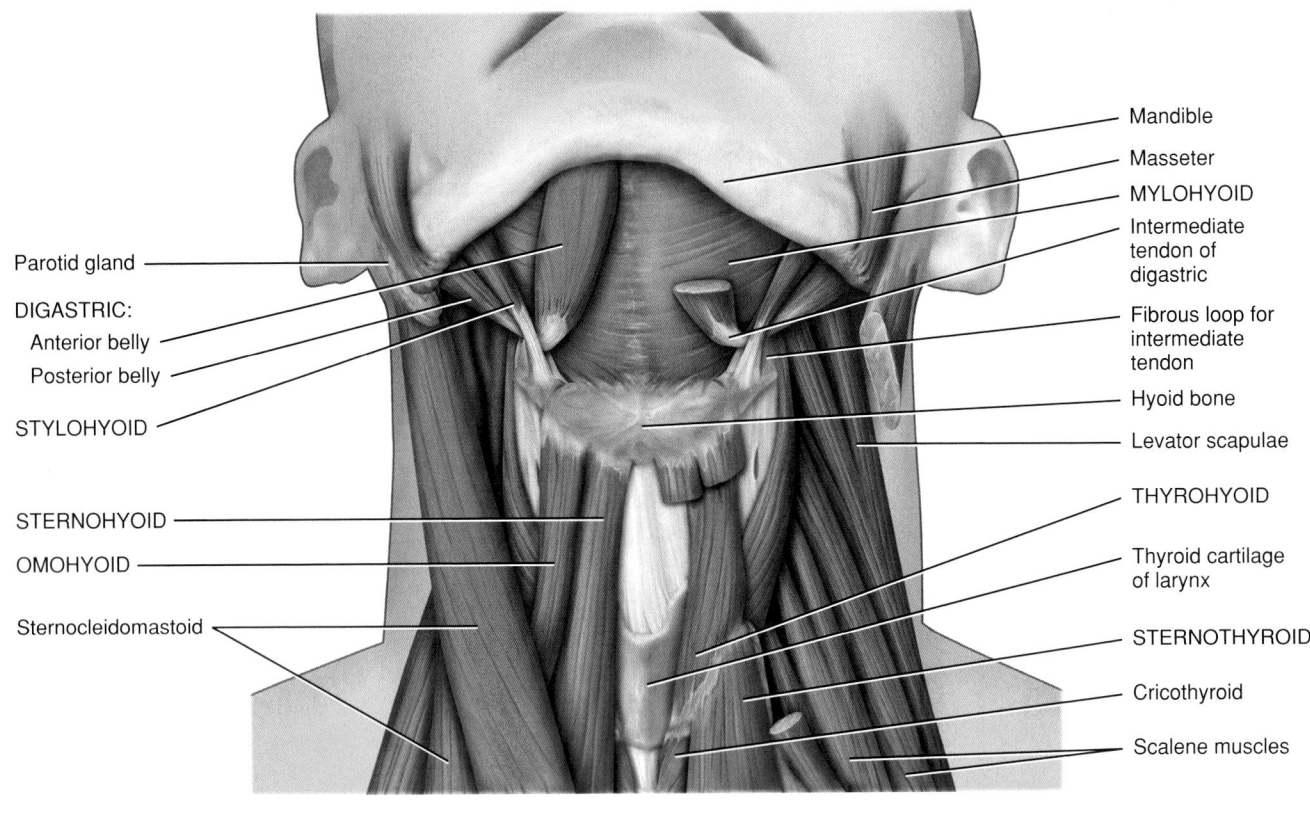

(a) Anterior superficial view   (b) Anterior deep view

RELATING MUSCLES TO MOVEMENTS

Arrange the muscles in this exhibit according to the following actions on the hyoid bone: (1) elevating it, (2) drawing it anteriorly, (3) drawing it posteriorly, and (4) depressing it; and on the thyroid cartilage: (1) elevating it and (2) depressing it. The same muscle may be mentioned more than once.

✔ CHECKPOINT

Which tongue, facial, and mandibular muscles do you use for chewing?

Hyoid bone

OMOHYOID:
Superior belly
Intermediate tendon
Inferior belly

Fascia

Clavicle

Coracoid process of scapula

Thyrohyoid membrane
Inferior constrictor
THYROHYOID
Thyroid cartilage of larynx
Cricothyroid
Cricoid cartilage of larynx
Tracheal cartilage
STERNOTHYROID
STERNOHYOID

Anterior superficial view   (c)   Anterior deep view

? **What is the combined action of the suprahyoid and infrahyoid muscles?**

EXHIBIT 11.E **387**

EXHIBIT 11.F — Muscles of the Neck That Move the Head (Figure 11.9)

OBJECTIVE
• Describe the origin, insertion, action, and innervation of the muscles that move the head.

The head is attached to the vertebral column at the atlanto-occipital joints formed by the atlas and occipital bone. Balance and movement of the head on the vertebral column involves the action of several neck muscles. For example, acting together (bilaterally), contraction of the two **sternocleidomastoid (SCM)** muscles flexes the cervical portion of the vertebral column and flexes the head. Acting singly (unilaterally), each sternocleidomastoid muscle laterally flexes and rotates the head. Each SCM consists of two bellies (Figure 11.9c); they are more evident near the anterior attachments. The separation of the two bellies is variable and thus more evident in some persons than in others. The two bellies insert as the **sternal head** and the **clavicular head** of the SCM. The bellies also function differently; muscular spasm in the two bellies causes somewhat different symptoms. Bilateral contraction of the **semispinalis capitis, splenius capitis,** and **longissimus capitis** muscles extends the head (Figure 11.9a, b). However, when these same muscles contract unilaterally, their actions are quite different, involving primarily rotation of the head.

The sternocleidomastoid muscle is an important landmark that divides the neck into two major triangles: anterior and posterior (Figure 11.9c). The triangles are important anatomically and surgically because of the structures that lie within their boundaries.

The **anterior triangle** is bordered superiorly by the mandible, medially by the cervical midline, and laterally by the anterior border of the sternocleidomastoid muscle. It has its apex at the sternum (Figure 11.9c).

The anterior triangle is subdivided into three paired triangles: *submandibular, carotid,* and *muscular.* An unpaired *submental triangle* is formed by the upper part of the combined right and left anterior triangles. The anterior triangle contains submental, submandibular, and deep cervical lymph nodes; the submandibular salivary gland and a portion of the parotid salivary gland; the facial artery and vein; carotid arteries and internal jugular vein; the thyroid gland and infrahyoid muscles; and the following cranial nerves: glossopharyngeal (IX), vagus (X), accessory (XI), and hypoglossal (XII).

The **posterior triangle** is bordered inferiorly by the clavicle, anteriorly by the posterior border of the sternocleidomastoid muscle, and posteriorly by the anterior border of the trapezius muscle (Figure 11.9c). The posterior triangle is subdivided into two triangles, *occipital* and *supraclavicular (omoclavicular),* by the inferior belly of the omohyoid muscle. The posterior triangle contains part of the subclavian artery, external jugular vein, cervical lymph nodes, brachial plexus, and the accessory (XI) nerve.

RELATING MUSCLES TO MOVEMENTS

Arrange the muscles in this exhibit according to the following actions on the head: (1) flexion, (2) lateral flexion, (3) extension, (4) rotation to side opposite contracting muscle, and (5) rotation to same side as contracting muscle. The same muscle may be mentioned more than once.

✔ CHECKPOINT
What muscles do you contract to signify "yes" and "no"?

MUSCLE	ORIGIN	INSERTION	ACTION	INNERVATION
Sternocleidomastoid (ster'-nō-klī'-dō-MAS-toyd; *sterno-* = breastbone; *-cleido-* = clavicle; *-mastoid* = mastoid process of temporal bone)	Sternal head: manubrium of sternum; clavicular head: medial third of clavicle.	Mastoid process of temporal bone and lateral half of superior nuchal line of occipital bone.	Acting together (bilaterally), flex cervical portion of vertebral column, extend head at atlanto-occipital joints; acting singly (unilaterally), laterally flex neck and head to same side and rotate head to side opposite contracting muscle. Laterally rotate and flex head to opposite side of contracting muscle. Posterior fibers of muscle can assist in extension of head. RMA: Elevate sternum during forced inhalation.	Accessory (XI) nerve, C2, and C3.
Semispinalis capitis (se'-mē-spi-NĀ-lis KAP-i-tis; *semi-* = half; *spine* = spinous process; *capit* = head)	Articular processes of C4–C6 and transverse processes of C7–T7.	Occipital bone between superior and inferior nuchal lines.	Acting together, extend head and vertebral column; acting singly, rotate head to side opposite contracting muscle.	Cervical spinal nerves.
Splenius capitis (SPLĒ-nē-us KAP-i-tis; *splenium* = bandage)	Ligamentum nuchae and spinous processes of C7–T4.	Occipital bone and mastoid process of temporal bone.	Acting together, extend head; acting singly, rotate head to side opposite contracting muscle.	Cervical spinal nerves.
Longissimus capitis (lon-JIS-i-mus KAP-i-tis; *longissimus* = longest)	Articular processes of T1–T4.	Mastoid process of temporal bone.	Acting together, extend head and vertebral column; acting singly, laterally flex and rotate head to same side as contracting muscle.	Cervical spinal nerves.
Spinalis capitis (spi-NĀ-lis KAP-i-tis; *spinal* = vertebral column)	Often absent or very small; arises with semispinalis capitis.	Occipital bone.	Extends head and vertebral column.	Cervical spinal nerves.

Figure 11.9 Muscles of the neck that move the head.

The sternocleidomastoid muscle divides the neck into two principal triangles: anterior and posterior.

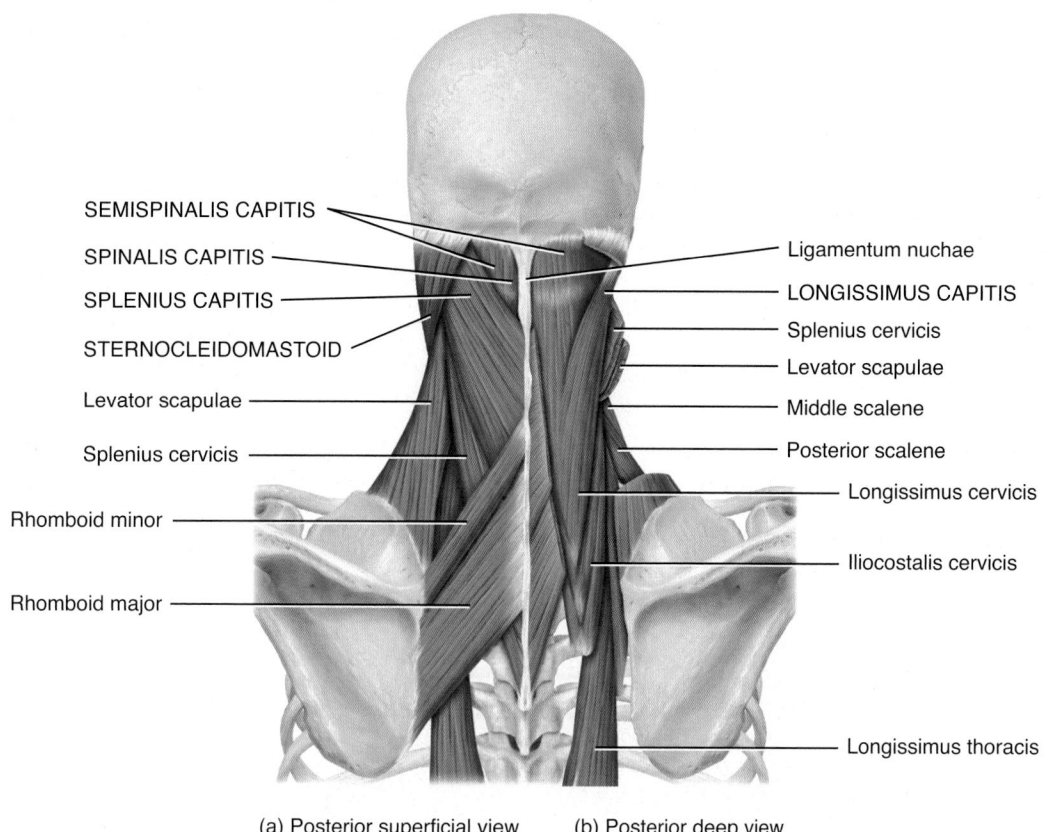

SEMISPINALIS CAPITIS

SPINALIS CAPITIS

SPLENIUS CAPITIS

STERNOCLEIDOMASTOID

Levator scapulae

Splenius cervicis

Rhomboid minor

Rhomboid major

Ligamentum nuchae

LONGISSIMUS CAPITIS

Splenius cervicis

Levator scapulae

Middle scalene

Posterior scalene

Longissimus cervicis

Iliocostalis cervicis

Longissimus thoracis

(a) Posterior superficial view   (b) Posterior deep view

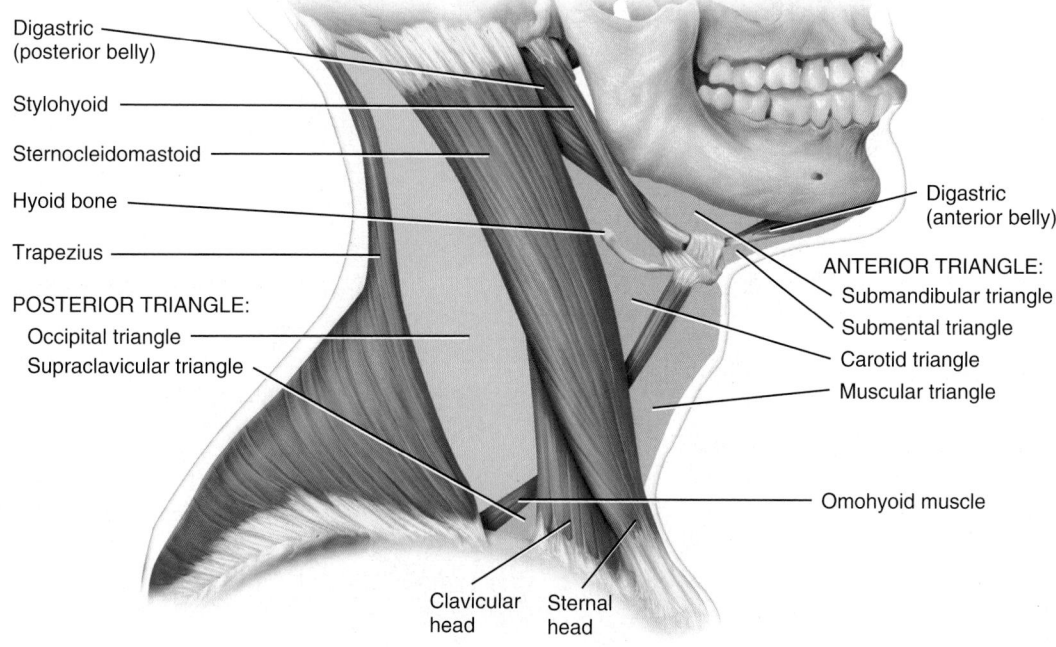

Digastric
(posterior belly)

Stylohyoid

Sternocleidomastoid

Hyoid bone

Trapezius

POSTERIOR TRIANGLE:

Occipital triangle

Supraclavicular triangle

Digastric
(anterior belly)

ANTERIOR TRIANGLE:

Submandibular triangle

Submental triangle

Carotid triangle

Muscular triangle

Omohyoid muscle

Clavicular   Sternal
head   head

(c) Right lateral view of triangles of neck

? **Why are triangles of the neck important?**

EXHIBIT 11.F **389**

EXHIBIT 11.G

Muscles of the Abdomen That Protect Abdominal Viscera and Move the Vertebral Column *(Figure 11.10)*

- Describe the origin, insertion, action, and innervation of the muscles that protect the abdominal viscera and move the vertebral column.

The anterolateral abdominal wall is composed of skin, fascia, and four pairs of muscles: the external oblique, internal oblique, transversus abdominis, and rectus abdominis (Figure 11.10). The first three muscles named are arranged from superficial to deep.

The **external oblique** is the superficial muscle. Its fascicles extend inferiorly and medially. The **internal oblique** is the intermediate flat muscle. Its fascicles extend at right angles to those of the external oblique. The **transversus abdominis** is the deep muscle, with most of its fascicles directed transversely around the abdominal wall. Together, the external oblique, internal oblique, and transversus abdominis form three layers of muscle around the abdomen. In each layer, the muscle fascicles extend in a different direction. This is a structural arrangement that affords considerable protection to the abdominal viscera, especially when the muscles have good tone.

The **rectus abdominis** muscle is a long muscle that extends the entire length of the anterior abdominal wall, originating at the pubic crest and pubic symphysis and inserting on the cartilages of ribs 5–7 and the xiphoid process of the sternum. The anterior surface of the muscle is interrupted by three transverse fibrous bands of tissue called **tendinous intersections,** believed to be remnants of septa that separated myotomes during embryological development (see Figure 10.17). There are usually three tendinous intersections, one at the level of the umbilicus, one near the xiphoid process, and one midway between the other two. A fourth intersection is sometimes found below the level of the umbilicus. These tendinous intersections are fused with the anterior wall of the rectus sheath but have no connections to the posterior abdominal wall. Muscular persons may possess easily demonstrated intersections as the result of exercise and the ensuing hypertrophy of the rectus muscle. Hypertrophy of the muscle tissue, of course, has no effect on the connective tissue of the intersections. Body

Figure 11.10 Muscles of the abdomen that protect abdominal viscera and move the vertebral column (backbone).

The anterolateral abdominal muscles protect the abdominal viscera, move the vertebral column, and assist in forced exhalation, defecation, urination, and childbirth.

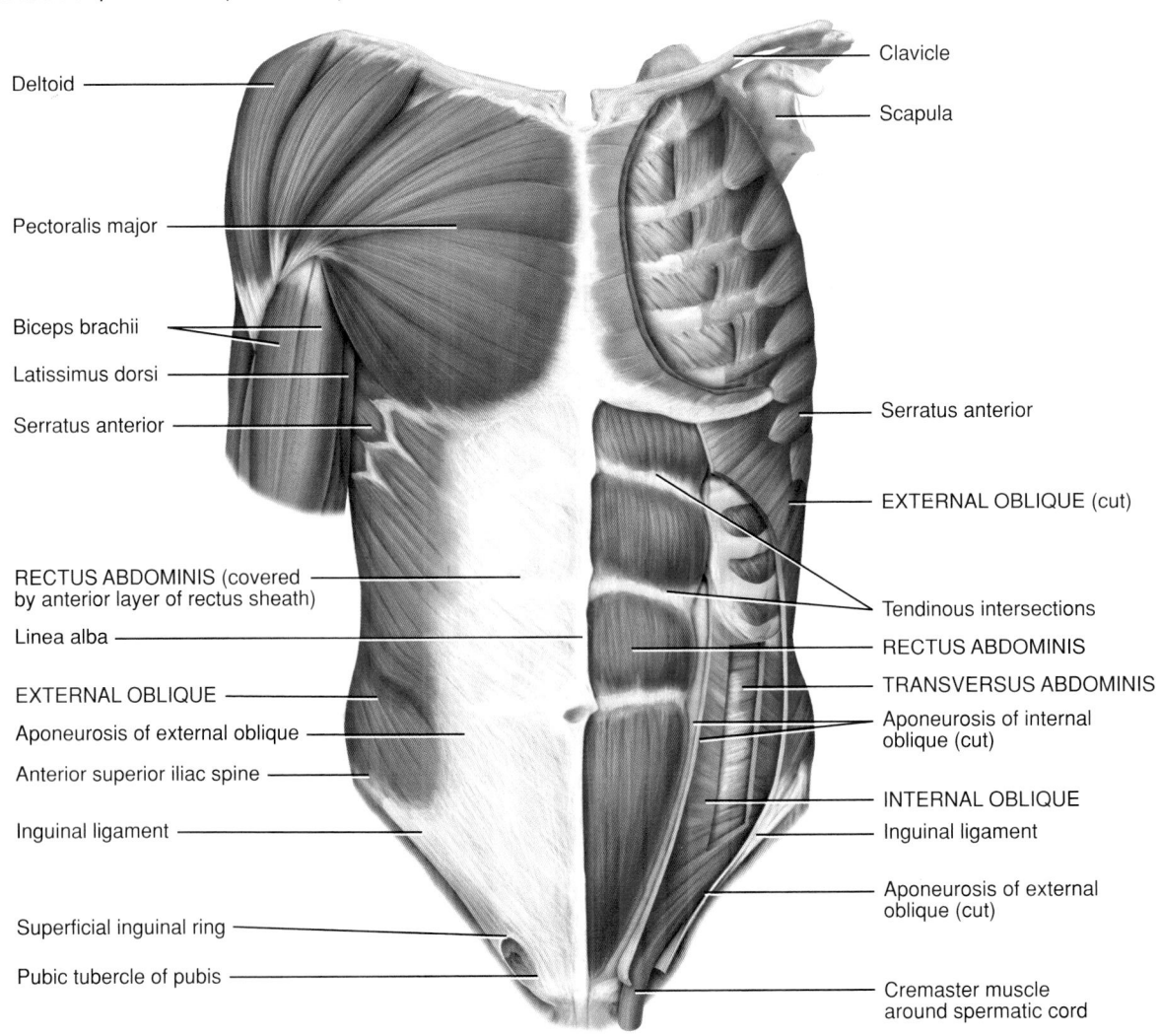

(a) Anterior superficial view   (b) Anterior deep view

builders focus on the development of the **"six-pack"** effect of the abdomen. Small percentages of the population have a variant of the intersections and are able to develop an "eight-pack."

As a group, the muscles of the anterolateral abdominal wall help contain and protect the abdominal viscera; flex, laterally flex, and rotate the vertebral column (backbone) at the intervertebral joints; compress

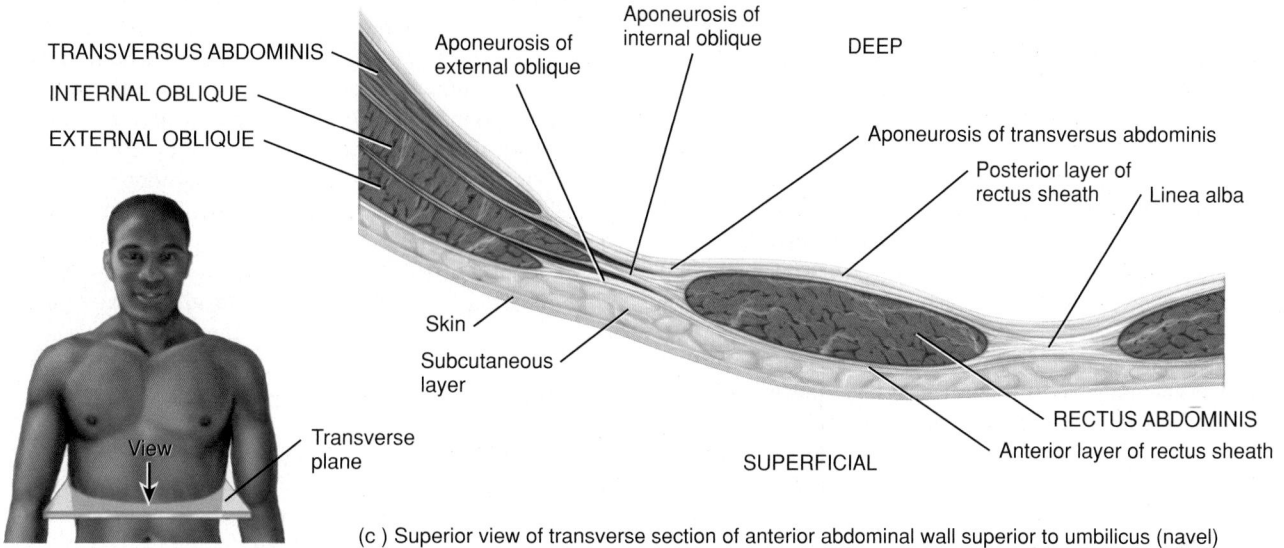

(c) Superior view of transverse section of anterior abdominal wall superior to umbilicus (navel)

(d) Anterior view

❓ **Which abdominal muscle aids in urination?**

EXHIBIT 11.G CONTINUES ▶

EXHIBIT 11.G **391**

the abdomen during forced exhalation; and produce the force required for defecation, urination, and childbirth.

The aponeuroses (sheathlike tendons) of the external oblique, internal oblique, and transversus abdominis muscles form the **rectus sheaths,** which enclose the rectus abdominis muscles. The sheaths meet at the midline to form the **linea alba** (= white line), a tough, fibrous band that extends from the xiphoid process of the sternum to the pubic symphysis. In the latter stages of pregnancy, the linea alba stretches to increase the distance between the rectus abdominis muscles. The inferior free border of the external oblique aponeurosis forms the **inguinal ligament,** which runs from the anterior superior iliac spine to the pubic tubercle (see Figure 11.20a). Just superior to the medial end of the inguinal ligament is a triangular slit in the aponeurosis referred to as the **superficial inguinal ring,** the outer opening of the inguinal canal (see Figure 28.2). The **inguinal canal** contains the spermatic cord and ilioinguinal nerve in males, and the round ligament of the uterus and ilioinguinal nerve in females.

The posterior abdominal wall is formed by the lumbar vertebrae, parts of the ilia of the hip bones, psoas major and iliacus muscles (described in Exhibit 11.Q), and quadratus lumborum muscle. The antero-lateral abdominal wall can contract and distend; the posterior abdominal wall is bulky and stable by comparison.

CLINICAL CONNECTION | *Inguinal Hernia*

A **hernia** (HER-nē-a) is a protrusion of an organ through a structure that normally contains it, which creates a lump that can be seen or felt through the skin's surface. The inguinal region is a weak area in the abdominal wall. It is often the site of an **inguinal hernia,** a rupture or separation of a portion of the inguinal area of the abdominal wall resulting in the protrusion of a part of the small intestine. A hernia is much more common in males than in females because the inguinal canals in males are larger to accommodate the spermatic cord and ilioinguinal nerve. Treatment of hernias most often involves surgery. The organ that protrudes is "tucked" back into the abdominal cavity and the defect in the abdominal muscles is repaired. In addition, a mesh is often applied to reinforce the area of weakness. •

RELATING MUSCLES TO MOVEMENTS

Arrange the muscles in this exhibit according to the following actions on the vertebral column: (1) flexion, (2) lateral flexion, (3) extension, and (4) rotation. The same muscle may be mentioned more than once.

✔ CHECKPOINT

Which muscles do you contract when you "suck in your gut," thereby compressing the anterior abdominal wall?

MUSCLE	ORIGIN	INSERTION	ACTION	INNERVATION
Rectus abdominis (REK-tus ab-DOM-in-is; *rectus* = fascicles parallel to midline: *abdominis* = abdomen)	Pubic crest and pubic symphysis.	Cartilage of ribs 5–7 and xiphoid process.	Flexes vertebral column, especially lumbar portion, and compresses abdomen to aid in defecation, urination, forced exhalation, and childbirth. RMA: Flexes pelvis on the vertebral column.	Thoracic spinal nerves T7–T12.
External oblique (ō-BLĒK; *external* = closer to surface; *oblique* = fascicles diagonal to midline)	Ribs 5–12.	Iliac crest and linea alba.	Acting together (bilaterally), compress abdomen and flex vertebral column; acting singly (unilaterally), laterally flex vertebral column, especially lumbar portion, and rotate vertebral column.	Thoracic spinal nerves T7–T12 and the iliohypogastric nerve.
Internal oblique (ō-BLĒK; *internal* = farther from surface)	Iliac crest, inguinal ligament, and thoracolumbar fascia.	Cartilage of ribs 7–10 and linea alba.	Acting together, compress abdomen and flex vertebral column; acting singly, laterally flex vertebral column, especially lumbar portion, and rotate vertebral column.	Thoracic spinal nerves T8–T12, the iliohypogastric nerve, and ilioinguinal nerve.
Transversus abdominis (tranz-VER-sus ab-DOM-in-is; *transverse* = fascicles perpendicular to midline)	Iliac crest, inguinal ligament, lumbar fascia, and cartilages of ribs 5–10.	Xiphoid process, linea alba, and pubis.	Compresses abdomen.	Thoracic spinal nerves T8–T12, iliohypogastric nerve, and ilioinguinal nerve.
Quadratus lumborum (kwod-RĀ-tus lum-BŌR-um; *quad* = four; *lumbo* = lumbar region) (see Figure 11.11)	Iliac crest and iliolumbar ligament.	Inferior border of rib 12 and L1–L4.	Acting together, pull twelfth ribs inferiorly during forced exhalation, fix twelfth ribs to prevent their elevation during deep inhalation, and held extend lumbar portion of vertebral column; acting singly, laterally flex vertebral column, especially lumbar portion. RMA: Elevates hip bone, commonly on one side.	Thoracic spinal nerves T12 and lumbar spinal nerves L1–L3 or L1–L4.

EXHIBIT 11.H

Muscles of the Thorax That Assist in Breathing *(Figure 11.11)*

⊙ **OBJECTIVE**

- Describe the origin, insertion, action, and innervation of the muscles of the thorax that assist in breathing.

The muscles of the thorax (chest) alter the size of the thoracic cavity so that breathing can occur. Inhalation (breathing in) occurs when the thoracic cavity increases in size, and exhalation (breathing out) occurs when the thoracic cavity decreases in size.

The dome-shaped **diaphragm** is the most important muscle that powers breathing. It also separates the thoracic and abdominal cavities. The diaphragm has a convex superior surface that forms the floor of the thoracic cavity (Figure 11.11b) and a concave, inferior surface that forms the roof of the abdominal cavity (Figure 11.11b). The **peripheral muscular portion** of the diaphragm originates on the xiphoid process of the sternum, the inferior six ribs and their costal cartilages, and the lumbar vertebrae and their intervertebral discs and the twelfth rib (Figure 11.11d). From their various origins, the fibers of the muscular portion converge and insert into the **central tendon,** a strong aponeurosis located near the center of the muscle (Figure 11.11a–d). The central tendon fuses with the inferior surface of the pericardium (covering of the heart) and the pleurae (coverings of the lungs).

The diaphragm has three major openings through which various structures pass between the thorax and abdomen. These structures include the aorta, along with the thoracic duct and azygous vein, which pass through the **aortic hiatus**; the esophagus with accompanying vagus (X) nerves, which pass through the **esophageal hiatus**; and the inferior vena cava, which passes through the **caval opening (foramen for the vena cava).** In a condition called a hiatus hernia, the stomach protrudes superiorly through the esophageal hiatus.

Movements of the diaphragm also help return venous blood passing through abdominal veins to the heart. Together with the anterolateral abdominal muscles, the diaphragm helps to increase intra-abdominal pressure to evacuate the pelvic contents during defecation, urination, and childbirth. This mechanism is further assisted when you take a deep breath and close the rima glottidis (the space between the vocal folds). The trapped air in the respiratory system prevents the diaphragm from elevating. The increase in intra-abdominal pressure also helps support the vertebral column and helps prevent flexion during weight lifting. This greatly assists the back muscles in lifting a heavy weight.

Other muscles involved in breathing, called **intercostal** muscles, span the intercostal spaces, the spaces between ribs. These muscles are arranged in three layers. The 11 pairs of **external intercostal** muscles occupy the superficial layer, and their fibers run in an oblique direction interiorly and anteriorly from the rib above to the rib below. They elevate the ribs during inhalation to help expand the thoracic cavity. The 11 pairs of **internal intercostal** muscles occupy the intermediate layer of the intercostal spaces. The fibers of these muscles run at right angles to the external intercostals, in an oblique direction interiorly and posteriorly from the inferior border of the rib above to the superior border of the rib below. They draw adjacent ribs together during forced exhalation to help decrease the size of the thoracic cavity.

Note: A mnemonic for the action of the intercostal muscles is singing "Old MacDonald had a farm, **E, I, E, I, O**" = **E**xternal **I**ntercostals **E**levate during **I**nhalation, **O**h!"

As you will see in Chapter 23, the diaphragm and external intercostal muscles are used during quiet inhalation and exhalation. However, during deep, forceful inhalation (during exercise or playing a wind instrument), the sternocleidomastoid, scalene, and pectoralis minor muscles are also used; during deep, forceful exhalation, the external oblique, internal oblique, transversus abdominis, rectus abdominis, and internal intercostals are also used.

MUSCLE	ORIGIN	INSERTION	ACTION	INNERVATION
Diaphragm (DĪ-a-fram; *dia-* = across; *-phragm* = wall)	Xiphoid process of sternum, costal cartilages and adjacent portions of inferior six ribs, lumbar vertebrae and their intervertebral discs.	Central tendon.	Contraction of diaphragm causes it to flatten and increases vertical dimension of thoracic cavity, resulting in inhalation; relaxation of diaphragm causes it to move superiorly and decreases vertical dimension of thoracic cavity, resulting in exhalation.	Phrenic nerve, which contains axons from cervical spinal nerves (C3–C5).
External intercostals (in'-ter-KOS-tals; *external* = closer to surface; *inter-* = between; *-costa* = rib)	Inferior border of rib above.	Superior border of rib below.	Contraction elevates ribs and increases anteroposterior and lateral dimensions of thoracic cavity, resulting in inhalation; relaxation depresses ribs and decreases anteroposterior and lateral dimensions of thoracic cavity, resulting in exhalation.	Thoracic spinal nerves T2–T12.
Internal intercostals (in'-ter-KOS-tals; *internal* = farther from surface)	Superior border of rib below.	Inferior border of rib above.	Contraction draws adjacent ribs together to further decrease anteroposterior and lateral dimensions of thoracic cavity during forced exhalation.	Thoracic spinal nerves T2–T12.

EXHIBIT 11.H CONTINUES ▶

EXHIBIT 11.H **393**

RELATING MUSCLES TO MOVEMENTS

Arrange the muscles in this exhibit according to the following actions: (1) increase in vertical length, (2) increase in lateral and anteroposterior dimensions, and (3) decrease in lateral and anteroposterior dimensions of the thorax.

✔CHECKPOINT

What are the names of the three openings in the diaphragm, and which structures pass through each?

Figure 11.11 **Muscles of the thorax (chest) that assist in breathing.**

Openings in the diaphragm permit the passage of the aorta, esophagus, and inferior vena cava.

Clavicle

INTERNAL
INTERCOSTALS

EXTERNAL
INTERCOSTALS

Pectoralis
minor (cut)

Ribs

External
oblique
(cut)

Rectus
abdominis
(cut)

Transversus abdominis
and aponeurosis

Rectus abdominis (covered
by anterior layer of rectus
sheath [cut])

Linea alba

Internal oblique

Anterior superior iliac spine

Aponeurosis of
internal oblique

Inguinal ligament

Spermatic cord

Ribs

INTERNAL
INTERCOSTALS

Innermost
intercostals

Sternum

Central tendon

DIAPHRAGM

Arcuate ligament

QUADRATUS LUMBORUM

Transversus abdominis

Fourth lumbar vertebra

Iliac crest

Sacrum

Pubic tubercle

Pubic symphysis

(a) Anterior superficial view   (b) Anterior deep view

Sternum
Fifth costal cartilage
Skin
Pectoralis major
Fifth rib
Pleura (cut)
Serratus anterior
Pleura (cut)
Inferior vena cava
in caval opening
DIAPHRAGM
Sixth rib
Central tendon
Vagus (X) nerve
EXTERNAL
INTERCOSTAL
Esophagus
In esophageal
hiatus
Central tendon
Seventh rib
Innermost
intercostal
INTERNAL
INTERCOSTAL
DIAPHRAGM
Eighth rib
Latissimus dorsi
Body of T9
Ninth rib
Azygos vein
Thoracic duct
In aortic hiatus
Erector spinae
Aorta
Spinal cord

(c) Superior view of diaphragm

Xiphoid process of sternum
DIAPHRAGM
Costal cartilages
Inferior vena cava
Vagus (X) nerve
Central tendon
Esophagus
Tenth rib
Aorta
Azygos vein
Twelfth rib
Quadratus
lumborum
Thoracic duct
Psoas major
Second lumbar vertebra
Third lumbar vertebra

(d) Inferior view of diaphragm

? **Which muscle associated with breathing is innervated by the phrenic nerve?**

EXHIBIT 11.H **395**

⊙ OBJECTIVE

- Describe the origin, insertion, action, and innervation of the muscles of the pelvic floor that support the pelvic viscera and function as sphincters.

The muscles of the pelvic floor are the levator ani and ischiococcygeus. Together with the fascia covering their internal and external surfaces, these muscles are referred to as the **pelvic diaphragm,** which stretches from the pubis anteriorly to the coccyx posteriorly, and from one lateral wall of the pelvis to the other. This arrangement gives the pelvic diaphragm the appearance of a funnel suspended from its attachments. The pelvic diaphragm separates the pelvic cavity above from the perineum below (see Exhibit 11.J). The anal canal and urethra pierce the pelvic diaphragm in both sexes, and the vagina also goes through it in females.

The two components of the **levator ani** muscle are the **pubococcygeus** and **iliococcygeus.** Figure 11.12 shows these muscles in the female and Figure 11.13 in Exhibit 11.J illustrates them in the male. The levator ani is the largest and most important muscle of the pelvic floor. It supports the pelvic viscera and resists the inferior thrust that accompanies increases in intra-abdominal pressure during functions such as forced exhalation, coughing, vomiting, urination, and defecation. The muscle also functions as a sphincter at the anorectal junction, urethra, and vagina. In addition to assisting the levator ani, the **ischiococcygeus** pulls the coccyx anteriorly after it has been pushed posteriorly during defecation or childbirth.

MUSCLE	ORIGIN	INSERTION	ACTION	INNERVATION
Levator ani (le-VĀ-tor Ā-nē; *levator* = raises; *ani* = anus)	Muscle is divisible into two parts: pubococcygeus muscle and iliococcygeus muscle.			
Pubococcygeus (pū′-bō-kok-SIJ-ē-us; *pubo-* = pubis; *-coccygeus* = coccyx)	Pubis and ischial spine.	Coccyx, urethra, anal canal, perineal body of perineum (wedge-shaped mass of fibrous tissue in center of perineum), and anococcygeal ligament (narrow fibrous band that extends from anus to coccyx).	Supports and maintains position of pelvic viscera; resists increase in intra-abdominal pressure during forced exhalation, coughing, vomiting, urination, and defecation; constricts anus, urethra, and vagina.	Sacral spinal nerves S2–S4.
Iliococcygeus (il′-ē-ō-kok-SIJ-ē-us; *ilio* = ilium)	Ischial spine.	Coccyx.	As above.	Sacral spinal nerves S2–S4.
Ischiococcygeus (is′-kē-ō-kok-SIJ-ē-us; *ischio* = hip)	Ischial spine.	Lower sacrum and upper coccyx.	Supports and maintains position of pelvic viscera; resists increase in intra-abdominal pressure during forced exhalation, coughing, vomiting, urination, and defecation; pulls coccyx anteriorly following defecation or childbirth.	Sacral spinal nerves S4–S5.

RELATING MUSCLES TO MOVEMENTS

Arrange the muscles in this exhibit according to the following actions: (1) supporting and maintaining the position of the pelvic viscera; (2) resisting an increase in intra-abdominal pressure; and (3) constriction of the anus, urethra, and vagina. The same muscle may be mentioned more than once.

✔**CHECKPOINT**
Which muscles are strengthened by Kegel exercises?

Figure 11.12 Muscles of the pelvic floor that support the pelvic viscera, assist in resisting intra-abdominal pressure, and function as sphincters.

The pelvic diaphragm supports the pelvic viscera.

Inferior superficial view of a female perineum

? What are the borders of the pelvic diaphragm?

EXHIBIT 11.I **397**

EXHIBIT 11.J   Muscles of the Perineum *(Figure 11.13)*

⦿ **OBJECTIVE**

• Describe the origin, insertion, action, and innervation of the muscles of the perineum.

The **perineum** is the region of the trunk inferior to the pelvic diaphragm. It is a diamond-shaped area that extends from the pubic symphysis anteriorly, to the coccyx posteriorly, and to the ischial tuberosities laterally. The female and the male perineums may be compared in Figures 11.12 and 11.13, respectively. A transverse line drawn between the ischial tuberosities divides the perineum into an **anterior urogenital triangle** that contains the external genitals and a posterior anal triangle that contains the anus (see Figure 28.21). Several perineal muscles insert into the perineal body of the perineum (described in Section 28.1). Clinically, the perineum is very important to physicians who care for women during

pregnancy and treat disorders related to the female genital tract, urogenital organs, and the anorectal region.

The muscles of the perineum are arranged in two layers: **superficial** and **deep.** The muscles of the superficial layer are the **superficial transverse perineal** muscle, the **bulbospongiosus,** and the **ischiocavernosus** (Figures 11.12 and 11.13). The deep muscles of the male perineum are the **deep transverse perineal** muscle and **external urethral sphincter** (Figure 11.13). The deep muscles of the female perineum are the **compressor urethrae, sphincter urethrovaginalis,** and **external urethral sphincter** (see Figure 11.12). The deep muscles of the perineum assist in urination and ejaculation in males and urination and compression of the vagina in females. The **external anal sphincter** closely adheres to the skin around the margin of the anus and keeps the anal canal and anus closed except during defecation.

MUSCLE	ORIGIN	INSERTION	ACTION	INNERVATION
SUPERFICIAL PERINEAL MUSCLES				
Superficial transverse perineal (per-i-NĒ-al; *superficial* = closer to surface; *transverse* = across; *perineus* = perineum)	Ischial tuberosity.	Perineal body of perineum.	Stabilizes perineal body of perineum.	Perineal branch of pudendal nerve of sacral plexus.
Bulbospongiosus (bul'-bō-spon'-jē-Ō-sus; *bulb-* = bulb; *-spongio-* = sponge)	Perineal body of perineum.	Perineal membrane of deep muscles of perineum, corpus spongiosum of penis, and deep fascia on dorsum of penis in male; pubic arch and root and dorsum of clitoris in female.	Helps expel urine during urination, helps propel semen along urethra, assists in erection of penis in male; constricts vaginal orifice and assists in erection of clitoris in female.	Perineal branch of pudendal nerve of sacral plexus.
Ischiocavernosus (is'-kē-ō-ka'-ver-NŌ-sus; *ischio* = hip)	Ischial tuberosity and ischial and pubic rami.	Corpora cavernosa of penis in male and clitoris in female and pubic symphysis.	Maintains erection of penis in male and clitoris in female by decreasing urine drainage.	Perineal branch of pudendal nerve of sacral plexus.
DEEP PERINEAL MUSCLES				
Deep transverse perineal (per-i-NĒ-al; *deep* = farther from surface)	Ischial ramus.	Perineal body of perineum.	Helps expel last drops of urine and semen in male.	Perineal branch of pudendal nerve of sacral plexus.
External urethral sphincter (ū-RĒ-thral SFINGK-ter)	Ischial and pubic rami.	Median raphe in male and vaginal wall in female.	Helps expel last drops of urine and semen in male and urine in female.	Sacral spinal nerve S4 and inferior rectal branch of pudendal nerve.
Compressor urethrae (ū-RĒ-thrē) (see Figure 11.12)	Ischiopubic ramus.	Blends with partner on other side anterior to urethra.	Serves as accessory sphincter of urethra.	Perineal branch of pudendal nerve of sacral plexus.
Sphincter urethrovaginalis (ū-RĒ-thrō-vaj-i-NAL-is) (see Figure 11.12)	Perineal body.	Blends with partner on other side anterior to urethra.	Serves as accessory sphincter of urethra and facilitates closing of vagina.	Perineal branch of pudendal nerve of sacral plexus.
External anal sphincter (Ā-nal)	Anococcygeal ligament.	Perineal body of perineum.	Keeps anal canal and anus closed.	Sacral spinal nerve S4 and inferior rectal branch of pudendal nerve.

RELATING MUSCLES TO MOVEMENTS

Arrange the muscles in this exhibit according to the following actions;
(1) expulsion of urine and semen, (2) erection of the clitoris and penis,
(3) closure of the anal orifice, and (4) constriction of the vaginal orifice.
The same muscle may be mentioned more than once.

✔CHECKPOINT

What are the borders and contents of the urogenital triangle
and the anal triangle?

Figure 11.13 **Muscles of the perineum.**

The urogenital diaphragm assists in urination in females and males, plays a part in ejaculation in males, and helps strengthen the pelvic floor.

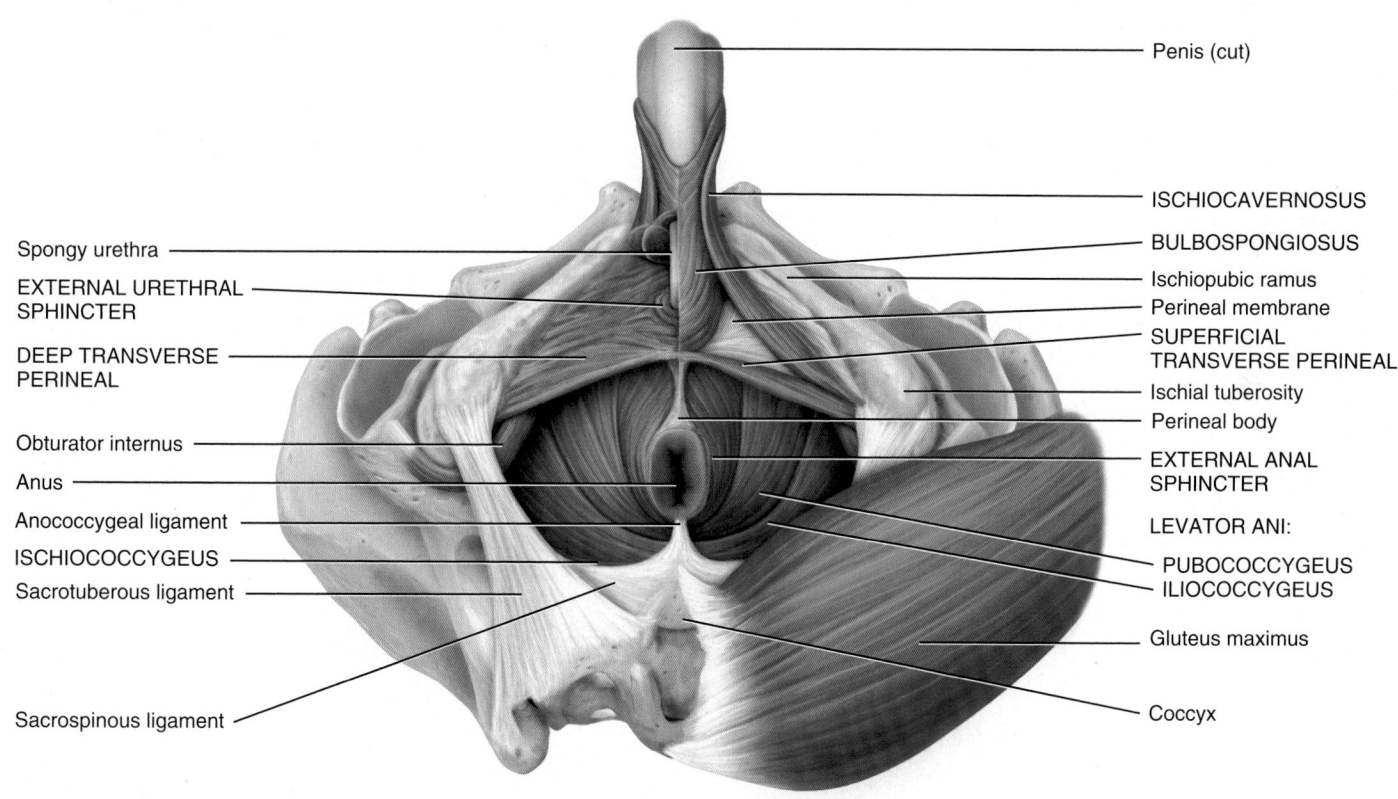

Inferior superficial view of male perineum

? **What are the borders of the perineum?**

EXHIBIT 11.J **399**

OBJECTIVE

- Describe the origin, insertion, action, and innervation of the muscles of the thorax that move the pectoral girdle.

The main action of the muscles that move the pectoral (shoulder) girdle (clavicle and scapula) is to stabilize the scapula so it can function as a steady origin for most of the muscles that move the humerus. Because scapular movements usually accompany humeral movements in the same direction, the muscles also move the scapula to increase the range of motion of the humerus. For example, it would not be possible to raise the arm above the head if the scapula did not move with the humerus. During abduction, the scapula follows the humerus by rotating upward.

Muscles that move the pectoral girdle can be classified into two groups based on their location in the thorax: **anterior** and **posterior thoracic muscles** (Figure 11.14). The anterior thoracic muscles are the subclavius, pectoralis minor, and serratus anterior. The **subclavius** is a small, cylindrical muscle under the clavicle that extends from the clavicle to the first rib. It steadies the clavicle during movements of the pectoral girdle. The **pectoralis minor** is a thin, flat, triangular muscle that is deep to the pectoralis major. Besides its role in movements of the scapula, the pectoralis

minor muscle assists in forced inhalation. The **serratus anterior** is a large, flat, fan-shaped muscle between the ribs and scapula. It is so named because of the saw-toothed appearance of its origins on the ribs.

The posterior thoracic muscles are the trapezius, levator scapulae, rhomboid major, and rhomboid minor. The **trapezius** is a large, flat, triangular sheet of muscle extending from the skull and vertebral column medially to the pectoral girdle laterally. It is the most superficial back muscle and covers the posterior neck region and superior portion of the trunk. The two trapezius muscles form a trapezoid (diamond-shaped quadrangle)—hence its name. The **levator scapulae** is a narrow, elongated muscle in the posterior portion of the neck. It is deep to the sternocleidomastoid and trapezius muscles. As its name suggests, one of its actions is to elevate the scapula (see Figure 11.15c). The **rhomboid major** and **rhomboid minor** lie deep to the trapezius and are not always distinct from each other. They appear as parallel bands that pass inferiorly and laterally from the vertebrae to the scapula (see Figure 11.15c). Their names are based on their shape—that is, a rhomboid (an oblique parallelogram). The rhomboid major is about two times wider than the rhomboid minor. Both muscles are used when forcibly lowering the raised upper limbs, as in driving a stake with a sledgehammer.

MUSCLE	ORIGIN	INSERTION	ACTION	INNERVATION
ANTERIOR THORACIC MUSCLES				
Subclavius (sub-KLĀ-vē-us; *sub-* = under; *-clavius* = clavicle)	First rib.	Clavicle.	Depresses and moves clavicle anteriorly and helps stabilize pectoral girdle.	Subclavian nerve.
Pectoralis minor (pek'-tō-RĀ-lis; *pector* = breast, chest, thorax; *minor* = lesser)	Second through fifth, third through fifth, or second through fourth ribs.	Coracoid process of scapula.	Abducts scapula and rotates it downward. RMA: Elevates third through fifth ribs during forced inhalation when scapula is fixed.	Medial pectoral nerve.
Serratus anterior (ser-Ā-tus; *serratus* = saw-toothed; *anterior* = front)	Superior eight or nine ribs.	Vertebral border and inferior angle of scapula.	Abducts scapula and rotates it upward. RMA: Elevates ribs when scapula is stabilized. Known as "boxer's muscle" because it is important in horizontal arm movements such as punching and pushing.	Long thoracic nerve.
POSTERIOR THORACIC MUSCLES				
Trapezius (tra-PĒ-zē-us; *trapezi* = trapezoid-shaped)	Superior nuchal line of occipital bone, ligamentum nuchae, and spines of C7–T12.	Clavicle and acromion and spine of scapula.	Superior fibers upward rotate scapula; middle fibers adduct scapula; inferior fibers depress and upward rotate scapula; superior and inferior fibers together rotate scapula upward; stabilizes scapula. RMA: Superior fibers can help extend head.	Accessory (XI) nerve and cervical spinal nerves C3–C5.
Levator scapulae (le-VĀ-tor SKA-pū-lē; *levator* = raises; *scapulae* = scapula)	Transverse processes of C1–C4.	Superior vertebral border of scapula.	Elevates scapula and rotates it downward.	Dorsal scapular nerve and cervical spinal nerves C3–C5.
Rhomboid major (rom-BOYD; *rhomboid* = rhomboid or diamond-shaped) (see Figure 11.15c)	Spines of T2–T5.	Vertebral border of scapula inferior to spine.	Elevates and adducts scapula and rotates it downward; stabilizes scapula.	Dorsal scapular nerve.
Rhomboid minor (rom-BOYD) (see Figure 11.15c)	Spines of C7–T1.	Vertebral border of scapula superior to spine.	Elevates and adducts scapula and rotates it downward; stabilizes scapula.	Dorsal scapular nerve.

To understand the actions of muscles that move the scapula, it is first helpful to review the various movements of the scapula:

- **Elevation:** superior movement of the scapula, such as shrugging the shoulders or lifting a weight over the head.
- **Depression:** inferior movement of the scapula, as in pulling down on a rope attached to a pulley.
- **Abduction (protraction):** movement of the scapula laterally and anteriorly, as in doing a "push-up" or punching.
- **Adduction (retraction):** movement of the scapula medially and posteriorly, as in pulling the oars in a rowboat.
- **Upward rotation:** movement of the inferior angle of the scapula laterally so that the glenoid cavity is moved upward. This movement is required to move the humerus past the horizontal, as in raising the arms in a "jumping jack."

- **Downward rotation:** movement of the inferior angle of the scapula medially so that the glenoid cavity is moved downward. This movement is seen when a gymnast on parallel bars supports the weight of the body on the hands.

RELATING MUSCLES TO MOVEMENTS

Arrange the muscles in this exhibit according to the following actions on the scapula: (1) depression, (2) elevation, (3) abduction, (4) adduction, (5) upward rotation, and (6) downward rotation. The same muscle may be mentioned more than once.

✔ CHECKPOINT

What muscles in this exhibit are used to raise your shoulders, lower your shoulders, join your hands behind your back, and join your hands in front of your chest?

Figure 11.14 Muscles of the thorax (chest) that move the pectoral (shoulder) girdle (clavicle and scapula).

Muscles that move the pectoral girdle originate on the axial skeleton and insert on the clavicle or scapula.

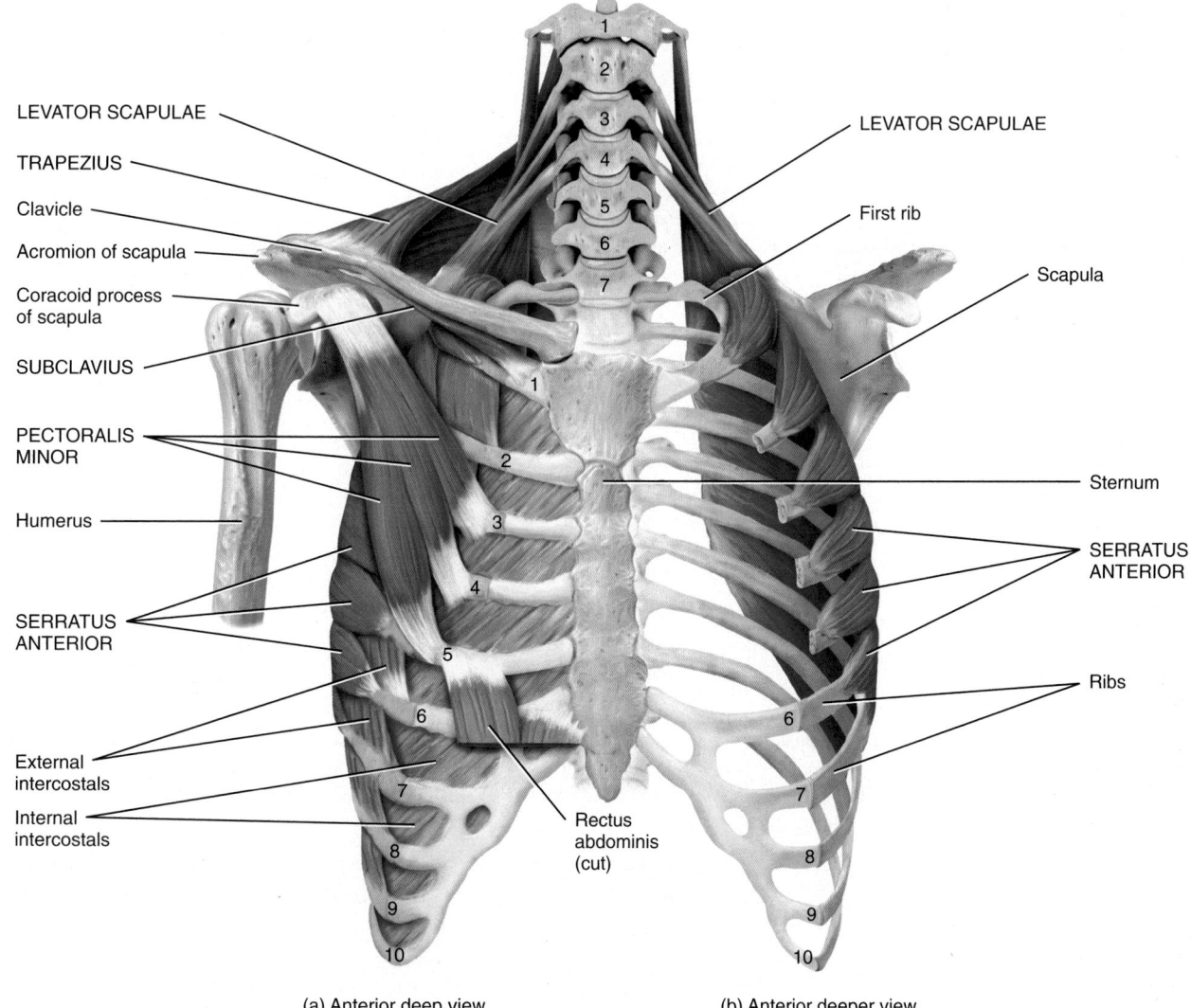

(a) Anterior deep view   (b) Anterior deeper view

❓ **What is the main action of the muscles that move the pectoral girdle?**

EXHIBIT 11.K **401**

EXHIBIT 11.L

Muscles of the Thorax and Shoulder That Move the Humerus *(Figure 11.15)*

● **OBJECTIVE**

- Describe the origin, insertion, action, and innervation of the muscles of the thorax that move the humerus.

Of the nine muscles that cross the shoulder joint, all except the pectoralis major and latissimus dorsi originate on the scapula (shoulder blade). The pectoralis major and latissimus dorsi thus are called **axial muscles,** because they originate on the axial skeleton. The remaining seven muscles, the **scapular muscles,** arise from the scapula (Figure 11.15).

Of the two axial muscles that move the humerus (arm bone), the **pectoralis major** is a large, thick, fan-shaped muscle that covers the superior part of the thorax and forms the anterior fold of the thorax. It has two origins: a smaller clavicular head and a larger sternocostal head. The **latissimus dorsi** is a broad, triangular muscle located on the interior part of the back that forms most of the posterior wall of the axilla. The reverse muscle action (RMA) of the latissimus dorsi enables the vertebral column and torso to be elevated, as in doing a pullup. It is commonly called the "swimmer's muscle" because its many actions are used while swimming; consequently, many competitive swimmers have well-developed "lats."

Among the scapular muscles, the **deltoid** is a thick, powerful shoulder muscle that covers the shoulder joint and forms the rounded contour

MUSCLE	ORIGIN	INSERTION	ACTION	INNERVATION
AXIAL MUSCLES THAT MOVE THE HUMERUS				
Pectoralis major (pek'-tō-RĀ-lis; *pector* = chest; *major* = larger) (see also Figure 11.10a)	Clavicle (clavicular head), sternum, and costal cartilages of second to sixth ribs and sometimes first to seventh ribs (sternocostal head).	Greater tubercle and lateral lip of intertubercular sulcus of humerus.	As a whole, adducts and medially rotates arm at shoulder joint; clavicular head flexes arm, and sternocostal head extends flexed arm to side of trunk.	Medial and lateral pectoral nerves.
Latissimus dorsi (la-TIS-i-mus DOR-sī; *latissimus* = widest; *dorsi* = of the back)	Spines of T7–L5, lumbar vertebrae, crests of sacrum and ilium, inferior four ribs via thoracolumbar fascia.	Intertubercular sulcus of humerus.	Extends, adducts, and medially rotates arm at shoulder joint; draws arm inferiorly and posteriorly. RMA: Elevates vertebral column and torso.	Thoracodorsal nerve.
SCAPULAR MUSCLES THAT MOVE THE HUMERUS				
Deltoid (DEL-toyd = triangularly shaped)	Acromial extremity of clavicle (anterior fibers), acromion of scapula (lateral fibers), and spine of scapula (posterior fibers).	Deltoid tuberosity of humerus.	Lateral fibers abduct arm at shoulder joint; anterior fibers flex and medially rotate arm at shoulder joint; posterior fibers extend and laterally rotate arm at shoulder joint.	Axillary nerve.
Subscapularis (sub-scap'-ū-LĀ-ris; *sub-* = below; *-scapularis* = scapula)	Subscapular fossa of scapula.	Lesser tubercle of humerus.	Medially rotates arm at shoulder joint.	Upper and lower subscapular nerve.
Supraspinatus (soo-pra-spī-NĀ-tus; *supra-* = above; *-spina-* = spine [of the scapula])	Supraspinous fossa of scapula.	Greater tubercle of humerus.	Assists deltoid muscle in abducting arm at shoulder joint.	Suprascapular nerve.
Infraspinatus (in'-fra-spī-NĀ-tus; *infra* = below)	Infraspinous fossa of scapula.	Greater tubercle of humerus.	Laterally rotates arm at shoulder joint.	Suprascapular nerve.
Teres major (TE-rēz; *teres* = long and round)	Inferior angle of scapula.	Medial lip of intertubercular sulcus of humerus.	Extends arm at shoulder joint and assists in adduction and medial rotation of arm at shoulder joint.	Lower subscapular nerve.
Teres minor (TE-rēz)	Inferior lateral border of scapula.	Greater tubercle of humerus.	Laterally rotates and extends arm at shoulder joint.	Axillary nerve.
Coracobrachialis (kor'-a-kō-brā-kē-Ā-lis; *coraco-* = coracoid process [of the scapula]; *-brachi-* = arm)	Coracoid process of scapula.	Middle of medial surface of shaft of humerus.	Flexes and adducts arm at shoulder joint.	Musculocutaneous nerve.

of the shoulder. This muscle is a frequent site of intramuscular injections. As you study the deltoid, note that its fascicles originate from three different points and that each group of fascicles moves the humerus differently. The **subscapularis** is a large triangular muscle that fills the subscapular fossa of the scapula and forms a small part in the apex of the posterior wall of the axilla. The **supraspinatus,** a rounded muscle named for its location in the supraspinous fossa of the scapula, lies deep to the trapezius. The **infraspinatus** is a triangular muscle, also named for its location in the infraspinous fossa of the scapula. The **teres major** is a thick, flattened muscle inferior to the teres minor that also helps form part of the posterior wall of the axilla. The **teres minor** is a cylindrical, elongated muscle, often inseparable from the infraspinatus, which lies

Figure 11.15 Muscles of the thorax (chest) and shoulder that move the humerus (arm bone).

The strength and stability of the shoulder joint are provided by the tendons that form the rotator cuff.

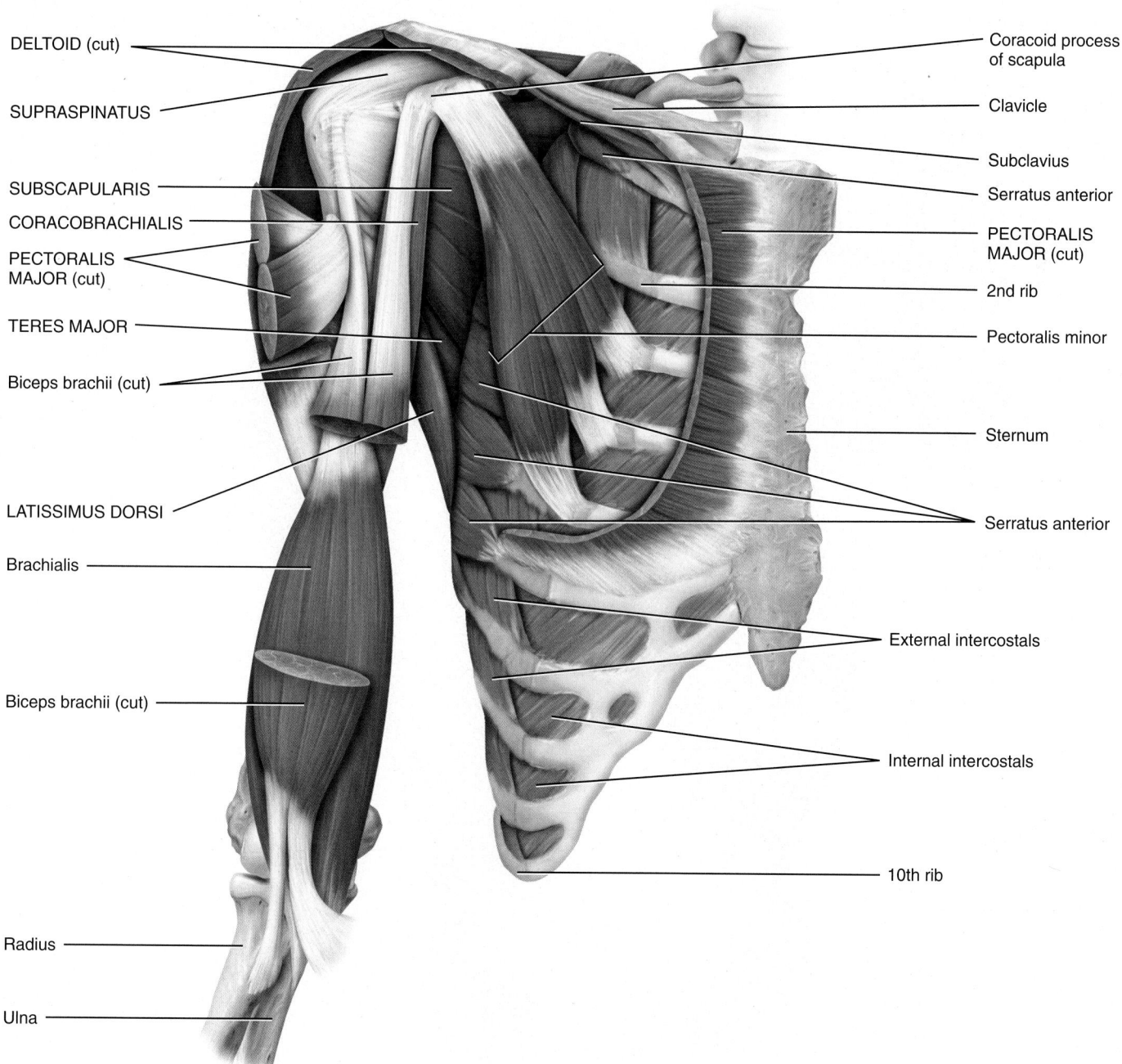

DELTOID (cut)
SUPRASPINATUS
SUBSCAPULARIS
CORACOBRACHIALIS
PECTORALIS MAJOR (cut)
TERES MAJOR
Biceps brachii (cut)
LATISSIMUS DORSI
Brachialis
Biceps brachii (cut)
Radius
Ulna

Coracoid process of scapula
Clavicle
Subclavius
Serratus anterior
PECTORALIS MAJOR (cut)
2nd rib
Pectoralis minor
Sternum
Serratus anterior
External intercostals
Internal intercostals
10th rib

(a) Anterior deep view (the intact pectoralis major muscle is shown in Figure 11.3a)

EXHIBIT 11.L CONTINUES ▶

EXHIBIT 11.L **403**

along its superior border. The **coracobrachialis** is an elongated, narrow muscle in the arm.

Four deep muscles of the shoulder—subscapularis, supraspinatus, infraspinatus, and teres minor—strengthen and stabilize the shoulder joint. These muscles join the scapula to the humerus. Their flat tendons fuse together to form the **rotator (musculotendinous) cuff,** a nearly com-

plete circle of tendons around the shoulder joint, like the cuff on a shirt-sleeve. The supraspinatus muscle is especially subject to wear and tear because of its location between the head of the humerus and acromion of the scapula, which compress its tendon during shoulder movements, especially abduction of the arm. This is further aggravated by poor posture with slouched shoulders.

FIGURE 11.15 CONTINUED ▶

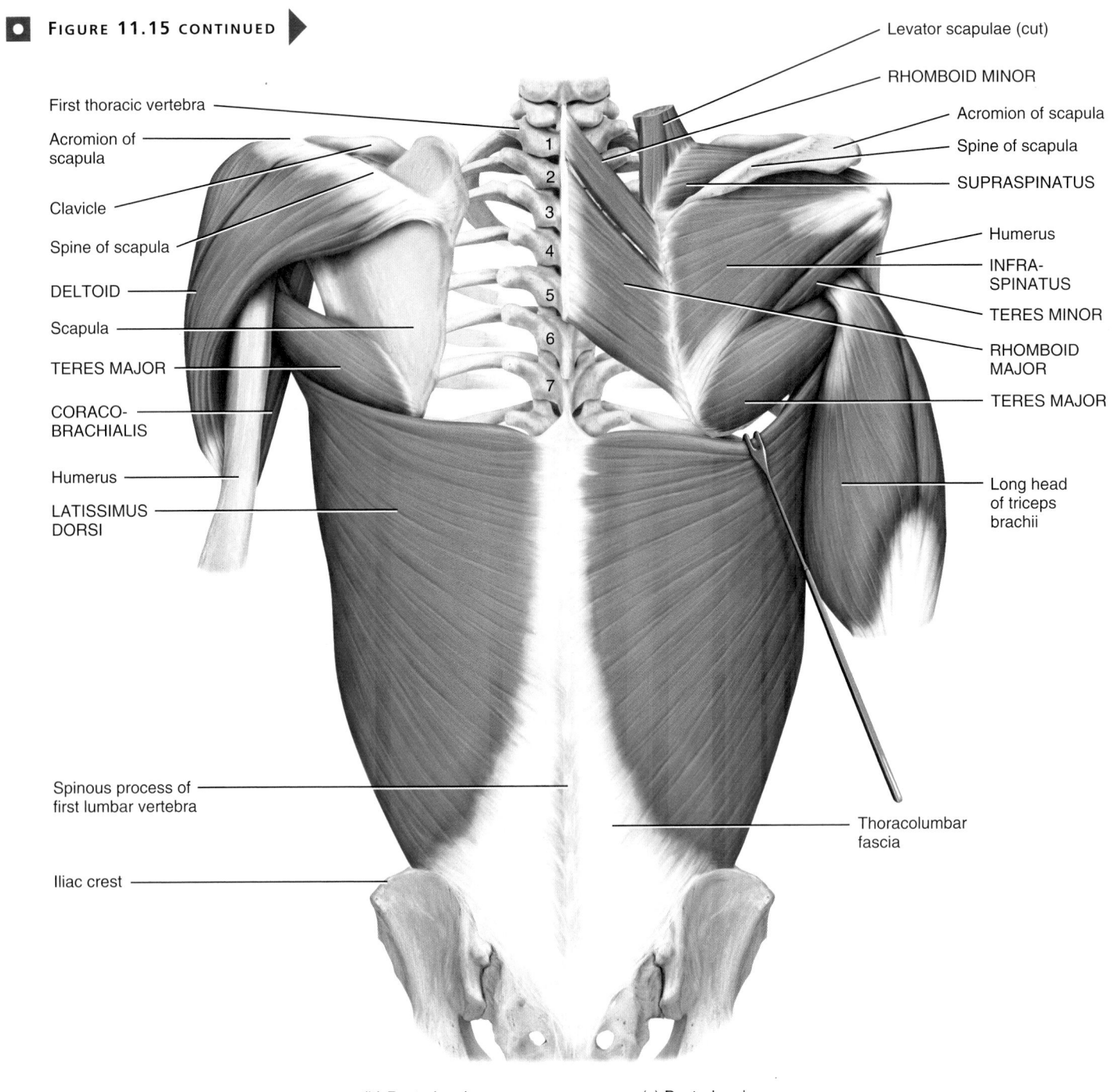

First thoracic vertebra

Acromion of scapula

Clavicle

Spine of scapula

DELTOID

Scapula

TERES MAJOR

CORACO-BRACHIALIS

Humerus

LATISSIMUS DORSI

Spinous process of first lumbar vertebra

Iliac crest

Levator scapulae (cut)

RHOMBOID MINOR

Acromion of scapula

Spine of scapula

SUPRASPINATUS

Humerus

INFRA-SPINATUS

TERES MINOR

RHOMBOID MAJOR

TERES MAJOR

Long head of triceps brachii

Thoracolumbar fascia

(b) Posterior view   (c) Posterior view

CLINICAL CONNECTION | *Rotator Cuff Injury and Impingement Syndrome*

Rotator cuff injury is a strain or tear in the rotator cuff muscles and is common among baseball pitchers, volleyball players, racquet sports players, and swimmers due to shoulder movements that involve vigorous circumduction. It also occurs as a result of wear and tear, aging, trauma, poor posture, improper lifting, and repetitive motions in certain jobs, such as placing items on a shelf above your head. Most often, there is tearing of the supraspinatus muscle tendon or the rotator cuff. This tendon is especially predisposed to wear and tear because of its location between the head of the humerus and acromion of the scapula, which compresses the tendon during shoulder movements. Poor posture and poor body mechanics also increase compression of the supraspinatus muscle tendon.

One of the most common causes of shoulder pain and dysfunction in athletes is known as **impingement syndrome,** which is sometimes confused with another common complaint, compartment syndrome, discussed in Disorders: Homeostatic Imbalances at the end of this chapter. The repetitive movement of the arm over the head that is common in baseball, overhead racquet sports, lifting weights over the head, spiking a volleyball, and swimming puts these athletes at risk. Impingement syndrome may also be caused by a direct blow or

stretch injury. Continual pinching of the supraspinatus tendon as a result of overhead motions causes it to become inflamed and results in pain. If movement is continued despite the pain, the tendon may degenerate near the attachment to the humerus and ultimately may tear away from the bone (rotator cuff injury). Treatment consists of resting the injured tendons, strengthening the shoulder through exercise, massage therapy, and surgery if the injury is particularly severe. During surgery, an inflamed bursa may be removed, bone may be trimmed, and/or the coracoacromial ligament may be detached. Torn rotator cuff tendons may be trimmed and then reattached with sutures, anchors, or surgical tacks. These steps make more space, thus relieving pressure and allowing the arm to move freely. •

RELATING MUSCLES TO MOVEMENTS

Arrange the muscles in this exhibit according to the following actions on the humerus at the shoulder joint: (1) flexion, (2) extension, (3) abduction, (4) adduction, (5) medial rotation, and (6) lateral rotation. The same muscle may be mentioned more than once.

✔ **CHECKPOINT**

Why are the two muscles that cross the shoulder joint called axial muscles, and the seven others called scapular muscles?

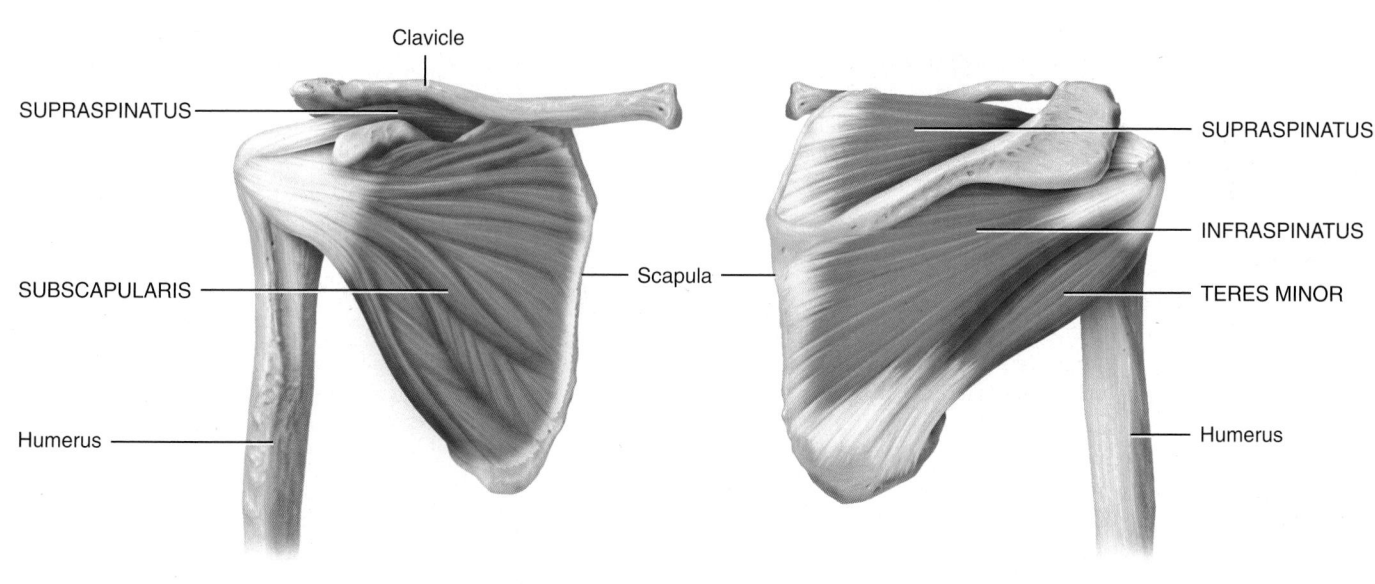

(d) Anterior deep view

(e) Posterior deep view

 Which tendons make up the rotator cuff?

EXHIBIT 11.L **405**

EXHIBIT 11.M

Muscles of the Arm That Move the Radius and Ulna *(Figure 11.16)*

● **OBJECTIVE**

- Describe the origin, insertion, action, and innervation of the muscles of the arm that move the radius and ulna.

Most of the muscles that move the radius and ulna (forearm bones) cause flexion and extension at the elbow, which is a hinge joint. The biceps brachii, brachialis, and brachioradialis muscles are the flexor muscles. The extensor muscles are the triceps brachii and the anconeus (Figure 11.16).

The **biceps brachii** is the large muscle located on the anterior surface of the arm. As indicated by its name, it has two heads of origin (long and short), both from the scapula. The muscle spans both the shoulder and elbow joints. In addition to its role in flexing the forearm at the elbow joint, it also supinates the forearm at the radioulnar joints and flexes the arm at the shoulder joint. The **brachialis** is deep to the biceps brachii muscle. It is the most powerful flexor of the forearm at the elbow joint. For this reason, it is called the "workhorse" of the elbow flexors. The **brachioradi-**

MUSCLE	ORIGIN	INSERTION	ACTION	INNERVATION
FOREARM FLEXORS				
Biceps brachii (BĪ-seps BRĀ-kē-ī; *biceps* = two heads of origin; *brachii* = arm)	Long head originates from tubercle above glenoid cavity of scapula (supraglenoid tubercle). Short head originates from coracoid process of scapula.	Radial tuberosity of radius and bicipital aponeurosis.*	Flexes forearm at elbow joint, supinates forearm at radioulnar joints, and flexes arm at shoulder joint.	Musculocutaneous nerve.
Brachialis (brā-kē-Ā-lis)	Distal, anterior surface of humerus.	Ulnar tuberosity and coronoid process of ulna.	Flexes forearm at elbow joint.	Musculocutaneous and radial nerves.
Brachioradialis (brā'-kē-ō-rā-dē-Ā-lis; *radi* = radius)	Lateral border of distal end of humerus.	Superior to styloid process of radius.	Flexes forearm at elbow joint; supinates and pronates forearm at radioulnar joints to neutral position.	Radial nerve.
FOREARM EXTENSORS				
Triceps brachii (TRĪ-seps BRĀ-kē-ī; *triceps* = three heads of origin)	Long head originates from infraglenoid tubercle, a projection inferior to glenoid cavity of scapula. Lateral head originates from lateral and posterior surface of humerus. Medial head originates from entire posterior surface of humerus inferior to a groove for the radial nerve.	Olecranon of ulna.	Extends forearm at elbow joint and extends arm at shoulder joint.	Radial nerve.
Anconeus (an-KŌ-nē-us; *ancon* = elbow)	Lateral epicondyle of humerus.	Olecranon and superior portion of shaft of ulna.	Extends forearm at elbow joint.	Radial nerve.
FOREARM PRONATORS				
Pronator teres (PRŌ-nā-tor TE-rēz; *pronator* = turns palm posteriorly; *tero* = round and long) (see also Figure 11.17a)	Medial epicondyle of humerus and coronoid process of ulna.	Midlateral surface of radius.	Pronates forearm at radioulnar joints and weakly flexes forearm at elbow joint.	Median nerve.
Pronator quadratus (PRŌ-nā-tor kwod-RĀ-tus; *quadratus* = square, four-sided) (see also Figure 11.17a–c)	Distal portion of shaft of ulna.	Distal portion of shaft of radius.	Pronates forearm at radioulnar joints.	Median nerve.
FOREARM SUPINATOR				
Supinator (SOO-pi-nā-tor; *supinator* = turns palm anteriorly) (see also Figure 11.17b,c)	Lateral epicondyle of humerus and ridge near radial notch of ulna (supinator crest).	Lateral surface of proximal one-third of radius.	Supinates forearm at radioulnar joints.	Deep radial nerve.

*The **bicipital aponeurosis** is a broad aponeurosis from the tendon of insertion of the biceps brachii muscle that descends medially across the brachial artery and fuses with deep fascia over the forearm flexor muscles. It also helps to protect the median nerve and brachial artery.

alis flexes the forearm at the elbow joint, especially when a quick movement is required or when a weight is lifted slowly during flexion of the forearm.

The **triceps brachii** is the large muscle located on the posterior surface of the arm. It is the more powerful of the extensors of the forearm at the elbow joint. As its name implies, it has three heads of origin, one from the scapula (long head) and two from the humerus (lateral and medial heads). The long head crosses the shoulder joint; the other heads do not. The **anconeus** is a small muscle located on the lateral part of the posterior aspect of the elbow that assists the triceps brachii in extending the forearm at the elbow joint.

Some muscles that move the radius and ulna are involved in pronation and supination at the radioulnar joints. The pronators, as suggested by their names, are the **pronator teres** and **pronator quadratus** muscles. The supinator of the forearm is aptly named the **supinator** muscle. You use the powerful action of the supinator when you twist a corkscrew or turn a screw with a screwdriver.

In the limbs, functionally related skeletal muscles and their associated blood vessels and nerves are grouped together by fascia into regions called **compartments.** In the arm, the biceps brachii, brachialis, and coracobrachialis muscles compose the **anterior (flexor) compartment.** The triceps brachii muscle forms the **posterior (extensor) compartment.**

Figure 11.16 **Muscles of the arm that move the radius and ulna (forearm bones).**

The anterior arm muscles flex the forearm, and the posterior arm muscles extend it.

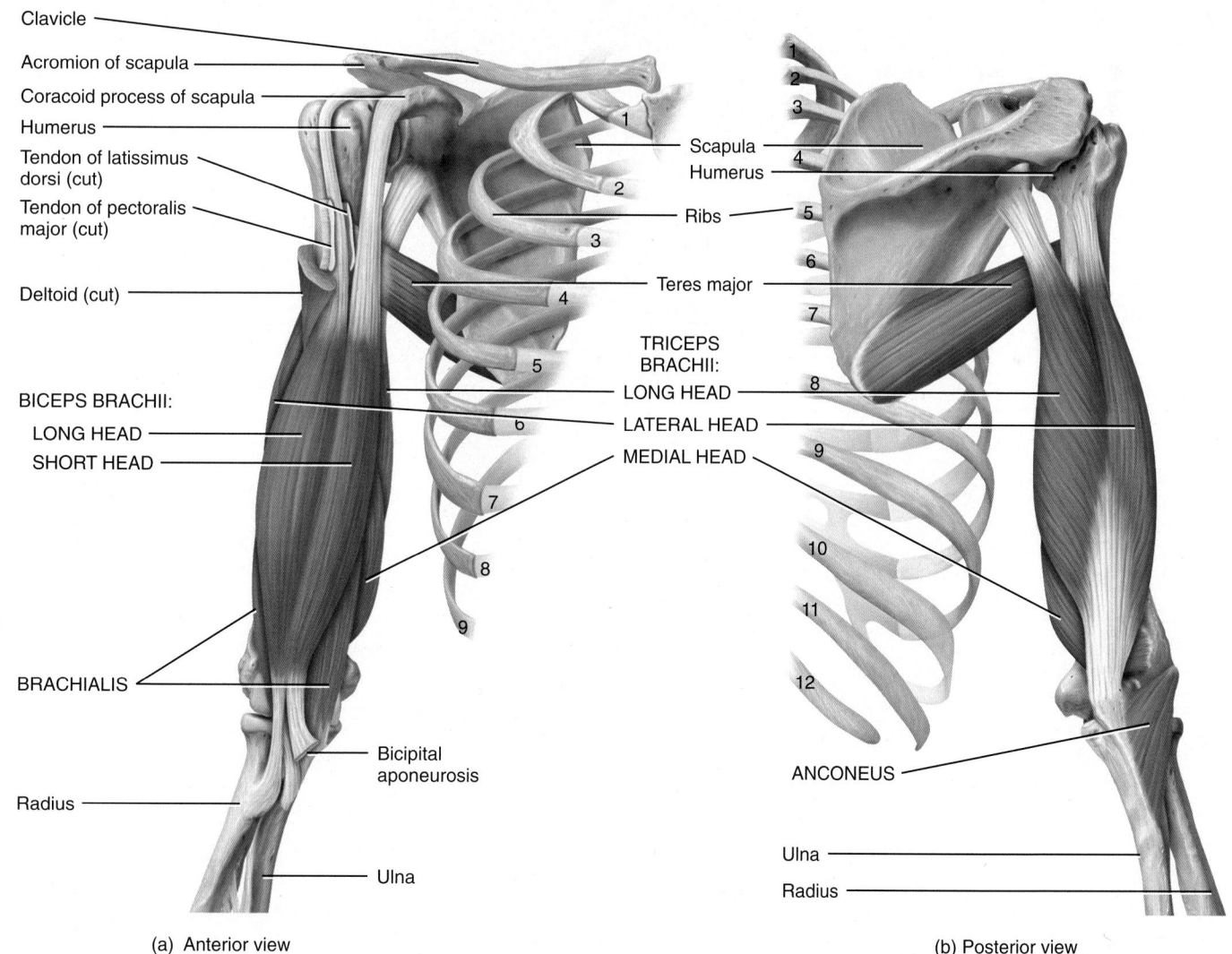

(a) Anterior view

(b) Posterior view

Exhibit 11.M continues ▶

EXHIBIT 11.M **407**

RELATING MUSCLES TO MOVEMENTS

Arrange the muscles in this exhibit according to the following actions on the elbow joint: (1) flexion and (2) extension; the following actions on the forearm at the radioulnar joints: (1) supination and (2) pronation; and the following actions on the humerus at the shoulder joint: (1) flexion and (2) extension. The same muscle may be mentioned more than once.

✔ CHECKPOINT

Flex your arm. Which group of muscles is contracting? Which group of muscles must relax so that you can flex your arm?

⬛ FIGURE **11.16** CONTINUED ▶

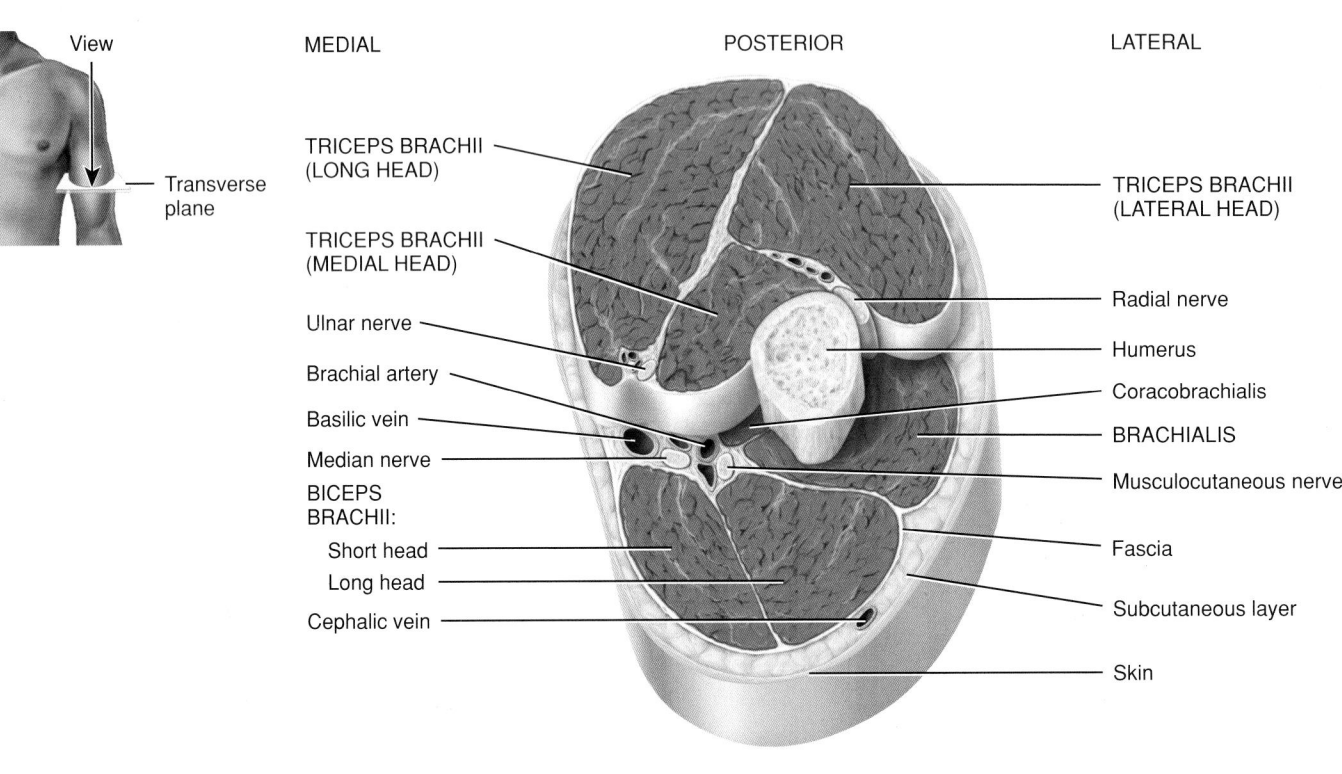

(c) Superior view of transverse section of arm

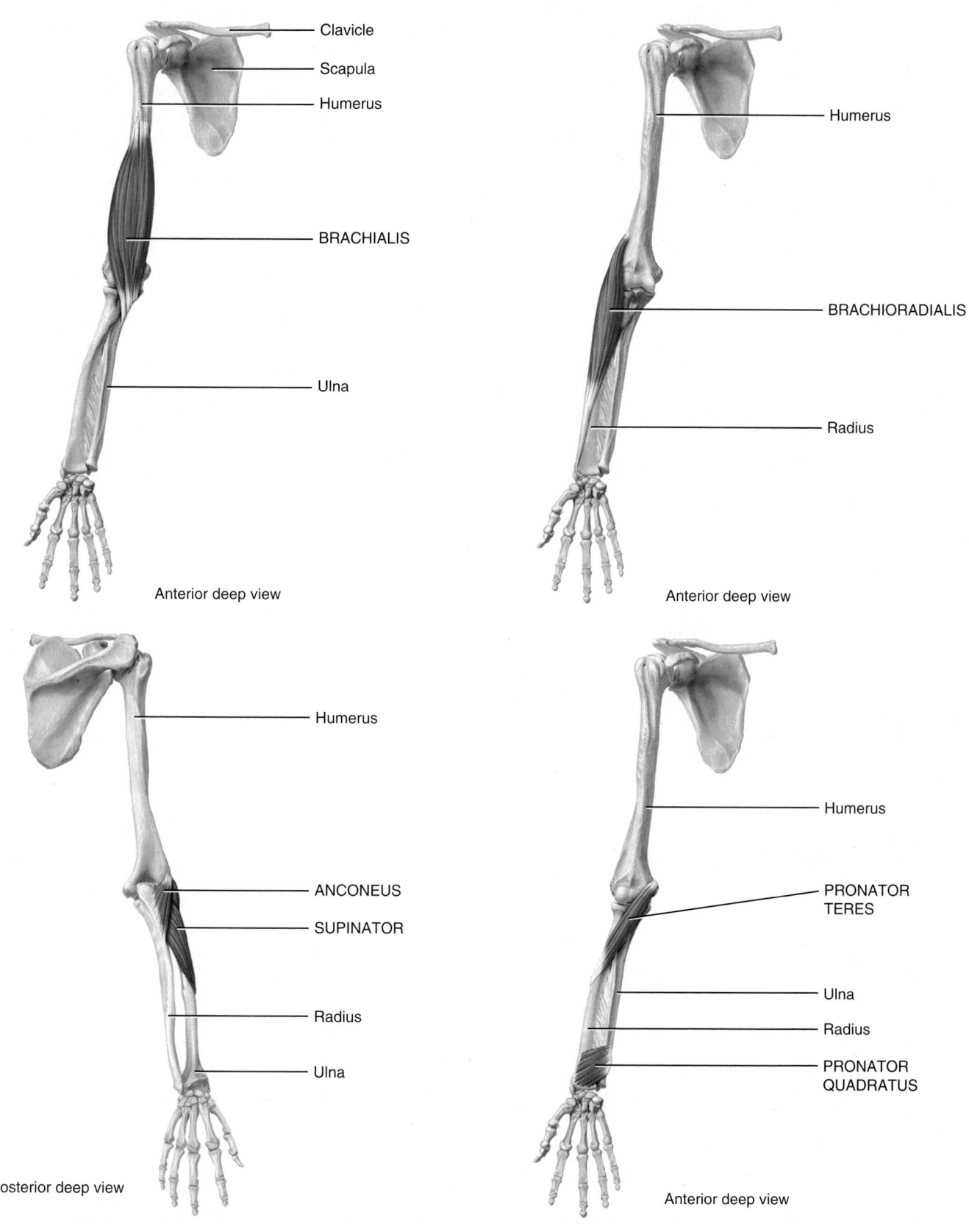

Clavicle

Scapula

Humerus

BRACHIALIS

Ulna

Anterior deep view

Humerus

BRACHIORADIALIS

Radius

Anterior deep view

Humerus

ANCONEUS

SUPINATOR

Radius

Ulna

Posterior deep view

Humerus

PRONATOR TERES

Ulna

Radius

PRONATOR QUADRATUS

Anterior deep view

(d) Isolated muscles

Which muscles are the most powerful flexor and the most powerful extensor of the forearm?

EXHIBIT 11.M **409**

◉ **OBJECTIVE**

- Describe the origin, insertion, action, and innervation of the muscles of the forearm that move the wrist, hand, and digits.

Muscles of the forearm that move the wrist, hand, and digits are many and varied (Figure 11.17). Those in this group that act on the digits are known as **extrinsic muscles of the hand** (*ex* = outside) because they originate *outside* the hand and insert within it. As you will see, the names for the muscles that move the wrist, hand, and digits give some indication of their origin, insertion, or action. Based on location and function, the muscles of the forearm are divided into two groups: (1) anterior compartment muscles and (2) posterior compartment muscles. The **anterior (flexor) compartment** muscles of the forearm originate on the humerus; typically insert on the carpals, metacarpals, and phalanges; and function primarily as flexors. The bellies of these muscles form the bulk of the forearm. One of the muscles in the superficial anterior compartment, the palmaris longus muscle, is missing in about 10% of individuals (usually in the left forearm) and is commonly used for tendon repair. The **posterior (extensor) compartment** muscles of the forearm originate on the humerus, insert on the metacarpals and phalanges, and function as extensors. Within each compartment, the muscles are grouped as superficial or deep.

The **superficial anterior compartment** muscles are arranged in the following order from lateral to medial: **flexor carpi radialis, palmaris longus,** and **flexor carpi ulnaris** (the ulnar nerve and artery are just lateral to the tendon of this muscle at the wrist). The **flexor digitorum** superficialis muscle is deep to the other three muscles and is the largest superficial muscle in the forearm.

The **deep anterior compartment** muscles are arranged in the following order from lateral to medial: **flexor pollicis longus** (the only flexor of the distal phalanx of the thumb) and **flexor digitorum profundus** (ends in four tendons that insert into the distal phalanges of the fingers).

The **superficial posterior compartment** muscles are arranged in the following order from lateral to medial: **extensor carpi radialis longus, extensor carpi radialis brevis, extensor digitorum** (occupies most of the posterior surface of the forearm and divides into four tendons that insert into the middle and distal phalanges of the fingers), **extensor digiti minimi** (a slender muscle usually connected to the extensor digitorum), and **extensor carpi ulnaris.**

The **deep posterior compartment** muscles are arranged in the following order from lateral to medial: **abductor pollicis longus, extensor pollicis brevis, extensor pollicis longus,** and **extensor indicis.**

The tendons of the muscles of the forearm that attach to the wrist or continue into the hand, along with blood vessels and nerves, are held close to bones by strong fasciae. The tendons are also surrounded by tendon sheaths. At the wrist, the deep fascia is thickened into fibrous bands called **retinacula** (*retinacul* = holdfast). The **flexor retinaculum** is located over the palmar surface of the carpal bones. The long flexor tendons of the digits and wrist and the median nerve pass deep to the flexor retinaculum. The **extensor retinaculum** is located over the dorsal surface of the carpal bones. The extensor tendons of the wrist and digits pass deep to it.

MUSCLE	ORIGIN	INSERTION	ACTION	INNERVATION
SUPERFICIAL ANTERIOR (FLEXOR) COMPARTMENT OF THE FOREARM				
Flexor carpi radialis (FLEK-sor KAR-pē-rā′-dē-Ā-lis; *flexor* = decreases angle at joint; *carpi* = wrist; *radi* = radius)	Medial epicondyle of humerus.	Second and third metacarpals.	Flexes and abducts hand (radial deviation) at wrist joint.	Median nerve.
Palmaris longus (pal-MA-ris LON-gus; *palma* = palm; *longus* = long)	Medial epicondyle of humerus.	Flexor retinaculum and palmar aponeurosis (fascia in center of palm).	Weakly flexes hand at wrist joint.	Median nerve.
Flexor carpi ulnaris (FLEK-sor KAR-pē ūl-NAR-is; *ulnar* = ulna)	Medial epicondyle of humerus and superior posterior border of ulna.	Pisiform, hamate, and base of fifth metacarpal.	Flexes and adducts hand (ulnar deviation) at wrist joint.	Ulnar nerve.
Flexor digitorum superficialis (FLEK-sor di-ji-TOR-um soo′-per-fish′-ē-Ā-lis; *digit* = finger or toe; *superficialis* = closer to surface)	Medial epicondyle of humerus, coronoid process of ulna, and ridge along lateral margin or anterior surface (anterior oblique line) of radius.	Middle phalanx of each finger.*	Flexes middle phalanx of each finger at proximal interphalangeal joint, proximal phalanx of each finger at metacarpophalangeal joint, and hand at wrist joint.	Median nerve.
DEEP ANTERIOR (FLEXOR) COMPARTMENT OF THE FOREARM				
Flexor pollicis longus (FLEK-sor POL-li-sis LON-gus; *pollicis* = thumb)	Anterior surface of radius and interosseous membrane (sheet of fibrous tissue that holds shafts of ulna and radius together).	Base of distal phalanx of thumb.	Flexes distal phalanx of thumb at interphalangeal joint.	Median nerve.
Flexor digitorum profundus (FLEK-sor di′-ji-TOR-um prō-FUN-dus; *profundus* = deep)	Anterior medial surface of body of ulna.	Base of distal phalanx of each finger.	Flexes distal and middle phalanges of each finger at interphalangeal joints, proximal phalanx of each finger at metacarpophalangeal joint, and hand at wrist joint.	Median and ulnar nerves.

MUSCLE	ORIGIN	INSERTION	ACTION	INNERVATION
SUPERFICIAL POSTERIOR (EXTENSOR) COMPARTMENT OF THE FOREARM				
Extensor carpi radialis longus (eks-TEN-sor KAR-pē rā'-dē-Ā-lis LON-gus; *extensor* = increases angle at joint)	Lateral supracondylar ridge of humerus.	Second metacarpal.	Extends and abducts hand at wrist joint (ulnar deviation).	Radial nerve.
Extensor carpi radialis brevis (eks-TEN-sor KAR-pē rā-dē-Ā-lis BREV-is; *brevis* = short)	Lateral epicondyle of humerus.	Third metacarpal.	Extends and abducts hand at wrist joint.	Radial nerve.
Extensor digitorum (eks-TEN-sor di'-ji-TOR-um)	Lateral epicondyle of humerus.	Distal and middle phalanges of each finger.	Extends distal and middle phalanges of each finger at interphalangeal joints, proximal phalanx of each finger at metacarpophalangeal joint, and hand at wrist joint.	Radial nerve.
Extensor digiti minimi (eks-TEN-sor DIJ-i-tē MIN-i-mē; *digit* = finger or toe; *minimi* = smallest)	Lateral epicondyle of humerus.	Tendon of extensor digitorum on fifth phalanx.	Extends proximal phalanx of little finger at metacarpophalangeal joint and hand at wrist joint.	Deep radial nerve.
Extensor carpi ulnaris (eks-TEN-sor KAR-pē ul-NAR-is)	Lateral epicondyle of humerus and posterior border of ulna.	Fifth metacarpal.	Extends and adducts hand at wrist joint (ulnar deviation).	Deep radial nerve.
DEEP POSTERIOR (EXTENSOR) COMPARTMENT OF THE FOREARM				
Abductor pollicis longus (ab-DUK-tor POL-li-sis LON-gus; *abductor* = moves part away from midline)	Posterior surface of middle of radius and ulna and interosseous membrane.	First metacarpal.	Abducts and extends thumb at carpometacarpal joint and abducts hand at wrist joint.	Deep radial nerve.
Extensor pollicis brevis (eks-TEN-sor POL-li-sis BREV-is)	Posterior surface of middle of radius and interosseous membrane.	Base of proximal phalanx of thumb.	Extends proximal phalanx of thumb at metacarpophalangeal joint, first metacarpal of thumb at carpometacarpal joint, and hand at wrist joint.	Deep radial nerve.
Extensor pollicis longus (eks-TEN-sor POL-li-sis LON-gus)	Posterior surface of middle of ulna and interosseous membrane.	Base of distal phalanx of thumb.	Extends distal phalanx of thumb at interphalangeal joint, extends first metacarpal of thumb at carpometacarpal joint, and abducts hand at wrist joint.	Deep radial nerve.
Extensor indicis (eks-TEN-sor IN-di-kis; *indicis* = index)	Posterior surface of ulna and interosseous membrane.	Tendon of extensor digitorum of index finger.	Extends distal and middle phalanges of index finger at interphalangeal joints, proximal phalanx of index finger at metacarpophalangeal joint, and hand at wrist joint.	Deep radial nerve.

*Reminder: The thumb or pollex is the first digit and has two phalanges: proximal and distal. The remaining digits, the fingers, are numbered II–V (2–5), and each has three phalanges: proximal, middle, and distal.

EXHIBIT 11.N CONTINUES

EXHIBIT 11.N   **411**

RELATING MUSCLES TO MOVEMENTS

Arrange the muscles in this exhibit according to the following actions on the wrist joint: (1) flexion, (2) extension, (3) abduction, and (4) adduction; the following actions on the fingers at the metacarpophalangeal joints: (1) flexion and (2) extension; the following actions on the fingers at the interphalangeal joints: (1) flexion and (2) extension; the following actions on the thumb at the carpometacarpal, metacarpophalangeal, and interphalangeal joints: (1) extension and (2) abduction; and the following action on the thumb at the interphalangeal joint: flexion. The same muscle may be mentioned more than once.

✔ CHECKPOINT

Which muscles and actions of the wrist, hand, thumb, and fingers are used when writing?

Figure 11.17 Muscles of the forearm that move the wrist, hand, thumb, and digits.

🔒 The anterior compartment muscles function as flexors, and the posterior compartment muscles function as extensors.

Biceps brachii
Brachialis
Medial epicondyle of humerus
Lateral epicondyle of humerus
Tendons of biceps brachii
Bicipital aponeurosis
PRONATOR TERES
BRACHIORADIALIS
PALMARIS LONGUS
FLEXOR CARPI RADIALIS
FLEXOR CARPI ULNARIS
EXTENSOR CARPI RADIALIS LONGUS
FLEXOR DIGITORUM SUPERFICIALIS
FLEXOR POLLICIS LONGUS
PRONATOR QUADRATUS
Radius
Tendon of abductor pollicis longus
Flexor retinaculum
Palmar aponeurosis
Metacarpal
Tendon of flexor pollicis longus
Tendons of flexor digitorum superficialis (cut)
Tendons of flexor digitorum profundus

Humerus
SUPINATOR
FLEXOR DIGITORUM PROFUNDUS
FLEXOR POLLICIS LONGUS
PRONATOR QUADRATUS
Tendon of flexor pollicis longus
Tendons of flexor digitorum profundus

(a) Anterior superficial view

(b) Anterior intermediate view

(c) Anterior deep view

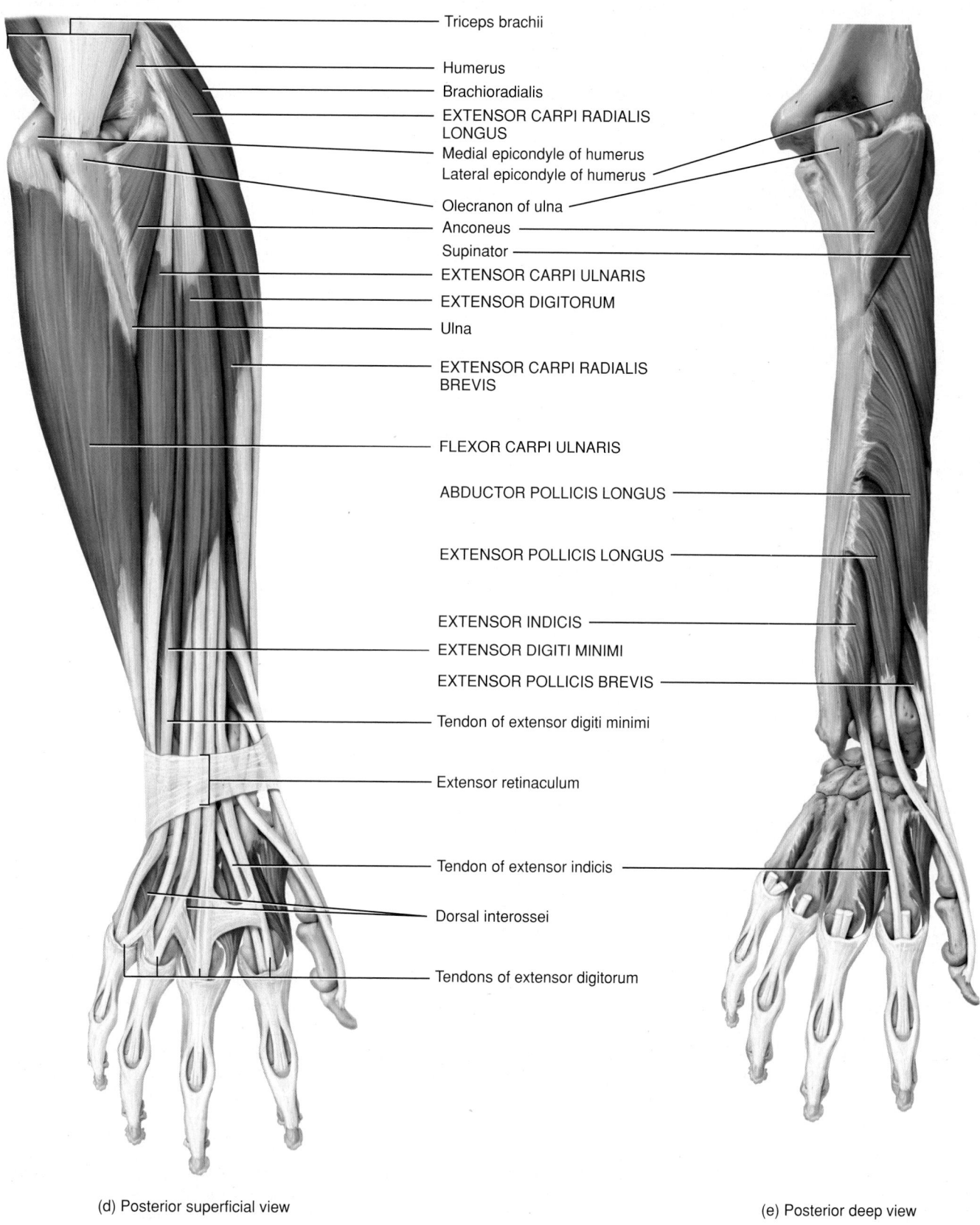

Triceps brachii
Humerus
Brachioradialis
EXTENSOR CARPI RADIALIS LONGUS
Medial epicondyle of humerus
Lateral epicondyle of humerus
Olecranon of ulna
Anconeus
Supinator
EXTENSOR CARPI ULNARIS
EXTENSOR DIGITORUM
Ulna
EXTENSOR CARPI RADIALIS BREVIS

FLEXOR CARPI ULNARIS

ABDUCTOR POLLICIS LONGUS

EXTENSOR POLLICIS LONGUS

EXTENSOR INDICIS
EXTENSOR DIGITI MINIMI
EXTENSOR POLLICIS BREVIS

Tendon of extensor digiti minimi

Extensor retinaculum

Tendon of extensor indicis

Dorsal interossei

Tendons of extensor digitorum

(d) Posterior superficial view

(e) Posterior deep view

? **What structures pass deep to the flexor retinaculum?**

EXHIBIT 11.N   **413**

EXHIBIT 11.O

Muscles of the Palm That Move the Digits—Intrinsic Muscles of the Hand (*Figure 11.18*)

OBJECTIVE

- Describe the origin, insertion, action, and innervation of the muscles of the palm that move the digits (the intrinsic muscles of the hand).

Several of the muscles discussed in Exhibit 11.N move the digits in various ways and are known as extrinsic muscles of the hand. They produce the powerful but crude movements of the digits. The **intrinsic muscles of the hand** in the palm produce the weak but intricate and precise movements of the digits that characterize the human hand (Figure 11.18). The muscles in this group are so named because their origins and insertions are *within* the hand.

The intrinsic muscles of the hand are divided into three groups: (1) **thenar,** (2) **hypothenar,** and (3) **intermediate.** The thenar muscles

MUSCLE	ORIGIN	INSERTION	ACTION	INNERVATION
THENAR (LATERAL ASPECT OF PALM)				
Abductor pollicis brevis (ab-DUK-tor POL-li-sis BREV-is; *abductor* = moves part away from middle; *pollicis* = thumb; *brevis* = short)	Flexor retinaculum, scaphoid, and trapezium.	Lateral side of proximal phalanx of thumb.	Abducts thumb at carpometacarpal joint.	Median nerve.
Opponens pollicis (op-PŌ-nenz POL-li-sis; *opponens* = opposes)	Flexor retinaculum and trapezium.	Lateral side of first metacarpal (thumb).	Moves thumb across palm to meet any finger (opposition) at carpometacarpal joint.	Median nerve.
Flexor pollicis brevis (FLEK-sor POL-li-sis BREV-is; *flexor* = decreases angle at joint)	Flexor retinaculum, trapezium, capitate, and trapezoid.	Lateral side of proximal phalanx of thumb.	Flexes thumb at carpometacarpal and metacarpophalangeal joints.	Median and ulnar nerves.
Adductor pollicis (ad-DUK-tor POL-li-sis; *adductor* = moves part toward midline)	Oblique head originates from capitate and second and third metacarpals. Transverse head originates from third metacarpal.	Medial side of proximal phalanx of thumb by tendon containing sesamoid bone.	Adducts thumb at carpometacarpal and metacarpophalangeal joints.	Ulnar nerve.
HYPOTHENAR (MEDIAL ASPECT OF PALM)				
Abductor digiti minimi (ab-DUK-tor DIJ-i-tē MIN-i-mē; *digit* = finger or toe; *minimi* = smallest)	Pisiform and tendon of flexor carpi ulnaris.	Medial side of proximal phalanx of little finger.	Abducts and flexes little finger at metacarpophalangeal joint.	Ulnar nerve.
Flexor digiti minimi brevis (FLEK-sor DIJ-i-tē MIN-i-mē BREV-is)	Flexor retinaculum and hamate.	Medial side of proximal phalanx of little finger.	Flexes little finger at carpometacarpal and metacarpophalangeal joints.	Ulnar nerve.
Opponens digiti minimi (op-PŌ-nenz DIJ-i-tē MIN-i-mē)	Flexor retinaculum and hamate.	Medial side of fifth metacarpal (little finger).	Moves little finger across palm to meet thumb (opposition) at carpometacarpal joint.	Ulnar nerve.
INTERMEDIATE (MIDPALMAR)				
Lumbricals (LUM-bri-kals; *lumbric* = earthworm) (four muscles)	Lateral sides of tendons and flexor digitorum profundus of each finger.	Lateral sides of tendons of extensor digitorum on proximal phalanges of each finger.	Flex each finger at metacarpophalangeal joints and extend each finger at interphalangeal joints.	Median and ulnar nerves.
Palmar interossei (PAL-mar in'-ter-OS-ē-ī; *palmar* = palm; *inter-* = between; *-ossei* = bones) (three muscles)	Sides of shafts of metacarpals of all digits (except middle one).	Sides of bases of proximal phalanges of all digits (except middle one).	Adduct and flex each finger (except middle finger) at metacarpophalangeal joints and extend these fingers at interphalangeal joints.	Ulnar nerve.
Dorsal interossei (DOR-sal in'-ter-OS-ē-ī; *dorsal* = back surface; *inter* = between; *ossei* = bones) (four muscles)	Adjacent sides of metacarpals.	Proximal phalanx of each finger.	Abduct fingers 2–4 at metacarpophalangeal joints, flex fingers 2–4 at metacarpophalangeal joints, and extend each finger at interphalangeal joints.	Ulnar nerve.

Figure 11.18 Muscles of the palm that move the digits—intrinsic muscles of the hand.

The intrinsic muscles of the hand produce the intricate and precise movements of the digits that characterize the human hand.

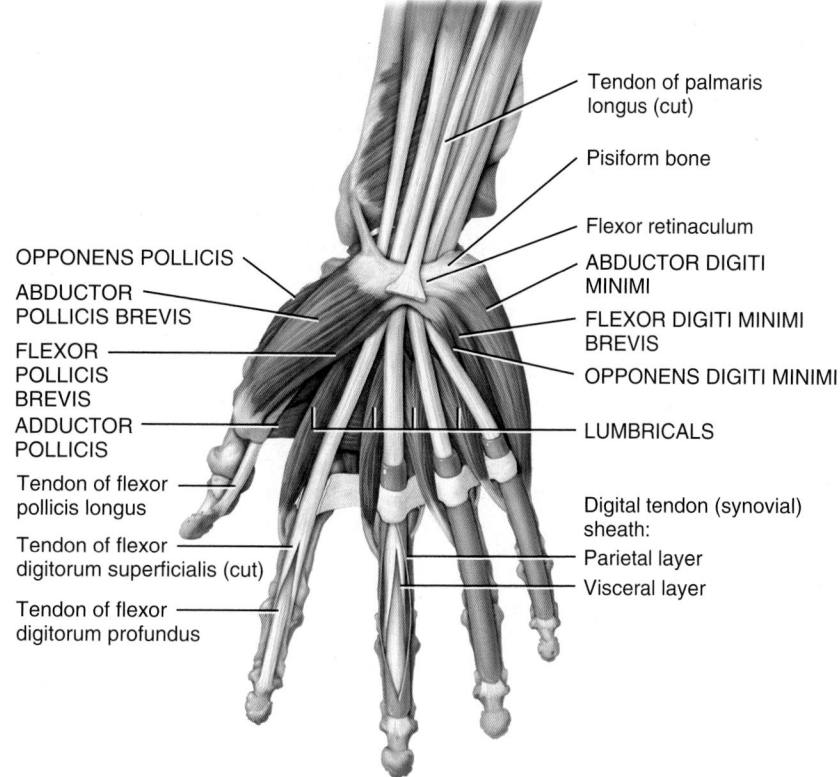

Tendon of palmaris longus (cut)

Pisiform bone

Flexor retinaculum

ABDUCTOR DIGITI MINIMI

FLEXOR DIGITI MINIMI BREVIS

OPPONENS DIGITI MINIMI

LUMBRICALS

Digital tendon (synovial) sheath:
Parietal layer
Visceral layer

OPPONENS POLLICIS

ABDUCTOR POLLICIS BREVIS

FLEXOR POLLICIS BREVIS

ADDUCTOR POLLICIS

Tendon of flexor pollicis longus

Tendon of flexor digitorum superficialis (cut)

Tendon of flexor digitorum profundus

(a) Anterior superficial view

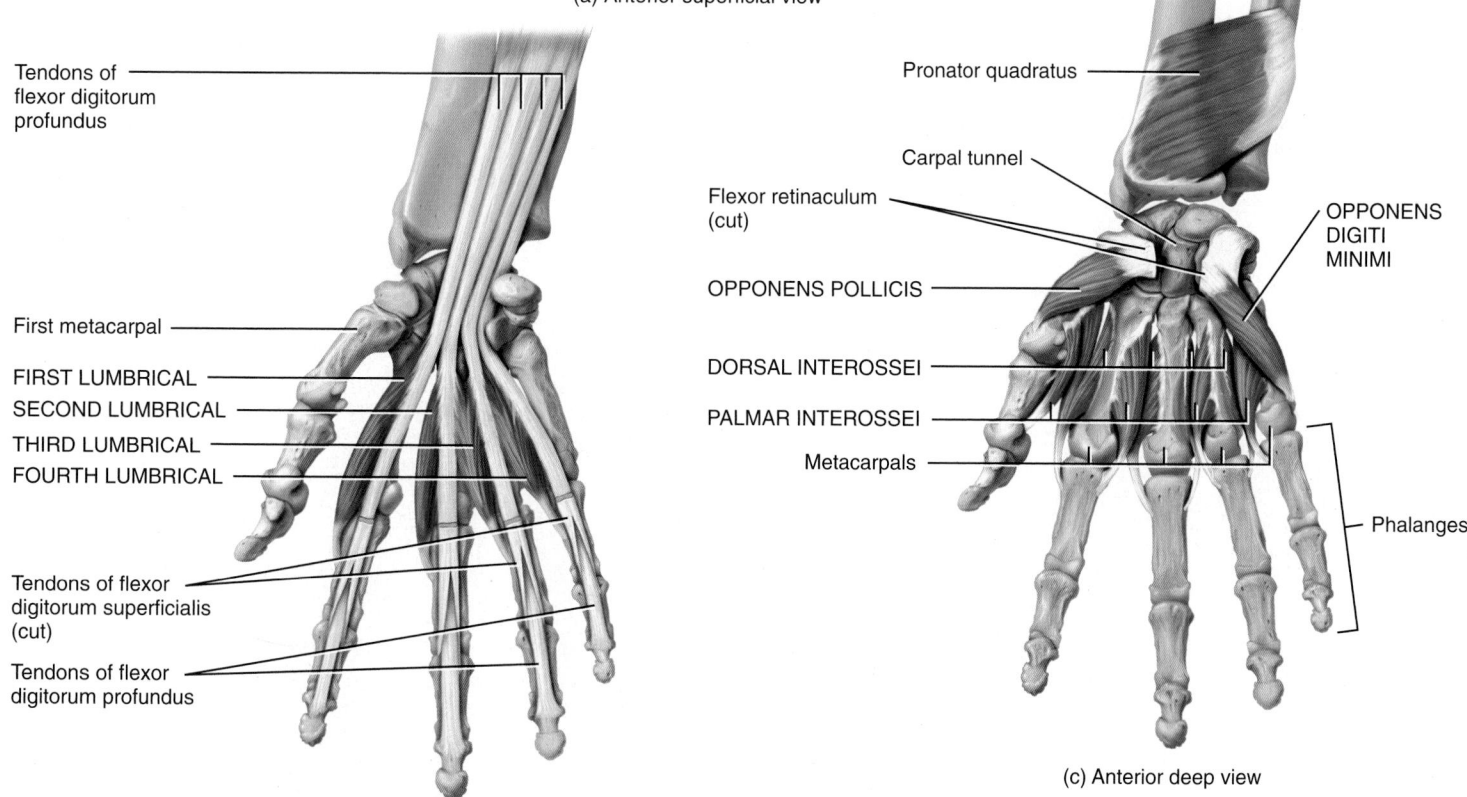

Tendons of flexor digitorum profundus

First metacarpal

FIRST LUMBRICAL

SECOND LUMBRICAL

THIRD LUMBRICAL

FOURTH LUMBRICAL

Tendons of flexor digitorum superficialis (cut)

Tendons of flexor digitorum profundus

(b) Anterior intermediate view of lumbricals

Pronator quadratus

Carpal tunnel

Flexor retinaculum (cut)

OPPONENS POLLICIS

DORSAL INTEROSSEI

PALMAR INTEROSSEI

Metacarpals

OPPONENS DIGITI MINIMI

Phalanges

(c) Anterior deep view

EXHIBIT 11.O CONTINUES ▶

EXHIBIT 11.O 415

include the abductor pollicis brevis, opponens pollicis, and flexor pollicis brevis. The **abductor pollicis brevis** is a thin, short, relatively broad superficial muscle on the lateral side of the thenar eminence. The **flexor pollicis brevis** is a short, wide muscle that is medial to the abductor pollicis brevis muscle. The **opponens pollicis** is a small, triangular muscle that is deep to the flexor pollicis brevis and abductor pollicis brevis muscles. The three thenar muscles plus the adductor pollicis form the **thenar eminence,** the lateral rounded contour on the palm that is also called the *ball of the thumb.* The **adductor pollicis** also acts on the thumb. The muscle is fan-shaped and has two heads (oblique and transverse) separated by a gap through which the radial artery passes.

The three hypothenar muscles act on the little finger and form the **hypothenar eminence,** the medial rounded contour on the palm that is also called the ball of the little finger. The hypothenar muscles are the abductor digiti minimi, flexor digiti minimi brevis, and opponens digiti minimi. The **abductor digiti minimi** is a short, wide muscle and is the most superficial of the hypothenar muscles. It is a powerful muscle that plays an important role in grasping an object with outspread fingers. The **flexor digiti minimi brevis** muscle is also short and wide and is lateral to the abductor digiti minimi muscle. The **opponens digiti minimi** muscle is triangular and deep to the other two hypothenar muscles.

The 11 intermediate (midpalmar) muscles include the lumbricals, palmar interossei, and dorsal interossei. The **lumbricals,** as their name indicates, are worm-shaped. They originate from and insert into the tendons of other muscles (flexor digitorum profundus and extensor digitorum). The **palmar interossei** are the smallest and more anterior of the

interossei muscles. The **dorsal interossei** are the most posterior of this series of muscles. Both sets of interossei muscles are located between the metacarpals and are important in abduction, adduction, flexion, and extension of the fingers, and in movements in skilled activities such as writing, typing, and playing a piano.

The functional importance of the hand is readily apparent when you consider that certain hand injuries can result in permanent disability. Most of the dexterity of the hand depends on movements of the thumb. The general activities of the hand are free motion, power grip (forcible movement of the fingers and thumb against the palm, as in squeezing), precision handling (a change in position of a handled object that requires exact control of finger and thumb positions, as in winding a watch or threading a needle), and pinch (compression between the thumb and index finger or between the thumb and first two fingers).

Movements of the thumb are very important in the precise activities of the hand, and they are defined in different planes from comparable movements of other digits because the thumb is positioned at a right angle to the other digits. The five principal movements of the thumb are illustrated in Figure 11.18g and include *flexion* (movement of the thumb medially across the palm), *extension* (movement of the thumb laterally away from the palm), *abduction* (movement of the thumb in an anteroposterior plane away from the palm), *adduction* (movement of the thumb in an anteroposterior plane toward the palm), and *opposition* (movement of the thumb across the palm so that the tip of the thumb meets the tip of a finger). Opposition is the single most distinctive digital movement that gives humans and other primates the ability to grasp and manipulate objects precisely.

⬤ FIGURE **11.18** CONTINUED ▶

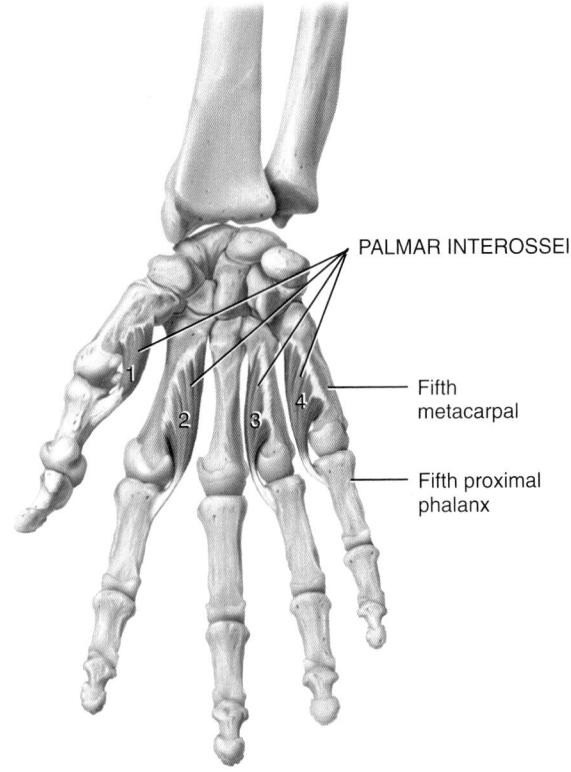

PALMAR INTEROSSEI

1

2   3   4

Fifth metacarpal

Fifth proximal phalanx

(d) Anterior deep view of palmar interossei

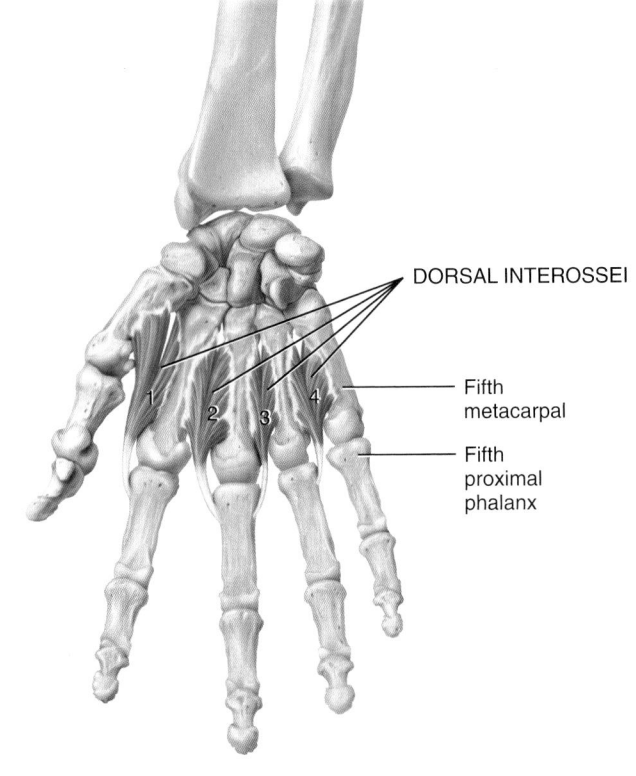

DORSAL INTEROSSEI

1   2   3   4

Fifth metacarpal

Fifth proximal phalanx

(e) Anterior deep view of dorsal interossei

RELATING MUSCLES TO MOVEMENTS

Arrange the muscles in this exhibit according to the following actions on the thumb at the carpometacarpal and metacarpophalangeal joints: (1) abduction, (2) adduction, (3) flexion, and (4) opposition; and the following actions on the fingers at the metacarpophalangeal and interphalangeal joints: (1) abduction, (2) adduction, (3) flexion, and (4) extension. The same muscle may be mentioned more than once.

✔ CHECKPOINT

How do the actions of the extrinsic and intrinsic muscles of the hand differ?

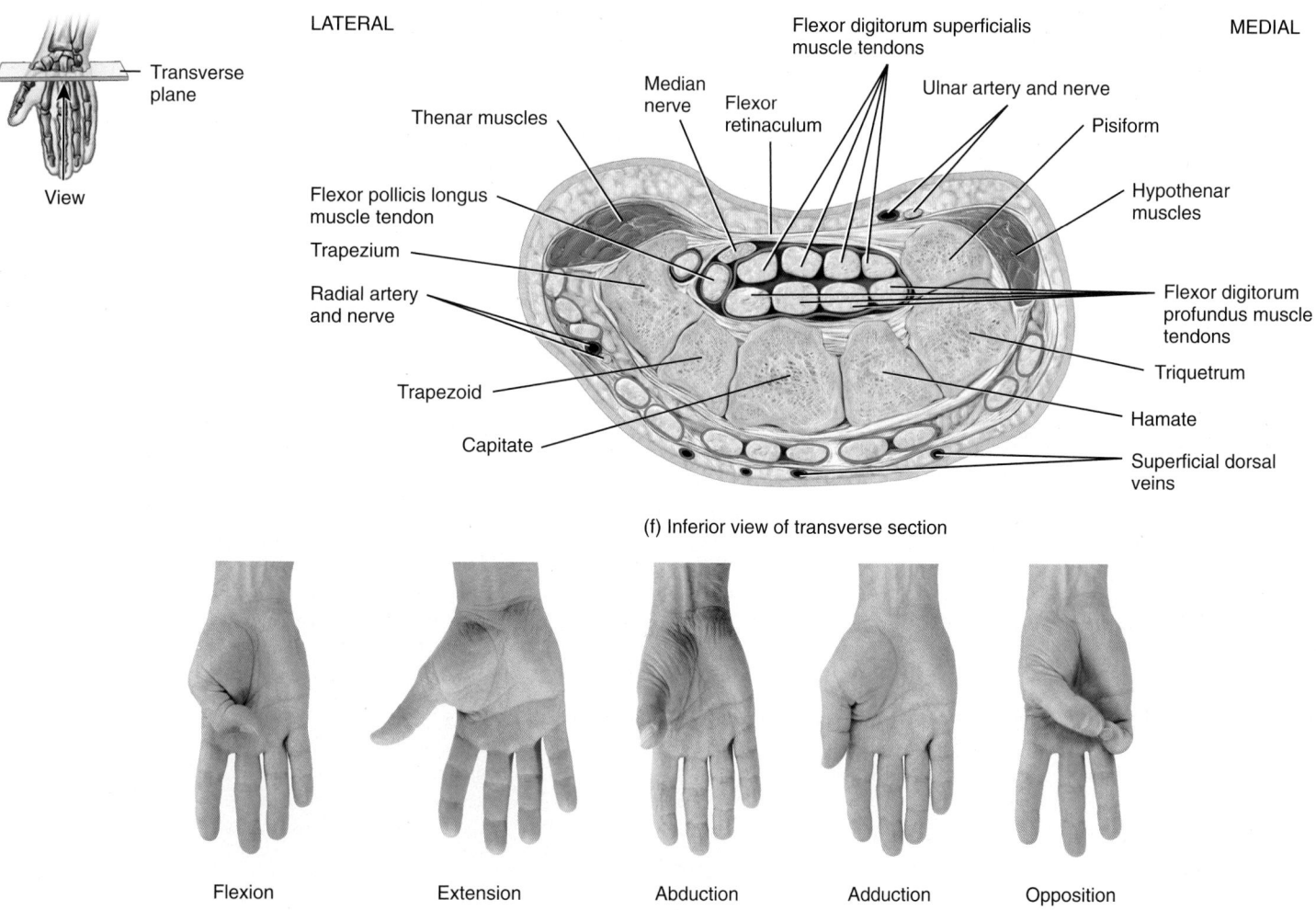

(f) Inferior view of transverse section

Flexion   Extension   Abduction   Adduction   Opposition

(g) Movements of the thumb

❓ **Muscles of the thenar eminence act on which digit?**

EXHIBIT 11.O **417**

OBJECTIVE

• Describe the origin, insertion, action, and innervation of the muscles that move the vertebral column.

The muscles that move the vertebral column (backbone) are quite complex because they have multiple origins and insertions and there is considerable overlap among them. One way to group the muscles is on the basis of the general direction of the muscle bundles and their approximate lengths. For example, the splenius muscles arise from the midline and extend laterally and superiorly to their insertions (Figure 11.19a). The erector spinae muscle group (consisting of the iliocostalis, longissimus, and spinalis muscles) arises either from the midline or more

MUSCLE	ORIGIN	INSERTION	ACTION	INNERVATION
SPLENIUS				
Splenius capitis (SPLĒ-nē-us KAP-i-tis; *splenium* = bandage; *capit* = head)	Ligamentum nuchae and spinous processes of C7–T4.	Occipital bone and mastoid process of temporal bone.	Acting together (bilaterally), extend head and extend vertebral column; acting singly (unilaterally), laterally flex and/or rotate head to same side as contracting muscle.	Middle cervical spinal nerves.
Splenius cervicis (SPLĒ-nē-us SER-vi-kis; *cervic* = neck)	Spinous processes of T3–T6.	Transverse processes of C1–C2 or C1–C4.	Acting together, extend head; acting singly, laterally flex and/or rotate head to same side as contracting muscle.	Inferior cervical spinal nerves.
ERECTOR SPINAE (e-REK-tor SPĪ-nē) Consists of iliocostalis muscles (lateral), longissimus muscles (intermediate), and spinalis muscles (medial).				
ILIOCOSTALIS GROUP (Lateral)				
Iliocostalis cervicis (il'-ē-ō-kos-TĀL-is SER-vi-kis; *ilio-* = flank; *-costa-* = rib)	Ribs 1–6.	Transverse processes of C4–C6.	Acting together, muscles of each region (cervical, thoracic, and lumbar) extend and maintain erect posture of vertebral column of their respective regions; acting singly, laterally flex vertebral column of their respective regions to same side as contracting muscle.	Cervical and thoracic spinal nerves.
Iliocostalis thoracis (il'-ē-ō-kos-TĀL-is thō-RĀ-sis; *thorac* = chest)	Ribs 7–12.	Ribs 1–6.		Thoracic spinal nerves.
Iliocostalis lumborum (il'-ē-ō-kos-TĀL-is lum-BOR-um)	Iliac crest.	Ribs 7–12.		Lumbar spinal nerves.
LONGISSIMUS GROUP (Intermediate)				
Longissimus capitis (lon-JIS-i-mus KAP-i-tis; *longissimus* = longest)	Articular processes of C4–C7 and transverse processes of T1–T4.	Mastoid process of temporal bone.	Acting together, both longissimus capitis muscles extend head and extend vertebral column; acting singly, rotate head to same side as contracting muscle.	Middle and inferior cervical spinal nerves.
Longissimus cervicis (lon-JIS-i-mus SER-vi-kis)	Transverse processes of T4–T5.	Transverse processes of C2–C6.	Acting together, longissimus cervicis and both longissimus thoracis muscles extend vertebral column of their respective regions; acting singly, laterally flex vertebral column of their respective regions.	Cervical and superior thoracic spinal nerves.
Longissimus thoracis (lon-JIS-i-mus thō-RĀ-sis)	Transverse processes of lumbar vertebrae.	Transverse processes of all thoracic and superior lumbar vertebrae and ribs 9 and 10.		Thoracic and lumbar spinal nerves.
SPINALIS GROUP (Medial)				
Spinalis capitis (spi-NĀ-lis KAP-i-tis; *spinal* = vertebral column)	Often absent or very small. Arises with semispinalis capitis.	Occipital bone.	Acting together, muscles of each region (cervical, thoracic, and lumbar) extend vertebral column of their respective regions and extend head.	Cervical spinal nerves.
Spinalis cervicis (spi-NĀ-lis SER-vi-kis)	Ligamentum nuchae and spinous process of C7.	Spinous process of axis.		Inferior cervical and thoracic spinal nerves.
Spinalis thoracis (spi-NĀ-lis thō-RĀ-sis)	Spinous processes of T10–L2.	Spinous processes of superior thoracic vertebrae.		Thoracic spinal nerves.

laterally but usually runs almost longitudinally, with neither a significant lateral nor medial direction as it is traced superiorly. The muscles of the transversospinalis group (semispinalis, multifidus, rotatores) arise laterally but extend toward the midline as they are traced superiorly. Deep to these three muscle groups are small segmental muscles that extend between spinous processes or transverse processes of vertebrae.

Note in Exhibit 11.G that the rectus abdominis, external oblique, internal oblique, and quadratus lumborum muscles also play a role in moving the vertebral column.

The bandagelike **splenius** muscles are attached to the sides and back of the neck. The two muscles in this group are named on the basis of their superior attachments (insertions): **splenius capitis** (head region)

MUSCLE	ORIGIN	INSERTION	ACTION	INNERVATION
TRANSVERSOSPINALES (trans-ver-sō-spi-NĀ-lēz)				
Semispinalis capitis (sem′-ē-spi-NĀ-lis KAP-i-tis; *semi* = partially or one half)	Articular processes of C4–C6 and transverse processes of C7–T7.	Occipital bone between superior and inferior nuchal lines.	Acting together, extend head and vertebral column; acting singly, rotate head to side opposite contracting muscle.	Cervical and thoracic spinal nerves.
Semispinalis cervicis (sem′-ē-spi-NĀ-lis SER-vi-kis)	Transverse processes of T1–T5.	Spinous processes of C1–C5.	Acting together, both semispinalis cervicis and both semispinalis thoracis muscles extend vertebral column of their respective regions; acting singly, rotate head to side opposite contracting muscle.	Cervical and thoracic spinal nerves.
Semispinalis thoracis (sem′-ē-spi-NĀ-lis thō-RĀ-sis)	Transverse processes of T6–T10.	Spinous processes of C6–T4.		Thoracic spinal nerves.
Multifidus (mul-TIF-i-dus; *multi-* = many; *-fid-* = segmented)	Sacrum; ilium; transverse processes of lumbar, thoracic, and C4–C7.	Spinous process of a more superior vertebra.	Acting together, extend vertebral column; acting singly, weakly laterally flex vertebral column and weakly rotate vertebral column to side opposite contracting muscle.	Cervical, thoracic, and lumbar spinal nerves.
Rotatores (rō′-ta-TŌ-rēz; singular is **rotatore**; *rotatore* = to rotate)	Transverse processes of all vertebrae.	Spinous process of vertebra superior to the one of origin.	Acting together, weakly extend vertebral column; acting singly, weakly rotate vertebral column to side opposite contracting muscle.	Cervical, thoracic, and lumbar spinal nerves.
SEGMENTAL (seg-MEN-tal)				
Interspinales (in-ter-spī-NĀ-lēz; *inter* = between)	Superior surface of all spinous processes.	Inferior surface of spinous process of vertebra superior to the one of origin.	Acting together, weakly extend, vertebral column; acting singly, stabilize vertebral column during movement.	Cervical, thoracic, and lumbar spinal nerves.
Intertransversarii (in′-ter-trans-vers-AR-ē-ī; singular is **intertransversarius**)	Transverse processes of all vertebrae.	Transverse process of vertebra superior to the one of origin.	Acting together, weakly extend vertebral column; acting singly, weakly laterally flex vertebral column and stabilize it during movements.	Cervical, thoracic, and lumbar spinal nerves.
SCALENES (SKĀ-lēnz)				
Anterior scalene (SKĀ-lēn; *anterior* = front; *scalene* = uneven)	Transverse processes of C3–C6.	First rib.	Acting together, right and left anterior scalene and middle scalene muscles elevate first ribs during deep inhalation. RMA: Flex cervical vertebrae; acting singly, laterally flex and slightly rotate cervical vertebrae.	Cervical spinal nerves.
Middle scalene	Transverse processes of C2–C7.	First rib.		Cervical spinal nerves.
Posterior scalene	Transverse processes of C4–C6.	Second rib.	Acting together, right and left posterior scalene elevate second ribs during deep inhalation. RMA: Flex cervical vertebrae; acting singly, laterally flex and slightly rotate cervical vertebrae.	Cervical spinal nerves.

EXHIBIT 11.P CONTINUES ▶

EXHIBIT 11.P **419**

and **splenius cervicis** (cervical region). They extend the head and laterally flex and rotate the head.

The **erector spinae** is the largest muscle mass of the back, forming a prominent bulge on either side of the vertebral column. It is the chief extensor of the vertebral column. It is also important in controlling flexion, lateral flexion, and rotation of the vertebral column and in maintaining the lumbar curve. As noted above, it consists of three groups: iliocostalis (laterally placed), longissimus (intermedially placed), and spinalis (medially placed). These groups, in turn, consist of a series of overlapping muscles, and the muscles within the groups are named according to the regions of the body with which they are associated. The **iliocostalis group** consists of three muscles: the **iliocostalis cervicis** (cervical region), **iliocostalis thoracis** (thoracic region), and **iliocostalis lumborum** (lumbar region). The **longissimus group** resembles a herringbone and consists of three muscles: the **longissimus capitis** (head region), **longissimus cervicis** (cervical region), and **longissimus thoracis** (thoracic region). The **spinalis group** also consists of three muscles: the **spinalis capitis**, **spinalis cervicis**, and **spinalis thoracis.**

The **transversospinales** are so named because their fibers run from the transverse processes to the spinous processes of the vertebrae. The semispinalis muscles in this group are also named according to the region of the body with which they are associated: **semispinalis capitis** (head region), **semispinalis cervicis** (cervical region), and **semispinalis thoracis** (thoracic region). These muscles extend the vertebral column and rotate the head. The **multifidus** muscle in this group, as its name implies, is segmented into several bundles. It extends and laterally flexes the vertebral column. This muscle is large and thick in the lumbar region and is important in maintaining the lumbar curve. The **rotatores** muscles of this group are short and are found along the entire length of the vertebral column. These small muscles contribute little to vertebral movement but play important roles in monitoring the position of the vertebral column and providing proprioceptive feedback to the stronger vertebral muscles.

Within the segmental muscle group (Figure 11.19b), the **interspinales** and **intertransversarii** muscles unite the spinous and transverse processes of consecutive vertebrae. They function primarily in stabilizing the vertebral column during its movements, and providing proprioceptive feedback.

Within the **scalene group** (Figure 11.19c), the **anterior scalene** muscle is anterior to the middle scalene muscle, the **middle scalene** muscle is intermediate in placement and is the longest and largest of the scalene muscles, and the **posterior scalene** muscle is posterior to the middle sca-

lene muscle and is the smallest of the scalene muscles. These muscles flex, laterally flex, and rotate the head and assist in deep inhalation.

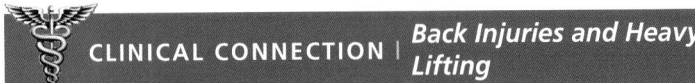

CLINICAL CONNECTION | *Back Injuries and Heavy Lifting*

The four factors associated with increased risk of **back injury** are amount of force, repetition, posture, and stress applied to the backbone. Poor physical condition, poor posture, lack of exercise, and excessive body weight contribute to the number and severity of sprains and strains. Back pain caused by a muscle strain or ligament sprain will normally heal within a short time and may never cause further problems. However, if ligaments and muscles are weak, discs in the lower back can become weakened and may herniate (rupture) with excessive lifting or a sudden fall, causing considerable pain.

Full flexion at the waist, as in touching your toes, overstretches the erector spinae muscles. Muscles that are overstretched cannot contract effectively. Straightening up from such a position is therefore initiated by the hamstring muscles on the back of the thigh and the gluteus maximus muscles of the buttocks. The erector spinae muscles join in as the degree of flexion decreases. Improperly lifting a heavy weight, however, can strain the erector spinae muscles. The result can be painful muscle spasms, tearing of tendons and ligaments of the lower back, and herniating of intervertebral discs. The lumbar muscles are adapted for maintaining posture, not for lifting. This is why it is important to bend at the knees and use the powerful extensor muscles of the thighs and buttocks while lifting a heavy load. •

RELATING MUSCLES TO MOVEMENTS

Arrange the muscles in this exhibit according to the following actions on the head at the atlanto-occipital and intervertebral joints: (1) extension, (2) lateral flexion, (3) rotation to same side as contracting muscle, and (4) rotation to opposite side as contracting muscle; and arrange the muscles according to the following actions on the vertebral column at the intervertebral joints: (1) flexion, (2) extension, (3) lateral flexion, (4) rotation, and (5) stabilization. The same muscle may be mentioned more than once.

✔ CHECKPOINT
What is the largest muscle group of the back?

Figure 11.19 **Muscles of the neck and back that move the vertebral column (backbone).** The trapezius and occipitofrontalis muscles have been removed.

The erector spinae group (iliocostalis, longissimus, and spinalis muscles) is the largest muscular mass of the body and is the chief extensor of the vertebral column.

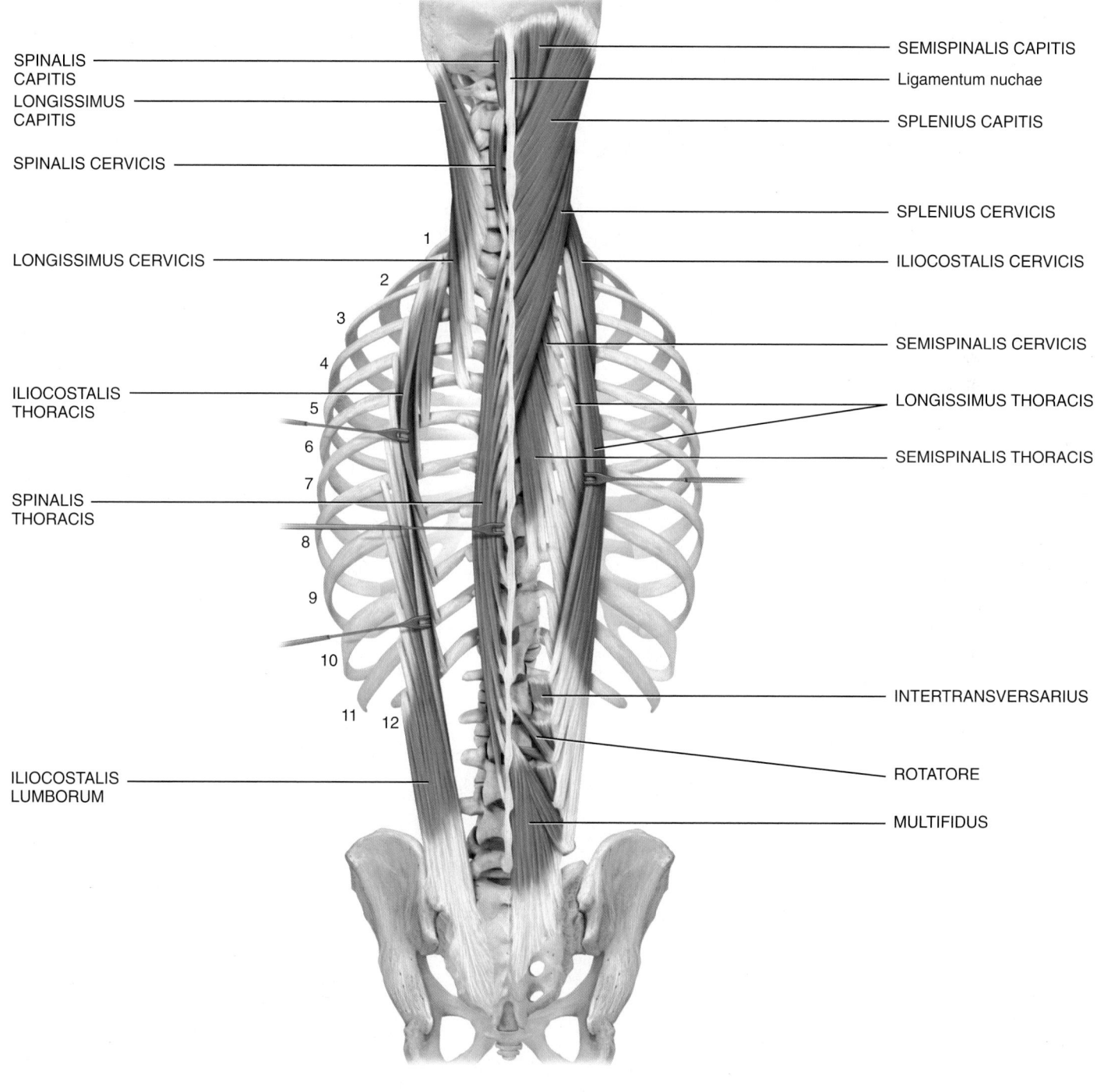

SPINALIS CAPITIS
LONGISSIMUS CAPITIS
SPINALIS CERVICIS
LONGISSIMUS CERVICIS
ILIOCOSTALIS THORACIS
SPINALIS THORACIS
ILIOCOSTALIS LUMBORUM

SEMISPINALIS CAPITIS
Ligamentum nuchae
SPLENIUS CAPITIS
SPLENIUS CERVICIS
ILIOCOSTALIS CERVICIS
SEMISPINALIS CERVICIS
LONGISSIMUS THORACIS
SEMISPINALIS THORACIS
INTERTRANSVERSARIUS
ROTATORE
MULTIFIDUS

(a) Posterior view

Exhibit 11.P CONTINUES ▶

EXHIBIT 11.P **421**

○ FIGURE 11.19 CONTINUED ▶

Transverse process of second lumbar vertebra

INTERTRANSVERSARII

ROTATORE

INTERSPINALES

Spinous process of fourth lumbar vertebra

(b) Posterolateral view

Atlas

Axis

C3
C4
C5
C6
C7
T1
T2

MIDDLE SCALENE (deep to anterior scalene)

ANTERIOR SCALENE (superficial to middle and posterior scalenes)

POSTERIOR SCALENE

First rib

Second rib

(c) Anterior view

❓ **Which muscles originate at the midline and extend laterally and upward to their insertion?**

● OBJECTIVE

- Describe the origin, insertion, action, and innervation of the muscles of the gluteal region that move the femur.

As you will see, muscles of the lower limbs are larger and more powerful than those of the upper limbs because of differences in function. While upper limb muscles are characterized by versatility of movement, lower limb muscles function in stability, locomotion, and maintenance of posture. In addition, muscles of the lower limbs often cross two joints and act equally on both.

The majority of muscles that move the femur (thigh bone) originate on the pelvic girdle and insert on the femur (Figure 11.20). The **psoas major** and **iliacus** muscles share a common insertion (lesser trochanter of femur) and are collectively known as the **iliopsoas** muscle. There are three gluteal muscles: gluteus maximus, gluteus medius, and gluteus minimus. The **gluteus maximus** is the largest and heaviest of the three muscles and is one of the largest muscles in the body. It is the chief extensor of the femur. In its reverse muscle action (RMA), it is a powerful extensor of the torso at the hip joint. The **gluteus medius** is mostly deep to the gluteus maximus and is a powerful abductor of the femur at the hip joint. It is a common site for intramuscular injection. The **gluteus minimus** is the smallest of the gluteal muscles and lies deep to the gluteus medius.

The **tensor fasciae latae** muscle is located on the lateral surface of the thigh. The *fascia lata* is a layer of deep fascia, composed of dense connective tissue, that encircles the entire thigh. It is well developed laterally where, together with the tendons of the tensor fasciae latae and gluteus maximus muscles, it forms a structure called the **iliotibial tract.** The tract inserts into the lateral condyle of the tibia.

The **piriformis, obturator internus, obturator externus, superior gemellus, inferior gemellus,** and **quadratus femoris** muscles are all deep to the gluteus maximus muscle and function as lateral rotators of the femur at the hip joint.

Three muscles on the medial aspect of the thigh are the **adductor longus, adductor brevis,** and **adductor magnus.** They originate on the pubic bone and insert on the femur. These three muscles adduct the thigh and are unique in their ability to both medially and laterally rotate the thigh. When the foot is on the ground, these muscles medially rotate the thigh, but when the foot is off the ground, they are lateral rotators of the thigh. This results from their oblique orientation from their anterior origin to their posterior insertion. In addition, the adductor longus flexes the thigh and the adductor magnus extends the thigh. The pectineus muscle also adducts and flexes the femur at the hip joint.

Technically, the adductor muscles and pectineus muscles are components of the medial compartment of the thigh and could be included in Exhibit 11.R. However, they are included here because they act on the femur.

At the junction between the trunk and lower limb is a space called the **femoral triangle.** The base is formed superiorly by the inguinal ligament, medially by the lateral border of the adductor longus muscle, and laterally by the medial border of the sartorius muscle. The apex is formed by the crossing of the adductor longus by the sartorius muscle (Figure 11.20a). The contents of the femoral triangle, from lateral to medial, are the femoral nerve and its branches, the femoral artery and several of its branches, the femoral vein and its proximal tributaries, and the deep inguinal lymph nodes. The femoral artery is easily accessible within the triangle and is the site for insertion of catheters that may extend into the aorta and ultimately into the coronary vessels of the heart. Such catheters are utilized during cardiac catheterization, coronary angiography, and other procedures involving the heart. Inguinal hernias frequently appear in this area.

☤ CLINICAL CONNECTION | *Groin Pull*

The five major muscles of the inner thigh function to move the legs medially. This muscle group is important in activities such as sprinting, hurdling, and horseback riding. A rupture or tear of one or more of these muscles can cause a **groin pull.** Groin pulls most often occur during sprinting or twisting, or from kicking a solid, perhaps stationary object. Symptoms of a groin pull may be sudden or may not surface until the day after the injury; they include sharp pain in the inguinal region, swelling, bruising, or inability to contract the muscles. As with most strain injuries, treatment involves PRICE therapy, which stands for *protection, rest, ice, compression,* and *elevation.* After the injured part is protected from further damage, ice should be applied immediately, and the injured part should be elevated and rested. An elastic bandage should be applied, if possible, to compress the injured tissue. •

RELATING MUSCLES TO MOVEMENTS

Arrange the muscles in this exhibit according to the following actions on the thigh at the hip joint: (1) flexion, (2) extension, (3) abduction, (4) adduction, (5) medial rotation, and (6) lateral rotation. The same muscle may be mentioned more than once.

✔ CHECKPOINT

What is the origin of most muscles that move the femur?

EXHIBIT 11.Q CONTINUES

EXHIBIT 11.Q **423**

MUSCLE	ORIGIN	INSERTION	ACTION	INNERVATION
Iliopsoas (il-ē-ō-SŌ-as)				
Psoas major (SŌ-as MĀ-jor; *psoa* = a muscle of the loin; *major* = larger)	Transverse processes and bodies of lumbar vertebrae.	With iliacus into lesser trochanter of femur.	Psoas major and iliacus muscles acting together flex thigh at hip joint, rotate thigh laterally, and flex trunk on hip as in sitting up from supine position.	Lumbar spinal nerves L2–L3.
Iliacus (il'-ē-A-cus; *iliac* = ilium)	Iliac fossa and sacrum.	With psoas major into lesser trochanter of femur.		Femoral nerve.
Gluteus maximus (GLOO-tē-us MAK-si-mus; *glute* = rump or buttock; *maximus* = largest)	Iliac crest, sacrum, coccyx, and aponeurosis of sacrospinalis.	Iliotibial tract of fascia lata and superior lateral part of linea aspera (gluteal tuberosity) under greater trochanter of femur.	Extends thigh at hip joint and laterally rotates thigh; helps lock knee in extension. RMA: Extends torso.	Inferior gluteal nerve.
Gluteus medius (GLOO-tē-us MĒ-dē-us; *medi* = middle)	Ilium.	Greater trochanter of femur.	Abducts thigh at hip joint and medially rotates thigh.	Superior gluteal nerve.
Gluteus minimus (GLOO-tē-us MIN-i-mus; *minimus* = smallest)	Ilium.	Greater trochanter of femur.	Abducts thigh at hip joint and medially rotates thigh.	Superior gluteal nerve.
Tensor fasciae latae (TEN-sor FA-shē-ē LĀ-tē; *tensor* = makes tense; *fasciae* = band; *lat* = wide)	Iliac crest.	Tibia by way of iliotibial tract.	Flexes and abducts thigh at hip joint.	Superior gluteal nerve.
Piriformis (pir-i-FOR-mis; *piri-* = pear; *-form-* = shape)	Anterior sacrum.	Superior border of greater trochanter of femur.	Laterally rotates and abducts thigh at hip joint.	Sacral spinal nerves S1 or S2, mainly S1.
Obturator internus (OB-too-rā'-tor in-TER-nus; *obturator* = obturator foramen; *intern* = inside)	Inner surface of obturator foramen, pubis, and ischium.	Medial surface of greater trochanter of femur.	Laterally rotates and abducts thigh at hip joint.	Nerve to obturator internus.
Obturator externus (OB-too-rā'-tor ex-TER-nus; *extern-* = outside)	Outer surface of obturator membrane.	Deep depression inferior to greater trochanter (trochanteric fossa) of femur.	Laterally rotates and abducts thigh at hip joint.	Obturator nerve.
Superior gemellus (jem-EL-lus; *superior* = above; *gemell* = twins)	Ischial spine.	Medial surface of greater trochanter of femur.	Laterally rotates and abducts thigh at hip joint.	Nerve to obturator internus.
Inferior gemellus (jem-EL-lus; *inferior* = below)	Ischial tuberosity.	Medial surface of greater trochanter of femur.	Laterally rotates and abducts thigh at hip joint.	Nerve to quadratus femoris.
Quadratus femoris (kwod-RĀ-tus FEM-or-is; *quad* = square, four-sided; *femoris* = femur)	Ischial tuberosity.	Elevation superior to mid-portion of intertrochanteric crest (quadrate tubercle) on posterior femur.	Laterally rotates and stabilizes hip joint.	Nerve to quadratus femoris.
Adductor longus (ad-DUK-tor LONG-us; *adductor* = moves part closer to midline; *longus* = long)	Pubic crest and pubic symphysis.	Linea aspera of femur.	Adducts and flexes thigh at hip joint and rotates thigh.* RMA: Extends thigh.	Obturator nerve.
Adductor brevis (BREV-is; *brevis* = short)	Inferior ramus of pubis.	Superior half of linea aspera of femur.	Adducts and flexes thigh at hip joint and rotates thigh.* RMA: Extends thigh.	Obturator nerve.
Adductor magnus (MAG-nus; *magnus* = large)	Inferior ramus of pubis and ischium to ischial tuberosity.	Linea aspera of femur.	Adducts thigh at hip joint and rotates thigh; anterior part flexes thigh at hip joint, and posterior part extends thigh at hip joint.*	Obturator and sciatic nerves.
Pectineus (pek-TIN-ē-us; *pectin* = comb)	Superior ramus of pubis.	Pectineal line of femur, between lesser trochanter and linea aspera.	Flexes and adducts thigh at hip joint.	Femoral nerve.

*All adductors are unique muscles that cross the thigh joint obliquely from an anterior origin to a posterior insertion. As a result they laterally rotate the hip joint when the foot is off the ground but medially rotate the hip joint when the foot is on the ground.

Figure 11.20 Muscles of the gluteal region that move the femur (thigh bone).

Most muscles that move the femur originate on the pelvic (hip) girdle and insert on the femur.

Twelfth rib

Quadratus lumborum

Iliac crest

ILIACUS

Anterior superior iliac spine

Femoral triangle

TENSOR FASCIAE LATAE

SARTORIUS

QUADRICEPS FEMORIS:

RECTUS FEMORIS (cut)

VASTUS LATERALIS

VASTUS INTERMEDIUS

VASTUS MEDIALIS

RECTUS FEMORIS (cut)

Iliotibial tract

Section of fascia lata (cut)

Tendon of quadriceps femoris

Patellar ligament

Psoas minor

PSOAS MAJOR

Sacrum

Inguinal ligament

Pubic tubercle

PECTINEUS

ADDUCTOR LONGUS

GRACILIS

ADDUCTOR MAGNUS

Patella

(a) Anterior superficial view (the femoral triangle is indicated by a dashed line)

EXHIBIT 11.Q CONTINUES ▶

EXHIBIT 11.Q **425**

● FIGURE 11.20 CONTINUED ▶

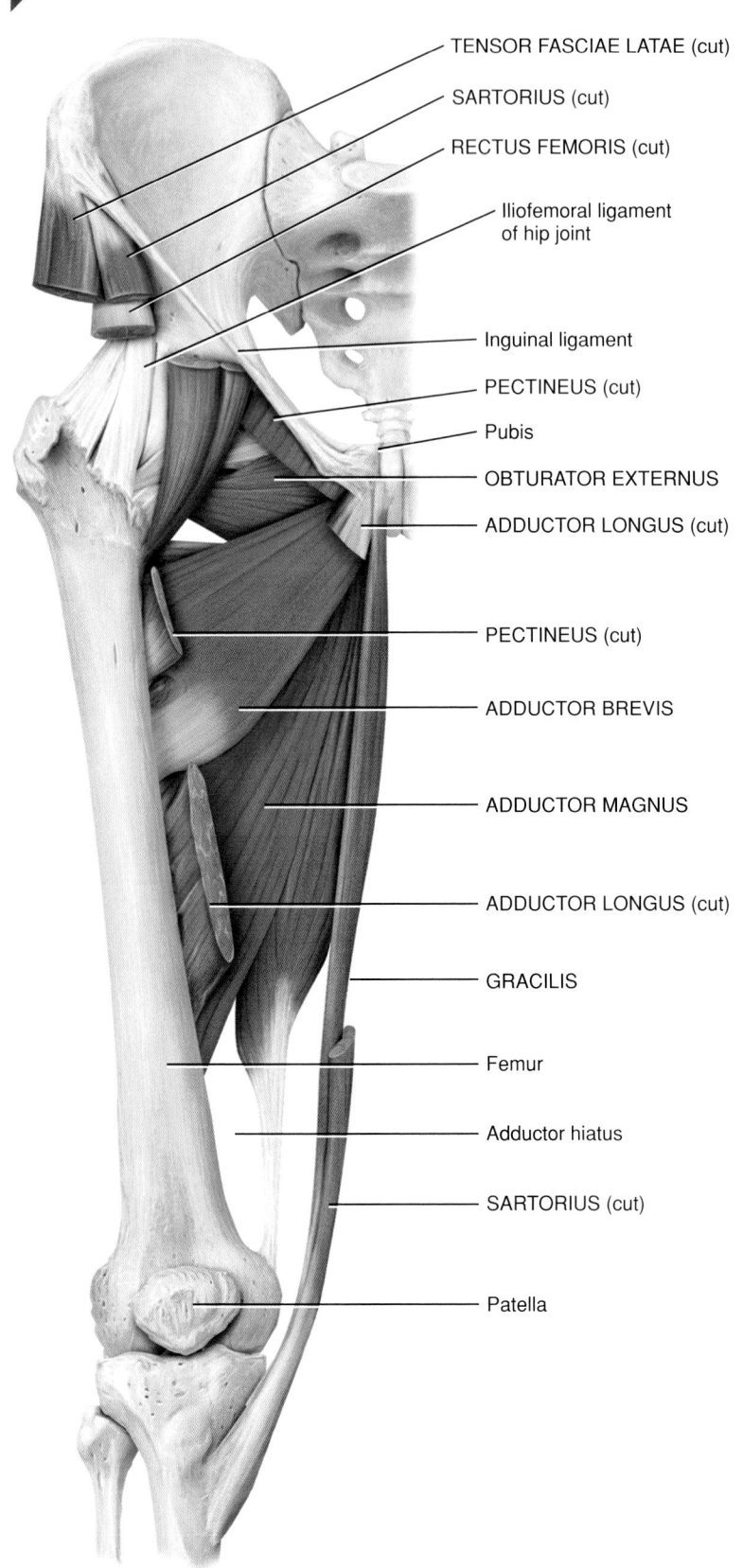

TENSOR FASCIAE LATAE (cut)

SARTORIUS (cut)

RECTUS FEMORIS (cut)

Iliofemoral ligament of hip joint

Inguinal ligament

PECTINEUS (cut)

Pubis

OBTURATOR EXTERNUS

ADDUCTOR LONGUS (cut)

PECTINEUS (cut)

ADDUCTOR BREVIS

ADDUCTOR MAGNUS

ADDUCTOR LONGUS (cut)

GRACILIS

Femur

Adductor hiatus

SARTORIUS (cut)

Patella

(b) Anterior deep view (femur rotated laterally)

GLUTEUS MEDIUS

GLUTEUS MAXIMUS

TENSOR FASCIAE LATAE

(c) Posterior superficial view

Iliac crest

Sacrum

Gluteus maximus (cut)

Obturator internus

Coccyx

Sacrotuberous ligament

Ischial tuberosity

Sciatic nerve

GRACILIS

SARTORIUS

Sartorius

Gracilis

Semitendinosus

Gluteus medius (cut)

Gluteus minimus

Piriformis

Superior gemellus

Greater trochanter

Inferior gemellus

Quadratus femoris

Gluteus maximus (cut)

Femur

ADDUCTOR MAGNUS

ADDUCTOR MINIMIS

HAMSTRINGS:
    SEMITENDINOSUS

BICEPS FEMORIS

SEMIMEMBRANOSUS

Iliotibial tract

VASTUS LATERALIS

Femur in floor of popliteal fossa

Tendon of biceps femoris

Plantaris

Gastrocnemius

(d) Posterior superficial view of thigh and deep view of gluteal region

EXHIBIT 11.Q CONTINUES ▶

EXHIBIT 11.Q **427**

FIGURE 11.20 CONTINUED ▶

Anterior deep view

Anterior views

Posterior deep views

(e) Isolated muscles

? What are the principal differences between the muscles of the free upper and lower limbs?

● OBJECTIVE

- Describe the origin, insertion, action, and innervation of the muscles that move the femur, tibia, and fibula.

Deep fascia separates the muscles of the thigh that act on the femur (thigh bone) and tibia and fibula (leg bones) into medial, anterior, and posterior compartments (Figure 11.21). Most of the muscles of the **medial (adductor) compartment of the thigh** have a similar orientation and adduct the femur at the hip joint. (See the adductor magnus, adductor longus, adductor brevis, and pectineus, which are components of the medial compartment, in Exhibit 11.Q.) The **gracilis,** the other muscle in the medial compartment, is a long, straplike muscle on the medial aspect of the thigh and knee. This muscle not only adducts the thigh, but also medially rotates the thigh and flexes the leg at the knee joint. For this reason, it is discussed here.

The muscles of the **anterior (extensor) compartment of the thigh** extend the leg (and flex the thigh). This compartment contains the quadriceps femoris and sartorius muscles. The **quadriceps femoris** muscle is the largest muscle in the body, covering most of the anterior surface and sides of the thigh. The muscle is actually a composite muscle, usually described as four separate muscles: (1) **rectus femoris,** on the anterior aspect of the thigh; (2) **vastus lateralis,** on the lateral aspect of the thigh; (3) **vastus medialis,** on the medial aspect of the thigh; and (4) **vastus intermedius,** located deep to the rectus femoris between the vastus lateralis and vastus medialis. The common tendon for the four muscles is known as the **quadriceps tendon,** which inserts into the patella. The tendon continues below the patella as the **patellar ligament,** which attaches to the tibial tuberosity. The quadriceps femoris muscle is the great extensor muscle of the leg. The **sartorius** is a long, narrow muscle that forms a band across the thigh from the ilium of the hip bone to the medial side of the tibia. The various movements it produces (flexion of the leg at the knee joint and flexion, abduction, and lateral rotation at the hip joint) help effect the cross-legged sitting position in which the heel of one limb is placed on the knee of the opposite limb. Its name means *tailor's muscle;* it was so called because tailors often assume a cross-legged sitting position. (Because the major action of the sartorius muscle is to move the thigh rather than the leg, it could have been included in Exhibit 11.Q.)

The muscles of the **posterior (flexor) compartment of the thigh** flex the leg (and extend the thigh). This compartment is composed of three muscles collectively called the **hamstrings:** (1) biceps femoris, (2) semitendinosus, and (3) semimembranosus. The hamstrings are so named because their tendons are long and stringlike in the popliteal area. Because the hamstrings span two joints (hip and knee), they are both extensors of the thigh and flexors of the leg. The **popliteal fossa** is a diamond-shaped space on the posterior aspect of the knee bordered laterally by the tendons of the biceps femoris muscle and medially by the tendons of the semitendinosus and semimembranosus muscles.

CLINICAL CONNECTION | *Pulled Hamstrings and Charley Horse*

A strain or partial tear of the proximal hamstring muscles is referred to as **pulled hamstrings** or **hamstring strains.** Like pulled groins (see Exhibit 11.Q), they are common sports injuries in individuals who run very hard and/or are required to perform quick starts and stops. Sometimes the violent muscular exertion required to perform a feat tears away a part of the tendinous origins of the hamstrings, especially the biceps femoris, from the ischial tuberosity. This is usually accompanied by a contusion (bruising), tearing of some of the muscle fibers, and rupture of blood vessels, producing a hematoma (collection of blood) and sharp pain. Adequate training with good balance between the quadriceps femoris and hamstrings and stretching exercises before running or competing are important in preventing this injury.

The term **charley horse** (CHAR-lē HŌRS), a slang term, is a popular name for a cramp or stiffness of muscles due to tearing of the muscle, followed by bleeding into the area. It is a common sports injury due to trauma or excessive activity and frequently occurs in the quadriceps femoris muscle, especially among football players. •

RELATING MUSCLES TO MOVEMENTS

Arrange the muscles in this exhibit according to the following actions on the thigh at the hip joint: (1) abduction, (2) adduction, (3) lateral rotation, (4) flexion, and (5) extension; and according to the following actions on the leg at the knee joint: (1) flexion and (2) extension. The same muscle may be mentioned more than once.

✔ CHECKPOINT

Which muscles are part of the medial, anterior, and posterior compartments of the thigh?

EXHIBIT 11.R CONTINUES ▶

EXHIBIT 11.R **429**

MUSCLE	ORIGIN	INSERTION	ACTION	INNERVATION
MEDIAL (ADDUCTOR) COMPARTMENT OF THE THIGH				
Adductor magnus (ad-DUK-tor MAG-nus)				
Adductor longus (LONG-us)	See Exhibit 11.Q			
Adductor brevis (BREV-is)				
Pectineus (pek-TIN-ē-us)				
Gracilis (GRAS-i-lis = slender) (see also Figure 11.20a)	Body and inferior ramus of pubis.	Medial surface of body of tibia.	Adducts thigh at hip joint, medially rotates thigh, and flexes leg at knee joint.	Obturator nerve.
ANTERIOR (EXTENSOR) COMPARTMENT OF THE THIGH (see also Figure 11.20a)				
Quadriceps femoris (KWOD-ri-seps FEM-or-is; *quadriceps* = four heads [of origin]; *femoris* = femur)				
Rectus femoris (REK-tus FEM-or-is; *rectus* = fascicles parallel to midline)	Anterior inferior iliac spine.	Patella via quadriceps tendon and then tibial tuberosity via patellar ligament.	All four heads extend leg at knee joint; rectus femoris muscle acting alone also flexes thigh at hip joint.	Femoral nerve.
Vastus lateralis (VAS-tus lat'-e-RA-lis; *vast* = huge; *lateralis* = lateral)	Greater trochanter and linea aspera of femur.			
Vastus medialis (VAS-tus mē-dē-Ā-lis; *medialis* = medial)	Linea aspera of femur.			
Vastus intermedius (VAS-tus in'-ter-MĒ-dē-us; *intermedius* = middle)	Anterior and lateral surfaces of body of femur.			
Sartorius (sar-TOR-ē-us; *sartor* = tailor; longest muscle in body)	Anterior superior iliac spine.	Medial surface of body of tibia.	Weakly flexes leg at knee joint; weakly flexes, abducts, and laterally rotates thigh at hip joint.	Femoral nerve.
POSTERIOR (FLEXOR) COMPARTMENT OF THE THIGH (see also Figure 11.20d)				
Hamstrings A collective designation for three separate muscles.				
Biceps femoris (BĪ-seps FEM-or-is; *biceps* = two heads of origin)	Long head arises from ischial tuberosity; short head arises from linea aspera of femur.	Head of fibula and lateral condyle of tibia.	Flexes leg at knee joint and extends thigh at hip joint.	Tibial and fibular nerves from sciatic nerve.
Semitendinosus (sem'-ē-ten-di-NŌ-sus; *semi-* = half; *-tendo-* = tendon)	Ischial tuberosity.	Proximal part of medial surface of shaft of tibia.	Flexes leg at knee joint and extends thigh at hip joint.	Tibial nerve from sciatic nerve.
Semimembranosus (sem'-ē-mem-bra-NŌ-sus; *membran-* = membrane)	Ischial tuberosity.	Medial condyle of tibia.	Flexes leg at knee joint and extends thigh at hip joint.	Tibial nerve from sciatic nerve.

Figure 11.21 Muscles of the thigh that move the femur (thigh bone) and tibia and fibula (leg bones).

Muscles that act on the leg originate in the hip and thigh and are separated into compartments by deep fascia.

(a) Superior view of transverse section of thigh

(b) Superior view of transverse section of thigh

Which muscles constitute the quadriceps femoris and hamstring muscles?

EXHIBIT 11.R **431**

EXHIBIT 11.S

Muscles of the Leg That Move the Foot and Toes (*Figure 11.22*)

● OBJECTIVE

• Describe the origin, insertion, action, and innervation of the muscles of leg that move the foot and toes.

Muscles that move the foot and toes are located in the leg (Figure 11.22). The muscles of the leg, like those of the thigh, are divided by deep fascia into three compartments: anterior, lateral, and posterior. The **anterior compartment of the leg** consists of muscles that dorsiflex the foot. In a situation analogous to the wrist, the tendons of the muscles of the anterior compartment are held firmly to the ankle by thickenings of deep fascia called the **superior extensor retinaculum** (*transverse ligament of the ankle*) and **inferior extensor retinaculum** (*cruciate ligament of the ankle*).

Within the anterior compartment, the **tibialis anterior** is a long, thick muscle against the lateral surface of the tibia, where it is easy to palpate (feel). The **extensor hallucis longus** is a thin muscle between and partly deep to the tibialis anterior and **extensor digitorum longus** muscles. This featherlike muscle is lateral to the tibialis anterior muscle, where it can also be palpated easily. The **fibularis (peroneus) tertius** muscle is part of the extensor digitorum longus, with which it shares a common origin.

The **lateral (fibular) compartment of the leg** contains two muscles that plantar flex and evert the foot: the **fibularis (peroneus) longus** and **fibularis (peroneus) brevis.**

The **posterior compartment of the leg** consists of muscles in superficial and deep groups. The superficial muscles share a common tendon of insertion, the **calcaneal (Achilles) tendon,** the strongest tendon of the body. It inserts into the calcaneal bone of the ankle. The superficial and most of the deep muscles plantar flex the foot at the ankle joint. The superficial muscles of the posterior compartment are the gastrocnemius, soleus, and plantaris—the so-called calf muscles. The large size of these muscles is directly related to the characteristic upright stance of humans. The **gastrocnemius** is the most superficial muscle and forms the prominence of the calf. The **soleus,** which lies deep to the gastrocnemius, is broad and flat. It derives its name from its resemblance to a flat fish (sole). The **plantaris** is a small muscle that may be absent; conversely, sometimes there are two of them in each leg. It runs obliquely between the gastrocnemius and soleus muscles.

The deep muscles of the posterior compartment are the popliteus, tibialis posterior, flexor digitorum longus, and flexor hallucis longus. The **popliteus** is a triangular muscle that forms the floor of the popliteal fossa. The **tibialis posterior** is the deepest muscle in the posterior compartment. It lies between the flexor digitorum longus and flexor hallucis longus muscles. The **flexor digitorum longus** is smaller than the **flexor hallucis longus,** even though the former flexes four toes and the latter flexes only the great toe at the interphalangeal joint.

MUSCLE	ORIGIN	INSERTION	ACTION	INNERVATION
ANTERIOR COMPARTMENT OF THE LEG				
Tibialis anterior (tib′-ē-Ā-lis an-TĒR-ē-or; *tibialis* = tibia; *anterior* = front)	Lateral condyle and body of tibia and interosseous membrane (sheet of fibrous tissue that holds shafts of tibia and fibula together).	First metatarsal and first (medial) cuneiform.	Dorsiflexes foot at ankle joint and inverts (supinates) foot at intertarsal joints.	Deep fibular (peroneal) nerve.
Extensor hallucis longus (eks-TEN-sor HAL-ū-sis LON-gus; *extensor* = increases angle at joint; *hallucis* = hallux or great toe; *longus* = long)	Anterior surface of middle third of fibula and interosseous membrane.	Distal phalanx of great toe.	Dorsiflexes foot at ankle joint and extends proximal phalanx of great toe at metatarsophalangeal joint.	Deep fibular (peroneal) nerve.
Extensor digitorum longus (di′-ji-TOR-um LON-gus; *digit-* = finger or toe)	Lateral condyle of tibia, anterior surface of fibula, and interosseous membrane.	Middle and distal phalanges of toes 2–5.*	Dorsiflexes foot at ankle joint and extends distal and middle phalanges of each toe at interphalangeal joints and proximal phalanx of each toe at metatarsophalangeal joint.	Deep fibular (peroneal) nerve.
Fibularis (peroneus) tertius (fib-ū-LĀ-ris; per′-Ō-nē-us; *person* = fibula; TER-shus; *tertius* = third)	Distal third of fibula and interosseous membrane.	Base of fifth metatarsal.	Dorsiflexes foot at ankle joint and everts (pronates) foot at intertarsal joints.	Deep fibular (peroneal) nerve.
LATERAL (FIBULAR) COMPARTMENT OF THE LEG				
Fibularis (peroneus) longus (LON-gus)	Head and body of fibula.	First metatarsal and first cuneiform.	Plantar flexes foot at ankle joint and everts (pronates) foot at intertarsal joints.	Superficial fibular (peroneal) nerve.
Fibularis (peroneus) brevis (BREV-is; *brevis* = short)	Distal half of body of fibula.	Base of fifth metatarsal.	Plantar flexes foot at ankle joint and everts (pronates) foot at intertarsal joints.	Superficial fibular (peroneal) nerve.

Shin splint syndrome, or simply **shin splints,** refers to pain or soreness along the tibia, specifically the medial, distal two-thirds. It may be caused by tendinitis of the anterior compartment muscles, especially the tibialis anterior muscle, inflammation of the periosteum (periostitis) around the tibia, or stress fractures of the tibia. The tendinitis usually occurs when poorly conditioned runners run on hard or banked surfaces with poorly supportive running shoes. The condition may also occur with vigorous activity of the legs following a period of relative inactivity or running in cold weather without proper warmup. The muscles in the anterior compartment (mainly the tibialis anterior) can be strengthened to balance the stronger posterior compartment muscles. •

RELATING MUSCLES TO MOVEMENTS

Arrange the muscles in this exhibit according to the following actions on the foot at the ankle joint: (1) dorsiflexion and (2) plantar flexion; according to the following actions on the foot at the intertarsal joints: (1) inversion and (2) eversion; and according to the following actions on the toes at the metatarsophalangeal and interphalangeal joints: (1) flexion and (2) extension. The same muscle may be mentioned more than once.

✔ CHECKPOINT

What are the superior extensor retinaculum and inferior extensor retinaculum?

MUSCLE	ORIGIN	INSERTION	ACTION	INNERVATION
SUPERFICIAL POSTERIOR COMPARTMENT OF THE LEG				
Gastrocnemius (gas'-trok-NĒ-mē-us; *gastro-* = belly; *cnem-* = leg)	Lateral and medial condyles of femur and capsule of knee.	Calcaneus by way of calcaneal (Achilles) tendon.	Plantar flexes foot at ankle joint and flexes leg at knee joint.	Tibial nerve.
Soleus (SŌ-lē-us; *sole* = type of flat fish)	Head of fibula and medial border of tibia.	Calcaneus by way of calcaneal (Achilles) tendon.	Plantar flexes foot at ankle joint.	Tibial nerve.
Plantaris (plan-TĀR-is = sole)	Lateral epicondyle of femur.	Calcaneus by way of calcaneal (Achilles) tendon.	Plantar flexes foot at ankle joint and flexes leg at knee joint.	Tibial nerve.
DEEP POSTERIOR COMPARTMENT OF THE LEG				
Popliteus (pop-LIT-ē-us; *poplit* = back of knee)	Lateral condyle of femur.	Proximal tibia.	Flexes leg at knee joint and medially rotates tibia to unlock the extended knee.	Tibial nerve.
Tibialis posterior (tib'-ē-Ā-lis; *posterior* = back)	Proximal tibia, fibula, and interosseous membrane.	Second, third, and fourth metatarsals; navicular; and all three cuneiforms.	Plantar flexes foot at ankle joint and inverts (supinates) foot at intertarsal joints.	Tibial nerve.
Flexor digitorum longus (FLEK-sor di'-ji-TOR-um LON-gus; *flexor* = decreases angle at point; *digit* = finger or toe)	Middle third of posterior surface of tibia.	Distal phalanges of toes 2–5.	Plantar flexes foot at ankle joint; flexes distal and middle proximal phalanx of each toe phalanges of each toe at interphalangeal joints and at metatarsophalangeal joint.	Tibial nerve.
Flexor hallucis longus (HAL-ū-sis)	Inferior two-thirds of posterior portion of fibula.	Distal phalanx of great toe.	Plantar flexes foot at ankle joint; flexes distal phalanx of great toe at interphalangeal joint and proximal phalanx of great toe at metatarsophalangeal joint.	Tibial nerve.

*Reminder: The great toe or hallux is the first toe and has two phalanges: proximal and distal. The remaining toes are numbered II–V (2–5), and each has three phalanges: proximal, middle, and distal.

EXHIBIT 11.S CONTINUES ▶

EXHIBIT 11.S   **433**

Figure 11.22 Muscles of the leg that move the foot and toes.

The superficial muscles of the posterior compartment share a common tendon of insertion, the calcaneal (Achilles) tendon, which inserts into the calcaneal bone of the ankle.

Quadriceps femoris

Tendon of quadriceps femoris

Iliotibial tract

Biceps femoris

Patella

Head of fibula

Patellar ligament

Tibia

TIBIALIS ANTERIOR

GASTROCNEMIUS

FIBULARIS LONGUS

SOLEUS

EXTENSOR DIGITORUM LONGUS

FLEXOR DIGITORUM LONGUS

FIBULARIS BREVIS

FIBULARIS TERTIUS

EXTENSOR HALLUCIS LONGUS

Calcaneal (Achilles) tendon

Fibula

EXTENSOR HALLUCIS BREVIS

Extensor digitorum brevis

Metatarsals

(a) Anterior superficial view

(b) Right lateral superficial view

Gracilis

Biceps femoris

Femur

Semitendinosus

Semimembranosus

Sartorius

Popliteal fossa

PLANTARIS

GASTROCNEMIUS
(cut)

Tibia

POPLITEUS

GASTROCNEMIUS

SOLEUS (cut)

Fibula

TIBIALIS POSTERIOR

SOLEUS

FIBULARIS LONGUS

FLEXOR DIGITORUM
LONGUS

FLEXOR HALLUCIS
LONGUS

FIBULARIS BREVIS

Tibia

Calcaneal (Achilles)
tendon (cut)

Tendon of
tibialis posterior

(c) Posterior superficial view

(d) Posterior deep view

EXHIBIT 11.S CONTINUES ▶

EXHIBIT 11.S **435**

⊙ **FIGURE 11.22 CONTINUED** ▶

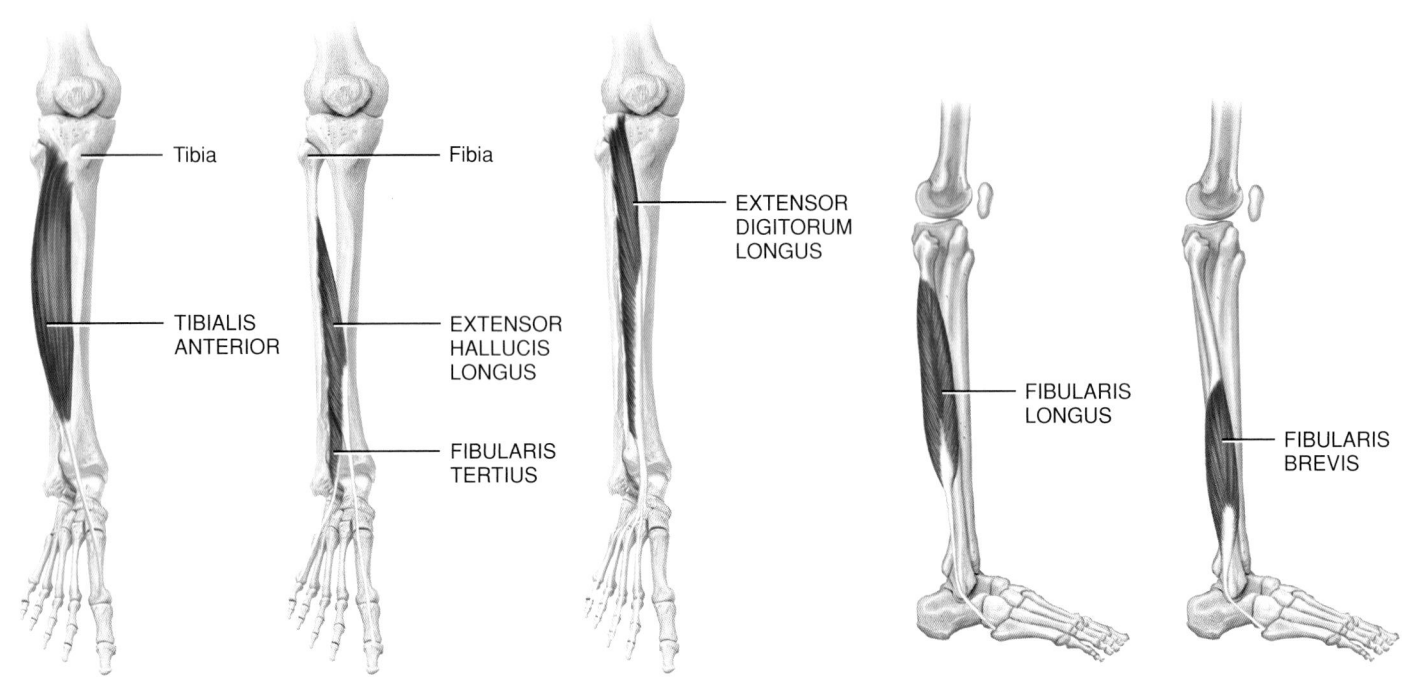

Anterior views

Right lateral view

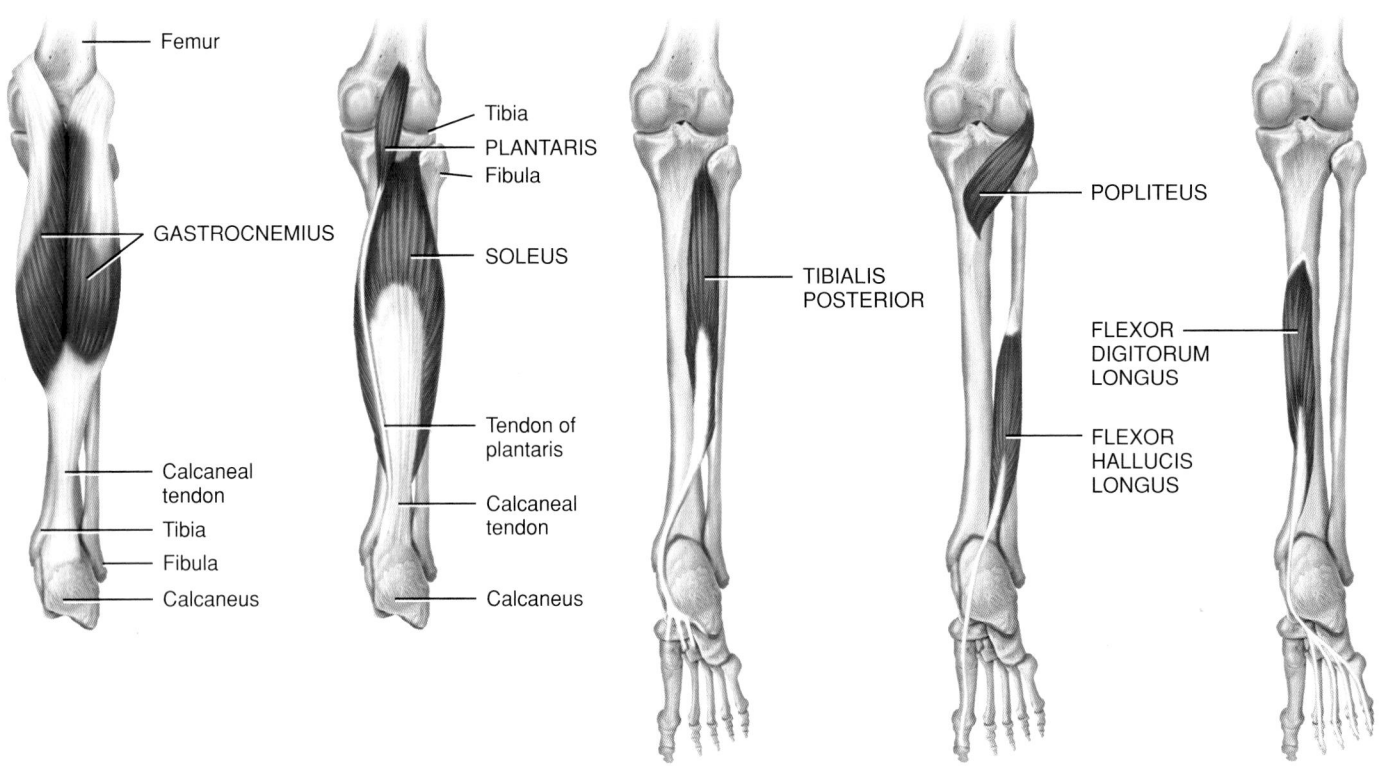

Posterior deep views

(e) Isolated muscles

❓ **What structures firmly hold the tendons of the anterior compartment muscles to the ankle?**

EXHIBIT 11.T

Intrinsic Muscles of the Foot That Move the Toes   *(Figure 11.23)*

● OBJECTIVE

- Describe the origin, insertion, action, and innervation of the intrinsic muscles of the foot that move the toes.

The muscles in this exhibit are termed **intrinsic muscles of the foot** because they originate and insert *within* the foot (Figure 11.23). The muscles of the hand are specialized for precise and intricate movements, but those of the foot are limited to support and locomotion. The deep fascia of the foot forms the **plantar aponeurosis (fascia)** that extends from the calcaneus bone to the phalanges of the toes. The aponeurosis supports the longitudinal arch of the foot and encloses the flexor tendons of the foot.

The intrinsic muscles of the foot are divided into two groups: **dorsal** and **plantar.** There are two dorsal muscles, the **extensor hallucis brevis** and the **extensor digitorum brevis.** The latter is a four-part muscle deep to the tendons of the extensor digitorum longus muscle, which extends toes 2–5 at the metatarsophalangeal joints.

The plantar muscles are arranged in four layers. The most superficial layer is called the first layer. Three muscles are in the first layer. The **abductor hallucis,** which lies along the medial border of the sole and is comparable to the abductor pollicis brevis in the hand, abducts the great toe at the metatarsophalangeal joint. The **flexor digitorum brevis,** which lies in the middle of the sole, flexes toes 2–5 at the interphalangeal and metatarsophalangeal joints. The **abductor digiti minimi,** which lies along the lateral border of the sole and is comparable to the same muscle in the hand, abducts the little toe.

The second layer consists of the **quadratus plantae,** a rectangular muscle that arises by two heads and flexes toes 2–5 at the metatarsophalangeal joints, and the **lumbricals,** four small muscles that are similar to the lumbricals in the hands. They flex the proximal phalanges and extend the distal phalanges of toes 2–5.

Three muscles compose the third layer. The **flexor hallucis brevis,** which lies adjacent to the plantar surface of the metatarsal of the great toe and is comparable to the same muscle in the hand, flexes the great toe. The **adductor hallucis,** which has an oblique and transverse head like the adductor pollicis in the hand, adducts the great toe. The **flexor digiti minimi brevis,** which lies superficial to the metatarsal of the little toe and is comparable to the same muscle in the hand, flexes the little toe.

The fourth layer is the deepest and consists of two muscle groups. The **dorsal interossei** are four muscles that abduct toes 2–4, flex the proximal phalanges, and extend the distal phalanges. The three **plantar interossei** abduct toes 3–5, flex the proximal phalanges, and extend the distal phalanges. The interossei of the feet are similar to those of the hand. However, their actions are relative to the midline of the second digit rather than the third digit as in the hand.

CLINICAL CONNECTION | *Plantar Fasciitis*

Plantar fasciitis (fas-ē-Ī-tis) or **painful heel syndrome** is an inflammatory reaction due to chronic irritation of the plantar aponeurosis (fascia) at its origin on the calcaneus (heel bone). The aponeurosis becomes less elastic with age. This condition is also related to weight-bearing activities (walking, jogging, lifting heavy objects), improperly constructed or fitting shoes, excess weight (puts pressure on the feet), and poor biomechanics (flat feet, high arches, and abnormalities in gait may cause uneven distribution of weight on the feet). Plantar fasciitis is the most common cause of heel pain in runners and arises in response to the repeated impact of running. Treatments include ice, deep heat, stretching exercises, weight loss, prosthetics (such as shoe inserts or heel lifts), steroid injections, and surgery. •

RELATING MUSCLES TO MOVEMENTS

Arrange the muscles in this exhibit according to the following actions on the great toe at the metatarsophalangeal joint: (1) flexion, (2) extension, (3) abduction, and (4) adduction; and according to the following actions on toes 2–5 at the metatarsophalangeal and interphalangeal joints: (1) flexion, (2) extension, (3) abduction, and (4) adduction. The same muscle may be mentioned more than once.

✔ CHECKPOINT

How do the intrinsic muscles of the hand and foot differ in function?

EXHIBIT 11.T CONTINUES ▶

EXHIBIT 11.T   **437**

EXHIBIT 11.T

Intrinsic Muscles of the Foot That Move the Toes *(Figure 11.23)* CONTINUED

MUSCLE	ORIGIN	INSERTION	ACTION	INNERVATION
DORSAL				
Extensor hallucis brevis (eks-TEN-sor HAL-ū-sis BREV-is; *extensor* = increases angle at joint; *hallucis* = hallux or great toe; *brevis* = short) (see Figure 11.22a)	Calcaneus and inferior extensor retinaculum.	Proximal phalanx of great toe.	Extends great toe at metatarsophalangeal joint.	Deep fibular (peroneal) nerve.
Extensor digitorum brevis *digit* = finger or toe; *brevis* = short) (see Figure 11.22a, b)	Calcaneus and inferior extensor retinaculum.	Middle phalanges of toes 2–4.	Extends toes 2–4 at interphalangeal joints.	Deep fibular (peroneal) nerve.
PLANTAR				
First layer (most superficial)				
Abductor hallucis (*abductor* = moves part away from midline; *hallucis* = hallux or great toe)	Calcaneus, plantar aponeurosis, and flexor retinaculum.	Medial side of proximal phalanx of great toe with the tendon of flexor hallucis brevis.	Abducts and flexes great toe at metatarsophalangeal joint.	Medial plantar nerve.
Flexor digitorum brevis (*flexor* = decreases angle at joint)	Calcaneus, plantar aponeurosis, and flexor retinaculum.	Sides of middle phalanx of toes 2–5.	Flexes toes 2–5 at proximal interphalangeal and metatarsophalangeal joints.	Medial plantar nerve.
Abductor digiti minimi (DIJ-i-tē MIN-i-mē; *minimi* = smallest)	Calcaneus, plantar aponeurosis, and flexor retinaculum.	Lateral side of proximal phalanx of little toe with tendon of flexor digiti minimi brevis.	Abducts and flexes little toe at metatarsophalangeal joint.	Lateral plantar nerve.
Second layer				
Quadratus plantae (kwod-RĀ-tus PLAN-tē; *quad* = square, four-sided; *planta* = sole)	Calcaneus.	Tendon of flexor digitorum longus.	Assists flexor digitorum longus to only flex toes 2–5 at interphalangeal and metatarsophalangeal joints.	Lateral plantar nerve.
Lumbricals (LUM-bri-kals; *lumbric* = earthworm)	Tendons of flexor digitorum longus.	Tendons of extensor digitorum longus on proximal phalanges of toes 2–5.	Extend toes 2–5 at interphalangeal joints and flex toes 2–5 at metatarsophalangeal joints.	Medial and lateral plantar nerves.
Third layer				
Flexor hallucis brevis	Cuboid and third (lateral) cuneiform.	Medial and lateral sides of proximal phalanx of great toe via tendon containing sesamoid bone.	Flexes great toe at metatarsophalangeal joint.	Medial plantar nerve.
Adductor hallucis (ad-DUK-tor = moves part closer to midline)	Metatarsals 2–4, ligaments of 3–5 metatarsophalangeal joints, and tendon of fibularis (peroneus) longus.	Lateral side of proximal phalanx of great toe.	Adducts and flexes great toe at metatarsophalangeal joint.	Lateral plantar nerve.
Flexor digiti minimi brevis	Metatarsal 5 and tendon of fibularis (peroneus) longus.	Lateral side of proximal phalanx of little toe.	Flexes little toe at metatarsophalangeal joint.	Lateral plantar nerve.
Fourth layer (deepest)				
Dorsal interossei (in-ter-OS-ē-ī) (not illustrated)	Adjacent side of all metatarsals.	Proximal phalanges: both sides of toe 2 and lateral side of toes 3 and 4.	Abduct and flex toes 2–4 at metatarsophalangeal joints and extend toes at interphalangeal joints.	Lateral plantar nerve.
Plantar interossei (PLAN-tar)	Metatarsals 3–5.	Medial side of proximal phalanges of toes 3–5.	Adduct and flex proximal metatarsophalangeal joints and extend toes at interphalangeal joints.	Lateral plantar nerve.

Figure 11.23 Intrinsic muscles of the foot that move the toes.

The muscles of the hand are specialized for precise and intricate movements; those of the foot are limited to support and movement.

(a) Plantar superficial and deep view

(b) Plantar deep view

EXHIBIT 11.T CONTINUES

EXHIBIT 11.T **439**

● FIGURE 11.23 CONTINUED ▶

Tendon of
flexor digitorum
brevis (cut)

Tendon of lumbrical (cut)

Tendons of flexor digitorum
longus (cut)

ADDUCTOR
HALLUCIS

PLANTAR
INTEROSSEI

FLEXOR HALLUCIS
BREVIS

FLEXOR
DIGITI
MINIMI
BREVIS

Long plantar ligament

Tendon of flexor digitorum
longus (cut)

QUADRATUS
PLANTAE
(cut)

Tendon of flexor
hallucis longus

ABDUCTOR
DIGITI
MINIMI
(cut)

(c) Plantar deeper view

Phalanges

Metatarsals

PLANTAR
INTEROSSEI

3 2 1

II   I

III

IV

V

Tarsals

Phalanges

Metatarsals

DORSAL
INTEROSSEI

4 3 2 1

II   I

III

IV

V

Tendon of
fibularis longus

Tendon of
fibularis brevis

Tarsals

(d) Plantar view

(e) Plantar view

❓ **What structure supports the longitudinal arch and encloses the flexor tendons of the foot?**

BODY SYSTEM	CONTRIBUTION OF THE MUSCULAR SYSTEM
For all body systems	The muscular system and muscle tissues produce body movements, stabilize body positions, move substances within the body, and produce heat that helps maintain normal body temperature.
Integumentary system	Pull of skeletal muscles on attachments to skin of face causes facial expressions; muscular exercise increases skin blood flow.
Skeletal system	Skeletal muscle causes movement of body parts by pulling on attachments to bones; skeletal muscle provides stability for bones and joints.
Nervous system	Smooth, cardiac, and skeletal muscles carry out commands for the nervous system; shivering—involuntary contraction of skeletal muscles that is regulated by the brain—generates heat to raise body temperature.
Endocrine system	Regular activity of skeletal muscles (exercise) improves action and signaling mechanisms of some hormones, such as insulin; muscles protect some endocrine glands.
Cardiovascular system	Cardiac muscle powers pumping action of heart; contraction and relaxation of smooth muscle in blood vessel walls help adjust the amount of blood flowing through various body tissues; contraction of skeletal muscles in the legs assists return of blood to the heart; regular exercise causes cardiac hypertrophy (enlargement) and increases heart's pumping efficiency; lactic acid produced by active skeletal muscles may be used for ATP production by the heart.
Lymphatic system and immunity	Skeletal muscles protect some lymph nodes and lymphatic vessels and promote flow of lymph inside lymphatic vessels; exercise may increase or decrease some immune responses.
Respiratory system	Skeletal muscles involved with breathing cause air to flow into and out of the lungs; smooth muscle fibers adjust size of airways; vibrations in skeletal muscles of larynx control air flowing past vocal cords, regulating voice production; coughing and sneezing, due to skeletal muscle contractions, help clear airways; regular exercise improves efficiency of breathing.
Digestive system	Skeletal muscles protect and support organs in the abdominal cavity; alternating contraction and relaxation of skeletal muscles power chewing and initiate swallowing; smooth muscle sphincters control volume of organs of the gastrointestinal (GI) tract; smooth muscles in walls of GI tract mix and move its contents through the tract.
Urinary system	Skeletal and smooth muscle sphincters and smooth muscle in wall of urinary bladder control whether urine is stored in the urinary bladder or voided (urination).
Reproductive systems	Skeletal and smooth muscle contractions eject semen from male; smooth muscle contractions propel oocyte along uterine tube, help regulate flow of menstrual blood from uterus, and force baby from uterus during childbirth; during intercourse, skeletal muscle contractions are associated with orgasm and pleasurable sensations in both sexes.

THE MUSCULAR SYSTEM

Running Injuries

Many individuals who jog or run sustain some type of **running-related injury.** Although such injuries may be minor, some can be quite serious. Untreated or inappropriately treated minor injuries may become chronic. Among runners, common sites of injury include the ankle, knee, calcaneal (Achilles) tendon, hip, groin, foot, and back. Of these, the knee often is the most severely injured area.

Running injuries are frequently related to faulty training techniques. This may involve improper or lack of sufficient warm-up routines, running too much, or running too soon after an injury. Or it might involve extended running on hard and/or uneven surfaces. Poorly constructed or worn-out running shoes can also contribute to injury, as can any biomechanical problem (such as a fallen arch) aggravated by running.

Most sports injuries should be treated initially with PRICE therapy, which stands for *protection, rest, ice, compression, and elevation.* Immediately protect the injured part, rest, and apply ice immediately after the injury, and elevate the injured part. Then apply an elastic bandage, if possible, to compress the injured tissue. Continue using PRICE for 2 to 3 days, and resist the temptation to apply heat, which may worsen the swelling. Follow-up treatment may include alternating moist heat and ice massage to enhance blood flow in the injured area. Sometimes it is helpful to take nonsteroidal anti-inflammatory drugs (NSAIDs) or to have local injections of corticosteroids. During the recovery period, it is important to keep active, using an alternative fitness program that does not worsen the original injury. This activity should be determined in consultation with a physician. Finally, careful exercise is needed to rehabilitate the injured area itself.

Massage therapy may also be used to prevent or treat many sports injuries.

Compartment Syndrome

As noted earlier in this chapter, skeletal muscles in the limbs are organized into functional units called *compartments.* In a disorder called **compartment syndrome,** some external or internal pressure constricts the structures within a compartment, resulting in damaged blood vessels and subsequent reduction of the blood supply (ischemia) to the structures within the compartment. Symptoms include pain, burning, pressure, pale skin, and paralysis. Common causes of compartment syndrome include crushing and penetrating injuries, contusion (damage to subcutaneous tissues without the skin being broken), muscle strain (overstretching of a muscle), or an improperly fitted cast. The pressure increase in the compartment can have serious consequences, such as hemorrhage, tissue injury, and edema (buildup of interstitial fluid). Because deep fasciae (connective tissue coverings) that enclose the compartments are very strong, accumulated blood and interstitial fluid cannot escape, and the increased pressure can literally choke off the blood flow and deprive nearby muscles and nerves of oxygen. One treatment option is **fasciotomy** (fash-ē-OT-ō-mē), a surgical procedure in which muscle fascia is cut to relieve the pressure. Without intervention, nerves can suffer damage, and muscles can develop scar tissue that results in permanent shortening of the muscles, a condition called *contracture.* If left untreated, tissues may die and the limb may no longer be able to function. Once the syndrome has reached this stage, amputation may be the only treatment option.

Muscle strain Tearing of fibers in a skeletal muscle or its tendon that attaches the muscle to bone. The tearing can also damage small blood vessels, causing local bleeding (bruising) and pain (caused by irritation of nerve endings in the region). Muscle strains usually occur when a muscle is stretched beyond its limit, for example, in response to sudden, quick heavy lifting; during sports activities; or while performing work tasks. A similar injury occurs if there is a direct blow to a muscle. Factors that contribute to muscle strain include muscle tightness due to insufficient stretching, not working antagonistic muscles equally, poor conditioning, muscle fatigue, and insufficient warm-up. The condition is treated by PRICE therapy: protection (P), rest (R), ice immediately after the injury (I), compression via a supportive wrap (C), and elevation of the limb (E). Also called **muscle pull** or **muscle tear.**

Paralysis (pa-RAL-i-sis; *para-* = departure from normal; *-lysis* = loosening) Loss of muscle function (voluntary movement) through injury, disease, or damage to its nerve supply. Most paralysis is due to stroke or spinal cord injury.

Repetitive strain or motion injuries (RSIs) Conditions resulting from overuse of equipment, poor posture, poor body mechanics, or activity that requires repeated movements; for example, various conditions of assembly line workers. Examples of overuse of equipment include overuse of a computer, hammer, guitar, or piano, to name a few.

Rhabdomyosarcoma (rab'-dō-mī'-ō-sar-KŌ-ma; *rhab-* = rod-shaped; *-myo-* = muscle; *-sarc-* = flesh; *-oma* = tumor) A tumor of skeletal muscle. Usually occurs in children and is highly malignant, with rapid metastasis.

Torticollis (tor-ti-KOL-is; *tortus-* = twisted; *-column* = neck) A contraction or shortening of the sternocleidomastoid muscle that causes the head to tilt toward the affected side and the chin to rotate toward the opposite side. It may be acquired or congenital. Also called **wryneck.**

Tic Spasmodic twitching made involuntarily by muscles that are usually under conscious control, for example, twitching of an eyelid.

CHAPTER REVIEW AND RESOURCE SUMMARY

Review

Resource

11.1 How Skeletal Muscles Produce Movements

1. Skeletal muscles that produce movement do so by pulling on bones.
2. The attachment to the more stationary bone is the origin; the attachment to the more movable bone is the insertion.
3. Bones serve as levers, and joints serve as fulcrums. Two different forces act on the lever: load (resistance) and effort.
4. Levers are categorized into three types—first-class, second-class, and third-class (most common)—according to the positions of the fulcrum, the effort, and the load on the lever.
5. Fascicular arrangements include parallel, fusiform, circular, triangular, and pennate. Fascicular arrangement affects a muscle's power and range of motion.
6. A prime mover produces the desired action; an antagonist produces an opposite action. Synergists assist a prime mover by reducing unnecessary movement. Fixators stabilize the origin of a prime mover so that it can act more efficiently.

Anatomy Overview - Skeletal Muscle
Animation - Contraction and
  Movement
Figure 11.1 - Relationship
  of Skeletal Muscles to
  Bones

11.2 How Skeletal Muscles Are Named

1. Distinctive features of different skeletal muscles include direction of muscle fascicles; size, shape, action, number of origins (or heads), and location of the muscle; and sites of origin and insertion of the muscle.
2. Most skeletal muscles are named based on combinations of features.

Anatomy Overview - The Muscular
  System

11.3 Principal Skeletal Muscles

1. Muscles of the head that produce facial expression move the skin rather than a joint when they contract, and they permit us to express a wide variety of emotions (see Exhibit 11.A). The muscles of the head that move the eyeballs are among the fastest contracting and most precisely controlled skeletal muscles in the body. They permit us to elevate, depress, abduct, adduct, and medially and laterally rotate the eyeballs. The muscles that move the eyelids open the eyes (see Exhibit 11.B).
2. Muscles that move the mandible play major roles in mastication (chewing) and speech (see Exhibit 11.C). The muscles of the head that move the tongue are important in mastication and speech, as well as deglutition (swallowing) (Exhibit 11.D).
3. Muscles of the anterior neck that assist in deglutition and speech, called suprahyoid muscles, are located above the hyoid bone (see Exhibit 11.E).
4. Muscles of the neck that move the head alter its position and help balance it on the vertebral column (see Exhibit 11.F).
5. Muscles of the abdomen help contain and protect the abdominal viscera, move the vertebral column, compress the abdomen, and produce the force required for defecation, urination, vomiting, and childbirth (see Exhibit 11.G).
6. Muscles of the thorax used in breathing alter the size of the thoracic cavity so that inhalation and exhalation can occur and assist in venous return of blood to the heart (see Exhibit 11.H).
7. Muscles of the pelvic floor support the pelvic viscera, resist the thrust that accompanies increases in intra-abdominal pressure, and function as sphincters at the anorectal junction, urethra, and vagina (see Exhibit 11.I). Muscles of the perineum assist in urination, erection of the penis and clitoris, ejaculation, and defecation (see Exhibit 11.J).
8. Muscles of the thorax that move the pectoral (shoulder) girdle stabilize the scapula so it can function as a stable point of origin for most of the muscles that move the humerus (see Exhibit 11.K).
9. Muscles of the thorax that move the humerus originate for the most part on the scapula (scapular muscles); the remaining muscles originate on the axial skeleton (axial muscles) (see Exhibit 11.L). Muscles of the arm that move the radius and ulna are involved in flexion and extension at the elbow joint and are organized into flexor and extensor compartments (see Exhibit 11.M).
10. Muscles of the forearm that move the wrist, hand, thumb, and digits (fingers) are many and varied; those muscles that act on the digits are called extrinsic muscles (see Exhibit 11.N). The muscles of the palm that move the digits (intrinsic muscles) are important in skilled activities and provide humans with the ability to grasp and manipulate objects precisely (see Exhibit 11.O).
11. Muscles of the neck and back that move the vertebral column are quite complex because they have multiple origins and insertions and because there is considerable overlap among them (see Exhibit 11.P).

Anatomy Overview - Muscles of Facial
  Expression
Anatomy Overview - Muscles Moving
  Eyeballs
Anatomy Overview - Muscles for
  Speech, Swallowing, and Chewing
Anatomy Overview - Muscles That
  Move the Head
Anatomy Overview - Muscles of the Torso
Anatomy Overview - Muscles for
  Breathing
Anatomy Overview - Muscles That
  Move the Pectoral Girdle
Anatomy Overview - Muscles That
  Move the Arm
Anatomy Overview - Muscles That
  Move the Forearm
Anatomy Overview - Muscles That
  Move the Hand
Anatomy Overview - Muscles That
  Move the Thigh
Anatomy Overview - Muscles of the Foot
Figure 11.3 - Principal
  Superficial Skeletal
  Muscles
Figure 11.11 - Muscles of
  the Thorax That Assist
  in Breathing
Figure 11.17 - Muscles of
  the Forearm That
  Move the Wrist, Hand,
  Thumb, and Digits
Exercise - Shoulder Movements Target
  Practice
Exercise - Lower Limb Movement Target
  Practice
Exercise - Manage Those Muscles

Review

12. Muscles of the gluteal region that move the femur originate for the most part on the pelvic girdle and insert on the femur; these muscles are larger and more powerful than comparable muscles in the upper limb (see Exhibit 11.Q). Muscles of the thigh that move the femur and the tibia and fibula are separated into medial (adductor), anterior (extensor), and posterior (flexor) compartments (see Exhibit 11.R).

13. Muscles of the leg that move the foot and toes are divided into anterior, lateral, and posterior compartments (see Exhibit 11.S).

14. Muscles of the foot that move the toes (intrinsic muscles), unlike those of the hand, are limited to the functions of support and locomotion (see Exhibit 11.T).

SELF-QUIZ QUESTIONS

Fill in the blanks in the following statements.

1. The muscle that forms the major portion of the cheek is the _____.

2. The three superficial posterior plantar flexors of the leg are the _____, _____, and _____. All three of these muscles insert on the _____ by way of the Achilles tendon.

Indicate whether the following statements are true or false.

3. Longer fibers in a muscle result in a greater range of motion.

4. When flexing the forearm, the biceps brachii acts as the prime mover and the triceps brachii acts as the antagonist.

Choose the one best answer to the following questions.

5. Which of the following muscles does *not* flex the thigh?
(a) rectus femoris
(b) gracilis
(c) sartorius
(d) iliacus
(e) tensor fascia latae

6. The iliotibial tract is composed of the tendon of the gluteus maximus muscle, the deep fascia that encircles the thigh, and the tendon of which of the following muscles?
(a) iliacus
(b) gluteus minimus
(c) tensor fascia latae
(d) adductor longus
(e) vastus lateralis

7. In order for movement to occur, (1) muscles generally need to cross a joint, (2) contraction of the muscle will pull on the origin, (3) muscles that move a body part cannot cover the moving part, (4) muscles need to exert force on tendons that pull on bones, (5) the insertion must act to stabilize the joint.
(a) 1, 2, 3, 4, and 5   (b) 1, 2, 3, and 4
(c) 1, 2, and 4   (d) 1, 3, and 4
(e) 3 and 4

8. Because you did not do well on your recent anatomy and physiology exam, you leave the classroom pouting. Which one of these muscles are you using?
(a) mentalis
(b) orbicularis oris
(c) risorius
(d) levator labii superioris
(e) zygomaticus minor

9. The rectus femoris has fascicles arranged on both sides of a centrally positioned tendon. This pattern of fascicle arrangement is
(a) unipennate.
(b) fusiform.
(c) multipennate.
(d) parallel.
(e) bipennate.

10. Which of the following muscle names and their naming descriptors are mismatched?
(a) adductor brevis: short muscle that moves a bone closer to the midline
(b) rectus abdominis: muscle with fibers parallel to the midline of the abdomen
(c) levator scapula: muscle that raises the scapula
(d) sternohyoid: muscle attached to the sternum and hyoid
(e) serratus anterior: comblike muscle located on the body's anterior surface

11. Match the following:
_____ (a) muscle that stabilizes the origin of the prime mover
_____ (b) site of muscle attachment to a stationary bone
_____ (c) muscle that stretches to allow desired motion
_____ (d) muscle that contracts to stabilize intermediate joints
_____ (e) site of muscle attachment to a movable bone
_____ (f) group of muscles, along with their blood and nerves, that have a common function
_____ (g) contracting muscle that produces the desired motion
_____ (h) fleshy part of the muscle

(1) compartment
(2) origin
(3) insertion
(4) belly
(5) synergist
(6) fixator
(7) prime mover (agonist)
(8) antagonist

12. Match the following:

_____ (a) compression of median nerve resulting in pain and numbness and tingling in the fingers

_____ (b) tendinitis of the anterior compartment muscles of the leg; inflammation of the tibial periosteum

_____ (c) improperly aligned eyeballs due to lesions in either the oculomotor or abducens nerves

_____ (d) stretching or tearing of distal attachments of adductor muscles

_____ (e) rupture of a portion of the inguinal area of the abdominal wall resulting in protrusion of part of the small intestine

_____ (f) caused by repetitive movement of the arm over the head that results in inflammation of the supraspinatus tendon

_____ (g) inflammation due to chronic irritation of the plantar aponeurosis at its origin on the calcaneus; most common cause of heel pain in runners

_____ (h) painful inflammation of tendons, tendon sheaths, and synovial membranes of joints

_____ (i) paralysis of facial muscles as a result of damage to the facial nerve

_____ (j) common in individuals who perform quick starts and stops; tearing away of part of the tendinous origins from the ischial tuberosity

_____ (k) permanent shortening of a muscle due to nerve damage and scar tissue development

_____ (l) may occur as a result of injury to levator ani muscle

_____ (m) external or internal pressure constricts structures in a compartment, causing a reduction of blood supply to the structures

(1) tenosynovitis
(2) Bell's palsy
(3) inguinal hernia
(4) urinary stress incontinence
(5) compartment syndrome
(6) groin pull
(7) pulled hamstrings
(8) strabismus
(9) shin splints
(10) plantar fasciitis
(11) impingement syndrome
(12) contracture
(13) carpal tunnel syndrome

13. Match the following:

_____ (a) rectus femoris, vastus lateralis, vastus medialis, vastus intermedius

_____ (b) biceps femoris, semitendinosus, semimembranosus

_____ (c) erector spinae; includes iliocostalis, longissimus, and spinalis groups

_____ (d) thenar, hypothenar, intermediate

_____ (e) biceps brachii, brachialis, coracobrachialis

_____ (f) latissimus dorsi

_____ (g) subscapularis, supraspinatus, infraspinatus, teres minor

_____ (h) diaphragm, external intercostals, internal intercostals

_____ (i) trapezius, levator scapulae, rhomboid major, rhomboid minor

(1) breathing muscles
(2) constitute flexor compartment of the arm
(3) hamstrings
(4) intrinsic muscle groups of the hand
(5) muscles that strengthen and stabilize the shoulder joint; the rotator cuff
(6) quadriceps femoris muscle
(7) largest muscle mass of the back
(8) posterior thoracic muscles
(9) swimmer's muscle

14. Match the following (some answers may be used more than once):

_____ (a) trapezius
_____ (b) orbicularis oculi
_____ (c) levator ani
_____ (d) rectus abdominis
_____ (e) triceps brachii
_____ (f) gastrocnemius
_____ (g) temporalis
_____ (h) external anal sphincter
_____ (i) external oblique
_____ (j) iliocostalis thoracis
_____ (k) digastric
_____ (l) styloglossus
_____ (m) masseter
_____ (n) adductor longus
_____ (o) zygomaticus major
_____ (p) latissimus dorsi
_____ (q) flexor carpi radialis
_____ (r) pronator teres
_____ (s) sternocleidomastoid
_____ (t) quadriceps femoris
_____ (u) deltoid
_____ (v) tibialis anterior
_____ (w) sartorius
_____ (x) gluteus maximus
_____ (y) superior rectus
_____ (z) trapezius

(1) muscle of facial expression
(2) muscle of mastication
(3) muscle that moves the eyeballs
(4) extrinsic muscle that moves the tongue
(5) suprahyoid muscle
(6) muscle of the perineum
(7) muscle that moves the head
(8) abdominal wall muscle
(9) pelvic floor muscle
(10) pectoral girdle muscle
(11) muscle that moves the humerus
(12) muscle that moves the radius and ulna
(13) muscle that moves the wrist, hand, and digits
(14) muscle that moves the vertebral column
(15) muscle that moves the femur
(16) muscle that acts on the femur, tibia, and fibula
(17) muscle that moves the foot and toes

15. Match the following (some answers may be used more than once):

_____ (a) most common lever in the body
_____ (b) lever formed by the head resting on the vertebral column
_____ (c) always produces a mechanical advantage
_____ (d) EFL
_____ (e) FLE
_____ (f) FEL
_____ (g) adduction of the thigh

(1) first-class lever
(2) second-class lever
(3) third-class lever

CRITICAL THINKING QUESTIONS

1. During a facelift, the cosmetic surgeon accidentally severs the facial nerve on the right side of the face. What are some of the effects this would have on the patient, and what muscles are involved?

2. While taking the bus to the supermarket, eleven-year-old Desmond informs his mother that he has to "go to the bathroom" (urinate). His mother tells him he must "hold it" until they arrive at the store. What muscles must remain contracted in order for him to prevent urination?

3. Minor-league pitcher José has been throwing a hundred pitches a day in order to perfect his curve ball. Lately he has experienced pain in his pitching arm. The doctor diagnosed a torn rotator cuff. José was confused because he thought cuffs were only found on shirt sleeves, not inside his shoulder. Explain to José what the doctor means and how this injury could affect his arm movement.

? ANSWERS TO FIGURE QUESTIONS

11.1 The belly of the muscle that extends the forearm, the triceps brachii, is located posterior to the humerus.

11.2 Second-class levers produce the most force.

11.3 For muscles named after their various characteristics, here are possible correct responses (for others, see Table 11.2): direction of fibers: external oblique; shape: deltoid; action: extensor digitorum; size: gluteus maximus; origin and insertion: sternocleidomastoid; location: tibialis anterior; number of tendons of origin: biceps brachii.

11.4 The corrugator supercilii muscle is involved in frowning; the zygomaticus major muscle contracts when you smile; the mentalis muscle contributes to pouting; the orbicularis oculi muscle contributes to squinting.

11.5 The inferior oblique muscle moves the eyeball superiorly and laterally because it originates at the anteromedial aspect of the floor of the orbit and inserts on the posterolateral aspect of the eyeball.

11.6 The masseter is the strongest muscle of mastication.

11.7 Functions of the tongue include chewing, detection of taste, swallowing, and speech.

11.8 The suprahyoid and infrahyoid muscles stabilize the hyoid bone to assist in tongue movements.

11.9 The triangles in the neck formed by the sternocleidomastoid muscles are important anatomically and surgically because of the structures that lie within their boundaries.

11.10 The rectus abdominis muscle aids in urination.

11.11 The diaphragm is innervated by the phrenic nerve.

11.12 The borders of the pelvic diaphragm are the pubic symphysis anteriorly, the coccyx posteriorly, and the walls of the pelvis laterally.

11.13 The borders of the perineum are the pubic symphysis anteriorly, the coccyx posteriorly, and the ischial tuberosities laterally.

11.14 The main action of the muscles that move the pectoral girdle is to stabilize the scapula to assist in movements of the humerus.

11.15 The rotator cuff consists of the flat tendons of the subscapularis, supraspinatus, infraspinatus, and teres minor muscles that form a nearly complete circle around the shoulder joint.

11.16 The brachialis is the most powerful forearm flexor; the triceps brachii is the most powerful forearm extensor.

11.17 Flexor tendons of the digits and wrist and the median nerve pass deep to the flexor retinaculum.

11.18 Muscles of the thenar eminence act on the thumb (pollex).

11.19 The splenius muscles arise from the midline and extend laterally and superiorly to their insertion.

11.20 Upper limb muscles exhibit diversity of movement; lower limb muscles function in stability, locomotion, and maintenance of posture. In addition, lower limb muscles usually cross two joints and act equally on both.

11.21 The quadriceps femoris consists of the rectus femoris, vastus lateralis, vastus medialis, and vastus intermedius; the hamstrings consist of the biceps femoris, semitendinosus, and semimembranosus.

11.22 The superior and inferior extensor retinacula firmly hold the tendons of the anterior compartment muscles to the ankle.

11.23 The plantar aponeurosis (fascia) supports the longitudinal arch and encloses the flexor tendons of the foot.

12 NERVOUS TISSUE

NERVOUS TISSUE AND HOMEOSTASIS *The excitable characteristic of nervous tissue allows for the generation of nerve impulses (action potentials) that provide communication with and regulation of most body tissues.*

Both the nervous and endocrine systems have the same objective: to keep controlled conditions within limits that maintain life. The nervous system regulates body activities by responding rapidly using nerve impulses; the endocrine system responds by releasing hormones. Chapter 18 compares the roles of both systems in maintaining homeostasis.

The nervous system is also responsible for our perceptions, behaviors, and memories, and it initiates all voluntary movements. Because this system is quite complex, we discuss its structure and function in several chapters. This chapter focuses on the organization of the nervous system and the properties of neurons (nerve cells) and neuroglia (cells that support the activities of neurons). We then examine the structure and functions of the spinal cord and spinal nerves (Chapter 13), and of the brain and cranial nerves (Chapter 14). The autonomic nervous system, which operates without voluntary control, will be covered in Chapter 15. Chapter 16 will discuss the somatic senses—touch, pressure, warmth, cold, pain, and others—and their sensory and motor pathways to show how nerve impulses pass into the spinal cord and brain or from the spinal cord and brain to muscles and glands. Exploration of the nervous system concludes with a discussion of the special senses: smell, taste, vision, hearing, and equilibrium (Chapter 17).

Neurology deals with normal functioning and disorders of the nervous system. A **neurologist** (noo-ROL-ō-jist) is a physician who diagnoses and treats disorders of the nervous system.

? *Did you ever wonder how the human nervous system coordinates and integrates all body systems so rapidly and efficiently?*

12.1 OVERVIEW OF THE NERVOUS SYSTEM

◉ OBJECTIVES

- Describe the organization of the nervous system.
- Describe the three basic functions of the nervous system.

Organization of the Nervous System

With a mass of only 2 kg (4.5 lb), about 3% of the total body weight, the **nervous system** is one of the smallest and yet the most complex of the 11 body systems. This intricate network of billions of neurons and even more neuroglia is organized into two main subdivisions: the central nervous system and the peripheral nervous system.

Central Nervous System

The **central nervous system (CNS)** consists of the brain and spinal cord (Figure 12.1a). The brain is the part of the CNS that is located in the skull and contains about 100 billion (10^{11}) neurons. The spinal cord is connected to the brain through the foramen magnum of the occipital bone and is encircled by the bones of the vertebral column. The spinal cord contains about 100 million neurons. The CNS processes many different kinds of incoming sensory information. It is also the source of thoughts, emotions, and memories. Most signals that stimulate muscles to contract and glands to secrete originate in the CNS.

Peripheral Nervous System

The **peripheral nervous system (PNS)** (pe-RIF-e-ral) consists of all nervous tissue outside the CNS (Figure 12.1a). Components of the PNS include nerves, ganglia, enteric plexuses, and sensory receptors. A **nerve** is a bundle of hundreds to thousands of axons plus associated connective tissue and blood vessels that lies outside the brain and spinal cord. Twelve pairs of **cranial nerves** emerge from the brain and thirty-one pairs of **spinal nerves** emerge from the spinal cord. Each nerve follows a defined path and serves a specific region of the body. **Ganglia** (GANG-glē-a = swelling or knot; singular is *ganglion*) are small masses of nervous tissue, consisting primarily of neuron cell bodies, that are located outside of the brain and spinal cord. Ganglia are closely associated with cranial and spinal nerves. **Enteric plexuses** (PLEK-sus-ez) are extensive networks of neurons located in the walls of organs of the gastrointestinal tract. The neurons of these plexuses help regulate the digestive system (see Section 24.3). The term **sensory receptor** refers to a structure of the nervous system that monitors changes in the external or internal environment. Examples of sensory receptors include touch receptors in the skin, photoreceptors in the eye, and olfactory receptors in the nose.

The PNS is divided into a **somatic nervous system (SNS)** (sō-MAT-ik; *soma* = body), an **autonomic nervous system (ANS)** (aw′-tō-NOM-ik; *auto-* = self; *-nomic* = law), and an **enteric nervous system (ENS)** (en-TER-ik; *enteron* = intestines). The SNS consists of (1) sensory neurons that convey information from somatic receptors in the head, body wall, and limbs and from recep-

tors for the special senses of vision, hearing, taste, and smell to the CNS and (2) motor neurons that conduct impulses from the CNS to *skeletal muscles* only. Because these motor responses can be consciously controlled, the action of this part of the PNS is *voluntary*.

The ANS consists of (1) sensory neurons that convey information from autonomic sensory receptors, located primarily in visceral organs such as the stomach and lungs, to the CNS and (2) motor neurons that conduct nerve impulses from the CNS to *smooth muscle, cardiac muscle,* and *glands*. Because its motor responses are not normally under conscious control, the action of the ANS is involuntary. The motor part of the ANS consists of two branches, the **sympathetic division** and the **parasympathetic division.** With a few exceptions, effectors receive nerves from both divisions, and usually the two divisions have opposing actions. For example, sympathetic neurons increase heart rate, and parasympathetic neurons slow it down. In general, the sympathetic division helps support exercise of emergency actions, the "fight-or-flight" responses, and the parasympathetic division takes care of "rest-and-digest" activities.

The operation of the ENS, the "brain of the gut," is involuntary. Once considered part of the ANS, the ENS consists of over 100 million neurons in enteric plexuses that extend most of the length of the gastrointestinal (GI) tract. Many of the neurons of the enteric plexuses function independently of the ANS and CNS to some extent, although they also communicate with the CNS via sympathetic and parasympathetic neurons. Sensory neurons of the ENS monitor chemical changes within the GI tract as well as the stretching of its walls. Enteric motor neurons govern contraction of GI tract smooth muscle to propel food through the GI tract, secretions of the GI tract organs such as acid from the stomach, and activity of GI tract endocrine cells, which secrete hormones.

Functions of the Nervous System

The nervous system carries out a complex array of tasks. It allows us to sense various smells, produce speech, and remember past events; in addition, it provides signals that control body movements and regulates the operation of internal organs. These diverse activities can be grouped into three basic functions: sensory (input), integrative (process), and motor (output).

- **Sensory function.** Sensory receptors *detect* internal stimuli, such as an increase in blood pressure, or external stimuli (for example, a raindrop landing on your arm). This sensory information is then carried into the brain and spinal cord through cranial and spinal nerves.

- **Integrative function.** The nervous system *processes* sensory information by analyzing it and making decisions for appropriate responses—an activity known as **integration.**

- **Motor function.** Once sensory information is integrated, the nervous system *may elicit an appropriate motor response* by activating **effectors** (muscles and glands) through cranial and spinal nerves. Stimulation of the effectors causes muscles to contract and glands to secrete.

Figure 12.1 Organization of the nervous system. (a) Subdivisions of the nervous system. (b) Nervous system organizational chart; blue boxes represent sensory components of the peripheral nervous system, red boxes represent motor components of the PNS, and green boxes represent effectors (muscles and glands).

🔑 The two main subdivisions of the nervous system are (1) the central nervous system (CNS), which consists of the brain and spinal cord, and (2) the peripheral nervous system (PNS), which consists of all nervous tissue outside the CNS.

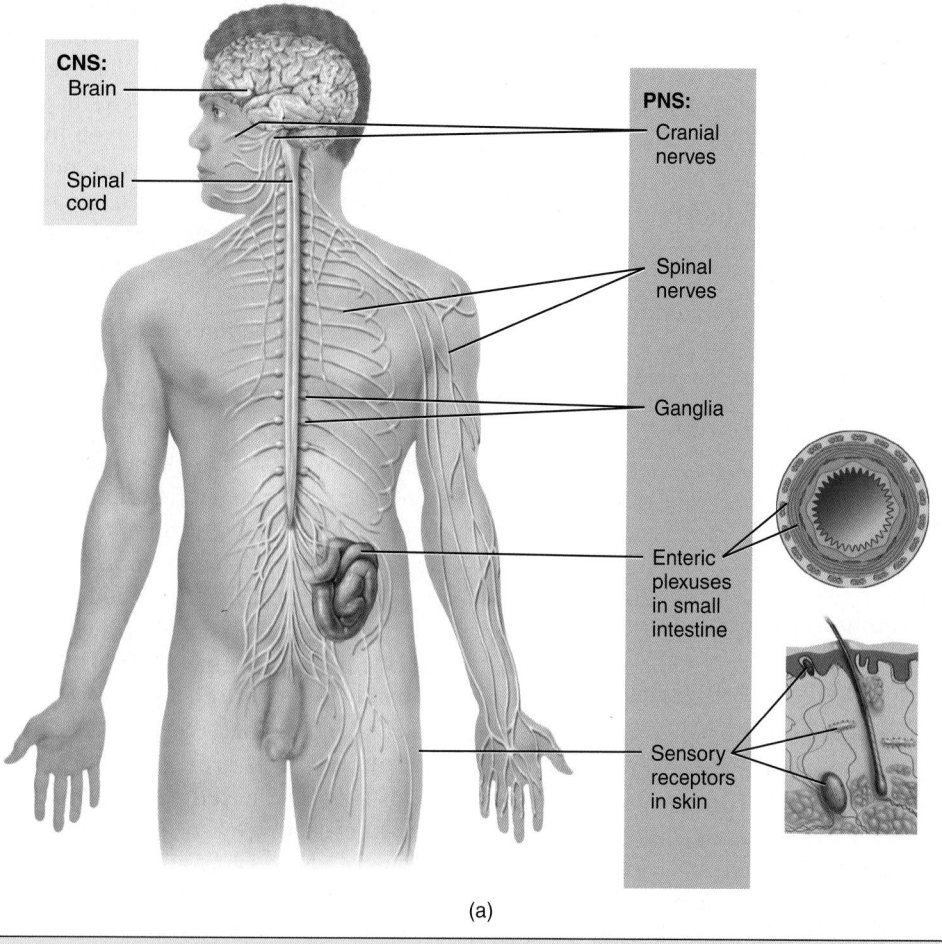

CNS:
Brain
Spinal cord

PNS:
Cranial nerves
Spinal nerves
Ganglia
Enteric plexuses in small intestine
Sensory receptors in skin

(a)

CENTRAL NERVOUS SYSTEM (CNS): brain and spinal cord

Somatic and special sensory receptors and somatic sensory neurons

Somatic motor neurons (voluntary)

Autonomic sensory receptors and autonomic sensory neurons

Autonomic motor neurons (involuntary): sympathetic and parasympathetic divisions

Enteric motor neurons (involuntary) in enteric plexuses

Enteric sensory receptors and enteric sensory neurons in GI tract and enteric plexuses

Skeletal muscle

Smooth muscle, cardiac muscle, and glands

Smooth muscle, glands, and endocrine cells of GI tract

Somatic Nervous System (SNS)

Autonomic Nervous System (ANS)

Enteric Nervous System (ENS)

PERIPHERAL NERVOUS SYSTEM (PNS): all nervous tissue outside the CNS

(b)

❓ **What are some of the functions of the CNS?**

The three basic functions of the nervous system occur, for example, when you answer your cell phone after hearing it ring. The sound of the ringing cell phone stimulates sensory receptors in your ears (sensory function). This auditory information is subsequently relayed into your brain where it is processed and the decision to answer the phone is made (integrative function). The brain then stimulates the contraction of specific muscles that will allow you to grab the phone and press the appropriate button to answer it (motor function).

✔ CHECKPOINT

1. What is the purpose of a sensory receptor?
2. What are the components and functions of the SNS, ANS, and ENS?
3. Which subdivisions of the PNS control voluntary actions? Involuntary actions?
4. Explain the concept of integration and provide an example.

12.2 HISTOLOGY OF NERVOUS TISSUE

◉ OBJECTIVES
- Contrast the histological characteristics and the functions of neurons and neuroglia.
- Distinguish between gray matter and white matter.

Nervous tissue comprises two types of cells—*neurons* and *neuroglia*. These cells combine in a variety of ways in different regions of the nervous system. In addition to forming the complex processing networks within the brain and spinal cord, neurons also connect all regions of the body to the brain and spinal cord. As highly specialized cells capable of reaching great lengths and making extremely intricate connections with other cells, neurons provide most of the unique functions of the nervous system, such as sensing, thinking, remembering, controlling muscle activity, and regulating glandular secretions. As a result of their specialization, they have lost the ability to undergo mitotic divisions. Neuroglia are smaller cells but they greatly outnumber neurons, perhaps by as much as 25 times. Neuroglia support, nourish, and protect neurons, and maintain the interstitial fluid that bathes them. Unlike neurons, neuroglia continue to divide throughout an individual's lifetime. Both neurons and neuroglia differ structurally depending on whether they are located in the central nervous system or the peripheral nervous system. These differences in structure correlate with the differences in function of the central nervous system and the peripheral nervous system.

Neurons

Like muscle cells, **neurons (nerve cells)** (NOO-rons) possess **electrical excitability** (ek-sīt′-a-BIL-i-tē), the ability to respond to a stimulus and convert it into an action potential. A **stimulus** is any change in the environment that is strong enough to initiate an action potential. An **action potential (nerve impulse)** is an electrical signal that propagates (travels) along the surface of the membrane of a neuron. It begins and travels due to the movement of ions (such as sodium and potassium) between interstitial fluid and the inside of a neuron through specific ion channels in its plasma membrane. Once begun, a nerve impulse travels rapidly and at a constant strength.

Some neurons are tiny and propagate impulses over a short distance (less than 1 mm) within the CNS. Others are the longest cells in the body. The neurons that enable you to wiggle your toes, for example, extend from the lumbar region of your spinal cord (just above waist level) to the muscles in your foot. Some neurons are even longer. Those that allow you to feel a feather tickling your toes stretch all the way from your foot to the lower portion of your brain. Nerve impulses travel these great distances at speeds ranging from 0.5 to 130 meters per second (1 to 290 mi/hr).

Parts of a Neuron

Most neurons have three parts: (1) a cell body, (2) dendrites, and (3) an axon (Figure 12.2). The **cell body,** also known as the *perikaryon* (per′-i-KAR-ē-on) or *soma,* contains a nucleus surrounded by cytoplasm that includes typical cellular organelles such as lysosomes, mitochondria, and a Golgi complex. Neuronal cell bodies also contain free ribosomes and prominent clusters of rough endoplasmic reticulum, termed **Nissl bodies** (NIS-el). The ribosomes are the sites of protein synthesis. Newly synthesized proteins produced by Nissl bodies are used to replace cellular components, as material for growth of neurons, and to regenerate damaged axons in the PNS. The cytoskeleton includes both **neurofibrils** (noo-rō-FĪ-brils), composed of bundles of intermediate filaments that provide the cell shape and support, and **microtubules** (mī-krō-TOO-būls′), which assist in moving materials between the cell body and axon. Aging neurons also contain **lipofuscin** (līp′-o-FYŪS-īn), a pigment that occurs as clumps of yellowish brown granules in the cytoplasm. Lipofuscin is a product of neuronal lysosomes that accumulates as the neuron ages, but does not seem to harm the neuron as it ages.

A **nerve fiber** is a general term for any neuronal process (extension) that emerges from the cell body of a neuron. Most neurons have two kinds of processes: multiple dendrites and a single axon. **Dendrites** (DEN-drīts = little trees) are the receiving or input portions of a neuron. The plasma membranes of dendrites (and cell bodies) contain numerous receptor sites for binding chemical messengers from other cells. Dendrites usually are short, tapering, and highly branched. In many neurons the dendrites form a tree-shaped array of processes extending from the cell body. Their cytoplasm contains Nissl bodies, mitochondria, and other organelles.

The single **axon** (= axis) of a neuron propagates nerve impulses toward another neuron, a muscle fiber, or a gland cell. An axon is a long, thin, cylindrical projection that often joins to the cell body at a cone-shaped elevation called the **axon hillock** (HIL-lok = small hill). The part of the axon closest to the axon hillock is the **initial segment.** In most neurons, nerve impulses arise at the junction of the axon hillock and the initial segment, an area called the **trigger zone,** from which they travel along the axon to their destination. An axon contains mitochondria, microtubules, and neurofibrils. Because rough endoplasmic reticulum is not present, protein synthesis does not occur in the axon. The cytoplasm of an axon, called **axoplasm,** is surrounded by a plasma membrane known as the **axolemma** (*lemma* = sheath or husk). Along the

Figure 12.2 **Structure of a multipolar neuron (a neuron with a large cell body, several short dendrites, and a single long axon).**
Arrows indicate the direction of information flow: dendrites → cell body → axon → axon terminals.

The basic parts of a neuron are dendrites, a cell body, and an axon.

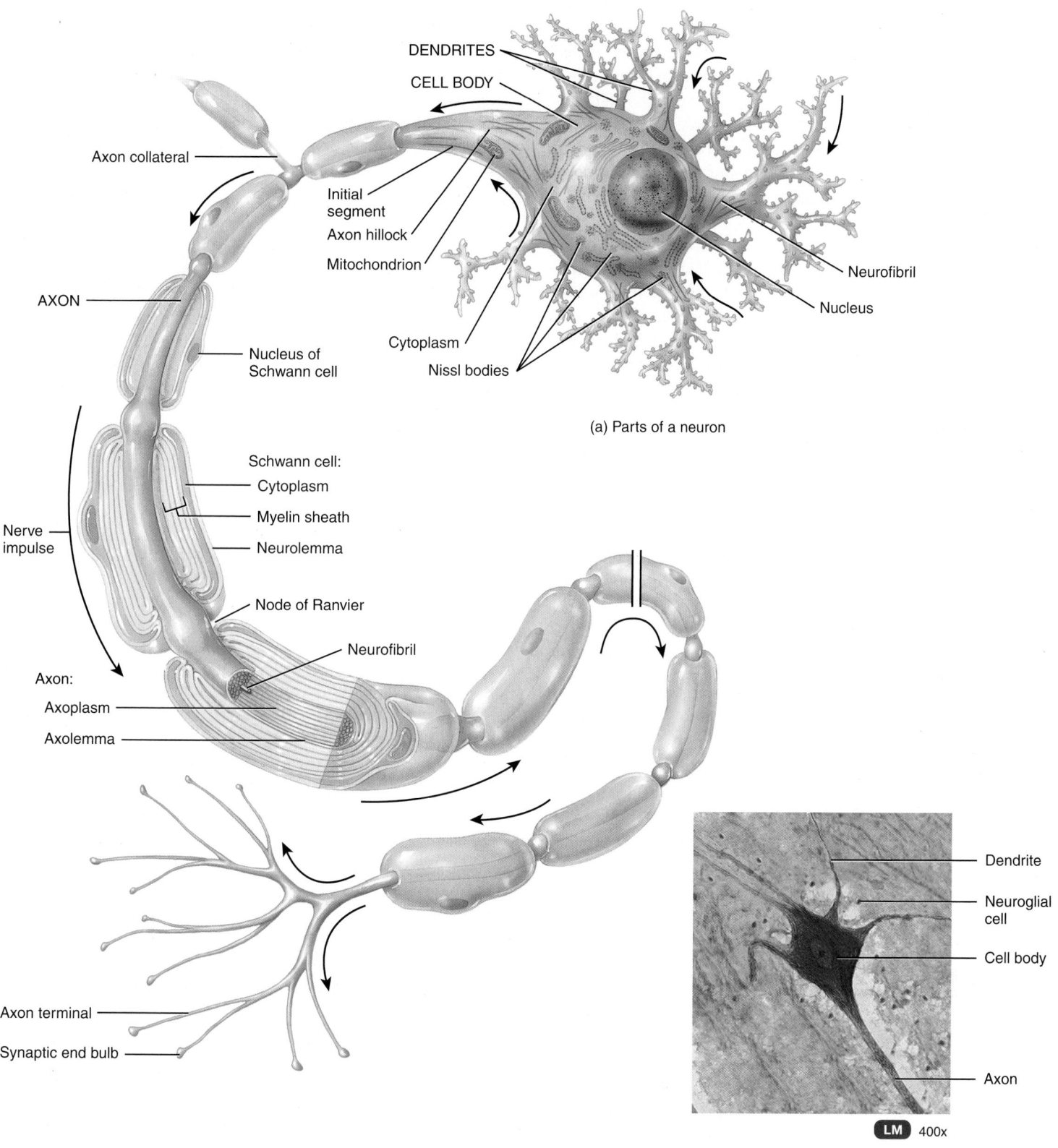

(a) Parts of a neuron

(b) Motor neuron

LM 400x

? **What roles do the dendrites, cell body, and axon play in communication of signals?**

length of an axon, side branches called **axon collaterals** may branch off, typically at a right angle to the axon. The axon and its collaterals end by dividing into many fine processes called **axon terminals (telodendria)** (tēl′-ō-DEN-drē-a).

The site of communication between two neurons or between a neuron and an effector cell is called a **synapse** (SIN-aps). The tips of some axon terminals swell into bulb-shaped structures called **synaptic end bulbs;** others exhibit a string of swollen bumps called **varicosities** (var′-i-KOS-i-tēz). Both synaptic end bulbs and varicosities contain many tiny membrane-enclosed sacs called **synaptic vesicles** that store a chemical called a **neurotransmitter** (noo′-rō-trans′-MIT-ter). A neurotransmitter is a molecule released from a synaptic vesicle that excites or inhibits another neuron, muscle fiber, or gland cell. Many neurons contain two or even three types of neurotransmitters, each with different effects on the postsynaptic cell.

Because some substances synthesized or recycled in the neuron cell body are needed in the axon or at the axon terminals, two types of transport systems carry materials from the cell body to the axon terminals and back. The slower system, which moves materials about 1–5 mm per day, is called **slow axonal transport.** It conveys axoplasm in one direction only—from the cell body toward the axon terminals. Slow axonal transport supplies new axoplasm to developing or regenerating axons and replenishes axoplasm in growing and mature axons.

Fast axonal transport, which is capable of moving materials a distance of 200–400 mm per day, uses proteins that function as "motors" to move materials along the surfaces of microtubules of the neuron's cytoskeleton. Fast axonal transport moves materials in both directions—away from and toward the cell body. Fast axonal transport that occurs in an **anterograde** (forward) direction moves organelles and synaptic vesicles from the cell body to the axon terminals. Fast axonal transport that occurs in a **retrograde** (backward) direction moves membrane vesicles and other cellular materials from the axon terminals to the cell body to be degraded or recycled. Substances that enter the neuron at the axon terminals are also moved to the cell body by fast retrograde transport. These substances include trophic chemicals such as nerve growth factor and harmful agents such as tetanus toxin and the viruses that cause rabies, herpes simplex, and polio.

Structural Diversity in Neurons

Neurons display great diversity in size and shape. For example, their cell bodies range in diameter from 5 micrometers (μm) (slightly smaller than a red blood cell) up to 135 μm (barely large enough to see with the unaided eye). The pattern of dendritic branching is varied and distinctive for neurons in different parts of the nervous system. A few small neurons lack an axon, and many others have very short axons. As we have already discussed, the longest axons are almost as long as a person is tall, extending from the toes to the lowest part of the brain.

Classification of Neurons

Both structural and functional features are used to classify the various neurons in the body.

STRUCTURAL CLASSIFICATION Structurally, neurons are classified according to the number of processes extending from the cell body (Figure 12.3):

Figure 12.3 Structural classification of neurons. Breaks indicate that axons are longer than shown.

A multipolar neuron has many processes extending from the cell body, a bipolar neuron has two, and a unipolar neuron has one.

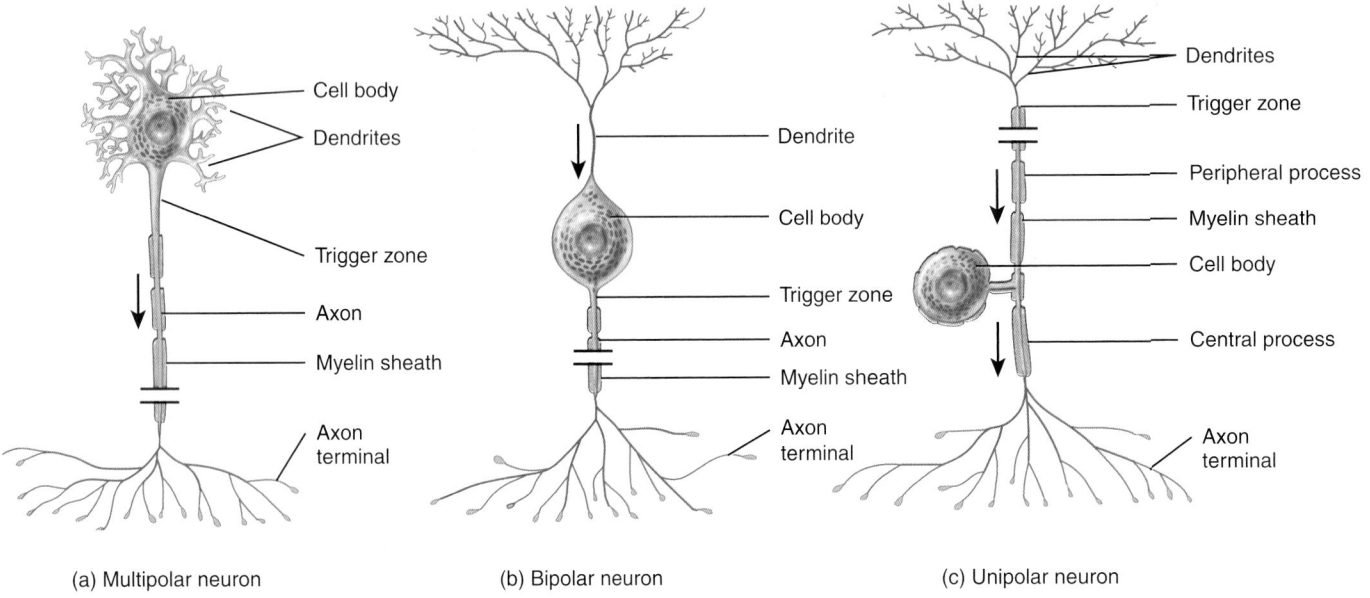

(a) Multipolar neuron

(b) Bipolar neuron

(c) Unipolar neuron

? **Which type of neuron shown in this figure is the most abundant type of neuron in the CNS?**

1. **Multipolar neurons** usually have several dendrites and one axon (Figure 12.3a). Most neurons in the brain and spinal cord are of this type, as well as all motor neurons (described shortly).

2. **Bipolar neurons** have one main dendrite and one axon (Figure 12.3b). They are found in the retina of the eye, the inner ear, and the olfactory (*olfact* = to smell) area of the brain.

3. **Unipolar neurons** have dendrites and one axon that are fused together to form a continuous process that emerges from the cell body (Figure 12.3c). These neurons are more appropriately called **pseudounipolar neurons** (soo′-dō-ū′-ni-PŌ-lar) because they begin in the embryo as bipolar neurons. During development, the dendrites and axon fuse together and become a single process. The dendrites of most unipolar neurons function as **sensory receptors** that detect a sensory stimulus such

as touch, pressure, pain, or thermal stimuli (see Figure 12.10). The trigger zone for nerve impulses in a unipolar neuron is at the junction of the dendrites and axon (Figure 12.3c). The impulses then propagate toward the synaptic end bulbs. The cell bodies of most unipolar neurons are located in the ganglia of spinal and cranial nerves. Several examples of sensory receptors that are dendrites of unipolar neurons are illustrated in Figure 12.4.

In addition to the structural classification scheme just described, some neurons are named for the histologist who first described them or for an aspect of their shape or appearance; examples include **Purkinje cells** (pur-KIN-jē) in the cerebellum and **pyramidal cells** (pi-RAM-i-dal), found in the cerebral cortex of the brain, which have pyramid-shaped cell bodies (Figure 12.5).

Figure 12.4 Examples of sensory receptors that are dendrites of unipolar neurons. Breaks indicate that axons are longer than shown. (a) A Meissner corpuscle is a touch receptor that consists of a mass of dendrites enclosed by a capsule of connective tissue. (b) A Merkel disc is a touch receptor that consists of free nerve endings (bare dendrites) that make contact with Merkel cells of the stratum basale of the skin. (c) A pacinian corpuscle is a pressure receptor composed of a multilayered connective tissue capsule that encloses a dendrite. (d) A nociceptor is a pain receptor that consists of free nerve endings (bare dendrites). Thermoreceptors (which detect thermal sensations), itch receptors, and tickle receptors resemble nociceptors in that they are unipolar neurons with free nerve endings that serve as sensory receptors.

The type of sensory receptor found in the dendrites of a unipolar neuron determines the type of sensory stimulus that the unipolar neuron can detect.

(a)

(b)

(c)

(d)

? **Where are most unipolar neurons located in the body?**

Figure 12.5 Two examples of CNS neurons. Arrows indicate the direction of information flow.

🔑 ⚏ The dendritic branching pattern often is distinctive for a particular type of neuron.

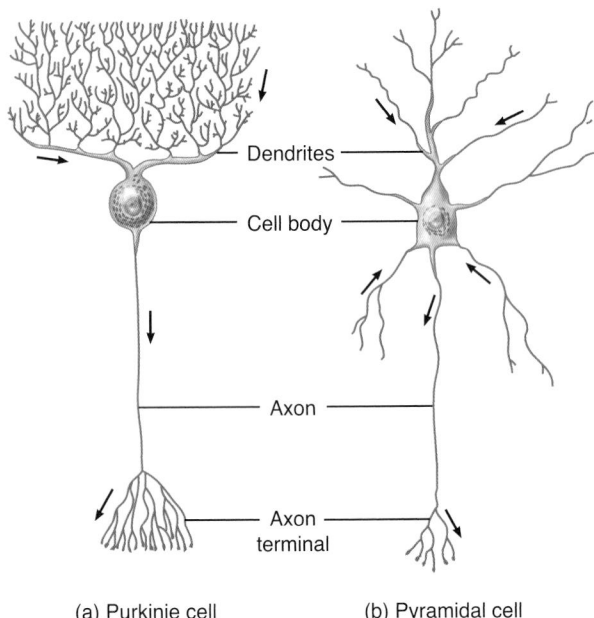

(a) Purkinje cell   (b) Pyramidal cell

❓ **How did the pyramidal cells get their name?**

FUNCTIONAL CLASSIFICATION Functionally, neurons are classified according to the direction in which the nerve impulse (action potential) is conveyed with respect to the CNS.

1. **Sensory** or **afferent neurons** (AF-er-ent NOO-ronz; *af-* = toward; *-ferrent* = carried) either contain sensory receptors at their distal ends (dendrites) (see Figure 12.10) or are located just after sensory receptors that are separate cells. Once an appropriate stimulus activates a sensory receptor, the sensory neuron forms an action potential in its axon and the action potential is conveyed *into* the CNS through cranial or spinal nerves. Most sensory neurons are unipolar in structure.

2. **Motor** or **efferent neurons** (EF-e-rent; *ef-* = away from) convey action potentials *away* from the CNS to **effectors** (muscles and glands) in the periphery (PNS) through cranial or spinal nerves (see Figure 12.10). Motor neurons are multipolar in structure.

3. **Interneurons** or **association neurons** are mainly located within the CNS between sensory and motor neurons (see Figure 12.10). Interneurons integrate (process) incoming sensory information from sensory neurons and then elicit a motor response by activating the appropriate motor neurons. Most interneurons are multipolar in structure.

Neuroglia

Neuroglia (noo-RŌG-lē-a; *-glia* = glue) or **glia** (GLĒ-a) make up about half the volume of the CNS. Their name derives from the idea of early histologists that they were the "glue" that held nervous tissue together. We now know that neuroglia are not merely passive bystanders but rather actively participate in the activities of nervous tissue. Generally, neuroglia are smaller than neurons, and they are 5 to 25 times more numerous. In contrast to neurons, glia do not generate or propagate action potentials, and they can multiply and divide in the mature nervous system. In cases of injury or disease, neuroglia multiply to fill in the spaces formerly occupied by neurons. Brain tumors derived from glia, called **gliomas** (glē-Ō-mas), tend to be highly malignant and to grow rapidly. Of the six types of neuroglia, four—astrocytes, oligodendrocytes, microglia, and ependymal cells—are found only in the CNS. The remaining two types—Schwann cells and satellite cells—are present in the PNS.

Neuroglia of the CNS

Neuroglia of the CNS can be classified on the basis of size, cytoplasmic processes, and intracellular organization into four types: astrocytes, oligodendrocytes, microglia, and ependymal cells (Figure 12.6).

ASTROCYTES (AS-trō-sīts; *astro-* = star; *-cyte* = cell) These star-shaped cells have many processes and are the largest and most numerous of the neuroglia. There are two types of **astrocytes.** *Protoplasmic astrocytes* have many short branching processes and are found in gray matter (described shortly). *Fibrous astrocytes* have many long unbranched processes and are located mainly in white matter (also described shortly). The processes of astrocytes make contact with blood capillaries, neurons, and the pia mater (a thin membrane around the brain and spinal cord).

The functions of astrocytes include the following: (1) Astrocytes contain microfilaments that give them considerable strength, which enables them to support neurons. (2) Processes of astrocytes wrapped around blood capillaries isolate neurons of the CNS from various potentially harmful substances in blood by secreting chemicals that maintain the unique selective permeability characteristics of the endothelial cells of the capillaries. In effect, the endothelial cells create a *blood–brain barrier,* which restricts the movement of substances between the blood and interstitial fluid of the CNS. Details of the blood–brain barrier are discussed in Chapter 14. (3) In the embryo, astrocytes secrete chemicals that appear to regulate the growth, migration, and interconnection among neurons in the brain. (4) Astrocytes help to maintain the appropriate chemical environment for the generation of nerve impulses. For example, they regulate the concentration of important ions such as K^+; take up excess neurotransmitters; and serve as a conduit for the passage of nutrients and other substances between blood capillaries and neurons. (5) Astrocytes may also play a role in learning and memory by influencing the formation of neural synapses (see Section 16.5).

OLIGODENDROCYTES (OL-i-gō-den′-drō-sīts; *oligo-* = few; *-dendro-* = tree) These resemble astrocytes but are smaller and contain fewer processes. **Oligodendrocyte** processes are responsible for forming and maintaining the myelin sheath around CNS axons. As you will see shortly, the **myelin sheath** is a multilayered lipid and protein covering around some axons that insulates them and increases the speed of nerve impulse conduction. Such axons are said to be **myelinated** (mī-e-li-NĀ-ted).

MICROGLIA   (mī-KROG-lē-a; *micro-* = small) These neuroglia are small cells with slender processes that give off numerous spinelike projections. **Microglia** function as phagocytes. Like tissue macrophages, they remove cellular debris formed during normal development of the nervous system and phagocytize microbes and damaged nervous tissue.

EPENDYMAL CELLS   (ep-EN-de-mal; *epen-* = above; *-dym-* = garment) **Ependymal cells** are cuboidal to columnar cells arranged in a single layer that possess microvilli and cilia. These cells line the ventricles of the brain and central canal of the spinal cord (spaces filled with cerebrospinal fluid, which protects and nourishes the brain and spinal cord). Functionally, ependymal cells produce, possibly monitor, and assist in the circulation of cerebrospinal fluid. They also form the blood–cerebrospinal fluid barrier, which is discussed in Chapter 14.

Figure 12.6 Neuroglia of the central nervous system.

Neuroglia of the CNS are distinguished on the basis of size, cytoplasmic processes, and intracellular organization.

Cells of pia mater
Oligodendrocyte
Microglial cell
Neuron
Blood capillary
Fibrous astrocytes
Protoplasmic astrocyte
Microglial cell
Ependymal cell
Microvillus
Cilia

Protoplasmic astrocyte
Node of Ranvier
Myelin sheath
Axon
Oligodendrocyte
Neurons

Ventricle

? **Which CNS neuroglia function as phagocytes?**

Neuroglia of the PNS

Neuroglia of the PNS completely surround axons and cell bodies. The two types of glial cells in the PNS are Schwann cells and satellite cells (Figure 12.7).

SCHWANN CELLS   (SCHVON or SCHWON) These cells encircle PNS axons. Like oligodendrocytes, they form the myelin sheath around axons. However, a single oligodendrocyte myelinates several axons, but each **Schwann cell** myelinates a single axon (Figure 12.7a; see also Figure 12.8a, c). A single Schwann cell can also enclose as many as 20 or more unmyelinated axons (axons that lack a myelin sheath) (Figure 12.7b). Schwann cells participate in axon regeneration, which is more easily accomplished in the PNS than in the CNS.

SATELLITE CELLS   (SAT-i-līt) These flat cells surround the cell bodies of neurons of PNS ganglia (Figure 12.7c). Besides providing structural support, **satellite cells** regulate the exchanges of materials between neuronal cell bodies and interstitial fluid.

Myelination

As you have already learned, axons surrounded by a multilayered lipid and protein covering, called the myelin sheath, are said to be **myelinated** (Figure 12.8a). The sheath electrically insulates the axon of a neuron and increases the speed of nerve impulse conduction. Axons without such a covering are said to be **unmyelinated** (Figure 12.8b).

Two types of neuroglia produce myelin sheaths: Schwann cells (in the PNS) and oligodendrocytes (in the CNS). Schwann cells begin to form myelin sheaths around axons during fetal development. Each Schwann cell wraps about 1 millimeter (1 mm = 0.04 in.) of a single axon's length by spiraling many times around the axon (Figure 12.8a). Eventually, multiple layers of glial plasma membrane surround the axon, with the Schwann cell's cytoplasm and nucleus forming the outermost layer. The inner portion, consisting of up to 100 layers of Schwann cell membrane, is the myelin sheath. The outer nucleated cytoplasmic layer of the Schwann cell, which encloses the myelin sheath, is the **neurolemma (sheath of Schwann)** (noo'-rō-LEM-ma). A neurolemma is found only around axons in the PNS. When an axon is injured, the neurolemma aids regeneration by forming a regeneration tube that guides and stimulates regrowth of the axon. Gaps in the myelin sheath, called **nodes of Ranvier** (RON-vē-ā), appear at intervals along the axon (Figure 12.8; see also Figure 12.2). Each Schwann cell wraps one axon segment between two nodes.

In the CNS, an oligodendrocyte myelinates parts of several axons. Each oligodendrocyte puts forth about 15 broad, flat processes that spiral around CNS axons, forming a myelin sheath. A neurolemma is not present, however, because the oligodendrocyte cell body and nucleus do not envelop the axon. Nodes of Ranvier are present, but they are fewer in number. Axons in the CNS display little regrowth after injury. This is thought to be due, in part, to the absence of a neurolemma, and in part to an inhibitory influence exerted by the oligodendrocytes on axon regrowth.

Figure 12.7 Neuroglia of the peripheral nervous system.

Neuroglia of the PNS completely surround axons and cell bodies of neurons.

(a)   (b)   (c)

? **How do Schwann cells and oligodendrocytes differ with respect to the number of axons they myelinate?**

The amount of myelin increases from birth to maturity, and its presence greatly increases the speed of nerve impulse conduction. An infant's responses to stimuli are neither as rapid nor as coordinated as those of an older child or an adult, in part because myelination is still in progress during infancy.

Collections of Nervous Tissue

The components of nervous tissue are grouped together in a variety of ways. Neuronal cell bodies are often grouped together in clusters. The axons of neurons are usually grouped together in

Figure 12.8 Myelinated and unmyelinated axons. Notice that one layer of Schwann cell plasma membrane surrounds unmyelinated axons.

Axons surrounded by a myelin sheath produced either by Schwann cells in the PNS or by oligodendrocytes in the CNS are said to be myelinated.

(a) Transverse sections of stages in the formation of a myelin sheath

(b) Transverse section of unmyelinated axons

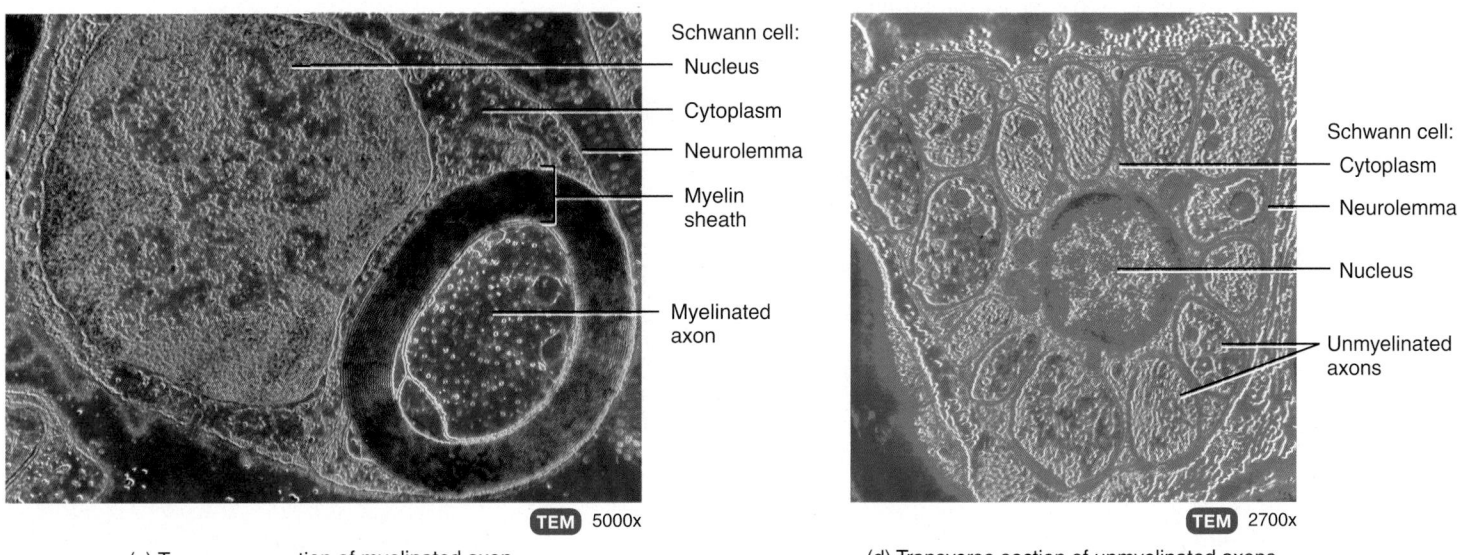

TEM 5000x

(c) Transverse section of myelinated axon

TEM 2700x

(d) Transverse section of unmyelinated axons

? **What is the functional advantage of myelination?**

bundles. In addition, widespread regions of nervous tissue are grouped together as either gray matter or white matter.

Clusters of Neuronal Cell Bodies

Recall that a **ganglion** (plural is *ganglia*) refers to a cluster of neuronal cell bodies located in the *PNS*. As mentioned earlier, ganglia are closely associated with cranial and spinal nerves. By contrast, a **nucleus** is a cluster of neuronal cell bodies located in the *CNS*.

Bundles of Axons

Recall that a **nerve** is a bundle of axons that is located in the *PNS*. Cranial nerves connect the brain to the periphery, whereas spinal nerves connect the spinal cord to the periphery. A **tract** is a bundle of axons that is located in the *CNS*. Tracts interconnect neurons in the spinal cord and brain.

Gray and White Matter

In a freshly dissected section of the brain or spinal cord, some regions look white and glistening, and others appear gray (Figure 12.9). **White matter** is composed primarily of myelinated axons. The whitish color of myelin gives white matter its name. The **gray matter** of the nervous system contains neuronal cell bodies, dendrites, unmyelinated axons, axon terminals, and neuroglia. It appears grayish, rather than white, because the Nissl bodies impart a gray color and there is little or no myelin in these areas. Blood vessels are present in both white and gray matter. In the spinal cord, the white matter surrounds an inner core of gray matter that, depending on how imaginative you are, is shaped like a butterfly or the letter H; in the brain, a thin shell of gray matter covers the surface of the largest portions of the brain, the cerebrum and cerebellum (Figure 12.9). The arrangement of gray matter

and white matter in the spinal cord and brain is discussed more extensively in Chapters 13 and 14, respectively.

✔ CHECKPOINT

5. Describe the parts of a neuron and the functions of each.
6. Give several examples of the structural and functional classifications of neurons.
7. What is a neurolemma and why is it important?
8. With reference to the nervous system, what is a nucleus?

12.3 ELECTRICAL SIGNALS IN NEURONS

◉ OBJECTIVES

• Describe the cellular properties that permit communication among neurons and effectors.
• Compare the basic types of ion channels, and explain how they relate to graded potentials and action potentials.
• Describe the factors that maintain a resting membrane potential.
• List the sequence of events that generate an action potential.

Like muscle fibers, neurons are electrically excitable. They communicate with one another using two types of electrical signals: (1) **Graded potentials** (described shortly) are used for short-distance communication only. (2) **Action potentials** (also described shortly) allow communication over long distances within the body. Recall that an action potential in a muscle fiber is called a **muscle action potential.** When an action potential occurs in a neuron (nerve cell), it is called a **nerve action potential (nerve**

Figure 12.9 Distribution of gray matter and white matter in the spinal cord and brain.

White matter primarily consists of myelinated axons of many neurons. Gray matter consists of neuron cell bodies, dendrites, unmyelinated axons, axon terminals, and neuroglia.

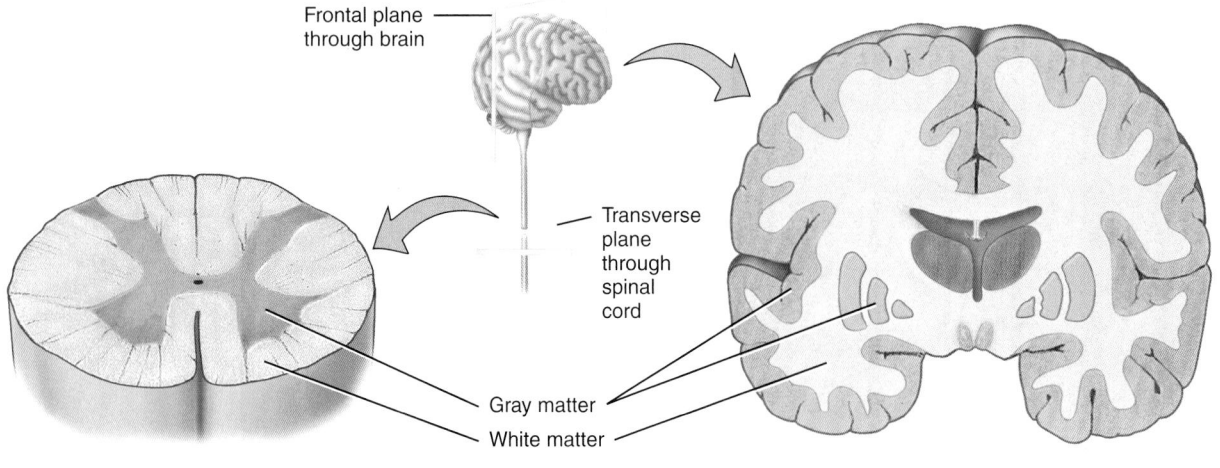

Frontal plane through brain

Transverse plane through spinal cord

Gray matter
White matter

(a) Transverse section of spinal cord

(b) Frontal section of brain

? **What is responsible for the white appearance of white matter?**

impulse). To understand the functions of graded potentials and action potentials, consider how the nervous system allows you to feel the smooth surface of a pen that you have picked up off of a table (Figure 12.10):

1 As you touch the pen, a graded potential develops in a sensory receptor in the skin of the fingers.

2 The graded potential triggers the axon of the sensory neuron to form a nerve action potential, which travels along the axon into the CNS and ultimately causes the release of neurotransmitter at a synapse with an interneuron.

Figure 12.10 Overview of nervous system functions.

Graded potentials and nerve and muscle action potentials are involved in the relay of sensory stimuli, integrative functions such as perception, and motor activities.

In which region of the brain does perception primarily occur?

③ The neurotransmitter stimulates the interneuron to form a graded potential in its dendrites and cell body.

④ In response to the graded potential, the axon of the interneuron forms a nerve action potential. The nerve action potential travels along the axon, which results in neurotransmitter release at the next synapse with another interneuron.

⑤ This process of neurotransmitter release at a synapse followed by the formation of a graded potential and then a nerve action potential occurs over and over as interneurons in higher parts of the brain (such as the thalamus and cerebral cortex) are activated. Once interneurons in the **cerebral cortex,** the outer part of the brain, are activated, perception occurs and you are able to feel the smooth surface of the pen touch your fingers. As you will learn in Chapter 14, perception, the conscious awareness of a sensation, is primarily a function of the cerebral cortex.

Suppose that you want to use the pen to write a letter. The nervous system would respond in the following way (Figure 12.10):

⑥ A stimulus in the brain causes a graded potential to form in the dendrites and cell body of an **upper motor neuron,** a type of motor neuron that synapses with a lower motor neuron farther down in the CNS in order to contract a skeletal muscle. The graded potential subsequently causes a nerve action potential to occur in the axon of the upper motor neuron, followed by neurotransmitter release.

⑦ The neurotransmitter generates a graded potential in a **lower motor neuron,** a type of motor neuron that directly supplies skeletal muscle fibers. The graded potential triggers the formation of a nerve action potential and then release of the neurotransmitter at neuromuscular junctions formed with skeletal muscle fibers that control movements of the fingers.

⑧ The neurotransmitter stimulates the muscle fibers that control finger movements to form muscle action potentials. The muscle action potentials cause these muscle fibers to contract, which allows you to write with the pen.

The production of graded potentials and action potentials depends on two basic features of the plasma membrane of excitable cells: the existence of a resting membrane potential and the presence of specific types of ion channels. Like most other cells in the body, the plasma membrane of excitable cells exhibits a **membrane potential,** an electrical potential difference (voltage) across the membrane. In excitable cells, this voltage is termed the **resting membrane potential.** The membrane potential is like voltage stored in a battery. If you connect the positive and negative terminals of a battery with a piece of wire, electrons will flow along the wire. This flow of charged particles is called **current.** In living cells, the flow of ions (rather than electrons) constitutes the electrical current.

Graded potentials and action potentials occur because the membranes of neurons contain many different kinds of ion channels that open or close in response to specific stimuli. Because the lipid bilayer of the plasma membrane is a good electrical insulator, the main paths for current to flow across the membrane are through the ion channels.

Ion Channels

When ion channels are open, they allow specific ions to move across the plasma membrane, down their **electrochemical gradient**—a concentration (chemical) difference plus an electrical difference. Recall that ions move from areas of higher concentration to areas of lower concentration (the chemical part of the gradient). Also, positively charged cations move toward a negatively charged area, and negatively charged anions move toward a positively charged area (the electrical aspect of the gradient). As ions move, they create a flow of electrical current that can change the membrane potential.

Ion channels open and close due to the presence of "gates." The gate is a part of the channel protein that can seal the channel pore shut or move aside to open the pore (see Figure 3.6). The electrical signals produced by neurons and muscle fibers rely on four types of ion channels: leak channels, ligand-gated channels, mechanically gated channels, and voltage-gated channels:

1. The gates of **leak channels** randomly alternate between open and closed positions (Figure 12.11a). Typically, plasma membranes have many more potassium ion (K^+) leak channels than sodium ion (Na^+) leak channels, and the potassium ion leak channels are leakier than the sodium ion leak channels. Thus, the membrane's permeability to K^+ is much higher than its permeability to Na^+. Leak channels are found in nearly all cells, including the dendrites, cell bodies, and axons of all types of neurons.

2. A **ligand-gated channel** opens and closes in response to the binding of a ligand (chemical) stimulus. A wide variety of chemical ligands—including neurotransmitters, hormones, and particular ions—can open or close ligand-gated channels. The neurotransmitter acetylcholine, for example, opens cation channels that allow Na^+ and Ca^{2+} to diffuse inward and K^+ to diffuse outward (Figure 12.11b). Ligand-gated channels are located in the dendrites of some sensory neurons, such as pain receptors, and in dendrites and cell bodies of interneurons and motor neurons.

3. A **mechanically gated channel** opens or closes in response to mechanical stimulation in the form of vibration (such as sound waves), touch, pressure, or tissue stretching (Figure 12.11c). The force distorts the channel from its resting position, opening the gate. Examples of mechanically gated channels are those found in auditory receptors in the ears, in receptors that monitor stretching of internal organs, and in touch receptors and pressure receptors in the skin.

4. A **voltage-gated channel** opens in response to a change in membrane potential (voltage) (Figure 12.11d). Voltage-gated channels participate in the generation and conduction of action potentials in the axons of all types of neurons.

Figure 12.11 Ion channels in the plasma membrane. (a) Leak channels randomly open and close. (b) A chemical stimulus—here, the neurotransmitter acetylcholine—opens a ligand-gated channel. (c) A mechanical stimulus opens a mechanically gated channel. (d) A change in membrane potential opens voltage-gated K$^+$ channels during an action potential.

The electrical signals produced by neurons and muscle fibers rely on four types of ion channels: leak channels, ligand-gated channels, mechanically gated channels, and voltage-gated channels.

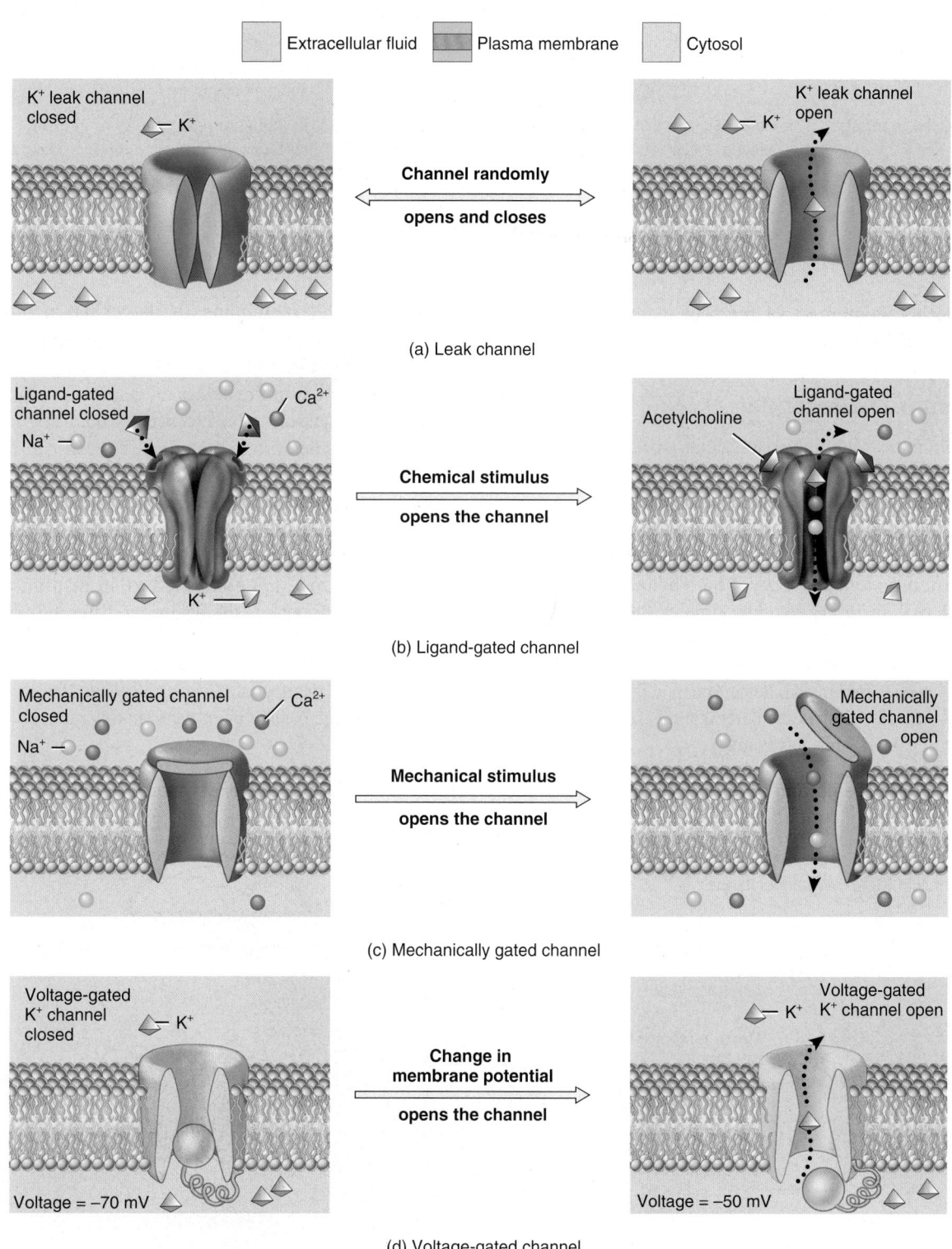

(a) Leak channel

(b) Ligand-gated channel

(c) Mechanically gated channel

(d) Voltage-gated channel

? **What type of gated channel is activated by a touch on the arm?**

TABLE 12.1

Ion Channels in Neurons

TYPE OF ION CHANNEL	DESCRIPTION	LOCATION
Leak channels	Gated channels that randomly open and close.	Found in nearly all cells, including dendrites, cell bodies, and axons of all types of neurons.
Ligand-gated channels	Gated channels that open in response to binding of ligand (chemical) stimulus.	Dendrites of some sensory neurons such as pain receptors and dendrites and cell bodies of interneurons and motor neurons.
Mechanically gated channels	Gated channels that open in response to mechanical stimulus (such as touch, pressure, vibration, or tissue stretching).	Dendrites of some sensory neurons such as touch receptors, pressure receptors, and some pain receptors.
Voltage-gated channels	Gated channels that open in response to voltage stimulus (change in membrane potential).	Axons of all types of neurons.

Table 12.1 presents a summary of the four major types of ion channels in neurons.

Resting Membrane Potential

The resting membrane potential exists because of a small buildup of negative ions in the cytosol along the inside of the membrane, and an equal buildup of positive ions in the extracellular fluid (ECF) along the outside surface of the membrane (Figure 12.12a). Such a separation of positive and negative electrical charges is a form of potential energy, which is measured in volts or millivolts (1 mV = 0.001 V). The greater the difference in charge across the membrane, the larger the membrane potential (voltage). Notice in Figure 12.12a that the buildup of charge occurs only very close to the membrane. The cytosol or extracellular fluid elsewhere in the cell contains equal numbers of positive and negative charges and is electrically neutral.

The resting membrane potential of a cell can be measured in the following way: The tip of a recording microelectrode is inserted inside the cell, and a reference electrode is placed outside the cell in the extracellular fluid. *Electrodes* are devices that conduct electrical charges. The recording microelectrode and the reference electrode are connected to an instrument known as a *voltmeter*, which detects the electrical difference (voltage) across

Figure 12.12 Resting membrane potential. To measure resting membrane potential, the tip of the recording microelectrode is inserted inside the neuron, and the reference electrode is placed in the extracellular fluid. The electrodes are connected to a voltmeter that measures the difference in charge across the plasma membrane (in this case −70 mV inside negative).

🔑 The resting membrane potential is an electrical potential difference (voltage) that exists across the plasma membrane of an excitable cell under resting conditions.

(a) Distribution of charges that produce the resting membrane potential of a neuron

(b) Measurement of the resting membrane potential of a neuron

❓ The resting membrane potential of a neuron typically is −70 mV. What does this mean?

the plasma membrane (Figure 12.12b). In neurons, the resting membrane potential ranges from −40 to −90 mV. A typical value is −70 mV. The minus sign indicates that the inside of the cell is negative relative to the outside. A cell that exhibits a membrane potential is said to be **polarized.** Most body cells are polarized; the membrane potential varies from +5 mV to −100 mV in different types of cells.

The resting membrane potential arises from three major factors:

1. *Unequal distribution of ions in the ECF and cytosol.* A major factor that contributes to the resting membrane potential is the unequal distribution of various ions in extracellular fluid and cytosol (Figure 12.13). Extracellular fluid is rich in Na$^+$

and chloride ions (Cl$^-$). In cytosol, however, the main cation is K$^+$, and the two dominant anions are phosphates attached to molecules, such as the three phosphates in ATP, and amino acids in proteins. Because the plasma membrane typically has more K$^+$ leak channels than Na$^+$ leak channels, the number of potassium ions that diffuse down their concentration gradient out of the cell into the ECF is greater than the number of sodium ions that diffuse down their concentration gradient from the ECF into the cell. As more and more positive potassium ions exit, the inside of the membrane becomes increasingly negative, and the outside of the membrane becomes increasingly positive.

Figure 12.13 Factors that contribute to the resting membrane potential. A major factor that contributes to the inside-negative resting membrane potential of a cell is the unequal distribution of ions in the ECF and cytosol. The ECF is rich in Na$^+$ and Cl$^-$ ions and the cytosol is rich in K$^+$ ions and anions such as phosphates in ATP and amino acids in proteins. Because the plasma membrane has more K$^+$ leak channels than Na$^+$ leak channels, the number of K$^+$ ions that leave the cell is greater than the number of Na$^+$ ions that enter the cell. As more and more K$^+$ ions leave the cell, the inside of the membrane becomes increasingly negative and the outside of the membrane becomes increasingly positive. The inability of most anions to leave the cell also contributes to the negativity of the resting membrane potential. These trapped anions cannot follow K$^+$ out of the cell because they are attached to nondiffusible molecules such as ATP and large proteins. The activity of the Na$^+$/K$^+$ ATPases is another factor that contributes to the resting membrane potential. The Na$^+$/K$^+$ ATPase expels 3 Na$^+$ ions for every 2 K$^+$ ions imported. Since the Na$^+$/K$^+$ ATPase removes more positive charges from the cell than it brings into the cell, it is electrogenic and helps establish the inside-negative resting membrane potential.

🔑 The resting membrane potential is determined by three major factors: (1) unequal distribution of ions in the ECF and cytosol, (2) inability of most anions to leave the cell, and (3) the electrogenic nature of the Na$^+$/K$^+$ ATPases.

? **Suppose that the plasma membrane of a neuron has more Na$^+$ leak channels than K$^+$ leak channels. What effect would this have on the resting membrane potential?**

2. ***Inability of most anions to leave the cell.*** Another factor contributes to the inside-negative resting membrane potential: Most anions inside the cell are not free to leave (Figure 12.13). They cannot follow the K⁺ out of the cell because they are attached to nondiffusible molecules such as ATP and large proteins.

3. ***Electrogenic nature of the Na⁺/K⁺ ATPases.*** Membrane permeability to Na⁺ is very low because there are only a few sodium leak channels. Nevertheless, sodium ions do slowly diffuse inward, down their concentration gradient. Left unchecked, such inward leakage of Na⁺ would eventually destroy the resting membrane potential. The small inward Na⁺ leak and outward K⁺ leak are offset by the Na⁺/K⁺ ATPases (sodium–potassium pumps) (Figure 12.13). These pumps help maintain the resting membrane potential by pumping out Na⁺ as fast as it leaks in. At the same time, the Na⁺/K⁺ ATPases bring in K⁺. However, the potassium ions eventually leak back out of the cell as they move down their concentration gradient. Recall that the Na⁺/K⁺ ATPases expel three Na⁺ for each two K⁺ imported (see Figure 3.10). Since these pumps remove more positive charges from the cell than they bring into the cell, they are *electrogenic,* which means they contribute to the negativity of the resting membrane potential. Their total contribution, however, is very small: only −3 mV of the total −70 mV resting membrane potential in a typical neuron.

Graded Potentials

A **graded potential** is a small deviation from the membrane potential that makes the membrane either more polarized (inside more negative) or less polarized (inside less negative). When the response makes the membrane more polarized (inside more negative), it is termed a **hyperpolarizing graded potential** (hī-per-PŌ-lar-ī′-zing) (Figure 12.14a). When the response makes the membrane less polarized (inside less negative), it is termed a **depolarizing graded potential** (Figure 12.14b).

A graded potential occurs when a stimulus causes mechanically gated or ligand-gated channels to open or close in an excitable cell's plasma membrane (Figure 12.15). Typically, mechanically gated channels and ligand-gated channels can be present in the dendrites of sensory neurons, and ligand-gated channels are numerous in the dendrites and cell bodies of interneurons and motor neurons. Hence, graded potentials occur mainly in the dendrites and cell body of a neuron.

To say that these electrical signals are *graded* means that they vary in amplitude (size), depending on the strength of the stimulus (Figure 12.16). They are larger or smaller depending on how many ligand-gated or mechanically gated channels have opened (or closed) and how long each remains open. The opening or closing of these ion channels alters the flow of specific ions across the membrane, producing a flow of current that is *localized,* which means that it spreads to adjacent regions along the plasma membrane in either direction from the stimulus source for a short distance and then gradually dies out as the charges are lost across the membrane

Figure 12.14 Graded potentials. Most graded potentials occur in the dendrites and cell body (areas colored blue).

 During a hyperpolarizing graded potential, the membrane potential is inside more negative than the resting level. During a depolarizing graded potential, the membrane potential is inside less negative than the resting level.

(a) Hyperpolarizing graded potential

(b) Depolarizing graded potential

? What kind of graded potential describes a change in membrane potential from −70 to −60 mV? From −70 to −80 mV?

Figure 12.15 **Generation of graded potentials in response to the opening of mechanically gated channels or ligand-gated channels.** (a) A mechanical stimulus (pressure) opens a mechanically gated channel that allows passage of cations (mainly Na^+ and Ca^{2+}) into the cell; a depolarizing graded potential occurs because the membrane potential becomes inside less negative than at rest. (b) The neurotransmitter acetylcholine (a ligand stimulus) opens a cation channel that allows passage of Na^+, K^+, and Ca^{2+}; Na^+ inflow is greater than either Ca^{2+} inflow or K^+ outflow, and a depolarizing graded potential occurs because the membrane potential becomes inside less negative than at rest. (c) The neurotransmitter glycine (a ligand stimulus) opens a Cl^- channel that allows passage of Cl^- ions into the cell; a hyperpolarizing graded potential occurs because the membrane potential becomes inside more negative than at rest.

🔑 A graded potential forms in response to the opening of mechanically gated channels or ligand-gated channels.

(a) Depolarizing graded potential caused by pressure, a mechanical stimulus

(b) Depolarizing graded potential caused by the neurotransmitter acetylcholine, a ligand stimulus

(c) Hyperpolarizing graded potential caused by the neurotransmitter glycine, a ligand stimulus

❓ **Which parts of a neuron contain mechanically gated channels? Ligand-gated channels?**

Figure 12.16 The graded nature of graded potentials. As stimulus strength increases (stimuli 1, 2, and 3), the amplitude (size) of each resulting depolarizing graded potential increases. Although not shown, a similar relationship exists between stimulus strength and the amplitude of a hyperpolarizing graded potential.

 The amplitude of a graded potential depends on the stimulus strength. The greater the stimulus strength, the larger the amplitude of the graded potential.

? Why does a stronger stimulus cause a larger graded potential than a weaker stimulus?

Figure 12.17 Summation of graded potentials. Summation of two depolarizing graded potentials happens in response to two stimuli of the same strength that occur very close together in time. The dotted lines represent the individual depolarizing graded potentials that would form if summation did not occur.

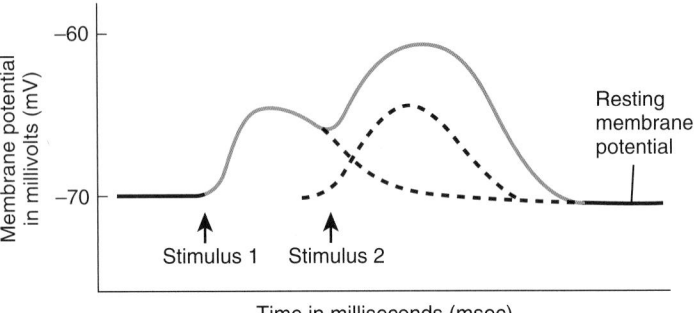 Summation occurs when two or more graded potentials add together to become larger in amplitude.

? What would happen if summation of graded potentials in a neuron did not occur?

through leak channels. This mode of travel by which graded potentials die out as they spread along the membrane is known as **decremental conduction** (dek-re-MENT-al). Because they die out within a few millimeters of their point of origin, graded potentials are useful for short-distance communication only.

Although an individual graded potential undergoes decremental conduction, it can become stronger and last longer by summating with other graded potentials. **Summation** is the process by which graded potentials add together. If two depolarizing graded potentials summate, the net result is a larger depolarizing graded potential (Figure 12.17). If two hyperpolarizing graded potentials summate, the net result is a larger hyperpolarizing graded potential. If two equal but opposite graded potentials summate (one depolarizing and the other hyperpolarizing), then they cancel each other out and the overall graded potential disappears. You will learn more about the process of summation later in this chapter.

Graded potentials have different names depending on which type of stimulus causes them and where they occur. For example, when a graded potential occurs in the dendrites or cell body of a neuron in response to a neurotransmitter, it is called a *postsynaptic potential* (explained shortly). On the other hand, the graded potentials that occur in sensory receptors and sensory neurons are termed *receptor potentials* and *generator potentials* (explained in Chapter 16).

Generation of Action Potentials

An **action potential (AP)** or **impulse** is a sequence of rapidly occurring events that decrease and reverse the membrane potential and then eventually restore it to the resting state. An action potential has two main phases: a depolarizing phase and a repolarizing phase (Figure 12.18). During the **depolarizing phase,** the negative membrane potential becomes less negative, reaches zero, and then becomes positive. During the **repolarizing phase,** the membrane potential is restored to the resting state of -70 mV. Following the repolarizing phase there may be an **after-hyperpolarizing phase,** during which the membrane potential temporarily becomes more negative than the resting level. Two types of voltage-gated channels open and then close during an action potential. These channels are present mainly in the axon plasma membrane and axon terminals. The first channels that open, the voltage-gated Na^+ channels, allow Na^+ to rush into the cell, which causes the depolarizing phase. Then voltage-gated K^+ channels open, allowing K^+ to flow out, which produces the repolarizing phase. The after-hyperpolarizing phase occurs when the voltage-gated K^+ channels remain open after the repolarizing phase ends.

An action potential occurs in the membrane of the axon of a neuron when depolarization reaches a certain level termed the **threshold** (about -55 mV in many neurons). Different neurons may have different thresholds for generation of an action potential, but the threshold in a particular neuron usually is constant. The generation of an action potential depends on whether a particular stimulus is able to bring the membrane potential to threshold (Figure 12.19). An action potential will not occur in response to a **subthreshold stimulus,** a stimulus that is a weak depolarization

Figure 12.18 Action potential (AP) or impulse. When a stimulus depolarizes the membrane to threshold (−55 mV), an AP is generated. The action potential arises at the trigger zone (here, at the junction of the axon hillock and the initial segment) and then propagates along the axon to the axon terminals. The green-colored regions of the neuron indicate parts that typically have voltage-gated Na^+ and K^+ channels (axon plasma membrane and axon terminals).

An action potential consists of a depolarizing phase and a repolarizing phase, which may be followed by an after-hyperpolarizing phase.

Key:

	Resting membrane potential: Voltage-gated Na^+ channels are in the resting state and voltage-gated K^+ channels are closed
	Stimulus causes depolarization to threshold
	Voltage-gated Na^+ channel activation gates are open — Absolute refractory period
	Voltage-gated K^+ channels are open; Na^+ channels are inactivating
	Voltage-gated K^+ channels are still open; Na^+ channels are in the resting state — Relative refractory period

? **Which channels are open during the depolarizing phase? During the repolarizing phase?**

Figure 12.19 Stimulus strength and action potential generation. A subthreshold stimulus does not cause an action potential because it does not bring the membrane potential to threshold. However, an action potential does occur in response to a threshold stimulus because a threshold stimulus is just strong enough to depolarize the membrane to threshold. Several action potentials form in response to a suprathreshold stimulus, which depolarizes the membrane above threshold. Each of the action potentials caused by the suprathreshold stimulus has the same amplitude (size) as the action potential caused by the threshold stimulus. For simplicity, the after-hyperpolarizing phase of the action potential is not shown.

An action potential will only occur once the membrane potential reaches threshold.

? **Will an action potential occur in response to a hyperpolarizing graded potential that spreads from the dendrites or cell body to the trigger zone of the axon of a neuron? Why or why not?**

that cannot bring the membrane potential to threshold. However, an action potential will occur in response to a **threshold stimulus,** a stimulus that is just strong enough to depolarize the membrane to threshold. Several action potentials will form in response to a **suprathreshold stimulus,** a stimulus that is strong enough to depolarize the membrane *above* threshold. Each of the action potentials caused by a suprathreshold stimulus has the same amplitude (size) as an action potential caused by a threshold stimulus. Therefore, once an action potential is generated, the amplitude of an action potential is always the same and does not depend on stimulus intensity. Instead, the greater the stimulus strength above threshold, the greater the frequency of the action potentials until a maximum frequency is reached as determined by the absolute refractory period (described shortly).

As you have just learned, an action potential is generated in response to a threshold stimulus but does not form when there is a subthreshold stimulus. In other words, an action potential either occurs completely or it does not occur at all. This characteristic of an action potential is known as the **all-or-none principle.** The all-or-none principle of the action potential is similar to pushing the first domino in a long row of standing dominoes. When the push on the first domino is strong enough (when depolarization reaches threshold), that domino falls against the second domino, and the *entire* row topples (an action potential occurs). Stronger pushes on the first domino produce the identical effect—toppling of the entire row. Thus, pushing on the first domino produces an all-or-none event: The dominoes all fall or none fall.

Depolarizing Phase

When a depolarizing graded potential or some other stimulus causes the membrane of the axon to depolarize to threshold, voltage-gated Na^+ channels open rapidly. Both the electrical and the chemical gradients favor inward movement of Na^+, and the resulting inrush of Na^+ causes the depolarizing phase of the action potential (see Figure 12.18). The inflow of Na^+ changes the membrane potential from -55 mV to $+30$ mV. At the peak of the action potential, the inside of the membrane is 30 mV more positive than the outside.

Each voltage-gated Na^+ channel has two separate gates, an *activation gate* and an *inactivation gate.* In the *resting state* of a voltage-gated Na^+ channel, the inactivation gate is open, but the activation gate is closed (step 1 in Figure 12.20). As a result, Na^+ cannot move into the cell through these channels. At threshold, voltage-gated Na^+ channels are activated. In the *activated state* of a voltage-gated Na^+ channel, both the activation and inactivation gates in the channel are open and Na^+ inflow begins (step 2 in Figure 12.20). As more channels open, Na^+ inflow increases, the membrane depolarizes further, and more Na^+ channels open. This is an example of a positive feedback mechanism. During the few ten-thousandths of a second that the voltage-gated Na^+ channel is open, about 20,000 Na^+ flow across the membrane and change the membrane potential considerably. But the concentration of Na^+ hardly changes because of the millions of Na^+ present in the extracellular fluid. The sodium–potassium pumps easily bail out the 20,000 or so Na^+ that enter the cell during a single action potential and maintain the low concentration of Na^+ inside the cell.

Repolarizing Phase

Shortly after the activation gates of the voltage-gated Na^+ channels open, the inactivation gates close (step 3 in Figure 12.20). Now the voltage-gated Na^+ channel is in an *inactivated state.* In addition to opening voltage-gated Na^+ channels, a threshold-level depolarization also opens voltage-gated K^+ channels (steps 3 and 4 in Figure 12.20). Because the voltage-gated K^+ channels open more slowly, their opening occurs at about the same time the voltage-gated Na^+ channels are closing. The slower opening of voltage-gated K^+ channels and the closing of previously open voltage-gated Na^+ channels produce the repolarizing phase of the action potential. As the Na^+ channels are inactivated, Na^+ inflow slows. At the same time, the K^+ channels are opening, accelerating K^+ outflow. Slowing of Na^+ inflow and acceleration of K^+ outflow cause the membrane potential to change from $+30$ mV to -70 mV. Repolarization also allows inactivated Na^+ channels to revert to the resting state.

After-hyperpolarizing Phase

While the voltage-gated K^+ channels are open, outflow of K^+ may be large enough to cause an after-hyperpolarizing phase of the action potential (see Figure 12.18). During this phase, the voltage-gated K^+ channels remain open and the membrane potential becomes even more negative (about -90 mV). As the voltage-gated K^+ channels close, the membrane potential returns to the resting level of -70 mV. Unlike voltage-gated Na^+ channels, most voltage-gated K^+ channels do not exhibit an inactivated state. Instead, they alternate between closed (resting) and open (activated) states.

Refractory Period

The period of time after an action potential begins during which an excitable cell cannot generate another action potential in response to a *normal* threshold stimulus is called the **refractory period** (rē-FRAK-tor-ē) (see key in Figure 12.18). During the **absolute refractory period,** even a very strong stimulus cannot initiate a second action potential. This period coincides with the period of Na^+ channel activation and inactivation (steps 2–4 in Figure 12.20). Inactivated Na^+ channels cannot reopen; they first must return to the resting state (step 1 in Figure 12.20). In contrast to action potentials, graded potentials do not exhibit a refractory period.

Large-diameter axons have a larger surface area and have a brief absolute refractory period of about 0.4 msec. Because a second nerve impulse can arise very quickly, up to 1000 impulses per second are possible. Small-diameter axons have absolute refractory periods as long as 4 msec, enabling them to transmit a maximum of 250 impulses per second. Under normal body conditions, the maximum frequency of nerve impulses in different axons ranges between 10 and 1000 per second.

The **relative refractory period** is the period of time during which a second action potential can be initiated, but only by a larger-than-normal stimulus. It coincides with the period when

Figure 12.20 **Changes in ion flow through voltage-gated channels during the depolarizing and repolarizing phases of an a‾ potential.** Leak channels and sodium–potassium pumps are not shown.

Inflow of sodium ions (Na⁺) causes the depolarizing phase, and outflow of potassium ions (K⁺) causes the repolarizing phase of an action potential.

Extracellular fluid   Plasma membrane   Cytosol

1. **Resting state:**
   All voltage-gated Na⁺ and K⁺ channels are closed. Axon plasma membrane is at resting membrane potential: small buildup of negative charges along inside surface of membrane and equal buildup of positive charges along outside surface of membrane.

2. **Depolarizing phase:**
   When membrane potential of axon reaches threshold, Na⁺ channel activation gates open. As Na⁺ ions move through these channels into neuron, buildup of positive charges forms along inside surface of membrane and membrane becomes depolarized.

4. **Repolarization phase continues:**
   K⁺ outflow continues. As more K⁺ ions leave neuron, more negative charges build up along inside surface of membrane. K⁺ outflow eventually restores resting membrane potential. Na⁺ channel inactivation gates open. Return to resting state occurs when K⁺ gates close.

3. **Repolarizing phase begins:**
   Na⁺ channel inactivation gates close and K⁺ channels open. Membrane starts to become repolarized as some K⁺ ions leave neuron and few negative charges begin to build up along inside surface of membrane.

? **Given the existence of leak channels for both K⁺ and Na⁺, could the membrane repolarize if the voltage-gated K⁺ channels did not exist?**

the voltage-gated K⁺ channels are still open after inactivated Na⁺ channels have returned to their resting state (see Figure 12.18).

Propagation of Action Potentials

To communicate information from one part of the body to another, action potentials in a neuron must travel from where they arise at the trigger zone of the axon to the axon terminals. In contrast to the graded potential, an action potential is not decremental (it does not die out). Instead, an action potential keeps its strength as it spreads along the membrane. This mode of conduction is called **propagation** (prop′-a-GĀ-shun), and it depends on positive feedback. As you have already learned, when sodium ions flow in, they cause voltage-gated Na⁺ channels in adjacent segments of the membrane to open. Thus, the action potential travels along the membrane rather like the activity of that long row of dominoes. In actuality, it is not the same action potential that propagates along the entire axon. Instead, the action potential regenerates over and over at adjacent regions of membrane from the trigger zone to the axon terminals. In a neuron, an action potential can propagate in this direction only—it cannot propagate back toward the cell body because any region of membrane that has just undergone an action potential is temporarily in the absolute refractory period and cannot generate another action potential. Because they can travel along a membrane without dying out, action potentials function in communication over long distances.

Continuous and Saltatory Conduction

There are two types of propagation: continuous conduction and saltatory conduction. The type of action potential propagation described so far is **continuous conduction,** which involves step-by-step depolarization and repolarization of each adjacent segment of the plasma membrane (Figure 12.21a). In continuous conduction, ions flow through their voltage-gated channels in each adja-

cent segment of the membrane. Note that the action potential propagates only a relatively short distance in a few milliseconds. Continuous conduction occurs in unmyelinated axons and in muscle fibers.

Action potentials propagate more rapidly along myelinated axons than along unmyelinated axons. If you compare parts a and b in Figure 12.21 you will see that the action potential propagates much farther along the myelinated axon in the same period of time. **Saltatory conduction** (SAL-ta-tō-rē; *saltat-* = leaping), the special mode of action potential propagation that occurs along myelinated axons, occurs because of the uneven distribution of voltage-gated channels. Few voltage-gated channels are present in regions where a myelin sheath covers the axolemma. By contrast, at the nodes of Ranvier (where there is no myelin sheath), the axolemma has many voltage-gated channels. Hence, current carried by Na⁺ and K⁺ flows across the membrane mainly at the nodes.

When an action potential propagates along a myelinated axon, an electric current (carried by ions) flows through the extracellular fluid surrounding the myelin sheath and through the cytosol from one node to the next. The action potential at the first node generates ionic currents in the cytosol and extracellular fluid that depolarize the membrane to threshold, opening voltage-gated Na⁺ channels at the second node. The resulting ionic flow through the opened channels constitutes an action potential at the second node. Then, the action potential at the second node generates an ionic current that opens voltage-gated Na⁺ channels at the third node, and so on. Each node repolarizes after it depolarizes.

The flow of current across the membrane only at the nodes of Ranvier has two consequences:

1. The action potential appears to "leap" from node to node as each nodal area depolarizes to threshold, thus the name "saltatory." Because an action potential leaps across long segments of the myelinated axolemma as current flows from one node to the next, it travels much faster than it would in an unmyelinated axon of the same diameter.

2. Opening a smaller number of channels only at the nodes, rather than many channels in each adjacent segment of membrane, represents a more energy-efficient mode of conduction. Because only small regions of the membrane depolarize and repolarize, minimal inflow of Na⁺ and outflow of K⁺ occurs each time an action potential passes by. Thus, less ATP is used by sodium–potassium pumps to maintain the low intracellular concentration of Na⁺ and the low extracellular concentration of K⁺.

Factors That Affect the Speed of Propagation

The speed of propagation of an action potential is affected by three major factors: amount of myelination, axon diameter, and temperature.

1. *Amount of myelination.* As you have just learned, action potentials propagate more rapidly along myelinated axons than along unmyelinated axons.

Figure 12.21 **Propagation of an action potential in a neuron after it arises at the trigger zone.** Dotted lines indicate ionic current flow. The insets show the path of current flow. (a) In continuous conduction along an unmyelinated axon, ionic currents flow across each adjacent segment of the membrane. (b) In saltatory conduction along a myelinated axon, the action potential (nerve impulse) at the first node generates ionic currents in the cytosol and interstitial fluid that open voltage-gated Na$^+$ channels at the second node, and so on at each subsequent node.

Unmyelinated axons exhibit continuous conduction; myelinated axons exhibit saltatory conduction.

(a) Continuous conduction

(b) Saltatory conduction

? **What factors determine the speed of propagation of an action potential?**

2. *Axon diameter.* Larger-diameter axons propagate action potentials faster than smaller ones due to their larger surface areas.

3. *Temperature.* Axons propagate action potentials at lower speeds when cooled.

Classification of Nerve Fibers

Axons can be classified into three major groups based on the amount of myelination, their diameters, and their propagation speeds:

• **A fibers** are the largest-diameter axons (5–20 μm) and are myelinated. A fibers have a brief absolute refractory period and conduct nerve impulses (action potentials) at speeds of 12 to 130 m/sec (27–290 mi/hr). The axons of sensory neurons that propagate impulses associated with touch, pressure, position

of joints, and some thermal and pain sensations are A fibers, as are the axons of motor neurons that conduct impulses to skeletal muscles.

• **B fibers** are axons with diameters of 2–3 μm. Like A fibers, B fibers are myelinated and exhibit saltatory conduction at speeds up to 15 m/sec (34 mi/hr). B fibers have a somewhat longer absolute refractory period than A fibers. B fibers conduct sensory nerve impulses from the viscera to the brain and spinal cord. They also constitute all the axons of the autonomic motor neurons that extend from the brain and spinal cord to the ANS relay stations called autonomic ganglia.

• **C fibers** are the smallest-diameter axons (0.5–1.5 μm) and all are unmyelinated. Nerve impulse propagation along a C fiber ranges from 0.5 to 2 m/sec (1–4 mi/hr). C fibers exhibit the longest absolute refractory periods. These unmyelinated axons conduct some sensory impulses for pain, touch, pressure, heat, and cold from the skin, and pain impulses from the viscera. Autonomic motor fibers that extend from autonomic ganglia to stimulate the heart, smooth muscle, and glands are C fibers. Examples of motor functions of B and C fibers are constricting and dilating the pupils, increasing and decreasing the heart rate, and contracting and relaxing the urinary bladder.

Encoding of Stimulus Intensity

How can your sensory systems detect stimuli of differing intensities if all nerve impulses are the same size? Why does a light touch feel different from firmer pressure? The main answer to this question is the *frequency of action potentials*—how often they are generated at the trigger zone. A light touch generates a low frequency of action potentials. A firmer pressure elicits action potentials that pass down the axon at a higher frequency. In addition to this "frequency code," a second factor is the number of sensory neurons recruited (activated) by the stimulus. A firm pressure stimulates a larger number of pressure-sensitive neurons than does a light touch.

Comparison of Electrical Signals Produced by Excitable Cells

We have seen that excitable cells—neurons and muscle fibers—produce two types of electrical signals: graded potentials and action potentials (impulses). One obvious difference between them is that the propagation of action potentials permits communication over long distances, but graded potentials can function only in short-distance communication because they are not propagated. Table 12.2 presents a summary of the differences between graded potentials and action potentials.

As we discussed in Chapter 10, propagation of a muscle action potential along the sarcolemma and into the T tubule system initiates the events of muscle contraction. Although action potentials in muscle fibers and in neurons are similar, there are some notable differences. The typical resting membrane potential of a neuron is −70 mV, but it is closer to −90 mV in skeletal and cardiac muscle fibers. The duration of a nerve impulse is 0.5–2 msec, but a muscle action potential is considerably longer—about 1.0–5.0 msec for skeletal muscle fibers and 10–300 msec for cardiac and smooth muscle fibers. Finally, the propagation speed of action potentials along the largest-diameter, myelinated axons is about 18 times faster than the propagation speed along the sarcolemma of a skeletal muscle fiber.

✔CHECKPOINT

9. Define the terms resting membrane potential, depolarization, repolarization, nerve impulse, and refractory period and identify the factors responsible for each.
10. How is saltatory conduction different from continuous conduction?
11. What effect does myelination have on the speed of propagation of an action potential?
12. How can you tell the difference between a stroke on the cheek and a slap across the face?

TABLE 12.2

Comparison of Graded Potentials and Action Potentials in Neurons

CHARACTERISTIC	GRADED POTENTIALS	ACTION POTENTIALS
Origin	Arise mainly in dendrites and cell body.	Arise at trigger zones and propagate along axon.
Types of channels	Ligand-gated or mechanically gated ion channels.	Voltage-gated channels for Na^+ and K^+.
Conduction	Decremental (not propagated); permit communication over short distances.	Propagate and thus permit communication over longer distances.
Amplitude (size)	Depending on strength of stimulus, varies from less than 1 mV to more than 50 mV.	All or none; typically about 100 mV.
Duration	Typically longer, ranging from several milliseconds to several minutes.	Shorter, ranging from 0.5 to 2 msec.
Polarity	May be hyperpolarizing (inhibitory to generation of action potential) or depolarizing (excitatory to generation of action potential).	Always consist of depolarizing followed by repolarizing phase and return to resting membrane potential.
Refractory period	Not present, thus summation can occur.	Present, thus summation cannot occur.

12.4 SIGNAL TRANSMISSION AT SYNAPSES

OBJECTIVES

- Explain the events of signal transmission at electrical and chemical synapses.
- Distinguish between spatial and temporal summation.
- Give examples of excitatory and inhibitory neurotransmitters, and describe how they act.

Recall from Chapter 10 that a synapse (SIN-aps) is a region where communication occurs between two neurons or between a neuron and an effector cell (muscle cell or glandular cell). The term **presynaptic neuron** (*pre-* = before) refers to a nerve cell that carries a nerve impulse toward a synapse. It is the cell that sends a signal. A **postsynaptic cell** is the cell that receives a signal. It may be a nerve cell called a **postsynaptic neuron** (*post-* = after) that carries a nerve impulse away from a synapse or an effector cell that responds to the impulse at the synapse.

Most synapses are either **axodendritic** (ak'-so-den-DRIT-ik = from axon to dendrite), while others are **axosomatic** (ak'-sō-sō-MAT-ik = from axon to cell body) or **axoaxonic** (ak'-so-ak-SON-ik = from axon to axon). In addition, synapses may be electrical or chemical and they differ both structurally and functionally.

In Chapter 10 we described the events occurring at one type of synapse, the neuromuscular junction. Our focus in this chapter is on synaptic communication among the billions of neurons in the nervous system. Synapses are essential for homeostasis because they allow information to be filtered and integrated. During learning, the structure and function of particular synapses change. The changes may allow some signals to be transmitted while others are blocked. For example, the changes in your synapses from studying will determine how well you do on your anatomy and physiology tests! Synapses are also important because some diseases and neurological disorders result from disruptions of synaptic communication, and many therapeutic and addictive chemicals affect the body at these junctions.

Electrical Synapses

At an **electrical synapse,** action potentials (impulses) conduct directly between the plasma membranes of adjacent neurons through structures called **gap junctions.** Each gap junction contains a hundred or so tubular *connexons,* which act like tunnels to connect the cytosol of the two cells directly (see Figure 4.2e). As ions flow from one cell to the next through the connexons, the action potential spreads from cell to cell. Gap junctions are common in visceral smooth muscle, cardiac muscle, and the developing embryo. They also occur in the brain.

Electrical synapses have two main advantages:

1. ***Faster communication.*** Because action potentials conduct directly through gap junctions, electrical synapses are faster than chemical synapses. At an electrical synapse, the action

potential passes directly from the presynaptic cell to the postsynaptic cell. The events that occur at a chemical synapse take some time and delay communication slightly.

2. ***Synchronization.*** Electrical synapses can synchronize (coordinate) the activity of a group of neurons or muscle fibers. In other words, a large number of neurons or muscle fibers can produce action potentials in unison if they are connected by gap junctions. The value of synchronized action potentials in the heart or in visceral smooth muscle is coordinated contraction of these fibers to produce a heartbeat or move food through the gastrointestinal tract.

Chemical Synapses

Although the plasma membranes of presynaptic and postsynaptic neurons in a **chemical synapse** are close, they do not touch. They are separated by the **synaptic cleft,** a space of 20–50 nm* that is filled with interstitial fluid. Nerve impulses cannot conduct across the synaptic cleft, so an alternative, indirect form of communication occurs. In response to a nerve impulse, the presynaptic neuron releases a neurotransmitter that diffuses through the fluid in the synaptic cleft and binds to receptors in the plasma membrane of the postsynaptic neuron. The postsynaptic neuron receives the chemical signal and in turn produces a **postsynaptic potential,** a type of graded potential. Thus, the presynaptic neuron converts an electrical signal (nerve impulse) into a chemical signal (released neurotransmitter). The postsynaptic neuron receives the chemical signal and in turn generates an electrical signal (postsynaptic potential). The time required for these processes at a chemical synapse, a **synaptic delay** of about 0.5 msec, is the reason that chemical synapses relay signals more slowly than electrical synapses.

A typical chemical synapse transmits a signal as follows (Figure 12.22):

① A nerve impulse arrives at a synaptic end bulb (or at a varicosity) of a presynaptic axon.

② The depolarizing phase of the nerve impulse opens **voltage-gated Ca^{2+} channels,** which are present in the membrane of synaptic end bulbs. Because calcium ions are more concentrated in the extracellular fluid, Ca^{2+} flows inward through the opened channels.

③ An increase in the concentration of Ca^{2+} inside the presynaptic neuron serves as a signal that triggers exocytosis of the synaptic vesicles. As vesicle membranes merge with the plasma membrane, neurotransmitter molecules within the vesicles are released into the synaptic cleft. Each synaptic vesicle contains several thousand molecules of neurotransmitter.

④ The neurotransmitter molecules diffuse across the synaptic cleft and bind to **neurotransmitter receptors** in the postsynaptic neuron's plasma membrane. The receptor shown in Figure 12.22 is part of a ligand-gated channel (see

*1 nanometer (nm) = 10^{-9} (0.000000001) meter.

Figure 12.22 Signal transmission at a chemical synapse. Through exocytosis of synaptic vesicles, a presynaptic neuron releases neurotransmitter molecules. After diffusing across the synaptic cleft, the neurotransmitter binds to receptors in the plasma membrane of the postsynaptic neuron and produces a postsynaptic potential.

At a chemical synapse, a presynaptic neuron converts an electrical signal (nerve impulse) into a chemical signal (neurotransmitter release). The postsynaptic neuron then converts the chemical signal back into an electrical signal (postsynaptic potential).

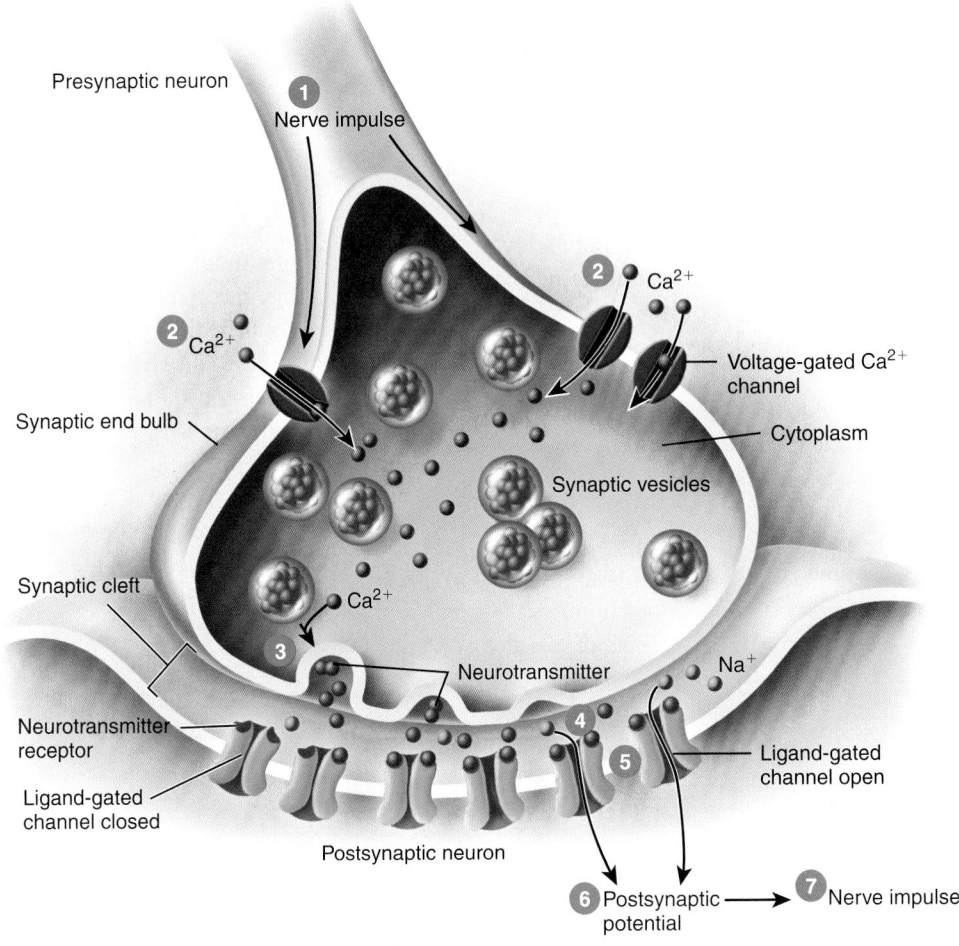

? Why may electrical synapses work in two directions, but chemical synapses can transmit a signal in only one direction?

Figure 12.11b); you will soon learn that this type of neurotransmitter receptor is called an *ionotropic receptor*. Not all neurotransmitters bind to ionotropic receptors; some bind to *metabotropic receptors* (described shortly).

5 Binding of neurotransmitter molecules to their receptors on ligand-gated channels opens the channels and allows particular ions to flow across the membrane.

6 As ions flow through the opened channels, the voltage across the membrane changes. This change in membrane voltage is a **postsynaptic potential.** Depending on which ions the channels admit, the postsynaptic potential may be a depolarization (excitation) or a hyperpolarization (inhibition). For example, opening of Na^+ channels allows inflow of Na^+, which causes depolarization. However, opening of Cl^- or K^+ channels causes hyperpolarization. Opening Cl^- channels

permits Cl^- to move into the cell, while opening the K^+ channels allows K^+ to move out—in either event, the inside of the cell becomes more negative.

7 When a depolarizing postsynaptic potential reaches threshold, it triggers an action potential in the axon of the postsynaptic neuron.

At most chemical synapses, only *one-way information transfer* can occur—from a presynaptic neuron to a postsynaptic neuron or an effector, such as a muscle fiber or a gland cell. For example, synaptic transmission at a neuromuscular junction (NMJ) proceeds from a somatic motor neuron to a skeletal muscle fiber (but not in the opposite direction). Only synaptic end bulbs of presynaptic neurons can release neurotransmitter, and only the postsynaptic neuron's membrane has the receptor proteins that can recognize and bind that neurotransmitter. As a result, action potentials move in one direction.

Excitatory and Inhibitory Postsynaptic Potentials

A neurotransmitter causes either an excitatory or an inhibitory graded potential. A neurotransmitter that causes *depolarization* of the postsynaptic membrane is excitatory because it brings the membrane closer to threshold (see Figure 12.14b). A depolarizing postsynaptic potential is called an **excitatory postsynaptic potential (EPSP).** Although a single EPSP normally does not initiate a nerve impulse, the postsynaptic cell does become more excitable. Because it is partially depolarized, it is more likely to reach threshold when the next EPSP occurs.

A neurotransmitter that causes *hyperpolarization* of the postsynaptic membrane (see Figure 12.14a) is inhibitory. During hyperpolarization, generation of an action potential is more difficult than usual because the membrane potential becomes inside more negative and thus even farther from threshold than in its resting state. A hyperpolarizing postsynaptic potential is termed an **inhibitory postsynaptic potential (IPSP).**

Structure of Neurotransmitter Receptors

As you have already learned, neurotransmitters released from a presynaptic neuron bind to **neurotransmitter receptors** in the plasma membrane of a postsynaptic cell. Each type of neurotransmitter receptor has one or more neurotransmitter binding sites where its specific neurotransmitter binds. When a neurotransmitter binds to the correct neurotransmitter receptor, an ion channel opens and a postsynaptic potential (either an EPSP or IPSP) forms in the membrane of the postsynaptic cell. Neurotransmitter receptors are classified as either ionotropic receptors or metabotropic receptors based on whether the neurotransmitter binding site and the ion channel are components of the same protein or are components of different proteins.

Ionotropic Receptors

An **ionotropic receptor** (ī-on-ō-TROP-ik) is a type of neurotransmitter receptor that contains a neurotransmitter binding site and an ion channel. In other words, the neurotransmitter binding site and the ion channel are components of the *same* protein. An ionotropic receptor is a type of ligand-gated channel (see Figure 12.11b). In the absence of neurotransmitter (the ligand), the ion channel component of the ionotropic receptor is closed. When the correct neurotransmitter binds to the ionotropic receptor, the ion channel opens, and an EPSP or IPSP occurs in the postsynaptic cell.

Many excitatory neurotransmitters bind to ionotropic receptors that contain cation channels (Figure 12.23a). EPSPs result from opening these cation channels. When cation channels open, they allow passage of the three most plentiful cations (Na^+, K^+, and Ca^{2+}) through the postsynaptic cell membrane, but Na^+ inflow is greater than either Ca^{2+} inflow or K^+ outflow and the inside of the postsynaptic cell becomes less negative (depolarized).

Many inhibitory neurotransmitters bind to ionotropic receptors that contain chloride channels (Figure 12.23b). IPSPs result from opening these Cl^- channels. When Cl^- channels open, a larger number of chloride ions diffuse inward. The inward flow of Cl^- ions causes the inside of the postsynaptic cell to become more negative (hyperpolarized).

Metabotropic Receptors

A **metabotropic receptor** (me-tab'-ō-TRO-pik) is a type of neurotransmitter receptor that contains a neurotransmitter binding site but lacks an ion channel as part of its structure. However, a metabotropic receptor is coupled to a separate ion channel by a type of membrane protein called a *G protein*. When a neurotransmitter binds to a metabotropic receptor, the G protein either directly opens (or closes) the ion channel or it may act indirectly by activating another molecule, a "second messenger," in the cytosol, which in turn opens (or closes) the ion channel (see Chapter 18 for a detailed discussion of G proteins). Thus, a metabotropic receptor differs from an ionotropic receptor in that the neurotransmitter binding site and the ion channel are components of *different* proteins.

Some inhibitory neurotransmitters bind to metabotropic receptors that are linked to K^+ channels (Figure 12.23c). IPSPs result from the opening of these K^+ channels. When K^+ channels open, a larger number of potassium ions diffuses outward. The outward flow of K^+ ions causes the inside of the postsynaptic cell to become more negative (hyperpolarized).

Different Postsynaptic Effects for the Same Neurotransmitter

The same neurotransmitter can be excitatory at some synapses and inhibitory at other synapses, depending on the structure of the neurotransmitter receptor to which it binds. For example, at some excitatory synapses acetylcholine (ACh) binds to ionotropic receptors that contain cation channels that open and subsequently generate EPSPs in the postsynaptic cell (Figure 12.23a). By contrast, at some inhibitory synapses ACh binds to metabotropic receptors coupled to G proteins that open K^+ channels, resulting in the formation of IPSPs in the postsynaptic cell (Figure 12.23c).

Removal of Neurotransmitter

Removal of the neurotransmitter from the synaptic cleft is essential for normal synaptic function. If a neurotransmitter could linger in the synaptic cleft, it would influence the postsynaptic neuron, muscle fiber, or gland cell indefinitely. Neurotransmitter is removed in three ways:

1. **Diffusion.** Some of the released neurotransmitter molecules diffuse away from the synaptic cleft. Once a neurotransmitter molecule is out of reach of its receptors, it can no longer exert an effect.

2. **Enzymatic degradation.** Certain neurotransmitters are inactivated through enzymatic degradation. For example, the enzyme acetylcholinesterase breaks down acetylcholine in the synaptic cleft.

3. **Uptake by cells.** Many neurotransmitters are actively transported back into the neuron that released them (reuptake).

Figure 12.23 Ionotropic and metabotropic neurotransmitter receptors. (a) The ionotropic acetylcholine (ACh) receptor contains two binding sites for the neurotransmitter ACh and a cation channel. Binding of ACh to this receptor causes the cation channel to open. Opening the cation channel allows passage of the three most plentiful cations (Na^+, K^+, and Ca^{2+}) through the postsynaptic cell membrane, but Na^+ inflow is greater than either Ca^{2+} inflow or K^+ outflow and an excitatory postsynaptic potential (EPSP) is generated. (b) The ionotropic gamma aminobutyric acid (GABA) receptor contains two binding sites for the neurotransmitter GABA and a Cl^- channel. Binding of GABA to this receptor causes the Cl^- channel to open. Opening the Cl^- channel allows a larger number of chloride ions to diffuse inward and an inhibitory postsynaptic potential (IPSP) is generated. (c) The metabotropic acetylcholine (ACh) receptor contains a binding site for the neurotransmitter ACh. Binding of ACh to this receptor activates a G protein, which in turn opens a K^+ channel. Opening the K^+ channel allows a larger number of potassium ions to diffuse out of the cell and an IPSP forms.

An ionotropic receptor is a type of neurotransmitter receptor that contains a neurotransmitter binding site and an ion channel; a metabotropic receptor is a type of neurotransmitter receptor that contains a neurotransmitter binding site and is coupled to a separate ion channel by a G protein.

(a) Ionotropic acetylcholine receptor

(b) Ionotropic GABA receptor

(c) Metabotropic acetylcholine receptor

The neurotransmitter acetylcholine (ACh) is excitatory at some synapses and inhibitory at other synapses. How is this possible?

Others are transported into neighboring neuroglia (uptake). The neurons that release norepinephrine, for example, rapidly take up the norepinephrine and recycle it into new synaptic vesicles. The membrane proteins that accomplish such uptake are called *neurotransmitter transporters*.

Spatial and Temporal Summation of Postsynaptic Potentials

A typical neuron in the CNS receives input from 1000 to 10,000 synapses. Integration of these inputs involves summation of the postsynaptic potentials that form in the postsynaptic neuron. Recall that summation is the process by which graded potentials add together. The greater the summation of EPSPs, the greater the chance that threshold will be reached. At threshold, one or more nerve impulses (action potentials) arise.

There are two types of summation: spatial summation and temporal summation. **Spatial summation** is summation of postsynaptic potentials in response to stimuli that occur at different *locations* in the membrane of a postsynaptic cell at the same

time. For example, spatial summation results from the buildup of neurotransmitter released simultaneously by *several* presynaptic end bulbs (Figure 12.24a). **Temporal summation** is summation of postsynaptic potentials in response to stimuli that occur at the same location in the membrane of the postsynaptic cell but at different *times*. For example, temporal summation results from buildup of neurotransmitter released by a *single* presynaptic end bulb two or more times in rapid succession (Figure 12.24b). Because a typical EPSP lasts about 15 msec, the second (and subsequent) release of neurotransmitter must occur soon after the first one if temporal summation is to occur. Summation is rather like a vote on the Internet. Many people voting "yes" or "no" on an issue at the same time can be compared to spatial summation. One person voting repeatedly and rapidly is like temporal summation. Most of the time, spatial and temporal summations are acting together to influence the chance that a neuron fires an action potential.

A single postsynaptic neuron receives input from many presynaptic neurons, some of which release excitatory neurotransmitters and some of which release inhibitory neurotransmitters

Figure 12.24 Spatial and temporal summation. (a) When presynaptic neurons 1 and 2 separately cause EPSPs (arrows) in postsynaptic neuron 3, the threshold level is not reached in neuron 3. Spatial summation occurs only when neurons 1 and 2 act simultaneously on neuron 3; their EPSPs sum to reach the threshold level and trigger a nerve impulse (action potential). (b) Temporal summation occurs when stimuli applied to the same axon in rapid succession (arrows) cause overlapping EPSPs that sum. When depolarization reaches the threshold level, a nerve impulse is triggered.

Spatial summation results from the buildup of neurotransmitter released simultaneously by several presynaptic end bulbs; temporal summation results from the buildup of neurotransmitter released by a single presynaptic end bulb two or more times in rapid succession.

(a) Spatial summation

(b) Temporal summation

? **Suppose that EPSPs summate in a postsynaptic neuron in response to simultaneous stimulation by the neurotransmitters glutamate, serotonin, and acetylcholine released by three separate presynaptic neurons. Is this an example of spatial or temporal summation?**

(Figure 12.25). The sum of all the excitatory and inhibitory effects at any given time determines the effect on the postsynaptic neuron, which may respond in the following ways:

1. **EPSP.** If the total excitatory effects are greater than the total inhibitory effects but less than the threshold level of stimulation, the result is an EPSP that does not reach threshold. Fol-

Figure 12.25 Summation of postsynaptic potentials at the trigger zone of a postsynaptic neuron. Presynaptic neurons 1, 3, and 5 release excitatory neurotransmitters (red dots) that generate excitatory postsynaptic potentials (EPSPs) (red arrows) in the membrane of a postsynaptic neuron. Presynaptic neurons 2 and 4 release inhibitory neurotransmitters (purple dots) that generate inhibitory postsynaptic potentials (IPSPs) (purple arrows) in the membrane of the postsynaptic neuron. The net summation of these EPSPs and IPSPs determines whether an action potential will be generated at the trigger zone of the postsynaptic neuron.

If the net summation of EPSPs and IPSPs is a depolarization that reaches threshold, then an action potential will occur at the trigger zone of a postsynaptic neuron.

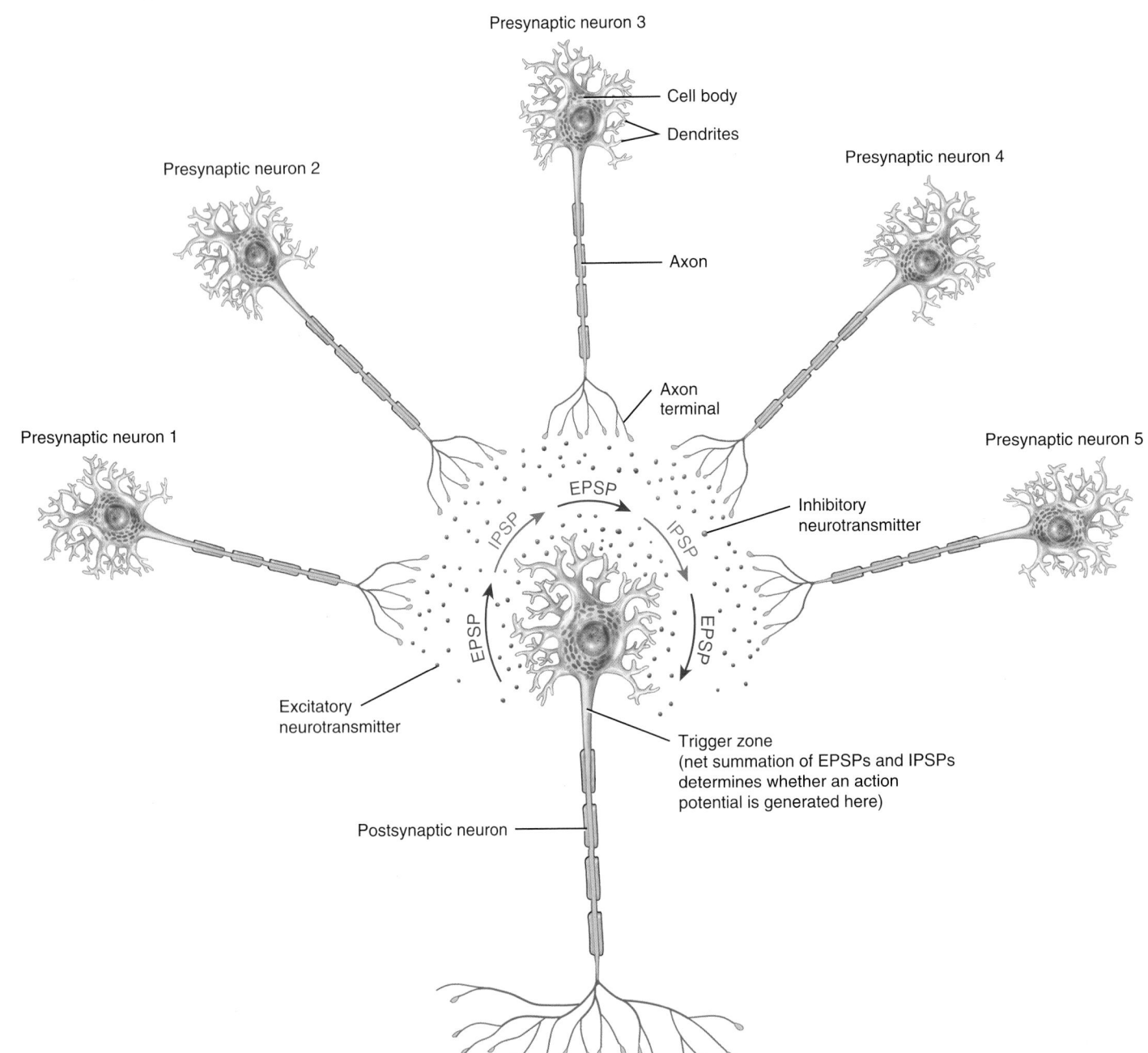

Suppose that the net summation of the EPSPs and IPSPs shown in this figure is a depolarization that brings the membrane potential of the trigger zone of the postsynaptic neuron to −60 mV. Will an action potential occur in the postsynaptic neuron?

The importance of inhibitory neurons can be appreciated by observing what happens when their activity is blocked. Normally, inhibitory neurons in the spinal cord called *Renshaw cells* release the neurotransmitter glycine at inhibitory synapses with somatic motor neurons. This inhibitory input to their motor neurons prevents excessive contraction of skeletal muscles. **Strychnine** (STRIK-nīn) is a lethal poison that is mainly used as a pesticide to control rats, moles, gophers,

and coyotes. When ingested, it binds to and blocks glycine receptors. The normal, delicate balance between excitation and inhibition in the CNS is disturbed, and motor neurons generate nerve impulses without restraint. All skeletal muscles, including the diaphragm, contract fully and remain contracted. Because the diaphragm cannot relax, the victim cannot inhale, and suffocation results. •

lowing an EPSP, subsequent stimuli can more easily generate a nerve impulse through summation because the neuron is partially depolarized.

2. ***Nerve impulse(s).*** If the total excitatory effects are greater than the total inhibitory effects and threshold is reached, one or more nerve impulses (action potentials) will be triggered. Impulses continue to be generated as long as the EPSP is at or above the threshold level.

3. ***IPSP.*** If the total inhibitory effects are greater than the excitatory effects, the membrane hyperpolarizes (IPSP). The result

is inhibition of the postsynaptic neuron and an inability to generate a nerve impulse.

Table 12.3 summarizes the structural and functional elements of a neuron.

✔CHECKPOINT

13. How is neurotransmitter removed from the synaptic cleft?
14. How are excitatory and inhibitory postsynaptic potentials similar and different?
15. Why are action potentials said to be "all-or-none," and EPSPs and IPSPs are described as "graded"?

TABLE 12.3

Summary of Neuronal Structure and Function

	STRUCTURE	FUNCTIONS
	Dendrites	Receive stimuli through activation of ligand-gated or mechanically gated ion channels; in sensory neurons, produce generator or receptor potentials; in motor neurons and interneurons, produce excitatory and inhibitory postsynaptic potentials (EPSPs and IPSPs).
	Cell body	Receives stimuli and produces EPSPs and IPSPs through activation of ligand-gated ion channels.
	Junction of axon hillock and initial segment of axon	Trigger zone in many neurons; integrates EPSPs and IPSPs and, if sum is depolarization that reaches threshold, initiates action potential (nerve impulse).
	Axon	Propagates nerve impulses from initial segment (or from dendrites of sensory neurons) to axon terminals in self-regenerating manner; impulse amplitude does not change as it propagates along axon.
	Axon terminals and synaptic end bulbs (or varicosities)	Inflow of Ca^{2+} caused by depolarizing phase of nerve impulse triggers exocytosis of neurotransmitter from synaptic vesicles.

Key:

Plasma membrane includes chemically gated channels

Plasma membrane includes voltage-gated Na^+ and K^+ channels

Plasma membrane includes voltage-gated Ca^{2+} channels

12.5 NEUROTRANSMITTERS

◉ OBJECTIVE

• Describe the classes and functions of neurotransmitters.

About 100 substances are either known or suspected neurotransmitters. Some neurotransmitters bind to their receptors and act quickly to open or close ion channels in the membrane. Others act more slowly via second-messenger systems to influence chemical reactions inside cells. The result of either process can be excitation or inhibition of postsynaptic neurons. Many neurotransmitters are also hormones released into the bloodstream by endocrine cells in organs throughout the body. Within the brain, certain neurons, called **neurosecretory cells,** also secrete hormones. Neurotransmitters can be divided into two classes based on size: small-molecule neurotransmitters and neuropeptides (Figure 12.26).

Small-Molecule Neurotransmitters

The small-molecule neurotransmitters include acetylcholine, amino acids, biogenic amines, ATP and other purines, nitric oxide, and carbon monoxide.

Acetylcholine

The best-studied neurotransmitter is **acetylcholine (ACh)** (a-sē′-til-KŌ-lēn), which is released by many PNS neurons and by some CNS neurons. ACh is an excitatory neurotransmitter at some synapses, such as the neuromuscular junction, where the binding of ACh to ionotropic receptors opens cation channels (see Figure 12.23a). It is also an inhibitory neurotransmitter at other synapses, where it binds to metabotropic receptors coupled to G proteins that open K^+ channels (see Figure 12.23c). For example, ACh slows heart rate at inhibitory synapses made by parasympathetic neurons of the vagus (X) nerve. The enzyme *acetylcholinesterase (AChE)* (a-sē′-til-kō′-lin-ES-ter-ās) inactivates ACh by splitting it into acetate and choline fragments.

Amino Acids

Several amino acids are neurotransmitters in the CNS. **Glutamate** (gloo-TA-māt) (glutamic acid) and **aspartate** (as-PAR-tāt) (aspartic acid) have powerful excitatory effects. Most excitatory neurons in the CNS and perhaps half of the synapses in the brain communicate via glutamate. At some glutamate synapses, binding of the neurotransmitter to ionotropic receptors opens cation channels. The consequent inflow of cations (mainly Na^+ ions) produces an EPSP. Inactivation of glutamate occurs via reuptake. Glutamate transporters actively transport glutamate back into the synaptic end bulbs and neighboring neuroglia.

Gamma aminobutyric (GAM-ma am-i-nō-bu-TIR-ik) **acid (GABA)** and **glycine** are important inhibitory neurotransmitters. At many synapses, the binding of GABA to ionotropic receptors opens Cl^- channels (see Figure 12.23b). GABA is found only in the CNS, where it is the most common inhibitory neurotransmitter. As many as one-third of all brain synapses use GABA. Antianxiety drugs such as diazepam (Valium®) enhance the action of GABA. Like GABA, the binding of glycine to ionotropic receptors opens Cl^- channels. About half of the inhibitory synapses in the spinal cord use the amino acid glycine; the rest use GABA.

A high level of glutamate in the interstitial fluid of the CNS causes **excitotoxicity** (ek-sī′-tō-tok-SIS-i-tē)—destruction of neurons through prolonged activation of excitatory synaptic transmission. The most common cause of excitotoxicity is oxygen deprivation of the brain due to ischemia (inadequate blood flow), as happens during a stroke. Lack of oxygen causes the glutamate transporters to fail, and glutamate accumulates in the interstitial spaces between neurons and glia, literally stimulating the neurons to death. Clinical trials are underway to see whether antiglutamate drugs administered after a stroke can offer some protection from excitotoxicity. •

Biogenic Amines

Certain amino acids are modified and decarboxylated (carboxyl group removed) to produce biogenic amines. Those that are prevalent in the nervous system include norepinephrine, epinephrine, dopamine, and serotonin. Most biogenic amines bind to metabotropic receptors; there are many different types of metabotropic receptors for each biogenic amine. Biogenic amines may cause either excitation or inhibition, depending on the type of metabotropic receptor at the synapse.

Norepinephrine (NE) (nor′-ep-i-NEF-rin) plays roles in arousal (awakening from deep sleep), dreaming, and regulating mood. A smaller number of neurons in the brain use **epinephrine** as a neurotransmitter. Both epinephrine and norepinephrine also serve as hormones. Cells of the adrenal medulla, the inner portion of the adrenal gland, release them into the blood.

Brain neurons containing the neurotransmitter **dopamine (DA)** (DŌ-pa-mēn) are active during emotional responses, addictive behaviors, and pleasurable experiences. In addition, dopamine-releasing neurons help regulate skeletal muscle tone and some aspects of movement due to contraction of skeletal muscles. The muscular stiffness that occurs in Parkinson disease is due to degeneration of neurons that release dopamine (see Disorders: Homeostatic Imbalances in Chapter 16). One form of schizophrenia is due to accumulation of excess dopamine.

Norepinephrine, dopamine, and epinephrine are classified chemically as **catecholamines** (kat-e-KŌL-a-mēns). They all include an amino group ($-NH_2$) and a catechol ring composed of six carbons and two adjacent hydroxyl ($-OH$) groups. Catecholamines are synthesized from the amino acid tyrosine. Inactivation of catecholamines occurs via reuptake into synaptic end bulbs. Then they are either recycled back into the synaptic vesicles or destroyed by enzymes. The two enzymes that break down catecholamines are **catechol-*O*-methyltransferase** (kat′-e-kōl-ō-meth-il-TRANS-fer-ās), or **COMT,** and **monoamine oxidase** (mon-ō-AM-īn OK-si-dās), or **MAO.**

Serotonin (ser′-ō-TŌ-nin), which is also known as **5-hydroxytryptamine (5-HT),** is concentrated in the neurons in a part of the brain called the raphe nucleus. It is thought to be involved in

SMALL-MOLECULE NEUROTRANSMITTERS

Acetylcholine

$$CH_3$$
$$|$$
$$H_3C - N^+ - CH_2 - CH_2 - O - C - CH_3$$
$$|$$
$$CH_3$$

(with O double-bonded to C)

Nitric oxide

$$N = O$$

Carbon monoxide

$$C \equiv O$$

Amino Acids

Glutamate

$$H$$
$$|$$
$$H_3N^+ - C - COO^-$$
$$|$$
$$CH_2$$
$$|$$
$$CH_2$$
$$|$$
$$COO^-$$

Aspartate

$$H$$
$$|$$
$$H_3N^+ - C - COO^-$$
$$|$$
$$CH_2$$
$$|$$
$$COO^-$$

Gamma aminobutyric acid (GABA)

$$H_3N^+ - CH_2 - CH_2 - CH_2 - COO^-$$

Glycine

$$H$$
$$|$$
$$H_3N^+ - C - COO^-$$
$$|$$
$$H$$

Biogenic Amines

Norepinephrine

$$HO$$
$$HO - \text{(ring)} - \overset{H}{\underset{OH}{C}} - CH_2 - NH_3^+$$

Epinephrine

$$HO$$
$$HO - \text{(ring)} - \overset{H}{\underset{OH}{C}} - CH_2 - NH_2^+ - CH_3$$

Dopamine

$$HO$$
$$HO - \text{(ring)} - CH_2 - CH_2 - NH_3^+$$

Serotonin

$$HO - \text{(ring)} - CH_2 - CH_2 - NH_3^+$$
(with N–H)

Purines

Example: ATP

Adenine

Adenosine

Ribose

Phosphate groups

$$NH_2$$

$$H_2C - O - P - O \sim P - O \sim P - O^-$$

OH   OH

NEUROPEPTIDES

Example: Substance P

Arg Pro Lys Pro Gln Gln Phe Phe Gly Leu Met

Figure 12.26 Neurotransmitters.
Neurotransmitters are divided into two major classes based on size: small-molecule neurotransmitters and neuropeptides. Small-molecule neurotransmitters include acetylcholine, amino acids, biogenic amines, purines, nitric oxide, and carbon monoxide. The neuropeptide shown is substance P, which consists of 11 amino acids linked by peptide bonds in the following order: arginine (Arg), proline (Pro), lysine (Lys), proline, glutamine (Gln), glutamine, phenylalanine (Phe), phenylalanine, glycine (Gly), leucine (Leu), and methionine (Met).

🔑 Neurotransmitters are chemical substances that neurons use to communicate with other neurons, muscle fibers, and glands.

❓ **Why are norepinephrine, epinephrine, dopamine, and serotonin classified as biogenic amines?**

sensory perception, temperature regulation, control of mood, appetite, and the induction of sleep.

Depression is a disorder that affects over 18 million people each year in the United States. People who are depressed feel sad and helpless, have a lack of interest in activities that they once enjoyed, and experience suicidal thoughts. There are several types of depression. A person with **major depression** experiences symptoms of depression that last for more than two weeks. A person with **dysthymia** (dis-THĪ-mē-a) experiences episodes of depression that alternate with periods of feeling normal. A person with **bipolar disorder,** or *manic-depressive illness,* experiences recurrent episodes of depression and extreme elation (mania). A person with **seasonal affective disorder (SAD)** experiences depression during the winter months, when day length is short (see Clinical Connection: Seasonal Affective Disorder and Jet Lag in Chapter 18). Although the exact cause of depression is unknown, research suggests that depression is linked to an imbalance of the neurotransmitters serotonin, norepinephrine, and dopamine in the brain. Factors that may contribute to depression include heredity, stress, chronic illnesses, certain personality traits (such as low self-esteem), and hormonal changes. Medication is the most common treatment for depression. For example, **selective serotonin reuptake inhibitors (SSRIs)** are drugs that provide relief from some forms of depression. By inhibiting reuptake of serotonin by serotonin transporters, SSRIs prolong the activity of this neurotransmitter at synapses in the brain. SSRIs include fluoxetine (Prozac®), paroxetine (Paxil®), and sertraline (Zoloft®). •

ATP and Other Purines

The characteristic ring structure of the adenosine portion of ATP (Figure 12.26) is called a purine ring. Adenosine itself, as well as its triphosphate, diphosphate, and monophosphate derivatives (ATP, ADP, and AMP), is an excitatory neurotransmitter in both the CNS and the PNS. Most of the synaptic vesicles that contain ATP also contain another neurotransmitter. In the PNS, ATP and norepinephrine are released together from some sympathetic neurons; some parasympathetic neurons release ATP and acetylcholine in the same vesicles.

Nitric Oxide

The simple gas **nitric oxide (NO)** is an important excitatory neurotransmitter secreted in the brain, spinal cord, adrenal glands, and nerves to the penis and has widespread effects throughout the body. NO contains a single nitrogen atom, in contrast to nitrous oxide (N_2O), or laughing gas, which has two nitrogen atoms. N_2O is sometimes used as an anesthetic during dental procedures.

The enzyme **nitric oxide synthase (NOS)** catalyzes formation of NO from the amino acid arginine. Based on the presence of NOS, it is estimated that more than 2% of the neurons in the brain produce NO. Unlike all previously known neurotransmitters, NO is not synthesized in advance and packaged into synaptic vesicles. Rather, it is formed on demand and acts immediately. Its action is brief because NO is a highly reactive free radical. It exists for less than 10 seconds before it combines with oxygen and water to form inactive nitrates

and nitrites. Because NO is lipid-soluble, it diffuses from cells that produce it into neighboring cells, where it activates an enzyme for production of a second messenger called cyclic GMP. Some research suggests that NO plays a role in memory and learning.

The first recognition of NO as a regulatory molecule was the discovery in 1987 that a chemical called EDRF (endothelium-derived relaxing factor) was actually NO. Endothelial cells in blood vessel walls release NO, which diffuses into neighboring smooth muscle cells and causes relaxation. The result is vasodilation, an increase in blood vessel diameter. The effects of such vasodilation range from a lowering of blood pressure to erection of the penis in males. Sildenafil (Viagra®) alleviates erectile dysfunction (impotence) by enhancing the effect of NO. In larger quantities, NO is highly toxic. Phagocytic cells, such as macrophages and certain white blood cells, produce NO to kill microbes and tumor cells.

Carbon Monoxide

Carbon monoxide (CO), like NO, is not produced in advance and packaged into synaptic vesicles. It too is formed as needed and diffuses out of cells that produce it into adjacent cells. CO is an excitatory neurotransmitter produced in the brain and in response to some neuromuscular and neuroglandular functions. CO might protect against excess neuronal activity and might be related to dilation of blood vessels, memory, olfaction (sense of smell), vision, thermoregulation, insulin release, and anti-inflammatory activity.

Neuropeptides

Neurotransmitters consisting of 3 to 40 amino acids linked by peptide bonds called **neuropeptides** (noor-ō-PEP-tīds) are numerous and widespread in both the CNS and PNS. Neuropeptides bind to metabotropic receptors and have excitatory or inhibitory actions, depending on the type of metabotropic receptor at the synapse. Neuropeptides are formed in the neuron cell body, packaged into vesicles, and transported to axon terminals. Besides their role as neurotransmitters, many neuropeptides serve as hormones that regulate physiological responses elsewhere in the body.

Scientists discovered that certain brain neurons have plasma membrane receptors for opiate drugs such as morphine and heroin. The quest to find the naturally occurring substances that use these receptors brought to light the first neuropeptides: two molecules, each a chain of five amino acids, named **enkephalins** (en-KEF-a-lins). Their potent analgesic (pain-relieving) effect is 200 times stronger than morphine. Other so-called *opioid peptides* include the **endorphins** (en-DOR-fins) and **dynorphins** (dī-NOR-fins). It is thought that opioid peptides are the body's natural painkillers. Acupuncture may produce analgesia (loss of pain sensation) by increasing the release of opioids. These neuropeptides have also been linked to improved memory and learning; feelings of pleasure or euphoria; control of body temperature; regulation of hormones that affect the onset of puberty, sexual drive, and reproduction; and mental illnesses such as depression and schizophrenia.

Another neuropeptide, **substance P,** is released by neurons that transmit pain-related input from peripheral pain receptors into the central nervous system, enhancing the perception of pain.

Enkephalin and endorphin suppress the release of substance P, thus decreasing the number of nerve impulses being relayed to the brain for pain sensations. Substance P has also been shown to counter the effects of certain nerve-damaging chemicals, prompting speculation that it might prove useful as a treatment for nerve degeneration.

Table 12.4 provides brief descriptions of these neuropeptides, as well as others that will be discussed in later chapters.

CLINICAL CONNECTION | *Modifying the Effects of Neurotransmitters*

Substances naturally present in the body as well as drugs and toxins can **modify the effects of neurotransmitters** in several ways:

1. Neurotransmitter synthesis can be stimulated or inhibited. For instance, many patients with Parkinson disease (see Disorders: Homeostatic Imbalances in Chapter 16) receive benefit from the drug L-dopa because it is a precursor of dopamine. For a limited period of time, taking L-dopa boosts dopamine production in affected brain areas.

2. Neurotransmitter release can be enhanced or blocked. Amphetamines promote release of dopamine and norepinephrine. Botulinum toxin causes paralysis by blocking release of acetylcholine from somatic motor neurons.

3. The neurotransmitter receptors can be activated or blocked. An agent that binds to receptors and enhances or mimics the effect of a natural neurotransmitter is an **agonist** (AG-ōn-ist). Isoproterenol (Isuprel®) is a powerful agonist of epinephrine and norepinephrine. It can be used to dilate the airways during an asthma attack. An agent that binds to and blocks neurotransmitter receptors is an **antagonist** (an-TAG-ō-nist). Zyprexa®, a drug prescribed for schizophrenia, is an antagonist of serotonin and dopamine.

4. Neurotransmitter removal can be stimulated or inhibited. For example, cocaine produces euphoria—intensely pleasurable feelings—by blocking transporters for dopamine reuptake. This action allows dopamine to linger longer in synaptic clefts, producing excessive stimulation of certain brain regions. •

✔ CHECKPOINT

16. Which neurotransmitters are excitatory and which are inhibitory? How do they exert their effects?
17. In what ways is nitric oxide different from all previously known neurotransmitters?

12.6 NEURAL CIRCUITS

◉ OBJECTIVE

• Identify the various types of neural circuits in the nervous system.

The CNS contains billions of neurons organized into complicated networks called **neural circuits,** functional groups of neurons that process specific types of information. In a **simple series circuit,** a presynaptic neuron stimulates a single postsynaptic neuron. The second neuron then stimulates another, and so on. However, most neural circuits are more complex.

A single presynaptic neuron may synapse with several postsynaptic neurons. Such an arrangement, called **divergence,** permits one presynaptic neuron to influence several postsynaptic neurons (or several muscle fibers or gland cells) at the same time. In a **diverging circuit,** the nerve impulse from a single presynaptic neuron causes the stimulation of increasing numbers of cells along the circuit (Figure 12.27a). For example, a small number of neurons in the brain that govern a particular body movement stimulate a much larger number of neurons in the spinal cord. Sensory signals are also arranged in diverging circuits, allowing a sensory impulse to be relayed to several regions of the brain. This arrangement amplifies the signal.

In another arrangement, called **convergence,** several presynaptic neurons synapse with a single postsynaptic neuron. This arrangement permits more effective stimulation or inhibition of the postsynaptic neuron. In a **converging circuit** (Figure 12.27b), the

TABLE 12.4

Neuropeptides

SUBSTANCE	DESCRIPTION
Substance P	Found in sensory neurons, spinal cord pathways, and parts of brain associated with pain; enhances perception of pain.
Enkephalins	Inhibit pain impulses by suppressing release of substance P; may have role in memory and learning, control of body temperature, sexual activity, and mental illness.
Endorphins	Inhibit pain by blocking release of substance P; may have role in memory and learning, sexual activity, control of body temperature, and mental illness.
Dynorphins	May be related to controlling pain and registering emotions.
Hypothalamic releasing and inhibiting hormones	Produced by hypothalamus; regulate release of hormones by anterior pituitary.
Angiotensin II	Stimulates thirst; may regulate blood pressure in brain. As a hormone, causes vasoconstriction and promotes release of aldosterone, which increases rate of salt and water reabsorption by kidneys.
Cholecystokinin (CCK)	Found in brain and small intestine; may regulate feeding as a "stop eating" signal. As a hormone, regulates pancreatic enzyme secretion during digestion, and contraction of smooth muscle in gastrointestinal tract.

Figure 12.27 Examples of neural circuits.

🔑 🔑 A neural circuit is a functional group of neurons that processes a specific kind of information.

(a) Diverging circuit

(b) Converging circuit

(c) Reverberating circuit

(d) Parallel after-discharge circuit

❓ **A motor neuron in the spinal cord typically receives input from neurons that originate in several different regions of the brain. Is this an example of convergence or divergence?**

postsynaptic neuron receives nerve impulses from several different sources. For example, a single motor neuron that synapses with skeletal muscle fibers at neuromuscular junctions receives input from several pathways that originate in different brain regions.

Some circuits are organized so that stimulation of the presynaptic cell causes the postsynaptic cell to transmit a series of nerve impulses. One such circuit is called a **reverberating circuit** (Figure 12.27c). In this pattern, the incoming impulse stimulates the first neuron, which stimulates the second, which stimulates the third, and so on. Branches from later neurons synapse with earlier ones. This arrangement sends impulses back through the circuit again and again. The output signal may last from a few seconds to many hours, depending on the number of synapses and the arrangement of neurons in the circuit. Inhibitory neurons may turn off a reverberating circuit after a period of time. Among the body responses thought to be the result of output signals from reverberating circuits are breathing, coordinated muscular activities, waking up, and short-term memory.

A fourth type of circuit is the **parallel after-discharge circuit** (Figure 12.27d). In this circuit, a single presynaptic cell stimulates a group of neurons, each of which synapses with a common postsynaptic cell. A differing number of synapses between the first and last neurons imposes varying synaptic delays, so that the last neuron exhibits multiple EPSPs or IPSPs. If the input is excitatory, the postsynaptic neuron then can send out a stream of impulses in quick succession. Parallel after-discharge circuits may be involved in precise activities such as mathematical calculations.

✔**CHECKPOINT**

18. What is a neural circuit?
19. What are the functions of diverging, converging, reverberating, and parallel after-discharge circuits?

12.7 REGENERATION AND REPAIR OF NERVOUS TISSUE

◉**OBJECTIVES**

• Define plasticity and neurogenesis.
• Describe the events involved in damage and repair of peripheral nerves.

Throughout your life, your nervous system exhibits **plasticity** (plas-TIS-i-tē), the capability to change based on experience. At the level of individual neurons, the changes that can occur include the sprouting of new dendrites, synthesis of new proteins, and changes in synaptic contacts with other neurons. Undoubtedly, both chemical and electrical signals drive the changes that occur. Despite this plasticity, mammalian neurons have very limited powers of **regeneration,** the capability to replicate or repair themselves. In the PNS, damage to dendrites and myelinated axons may be repaired if the cell body remains intact and if the Schwann cells that produce myelination remain active. In the CNS, little or no repair of damage to neurons occurs. Even when the cell body remains intact, a severed axon cannot be repaired or regrown.

Neurogenesis in the CNS

Neurogenesis (noo′-rō-JEN-e-sis)—the birth of new neurons from undifferentiated stem cells—occurs regularly in some animals. For example, new neurons appear and disappear every year in some songbirds. Until recently, the dogma in humans and other primates was "no new neurons" in the adult brain. Then, in 1992, Canadian researchers published their unexpected finding that **epidermal growth factor (EGF)** stimulated cells taken from the brains of adult mice to proliferate into both neurons and astrocytes. Previously, EGF was known to trigger mitosis in a variety of nonneuronal cells and to promote wound healing and tissue regeneration. In 1998, scientists discovered that significant numbers of new neurons do arise in the adult human hippocampus, an area of the brain that is crucial for learning.

The nearly complete lack of neurogenesis in other regions of the brain and spinal cord seems to result from two factors: (1) inhibitory influences from neuroglia, particularly oligodendrocytes, and (2) absence of growth-stimulating cues that were present during fetal development. Axons in the CNS are myelinated by oligodendrocytes rather than Schwann cells, and this CNS myelin is one of the factors inhibiting regeneration of neurons. Perhaps this same mechanism stops axonal growth once a target region has been reached during development. Also, after axonal damage, nearby astrocytes proliferate rapidly, forming a type of scar tissue that acts as a physical barrier to regeneration. Thus, injury of the brain or spinal cord usually is permanent. Ongoing research seeks ways to improve the environment for existing spinal cord axons to bridge the injury gap. Scientists also are trying to find ways to stimulate dormant stem cells to replace neurons lost through damage or disease and to develop tissue-cultured neurons that can be used for transplantation purposes.

Damage and Repair in the PNS

Axons and dendrites that are associated with a neurolemma may undergo repair if the cell body is intact, if the Schwann cells are functional, and if scar tissue formation does not occur too rapidly (Figure 12.28). Most nerves in the PNS consist of processes that are covered with a neurolemma. A person who injures axons of a nerve in an upper limb, for example, has a good chance of regaining nerve function.

When there is damage to an axon, changes usually occur both in the cell body of the affected neuron and in the portion of the

Figure 12.28 Damage and repair of a neuron in the PNS.

 Myelinated axons in the peripheral nervous system may be repaired if the cell body remains intact and if Schwann cells remain active.

(a) Normal neuron

(b) Chromatolysis and Wallerian degeneration

(c) Regeneration

? What is the role of the neurolemma in regeneration?

axon distal to the site of injury. Changes also may occur in the portion of the axon proximal to the site of injury.

About 24 to 48 hours after injury to a process of a normal peripheral neuron (Figure 12.28a), the Nissl bodies break up into fine granular masses. This alteration is called **chromatolysis** (krō′-ma-TOL-i-sis; *chromato-* = color; *-lysis* = destruction). By the third to fifth day, the part of the axon distal to the damaged region becomes slightly swollen and then breaks up into fragments; the myelin sheath also deteriorates (Figure 12.28b). Even though the axon and myelin sheath degenerate, the neurolemma remains. Degeneration of the distal portion of the axon and myelin sheath is called **Wallerian degeneration** (waw-LE-rē′-an).

Following chromatolysis, signs of recovery in the cell body become evident. Macrophages phagocytize the debris. Synthesis of RNA and protein accelerates, which favors rebuilding or **regeneration** of the axon. The Schwann cells on either side of the injured site multiply by mitosis, grow toward each other, and may form a **regeneration tube** across the injured area (Figure 12.28c). The tube guides growth of a new axon from the proximal area across the injured area into the distal area previously occupied by the original axon. However, new axons cannot grow if the gap at the site of injury is too large or if the gap becomes filled with collagen fibers.

During the first few days following damage, buds of regenerating axons begin to invade the tube formed by the Schwann cells (Figure 12.28b). Axons from the proximal area grow at a rate of about 1.5 mm (0.06 in.) per day across the area of damage, find their way into the distal regeneration tubes, and grow toward the distally located receptors and effectors. Thus, some sensory and motor connections are reestablished and some functions restored. In time, the Schwann cells form a new myelin sheath.

✔ **CHECKPOINT**

20. What factors contribute to a lack of neurogenesis in most parts of the brain?
21. What is the function of the regeneration tube in repair of neurons?

DISORDERS: HOMEOSTATIC IMBALANCES

Multiple Sclerosis

Multiple sclerosis (MS) is a disease that causes a progressive destruction of myelin sheaths surrounding neurons in the CNS. It afflicts about 350,000 people in the United States and 2 million people worldwide. It usually appears between the ages of 20 and 40, affecting females twice as often as males. MS is most common in whites, less common in blacks, and rare in Asians. MS is an autoimmune disease—the body's own immune system spearheads the attack. The condition's name describes the anatomical pathology: In *multiple* regions the myelin sheaths deteriorate to *scleroses,* which are hardened scars or plaques. Magnetic resonance imaging (MRI) studies reveal numerous plaques in the white matter of the brain and spinal cord. The destruction of myelin sheaths slows and then short-circuits propagation of nerve impulses.

The most common form of the condition is relapsing–remitting MS, which usually appears in early adulthood. The first symptoms may include a feeling of heaviness or weakness in the muscles, abnormal sensations, or double vision. An attack is followed by a period of remission during which the symptoms temporarily disappear. One attack follows another over the years, usually every year or two. The result is a progressive loss of function interspersed with remission periods, during which symptoms abate.

Although the cause of MS is unclear, both genetic susceptibility and exposure to some environmental factor (perhaps a herpes virus) appear to contribute. Since 1993, many patients with relapsing–remitting MS have been treated with injections of beta interferon. This treatment lengthens the time between relapses, decreases the severity of relapses, and slows formation of new lesions in some cases. Unfortunately, not all MS patients can tolerate beta interferon, and therapy becomes less effective as the disease progresses.

Epilepsy

Epilepsy (ep'-i-LEP-sē) is characterized by short, recurrent attacks of motor, sensory, or psychological malfunction, although it almost never affects intelligence. The attacks, called *epileptic seizures,* afflict about 1% of the world's population. They are initiated by abnormal, synchronous electrical discharges from millions of neurons in the brain, perhaps resulting from abnormal reverberating circuits. The discharges stimulate many of the neurons to send nerve impulses over their conduction pathways. As a result, lights, noise, or smells may be sensed when the eyes, ears, and nose have not been stimulated. Moreover, the skeletal muscles of a person having a seizure may contract involuntarily. *Partial seizures* begin in a small area on one side of the brain and produce milder symptoms; *generalized seizures* involve larger areas on both sides of the brain and loss of consciousness.

Epilepsy has many causes, including brain damage at birth (the most common cause); metabolic disturbances (hypoglycemia, hypocalcemia, uremia, hypoxia); infections (encephalitis or meningitis); toxins (alcohol, tranquilizers, hallucinogens); vascular disturbances (hemorrhage, hypotension); head injuries; and tumors and abscesses of the brain. Seizures associated with fever are most common in children under the age of two. However, most epileptic seizures have no demonstrable cause.

Epileptic seizures often can be eliminated or alleviated by antiepileptic drugs, such as phenytoin, carbamazepine, and valproate sodium. An implantable device that stimulates the vagus (X) nerve has produced dramatic results in reducing seizures in some patients whose epilepsy was not well-controlled by drugs. In very severe cases, surgical intervention may be an option.

MEDICAL TERMINOLOGY

Guillain-Barré syndrome (GBS) (GHE-an ba-RĀ) An acute demyelinating disorder in which macrophages strip myelin from axons in the PNS. It is the most common cause of acute paralysis in North America and Europe and may result from the immune system's response to a bacterial infection. Most patients recover completely or partially, but about 15% remain paralyzed.

Neuroblastoma (noor-ō-blas-TŌ-ma) A malignant tumor that consists of immature nerve cells (neuroblasts); occurs most commonly in the abdomen and most frequently in the adrenal glands. Although rare, it is the most common tumor in infants.

Neuropathy (noo-ROP-a-thē; *neuro-* = a nerve; *-pathy* = disease) Any disorder that affects the nervous system but particularly a disorder of a cranial or spinal nerve. An example is *facial neuropathy* (Bell's palsy), a disorder of the facial (VII) nerve.

Rabies (RĀ-bēz; *rabi-* = mad, raving) A fatal disease caused by a virus that reaches the CNS via fast axonal transport. It is usually transmitted by the bite of an infected dog or other meat-eating animal. The symptoms are excitement, aggressiveness, and madness, followed by paralysis and death.

CHAPTER REVIEW AND RESOURCE SUMMARY

Review	Resource

12.1 Overview of the Nervous System

1. The central nervous system (CNS) consists of the brain and spinal cord.
2. The peripheral nervous system (PNS) consists of all nervous tissue outside the CNS. Components of the PNS include nerves, ganglia, enteric plexuses, and sensory receptors.
3. The PNS is divided into a somatic nervous system (SNS), autonomic nervous system (ANS), and enteric nervous system (ENS).
4. The SNS consists of sensory neurons that conduct impulses from somatic and special sense receptors to the CNS and motor neurons from the CNS to skeletal muscles.
5. The ANS contains sensory neurons from visceral organs and motor neurons that convey impulses from the CNS to smooth muscle tissue, cardiac muscle tissue, and glands.
6. The ENS consists of neurons in enteric plexuses in the gastrointestinal (GI) tract that function somewhat independently of the ANS and CNS. The ENS monitors chemical changes within the GI tract and stretching of its walls; the ENS also controls contraction of GI tract smooth muscle.
7. The nervous system helps maintain homeostasis and integrates all body activities by sensing changes (sensory function), interpreting them (integrative function), and reacting to them (motor function).

Anatomy Overview - Nervous System
Anatomy Overview - Nerve
Animation - Structure and Function of the Nervous System: System Organization

12.2 Histology of Nervous Tissue

1. Nervous tissue consists of neurons (nerve cells) and neuroglia. Neurons have the property of electrical excitability and are responsible for most unique functions of the nervous system: sensing, thinking, remembering, controlling muscle activity, and regulating glandular secretions.
2. Most neurons have three parts. The dendrites are the main receiving or input region. Integration occurs in the cell body, which includes typical cellular organelles. The output part typically is a single axon, which propagates nerve impulses toward another neuron, a muscle fiber, or a gland cell.
3. Synapses are the site of functional contact between two excitable cells. Axon terminals contain synaptic vesicles filled with neurotransmitter molecules.
4. Slow axonal transport and fast axonal transport are systems for conveying materials to and from the cell body and axon terminals.
5. On the basis of their structure, neurons are classified as multipolar, bipolar, or unipolar.
6. Neurons are functionally classified as sensory (afferent) neurons, motor (efferent) neurons, and interneurons. Sensory neurons carry sensory information into the CNS. Motor neurons carry information out of the CNS to effectors (muscles and glands). Interneurons are located within the CNS between sensory and motor neurons.
7. Neuroglia support, nurture, and protect neurons and maintain the interstitial fluid that bathes them. Neuroglia in the CNS include astrocytes, oligodendrocytes, microglia, and ependymal cells. Neuroglia in the PNS include Schwann cells and satellite cells.
8. Two types of neuroglia produce myelin sheaths: Oligodendrocytes myelinate axons in the CNS, and Schwann cells myelinate axons in the PNS.
9. White matter consists of aggregates of myelinated axons; gray matter contains cell bodies, dendrites, and axon terminals of neurons, unmyelinated axons, and neuroglia.
10. In the spinal cord, gray matter forms an H-shaped inner core that is surrounded by white matter. In the brain, a thin, superficial shell of gray matter covers the cerebral and cerebellar hemispheres.

Anatomy Overview - Nervous Tissue: Neuron
Anatomy Overview - Nervous Tissue: Neuroglia
Animation - Neuron Structure and Function
Animation - Factors That Affect Conduction Rates: Myelination
Figure 12.6 - Neuroglia of the Central Nervous System
Exercise - Paint a Neuron

12.3 Electrical Signals in Neurons

1. Neurons communicate with one another using graded potentials, which are used for short-distance communication only, and action potentials, which allow communication over long distances within the body.
2. The electrical signals produced by neurons and muscle fibers rely on four kinds of ion channels: leak channels, ligand-gated channels, mechanically gated channels, and voltage-gated channels. Table 12.1 summarizes the different types of ion channels in neurons.
3. A resting membrane potential exists across the plasma membrane of excitable cells that are unstimulated (at rest). The resting membrane potential exists because of a small buildup of negative ions in the cytosol along the inside surface of the membrane, and an equal buildup of positive ions in the extracellular fluid (ECF) along the outside surface of the membrane. A typical value for the resting membrane potential of a neuron is -70 mV. A cell that exhibits a membrane potential is polarized. The resting membrane potential is determined by three major factors: (1) unequal distribution of ions

Animation - Introduction to Membrane Potentials
Animation - Resting Membrane Potentials
Animation - Action Potentials
Animation - Graded Potentials
Figure 12.12 - Resting Membrane Potential
Figure 12.18 - Action Potential or Impulse
Exercise - Ruling the Gated Channels
Exercise - Keep the Resting Potential
Exercise - Nervous Race

Review	**Resource**

in the ECF and cytosol; (2) inability of most cytosolic anions to leave the cell; and (3) electrogenic nature of the Na^+/K^+ ATPases.

4. A graded potential is a small deviation from the resting membrane potential that occurs because ligand-gated or mechanically gated channels open or close. A hyperpolarizing graded potential makes the membrane potential more negative (more polarized); a depolarizing graded potential makes the membrane potential less negative (less polarized). The amplitude of a graded potential varies, depending on the strength of the stimulus.

5. According to the all-or-none principle, if a stimulus is strong enough to generate an action potential, the impulse generated is of a constant size. A stronger stimulus does not generate a larger action potential. Instead, the greater the stimulus strength above threshold, the greater the frequency of the action potentials.

6. During an action potential, voltage-gated Na^+ and K^+ channels open and close in sequence. This results first in depolarization, the reversal of membrane polarization (from -70 mV to $+30$ mV). Then repolarization, the recovery of the resting membrane potential (from $+30$ mV to -70 mV), occurs.

7. During the first part of the refractory period (RP), another impulse cannot be generated at all (absolute RP); a little later, it can be triggered only by a larger-than-normal stimulus (relative RP).

8. Because an action potential travels from point to point along the membrane without getting smaller, it is useful for long-distance communication. Nerve impulse propagation in which the impulse "leaps" from one node of Ranvier to the next along a myelinated axon is saltatory conduction. Saltatory conduction is faster than continuous conduction.

9. Axons with larger diameters conduct impulses at higher speeds than do axons with smaller diameters.

10. The intensity of a stimulus is encoded in the frequency of action potentials and in the number of sensory neurons that are recruited. Table 12.2 compares graded potentials and action potentials.

Resource

Exercise - Define Stages of the Action Potential
Exercise - Investigate Types of Impulse Conduction
Concepts and Connections - Membrane Potentials

12.4 Signal Transmission at Synapses

1. A synapse is the functional junction between one neuron and another, or between a neuron and an effector such as a muscle or a gland. The two types of synapses are electrical and chemical.

2. A chemical synapse produces one-way information transfer—from a presynaptic neuron to a postsynaptic neuron.

3. An excitatory neurotransmitter is one that can depolarize the postsynaptic neuron's membrane, bringing the membrane potential closer to threshold. An inhibitory neurotransmitter hyperpolarizes the membrane of the postsynaptic neuron, moving it further from threshold.

4. There are two major types of neurotransmitter receptors: ionotropic receptors and metabotropic receptors. An ionotropic receptor contains a neurotransmitter binding site and an ion channel. A metabotropic receptor contains a neurotransmitter binding site and is coupled to a separate ion channel by a G protein.

5. Neurotransmitter is removed from the synaptic cleft in three ways: diffusion, enzymatic degradation, and uptake by cells (neurons and neuroglia).

6. If several presynaptic end bulbs release their neurotransmitter at about the same time, the combined effect may generate a nerve impulse, due to summation. Summation may be spatial or temporal.

7. The postsynaptic neuron is an integrator. It receives excitatory and inhibitory signals, integrates them, and then responds accordingly.

8. Table 12.3 summarizes the structural and functional elements of a neuron.

Animation - Events at the Synapse
Exercise - Are Your Synapses Working?
Figure 12.22 - Signal Transmission at a Chemical Synapse

12.5 Neurotransmitters

1. Both excitatory and inhibitory neurotransmitters are present in the CNS and the PNS. A given neurotransmitter may be excitatory in some locations and inhibitory in others.

2. Neurotransmitters can be divided into two classes based on size: (1) small-molecule neurotransmitters (acetylcholine, amino acids, biogenic amines, ATP and other purines, nitric oxide, and carbon monoxide), and (2) neuropeptides, which are composed of 3 to 40 amino acids.

3. Chemical synaptic transmission may be modified by affecting synthesis, release, or removal of a neurotransmitter or by blocking or stimulating neurotransmitter receptors.

4. Table 12.4 describes several important neuropeptides.

Anatomy Overview - Neurotransmitters

12.6 Neural Circuits

1. Neurons in the central nervous system are organized into networks called neural circuits.

2. Neural circuits include simple series, diverging, converging, reverberating, and parallel after-discharge circuits.

Animation - Introduction to Membrane Potentials
Figure 12.27 - Neural Circuits

Review   **Resource**

12.7 Regeneration and Repair of Nervous Tissue

1. The nervous system exhibits plasticity (the capability to change based on experience), but it has very limited powers of regeneration (the capability to replicate or repair damaged neurons).
2. Neurogenesis, the birth of new neurons from undifferentiated stem cells, is normally very limited. Repair of damaged axons does not occur in most regions of the CNS.
3. Axons and dendrites that are associated with a neurolemma in the PNS may undergo repair if the cell body is intact, the Schwann cells are functional, and scar tissue formation does not occur too rapidly.

SELF-QUIZ QUESTIONS

Fill in the blanks in the following statements.

1. The subdivisions of the PNS are the _somatic_, _autonomic_, and _enteric_.
2. The two divisions of the autonomic nervous system are the _sympathetic_ division and the _parasympathetic_ division.

Indicate whether the following statements are true or false.

3. At a chemical synapse between two neurons, the neuron receiving the signal is called the presynaptic neuron, and the neuron sending the signal is called the postsynaptic neuron. _false_

4. Neurons in the PNS are always capable of repair while those in the CNS are not. _false_

Choose the one best answer to the following questions.

5. Which of the following statements are *true*? (1) The sensory function of the nervous system involves sensory receptors sensing certain changes in the internal and external environments. (2) Sensory neurons receive electrical signals from sensory receptors. (3) The integrative function of the nervous system involves analyzing sensory information, storing some of it, and making decisions regarding appropriate responses. (4) Interneurons are located primarily in the PNS. (5) Motor function involves the activation of effectors (muscles and glands).
(a) 1, 2, 3, and 4   (b) 2, 4, and 5   (c) 1, 2, 3, and 5
(d) 1, 2, and 4   (e) 2, 3, 4, and 5

6. A neuron's resting membrane potential is established and maintained by (1) a high concentration of K^+ in the extracellular fluid and a high concentration of Na^+ in the cytosol, (2) the plasma membrane's higher permeability to Na^+ because of the presence of numerous Na^+ leakage channels, (3) differences in both ion concentrations and electrical gradients, (4) the fact that there are numerous large, nondiffusible anions in the cytosol, (5) sodium–potassium pumps that help to maintain the proper distribution of sodium and potassium.
(a) 1, 2, and 5   (b) 1, 2, and 3   (c) 2, 3, and 4
(d) 3, 4, and 5   (e) 1, 2, 3, 4, and 5

7. Place the following events in a chemical synapse in the correct order: (1) release of neurotransmitters into the synaptic cleft, (2) arrival of nerve impulse at the presynaptic neuron's synaptic end bulb (or varicosity), (3) either depolarization or hyperpolarization of postsynaptic membrane, (4) inward flow of Ca^{2+} through activated voltage-gated Ca^{2+} channels in the synaptic end bulb membrane, (5) exocytosis of synaptic vesicles, (6) opening of ligand-gated channels on the postsynaptic plasma membrane, (7) binding of neurotransmitters to receptors in the postsynaptic neuron's plasma membrane.
(a) 2, 1, 5, 4, 7, 6, 3   (b) 1, 2, 4, 5, 7, 6, 3
(c) 2, 4, 5, 1, 7, 6, 3   (d) 4, 5, 1, 7, 6, 3, 2
(e) 2, 5, 1, 4, 6, 7, 3

8. Several neurons in the brain sending impulses to a single motor neuron that terminates at a neuromuscular junction is an example of a _____ circuit.
(a) reverberating   (b) simple series
(c) parallel after-discharge   (d) diverging
(e) converging

9. Which of the following statements are *true*? (1) If the excitatory effect is greater than the inhibitory effect but less than the threshold of stimulation, the result is a subthreshold EPSP. (2) If the excitatory effect is greater than the inhibitory effect and reaches or surpasses the threshold level of stimulation, the result is a threshold or suprathreshold EPSP and one or more nerve impulses. (3) If the inhibitory effect is greater than the excitatory effect, the membrane hyperpolarizes, resulting in inhibition of the postsynaptic neuron and the inability of the neuron to generate a nerve impulse. (4) The greater the summation of hyperpolarizations, the more likely a nerve impulse will be initiated.
(a) 1 and 4   (b) 2 and 4   (c) 1, 3, and 4
(d) 2, 3, and 4   (e) 1, 2, and 3

10. Which of the following statements are *true*? (1) The basic types of ion channels are gated, leakage, and electrical. (2) Ion channels allow for the development of graded potentials and action potentials. (3) Voltage-gated channels open in response to changes in membrane potential. (4) Ligand-gated channels open due to the presence of specific chemicals. (5) A graded potential is useful for communication over long distances.
(a) 1, 2, and 3   (b) 2, 3, and 4   (c) 2, 3, and 5
(d) 2, 3, 4, and 5   (e) 1, 3, and 5

11. Which of the following statements are *true*? (1) The frequency of impulses and number of activated sensory neurons encode differences in stimuli intensity. (2) Larger-diameter axons conduct nerve impulses faster than smaller-diameter ones. (3) Continuous conduction is faster than saltatory conduction. (4) The presence or absence of a myelin sheath is an important factor that determines the speed of nerve impulse propagation. (5) Action potentials are localized, but graded potentials are propagated.
(a) 1, 3, and 5   (b) 3 and 4   (c) 2, 4, and 5
(d) 2 and 4   (e) 1, 2, and 4

12. Neurotransmitters are removed from the synaptic cleft by (1) axonal transport, (2) diffusion away from the cleft, (3) neurosecretory cells, (4) enzymatic breakdown, (5) cellular uptake.
(a) 1, 2, 3, and 4   (b) 2, 4, and 5   (c) 2, 3, and 4
(d) 1, 4, and 5   (e) 1, 2, 3, 4, and 5

13. Match the following:

___6___ (a) neurons with just one process extending from the cell body; are always sensory neurons

___6___ (b) small phagocytic neuroglia

___4___ (c) help maintain an appropriate chemical environment for generation of action potentials by neurons; part of the blood–brain barrier

___2___ (d) provide myelin sheath for CNS axons

___11___ (e) contains neuronal cell bodies, dendrites, axon terminals, unmyelinated axons, and neuroglia

___1___ (f) a cluster of cell bodies within the CNS

___4___ (g) form CSF and assist in its circulation; form blood–cerebrospinal barrier

___8___ (h) neurons having several dendrites and one axon; most common neuronal type

___3___ (i) neurons with one main dendrite and one axon; found in the retina of the eye

___9___ (j) provide myelin sheath for PNS axons

_____ (k) support neurons in PNS ganglia

_____ (l) a cluster of neuronal cell bodies located outside the brain and spinal cord

_____(m) composed primarily of myelinated axons

_____ (n) bundles of axons and associated connective tissue and blood vessels in the PNS

_____ (o) extensive neuronal networks that help regulate the digestive system

(1) astrocytes
(2) oligodendrocytes
(3) ganglion
(4) ependymal cells
(5) satellite cells
(6) unipolar neurons
(7) bipolar neurons
(8) multipolar neurons
(9) gray matter
(10) white matter
(11) enteric plexuses
(12) microglia
(13) Schwann cells
(14) nucleus
(15) nerves

14. Match the following:

_____ (a) a sequence of rapidly occurring events that decreases and eventually reverses the membrane potential and then restores it to the resting state; a nerve impulse

_____ (b) a small deviation from the resting membrane potential that makes the membrane either more or less polarized

_____ (c) period of time when a second action potential can be initiated with a very strong stimulus

_____ (d) the minimum level of depolarization required for a nerve impulse to be generated

_____ (e) the recovery of the resting membrane potential

_____ (f) a neurotransmitter-caused depolarization of the postsynaptic membrane

_____ (g) a neurotransmitter-caused hyperpolarization of the postsynaptic membrane

_____ (h) time during which a neuron cannot produce an action potential even with a very strong stimulus

_____ (i) polarization that is less negative than the resting level

_____ (j) results from the buildup of neurotransmitter released simultaneously by several presynaptic end bulbs

_____ (k) the hyperpolarization that occurs after the repolarizing phase of an action potential

_____ (l) polarization that is more negative than the resting level

_____(m) results from the buildup of neurotransmitter from the rapid, successive release by a single presynaptic end bulb

(1) graded potential
(2) action potential
(3) excitatory postsynaptic potential
(4) inhibitory postsynaptic potential
(5) absolute refractory period
(6) repolarization
(7) after-hyperpolarizing phase
(8) spatial summation
(9) threshold
(10) relative refractory period
(11) temporal summation
(12) depolarizing graded potential
(13) hyperpolarizing graded potential

15. Match the following:

_____ (a) the part of the neuron that contains the nucleus and organelles

_____ (b) rough endoplasmic reticulum in neurons; site of protein synthesis

_____ (c) store neurotransmitter

_____ (d) the process that propagates nerve impulses toward another neuron, muscle fiber, or gland cell

_____ (e) the highly branched receiving or input portions of a neuron

_____ (f) a multilayered lipid and protein covering for axons produced by neuroglia

_____ (g) the outer nucleated cytoplasmic layer of the Schwann cell

_____ (h) first portion of the axon, closest to the axon hillock

_____ (i) site of communication between two neurons or between a neuron and an effector cell

_____ (j) form the cytoskeleton of a neuron

_____ (k) gaps in the myelin sheath of an axon

_____ (l) general term for any neuronal process

_____(m) area where the axon joins the cell body

_____ (n) area where nerve impulses arise

_____ (o) the numerous fine processes at the ends of an axon and its collaterals

_____ (p) interstitial fluid-filled space separating two neurons

(1) myelin sheath
(2) neurolemma
(3) nodes of Ranvier
(4) cell body
(5) Nissl bodies
(6) neurofibrils
(7) dendrites
(8) axon
(9) axon hillock
(10) initial segment
(11) trigger zone
(12) synaptic cleft
(13) nerve fiber
(14) axon terminals
(15) synapse
(16) synaptic vesicles

CRITICAL THINKING QUESTIONS

1. The buzzing of the alarm clock woke Carrie. She stretched, yawned, and started to salivate as she smelled the brewing coffee. She could feel her stomach rumble. List the divisions of the nervous system that are involved in each of these actions.

2. Baby Ming is learning to crawl. He also likes to pull himself onto window sills, gnawing on the painted wood of his century-old home as he looks out the windows. Lately his mother, an anatomy and physiology student, has noticed some odd behavior and took Ming to the pediatrician. Blood work determined that Ming had a high level of lead in his blood, ingested from the old leaded paint on the window sill. The doctor indicated that lead poisoning is a type of demyelination disorder. Why should Ming's mother be concerned?

3. As a torture procedure for his enemies, mad scientist Dr. Moro is trying to develop a drug that will enhance the effects of substance P. What cellular mechanisms could he enlist to design such a drug?

? ANSWERS TO FIGURE QUESTIONS

12.1 The CNS processes many different kinds of sensory information; is the source of thoughts, emotions, and memories; and gives rise to signals that stimulate muscles to contract and glands to secrete.

12.2 Dendrites and the cell body receive input; the axon conducts nerve impulses (action potentials) and transmits the message to another neuron or effector cell by releasing a neurotransmitter at its synaptic end bulbs.

12.3 Most neurons in the CNS are multipolar neurons.

12.4 Most unipolar neurons are located in the PNS; the cell bodies of these neurons are mainly in the ganglia of cranial and spinal nerves.

12.5 The cell body of a pyramidal cell is shaped like a pyramid.

12.6 Microglia function as phagocytes in the central nervous system.

12.7 One Schwann cell myelinates a single axon; one oligodendrocyte myelinates several axons.

12.8 Myelination increases the speed of nerve impulse conduction.

12.9 Myelin makes white matter look shiny and white.

12.10 Perception primarily occurs in the cerebral cortex.

12.11 A touch on the arm activates mechanically gated channels.

12.12 The -70 mV resting membrane potential of a neuron means that the inside of the neuron is 70 mV more negative than the outside when the neuron is at rest (not excited by a stimulus).

12.13 More Na^+ ions would leak into the cell and fewer K^+ ions would leak out of the cell, which would make the resting membrane potential more inside-positive.

12.14 A change in membrane potential from -70 to -60 mV is a depolarizing graded potential since the membrane potential is inside less negative than at rest. A change in membrane potential from -70 to -80 mV is a hyperpolarizing graded potential since the membrane potential is inside more negative than at rest.

12.15 Ligand-gated channels and mechanically gated channels can be present in the dendrites of sensory neurons, and ligand-gated channels are numerous in the dendrites and cell bodies of interneurons and motor neurons.

12.16 A stronger stimulus opens more mechanically gated channels or ligand-gated channels than a weaker stimulus.

12.17 Since individual graded potentials undergo decremental conduction, they would die out as they spread through the dendrites and cell body if summation did not occur and an action potential would not be generated at the trigger zone of the axon.

12.18 Voltage-gated Na^+ channels are open during the depolarizing phase, and voltage-gated K^+ channels are open during the repolarizing phase.

12.19 An action potential will not occur in response to a hyperpolarizing graded potential because a hyperpolarizing graded potential causes the membrane potential to become inside more negative and, therefore, farther away from threshold (-55 mV).

12.20 Yes, because the leak channels would still allow K^+ to exit more rapidly than Na^+ could enter the axon. Some mammalian myelinated axons have only a few voltage-gated K^+ channels.

12.21 The diameter of an axon, presence or absence of a myelin sheath, and temperature determine the speed of propagation of an action potential.

12.22 In some electrical synapses (gap junctions), ions may flow equally well in either direction, so either neuron may be the presynaptic one. At a chemical synapse, one neuron releases neurotransmitter and the other neuron has receptors that bind this chemical. Thus, the signal can proceed in only one direction.

12.23 At some excitatory synapses, ACh binds to ionotropic receptors with cation channels that open and subsequently generate EPSPs in the postsynaptic cell. At some inhibitory synapses, ACh binds to metabotropic receptors coupled to G proteins that open K^+ channels, resulting in the formation of IPSPs in the postsynaptic cell.

12.24 This is an example of spatial summation since the summation results from the buildup of neurotransmitter released simultaneously by several presynaptic end bulbs.

12.25 Since -60 mV is below threshold, an action potential will not occur in the postsynaptic neuron.

12.26 Norepinephrine, epinephrine, dopamine, and serotonin are classified as biogenic amines because they are derived from amino acids that have been chemically modified.

12.27 A motor neuron receiving input from several other neurons is an example of convergence.

12.28 The neurolemma provides a regeneration tube that guides regrowth of a severed axon.

13 THE SPINAL CORD AND SPINAL NERVES

THE SPINAL CORD AND SPINAL NERVES AND HOMEOSTASIS *The spinal cord and spinal nerves contribute to homeostasis by providing quick, reflexive responses to many stimuli. The spinal cord is the pathway for sensory input to the brain and motor output from the brain.*

About 100 million neurons and even more neuroglia compose the **spinal cord,** the part of the central nervous system that extends from the brain. The spinal cord and its associated spinal nerves contain neural circuits that control some of your most rapid reactions to environmental changes. If you pick up something hot, the grasping muscles may relax and you may drop the hot object even before you are consciously aware of the extreme heat or pain. This is an example of a spinal cord reflex—a quick, automatic response to certain kinds of stimuli that involves neurons only in the spinal nerves and spinal cord. Besides processing reflexes, the gray matter of the spinal cord also is a site for integration (summing) of excitatory postsynaptic potentials (EPSPs) and inhibitory postsynaptic potentials (IPSPs), which you learned about in Chapter 12. These graded potentials arise as neurotransmitter molecules interact with their receptors at synapses in the spinal cord. The white matter of the spinal cord contains a dozen major sensory and motor tracts, which function as the "highways" along which sensory input travels to the brain and motor output travels from the brain to skeletal muscles and other effectors. Recall that the spinal cord is continuous with the brain and that together they make up the central nervous system (CNS).

? *Did you ever wonder why spinal cord injuries can have such widespread effects on the body?*

492

13.1 SPINAL CORD ANATOMY

⊙ OBJECTIVES

- Describe the protective structures and the gross anatomical features of the spinal cord.
- Describe how spinal nerves are connected to the spinal cord.

Protective Structures

Recall from the previous chapter that the nervous tissue of the central nervous system is very delicate and does not respond well to injury or damage. Accordingly, nervous tissue requires considerable protection. The first layer of protection for the central nervous system is the hard bony skull and vertebral column. The skull encases the brain and the vertebral column surrounds the spinal cord, providing strong protective defenses against damaging blows or bumps. The second protective layer is the meninges, three membranes that lie between the bony encasement and the nervous tissue in both the brain and spinal cord. Finally, a space between two of the meningeal membranes contains cerebrospinal fluid, a buoyant liquid that suspends the central nervous tissue in a weightless environment while surrounding it with a shock-absorbing, hydraulic cushion.

Vertebral Column

The spinal cord is located within the vertebral canal of the vertebral column. As you learned in Chapter 7, the vertebral foramina of all the vertebrae, stacked one on top of the other, form the vertebral canal. The surrounding vertebrae provide a sturdy shelter for the enclosed spinal cord (see Figure 13.1c). The vertebral ligaments, meninges, and cerebrospinal fluid provide additional protection.

Meninges

The **meninges** (me-NIN-jē-z; singular is **meninx** [ME-ninks]) are three protective, connective tissue coverings that encircle the spinal cord and brain. From superficial to deep they are the (1) dura mater, (2) arachnoid mater, and (3) pia mater. The **spinal meninges** surround the spinal cord (Figure 13.1a) and are continuous with the **cranial meninges,** which encircle the brain (shown in Figure 14.2a). All three spinal meninges cover the spinal nerves up to the point where they exit the spinal column through the intervertebral foramina. The spinal cord is also protected by a cushion of fat and connective tissue located in the **epidural space** (ep′-i-DOO-ral), a space between the dura mater and the wall of the vertebral canal (Figure 13.1c). Following is a description of each meningeal layer.

1. **Dura mater** (DOO-ra MĀ-ter = tough mother). The most superficial of the three spinal meninges is a thick strong layer composed of dense, irregular connective tissue. The dura mater forms a sac from the level of the foramen magnum in the occipital bone, where it is continuous with the meningeal dura mater of the brain, to the second sacral vertebra. The dura mater is also continuous with the epineurium, the outer covering of spinal and cranial nerves.

2. **Arachnoid mater** (a-RAK-noyd MĀ-ter; *arachn-* = spider; *-oid* = similar to). This layer, the middle of the meningeal membranes, is a thin, avascular covering comprised of cells and thin, loosely arranged collagen and elastic fibers. It is called the arachnoid mater because of its spider's web arrangement of delicate collagen fibers and some elastic fibers. It is deep to the dura mater and is continuous through the foramen magnum with the arachnoid mater of the brain. Between the dura mater and the arachnoid mater is a thin **subdural space,** which contains interstitial fluid.

3. **Pia mater** (PĒ-a MĀ-ter; *pia* = delicate). This innermost meninx is a thin transparent connective tissue layer that adheres to the surface of the spinal cord and brain. It consists of thin squamous to cuboidal cells within interlacing bundles of collagen fibers and some fine elastic fibers. Within the pia mater are many blood vessels that supply oxygen and nutrients to the spinal cord. Triangular-shaped membranous extensions of the pia mater suspend the spinal cord in the middle of its dural sheath. These extensions, called **denticulate ligaments** (den-TIK-ū-lāt = small tooth), are thickenings of the pia mater. They project laterally and fuse with the arachnoid mater and inner surface of the dura mater between the anterior and posterior nerve roots of spinal nerves on either side (Figure 13.1a, b). Extending along the entire length of the spinal cord, the denticulate ligaments protect the spinal cord against sudden displacement that could result in shock. Between the arachnoid mater and pia mater is a space, the **subarachnoid space,** which contains shock-absorbing cerebrospinal fluid.

⚕ CLINICAL CONNECTION | *Spinal Tap*

In a **spinal tap (lumbar puncture),** a local anesthetic is given, and a long hollow needle is inserted into the subarachnoid space to withdraw cerebrospinal fluid (CSF) for diagnostic purposes; to introduce antibiotics, contrast media for myelography, or anesthetics; to administer chemotherapy; to measure CSF pressure; and/or to evaluate the effects of treatment for diseases such as meningitis. During this procedure, the patient lies on his or her side with the vertebral column flexed. Flexion of the vertebral column increases the distance between the spinous processes of the vertebrae, which allows easy access to the subarachnoid space. The spinal cord ends around the second lumbar vertebra (L2); however, the spinal meninges and circulating cerebrospinal fluid extend to the second sacral vertebra (S2). Between vertebrae L2 and S2 the spinal meninges are present, but the spinal cord is absent. Consequently, a spinal tap is normally performed in adults between the L3 and L4 or L4 and L5 lumbar vertebrae because this region provides safe access to the subarachnoid space without the risk of damaging the spinal cord. (A line drawn across the highest points of the iliac crests, called the *supracristal line,* passes through the spinous process of the fourth lumbar vertebra and is used as a landmark for administering a spinal tap.) •

External Anatomy of the Spinal Cord

The **spinal cord** is roughly oval in shape, being flattened slightly anteriorly and posteriorly. In adults, it extends from the medulla oblongata, the inferior part of the brain, to the superior border of the second lumbar vertebra (Figure 13.2). In newborn infants, it extends to the third or fourth lumbar vertebra. During early childhood, both the spinal cord and the vertebral column grow longer as part of overall body growth. Elongation of the spinal cord stops around age 4 or 5, but growth of the vertebral column continues. Thus, the spinal cord does not extend the entire length of the adult vertebral column. The length of the adult spinal cord ranges from 42 to 45 cm (16–18 in.). Its diameter is about 2 cm (0.75 in.) in the midthoracic region, somewhat larger in the lower cervical and midlumbar regions, and smallest at the inferior tip.

When the spinal cord is viewed externally, two conspicuous enlargements can be seen. The superior enlargement, the **cervical enlargement,** extends from the fourth cervical vertebra (C4) to the first thoracic vertebra (T1). Nerves to and from the upper limbs arise from the cervical enlargement. The inferior enlargement, called the **lumbar enlargement,** extends from the ninth to the twelfth thoracic vertebra. Nerves to and from the lower limbs arise from the lumbar enlargement.

Inferior to the lumbar enlargement, the spinal cord terminates as a tapering, conical structure called the **conus medullaris** (KŌ-nus med-ū-LAR-is; *conus* = cone), which ends at the level of the intervertebral disc between the first and second lumbar vertebrae (L1–L2) in adults. Arising from the conus medullaris is the **filum terminale** (FĪ-lum ter-mi-NAL-ē = terminal filament), an extension of the pia mater that extends inferiorly and fuses with the arachnoid mater and dura mater and anchors the spinal cord to the coccyx.

Spinal nerves are the paths of communication between the spinal cord and specific regions of the body. The spinal cord appears to be segmented because the 31 pairs of spinal nerves emerge at regular intervals from intervertebral foramina (Figure 13.2). Indeed, each pair of spinal nerves is said to arise from a *spinal segment*. Within the spinal cord there is no obvious segmentation but, for convenience, the naming of spinal nerves is based on the segment in

Figure 13.1 Gross anatomy of the spinal cord. The spinal meninges are evident in parts (a) and (c).

Meninges are connective tissue coverings that surround the spinal cord and brain.

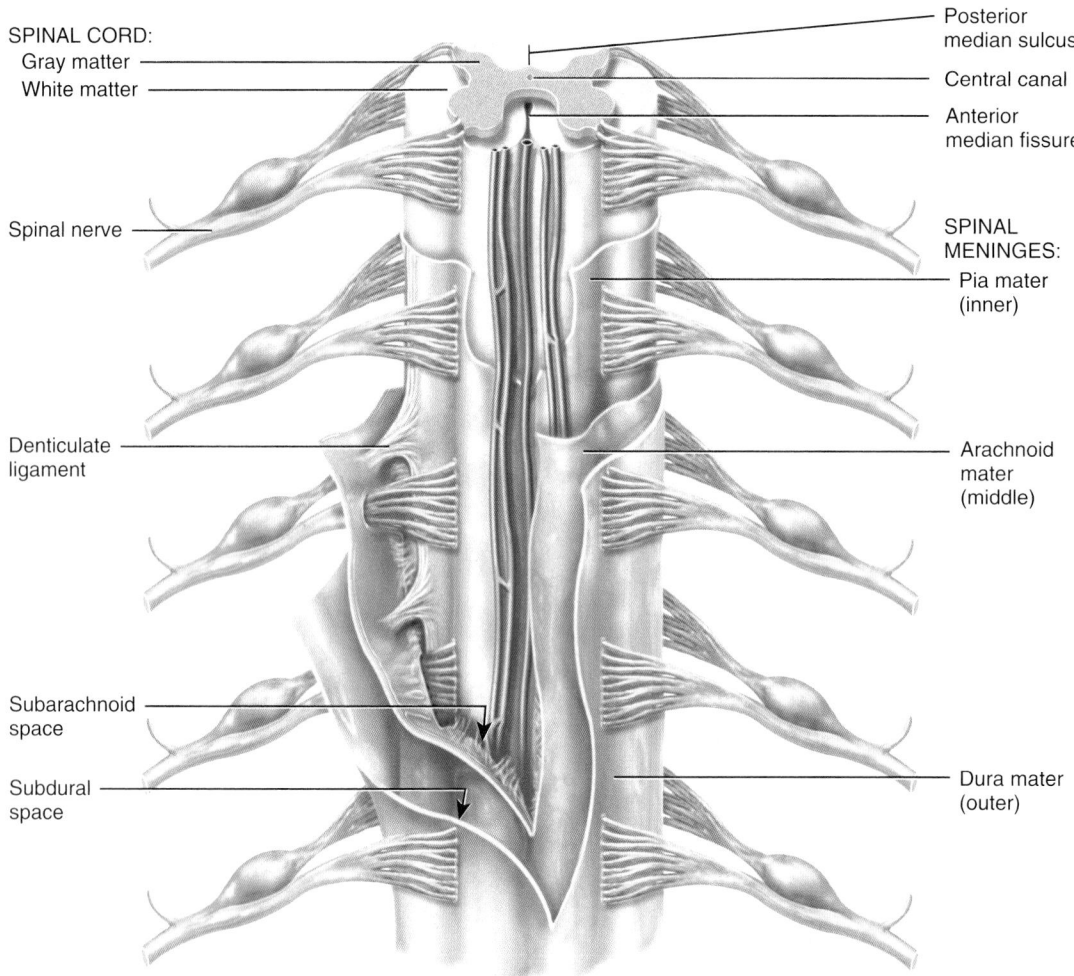

(a) Anterior view and transverse section through spinal cord

SUPERIOR

Fourth ventricle

Cerebellum of brain (cut)

Glossopharyngeal (IX) and vagus (X) nerves

Accessory (XI) nerve

Gracile fasciculus

Occipital bone (cut)

Posterior median sulcus

Cuneate fasciculus

Vertebral artery

Denticulate ligament

Posterior (dorsal) rootlets of spinal nerve

Dura mater and arachnoid mater (cut)

INFERIOR

(b) Posterior view of cervical region of spinal cord

POSTERIOR

View

Transverse plane

Spinous process of vertebra

Subarachnoid space

Posterior (dorsal) root of spinal nerve

Denticulate ligament

Anterior (ventral) root of spinal nerve

Transverse foramen

Body of vertebra

Dura mater and arachnoid mater

Spinal cord

Pia mater

Epidural space

Superior articular facet of vertebra

Posterior (dorsal) ramus of spinal nerve

Spinal nerve

Anterior (ventral) ramus of spinal nerve

Vertebral artery in transverse foramen

ANTERIOR

(c) Transverse section of the spinal cord within a cervical vertebra

? **What are the superior and inferior boundaries of the spinal dura mater?**

Figure 13.2 External anatomy of the spinal cord and the spinal nerves.

The spinal cord extends from the medulla oblongata of the brain to the superior border of the second lumbar vertebra.

CERVICAL PLEXUS (C1–C5):
- Lesser occipital nerve
- Great auricular
- Ansa cervicalis
- Transverse cervical nerve
- Supraclavicular nerve
- Phrenic nerve

BRACHIAL PLEXUS (C5–T1):
- Musculocutaneous nerve
- Axillary nerve
- Median nerve
- Radial nerve
- Ulnar nerve

Intercostal (thoracic) nerves

Subcostal nerve (intercostal nerve 12)

LUMBAR PLEXUS (L1–L4):
- Iliohypogastric nerve
- Ilioinguinal nerve
- Genitofemoral nerve
- Lateral femoral cutaneous nerve

Femoral nerve

Obturator nerve

SACRAL PLEXUS (L4–S4):
- Superior gluteal nerve
- Inferior gluteal nerve

Sciatic nerve:
- Common fibular nerve
- Tibial nerve

Posterior cutaneous nerve of thigh

Pudendal nerve

Medulla oblongata

Atlas (first cervical vertebra)

CERVICAL NERVES (8 pairs)

Cervical enlargement

First thoracic vertebra

THORACIC NERVES (12 pairs)

Lumbar enlargement

First lumbar vertebra
Conus medullaris

LUMBAR NERVES (5 pairs)

Cauda equina

Ilium of hip bone

Sacrum

SACRAL NERVES (5 pairs)

COCCYGEAL NERVES (1 pair)

Filum terminale

C1, C2, C3, C4, C5, C6, C7, C8, T1, T2, T3, T4, T5, T6, T7, T8, T9, T10, T11, T12, L1, L2, L3, L4, L5, S1, S2, S3, S4, S5

Posterior view of entire spinal cord and portions of spinal nerves

What portion of the spinal cord connects with nerves of the upper limbs?

which they are located. There are 8 pairs of *cervical nerves* (represented in Figure 13.2 as C1–C8), 12 pairs of *thoracic nerves* (T1–T12), 5 pairs of *lumbar nerves* (L1–L5), 5 pairs of *sacral nerves* (S1–S5), and 1 pair of *coccygeal nerves* (Co1).

Two bundles of axons, called **roots,** connect each spinal nerve to a segment of the cord by even smaller bundles of axons called **rootlets** (see Figure 13.3a). The **posterior (dorsal) root** and rootlets contain only sensory axons, which conduct nerve

Figure 13.3 **Internal anatomy of the spinal cord: the organization of gray matter and white matter.** For simplicity, dendrites are not shown in this and several other illustrations of transverse sections of the spinal cord. Blue, red, and green arrows indicate the direction of nerve impulse propagation.

🔑 The posterior gray horn contains axons of sensory neurons and cell bodies of interneurons; the lateral gray horn contains cell bodies of autonomic motor neurons; and the anterior gray horn contains cell bodies of somatic motor neurons.

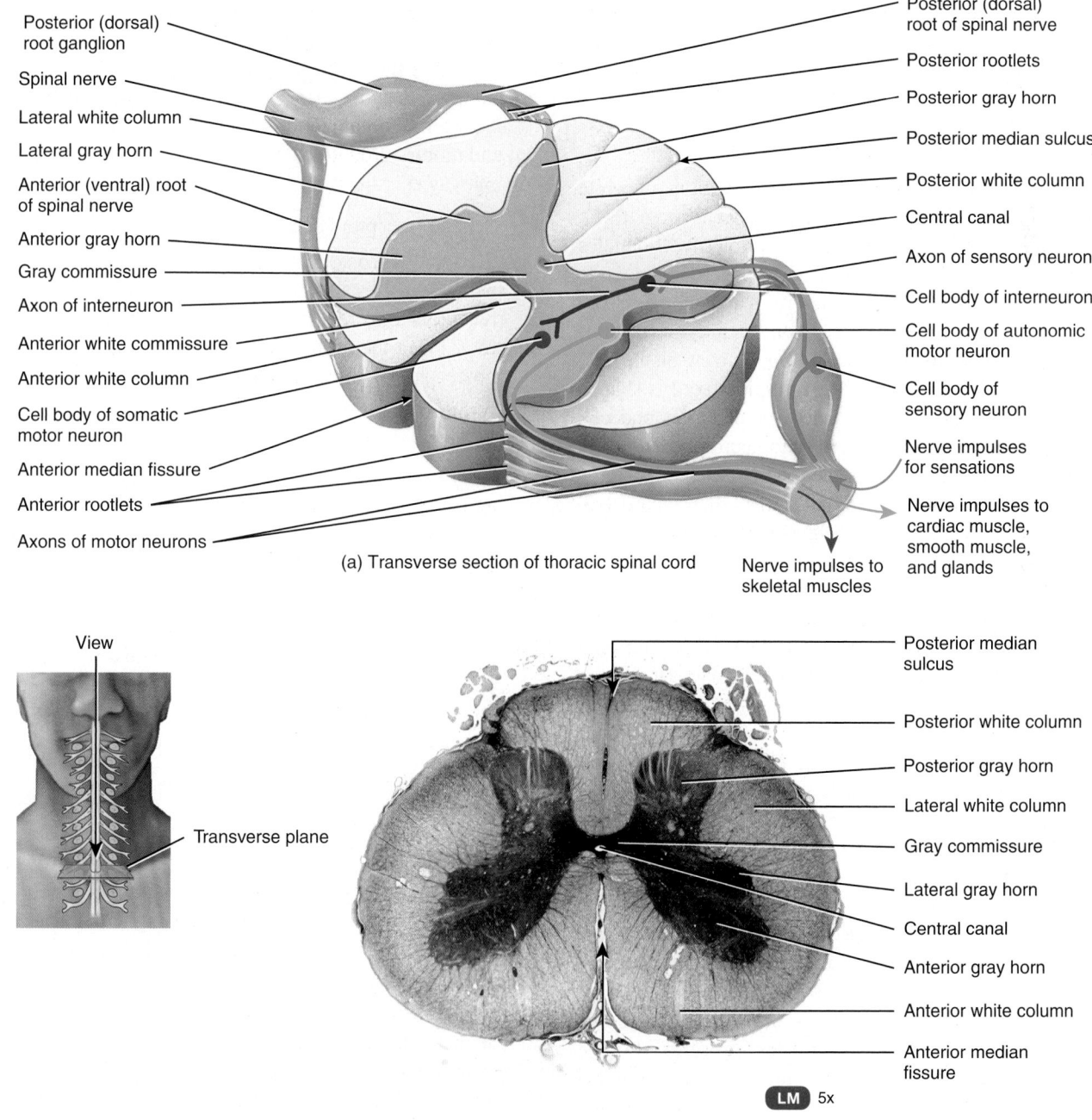

(a) Transverse section of thoracic spinal cord

(b) Transverse section of thoracic spinal cord

LM 5x

❓ **What is the difference between a horn and a column in the spinal cord?**

impulses from sensory receptors in the skin, muscles, and internal organs into the central nervous system. Each posterior root has a swelling, the **posterior (dorsal) root ganglion,** which contains the cell bodies of sensory neurons. The **anterior (ventral) root** and rootlets contain axons of motor neurons, which conduct nerve impulses from the CNS to effectors (muscles and glands).

As spinal nerves branch from the spinal cord, they pass laterally to exit the vertebral canal through the intervertebral foramina between adjacent vertebrae. However, because the spinal cord is shorter than the vertebral column, nerves that arise from the lumbar, sacral, and coccygeal regions of the spinal cord do not leave the vertebral column at the same level they exit the cord. The roots of these lower spinal nerves angle inferiorly alongside the filum terminale in the vertebral canal like wisps of hair. Accordingly, the roots of these nerves are collectively named the **cauda equina** (KAW-da ē-KWĪ-na), meaning "horse's tail" (Figure 13.2).

Internal Anatomy of the Spinal Cord

A transverse section of the spinal cord reveals regions of white matter that surround an inner core of gray matter (Figure 13.3). The white matter of the spinal cord consists primarily of bundles of myelinated axons of neurons. Two grooves penetrate the white matter of the spinal cord and divide it into right and left sides. The **anterior median fissure** is a wide groove on the anterior (ventral) side. The **posterior median sulcus** is a narrow furrow on the posterior (dorsal) side. The gray matter of the spinal cord is shaped like the letter H or a butterfly; it consists of dendrites and cell bodies of neurons, unmyelinated axons, and neuroglia. The **gray commissure** (KOM-mi-shur) forms the crossbar of the H. In the center of the gray commissure is a small space called the **central canal;** it extends the entire length of the spinal cord and is filled with cerebrospinal fluid. At its superior end, the central canal is continuous with the fourth ventricle (a space that contains cerebrospinal fluid) in the medulla oblongata of the brain. Anterior to the gray commissure is the **anterior (ventral) white commissure,** which connects the white matter of the right and left sides of the spinal cord.

In the gray matter of the spinal cord and brain, clusters of neuronal cell bodies form functional groups called **nuclei.** *Sensory nuclei* receive input from receptors via sensory neurons, and *motor nuclei* provide output to effector tissues via motor neurons. The gray matter on each side of the spinal cord is subdivided into regions called **horns** (Figure 13.3). The **posterior (dorsal) gray horns** contain cell bodies and axons of interneurons as well as axons of incoming sensory neurons. Recall that cell bodies of sensory neurons are located in the posterior (dorsal) root ganglion of a spinal nerve. The **anterior (ventral) gray horns** contain *somatic motor nuclei,* which are clusters of cell bodies of somatic motor neurons that provide nerve impulses for contraction of skeletal muscles. Between the posterior and anterior gray horns are the **lateral gray horns,** which are present only in thoracic and upper lumbar segments of the spinal cord. The lateral gray horns contain *autonomic motor nuclei,* which are clusters of cell bodies

of autonomic motor neurons that regulate the activity of cardiac muscle, smooth muscle, and glands.

The white matter of the spinal cord, like the gray matter, is organized into regions. The anterior and posterior gray horns divide the white matter on each side into three broad areas called **columns:** (1) **anterior (ventral) white columns,** (2) **posterior (dorsal) white columns,** and (3) **lateral white columns** (Figure 13.3). Each column in turn contains distinct bundles of axons having a common origin or destination and carrying similar information. These bundles, which may extend long distances up or down the spinal cord, are called **tracts.** Recall that tracts are bundles of axons in the CNS, whereas nerves are bundles of axons in the PNS. **Sensory (ascending) tracts** consist of axons that conduct nerve impulses toward the brain. Tracts consisting of axons that carry nerve impulses from the brain are called **motor (descending) tracts.** Sensory and motor tracts of the spinal cord are continuous with sensory and motor tracts in the brain.

The internal organization of the spinal cord allows sensory input and motor output to be processed by the spinal cord in the following way (Figure 13.4):

1 Sensory receptors detect a sensory stimulus.

2 Sensory neurons convey this sensory input in the form of nerve impulses along their axons, which extend from sensory receptors into the spinal nerve and then into the posterior root. From the posterior root, axons of sensory neurons may proceed along three possible paths (see steps **3**, **4**, and **5**).

3 Axons of sensory neurons may extend into the white matter of the spinal cord and ascend to the brain as part of a sensory tract.

4 Axons of sensory neurons may enter the posterior gray horn and synapse with interneurons whose axons extend into the white matter of the spinal cord and then ascend to the brain as part of a sensory tract.

5 Axons of sensory neurons may enter the posterior gray horn and synapse with interneurons that in turn synapse with somatic motor neurons that are involved in spinal reflex pathways. Spinal cord reflexes are described in more detail later in this chapter.

6 Motor output from the spinal cord to skeletal muscles involves somatic motor neurons of the anterior gray horn. Many somatic motor neurons are regulated by the brain. Axons from higher brain centers form motor tracts that descend from the brain into the white matter of the spinal cord. There they synapse with somatic motor neurons either directly or indirectly by first synapsing with interneurons that in turn synapse with somatic motor neurons.

7 When activated, somatic motor neurons convey motor output in the form of nerve impulses along their axons, which sequentially pass through the anterior gray horn and anterior root to enter the spinal nerve. From the spinal nerve, axons of somatic motor neurons extend to skeletal muscles of the body.

8 Motor output from the spinal cord to cardiac muscle, smooth muscle, and glands involves autonomic motor neurons of the

Figure 13.4 Processing of sensory input and motor output by the spinal cord.

🔑 Sensory input is conveyed from sensory receptors to the posterior gray horns of the spinal cord, whereas motor output is conveyed from the anterior and lateral gray horns of the spinal cord to effectors (muscles and glands).

Sensory (ascending) tract

Motor (descending) tract

Sensory (ascending) tract

Posterior (dorsal) root ganglion

Spinal nerve

Lateral gray horn

Anterior gray horn

Anterior (ventral) root of spinal nerve

Interneuron

Somatic motor neuron

Posterior (dorsal) root of spinal nerve

Posterior gray horn

Interneuron

Autonomic motor neuron

Sensory neuron

Sensory receptors in the skin

Skeletal muscle

Autonomic motor neuron

Cardiac muscle, smooth muscle, and glands

❓ **Lateral gray horns are found in which segments of the spinal cord?**

lateral gray horn. When activated, autonomic motor neurons convey motor output in the form of nerve impulses along their axons, which sequentially pass through the lateral gray horn, anterior gray horn, and anterior root to enter the spinal nerve.

9 From the spinal nerve, axons of autonomic motor neurons from the spinal cord synapse with another group of autonomic motor neurons located in the peripheral nervous system (PNS). The axons of this second group of autonomic motor neurons in turn synapse with cardiac muscle, smooth muscle, and glands. You will learn more about autonomic motor neurons when the autonomic nervous system is described in Chapter 15.

The various spinal cord segments vary in size, shape, relative amounts of gray and white matter, and distribution and shape of gray matter. For example, the amount of gray matter is largest in the cervical and lumbar segments of the spinal cord because these segments are responsible for sensory and motor innervation of the limbs. In addition, more sensory and motor tracts are present in the upper segments of the spinal cord than in the lower segments. Therefore, the amount of white matter decreases from cervical to sacral segments of the spinal cord. There are two major reasons for this variation in spinal cord white matter: (1) As the spinal cord ascends from sacral to cervical segments, more ascending axons are added to spinal cord white matter to form more sensory

tracts. (2) As the spinal cord descends from cervical to sacral segments, the motor tracts decrease in thickness as more descending axons leave the motor tracts to synapse with neurons in the gray matter of the spinal cord. Table 13.1 summarizes the variations in spinal cord segments.

✔ **CHECKPOINT**

1. Where are the spinal meninges located? Where are the epidural, subdural, and subarachnoid spaces located?
2. What are the cervical and lumbar enlargements?
3. Define conus medullaris, filum terminale, and cauda equina. What is a spinal segment? How is the spinal cord partially divided into right and left sides?
4. What does each of the following terms mean? Gray commissure, central canal, anterior gray horn, lateral gray horn, posterior gray horn, anterior white column, lateral white column, posterior white column, ascending tract, and descending tract.

13.2 SPINAL NERVES

◉ **OBJECTIVES**

• Describe the components, connective tissue coverings, and branching of a spinal nerve.
• Define plexus, and identify the distribution of nerves of the cervical, brachial, lumbar, and sacral plexuses.
• Describe the clinical significance of dermatomes.

Spinal nerves are associated with the spinal cord and, like all nerves of the peripheral nervous system (PNS), are parallel bundles of axons and their associated neuroglial cells wrapped in several layers of connective tissue. Spinal nerves connect the CNS to sensory receptors, muscles, and glands in all parts of the body. The 31 pairs of spinal nerves are named and numbered according to the region and level of the vertebral column from which they emerge (see Figure 13.2). Not all spinal cord segments are aligned with their corresponding vertebrae. Recall that the spinal cord ends near the level of the superior border of the second lumbar vertebra (L2), and that the roots of the lumbar, sacral, and coccygeal nerves descend at an angle to reach their respective foramina before emerging from the vertebral column. This arrangement constitutes the cauda equina.

The first cervical pair of spinal nerves emerges from the spinal cord between the occipital bone and the atlas (first cervical vertebra, or C1). Most of the remaining spinal nerves emerge from the spinal cord through the intervertebral foramina between adjoining vertebrae. Spinal nerves C1–C7 exit the vertebral canal *above* their corresponding vertebrae. Spinal nerve C8 exits the vertebral canal between vertebrae C7 and T1. Spinal nerves T1–L5 exit the vertebral canal *below* their corresponding vertebrae. From the spinal cord, the roots of the sacral spinal nerves (S1–S5) and the coccygeal spinal nerves (Co1) enter the sacral canal, the part of the vertebral canal in the sacrum (see Figure 7.21). Subsequently, spinal nerves S1–S4 exit the sacral canal via the four pairs of anterior and posterior sacral foramina, and spinal nerves S5 and Co1 exit the sacral canal via the sacral hiatus.

TABLE 13.1	
Comparison of Various Spinal Cord Segments	
SEGMENT	**DISTINGUISHING CHARACTERISTICS**
Cervical (Segment C6)	Relatively large diameter, relatively large amounts of white matter, oval; in upper cervical segments (C1–C4), posterior gray horn is large but anterior gray horn is relatively small; in lower cervical segments (C5 and below), posterior gray horns are enlarged and anterior gray horns are well developed.
Thoracic (Segment T5)	Small diameter due to relatively small amounts of gray matter; except for first thoracic segment, anterior and posterior gray horns are relatively small; small lateral gray horn is present.
Lumbar (Segment L4)	Nearly circular; very large anterior and posterior gray horns; small lateral gray horn is present in upper segments; relatively less white matter than cervical segments.
Sacral (Segment S3)	Relatively small, but relatively large amounts of gray matter; relatively small amounts of white matter; anterior and posterior gray horns are large and thick.
Coccygeal	Resembles lower sacral spinal segments, but much smaller.

As noted earlier, a typical spinal nerve has two connections to the cord: a posterior root and an anterior root (see Figure 13.3a). The posterior and anterior roots unite to form a spinal nerve at the intervertebral foramen. Because the posterior root contains sensory axons and the anterior root contains motor axons, a spinal nerve is classified as a **mixed nerve.** The posterior root contains a posterior root ganglion in which cell bodies of sensory neurons are located.

Connective Tissue Coverings of Spinal Nerves

Each spinal nerve and cranial nerve consists of many individual axons and contains layers of protective connective tissue coverings (Figure 13.5). Individual axons within a nerve, whether myelinated or unmyelinated, are wrapped in **endoneurium** (en′-dō-NOO-rē-um; *endo-* = within or inner; *-neurium* = nerve), the innermost layer. The endoneurium consists of a mesh of collagen fibers, fibroblasts, and macrophages. Groups of axons with their endoneurium are held together in bundles called **fascicles,** each of which is wrapped in **perineurium** (per′-i-NOO-rē-um; *peri-* = around), the middle layer. The perineurium is a thicker layer of connective tissue. It consists of up to 15 layers of fibroblasts within a network of collagen fibers. The outermost covering over the entire nerve is the **epineurium** (ep′-i-NOO-rē-um; *epi-* = over). It consists of fibroblasts and thick collagen fibers. Exten-

sions of the epineurium also fill the spaces between fascicles. The dura mater of the spinal meninges fuses with the epineurium as the nerve passes through the intervertebral foramen. Note the presence of many blood vessels, which nourish the spinal meninges (Figure 13.5b). You may recall from Chapter 10 that the connective tissue coverings of skeletal muscles—endomysium, perimysium, and epimysium—are similar in organization to those of nerves.

Distribution of Spinal Nerves

Branches

A short distance after passing through its intervertebral foramen, a spinal nerve divides into several branches (Figure 13.6). These branches are known as **rami** (RĀ-mī = branches). The **posterior (dorsal) ramus** (RĀ-mus; singular form) serves the deep muscles and skin of the posterior surface of the trunk. The **anterior (ventral) ramus** serves the muscles and structures of the upper and lower limbs and the skin of the lateral and anterior surfaces of the trunk. In addition to posterior and anterior rami, spinal nerves also give off a **meningeal branch** (me-NIN-jē′-al). This branch reenters the vertebral cavity through the intervertebral foramen and supplies the vertebrae, vertebral ligaments, blood vessels of the spinal cord, and meninges. Other branches of a spinal nerve

Figure 13.5 Organization and connective tissue coverings of a spinal nerve. (Part b © Dr. Richard Kessel and Dr. Randy Kardon/Visuals Unlimited.)

Three layers of connective tissue wrappings protect axons: Endoneurium surrounds individual axons, perineurium surrounds bundles of axons (fascicles), and epineurium surrounds an entire nerve.

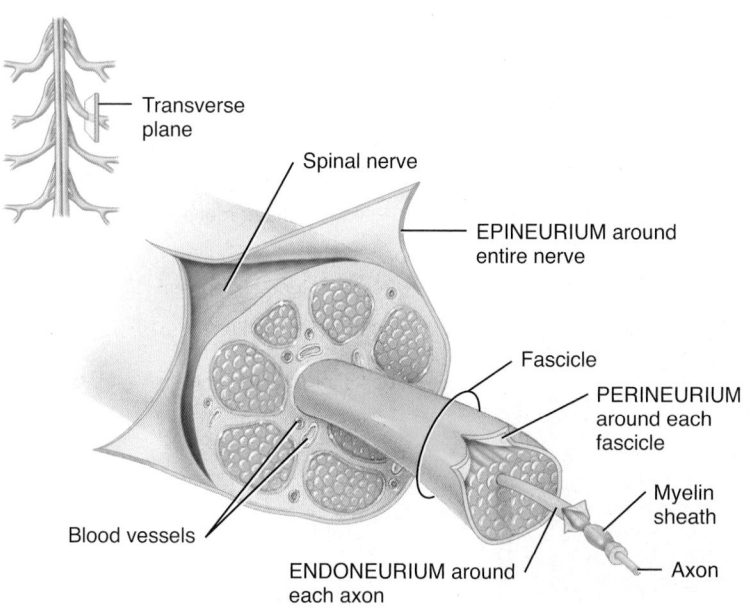

(a) Transverse section showing the coverings of a spinal nerve

(b) Transverse section of 12 nerve fascicles

? Why are all spinal nerves classified as mixed nerves?

Figure 13.6 **Branches of a typical spinal nerve, shown in transverse section through the thoracic portion of the spinal cord.** (See also Figure 13.1c.)

The branches of a spinal nerve are the posterior ramus, the anterior ramus, the meningeal branch, and the rami communicantes.

View

Transverse plane

POSTERIOR

Spinous process of vertebra

Deep muscles of back

POSTERIOR (DORSAL) RAMUS

ANTERIOR (VENTRAL) RAMUS

MENINGEAL BRANCH

Denticulate ligament

Subarachnoid space (contains CSF)

Body of vertebra

Spinal cord

Posterior (dorsal) root

Posterior (dorsal) root ganglion

Anterior (ventral) root

RAMI COMMUNICANTES

Dura mater and arachnoid mater

Sympathetic ganglion on sympathetic trunk

Epidural space (contains fat and blood vessels)

ANTERIOR

(a) Superior view

Denticulate ligament

Pedicle of vertebra (cut)

Dura mater and arachnoid mater

Spinal nerve

ANTERIOR (VENTRAL) RAMUS

POSTERIOR (DORSAL) RAMUS

Anterior (ventral) rootlets

Posterior (dorsal) rootlets

Anterior (ventral) root

Posterior (dorsal) root

(b) Anterior view and oblique section of spinal cord

Which spinal nerve branches serve the upper and lower limbs?

are the **rami communicantes** (kō-mū-ni-KAN-tēz), components of the autonomic nervous system that will be discussed in Chapter 15.

Plexuses

Axons from the anterior rami of spinal nerves, except for thoracic nerves T2–T12, do not go directly to the body structures they supply. Instead, they form networks on both the left and right sides of the body by joining with various numbers of axons from anterior rami of adjacent nerves. Such a network of axons is called a **plexus** (PLEK-sus = braid or network). The principal plexuses are the **cervical plexus, brachial plexus, lumbar plexus,** and **sacral plexus.** A smaller **coccygeal plexus** is also present. Refer to Figure 13.2 to see their relationships to one another. Emerging from the plexuses are nerves bearing names that are often descriptive of the general regions they serve or the course they take. Each of the nerves in turn may have several branches named for the specific structures they innervate.

Exhibits 13.A–13.D (Figures 13.7–13.10) summarize the principal plexuses. The anterior rami of spinal nerves T2–T12 are called intercostal nerves and will be discussed next.

Intercostal Nerves

The anterior rami of spinal nerves T2–T12 do not enter into the formation of plexuses and are known as **intercostal** or **thoracic nerves.** These nerves directly connect to the structures they supply in the intercostal spaces. After leaving its intervertebral foramen, the anterior ramus of nerve T2 innervates the intercostal muscles of the second intercostal space and supplies the skin of the axilla and posteromedial aspect of the arm. Nerves T3–T6 extend along the costal grooves of the ribs and then to the intercostal muscles and skin of the anterior and lateral chest wall. Nerves T7–T12 supply the intercostal muscles and abdominal muscles, along with the overlying skin. The posterior rami of the intercostal nerves supply the deep back muscles and skin of the posterior aspect of the thorax.

EXHIBIT 13.A   Cervical Plexus   *(Figure 13.7)*

◉ OBJECTIVE
• Describe the origin and distribution of the cervical plexus.

The **cervical plexus** (SER-vi-kul) is formed by the roots (anterior rami) of the first four cervical nerves (C1–C4), with contributions from C5 (Figure 13.7). There is one on each side of the neck alongside the first four cervical vertebrae.

The cervical plexus supplies the skin and muscles of the head, neck, and superior part of the shoulders and chest. The phrenic nerves arise from the cervical plexuses and supply motor fibers to the diaphragm. Branches of the cervical plexus also run parallel to two cranial nerves, the accessory (XI) nerve and hypoglossal (XII) nerve.

NERVE	ORIGIN	DISTRIBUTION
SUPERFICIAL (SENSORY) BRANCHES		
Lesser occipital	C2	Skin of scalp posterior and superior to ear.
Great auricular (aw-RIK-ū-lar)	C2–C3	Skin anterior, inferior, and over ear, and over parotid glands.
Transverse cervical	C2–C3	Skin over anterior aspect of neck.
Supraclavicular	C3–C4	Skin over superior portion of chest and shoulder.
DEEP (LARGELY MOTOR) BRANCHES		
Ansa cervicalis (AN-sa ser-vi-KAL-is)		Divides into superior and inferior roots.
Superior root	C1	Infrahyoid and geniohyoid muscles of neck.
Inferior root	C2–C3	Infrahyoid muscles of neck.
Phrenic (FREN-ik)	C3–C5	Diaphragm.
Segmental branches	C1–C5	Prevertebral (deep) muscles of neck, levator scapulae, and middle scalene muscles.

EXHIBIT 13.A CONTINUES ▶

EXHIBIT 13.A Cervical Plexus *(Figure 13.7)* CONTINUED

🔱 **CLINICAL CONNECTION** | *Injuries to the Phrenic Nerves*

The phrenic nerves originate from C3, C4, and C5 and supply the diaphragm. Complete severing of the spinal cord above the origin of the phrenic nerves (C3, C4, and C5) causes respiratory arrest. In **injuries to the phrenic nerves,** breathing stops because the phrenic nerves no longer send nerve impulses to the diaphragm. The phrenic nerves may also be damaged due to pressure from malignant tracheal or esophageal tumors in the mediastinum. •

✔**CHECKPOINT**

Which nerve from the cervical plexus causes contraction of the diaphragm?

Figure 13.7 Cervical plexus in anterior view.

🔑 The cervical plexus supplies the skin and muscles of the head, neck, superior portion of the shoulders and chest, and diaphragm.

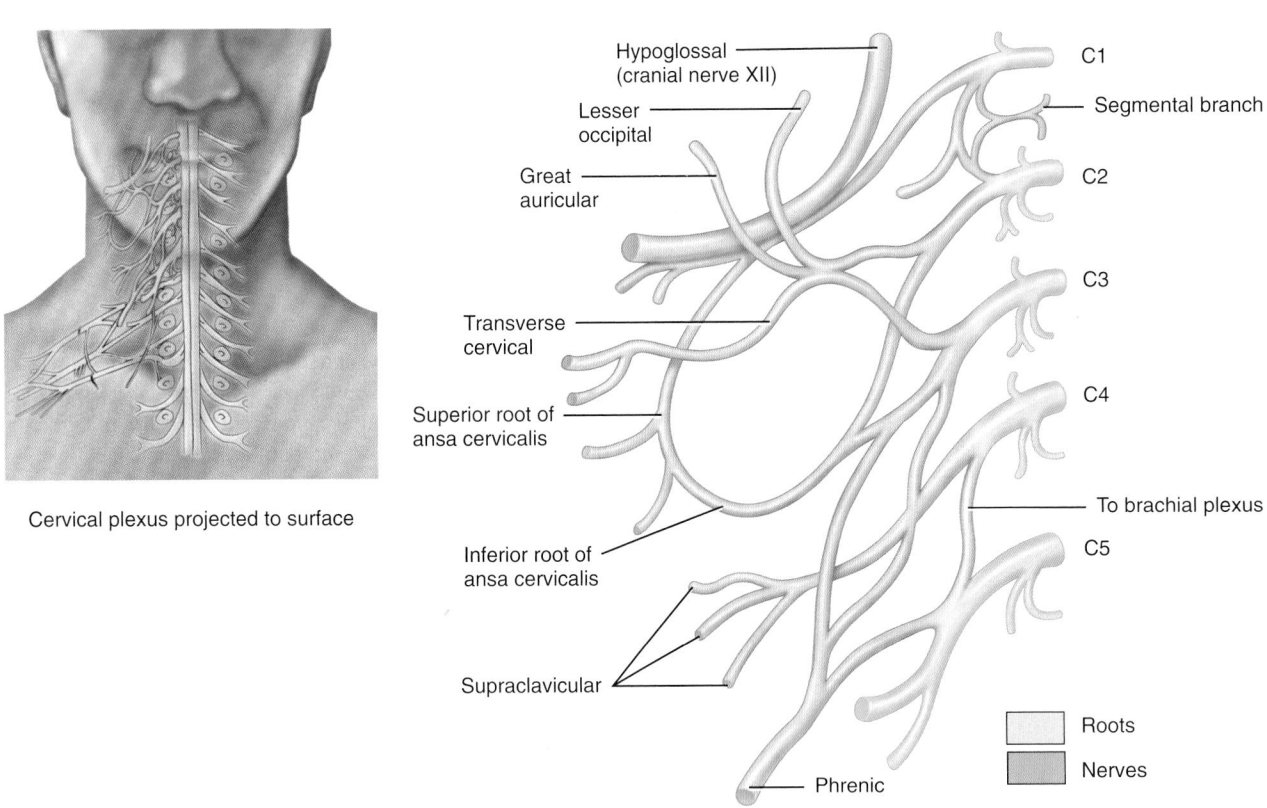

Cervical plexus projected to surface

Origin of cervical plexus

❓ **Why does complete severing of the spinal cord at level C2 cause respiratory arrest?**

EXHIBIT 13.B   Brachial Plexus   *(Figure 13.8)*

OBJECTIVE

- Describe the origin, distribution, and effects of damage to the brachial plexus.

The roots (anterior rami) of spinal nerves C5–C8 and T1 form the **brachial plexus** (BRĀ-kē-al), which extends inferiorly and laterally on either side of the last four cervical and first thoracic vertebrae (Figure 13.8a). It passes above the first rib posterior to the clavicle and then enters the axilla.

Since the brachial plexus is so complex, an explanation of its various parts is helpful. As with the cervical and other plexuses, the **roots** are the anterior rami of the spinal nerves. The roots of several spinal nerves unite to form **trunks** in the inferior part of the neck. These are the *superior, middle,* and *inferior trunks.* Posterior to the clavicles, the trunks divide into **divisions,** called the *anterior* and *posterior divisions.* In the axillae, the divisions unite to form **cords** called the *lateral, medial,* and *posterior cords.* The cords are named for their relationship to the axillary artery, a large artery that supplies blood to the upper limb. The principal **nerves** of the brachial plexus branch from the cords.

The brachial plexus provides almost the entire nerve supply of the shoulders and upper limbs (Figure 13.8b). Five important nerves arise from the brachial plexus: (1) The **axillary nerve** supplies the deltoid and teres minor muscles. (2) The **musculocutaneous nerve** supplies the anterior muscles of the arm. (3) The **radial nerve** supplies the muscles on the posterior aspect of the arm and forearm. (4) The **median nerve** supplies most of the muscles of the anterior forearm and some of the muscles of the hand. (5) The **ulnar nerve** supplies the anteromedial muscles of the forearm and most of the muscles of the hand.

CLINICAL CONNECTION | Injuries to Nerves Emerging from the Brachial Plexus

Injury to the superior roots of the brachial plexus (C5–C6) may result from forceful pulling away of the head from the shoulder, as might occur from a heavy fall on the shoulder or excessive stretching of an infant's neck during childbirth. The presentation of this injury is characterized by an upper limb in which the shoulder is adducted, the arm is medially rotated, the elbow is extended, the forearm is pronated, and the wrist is flexed (Figure 13.8c). This condition is called **Erb-Duchenne palsy** or **waiter's tip position.** There is loss of sensation along the lateral side of the arm.

Injury to the radial (and axillary) **nerve** can be caused by improperly administered intramuscular injections into the deltoid muscle. The radial nerve may also be injured when a cast is applied too tightly around the mid-humerus. Radial nerve injury is indicated by **wrist drop,** the inability to extend the wrist and fingers (Figure 13.8c). Sensory loss is minimal due to the overlap of sensory innervation by adjacent nerves.

NERVE	ORIGIN	DISTRIBUTION
Dorsal scapular (SKAP-ū-lar)	C5	Levator scapulae, rhomboid major, and rhomboid minor muscles.
Long thoracic (thor-RAS-ik)	C5–C7	Serratus anterior muscle.
Nerve to subclavius (sub-KLĀ-vē-us)	C5–C6	Subclavius muscle.
Suprascapular	C5–C6	Supraspinatus and infraspinatus muscles.
Musculocutaneous (mus′-kū-lō-kū-TĀN-ē-us)	C5–C7	Coracobrachialis, biceps brachii, and brachialis muscles.
Lateral pectoral (PEK-to-ral)	C5–C7	Pectoralis major muscle.
Upper subscapular	C5–C6	Subscapularis muscle.
Thoracodorsal (tho-RĀ-kō-dor-sal)	C6–C8	Latissimus dorsi muscle.
Lower subscapular	C5–C6	Subscapularis and teres major muscles.
Axillary (AK-si-lar-ē)	C5–C6	Deltoid and teres minor muscles; skin over deltoid and superior posterior aspect of arm.
Median	C5–T1	Flexors of forearm, except flexor carpi ulnaris; ulnar half of flexor digitorum profundus, and some muscles of hand (lateral palm); skin of lateral two-thirds of palm of hand and fingers.
Radial		Triceps brachii, anconeus, and extensor muscles of forearm; skin of posterior arm and forearm, lateral two-thirds of dorsum of hand, and fingers over proximal and middle phalanges.
Medial pectoral	C8–T1	Pectoralis major and pectoralis minor muscles.
Medial cutaneous nerve of arm (kū′-TĀ-nē-us)	C8–T1	Skin of medial and posterior aspects of distal third of arm.
Medial cutaneous nerve of forearm	C8–T1	Skin of medial and posterior aspects of forearm.
Ulnar	C8–T1	Flexor carpi ulnaris, ulnar half of flexor digitorum profundus, and most muscles of hand; skin of medial side of hand, little finger, and medial half of ring finger.

EXHIBIT 13.B CONTINUES

EXHIBIT 13.B **505**

Injury to the median nerve may result in **median nerve palsy,** which is indicated by numbness, tingling, and pain in the palm and fingers. There is also inability to pronate the forearm and flex the proximal interphalangeal joints of all digits and the distal interphalangeal joints of the second and third digits (Figure 13.8c). In addition, wrist flexion is weak and is accompanied by adduction, and thumb movements are weak.

Injury to the ulnar nerve may result in **ulnar nerve palsy,** which is indicated by an inability to abduct or adduct the fingers, atrophy of the interosseous muscles of the hand, hyperextension of the metacarpophalangeal joints, and flexion of the interphalangeal joints, a condition called **clawhand** (Figure 13.8c). There is also loss of sensation over the little finger.

Injury to the long thoracic nerve results in paralysis of the serratus anterior muscle. The medial border of the scapula protrudes, giving it the appearance of a wing. When the arm is raised, the vertebral border and inferior angle of the scapula pull away from the thoracic wall and protrude outward, causing the medial border of the

scapula to protrude; because the scapula looks like a wing, this condition is called **winged scapula** (Figure 13.8c). The arm cannot be abducted beyond the horizontal position.

Compression of the brachial plexus on one or more of its nerves is sometimes known as **thoracic outlet syndrome.** The subclavian artery and subclavian vein may also be compressed. The compression may result from spasm of the scalene or pectoralis minor muscles, the presence of a cervical rib (an embryological anomaly), or misaligned ribs. The patient may experience pain, numbness, weakness, or tingling in the upper limb, across the upper thoracic area, and over the scapula on the affected side. The symptoms of thoracic outlet syndrome are exaggerated during physical or emotional stress because the added stress increases the contraction of the involved muscles. •

✔ CHECKPOINT

Injury of which nerve could cause paralysis of the serratus anterior muscle?

Figure 13.8 Brachial plexus in anterior view.

The brachial plexus supplies the shoulders and upper limbs.

Brachial plexus projected to surface

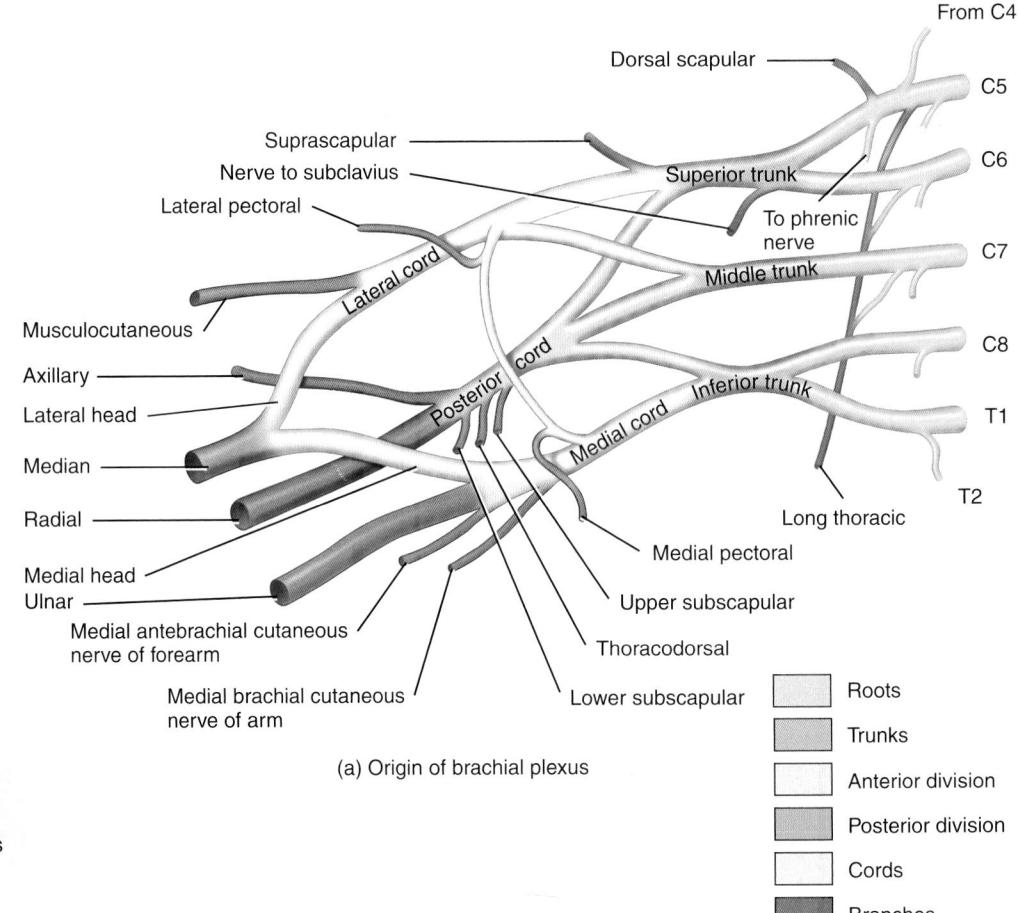

(a) Origin of brachial plexus

MNEMONIC for subunits of the brachial plexus:
Risk **T**akers **D**on't **C**autiously **B**ehave.
Roots, **T**runks, **D**ivisions, **C**ords, **B**ranches

Roots
Trunks
Anterior division
Posterior division
Cords
Branches

Dorsal scapular nerve
Nerve to subclavius
Suprascapular nerve

Superior trunk
Middle trunk
Inferior trunk

From C4
C5
C6
C7
C8
T1

Clavicle
Long thoracic nerve

Lateral cord
Posterior cord
Lateral pectoral nerve
Axillary nerve
Medial cord
Musculocutaneous nerve
Radial nerve
Median nerve

Medial pectoral
Medial brachial cutaneous
Medial antebrachial cutaneous
Scapula
Ulnar nerve

Humerus

Deep branch of radial nerve

Radius

Superficial branch of radial nerve
Median nerve

Ulna
Ulnar nerve

Superficial branch of ulnar nerve
Radial nerve

Digital branch of median nerve

Digital branch of ulnar nerve

(b) Distribution of nerves from the brachial plexus

Wrist drop

Erb-Duchenne palsy (waiter's tip)

Median nerve palsy

Ulnar nerve palsy

Winging of right scapula

(c) Injuries to the brachial plexus

? **What five important nerves arise from the brachial plexus?**

EXHIBIT 13.B   **507**

EXHIBIT 13.C Lumbar Plexus *(Figure 13.9)*

◉ OBJECTIVE

• Describe the origin and distribution of the lumbar plexus.

The roots (anterior rami) of spinal nerves L1–L4 form the **lumbar plexus** (LUM-bar) (Figure 13.9). Unlike the brachial plexus, there is minimal intermingling of fibers in the lumbar plexus. On either side of the first four lumbar vertebrae, the lumbar plexus passes obliquely outward, between the superficial and deep heads of the psoas major muscle and anterior to the quadratus lumborum muscle. It then gives rise to its peripheral nerves.

The lumbar plexus supplies the anterolateral abdominal wall, external genitals, and part of the lower limbs.

NERVE	ORIGIN	DISTRIBUTION
Iliohypogastric (il'-ē-ō-hī-pō-GAS-trik)	L1	Muscles of anterolateral abdominal wall; skin of inferior abdomen and buttock.
Ilioinguinal (il'-ē-ō-IN-gwi-nal)	L1	Muscles of anterolateral abdominal wall; skin of superior and medial aspect of thigh, root of penis and scrotum in male, and labia majora and mons pubis in female.
Genitofemoral (jen'-i-tō-FEM-or-al)	L1–L2	Cremaster muscle; skin over middle anterior surface of thigh, scrotum in male, and labia majora in female.
Lateral cutaneous nerve of thigh	L2–L3	Skin over lateral, anterior, and posterior aspects of thigh.
Femoral	L2–L4	Largest nerve arising from lumbar plexus; distributed to flexor muscles of hip joint and extensor muscles of knee joint, skin over anterior and medial aspect of thigh and medial side of leg and foot.
Obturator (OB-too-rā-tor)	L2–L4	Adductor muscles of hip joint; skin over medial aspect of thigh.

Figure 13.9 **Lumbar plexus in anterior view.**

🔑 The lumbar plexus supplies the anterolateral abdominal wall, external genitals, and part of the lower limbs.

Lumbar plexus

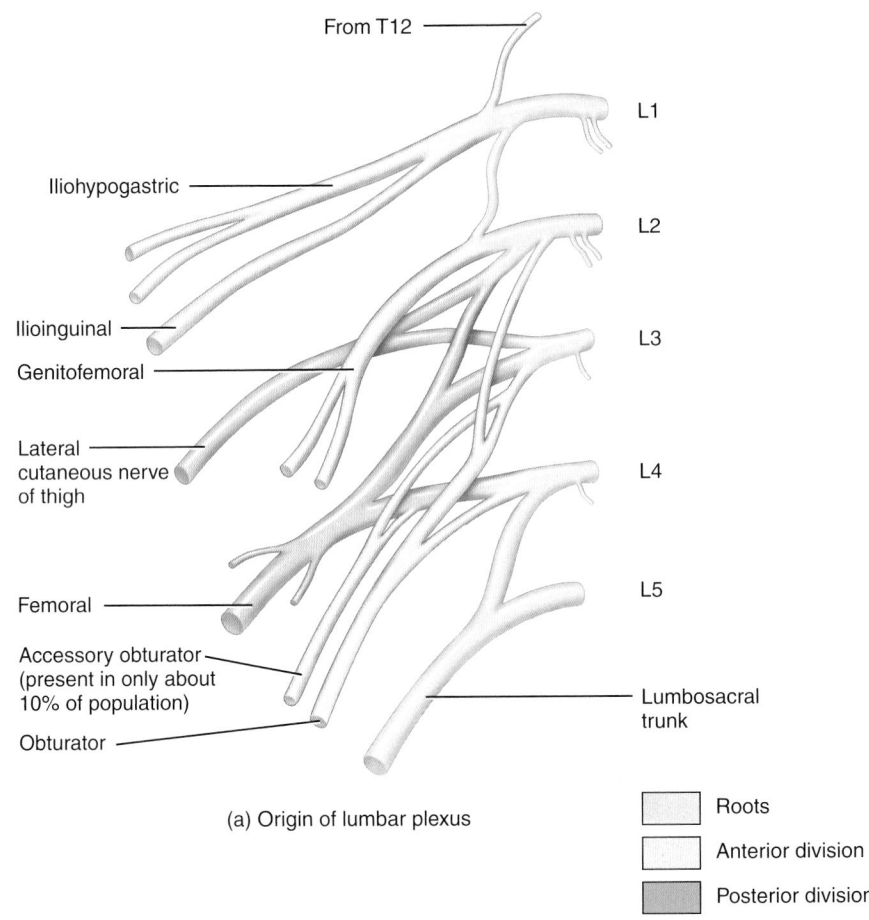

(a) Origin of lumbar plexus

	Roots
	Anterior division
	Posterior division

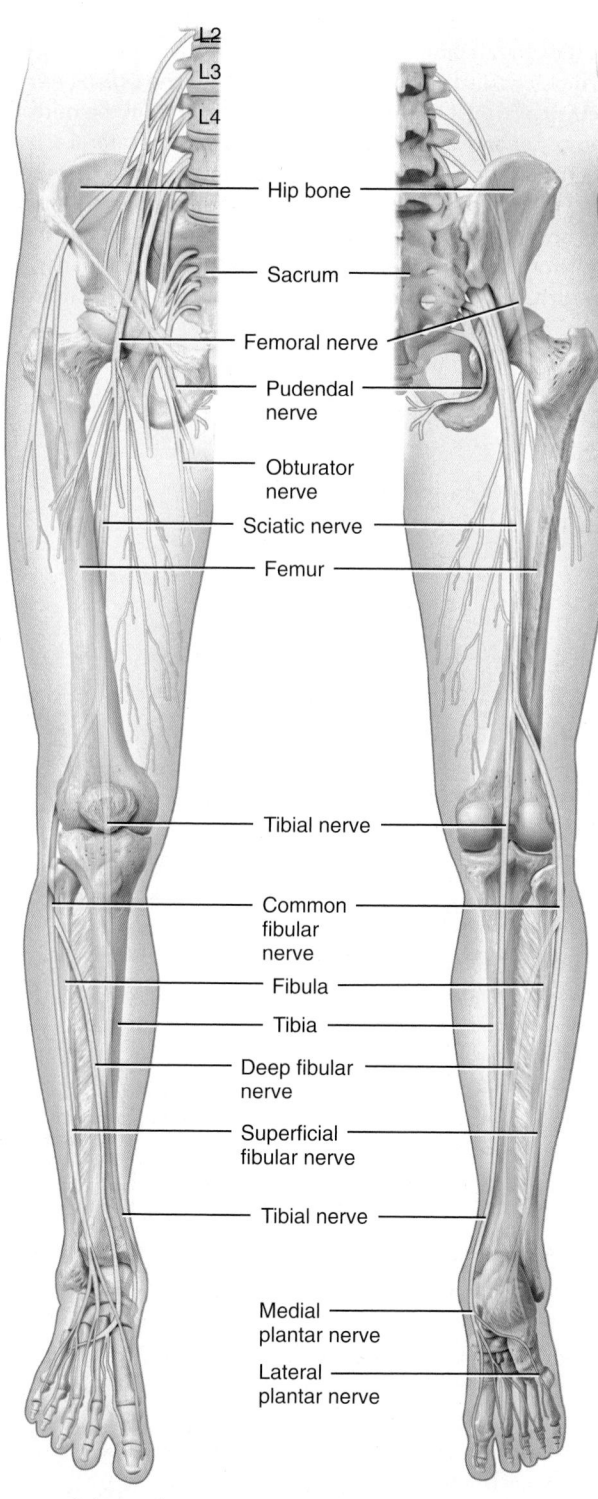

L2
L3
L4

Hip bone

Sacrum

Femoral nerve

Pudendal
nerve

Obturator
nerve

Sciatic nerve

Femur

Tibial nerve

Common
fibular
nerve

Fibula

Tibia

Deep fibular
nerve

Superficial
fibular nerve

Tibial nerve

Medial
plantar nerve

Lateral
plantar nerve

Anterior view   Posterior view

(b) Distribution of nerves from the lumbar and sacral plexuses

 What are the signs of femoral nerve injury?

CLINICAL CONNECTION | *Injuries to the Lumbar Plexus*

The largest nerve arising from the lumbar plexus is the femoral nerve. **Injuries to the femoral nerve,** which can occur in stab or gunshot wounds, are indicated by an inability to extend the leg and by loss of sensation in the skin over the anteromedial aspect of the thigh.

Injuries to the obturator nerve result in paralysis of the adductor muscles of the thigh and loss of sensation over the medial aspect of the thigh. It may result from pressure on the nerve by the fetal head during pregnancy. •

✔CHECKPOINT

What is the largest nerve arising from the lumbar plexus?

EXHIBIT 13.C **509**

EXHIBIT 13.D Sacral and Coccygeal Plexuses *(Figure 13.10)*

◉ OBJECTIVE

- Describe the origin and distribution of the sacral and coccygeal plexuses.

The roots (anterior rami) of spinal nerves L4–L5 and S1–S4 form the **sacral plexus** (SĀ-kral) (Figure 13.10). This plexus is situated largely anterior to the sacrum. The sacral plexus supplies the buttocks, perineum, and lower limbs. The largest nerve in the body—the sciatic nerve—arises from the sacral plexus.

The roots (anterior rami) of spinal nerves S4–S5 and the coccygeal nerves form a small **coccygeal plexus** (kok-SIG-ē-al), which supplies a small area of skin in the coccygeal region.

⚕ CLINICAL CONNECTION | *Injury to the Sciatic Nerve*

The most common form of back pain is caused by compression or irritation of the sciatic nerve, the longest nerve in the human body. The sciatic nerve is actually two nerves—tibial and common fibular—bound together by a common sheath of connective tissue. It splits into its two divisions, usually at the knee. **Injury to the sciatic nerve** results in **sciatica** (sī-AT-i-ka), pain that may extend from the buttock down the posterior and lateral aspect of the leg and the lateral aspect of the foot. The sciatic nerve may be injured because of a herniated (slipped) disc, dislocated hip, osteoarthritis of the lumbosacral spine, pathological shortening of the lateral rotator muscles of the thigh (especially piriformis), pressure from the uterus during pregnancy, inflammation, irritation, or an improperly administered gluteal intra-

NERVE	ORIGIN	DISTRIBUTION
Superior gluteal (GLOO-tē-al)	L4–L5 and S1	Gluteus minimus, gluteus medius, and tensor fasciae latae muscles.
Inferior gluteal	L5–S2	Gluteus maximus muscle.
Nerve to piriformis (pir-i-FORM-is)	S1–S2	Piriformis muscle.
Nerve to quadratus femoris (quad-RĀ-tus FEM-or-is) and **inferior gemellus** (jem-EL-us)	L4–L5 and S1	Quadratus femoris and inferior gemellus muscles.
Nerve to obturator internus (OB-too-rā′-tor in-TER-nus) and **superior gemellus**	L5–S2	Obturator internus and superior gemellus muscles.
Perforating cutaneous (kū′-TĀ-nē-us)	S2–S3	Skin over inferior medial aspect of buttock.
Posterior cutaneous nerve of thigh	S1–S3	Skin over anal region, inferior lateral aspect of buttock, superior posterior aspect of thigh, superior part of calf, scrotum in male, and labia majora in female.
Sciatic (sī-AT-ik)	L4–S3	Actually two nerves—tibial and common fibular—bound together by common sheath of connective tissue; splits into its two divisions, usually at the knee. (See below for distributions.) As sciatic nerve descends through thigh, it sends branches to hamstring muscles and adductor magnus.
Tibial (TIB-ē-al)	L4–S3	Gastrocnemius, plantaris, soleus, popliteus, tibialis posterior, flexor digitorum longus, and flexor hallucis longus muscles. Branches of tibial nerve in foot are medial plantar nerve and lateral plantar nerve.
Medial plantar (PLAN-tar) (see Figure 13.9b)		Abductor hallucis, flexor digitorum brevis, and flexor hallucis brevis muscles; skin over medial two-thirds of plantar surface of foot.
Lateral plantar (see Figure 13.9b)		Remaining muscles of foot not supplied by medial plantar nerve; skin over lateral third of plantar surface of foot.
Common fibular (FIB-ū-lar)	L4–S2	Divides into superficial fibular and deep fibular branch.
Superficial fibular		Fibularis longus and fibularis brevis muscles; skin over distal third of anterior aspect of leg and dorsum of foot.
Deep fibular		Tibialis anterior, extensor hallucis longus, fibularis tertius, and extensor digitorum longus and extensor digitorum brevis muscles; skin on adjacent sides of great and second toes.
Pudendal (pū-DEN-dal)	S2–S4	Muscles of perineum; skin of penis and scrotum in male and clitoris, labia majora, labia minora, and vagina in female.

muscular injection. In addition, sitting on a wallet or other object for a long period of time can compress the nerve and induce pain.

In many sciatic nerve injuries, the common fibular portion is the most affected, frequently from fractures of the fibula or by pressure from casts or splints over the thigh or leg. Damage to the common fibular nerve causes the foot to be plantar flexed, a condition called **foot drop**, and inverted, a condition called **equinovarus** (e-KWĪ-nō-va-rus). There is also loss of function along the anterolateral apects of the leg and dorsum of the foot and toes. Injury to the tibial portion of the sciatic nerve results in dorsiflexion of the foot plus eversion, a condition called **calcaneovalgus** (kal-KĀ-nē-ō-valʹ-gus). Loss of sensation on the sole also occurs. Treatments for sciatica are similar to those for a herniated (slipped) disc—rest, pain medications, exercises, ice or heat, and massage. •

✔ CHECKPOINT

Injury of which nerve causes footdrop?

Figure 13.10 Sacral and coccygeal plexuses in anterior view. The distribution of the nerves of the sacral plexus is shown in Figure 13.9b.

🔑 The sacral plexus supplies the buttocks, perineum, and lower limbs.

Sacral and coccygeal plexuses

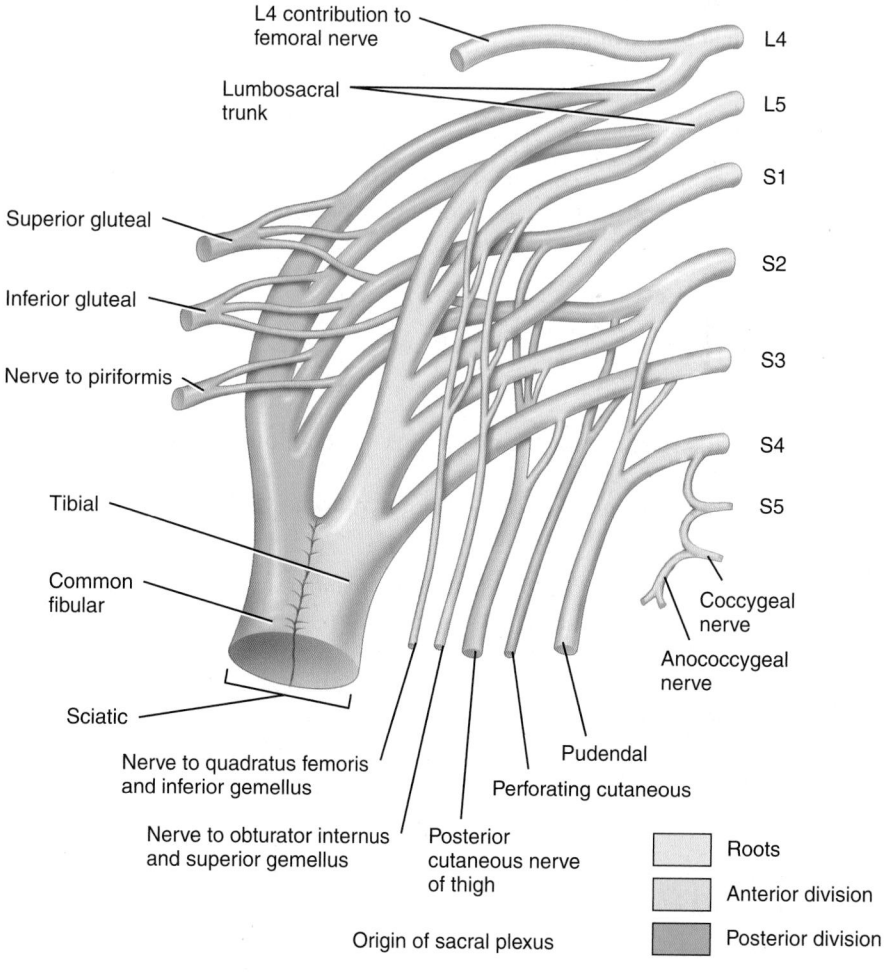

L4 contribution to femoral nerve

Lumbosacral trunk

Superior gluteal

Inferior gluteal

Nerve to piriformis

Tibial

Common fibular

Sciatic

Nerve to quadratus femoris and inferior gemellus

Nerve to obturator internus and superior gemellus

Posterior cutaneous nerve of thigh

Perforating cutaneous

Pudendal

Anococcygeal nerve

Coccygeal nerve

L4
L5
S1
S2
S3
S4
S5

Origin of sacral plexus

Roots
Anterior division
Posterior division

? **What is the origin of the sacral plexus?**

EXHIBIT 13.D **511**

Dermatomes

The skin over the entire body is supplied by somatic sensory neurons that carry nerve impulses from the skin into the spinal cord and brain. Each spinal nerve contains sensory neurons that serve a specific, predictable segment of the body. One of the cranial nerves, the trigeminal (V) nerve, serves most of the skin of the face and scalp. The area of the skin that provides sensory input to the CNS via one pair of spinal nerves or the trigeminal (V) nerve is called a **dermatome** (DER-ma-tōm; *derma-* = skin; *-tome* = thin segment) (Figure 13.11). The nerve supply in adjacent dermatomes overlaps somewhat. Knowing which spinal cord segments supply each dermatome makes it possible to locate damaged regions of the spinal cord. If the skin in a particular region is stimulated but the sensation is not perceived, the nerves supplying that dermatome are probably damaged. In regions where the overlap is considerable, little loss of sensation may result if only one of the nerves supplying the dermatome is damaged. Information about the innervation patterns of spinal nerves can also be used therapeutically. Cutting posterior roots or infusing local anesthetics can block pain either permanently or transiently. Because dermatomes overlap, deliberate production of a region of complete anesthesia may require that at least three adjacent spinal nerves be cut or blocked by an anesthetic drug.

✔ **CHECKPOINT**

5. How are spinal nerves named and numbered? Why are all spinal nerves classified as mixed nerves?
6. How do spinal nerves connect to the spinal cord?
7. Which regions of the body are supplied by plexuses and by intercostal nerves?

13.3 SPINAL CORD PHYSIOLOGY

◉ **OBJECTIVES**

• Describe the functions of the major sensory and motor tracts of the spinal cord.
• Describe the functional components of a reflex arc and the ways reflexes maintain homeostasis.

The spinal cord has two principal functions in maintaining homeostasis: nerve impulse propagation and integration of information. The *white matter tracts* in the spinal cord are highways for nerve impulse propagation. Sensory input travels along these tracts toward the brain, and motor output travels from the brain along these tracts toward skeletal muscles and other effector tissues. The *gray matter* of the spinal cord receives and integrates incoming and outgoing information.

Sensory and Motor Tracts

As noted previously, one of the ways the spinal cord promotes homeostasis is by conducting nerve impulses along tracts. Often, the name of a tract indicates its position in the white matter and where it begins and ends. For example, the anterior corticospinal tract is located in the *anterior* white column; it begins in the *cerebral*

Figure 13.11 Distribution of dermatomes.

 A dermatome is an area of skin that provides sensory input to the CNS via the posterior roots of one pair of spinal nerves or via the trigeminal (V) nerve.

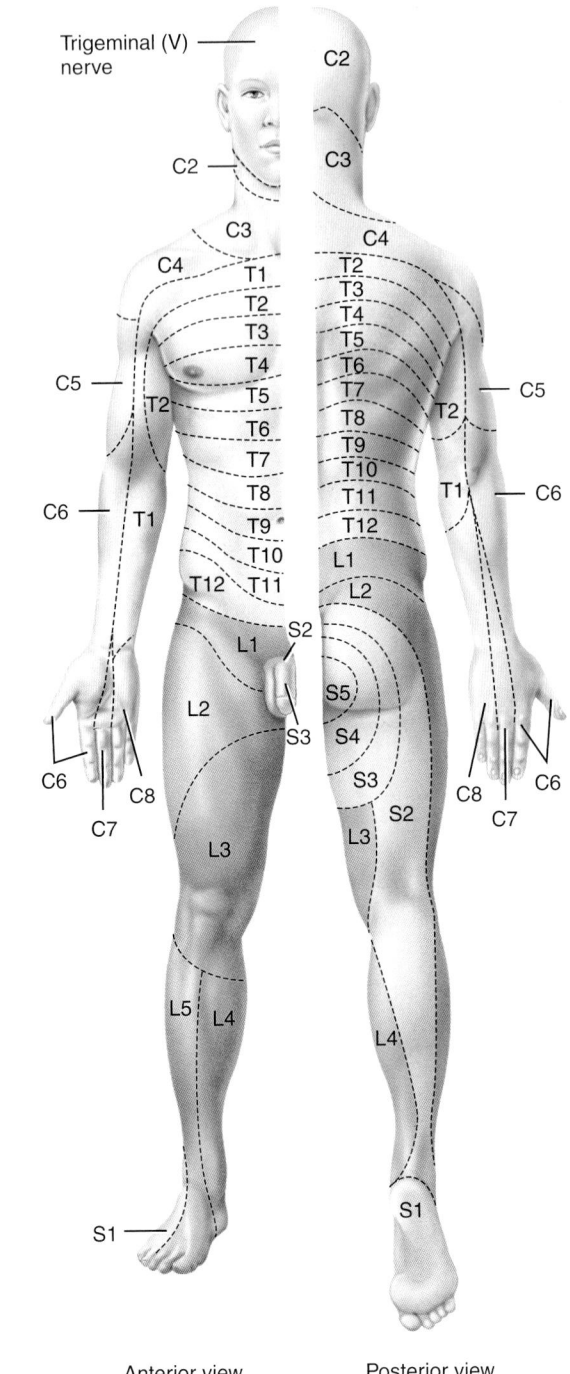

Anterior view   Posterior view

❓ **Which is the only spinal nerve that does not have a corresponding dermatome?**

cortex (superficial gray matter of the cerebrum of the brain) and ends in the *spinal cord*. Notice that the location of the axon terminals comes last in the name. This regularity in naming allows you to determine the direction of information flow along any tract named according to this convention. Because the anterior corticospinal tract conveys nerve impulses from the brain toward the spinal cord, it is a motor (descending) tract. Figure 13.12 highlights the major sensory and motor tracts in the spinal cord. These tracts are described in detail in Chapter 16 and summarized in Tables 16.3 and 16.4.

Nerve impulses from sensory receptors propagate up the spinal cord to the brain along two main routes on each side: the spinothalamic tract and the posterior column. The **spinothalamic tract** (spī′-nō-tha-LAM-ik) conveys nerve impulses for sensing pain, warmth, coolness, itching, tickling, deep pressure, and crude touch. The **posterior column** consists of two tracts: the **gracile fasciculus** (GRAS-īl fa-SIK-ū-lus) and the **cuneate fasciculus** (KŪ-nē-āt). The posterior column tracts convey nerve impulses for discriminative touch, light pressure, vibration, and conscious proprioception (the awareness of the positions and movements of muscles, tendons, and joints).

The sensory systems keep the CNS informed of changes in the external and internal environments. The sensory information is integrated (processed) by interneurons in the spinal cord and brain. Responses to the integrative decisions are brought about by motor activities (muscular contractions and glandular secretions). The cerebral cortex, the outer part of the brain, plays a major role in controlling precise voluntary muscular movements. Other brain regions provide important integration for regulation of automatic movements. Motor output to skeletal muscles travels down

Figure 13.12 Locations of major sensory and motor tracts, shown in a transverse section of the spinal cord. Sensory tracts are indicated on one half and motor tracts on the other half of the cord, but actually all tracts are present on both sides.

The name of a tract often indicates its location in the white matter and where it begins and ends.

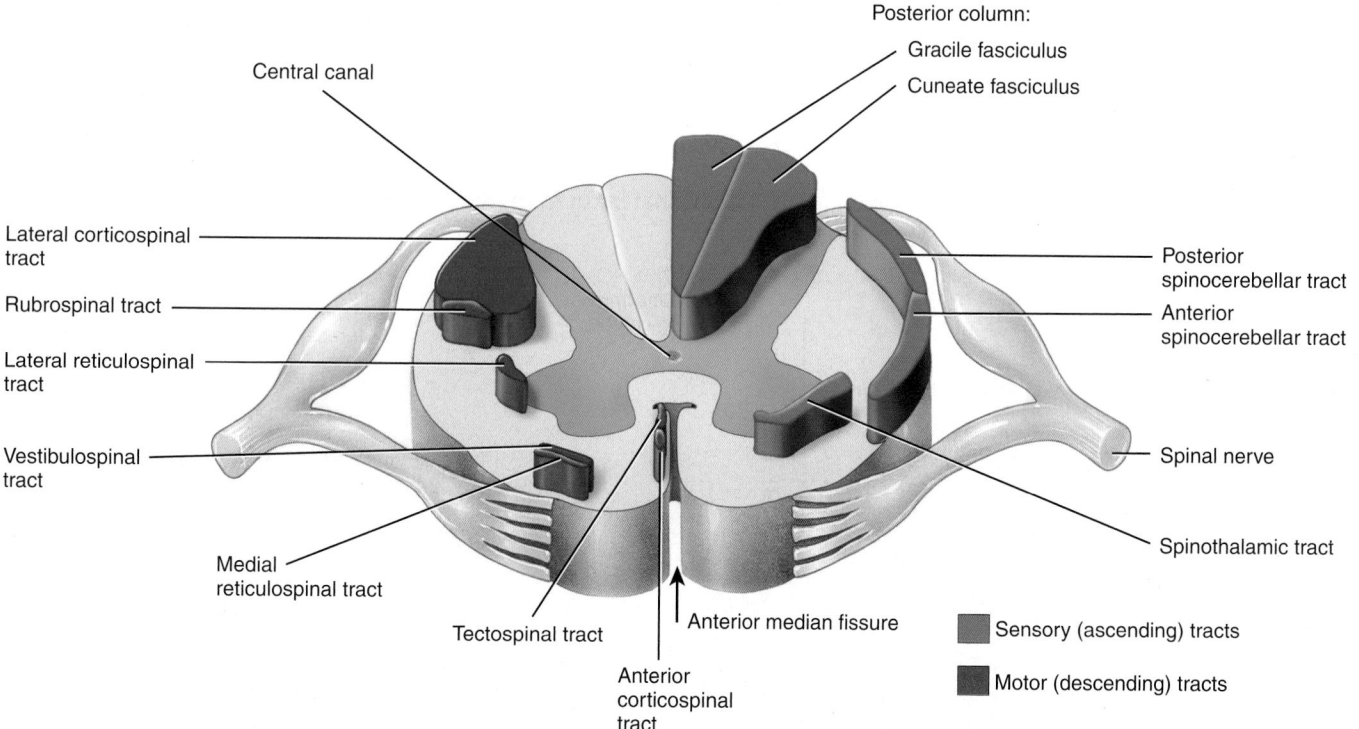

FUNCTIONS OF THE SPINAL CORD AND SPINAL NERVES

1. The white matter of the spinal cord contains sensory and motor tracts, the "highways" for conduction of sensory nerve impulses toward the brain and motor nerve impulses from the brain toward effector tissues.
2. The spinal cord gray matter is a site for integration (summing) of excitatory postsynaptic potentials (EPSPs) and inhibitory postsynaptic potentials (IPSPs).
3. Spinal nerves and the nerves that branch from them connect the CNS to the sensory receptors, muscles, and glands in all parts of the body.

? **Based on its name, list the origin and destination of the spinothalamic tract. Is this a sensory or a motor tract?**

the spinal cord in two types of descending pathways: direct and indirect. The **direct pathways** include the **lateral corticospinal** (kor′-ti-kō-SPĪ-nal), **anterior corticospinal,** and **corticobulbar tracts** (kor′-ti-kō-BUL-bar). They convey nerve impulses that originate in the cerebral cortex and are destined to cause *voluntary* movements of skeletal muscles. **Indirect pathways** include the **rubrospinal** (ROO-brō-spī-nal), **tectospinal** (TEK-tō-spī-nal), **vestibulospinal** (ves-TIB-ū-lō-spī-nal), **lateral reticulospinal** (re-TIK-ū-lō-spī-nal), and **medial reticulospinal tracts.** These tracts convey nerve impulses from the brain stem to cause *automatic movements* and help coordinate body movements with visual stimuli. Indirect pathways also maintain skeletal muscle tone, sustain contraction of postural muscles, and play a major role in equilibrium by regulating muscle tone in response to movements of the head.

Reflexes and Reflex Arcs

The second way the spinal cord promotes homeostasis is by serving as an integrating center for some reflexes. A **reflex** is a fast, involuntary, unplanned sequence of actions that occurs in response to a particular stimulus. Some reflexes are inborn, such as pulling your hand away from a hot surface before you even feel that it is

hot. Other reflexes are learned or acquired. For instance, you learn many reflexes while acquiring driving expertise. Slamming on the brakes in an emergency is one example. When integration takes place in the spinal cord gray matter, the reflex is a **spinal reflex.** An example is the familiar patellar reflex (knee jerk). If integration occurs in the brain stem rather than the spinal cord, the reflex is called a **cranial reflex.** An example is the tracking movements of your eyes as you read this sentence. You are probably most aware of **somatic reflexes,** which involve contraction of skeletal muscles. Equally important, however, are the **autonomic (visceral) reflexes,** which generally are not consciously perceived. They involve responses of smooth muscle, cardiac muscle, and glands. As you will see in Chapter 15, body functions such as heart rate, digestion, urination, and defecation are controlled by the autonomic nervous system through autonomic reflexes.

Nerve impulses propagating into, through, and out of the CNS follow specific pathways, depending on the kind of information, its origin, and its destination. The pathway followed by nerve impulses that produce a reflex is a **reflex arc (reflex circuit).** A reflex arc includes the following five functional components (Figure 13.13):

① Sensory receptor. The distal end of a sensory neuron (dendrite) or an associated sensory structure serves as a sensory

Figure 13.13 General components of a reflex arc. Arrows show the direction of nerve impulse propagation.

A reflex is a fast, predictable sequence of involuntary actions that occur in response to certain changes in the environment.

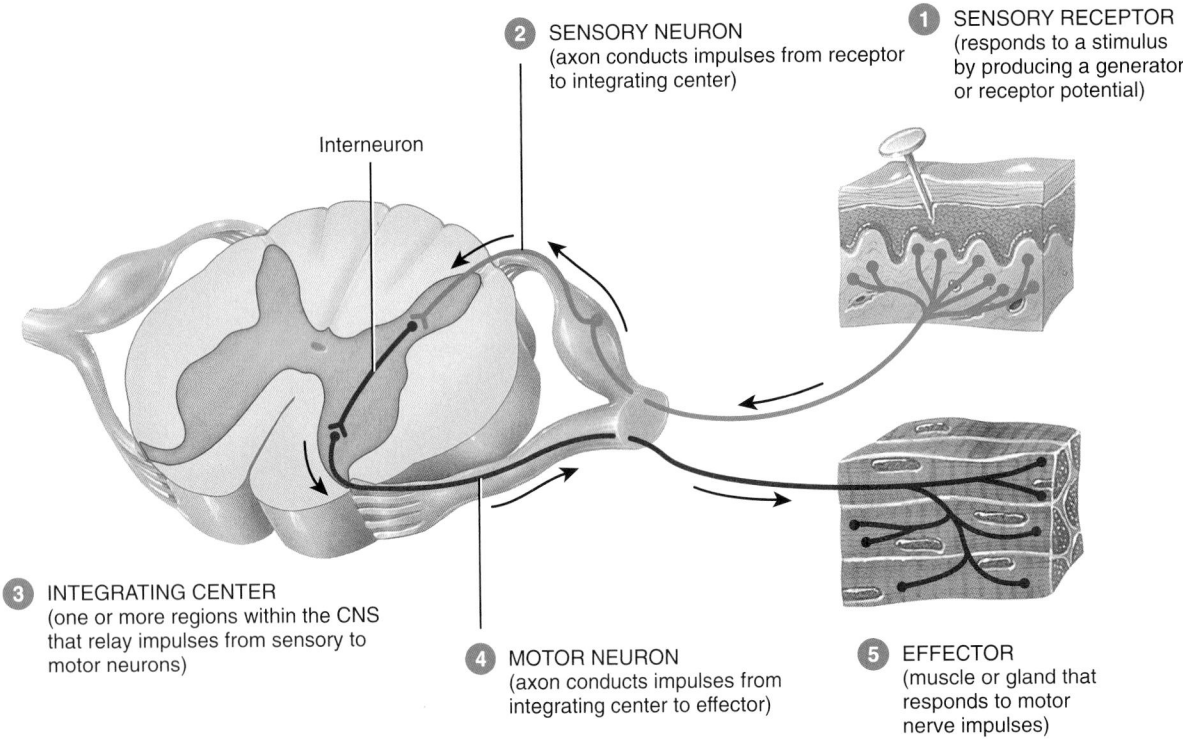

2 SENSORY NEURON (axon conducts impulses from receptor to integrating center)

1 SENSORY RECEPTOR (responds to a stimulus by producing a generator or receptor potential)

Interneuron

3 INTEGRATING CENTER (one or more regions within the CNS that relay impulses from sensory to motor neurons)

4 MOTOR NEURON (axon conducts impulses from integrating center to effector)

5 EFFECTOR (muscle or gland that responds to motor nerve impulses)

❓ What initiates a nerve impulse in a sensory neuron? Which branch of the nervous system includes all integrating centers for reflexes?

receptor. It responds to a specific **stimulus**—a change in the internal or external environment—by producing a graded potential called a generator (or receptor) potential (described in Section 16.1). If a generator potential reaches the threshold level of depolarization, it will trigger one or more nerve impulses in the sensory neuron.

2 **Sensory neuron.** The nerve impulses propagate from the sensory receptor along the axon of the sensory neuron to the axon terminals, which are located in the gray matter of the spinal cord or brain stem. From here, relay neurons send nerve impulses to the area of the brain that allows conscious awareness that the reflex has occurred.

3 **Integrating center.** One or more regions of gray matter within the CNS act as an integrating center. In the simplest type of reflex, the integrating center is a single synapse between a sensory neuron and a motor neuron. A reflex pathway having only one synapse in the CNS is termed a **monosynaptic reflex arc** (mon′-ō-si-NAP-tik; *mono-* = one). More often, the integrating center consists of one or more interneurons, which may relay impulses to other interneurons as well as to a motor neuron. A **polysynaptic reflex arc** (*poly-* = many) involves more than two types of neurons and more than one CNS synapse.

4 **Motor neuron.** Impulses triggered by the integrating center propagate out of the CNS along a motor neuron to the part of the body that will respond.

5 **Effector.** The part of the body that responds to the motor nerve impulse, such as a muscle or gland, is the effector. Its action is called a reflex. If the effector is skeletal muscle, the reflex is a **somatic reflex.** If the effector is smooth muscle, cardiac muscle, or a gland, the reflex is an **autonomic (visceral) reflex.**

Because reflexes are normally so predictable, they provide useful information about the health of the nervous system and can greatly aid diagnosis of disease. Damage or disease anywhere along its reflex arc can cause a reflex to be absent or abnormal. For example, tapping the patellar ligament normally causes reflex extension of the knee joint. Absence of the patellar reflex could indicate damage of the sensory or motor neurons, or a spinal cord injury in the lumbar region. Somatic reflexes generally can be tested simply by tapping or stroking the body surface.

Next, we examine four important somatic spinal reflexes: the stretch reflex, the tendon reflex, the flexor (withdrawal) reflex, and the crossed extensor reflex.

The Stretch Reflex

A **stretch reflex** causes contraction of a skeletal muscle (the effector) in response to stretching of the muscle. This type of reflex occurs via a monosynaptic reflex arc. The reflex can occur by activation of a single sensory neuron that forms one synapse in the CNS with a single motor neuron. Stretch reflexes can be elicited by tapping on tendons attached to muscles at the elbow, wrist,

knee, and ankle joints. An example of a stretch reflex is the patellar reflex (knee jerk), which is described in Clinical Connection: Reflexes and Diagnosis later in the chapter.

A stretch reflex operates as follows (Figure 13.14):

1 Slight stretching of a muscle stimulates sensory receptors in the muscle called **muscle spindles** (shown in more detail in Figure 16.4). The spindles monitor changes in the length of the muscle.

2 In response to being stretched, a muscle spindle generates one or more nerve impulses that propagate along a somatic sensory neuron through the posterior root of the spinal nerve and into the spinal cord.

3 In the spinal cord (integrating center), the sensory neuron makes an excitatory synapse with and thereby activates a motor neuron in the anterior gray horn.

4 If the excitation is strong enough, one or more nerve impulses arise in the motor neuron and propagate along its axon, which extends from the spinal cord into the anterior root and through peripheral nerves to the stimulated muscle. The axon terminals of the motor neuron form neuromuscular junctions (NMJs) with skeletal muscle fibers of the stretched muscle.

5 Acetylcholine released by nerve impulses at the NMJs triggers one or more muscle action potentials in the stretched muscle (effector), and the muscle contracts. Thus, muscle stretch is followed by muscle contraction, which relieves the stretching.

In the reflex arc just described, sensory nerve impulses enter the spinal cord on the same side from which motor nerve impulses leave it. This arrangement is called an **ipsilateral reflex** (ip′-si-LAT-er-al = same side). All monosynaptic reflexes are ipsilateral.

In addition to the large-diameter motor neurons that innervate typical skeletal muscle fibers, smaller-diameter motor neurons innervate smaller, specialized muscle fibers within the muscle spindles themselves. The brain regulates muscle spindle sensitivity through pathways to these smaller motor neurons. This regulation ensures proper muscle spindle signaling over a wide range of muscle lengths during voluntary and reflex contractions. By adjusting how vigorously a muscle spindle responds to stretching, the brain sets an overall level of **muscle tone,** which is the small degree of contraction present while the muscle is at rest. Because the stimulus for the stretch reflex is stretching of muscle, this reflex helps avert injury by preventing overstretching of muscles.

Although the stretch reflex pathway itself is monosynaptic (just two neurons and one synapse), a polysynaptic reflex arc to the antagonistic muscles operates at the same time. This arc involves three neurons and two synapses. An axon collateral (branch) from the muscle spindle sensory neuron also synapses with an inhibitory interneuron in the integrating center. In turn, the interneuron synapses with and inhibits a motor neuron that

Figure 13.14 Stretch reflex. This monosynaptic reflex arc has only one synapse in the CNS—between a single sensory neuron and a single motor neuron. A polysynaptic reflex arc to antagonistic muscles that includes two synapses in the CNS and one interneuron is also illustrated. Plus signs (+) indicate excitatory synapses; the minus sign (−) indicates an inhibitory synapse.

The stretch reflex causes contraction of a muscle that has been stretched.

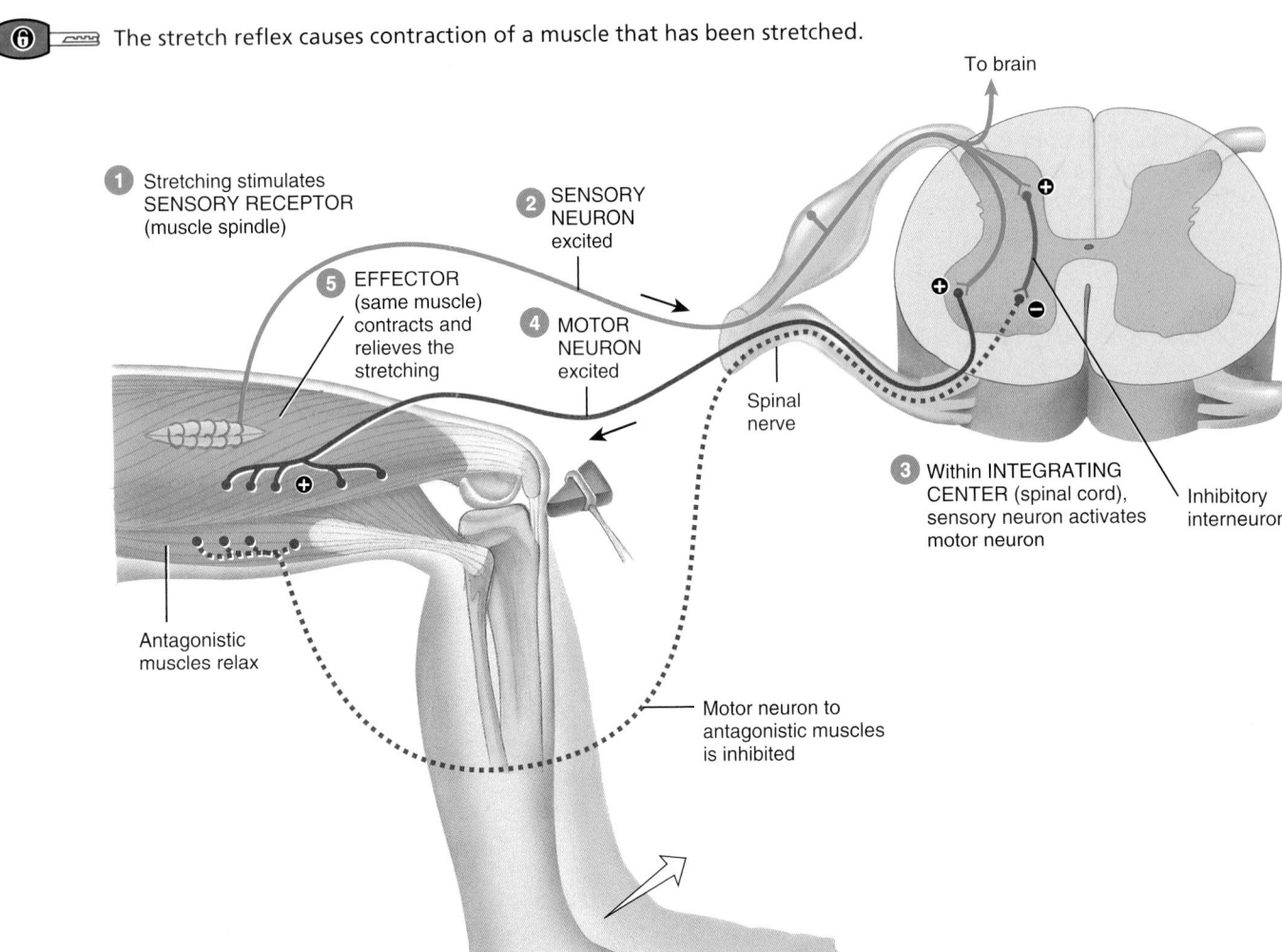

? What makes this an ipsilateral reflex?

normally excites the antagonistic muscles (Figure 13.14). Thus, when the stretched muscle contracts during a stretch reflex, antagonistic muscles that oppose the contraction relax. This type of arrangement, in which the components of a neural circuit simultaneously cause contraction of one muscle and relaxation of its antagonists, is termed **reciprocal innervation** (rē-SIP-ro′-kal in′-er-VĀ-shun). Reciprocal innervation prevents conflict between opposing muscles and is vital in coordinating body movements.

Axon collaterals of the muscle spindle sensory neuron also relay nerve impulses to the brain over specific ascending pathways. In this way, the brain receives input about the state of stretch or contraction of skeletal muscles, enabling it to coordinate muscular movements. The nerve impulses that pass to the brain also allow conscious awareness that the reflex has occurred.

The stretch reflex can also help maintain posture. For example, if a standing person begins to lean forward, the gastrocnemius and other calf muscles are stretched. Consequently, stretch reflexes are initiated in these muscles, which cause them to contract and reestablish the body's upright posture. Similar types of stretch reflexes occur in the muscles of the shin when a standing person begins to lean backward.

The Tendon Reflex

The stretch reflex operates as a feedback mechanism to control muscle *length* by causing muscle contraction. In contrast, the **tendon reflex** operates as a feedback mechanism to control muscle *tension* by causing muscle relaxation before muscle force becomes so great that tendons might be torn. Although the tendon reflex is less sensitive than the stretch reflex, it can override the stretch reflex when tension is great, making you drop a very heavy weight, for example. Like the stretch reflex, the tendon reflex is ipsilateral. The sensory receptors for this reflex are called **tendon (Golgi tendon) organs** (shown in more detail in Figure 16.4), which lie within a tendon near its junction with a muscle. In contrast to muscle spindles, which are sensitive to changes in muscle length, tendon organs detect and respond to changes in muscle tension that are caused by passive stretch or muscular contraction.

A tendon reflex operates as follows (Figure 13.15):

1 As the tension applied to a tendon increases, the tendon organ (sensory receptor) is stimulated (depolarized to threshold).

2 Nerve impulses arise and propagate into the spinal cord along a sensory neuron.

3 Within the spinal cord (integrating center), the sensory neuron activates an inhibitory interneuron that synapses with a motor neuron.

4 The inhibitory neurotransmitter inhibits (hyperpolarizes) the motor neuron, which then generates fewer nerve impulses.

5 The muscle relaxes and relieves excess tension.

Thus, as tension on the tendon organ increases, the frequency of inhibitory impulses increases; inhibition of the motor neurons to the muscle developing excess tension (effector) causes relaxation

Figure 13.15 Tendon reflex. This reflex arc is polysynaptic—more than one CNS synapse and more than two different neurons are involved in the pathway. The sensory neuron synapses with two interneurons. An inhibitory interneuron causes relaxation of the effector, and a stimulatory interneuron causes contraction of the antagonistic muscle. Plus signs (+) indicate excitatory synapses; the minus sign (−) indicates an inhibitory synapse.

The tendon reflex causes relaxation of the muscle attached to the stimulated tendon organ.

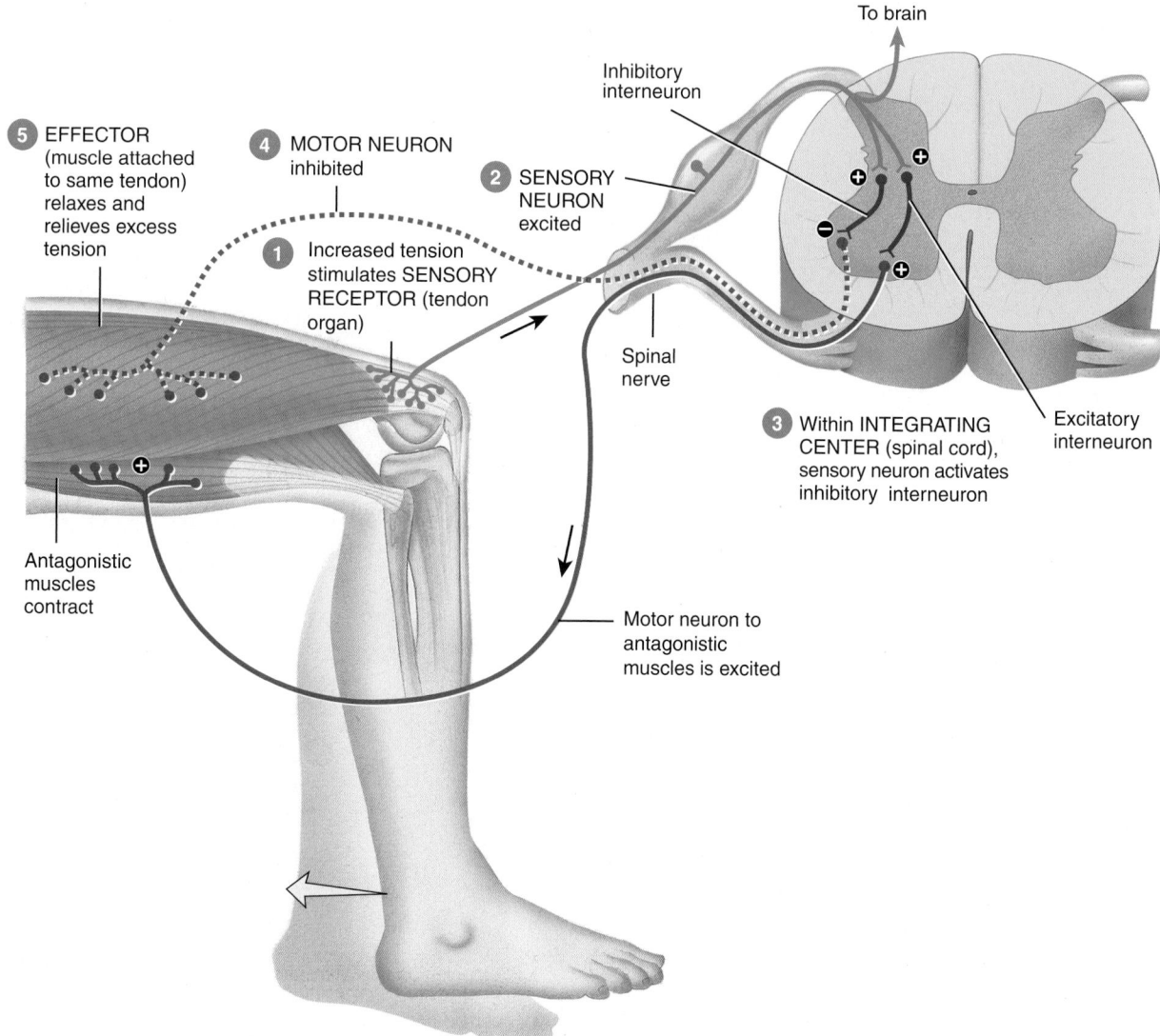

? **What is reciprocal innervation?**

of the muscle. In this way, the tendon reflex protects the tendon and muscle from damage due to excessive tension.

Note in Figure 13.15 that the sensory neuron from the tendon organ also synapses with an excitatory interneuron in the spinal cord. The excitatory interneuron in turn synapses with motor neurons controlling antagonistic muscles. Thus, while the tendon reflex brings about relaxation of the muscle attached to the tendon organ, it also triggers contraction of antagonists. Here we have another example of reciprocal innervation. The sensory neuron also relays nerve impulses to the brain by way of sensory tracts, thus informing the brain about the state of muscle tension throughout the body.

The Flexor and Crossed Extensor Reflexes

Another reflex involving a polysynaptic reflex arc results when, for instance, you step on a tack. In response to such a painful stimulus, you immediately withdraw your leg. This reflex, called the **flexor** or **withdrawal reflex,** operates as follows (Figure 13.16):

1 Stepping on a tack stimulates the dendrites (sensory receptor) of a pain-sensitive neuron.

2 This sensory neuron then generates nerve impulses, which propagate into the spinal cord.

Figure 13.16 Flexor (withdrawal) reflex. Plus signs (+) indicate excitatory synapses.

The flexor reflex causes withdrawal of a part of the body in response to a painful stimulus.

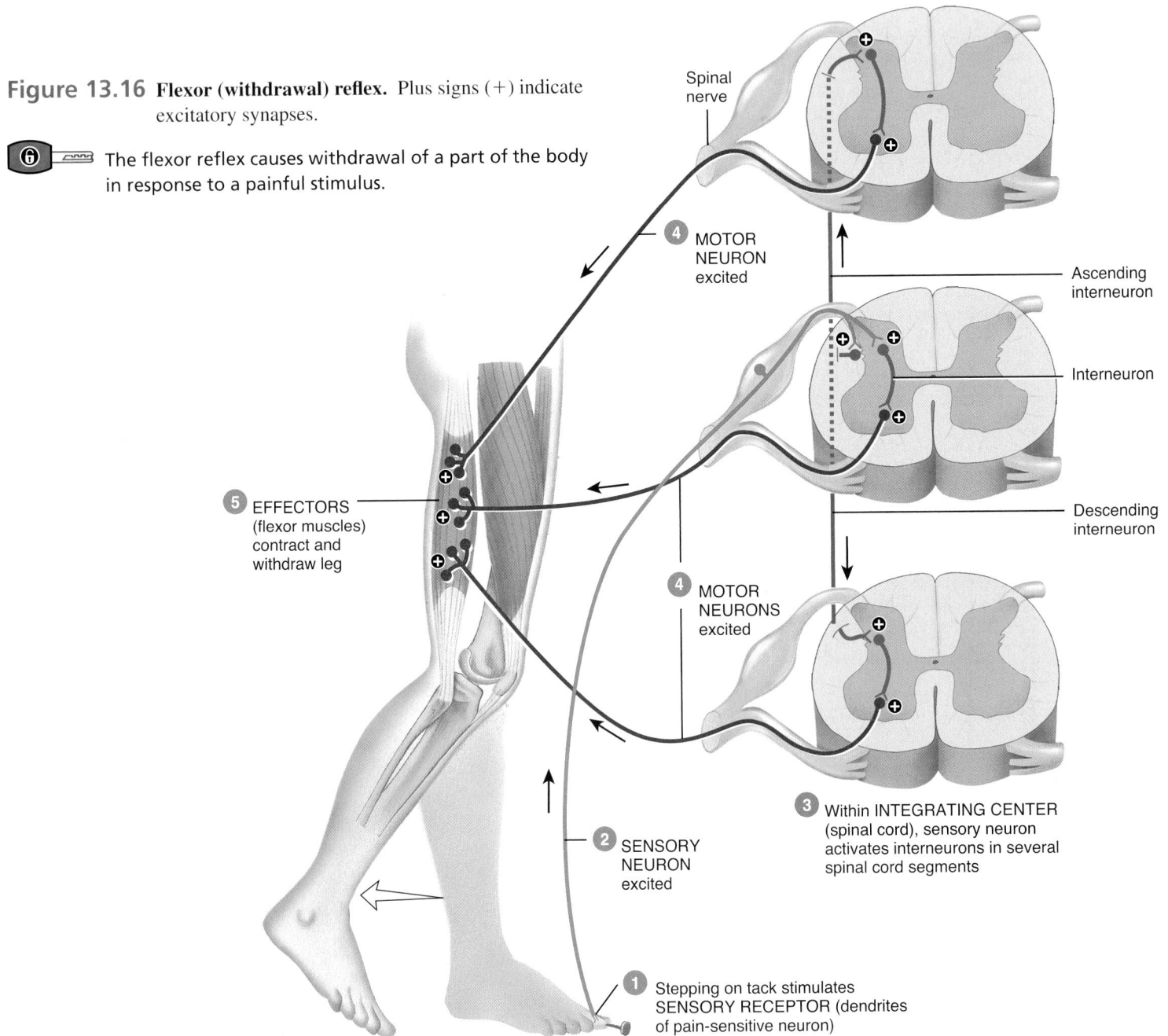

4 MOTOR NEURON excited

Spinal nerve

Ascending interneuron

Interneuron

Descending interneuron

5 EFFECTORS (flexor muscles) contract and withdraw leg

4 MOTOR NEURONS excited

3 Within INTEGRATING CENTER (spinal cord), sensory neuron activates interneurons in several spinal cord segments

2 SENSORY NEURON excited

1 Stepping on tack stimulates SENSORY RECEPTOR (dendrites of pain-sensitive neuron)

? Why is the flexor reflex classified as an intersegmental reflex arc?

3 Within the spinal cord (integrating center), the sensory neuron activates interneurons that extend to several spinal cord segments.

4 The interneurons activate motor neurons in several spinal cord segments. As a result, the motor neurons generate nerve impulses, which propagate toward the axon terminals.

5 Acetylcholine released by the motor neurons causes the flexor muscles in the thigh (effectors) to contract, producing withdrawal of the leg. This reflex is protective because contraction of flexor muscles moves a limb away from the source of a possibly damaging stimulus.

The flexor reflex, like the stretch reflex, is ipsilateral—the incoming and outgoing impulses propagate into and out of the same side of the spinal cord. The flexor reflex also illustrates another feature of polysynaptic reflex arcs. Moving your entire lower or upper limb away from a painful stimulus involves contraction of more than one muscle group. Hence, several motor neurons must simultaneously convey impulses to several limb muscles. Because nerve impulses from one sensory neuron ascend and descend in the spinal cord and activate interneurons in several segments of the spinal cord, this type of reflex is called an **intersegmental reflex arc** (in′-ter-seg-MEN-tal; *inter-* = between). Through intersegmental reflex arcs, a single sensory neuron can activate several motor neurons, thereby stimulating more than one effector. The monosynaptic stretch reflex, in contrast, involves muscles receiving nerve impulses from one spinal cord segment only.

Something else may happen when you step on a tack: You may start to lose your balance as your body weight shifts to the other foot. Besides initiating the flexor reflex that causes you to withdraw the limb, the pain impulses from stepping on the tack also initiate a **crossed extensor reflex** to help you maintain your balance; it operates as follows (Figure 13.17):

1 Stepping on a tack stimulates the sensory receptor of a pain-sensitive neuron in the right foot.

2 This sensory neuron then generates nerve impulses, which propagate into the spinal cord.

3 Within the spinal cord (integrating center), the sensory neuron activates several interneurons that synapse with motor neurons on the left side of the spinal cord in several spinal cord segments. Thus, incoming pain signals cross to the opposite side through interneurons at that level, and at several levels above and below the point of entry into the spinal cord.

4 The interneurons excite motor neurons in several spinal cord segments that innervate extensor muscles. The motor neurons in turn generate more nerve impulses, which propagate toward the axon terminals.

5 Acetylcholine released by the motor neurons causes extensor muscles in the thigh (effectors) of the unstimulated left limb to contract, producing extension of the left leg. In this way, weight can be placed on the foot that must now support the entire body. A comparable reflex occurs with painful stimulation of the left lower limb or either upper limb.

Unlike the flexor reflex, which is an ipsilateral reflex, the crossed extensor reflex involves a **contralateral reflex arc** (kon′-tra-LAT-er-al = opposite side): Sensory impulses enter one side of the spinal cord and motor impulses exit on the opposite side. Thus, a crossed extensor reflex synchronizes the extension of the contralateral limb with the withdrawal (flexion) of the stimulated limb. Reciprocal innervation also occurs in both the flexor reflex and the crossed extensor reflex. In the flexor reflex, when the flexor muscles of a painfully stimulated lower limb are contracting, the extensor muscles of the same limb are relaxing to some degree. If both sets of muscles contracted at the same time, the two sets of muscles would pull on the bones in opposite directions, which might immobilize the limb. Because of reciprocal innervation, one set of muscles contracts while the other relaxes.

CLINICAL CONNECTION | *Reflexes and Diagnosis*

Reflexes are often used for diagnosing disorders of the nervous system and locating injured tissue. If a reflex ceases to function or functions abnormally, the physician may suspect that the damage lies somewhere along a particular conduction pathway. Many somatic reflexes can be tested simply by tapping or stroking the body. Among the somatic reflexes of clinical significance are the following:

- **Patellar reflex (knee jerk).** This stretch reflex involves extension of the leg at the knee joint by contraction of the quadriceps femoris muscle in response to tapping the patellar ligament (see Figure 13.14). This reflex is blocked by damage to the sensory or motor nerves supplying the muscle or to the integrating centers in the second, third, or fourth lumbar segments of the spinal cord. It is often absent in people with chronic diabetes mellitus or neurosyphilis, both of which cause degeneration of nerves. It is exaggerated in disease or injury involving certain motor tracts descending from the higher centers of the brain to the spinal cord.

- **Achilles reflex (a-KIL-ēz) (ankle jerk).** This stretch reflex involves plantar flexion of the foot by contraction of the gastrocnemius and soleus muscles in response to tapping the calcaneal (Achilles) tendon. Absence of the Achilles reflex indicates damage to the nerves supplying the posterior leg muscles or to neurons in the lumbosacral region of the spinal cord. This reflex may also disappear in people with chronic diabetes, neurosyphilis, alcoholism, and subarachnoid hemorrhages. An exaggerated Achilles reflex indicates cervical cord compression or a lesion of the motor tracts of the first or second sacral segments of the cord.

- **Babinski sign (ba-BIN-skē).** This reflex results from gentle stroking of the lateral outer margin of the sole. The great toe extends, with or without a lateral fanning of the other toes. This phenomenon normally occurs in children under 1$\frac{1}{2}$ years of age and is due to incomplete myelination of fibers in the corticospinal tract. A positive Babinski sign after age 1$\frac{1}{2}$ is abnormal and indicates an interruption of the corticospinal tract as the result of a lesion of the tract, usually in the upper portion. The normal response after age 1$\frac{1}{2}$ is the **plantar flexion reflex,** or **negative Babinski**—a curling under of all the toes.

- **Abdominal reflex.** This reflex involves contraction of the muscles that compress the abdominal wall in response to stroking the side

Figure 13.17 Crossed extensor reflex. The flexor reflex arc is shown (at left) for comparison with the crossed extensor reflex arc. Plus signs (+) indicate excitatory synapses.

🔑 A crossed extensor reflex causes contraction of muscles that extend joints in the limb opposite a painful stimulus.

Spinal nerve

Ascending interneurons

4 MOTOR NEURON excited

5 EFFECTORS (extensor muscles) contract, and extend **left** leg

Interneurons from other side

Flexor muscles contract and withdraw **right** leg

Descending interneurons

4 MOTOR NEURONS excited

3 Within INTEGRATING CENTER (spinal cord), sensory neuron activates several interneurons

2 SENSORY NEURON excited

1 Stepping on a tack stimulates SENSORY RECEPTOR (dendrites of pain-sensitive neuron) in **right** foot

Withdrawal of right leg (flexor reflex)

Extension of left leg (crossed extensor reflex)

❓ **Why is the crossed extensor reflex classified as a contralateral reflex arc?**

of the abdomen. The response is an abdominal muscle contraction that causes the umbilicus to move in the direction of the stimulus. Absence of this reflex is associated with lesions of the corticospinal tracts. It may also be absent because of lesions of the peripheral nerves, lesions of integrating centers in the thoracic part of the cord, or multiple sclerosis.

Most autonomic reflexes are not practical diagnostic tools because it is difficult to stimulate visceral effectors, which are deep inside the body. An exception is the pupillary light reflex, in which the pupils of both eyes decrease in diameter when either eye is exposed to light. Because the reflex arc includes synapses in lower parts of the brain, the **absence of a normal pupillary light reflex** may indicate brain damage or injury. •

✔CHECKPOINT

8. Which spinal cord tracts are ascending tracts? Which are descending tracts?

9. How are somatic and autonomic reflexes similar and different?

10. Describe the mechanism and function of a stretch reflex, tendon reflex, flexor (withdrawal) reflex, and crossed extensor reflex.

11. What does each of the following terms mean in relation to reflex arcs? Monosynaptic, ipsilateral, polysynaptic, intersegmental, contralateral, and reciprocal innervation.

DISORDERS: HOMEOSTATIC IMBALANCES

The spinal cord can be damaged in several ways. Outcomes range from little or no long-term neurological deficits to severe deficits and even death.

Traumatic Injuries

Most **spinal cord injuries** are due to trauma as a result of factors such as automobile accidents, falls, contact sports, diving, and acts of violence. The effects of the injury depend on the extent of direct trauma to the spinal cord or compression of the cord by fractured or displaced vertebrae or blood clots. Although any segment of the spinal cord may be involved, the most common sites of injury are in the cervical, lower thoracic, and upper lumbar regions. Depending on the location and extent of spinal cord damage, paralysis may occur. **Monoplegia** (mon′-ō-PLĒ-jē-a; *mono-* = one; *-plegia* = blow or strike) is paralysis of one limb only. **Diplegia** (*di-* = two) is paralysis of both upper limbs or both lower limbs. **Paraplegia** (*para-* = beyond) is paralysis of both lower limbs. **Hemiplegia** (*hemi-* = half) is paralysis of the upper limb, trunk, and lower limb on one side of the body, and **quadriplegia** (*quad-* = four) is paralysis of all four limbs.

Complete transection (tran-SEK-shun; *trans-* = across; *-section* = a cut) of the spinal cord means that the cord is severed from one side to the other, thus cutting all sensory and motor tracts. It results in a loss of all sensations and voluntary movement below the level of the transection. A person will have permanent loss of all sensations in dermatomes below the injury because ascending nerve impulses cannot propagate past the transection to reach the brain. A **dermatome** (DER-ma-tōm; *derma* = skin, *-tome* = thin segment) is an area of skin that provides sensory input to the CNS via a pair of spinal nerves. At the same time, all voluntary muscle contractions will be lost below the transection because nerve impulses descending from the brain also cannot pass. The extent of paralysis of skeletal muscles depends on the level of injury. The closer the injury is to the head, the greater the area of the body that may be affected. The following list outlines which muscle functions may be *retained* at progressively lower levels of spinal cord transection. (These are spinal cord levels and not vertebral column levels. Recall that spinal cord levels differ from vertebral column levels because of the differential growth of the cord versus the column, especially as you progress inferiorly.)

- C1–C3: no function maintained from the neck down; ventilator needed for breathing; electric wheelchair with breath, head, or shoulder-controlled device required (see Figure A)
- C4–C5: diaphragm, which allows breathing
- C6–C7: some arm and chest muscles, which allows feeding, some dressing, and manual wheelchair required (see Figure B)
- T1–T3: intact arm function
- T4–T9: control of trunk above the umbilicus
- T10–L1: most thigh muscles, which allows walking with long leg braces (see Figure C)
- L1–L2: most leg muscles, which allows walking with short leg braces (see Figure D)

Hemisection is a partial transection of the cord on either the right or the left side. After hemisection, three main symptoms, known together as *Brown-Séquard syndrome* (sē-KAR), occur below the level of the injury: (1) Damage of the posterior column (sensory tracts)

C1

Cervical segment

(A)

C7
T1

Thoracic segment

(B)

T8
T9

Lumbar segment

T11
T12

Sacral segment

L2

(C)

(D)

causes loss of proprioception and fine touch sensations on the *ipsilateral* (same) side as the injury. (2) Damage of the lateral corticospinal tract (motor tract) causes ipsilateral paralysis. (3) Damage of the spinothalamic tracts (sensory tracts) causes loss of pain and temperature sensations on the *contralateral* (opposite) side.

Following complete transection, and to varying degrees after hemisection, spinal shock occurs. **Spinal shock** is an immediate response to spinal cord injury characterized by temporary **areflexia** (a'-re-FLEK-sē-a), loss of reflex function. The areflexia occurs in parts of the body served by spinal nerves below the level of the injury. Signs of acute spinal shock include slow heart rate, low blood pressure, flaccid paralysis of skeletal muscles, loss of somatic sensations, and urinary bladder dysfunction. Spinal shock may begin within 1 hour after injury and may last from several minutes to several months, after which reflex activity gradually returns.

In many cases of traumatic injury of the spinal cord, the patient may have an improved outcome if an anti-inflammatory corticosteroid drug called methylprednisolone is given within 8 hours of the injury. This is because the degree of neurologic deficit is greatest immediately following traumatic injury as a result of *edema* (collection of fluid within tissues) as the immune system responds to the injury.

Spinal Cord Compression

Although the spinal cord is normally protected by the vertebral column, certain disorders may put pressure on it and disrupt its normal functions. Spinal cord compression may result from fractured vertebrae, herniated intervertebral discs, tumors, osteoporosis, or infections. If the source of the compression is determined before neural tissue is destroyed, spinal cord function usually returns to normal. Depending on the location and degree of compression, symptoms include pain, weakness or paralysis, and either decreased or complete loss of sensation below the level of the injury.

Degenerative Diseases

A number of degenerative diseases affect the functions of the spinal cord. One of these is multiple sclerosis, the details of which were presented in Disorders: Homeostatic Imbalances at the end of Chapter 12. Another progressive degenerative disease is amyotrophic lateral sclerosis (Lou Gehrig's disease), which affects motor neurons of the brain and spinal cord and results in muscle weakness and atrophy. Details are presented in Clinical Connection: Amyotrophic Lateral Sclerosis in Chapter 16.

Shingles

Shingles is an acute infection of the peripheral nervous system caused by herpes zoster (HER-pēz ZOS-ter), the virus that also causes chickenpox. After a person recovers from chickenpox, the virus retreats to a posterior root ganglion. If the virus is reactivated, the immune system usually prevents it from spreading. From time to time, however, the reactivated virus overcomes a weakened immune system, leaves the ganglion, and travels down sensory neurons of the skin by fast axonal transport (described in Section 12.2). The result is pain, discoloration of the skin, and a characteristic line of skin blisters. The line of blisters marks the distribution (dermatome) of the particular cutaneous sensory nerve belonging to the infected posterior root ganglion.

Poliomyelitis

Poliomyelitis (pō'-lē-ō-mī-e-LĪ-tus), or simply **polio,** is caused by a virus called poliovirus. The onset of the disease is marked by fever, severe headache, a stiff neck and back, deep muscle pain and weakness, and loss of certain somatic reflexes. In its most serious form, the virus produces paralysis by destroying cell bodies of motor neurons, specifically those in the anterior horns of the spinal cord and in the nuclei of the cranial nerves. Polio can cause death from respiratory or heart failure if the virus invades neurons in vital centers that control breathing and heart functions in the brain stem. Even though polio vaccines have virtually eradicated polio in the United States, outbreaks of polio continue throughout the world. Due to international travel, polio could be easily reintroduced into North America if individuals are not vaccinated appropriately.

Several decades after suffering a severe attack of polio and following their recovery from it, some individuals develop a condition called **post-polio syndrome.** This neurological disorder is characterized by progressive muscle weakness, extreme fatigue, loss of function, and pain, especially in muscles and joints. Post-polio syndrome seems to involve a slow degeneration of motor neurons that innervate muscle fibers. Triggering factors appear to be a fall, a minor accident, surgery, or prolonged bed rest. Possible causes include overuse of surviving motor neurons over time, smaller motor neurons because of the initial infection by the virus, reactivation of dormant polio viral particles, immune-mediated responses, hormone deficiencies, and environmental toxins. Treatment consists of muscle-strengthening exercises, administration of pyridostigmine to enhance the action of acetylcholine in stimulating muscle contraction, and administration of nerve growth factors to stimulate both nerve and muscle growth.

MEDICAL TERMINOLOGY

Epidural block (ep'-i-DOO-ral) Injection of an anesthetic drug into the epidural space, the space between the dura mater and the vertebral column, in order to cause a temporary loss of sensation. Such injections in the lower lumbar region are used to control pain during childbirth.

Meningitis (men-in-JĪ-tis; *-itis* = inflammation) Inflammation of the meninges due to an infection, usually caused by a bacterium or virus. Symptoms include fever, headache, stiff neck, vomiting, confusion, lethargy, and drowsiness. Bacterial meningitis is much more serious and is treated with antibiotics. Viral meningitis has no specific treatment. Bacterial meningitis may be fatal if not treated promptly; viral meningitis usually resolves on its own in 1–2 weeks. A vaccine is available to help protect against some types of bacterial meningitis.

Myelitis (mī-e-LĪ-tis; *myel-* = spinal cord) Inflammation of the spinal cord.

Nerve block Loss of sensation in a region due to injection of a local anesthetic; an example is local dental anesthesia.

Neuralgia (noo-RAL-jē-a; *neur-* = nerve; *-algia* = pain) Attacks of pain along the entire course or a branch of a sensory nerve.

Neuritis (*neur-* = nerve; *-itis* = inflammation) Inflammation of one or several nerves that may result from irritation to the nerve produced by direct blows, bone fractures, contusions, or penetrating injuries. Additional causes include infections, vitamin deficiency (usually thiamine), and poisons such as carbon monoxide, carbon tetrachloride, heavy metals, and some drugs.

Paresthesia (par-es-THĒ-zē-a; *par-* = departure from normal; *-esthesia* = sensation) An abnormal sensation such as burning, pricking, tickling, or tingling resulting from a disorder of a sensory nerve.

CHAPTER REVIEW AND RESOURCE SUMMARY

Review

Resource

13.1 Spinal Cord Anatomy

1. The spinal cord is protected by the vertebral column, the meninges, cerebrospinal fluid, and denticulate ligaments.
2. The three meninges are coverings that run continuously around the spinal cord and brain. They are the dura mater, arachnoid mater, and pia mater.
3. The spinal cord begins as a continuation of the medulla oblongata and ends at about the second lumbar vertebra in an adult.
4. The spinal cord contains cervical and lumbar enlargements that serve as points of origin for nerves to the limbs.
5. The tapered inferior portion of the spinal cord is the conus medullaris, from which arise the filum terminale and cauda equina.
6. Spinal nerves connect to each segment of the spinal cord by two roots. The posterior or dorsal root contains sensory axons, and the anterior or ventral root contains motor neuron axons.
7. The anterior median fissure and the posterior median sulcus partially divide the spinal cord into right and left sides.
8. The gray matter in the spinal cord is divided into horns, and the white matter into columns. In the center of the spinal cord is the central canal, which runs the length of the spinal cord.
9. Parts of the spinal cord observed in transverse section are the gray commissure; central canal; anterior, posterior, and lateral gray horns; and anterior, posterior, and lateral white columns, which contain ascending and descending tracts. Each part has specific functions.
10. The spinal cord conveys sensory and motor information by way of ascending and descending tracts, respectively.

Anatomy Overview - The Spinal Cord
Figure 13.1 - Gross Anatomy of the Spinal Cord
Figure 13.2 - External Anatomy of the Spinal Cord and Spinal Nerves
Figure 13.4 - Processing of Sensory and Motor Input

13.2 Spinal Nerves

1. The 31 pairs of spinal nerves are named and numbered according to the region and level of the spinal cord from which they emerge. There are 8 pairs of cervical, 12 pairs of thoracic, 5 pairs of lumbar, 5 pairs of sacral, and 1 pair of coccygeal nerves.
2. Spinal nerves typically are connected with the spinal cord by a posterior root and an anterior root. All spinal nerves contain both sensory and motor axons (are mixed nerves).
3. Three connective tissue coverings associated with spinal nerves are the endoneurium, perineurium, and epineurium.
4. Branches of a spinal nerve include the posterior ramus, anterior ramus, meningeal branch, and rami communicantes.
5. The anterior rami of spinal nerves, except for T2–T12, form networks of nerves called plexuses.
6. Emerging from the plexuses are nerves bearing names that typically describe the general regions they supply or the route they follow.
7. Nerves of the cervical plexus supply the skin and muscles of the head, neck, and upper part of the shoulders; they connect with some cranial nerves and innervate the diaphragm. Nerves of the brachial plexus supply the upper limbs and several neck and shoulder muscles. Nerves of the lumbar plexus supply the anterolateral abdominal wall, external genitals, and part of the lower limbs. Nerves of the sacral plexus supply the buttocks, perineum, and part of the lower limbs. Nerves of the coccygeal plexus supply the skin of the coccygeal region.
8. Anterior rami of nerves T2–T12 do not form plexuses and are called intercostal (thoracic) nerves. They are distributed directly to the structures they supply in intercostal spaces.
9. Sensory neurons within spinal nerves and the trigeminal (V) nerve serve specific, constant segments of the skin called dermatomes.
10. Knowledge of dermatomes helps health care providers determine which segment of the spinal cord or which spinal nerve is damaged.

Anatomy Overview - Spinal Nerves

13.3 Spinal Cord Physiology

1. The white matter tracts in the spinal cord are highways for nerve impulse propagation. Along these tracts, sensory input travels toward the brain, and motor output travels from the brain toward skeletal muscles and other effector tissues. Sensory input travels along two main routes in the white matter of the spinal cord: the posterior column and the spinothalamic tract. Motor output travels along two main routes in the white matter of the spinal cord: direct pathways and indirect pathways.

Animation - Somatic Sensory and Motor Pathways
Animation - Reflex Arcs
Animation - Reflexes
Figure 13.12 - Locations of Major Sensory and Motor Tracts

Review

2. A second major function of the spinal cord is to serve as an integrating center for spinal reflexes. This integration occurs in the gray matter.

3. A reflex is a fast, predictable sequence of involuntary actions, such as muscle contractions or glandular secretions, which occurs in response to certain changes in the environment. Reflexes may be spinal or cranial and somatic or autonomic (visceral).

4. The components of a reflex arc are sensory receptor, sensory neuron, integrating center, motor neuron, and effector.

5. Somatic spinal reflexes include the stretch reflex, the tendon reflex, the flexor (withdrawal) reflex, and the crossed extensor reflex; all exhibit reciprocal innervation.

6. A two-neuron or monosynaptic reflex arc consists of one sensory neuron and one motor neuron. A stretch reflex, such as the patellar reflex, is an example.

7. The stretch reflex is ipsilateral and is important in maintaining muscle tone.

8. A polysynaptic reflex arc contains sensory neurons, interneurons, and motor neurons. The tendon reflex, flexor (withdrawal) reflex, and crossed extensor reflexes are examples.

9. The tendon reflex is ipsilateral and prevents damage to muscles and tendons when muscle force becomes too extreme. The flexor reflex is ipsilateral and moves a limb away from the source of a painful stimulus. The crossed extensor reflex extends the limb contralateral to a painfully stimulated limb, allowing the weight of the body to shift when a supporting limb is withdrawn.

10. Several important somatic reflexes are used to diagnose various disorders. These include the patellar reflex, Achilles reflex, Babinski sign, and abdominal reflex.

Resource

Figure 13.17 - Crossed Extensor Reflex
Exercise - Assemble an Arc
Exercise - Stretch Reflex

SELF-QUIZ QUESTIONS

Fill in the blanks in the following statements.

1. Because they contain both sensory and motor axons, spinal nerves are considered to be _____ nerves.

2. The five components of a reflex arc, in order from the beginning to the end, are (1) _____, (2) _____, (3) _____, (4) _____, and (5) _____.

Indicate whether the following statements are true or false.

3. Gray matter of the spinal cord contains somatic motor and sensory nuclei, autonomic motor and sensory nuclei, and functions to receive and integrate both incoming and outgoing information.

4. The epidural space is located between the wall of the vertebral canal and the pia mater.

Choose the one best answer to the following questions.

5. Which of the following is *not* true? (1) Dermatomes are areas of the body that are stimulated by motor neurons exiting specific spinal nerves. (2) The stretch reflex helps to maintain muscle tone. (3) The Achilles reflex is an example of a stretch reflex. (4) The abdominal reflex is used to diagnose problems with autonomic reflexes. (5) Spinal nerves T2–T12 do not enter into the formation of a plexus.
   (a) 1, 2, and 4   (b) 2 and 5   (c) 1 and 4
   (d) 1, 3, and 5   (e) 1, 3, and 4

6. While identifying and labeling cadaver muscles, your lab partner accidentally pokes your finger with a pin. Place the following steps in the correct order from beginning to end of your body's response. (1) Impulses travel through anterior (ventral) root of spinal nerve(s). (2) Sensory neuron relays impulse to spinal cord. (3) Motor impulses reach muscles, causing withdrawal of the affected limb. (4) Integrating centers interpret sensory impulses, and then generate motor impulses. (5) Sensory receptor activated by stimulus. (6) Impulse travels through posterior (dorsal) root of spinal nerve.
   (a) 5, 3, 6, 4, 1, 2   (b) 5, 2, 1, 4, 6, 3   (c) 5, 2, 6, 4, 1, 3
   (d) 3, 5, 1, 2, 4, 6   (e) 2, 1, 5, 4, 6, 3

7. The connective tissue surrounding each individual axon is
   (a) endoneurium.   (b) epineurium.   (c) perineurium.
   (d) fascicle.   (e) arachnoid mater.

8. The tracts of the posterior column are involved in (1) conscious proprioception. (2) touch, (3) pain, (4) thermal sensations, (5) pressure, (6) vibration.
   (a) 1, 2, 4, and 5   (b) 2, 4, and 6   (c) 1, 2, 5, and 6
   (d) 3, 4, 5, and 6   (e) 1, 3, 5, and 6

9. Which of the following is a motor tract?
   (a) posterior spinocerebellar   (b) lateral spinothalamic
   (c) anterior spinocerebellar   (d) lateral corticospinal
   (e) posterior column

10. Cutting the posterior root of a spinal nerve would
   (a) interfere with the circulation of cerebrospinal fluid.
   (b) impair motor control of skeletal muscles.
   (c) interfere with the ability of the brain to transmit motor impulses.
   (d) impair motor control of organs.
   (e) interfere with the flow of sensory impulses.

11. Which of the following statements is *false*?
   (a) The two main spinal cord sensory paths are the spinothalamic and anterior columns.
   (b) The spinothalamic tracts convey impulses for sensing pain, temperature, itching, and tickling.
   (c) Direct pathways convey nerve impulses destined to cause precise, voluntary movements of skeletal muscles.
   (d) Indirect pathways convey nerve impulses that program automatic movements, help coordinate body movements with visual stimuli, maintain skeletal muscle tone and posture, and contribute to equilibrium.
   (e) The direct pathways are motor pathways.

12. Which of the following are *true*?
(1) The anterior (ventral) gray horns contain cell bodies of neurons that cause skeletal muscle contraction. (2) The gray commissure connects the white matter of the right and left sides of the spinal cord. (3) Cell bodies of autonomic motor neurons are located in the lateral gray horns. (4) Sensory (ascending) tracts conduct motor impulses down the spinal cord. (5) Gray matter in the spinal cord consists of cell bodies of neurons, neuroglia, unmyelinated axons, and dendrites of interneurons and motor neurons.

(a) 1, 2, 3, and 5   (b) 2 and 4   (c) 2, 3, 4, and 5
(d) 1, 3, and 5   (e) 1, 2, 3, and 4

13. Match the following (some answers may be used more than once):
_____ (a) a reflex resulting in the contraction of a skeletal muscle when it is stretched
_____ (b) receptors that monitor changes in muscle length
_____ (c) a balance-maintaining reflex
_____ (d) operates as a feedback mechanism to control muscle tension by causing muscle relaxation when muscle force becomes too extreme
_____ (e) reflex arc that consists of one sensory and one motor neuron
_____ (f) acts as a feedback mechanism to control muscle length by causing muscle contraction
_____ (g) sensory impulses enter on one side of the spinal cord and motor impulses exit on the opposite side
_____ (h) occurs when sensory nerve impulse travels up and down the spinal cord, thereby activating several motor neurons and more than one effector
_____ (i) polysynaptic reflex initiated in response to a painful stimulus
_____ (j) receptors that monitor changes in muscle tension
_____ (k) maintains proper muscle tone
_____ (l) reflex pathway that contains sensory neurons, interneurons, and motor neurons
_____ (m) motor nerve impulses exit the spinal cord on the same side that sensory impulses entered the spinal cord
_____ (n) protects the tendon and muscle from damage due to excessive tension
_____ (o) a neural circuit that coordinates body movements by causing contraction of one muscle and relaxation of antagonistic muscles or relaxation of a muscle and contraction of the antagonists

(1) stretch reflex
(2) tendon reflex
(3) flexor (withdrawal) reflex
(4) crossed extensor reflex
(5) intersegmental reflex arc
(6) contralateral reflex arc
(7) ipsilateral reflex arc
(8) muscle spindles
(9) tendon (Golgi tendon) organs
(10) reciprocal innervation
(11) monosynaptic reflex
(12) polysynaptic reflex

14. Match the following:
_____ (a) the joining together of the anterior rami of adjacent nerves
_____ (b) spinal nerve branches that serve the deep muscles and skin of the posterior surface of the trunk
_____ (c) spinal nerve branches that serve the muscles and structures of the upper and lower limbs and the lateral and ventral trunk
_____ (d) area of the spinal cord from which nerves to and from the upper limbs arise
_____ (e) area of the spinal cord from which nerves to and from the lower limbs arise
_____ (f) the roots form the nerves that arise from the inferior part of the spinal cord but do not leave the vertebral column at the same level as they exit the cord
_____ (g) contains motor neuron axons and conducts impulses from the spinal cord to the peripheral organs and cells
_____ (h) avascular covering of spinal cord composed of delicate collagen fibers and some elastic fibers
_____ (i) contains sensory neuron axons and conducts impulses from the peripheral receptors into the spinal cord
_____ (j) superficial spinal cord covering of dense, irregular connective tissue
_____ (k) an extension of the pia mater that anchors the spinal cord to the coccyx
_____ (l) extending the length of the spinal cord, these pia mater thickenings fuse with the arachnoid mater and dura mater and help to protect the spinal cord from shock and sudden displacement
_____ (m) thin transparent connective tissue composed of interlacing bundles of collagen fibers and some elastic fibers adhering to the spinal cord's surface
_____ (n) space within the spinal cord filled with cerebrospinal fluid
_____ (o) spinal nerve branch that supplies vertebrae, vertebral ligaments, blood vessels of the spinal cord, and meninges

(1) cervical enlargement
(2) lumbar enlargement
(3) central canal
(4) denticulate ligaments
(5) cauda equina
(6) meningeal branch
(7) pia mater
(8) arachnoid mater
(9) dura mater
(10) posterior (dorsal) root
(11) anterior (ventral) root
(12) posterior (dorsal) ramus
(13) anterior (ventral) ramus
(14) plexus
(15) filum terminale

15. Match the following:
_____ (a) provides the entire nerve supply of the shoulders and upper limbs
_____ (b) provides the nerve supply of the skin and muscles of the head, neck, and superior part of the shoulders and chest
_____ (c) provides the nerve supply of the anterolateral abdominal wall, external genitals, and part of the lower limbs
_____ (d) supplies the buttocks, perineum, and lower limbs
_____ (e) formed by the anterior rami of C1–C4 with some contribution by C5
_____ (f) formed by anterior rami of S4–S5 and coccygeal nerves
_____ (g) formed by the anterior rami of L1–L4
_____ (h) formed by the anterior rami of C5–C8 and T1
_____ (i) formed by the anterior rami of L4–L5 and S1–S4
_____ (j) phrenic nerve arises from this plexus
_____ (k) median nerve arises from this plexus
_____ (l) sciatic nerve arises from this plexus
_____(m) femoral nerve arises from this plexus
_____ (n) supplies a small area of skin in coccygeal region
_____ (o) injury to this plexus can affect breathing

(1) cervical plexus
(2) brachial plexus
(3) lumbar plexus
(4) sacral plexus
(5) coccygeal plexus

CRITICAL THINKING QUESTIONS

1. Evalina's severe headaches and other symptoms were suggestive of meningitis, so her physician ordered a spinal tap. List the structures that the needle will pierce from the most superficial to the deepest. Why would the physician order a test in the spinal region to check a problem in Evalina's head?

2. Sunil has developed an infection that is destroying cells in the anterior gray horns in the lower cervical region of the spinal cord. What kinds of symptoms would you expect to occur?

3. Allyson is in a car accident and suffers spinal cord compression in the lower spinal cord. Although she is in pain, she cannot distinguish when the doctor is touching her calf or her toes and she is having trouble telling how her lower limbs are positioned. What part of the spinal cord has been affected by the accident?

❓ ANSWERS TO FIGURE QUESTIONS

13.1 The superior boundary of the spinal dura mater is the foramen magnum of the occipital bone. The inferior boundary is the second sacral vertebra.

13.2 The cervical enlargement connects with sensory and motor nerves of the upper limbs.

13.3 A horn is an area of gray matter, and a column is a region of white matter in the spinal cord.

13.4 Lateral gray horns are found in the thoracic and upper lumbar segments of the spinal cord.

13.5 All spinal nerves are classified as mixed because their posterior roots contain sensory axons and their anterior roots contain motor axons.

13.6 The anterior rami serve the upper and lower limbs.

13.7 Severing the spinal cord at level C2 causes respiratory arrest because it prevents descending nerve impulses from reaching the phrenic nerve, which stimulates contraction of the diaphragm, the main muscle needed for breathing.

13.8 The axillary, musculocutaneous, radial, median, and ulnar nerves are five important nerves that arise from the brachial plexus.

13.9 Signs of femoral nerve injury include inability to extend the leg and loss of sensation in the skin over the anterolateral aspect of the thigh.

13.10 The origin of the sacral plexus is the anterior rami of spinal nerves L4–L5 and S1–S4.

13.11 The only spinal nerve without a corresponding dermatome is C1.

13.12 The spinothalamic tract originates in the spinal cord and ends in the thalamus (a region of the brain). Because "spinal" comes first in the name, you know it contains ascending axons and thus is a sensory tract.

13.13 A sensory receptor produces a generator potential, which triggers a nerve impulse if the generator potential reaches threshold. Reflex integrating centers are in the CNS.

13.14 In an ipsilateral reflex, the sensory and motor neurons are on the same side of the spinal cord.

13.15 Reciprocal innervation is a type of arrangement of a neural circuit involving simultaneous contraction of one muscle and relaxation of its antagonist.

13.16 The flexor reflex is intersegmental because impulses go out over motor neurons located in several spinal nerves, each arising from a different segment of the spinal cord.

13.17 The crossed extensor reflex is a contralateral reflex arc because the motor impulses leave the spinal cord on the side opposite the entry of sensory impulses.

14

THE BRAIN AND CRANIAL NERVES

THE BRAIN, CRANIAL NERVES, AND HOMEOSTASIS *Your brain contributes to homeostasis by receiving sensory input, integrating new and stored information, making decisions, and executing responses through motor activities.*

Solving an equation, feeling hungry, laughing—the neural processes needed for each of these activities occur in different regions of the **brain,** that portion of the central nervous system contained within the cranium. About 100 billion neurons and 10 trillion to 50 trillion neuroglia make up the brain, which has a mass of about 1300 g (almost 3 lb) in adults. On average, each neuron forms 1000 synapses with other neurons. Thus, the total number of synapses, about a thousand trillion or 10^{15}, is larger than the number of stars in our galaxy.

The brain is the control center for registering sensations, correlating them with one another and with stored information, making decisions, and taking actions. It also is the center for the intellect, emotions, behavior, and memory. But the brain encompasses yet a larger domain: It directs our behavior toward others. With ideas that excite, artistry that dazzles, or rhetoric that mesmerizes, one person's thoughts and actions may influence and shape the lives of many others. As you will see shortly, different regions of the brain are specialized for different functions. Different parts of the brain also work together to accomplish certain shared functions. This chapter explores how the brain is protected and nourished, what functions occur in the major regions of the brain, and how the spinal cord and the 12 pairs of cranial nerves connect with the brain to form the control center of the human body.

? *Did you ever wonder how cerebrovascular accidents (strokes) occur and how they are treated?*

14.1 BRAIN ORGANIZATION, PROTECTION, AND BLOOD SUPPLY

◉ **OBJECTIVES**

- Identify the major parts of the brain.
- Describe how the brain is protected.
- Describe the blood supply of the brain.

In order to understand the terminology used for the principal parts of the adult brain, it will be helpful to know how the brain develops. The brain and spinal cord develop from the ectodermal **neural tube** (see Figure 14.27). The anterior part of the neural tube expands, along with the associated neural crest tissue. Constrictions in this expanded tube soon appear, creating three regions called **primary brain vesicles**: *prosencephalon, mesencephalon,* and *rhombencephalon* (see Figure 14.28). Both the prosencephalon and rhombencephalon subdivide further, forming **secondary brain vesicles.** The *prosencephalon* (PROS-en-sef′-a-lon), or forebrain, gives rise to the telencephalon and diencephalon, and the *rhombencephalon* (ROM-ben-sef′-a-lon), or hindbrain, develops into the metencephalon and myelencephalon. The various brain vesicles give rise to the following adult structures:

- The **telencephalon** develops into the *cerebrum* and *lateral ventricles.*
- The **diencephalon** forms the *thalamus, hypothalamus, epithalamus,* and *third ventricle.*
- The **metencephalon** becomes the *pons, cerebellum,* and *upper part of the fourth ventricle.*
- The **myelencephalon** forms the *medulla oblongata* and *lower part of the fourth ventricle.*

- The **mesencephalon** (mes′-en-SEF-a-lon), or midbrain, gives rise to the *midbrain* and *aqueduct of the midbrain (cerebral aqueduct).*

The walls of these brain regions develop into the nervous tissue of the brain, while the hollow interior of the tube is transformed into the various ventricles (fluid-filled spaces) of the brain. The expanded neural crest tissue becomes prominent in head development. Most of the protective structures of the brain—that is, most of the bones of the skull, associated connective tissues, and meningeal membranes—arise from this expanded neural crest tissue.

These relationships are summarized in Table 14.1.

Major Parts of the Brain

The adult brain consists of four major parts: brain stem, cerebellum, diencephalon, and cerebrum (Figure 14.1). The **brain stem** is continuous with the spinal cord and consists of the medulla oblongata, pons, and midbrain. Posterior to the brain stem is the **cerebellum** (ser′-e-BEL-um = little brain). Superior to the brain stem is the **diencephalon** (dī′-en-SEF-a-lon; *di-* = through; *-encephalon* = brain), which consists of the thalamus, hypothalamus, and epithalamus. Supported on the diencephalon and brain stem is the **cerebrum** (se-RĒ-brum = brain), the largest part of the brain.

Protective Coverings of the Brain

The cranium (see Figure 7.4) and the cranial meninges surround and protect the brain. The **cranial meninges** (me-NIN-jēz) are continuous with the spinal meninges, have the same basic structure, and bear the same names: the outer **dura mater** (DOO-ra MĀ-ter), the middle **arachnoid mater** (a-RAK-noyd), and the inner **pia mater** (PĪ-a or PĒ-a) (Figure 14.2). However, the cranial

---

TABLE 14.1

Development of the Brain

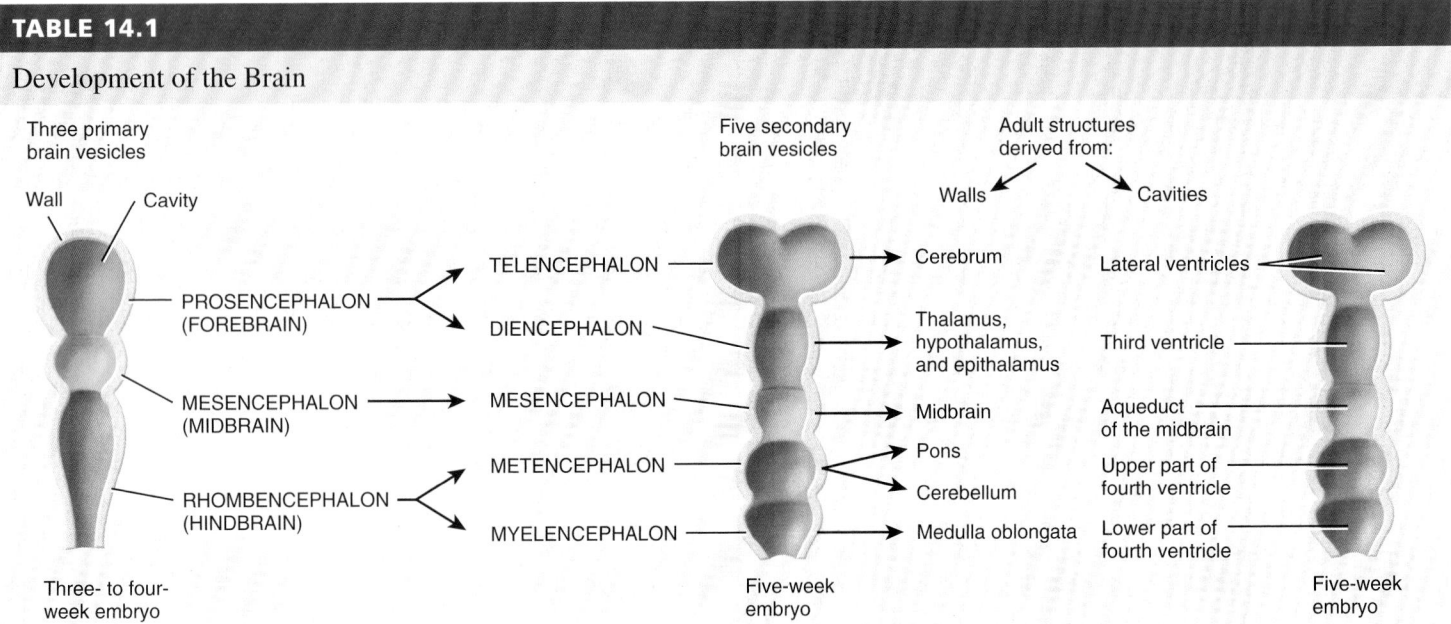

Three primary brain vesicles	Five secondary brain vesicles	Adult structures derived from: Walls	Adult structures derived from: Cavities
PROSENCEPHALON (FOREBRAIN)	TELENCEPHALON	Cerebrum	Lateral ventricles
	DIENCEPHALON	Thalamus, hypothalamus, and epithalamus	Third ventricle
MESENCEPHALON (MIDBRAIN)	MESENCEPHALON	Midbrain	Aqueduct of the midbrain
RHOMBENCEPHALON (HINDBRAIN)	METENCEPHALON	Pons / Cerebellum	Upper part of fourth ventricle
	MYELENCEPHALON	Medulla oblongata	Lower part of fourth ventricle

Three- to four-week embryo · Five-week embryo · Five-week embryo

Figure 14.1 The brain. The pituitary gland is discussed with the endocrine system in Chapter 18.

The four principal parts of the brain are the brain stem, cerebellum, diencephalon, and cerebrum.

(a) Sagittal section, medial view

(b) Sagittal section, medial view

? **Which part of the brain is the largest?**

Figure 14.2 **The protective coverings of the brain.**

Cranial bones and cranial meninges protect the brain.

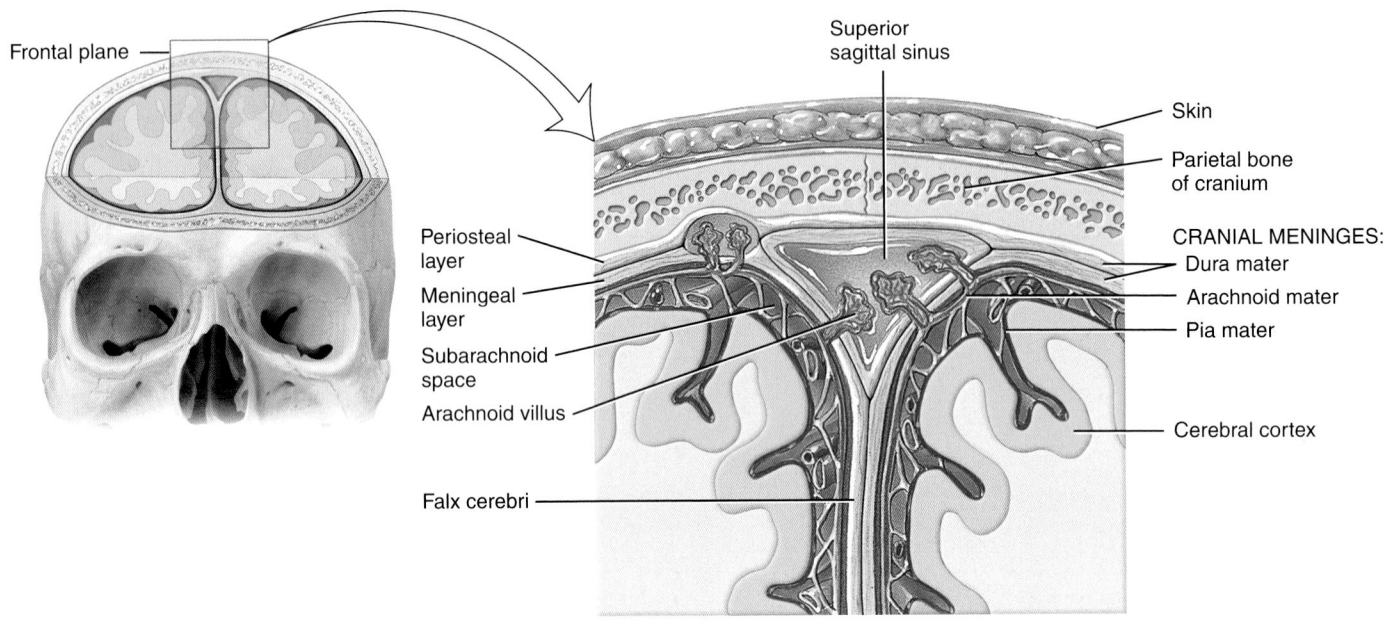

(a) Anterior view of frontal section through skull showing the cranial meninges

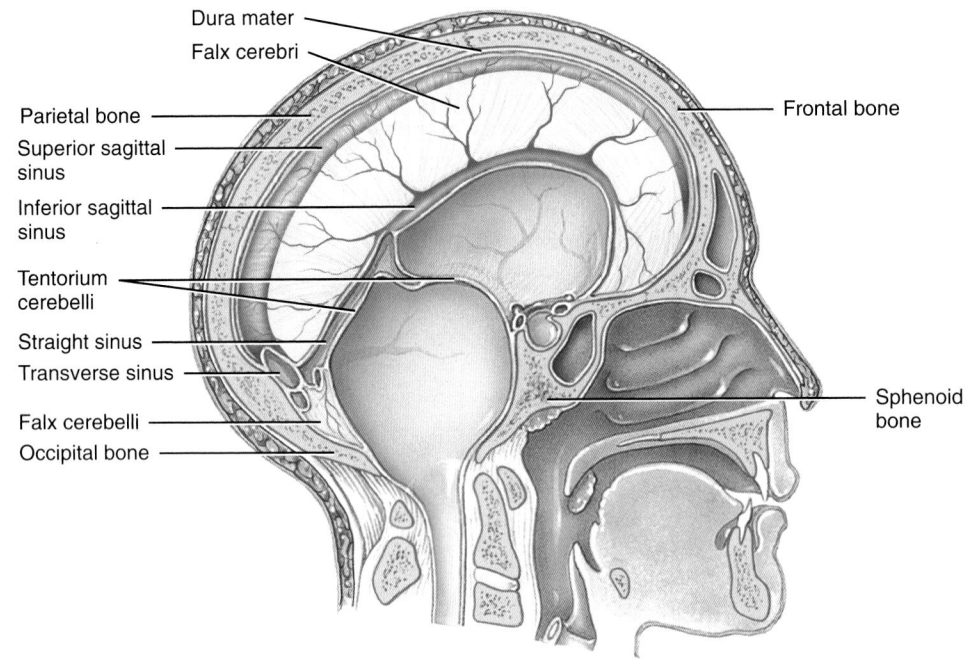

(b) Sagittal section of extensions of the dura mater

? **What are the three layers of the cranial meninges, from superficial to deep?**

dura mater has two layers; the spinal dura mater has only one. The two dural layers are called the *periosteal layer* (which is external) and the *meningeal layer* (which is internal). The dural layers around the brain are fused together except where they separate to enclose the dural venous sinuses (endothelial-lined venous channels) that drain venous blood from the brain and deliver it into the internal jugular veins. Also, there is no epidural space around the brain. Blood vessels that enter brain tissue pass along the surface of the brain, and as they penetrate inward, they are sheathed by a loose-fitting sleeve of pia mater. Three extensions of the dura mater separate parts of the brain: (1) The **falx cerebri** (FALKS ser-i-BRĒ; *falx* = sickle-shaped) separates the two hemispheres (sides) of the cerebrum. (2) The **falx cerebelli** (ser'-e-BEL-ī) separates the two hemispheres of the cerebellum. (3) The **tentorium cerebelli** (ten-TŌ-rē-um = tent) separates the cerebrum from the cerebellum.

Brain Blood Flow and the Blood–Brain Barrier

Blood flows to the brain mainly via the internal carotid and vertebral arteries (see Figure 21.19); the dural venous sinuses drain into the internal jugular veins to return blood from the head to the heart (see Figure 21.24).

In an adult, the brain represents only 2% of total body weight, but it consumes about 20% of the oxygen and glucose used by the body, even when you are resting. Neurons synthesize ATP almost exclusively from glucose via reactions that use oxygen. When the activity of neurons and neuroglia increases in a particular region of the brain, blood flow to that area also increases. Even a brief slowing of brain blood flow may cause disorientation or a lack of consciousness, such as when you stand up too quickly after sitting for a long period of time. Typically, an interruption in blood flow for 1 or 2 minutes impairs neuronal function, and total deprivation of oxygen for about 4 minutes causes permanent injury. Because virtually no glucose is stored in the brain, the supply of glucose also must be continuous. If blood entering the brain has a low level of glucose, mental confusion, dizziness, convulsions, and loss of consciousness may occur. People with diabetes must be vigilant about their blood sugar levels because these levels can drop quickly, leading to diabetic shock, which is characterized by seizure, coma, and possibly death.

The **blood–brain barrier (BBB)** consists mainly of tight junctions that seal together the endothelial cells of brain blood capillaries and a thick basement membrane that surrounds the capillaries. As you learned in Chapter 12, astrocytes are one type of neuroglia; the processes of many astrocytes press up against the capillaries and secrete chemicals that maintain the permeability characteristics of the tight junctions. A few water-soluble substances, such as glucose, cross the BBB by active transport. Other substances, such as creatinine, urea, and most ions, cross the BBB very slowly. Still other substances—proteins and most antibiotic drugs—do not pass at all from the blood into brain tissue. However, lipid-soluble substances, such as oxygen, carbon dioxide, alcohol, and most anesthetic agents, are able to access brain tissue freely. Trauma, certain toxins, and inflammation can cause a breakdown of the blood–brain barrier.

CLINICAL CONNECTION | Breaching the Blood–Brain Barrier

Because it is so effective, the BBB prevents the passage of helpful substances as well as those that are potentially harmful. Researchers are exploring ways to move drugs that could be therapeutic for brain cancer or other CNS disorders past the BBB. In one method, the drug is injected in a concentrated sugar solution. The high osmotic pressure of the sugar solution causes the endothelial cells of the capillaries to shrink, which opens gaps between their tight junctions, making the BBB more leaky and allowing the drug to enter the brain tissue. •

✔ CHECKPOINT

1. Compare the sizes and locations of the cerebrum and cerebellum.
2. Describe the locations of the cranial meninges.
3. Explain the blood supply to the brain and the importance of the blood–brain barrier.

14.2 CEREBROSPINAL FLUID

◉ OBJECTIVE

• Explain the formation and circulation of cerebrospinal fluid.

Cerebrospinal fluid (CSF) is a clear, colorless liquid composed primarily of water that protects the brain and spinal cord from chemical and physical injuries. It also carries small amounts of oxygen, glucose, and other needed chemicals from the blood to neurons and neuroglia. CSF continuously circulates through cavities in the brain and spinal cord and around the brain and spinal cord in the subarachnoid space (the space between the arachnoid mater and pia mater). The total volume of CSF is 80 to 150 mL (3 to 5 oz) in an adult. CSF contains small amounts of glucose, proteins, lactic acid, urea, cations (Na^+, K^+, Ca^{2+}, Mg^{2+}), and anions (Cl^- and HCO_3^-); it also contains some white blood cells.

Figure 14.3 shows the four CSF-filled cavities within the brain, which are called **ventricles** (VEN-tri-kuls = little cavities). There is one **lateral ventricle** in each hemisphere of the cerebrum. (Think of them as ventricles 1 and 2.) Anteriorly, the lateral ventricles are separated by a thin membrane, the **septum pellucidum** (SEP-tum pe-LOO-si-dum; *pellucid* = transparent). The **third ventricle** is a narrow slitlike cavity along the midline superior to the hypothalamus and between the right and left halves of the thalamus. The **fourth ventricle** lies between the brain stem and the cerebellum.

Functions of CSF

The CSF has three basic functions:

1. *Mechanical protection.* CSF serves as a shock-absorbing medium that protects the delicate tissues of the brain and spinal

Figure 14.3 Locations of ventricles within a "transparent" brain. One interventricular foramen on each side connects a lateral ventricle to the third ventricle, and the aqueduct of the midbrain connects the third ventricle to the fourth ventricle.

🔑 Ventricles are cavities within the brain that are filled with cerebrospinal fluid.

Right lateral view of brain

❓ **Which brain region is anterior to the fourth ventricle? Which is posterior to it?**

cord from jolts that would otherwise cause them to hit the bony walls of the cranial cavity and vertebral canal. The fluid also buoys the brain so that it "floats" in the cranial cavity.

2. **Homeostatic function.** The pH of the CSF affects pulmonary ventilation and cerebral blood flow, which is important in maintaining homeostatic controls for brain tissue. CSF also serves as a transport system for polypeptide hormones secreted by hypothalamic neurons that act at remote sites in the brain.

3. **Circulation.** CSF is a medium for minor exchange of nutrients and waste products between the blood and adjacent nervous tissue.

Formation of CSF in the Ventricles

The majority of CSF production is from the **choroid plexuses** (KŌ-royd = membranelike), networks of blood capillaries in the walls of the ventricles (Figure 14.4a). Ependymal cells joined by tight junctions cover the capillaries of the choroid plexuses. Selected substances (mostly water) from the blood plasma, which are filtered from the capillaries, are secreted by the ependymal cells to produce the cerebrospinal fluid. This secretory capacity is bidirectional and accounts for continuous production of CSF and transport of metabolites from the nervous tissue back to the blood.

Because of the tight junctions between ependymal cells, materials entering CSF from choroid capillaries cannot leak between these cells; instead, they must pass through the ependymal cells. This **blood–cerebrospinal fluid barrier** permits certain substances to enter the CSF but excludes others, protecting the brain and spinal cord from potentially harmful blood-borne substances. In contrast to the blood–brain barrier, which is formed mainly by tight junctions of brain capillary endothelial cells, the blood–cerebrospinal fluid barrier is formed by tight junctions of ependymal cells.

Circulation of CSF

The CSF formed in the choroid plexuses of each lateral ventricle flows into the third ventricle through two narrow, oval openings, the **interventricular foramina** (in′-ter-ven-TRIK-ū-lar) (singular is *foramen;* Figure 14.4b). More CSF is added by the choroid plexus in the roof of the third ventricle. The fluid then flows through the **aqueduct of the midbrain (cerebral aqueduct)** (AK-we-dukt), which passes through the midbrain, into the fourth ventricle. The choroid plexus of the fourth ventricle contributes more fluid. CSF enters the subarachnoid space through three openings in the roof of the fourth ventricle: a single **median aperture** (AP-er-chur) and paired **lateral apertures,** one on each side. CSF then circulates in the central canal of the spinal cord and in the subarachnoid space around the surface of the brain and spinal cord.

CSF is gradually reabsorbed into the blood through **arachnoid villi,** fingerlike extensions of the arachnoid that project into the dural venous sinuses, especially the **superior sagittal sinus** (see Figure 14.2). (A cluster of arachnoid villi is called an **arachnoid granulation.**) Normally, CSF is reabsorbed as rapidly as it is formed by the choroid plexuses, at a rate of about 20 mL/hr (480 mL/day). Because the rates of formation and reabsorption are the same, the pressure of CSF normally is constant. For the same reason, the volume of CSF remains constant. Figure 14.4d summarizes the production and flow of CSF.

mulates in the ventricles, the CSF pressure rises. Elevated CSF pressure causes a condition called **hydrocephalus** (hī'-drō-SEF-a-lus; *hydro-* = water; *-cephal-* = head). The abnormal accumulation of CSF may be due to an obstruction to CSF flow or an abnormal rate of CSF production and/or reabsorption. In a baby whose fontanels have not yet closed, the head bulges due to the increased pressure. If the condition persists, the fluid buildup compresses and damages the delicate nervous tissue. Hydrocephalus is relieved by draining the excess CSF. In one procedure, called *endoscopic third ventriculostomy (ETV),* a neurosurgeon makes a hole in the floor of the third ventricle and the CSF drains directly into the subarachnoid space. In adults, hydrocephalus may occur after head injury, meningitis, or subarachnoid hemorrhage. Because the adult skull bones are fused together, this condition can quickly become life-threatening and requires immediate intervention. •

CLINICAL CONNECTION | *Hydrocephalus*

Abnormalities in the brain—tumors, inflammation, or developmental malformations—can interfere with the circulation of CSF from the ventricles into the subarachnoid space. When excess CSF accu-

✔CHECKPOINT

4. What structures produce CSF, and where are they located?
5. What is the difference between the blood–brain barrier and the blood–cerebrospinal fluid barrier?

Figure 14.4 Pathways of circulating cerebrospinal fluid.

🔑 CSF is formed from blood plasma by ependymal cells that cover the choroid plexuses of the ventricles.

(a) Superior view of transverse section of brain showing choroid plexuses

FIGURE 14.4 CONTINUES ▶

⬤ **FIGURE 14.4 CONTINUED** ▶

POSTERIOR

ANTERIOR

CHOROID PLEXUS OF
LATERAL VENTRICLE

CHOROID PLEXUS OF
THIRD VENTRICLE

Cerebrum

Intermediate mass
of thalamus

Posterior
commissure

Great cerebral
vein

Straight sinus

AQUEDUCT OF THE
MIDBRAIN (CEREBRAL
AQUEDUCT)

Cerebellum

CHOROID PLEXUS OF
FOURTH VENTRICLE

MEDIAN APERTURE

Superior cerebral vein

ARACHNOID VILLUS

SUBARACHNOID SPACE

SUPERIOR SAGITTAL
SINUS

Corpus callosum

LATERAL VENTRICLE

INTERVENTRICULAR
FORAMEN

Anterior commissure

THIRD VENTRICLE

Hypothalamus

Cranial meninges:

Pia mater

Arachnoid mater

Dura mater

Midbrain

Pons

LATERAL APERTURE

FOURTH VENTRICLE

Medulla oblongata

Spinal cord

CENTRAL CANAL

SUBARACHNOID SPACE

Filum terminale

Sagittal
plane

View

Path of:

⟶ CSF

⟶ Venous blood

(b) Sagittal section of brain and spinal cord

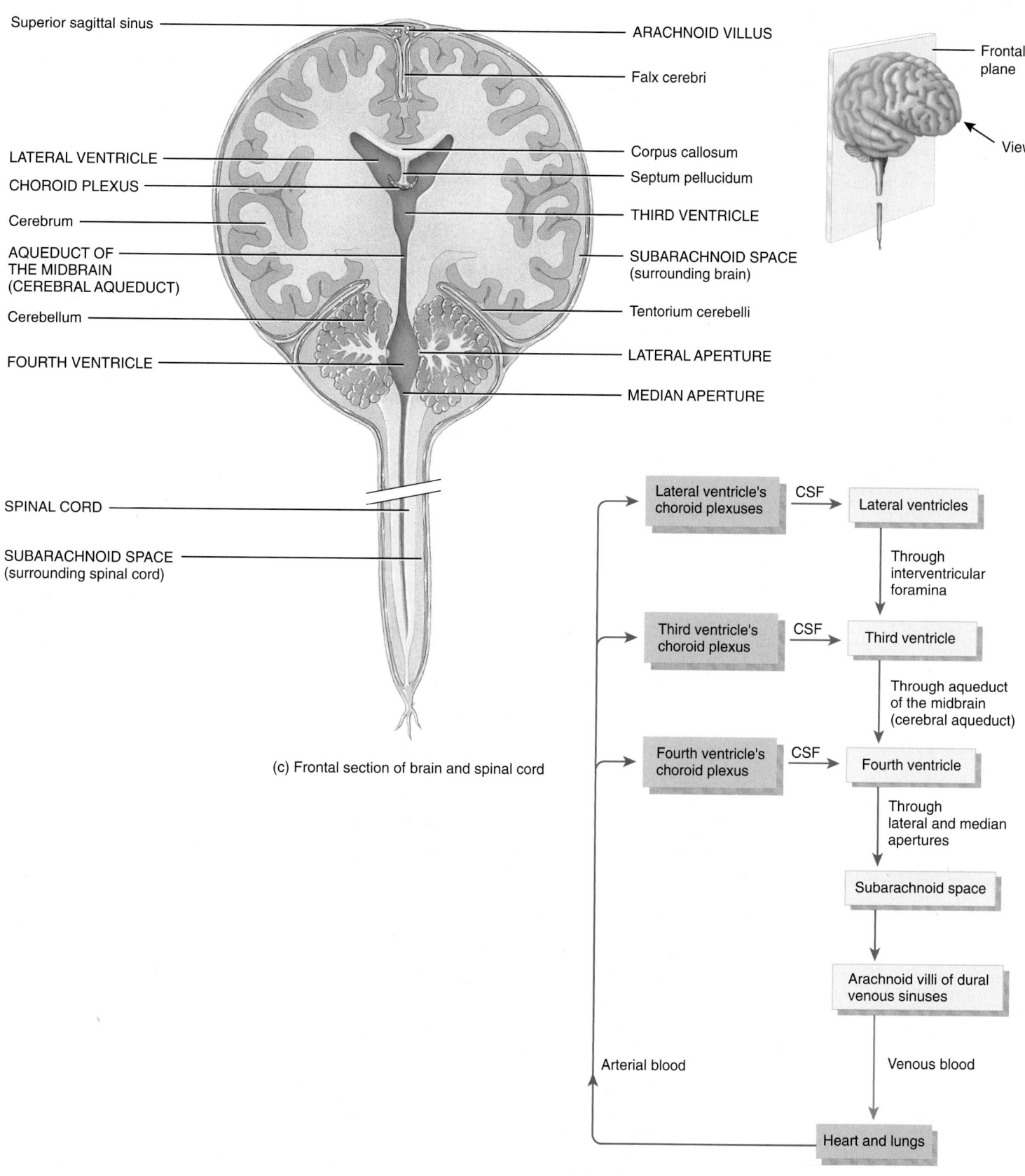

Superior sagittal sinus

ARACHNOID VILLUS

Falx cerebri

LATERAL VENTRICLE

CHOROID PLEXUS

Cerebrum

AQUEDUCT OF THE MIDBRAIN (CEREBRAL AQUEDUCT)

Cerebellum

FOURTH VENTRICLE

Corpus callosum

Septum pellucidum

THIRD VENTRICLE

SUBARACHNOID SPACE (surrounding brain)

Tentorium cerebelli

LATERAL APERTURE

MEDIAN APERTURE

SPINAL CORD

SUBARACHNOID SPACE (surrounding spinal cord)

Frontal plane

View

(c) Frontal section of brain and spinal cord

Lateral ventricle's choroid plexuses — CSF → Lateral ventricles

Through interventricular foramina

Third ventricle's choroid plexus — CSF → Third ventricle

Through aqueduct of the midbrain (cerebral aqueduct)

Fourth ventricle's choroid plexus — CSF → Fourth ventricle

Through lateral and median apertures

Subarachnoid space

Arachnoid villi of dural venous sinuses

Arterial blood

Venous blood

Heart and lungs

(d) Summary of the formation, circulation, and absorption of cerebrospinal fluid (CSF)

? Where is CSF reabsorbed?

14.3 THE BRAIN STEM AND RETICULAR FORMATION

◉ OBJECTIVE

• Describe the structures and functions of the brain stem and reticular formation.

The brain stem is the part of the brain between the spinal cord and the diencephalon. It consists of three structures: (1) medulla oblongata, (2) pons, and (3) midbrain. Extending through the brain stem is the reticular formation, a netlike region of interspersed gray and white matter.

Medulla Oblongata

The **medulla oblongata** (me-DOOL-la ob'-long-GA-ta), or more simply the **medulla,** is continuous with the superior part of the spinal cord; it forms the inferior part of the brain stem (Figure 14.5; see also Figure 14.1). The medulla begins at the foramen magnum and extends to the inferior border of the pons, a distance of about 3 cm (1.2 in.).

The medulla's white matter contains all sensory (ascending) tracts and motor (descending) tracts that extend between the spinal cord and other parts of the brain. Some of the white matter forms bulges on the anterior aspect of the medulla. These protru-

Figure 14.5 Medulla oblongata in relation to the rest of the brain stem.

🔑 The brain stem consists of the medulla oblongata, pons, and midbrain.

View

Olfactory bulb
Olfactory tract
Pituitary gland
Optic tract
Mammillary body
Cerebellar peduncles
Olive
Pyramids
Spinal nerve C1
Spinal cord

ANTERIOR

Cerebrum

CEREBRAL PEDUNCLE OF MIDBRAIN

PONS

MEDULLA OBLONGATA

Cerebellum

POSTERIOR

Inferior aspect of brain

❓ **What part of the brain stem contains the pyramids? The cerebral peduncles? Literally means "bridge"?**

sions, called the **pyramids** (Figure 14.6; see also Figure 14.5), are formed by the large corticospinal tracts that pass from the cerebrum to the spinal cord. The corticospinal tracts control voluntary movements of the limbs and trunk (see Figure 16.10). Just superior to the junction of the medulla with the spinal cord, 90% of the axons in the left pyramid cross to the right side, and 90% of the axons in the right pyramid cross to the left side. This crossing is called the **decussation of pyramids** (dē′-ku-SĀ-shun; *decuss* = crossing) and explains why each side of the brain controls voluntary movements on the opposite side of the body.

The medulla also contains several **nuclei.** (Recall that a nucleus is a collection of neuronal cell bodies within the CNS.) Some of these nuclei control vital body functions. Examples of nuclei in the medulla that regulate vital activities include the cardiovascular center and the medullary rhythmicity area. The **cardiovascular center** regulates the rate and force of the heartbeat and the diameter of blood vessels (see Figure 21.13). The **medullary rhythmicity area** (rith-MIS-i-tē) of the **respiratory center** adjusts the basic rhythm of breathing (see Figure 23.25).

Besides regulating heartbeat, blood vessel diameter, and the normal breathing rhythm, nuclei in the medulla also control reflexes for vomiting, swallowing, sneezing, coughing, and hiccupping. The **vomiting center** of the medulla causes **vomiting,** the forcible expulsion of the contents of the upper gastrointestinal (GI) tract through the mouth (see Section 24.8). The **deglutition center** (dē-gloo-TISH-un) of the medulla promotes **deglutition** (swallowing) of a mass of food that has moved from the oral cavity of the mouth into the pharynx (throat) (see Section 24.8). **Sneezing** involves spasmodic contraction of breathing muscles that forcefully expel air through the nose and mouth. **Coughing** involves a long-drawn and deep inhalation and then a strong exhalation that suddenly sends a blast of air through the upper respiratory passages. **Hiccupping** is caused by spasmodic contractions of the diaphragm (a muscle of breathing) that ultimately result in the production of a sharp sound on inhalation. Sneezing, coughing, and hiccupping are described in more detail in Table 23.1.

Just lateral to each pyramid is an oval-shaped swelling called an **olive** (see Figures 14.5, 14.6). Within the olive is the **inferior olivary nucleus,** which receives input from the cerebral cortex, red nucleus of the midbrain, and spinal cord. Neurons of the inferior olivary nucleus extend their axons into the cerebellum, where they regulate the activity of cerebellar neurons. By influencing cerebellar neuron activity, the inferior olivary nucleus provides instructions that the cerebellum uses to make adjustments to muscle activity as you learn new motor skills.

Nuclei associated with sensations of touch, pressure, vibration, and conscious proprioception are located in the posterior part of

Figure 14.6 Internal anatomy of the medulla oblongata.

The pyramids of the medulla contain the large motor tracts that run from the cerebrum to the spinal cord.

View

Transverse plane

Fourth ventricle

VAGUS NUCLEUS (dorsal motor)

HYPOGLOSSAL NUCLEUS

INFERIOR OLIVARY NUCLEUS

DECUSSATION OF PYRAMIDS

Lateral corticospinal tract axons

Anterior corticospinal tract axons

Choroid plexus

Vagus (X) nerve

OLIVE

Hypoglossal (XII) nerve

PYRAMIDS

Spinal nerve C1

Spinal cord

Transverse section and anterior surface of medulla oblongata

? **What does decussation mean? What is the functional consequence of decussation of the pyramids?**

the medulla. These nuclei are the right and left **gracile nucleus** (GRAS-il = slender) and **cuneate nucleus** (KŪ-nē-āt = wedge). Ascending sensory axons of the **gracile fasciculus** (fa-SIK-ū-lus) and the **cuneate fasciculus,** which are two tracts in the posterior columns of the spinal cord, form synapses in these nuclei (see Figure 16.5). Postsynaptic neurons then relay the sensory information to the thalamus on the opposite side of the brain. The axons ascend to the thalamus in a band of white matter called the **medial lemniscus** (lem-NIS-kus = ribbon), which extends through the medulla, pons, and midbrain (see Figure 14.7b). The tracts of the posterior columns and the axons of the medial lemniscus are collectively known as the **posterior column–medial lemniscus pathway.**

The medulla also contains nuclei that are components of sensory pathways for gustation (taste), audition (hearing), and equilibrium (balance). The **gustatory nucleus** (GUS-ta-tō′-rē) of the medulla is part of the gustatory pathway from the tongue to the brain; it receives gustatory input from the taste buds of the tongue (see Figure 17.3e). The **cochlear nuclei** (KOK-lē-ar) of the medulla are part of the auditory pathway from the inner ear to the brain; they receive auditory input from the cochlea of the inner ear (see Figure 17.23). The **vestibular nuclei** (ves-TIB-ū-lar) of the medulla and pons are components of the equilibrium pathway from the inner ear to the brain; they receive sensory information associated with equilibrium from proprioceptors in the vestibular apparatus of the inner ear (see Figure 17.26).

Finally, the medulla contains nuclei associated with the following five pairs of cranial nerves (see Figure 14.5):

1. **Vestibulocochlear (VIII) nerves.** Several nuclei in the medulla receive sensory input from and provide motor output to the cochlea of the internal ear via the vestibulocochlear nerves. These nerves convey impulses related to hearing.

2. **Glossopharyngeal (IX) nerves.** Nuclei in the medulla relay sensory and motor impulses related to taste, swallowing, and salivation via the glossopharyngeal nerves.

3. **Vagus (X) nerves.** Nuclei in the medulla receive sensory impulses from and provide motor impulses to the pharynx and larynx and many thoracic and abdominal viscera via the vagus nerves.

4. **Accessory (XI) nerves (cranial portion).** These fibers are actually part of the vagus (X) nerves. Nuclei in the medulla are the origin for nerve impulses that control swallowing via the vagus nerves (cranial portion of the accessory nerves).

5. **Hypoglossal (XII) nerves.** Nuclei in the medulla are the origin for nerve impulses that control tongue movements during speech and swallowing via the hypoglossal nerves.

CLINICAL CONNECTION | Injury of the Medulla

Given the vital activities controlled by the medulla, it is not surprising that **injury to the medulla** from a hard blow to the back of the head or upper neck such as falling back on ice can be fatal. Damage to the medullary rhythmicity area is particularly serious and can rapidly lead to death. Symptoms of nonfatal injury to the medulla may include cranial nerve malfunctions on the same side of the body as the injury, paralysis and loss of sensation on the opposite side of the body, and irregularities in breathing or heart rhythm. Alcohol overdose also suppresses the medullary rhythmicity area and may result in death. •

Pons

The **pons** (= bridge) lies directly superior to the medulla and anterior to the cerebellum and is about 2.5 cm (1 in.) long (see Figures 14.1, 14.5). Like the medulla, the pons consists of both nuclei and tracts. As its name implies, the pons is a bridge that connects parts of the brain with one another. These connections are provided by bundles of axons. Some axons of the pons connect the right and left sides of the cerebellum. Others are part of ascending sensory tracts and descending motor tracts.

The pons has two major structural components: a ventral region and a dorsal region. The ventral region of the pons forms a large synaptic relay station consisting of scattered gray centers called the **pontine nuclei** (PON-tīn). Entering and exiting these nuclei are numerous white matter tracts, each of which provides a connection between the cortex (outer layer) of a cerebral hemisphere and that of the opposite hemisphere of the cerebellum. This complex circuitry plays an essential role in coordinating and maximizing the efficiency of voluntary motor output throughout the body. The dorsal region of the pons is more like the other regions of the brain stem, the medulla and midbrain. It contains ascending and descending tracts along with the nuclei of cranial nerves.

Other nuclei located in the pons are the **pneumotaxic area** (noo-mō-TAK-sik) and the **apneustic area** (ap-NOO-stik), shown in Figure 23.24. Together with the medullary rhythmicity area, the pneumotaxic and apneustic areas help control breathing.

The pons also contains nuclei associated with the following four pairs of cranial nerves (see Figure 14.5):

1. **Trigeminal (V) nerves.** Nuclei in the pons receive sensory impulses for somatic sensations from the head and face and provide motor impulses that govern chewing via the trigeminal nerves.

2. **Abducens (VI) nerves.** Nuclei in the pons provide motor impulses that control eyeball movement via the abducens nerves.

3. **Facial (VII) nerves.** Nuclei in the pons receive sensory impulses for taste and provide motor impulses to regulate secretion of saliva and tears and contraction of muscles of facial expression via the facial nerves.

4. **Vestibulocochlear (VIII) nerves.** Nuclei in the pons receive sensory impulses from and provide motor impulses to the vestibular apparatus via the vestibulocochlear nerves. These nerves convey impulses related to balance and equilibrium.

Midbrain

The **midbrain** or **mesencephalon** extends from the pons to the diencephalon (see Figures 14.1, 14.5) and is about 2.5 cm (1 in.)

long. The aqueduct of the midbrain (cerebral aqueduct) passes through the midbrain, connecting the third ventricle above with the fourth ventricle below. Like the medulla and the pons, the midbrain contains both nuclei and tracts (Figure 14.7).

The anterior part of the midbrain contains paired bundles of axons known as the **cerebral peduncles** (pe-DUNK-kuls or PĒ-dung-kuls = little feet; see Figures 14.5, 14.7b). The cerebral peduncles consist of axons of the corticospinal, corticobulbar, and corticopontine tracts, which conduct nerve impulses from motor areas in the cerebral cortex to the spinal cord, medulla, and pons, respectively.

The posterior part of the midbrain, called the **tectum** (TEK-tum = roof), contains four rounded elevations (Figure 14.7a). The two superior elevations, nuclei known as the **superior colliculi** (ko-LIK-ū-lī = little hills; singular is *colliculus*), serve as reflex centers for certain visual activities. Through neural circuits from the retina of the eye to the superior colliculi to the extrinsic eye muscles, visual stimuli elicit eye movements for tracking moving images (such as a moving car) and scanning stationary images (as you are doing to read this sentence). The superior colliculi are also responsible for reflexes that govern movements of the head, eyes, and trunk in response to visual stimuli. The two inferior

elevations, the **inferior colliculi,** are part of the auditory pathway, relaying impulses from the receptors for hearing in the inner ear to the brain. These two nuclei are also reflex centers for the *startle reflex,* sudden movements of the head, eyes, and trunk that occur when you are surprised by a loud noise such as a gunshot.

The midbrain contains several other nuclei, including the left and right **substantia nigra** (sub-STAN-shē-a = substance; NĪ-gra = black), which are large and darkly pigmented (Figure 14.7b). Neurons that release dopamine, extending from the substantia nigra to the basal nuclei, help control subconscious muscle activities. Loss of these neurons is associated with Parkinson disease (see Disorders: Homeostatic Imbalances at the end of Chapter 16). Also present are the left and right **red nuclei,** which look reddish due to their rich blood supply and an iron-containing pigment in their neuronal cell bodies. Axons from the cerebellum and cerebral cortex form synapses in the red nuclei, which help control muscular movements.

Still other nuclei in the midbrain are associated with two pairs of cranial nerves (see Figure 14.5):

1. **Oculomotor (III) nerves.** Nuclei in the midbrain provide motor impulses that control movements of the eyeball, while

Figure 14.7 Midbrain.

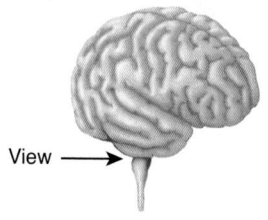 The midbrain connects the pons to the diencephalon.

(a) Posterior view of midbrain in relation to brain stem

FIGURE 14.7 CONTINUES ▶

⬤ **FIGURE 14.7 CONTINUED** ▶

View

Transverse plane

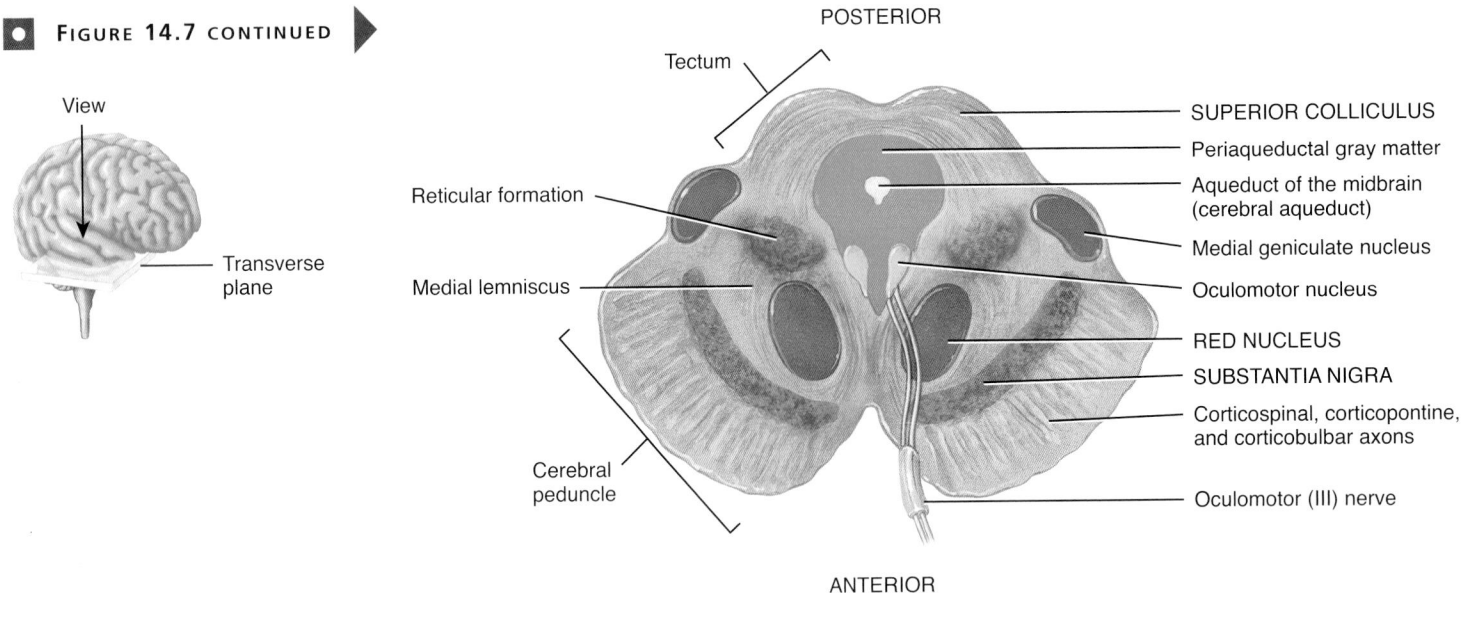

POSTERIOR

Tectum

SUPERIOR COLLICULUS

Periaqueductal gray matter

Aqueduct of the midbrain (cerebral aqueduct)

Medial geniculate nucleus

Oculomotor nucleus

RED NUCLEUS

SUBSTANTIA NIGRA

Corticospinal, corticopontine, and corticobulbar axons

Oculomotor (III) nerve

Reticular formation

Medial lemniscus

Cerebral peduncle

ANTERIOR

(b) Transverse section of midbrain

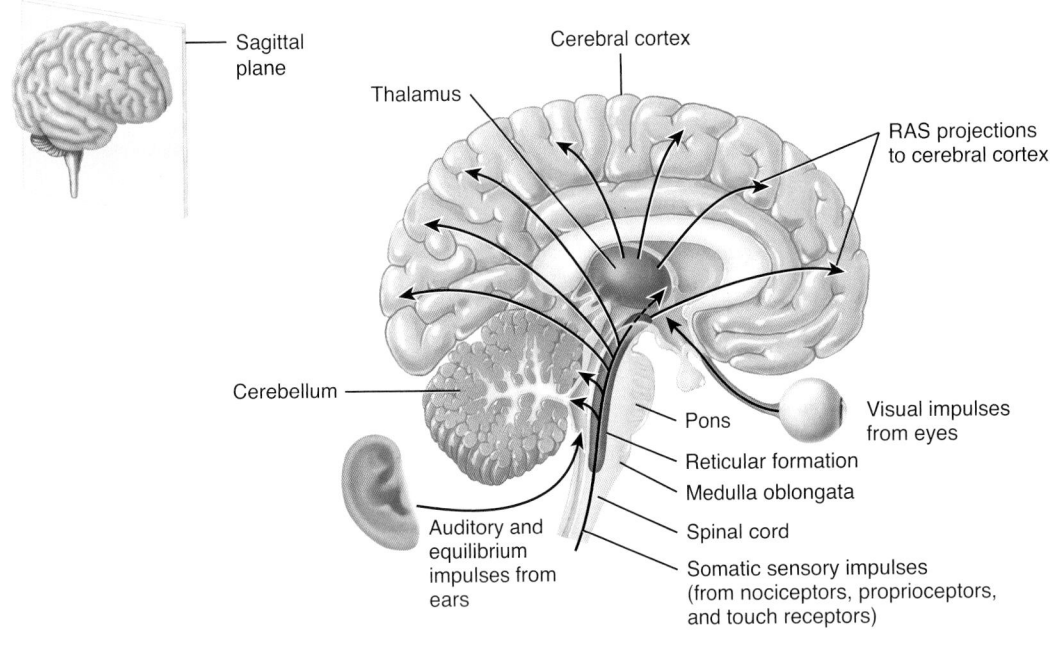

Sagittal plane

Cerebral cortex

Thalamus

RAS projections to cerebral cortex

Cerebellum

Visual impulses from eyes

Pons

Reticular formation

Medulla oblongata

Spinal cord

Auditory and equilibrium impulses from ears

Somatic sensory impulses (from nociceptors, proprioceptors, and touch receptors)

(c) Sagittal section through brain and spinal cord showing the reticular formation

❓ **What is the importance of the cerebral peduncles?**

accessory oculomotor nuclei provide motor control to the smooth muscles that regulate constriction of the pupil and changes in shape of the lens via the oculomotor nerves.

2. **Trochlear (IV) nerves.** Nuclei in the midbrain provide motor impulses that control movements of the eyeball via the trochlear nerves.

Reticular Formation

In addition to the well-defined nuclei already described, much of the brain stem consists of small clusters of neuronal cell bodies (gray matter) interspersed among small bundles of myelinated axons (white matter). The broad region where white matter and gray matter exhibit a netlike arrangement is known as the **reticular** formation (re-TIK-ū-lar; *ret-* = net; Figure 14.7c). It extends from the superior part of the spinal cord, throughout the brain stem, and into the inferior part of the diencephalon. Neurons within the reticular formation have both ascending (sensory) and descending (motor) functions.

The ascending portion of the reticular formation is called the **reticular activating system (RAS),** which consists of sensory axons that project to the cerebral cortex, both directly and through the thalamus. Many sensory stimuli can activate the ascending portion of the RAS. Among these are visual and auditory stimuli; mental activities; stimuli from pain, touch, and pressure receptors; and receptors in our limbs and head that keep us aware of the position of our body parts. Perhaps the most important function of the RAS is **consciousness,** a state of wakefulness in which an individual is fully alert, aware, and oriented. Visual and auditory stimuli and mental activities can stimulate the RAS to help maintain consciousness. The RAS is also active during **arousal,** or awakening from sleep. Another function of the RAS is to help maintain **attention** and *alertness.* The RAS also prevents **sensory overload** by filtering out insignificant information so that it does not reach consciousness. For example, while waiting in the hallway for your anatomy class to begin, you may be unaware of all the noise around you while reviewing your notes for class. Inactivation of the RAS produces **sleep,** a state of partial consciousness from which an individual can be aroused. Damage to the RAS, on the other hand, results in **coma,** a state of unconsciousness from which an individual cannot be aroused. In the lightest stages of coma, brain stem and spinal cord reflexes persist, but in the deepest states even those reflexes are lost, and if respiratory and cardiovascular controls are lost, the patient dies. Drugs such as melatonin affect the RAS by helping to induce sleep, and general anesthetics turn off consciousness via the RAS. The descending portion of the RAS has connections to the cerebellum and spinal cord and helps regulate **muscle tone,** the slight degree of involuntary contraction in normal resting skeletal muscles. This portion of the RAS also assists in the regulation of heart rate, blood pressure, and respiratory rate.

Even though the RAS receives input from the eyes, ears, and other sensory receptors, there is no input from receptors for the sense of smell; even strong odors may fail to cause arousal. People who die in house fires usually succumb to smoke inhalation without awakening. For this reason, all sleeping areas should have a nearby smoke detector that emits a loud alarm. A vibrating pillow or flashing light can serve the same purpose for those who are hearing impaired.

The functions of the brain stem are summarized in Table 14.2.

✔**CHECKPOINT**

6. Where are the medulla, pons, and midbrain located relative to one another?
7. What body functions are governed by nuclei in the brain stem?
8. List the functions of the reticular formation.

14.4 THE CEREBELLUM

⦿ OBJECTIVE

• Describe the structure and functions of the cerebellum.

The **cerebellum** (ser′-e-BEL-um), second only to the cerebrum in size, occupies the inferior and posterior aspects of the cranial cavity. Like the cerebrum, the cerebellum has a highly folded surface that greatly increases the surface area of its outer gray matter cortex, allowing for a greater number of neurons. The cerebellum accounts for about a tenth of the brain mass yet contains nearly half of the neurons in the brain. The cerebellum is posterior to the medulla and pons and inferior to the posterior portion of the cerebrum (see Figure 14.1). A deep groove known as the **transverse fissure,** along with the **tentorium cerebelli** (ten-TŌ-rē-um), which supports the posterior part of the cerebrum, separates the cerebellum from the cerebrum (see Figures 14.2b, 14.11b).

In superior or inferior views, the shape of the cerebellum resembles a butterfly. The central constricted area is the **vermis** (= worm), and the lateral "wings" or lobes are the **cerebellar hemispheres** (Figure 14.8a, b). Each hemisphere consists of lobes separated by deep and distinct fissures. The **anterior lobe** and **posterior lobe** govern subconscious aspects of skeletal muscle movements. The **flocculonodular lobe** (flok-ū-lō-NOD-ū-lar; *flocculo-* = wool-like tuft) on the inferior surface contributes to equilibrium and balance.

The superficial layer of the cerebellum, called the **cerebellar cortex,** consists of gray matter in a series of slender, parallel folds called **folia** (= leaves). Deep to the gray matter are tracts of white matter called **arbor vitae** (AR-bor VĪ-tē = tree of life) that resemble branches of a tree. Even deeper, within the white matter, are the **cerebellar nuclei,** regions of gray matter that give rise to axons carrying impulses from the cerebellum to other brain centers.

Three paired **cerebellar peduncles** (pe-DUNG-kuls) attach the cerebellum to the brain stem (see Figures 14.7a, 14.8b). These bundles of white matter consist of axons that conduct impulses between the cerebellum and other parts of the brain. The **superior cerebellar peduncles** contain axons that extend from the cerebellum to the red nuclei of the midbrain and to several nuclei of the thalamus. The **middle cerebellar peduncles** are the largest peduncles; their axons carry impulses for voluntary movements from the pontine nuclei (which receive input from motor areas of the cerebral cortex) into the cerebellum. The **inferior cerebellar peduncles** consist of (1) axons of the spinocerebellar tracts that carry sensory information into the cerebellum from proprioceptors in the trunk and limbs; (2) axons from the vestibular apparatus of the inner ear and from the vestibular nuclei of the medulla and pons that carry sensory information into the cerebellum from proprioceptors in the head; (3) axons from the inferior olivary nucleus of the medulla that enter the cerebellum and regulate the activity of cerebellar neurons; (4) axons that extend from the cerebellum to the vestibular nuclei of the medulla and pons; and (5) axons that extend from the cerebellum to the reticular formation.

Figure 14.8 Cerebellum.

The cerebellum coordinates skilled movements and regulates posture and balance.

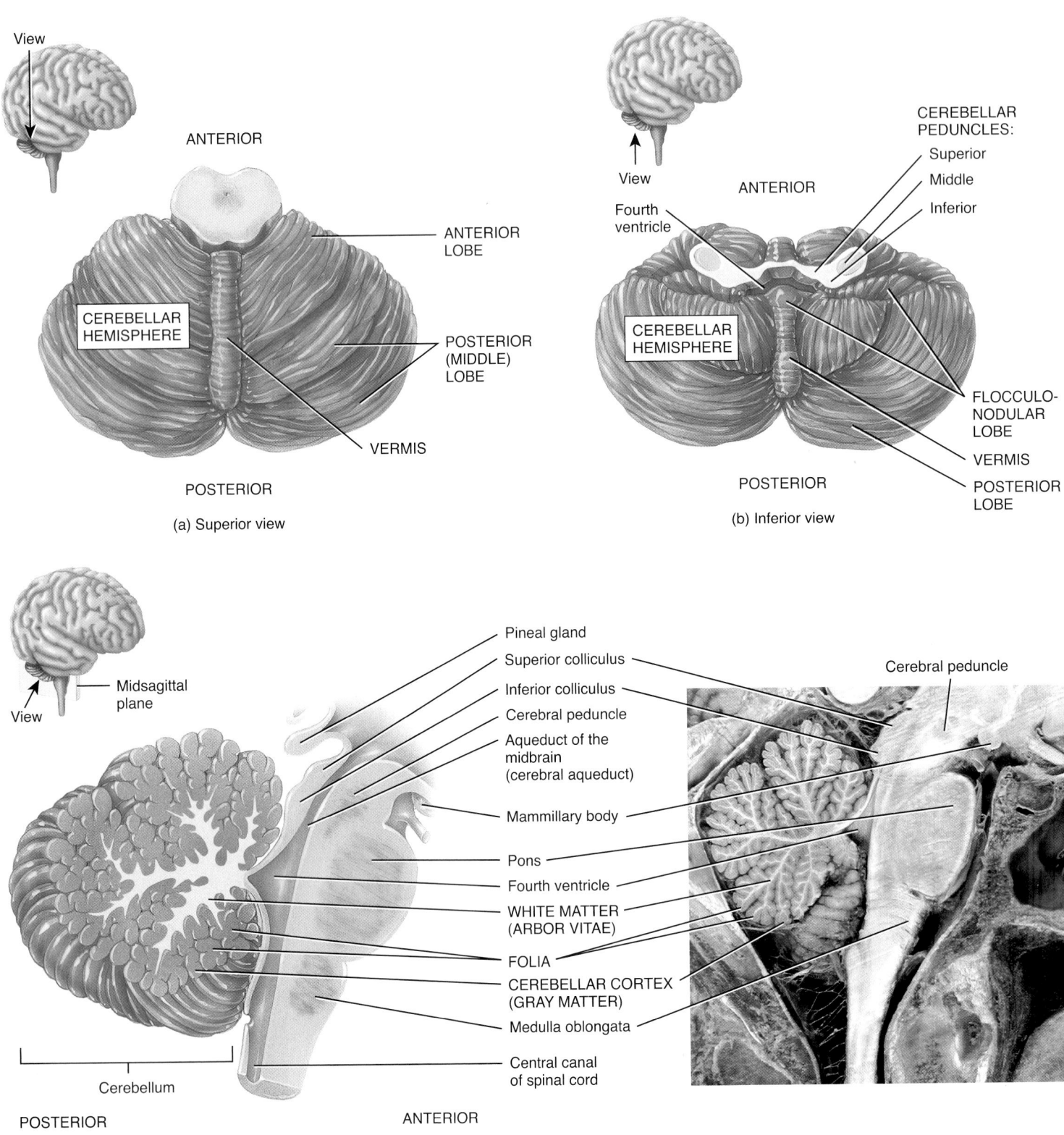

(a) Superior view

(b) Inferior view

(c) Midsagittal section of cerebellum and brain stem

(d) Midsagittal section

Which structures contain the axons that carry information into and out of the cerebellum?

The primary function of the cerebellum is to evaluate how well movements initiated by motor areas in the cerebrum are actually being carried out. When movements initiated by the cerebral motor areas are not being carried out correctly, the cerebellum detects the discrepancies. It then sends feedback signals to motor areas of the cerebral cortex, via its connections to the thalamus. The feedback signals help correct the errors, smooth the movements, and coordinate complex sequences of skeletal muscle contractions. Aside from this coordination of skilled movements, the cerebellum is the main brain region that regulates posture and balance. These aspects of cerebellar function make possible all skilled muscular activities, from catching a baseball to dancing to speaking. The presence of reciprocal connections between the cerebellum and association areas of the cerebral cortex suggests that the cerebellum may also have nonmotor functions such as cognition (acquisition of knowledge) and language processing. This view is supported by imaging studies using MRI and PET. Studies also suggest that the cerebellum may play a role in processing sensory information.

The functions of the cerebellum are summarized in Table 14.2.

CLINICAL CONNECTION | *Ataxia*

Damage to the cerebellum can result in a loss of ability to coordinate muscular movements, a condition called **ataxia** (a-TAK-sē-a; *a-* = without; *-taxia* = order). Blindfolded people with ataxia cannot touch the tip of their nose with a finger because they cannot coordinate movement with their sense of where a body part is located. Another sign of ataxia is a changed speech pattern due to uncoordinated speech muscles. Cerebellar damage may also result in staggering or abnormal walking movements. People who consume too much alcohol show signs of ataxia because alcohol inhibits activity of the cerebellum. Such individuals have difficulty in passing sobriety tests. Ataxia can also occur as a result of degenerative diseases (multiple sclerosis and Parkinson disease), trauma, brain tumors, and genetic factors, and as a side effect of medication prescribed for bipolar disorder. •

✔ CHECKPOINT

9. Describe the location and principal parts of the cerebellum.
10. Where do the axons of each of the three pairs of cerebellar peduncles begin and end? What are their functions?

14.5 THE DIENCEPHALON

◉ OBJECTIVE

• Describe the components and functions of the diencephalon (thalamus, hypothalamus, and epithalamus).

The **diencephalon** forms a central core of brain tissue just superior to the midbrain. It is almost completely surrounded by the cerebral hemispheres and contains numerous nuclei involved in a wide variety of sensory and motor processing between higher and lower brain centers. The diencephalon extends from the brain stem to the cerebrum and surrounds the third ventricle; it includes the thalamus, hypothalamus, and epithalamus. Projecting from the hypothalamus is the hypophysis, or pituitary gland. Portions of the diencephalon in the wall of the third ventricle are called circumventricular organs and will be discussed shortly. The optic tracts carrying neurons from the retina enter the diencephalon.

Thalamus

The **thalamus** (THAL-a-mus = inner chamber), which measures about 3 cm (1.2 in.) in length and makes up 80% of the diencephalon, consists of paired oval masses of gray matter organized into nuclei with interspersed tracts of white matter (Figure 14.9). A bridge of gray matter called the **intermediate mass (interthalamic adhesion)** joins the right and left halves of the thalamus in about 70% of human brains. A vertical Y-shaped sheet of white matter called the **internal medullary lamina** divides the gray matter of the right and left sides of the thalamus (Figure 14.9c). It consists of myelinated axons that enter and leave the various thalamic nuclei. Axons that connect the thalamus and cerebral cortex pass through the **internal capsule,** a thick band of white matter lateral to the thalamus (see Figure 14.13b).

The thalamus is the major relay station for most sensory impulses that reach the primary sensory areas of the cerebral cortex from the spinal cord and brain stem. In addition, the thalamus contributes to motor functions by transmitting information from the cerebellum and basal nuclei to the primary motor area of the cerebral cortex. The thalamus also relays nerve impulses between different areas of the cerebrum and plays a role in the maintenance of consciousness.

Based on their positions and functions, there are seven major groups of nuclei on each side of the thalamus (Figure 14.9c, d):

1. The **anterior nucleus** receives input from the hypothalamus and sends output to the limbic system (described in Section 14.6). It functions in emotions and memory.

2. The **medial nuclei** receive input from the limbic system and basal nuclei and send output to the cerebral cortex. They function in emotions, learning, memory, and cognition (thinking and knowing).

3. Nuclei in the **lateral group** receive input from the limbic system, superior colliculi, and cerebral cortex and send output to the cerebral cortex. The **lateral dorsal nucleus** functions in the expression of emotions. The **lateral posterior nucleus** and **pulvinar nucleus** help integrate sensory information.

4. Five nuclei are part of the **ventral group.** The **ventral anterior nucleus** receives input from the basal nuclei and sends output to motor areas of the cerebral cortex; it plays a role in movement control. The **ventral lateral nucleus** receives input from the cerebellum and basal nuclei and sends output to motor areas of the cerebral cortex; it also plays a role in movement control. The **ventral posterior nucleus** relays impulses for somatic sensations such as touch, pressure, vibration, itch,

Figure 14.9 Thalamus. Note the position of the thalamus in the lateral view (a) and in the medial view (b). The various thalamic nuclei shown in (c) and (d) are correlated by color to the cortical regions to which they project in (a) and (b).

The thalamus is the principal relay station for sensory impulses that reach the cerebral cortex from other parts of the brain and the spinal cord.

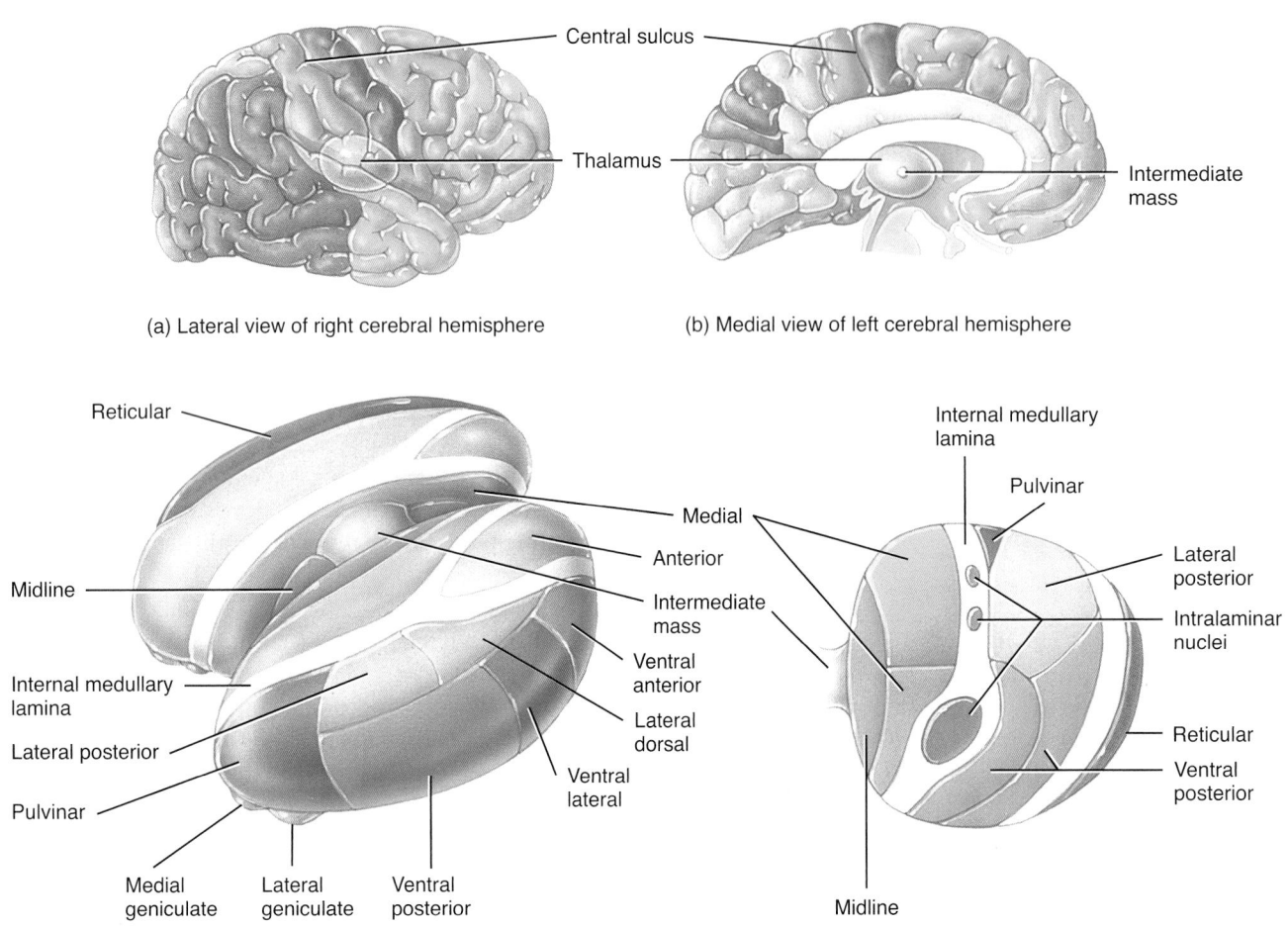

(a) Lateral view of right cerebral hemisphere

(b) Medial view of left cerebral hemisphere

(c) Superolateral view of thalamus showing locations of thalamic nuclei (reticular nucleus is shown on the left side only; all other nuclei are shown on the right side)

(d) Transverse section of right side of thalamus showing locations of thalamic nuclei

? **What structure usually connects the right and left halves of the thalamus?**

tickle, temperature, pain, and proprioception from the face and body to the cerebral cortex. The **lateral geniculate nucleus** (je-NIK-ū-lat = bent like a knee) relays visual impulses for sight from the retina to the primary visual area of the cerebral cortex. The **medial geniculate nucleus** relays auditory impulses for hearing from the ear to the primary auditory area of the cerebral cortex.

5. **Intralaminar nuclei** (in′-tra-LA-mi′-nar) lie within the internal medullary lamina and make connections with the reticular formation, cerebellum, basal nuclei, and wide areas of the cerebral cortex. They function in arousal (activation of the cerebral cortex from the brain stem reticular formation) and integration of sensory and motor information.

6. The **midline nucleus** forms a thin band adjacent to the third ventricle and has a presumed function in memory and olfaction.

7. The **reticular nucleus** surrounds the lateral aspect of the thalamus, next to the internal capsule. This nucleus monitors, filters, and integrates activities of other thalamic nuclei.

Hypothalamus

The **hypothalamus** (hī′-pō-THAL-a-mus; *hypo-* = under) is a small part of the diencephalon located inferior to the thalamus. It is composed of a dozen or so nuclei in four major regions:

1. The **mammillary region** (MAM-i-ler-ē; *mammill-* = nipple-shaped), adjacent to the midbrain, is the most posterior part of

the hypothalamus. It includes the *mammillary bodies* and *posterior hypothalamic nuclei* (Figure 14.10). The mammillary bodies are two small, rounded projections that serve as relay stations for reflexes related to the sense of smell.

2. The **tuberal region** (TOO-ber-al), the widest part of the hypothalamus, includes the *dorsomedial nucleus, ventromedial nucleus,* and *arcuate nucleus* (AR-kū-āt), plus the stalklike **infundibulum** (in-fun-DIB-ū-lum = funnel), which connects the pituitary gland to the hypothalamus (Figure 14.10). The **median eminence** is a slightly raised region that encircles the infundibulum (see Figure 14.7a).

3. The **supraoptic region** (*supra-* = above; *-optic* = eye) lies superior to the optic chiasm (point of crossing of optic nerves) and contains the *paraventricular nucleus, supraoptic nucleus, anterior hypothalamic nucleus,* and *suprachiasmatic nucleus* (soo′-pra-kī′-az-MA-tik) (Figure 14.10). Axons from the paraventricular and supraoptic nuclei form the hypothalamo-hypophyseal tract (hī′-pō-thal′-a-mō-hī-pō-FIZ-ē-al), which extends through the infundibulum to the posterior lobe of the pituitary (see Figure 18.8).

4. The **preoptic region** anterior to the supraoptic region is usually considered part of the hypothalamus because it participates with the hypothalamus in regulating certain autonomic activities. The preoptic region contains the *medial* and *lateral preoptic nuclei* (Figure 14.10).

The hypothalamus controls many body activities and is one of the major regulators of homeostasis. Sensory impulses related to both somatic and visceral senses arrive at the hypothalamus, as do impulses from receptors for vision, taste, and smell. Other receptors within the hypothalamus itself continually monitor osmotic pressure, blood glucose level, certain hormone concentrations, and the temperature of blood. The hypothalamus has several very important connections with the pituitary gland and produces a variety of hormones, which are described in more detail in Chapter 18. Some functions can be attributed to specific hypothalamic nuclei, but others are not so precisely localized. Important functions of the hypothalamus include the following:

• *Control of the ANS.* The hypothalamus controls and integrates activities of the autonomic nervous system, which regulates contraction of smooth muscle and cardiac muscle and the secretions of many glands. Axons extend from the hypothalamus to parasympathetic and sympathetic nuclei in the brain stem and spinal cord. Through the ANS, the hypothalamus is a major regulator of visceral activities, including regulation of heart rate, movement of food through the gastrointestinal tract, and contraction of the urinary bladder.

Figure 14.10 Hypothalamus. Selected portions of the hypothalamus and a three-dimensional representation of hypothalamic nuclei are shown (after Netter).

The hypothalamus controls many body activities and is an important regulator of homeostasis.

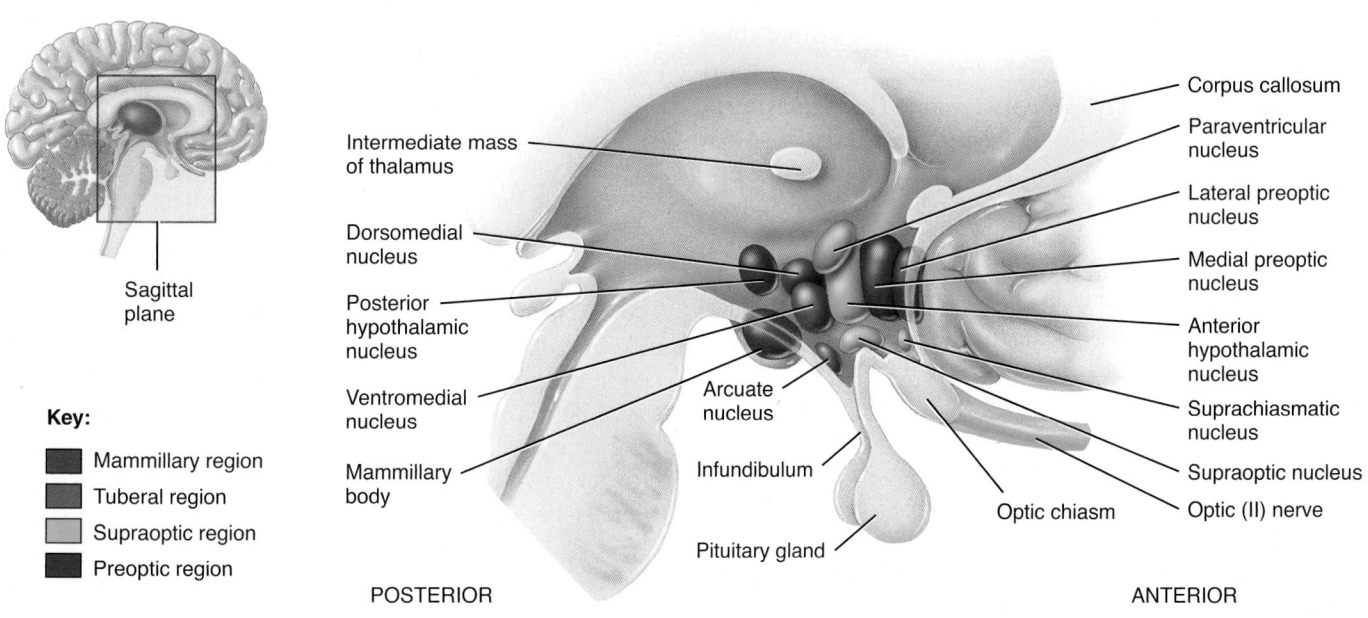

Sagittal section of brain showing hypothalamic nuclei

? **What are the four major regions of the hypothalamus, from posterior to anterior?**

- *Production of hormones.* The hypothalamus produces several hormones and has two types of important connections with the pituitary gland, an endocrine gland located inferior to the hypothalamus (see Figure 14.1). First, hypothalamic hormones known as *releasing hormones* and *inhibiting hormones* are released into capillary networks in the median eminence (see Figure 18.5). The bloodstream carries these hormones directly to the anterior lobe of the pituitary, where they stimulate or inhibit secretion of anterior pituitary hormones. Second, axons extend from the paraventricular and supraoptic nuclei through the infundibulum into the posterior lobe of the pituitary (see Figure 18.8). The cell bodies of these neurons make one of two hormones (*oxytocin* or *antidiuretic hormone*). Their axons transport the hormones to the posterior pituitary, where they are released.

- *Regulation of emotional and behavioral patterns.* Together with the limbic system (described shortly), the hypothalamus participates in expressions of rage, aggression, pain, and pleasure, and the behavioral patterns related to sexual arousal.

- *Regulation of eating and drinking.* The hypothalamus regulates food intake. It contains a **feeding center,** which promotes eating, and a **satiety center,** which causes a sensation of fullness and cessation of eating. The hypothalamus also contains a **thirst center.** When certain cells in the hypothalamus are stimulated by rising osmotic pressure of the extracellular fluid, they cause the sensation of thirst. The intake of water by drinking restores the osmotic pressure to normal, removing the stimulation and relieving the thirst.

- *Control of body temperature.* The hypothalamus also functions as the body's **thermostat.** If the temperature of blood flowing through the hypothalamus is above normal, the hypothalamus directs the autonomic nervous system to stimulate activities that promote heat loss. When blood temperature is below normal, by contrast, the hypothalamus generates impulses that promote heat production and retention.

- *Regulation of circadian rhythms and states of consciousness.* The suprachiasmatic nucleus of the hypothalamus serves as the body's internal biological clock because it establishes **circadian (daily) rhythms** (ser-KĀ-dē-an), patterns of biological activity (such as the sleep–wake cycle) that occur on a circadian schedule (cycle of about 24 hours). This nucleus receives input from the eyes (retina) and sends output to other hypothalamic nuclei, the reticular formation, and the pineal gland.

Epithalamus

The **epithalamus** (ep'-i-THAL-a-mus; *epi-* = above), a small region superior and posterior to the thalamus, consists of the pineal gland and habenular nuclei. The **pineal gland** (PĪN-ē-al = pineconelike) is about the size of a small pea and protrudes from the posterior midline of the third ventricle (see Figure 14.1). The pineal gland is part of the endocrine system because it secretes the hormone **melatonin.** As more melatonin is liberated during darkness than in light, this hormone is thought to promote sleepiness.

When taken orally, melatonin also appears to contribute to the setting of the body's biological clock by inducing sleep and helping the body to adjust to jet lag. The **habenular nuclei** (ha-BEN-ū-lar), shown in Figure 14.7a, are involved in olfaction, especially emotional responses to odors such as a loved one's cologne or Mom's chocolate chip cookies baking in the oven.

The functions of the three parts of the diencephalon are summarized in Table 14.2.

Circumventricular Organs

Parts of the diencephalon, called **circumventricular** (ser'-kum-ven-TRIK-ū-lar) **organs (CVOs)** because they lie in the wall of the third ventricle, can monitor chemical changes in the blood because they lack a blood–brain barrier. CVOs include part of the hypothalamus, the pineal gland, the pituitary gland, and a few other nearby structures. Functionally, these regions coordinate homeostatic activities of the endocrine and nervous systems, such as the regulation of blood pressure, fluid balance, hunger, and thirst. CVOs are also thought to be the sites of entry into the brain of HIV, the virus that causes AIDS. Once in the brain, HIV may cause dementia (irreversible deterioration of mental state) and other neurological disorders.

✔CHECKPOINT

11. Why is the thalamus considered a "relay station" in the brain?
12. Why is the hypothalamus considered part of both the nervous system and the endocrine system?
13. What are the functions of the epithalamus?
14. Define a circumventricular organ.

14.6 THE CEREBRUM

◉ OBJECTIVES

- Describe the cortex, gyri, fissures, and sulci of the cerebrum.
- List and locate the lobes of the cerebrum.
- Describe the tracts that compose the cerebral white matter.
- Describe the nuclei that compose the basal nuclei.
- List the structures and describe the functions of the limbic system.

The **cerebrum** (se-RĒ-brum) is the "seat of intelligence." It provides us with the ability to read, write, and speak; to make calculations and compose music; and to remember the past, plan for the future, and imagine things that have never existed before. The cerebrum consists of an outer cerebral cortex, an internal region of cerebral white matter, and gray matter nuclei deep within the white matter.

Cerebral Cortex

The **cerebral cortex** (*cortex* = rind or bark) is a region of gray matter that forms the outer rim of the cerebrum (Figure 14.11a). Although only 2–4 mm (0.08–0.16 in.) thick, the cerebral cortex contains billions of neurons arranged in layers. During embryonic development, when brain size increases rapidly, the gray matter of the cortex enlarges much faster than the deeper white matter. As a result, the cortical region rolls and folds on itself. The folds are called **gyri** (JĪ-rī = circles; singular is *gyrus*) or **convolutions** (kon′-vō-LOO-shuns) (Figure 14.11a, b). The deepest grooves between folds are known as **fissures;** the shallower grooves between folds are termed **sulci** (SUL-sī = grooves; singular is *sulcus*). The most prominent fissure, the **longitudinal fissure,** separates the cerebrum into right and left halves called **cerebral**

Figure 14.11 Cerebrum. Because the insula cannot be seen externally, it has been projected to the surface in (b).

The cerebrum is the "seat of intelligence"; it provides us with the ability to read, write, and speak; to make calculations and compose music; to remember the past and plan for the future; and to create.

ANTERIOR

Frontal lobe

Longitudinal fissure

Precentral gyrus

Central sulcus

Postcentral gyrus

Parietal lobe

Occipital lobe

Left hemisphere

Right hemisphere

POSTERIOR

(a) Superior view

Gyrus

Sulcus

Cerebral cortex

Cerebral white matter

Fissure

Details of a gyrus, sulcus, and fissure

Postcentral gyrus

Parietal lobe

Parieto-occipital sulcus

Occipital lobe

Transverse fissure

Cerebellum

Central sulcus

Precentral gyrus

Frontal lobe

Insula (projected to surface)

Lateral cerebral sulcus

Temporal lobe

(b) Right lateral view

FIGURE 14.11 CONTINUES ▶

FIGURE 14.11 CONTINUED ▶

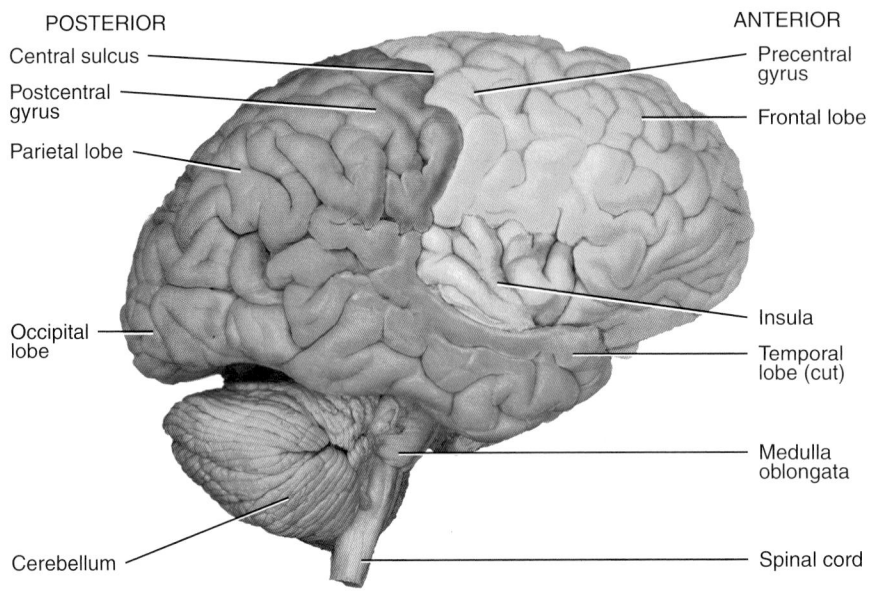

POSTERIOR · ANTERIOR

Central sulcus
Postcentral gyrus
Parietal lobe
Occipital lobe
Cerebellum

Precentral gyrus
Frontal lobe
Insula
Temporal lobe (cut)
Medulla oblongata
Spinal cord

(c) Right lateral view with temporal lobe cut away

During development, does the gray matter or the white matter enlarge more rapidly? What are the brain folds, shallow grooves, and deep grooves called?

hemispheres. Within the longitudinal fissure between the cerebral hemispheres is the falx cerebri. The cerebral hemispheres are connected internally by the **corpus callosum** (kal-LŌ-sum; *corpus* = body; *callosum* = hard), a broad band of white matter containing axons that extend between the hemispheres (Figure 14.12).

Figure 14.12 Organization of white matter tracts of the left cerebral hemisphere.

Association tracts, commissural tracts, and projection tracts form white matter tracts in the cerebral hemispheres.

Midsagittal plane

View

ASSOCIATION TRACTS
Septum pellucidum
Mammillary body
POSTERIOR

Cerebral cortex
COMMISSURAL AND PROJECTION TRACTS
COMMISSURAL TRACTS:
CORPUS CALLOSUM
ANTERIOR COMMISSURE
ANTERIOR

Medial view of tracts revealed by removing gray matter from a midsagittal section

Which tracts carry impulses between gyri of the same hemisphere? Between gyri in opposite hemispheres? Between the cerebrum and thalamus, brain stem, and spinal cord?

Lobes of the Cerebrum

Each cerebral hemisphere can be further subdivided into several lobes. The lobes are named after the bones that cover them: frontal, parietal, temporal, and occipital lobes (see Figure 14.11a, b). The **central sulcus** (SUL-kus) separates the **frontal lobe** from the **parietal lobe.** A major gyrus, the **precentral gyrus**—located immediately anterior to the central sulcus—contains the primary motor area of the cerebral cortex. Another major gyrus, the **postcentral gyrus,** which is located immediately posterior to the central sulcus, contains the primary somatosensory area of the cerebral cortex. The **lateral cerebral sulcus (fissure)** separates the **frontal lobe** from the **temporal lobe.** The **parieto-occipital sulcus** separates the **parietal lobe** from the **occipital lobe.** A fifth part of the cerebrum, the **insula,** cannot be seen at the surface of the brain because it lies within the lateral cerebral sulcus, deep to the parietal, frontal, and temporal lobes (Figure 14.11b).

Cerebral White Matter

The **cerebral white matter** consists primarily of myelinated axons in three types of tracts (Figure 14.12):

1. **Association tracts** contain axons that conduct nerve impulses between gyri in the same hemisphere.

2. **Commissural tracts** (kom′-i-SYUR-al) contain axons that conduct nerve impulses from gyri in one cerebral hemisphere to corresponding gyri in the other cerebral hemisphere. Three important groups of commissural tracts are the **corpus callosum** (the largest fiber bundle in the brain, containing about 300 million fibers), **anterior commissure,** and **posterior commissure.**

3. **Projection tracts** contain axons that conduct nerve impulses from the cerebrum to lower parts of the CNS (thalamus, brain stem, or spinal cord) or from lower parts of the CNS to the cerebrum. An example is the **internal capsule,** a thick band of white matter that contains both ascending and descending axons (see Figure 14.13).

Basal Nuclei

Deep within each cerebral hemisphere are three nuclei (masses of gray matter) that are collectively termed the **basal nuclei** (Figure 14.13). (Historically, these nuclei have been called the basal *ganglia.* However, this is a misnomer because a *ganglion* is an aggregate of neuronal cell bodies in the peripheral nervous system. While both terms still appear in the literature, we use *nuclei,* as this is the correct term as determined by the *Terminologia Anatomica,* the final say on correct anatomical terminology.)

Two of the basal nuclei are side-by-side, just lateral to the thalamus. They are the **globus pallidus** (GLŌ-bus PAL-i-dus; *globus* = ball; *pallidus* = pale), which is closer to the thalamus, and the **putamen** (pū-TĀ-men = shell), which is closer to the cerebral cortex. Together, the globus pallidus and putamen are referred to as the **lentiform nucleus** (LEN-ti-form = shaped like a lens). The third of the basal nuclei is the **caudate nucleus** (KAW-dāt;

caud- = tail), which has a large "head" connected to a smaller "tail" by a long comma-shaped "body." Together, the lentiform and caudate nuclei are known as the **corpus striatum** (strī-Ā-tum; *corpus* = body; *striatum* = striated). The term corpus striatum refers to the striated (striped) appearance of the internal capsule as it passes among the basal nuclei. Nearby structures that are functionally linked to the basal nuclei are the *substantia nigra* of the midbrain and the *subthalamic nuclei* of the diencephalon (see Figures 14.7b, 14.13b). Axons from the substantia nigra terminate in the caudate nucleus and putamen. The subthalamic nuclei interconnect with the globus pallidus.

The **claustrum** (KLAWS-trum) is a thin sheet of gray matter situated lateral to the putamen. It is considered by some to be a subdivision of the basal nuclei. The function of the claustrum in humans has not been clearly defined, but it may be involved in visual attention.

The basal nuclei receive input from the cerebral cortex and provide output to motor parts of the cortex via the medial and ventral group nuclei of the thalamus. In addition, the basal nuclei have extensive connections with one another. A major function of the basal nuclei is to help regulate initiation and termination of movements. Activity of neurons in the putamen precedes or anticipates body movements; activity of neurons in the caudate nucleus occurs prior to eye movements. The globus pallidus helps regulate the muscle tone required for specific body movements. The basal nuclei also control subconscious contractions of skeletal muscles. Examples include automatic arm swings while walking and true laughter in response to a joke (not the kind you consciously initiate to humor your A&P instructor).

In addition to influencing motor functions, the basal nuclei have other roles. They help initiate and terminate some cognitive processes, such as attention, memory, and planning, and may act with the limbic system to regulate emotional behaviors. Disorders such as Parkinson disease, obsessive–compulsive disorder, schizophrenia, and chronic anxiety are thought to involve dysfunction of circuits between the basal nuclei and the limbic system and are described in more detail in Chapter 16.

The Limbic System

Encircling the upper part of the brain stem and the corpus callosum is a ring of structures on the inner border of the cerebrum and floor of the diencephalon that constitutes the **limbic system** (*limbic* = border). The main components of the limbic system are as follows (Figure 14.14):

- The so-called **limbic lobe** is a rim of cerebral cortex on the medial surface of each hemisphere. It includes the **cingulate gyrus** (SIN-gyu-lat; *cingul-* = belt), which lies above the corpus callosum, and the **parahippocampal gyrus** (par′-a-hip-ō-KAM-pal), which is in the temporal lobe below. The **hippocampus** (hip′-ō-KAM-pus = seahorse) is a portion of the parahippocampal gyrus that extends into the floor of the lateral ventricle.

- The **dentate gyrus** (*dentate* = toothed) lies between the hippocampus and parahippocampal gyrus.

Figure 14.13 Basal nuclei. In (a) the basal nuclei have been projected to the surface; in both (a) and (b) they are shown in purple.

The basal nuclei help initiate and terminate movements, suppress unwanted movements, and regulate muscle tone.

(a) Lateral view of right side of brain

(b) Anterior view of frontal section

Where are the basal nuclei located relative to the thalamus?

- The **amygdala** (a-MIG-da-la; *amygda-* = almond-shaped) is composed of several groups of neurons located close to the tail of the caudate nucleus.

- The **septal nuclei** are located within the septal area formed by the regions under the corpus callosum and the paraterminal gyrus (a cerebral gyrus).

- The **mammillary bodies** of the hypothalamus are two round masses close to the midline near the cerebral peduncles.

- Two nuclei of the thalamus, the **anterior nucleus** and the **medial nucleus,** participate in limbic circuits (see Figure 14.9c, d).

- The **olfactory bulbs** are flattened bodies of the olfactory pathway that rest on the cribriform plate.

- The **fornix, stria terminalis, stria medullaris, medial forebrain bundle,** and **mammillothalamic tract** (mam-i-lō-tha-LAM-ik) are linked by bundles of interconnecting myelinated axons.

The limbic system is sometimes called the "emotional brain" because it plays a primary role in a range of emotions, including pain, pleasure, docility, affection, and anger. It also is involved in olfaction (smell) and memory. Experiments have shown that when different areas of animals' limbic systems are stimulated, the animals' reactions indicate that they are experiencing intense pain or extreme pleasure. Stimulation of other limbic system areas in animals produces tameness and signs of affection.

Figure 14.14 Components of the limbic system (shaded green) and surrounding structures.

The limbic system governs emotional aspects of behavior.

Sagittal plane

View

Fornix

Stria medullaris

Stria terminalis

Hippocampus (in temporal lobe)

Dentate gyrus

Anterior nucleus of thalamus

Mammillothalamic tract

Corpus callosum

Cingulate gyrus (in frontal lobe)

Anterior commissure

Septal nuclei

Mammillary body in hypothalamus

Olfactory bulb

Amygdala

Parahippocampal gyrus (in temporal lobe)

POSTERIOR

Sagittal section

ANTERIOR

? **Which part of the limbic system functions with the cerebrum in memory?**

Stimulation of a cat's amygdala or certain nuclei of the hypothalamus produces a behavioral pattern called rage—the cat extends its claws, raises its tail, opens its eyes wide, hisses, and spits. By contrast, removal of the amygdala produces an animal that lacks fear and aggression. Likewise, a person whose amygdala is damaged fails to recognize fearful expressions in others or to express fear in situations where this emotion would normally be appropriate, for example, while being attacked by an animal.

Together with parts of the cerebrum, the limbic system also functions in memory; damage to the limbic system causes memory impairment. One portion of the limbic system, the hippocampus, is seemingly unique among structures of the central nervous system—it has cells reported to be capable of mitosis. Thus, the portion of the brain that is responsible for some aspects of memory may develop new neurons, even in the elderly.

The functions of the cerebrum are summarized in Table 14.2.

CLINICAL CONNECTION | Brain Injuries

Brain injuries are commonly associated with head trauma and result in part from displacement and distortion of neural tissue at the moment of impact. Additional tissue damage may occur when normal blood flow is restored after a period of ischemia (reduced blood flow). The sudden increase in oxygen level produces large numbers of oxygen free radicals (charged oxygen molecules with an unpaired electron).

Brain cells recovering from the effects of a stroke or cardiac arrest also release free radicals. Free radicals cause damage by disrupting cellular DNA and enzymes and by altering plasma membrane permeability. Brain injuries can also result from hypoxia (cellular oxygen deficiency).

Various degrees of brain injury are described by specific terms. A **concussion** (kon-KUSH-un) is an injury characterized by an abrupt, but temporary, loss of consciousness (from seconds to hours), disturbances of vision, and problems with equilibrium. It is caused by a blow to the head or the sudden stopping of a moving head (as in an automobile accident) and is the most common brain injury. A concussion produces no obvious bruising of the brain. Signs of a concussion are headache, drowsiness, nausea and/or vomiting, lack of concentration, confusion, or post-traumatic amnesia (memory loss).

A brain **contusion** (kon-TOO-zhun) is bruising due to trauma and includes the leakage of blood from microscopic vessels. It is usually associated with a concussion. In a contusion, the pia mater may be torn, allowing blood to enter the subarachnoid space. The area most commonly affected is the frontal lobe. A contusion usually results in an immediate loss of consciousness (generally lasting no longer than 5 minutes), loss of reflexes, transient cessation of respiration, and decreased blood pressure. Vital signs typically stabilize in a few seconds.

A **laceration** (las-er-Ā-shun) is a tear of the brain, usually from a skull fracture or a gunshot wound. A laceration results in rupture of large blood vessels, with bleeding into the brain and subarachnoid space. Consequences include cerebral hematoma (localized pool of blood, usually clotted, that swells against the brain tissue), edema, and increased intracranial pressure. If the blood clot is small enough, it may pose no major threat and may be absorbed. If the blood clot is large, it may require surgical removal. Swelling infringes on the limited space that the brain occupies in the cranial cavity. Swelling causes excruciating headaches. Brain tissue can also undergo *necrosis* (cellular death) due to the swelling; if the swelling is severe enough, the brain can herniate through the foramen magnum, resulting in death. •

✔ CHECKPOINT

15. List and locate the lobes of the cerebrum. How are they separated from one another? What is the insula?
16. Distinguish between the precentral gyrus and the postcentral gyrus.
17. Describe the organization of cerebral white matter and indicate the function of each major group of fibers.
18. List the basal nuclei. What are the functions of the basal nuclei?
19. Define the limbic system and list several of its functions.

14.7 FUNCTIONAL ORGANIZATION OF THE CEREBRAL CORTEX

⦿ OBJECTIVES

• Describe the locations and functions of the sensory, association, and motor areas of the cerebral cortex.
• Explain the significance of hemispheric lateralization.
• Define brain waves and indicate their significance.

Specific types of sensory, motor, and integrative signals are processed in certain regions of the cerebral cortex (Figure 14.15). Generally, **sensory areas** receive sensory information and are involved in **perception,** the conscious awareness of a sensation; **motor areas** control the execution of voluntary movements; and **association areas** deal with more complex integrative functions such as memory, emotions, reasoning, will, judgment, personality traits, and intelligence. In this section we will also discuss hemispheric lateralization and brain waves.

Figure 14.15 Functional areas of the cerebrum. Broca's speech area and Wernicke's area are in the left cerebral hemisphere of most people; they are shown here to indicate their relative locations. The numbers, still used today, are from K. Brodmann's map of the cerebral cortex, first published in 1909.

Particular areas of the cerebral cortex process sensory, motor, and integrative signals.

Lateral view of right cerebral hemisphere

❓ What area(s) of the cerebrum integrate(s) interpretation of visual, auditory, and somatic sensations? Translates thoughts into speech? Controls skilled muscular movements? Interprets sensations related to taste? Interprets pitch and rhythm? Interprets shape, color, and movement of objects? Controls voluntary scanning movements of the eyes?

Sensory Areas

Sensory impulses arrive mainly in the posterior half of both cerebral hemispheres, in regions behind the central sulci. In the cerebral cortex, primary sensory areas receive sensory information that has been relayed from peripheral sensory receptors through lower regions of the brain. Sensory association areas often are adjacent to the primary areas. They usually receive input both from the primary areas and from other brain regions. Sensory association areas integrate sensory experiences to generate meaningful patterns of recognition and awareness. For example, a person with damage in the *primary* visual area would be blind in at least part of his visual field, but a person with damage to a visual *association* area might see normally yet be unable to recognize ordinary objects such as a lamp or a toothbrush just by looking at them.

The following are some important sensory areas (Figure 14.15; the significance of the numbers in parentheses is explained in the figure caption):

- The **primary somatosensory area** (areas 1, 2, and 3) is located directly posterior to the central sulcus of each cerebral hemisphere in the postcentral gyrus of each parietal lobe. It extends from the lateral cerebral sulcus, along the lateral surface of the parietal lobe to the longitudinal fissure, and then along the medial surface of the parietal lobe within the longitudinal fissure. The primary somatosensory area receives nerve impulses for touch, pressure, vibration, itch, tickle, temperature (coldness and warmth), pain, and proprioception (joint and muscle position) and is involved in the perception of these somatic sensations. A "map" of the entire body is present in the primary somatosensory area: Each point within the area receives impulses from a specific part of the body (see Figure 16.8a). The size of the cortical area receiving impulses from a particular part of the body depends on the number of receptors present there rather than on the size of the body part. For example, a larger region of the somatosensory area receives impulses from the lips and fingertips than from the thorax or hip. This distorted somatic sensory map of the body is known as the **sensory homunculus** (*homunculus* = little man). The primary somatosensory area allows you to pinpoint where somatic sensations originate, so that you know exactly where on your body to swat that mosquito.
- The **primary visual area** (area 17), located at the posterior tip of the occipital lobe mainly on the medial surface (next to the longitudinal fissure), receives visual information and is involved in visual perception.
- The **primary auditory area** (areas 41 and 42), located in the superior part of the temporal lobe near the lateral cerebral sulcus, receives information for sound and is involved in auditory perception.
- The **primary gustatory area** (area 43), located at the base of the postcentral gyrus superior to the lateral cerebral sulcus in the parietal cortex, receives impulses for taste and is involved in gustatory perception and taste discrimination.
- The **primary olfactory area** (area 28), located in the temporal lobe on the medial aspect (and thus not visible in Figure 14.15), receives impulses for smell and is involved in olfactory perception.

Motor Areas

Motor output from the cerebral cortex flows mainly from the anterior part of each hemisphere. Among the most important motor areas are the following (Figure 14.15):

- The **primary motor area** (area 4) is located in the precentral gyrus of the frontal lobe. As is true for the primary somatosensory area, a "map" of the entire body is present in the primary motor area: Each region within the area controls voluntary contractions of specific muscles or groups of muscles (see Figure 16.8b). Electrical stimulation of any point in the primary motor area causes contraction of specific skeletal muscle fibers on the opposite side of the body. Different muscles are represented unequally in the primary motor area. More cortical area is devoted to those muscles involved in skilled, complex, or delicate movement. For instance, the cortical region devoted to muscles that move the fingers is much larger than the region for muscles that move the toes. This distorted muscle map of the body is called the **motor homunculus.**
- **Broca's speech area** (BRŌ-kaz) (areas 44 and 45) is located in the frontal lobe close to the lateral cerebral sulcus. Speaking and understanding language are complex activities that involve several sensory, association, and motor areas of the cortex. In about 97% of the population, these language areas are localized in the *left* hemisphere. The planning and production of speech occur in the *left* frontal lobe in most people. From Broca's speech area, nerve impulses pass to the premotor regions that control the muscles of the larynx, pharynx, and mouth. The impulses from the premotor area result in specific, coordinated muscle contractions. Simultaneously, impulses propagate from Broca's speech area to the primary motor area. From here, impulses also control the breathing muscles to regulate the proper flow of air past the vocal cords. The coordinated contractions of your speech and breathing muscles enable you to speak your thoughts. People who suffer a cerebrovascular accident (CVA) or stroke in this area can still have clear thoughts but are unable to form words, a phenomenon referred to as *nonfluent aphasia*; see the next Clinical Connection.

Association Areas

The association areas of the cerebrum consist of large areas of the occipital, parietal, and temporal lobes and of the frontal lobes anterior to the motor areas. Association areas are connected with one another by association tracts and include the following (Figure 14.15):

- The **somatosensory association area** (areas 5 and 7) is just posterior to and receives input from the primary somatosensory area, as well as from the thalamus and other parts of the brain. This area permits you to determine the exact shape and texture of an object by feeling it, to determine the orientation of one object with respect to another as they are felt, and to sense the relationship of one body part to another. Another role of the somatosensory association area is the storage of memories of past somatic sensory experiences, enabling you to compare current sensations with previous experiences. For example, the somatosensory association area allows you to recognize objects such as a pencil and a paperclip simply by touching them.

- The **visual association area** (areas 18 and 19), located in the occipital lobe, receives sensory impulses from the primary visual area and the thalamus. It relates present and past visual experiences and is essential for recognizing and evaluating what is seen. For example, the visual association area allows you to recognize an object such as a spoon simply by looking at it.

- The **facial recognition area,** corresponding roughly to areas 20, 21, and 37 in the inferior temporal lobe, receives nerve impulses from the visual association area. This area stores information about faces, and it allows you to recognize people by their faces. The facial recognition area in the *right* hemisphere is usually more dominant than the corresponding region in the left hemisphere.

- The **auditory association area** (area 22), located inferior and posterior to the primary auditory area in the temporal cortex, allows you to recognize a particular sound as speech, music, or noise.

- The **orbitofrontal cortex,** corresponding roughly to area 11 along the lateral part of the frontal lobe, receives sensory impulses from the primary olfactory area. This area allows you to identify odors and to discriminate among different odors. During olfactory processing, the orbitofrontal cortex of the *right* hemisphere exhibits greater activity than the corresponding region in the left hemisphere.

- **Wernicke's (posterior language) area** (VER-ni-kēz) (area 22, and possibly areas 39 and 40), a broad region in the *left* temporal and parietal lobes, interprets the meaning of speech by recognizing spoken words. It is active as you translate words into thoughts. The regions in the *right* hemisphere that correspond to Broca's and Wernicke's areas in the left hemisphere also contribute to verbal communication by adding emotional content, such as anger or joy, to spoken words. Unlike those who have CVAs in Broca's area, people who suffer strokes in Wernicke's area can still speak, but cannot arrange words in a coherent fashion (fluent aphasia, or "word salad"; see the Clinical Connection on this page).

- The **common integrative area** (areas 5, 7, 39, and 40) is bordered by somatosensory, visual, and auditory association areas. It receives nerve impulses from these areas and from the primary gustatory area, the primary olfactory area, the thalamus, and parts of the brain stem. This area integrates sensory interpretations from the association areas and impulses from other areas, allowing the formation of thoughts based on a variety of sensory inputs. It then transmits signals to other parts of the brain for the appropriate response to the sensory signals it has interpreted.

- The **prefrontal cortex (frontal association area)** is an extensive area in the anterior portion of the frontal lobe that is well developed in primates, especially humans (areas 9, 10, 11, and 12; area 12 is not illustrated since it can be seen only in a medial view). This area has numerous connections with other areas of the cerebral cortex, thalamus, hypothalamus, limbic system, and cerebellum. The prefrontal cortex is concerned with the makeup of a person's personality, intellect, complex learning abilities, recall of information, initiative, judgment, foresight, reasoning, conscience, intuition, mood, planning for the future, and development of abstract ideas. A person with bilateral damage to the prefrontal cortices typically becomes rude, inconsiderate, incapable of accepting advice, moody, inattentive, less creative, unable to plan for the future, and incapable of anticipating the consequences of rash or reckless words or behavior.

- The **premotor area** (area 6) is a motor association area that is immediately anterior to the primary motor area. Neurons in this area communicate with the primary motor cortex, the sensory association areas in the parietal lobe, the basal nuclei, and the thalamus. The premotor area deals with learned motor activities of a complex and sequential nature. It generates nerve impulses that cause specific groups of muscles to contract in a specific sequence, as when you write your name. The premotor area also serves as a memory bank for such movements.

- The **frontal eye field area** (area 8) in the frontal cortex is sometimes included in the premotor area. It controls voluntary scanning movements of the eyes—like those you just used in reading this sentence.

☤ CLINICAL CONNECTION | *Aphasia*

Much of what we know about language areas comes from studies of patients with language or speech disturbances that have resulted from brain damage. Broca's speech area, Wernicke's area, and other language areas are located in the left cerebral hemisphere of most people, regardless of whether they are left-handed or right-handed. Injury to language areas of the cerebral cortex results in **aphasia** (a-FĀ-zē-a; *a-* = without; *-phasia* = speech), an inability to use or comprehend words. Damage to Broca's speech area results in *nonfluent aphasia,* an inability to properly articulate or form words; people with nonfluent aphasia know what they wish to say but cannot speak. Damage to Wernicke's area, the common integrative area, or auditory association area results in *fluent aphasia,* characterized by faulty understanding of spoken or written words. A person experiencing this type of aphasia may fluently produce strings of words that have no meaning ("word salad"). For example, someone with fluent aphasia might say, "I rang car porch dinner light river pencil." The underlying deficit may be **word deafness** (an inability to understand spoken words), **word blindness** (an inability to understand written words), or both. •

The functions of the various parts of the brain are summarized in Table 14.2.

TABLE 14.2

Summary of Functions of Principal Parts of the Brain

PART	FUNCTION	PART	FUNCTION

BRAIN STEM

Medulla oblongata

Medulla oblongata: Contains sensory (ascending) and motor (descending) tracts. Cardiovascular center regulates heartbeat and blood vessel diameter. Medullary rhythmicity area (together with pons) regulates breathing. Contains gracile nucleus, cuneate nucleus, gustatory nucleus, cochlear nuclei, and vestibular nuclei (components of sensory pathways to brain). Inferior olivary nucleus provides instructions that cerebellum uses to adjust muscle activity when learning new motor skills. Other nuclei coordinate vomiting, swallowing, sneezing, coughing, and hiccupping. Contains nuclei of origin for cranial nerves VIII, IX, X, XI, and XII. Reticular formation (also in pons, midbrain, and diencephalon) functions in consciousness and arousal.

Pons: Contains sensory and motor tracts. Pontine nuclei relay nerve impulses from motor areas of cerebral cortex to cerebellum. Contains vestibular nuclei (along with medulla) that are part of equilibrium pathway to brain. Pneumotaxic area and apneustic area (together with the medulla) help control breathing. Contains nuclei of origin for cranial nerves V, VI, VII, and VIII.

Pons

Midbrain: Contains sensory and motor tracts. Superior colliculi coordinate movements of head, eyes, and trunk in response to visual stimuli. Inferior colliculi coordinate movements of head, eyes, and trunk in response to auditory stimuli. Substantia nigra and red nucleus contribute to control of movement. Contains nuclei of origin for cranial nerves III and IV.

Midbrain

CEREBELLUM

Cerebellum

Smooths and coordinates contractions of skeletal muscles. Regulates posture and balance. May have role in cognition and language processing.

DIENCEPHALON

Epithalamus   Thalamus

Hypothalamus

Thalamus: Relays almost all sensory input to cerebral cortex. Contributes to motor functions by transmitting information from cerebellum and basal nuclei to primary motor area of cerebral cortex. Plays role in maintenance of consciousness.

Hypothalamus: Controls and integrates activities of autonomic nervous system. Produces hormones, including releasing hormones, inhibiting hormones, oxytocin, and antidiuretic hormone (ADH). Regulates emotional and behavioral patterns (together with limbic system). Contains feeding and satiety centers (regulate eating), thirst center (regulates drinking), and suprachiasmatic nucleus (regulates circadian rhythms). Controls body temperature by serving as body's thermostat.

Epithalamus: Consists of pineal gland (secretes melatonin) and habenular nuclei.

CEREBRUM

Cerebrum

Sensory areas of cerebral cortex are involved in perception of sensory information; motor areas control execution of voluntary movements; association areas deal with more complex integrative functions such as memory, personality traits, and intelligence. Basal nuclei help initiate and terminate movements, suppress unwanted movements, and regulate muscle tone. Limbic system promotes range of emotions, including pleasure, pain, docility, affection, fear, and anger.

Hemispheric Lateralization

Although the brain is almost symmetrical on its right and left sides, subtle anatomical differences between the two hemispheres exist. For example, in about two-thirds of the population, the planum temporale, a region of the temporal lobe that includes Wernicke's area, is 50% larger on the left side than on the right side. This asymmetry appears in the human fetus at about 30 weeks of gestation. Physiological differences also exist; although the two hemispheres share performance of many functions, each hemisphere also specializes in performing certain unique functions. This functional asymmetry is termed **hemispheric lateralization.**

In most people the left hemisphere is more important for reasoning, numerical and scientific skills, spoken and written language, and the ability to use and understand sign language. Patients with damage in the left hemisphere often exhibit aphasia. Conversely, the right hemisphere is more specialized for musical and artistic awareness; spatial and pattern perception; recognition of faces and emotional content of language; discrimination of different smells; and generating mental images of sight, sound, touch, taste, and smell to compare relationships among them. Patients with damage in right hemisphere regions that correspond to Broca's and Wernicke's areas in the left hemisphere speak in a monotonous voice, having lost the ability to impart emotional inflection to what they say.

Despite some dramatic differences in functions of the two hemispheres, there is considerable variation from one person to another. Also, lateralization seems less pronounced in females than in males, both for language (left hemisphere) and for visual and spatial skills (right hemisphere). For instance, females are less likely than males to suffer aphasia after damage to the left hemisphere. A possibly related observation is that the anterior commissure is 12% larger and the corpus callosum has a broader posterior portion in females. Recall that both the anterior commissure and the corpus callosum are commissural tracts that provide communication between the two hemispheres.

Table 14.3 summarizes some of the functional differences between the two cerebral hemispheres.

Brain Waves

At any instant, brain neurons are generating millions of nerve impulses (action potentials). Taken together, these electrical signals are called **brain waves.** Brain waves generated by neurons close to the brain surface, mainly neurons in the cerebral cortex, can be detected by sensors called electrodes placed on the forehead and scalp. A record of such waves is called an **electroencephalogram** (e-lek'-trō-en-SEF-a-lō-gram; *electro-* = electricity; *-gram* = recording) or **EEG.** Electroencephalograms are useful both in studying normal brain functions, such as changes that occur during sleep, and in diagnosing a variety of brain disorders, such as epilepsy, tumors, trauma, hematomas, metabolic abnormalities, sites of trauma, and degenerative diseases. The EEG is also utilized to determine if "life" is present, that is, to establish or confirm that brain death has occurred.

Patterns of activation of brain neurons produce four types of brain waves (Figure 14.16):

1. **Alpha waves.** These rhythmic waves occur at a frequency of about 8–13 cycles per second. (The unit commonly used to express frequency is the hertz, Hz. One hertz is one cycle per second.) Alpha waves are present in the EEGs of nearly all normal individuals when they are awake and resting with their eyes closed. These waves disappear entirely during sleep.

2. **Beta waves.** The frequency of these waves is between 14 and 30 Hz. Beta waves generally appear when the nervous system is active—that is, during periods of sensory input and mental activity.

3. **Theta waves.** Theta waves (THĀ-ta) have frequencies of 4–7 Hz. These waves normally occur in children and adults experiencing emotional stress. They also occur in many disorders of the brain.

TABLE 14.3

Functional Differences between Right and Left Hemispheres

RIGHT HEMISPHERE FUNCTIONS

Receives somatic sensory signals from, and controls muscles on, left side of body.

Musical and artistic awareness.

Space and pattern perception.

Recognition of faces and emotional content of facial expressions.

Generating emotional content of language.

Generating mental images to compare spatial relationships.

Identifying and discriminating among odors.

LEFT HEMISPHERE FUNCTIONS

Receives somatic sensory signals from, and controls muscles on, right side of body.

Reasoning.

Numerical and scientific skills.

Ability to use and understand sign language.

Spoken and written language.

Figure 14.16 Types of brain waves recorded in an electroencephalogram (EEG).

Brain waves indicate electrical activity of the cerebral cortex.

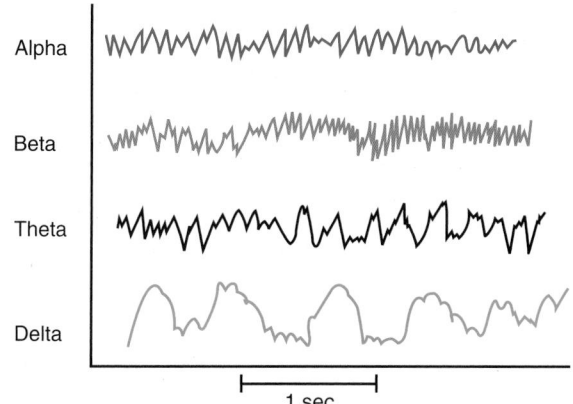

Alpha

Beta

Theta

Delta

1 sec

? Which type of brain wave indicates emotional stress?

4. **Delta waves.** The frequency of these waves is 1–5 Hz. Delta waves occur during deep sleep in adults, but they are normal in awake infants. When produced by an awake adult, they indicate brain damage.

✔ **CHECKPOINT**

20. Compare the functions of the sensory, motor, and association areas of the cerebral cortex.
21. What is hemispheric lateralization?
22. What is the diagnostic value of an EEG?

14.8 CRANIAL NERVES

◉ **OBJECTIVE**

- Identify the cranial nerves by name, number, and type, and give the function of each.

The 12 pairs of **cranial nerves** are so named because they pass through various foramina in the bones of the cranium and arise from the brain inside the cranial cavity. Like the 31 pairs of spinal nerves, they are part of the peripheral nervous system (PNS). Each cranial nerve has both a number, designated by a roman numeral, and a name. The numbers indicate the order, from anterior to posterior, in which the nerves arise from the brain. The names designate a nerve's distribution or function.

Three cranial nerves (I, II, and VIII) carry axons of sensory neurons and thus are called **special sensory nerves.** These nerves are unique to the head and are associated with the special senses of smelling, seeing, and hearing. The cell bodies of most sensory neurons are located in ganglia outside the brain.

Five cranial nerves (III, IV, VI, XI, and XII) are classified as **motor nerves** because they contain only axons of motor neurons as they leave the brain stem. The cell bodies of motor neurons lie in nuclei within the brain. Motor axons that innervate skeletal muscles are of two types:

1. *Branchial motor axons* innervate skeletal muscles that develop from the pharyngeal (branchial) arches (see Figure 14.28). These neurons leave the brain through the mixed cranial nerves and the accessory nerve.

2. *Somatic motor axons* innervate skeletal muscles that develop from head somites (eye muscles and tongue muscles). These neurons exit the brain through five motor cranial nerves (III, IV, VI, XI, and XII). Motor axons that innervate smooth muscle, cardiac muscle, and glands are called *autonomic motor axons* and are part of the parasympathetic division.

The remaining four cranial nerves (V, VII, IX, and X) are **mixed nerves**—they contain axons of both sensory neurons entering the brain stem and motor neurons leaving the brain stem.

Each cranial nerve is covered in detail in Exhibits 14.A through 14.J. Although the cranial nerves are mentioned singly in the exhibits with regard to their type, location, and function, remember that they are paired structures.

Table 14.4 presents a summary of the components and principal functions of the cranial nerves, including a mnemonic to help you remember their names.

⚕ **CLINICAL CONNECTION | *Dental Anesthesia***

The inferior alveolar nerve, a branch of the mandibular nerve, supplies all the teeth in one half of the mandible; it is often anesthetized in dental procedures. The same procedure will anesthetize the lower lip because the mental nerve is a branch of the inferior alveolar nerve. Because the lingual nerve runs very close to the inferior alveolar nerve near the mental foramen, it too is often anesthetized at the same time. For anesthesia to the upper teeth, the superior alveolar nerve endings, which are branches of the maxillary nerve, are blocked by inserting the needle beneath the mucous membrane. The anesthetic solution is then infiltrated slowly throughout the area of the roots of the teeth to be treated. •

✔ **CHECKPOINT**

23. How are cranial nerves named and numbered?
24. What is the difference between a special sensory, motor, and mixed cranial nerve?
25. Briefly list the function for each cranial nerve.

| EXHIBIT 14.A | Olfactory (I) Nerve *(Figure 14.17)* |

● OBJECTIVE

• Identify the termination of the olfactory nerve in the brain, the foramen through which it passes, and its function.

The **olfactory (I) nerve** (ōl-FAK-tō-rē; *olfact-* = to smell) is entirely sensory; it contains axons that conduct nerve impulses for olfaction, the sense of smell (Figure 14.17). The olfactory epithelium occupies the superior part of the nasal cavity, covering the inferior surface of the cribriform plate and extending down along the superior nasal concha. The olfactory receptors within the olfactory epithelium are bipolar neurons. Each has a single odor-sensitive, knob-shaped dendrite projecting from one side of the cell body and an unmyelinated axon extending from the other side. Bundles of axons of olfactory receptors extend through about 20 olfactory foramina in the cribriform plate of the ethmoid bone on each side of the nose. These 40 or so bundles of axons collectively form the right and left olfactory nerves.

Olfactory nerves end in the brain in paired masses of gray matter called the **olfactory bulbs,** two extensions of the brain that rest on the cribriform plate. Within the olfactory bulbs, the axon terminals of olfactory receptors form synapses with the dendrites and cell bodies of the next neurons in the olfactory pathway. The axons of these neurons make up the **olfactory tracts,** which extend posteriorly from the olfactory bulbs (Figure 14.17). Axons in the olfactory tracts end in the primary olfactory area in the temporal lobe of the cerebral cortex.

✔ CHECKPOINT

Where is the olfactory epithelium located?

Figure 14.17 Olfactory (I) nerve.

🔑 The olfactory epithelium is located on the inferior surface of the cribriform plate and superior nasal conchae.

CLINICAL CONNECTION | *Anosmia*

Loss of the sense of smell, called **anosmia** (an-OZ-mē-a), may result from infections of the nasal mucosa, head injuries in which the cribriform plate of the ethmoid bone is fractured, lesions along the olfactory pathway or in the brain, meningitis, smoking, or cocaine use. •

❓ Where do axons in the olfactory tracts terminate?

EXHIBIT 14.B   Optic (II) Nerve   *(Figure 14.18)*

⊙ OBJECTIVE

- Identify the termination of the optic nerve in the brain, the foramen through which it exits the skull, and its function.

The **optic (II) nerve** (OP-tik; *opti-* = the eye, vision) is entirely sensory; it contains axons that conduct nerve impulses for vision (Figure 14.18). In the retina, rods and cones initiate visual signals and relay them to bipolar cells, which transmit the signals to ganglion cells. Axons of all the ganglion cells in the retina of each eye join to form an optic nerve, which passes through the optic foramen. About 10 mm (0.4 in.) posterior to the eyeball, the two optic nerves merge to form the **optic chiasm** (KĪ-azm = a crossover, as in the letter **X**). Within the chiasm, axons from the medial half of each eye cross to the opposite side; axons from the lateral half remain on the same side. Posterior to the chiasm, the regrouped axons, some from each eye, form the **optic tracts.** Most axons in the optic tracts end in the lateral geniculate nucleus of the thalamus. There they synapse with neurons whose axons extend to the primary visual area in the occipital lobe of the cerebral cortex (area 17 in Figure 14.15). A few axons pass through the lateral geniculate nucleus and then extend to the superior colliculi of the midbrain. They synapse with motor neurons that control the extrinsic and intrinsic eye muscles.

✔ CHECKPOINT

Trace the sequence of nerve cells that process visual impulses within the retina.

Figure 14.18 Optic (II) nerve.

🔑 In sequence, visual signals are relayed from rods and cones to bipolar cells to ganglion cells.

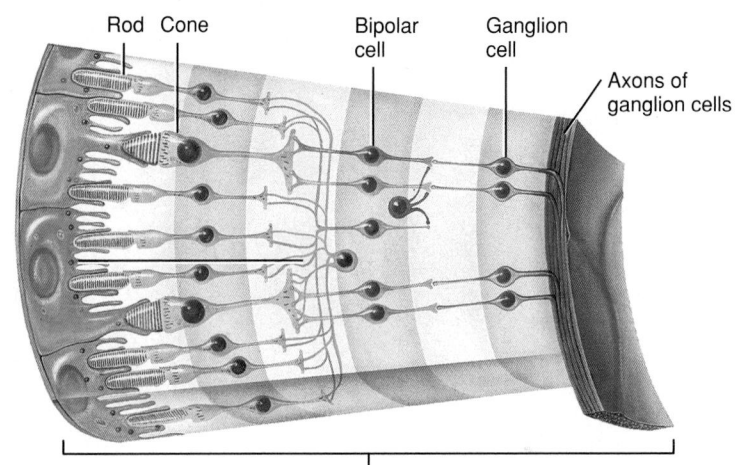

Rod Cone   Bipolar cell   Ganglion cell   Axons of ganglion cells

Retina

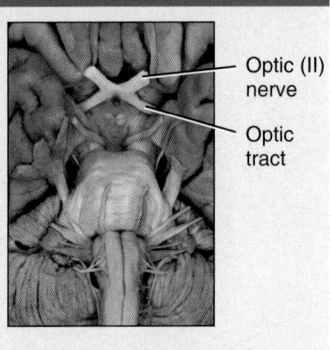

CLINICAL CONNECTION | *Anopia*

Fractures in the orbit, brain lesions, damage along the visual pathway, diseases of the nervous system (such as multiple sclerosis), pituitary gland tumors, or cerebral aneurysms (enlargements of blood vessels due to weakening of their walls) may result in visual field defects and loss of visual acuity. Blindness due to a defect in or loss of one or both eyes is called **anopia** (an-Ō-pē-a). •

Optic (II) nerve

Optic tract

Anterior

Optic chiasm

Optic tract

Optic (II) nerve

Retina

Eyeball

Posterior

❓ **Where do most axons in the optic tracts terminate?**

EXHIBIT 14.B **559**

■ **OBJECTIVE**

- Identify the origins of the oculomotor, trochlear, and abducens nerves in the brain, the foramen through which each exits the skull, and their functions.

The oculomotor, trochlear, and abducens nerves are the cranial nerves that control the muscles that move the eyeballs. They are all motor nerves that contain only motor axons as they exit the brain stem. Sensory axons from the extrinsic eyeball muscles begin their course toward the brain in each of these nerves, but eventually these sensory axons leave these nerves to join the ophthalmic branch of the trigeminal nerve. The sensory axons *do not* return to the brain in the oculomotor, trochlear, or abducens nerves. The cell bodies of the unipolar sensory neurons reside in the mesencephalic nucleus and they

enter the midbrain via the trigeminal nerve. These axons convey nerve impulses from the extrinsic eyeball muscles for *proprioception,* the perception of the movements and position of the body independent of vision.

The **oculomotor (III) nerve** (ok'-ū-lō-MŌ-tor; *oculo-* = eye; *-motor* = a mover) has its motor nucleus in the anterior part of the midbrain. The oculomotor nerve extends anteriorly and divides into superior and inferior branches, both of which pass through the superior orbital fissure into the orbit (Figure 14.19a). Axons in the superior branch innervate the superior rectus (an extrinsic eyeball muscle) and the levator palpebrae superioris (the muscle of the upper eyelid). Axons in the inferior branch supply the medial rectus, inferior rectus, and inferior oblique muscles—all extrinsic eyeball muscles. These somatic motor neurons control movements of the eyeball and upper eyelid.

Figure 14.19 Oculomotor (III), trochlear (IV), and abducens (VI) nerves.

The oculomotor nerve has the widest distribution among extrinsic eye muscles.

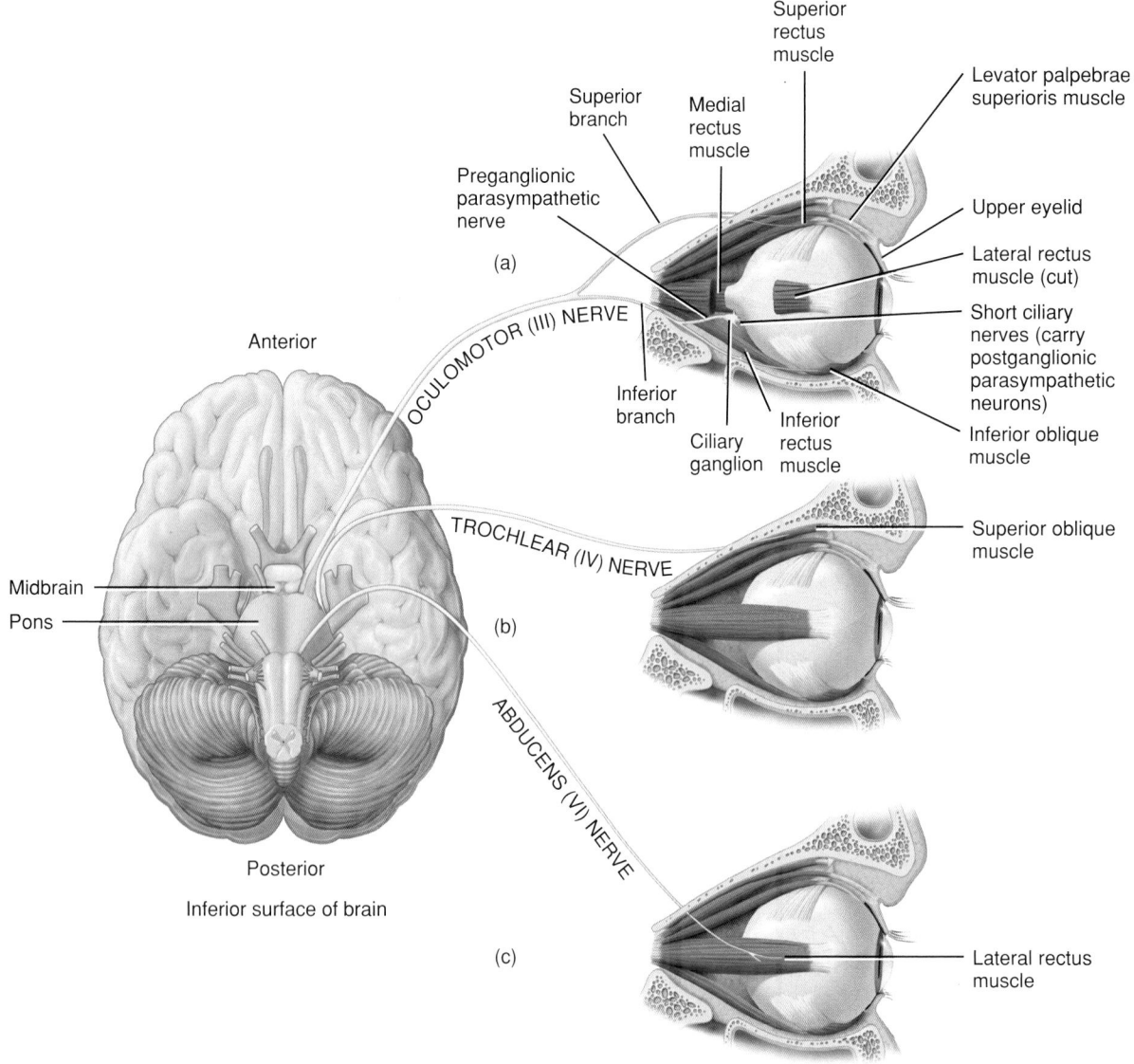

Which branch of the oculomotor nerve is distributed to the superior rectus muscle? Which is the smallest cranial nerve?

The inferior branch of the oculomotor nerve also supplies autonomic motor axons to intrinsic eyeball muscles, which consist of smooth muscle. They include the ciliary muscle of the eyeball and the circular muscles (sphincter pupillae) of the iris. Autonomic impulses propagate from a nucleus in the midbrain (accessory oculomotor) to the **ciliary ganglion,** a synaptic relay center for the two motor neurons of the autonomic nervous system. From the ciliary ganglion, autonomic motor axons extend to the ciliary muscle, which adjusts the lens for near vision *(accommodation)*. Other autonomic motor axons stimulate the circular muscles of the iris to contract when bright light stimulates the eye, causing a decrease in the size of the pupil *(constriction)*.

The **trochlear (IV) nerve** (TRŌK-lē-ar; *trochle-* = a pulley) is the smallest of the 12 cranial nerves and is the only one that arises from the posterior aspect of the brain stem. The somatic motor neurons originate in a nucleus in the midbrain (trochlear nucleus), and axons from the nucleus cross to the opposite side as they exit the brain on its posterior aspect. The nerve then wraps around the pons and exits through the superior orbital fissure into the orbit. These somatic motor axons innervate the superior oblique muscle of the eyeball, another extrinsic eyeball muscle that controls movement of the eyeball (Figure 14.19b).

Neurons of the **abducens (VI) nerve** (ab-DOO-senz; *ab-* = away; *-ducens* = to lead) originate from a nucleus in the pons (abducens nucleus). Somatic motor axons extend from the nucleus to the lateral rectus muscle of the eyeball, an extrinsic eyeball muscle, through the superior orbital fissure of the orbit (Figure 14.19c). The abducens nerve is so named because nerve impulses cause abduction (lateral rotation) of the eyeball.

✔CHECKPOINT

How are the oculomotor, trochlear, and abducens nerves related functionally?

CLINICAL CONNECTION | *Strabismus, Ptosis, and Diplopia*

Damage to the oculomotor (III) nerve causes **strabismus** (stra-BIZ-mus) (a condition in which both eyes do not fix on the same object, since one or both eyes may turn inward or outward), **ptosis** (TŌ-sis) (drooping) of the upper eyelid, dilation of the pupil, movement of the eyeball downward and outward on the damaged side, loss of accommodation for near vision, and **diplopia** (di-PLŌ-pē-a) (double vision).

Trochlear (IV) nerve damage can also result in strabismus and diplopia.

With damage to the abducens (VI) nerve, the affected eyeball cannot move laterally beyond the midpoint, and the eyeball usually is directed medially. This leads to strabismus and diplopia.

Causes of damage to the oculomotor, trochlear, and abducens nerves include trauma to the skull or brain, compression resulting from aneurysms, and lesions of the superior orbital fissure. Individuals with damage to these nerves are forced to tilt their heads in various directions to help bring the affected eyeball into the correct frontal plane. •

Oculomotor (III) nerve

Trochlear (IV) nerve

Abducens (VI) nerve

EXHIBIT 14.C **561**

EXHIBIT 14.D   Trigeminal (V) Nerve   *(Figure 14.20)*

OBJECTIVE

• Identify the origin of the trigeminal nerve in the brain, describe the foramina through which each of its three major branches exits the skull, and explain the function of each branch.

The **trigeminal (V) nerve** (trī-JEM-i-nal = triple, for its three branches) is a mixed cranial nerve and the largest of the cranial nerves. The trigeminal nerve emerges from two roots on the anterolateral surface of the pons. The large sensory root has a swelling called the **trigeminal (semilunar) ganglion,** which is located in a fossa on the inner surface of the

petrous portion of the temporal bone. The ganglion contains cell bodies of most of the primary sensory neurons. Neurons of the smaller motor root originate in a nucleus in the pons.

As indicated by its name, the trigeminal nerve has three branches: ophthalmic, maxillary, and mandibular (Figure 14.20). The **ophthalmic nerve** (of-THAL-mik; *ophthalm-* = the eye), the smallest branch, passes into the orbit via the superior orbital fissure. The **maxillary nerve** (*maxilla* = upper jawbone) is intermediate in size between the ophthalmic and mandibular nerves and passes through the foramen rotundum. The **mandibular nerve** (*mandibula* = lower jawbone), the largest branch, passes through the foramen ovale.

Figure 14.20 **Trigeminal (V) nerve.**

The three branches of the trigeminal nerve leave the cranium through the superior orbital fissure, foramen rotundum, and foramen ovale.

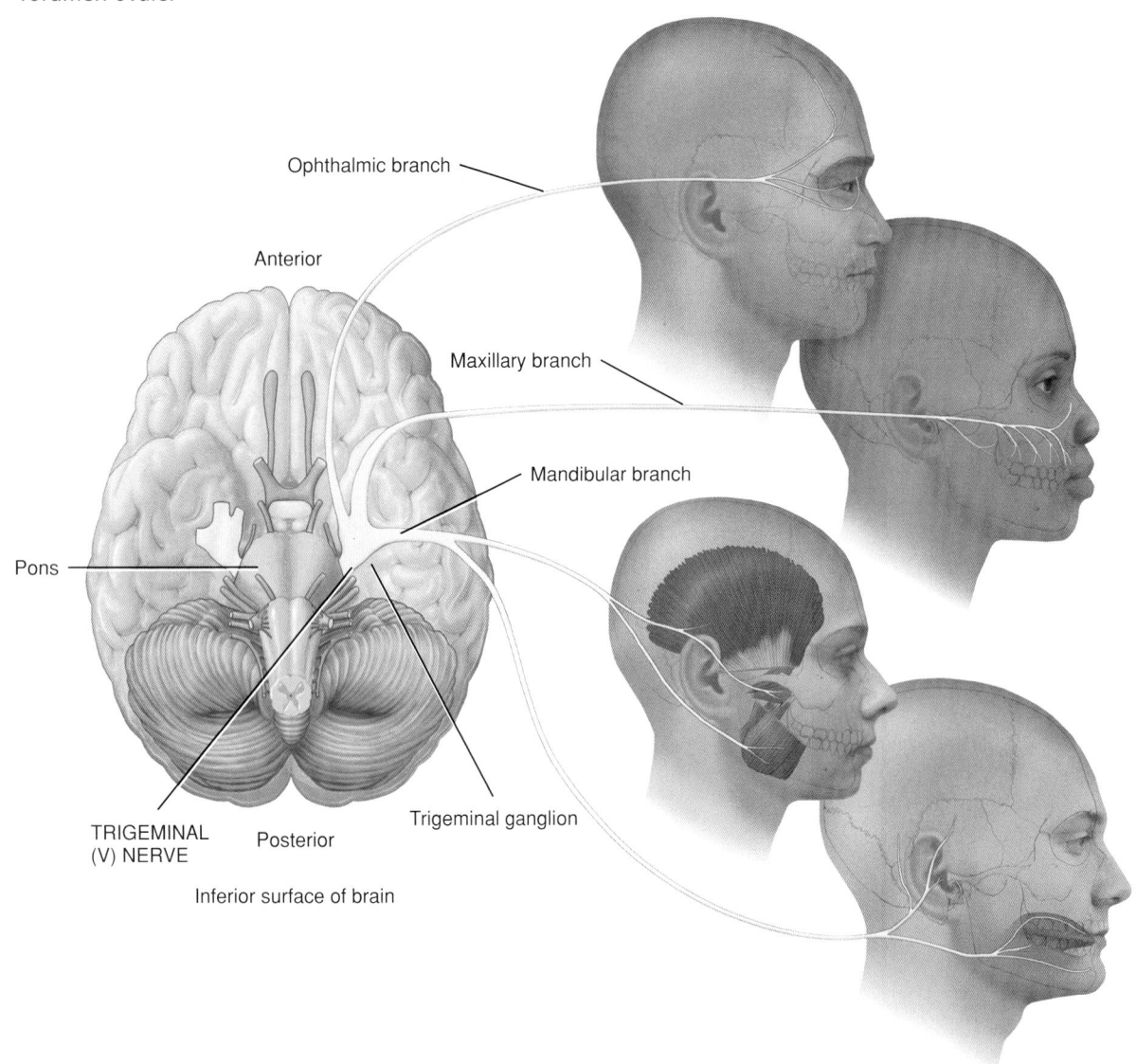

Ophthalmic branch

Anterior

Maxillary branch

Mandibular branch

Pons

TRIGEMINAL (V) NERVE   Posterior   Trigeminal ganglion

Inferior surface of brain

? **How does the trigeminal nerve compare in size with the other cranial nerves?**

Sensory axons in the trigeminal nerve carry nerve impulses for touch, pain, and thermal sensations (heat and cold). The ophthalmic nerve contains sensory axons from the skin over the upper eyelid, cornea, lacrimal glands, upper part of the nasal cavity, side of the nose, forehead, and anterior half of the scalp. The maxillary nerve includes sensory axons from the mucosa of the nose, palate, part of the pharynx, upper teeth, upper lip, and lower eyelid. The mandibular nerve contains sensory axons from the anterior two-thirds of the tongue (not taste), cheek and mucosa deep to it, lower teeth, skin over the mandible and side of the head anterior to the ear, and mucosa of the floor of the mouth. The sensory axons from the three branches enter the trigeminal ganglion, where their cell bodies are located, and terminate in nuclei in the pons. The trigeminal nerve also contains sensory axons from proprioceptors (receptors that provide information regarding body position and movements) located in the muscles of mastication and extrinsic muscles of the eyeball, but the cell bodies of these neurons are located in the mesencephalic nucleus.

Branchial motor neurons of the trigeminal nerve are part of the mandibular nerve and supply muscles of mastication (masseter, temporalis, medial pterygoid, lateral pterygoid, anterior belly of digastric, and mylohyoid muscles, as well as the tensor veli palatini muscle in the soft palate and tensor tympani muscle in the middle ear). These motor neurons mainly control chewing movements.

✔CHECKPOINT

What are the three branches of the trigeminal nerve, and which branch is the largest?

CLINICAL CONNECTION | *Trigeminal Neuralgia*

Neuralgia (pain) relayed via one or more branches of the trigeminal (V) nerve, caused by conditions such as inflammation or lesions, is called **trigeminal neuralgia (tic douloureux).** This is a sharp cutting or tearing pain that lasts for a few seconds to a minute and is caused by anything that presses on the trigeminal nerve or its branches. It occurs almost exclusively in people over 60 and can be the first sign of a disease, such as multiple sclerosis or diabetes, or lack of vitamin B_{12}, which damage the nerves. Injury of the mandibular nerve may cause paralysis of the chewing muscles and a loss of the sensations of touch, temperature, and proprioception in the lower part of the face. •

Trigeminal (V) nerve

EXHIBIT 14.D **563**

EXHIBIT 14.E   Facial (VII) Nerve   *(Figure 14.21)*

■ **OBJECTIVE**

• Identify the origins of the facial nerve in the brain, the foramen through which it exits the skull, and its function.

The **facial (VII) nerve** (FĀ-shal = face) is a mixed cranial nerve. Its sensory axons extend from the taste buds of the anterior two-thirds of the tongue, which enter the temporal bone to join the facial nerve. From here the sensory axons pass to the **geniculate ganglion** (je-NIK-ū-lat), a cluster of cell bodies of sensory neurons of the facial nerve within the temporal bone, and ends in the pons. From the pons, axons extend to the thalamus, and then to the gustatory areas of the cerebral cortex (Figure 14.21). The sensory portion of the facial nerve also contains axons from skin in the ear canal that relay touch, pain, and thermal sensations. Additionally, proprioceptors from muscles of the face and scalp relay information through their cell bodies in a nucleus in the midbrain (mesencephalic nucleus).

Axons of branchial motor neurons arise from a nucleus in the pons and exit the stylomastoid foramen to innervate middle ear, facial, scalp, and neck muscles. Nerve impulses propagating along these axons cause contraction of the muscles of facial expression plus the stylohyoid muscle, the posterior belly of the digastric muscle, and the stapedius muscle. The facial nerve innervates more named muscles than any other nerve in the body.

Axons of the autonomic motor neurons run in branches of the facial nerve and end in two ganglia: the **pterygopalatine ganglion** (ter'-i-gō-PAL-a-tīn) and the **submandibular ganglion,** respectively. From synaptic relays in the two ganglia, postganglionic autonomic motor axons extend to lacrimal glands (which secrete tears), nasal glands, palatine glands, and saliva-producing sublingual and submandibular glands.

✔ **CHECKPOINT**

Why is the facial nerve considered the major motor nerve of the head?

Figure 14.21 **Facial (VII) nerve.**

 The facial nerve causes contraction of the muscles of facial expression.

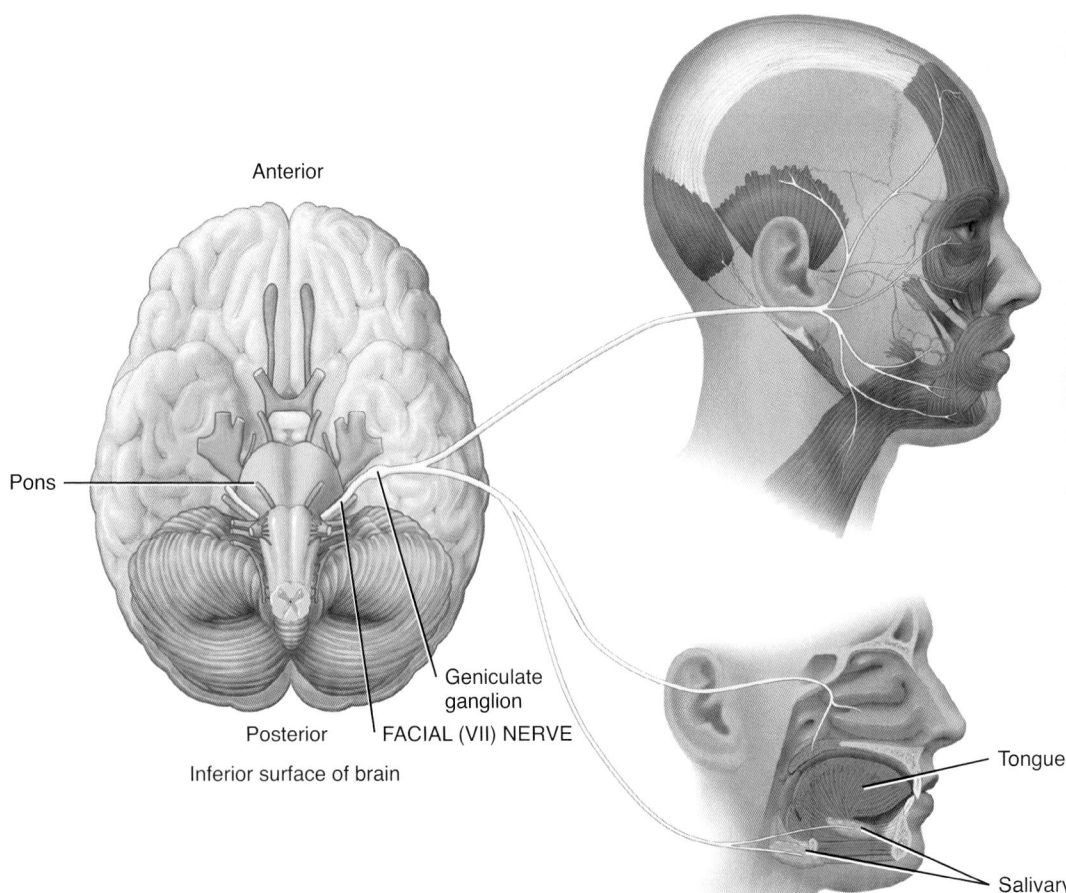

Anterior

Pons

Geniculate ganglion

Posterior   FACIAL (VII) NERVE

Inferior surface of brain

Tongue

Salivary glands

 CLINICAL CONNECTION | *Bell's Palsy*

Damage to the facial (VII) nerve due to conditions such as viral infection (shingles) or a bacterial infection (Lyme disease) produces **Bell's palsy** (paralysis of the facial muscles), loss of taste, decreased salivation, and loss of ability to close the eyes, even during sleep. The nerve can also be damaged by trauma, tumors, and stroke. •

Facial (VII) nerve

❓ **Where do the motor axons of the facial nerve originate?**

EXHIBIT 14.F Vestibulocochlear (VIII) Nerve *(Figure 14.22)*

OBJECTIVE

• Identify the origin of the vestibulocochlear nerve in the brain, the foramen through which it exits the skull, and the functions of each of its branches.

The **vestibulocochlear (VIII) nerve** (ves-tib-ū-lō-KOK-lē-ar; *vestibulo-* = small cavity; *-cochlear* = spiral, snail-like) was formerly known as the *acoustic* or *auditory nerve*. It is a sensory cranial nerve and has two branches, the vestibular branch and the cochlear branch (Figure 14.22). The **vestibular branch** carries impulses for equilibrium and the **cochlear branch** carries impulses for hearing.

Sensory axons in the vestibular branch extend from the semicircular canals, the saccule, and the utricle of the inner ear to the **vestibular ganglion,** where the cell bodies of the neurons are located (see Fig-

ure 17.21b), and end in vestibular nuclei in the pons and cerebellum. Some sensory axons also enter the cerebellum via the inferior cerebellar peduncle.

Sensory axons in the cochlear branch arise in the spiral organ (organ of Corti) in the cochlea of the internal ear. The cell bodies of cochlear branch sensory neurons are located in the **spiral ganglion** of the cochlea (see Figure 17.21b). From there, axons extend to nuclei in the medulla oblongata and end in the thalamus.

The nerve contains some motor fibers, but they do not innervate muscle tissue. Instead, they modulate the hair cells in the inner ear.

✔ CHECKPOINT

What are the functions of each of the two branches of the vestibulocochlear nerve?

Figure 14.22 Vestibulocochlear (VIII) nerve.

The vestibular branch of the vestibulocochlear nerve carries impulses for equilibrium, while the cochlear branch carries impulses for hearing.

CLINICAL CONNECTION | Vertigo, Ataxia, and Nystagmus

Injury to the vestibular branch of the vestibulocochlear (VIII) nerve may cause **vertigo** (ver-TI-gō) (a subjective feeling that one's own body or the environment is rotating), **ataxia** (a-TAK-sē-a) (muscular incoordination), and **nystagmus** (nis-TAG-mus) (involuntary rapid movement of the eyeball). Injury to the cochlear branch may cause **tinnitus** (ringing in the ears) or deafness. The vestibulocochlear nerve may be injured as a result of conditions such as trauma, lesions, or middle ear infections. •

Vestibulocochlear (VIII) nerve

? What structures are found in the vestibular and spiral ganglia?

EXHIBIT 14.F **565**

EXHIBIT 14.G   Glossopharyngeal (IX) Nerve   *(Figure 14.23)*

OBJECTIVE

• Identify the origin of the glossopharyngeal nerve in the brain, the foramen through which it exits the skull, and its function.

The **glossopharyngeal (IX) nerve** (glos'-ō-fa-RIN-jē-al; *glosso-* = tongue; *-pharyngeal* = throat) is a mixed cranial nerve (Figure 14.23). Sensory axons of the glossopharyngeal nerve arise from (1) taste buds on the posterior one-third of the tongue, (2) proprioceptors from some swallowing muscles supplied by the motor portion, (3) baroreceptors (pressure-monitoring receptors) in the carotid sinus that monitor blood pressure, (4) chemoreceptors (receptors that monitor blood levels of oxygen and carbon dioxide) in the carotid bodies near the carotid arteries (see Figure 23.26) and aortic bodies near the arch of the aorta (see Figure 23.26),

and (5) the external ear to convey touch, pain, and thermal (heat and cold) sensations. The cell bodies of these sensory neurons are located in the **superior** and **inferior ganglia.** From the ganglia, sensory axons pass through the jugular foramen and end in the medulla.

Axons of motor neurons in the glossopharyngeal nerve arise in nuclei of the medulla and exit the skull through the jugular foramen. Branchial motor neurons innervate the stylopharyngeus muscle, which assists in swallowing, and axons of autonomic motor neurons stimulate the parotid gland to secrete saliva. The postganglionic cell bodies of autonomic motor neurons are located in the **otic ganglion.**

✔ CHECKPOINT

Which other cranial nerves are also distributed to the tongue?

Figure 14.23 Glossopharyngeal (IX) nerve.

Sensory axons in the glossopharyngeal nerve supply the taste buds.

Inferior surface of brain

CLINICAL CONNECTION | *Dysphagia, Aptyalia, and Ageusia*

Injury to the glossopharyngeal (IX) nerve causes **dysphagia** (dis-FĀ-gē-a), or difficulty in swallowing; **aptyalia** (ap-tē-Ā-lē-a), or reduced secretion of saliva; loss of sensation in the throat; and **ageusia** (a-GOO-sē-a), or loss of taste sensation. The glossopharyngeal nerve may be injured as a result of conditions such as trauma or lesions. •

Glossopharyngeal (IX) nerve

? Through which foramen does the glossopharyngeal nerve exit the skull?

EXHIBIT 14.H   Vagus (X) Nerve   *(Figure 14.24)*

OBJECTIVE

- Identify the origin of the vagus nerve in the brain, the foramen through which it exits the skull, and its function.

The **vagus (X) nerve** (VĀ-gus = vagrant or wandering) is a mixed cranial nerve that is distributed from the head and neck into the thorax and abdomen (Figure 14.24). The nerve derives its name from its wide distribution. In the neck, it lies medial and posterior to the internal jugular vein and common carotid artery.

Sensory axons in the vagus nerve arise from the skin of the external ear for touch, pain, and thermal sensations; a few taste buds in the epiglottis and pharynx; and proprioceptors in muscles of the neck and throat. Also, sensory axons come from baroreceptors in the carotid sinus and chemoreceptors in the carotid and aortic bodies. The majority of sensory neurons come from visceral sensory receptors in most organs of the thoracic and abdominal cavities that convey sensations (such as hunger, fullness, and discomfort) from these organs. The sensory neurons have cell bodies in the **superior** and **inferior ganglia** and then pass through the jugular foramen to end in the medulla and pons.

The branchial motor neurons, which run briefly with the accessory nerve, arise from nuclei in the medulla oblongata and supply muscles of the pharynx, larynx, and soft palate that are used in swallowing, vocalization, and coughing. Historically these motor neurons have been called the cranial accessory nerve, but these fibers actually belong to the vagus (X) nerve.

Axons of autonomic motor neurons in the vagus nerve originate in nuclei of the medulla and supply the lungs, heart, glands of the gastrointestinal (GI) tract, and smooth muscle of the respiratory passageways, esophagus, stomach, gallbladder, small intestine, and most of the large intestine (see Figure 15.3). Autonomic motor axons initiate smooth muscle contractions in the gastrointestinal tract to aid motility and stimulate secretion by digestive glands; activate smooth muscle to constrict respiratory passageways; and decrease heart rate.

✔ CHECKPOINT

On what basis is the vagus nerve named?

Figure 14.24 Vagus (X) nerve.

The vagus nerve is widely distributed in the head, neck, thorax, and abdomen.

Inferior surface of brain

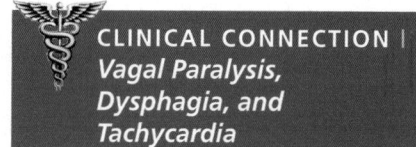

CLINICAL CONNECTION |
Vagal Paralysis, Dysphagia, and Tachycardia

Injury to the vagus (X) nerve due to conditions such as trauma or lesions causes **vagal paralysis,** or interruptions of sensations from many organs in the thoracic and abdominal cavities; **dysphagia** (dis-FĀ-gē-a), or difficulty in swallowing; and **tachycardia** (tak'-i-KAR-dē-a), or increased heart rate. •

Vagus (X) nerve

? **Where is the vagus nerve located in the neck region?**

EXHIBIT 14.H   **567**

EXHIBIT 14.I   Accessory (XI) Nerve   *(Figure 14.25)*

⊙ **OBJECTIVE**

• Identify the origin of the accessory nerve in the spinal cord, the foramina through which it first enters and then exits the skull, and its function.

The **accessory (XI) nerve** (ak-SES-ō-rē = assisting) is a motor cranial nerve (Figure 14.25). Historically it has been divided into two parts, a cranial accessory nerve and a spinal accessory nerve. The cranial accessory nerve actually is part of the vagus (X) nerve (see Exhibit 14.H). The "old" spinal accessory nerve is the accessory nerve we discuss in this exhibit. Its motor axons arise in the anterior gray horn of the first five segments of the cervical portion of the spinal cord. The axons from the segments exit the spinal cord laterally and come together, ascend through the foramen mag-

num, and then exit through the jugular foramen along with the vagus and glossopharyngeal nerves. The accessory nerve conveys motor impulses to the sternocleidomastoid and trapezius muscles to coordinate head movements. Sensory axons in the accessory nerve, which originate from proprioceptors in the sternocleidomastoid and trapezius muscles, begin their course toward the brain in the accessory nerve, but eventually leave the nerve to join nerves of the cervical plexus. From the cervical plexus they enter the spinal cord via the posterior roots of cervical spinal nerves; their cell bodies are located in the posterior root ganglia of those nerves. In the spinal cord the axons ascend to nuclei in the medulla oblongata.

✔ **CHECKPOINT**

Where do the motor axons of the accessory nerve originate?

Figure 14.25 Accessory (XI) nerve.

🔑 The accessory nerve exits the cranium through the jugular foramen.

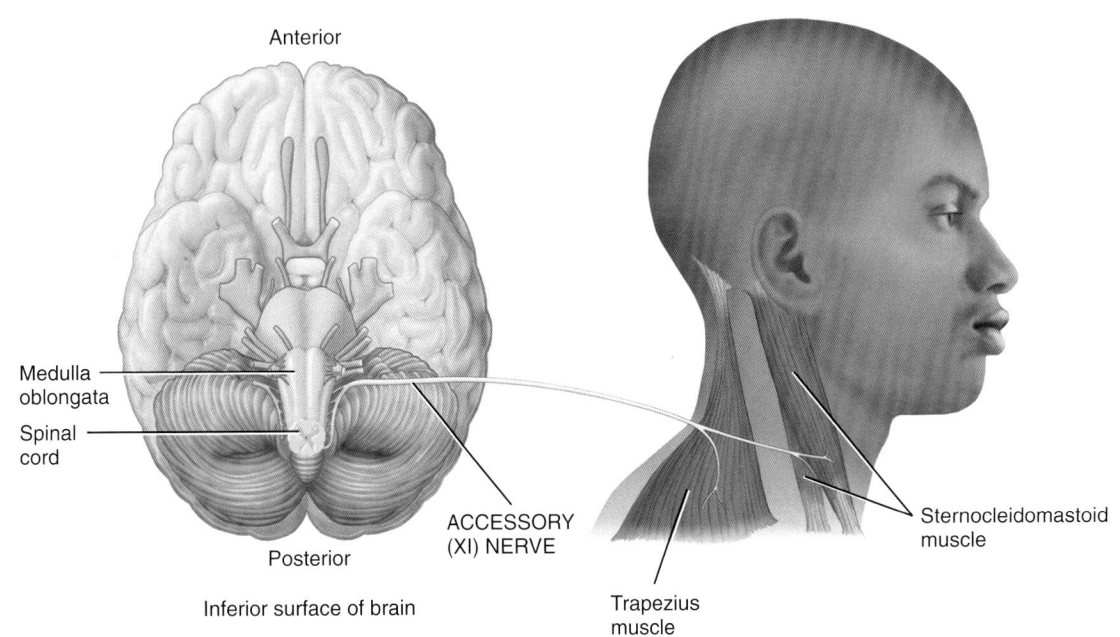

Anterior

Medulla oblongata

Spinal cord

ACCESSORY (XI) NERVE

Posterior

Inferior surface of brain

Trapezius muscle

Sternocleidomastoid muscle

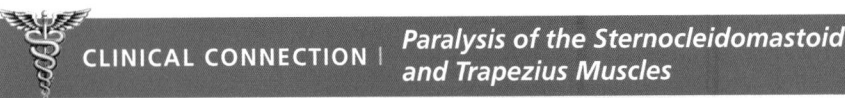

⚕ **CLINICAL CONNECTION |** *Paralysis of the Sternocleidomastoid and Trapezius Muscles*

If the accessory (XI) nerve is damaged due to conditions such as trauma, lesions, or stroke, the result is **paralysis of the sternocleidomastoid and trapezius muscles** so that the person is unable to raise the shoulders and has difficulty in turning the head. •

Accessory (XI) nerve

❓ **How does the accessory nerve differ from the other cranial nerves?**

EXHIBIT 14.J | Hypoglossal (XII) Nerve *(Figure 14.26)*

⊙ OBJECTIVE

• Identify the origin of the hypoglossal nerve in the brain, the foramen through which it exits the skull, and its function.

The **hypoglossal (XII) nerve** (hī′-pō-GLOS-al; *hypo-* = below; *-glossal* = tongue) is a motor cranial nerve. The somatic motor axons originate in a nucleus in the medulla oblongata (hypoglossal nucleus), exit the medulla on its anterior surface, and pass through the hypoglossal canal to supply the muscles of the tongue (Figure 14.26). These axons conduct nerve impulses for speech and swallowing. The sensory axons do not return to the brain in the hypoglossal nerve. Instead, sensory axons that originate from proprioceptors in the tongue muscles begin their course toward the brain in the hypoglossal nerve but leave the nerve to join cervical spinal nerves and end in the medulla oblongata, again entering the central nervous system via posterior roots of cervical spinal nerves.

✔ CHECKPOINT

In what portion of the brain does the hypoglossal nucleus originate?

Figure 14.26 Hypoglossal (XII) nerve.

🔑 The hypoglossal nerve exits the cranium through the hypoglossal canal.

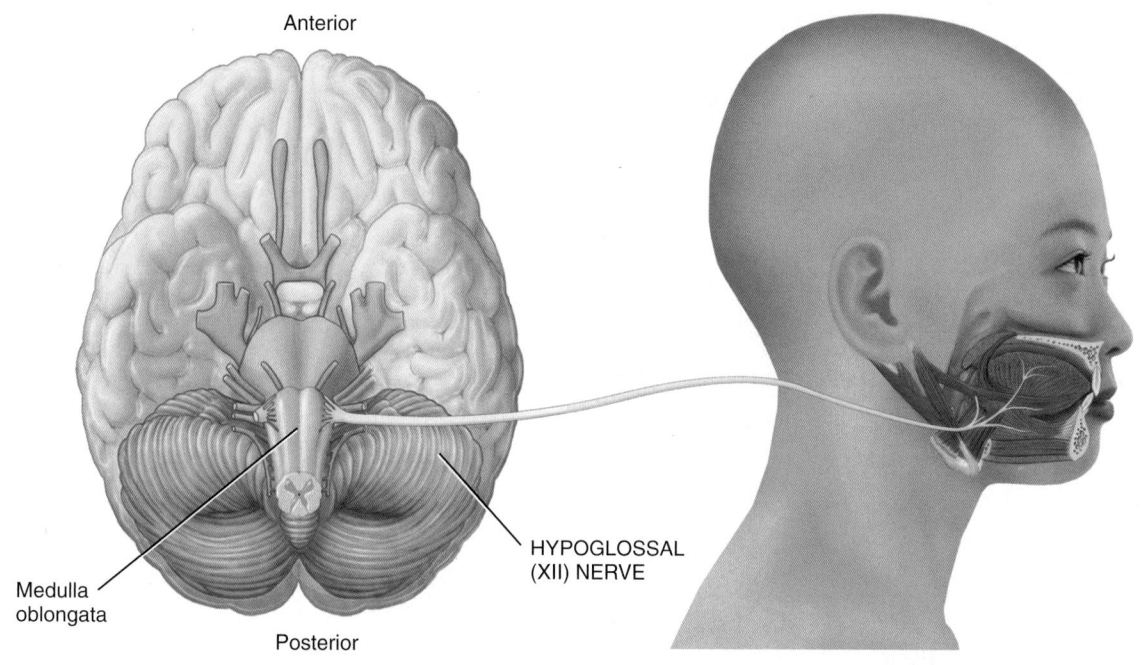

Anterior

Medulla oblongata

HYPOGLOSSAL (XII) NERVE

Posterior

Inferior surface of brain

⚕ CLINICAL CONNECTION | *Dysarthria and Dysphagia*

Injury to the hypoglossal (XII) nerve results in difficulty in chewing; **dysarthria** (dis-AR-thrē-a), or difficulty in speaking; and **dysphagia** (dis-FĀ-gē-a), or difficulty in swallowing. The tongue, when protruded, curls toward the affected side, and the affected side atrophies. The hypoglossal nerve may be injured as a result of conditions such as trauma, lesions, stroke, amyotrophic lateral sclerosis (Lou Gehrig's disease), or infections in the brain stem. •

Hypoglossal (XII) nerve

❓ **What important motor functions are related to the hypoglossal nerve?**

EXHIBIT 14.J **569**

TABLE 14.4

Summary of Cranial Nerves*

CRANIAL NERVE	COMPONENTS	PRINCIPAL FUNCTIONS
Olfactory (I)	*Special sensory*	Olfaction (smell).
Optic (II)	*Special sensory*	Vision (sight).
Oculomotor (III)	*Motor*	
	Somatic	Movement of eyeballs and upper eyelid.
	Motor (autonomic)	Adjusts lens for near vision (accommodation).
		Constriction of pupil.
Trochlear (IV)	*Motor*	
	Somatic	Movement of eyeballs.
Trigeminal (V)	*Mixed*	
	Sensory	Touch, pain, and thermal sensations from scalp, face, and oral cavity (including teeth and anterior two-thirds of tongue).
	Motor (branchial)	Chewing and controls middle ear muscle.
Abducens (VI)	*Motor*	
	Somatic	Movement of eyeballs.
Facial (VII)	*Mixed*	
	Sensory	Taste from anterior two-thirds of tongue.
		Touch, pain, and thermal sensations from skin in external ear canal.
	Motor (branchial)	Control of muscles of facial expression and middle ear muscle.
	Motor (autonomic)	Secretion of tears and saliva.
Vestibulocochlear (VIII)	*Special sensory*	Hearing and equilibrium.
Glossopharyngeal (IX)	*Mixed*	
	Sensory	Taste from posterior one-third of tongue.
		Proprioception in some swallowing muscles.
		Monitors blood pressure and oxygen and carbon dioxide levels in blood.
		Touch, pain, and thermal sensations from skin of external ear and upper pharynx.
	Motor (branchial)	Assists in swallowing.
	Motor (autonomic)	Secretion of saliva.
Vagus (X)	*Mixed*	
	Sensory	Taste from epiglottis.
		Proprioception from throat and voice box muscles.
		Monitors blood pressure and oxygen and carbon dioxide levels in blood.
		Touch, pain, and thermal sensations from skin of external ear.
		Sensations from thoracic and abdominal organs.
	Motor (branchial)	Swallowing, vocalization, and coughing.
	Motor (autonomic)	Motility and secretion of gastrointestinal organs.
		Constriction of respiratory passageways.
		Decreases heart rate.
Accessory (XI)	*Motor*	
	Branchial	Movement of head and pectoral girdle.
Hypoglossal (XII)	*Motor*	
	Somatic	Speech, manipulation of food, and swallowing.

***MNEMONIC FOR CRANIAL NERVES:**

Oh	Oh	Oh	To	Touch	And	Feel	Very	Green	Vegetables	AH!	
Olfactory	Optic	Oculomotor	Trochlear	Trigeminal	Abducens	Facial	Vestibulocochlear	Glossopharyngeal	Vagus	Accessory	Hypoglossal

14.9 DEVELOPMENT OF THE NERVOUS SYSTEM

◉ **OBJECTIVE**

• Describe how the parts of the brain develop.

Development of the nervous system begins in the third week of gestation with a thickening of the **ectoderm** called the **neural plate** (Figure 14.27). The plate folds inward and forms a longitudinal groove, the **neural groove.** The raised edges of the neural plate are called **neural folds.** As development continues, the neural folds increase in height and meet to form a tube called the **neural tube.**

Three layers of cells differentiate from the wall that encloses the neural tube. The outer or **marginal layer** cells develop into the *white matter* of the nervous system. The middle or **mantle layer** cells develop into the *gray matter.* The inner or **ependymal layer** (e-PEN-di-mal) cells eventually form the *lining of the central canal of the spinal cord* and *ventricles of the brain.*

The **neural crest** is a mass of tissue between the neural tube and the skin ectoderm (Figure 14.27b). It differentiates and eventually forms the *posterior (dorsal) root ganglia of spinal nerves, spinal nerves, ganglia of cranial nerves, cranial nerves, ganglia of the autonomic nervous system, adrenal medulla,* and *meninges.*

Figure 14.27 Origin of the nervous system. (a) Dorsal view of an embryo in which the neural folds have partially united, forming the early neural tube. (b) Transverse sections through the embryo showing the formation of the neural tube.

🔑 The nervous system begins developing in the third week from a thickening of ectoderm called the neural plate.

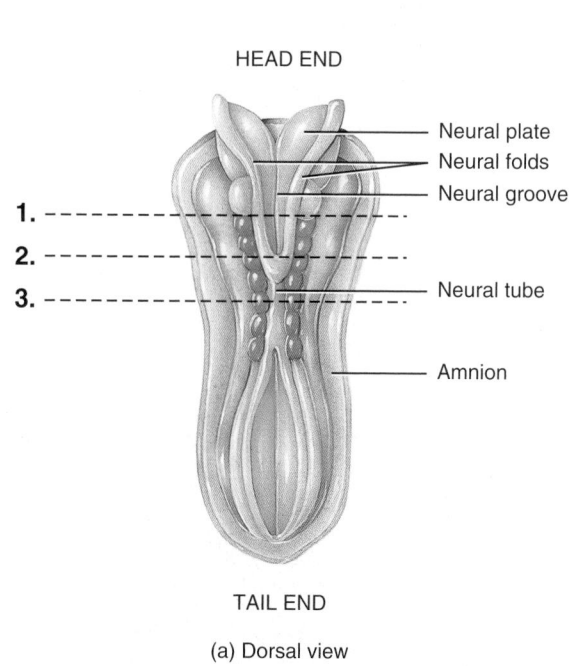

HEAD END

- Neural plate
- Neural folds
- Neural groove

1. - - - - - - - - - - - - - -
2. - - - - - - - - - - - - - -
3. - - - - - - - - - - - - - - — Neural tube

— Amnion

TAIL END

(a) Dorsal view

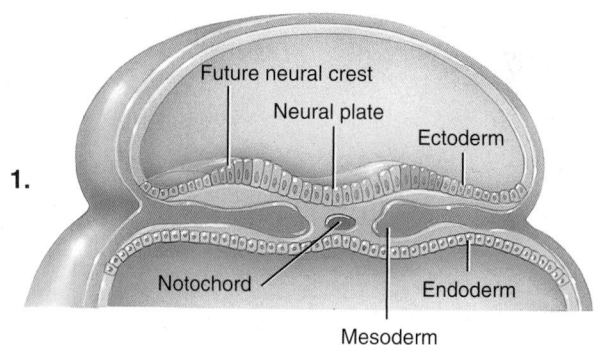

1.
- Future neural crest
- Neural plate
- Ectoderm
- Notochord
- Endoderm
- Mesoderm

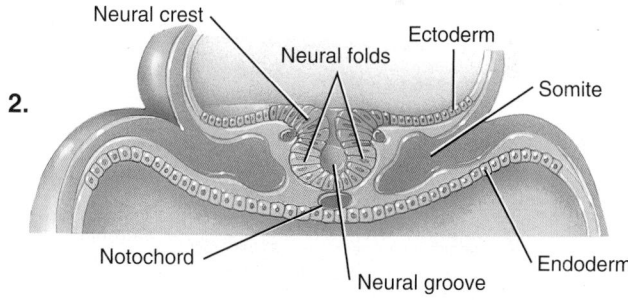

2.
- Neural crest
- Ectoderm
- Neural folds
- Somite
- Notochord
- Endoderm
- Neural groove

3.
- Neural crest
- Neural tube
- Somite
- Notochord
- Ectoderm
- Endoderm

(b) Transverse sections

❓ **What is the origin of the gray matter of the nervous system?**

As discussed at the beginning of this chapter, during the third to fourth week of embryonic development, the anterior part of the neural tube develops into three enlarged areas called **primary brain vesicles** that are named for their relative positions. These are the **prosencephalon** (pros'-en-SEF-a-lon; *pros-* = before) or forebrain, **mesencephalon** (mes'-en-SEF-a-lon; *mes-* = middle) or midbrain, and **rhombencephalon** (rom'-ben-SEF-a-lon; *rhomb-* = behind) or hindbrain (Figure 14.28a; see also Table 14.1). During the fifth week of development, **secondary brain vesicles** begin to develop. The prosencephalon develops into two secondary brain vesicles called the **telencephalon** (tel'-en-SEF-a-lon; *tel-* = distant) and the **diencephalon** (dī-en-

SEF-a-lon; *di-* = through) (Figure 14.28b). The rhombencephalon also develops into two secondary brain vesicles called the **metencephalon** (met'-en-SEF-a-lon; *met-* = after) and the **myelencephalon** (mī-el-en-SEF-a-lon; *myel-* = marrow). The area of the neural tube inferior to the myelencephalon gives rise to the *spinal cord.*

The brain vesicles continue to develop as follows (Figure 14.28c, d; see also Table 14.1):

- The telencephalon develops into the *cerebral hemispheres,* including the *basal nuclei,* and houses the paired *lateral ventricles.*

Figure 14.28 Development of the brain and spinal cord.

The various parts of the brain develop from the primary brain vesicles.

Lateral view of right side

(a) Three- to four-week embryo showing primary brain vesicles

(b) Seven-week embryo showing secondary brain vesicles

(c) Eleven-week fetus showing expanding cerebral hemispheres overgrowing the diencephalon

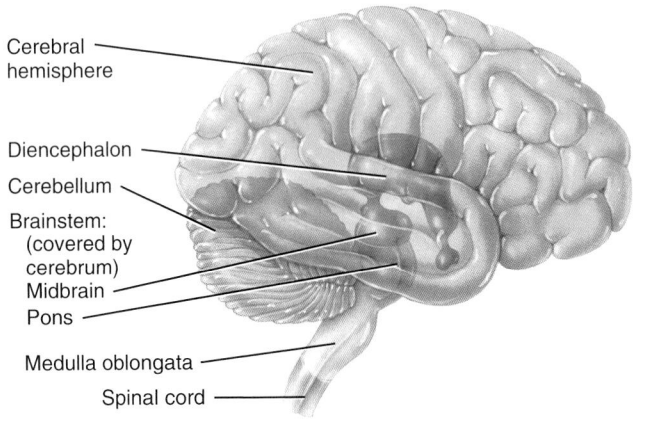

(d) Brain at birth (diencephalon and superior portion of brain stem projected to surface)

? **Which primary brain vesicle does not develop into a secondary brain vesicle?**

- The diencephalon develops into the *thalamus, hypothalamus,* and *epithalamus.*

- The mesencephalon develops into the *midbrain,* which surrounds the *aqueduct of the midbrain (cerebral aqueduct).*

- The metencephalon becomes the *pons* and *cerebellum* and houses part of the *fourth ventricle.*

- The myelencephalon develops into the *medulla oblongata* and houses the remainder of the *fourth ventricle.*

Two **neural tube defects**—spina bifida (see Disorders: Homeostatic Imbalances at the end of Chapter 7) and anencephaly (absence of the skull and cerebral hemispheres, discussed in Section 29.1)—are associated with low levels of folic acid (folate), one of the B vitamins, in the first few weeks of development. Many foods, especially grain products such as cereals and bread, are now fortified with folic acid; however, the incidence of both disorders is greatly reduced when women who are or may become pregnant take folic acid supplements.

✔ CHECKPOINT

26. What parts of the brain develop from each primary brain vesicle?

14.10 AGING AND THE NERVOUS SYSTEM

◉ OBJECTIVE

- Describe the effects of aging on the nervous system.

The brain grows rapidly during the first few years of life. Growth is due mainly to an increase in the size of neurons already present, the proliferation and growth of neuroglia, the development of dendritic branches and synaptic contacts, and continuing myelination of axons. From early adulthood onward, brain mass declines. By the time a person reaches 80, the brain weighs about 7% less than it did in young adulthood. Although the number of neurons present does not decrease very much, the number of synaptic contacts declines. Associated with the decrease in brain mass is a decreased capacity for sending nerve impulses to and from the brain. As a result, processing of information diminishes. Conduction velocity decreases, voluntary motor movements slow down, and reflex times increase.

✔ CHECKPOINT

27. How is brain mass related to age?

DISORDERS: HOMEOSTATIC IMBALANCES

Cerebrovascular Accident

The most common brain disorder is a **cerebrovascular accident (CVA),** also called a **stroke** or **brain attack.** CVAs affect 500,000 people each year in the United States and represent the third leading cause of death, behind heart attacks and cancer. A CVA is characterized by abrupt onset of persisting neurological symptoms, such as paralysis or loss of sensation, that arise from destruction of brain tissue. Common causes of CVAs are intracerebral hemorrhage (bleeding from a blood vessel in the pia mater or brain), emboli (blood clots), and atherosclerosis (formation of cholesterol-containing plaques that block blood flow) of the cerebral arteries.

Among the risk factors implicated in CVAs are high blood pressure, high blood cholesterol, heart disease, narrowed carotid arteries, transient ischemic attacks (TIAs; discussed next), diabetes, smoking, obesity, and excessive alcohol intake.

A clot-dissolving drug called *tissue plasminogen activator (t-PA)* is used to open up blocked blood vessels in the brain. The drug is most effective when administered *within three hours* of the onset of the CVA, however, and is helpful only for CVAs due to a blood clot *(ischemic CVAs).* Use of t-PA can decrease the permanent disability associated with these types of CVAs by 50%. However, t-PA should not be administered to individuals with strokes caused by hemorrhaging *(hemorrhagic CVAs)* since it can cause further injury or even death. The distinction between the types of CVA is made on the basis of a CT scan.

New studies show that "cold therapy" might be successful in limiting the amount of residual damage from a CVA. States of hypothermia, such as those experienced by cold-water drowning victims, seem to trigger a survival response in which the body requires less oxygen; application of this principle to stroke victims has showed promise. Some commercial companies now provide "CVA survival kits" including cooling blankets that can be kept in the home.

Transient Ischemic Attacks

A **transient ischemic attack (TIA)** (is-KĒ-mik) is an episode of temporary cerebral dysfunction caused by impaired blood flow to part of the brain. Symptoms include dizziness, weakness, numbness, or paralysis in a limb or on one side of the body; drooping of one side of the face; headache; slurred speech or difficulty understanding speech; and/or a partial loss of vision or double vision. Sometimes nausea or vomiting also occurs. The onset of symptoms is sudden and reaches maximum intensity almost immediately. A TIA usually persists for 5 to 10 minutes and only rarely lasts as long as 24 hours. It leaves no permanent neurological deficits. The causes of TIA include blood clots, atherosclerosis, and certain blood disorders. About one-third of patients who experience a TIA will have a CVA eventually. Therapy for TIAs includes drugs such as aspirin, which blocks the aggregation of blood platelets, and anticoagulants; cerebral artery bypass grafting; and carotid endarterectomy (removal of the cholesterol-containing plaques and inner lining of an artery).

Alzheimer Disease

Alzheimer disease (ALTZ-hī-mer) or **AD** is a disabling senile dementia, the loss of reasoning and ability to care for oneself, that afflicts about 11% of the population over age 65. In the United States, about 4 million people suffer from AD. Claiming over 100,000 lives a year, AD is the fourth leading cause of death among the elderly, after heart disease, cancer, and stroke. The cause of most AD cases is still unknown, but evidence suggests it is due to a combination of genetic factors, environmental or lifestyle factors, and the aging process. Mutations in three different genes (coding for presenilin-1, presenilin-2, and amyloid precursor protein) lead to early-onset forms of AD in afflicted families but account for less than 1% of all cases. An environmental risk factor for developing AD is a history of head injury. A similar dementia occurs in boxers, probably caused by repeated blows to the head.

Individuals with AD initially have trouble remembering recent events. They then become confused and forgetful, often repeating questions or getting lost while traveling to familiar places. Disorientation grows, memories of past events disappear, and episodes of paranoia, hallucination, or violent changes in mood may occur. As their minds continue to deteriorate, they lose their ability to read, write, talk, eat, or walk. The disease culminates in dementia. A person with AD usually dies of some complication that afflicts bedridden patients, such as pneumonia.

At autopsy, brains of AD victims show three distinct structural abnormalities:

1. *Loss of neurons that liberate acetylcholine.* A major center of neurons that liberate ACh is the *nucleus basalis*, a group of large cells below the globus pallidus. Axons of these neurons project widely throughout the cerebral cortex and limbic system. Their destruction is a hallmark of Alzheimer disease.

2. *Beta-amyloid plaques* (bā'-ta-AM-i-loyd), clusters of abnormal proteins deposited outside neurons.

3. *Neurofibrillary tangles* (noo-rō-FĪ-bril-ler'-ē), abnormal bundles of filaments inside neurons in affected brain regions. These filaments consist of a protein called *tau* (TAW) that has been *hyperphosphorylated* (hī-per-fos-FOR-i-lā-ted, meaning too many phosphate groups have been added to it).

Drugs that inhibit acetylcholinesterase (AChE), the enzyme that inactivates ACh, improve alertness and behavior in about 5% of AD patients. Tacrine®, the first anticholinesterase inhibitor approved for treatment of AD in the United States, has significant side effects and requires dosing four times a day. Donepezil®, approved in 1998, is less toxic to the liver and has the advantage of once-a-day dosing. Some evidence suggests that vitamin E (an antioxidant), estrogen, ibuprofen, and ginkgo biloba extract may have slight beneficial effects in AD patients. In addition, researchers are currently exploring ways to develop drugs that will prevent beta-amyloid plaque formation by inhibiting the enzymes involved in beta-amyloid synthesis and by increasing the activity of the enzymes involved in beta-amyloid degradation. Researchers are also trying to develop drugs that will reduce the formation of neurofibrillary tangles by inhibiting the enzymes that hyperphosphorylate tau.

Brain Tumors

A **brain tumor** is an abnormal growth of tissue in the brain that may be malignant or benign. Unlike most other tumors in the body, malignant and benign brain tumors may be equally serious, compressing adjacent tissues and causing a buildup of pressure in the skull. The most common malignant tumors are secondary tumors that metastasize from other cancers in the body, such as those in the lungs, breasts, skin (malignant melanoma), blood (leukemia), and lymphatic organs (lymphoma). Most primary brain tumors (those that originate within the brain) are gliomas, which develop in neuroglia. The symptoms of a brain tumor depend on its size, location, and rate of growth. Among the symptoms are headache, poor balance and coordination, dizziness, double vision, slurred speech, nausea and vomiting, fever, abnormal pulse and breathing rates, personality changes, numbness and weakness of the limbs, and seizures. Treatment options for brain tumors vary with their size, location, and type and may include surgery, radiation therapy, and/or chemotherapy. Unfortunately, chemotherapeutic agents do not readily cross the blood–brain barrier.

Attention Deficit Hyperactivity Disorder

Attention deficit hyperactivity disorder (ADHD) is a learning disorder characterized by poor or short attention span, a consistent level of hyperactivity, and a level of impulsiveness inappropriate for the child's age. ADHD is believed to affect about 5% of children and is diagnosed 10 times more often in boys than in girls. The condition typically begins in childhood and continues into adolescence and adulthood. Symptoms of ADHD develop in early childhood, often before age four, and include difficulty in organizing and finishing tasks, lack of attention to details, short attention span and inability to concentrate, difficulty following instructions, talking excessively and frequently interrupting others, frequent running or excessive climbing, inability to play quietly alone, and difficulty waiting or taking turns.

The causes of ADHD are not fully understood, but it does have a strong genetic component. Some evidence also suggests that ADHD is related to problems with neurotransmitters. In addition, recent imaging studies have demonstrated that people with ADHD have less nervous tissue in specific regions of the brain such as the frontal and temporal lobes, caudate nucleus, and cerebellum. Treatment may involve remedial education, behavioral modification techniques, restructuring routines, and drugs that calm the child and help focus attention.

MEDICAL TERMINOLOGY

Agnosia (ag-NŌ-zē-a; *a-* = without; *-gnosia* = knowledge) Inability to recognize the significance of sensory stimuli such as sounds, sights, smells, tastes, and touch.

Apraxia (a-PRAK-sē-a; *-praxia* = coordinated) Inability to carry out purposeful movements in the absence of paralysis.

Consciousness (KON-shus-nes) A state of wakefulness in which an individual is fully alert, aware, and oriented, partly as a result of feedback between the cerebral cortex and reticular activating system.

Delirium (dē-LIR-ē-um = off the track) A transient disorder of abnormal cognition and disordered attention accompanied by disturbances of the sleep–wake cycle and psychomotor behavior (hyperactivity or hypoactivity of movements and speech). Also called **acute confusional state (ACS).**

Dementia (de-MEN-shē-a; *de-* = away from; *-mentia* = mind) Permanent or progressive general loss of intellectual abilities, including impairment of memory, judgment, and abstract thinking and changes in personality.

Encephalitis (en'-sef-a-LĪ-tis) An acute inflammation of the brain caused by either a direct attack by any of several viruses or an allergic reaction to any of the many viruses that are normally harmless to the central nervous system. If the virus affects the spinal cord as well, the condition is called **encephalomyelitis.**

Encephalopathy (en-sef'-a-LOP-a-thē; *encephalo-* = brain; *-pathos* = disease) Any disorder of the brain.

Lethargy (LETH-ar-jē) A condition of functional sluggishness.

Microcephaly (mī-krō-SEF-a-lē; *micro-* = small; *-cephal* = head) A congenital condition that involves the development of a small brain and skull and frequently results in mental retardation.

Prosopagnosia (pros'-ō-pag-NŌ-sē-a; *a-* = without; *-gnosia* = knowledge) Inability to recognize faces, usually caused by damage to the facial recognition area in the inferior temporal lobe of both cerebral hemispheres.

Reye's syndrome (RĪZ) Occurs after a viral infection, particularly chickenpox or influenza, most often in children or teens who have taken aspirin; characterized by vomiting and brain dysfunction (disorientation, lethargy, and personality changes) that may progress to coma and death.

Stupor (STOO-por) Unresponsiveness from which a patient can be aroused only briefly and only by vigorous and repeated stimulation.

CHAPTER REVIEW AND RESOURCE SUMMARY

Review

Resource

14.1 Brain Organization, Protection, and Blood Supply

1. The major parts of the brain are the brain stem, cerebellum, diencephalon, and cerebrum.
2. The brain is protected by cranial bones and the cranial meninges.
3. The cranial meninges are continuous with the spinal meninges. From superficial to deep they are the dura mater, arachnoid mater, and pia mater.
4. Blood flow to the brain is mainly via the internal carotid and vertebral arteries.
5. Any interruption of the oxygen or glucose supply to the brain can result in weakening of, permanent damage to, or death of brain cells.
6. The blood–brain barrier (BBB) causes different substances to move between the blood and the brain tissue at different rates and prevents the movement of some substances from blood into the brain.

Anatomy Overview - Nervous System
Anatomy Overview - The Brain
Figure 14.1 - The Brain
Exercise - Concentrate on Your Brain

14.2 Cerebrospinal Fluid

1. Cerebrospinal fluid (CSF) is formed in the choroid plexuses and circulates through the lateral ventricles, third ventricle, fourth ventricle, subarachnoid space, and central canal. Most of the fluid is absorbed into the blood across the arachnoid villi of the superior sagittal sinus.
2. Cerebrospinal fluid provides mechanical protection, chemical protection, and circulation of nutrients.

14.3 The Brain Stem and Reticular Formation

1. The medulla oblongata is continuous with the superior part of the spinal cord and contains both sensory tracts and motor tracts. It contains a cardiovascular center, which regulates heart rate and blood vessel diameter, and a medullary rhythmicity area, which helps control breathing. It also contains the gracile nucleus, cuneate nucleus, gustatory nucleus, cochlear nuclei, and vestibular nuclei, which are components of sensory pathways to the brain. Also present in the medulla is the inferior olivary nucleus, which provides instructions that the cerebellum uses to adjust muscle activity when you learn new motor skills. Other nuclei of the medulla coordinate vomiting, swallowing, sneezing, coughing, and hiccupping. The medulla also contains nuclei associated with cranial nerves VIII–XII.
2. The pons is superior to the medulla. It contains both sensory tracts and motor tracts. Pontine nuclei relay nerve impulses related to voluntary skeletal movements from the cerebral cortex to the cerebellum. The pons also contains the pneumotaxic and apneustic centers, which help control breathing. Vestibular nuclei, which are present in the pons and medulla, are part of the equilibrium pathway to the brain. Also present in the pons are nuclei associated with cranial nerves V–VIII.
3. The midbrain connects the pons and diencephalon and surrounds the cerebral aqueduct. It contains both sensory tracts and motor tracts. The superior colliculi coordinate movements of the head, eye, and trunk in response to visual stimuli; the inferior colliculi coordinate movements of the head, eyes, and trunk in response to auditory stimuli. The midbrain also contains nuclei associated with cranial nerves III and IV.
4. A large part of the brain stem consists of small areas of gray matter and white matter called the reticular formation, which helps maintain consciousness, causes awakening from sleep, and contributes to regulating muscle tone.

Anatomy Overview - Brain Stem
Figure 14.5 - Medulla Oblongata in Relation to the Rest of the Brain Stem

14.4 The Cerebellum

1. The cerebellum occupies the inferior and posterior aspects of the cranial cavity. It consists of two lateral hemispheres and a medial, constricted vermis.
2. It connects to the brain stem by three pairs of cerebellar peduncles.
3. The cerebellum smooths and coordinates the contractions of skeletal muscles. It also maintains posture and balance.

Anatomy Overview - Cerebellum

Review

14.5 The Diencephalon

1. The diencephalon surrounds the third ventricle and consists of the thalamus, hypothalamus, and epithalamus.
2. The thalamus is superior to the midbrain and contains nuclei that serve as relay stations for most sensory input to the cerebral cortex. It also contributes to motor functions by transmitting information from the cerebellum and basal nuclei to the primary motor area of the cerebral cortex. In addition, the thalamus plays a role in maintenance of consciousness.
3. The hypothalamus is inferior to the thalamus. It controls the autonomic nervous system, produces hormones, and regulates emotional and behavioral patterns (along with the limbic system). The hypothalamus also contains a feeding center and satiety center, which regulate eating, and a thirst center, which regulates drinking. In addition, the hypothalamus controls body temperature by serving as the body's thermostat. Also present in the hypothalamus is the suprachiasmatic nucleus, which regulates circadian rhythms and functions as the body's internal biological clock.
4. The epithalamus consists of the pineal gland and the habenular nuclei. The pineal gland secretes melatonin, which is thought to promote sleep and to help set the body's biological clock.
5. Circumventricular organs (CVOs) can monitor chemical changes in the blood because they lack the blood–brain barrier.

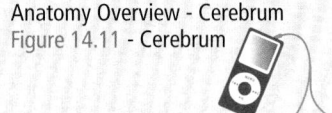

Anatomy Overview - Diencephalon
Figure 14.10 -
  Hypothalamus

14.6 The Cerebrum

1. The cerebrum is the largest part of the brain. Its cortex contains gyri (convolutions), fissures, and sulci.
2. The cerebral hemispheres are divided into four lobes: frontal, parietal, temporal, and occipital.
3. The white matter of the cerebrum is deep to the cortex and consists primarily of myelinated axons extending to other regions as association, commissural, and projection fibers.
4. The basal nuclei are several groups of nuclei in each cerebral hemisphere. They help initiate and terminate movements, suppress unwanted movements, and regulate muscle tone.
5. The limbic system encircles the upper part of the brain stem and the corpus callosum. It functions in emotional aspects of behavior and memory.
6. Table 14.2 summarizes the functions of various parts of the brain.

Anatomy Overview - Cerebrum
Figure 14.11 - Cerebrum

14.7 Functional Organization of the Cerebral Cortex

1. The sensory areas of the cerebral cortex allow perception of sensory information. The motor areas control the execution of voluntary movements. The association areas are concerned with more complex integrative functions such as memory, personality traits, and intelligence.
2. The primary somatosensory area (areas 1, 2, and 3) receives nerve impulses from somatic sensory receptors for touch, pressure, vibration, itch, tickle, temperature, pain, and proprioception and is involved in the perception of these sensations. Each point within the area receives impulses from a specific part of the face or body. The primary visual area (area 17) receives visual information and is involved in visual perception. The primary auditory area (areas 41 and 42) receives information for sound and is involved in auditory perception. The primary gustatory area (area 43) receives impulses for taste and is involved in gustatory perception and taste discrimination. The primary olfactory area (area 28) receives impulses for smell and is involved in olfactory perception.
3. Motor areas include the primary motor area (area 4), which controls voluntary contractions of specific muscles or groups of muscles, and Broca's speech area (areas 44 and 45), which controls production of speech.
4. The somatosensory association area (areas 5 and 7) permits you to determine the exact shape and texture of an object simply by touching it and to sense the relationship of one body part to another. It also stores memories of past somatic sensory experiences.
5. The visual association area (areas 18 and 19) relates present to past visual experiences and is essential for recognizing and evaluating what is seen. The facial recognition area (areas 20, 21, and 37) stores information about faces and allows you to recognize people by their faces. The auditory association area (area 22) allows you to recognize a particular sound as speech, music, or noise.
6. The orbitofrontal cortex (area 11) allows you to identify odors and discriminate among different odors. Wernicke's area (area 22 and possibly 39 and 40) interprets the meaning of speech by translating words into thoughts. The common integrative area (areas 5, 7, 39, and 40) integrates sensory interpretations from the association areas and impulses from other areas, allowing thoughts based on sensory inputs.
7. The prefrontal cortex (areas 9, 10, 11, and 12) is concerned with personality, intellect, complex learning abilities, judgment, reasoning, conscience, intuition, and development of abstract ideas. The premotor area (area 6) generates nerve impulses that cause specific groups of muscles to contract in

Review **Resource**

specific sequences. It also serves as a memory bank for complex movements. The frontal eye field area (area 8) controls voluntary scanning movements of the eyes.

8. Subtle anatomical differences exist between the two hemispheres, and each has unique functions. Each hemisphere receives sensory signals from and controls movements of the opposite side of the body. The left hemisphere is more important for language, numerical and scientific skills, and reasoning. The right hemisphere is more important for musical and artistic awareness, spatial and pattern perception, recognition of faces, emotional content of language, identifying odors, and generating mental images of sight, sound, touch, taste, and smell.

9. Brain waves generated by the cerebral cortex are recorded from the surface of the head in an electroencephalogram (EEG). The EEG may be used to diagnose epilepsy, infections, and tumors.

14.8 Cranial Nerves

1. Twelve pairs of cranial nerves originate from the nose, eyes, inner ear, brain stem, and spinal cord.
2. They are named primarily based on their distribution and are numbered I–XII in order of attachment to the brain. Exhibits 14.A–14.J and Table 14.4 summarize the types, locations, functions, and disorders of the cranial nerves.

Anatomy Overview - Cranial Nerves
Exercise - Cranial Nerve Target Practice: Motor Functions
Exercise - Cranial Nerve Target Practice: Sensory Functions

14.9 Development of the Nervous System

1. The development of the nervous system begins with a thickening of a region of the ectoderm called the neural plate.
2. During embryological development, primary brain vesicles form from the neural tube and serve as forerunners of various parts of the brain.
3. The telencephalon forms the cerebrum, the diencephalon develops into the thalamus and hypothalamus, the mesencephalon develops into the midbrain, the metencephalon develops into the pons and cerebellum, and the myelencephalon forms the medulla.

14.10 Aging and the Nervous System

1. The brain grows rapidly during the first few years of life.
2. Age-related effects involve loss of brain mass and decreased capacity for sending nerve impulses.

SELF-QUIZ QUESTIONS

Fill in the blanks in the following statements.

1. The cerebral hemispheres are connected internally by a broad band of white matter known as the _____.
2. List the five lobes of the cerebrum: _____, _____, _____, _____, _____.
3. The _____ separates the cerebrum into right and left halves.

Indicate whether the following statements are true or false.

4. The brain stem consists of the medulla oblongata, pons, and diencephalon.
5. You are the greatest student of anatomy and physiology, and you are well-prepared for your exam on the brain. As you confidently answer the questions, your brain is exhibiting beta waves.

Choose the one best answer to the following questions.

6. Which of the following is *not* a function of the thalamus?
   (a) relaying information from the cerebellum and basal nuclei to primary motor areas of the cerebral cortex
   (b) helping maintain consciousness
   (c) playing a role in emotions and memory
   (d) regulating body temperature
   (e) relaying sensory impulses to the cerebral cortex

7. Which of the following statements is *false*?
   (a) The blood supply to the brain is provided mainly by the internal carotid and vertebral arteries.
   (b) Neurons in the brain rely almost exclusively on aerobic respiration to produce ATP.
   (c) An interruption of blood flow to the brain for even 20 seconds may impair brain function.
   (d) Glucose supply to the brain must be continuous.
   (e) Low levels of glucose in the blood to the brain may result in unconsciousness.

8. In which of the following ways does cerebrospinal fluid contribute to homeostasis? (1) mechanical protection, (2) chemical protection, (3) electrical protection, (4) circulation, (5) immunity.
   (a) 1, 2, and 3   (b) 2, 3, and 4   (c) 3, 4, and 5
   (d) 1, 2, and 4   (e) 2, 4, and 5

9. Which of the following are functions of the hypothalamus? (1) control of the ANS, (2) production of hormones, (3) regulation of emotional and behavioral patterns, (4) regulation of eating and drinking, (5) control of body temperature, (6) regulation of circadian rhythms.
   (a) 1, 2, 4, and 6   (b) 2, 3, 5, and 6   (c) 1, 3, 5, and 6
   (d) 1, 4, 5, and 6   (e) 1, 2, 3, 4, 5, and 6

10. Which of the following statements is *false*?
(a) Association tracts transmit nerve impulses between gyri in the same hemisphere.
(b) Commissural tracts transmit impulses from the gyri in one cerebral hemisphere to the corresponding gyri in the other hemisphere.
(c) Projection tracts form descending and ascending tracts that transmit impulses from the cerebrum and other parts of the brain to the spinal cord, or from the spinal cord to the brain.
(d) The internal capsule is an example of a commissural tract.
(e) The corpus callosum is an example of a commissural tract.

11. Which of the following statements is *true*?
(a) The right and left hemispheres of the cerebrum are completely symmetrical.
(b) The left hemisphere controls the left side of the body.
(c) The right hemisphere is more important for spoken and written language.
(d) The left hemisphere is more important for musical and artistic awareness.
(e) Hemispheric lateralization is more pronounced in males than in females.

12. Match the following (some answers will be used more than once):
_____ (a) oculomotor
_____ (b) trigeminal
_____ (c) abducens
_____ (d) vestibulocochlear
_____ (e) accessory
_____ (f) vagus
_____ (g) facial
_____ (h) glossopharyngeal
_____ (i) olfactory
_____ (j) trochlear
_____ (k) optic
_____ (l) hypoglossal
_____ (m) functions in sense of smell
_____ (n) functions in hearing and equilibrium
_____ (o) functions in chewing
_____ (p) functions in facial expression and secretion of saliva and tears
_____ (q) functions in movement of tongue during speech and swallowing
_____ (r) functions in secretion of digestive fluids
_____ (s) functions in secretion of saliva, taste, regulation of blood pressure, and muscle sense
_____ (t) sensory only
_____ (u) functions in eye movement by controlling extrinsic eye muscles
_____ (v) functions in swallowing and head movements

(1) cranial nerve I
(2) cranial nerve II
(3) cranial nerve III
(4) cranial nerve IV
(5) cranial nerve V
(6) cranial nerve VI
(7) cranial nerve VII
(8) cranial nerve VIII
(9) cranial nerve IX
(10) cranial nerve X
(11) cranial nerve XI
(12) cranial nerve XII

13. Match the following (some answers may be used more than once):
_____ (a) emotional brain; involved in olfaction and memory
_____ (b) bridge connecting parts of the brain with each other
_____ (c) sensory relay area
_____ (d) alerts the cerebral cortex to incoming sensory signals
_____ (e) regulates posture and balance
_____ (f) lacks a blood-brain barrier; can monitor chemical changes in the blood
_____ (g) site of decussation of pyramids
_____ (h) site of pneumotaxic and apneustic areas
_____ (i) secretes melatonin
_____ (j) contains sensory, motor, and association areas
_____ (k) responsible for maintaining consciousness and awakening from sleep
_____ (l) controls ANS
_____ (m) contains reflex centers for movements of the eyes, head, and neck in response to visual and other stimuli, and reflex center for movements of the head and trunk in response to auditory stimuli
_____ (n) plays an essential role in awareness and in the acquisition of knowledge; cognition
_____ (o) several groups of nuclei that control large autonomic movements of skeletal muscles and help regulate muscle tone required for specific body movements
_____ (p) produces hormones that regulate endocrine gland function
_____ (q) contains the vital cardiovascular center and medullary rhythmicity center

(1) medulla oblongata
(2) pons
(3) midbrain
(4) cerebellum
(5) pineal gland
(6) thalamus
(7) hypothalamus
(8) cerebrum
(9) limbic system
(10) reticular formation
(11) circumventricular organs
(12) reticular activating system
(13) basal nuclei

14. Match the following:

_____ (a) protrusions in the medulla formed by the large corticospinal tracts

_____ (b) dura mater extension that separates the two cerebral hemispheres

_____ (c) fingerlike extensions of arachnoid mater where CSF is reabsorbed

_____ (d) dura mater extension that separates the two cerebellar hemispheres

_____ (e) located in the hypothalamus; relay stations for reflexes related to smell

_____ (f) folds in the cerebral cortex

_____ (g) shallow grooves in the cerebral cortex

_____ (h) bundles of white matter that relay information between the cerebellum and other parts of the brain

_____ (i) a thick band of sensory and motor tracts that connect the cerebral cortex with the brain stem and spinal cord

_____ (j) dura mater extension that separates the cerebrum from the cerebellum

_____ (k) thin membranous partition between the lateral ventricles

(1) gyri
(2) internal capsule
(3) mammillary bodies
(4) tentorium cerebelli
(5) pyramids
(6) falx cerebelli
(7) septum pellucidum
(8) cerebellar peduncles
(9) falx cerebri
(10) sulci
(11) arachnoid villi

15. Match the following:

_____ (a) allows planning and production of speech

_____ (b) receives impulses for sound

_____ (c) controls voluntary contraction of muscles

_____ (d) allows recognition and evaluation of visual experiences

_____ (e) integration and interpretation of somatic sensations; comparison of past to present sensations

_____ (f) receives impulses for touch, proprioception, pain, and temperature

_____ (g) receives impulses for taste

_____ (h) interpretation of sounds as speech, music, or noise

_____ (i) receives impulses from many sensory and association areas as well as the thalamus and brain stem; allows formation of thoughts so appropriate action can occur

_____ (j) translates words into thoughts

_____ (k) receives impulses for smell

_____ (l) allows awareness of shape, color, and movement

_____ (m) coordinates muscle movement for complex, learned sequential motor activities

_____ (n) involved in scanning eye movements

_____ (o) allows discrimination among different odors

(1) primary visual area
(2) primary auditory area
(3) primary gustatory area
(4) primary olfactory area
(5) primary somatosensory area
(6) primary motor area
(7) somatosensory association area
(8) visual association area
(9) frontal eye field
(10) Broca's area
(11) auditory association area
(12) premotor area
(13) Wernicke's area
(14) common integrative area
(15) orbitofrontal cortex

CRITICAL THINKING QUESTIONS

1. An elderly relative suffered a CVA (stroke) and now has difficulty moving her right arm, and she also has speech problems. What areas of the brain were damaged by the stroke?

2. Nicky has recently had a viral infection and now she cannot move the muscles on the right side of her face. In addition, she is experiencing a loss of taste and a dry mouth, and she cannot close her right eye. What cranial nerve has been affected by the viral infection?

3. You have been hired by a pharmaceutical company to develop a drug to regulate a specific brain disorder. What is a major physiological roadblock to developing such a drug, and how can you design a drug to bypass that roadblock so that the drug is delivered to the brain where it is needed?

? ANSWERS TO FIGURE QUESTIONS

14.1 The largest part of the brain is the cerebrum.

14.2 From superficial to deep, the three cranial meninges are the dura mater, arachnoid, and pia mater.

14.3 The brain stem is anterior to the fourth ventricle, and the cerebellum is posterior to it.

14.4 Cerebrospinal fluid is reabsorbed by the arachnoid villi that project into the dural venous sinuses.

14.5 The medulla oblongata contains the pyramids; the midbrain contains the cerebral peduncles; *pons* means "bridge."

14.6 Decussation means crossing to the opposite side. The functional consequence of decussation of the pyramids is that each side of the cerebrum controls muscles on the opposite side of the body.

14.7 The cerebral peduncles are the main sites through which tracts extend and nerve impulses are conducted between the superior parts of the brain and the inferior parts of the brain and the spinal cord.

14.8 The cerebellar peduncles carry information into and out of the cerebellum.

14.9 In about 70% of human brains, the intermediate mass connects the right and left halves of the thalamus.

14.10 From posterior to anterior, the four major regions of the hypothalamus are the mammillary, tuberal, supraoptic, and preoptic regions.

14.11 The gray matter enlarges more rapidly during development, in the process producing convolutions or gyri (folds), sulci (shallow grooves), and fissures (deep grooves).

14.12 Association tracts connect gyri of the same hemisphere; commissural tracts connect gyri in opposite hemispheres; projection tracts connect the cerebrum with the thalamus, brain stem, and spinal cord.

14.13 The basal nuclei are lateral, superior, and inferior to the thalamus.

14.14 The hippocampus is the part of the limbic system that functions with the cerebrum in memory.

14.15 The common integrative association area integrates interpretation of visual, auditory, and somatic sensations; Broca's speech area translates thoughts into speech; the premotor area controls skilled muscular movements; the primary gustatory area interprets sensations related to taste; the primary auditory area allows you to interpret pitch and rhythm; the primary visual area allows you to interpret shape, color, and movement of objects; and the frontal eye field area controls voluntary scanning movements of the eyes.

14.16 In an EEG, theta waves indicate emotional stress.

14.17 Axons in the olfactory tracts terminate in the primary olfactory area in the temporal lobe of the cerebral cortex.

14.18 Most axons in the optic tracts terminate in the lateral geniculate nucleus of the thalamus.

14.19 The superior branch of the oculomotor nerve is distributed to the superior rectus muscle; the trochlear nerve is the smallest cranial nerve.

14.20 The trigeminal nerve is the largest cranial nerve.

14.21 Motor axons of the facial nerve originate in the pons.

14.22 The vestibular ganglion contains cell bodies from sensory axons that arise in the semicircular canals, saccule, and utricle; the spiral ganglion contains cell bodies from axons that arise in the spiral organ of the cochlea.

14.23 The glossopharyngeal nerve exits the skull through the jugular foramen.

14.24 The vagus nerve is located medial and posterior to the internal jugular vein and common carotid artery in the neck.

14.25 The accessory nerve is the only cranial nerve that originates from both the brain and spinal cord.

14.26 Two important motor functions of the hypoglossal nerve are speech and swallowing.

14.27 The gray matter of the nervous system derives from the mantle layer cells of the neural tube.

14.28 The mesencephalon does not develop into a secondary brain vesicle.

15

THE AUTONOMIC NERVOUS SYSTEM

THE AUTONOMIC NERVOUS SYSTEM AND HOMEOSTASIS *The autonomic nervous system contributes to homeostasis by responding to subconscious visceral sensations and exciting or inhibiting smooth muscle, cardiac muscle, and glands.*

As you learned in Chapter 12, the peripheral nervous system (PNS) includes cranial and spinal nerves and is divided into a somatic nervous system (SNS), autonomic nervous system (ANS), and enteric nervous system (ENS). Like the somatic nervous system, the **autonomic nervous system (ANS)** (aw′-tō-NOM-ik) operates via reflex arcs. (The derivation is *auto-* = self; *-nomic* = law.) Structurally, the ANS includes autonomic sensory neurons, integrating centers in the central nervous system (CNS), autonomic motor neurons, and the enteric division. A continual flow of nerve impulses from (1) *autonomic sensory neurons* in visceral organs and blood vessels propagate into (2) *integrating centers* in the CNS. Then, impulses in (3) *autonomic motor neurons* propagate to various effector tissues, thereby regulating the activity of smooth muscle, cardiac muscle, and many glands. (4) The *enteric division* is a specialized network of nerves and ganglia forming an independent nerve network within the wall of the gastrointestinal (GI) tract. The ANS usually operates without conscious control. However, centers in the hypothalamus and brain stem do regulate ANS reflexes. In this chapter, we compare structural and functional features of the somatic and autonomic nervous systems. Then we discuss the anatomy of the motor portion of the ANS and compare the organization and actions of its two major parts, the sympathetic and parasympathetic divisions.

? *Did you ever wonder how some blood pressure medications exert their effects through the autonomic nervous system?*

15.1 COMPARISON OF SOMATIC AND AUTONOMIC NERVOUS SYSTEMS

◉ **OBJECTIVE**

• Compare the structural and functional differences between the somatic and autonomic parts of the nervous system.

Somatic Nervous System

The somatic nervous system includes both sensory and motor neurons. Sensory neurons convey input from receptors for somatic senses (tactile, thermal, pain, and proprioceptive sensations; see Chapter 16) and from receptors for the special senses (sight, hearing, taste, smell, and equilibrium; see Chapter 17). All these sensations normally are consciously perceived. In turn, somatic motor neurons innervate skeletal muscles—the effectors of the somatic nervous system—and produce both reflexive and voluntary movements. When a somatic motor neuron stimulates the muscle, it contracts; the effect always is excitation. If somatic motor neurons cease to stimulate a muscle, the result is a paralyzed, limp muscle that has no muscle tone. Although we are generally not conscious of breathing, the muscles that generate respiratory movements are also skeletal muscles controlled by somatic motor neurons. If the respiratory motor neurons become inactive, breathing stops. A few skeletal muscles, such as those in the middle ear, are controlled by reflexes and cannot be contracted voluntarily.

Autonomic Nervous System

The main input to the ANS comes from **autonomic (visceral) sensory neurons.** Mostly, these neurons are associated with **interoceptors** (IN-ter-ō-sep′-tors), which are sensory receptors located in blood vessels, visceral organs, muscles, and the nervous system that monitor conditions in the *internal* environment. Examples of interoceptors are chemoreceptors that monitor blood CO_2 level and mechanoreceptors that detect the degree of stretch in the walls of organs or blood vessels. Unlike those triggered by a flower's perfume, a beautiful painting, or a delicious meal, these sensory signals are not consciously perceived most of the time, although intense activation of interoceptors may produce conscious sensations. Two examples of perceived visceral sensations are pain sensations from damaged viscera and angina pectoris (chest pain) from inadequate blood flow to the heart. Input that influences the ANS also includes some sensations monitored by somatic sensory and special sensory neurons. For example, pain can produce dramatic changes in some autonomic activities.

Autonomic motor neurons regulate visceral activities by either increasing (exciting) or decreasing (inhibiting) ongoing activities in their effector tissues (cardiac muscle, smooth muscle, and glands). Changes in the diameter of the pupils, dilation and constriction of blood vessels, and adjustment of the rate and force of the heartbeat are examples of autonomic motor responses. Unlike skeletal muscle, tissues innervated by the ANS often

function to some extent even if their nerve supply is damaged. For example, the heart continues to beat when it is removed for transplantation into another person, smooth muscle in the lining of the gastrointestinal tract contracts rhythmically on its own, and glands produce some secretions in the absence of ANS control.

Most autonomic responses cannot be consciously altered to any great degree. You probably cannot voluntarily slow your heartbeat to half its normal rate. For this reason, some autonomic responses are the basis for polygraph ("lie detector") tests. However, practitioners of yoga or other techniques of meditation may learn how to regulate at least some of their autonomic activities through long practice. **Biofeedback,** in which monitoring devices display information about a body function such as heart rate or blood pressure, enhances the ability to learn such conscious control. Signals from the general somatic and special senses, acting via the limbic system, also influence responses of autonomic motor neurons. Seeing a bike about to hit you, hearing squealing brakes of a nearby car, or being grabbed by an attacker would all increase the rate and force of your heartbeat.

Comparison of Somatic and Autonomic Motor Neurons

Recall from Chapter 10 that the axon of a single, myelinated somatic motor neuron extends from the central nervous system (CNS) all the way to the skeletal muscle fibers in its motor unit (Figure 15.1a). By contrast, most autonomic motor pathways consist of two motor neurons in series, that is, one following the other (Figure 15.1b). The first neuron (preganglionic neuron) has its cell body in the CNS; its myelinated axon extends from the CNS to an **autonomic ganglion.** (Recall that a *ganglion* is a collection of neuronal cell bodies in the PNS.) The cell body of the second neuron (postganglionic neuron) is also in that same autonomic ganglion; its unmyelinated axon extends directly from the ganglion to the effector (smooth muscle, cardiac muscle, or a gland). Alternatively, in some autonomic pathways, the first motor neuron extends to specialized cells called *chromaffin cells* in the adrenal medullae (inner portions of the adrenal glands) rather than an autonomic ganglion. In addition, all somatic motor neurons release only acetylcholine (ACh) as their neurotransmitter, but autonomic motor neurons release either ACh or norepinephrine (NE).

Unlike somatic output (motor), the output part of the ANS has two divisions: the **sympathetic division** and the **parasympathetic division.** Most organs have **dual innervation;** that is, they receive impulses from both sympathetic and parasympathetic neurons. In some organs, nerve impulses from one division of the ANS stimulate the organ to increase its activity (excitation), and impulses from the other division decrease the organ's activity (inhibition). For example, an increased rate of nerve impulses from the *sympathetic* division increases heart rate, and an increased rate of nerve impulses from the *parasympathetic* division decreases heart rate. The sympathetic division is often called the *fight-or-flight division.* Sympathetic activities result in increased alertness and metabolic activities in order to prepare the body for an emergency situation. Responses to such situations, which may

Figure 15.1 **Motor neuron pathways in the (a) somatic nervous system and (b) autonomic nervous system (ANS).**
Note that autonomic motor neurons release either acetylcholine (ACh) or norepinephrine (NE); somatic motor neurons release ACh.

🔑 Somatic nervous system stimulation always excites its effectors (skeletal muscle fibers); stimulation by the autonomic nervous system either excites or inhibits visceral effectors.

Somatic
motor neuron
(myelinated)

ACh

Spinal cord

Effector: skeletal muscle

(a) Somatic nervous system

Autonomic
motor neurons

ACh

NE

Spinal cord

Sympathetic
preganglionic
neuron
(myelinated)

Autonomic
ganglion

Sympathetic
postganglionic
neuron
(unmyelinated)

Effectors: glands, cardiac
muscle (in heart), and
smooth muscle (e.g., in
urinary bladder)

Adrenal cortex

Adrenal medulla

Chromaffin cell

ACh

Epinephrine
and NE

Spinal cord

Sympathetic
preganglionic
neuron
(myelinated)

Adrenal medulla

Blood vessel

ACh

ACh

Spinal cord

Parasympathetic
preganglionic
neuron
(myelinated)

Autonomic
ganglion

Parasympathetic
postganglionic
neuron
(unmyelinated)

Effectors: glands, cardiac
muscle (in heart), and
smooth muscle (e.g., in
urinary bladder)

(b) Autonomic nervous system

? **What does dual innervation mean?**

occur during physical activity or emotional stress, include a rapid heart rate, faster breathing rate, dilation of the pupils, dry mouth, sweaty but cool skin, dilation of blood vessels to organs involved in combating stress (such as the heart and skeletal muscles), constriction of blood vessels to organs not involved in combating

stress (for example, the gastrointestinal tract and kidneys), and release of glucose from the liver.

The parasympathetic division is often referred to as the *rest-and-digest division* because its activities conserve and restore body energy during times of rest or digesting a meal; the majority of its output is directed to the smooth muscle and glandular tissue of the gastrointestinal and respiratory tracts. The parasympathetic division conserves energy and replenishes nutrient stores. Although both the sympathetic and parasympathetic divisions are concerned with maintaining homeostasis, they do so in dramatically different ways.

Table 15.1 compares the somatic and autonomic nervous systems.

✔ CHECKPOINT

1. How do the autonomic nervous system and somatic nervous system compare in structure and function?
2. What are the main input and output components of the autonomic nervous system?

15.2 ANATOMY OF AUTONOMIC MOTOR PATHWAYS

◉ OBJECTIVES

• Describe preganglionic and postganglionic neurons of the autonomic nervous system.
• Compare the anatomical components of the sympathetic and parasympathetic divisions of the autonomic nervous system.

Anatomical Components

Each division of the ANS has two motor neurons. The first of the two motor neurons in any autonomic motor pathway is called a **preganglionic neuron** (Figure 15.1b). Its cell body is in the brain or spinal cord; its axon exits the CNS as part of a cranial or spinal nerve. The axon of a preganglionic neuron is a small-diameter, myelinated type B fiber that usually extends to an autonomic ganglion, where it synapses with a **postganglionic neuron,** the second neuron in the autonomic motor pathway. Note that the postganglionic neuron lies entirely outside the CNS in the PNS. Its cell body and dendrites are located in an **autonomic ganglion,** where it forms synapses with one or more preganglionic axons. The axon of a postganglionic neuron is a small-diameter, unmyelinated type C fiber that terminates in a visceral effector. Thus, preganglionic neurons convey nerve impulses from the CNS to autonomic ganglia, and postganglionic neurons relay the impulses from autonomic ganglia to visceral effectors.

Preganglionic Neurons

In the sympathetic division, the preganglionic neurons have their cell bodies in the lateral horns of the gray matter in the 12 thoracic segments and the first two (and sometimes three) lumbar segments of the spinal cord (Figure 15.2). For this reason, the sympathetic division is also called the **thoracolumbar division** (thōr′-a-kō-LUM-bar), and the axons of the sympathetic preganglionic neurons are known as the **thoracolumbar outflow.**

Cell bodies of preganglionic neurons of the parasympathetic division are located in the nuclei of four cranial nerves in the brain stem (III, VII, IX, and X) and in the lateral gray matter of the second through fourth sacral segments of the spinal cord (Figure 15.3). Hence, the parasympathetic division is also known

TABLE 15.1

Comparison of the Somatic and Autonomic Nervous Systems

	SOMATIC NERVOUS SYSTEM	AUTONOMIC NERVOUS SYSTEM
Sensory input	From somatic senses and special senses.	Mainly from interoceptors; some from somatic senses and special senses.
Control of motor output	Voluntary control from cerebral cortex, with contributions from basal ganglia, cerebellum, brain stem, and spinal cord.	Involuntary control from hypothalamus, limbic system, brain stem, and spinal cord; limited control from cerebral cortex.
Motor neuron pathway	One-neuron pathway: Somatic motor neurons extending from CNS synapse directly with effector.	Usually two-neuron pathway: Preganglionic neurons extending from CNS synapse with postganglionic neurons in autonomic ganglion, and postganglionic neurons extending from ganglion synapse with visceral effector. Alternatively, preganglionic neurons may extend from CNS to synapse with chromaffin cells of adrenal medullae.
Neurotransmitters and hormones	All somatic motor neurons release ACh.	All sympathetic and parasympathetic preganglionic neurons release ACh. Most sympathetic postganglionic neurons release NE; those to most sweat glands release ACh. All parasympathetic postganglionic neurons release ACh. Chromaffin cells of adrenal medullae release epinephrine and norepinephrine.
Effectors	Skeletal muscle.	Smooth muscle, cardiac muscle, and glands.
Responses	Contraction of skeletal muscle.	Contraction or relaxation of smooth muscle; increased or decreased rate and force of contraction of cardiac muscle; increased or decreased secretions of glands.

Figure 15.2 Structure of the sympathetic division of the autonomic nervous system. Solid lines represent preganglionic axons; dashed lines represent postganglionic axons. Although the innervated structures are shown for only one side of the body for diagrammatic purposes, the sympathetic division actually innervates tissues and organs on both sides.

 Cell bodies of sympathetic preganglionic neurons are located in the lateral horns of gray matter in the 12 thoracic and first two lumbar segments of the spinal cord.

SYMPATHETIC DIVISION (thoracolumbar)

Key:
— Preganglionic neurons
— Postganglionic neurons

Distributed primarily to smooth muscle of blood vessels of these organs:

Brain

Eye
Pineal gland
Lacrimal gland
Mucous membrane of nose and palate
Sublingual and submandibular glands
Parotid gland

Spinal cord

C1
C2
C3
C4
C5
C6
C7
C8

Superior cervical ganglion

Middle cervical ganglion

Inferior cervical ganglion

Heart
Atrial muscle fibers
SA/AV nodes
Ventricular muscle fibers

Cardiac plexus

Trachea
Bronchi
Lungs

T1
T2
T3
T4
T5
T6
T7
T8
T9
T10
T11
T12

Pulmonary plexus

Liver, gallbladder, and bile ducts

Greater splanchnic nerve

Celiac ganglion

Stomach
Spleen
Pancreas

Transverse colon

Aorticorenal ganglion

Lesser splanchnic nerve

Small intestine

Ascending colon

Descending colon

Sigmoid colon

Least splanchnic nerve

Rectum

Skin

Sweat gland
Hair follicle smooth muscle
Blood vessels (each sympathetic trunk innervates skin and viscera)

Superior mesenteric ganglion

Adrenal gland
Kidney
Ureter

L1
L2
L3
L4
L5

Renal ganglion

S1
S2
S3
S4
S5

Sympathetic trunk ganglia (on both sides)

Lumbar splanchnic nerve

Inferior mesenteric ganglion

Prevertebral ganglia

Coccygeal (fused together)

Urinary bladder   External genitals   Uterus

Which division, sympathetic or parasympathetic, has longer preganglionic axons? Why?

585

Figure 15.3 Structure of the parasympathetic division of the autonomic nervous system. Solid lines represent preganglionic axons; dashed lines represent postganglionic axons. Although the innervated structures are shown only for one side of the body for diagrammatic purposes, the parasympathetic division actually innervates tissues and organs on both sides.

6 Cell bodies of parasympathetic preganglionic neurons are located in brain stem nuclei and in the lateral gray matter in the second through fourth sacral segments of the spinal cord.

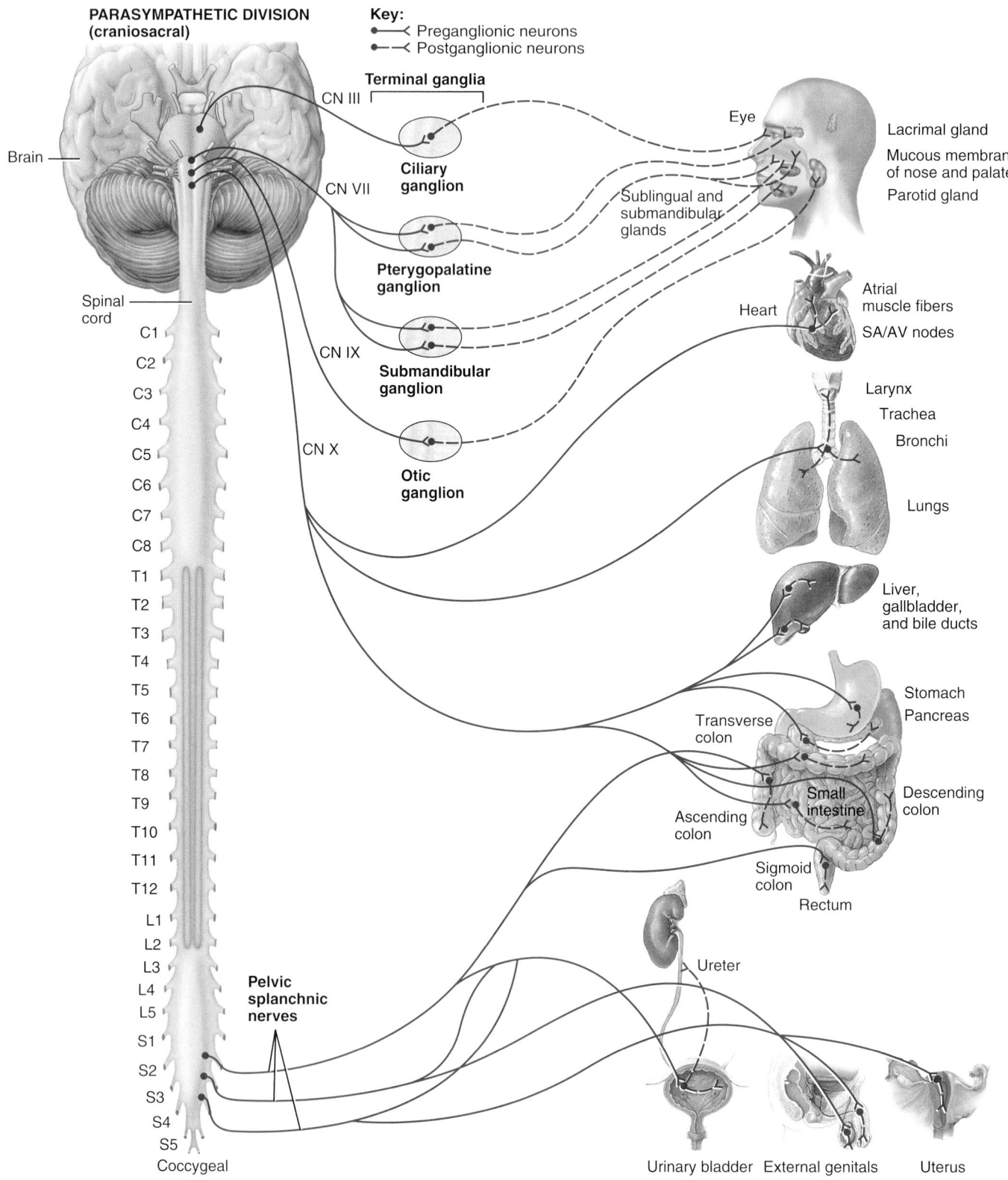

? Which ganglia are associated with the parasympathetic division? Sympathetic division?

as the **craniosacral division** (krā′-nē-ō-SĀK-ral), and the axons of the parasympathetic preganglionic neurons are referred to as the **craniosacral outflow.**

Autonomic Ganglia

There are two major groups of autonomic ganglia: (1) sympathetic ganglia, which are components of the sympathetic division of the ANS, and (2) parasympathetic ganglia, which are components of the parasympathetic division of the ANS.

SYMPATHETIC GANGLIA The sympathetic ganglia are the sites of synapses between sympathetic preganglionic and postganglionic neurons. There are two major types of sympathetic ganglia: sympathetic trunk ganglia and prevertebral ganglia. **Sympathetic trunk ganglia** (also called *vertebral chain ganglia* or *paravertebral ganglia*) lie in a vertical row on either side of the vertebral column. These ganglia extend from the base of the skull to the coccyx (Figure 15.2). Postganglionic axons from sympathetic trunk ganglia primarily innervate organs above the diaphragm, such as the head, neck, shoulders, and heart. Sympathetic trunk ganglia in the neck have specific names. They are the **superior, middle,** and **inferior cervical ganglia.** The remaining sympathetic trunk ganglia do not have individual names. Because the sympathetic trunk ganglia are near the spinal cord, most sympathetic preganglionic axons are short and most sympathetic postganglionic axons are long.

The second group of sympathetic ganglia, the **prevertebral (collateral) ganglia,** lies anterior to the vertebral column and close to the large abdominal arteries. In general, postganglionic axons from prevertebral ganglia innervate organs below the diaphragm. There are five major prevertebral ganglia (Figure 15.2; see also Figure 15.5): (1) The **celiac ganglion** (SĒ-lē-ak) is on either side of the celiac trunk, an artery that is just inferior to the diaphragm. (2) The **superior mesenteric ganglion** (MEZ-en-ter′-ik) is near the beginning of the superior mesenteric artery in the upper abdomen. (3) The **inferior mesenteric ganglion** is near the beginning of the inferior mesenteric artery in the middle of the abdomen. (4) The **aorticorenal ganglion** (ā-or′-ti-kō-RĒ-nal) and (5) the **renal ganglion** are near the renal artery of each kidney.

PARASYMPATHETIC GANGLIA Preganglionic axons of the parasympathetic division synapse with postganglionic neurons in **terminal (intramural) ganglia.** Most of these ganglia are located close to or actually within the wall of a visceral organ. Terminal ganglia in the head have specific names. They are the **ciliary ganglion, pterygopalatine ganglion** (ter′-i-gō-PAL-a-tīn), **submandibular ganglion,** and **otic ganglion** (Figure 15.3). The remaining terminal ganglia do not have specific names. Because terminal ganglia are located either close to or in the wall of the visceral organ, parasympathetic preganglionic axons are long, in contrast to parasympathetic postganglionic axons, which are short.

Postganglionic Neurons

Once axons of sympathetic preganglionic neurons pass to sympathetic trunk ganglia, they may connect with postganglionic neurons in one of the following ways (Figure 15.4):

1 An axon may synapse with postganglionic neurons in the ganglion it first reaches.

2 An axon may ascend or descend to a higher or lower ganglion before synapsing with postganglionic neurons. The axons of incoming sympathetic preganglionic neurons that pass up or down the sympathetic trunk collectively form the **sympathetic chains,** the fibers on which the ganglia are strung.

3 An axon may continue, without synapsing, through the sympathetic trunk ganglion to end at a prevertebral ganglion and synapse with postganglionic neurons there.

4 An axon may also pass, without synapsing, through the sympathetic trunk ganglion and a prevertebral ganglion and then extend to chromaffin cells of the adrenal medullae that are functionally similar to sympathetic postganglionic neurons.

A single sympathetic preganglionic fiber has many axon collaterals (branches) and may synapse with 20 or more postganglionic neurons. This pattern of projection is an example of divergence and helps explain why many sympathetic responses affect almost the entire body simultaneously. After exiting their ganglia, the postganglionic axons typically terminate in several visceral effectors (see Figure 15.2).

Axons of preganglionic neurons of the parasympathetic division pass to terminal ganglia near or within a visceral effector (see Figure 15.3). In the ganglion, the presynaptic neuron usually synapses with only four or five postsynaptic neurons, all of which supply a single visceral effector, allowing parasympathetic responses to be localized to a single effector.

Autonomic Plexuses

In the thorax, abdomen, and pelvis, axons of both sympathetic and parasympathetic neurons form tangled networks called **autonomic plexuses,** many of which lie along major arteries. The autonomic plexuses also may contain sympathetic ganglia and axons of autonomic sensory neurons. The major plexuses in the thorax are the **cardiac plexus,** which supplies the heart, and the **pulmonary plexus,** which supplies the bronchial tree (Figure 15.5).

The abdomen and pelvis also contain major autonomic plexuses (Figure 15.5), and often the plexuses are named after the artery along which they are distributed. The **celiac (solar) plexus** is the largest autonomic plexus and surrounds the celiac trunk. It contains two large celiac ganglia, two aorticorenal ganglia, and a dense network of autonomic axons and is distributed to the stomach, spleen, pancreas, liver, gallbladder, kidneys, adrenal medullae, testes, and ovaries. The **superior mesenteric plexus** contains the superior mesenteric ganglion and supplies the small and large intestines. The **inferior mesenteric plexus** contains the inferior mesenteric ganglion, which innervates the large intestine. Axons of some sympathetic postganglionic neurons from the inferior

Figure 15.4 Types of connections between ganglia and postganglionic neurons in the sympathetic division of the ANS.
Numbers correspond to descriptions in the text. Also illustrated are the gray and white rami communicantes.

Sympathetic ganglia lie in two chains on either side of the vertebral column (sympathetic trunk ganglia) and near large abdominal arteries anterior to the vertebral column (prevertebral ganglia).

Cephalic periarterial nerve

Carotid artery

Visceral effectors: eyes, lacrimal glands, salivary glands, pineal gland, nasal mucosa, and sweat glands, blood vessels, and arrector pili muscles of skin of face

Posterior ramus of spinal nerve

Anterior ramus of spinal nerve

Sympathetic nerve

Visceral effector: heart

Posterior horn

Posterior root

Posterior root ganglion

Lateral horn

Above T1

Sympathetic chain

Skin

Anterior horn

Spinal nerve

Spinal cord (thoracic or upper lumbar segment)

Anterior root

Visceral effectors: sweat glands, blood vessels, and arrector pili muscles of skin of neck, trunk, and limbs

Splanchnic nerve

Sympathetic trunk ganglion

Gray ramus communicans

Prevertebral ganglion (celiac ganglion)

White ramus communicans

Visceral effector: stomach

Adrenal cortex
Adrenal medulla

Chromaffin cell

Below L2

Adrenal gland

Key:
—< Sympathetic preganglionic neurons
—< Sympathetic postganglionic neurons

Anterior view

? **What is the significance of the sympathetic trunk ganglia?**

Figure 15.5 Autonomic plexuses in the thorax, abdomen, and pelvis.

 An autonomic plexus is a network of sympathetic and parasympathetic axons that sometimes also includes autonomic sensory axons and sympathetic ganglia.

Right vagus (X) nerve

Arch of aorta

Right primary bronchus

Right sympathetic trunk ganglion

Greater splanchnic nerve

Lesser splanchnic nerve

Inferior vena cava (cut)

Celiac trunk (artery)

AORTICORENAL GANGLION

Right kidney

Superior mesenteric artery

Right sympathetic trunk ganglion

Trachea

Left vagus (X) nerve

CARDIAC PLEXUS

PULMONARY PLEXUS

Esophagus

Thoracic aorta

ESOPHAGEAL PLEXUS

Diaphragm

CELIAC GANGLION AND PLEXUS

SUPERIOR MESENTERIC GANGLION AND PLEXUS

RENAL GANGLION AND RENAL PLEXUS

INFERIOR MESENTERIC GANGLION AND PLEXUS

Inferior mesenteric artery

HYPOGASTRIC PLEXUS

Anterior view

 Which is the largest autonomic plexus?

mesenteric ganglion also extend through the **hypogastric plexus,** which is anterior to the fifth lumbar vertebra, to supply the pelvic viscera. The **renal plexus** contains the renal ganglion and supplies the renal arteries within the kidneys and ureters.

With this background in mind, we can now examine some of the specific structural features of the sympathetic and parasympathetic divisions of the ANS in more detail.

Structure of the Sympathetic Division

Pathway from Spinal Cord to Sympathetic Trunk Ganglia

Cell bodies of sympathetic preganglionic neurons are part of the lateral gray horns of all thoracic segments and of the first two lumbar segments of the spinal cord (see Figure 15.2). The preganglionic axons leave the spinal cord along with the somatic motor neurons at the same segmental level. After exiting through the intervertebral foramina, the myelinated preganglionic sympathetic axons pass into the anterior root of a spinal nerve and enter a short pathway called a **white ramus** (RĀ-mus) before passing to the nearest sympathetic trunk ganglion on the same side (see Figure 15.4). Collectively, the white rami are called the **white rami communicantes** (kō-mū-ni-KAN-tēz; singular is **ramus communicans**). Thus, white rami communicantes are structures containing sympathetic preganglionic axons that connect the anterior ramus of the spinal nerve with the ganglia of the sympathetic trunk. The "white" in their name indicates that they contain myelinated axons. Only the thoracic and first two or three lumbar nerves have white rami communicantes.

Organization of Sympathetic Trunk Ganglia

The paired **sympathetic trunk ganglia** are arranged anterior and lateral to the vertebral column, one on either side. Typically, there are 3 cervical, 11 or 12 thoracic, 4 or 5 lumbar, 4 or 5 sacral sympathetic trunk ganglia, and 1 coccygeal ganglion. The right and left coccygeal ganglia are fused together and usually lie at the midline. Although the sympathetic trunk ganglia extend inferiorly from the neck, chest, and abdomen to the coccyx, they receive preganglionic axons only from the thoracic and lumbar segments of the spinal cord (see Figure 15.2).

The cervical portion of each sympathetic trunk is located in the neck and is subdivided into superior, middle, and inferior ganglia (see Figure 15.2). Postganglionic neurons leaving the **superior cervical ganglion** serve the head and heart. They are distributed to the sweat glands, smooth muscle of the eye, blood vessels of the face, lacrimal glands, pineal gland, nasal mucosa, salivary glands (which include the submandibular, sublingual, and parotid glands), and heart. Postganglionic neurons leaving the **middle cervical ganglion** and the **inferior cervical ganglion** innervate the heart.

The thoracic portion of each sympathetic trunk lies anterior to the necks of the corresponding ribs. This region of the sympathetic trunk receives most of the sympathetic preganglionic axons. Postganglionic neurons from the thoracic sympathetic trunk innervate the heart, lungs, bronchi, and other thoracic viscera. In the skin, these neurons also innervate sweat glands, blood vessels, and arrector pili muscles of hair follicles. The lumbar portion of each sympathetic trunk lies lateral to the corresponding lumbar vertebrae. The sacral region of the sympathetic trunk lies in the pelvic cavity on the medial side of the anterior sacral foramina.

Pathways from Sympathetic Trunk Ganglia to Visceral Effectors

Axons leave the sympathetic trunk in four possible ways: (1) They can enter spinal nerves; (2) they can form cephalic periarterial nerves; (3) they can form sympathetic nerves; and (4) they can form splanchnic nerves.

SPINAL NERVES   Recall that some of the incoming sympathetic preganglionic neurons synapse with postganglionic neurons in the sympathetic trunk, either in the ganglion at the level of entry or in a ganglion farther up or down the sympathetic trunk. The axons of some of these postganglionic neurons leave the sympathetic trunk by entering a short pathway called a **gray ramus** and then merge with the anterior ramus of a spinal nerve. Therefore, **gray rami communicantes** are structures containing sympathetic postganglionic axons that connect the ganglia of the sympathetic trunk to spinal nerves (see Figure 15.4). The "gray" in their name indicates that they contain unmyelinated axons. Gray rami communicantes outnumber the white rami because there is a gray ramus leading to each of the 31 pairs of spinal nerves. The axons of the postganglionic neurons that leave the sympathetic trunk to enter spinal nerves provide sympathetic innervation to the visceral effectors in the skin of the neck, trunk, and limbs, including sweat glands, smooth muscle in blood vessels, and arrector pili muscles of hair follicles.

CEPHALIC PERIARTERIAL NERVES   Some sympathetic preganglionic neurons that enter the sympathetic trunk ascend to the superior cervical ganglion, where they synapse with postganglionic neurons. The axons of some of these postganglionic neurons leave the sympathetic trunk by forming **cephalic periarterial nerves** (per'-ē-ar-TĒ-rē-al), nerves that extend to the head by wrapping around and following the course of various arteries (such as the carotid arteries) that pass from the neck to the head (see Figure 15.4). Cephalic periarterial nerves provide sympathetic innervation to visceral effectors in the skin of the face (sweat glands, smooth muscle of blood vessels, and arrector pili muscles of hair follicles), as well as other visceral effectors of the head (smooth muscle of the eye, lacrimal glands, pineal gland, nasal mucosa, and salivary glands).

SYMPATHETIC NERVES   Some of the incoming sympathetic preganglionic neurons synapse with postganglionic neurons in one or more ganglia of the sympathetic trunk. Then, the axons of the postganglionic neurons leave the trunk by forming **sympathetic nerves** that extend to visceral effectors in the thoracic cavity (Figure 15.4). Sympathetic nerves provide sympathetic innervation to the heart and lungs.

• **Sympathetic nerves to the heart.** Sympathetic innervation of the heart consists of axons of preganglionic neurons that enter the sympathetic trunk and then form synapses with postganglionic neurons in the superior, middle, and inferior cervical ganglia and first through fourth thoracic ganglia (T1–T4). From these ganglia, axons of postganglionic neurons exit the sympathetic trunk by forming sympathetic nerves that enter the cardiac plexus to supply the heart (see Figure 15.2).

- *Sympathetic nerves to the lungs.* Sympathetic innervation of the lungs consists of axons of preganglionic neurons that enter the sympathetic trunk and then form synapses with postganglionic neurons in the second through fourth thoracic ganglia (T2–T4). From these ganglia, axons of sympathetic postganglionic neurons exit the trunk by forming sympathetic nerves that enter the pulmonary plexus to supply the smooth muscle of the bronchi and bronchioles of the lungs (see Figure 15.2).

SPLANCHNIC NERVES Recall that some sympathetic preganglionic axons pass through the sympathetic trunk without terminating in it. Beyond the trunk, they form nerves known as **splanchnic nerves** (SPLANK-nik; see Figures 15.2 and 15.4), which extend to outlying prevertebral ganglia.

- *Splanchnic nerves to abdominopelvic organs.* Most sympathetic preganglionic axons that enter splanchnic nerves are destined to synapse with sympathetic postganglionic neurons in the prevertebral ganglia that supply the organs of the abdominopelvic cavity. Preganglionic axons from the fifth through ninth or tenth thoracic ganglia (T5–T9 or T10) form the **greater splanchnic nerve.** It pierces the diaphragm and enters the **celiac ganglion** of the celiac plexus. From there, postganglionic neurons follow and innervate blood vessels to the stomach, spleen, liver, kidneys, and small intestine. Preganglionic axons from the tenth and eleventh thoracic ganglia (T10–T11) form the **lesser splanchnic nerve.** It pierces the diaphragm and passes through the celiac plexus to enter the aorticorenal ganglion and superior mesenteric ganglion of the superior mesenteric plexus. Postganglionic neurons from the superior mesenteric ganglion follow and innervate blood vessels of the small intestine and proximal colon. The **least (lowest) splanchnic nerve,** which is not always present, is formed by preganglionic axons from the twelfth thoracic ganglia (T12) or a branch of the lesser splanchnic nerve. It pierces the diaphragm and enters the renal plexus near the kidney. Postganglionic neurons from the renal plexus supply kidney arterioles and the ureters. Preganglionic axons that form the **lumbar splanchnic nerve** from the first through fourth lumbar ganglia (L1–L4) enter the inferior mesenteric plexus and terminate in the **inferior mesenteric ganglion,** where they synapse with postganglionic neurons. Axons of postganglionic neurons extend through the inferior mesenteric plexus to supply the distal colon and rectum; they also extend through the hypogastric plexus to supply blood vessels of the distal colon, rectum, urinary bladder, and genital organs. Postganglionic axons leaving the prevertebral ganglia follow the course of various arteries to abdominal and pelvic visceral effectors.

- *Splanchnic nerves to the adrenal medulla.* Some sympathetic preganglionic axons pass, without synapsing, through the sympathetic trunk, greater splanchnic nerves, and celiac ganglion, and then extend to **chromaffin cells** in the adrenal medullae of the adrenal glands (see Figures 15.2 and 15.4). Developmentally, the adrenal medullae and sympathetic ganglia are derived from the same tissue, the neural crest (see Figure 14.27). The adrenal medullae are modified sympathetic ganglia, and the chromaffin cells are similar to sympathetic postganglionic neurons except they lack dendrites and axons. Rather than extending to another organ, however, these cells release hormones into the blood. Upon stimulation by sympathetic preganglionic neurons, the chromaffin cells of the adrenal medullae release a mixture of catecholamine hormones—about 80% **epinephrine,** 20% **norepinephrine,** and a trace amount of **dopamine.** These hormones circulate throughout the body and intensify responses elicited by sympathetic postganglionic neurons.

CLINICAL CONNECTION | *Horner's Syndrome*

In **Horner's syndrome,** the sympathetic innervation to one side of the face is lost due to an inherited mutation, an injury, or a disease that affects sympathetic outflow through the superior cervical ganglion. Symptoms occur on the affected side and include ptosis (drooping of the upper eyelid), miosis (constricted pupil), and anhidrosis (lack of sweating). •

Structure of the Parasympathetic Division

Cell bodies of parasympathetic preganglionic neurons are found in nuclei in the brain stem and in the lateral gray matter of the second through fourth sacral segments of the spinal cord (see Figure 15.3). Their axons emerge as part of a cranial nerve or as part of the anterior root of a spinal nerve. The **cranial parasympathetic outflow** consists of preganglionic axons that extend from the brain stem in four cranial nerves. The **sacral parasympathetic outflow** consists of preganglionic axons in anterior roots of the second through fourth sacral spinal nerves. The preganglionic axons of both the cranial and sacral outflows end in terminal ganglia, where they synapse with postganglionic neurons.

The cranial outflow has four pairs of ganglia and the ganglia associated with the vagus (X) nerve. The four pairs of cranial parasympathetic ganglia innervate structures in the head and are located close to the organs they innervate (see Figure 15.3).

1. The **ciliary ganglia** lie lateral to each optic (II) nerve near the posterior aspect of the orbit. Preganglionic axons pass with the oculomotor (III) nerves to the ciliary ganglia. Postganglionic axons from the ganglia innervate smooth muscle fibers in the eyeball.

2. The **pterygopalatine ganglia** (ter′-i-gō-PAL-a-tīn) are located lateral to the sphenopalatine foramen, between the sphenoid and palatine bones. They receive preganglionic axons from the facial (VII) nerve and send postganglionic axons to the nasal mucosa, palate, pharynx, and lacrimal glands.

3. The **submandibular ganglia** are found near the ducts of the submandibular salivary glands. They receive preganglionic axons from the facial nerves and send postganglionic axons to the submandibular and sublingual salivary glands.

4. The **otic ganglia** are situated just inferior to each foramen ovale. They receive preganglionic axons from the glossopharyngeal (IX) nerves and send postganglionic axons to the parotid salivary glands.

Preganglionic axons that leave the brain as part of the vagus (X) nerves carry nearly 80% of the total craniosacral outflow. Vagal axons extend to many terminal ganglia in the thorax and abdomen. As the vagus nerve passes through the thorax, it sends axons to the heart and the airways of the lungs. In the abdomen, it supplies the liver, gallbladder, stomach, pancreas, small intestine, and part of the large intestine.

The sacral parasympathetic outflow consists of preganglionic axons from the anterior roots of the second through fourth sacral spinal nerves (S2–S4). As the preganglionic axons course through the sacral spinal nerves, they branch off these nerves to form **pelvic splanchnic nerves** (Figure 15.6). Pelvic splanchnic nerves synapse with parasympathetic postganglionic neurons located in terminal ganglia in the walls of the innervated viscera. From the terminal ganglia, parasympathetic postganglionic axons innervate

Figure 15.6 **Pelvic splanchnic nerves.**

Through pelvic splanchnic nerves, axons of parasympathetic preganglionic neurons extend to parasympathetic postganglionic neurons in terminal ganglia in the walls of the colon, ureters, urinary bladder, and reproductive organs.

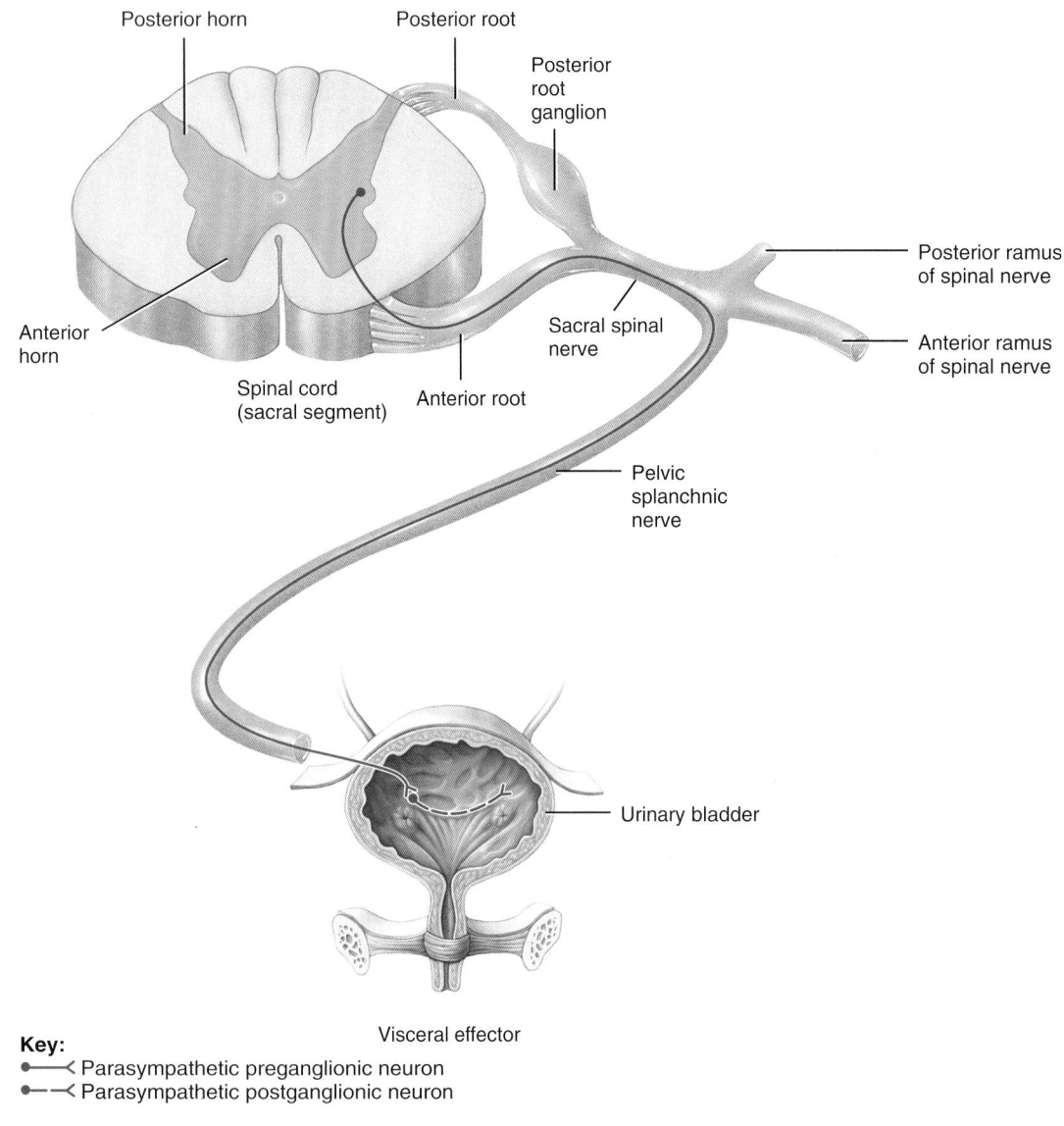

Posterior horn

Posterior root

Posterior root ganglion

Anterior horn

Spinal cord (sacral segment)

Anterior root

Sacral spinal nerve

Posterior ramus of spinal nerve

Anterior ramus of spinal nerve

Pelvic splanchnic nerve

Urinary bladder

Visceral effector

Key:
⚫━< Parasympathetic preganglionic neuron
⚫━< Parasympathetic postganglionic neuron

❓ **Pelvic splanchnic nerves branch from which spinal nerves?**

smooth muscle and glands in the walls of the colon, ureters, urinary bladder, and reproductive organs.

Structure of the Enteric Division

To understand and appreciate the **enteric division** (en-TER-ik) of the autonomic nervous system, it is important to realize that the gastrointestinal tract, like the surface of the body, forms an extensive area of contact with the environment. Although this gastrointestinal environment is inside the body, it is still considered part of the external environment. Just as the surface of the body must respond to important environmental stimuli in order to function properly, the surface of the gastrointestinal tract must respond to surrounding stimuli to generate proper homeostatic controls. In fact, these responses and controls are so important that the gastrointestinal tract has its own nervous system with intrinsic input, processing, and output. The enteric division can and does function independently of central nervous system activity, but it can also receive controlling input from the central nervous system.

The enteric division is the specialized collection of nerves and ganglia forming a complex, integrated neuronal network within the wall of the gastrointestinal tract, pancreas, and gallbladder. This incredible nerve network contains in the neighborhood of 100 million neurons, approximately the same number as the spinal cord, and is capable of continued function without input from the central nervous system. The enteric network of nerves and ganglia contains sensory neurons capable of monitoring tension in the intestinal wall and accessing the composition of the intestinal contents. These sensory neurons relay their input signals to interneurons within the enteric ganglia. The interneurons establish an integrative network that processes the incoming signals and generates regulatory output signals to motor neurons throughout plexuses within the wall of the digestive organs. The motor neurons carry the output signals to the smooth muscle and glands of the gastrointestinal tract to exert control over its motility and secretory activities.

Most of the nerve fibers that innervate the digestive organs arise from two plexuses within the enteric system. The largest, the **myenteric plexus** (mī-en-TER-ik), is positioned between the outer longitudinal and circular muscle layers from the upper esophagus to the anus. The myenteric plexus communicates extensively with a somewhat smaller plexus, the **submucosal plexus,** which occupies the gut wall between the circular muscle layer and the muscularis mucosae (see Section 24.3) and runs from the stomach to the anus. Neurons emerge from the ganglia of these two plexuses to form smaller plexuses around blood vessels and within the muscle layers and mucosa of the gut wall. It is this system of nerves that makes possible the normal motility and secretory functions of the gastrointestinal tract.

✔ CHECKPOINT

3. Why is the sympathetic division called the thoracolumbar division even though its ganglia extend from the cervical to the sacral region?
4. List the organs served by each sympathetic and parasympathetic ganglion.
5. Describe the locations of sympathetic trunk ganglia, prevertebral ganglia, and terminal ganglia. Which types of autonomic neurons synapse in each type of ganglion?
6. Why does the sympathetic division produce simultaneous effects throughout the body, in contrast to parasympathetic effects, which typically are localized to specific organs?
7. What are the functions of the enteric division of the ANS?

15.3 ANS NEUROTRANSMITTERS AND RECEPTORS

● OBJECTIVE

• Describe the neurotransmitters and receptors involved in autonomic responses.

Based on the neurotransmitter they produce and release, autonomic neurons are classified as either cholinergic or adrenergic. The receptors for the neurotransmitters are integral membrane proteins located in the plasma membrane of the postsynaptic neuron or effector cell.

Cholinergic Neurons and Receptors

Cholinergic neurons (kō′-lin-ER-jik) release the neurotransmitter **acetylcholine (ACh).** In the ANS, the cholinergic neurons include (1) all sympathetic and parasympathetic preganglionic neurons, (2) sympathetic postganglionic neurons that innervate most sweat glands, and (3) all parasympathetic postganglionic neurons (Figure 15.7).

ACh is stored in synaptic vesicles and released by exocytosis. It then diffuses across the synaptic cleft and binds with specific **cholinergic receptors,** integral membrane proteins in the *postsynaptic* plasma membrane. The two types of cholinergic receptors, both of which bind ACh, are nicotinic receptors and muscarinic receptors. **Nicotinic receptors** (nik′-ō-TIN-ik) are present in the plasma membrane of dendrites and cell bodies of both sympathetic and parasympathetic postganglionic neurons (Figure 15.7a, b), the plasma membranes of chromaffin cells of the adrenal medullae, and in the motor end plate at the neuromuscular junction. They are so named because nicotine mimics the action of ACh by binding to these receptors. (Nicotine, a natural substance in tobacco leaves, is not a naturally occurring substance in humans and is not normally present in nonsmokers.) **Muscarinic receptors** (mus′-ka-RIN-ik) are present in the plasma membranes of all effectors (smooth muscle, cardiac muscle, and glands) innervated by parasympathetic postganglionic axons. In addition, most sweat glands receive their innervation from *cholinergic* sympathetic postganglionic neurons and possess muscarinic receptors (see Figure 15.7b). These receptors are so named because a mushroom poison called muscarine mimics the actions of ACh by binding to them. Nicotine does not activate muscarinic receptors, and muscarine does not activate nicotinic receptors, but ACh does activate both types of cholinergic receptors.

Figure 15.7 Cholinergic neurons and adrenergic neurons in the sympathetic and parasympathetic divisions.

 Cholinergic neurons release acetylcholine; adrenergic neurons release norepinephrine. Cholinergic receptors (nicotinic or muscarinic) and adrenergic receptors are integral membrane proteins located in the plasma membrane of a postsynaptic neuron or an effector cell.

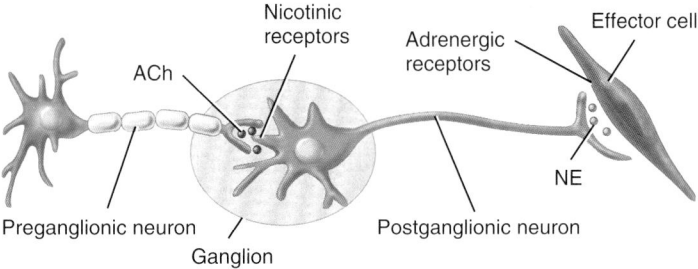

(a) Sympathetic division—innervation to most effector tissues

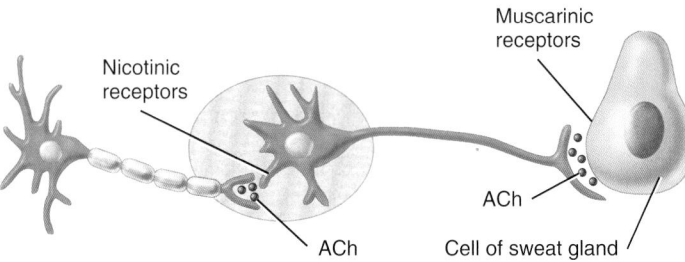

(b) Sympathetic division—innervation to most sweat glands

(c) Parasympathetic division

? **Which ANS neurons are adrenergic? What types of effector tissues contain muscarinic receptors?**

Activation of nicotinic receptors by ACh causes depolarization and thus excitation of the postsynaptic cell, which can be a postganglionic neuron, an autonomic effector, or a skeletal muscle fiber. Activation of muscarinic receptors by ACh sometimes causes depolarization (excitation) and sometimes causes hyperpolarization (inhibition), depending on which particular cell bears the muscarinic receptors. For example, binding of ACh to muscarinic receptors inhibits (relaxes) smooth muscle sphincters in the gastrointestinal tract. By contrast, ACh excites muscarinic receptors in smooth muscle fibers in the circular muscles of the iris of the eye, causing them to contract. Because acetylcholine is quickly inactivated by the enzyme **acetylcholinesterase (AChE),** effects triggered by cholinergic neurons are brief.

Adrenergic Neurons and Receptors

In the ANS, **adrenergic neurons** (ad′-ren-ER-jik) release **norepinephrine (NE),** also known as **noradrenalin** (Figure 15.7a). Most sympathetic postganglionic neurons are adrenergic. Like ACh, NE is stored in synaptic vesicles and released by exocytosis. Molecules of NE diffuse across the synaptic cleft and bind to specific adrenergic receptors on the postsynaptic membrane, causing either excitation or inhibition of the effector cell.

Adrenergic receptors bind both norepinephrine and epinephrine. The norepinephrine can be either released as a neurotransmitter by sympathetic postganglionic neurons or released as a hormone into the blood by chromaffin cells of the adrenal medullae; epinephrine is released as a hormone. The two main types of adrenergic receptors are **alpha (α) receptors** and **beta (β) receptors,** which are found on visceral effectors innervated by most sympathetic postganglionic axons. These receptors are further classified into subtypes—α_1, α_2, β_1, β_2, and β_3—based on the specific responses they elicit and by their selective binding of drugs that activate or block them. Although there are some exceptions, activation of α_1 and β_1 receptors generally produces excitation, and activation of α_2 and β_2 receptors causes inhibition of effector tissues. β_3 receptors are present only on cells of brown adipose tissue, where their activation causes *thermogenesis* (heat production). Cells of most effectors contain either alpha or beta receptors; some visceral effector cells contain both. Norepinephrine stimulates alpha receptors more strongly than beta receptors; epinephrine is a potent stimulator of both alpha and beta receptors.

The activity of norepinephrine at a synapse is terminated either when the NE is taken up by the axon that released it or when the NE is enzymatically inactivated by either **catechol-*O*-methyltransferase (COMT)** (kat′-e-kol′-ō-meth′-il-TRANS-fer-ās) or **monoamine oxidase (MAO)** (mon′-ō-AM-ēn OK-si-dās′). Compared to ACh, norepinephrine lingers in the synaptic cleft for a longer time. Thus, effects triggered by adrenergic neurons typically are longer lasting than those triggered by cholinergic neurons.

Table 15.2 describes the locations of cholinergic and adrenergic receptors and summarizes the responses that occur when each type of receptor is activated.

Receptor Agonists and Antagonists

A large variety of drugs and natural products can selectively activate or block specific cholinergic or adrenergic receptors. An **agonist** (*agon* = a contest) is a substance that binds to and activates a receptor, in the process mimicking the effect of a natural

TABLE 15.2

Location and Responses of Adrenergic and Cholinergic Receptors

TYPE OF RECEPTOR	MAJOR LOCATIONS	EFFECTS OF RECEPTOR ACTIVATION
CHOLINERGIC	Integral proteins in postsynaptic plasma membranes; activated by the neurotransmitter acetylcholine.	
Nicotinic	Plasma membrane of postganglionic sympathetic and parasympathetic neurons.	Excitation → impulses in postganglionic neurons.
	Chromaffin cells of adrenal medullae.	Epinephrine and norepinephrine secretion.
	Sarcolemma of skeletal muscle fibers (motor end plate).	Excitation → contraction.
Muscarinic	Effectors innervated by parasympathetic postganglionic neurons.	In some receptors, excitation; in others, inhibition.
	Sweat glands innervated by cholinergic sympathetic postganglionic neurons.	Increased sweating.
	Skeletal muscle blood vessels innervated by cholinergic sympathetic postganglionic neurons.	Inhibition → relaxation → vasodilation.
ADRENERGIC	Integral proteins in postsynaptic plasma membranes; activated by the neurotransmitter norepinephrine and the hormones norepinephrine and epinephrine.	
α_1	Smooth muscle fibers in blood vessels that serve salivary glands, skin, mucosal membranes, kidneys, and abdominal viscera; radial muscle in iris of eye; sphincter muscles of stomach and urinary bladder.	Excitation → contraction, which causes vasoconstriction, dilation of pupil, and closing of sphincters.
	Salivary gland cells.	Secretion of K^+ and water.
	Sweat glands on palms and soles.	Increased sweating.
α_2	Smooth muscle fibers in some blood vessels.	Inhibition → relaxation → vasodilation.
	Cells of pancreatic islets that secrete the hormone insulin (beta cells).	Decreased insulin secretion.
	Pancreatic acinar cells.	Inhibition of digestive enzyme secretion.
	Platelets in blood.	Aggregation to form platelet plug.
β_1	Cardiac muscle fibers.	Excitation → increased force and rate of contraction.
	Juxtaglomerular cells of kidneys.	Renin secretion.
	Posterior pituitary.	Antidiuretic hormone (ADH) secretion.
	Adipose cells.	Breakdown of triglycerides → release of fatty acids into blood.
β_2	Smooth muscle in walls of airways; in blood vessels that serve heart, skeletal muscle, adipose tissue, and liver; and in walls of visceral organs, such as urinary bladder.	Inhibition → relaxation, which causes dilation of airways, vasodilation, and relaxation of organ walls.
	Ciliary muscle in eye.	Inhibition → relaxation.
	Hepatocytes in liver.	Glycogenolysis (breakdown of glycogen into glucose).
β_3	Brown adipose tissue.	Thermogenesis (heat production).

neurotransmitter or hormone. Phenylephrine, an adrenergic agonist at α_1 receptors, is a common ingredient in cold and sinus medications. Because it constricts blood vessels in the nasal mucosa, phenylephrine reduces production of mucus, thus relieving nasal congestion. An **antagonist** (*anti* = against) is a substance that binds to and blocks a receptor, thereby preventing a natural neurotransmitter or hormone from exerting its effect. For example, atropine blocks muscarinic ACh receptors, dilates the pupils, reduces glandular secretions, and relaxes smooth muscle in the

gastrointestinal tract. As a result, it is used to dilate the pupils during eye examinations, in the treatment of smooth muscle disorders such as iritis and intestinal hypermotility, and as an antidote for chemical warfare agents that inactivate acetylcholinesterase.

Propranolol (Inderal®) often is prescribed for patients with hypertension (high blood pressure). It is a nonselective beta blocker, meaning it binds to all types of beta receptors and prevents their activation by epinephrine and norepinephrine. The desired effects of propranolol are due to its *blockade* of β_1 receptors—namely, decreased heart rate and force of contraction and a consequent decrease in blood pressure. Undesired effects due to blockade of β_2 receptors may include hypoglycemia (low blood glucose), resulting from decreased glycogen breakdown and decreased gluconeogenesis (the conversion of a noncarbohydrate into glucose in the liver), and mild bronchoconstriction (narrowing of the airways). If these side effects pose a threat to the patient, a selective β_1 blocker (which binds only to specific beta receptors) such as metoprolol (Lopressor®) can be prescribed instead of propranolol.

✔**CHECKPOINT**

8. Why are cholinergic and adrenergic neurons so named?
9. What neurotransmitters and hormones bind to adrenergic receptors?
10. What do the terms agonist and antagonist mean?

15.4 PHYSIOLOGY OF THE ANS

◉ **OBJECTIVE**

• Describe the major responses of the body to stimulation by the sympathetic and parasympathetic divisions of the ANS.

Autonomic Tone

As noted earlier, most body organs receive innervation from both divisions of the ANS, which typically work in opposition to one another. The balance between sympathetic and parasympathetic activity, called **autonomic tone,** is regulated by the hypothalamus. Typically, the hypothalamus turns up sympathetic tone at the same time it turns down parasympathetic tone, and vice versa. The two divisions can affect body organs differently because their postganglionic neurons release different neurotransmitters and because the effector organs possess different adrenergic and cholinergic receptors. A few structures receive only sympathetic innervation—sweat glands, arrector pili muscles attached to hair follicles in the skin, the kidneys, the spleen, most blood vessels, and the adrenal medullae (see Figure 15.2). In these structures there is no opposition from the parasympathetic division. Still, an increase in sympathetic tone has one effect, and a decrease in sympathetic tone produces the opposite effect.

Sympathetic Responses

During physical or emotional stress, the sympathetic division dominates the parasympathetic division. High sympathetic tone favors body functions that can support vigorous physical activity and rapid production of ATP. At the same time, the sympathetic division reduces body functions that favor the storage of energy. Besides physical exertion, various emotions—such as fear, embarrassment, or rage—stimulate the sympathetic division. Visualizing body changes that occur during "E situations" such as exercise, emergency, excitement, and embarrassment will help you

remember most of the sympathetic responses. Activation of the sympathetic division and release of hormones by the adrenal medullae set in motion a series of physiological responses collectively called the **fight-or-flight response,** which includes the following effects:

• The pupils of the eyes dilate.

• Heart rate, force of heart contraction, and blood pressure increase.

• The airways dilate, allowing faster movement of air into and out of the lungs.

• The blood vessels that supply the kidneys and gastrointestinal tract constrict, which decreases blood flow through these tissues. The result is a slowing of urine formation and digestive activities, which are not essential during exercise.

• Blood vessels that supply organs involved in exercise or fighting off danger—skeletal muscles, cardiac muscle, liver, and adipose tissue—dilate, allowing greater blood flow through these tissues.

• Liver cells perform glycogenolysis (breakdown of glycogen to glucose), and adipose tissue cells perform lipolysis (breakdown of triglycerides to fatty acids and glycerol).

• Release of glucose by the liver increases blood glucose level.

• Processes that are not essential for meeting the stressful situation are inhibited. For example, muscular movements of the gastrointestinal tract and digestive secretions slow down or even stop.

The effects of sympathetic stimulation are longer lasting and more widespread than the effects of parasympathetic stimulation for three reasons: (1) Sympathetic postganglionic axons diverge more extensively; as a result, many tissues are activated simultaneously. (2) Acetylcholinesterase quickly inactivates acetylcholine, but norepinephrine lingers in the synaptic cleft for a longer period. (3) Epinephrine and norepinephrine secreted into the blood from the adrenal medullae intensify and prolong the responses caused by NE liberated from sympathetic postganglionic axons. These blood-borne hormones circulate throughout the body, affecting all tissues that have alpha and beta receptors. In time, blood-borne NE and epinephrine are destroyed by enzymes in the liver.

Parasympathetic Responses

In contrast to the fight-or-flight activities of the sympathetic division, the parasympathetic division enhances **rest-and-digest** activities. Parasympathetic responses support body functions that conserve and restore body energy during times of rest and recovery. In the quiet intervals between periods of exercise, parasympathetic impulses to the digestive glands and the smooth muscle of the gastrointestinal tract predominate over sympathetic impulses. This allows energy-supplying food to be digested and absorbed. At the same time, parasympathetic responses reduce body functions that support physical activity.

The acronym *SLUDD* can be helpful in remembering five parasympathetic responses. It stands for salivation (S), lacrimation (L), urination (U), digestion (D), and defecation (D). All of these activities are stimulated mainly by the parasympathetic division. Besides the increasing SLUDD responses, other important parasympathetic responses are "three decreases": decreased heart rate, decreased diameter of airways (bronchoconstriction), and decreased diameter (constriction) of the pupils.

Table 15.3 compares the structural and functional features of the sympathetic and parasympathetic divisions of the ANS. Table 15.4 lists the responses of glands, cardiac muscle, and smooth muscle to stimulation by the sympathetic and parasympathetic divisions of the ANS.

✔ CHECKPOINT

11. Define autonomic tone.
12. What are some examples of the antagonistic effects of the sympathetic and parasympathetic divisions of the autonomic nervous system?
13. What happens during the fight-or-flight response?
14. Why is the parasympathetic division of the ANS called an energy conservation/restoration system?
15. Describe the sympathetic response in a frightening situation for each of the following body parts: hair follicles, iris of eye, lungs, spleen, adrenal medullae, urinary bladder, stomach, intestines, gallbladder, liver, heart, arterioles of the abdominal viscera, and arterioles of skeletal muscles.

TABLE 15.3

Comparison of Sympathetic and Parasympathetic Divisions of the ANS

	SYMPATHETIC (THORACOLUMBAR)	PARASYMPATHETIC (CRANIOSACRAL)
Distribution	Wide regions of body: skin, sweat glands, arrector pili muscles of hair follicles, adipose tissue, smooth muscle of blood vessels.	Limited mainly to head and to viscera of thorax, abdomen, and pelvis; some blood vessels.
Location of preganglionic neuron cell bodies and site of outflow	Lateral gray horns of spinal cord segments T1–L2. Axons of preganglionic neurons constitute thoracolumbar outflow.	Nuclei of cranial nerves III, VII, IX, and X and lateral gray matter of spinal cord segments S2–S4. Axons of preganglionic neurons constitute craniosacral outflow.
Associated ganglia	Sympathetic trunk ganglia and prevertebral ganglia.	Terminal ganglia.
Ganglia locations	Close to CNS and distant from visceral effectors.	Typically near or within wall of visceral effectors.
Axon length and divergence	Preganglionic neurons with short axons synapse with many postganglionic neurons with long axons that pass to many visceral effectors.	Preganglionic neurons with long axons usually synapse with four to five postganglionic neurons with short axons that pass to single visceral effector.
White and gray rami communicantes	Both present; white rami communicantes contain myelinated preganglionic axons; gray rami communicantes contain unmyelinated postganglionic axons.	Neither present.
Neurotransmitters	Preganglionic neurons release acetylcholine (ACh), which is excitatory and stimulates postganglionic neurons; most postganglionic neurons release norepinephrine (NE); postganglionic neurons that innervate most sweat glands and some blood vessels in skeletal muscle release ACh.	Preganglionic neurons release ACh, which is excitatory and stimulates postganglionic neurons; postganglionic neurons release ACh.
Physiological effects	Fight-or-flight responses.	Rest-and-digest activities.

TABLE 15.4

Effects of Sympathetic and Parasympathetic Divisions of the ANS

VISCERAL EFFECTOR	EFFECT OF SYMPATHETIC STIMULATION (α OR β ADRENERGIC RECEPTORS, EXCEPT AS NOTED)*	EFFECT OF PARASYMPATHETIC STIMULATION (MUSCARINIC ACh RECEPTORS)
GLANDS		
Adrenal medullae	Secretion of epinephrine and norepinephrine (nicotinic ACh receptors).	No known effect.
Lacrimal (tear)	Slight secretion of tears (α).	Secretion of tears.
Pancreas	Inhibits secretion of digestive enzymes and the hormone insulin (α_2); promotes secretion of the hormone glucagon (β_2).	Secretion of digestive enzymes and the hormone insulin.
Posterior pituitary	Secretion of antidiuretic hormone (ADH) (β_1).	No known effect.
Pineal	Increases synthesis and release of melatonin (β).	No known effect.
Sweat	Increases sweating in most body regions (muscarinic ACh receptors); sweating on palms and soles (α_1).	No known effect.
Adipose tissue[†]	Lipolysis (breakdown of triglycerides into fatty acids and glycerol) (β_1); release of fatty acids into blood (β_1 and β_3).	No known effect.
Liver[†]	Glycogenolysis (conversion of glycogen into glucose); gluconeogenesis (conversion of noncarbohydrates into glucose); decreased bile secretion (α and β_2).	Glycogen synthesis; increased bile secretion.
Kidney, juxtaglomerular cells[†]	Secretion of renin (β_1).	No known effect.
CARDIAC (HEART) MUSCLE		
	Increased heart rate and force of atrial and ventricular contractions (β_1).	Decreased heart rate; decreased force of atrial contraction.
SMOOTH MUSCLE		
Iris, radial muscle	Contraction → dilation of pupil (α_1).	No known effect.
Iris, circular muscle	No known effect.	Contraction → constriction of pupil.
Ciliary muscle of eye	Relaxation to adjust shape of lens for distant vision (β_2).	Contraction for close vision.
Lungs, bronchial muscle	Relaxation → airway dilation (β_2).	Contraction → airway constriction.
Gallbladder and ducts	Relaxation to facilitate storage of bile in the gallbladder ($\beta2$).	Contraction → release of bile into small intestine.
Stomach and intestines	Decreased motility and tone (α_1, α_2, β_2); contraction of sphincters (α_1).	Increased motility and tone; relaxation of sphincters.
Spleen	Contraction and discharge of stored blood into general circulation (α_1).	No known effect.
Ureter	Increases motility (α_1).	Increases motility (?).
Urinary bladder	Relaxation of muscular wall (β_2); contraction of internal urethral sphincter (α_1).	Contraction of muscular wall; relaxation of internal urethral sphincter.
Uterus	Inhibits contraction in nonpregnant women (β_2); promotes contraction in pregnant women (α_1).	Minimal effect.
Sex organs	In males: contraction of smooth muscle of ductus (vas) deferens, prostate, and seminal vesicle resulting in ejaculation (α_1).	Vasodilation; erection of clitoris (females) and penis (males).
Hair follicles, arrector pili muscle	Contraction → erection of hairs resulting in goosebumps (α_1).	No known effect.

TABLE 15.4 CONTINUED

Effects of Sympathetic and Parasympathetic Divisions of the ANS

VISCERAL EFFECTOR	EFFECT OF SYMPATHETIC STIMULATION (α OR β ADRENERGIC RECEPTORS, EXCEPT AS NOTED)*	EFFECT OF PARASYMPATHETIC STIMULATION (MUSCARINIC ACh RECEPTORS)
VASCULAR SMOOTH MUSCLE		
Salivary gland arterioles	Vasoconstriction, which decreases secretion of saliva (α_1).	Vasodilation, which increases secretion of saliva.
Gastric gland arterioles	Vasoconstriction, which inhibits secretion (α_1).	Secretion of gastric juice.
Intestinal gland arterioles	Vasoconstriction, which inhibits secretion (α_1).	Secretion of intestinal juice.
Coronary (heart) arterioles	Relaxation → vasodilation (β_2); contraction → vasoconstriction (α_1, α_2); contraction → vasoconstriction (muscarinic ACh receptors).	Contraction → vasoconstriction.
Skin and mucosal arterioles	Contraction → vasoconstriction (α_1).	Vasodilation, which may not be physiologically significant.
Skeletal muscle arterioles	Contraction → vasoconstriction (α_1); relaxation → vasodilation (β_2); relaxation → vasodilation (muscarinic ACh receptors).	No known effect.
Abdominal viscera arterioles	Contraction → vasoconstriction (α_1, β_2).	No known effect.
Brain arterioles	Slight contraction → vasoconstriction (α_1).	No known effect.
Kidney arterioles	Constriction of blood vessels → decreased urine volume (α_1).	No known effect.
Systemic veins	Contraction → constriction (α_1); relaxation → dilation (β_2).	No known effect.

*Subcategories of α and β receptors are listed if known.   †Grouped with glands because they release substances into the blood.

15.5 INTEGRATION AND CONTROL OF AUTONOMIC FUNCTIONS

⊙ **OBJECTIVES**

- Describe the components of an autonomic reflex.
- Explain the relationship of the hypothalamus to the ANS.

Autonomic Reflexes

Autonomic reflexes are responses that occur when nerve impulses pass through an autonomic reflex arc. These reflexes play a key role in regulating controlled conditions in the body, such as *blood pressure,* by adjusting heart rate, force of ventricular contraction, and blood vessel diameter; *digestion,* by adjusting the motility (movement) and muscle tone of the gastrointestinal tract; and *defecation* and *urination,* by regulating the opening and closing of sphincters.

The components of an autonomic reflex arc are as follows:

- **Receptor.** Like the receptor in a somatic reflex arc (see Figure 13.14), the receptor in an autonomic reflex arc is the distal end of a sensory neuron, which responds to a stimulus and produces a change that will ultimately trigger nerve impulses. Autonomic sensory receptors are mostly associated with interoceptors.

- **Sensory neuron.** Conducts nerve impulses from receptors to the CNS.

- **Integrating center.** Interneurons within the CNS relay signals from sensory neurons to motor neurons. The main integrating centers for most autonomic reflexes are located in the hypothalamus and brain stem. Some autonomic reflexes, such as those for urination and defecation, have integrating centers in the spinal cord.

- **Motor neurons.** Nerve impulses triggered by the integrating center propagate out of the CNS along motor neurons to an effector. In an autonomic reflex arc, two motor neurons connect the CNS to an effector: The preganglionic neuron conducts motor impulses from the CNS to an autonomic ganglion, and the postganglionic neuron conducts motor impulses from an autonomic ganglion to an effector (see Figure 15.1).

- **Effector.** In an autonomic reflex arc, the effectors are smooth muscle, cardiac muscle, and glands, and the reflex is called an autonomic reflex.

Focus on Homeostasis

BODY SYSTEM	CONTRIBUTION OF THE NERVOUS SYSTEM

 For all body systems
Together with hormones from the endocrine system, nerve impulses provide communication and regulation of most body tissues.

 Integumentary system
Sympathetic nerves of the autonomic nervous system (ANS) control contraction of smooth muscles attached to hair follicles and secretion of perspiration from sweat glands.

 Skeletal system
Pain receptors in bone tissue warn of bone trauma or damage.

 Muscular system
Somatic motor neurons receive instructions from motor areas of the brain and stimulate contraction of skeletal muscles to bring about body movements; basal nuclei and reticular formation set level of muscle tone; cerebellum coordinates skilled movements.

 Endocrine system
Hypothalamus regulates secretion of hormones from anterior and posterior pituitary; ANS regulates secretion of hormones from adrenal medulla and pancreas.

 Cardiovascular system
Cardiovascular center in the medulla oblongata provides nerve impulses to ANS that govern heart rate and the forcefulness of the heartbeat; nerve impulses from ANS also regulate blood pressure and blood flow through blood vessels.

  Lymphatic system and immunity
Certain neurotransmitters help regulate immune responses; activity in nervous system may increase or decrease immune responses.

 Respiratory system
Respiratory areas in brain stem control breathing rate and depth; ANS helps regulate diameter of airways.

 Digestive system
ANS and enteric nervous system (ENS) help regulate digestion; parasympathetic division of ANS stimulates many digestive processes.

 Urinary system
ANS helps regulate blood flow to kidneys, thereby influencing the rate of urine formation; brain and spinal cord centers govern emptying of the urinary bladder.

Reproductive systems
Hypothalamus and limbic system govern a variety of sexual behaviors; ANS brings about erection of penis in males and clitoris in females and ejaculation of semen in males; hypothalamus regulates release of anterior pituitary hormones that control gonads (ovaries and testes); nerve impulses elicited by touch stimuli from suckling infant cause release of oxytocin and milk ejection in nursing mothers.

THE NERVOUS SYSTEM

Autonomic Control by Higher Centers

Normally, we are not aware of muscular contractions of our digestive organs, our heartbeat, changes in the diameter of our blood vessels, and pupil dilation and constriction because the integrating centers for these autonomic responses are in the spinal cord or the lower regions of the brain. Somatic or autonomic sensory neurons deliver input to these centers, and autonomic motor neurons provide output that adjusts activity in the visceral effector, usually without our conscious perception.

The hypothalamus is the major control and integration center of the ANS. The hypothalamus receives sensory input related to visceral functions, olfaction (smell), and gustation (taste), as well as changes in temperature, osmolarity, and levels of various substances in blood. It also receives input relating to emotions from the limbic system. Output from the hypothalamus influences autonomic centers in both the brain stem (such as the cardiovascular, salivation, swallowing, and vomiting centers) and the spinal cord (such as the defecation and urination reflex centers in the sacral spinal cord).

Anatomically, the hypothalamus is connected to both the sympathetic and parasympathetic divisions of the ANS by axons of neurons with dendrites and cell bodies in various hypothalamic nuclei. The axons form tracts from the hypothalamus to parasympathetic and sympathetic nuclei in the brain stem and spinal cord through relays in the reticular formation. The posterior and lateral parts of the hypothalamus control the sympathetic division. Stimulation of these areas produces an increase in heart rate and force of contraction, a rise in blood pressure due to constriction of blood vessels, an increase in body temperature, dilation of the pupils, and inhibition of the gastrointestinal tract. In contrast, the anterior and medial parts of the hypothalamus control the parasympathetic division. Stimulation of these areas results in a decrease in heart rate, lowering of blood pressure, constriction of the pupils, and increased secretion and motility of the gastrointestinal tract.

✔ CHECKPOINT

16. Give three examples of controlled conditions in the body that are kept in homeostatic balance by autonomic reflexes.
17. How does an autonomic reflex arc differ from a somatic reflex arc?

• • •

Now that we have discussed the structure and function of the nervous system, you can appreciate the many ways that this system contributes to homeostasis of other body systems by examining *Focus on Homeostasis: The Nervous System.*

DISORDERS: HOMEOSTATIC IMBALANCES

Autonomic Dysreflexia

Autonomic dysreflexia (dis'-rē-FLEKS-sē-a) is an exaggerated response of the sympathetic division of the ANS that occurs in about 85% of individuals with spinal cord injury at or above the level of T6. The condition is seen after recovery from spinal shock (see Disorders: Homeostatic Imbalances in Chapter 13) and occurs due to interruption of the control of ANS neurons by higher centers. When certain sensory impulses, such as those resulting from stretching of a full urinary bladder, are unable to ascend the spinal cord, mass stimulation of the sympathetic nerves inferior to the level of injury occurs. Other triggers include stimulation of pain receptors and the visceral contractions resulting from sexual stimulation, labor/delivery, and bowel stimulation. Among the effects of increased sympathetic activity is severe vasoconstriction, which elevates blood pressure. In response, the cardiovascular center in the medulla oblongata (1) increases parasympathetic output via the vagus (X) nerve, which decreases heart rate, and (2) decreases sympathetic output, which causes dilation of blood vessels superior to the level of the injury.

The condition is characterized by a pounding headache; hypertension; flushed, warm skin with profuse sweating above the injury level; pale, cold, dry skin below the injury level; and anxiety. This emergency condition requires immediate intervention. The first approach is to quickly identify and remove the problematic stimulus. If this does not relieve the symptoms, an antihypertensive drug such as clonidine or nitroglycerin can be administered. Untreated autonomic dysreflexia can cause seizures, stroke, or heart attack.

Raynaud Phenomenon

In **Raynaud phenomenon** (rā-NŌ) the digits (fingers and toes) become ischemic (lack blood) after exposure to cold or with emotional stress. The condition is due to excessive sympathetic stimulation of smooth muscle in the arterioles of the digits and a heightened response to stimuli that cause vasoconstriction. When arterioles in the digits vasoconstrict in response to sympathetic stimulation, blood flow is greatly diminished. As a result, the digits may blanch (look white due to blockage of blood flow) or become cyanotic (look blue due to deoxygenated blood in capillaries). In extreme cases, the digits may become necrotic from lack of oxygen and nutrients. With rewarming after cold exposure, the arterioles may dilate, causing the fingers and toes to look red. Many patients with Raynaud phenomenon have low blood pressure. Some have increased numbers of alpha adrenergic receptors. Raynaud is most common in young women and occurs more often in cold climates. Patients with Raynaud phenomenon should avoid exposure to cold, wear warm clothing, and keep the hands and feet warm. Drugs used to treat Raynaud include nifedipine, a calcium channel blocker that relaxes vascular smooth muscle, and prazosin, which relaxes smooth muscle by blocking alpha receptors. Smoking and the use of alcohol or illicit drugs can exacerbate the symptoms of this condition.

MEDICAL TERMINOLOGY

Autonomic nerve neuropathy (noo-ROP-a-thē) A *neuropathy* (disorder of a cranial or spinal nerve) that affects one or more autonomic nerves, with multiple effects on the autonomic nervous system, including constipation, urinary incontinence, impotence, and fainting and low blood pressure when standing (*orthostatic hypotension*) due to decreased sympathetic control of the cardiovascular system. Often caused by long-term diabetes mellitus (*diabetic neuropathy*).

Biofeedback A technique in which an individual is provided with information regarding an autonomic response such as heart rate, blood pressure, or skin temperature. Various electronic monitoring devices provide visual or auditory signals about the autonomic responses. By concentrating on positive thoughts, individuals learn to alter autonomic responses. For example, biofeedback has been used to decrease heart rate and blood pressure and increase skin temperature in order to decrease the severity of migraine headaches.

Dysautonomia (dis-aw-tō-NŌ-mē-a; *dys-* = difficult; *-autonomia* = self-governing) An inherited disorder in which the autonomic nervous system functions abnormally, resulting in reduced tear gland secretions, poor vasomotor control, motor incoordination, skin blotching, absence of pain sensation, difficulty in swallowing, hyporeflexia, excessive vomiting, and emotional instability.

Hyperhidrosis (hī′-per-hī-DRŌ-sis; *hyper-* = above or too much; *-hidrosis* = sweat) Excessive or profuse sweating due to intense stimulation of sweat glands.

Mass reflex In cases of severe spinal cord injury above the level of the sixth thoracic vertebra, stimulation of the skin or overfilling of a visceral organ (such as the urinary bladder or colon) below the level of the injury results in intense activation of autonomic and somatic output from the spinal cord as reflex activity returns. The exaggerated response occurs because there is no inhibitory input from the brain. The mass reflex consists of flexor spasms of the lower limbs, evacuation of the urinary bladder and colon, and profuse sweating below the level of the lesion.

Megacolon (*mega-* = big) An abnormally large colon. In congenital megacolon, parasympathetic nerves to the distal segment of the colon do not develop properly. Loss of motor function in the segment causes massive dilation of the normal proximal colon. The condition results in extreme constipation, abdominal distension, and occasionally, vomiting. Surgical removal of the affected segment of the colon corrects the disorder.

Reflex sympathetic dystrophy (RSD) A syndrome that includes spontaneous pain, painful hypersensitivity to stimuli such as light touch, and excessive coldness and sweating in the involved body part. The disorder frequently involves the forearms, hands, knees, and feet. It appears that activation of the sympathetic division of the autonomic nervous system due to traumatized nociceptors as a result of trauma or surgery on bones or joints is involved. Treatment consists of anesthetics and physical therapy. Recent clinical studies also suggest that the drug baclofen can be used to reduce pain and restore normal function to the affected body part. Also called **complex regional pain syndrome type 1**.

Vagotomy (vā-GOT-ō-mē; *-tome* = incision) Cutting the vagus (X) nerve. It is frequently done to decrease the production of hydrochloric acid in persons with ulcers.

CHAPTER REVIEW AND RESOURCE SUMMARY

Review	Resource
15.1 Comparison of Somatic and Autonomic Nervous Systems	Anatomy Overview - The Nervous System: Overview
1. The somatic nervous system operates under conscious control; the ANS usually operates without conscious control.	Anatomy Overview - Organization of the ANS
2. Sensory input for the somatic nervous system is mainly from the somatic senses and special senses; sensory input for the ANS is from interoceptors, in addition to somatic senses and special senses.	Animation - Somatic Sensory and Motor Pathways
3. The axons of somatic motor neurons extend from the CNS and synapse directly with an effector. Autonomic motor pathways consist of two motor neurons in series. The axon of the first motor neuron extends from the CNS and synapses in an autonomic ganglion with the second motor neuron; the second neuron synapses with an effector.	Figure 15.1 - Motor Neuron Pathways in the Somatic and Autonomic Nervous Systems
4. The output (motor) portion of the ANS has two major divisions: sympathetic and parasympathetic. Most body organs receive dual innervation; usually one ANS division causes excitation and the other causes inhibition. The enteric division consists of nerves and ganglia within the wall of the GI tract.	Exercise - Assemble the Structure of the ANS
5. Somatic nervous system effectors are skeletal muscles; ANS effectors include cardiac muscle, smooth muscle, and glands.	Exercise - Sort ANS Functions
6. Table 15.1 compares the somatic and autonomic nervous systems.	Exercise - What Is Your ANS Status?
15.2 Anatomy of Autonomic Motor Pathways	Animation - ANS: Motor Pathways
1. A preganglionic neuron is the first of the two motor neurons in any autonomic motor pathway; the axon of the preganglionic neuron extends to an autonomic ganglion, where it synapses with a postganglionic neuron, the second neuron in the autonomic motor pathway. Preganglionic neurons are myelinated; postganglionic neurons are unmyelinated.	Anatomy Overview - Visceral Receptors Anatomy Overview - Visceral Effectors
2. The cell bodies of sympathetic preganglionic neurons are in the lateral gray horns of the 12 thoracic and the first two or three lumbar segments of the spinal cord; the cell bodies of parasympathetic	

Review   **Resource**

preganglionic neurons are in four cranial nerve nuclei (III, VII, IX, and X) in the brain stem and lateral gray matter of the second through fourth sacral segments of the spinal cord.

3. There are two major groups of autonomic ganglia: sympathetic ganglia and parasympathetic ganglia. Sympathetic ganglia include sympathetic trunk ganglia (on both sides of vertebral column) and prevertebral ganglia (anterior to vertebral column). Parasympathetic ganglia are known as terminal ganglia (near or inside visceral effectors).

4. Sympathetic preganglionic neurons synapse with postganglionic neurons in ganglia of the sympathetic trunk or in prevertebral ganglia; parasympathetic preganglionic neurons synapse with postganglionic neurons in terminal ganglia.

15.3 ANS Neurotransmitters and Receptors

1. Cholinergic neurons release acetylcholine. In the ANS, cholinergic neurons include all sympathetic and parasympathetic preganglionic neurons, sympathetic postganglionic neurons that innervate most sweat glands, and all parasympathetic postganglionic neurons.

2. Acetylcholine binds to cholinergic receptors. The two types of cholinergic receptors, both of which bind acetylcholine, are nicotinic receptors and muscarinic receptors. Nicotinic receptors are present in the plasma membranes of dendrites and cell bodies of both sympathetic and parasympathetic postganglionic neurons, in the plasma membranes of chromaffin cells of the adrenal medullae, and in the motor end plate at the neuromuscular junction. Muscarinic receptors are present in the plasma membranes of all effectors innervated by parasympathetic postganglionic neurons and in most sweat glands innervated by cholinergic sympathetic postganglionic neurons.

3. In the ANS, adrenergic neurons release norepinephrine. Most sympathetic postganglionic neurons are adrenergic.

4. Both epinephrine and norepinephrine bind to adrenergic receptors, which are found on visceral effectors innervated by most sympathetic postganglionic neurons. The two main types of adrenergic receptors are alpha receptors and beta receptors.

5. Table 15.2 summarizes the types of cholinergic and adrenergic receptors.

6. An agonist is a substance that binds to and activates a receptor, mimicking the effect of a natural neurotransmitter or hormone. An antagonist is a substance that binds to and blocks a receptor, thereby preventing a natural neurotransmitter or hormone from exerting its effect.

Anatomy Overview - Neurotransmitters
Animation - The ANS: Types of Neurotransmitters and Neurons
Figure 15.7 - Cholinergic and Adrenergic Neurons in the Sympathetic and Parasympathetic Divisions

15.4 Physiology of the ANS

1. The sympathetic division favors body functions that can support vigorous physical activity and rapid production of ATP (fight-or-flight response); the parasympathetic division regulates activities that conserve and restore body energy.

2. The effects of sympathetic stimulation are longer lasting and more widespread than the effects of parasympathetic stimulation.

3. Table 15.3 compares structural and functional features of the sympathetic and parasympathetic divisions.

4. Table 15.4 lists sympathetic and parasympathetic responses.

Anatomy Overview - Effectors
Animation - Physiological Effects of the ANS
Animation - The Alarm Reaction

15.5 Integration and Control of Autonomic Functions

1. An autonomic reflex adjusts the activities of smooth muscle, cardiac muscle, and glands.

2. An autonomic reflex arc consists of a receptor, a sensory neuron, an integrating center, two autonomic motor neurons, and a visceral effector.

3. The hypothalamus is the major control and integration center of the ANS. It is connected to both the sympathetic and the parasympathetic divisions.

Anatomy Overview - The ANS Control Centers
Animation - Reflexes

SELF-QUIZ QUESTIONS

Fill in the blanks in the following statements.

1. Cholinergic neurons release _____ and adrenergic neurons release _____.

2. Because of the location of the preganglionic cell bodies, the sympathetic division of the ANS is also called the _____ division; the parasympathetic division is also called the _____ division.

Indicate whether the following statements are true or false.

3. The vagus nerves transmit 80% of the outflow of the parasympathetic preganglionic axons.

4. Organs that receive both sympathetic and parasympathetic motor impulses are said to have dual innervation.

Choose the one best answer to the following questions.

5. Which of the following statements is *false*?
   (a) A single sympathetic preganglionic fiber may synapse with 20 or more postganglionic fibers, which partly explains why sympathetic responses are widespread throughout the body.
   (b) Parasympathetic effects tend to be localized because parasympathetic neurons usually synapse in the terminal ganglia with only four or five postsynaptic neurons (all of which supply a single effector).
   (c) Some sympathetic preganglionic neurons extend to and terminate in the adrenal medullae.
   (d) The parasympathetic preganglionic neurons synapse with the postganglionic axons in the prevertebral ganglia.
   (e) Parasympathetic preganglionic neurons emerge from the CNS as part of a cranial nerve or anterior root of a spinal nerve.

6. Which autonomic plexus supplies the large intestine? (1) renal, (2) inferior mesenteric, (3) hypogastric, (4) superior mesenteric, (5) celiac.
   (a) 2, 3, and 4   (b) 1, 2, 3, 4, and 5   (c) 3 and 4
   (d) 4 and 5   (e) 2 and 4

7. Which of the following statements are *true*? (1) The somatic nervous system and the ANS both include sensory and motor neurons. (2) Somatic motor neurons release the neurotransmitter norepinephrine. (3) The effect of an autonomic motor neuron is either excitation or inhibition, but that of a somatic motor neuron is always excitation. (4) Autonomic sensory neurons are mostly associated with interoceptors. (5) Autonomic motor pathways consist of two motor neurons in series. (6) Somatic motor pathways consist of two motor neurons in series.
   (a) 1, 2, 3, 4, and 5   (b) 1, 3, 4, and 5   (c) 2, 3, 5, and 6
   (d) 1, 3, 5, and 6   (e) 2, 4, 5, and 6

8. Which of the following statements is *false*?
   (a) The first neuron in an autonomic pathway is the preganglionic neuron.
   (b) The axons of preganglionic neurons are located in spinal or cranial nerves.
   (c) The postganglionic neuron's cell body is within the CNS.
   (d) Postganglionic neurons relay impulses from autonomic ganglia to visceral effectors.
   (e) All somatic motor neurons release acetylcholine.

9. Which of the following are *true*? (1) Monoamine oxidase enzymatically breaks down norepinephrine. (2) Activation of α_2 and β_2 receptors generally produces excitation in the effectors. (3) A beta blocker works by preventing activation of β receptors by epinephrine and norepinephrine. (4) An agonist is a substance that binds to a receptor and prevents the natural neurotransmitter from exerting its effect. (5) Activation of nicotinic receptors always causes excitation of the postsynaptic cell.
   (a) 2 and 3   (b) 1, 2, and 3   (c) 2, 4, and 5
   (d) 1, 2, 3, 4, and 5   (e) 1, 3, and 5

10. Which of the following are cholinergic neurons? (1) all sympathetic preganglionic neurons, (2) all parasympathetic preganglionic neurons, (3) all parasympathetic postganglionic neurons, (4) all sympathetic postganglionic neurons, (5) some sympathetic postganglionic neurons.
    (a) 1, 2, 3, and 5   (b) 1, 2, 3, and 4   (c) 2, 3, and 5
    (d) 2 and 5   (e) 1, 3, and 5

11. Which of the following statements are *true*? (1) Most sympathetic postganglionic axons are adrenergic. (2) Cholinergic receptors are classified as nicotinic and muscarinic. (3) Adrenergic receptors are classified as alpha and beta. (4) Muscarinic receptors are present on all effectors innervated by parasympathetic postganglionic axons. (5) In general, norepinephrine stimulates alpha receptors more vigorously than beta receptors; epinephrine is a potent stimulator of both alpha and beta receptors.
    (a) 1, 2, 3, 4, and 5   (b) 2, 3, 4 and 5   (c) 1, 3, 4, and 5
    (d) 3, 4, and 5   (e) 1, 2, 3, and 4

12. Which of the following are reasons why the effects of sympathetic stimulation are longer lasting and more widespread than those of parasympathetic stimulation? (1) There is greater divergence of sympathetic postganglionic fibers. (2) There is less divergence of sympathetic postganglionic fibers. (3) Acetylcholinesterase quickly inactivates ACh, but norepinephrine lingers in the synaptic cleft for a longer time. (4) Norepinephrine and epinephrine secreted into the blood by the adrenal medullae intensify the actions of the sympathetic division. (5) ACh remains in the synaptic cleft until norepinephrine is produced.
    (a) 1 and 3   (b) 1, 3, and 5   (c) 1, 3, and 4
    (d) 2, 3, and 4   (e) 2, 3, and 5

13. Place the following components of an autonomic reflex arc in the correct order from beginning to end.
    (a) postganglionic neuron   (b) sensory neuron
    (c) effector   (d) autonomic ganglion
    (e) receptor   (f) preganglionic neuron
    (g) integrating center

14. Match the following:
    ___ (a) also known as intramural ganglia
    ___ (b) includes the celiac, superior mesenteric, and inferior mesenteric ganglia
    ___ (c) also called vertebral chain or paravertebral ganglia
    ___ (d) lie in a vertical row on either side of the vertebral column
    ___ (e) postganglionic fibers, in general, innervate organs below the diaphragm
    ___ (f) ganglia located at the end of an autonomic motor pathway close to or actually within the wall of a visceral organ
    ___ (g) includes ciliary, pterygopalatine, submandibular, and otic ganglia
    ___ (h) extend from base of the skull to the coccyx
    ___ (i) myelinated preganglionic fibers that connect the anterior rami of spinal nerves with the ganglia of the sympathetic trunk
    ___ (j) also known as collateral ganglia
    ___ (k) unmyelinated postganglionic axons that connect the ganglia of the sympathetic trunk to spinal nerves

    (1) sympathetic trunk ganglia
    (2) prevertebral ganglia
    (3) terminal ganglia
    (4) white rami communicantes
    (5) gray rami communicantes

15. Match the following:

    ____ (a) stimulates urination and defecation

    ____ (b) prepares the body for emergency situations

    ____ (c) fight-or-flight response

    ____ (d) promotes digestion and absorption of food

    ____ (e) concerned primarily with processes involving the expenditure of energy

    ____ (f) controlled by the posterior and lateral portions of the hypothalamus

    ____ (g) controlled by the anterior and medial portions of the hypothalamus

    ____ (h) causes a decrease in heart rate

(1) increased activity of the sympathetic division of the ANS

(2) increased activity of the parasympathetic division of the ANS

CRITICAL THINKING QUESTIONS

1. You've been to the "all-you-can-eat" buffet and have consumed large amounts of food. After returning home, you recline on the couch to watch television. Which division of the nervous system will be handling your body's after-dinner activities? List several organs involved, the major nerve supply to each organ, and the effects of the nervous system on their functions.

2. Ciara is driving home from school, listening to her favorite music, when a dog darts into the street in front of her car. She manages to swerve to avoid hitting the dog. As she continues on her way, she notices her heart is racing, she has "goose-bumps," and her hands are sweaty. Why is she experiencing these effects?

3. Mrs. Young is experiencing a bout of diarrhea that is keeping her house-bound. She would like to attend a birthday party for her brother but is afraid to attend because of her diarrhea. What type of drug, related to the autonomic nervous system function, could she take to help relieve her diarrhea?

? ANSWERS TO FIGURE QUESTIONS

15.1 Dual innervation means that a body organ receives neural innervation from both sympathetic and parasympathetic neurons of the ANS.

15.2 Most parasympathetic preganglionic axons are longer than most sympathetic preganglionic axons because most parasympathetic ganglia are in the walls of visceral organs, but most sympathetic ganglia are close to the spinal cord in the sympathetic trunk.

15.3 Terminal ganglia are associated with the parasympathetic division; sympathetic trunk and prevertebral ganglia are associated with the sympathetic division.

15.4 Sympathetic trunk ganglia contain sympathetic postganglionic neurons that lie in a vertical row on either side of the vertebral column.

15.5 The largest autonomic plexus is the celiac (solar) plexus.

15.6 Pelvic splanchnic nerves branch from the second through fourth sacral spinal nerves.

15.7 Most (but not all) sympathetic postganglionic neurons are adrenergic. Muscarinic receptors are present in the plasma membranes of all effectors (smooth muscle, cardiac muscle, and glands) innervated by parasympathetic postganglionic neurons and in sweat glands innervated by cholinergic sympathetic postganglionic neurons.

16 SENSORY, MOTOR, AND INTEGRATIVE SYSTEMS

SENSORY, MOTOR, AND INTEGRATIVE SYSTEMS AND HOMEOSTASIS *The sensory and motor pathways of the body provide routes for input into the brain and spinal cord and for output to targeted organs for responses, such as muscle contraction.*

In the previous four chapters we described the organization of the nervous system. In this chapter, we explore the levels and components of sensation. We also examine the pathways that convey somatic sensory nerve impulses from the body to the brain, and the pathways that carry impulses from the brain to skeletal muscles to produce movements. As sensory impulses reach the CNS, they become part of a large pool of sensory input. However, not every nerve impulse transmitted to the CNS elicits a response. Rather, each piece of incoming information is combined with other arriving and previously stored information in a process called *integration*. Integration occurs at many places along pathways in the CNS, such as the spinal cord, brain stem, cerebellum, basal nuclei, and cerebral cortex. You will also learn how the motor responses that govern muscle contraction are modified at several of these levels. To conclude this chapter, we introduce two complex integrative functions of the brain: (1) wakefulness and sleep and (2) learning and memory.

? *Did you ever wonder how drugs such as aspirin and ibuprofen relieve pain?*

16.1 SENSATION

◉ OBJECTIVES

- Define sensation, and discuss the components of sensation.
- Describe the different ways to classify sensory receptors.

In its broadest definition, **sensation** is the conscious or subconscious awareness of changes in the external or internal environment. The nature of the sensation and the type of reaction generated vary according to the ultimate destination of nerve impulses that convey sensory information to the CNS. Sensory impulses that reach the spinal cord may serve as input for spinal reflexes, such as the stretch reflex you learned about in Chapter 13. Sensory impulses that reach the lower brain stem elicit more complex reflexes, such as changes in heart rate or breathing rate. When sensory impulses reach the cerebral cortex, we become consciously aware of the sensory stimuli and can precisely locate and identify specific sensations such as touch, pain, hearing, or taste. As you learned in Chapter 14, **perception** is the conscious awareness and interpretation of sensations and is primarily a function of the cerebral cortex. We have no perception of some sensory information because it never reaches the cerebral cortex. For example, certain sensory receptors constantly monitor the pressure of blood in blood vessels. Because the nerve impulses conveying blood pressure information propagate to the cardiovascular center in the medulla oblongata rather than to the cerebral cortex, blood pressure is not consciously perceived.

Sensory Modalities

Each unique type of sensation—such as touch, pain, vision, or hearing—is called a **sensory modality** (mō-DAL-i-tē). A given sensory neuron carries information for only one sensory modality. Neurons relaying impulses for touch to the somatosensory area of the cerebral cortex do not transmit impulses for pain. Likewise, nerve impulses from the eyes are perceived as sight, and those from the ears are perceived as sounds.

The different sensory modalities can be grouped into two classes: general senses and special senses.

1. The **general senses** refer to both somatic senses and visceral senses. **Somatic senses** (*somat-* = of the body) include tactile sensations (touch, pressure, vibration, itch, and tickle), thermal sensations (warm and cold), pain sensations, and proprioceptive sensations. Proprioceptive sensations allow perception of both the static (nonmoving) positions of limbs and body parts (joint and muscle position sense) and movements of the limbs and head. **Visceral senses** provide information about conditions within internal organs, for example, pressure, stretch, chemicals, nausea, hunger, and temperature.

2. The **special senses** include the sensory modalities of smell, taste, vision, hearing, and equilibrium or balance.

In this chapter we discuss the somatic senses and visceral pain. The special senses are the focus of Chapter 17. Visceral senses were discussed in Chapter 15 and will be discussed in association with individual organs in later chapters.

The Process of Sensation

The process of sensation begins in a **sensory receptor,** which can be either a specialized cell or the dendrites of a sensory neuron. As previously noted, a given sensory receptor responds vigorously to one particular kind of **stimulus,** a change in the environment that can activate certain sensory receptors. A sensory receptor responds only weakly or not at all to other stimuli. This characteristic of sensory receptors is known as **selectivity.**

For a sensation to arise, the following four events typically occur:

1. *Stimulation of the sensory receptor.* An appropriate stimulus must occur within the sensory receptor's *receptive field,* that is, the body region where stimulation activates the receptor and produces a response.

2. *Transduction of the stimulus.* A sensory receptor *transduces* (converts) energy in a stimulus into a graded potential. Recall that graded potentials vary in amplitude (size), depending on the strength of the stimulus that causes them, and are not propagated. (See Section 12.3 to review the differences between action potentials and graded potentials.) Each type of sensory receptor exhibits selectivity: It can transduce only one kind of stimulus. For example, odorant molecules in the air stimulate olfactory (smell) receptors in the nose, which transduce the molecules' chemical energy into electrical energy in the form of a graded potential.

3. *Generation of nerve impulses.* When a graded potential in a sensory neuron reaches threshold, it triggers one or more nerve impulses, which then propagate toward the CNS. Sensory neurons that conduct impulses from the PNS into the CNS are called **first-order neurons.**

4. *Integration of sensory input.* A particular region of the CNS receives and integrates the sensory nerve impulses. Conscious sensations or perceptions are integrated in the cerebral cortex. You seem to see with your eyes, hear with your ears, and feel pain in an injured part of your body because sensory impulses from each part of the body arrive in a specific region of the cerebral cortex, which interprets the sensation as coming from the stimulated sensory receptors.

Sensory Receptors

Types of Sensory Receptors

Several structural and functional characteristics of sensory receptors can be used to group them into different classes. These include (1) microscopic structure, (2) location of the receptors and the origin of stimuli that activate them, and (3) type of stimulus detected.

MICROSCOPIC STRUCTURE On a microscopic level, sensory receptors may be (1) free nerve endings of first-order sensory neurons, (2) encapsulated nerve endings of first-order sensory neurons, or (3) separate cells that synapse with first-order sensory neurons. **Free nerve endings** are bare dendrites; they lack any structural specializations that can be seen under a light microscope (Figure 16.1a). Receptors for pain, temperature, tickle, itch, and some touch sensations are free nerve endings. Receptors for other somatic and visceral sensations, such as pressure, vibration, and some touch sensations, are **encapsulated nerve endings.** Their dendrites are enclosed in a connective tissue capsule that has a distinctive microscopic structure—for example, Pacinian corpuscles (Figure 16.1b). The different types of capsules enhance the sensitivity or specificity of the receptor. Sensory receptors for some special senses are specialized, **separate cells** that synapse with sensory neurons. These include *hair cells* for hearing and equilibrium in the inner ear, *gustatory receptor cells* in taste buds (Figure 16.1c), and *photoreceptors* in the retina of the eye for vision; you will learn more about separate cells in Chapter 17.

Sensory receptors produce two different kinds of graded potentials—generator potentials and receptor potentials—in response to a stimulus. When stimulated, the dendrites of free nerve endings, encapsulated nerve endings, and the receptive part of olfactory receptors produce a **generator potential** (Figure 16.1a, b). When a generator potential is large enough to reach threshold, it triggers one or more nerve impulses in the axon of a first-order sensory neuron. The resulting nerve impulse propagates along the axon into the CNS. Thus, generator potentials generate action potentials.

By contrast, sensory receptors that are separate cells produce graded potentials termed **receptor potentials.** Receptor potentials trigger release of neurotransmitter through exocytosis of

Figure 16.1 **Types of sensory receptors and their relationship to first-order sensory neurons.** (a) Free nerve endings, in this case, a cold-sensitive receptor. These endings are bare dendrites of first-order neurons with no apparent structural specialization. (b) An encapsulated nerve ending, in this case a pressure-sensitive receptor. Encapsulated nerve endings are dendrites of first-order neurons. (c) A separate receptor cell—here, a gustatory (taste) receptor—and its synapse with a first-order neuron.

Free nerve endings and encapsulated nerve endings produce generator potentials that trigger nerve impulses in first-order neurons. Separate sensory receptors produce a receptor potential that causes the release of a neurotransmitter. The neurotransmitter then triggers nerve impulses in a first-order neuron.

? **Which senses are served by receptors that are separate cells?**

synaptic vesicles (Figure 16.1c). The neurotransmitter molecules liberated from synaptic vesicles diffuse across the synaptic cleft and produce a postsynaptic potential (PSP) in the first-order neuron. In turn, the PSPs may trigger one or more nerve impulses, which propagate along the axon into the CNS.

The amplitude of both generator potentials and receptor potentials varies with the intensity of the stimulus, with an intense stimulus producing a large potential and a weak stimulus eliciting a small one. Similarly, large generator potentials or receptor potentials trigger nerve impulses at high frequencies in the first-order neuron, in contrast to small generator potentials or receptor potentials, which trigger nerve impulses at lower frequencies.

LOCATION OF RECEPTORS AND ORIGIN OF ACTIVATING STIMULI
Another way to group sensory receptors is based on the location of the receptors and the origin of the stimuli that activate them.

- **Exteroceptors** (EKS-ter-ō-sep′-tors) are located at or near the external surface of the body; they are sensitive to stimuli originating outside the body and provide information about the *external* environment. The sensations of hearing, vision, smell, taste, touch, pressure, vibration, temperature, and pain are conveyed by exteroceptors.

- **Interoceptors** (IN-ter-ō-sep′-tors) or **visceroceptors** are located in blood vessels, visceral organs, muscles, and the nervous system and monitor conditions in the *internal* environment. The nerve impulses produced by interoceptors usually are not consciously perceived; occasionally, however, activation of interoceptors by strong stimuli may be felt as pain or pressure.

- **Proprioceptors** (PRŌ-prē-ō-sep′-tors; *proprio-* = one's own) are located in muscles, tendons, joints, and the inner ear. They provide information about body position, muscle length and tension, and the position and movement of your joints.

TYPE OF STIMULUS DETECTED   A third way to group sensory receptors is according to the type of stimulus they detect. Most stimuli are in the form of mechanical energy, such as sound waves or pressure changes; electromagnetic energy, such as light or heat; or chemical energy, such as in a molecule of glucose.

- **Mechanoreceptors** are sensitive to mechanical stimuli such as the deformation, stretching, or bending of cells. Mechanoreceptors provide sensations of touch, pressure, vibration, proprioception, and hearing and equilibrium. They also monitor the stretching of blood vessels and internal organs.

- **Thermoreceptors** detect changes in temperature.

- **Nociceptors** (nō′-sē-SEP-tors) respond to painful stimuli resulting from physical or chemical damage to tissue.

- **Photoreceptors** detect light that strikes the retina of the eye.

- **Chemoreceptors** detect chemicals in the mouth (taste), nose (smell), and body fluids.

- **Osmoreceptors** detect the osmotic pressure of body fluids.

Table 16.1 summarizes the classification of sensory receptors.

Adaptation in Sensory Receptors

A characteristic of most sensory receptors is **adaptation,** in which the generator potential or receptor potential decreases in amplitude during a maintained, constant stimulus. As you may already have guessed, this causes the frequency of nerve impulses in the first-order neuron to decrease. Because of adaptation, the perception of a sensation may fade or disappear even though the stimulus persists. For example, when you first step into a hot shower, the water may feel very hot, but soon the sensation decreases to one of comfortable warmth even though the stimulus (the high temperature of the water) does not change.

TABLE 16.1

Classification of Sensory Receptors

BASIS OF CLASSIFICATION	DESCRIPTION
MICROSCOPIC STRUCTURE	
Free nerve endings	Bare dendrites associated with pain, thermal, tickle, itch, and some touch sensations.
Encapsulated nerve endings	Dendrites enclosed in connective tissue capsule for pressure, vibration, and some touch sensations.
Separate cells	Receptor cells synapse with first-order sensory neurons; located in retina of eye (photoreceptors), inner ear (hair cells), and taste buds of tongue (gustatory receptor cells).
RECEPTOR LOCATION AND ACTIVATING STIMULI	
Exteroceptors	Located at or near body surface; sensitive to stimuli originating outside body; provide information about external environment; convey visual, smell, taste, touch, pressure, vibration, thermal, and pain sensations.
Interoceptors	Located in blood vessels, visceral organs, and nervous system; provide information about internal environment; impulses usually are not consciously perceived but occasionally may be felt as pain or pressure.
Proprioceptors	Located in muscles, tendons, joints, and inner ear; provide information about body position, muscle length and tension, position and motion of joints, and equilibrium (balance).
TYPE OF STIMULUS DETECTED	
Mechanoreceptors	Detect mechanical stimuli; provide sensations of touch, pressure, vibration, proprioception, and hearing and equilibrium; also monitor stretching of blood vessels and internal organs.
Thermoreceptors	Detect changes in temperature.
Nociceptors	Respond to painful stimuli resulting from physical or chemical damage to tissue.
Photoreceptors	Detect light that strikes the retina of the eye.
Chemoreceptors	Detect chemicals in mouth (taste), nose (smell), and body fluids.
Osmoreceptors	Sense osmotic pressure of body fluids.

Receptors vary in how quickly they adapt. **Rapidly adapting receptors** adapt very quickly. They are specialized for signaling *changes* in a stimulus. Receptors associated with pressure, touch, and smell are rapidly adapting. **Slowly adapting receptors,** by contrast, adapt slowly and continue to trigger nerve impulses as long as the stimulus persists. Slowly adapting receptors monitor stimuli associated with pain, body position, and chemical composition of the blood.

✔ CHECKPOINT

1. How is sensation different from perception?
2. What is a sensory modality?
3. How are generator potentials and receptor potentials similar? How are they different?
4. What is the difference between rapidly adapting and slowly adapting receptors?

16.2 SOMATIC SENSATIONS

◉ OBJECTIVES

- Describe the location and function of the somatic sensory receptors for tactile, thermal, and pain sensations.
- Identify the receptors for proprioception and describe their functions.

Somatic sensations arise from stimulation of sensory receptors embedded in the skin or subcutaneous layer; in mucous membranes of the mouth, vagina, and anus; in muscles, tendons, and joints; and in the inner ear. The sensory receptors for somatic sensations are distributed unevenly—some parts of the body surface are densely populated with receptors, and others contain only a few. The areas with the highest density of somatic sensory receptors are the tip of the tongue, the lips, and the fingertips. Somatic sensations that arise from stimulating the skin surface are **cutaneous sensations** (kū-TĀ-nē-us; *cutane-* = skin). There are four modalities of somatic sensation: tactile, thermal, pain, and proprioceptive.

Tactile Sensations

The **tactile sensations** (TAK-tĭl; *tact-* = touch) include touch, pressure, vibration, itch, and tickle. Although we perceive differences among these sensations, they arise by activation of some of the same types of receptors. Several types of encapsulated mechanoreceptors attached to large-diameter myelinated A fibers mediate sensations of touch, pressure, and vibration. Other tactile sensations, such as itch and tickle sensations, are detected by free nerve endings attached to small-diameter, unmyelinated C fibers. Recall that larger-diameter, myelinated axons propagate nerve impulses more rapidly than do smaller-diameter, unmyelinated axons. Tactile receptors in the skin or subcutaneous layer include Meissner corpuscles, hair root plexuses, Merkel discs, Ruffini corpuscles, Pacinian corpuscles, and free nerve endings (Figure 16.2).

Touch

Sensations of **touch** generally result from stimulation of tactile receptors in the skin or subcutaneous layer. There are two types of

rapidly adapting touch receptors. **Meissner corpuscles** (MĪS-ner) or *corpuscles of touch* are touch receptors that are located in the dermal papillae of hairless skin. Each corpuscle is an egg-shaped mass of dendrites enclosed by a capsule of connective tissue. Because Meissner corpuscles are rapidly adapting receptors, they generate nerve impulses mainly at the onset of a touch. They are abundant in the fingertips, hands, eyelids, tip of the tongue, lips, nipples, soles, clitoris, and tip of the penis. **Hair root plexuses** are rapidly adapting touch receptors found in hairy skin; they consist of free nerve endings wrapped around hair follicles. Hair root plexuses detect movements on the skin surface that disturb hairs. For example, an insect landing on a hair causes movement of the hair shaft that stimulates the free nerve endings.

There are also two types of slowly adapting touch receptors. **Merkel discs,** also known as *tactile discs* or *type I cutaneous mechanoreceptors,* are saucer-shaped, flattened free nerve endings that make contact with Merkel cells of the stratum basale (see Figure 5.2d). These touch receptors are plentiful in the fingertips, hands, lips, and external genitalia. **Ruffini corpuscles** or *type II cutaneous mechanoreceptors* are elongated, encapsulated receptors located deep in the dermis, and in ligaments and tendons. Present in the hands and abundant on the soles, they are most sensitive to stretching that occurs as digits or limbs are moved.

Pressure

Pressure, a sustained sensation that is felt over a larger area than touch, occurs with deformation of deeper tissues. Receptors that contribute to sensations of pressure include Meissner corpuscles, Merkel discs, and Pacinian corpuscles. A **Pacinian** (pa-SIN-ē-an), or **lamellated, corpuscle** is a large oval structure composed of a multilayered connective tissue capsule that encloses a dendrite. Like Meissner corpuscles, Pacinian corpuscles adapt rapidly. They are widely distributed in the body: in the dermis and subcutaneous layer; in submucosal tissues that underlie mucous and serous membranes; around joints, tendons, and muscles; in the periosteum; and in the mammary glands, external genitalia, and certain viscera, such as the pancreas and urinary bladder.

Vibration

Sensations of **vibration,** such as using an electric knife to carve a turkey, result from rapidly repetitive sensory signals from tactile receptors. The receptors for vibration sensations are Meissner corpuscles and Pacinian corpuscles. Meissner corpuscles can detect lower-frequency vibrations, and Pacinian corpuscles detect higher-frequency vibrations.

Itch

The **itch** sensation results from stimulation of free nerve endings by certain chemicals, such as bradykinin or antigens in mosquito saliva injected from a bite, often because of a local inflammatory response (bradykinin, a kinin, is a potent vasodilator).

Tickle

Free nerve endings are thought to mediate the **tickle** sensation. This intriguing sensation typically arises only when someone else touches you, not when you touch yourself. The solution to this puzzle seems to lie in the impulses that conduct to and from the cerebellum when you are moving your fingers and touching yourself that don't occur when someone else is tickling you.

CLINICAL CONNECTION | *Phantom Limb Sensation*

Patients who have had a limb amputated may still experience sensations such as itching, pressure, tingling, or pain as if the limb were still there. This phenomenon is called **phantom limb sensation.** Although the limb has been removed, severed endings of sensory axons are still present in the remaining stump. If these severed endings are activated, the cerebral cortex interprets the sensation as coming from the sensory receptors in the nonexisting (phantom) limb. Another explanation for phantom limb sensation is that the area of the cerebral cortex that previously received sensory input from the missing limb undergoes extensive functional reorganization that allows it to respond to stimuli from another body part. The remodeling of this cortical area is thought to give rise to false sensory perceptions from the missing limb. Phantom limb pain can be very distressing to an amputee. Many report that the pain is severe or extremely intense, and that it often does not respond to traditional pain medication therapy. In such cases, alternative treatments may include electrical nerve stimulation, acupuncture, and biofeedback. •

Thermal Sensations

Thermoreceptors are free nerve endings that have receptive fields about 1 mm in diameter on the skin surface. Two distinct **thermal sensations**—coldness and warmth—are detected by different receptors. **Cold receptors** are located in the stratum basale of the epidermis and are attached to medium-diameter, myelinated A fibers, although a few connect to small-diameter, unmyelinated C fibers. Temperatures between 10° and 40°C (50–105°F) activate cold receptors. **Warm receptors,** which are not as abundant as cold receptors, are located in the dermis and are attached to small-diameter, unmyelinated C fibers; they are activated by temperatures between 32° and 48°C (90–118°F). Cold and warm receptors both adapt rapidly at the onset of a stimulus, but as noted earlier in the chapter they continue to generate impulses at a lower frequency throughout a prolonged stimulus. Temperatures below 10°C and above 48°C primarily stimulate pain receptors, rather than thermoreceptors, producing painful sensations, which we discuss next.

Pain Sensations

Pain is indispensable for survival. It serves a protective function by signaling the presence of noxious, tissue-damaging conditions.

Figure 16.2 Structure and location of sensory receptors in the skin and subcutaneous layer.

The somatic sensations of touch, pressure, vibration, warmth, cold, and pain arise from sensory receptors in the skin, subcutaneous layer, and mucous membranes.

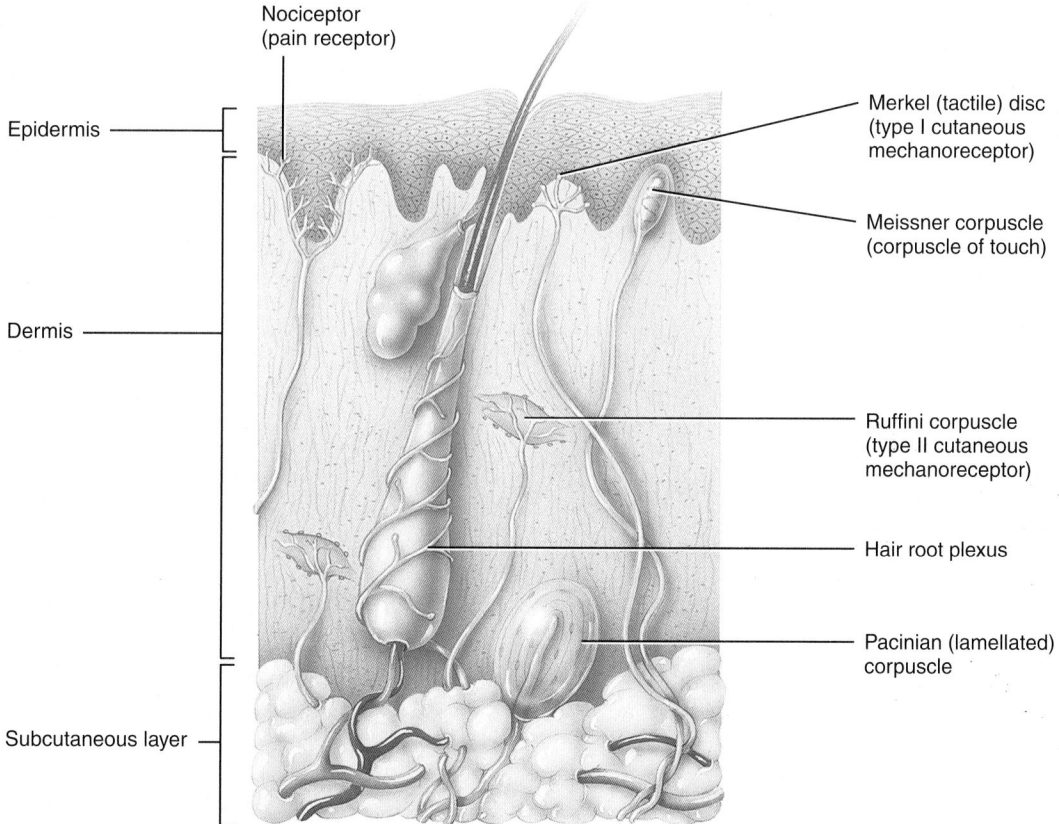

Which sensations can arise when free nerve endings are stimulated?

From a medical standpoint, the subjective description and indication of the location of pain may help pinpoint the underlying cause of disease.

Nociceptors (*noci-* = harmful), the receptors for pain, are free nerve endings found in every tissue of the body except the brain (Figure 16.2). Intense thermal, mechanical, or chemical stimuli can activate nociceptors. Tissue irritation or injury releases chemicals such as prostaglandins, kinins, and potassium ions (K^+) that stimulate nociceptors. Pain may persist even after a pain-producing stimulus is removed because pain-mediating chemicals linger, and because nociceptors exhibit very little adaptation. Conditions that elicit pain include excessive distention (stretching) of a structure, prolonged muscular contractions, muscle spasms, or ischemia (inadequate blood flow to an organ).

Types of Pain

There are two types of pain: fast and slow. The perception of **fast pain** occurs very rapidly, usually within 0.1 second after a stimulus is applied, because the nerve impulses propagate along medium-diameter, myelinated A fibers. This type of pain is also known as acute, sharp, or pricking pain. The pain felt from a needle puncture or knife cut to the skin is fast pain. Fast pain is not felt in deeper tissues of the body. The perception of **slow pain,** by contrast, begins a second or more after a stimulus is applied. It then gradually increases in intensity over a period of several seconds or minutes. Impulses for slow pain conduct along small-diameter, unmyelinated C fibers. This type of pain, which may be excruciating, is also referred to as chronic, burning, aching, or throbbing pain. Slow pain can occur both in the skin and in deeper

tissues or internal organs. An example is the pain associated with a toothache. You can perceive the difference in onset of these two types of pain best when you injure a body part that is far from the brain because the conduction distance is long. When you stub your toe, for example, you first feel the sharp sensation of fast pain and then feel the slower, aching sensation of slow pain.

Pain that arises from stimulation of receptors in the skin is called **superficial somatic pain;** stimulation of receptors in skeletal muscles, joints, tendons, and fascia causes **deep somatic pain. Visceral pain** results from stimulation of nociceptors in visceral organs. If stimulation is *diffuse* (involves large areas), visceral pain can be severe. Diffuse stimulation of visceral nociceptors might result from distention or ischemia of an internal organ. For example, a kidney stone or a gallstone might cause severe pain by obstructing and distending a ureter or bile duct.

Localization of Pain

Fast pain is very precisely localized to the stimulated area. For example, if someone pricks you with a pin, you know exactly which part of your body was stimulated. Somatic slow pain also is well localized but more diffuse (involves large areas); it usually appears to come from a larger area of the skin. In some instances of visceral slow pain, the affected area is where the pain is felt. If the pleural membranes around the lungs are inflamed, for example, you experience chest pain.

However, in many instances of visceral pain, the pain is felt in or just deep to the skin that overlies the stimulated organ, or in a surface area far from the stimulated organ. This phenomenon is called **referred pain.** Figure 16.3 shows skin regions to which

Figure 16.3 Distribution of referred pain. Colored parts indicate skin areas to which visceral pain is referred.

🔒 ▭ Nociceptors are present in almost every tissue of the body.

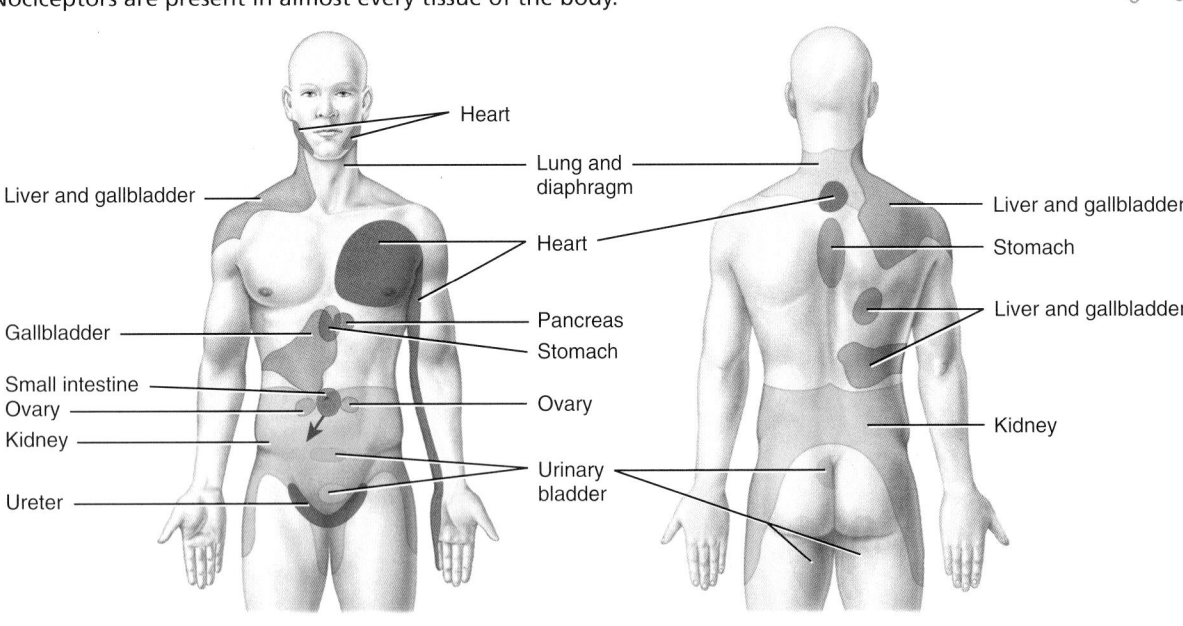

(a) Anterior view   (b) Posterior view

❓ **Which visceral organ has the broadest area for referred pain?**

visceral pain may be referred. In general, the visceral organ involved and the area to which the pain is referred are served by the same segment of the spinal cord. For example, sensory fibers from the heart, the skin superficial to the heart, and the skin along the medial aspect of the left arm enter spinal cord segments T1 to T5. Thus, the pain of a heart attack typically is felt in the skin over the heart and along the left arm.

CLINICAL CONNECTION | *Analgesia: Relief from Pain*

Pain sensations sometimes occur out of proportion to minor damage, persist chronically due to an injury, or even appear for no obvious reason. In such cases, **analgesia** (an-al-JĒ-zē-a; *an-* = without; *-algesia* = pain) or pain relief is needed. Analgesic drugs such as aspirin and ibuprofen (for example, Advil® or Motrin®) block formation of prostaglandins, which stimulate nociceptors. Local anesthetics, such as Novocaine®, provide short-term pain relief by blocking conduction of nerve impulses along the axons of first-order pain neurons. Morphine and other opiate drugs alter the quality of pain perception in the brain; pain is still sensed but it is no longer perceived as being so noxious. Many pain clinics use anticonvulsant and antidepressant medications to treat those suffering from chronic pain. •

Proprioceptive Sensations

Proprioceptive sensations allow us to know where our head and limbs are located and how they are moving even if we are not looking at them, so that we can walk, type, or dress without using our eyes. **Kinesthesia** (kin′-es-THĒ-zē-a; *kin-* = motion; *-esthesia* = perception) is the perception of body movements. Proprioceptive sensations arise in receptors termed **proprioceptors.** Those proprioceptors embedded in muscles (especially postural muscles) and tendons inform us of the degree to which muscles are contracted, the amount of tension on tendons, and the positions of joints. Hair cells of the inner ear monitor the orientation of the head relative to the ground and head position during movements. The way they provide information for maintaining balance and equilibrium will be described in Chapter 17. Because proprioceptors adapt slowly and only slightly, the brain continually receives nerve impulses related to the position of different body parts and makes adjustments to ensure coordination.

Proprioceptors also allow **weight discrimination,** the ability to assess the weight of an object. This type of information helps you to determine the muscular effort necessary to perform a task. For example, as you pick up a shopping bag, you quickly realize whether it contains books or feathers, and you then exert the correct amount of effort needed to lift it.

Here we discuss three types of proprioceptors: muscle spindles within skeletal muscles, tendon organs within tendons, and joint kinesthetic receptors within synovial joint capsules.

Muscle Spindles

Muscle spindles are the proprioceptors in skeletal muscles that monitor changes in the length of skeletal muscles and participate in stretch reflexes (shown in Figure 13.15). By adjusting how vigorously a muscle spindle responds to stretching of a skeletal muscle, the brain sets an overall level of **muscle tone,** the small degree of contraction that is present while the muscle is at rest.

Each **muscle spindle** consists of several slowly adapting sensory nerve endings that wrap around 3 to 10 specialized muscle fibers, called **intrafusal muscle fibers** (in′-tra-FŪ-sal = within a spindle). A connective tissue capsule encloses the sensory nerve endings and intrafusal fibers and anchors the spindle to the endomysium and perimysium (Figure 16.4). Muscle spindles are interspersed among most skeletal muscle fibers and aligned parallel to them. In muscles that produce finely controlled movements, such as those of the fingers or eyes as you read music and play a musical instrument, muscle spindles are plentiful. Muscles involved in coarser but more forceful movements, like the quadriceps femoris and hamstring muscles of the thigh, have fewer muscle spindles. The only skeletal muscles that lack spindles are the tiny muscles of the middle ear.

The main function of muscle spindles is to measure *muscle length*—how much a muscle is being stretched. Either sudden or prolonged stretching of the central areas of the intrafusal muscle fibers stimulates the sensory nerve endings. The resulting nerve impulses propagate into the CNS. Information from muscle spindles arrives quickly at the somatic sensory areas of the cerebral cortex, which allows conscious perception of limb positions and movements. At the same time, impulses from muscle spindles pass to the cerebellum, where the input is used to coordinate muscle contractions.

In addition to their sensory nerve endings near the middle of intrafusal fibers, muscle spindles contain motor neurons called **gamma motor neurons.** These motor neurons terminate near both ends of the intrafusal fibers and adjust the tension in a muscle spindle to variations in the length of the muscle. For example, when your biceps muscle shortens in response to lifting a weight, gamma motor neurons stimulate the ends of the intrafusal fibers to contract slightly. This keeps the intrafusal fibers taut even though the contractile muscle fibers surrounding the spindle are reducing spindle tension. This maintains the sensitivity of the muscle spindle to stretching of the muscle. As the frequency of impulses in its gamma motor neuron increases, a muscle spindle becomes more sensitive to stretching of its midregion.

Surrounding muscle spindles are ordinary skeletal muscle fibers, called **extrafusal muscle fibers** (*extrafusal* = outside a spindle), which are supplied by large-diameter A fibers called **alpha motor neurons.** The cell bodies of both gamma and alpha motor neurons are located in the anterior gray horn of the spinal cord (or in the brain stem for muscles in the head). During the stretch reflex, impulses in muscle spindle sensory axons propagate into the spinal cord and brain stem and activate alpha motor neurons that connect to extrafusal muscle fibers in the same muscle. In this way, activation of its muscle spindles causes contraction of a skeletal muscle, which relieves the stretching.

Tendon Organs

Tendon organs are located at the junction of a tendon and a muscle. By initiating tendon reflexes (see Figure 13.16), tendon

Figure 16.4 **Two types of proprioceptors: a muscle spindle and a tendon organ.** In muscle spindles, which monitor changes in skeletal muscle length, sensory nerve endings wrap around the central portion of intrafusal muscle fibers. In tendon organs, which monitor the force of muscle contraction, sensory nerve endings are activated by increasing tension on a tendon. If you examine Figure 13.15 you can see the relationship of a muscle spindle to the spinal cord as a component of a stretch reflex. In Figure 13.16, you can see the relationship of a tendon organ to the spinal cord as a component of a tendon reflex.

Proprioceptors provide information about body position and movement.

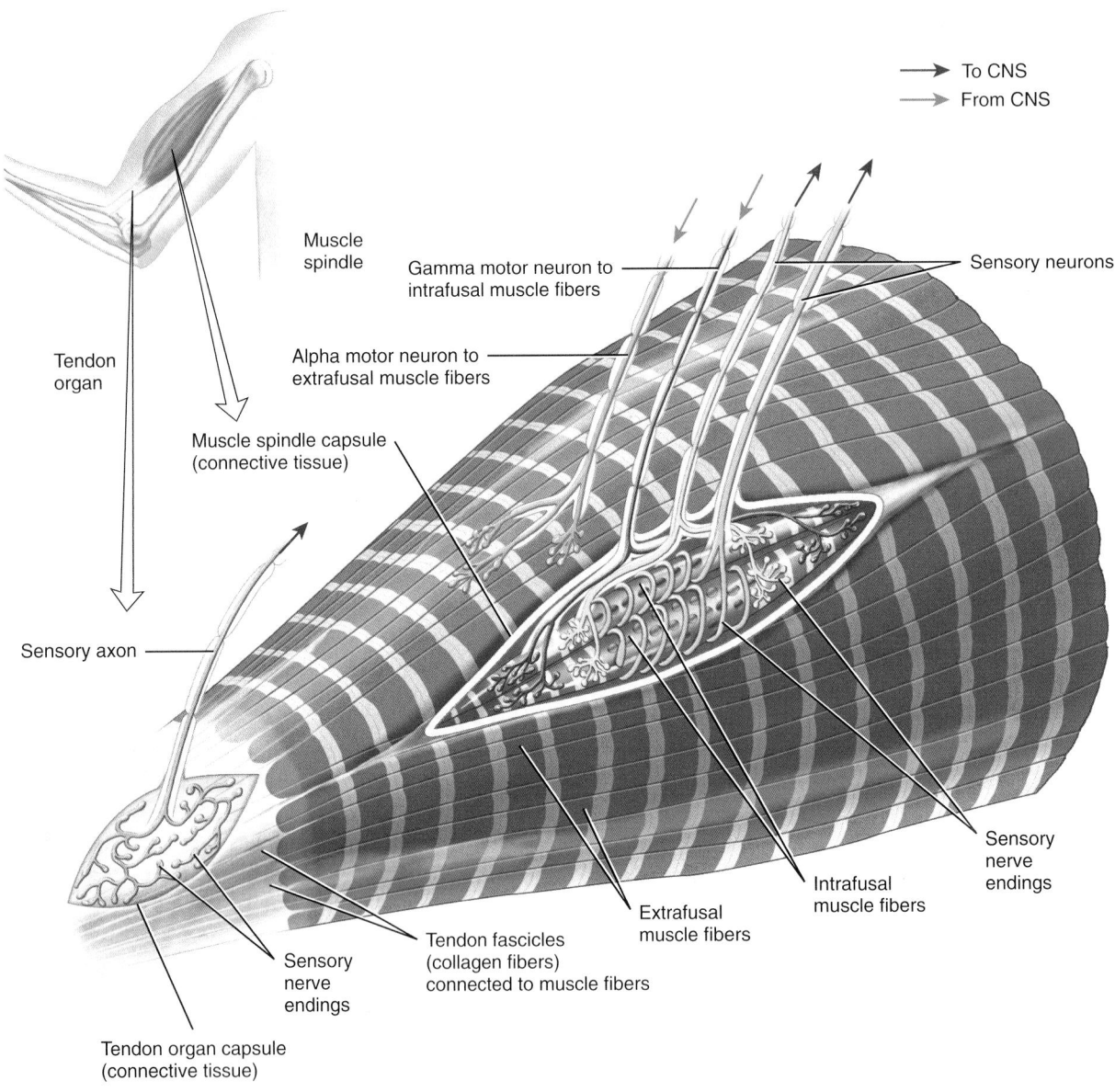

How is a muscle spindle activated?

organs protect tendons and their associated muscles from damage due to excessive tension. (When a muscle contracts, it exerts a force that pulls the points of attachment of the muscle at either end toward each other. This force is the muscle tension.) Each tendon organ consists of a thin capsule of connective tissue that encloses a few tendon fascicles (bundles of collagen fibers) (Figure 16.4). Penetrating the capsule are one or more sensory nerve endings that entwine among and around the collagen fibers of the tendon. When tension is applied to a muscle, the tendon

organs generate nerve impulses that propagate into the CNS, providing information about changes in muscle tension. The resulting tendon reflexes decrease muscle tension by causing muscle relaxation.

Joint Kinesthetic Receptors

Several types of **joint kinesthetic receptors** (kin′-es-THET-ik) are present within and around the articular capsules of synovial joints. Free nerve endings and Ruffini corpuscles in the capsules

TABLE 16.2

Summary of Receptors for Somatic Sensations

RECEPTOR TYPE	RECEPTOR STRUCTURE AND LOCATION	SENSATIONS	ADAPTATION RATE
TACTILE RECEPTORS			
Meissner corpuscles (corpuscles of touch)	Capsule surrounds mass of dendrites in dermal papillae of hairless skin.	Touch, pressure, and slow vibrations.	Rapid.
Hair root plexuses	Free nerve endings wrapped around hair follicles in skin.	Touch.	Rapid.
Merkel (tactile) discs (type I cutaneous mechanoreceptors)	Saucer-shaped free nerve endings make contact with Merkel cells in epidermis.	Touch and pressure.	Slow.
Ruffini corpuscles (type II cutaneous mechanoreceptors)	Elongated capsule surrounds dendrites deep in dermis and in ligaments and tendons.	Stretching of skin.	Slow.
Pacinian (lamellated) corpuscles	Oval, layered capsule surrounds dendrites; present in dermis and subcutaneous layer, submucosal tissues, joints, periosteum, and some viscera.	Pressure and fast vibrations.	Rapid.
Itch and tickle receptors	Free nerve endings in skin and mucous membranes.	Itching and tickling.	Both slow and rapid.
THERMORECEPTORS			
Warm receptors and cold receptors	Free nerve endings in skin and mucous membranes of mouth, vagina, and anus.	Warmth or cold.	Initially rapid, then slow.
PAIN RECEPTORS			
Nociceptors	Free nerve endings in every body tissue except brain.	Pain.	Slow.
PROPRIOCEPTORS			
Muscle spindles	Sensory nerve endings wrap around central area of encapsulated intrafusal muscle fibers within most skeletal muscles.	Muscle length.	Slow.
Tendon organs	Capsule encloses collagen fibers and sensory nerve endings at junction of tendon and muscle.	Muscle tension.	Slow.
Joint kinesthetic receptors	Pacinian corpuscles, Ruffini corpuscles, tendon organs, and free nerve endings.	Joint position and movement.	Rapid.

of joints respond to pressure. Small Pacinian corpuscles in the connective tissue outside articular capsules respond to acceleration and deceleration of joints during movement. Joint ligaments contain receptors similar to tendon organs that adjust reflex inhibition of the adjacent muscles when excessive strain is placed on the joint.

Table 16.2 summarizes the types of somatic sensory receptors and the sensations they convey.

✔ CHECKPOINT

5. Which somatic sensory receptors are encapsulated?
6. Why do some receptors adapt slowly, and others adapt rapidly?
7. Which somatic sensory receptors mediate touch sensations?
8. How does fast pain differ from slow pain?
9. What is referred pain, and how is it useful in diagnosing internal disorders?
10. What aspects of muscle function are monitored by muscle spindles and tendon organs?

16.3 SOMATIC SENSORY PATHWAYS

◉ OBJECTIVE

• Describe the neuronal components and functions of the posterior column–medial lemniscus pathway, the anterolateral pathway, and the spinocerebellar pathway.

Somatic sensory pathways relay information from the somatic sensory receptors just described to the primary somatosensory area in the cerebral cortex and to the cerebellum. The pathways to the cerebral cortex consist of thousands of sets of three neurons: a first-order neuron, a second-order neuron, and a third-order neuron.

1. **First-order neurons** conduct impulses from somatic receptors into the brain stem or spinal cord. From the face, mouth, teeth, and eyes, somatic sensory impulses propagate along *cranial nerves* into the brain stem. From the neck, trunk, limbs, and posterior aspect of the head, somatic sensory impulses propagate along *spinal nerves* into the spinal cord.

2. **Second-order neurons** conduct impulses from the brain stem and spinal cord to the thalamus. Axons of second-order neurons *decussate* (cross over to the opposite side) in the brain stem or spinal cord before ascending to the ventral posterior nucleus of the thalamus. Thus, all somatic sensory information from one side of the body reaches the thalamus on the opposite side.

3. **Third-order neurons** conduct impulses from the thalamus to the primary somatosensory area of the cortex on the same side.

Regions within the CNS where neurons synapse with other neurons that are a part of a particular sensory or motor pathway are known as **relay stations** because neural signals are being relayed from one region of the CNS to another. For example, the neurons of many sensory pathways synapse with neurons in the thalamus; therefore the thalamus functions as a major relay station. In addition to the thalamus, many other regions of the CNS, including the spinal cord and brain stem, can function as relay stations.

Somatic sensory impulses ascend to the cerebral cortex via three general pathways: (1) the posterior column–medial lemniscus pathway, (2) the anterolateral (spinothalamic) pathway, and (3) the trigeminothalamic pathway. Somatic sensory impulses reach the cerebellum via the spinocerebellar tracts.

Posterior Column–Medial Lemniscus Pathway to the Cortex

Nerve impulses for touch, pressure, vibration, and conscious proprioception from the limbs, trunk, neck, and posterior head ascend to the cerebral cortex along the **posterior column–medial lemniscus pathway** (lem-NIS-kus = ribbon) (Figure 16.5). The name of the pathway comes from the names of two white-matter tracts that convey the impulses: the posterior column of the spinal cord and the medial lemniscus of the brain stem.

First-order neurons in the posterior column–medial lemniscus pathway extend from sensory receptors in the limbs, trunk, neck, and posterior head into the spinal cord and ascend to the medulla oblongata on the same side of the body. The cell bodies of these first-order neurons are in the posterior (dorsal) root ganglia of spinal nerves. In the spinal cord, their axons form the **posterior (dorsal) columns,** which consist of two parts: the **gracile fasciculus** (GRAS-īl fa-SIK-ū-lus) and the **cuneate fasciculus** (KŪ-nē-āt). The axons synapse with the dendrites of second-order neurons whose cell bodies are located in the gracile nucleus or cuneate nucleus of the medulla. Nerve impulses for touch, pressure, vibration, and conscious proprioception from the upper limbs, upper trunk, neck, and posterior head propagate along axons in the cuneate fasciculus and arrive at the cuneate nucleus. Nerve impulses for touch, pressure, and vibration from the lower limbs and lower trunk propagate along axons in the gracile fasciculus and arrive at the gracile nucleus.

The axons of the second-order neurons cross to the opposite side of the medulla and enter the **medial lemniscus,** a thin ribbon-like projection tract that extends from the medulla to the ventral posterior nucleus of the thalamus. In the thalamus, the axon terminals of second-order neurons synapse with third-order neurons,

Figure 16.5 The posterior column–medial lemniscus pathway.

 The posterior column–medial lemniscus pathway conveys nerve impulses for touch, pressure, vibration, and conscious proprioception from the limbs, trunk, neck, and posterior head to the cerebral cortex.

RIGHT SIDE OF BODY

LEFT SIDE OF BODY

Primary somatosensory area of cerebral cortex

THIRD-ORDER NEURONS

Thalamus (ventral posterior nucleus)

Medial lemniscus

Midbrain

SECOND-ORDER NEURONS

Gracile nucleus

Cuneate nucleus

FIRST-ORDER NEURONS

Medulla

POSTERIOR COLUMN:

Posterior root ganglion

Gracile fasciculus

Cuneate fasciculus

Receptors for touch, pressure, vibration, and proprioception in the upper limbs, upper trunk, neck, and posterior head

Spinal nerve

Cervical spinal cord

Receptors for touch, pressure, and vibration in the lower limbs and lower trunk

Lumbar spinal cord

? **What are the two major tracts that form the posterior columns?**

which project their axons to the primary somatosensory area of the cerebral cortex.

Anterolateral Pathway to the Cortex

Nerve impulses for pain, temperature, itch, and tickle from the limbs, trunk, neck, and posterior head ascend to the cerebral cortex along the **anterolateral** or **spinothalamic pathway** (spī-nō-tha-

LAM-ik). Like the posterior column–medial lemniscus pathway, the anterolateral pathway is composed of three-neuron sets (Figure 16.6). The first-order neurons connect a receptor of the limbs, trunk, neck, or posterior head with the spinal cord. The cell bodies of the first-order neurons are in the posterior root ganglion. The axon terminals of the first-order neurons synapse with second-order neurons, whose cell bodies are located in the posterior gray horn of the spinal cord. The axons of the second-order neurons cross to the opposite side of the spinal cord. Then, they pass upward to the brain stem as the **spinothalamic tract.** The axons of the second-order neurons end in the ventral posterior nucleus of the thalamus, where they synapse with the third-order neurons. The axons of the third-order neurons project to the primary somatosensory area on the same side of the cerebral cortex as the thalamus.

Trigeminothalamic Pathway to the Cortex

Nerve impulses for most somatic sensations (tactile, thermal, and pain) from the face, nasal cavity, oral cavity, and teeth ascend to the cerebral cortex along the **trigeminothalamic pathway** (trī-JEM-i-nō-tha-LAM-ik). Like the other somatosensory pathways just described, the trigeminothalamic pathway consists of three-neuron sets (Figure 16.7). First-order neurons extend from somatic sensory receptors in the face, nasal cavity, oral cavity, and teeth into the pons through the trigeminal (V) nerves. The cell

Figure 16.6 The anterolateral (spinothalamic) pathway.

The anterolateral pathway conveys nerve impulses for pain, cold, warmth, itch, and tickle from the limbs, trunk, neck, and posterior head to the cerebral cortex.

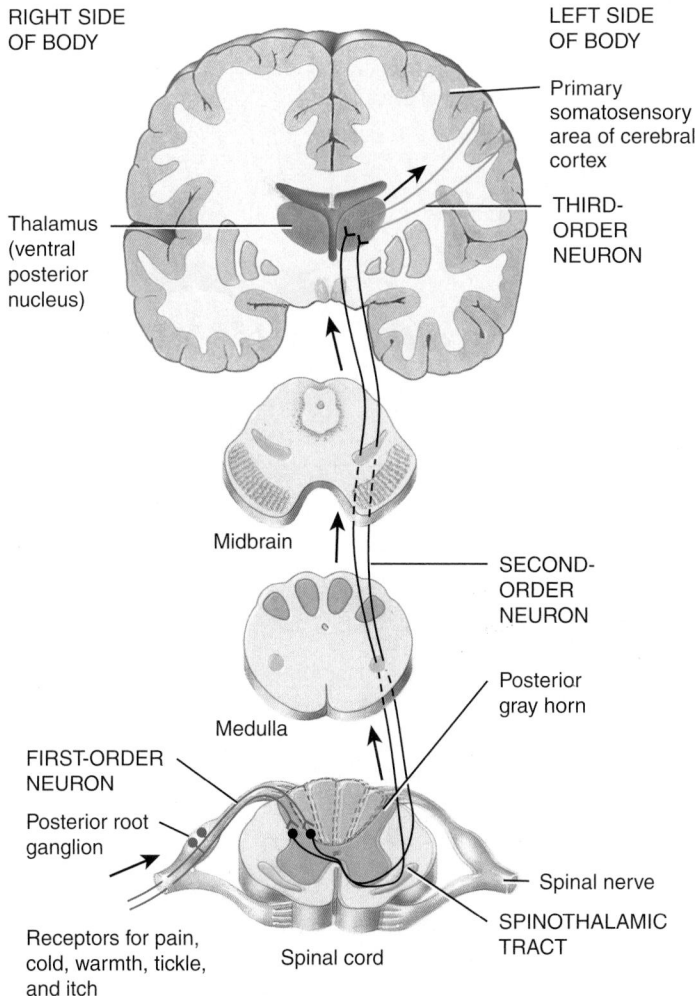

Figure 16.7 The trigeminothalamic pathway.

The trigeminothalamic pathway conveys nerve impulses for most somatic sensations (tactile, thermal, pain, and proprioceptive) from the face, nasal cavity, oral cavity, and teeth to the cerebral cortex.

? What types of sensory deficits could be produced by damage to the right spinothalamic tract?

? Which cranial nerve conveys impulses for most somatic sensations from the left side of the face into the pons?

bodies of these first-order neurons are in the trigeminal ganglion. The axon terminals of some first-order neurons synapse with second-order neurons in the pons. The axons of other first-order neurons descend into the medulla to synapse with second-order neurons. The axons of the second-order neurons cross to the opposite side of the pons and medulla and then ascend as the **trigeminothalamic tract** to the ventral posterior nucleus of the thalamus. In the thalamus, the axon terminals of the second-order neurons synapse with third-order neurons, which project their axons to the primary somatosensory area on the same side of the cerebral cortex as the thalamus.

Mapping the Primary Somatosensory Area

Specific areas of the cerebral cortex receive somatic sensory input from particular parts of the body. Other areas of the cerebral cor-

tex provide output in the form of instructions for movement of particular parts of the body. The *somatic sensory map* and the *somatic motor map* relate body parts to these cortical areas.

Precise localization of somatic sensations occurs when nerve impulses arrive at the **primary somatosensory area** (areas 1, 2, and 3 in Figure 14.15), which occupies the postcentral gyri of the parietal lobes of the cerebral cortex. Each region in this area receives sensory input from a different part of the body. Figure 16.8a maps the destination of somatic sensory signals from different parts of the left side of the body in the somatosensory area of the right cerebral hemisphere. The left cerebral hemisphere has a similar primary somatosensory area that receives sensory input from the right side of the body.

Note that some parts of the body—chiefly the lips, face, tongue, and hand—provide input to large regions in the somatosensory area. Other parts of the body, such as the trunk and lower limbs,

Figure 16.8 Somatic sensory and somatic motor maps in the cerebral cortex, right hemisphere. (a) Primary somatosensory area (postcentral gyrus) and (b) primary motor area (precentral gyrus) of the right cerebral hemisphere. The left hemisphere has similar representation. (After Penfield and Rasmussen.)

Each point on the body surface maps to a specific region in both the primary somatosensory area and the primary motor area.

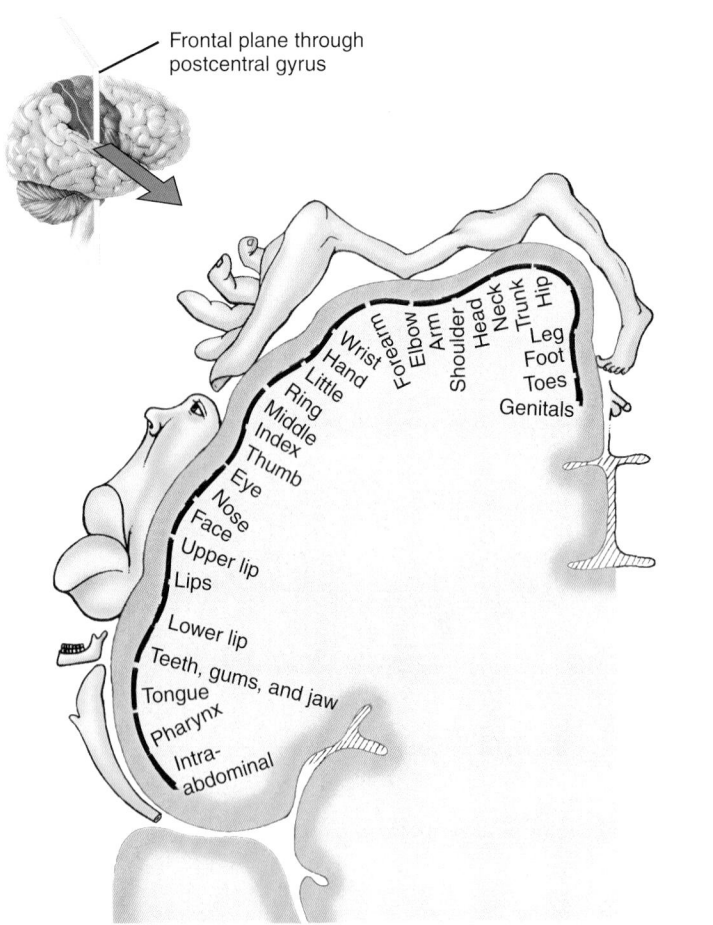

(a) Frontal section of primary somatosensory area in right cerebral hemisphere

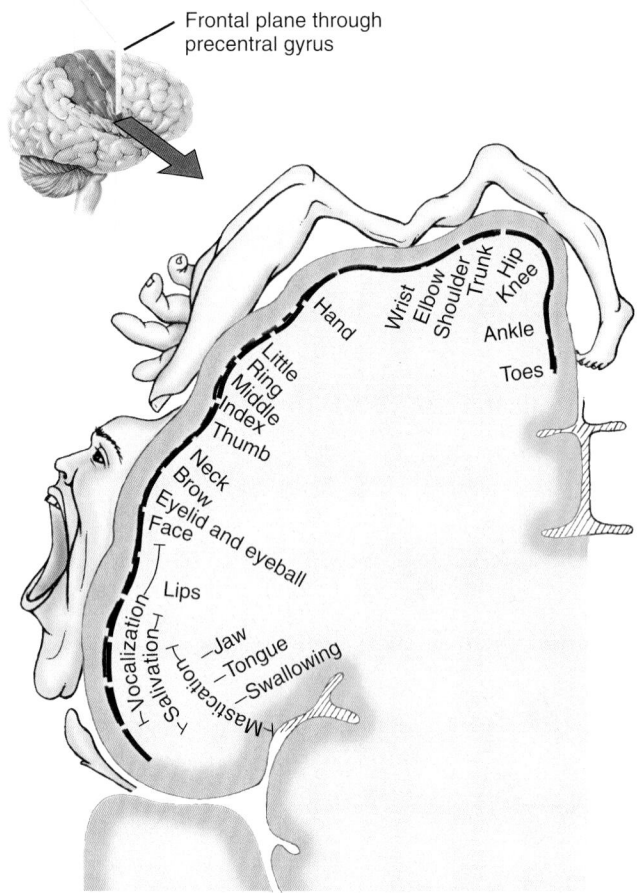

(b) Frontal section of primary motor area in right cerebral hemisphere

How do the somatosensory and motor representations compare for the hand, and what does this difference imply?

project to much smaller cortical regions. The relative sizes of these regions in the somatosensory area are proportional to the number of specialized sensory receptors within the corresponding part of the body. For example, there are many sensory receptors in the skin of the lips but few in the skin of the trunk. This distorted somatic sensory map of the body is known as the **sensory homunculus.** The size of the cortical region that represents a body part may expand or shrink somewhat, depending on the quantity of sensory impulses received from that body part. For example, people who learn to read Braille eventually have a larger cortical region in the somatosensory area to represent the fingertips.

Somatic Sensory Pathways to the Cerebellum

Two tracts in the spinal cord—the **posterior spinocerebellar tract** (spī-nō-ser-e-BEL-ar) and the **anterior spinocerebellar tract**—are the major routes proprioceptive impulses take to reach the cerebellum. Although they are not consciously perceived, sensory impulses conveyed to the cerebellum along these two pathways are critical for posture, balance, and coordination of skilled movements.

Table 16.3 summarizes the major somatic sensory tracts and pathways.

TABLE 16.3

Major Somatic Sensory Tracts and Pathways

TRACTS AND LOCATIONS	PATHWAY FUNCTIONS
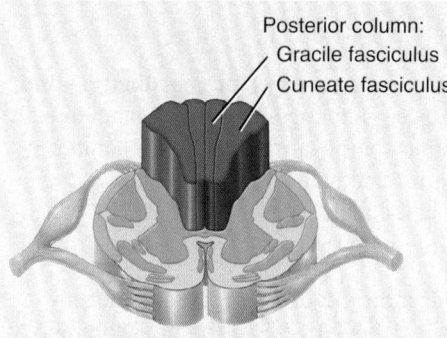 Posterior column: Gracile fasciculus Cuneate fasciculus Spinal cord	**Posterior column:** Consists of two tracts: (1) **cuneate fasciculus,** which conveys nerve impulses for touch, pressure, vibration, and conscious proprioception from upper limbs, upper trunk, neck, and posterior head, and (2) **gracile fasciculus,** which conveys nerve impulses for touch, pressure, and vibration from lower limbs and lower trunk. Axons of first-order neurons from one side of body form posterior column on same side and end in medulla, where they synapse with dendrites and cell bodies of second-order neurons. Axons of second-order neurons decussate, enter medial lemniscus on opposite side, and extend to thalamus. Third-order neurons transmit nerve impulses from thalamus to primary somatosensory cortex on side opposite the site of stimulation.
 Spinothalamic tract Spinal cord	**Spinothalamic:** Conveys nerve impulses for pain, cold, warmth, itch, and tickle from limbs, trunk, neck, and posterior head. Axons of first-order neurons from one side of body synapse with dendrites and cell bodies of second-order neurons in posterior gray horn on same side of body. Axons of second-order neurons decussate, enter spinothalamic tract on opposite side, and extend to thalamus. Third-order neurons transmit nerve impulses from thalamus to primary somatosensory cortex on side opposite the site of stimulation.
 Trigeminothalamic tract Pons	**Trigeminothalamic:** Conveys nerve impulses for touch, pressure, vibration, pain, cold, warmth, itch, and tickle from face, nasal cavity, oral cavity, and teeth. Axons of first-order neurons from one side of head synapse with dendrites and cell bodies of second-order neurons in pons and medulla on same side of head. Axons of second-order neurons decussate, enter trigeminothalamic tract on opposite side, and extend to thalamus. Third-order neurons transmit nerve impulses from thalamus to primary somatosensory cortex on side opposite the site of stimulation.
 Posterior spinocerebellar tract Spinal cord Anterior spinothalamic tract	**Anterior and posterior spinocerebellar:** Convey nerve impulses from proprioceptors in trunk and lower limb of one side of body to same side of cerebellum. Proprioceptive input informs cerebellum of actual movements, allowing it to coordinate, smooth, and refine skilled movements and maintain posture and balance.

Syphilis is a sexually transmitted disease caused by the bacterium *Treponema pallidum.* Because it is a bacterial infection, it can be treated with antibiotics. However, if the infection is not treated, the third stage of syphilis typically causes debilitating neurological symptoms. A common outcome is progressive degeneration of the posterior portions of the spinal cord, including the posterior columns, posterior spinocerebellar tracts, and posterior roots. Somatic sensations are lost, and the person's gait becomes uncoordinated and jerky because proprioceptive impulses fail to reach the cerebellum. •

✔ CHECKPOINT

11. What are the functional differences between the posterior column–medial lemniscus pathway, the anterolateral pathway, and the trigeminothalamic pathway?
12. Which body parts have the largest representation in the primary somatosensory area?
13. What type of sensory information is carried in the spinocerebellar tracts, and what is its function?

16.4 SOMATIC MOTOR PATHWAYS

◉ OBJECTIVES

• Identify the locations and functions of the different types of neurons in the somatic motor pathways.
• Compare the locations and functions of the direct and indirect motor pathways.
• Explain how the basal nuclei and cerebellum contribute to movements.

Neural circuits in the brain and spinal cord orchestrate all voluntary and involuntary movements. Ultimately, all excitatory and inhibitory signals that control movement converge on the motor neurons that extend out of the brain stem and spinal cord to innervate skeletal muscles in the body. These neurons, also known as **lower motor neurons (LMNs),** have their cell bodies in the brain stem and spinal cord. From the brain stem, axons of LMNs extend through *cranial nerves* to innervate skeletal muscles of the face and head. From the spinal cord, axons of LMNs extend through *spinal nerves* to innervate skeletal muscles of the limbs and trunk. Only LMNs provide output from the CNS to skeletal muscle fibers. For this reason, they are also called the *final common pathway.*

Neurons in four distinct but highly interactive neural circuits, collectively termed the **somatic motor pathways,** participate in control of movement by providing input to lower motor neurons (Figure 16.9):

❶ *Local circuit neurons.* Input arrives at lower motor neurons from nearby interneurons called **local circuit neurons.** These neurons are located close to the lower motor neuron cell bodies in the brain stem and spinal cord. Local circuit neurons receive input from somatic sensory receptors, such as nociceptors and muscle spindles, as well as from higher

centers in the brain. They help coordinate rhythmic activity in specific muscle groups, such as alternating flexion and extension of the lower limbs during walking.

❷ *Upper motor neurons.* Both local circuit neurons and lower motor neurons receive input from **upper motor neurons (UMNs).** Most upper motor neurons synapse with local circuit neurons, which in turn synapse with lower motor neurons. (A few upper motor neurons synapse directly with lower motor neurons.) UMNs from the cerebral cortex are essential for the execution of voluntary movements of the body. Other UMNs originate in motor centers of the brain stem: the red nucleus, the vestibular nucleus, the superior colliculus, and the reticular formation. UMNs from the brain stem regulate muscle tone, control postural muscles, and help maintain balance and orientation of the head and body. Both the basal nuclei and cerebellum exert influence on upper motor neurons.

Figure 16.9 Somatic motor pathways for coordination and control of movement. Lower motor neurons receive input directly from ❶ local circuit neurons (purple arrow) and ❷ upper motor neurons in the cerebral cortex and brain stem (green arrows). Neural circuits involving basal nuclei neurons ❸ and cerebellar neurons ❹ regulate activity of upper motor neurons (red arrows).

🔑 Because lower motor neurons provide all output to skeletal muscles, they are called the final common pathway.

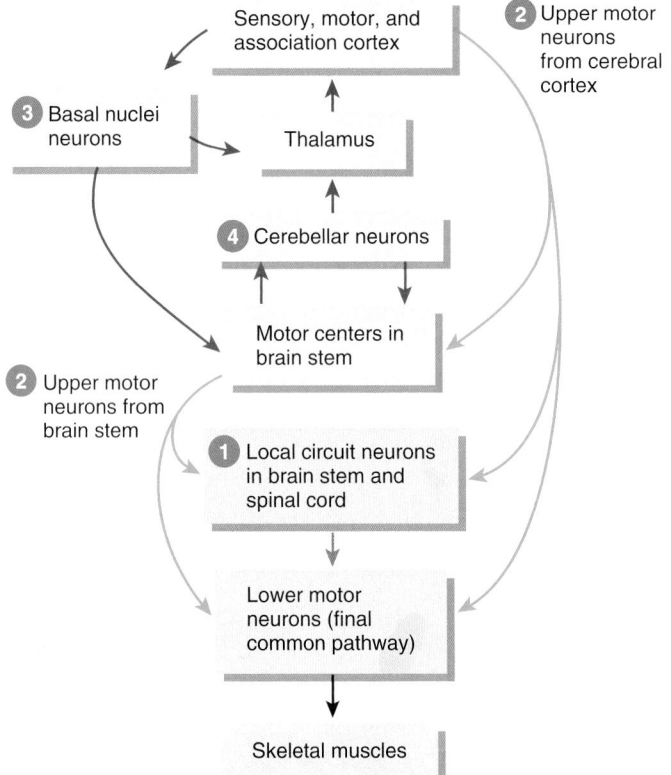

❓ How do the functions of upper motor neurons from the cerebral cortex and from the brain stem differ?

3 *Basal nuclei neurons.* Basal nuclei neurons assist movement by providing input to upper motor neurons. Neural circuits interconnect the basal nuclei with motor areas of the cerebral cortex (via the thalamus) and the brain stem. These circuits help initiate and terminate movements, suppress unwanted movements, and establish a normal level of muscle tone.

4 *Cerebellar neurons.* Cerebellar neurons also aid movement by controlling the activity of upper motor neurons. Neural circuits interconnect the cerebellum with motor areas of the cerebral cortex (via the thalamus) and the brain stem. A prime function of the cerebellum is to monitor differences between intended movements and movements actually performed. Then, it issues commands to upper motor neurons to reduce errors in movement. The cerebellum thus coordinates body movements and helps maintain normal posture and balance.

CLINICAL CONNECTION | Paralysis

Damage or disease of *lower* motor neurons produces **flaccid paralysis** (FLAS-id or FLAK-sid) of muscles on the same side of the body. There is neither voluntary nor reflex action of the innervated muscle fibers, muscle tone is decreased or lost, and the muscle remains limp or flaccid. Injury or disease of *upper* motor neurons in the cerebral cortex removes inhibitory influences that some of these neurons have on lower motor neurons, which causes **spastic paralysis** of muscles on the opposite side of the body. In this condition muscle tone is increased, reflexes are exaggerated, and pathological reflexes such as the Babinski sign appear (see Clinical Connection: Reflexes and Diagnosis in Chapter 13). •

Organization of Upper Motor Neuron Pathways

The axons of upper motor neurons extend from the brain to lower motor neurons via two types of somatic motor pathways—direct and indirect. **Direct motor pathways** provide input to lower motor neurons via axons that extend directly from the cerebral cortex. **Indirect motor pathways** provide input to lower motor neurons from motor centers in the basal nuclei, cerebellum, and cerebral cortex. Direct and indirect pathways both govern generation of nerve impulses in the lower motor neurons, the neurons that stimulate contraction of skeletal muscles.

Before we examine these pathways we consider the role of the motor cortex in voluntary movement.

Mapping the Motor Areas

Control of body movements occurs via neural circuits in several regions of the brain. The **primary motor area** (area 4 in Figure 14.15), located in the precentral gyrus of the frontal lobe (see Figure 16.8b) of the cerebral cortex, is a major control region for the execution of voluntary movements. The adjacent **premotor area** (area 6) also contributes axons to the descending motor pathways. As is true for somatic sensory representation in the somatosensory area, different muscles are represented unequally in the primary motor area. More cortical area is devoted to those muscles involved in skilled, complex, or delicate movements.

Muscles in the thumb, fingers, lips, tongue, and vocal cords have large representations; the trunk has a much smaller representation. This distorted muscle map of the body is referred to as the **motor homunculus.** By comparing Figures 16.8a and 16.8b, you can see that somatosensory and somatic motor representations are similar but not identical for most parts of the body.

Direct Motor Pathways

Nerve impulses for voluntary movements propagate from the cerebral cortex to lower motor neurons via the direct motor pathways. The direct motor pathways, which are also known as the *pyramidal pathways,* consist of axons that descend from pyramidal cells. *Pyramidal cells* are upper motor neurons with pyramid-shaped cell bodies (see Figure 12.5b) located in the primary motor area and the premotor area of the cerebral cortex (areas 4 and 6, respectively, in Figure 14.15). The direct motor pathways consist of the corticospinal pathways and the corticobulbar pathway.

CORTICOSPINAL PATHWAYS The **corticospinal pathways** (kor′-ti-kō-SPĪ-nal) conduct impulses for the control of muscles of the limbs and trunk. Axons of upper motor neurons in the cerebral cortex form the **corticospinal tracts,** which descend through the internal capsule of the cerebrum and the cerebral peduncle of the midbrain. In the medulla oblongata, the axon bundles of the corticospinal tracts form the ventral bulges known as the *pyramids.* About 90% of the corticospinal axons *decussate* (cross over) to the *contralateral* (opposite) side in the medulla oblongata and then descend into the spinal cord where they synapse with a local circuit neuron or a lower motor neuron. The 10% that remain on the *ipsilateral* (same) side eventually decussate at the spinal cord levels where they synapse with a local circuit neuron or lower motor neuron. Thus, the right cerebral cortex controls most of the muscles on the left side of the body, and the left cerebral cortex controls most of the muscles on the right side of the body. There are two types of corticospinal tracts: the lateral corticospinal tract and the anterior corticospinal tract.

1. *Lateral corticospinal tract.* Corticospinal axons that decussate in the medulla form the **lateral corticospinal tract** in the lateral white column of the spinal cord (Figure 16.10a). These axons synapse with local circuit neurons or lower motor neurons in the anterior gray horn of the spinal cord. Axons of these lower motor neurons exit the cord in the anterior roots of spinal nerves and terminate in skeletal muscles that control movements of the distal parts of the limbs. The distal muscles are responsible for precise, agile, and highly skilled movements. Examples include the movements needed to button a shirt or play the piano.

2. *Anterior corticospinal tract.* Corticospinal axons that do not decussate in the medulla form the **anterior corticospinal tract** in the anterior white column of the spinal cord (Figure 16.10b). At each spinal cord level, some of these axons decussate via the anterior white commissure. Then, they synapse with local circuit neurons or lower motor neurons in the anterior gray horn. Axons of these lower motor neurons exit the cord in the

Figure 16.10 The corticospinal pathways.

🔒 ⚏ The corticospinal pathways conduct nerve impulses for the control of muscles of the limbs and trunk.

Primary motor area of cerebral cortex

RIGHT SIDE
OF BODY

LEFT SIDE
OF BODY

Internal
capsule

Midbrain

UPPER
MOTOR
NEURON

Cerebral
peduncle

Pons

Medulla

Pyramid

Decussation
(crossing)
in medulla

Spinal cord

LATERAL
CORTICOSPINAL
TRACT

Spinal nerve

LOWER
MOTOR
NEURON

To skeletal
muscles in
the distal
parts of the
limbs

Spinal cord

(a) The lateral corticospinal pathway

Primary motor area of cerebral cortex

RIGHT SIDE
OF BODY

LEFT SIDE
OF BODY

Internal
capsule

Midbrain

UPPER
MOTOR
NEURON

Cerebral
peduncle

Pons

Medulla

Pyramid

Spinal cord

ANTERIOR
CORTICOSPINAL
TRACT

Spinal nerve

LOWER
MOTOR
NEURON

To skeletal
muscles in
the trunk
and
proximal parts
of the limbs

Spinal cord

(b) The anterior corticospinal pathway

❓ **Which tract conveys nerve impulses that result in contractions of muscles in the distal parts of the limbs?**

anterior roots of spinal nerves. They terminate in skeletal muscles that control movements of the trunk and proximal parts of the limbs.

CORTICOBULBAR PATHWAY The **corticobulbar pathway** (kor′-ti-kō-BUL-bar) conducts impulses for the control of skeletal muscles in the head. Axons of upper motor neurons from the cerebral cortex form the **corticobulbar tract**, which descends along with the corticospinal tracts through the internal capsule of the cerebrum and cerebral peduncle of the midbrain (Figure 16.11). Some of the axons of the corticobulbar tract decussate; others do not. The axons terminate in the motor nuclei of nine pairs of cranial nerves in the brain stem: the oculomotor (III), trochlear (IV), trigeminal (V), abducens (VI), facial (VII), glossopharyngeal (IX),

Figure 16.11 The corticobulbar pathway. For simplicity, only two cranial nerves are illustrated.

The corticobulbar pathway conducts nerve impulses for the control of skeletal muscles in the head.

The axons of the corticobulbar tract terminate in the motor nuclei of which cranial nerves?

vagus (X), accessory (XI), and hypoglossal (XII). The lower motor neurons of the cranial nerves convey impulses that control precise, voluntary movements of the eyes, tongue, and neck, plus chewing, facial expression, speech, and swallowing.

Table 16.4 summarizes the major somatic motor tracts and pathways.

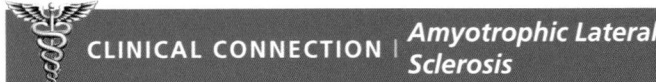

CLINICAL CONNECTION | *Amyotrophic Lateral Sclerosis*

Amyotrophic lateral sclerosis (ALS) (ā'-mī-ō-TRŌF-ik; *a*- = without; *-myo-* = muscle; *-trophic* = nourishment) is a progressive degenerative disease that attacks motor areas of the cerebral cortex, axons of upper motor neurons in the lateral white columns (corticospinal and rubrospinal tracts), and lower motor neuron cell bodies. It causes progressive muscle weakness and atrophy. ALS often begins in sections of the spinal cord that serve the hands and arms but rapidly spreads to involve the whole body and face, without affecting intellect or

TABLE 16.4

Major Somatic Motor Tracts and Pathways

TRACTS AND LOCATIONS

PATHWAY FUNCTIONS

DIRECT (PYRAMIDAL) PATHWAYS

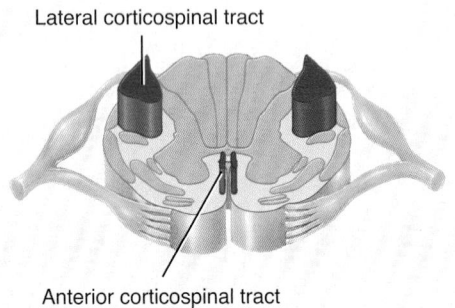

Lateral corticospinal tract

Anterior corticospinal tract

Spinal cord

Lateral corticospinal: Conveys nerve impulses from motor cortex to skeletal muscles on opposite side of body for precise, voluntary movements of distal parts of limbs. Axons of upper motor neurons (UMNs) descend from precentral gyrus of cortex into medulla. Here 90% decussate (cross over to opposite side) and then enter contralateral side of spinal cord to form this tract. At their level of termination, UMNs end in anterior gray horn on same side. They provide input to lower motor neurons, which innervate skeletal muscles.

Anterior corticospinal: Conveys nerve impulses from motor cortex to skeletal muscles on opposite side of body for movements of trunk and proximal parts of limbs. Axons of UMNs descend from cortex into medulla. Here the 10% that do not decussate enter spinal cord and form this tract. At their level of termination, UMNs decussate and end in anterior gray horn on opposite side of body. They provide input to lower motor neurons, which innervate skeletal muscles.

Cerebral peduncle

Corticobulbar tract

Midbrain of brain stem

Corticobulbar: Conveys nerve impulses from motor cortex to skeletal muscles of head and neck to coordinate precise, voluntary movements. Axons of UMNs descend from cortex into brain stem, where some decussate and others do not. They provide input to lower motor neurons in nuclei of cranial nerves III, IV, V, VI, VII, IX, X, XI, and XII, which control voluntary movements of eyes, tongue, and neck; chewing; facial expression; and speech.

INDIRECT (EXTRAPYRAMIDAL) PATHWAYS

Rubrospinal tract

Tectospinal tract

Medial reticulospinal tract

Lateral reticulospinal tract

Vestibulospinal tract

Spinal cord

Rubrospinal: Conveys nerve impulses from red nucleus (which receives input from cerebral cortex and cerebellum) to contralateral skeletal muscles that govern precise, voluntary movements of distal parts of upper limbs.

Tectospinal: Conveys nerve impulses from superior colliculus to contralateral skeletal muscles that reflexively move head, eyes, and trunk in response to visual or auditory stimuli.

Vestibulospinal: Conveys nerve impulses from vestibular nucleus (which receives input about head movements from inner ear) to ipsilateral skeletal muscles of trunk and proximal parts of limbs for maintaining posture and balance in response to head movements.

Medial and lateral reticulospinal: Conveys nerve impulses from reticular formation to ipsilateral skeletal muscles of trunk and proximal parts of limbs for maintaining posture and regulating muscle tone in response to ongoing body movements.

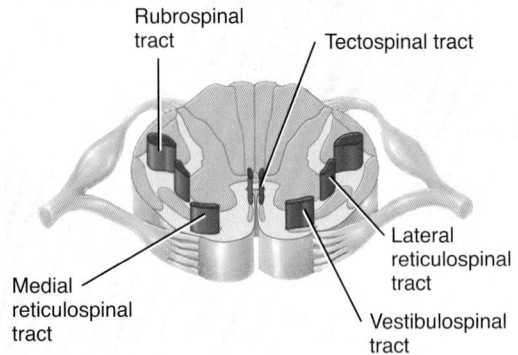

sensations. Death typically occurs in 2 to 5 years. ALS is commonly known as *Lou Gehrig's disease* after the New York Yankees baseball player who died from it at age 37 in 1941.

Inherited mutations account for about 15% of all cases of ALS (familial ALS). Noninherited (sporadic) cases of ALS appear to have several implicating factors. According to one theory there is a buildup in the synaptic cleft of the neurotransmitter glutamate released by motor neurons due to a mutation of the protein that normally deactivates and recycles the neurotransmitter. The excess glutamate causes motor neurons to malfunction and eventually die. The drug riluzole, which is used to treat ALS, reduces damage to motor neurons by decreasing the release of glutamate. Other factors may include damage to motor neurons by free radicals, autoimmune responses, viral infections, deficiency of nerve growth factor, apoptosis (programmed cell death), environmental toxins, and trauma.

In addition to riluzole, ALS is treated with drugs that relieve symptoms such as fatigue, muscle pain and spasticity, excessive saliva, and difficulty sleeping. The only other treatment is supportive care provided by physical, occupational, and speech therapists; nutritionists; social workers; and home care and hospice nurses. •

Indirect Motor Pathways

The **indirect motor pathways** or **extrapyramidal pathways** include all somatic motor tracts other than the corticospinal and corticobulbar tracts. Axons of upper motor neurons that give rise to the indirect motor pathways descend from various nuclei of the brain stem into five major tracts of the spinal cord and terminate on local circuit neurons or lower motor neurons. These tracts are the **rubrospinal** (ROO-brō-spī-nal), **tectospinal** (TEK-tō-spī-nal), **vestibulospinal** (ves-TIB-ū-lō-spī-nal), **lateral reticulospinal** (re-TIK-ū-lō-spī-nal), and **medial reticulospinal tracts.**
Table 16.4 summarizes the tracts of the indirect motor pathways.

Roles of the Basal Nuclei

As previously noted, the basal nuclei and cerebellum influence movement through their effects on upper motor neurons. The functions of the basal nuclei include the following:

1. The basal nuclei play a major role in the initiation and termination of movements. Two parts of the basal nuclei, the caudate nucleus and the putamen, receive input from sensory, association, and motor areas of the cerebral cortex and from the substantia nigra. Output from the basal nuclei comes from the globus pallidus and substantia nigra, which send feedback signals to the upper motor cortex by way of the thalamus. (Figure 14.13b shows these parts of the basal nuclei.) This circuit—from cortex to basal nuclei to thalamus to cortex—appears to function in initiating and terminating movements. Neurons in the putamen generate impulses just before body movements occur, and neurons in the caudate nucleus generate impulses just before eye movements occur.

2. The basal nuclei suppress unwanted movements by their inhibitory effects on the thalamus and superior colliculus.

3. The basal nuclei influence muscle tone. The globus pallidus sends impulses into the reticular formation that reduce muscle tone. Damage or destruction of some basal nuclei connections causes a generalized increase in muscle tone.

4. The basal nuclei influence many aspects of cortical function, including sensory, limbic, cognitive, and linguistic functions. For example, the basal nuclei help initiate and terminate some cognitive processes, such as attention, memory, and planning. In addition, the basal nuclei may act with the limbic system to regulate emotional behaviors.

CLINICAL CONNECTION | *Disorders of the Basal Nuclei*

Disorders of the basal nuclei can affect body movements, cognition, and behavior. Uncontrollable shaking (tremor) and muscle rigidity (stiffness) are hallmark signs of **Parkinson disease (PD)** (see Disorders: Homeostatic Imbalances at the end of this chapter). In this disorder, dopamine-releasing neurons that extend from the substantia nigra to the putamen and caudate nucleus degenerate.

Huntington disease (HD) is an inherited disorder in which the caudate nucleus and putamen degenerate, with loss of neurons that normally release GABA or acetylcholine. A key sign of HD is **chorea** (KŌ-rē-a = a dance), in which rapid, jerky movements occur involuntarily and without purpose. Progressive mental deterioration also occurs. Symptoms of HD often do not appear until age 30 or 40. Death occurs 10 to 20 years after symptoms first appear.

Tourette syndrome is a disorder that is characterized by involuntary body movements (motor tics) and the use of inappropriate or unnecessary sounds or words (vocal tics). Although the cause is unknown, research suggests that this disorder involves a dysfunction of the cognitive neural circuits between the basal nuclei and the prefrontal cortex.

Some psychiatric disorders, such as schizophrenia and obsessive–compulsive disorder, are thought to involve dysfunction of the behavioral neural circuits between the basal nuclei and the limbic system. In **schizophrenia,** excess dopamine activity in the brain causes a person to experience delusions, distortions of reality, paranoia, and hallucinations. People who have **obsessive–compulsive disorder (OCD)** experience repetitive thoughts (obsessions) that cause repetitive behaviors (compulsions) that they feel obligated to perform. For example, a person with OCD might have repetitive thoughts about someone breaking into the house; these thoughts might drive that person to check the doors of the house over and over again (for minutes or hours at a time) to make sure that they are locked. •

Modulation of Movement by the Cerebellum

In addition to maintaining proper posture and balance, the cerebellum is active in both learning and performing rapid, coordinated, highly skilled movements such as hitting a golf ball, speaking, and swimming. Cerebellar function involves four activities (Figure 16.12):

① **Monitoring intentions for movement.** The cerebellum receives impulses from the motor cortex and basal nuclei via

Figure 16.12 Input to and output from the cerebellum.

The cerebellum coordinates and smooths contractions of skeletal muscles during skilled movements and helps maintain posture and balance.

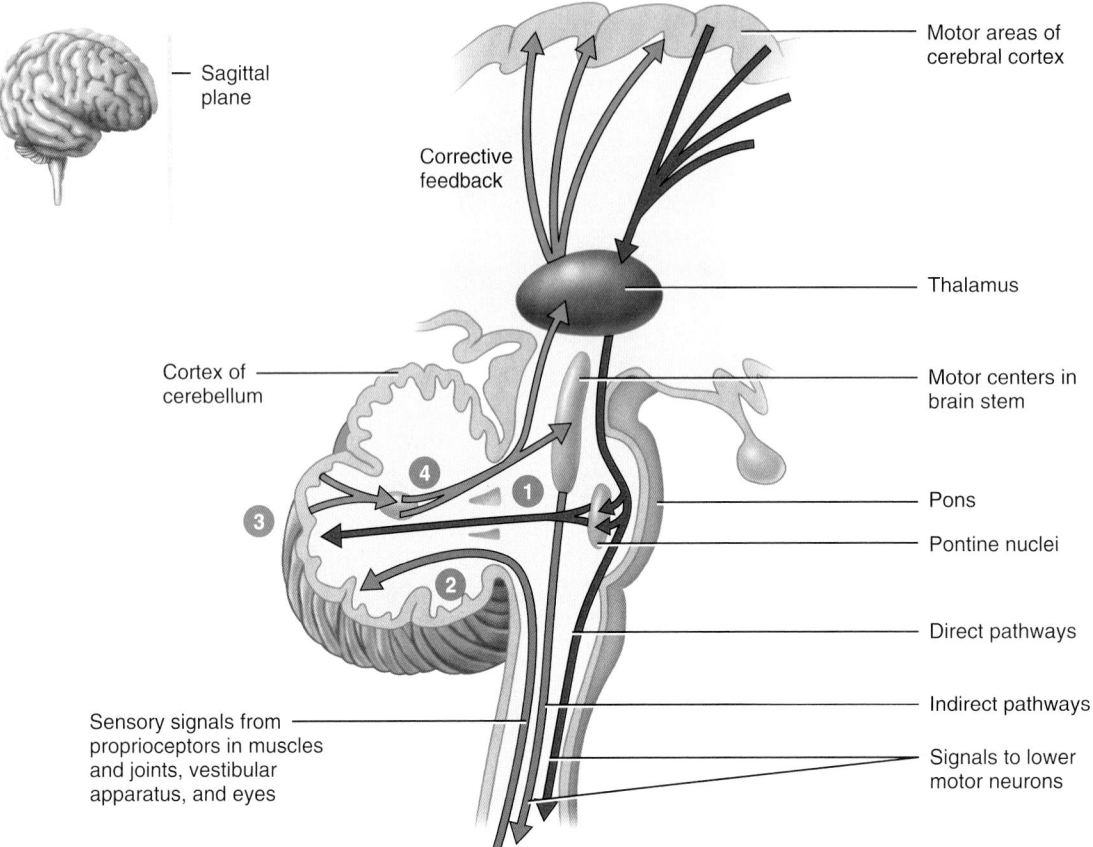

Sagittal section through brain and spinal cord

Which tracts carry information from proprioceptors in joints and muscles to the cerebellum?

the pontine nuclei in the pons regarding what movements are planned (red arrows).

2 Monitoring actual movement. The cerebellum receives input from proprioceptors in joints and muscles that reveals what actually is happening. These nerve impulses travel in the anterior and posterior spinocerebellar tracts. Nerve impulses from the vestibular (equilibrium-sensing) apparatus in the inner ear and from the eyes also enter the cerebellum.

3 Comparing command signals with sensory information. The cerebellum compares intentions for movement with the actual movement performed.

4 Sending out corrective feedback. If there is a discrepancy between intended and actual movement, the cerebellum sends feedback to upper motor neurons. This information travels via the thalamus to UMNs in the cerebral cortex but goes directly to UMNs in brain stem motor centers (green arrows). As movements occur, the cerebellum continuously provides error corrections to upper motor neurons, which decreases errors and smoothes the motion.

It also contributes over longer periods to the learning of new motor skills.

Skilled activities such as tennis or volleyball provide good examples of the contribution of the cerebellum to movement. To make a good serve or to block a spike, you must bring your racket or arms forward just far enough to make solid contact. How do you stop at exactly the right point? Before you even hit the ball, the cerebellum has sent nerve impulses to the cerebral cortex and basal nuclei informing them where your swing must stop. In response to impulses from the cerebellum, the cortex and basal nuclei transmit motor impulses to opposing body muscles to stop the swing.

✔CHECKPOINT
14. Trace the path of a motor impulse from the upper motor neurons through the final common pathway.
15. Which parts of the body have the largest representation in the motor cortex? Which have the smallest?
16. Explain why the two main somatic motor pathways are called "direct" and "indirect."
17. Explain the role of the cerebellum in performing rapid, coordinated, highly skilled movements.

16.5 INTEGRATIVE FUNCTIONS OF THE CEREBRUM

◉ **OBJECTIVES**

• Compare the integrative cerebral functions of wakefulness and sleep, and learning and memory.
• Describe the four stages of sleep.
• Explain the factors that contribute to memory.

We turn now to a fascinating, though incompletely understood, function of the cerebrum: integration, the processing of sensory information by analyzing and storing it and making decisions for various responses. The **integrative functions** include cerebral activities such as sleep and wakefulness, learning and memory, and emotional responses. (The role of the limbic system in emotional behavior was discussed in Chapter 14.)

Wakefulness and Sleep

Humans sleep and awaken in a 24-hour cycle called a **circadian rhythm** (ser-KĀ-dē-an; *circa-* = about; *-dia* = a day) that is established by the suprachiasmatic nucleus of the hypothalamus (see Figure 14.10). A person who is awake is in a state of readiness and is able to react consciously to various stimuli. EEG recordings show that the cerebral cortex is very active during wakefulness; fewer impulses arise during most stages of sleep.

The Role of the Reticular Activating System in Awakening

How does your nervous system make the transition between these two states? Because stimulation of some of its parts increases activity of the cerebral cortex, a portion of the reticular formation is known as the **reticular activating system (RAS)** (Figure 16.13). When this area is active, many nerve impulses are transmitted to widespread areas of the cerebral cortex, both directly and via the thalamus. The effect is a generalized increase in cortical activity.

Arousal, or awakening from sleep, also involves increased activity in the RAS. For arousal to occur, the RAS must be stimulated. Many sensory stimuli can activate the RAS: painful stimuli detected by nociceptors, touch and pressure on the skin, movement of the limbs, bright light, or the buzz of an alarm clock. Once the RAS is activated, the cerebral cortex is also activated, and arousal occurs. The result is a state of wakefulness called **consciousness** (KON-shus-nes). Notice in Figure 16.13 that even though the RAS receives input from somatic sensory receptors, the eyes, and the ears, there is no input from olfactory receptors; even strong odors may fail to cause arousal. People who die in house fires usually succumb to smoke inhalation without awakening. For this reason, all sleeping areas should have a nearby

Figure 16.13 The reticular activating system (RAS) consists of neurons whose axons project from the reticular formation through the thalamus to the cerebral cortex.

🔑 Increased activity of the RAS causes awakening from sleep (arousal).

Sagittal section through brain and spinal cord showing the reticular formation

 Why should every sleeping room have a smoke detector?

smoke detector that emits a loud alarm. A vibrating pillow or flashing light can serve the same purpose for those who are hearing impaired.

Sleep

Sleep is a state of altered consciousness or partial unconsciousness from which an individual can be aroused. Although it is essential, the exact functions of sleep are still unclear. Sleep deprivation impairs attention, learning, and performance. Normal sleep consists of two components: non-rapid eye movement (NREM) sleep and rapid eye movement (REM) sleep. **NREM sleep** consists of four gradually merging stages:

1. *Stage 1* is a transition stage between wakefulness and sleep that normally lasts 1–7 minutes. The person is relaxed with eyes closed and has fleeting thoughts. People awakened during this stage often say they have not been sleeping.

2. *Stage 2* or *light sleep* is the first stage of true sleep. In it, a person is a little more difficult to awaken. Fragments of dreams may be experienced, and the eyes may slowly roll from side to side.

3. *Stage 3* is a period of moderately deep sleep. Body temperature and blood pressure decrease, and it is difficult to awaken the person. This stage occurs about 20 minutes after falling asleep.

4. *Stage 4* is the deepest level of sleep. Although brain metabolism decreases significantly and body temperature drops slightly at this time, most reflexes are intact, and muscle tone is decreased only slightly. When sleepwalking occurs, it does so during this stage.

Typically, a person goes from stage 1 to stage 4 of NREM sleep in less than an hour. During a typical 7- or 8-hour sleep period, there are three to five episodes of **REM sleep,** during which the eyes move rapidly back and forth under closed eyelids. The person may rapidly ascend through stages 3 and 2 before entering REM sleep. The first episode of REM sleep lasts 10–20 minutes. Then, another interval of NREM sleep follows.

REM and NREM sleep alternate throughout the night. REM periods, which occur approximately every 90 minutes, gradually lengthen, until the final one lasts about 50 minutes. In adults, REM sleep totals 90–120 minutes during a typical sleep period. As a person ages, the average total time spent sleeping decreases, and the percentage of REM sleep declines. As much as 50% of an infant's sleep is REM sleep, as opposed to 35% for 2-year-olds and 25% for adults. Although we do not yet understand the function of REM sleep, the high percentage of REM sleep in infants and children is thought to be important for the maturation of the brain. Neuronal activity is high during REM sleep—brain blood flow and oxygen use are higher during REM sleep than during intense mental or physical activity while awake.

Different parts of the brain mediate NREM and REM sleep. Neurons in the preoptic area of the hypothalamus, the basal forebrain, and the medulla oblongata govern NREM sleep; neurons in the pons and midbrain turn REM sleep on and off. Several lines of evidence suggest the existence of sleep-inducing chemicals in the brain. One apparent sleep-inducer is adenosine, which accumulates during periods of high ATP (adenosine triphosphate) use by the nervous system. Adenosine binds to specific receptors, called A1 receptors, and inhibits certain cholinergic (acetylcholine-releasing) neurons of the RAS that participate in arousal. Thus, activity in the RAS during sleep is low due to the inhibitory effect of adenosine. Caffeine (in coffee) and theophylline (in tea)—substances known for their ability to maintain wakefulness—bind to and block the A1 receptors, preventing adenosine from binding and inducing sleep.

Several physiological changes occur during sleep. Most dreaming occurs during REM sleep, and the EEG readings are similar to those of a person who is awake. With the exception of motor neurons that govern breathing and eye movements, most somatic motor neurons are inhibited during REM sleep, which decreases muscle tone and even paralyzes the skeletal muscles. Many people experience a momentary feeling of paralysis if they are awakened during REM sleep. During sleep, activity in the parasympathetic division of the autonomic nervous system (ANS) increases while sympathetic activity decreases. Heart rate and blood pressure decrease during NREM sleep and decrease further during REM sleep. Increased parasympathetic activity during REM sleep sometimes causes erection of the penis, even when dream content is not sexual. The presence of penile erections during REM sleep in a man with erectile dysfunction (inability to attain an erection while awake) indicates that his problem has a psychological, rather than a physical, cause.

CLINICAL CONNECTION | Sleep Disorders

Sleep disorders affect over 70 million Americans each year. Common sleep disorders include insomnia, sleep apnea, and narcolepsy. A person with **insomnia** (in-SOM-nē-a) has difficulty in falling asleep or staying asleep. Possible causes of insomnia include stress, excessive caffeine intake, disruption of circadian rhythms (for example, working the night shift instead of the day shift at your job), and depression. **Sleep apnea** (AP-nē-a) is a disorder in which a person repeatedly stops breathing for 10 or more seconds while sleeping. Most often, it occurs because a loss of muscle tone in pharyngeal muscles allows the airway to collapse. **Narcolepsy** (NAR-kō-lep-sē) is a condition in which REM sleep cannot be inhibited during waking periods. As a result, involuntary periods of sleep that last about 15 minutes occur throughout the day. Recent studies have revealed that people with narcolepsy have a deficiency of the neuropeptide *orexin*, which is also known as **hypocretin.** Orexin is released from certain neurons of the hypothalamus and has a role in promoting wakefulness. •

Learning and Memory

Without memory, we would repeat mistakes and be unable to learn. Similarly, we would not be able to repeat our successes or accomplishments, except by chance. Although both learning and memory have been studied extensively, we still have no completely satisfactory explanation for how we recall information or how we remember events. However, we do know something about how information is acquired and stored, and it is clear that there are different categories of memory.

Learning is the ability to acquire new information or skills through instruction or experience. **Memory** is the process by which information acquired through learning is stored and retrieved. For an experience to become part of memory, it must produce persistent structural and functional changes that represent the experience in the brain. This capability for change associated with learning is termed **plasticity** (plas-TIS-i-tē). Nervous system plasticity underlies our ability to change our behavior in response to stimuli from the external and internal environments. It involves changes in individual neurons—for example, synthesis of different proteins or sprouting of new dendrites—as well as changes in the strengths of synaptic connections among neurons. The parts of the brain known to be involved with memory include the association areas of the frontal, parietal, occipital, and temporal lobes; parts of the limbic system, especially the hippocampus and amygdala; and the diencephalon. The primary somatosensory and primary motor areas in the brain also exhibit plasticity. If a particular body part is used more intensively or in a newly learned activity, such as reading Braille, the cortical areas devoted to that body part gradually expand.

Memory occurs in stages over a period of time. **Immediate memory** is the ability to recall ongoing experiences for a few seconds. It provides a perspective to the present time that allows us to know where we are and what we are doing. **Short-term memory** is the temporary ability to recall a few pieces of information for seconds to minutes. One example is when you look up an unfamiliar telephone number, cross the room to the phone, and then dial the new number. If the number has no special significance, it is usually forgotten within a few seconds. Brain areas involved in immediate and short-term memory include the hippocampus, the mammillary bodies, and two nuclei of the thalamus (anterior and medial nuclei). Some evidence supports the notion that short-term memory depends more on electrical and chemical events in the brain than on structural changes, such as the formation of new synapses.

Information in short-term memory may later be transformed into a more permanent type of memory, called **long-term memory,** which lasts from days to years. If you use that new telephone number often enough, it becomes part of long-term memory. Information in long-term memory usually can be retrieved for use whenever needed. The reinforcement that results from the frequent retrieval of a piece of information is called **memory consolidation.** Long-term memories for information that can be expressed by language, such as a telephone number, apparently are stored in wide regions of the cerebral cortex. Memories for motor skills, such as how to serve a tennis ball, are stored in the basal nuclei and cerebellum as well as in the cerebral cortex.

Although the brain receives many stimuli, we pay attention to only a few of them at a time. It has been estimated that only 1% of all the information that comes to our consciousness is stored as long-term memory. Moreover, much of what goes into long-term memory is eventually forgotten. Memory does not record every detail as if it were magnetic tape. Even when details are lost, we can often explain the idea or concept using our own words and ways of viewing things.

CLINICAL CONNECTION | *Amnesia*

Amnesia (am-NĒ-zē-a = forgetfulness) refers to the lack or loss of memory. It is a total or partial inability to remember past experiences. In *anterograde amnesia,* there is memory loss for events that occur *after* the trauma or disease that caused the condition. In other words, it is an inability to form new memories. In *retrograde amnesia,* there is a memory loss for events that occurred *before* the trauma or disease that caused the condition. In other words, it is an inability to recall past events. •

Several conditions that inhibit the electrical activity of the brain, such as anesthesia, coma, electroconvulsive therapy (ECT), and ischemia of the brain, disrupt retention of recently acquired information without altering previously established long-term memories. People who suffer retrograde amnesia cannot remember anything that occurred during the 30 minutes or so before the amnesia developed. As a person recovers from amnesia, the most recent memories return last.

Anatomical changes occur in neurons when they are stimulated. For example, electron micrographs of neurons subjected to prolonged, intense activity reveal an increase in the number of presynaptic terminals and enlargement of synaptic end bulbs in presynaptic neurons, as well as an increase in the number of dendritic branches in postsynaptic neurons. Moreover, neurons grow new synaptic end bulbs with increasing age, presumably because of increased use. Opposite changes occur when neurons are inactive. For example, the cerebral cortex in the visual area of animals that have lost their eyesight becomes thinner.

A phenomenon called **long-term potentiation (LTP)** (pō-ten′-shē-Ā-shun) is believed to underlie some aspects of memory; transmission at some synapses within the hippocampus is enhanced (potentiated) for hours or weeks after a brief period of high-frequency stimulation. The neurotransmitter released is glutamate, which acts on NMDA* glutamate receptors on the postsynaptic neurons. In some cases, induction of LTP depends on the release of nitric oxide (NO) from the postsynaptic neurons after they have been activated by glutamate. The NO in turn diffuses into the presynaptic neurons and causes LTP.

✔ CHECKPOINT

18. Describe how sleep and wakefulness are related to the reticular activating system (RAS).
19. What are the four stages of non-rapid eye movement (NREM) sleep? How is NREM sleep distinguished from rapid eye movement (REM) sleep?
20. Define memory. What are the three kinds of memory? What is memory consolidation?
21. What is long-term potentiation?

*Named after the chemical *N*-methyl-D-aspartate, which is used to detect this type of glutamate receptor.

DISORDERS: HOMEOSTATIC IMBALANCES

Parkinson Disease

Parkinson disease (PD) is a progressive disorder of the CNS that typically affects its victims around age 60. Neurons that extend from the substantia nigra to the putamen and caudate nucleus, where they release the neurotransmitter dopamine (DA), degenerate in PD. The caudate nucleus of the basal nuclei contains neurons that liberate the neurotransmitter acetylcholine (ACh). Although the level of ACh does not change as the level of DA declines, the imbalance of neurotransmitter activity—too little DA and too much ACh—is thought to cause most of the symptoms. The cause of PD is unknown, but toxic environmental chemicals, such as pesticides, herbicides, and carbon monoxide, are suspected contributing agents. Only 5% of PD patients have a family history of the disease.

In PD patients, involuntary skeletal muscle contractions often interfere with voluntary movement. For instance, the muscles of the upper limb may alternately contract and relax, causing the hand to shake. This shaking, called **tremor,** is the most common symptom of PD. Also, muscle tone may increase greatly, causing rigidity of the involved body part. Rigidity of the facial muscles gives the face a masklike appearance. The expression is characterized by a wide-eyed, unblinking stare and a slightly open mouth with uncontrolled drooling.

Motor performance is also impaired by **bradykinesia** (*brady-* = slow), slowness of movements. Activities such as shaving, cutting food, and buttoning a shirt take longer and become increasingly more difficult as the disease progresses. Muscular movements also exhibit **hypokinesia** (*hypo-* = under), decreasing range of motion. For example, words are written smaller, letters are poorly formed, and eventually handwriting becomes illegible. Often, walking is impaired; steps become shorter and shuffling, and arm swing diminishes. Even speech may be affected.

Treatment of PD is directed toward increasing levels of DA and decreasing levels of ACh. Although people with PD do not manufacture enough dopamine, taking it orally is useless because DA cannot cross the blood–brain barrier. Even though symptoms are partially relieved by a drug developed in the 1960s called levodopa (L-dopa), a precursor of DA, the drug does not slow the progression of the disease. As more and more affected brain cells die, the drug becomes useless. Another drug, called selegiline (Deprenyl®), is used to inhibit monoamine oxidase, an enzyme that degrades dopamine. This drug slows progression of PD and may be used together with levodopa. Anticholinergic drugs such as benztropine and trihexyphenidyl can also be used to block the effects of ACh at some of the synapses between basal nuclei neurons. This helps to restore the balance between ACh and DA. Anticholinergic drugs effectively reduce symptomatic tremor, rigidity, and drooling.

For more than a decade, surgeons have sought to reverse the effects of Parkinson disease by transplanting dopamine-rich fetal nervous tissue into the basal nuclei (usually the putamen) of patients with severe PD. Only a few postsurgical patients have shown any degree of improvement, such as less rigidity and improved quickness of motion. Another surgical technique that has produced improvement for some patients is *pallidotomy,* in which a part of the globus pallidus that generates tremors and produces muscle rigidity is destroyed. In addition, some patients are being treated with a surgical procedure called *deep-brain stimulation (DBS),* which involves the implantation of electrodes into the subthalamic nucleus. The electrical currents released by the implanted electrodes reduce many of the symptoms of PD.

MEDICAL TERMINOLOGY

Acupuncture (ak-ū-PUNK-chur) The use of fine needles (lasers, ultrasound, or electricity) inserted into specific exterior body locations (acupoints) and manipulated to relieve pain and provide therapy for various conditions. The placement of needles may cause the release of neurotransmitters such as endorphins, painkillers that may inhibit pain pathways.

Cerebral palsy (CP) A motor disorder that results in the loss of muscle control and coordination; caused by damage of the motor areas of the brain during fetal life, birth, or infancy. Radiation during fetal life, temporary lack of oxygen during birth, and hydrocephalus during infancy may also cause cerebral palsy.

Coma (KŌ-ma) A state of unconsciousness in which a person's responses to stimuli are reduced or absent. In a *light coma,* an individual may respond to certain stimuli, such as sound, touch, or light, and move his or her eyes, cough, and even murmur. In a *deep coma,* a person does not respond to any stimuli and does not make any movements. Causes of coma include head injuries, cardiac arrest, stroke, brain tumors, infections (encephalitis and meningitis), seizures, alcoholic intoxication, drug overdose, severe lung disorders (chronic obstructive pulmonary disease, pulmonary edema, pulmonary embolism), inhalation of large amounts of carbon monoxide, liver or kidney failure, low or high blood sugar or sodium levels, and low or high body temperature. If brain damage is minor or reversible, a person may come out of a coma and recover fully; if brain damage is severe and irreversible, recovery is unlikely.

Pain threshold The smallest intensity of a painful stimulus at which a person perceives pain. All individuals have the same pain threshold.

Pain tolerance The greatest intensity of painful stimulation that a person is able to tolerate. Individuals vary in their tolerance to pain.

Synesthesia (sin-es-THĒ-zē-a; *syn-* = together; *-aisthesis* = sensation) A condition in which sensations of two or more modalities accompany one another. In some cases, a stimulus for one sensation is perceived as a stimulus for another; for example, a sound produces a sensation of color. In other cases, a stimulus from one part of the body is experienced as coming from a different part.

Review	Resource

16.1 Sensation

1. Sensation is the conscious or subconscious awareness of changes in the external or internal environment. Perception is the conscious awareness and interpretation of sensations and is primarily a function of the cerebral cortex.
2. The nature of a sensation and the type of reaction generated vary according to the destination of sensory impulses in the CNS.
3. Each different type of sensation is a sensory modality; usually, a given sensory neuron serves only one modality.
4. General senses include somatic senses (touch, pressure, vibration, warmth, cold, pain, itch, tickle, and proprioception) and visceral senses; special senses include the modalities of smell, taste, vision, hearing, and equilibrium.
5. For a sensation to arise, four events typically occur: stimulation, transduction, generation of impulses, and integration.
6. Simple receptors, consisting of free nerve endings and encapsulated nerve endings, are associated with the general senses; complex receptors are associated with the special senses.
7. Sensory receptors respond to stimuli by producing receptor or generator potentials.
8. Table 16.1 summarizes the classification of sensory receptors.
9. Adaptation is a decrease in sensitivity during a long-lasting stimulus. Receptors are either rapidly adapting or slowly adapting.

Resource:
Anatomy Overview - Chemoreceptors
Anatomy Overview - Baroreceptors
Anatomy Overview - Proprioceptors

16.2 Somatic Sensations

1. Somatic sensations include tactile sensations (touch, pressure, vibration, itch, and tickle), thermal sensations (warmth and cold), pain, and proprioception.
2. Receptors for tactile, thermal, and pain sensations are located in the skin, subcutaneous layer, and mucous membranes of the mouth, vagina, and anus.
3. Receptors for touch are (a) Meissner corpuscles and hair root plexuses, which are rapidly adapting, and (b) slowly adapting Merkel discs. Ruffini corpuscles, which are slowly adapting, are sensitive to stretching.
4. Receptors for pressure include Meissner corpuscles, Merkel discs, and Pacinian corpuscles.
5. Receptors for vibration are Meissner corpuscles and Pacinian corpuscles.
6. Itch receptors, tickle receptors, and thermoreceptors are free nerve endings. Cold receptors are located in the stratum basale of the epidermis; warm receptors are located in the dermis.
7. Pain receptors (nociceptors) are free nerve endings that are located in nearly every body tissue.
8. Nerve impulses for fast pain propagate along medium-diameter, myelinated A fibers, whereas those for slow pain conduct along small-diameter, unmyelinated C fibers.
9. Receptors for proprioceptive sensations (position and movement of body parts) are located in muscles, tendons, joints, and the inner ear. Proprioceptors include muscle spindles, tendon organs, joint kinesthetic receptors, and hair cells of the inner ear.
10. Table 16.2 summarizes the somatic sensory receptors and the sensations they convey.

Resource:
Anatomy Overview - Sensory Receptors of the Skin
Anatomy Overview - Proprioceptors
Figure 16.3 - Referred Pain
Figure 16.4 - Two Types of Proprioceptors

16.3 Somatic Sensory Pathways

1. Somatic sensory pathways from receptors to the cerebral cortex involve three-neuron sets: first-order, second-order, and third-order neurons.
2. Axon collaterals (branches) of somatic sensory neurons simultaneously carry signals into the cerebellum and the reticular formation of the brain stem.
3. Nerve impulses for touch, pressure, vibration, and conscious proprioception in the limbs, trunk, neck, and posterior head ascend to the cerebral cortex along the posterior column–medial lemniscus pathway.
4. Nerve impulses for pain, temperature, itch, and tickle from the limbs, trunk, neck, and posterior head ascend to the cerebral cortex along the anterolateral (spinothalamic) pathway.
5. Nerve impulses for most somatic sensations (tactile, thermal, pain, and proprioceptive) from the face, nasal cavity, oral cavity, and teeth ascend to the cerebral cortex along the trigeminothalamic pathway.
6. Specific regions of the primary somatosensory area (postcentral gyrus) of the cerebral cortex receive somatic sensory input from different parts of the body.
7. The neural pathways to the cerebellum are the anterior and posterior spinocerebellar tracts, which transmit impulses for subconscious proprioception from the trunk and lower limbs.
8. Table 16.3 summarizes the major somatic sensory pathways.

Resource:
Anatomy Overview - Cerebrum
Animation - Somatic Sensory Pathways
Figure 16.6 - The Anterolateral Pathway
Figure 16.8 - Somatic Sensory and Somatic Motor Maps in the Cerebral Cortex

Review

16.4 Somatic Motor Pathways

Animation - Somatic Motor Pathways
Figure 16.10 - The Corticospinal Pathways

1. All excitatory and inhibitory signals that control movement converge on motor neurons, also known as lower motor neurons (LMNs) or the final common pathway.
2. Neurons in four neural circuits, collectively termed the somatic motor pathways, participate in control of movement by providing input to lower motor neurons: local circuit neurons, upper motor neurons, basal nuclei neurons, and cerebellar neurons.
3. The primary motor area (precentral gyrus) of the cortex is a major control region for executing voluntary movements.
4. The axons of upper motor neurons (UMNs) extend from the brain to lower motor neurons via direct and indirect motor pathways.
5. The direct (pyramidal) pathways include the corticospinal pathways and the corticobulbar pathway. The corticospinal pathways convey nerve impulses from the motor cortex to skeletal muscles in the limbs and trunk. The corticobulbar pathway conveys nerve impulses from the motor cortex to skeletal muscles in the head.
6. Indirect (extrapyramidal) pathways extend from several motor centers of the brain stem into the spinal cord. Indirect pathways include the rubrospinal, tectospinal, vestibulospinal, and medial and lateral reticulospinal tracts.
7. Table 16.4 summarizes the major somatic motor pathways.
8. Neurons of the basal nuclei assist movement by providing input to the upper motor neurons. They help initiate and terminate movements, suppress unwanted movements, and establish a normal level of muscle tone.
9. The cerebellum is active in learning and performing rapid, coordinated, highly skilled movements. It also contributes to maintaining balance and posture.

16.5 Integrative Functions of the Cerebrum

1. Sleep and wakefulness are integrative functions that are controlled by the suprachiasmatic nucleus and the reticular activating system (RAS).
2. Non-rapid eye movement (NREM) sleep consists of four stages.
3. Most dreaming occurs during rapid eye movement (REM) sleep.
4. Memory, the ability to store and recall thoughts, involves persistent changes in the brain, a capability called plasticity. Three types are immediate, short-term, and long-term memory.

SELF-QUIZ QUESTIONS

Fill in the blanks in the following statements.

1. _____ is the conscious or subconscious awareness of external or internal stimuli; _____ is the conscious awareness and interpretation of sensory input.

2. The term used to describe the crossing over of axons from one side of the brain or spinal cord to the other side is _____.

Indicate whether the following statements are true or false.

3. Touch, pressure, and pain are all classified as tactile sensations.

4. Awakening from sleep involves increased activity in the reticular activating system.

Choose the one best answer to the following questions.

5. A nurse touches the lower back of a patient, but the patient does not feel the sensation. Which of the following could explain the lack of sensation? (1) The stimulus was not in the receptive field. (2) The generator potential has not reached threshold. (3) There is damage to the somatosensory region of the cerebral cortex. (4) The nurse was stimulating a proprioceptor. (5) A slowly adapting receptor has been stimulated.
   (a) 1, 3 and 5   (b) 3, 4, and 5   (c) 1, 2, and 3
   (d) 2, 3, and 4   (e) 1 only

6. Which of the following statements is *false*? (1) Upper motor neurons transmit impulses from the CNS to skeletal muscle fibers. (2) Lower motor neurons have their cell bodies in the brain stem and spinal cord. (3) Local circuit neurons receive input from somatic sensory receptors and help coordinate rhythmic activity in specific muscle groups. (4) The activity of upper motor neurons is influenced by both the basal ganglia and cerebellum. (5) The cerebellum helps to monitor differences between intended movements and actual movements for coordination, posture, and balance.

7. Which of the following statements are *true*? (1) Slow pain is a result of impulse propagation along myelinated A nerve fibers. (2) Visceral pain occurs when nociceptors in the skin are stimulated. (3) Referred pain is pain felt in an area far from the stimulated organ. (4) Nociceptors exhibit very little adaptation. (5) Nociceptors are located in every body tissue.
   (a) 1, 3, 4, and 5   (b) 2, 3, and 5   (c) 1 and 5
   (d) 3 and 4   (e) 3, 4, and 5

8. You cannot "hear" with your eyes because
   (a) hearing is a somatic sense and vision is a special sense.
   (b) the sensory neurons for sight carry information only for the modality of vision.

(c) the impulses for hearing are transmitted to the somatosensory area of the cerebral cortex.

(d) hearing receptors are selective and vision receptors are not.

(e) hearing receptors produce a generator potential and vision receptors produce a receptor potential.

9. Which of the following statements is *false*?
   (a) First-order sensory neurons carry signals from the somatic receptors into either the brain stem or spinal cord.
   (b) Second-order neurons carry signals from the spinal cord and brain stem to the thalamus.
   (c) Third-order neurons project to the primary somatosensory area of the cortex, where conscious perception of the sensation results.
   (d) The somatic sensory pathways to the cerebellum are the posterior column–medial lemniscus pathway and the anterolateral pathway.
   (e) Axons of second-order neurons decussate (cross) in the spinal cord or brain stem before ascending to the thalamus.

10. Which of the following is not an aspect of cerebellar function?
    (a) monitoring movement intention
    (b) monitoring actual movement
    (c) comparing intent with actual performance
    (d) sending out corrective signals
    (e) directing sensory input to effectors

11. During REM sleep (1) neuronal activity in the pons and midbrain is high, (2) most somatic motor neurons are inhibited, (3) most dreaming occurs, (4) sleepwalking can occur, (5) there is an increase in heart rate and blood pressure.
    (a) 1, 2, 4, and 5   (b) 2, 3, and 5   (c) 1, 2, 3, 4, and 5
    (d) 2, 3, and 4   (e) 1, 2, and 3

12. Which of the following statements is *incorrect*?
    (a) The graded potentials produced by receptors that serve the senses of touch, pressure, stretching, vibration, pain, proprioception, and smell are generator potentials.
    (b) The graded potentials produced by receptors that serve the special senses of vision, hearing, equilibrium, and taste are receptor potentials.
    (c) When a generator potential is large enough to reach threshold, it generates one or more nerve impulses in its first-order sensory neuron.
    (d) A receptor potential generates nerve impulses in a second-order neuron.
    (e) The amplitude of both generator and receptor potentials varies with the intensity of the stimulus.

13. Match the following (some answers may be used more than once):
    ____ (a) receptors located in muscles, tendons, joints, and the inner ear
    ____ (b) receptors located in blood vessels, visceral organs, muscles, and the nervous system
    ____ (c) receptors that detect temperature changes
    ____ (d) receptors that detect light that strikes the retina of the eye
    ____ (e) receptors located at or near the external surface of the body
    ____ (f) bare dendrites associated with pain, thermal, tickle, itch, and some touch sensations
    ____ (g) receptors that provide information about body position, muscle tension, and position and activity of joints
    ____ (h) receptors that sense osmotic pressures of body fluids
    ____ (i) receptors that detect chemicals in the mouth, nose, and body fluids
    ____ (j) receptors that detect mechanical pressure or stretching
    ____ (k) receptors that respond to stimuli resulting from physical or chemical damage to tissues
    ____ (l) dendrites enclosed in a connective tissue capsule

    (1) exteroceptors
    (2) interoceptors
    (3) proprioceptors
    (4) mechanoreceptors
    (5) thermoreceptors
    (6) nociceptors
    (7) photoreceptors
    (8) chemoreceptors
    (9) free nerve endings
    (10) encapsulated nerve endings
    (11) osmoreceptors

14. Match the following:
    ____ (a) specialized groupings of muscle fibers interspersed among regular skeletal muscle fibers and oriented parallel to them; monitor changes in the length of a skeletal muscle
    ____ (b) inform the CNS about changes in muscle tension
    ____ (c) widely distributed free nerve ending receptors for pain
    ____ (d) encapsulated receptors for touch located in the dermal papillae; found in hairless skin, eyelids, tip of the tongue, and lips
    ____ (e) lamellated corpuscles that detect pressure
    ____ (f) type II cutaneous mechanoreceptors; most sensitive to stretching that occurs as digits or limbs are moved
    ____ (g) located in the stratum basale and activated by low temperatures
    ____ (h) located in the dermis and activated by high temperatures
    ____ (i) found within and around the articular capsules of synovial joints; respond to pressure and acceleration and deceleration of joints
    ____ (j) type I cutaneous mechanoreceptors that function in touch

    (1) Meissner corpuscles
    (2) Merkel discs
    (3) Ruffini corpuscles
    (4) Pacinian corpuscles
    (5) cold receptors
    (6) warm receptors
    (7) nociceptors
    (8) tendon organs
    (9) joint kinesthetic receptors
    (10) muscle spindles

15. Match the following:
    _____ (a) located in the precentral gyrus, this is the major control region of the cerebral cortex for initiation of voluntary movements
    _____ (b) direct pathways conveying impulses from the cerebral cortex to the spinal cord that result in precise, voluntary movements
    _____ (c) contains motor neurons that control skilled movements of the hands and feet
    _____ (d) tracts include rubrospinal, tectospinal, vestibulospinal, lateral reticulospinal, and medial reticulospinal
    _____ (e) contain neurons that help initiate and terminate movements; can suppress unwanted movements; influence muscle tone
    _____ (f) carries impulses for pain, temperature, tickle, and itch
    _____ (g) the major routes relaying proprioceptive input to the cerebellum; critical for posture, balance, and coordination of skilled movements
    _____ (h) composed of axons of first-order neurons; include the gracile fasciculus and cuneate fasciculus
    _____ (i) contains motor neurons that coordinate movements of the axial skeleton
    _____ (j) contains axons that convey impulses for precise, voluntary movements of the eyes, tongue, and neck, plus chewing, facial expression, and speech
    _____ (k) conveys sensations of touch, conscious proprioception, pressure, and vibration to the cerebral cortex
    _____ (l) carries impulses for most somatic sensations from the face, nasal cavity, oral cavity, and teeth

    (1) posterior column
    (2) anterolateral (spinothalamic) pathway
    (3) spinocerebellar tracts
    (4) lateral corticospinal tract
    (5) anterior corticospinal tract
    (6) corticobulbar tracts
    (7) extrapyramidal pathways
    (8) pyramidal pathways
    (9) primary motor area
    (10) basal nuclei
    (11) posterior column–medial lemniscus pathway
    (12) trigeminothalamic pathway

CRITICAL THINKING QUESTIONS

1. When Joni first stepped onto the sailboat, she smelled the tangy sea air and felt the motion of water beneath her feet. After a few minutes, she no longer noticed the smell, but unfortunately she was aware of the rolling motion for hours. What types of receptors are involved in smell and detection of motion? Why did her sensation of smell fade but the rolling sensation remain?

2. Monique sticks her left hand into a hot tub heated to about 43°C (110°F) in order to decide if she wants to enter the hot tub. Trace the pathway involved in transmitting the sensation of heat from her left hand to the somatosensory area in the cerebral cortex.

3. Marvin has had trouble sleeping. Last night his mother found him sleepwalking and gently led him back to his bed. When Marvin was awakened by his alarm clock the next day, he had no recollection of sleepwalking and, in fact, told his mother about the vivid dreams he had. What specific stages of sleep did Marvin undergo during the night? What neurological mechanism awakened Marvin in the morning?

? ANSWERS TO FIGURE QUESTIONS

16.1 The special senses of vision, taste, hearing, and equilibrium are served by separate sensory cells.

16.2 Pain, thermal sensations, and tickle and itch arise with activation of different free nerve endings.

16.3 The kidneys have the broadest area for referred pain.

16.4 Muscle spindles are activated when the central areas of the intrafusal fibers are stretched.

16.5 The posterior columns consist of the cuneate fasciculus and the gracile fasciculus.

16.6 Damage to the right spinothalamic tract could result in loss of pain, thermal, itch, and tickle sensations on the left side of the body.

16.7 The left trigeminal (V) nerve conveys nerve impulses for most somatic sensations from the left side of the face into the pons.

16.8 The hand has a larger representation in the motor area than in the somatosensory area, which implies greater precision in the hand's movement control than fine ability in its sensation.

16.9 Cerebral cortex UMNs are essential for the execution of voluntary movements of the body. Brain stem UMNs regulate muscle tone, control postural muscles, and help maintain balance and orientation of the head and body.

16.10 The lateral corticospinal tract conducts impulses that result in contractions of the muscles in the distal parts of the limbs.

16.11 The axons of the corticobulbar tract terminate in the motor nuclei of the following cranial nerves: oculomotor (III), trochlear (IV), trigeminal (V), abducens (VI), facial (VII), glossopharyngeal (IX), vagus (X), accessory (XI), and hypoglossal (XII).

16.12 The anterior and posterior spinocerebellar tracts carry information from proprioceptors in joints and muscles to the cerebellum.

16.13 Olfactory input does not stimulate the RAS; a smoke detector responds to smoke by sounding a loud bell or buzzer, which awakens sleepers by providing auditory input that stimulates the RAS.

17 | THE SPECIAL SENSES

THE SPECIAL SENSES AND HOMEOSTASIS *Sensory organs have special receptors that allow us to smell, taste, see, hear, and maintain equilibrium or balance. Information conveyed from these receptors to the central nervous system is used to help maintain homeostasis.*

Recall from Chapter 16 that the general senses include somatic senses (tactile, thermal, pain, and proprioceptive) and visceral sensations. As you learned in that chapter, receptors for the general senses are scattered throughout the body and are relatively simple in structure. Receptors for the special senses—smell, taste, vision, hearing, and equilibrium—are anatomically distinct from one another and are concentrated in specific locations in the head. They are usually embedded in the epithelial tissue within complex sensory organs such as the eyes and ears. Neural pathways for the special senses are also more complex than those for the general senses.

In this chapter we examine the structure and function of the special sense organs, and the pathways involved in conveying their information to the central nervous system. **Ophthalmology** (of-thal-MOL-ō-jē; *ophthalmo-* = eye; *-logy* = study of) is the science that deals with the eyes and their disorders. The other special senses are, in large part, the concern of **otorhinolaryngology** (ō-tō-rī′-nō-lar-in-GOL-ō-jē; *oto-* = ear; *-rhino-* = nose; *-laryngo-* = larynx), the science that deals with the ears, nose, pharynx (throat), and larynx (voice box) and their disorders.

? *Did you ever wonder how LASIK is performed?*

17.1 OLFACTION: SENSE OF SMELL

◉ OBJECTIVES

- Describe the structure of the olfactory receptors and other cells involved in olfaction.
- Outline the neural pathway for olfaction.

Last night as you were studying anatomy in the lounge, all of a sudden you were surrounded by the smell of freshly baked brownies. When you followed your nose and begged for one, biting into the moist, flavorful treat transported you back 10 years into your mother's kitchen. Both smell and taste are chemical senses; the sensations arise from the interaction of molecules with smell or taste receptors. To be detected by either sense, the stimulating molecules must be dissolved. Because impulses for smell and taste propagate to the limbic system (and to higher cortical areas as well), certain odors and tastes can evoke strong emotional responses or a flood of memories.

Anatomy of Olfactory Receptors

It is estimated that humans can recognize about 10,000 different odors. To make this possible, the nose contains 10 million to 100 million receptors for the sense of smell or **olfaction** (ōl-FAK-shun; *olfact-* = smell), contained within a region called the olfactory epithelium. With a total area of 5 cm² (a little less than 1 in.²), the **olfactory epithelium** (ōl-FAK-tō-rē) occupies the superior part of the nasal cavity, covering the inferior surface of the cribriform plate and extending along the superior nasal concha (Figure 17.1a). The olfactory epithelium consists of three kinds of cells: olfactory receptors, supporting cells, and basal cells (Figure 17.1b).

Olfactory receptors are the first-order neurons of the olfactory pathway. Each olfactory receptor is a bipolar neuron with an exposed knob-shaped dendrite and an axon projecting through the cribriform plate and ending in the olfactory bulb. The parts of the olfactory receptors that respond to inhaled chemicals are the **olfactory hairs,** cilia that project from the dendrite. (Recall that

Figure 17.1 Olfactory epithelium and olfactory pathway. (a) Location of olfactory epithelium in nasal cavity. (b) Anatomy of olfactory receptors, consisting of first-order neurons whose axons extend through the cribriform plate and terminate in the olfactory bulb. (c) Histology of the olfactory epithelium. (d) Olfactory pathway.

🔑 The olfactory epithelium consists of olfactory receptors, supporting cells, and basal cells.

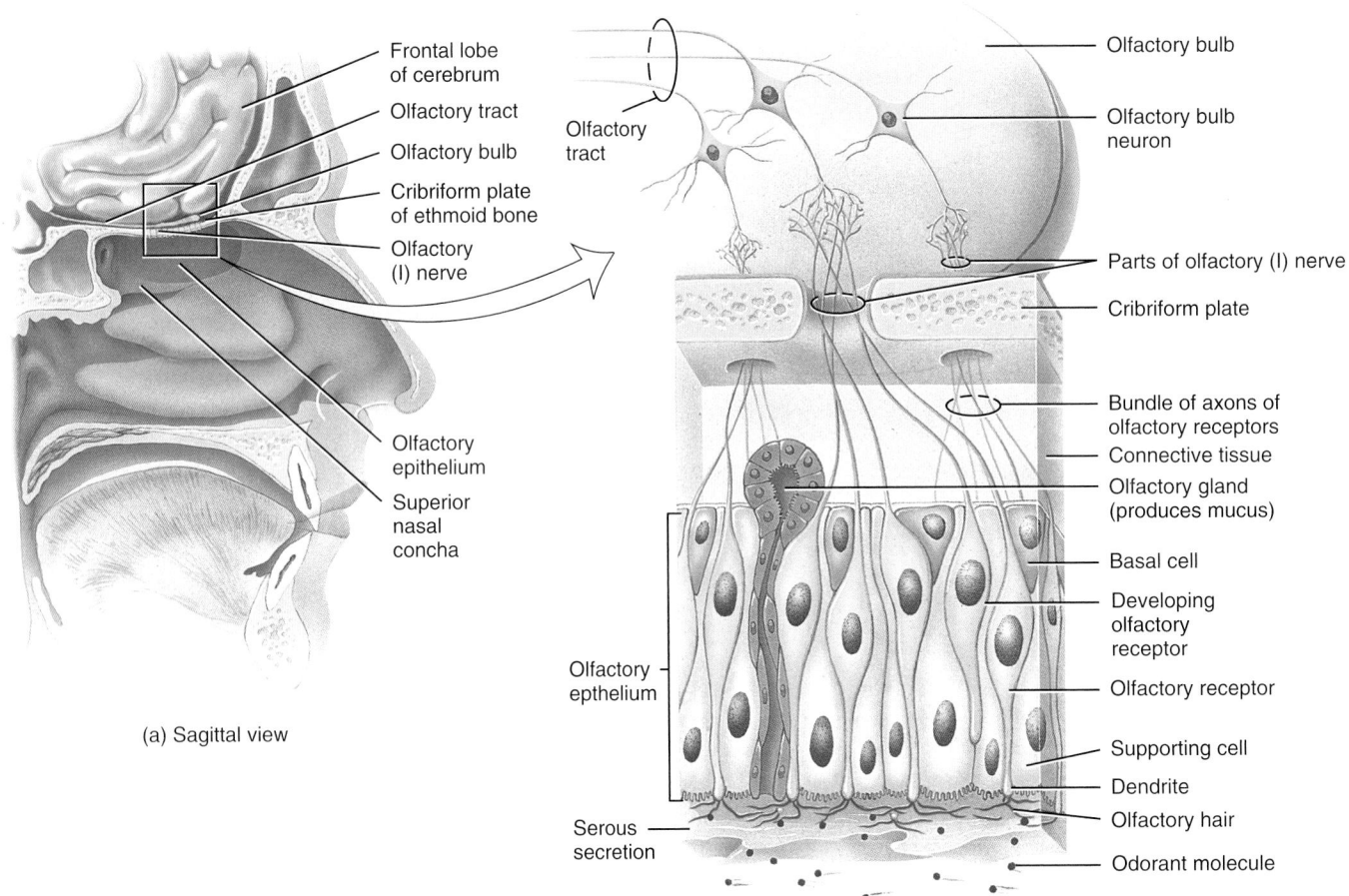

(a) Sagittal view

(b) Enlarged aspect of olfactory receptors

transduction is the conversion of stimulus energy into a graded potential in a sensory receptor.) Chemicals that have an odor and can therefore stimulate the olfactory hairs are called **odorants.** Olfactory receptors respond to the chemical stimulation of an odorant molecule by producing a generator potential, thus initiating the olfactory response.

Supporting cells are columnar epithelial cells of the mucous membrane lining the nose. They provide physical support, nourishment, and electrical insulation for the olfactory receptors, and they help detoxify chemicals that come in contact with the olfac-

tory epithelium. **Basal cells** are stem cells located between the bases of the supporting cells. They continually undergo cell division to produce new olfactory receptors, which live for only a month or so before being replaced. This process is remarkable considering that olfactory receptors are neurons, and as you have already learned, mature neurons are generally not replaced.

Within the connective tissue that supports the olfactory epithelium are **olfactory (Bowman's) glands,** which produce mucus that is carried to the surface of the epithelium by ducts. The secretion moistens the surface of the olfactory epithelium and dissolves odorants so that transduction can occur. Both supporting cells of the nasal epithelium and olfactory glands are innervated by autonomic neurons within branches of the facial (VII) nerve, which can be stimulated by certain chemicals. Impulses in these nerves in turn stimulate the lacrimal glands in the eyes and nasal mucous glands. The result is tears and a runny nose after inhaling substances such as pepper or the vapors of household ammonia.

Physiology of Olfaction

Many attempts have been made to distinguish among and classify "primary" sensations of smell. Genetic evidence now suggests the existence of hundreds of primary odors. Our ability to recognize about 10,000 different odors probably depends on patterns of activity in the brain that arise from activation of many different combinations of olfactory receptors.

Connective tissue
Basal cell
Olfactory gland
Olfactory receptor
Duct
Supporting cell
Olfactory hairs

LM 300x

(c) Histology of olfactory epithelium

Frontal plane

View

Primary olfactory area in cerebral cortex

Olfactory tract

Part of olfactory (I) nerve

Olfactory epithelium

Olfactory bulb

Cribriform plate

Olfactory receptor

Olfactory tract

Olfactory bulb

Cribriform plate of ethmoid bone

Olfactory epithelium

(d) Olfactory pathway

? **What is the life span of an olfactory receptor?**

Olfactory receptors react to odorant molecules in the same way that most sensory receptors react to their specific stimuli: A generator potential (depolarization) develops and triggers one or more nerve impulses. In some cases, an odorant binds to an *olfactory receptor protein* in the plasma membrane of an olfactory hair (Figure 17.2). The olfactory receptor protein is coupled to a membrane protein called a G protein, which in turn activates the enzyme adenylate cyclase (see Section 18.4). The result is the following chain of events: production of cyclic adenosine monophosphate (cAMP) → opening of sodium ion (Na$^+$) channels → inflow of Na$^+$ → depolarizing generator potential → generation of nerve impulse and propagation along an axon of the olfactory receptor.

Odor Thresholds and Adaptation

Olfaction, like all the special senses, has a low threshold. Only a few molecules of certain substances need be present in air to be perceived as an odor. A good example is the chemical methyl mercaptan, which smells like rotten cabbage and can be detected in concentrations as low as 1/25 billionth of a milligram per milliliter of air. Because the natural gas used for cooking and heating is odorless but lethal and potentially explosive if it accumulates, a small amount of methyl mercaptan is added to natural gas to provide olfactory warning of gas leaks.

Adaptation (decreasing sensitivity) to odors occurs rapidly. Olfactory receptors adapt by about 50% in the first second or so after stimulation but adapt very slowly thereafter. Still, complete insensitivity to certain strong odors occurs about a minute after exposure. Apparently, reduced sensitivity involves an adaptation process in the central nervous system as well.

The Olfactory Pathway

On each side of the nose, about 40 bundles of the slender, unmyelinated axons of olfactory receptors extend through about 20 olfactory foramina in the cribriform plate of the ethmoid bone (see Figure 17.1b). These 40 or so bundles of axons collectively form the right and left **olfactory (I) nerves.** The olfactory nerves terminate in the brain in paired masses of gray matter called the **olfactory bulbs,** which are located below the frontal lobes of the cerebrum and lateral to the crista galli of the ethmoid bone. Within the olfactory bulbs, the axon terminals of olfactory receptors form synapses with the dendrites and cell bodies of olfactory bulb neurons in the olfactory pathway.

Axons of olfactory bulb neurons extend posteriorly and form the **olfactory tract** (see Figure 17.1b). Some of the axons of the olfactory tract project to the **primary olfactory area** of the cerebral cortex; located at the inferior and medial surface of the temporal lobe, the primary olfactory area is where conscious awareness

Figure 17.2 Olfactory transduction. Binding of an odorant molecule to an olfactory receptor protein activates a G protein and adenylate cyclase, resulting in the production of cAMP. Cyclic AMP opens sodium ion (Na$^+$) channels, and Na$^+$ ions enter the olfactory receptor. The resulting depolarization may generate an action potential, which propagates along the axon of the olfactory receptor.

Odorants can produce depolarizing generator potentials, which can lead to action potentials.

In which part of an olfactory receptor does olfactory transduction occur?

of smell begins (see Figure 17.1d). Olfactory sensations are the only sensations that reach the cerebral cortex without first synapsing in the thalamus. Other axons of the olfactory tract project to the limbic system and hypothalamus; these connections account for our emotional and memory-evoked responses to odors. Examples include sexual excitement on smelling a certain perfume, nausea on smelling a food that once made you violently ill, or an odor-evoked memory of a childhood experience.

From the primary olfactory area, pathways also extend to the **frontal lobe.** An important region for odor identification and discrimination is the **orbitofrontal area** (area 11 in Figure 14.15). People who suffer damage in this area have difficulty identifying different odors. Positron emission tomography (PET) studies suggest some degree of hemispheric lateralization: The orbitofrontal area of the *right* hemisphere exhibits greater activity during olfactory processing.

CLINICAL CONNECTION | *Hyposmia*

Women often have a keener sense of smell than men do, especially at the time of ovulation. Smoking seriously impairs the sense of smell in the short term and may cause long-term damage to olfactory receptors. With aging the sense of smell deteriorates. **Hyposmia** (hī-POZ-mē-a; *osmi* = smell, odor), a reduced ability to smell, affects half of those over age 65 and 75% of those over age 80. Hyposmia also can be caused by neurological changes, such as a head injury, Alzheimer disease, or Parkinson disease; certain drugs, such as antihistamines, analgesics, or steroids; and the damaging effects of smoking. •

✔CHECKPOINT

1. How do basal cells contribute to olfaction?
2. What is the sequence of events from the binding of an odorant molecule to an olfactory hair to the arrival of a nerve impulse in the orbitofrontal area?

17.2 GUSTATION: SENSE OF TASTE

◉ OBJECTIVE

• Describe the structure of the gustatory receptors and the neural pathway for gustation.

Taste or **gustation** (gus-TĀ-shun; *gust-* = taste), like olfaction, is a chemical sense. However, it is much simpler than olfaction in that only five primary tastes can be distinguished: *sour, sweet, bitter, salty,* and *umami* (ū-MAM-ē). The umami taste, more recently discovered than the other tastes, was first reported by Japanese scientists and is described as "meaty" or "savory." Umami is believed to arise from taste receptors that are stimulated by monosodium glutamate (MSG), a substance naturally present in many foods and added to others as a flavor enhancer. All other flavors, such as chocolate, pepper, and coffee, are combinations of the five primary tastes, plus accompanying olfactory and tactile (touch) sensations. Odors from food can pass upward from the mouth into the nasal cavity, where they stimulate olfactory receptors. Because olfaction is much more sensitive than taste, a given concentration of a food substance may stimulate the olfactory

system thousands of times more strongly than it stimulates the gustatory system. When you have a cold or are suffering from allergies and cannot taste your food, it is actually olfaction that is blocked, not taste.

Anatomy of Taste Buds and Papillae

The receptors for sensations of taste are located in the taste buds (Figure 17.3). Most of the nearly 10,000 taste buds of a young adult are on the tongue, but some are found on the soft palate (posterior portion of the roof of the mouth), pharynx (throat), and epiglottis (cartilage lid over voice box). The number of taste buds declines with age. Each **taste bud** is an oval body consisting of three kinds of epithelial cells: supporting cells, gustatory receptor cells, and basal cells (see Figure 17.3c). The **supporting cells** surround about 50 **gustatory receptor cells** (GUS-ta-tōr-ē) in each taste bud. A single, long microvillus, called a **gustatory hair,** projects from each gustatory receptor cell to the external surface through the **taste pore,** an opening in the taste bud. **Basal cells,** stem cells found at the periphery of the taste bud near the connective tissue layer, produce supporting cells, which then develop into gustatory receptor cells. Each gustatory receptor cell has a life span of about 10 days. This is why it doesn't take taste sensors on the tongue too long to recover from being burned by that too hot cup of coffee or cocoa. At their base, the gustatory receptor cells synapse with dendrites of the first-order neurons that form the first part of the gustatory pathway. The dendrites of each first-order neuron branch profusely and contact many gustatory receptor cells in several taste buds.

Taste buds are found in elevations on the tongue called **papillae** (pa-PIL-ē; singular is *papilla*), which increase the surface area and provide a rough texture to the upper surface of the tongue (Figure 17.3a, b). Three types of papillae contain taste buds:

1. About 12 very large, circular **vallate (circumvallate) papillae** (VAL-āt = wall-like) form an inverted V-shaped row at the back of the tongue. Each of these papillae houses 100–300 taste buds.

2. **Fungiform papillae** (FUN-ji-form = mushroomlike) are mushroom-shaped elevations scattered over the entire surface of the tongue that contain about five taste buds each.

3. **Foliate papillae** (FO-lē-āt = leaflike) are located in small trenches on the lateral margins of the tongue, but most of their taste buds degenerate in early childhood.

In addition, the entire surface of the tongue has **filiform papillae** (FIL-i-form = threadlike). These pointed, threadlike structures contain tactile receptors but no taste buds. They increase friction between the tongue and food, making it easier for the tongue to move food in the oral cavity.

Physiology of Gustation

Chemicals that stimulate gustatory receptor cells are known as **tastants.** Once a tastant is dissolved in saliva, it can make contact with the plasma membrane of the gustatory hairs, which are the

Figure 17.3 **The relationship of gustatory receptor cells in taste buds to tongue papillae.**

Gustatory receptor cells are located in taste buds.

Epiglottis

Palatine tonsil

Lingual tonsil

Vallate papilla

Fungiform papilla

Filiform papilla

Foliate papilla

(a) Dorsum of tongue showing location of papillae

Vallate papilla

Filiform papilla

Fungiform papilla

Taste bud

(b) Details of papillae

Taste pore

Gustatory hair

Stratified squamous epithelium

Gustatory receptor cell

Supporting cell

Basal cell

Sensory neurons

Connective tissue

(c) Structure of a taste bud

Taste pore

Gustatory hair

Stratified squamous epithelium

Gustatory receptor cell

Supporting cell

Basal cell

Connective tissue

LM 200x

LM 700x

(d) Histology of a taste bud from a vallate papilla

sites of taste transduction. The result is a receptor potential that stimulates exocytosis of synaptic vesicles from the gustatory receptor cell. In turn, the liberated neurotransmitter molecules trigger nerve impulses in the first-order sensory neurons that synapse with gustatory receptor cells.

The receptor potential arises differently for different tastants. The sodium ions (Na^+) in a salty food enter gustatory receptor cells via Na^+ channels in the plasma membrane. The accumulation of Na^+ inside causes depolarization that leads to release of neurotransmitter. The hydrogen ions (H^+) in sour tastants may flow into gustatory receptor cells via H^+ channels. They also influence opening and closing of other types of ion channels. Again, the result is depolarization and the liberation of neurotransmitter.

Other tastants, responsible for stimulating sweet, bitter, and umami tastes, do not themselves enter gustatory receptor cells. Rather, they bind to receptors on the plasma membrane that are linked to G proteins. The G proteins then activate several different chemicals known as second messengers inside the gustatory receptor cell. Different second messengers cause depolarization in different ways, but the result is the same—release of neurotransmitter.

If all tastants cause release of neurotransmitter from many gustatory receptor cells, why do foods taste different? The answer to this question is thought to lie in the patterns of nerve impulses in groups of first-order taste neurons that synapse with the gustatory receptor cells. Different tastes arise from activation of different groups of taste neurons. In addition, although each individual gustatory receptor cell responds to more than one of the five primary tastes, it may respond more strongly to some tastants than to others.

Taste Thresholds and Adaptation

The threshold for taste varies for each of the primary tastes. The threshold for bitter substances, such as quinine, is lowest. Because poisonous substances often are bitter, the low threshold (or high sensitivity) may have a protective function. The threshold for sour substances, such as lemon, as measured by using hydrochloric acid, is somewhat higher. The thresholds for salty substances, represented by sodium chloride, and for sweet substances, as measured by using sucrose, are similar, and are higher than those for bitter or sour substances.

Complete adaptation to a specific taste can occur in 1–5 minutes of continuous stimulation. Taste adaptation is due to changes that occur in the taste receptors, in olfactory receptors, and in neurons of the gustatory pathway in the CNS.

The Gustatory Pathway

Three cranial nerves contain axons of the first-order gustatory neurons that innervate the taste buds. The **facial (VII) nerve** serves taste buds in the anterior two-thirds of the tongue; the **glossopharyngeal (IX) nerve** serves taste buds in the posterior one-third of the tongue; and the **vagus (X) nerve** serves taste buds in the throat and epiglottis (Figure 17.3e). From the taste buds, nerve impulses propagate along these cranial nerves to the **gustatory nucleus** in the medulla oblongata. From the medulla, some axons carrying taste signals project to the **limbic system** and the **hypothalamus**; others project to the **thalamus.** Taste signals that project from the thalamus to the **primary gustatory area** in the parietal lobe of the cerebral cortex (see area 43 in Figure 14.15) give rise to the conscious perception of taste.

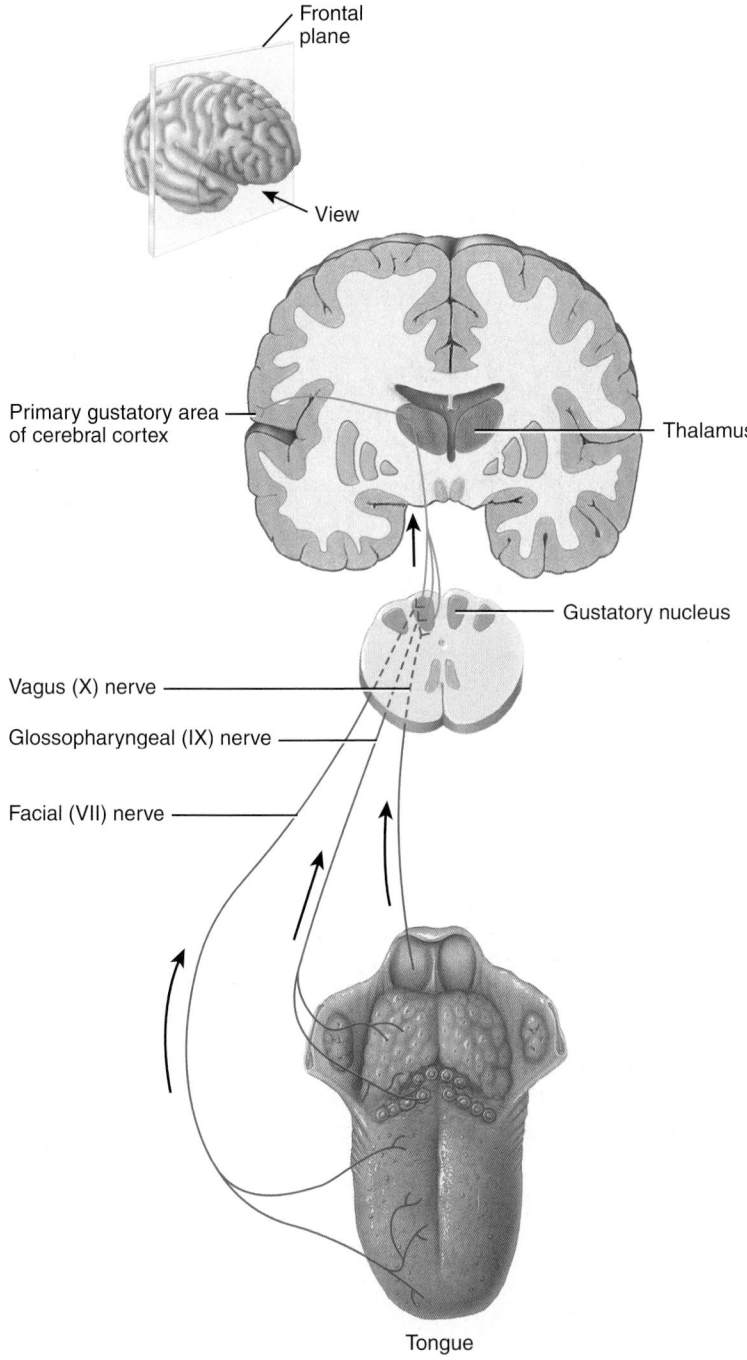

(e) Gustatory pathway

❓ **What role do basal cells play in taste buds?**

CLINICAL CONNECTION | *Taste Aversion*

Probably because of taste projections to the hypothalamus and limbic system, there is a strong link between taste and pleasant or unpleasant emotions. Sweet foods evoke reactions of pleasure while bitter ones cause expressions of disgust even in newborn babies. This phenomenon is the basis for **taste aversion,** in which people and animals quickly learn to avoid a food if it upsets the digestive system. The advantage of avoiding foods that cause such illness is longer survival. However, the drugs and radiation treatments used to combat cancer often cause nausea and gastrointestinal upset regardless of what foods are consumed. Thus, cancer patients may lose their appetite because they develop taste aversions for most foods. •

✔ CHECKPOINT

3. How do olfactory receptors and gustatory receptor cells differ in structure and function?
4. Trace the path of a gustatory stimulus from contact of a tastant with saliva to the primary gustatory area in the cerebral cortex.
5. Compare the olfactory and gustatory pathways.

17.3 VISION

◉ OBJECTIVES

- List and describe the accessory structures of the eye and the structural components of the eyeball.
- Discuss image formation by describing refraction, accommodation, and constriction of the pupil.
- Describe the processing of visual signals in the retina and the neural pathway for vision.

Vision, the act of seeing, is extremely important to human survival. More than half the sensory receptors in the human body are located in the eyes, and a large part of the cerebral cortex is devoted to processing visual information. In this section, we examine electromagnetic radiation, the accessory structures of the eye, the eyeball itself, the formation of visual images, the physiology of vision, and the visual pathway from the eye to the brain.

Electromagnetic Radiation

Electromagnetic radiation (e-lek′-trō-mag′-NET-ik) is energy in the form of waves that radiates from the sun. There are many types of electromagnetic radiation, including gamma rays, x-rays, UV rays, visible light, infrared radiation, microwaves, and radio waves. This range of electromagnetic radiation is known as the **electromagnetic spectrum** (Figure 17.4). The distance between two consecutive peaks of an electromagnetic wave is the *wavelength.* Wavelengths range from short to long; for example, gamma rays have wavelengths smaller than a nanometer, and most radio waves have wavelengths greater than a meter.

The eyes are responsible for the detection of **visible light,** the part of the electromagnetic spectrum with wavelengths ranging from about 400 to 700 nm. Visible light exhibits colors: The color

Figure 17.4 The electromagnetic spectrum.

Visible light is the part of the electromagnetic spectrum with wavelengths ranging from about 400 to 700 nm.

(a) Electromagnetic spectrum

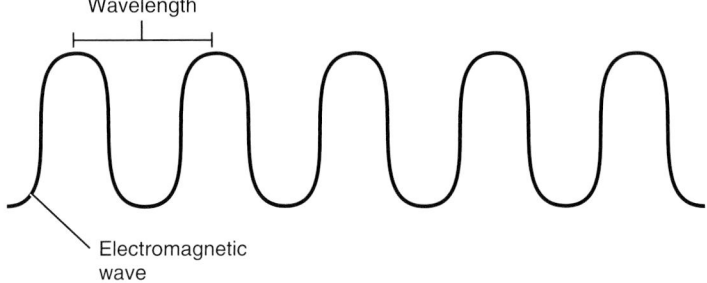

(b) An electromagnetic wave

? **Visible light that has a wavelength of 700 nm is what color?**

of visible light depends on its wavelength. For example, light that has a wavelength of 400 nm is violet, and light that has a wavelength of 700 nm is red. An object can absorb certain wavelengths of visible light and reflect others; the object will appear the color of the wavelength that is reflected. For example, a green apple appears green because it reflects mostly green light and absorbs most other wavelengths of visible light. An object appears white because it reflects all wavelengths of visible light. An object appears black because it absorbs all wavelengths of visible light.

Accessory Structures of the Eye

The **accessory structures** of the eye include the eyelids, eyelashes, eyebrows, the lacrimal (tear-producing) apparatus, and extrinsic eye muscles.

Eyelids

The upper and lower **eyelids,** or **palpebrae** (PAL-pe-brē; singular is *palpebra*), shade the eyes during sleep, protect the eyes from

Figure 17.5 Surface anatomy of the right eye.

 The palpebral fissure is the space between the upper and lower eyelids that exposes the eyeball.

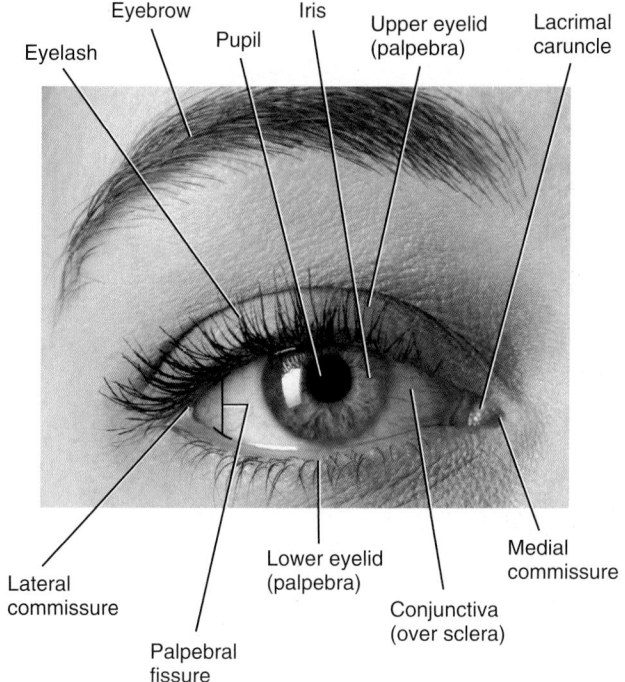

Which structure shown here is continuous with the inner lining of the eyelids?

excessive light and foreign objects, and spread lubricating secretions over the eyeballs (Figure 17.5). The upper eyelid is more movable than the lower and contains in its superior region the **levator palpebrae superioris muscle** (see Figure 17.6a). Sometimes a person may experience an annoying *twitch* in an eyelid, an involuntary quivering similar to muscle twitches in the hand, forearm, leg, or foot. Twitches are almost always harmless and usually last for only a few seconds. They are often associated with stress and fatigue. The space between the upper and lower eyelids that exposes the eyeball is the **palpebral fissure** (PAL-pe-bral). Its angles are known as the **lateral commissure** (KOM-i-shur), which is narrower and closer to the temporal bone, and the **medial commissure,** which is broader and nearer the nasal bone. In the medial commissure is a small, reddish elevation, the **lacrimal caruncle** (KAR-ung-kul), which contains sebaceous (oil) glands and sudoriferous (sweat) glands. The whitish material that sometimes collects in the medial commissure comes from these glands.

From superficial to deep, each eyelid consists of epidermis, dermis, subcutaneous tissue, fibers of the orbicularis oculi muscle, a tarsal plate, tarsal glands, and conjunctiva (Figure 17.6a). The **tarsal plate** is a thick fold of connective tissue that gives form and support to the eyelids. Embedded in each tarsal plate is a row of elongated modified sebaceous glands, known as **tarsal** or

Meibomian glands (mī-BŌ-mē-an), that secrete a fluid that helps keep the eyelids from adhering to each other. Infection of the tarsal glands produces a tumor or cyst on the eyelid called a **chalazion** (ka-LĀ-zē-on = small bump). The **conjunctiva** (kon′-junk-TĪ-va) is a thin, protective mucous membrane composed of nonkeratinized stratified squamous epithelium with numerous goblet cells that is supported by areolar connective tissue. The **palpebral conjunctiva** lines the inner aspect of the eyelids, and the **bulbar conjunctiva** passes from the eyelids onto the surface of the eyeball, where it covers the sclera (the "white" of the eye) but not the cornea, which is a transparent region that forms the outer anterior surface of the eyeball. Over the sclera, the conjunctiva is vascular. Both the sclera and the cornea will be discussed in more detail shortly. Dilation and congestion of the blood vessels of the bulbar conjunctiva due to local irritation or infection are the cause of **bloodshot eyes.**

Eyelashes and Eyebrows

The **eyelashes,** which project from the border of each eyelid, and the **eyebrows,** which arch transversely above the upper eyelids, help protect the eyeballs from foreign objects, perspiration, and the direct rays of the sun. Sebaceous glands at the base of the hair follicles of the eyelashes, called **sebaceous ciliary glands,** release a lubricating fluid into the follicles. Infection of these glands, usually by bacteria, causes a painful, pus-filled swelling called a **sty.**

The Lacrimal Apparatus

The **lacrimal apparatus** (LAK-ri-mal; *lacrim-* = tears) is a group of structures that produces and drains **lacrimal fluid** or **tears.** The **lacrimal glands,** each about the size and shape of an almond, secrete lacrimal fluid, which drains into 6–12 **excretory lacrimal ducts** that empty tears onto the surface of the conjunctiva of the upper lid (Figure 17.6b). From here the tears pass medially over the anterior surface of the eyeball to enter two small openings called **lacrimal puncta** (singular is *punctum*). Tears then pass into two ducts, the **lacrimal canals,** which lead into the **lacrimal sac** and then into the **nasolacrimal duct.** This duct carries the lacrimal fluid into the nasal cavity just inferior to the inferior nasal concha. An infection of the lacrimal sacs is called **dacryocystitis** (dak′-rē-ō-sis-TĪ-tis; *dacryo-* = lacrimal sac; *-itis* = inflammation of). It is usually caused by a bacterial infection and results in blockage of the nasolacrimal ducts.

The lacrimal glands are supplied by parasympathetic fibers of the facial (VII) nerves. The lacrimal fluid produced by these glands is a watery solution containing salts, some mucus, and **lysozyme** (LĪ-sō-zīm), a protective bactericidal enzyme. The fluid protects, cleans, lubricates, and moistens the eyeball. After being secreted from the lacrimal gland, lacrimal fluid is spread medially over the surface of the eyeball by the blinking of the eyelids. Each gland produces about 1 mL of lacrimal fluid per day.

Normally, tears are cleared away as fast as they are produced, either by evaporation or by passing into the lacrimal canals and then into the nasal cavity. If an irritating substance makes contact with the conjunctiva, however, the lacrimal glands are stimulated

Figure 17.6 Accessory structures of the eye.

Accessory structures of the eye include the eyelids, eyelashes, eyebrows, lacrimal apparatus, and extrinsic eye muscles.

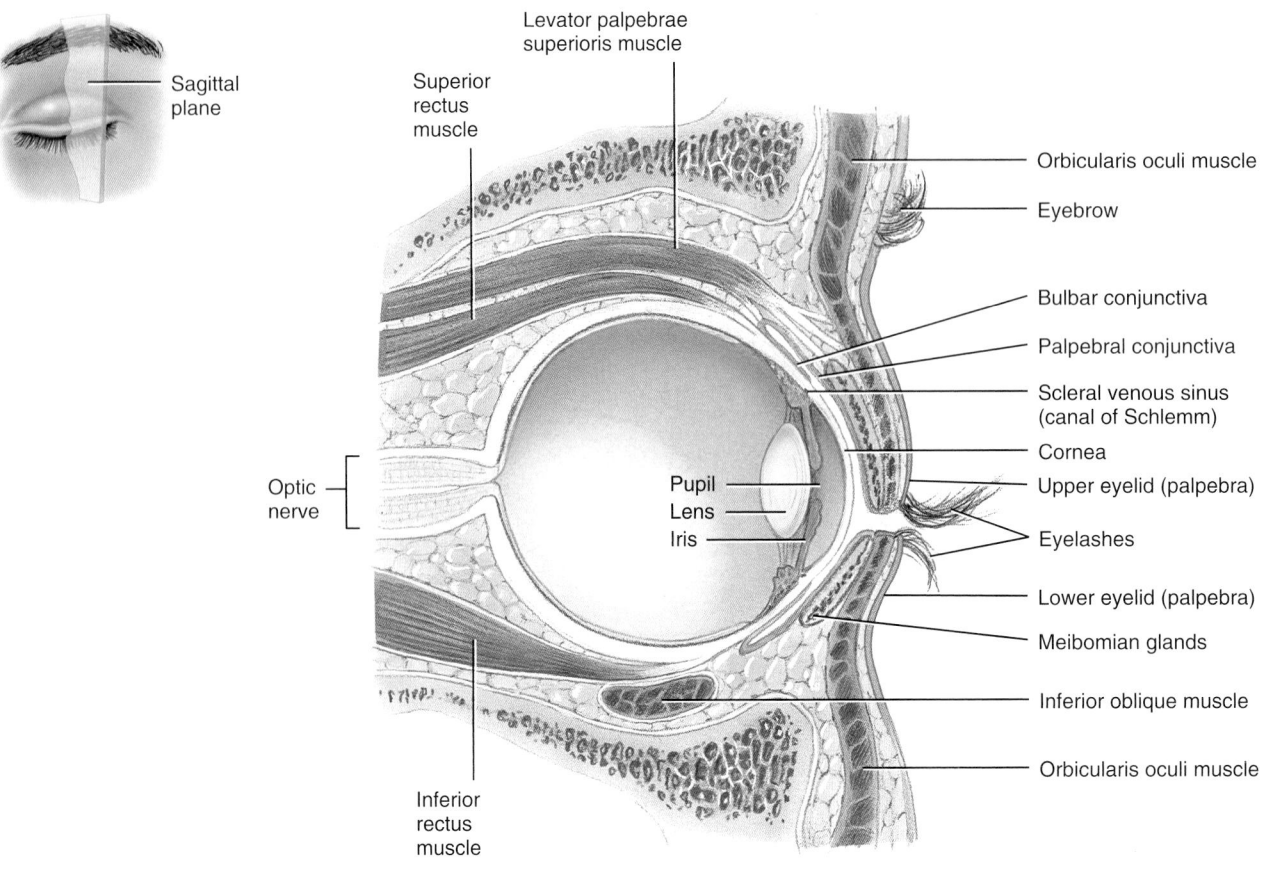

(a) Sagittal section of eye and its accessory structures

(b) Anterior view of the lacrimal apparatus

FLOW OF TEARS

Lacrimal gland
↓
Lacrimal ducts
↓
Superior or inferior lacrimal canal
↓
Lacrimal sac
↓
Nasolacrimal duct
↓
Nasal cavity

? **What is lacrimal fluid, and what are its functions?**

to oversecrete, and tears accumulate (watery eyes). Lacrimation is a protective mechanism, as the tears dilute and wash away the irritating substance. Watery eyes also occur when an inflamma-

tion of the nasal mucosa, such as occurs with a cold, obstructs the nasolacrimal ducts and blocks drainage of tears. Only humans express emotions, both happiness and sadness, by **crying.** In response to parasympathetic stimulation, the lacrimal glands produce excessive lacrimal fluid that may spill over the edges of the eyelids and even fill the nasal cavity with fluid. This is how crying produces a runny nose.

Extrinsic Eye Muscles

The eyes sit in the bony depressions of the skull called the *orbits.* The orbits help protect the eyes, stabilize them in three-dimensional space, and anchor them to the muscles that produce their essential movements. The extrinsic eye muscles extend from the walls of the bony orbit to the sclera (white) of the eye and are surrounded in the orbit by a significant quantity of **periorbital fat** (per′-ē-OR-bi-tal). These muscles are capable of moving the eye in al-

most any direction. Six extrinsic eye muscles move each eye: the **superior rectus, inferior rectus, lateral rectus, medial rectus, superior oblique,** and **inferior oblique** (Figures 17.6a and 17.7). They are supplied by cranial nerves III, IV, or VI. In general, the motor units in these muscles are small. Some motor neurons serve only two or three muscle fibers—fewer than in any other part of the body except the larynx (voice box). Such small motor units permit smooth, precise, and rapid movement of the eyes. As indicated in Exhibit 11.B, the extrinsic eye muscles move the eyeball laterally, medially, superiorly, and inferiorly. For example, looking to the right requires simultaneous contraction of the right lateral rectus and left medial rectus muscles of the eyeball and relaxation of the left lateral rectus and right medial rectus of the

Figure 17.7 Anatomy of the eyeball.

The wall of the eyeball consists of three layers: the fibrous tunic, the vascular tunic, and the retina.

Superior view of transverse section of right eyeball

? What are the components of the fibrous tunic and vascular tunic?

eyeball. The oblique muscles preserve rotational stability of the eyeball. Neural circuits in the brain stem and cerebellum coordinate and synchronize the movements of the eyes.

Anatomy of the Eyeball

The adult **eyeball** measures about 2.5 cm (1 in.) in diameter. Of its total surface area, only the anterior one-sixth is exposed; the remainder is recessed and protected by the orbit, into which it fits. Anatomically, the wall of the eyeball consists of three layers: (1) fibrous tunic, (2) vascular tunic, and (3) retina.

Fibrous Tunic

The **fibrous tunic** (TOO-nik) is the superficial layer of the eyeball and consists of the anterior cornea and posterior sclera (Figure 17.7). The **cornea** (KOR-nē-a) is a transparent coat that covers the colored iris. Because it is curved, the cornea helps focus light onto the retina. Its outer surface consists of nonkeratinized stratified squamous epithelium. The middle coat of the cornea consists of collagen fibers and fibroblasts, and the inner surface is simple squamous epithelium. Since the central part of the cornea receives oxygen from the outside air, contact lenses that are worn for long periods of time must be permeable to permit oxygen to pass through them. The **sclera** (SKLE-ra; *scler-* = hard), the "white" of the eye, is a layer of dense connective tissue made up mostly of collagen fibers and fibroblasts. The sclera covers the entire eyeball except the cornea; it gives shape to the eyeball, makes it more rigid, protects its inner parts, and serves as a site of attachment for the extrinsic eye muscles. At the junction of the sclera and cornea is an opening known as the **scleral venous sinus (canal of Schlemm).** A fluid called *aqueous humor,* which will be described later, drains into this sinus (Figure 17.7).

Vascular Tunic

The **vascular tunic** or **uvea** (Ū-ve-a) is the middle layer of the eyeball. It is composed of three parts: choroid, ciliary body, and iris (Figure 17.7). The highly vascularized **choroid** (KŌ-royd), which is the posterior portion of the vascular tunic, lines most of the internal surface of the sclera. Its numerous blood vessels provide nutrients to the posterior surface of the retina. The choroid also contains melanocytes that produce the pigment melanin, which causes this layer to appear dark brown in color. Melanin in the choroid absorbs stray light rays, which prevents reflection and scattering of light within the eyeball. As a result, the image cast on the retina by the cornea and lens remains sharp and clear. Albinos lack melanin in all parts of the body, including the eye. They often need to wear sunglasses, even indoors, because even moderately bright light is perceived as bright glare due to light scattering.

In the anterior portion of the vascular tunic, the choroid becomes the **ciliary body** (SIL-ē-ar′-ē). It extends from the **ora serrata** (Ō-ra ser-RĀ-ta), the jagged anterior margin of the retina, to a point just posterior to the junction of the sclera and cornea. Like the choroid, the ciliary body appears dark brown in color because it contains melanin-producing melanocytes. In addition, the cil-

iary body consists of ciliary processes and ciliary muscle. The **ciliary processes** are protrusions or folds on the internal surface of the ciliary body. They contain blood capillaries that secrete aqueous humor. Extending from the ciliary process are **zonular fibers (suspensory ligaments)** that attach to the lens. The fibers consist of thin, hollow fibrils that resemble elastic connective tissue fibers. The **ciliary muscle** is a circular band of smooth muscle. Contraction or relaxation of the ciliary muscle changes the tightness of the zonular fibers, which alters the shape of the lens, adapting it for near or far vision.

The **iris** (= rainbow), the colored portion of the eyeball, is shaped like a flattened donut. It is suspended between the cornea and the lens and is attached at its outer margin to the ciliary processes. It consists of melanocytes and circular and radial smooth muscle fibers. The amount of melanin in the iris determines the eye color. The eyes appear brown to black when the iris contains a large amount of melanin, blue when its melanin concentration is very low, and green when its melanin concentration is moderate.

A principal function of the iris is to regulate the amount of light entering the eyeball through the **pupil** (*pupil* = little person; because this is where you see a reflection of yourself when looking into someone's eyes), the hole in the center of the iris. The pupil appears black because, as you look through the lens, you see the heavily pigmented back of the eye (choroid and retina). However, if bright light is directed into the pupil, the reflected light is red because of the blood vessels on the surface of the retina. It is for this reason that a person's eyes appear red in a photograph ("red eye") when the flash is directed into the pupil. Autonomic reflexes regulate pupil diameter in response to light levels (Figure 17.8). When bright light stimulates the eye, parasympathetic

Figure 17.8 Responses of the pupil to light of varying brightness.

Contraction of the circular muscles causes constriction of the pupil; contraction of the radial muscles causes dilation of the pupil.

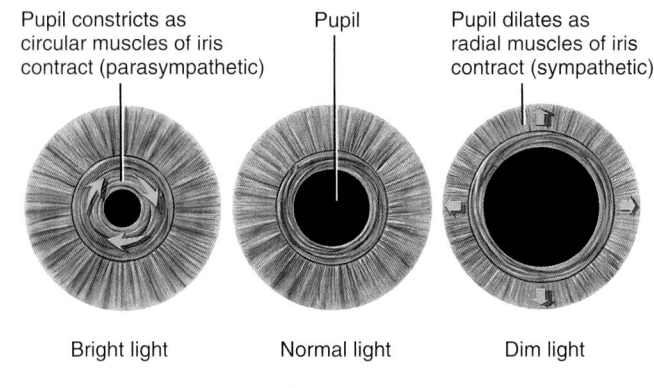

Pupil constricts as circular muscles of iris contract (parasympathetic)   Pupil   Pupil dilates as radial muscles of iris contract (sympathetic)

Bright light   Normal light   Dim light

Anterior views

Which division of the autonomic nervous system causes pupillary constriction? Which causes pupillary dilation?

fibers of the oculomotor (III) nerve stimulate the **circular muscles** or **sphincter pupillae** (pu-PIL-ē) of the iris to contract, causing a decrease in the size of the pupil (constriction). In dim light, sympathetic neurons stimulate the **radial muscles** or **dilator pupillae** of the iris to contract, causing an increase in the pupil's size (dilation).

Retina

The third and inner layer of the eyeball, the **retina,** lines the posterior three-quarters of the eyeball and is the beginning of the visual pathway (see Figure 17.7). This layer's anatomy can be viewed with an *ophthalmoscope* (of-THAL-mō-skōp; *ophthalmos-* = eye; *-skopeo* = to examine), an instrument that shines light into the eye and allows an observer to peer through the pupil, providing a magnified image of the retina and its blood vessels as well as the optic (II) nerve (Figure 17.9). The surface of the retina is the only place in the body where blood vessels can be viewed directly and examined for pathological changes, such as those that occur with hypertension, diabetes mellitus, cataracts, and age-related macular disease. Several landmarks are visible through an ophthalmoscope. The **optic disc** is the site where the optic (II) nerve exits the eyeball. Bundled together with the optic nerve are the **central retinal artery,** a branch of the ophthalmic artery, and the **central retinal vein** (see Figure 17.7). Branches of the central retinal artery fan out to nourish the anterior surface of the retina; the central retinal vein drains blood from the retina through the optic disc. Also visible are the macula lutea and fovea centralis, which are described shortly.

The retina consists of a pigmented layer and a neural layer. The **pigmented layer** is a sheet of melanin-containing epithelial cells located between the choroid and the neural part of the retina. The melanin in the pigmented layer of the retina, as in the choroid, also helps to absorb stray light rays. The **neural (sensory) layer** of the retina is a multilayered outgrowth of the brain that processes visual data extensively before sending nerve impulses into axons that form the optic nerve. Three distinct layers of retinal neurons—the **photoreceptor layer,** the **bipolar cell layer,** and the **ganglion cell layer**—are separated by two zones, the *outer* and *inner synaptic layers,* where synaptic contacts are made (Figure 17.10). Note that light passes through the ganglion and bipolar cell layers and both synaptic layers before it reaches the photoreceptor layer. Two other types of cells present in the bipolar cell layer of the retina are called **horizontal cells** and **amacrine cells** (AM-a-krin). These cells form laterally directed neural circuits that modify the signals being transmitted along the pathway from photoreceptors to bipolar cells to ganglion cells.

CLINICAL CONNECTION | *Detached Retina*

A **detached retina** may occur due to trauma, such as a blow to the head, in various eye disorders, or as a result of age-related degeneration. The detachment occurs between the neural portion of the retina and the pigmented epithelium. Fluid accumulates between these layers, forcing the thin, pliable retina to billow outward. The result is distorted vision and blindness in the corresponding field of vision. The retina may be reattached by laser surgery or cryosurgery (localized application of extreme cold), and reattachment must be accomplished quickly to avoid permanent damage to the retina. •

Photoreceptors are specialized cells that begin the process by which light rays are ultimately converted to nerve impulses. There are two types of photoreceptors: rods and cones. Each retina has about 6 million cones and 120 million rods. **Rods** allow us to see in dim light, such as moonlight. Because rods do not provide color vision, in dim light we can see only black, white, and all shades of gray in between. Brighter lights stimulate **cones,** which produce color vision. Three types of cones are present in the retina: (1) *blue cones,* which are sensitive to blue light, (2) *green cones,* which are sensitive to green light, and (3) *red cones,* which are sensitive to red light. Color vision results from the stimulation of various combinations of these three types of cones. Most of our experiences are mediated by the cone system, the loss of which produces legal blindness. A person who loses rod vision mainly has difficulty seeing in dim light and thus should not drive at night.

From photoreceptors, information flows through the outer synaptic layer to bipolar cells and then from bipolar cells through the inner synaptic layer to ganglion cells. The axons of ganglion cells extend posteriorly to the optic disc and exit the eyeball as the optic (II) nerve. The optic disc is also called the **blind spot.** Because it contains no rods or cones, we cannot see images that strike the blind spot. Normally, you are not aware of having a blind spot, but you can easily demonstrate its presence. Hold this page about 20 in. from your face with the cross shown at the end

---

Figure 17.9 A normal retina, as seen through an ophthalmoscope. Blood vessels in the retina can be viewed directly and examined for pathological changes.

The optic disc is the site where the optic nerve exits the eyeball. The fovea centralis is the area of highest visual acuity.

NASAL SIDE

TEMPORAL SIDE

Optic disc

Macula lutea

Fovea centralis

Retinal blood vessels

Left eye

? Evidence of what diseases may be seen through an ophthalmoscope?

Figure 17.10 Microscopic structure of the retina. The downward blue arrow at left indicates the direction of the signals passing through the neural layer of the retina. Eventually, nerve impulses arise in ganglion cells and propagate along their axons, which make up the optic (II) nerve.

In the retina, visual signals pass from photoreceptors to bipolar cells to ganglion cells.

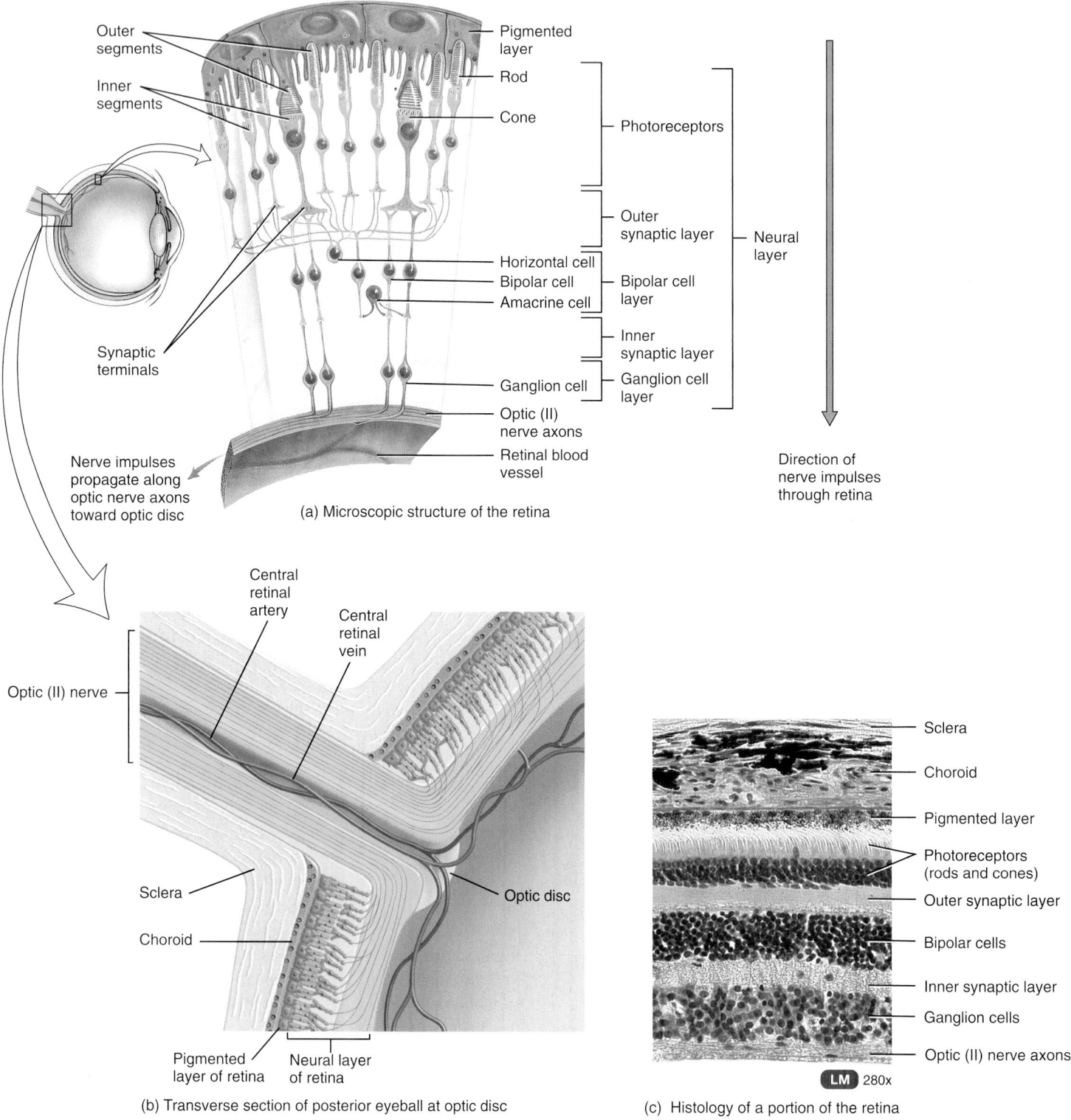

(a) Microscopic structure of the retina

(b) Transverse section of posterior eyeball at optic disc

(c) Histology of a portion of the retina

? **What are the two types of photoreceptors, and how do their functions differ?**

of this paragraph directly in front of your right eye. You should be able to see the cross and the square when you close your left eye. Now, keeping the left eye closed, slowly bring the page closer to your face while keeping the right eye on the cross. At a certain distance the square will disappear from your field of vision because its image falls on the blind spot.

The **macula lutea** (MAK-ū-la LOO-tē-a; *macula* = a small, flat spot; *lute-* = yellowish) is in the exact center of the posterior portion of the retina, at the visual axis of the eye (see Figure 17.9). The **fovea centralis** (FŌ-vē-a) (see Figures 17.7 and 17.9), a small depression in the center of the macula lutea, contains only cones. In addition, the layers of bipolar and ganglion cells, which scatter light to some extent, do not cover the cones here; these layers are displaced to the periphery of the fovea centralis. As a result, the fovea centralis is the area of highest **visual acuity** (a-KU-i-tē) or **resolution** (sharpness of vision). A main reason that you move your head and eyes while looking at something is to place images of interest on your fovea centralis—as you do to read the words in this sentence! Rods are absent from the fovea centralis and are more plentiful toward the periphery of the retina. Because rod vision is more sensitive than cone vision, you can see a faint object (such as a dim star) better if you gaze slightly to one side rather than looking directly at it.

CLINICAL CONNECTION | Age-related Macular Disease

Age-related macular disease (AMD), also known as *macular degeneration,* is a degenerative disorder of the retina in persons 50 years of age and older. In AMD, abnormalities occur in the region of the macula lutea, which is ordinarily the area of most acute vision. Victims of advanced AMD retain their peripheral vision but lose the ability to see straight ahead. For instance, they cannot see facial features to identify a person in front of them. AMD is the leading cause of blindness in those over age 75, afflicting 13 million Americans, and is 2.5 times more common in pack-a-day smokers than in nonsmokers. Initially, a person may experience blurring and distortion at the center of the visual field. In "dry" AMD, central vision gradually diminishes because the pigmented layer atrophies and degenerates. There is no effective treatment. In about 10% of cases, dry AMD progresses to "wet" AMD, in which new blood vessels form in the choroid and leak plasma or blood under the retina. Vision loss can be slowed by using laser surgery to destroy the leaking blood vessels. •

Lens

Behind the pupil and iris, within the cavity of the eyeball, is the **lens** (see Figure 17.7). Within the cells of the lens, proteins called **crystallins** (KRIS-ta-lins), arranged like the layers of an onion, make up the refractive media of the lens, which normally is perfectly transparent and lacks blood vessels. It is enclosed by a clear connective tissue capsule and held in position by encircling zonular fibers, which attach to the ciliary processes. The lens helps focus images on the retina to facilitate clear vision.

Interior of the Eyeball

The lens divides the interior of the eyeball into two cavities: the anterior cavity and vitreous chamber. The **anterior cavity**—the space anterior to the lens—consists of two chambers. The **anterior chamber** lies between the cornea and the iris. The **posterior chamber** lies behind the iris and in front of the zonular fibers and lens (Figure 17.11). Both chambers of the anterior cavity are filled with **aqueous humor** (AK-wē-us HŪ-mer; *aqua* = water), a transparent watery fluid that nourishes the lens and cornea. Aqueous humor continually filters out of blood capillaries in the ciliary processes of the ciliary body and enters the posterior chamber. It then flows forward between the iris and the lens, through the pupil, and into the anterior chamber. From the anterior chamber, aqueous humor drains into the scleral venous sinus (canal of Schlemm) and then into the blood. Normally, aqueous humor is completely replaced about every 90 minutes.

The larger posterior cavity of the eyeball is the **vitreous chamber** (VIT-rē-us), which lies between the lens and the retina. Within the vitreous chamber is the **vitreous body,** a transparent jellylike substance that holds the retina flush against the choroid, giving the retina an even surface for the reception of clear images. It occupies about four-fifths of the eyeball. Unlike the aqueous humor, the vitreous body does not undergo constant replacement. It is formed during embryonic life and consists of mostly water plus collagen fibers and hyaluronic acid. The vitreous body also contains phagocytic cells that remove debris, keeping this part of the eye clear for unobstructed vision. Occasionally, collections of debris may cast a shadow on the retina and create the appearance of specks that dart in and out of the field of vision. These *vitreal floaters,* which are more common in older individuals, are usually harmless and do not require treatment. The **hyaloid canal** (HĪ-a-loyd) is a narrow channel that is inconspicuous in adults and runs through the vitreous body from the optic disc to the posterior aspect of the lens. In the fetus, it is occupied by the hyaloid artery (see Figure 17.27d).

The pressure in the eye, called **intraocular pressure,** is produced mainly by the aqueous humor and partly by the vitreous body; normally it is about 16 mmHg (millimeters of mercury). The intraocular pressure maintains the shape of the eyeball and prevents it from collapsing. Puncture wounds to the eyeball may cause the loss of aqueous humor and the vitreous body. This in turn causes a decrease in intraocular pressure, a detached retina, and in some cases blindness.

Table 17.1 summarizes the structures associated with the eyeball.

Image Formation

In some ways the eye is like a camera: Its optical elements focus an image of some object on a light-sensitive "film"—the retina—while ensuring the correct amount of light to make the proper "exposure." To understand how the eye forms clear images of objects on the retina, we must examine three processes: (1) the refraction or bending of light by the lens and cornea; (2) accommodation, the change in shape of the lens; and (3) constriction or narrowing of the pupil.

Figure 17.11 **The iris separating the anterior and posterior chambers of the eye.** The section is through the anterior portion of the eyeball at the junction of the cornea and sclera. Arrows indicate the flow of aqueous humor.

🔑 ▭ The lens separates the posterior chamber of the anterior cavity from the vitreous chamber.

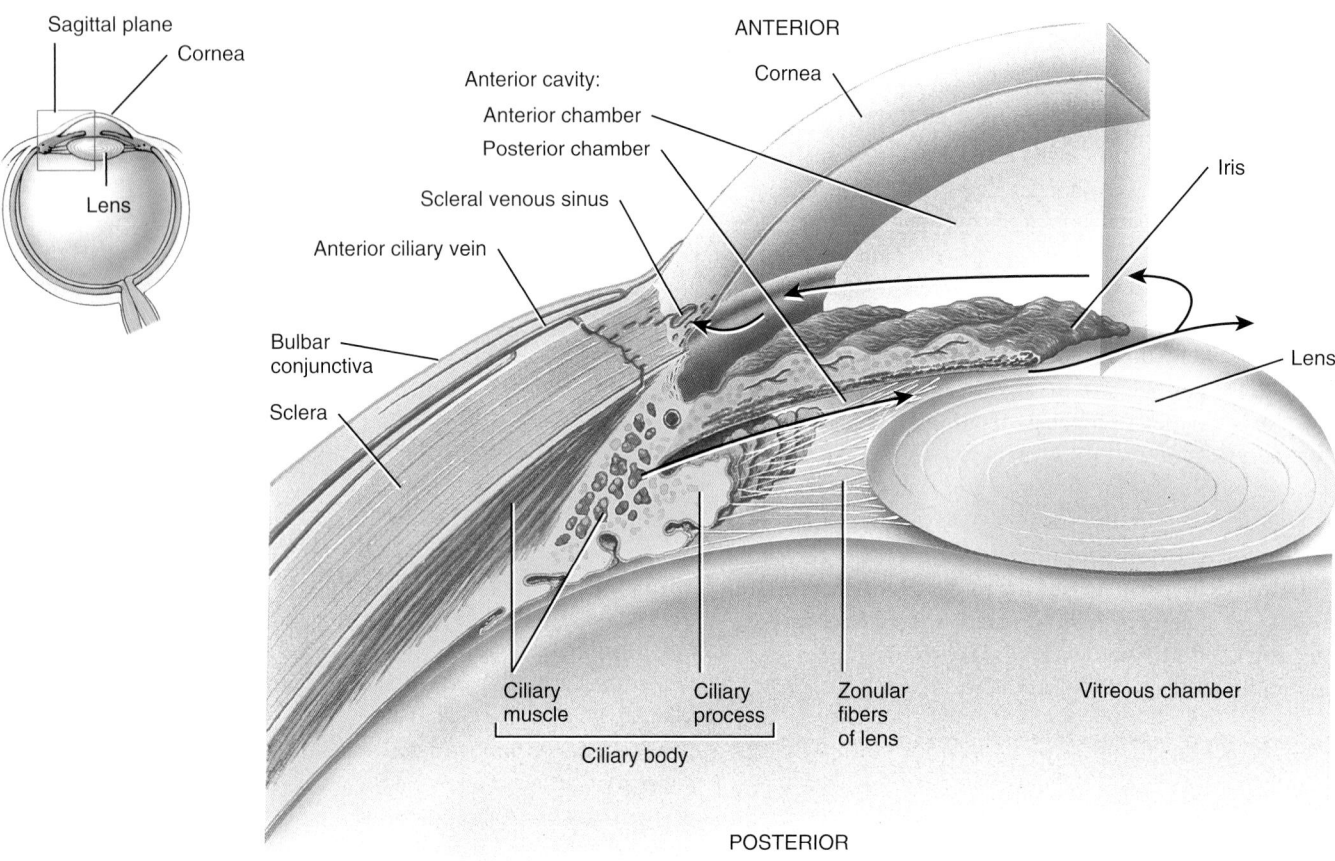

Where is aqueous humor produced, what is its circulation path, and where does it drain from the eyeball?

Refraction of Light Rays

When light rays traveling through a transparent substance (such as air) pass into a second transparent substance with a different density (such as water), they bend at the junction between the two substances. This bending is called **refraction** (re-FRAK-shun) (Figure 17.12a). As light rays enter the eye, they are refracted at the anterior and posterior surfaces of the cornea. Both surfaces of the lens of the eye further refract the light rays so they come into exact focus on the retina.

Images focused on the retina are inverted (upside down) (Figure 17.12b, c). They also undergo right-to-left reversal; that is, light from the right side of an object strikes the left side of the retina, and vice versa. The reason the world does not look inverted and reversed is that the brain "learns" early in life to coordinate visual images with the orientations of objects. The brain stores the inverted and reversed images we acquired when we first reached for and touched objects and interprets those visual images as being correctly oriented in space.

About 75% of the total refraction of light occurs at the cornea. The lens provides the remaining 25% of focusing power and also changes the focus to view near or distant objects. When an object

is 6 m (20 ft) or more away from the viewer, the light rays reflected from the object are nearly parallel to one another (Figure 17.12b). The lens must bend these parallel rays just enough so that they fall exactly focused on the central fovea, where vision is sharpest. Because light rays that are reflected from objects closer than 6 m (20 ft) are divergent rather than parallel (Figure 17.12c), the rays must be refracted more if they are to be focused on the retina. This additional refraction is accomplished through a process called accommodation.

Accommodation and the Near Point of Vision

A surface that curves outward, like the surface of a ball, is said to be *convex*. When the surface of a lens is convex, that lens will refract incoming light rays toward each other, so that they eventually intersect. If the surface of a lens curves inward, like the inside of a hollow ball, the lens is said to be *concave* and causes light rays to refract away from each other. The lens of the eye is convex on both its anterior and posterior surfaces, and its focusing power increases as its curvature becomes greater. When the eye is focusing on a close object, the lens becomes more curved, causing greater refraction of the light rays. This increase in the curvature

TABLE 17.1

Summary of the Structures of the Eyeball

STRUCTURE	FUNCTION
Fibrous tunic	*Cornea:* Admits and refracts (bends) light. *Sclera:* Provides shape and protects inner parts.
Vascular tunic	*Iris:* Regulates amount of light that enters eyeball. *Ciliary body:* Secretes aqueous humor and alters shape of lens for near or far vision (accommodation). *Choroid:* Provides blood supply and absorbs scattered light.
Retina	Receives light and converts it into receptor potentials and nerve impulses. Output to brain via axons of ganglion cells, which form optic (II) nerve.
Lens	Refracts light.
Anterior cavity	Contains aqueous humor that helps maintain shape of eyeball and supplies oxygen and nutrients to lens and cornea.
Vitreous chamber	Contains vitreous body that helps maintain shape of eyeball and keeps retina attached to choroid.

Figure 17.12 Refraction of light rays. (a) Refraction is the bending of light rays at the junction of two transparent substances with different densities. (b) The cornea and lens refract light rays from distant objects so the image is focused on the retina. (c) In accommodation, the lens becomes more spherical, which increases the refraction of light.

🔑 Images focused on the retina are inverted and left-to-right reversed.

(a) Refraction of light rays

(b) Viewing distant object

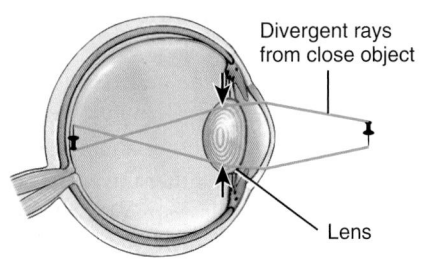

(c) Accommodation

❓ **What sequence of events occurs during accommodation?**

of the lens for near vision is called **accommodation** (a-kom-a-DĀ-shun) (Figure 17.12c). The **near point of vision** is the minimum distance from the eye that an object can be clearly focused with maximum accommodation. This distance is about 10 cm (4 in.) in a young adult.

How does accommodation occur? When you are viewing distant objects, the ciliary muscle of the ciliary body is relaxed and the lens is flatter because it is stretched in all directions by taut zonular fibers. When you view a close object, the ciliary muscle contracts, which pulls the ciliary process and choroid forward toward the lens. This action releases tension on the lens and zonular fibers. Because it is elastic, the lens becomes more spherical (more convex), which increases its focusing power and causes greater convergence of the light rays. Parasympathetic fibers of the oculomotor (III) nerve innervate the ciliary muscle of the ciliary body and, therefore, mediate the process of accommodation.

With aging, the lens loses elasticity and thus its ability to curve to focus on objects that are close. Therefore, older people cannot read print at the same close range as can younger people. This condition is called **presbyopia** (prez-bē-Ō-pē-a; *presby-* = old; *-opia* = pertaining to the eye or vision). By age 40 the near point of vision may have increased to 20 cm (8 in.), and at age 60 it may be as much as 80 cm (31 in.). Presbyopia usually begins in the mid-forties. At about that age, people who have not previously worn glasses begin to need them for reading. Those who already wear glasses typically start to need bifocals, lenses that can focus for both distant and close vision. •

Refraction Abnormalities

The normal eye, known as an **emmetropic eye** (em′-e-TROP-ik), can sufficiently refract light rays from an object 6 m (20 ft) away so that a clear image is focused on the retina. Many people, however, lack this ability because of refraction abnormalities. Among these abnormalities are **myopia** (mī-Ō-pē-a), or *nearsightedness,* which occurs when the eyeball is too long relative to the focusing power of the cornea and lens, or when the lens is thicker than normal, so an image converges in front of the retina. Myopic individuals can see close objects clearly, but not distant objects. In **hyperopia** (hī-per-Ō-pē-a) or *farsightedness,* also known as **hypermetropia** (hī′-per-me-TRŌ-pē-a), the eyeball length is short relative to the focusing power of the cornea and lens, or the lens is thinner than normal, so an image converges behind the retina. Hyperopic individuals can see distant objects clearly, but not close ones. Figure 17.13 illustrates these conditions and explains how they are corrected. Another refraction abnormality is **astigmatism** (a-STIG-ma-tizm), in which either the cornea or the lens has an irregular curvature. As a result, parts of the image are out of focus, and thus vision is blurred or distorted.

Most errors of vision can be corrected by eyeglasses, contact lenses, or surgical procedures. A contact lens floats on a film of tears over the cornea. The anterior outer surface of the contact lens corrects the visual defect, and its posterior surface matches the curvature of the cornea. LASIK involves reshaping the cornea to permanently correct refraction abnormalities.

An increasingly popular alternative to wearing glasses or contact lenses is refractive surgery to correct the curvature of the cornea for conditions such as farsightedness, nearsightedness, and astigmatism. The most common type of refractive surgery is **LASIK** (laser-assisted in-situ keratomileusis). After anesthetic drops are placed in the eye, a circular flap of tissue from the center of the cornea is cut. The flap is folded out of the way, and the underlying layer of cornea is reshaped with a laser, one microscopic layer at a time. A computer assists the physician in removing very precise layers of the cornea. After the sculpting is complete, the corneal flap is repositioned over the treated area. A patch is placed over the eye overnight and the flap quickly reattaches to the rest of the cornea. •

Figure 17.13 Refraction abnormalities in the eyeball and their correction. (a) Normal (emmetropic) eye. (b) In the nearsighted or myopic eye, the image is focused in front of the retina. The condition may result from an elongated eyeball or thickened lens. (c) Correction of myopia is by use of a concave lens that diverges entering light rays so that they come into focus directly on the retina. (d) In the farsighted or hyperopic eye, the image is focused behind the retina. The condition results from a shortened eyeball or a thin lens. (e) Correction of hyperopia is by a convex lens that converges entering light rays so that they focus directly on the retina.

In myopia (nearsightedness), only close objects can be seen clearly; in hyperopia (farsightedness), only distant objects can be seen clearly.

(a) Normal (emmetropic) eye

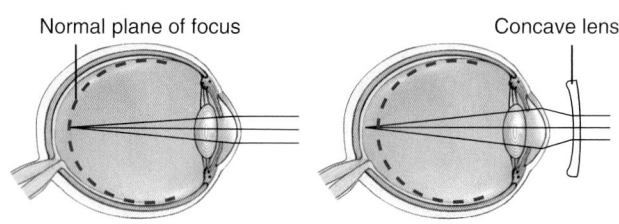

(b) Nearsighted (myopic) eye, uncorrected

(c) Nearsighted (myopic) eye, corrected

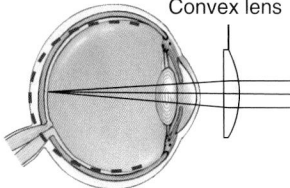

(d) Farsighted (hyperopic) eye, uncorrected

(e) Farsighted (hyperopic) eye, corrected

 What is presbyopia?

Constriction of the Pupil

The circular muscle fibers of the iris also have a role in the formation of clear retinal images. As you have already learned, **constriction of the pupil** is a narrowing of the diameter of the hole through which light enters the eye due to the contraction of the circular muscles of the iris. This autonomic reflex occurs simultaneously with accommodation and prevents light rays from entering the eye through the periphery of the lens. Light rays entering at the periphery would not be brought to focus on the retina and would result in blurred vision. The pupil, as noted earlier, also constricts in bright light.

Convergence

Because of the position of their eyes in their heads, many animals, such as horses and goats, see one set of objects off to the left through one eye, and an entirely different set of objects off to the right through the other. In humans, both eyes focus on only one set of objects—a characteristic called **binocular vision.** This feature of our visual system allows the perception of depth and an appreciation of the three-dimensional nature of objects.

Binocular vision occurs when light rays from an object strike corresponding points on the two retinas. When we stare straight ahead at a distant object, the incoming light rays are aimed directly at both pupils and are refracted to comparable spots on the retinas of both eyes. As we move closer to an object, however, the eyes must rotate medially if the light rays from the object are to strike the same points on both retinas. The term **convergence** refers to this medial movement of the two eyeballs so that both are directed toward the object being viewed, for example, tracking a pencil moving toward your eyes. The nearer the object, the greater the degree of convergence needed to maintain binocular vision. The coordinated action of the extrinsic eye muscles brings about convergence.

Physiology of Vision

Photoreceptors and Photopigments

Rods and cones were named for the different appearance of the *outer segment*—the distal end next to the pigmented layer—of each of these types of photoreceptors. The outer segments of rods are cylindrical or rod-shaped; those of cones are tapered or cone-shaped (Figure 17.14). Transduction of light energy into a receptor potential occurs in the outer segment of both rods and cones. The photopigments are integral proteins in the plasma membrane of the outer segment. In cones the plasma membrane is folded back and forth in a pleated fashion; in rods the pleats pinch off from the plasma membrane to form discs. The outer segment of each rod contains a stack of about 1000 discs, piled up like coins inside a wrapper.

Photoreceptor outer segments are renewed at an astonishingly rapid pace. In rods, one to three new discs are added to the base of the outer segment every hour while old discs slough off at the tip and are phagocytized by pigment epithelial cells. The *inner segment* contains the cell nucleus, Golgi complex, and many mitochondria. At its proximal end, the photoreceptor expands into bulblike synaptic terminals filled with synaptic vesicles.

Figure 17.14 Structure of rod and cone photoreceptors.
The inner segments contain the metabolic machinery for synthesis of photopigments and production of ATP. The photopigments are embedded in the membrane discs or folds of the outer segments. New discs in rods and new folds in cones form at the base of the outer segment. Pigmented epithelial cells phagocytize the old discs and folds that slough off the distal tip of the outer segments.

 Transduction of light energy into a receptor potential occurs in the outer segments of rods and cones.

? **What are the functional similarities between rods and cones?**

The first step in visual transduction is absorption of light by a **photopigment,** a colored protein that undergoes structural changes when it absorbs light, in the outer segment of a photoreceptor. Light absorption initiates the events that lead to the production of a receptor potential. The single type of photopigment in rods is **rhodopsin** (rō-DOP-sin; *rhod-* = rose; *-opsin* = related to vision). Three different **cone photopigments** are present in the retina, one in each of the three types of cones. Color vision results from different colors of light selectively activating the different cone photopigments.

All photopigments associated with vision contain two parts: a glycoprotein known as **opsin** and a derivative of vitamin A called **retinal.** Vitamin A derivatives are formed from carotene, the plant pigment that gives carrots their orange color. Good vision depends on adequate dietary intake of carotene-rich vegetables such as carrots, spinach, broccoli, and yellow squash, or foods that contain vitamin A, such as liver.

Retinal is the light-absorbing part of all visual photopigments. In the human retina, there are four different opsins, three in the cones and one in the rods (rhodopsin). Small variations in the amino acid sequences of the different opsins permit the rods and cones to absorb different colors (wavelengths) of incoming light.

Photopigments respond to light in the following cyclical process (Figure 17.15):

❶ In darkness, retinal has a bent shape, called *cis*-retinal, which fits snugly into the opsin portion of the photopigment. When *cis*-retinal absorbs a photon of light, it straightens out to a shape called *trans*-retinal. This cis-to-trans conversion is called **isomerization** (ī-som′-er-i-ZĀ-shun) and is the first step in visual transduction. After retinal isomerizes, several unstable chemical intermediates form and disappear. These chemical changes lead to production of a receptor potential (see Figure 17.16).

❷ In about a minute, *trans*-retinal completely separates from opsin. The final products look colorless, so this part of the cycle is termed **bleaching** of photopigment.

❸ An enzyme called **retinal isomerase** converts *trans*-retinal back to *cis*-retinal.

❹ The *cis*-retinal then can bind to opsin, reforming a functional photopigment. This part of the cycle—resynthesis of a photopigment—is called **regeneration.**

The pigmented layer of the retina adjacent to the photoreceptors stores a large quantity of vitamin A and contributes to the regeneration process in rods. The extent of rhodopsin regeneration decreases drastically if the retina detaches from the pigmented layer. Cone photopigments regenerate much more quickly than the rhodopsin in rods and are less dependent on the pigmented layer. After complete bleaching, regeneration of half of the rhodopsin takes 5 minutes; half of the cone photopigments regenerate in only 90 seconds. Full regeneration of bleached rhodopsin takes 30 to 40 minutes.

Light and Dark Adaptation

When you emerge from dark surroundings (say, a tunnel) into the sunshine, **light adaptation** occurs—your visual system adjusts in seconds to the brighter environment by decreasing its sensitivity. On the other hand, when you enter a darkened room such as a theater, your visual system undergoes **dark adaptation**—its sensitivity increases slowly over many minutes. The difference in the rates of bleaching and regeneration of the photopigments in the rods and cones accounts for some (but not all) of the sensitivity changes during light and dark adaptation.

Figure 17.15 The cyclical bleaching and regeneration of photopigment. Blue arrows indicate bleaching steps; black arrows indicate regeneration steps.

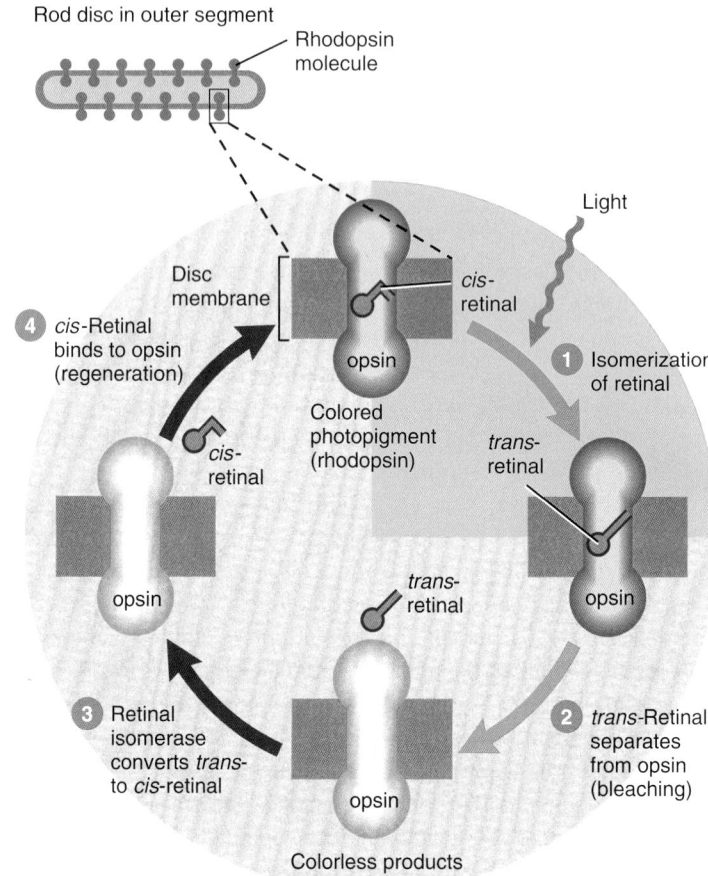

Retinal, a derivative of vitamin A, is the light-absorbing part of all visual photopigments.

Rod disc in outer segment

Rhodopsin molecule

Light

Disc membrane

❹ *cis*-Retinal binds to opsin (regeneration)

cis-retinal

Colored photopigment (rhodopsin)

opsin

cis-retinal

opsin

cis-retinal

❶ Isomerization of retinal

trans-retinal

opsin

trans-retinal

❷ *trans*-Retinal separates from opsin (bleaching)

❸ Retinal isomerase converts *trans*-to *cis*-retinal

trans-retinal

opsin

Colorless products

? **What is the conversion of *cis*-retinal to *trans*-retinal called?**

As the light level increases, more and more photopigment is bleached. While light is bleaching some photopigment molecules, however, others are being regenerated. In daylight, regeneration of rhodopsin cannot keep up with the bleaching process, so rods contribute little to daylight vision. In contrast, cone photopigments regenerate rapidly enough that some of the cis form is always present, even in very bright light.

If the light level decreases abruptly, sensitivity increases rapidly at first and then more slowly. In complete darkness, full regeneration of cone photopigments occurs during the first 8 minutes of dark adaptation. During this time, a threshold (barely perceptible) light flash is seen as having color. Rhodopsin regenerates more slowly, and our visual sensitivity increases until even a single photon (the smallest unit of light) can be detected. In that situation, although much dimmer light can be detected, threshold flashes appear gray-white, regardless of their color. At very low light levels, such as starlight, objects appear as shades of gray because only the rods are functioning.

Figure 17.16 Operation of rod photoreceptors.

 Light causes a hyperpolarizing receptor potential in photoreceptors, which decreases release of an inhibitory neurotransmitter (glutamate).

cGMP-gated
Na$^+$ channels open
↓
Inflow of Na$^+$
(dark current)
↓
Membrane potential
of −30 mV
↓
Glutamate released
at synaptic terminals
inhibits bipolar cell

cGMP

Na$^+$

Glutamate

(a) In darkness

Isomerization of retinal
activates enzyme that
breaks down cGMP
↓
cGMP-gated
Na$^+$ channels close
↓
Inflow of Na$^+$ slows
↓
Hyperpolarizing
receptor potential
↓
Glutamate release
turned off, which excites
bipolar cell

(b) In light

? **What is the function of cyclic GMP in photoreceptors?**

Release of Neurotransmitter by Photoreceptors

As mentioned previously, the absorption of light and isomerization of retinal initiates chemical changes in the photoreceptor outer segment that lead to production of a receptor potential. To understand how the receptor potential arises, however, we first need to

examine the operation of photoreceptors in the absence of light. In darkness, sodium ions (Na$^+$) flow into photoreceptor outer segments through ligand-gated Na$^+$ channels (Figure 17.16a). The ligand that holds these channels open is **cyclic GMP (guanosine monophosphate)** or **cGMP.** The inflow of Na$^+$, called the "dark current," partially depolarizes the photoreceptor. As a result, in darkness the membrane potential of a photoreceptor is about −30 mV. This is much closer to zero than a typical neuron's resting membrane potential of −70 mV. The partial depolarization during darkness triggers continual release of neurotransmitter at the synaptic terminals. The neurotransmitter in rods, and perhaps in cones, is the amino acid glutamate (glutamic acid). At synapses between rods and some bipolar cells, glutamate is an inhibitory neurotransmitter: It triggers inhibitory postsynaptic potentials (IPSPs) that hyperpolarize the bipolar cells and prevent them from sending signals on to the ganglion cells.

When light strikes the retina and *cis*-retinal undergoes isomerization, enzymes are activated that break down cGMP. As a result, some cGMP-gated Na$^+$ channels close, Na$^+$ inflow decreases, and the membrane potential becomes more negative, approaching −70 mV (Figure 17.16b). This sequence of events produces a hyperpolarizing receptor potential that decreases the release of glutamate. Dim lights cause small and brief receptor potentials that partially turn off glutamate release; brighter lights elicit larger and longer receptor potentials that more completely shut down neurotransmitter release. Thus, light excites the bipolar cells that synapse with rods by turning off the release of an inhibitory neurotransmitter. The excited bipolar cells subsequently stimulate the ganglion cells to form action potentials in their axons.

CLINICAL CONNECTION | *Color Blindness and Night Blindness*

Most forms of **color blindness,** an inherited inability to distinguish between certain colors, result from the absence or deficiency of one of the three types of cones. The most common type is *red–green color blindness,* in which red cones or green cones are missing. As a result, the person cannot distinguish between red and green. Prolonged vitamin A deficiency and the resulting below-normal amount of rhodopsin may cause **night blindness** or **nyctalopia** (nik′-ta-LŌ-pē-a), an inability to see well at low light levels. •

The Visual Pathway

Visual signals in the retina undergo considerable processing at synapses among the various types of neurons in the retina (horizontal cells, bipolar cells, and amacrine cells; see Figure 17.10). Then, the axons of retinal ganglion cells provide output from the retina to the brain, exiting the eyeball as the **optic (II) nerve.**

Processing of Visual Input in the Retina

Within the neural layer of the retina, certain features of visual input are enhanced while other features may be discarded. Input from several cells may either converge on a smaller number of postsynaptic neurons or diverge to a large number. Overall, convergence predominates: There are only 1 million ganglion cells, but 126 million photoreceptors in the human eye.

Once receptor potentials arise in the outer segments of rods and cones, they spread through the inner segments to the synaptic terminals. Neurotransmitter molecules released by rods and cones induce local graded potentials in both bipolar cells and horizontal cells. Between 6 and 600 rods synapse with a single bipolar cell in the outer synaptic layer of the retina; a cone more often synapses with a single bipolar cell. The convergence of many rods onto a single bipolar cell increases the light sensitivity of rod vision but slightly blurs the image that is perceived. Cone vision, although less sensitive, is sharper because of the one-to-one synapses between cones and their bipolar cells. Stimulation of rods by light excites bipolar cells; cone bipolar cells may be either excited or inhibited when a light is turned on.

Horizontal cells transmit inhibitory signals to bipolar cells in the areas lateral to excited rods and cones. This lateral inhibition enhances contrasts in the visual scene between areas of the retina that are strongly stimulated and adjacent areas that are more weakly stimulated. Horizontal cells also assist in the differentiation of various colors. Amacrine cells, which are excited by bipolar cells, synapse with ganglion cells and transmit information to them that signals a change in the level of illumination of the retina. When bipolar or amacrine cells transmit excitatory signals to ganglion cells, the ganglion cells become depolarized and initiate nerve impulses.

Brain Pathway and Visual Fields

The axons within the **optic (II) nerve** pass through the **optic chiasm** (kī-AZ-em = a crossover, as in the letter X), a crossing point of the optic nerves (Figure 17.17a, b). Some axons cross to the opposite side, but others remain uncrossed. After passing through the optic chiasm, the axons, now part of the **optic tract,** enter the brain and most of them terminate in the **lateral geniculate nucleus** of the thalamus. Here they synapse with neurons whose axons form the **optic radiations,** which project to the **primary visual areas** in the occipital lobes of the cerebral cortex (area 17 in Figure 14.15), and visual perception begins. Some of the fibers in the optic tracts terminate in the **superior colliculi,** which control the extrinsic eye muscles, and the **pretectal nuclei,** which control pupillary and accommodation reflexes.

Everything that can be seen by one eye is that eye's **visual field.** As noted earlier, because our eyes are located anteriorly in our heads, the visual fields overlap considerably (Figure 17.17b). We have binocular vision due to the large region where the visual fields of the two eyes overlap—the **binocular visual field.** The visual field of each eye is divided into two regions: the **nasal** or **central half** and the **temporal** or **peripheral half.** For each eye, light rays from an object in the nasal half of the visual field fall on the temporal half of the retina, and light rays from an object in the temporal half of the visual field fall on the nasal half of the retina. Visual information from the *right* half of each visual field is conveyed to the *left* side of the brain, and visual information from the *left* half of each visual field is conveyed to the *right* side of the brain, as follows (Figure 17.17c, d):

❶ The axons of all retinal ganglion cells in one eye exit the eyeball at the optic disc and form the optic nerve on that side.

❷ At the optic chiasm, axons from the temporal half of each retina do not cross but continue directly to the lateral geniculate nucleus of the thalamus on the same side.

❸ In contrast, axons from the nasal half of each retina cross the optic chiasm and continue to the opposite thalamus.

❹ Each optic tract consists of crossed and uncrossed axons that project from the optic chiasm to the thalamus on one side.

❺ Axon collaterals (branches) of the retinal ganglion cells project to the midbrain, where they participate in neural circuits that govern constriction of the pupils in response to light and coordination of head and eye movements. Collaterals also extend to the suprachiasmatic nucleus of the hypothalamus, which establishes patterns of sleep and other activities that occur on a circadian or daily schedule in response to intervals of light and darkness.

❻ The axons of thalamic neurons form the optic radiations as they project from the thalamus to the primary visual area of the cortex on the same side.

Although we have just described the visual pathway as a single pathway, visual signals are thought to be processed by at least three separate systems in the cerebral cortex, each with its own function. One system processes information related to the shape of objects, another system processes information regarding color of objects, and a third system processes information about movement, location, and spatial organization.

✔ **CHECKPOINT**

6. What is the function of the lacrimal apparatus?
7. What types of cells make up the neural layer and the pigmented layer of the retina?
8. How do photopigments respond to light and recover in darkness?
9. How do receptor potentials arise in photoreceptors?
10. By what pathway do nerve impulses triggered by an object in the nasal half of the visual field of the left eye reach the primary visual area of the cortex?

17.4 HEARING AND EQUILIBRIUM

◉ **OBJECTIVES**

• Describe the anatomy of the structures in the three main regions of the ear.
• List the major events in the physiology of hearing.
• Identify the receptor organs for equilibrium, and describe how they function.
• Describe the auditory and equilibrium pathways.

Hearing is the ability to perceive sounds. The ear is an engineering marvel because its sensory receptors can transduce sound vibrations with amplitudes as small as the diameter of an atom of gold (0.3 nm) into electrical signals 1000 times faster than photoreceptors can respond to light. The ear also contains receptors

Figure 17.17 The visual pathway. (a) Partial dissection of the brain reveals the optic radiations (axons extending from the thalamus to the occipital lobe). (b) An object in the binocular visual field can be seen with both eyes. In (c) and (d), note that information from the right side of the visual field of each eye projects to the left side of the brain, and information from the left side of the visual field of each eye projects to the right side of the brain.

🔑 The axons of ganglion cells in the temporal half of each retina extend to the thalamus on the same side; the axons of ganglion cells in the nasal half of each retina extend to the thalamus on the opposite side.

View

ANTERIOR

Visual pathway:

Optic (II) nerve

↓

Optic chiasm

↓

Optic tract

↓

Lateral geniculate nucleus of thalamus

↓

Optic radiation

↓

Primary visual area of cerebral cortex (area 17) in occipital lobe

POSTERIOR

(a) Inferior view

View

Transverse plane

Visual field of left eye

Visual field of right eye

Binocular visual field

Optic (II) nerves

Optic chiasm

Midbrain

Optic tracts

Lateral geniculate nucleus of thalamus

Optic radiation

Primary visual area of cerebral cortex (area 17) of occipital lobe

(b) Superior view of transverse section through eyeballs and brain

Visual field of left eye

Temporal half

Nasal half

Left eye

Temporal retina

Nasal retina

1

2

3

4 Optic tract

5

Midbrain

Optic radiations

6

Lateral geniculate nucleus of the thalamus

Primary visual area of cerebral cortex (area 17) in occipital lobe

(c) Left eye and its pathways

Visual field of right eye

Nasal half

Temporal half

Nasal retina

Right eye

Temporal retina

1

2

3

4

5

Midbrain

Optic radiations

6

(d) Right eye and its pathways

❓ **Light rays from an object in the temporal half of the visual field strike which half of the retina?**

for equilibrium, the sense that helps you maintain your balance and be aware of your orientation in space.

Anatomy of the Ear

The ear is divided into three main regions: (1) the external ear, which collects sound waves and channels them inward; (2) the middle ear, which conveys sound vibrations to the oval window; and (3) the internal ear, which houses the receptors for hearing and equilibrium.

External (Outer) Ear

The **external (outer) ear** consists of the auricle, external auditory canal, and eardrum (Figure 17.18). The **auricle** (AW-ri-kul) or **pinna** is a flap of elastic cartilage shaped like the flared end of a trumpet and covered by skin. The rim of the auricle is the **helix;** the inferior portion is the **lobule.** Ligaments and muscles attach the auricle to the head. The **external auditory canal** (*audit-* = hearing) is a curved tube about 2.5 cm (1 in.) long that lies in the temporal bone and leads to the eardrum. The **tympanic mem-brane** (tim-PAN-ik; *tympan-* = a drum) or **eardrum** is a thin, semitransparent partition between the external auditory canal and middle ear. The tympanic membrane is covered by epidermis and lined by simple cuboidal epithelium. Between the epithelial layers is connective tissue composed of collagen, elastic fibers, and fibroblasts. Tearing of the tympanic membrane is called a **perforated eardrum.** It may be due to pressure from a cotton swab, trauma, or a middle ear infection, and usually heals within a month. The tympanic membrane may be examined directly by an **otoscope** (Ō-tō-skōp; *oto-* = ear; *-skopeo* = to view), a viewing instrument that illuminates and magnifies the external auditory canal and tympanic membrane.

Near the exterior opening, the external auditory canal contains a few hairs and specialized sweat glands called **ceruminous glands** (se-RŪ-mi-nus) that secrete **earwax** or **cerumen** (se-RŪ-men). The combination of hairs and cerumen helps prevent dust and foreign objects from entering the ear. Cerumen also prevents damage to the delicate skin of the external ear canal by water and insects. Cerumen usually dries up and falls out of the ear canal. However, some people produce a large amount of cerumen, which can become

Figure 17.18 Anatomy of the ear.

The ear has three principal regions: the external (outer) ear, the middle ear, and the internal (inner) ear.

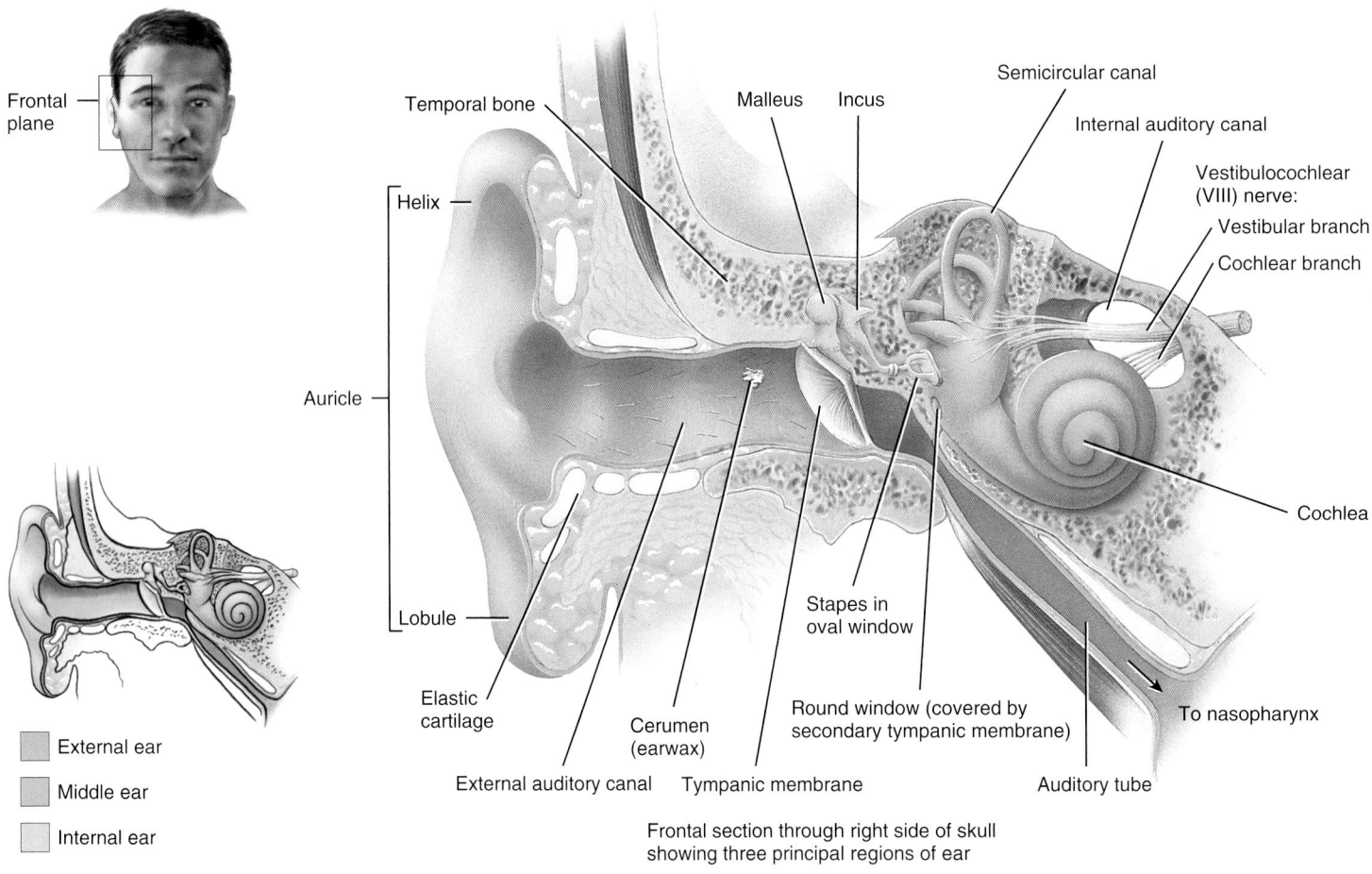

Frontal section through right side of skull showing three principal regions of ear

External ear

Middle ear

Internal ear

? **To which structure of the external ear does the malleus of the middle ear attach?**

impacted and can muffle incoming sounds. The treatment for **impacted cerumen** is usually periodic ear irrigation or removal of wax with a blunt instrument by trained medical personnel.

Middle Ear

The **middle ear** is a small, air-filled cavity in the petrous portion of the temporal bone that is lined by epithelium (Figure 17.19). It is separated from the external ear by the tympanic membrane and from the internal ear by a thin bony partition that contains two small membrane-covered openings: the oval window and the round window. Extending across the middle ear and attached to it by ligaments are the three smallest bones in the body, the **auditory ossicles** (OS-si-kuls), which are connected by synovial joints. The bones, named for their shapes, are the malleus, incus, and stapes—commonly called the hammer, anvil, and stirrup, respectively. The "handle" of the **malleus** (MAL-ē-us) attaches to the internal surface of the tympanic membrane. The head of the malleus articulates with the body of the incus. The **incus** (ING-kus), the middle bone in the series, articulates with the head of the stapes. The base or footplate of the **stapes** (STĀ-pēz) fits into the

oval window. Directly below the oval window is another opening, the **round window,** which is enclosed by a membrane called the **secondary tympanic membrane.**

Besides the ligaments, two tiny skeletal muscles also attach to the ossicles (Figure 17.19). The **tensor tympani** (TIM-pan-ē) **muscle,** which is supplied by the mandibular branch of the trigeminal (V) nerve, limits movement and increases tension on the eardrum to prevent damage to the inner ear from loud noises. The **stapedius** (sta-PĒ-de-us) **muscle,** which is supplied by the facial (VII) nerve, is the smallest skeletal muscle in the human body. By dampening large vibrations of the stapes due to loud noises, it protects the oval window, but it also decreases the sensitivity of hearing. For this reason, paralysis of the stapedius muscle is associated with **hyperacusia** (hī′-per-a-KŪ-sē-a), which is abnormally sensitive hearing. Because it takes a fraction of a second for the tensor tympani and stapedius muscles to contract, they can protect the inner ear from prolonged loud noises but not from brief ones such as a gunshot.

The anterior wall of the middle ear contains an opening that leads directly into the **auditory (pharyngotympanic) tube,**

Figure 17.19 **The right middle ear and the auditory ossicles.**

Common names for the malleus, incus, and stapes are the hammer, anvil, and stirrup, respectively.

Auditory ossicles

Auditory tube

- External ear
- Middle ear
- Internal ear

Superior ligament of malleus

Malleus

Incus

Posterior ligament of incus

Stapes in oval window (fenestra vestibuli)

Facial (VII) nerve

Lateral ligament of malleus

Anterior ligament of malleus (cut)

Round window (fenestra cochlea)

Tensor tympani muscle

LATERAL

MEDIAL

Tympanic membrane

External auditory canal

Stapedius muscle

Middle ear

Auditory tube

Frontal section showing location of auditory ossicles in the middle ear

? **What structures separate the middle ear from the internal ear?**

commonly known as the **eustachian tube** (ū′-STĀ-kē-an, ū-STĀ-shun). The auditory tube, which consists of both bone and elastic cartilage, connects the middle ear with the nasopharynx (superior portion of the throat). It is normally closed at its medial (pharyngeal) end. During swallowing and yawning, it opens, allowing air to enter or leave the middle ear until the pressure in the middle ear equals the atmospheric pressure. Most of us have experienced our ears popping as the pressures equalize. When the pressures are balanced, the tympanic membrane vibrates freely as sound waves strike it. If the pressure is not equalized, intense pain, hearing impairment, ringing in the ears, and vertigo could develop. The auditory tube also is a route for pathogens to travel from the nose and throat to the middle ear, causing the most common type of ear infection (see *otitis media* in Disorders: Homeostatic Imbalances at the end of this chapter).

Internal (Inner) Ear

The **internal (inner) ear** is also called the **labyrinth** (LAB-i-rinth) because of its complicated series of canals (Figure 17.20). Structurally, it consists of two main divisions: an outer bony labyrinth that encloses an inner membranous labyrinth. It is like long balloons put inside a rigid tube. The **bony labyrinth** is a series of cavities in the petrous portion of the temporal bone divided into three areas: (1) the semicircular canals, (2) the vestibule, and (3) the cochlea. The bony labyrinth is lined with periosteum and contains **perilymph.** This fluid, which is chemically similar to cerebrospinal fluid, surrounds the **membranous labyrinth,** a series of epithelial sacs and tubes inside the bony labyrinth that have the same general form as the bony labyrinth and house the receptors for hearing and equilibrium. The epithelial membranous labyrinth contains **endolymph.** The level of potassium ions (K^+) in endolymph is unusually high for an extracellular fluid, and potassium ions play a role in the generation of auditory signals (described shortly).

The **vestibule** (VES-ti-būl) is the oval central portion of the bony labyrinth. The membranous labyrinth in the vestibule consists of two sacs called the **utricle** (Ū-tri-kul = little bag) and the **saccule** (SAK-ūl = little sac), which are connected by a small duct. Projecting superiorly and posteriorly from the vestibule are the three bony **semicircular canals,** each of which lies at approximately right angles to the other two. Based on their positions,

Figure 17.20 The right internal ear. The outer, cream-colored area is part of the bony labyrinth; the inner, pink-colored area is the membranous labyrinth.

The bony labyrinth contains perilymph, and the membranous labyrinth contains endolymph.

- External ear
- Middle ear
- Internal ear

Semicircular canals (contain semicircular ducts):
- Anterior
- Posterior
- Lateral

LATERAL

Ampulla of semicircular duct

Stapes in oval window

Round window

Bony labyrinth (contains perilymph)
Membranous labyrinth (contains endolymph)
Ampulla of semicircular canal
Utricle
Vestibule
Oval window
Saccule
Cochlea
Cochlear duct

MEDIAL

Components of right internal ear

? **What are the names of the two sacs that lie in the membranous labyrinth of the vestibule?**

they are named the anterior, posterior, and lateral semicircular canals. The anterior and posterior semicircular canals are vertically oriented; the lateral one is horizontally oriented. At one end of each canal is a swollen enlargement called the **ampulla** (am-PUL-la = saclike duct). The portions of the membranous labyrinth that lie inside the bony semicircular canals are called the **semicircular ducts.** These structures connect with the utricle of the vestibule.

The vestibular branch of the vestibulocochlear (VIII) nerve consists of *ampullary, utricular,* and *saccular nerves.* These nerves contain both first-order sensory neurons and motor neurons that synapse with receptors for equilibrium. The first-order sensory neurons carry sensory information from the receptors, and the motor neurons carry feedback signals to the receptors, apparently to modify their sensitivity. Cell bodies of the sensory neurons are located in the **vestibular ganglia** (ves-TIB-ū-lar) (see Figure 17.21b).

Anterior to the vestibule is the **cochlea** (KOK-lē-a = snail-shaped), a bony spiral canal (Figure 17.21a) that resembles a snail's shell and makes almost three turns around a central bony core called the **modiolus** (mō-DĪ-ō′-lus; Figure 17.21b). Sections through the cochlea reveal that it is divided into three channels: cochlear duct, scala vestibuli, and scala tympani (Figure 17.21a–c). The **cochlear duct (scala media)** (KOK-lē-ar) is a continuation of the membranous labyrinth into the cochlea; it is filled with endolymph. The channel above the cochlear duct is the **scala vestibuli,** which ends at the oval window. The channel below is the **scala tympani,** which ends at the round window. Both the scala vestibuli and scala tympani are part of the bony labyrinth of the cochlea; therefore, these chambers are filled with perilymph. The scala vestibuli and scala tympani are completely separated by the cochlear duct, except for an opening at the apex of the cochlea, the **helicotrema** (hel-i-kō-TRĒ-ma; Figure 17.22). The cochlea adjoins the wall of the vestibule, into which the scala vestibuli opens. The perilymph in the vestibule is continuous with that of the scala vestibuli.

The **vestibular membrane** separates the cochlear duct from the scala vestibuli, and the **basilar membrane** (BĀS-i-lar) separates the cochlear duct from the scala tympani. Resting on the basilar membrane is the **spiral organ** or **organ of Corti** (KOR-tē) (Figure 17.21c, d). The spiral organ is a coiled sheet of epithelial cells, including supporting cells and about 16,000 **hair cells,** which are the receptors for hearing. There are two groups of hair cells: The *inner hair cells* are arranged in a single row, whereas the *outer hair cells* are arranged in three rows. At the apical tip of each hair cell are 40–80 stereocilia that extend into the endolymph of the cochlear duct. Despite their name, stereocilia are actually long, hairlike microvilli arranged in several rows of graded height.

At their basal ends, inner and outer hair cells synapse both with first-order sensory neurons and with motor neurons from the cochlear branch of the vestibulocochlear (VIII) nerve. Cell bodies of the sensory neurons are located in the **spiral ganglion** (Figure 17.21b, c). Although outer hair cells outnumber them by 3 to 1, the inner hair cells synapse with 90–95% of the first-order sensory neurons in the cochlear nerve that relay auditory information to the brain. By contrast, 90% of the motor neurons in the cochlear nerve synapse with outer hair cells. The **tectorial membrane** (tek-TŌ-rē-al; *tector-* = covering), a flexible gelatinous membrane, covers the hair cells of the spiral organ (Figure 17.21d). In fact, the ends of the stereocilia of the hair cells are embedded in the tectorial membrane while the bodies of the hair cells rest on the basilar membrane.

The Nature of Sound Waves

In order to understand the physiology of hearing, it is necessary to learn something about its input, which occurs in the form of sound waves. **Sound waves** are alternating high- and low-pressure regions traveling in the same direction through some medium (such as air). They originate from a vibrating object in much the same way that ripples arise and travel over the surface of a pond when you toss a stone into it. The *frequency* of a sound vibration is its *pitch.* The higher the frequency of vibration, the higher is the pitch. The sounds heard most acutely by the human ear are those from sources that vibrate at frequencies between 500 and 5000 hertz (Hz; 1 Hz = 1 cycle per second). The entire audible range extends from 20 to 20,000 Hz. Sounds of speech primarily contain frequencies between 100 and 3000 Hz, and the "high C" sung by a soprano has a dominant frequency at 1048 Hz. The sounds from a jet plane several miles away range from 20 to 100 Hz.

The larger the *intensity* (size or amplitude) of the vibration, the *louder* is the sound. Sound intensity is measured in units called **decibels (dB).** An increase of one decibel represents a tenfold increase in sound intensity. The hearing threshold—the point at which an average young adult can just distinguish sound from silence—is defined as 0 dB at 1000 Hz. Rustling leaves have a decibel level of 15; whispered speech, 30; normal conversation, 60; a vacuum cleaner, 75; shouting, 80; and a nearby motorcycle or jackhammer, 90. Sound becomes uncomfortable to a normal ear at about 120 dB, and painful above 140 dB.

CLINICAL CONNECTION | Loud Sounds and Hair Cell Damage

Exposure to loud music and the engine roar of jet planes, revved-up motorcycles, lawn mowers, and vacuum cleaners damages hair cells of the cochlea. Because prolonged noise exposure causes hearing loss, employers in the United States must require workers to use hearing protectors when occupational noise levels exceed 90 dB. Rock concerts and even inexpensive headphones can easily produce sounds over 110 dB. Continued exposure to high-intensity sounds is one cause of **deafness,** a significant or total hearing loss. The louder the sounds, the more rapid is the hearing loss. Deafness usually begins with loss of sensitivity for high-pitched sounds. If you are listening to music through ear buds and bystanders can hear it, the decibel level is in the damaging range. Most people fail to notice their progressive hearing loss until destruction is extensive and they begin having difficulty understanding speech. Wearing earplugs with a noise-reduction rating of 30 dB while engaging in noisy activities can protect the sensitivity of your ears. •

Figure 17.21 Semicircular canals, vestibule, and cochlea of the right ear. Note that the cochlea makes nearly three complete turns.

The three channels in the cochlea are the scala vestibuli, the scala tympani, and the cochlear duct.

(a) Sections through cochlea

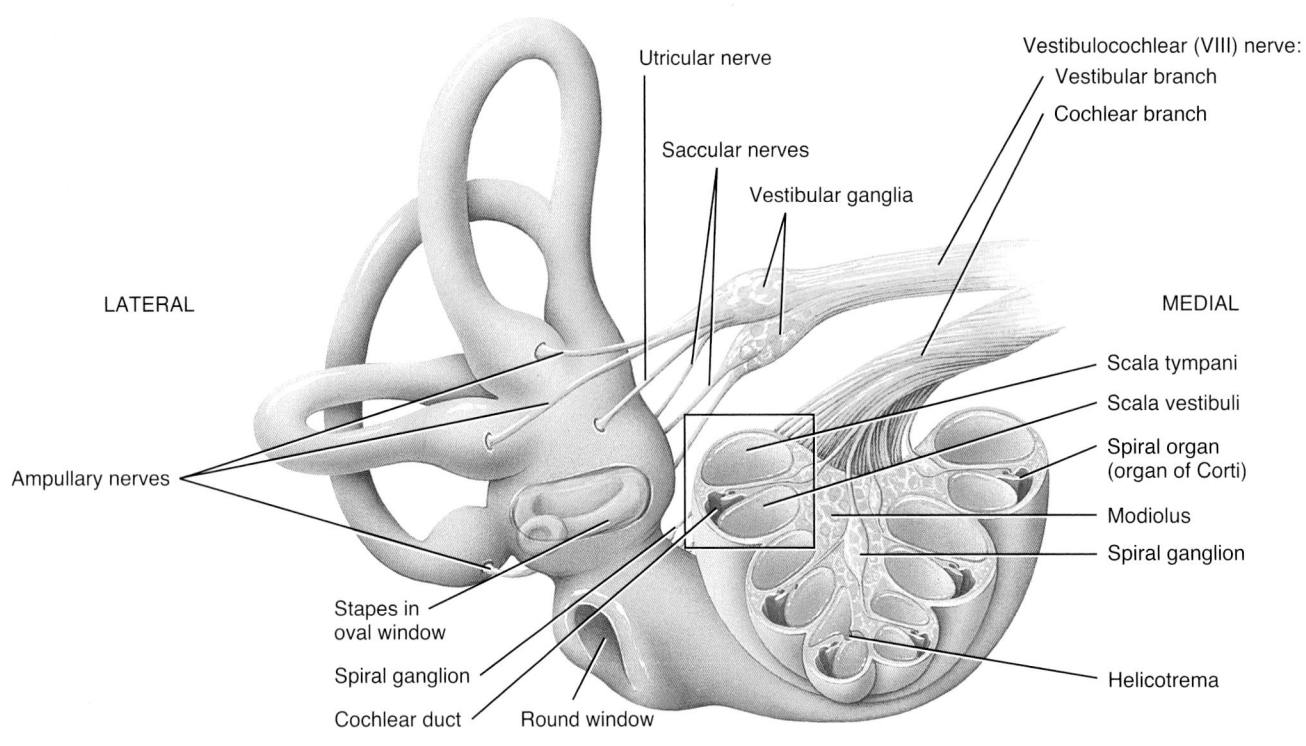

(b) Components of vestibulocochlear nerve (cranial nerve VIII)

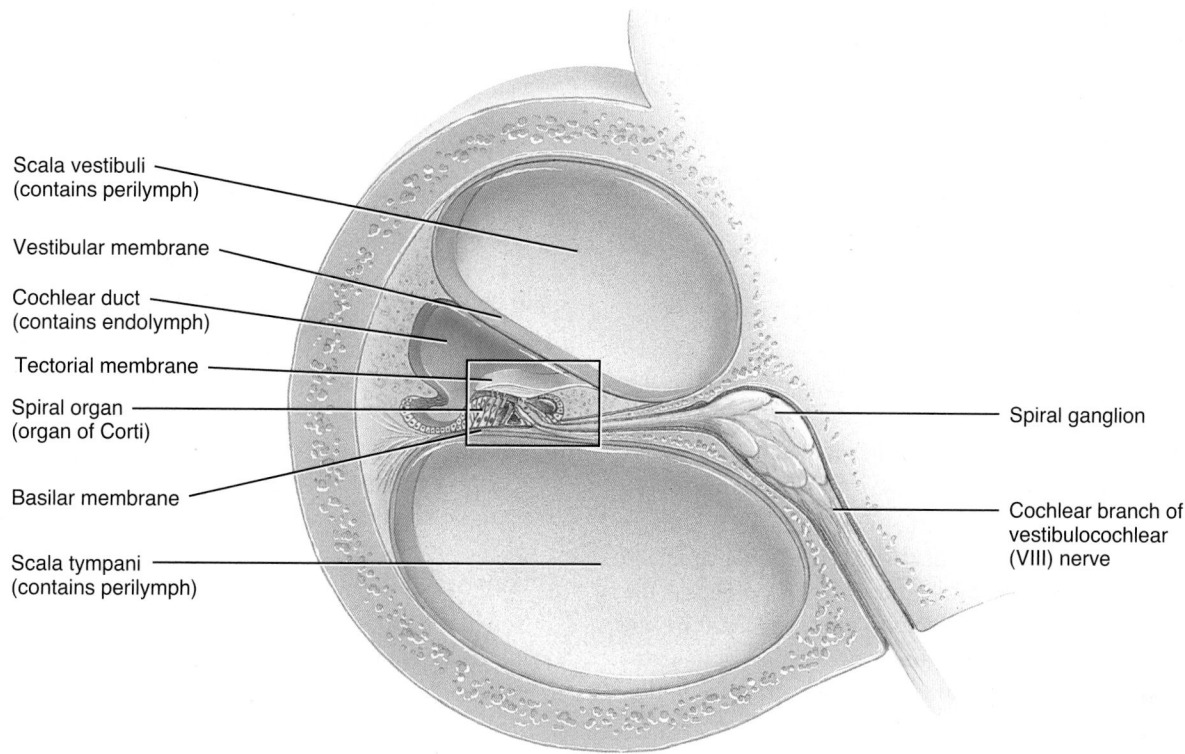

Scala vestibuli
(contains perilymph)

Vestibular membrane

Cochlear duct
(contains endolymph)

Tectorial membrane

Spiral organ
(organ of Corti)

Basilar membrane

Scala tympani
(contains perilymph)

Spiral ganglion

Cochlear branch of
vestibulocochlear
(VIII) nerve

(c) Section through one turn of cochlea

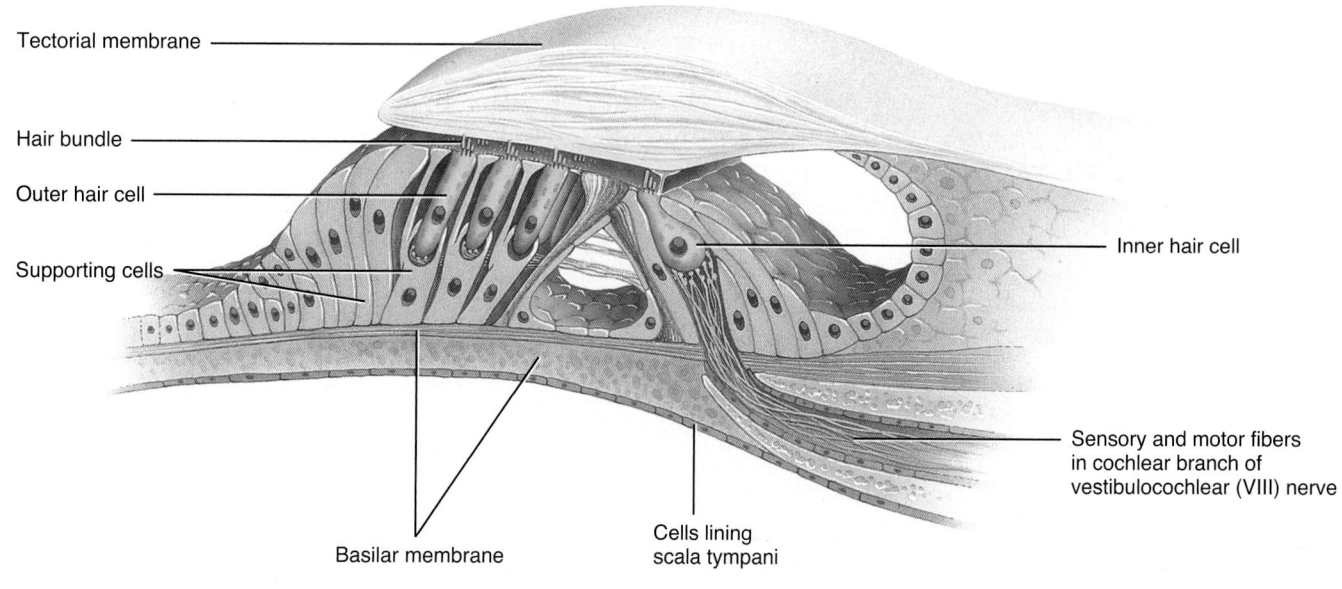

Tectorial membrane

Hair bundle

Outer hair cell

Supporting cells

Inner hair cell

Sensory and motor fibers
in cochlear branch of
vestibulocochlear (VIII) nerve

Basilar membrane

Cells lining
scala tympani

(d) Enlargement of spiral organ (organ of Corti)

? **What are the three subdivisions of the bony labyrinth?**

Physiology of Hearing

The following events are involved in hearing (Figure 17.22):

1 The auricle directs sound waves into the external auditory canal.

2 When sound waves strike the tympanic membrane, the alternating waves of high and low pressure in the air cause the tympanic membrane to vibrate back and forth. The tympanic membrane vibrates slowly in response to low-frequency (low-pitched) sounds and rapidly in response to high-frequency (high-pitched) sounds.

3 The central area of the tympanic membrane connects to the malleus, which vibrates along with the tympanic membrane. This vibration is transmitted from the malleus to the incus and then to the stapes.

4 As the stapes moves back and forth, its oval-shaped footplate, which is attached via a ligament to the circumference of the oval window, vibrates in the oval window. The vibrations at the oval window are about 20 times more vigorous than the tympanic membrane because the auditory ossicles efficiently transmit small vibrations spread over a large surface area (the tympanic membrane) into larger vibrations at a smaller surface (the oval window).

5 The movement of the stapes at the oval window sets up fluid pressure waves in the perilymph of the cochlea. As the oval window bulges inward, it pushes on the perilymph of the scala vestibuli.

6 Pressure waves are transmitted from the scala vestibuli to the scala tympani and eventually to the round window, causing it to bulge outward into the middle ear. (See **9** in the figure.)

7 The pressure waves travel through the perilymph of the scala vestibuli, then the vestibular membrane, and then move into the endolymph inside the cochlear duct.

8 The pressure waves in the endolymph cause the basilar membrane to vibrate, which moves the hair cells of the spiral organ against the tectorial membrane. This leads to bending of the stereocilia and ultimately to the generation of nerve impulses in first-order neurons in cochlear nerve fibers.

9 Sound waves of various frequencies cause certain regions of the basilar membrane to vibrate more intensely than other regions. Each segment of the basilar membrane is "tuned" for

Figure 17.22 Events in the stimulation of auditory receptors in the right ear. The numbers correspond to the events listed in the text. The cochlea has been uncoiled to more easily visualize the transmission of sound waves and their distortion of the vestibular and basilar membranes of the cochlear duct.

Hair cells of the spiral organ (organ of Corti) convert a mechanical vibration (stimulus) into an electrical signal (receptor potential).

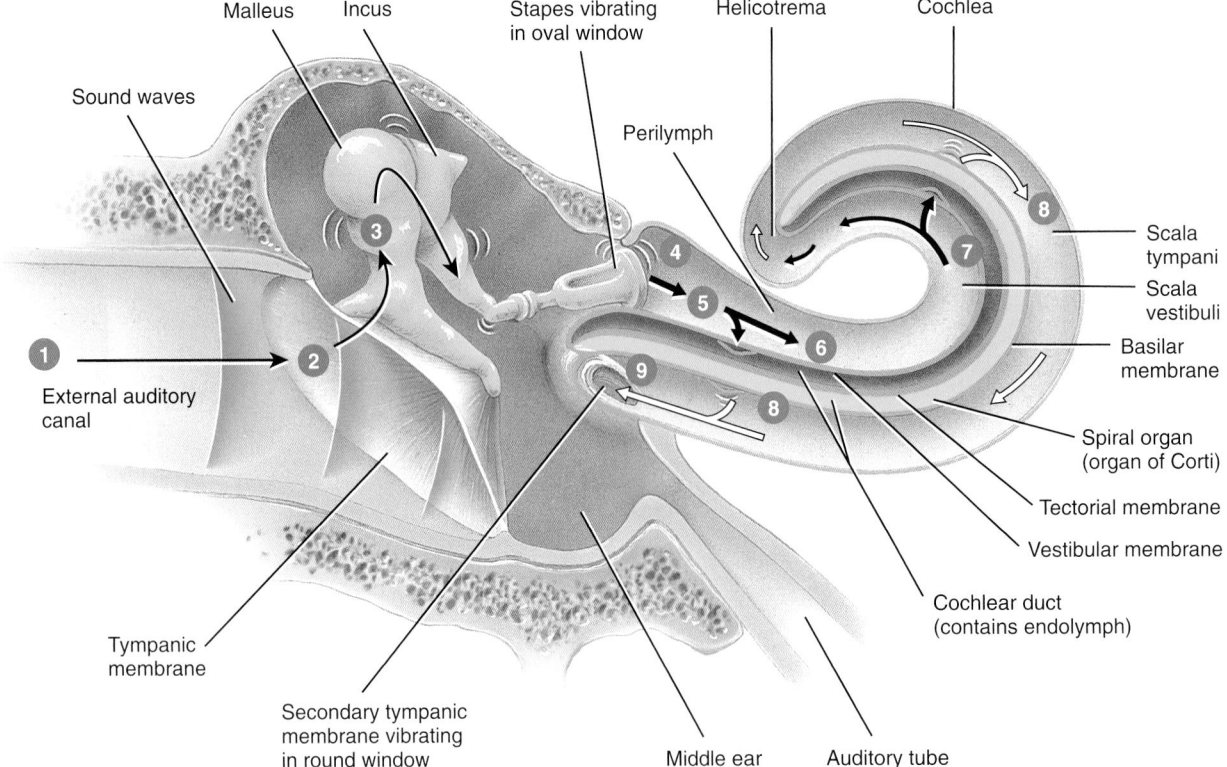

? **Which part of the basilar membrane vibrates most vigorously in response to high-frequency (high-pitched) sounds?**

a particular pitch. Because the membrane is narrower and stiffer at the base of the cochlea (closer to the oval window), high-frequency (high-pitched) sounds induce maximal vibrations in this region. Toward the apex of the cochlea, the basilar membrane is wider and more flexible; low-frequency (low-pitched) sounds cause maximal vibration of the basilar membrane there. Loudness is determined by the intensity of sound waves. High-intensity sound waves cause larger vibrations of the basilar membrane, which leads to a higher frequency of nerve impulses reaching the brain. Louder sounds also may stimulate a larger number of hair cells.

The hair cells transduce mechanical vibrations into electrical signals. As the basilar membrane vibrates, the hair bundles at the apex of the hair cell bend back and forth and slide against one another. A *tip link* protein connects the tip of each stereocilium to a mechanically gated ion channel called the **transduction channel** in its taller stereocilium neighbor. As the stereocilia bend in the direction of the taller stereocilia, the tip links tug on the transduction channels and open them. These channels allow cations in the endolymph, primarily K^+, to enter the hair cell cytosol. As cations enter, they produce a depolarizing receptor potential. Depolarization quickly spreads along the plasma membrane and opens voltage-gated Ca^{2+} channels in the base of the hair cell. The resulting inflow of Ca^{2+} triggers exocytosis of synaptic vesicles containing a neurotransmitter, which is probably glutamate. As more neurotransmitter is released, the frequency of nerve impulses in the first-order sensory neurons that synapse with the base of the hair cell increases. Bending of the stereocilia in the opposite direction closes the transduction channels, allows hyperpolarization to occur, and reduces neurotransmitter release from the hair cells. This decreases the frequency of nerve impulses in the sensory neurons.

Besides its role in detecting sounds, the cochlea has the surprising ability to produce sounds. These usually inaudible sounds, called **otoacoustic emissions** (ō-tō-a-KOO-stik), can be picked up by placing a sensitive microphone next to the eardrum. They are caused by vibrations of the outer hair cells that occur in response to sound waves and to signals from motor neurons. As they depolarize and repolarize, the outer hair cells rapidly shorten and lengthen. This vibratory behavior appears to change the stiffness of the tectorial membrane and is thought to enhance the movement of the basilar membrane, which amplifies the responses of the inner hair cells. At the same time, the outer hair cell vibrations set up a traveling wave that goes back toward the stapes and leaves the ear as an otoacoustic emission. Detection of these inner ear–produced sounds is a fast, inexpensive, and noninvasive way to screen newborns for hearing defects. In deaf babies, otoacoustic emissions are not produced or are greatly reduced in size.

The Auditory Pathway

Bending of the stereocilia of the hair cells of the spiral organ causes the release of a neurotransmitter (probably glutamate), which generates nerve impulses in the sensory neurons that innervate the hair cells. The cell bodies of the sensory neurons are located in the **spiral ganglia.** Nerve impulses pass along the axons of these neurons, which form the cochlear branch of the vestibulocochlear (VIII) nerve (Figure 17.23). These axons synapse with neurons in the **cochlear nuclei** in the medulla oblongata on the same side. Some of the axons from the cochlear nuclei decussate (cross over) in the medulla, ascend in a tract called the **lateral meniscus** on the opposite side, and terminate in the **inferior colliculus** in the midbrain. Other axons from the cochlear nuclei end in the **superior olivary nucleus** in the pons on each side. Slight differences in the timing of nerve impulses arriving from the two ears at the superior olivary nuclei allow us to locate the source of a sound. Axons from the superior olivary nuclei also ascend in the lateral meniscus tracts on both sides and end in the inferior colliculi. From each inferior colliculus, nerve impulses are conveyed to the **medial geniculate nucleus** in the thalamus and finally to the **primary auditory area** of the cerebral cortex in the temporal lobe of the cerebrum (see areas 41 and 42 in Figure 14.15). Because many auditory axons decussate in the medulla while others remain on the same side, the right and left primary auditory areas receive nerve impulses from both ears.

CLINICAL CONNECTION | *Cochlear Implants*

A **cochlear implant** is a device that translates sounds into electrical signals that can be interpreted by the brain. Such a device is useful for people with deafness that is caused by damage to hair cells in the cochlea. The external parts of a cochlear implant consist of (1) a *microphone* worn around the ear that picks up sound waves, (2) a *sound processor,* which may be placed in a shirt pocket, that converts sound waves into electrical signals, and (3) a *transmitter,* worn behind the ear, which receives signals from the sound processor and passes them to an internal receiver. The internal parts of a cochlear implant are the (1) *internal receiver,* which relays signals to (2) *electrodes* implanted in the cochlea, where they trigger nerve impulses in sensory neurons in the cochlear branch of the vestibulocochlear (VIII) nerve. These artificially induced nerve impulses propagate over their normal pathways to the brain. The perceived sounds are crude compared to normal hearing, but they provide a sense of rhythm and loudness; information about certain noises, such as those made by telephones and automobiles; and the pitch and cadence of speech. Some patients hear well enough with a cochlear implant to use the telephone. •

Physiology of Equilibrium

There are two types of **equilibrium** (ē-kwi-LIB-rē-um) or balance. **Static equilibrium** refers to the maintenance of the position of the body (mainly the head) relative to the force of gravity. Body movements that stimulate the receptors for static equilibrium include tilting the head and *linear* acceleration or deceleration, such as when the body is being moved in an elevator or in a car that is speeding up or slowing down. **Dynamic equilibrium** is the maintenance of body position (mainly the head) in response to sudden movements such as *rotational* acceleration or deceleration.

Figure 17.23 **The auditory pathway.**

From hair cells of the cochlea, auditory information is conveyed along the cochlear branch of the vestibulocochlear (VIII) nerve and then to the brain stem, thalamus, and cerebral cortex.

Primary auditory area in cerebral cortex

Medial geniculate nucleus in thalamus

Cochlear branch of vestibulocochlear (VIII) nerve

Inferior colliculus in midbrain

Lateral menisci

Superior olivary nucleus in pons

Cerebellum

Cochlear nuclei in medulla oblongata

? **What is the function of the superior olivary nucleus of the pons?**

Collectively, the receptor organs for equilibrium are called the **vestibular apparatus** (ves-TIB-ū-lar); these include the saccule, utricle, and semicircular ducts.

Otolithic Organs: Saccule and Utricle

The walls of both the utricle and the saccule contain a small, thickened region called a **macula** (MAK-ū-la; Figure 17.24). The two *maculae* (plural) (MAK-u-lē), which are perpendicular to one another, are the receptors for static equilibrium. They provide sensory information on the position of the head in space and are essential for maintaining appropriate posture and balance. The maculae also detect linear acceleration and deceleration—for example, the sensations you feel while in an elevator or car that is speeding up or slowing down.

The maculae consist of two kinds of cells: **hair cells,** which are the sensory receptors, and **supporting cells.** Hair cells have on their surface 40–80 stereocilia (which are actually microvilli) of graduated height, plus one *kinocilium,* a conventional cilium anchored firmly to its basal body and extending beyond the longest stereocilium. As in the cochlea, the stereocilia are connected by

tip links. Collectively, the stereocilia and kinocilium are called a **hair bundle.** Scattered among the hair cells are columnar supporting cells that probably secrete the thick, gelatinous, glycoprotein layer, called the **otolithic membrane** (ō-tō-LITH-ik), that rests on the hair cells. A layer of dense calcium carbonate crystals, called **otoliths** (Ō-tō-liths; *oto-* = ear; *-liths* = stones), extends over the entire surface of the otolithic membrane.

Because the otolithic membrane sits on top of the macula, if you tilt your head forward, the otolithic membrane (along with the otoliths as well) is pulled by gravity. It slides "downhill" over the hair cells in the direction of the tilt, bending the hair bundles. However, if you are sitting upright in a car that suddenly jerks forward, the otolithic membrane lags behind the head movement, pulls on the hair bundles, and makes them bend in the other direction. Bending of the hair bundles in one direction stretches the tip links, which pull open transduction channels, producing depolarizing receptor potentials; bending in the opposite direction closes the transduction channels and produces hyperpolarization.

As the hair cells depolarize and repolarize, they release neurotransmitter at a faster or slower rate. The hair cells synapse with

Figure 17.24 Location and structure of receptors in the maculae of the right ear. Both first-order sensory neurons (blue) and motor neurons (red) synapse with the hair cells.

The movement of stereocilia initiates depolarizing receptor potentials.

(a) Overall structure of section of macula

(c) Position of macula with head upright (left) and tilted forward (right)

(b) Details of two hair cells

? **With which type of equilibrium are the maculae associated?**

Figure 17.25 Location and structure of the semicircular ducts of the right ear. Both first-order sensory neurons (blue) and motor neurons (red) synapse with the hair cells. The ampullary nerves are branches of the vestibular division of the vestibulocochlear (VIII) nerve.

🔑 The positions of the semicircular ducts permit detection of rotational movements.

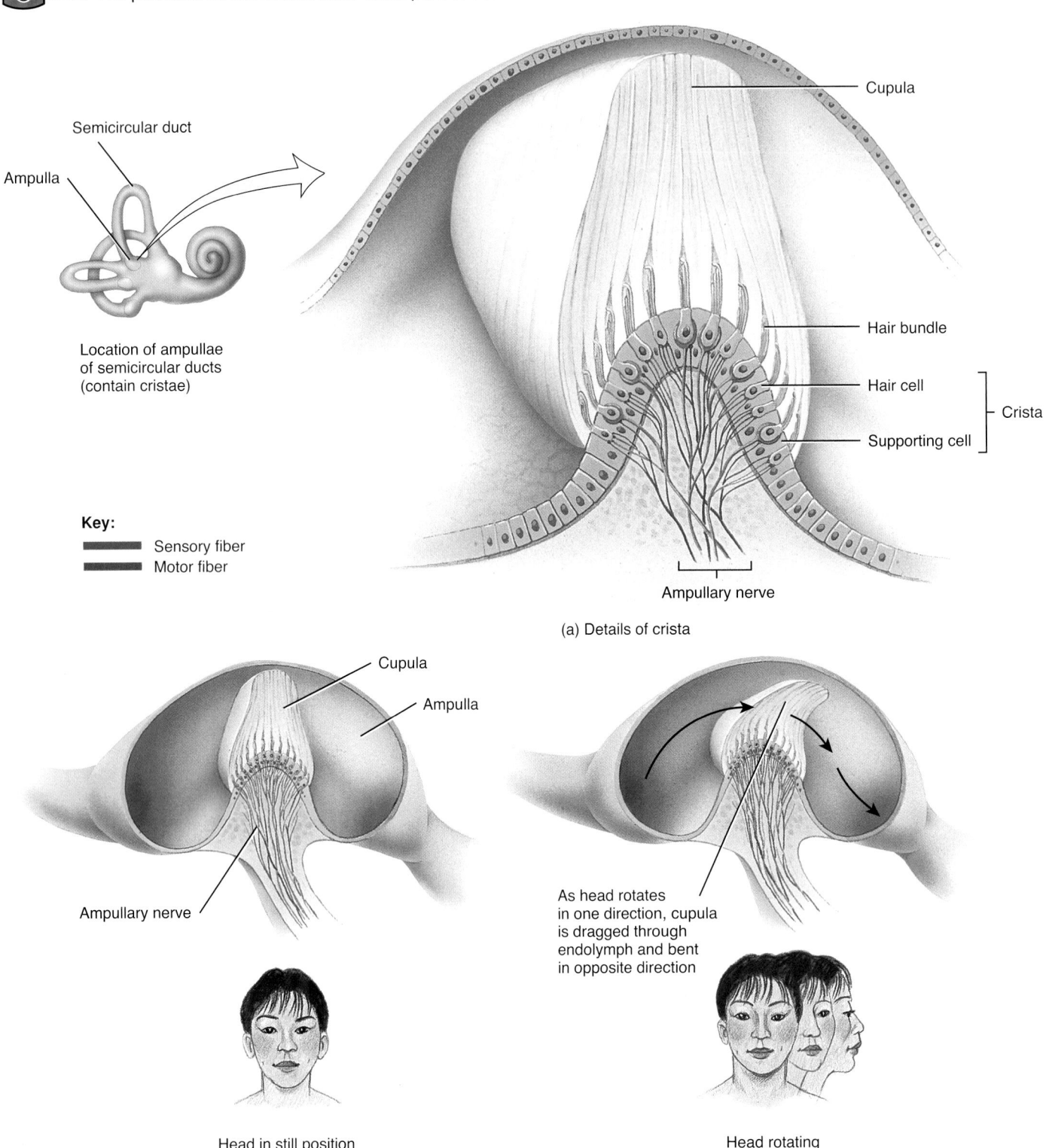

Semicircular duct

Ampulla

Location of ampullae of semicircular ducts (contain cristae)

Cupula

Hair bundle

Hair cell

Supporting cell

Crista

Key:
Sensory fiber
Motor fiber

Ampullary nerve

(a) Details of crista

Cupula

Ampulla

Ampullary nerve

As head rotates in one direction, cupula is dragged through endolymph and bent in opposite direction

Head in still position

Head rotating

(b) Position of cupula with head in still position (left) and when head rotates (right)

❓ **With which type of equilibrium are the semicircular ducts associated?**

first-order sensory neurons in the vestibular branch of the vestibulocochlear (VIII) nerve (see Figure 17.21b). These neurons fire impulses at a slow or rapid pace depending on the amount of neurotransmitter present. Motor neurons also synapse with the hair cells and sensory neurons. Evidently, the motor neurons regulate the sensitivity of the hair cells and sensory neurons.

Semicircular Ducts

The three semicircular ducts function in dynamic equilibrium. The ducts lie at right angles to one another in three planes (Figure 17.25): The two vertical ducts are the anterior and posterior semicircular ducts, and the horizontal one is the lateral semicircular duct (see also Figure 17.20). This positioning permits detection of rotational acceleration or deceleration. In the **ampulla,** the dilated portion of each duct, is a small elevation called the **crista** (KRIS-ta = crest; plural is *cristae*). Each crista contains a group of hair cells and supporting cells. Covering the crista is a mass of gelatinous material called the **cupula** (KU-pū-la). When you move your head, the attached semicircular ducts and hair cells move with it. The endolymph within the ampulla, however, is not

attached and lags behind. As the moving hair cells drag along the stationary endolymph, the hair bundles bend. Bending of the hair bundles produces receptor potentials. In turn, the receptor potentials lead to nerve impulses that pass along the vestibular branch of the vestibulocochlear (VIII) nerve.

Equilibrium Pathways

Bending of **hair bundles of the hair cells** in the semicircular ducts, utricle, or saccule causes the release of a neurotransmitter (probably glutamate), which generates nerve impulses in the sensory neurons that innervate the hair cells. The cell bodies of sensory neurons are located in the **vestibular ganglia.** Nerve impulses pass along the axons of these neurons, which form the **vestibular branch of the vestibulocochlear (VIII) nerve** (Figure 17.26). Most of these axons synapse with sensory neurons in **vestibular nuclei,** the major integrating centers for equilibrium, in the medulla oblongata and pons. The vestibular nuclei also receive input from the eyes and proprioceptors, especially proprioceptors in the neck and limb muscles that indicate the position of the head

Figure 17.26 The equilibrium pathway.

From hair cells of the semicircular ducts, utricle, and saccule, vestibular information is conveyed along the vestibular branch of the vestibulocochlear (VIII) nerve and then to the brain stem, cerebellum, thalamus, and cerebral cortex.

? Where are the vestibular nuclei located?

and limbs. The remaining axons enter the cerebellum through the **inferior cerebellar peduncles** (see Figure 14.8b). Bidirectional pathways connect the cerebellum and vestibular nuclei.

The vestibular nuclei integrate information from vestibular, visual, and somatic receptors and then send commands to (1) the **nuclei of cranial nerves**—oculomotor (III), trochlear (IV), and abducens (VI)—that control coupled movements of the eyes with those of the head to help maintain focus on the visual field; (2) **nuclei of the accessory (XI) nerves** to help control head and neck movements to assist in maintaining equilibrium; (3) the **vestibulospinal tract,** which conveys impulses down the spinal cord to maintain muscle tone in skeletal muscles to help maintain equilibrium; and (4) the **ventral posterior nucleus** in the thalamus and then to the **vestibular area** in the parietal lobe of the cerebral cortex (which is part of the primary somatosensory area; see areas 1, 2, and 3 in Figure 14.15) to provide us with the conscious awareness of the position and movements of the head and limbs.

Table 17.2 summarizes the structures of the ear related to hearing and equilibrium.

TABLE 17.2

Summary of Structures of the Ear

REGIONS OF THE EAR AND KEY STRUCTURES | **FUNCTION**

External (outer) ear

Auricle (pinna): Collects sound waves.

External auditory canal (meatus): Directs sound waves to eardrum.

Tympanic membrane (eardrum): Sound waves cause it to vibrate, which in turn causes malleus to vibrate.

Middle ear

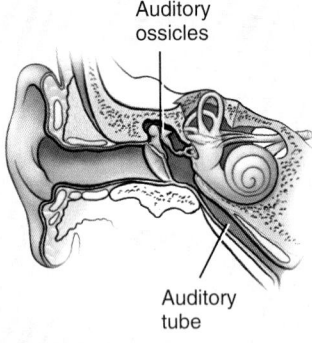

Auditory ossicles: Transmit and amplify vibrations from tympanic membrane to oval window.

Auditory (eustachian) tube: Equalizes air pressure on both sides of tympanic membrane.

Internal (inner) ear

Cochlea: Contains a series of fluids, channels, and membranes that transmit vibrations to spiral organ (organ of Corti), the organ of hearing; hair cells in spiral organ produce receptor potentials, which elicit nerve impulses in cochlear branch of vestibulocochlear (VIII) nerve.

Vestibular apparatus: Includes semicircular ducts, utricle, and saccule, which generate nerve impulses that propagate along vestibular branch of vestibulocochlear (VIII) nerve.

Semicircular ducts: Contain cristae, site of hair cells for dynamic equilibrium (maintenance of body position, mainly the head, in response to rotational acceleration and deceleration movements).

Utricle: Contains macula, site of hair cells for static equilibrium (maintenance of body position, mainly the head, relative to force of gravity).

Saccule: Contains macula, site of hair cells for static equilibrium (maintenance of body position, mainly the head, relative to force of gravity).

✔CHECKPOINT

11. How are sound waves transmitted from the auricle to the spiral organ of Corti?
12. How do hair cells in the cochlea and vestibular apparatus transduce mechanical vibrations into electrical signals?
13. What is the pathway for auditory impulses from the cochlea to the cerebral cortex?
14. Compare the function of the maculae in maintaining static equilibrium with the role of the cristae in maintaining dynamic equilibrium.
15. What is the role of vestibular input to the cerebellum?
16. Describe the equilibrium pathways.

17.5 DEVELOPMENT OF THE EYES AND EARS

⊙ OBJECTIVE

• Describe the development of the eyes and the ears.

Eyes

The *eyes* begin to develop about 22 days after fertilization when the **ectoderm** of the lateral walls of the prosencephalon (forebrain) bulges out to form a pair of shallow grooves called the **optic grooves.** Within a few days, as the neural tube is closing, the optic grooves enlarge and grow toward the surface ectoderm and become known as the **optic vesicles.** When the optic vesicles reach the surface ectoderm, the surface ectoderm thickens to form the **lens placodes** (PLAK-ods). In addition, the distal portions of the optic vesicles invaginate, forming the **optic cups;** they remain attached to the prosencephalon by narrow, hollow proximal structures called **optic stalks.** Figure 17.27 shows the stages in the development of the eyes.

The lens placodes also invaginate and develop into lens vesicles that sit in the optic cups. The lens vesicles eventually develop into the *lenses.* Blood is supplied to the developing lenses (and retina) by the hyaloid arteries. These arteries gain access to the developing eyes through a groove on the inferior surface of the optic cup and optic stalk called the **choroid fissure.** As the lenses mature, part of the hyaloid arteries that pass through the vitreous chamber degenerate; the remaining portions of the hyaloid arteries become the *central retinal arteries.*

The inner wall of the optic cup forms the *neural layer* of the retina, while the outer layer forms the *pigmented layer* of the retina. Axons from the neural layer grow through the optic stalk to the brain, converting the optic stalk to the *optic (II) nerve.* Although myelination of the optic nerves begins late in fetal life, it is not completed until the tenth week after birth.

The anterior portion of the optic cup forms the epithelium of the *ciliary body, iris,* and *circular and radial muscles* of the iris. The connective tissue of the ciliary body, *ciliary muscle,* and *zonular fibers* of the lens develop from **mesenchyme** around the anterior portion of the optic cup.

Mesenchyme surrounding the optic cup and optic stalk differentiates into an inner layer that gives rise to the *choroid* and an outer layer that develops into the *sclera* and part of the *cornea.* The remainder of the cornea is derived from surface ectoderm.

The *anterior chamber* develops from a cavity that forms in the mesenchyme between the iris and cornea; the *posterior chamber* develops from a cavity that forms in the mesenchyme between the iris and lens.

Some mesenchyme around the developing eye enters the optic cup through the choroid fissure. This mesenchyme occupies the space between the lens and retina and differentiates into a delicate network of fibers. Later the spaces between the fibers fill with a jellylike substance, thus forming the *vitreous body* in the vitreous chamber.

The *eyelids* form from surface ectoderm and mesenchyme. The upper and lower eyelids meet and fuse at about eight weeks of development and remain closed until about 26 weeks of development.

Ears

The first portion of the ear to develop is the *internal ear.* It begins to form about 22 days after fertilization as a thickening of the surface ectoderm, called **otic placodes** (Figure 17.28a), that appears on either side of the rhombencephalon (hindbrain). The otic placodes invaginate quickly (Figure 17.28b) to form the **otic pits** (Figure 17.28c). Next, the otic pits pinch off from the surface ectoderm to form the **otic vesicles** within the mesenchyme of the head (Figure 17.28d). During later development, the otic vesicles will form the structures associated with the *membranous labyrinth* of the internal ear. Mesenchyme around the otic vesicles produces cartilage that later ossifies to form the bone associated with the *bony labyrinth* of the internal ear.

The *middle ear* develops from a structure called the first **pharyngeal (branchial) pouch,** an **endoderm**-lined outgrowth of the primitive pharynx (see Figure 18.21a). The pharyngeal

Figure 17.27 Development of the eyes.

🔑 The eyes begin to develop about 22 days after fertilization from ectoderm of the prosencephalon.

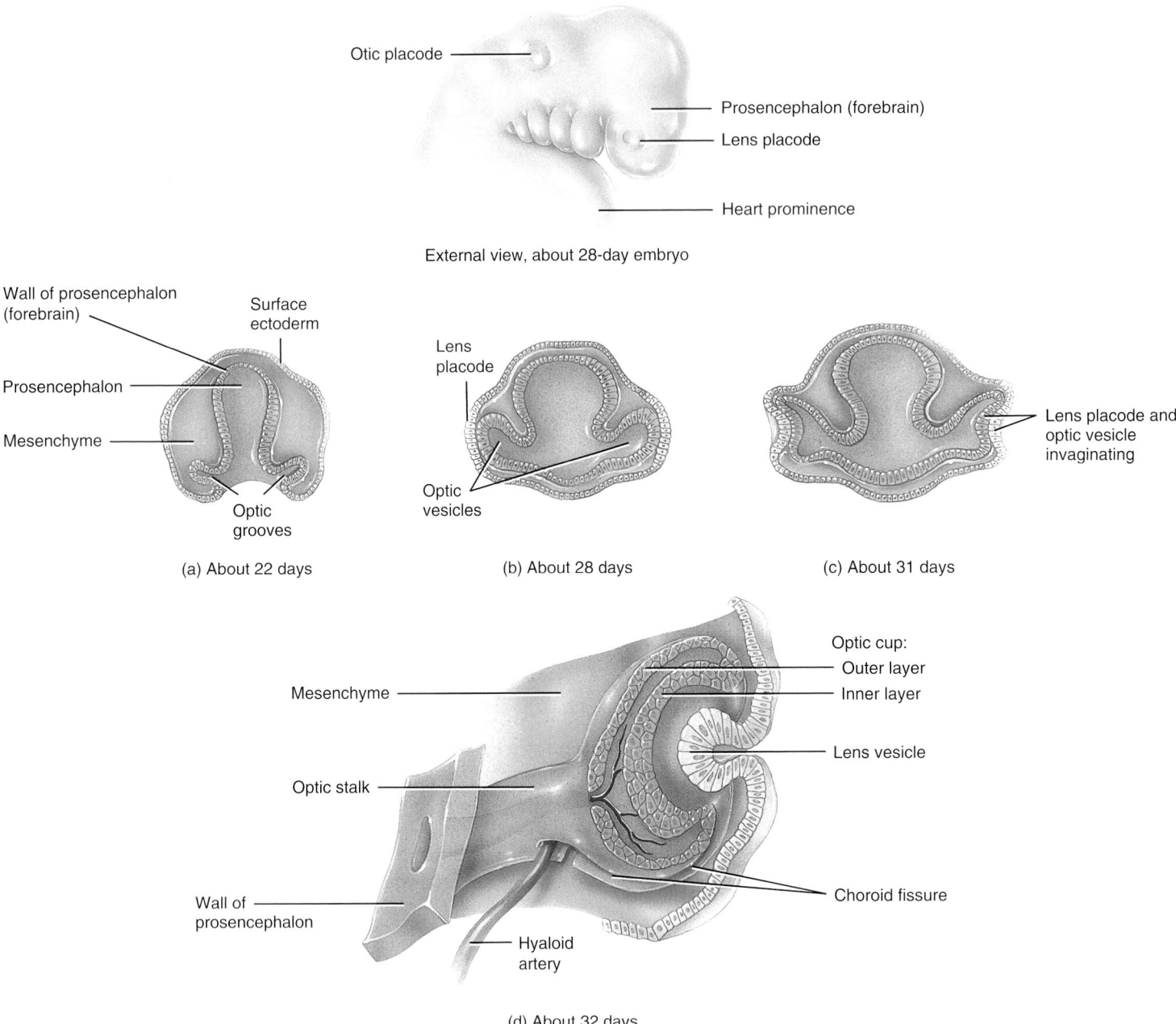

External view, about 28-day embryo

(a) About 22 days

(b) About 28 days

(c) About 31 days

(d) About 32 days

❓ **Which structure gives rise to the neural and pigmented layers of the retina?**

pouches are discussed in detail in Section 29.1. The *auditory ossicles* develop from the first and second pharyngeal arches.

The *external ear* develops from the first **pharyngeal cleft,** an endoderm-lined groove between the first and second pharyngeal arches (see the inset in Figure 17.28). The pharyngeal clefts are discussed in detail in Section 29.1.

✔ CHECKPOINT

17. How do the origins of the eyes and ears differ?

Figure 17.28 Development of the ears.

 The first parts of the ears to develop are the internal ears, which begin to form about 22 days after fertilization as thickenings of surface ectoderm.

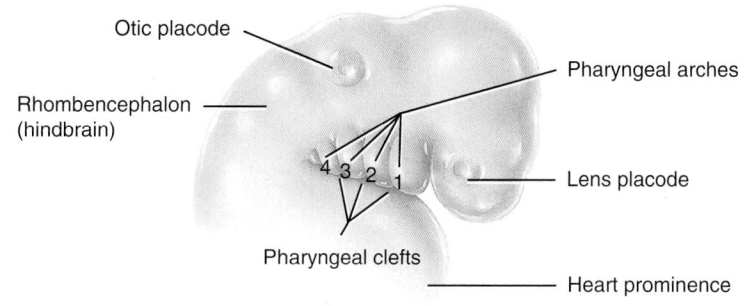

External view, about 28-day embryo

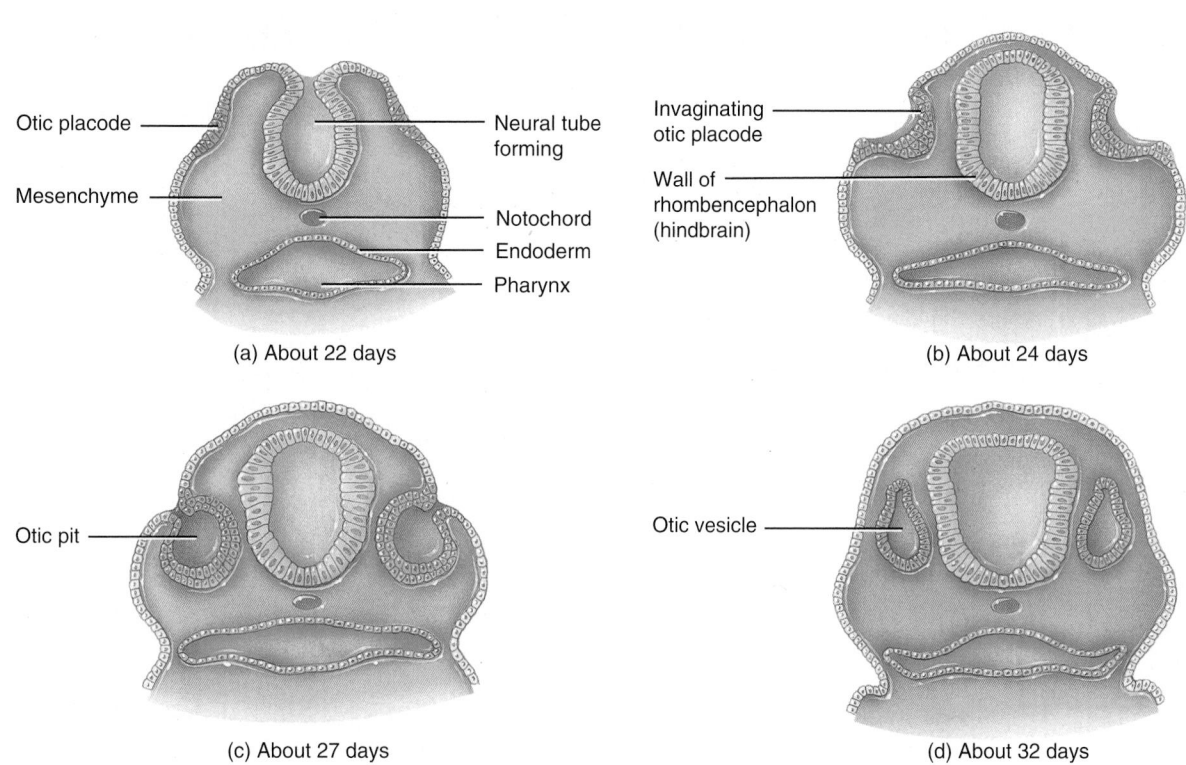

(a) About 22 days

(b) About 24 days

(c) About 27 days

(d) About 32 days

❓ **How do the three parts of the ear differ in origin?**

17.6 AGING AND THE SPECIAL SENSES

◉ **OBJECTIVE**

• Describe the age-related changes that occur in the eyes and ears.

Most people do not experience any problems with the senses of smell and taste until about age 50. This is due to a gradual loss of olfactory receptors and gustatory receptor cells coupled with their slower rate of replacement as we age.

Several age-related changes occur in the eyes. As noted earlier, the lens loses some of its elasticity and thus cannot change shape as easily, resulting in presbyopia (see Section 17.3). Cataracts (loss of transparency of the lenses) also occur with aging (see Disorders: Homeostatic Imbalances at the end of the chapter). In old age, the sclera ("white" of the eye) becomes thick and rigid and develops a yellowish or brownish coloration due to many years of exposure to ultraviolet light, wind, and dust. The sclera may also develop random splotches of pigment, especially in people with dark complexions. The iris fades or develops irregular pigment. The muscles that regulate the size of the pupil weaken with age and the pupils become smaller, react more slowly to light, and dilate more slowly in the dark. For these reasons, elderly people find that objects are not as bright, their eyes may adjust more slowly when going outdoors, and they have problems going from brightly lit to darkly lit places. Some diseases of the retina are more likely to occur in old age, including age-related macular disease and detached retina (see the Clinical Connections in Section 17.3). A disorder called glaucoma (see below) develops in the eyes of aging people as a result of the buildup of aqueous humor. Tear production and the number of mucous cells in the conjunctiva may decrease with age, resulting in dry eyes. The eyelids lose their elasticity, becoming baggy and wrinkled. The amount of fat around the orbits may decrease, causing the eyeballs to sink into the orbits. Finally, as we

age the sharpness of vision decreases, color and depth perception are reduced, and "vitreal floaters" increase.

By about age 60, around 25% of individuals experience a noticeable hearing loss, especially for higher-pitched sounds. The age-associated progressive loss of hearing in both ears is called **presbycusis** (pres'-bī-KŪ-sis; *presby-* = old; *-acou-* = hearing; *-sis* = condition). It may be related to damaged and lost hair cells in the spiral organ or degeneration of the nerve pathway for hearing. Tinnitus (ringing in the ears) and vestibular imbalance also occur more frequently in the elderly.

✔CHECKPOINT

18. What changes in the eyes and ears are related to the aging process, and how do they take place?

DISORDERS: HOMEOSTATIC IMBALANCES

Cataracts

A common cause of blindness is a loss of transparency of the lens known as a **cataract** (KAT-a-rakt = waterfall). The lens becomes cloudy (less transparent) due to changes in the structure of the lens proteins. Cataracts often occur with aging but may also be caused by injury, excessive exposure to ultraviolet rays, certain medications (such as long-term use of steroids), or complications of other diseases (for example, diabetes). People who smoke also have increased risk of developing cataracts. Fortunately, sight can usually be restored by surgical removal of the old lens and implantation of a new artificial one.

Glaucoma

Glaucoma (glaw-KŌ-ma) is the most common cause of blindness in the United States, afflicting about 2 percent of the population over age 40. In many cases, glaucoma is due to an abnormally high intraocular pressure as a result of a buildup of aqueous humor within the anterior cavity. The fluid compresses the lens into the vitreous body and puts pressure on the neurons of the retina. Persistent pressure results in a progression from mild visual impairment to irreversible destruction of neurons of the retina, damage to the optic nerve, and blindness. Glaucoma is painless, and the other eye compensates largely, so a person may experience considerable retinal damage and loss of vision before the condition is diagnosed. Because glaucoma occurs more often with advancing age, regular measurement of intraocular pressure is an increasingly important part of an eye exam as people grow older. Risk factors include race (blacks are more susceptible), increasing age, family history, and past eye injuries and disorders.

Some individuals have another form of glaucoma called **normal-tension (low-tension) glaucoma.** In this condition, there is damage to the optic nerve with a corresponding loss of vision, even though intraocular pressure is normal. Although the cause is unknown, it appears to be related to a fragile optic nerve, vasospasm of blood vessels around the optic nerve, and ischemia due to narrowed or obstructed blood vessels around the optic nerve. The incidence of normal-tension glaucoma is higher among Japanese and Koreans and females.

Deafness

Deafness is significant or total hearing loss. **Sensorineural deafness** (sen'-so-rē-NOO-ral) is caused by either impairment of hair cells

in the cochlea or damage of the cochlear branch of the vestibulocochlear (VIII) nerve. This type of deafness may be caused by atherosclerosis, which reduces blood supply to the ears; by repeated exposure to loud noise, which destroys hair cells of the spiral organ; by certain drugs such as aspirin and streptomycin; and/or by genetic factors. **Conduction deafness** is caused by impairment of the external and middle ear mechanisms for transmitting sounds to the cochlea. Causes of conduction deafness include otosclerosis, the deposition of new bone around the oval window; impacted cerumen; injury to the eardrum; and aging, which often results in thickening of the eardrum and stiffening of the joints of the auditory ossicles. A hearing test called *Weber's test* is used to distinguish between sensorineural and conduction deafness. In the test, the stem of a vibrating fork is held to the forehead. In people with normal hearing, the sound is heard equally in both ears. If the sound is heard best in the affected ear, the deafness is probably of the conduction type; if the sound is heard best in the normal ear, it is probably of the sensorineural type.

Ménière's Disease

Ménière's disease (men'-ē-ĀRZ) results from an increased amount of endolymph that enlarges the membranous labyrinth. Among the symptoms are fluctuating hearing loss (caused by distortion of the basilar membrane of the cochlea) and roaring tinnitus (ringing). Spinning or whirling vertigo (dizziness) is characteristic of Ménière's disease. Almost total destruction of hearing may occur over a period of years.

Otitis Media

Otitis media (ō-TĪ-tis MĒ-dē-a) is an acute infection of the middle ear caused mainly by bacteria and associated with infections of the nose and throat. Symptoms include pain, malaise, fever, and a reddening and outward bulging of the eardrum, which may rupture unless prompt treatment is received. (This may involve draining pus from the middle ear.) Bacteria passing into the auditory tube from the nasopharynx are the primary cause of middle ear infections. Children are more susceptible than adults to middle ear infections because their auditory tubes are almost horizontal, which decreases drainage. If otitis media occurs frequently, a surgical procedure called **tympanotomy** (tim'-pa-NOT-ō-mē; *tympano-* = drum; *-tome* = incision) is often employed. This consists of the insertion of a small tube into the eardrum to provide a pathway for the drainage of fluid from the middle ear.

MEDICAL TERMINOLOGY

Ageusia (a-GOO-sē-a; *a-* = without; *-geusis* = taste) Loss of the sense of taste.

Amblyopia (am′-blē-Ō-pē-a; *ambly-* = dull or dim) Term used to describe the loss of vision in an otherwise normal eye that, because of muscle imbalance, cannot focus in synchrony with the other eye. Sometimes called "wandering eyeball" or a "lazy eye."

Anosmia (an-OZ-mē-a; *a-* = without; *-osmi* = smell, odor) Total lack of the sense of smell.

Barotrauma (bar′-ō-TRAW-ma; *baros-* = weight) Damage or pain, mainly affecting the middle ear, as a result of pressure changes. It occurs when pressure on the outer side of the tympanic membrane is higher than on the inner side, for example, when flying in an airplane or diving. Swallowing or holding your nose and exhaling with your mouth closed usually opens the auditory tubes, allowing air into the middle ear to equalize the pressure.

Blepharitis (blef-a-RĪ-tis; *blephar-* = eyelid; *-itis* = inflammation of) An inflammation of the eyelid.

Conjunctivitis (pinkeye) An inflammation of the conjunctiva; when caused by bacteria such as pneumococci, staphylococci, or *Haemophilus influenzae,* it is very contagious and more common in children. Conjunctivitis may also be caused by irritants, such as dust, smoke, or pollutants in the air, in which case it is not contagious.

Corneal abrasion (KOR-nē-al a-BRĀ-zhun) A scratch on the surface of the cornea, for example, from a speck of dirt or damaged contact lenses. Symptoms include pain, redness, watering, blurry vision, sensitivity to bright light, and frequent blinking.

Corneal transplant A procedure in which a defective cornea is removed and a donor cornea of similar diameter is sewn in. It is the most common and most successful transplant operation. Since the cornea is avascular, antibodies in the blood that might cause rejection do not enter the transplanted tissue, and rejection rarely occurs. The shortage of donor corneas has been partially overcome by the development of artificial corneas made of plastic.

Diabetic retinopathy (ret-i-NOP-a-thē; *retino-* = retina; *-pathos* = suffering) Degenerative disease of the retina due to diabetes mellitus, in which blood vessels in the retina are damaged or new ones grow and interfere with vision.

Exotropia (ek′-sō-TRŌ-pē-a; *ex-* = out; *-tropia* = turning) Turning outward of the eyes.

Keratitis (ker′-a-TĪ-tis; *kerat-* = cornea) An inflammation or infection of the cornea.

Miosis (mī-Ō-sis) Constriction of the pupil.

Mydriasis (mi-DRĪ-a-sis) Dilation of the pupil.

Nystagmus (nis-TAG-mus; *nystagm-* = nodding or drowsy) A rapid involuntary movement of the eyeballs, possibly caused by a disease of the central nervous system. It is associated with conditions that cause vertigo.

Otalgia (ō-TAL-jē-a; *oto-* = ear; *-algia* = pain) Earache.

Photophobia (fō′-tō-FŌ-bē-a; *photo-* = light; *-phobia* = fear) Abnormal visual intolerance to light.

Ptosis (TŌ-sis = fall) Falling or drooping of the eyelid (or slippage of any organ below its normal position).

Retinoblastoma (ret-i-nō-blas-TŌ-ma; *-oma* = tumor) A tumor arising from immature retinal cells; it accounts for 2% of childhood cancers.

Scotoma (skō-TŌ-ma = darkness) An area of reduced or lost vision in the visual field.

Strabismus (stra-BIZ-mus; *strabismos* = squinting) Misalignment of the eyeballs so that the eyes do not move in unison when viewing an object; the affected eye turns either medially or laterally with respect to the normal eye and the result is double vision (diplopia). It may be caused by physical trauma, vascular injuries, or tumors of the extrinsic eye muscle or the oculomotor (III), trochlear (IV), or abducens (VI) cranial nerves.

Tinnitus (ti-NĪ-tus) A ringing, roaring, or clicking in the ears.

Tonometer (tō-NOM-ē-ter; *tono-* = tension or pressure; *-metron* = measure) An instrument for measuring pressure, especially intraocular pressure.

Trachoma (tra-KŌ-ma) A serious form of conjunctivitis and the greatest single cause of blindness in the world. It is caused by the bacterium *Chlamydia trachomatis.* The disease produces an excessive growth of subconjunctival tissue and invasion of blood vessels into the cornea, which progresses until the entire cornea is opaque.

Vertigo (VER-ti-gō = dizziness) A sensation of spinning or movement in which the world seems to revolve or the person seems to revolve in space, often associated with nausea and, in some cases, vomiting. It may be caused by arthritis of the neck or an infection of the vestibular apparatus.

CHAPTER REVIEW AND RESOURCE SUMMARY

Review	Resource
17.1 Olfaction: Sense of Smell	Anatomy Overview - The Special Senses
1. The receptors for olfaction, which are bipolar neurons, are in the nasal epithelium along with olfactory glands, which produce mucus that dissolves odorants.	Anatomy Overview - Chemoreceptors Anatomy Overview - Olfactory Nerve
2. In olfactory reception, a generator potential develops and triggers one or more nerve impulses.	
3. The threshold of smell is low, and adaptation to odors occurs quickly.	
4. Axons of olfactory receptors form the olfactory (I) nerves, which convey nerve impulses to the olfactory bulbs, olfactory tracts, limbic system, and cerebral cortex (temporal and frontal lobes).	
17.2 Gustation: Sensation of Taste	Anatomy Overview - Chemoreceptors
1. The receptors for gustation, the gustatory receptor cells, are located in taste buds.	Anatomy Overview - Facial Nerve Anatomy Overview - Glossopharyngeal Nerve
2. Dissolved chemicals, called tastants, stimulate gustatory receptor cells by flowing through ion channels in the plasma membrane or by binding to receptors attached to G proteins in the membrane.	Anatomy Overview - Vagus Nerve
3. Receptor potentials developed in gustatory receptor cells cause the release of neurotransmitter, which can generate nerve impulses in first-order sensory neurons.	

Review

4. The threshold varies with the taste involved, and adaptation to taste occurs quickly.
5. Gustatory receptor cells trigger nerve impulses in cranial nerves VII, IX, and X. Taste signals then pass to the medulla oblongata, thalamus, and cerebral cortex (parietal lobe).

17.3 Vision

1. Accessory structures of the eyes include the eyebrows, eyelids, eyelashes, lacrimal apparatus, and extrinsic eye muscles. The lacrimal apparatus consists of structures that produce and drain tears.
2. The eye is constructed of three layers: (a) fibrous tunic (sclera and cornea), (b) vascular tunic (choroid, ciliary body, and iris), and (c) retina. .
3. The retina consists of a pigmented layer and a neural layer that includes a photoreceptor layer, bipolar cell layer, ganglion cell layer, horizontal cells, and amacrine cells.
4. The anterior cavity contains aqueous humor; the vitreous chamber contains the vitreous body.
5. Image formation on the retina involves refraction of light rays by the cornea and lens, which focus an inverted image on the fovea centralis of the retina. For viewing close objects, the lens increases its curvature (accommodation) and the pupil constricts to prevent light rays from entering the eye through the periphery of the lens.
6. The near point of vision is the minimum distance from the eye at which an object can be clearly focused with maximum accommodation.
7. In convergence, the eyeballs move medially so they are both directed toward an object being viewed.
8. The first step in vision is the absorption of light by photopigments in rods and cones and isomerization of *cis*-retinal. Receptor potentials in rods and cones decrease the release of inhibitory neurotransmitter, which induces graded potentials in bipolar cells and horizontal cells.
9. Horizontal cells transmit inhibitory signals to bipolar cells; bipolar or amacrine cells transmit excitatory signals to ganglion cells, which depolarize and initiate nerve impulses.
10. Impulses from ganglion cells are conveyed into the optic (II) nerve, through the optic chiasm and optic tract, to the thalamus. From the thalamus, impulses for vision propagate to the cerebral cortex (occipital lobe). Axon collaterals of retinal ganglion cells extend to the midbrain and hypothalamus.

Anatomy Overview - Oculomotor Nerve
Anatomy Overview - Trochlear Nerve
Anatomy Overview - Abducens Nerve
Anatomy Overview - Optic Nerve
Figure 17.7 - Anatomy of the Eyeball
Figure 17.17 - The Visual Pathway

17.4 Hearing and Equilibrium

1. The external (outer) ear consists of the auricle, external auditory canal, and tympanic membrane (eardrum).
2. The middle ear consists of the auditory tube, ossicles, oval window, and round window.
3. The internal (inner) ear consists of the bony labyrinth and membranous labyrinth. The internal ear contains the spiral organ (organ of Corti), the organ of hearing.
4. Sound waves enter the external auditory canal, strike the tympanic membrane, pass through the ossicles, strike the oval window, set up waves in the perilymph, strike the vestibular membrane and scala tympani, increase pressure in the endolymph, vibrate the basilar membrane, and stimulate hair bundles on the spiral organ (organ of Corti).
5. Hair cells convert mechanical vibrations into a receptor potential, which releases neurotransmitter that can initiate nerve impulses in first-order sensory neurons.
6. Sensory axons in the cochlear branch of the vestibulocochlear (VIII) nerve terminate in the medulla oblongata. Auditory signals then pass to the inferior colliculus, thalamus, and temporal lobes of the cerebral cortex.
7. Static equilibrium is the orientation of the body relative to the pull of gravity. The maculae of the utricle and saccule are the sense organs of static equilibrium. Body movements that stimulate the receptors for static equilibrium include tilting the head and linear acceleration or deceleration.
8. Dynamic equilibrium is the maintenance of body position in response to rotational acceleration or deceleration. The cristae in the semicircular ducts are the main sense organs of dynamic equilibrium.
9. Most vestibular branch axons of the vestibulocochlear nerve enter the brain stem and terminate in the medulla and pons; other axons enter the cerebellum.

Anatomy Overview - Vestibulocochlear Nerve
Figure 17.19 - The Right Middle Ear and Auditory Ossicles
Figure 17.20 - The Right Internal Ear
Figure 17.22 - Stimulation of Auditory Receptors in the Right Ear

17.5 Development of the Eyes and Ears

1. The eyes begin their development about 22 days after fertilization from ectoderm of the lateral walls of the prosencephalon (forebrain).
2. The ears begin their development about 22 days after fertilization from a thickening of ectoderm on either side of the rhombencephalon (hindbrain). The sequence of development of the ear is internal ear, middle ear, and external ear.

Review   **Resource**

17.6 Aging and the Special Senses

1. Most people do not experience problems with the senses of smell and taste until about age 50.
2. Among the age-related changes to the eyes are presbyopia, cataracts, difficulty adjusting to light, macular disease, glaucoma, dry eyes, and decreased sharpness of vision.
3. With age there is a progressive loss of hearing, and tinnitus occurs more frequently.

SELF-QUIZ QUESTIONS

Fill in the blanks in the following statements.

1. The five primary taste sensations are _____, _____, _____, _____, and _____.

2. _____ equilibrium refers to the maintenance of the position of the body relative to the force of gravity; _____ equilibrium refers to the maintenance of body position in response to rotational acceleration or deceleration.

Indicate whether the following statements are true or false.

3. Of all of the special senses, only smell and taste sensations project both to higher cortical areas and to the limbic system.

4. The ability to change the curvature of the lens for near vision is convergence.

Choose the one best answer to the following questions.

5. Which of the following are *true*? (1) The sites of olfactory transduction are the olfactory hairs. (2) The olfactory bulbs transmit impulses to the temporal lobe of the cerebral cortex. (3) The axons of olfactory receptors pass through the olfactory foramina in the cribriform plate of the ethmoid bone. (4) The olfactory nerves are bundles of axons that terminate in the olfactory tracts. (5) Within the olfactory bulbs, the first-order neurons synapse with the second-order neurons.
(a) 1, 2, and 4   (b) 2, 3, 4, and 5   (c) 1, 2, 3, 4, and 5
(d) 1, 3, and 5   (e) 1, 2, 3, and 5

6. Which of the following statements is *incorrect*?
(a) Olfactory receptors respond to the chemical stimulation of an odorant molecule by producing a receptor potential.
(b) Basal stem cells continually produce new olfactory receptors.
(c) Adaptation to odors is rapid and occurs in both olfactory receptors and the CNS.
(d) Production of nasal mucus by olfactory glands serves to moisten the olfactory epithelium and dissolve odorants.
(e) The orbitofrontal area is an important region for odor identification and discrimination.

7. Which of the following statements is *incorrect*?
(a) Taste is a chemical sense.
(b) The receptors for taste sensations are found in taste buds located on the tongue, the soft palate, the pharynx, and the epiglottis.
(c) Gustatory hairs are the sites of taste transduction.
(d) The threshold for bitter substances is the highest.
(e) Complete adaptation to taste can occur in 1 to 5 minutes.

8. When viewing an object close to your eyes, which of the following are required for proper image formation on the retina? (1) increased curvature of the lens, (2) contraction of the ciliary muscle, (3) divergence of the eyeballs, (4) refraction of light at the anterior and posterior surfaces of the cornea, (5) constriction of the pupil by contraction of the extrinsic eye muscles.
(a) 1, 2, 3, 4, and 5   (b) 1, 2, and 4   (c) 1, 2, 3, and 4
(d) 2, 4, and 5   (e) 2, 3, and 4

9. Which of the following are *mismatched*?
(a) fungiform papillae: scattered over the entire tongue's surface.
(b) filiform papillae: contain taste buds in early childhood.
(c) vallate papillae: each houses 100–300 taste buds.
(d) foliate papillae: located in trenches on the lateral margins of the tongue.
(e) fungiform papillae: each houses about five taste buds.

10. Place in order the structures involved in the visual pathway.
(a) optic tract   (b) ganglion cells   (c) cornea
(d) lens   (e) bipolar cells   (f) optic nerve
(g) visual cortex   (h) vitreous body   (i) optic chiasm
(j) aqueous humor   (k) pupil   (l) photoreceptors
(m) thalamus

11. Which of the following statements is *incorrect*?
(a) Retinal is the light-absorbing portion of all visual photopigments.
(b) The only photopigment in rods is rhodopsin, but three different cone photopigments are present in the retina.
(c) Retinal is a derivative of vitamin C.
(d) Color vision results from different colors of light selectively activating different cone photopigments.
(e) Bleaching and regeneration of the photopigments account for much but not all of the sensitivity changes during light and dark adaptation.

12. Which of the following is the *correct* sequence for the auditory pathway?
(a) external auditory canal, tympanic membrane, auditory ossicles, oval window, cochlea and spiral organ
(b) tympanic membrane, external auditory canal, auditory ossicles, cochlea and spiral organ, round window
(c) auditory ossicles, tympanic membrane, cochlea and spiral organ, round window, oval window, external auditory canal
(d) auricle, tympanic membrane, round window, cochlea and spiral organ, oval window
(e) external auditory canal, tympanic membrane, auditory ossicles, internal auditory canal, spiral organ, oval window

13. Match the following:

_____ (a) upper and lower eyelids; shade the eyes during sleep, spread lubricating secretions over the eyeballs

_____ (b) produces and drains tears

_____ (c) arch transversely above the eyeballs and help protect the eyeballs from foreign objects, perspiration, and the direct rays of the sun

_____ (d) move the eyeball medially, laterally, superiorly, or inferiorly

_____ (e) a thick fold of connective tissue that gives form and support to the eyelids

_____ (f) modified sebaceous glands; secretion helps keep eyelids from adhering to one another

_____ (g) project from the border of each eyelid; help protect the eyeballs from foreign objects, perspiration, and direct rays of the sun

_____ (h) a thin, protective mucous membrane that lines the inner aspect of the eyelids and passes from the eyelids onto the surface of the eyeball, where it covers the sclera

(1) palpebrae
(2) tarsal or Meibomian glands
(3) conjunctiva
(4) eyelashes
(5) lacrimal apparatus
(6) extrinsic eye muscles
(7) eyebrows
(8) tarsal plate

14. Match the following:

_____ (a) colored portion of the eyeball; regulates the amount of light entering the posterior part of the eyeball

_____ (b) innermost layer of the eyeball; beginning of the visual pathway; contains rods and cones

_____ (c) biconvex transparent structure that fine tunes focusing of light rays for clear vision

_____ (d) circular band of smooth muscle that alters the shape of the lens for near or far vision

_____ (e) watery fluid in the anterior cavity that helps nourish the lens and cornea; helps maintain shape of the eyeball

_____ (f) the hole in the center of the iris

_____ (g) white of the eye; gives shape to the eyeball, makes it more rigid, protects its inner parts

_____ (h) avascular superficial layer of the eyeball; includes cornea and sclera

_____ (i) middle, vascularized layer of the eyeball; includes choroid, ciliary body, and iris

(1) sclera
(2) ciliary muscle
(3) iris
(4) pupil
(5) retina
(6) fibrous tunic
(7) aqueous humor
(8) lens
(9) vascular tunic

15. Match the following:

_____ (a) partition between external auditory canal and middle ear

_____ (b) oval central portion of the bony labyrinth; contains utricle and saccule

_____ (c) receptor for static equilibrium; also contributes to some aspects of dynamic equilibrium; consists of hair cells and supporting cells

_____ (d) contains hair cells which are the receptors for hearing

_____ (e) ear bones: malleus, incus, stapes

_____ (f) the pressure equalization tube that connects the middle ear to the nasopharynx

_____ (g) contains the spiral organ

_____ (h) fluid found within the membranous labyrinth; pressure waves in this fluid cause vibration of the basilar membrane

_____ (i) receptor organs for equilibrium; the saccule, utricle, and semicircular canals

_____ (j) swollen enlargement in semicircular canals; contains structures involved in dynamic equilibrium

_____ (k) opening between the middle ear and internal ear; is enclosed by a membrane called the secondary tympanic membrane

_____ (l) the flap of elastic cartilage covered by skin that captures sound waves; the pinna

_____ (m) fluid found inside bony labyrinth; bulging of the oval window causes pressure waves in this fluid

_____ (n) opening between the middle and inner ear; receives base of stapes

(1) auricle
(2) tympanic membrane
(3) auditory ossicles
(4) vestibular apparatus
(5) ampulla
(6) cochlea
(7) perilymph
(8) oval window
(9) round window
(10) auditory or eustachian tube
(11) vestibule
(12) endolymph
(13) spiral organ
(14) macula

CRITICAL THINKING QUESTIONS

1. Mario has experienced damage to his facial nerve. How would this affect his special senses?

2. The shift nurse brings ailing eighty-year-old Gertrude her dinner. As Gertrude eats a small amount of her food, she comments that she isn't hungry and that "hospital food just doesn't taste good!" The nurse gives Gertrude a menu so she can choose her morning breakfast. Gertrude complains that she is having trouble reading the menu and asks the nurse to read it to her. As the nurse begins to read, Gertrude loudly asks her to "speak up and turn off the buzzing." What does the nurse know about aging and the special senses that help to explain Gertrude's comments?

3. As you help your neighbor put drops in her six-year-old daughter's eyes, the daughter states, "That medicine tastes bad." How do you explain to the neighbor how her daughter can "taste" the eyedrops?

? ANSWERS TO FIGURE QUESTIONS

17.1 An olfactory receptor has a life span of about one month.

17.2 Olfactory transduction occurs in the olfactory hairs of an olfactory receptor.

17.3 Basal cells develop into gustatory receptor cells.

17.4 Visible light that has a wavelength of 700 nm is red.

17.5 The conjunctiva is continuous with the inner lining of the eyelids.

17.6 Lacrimal fluid, or tears, is a watery solution containing salts, some mucus, and lysozyme that protects, cleans, lubricates, and moistens the eyeball.

17.7 The fibrous tunic consists of the cornea and sclera; the vascular tunic consists of the choroid, ciliary body, and iris.

17.8 The parasympathetic division of the ANS causes pupillary constriction; the sympathetic division causes pupillary dilation.

17.9 An ophthalmoscopic examination of the blood vessels of the eye can reveal evidence of hypertension, diabetes mellitus, cataracts, and age-related macular disease.

17.10 The two types of photoreceptors are rods and cones. Rods provide black-and-white vision in dim light; cones provide high visual acuity and color vision in bright light.

17.11 After its secretion by the ciliary process, aqueous humor flows into the posterior chamber, around the iris, into the anterior chamber, and out of the eyeball through the scleral venous sinus.

17.12 During accommodation the ciliary muscle contracts, causing the zonular fibers to slacken. The lens then becomes more convex, increasing its focusing power.

17.13 Presbyopia is the loss of lens elasticity that occurs with aging.

17.14 Both rods and cones transduce light into receptor potentials, use a photopigment embedded in outer segment discs or folds, and release neurotransmitter at synapses with bipolar cells and horizontal cells.

17.15 The conversion of *cis*-retinal to *trans*-retinal is called isomerization.

17.16 Cyclic GMP is the ligand that opens Na^+ channels in photoreceptors, causing the dark current to flow.

17.17 Light rays from an object in the temporal half of the visual field fall on the nasal half of the retina.

17.18 The malleus of the middle ear is attached to the eardrum, which is part of the external ear.

17.19 The oval and round windows separate the middle ear from the internal ear.

17.20 The two sacs in the membranous labyrinth of the vestibule are the utricle and saccule.

17.21 The three subdivisions of the bony labyrinth are the semicircular canals, vestibule, and cochlea.

17.22 The region of the basilar membrane close to the oval and round windows vibrates most vigorously in response to high-frequency sounds.

17.23 The superior olivary nucleus of the pons is the part of the auditory pathway that allows a person to locate the source of a sound.

17.24 The maculae are associated with static equilibrium; they provide sensory information about the position of the head in space.

17.25 The semicircular ducts are associated with dynamic equilibrium.

17.26 The vestibular nuclei are located in the medulla and the pons.

17.27 The optic cup forms the neural and pigmented layers of the retina.

17.28 The internal ear develops from surface ectoderm, the middle ear develops from pharyngeal pouches, and the external ear develops from a pharyngeal cleft.

18 THE ENDOCRINE SYSTEM

THE ENDOCRINE SYSTEM AND HOMEOSTASIS *Circulating or local hormones of the endocrine system contribute to homeostasis by regulating the activity and growth of target cells in your body. Hormones also regulate your metabolism.*

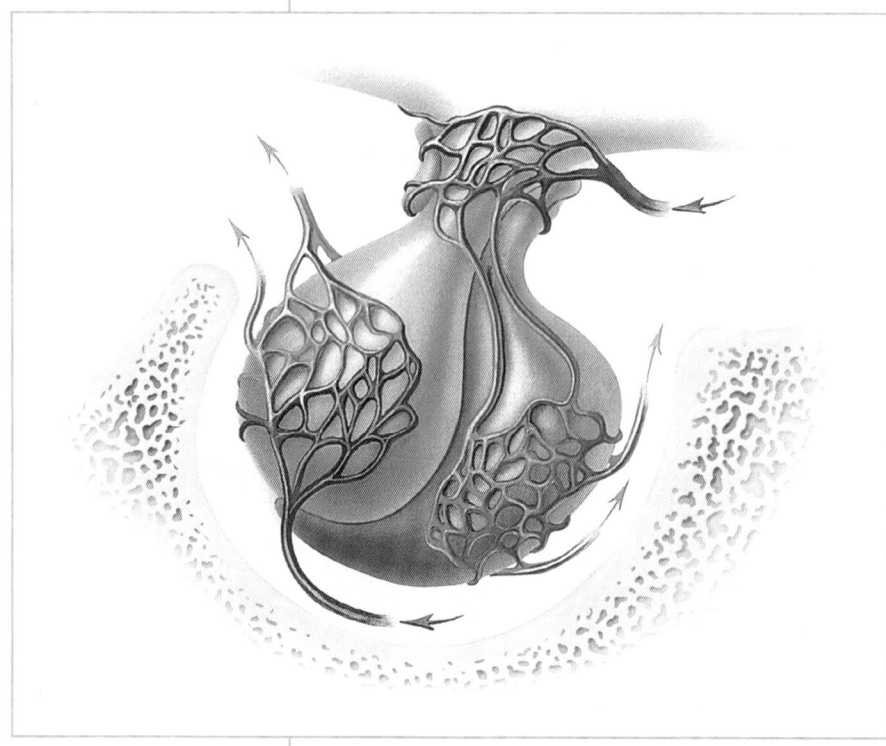

As girls and boys enter puberty, they start to develop striking differences in physical appearance and behavior. Perhaps no other period in life so dramatically shows the impact of the endocrine system in directing development and regulating body functions. In girls, estrogens promote accumulation of adipose tissue in the breasts and hips, sculpting a feminine shape. At the same time or a little later, increasing levels of testosterone in boys begin to help build muscle mass and enlarge the vocal cords, producing a lower-pitched voice. These changes are just a few examples of the powerful influence of endocrine secretions. Less dramatically, perhaps, multitudes of hormones help maintain homeostasis on a daily basis. They regulate the activity of smooth muscle, cardiac muscle, and some glands; alter metabolism; spur growth and development; influence reproductive processes; and participate in circadian (daily) rhythms established by the suprachiasmatic nucleus of the hypothalamus.

Did you ever wonder why thyroid gland disorders affect all major body systems?

18.1 COMPARISON OF CONTROL BY THE NERVOUS AND ENDOCRINE SYSTEMS

◉ **OBJECTIVE**

• Compare control of body functions by the nervous system and endocrine system.

The nervous and endocrine systems act together to coordinate functions of all body systems. Recall that the nervous system acts through nerve impulses (action potentials) conducted along axons of neurons. At synapses, nerve impulses trigger the release of mediator (messenger) molecules called *neurotransmitters* (shown in Figure 12.22). The endocrine system also controls body activities by releasing mediators, called *hormones,* but the means of control of the two systems are very different.

A **hormone** (*hormon* = to excite or get moving) is a mediator molecule that is released in one part of the body but regulates the activity of cells in other parts of the body. Most hormones enter interstitial fluid and then the bloodstream. The circulating blood delivers hormones to cells throughout the body. Both neurotransmitters and hormones exert their effects by binding to receptors on or in their "target" cells. Several mediators act as both neurotransmitters and hormones. One familiar example is norepinephrine, which is released as a neurotransmitter by sympathetic postganglionic neurons and as a hormone by chromaffin cells of the adrenal medullae.

Responses of the endocrine system often are slower than responses of the nervous system; although some hormones act within seconds, most take several minutes or more to cause a response. The effects of nervous system activation are generally briefer than those of the endocrine system. The nervous system acts on specific muscles and glands. The influence of the endocrine system is much broader; it helps regulate virtually all types of body cells.

We will also have several opportunities to see how the nervous and endocrine systems function together as an interlocking "supersystem." For example, certain parts of the nervous system stimulate or inhibit the release of hormones by the endocrine system.

Table 18.1 compares the characteristics of the nervous and endocrine systems. In this chapter, we focus on the major endocrine glands and hormone-producing tissues and examine how their hormones govern body activities.

✔**CHECKPOINT**

1. List the similarities among and differences between the nervous and endocrine systems with regard to the control of homeostasis.

18.2 ENDOCRINE GLANDS

◉ **OBJECTIVE**

• Distinguish between exocrine and endocrine glands.

Recall from Chapter 4 that the body contains two kinds of glands: exocrine glands and endocrine glands. **Exocrine glands** (EK-sō-krin; *exo-* = outside) secrete their products into ducts that carry the secretions into body cavities, into the lumen of an organ, or to the outer surface of the body. Exocrine glands include sudoriferous (sweat), sebaceous (oil), mucous, and digestive glands. **Endocrine glands** (EN-dō-krin; *endo-* = within) secrete their products (hormones) into the interstitial fluid surrounding the secretory cells rather than into ducts. From the interstitial fluid, hormones diffuse into blood capillaries and blood carries them to target cells throughout the body. Because of their dependence on the cardiovascular system to distribute their products, endocrine glands are some of the most vascular tissues of the body. Considering that most hormones are required in very small amounts, circulating levels typically are low.

The endocrine glands include the pituitary, thyroid, parathyroid, adrenal, and pineal glands (Figure 18.1). In addition, several organs and tissues are not exclusively classified as endocrine glands but contain cells that secrete hormones. These include the hypothalamus, thymus, pancreas, ovaries, testes, kidneys, stomach, liver, small intestine, skin, heart, adipose tissue, and placenta. Taken together, all endocrine glands and hormone-secreting cells constitute the **endocrine system.** The science of the structure and function of

TABLE 18.1		
Comparison of Control by the Nervous and Endocrine Systems		
CHARACTERISTIC	**NERVOUS SYSTEM**	**ENDOCRINE SYSTEM**
Mediator molecules	Neurotransmitters released locally in response to nerve impulses.	Hormones delivered to tissues throughout body by blood.
Site of mediator action	Close to site of release, at synapse; binds to receptors in postsynaptic membrane.	Far from site of release (usually); binds to receptors on or in target cells.
Types of target cells	Muscle (smooth, cardiac, and skeletal) cells, gland cells, other neurons.	Cells throughout body.
Time to onset of action	Typically within milliseconds (thousandths of a second).	Seconds to hours or days.
Duration of action	Generally briefer (milliseconds).	Generally longer (seconds to days).

Figure 18.1 Location of many endocrine glands. Also shown are other organs that contain endocrine cells and associated structures.

Endocrine glands secrete hormones, which circulating blood delivers to target tissues.

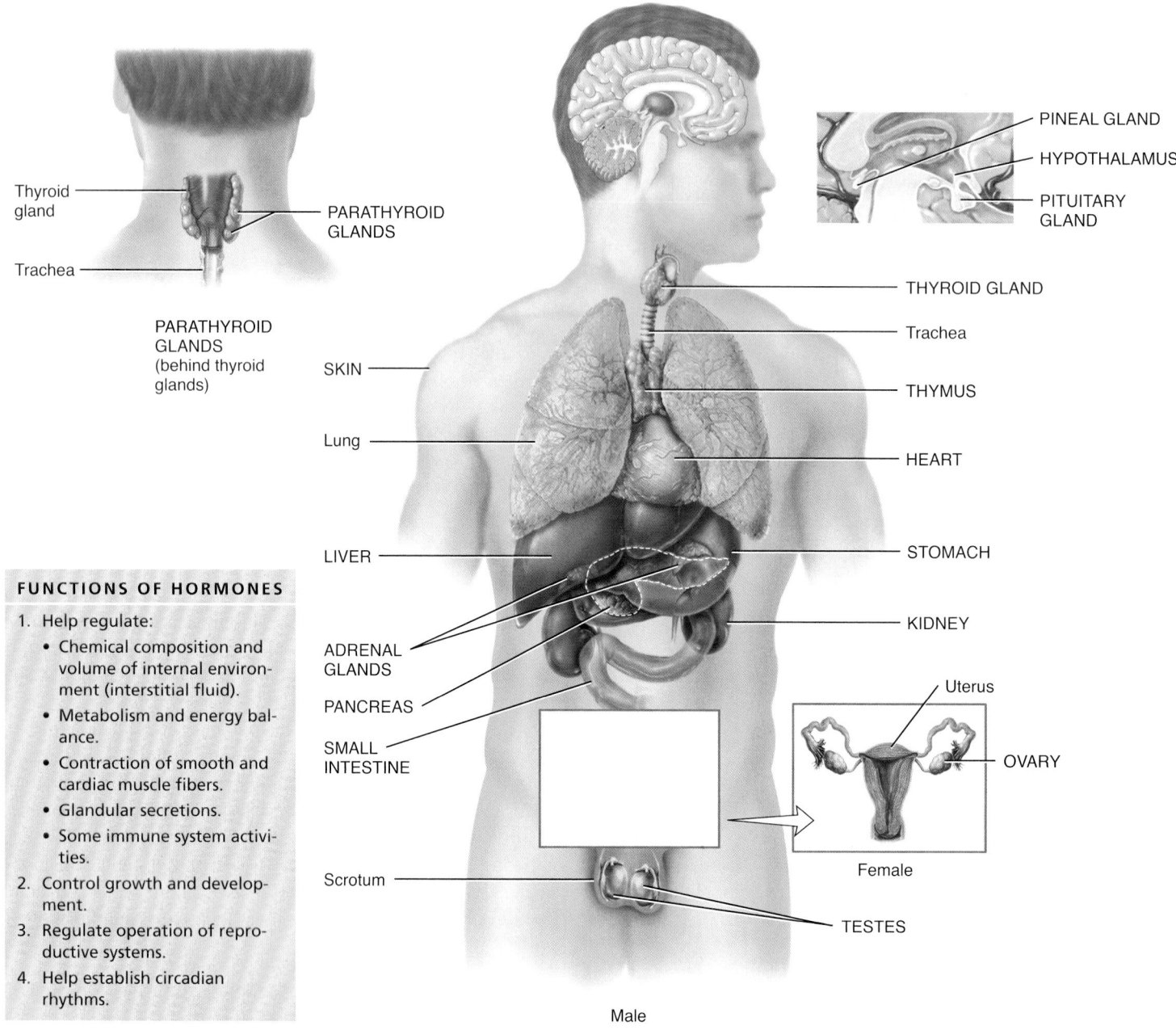

Thyroid gland

PARATHYROID GLANDS

Trachea

PARATHYROID GLANDS (behind thyroid glands)

PINEAL GLAND

HYPOTHALAMUS

PITUITARY GLAND

SKIN

Lung

LIVER

THYROID GLAND

Trachea

THYMUS

HEART

STOMACH

KIDNEY

Uterus

ADRENAL GLANDS

PANCREAS

SMALL INTESTINE

OVARY

Female

Scrotum

TESTES

Male

FUNCTIONS OF HORMONES

1. Help regulate:
   • Chemical composition and volume of internal environment (interstitial fluid).
   • Metabolism and energy balance.
   • Contraction of smooth and cardiac muscle fibers.
   • Glandular secretions.
   • Some immune system activities.
2. Control growth and development.
3. Regulate operation of reproductive systems.
4. Help establish circadian rhythms.

What is the basic difference between endocrine glands and exocrine glands?

the endocrine glands and the diagnosis and treatment of disorders of the endocrine system is **endocrinology** (en′-dō-kri-NOL-ō-jē; *endo-* = within; *-crino-* = to secrete; *-logy* = study of).

✔**CHECKPOINT**

2. List three organs or tissues that are not exclusively classified as endocrine glands but contain cells that secrete hormones.

18.3 HORMONE ACTIVITY

◉ **OBJECTIVES**

• Describe how hormones interact with target-cell receptors.
• Compare the two chemical classes of hormones based on their solubility.

The Role of Hormone Receptors

Although a given hormone travels throughout the body in the blood, it affects only specific target cells. Hormones, like neurotransmitters, influence their target cells by chemically binding to specific protein **receptors.** Only the target cells for a given hormone have receptors that bind and recognize that hormone. For example, thyroid-stimulating hormone (TSH) binds to receptors on cells of the thyroid gland, but it does not bind to cells of the ovaries because ovarian cells do not have TSH receptors.

Receptors, like other cellular proteins, are constantly being synthesized and broken down. Generally, a target cell has 2000 to 100,000 receptors for a particular hormone. If a hormone is present in excess, the number of target-cell receptors may decrease—an effect called **down-regulation.** For example, when certain cells of the testes are exposed to a high concentration of luteinizing hormone (LH), the number of LH receptors decreases. Down-regulation makes a target cell *less sensitive* to a hormone. In contrast, when a hormone is deficient, the number of receptors may increase. This phenomenon, known as **up-regulation,** makes a target cell *more sensitive* to a hormone.

Circulating and Local Hormones

Most endocrine hormones are **circulating hormones**—they pass from the secretory cells that make them into interstitial fluid and then into the blood (Figure 18.2a). Other hormones, termed **local hormones,** act locally on neighboring cells or on the same cell that secreted them without first entering the bloodstream (Figure 18.2b). Local hormones that act on neighboring cells are called **paracrines** (PAR-a-krins; *para-* = beside or near), and those that act on the same cell that secreted them are called **autocrines** (AW-tō-krins; *auto-* = self). One example of a local hormone is interleukin-2 (IL-2), which is released by helper T cells (a type of white blood cell) during immune responses (see Chapter 22). IL-2 helps activate other nearby immune cells, a paracrine effect. But it also acts as an autocrine by stimulating the same cell that released it to proliferate. This action generates more helper T cells that can secrete even more IL-2 and thus strengthen the immune response. Another example of a local hormone is the gas nitric oxide (NO), which is released by endothelial cells lining blood vessels. NO causes relaxation of nearby smooth muscle fibers in blood vessels, which in turn causes vasodilation (increase in blood vessel diameter). The effects of such vasodilation

Figure 18.2 Comparison between circulating hormones and local hormones (autocrines and paracrines).

 Circulating hormones are carried through the bloodstream to act on distant target cells. Paracrines act on neighboring cells, and autocrines act on the same cells that produced them.

(a) Circulating hormones

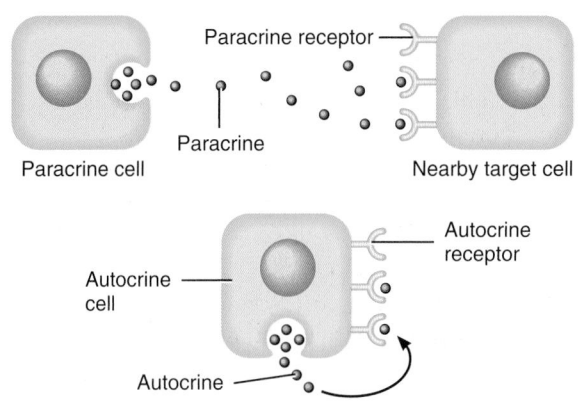

(b) Local hormones (paracrines and autocrines)

 In the stomach, one stimulus for secretion of hydrochloric acid by parietal cells is the release of histamine by neighboring mast cells. Is histamine an autocrine or a paracrine in this situation?

range from a lowering of blood pressure to erection of the penis in males. The drug *Viagra*® (sildenafil) enhances the effects stimulated by nitric oxide in the penis.

Local hormones usually are inactivated quickly; circulating hormones may linger in the blood and exert their effects for a few minutes or occasionally for a few hours. In time, circulating hormones are inactivated by the liver and excreted by the kidneys. In cases of kidney or liver failure, excessive levels of hormones may build up in the blood.

Chemical Classes of Hormones

Chemically, hormones can be divided into two broad classes: those that are soluble in lipids, and those that are soluble in water. This chemical classification is also useful functionally because the two classes exert their effects differently.

Lipid-Soluble Hormones

The lipid-soluble hormones include steroid hormones, thyroid hormones, and nitric oxide.

1. **Steroid hormones** are derived from cholesterol. Each steroid hormone is unique due to the presence of different chemical groups attached at various sites on the four rings at the core of its structure. These small differences allow for a large diversity of functions.

2. Two **thyroid hormones** (T_3 and T_4) are synthesized by attaching iodine to the amino acid tyrosine. The presence of two benzene rings within a T_3 or T_4 molecule makes these molecules very lipid-soluble.

3. The gas **nitric oxide (NO)** is both a hormone and a neurotransmitter. Its synthesis is catalyzed by the enzyme nitric oxide synthase.

Water-Soluble Hormones

The water-soluble hormones include amine hormones, peptide and protein hormones, and eicosanoid hormones.

1. **Amine hormones** (a-MĒN) are synthesized by decarboxylating (removing a molecule of CO_2) and otherwise modifying certain amino acids. They are called amines because they retain an amino group ($—NH_3^+$). The catecholamines—epinephrine, norepinephrine, and dopamine—are synthesized by modifying the amino acid tyrosine. Histamine is synthesized from the amino acid histidine by mast cells and platelets. Serotonin and melatonin are derived from tryptophan.

2. **Peptide hormones** and **protein hormones** are amino acid polymers. The smaller peptide hormones consist of chains of 3 to 49 amino acids; the larger protein hormones include 50 to 200 amino acids. Examples of peptide hormones are antidiuretic hormone and oxytocin; protein hormones include human growth hormone and insulin. Several of the protein hormones, such as thyroid-stimulating hormone, have attached carbohydrate groups and thus are **glycoprotein hormones.**

3. The **eicosanoid hormones** (ī-KŌ-sa-noyd; *eicos-* = twenty forms; *-oid* = resembling) are derived from arachidonic acid, a 20-carbon fatty acid. The two major types of eicosanoids are **prostaglandins** and **leukotrienes.** The eicosanoids are important local hormones, and they may act as circulating hormones as well.

Table 18.2 summarizes the classes of lipid-soluble and water-soluble hormones and provides an overview of the major hormones and their sites of secretion.

Hormone Transport in the Blood

Most water-soluble hormone molecules circulate in the watery blood plasma in a "free" form (not attached to other molecules), but most lipid-soluble hormone molecules are bound to **transport proteins.** The transport proteins, which are synthesized by cells in the liver, have three functions:

1. They make lipid-soluble hormones temporarily water-soluble, thus increasing their solubility in blood.

2. They retard passage of small hormone molecules through the filtering mechanism in the kidneys, thus slowing the rate of hormone loss in the urine.

3. They provide a ready reserve of hormone, already present in the bloodstream.

In general, 0.1–10% of the molecules of a lipid-soluble hormone are not bound to a transport protein. This **free fraction** diffuses out of capillaries, binds to receptors, and triggers responses. As free hormone molecules leave the blood and bind to their receptors, transport proteins release new ones to replenish the free fraction.

CLINICAL CONNECTION | *Administering Hormones*

Both steroid hormones and thyroid hormones are effective when taken by mouth. They are not split apart during digestion and easily cross the intestinal lining because they are lipid-soluble. By contrast, peptide and protein hormones, such as insulin, are not effective oral medications because digestive enzymes destroy them by breaking their peptide bonds. This is why people who need insulin must take it by injection. •

✔ CHECKPOINT

3. What is the difference between down-regulation and up-regulation?
4. Identify the chemical classes of hormones, and give an example of each.
5. How are hormones transported in the blood?

18.4 MECHANISMS OF HORMONE ACTION

⦿ **OBJECTIVE**

• Describe the two general mechanisms of hormone action.

The response to a hormone depends on both the hormone and the target cell. Various target cells respond differently to the same hormone. Insulin, for example, stimulates synthesis of glycogen in liver cells and synthesis of triglycerides in adipose cells.

The response to a hormone is not always the synthesis of new molecules, as is the case for insulin. Other hormonal effects include changing the permeability of the plasma mem-

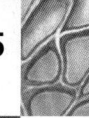

TABLE 18.2

Summary of Hormones by Chemical Class

CHEMICAL CLASS	HORMONES	SITE OF SECRETION
LIPID-SOLUBLE		
Steroid hormones Aldosterone	Aldosterone, cortisol, androgens.	Adrenal cortex.
	Calcitriol.	Kidneys.
	Testosterone.	Testes.
	Estrogens, progesterone.	Ovaries.
Thyroid hormones Triiodothyronine (T_3)	T_3 (triiodothyronine), T_4 (thyroxine).	Thyroid gland (follicular cells).
Gas	Nitric oxide (NO).	Endothelial cells lining blood vessels.
WATER-SOLUBLE		
Amines Norepinephrine	Epinephrine, norepinephrine (catecholamines).	Adrenal medulla.
	Melatonin.	Pineal gland.
	Histamine.	Mast cells in connective tissues.
	Serotonin.	Platelets in blood.
Peptides and proteins Oxytocin	All hypothalamic releasing and inhibiting hormones.	Hypothalamus.
	Oxytocin, antidiuretic hormone.	Posterior pituitary.
	Human growth hormone, thyroid-stimulating hormone, adrenocorticotropic hormone, follicle-stimulating hormone, luteinizing hormone, prolactin, melanocyte-stimulating hormone.	Anterior pituitary.
	Insulin, glucagon, somatostatin, pancreatic polypeptide.	Pancreas.
	Parathyroid hormone.	Parathyroid glands.
	Calcitonin.	Thyroid gland (parafollicular cells).
	Gastrin, secretin, cholecystokinin, GIP (glucose-dependent insulinotropic peptide).	Stomach and small intestine (enteroendocrine cells).
	Erythropoietin.	Kidneys.
	Leptin.	Adipose tissue.
Eicosanoids A leukotriene (LTB_4)	Prostaglandins, leukotrienes.	All cells except red blood cells.

brane, stimulating transport of a substance into or out of the target cells, altering the rate of specific metabolic reactions, or causing contraction of smooth muscle or cardiac muscle. In part, these varied effects of hormones are possible because a single hormone can set in motion several different cellular responses. However, a hormone must first "announce its arrival" to a target cell by binding to its receptors. The receptors for lipid-soluble hormones are located inside target cells. The receptors for water-soluble hormones are part of the plasma membrane of target cells.

Action of Lipid-Soluble Hormones

As you just learned, lipid-soluble hormones, including steroid hormones and thyroid hormones, bind to receptors within target cells. Their mechanism of action is as follows (Figure 18.3):

1 A free lipid-soluble hormone molecule diffuses from the blood, through interstitial fluid, and through the lipid bilayer of the plasma membrane into a cell.

2 If the cell is a target cell, the hormone binds to and activates receptors located within the cytosol or nucleus. The activated receptor–hormone complex then alters gene expression: It turns specific genes of the nuclear DNA on or off.

3 As the DNA is transcribed, new messenger RNA (mRNA) forms, leaves the nucleus, and enters the cytosol. There, it directs synthesis of a new protein, often an enzyme, on the ribosomes.

4 The new proteins alter the cell's activity and cause the responses typical of that hormone.

Action of Water-Soluble Hormones

Because amine, peptide, protein, and eicosanoid hormones are not lipid-soluble, they cannot diffuse through the lipid bilayer of the plasma membrane and bind to receptors inside target cells. Instead, water-soluble hormones bind to receptors that protrude from the target-cell surface. The receptors are integral transmembrane proteins in the plasma membrane. When a water-soluble hormone binds to its receptor at the outer surface of the plasma membrane, it acts as the **first messenger.** The first messenger (the hormone) then causes production of a **second messenger** inside the cell, where specific hormone-stimulated responses take place. One common second messenger is **cyclic AMP (cAMP).** Neurotransmitters, neuropeptides, and several sensory transduction mechanisms (for example, vision; see Figure 17.16) also act via second-messenger systems.

The action of a typical water-soluble hormone occurs as follows (Figure 18.4):

1 A water-soluble hormone (the first messenger) diffuses from the blood through interstitial fluid and then binds to its receptor at the exterior surface of a target cell's plasma membrane. The hormone–receptor complex activates a membrane protein called a **G protein.** The activated G protein in turn activates **adenylate cyclase** (a-DEN-i-lāt SĪ-klās).

2 Adenylate cyclase converts ATP into cyclic AMP (cAMP). Because the enzyme's active site is on the inner surface of the plasma membrane, this reaction occurs in the cytosol of the cell.

3 Cyclic AMP (the second messenger) activates one or more protein kinases, which may be free in the cytosol or bound to the plasma membrane. A **protein kinase** is an enzyme that phosphorylates (adds a phosphate group to) other cellular proteins (such as enzymes). The donor of the phosphate group is ATP, which is converted to ADP.

4 Activated protein kinases phosphorylate one or more cellular proteins. Phosphorylation activates some of these proteins and inactivates others, rather like turning a switch on or off.

5 Phosphorylated proteins in turn cause reactions that produce physiological responses. Different protein kinases exist within different target cells and within different organelles of the same target cell. Thus, one protein kinase might trigger glycogen synthesis, a second might cause the breakdown of triglyceride, a third may promote protein synthesis, and so forth. As noted in step **4**, phosphorylation by a protein kinase can also inhibit certain proteins. For example, some of the kinases unleashed when epinephrine binds to liver cells inactivate an enzyme needed for glycogen synthesis.

6 After a brief period, an enzyme called **phosphodiesterase** (fos'-fō-dī-ES-ter'-ās) inactivates cAMP. Thus, the cell's response is turned off unless new hormone molecules continue to bind to their receptors in the plasma membrane.

The binding of a hormone to its receptor activates many G-protein molecules, which in turn activate molecules of adenylate cyclase (as noted in step **1**). Unless they are further stimulated by the binding of more hormone molecules to receptors, G proteins

Figure 18.3 Mechanism of action of the lipid-soluble steroid hormones and thyroid hormones.

🔑 Lipid-soluble hormones bind to receptors inside target cells.

Free hormone

Blood capillary

Transport protein

1 Lipid-soluble hormone diffuses into cell

2 Activated receptor–hormone complex alters gene expression

Nucleus

Receptor

Cytosol

DNA

mRNA

3 Newly formed mRNA directs synthesis of specific proteins on ribosomes

Ribosome

New protein

4 New proteins alter cell's activity

Target cell

? What is the action of the receptor–hormone complex?

slowly inactivate, thus decreasing the activity of adenylate cyclase and helping to stop the hormone response. G proteins are a common feature of most second-messenger systems.

Many hormones exert at least some of their physiological effects through the *increased* synthesis of cAMP. Examples include antidiuretic hormone (ADH), thyroid-stimulating hormone (TSH), adrenocorticotropic hormone (ACTH), glucagon, epinephrine, and hypothalamic releasing hormones. In other cases, such as growth hormone–inhibiting hormone (GHIH), the level of cyclic AMP *decreases* in response to the binding of a hormone to its receptor. Besides cAMP, other second messengers include calcium ions (Ca^{2+}), cGMP (cyclic guanosine monophosphate, a cyclic nucleotide similar to cAMP), inositol trisphosphate (IP_3), and diacylglycerol (DAG). A given hormone may use different second messengers in different target cells.

Hormones that bind to plasma membrane receptors can induce their effects at very low concentrations because they initiate a cascade or chain reaction, each step of which multiplies or amplifies the initial effect. For example, the binding of a single molecule of epinephrine to its receptor on a liver cell may activate a hundred or so G proteins, each of which activates an adenylate cyclase molecule. If each adenylate cyclase produces even 1000 cAMP, then 100,000 of these second messengers will be liberated inside the cell. Each cAMP may activate a protein kinase, which in turn can act on hundreds or thousands of substrate molecules. Some of the kinases phosphorylate and activate a key enzyme needed for glycogen breakdown. The end result of the binding of a single molecule of epinephrine to its receptor is the breakdown of millions of glycogen molecules into glucose monomers.

Hormone Interactions

The responsiveness of a target cell to a hormone depends on (1) the hormone's concentration in the blood, (2) the abundance of the target cell's hormone receptors, and (3) influences exerted by other hormones. A target cell responds more vigorously when the level of a hormone rises or when it has more receptors (upregulation). In addition, the actions of some hormones on target cells require a simultaneous or recent exposure to a second hormone. In such cases, the second hormone is said to have a **permissive effect.** For example, epinephrine alone only weakly stimulates lipolysis (the breakdown of triglycerides), but when small amounts of thyroid hormones (T_3 and T_4) are present, the same amount of epinephrine stimulates lipolysis much more powerfully. Sometimes the permissive hormone increases the number of receptors for the other hormone, and sometimes it promotes the synthesis of an enzyme required for the expression of the other hormone's effects.

When the effect of two hormones acting together is greater or more extensive than the effect of each hormone acting alone, the two hormones are said to have a **synergistic effect.** For example, normal development of oocytes in the ovaries requires both follicle-stimulating hormone from the anterior pituitary and estrogens from the ovaries. Neither hormone alone is sufficient.

When one hormone opposes the actions of another hormone, the two hormones are said to have **antagonistic effects.** An example of an antagonistic pair of hormones is insulin, which promotes synthesis of glycogen by liver cells, and glucagon, which stimulates breakdown of glycogen in the liver.

Figure 18.4 Mechanism of action of the water-soluble hormones (amines, peptides, proteins, and eicosanoids).

Water-soluble hormones bind to receptors embedded in the plasma membranes of target cells.

Blood capillary

1 Binding of hormone (first messenger) to its receptor activates G protein, which activates adenylate cyclase

Water-soluble hormone

Receptor

Adenylate cyclase

G protein

Second messenger

ATP

cAMP

2 Activated adenylate cyclase converts ATP to cAMP

Protein kinases

6 Phosphodiesterase inactivates cAMP

3 cAMP serves as a second messenger to activate protein kinases

Activated protein kinases

Protein

4 Activated protein kinases phosphorylate cellular proteins

ATP

ADP

Protein – P

5 Millions of phosphorylated proteins cause reactions that produce physiological responses

Target cell

? Why is cAMP called a "second messenger"?

✔ CHECKPOINT

6. What factors determine the responsiveness of a target cell to a hormone?

7. What are the differences among permissive effects, synergistic effects, and antagonistic effects of hormones?

18.5 CONTROL OF HORMONE SECRETION

● OBJECTIVE

• Describe the mechanisms of control of hormone secretion.

The release of most hormones occurs in short bursts, with little or no secretion between bursts. When stimulated, an endocrine gland will release its hormone in more frequent bursts, increasing the concentration of the hormone in the blood. In the absence of stimulation, the blood level of the hormone decreases. Regulation of secretion normally prevents overproduction or underproduction of any given hormone.

Hormone secretion is regulated by (1) signals from the nervous system, (2) chemical changes in the blood, and (3) other hormones. For example, nerve impulses to the adrenal medullae regulate the release of epinephrine; blood Ca^{2+} level regulates the secretion of parathyroid hormone; and a hormone from the anterior pituitary (adrenocorticotropic hormone) stimulates the release of cortisol by the adrenal cortex. Most hormonal regulatory systems work via negative feedback (see Figure 1.3), but a few operate via positive feedback (see Figure 1.4). For example, during childbirth, the hormone oxytocin stimulates contractions of the uterus, and uterine contractions in turn stimulate more oxytocin release, a positive feedback effect.

Now that you have a general understanding of the roles of hormones in the endocrine system, we turn to discussions of the various endocrine glands and the hormones they secrete.

✔ CHECKPOINT

8. What three types of signals control hormone secretion?

18.6 HYPOTHALAMUS AND PITUITARY GLAND

● OBJECTIVES

• Describe the locations of and relationships between the hypothalamus and pituitary gland.
• Describe the location, histology, hormones, and functions of the anterior and posterior pituitary.

For many years, the **pituitary gland** (pi-TOO-i-tār-ē) or **hypophysis** (hī-POF-i-sis) was called the "master" endocrine gland because it secretes several hormones that control other endocrine glands. We now know that the pituitary gland itself has a master—the **hypothalamus.** This small region of the brain below the thalamus is the major link between the nervous and endocrine systems. Cells in the hypothalamus synthesize at least nine different hormones, and the pituitary gland secretes seven. Together, these hormones play important roles in the regulation of virtually all aspects of growth, development, metabolism, and homeostasis.

The pituitary gland is a pea-shaped structure that measures 1–1.5 cm (0.5 in.) in diameter and lies in the hypophyseal fossa of the sella turcica of the sphenoid bone. It attaches to the hypothalamus by a stalk, the **infundibulum** (in'-fun-DIB-ū-lum = a funnel;

Figure 18.5), and has two anatomically and functionally separate portions: the anterior pituitary and the posterior pituitary. The **anterior pituitary (anterior lobe),** also called the **adenohypophysis,** accounts for about 75% of the total weight of the gland and is composed of epithelial tissue. The anterior pituitary consists of two parts in an adult: The **pars distalis** is the larger portion, and the **pars tuberalis** (PARS too-be'-RAL-is) forms a sheath around the infundibulum. The **posterior pituitary (posterior lobe),** also called the **neurohypophysis,** is composed of neural tissue. It also consists of two parts: the **pars nervosa** (ner-VŌ-sa), the larger bulbar portion, and the infundibulum. A third region of the pituitary gland called the **pars intermedia** atrophies during human fetal development and ceases to exist as a separate lobe in adults (see Figure 18.21b). However, some of its cells migrate into adjacent parts of the anterior pituitary, where they persist.

Anterior Pituitary

The **anterior pituitary** or **adenohypophysis** (ad'-e-nō-hī-POF-i-sis; *adeno-* = gland; *-hypophysis* = undergrowth) secretes hormones that regulate a wide range of bodily activities, from growth to reproduction. Release of anterior pituitary hormones is stimulated by **releasing hormones** and suppressed by **inhibiting hormones** from the hypothalamus. Thus, the hypothalamic hormones are an important link between the nervous and endocrine systems.

Hypophyseal Portal System

Hypothalamic hormones that release or inhibit anterior pituitary hormones reach the anterior pituitary through a portal system. Usually, blood passes from the heart through an artery to a capillary to a vein and back to the heart. In a *portal system,* blood flows from one capillary network into a portal vein, and then into a second capillary network before returning to the heart. The name of the portal system indicates the location of the second capillary network. In the **hypophyseal portal system** (hī'-pō-FIZ-ē-al), blood flows from capillaries in the hypothalamus into portal veins that carry blood to capillaries of the anterior pituitary.

The **superior hypophyseal arteries,** branches of the internal carotid arteries, bring blood into the hypothalamus (Figure 18.5a). At the junction of the median eminence of the hypothalamus and the infundibulum, these arteries divide into a capillary network called the **primary plexus of the hypophyseal portal system.** From the primary plexus, blood drains into the **hypophyseal portal veins** that pass down the outside of the infundibulum. In the anterior pituitary, the hypophyseal portal veins divide again and form another capillary network called the **secondary plexus of the hypophyseal portal system.**

Above the optic chiasm are clusters of specialized neurons called **neurosecretory cells** (Figure 18.5b). They synthesize the hypothalamic releasing and inhibiting hormones in their cell bodies and package the hormones inside vesicles, which reach the axon terminals by axonal transport. Nerve impulses stimulate the vesicles to undergo exocytosis. The hormones then diffuse into the primary plexus of the hypophyseal portal system. Quickly, the hypothalamic hormones flow with the blood

Figure 18.5 Hypothalamus and pituitary gland, and their blood supply. Releasing and inhibiting hormones synthesized by hypothalamic neurosecretory cells are transported within axons and released at the axon terminals. The hormones diffuse into capillaries of the primary plexus of the hypophyseal portal system and are carried by the hypophyseal portal veins to the secondary plexus of the hypophyseal portal system for distribution to target cells in the anterior pituitary.

Hypothalamic hormones are an important link between the nervous and endocrine systems.

Infundibulum

POSTERIOR PITUITARY

ANTERIOR PITUITARY

Sagittal section of pituitary gland

Hypothalamus

Pituitary gland

Hypothalamic neurosecretory cell

Hypophyseal portal veins

(b) Path of releasing and inhibiting hormones

Primary plexus of hypophyseal portal system

HYPOTHALAMUS

Infundibulum

Median eminence

Superior hypophyseal artery

Posterior hypophyseal veins

Hypophyseal portal veins

Sphenoid bone

POSTERIOR PITUITARY

ANTERIOR PITUITARY

Capillary plexus of infundibular process

Secondary plexus of hypophyseal portal system

Anterior hypophyseal veins

Hypophyseal fossa

POSTERIOR

ANTERIOR

Inferior hypophyseal artery

(a) Relationship of hypothalamus to pituitary gland

Somatotroph

Thyrotroph

Gonadotroph

Corticotroph

Lactotroph

LM   all about 65x

(c) Histology of anterior pituitary

What is the functional importance of the hypophyseal portal veins?

through the portal veins and into the secondary plexus. This direct route permits hypothalamic hormones to act immediately on anterior pituitary cells, before the hormones are diluted or destroyed in the general circulation. Hormones secreted by anterior pituitary cells pass into the secondary plexus capillaries, which drain into the anterior hypophyseal veins and out into the general circulation. Anterior pituitary hormones then travel to target tissues throughout the body. Those anterior pituitary hormones that act on other endocrine glands are called **tropic hormones** (TRŌ-pik) or tropins.

Types of Anterior Pituitary Cells and Their Hormones

Five types of anterior pituitary cells—somatotrophs, thyrotrophs, gonadotrophs, lactotrophs, and corticotrophs—secrete seven hormones (Table 18.3):

1. **Somatotrophs** (sō-MAT-ō-trōfs) secrete **human growth hormone (hGH),** also known as **somatotropin** (sō′-ma-tō-TRŌ-pin; *somato-* = body; *-tropin* = change). Human growth hormone in turn stimulates several tissues to secrete **insulinlike growth factors,** hormones that stimulate general body growth and regulate aspects of metabolism.

2. **Thyrotrophs** (THĪ-rō-trōfs) secrete **thyroid-stimulating hormone (TSH),** also known as **thyrotropin** (thī-rō-TRŌ-pin; *thyro-* = pertaining to the thyroid gland). TSH controls the secretions and other activities of the thyroid gland.

3. **Gonadotrophs** (gō-NAD-ō-trōfs; *gonado-* = seed) secrete two **gonadotropins: follicle-stimulating hormone (FSH)** and **luteinizing hormone (LH)** (LOO-tē-in′-īz-ing). FSH and LH both act on the gonads. They stimulate secretion of estrogens and progesterone and the maturation of oocytes in the ovaries, and they stimulate sperm production and secretion of testosterone in the testes.

4. **Lactotrophs** (LAK-tō-trōfs; *lacto-* = milk) secrete **prolactin (PRL),** which initiates milk production in the mammary glands.

5. **Corticotrophs** (KOR-ti-kō-trōfs) secrete **adrenocorticotropic hormone (ACTH),** also known as **corticotropin** (kor′-ti-kō-TRŌ-pin; *cortico-* = rind or bark), which stimulates the adrenal cortex to secrete glucocorticoids such as cortisol. Some corticotrophs, remnants of the pars intermedia, also secrete **melanocyte-stimulating hormone (MSH).**

Control of Secretion by the Anterior Pituitary

Secretion of anterior pituitary hormones is regulated in two ways. First, neurosecretory cells in the hypothalamus secrete five releasing hormones, which stimulate secretion of anterior pituitary hormones, and two inhibiting hormones, which suppress secretion of anterior pituitary hormones (Table 18.3). Second, negative feedback in the form of hormones released by target glands decreases secretions of three types of anterior pituitary cells (Figure 18.6). In such negative feedback loops, the secretory activity of thyrotrophs, gonadotrophs, and corticotrophs decreases when blood levels of their target gland hormones rise. For example, adrenocorticotropic hormone (ACTH) stimulates the cortex of the adrenal gland to secrete glucocorticoids, mainly cortisol. In turn, an elevated blood level of cortisol decreases secretion of both corticotropin and corticotropin-releasing hormone (CRH) by suppressing the activity of the anterior pituitary corticotrophs and hypothalamic neurosecretory cells.

Human Growth Hormone and Insulinlike Growth Factors

Somatotrophs are the most numerous cells in the anterior pituitary, and human growth hormone (hGH) is the most plentiful anterior pituitary hormone. The main function of hGH is to promote synthesis and secretion of small protein hormones called **insulinlike growth factors (IGFs)** or **somatomedins** (sō′-ma-tō-MĒ-

---

TABLE 18.3

Hormones of the Anterior Pituitary

HORMONE	SECRETED BY	HYPOTHALAMIC-RELEASING HORMONE (STIMULATES SECRETION)	HYPOTHALAMIC-INHIBITING HORMONE (SUPPRESSES SECRETION)
Human growth hormone (hGH), also known as somatotropin	Somatotrophs.	Growth hormone–releasing hormone (GHRH), also known as somatocrinin.	Growth hormone–inhibiting hormone (GHIH), also known as somatostatin.
Thyroid-stimulating hormone (TSH), also known as thyrotropin	Thyrotrophs.	Thyrotropin-releasing hormone (TRH).	Growth hormone–inhibiting hormone (GHIH).
Follicle-stimulating hormone (FSH)	Gonadotrophs.	Gonadotropin-releasing hormone (GnRH).	—
Luteinizing hormone (LH)	Gonadotrophs.	Gonadotropin-releasing hormone (GnRH).	—
Prolactin (PRL)	Lactotrophs.	Prolactin-releasing hormone (PRH).*	Prolactin-inhibiting hormone (PIH), which is dopamine.
Adrenocorticotropic hormone (ACTH), also known as corticotropin	Corticotrophs.	Corticotropin-releasing hormone (CRH).	—
Melanocyte-stimulating hormone (MSH)	Corticotrophs.	Corticotropin-releasing hormone (CRH).	Dopamine.

*Thought to exist, but exact nature is uncertain.

Figure 18.6 Negative feedback regulation of hypothalamic neurosecretory cells and anterior pituitary corticotrophs. Solid arrows indicate stimulation of secretions; dashed arrows indicate inhibition of secretion via negative feedback.

 Cortisol secreted by the adrenal cortex suppresses secretion of CRH and ACTH.

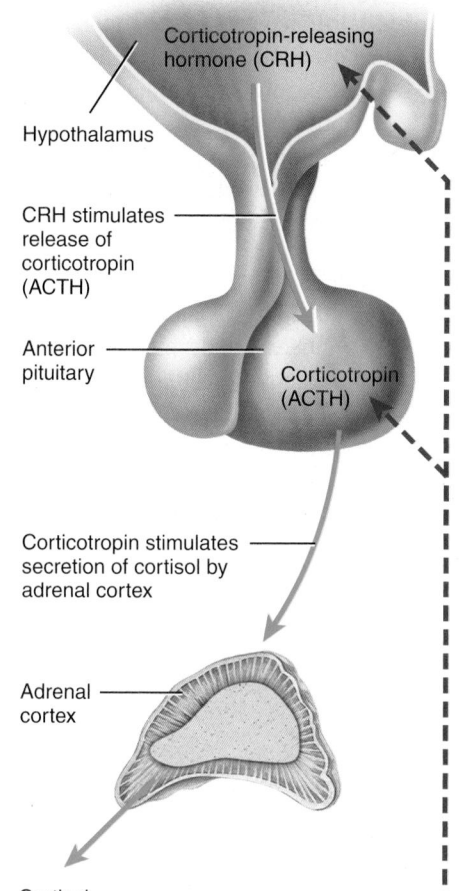

Corticotropin-releasing hormone (CRH)

Hypothalamus

CRH stimulates release of corticotropin (ACTH)

Anterior pituitary

Corticotropin (ACTH)

Elevated cortisol inhibits release of CRH by hypothalamic neurosecretory cells

Corticotropin stimulates secretion of cortisol by adrenal cortex

Elevated cortisol inhibits release of corticotropin by anterior pituitary corticotrophs

Adrenal cortex

Cortisol

? Which other target gland hormones suppress secretion of hypothalamic and anterior pituitary hormones by negative feedback?

dins). In response to human growth hormone, cells in the liver, skeletal muscles, cartilage, bones, and other tissues secrete IGFs, which may either enter the bloodstream from the liver or act locally in other tissues as autocrines or paracrines.

The functions of IGFs include the following:

1. IGFs cause cells to grow and multiply by increasing uptake of amino acids into cells and accelerating protein synthesis. IGFs also decrease the breakdown of proteins and the use of amino acids for ATP production. Due to these effects of the IGFs, human growth hormone increases the growth rate of the skeleton and skeletal muscles during childhood and the teenage years. In adults, human growth hormone and IGFs help maintain the mass of muscles and bones and promote healing of injuries and tissue repair.

2. IGFs also enhance lipolysis in adipose tissue, which results in increased use of the released fatty acids for ATP production by body cells.

3. In addition to affecting protein and lipid metabolism, human growth hormone and IGFs influence carbohydrate metabolism by decreasing glucose uptake, which decreases the use of glucose for ATP production by most body cells. This action spares glucose so that it is available to neurons for ATP production in times of glucose scarcity. IGFs and human growth hormone may also stimulate liver cells to release glucose into the blood.

Somatotrophs in the anterior pituitary release bursts of human growth hormone every few hours, especially during sleep. Their secretory activity is controlled mainly by two hypothalamic hormones: (1) growth hormone–releasing hormone (GHRH) promotes secretion of human growth hormone, and (2) growth hormone–inhibiting hormone (GHIH) suppresses it. A major regulator of GHRH and GHIH secretion is the blood glucose level (Figure 18.7):

1 **Hypoglycemia** (hī′-pō-glī-SĒ-mē-a), an abnormally low blood glucose concentration, stimulates the hypothalamus to secrete GHRH, which flows toward the anterior pituitary in the hypophyseal portal veins.

2 On reaching the anterior pituitary, GHRH stimulates somatotrophs to release human growth hormone.

3 Human growth hormone stimulates secretion of insulinlike growth factors, which speed breakdown of liver glycogen into glucose, causing glucose to enter the blood more rapidly.

4 As a result, blood glucose rises to the normal level (about 90 mg/100 mL of blood plasma).

5 An increase in blood glucose above the normal level inhibits release of GHRH.

6 **Hyperglycemia** (hī′-per-glī-SĒ-mē-a), an abnormally high blood glucose concentration, stimulates the hypothalamus to secrete GHIH (while inhibiting the secretion of GHRH).

7 On reaching the anterior pituitary in portal blood, GHIH inhibits secretion of human growth hormone by somatotrophs.

8 A low level of human growth hormone and IGFs slows breakdown of glycogen in the liver, and glucose is released into the blood more slowly.

9 Blood glucose falls to the normal level.

10 A decrease in blood glucose below the normal level (hypoglycemia) inhibits release of GHIH.

Other stimuli that promote secretion of human growth hormone include decreased fatty acids and increased amino acids in the blood; deep sleep (stages 3 and 4 of non-rapid eye movement sleep); increased activity of the sympathetic division of the autonomic nervous system, such as might occur with stress or vigorous physical exercise; and other hormones, including glucagon, estrogens, cortisol, and insulin. Factors that inhibit human growth hormone secretion are increased levels of fatty acids and decreased levels of amino acids in the blood; rapid eye movement sleep;

Figure 18.7 **Effects of human growth hormone (hGH) and insulinlike growth factors (IGFs).** Solid arrows indicate stimulation of secretion; dashed arrows indicate inhibition of secretion via negative feedback.

🔑 Secretion of hGH is stimulated by growth hormone–releasing hormone (GHRH) and inhibited by growth hormone–inhibiting hormone (GHIH).

1 Low blood glucose (hypoglycemia) stimulates release of

6 High blood glucose (hyperglycemia) stimulates release of

GHRH GHIH

2 GHRH stimulates secretion of hGH by somatotrophs

7 GHIH inhibits secretion of hGH by somatotrophs

hGH

Anterior pituitary

3 hGH and IGFs speed up breakdown of liver glycogen into glucose, which enters the blood more rapidly

8 A low level of hGH and IGFs decreases the rate of glycogen breakdown in the liver and glucose enters the blood more slowly

4 Blood glucose level rises to normal (about 90 mg/100 mL)

9 Blood glucose level falls to normal (about 90 mg/100 mL)

5 If blood glucose continues to increase, hyperglycemia inhibits release of GHRH

10 If blood glucose continues to decrease, hypoglycemia inhibits release of GHIH

❓ **If a person has a pituitary tumor that secretes a large amount of hGH and the tumor cells are not responsive to regulation by GHRH and GHIH, will hyperglycemia or hypoglycemia be more likely?**

emotional deprivation; obesity; low levels of thyroid hormones; and human growth hormone itself (through negative feedback). Growth hormone–inhibiting hormone (GHIH), alternatively known as somatostatin, also inhibits the secretion of human growth hormone.

⚕ **CLINICAL CONNECTION** | *Diabetogenic Effect of hGH*

One symptom of excess human growth hormone (hGH) is hyperglycemia. Persistent hyperglycemia in turn stimulates the pancreas to secrete insulin continually. Such excessive stimulation, if it lasts for weeks or months, may cause "beta-cell burnout," a greatly decreased capacity of pancreatic beta cells to synthesize and secrete insulin. Thus, excess secretion of human growth hormone may have a **diabetogenic effect** (dī′-a-bet′-o-JEN-ik); that is, it causes diabetes mellitus (lack of insulin activity). •

Thyroid-Stimulating Hormone

Thyroid-stimulating hormone (TSH) stimulates the synthesis and secretion of the two thyroid hormones, triiodothyronine (T_3) and thyroxine (T_4), both produced by the thyroid gland. Thyrotropin-releasing hormone (TRH) from the hypothalamus controls TSH secretion. Release of TRH in turn depends on blood levels of T_3 and T_4; high levels of T_3 and T_4 inhibit secretion of TRH via negative feedback. There is no thyrotropin-inhibiting hormone. The release of TRH is explained later in the chapter (see Figure 18.12).

Follicle-Stimulating Hormone

In females, the ovaries are the targets for follicle-stimulating hormone (FSH). Each month FSH initiates the development of several ovarian follicles, saclike arrangements of secretory cells that surround a developing oocyte. FSH also stimulates follicular cells to secrete estrogens (female sex hormones). In males, FSH stimulates sperm production in the testes. Gonadotropin-releasing hormone (GnRH) from the hypothalamus stimulates FSH release. Release of GnRH and FSH is suppressed by estrogens in females and by testosterone (the principal male sex hormone) in males through negative feedback systems. There is no gonadotropin-inhibiting hormone.

Luteinizing Hormone

In females, luteinizing hormone (LH) triggers **ovulation,** the release of a secondary oocyte (future ovum) by an ovary. LH stimulates formation of the corpus luteum (structure formed after ovulation) in the ovary and the secretion of progesterone (another female sex hormone) by the corpus luteum. Together, FSH and LH also stimulate secretion of estrogens by ovarian cells. Estrogens and progesterone prepare the uterus for implantation of a fertilized ovum and help prepare the mammary glands for milk secretion. In males, LH stimulates cells in the testes to secrete testosterone. Secretion of LH, like that of FSH, is controlled by gonadotropin-releasing hormone (GnRH).

Prolactin

Prolactin (PRL), together with other hormones, initiates and maintains milk production by the mammary glands. By itself, prolactin has only a weak effect. Only after the mammary glands have been primed by estrogens, progesterone, glucocorticoids, human growth hormone, thyroxine, and insulin, which exert permissive effects, does PRL bring about milk production. Ejection of milk from the mammary glands depends on the hormone oxytocin, which is released from the posterior pituitary. Together, milk production and ejection constitute *lactation.*

The hypothalamus secretes both inhibitory and excitatory hormones that regulate prolactin secretion. In females, prolactin-inhibiting hormone (PIH), which is dopamine, inhibits the release of prolactin from the anterior pituitary most of the time. Each month, just before menstruation begins, the secretion of PIH diminishes and

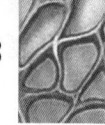

the blood level of prolactin rises, but not enough to stimulate milk production. Breast tenderness just before menstruation may be caused by elevated prolactin. As the menstrual cycle begins anew, PIH is again secreted and the prolactin level drops. During pregnancy, the prolactin level rises, stimulated by prolactin-releasing hormone (PRH) from the hypothalamus. The sucking action of a nursing infant causes a reduction in hypothalamic secretion of PIH.

The function of prolactin is not known in males, but its hypersecretion causes erectile dysfunction (impotence, the inability to have an erection of the penis). In females, hypersecretion of prolactin causes galactorrhea (inappropriate lactation) and amenorrhea (absence of menstrual cycles).

Adrenocorticotropic Hormone

Corticotrophs secrete mainly adrenocorticotropic hormone (ACTH). ACTH controls the production and secretion of cortisol and other glucocorticoids by the cortex (outer portion) of the adrenal glands. Corticotropin-releasing hormone (CRH) from the hypothalamus stimulates secretion of ACTH by corticotrophs. Stress-related stimuli, such as low blood glucose or physical trauma, and interleukin-1, a substance produced by macrophages, also stimulate release of ACTH. Glucocorticoids inhibit CRH and ACTH release via negative feedback.

Melanocyte-Stimulating Hormone

Melanocyte-stimulating hormone (MSH) increases skin pigmentation in amphibians by stimulating the dispersion of melanin granules in melanocytes. Its exact role in humans is unknown, but the presence of MSH receptors in the brain suggests it may influence brain activity. There is little circulating MSH in humans. However, continued administration of MSH for several days does produce a darkening of the skin. Excessive levels of corticotropin-releasing hormone (CRH) can stimulate MSH release; dopamine inhibits MSH release.

Table 18.4 summarizes the principal actions of the anterior pituitary hormones.

TABLE 18.4

Summary of the Principal Actions of Anterior Pituitary Hormones

HORMONE	TARGET TISSUES	PRINCIPAL ACTIONS
Human growth hormone (hGH), also known as somatotropin	Liver (and other tissues)	Stimulates liver, muscle, cartilage, bone, and other tissues to synthesize and secrete insulinlike growth factors (IGFs); IGFs promote growth of body cells, protein synthesis, tissue repair, lipolysis, and elevation of blood glucose concentration.
Thyroid-stimulating hormone (TSH), also known as thyrotropin	Thyroid gland	Stimulates synthesis and secretion of thyroid hormones by thyroid gland.
Follicle-stimulating hormone (FSH)	Ovaries Testes	In females, initiates development of oocytes and induces ovarian secretion of estrogens. In males, stimulates testes to produce sperm.
Luteinizing hormone (LH)	Ovaries Testes	In females, stimulates secretion of estrogens and progesterone, ovulation, and formation of corpus luteum. In males, stimulates testes to produce testosterone.
Prolactin (PRL)	Mammary glands	Together with other hormones, promotes milk production by mammary glands.
Adrenocorticotropic hormone (ACTH), also known as corticotropin	Adrenal cortex	Stimulates secretion of glucocorticoids (mainly cortisol) by adrenal cortex.
Melanocyte-stimulating hormone (MSH)	Brain	Exact role in humans is unknown but may influence brain activity; when present in excess, can cause darkening of skin.

Posterior Pituitary

Although the **posterior pituitary** or **neurohypophysis** (noo′-rō-hī-POF-i-sis; *neuro-* = nerve; *-hypophysis* = undergrowth) does not *synthesize* hormones, it does *store* and *release* two hormones. It consists of axons and axon terminals of more than 10,000 hypothalamic neurosecretory cells. The cell bodies of the neurosecretory cells are in the paraventricular and supraoptic nuclei of the hypothalamus; their axons form the **hypothalamohypophyseal tract** (hī′-pō-thal′-a-mō-hī-pō-FIZ-ē-al). This tract begins in the hypothalamus and ends near blood capillaries in the posterior pituitary (Figure 18.8). The neuronal cell bodies in the paraventricular nucleus synthesize the hormone **oxytocin** (**OT;** ok′-sē-TŌ-sin; *okytoc* = quick birth) and the neuronal cell bodies in the supraoptic nucleus produce **antidiuretic hormone** (**ADH**), also called **vasopressin.** The axon terminals in the posterior pituitary are associated with specialized neuroglia called **pituicytes** (pi-TOO-i-sītz). These cells have a supporting role similar to that of astrocytes (see Chapter 12).

After their production in the cell bodies of neurosecretory cells, oxytocin and antidiuretic hormone are packaged into secretory vesicles, which move by fast axonal transport (described in Section 12.2) to the axon terminals in the posterior pituitary, where they are stored until nerve impulses trigger exocytosis and release of the hormone.

Blood is supplied to the posterior pituitary by the **inferior hypophyseal arteries,** which branch from the internal carotid arteries. In the posterior pituitary, the inferior hypophyseal arteries drain into the **capillary plexus of the infundibular process,** a capillary network that receives secreted oxytocin and antidiuretic hormone (see Figure 18.5). From this plexus, hormones pass into the **posterior hypophyseal veins** for distribution to target cells in other tissues.

Oxytocin

During and after delivery of a baby, oxytocin affects two target tissues: the mother's uterus and breasts. During delivery, stretching of the cervix of the uterus stimulates the release of oxytocin which, in turn, enhances contraction of smooth muscle cells in the wall of the uterus (see Figure 1.4); after delivery, it stimulates milk ejection ("letdown") from the mammary glands in response to the mechanical stimulus provided by a suckling infant. The

Figure 18.8 **Axons of hypothalamic neurosecretory cells form the hypothalamohypophyseal tract, which extends from the paraventricular and supraoptic nuclei to the posterior pituitary.** Hormone molecules synthesized in the cell body of a neurosecretory cell are packaged into secretory vesicles that move down to the axon terminals. Nerve impulses trigger exocytosis of the vesicles, thereby releasing the hormone.

Oxytocin and antidiuretic hormone are synthesized in the hypothalamus and released into the capillary plexus of the infundibular process in the posterior pituitary.

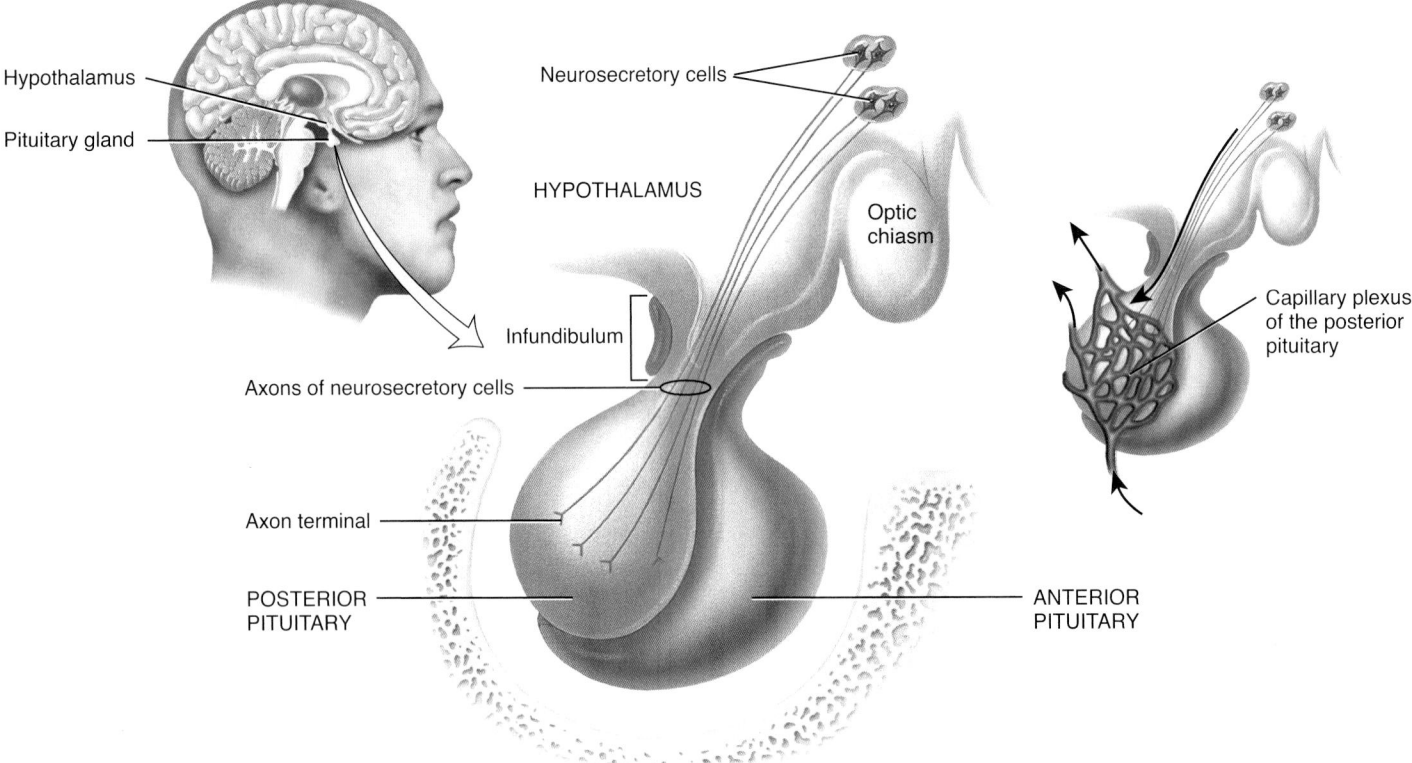

? Functionally, how are the hypothalamohypophyseal tract and the hypophyseal portal veins similar? Structurally, how are they different?

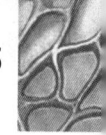

function of oxytocin in males and in nonpregnant females is not clear. Experiments with animals have suggested that it has actions within the brain that foster parental caretaking behavior toward young offspring. It may also be responsible, in part, for the feelings of sexual pleasure during and after intercourse.

CLINICAL CONNECTION | *Oxytocin and Childbirth*

Years before oxytocin was discovered, it was common practice in midwifery to let a first-born twin nurse at the mother's breast to speed the birth of the second child. Now we know why this practice is helpful—it stimulates the release of oxytocin. Even after a single birth, nursing promotes expulsion of the placenta (afterbirth) and helps the uterus regain its smaller size. Synthetic oxytocin (Pitocin) often is given to induce labor or to increase uterine tone and control hemorrhage just after giving birth. •

Antidiuretic Hormone

As its name implies, an **antidiuretic** (an-ti-dī-ū-RET-ik; *anti-* = against; *-dia-* = throughout; *-ouresis* = urination) is a substance that decreases urine production. ADH causes the kidneys to return more water to the blood, thus decreasing urine volume. In the absence of ADH, urine output increases more than tenfold, from the normal 1 to 2 liters to about 20 liters a day. Drinking alcohol often causes frequent and copious urination because alcohol inhibits secretion of ADH. ADH also decreases the water lost through sweating and causes constriction of arterioles, which increases blood pressure. This hormone's other name, **vasopressin** (vā-sō-PRESS-in; *vaso-* = blood; *-pressus* = to press), reflects this effect on blood pressure.

The amount of ADH secreted varies with blood osmotic pressure and blood volume. Figure 18.9 shows regulation of ADH secretion and the actions of ADH:

1 High blood osmotic pressure (or decreased blood volume)—due to dehydration or a decline in blood volume because of hemorrhage, diarrhea, or excessive sweating—stimulates **osmoreceptors,** neurons in the hypothalamus that monitor blood osmotic pressure. Elevated blood osmotic pressure activates the osmoreceptors directly; they also receive excitatory input from other brain areas when blood volume decreases.

2 Osmoreceptors activate the hypothalamic neurosecretory cells that synthesize and release ADH.

3 When neurosecretory cells receive excitatory input from the osmoreceptors, they generate nerve impulses that cause exocytosis of ADH-containing vesicles from their axon terminals in the posterior pituitary. This liberates ADH, which diffuses into blood capillaries of the posterior pituitary.

4 The blood carries ADH to three target tissues: the kidneys, sudoriferous (sweat) glands, and smooth muscle in blood vessel walls. The kidneys respond by retaining more water, which decreases urine output. Secretory activity of sweat glands decreases, which lowers the rate of water loss by perspiration from the skin. Smooth muscle in the walls of ar-

Figure 18.9 Regulation of secretion and actions of antidiuretic hormone (ADH).

ADH acts to retain body water and increase blood pressure.

1 High blood osmotic pressure stimulates hypothalamic osmoreceptors

5 Low blood osmotic pressure inhibits hypothalamic osmoreceptors

Osmoreceptors

2 Osmoreceptors activate neurosecretory cells that synthesize and release ADH

6 Inhibition of osmoreceptors reduces or stops ADH secretion

Hypothalamus

3 Nerve impulses liberate ADH from axon terminals in posterior pituitary into bloodstream

ADH

Target tissues

4 Kidneys retain more water, which decreases urine output

Sudoriferous (sweat) glands decrease water loss by perspiration from skin

Arterioles constrict, which increases blood pressure

? **If you drank a liter of water, what effect would this have on the osmotic pressure of your blood, and how would the level of ADH change in your blood?**

terioles (small arteries) contracts in response to high levels of ADH, which constricts (narrows) the lumen of these blood vessels and increases blood pressure.

5 Low osmotic pressure of blood (or increased blood volume) inhibits the osmoreceptors.

6 Inhibition of osmoreceptors reduces or stops ADH secretion. The kidneys then retain less water by forming a larger volume of urine, secretory activity of sweat glands increases, and arterioles dilate. The blood volume and osmotic pressure of body fluids return to normal.

Secretion of ADH can also be altered in other ways. Pain, stress, trauma, anxiety, acetylcholine, nicotine, and drugs such as morphine, tranquilizers, and some anesthetics stimulate ADH secretion. The dehydrating effect of alcohol, which has already been mentioned, may cause both the thirst and the headache typical of a hangover. Hyposecretion of ADH or nonfunctional ADH receptors cause diabetes insipidus (see Disorders: Homeostatic Imbalances at the end of the chapter).

Table 18.5 lists the posterior pituitary hormones, control of their secretion, and their principal actions.

✔ **CHECKPOINT**

9. In what respect is the pituitary gland actually two glands?
10. How do hypothalamic releasing and inhibiting hormones influence secretions of the anterior pituitary?
11. Describe the structure and importance of the hypothalamohypophyseal tract.

18.7 THYROID GLAND

◉ **OBJECTIVE**

• Describe the location, histology, hormones, and functions of the thyroid gland.

The butterfly-shaped **thyroid gland** is located just inferior to the larynx (voice box). It is composed of right and left **lateral lobes,** one on either side of the trachea, that are connected by an **isthmus** (IS-mus = a narrow passage) anterior to the trachea (Figure 18.10a). About 50% of thyroid glands have a small third lobe, called the *pyramidal lobe.* It extends superiorly from the isthmus. The normal mass of the thyroid is about 30 g (1 oz).

Microscopic spherical sacs called **thyroid follicles** (Figure 18.10b) make up most of the thyroid gland. The wall of each

follicle consists primarily of cells called **follicular cells** (fo-LIK-u-lar), most of which extend to the lumen (internal space) of the follicle. A **basement membrane** surrounds each follicle. When the follicular cells are inactive, their shape is low cuboidal to squamous, but under the influence of TSH they become active in secretion and range from cuboidal to low columnar in shape. The follicular cells produce two hormones: **thyroxine** (thī-ROK-sēn), which is also called **tetraiodothyronine** (tet-ra-ī-ō-dō-THĪ-rō-nēn) or T_4 because it contains four atoms of iodine, and **triiodothyronine** (trī-ī'-ō-dō-THĪ-rō-nēn) or T_3, which contains three atoms of iodine. T_3 and T_4 together are also known as **thyroid hormones.** A few cells called **parafollicular cells** (par'-a-fo-LIK-u-lar) or **C cells** lie between follicles. They produce the hormone **calcitonin** (kal-si-TŌ-nin), which helps regulate calcium homeostasis.

Formation, Storage, and Release of Thyroid Hormones

The thyroid gland is the only endocrine gland that stores its secretory product in large quantities—normally about a 100-day supply. Synthesis and secretion of T_3 and T_4 occurs as follows (Figure 18.11):

❶ *Iodide trapping.* Thyroid follicular cells trap iodide ions (I^-) by actively transporting them from the blood into the cytosol. As a result, the thyroid gland normally contains most of the iodide in the body.

❷ *Synthesis of thyroglobulin.* While the follicular cells are trapping I^-, they are also synthesizing **thyroglobulin (TGB)**

TABLE 18.5		
Summary of Posterior Pituitary Hormones		
HORMONE AND TARGET TISSUES	**CONTROL OF SECRETION**	**PRINCIPAL ACTIONS**
Oxytocin (OT) Uterus Mammary glands	Neurosecretory cells of hypothalamus secrete OT in response to uterine distention and stimulation of nipples.	Stimulates contraction of smooth muscle cells of uterus during childbirth; stimulates contraction of myoepithelial cells in mammary glands to cause milk ejection.
Antidiuretic hormone (ADH) or vasopressin Kidneys Sudoriferous (sweat) glands Arterioles	Neurosecretory cells of hypothalamus secrete ADH in response to elevated blood osmotic pressure, dehydration, loss of blood volume, pain, or stress; inhibitors of ADH secretion include low blood osmotic pressure, high blood volume, and alcohol.	Conserves body water by decreasing urine volume; decreases water loss through perspiration; raises blood pressure by constricting arterioles.

Figure 18.10 Location, blood supply, and histology of the thyroid gland.

Thyroid hormones regulate (1) oxygen use and basal metabolic rate, (2) cellular metabolism, and (3) growth and development.

Trachea

Thyroid gland

Pyramidal lobe of thyroid gland

RIGHT LATERAL LOBE OF THYROID GLAND

Middle thyroid vein

Inferior thyroid artery

Subclavian artery

Hyoid bone
Superior thyroid artery
Superior thyroid vein
Thyroid cartilage of larynx
Internal jugular vein
LEFT LATERAL LOBE OF THYROID GLAND
Common carotid artery
ISTHMUS OF THYROID GLAND
Vagus (X) nerve
Trachea
Inferior thyroid veins
Sternum

(a) Anterior view of thyroid gland

Parafollicular (C) cell

Follicular cell

Thyroid follicle

Thyroglobulin (TGB) (colloid)

Basement membrane

LM 500x

(b) Thyroid follicles

RIGHT LATERAL LOBE

ISTHMUS

LEFT LATERAL LOBE

(c) Anterior view of thyroid gland

Thyroid cartilage of larynx

Cricoid cartilage of larynx
RIGHT LATERAL LOBE OF THYROID GLAND
LEFT LATERAL LOBE OF THYROID GLAND
ISTHMUS OF THYROID GLAND

Trachea

Right lung

Arch of aorta

(d) Anterior view

? Which cells secrete T_3 and T_4? Which secrete calcitonin? Which of these hormones are also called thyroid hormones?

Figure 18.11 Steps in the synthesis and secretion of thyroid hormones.

Thyroid hormones are synthesized by attaching iodine atoms to the amino acid tyrosine.

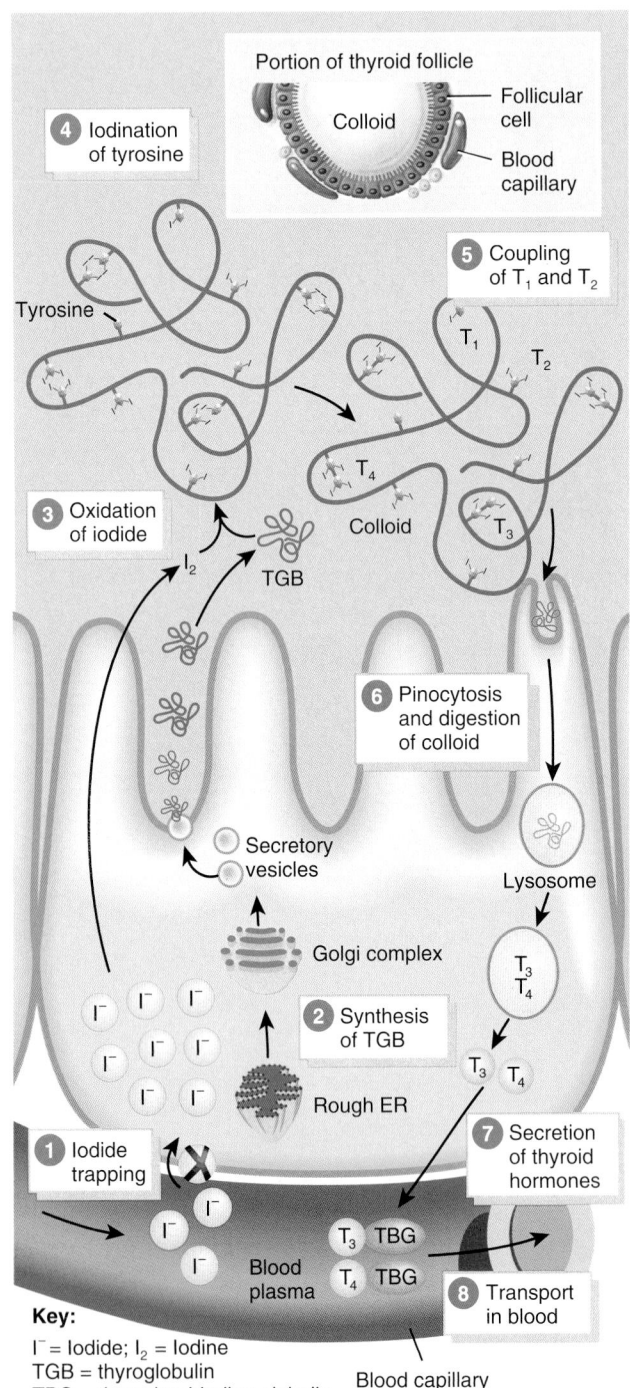

Key:
I⁻ = Iodide; I₂ = Iodine
TGB = thyroglobulin
TBG = thyroxine-binding globulin

? **What is the storage form of thyroid hormones?**

(thī′-rō-GLOB-u-lin), a large glycoprotein that is produced in the rough endoplasmic reticulum, modified in the Golgi complex, and packaged into secretory vesicles. The vesicles then undergo exocytosis, which releases TGB into the lumen of the follicle.

③ *Oxidation of iodide.* Some of the amino acids in TGB are tyrosines that will become iodinated. However, negatively charged iodide ions cannot bind to tyrosine until they undergo oxidation (removal of electrons) to iodine: $2\ I^- \rightarrow I_2$. As the iodide ions are being oxidized, they pass through the membrane into the lumen of the follicle.

④ *Iodination of tyrosine.* As iodine molecules (I_2) form, they react with tyrosines that are part of thyroglobulin molecules. Binding of one iodine atom yields monoiodotyrosine (T_1), and a second iodination produces diiodotyrosine (T_2). The TGB with attached iodine atoms, a sticky material that accumulates and is stored in the lumen of the thyroid follicle, is termed **colloid.**

⑤ *Coupling of T_1 and T_2.* During the last step in the synthesis of thyroid hormone, two T_2 molecules join to form T_4, or one T_1 and one T_2 join to form T_3.

⑥ *Pinocytosis and digestion of colloid.* Droplets of colloid reenter follicular cells by pinocytosis and merge with lysosomes. Digestive enzymes in the lysosomes break down TGB, cleaving off molecules of T_3 and T_4.

⑦ *Secretion of thyroid hormones.* Because T_3 and T_4 are lipid-soluble, they diffuse through the plasma membrane into interstitial fluid and then into the blood. T_4 normally is secreted in greater quantity than T_3, but T_3 is several times more potent. Moreover, after T_4 enters a body cell, most of it is converted to T_3 by removal of one iodine.

⑧ *Transport in the blood.* More than 99% of both the T_3 and the T_4 combine with transport proteins in the blood, mainly **thyroxine-binding globulin (TBG).**

Actions of Thyroid Hormones

Because most body cells have receptors for thyroid hormones, T_3 and T_4 exert their effects throughout the body.

1. Thyroid hormones increase **basal metabolic rate (BMR),** the rate of oxygen consumption under standard or basal conditions (awake, at rest, and fasting), by stimulating the use of cellular oxygen to produce ATP. When the basal metabolic rate increases, cellular metabolism of carbohydrates, lipids, and proteins increases.

2. A second major effect of thyroid hormones is to stimulate synthesis of additional sodium–potassium pumps (Na^+/K^+ ATPase), which use large amounts of ATP to continually eject sodium ions (Na^+) from the cytosol into the extracellular fluid and potassium ions (K^+) from the extracellular fluid into the cytosol. As cells produce and use more ATP, more heat is given off, and body temperature rises. This phenomenon is called the **calorigenic effect** (ka-lor′-i-JEN-ik). In this way,

thyroid hormones play an important role in the maintenance of normal body temperature. Normal mammals can survive in freezing temperatures, but those whose thyroid glands have been removed cannot.

3. In the regulation of metabolism, the thyroid hormones stimulate protein synthesis and increase the use of glucose and fatty acids for ATP production. They also increase lipolysis and enhance cholesterol excretion, thus reducing blood cholesterol level.

4. The thyroid hormones enhance some actions of the catecholamines (norepinephrine and epinephrine) because they up-regulate beta (β) receptors. For this reason, symptoms of hyperthyroidism include increased heart rate, more forceful heartbeats, and increased blood pressure.

5. Together with human growth hormone and insulin, thyroid hormones accelerate body growth, particularly the growth of the nervous and skeletal systems. Deficiency of thyroid hormones during fetal development, infancy, or childhood causes severe mental retardation and stunted bone growth.

Control of Thyroid Hormone Secretion

Thyrotropin-releasing hormone (TRH) from the hypothalamus and thyroid-stimulating hormone (TSH) from the anterior pituitary stimulate synthesis and release of thyroid hormones, as shown in Figure 18.12:

1. Low blood levels of T_3 and T_4 or low metabolic rate stimulates the hypothalamus to secrete TRH.

2. TRH enters the hypophyseal portal veins and flows to the anterior pituitary, where it stimulates thyrotrophs to secrete TSH.

3. TSH stimulates virtually all aspects of thyroid follicular cell activity, including iodide trapping (1 in Figure 18.11), hormone synthesis and secretion (2 and 7 in Figure 18.11), and growth of the follicular cells.

4. The thyroid follicular cells release T_3 and T_4 into the blood until the metabolic rate returns to normal.

5. An elevated level of T_3 inhibits release of TRH and TSH (negative feedback inhibition).

Conditions that increase ATP demand—a cold environment, hypoglycemia, high altitude, and pregnancy—also increase the secretion of the thyroid hormones.

Calcitonin

The hormone produced by the **parafollicular cells** of the thyroid gland (see Figure 18.10b) is **calcitonin (CT).** CT can decrease the level of calcium in the blood by inhibiting the action of osteoclasts, the cells that break down bone extracellular matrix. The secretion of CT is controlled by a negative feedback system (see Figure 18.14).

When its blood level is high, calcitonin lowers the amount of blood calcium and phosphates by inhibiting bone resorption (breakdown of bone extracellular matrix) by osteoclasts and by accelerating uptake of calcium and phosphates into bone extracellular

Figure 18.12 Regulation of secretion and actions of thyroid hormones. TRH = thyrotropin-releasing hormone, TSH = thyroid-stimulating hormone, T_3 = triiodothyronine, and T_4 = thyroxine (tetraiodothyronine).

🔑 TSH promotes release of thyroid hormones (T_3 and T_4) by the thyroid gland.

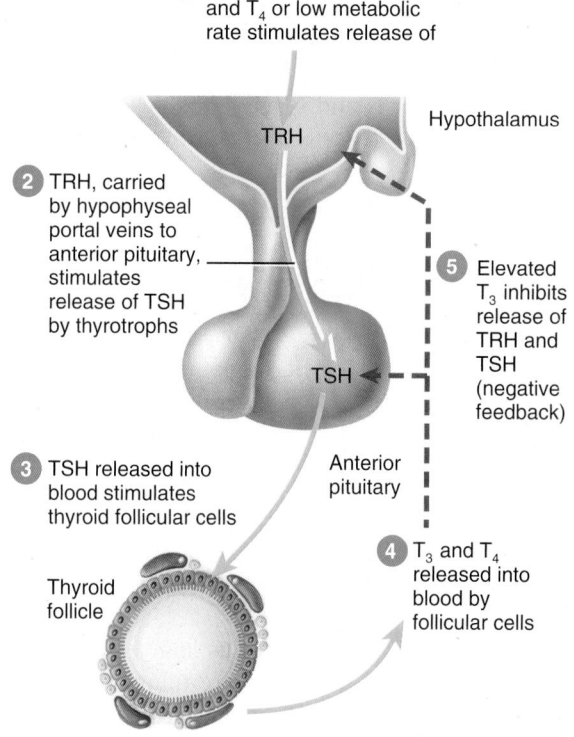

1. Low blood levels of T_3 and T_4 or low metabolic rate stimulates release of

Hypothalamus

TRH

2. TRH, carried by hypophyseal portal veins to anterior pituitary, stimulates release of TSH by thyrotrophs

5. Elevated T_3 inhibits release of TRH and TSH (negative feedback)

TSH

Anterior pituitary

3. TSH released into blood stimulates thyroid follicular cells

Thyroid follicle

4. T_3 and T_4 released into blood by follicular cells

Actions of Thyroid Hormones:

Increase basal metabolic rate

Stimulate synthesis of Na$^+$/K$^+$ ATPase

Increase body temperature (calorigenic effect)

Stimulate protein synthesis

Increase the use of glucose and fatty acids for ATP production

Stimulate lipolysis

Enhance some actions of catecholamines

Regulate development and growth of nervous tissue and bones

❓ **How could an iodine-deficient diet lead to goiter, which is an enlargement of the thyroid gland?**

matrix. Miacalcin, a calcitonin extract derived from salmon that is 10 times more potent than human calcitonin, is prescribed to treat osteoporosis.

Table 18.6 summarizes the hormones produced by the thyroid gland, control of their secretion, and their principal actions.

✔CHECKPOINT

12. Explain how blood levels of T_3/T_4, TSH, and TRH would change in a laboratory animal that has undergone a thyroidectomy (complete removal of its thyroid gland).
13. How are the thyroid hormones synthesized, stored, and secreted?
14. How is the secretion of T_3 and T_4 regulated?
15. What are the physiological effects of the thyroid hormones?

18.8 PARATHYROID GLANDS

◉ OBJECTIVE

- Describe the location, histology, hormone, and functions of the parathyroid glands.

Partially embedded in the posterior surface of the lateral lobes of the thyroid gland are several small, round masses of tissue called the **parathyroid glands** (*para-* = beside). Each has a mass of about 40 mg (0.04 g). Usually, one superior and one inferior parathyroid gland are attached to each lateral thyroid lobe (Figure 18.13a), for a total of four.

Microscopically, the parathyroid glands contain two kinds of epithelial cells (Figure 18.13b, c). The more numerous cells, called **chief (principal) cells,** produce **parathyroid hormone (PTH),** also called **parathormone.** The function of the other kind of cell, called an **oxyphil cell,** is not known in a normal parathyroid gland. However, its presence clearly helps to identify the parathyroid gland histologically due to its unique staining characteristics. Furthermore, in a cancer of the parathyroid glands, oxyphil cells secrete excess PTH.

Parathyroid Hormone

Parathyroid hormone is the major regulator of the levels of calcium (Ca^{2+}), magnesium (Mg^{2+}), and phosphate (HPO_4^{2-}) ions in the blood. The specific action of PTH is to increase the number and activity of osteoclasts. The result is elevated bone *resorption,* which releases ionic calcium (Ca^{2+}) and phosphates (HPO_4^{2-}) into the blood. PTH also acts on the kidneys. First, it slows the rate at which Ca^{2+} and Mg^{2+} are lost from blood into the urine. Second, it increases loss of HPO_4^{2-} from blood into the urine. Because more HPO_4^{2-} is lost in the urine than is gained from the bones, PTH decreases blood HPO_4^{2-} level and increases blood Ca^{2+} and Mg^{2+} levels. A third effect of PTH on the kidneys is to promote formation of the hormone **calcitriol** (kal′-si-TRĪ-ol), the active form of vitamin D. Calcitriol, also known as *1,25-dihydroxyvitamin D_3,* increases the rate of Ca^{2+}, HPO_4^{2-}, and Mg^{2+} *absorption* from the gastrointestinal tract into the blood.

The blood calcium level directly controls the secretion of both calcitonin and parathyroid hormone via negative feedback loops that do not involve the pituitary gland (Figure 18.14):

1 A higher-than-normal level of calcium ions (Ca^{2+}) in the blood stimulates parafollicular cells of the thyroid gland to release more calcitonin.

2 Calcitonin inhibits the activity of osteoclasts, thereby decreasing the blood Ca^{2+} level.

TABLE 18.6

Summary of Thyroid Gland Hormones

HORMONE AND SOURCE	CONTROL OF SECRETION	PRINCIPAL ACTIONS
T_3 (triiodothyronine) and **T_4 (thyroxine)** or thyroid hormones from follicular cells	Secretion is increased by thyrotropin-releasing hormone (TRH), which stimulates release of thyroid-stimulating hormone (TSH) in response to low thyroid hormone levels, low metabolic rate, cold, pregnancy, and high altitudes; TRH and TSH secretions are inhibited in response to high thyroid hormone levels; high iodine level suppresses T_3/T_4 secretion.	Increase basal metabolic rate; stimulate synthesis of proteins; increase use of glucose and fatty acids for ATP production; increase lipolysis; enhance cholesterol excretion; accelerate body growth; contribute to development of nervous system.
Calcitonin (CT) from parafollicular cells	High blood Ca^{2+} levels stimulate secretion; low blood Ca^{2+} levels inhibit secretion.	Lowers blood levels of Ca^{2+} and HPO_4^{2-} by inhibiting bone resorption by osteoclasts and by accelerating uptake of calcium and phosphates into bone extracellular matrix.

Figure 18.13 **Location, blood supply, and histology of the parathyroid glands.**

🔑 ⚊⚊ The parathyroid glands, normally four in number, are embedded in the posterior surface of the thyroid gland.

Parathyroid glands (behind thyroid gland)

Trachea

LEFT SUPERIOR PARATHYROID GLAND

Esophagus

LEFT INFERIOR PARATHYROID GLAND

Left inferior thyroid artery

Left subclavian artery

Left subclavian vein

Left common carotid artery

Right internal jugular vein

Right common carotid artery

Middle cervical sympathetic ganglion

Thyroid gland

RIGHT SUPERIOR PARATHYROID GLAND

Inferior cervical sympathetic ganglion

RIGHT INFERIOR PARATHYROID GLAND

Vagus (X) nerve

Right brachiocephalic vein

Brachiocephalic trunk

Trachea

C.KELLEY

(a) Posterior view

Venule

Capillary

Arteriole

Chief cell

Oxyphil cell

LM 240x

(b) Parathyroid gland

Capsule { Parathyroid / Thyroid }

Chief cell / Oxyphil cell } Parathyroid gland

Follicular cell / Parafollicular cell } Thyroid gland

Blood vessel

(c) Portion of thyroid gland (left) and parathyroid gland (right)

PARATHYROID GLAND

Thyroid gland

PARATHYROID GLAND

(d) Posterior view of parathyroid glands

❓ **What are the secretory products of (1) parafollicular cells of the thyroid gland and (2) chief (principal) cells of the parathyroid glands?**

Figure 18.14 The roles of calcitonin (green arrows), parathyroid hormone (blue arrows), and calcitriol (orange arrows) in calcium homeostasis.

With respect to regulation of blood Ca^{2+} level, calcitonin and PTH are antagonists.

1 High level of Ca^{2+} in blood stimulates thyroid gland parafollicular cells to release more CT.

3 Low level of Ca^{2+} in blood stimulates parathyroid gland chief cells to release more PTH.

6 CALCITRIOL stimulates increased absorption of Ca^{2+} from foods, which increases blood Ca^{2+} level.

5 PTH also stimulates the kidneys to release CALCITRIOL.

4 PARATHYROID HORMONE (PTH) promotes release of Ca^{2+} from bone extracellular matrix into blood and slows loss of Ca^{2+} in urine, thus increasing blood Ca^{2+} level.

2 CALCITONIN inhibits osteoclasts, thus decreasing blood Ca^{2+} level.

What are the primary target tissues for PTH, CT, and calcitriol?

3 A lower-than-normal level of Ca^{2+} in the blood stimulates chief cells of the parathyroid gland to release more PTH.

4 PTH promotes resorption of bone extracellular matrix, which releases Ca^{2+} into the blood and slows loss of Ca^{2+} in the urine, raising the blood level of Ca^{2+}.

5 PTH also stimulates the kidneys to synthesize calcitriol, the active form of vitamin D.

6 Calcitriol stimulates increased absorption of Ca^{2+} from foods in the gastrointestinal tract, which helps increase the blood level of Ca^{2+}.

TABLE 18.7

Summary of Parathyroid Gland Hormone

HORMONE AND SOURCE	CONTROL OF SECRETION	PRINCIPAL ACTIONS
Parathyroid hormone (PTH) from chief cells Chief cell	Low blood Ca^{2+} levels stimulate secretion; high blood Ca^{2+} levels inhibit secretion.	Increases blood Ca^{2+} and Mg^{2+} levels and decreases blood HPO_4^{2-} level; increases bone resorption by osteoclasts; increases Ca^{2+} reabsorption and HPO_4^{2-} excretion by kidneys; promotes formation of calcitriol (active form of vitamin D), which increases rate of dietary Ca^{2+} and Mg^{2+} absorption.

Table 18.7 summarizes control of secretion and the principal actions of parathyroid hormone.

✔CHECKPOINT

16. How is secretion of parathyroid hormone regulated?
17. In what ways are the actions of PTH and calcitriol similar? How are they different?

18.9 ADRENAL GLANDS

◉ **OBJECTIVE**

• Describe the location, histology, hormones, and functions of the adrenal glands.

The paired **adrenal (suprarenal) glands,** one of which lies superior to each kidney in the retroperitoneal space (Figure 18.15a), have a flattened pyramidal shape. In an adult, each adrenal gland is 3–5 cm in height, 2–3 cm in width, and a little less than 1 cm thick, with a mass of 3.5–5 g, only half its size at birth. During embryonic development, the adrenal glands differentiate into two structurally and functionally distinct regions: a large, peripherally located **adrenal cortex,** comprising 80–90% of the gland, and a small, centrally located **adrenal medulla** (Figure 18.15b). A connective tissue capsule covers the gland. The adrenal glands, like the thyroid gland, are highly vascularized.

The adrenal cortex produces steroid hormones that are essential for life. Complete loss of adrenocortical hormones leads to death due to dehydration and electrolyte imbalances in a few days to a week, unless hormone replacement therapy begins promptly. The adrenal medulla produces three catecholamine hormones—norepinephrine, epinephrine, and a small amount of dopamine.

Adrenal Cortex

The adrenal cortex is subdivided into three zones, each of which secretes different hormones (Figure 18.15d). The outer zone, just deep to the connective tissue capsule, is the **zona glomerulosa** (glo-mer′-ū-LŌ-sa; *zona* = belt; *glomerul-* = little ball). Its cells, which are closely packed and arranged in spherical clusters and arched columns, secrete hormones called **mineralocorticoids** (min′-er-al-ō-KOR-ti-koyds) because they affect mineral homeostasis. The middle zone, or **zona fasciculata** (fa-sik′-ū-LA-ta; *fascicul-* = little bundle), is the widest of the three zones and consists of cells arranged in long, straight columns. The cells of the

Figure 18.15 **Location, blood supply, and histology of the adrenal (suprarenal) glands.**

🔑 ⚏ The adrenal cortex secretes steroid hormones that are essential for life; the adrenal medulla secretes norepinephrine and epinephrine.

(a) Anterior view

FIGURE 18.15 CONTINUES ▶

FIGURE 18.15 CONTINUED ▶

(b) Section through left adrenal gland

(c) Anterior view of adrenal gland and kidney

(d) Subdivisions of adrenal gland

? **What is the position of the adrenal glands relative to the kidneys?**

zona fasciculata secrete mainly **glucocorticoids** (gloo′-kō-KOR-ti-koyds), primarily cortisol, so named because they affect glucose homeostasis. The cells of the inner zone, the **zona reticularis** (re-tik′-ū-LAR-is; *reticul-* = network), are arranged in branching cords. They synthesize small amounts of weak **androgens** (*andro-* = a man), steroid hormones that have masculinizing effects.

Mineralocorticoids

Aldosterone (al-DOS-ter-ōn) is the major mineralocorticoid. It regulates homeostasis of two mineral ions—namely, sodium ions (Na^+) and potassium ions (K^+)—and helps adjust blood pressure and blood volume. Aldosterone also promotes excretion of H^+ in the urine; this removal of acids from the body can help prevent acidosis (blood pH below 7.35), which is discussed in Chapter 27.

The **renin–angiotensin–aldosterone** or **RAA pathway** (RĒ-nin an′-jē-ō-TEN-sin) controls secretion of aldosterone (Figure 18.16):

1 Stimuli that initiate the renin–angiotensin–aldosterone pathway include dehydration, Na^+ deficiency, or hemorrhage.

2 These conditions cause a decrease in blood volume.

3 Decreased blood volume leads to decreased blood pressure.

4 Lowered blood pressure stimulates certain cells of the kidneys, called juxtaglomerular cells, to secrete the enzyme **renin.**

5 The level of renin in the blood increases.

6 Renin converts **angiotensinogen** (an′-jē-ō-ten-SIN-ō-jen), a plasma protein produced by the liver, into **angiotensin I.**

7 Blood containing increased levels of angiotensin I circulates in the body.

8 As blood flows through capillaries, particularly those of the lungs, the enzyme **angiotensin-converting enzyme (ACE)** converts angiotensin I into the hormone **angiotensin II.**

9 Blood level of angiotensin II increases.

10 Angiotensin II stimulates the adrenal cortex to secrete aldosterone.

11 Blood containing increased levels of aldosterone circulates to the kidneys.

12 In the kidneys, aldosterone increases reabsorption of Na^+, which in turn causes reabsorption of water by osmosis. As a result, less water is lost in the urine. Aldosterone also stimulates the kidneys to increase secretion of K^+ and H^+ into the urine.

13 With increased water reabsorption by the kidneys, blood volume increases.

14 As blood volume increases, blood pressure increases to normal.

15 Angiotensin II also stimulates contraction of smooth muscle in the walls of arterioles. The resulting vasoconstriction of

Figure 18.16 Regulation of aldosterone secretion by the renin–angiotensin–aldosterone (RAA) pathway.

🔑 🖴 Aldosterone helps regulate blood volume, blood pressure, and levels of Na⁺, K⁺, and H⁺ in the blood.

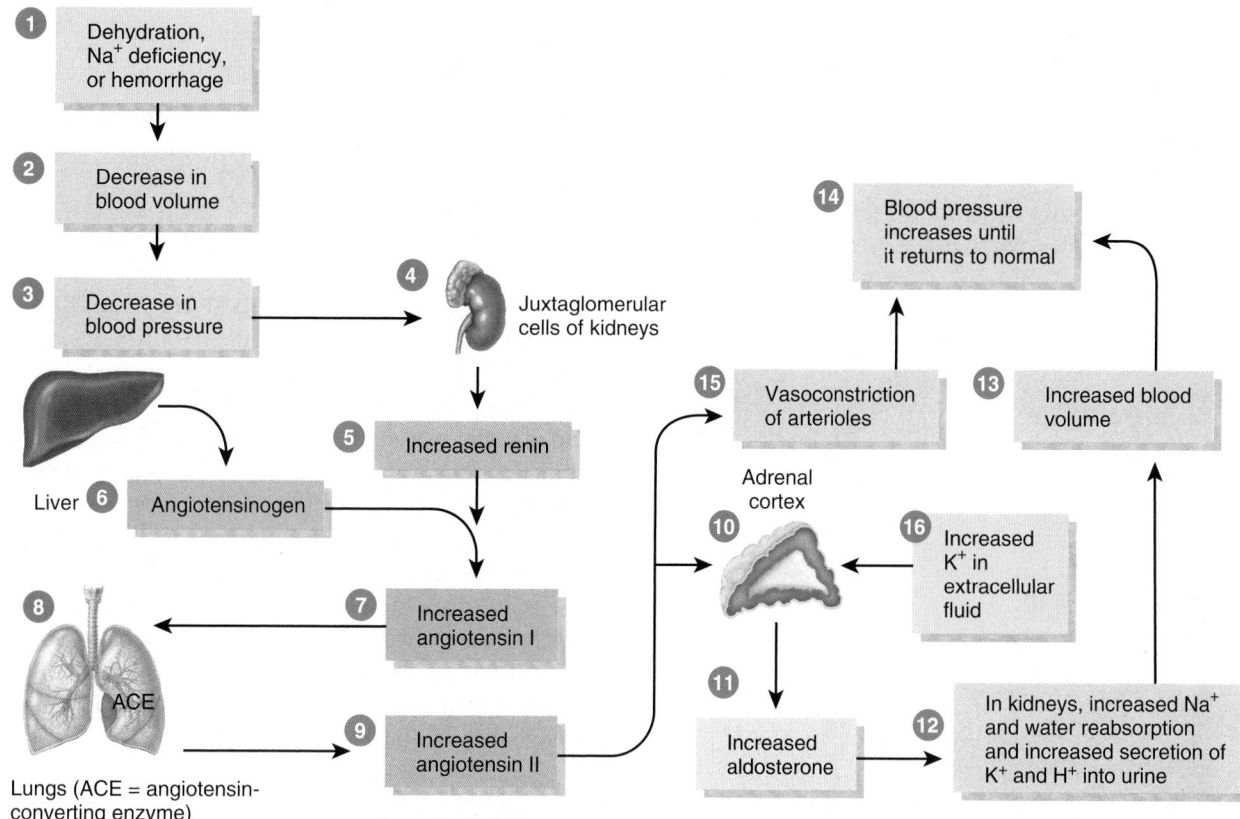

? In what two ways can angiotensin II increase blood pressure, and what are its target tissues in each case?

the arterioles increases blood pressure and thus helps raise blood pressure to normal.

16 Besides angiotensin II, a second stimulator of aldosterone secretion is an increase in the K⁺ concentration of blood (or interstitial fluid). A decrease in the blood K⁺ level has the opposite effect.

Glucocorticoids

The glucocorticoids, which regulate metabolism and resistance to stress, include **cortisol** (KOR-ti-sol; also called **hydrocortisone**), **corticosterone** (kor′-ti-KOS-ter-ōn), and **cortisone** (KOR-ti-sōn). Of these three hormones secreted by the zona fasciculata, cortisol is the most abundant, accounting for about 95% of glucocorticoid activity.

Control of glucocorticoid secretion occurs via a typical negative feedback system (Figure 18.17). Low blood levels of glucocorticoids, mainly cortisol, stimulate neurosecretory cells in the hypothalamus to secrete **corticotropin-releasing hormone (CRH).** CRH (together with a low level of cortisol) promotes the release of ACTH from the anterior pituitary. ACTH flows in the blood to the adrenal cortex, where it stimulates glucocorticoid secretion. (To a much smaller extent, ACTH also stimulates secretion of aldosterone.) The discussion of stress at the end of the chapter describes

how the hypothalamus also increases CRH release in response to a variety of physical and emotional stresses (see Section 18.15).

Glucocorticoids have the following effects:

1. **Protein breakdown.** Glucocorticoids increase the rate of protein breakdown, mainly in muscle fibers, and thus increase the liberation of amino acids into the bloodstream. The amino acids may be used by body cells for synthesis of new proteins or for ATP production.

2. **Glucose formation.** On stimulation by glucocorticoids, liver cells may convert certain amino acids or lactic acid to glucose, which neurons and other cells can use for ATP production. Such conversion of a substance other than glycogen or another monosaccharide into glucose is called **gluconeogenesis** (gloo′-ko-nē′-ō-JEN-e-sis).

3. **Lipolysis.** Glucocorticoids stimulate **lipolysis** (lī-POL-i-sis), the breakdown of triglycerides and release of fatty acids from adipose tissue into the blood.

4. **Resistance to stress.** Glucocorticoids work in many ways to provide resistance to stress. The additional glucose supplied by the liver cells provides tissues with a ready source of ATP to combat a range of stresses, including exercise, fasting, fright, temperature extremes, high altitude, bleeding, infection, surgery, trauma, and

Figure 18.17 Negative feedback regulation of glucocorticoid secretion.

A high level of CRH and a low level of glucocorticoids promote the release of ACTH, which stimulates glucocorticoid secretion by the adrenal cortex.

If a heart transplant patient receives prednisone (a glucocorticoid) to help prevent rejection of the transplanted tissue, will blood levels of ACTH and CRH be high or low? Explain.

disease. Because glucocorticoids make blood vessels more sensitive to other hormones that cause vasoconstriction, they raise blood pressure. This effect would be an advantage in cases of severe blood loss, which causes blood pressure to drop.

5. *Anti-inflammatory effects.* Glucocorticoids inhibit white blood cells that participate in inflammatory responses. Unfortunately,

glucocorticoids also retard tissue repair, and as a result, they slow wound healing. Although high doses can cause severe mental disturbances, glucocorticoids are very useful in the treatment of chronic inflammatory disorders such as rheumatoid arthritis.

6. *Depression of immune responses.* High doses of glucocorticoids depress immune responses. For this reason, glucocorticoids are prescribed for organ transplant recipients to retard tissue rejection by the immune system.

Androgens

In both males and females, the adrenal cortex secretes small amounts of weak androgens. The major androgen secreted by the adrenal gland is **dehydroepiandrosterone (DHEA)** (dē-hī-drō-ep′-ē-an-DROS-ter-ōn). After puberty in males, the androgen testosterone is also released in much greater quantity by the testes. Thus, the amount of androgens secreted by the adrenal gland in males is usually so low that their effects are insignificant. In females, however, adrenal androgens play important roles. They promote libido (sex drive) and are converted into estrogens (feminizing sex steroids) by other body tissues. After menopause, when ovarian secretion of estrogens ceases, all female estrogens come from conversion of adrenal androgens. Adrenal androgens also stimulate growth of axillary and pubic hair in boys and girls and contribute to the prepubertal growth spurt. Although control of adrenal androgen secretion is not fully understood, the main hormone that stimulates its secretion is ACTH.

CLINICAL CONNECTION | *Congenital Adrenal Hyperplasia*

Congenital adrenal hyperplasia (CAH) (hī-per-PLĀ-zhē-a) is a genetic disorder in which one or more enzymes needed for synthesis of cortisol are absent. Because the cortisol level is low, secretion of ACTH by the anterior pituitary is high due to lack of negative feedback inhibition. ACTH in turn stimulates growth and secretory activity of the adrenal cortex. As a result, both adrenal glands are enlarged. However, certain steps leading to synthesis of cortisol are blocked. Thus, precursor molecules accumulate, and some of these are weak androgens that can undergo conversion to testosterone. The result is **virilism** (VIR-i-lizm), or masculinization. In a female, virile characteristics include growth of a beard, development of a much deeper voice and a masculine distribution of body hair, growth of the clitoris so it may resemble a penis, atrophy of the breasts, and increased muscularity that produces a masculine physique. In prepubertal males, the syndrome causes the same characteristics as in females, plus rapid development of the male sexual organs and emergence of male sexual desires. In adult males, the virilizing effects of CAH are usually completely obscured by the normal virilizing effects of the testosterone secreted by the testes. As a result, CAH is often difficult to diagnose in adult males. Treatment involves cortisol therapy, which inhibits ACTH secretion and thus reduces production of adrenal androgens. •

Adrenal Medulla

The inner region of the adrenal gland, the **adrenal medulla,** is a modified sympathetic ganglion of the autonomic nervous system (ANS). It develops from the same embryonic tissue as all other sympathetic ganglia, but its cells, which lack axons, form clusters around large blood vessels. Rather than releasing a neurotransmitter, the cells of the adrenal medulla secrete hormones. The hormone-producing cells, called **chromaffin cells** (KRŌ-maf-in; *chrom-* = color; *-affin* = affinity for; see Figure 18.15d), are innervated by sympathetic preganglionic neurons of the ANS. Because the ANS exerts direct control over the chromaffin cells, hormone release can occur very quickly.

The two major hormones synthesized by the adrenal medulla are **epinephrine** (ep'-i-NEF-rin) and **norepinephrine (NE),** also called *adrenaline* and *noradrenaline,* respectively. The chromaffin cells of the adrenal medulla secrete an unequal amount of these hormones—about 80% epinephrine and 20% norepinephrine. The hormones of the adrenal medulla intensify sympathetic responses that occur in other parts of the body.

In stressful situations and during exercise, impulses from the hypothalamus stimulate sympathetic preganglionic neurons, which in turn stimulate the chromaffin cells to secrete epinephrine and norepinephrine. These two hormones greatly augment the fight-or-flight response that you learned about in Chapter 15. By increasing heart rate and force of contraction, epinephrine and norepinephrine increase the output of the heart, which increases blood pressure. They also increase blood flow to the heart, liver, skeletal muscles, and adipose tissue; dilate airways to the lungs; and increase blood levels of glucose and fatty acids.

Table 18.8 summarizes the hormones produced by the adrenal glands, control of their secretion, and their principal actions.

✔ CHECKPOINT

18. How do the adrenal cortex and adrenal medulla compare with regard to location and histology?
19. How is secretion of adrenal cortex hormones regulated?
20. How is the adrenal medulla related to the autonomic nervous system?

18.10 PANCREATIC ISLETS

◉ OBJECTIVE

• Describe the location, histology, hormones, and functions of the pancreatic islets.

The **pancreas** (*pan-* = all; *-creas* = flesh) is both an endocrine gland and an exocrine gland. We discuss its endocrine functions here and include its exocrine functions in Chapter 24 in the coverage of the digestive system. A flattened organ that measures about 12.5–15 cm (5–6 in.) in length, the pancreas is located in the curve of the duodenum, the first part of the small intestine, and consists of a head, a body, and a tail (Figure 18.18a). Roughly 99% of the exocrine cells of the pancreas are arranged in clusters called **acini** (AS-i-nī). The acini produce digestive enzymes,

TABLE 18.8

Summary of Adrenal Gland Hormones

HORMONES AND SOURCE	CONTROL OF SECRETION	PRINCIPAL ACTIONS
ADRENAL CORTEX HORMONES		
Mineralocorticoids (mainly aldosterone) from zona glomerulosa cells	Increased blood K^+ level and angiotensin II stimulate secretion.	Increase blood levels of Na^+ and water; decrease blood level of K^+.
Glucocorticoids (mainly cortisol) from zona fasciculata cells	ACTH stimulates release; corticotropin-releasing hormone (CRH) promotes ACTH secretion in response to stress and low blood levels of glucocorticoids.	Increase protein breakdown (except in liver), stimulate gluconeogenesis and lipolysis, provide resistance to stress, dampen inflammation, depress immune responses.
Androgens (mainly dehydroepiandrosterone, or DHEA) from zona reticularis cells	ACTH stimulates secretion.	Assist in early growth of axillary and pubic hair in both sexes; in females, contribute to libido and are source of estrogens after menopause.
—Adrenal cortex		
ADRENAL MEDULLA HORMONES		
Epinephrine and **norepinephrine** from chromaffin cells	Sympathetic preganglionic neurons release acetylcholine, which stimulates secretion.	Enhance effects of sympathetic division of autonomic nervous system (ANS) during stress.
—Adrenal medulla		

Figure 18.18 Location, blood supply, and histology of the pancreas.

🔑 Pancreatic hormones regulate blood glucose level.

Pancreas
Kidney

Common hepatic artery
Abdominal aorta
Celiac trunk
Splenic artery

Gastroduodenal artery
Dorsal pancreatic artery
Anterior pancreaticoduodenal artery
Duodenum of small intestine

Spleen (elevated)

TAIL OF PANCREAS
BODY OF PANCREAS
Inferior pancreatic artery
Superior mesenteric artery
Inferior pancreaticoduodenal artery

HEAD OF PANCREAS

(a) Anterior view

Blood capillary
Exocrine acini
Alpha cell (secretes glucagon)
Beta cell (secretes insulin)
Delta cell (secretes somatostatin)
F cell (secretes pancreatic polypeptide)

(b) Pancreatic islet and surrounding acini

Exocrine acinus
Pancreatic islet
Beta cell
Alpha cell

LM 200x

LM 40x

Pancreatic duct

(c) Pancreatic islet and surrounding acini

Pancreas

Duodenum (cut open)

(d) Anterior view of pancreas dissected to reveal pancreatic duct

❓ **Is the pancreas an exocrine gland or an endocrine gland?**

708

which flow into the gastrointestinal tract through a network of ducts. Scattered among the exocrine acini are 1–2 million tiny clusters of endocrine tissue called **pancreatic islets** (Ī-lets) or **islets of Langerhans** (LAHNG-er-hanz; Figure 18.18b, c). Abundant capillaries serve both the exocrine and endocrine portions of the pancreas.

Cell Types in the Pancreatic Islets

Each pancreatic islet includes four types of hormone-secreting cells:

1. **Alpha** or **A cells** constitute about 17% of pancreatic islet cells and secrete **glucagon** (GLOO-ka-gon).

2. **Beta** or **B cells** constitute about 70% of pancreatic islet cells and secrete **insulin** (IN-soo-lin).

3. **Delta** or **D cells** constitute about 7% of pancreatic islet cells and secrete **somatostatin** (sō-ma-to-STAT-in).

4. **F cells** constitute the remainder of pancreatic islet cells and secrete **pancreatic polypeptide.**

The interactions of the four pancreatic hormones are complex and not completely understood. We do know that glucagon raises blood glucose level, and insulin lowers it. Somatostatin acts in a paracrine manner to inhibit both insulin and glucagon release from neighboring beta and alpha cells. It may also act as a circulating hormone to slow absorption of nutrients from the gastrointestinal tract. In addition, somatostatin inhibits the secretion of growth hormone. Pancreatic polypeptide inhibits somatostatin secretion, gallbladder contraction, and secretion of digestive enzymes by the pancreas.

Regulation of Glucagon and Insulin Secretion

The principal action of glucagon is to increase blood glucose level when it falls below normal. Insulin, on the other hand, helps lower blood glucose level when it is too high. The level of blood glucose controls secretion of glucagon and insulin via negative feedback (Figure 18.19):

1 Low blood glucose level (hypoglycemia) stimulates secretion of glucagon from alpha cells of the pancreatic islets.

2 Glucagon acts on hepatocytes (liver cells) to accelerate the conversion of glycogen into glucose (glycogenolysis) and to promote formation of glucose from lactic acid and certain amino acids (gluconeogenesis).

3 As a result, hepatocytes release glucose into the blood more rapidly, and blood glucose level rises.

4 If blood glucose continues to rise, high blood glucose level (hyperglycemia) inhibits release of glucagon (negative feedback).

5 High blood glucose (hyperglycemia) stimulates secretion of insulin by beta cells of the pancreatic islets.

6 Insulin acts on various cells in the body to accelerate facilitated diffusion of glucose into cells; to speed conversion of glucose into glycogen (glycogenesis); to increase uptake of amino acids

by cells and to increase protein synthesis; to speed synthesis of fatty acids (lipogenesis); to slow the conversion of glycogen to glucose (glycogenolysis); and to slow the formation of glucose from lactic acid and amino acids (gluconeogenesis).

7 As a result, blood glucose level falls.

8 If blood glucose level drops below normal, low blood glucose inhibits release of insulin (negative feedback) and stimulates release of glucagon.

Although blood glucose level is the most important regulator of insulin and glucagon, several hormones and neurotransmitters also regulate the release of these two hormones. In addition to the responses to blood glucose level just described, glucagon

Figure 18.19 Negative feedback regulation of the secretion of glucagon (blue arrows) and insulin (orange arrows).

🔑 Low blood glucose stimulates release of glucagon; high blood glucose stimulates secretion of insulin.

1 Low blood glucose (hypoglycemia) stimulates alpha cells to secrete

5 High blood glucose (hyperglycemia) stimulates beta cells to secrete

GLUCAGON

INSULIN

2 Glucagon acts on hepatocytes (liver cells) to:
- convert glycogen into glucose (glycogenolysis)
- form glucose from lactic acid and certain amino acids (gluconeogenesis)

3 Glucose released by hepatocytes raises blood glucose level to normal

4 If blood glucose continues to rise, hyperglycemia inhibits release of glucagon

6 Insulin acts on various body cells to:
- accelerate facilitated diffusion of glucose into cells
- speed conversion of glucose into glycogen (glycogenesis)
- increase uptake of amino acids and increase protein synthesis
- speed synthesis of fatty acids (lipogenesis)
- slow glycogenolysis
- slow gluconeogenesis

7 Blood glucose level falls

8 If blood glucose continues to fall, hypoglycemia inhibits release of insulin

❓ **Does glycogenolysis increase or decrease blood glucose level?**

stimulates insulin release directly; insulin has the opposite effect, suppressing glucagon secretion. As blood glucose level declines and less insulin is secreted, the alpha cells of the pancreas are released from the inhibitory effect of insulin so they can secrete more glucagon. Indirectly, human growth hormone (hGH) and adrenocorticotropic hormone (ACTH) stimulate secretion of insulin because they act to elevate blood glucose.

Insulin secretion is also stimulated by:

- Acetylcholine, the neurotransmitter liberated from axon terminals of parasympathetic vagus nerve fibers that innervate the pancreatic islets
- The amino acids arginine and leucine, which would be present in the blood at higher levels after a protein-containing meal

- Glucose-dependent insulinotropic peptide (GIP),* a hormone released by enteroendocrine cells of the small intestine in response to the presence of glucose in the gastrointestinal tract

Thus, digestion and absorption of food containing both carbohydrates and proteins provide strong stimulation for insulin release.

Glucagon secretion is stimulated by:

- Increased activity of the sympathetic division of the ANS, as occurs during exercise

*GIP—previously called gastric inhibitory peptide—was renamed because at physiological concentration its inhibitory effect on stomach function is negligible.

TABLE 18.9

Summary of Pancreatic Islet Hormones

HORMONE AND SOURCE	CONTROL OF SECRETION	PRINCIPAL ACTIONS
Glucagon from alpha cells of pancreatic islets Alpha cell	Decreased blood level of glucose, exercise, and mainly protein meals stimulate secretion; somatostatin and insulin inhibit secretion.	Raises blood glucose level by accelerating breakdown of glycogen into glucose in liver (glycogenolysis), converting other nutrients into glucose in liver (gluconeogenesis), and releasing glucose into blood.
Insulin from beta cells of pancreatic islets 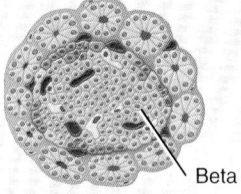 Beta cell	Increased blood level of glucose, acetylcholine (released by parasympathetic vagus nerve fibers), arginine and leucine (two amino acids), glucagon GIP, hGH, and ACTH stimulate secretion; somatostatin inhibits secretion.	Lowers blood glucose level by accelerating transport of glucose into cells, converting glucose into glycogen (glycogenesis), and decreasing glycogenolysis and gluconeogenesis; increases lipogenesis and stimulates protein synthesis.
Somatostatin from delta cells of pancreatic islets 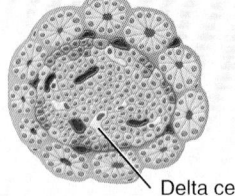 Delta cell	Pancreatic polypeptide inhibits secretion.	Inhibits secretion of insulin and glucagon; slows absorption of nutrients from gastrointestinal tract.
Pancreatic polypeptide from F cells of pancreatic islets 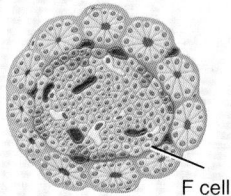 F cell	Meals containing protein, fasting, exercise, and acute hypoglycemia stimulate secretion; somatostatin and elevated blood glucose level inhibit secretion.	Inhibits somatostatin secretion, gallbladder contraction, and secretion of pancreatic digestive enzymes.

- A rise in blood amino acids if blood glucose level is low, which could occur after a meal that contained mainly protein

Table 18.9 summarizes the hormones produced by the pancreas, control of their secretion, and their principal actions.

✔CHECKPOINT

21. How are blood levels of glucagon and insulin controlled?
22. What are the effects on secretion of insulin and glucagon of exercise versus eating a carbohydrate- and protein-rich meal?

18.11 OVARIES AND TESTES

◉ OBJECTIVE

- Describe the location, hormones, and functions of the male and female gonads.

Gonads are the organs that produce gametes—sperm in males and oocytes in females. In addition to their reproductive function, the gonads secrete hormones. The **ovaries,** paired oval bodies located in the female pelvic cavity, produce several steroid hormones including two **estrogens** (estradiol and estrone) and **progesterone.** These female sex hormones, along with FSH and LH from the anterior pituitary, regulate the menstrual cycle, maintain pregnancy, and prepare the mammary glands for lactation. They also promote enlargement of the breasts and widening of the hips at puberty, and help maintain these female secondary sex characteristics. The ovaries also produce **inhibin,** a protein hormone that inhibits secretion of follicle-stimulating hormone (FSH). During pregnancy, the ovaries and placenta produce a peptide hormone called **relaxin (RLX),** which increases the flexibility of the pubic symphysis during pregnancy and helps dilate the uterine cervix during labor and delivery. These actions help ease the baby's passage by enlarging the birth canal.

The male gonads, the **testes,** are oval glands that lie in the scrotum. The main hormone produced and secreted by the testes is **testosterone,** an **androgen** or male sex hormone. Testosterone stimulates descent of the testes before birth, regulates production of sperm, and stimulates the development and maintenance of male secondary sex characteristics, such as beard growth and deepening of the voice. The testes also produce inhibin, which inhibits secretion of FSH. The detailed structure of the ovaries and testes and the specific roles of sex hormones are discussed in Chapter 28.

Table 18.10 summarizes the hormones produced by the ovaries and testes and their principal actions.

✔CHECKPOINT

23. Why are the ovaries and testes classified as endocrine glands as well as reproductive organs?

18.12 PINEAL GLAND AND THYMUS

◉ OBJECTIVE

- Describe the location, histology, hormone, and functions of the pineal gland.
- Describe the role of the thymus in immunity.

TABLE 18.10

Summary of Hormones of the Ovaries and Testes

HORMONE	PRINCIPAL ACTIONS
OVARIAN HORMONES	
Estrogens and progesterone Ovaries	Together with gonadotropic hormones of anterior pituitary, regulate female reproductive cycle and oogenesis, maintain pregnancy, prepare mammary glands for lactation, and promote development and maintenance of female secondary sex characteristics.
Relaxin	Increases flexibility of pubic symphysis during pregnancy; helps dilate uterine cervix during labor and delivery.
Inhibin	Inhibits secretion of FSH from anterior pituitary.
TESTICULAR HORMONES	
Testosterone Testes	Stimulates descent of testes before birth; regulates spermatogenesis; promotes development and maintenance of male secondary sex characteristics.
Inhibin	Inhibits secretion of FSH from anterior pituitary.

The **pineal gland** (PĪN-ē-al = pinecone shape) is a small endocrine gland attached to the roof of the third ventricle of the brain at the midline (see Figure 18.1). Part of the epithalamus, it is positioned between the two superior colliculi, has a mass of 0.1–0.2 g, and is covered by a capsule formed by the pia mater. The gland consists of masses of neuroglia and secretory cells called **pinealocytes** (pīn-ē-AL-ō-sītz).

The pineal gland secretes **melatonin,** an amine hormone derived from serotonin. Melatonin appears to contribute to the setting of the body's biological clock, which is controlled by the suprachiasmatic nucleus of the hypothalamus. As more melatonin is liberated during darkness than in light, this hormone is thought to promote sleepiness. In response to visual input from the eyes (retina), the suprachiasmatic nucleus stimulates sympathetic postganglionic neurons of the superior cervical ganglion, which in turn stimulate the pinealocytes of the pineal gland to secrete melatonin in a rhythmic pattern, with low levels of melatonin secreted during the day and significantly higher levels secreted at night. During sleep, plasma levels of melatonin increase tenfold and then decline to a low level again before awakening. Small doses of melatonin given orally can induce sleep and reset daily rhythms, which might benefit workers whose shifts alternate between daylight and nighttime hours. Melatonin also is a potent antioxidant that may provide some protection against damaging oxygen free radicals.

In animals that breed during specific seasons, melatonin inhibits reproductive functions, but it is unclear whether melatonin influences human reproductive function. Melatonin levels are higher in children and decline with age into adulthood, but there is no evidence that changes in melatonin secretion correlate with the onset of puberty and sexual maturation. Nevertheless, because melatonin causes atrophy of the gonads in several animal species, the possibility of adverse effects on human reproduction must be studied before its use to reset daily rhythms can be recommended.

CLINICAL CONNECTION | *Seasonal Affective Disorder and Jet Lag*

Seasonal affective disorder (SAD) is a type of depression that afflicts some people during the winter months, when day length is short. It is thought to be due, in part, to overproduction of melatonin. Full-spectrum bright-light therapy—repeated doses of several hours of exposure to artificial light as bright as sunlight—provides relief for some people. Three to six hours of exposure to bright light also appears to speed recovery from jet lag, the fatigue suffered by travelers who quickly cross several time zones. •

The **thymus** is located behind the sternum between the lungs. Because of the role of the thymus in immunity, the details of its structure and functions are discussed in Chapter 22. The hormones produced by the thymus—**thymosin, thymic humoral factor (THF), thymic factor (TF),** and **thymopoietin** (thī-mō-poy-Ē-tin)—promote the maturation of T cells (a type of white blood cell that destroys microbes and foreign substances) and may retard the aging process.

✔ CHECKPOINT

24. What is the relationship between melatonin and sleep?
25. Which thymic hormones play a role in immunity?

18.13 OTHER ENDOCRINE TISSUES AND ORGANS, EICOSANOIDS, AND GROWTH FACTORS

⦿ OBJECTIVES

• List the hormones secreted by cells in tissues and organs other than endocrine glands, and describe their functions.
• Describe the actions of eicosanoids and growth factors.

Hormones from Other Endocrine Tissues and Organs

As you learned at the beginning of this chapter, cells in organs other than those usually classified as endocrine glands have an endocrine function and secrete hormones. You learned about several of these in this chapter: the hypothalamus, thymus, pancreas, ovaries, and testes. Table 18.11 provides an overview of these organs and tissues and their hormones and actions.

Eicosanoids

Two families of eicosanoid molecules—the **prostaglandins** (pros'-ta-GLAN-dins), or **PGs,** and the **leukotrienes** (loo-kō-TRĪ-ēns), or **LTs**—are found in virtually all body cells except red blood cells, where they act as local hormones (paracrines or autocrines) in response to chemical or mechanical stimuli. They are synthesized by clipping a 20-carbon fatty acid called **arachidonic acid** (a-rak-i-DON-ik) from membrane phospholipid molecules. From arachidonic acid, different enzymatic reactions produce PGs or LTs. **Thromboxane (TX)** (throm-BOK-sān) is a modified PG that constricts blood vessels and promotes platelet activation.

TABLE 18.11

Summary of Hormones Produced by Other Organs and Tissues That Contain Endocrine Cells

HORMONE	PRINCIPAL ACTIONS
GASTROINTESTINAL TRACT	
Gastrin	Promotes secretion of gastric juice; increases movements of the stomach.
Glucose-dependent insulinotropic peptide (GIP)	Stimulates release of insulin by pancreatic beta cells.
Secretin	Stimulates secretion of pancreatic juice and bile.
Cholecystokinin (CCK)	Stimulates secretion of pancreatic juice; regulates release of bile from gallbladder; causes feeling of fullness after eating.
PLACENTA	
Human chorionic gonadotropin (hCG)	Stimulates corpus luteum in ovary to continue production of estrogens and progesterone to maintain pregnancy.
Estrogens and progesterone	Maintain pregnancy; help prepare mammary glands to secrete milk.
Human chorionic somatomammotropin (hCS)	Stimulates development of mammary glands for lactation.
KIDNEYS	
Renin	Part of reaction sequence that raises blood pressure by bringing about vasoconstriction and secretion of aldosterone.
Erythropoietin (EPO)	Increases rate of red blood cell formation.
Calcitriol* (active form of vitamin D)	Aids in absorption of dietary calcium and phosphorus.
HEART	
Atrial natriuretic peptide (ANP)	Decreases blood pressure.
ADIPOSE TISSUE	
Leptin	Suppresses appetite; may increase FSH and LH activity.

*Synthesis begins in the skin, continues in the liver, and ends in the kidneys.

Appearing in the blood in minute quantities, eicosanoids are present only briefly due to rapid inactivation.

To exert their effects, eicosanoids bind to receptors on target-cell plasma membranes and stimulate or inhibit the synthesis of second messengers such as cyclic AMP. Leukotrienes stimulate chemotaxis (attraction to a chemical stimulus) of white blood cells and mediate inflammation. The prostaglandins alter smooth muscle contraction, glandular secretions, blood flow, reproductive processes, platelet function, respiration, nerve impulse transmission, lipid metabolism, and immune responses. They also have roles in promoting inflammation and fever, and in intensifying pain.

CLINICAL CONNECTION | *Nonsteroidal Anti-inflammatory Drugs*

In 1971, scientists solved the long-standing puzzle of how aspirin works. Aspirin and related **nonsteroidal anti-inflammatory drugs (NSAIDs),** such as ibuprofen (Motrin®), inhibit *cyclooxygenase (COX),* a key enzyme involved in prostaglandin synthesis. NSAIDs are used to treat a wide variety of inflammatory disorders, from rheumatoid arthritis to tennis elbow. The success of NSAIDs in reducing fever, pain, and inflammation shows how prostaglandins contribute to these woes. •

Growth Factors

Several of the hormones we have described—insulinlike growth factor, thymosin, insulin, thyroid hormones, human growth hormone, and prolactin—stimulate cell growth and division. In addition, several more recently discovered hormones called **growth factors** play important roles in tissue development, growth, and repair. Growth factors are *mitogenic* substances—they cause growth by stimulating cell division. Many growth factors act locally, as autocrines or paracrines.

A summary of sources and actions of six important growth factors is presented in Table 18.12.

✔CHECKPOINT

26. What hormones are secreted by the gastrointestinal tract, placenta, kidneys, skin, adipose tissue, and heart?
27. What are some functions of prostaglandins, leukotrienes, and growth factors?

18.14 THE STRESS RESPONSE

◉ OBJECTIVE

• Describe how the body responds to stress.

It is impossible to remove all of the stress from our everyday lives. Some stress, called **eustress,** prepares us to meet certain challenges and thus is helpful. Other stress, called **distress,** is harmful. Any stimulus that produces a stress response is called a **stressor.** A stressor may be almost any disturbance of the human body—heat or cold, environmental poisons, toxins given off by bacteria, heavy bleeding from a wound or surgery, or a strong

TABLE 18.12

Summary of Selected Growth Factors

GROWTH FACTOR	COMMENT
Epidermal growth factor (EGF)	Produced in submaxillary (salivary) glands; stimulates proliferation of epithelial cells, fibroblasts, neurons, and astrocytes; suppresses some cancer cells and secretion of gastric juice by stomach.
Platelet-derived growth factor (PDGF)	Produced in blood platelets; stimulates proliferation of neuroglia, smooth muscle fibers, and fibroblasts; appears to have role in wound healing; may contribute to atherosclerosis development.
Fibroblast growth factor (FGF)	Found in pituitary gland and brain; stimulates proliferation of many cells derived from embryonic mesoderm (fibroblasts, adrenocortical cells, smooth muscle fibers, chondrocytes, and endothelial cells); stimulates formation of new blood vessels (angiogenesis).
Nerve growth factor (NGF)	Produced in submandibular (salivary) glands and hippocampus of brain; stimulates growth of ganglia in embryo; maintains sympathetic nervous system; stimulates hypertrophy and differentiation of neurons.
Tumor angiogenesis factors (TAFs)	Produced by normal and tumor cells; stimulate growth of new capillaries, organ regeneration, and wound healing.
Transforming growth factors (TGFs)	Produced by various cells as separate molecules: TGF-alpha has activities similar to epidermal growth factor; TGF-beta inhibits proliferation of many cell types.

emotional reaction. The responses to stressors may be pleasant or unpleasant, and they vary among people and even within the same person at different times.

Your body's homeostatic mechanisms attempt to counteract stress. When they are successful, the internal environment remains within normal physiological limits. If stress is extreme, unusual, or long lasting, the normal mechanisms may not be enough. In 1936, Hans Selye, a pioneer in stress research, showed that a variety of stressful conditions or noxious agents elicit a similar sequence of bodily changes. These changes, called the **stress response** or **general adaptation syndrome (GAS),** are controlled mainly by the hypothalamus. The stress response occurs in three stages: (1) an initial fight-or-flight response, (2) a slower resistance reaction, and eventually (3) exhaustion.

The Fight-or-Flight Response

The **fight-or-flight response,** initiated by nerve impulses from the hypothalamus to the sympathetic division of the autonomic nervous system (ANS), including the adrenal medulla, quickly mobilizes the body's resources for immediate physical activity

(Figure 18.20a). It brings huge amounts of glucose and oxygen to the organs that are most active in warding off danger: the brain, which must become highly alert; the skeletal muscles, which may have to fight off an attacker or flee; and the heart, which must work vigorously to pump enough blood to the brain and muscles. During the fight-or-flight response, nonessential body functions such as digestive, urinary, and reproductive activities are inhibited. Reduction of blood flow to the kidneys promotes release of renin, which sets into motion the renin–angiotensin–aldosterone pathway (see Figure 18.16). Aldosterone causes the kidneys to retain Na$^+$, which leads to water retention and elevated blood pressure. Water retention also helps preserve body fluid volume in the case of severe bleeding.

The Resistance Reaction

The second stage in the stress response is the **resistance reaction** (Figure 18.20b). Unlike the short-lived fight-or-flight response, which is initiated by nerve impulses from the hypothalamus, the resistance reaction is initiated in large part by hypothalamic releasing hormones and is a longer-lasting response. The hormones involved are corticotropin-releasing hormone (CRH), growth hormone–releasing hormone (GHRH), and thyrotropin-releasing hormone (TRH).

CRH stimulates the anterior pituitary to secrete ACTH, which in turn stimulates the adrenal cortex to increase release of cortisol. Cortisol then stimulates gluconeogenesis by liver cells, breakdown of triglycerides into fatty acids (lipolysis), and catabolism of proteins into amino acids. Tissues throughout the body can use the resulting glucose, fatty acids, and amino acids to produce ATP or to repair damaged cells. Cortisol also reduces inflammation.

A second hypothalamic releasing hormone, GHRH, causes the anterior pituitary to secrete human growth hormone (hGH). Acting via insulinlike growth factors, hGH stimulates lipolysis and glycogenolysis, the breakdown of glycogen to glucose, in the liver. A third hypothalamic releasing hormone, TRH, stimulates the anterior pituitary to secrete thyroid-stimulating hormone (TSH). TSH promotes secretion of thyroid hormones, which stimulate the increased use of glucose for ATP production. The combined actions of hGH and TSH supply additional ATP for metabolically active cells throughout the body.

The resistance stage helps the body continue fighting a stressor long after the fight-or-flight response dissipates. This is why your heart continues to pound for several minutes even after the stressor is removed. Generally, it is successful in seeing us through a stressful episode, and our bodies then return to normal. Occasionally, however, the resistance stage fails to combat the stressor, and the body moves into the state of exhaustion.

Exhaustion

The resources of the body may eventually become so depleted that they cannot sustain the resistance stage, and **exhaustion**

ensues. Prolonged exposure to high levels of cortisol and other hormones involved in the resistance reaction causes wasting of muscle, suppression of the immune system, ulceration of the gastrointestinal tract, and failure of pancreatic beta cells. In addition, pathological changes may occur because resistance reactions persist after the stressor has been removed.

Stress and Disease

Although the exact role of stress in human diseases is not known, it is clear that stress can lead to particular diseases by temporarily inhibiting certain components of the immune system. Stress-related disorders include gastritis, ulcerative colitis, irritable bowel syndrome, hypertension, asthma, rheumatoid arthritis (RA), migraine headaches, anxiety, and depression. People under stress are at a greater risk of developing chronic disease or dying prematurely.

Interleukin-1, a substance secreted by macrophages of the immune system (see the discussion of ACTH in Section 18.6), is an important link between stress and immunity. One action of interleukin-1 is to stimulate secretion of ACTH, which in turn stimulates the production of cortisol. Not only does cortisol provide resistance to stress and inflammation, but it also suppresses further production of interleukin-1. Thus, the immune system turns on the stress response, and the resulting cortisol then turns off one immune system mediator. This negative feedback system keeps the immune response in check once it has accomplished its goal. Because of this activity, cortisol and other glucocorticoids are used as immunosuppressive drugs for organ transplant recipients.

CLINICAL CONNECTION | *Posttraumatic Stress Disorder*

Posttraumatic stress disorder (PTSD) is an anxiety disorder that may develop in an individual who has experienced, witnessed, or learned about a physically or psychologically distressing event. The immediate cause of PTSD appears to be the specific stressors associated with the events. Among the stressors are terrorism, hostage taking, imprisonment, military duty, serious accidents, torture, sexual or physical abuse, violent crimes, school shootings, massacres, and natural disasters. In the United States, PTSD affects 10% of females and 5% of males. Symptoms of PTSD include reexperiencing the event through nightmares or flashbacks; avoidance of any activity, person, place, or event associated with the stressors; loss of interest and lack of motivation; poor concentration; irritability; and insomnia. Treatment may include the use of antidepressants, mood stabilizers, and antianxiety and antipsychotic agents. •

✔ CHECKPOINT

28. What is the central role of the hypothalamus during stress?
29. What body reactions occur during the fight-or-flight response, the resistance reaction, and exhaustion?
30. What is the relationship between stress and immunity?

Figure 18.20 **Responses to stressors during the stress response.** Red arrows (hormonal responses) and green arrows (neural responses) in (a) indicate immediate fight-or-flight reactions; black arrows in (b) indicate long-term resistance reactions.

Stressors stimulate the hypothalamus to initiate the stress response through the fight-or-flight response and the resistance reaction.

STRESSORS
stimulate

Key:
CRH = Corticotropin-releasing hormone
ACTH = Adrenocorticotropic hormone
GHRH = Growth hormone–releasing hormone
hGH = Human growth hormone
TRH = Thyrotropin-releasing hormone
TSH = Thyroid-stimulating hormone

CRH
GHRH
TRH

Hypothalamus

Nerve impulses

Sympathetic centers in spinal cord

Anterior pituitary

TSH
hGH
ACTH

Sympathetic nerves

ACTH

hGH

TSH

Adrenal medulla

Adrenal cortex

Liver

Thyroid gland

Visceral effectors

Epinephrine and norepinephrine

Supplement and prolong "fight-or-flight" responses

Cortisol

IGFs

Thyroid hormones (T_3 and T_4)

STRESS RESPONSES
1. Increased heart rate and force of beat
2. Constriction of blood vessels of most viscera and skin
3. Dilation of blood vessels of heart, lungs, brain, and skeletal muscles
4. Contraction of spleen
5. Conversion of glycogen into glucose in liver
6. Sweating
7. Dilation of airways
8. Decrease in digestive activities
9. Water retention and elevated blood pressure

STRESS RESPONSES
Lipolysis
Gluconeogenesis
Protein catabolism
Sensitized blood vessels
Reduced inflammation

STRESS RESPONSES
Lipolysis
Glycogenolysis

STRESS RESPONSES
Increased use of glucose to produce ATP

(a) Fight-or-flight responses

(b) Resistance reaction

? **What is the basic difference between the stress response and homeostasis?**

18.15 DEVELOPMENT OF THE ENDOCRINE SYSTEM

● **OBJECTIVE**

• Describe the development of endocrine glands.

The development of the endocrine system is not as localized as the development of other systems because, as you have already learned, endocrine organs are distributed throughout the body.

About 3 weeks after fertilization, the *pituitary gland (hypophysis)* begins to develop from two different regions of the **ectoderm.** The *posterior pituitary (neurohypophysis)* is derived from an outgrowth of ectoderm called the **neurohypophyseal bud** (noor′-ō-hī-pō-FIZ-ē-al), located on the floor of the hypothalamus (Figure 18.21). The *infundibulum,* also an outgrowth of the neurohypophyseal bud, connects the posterior pituitary to the hypothalamus. The *anterior pituitary (adenohypophysis)* is derived from an outgrowth of ectoderm from the roof of the mouth called the **hypophyseal (Rathke's) pouch.** The pouch grows toward the neurohypophyseal bud and eventually loses its connection with the roof of the mouth.

The *thyroid gland* develops during the fourth week as a midventral outgrowth of **endoderm,** called the **thyroid diverticulum** (dī′-ver-TIK-ū-lum), from the floor of the pharynx at the level of the second pair of pharyngeal pouches (Figure 18.21a). The outgrowth projects inferiorly and differentiates into the right and left lateral lobes and the isthmus of the gland.

The *parathyroid glands* develop during the fourth week from **endoderm** as outgrowths from the third and fourth **pharyngeal pouches** (fa-RIN-jē-al), which help to form structures of the head and neck.

Figure 18.21 Development of the endocrine system.

🔑 Glands of the endocrine system develop from all three primary germ layers: ectoderm, mesoderm, and endoderm.

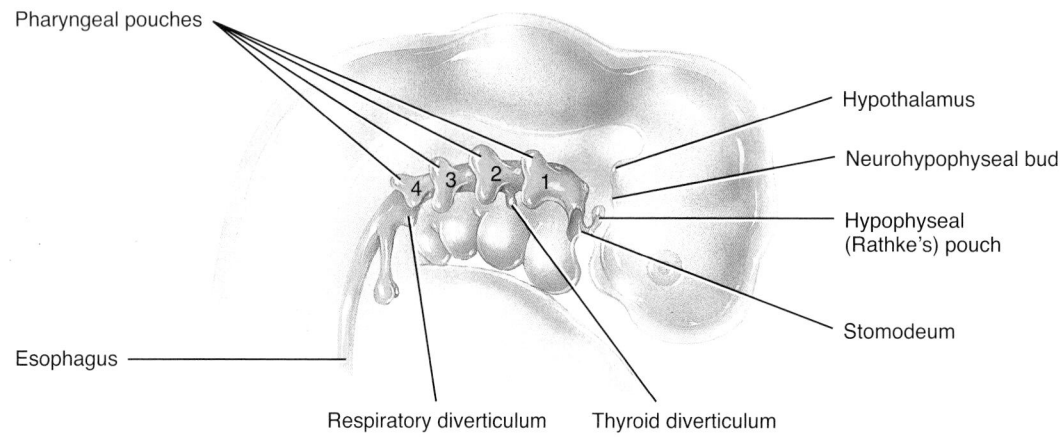

(a) Location of neurohypophyseal bud, hypophyseal (Rathke's) pouch, thyroid diverticulum, and pharyngeal pouches in 28-day embryo

(b) Development of pituitary gland between 5 and 16 weeks

❓ **Which endocrine gland develops from tissues with two different embryological origins?**

The adrenal cortex and adrenal medulla develop during the fifth week and have completely different embryological origins. The *adrenal cortex* is derived from the same region of **mesoderm** that produces the gonads. Endocrine tissues that secrete steroid hormones all are derived from mesoderm. The *adrenal medulla* is derived from **ectoderm** from **neural crest** cells that migrate to the superior pole of the kidney. Recall that neural crest cells also give rise to sympathetic ganglia and other structures of the nervous system (see Figure 14.27b).

The *pancreas* develops during the fifth through seventh weeks from two outgrowths of **endoderm** from the part of the **foregut** that later becomes the duodenum (see Figure 29.12c). The two outgrowths eventually fuse to form the pancreas. The origin of the ovaries and testes is discussed in Section 28.5.

The *pineal gland* arises during the seventh week as an outgrowth between the thalamus and colliculi of the midbrain from **ectoderm** associated with the **diencephalon** (see Figure 14.28).

The *thymus* arises during the fifth week from **endoderm** of the third pharyngeal pouches.

✔**CHECKPOINT**

31. Compare the origins of the adrenal cortex and adrenal medulla.

18.16 AGING AND THE ENDOCRINE SYSTEM

◉ **OBJECTIVE**

• Describe the effects of aging on the endocrine system.

Although some endocrine glands shrink as we get older, their performance may or may not be compromised. Production of human growth hormone by the anterior pituitary decreases, which is one cause of muscle atrophy as aging proceeds. The thyroid gland often decreases its output of thyroid hormones with age, causing a decrease in metabolic rate, an increase in body fat, and hypothyroidism, which is seen more often in older people. Because there is less negative feedback (lower levels of thyroid hormones), the level of thyroid-stimulating hormone increases with age (see Figure 18.12).

With aging, the blood level of PTH rises, perhaps due to inadequate dietary intake of calcium. In a study of older women who took 2400 mg/day of supplemental calcium, blood levels of PTH were as low as those of younger women. Both calcitriol and calcitonin levels are lower in older persons. Together, the rise in PTH and the fall in calcitonin level heighten the age-related decrease in bone mass that leads to osteoporosis and increased risk of fractures (see Figure 18.14).

The adrenal glands contain increasingly more fibrous tissue and produce less cortisol and aldosterone with advancing age. However, production of epinephrine and norepinephrine remains normal. The pancreas releases insulin more slowly with age, and receptor sensitivity to glucose declines. As a result, blood glucose levels in older people increase faster and return to normal more slowly than in younger individuals.

The thymus is largest in infancy. After puberty, its size begins to decrease, and thymic tissue is replaced by adipose and areolar connective tissue. In older adults, the thymus has atrophied significantly. However, it still produces new T cells for immune responses.

The ovaries decrease in size with age, and they no longer respond to gonadotropins. The resultant decreased output of estrogens leads to conditions such as osteoporosis, high blood cholesterol, and atherosclerosis. FSH and LH levels are high due to less negative feedback inhibition of estrogens. Although testosterone production by the testes decreases with age, the effects are not usually apparent until very old age; and many elderly males can still produce active sperm in normal numbers, even though there are higher numbers of morphologically abnormal sperm and decreased sperm motility.

✔**CHECKPOINT**

32. Which hormone is related to the muscle atrophy that occurs with aging?

• • •

To appreciate the many ways the endocrine system contributes to homeostasis of other body systems, examine *Focus on Homeostasis: The Endocrine System*. Next, in Chapter 19, we will begin to explore the cardiovascular system, starting with a description of the composition and functions of blood.

Focus on Homeostasis

BODY SYSTEM	CONTRIBUTION OF THE ENDOCRINE SYSTEM
For all body systems	Together with the nervous system, circulating and local hormones of the endocrine system regulate activity and growth of target cells throughout the body; several hormones regulate metabolism, uptake of glucose, and molecules used for ATP production by body cells.
Integumentary system	Androgens stimulate growth of axillary and pubic hair and activation of sebaceous glands; excess melanocyte-stimulating hormone (MSH) causes darkening of skin.
Skeletal system	Human growth hormone (hGH) and insulinlike growth factors (IGFs) stimulate bone growth; estrogens cause closure of the epiphyseal plates at the end of puberty and help maintain bone mass in adults; parathyroid hormone (PTH) and calcitonin regulate levels of calcium and other minerals in bone matrix and blood; thyroid hormones are needed for normal development and growth of the skeleton.
Muscular system	Epinephrine and norepinephrine help increase blood flow to exercising muscle; PTH maintains proper level of Ca^{2+}, needed for muscle contraction; glucagon, insulin, and other hormones regulate metabolism in muscle fibers; hGH, IGFs, and thyroid hormones help maintain muscle mass.
Nervous system	Several hormones, especially thyroid hormones, insulin, and growth hormone, influence growth and development of the nervous system; PTH maintains proper level of Ca^{2+}, needed for generation and conduction of nerve impulses.
Cardiovascular system	Erythropoietin (EPO) promotes formation of red blood cells; aldosterone and antidiuretic hormone (ADH) increase blood volume; epinephrine and norepinephrine increase heart rate and force of contraction; several hormones elevate blood pressure during exercise and other stresses.
Lymphatic system and immunity	Glucocorticoids such as cortisol depress inflammation and immune responses; thymic hormones promote maturation of T cells (a type of white blood cell).
Respiratory system	Epinephrine and norepinephrine dilate (widen) airways during exercise and other stresses; erythropoietin regulates amount of oxygen carried in blood by adjusting number of red blood cells.
Digestive system	Epinephrine and norepinephrine depress activity of the digestive system; gastrin, cholecystokinin, secretin, and GIP help regulate digestion; calcitriol promotes absorption of dietary calcium; leptin suppresses appetite.
Urinary system	ADH, aldosterone, and atrial natriuretic peptide (ANP) adjust the rate of loss of water and ions in the urine, thereby regulating blood volume and ion content of the blood.
Reproductive systems	Hypothalamic releasing and inhibiting hormones, follicle-stimulating hormone (FSH), and luteinizing hormone (LH) regulate development, growth, and secretions of the gonads (ovaries and testes); estrogens and testosterone contribute to development of oocytes and sperm and stimulate development of secondary sex characteristics; prolactin promotes milk secretion in mammary glands; oxytocin causes contraction of the uterus and ejection of milk from the mammary glands.

THE ENDOCRINE SYSTEM

DISORDERS: HOMEOSTATIC IMBALANCES

Disorders of the endocrine system often involve either **hyposecretion** (hypo- = too little or under), inadequate release of a hormone, or **hypersecretion** (hyper- = too much or above), excessive release of a hormone. In other cases, the problem is faulty hormone receptors, an inadequate number of receptors, or defects in second-messenger systems. Because hormones are distributed in the blood to target tissues throughout the body, problems associated with endocrine dysfunction may also be widespread.

Pituitary Gland Disorders

Pituitary Dwarfism, Giantism, and Acromegaly

Several disorders of the anterior pituitary involve human growth hormone (hGH). Hyposecretion of hGH during the growth years slows bone growth, and the epiphyseal plates close before normal height is reached. This condition is called **pituitary dwarfism** (see Clinical Connection: Hormonal Abnormalities That Affect Height in Chapter 6). Other organs of the body also fail to grow, and the body proportions are childlike. Treatment requires administration of hGH during childhood, before the epiphyseal plates close.

Hypersecretion of hGH during childhood causes **giantism,** an abnormal increase in the length of long bones. The person grows to be very tall, but body proportions are about normal. Figure 18.22a shows identical twins; one brother developed giantism due to a pituitary tumor. Hypersecretion of hGH during adulthood is called **acromegaly** (ak′-rō-MEG-a-lē). Although hGH cannot produce further lengthening of the long bones because the epiphyseal plates are already closed, the bones of the hands, feet, cheeks, and jaws thicken and other tissues enlarge. In addition, the eyelids, lips, tongue, and nose enlarge, and the skin thickens and develops furrows, especially on the forehead and soles (Figure 18.22b).

Diabetes Insipidus

The most common abnormality associated with dysfunction of the posterior pituitary is **diabetes insipidus** (dī-a-BĒ-tēs in-SIP-i-dus;

Figure 18.22 Various endocrine disorders.

 Disorders of the endocrine system often involve hyposecretion or hypersecretion of hormones.

(a) A 22-year-old man with pituitary giantism shown beside his identical twin

(b) Acromegaly (excess hGH during adulthood)

(c) Goiter (enlargement of thyroid gland)

(d) Exophthalmos (excess thyroid hormones, as in Graves' disease)

(e) Cushing's syndrome (excess glucocorticoids)

 Which endocrine disorder is due to antibodies that mimic the action of TSH?

diabetes = overflow; *insipidus* = tasteless) or **DI.** This disorder is due to defects in antidiuretic hormone (ADH) receptors or an inability to secrete ADH. *Neurogenic diabetes insipidus* results from hyposecretion of ADH, usually caused by a brain tumor, head trauma, or brain surgery that damages the posterior pituitary or the hypothalamus. In *nephrogenic diabetes insipidus,* the kidneys do not respond to ADH. The ADH receptors may be nonfunctional, or the kidneys may be damaged. A common symptom of both forms of DI is excretion of large volumes of urine, with resulting dehydration and thirst. Bed-wetting is common in afflicted children. Because so much water is lost in the urine, a person with DI may die of dehydration if deprived of water for only a day or so.

Treatment of neurogenic diabetes insipidus involves hormone replacement, usually for life. Either subcutaneous injection or nasal spray application of ADH analogs is effective. Treatment of nephrogenic diabetes insipidus is more complex and depends on the nature of the kidney dysfunction. Restriction of salt in the diet and, paradoxically, the use of certain diuretic drugs, are helpful.

Thyroid Gland Disorders

Thyroid gland disorders affect all major body systems and are among the most common endocrine disorders. **Congenital hypothyroidism,** hyposecretion of thyroid hormones that is present at birth, has devastating consequences if not treated promptly. Previously termed *cretinism,* this condition causes severe mental retardation and stunted bone growth. At birth, the baby typically is normal because lipid-soluble maternal thyroid hormones crossed the placenta during pregnancy and allowed normal development. Most states require testing of all newborns to ensure adequate thyroid function. If congenital hypothyroidism exists, oral thyroid hormone treatment must be started soon after birth and continued for life.

Hypothyroidism during the adult years produces **myxedema** (mix-e-DE-ma), which occurs about five times more often in females than in males. A hallmark of this disorder is edema (accumulation of interstitial fluid) that causes the facial tissues to swell and look puffy. A person with myxedema has a slow heart rate, low body temperature, sensitivity to cold, dry hair and skin, muscular weakness, general lethargy, and a tendency to gain weight easily. Because the brain has already reached maturity, mental retardation does not occur, but the person may be less alert. Oral thyroid hormones reduce the symptoms.

The most common form of hyperthyroidism is **Graves' disease,** which also occurs seven to ten times more often in females than in males, usually before age 40. Graves' disease is an autoimmune disorder in which the person produces antibodies that mimic the action of thyroid-stimulating hormone (TSH). The antibodies continually stimulate the thyroid gland to grow and produce thyroid hormones. A primary sign is an enlarged thyroid, which may be two to three times its normal size. Graves' patients often have a peculiar edema behind the eyes, called **exophthalmos** (ek'-sof-THAL-mos), which causes the eyes to protrude (Figure 18.22d). Treatment may include surgical removal of part or all of the thyroid gland (thyroidectomy), the use of radioactive iodine (131I) to selectively destroy thyroid tissue, and the use of antithyroid drugs to block synthesis of thyroid hormones.

A **goiter** (GOY-ter; *guttur* = throat) is simply an enlarged thyroid gland. It may be associated with hyperthyroidism, hypothyroidism, or **euthyroidism** (ū-THĪ-royd-izm; *eu* = good), which means normal secretion of thyroid hormone. In some places in the world, dietary iodine intake is inadequate; the resultant low level of thyroid hormone in the blood stimulates secretion of TSH, which causes thyroid gland enlargement (Figure 18.22c).

Parathyroid Gland Disorders

Hypoparathyroidism (hī-pō-par'-a-THĪ-royd-izm)—too little parathyroid hormone—leads to a deficiency of blood Ca^{2+}, which causes neurons and muscle fibers to depolarize and produce action potentials spontaneously. This leads to twitches, spasms, and **tetany** (maintained contraction) of skeletal muscle. The leading cause of hypoparathyroidism is accidental damage to the parathyroid glands or to their blood supply during thyroidectomy surgery.

Hyperparathyroidism, an elevated level of parathyroid hormone, most often is due to a tumor of one of the parathyroid glands. An elevated level of PTH causes excessive resorption of bone matrix, raising the blood levels of calcium and phosphate ions and causing bones to become soft and easily fractured. High blood calcium level promotes formation of kidney stones. Fatigue, personality changes, and lethargy are also seen in patients with hyperparathyroidism.

Adrenal Gland Disorders

Cushing's Syndrome

Hypersecretion of cortisol by the adrenal cortex produces **Cushing's syndrome** (Figure 18.22e). Causes include a tumor of the adrenal gland that secretes cortisol, or a tumor elsewhere that secretes adrenocorticotropic hormone (ACTH), which in turn stimulates excessive secretion of cortisol. The condition is characterized by breakdown of muscle proteins and redistribution of body fat, resulting in spindly arms and legs accompanied by a rounded "moon face," "buffalo hump" on the back, and pendulous (hanging) abdomen. Facial skin is flushed, and the skin covering the abdomen develops stretch marks. The person also bruises easily, and wound healing is poor. The elevated level of cortisol causes hyperglycemia, osteoporosis, weakness, hypertension, increased susceptibility to infection, decreased resistance to stress, and mood swings. People who need long-term glucocorticoid therapy—for instance, to prevent rejection of a transplanted organ—may develop a cushingoid appearance.

Addison's Disease

Hyposecretion of glucocorticoids and aldosterone causes **Addison's disease (chronic adrenocortical insufficiency).** The majority of cases are autoimmune disorders in which antibodies cause adrenal cortex destruction or block binding of ACTH to its receptors. Pathogens, such as the bacterium that causes tuberculosis, also may trigger adrenal cortex destruction. Symptoms, which typically do not appear until 90% of the adrenal cortex has been destroyed, include mental lethargy, anorexia, nausea and vomiting, weight loss, hypoglycemia, and muscular weakness. Loss of aldosterone leads to elevated potassium and decreased sodium in the blood, low blood pressure, dehydration, decreased cardiac output, arrhythmias, and even cardiac arrest. The skin may have a "bronzed" appearance that often is mistaken for a suntan. Such was true in the case of President John F. Kennedy, whose Addison's disease was known to only a few while he was alive. Treatment consists of replacing glucocorticoids and mineralocorticoids and increasing sodium in the diet.

Pheochromocytomas

Usually benign tumors of the chromaffin cells of the adrenal medulla, called **pheochromocytomas** (fē-ō-krō'-mō-si-TŌ-mas; *pheo-* = dusky; *-chromo-* = color; *-cyto-* = cell), cause hypersecretion of epinephrine and norepinephrine. The result is a prolonged version of the fight-or-flight response: rapid heart rate, high blood pressure, high levels of glucose in blood and urine, an elevated basal metabolic rate

(BMR), flushed face, nervousness, sweating, and decreased gastrointestinal motility. Treatment is surgical removal of the tumor.

Pancreatic Islet Disorders

The most common endocrine disorder is **diabetes mellitus** (MEL-i-tus; *melli-* = honey sweetened), caused by an inability to produce or use insulin. Diabetes mellitus is the fourth leading cause of death by disease in the United States, primarily because of its damage to the cardiovascular system. Because insulin is unavailable to aid transport of glucose into body cells, blood glucose level is high and glucose "spills" into the urine (glucosuria). Hallmarks of diabetes mellitus are the three "polys": *polyuria,* excessive urine production due to an inability of the kidneys to reabsorb water; *polydipsia,* excessive thirst; and *polyphagia,* excessive eating.

Both genetic and environmental factors contribute to onset of the two types of diabetes mellitus—type 1 and type 2—but the exact mechanisms are still unknown. **Type 1 diabetes,** previously known as **insulin-dependent diabetes mellitus (IDDM),** occurs because the person's immune system destroys the pancreatic beta cells. As a result, the pancreas produces little or no insulin. Type 1 diabetes usually develops in people younger than age 20 and it persists throughout life. By the time symptoms of type 1 diabetes arise, 80–90% of the islet beta cells have been destroyed. Type 1 diabetes is most common in northern Europe, especially in Finland, where nearly 1% of the population develops type 1 diabetes by 15 years of age. In the United States, type 1 diabetes is 1.5–2.0 times more common in whites than in African American or Asian populations.

The cellular metabolism of an untreated type 1 diabetic is similar to that of a starving person. Because insulin is not present to aid the entry of glucose into body cells, most cells use fatty acids to produce ATP. Stores of triglycerides in adipose tissue are catabolized to yield fatty acids and glycerol. The by-products of fatty acid breakdown—organic acids called ketones or ketone bodies—accumulate. Buildup of ketones causes blood pH to fall, a condition known as **ketoacidosis** (kē′-tō-as-i-DŌ-sis). Unless treated quickly, ketoacidosis can cause death.

The breakdown of stored triglycerides also causes weight loss. As lipids are transported by the blood from storage depots to cells, lipid particles are deposited on the walls of blood vessels, leading to atherosclerosis and a multitude of cardiovascular problems, including cerebrovascular insufficiency, ischemic heart disease, peripheral vascular disease, and gangrene. A major complication of diabetes is loss of vision due either to cataracts (excessive glucose attaches to lens proteins, causing cloudiness) or to damage to blood vessels of the retina. Severe kidney problems also may result from damage to renal blood vessels.

Type 1 diabetes is treated through self-monitoring of blood glucose level (up to 7 times daily), regular meals containing 45–50% carbohydrates and less than 30% fats, exercise, and periodic insulin injections (up to 3 times a day). Several implantable pumps are available to provide insulin without the need for repeated injections. Because they lack a reliable glucose sensor, however, the person must self-monitor blood glucose level to determine insulin doses. It is also possible to successfully transplant a pancreas, but immunosuppressive drugs must then be taken for life. Another promising approach under investigation is transplantation of isolated islets in semipermeable hollow tubes. The tubes allow glucose and insulin to enter and leave but prevent entry of immune system cells that might attack the islet cells.

Type 2 diabetes, formerly known as **non-insulin-dependent diabetes mellitus (NIDDM),** is much more common than type 1, representing more than 90% of all cases. Type 2 diabetes most often occurs in obese people who are over age 35. However, the number of obese children and teenagers with type 2 diabetes is increasing. Clinical symptoms are mild, and the high glucose levels in the blood often can be controlled by diet, exercise, and weight loss. Sometimes, drugs such as *glyburide* (DiaBeta®) and metformin (Fortamet®) are used to stimulate secretion of insulin by pancreatic beta cells. Although some type 2 diabetics need insulin, many have a sufficient amount (or even a surplus) of insulin in the blood. For these people, diabetes arises not from a shortage of insulin but because target cells become less sensitive to it due to down-regulation of insulin receptors.

Hyperinsulinism most often results when a diabetic injects too much insulin. The main symptom is **hypoglycemia,** decreased blood glucose level, which occurs because the excess insulin stimulates too much uptake of glucose by body cells. The resulting hypoglycemia stimulates the secretion of epinephrine, glucagon, and human growth hormone. As a consequence, anxiety, sweating, tremor, increased heart rate, hunger, and weakness occur. When blood glucose falls, brain cells are deprived of the steady supply of glucose they need to function effectively. Severe hypoglycemia leads to mental disorientation, convulsions, unconsciousness, and shock. Shock due to an insulin overdose is termed **insulin shock.** Death can occur quickly unless blood glucose is restored to normal levels. From a clinical standpoint, a diabetic suffering from either a hyperglycemia or a hypoglycemia crisis can have very similar symptoms—mental changes, coma, seizures, and so on. It is important to quickly and correctly identify the cause of the underlying symptoms and treat them appropriately.

MEDICAL TERMINOLOGY

Gynecomastia (gī-ne′-kō-MAS-tē-a; *gyneco-* = woman; *-mast-* = breast) Excessive development of mammary glands in a male. Sometimes a tumor of the adrenal gland may secrete sufficient amounts of estrogen to cause the condition.

Hirsutism (HER-soo-tizm; *hirsut-* = shaggy) Presence of excessive body and facial hair in a male pattern, especially in women; may be due to excess androgen production due to tumors or drugs.

Thyroid crisis (storm) A severe state of hyperthyroidism that can be life-threatening. It is characterized by high body temperature, rapid heart rate, high blood pressure, gastrointestinal symptoms (abdominal pain, vomiting, diarrhea), agitation, tremors, confusion, seizures, and possibly coma.

Virilizing adenoma (*aden-* = gland; *-oma* = tumor) Tumor of the adrenal gland that liberates excessive androgens, causing virilism (masculinization) in females. Occasionally, adrenal tumor cells liberate estrogens to the extent that a male patient develops gynecomastia. Such a tumor is called a **feminizing adenoma.**

Review

Introduction

1. Hormones regulate the activity of smooth muscle, cardiac muscle, and some glands; alter metabolism; spur growth and development; influence reproductive processes; and participate in circadian (daily) rhythms.

18.1 Comparison of Control by the Nervous and Endocrine Systems

1. The nervous system controls homeostasis through nerve impulses and neurotransmitters, which act locally and quickly. The endocrine system uses hormones, which act more slowly in distant parts of the body. (See Table 18.1.)
2. The nervous system controls neurons, muscle cells, and glandular cells; the endocrine system regulates virtually all body cells.

18.2 Endocrine Glands

1. Exocrine glands (sudoriferous, sebaceous, mucous, and digestive) secrete their products through ducts into body cavities or onto body surfaces. Endocrine glands secrete hormones into interstitial fluid. Then, the hormones diffuse into the blood.
2. The endocrine system consists of endocrine glands (pituitary, thyroid, parathyroid, adrenal, and pineal glands) and other hormone-secreting tissues (hypothalamus, thymus, pancreas, ovaries, testes, kidneys, stomach, liver, small intestine, skin, heart, adipose tissue, and placenta).

18.3 Hormone Activity

1. Hormones affect only specific target cells that have receptors to recognize (bind) a given hormone. The number of hormone receptors may decrease (down-regulation) or increase (up-regulation).
2. Circulating hormones enter the bloodstream; local hormones (paracrines and autocrines) act locally on neighboring cells.
3. Chemically, hormones are either lipid-soluble (steroids, thyroid hormones, and nitric oxide) or water-soluble (amines; peptides, proteins, and glycoproteins; and eicosanoids). (See Table 18.2.)
4. Water-soluble hormone molecules circulate in the watery blood plasma in a "free" form (not attached to plasma proteins); most lipid-soluble hormones are bound to transport proteins synthesized by the liver.

18.4 Mechanisms of Hormone Action

1. Lipid-soluble steroid hormones and thyroid hormones affect cell function by altering gene expression.
2. Water-soluble hormones alter cell function by activating plasma membrane receptors, which elicit production of a second messenger that activates various enzymes inside the cell.
3. Hormonal interactions can have three types of effects: permissive, synergistic, or antagonistic.

18.5 Control of Hormone Secretion

1. Hormone secretion is controlled by signals from the nervous system, chemical changes in blood, and other hormones.
2. Negative feedback systems regulate the secretion of many hormones.

18.6 Hypothalamus and Pituitary Gland

1. The hypothalamus is the major integrating link between the nervous and endocrine systems. The hypothalamus and pituitary gland regulate virtually all aspects of growth, development, metabolism, and homeostasis. The pituitary gland is located in the hypophyseal fossa and is divided into two main portions: the anterior pituitary (glandular portion) and the posterior pituitary (nervous portion).
2. Secretion of anterior pituitary hormones is stimulated by releasing hormones and suppressed by inhibiting hormones from the hypothalamus.
3. The blood supply to the anterior pituitary is from the superior hypophyseal arteries. Hypothalamic releasing and inhibiting hormones enter the primary plexus and flow to the secondary plexus in the anterior pituitary by the hypophyseal portal veins.
4. The anterior pituitary consists of somatotrophs that produce human growth hormone (hGH), lactotrophs that produce prolactin (PRL), corticotrophs that secrete adrenocorticotropic hormone (ACTH) and

Resource

Anatomy Overview - The Endocrine System

Anatomy Overview - The Nervous System

Anatomy Overview - The Endocrine System

Anatomy Overview - Hormones
Anatomy Overview - Lipid-Soluble Hormones
Anatomy Overview - Water-Soluble Hormones
Anatomy Overview - Local Hormones

Animation - Introduction to Hormonal Regulation, Secretion, and Concentration
Animation - Mechanisms of Hormone Action
Exercise - Hormone Actions

Animation - Introduction to Hormone Feedback Loops
Exercise - Produce That Hormone

Anatomy Overview - The Hypothalamus and Pituitary Gland
Anatomy Overview - Hormones of the Anterior Pituitary Gland
Anatomy Overview - The Hypothalamus
Anatomy Overview - Hormones of the Hypothalamus
Animation - hGH: Growth and Development
Animation - GHRH/hGR

Review	Resource

melanocyte-stimulating hormone (MSH), thyrotrophs that secrete thyroid-stimulating hormone (TSH), and gonadotrophs that synthesize follicle-stimulating hormone (FSH) and luteinizing hormone (LH). (See Tables 18.3 and 18.4.)

5. Human growth hormone (hGH) stimulates body growth through insulinlike growth factors (IGFs). Secretion of hGH is inhibited by GHIH (growth hormone–inhibiting hormone, or somatostatin) and promoted by GHRH (growth hormone–releasing hormone).

6. TSH regulates thyroid gland activities. Its secretion is stimulated by TRH (thyrotropin-releasing hormone) and suppressed by GHIH.

7. FSH and LH regulate the activities of the gonads—ovaries and testes. Their secretion is controlled by GnRH (gonadotropin-releasing hormone).

8. Prolactin (PRL) helps initiate milk secretion. Prolactin-inhibiting hormone (PIH) suppresses secretion of PRL; prolactin-releasing hormone (PRH) stimulates PRL secretion.

9. ACTH regulates the activities of the adrenal cortex and is controlled by CRH (corticotropin-releasing hormone). Dopamine inhibits secretion of MSH.

10. The posterior pituitary contains axon terminals of neurosecretory cells whose cell bodies are in the hypothalamus. Hormones made by the hypothalamus and stored in the posterior pituitary are oxytocin (OT), which stimulates contraction of the uterus and ejection of milk from the breasts, and antidiuretic hormone (ADH), which stimulates water reabsorption by the kidneys and constriction of arterioles. (See Table 18.5.) Oxytocin secretion is stimulated by uterine stretching and suckling during nursing; ADH secretion is controlled by osmotic pressure of the blood and blood volume.

Resource (right column, top):
Animation - ACTH/Cortisol: Glycogenolysis
Animation - TRH/TSH: Production
Animation - hGH: Glycogenolysis and Lipolysis
Animation - Antidiuretic Hormone
Figure 18.5 - Hypothalamus and Pituitary Gland and Their Blood Supply

18.7 Thyroid Gland

1. The thyroid gland is located inferior to the larynx.

2. It consists of thyroid follicles composed of follicular cells, which secrete the thyroid hormones thyroxine (T_4) and triiodothyronine (T_3), and parafollicular cells, which secrete calcitonin (CT).

3. Thyroid hormones are synthesized from iodine and tyrosine within thyroglobulin (TGB). They are transported in the blood bound to plasma proteins, mostly thyroxine-binding globulin (TBG).

4. Secretion is controlled by TRH from the hypothalamus and thyroid-stimulating hormone (TSH) from the anterior pituitary.

5. Thyroid hormones regulate oxygen use and metabolic rate, cellular metabolism, and growth and development.

6. Calcitonin (CT) can lower the blood level of calcium ions (Ca^{2+}) and promote deposition of Ca^{2+} into bone matrix. Secretion of CT is controlled by the Ca^{2+} level in the blood. (See Table 18.6.)

Resource (right column):
Anatomy Overview - The Thyroid
Animation - Thyroid Hormones and Glucose and Lipid Catabolism
Animation - TRH/TSH
Animation - Calcitonin
Figure 18.11 - Steps in the Synthesis and Secretion of Thyroid Hormones
Exercise - Calcium Homeostasis

18.8 Parathyroid Glands

1. The parathyroid glands are embedded in the posterior surfaces of the lateral lobes of the thyroid gland. They consist of chief cells and oxyphil cells.

2. Parathyroid hormone (PTH) regulates the homeostasis of calcium, magnesium, and phosphate ions by increasing blood calcium and magnesium levels and decreasing blood phosphate levels. PTH secretion is controlled by the level of calcium in the blood. (See Table 18.7.)

Resource (right column):
Anatomy Overview - Parathyroid Glands
Anatomy Overview - Hormones of the Parathyroid
Animation - Parathyroid Hormone

18.9 Adrenal Glands

1. The adrenal glands are located superior to the kidneys. They consist of an outer adrenal cortex and inner adrenal medulla.

2. The adrenal cortex is divided into a zona glomerulosa, a zona fasciculata, and a zona reticularis; the adrenal medulla consists of chromaffin cells and large blood vessels.

3. Cortical secretions include mineralocorticoids, glucocorticoids, and androgens.

4. Mineralocorticoids (mainly aldosterone) increase sodium and water reabsorption and decrease potassium reabsorption. Secretion is controlled by the renin–angiotensin–aldosterone (RAA) pathway and by K^+ level in the blood.

5. Glucocorticoids (mainly cortisol) promote protein breakdown, gluconeogenesis, and lipolysis; help resist stress; and serve as anti-inflammatory substances. Their secretion is controlled by ACTH.

6. Androgens secreted by the adrenal cortex stimulate growth of axillary and pubic hair, aid the prepubertal growth spurt, and contribute to libido.

7. The adrenal medulla secretes epinephrine and norepinephrine (NE), which are released during stress and produce effects similar to sympathetic responses. (See Table 18.8.)

Resource (right column):
Anatomy Overview - Adrenal Glands
Anatomy Overview - Hormones of the Adrenal Glands
Animation - ACTH/Cortisol
Animation - Epinephrine/NE
Figure 18.15 - The Adrenal Glands
Figure 18.16 - Regulation of Aldosterone Secretion

Review	Resource

Review

18.10 Pancreatic Islets

1. The pancreas lies in the curve of the duodenum. It has both endocrine and exocrine functions.
2. The endocrine portion consists of pancreatic islets or islets of Langerhans, made up of four types of cells: alpha, beta, delta, and F cells.
3. Alpha cells secrete glucagon, beta cells secrete insulin, delta cells secrete somatostatin, and F cells secrete pancreatic polypeptide.
4. Glucagon increases blood glucose level; insulin decreases blood glucose level. Secretion of both hormones is controlled by the level of glucose in the blood. (See Table 18.9.)

18.11 Ovaries and Testes

1. The ovaries are located in the pelvic cavity and produce estrogens, progesterone, and inhibin. These sex hormones govern the development and maintenance of female secondary sex characteristics, reproductive cycles, pregnancy, lactation, and normal female reproductive functions. (See Table 18.10.)
2. The testes lie inside the scrotum and produce testosterone and inhibin. These sex hormones govern the development and maintenance of male secondary sex characteristics and normal male reproductive functions. (See Table 18.10.)

18.12 Pineal Gland and Thymus

1. The pineal gland is attached to the roof of the third ventricle of the brain. It consists of secretory cells called pinealocytes, neuroglia, and endings of sympathetic postganglionic axons.
2. The pineal gland secretes melatonin, which contributes to setting the body's biological clock (controlled in the suprachiasmatic nucleus). During sleep, plasma levels of melatonin increase.
3. The thymus secretes several hormones related to immunity.
4. Thymosin, thymic humoral factor (THF), thymic factor (TF), and thymopoietin promote the maturation of T cells.

18.13 Other Endocrine Tissues and Organs, Eicosanoids, and Growth Factors

1. Body tissues other than those normally classified as endocrine glands contain endocrine tissue and secrete hormones; they include the gastrointestinal tract, placenta, kidneys, skin, and heart. (See Table 18.11.)
2. Prostaglandins and leukotrienes are eicosanoids that act as local hormones in most body tissues.
3. Growth factors are local hormones that stimulate cell growth and division. (See Table 18.12.)

18.14 The Stress Response

1. Productive stress is termed eustress, and harmful stress is termed distress.
2. If stress is extreme, it triggers the stress response (general adaptation syndrome), which occurs in three stages: the fight-or-flight response, resistance reaction, and exhaustion.
3. The stimuli that produce the stress response are called stressors. Stressors include surgery, poisons, infections, fever, and strong emotional responses.
4. The fight-or-flight response is initiated by nerve impulses from the hypothalamus to the sympathetic division of the autonomic nervous system and the adrenal medulla. This response rapidly increases circulation, promotes ATP production, and decreases nonessential activities.
5. The resistance reaction is initiated by releasing hormones secreted by the hypothalamus, most importantly CRH, TRH, and GHRH. Resistance reactions are longer lasting and accelerate breakdown reactions to provide ATP for counteracting stress.
6. Exhaustion results from depletion of body resources during the resistance stage.
7. Stress may trigger certain diseases by inhibiting the immune system. An important link between stress and immunity is interleukin-l, produced by macrophages; it stimulates secretion of ACTH.

18.15 Development of the Endocrine System

1. The development of the endocrine system is not as localized as in other systems because endocrine organs develop in widely separated parts of the embryo.
2. The pituitary gland, adrenal medulla, and pineal gland develop from ectoderm; the adrenal cortex develops from mesoderm; and the thyroid gland, parathyroid glands, pancreas, and thymus develop from endoderm.

Resource

Anatomy Overviews: The Pancreas; Hormones of the Pancreas
Animations: Glucagon; Insulin
Figure 18.18 - The Pancreas
Figure 18.19 - Secretion of Glucagon and Insulin
Exercise - Glucose Regulation Feedback Loop
Concepts and Connections - Blood Glucose Regulation

Anatomy Overviews: The Ovaries; Ovarian Hormones; The Testes; Testicular Hormones
Animations: Hormonal Control of Male Reproductive Function; Hormonal Regulation of Female Reproductive System
Exercises: Match the Female Hormones; Match the Male Hormones

Anatomy Overview - The Pineal Gland

Animations: Hormonal Control of Digestive Activities; Hormonal Regulation of Pregnancy
Exercise - Concentrate on Endocrine Activities

Animation - General Adaptation Syndrome
Concepts and Connections - General Adaptation Syndrome

Review   **Resource**

18.16 Aging and the Endocrine System

1. Although some endocrine glands shrink as we get older, their performance may or may not be compromised.
2. Production of human growth hormone, thyroid hormones, cortisol, aldosterone, and estrogens decreases with advancing age.
3. With aging, the blood levels of TSH, LH, FSH, and PTH rise.
4. The pancreas releases insulin more slowly with age, and receptor sensitivity to glucose declines.
5. After puberty, thymus size begins to decrease, and thymic tissue is replaced by adipose and areolar connective tissue.

SELF-QUIZ QUESTIONS

Fill in the blanks in the following statements.

1. The three stages of the stress response or general adaptation syndrome, in order of occurrence, are _____, _____ , and _____.

2. The _____ is the major integrating link between the nervous and endocrine systems, acts as an endocrine gland itself, and helps control the stress response.

3. Down-regulation makes a target cell _____ sensitive to a hormone while up-regulation makes a target cell _____ sensitive to a hormone.

Indicate whether the following statements are true or false.

4. If the effect of two or more hormones acting together is greater than the sum of each acting alone, then the two hormones are said to have a permissive effect.

5. In the direct gene activation method of hormone action, the hormone enters the target cell and binds to an intracellular receptor. The activated receptor–hormone complex then alters gene expression to produce the protein that causes the physiological responses that are characteristic of the hormone.

Choose the one best answer to the following questions.

6. Which of the following comparisons are *true*? (1) Nerve impulses produce their effects quickly; hormonal responses generally are slower. (2) Nervous system effects are brief; endocrine system effects are longer lasting. (3) The nervous system controls homeostasis through mediator molecules called neurotransmitters; the endocrine system works through mediator molecules called hormones. (4) The nervous system can stimulate or inhibit the release of hormones; some hormones are released by neurons as neurotransmitters. (5) Neurotransmitters transmit impulses directly; hormones must bind to receptors on or in target cells in order to exert their effects.
   (a) 1, 2, 3, 4, and 5   (b) 1, 2, 3, and 4   (c) 2, 3, 4, and 5
   (d) 2, 4, and 5   (e) 1, 4, and 5

7. Insulin and thyroxine arrive at an organ at the same time. Thyroxine causes an effect on the organ but insulin does not. Why?
   (a) Insulin is a lipid-soluble hormone and thyroxine is not.
   (b) The target cells in the organ have up-regulated for thyroxine.
   (c) Thyroxine is a local hormone and insulin is a circulating hormone.
   (d) Thyroxine inhibits the action of insulin.
   (e) The organ's cells have receptors for thyroxine but not for insulin.

8. Which of the following is not a category of water-soluble hormones?
   (a) peptides   (b) amines   (c) eicosanoids
   (d) steroids   (e) proteins

9. Place in correct order the action of a water-soluble hormone on its target cell. (1) Adenylate cyclase is activated, catalyzing the conversion of ATP to cAMP. (2) Enzymes catalyze reactions that produce a physiological response attributed to the hormone. (3) The hormone binds to a membrane receptor. (4) Activated protein kinases phosphorylate cellular proteins. (5) The hormone–receptor complex activates G proteins. (6) Cyclic AMP activates protein kinases.
   (a) 3, 5, 1, 6, 4, 2   (b) 3, 1, 5, 6, 4, 2   (c) 5, 1, 4, 2, 3, 6
   (d) 3, 4, 5, 1, 6, 2   (e) 6, 3, 5, 1, 4, 2

10. Hormones (1) generally utilize negative feedback mechanisms to regulate their secretion; (2) will only affect target cells far removed from the hormone-producing secretory cells; (3) must bind to transport proteins in order to circulate in the blood; (4) may be released in low concentrations but can produce large effects in the target cells because of amplification; (5) can regulate the responsiveness of the target tissue by controlling the number of receptor sites for the hormone.
   (a) 1, 2, and 3   (b) 1, 2, 4, and 5   (c) 2, 3, and 4
   (d) 2, 3, 4, and 5   (e) 1, 4, and 5

11. The pituitary gland (1) is located in the cribriform plate of the ethmoid bone, (2) is linked to the hypothalamus by the infundibulum, (3) has a posterior portion that contains axon terminals from hypothalamic neurosecretory cells, (4) produces releasing and inhibiting hormones, (5) has a vascular connection with the hypothalamus known as the hypophyseal portal system.
   (a) 1, 2, and 4   (b) 2, 3, 4, and 5   (c) 2, 3, and 5
   (d) 1, 2, 3, 4, and 5   (e) 2, 4, and 5

12. The class of adrenal gland hormones that provides resistance to stress, produces anti-inflammatory effects, and promotes normal metabolism to ensure adequate quantities of ATP is
   (a) glucocorticoids.
   (b) mineralocorticoids.
   (c) androgens.
   (d) catecholamines.
   (e) gonadocorticoids.

13. Match the following:

_____ (a) increases blood Ca²⁺ level
_____ (b) increases blood glucose level
_____ (c) decreases blood Ca²⁺ level
_____ (d) decreases blood glucose level
_____ (e) initiates and maintains milk secretion by the mammary glands
_____ (f) regulates the body's biological clock
_____ (g) stimulates sex hormone production; triggers ovulation
_____ (h) augments fight-or-flight responses
_____ (i) regulates metabolism and resistance to stress
_____ (j) helps control water and electrolyte homeostasis
_____ (k) suppresses release of FSH
_____ (l) stimulates growth of axillary and pubic hair
_____(m) promotes T cell maturation
_____ (n) regulates oxygen use, basal metabolic rate, cellular metabolism, and growth and development
_____(o) stimulates protein synthesis, inhibits protein breakdown, stimulates lipolysis, and retards use of glucose for ATP production
_____ (p) inhibits water loss through the kidneys
_____ (q) stimulates egg and sperm formation
_____ (r) enhances uterine contractions during labor; stimulates milk ejection
_____ (s) stimulates and inhibits secretion of anterior pituitary hormones
_____ (t) increases skin pigmentation when present in excess
_____ (u) stimulates synthesis and release of T₃ and T₄
_____ (v) local hormones involved in inflammation, smooth muscle contraction, blood flow, and inflammation

(1) insulin
(2) glucagon
(3) inhibin
(4) follicle-stimulating hormone
(5) luteinizing hormone
(6) thyroxine and tri-iodothyronine
(7) calcitonin
(8) parathormone
(9) melanocyte-stimulating hormone
(10) oxytocin
(11) antidiuretic hormone
(12) prolactin
(13) human growth hormone
(14) hypothalamic regulating hormones
(15) aldosterone
(16) thyroid-stimulating hormone
(17) androgens
(18) epinephrine and norepinephrine
(19) prostaglandins
(20) melatonin
(21) thymosin
(22) cortisol

14. Match the following hormone-secreting cells to the hormone(s) they release:

_____ (a) ACTH and MSH
_____ (b) TSH
_____ (c) glucagon
_____ (d) PTH
_____ (e) glucocorticoids
_____ (f) calcitonin
_____ (g) insulin
_____ (h) androgens
_____ (i) progesterone
_____ (j) FSH and LH
_____ (k) epinephrine and norepinephrine
_____ (l) hGH
_____(m) testosterone
_____ (n) mineralocorticoids
_____ (o) thyroxine and triiodothyronine
_____ (p) PRL

(1) beta cells of pancreatic islet
(2) alpha cells of pancreatic islet
(3) follicular cells of thyroid gland
(4) parafollicular cells of thyroid gland
(5) testes
(6) ovary
(7) somatotrophs
(8) thyrotrophs
(9) gonadotrophs
(10) corticotrophs
(11) lactotrophs
(12) chief cells
(13) chromaffin cells
(14) zona glomerulosa cells
(15) zona fasciculata cells
(16) zona reticularis cells

15. Match the endocrine disorder to the problem that produced the disorder:

_____ (a) hyposecretion of insulin or down-regulation of insulin receptors
_____ (b) hypersecretion of hGH before closure of epiphyseal plates
_____ (c) hyposecretion of thyroid hormone that is present at birth
_____ (d) hypersecretion of glucocorticoids
_____ (e) hyposecretion of hGH before closure of epiphyseal plates
_____ (f) hypersecretion of epinephrine and norepinephrine
_____ (g) hypersecretion of hGH after closure of epiphyseal plates
_____ (h) hyposecretion of glucocorticoids and aldosterone
_____ (i) hyposecretion of ADH
_____ (j) hypersecretion of melatonin
_____ (k) hyposecretion of thyroid hormone in adults
_____ (l) hyperthyroidism, an autoimmune disease

(1) giantism
(2) acromegaly
(3) pituitary dwarfism
(4) diabetes insipidus
(5) myxedema
(6) Graves' disease
(7) Cushing's syndrome
(8) seasonal affective disorder
(9) Addison's disease
(10) pheochromocytomas
(11) congenital hypothyroidism
(12) diabetes mellitus

CRITICAL THINKING QUESTIONS

1. Amanda hates her new student ID photo. Her hair looks dry, the extra weight that she's gained shows, and her neck looks fat. In fact, there's an odd butterfly-shaped swelling across the front of her neck, under her chin. Amanda's also been feeling very tired and mentally "dull" lately, but she figures all new A&P students feel that way. Should she visit the clinic or just wear turtlenecks?

2. Amanda (from question 1 above) goes to the clinic and blood is drawn. The results show that her T_4 levels are low and her TSH levels are low. Later she is given a TSH stimulation test in which TSH is injected and the T_4 levels are monitored. After TSH injection, her T_4 levels rise. Does Amanda have problems with her pituitary or with her thyroid gland? How did you come to your conclusion?

3. Mr. Hernandez visited his doctor complaining that he is constantly thirsty and is "in the bathroom day and night" relieving his bladder. The doctor ordered blood and urine tests to check for glucose and ketones, which were all negative. What is the doctor's diagnosis of Mr. Hernandez, and what gland(s) or organ(s) is(are) involved?

? ANSWERS TO FIGURE QUESTIONS

18.1 Secretions of endocrine glands diffuse into interstitial fluid and then into the blood; exocrine secretions flow into ducts that lead into body cavities or to the body surface.

18.2 In the stomach, histamine is a paracrine because it acts on nearby parietal cells without entering the blood.

18.3 The receptor–hormone complex alters gene expression by turning specific genes of nuclear DNA on or off.

18.4 Cyclic AMP is termed a second messenger because it translates the presence of the first messenger, the water-soluble hormone, into a response inside the cell.

18.5 The hypophyseal portal veins carry blood from the median eminence of the hypothalamus, where hypothalamic releasing and inhibiting hormones are secreted, to the anterior pituitary, where these hormones act.

18.6 Thyroid hormones suppress secretion of TSH by thyrotrophs and of TRH by hypothalamic neurosecretory cells; gonadal hormones suppress secretion of FSH and LH by gonadotrophs and of GnRH by hypothalamic neurosecretory cells.

18.7 Excess levels of hGH would cause hyperglycemia.

18.8 Functionally, both the hypothalamohypophyseal tract and the hypophyseal portal veins carry hypothalamic hormones to the pituitary gland. Structurally, the tract is composed of axons of neurons that extend from the hypothalamus to the posterior pituitary; the portal veins are blood vessels that extend from the hypothalamus to the anterior pituitary.

18.9 Absorption of a liter of water in the intestines would decrease the osmotic pressure of your blood plasma, turning off secretion of ADH and decreasing the ADH level in your blood.

18.10 Follicular cells secrete T_3 and T_4, also known as thyroid hormones. Parafollicular cells secrete calcitonin.

18.11 The storage form of thyroid hormones is thyroglobulin.

18.12 Lack of iodine in the diet → diminished production of T_3 and T_4 → increased release of TSH → growth (enlargement) of the thyroid gland → goiter.

18.13 Parafollicular cells of the thyroid gland secrete calcitonin; chief (principal) cells of the parathyroid gland secrete PTH.

18.14 Target tissues for PTH are bones and the kidneys; target tissue for CT is bone; target tissue for calcitriol is the GI tract.

18.15 The adrenal glands are superior to the kidneys in the retroperitoneal space.

18.16 Angiotensin II acts to constrict blood vessels by causing contraction of vascular smooth muscle, and it stimulates secretion of aldosterone (by zona glomerulosa cells of the adrenal cortex), which in turn causes the kidneys to conserve water and thereby increase blood volume.

18.17 A transplant recipient who takes prednisone will have low blood levels of ACTH and CRH due to negative feedback suppression of the anterior pituitary and hypothalamus by the prednisone.

18.18 The pancreas is both an endocrine and an exocrine gland.

18.19 Glycogenolysis is the conversion of glycogen into glucose and therefore it increases blood glucose level.

18.20 Homeostasis maintains controlled conditions typical of a normal internal environment; the stress response resets controlled conditions at a different level to cope with various stressors.

18.21 The adrenal cortex of the adrenal gland is derived from mesoderm, while the adrenal medulla of the adrenal gland is derived from ectoderm.

18.22 Antibodies that mimic the action of TSH are produced in Graves' disease.

19

THE CARDIOVASCULAR SYSTEM: THE BLOOD

BLOOD AND HOMEOSTASIS *Blood contributes to homeostasis by transporting oxygen, carbon dioxide, nutrients, and hormones to and from your body's cells. It helps regulate body pH and temperature, and provides protection against disease through phagocytosis and the production of antibodies.*

The **cardiovascular system** (*cardio-* = heart; *vascular* = blood or blood vessels) consists of three interrelated components: blood, the heart, and blood vessels. The focus of this chapter is blood; the next two chapters will examine the heart and blood vessels, respectively. Blood transports various substances, helps regulate several life processes, and affords protection against disease. For all of its similarities in origin, composition, and functions, blood is as unique from one person to another as are skin, bone, and hair. Health-care professionals routinely examine and analyze its differences through various blood tests when trying to determine the cause of different diseases. The branch of science concerned with the study of blood, blood-forming tissues, and the disorders associated with them is **hematology** (hēm-a-TOL-ō-jē; *hema-* or *hemato-* = blood; *-logy* = study of).

Did you ever wonder why blood is such a unique substance that can be analyzed to determine if we are healthy, to detect a multitude of infections, and to detect or rule out various diseases and injuries?

19.1 FUNCTIONS AND PROPERTIES OF BLOOD

◉ OBJECTIVES

- Describe the functions of blood.
- Describe the physical characteristics and principal components of blood.

Most cells of a multicellular organism cannot move around to obtain oxygen and nutrients or eliminate carbon dioxide and other wastes. Instead, these needs are met by two fluids: blood and interstitial fluid. **Blood** is a liquid connective tissue that consists of cells surrounded by a liquid extracellular matrix. The extracellular matrix is called blood plasma, and it suspends various cells and cell fragments. **Interstitial fluid** is the fluid that bathes body cells (see Figure 27.1) and is constantly renewed by the blood. Blood transports oxygen from the lungs and nutrients from the gastrointestinal tract, which diffuse from the blood into the interstitial fluid and then into body cells. Carbon dioxide and other wastes move in the reverse direction, from body cells to interstitial fluid to blood. Blood then transports the wastes to various organs—the lungs, kidneys, and skin—for elimination from the body.

Functions of Blood

Blood has three general functions:

1. *Transportation.* As you just learned, blood transports oxygen from the lungs to the cells of the body and carbon dioxide from the body cells to the lungs for exhalation. It carries nutrients from the gastrointestinal tract to body cells and hormones from endocrine glands to other body cells. Blood also transports heat and waste products to various organs for elimination from the body.

2. *Regulation.* Circulating blood helps maintain homeostasis of all body fluids. Blood helps regulate pH through the use of buffers (chemicals that convert strong acids or bases into weak ones). It also helps adjust body temperature through the heat-absorbing and coolant properties of the water (see Section 2.4) in blood plasma and its variable rate of flow through the skin, where excess heat can be lost from the blood to the environment. In addition, blood osmotic pressure influences the water content of cells, mainly through interactions of dissolved ions and proteins.

3. *Protection.* Blood can clot (become gel-like), which protects against its excessive loss from the cardiovascular system after an injury. In addition, its white blood cells protect against disease by carrying on phagocytosis. Several types of blood proteins, including antibodies, interferons, and complement, help protect against disease in a variety of ways.

Physical Characteristics of Blood

Blood is denser and more viscous (thicker) than water and feels slightly sticky. The temperature of blood is 38°C (100.4°F), about

1°C higher than oral or rectal body temperature, and it has a slightly alkaline pH ranging from 7.35 to 7.45. The color of blood varies with its oxygen content. When saturated with oxygen, it is bright red. When unsaturated with oxygen, it is dark red. Blood constitutes about 20% of extracellular fluid, amounting to 8% of the total body mass. The blood volume is 5 to 6 liters (1.5 gal) in an average-sized adult male and 4 to 5 liters (1.2 gal) in an average-sized adult female. The gender difference in volume is due to differences in body size. Several hormones, regulated by negative feedback, ensure that blood volume and osmotic pressure remain relatively constant. Especially important are the hormones aldosterone, antidiuretic hormone, and atrial natriuretic peptide, which regulate how much water is excreted in the urine (see Section 27.1).

⚕ CLINICAL CONNECTION | *Withdrawing Blood*

Blood samples for laboratory testing may be obtained in several ways. The most common procedure is **venipuncture** (vēn′-i-PUNK-chur), withdrawal of blood from a vein using a needle and collecting tube, which contains various additives. A tourniquet is wrapped around the arm above the venipuncture site, which causes blood to accumulate in the vein. This increased blood volume makes the vein stand out. Opening and closing the fist further causes it to stand out, making the venipuncture more successful. A common site for venipuncture is the median cubital vein anterior to the elbow (see Figure 21.25b). Another method of withdrawing blood is through a **finger** or **heel stick.** Diabetic patients who monitor their daily blood sugar typically perform a finger stick, and it is often used for drawing blood from infants and children. In an **arterial stick,** blood is withdrawn from an artery; this test is used to determine the level of oxygen in oxygenated blood. •

Components of Blood

Whole blood has two components: (1) blood plasma, a watery liquid extracellular matrix that contains dissolved substances, and (2) formed elements, which are cells and cell fragments. If a sample of blood is centrifuged (spun) in a small glass tube, the cells (which are more dense) sink to the bottom of the tube while the plasma (which is less dense) forms a layer on top (Figure 19.1a). Blood is about 45% formed elements and 55% blood plasma. Normally, more than 99% of the formed elements are cells named for their red color—red blood cells (RBCs). Pale, colorless white blood cells (WBCs) and platelets occupy less than 1% of the formed elements. Because they are less dense than red blood cells but more dense than blood plasma, they form a very thin **buffy coat** layer between the packed RBCs and plasma in centrifuged blood. Figure 19.1b shows the composition of blood plasma and the numbers of the various types of formed elements in blood.

Blood Plasma

When the formed elements are removed from blood, a straw-colored liquid called **blood plasma** (or simply **plasma**) is left. Blood plasma is about 91.5% water and 8.5% solutes, most of which

Figure 19.1 Components of blood in a normal adult.

Blood is a connective tissue that consists of blood plasma (liquid) plus formed elements (red blood cells, white blood cells, and platelets).

Plasma (55%)

Buffy coat, composed of white blood cells and platelets

Red blood cells (45%)

(a) Appearance of centrifuged blood

FUNCTIONS OF BLOOD

1. Transports oxygen, carbon dioxide, nutrients, hormones, heat, and wastes.
2. Regulates pH, body temperature, and water content of cells.
3. Protects against blood loss through clotting, and against disease through phagocytic white blood cells and proteins such as antibodies, interferons, and complement.

BODY WEIGHT	VOLUME	FORMED ELEMENTS (number per µL)	WHITE BLOOD CELLS
Whole blood 8%	Blood plasma 55% → Proteins 7%		Albumins 54%
			Globulins 38%
	Water 91.5%		Fibrinogen 7%
Other fluids and tissues 92%			All others 1%
			Electrolytes
			Nutrients
			Gases
			Regulatory substances
	Other solutes 1.5%		Waste products
		PLASMA (weight)	SOLUTES
	Formed elements 45%	Platelets 150,000–400,000	Neutrophils 60–70%
		White blood cells 5000–10,000	Lymphocytes 20–25%
		Red blood cells 4.8–5.4 million	Monocytes 3–8%
			Eosinophils 2–4%
			Basophils 0.5–1.0%

(b) Components of blood

? What is the approximate volume of blood in your body?

(7% by weight) are proteins. Some of the proteins in blood plasma are also found elsewhere in the body, but those confined to blood are called **plasma proteins.** Hepatocytes (liver cells) synthesize most of the plasma proteins, which include the **albumins** (al′-BŪ-mins) (54% of plasma proteins), **globulins** (GLOB-ū-lins) (38%), and **fibrinogen** (fī-BRIN-ō-jen) (7%). Certain blood cells develop into cells that produce gamma globulins, an important type of globulin. These plasma proteins are also called **antibodies** or **immunoglobulins** (im′-ū-nō-GLOB-ū-lins) because they are produced during certain immune responses. Foreign substances (antigens) such as bacteria and viruses stimulate production of millions of different antibodies. An antibody binds specifically to the antigen that stimulated its production and thus disables the invading antigen.

Besides proteins, other solutes in plasma include electrolytes, nutrients, regulatory substances such as enzymes and hormones, gases, and waste products such as urea, uric acid, creatinine, ammonia, and bilirubin.

Table 19.1 describes the chemical composition of blood plasma.

Formed Elements

The **formed elements** of the blood include three principal components: **red blood cells (RBCs), white blood cells (WBCs),** and **platelets** (Figure 19.2). RBCs and WBCs are whole cells; platelets are cell fragments. RBCs and platelets have just a few roles, but WBCs have a number of specialized functions. Several distinct types of WBCs—neutrophils, lymphocytes, monocytes, eosinophils, and basophils—each with a unique microscopic appearance, carry out these functions, which are discussed later in this chapter.

Following is the classification of the formed elements in blood:

I. Red blood cells
II. White blood cells
    A. Granular leukocytes (contain conspicuous granules that are visible under a light microscope after staining)
        1. Neutrophils
        2. Eosinophils
        3. Basophils
    B. Agranular leukocytes (no granules are visible under a light microscope after staining)
        1. T and B lymphocytes and natural killer (NK) cells
        2. Monocytes
III. Platelets

The percentage of total blood volume occupied by RBCs is called the **hematocrit** (hē-MAT-ō-krit); a hematocrit of 40 indicates that 40% of the volume of blood is composed of RBCs. The

---

TABLE 19.1

Substances in Blood Plasma

CONSTITUENT	DESCRIPTION	FUNCTION
Water (91.5%)	Liquid portion of blood.	Solvent and suspending medium. Absorbs, transports, and releases heat.
Plasma proteins (7%)	Most produced by liver.	Responsible for colloid osmotic pressure. Major contributors to blood viscosity. Transport hormones (steroid), fatty acids, and calcium. Help regulate blood pH.
Albumin	Smallest and most numerous of proteins.	
Globulins	Large proteins (plasma cells produce immunoglobulins).	Immunoglobulins help attack viruses and bacteria. Alpha and beta globulins transport iron, lipids, and fat-soluble vitamins.
Fibrinogen	Large protein.	Plays essential role in blood clotting.
Other solutes (1.5%)		
Electrolytes	Inorganic salts; positively charged (cations) Na^+, K^+, Ca^{2+}, Mg^{2+}; negatively charged (anions) Cl^-, HPO_4^{2-}, SO_4^{2-}, HCO_3^-.	Help maintain osmotic pressure and essential roles in cell functions.
Nutrients	Products of digestion, such as amino acids, glucose, fatty acids, glycerol, vitamins, and minerals.	Essential roles in cell functions, growth, and development.
Gases	Oxygen (O_2).	Oxygen is important in many cellular functions.
	Carbon dioxide (CO_2).	Carbon dioxide is involved in the regulation of blood pH.
	Nitrogen (N_2).	Nitrogen has no known function.
Regulatory substances	Enzymes.	Catalyze chemical reactions.
	Hormones.	Regulate metabolism, growth, and development.
	Vitamins.	Cofactors for enzymatic reactions.
Waste products	Urea, uric acid, creatine, creatinine, bilirubin, Ammonia.	Most are breakdown products of protein metabolism that are carried by the blood to organs of excretion.

Figure 19.2 Scanning electron micrograph and photomicrograph of the formed elements of blood.

The formed elements of blood are red blood cells (RBCs), white blood cells (WBCs), and platelets.

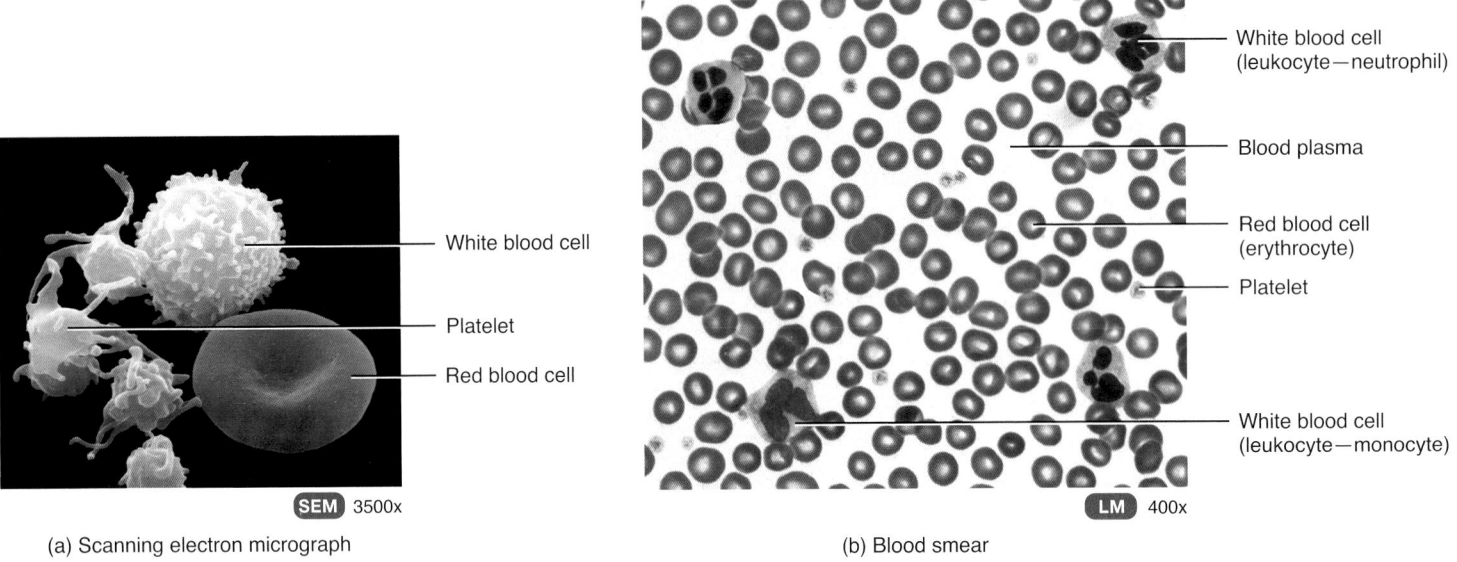

(a) Scanning electron micrograph   SEM 3500x

(b) Blood smear   LM 400x

Which formed elements of the blood are cell fragments?

normal range of hematocrit for adult females is 38–46% (average = 42); for adult males, it is 40–54% (average = 47). The hormone testosterone, present in much higher concentration in males than in females, stimulates synthesis of erythropoietin (EPO), the hormone that in turn stimulates production of RBCs. Thus, testosterone contributes to higher hematocrits in males. Lower values in women during their reproductive years also may be due to excessive loss of blood during menstruation. A significant drop in hematocrit indicates *anemia,* a lower-than-normal number of RBCs. In *polycythemia* (pol′-ē-sī-THĒ-mē-a) the percentage of RBCs is abnormally high, and the hematocrit may be 65% or higher. This raises the viscosity of blood, which increases the resistance to flow and makes the blood more difficult for the heart to pump. Increased viscosity also contributes to high blood pressure and increased risk of stroke. Causes of polycythemia include abnormal increases in RBC production, tissue hypoxia, dehydration, and blood doping or the use of EPO by athletes.

✔ **CHECKPOINT**

1. In what ways is blood plasma similar to interstitial fluid? How does it differ?
2. What substances does blood transport?
3. How many kilograms or pounds of blood are there in your body?
4. How does the volume of blood plasma in your body compare to the volume of fluid in a 2-liter bottle of Coke?
5. List the formed elements in blood plasma and describe their functions.
6. What is the significance of lower-than-normal or higher-than-normal hematocrit?

19.2 FORMATION OF BLOOD CELLS

◉ **OBJECTIVE**
• Explain the origin of blood cells.

Although some lymphocytes have a lifetime measured in years, most formed elements of the blood last only hours, days, or weeks, and must be replaced continually. Negative feedback systems regulate the total number of RBCs and platelets in circulation, and their numbers normally remain steady. The abundance of the different types of WBCs, however, varies in response to challenges by invading pathogens and other foreign antigens.

The process by which the formed elements of blood develop is called **hemopoiesis** (hē-mō-poy-Ē-sis; *-poiesis* = making) or *hematopoiesis.* Before birth, hemopoiesis first occurs in the yolk sac of an embryo and later in the liver, spleen, thymus, and lymph nodes of a fetus. Red bone marrow becomes the primary site of hemopoiesis in the last 3 months before birth, and continues as the source of blood cells after birth and throughout life.

Red bone marrow is a highly vascularized connective tissue located in the microscopic spaces between trabeculae of spongy bone tissue. It is present chiefly in bones of the axial skeleton, pectoral and pelvic girdles, and the proximal epiphyses of the humerus and femur. About 0.05–0.1% of red bone marrow cells are derived from mesenchyme (tissue from which almost all connective tissues develop) and are called **pluripotent stem cells** (ploo-RIP-ō-tent; *pluri-* = several) or *hemocytoblasts.* These cells have the capacity to develop into many different types of cells (Figure 19.3). In newborns, all bone marrow is red and thus

Figure 19.3 Origin, development, and structure of blood cells. A few of the generations of some cell lines have been omitted.

Blood cell production, called hemopoiesis, occurs mainly in red bone marrow after birth.

Key:

▓	Progenitor cells
☐	Precursor cells or "blasts"
☐	Formed elements of circulating blood
▒	Tissue cells

Key:

CFU–E	Colony-forming unit—erythrocyte
CFU–Meg	Colony-forming unit—megakaryocyte
CFU–GM	Colony-forming unit—granulocyte macrophage

Pluripotent stem cell

Myeloid stem cell

Lymphoid stem cell

CFU–E → Proerythroblast → Nucleus ejected → Reticulocyte → Red blood cell (erythrocyte)

CFU–Meg → Megakaryoblast → Megakaryocyte → Platelets (thrombocytes)

CFU–GM → Eosinophilic myeloblast → Eosinophil

Basophilic myeloblast → Basophil

Myeloblast → Neutrophil

Granular leukocytes

Monoblast → Monocyte

T lymphoblast → T lymphocyte (T cell)

B lymphoblast → B lymphocyte (B cell)

NK lymphoblast → Natural killer (NK) cell

Agranular leukocytes

Mast cell

Macrophage

Plasma cell

? **From which connective tissue cells do pluripotent stem cells develop?**

active in blood cell production. As an individual ages, the rate of blood cell formation decreases; the red bone marrow in the medullary (marrow) cavity of long bones becomes inactive and is replaced by yellow bone marrow, which consists largely of fat cells. Under certain conditions, such as severe bleeding, yellow bone marrow can revert to red bone marrow; this occurs as blood-forming stem cells from the red bone marrow move into the yellow bone marrow, which is then repopulated by pluripotent stem cells.

CLINICAL CONNECTION | *Bone Marrow Examination*

Sometimes a sample of red bone marrow must be obtained in order to diagnose certain blood disorders, such as leukemia and severe anemias. **Bone marrow examination** may involve *bone marrow aspiration* (withdrawal of a small amount of red bone marrow with a fine needle and syringe) or a *bone marrow biopsy* (removal of a core of red bone marrow with a larger needle).

Both types of samples are usually taken from the iliac crest of the hip bone, although samples are sometimes aspirated from the sternum. In young children, bone marrow samples are taken from a vertebra or tibia (shin bone). The tissue or cell sample is then sent to a pathology lab for analysis. Specifically, laboratory technicians look for signs of neoplastic (cancer) cells or other diseased cells to assist in diagnosis. •

Stem cells in red bone marrow reproduce themselves, proliferate, and differentiate into cells that give rise to blood cells, macrophages, reticular cells, mast cells, and adipocytes. Some stem cells can also form osteoblasts, chondroblasts, and muscle cells, and may be destined for use as a source of bone, cartilage, and muscular tissue for tissue and organ replacement. The reticular cells produce reticular fibers, which form the stroma (framework) that supports red bone marrow cells. Blood from nutrient and metaphyseal arteries (see Figure 6.4) enters a bone and passes into the enlarged and leaky capillaries, called *sinuses,* that surround red bone marrow cells and fibers. After blood cells form, they enter the sinuses and other blood vessels and leave the bone through nutrient and periosteal veins (see Figure 6.4). With the exception of lymphocytes, formed elements do not divide once they leave red bone marrow.

In order to form blood cells, pluripotent stem cells in red bone marrow produce two further types of stem cells, which have the capacity to develop into several types of cells. These stem cells are called *myeloid stem cells* and *lymphoid stem cells.* Myeloid stem cells begin their development in red bone marrow and give rise to red blood cells, platelets, monocytes, neutrophils, eosinophils, and basophils. Lymphoid stem cells begin their development in red bone marrow but complete it in lymphatic tissues; they give rise to lymphocytes. Although the various stem cells have distinctive cell identity markers in their plasma membranes, they cannot be distinguished histologically and resemble lymphocytes.

During hemopoiesis, some of the myeloid stem cells differentiate into **progenitor cells** (prō-JEN-i-tor). Other myeloid stem cells and the lymphoid stem cells develop directly into precursor cells (described shortly). Progenitor cells are no longer capable of reproducing themselves and are committed to giving rise to more specific elements of blood. Some progenitor cells are known as *colony-forming units (CFUs).* Following the CFU designation is an abbreviation that indicates the mature elements in blood that they will produce: CFU–E ultimately produces erythrocytes (red blood cells); CFU–Meg produces megakaryocytes, the source of platelets; and CFU–GM ultimately produces granulocytes (specifically, neutrophils) and monocytes (see Figure 19.3). Progenitor cells, like stem cells, resemble lymphocytes and cannot be distinguished by their microscopic appearance alone.

In the next generation, the cells are called **precursor cells,** also known as **blasts.** Over several cell divisions they develop into the actual formed elements of blood. For example, monoblasts develop into monocytes, eosinophilic myeloblasts develop into

eosinophils, and so on. Precursor cells have recognizable microscopic appearances.

Several hormones called **hemopoietic growth factors** (hē-mō-poy-ET-ik) regulate the differentiation and proliferation of particular progenitor cells. **Erythropoietin** (e-rith'-rō-POY-e-tin) or **EPO** increases the number of red blood cell precursors. EPO is produced primarily by cells in the kidneys that lie between the kidney tubules (peritubular interstitial cells). With renal failure, EPO release slows and RBC production is inadequate. This leads to a decreased hematocrit, which leads to a decreased ability to deliver oxygen to body tissues. **Thrombopoietin** (throm'-bō-POY-e-tin) or **TPO** is a hormone produced by the liver that stimulates the formation of platelets (thrombocytes) from megakaryocytes. Several different cytokines regulate development of different blood cell types. **Cytokines** (SĪ-tō-kīns) are small glycoproteins that are typically produced by cells such as red bone marrow cells, leukocytes, macrophages, fibroblasts, and endothelial cells. They generally act as local hormones (autocrines or paracrines; see Chapter 18). Cytokines stimulate proliferation of progenitor cells in red bone marrow and regulate the activities of cells involved in nonspecific defenses (such as phagocytes) and immune responses (such as B cells and T cells). Two important families of cytokines that stimulate white blood cell formation are **colony-stimulating factors (CSFs)** and **interleukins** (in'-ter-LOO-kins).

CLINICAL CONNECTION | *Medical Uses of Hemopoietic Growth Factors*

Hemopoietic growth factors made available through recombinant DNA technology hold tremendous potential for medical uses when a person's natural ability to form new blood cells is diminished or defective. The artificial form of erythropoietin (epoetin alfa) is very effective in treating the diminished red blood cell production that accompanies end-stage kidney disease. Granulocyte–macrophage colony-stimulating factor and granulocyte CSF are given to stimulate white blood cell formation in cancer patients who are undergoing chemotherapy, which kills red bone marrow cells as well as cancer cells because both cell types are undergoing mitosis. (Recall that white blood cells help protect against disease.) Thrombopoietin shows great promise for preventing the depletion of platelets, which are needed to help blood clot, during chemotherapy. CSFs and thrombopoietin also improve the outcome of patients who receive bone marrow transplants. Hemopoietic growth factors are also used to treat thrombocytopenia in neonates, other clotting disorders, and various types of anemia. Research on these medications is ongoing and shows a great deal of promise. •

✔**CHECKPOINT**

7. Which hemopoietic growth factors regulate differentiation and proliferation of CFU–E and formation of platelets from megakaryocytes?

8. Describe the formation of platelets from pluripotent stem cells, including the influence of hormones.

19.3 RED BLOOD CELLS

◉ OBJECTIVE

• Describe the structure, functions, life cycle, and production of red blood cells.

Red blood cells (RBCs) or **erythrocytes** (e-RITH-rō-sīts; *erythro-* = red; *-cyte* = cell) contain the oxygen-carrying protein **hemoglobin,** which is a pigment that gives whole blood its red color. A healthy adult male has about 5.4 million red blood cells per microliter (μL) of blood,* and a healthy adult female has about 4.8 million. (One drop of blood is about 50 μL.) To maintain normal numbers of RBCs, new mature cells must enter the circulation at the astonishing rate of at least 2 million per second, a pace that balances the equally high rate of RBC destruction.

RBC Anatomy

RBCs are biconcave discs with a diameter of 7–8 μm (Figure 19.4a). Recall that 1 μm = 1/25,000 of an inch or 1/10,000 of a centimeter or 1/1000 of a millimeter. Mature red blood cells have a simple structure. Their plasma membrane is both strong and flexible, which allows them to deform without rupturing as they squeeze through narrow capillaries. As you will see later, certain glycolipids in the plasma membrane of RBCs are antigens that account for the various blood groups such as the ABO and Rh groups. RBCs lack a nucleus and other organelles and can neither reproduce nor carry on extensive metabolic activities. The cytosol of RBCs contains hemoglobin molecules; these important mole-

*1 μL = 1 mm^3 = 10^{-6} liter.

cules are synthesized before loss of the nucleus during RBC production and constitute about 33% of the cell's weight.

RBC Physiology

Red blood cells are highly specialized for their oxygen transport function. Because mature RBCs have no nucleus, all their internal space is available for oxygen transport. Because RBCs lack mitochondria and generate ATP anaerobically (without oxygen), they do not use up any of the oxygen they transport. Even the shape of an RBC facilitates its function. A biconcave disc has a much greater surface area for the diffusion of gas molecules into and out of the RBC than would, say, a sphere or a cube.

Each RBC contains about 280 million hemoglobin molecules. A hemoglobin molecule consists of a protein called **globin,** composed of four polypeptide chains (two alpha and two beta chains); a ring-like nonprotein pigment called a **heme** (Figure 19.4b) is bound to each of the four chains. At the center of each heme ring is an iron ion (Fe^{2+}) that can combine reversibly with one oxygen molecule (Figure 19.4c), allowing each hemoglobin molecule to bind four oxygen molecules. Each oxygen molecule picked up from the lungs is bound to an iron ion. As blood flows through tissue capillaries, the iron–oxygen reaction reverses. Hemoglobin releases oxygen, which diffuses first into the interstitial fluid and then into cells.

Hemoglobin also transports about 23% of the total carbon dioxide, a waste product of metabolism. (The remaining carbon dioxide is dissolved in plasma or carried as bicarbonate ions.) Blood flowing through tissue capillaries picks up carbon dioxide, some of which combines with amino acids in the globin part of hemoglobin. As blood flows through the lungs, the carbon dioxide is released from hemoglobin and then exhaled.

Figure 19.4 **The shapes of a red blood cell (RBC) and a hemoglobin molecule.** In (b), note that each of the four polypeptide chains of a hemoglobin molecule (blue) has one heme group (gold), which contains an iron ion (Fe^{2+}), shown in red.

🔑 The iron portion of a heme group binds oxygen for transport by hemoglobin.

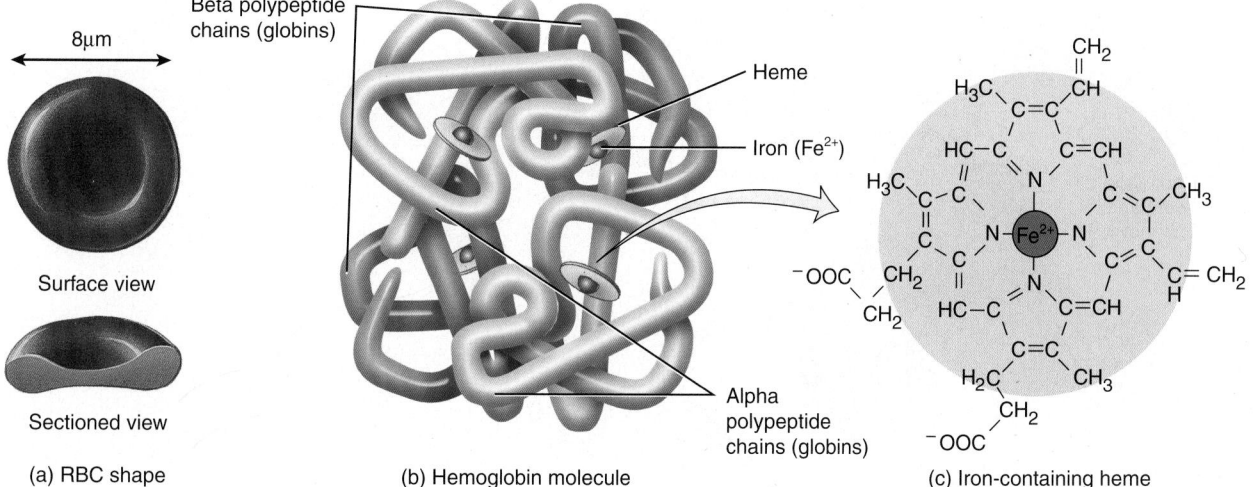

(a) RBC shape

(b) Hemoglobin molecule

(c) Iron-containing heme

❓ **How many molecules of O$_2$ can one hemoglobin molecule transport?**

In addition to its key role in transporting oxygen and carbon dioxide, hemoglobin also plays a role in the regulation of blood flow and blood pressure. The gaseous hormone **nitric oxide (NO),** produced by the endothelial cells that line blood vessels, binds to hemoglobin. Under some circumstances, hemoglobin releases NO. The released NO causes *vasodilation,* an increase in blood vessel diameter that occurs when the smooth muscle in the vessel wall relaxes. Vasodilation improves blood flow and enhances oxygen delivery to cells near the site of NO release.

Red blood cells also contain the enzyme carbonic anhydrase (CA), which catalyzes the conversion of carbon dioxide and water to carbonic acid, which in turn dissociates into H^+ and HCO_3^-. The entire reaction is reversible and is summarized as follows:

$$CO_2 + H_2O \xrightleftharpoons{CA} H_2CO_3 \rightleftharpoons H^+ + HCO_3^-$$

Carbon dioxide | Water | | Carbonic acid | | Hydrogen ion | Bicarbonate ion

This reaction is significant for two reasons: (1) It allows about 70% of CO_2 to be transported in blood plasma from tissue cells to the lungs in the form of HCO_3^- (see Chapter 23). (2) It also serves as an important buffer in extracellular fluid (see Chapter 27).

RBC Life Cycle

Red blood cells live only about 120 days because of the wear and tear their plasma membranes undergo as they squeeze through blood capillaries. Without a nucleus and other organelles, RBCs cannot synthesize new components to replace damaged ones. The plasma membrane becomes more fragile with age, and the cells are more likely to burst, especially as they squeeze through narrow channels in the spleen. Ruptured red blood cells are removed from circulation and destroyed by fixed phagocytic macrophages in the spleen and liver, and the breakdown products are recycled and used in numerous metabolic processes, including the formation of new red blood cells. The recycling occurs as follows (Figure 19.5):

1 Macrophages in the spleen, liver, or red bone marrow phagocytize ruptured and worn-out red blood cells.

2 The globin and heme portions of hemoglobin are split apart.

3 Globin is broken down into amino acids, which can be reused to synthesize other proteins.

4 Iron is removed from the heme portion in the form of Fe^{3+}, which associates with the plasma protein **transferrin**

Figure 19.5 Formation and destruction of red blood cells, and the recycling of hemoglobin components. RBCs circulate for about 120 days after leaving red bone marrow before they are phagocytized by macrophages.

🔑 The rate of RBC formation by red bone marrow equals the rate of RBC destruction by macrophages.

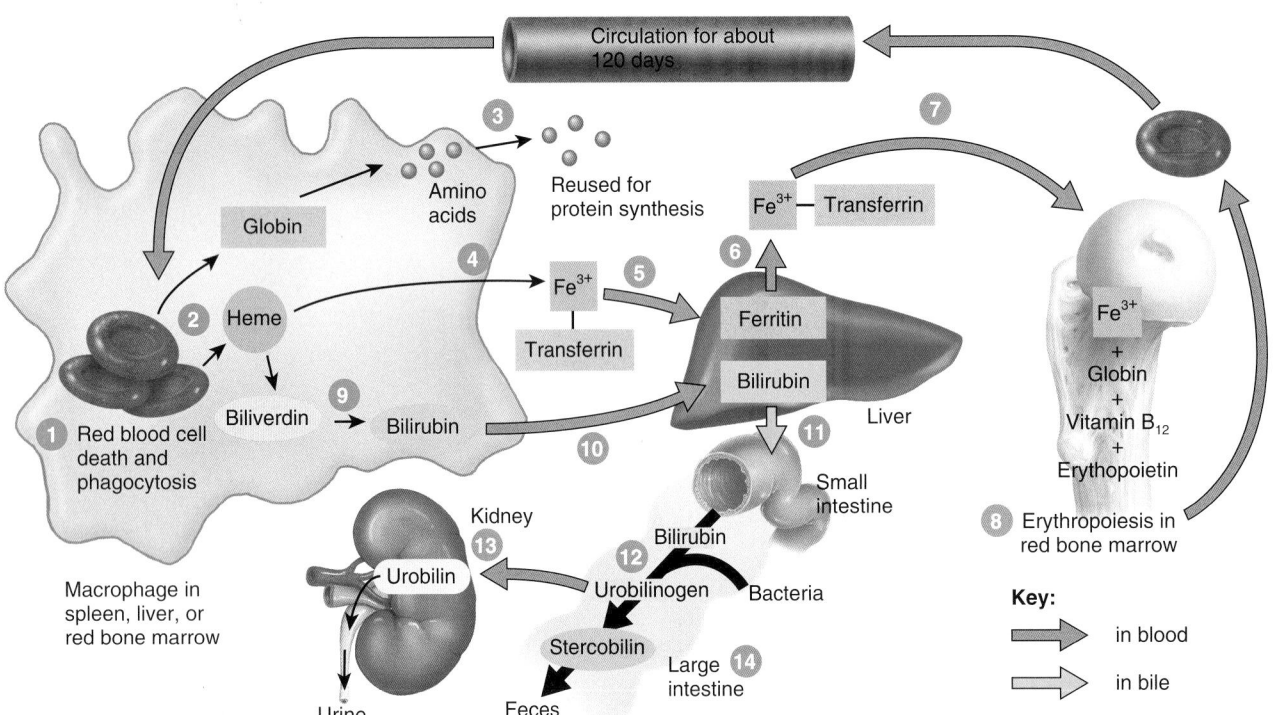

? **What is the function of transferrin?**

(trans-FER-in; *trans-* = across; *-ferr-* = iron), a transporter for Fe^{3+} in the bloodstream.

⑤ In muscle fibers, liver cells, and macrophages of the spleen and liver, Fe^{3+} detaches from transferrin and attaches to an iron-storage protein called **ferritin** (FER-i-tin).

⑥ On release from a storage site or absorption from the gastrointestinal tract, Fe^{3+} reattaches to transferrin.

⑦ The Fe^{3+}–transferrin complex is then carried to red bone marrow, where RBC precursor cells take it up through receptor-mediated endocytosis (see Figure 3.12) for use in hemoglobin synthesis. Iron is needed for the heme portion of the hemoglobin molecule, and amino acids are needed for the globin portion. Vitamin B_{12} is also needed for the synthesis of hemoglobin.

⑧ Erythropoiesis in red bone marrow results in the production of red blood cells, which enter the circulation.

⑨ When iron is removed from heme, the non-iron portion of heme is converted to **biliverdin** (bil′-i-VER-din), a green pigment, and then into **bilirubin** (bil′-i-ROO-bin), a yellow-orange pigment.

⑩ Bilirubin enters the blood and is transported to the liver.

⑪ Within the liver, bilirubin is released by liver cells into bile, which passes into the small intestine and then into the large intestine.

⑫ In the large intestine, bacteria convert bilirubin into **urobilinogen** (ūr-ō-bī-LIN-ō-jen).

⑬ Some urobilinogen is absorbed back into the blood, converted to a yellow pigment called **urobilin** (ūr-ō-BĪ-lin), and excreted in urine.

⑭ Most urobilinogen is eliminated in feces in the form of a brown pigment called **stercobilin** (ster′-kō-BĪ-lin), which gives feces its characteristic color.

CLINICAL CONNECTION | *Iron Overload and Tissue Damage*

Because free iron ions (Fe^{2+} and Fe^{3+}) bind to and damage molecules in cells or in the blood, transferrin and ferritin act as protective "protein escorts" during transport and storage of iron ions. As a result, plasma contains virtually no free iron. Furthermore, only small amounts are available inside body cells for use in synthesis of iron-containing molecules such as the cytochrome pigments needed for ATP production in mitochondria (see Figure 25.9). In cases of **iron overload,** the amount of iron present in the body builds up. Because we have no method for eliminating excess iron, any condition that increases dietary iron absorption can cause iron overload. At some point, the proteins transferrin and ferritin become saturated with iron ions, and free iron level rises. Common consequences of iron overload are diseases of the liver, heart, pancreatic islets, and gonads. Iron overload also allows certain iron-dependent microbes to flourish. Such microbes normally are not pathogenic, but they multiply rapidly and can cause lethal effects in a short time when free iron is present. •

Erythropoiesis: Production of RBCs

Erythropoiesis (e-rith′-rō-poy-Ē-sis), the production of RBCs, starts in the red bone marrow with a precursor cell called a **proerythroblast** (pro-e-RITH-ro-blast) (see Figure 19.3). The proerythroblast divides several times, producing cells that begin to synthesize hemoglobin. Ultimately, a cell near the end of the development sequence ejects its nucleus and becomes a **reticulocyte** (re-TIK-ū-lō-sīt). Loss of the nucleus causes the center of the cell to indent, producing the red blood cell's distinctive biconcave shape. Reticulocytes retain some mitochondria, ribosomes, and endoplasmic reticulum. They pass from red bone marrow into the bloodstream by squeezing between the endothelial cells of blood capillaries. Reticulocytes develop into mature red blood cells within 1 to 2 days after their release from red bone marrow.

CLINICAL CONNECTION | *Reticulocyte Count*

The rate of erythropoiesis is measured by a **reticulocyte count.** Normally, a little less than 1% of the oldest RBCs are replaced by newcomer reticulocytes on any given day. It then takes 1 to 2 days for the reticulocytes to lose the last vestiges of endoplasmic reticulum and become mature RBCs. Thus, reticulocytes account for about 0.5–1.5% of all RBCs in a normal blood sample. A low "retic" count in a person who is anemic might indicate a shortage of erythropoietin or an inability of the red bone marrow to respond to EPO, perhaps because of a nutritional deficiency or leukemia. A high "retic" count might indicate a good red bone marrow response to previous blood loss or to iron therapy in someone who had been iron deficient. It could also point to illegal use of epoetin alfa by an athlete. •

Normally, erythropoiesis and red blood cell destruction proceed at roughly the same pace. If the oxygen-carrying capacity of the blood falls because erythropoiesis is not keeping up with RBC destruction, a negative feedback system steps up RBC production (Figure 19.6). The controlled condition is the amount of oxygen delivered to body tissues. Cellular oxygen deficiency, called **hypoxia** (hī-POKS-ē-a), may occur if too little oxygen enters the blood. For example, the lower oxygen content of air at high altitudes reduces the amount of oxygen in the blood. Oxygen delivery may also fall due to anemia, which has many causes: Lack of iron, lack of certain amino acids, and lack of vitamin B_{12} are but a few (see Disorders: Homeostatic Imbalances at the end of this chapter). Circulatory problems that reduce blood flow to tissues may also reduce oxygen delivery. Whatever the cause, hypoxia stimulates the kidneys to step up the release of erythropoietin, which speeds the development of proerythroblasts into reticulocytes in the red bone marrow. As the number of circulating RBCs increases, more oxygen can be delivered to body tissues.

Premature newborns often exhibit anemia, due in part to inadequate production of erythropoietin. During the first weeks after birth, the liver, not the kidneys, produces most EPO. Because the liver is less sensitive than the kidneys to hypoxia, newborns have a smaller EPO response to anemia than do adults. Because fetal hemoglobin (hemoglobin present at birth) carries up to 30% more oxygen, the loss of fetal hemoglobin, due to insufficient erythropoietin production, makes the anemia worse.

Figure 19.6 Negative feedback regulation of erythropoiesis (red blood cell formation). Lower oxygen content of air at high altitudes, anemia, and circulatory problems may reduce oxygen delivery to body tissues.

🔑 The main stimulus for erythropoiesis is hypoxia, a decrease in the oxygen-carrying capacity of the blood.

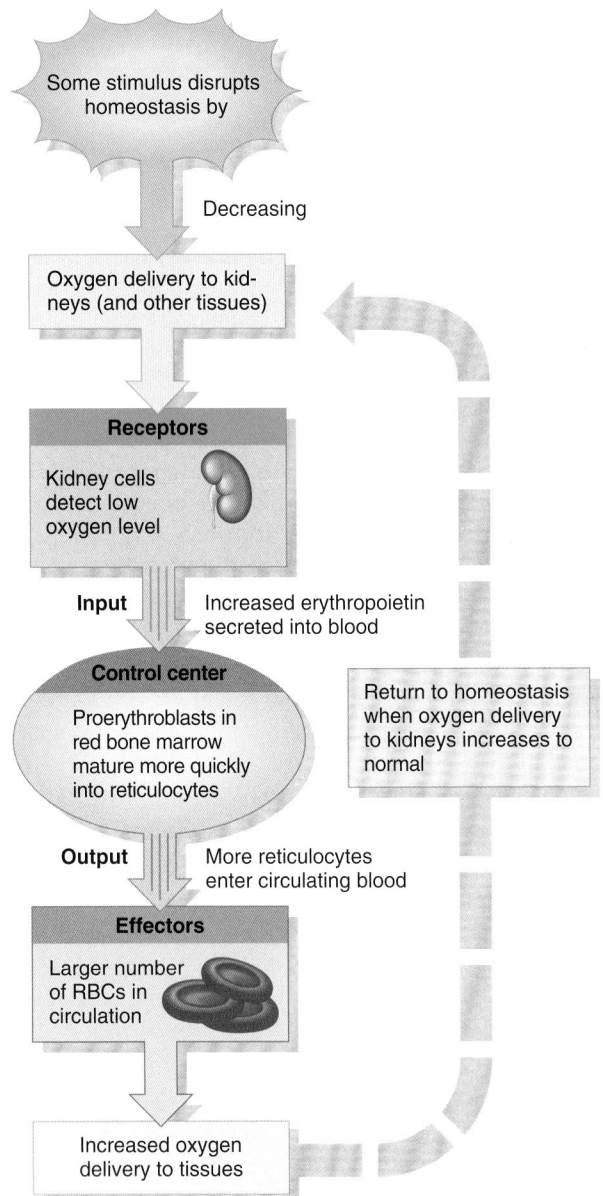

How might your hematocrit change if you moved from a town at sea level to a high mountain village?

CLINICAL CONNECTION | *Blood Doping*

Delivery of oxygen to muscles is a limiting factor in muscular feats from weightlifting to running a marathon. As a result, increasing the oxygen-carrying capacity of the blood enhances athletic performance, especially in endurance events. Because RBCs transport oxygen, athletes have tried several means of increasing their RBC count, known as **blood doping** or **artificially induced polycythemia** (an abnormally high number of RBCs), to gain a competitive edge. Athletes have enhanced their RBC production by injecting epoetin alfa (Procrit® or Epogen®), a drug that is used to treat anemia by stimulating the production of RBCs by red bone marrow. Practices that increase the number of RBCs are dangerous because they raise the viscosity of the blood, which increases the resistance to blood flow and makes the blood more difficult for the heart to pump. Increased viscosity also contributes to high blood pressure and increased risk of stroke. During the 1980s, at least 15 competitive cyclists died from heart attacks or strokes linked to suspected use of epoetin alfa. Although the International Olympics Committee bans the use of epoetin alfa, enforcement is difficult because the drug is identical to naturally occurring erythropoietin (EPO).

So-called **natural blood doping** is seemingly the key to the success of marathon runners from Kenya. The average altitude throughout Kenya's highlands is about 2000 meters (6600 feet) above sea level; other areas of Kenya are even higher. Altitude training greatly improves fitness, endurance, and performance. At these higher altitudes, the body increases the production of red blood cells, which means that exercise greatly oxygenates the blood. When these runners compete in Boston, for example, at an altitude just above sea level, the bodies of these runners contain more erythrocytes than do the bodies of competitors who trained in Boston. A number of training camps have been established in Kenya and now attract endurance athletes from all over the world. •

✔ CHECKPOINT

9. Describe the size, microscopic appearance, and functions of RBCs.
10. How is hemoglobin recycled?
11. What is erythropoiesis? How does erythropoiesis affect hematocrit? What factors speed up and slow down erythropoiesis?

19.4 WHITE BLOOD CELLS

◉ OBJECTIVE

• Describe the structure, functions, and production of white blood cells (WBCs).

Types of WBCs

Unlike red blood cells, white blood cells or **leukocytes** (LOO-kō-sīts; *leuko-* = white) have nuclei and a full complement of other organelles but they do not contain hemoglobin (Figure 19.7). WBCs are classified as either granular or agranular, depending on whether they contain conspicuous chemical-filled cytoplasmic granules (vesicles) that are made visible by staining when viewed through a light microscope. *Granular leukocytes* include neutrophils, eosinophils, and basophils; *agranular leukocytes* include lymphocytes and monocytes. As shown in Figure 19.3, monocytes and granular leukocytes develop from myeloid stem cells. In contrast, lymphocytes develop from lymphoid stem cells.

Granular Leukocytes

After staining, each of the three types of granular leukocytes displays conspicuous granules with distinctive coloration that can be recognized under a light microscope. Granular leukocytes can be distinguished as follows:

* **Neutrophil** (NOO-trō-fil). The granules of a neutrophil are smaller than those of other granular leukocytes, evenly distributed, and pale lilac (Figure 19.7a). Because the granules do not strongly attract either the acidic (red) or basic (blue) stain, these WBCs are neutrophilic (= neutral loving). The nucleus has two to five lobes, connected by very thin strands of nuclear material. As the cells age, the number of nuclear lobes increases. Because older neutrophils thus have several differently shaped nuclear lobes, they are often called *polymorphonuclear leukocytes (PMNs),* polymorphs, or "polys."

* **Eosinophil** (ē-ō-SIN-ō-fil). The large, uniform-sized granules within an eosinophil are *eosinophilic* (= eosin-loving)—they stain red-orange with acidic dyes (Figure 19.7b). The granules usually do not cover or obscure the nucleus, which most often has two lobes connected by either a thin strand or a thick strand of nuclear material.

* **Basophil** (BĀ-sō-fil). The round, variable-sized granules of a basophil are *basophilic* (= basic loving)—they stain blue-purple with basic dyes (Figure 19.7c). The granules commonly obscure the nucleus, which has two lobes.

Agranular Leukocytes

Even though so-called agranular leukocytes possess cytoplasmic granules, the granules are not visible under a light microscope because of their small size and poor staining qualities.

* **Lymphocyte** (LIM-fō-sīt). The nucleus of a lymphocyte stains dark and is round or slightly indented (Figure 19.7d). The cytoplasm stains sky blue and forms a rim around the nucleus. The larger the cell, the more cytoplasm is visible. Lymphocytes are classified by cell diameter as large lymphocytes (10–14 μm) or small lymphocytes (6–9 μm). Although the functional significance of the size difference between small and large lymphocytes is unclear, the distinction is still clinically useful because an increase in the number of large lymphocytes has diagnostic significance in acute viral infections and in some immunodeficiency diseases.

* **Monocyte** (MON-ō-sīt). The nucleus of a monocyte is usually kidney-shaped or horseshoe-shaped, and the cytoplasm is blue-gray and has a foamy appearance (Figure 19.7e). The cytoplasm's color and appearance are due to very fine *azurophilic granules* (az'-ū-rō-FIL-ik; *azur-* = blue; *-philic* = loving), which are lysosomes. Blood is merely a conduit for monocytes, which migrate from the blood into the tissues, where they enlarge and differentiate into **macrophages** (MAK-rō-fā-jez = large eaters). Some become **fixed (tissue) macrophages,** which means they reside in a particular tissue; examples are alveolar macrophages in the lungs or macrophages in the spleen. Others become **wandering macrophages,** which roam the tissues and gather at sites of infection or inflammation.

White blood cells and all other nucleated cells in the body have proteins, called **major histocompatibility (MHC) antigens,** protruding from their plasma membrane into the extracellular fluid. These "cell identity markers" are unique for each person (except identical twins). Although RBCs possess blood group antigens, they lack the MHC antigens.

Functions of WBCs

In a healthy body, some WBCs, especially lymphocytes, can live for several months or years, but most live only a few days. During

Figure 19.7 Types of white blood cells.

 The shapes of their nuclei and the staining properties of their cytoplasmic granules distinguish white blood cells from one another.

    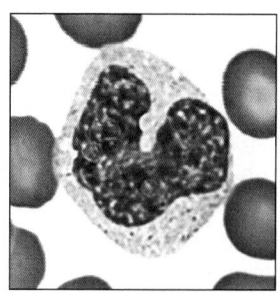

LM all 1600x

| (a) Neutrophil | (b) Eosinophil | (c) Basophil | (d) Lymphocyte | (e) Monocyte |

? **Which WBCs are called granular leukocytes? Why?**

a period of infection, phagocytic WBCs may live only a few hours. WBCs are far less numerous than red blood cells; at about 5000–10,000 cells per microliter of blood, they are outnumbered by RBCs by about 700:1. **Leukocytosis** (loo′-kō-sī-TŌ-sis), an increase in the number of WBCs above 10,000/μL, is a normal, protective response to stresses such as invading microbes, strenuous exercise, anesthesia, and surgery. An abnormally low level of white blood cells (below 5000/μL) is termed **leukopenia** (loo′-kō-PĒ-nē-a). It is never beneficial and may be caused by radiation, shock, and certain chemotherapeutic agents.

The skin and mucous membranes of the body are continuously exposed to microbes and their toxins. Some of these microbes can invade deeper tissues to cause disease. Once pathogens enter the body, the general function of white blood cells is to combat them by phagocytosis or immune responses. To accomplish these tasks, many WBCs leave the bloodstream and collect at sites of pathogen invasion or inflammation. Once granular leukocytes and monocytes leave the bloodstream to fight injury or infection, they never return to it. Lymphocytes, on the other hand, continually recirculate—from blood to interstitial spaces of tissues to lymphatic fluid and back to blood. Only 2% of the total lymphocyte population is circulating in the blood at any given time; the rest is in lymphatic fluid and organs such as the skin, lungs, lymph nodes, and spleen.

RBCs are contained within the bloodstream, but WBCs leave the bloodstream by a process termed **emigration** (em′-i-GRĀ-shun; e- = out; -migra- = wander), also called *diapedesis* (dī-a-pe-DĒ-sis), in which they roll along the endothelium, stick to it, and then squeeze between endothelial cells (Figure 19.8). The precise signals that stimulate emigration through a particular blood vessel vary for the different types of WBCs. Molecules known as **adhesion molecules** help WBCs stick to the endothelium. For example, endothelial cells display adhesion molecules called *selectins* in response to nearby injury and inflammation. Selectins stick to carbohydrates on the surface of neutrophils, causing them to slow down and roll along the endothelial surface. On the neutrophil surface are other adhesion molecules called *integrins,* which tether neutrophils to the endothelium and assist their movement through the blood vessel wall and into the interstitial fluid of the injured tissue.

Neutrophils and macrophages are active in **phagocytosis** (fag′-ō-sī-TŌ-sis); they can ingest bacteria and dispose of dead matter (see Figure 3.13). Several different chemicals released by microbes and inflamed tissues attract phagocytes, a phenomenon called **chemotaxis** (kē-mō-TAK-sis). The substances that provide stimuli for chemotaxis include toxins produced by microbes; kinins, which are specialized products of damaged tissues; and some of the colony-stimulating factors (CSFs). The CSFs also enhance the phagocytic activity of neutrophils and macrophages.

Among WBCs, neutrophils respond most quickly to tissue destruction by bacteria. After engulfing a pathogen during phagocytosis, a neutrophil unleashes several chemicals to destroy the pathogen. These chemicals include the enzyme **lysozyme** (LĪ-sō-zīm′), which destroys certain bacteria, and **strong oxidants,** such as the superoxide anion (O_2^-), hydrogen peroxide (H_2O_2), and the hypochlorite anion (OCl^-), which is similar to household bleach.

Figure 19.8 **Emigration of white blood cells.**

 Adhesion molecules (selectins and integrins) assist the emigration of WBCs from the bloodstream into interstitial fluid.

Key:

⌇ Selectins on endothelial cells

▪ Integrins on neutrophil

❓ **In what way is the "traffic pattern" of lymphocytes in the body different from that of other WBCs?**

Neutrophils also contain **defensins,** proteins that exhibit a broad range of antibiotic activity against bacteria and fungi. Within a neutrophil, vesicles containing defensins merge with phagosomes containing microbes. Defensins form peptide "spears" that poke holes in microbe membranes; the resulting loss of cellular contents kills the invader.

Eosinophils leave the capillaries and enter tissue fluid. They are believed to release enzymes, such as histaminase, that combat the effects of histamine and other substances involved in inflammation during allergic reactions. Eosinophils also phagocytize antigen–antibody complexes and are effective against certain parasitic worms. A high eosinophil count often indicates an allergic condition or a parasitic infection.

At sites of inflammation, basophils leave capillaries, enter tissues, and release granules that contain heparin, histamine, and serotonin. These substances intensify the inflammatory reaction and are involved in hypersensitivity (allergic) reactions. Basophils are similar in function to mast cells, connective tissue cells that originate from pluripotent stem cells in red bone marrow. Like basophils, mast cells release substances involved in inflammation, including heparin, histamine, and proteases. Mast cells are widely dispersed in the body, particularly in connective tissues of the skin and mucous membranes of the respiratory and gastrointestinal tracts.

Lymphocytes are the major soldiers in immune system battles (described in detail in Chapter 22). Most lymphocytes continually move among lymphoid tissues, lymph, and blood, spending only a few hours at a time in blood. Thus, only a small proportion of the total lymphocytes are present in the blood at any given time. Three main types of lymphocytes are B cells, T cells, and natural killer (NK) cells. B cells are particularly effective in destroying bacteria and inactivating their toxins. T cells attack viruses, fungi, transplanted cells, cancer cells, and some bacteria, and are responsible for transfusion reactions, allergies, and the rejection of transplanted organs. Immune responses carried out by both B cells and T cells help combat infection and provide protection against some diseases. Natural killer cells attack a wide variety of infectious microbes and certain spontaneously arising tumor cells.

Monocytes take longer to reach a site of infection than neutrophils, but they arrive in larger numbers and destroy more microbes. On their arrival, monocytes enlarge and differentiate into wandering macrophages, which clean up cellular debris and microbes by phagocytosis after an infection.

As you have already learned, an increase in the number of circulating WBCs usually indicates inflammation or infection. A physician may order a **differential white blood cell count,** a count of each of the five types of white blood cells, to detect infection or inflammation, determine the effects of possible poisoning by chemicals or drugs, monitor blood disorders (for example, leukemia) and the effects of chemotherapy, or detect allergic reactions and parasitic infections. Because each type of white blood cell plays a different role, determining the *percentage* of each type in the blood assists in diagnosing the condition. Table 19.2 lists the significance of both high and low WBC counts.

✔ CHECKPOINT

12. What is the importance of emigration, chemotaxis, and phagocytosis in fighting bacterial invaders?
13. How are leukocytosis and leukopenia different?
14. What is a differential white blood cell count?
15. What functions do granular leukocytes, macrophages, B cells, T cells, and natural killer cells perform?

19.5 PLATELETS

◉ OBJECTIVE

• Describe the structure, function, and origin of platelets.

Besides the immature cell types that develop into erythrocytes and leukocytes, hemopoietic stem cells also differentiate into cells that produce platelets. Under the influence of the hormone **thrombopoietin** (throm′-bo-POY-e-tin), myeloid stem cells develop into megakaryocyte colony-forming cells that in turn develop into precursor cells called *megakaryoblasts* (see Figure 19.3). Megakaryoblasts transform into megakaryocytes, huge cells that splinter into 2000 to 3000 fragments. Each fragment,

TABLE 19.2

Significance of High and Low White Blood Cell Counts

WBC TYPE	HIGH COUNT MAY INDICATE	LOW COUNT MAY INDICATE
Neutrophils	Bacterial infection, burns, stress, inflammation.	Radiation exposure, drug toxicity, vitamin B_{12} deficiency, systemic lupus erythematosus (SLE).
Lymphocytes	Viral infections, some leukemias.	Prolonged illness, immunosuppression, treatment with cortisol.
Monocytes	Viral or fungal infections, tuberculosis, some leukemias, other chronic diseases.	Bone marrow suppression, treatment with cortisol.
Eosinophils	Allergic reactions, parasitic infections, autoimmune diseases.	Drug toxicity, stress.
Basophils	Allergic reactions, leukemias, cancers, hypothyroidism.	Pregnancy, ovulation, stress, hyperthyroidism.

enclosed by a piece of the plasma membrane, is a **platelet (thrombocyte).** Platelets break off from the megakaryocytes in red bone marrow and then enter the blood circulation. Between 150,000 and 400,000 platelets are present in each microliter of blood. Each is irregularly disc-shaped, 2–4 μm in diameter, and has many vesicles but no nucleus.

Their granules contain chemicals that, once released, promote blood clotting. Platelets help stop blood loss from damaged blood vessels by forming a platelet plug. Platelets have a short life span, normally just 5 to 9 days. Aged and dead platelets are removed by fixed macrophages in the spleen and liver.

Table 19.3 summarizes the formed elements in blood.

⚕ **CLINICAL CONNECTION** | *Complete Blood Count*

A **complete blood count (CBC)** is a very valuable test that screens for anemia and various infections. Usually included are counts of RBCs, WBCs, and platelets per microliter of whole blood; hematocrit; and differential white blood cell count. The amount of hemoglobin in grams per milliliter of blood also is determined. Normal hemoglobin ranges are as follows: infants, 14–20 g/100 mL of blood; adult females, 12–16 g/100 mL of blood; and adult males, 13.5–18 g/100 mL of blood. •

✔ CHECKPOINT

16. How do RBCs, WBCs, and platelets compare with respect to size, number per microliter of blood, and life span?

TABLE 19.3

Summary of Formed Elements in Blood

NAME AND APPEARANCE	NUMBER	CHARACTERISTICS*	FUNCTIONS
Red Blood Cells (RBCs) or **Erythrocytes**	4.8 million/μL in females; 5.4 million/μL in males.	7–8 μm diameter, biconcave discs, without nuclei; live for about 120 days.	Hemoglobin within RBCs transports most oxygen and part of carbon dioxide in blood.
White Blood Cells (WBCs) or **Leukocytes**	5000–10,000/μL.	Most live for a few hours to a few days.[†]	Combat pathogens and other foreign substances that enter body.
Granular leukocytes			
Neutrophils	60–70% of all WBCs.	10–12 μm diameter; nucleus has 2–5 lobes connected by thin strands of chromatin; cytoplasm has very fine, pale lilac granules.	Phagocytosis. Destruction of bacteria with lysozyme, defensins, and strong oxidants, such as superoxide anion, hydrogen peroxide, and hypochlorite anion.
Eosinophils	2–4% of all WBCs.	10–12 μm diameter; nucleus usually has 2 lobes connected by thick strand of chromatin; large, red-orange granules fill cytoplasm.	Combat effects of histamine in allergic reactions, phagocytize antigen–antibody complexes, and destroy certain parasitic worms.
Basophils	0.5–1% of all WBCs.	8–10 μm diameter; nucleus has 2 lobes; large cytoplasmic granules appear deep blue-purple.	Liberate heparin, histamine, and serotonin in allergic reactions that intensify overall inflammatory response.
Agranular leukocytes			
Lymphocytes (T cells, B cells, and natural killer cells)	20–25% of all WBCs.	Small lymphocytes are 6–9 μm in diameter; large lymphocytes are 10–14 μm in diameter; nucleus is round or slightly indented; cytoplasm forms rim around nucleus that looks sky blue; the larger the cell, the more cytoplasm is visible.	Mediate immune responses, including antigen–antibody reactions. B cells develop into plasma cells, which secrete antibodies. T cells attack invading viruses, cancer cells, and transplanted tissue cells. Natural killer cells attack wide variety of infectious microbes and certain spontaneously arising tumor cells.
Monocytes	3–8% of all WBCs.	12–20 μm diameter; nucleus is kidney or horseshoe shaped; cytoplasm is blue-gray and appears foamy.	Phagocytosis (after transforming into fixed or wandering macrophages).
Platelets (thrombocytes)	150,000–400,000/μL.	2–4 μm diameter cell fragments that live for 5–9 days; contain many vesicles but no nucleus.	Form platelet plug in hemostasis; release chemicals that promote vascular spasm and blood clotting.

*Colors are those seen when using Wright's stain.
[†]Some lymphocytes, called T and B memory cells, can live for many years once they are established.

19.6 STEM CELL TRANSPLANTS FROM BONE MARROW AND CORD BLOOD

◉ **OBJECTIVE**

• Explain the importance of bone marrow transplants and stem cell transplants.

A **bone marrow transplant** is the replacement of cancerous or abnormal red bone marrow with healthy red bone marrow in order to establish normal blood cell counts. In patients with cancer or certain genetic diseases, the defective red bone marrow is destroyed by high doses of chemotherapy and whole body radiation just before the transplant takes place. These treatments kill the cancer cells and destroy the patient's immune system in order to decrease the chance of transplant rejection.

Healthy red bone marrow for transplanting may be supplied by a donor or by the patient when the underlying disease is inactive, as when leukemia is in remission. The red bone marrow from a donor is usually removed from the iliac crest of the hip bone under general anesthesia with a syringe and is then injected into the recipient's vein, much like a blood transfusion. The injected marrow migrates to the recipient's red bone marrow cavities, where the donor's stem cells multiply. If all goes well, the recipient's red bone marrow is replaced entirely by healthy, noncancerous cells.

Bone marrow transplants have been used to treat aplastic anemia, certain types of leukemia, severe combined immunodeficiency disease (SCID), Hodgkin's disease, non-Hodgkin's lymphoma, multiple myeloma, thalassemia, sickle-cell disease, breast cancer, ovarian cancer, testicular cancer, and hemolytic anemia. However, there are some drawbacks. Since the recipient's white blood cells have been completely destroyed by chemotherapy and radiation, the patient is extremely vulnerable to infection. (It takes about 2–3 weeks for transplanted bone marrow to produce enough white blood cells to protect against infection.) In addition, transplanted red bone marrow may produce T cells that attack the recipient's tissues, a reaction called *graft-versus-host disease.* Similarly, any of the recipient's T cells that survived the chemotherapy and radiation can attack donor transplant cells. Another drawback is that patients must take immunosuppressive drugs for life. Because these drugs reduce the level of immune system activity, they increase the risk of infection. Immunosuppressive drugs also have side effects such as fever, muscle aches, headache, nausea, fatigue, depression, high blood pressure, and kidney and liver damage.

A more recent advance for obtaining stem cells involves a **cord-blood transplant.** The connection between the mother and embryo (and later the fetus) is the umbilical cord. Stem cells may be obtained from the umbilical cord shortly after birth. The stem cells are removed from the cord with a syringe and then frozen. Stem cells from the cord have several advantages over those obtained from red bone marrow:

1. They are easily collected following permission of the newborn's parents.

2. They are more abundant than stem cells in red bone marrow.

3. They are less likely to cause graft-versus-host disease, so the match between donor and recipient does not have to be as close as in a bone marrow transplant. This provides a larger number of potential donors.

4. They are less likely to transmit infections.

5. They can be stored indefinitely in cord-blood banks.

✔ **CHECKPOINT**

17. How are cord-blood transplants and bone marrow transplants similar? How do they differ?

19.7 HEMOSTASIS

◉ **OBJECTIVES**

• Describe the three mechanisms that contribute to hemostasis.
• Identify the stages of blood clotting and explain the various factors that promote and inhibit blood clotting.

Hemostasis (hē-mō-STĀ-sis), not to be confused with the very similar term homeostasis, is a sequence of responses that stops bleeding. When blood vessels are damaged or ruptured, the hemostatic response must be quick, localized to the region of damage, and carefully controlled in order to be effective. Three mechanisms reduce blood loss: (1) vascular spasm, (2) platelet plug formation, and (3) blood clotting (coagulation). When successful, hemostasis prevents **hemorrhage** (HEM-o-rij; *-rhage* = burst forth), the loss of a large amount of blood from the vessels. Hemostatic mechanisms can prevent hemorrhage from smaller blood vessels, but extensive hemorrhage from larger vessels usually requires medical intervention.

Vascular Spasm

When arteries or arterioles are damaged, the circularly arranged smooth muscle in their walls contracts immediately, a reaction called **vascular spasm.** This reduces blood loss for several minutes to several hours, during which time the other hemostatic mechanisms go into operation. The spasm is probably caused by damage to the smooth muscle, by substances released from activated platelets, and by reflexes initiated by pain receptors.

Platelet Plug Formation

Considering their small size, platelets store an impressive array of chemicals. Within many vesicles are clotting factors, ADP, ATP, Ca^{2+}, and serotonin. Also present are enzymes that produce thromboxane A2, a prostaglandin; *fibrin-stabilizing factor,* which helps to strengthen a blood clot; lysosomes; some mitochondria; membrane systems that take up and store calcium and provide channels for release of the contents of granules; and glycogen. Also within platelets is **platelet-derived growth factor (PDGF),** a hormone that can cause proliferation of vascular endothelial cells, vascular smooth muscle fibers, and fibroblasts to help repair damaged blood vessel walls.

Platelet plug formation occurs as follows (Figure 19.9):

❶ Initially, platelets contact and stick to parts of a damaged blood vessel, such as collagen fibers of the connective tissue

Figure 19.9 **Platelet plug formation.**

🔒 ⊏══ A platelet plug can stop blood loss completely if the hole in a blood vessel is small enough.

❶ Platelet adhesion

Collagen fibers

Liberated ADP, serotonin, and thromboxane A2

❷ Platelet release reaction

Collagen fibers

Platelet plug

❸ Platelet aggregation

❓ **Along with platelet plug formation, which two mechanisms contribute to hemostasis?**

underlying the damaged endothelial cells. This process is called **platelet adhesion.**

❷ Due to adhesion, the platelets become activated, and their characteristics change dramatically. They extend many projections that enable them to contact and interact with one another, and they begin to liberate the contents of their vesicles. This phase is called the **platelet release reaction.** Liberated ADP and thromboxane A2 play a major role by activating nearby platelets. Serotonin and thromboxane A2 function as vasoconstrictors, causing and sustaining contraction of vascular smooth muscle, which decreases blood flow through the injured vessel.

❸ The release of ADP makes other platelets in the area sticky, and the stickiness of the newly recruited and activated platelets causes them to adhere to the originally activated platelets. This gathering of platelets is called **platelet aggregation.** Eventually, the accumulation and attachment of large numbers of platelets form a mass called a **platelet plug.**

A platelet plug is very effective in preventing blood loss in a small vessel. Although initially the platelet plug is loose, it becomes quite tight when reinforced by fibrin threads formed during clotting (see Figure 19.10). A platelet plug can stop blood loss completely if the hole in a blood vessel is not too large.

Blood Clotting

Normally, blood remains in its liquid form as long as it stays within its vessels. If it is drawn from the body, however, it thickens and forms a gel. Eventually, the gel separates from the liquid. The straw-colored liquid, called **serum,** is simply blood plasma minus the clotting proteins. The gel is called a **clot.** It consists of a network of insoluble protein fibers called fibrin in which the formed elements of blood are trapped (Figure 19.10).

The process of gel formation, called **clotting** or **coagulation** (kō-ag-u-LĀ-shun), is a series of chemical reactions that culminates in formation of fibrin threads. If blood clots too easily, the result can be **thrombosis**—clotting in an undamaged blood vessel. If the blood takes too long to clot, hemorrhage can occur.

Clotting involves several substances known as **clotting (coagulation) factors.** These factors include calcium ions (Ca^{2+}), several inactive enzymes that are synthesized by hepatocytes (liver cells) and released into the bloodstream, and various molecules associated with platelets or released by damaged tissues. Most clotting factors are identified by Roman numerals that indicate the order of their discovery (not necessarily the order of their participation in the clotting process).

Clotting is a complex cascade of enzymatic reactions in which each clotting factor activates many molecules of the next one in a fixed sequence. Finally, a large quantity of product (the insoluble protein fibrin) is formed. Clotting can be divided into three stages (Figure 19.11):

❶ Two pathways, called the extrinsic pathway and the intrinsic pathway (Figures 19.11a, b), which will be described shortly, lead to the formation of prothrombinase. Once prothrombi-

Figure 19.10 Blood clot formation. Notice the platelet and red blood cells entrapped in fibrin threads.

A blood clot is a gel that contains formed elements of the blood entangled in fibrin threads.

Platelet

Red blood cell

Fibrin threads

SEM 900x

(a) Early stage

SEM 900x

(b) Intermediate stage

Fibrin threads

Red blood cell

SEM 900x

(c) Late stage showing red blood cells trapped in fibrin threads

? **What is serum?**

nase is formed, the steps involved in the next two stages of clotting are the same for both the extrinsic and intrinsic pathways, and together these two stages are referred to as the common pathway.

2 Prothrombinase converts prothrombin (a plasma protein formed by the liver) into the enzyme thrombin.

3 Thrombin converts soluble fibrinogen (another plasma protein formed by the liver) into insoluble fibrin. Fibrin forms the threads of the clot.

The Extrinsic Pathway

The **extrinsic pathway** of blood clotting has fewer steps than the intrinsic pathway and occurs rapidly—within a matter of seconds if trauma is severe. It is so named because a tissue protein called **tissue factor (TF),** also known as **thromboplastin** (throm′-bō-PLAS-tin), leaks into the blood from cells *outside (extrinsic to)* blood vessels and initiates the formation of prothrombinase. TF is a complex mixture of lipoproteins and phospholipids released from the surfaces of damaged cells. In the presence of Ca^{2+}, TF

Figure 19.11 The blood-clotting cascade.

In blood clotting, coagulation factors are activated in sequence, resulting in a cascade of reactions that includes positive feedback cycles.

(a) Extrinsic pathway

Tissue trauma

(b) Intrinsic pathway

Blood trauma

Damaged endothelial cells expose collagen fibers

Tissue factor (TF)

Damaged platelets

Activated XII

Ca^{2+} Ca^{2+}

Activated platelets

Platelet phospholipids

Activated X   Activated X

V
Ca^{2+} V Ca^{2+}

1 **PROTHROMBINASE**

(c) Common pathway

Ca^{2+}

Prothrombin (II)

THROMBIN **2**

Ca^{2+} XIII

Fibrinogen (I)

Activated XIII

Loose fibrin threads → **STRENGTHENED FIBRIN THREADS** **3**

? **What is the outcome of the first stage of blood clotting?**

begins a sequence of reactions that ultimately activates clotting factor X (Figure 19.11a). Once factor X is activated, it combines with factor V in the presence of Ca^{2+} to form the active enzyme prothrombinase, completing the extrinsic pathway.

The Intrinsic Pathway

The **intrinsic pathway** of blood clotting is more complex than the extrinsic pathway, and it occurs more slowly, usually requiring several minutes. The intrinsic pathway is so named because its activators are either in direct contact with blood or contained *within (intrinsic to)* the blood; outside tissue damage is not needed. If endothelial cells become roughened or damaged, blood can come in contact with collagen fibers in the connective tissue around the endothelium of the blood vessel. In addition, trauma to endothelial cells causes damage to platelets, resulting in the release of phospholipids by the platelets. Contact with collagen fibers (or with the glass sides of a blood collection tube) activates clotting factor XII (Figure 19.11b), which begins a sequence of reactions that eventually activates clotting factor X. Platelet phospholipids and Ca^{2+} can also participate in the activation of factor X. Once factor X is activated, it combines with factor V to form the active enzyme prothrombinase (just as occurs in the extrinsic pathway), completing the intrinsic pathway.

The Common Pathway

The formation of prothrombinase marks the beginning of the common pathway. In the second stage of blood clotting (Figure 19.11c), prothrombinase and Ca^{2+} catalyze the conversion of prothrombin to thrombin. In the third stage, thrombin, in the presence of Ca^{2+}, converts fibrinogen, which is soluble, to loose fibrin threads, which are insoluble. Thrombin also activates factor XIII (fibrin stabilizing factor), which strengthens and stabilizes the fibrin threads into a sturdy clot. Plasma contains some factor XIII, which is also released by platelets trapped in the clot.

Thrombin has two positive feedback effects. In the first positive feedback loop, which involves factor V, it accelerates the formation of prothrombinase. Prothrombinase in turn accelerates the production of more thrombin, and so on. In the second positive feedback loop, thrombin activates platelets, which reinforces their aggregation and the release of platelet phospholipids.

Clot Retraction

Once a clot is formed, it plugs the ruptured area of the blood vessel and thus stops blood loss. **Clot retraction** is the consolidation or tightening of the fibrin clot. The fibrin threads attached to the damaged surfaces of the blood vessel gradually contract as platelets pull on them. As the clot retracts, it pulls the edges of the damaged vessel closer together, decreasing the risk of further damage. During retraction, some serum can escape between the fibrin threads, but the formed elements in blood cannot. Normal retraction depends on an adequate number of platelets in the clot, which release factor XIII and other factors, thereby strengthening and stabilizing the clot. Permanent repair of the blood vessel can then take place. In time, fibroblasts form connective tissue in the ruptured area, and new endothelial cells repair the vessel lining.

Role of Vitamin K in Clotting

Normal clotting depends on adequate levels of vitamin K in the body. Although vitamin K is not involved in actual clot formation,

it is required for the synthesis of four clotting factors. Normally produced by bacteria that inhabit the large intestine, vitamin K is a fat-soluble vitamin that can be absorbed through the lining of the intestine and into the blood if absorption of lipids is normal. People suffering from disorders that slow absorption of lipids (for example, inadequate release of bile into the small intestine) often experience uncontrolled bleeding as a consequence of vitamin K deficiency.

The various clotting factors, their sources, and the pathways of activation are summarized in Table 19.4.

Hemostatic Control Mechanisms

Many times a day little clots start to form, often at a site of minor roughness or at a developing atherosclerotic plaque inside a blood vessel. Because blood clotting involves amplification and positive feedback cycles, a clot has a tendency to enlarge, creating the potential for impairment of blood flow through undamaged vessels. The **fibrinolytic system** (fī-bri-nō-LIT-ik) dissolves small, inappropriate clots; it also dissolves clots at a site of damage once the damage is repaired. Dissolution of a clot is called **fibrinolysis** (fī-bri-NOL-i-sis). When a clot is formed, an inactive plasma enzyme called **plasminogen** (plaz-MIN-o-jen) is incorporated into the clot. Both body tissues and blood contain substances that can activate plasminogen to **plasmin** or **fibrinolysin** (fī-brin-ō-LĪ-sin), an active plasma enzyme. Among these substances are thrombin, activated factor XII, and tissue plasminogen activator (t-PA), which is synthesized in endothelial cells of most tissues and liberated into the blood. Once plasmin is formed, it can dissolve the clot by digesting fibrin threads and inactivating substances such as fibrinogen, prothrombin, and factors V and XII.

Even though thrombin has a positive feedback effect on blood clotting, clot formation normally remains localized at the site of damage. A clot does not extend beyond a wound site into the general circulation, in part because fibrin absorbs thrombin into the clot. Another reason for localized clot formation is that because of the dispersal of some of the clotting factors by the blood, their concentrations are not high enough to bring about widespread clotting.

Several other mechanisms also control blood clotting. For example, endothelial cells and white blood cells produce a prostaglandin called **prostacyclin** (pros-ta-SĪ-klin) that opposes the actions of thromboxane A2. Prostacyclin is a powerful inhibitor of platelet adhesion and release.

In addition, substances that delay, suppress, or prevent blood clotting, called **anticoagulants** (an′-tī-kō-AG-ū-lants), are present in blood. These include **antithrombin,** which blocks the action of several factors, including XII, X, and II (prothrombin). **Heparin,** an anticoagulant that is produced by mast cells and basophils, combines with antithrombin and increases its effectiveness in blocking thrombin. Another anticoagulant, **activated protein C (APC),** inactivates the two major clotting factors not blocked by antithrombin and enhances activity of plasminogen activators. Babies that lack the ability to produce APC due to a genetic mutation usually die of blood clots in infancy.

TABLE 19.4

Clotting (Coagulation) Factors

NUMBER*	NAME(S)	SOURCE	PATHWAY(S) OF ACTIVATION
I	Fibrinogen.	Liver.	Common.
II	Prothrombin.	Liver.	Common.
III	Tissue factor (thromboplastin).	Damaged tissues and activated platelets.	Extrinsic.
IV	Calcium ions (Ca^{2+}).	Diet, bones, and platelets.	All.
V	Proaccelerin, labile factor, or accelerator globulin (AcG).	Liver and platelets.	Extrinsic and intrinsic.
VII	Serum prothrombin conversion accelerator (SPCA), stable factor, or proconvertin.	Liver.	Extrinsic.
VIII	Antihemophilic factor (AHF), antihemophilic factor A, or antihemophilic globulin (AHG).	Liver.	Intrinsic.
IX	Christmas factor, plasma thromboplastin component (PTC), or antihemophilic factor B.	Liver.	Intrinsic.
X	Stuart factor, Prower factor, or thrombokinase.	Liver.	Extrinsic and intrinsic.
XI	Plasma thromboplastin antecedent (PTA) or antihemophilic factor C.	Liver.	Intrinsic.
XII	Hageman factor, glass factor, contact factor, or antihemophilic factor D.	Liver.	Intrinsic.
XIII	Fibrin-stabilizing factor (FSF).	Liver and platelets.	Common.

*There is no factor VI. Prothrombinase (prothrombin activator) is a combination of activated factors V and X.

CLINICAL CONNECTION | Anticoagulants

Patients who are at increased risk of forming blood clots may receive anticoagulants. Examples are heparin or warfarin. Heparin is often administered during hemodialysis and open-heart surgery. **Warfarin (Coumadin®)** acts as an antagonist to vitamin K and thus blocks synthesis of four clotting factors. Warfarin is slower acting than heparin. To prevent clotting in donated blood, blood banks and laboratories often add substances that remove Ca^{2+}; examples are EDTA (ethylenediaminetetraacetic acid) and CPD (citrate phosphate dextrose). •

Intravascular Clotting

Despite the anticoagulating and fibrinolytic mechanisms, blood clots sometimes form within the cardiovascular system. Such clots may be initiated by roughened endothelial surfaces of a blood vessel resulting from atherosclerosis, trauma, or infection. These conditions induce adhesion of platelets. Intravascular clots may also form when blood flows too slowly (stasis), allowing clotting factors to accumulate locally in high enough concentrations to initiate coagulation. Clotting in an unbroken blood vessel (usually a vein) is called **thrombosis** (throm-BŌ-sis; *thromb-* = clot; *-osis* = a condition of). The clot itself, called a **thrombus** (THROM-bus), may dissolve spontaneously. If it remains intact, however, the thrombus may become dislodged and be swept away in the blood. A blood clot, bubble of air, fat from broken bones, or a piece of debris transported by the bloodstream is called an **embolus** (EM-bō-lus; *em-* = in; *-bolus* = a mass). An embolus that breaks away from an arterial wall may lodge in a smaller-diameter artery downstream and block blood flow to a vital organ. When an embolus lodges in the lungs, the condition is called **pulmonary embolism.**

CLINICAL CONNECTION | Aspirin and Thrombolytic Agents

In patients with heart and blood vessel disease, the events of hemostasis may occur even without external injury to a blood vessel. At low doses, **aspirin** inhibits vasoconstriction and platelet aggregation by blocking synthesis of thromboxane A2. It also reduces the chance of thrombus formation. Due to these effects, aspirin reduces the risk of transient ischemic attacks (TIA), strokes, myocardial infarction, and blockage of peripheral arteries.

Thrombolytic agents (throm'-bō-LIT-ik) are chemical substances that are injected into the body to dissolve blood clots that have already formed to restore circulation. They either directly or indirectly activate plasminogen. The first thrombolytic agent, approved in 1982 for dissolving clots in the coronary arteries of the heart, was **streptokinase,** which is produced by streptococcal bacteria. A genetically engineered version of human **tissue plasminogen activator (t-PA)** is now used to treat victims of both heart attacks and brain attacks (strokes) that are caused by blood clots. •

✔ CHECKPOINT

18. What is hemostasis?
19. How do vascular spasm and platelet plug formation occur?
20. What is fibrinolysis? Why does blood rarely remain clotted inside blood vessels?
21. How do the extrinsic and intrinsic pathways of blood clotting differ?
22. Define each of the following terms: anticoagulant, thrombus, embolus, and thrombolytic agent.

19.8 BLOOD GROUPS AND BLOOD TYPES

◉ OBJECTIVES

• Distinguish between the ABO and Rh blood groups.
• Explain why it is so important to match donor and recipient blood types before administering a transfusion.

The surfaces of erythrocytes contain a genetically determined assortment of **antigens** composed of glycoproteins and glycolipids. These antigens, called **agglutinogens** (a-gloo-TIN-ō-jens), occur in characteristic combinations. Based on the presence or absence of various antigens, blood is categorized into different **blood groups.** Within a given blood group, there may be two or more different **blood types.** There are at least 24 blood groups and more than 100 antigens that can be detected on the surface of red blood cells. Here we discuss two major blood groups—ABO and Rh. Other blood groups include the Lewis, Kell, Kidd, and Duffy

systems. The incidence of ABO and Rh blood types varies among different population groups, as indicated in Table 19.5.

TABLE 19.5

Blood Types in the United States

POPULATION GROUP	BLOOD TYPE (PERCENTAGE)				
	O	A	B	AB	Rh+
European-American	45	40	11	4	85
African-American	49	27	20	4	95
Korean-American	32	28	30	10	100
Japanese-American	31	38	21	10	100
Chinese-American	42	27	25	6	100
Native American	79	16	4	1	100

ABO Blood Group

The **ABO blood group** is based on two glycolipid antigens called A and B (Figure 19.12). People whose RBCs display *only antigen A* have **type A** blood. Those who have *only antigen B* are **type B.** Individuals who have *both A and B antigens* are **type AB;** those who have *neither antigen A nor B* are **type O.**

Blood plasma usually contains **antibodies** called **agglutinins** (a-GLOO-ti-nins) that react with the A or B antigens if the two are mixed. These are the **anti-A antibody,** which reacts with antigen A, and the **anti-B antibody,** which reacts with antigen B. The

Figure 19.12 Antigens and antibodies of the ABO blood types.

🔑 The antibodies in your plasma do not react with the antigens on your red blood cells.

BLOOD TYPE	TYPE A	TYPE B	TYPE AB	TYPE O
Red blood cells	A antigen	B antigen	Both A and B antigens	Neither A nor B antigen
Plasma	Anti-B antibody	Anti-A antibody	Neither antibody	Both anti-A and anti-B antibodies

 Which antibodies are usually present in type O blood?

antibodies present in each of the four blood types are shown in Figure 19.12. You do not have antibodies that react with the antigens of your own RBCs, but you do have antibodies for any antigens that your RBCs lack. For example, if your blood type is B, you have B antigens on your red blood cells, and you have anti-A antibodies in your blood plasma. Although agglutinins start to appear in the blood within a few months after birth, the reason for their presence is not clear. Perhaps they are formed in response to bacteria that normally inhabit the gastrointestinal tract. Because the antibodies are large IgM-type antibodies (see Table 22.3) that do not cross the placenta, ABO incompatibility between a mother and her fetus rarely causes problems.

Transfusions

Despite the differences in RBC antigens reflected in the blood group systems, blood is the most easily shared of human tissues, saving many thousands of lives every year through transfusions. A **transfusion** (trans-FŪ-zhun) is the transfer of whole blood or blood components (red blood cells only or blood plasma only) into the bloodstream or directly into the red bone marrow. A transfusion is most often given to alleviate anemia, to increase blood volume (for example, after a severe hemorrhage), or to improve immunity. However, the normal components of one person's RBC plasma membrane can trigger damaging antigen–antibody responses in a transfusion recipient. In an incompatible blood transfusion, antibodies in the recipient's plasma bind to the antigens on the donated RBCs, which causes **agglutination** (a-gloo-ti-NĀ-shun), or clumping, of the RBCs. Agglutination is an antigen–antibody response in which RBCs become cross-linked to one another. (Note that agglutination is not the same as blood clotting.) When these antigen–antibody complexes form, they activate plasma proteins of the complement family (described in Section 22.6). In essence, complement molecules make the plasma membrane of the donated RBCs leaky, causing **hemolysis** (hē-MOL-i-sis) or rupture of the RBCs and the release of hemoglobin into the blood plasma. The liberated hemoglobin may cause kidney damage by clogging the filtration membranes. Although quite rare, it is possible for the viruses that cause AIDS and hepatitis B and C to be transmitted through transfusion of contaminated blood products.

Consider what happens if a person with type A blood receives a transfusion of type B blood. The recipient's blood (type A) contains A antigens on the red blood cells and anti-B antibodies in the plasma. The donor's blood (type B) contains B antigens and anti-A antibodies. In this situation, two things can happen. First, the anti-B antibodies in the recipient's plasma can bind to the B antigens on the donor's erythrocytes, causing agglutination and hemolysis of the red blood cells. Second, the anti-A antibodies in the donor's plasma can bind to the A antigens on the recipient's red blood cells, a less serious reaction because the donor's anti-A antibodies become so diluted in the recipient's plasma that they do not cause significant agglutination and hemolysis of the recipient's RBCs.

Table 19.6 summarizes the interactions of the four blood types of the ABO system.

People with type AB blood do not have anti-A or anti-B antibodies in their blood plasma. They are sometimes called *universal recipients* because theoretically they can receive blood from donors of all four blood types. They have no antibodies to attack antigens on donated RBCs. People with type O blood have neither A nor B antigens on their RBCs and are sometimes called *universal donors* because theoretically they can donate blood to all four ABO blood types. Type O persons requiring blood may receive only type O blood (Table 19.6). In practice, use of the terms universal recipient and universal donor is misleading and dangerous. Blood contains antigens and antibodies other than those associated with the ABO system that can cause transfusion problems. Thus, blood should be carefully cross-matched or screened before transfusion. In about 80% of the population, soluble antigens of the ABO type appear in saliva and other body fluids, in which case blood type can be identified from a sample of saliva.

TABLE 19.6				
Summary of ABO Blood Group Interactions				
	BLOOD TYPE			
CHARACTERISTIC	A	B	AB	O
Agglutinogen (antigen) on RBCs	A	B	Both A and B	Neither A nor B
Agglutinin (antibody) in plasma	Anti-B	Anti-A	Neither anti-A nor anti-B	Both anti-A and anti-B
Compatible donor blood types (no hemolysis)	A, O	B, O	A, B, AB, O	O
Incompatible donor blood types (hemolysis)	B, AB	A, AB	—	A, B, AB

Rh Blood Group

The **Rh blood group** is so named because the Rh antigen, called **Rh factor,** was first found in the blood of the *Rhesus* monkey. The alleles of three genes may code for the Rh antigen. People whose RBCs have Rh antigens are designated Rh^+ (Rh positive); those who lack Rh antigens are designated Rh^- (Rh negative). Table 19.5 shows the incidence of Rh^+ and Rh^- in various populations. Normally, blood plasma does not contain anti-Rh antibodies. If an Rh^- person receives an Rh^+ blood transfusion, however, the immune system starts to make anti-Rh antibodies that will remain in the blood. If a second transfusion of Rh^+ blood is given later, the previously formed anti-Rh antibodies will cause agglutination and hemolysis of the RBCs in the donated blood, and a severe reaction may occur.

CLINICAL CONNECTION | *Hemolytic Disease of the Newborn*

The most common problem with Rh incompatibility, **hemolytic disease of the newborn (HDN),** may arise during pregnancy (Figure 19.13). Normally, no direct contact occurs between maternal and fetal blood while a woman is pregnant. However, if a small amount of Rh⁺ blood leaks from the fetus through the placenta into the bloodstream of an Rh⁻ mother, the mother will start to make anti-Rh antibodies. Because the greatest possibility of fetal blood leakage into the maternal circulation occurs at delivery, the firstborn baby usually is not affected. If the mother becomes pregnant again, however, her anti-Rh antibodies can cross the placenta and enter the bloodstream of the fetus. If the fetus is Rh⁻, there is no problem, because Rh⁻ blood does not have the Rh antigen. If the fetus is Rh⁺, however, agglutination and hemolysis brought on by fetal–maternal incompatibility may occur in the fetal blood.

An injection of anti-Rh antibodies called anti-Rh gamma globulin (RhoGAM®) can be given to prevent HDN. Rh⁻ women should receive RhoGAM® before delivery, and soon after every delivery, miscarriage, or abortion. These antibodies bind to and inactivate the fetal Rh antigens before the mother's immune system can respond to the foreign antigens by producing her own anti-Rh antibodies. •

Figure 19.13 Development of hemolytic disease of the newborn (HDN). (a) At birth, a small quantity of fetal blood usually leaks across the placenta into the maternal bloodstream. A problem can arise when the mother is Rh⁻ and the baby is Rh⁺, having inherited an allele for the Rh antigens from the father. (b) On exposure to Rh antigen, the mother's immune system responds by making anti-Rh antibodies. (c) During a subsequent pregnancy, the maternal antibodies cross the placenta into the fetal blood. If the second fetus is Rh⁺, the ensuing antigen–antibody reaction causes agglutination and hemolysis of fetal RBCs. The result is HDN.

HDN occurs when maternal anti-Rh antibodies cross the placenta and cause hemolysis of fetal RBCs.

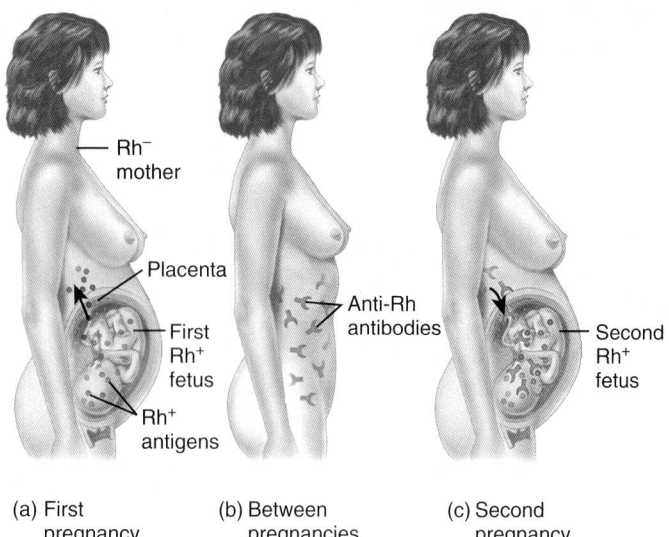

(a) First pregnancy   (b) Between pregnancies   (c) Second pregnancy

? Why is the firstborn baby unlikely to have HDN?

Typing and Cross-Matching Blood for Transfusion

To avoid blood-type mismatches, laboratory technicians type the patient's blood and then either cross-match it to potential donor blood or screen it for the presence of antibodies. In the procedure for ABO blood typing, single drops of blood are mixed with different *antisera*, solutions that contain antibodies (Figure 19.14). One drop of blood is mixed with anti-A serum, which contains anti-A antibodies that will agglutinate red blood cells that possess A antigens. Another drop is mixed with anti-B serum, which contains anti-B antibodies that will agglutinate red blood cells that possess B antigens. If the red blood cells agglutinate only when

Figure 19.14 ABO blood typing.

In the procedure for ABO blood typing, blood is mixed with anti-A serum and anti-B serum.

? What is agglutination?

mixed with anti-A serum, the blood is type A. If the red blood cells agglutinate only when mixed with anti-B serum, the blood is type B. The blood is type AB if both drops agglutinate; if neither drop agglutinates, the blood is type O.

In the procedure for determining Rh factor, a drop of blood is mixed with antiserum containing antibodies that will agglutinate RBCs displaying Rh antigens. If the blood agglutinates, it is Rh^+; no agglutination indicates Rh^-.

Once the patient's blood type is known, donor blood of the same ABO and Rh type is selected. In a **cross-match,** the possible donor RBCs are mixed with the recipient's serum. If agglutination does not occur, the recipient does not have antibodies that

will attack the donor RBCs. Alternatively, the recipient's serum can be **screened** against a test panel of RBCs having antigens known to cause blood transfusion reactions to detect any antibodies that may be present.

✔ CHECKPOINT

23. What precautions must be taken before giving a blood transfusion?
24. What is hemolysis, and how can it occur after a mismatched blood transfusion?
25. Explain the conditions that may cause hemolytic disease of the newborn.

 ## DISORDERS: HOMEOSTATIC IMBALANCES

Anemia

Anemia (a-NĒ-mē-a) is a condition in which the oxygen-carrying capacity of blood is reduced. All of the many types of anemia are characterized by reduced numbers of RBCs or a decreased amount of hemoglobin in the blood. The person feels fatigued and is intolerant of cold, both of which are related to lack of oxygen needed for ATP and heat production. Also, the skin appears pale, due to the low content of red-colored hemoglobin circulating in skin blood vessels. Among the most important causes and types of anemia are the following:

- *Inadequate absorption of iron, excessive loss of iron, increased iron requirement, or insufficient intake of iron* causes **iron-deficiency anemia,** the most common type of anemia. Women are at greater risk for iron-deficiency anemia due to menstrual blood losses and increased iron demands of the growing fetus during pregnancy. Gastrointestinal losses, such as those that occur with malignancy or ulceration, also contribute to this type of anemia.

- *Inadequate intake of vitamin B_{12} or folic acid* causes **megaloblastic anemia** in which red bone marrow produces large, abnormal red blood cells (megaloblasts). It may also be caused by drugs that alter gastric secretion or are used to treat cancer.

- *Insufficient hemopoiesis* resulting from an inability of the stomach to produce intrinsic factor, which is needed for absorption of vitamin B_{12} in the small intestine, causes **pernicious anemia.**

- *Excessive loss of RBCs* through bleeding resulting from large wounds, stomach ulcers, or especially heavy menstruation leads to **hemorrhagic anemia.**

- *RBC plasma membranes rupture prematurely* in **hemolytic anemia.** The released hemoglobin pours into the plasma and may damage the filtering units (glomeruli) in the kidneys. The condition may result from inherited defects such as abnormal red blood cell enzymes, or from outside agents such as parasites, toxins, or antibodies from incompatible transfused blood.

- *Deficient synthesis of hemoglobin* occurs in **thalassemia** (thal'-a-SĒ-mē-a), a group of hereditary hemolytic anemias. The RBCs are small (microcytic), pale (hypochromic), and short-lived. Thalassemia occurs primarily in populations from countries bordering the Mediterranean Sea.

- *Destruction of red bone marrow* results in **aplastic anemia.** It is caused by toxins, gamma radiation, and certain medications that inhibit enzymes needed for hemopoiesis.

Sickle-Cell Disease

The RBCs of a person with **sickle-cell disease (SCD)** contain Hb-S, an abnormal kind of hemoglobin. When Hb-S gives up oxygen to the interstitial fluid, it forms long, stiff, rodlike structures that bend the erythrocyte into a sickle shape (Figure 19.15). The sickled cells rupture easily. Even though erythropoiesis is stimulated by the loss of the cells, it cannot keep pace with hemolysis. Signs and symptoms of SCD are caused by the sickling of red blood cells. When red blood cells sickle, they break down prematurely (sickled cells die in about 10 to 20 days). This leads to anemia, which can cause shortness of breath, fatigue, paleness, and delayed growth and development in children. The rapid breakdown and loss of blood cells may also cause jaundice, yellowing of the eyes and skin. Sickled cells do not move easily through blood vessels and they tend to stick together and form clumps that cause blockages in blood vessels. This deprives body

Figure 19.15 Red blood cells from a person with sickle-cell disease.

 The red blood cells of a person with sickle-cell disease contain an abnormal type of hemoglobin.

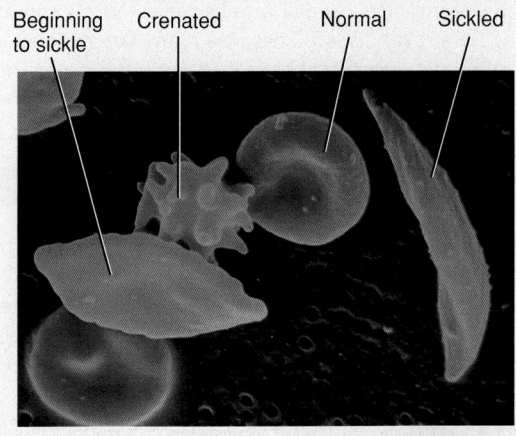

Beginning to sickle   Crenated   Normal   Sickled

SEM 3310x

Red blood cells

 What are some symptoms of sickle-cell disease?

organs of sufficient oxygen and causes pain, for example, in bones and the abdomen; serious infections; and organ damage, especially in the lungs, brain, spleen, and kidneys. Other symptoms of SCD include fever, rapid heart rate, swelling and inflammation of the hands and/or feel, leg ulcers, eye damage, excessive thirst, frequent urination, and painful and prolonged erections in males. Almost all individuals with SCD have painful episodes that can last from hours to days. Some people have one episode every few years; others have several episodes a year. The episodes may range from mild to those that require hospitalization. Any activity that reduces the amount of oxygen in the blood, such as vigorous exercise, may produce a **sickle-cell crisis** (worsening of the anemia, pain in the abdomen and long bones of the limbs, fever, and shortness of breath).

Sickle-cell disease is inherited. People with two sickle-cell genes have severe anemia; those with only one defective gene have the sickle cell trait. Sickle-cell genes are found primarily among populations (or their descendants) that live in the malaria belt around the world, including parts of Mediterranean Europe, sub-Saharan Africa, and tropical Asia. The genes responsible for the tendency of the RBCs to sickle also alter the permeability of the plasma membranes of sickled cells, causing potassium ions to leak out. Low levels of potassium kill the malaria parasites that may infect sickled cells. Because of this effect, a person with one normal gene and one sickle-cell gene has a higher-than-average resistance to malaria. The possession of a single sickle-cell gene thus confers a survival benefit.

Treatment of SCD consists of administration of analgesics to relieve pain, fluid therapy to maintain hydration, oxygen to reduce oxygen deficiency, antibiotics to counter infections, and blood transfusions. People who suffer from SCD have normal fetal hemoglobin (Hb-F), a slightly different form of hemoglobin that predominates at birth and is present in small amounts after birth. In some patients with sickle-cell disease, a drug called hydroxyurea promotes transcription of the normal Hb-F gene, elevates the level of Hb-F, and reduces the chance that the RBCs will sickle. Unfortunately, this drug also has toxic effects on the bone marrow; thus, its safety for long-term use is questionable.

Hemophilia

Hemophilia (hē-mō-FIL-ē-a; *-philia* = loving) is an inherited deficiency of clotting in which bleeding may occur spontaneously or after only minor trauma. It is the oldest known hereditary bleeding disorder; descriptions of the disease are found as early as the second century A.D. Hemophilia usually affects males and is sometimes referred to as "the royal disease" because many descendants of Queen Victoria, beginning with one of her sons, were affected by the disease. Different types of hemophilia are due to deficiencies of different blood clotting factors and exhibit varying degrees of severity, ranging from mild to severe bleeding tendencies. Hemophilia is characterized by spontaneous or traumatic subcutaneous and intramuscular hemorrhaging, nosebleeds, blood in the urine, and hemorrhages in joints that produce pain and tissue damage. Treatment involves transfusions of fresh blood plasma or concentrates of the deficient clotting factor to relieve the tendency to bleed. Another treatment is the drug desmopressin (DDAVP), which can boost the levels of the clotting factors.

Leukemia

The term **leukemia** (loo-KĒ-mē-a; *leuko-* = white) refers to a group of red bone marrow cancers in which abnormal white blood cells multiply uncontrollably. The accumulation of the cancerous white blood cells in red bone marrow interferes with the production of red blood cells, white blood cells, and platelets. As a result the oxygen-carrying capacity of the blood is reduced, an individual is more susceptible to infection, and blood clotting is abnormal. In most leukemias, the cancerous white blood cells spread to the lymph nodes, liver, and spleen, causing them to enlarge. All leukemias produce the usual symptoms of anemia (fatigue, intolerance to cold, and pale skin). In addition, weight loss, fever, night sweats, excessive bleeding, and recurrent infections may occur.

In general, leukemias are classified as **acute** (symptoms develop rapidly) and **chronic** (symptoms may take years to develop). Leukemias are also classified on the basis of the type of white blood cell that becomes malignant. **Lymphoblastic leukemia** (lim-fō-BLAS-tik) involves cells derived from lymphoid stem cells (lymphoblasts) and/or lymphocytes. **Myelogenous leukemia** (mī-e-LOJ-e-nus) involves cells derived from myeloid stem cells (myeloblasts). Combining onset of symptoms and cells involved, there are four types of leukemia:

1. **Acute lymphoblastic leukemia (ALL)** is the most common leukemia in children, but adults can also get it.
2. **Acute myelogenous leukemia (AML)** affects both children and adults.
3. **Chronic lymphoblastic anemia (CLA)** is the most common leukemia in adults, usually those older than 55.
4. **Chronic myelogenous leukemia (CML)** occurs mostly in adults.

The cause of most types of leukemia is unknown. However, certain risk factors have been implicated. These include exposure to radiation or chemotherapy for other cancers, genetics (some genetic disorders such as Down syndrome), environmental factors (smoking and benzene), and microbes such as the human T cell leukemia–lymphoma virus-1 (HTLV-1) and the Epstein-Barr virus.

Treatment options include chemotherapy, radiation, stem cell transplantation, interferon, antibodies, and blood transfusion.

MEDICAL TERMINOLOGY

Acute normovolemic hemodilution (nor-mō-vō-LĒ-mik hē-mō-di-LOO-shun) Removal of blood immediately before surgery and its replacement with a cell-free solution to maintain sufficient blood volume for adequate circulation. At the end of surgery, once bleeding has been controlled, the collected blood is returned to the body.

Autologous preoperative transfusion (aw-TOL-o-gus trans-FŪ-zhun; *auto-* = self) Donating one's own blood; can be done up to 6 weeks before elective surgery. Also called **predonation.** This procedure eliminates the risk of incompatibility and blood-borne disease.

Blood bank A facility that collects and stores a supply of blood for future use by the donor or others. Because blood banks have additional and diverse functions (immunohematology reference work, continuing medical education, bone and tissue storage, and clinical consultation), they are more appropriately referred to as **centers of transfusion medicine.**

Cyanosis (sī-a-NŌ-sis; *cyano-* = blue) Slightly bluish/dark-purple skin discoloration, most easily seen in the nail beds and mucous membranes, due to an increased quantity of *methemoglobin,* hemoglobin not combined with oxygen in systemic blood.

Gamma globulin (GLOB-ū-lin) Solution of immunoglobulins from blood consisting of antibodies that react with specific pathogens, such as viruses. It is prepared by injecting the specific virus into animals, removing blood from the animals after antibodies have accumulated, isolating the antibodies, and injecting them into a human to provide short-term immunity.

Hemochromatosis (hē-mō-krō-ma-TŌ-sis; *chroma* = color) Disorder of iron metabolism characterized by excessive absorption of ingested iron and excess deposits of iron in tissues (especially the liver, heart, pituitary gland, gonads, and pancreas) that result in bronze discoloration of the skin, cirrhosis, diabetes mellitus, and bone and joint abnormalities.

Hemorrhage (HEM-or-ij; *rhegnynai* = bursting forth) Loss of a large amount of blood; can be either internal (from blood vessels into tissues) or external (from blood vessels directly to the surface of the body).

Jaundice (*jaund-* = yellow) An abnormal yellowish discoloration of the sclerae of the eyes, skin, and mucous membranes due to excess bilirubin (yellow-orange pigment) in the blood. The three main categories of jaundice are *prehepatic jaundice*, due to excess pro-

duction of bilirubin; *hepatic jaundice*, abnormal bilirubin processing by the liver caused by congenital liver disease, cirrhosis (scar tissue formation) of the liver, or hepatitis (liver inflammation); and *extrahepatic jaundice*, due to blockage of bile drainage by gallstones or cancer of the bowel or pancreas.

Phlebotomist (fle-BOT-ō-mist; *phlebo-* = vein; *-tom* = cut) A technician who specializes in withdrawing blood.

Septicemia (sep'-ti-SĒ-mē-a; *septic-* = decay; *-emia* = condition of blood) Toxins or disease-causing bacteria in the blood. Also called "blood poisoning."

Thrombocytopenia (throm'-bō-sī-tō-PĒ-nē-a; *-penia* = poverty) Very low platelet count that results in a tendency to bleed from capillaries.

Venesection (vē'-ne-SEK-shun; *ven-* = vein) Opening of a vein for withdrawal of blood. Although **phlebotomy** (fle-BOT-ō-mē) is a synonym for venesection, in clinical practice phlebotomy refers to therapeutic bloodletting, such as the removal of some blood to lower its viscosity in a patient with polycythemia.

Whole blood Blood containing all formed elements, plasma, and plasma solutes in natural concentrations.

CHAPTER REVIEW AND RESOURCE SUMMARY

Review	Resource

Introduction

1. The cardiovascular system consists of the blood, heart, and blood vessels.
2. Blood is a liquid connective tissue that consists of cells and cell fragments surrounded by a liquid extracellular matrix (blood plasma).

Resource: Anatomy Overview - The Cardiovascular System

19.1 Functions and Properties of Blood

1. Blood transports oxygen, carbon dioxide, nutrients, wastes, and hormones.
2. It helps regulate pH, body temperature, and water content of cells.
3. It provides protection through clotting and by combating toxins and microbes through certain phagocytic white blood cells or specialized blood plasma proteins.
4. Physical characteristics of blood include a viscosity greater than that of water; a temperature of 38°C (100.4°F); and a pH of 7.35–7.45.
5. Blood constitutes about 8% of body weight, and its volume is 4–6 liters in adults.
6. Blood is about 55% blood plasma and 45% formed elements.
7. The hematocrit is the percentage of total blood volume occupied by red blood cells.
8. Blood plasma consists of 91.5% water and 8.5% solutes. Principal solutes include proteins (albumins, globulins, fibrinogen), nutrients, vitamins, hormones, respiratory gases, electrolytes, and waste products.
9. The formed elements in blood include red blood cells (erythrocytes), white blood cells (leukocytes), and platelets.

Resource: Anatomy Overview - Blood Concepts and Connections - Blood

19.2 Formation of Blood Cells

1. Hemopoiesis is the formation of blood cells from hemopoietic stem cells in red bone marrow.
2. Myeloid stem cells form RBCs, platelets, granulocytes, and monocytes. Lymphoid stem cells give rise to lymphocytes.
3. Several hemopoietic growth factors stimulate differentiation and proliferation of the various blood cells.

Resource: Anatomy Overview - Erythrocytes
Figure 19.3 - Origin, Development, and Structure of Blood Cells

19.3 Red Blood Cells

1. Mature RBCs are biconcave discs that lack nuclei and contain hemoglobin.
2. The function of the hemoglobin in red blood cells is to transport oxygen and some carbon dioxide.
3. RBCs live about 120 days. A healthy male has about 5.4 million RBCs/μL of blood; a healthy female has about 4.8 million/μL.
4. After phagocytosis of aged RBCs by macrophages, hemoglobin is recycled.

Resource: Anatomy Overview - Erythrocytes
Animation - Erythropoietin
Figure 19.4 - The Shapes of a Red Blood Cell and a Hemoglobin Molecule

Review	Resource

5. RBC formation, called erythropoiesis, occurs in adult red bone marrow of certain bones. It is stimulated by hypoxia, which stimulates the release of erythropoietin by the kidneys.

6. A reticulocyte count is a diagnostic test that indicates the rate of erythropoiesis.

19.4 White Blood Cells

1. WBCs are nucleated cells. The two principal types are granulocytes (neutrophils, eosinophils, and basophils) and agranulocytes (lymphocytes and monocytes).

2. The general function of WBCs is to combat inflammation and infection. Neutrophils and macrophages (which develop from monocytes) do so through phagocytosis.

3. Eosinophils combat the effects of histamine in allergic reactions, phagocytize antigen–antibody complexes, and combat parasitic worms. Basophils liberate heparin, histamine, and serotonin in allergic reactions that intensify the inflammatory response.

4. B lymphocytes, in response to the presence of foreign substances called antigens, differentiate into plasma cells that produce antibodies. Antibodies attach to the antigens and render them harmless. This antigen–antibody response combats infection and provides immunity. T lymphocytes destroy foreign invaders directly. Natural killer cells attack infectious microbes and tumor cells.

5. Except for lymphocytes, which may live for years, WBCs usually live for only a few hours or a few days. Normal blood contains 5000–10,000 WBCs/μL.

Anatomy Overview - White Blood Cells

Figure 19.7 - Types of White Blood Cells

19.5 Platelets

1. Platelets (thrombocytes) are disc-shaped cell fragments that splinter from megakaryocytes. Normal blood contains 150,000–400,000 platelets/μL.

2. Platelets help stop blood loss from damaged blood vessels by forming a platelet plug.

Anatomy Overview - Platelets

19.6 Stem Cell Transplants from Bone Marrow and Cord Blood

1. Bone marrow transplants involve removal of red bone marrow as a source of stem cells from the iliac crest.

2. In a cord-blood transplant, stem cells from the placenta are removed from the umbilical cord.

3. Cord-blood transplants have several advantages over bone marrow transplants.

19.7 Hemostasis

1. Hemostasis refers to the stoppage of bleeding.

2. It involves vascular spasm, platelet plug formation, and blood clotting (coagulation).

3. In vascular spasm, the smooth muscle of a blood vessel wall contracts, which slows blood loss.

4. Platelet plug formation involves the aggregation of platelets to stop bleeding.

5. A clot is a network of insoluble protein fibers (fibrin) in which formed elements of blood are trapped.

6. The chemicals involved in clotting are known as clotting (coagulation) factors.

7. Blood clotting involves a cascade of reactions that may be divided into three stages: formation of prothrombinase, conversion of prothrombin into thrombin, and conversion of soluble fibrinogen into insoluble fibrin.

8. Clotting is initiated by the interplay of the extrinsic and intrinsic pathways of blood clotting.

9. Normal coagulation requires vitamin K and is followed by clot retraction (tightening of the clot) and ultimately fibrinolysis (dissolution of the clot).

10. Clotting in an unbroken blood vessel is called thrombosis. A thrombus that moves from its site of origin is called an embolus.

Figure 19.11 - The Blood-Clotting Cascade

19.8 Blood Groups and Blood Types

1. ABO and Rh blood groups are genetically determined and based on antigen–antibody responses.

2. In the ABO blood group, the presence or absence of A and B antigens on the surface of RBCs determines blood type.

3. In the Rh system, individuals whose RBCs have Rh antigens are classified as Rh$^+$; those who lack the antigen are Rh$^-$.

4. Hemolytic disease of the newborn (HDN) can occur when an Rh$^-$ mother is pregnant with an Rh$^+$ fetus.

5. Before blood is transfused, a recipient's blood is typed and then either cross-matched to potential donor blood or screened for the presence of antibodies.

Fill in the blanks in the following statements.

1. Plasma minus its clotting proteins is termed _____.

2. _____ is the consolidation or tightening of the fibrin clot that helps to bring the edges of a damaged vessel closer together.

Indicate whether the following statements are true or false.

3. Hemoglobin functions in transporting both oxygen and carbon dioxide and in regulating blood pressure.

4. The most numerous white blood cells in a differential white blood cell count of a healthy individual are the neutrophils.

Choose the one best answer to the following questions.

5. Which of the following are *not* required for clot formation? (1) vitamin K, (2) calcium, (3) prostacyclin, (4) plasmin, (5) fibrinogen.
   (a) 1, 2, and 5 (b) 3, 4, and 5 (c) 4 and 5
   (d) 1, 2, and 3 (e) 3 and 4

6. Place the steps involved in hemostasis in the correct order. (1) conversion of fibrinogen into fibrin, (2) conversion of prothrombin into thrombin, (3) adhesion and aggregation of platelets on damaged vessel, (4) prothrombinase formed by extrinsic or intrinsic pathway, (5) reduction of blood loss by initiation of a vascular spasm.
   (a) 5, 3, 4, 2, 1 (b) 5, 4, 3, 1, 2 (c) 3, 5, 4, 2, 1
   (d) 5, 3, 2, 1, 4 (e) 5, 3, 2, 4, 1

7. Which of the following statements explain why red blood cells (RBCs) are highly specialized for oxygen transport? (1) RBCs contain hemoglobin. (2) RBCs lack a nucleus. (3) RBCs have many mitochondria and thus generate ATP aerobically. (4) The biconcave shape of RBCs provides a large surface area for the inward and outward diffusion of gas molecules. (5) RBCs can carry up to four oxygen molecules for each hemoglobin molecule.
   (a) 1, 2, 3, and 5 (b) 1, 2, 4, and 5 (c) 2, 3, 4, and 5
   (d) 1, 3, and 5 (e) 2, 4, and 5

8. Which of the following are *true?* (1) White blood cells leave the bloodstream by emigration. (2) Adhesion molecules help white blood cells stick to the endothelium, which aids emigration. (3) Neutrophils and macrophages are active in phagocytosis. (4) The attraction of phagocytes to microbes and inflamed tissue is termed chemotaxis. (5) Leukopenia is an increase in white blood cell count that occurs during infection.
   (a) 1, 2, 4, and 5 (b) 2, 3, 4, and 5 (c) 1, 2, 3, and 4
   (d) 1, 3, and 5 (e) 1, 2, and 4

9. A person with type A Rh$^-$ blood can receive a blood transfusion from which of the following types? (1) A Rh$^+$, (2) B Rh$^-$, (3) AB Rh$^-$, (4) O Rh$^-$, (5) A Rh$^-$.
   (a) 1 only (b) 3 only (c) 4 only
   (d) 4 and 5 (e) 1 and 5

10. A person with type B positive blood receives a transfusion of type AB positive blood. What will happen?
    (a) The recipient's antibodies will react with the donor's red blood cells.
    (b) The donor's antigens will destroy the recipient's antibodies.
    (c) The donor's antibodies will react with and destroy all of the recipient's red blood cells.
    (d) The recipient's blood type will change from Rh$^+$ to Rh$^-$.
    (e) These blood types are compatible, and the transfusion will be accepted.

11. What happens to the iron (Fe^{3+}) that is released during the breakdown of damaged red blood cells?
    (a) It is used to synthesize proteins.
    (b) It is transported to the liver where it becomes part of bile.
    (c) It is converted into urobilin and excreted in urine.
    (d) It attaches to transferrin and is transported to bone marrow for use in hemoglobin synthesis.
    (e) It is utilized by intestinal bacteria to convert bilirubin into urobilinogen.

12. Which of the following would *not* cause an increase in erythropoietin?
    (a) anemia
    (b) high altitude
    (c) hemorrhage
    (d) donating blood to a blood bank
    (e) polycythemia

13. Match the following:
    _____ (a) the percentage of total blood volume occupied by red blood cells
    _____ (b) the percentage of each type of white blood cell
    _____ (c) measures numbers of RBCs, WBCs, platelets per μ of blood; hematocrit; and differential WBC count
    _____ (d) measures the rate of erythropoiesis
    _____ (e) withdrawal of blood from a vein using a needle and collecting tube
    _____ (f) withdrawal of a small amount of red bone marrow with a fine needle and syringe
    _____ (g) removal of a core of red bone marrow with a large needle

    (1) reticulocyte count
    (2) bone marrow biopsy
    (3) venipuncture
    (4) hematocrit
    (5) bone marrow aspiration
    (6) complete blood count
    (7) differential white blood cell count

14. Match the following:

_____ (a) contain hemoglobin and function in gas transport

_____ (b) cell fragments enclosed by a piece of the cell membrane of megakaryocytes; contain clotting factors

_____ (c) white blood cell showing a kidney-shaped nucleus; capable of phagocytosis

_____ (d) monocytes that roam the tissues and gather at sites of infection or inflammation

_____ (e) occur as B cells, T cells, and natural killer cells

_____ (f) give rise to red blood cells, monocytes, neutrophils, eosinophils, basophils, and platelets

_____ (g) cells that give rise to all the formed elements of blood; derived from mesenchyme

(1) lymphocytes
(2) monocytes
(3) pluripotent stem cells
(4) red blood cells
(5) myeloid stem cells
(6) platelets
(7) wandering macrophages

15. Match the following:

_____ (a) tissue protein that leaks into the blood from cells outside blood vessels and initiates the formation of prothrombinase

_____ (b) an anticoagulant

_____ (c) its formation is initiated by either the extrinsic or intrinsic pathway or both; catalyzes the conversion of prothrombin to thrombin

_____ (d) forms the threads of a clot; produced from fibrinogen

_____ (e) can dissolve a clot by digesting fibrin threads

_____ (f) serves as the catalyst to form fibrin; formed from prothrombin

(1) prothrombinase
(2) thrombin
(3) fibrin
(4) thromboplastin
(5) plasmin
(6) heparin

CRITICAL THINKING QUESTIONS

1. Shilpa has recently been on broad-spectrum antibiotics for a recurrent urinary bladder infection. While slicing vegetables, she cut herself and had difficulty stopping the bleeding. How could the antibiotics have played a role in her bleeding?

2. Mrs. Brown is in renal failure. Her recent blood tests indicated a hematocrit of 22. Why is her hematocrit low? What can she be given to raise her hematocrit?

3. Thomas has hepatitis, which is disrupting his liver functions. What kinds of symptoms would he be experiencing based on the role(s) of the liver related to blood?

❓ ANSWERS TO FIGURE QUESTIONS

19.1 Blood volume is about 8% of your body mass, roughly 5–6 liters in males and 4–5 liters in females. For instance, a 70-kg (150-lb) person has a blood volume of 5.6 liters (70 kg × 8% × 1 liter/kg).

19.2 Platelets are cell fragments.

19.3 Pluripotent stem cells develop from mesenchyme.

19.4 One hemoglobin molecule can transport a maximum of four O_2 molecules, one O_2 bound to each heme group.

19.5 Transferrin is a plasma protein that transports iron in the blood.

19.6 Once you moved to high altitude, your hematocrit would increase due to increased secretion of erythropoietin.

19.7 Neutrophils, eosinophils, and basophils are called granular leukocytes because all have cytoplasmic granules that are visible through a light microscope when stained.

19.8 Lymphocytes recirculate from blood to tissues and back to blood. After leaving the blood, other WBCs remain in the tissues until they die.

19.9 Along with platelet plug formation, vascular spasm and blood clotting contribute to hemostasis.

19.10 Serum is blood plasma minus the clotting proteins.

19.11 The outcome of the first stage of clotting is the formation of prothrombinase.

19.12 Type O blood usually contains both anti-A and anti-B antibodies.

19.13 Because the mother is most likely to start making anti-Rh antibodies after the first baby is already born, that baby suffers no damage.

19.14 Agglutination refers to clumping of red blood cells.

19.15 Some symptoms of sickle-cell disease are anemia, jaundice, bone pain, shortness of breath, rapid heart rate, abdominal pain, fever, and fatigue.

20 THE CARDIOVASCULAR SYSTEM: THE HEART

THE HEART AND HOMEOSTASIS *The heart pumps blood through blood vessels to all body tissues.*

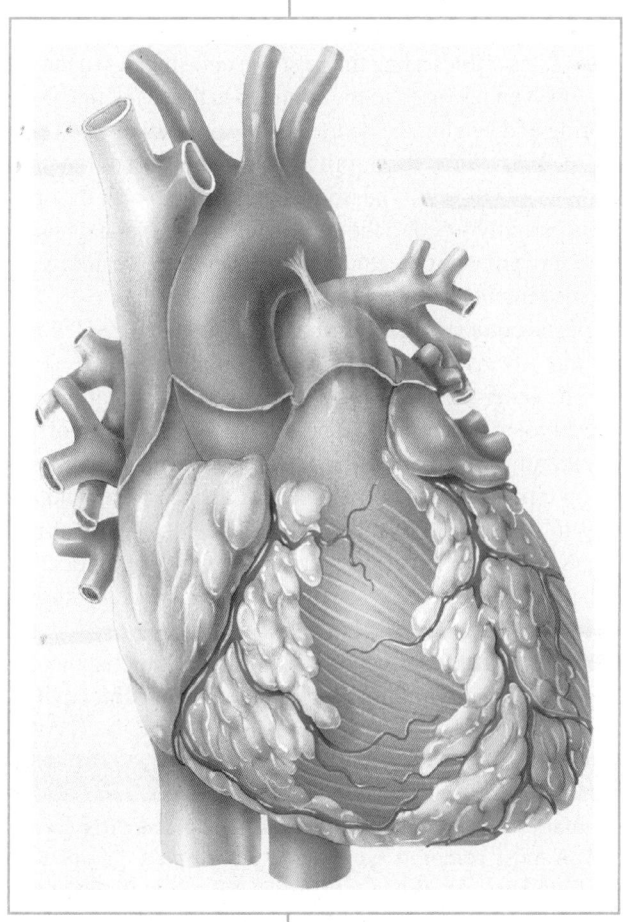

As you learned in the previous chapter, the **cardiovascular system** consists of the blood, the heart, and blood vessels. We also examined the composition and functions of blood, and in this chapter you will learn about the pump that circulates it throughout the body—the heart. For blood to reach body cells and exchange materials with them, it must be pumped continuously by the heart through the body's blood vessels. The heart beats about 100,000 times every day, which adds up to about 35 million beats in a year, and approximately 2.5 billion times in an average lifetime. The left side of the heart pumps blood through an estimated 120,000 km (75,000 mi) of blood vessels, which is equivalent to traveling around the earth's equator about 3 times. The right side of the heart pumps blood through the lungs, enabling blood to pick up oxygen and unload carbon dioxide. Even while you are sleeping, your heart pumps 30 times its own weight each minute, which amounts to about 5 liters (5.3 qt) to the lungs and the same volume to the rest of the body. At this rate, your heart pumps more than about 14,000 liters (3600 gal) of blood in a day, or 5 million liters (1.3 million gal) in a year. You don't spend all your time sleeping, however, and your heart pumps more vigorously when you are active. Thus, the actual blood volume your heart pumps in a single day is much larger.

The scientific study of the normal heart and the diseases associated with it is **cardiology** (kar-dē-OL-ō-jē; *cardio-* = heart; *-logy* = study of). This chapter explores the structure of the heart and the unique properties that permit it to pump for a lifetime without rest.

? *Did you ever wonder about the difference between "good" and "bad" cholesterol?*

20.1 ANATOMY OF THE HEART

⊙ **OBJECTIVES**

- Describe the location of the heart.
- Describe the structure of the pericardium and the heart wall.
- Discuss the external and internal anatomy of the chambers of the heart.
- Relate the thickness of the chambers of the heart to their functions.

Location of the Heart

For all its might, the **heart** is relatively small, roughly the same size (but not the same shape) as your closed fist. It is about 12 cm (5 in.) long, 9 cm (3.5 in.) wide at its broadest point, and 6 cm (2.5 in.) thick, with an average mass of 250 g (8 oz) in adult females and 300 g (10 oz) in adult males. The heart rests on the diaphragm, near the midline of the thoracic cavity. Recall that the midline is an imaginary vertical line that divides the body into unequal left and right sides. The heart lies in the **mediastinum** (mē'-dē-as-TĪ-num), an anatomical region that extends from the sternum to the vertebral column, from the first rib to the diaphragm, and between the lungs (Figure 20.1a). About two-thirds of the mass of the heart lies to the left of the body's midline (Figure 20.1b). You can visualize the heart as a cone lying on its side. The pointed **apex** is formed by the tip of the left ventricle (a lower chamber of the heart) and rests on the diaphragm. It is directed anteriorly, inferiorly, and to the left. The **base** of the heart is its posterior surface. It is formed by the atria (upper chambers) of the heart, mostly the left atrium.

In addition to the apex and base, the heart has several distinct surfaces and borders (margins). The **anterior surface** is deep to the sternum and ribs. The **inferior surface** is the part of the heart between the apex and right border and rests mostly on the diaphragm (Figure 20.1b). The **right border** faces the right lung and extends from the inferior surface to the base. The **left border,** also called the *pulmonary border,* faces the left lung and extends from the base to the apex.

✚ **CLINICAL CONNECTION | *Cardiopulmonary Resuscitation***

Because the heart lies between two rigid structures—the vertebral column and the sternum (Figure 20.1a)—external pressure on the chest (compression) can be used to force blood out of the heart and into the circulation. In cases in which the heart suddenly stops beating, **cardiopulmonary resuscitation** (kar-dē-ō-PUL-mo-nar'-ē re-sus-i-TĀ-shun) or **CPR**—properly applied cardiac compressions, performed with artificial ventilation of the lungs via mouth-to-mouth respiration—saves lives. CPR keeps oxygenated blood circulating until the heart can be restarted.

Researchers have found that chest compressions alone are equally as effective as, if not better than, traditional CPR with lung ventilation. This is good news because it is easier for an emergency dispatcher to give instructions limited to chest compressions to frightened, non-medical bystanders. As public fear of contracting contagious diseases such as hepatitis, HIV, and tuberculosis continues to rise, bystanders are much more likely to perform chest compressions alone than treatment involving mouth-to-mouth rescue breathing. •

Pericardium

The membrane that surrounds and protects the heart is the **pericardium** (per'-i-KAR-dē-um; *peri-* = around). It confines the heart to its position in the mediastinum, while allowing sufficient freedom of movement for vigorous and rapid contraction. The pericardium consists of two main parts: (1) the fibrous pericardium and (2) the serous pericardium (Figure 20.2a). The superficial **fibrous pericardium** is composed of tough, inelastic, dense irregular connective tissue. It resembles a bag that rests on and attaches to the diaphragm; its open end is fused to the connective tissues of the blood vessels entering and leaving the heart. The fibrous pericardium prevents overstretching of the heart, provides protection, and anchors the heart in the mediastinum. The fibrous pericardium near the apex of the heart is partially fused to the central tendon of the diaphragm and therefore movement of the diaphragm, as in deep breathing, facilitates the movement of blood by the heart.

The deeper **serous pericardium** is a thinner, more delicate membrane that forms a double layer around the heart (Figure 20.2a). The outer **parietal layer** of the serous pericardium is fused to the fibrous pericardium. The inner **visceral layer** of the serous pericardium, also called the **epicardium** (ep'-i-KAR-dē-um; *epi-* = on top of), is one of the layers of the heart wall and adheres tightly to the surface of the heart. Between the parietal and visceral layers of the serous pericardium is a thin film of lubricating serous fluid. This slippery secretion of the pericardial cells, known as **pericardial fluid,** reduces friction between the layers of the serous pericardium as the heart moves. The space that contains the few milliliters of pericardial fluid is called the **pericardial cavity.**

✚ **CLINICAL CONNECTION | *Pericarditis***

Inflammation of the pericardium is called **pericarditis** (per'-i-kar-DĪ-tis). The most common type, *acute pericarditis,* begins suddenly and has no known cause in most cases but is sometimes linked to a viral infection. As a result of irritation to the pericardium, there is chest pain that may extend to the left shoulder and down the left arm (often mistaken for a heart attack) and *pericardial friction rub* (a scratchy or creaking sound heard through a stethoscope as the visceral layer of the serous pericardium rubs against the parietal layer of the serous pericardium). Acute pericarditis usually lasts for about 1 week and is treated with drugs that reduce inflammation and pain, such as ibuprofen or aspirin.

Chronic pericarditis begins gradually and is long-lasting. In one form of this condition, there is a buildup of pericardial fluid. If a great deal of fluid accumulates, this is a life-threatening condition because the fluid compresses the heart, a condition called *cardiac tamponade* (tam'-pon-ĀD). As a result of the compression, ventricular filling is decreased, cardiac output is reduced, venous return to the heart is diminished, blood pressure falls, and breathing is difficult. Most causes of chronic pericarditis involving cardiac tamponade are unknown, but

Figure 20.1 **Position of the heart and associated structures in the mediastinum (dashed outline).**

The heart is located in the mediastinum, with two-thirds of its mass to the left of the midline.

Transverse plane

View

ANTERIOR

Heart

PERICARDIAL CAVITY

Right lung

Aorta

RIGHT PLEURAL CAVITY

Sternum

Muscle

Left lung

Esophagus

Sixth thoracic vertebra

LEFT PLEURAL CAVITY

POSTERIOR

(a) Inferior view of transverse section of thoracic cavity showing heart in mediastinum

Superior vena cava

SUPERIOR BORDER

RIGHT BORDER

Right lung

Pleura (cut to reveal lung inside)

Pericardium (cut)

Diaphragm

INFERIOR SURFACE

Arch of aorta

Pulmonary trunk

Left lung

Heart

LEFT BORDER

APEX OF HEART

(b) Anterior view of heart in thoracic cavity

Superior vena cava

Rib (cut)

Right lung

Arch of aorta

Pulmonary trunk

Left lung

Heart

Apex of heart

Diaphragm

(c) Anterior view

? **What is the mediastinum?**

Figure 20.2 Pericardium and heart wall.

The pericardium is a triple-layered sac that surrounds and protects the heart.

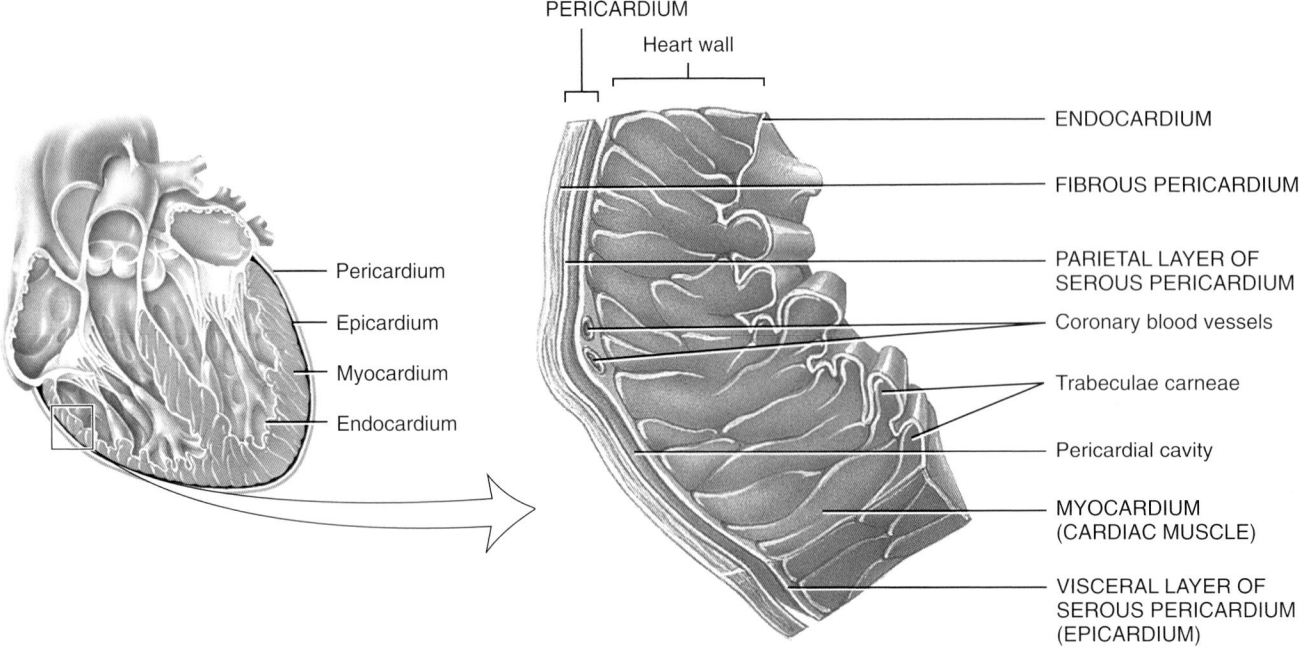

(a) Portion of pericardium and right ventricular heart wall showing divisions of pericardium and layers of heart wall

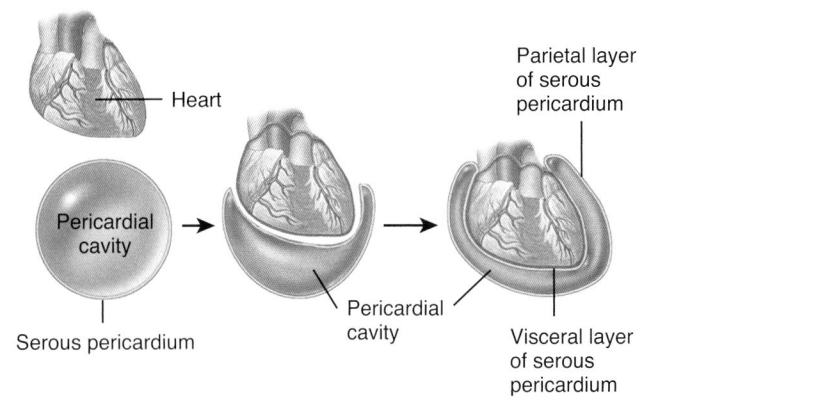

(b) Simplified relationship of serous pericardium to heart

(c) Cardiac muscle bundles of myocardium

? Which layer is both a part of the pericardium and a part of the heart wall?

---

it is sometimes caused by conditions such as cancer and tuberculosis. Treatment consists of draining the excess fluid through a needle passed into the pericardial cavity. •

Layers of the Heart Wall

The wall of the heart consists of three layers (Figure 20.2a): the epicardium (external layer), the myocardium (middle layer), and the endocardium (inner layer). The **epicardium** is composed of two tissue layers. The outermost, as you just learned, is called the *visceral layer of the serous pericardium.* This thin, transparent outer layer of the heart wall is composed of mesothelium. Beneath the mesothelium is a variable layer of delicate fibroelastic tissue and adipose tissue. The adipose tissue predominates and becomes thickest over the ventricular surfaces, where it houses the major coronary and cardiac vessels of the heart. The amount of fat varies from person to person, corresponds to the general extent of body fat in an individual, and typically increases with age. The epicardium imparts a smooth, slippery texture to the outermost surface of the heart. The epicardium contains blood vessels, lymphatics, and vessels that supply the myocardium.

The middle **myocardium** (mī′-ō-KAR-dē-um; *myo-* = muscle) is responsible for the pumping action of the heart and is composed of cardiac muscle tissue. It makes up approximately 95% of the heart wall. The muscle fibers (cells), like those of striated skeletal muscle tissue, are wrapped and bundled with connective tissue sheaths composed of endomysium and perimysium. The cardiac muscle fibers are organized in bundles that swirl diagonally around the heart and generate the strong pumping actions of the heart (Figure 20.2c). Although it is striated like skeletal muscle, recall that cardiac muscle is involuntary like smooth muscle.

The innermost **endocardium** (en′-dō-KAR-dē-um; *endo-* = within) is a thin layer of endothelium overlying a thin layer of connective tissue. It provides a smooth lining for the chambers of the heart and covers the valves of the heart. The smooth endothelial lining minimizes the surface friction as blood passes through the heart. The endocardium is continuous with the endothelial lining of the large blood vessels attached to the heart.

CLINICAL CONNECTION | Myocarditis and Endocarditis

Myocarditis (mī-ō-kar-DĪ-tis) is an inflammation of the myocardium that usually occurs as a complication of a viral infection, rheumatic fever, or exposure to radiation or certain chemicals or medications. Myocarditis often has no symptoms. However, if they do occur, they may include fever, fatigue, vague chest pain, irregular or rapid heartbeat, joint pain, and breathlessness. Myocarditis is usually mild and recovery occurs within 2 weeks. Severe cases can lead to cardiac failure and death. Treatment consists of avoiding vigorous exercise, a low-salt diet, electrocardiographic monitoring, and treatment of the cardiac failure. **Endocarditis** (en′-dō-kar-DĪ-tis) refers to an inflammation of the endocardium and typically involves the heart valves. Most cases are caused by bacteria (bacterial endocarditis). Signs and symptoms of endocarditis include fever, heart murmur, irregular or rapid heartbeat, fatigue, loss of appetite, night sweats, and chills. Treatment is with intravenous antibiotics. •

Chambers of the Heart

The heart has four chambers. The two superior receiving chambers are the **atria** (= entry halls or chambers), and the two inferior pumping chambers are the **ventricles** (= little bellies). The paired atria receive blood from blood vessels returning blood to the heart, called veins, while the ventricles eject the blood from the heart into blood vessels called arteries. On the anterior surface of each atrium is a wrinkled pouchlike structure called an **auricle** (OR-i-kul; *auri-* = ear), so named because of its resemblance to a dog's ear (Figure 20.3). Each auricle slightly increases the

---

Figure 20.3 Structure of the heart: surface features. Throughout this book, blood vessels that carry oxygenated blood (which looks bright red) are colored red, whereas those that carry deoxygenated blood (which looks dark red) are colored blue.

Sulci are grooves that contain blood vessels and fat and that mark the external boundaries between the various chambers.

Brachiocephalic trunk
Left common carotid artery
Left subclavian artery
Superior vena cava
Ascending aorta
Arch of aorta
Ligamentum arteriosum
Left pulmonary artery
Pulmonary trunk
Right pulmonary artery
Fibrous pericardium (cut)
Left pulmonary veins
Right pulmonary veins
AURICLE OF LEFT ATRIUM
Branch of left coronary artery
AURICLE OF RIGHT ATRIUM
LEFT VENTRICLE
Right coronary artery
RIGHT ATRIUM
CORONARY SULCUS (deep to fat)
ANTERIOR INTERVENTRICULAR SULCUS (deep to fat)
RIGHT VENTRICLE
Inferior vena cava
Descending aorta

(a) Anterior external view showing surface features

FIGURE 20.3 CONTINUES ▶

◉ **FIGURE 20.3** CONTINUED ▶

Left subclavian artery
Brachiocephalic trunk
Left common carotid artery
Arch of aorta
Superior vena cava
Ligamentum arteriosum
Ascending aorta
Left pulmonary artery
Left pulmonary veins
Right pulmonary veins
Pulmonary trunk
RIGHT AURICLE OF RIGHT ATRIUM
LEFT AURICLE OF LEFT ATRIUM
RIGHT ATRIUM
CORONARY SULCUS
LEFT VENTRICLE
RIGHT VENTRICLE
ANTERIOR INTERVENTRICULAR SULCUS

(b) Anterior external view

Left common carotid artery
Left subclavian artery
Brachiocephalic trunk
Arch of aorta
Superior vena cava
Ascending aorta
Descending aorta
Left pulmonary artery
Right pulmonary artery
Left pulmonary veins
Right pulmonary veins
LEFT ATRIUM
RIGHT ATRIUM
Coronary sinus (in coronary sulcus)
Right coronary artery
Inferior vena cava
LEFT VENTRICLE
Middle cardiac vein
POSTERIOR INTERVENTRICULAR SULCUS (deep to fat)
RIGHT VENTRICLE

(c) Posterior external view showing surface features

❓ **The coronary sulcus forms an external boundary between which chambers of the heart?**

capacity of an atrium so that it can hold a greater volume of blood. Also on the surface of the heart are a series of grooves, called **sulci** (SUL-sī), that contain coronary blood vessels and a variable amount of fat. Each *sulcus* (SUL-kus; singular) marks the exter-

nal boundary between two chambers of the heart. The deep **coronary sulcus** (*coron-* = resembling a crown) encircles most of the heart and marks the external boundary between the superior atria and inferior ventricles. The **anterior interventricular sulcus** (in′-ter-ven-TRIK-ū-lar) is a shallow groove on the anterior surface of the heart that marks the external boundary between the right and left ventricles. This sulcus continues around to the posterior surface of the heart as the **posterior interventricular sulcus,** which marks the external boundary between the ventricles on the posterior aspect of the heart (Figure 20.3c).

Right Atrium

The **right atrium** forms the right border of the heart and receives blood from three veins: the *superior vena cava, inferior vena cava,* and *coronary sinus* (Figure 20.4a). (Veins always carry blood toward the heart.) The right atrium is about 2–3 mm (0.08–0.12 in.) in average thickness. The anterior and posterior walls of the right atrium are very different. The inside of the posterior wall is smooth; the inside of the anterior wall is rough due to the presence of muscular ridges called **pectinate muscles** (PEK-ti-nāt; *pectin* = comb), which also extend into the auricle (Figure 20.4b). Between the right atrium and left atrium is a thin partition called the **interatrial septum** (*inter-* = between; *septum* = a dividing wall or partition). A prominent feature of this septum is

an oval depression called the **fossa ovalis,** the remnant of the *foramen ovale,* an opening in the interatrial septum of the fetal heart that normally closes soon after birth (see Figure 21.30). Blood passes from the right atrium into the right ventricle through a valve that is called the **tricuspid valve** (trī-KUS-pid; *tri-* = three; *-cuspid* = point) because it consists of three leaflets or cusps (Figure 20.4a). It is also called the **right atrioventricular valve** (ā′-trē-ō-ven-TRIK-ū-lar). The valves of the heart are composed of dense connective tissue covered by endocardium.

Right Ventricle

The **right ventricle** is about 4–5 mm (0.16–0.2 in.) in average thickness and forms most of the anterior surface of the heart. The inside of the right ventricle contains a series of ridges formed by raised bundles of cardiac muscle fibers called **trabeculae carneae** (tra-BEK-ū-lē KAR-nē-ē; *trabeculae* = little beams; *carneae* = fleshy; see Figure 20.2a). Some of the trabeculae carneae convey part of the conduction system of the heart, which you will learn about later in this chapter (see Section 20.3). The cusps of the tricuspid valve are connected to tendonlike cords, the

Figure 20.4 Structure of the heart: internal anatomy.

⑥ 🔑 Blood flows into the right atrium through the superior vena cava, inferior vena cava, and coronary sinus and into the left atrium through four pulmonary veins.

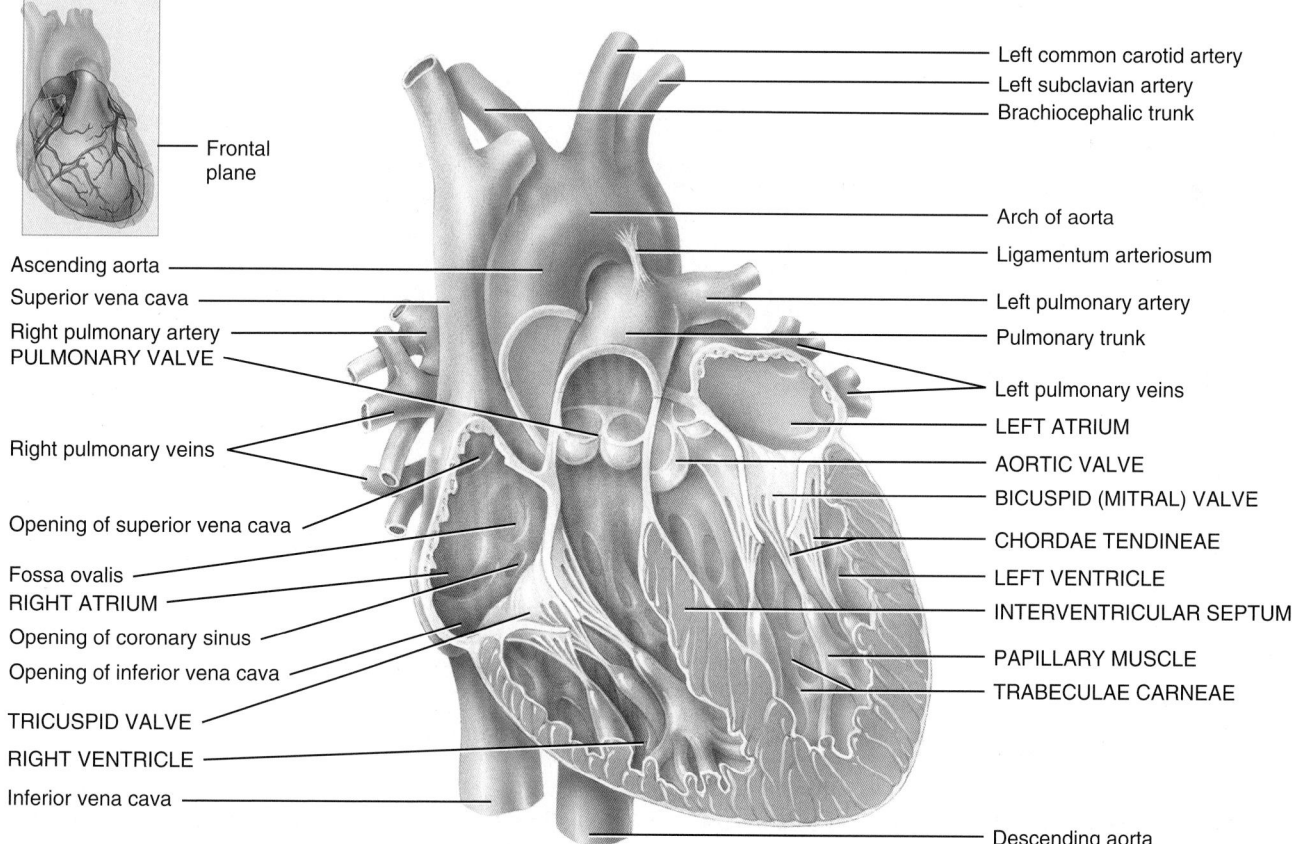

Frontal plane

Ascending aorta
Superior vena cava
Right pulmonary artery
PULMONARY VALVE

Right pulmonary veins

Opening of superior vena cava

Fossa ovalis
RIGHT ATRIUM
Opening of coronary sinus
Opening of inferior vena cava

TRICUSPID VALVE
RIGHT VENTRICLE
Inferior vena cava

Left common carotid artery
Left subclavian artery
Brachiocephalic trunk

Arch of aorta
Ligamentum arteriosum
Left pulmonary artery
Pulmonary trunk
Left pulmonary veins
LEFT ATRIUM
AORTIC VALVE
BICUSPID (MITRAL) VALVE
CHORDAE TENDINEAE
LEFT VENTRICLE
INTERVENTRICULAR SEPTUM
PAPILLARY MUSCLE
TRABECULAE CARNEAE

Descending aorta

(a) Anterior view of frontal section showing internal anatomy

FIGURE 20.4 CONTINUES ▶

FIGURE 20.4 CONTINUED ▶

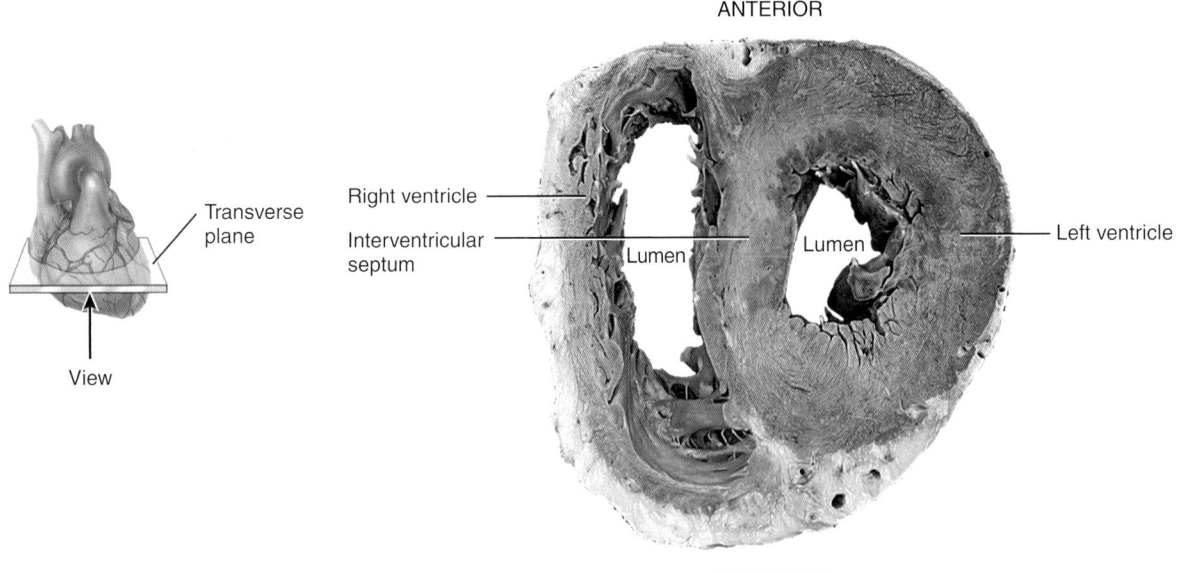

Brachiocephalic trunk

Superior vena cava

Right pulmonary vein

Ascending aorta

RIGHT AURICLE (cut open)

Pectinate muscles

RIGHT ATRIUM

Cusp of tricuspid valve

Chordae tendineae

Papillary muscle

RIGHT VENTRICLE

Left subclavian artery

Left common carotid artery

Arch of aorta

Ligamentum arteriosum

Pulmonary trunk

Left pulmonary vein

LEFT AURICLE

LEFT VENTRICLE

INTERVENTRICULAR SEPTUM

TRABECULAE CARNEAE

(b) Anterior view of partially sectioned heart

ANTERIOR

Transverse plane

View

Right ventricle

Interventricular septum

Lumen

Lumen

Left ventricle

POSTERIOR

(c) Inferior view of transverse section showing differences in thickness of ventricular walls

? **How does thickness of the myocardium relate to the workload of a cardiac chamber?**

chordae tendineae (KOR-dē ten-DI-nē-ē; *chord-* = cord; *tend-* = tendon), which in turn are connected to cone-shaped trabeculae carneae called **papillary muscles** (*papill-* = nipple). Internally, the right ventricle is separated from the left ventricle by a parti-

tion called the **interventricular septum.** Blood passes from the right ventricle through the **pulmonary valve** (*pulmonary semilunar valve*) into a large artery called the *pulmonary trunk,* which divides into right and left *pulmonary arteries* and carries blood to the lungs. Arteries always take blood away from the heart (a mnemonic to help you: artery = away).

Left Atrium

The **left atrium** is about the same thickness as the right atrium and forms most of the base of the heart (Figure 20.4a). It receives blood from the lungs through four *pulmonary veins.* Like the right atrium, the inside of the left atrium has a smooth posterior wall. Because pectinate muscles are confined to the auricle of the left atrium, the anterior wall of the left atrium also is smooth. Blood passes from the left atrium into the left ventricle through the **bicuspid (mitral) valve** (*bi-* = two), which, as its name implies, has two cusps. The term *mitral* refers to the resemblance of the bicuspid valve to a bishop's miter (hat), which is two-sided. It is also called the **left atrioventricular valve.**

Left Ventricle

The **left ventricle** is the thickest chamber of the heart, averaging 10–15 mm (0.4–0.6 in.) and forms the apex of the heart (see Figure 20.1b). Like the right ventricle, the left ventricle contains trabeculae carneae and has chordae tendineae that anchor the cusps of the bicuspid valve to papillary muscles. Blood passes from the left ventricle through the **aortic valve** (*aortic semilunar valve*) into the *ascending aorta* (*aorte* = to suspend, because the aorta once was believed to lift up the heart). Some of the blood in the aorta flows into the *coronary arteries,* which branch from the ascending aorta and carry blood to the heart wall. The remainder of the blood passes into the *arch of the aorta* and *descending aorta* (*thoracic aorta* and *abdominal aorta*). Branches of the arch of the aorta and descending aorta carry blood throughout the body.

During fetal life, a temporary blood vessel, called the *ductus arteriosus,* shunts blood from the pulmonary trunk into the aorta. Hence, only a small amount of blood enters the nonfunctioning fetal lungs (see Figure 21.30). The ductus arteriosus normally closes shortly after birth, leaving a remnant known as the **ligamentum arteriosum** (lig'-a-MEN-tum ar-ter-ē-Ō-sum), which connects the arch of the aorta and pulmonary trunk (Figure 20.4a).

Myocardial Thickness and Function

The thickness of the myocardium of the four chambers varies according to each chamber's function. The thin-walled atria deliver blood under less pressure into the adjacent ventricles. Because the ventricles pump blood under higher pressure over greater distances, their walls are thicker (Figure 20.4a). Although the right and left ventricles act as two separate pumps that simultaneously eject equal volumes of blood, the right side has a much smaller workload. It pumps blood a short distance to the lungs at lower pressure, and the resistance to blood flow is small. The left ventricle pumps blood great distances to all other parts of the body at higher pressure, and the resistance to blood flow is larger. Therefore, the left ventricle works much harder than the right ventricle to maintain the same rate of blood flow. The anatomy of the two ventricles confirms this functional difference—the muscular wall of the left ventricle is considerably thicker than the wall of the right ventricle (Figure 20.4c). Note also that the perimeter of the lumen (space) of the left ventricle is roughly circular, in contrast to that of the right ventricle, which is somewhat crescent-shaped.

Fibrous Skeleton of the Heart

In addition to cardiac muscle tissue, the heart wall also contains dense connective tissue that forms the **fibrous skeleton of the heart** (Figure 20.5). Essentially, the fibrous skeleton consists of four dense connective tissue rings that surround the valves of the heart, fuse with one another, and merge with the interventricular septum. In addition to forming a structural foundation for the heart valves, the fibrous skeleton prevents overstretching of the valves as blood passes through them. It also serves as a point of

Figure 20.5 Fibrous skeleton of the heart. Elements of the fibrous skeleton are shown in capital letters.

 Fibrous rings support the four valves of the heart and are fused to one another.

Superior view (the atria have been removed)

? In what two ways does the fibrous skeleton contribute to the functioning of heart valves?

insertion for bundles of cardiac muscle fibers and acts as an electrical insulator between the atria and ventricles.

✔CHECKPOINT

1. Define each of the following external features of the heart: auricle, coronary sulcus, anterior interventricular sulcus, and posterior interventricular sulcus.
2. Describe the structure of the pericardium and the layers of the wall of the heart.
3. What are the characteristic internal features of each chamber of the heart?
4. Which blood vessels deliver blood to the right and left atria?
5. What is the relationship between wall thickness and function among the various chambers of the heart?
6. What type of tissue composes the fibrous skeleton of the heart, and how is it organized?

20.2 HEART VALVES AND CIRCULATION OF BLOOD

◉ OBJECTIVES

• Describe the structure and function of the valves of the heart.
• Outline the flow of blood through the chambers of the heart and through the systemic and pulmonary circulations.
• Discuss the coronary circulation.

As each chamber of the heart contracts, it pushes a volume of blood into a ventricle or out of the heart into an artery. Valves open and close in response to *pressure changes* as the heart contracts and relaxes. Each of the four valves helps ensure the one-way flow of blood by opening to let blood through and then closing to prevent its backflow.

Figure 20.6 Responses of the valves to the pumping of the heart.

⑥ ⊏⊐ Heart valves prevent the backflow of blood.

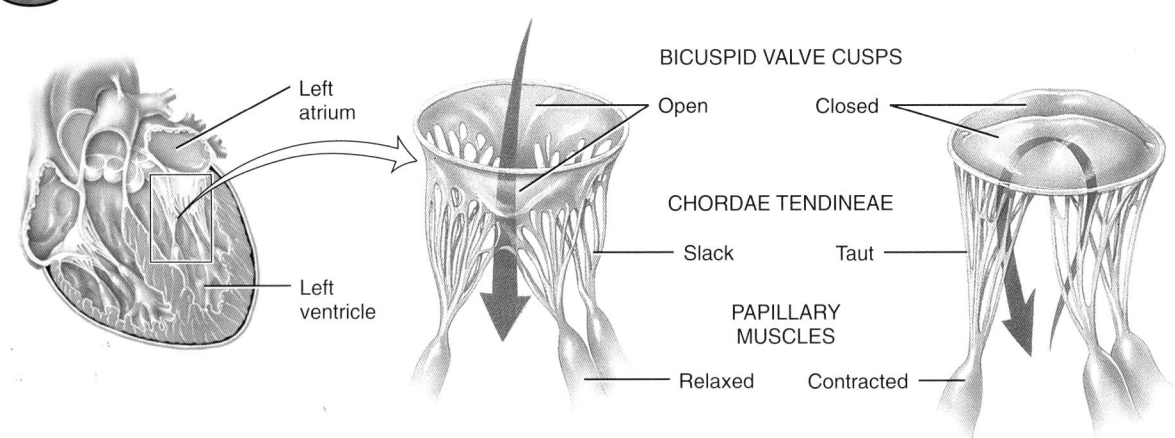

(a) Bicuspid valve open   (b) Bicuspid valve closed

(c) Tricuspid valve open

Operation of the Atrioventricular Valves

Because they are located between an atrium and a ventricle, the tricuspid and bicuspid valves are termed **atrioventricular (AV) valves.** When an AV valve is open, the rounded ends of the cusps project into the ventricle. When the ventricles are relaxed, the papillary muscles are relaxed, the chordae tendineae are slack, and blood moves from a higher pressure in the atria to a lower pressure in the ventricles through open AV valves (Figure 20.6a, d). When the ventricles contract, the pressure of the blood drives the cusps upward until their edges meet and close the opening (Figure 20.6b, e). At the same time, the papillary muscles contract, which pulls on and tightens the chordae tendineae. This prevents the valve cusps from everting (opening into the atria) in response to the high ventricular pressure. If the AV valves or chordae tendineae are damaged, blood may regurgitate (flow back) into the atria when the ventricles contract.

Operation of the Semilunar Valves

The aortic and pulmonary valves are known as the **semilunar (SL) valves** (sem-ē-LOO-nar; *semi-* = half; *-lunar* = moon-shaped) because they are made up of three crescent moon–shaped

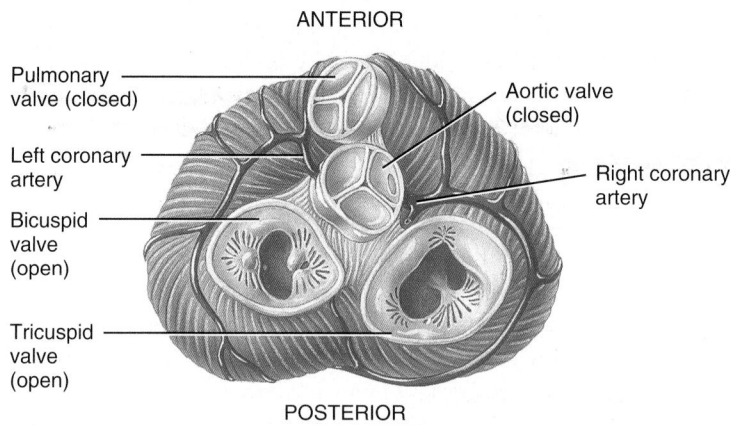

Pulmonary valve (closed)
Left coronary artery
Bicuspid valve (open)
Tricuspid valve (open)
Aortic valve (closed)
Right coronary artery

ANTERIOR

POSTERIOR

(d) Superior view with atria removed: pulmonary and aortic valves closed, bicuspid and tricuspid valves open

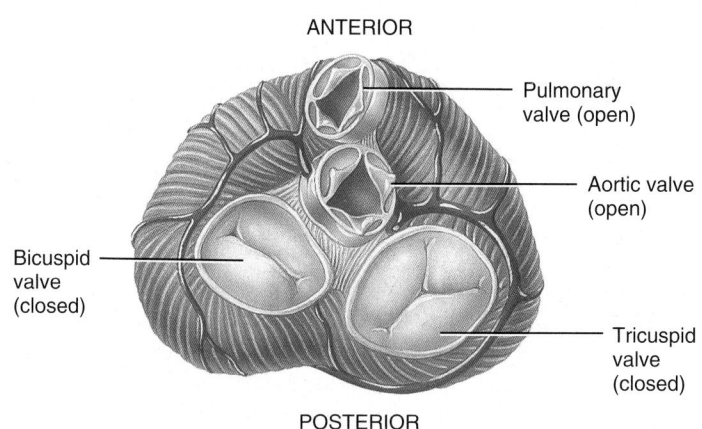

ANTERIOR

Pulmonary valve (open)
Aortic valve (open)
Bicuspid valve (closed)
Tricuspid valve (closed)

POSTERIOR

(e) Superior view with atria removed: pulmonary and aortic valves open, bicuspid and tricuspid valves closed

ANTERIOR

Pulmonary trunk
PULMONARY VALVE
Pectinate muscle of left atrium
Left coronary artery
BICUSPID (MITRAL) VALVE

Ascending aorta
Right coronary artery
Pectinate muscle of right atrium
AORTIC VALVE
TRICUSPID VALVE
Coronary sinus

POSTERIOR

(f) Superior view of atrioventricular and semilunar valves

Semilunar cusp of aortic valve

(g) Superior view of aortic valve

How do papillary muscles prevent atrioventricular valve cusps from everting (swinging upward) into the atria?

cusps (Figure 20.6d). Each cusp attaches to the arterial wall by its convex outer margin. The SL valves allow ejection of blood from the heart into arteries but prevent backflow of blood into the ventricles. The free borders of the cusps project into the lumen of the artery. When the ventricles contract, pressure builds up within the chambers. The semilunar valves open when pressure in the ventricles exceeds the pressure in the arteries, permitting ejection of blood from the ventricles into the pulmonary trunk and aorta (Figure 20.6e). As the ventricles relax, blood starts to flow back toward the heart. This backflowing blood fills the valve cusps, which causes the free edges of the semilunar valves to contact each other tightly and close the opening between the ventricle and artery (Figure 20.6d).

Surprisingly perhaps, there are no valves guarding the junctions between the venae cavae and the right atrium or the pulmonary veins and the left atrium. As the atria contract, a small amount of blood does flow backward from the atria into these vessels. However, backflow is minimized by a different mechanism; as the atrial muscle contracts, it compresses and nearly collapses the venous entry points.

CLINICAL CONNECTION | *Heart Valve Disorders*

When heart valves operate normally, they open fully and close completely at the proper times. A narrowing of a heart valve opening that restricts blood flow is known as **stenosis** (ste-NŌ-sis = a narrowing); failure of a valve to close completely is termed **insufficiency** (in'-su-FISH-en-sē) or **incompetence.** In **mitral stenosis,** scar formation or a congenital defect causes narrowing of the mitral valve. One cause of **mitral insufficiency,** in which there is backflow of blood from the left ventricle into the left atrium, is **mitral valve prolapse (MVP).** In MVP one or both cusps of the mitral valve protrude into the left atrium during ventricular contraction. Mitral valve prolapse is one of the most common valvular disorders, affecting as much as 30% of the population. It is more prevalent in women than in men, and does not always pose a serious threat. In **aortic stenosis** the aortic valve is narrowed, and in **aortic insufficiency** there is backflow of blood from the aorta into the left ventricle.

Certain infectious diseases can damage or destroy the heart valves. One example is **rheumatic fever,** an acute systemic inflammatory disease that usually occurs after a streptococcal infection of the throat. The bacteria trigger an immune response in which antibodies produced to destroy the bacteria instead attack and inflame the connective tissues in joints, heart valves, and other organs. Even though rheumatic fever may weaken the entire heart wall, most often it damages the mitral and aortic valves.

If daily activities are affected by symptoms and if a heart valve cannot be repaired surgically, then the valve must be replaced. Tissue valves may be provided by human donors or pigs; sometimes, mechanical replacements are used. In any case, valve replacement involves open heart surgery. The aortic valve is the most commonly replaced heart valve. •

Systemic and Pulmonary Circulations

In postnatal (after birth) circulation, the heart pumps blood into two closed circuits with each beat—**systemic circulation** and **pulmonary circulation** (*pulmon-* = lung). The two circuits are arranged in series: The output of one becomes the input of the other, as would happen if you attached two garden hoses (see Figure 21.17). The left side of the heart is the pump for systemic circulation; it receives bright red *oxygenated* (oxygen-rich) *blood* from the lungs. The left ventricle ejects blood into the *aorta* (Figure 20.7). From the aorta, the blood divides into separate streams, entering progressively smaller *systemic arteries* that carry it to all organs throughout the body—except for the air sacs (alveoli) of the lungs, which are supplied by pulmonary circulation. In systemic tissues, arteries give rise to smaller-diameter *arterioles,* which finally lead into extensive beds of *systemic capillaries.* Exchange of nutrients and gases occurs across the thin capillary walls. Blood unloads O_2 (oxygen) and picks up CO_2 (carbon dioxide). In most cases, blood flows through only one capillary and then enters a *systemic venule.* Venules carry *deoxygenated* (oxygen-poor) *blood* away from tissues and merge to form larger *systemic veins.* Ultimately the blood flows back to the right atrium.

The right side of the heart is the pump for pulmonary circulation; it receives all the dark-red deoxygenated blood returning from the systemic circulation. Blood ejected from the right ventricle flows into the *pulmonary trunk,* which branches into *pulmonary arteries* that carry blood to the right and left lungs. In pulmonary capillaries, blood unloads CO_2, which is exhaled, and picks up O_2 from inhaled air. The freshly oxygenated blood then flows into pulmonary veins and returns to the left atrium.

Coronary Circulation

Nutrients are not able to diffuse quickly enough from blood in the chambers of the heart to supply all the layers of cells that make up the heart wall. For this reason, the myocardium has its own network of blood vessels, the **coronary** or **cardiac circulation** (*coron-* = crown). The **coronary arteries** branch from the ascending aorta and encircle the heart like a crown encircles the head (Figure 20.8a). While the heart is contracting, little blood flows in the coronary arteries because they are squeezed shut. When the heart relaxes, however, the high pressure of blood in the aorta propels blood through the coronary arteries, into capillaries, and then into **coronary veins** (Figure 20.8b).

Coronary Arteries

Two coronary arteries, the right and left coronary arteries, branch from the ascending aorta and supply oxygenated blood to the myocardium (Figure 20.8a). The **left coronary artery** passes inferior to the left auricle and divides into the anterior interventricular and circumflex branches. The **anterior interventricular branch** or *left anterior descending (LAD) artery* is in the anterior interventricular sulcus and supplies oxygenated blood to the walls of both ventricles. The **circumflex branch** (SER-kum-fleks) lies in the coronary sulcus and distributes oxygenated blood to the walls of the left ventricle and left atrium.

The **right coronary artery** supplies small branches (*atrial branches*) to the right atrium. It continues inferior to the right auricle and ultimately divides into the posterior interventricular and marginal branches. The **posterior interventricular branch** follows the posterior interventricular sulcus and supplies the walls of the two ventricles with oxygenated blood. The **marginal branch** beyond the coronary sulcus runs along the right margin of the heart and transports oxygenated blood to the myocardium of the right ventricle.

Most parts of the body receive blood from branches of more than one artery, and where two or more arteries supply the same region, they usually connect. These connections, called **anastomoses** (a-nas´-tō-MŌ-sēs), provide alternate routes, called **collateral circulation,** for blood to reach a particular organ or tissue. The myocardium contains many anastomoses that connect branches of a given coronary artery or extend between branches of different coronary arteries. They provide detours for arterial blood if a main route becomes obstructed. Thus, heart muscle may receive sufficient oxygen even if one of its coronary arteries is partially blocked.

Figure 20.7 Systemic and pulmonary circulations.

The left side of the heart pumps oxygenated blood into the systemic circulation to all tissues of the body except the air sacs (alveoli) of the lungs. The right side of the heart pumps deoxygenated blood into the pulmonary circulation to the air sacs (alveoli) of the lungs.

9. Systemic capillaries of head and upper limbs

4. Pulmonary capillaries of right lung

4. Pulmonary capillaries of left lung

Key:
- Oxygen-rich blood
- Oxygen-poor blood

(a) Path of blood flow through heart

9. Systemic capillaries of trunk and lower limbs

4. In pulmonary capillaries, blood loses CO_2 and gains O_2

3. Pulmonary trunk and pulmonary arteries

5. Pulmonary veins (oxygenated blood)

Pulmonary valve

2. Right ventricle

6. Left atrium

Tricuspid valve

Bicuspid valve

1. Right atrium (deoxygenated blood)

7. Left ventricle

Aortic valve

10. Superior vena cava | Inferior vena cava | Coronary sinus

8. Aorta and systemic arteries

9. In systemic capillaries, blood loses O_2 and gains CO_2

(b) Diagram of blood flow

Which numbers constitute the pulmonary circulation? Which constitute the systemic circulation?

Figure 20.8   The coronary circulation. The views of the heart from the anterior aspect in (a) and (b) are drawn as if the heart were transparent to reveal blood vessels on the posterior aspect.

The right and left coronary arteries deliver blood to the heart; the coronary veins drain blood from the heart into the coronary sinus.

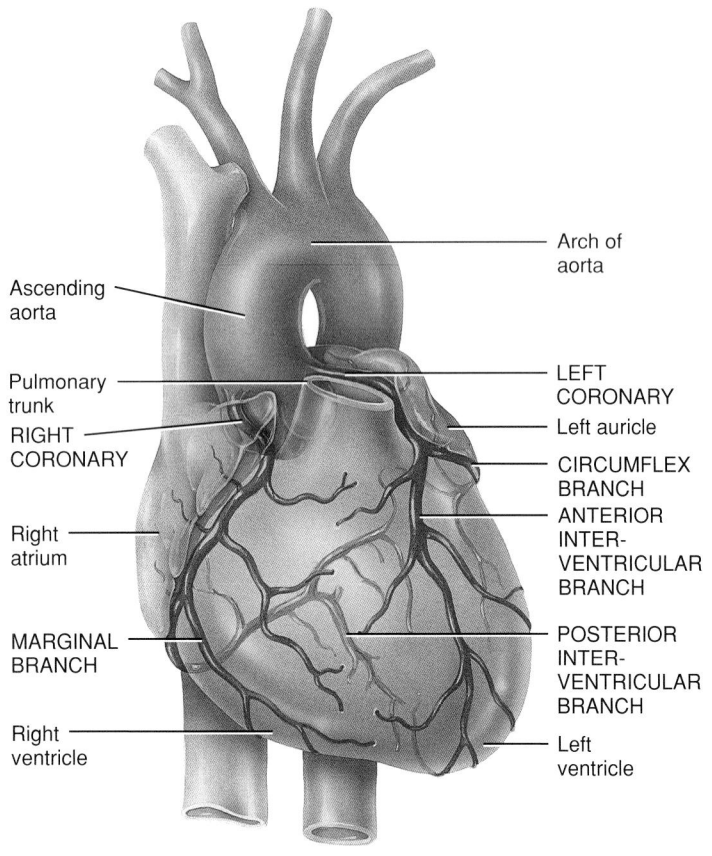

(a) Anterior view of coronary arteries

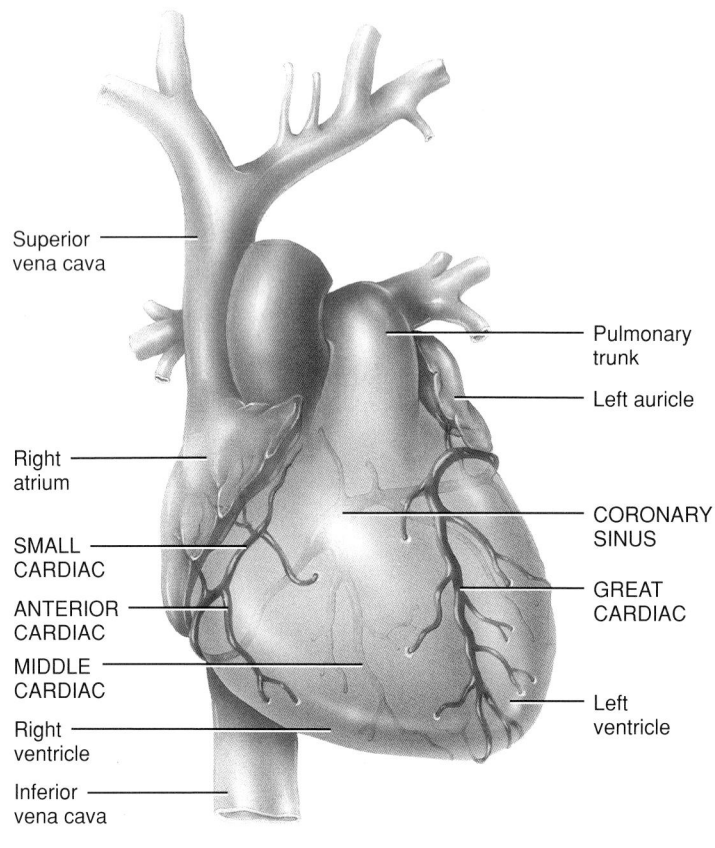

(b) Anterior view of coronary veins

Coronary Veins

After blood passes through the arteries of the coronary circulation, it flows into capillaries, where it delivers oxygen and nutrients to the heart muscle and collects carbon dioxide and waste, and then moves into coronary veins. Most of the deoxygenated blood from the myocardium drains into a large *vascular sinus* in the coronary sulcus on the posterior surface of the heart, called the **coronary sinus** (Figure 20.8b). (A *vascular sinus* is a thin-walled vein that has no smooth muscle to alter its diameter.) The deoxygenated blood in the coronary sinus empties into the right atrium. The principal tributaries carrying blood into the coronary sinus are the following:

- **Great cardiac vein** in the anterior interventricular sulcus, which drains the areas of the heart supplied by the left coronary artery (left and right ventricles and left atrium)
- **Middle cardiac vein** in the posterior interventricular sulcus, which drains the areas supplied by the posterior interventricular branch of the right coronary artery (left and right ventricles)

- **Small cardiac vein** in the coronary sulcus, which drains the right atrium and right ventricle
- **Anterior cardiac veins,** which drain the right ventricle and open directly into the right atrium

When blockage of a coronary artery deprives the heart muscle of oxygen, **reperfusion** (re′-per-FYŪ-zhun), the reestablishment of blood flow, may damage the tissue further. This surprising effect is due to the formation of oxygen **free radicals** from the reintroduced oxygen. As you learned in Chapter 2, free radicals are molecules that have an unpaired electron (see Figure 2.3b). These unstable, highly reactive molecules cause chain reactions that lead to cellular damage and death. To counter the effects of oxygen free radicals, body cells produce enzymes that convert free radicals to less reactive substances. Two such enzymes are *superoxide dismutase* (dis-MŪ-tās) and *catalase* (KAT-a-lās). In addition, nutrients such as vitamin E, vitamin C, beta-carotene, zinc, and selenium serve as antioxidants, which remove oxygen free radicals from circulation. Drugs that lessen reperfusion damage after a heart attack or stroke are currently under development.

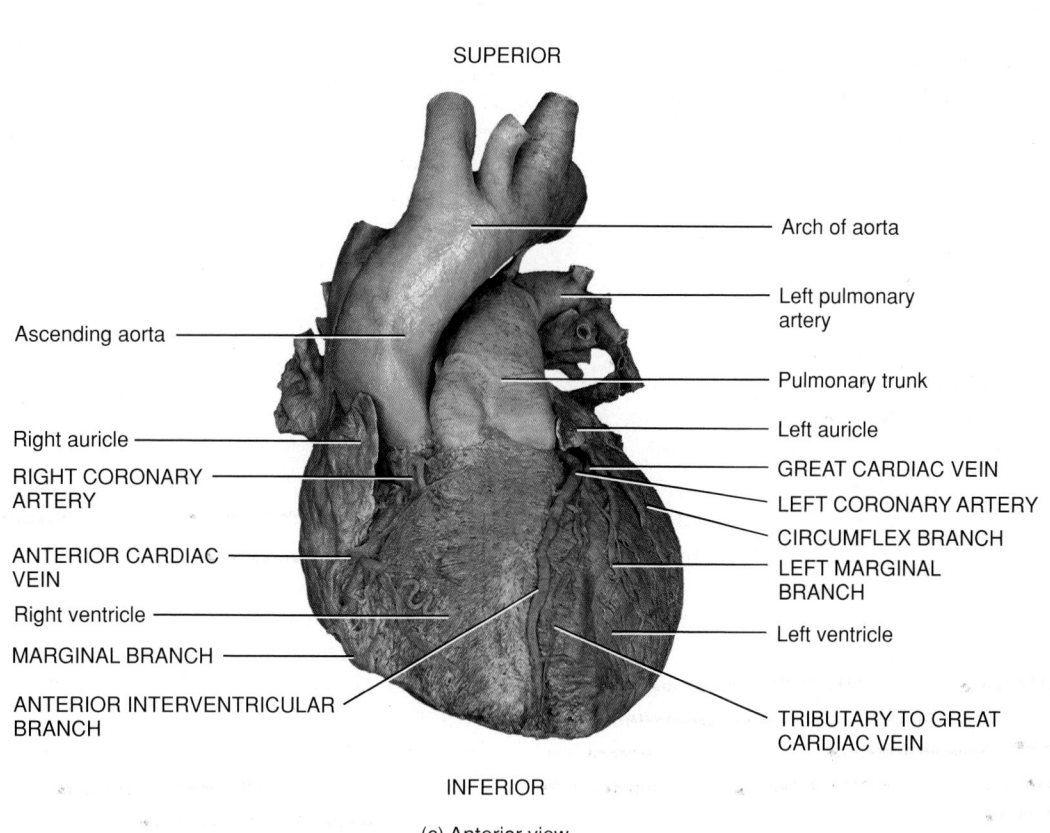

SUPERIOR

Arch of aorta

Left pulmonary artery

Ascending aorta

Pulmonary trunk

Left auricle

Right auricle

GREAT CARDIAC VEIN

RIGHT CORONARY ARTERY

LEFT CORONARY ARTERY

CIRCUMFLEX BRANCH

ANTERIOR CARDIAC VEIN

LEFT MARGINAL BRANCH

Right ventricle

Left ventricle

MARGINAL BRANCH

ANTERIOR INTERVENTRICULAR BRANCH

TRIBUTARY TO GREAT CARDIAC VEIN

INFERIOR

(c) Anterior view

? **Which coronary blood vessel delivers oxygenated blood to the walls of the left atrium and left ventricle?**

CLINICAL CONNECTION | *Myocardial Ischemia and Infarction*

Partial obstruction of blood flow in the coronary arteries may cause **myocardial ischemia** (is-KĒ-mē-a; *ische-* = to obstruct; *-emia* = in the blood), a condition of reduced blood flow to the myocardium. Usually, ischemia causes **hypoxia** (hī-POKS-ē-a = reduced oxygen supply), which may weaken cells without killing them. **Angina pectoris** (an-JĪ-na, or AN-ji-na, PEK-tō-ris), which literally means "strangled chest," is a severe pain that usually accompanies myocardial ischemia. Typically, sufferers describe it as a tightness or squeezing sensation, as though the chest were in a vise. The pain associated with angina pectoris is often referred to the neck, chin, or down the left arm to the elbow. **Silent myocardial ischemia,** ischemic episodes without pain, is particularly dangerous because the person has no forewarning of an impending heart attack.

A complete obstruction to blood flow in a coronary artery may result in a **myocardial infarction (MI)** (in-FARK-shun), commonly called a *heart attack. Infarction* means the death of an area of tissue because of interrupted blood supply. Because the heart tissue distal to the obstruction dies and is replaced by noncontractile scar tissue,

the heart muscle loses some of its strength. Depending on the size and location of the infarcted (dead) area, an infarction may disrupt the conduction system of the heart and cause sudden death by triggering ventricular fibrillation. Treatment for a myocardial infarction may involve injection of a thrombolytic (clot-dissolving) agent such as streptokinase or t-PA, plus heparin (an anticoagulant), or performing coronary angioplasty or coronary artery bypass grafting. Fortunately, heart muscle can remain alive in a resting person if it receives as little as 10–15% of its normal blood supply. •

✔ CHECKPOINT

7. What causes the heart valves to open and to close? What supporting structures ensure that the valves operate properly?

8. In correct sequence, which heart chambers, heart valves, and blood vessels would a drop of blood encounter as it flows from the right atrium to the aorta?

9. Which arteries deliver oxygenated blood to the myocardium of the left and right ventricles?

20.3 CARDIAC MUSCLE TISSUE AND THE CARDIAC CONDUCTION SYSTEM

⊙ **OBJECTIVES**

• Describe the structural and functional characteristics of cardiac muscle tissue and the cardiac conduction system.
• Explain how an action potential occurs in cardiac contractile fibers.
• Describe the electrical events of a normal electrocardiogram (ECG).

Histology of Cardiac Muscle Tissue

Compared with skeletal muscle fibers, cardiac muscle fibers are shorter in length and less circular in transverse section (Figure 20.9). They also exhibit branching, which gives individual cardiac muscle fibers a "stair-step" appearance (see Table 4.9). A typical cardiac muscle fiber is 50–100 μm long and has a diameter of about 14 μm. Usually one centrally located nucleus is present, although an occasional cell may have two nuclei. The ends of cardiac muscle fibers connect to neighboring fibers by irregular transverse thickenings of the sarcolemma called **intercalated discs** (in-TER-ka-lāt-ed; *intercalat-* = to insert between). The discs contain **desmosomes,** which hold the fibers together, and **gap junctions,** which allow muscle action potentials to conduct from one muscle fiber to its neighbors. Gap junctions allow the entire myocardium of the atria or the ventricles to contract as a single, coordinated unit.

Mitochondria are larger and more numerous in cardiac muscle fibers than in skeletal muscle fibers. In a cardiac muscle fiber, they take up 25% of the cytosolic space; in a skeletal muscle fiber only 2% of the cytosolic space is occupied by mitochondria. Cardiac muscle fibers have the same arrangement of actin and myosin, and the same bands, zones, and Z discs, as skeletal muscle fibers. The transverse tubules of cardiac muscle are wider but less abundant than those of skeletal muscle; the one transverse tubule per sarcomere is located at the Z disc. The sarcoplasmic reticulum of cardiac muscle fibers is somewhat smaller than the SR of skeletal muscle fibers. As a result, cardiac muscle has a smaller intracellular reserve of Ca^{2+}.

| CLINICAL CONNECTION | Regeneration of Heart Cells |

As noted earlier in the chapter, the heart of a heart attack survivor often has regions of infarcted (dead) cardiac muscle tissue that typically are replaced with noncontractile fibrous scar tissue over time. Our inability to repair damage from a heart attack has been attributed to a lack of stem cells in cardiac muscle and to the absence of mitosis in mature cardiac muscle fibers. A recent study of heart transplant recipients by American and Italian scientists, however, provides evidence for significant replacement of heart cells. The researchers studied men who had received a heart from a female, and then looked for the presence of a Y chromosome in heart cells. (All female cells except gametes have two X chromosomes and lack the Y chromosome.) Several years after the transplant surgery, between 7% and 16% of the heart cells in the transplanted tissue, including cardiac muscle fibers and endothelial cells in coronary arterioles and capillaries, had been replaced by the recipient's own cells, as evidenced by the presence of a Y chromosome. The study also revealed cells with some of the characteristics of stem cells in both transplanted hearts and control hearts. Evidently, stem cells can migrate from the blood into the heart and differentiate into functional muscle and endothelial cells. The hope is that researchers can learn how to "turn on" such regeneration of heart cells to treat people with heart failure or cardiomyopathy (diseased heart). •

Autorhythmic Fibers: The Conduction System

An inherent and rhythmical electrical activity is the reason for the heart's lifelong beat. The source of this electrical activity is a network of specialized cardiac muscle fibers called **autorhythmic fibers** (aw'-tō-RITH-mik; *auto-* = self) because they are self-excitable. Autorhythmic fibers repeatedly generate action potentials that trigger heart contractions. They continue to stimulate a heart to beat even after it is removed from the body—for example, to be transplanted into another person—and all of its nerves have been cut. (Note: Surgeons do not attempt to reattach heart nerves during heart transplant operations. For this reason, it has been said that heart surgeons are better "plumbers" than they are "electricians.")

During embryonic development, only about 1% of the cardiac muscle fibers become autorhythmic fibers; these relatively rare fibers have two important functions:

1. They act as a **pacemaker,** setting the rhythm of electrical excitation that causes contraction of the heart.

2. They form the **cardiac conduction system,** a network of specialized cardiac muscle fibers that provide a path for each cycle of cardiac excitation to progress through the heart. The conduction system ensures that cardiac chambers become stimulated to contract in a coordinated manner, which makes the heart an effective pump. As you will see later in the chapter, problems with autorhythmic fibers can result in arrhythmias (abnormal rhythms) in which the heart beats irregularly, too fast, or too slow.

Cardiac action potentials propagate through the conduction system in the following sequence (Figure 20.10a):

❶ Cardiac excitation normally begins in the **sinoatrial (SA) node,** located in the right atrial wall just inferior and lateral to the opening of the superior vena cava. SA node cells do not have a stable resting potential. Rather, they repeatedly depolarize to threshold spontaneously. The spontaneous depolarization is a **pacemaker potential.** When the pacemaker potential reaches threshold, it triggers an action potential (Figure 20.10b). Each action potential from the SA node propagates throughout both atria via gap junctions in the intercalated discs of atrial muscle fibers. Following the action potential, the two atria contract at the same time.

Figure 20.9 **Histology of cardiac muscle tissue.** (See Table 4.9 for a light micrograph of cardiac muscle.)

Cardiac muscle fibers connect to neighboring fibers by intercalated discs, which contain desmosomes and gap junctions.

Desmosomes

Mitochondrion

Intercalated discs

Opening of transverse tubule

Gap junctions

Cardiac muscle fiber (cell)

Nucleus

Sarcolemma

(a) Cardiac muscle fibers

Sarcolemma

Transverse tubule

Mitochondrion

Sarcoplasmic reticulum

Nucleus

Thin filament

Thick filament

Z disc

M line

H zone

Z disc

I band

A band

I band

Sarcomere

(b) Arrangement of components in a cardiac muscle fiber

What are the functions of intercalated discs in cardiac muscle fibers?

Figure 20.10 **The conduction system of the heart.** Autorhythmic fibers in the SA node, located in the right atrial wall (a), act as the heart's pacemaker, initiating cardiac action potentials (b) that cause contraction of the heart's chambers.

🔑 The conduction system ensures that the chambers of the heart contract in a coordinated manner.

- Frontal plane

- Right atrium
- ① SINOATRIAL (SA) NODE
- ② ATRIOVENTRICULAR (AV) NODE
- ③ ATRIOVENTRICULAR (AV) BUNDLE (BUNDLE OF HIS)
- ④ RIGHT AND LEFT BUNDLE BRANCHES
- Right ventricle
- ⑤ PURKINJE FIBERS

- Left atrium
- Left ventricle

(a) Anterior view of frontal section

(b) Pacemaker potentials and action potentials in autorhythmic fibers of SA node

❓ **Which component of the conduction system provides the only electrical connection between the atria and the ventricles?**

structure in the AV node. This delay provides time for the atria to empty their blood into the ventricles.

③ From the AV node, the action potential enters the **atrioventricular (AV) bundle** (also known as the **bundle of His,** pronounced HIZ). This bundle is the only site where action potentials can conduct from the atria to the ventricles. (Elsewhere, the fibrous skeleton of the heart electrically insulates the atria from the ventricles.)

④ After propagating along the AV bundle, the action potential enters both the **right** and **left bundle branches.** The bundle branches extend through the interventricular septum toward the apex of the heart.

⑤ Finally, the large-diameter **Purkinje fibers** (pur-KIN-jē) rapidly conduct the action potential beginning at the apex of the heart upward to the remainder of the ventricular myocardium. Then the ventricles contract, pushing the blood upward toward the semilunar valves.

On their own, autorhythmic fibers in the SA node would initiate an action potential about every 0.6 second, or 100 times per minute. Thus, the SA node sets the rhythm for contraction of the heart—it is the *natural pacemaker.* This rate is faster than that of any other autorhythmic fibers. Because action potentials from the

② By conducting along atrial muscle fibers, the action potential reaches the **atrioventricular (AV) node,** located in the interatrial septum, just anterior to the opening of the coronary sinus (Figure 20.10a). At the AV node, the action potential slows considerably as a result of various differences in cell

SA node spread through the conduction system and stimulate other areas before the other areas are able to generate an action potential at their own, slower rate, the SA node acts as the natural pacemaker of the heart. Nerve impulses from the autonomic nervous system (ANS) and blood-borne hormones (such as epinephrine) *modify the timing and strength* of each heartbeat, but they *do not establish the fundamental rhythm.* In a person at rest, for example, acetylcholine released by the parasympathetic division of the ANS slows SA node pacing to about every 0.8 second or 75 action potentials per minute (Figure 20.10b).

CLINICAL CONNECTION | *Artificial Pacemakers*

If the SA node becomes damaged or diseased, the slower AV node can pick up the pacemaking task. Its spontaneous pacing rate is 40 to 60 times per minute. If the activity of both nodes is suppressed, the heartbeat may still be maintained by autorhythmic fibers in the ventricles—the AV bundle, a bundle branch, or Purkinje fibers. However, the pacing rate is so slow (20–35 beats per minute) that blood flow to the brain is inadequate. When this condition occurs, normal heart rhythm can be restored and maintained by surgically implanting an **artificial pacemaker,** a device that sends out small electrical currents to stimulate the heart to contract. A pacemaker consists of a battery and impulse generator and is usually implanted beneath the skin just inferior to the clavicle. The pacemaker is connected to one or two flexible leads (wires) that are threaded through the superior vena cava and then passed into the various chambers of the heart. Many of the newer pacemakers, referred to as *activity-adjusted pacemakers,* automatically speed up the heartbeat during exercise. •

Action Potential and Contraction of Contractile Fibers

The action potential initiated by the SA node travels along the conduction system and spreads out to excite the "working" atrial and ventricular muscle fibers, called **contractile fibers.** An action potential occurs in a contractile fiber as follows (Figure 20.11):

1 *Depolarization.* Unlike autorhythmic fibers, contractile fibers have a stable resting membrane potential that is close to −90 mV. When a contractile fiber is brought to threshold by an action potential from neighboring fibers, its **voltage-gated fast Na^+ channels** open. These sodium ion channels are referred to as "fast" because they open very rapidly in response to a threshold-level depolarization. Opening of these channels allows Na^+ inflow because the cytosol of contractile fibers is electrically more negative than interstitial fluid and Na^+ concentration is higher in interstitial fluid. Inflow of Na^+ down the electrochemical gradient produces a **rapid depolarization** (dē′-pō-lar-i-ZĀ-shun). Within a few milliseconds, the fast Na^+ channels automatically inactivate and Na^+ inflow decreases.

2 *Plateau.* The next phase of an action potential in a contractile fiber is the **plateau,** a period of maintained depolarization. It is due in part to opening of **voltage-gated slow Ca^{2+} channels** in the sarcolemma. When these channels open, calcium ions move from the interstitial fluid (which has a higher Ca^{2+} concentration) into the cytosol. This inflow of Ca^{2+} causes even more Ca^{2+} to pour out of the sarcoplasmic reticulum into the cytosol through additional Ca^{2+} channels in the sarcoplasmic reticulum membrane. The increased Ca^{2+}

Figure 20.11 Action potential in a ventricular contractile fiber. The resting membrane potential is about −90 mV.

A long refractory period prevents tetanus in cardiac muscle fibers.

? How does the duration of an action potential in a ventricular contractile fiber compare with that in a skeletal muscle fiber?

concentration in the cytosol ultimately triggers contraction. Several different types of **voltage-gated K$^+$ channels** are also found in the sarcolemma of a contractile fiber. Just before the plateau phase begins, some of these K$^+$ channels open, allowing potassium ions to leave the contractile fiber. Therefore, depolarization is sustained during the plateau phase because Ca^{2+} inflow just balances K$^+$ outflow. The plateau phase lasts for about 0.25 sec, and the membrane potential of the contractile fiber is close to 0 mV. By comparison, depolarization in a neuron or skeletal muscle fiber is much briefer, about 1 msec (0.001 sec), because it lacks a plateau phase.

③ *Repolarization.* The recovery of the resting membrane potential during the **repolarization** (rē′-pō-lar-i-ZĀ-shun) phase of a cardiac action potential resembles that in other excitable cells. After a delay (which is particularly prolonged in cardiac muscle), additional voltage-gated K$^+$ channels open. Outflow of K$^+$ restores the negative resting membrane potential (−90 mV). At the same time, the calcium channels in the sarcolemma and the sarcoplasmic reticulum are closing, which also contributes to repolarization.

The mechanism of contraction is similar in cardiac and skeletal muscle: The electrical activity (action potential) leads to the mechanical response (contraction) after a short delay. As Ca^{2+} concentration rises inside a contractile fiber, Ca^{2+} binds to the regulatory protein troponin, which allows the actin and myosin filaments to begin sliding past one another, and tension starts to develop. Substances that alter the movement of Ca^{2+} through slow Ca^{2+} channels influence the strength of heart contractions. Epinephrine, for example, increases contraction force by enhancing Ca^{2+} flow into the cytosol.

In muscle, the **refractory period** (re-FRAK-to-rē) is the time interval during which a second contraction cannot be triggered. The refractory period of a cardiac muscle fiber lasts longer than the contraction itself (Figure 20.11). As a result, another contraction cannot begin until relaxation is well under way. For this reason, tetanus (maintained contraction) cannot occur in cardiac muscle as it can in skeletal muscle. The advantage is apparent if you consider how the ventricles work. Their pumping function depends on alternating contraction (when they eject blood) and relaxation (when they refill). If heart muscle could undergo tetanus, blood flow would cease.

ATP Production in Cardiac Muscle

In contrast to skeletal muscle, cardiac muscle produces little of the ATP it needs by anaerobic cellular respiration (see Figure 10.11). Instead, it relies almost exclusively on aerobic cellular respiration in its numerous mitochondria. The needed oxygen diffuses from blood in the coronary circulation and is released from myoglobin inside cardiac muscle fibers. Cardiac muscle fibers use several fuels to power mitochondrial ATP production. In a person at rest, the heart's ATP comes mainly from oxidation of fatty acids (60%) and glucose (35%), with smaller contributions from lactic acid, amino acids, and ketone bodies. During exercise,

the heart's use of lactic acid, produced by actively contracting skeletal muscles, rises.

Like skeletal muscle, cardiac muscle also produces some ATP from creatine phosphate. One sign that a myocardial infarction (heart attack, see Clinical Connection: Myocardial Ischemia and Infarction) has occurred is the presence in blood of creatine kinase (CK), the enzyme that catalyzes transfer of a phosphate group from creatine phosphate to ADP to make ATP. Normally, CK and other enzymes are confined within cells. Injured or dying cardiac or skeletal muscle fibers release CK into the blood.

Electrocardiogram

As action potentials propagate through the heart, they generate electrical currents that can be detected at the surface of the body. An **electrocardiogram** (e-lek′-trō-KAR-dē-ō-gram), abbreviated either **ECG** or **EKG** (from the German word *Elektrokardiogram*), is a recording of these electrical signals. The ECG is a composite record of action potentials produced by all the heart muscle fibers during each heartbeat. The instrument used to record the changes is an **electrocardiograph.**

In clinical practice, electrodes are positioned on the arms and legs (limb leads) and at six positions on the chest (chest leads) to record the ECG. The electrocardiograph amplifies the heart's electrical signals and produces 12 different tracings from different combinations of limb and chest leads. Each limb and chest electrode records slightly different electrical activity because of the difference in its position relative to the heart. By comparing these records with one another and with normal records, it is possible to determine (1) if the conducting pathway is abnormal, (2) if the heart is enlarged, (3) if certain regions of the heart are damaged, and (4) the cause of chest pain.

In a typical record, three clearly recognizable waves appear with each heartbeat (Figure 20.12). The first, called the **P wave,** is a small upward deflection on the ECG. The P wave represents **atrial depolarization,** which spreads from the SA node through contractile fibers in both atria. The second wave, called the **QRS complex,** begins as a downward deflection, continues as a large, upright, triangular wave, and ends as a downward wave. The QRS complex represents **rapid ventricular depolarization,** as the action potential spreads through ventricular contractile fibers. The third wave is a dome-shaped upward deflection called the **T wave.** It indicates **ventricular repolarization** and occurs just as the ventricles are starting to relax. The T wave is smaller and wider than the QRS complex because repolarization occurs more slowly than depolarization. During the plateau period of steady depolarization, the ECG tracing is flat.

In reading an ECG, the size of the waves can provide clues to abnormalities. Larger P waves indicate enlargement of an atrium; an enlarged Q wave may indicate a myocardial infarction; and an enlarged R wave generally indicates enlarged ventricles. The T wave is flatter than normal when the heart muscle is receiving insufficient oxygen—as, for example, in coronary artery disease. The T wave may be elevated in hyperkalemia (high blood K$^+$ level).

Figure 20.12 **Normal electrocardiogram or ECG (lead II).** P wave = atrial depolarization; QRS complex = onset of ventricular depolarization; T wave = ventricular repolarization.

An ECG is a recording of the electrical activity that initiates each heartbeat.

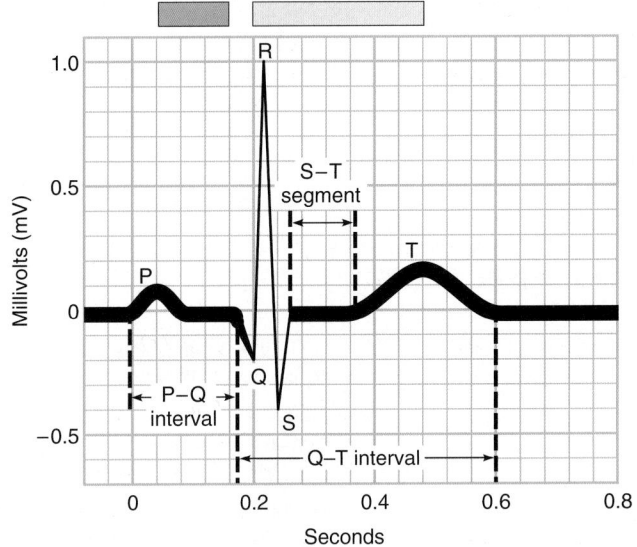

Key:

Atrial contraction

Ventricular contraction

? **What is the significance of an enlarged Q wave?**

Analysis of an ECG also involves measuring the time spans between waves, which are called **intervals** or **segments.** For example, the **P–Q interval** is the time from the beginning of the P wave to the beginning of the QRS complex. It represents the conduction time from the beginning of atrial excitation to the beginning of ventricular excitation. Put another way, the P–Q interval is the time required for the action potential to travel through the atria, atrioventricular node, and the remaining fibers of the conduction system. As the action potential is forced to detour around scar tissue caused by disorders such as coronary artery disease and rheumatic fever, the P–Q interval lengthens.

The **S–T segment,** which begins at the end of the S wave and ends at the beginning of the T wave, represents the time when the ventricular contractile fibers are depolarized during the plateau phase of the action potential. The S–T segment is elevated (above the baseline) in acute myocardial infarction and depressed (below the baseline) when the heart muscle receives insufficient oxygen. The **Q–T interval** extends from the start of the QRS complex to the end of the T wave. It is the time from the beginning of ventricular depolarization to the end of ventricular

repolarization. The Q–T interval may be lengthened by myocardial damage, myocardial ischemia (decreased blood flow), or conduction abnormalities.

Sometimes it is helpful to evaluate the heart's response to the stress of physical exercise (stress testing) (see Disorders: Homeostatic Imbalances at the end of this chapter). Although narrowed coronary arteries may carry adequate oxygenated blood while a person is at rest, they will not be able to meet the heart's increased need for oxygen during strenuous exercise. This situation creates changes that can be seen on an electrocardiogram.

Abnormal heart rhythms and inadequate blood flow to the heart may occur only briefly or unpredictably. To detect these problems, **continuous ambulatory electrocardiographs** are used. With this procedure, a person wears a battery-operated monitor (Holter monitor) that records an ECG continuously for 24 hours. Electrodes attached to the chest are connected to the monitor, and information on the heart's activity is stored in the monitor and retrieved later by medical personnel.

Correlation of ECG Waves with Atrial and Ventricular Systole

As we have seen, the atria and ventricles depolarize and then contract at different times because the conduction system routes cardiac action potentials along a specific pathway. The term **systole** (SIS-tō-lē = contraction) refers to the phase of contraction; the phase of relaxation is **diastole** (dī-AS-tō-lē = dilation or expansion). The ECG waves predict the timing of atrial and ventricular systole and diastole. At a heart rate of 75 beats per minute, the timing is as follows (Figure 20.13):

1 A cardiac action potential arises in the SA node. It propagates throughout the atrial muscle and down to the AV node in about 0.03 sec. As the atrial contractile fibers depolarize, the P wave appears in the ECG.

2 After the P wave begins, the atria contract (atrial systole). Conduction of the action potential slows at the AV node because the fibers there have much smaller diameters and fewer gap junctions. (Traffic slows in a similar way where a four-lane highway narrows to one lane in a construction zone!) The resulting 0.1-sec delay gives the atria time to contract, thus adding to the volume of blood in the ventricles, before ventricular systole begins.

3 The action potential propagates rapidly again after entering the AV bundle. About 0.2 sec after onset of the P wave, it has propagated through the bundle branches, Purkinje fibers, and the entire ventricular myocardium. Depolarization progresses down the septum, upward from the apex, and outward from the endocardial surface, producing the QRS complex. At the same time, atrial repolarization is occurring, but it is not usually evident in an ECG because the larger QRS complex masks it.

4 Contraction of ventricular contractile fibers (ventricular systole) begins shortly after the QRS complex appears and continues during the S–T segment. As contraction proceeds from

Figure 20.13 **Timing and route of action potential depolarization and repolarization through the conduction system and myocardium.** Green indicates depolarization, and red indicates repolarization.

Depolarization causes contraction and repolarization causes relaxation of cardiac muscle fibers.

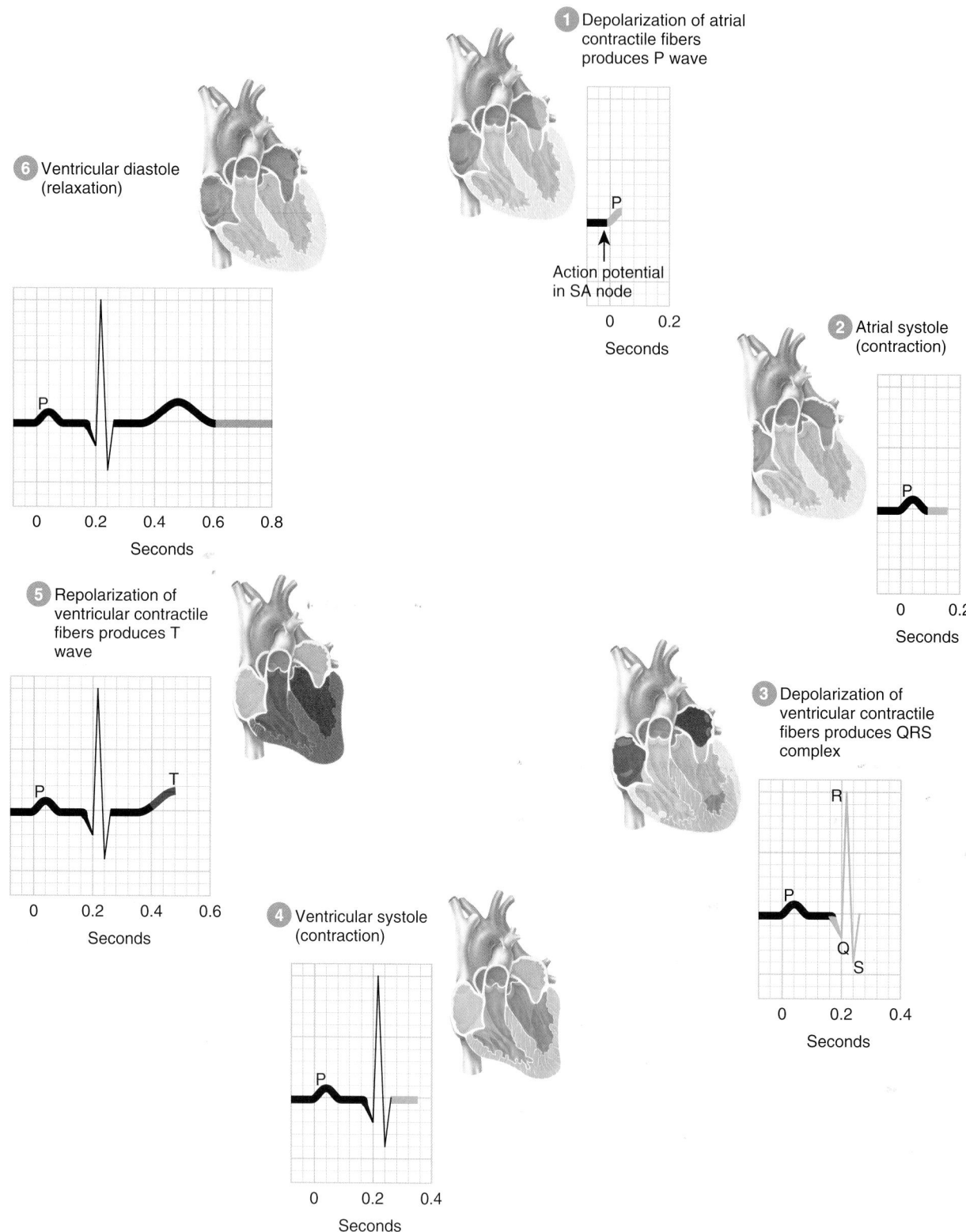

1 Depolarization of atrial contractile fibers produces P wave

Action potential in SA node

6 Ventricular diastole (relaxation)

2 Atrial systole (contraction)

5 Repolarization of ventricular contractile fibers produces T wave

3 Depolarization of ventricular contractile fibers produces QRS complex

4 Ventricular systole (contraction)

Where in the conduction system do action potentials propagate most slowly?

the apex toward the base of the heart, blood is squeezed upward toward the semilunar valves.

5 Repolarization of ventricular contractile fibers begins at the apex and spreads throughout the ventricular myocardium. This produces the T wave in the ECG about 0.4 sec after the onset of the P wave.

6 Shortly after the T wave begins, the ventricles start to relax (ventricular diastole). By 0.6 sec, ventricular repolarization is complete and ventricular contractile fibers are relaxed.

During the next 0.2 sec, contractile fibers in both the atria and ventricles are relaxed. At 0.8 sec, the P wave appears again in the ECG, the atria begin to contract, and the cycle repeats.

As you have just learned, events in the heart occur in cycles that repeat for as long as you live. Next, we will see how the pressure changes associated with relaxation and contraction of the heart chambers allow the heart to alternately fill with blood and then eject blood into the aorta and pulmonary trunk.

✔ CHECKPOINT

10. How do cardiac muscle fibers differ structurally and functionally from skeletal muscle fibers?
11. In what ways are autorhythmic fibers similar to and different from contractile fibers?
12. What happens during each of the three phases of an action potential in ventricular contractile fibers?
13. In what ways are ECGs helpful in diagnosing cardiac problems?
14. How does each ECG wave, interval, and segment relate to contraction (systole) and relaxation (diastole) of the atria and ventricles?

20.4 THE CARDIAC CYCLE

◉ OBJECTIVES

- Describe the pressure and volume changes that occur during a cardiac cycle.
- Relate the timing of heart sounds to the ECG waves and pressure changes during systole and diastole.

A single **cardiac cycle** includes all the events associated with one heartbeat. Thus, a cardiac cycle consists of systole and diastole of the atria plus systole and diastole of the ventricles.

Pressure and Volume Changes during the Cardiac Cycle

In each cardiac cycle, the atria and ventricles alternately contract and relax, forcing blood from areas of higher pressure to areas of lower pressure. As a chamber of the heart contracts, blood pressure within it increases. Figure 20.14 shows the relation between the heart's electrical signals (ECG) and changes in atrial pressure, ventricular pressure, aortic pressure, and ventricular volume during the cardiac cycle. The pressures given in the figure apply to the left side of the heart; pressures on the right side are consider-

ably lower. Each ventricle, however, expels the same volume of blood per beat, and the same pattern exists for both pumping chambers. When heart rate is 75 beats/min, a cardiac cycle lasts 0.8 sec. To examine and correlate the events taking place during a cardiac cycle, we will begin with atrial systole.

Atrial Systole

During **atrial systole,** which lasts about 0.1 sec, the atria are contracting. At the same time, the ventricles are relaxed.

1 Depolarization of the SA node causes atrial depolarization, marked by the P wave in the ECG.

2 Atrial depolarization causes atrial systole. As the atria contract, they exert pressure on the blood within, which forces blood through the open AV valves into the ventricles.

3 Atrial systole contributes a final 25 mL of blood to the volume already in each ventricle (about 105 mL). The end of atrial systole is also the end of ventricular diastole (relaxation). Thus, each ventricle contains about 130 mL at the end of its relaxation period (diastole). This blood volume is called the **end-diastolic volume (EDV).**

4 The QRS complex in the ECG marks the onset of ventricular depolarization.

Ventricular Systole

During **ventricular systole,** which lasts about 0.3 sec, the ventricles are contracting. At the same time, the atria are relaxed in **atrial diastole.**

5 Ventricular depolarization causes ventricular systole. As ventricular systole begins, pressure rises inside the ventricles and pushes blood up against the atrioventricular (AV) valves, forcing them shut. For about 0.05 seconds, both the SL (semilunar) and AV valves are closed. This is the period of **isovolumetric contraction** (ī-sō-VOL-u-met′-rik; *iso-* = same). During this interval, cardiac muscle fibers are contracting and exerting force but are not yet shortening. Thus, the muscle contraction is isometric (same length). Moreover, because all four valves are closed, ventricular volume remains the same (isovolumic).

6 Continued contraction of the ventricles causes pressure inside the chambers to rise sharply. When left ventricular pressure surpasses aortic pressure at about 80 millimeters of mercury (mmHg) and right ventricular pressure rises above the pressure in the pulmonary trunk (about 20 mmHg), both SL valves open. At this point, ejection of blood from the heart begins. The period when the SL valves are open is **ventricular ejection** and lasts for about 0.25 sec. The pressure in the left ventricle continues to rise to about 120 mmHg, whereas the pressure in the right ventricle climbs to about 25–30 mmHg.

7 The left ventricle ejects about 70 mL of blood into the aorta and the right ventricle ejects the same volume of blood into the pulmonary trunk. The volume remaining in each ventricle at the end of systole, about 60 mL, is the **end-systolic volume (ESV). Stroke volume,** the volume ejected per beat from

Figure 20.14 Cardiac cycle. (a) ECG. (b) Changes in left atrial pressure (green line), left ventricular pressure (blue line), and aortic pressure (red line) as they relate to the opening and closing of heart valves. (c) Heart sounds. (d) Changes in left ventricular volume. (e) Phases of the cardiac cycle.

🔑 A cardiac cycle is composed of all the events associated with one heartbeat.

(a) ECG

P, R, Q, S, T — ① ④ ⑧

0.1 sec | 0.3 sec | 0.4 sec
Atrial systole | Ventricular systole | Relaxation period

(b) Pressure (mmHg)

120, 100, 80, 60, 40, 20, 0

⑨ Aortic valve closes
Dicrotic wave
Aortic pressure
⑥ Aortic valve opens
Left ventricular pressure
⑤ Bicuspid valve closes
②
⑩ Bicuspid valve opens
Left atrial pressure

(c) Heart sounds

S1, S2, S3, S4

(d) Volume in ventricle (mL)

130, 60, 0

③ End-diastolic volume
Stroke volume
⑦ End-systolic volume

(e) Phases of the cardiac cycle

Atrial contraction | Isovolumetric contraction | Ventricular ejection | Isovolumetric relaxation | Ventricular filling | Atrial contraction

❓ How much blood remains in each ventricle at the end of ventricular diastole in a resting person? What is this volume called?

each ventricle, equals end-diastolic volume minus end-systolic volume: SV = EDV − ESV. At rest, the stroke volume is about 130 mL − 60 mL = 70 mL (a little more than 2 oz).

8 The T wave in the ECG marks the onset of ventricular repolarization.

Relaxation Period

During the **relaxation period,** which lasts about 0.4 sec, the atria and the ventricles are both relaxed. As the heart beats faster and faster, the relaxation period becomes shorter and shorter, whereas the durations of atrial systole and ventricular systole shorten only slightly.

9 Ventricular repolarization causes **ventricular diastole.** As the ventricles relax, pressure within the chambers falls, and blood in the aorta and pulmonary trunk begins to flow backward toward the regions of lower pressure in the ventricles. Backflowing blood catches in the valve cusps and closes the SL valves. The aortic valve closes at a pressure of about 100 mmHg. Rebound of blood off the closed cusps of the aortic valve produces the **dicrotic wave** on the aortic pressure curve. After the SL valves close, there is a brief interval when ventricular blood volume does not change because all four valves are closed. This is the period of **isovolumetric relaxation.**

10 As the ventricles continue to relax, the pressure falls quickly. When ventricular pressure drops below atrial pressure, the AV valves open, and **ventricular filling** begins. The major part of ventricular filling occurs just after the AV valves open. Blood that has been flowing into and building up in the atria during ventricular systole then rushes rapidly into the ventricles. At the end of the relaxation period, the ventricles are about three-quarters full. The P wave appears in the ECG, signaling the start of another cardiac cycle.

Heart Sounds

Auscultation (aws-kul-TĀ-shun; *ausculta-* = listening), the act of listening to sounds within the body, is usually done with a stethoscope. The sound of the heartbeat comes primarily from blood turbulence caused by the closing of the heart valves. Smoothly flowing blood is silent. Compare the sounds made by white-water rapids or a waterfall with the silence of a smoothly flowing river. During each cardiac cycle, there are four **heart sounds,** but in a normal heart only the first and second heart sounds (S1 and S2) are loud enough to be heard through a stethoscope. Figure 20.14c shows the timing of heart sounds relative to other events in the cardiac cycle.

The first sound (S1), which can be described as a **lubb** sound, is louder and a bit longer than the second sound. S1 is caused by blood turbulence associated with closure of the AV valves soon after ventricular systole begins. The second sound (S2), which is shorter and not as loud as the first, can be described as a **dupp** sound. S2 is caused by blood turbulence associated with closure of the SL valves at the beginning of ventricular diastole. Although S1 and S2 are due to blood turbulence associated with the closure

of valves, they are best heard at the surface of the chest in locations that are slightly different from the locations of the valves (Figure 20.15). This is because the sound is carried by the blood flow away from the valves. Normally not loud enough to be heard, S3 is due to blood turbulence during rapid ventricular filling, and S4 is due to blood turbulence during atrial systole.

CLINICAL CONNECTION | *Heart Murmurs*

Heart sounds provide valuable information about the mechanical operation of the heart. A **heart murmur** is an abnormal sound consisting of a clicking, rushing, or gurgling noise that either is heard before, between, or after the normal heart sounds, or may mask the normal heart sounds. Heart murmurs in children are extremely common and usually do not represent a health condition. Murmurs are most frequently discovered in children between the ages of 2 and 4. These types of heart murmurs are referred to as *innocent* or *functional heart murmurs;* they often subside or disappear with growth. Although some heart murmurs in adults are innocent, most often an adult murmur indicates a valve disorder. When a heart valve exhibits stenosis, the heart murmur is heard while the valve should be fully open but is not. For example, mitral stenosis (see Clinical Connection: Heart Valve Disorders) produces a murmur during the relaxation period, between S2 and the next S1. An incompetent heart valve, by

Figure 20.15 Heart sounds. Location of valves (blue) and auscultation sites (red) for heart sounds.

Listening to sounds within the body is called auscultation; it is usually done with a stethoscope.

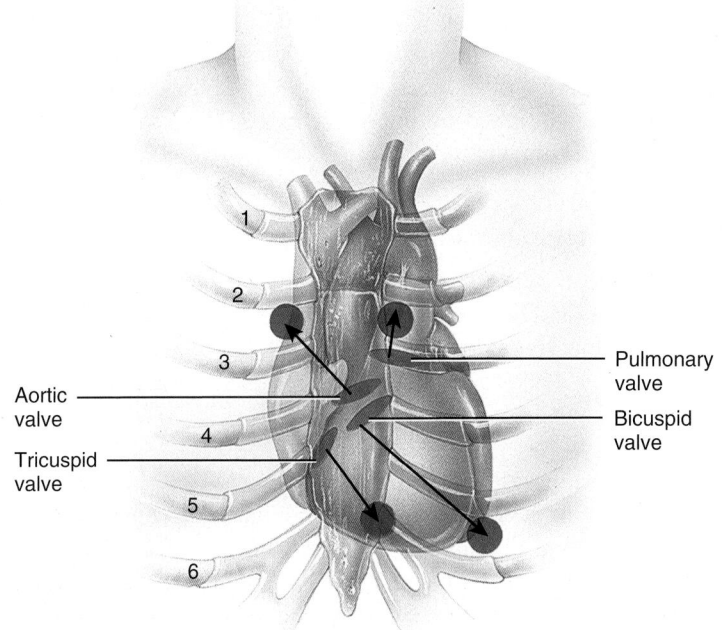

Anterior view of heart valve locations and auscultation sites

? Which heart sound is related to blood turbulence associated with closure of the atrioventricular valves?

contrast, causes a murmur to appear when the valve should be fully closed but is not. So, a murmur due to mitral incompetence (see Clinical Connection: Heart Valve Disorders) occurs during ventricular systole, between S1 and S2. •

✔CHECKPOINT

15. Why must left ventricular pressure be greater than aortic pressure during ventricular ejection?
16. Does more blood flow through the coronary arteries during ventricular diastole or ventricular systole? Explain your answer.
17. During which two periods of the cardiac cycle do the heart muscle fibers exhibit isometric contractions?
18. What events produce the four normal heart sounds? Which ones can usually be heard through a stethoscope?

20.5 CARDIAC OUTPUT

◉ OBJECTIVES

• Define cardiac output.
• Describe the factors that affect regulation of stroke volume.
• Outline the factors that affect the regulation of heart rate.

Although the heart has autorhythmic fibers that enable it to beat independently, its operation is governed by events occurring throughout the body. Body cells must receive a certain amount of oxygen from blood each minute to maintain health and life. When cells are metabolically active, as during exercise, they take up even more oxygen from the blood. During rest periods, cellular metabolic need is reduced, and the workload of the heart decreases.

Cardiac output (CO) is the volume of blood ejected from the left ventricle (or the right ventricle) into the aorta (or pulmonary trunk) each minute. Cardiac output equals the **stroke volume (SV),** the volume of blood ejected by the ventricle during each contraction, multiplied by the **heart rate (HR),** the number of heartbeats per minute:

$$\underset{\text{(mL/min)}}{\text{CO}} = \underset{\text{(mL/beat)}}{\text{SV}} \times \underset{\text{(beats/min)}}{\text{HR}}$$

In a typical resting adult male, stroke volume averages 70 mL/beat, and heart rate is about 75 beats/min. Thus, average cardiac output is

$$\begin{aligned} \text{CO} &= 70 \text{ mL/beat} \times 75 \text{ beats/min} \\ &= 5250 \text{ mL/min} \\ &= 5.25 \text{ L/min} \end{aligned}$$

This volume is close to the total blood volume, which is about 5 liters in a typical adult male. Thus, your entire blood volume flows through your pulmonary and systemic circulations each minute. Factors that increase stroke volume or heart rate normally increase CO. During mild exercise, for example, stroke volume may increase to 100 mL/beat, and heart rate to 100 beats/min. Cardiac output then would be 10 L/min. During intense (but still not maximal) exercise, the heart rate may accelerate to

150 beats/min, and stroke volume may rise to 130 mL/beat, resulting in a cardiac output of 19.5 L/min.

Cardiac reserve is the difference between a person's maximum cardiac output and cardiac output at rest. The average person has a cardiac reserve of four or five times the resting value. Top endurance athletes may have a cardiac reserve seven or eight times their resting CO. People with severe heart disease may have little or no cardiac reserve, which limits their ability to carry out even the simple tasks of daily living.

Regulation of Stroke Volume

A healthy heart will pump out the blood that entered its chambers during the previous diastole. In other words, if more blood returns to the heart during diastole, then more blood is ejected during the next systole. At rest, the stroke volume is 50–60% of the end-diastolic volume because 40–50% of the blood remains in the ventricles after each contraction (end-systolic volume). Three factors regulate stroke volume and ensure that the left and right ventricles pump equal volumes of blood: (1) **preload,** the degree of stretch on the heart before it contracts; (2) **contractility,** the forcefulness of contraction of individual ventricular muscle fibers; and (3) **afterload,** the pressure that must be exceeded before ejection of blood from the ventricles can occur.

Preload: Effect of Stretching

A greater preload (stretch) on cardiac muscle fibers prior to contraction increases their force of contraction. Preload can be compared to the stretching of a rubber band. The more the rubber band is stretched, the more forcefully it will snap back. Within limits, the more the heart fills with blood during diastole, the greater the force of contraction during systole. This relationship is known as the **Frank–Starling law of the heart.** The preload is proportional to the end-diastolic volume (EDV) (the volume of blood that fills the ventricles at the end of diastole). Normally, the greater the EDV, the more forceful the next contraction.

Two key factors determine EDV: (1) the duration of ventricular diastole and (2) **venous return,** the volume of blood returning to the right ventricle. When heart rate increases, the duration of diastole is shorter. Less filling time means a smaller EDV, and the ventricles may contract before they are adequately filled. By contrast, when venous return increases, a greater volume of blood flows into the ventricles, and the EDV is increased.

When heart rate exceeds about 160 beats/min, stroke volume usually declines due to the short filling time. At such rapid heart rates, EDV is less, and the preload is lower. People who have slow resting heart rates usually have large resting stroke volumes because filling time is prolonged and preload is larger.

The Frank–Starling law of the heart equalizes the output of the right and left ventricles and keeps the same volume of blood flowing to both the systemic and pulmonary circulations. If the left side of the heart pumps a little more blood than the right side, the volume of blood returning to the right ventricle (venous return) increases. The increased EDV causes the right ventricle to contract more forcefully on the next beat, bringing the two sides back into balance.

Contractility

The second factor that influences stroke volume is myocardial **contractility,** the strength of contraction at any given preload. Substances that increase contractility are **positive inotropic agents** (īn′-ō-TRŌ-pik); those that decrease contractility are **negative inotropic agents.** Thus, for a constant preload, the stroke volume increases when a positive inotropic substance is present. Positive inotropic agents often promote Ca^{2+} inflow during cardiac action potentials, which strengthens the force of the next contraction. Stimulation of the sympathetic division of the autonomic nervous system (ANS), hormones such as epinephrine and norepinephrine, increased Ca^{2+} level in the interstitial fluid, and the drug digitalis all have positive inotropic effects. In contrast, inhibition of the sympathetic division of the ANS, anoxia, acidosis, some anesthetics, and increased K^+ level in the interstitial fluid have negative inotropic effects. *Calcium channel blockers* are drugs that can have a negative inotropic effect by reducing Ca^{2+} inflow, thereby decreasing the strength of the heartbeat.

Afterload

Ejection of blood from the heart begins when pressure in the right ventricle exceeds the pressure in the pulmonary trunk (about 20 mmHg), and when the pressure in the left ventricle exceeds the pressure in the aorta (about 80 mmHg). At that point, the higher pressure in the ventricles causes blood to push the semilunar valves open. The pressure that must be overcome before a semilunar valve can open is termed the **afterload.** An increase in afterload causes stroke volume to decrease, so that more blood remains in the ventricles at the end of systole. Conditions that can increase afterload include hypertension (elevated blood pressure) and narrowing of arteries by atherosclerosis (see Disorders: Homoeostatic Imbalances at the end of this chapter).

CLINICAL CONNECTION | Congestive Heart Failure

In **congestive heart failure (CHF),** there is a loss of pumping efficiency by the heart. Causes of CHF include coronary artery disease (see Disorders: Homeostatic Imbalances at the end of this chapter), congenital defects, long-term high blood pressure (which increases the afterload), myocardial infarctions (regions of dead heart tissue due to a previous heart attack), and valve disorders. As the pump becomes less effective, more blood remains in the ventricles at the end of each cycle, and gradually the end-diastolic volume (preload) increases. Initially, increased preload may promote increased force of contraction (the Frank–Starling law of the heart), but as the preload increases further, the heart is overstretched and contracts less forcefully. The result is a potentially lethal positive feedback loop: Less-effective pumping leads to even lower pumping capability.

Often, one side of the heart starts to fail before the other. If the left ventricle fails first, it can't pump out all the blood it receives. As a result, blood backs up in the lungs and causes *pulmonary edema,* fluid accumulation in the lungs that can cause suffocation if left untreated. If the right ventricle fails first, blood backs up in the systemic veins and, over time, the kidneys cause an increase in blood volume. In this case, the resulting *peripheral edema* usually is most noticeable in the feet and ankles. •

Regulation of Heart Rate

As you have just learned, cardiac output depends on both heart rate and stroke volume. Adjustments in heart rate are important in the short-term control of cardiac output and blood pressure. The sinoatrial (SA) node initiates contraction and, if left to itself, would set a constant heart rate of about 100 beats/min. However, tissues require different volumes of blood flow under different conditions. During exercise, for example, cardiac output rises to supply working tissues with increased amounts of oxygen and nutrients. Stroke volume may fall if the ventricular myocardium is damaged or if blood volume is reduced by bleeding. In these cases, homeostatic mechanisms maintain adequate cardiac output by increasing the heart rate and contractility. Among the several factors that contribute to regulation of heart rate, the most important are the autonomic nervous system and hormones released by the adrenal medullae (epinephrine and norepinephrine).

Autonomic Regulation of Heart Rate

Nervous system regulation of the heart originates in the **cardiovascular center** in the medulla oblongata. This region of the brain stem receives input from a variety of sensory receptors and from higher brain centers, such as the limbic system and cerebral cortex. The cardiovascular center then directs appropriate output by increasing or decreasing the frequency of nerve impulses in both the sympathetic and parasympathetic branches of the ANS (Figure 20.16).

Even before physical activity begins, especially in competitive situations, heart rate may climb. This anticipatory increase occurs because the limbic system sends nerve impulses to the cardiovascular center in the medulla. As physical activity begins, **proprioceptors** that are monitoring the position of limbs and muscles send nerve impulses at an increased frequency to the cardiovascular center. Proprioceptor input is a major stimulus for the quick rise in heart rate that occurs at the onset of physical activity. Other sensory receptors that provide input to the cardiovascular center include **chemoreceptors,** which monitor chemical changes in the blood, and **baroreceptors,** which monitor the stretching of major arteries and veins caused by the pressure of the blood flowing through them. Important baroreceptors located in the arch of the aorta and in the carotid arteries (see Figure 21.13) detect changes in blood pressure and provide input to the cardiovascular center when it changes. The role of baroreceptors in the regulation of blood pressure is discussed in detail in Chapter 21. Here we focus on the innervation of the heart by the sympathetic and parasympathetic branches of the ANS.

Sympathetic neurons extend from the medulla oblongata into the spinal cord. From the thoracic region of the spinal cord, sympathetic **cardiac accelerator nerves** extend out to the SA node, AV node, and most portions of the myocardium. Impulses in the cardiac accelerator nerves trigger the release of norepinephrine, which binds to beta-1 (β_1) receptors on cardiac muscle fibers. This interaction has two separate effects: (1) In SA (and AV) node fibers, norepinephrine speeds the rate of spontaneous depolarization so that these pacemakers fire impulses more rapidly and heart rate increases; (2) in

Figure 20.16 Nervous system control of the heart.

The cardiovascular center in the medulla oblongata controls both sympathetic and parasympathetic nerves that innervate the heart.

INPUT TO CARDIOVASCULAR CENTER

From higher brain centers: cerebral cortex, limbic system, and hypothalamus

From sensory receptors:
Proprioceptors—monitor movements
Chemoreceptors—monitor blood chemistry
Baroreceptors—monitor blood pressure

OUTPUT TO HEART

Increased rate of spontaneous depolarization in SA node (and AV node) increases heart rate

Increased contractility of atria and ventricles increases stroke volume

Decreased rate of spontaneous depolarization in SA node (and AV node) decreases heart rate

Cardiac accelerator nerves (sympathetic)

Vagus nerves (cranial nerve X, parasympathetic)

Cardiovascular (CV) center

What region of the heart is innervated by the sympathetic division but not by the parasympathetic division?

contractile fibers throughout the atria and ventricles, norepinephrine enhances Ca^{2+} entry through the voltage-gated slow Ca^{2+} channels, thereby increasing contractility. As a result, a greater volume of blood is ejected during systole. With a moderate increase in heart rate, stroke volume does not decline because the increased contractility offsets the decreased preload. With maximal sympathetic stimulation, however, heart rate may reach 200 beats/min in a 20-year-old person. At such a high heart rate, stroke volume is lower than at rest due to the very short filling time. The maximal heart rate declines with age; as a rule, subtracting your age from 220 provides a good estimate of your maximal heart rate in beats per minute.

Parasympathetic nerve impulses reach the heart via the right and left **vagus (X) nerves.** Vagal axons terminate in the SA node, AV node, and atrial myocardium. They release acetylcholine, which decreases heart rate by slowing the rate of spontaneous depolarization in autorhythmic fibers. As only a few vagal fibers innervate ventricular muscle, changes in parasympathetic activity have little effect on contractility of the ventricles.

A continually shifting balance exists between sympathetic and parasympathetic stimulation of the heart. At rest, parasympathetic stimulation predominates. The resting heart rate—about 75 beats/min—is usually lower than the autorhythmic rate of the SA node (about 100 beats/min). With maximal stimulation by the parasympathetic division, the heart can slow to 20 or 30 beats/min, or can even stop momentarily.

Chemical Regulation of Heart Rate

Certain chemicals influence both the basic physiology of cardiac muscle and the heart rate. For example, hypoxia (lowered oxygen level), acidosis (low pH), and alkalosis (high pH) all depress

cardiac activity. Several hormones and cations have major effects on the heart:

1. ***Hormones.*** Epinephrine and norepinephrine (from the adrenal medullae) enhance the heart's pumping effectiveness. These hormones affect cardiac muscle fibers in much the same way as does norepinephrine released by cardiac accelerator nerves—they increase both heart rate and contractility. Exercise, stress, and excitement cause the adrenal medullae to release more hormones. Thyroid hormones also enhance cardiac contractility and increase heart rate. One sign of hyperthyroidism (excessive thyroid hormone) is **tachycardia** (tak′-i-KAR-dē-a), an elevated resting heart rate.

2. ***Cations.*** Given that differences between intracellular and extracellular concentrations of several cations (for example, Na^+ and K^+) are crucial for the production of action potentials in all nerve and muscle fibers, it is not surprising that ionic imbalances can quickly compromise the pumping effectiveness of the heart. In particular, the relative concentrations of three cations—K^+, Ca^{2+}, and Na^+—have a large effect on cardiac function. Elevated blood levels of K^+ or Na^+ decrease heart rate and contractility. Excess Na^+ blocks Ca^{2+} inflow during cardiac action potentials, thereby decreasing the force of contraction, whereas excess K^+ blocks generation of action potentials. A moderate increase in interstitial (and thus intracellular) Ca^{2+} level speeds heart rate and strengthens the heartbeat.

Other Factors in Heart Rate Regulation

Age, gender, physical fitness, and body temperature also influence resting heart rate. A newborn baby is likely to have a resting heart rate over 120 beats/min; the rate then gradually declines throughout life. Adult females often have slightly higher resting heart rates than adult males, although regular exercise tends to bring resting heart rate down in both sexes. A physically fit person may even exhibit **bradycardia** (brād′-i-KAR-dē-a; *bradys-* = slow), a resting heart rate under 50 beats/min. This is a beneficial effect of endurance-type training because a slowly beating heart is more energy efficient than one that beats more rapidly.

Increased body temperature, as occurs during a fever or strenuous exercise, causes the SA node to discharge impulses more quickly, thereby increasing heart rate. Decreased body temperature decreases heart rate and strength of contraction.

During surgical repair of certain heart abnormalities, it is helpful to slow a patient's heart rate by **hypothermia** (hī-pō-THER-mē-a), in which the person's body is deliberately cooled to a low core temperature. Hypothermia slows metabolism, which reduces the oxygen needs of the tissues, allowing the heart and brain to withstand short periods of interrupted or reduced blood flow during the procedure.

Figure 20.17 summarizes the factors that can increase stroke volume and heart rate to achieve an increase in cardiac output.

Figure 20.17 Factors that increase cardiac output.

Cardiac output equals stroke volume multiplied by heart rate.

Increased end-diastolic volume (stretches the heart)

Positive inotropic agents such as increased sympathetic stimulation; catecholamines, glucagon, or thyroid hormones in the blood; increased Ca^{2+} in extracellular fluid

Decreased arterial blood pressure during diastole

Increased PRELOAD

Increased CONTRACTILITY

Decreased AFTERLOAD

Within limits, cardiac muscle fibers contract more forcefully with stretching (Frank–Starling law of the heart)

Positive inotropic agents increase force of contraction at all physiological levels of stretch

Semilunar valves open sooner when blood pressure in aorta and pulmonary artery is lower

Increased STROKE VOLUME

Increased CARDIAC OUTPUT

Increased HEART RATE

Increased sympathetic stimulation and decreased parasympathetic stimulation

Catecholamine or thyroid hormones in the blood; moderate increase in extracellular Ca^{2+}

Infants and senior citizens, females, low physical fitness, increased body temperature

NERVOUS SYSTEM
Cardiovascular center in medulla oblongata receives input from cerebral cortex, limbic system, proprioceptors, baroreceptors, and chemoreceptors

CHEMICALS

OTHER FACTORS

When you are exercising, contraction of skeletal muscles helps return blood to the heart more rapidly. Would this tend to increase or decrease stroke volume?

19. How is cardiac output calculated?
20. Define stroke volume (SV), and explain the factors that regulate it.
21. What is the Frank–Starling law of the heart? What is its significance?
22. Define cardiac reserve. How does it change with training or with heart failure?
23. How do the sympathetic and parasympathetic divisions of the autonomic nervous system adjust heart rate?

20.6 EXERCISE AND THE HEART

◉ **OBJECTIVE**

• Explain how the heart is affected by exercise.

A person's cardiovascular fitness can be improved at any age with regular exercise. Some types of exercise are more effective than others for improving the health of the cardiovascular system. **Aerobics,** any activity that works large body muscles for at least 20 minutes, elevates cardiac output and accelerates metabolic rate. Three to five such sessions a week are usually recommended for improving the health of the cardiovascular system. Brisk walking, running, bicycling, cross-country skiing, and swimming are examples of aerobic activities.

Sustained exercise increases the oxygen demand of the muscles. Whether the demand is met depends mainly on the adequacy of cardiac output and proper functioning of the respiratory system. After several weeks of training, a healthy person increases maximal cardiac output (the amount of blood ejected from the ventricles into their respective arteries per minute), thereby increasing the maximal rate of oxygen delivery to the tissues. Oxygen delivery also rises because skeletal muscles develop more capillary networks in response to long-term training.

During strenuous activity, a well-trained athlete can achieve a cardiac output double that of a sedentary person, in part because training causes hypertrophy (enlargement) of the heart. This condition is referred to as **physiological cardiomegaly** (kar′-dē-ō-MEG-a-lē; *mega* = large). A **pathological cardiomegaly** is related to significant heart disease. Even though the heart of a well-trained athlete is larger, *resting* cardiac output is about the same as in a healthy untrained person, because *stroke volume* (volume of blood pumped by each beat of a ventricle) is increased while heart rate is decreased. The resting heart rate of a trained athlete often is only 40–60 beats per minute (*resting bradycardia*). Regular exercise also helps to reduce blood pressure, anxiety, and depression; control weight; and increase the body's ability to dissolve blood clots.

✔**CHECKPOINT**

24. What are some of the cardiovascular benefits of regular exercise?

20.7 HELP FOR FAILING HEARTS

◉ **OBJECTIVE**

• Describe several techniques used for failing hearts.

As the heart fails, a person has decreasing ability to exercise or even to move around. A variety of surgical techniques and medical devices exist to aid a failing heart. For some patients, even a 10% increase in the volume of blood ejected from the ventricles can mean the difference between being bedridden and having limited mobility.

A **cardiac (heart) transplant** is the replacement of a severely damaged heart with a normal heart from a brain-dead or recently deceased donor. Cardiac transplants are performed on patients with end-stage heart failure or severe coronary artery disease. Once a suitable heart is located, the chest cavity is exposed through a midsternal cut. After the patient is placed on a heart–lung bypass machine, which oxygenates and circulates blood, the pericardium is cut to expose the heart. Next, the diseased heart is removed (usually except for the posterior wall of the left atrium) (Figure 20.18) and the donor heart is trimmed and sutured into position so that the remaining left atrium and great vessels are connected to the donor heart. The new heart is started as blood flows through it (an electrical shock may be used to correct an abnormal rhythm), the patient is weaned from the heart–lung bypass machine, and the chest is closed. The patient must remain on immunosuppressant drugs for a lifetime to prevent rejection. Since the vagus (X) nerve is severed during the surgery, the new heart will beat at about 100 times per minute (compared with a normal rate of about 75 times per minute).

Usually, a donor heart is perfused with a cold solution and then preserved in sterile ice. This can keep the heart viable for about 4–5 hours. In May 2007, surgeons in the United States performed the first beating-heart transplant. The donor heart was maintained at normal body temperature and hooked up to an organ care system that allowed it to keep beating with warm, oxygenated blood flowing through it. This greatly prolongs the time between removal of the heart from the donor and transplantation into a recipient and also decreases injury to the heart while being deprived of blood, which can lead to rejection. The safety and benefits of the oxygen care system are still being evaluated.

Cardiac transplants are common today and produce good results, but the availability of donor hearts is very limited. Another approach is the use of cardiac assist devices and other surgical procedures that assist heart function without removing the heart. Table 20.1 describes several of these devices and procedures.

✔**CHECKPOINT**

25. Describe how a heart transplant is performed.
26. Explain four different cardiac assist devices and procedures.

Figure 20.18 Cardiac transplantation.

Cardiac transplantation is the replacement of a severely damaged heart with a normal heart from a brain-dead or recently deceased donor.

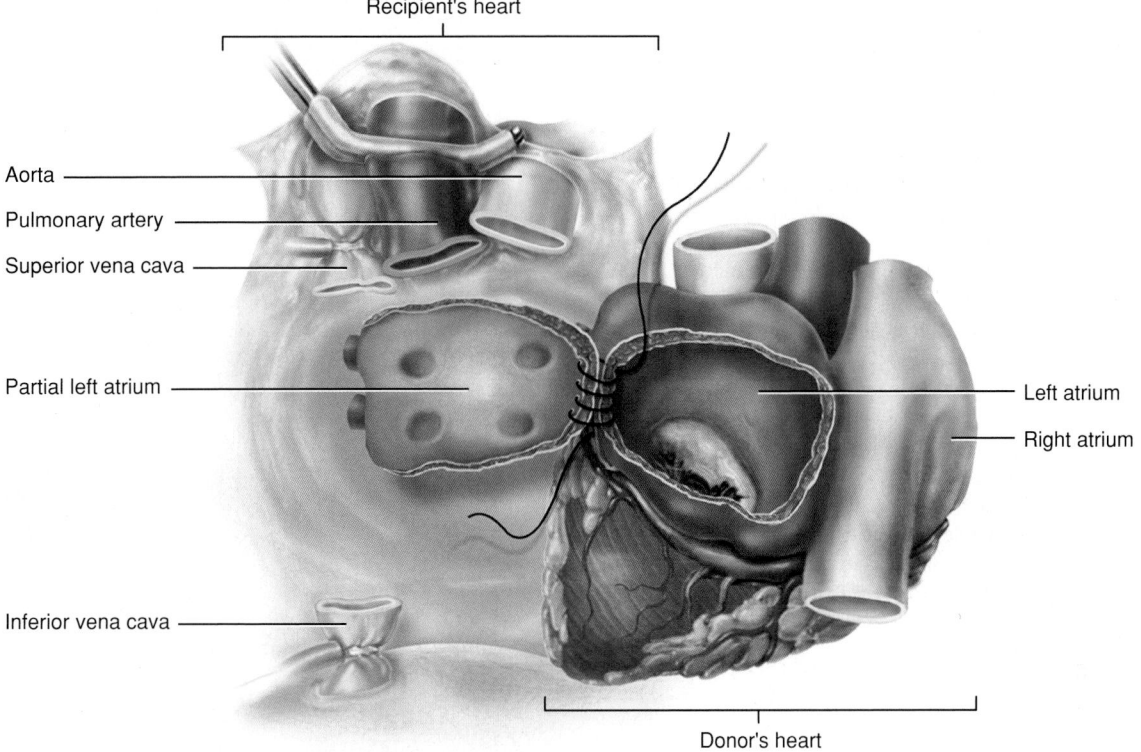

Recipient's heart

- Aorta
- Pulmonary artery
- Superior vena cava
- Partial left atrium
- Inferior vena cava

- Left atrium
- Right atrium

Donor's heart

(a) Donor's left atrium is sutured to recipient's left atrium

- Recipient's superior vena cava
- Donor's right atrium
- Recipient's inferior vena cava

(b) Donor's right atrium is sutured to recipient's superior and inferior venae cavae

(c) Transplanted heart with sutures

? **Which patients are candidates for cardiac transplantation?**

TABLE 20.1

Cardiac Assist Devices and Procedures

DEVICE	DESCRIPTION
Intra-aortic balloon pump (IABP)	A 40-mL polyurethane balloon mounted on a catheter is inserted into an artery in the groin and threaded through the femoral artery into the thoracic aorta (see a). An external pump inflates the balloon with helium gas at the beginning of ventricular diastole. As the balloon inflates, it pushes blood both backward toward the heart (improves coronary blood flow) and forward toward peripheral tissues. The balloon then is rapidly deflated just before the next ventricular systole, drawing blood out of the left ventricle (making it easier for the left ventricle to eject blood). Because the balloon is inflated between heartbeats, this technique is called *intra-aortic balloon counterpulsation.*
Ventricular assist device (VAD)	A mechanical pump helps a weakened ventricle pump blood throughout the body so the heart does not have to work as hard. A VAD may be used to help a patient survive until a heart transplant can be performed *(bridge to transplant)* or provide an alternative to heart transplantation *(destination therapy).*

VADs are classified according to the ventricle that requires support. A *left ventricular assist device (LVAD),* the most common, helps the left ventricle pump blood into the aorta (see b). A *right ventricular assist device (RVAD)* helps the right ventricle pump blood into the pulmonary trunk. A *biventricular assist device (BVAD)* helps both the left and right ventricles perform.

The kind of VAD used depends on the patient's specific needs. To help you understand how a VAD works, see the LVAD (b). An inflow tube attached to the apex of the left ventricle takes blood from the ventricle through a one-way valve into the pump unit. Once the pump fills with blood, an external control system triggers pumping, and blood flows through a one-way valve into an outflow tube that delivers blood into the aorta. The external control system is on a belt around the waist or on a shoulder strap. Some VADs pump at a constant rate; others are coordinated with the person's heartbeat.

Thoracic aorta

Catheter

Anterior view

(a) Intra-aortic balloon pump

Posterior view

Outflow tube

Inflow tube

Outflow one-way valve

Inflow one-way valve

Pump unit

Parts of left ventricular assist device (LVAD)

Driveline

(b) Left ventricular assist device (LVAD)

Aorta

Left ventricle

Implanted left ventricular assist device (LVAD)

TABLE 20.1 CONTINUED

Cardiac Assist Devices and Procedures

DEVICE	DESCRIPTION
Cardiomyoplasty	A large piece of the patient's own skeletal muscle (left latissimus dorsi) is partially freed from connective tissue attachments and wrapped around the heart, leaving the blood and nerve supply intact. An implanted pacemaker stimulates skeletal muscle's motor neurons to cause contraction 10–20 times per minute, in synchrony with some of the heartbeats.
Skeletal muscle assist device	A piece of the patient's own skeletal muscle is used to fashion a pouch inserted between the heart and aorta, functioning as a booster heart. A pacemaker stimulates the muscle's motor neurons to elicit contraction.

20.8 DEVELOPMENT OF THE HEART

◉ OBJECTIVE

- Describe the development of the heart.

Listening to a fetal heartbeat for the first time is an exciting moment for prospective parents, but it is also an important diagnostic tool. The cardiovascular system is one of the first systems to form in an embryo, and the heart is the first functional organ. This order of development is essential because of the need of the rapidly growing embryo to obtain oxygen and nutrients and get rid of wastes. As you will learn shortly, the development of the heart is a complex process, and any disruptions along the way can result in congenital (present at birth) disorders of the heart. Such disorders, described in Disorders: Homeostatic Imbalances at the end of the chapter, are responsible for almost half of all deaths from birth defects.

The *heart* begins its development from **mesoderm** on day 18 or 19 following fertilization. In the head end of the embryo, the heart develops from a group of mesodermal cells called the **cardiogenic area** (kar-dē-ō-JEN-ik; *cardio-* = heart; *-genic* = producing) (Figure 20.19a). In response to signals from the underlying endoderm, the mesoderm in the cardiogenic area forms a pair of elongated strands called **cardiogenic cords.** Shortly after, these cords develop a hollow center and then become known as **endocardial tubes** (Figure 20.19b). With lateral folding of the embryo, the paired endocardial tubes approach each other and fuse into a single tube called the **primitive heart tube** on day 21 following fertilization (Figure 20.19c).

On the twenty-second day, the primitive heart tube develops into five distinct regions and begins to pump blood. From tail end to head end (and the direction of blood flow) they are the (1) **sinus venosus,** (2) **primitive atrium,** (3) **primitive ventricle,** (4) **bulbus cordis,** and (5) **truncus arteriosus.** The sinus venosus initially receives blood from all the veins in the embryo; contractions of the heart begin in this region and follow sequentially in the other regions. Thus, at this stage, the heart consists of a series of unpaired regions. The fates of the regions are as follows:

1. The sinus venosus develops into *part of the right atrium (posterior wall), coronary sinus,* and *sinoatrial (SA) node.*

2. The primitive atrium develops into *part of the right atrium (anterior wall), right auricle, part of the left atrium (anterior wall),* and the *left auricle.*

3. The primitive ventricle gives rise to the *left ventricle.*

4. The bulbus cordis develops into the *right ventricle.*

5. The truncus arteriosus gives rise to the *ascending aorta* and *pulmonary trunk.*

On day 23, the primitive heart tube elongates. Because the bulbus cordis and primitive ventricle grow more rapidly than other parts of the tube and because the atrial and venous ends of the tube are confined by the pericardium, the tube begins to loop and fold. At first, the primitive heart tube assumes a U-shape; later it becomes S-shaped (Figure 20.19e). As a result of these movements, which are completed by day 28, the primitive atria and ventricles of the future heart are reoriented to assume their final adult positions. The remainder of heart development consists of remodeling of the chambers and the formation of septa and valves to form a four-chambered heart.

On about day 28, thickenings of mesoderm of the inner lining of the heart wall, called **endocardial cushions,** appear (Figure 20.20). They grow toward each other, fuse, and divide the single **atrioventricular canal** (region between atria and ventricles) into smaller, separate left and right atrioventricular canals. Also, the *interatrial septum* begins its growth toward the fused endocardial cushions. Ultimately, the interatrial septum and endocardial cushions unite and an opening in the septum, the **foramen ovale** (ō-VAL-ē), develops. The interatrial septum divides the atrial region into a *right atrium* and a *left atrium.* Before birth, the foramen ovale allows most blood entering the right atrium to pass into the left atrium. After birth, it normally closes so that the interatrial septum is a complete partition. The remnant of the foramen ovale is the fossa ovalis (Figure 20.4a). Formation of the *interventricular septum* partitions the ventricular region into a *right ventricle* and a *left ventricle.* Partitioning of the atrioventricular canal, atrial region, and ventricular region is basically complete by the end of the fifth week. The *atrioventricular valves* form between the fifth and eighth weeks. The *semilunar valves* form between the fifth and ninth weeks.

✔CHECKPOINT

27. Why is the cardiovascular system one of the first systems to develop?

28. From which tissue does the heart develop?

Figure 20.19 Development of the heart. Arrows within the structures indicate the direction of blood flow.

The heart begins its development from a group of mesodermal cells called the cardiogenic area during the third week after fertilization.

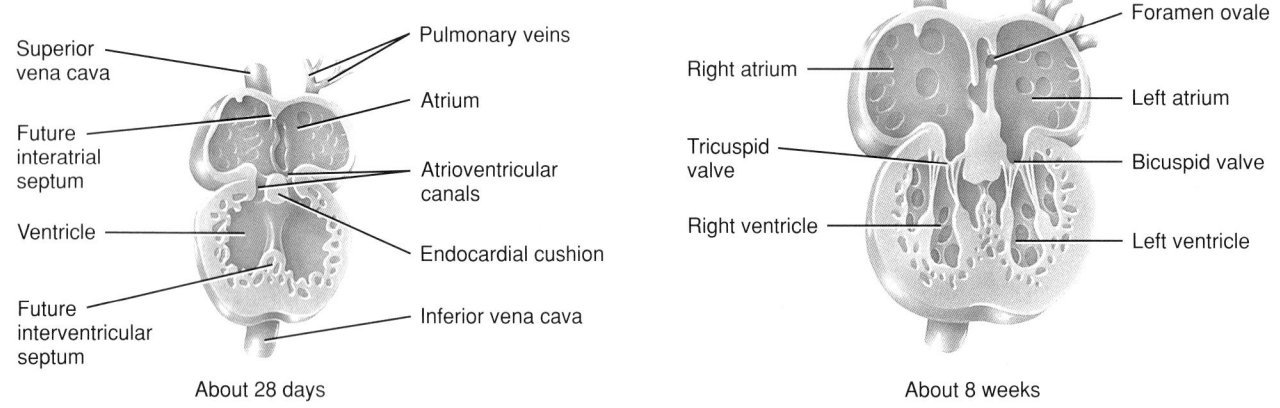

Head end

Cardiogenic area

Neural plate

19 days

(a) Location of cardiogenic area

Arterial end of heart

Endocardial tubes

Venous end of heart

20 days

(b) Formation of endocardial tubes

Fusion of endocardial tubes into primitive heart tube

21 days

(c) Formation of primitive heart tube

Truncus arteriosus

Bulbus cordis

Primitive ventricle

Primitive atrium

Sinus venosus

22 days

(d) Development of regions in primitive heart tube

Truncus arteriosus

Bulbus cordis

Primitive ventricle

Primitive atrium

Sinus venosus

23 days

Superior vena cava

Inferior vena cava

24 days

(e) Bending of primitive heart

Aorta

Pulmonary trunk

Atrium

Ventricle

28 days

(f) Orientation of atria and ventricles to their final adult position

When during embryonic development does the primitive heart begin to contract?

Figure 20.20 Partitioning of the heart into four chambers.

Partitioning of the heart begins on about the twenty-eighth day after fertilization.

Superior vena cava

Future interatrial septum

Ventricle

Future interventricular septum

Pulmonary veins

Atrium

Atrioventricular canals

Endocardial cushion

Inferior vena cava

About 28 days

Foramen ovale

Right atrium

Left atrium

Tricuspid valve

Bicuspid valve

Right ventricle

Left ventricle

About 8 weeks

When is the partitioning of the heart complete?

Coronary Artery Disease

Coronary artery disease (CAD) is a serious medical problem that affects about 7 million people annually. Responsible for nearly three-quarters of a million deaths in the United States each year, it is the leading cause of death for both men and women. CAD results from the effects of the accumulation of atherosclerotic plaques (described shortly) in coronary arteries, which leads to a reduction in blood flow to the myocardium. Some individuals have no signs or symptoms; others experience angina pectoris (chest pain), and still others suffer heart attacks.

Risk Factors for CAD

People who possess combinations of certain risk factors are more likely to develop CAD. *Risk factors* (characteristics, symptoms, or signs present in a disease-free person that are statistically associated with a greater chance of developing a disease) include smoking, high blood pressure, diabetes, high cholesterol levels, obesity, "type A" personality, sedentary lifestyle, and a family history of CAD. Most of these can be modified by changing diet and other habits or can be controlled by taking medications. However, other risk factors are unmodifiable (beyond our control), including genetic predisposition (family history of CAD at an early age), age, and gender. For example, adult males are more likely than adult females to develop CAD; after age 70 the risks are roughly equal. Smoking is undoubtedly the number-one risk factor in all CAD-associated diseases, roughly doubling the risk of morbidity and mortality.

Development of Atherosclerotic Plaques

Although the following discussion applies to coronary arteries, the process can also occur in arteries outside the heart. Thickening of the walls of arteries and loss of elasticity are the main characteristics of a group of diseases called **arteriosclerosis** (ar-tē-rē-ō-skle-RŌ-sis; *sclero-* = hardening). One form of arteriosclerosis is **atherosclerosis** (ath-er-ō-skle-RŌ-sis), a progressive disease characterized by the formation in the walls of large and medium-sized arteries of lesions called **atherosclerotic plaques** (ath-er-ō-skle-RO-tik) (Figure 20.21).

To understand how atherosclerotic plaques develop, you will need to learn about the role of molecules produced by the liver and small intestine called **lipoproteins.** These spherical particles consist of an inner core of triglycerides and other lipids and an outer shell of proteins, phospholipids, and cholesterol. Like most lipids, cholesterol does not dissolve in water and must be made water-soluble in order to be transported in the blood. This is accomplished by combining it with lipoproteins. Two major lipoproteins are **low-density lipoproteins (LDLs)** and **high-density lipoproteins (HDLs).** LDLs transport cholesterol from the liver to body cells for use in cell membrane repair and the production of steroid hormones and bile salts. However, excessive amounts of LDLs promote atherosclerosis, so the cholesterol in these particles is commonly known as "bad cholesterol." HDLs, on the other hand, remove excess cholesterol from body cells and transport it to the liver for elimination. Because HDLs decrease blood cholesterol level, the cholesterol in HDLs is commonly referred to as "good cholesterol." Basically, you want your LDL concentration to be low and your HDL concentration to be high.

Inflammation, a defensive response of the body to tissue damage, plays a key role in the development of atherosclerotic plaques. As a result of tissue damage, blood vessels dilate and increase their permeability, and phagocytes, including macrophages, appear in large numbers. The formation of atherosclerotic plaques begins when excess LDLs from the blood accumulate in the inner layer of an artery wall (layer closest to the bloodstream) and the lipids and proteins in the LDLs undergo oxidation (removal of electrons) and the proteins also bind to sugars. In response, endothelial and smooth muscle cells of the artery secrete substances that attract monocytes from the blood and convert them into macrophages. The macrophages then ingest and become so filled with oxidized LDL particles that they have a foamy appearance when viewed microscopically **(foam cells).** T cells (lymphocytes) follow monocytes into the inner lining of an artery, where they release chemicals that intensify the inflammatory response. Together, the foam cells, macrophages, and T cells form a **fatty streak,** the beginning of an atherosclerotic plaque.

Macrophages secrete chemicals that cause smooth muscle cells of the middle layer of an artery to migrate to the top of the atherosclerotic plaque, forming a cap over it and thus walling it off from the blood.

Because most atherosclerotic plaques expand away from the bloodstream rather than into it, blood can still flow through the

Figure 20.21 Photomicrographs of a transverse section of a normal artery and one partially obstructed by an atherosclerotic plaque.

 Inflammation plays a key role in the development of atherosclerotic plaques.

LM 20x — Normal artery

LM 20x — Obstructed artery

Partially obstructed lumen (space through which blood flows)

Atherosclerotic plaque

? **What is the role of HDL?**

affected artery with relative ease, often for decades. Relatively few heart attacks occur when plaque in a coronary artery expands into the bloodstream and restricts blood flow. Most heart attacks occur when the cap over the plaque breaks open in response to chemicals produced by foam cells. In addition, T cells induce foam cells to produce tissue factor (TF), a chemical that begins the cascade of reactions that result in blood clot formation. If the clot in a coronary artery is large enough, it can significantly decrease or stop the flow of blood and result in a heart attack.

A number of other risk factors (all modifiable) have also been identified as significant predictors of CAD when their levels are elevated. **C-reactive proteins (CRPs)** are proteins produced by the liver or present in blood in an inactive form that are converted to an active form during inflammation. CRPs may play a direct role in the development of atherosclerosis by promoting the uptake of LDLs by macrophages. **Lipoprotein (a)** is an LDL-like particle that binds to endothelial cells, macrophages, and blood platelets; may promote the proliferation of smooth muscle fibers; and inhibits the breakdown of blood clots. **Fibrinogen** is a glycoprotein involved in blood clotting that may help regulate cellular proliferation, vasoconstriction, and platelet aggregation. **Homocysteine** (hō′-mō-SIS-tē-ēn) is an amino acid that may induce blood vessel damage by promoting platelet aggregation and smooth muscle fiber proliferation.

Diagnosis of CAD

Many procedures may be employed to diagnose CAD; the specific procedure used will depend on the signs and symptoms of the individual.

A resting electrocardiogram (see Section 20.3) is the standard test employed to diagnose CAD. **Stress testing** can also be performed. In an *exercise stress test,* the functioning of the heart is monitored when placed under physical stress by exercising using a treadmill, an exercise bicycle, or arm exercises. During the procedure, ECG recordings are monitored continuously and blood pressure is monitored at intervals. A *nonexercise (pharmacologic) stress test* is used for individuals who cannot exercise due to conditions such as arthritis. A medication is injected that stresses the heart to mimic the effects of exercise. During both exercise and nonexercise stress testing, **radionuclide imaging** may be performed to evaluate blood flow through heart muscle (see Table 1.3).

Diagnosis of CAD may also involve **echocardiography** (ek′-ō-kar-dē-OG-ra-fē), a technique that uses ultrasound waves to image the interior of the heart. Echocardiography allows the heart to be seen in motion and can be used to determine the size, shape, and functions of the chambers of the heart; the volume and velocity of blood pumped from the heart; the status of heart valves; the presence of birth defects; and abnormalities of the pericardium. A fairly recent technique for evaluating CAD is **electron beam computerized tomography (EBCT),** which detects calcium deposits in coronary arteries. These calcium deposits are indicators of atherosclerosis.

Coronary (cardiac) computed tomography radiography (CCTA) is a computer-assisted radiography procedure in which a contrast medium is injected into a vein and a beta-blocker is given to decrease heart rate. Then x-ray beams trace an arc around the heart and ultimately produce an image called a *CCTA scan.* This procedure is used primarily to detect blockages such as atherosclerotic plaque or calcium (see Table 1.3).

Cardiac catheterization (kath′-e-ter-i-ZĀ-shun) is an invasive procedure used to visualize the heart's chambers, valves, and great vessels in order to diagnose and treat disease not related to abnormalities of the coronary arteries. It may also be used to measure pressure in the heart and great vessels; to assess cardiac output; to measure the flow of blood through the heart and great vessels; to identify

the location of septal and valvular defects; and to take tissue and blood samples. The basic procedure involves inserting a long, flexible, radiopaque **catheter** (plastic tube) into a peripheral vein (for *right heart catheterization*) or a peripheral artery (for *left heart catheterization*) and guiding it under fluoroscopy (x-ray observation).

Coronary angiography (an′-jē-OG-ra-fē; *angio-* = blood vessel; *-grapho* = to write) is an invasive procedure used to obtain information about the coronary arteries. In the procedure, a catheter is inserted into an artery in the groin or wrist and threaded under fluoroscopy toward the heart and then into the coronary arteries. After the tip of the catheter is in place, a radiopaque contrast medium is injected into the coronary arteries. The radiographs of the arteries, called *angiograms,* appear in motion on a monitor and the information is recorded on a videotape or computer disc. Coronary angiography may be used to visualize coronary arteries (see Table 1.3) and to inject clot-dissolving drugs, such as streptokinase or tissue plasminogen activator (t-PA), into a coronary artery to dissolve an obstructing thrombus.

Treatment of CAD

Treatment options for CAD include **drugs** (antihypertensives, nitroglycerin, beta-blockers, cholesterol-lowering drugs, and clot-dissolving agents) and various surgical and nonsurgical procedures designed to increase the blood supply to the heart.

Coronary artery bypass grafting (CABG) is a surgical procedure in which a blood vessel from another part of the body is attached ("grafted") to a coronary artery to bypass an area of blockage. A piece of the grafted blood vessel is sutured between the aorta and the unblocked portion of the coronary artery (Figure 20.22a). Sometimes multiple blood vessels have to be grafted.

A nonsurgical procedure used to treat CAD is **percutaneous transluminal coronary angioplasty (PTCA)** (*percutaneous* = through the skin; *trans-* = across; *-lumen* = an opening or channel in a tube; *angio-* = blood vessel; *-plasty* = to mold or to shape). In one variation of this procedure, a balloon catheter is inserted into an artery of an arm or leg and gently guided into a coronary artery (Figure 20.22b). While dye is released, angiograms (x-rays of blood vessels) are taken to locate the plaques. Next, the catheter is advanced to the point of obstruction, and a balloonlike device is inflated with air to squash the plaque against the blood vessel wall. Because 30–50% of PTCA-opened arteries fail due to restenosis (renarrowing) within 6 months after the procedure is done, a stent may be inserted via a catheter. A **stent** is a metallic, fine wire tube that is permanently placed in an artery to keep the artery *patent* (open), permitting blood to circulate (Figure 20.22c, d). Restenosis may be due to damage from the procedure itself, for PTCA may damage the arterial wall, leading to platelet activation, proliferation of smooth muscle fibers, and plaque formation. Recently, *drug-coated (drug-eluting) coronary stents* have been used to prevent restenosis. The stents are coated with one of several antiproliferative drugs (drugs that inhibit the proliferation of smooth muscle fibers of the middle layer of an artery) and anti-inflammatory drugs. It has been shown that drug-coated stents reduce the rate of restenosis when compared with bare-metal (noncoated) stents. In addition to balloon and stent angioplasty, laser-emitting catheters are used to vaporize plaques (excimer laser coronary angioplasty or ELCA) and small blades inside catheters are used to remove part of the plaque (directional coronary atherectomy).

One area of current research involves cooling the body's core temperature during procedures such as coronary artery bypass grafting (CABG). There have been some promising results from the application of cold therapy during a cerebral vascular accident (CVA or stroke). This research stemmed from observations of people who had suffered

Figure 20.22 Procedures for reestablishing blood flow in occluded coronary arteries.

Treatment options for CAD include drugs and various nonsurgical and surgical procedures.

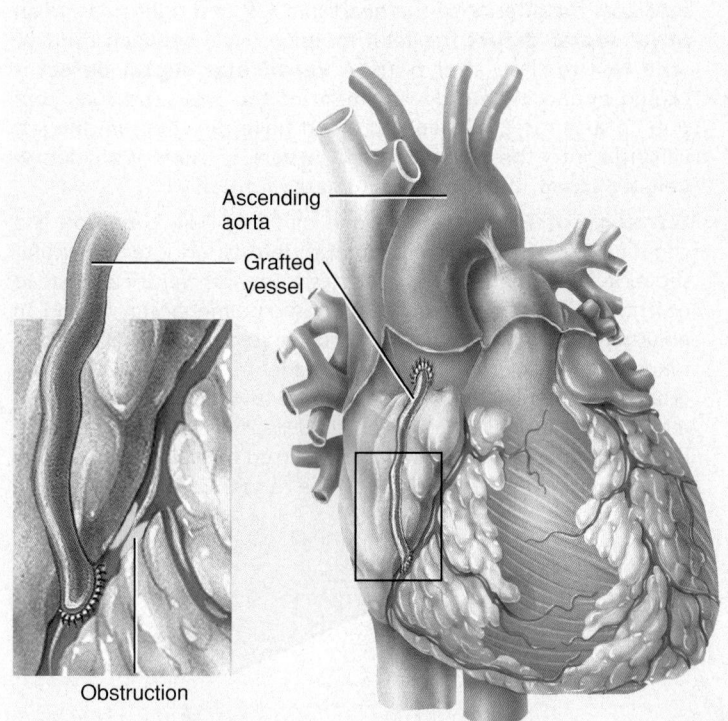

Ascending aorta

Grafted vessel

Obstruction

(a) Coronary artery bypass grafting (CABG)

Balloon | Atherosclerotic plaque | Narrowed lumen of artery | Coronary artery

Balloon catheter with uninflated balloon is threaded to obstructed area in artery

When balloon is inflated, it stretches arterial wall and squashes atherosclerotic plaque

After lumen is widened, balloon is deflated and catheter is withdrawn

(b) Percutaneous transluminal coronary angioplasty (PTCA)

Stent

Lumen of artery

(c) Stent in an artery

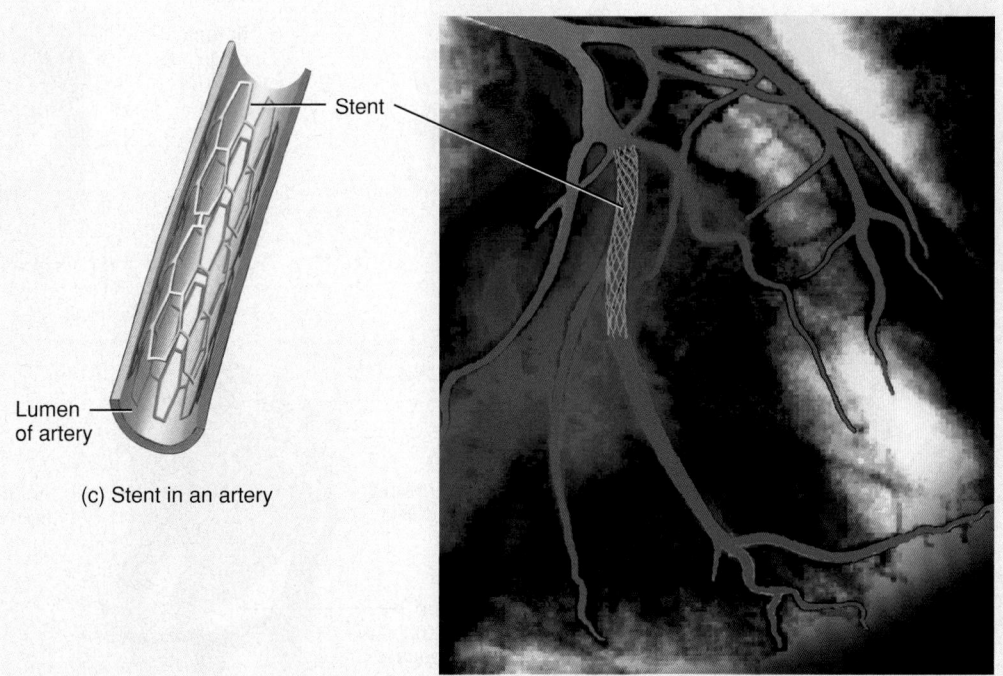

(d) Angiogram showing stent in circumflex artery

Which diagnostic procedure for CAD is used to visualize coronary blood vessels?

a hypothermic incident (such as cold-water drowning) and recovered with relatively minimal neurologic deficits.

Congenital Heart Defects

A defect that is present at birth, and usually before, is called a **congenital defect** (kon-JEN-i-tal). Many such defects are not serious and may go unnoticed for a lifetime. Others are life-threatening and must be surgically repaired. Among the several congenital defects that affect the heart are the following (Figure 20.23):

- **Coarctation of the aorta** (kō′-ark-TĀ-shun). In this condition, a segment of the aorta is too narrow, and thus the flow of oxygenated blood to the body is reduced, the left ventricle is forced to pump harder, and high blood pressure develops. Coarctation is usually repaired surgically by removing the area of obstruction. Surgical interventions that are done in childhood may require revisions in adulthood. Another surgical procedure is balloon dilation, insertion and inflation of a device in the aorta to stretch the vessel. A stent can be inserted and left in place to hold the vessel open.

- **Patent ductus arteriosus (PDA)** (PĀ-tent). In some babies, the ductus arteriosus, a temporary blood vessel between the aorta and the pulmonary trunk, remains open rather than closing shortly after birth. As a result, aortic blood flows into the lower-pressure pulmonary trunk, thus increasing the pulmonary trunk blood pressure and overworking both ventricles. In uncomplicated PDA, medication can be used to facilitate the closure of the defect. In more severe cases, surgical intervention may be required.

- **Septal defect.** A septal defect is an opening in the septum that separates the interior of the heart into left and right sides. In an **atrial septal defect** the fetal foramen ovale between the two atria fails to close after birth. A **ventricular septal defect** is caused by incomplete development of the interventricular septum. In such cases, oxygenated blood flows directly from the left ventricle into the right ventricle, where it mixes with deoxygenated blood. The condition is treated surgically.

- **Tetralogy of Fallot** (tet-RAL-ō-jē of fal-Ō). This condition is a combination of four developmental defects: an interventricular septal defect, an aorta that emerges from both ventricles instead of from the left ventricle only, a stenosed pulmonary valve, and an enlarged right ventricle. There is a decreased flow of blood to the lungs and mixing of blood from both sides of the heart. This causes cyanosis, the bluish discoloration most easily seen in nail beds and mucous membranes when the level of deoxygenated hemoglobin is high; in infants, this condition is referred to as "blue baby." Despite the apparent complexity of this condition, surgical repair is usually successful.

Figure 20.23 Congenital heart defects.

A congenital defect is one that is present at birth, and usually before.

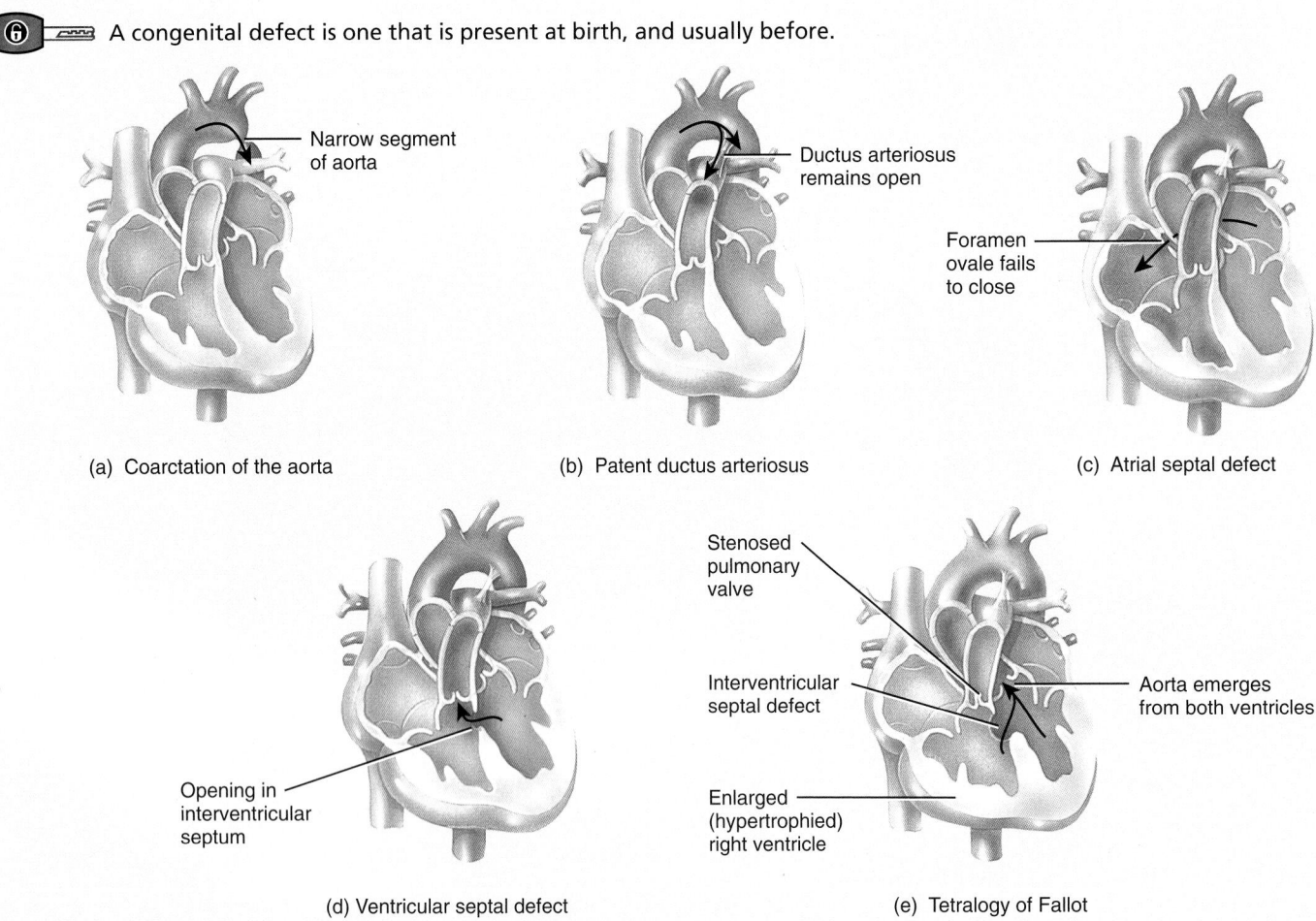

(a) Coarctation of the aorta

(b) Patent ductus arteriosus

(c) Atrial septal defect

(d) Ventricular septal defect

(e) Tetralogy of Fallot

Which four developmental defects occur in tetralogy of Fallot?

Arrhythmias

The usual rhythm of heartbeats, established by the SA node, is called **normal sinus rhythm.** The term **arrhythmia** (a-RITH-mē-a) or **dysrhythmia** refers to an abnormal rhythm as a result of a defect in the conduction system of the heart. The heart may beat irregularly, too quickly, or too slowly. Symptoms include chest pain, shortness of breath, lightheadedness, dizziness, and fainting. Arrhythmias may be caused by factors that stimulate the heart such as stress, caffeine, alcohol, nicotine, cocaine, and certain drugs that contain caffeine or other stimulants. Arrhythmias may also be caused by a congenital defect, coronary artery disease, myocardial infarction, hypertension, defective heart valves, rheumatic heart disease, hyperthyroidism, and potassium deficiency.

Arrhythmias are categorized by their speed, rhythm, and origination of the problem. **Bradycardia** (brād′-i-KAR-dē-a; brady- = slow) refers to a slow heart rate (below 50 beats per minute); **tachycardia** (tak′-i-KAR-dē-a; tachy- = quick) refers to a rapid heart rate (over 100 beats per minute); and **fibrillation** (fi-bri-LĀ-shun) refers to rapid, uncoordinated heartbeats. Arrhythmias that begin in the atria are called **supraventricular** or **atrial arrhythmias;** those that originate in the ventricles are called **ventricular arrhythmias.**

- **Supraventricular tachycardia (SVT)** is a rapid but regular heart rate (160–200 beats per minute) that originates in the atria. The episodes begin and end suddenly and may last from a few minutes to many hours. SVTs can sometimes be stopped by maneuvers that stimulate the vagus (X) nerve and decrease heart rate. These include straining as if having a difficult bowel movement, rubbing the area over the carotid artery in the neck to stimulate the carotid sinus (not recommended for people over 50 since it may cause a stroke), and plunging the face into a bowl of ice-cold water. Treatment may also involve antiarrhythmic drugs and destruction of the abnormal pathway by radiofrequency ablation.

- **Heart block** is an arrhythmia that occurs when the electrical pathways between the atria and ventricles are blocked, slowing the transmission of nerve impulses. The most common site of blockage is the atrioventricular node, a condition called *atrioventricular (AV) block.* In *first-degree AV block,* the P–Q interval is prolonged, usually because conduction through the AV node is slower than normal (Figure 20.24b). In *second-degree AV block,* some of the action potentials from the SA node are not conducted through the AV node. The result is "dropped" beats because excitation doesn't always reach the ventricles. Consequently, there are fewer QRS

Figure 20.24 Representative arrhythmias.

An arrhythmia is an abnormal rhythm as a result of a defect in the cardiac conduction system.

(a) Normal electrocardiogram (ECG)

(b) First-degree AV block

(c) Atrial fibrillation

(d) Ventricular tachycardia

(e) Ventricular fibrillation

 Why is ventricular fibrillation such a serious arrhythmia?

complexes than P waves on the ECG. In *third-degree (complete) AV block,* no SA node action potentials get through the AV node. Autorhythmic fibers in the atria and ventricles pace the upper and lower chambers separately. With complete AV block, the ventricular contraction rate is less than 40 beats/min.

- **Atrial premature contraction (APC)** is a heartbeat that occurs earlier than expected and briefly interrupts the normal heart rhythm. It often causes a sensation of a skipped heartbeat followed by a more forceful heartbeat. APCs originate in the atrial myocardium and are common in healthy individuals.

- **Atrial flutter** consists of rapid, regular atrial contractions (240–360 beats/min) accompanied by an atrioventricular (AV) block in which some of the nerve impulses from the SA node are not conducted through the AV node.

- **Atrial fibrillation (AF)** is a common arrhythmia, affecting mostly older adults, in which contraction of the atrial fibers is asynchronous (not in unison) so that atrial pumping ceases altogether. The atria may beat 300–600 beats/min. The ventricles may also speed up, resulting in a rapid heartbeat (up to 160 beats/min). The ECG of an individual with atrial fibrillation typically has no clearly defined P waves and irregularly spaced QRS complexes (and P–R intervals) (Figure 20.24c). Since the atria and ventricles do not beat in rhythm, the heartbeat is irregular in timing and strength. In an otherwise strong heart, atrial fibrillation reduces the pumping effectiveness of the heart by 20–30%. The most dangerous complication of atrial fibrillation is stroke since blood may stagnate in the atria and form blood clots. A stroke occurs when part of a blood clot occludes an artery supplying the brain.

- **Ventricular premature contraction,** another form of arrhythmia, arises when an *ectopic focus* (ek-TOP-ik), a region of the heart other than the conduction system, becomes more excitable than normal and causes an occasional abnormal action potential to occur. As a wave of depolarization spreads outward from the ectopic focus, it causes a ventricular premature contraction (beat). The contraction occurs early in diastole before the SA node is normally scheduled to discharge its action potential. Ventricular premature contractions may be relatively benign and may be caused by emotional stress, excessive intake of stimulants such as caffeine, alcohol, or nicotine, and lack of sleep. In other cases, the premature beats may reflect an underlying pathology.

- **Ventricular tachycardia (VT or V-tach)** is an arrhythmia that originates in the ventricles and is characterized by four or more ventricular premature contractions. It causes the ventricles to beat too fast (at least 120 beats/min) (Figure 20.24d). VT is almost always associated with heart disease or a recent myocardial infarction and may develop into a very serious arrhythmia called ventricular fibrillation (described shortly). Sustained VT is dangerous because the ventricles do not fill properly and thus do not pump sufficient blood. The result may be low blood pressure and heart failure.

- **Ventricular fibrillation (VF or V-fib)** is the most deadly arrhythmia, in which contractions of the ventricular fibers are completely asynchronous so that the ventricles quiver rather than contract in a coordinated way. As a result, ventricular pumping stops, blood ejection ceases, and circulatory failure and death occur unless there is immediate medical intervention. During ventricular fibrillation, the ECG has no clearly defined P waves, QRS complexes, or T waves (Figure 20.24e). The most common cause of ventricular fibrillation is inadequate blood flow to the heart due to coronary artery disease, as occurs during a myocardial infarction. Other causes are cardiovascular shock, electrical shock, drowning, and very low potassium levels. Ventricular fibrillation causes unconsciousness in seconds and, if untreated, seizures occur and irreversible brain damage may occur after 5 minutes. Death soon follows. Treatment involves cardiopulmonary resuscitation (CPR) and defibrillation. In **defibrillation** (dē-fib-re-LĀ-shun), also called **cardioversion** (kar'-dē-ō-VER-shun), a strong, brief electrical current is passed to the heart and often can stop the ventricular fibrillation. The electrical shock is generated by a device called a **defibrillator** (de-FIB-ri-lā-tor) and applied via two large paddle-shaped electrodes pressed against the skin of the chest. Patients who face a high risk of dying from heart rhythm disorders now can receive an **automatic implantable cardioverter defibrillator (AICD),** an implanted device that monitors their heart rhythm and delivers a small shock directly to the heart when a life-threatening rhythm disturbance occurs. Thousands of patients around the world have AICDs. Also available are **automated external defibrillators (AEDs)** that function like AICDs, except that they are external devices. About the size of a laptop computer, AEDs are used by emergency response teams and are found increasingly in public places such as stadiums, casinos, airports, hotels, and shopping malls. Defibrillation may also be used as an emergency treatment for cardiac arrest.

MEDICAL TERMINOLOGY

Asystole (ā-SIS-tō-lē; *a-* = without) Failure of the myocardium to contract.

Cardiac arrest (KAR-dē-ak a-REST) Cessation of an effective heartbeat. The heart may be completely stopped or in ventricular fibrillation.

Cardiac rehabilitation (rē-ha-bil-i-TĀ-shun) A supervised program of progressive exercise, psychological support, education, and training to enable a patient to resume normal activities following a myocardial infarction.

Cardiomegaly (kar'-dē-ō-MEG-a-lē; *mega* = large) Heart enlargement.

Cardiomyopathy (kar'-dē-ō-mī-OP-a-thē; *myo-* = muscle; *-pathos* = disease) A progressive disorder in which ventricular structure or function is impaired. In dilated cardiomyopathy, the ventricles enlarge (stretch) and become weaker and reduce the heart's pumping action. In hypertrophic cardiomyopathy, the ventricular walls thicken and the pumping efficiency of the ventricles is reduced.

Commotio cordis (kō-MŌ-shē-ō KOR-dis; *commotio* = disturbance; *cordis* = heart) Damage to the heart, frequently fatal, as a result of a sharp, nonpenetrating blow to the chest while the ventricles are repolarizing.

Cor pulmonale (CP) (KOR pul-mōn-AL-ē; *cor* = heart; *pulmon-* = lung) A term referring to right ventricular hypertrophy from disorders that bring about hypertension (high blood pressure) in the pulmonary circulation.

Ejection fraction The fraction of the end-diastolic volume (EDV) that is ejected during an average heartbeat. Equal to stroke volume (SV) divided by EDV.

Electrophysiological testing (e-lek'-trō-fiz'-ē-ō-LOJ-i-kal) A procedure in which a catheter with an electrode is passed through blood vessels and introduced into the heart. It is used to detect the exact locations of abnormal electrical conduction pathways. Once an abnormal pathway is located, it can be destroyed by sending a current through the electrode, a procedure called *radiofrequency ablation.*

Palpitation (pal′-pi-TĀ-shun) A fluttering of the heart or an abnormal rate or rhythm of the heart about which an individual is aware.

Paroxysmal tachycardia (par′-ok-SIZ-mal tak′-i-KAR-dē-a; *tachy-* = quick) A period of rapid heartbeats that begins and ends suddenly.

Sick sinus syndrome An abnormally functioning SA node that initiates heartbeats too slowly or rapidly, pauses too long between heartbeats, or stops producing heartbeats. Symptoms include lightheadedness, shortness of breath, loss of consciousness, and palpitations. It is caused by degeneration of cells in the SA node and is common in elderly persons. It is also related to coronary artery disease. Treatment consists of drugs to speed up or slow down the heart or implantation of an artificial pacemaker.

Sudden cardiac death The unexpected cessation of circulation and breathing due to an underlying heart disease such as ischemia, myocardial infarction, or a disturbance in cardiac rhythm.

CHAPTER REVIEW AND RESOURCE SUMMARY

Review

Resource

20.1 Anatomy of the Heart

1. The heart is located in the mediastinum; about two-thirds of its mass is to the left of the midline. It is shaped like a cone lying on its side. Its apex is the pointed, inferior part; its base is the broad, superior part.
2. The pericardium is the membrane that surrounds and protects the heart; it consists of an outer fibrous layer and an inner serous pericardium, which is composed of a parietal and a visceral layer. Between the parietal and visceral layers of the serous pericardium is the pericardial cavity, a potential space filled with a few milliliters of lubricating pericardial fluid that reduces friction between the two membranes.
3. Three layers make up the wall of the heart: epicardium (visceral layer of the serous pericardium), myocardium, and endocardium. The epicardium consists of mesothelium and connective tissue, the myocardium is composed of cardiac muscle tissue, and the endocardium consists of endothelium and connective tissue.
4. The heart chambers include two superior chambers, the right and left atria, and two inferior chambers, the right and left ventricles. External features of the heart include the auricles (flaps of each atrium that slightly increase their volume), the coronary sulcus between the atria and ventricles, and the anterior and posterior sulci between the ventricles on the anterior and posterior surfaces of the heart, respectively.
5. The right atrium receives blood from the superior vena cava, inferior vena cava, and coronary sinus. It is separated internally from the left atrium by the interatrial septum, which contains the fossa ovalis. Blood exits the right atrium through the tricuspid valve.
6. The right ventricle receives blood from the right atrium. It is separated internally from the left ventricle by the interventricular septum and pumps blood through the pulmonary valve into the pulmonary trunk.
7. Oxygenated blood enters the left atrium from the pulmonary veins and exits through the bicuspid (mitral) valve.
8. The left ventricle pumps oxygenated blood through the aortic valve into the aorta.
9. The thickness of the myocardium of the four chambers varies according to the chamber's function. The left ventricle, with the highest workload, has the thickest wall.
10. The fibrous skeleton of the heart is dense connective tissue that surrounds and supports the valves of the heart.

Anatomy Overview - The Cardiovascular System
Anatomy Overview - The Heart
Figure 20.3 - Structure of the Heart: Surface Features
Exercise - Paint the Heart

20.2 Heart Valves and Circulation of Blood

1. Heart valves prevent backflow of blood within the heart. The atrioventricular (AV) valves, which lie between atria and ventricles, are the tricuspid valve on the right side of the heart and the bicuspid (mitral) valve on the left. The semilunar (SL) valves are the aortic valve, at the entrance to the aorta, and the pulmonary valve, at the entrance to the pulmonary trunk.
2. The left side of the heart is the pump for systemic circulation, the circulation of blood throughout the body except for the air sacs of the lungs. The left ventricle ejects blood into the aorta, and blood then flows into systemic arteries, arterioles, capillaries, venules, and veins, which carry it back to the right atrium.
3. The right side of the heart is the pump for pulmonary circulation, the circulation of blood through the lungs. The right ventricle ejects blood into the pulmonary trunk, and blood then flows into pulmonary arteries, pulmonary capillaries, and pulmonary veins, which carry it back to the left atrium.

Anatomy Overview - Heart Structures
Anatomy Overview - Pulmonary Circulation
Figure 20.7 - Systemic and Pulmonary Circulations
Exercise - Drag and Drop Blood Flow
Negative Feedback Loop Exercise - Homeostasis of Blood Circulation

Review

4. The coronary circulation provides blood flow to the myocardium. The main arteries of the coronary circulation are the left and right coronary arteries; the main veins are the cardiac veins and the coronary sinus.

20.3 Cardiac Muscle Tissue and the Cardiac Conduction System

1. Cardiac muscle fibers usually contain a single centrally located nucleus. Compared with skeletal muscle fibers, cardiac muscle fibers have more and larger mitochondria, slightly smaller sarcoplasmic reticulum, and wider transverse tubules, which are located at Z discs.
2. Cardiac muscle fibers are connected end-to-end via intercalated discs. Desmosomes in the discs provide strength, and gap junctions allow muscle action potentials to conduct from one muscle fiber to its neighbors.
3. Autorhythmic fibers form the conduction system, cardiac muscle fibers that spontaneously depolarize and generate action potentials.
4. Components of the conduction system are the sinoatrial (SA) node (pacemaker), atrioventricular (AV) node, atrioventricular (AV) bundle (bundle of His), bundle branches, and Purkinje fibers.
5. Phases of an action potential in a ventricular contractile fiber include rapid depolarization, a long plateau, and repolarization.
6. Cardiac muscle tissue has a long refractory period, which prevents tetanus.
7. The record of electrical changes during each cardiac cycle is called an electrocardiogram (ECG). A normal ECG consists of a P wave (atrial depolarization), a QRS complex (onset of ventricular depolarization), and a T wave (ventricular repolarization).
8. The P–Q interval represents the conduction time from the beginning of atrial excitation to the beginning of ventricular excitation. The S–T segment represents the time when ventricular contractile fibers are fully depolarized.

20.4 The Cardiac Cycle

1. A cardiac cycle consists of the systole (contraction) and diastole (relaxation) of both atria, plus the systole and diastole of both ventricles. With an average heartbeat of 75 beats/min, a complete cardiac cycle requires 0.8 sec.
2. The phases of the cardiac cycle are (a) atrial systole, (b) ventricular systole, and (c) relaxation period.
3. S1, the first heart sound (lubb), is caused by blood turbulence associated with the closing of the atrioventricular valves. S2, the second sound (dupp), is caused by blood turbulence associated with the closing of semilunar valves.

20.5 Cardiac Output

1. Cardiac output (CO) is the amount of blood ejected per minute by the left ventricle into the aorta (or by the right ventricle into the pulmonary trunk). It is calculated as follows: CO (mL/min) = stroke volume (SV) in mL/beat × heart rate (HR) in beats/min.
2. Stroke volume (SV) is the amount of blood ejected by a ventricle during each systole.
3. Cardiac reserve is the difference between a person's maximum cardiac output and his or her cardiac output at rest.
4. Stroke volume is related to preload (stretch on the heart before it contracts), contractility (forcefulness of contraction), and afterload (pressure that must be exceeded before ventricular ejection can begin).
5. According to the Frank–Starling law of the heart, a greater preload (end-diastolic volume) stretching cardiac muscle fibers just before they contract increases their force of contraction until the stretching becomes excessive.
6. Nervous control of the cardiovascular system originates in the cardiovascular center in the medulla oblongata.
7. Sympathetic impulses increase heart rate and force of contraction; parasympathetic impulses decrease heart rate.
8. Heart rate is affected by hormones (epinephrine, norepinephrine, thyroid hormones), ions (Na^+, K^+, Ca^{2+}), age, gender, physical fitness, and body temperature.

20.6 Exercise and the Heart

1. Sustained exercise increases oxygen demand on muscles.
2. Among the benefits of aerobic exercise are increased cardiac output, decreased blood pressure, weight control, and increased fibrinolytic activity.

Resource

Anatomy Overview - Cardiac Muscle
Animation - Cardiac Conduction
Figure 20.10 - The
   Conduction System
   of the Heart
Exercise - Sequence
   Cardiac Conduction
Exercise - ECG Jigsaw Puzzle

Animation - Cardiac Cycle and ECG
Animation - Cardiac Cycle
Exercise - Cardiac Cycle
Concepts and Connections - Cardiac
   Cycle

Animation - Cardiac Output
Animation - ANS Effects on Cardiac
   Conduction
Exercise - Cardiac Output Factors
Concepts and Connections - Cardiac
   Output

Review

Resource

20.7 Help for Failing Hearts

1. A cardiac (heart) transplant is the replacement of a severely damaged heart with a normal one.
2. Cardiac assist devices and procedures include the intra-aortic balloon pump, the ventricular assist device, cardiomyoplasty, and a skeletal muscle assist device.

20.8 Development of the Heart

Partitioning of the Heart

1. The heart develops from mesoderm.
2. The endocardial tubes develop into the four-chambered heart and great vessels of the heart.

SELF-QUIZ QUESTIONS

Fill in the blanks in the following statements.

1. The chamber of the heart with the thickest myocardium is the _____. *left ventricle*

2. The phase of heart contraction is called _____; *systole* the phase of relaxation is called _____. *diastole*

Indicate whether the following statements are true or false.

3. In auscultation, the lubb represents closing of the semilunar valves and the dupp represents closing of the atrioventricular valves. *F*

4. The Frank–Starling law of the heart equalizes the output of the right and left ventricles and keeps the same volume of blood flowing to both the systemic and pulmonary circulations. *T*

Choose the one best answer to the following questions.

5. Which of the following is the correct route of blood through the heart from the systemic circulation to the pulmonary circulation and back to the systemic circulation?
   (a) right atrium, tricuspid valve, right ventricle, pulmonary semilunar valve, left atrium, mitral valve, left ventricle, aortic semilunar valve
   (b) left atrium, tricuspid valve, left ventricle, pulmonary semilunar valve, right atrium, mitral valve, right ventricle, aortic semilunar valve
   (c) left atrium, pulmonary semilunar valve, right atrium, tricuspid valve, left ventricle, aortic semilunar valve, right ventricle, mitral valve
   (d) left ventricle, mitral valve, left atrium, pulmonary semilunar valve, right ventricle, tricuspid valve, right atrium, aortic semilunar valve
   (e) right atrium, mitral valve, right ventricle, pulmonary semilunar valve, left atrium, tricuspid valve, left ventricle, aortic semilunar valve

6. Which of the following represents the correct pathway for conduction of an action potential through the heart?
   (a) AV node, AV bundle, SA node, Purkinje fibers, bundle branches
   (b) AV node, bundle branches, AV bundle, SA node, Purkinje fibers
   (c) SA node, AV node, AV bundle, bundle branches, Purkinje fibers
   (d) SA node, AV bundle, bundle branches, AV node, Purkinje fibers
   (e) SA node, AV node, Purkinje fibers, bundle branches, AV bundle

7. The external boundary between the atria and ventricles is the
   (a) anterior interventricular sulcus.
   (b) interventricular septum.
   (c) interatrial septum.
   (d) coronary sulcus.
   (e) posterior interventricular sulcus.

8. A softball player is found to have a resting cardiac output of 5.0 liters per minute and a heart rate of 50 beats per minute. What is her stroke volume? $CO = SV \times HR$
   (a) 10 mL   (b) 100 mL $5.0 = SV \times 50$
   (c) 1000 mL   (d) 250 mL
   (e) The information given is insufficient to calculate stroke volume.

9. Which of the following are true? (1) ANS regulation of heart rate originates in the cardiovascular center of the medulla oblongata. (2) Proprioceptor input is a major stimulus that accounts for the rapid rise in the heart rate at the onset of physical activity. (3) The vagus nerves release norepinephrine, causing the heart rate to increase. (4) Hormones from the adrenal medulla and the thyroid gland can increase the heart rate. (5) Hypothermia increases the heart rate.
   (a) 1, 2, 3, and 4   (b) 1, 2, and 4
   (c) 2, 3, 4, and 5   (d) 3, 5, and 6
   (e) 1, 2, 4, and 5

10. Which of the following are true concerning action potentials and contraction in the myocardium? (1) The refractory period in a cardiac muscle fiber is very brief. (2) The binding of Ca^{2+} to troponin allows the interaction of actin and myosin filaments, resulting in contraction. (3) Repolarization occurs when the voltage-gated K^+ channels open and calcium channels are closing. (4) Opening of voltage-gated fast Na^+ channels results in depolarization. (5) Opening of voltage-gated slow Ca^{2+} channels results in a period of maintained depolarization, known as the plateau.
   (a) 1, 3, and 5   (b) 2, 3, and 4
   (c) 2 and 5   (d) 3, 4, and 5
   (e) 2, 3, 4, and 5

11. Which of the following would not increase stroke volume?
   (a) increased Ca^{2+} in the interstitial fluid
   (b) epinephrine
   (c) increased K^+ in the interstitial fluid
   (d) increase in venous return
   (e) slow resting heart rate

12. Match the following:

_____ (a) indicates ventricular repolarization

_____ (b) represents the time from the beginning of ventricular depolarization to the end of ventricular repolarization

_____ (c) represents atrial depolarization

_____ (d) represents the time when the ventricular contractile fibers are fully depolarized; occurs during the plateau phase of the action potential

_____ (e) represents the onset of ventricular depolarization

_____ (f) represents the conduction time from the beginning of atrial excitation to the beginning of ventricular excitation

(1) P wave
(2) QRS complex
(3) T wave
(4) P–Q interval
(5) S–T segment
(6) Q–T interval

13. Match the following:

_____ (a) major branch from the ascending aorta; passes inferior to the left auricle

_____ (b) lies in the posterior interventricular sulcus; supplies the walls of the ventricles with oxygenated blood

_____ (c) located in the coronary sulcus on the posterior surface of the heart; receives most of the deoxygenated blood from the myocardium

_____ (d) lies in the coronary sulcus; supplies oxygenated blood to the walls of the right ventricle

_____ (e) lies in the coronary sulcus; drains the right atrium and right ventricle

_____ (f) major branch from the ascending aorta; lies inferior to the right auricle

_____ (g) lies in the posterior interventricular sulcus; drains the right and left ventricles

_____ (h) lies in the anterior interventricular sulcus; supplies oxygenated blood to the walls of both ventricles

_____ (i) lies in the anterior interventricular sulcus; drains the walls of both ventricles and the left atrium

_____ (j) lies in the coronary sulcus; supplies oxygenated blood to the walls of the left ventricle and left atrium

_____ (k) drain the right ventricle and open directly into the right atrium

(1) small cardiac vein
(2) anterior interventricular branch (left anterior descending artery)
(3) anterior cardiac veins
(4) posterior interventricular branch
(5) marginal branch
(6) circumflex branch
(7) middle cardiac vein
(8) left coronary artery
(9) right coronary artery
(10) great cardiac vein
(11) coronary sinus

14. Match the following:

_____ (a) collects oxygenated blood from the pulmonary circulation

_____ (b) pumps deoxygenated blood to the lungs for oxygenation

_____ (c) their contraction pulls on and tightens the chordae tendineae, preventing the valve cusps from everting

_____ (d) cardiac muscle tissue

_____ (e) increase blood-holding capacity of the atria

_____ (f) tendonlike cords connected to the atrioventricular valve cusps which, along with the papillary muscles, prevent valve eversion

_____ (g) the superficial dense irregular connective tissue covering of the heart

_____ (h) outer layer of the serous pericardium; is fused to the fibrous pericardium

_____ (i) endothelial cells lining the interior of the heart; are continuous with the endothelium of the blood vessels

_____ (j) pumps oxygenated blood to all body cells, except the air sacs of the lungs

_____ (k) prevents backflow of blood from the right ventricle into the right atrium

_____ (l) collects deoxygenated blood from the systemic circulation

_____ (m) left atrioventricular valve

_____ (n) the remnant of the foramen ovale, an opening in the interatrial septum of the fetal heart

_____ (o) blood vessels that pierce the heart muscle and supply blood to the cardiac muscle fibers

_____ (p) grooves on the surface of the heart which delineate the external boundaries between the chambers

_____ (q) prevent backflow of blood from the arteries into the ventricles

_____ (r) the gap junction and desmosome connections between individual cardiac muscle fibers

_____ (s) internal wall dividing the chambers of the heart

_____ (t) separate the upper and lower heart chambers, preventing backflow of blood from the ventricles back into the atria

_____ (u) inner visceral layer of the pericardium; adheres tightly to the surface of the heart

_____ (v) ridges formed by raised bundles of cardiac muscle fibers

(1) right atrium
(2) right ventricle
(3) left atrium
(4) left ventricle
(5) tricuspid valve
(6) bicuspid (mitral) valve
(7) chordae tendineae
(8) auricles
(9) papillary muscles
(10) trabeculae carneae
(11) fibrous pericardium
(12) parietal pericardium
(13) epicardium
(14) myocardium
(15) endocardium
(16) atrioventricular valves
(17) semilunar valves
(18) intercalated discs
(19) sulci
(20) septum
(21) fossa ovalis
(22) coronary circulation

15. Match the following:

3 _3_ (a) amount of blood contained in the ventricles at the end of ventricular relaxation

7 _7_ (b) period of time when cardiac muscle fibers are contracting and exerting force but not shortening

2 _6_ (c) amount of blood ejected per beat by each ventricle

5 _3_ (d) amount of blood remaining in the ventricles following ventricular contraction

1 _1_ (e) difference between a person's maximum cardiac output and cardiac output at rest

6 _2_ (f) period of time when semilunar valves are open and blood flows out of the ventricles

4&7 _4_ (g) period when all four valves are closed and ventricular blood volume does not change

(1) cardiac reserve
(2) stroke volume
(3) end-diastolic volume (EDV)
(4) isovolumetric relaxation
(5) end-systolic volume (ESV)
(6) ventricular ejection
(7) isovolumetric contraction

CRITICAL THINKING QUESTIONS

1. Gerald recently visited the dentist to have his teeth cleaned and checked. During the cleaning process, Gerald had some bleeding from his gums. A couple of days later, Gerald developed a fever, rapid heartbeat, sweating, and chills. He visited his family physician, who detected a slight heart murmur. Gerald was given antibiotics and continued to have his heart monitored. How was Gerald's dental visit related to his illness?

2. Unathletic Sylvia makes a resolution to begin an exercise program. She tells you that she wants to make her heart "beat as fast as it can" during exercise. Explain to her why that may not be a good idea.

3. Mr. Perkins is a large, 62-year-old man with a weakness for sweets and fried foods. His idea of exercise is walking to the kitchen for more potato chips to eat while he is watching sports on television. Lately, he's been troubled by chest pains when he walks up stairs. His doctor told him to quit smoking and scheduled cardiac angiography for the next week. What is involved in performing this procedure? Why did the doctor order this test?

? ANSWERS TO FIGURE QUESTIONS

20.1 The mediastinum is the anatomical region that extends from the sternum to the vertebral column, from the first rib to the diaphragm, and between the lungs.

20.2 The visceral layer of the serous pericardium (epicardium) is both a part of the pericardium and a part of the heart wall.

20.3 The coronary sulcus forms a boundary between the atria and ventricles.

20.4 The greater the workload of a heart chamber, the thicker its myocardium.

20.5 The fibrous skeleton attaches to the heart valves and prevents overstretching of the valves as blood passes through them.

20.6 The papillary muscles contract, which pulls on the chordae tendineae and prevents cusps of the atrioventricular valves from everting and letting blood flow back into the atria.

20.7 Numbers 2 (right ventricle) through 6 (left atrium) depict the pulmonary circulation; numbers 7 (left ventricle) through 1 (right atrium) depict the systemic circulation.

20.8 The circumflex artery delivers oxygenated blood to the left atrium and left ventricle.

20.9 The intercalated discs hold the cardiac muscle fibers together and enable action potentials to propagate from one muscle fiber to another.

20.10 The only electrical connection between the atria and the ventricles is the atrioventricular bundle.

20.11 The duration of an action potential is much longer in a ventricular contractile fiber (0.3 sec = 300 msec) than in a skeletal muscle fiber (1–2 msec).

20.12 An enlarged Q wave may indicate a myocardial infarction (heart attack).

20.13 Action potentials propagate most slowly through the AV node.

20.14 The amount of blood in each ventricle at the end of ventricular diastole—called the end-diastolic volume—is about 130 mL in a resting person.

20.15 The first heart sound (S1), or lubb, is associated with closure of the atrioventricular valves.

20.16 The ventricular myocardium receives innervation from the sympathetic division only.

20.17 The skeletal muscle "pump" increases stroke volume by increasing preload (end-diastolic volume).

20.18 Individuals with end-stage heart failure or severe coronary artery disease are candidates for cardiac transplantation.

20.19 The heart begins to contract by the twenty-second day of gestation.

20.20 Partitioning of the heart is complete by the end of the fifth week.

20.21 HDL removes excess cholesterol from body cells and transports it to the liver for elimination.

20.22 Coronary angiography is used to visualize many blood vessels.

20.23 Tetralogy of Fallot involves an interventricular septal defect, an aorta that emerges from both ventricles, a stenosed pulmonary valve, and an enlarged right ventricle.

20.24 Ventricular fibrillation is such a serious arrhythmia because ventricular pumping stops, blood ejection ceases, and circulatory failure and death can occur without immediate medical intervention.

21 THE CARDIOVASCULAR SYSTEM: BLOOD VESSELS AND HEMODYNAMICS

BLOOD VESSELS, HEMODYNAMICS, AND HOMEOSTASIS *Blood vessels contribute to homeostasis by providing the structures for the flow of blood to and from the heart and the exchange of nutrients and wastes in tissues. They also play an important role in adjusting the velocity and volume of blood flow.*

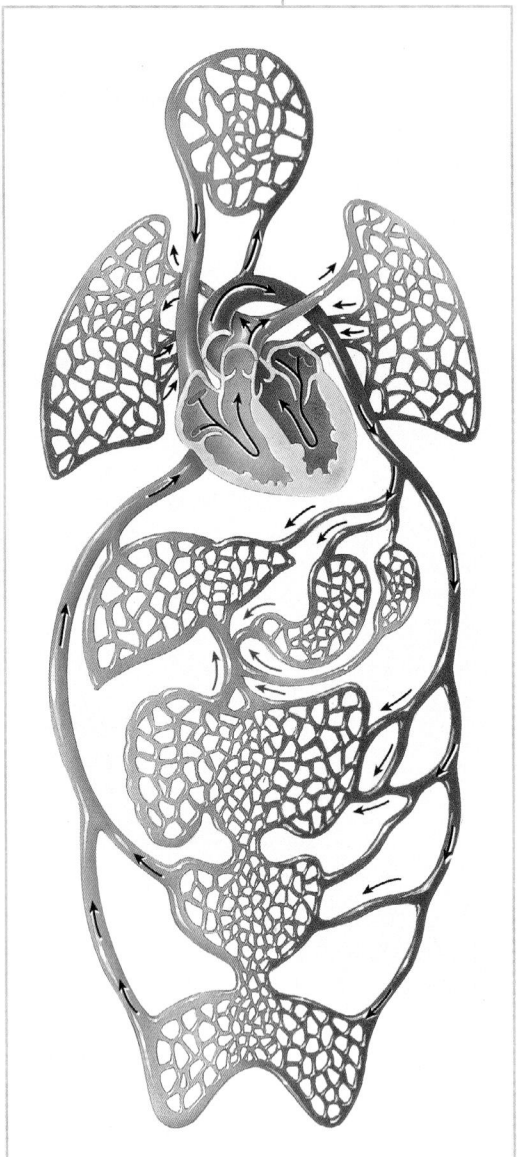

The cardiovascular system contributes to homeostasis of other body systems by transporting and distributing blood throughout the body to deliver materials (such as oxygen, nutrients, and hormones) and carry away wastes. The structures involved in these important tasks are the blood vessels, which form a closed system of tubes that carries blood away from the heart, transports it to the tissues of the body, and then returns it to the heart. The left side of the heart pumps blood through an estimated 100,000 km (60,000 mi) of blood vessels. The right side of the heart pumps blood through the lungs, enabling blood to pick up oxygen and unload carbon dioxide. Chapters 19 and 20 described the composition and functions of blood and the structure and function of the heart. In this chapter, we focus on the structure and functions of the various types of blood vessels; on **hemodynamics** (hē-mō-dī-NAM-iks; *hemo-* = blood; *-dynamics* = power), the forces involved in circulating blood throughout the body; and on the blood vessels that constitute the major circulatory routes.

Did you ever wonder why untreated hypertension has so many damaging effects?

21.1 STRUCTURE AND FUNCTION OF BLOOD VESSELS

⊙ **OBJECTIVES**

- Contrast the structure and function of arteries, arterioles, capillaries, venules, and veins.
- Outline the vessels through which the blood moves in its passage from the heart to the capillaries and back.
- Distinguish between pressure reservoirs and blood reservoirs.

The five main types of blood vessels are arteries, arterioles, capillaries, venules, and veins (see Figure 20.7). **Arteries** (AR-ter-ēz) carry blood *away from the heart* to other organs. Large, elastic arteries leave the heart and divide into medium-sized, muscular arteries that branch out into the various regions of the body. Medium-sized arteries then divide into small arteries, which in turn divide into still smaller arteries called **arterioles** (ar-TĒR-ē-ōls). As the arterioles enter a tissue, they branch into numerous tiny vessels called **capillaries** (KAP-i-lar′-ēz = hairlike). The thin walls of capillaries allow the exchange of substances between the blood and body tissues. Groups of capillaries within a tissue reunite to form small veins called **venules** (VEN-ūls). These in turn merge to form progressively larger blood vessels called veins. **Veins** (VĀNZ) are the blood vessels that convey blood from the tissues *back to the heart.*

CLINICAL CONNECTION | *Angiogenesis and Disease*

Angiogenesis (an′-jē-ō-JEN-e-sis; *angio-* = blood vessel; *-genesis* = production) refers to the growth of new blood vessels. It is an important process in embryonic and fetal development, and in postnatal life serves important functions such as wound healing, formation of a new uterine lining after menstruation, formation of the corpus luteum after ovulation, and development of blood vessels around obstructed arteries in the coronary circulation. Several proteins (peptides) are known to promote and inhibit angiogenesis.

Clinically angiogenesis is important because cells of a malignant tumor secrete proteins called *tumor angiogenesis factors (TAFs)* that stimulate blood vessel growth to provide nourishment for the tumor cells. Scientists are seeking chemicals that would inhibit angiogenesis and thus stop the growth of tumors. In *diabetic retinopathy* (ret-i-NOP-a-thē), angiogenesis may be important in the development of blood vessels that actually cause blindness, so finding inhibitors of angiogenesis may also prevent the blindness associated with diabetes. •

Basic Structure of a Blood Vessel

The wall of a blood vessel consists of three layers, or tunics, of different tissues: an epithelial inner lining, a middle layer consisting of smooth muscle and elastic connective tissue, and a connective tissue outer covering. The three structural layers of a generalized blood vessel from innermost to outermost are the tunica interna (intima), tunica media, and tunica externa (adventia) (Figure 21.1).

Modifications of this basic design account for the five types of blood vessels and the structural and functional differences among the various vessel types. Always remember that structural variations correlate to the differences in function that occur throughout the cardiovascular system.

Tunica Interna (Intima)

The **tunica interna (intima)** (TOO-ni-ka; *tunic* = garment or coat; *interna or intima* = innermost) forms the inner lining of a blood vessel and is in direct contact with the blood as it flows through the **lumen** (LOO-men), or interior opening, of the vessel (Figure 21.1a, b). Although this layer has multiple parts, these tissue components contribute minimally to the thickness of the vessel wall. Its innermost layer is called *endothelium,* which is continuous with the endocardial lining of the heart. The endothelium is a thin layer of flattened cells that lines the inner surface of the entire cardiovascular system (heart and blood vessels). Until recently, endothelial cells were regarded as little more than a passive barrier between the blood and the remainder of the vessel wall. It is now known that endothelial cells are active participants in a variety of vessel-related activities, including physical influences on blood flow, secretion of locally acting chemical mediators that influence the contractile state of the vessel's overlying smooth muscle, and assistance with capillary permeability. In addition, their smooth luminal surface facilitates efficient blood flow by reducing surface friction.

The second component of the tunica interna is a *basement membrane* deep to the endothelium. It provides a physical support base for the epithelial layer. Its framework of collagen fibers affords the basement membrane significant tensile strength, yet its properties also provide resilience for stretching and recoil. The basement membrane anchors the endothelium to the underlying connective tissue while also regulating molecular movement. It appears to play an important role in guiding cell movements during tissue repair of blood vessel walls. The outermost part of the tunica interna, which forms the boundary between the tunica interna and tunica media, is the *internal elastic lamina (lamina* = thin plate). The internal elastic lamina is a thin sheet of elastic fibers with a variable number of window-like openings that give it the look of Swiss cheese. These openings facilitate diffusion of materials through the tunica interna to the thicker tunica media.

Tunica Media

The **tunica media** (*media* = middle) is a muscular and connective tissue layer that displays the greatest variation among the different vessel types (Figure 21.1a, b). In most vessels, it is a relatively thick layer comprising mainly smooth muscle cells and substantial amounts of elastic fibers. The primary role of the smooth muscle cells, which extend circularly around the lumen like a ring encircles your finger, is to regulate the diameter of the lumen. An increase in sympathetic stimulation typically stimulates the smooth muscle to contract, squeezing the vessel wall and narrowing the lumen. Such a decrease in the diameter of the lumen of a blood vessel is called **vasoconstriction** (vā-sō-kon-STRIK-shun). In contrast, when sympathetic stimulation decreases, or in

Figure 21.1 Comparative structure of blood vessels. The capillary (c) is enlarged relative to the artery (a) and vein (b).

Arteries carry blood from the heart to tissues; veins carry blood from tissues to the heart.

TUNICA INTERNA:
Endothelium

Basement membrane

Internal elastic lamina

TUNICA MEDIA:
Smooth muscle

External elastic lamina

TUNICA EXTERNA

Valve

Lumen

(a) Artery

Lumen

(b) Vein

Lumen

Endothelium

Basement membrane

(c) Capillary

Internal elastic lamina

External elastic lamina

Tunica externa

Lumen with blood cells

Tunica interna

Tunica media

Connective tissue

LM 200x

(d) Transverse section through an artery

Connective tissue

Red blood cell

Capillary endothelial cells

LM 600x

(e) Red blood cells passing through a capillary

? Which vessel—the femoral artery or the femoral vein—has a thicker wall? Which has a wider lumen?

the presence of certain chemicals (such as nitric oxide, H^+, and lactic acid) or in response to blood pressure, smooth muscle fibers relax. The resulting increase in lumen diameter is called **vasodilation** (vā-sō-dī-LĀ-shun). As you will learn in more detail shortly, the rate of blood flow through different parts of the body is regulated by the extent of smooth muscle contraction in the walls of particular vessels. Furthermore, the extent of smooth muscle contraction in particular vessel types is crucial in the regulation of blood pressure.

In addition to regulating blood flow and blood pressure, smooth muscle contracts when an artery or arteriole is damaged *(vascular spasm)* to help limit loss of blood through the injured vessel if it is small. Smooth muscle cells also help produce the elastic fibers within the tunica media that allow the vessels to stretch and recoil under the applied pressure of the blood.

The tunica media is the most variable of the tunics. As you study the different types of blood vessels in the remainder of this chapter, you will see that the structural differences in this layer account for the many variations in function among the different vessel types. Separating the tunica media from the tunica externa is a network of elastic fibers, the *external elastic lamina,* which is part of the tunica media.

Tunica Externa

The outer covering of a blood vessel, the **tunica externa** *(externa = outermost),* consists of elastic and collagen fibers (Figure 21.1a, b). The tunica externa contains numerous nerves and, especially in larger vessels, tiny blood vessels that supply the tissue of the vessel wall. These small vessels that supply blood to the tissues of the vessel are called **vasa vasorum** (VĀ-sa va-SŌ-rum; *vas* = vessel), or vessels to the vessels. They are easily seen on large vessels such as the aorta. In addition to the important role of supplying the vessel wall with nerves and self-vessels, the tunica externa helps anchor the vessels to surrounding tissues.

Arteries

Because **arteries** *(ar- = air; -ter- = to carry)* were found empty at death, in ancient times they were thought to contain only air. The wall of an artery has the three layers of a typical blood vessel, but has a thick muscular-to-elastic tunica media (Figure 21.1a). Due to their plentiful elastic fibers, arteries normally have high *compliance,* which means that their walls stretch easily or expand without tearing in response to a small increase in pressure.

Elastic Arteries

Elastic arteries are the largest arteries in the body, ranging from the garden hose–sized aorta and pulmonary trunk to the finger-sized branches of the aorta. They have the largest diameter among arteries, but their vessel walls (approximately one-tenth of the vessel's total diameter) are relatively thin compared with the overall size of the vessel. These vessels are characterized by well-defined internal and external elastic laminae, along with a thick tunica media that is dominated by elastic fibers, called the **elastic lamellae** (la-MEL-ē = little plates). Elastic arteries

include the two major trunks that exit the heart (the aorta and the pulmonary trunk), along with the aorta's major initial branches, such as the brachiocephalic, subclavian, common carotid, and common iliac arteries (see Figure 21.19a). Elastic arteries perform an important function: They help propel blood onward while the ventricles are relaxing. As blood is ejected from the heart into elastic arteries, their walls stretch, easily accommodating the surge of blood. As they stretch, the elastic fibers momentarily store mechanical energy, functioning as a **pressure reservoir** (REZ-er-vwar) (Figure 21.2a). Then, the elastic fibers recoil and convert stored (potential) energy in the vessel into kinetic energy of the blood. Thus, blood continues to move through the arteries even while the ventricles are relaxed (Figure 21.2b). Because they conduct blood from the heart to medium-sized, more muscular arteries, elastic arteries also are called *conducting arteries.*

Figure 21.2 Pressure reservoir function of elastic arteries.

Recoil of elastic arteries keeps blood flowing during ventricular relaxation (diastole).

Aorta and elastic arteries

Blood flows toward capillaries

Left atrium

Left ventricle contracts (systole) and ejects blood

(a) Elastic aorta and arteries stretch during ventricular contraction

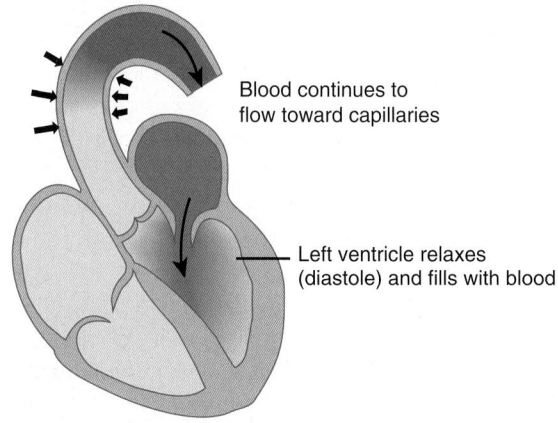

Blood continues to flow toward capillaries

Left ventricle relaxes (diastole) and fills with blood

(b) Elastic aorta and arteries recoil during ventricular relaxation

? **In atherosclerosis, the walls of elastic arteries become less compliant (stiffer). What effect does reduced compliance have on the pressure reservoir function of arteries?**

Muscular Arteries

Medium-sized arteries are called **muscular arteries** because their tunica media contains more smooth muscle and fewer elastic fibers than elastic arteries. The large amount of smooth muscle, approximately three-quarters of the total mass, makes the walls of muscular arteries relatively thick. Thus, muscular arteries are capable of greater vasoconstriction and vasodilation to adjust the rate of blood flow. Muscular arteries have a well-defined internal elastic lamina but a thin external elastic lamina. These two elastic laminae form the inner and outer boundaries of the muscular tunica media. In large arteries, the thick tunica media can have as many as 40 layers of circumferentially arranged smooth muscle cells; in smaller arteries there are as few as three layers.

Muscular arteries span a range of sizes from the pencil-sized femoral and axillary arteries to string-sized arteries that enter organs, measuring as little as 0.5 mm in diameter. Compared to elastic arteries, the vessel wall of muscular arteries comprises a larger percentage (25%) of the total vessel diameter. Because the muscular arteries continue to branch and ultimately distribute blood to each of the various organs, they are called **distributing arteries.** Examples include the brachial artery in the arm and radial artery in the forearm (see Figure 21.19a).

The tunica externa is often thicker than the tunica media in muscular arteries. This outer layer contains fibroblasts, collagen fibers, and elastic fibers all oriented longitudinally. The loose structure of this layer permits changes in the diameter of the vessel to take place but also prevents shortening or retraction of the vessel when it is cut.

Because of the reduced amount of elastic tissue in the walls of muscular arteries, these vessels do not have the ability to recoil and help propel the blood like the elastic arteries. Instead, the thick, muscular tunica media is primarily responsible for the functions of the muscular arteries. The ability of the muscle to contract and maintain a state of partial contraction is referred to as *vascular tone.* Vascular tone stiffens the vessel wall and is important in maintaining vessel pressure and efficient blood flow.

Anastomoses

Most tissues of the body receive blood from more than one artery. The union of the branches of two or more arteries supplying the same body region is called an **anastomosis** (a-nas'-tō-MŌ-sis = connecting; plural is *anastomoses*) (see Figure 21.21c). Anastomoses between arteries provide alternative routes for blood to reach a tissue or organ. If blood flow stops for a short time when normal movements compress a vessel, or if a vessel is blocked by disease, injury, or surgery, then circulation to a part of the body is not necessarily stopped. The alternative route of blood flow to a body part through an anastomosis is known as **collateral circulation.** Anastomoses may also occur between veins and between arterioles and venules. Arteries that do not anastomose are known as **end arteries.** Obstruction of an end artery interrupts the blood supply to a whole segment of an organ, producing necrosis (death) of that segment. Alternative blood routes may also be provided by nonanastomosing vessels that supply the same region of the body.

Arterioles

Literally meaning small arteries, **arterioles** are abundant microscopic vessels that regulate the flow of blood into the capillary networks of the body's tissues (see Figure 21.3). The approximately 400 million arterioles have diameters that range in size from 15 μm to 300 μm. The wall thickness of arterioles is one-half of the total vessel diameter.

Arterioles have a thin tunica interna with a thin, fenestrated (with small pores) internal elastic lamina that disappears at the terminal end. The tunica media consists of one to two layers of smooth muscle cells having a circular orientation in the vessel wall. The terminal end of the arteriole, the region called the **metarteriole** (met'-ar-TER-ē-ōl; *meta* = after), tapers toward the capillary junction. At the metarteriole–capillary junction, the distal-most muscle cell forms the **precapillary sphincter** (SFINGK-ter = to bind tight), which monitors the blood flow into the capillary; the other muscle cells in the arteriole regulate the resistance (opposition) to blood flow (see Figure 21.3).

The tunica externa of the arteriole consists of areolar connective tissue containing abundant unmyelinated sympathetic nerves. This sympathetic nerve supply, along with the actions of local chemical mediators, can alter the diameter of arterioles and thus vary the rate of blood flow and resistance through these vessels.

Arterioles play a key role in regulating blood flow from arteries into capillaries by regulating **resistance,** the opposition to blood flow. Because of this they are known as *resistance vessels.* In a blood vessel, resistance is due mainly to friction between blood and the inner walls of blood vessels. When the blood vessel diameter is smaller, the friction is greater, so there is more resistance. Contraction of the smooth muscle of an arteriole causes vasoconstriction, which increases resistance even more and decreases blood flow into capillaries supplied by that arteriole. By contrast, relaxation of the smooth muscle of an arteriole causes vasodilation, which decreases resistance and increases blood flow into capillaries. A change in arteriole diameter can also affect blood pressure: Vasoconstriction of arterioles increases blood pressure, and vasodilation of arterioles decreases blood pressure.

Capillaries

Capillaries (*capillus* = little hair), the smallest of blood vessels, have diameters of 5–10 μm, and form the U-turns that connect the arterial outflow to the venous return (Figure 21.3). Since red blood cells have a diameter of 8 μm, they must often fold on themselves in order to pass single file through the lumens of these vessels. Capillaries form an extensive network, approximately 20 billion in number, of short (hundreds of micrometers in length), branched, interconnecting vessels that course among the individual cells of the body. This network forms an enormous surface area to make contact with the body's cells. The flow of blood from a metarteriole through capillaries and into a **postcapillary venule** (venule that receives blood from a capillary) is called the **microcirculation** (*micro* = small) of the body. The primary function of capillaries is the exchange of substances between the blood and

Figure 21.3 Arteriole, capillaries, and venule. Precapillary sphincters regulate the flow of blood through capillary beds.

In capillaries, nutrients, gases, and wastes are exchanged between the blood and interstitial fluid.

(a) Sphincters relaxed: blood flowing through capillaries

(b) Sphincters contracted: blood flowing through thoroughfare channel

? **Why do metabolically active tissues have extensive capillary networks?**

interstitial fluid. Because of this, these thin-walled vessels are referred to as *exchange vessels.*

Capillaries are found near almost every cell in the body, but their number varies with the metabolic activity of the tissue they serve. Body tissues with high metabolic requirements, such as muscles, the brain, the liver, the kidneys, and the nervous system, use more O_2 and nutrients and thus have extensive capillary networks. Tissues with lower metabolic requirements, such as tendons and ligaments, contain fewer capillaries. Capillaries are absent in a few tissues, such as all covering and lining epithelia, the cornea and lens of the eye, and cartilage.

The structure of capillaries is well suited to their function as exchange vessels because they lack both a tunica media and a tunica externa. Because capillary walls are composed of only a single layer of endothelial cells (see Figure 21.1e) and a basement membrane, a substance in the blood must pass through just one cell layer to reach the interstitial fluid and tissue cells. Exchange of materials occurs only through the walls of capillaries and the beginning of venules; the walls of arteries, arterioles, most venules, and veins present too thick a barrier. Capillaries form extensive branching networks that increase the surface area available for rapid exchange of materials. In most tissues, blood flows through only a small part of the capillary network when metabolic

needs are low. However, when a tissue is active, such as contracting muscle, the entire capillary network fills with blood.

Throughout the body, capillaries function as part of a **capillary bed** (Figure 21.3), a network of 10–100 capillaries that arises from a single metarteriole. In most parts of the body, blood can flow through a capillary network from an arteriole into a venule as follows:

1. *Capillaries.* In this route, blood flows from an arteriole into capillaries and then into venules (postcapillary venules). As noted earlier, at the junctions between the metarteriole and the capillaries are rings of smooth muscle fibers called precapillary sphincters that control the flow of blood through the capillaries. When the precapillary sphincters are relaxed (open), blood flows into the capillaries (Figure 21.3a); when precapillary sphincters contract (close or partially close), blood flow through the capillaries ceases or decreases (Figure 21.3b). Typically, blood flows intermittently through capillaries due to alternating contraction and relaxation of the smooth muscle of metarterioles and the precapillary sphincters. This intermittent contraction and relaxation, which may occur 5 to 10 times per minute, is called **vasomotion** (vā-sō-MŌ-shun). In part, vasomotion is due to chemicals released by the endothelial cells;

nitric oxide is one example. At any given time, blood flows through only about 25% of the capillaries.

2. *Thoroughfare channel.* The proximal end of a metarteriole is surrounded by scattered smooth muscle fibers whose contraction and relaxation help regulate blood flow. The distal end of the vessel has no smooth muscle; it resembles a capillary and is called a **thoroughfare channel.** Such a channel provides a direct route for blood from an arteriole to a venule, thus bypassing capillaries.

The body contains three different types of capillaries: continuous capillaries, fenestrated capillaries, and sinusoids (Figure 21.4). Most capillaries are **continuous capillaries,** in which the plasma membranes of endothelial cells form a continuous tube that is interrupted only by **intercellular clefts,** gaps between neighboring endothelial cells (Figure 21.4a). Continuous capillaries are found in the central nervous system, lungs, skin, muscle tissue, and the skin.

Other capillaries of the body are **fenestrated capillaries** (fen′-es-TRĀ-ted; *fenestr-* = window). The plasma membranes of the endothelial cells in these capillaries have many **fenestrations** (fen′-es-TRĀ-shuns), small pores (holes) ranging from 70 to 100 nm in diameter (Figure 21.4b). Fenestrated capillaries are found in the kidneys, villi of the small intestine, choroid plexuses of the ventricles in the brain, ciliary processes of the eyes, and most endocrine glands.

Sinusoids (SĪ-nū-soyds; *sinus* = curve) are wider and more winding than other capillaries. Their endothelial cells may have unusually large fenestrations. In addition to having an incomplete or absent basement membrane (Figure 21.4c), sinusoids have very large intercellular clefts that allow proteins and in some cases even blood cells to pass from a tissue into the bloodstream. For example, newly formed blood cells enter the bloodstream through the sinusoids of red bone marrow. In addition, sinusoids contain specialized lining cells that are adapted to the function of the tissue. Sinusoids in the liver, for example, contain phagocytic cells that remove bacteria and other debris from the blood. The spleen, anterior pituitary, and parathyroid and adrenal glands also have sinusoids.

Usually blood passes from the heart and then in sequence through arteries, arterioles, capillaries, venules, and veins and then back to the heart. In some parts of the body, however, blood passes from one capillary network into another through a vein called a *portal vein.* Such a circulation of blood is called a **portal system.** The name of the portal system gives the name of the second capillary location. For example, there are portal systems associated with the liver (hepatic portal circulation; see Figure 21.28) and the pituitary gland (hypophyseal portal system; see Figure 18.5).

Venules

Unlike their thick-walled arterial counterparts, **venules** (= little vein) and veins have thin walls that do not readily maintain their shape. Venules drain the capillary blood and begin the return flow of blood back toward the heart (see Figure 21.3).

Figure 21.4 Types of capillaries.

 Capillaries are microscopic blood vessels that connect arterioles and venules.

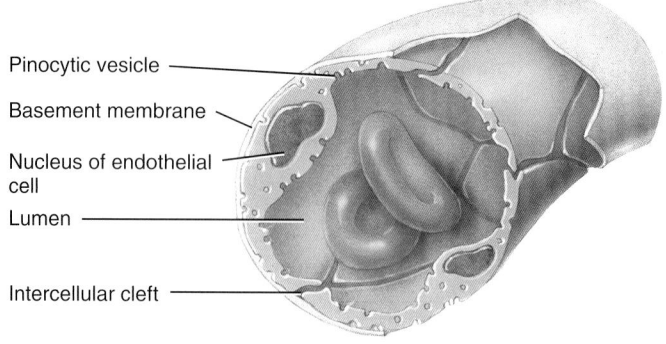

(a) Continuous capillary formed by endothelial cells

Labels: Pinocytic vesicle; Basement membrane; Nucleus of endothelial cell; Lumen; Intercellular cleft

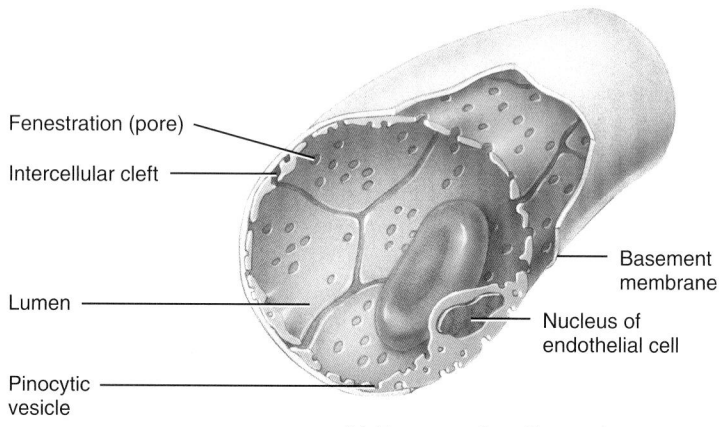

(b) Fenestrated capillary

Labels: Fenestration (pore); Intercellular cleft; Lumen; Pinocytic vesicle; Basement membrane; Nucleus of endothelial cell

(c) Sinusoid

Labels: Incomplete basement membrane; Lumen; Nucleus of endothelial cell; Intercellular cleft

? How do materials move through capillary walls?

As noted earlier, venules that initially receive blood from capillaries are called **postcapillary venules.** They are the smallest venules, measuring 10 μm to 50 μm in diameter, and have loosely organized intercellular junctions (the weakest endothelial contacts encountered along the entire vascular tree) and thus are very porous. They function as significant sites of exchange of nutrients and wastes and white blood cell emigration, and for this reason form part of the microcirculatory exchange unit along with the capillaries.

As the postcapillary venules move away from capillaries, they acquire one or two layers of circularly arranged smooth muscle cells. These **muscular venules** (50 μm to 200 μm) have thicker walls across which exchanges with the interstitial fluid can no longer occur. The thin walls of the postcapillary and muscular venules are the most distensible elements of the vascular system; this allows them to expand and serve as excellent reservoirs for accumulating large volumes of blood. Blood volume increases of 360% have been measured in the postcapillary and muscular venules.

Veins

While **veins** do show structural changes as they increase in size from small to medium to large, the structural changes are not as distinct as they are in arteries. Veins, in general, have very thin walls relative to their total diameter (average thickness is less than one-tenth of the vessel diameter). They range in size from 0.5 mm in diameter for small veins to 3 cm in the large superior and interior venae cavae entering the heart.

Although veins are composed of essentially the same three layers as arteries, the relative thicknesses of the layers are different. The tunica interna of veins is thinner than that of arteries; the tunica media of veins is much thinner than in arteries, with relatively little smooth muscle and elastic fibers. The tunica externa of veins is the thickest layer and consists of collagen and elastic fibers. Veins lack the internal or external elastic laminae found in arteries (see Figure 22.1b). They are distensible enough to adapt to variations in the volume and pressure of blood passing through them, but are not designed to withstand high pressure. The lumen of a vein is larger than that of a comparable artery, and veins often appear collapsed (flattened) when sectioned.

The pumping action of the heart is a major factor in moving venous blood back to the heart. The contraction of skeletal muscles in the lower limbs also helps boost venous return to the heart (Figure 21.9). The average blood pressure in veins is considerably lower than in arteries. The difference in pressure can be noticed when blood flows from a cut vessel. Blood leaves a cut vein in an even, slow flow but spurts rapidly from a cut artery. Most of the structural differences between arteries and veins reflect this pressure difference. For example, the walls of veins are not as strong as those of arteries.

Many veins, especially those in the limbs, also contain **valves**, thin folds of tunica interna that form flaplike cusps. The valve cusps project into the lumen, pointing toward the heart (see Figure 21.5). The low blood pressure in veins allows blood returning to the heart to slow and even back up; the valves aid in venous return by preventing the backflow of blood.

A **vascular (venous) sinus** is a vein with a thin endothelial wall that has no smooth muscle to alter its diameter. In a vascular sinus, the surrounding dense connective tissue replaces the tunica media

and tunica externa in providing support. For example, dural venous sinuses, which are supported by the dura mater, convey deoxygenated blood from the brain to the heart. Another example of a vascular sinus is the coronary sinus of the heart (see Figure 20.3c).

While veins follow paths similar to those of their arterial counterparts, they differ from arteries in a number of ways, aside from the structures of their walls. First, veins are more numerous than arteries for several reasons. Some veins are paired and accompany medium- to small-sized muscular arteries. These double sets of veins escort the arteries and connect with one another via venous channels called **anastomotic veins** (a-nas'-tō-MOT-ik). The anastomotic veins cross the accompanying artery to form ladderlike rungs between the paired veins (see Figure 21.25c). The greatest number of paired veins occurs within the limbs. The subcutaneous layer deep to the skin is another source of veins. These veins, called **superficial veins,** course through the subcutaneous layer unaccompanied by parallel arteries. Along their course, the superficial veins form small connections (anastomoses) with the **deep veins** that travel between the skeletal muscles. These connections allow communication between the deep and superficial flow of blood. The amount of blood flow through superficial veins varies from location

Figure 21.5 Venous valves.

Valves in veins allow blood to flow in one direction only—toward the heart.

Transverse plane

Vein

Frontal plane

Cusps of valve

Transverse section, superior view

Cusps of valve

Longitudinally cut

Photographs of valve in vein

? **Why are valves more important in arm veins and leg veins than in neck veins?**

to location within the body. In the upper limb, the superficial veins are much larger than the deep veins and serve as the major pathways from the capillaries of the upper limb back to the heart. In the lower limb, the opposite is true; the deep veins serve as the principal return pathways. In fact, one-way valves in small anastomosing vessels allow blood to pass from the superficial to the deep veins, but prevent the blood from passing in the reverse direction. This design has important implications in the development of varicose veins.

In some individuals the superficial veins can be seen as blue-colored tubes passing under the skin. While the venous blood is a deep dark red, the veins appear blue because their thin walls and the tissues of the skin absorb the red-light wavelengths, allowing the blue light to pass through the surface to our eyes where we see them as blue.

A summary of the distinguishing features of blood vessels is presented in Table 21.1.

CLINICAL CONNECTION | *Varicose Veins*

Leaky venous valves can cause veins to become dilated and twisted in appearance, a condition called **varicose veins** (VAR-i-kōs) or **varices** (VAR-i-sēz; *varic-* = a swollen vein). The singular is *varix* (VAR-iks). The condition may occur in the veins of almost any body part, but it is most common in the esophagus, anal canal, and superficial veins of the lower limbs. Those in the lower limbs can range from cosmetic problems to serious medical conditions. The valvular defect may be congenital or may result from mechanical stress (prolonged standing or pregnancy) or aging. The leaking venous valves allow the backflow of blood from the deep veins to the less efficient superficial veins, where the blood pools. This creates pressure that distends the vein and allows fluid to leak into surrounding tissue. As a result, the affected vein and the tissue around it may become inflamed and painfully tender. Veins close to the surface of the legs, especially the

TABLE 21.1

Distinguishing Features of Blood Vessels

BLOOD VESSEL	SIZE	TUNICA INTERNA	TUNICA MEDIA	TUNICA EXTERNA	FUNCTION
Elastic arteries	Largest arteries in the body.	Well-defined internal elastic lamina.	Thick and dominated by elastic fibers; well-defined external elastic lamina.	Thinner than tunica media.	Conduct blood from heart to muscular arteries.
Muscular arteries	Medium-sized arteries.	Well-defined internal elastic lamina.	Thick and dominated by smooth muscle; thin external elastic lamina.	Thicker than tunica media.	Distribute blood to arterioles.
Arterioles	Microscopic (15–300 μm in diameter).	Thin with a fenestrated internal elastic lamina that disappears distally.	One or two layers of circularly oriented smooth muscle; distalmost smooth muscle cell forms a precapillary sphincter.	Loose collagenous connective tissue and sympathetic nerves.	Deliver blood to capillaries and help regulate blood flow from arteries to capillaries.
Capillaries	Microscopic; smallest blood vessels (5–10 μm in diameter).	Endothelium and basement membrane.	None.	None.	Permit exchange of nutrients and wastes between blood and interstitial fluid; distribute blood to postcapillary venules.
Postcapillary venules	Microscopic (10–50 μm in diameter).	Endothelium and basement membrane.	None.	Sparse.	Pass blood into muscular venules; permit exchange of nutrients and wastes between blood and interstitial fluid and function in white blood cell emigration.
Muscular venules	Microscopic (50–200 μm in diameter).	Endothelium and basement membrane.	One or two layers of circularly oriented smooth muscle.	Sparse.	Pass blood into vein; act as reservoirs for accumulating large volumes of blood (along with postcapillary venules).
Veins	Range from 0.5 mm to 3 cm in diameter.	Endothelium and basement membrane; no internal elastic lamina; contain valves; lumen much larger than in accompanying artery.	Much thinner than in arteries; no external elastic lamina.	Thickest of the three layers.	Return blood to heart, facilitated by valves in veins in limbs.

saphenous vein, are highly susceptible to varicosities; deeper veins are not as vulnerable because surrounding skeletal muscles prevent their walls from stretching excessively. Varicose veins in the anal canal are referred to as *hemorrhoids* (HEM-o-royds). Esophageal varices result from dilated veins in the walls of the lower part of the esophagus and sometimes the upper part of the stomach. Bleeding esophageal varices are life-threatening and are usually a result of chronic liver disease.

Several treatment options are available for varicose veins in the lower limbs. *Elastic stockings* (support hose) may be used for individuals with mild symptoms or for whom other options are not recommended. *Sclerotherapy* (skle-rō-THER-a-pē) involves injection of a solution into varicose veins that damages the tunica interna by producing a harmless superficial thrombophlebitis (inflammation involving a blood clot). Healing of the damaged part leads to scar formation that occludes the vein. *Radiofrequency endovenous occlusion* (ō-KLOO-zhun) involves the application of radiofrequency energy to heat up and close off varicose veins. *Laser occlusion* uses laser therapy to shut down veins. In a surgical procedure called *stripping,* veins may be removed. In this procedure, a flexible wire is threaded through the vein and then pulled out to strip (remove) it from the body.

In some individuals the superficial veins can be seen as blue-colored tubes passing under the skin. While the venous blood is a deep dark red, the veins appear blue because their thin walls and the tissue of the skin absorb the red-light wavelengths, allowing the blue light to pass through the surface to our eyes, where we see them as blue. •

Figure 21.6 Blood distribution in the cardiovascular system at rest.

 Because systemic veins and venules contain more than half the total blood volume, they are called blood reservoirs.

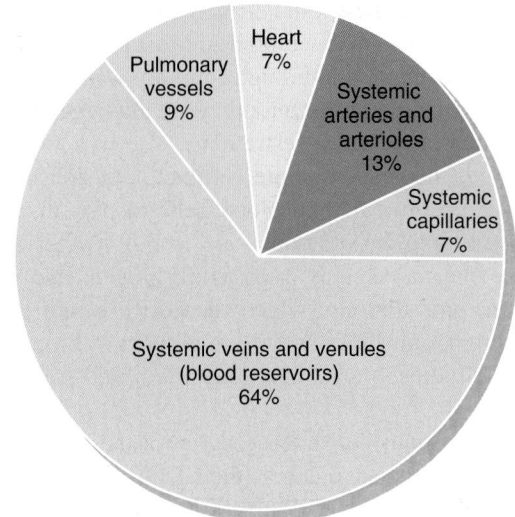

? If your total blood volume is 5 liters, what volume is in your venules and veins right now? In your capillaries?

Blood Distribution

The largest portion of your blood volume at rest—about 64%—is in systemic veins and venules (Figure 21.6). Systemic arteries and arterioles hold about 13% of the blood volume, systemic capillaries hold about 7%, pulmonary blood vessels hold about 9%, and the heart holds about 7%. Because systemic veins and venules contain a large percentage of the blood volume, they function as **blood reservoirs** from which blood can be diverted quickly if the need arises. For example, during increased muscular activity, the cardiovascular center in the brain stem sends a larger number of sympathetic impulses to veins. The result is *venoconstriction,* constriction of veins, which reduces the volume of blood in reservoirs and allows a greater blood volume to flow to skeletal muscles, where it is needed most. A similar mechanism operates in cases of hemorrhage, when blood volume and pressure decrease; in this case, venoconstriction helps counteract the drop in blood pressure. Among the principal blood reservoirs are the veins of the abdominal organs (especially the liver and spleen) and the veins of the skin.

✔ CHECKPOINT

1. What is the function of elastic fibers and smooth muscle in the tunica media of arteries?
2. How are elastic arteries and muscular arteries different?
3. What structural features of capillaries allow the exchange of materials between blood and body cells?
4. What is the difference between pressure reservoirs and blood reservoirs? Why is each important?
5. What is the relationship between anastomoses and collateral circulation?

21.2 CAPILLARY EXCHANGE

⊙ OBJECTIVE

• Discuss the pressures that cause movement of fluids between capillaries and interstitial spaces.

The mission of the entire cardiovascular system is to keep blood flowing through capillaries to allow **capillary exchange,** the movement of substances between blood and interstitial fluid. The 7% of the blood in systemic capillaries at any given time is continually exchanging materials with interstitial fluid. Substances enter and leave capillaries by three basic mechanisms: diffusion, transcytosis, and bulk flow.

Diffusion

The most important method of capillary exchange is simple diffusion. Many substances, such as oxygen (O_2), carbon dioxide (CO_2), glucose, amino acids, and hormones, enter and leave capillaries by simple diffusion. Because O_2 and nutrients normally are present in higher concentrations in blood, they diffuse down their concentration gradients into interstitial fluid and then into body cells. CO_2 and other wastes released by body cells are present in higher concentrations in interstitial fluid, so they diffuse into blood.

Substances in blood or interstitial fluid can cross the walls of a capillary by diffusing through the intercellular clefts or fenestrations or by diffusing through the endothelial cells (see Figure 21.4). Water-soluble substances such as glucose and amino acids pass across capillary walls through intercellular clefts or fenestrations. Lipid-soluble materials, such as O_2, CO_2, and steroid hormones, may pass across capillary walls directly through the lipid bilayer of endothelial cell plasma membranes. Most plasma proteins and red blood cells cannot pass through capillary walls of continuous and fenestrated capillaries because they are too large to fit through the intercellular clefts and fenestrations.

In sinusoids, however, the intercellular clefts are so large that they allow even proteins and blood cells to pass through their walls. For example, hepatocytes (liver cells) synthesize and release many plasma proteins, such as fibrinogen (the main clotting protein) and albumin, which then diffuse into the bloodstream through sinusoids. In red bone marrow, blood cells are formed (hemopoiesis) and then enter the bloodstream through sinusoids.

In contrast to sinusoids, the capillaries of the brain allow only a few substances to move across their walls. Most areas of the brain contain continuous capillaries; however, these capillaries are very "tight." The endothelial cells of most brain capillaries are sealed together by tight junctions. The resulting blockade to movement of materials into and out of brain capillaries is known as the *blood–brain barrier* (see Section 14.1). In brain areas that lack the blood–brain barrier, for example, the hypothalamus, pineal gland, and pituitary gland, materials undergo capillary exchange more freely.

Transcytosis

A small quantity of material crosses capillary walls by **transcytosis** (tranz'-sī-TŌ-sis; *trans-* = across; *-cyt-* = cell; *-osis* = process). In this process, substances in blood plasma become enclosed within tiny pinocytic vesicles that first enter endothelial cells by endocytosis, then move across the cell and exit on the other side by exocytosis. This method of transport is important mainly for large, lipid-insoluble molecules that cannot cross capillary walls in any other way. For example, the hormone insulin (a small protein) enters the bloodstream by transcytosis, and certain antibodies (also proteins) pass from the maternal circulation into the fetal circulation by transcytosis.

Bulk Flow: Filtration and Reabsorption

Bulk flow is a passive process in which *large* numbers of ions, molecules, or particles in a fluid move together in the same direction. The substances move at rates far greater than can be accounted for by diffusion alone. Bulk flow occurs from an area of higher pressure to an area of lower pressure, and it continues as long as a pressure difference exists. Diffusion is more important for *solute exchange* between blood and interstitial fluid, but bulk flow is more important for regulation of the *relative volumes of blood and interstitial fluid.* Pressure-driven movement of fluid

and solutes *from* blood capillaries *into* interstitial fluid is called **filtration.** Pressure-driven movement *from* interstitial fluid *into* blood capillaries is called **reabsorption.**

Two pressures promote filtration: blood hydrostatic pressure (BHP), the pressure generated by the pumping action of the heart, and interstitial fluid osmotic pressure (in'-ter-STISH-al). The main pressure promoting reabsorption of fluid is blood colloid osmotic pressure. The balance of these pressures, called **net filtration pressure (NFP),** determines whether the volumes of blood and interstitial fluid remain steady or change. Overall, the volume of fluid and solutes reabsorbed normally is almost as large as the volume filtered. This near equilibrium is known as **Starling's law of the capillaries.** Let's see how these hydrostatic and osmotic pressures balance.

Within vessels, the hydrostatic pressure is due to the pressure that water in blood plasma exerts against blood vessel walls. The **blood hydrostatic pressure (BHP)** is about 35 millimeters of mercury (mmHg) at the arterial end of a capillary, and about 16 mmHg at the capillary's venous end (Figure 21.7). BHP "pushes" fluid out of capillaries into interstitial fluid. The opposing pressure of the interstitial fluid, called **interstitial fluid hydrostatic pressure (IFHP),** "pushes" fluid from interstitial spaces back into capillaries. However, IFHP is close to zero. (IFHP is difficult to measure, and its reported values vary from small positive values to small negative values.) For our discussion we assume that IFHP equals 0 mmHg all along the capillaries.

The difference in osmotic pressure across a capillary wall is due almost entirely to the presence in blood of plasma proteins, which are too large to pass through either fenestrations or gaps between endothelial cells. **Blood colloid osmotic pressure (BCOP)** is a force caused by the colloidal suspension of these large proteins in plasma that averages 26 mmHg in most capillaries. The effect of BCOP is to "pull" fluid from interstitial spaces into capillaries. Opposing BCOP is **interstitial fluid osmotic pressure (IFOP),** which "pulls" fluid out of capillaries into interstitial fluid. Normally, IFOP is very small—0.1–5 mmHg—because only tiny amounts of protein are present in interstitial fluid. The small amount of protein that leaks from blood plasma into interstitial fluid does not accumulate there because it passes into lymph in lymphatic capillaries and is eventually returned to the blood. For discussion, we can use a value of 1 mmHg for IFOP.

Whether fluids leave or enter capillaries depends on the balance of pressures. If the pressures that push fluid out of capillaries exceed the pressures that pull fluid into capillaries, fluid will move from capillaries into interstitial spaces (filtration). If, however, the pressures that push fluid out of interstitial spaces into capillaries exceed the pressures that pull fluid out of capillaries, then fluid will move from interstitial spaces into capillaries (reabsorption).

The net filtration pressure (NFP), which indicates the direction of fluid movement, is calculated as follows:

$$NFP = (BHP + IFOP) - (BCOP + IFHP)$$

Pressures that promote filtration	Pressures that promote reabsorption

At the arterial end of a capillary,

$$NFP = (35 + 1)\text{ mmHg} - (26 + 0)\text{ mmHg}$$
$$= 36 - 26\text{ mmHg} = 10\text{ mmHg}$$

Thus, at the arterial end of a capillary, there is a *net outward pressure* of 10 mmHg, and fluid moves out of the capillary into interstitial spaces (filtration).

At the venous end of a capillary,

$$NFP = (16 + 1)\text{ mmHg} - (26 + 0)\text{ mmHg}$$
$$= 17 - 26\text{ mmHg} = -9\text{ mmHg}$$

At the venous end of a capillary, the negative value (−9 mmHg) represents a *net inward pressure,* and fluid moves into the capillary from tissue spaces (reabsorption).

On average, about 85% of the fluid filtered out of capillaries is reabsorbed. The excess filtered fluid and the few plasma proteins that do escape from blood into interstitial fluid enter lymphatic capillaries (see Figure 22.2). As lymph drains into the junction of the jugular and subclavian veins in the upper thorax (see Figure 22.3), these materials return to the blood. Every day about 20 liters of fluid filter out of capillaries in tissues throughout the body. Of this fluid, 17 liters are reabsorbed and 3 liters enter lymphatic capillaries (excluding filtration during urine formation).

Figure 21.7 Dynamics of capillary exchange (Starling's law of the capillaries). Excess filtered fluid drains into lymphatic capillaries.

Blood hydrostatic pressure pushes fluid out of capillaries (filtration), and blood colloid osmotic pressure pulls fluid into capillaries (reabsorption).

Key:
BHP = Blood hydrostatic pressure
IFHP = Interstitial fluid hydrostatic pressure
BCOP = Blood colloid osmotic pressure
IFOP = Interstitial fluid osmotic pressure
NFP = Net filtration pressure

Net filtration at arterial end of capillaries (20 liters per day)

Net reabsorption at venous end of capillaries (17 liters per day)

Net filtration pressure (NFP) = (BHP + IFOP) [Pressures promoting filtration] − (BCOP + IFHP) [Pressures promoting reabsorption]

Arterial end	Venous end
NFP = (35 + 1) − (26 + 0) = 10 mmHg	**NFP** = (16 + 1) − (26 + 0) = −9 mmHg
Net filtration	Net reabsorption

? **A person who has liver failure cannot synthesize the normal amount of plasma proteins. How does a deficit of plasma proteins affect blood colloid osmotic pressure, and what is the effect on capillary exchange?**

venous end of capillaries, blood pressure has dropped to about 16 mmHg. Blood pressure continues to drop as blood enters systemic venules and then veins because these vessels are farthest from the left ventricle. Finally, blood pressure reaches 0 mmHg as blood flows into the right ventricle.

Mean arterial pressure (MAP), the average blood pressure in arteries, is roughly one-third of the way between the diastolic and systolic pressures. It can be estimated as follows:

MAP = diastolic BP + 1/3 (systolic BP − diastolic BP)

Thus, in a person whose BP is 110/70 mmHg, MAP is about 83 mmHg [70 + 1/3(110 − 70)].

We have already seen that cardiac output equals heart rate multiplied by stroke volume. Another way to calculate cardiac output is to divide mean arterial pressure (MAP) by resistance (R): CO = MAP ÷ R. By rearranging the terms of this equation, you can see that MAP = CO × R. If cardiac output rises due to an increase in stroke volume or heart rate, then the mean arterial pressure rises as long as resistance remains steady. Likewise, a decrease in cardiac output causes a decrease in mean arterial pressure if resistance does not change.

Blood pressure also depends on the total volume of blood in the cardiovascular system. The normal volume of blood in an adult is about 5 liters (5.3 qt). Any decrease in this volume, as from hemorrhage, decreases the amount of blood that is circulated through the arteries each minute. A modest decrease can be compensated for by homeostatic mechanisms that help maintain blood pressure (described in Section 21.4), but if the decrease in blood volume is greater than 10% of the total, blood pressure drops. Conversely, anything that increases blood volume, such as water retention in the body, tends to increase blood pressure.

Vascular Resistance

As noted earlier, **vascular resistance** is the opposition to blood flow due to friction between blood and the walls of blood vessels. Vascular resistance depends on (1) size of the blood vessel lumen, (2) blood viscosity, and (3) total blood vessel length.

1. *Size of the lumen.* The smaller the lumen of a blood vessel, the greater its resistance to blood flow. Resistance is inversely proportional to the fourth power of the diameter (d) of the blood vessel's lumen ($R \propto 1/d^4$). The smaller the diameter of the blood vessel, the greater the resistance it offers to blood flow. For example, if the diameter of a blood vessel decreases by one-half, its resistance to blood flow increases 16 times. Vasoconstriction narrows the lumen, and vasodilation widens it. Normally, moment-to-moment fluctuations in blood flow through a given tissue are due to vasoconstriction and vasodilation of the tissue's arterioles. As arterioles dilate, resistance decreases, and blood pressure falls. As arterioles constrict, resistance increases, and blood pressure rises.

2. *Blood viscosity.* The viscosity (vis-KOS-i-tē = thickness) of blood depends mostly on the ratio of red blood cells to plasma (fluid) volume, and to a smaller extent on the concentration of proteins in plasma. The higher the blood's viscosity, the higher the resistance. Any condition that increases the viscosity of blood, such as dehydration or polycythemia (an unusually high number of red blood cells), thus increases blood pressure. A depletion of plasma proteins or red blood cells, due to anemia or hemorrhage, decreases viscosity and thus decreases blood pressure.

3. *Total blood vessel length.* Resistance to blood flow through a vessel is directly proportional to the length of the blood vessel. The longer a blood vessel, the greater the resistance. Obese people often have hypertension (elevated blood pressure) because the additional blood vessels in their adipose tissue increase their total blood vessel length. An estimated 650 km (about 400 miles) of additional blood vessels develop for each extra kilogram (2.2 lb) of fat.

Systemic vascular resistance (SVR), also known as *total peripheral resistance (TPR),* refers to all the vascular resistances offered by systemic blood vessels. The diameters of arteries and veins are large, so their resistance is very small because most of the blood does not come into physical contact with the walls of the blood vessel. The smallest vessels—arterioles, capillaries, and venules—contribute the most resistance. A major function of arterioles is to control SVR—and therefore blood pressure and blood flow to particular tissues—by changing their diameters. Arterioles need to vasodilate or vasoconstrict only slightly to have a large effect on SVR. The main center for regulation of SVR is the vasomotor center in the brain stem (described shortly).

Venous Return

Venous return, the volume of blood flowing back to the heart through the systemic veins, occurs due to the pressure generated by contractions of the heart's left ventricle. The pressure difference from venules (averaging about 16 mmHg) to the right ventricle (0 mmHg), although small, normally is sufficient to cause venous return to the heart. If pressure increases in the right atrium or ventricle, venous return will decrease. One cause of increased pressure in the right atrium is an incompetent (leaky) tricuspid valve, which lets blood regurgitate (flow backward) as the ventricles contract. The result is decreased venous return and buildup of blood on the venous side of the systemic circulation.

When you stand up, for example, at the end of an anatomy and physiology lecture, the pressure pushing blood up the veins in your lower limbs is barely enough to overcome the force of gravity pushing it back down. Besides the heart, two other mechanisms "pump" blood from the lower body back to the heart: (1) the skeletal muscle pump and (2) the respiratory pump. Both pumps depend on the presence of valves in veins.

The **skeletal muscle pump** operates as follows (Figure 21.9):

1 While you are standing at rest, both the venous valve closer to the heart (proximal valve) and the one farther from the heart (distal valve) in this part of the leg are open, and blood flows upward toward the heart.

Figure 21.9 Action of the skeletal muscle pump in returning blood to the heart.

⊙ ⟐ Milking refers to skeletal muscle contractions that drive venous blood toward the heart.

Proximal valve

Distal valve

① ② ③

❓ **Aside from cardiac contractions, what mechanisms act as pumps to boost venous return?**

② Contraction of leg muscles, such as when you stand on tiptoes or take a step, compresses the vein. The compression pushes blood through the proximal valve, an action called *milking*. At the same time, the distal valve in the uncompressed segment of the vein closes as some blood is pushed against it. People who are immobilized through injury or disease lack these contractions of leg muscles. As a result, their venous return is slower and they may develop circulation problems.

③ Just after muscle relaxation, pressure falls in the previously compressed section of vein, which causes the proximal valve to close. The distal valve now opens because blood pressure in the foot is higher than in the leg, and the vein fills with blood from the foot. The proximal valve then reopens.

The **respiratory pump** is also based on alternating compression and decompression of veins. During inhalation, the diaphragm moves downward, which causes a decrease in pressure in the thoracic cavity and an increase in pressure in the abdominal cavity. As a result, abdominal veins are compressed, and a greater volume of blood moves from the compressed abdominal veins into the decompressed thoracic veins and then into the right atrium. When the pressures reverse during exhalation, the valves in the veins prevent backflow of blood from the thoracic veins to the abdominal veins.

Figure 21.10 summarizes the factors that increase blood pressure through increasing cardiac output or systemic vascular resistance.

Velocity of Blood Flow

Earlier we saw that blood flow is the *volume* of blood that flows through any tissue in a given time period (in mL/min). The speed or *velocity* of blood flow (in cm/sec) is inversely related to the cross-sectional area. Velocity is slowest where the total cross-sectional area is greatest (Figure 21.11). Each time an artery branches, the total cross-sectional area of all its branches is greater than the cross-sectional area of the original vessel, so blood flow becomes slower and slower as blood moves further away from the heart, and is slowest in the capillaries. Conversely, when venules unite to form veins, the total cross-sectional area becomes smaller and flow becomes faster. In an adult, the cross-sectional area of the aorta is only 3–5 cm^2, and the average velocity of the blood there is 40 cm/sec. In capillaries, the total cross-sectional area is 4500–6000 cm^2, and the velocity of blood flow is less than 0.1 cm/sec. In the two venae cavae combined, the cross-sectional area is about 14 cm^2, and the velocity is about 15 cm/sec. Thus, the velocity of blood flow decreases as blood flows from the aorta to arteries to arterioles to capillaries, and increases as it leaves capillaries and returns to the heart. The relatively slow rate of flow through capillaries aids the exchange of materials between blood and interstitial fluid.

Circulation time is the time required for a drop of blood to pass from the right atrium, through the pulmonary circulation, back to the left atrium, through the systemic circulation down to the foot, and back again to the right atrium. In a resting person, circulation time normally is about 1 minute.

⚕ **CLINICAL CONNECTION | *Syncope***

Syncope (SIN-kō-pē), or fainting, is a sudden, temporary loss of consciousness that is not due to head trauma, followed by spontaneous recovery. It is most commonly due to cerebral ischemia, lack of sufficient blood flow to the brain. Syncope may occur for several reasons:

- *Vasodepressor syncope* is due to sudden emotional stress or real, threatened, or fantasized injury.
- *Situational syncope* is caused by pressure stress associated with urination, defecation, or severe coughing.
- *Drug-induced syncope* may be caused by drugs such as antihypertensives, diuretics, vasodilators, and tranquilizers.
- *Orthostatic hypotension,* an excessive decrease in blood pressure that occurs on standing up, may cause fainting. •

✔ CHECKPOINT

9. Explain how blood pressure and resistance determine volume of blood flow.
10. What is systemic vascular resistance and what factors contribute to it?
11. How is the return of venous blood to the heart accomplished?
12. Why is the velocity of blood flow faster in arteries and veins than in capillaries?

Figure 21.10 **Summary of factors that increase blood pressure.** Changes noted within green boxes increase cardiac output; changes noted within blue boxes increase systemic vascular resistance.

Increases in cardiac output and increases in systemic vascular resistance will increase mean arterial pressure.

Which type of blood vessel exerts the major control of systemic vascular resistance, and how does it achieve this?

Figure 21.11 **Relationship between velocity (speed) of blood flow and total cross-sectional area in different types of blood vessels.**

Velocity of blood flow is slowest in the capillaries because they have the largest total cross-sectional area.

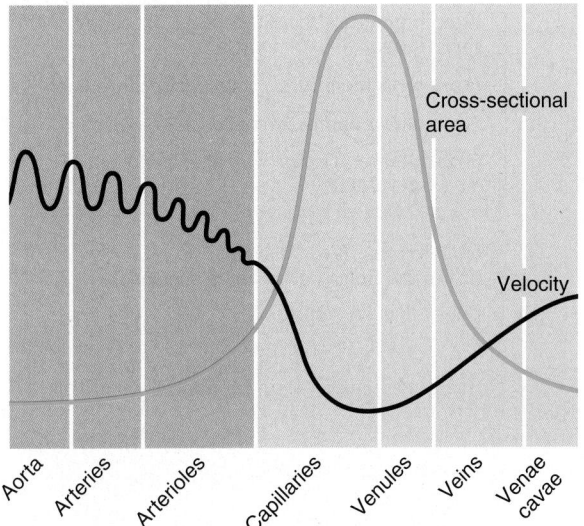

In which blood vessels is the velocity of flow fastest?

21.4 CONTROL OF BLOOD PRESSURE AND BLOOD FLOW

◉ OBJECTIVE

- Describe how blood pressure is regulated.

Several interconnected negative feedback systems control blood pressure by adjusting heart rate, stroke volume, systemic vascular resistance, and blood volume. Some systems allow rapid adjustments to cope with sudden changes, such as the drop in blood pressure in the brain that occurs when you get out of bed; others act more slowly to provide long-term regulation of blood pressure. The body may also require adjustments to the distribution of blood flow. During exercise, for example, a greater percentage of the total blood flow is diverted to skeletal muscles.

Role of the Cardiovascular Center

In Chapter 20, we noted how the **cardiovascular (CV) center** in the medulla oblongata helps regulate heart rate and stroke volume. The CV center also controls neural, hormonal, and local negative feedback systems that regulate blood pressure and blood flow to specific tissues. Groups of neurons scattered within the

CV center regulate heart rate, contractility (force of contraction) of the ventricles, and blood vessel diameter. Some neurons stimulate the heart (cardiostimulatory center); others inhibit the heart (cardioinhibitory center). Still others control blood vessel diameter by causing constriction (vasoconstrictor center) or dilation (vasodilator center); these neurons are referred to collectively as the vasomotor center. Because the CV center neurons communicate with one another, function together, and are not clearly separated anatomically, we discuss them here as a group.

The cardiovascular center receives input both from higher brain regions and from sensory receptors (Figure 21.12). Nerve impulses descend from the cerebral cortex, limbic system, and hypothalamus to affect the cardiovascular center. For example, even before you start to run a race, your heart rate may increase due to nerve impulses conveyed from the limbic system to the CV center. If your body temperature rises during a race, the hypothalamus sends nerve impulses to the CV center. The resulting vasodilation of skin blood vessels allows heat to dissipate more rapidly from the surface of the skin. The three main types of sensory receptors that provide input to the cardiovascular center are proprioceptors, baroreceptors, and chemoreceptors. *Proprioceptors* (PRŌ-prē-ō-sep′-tors) monitor movements of joints and muscles and provide input to the cardiovascular center during physical activity. Their activity accounts for the rapid increase in heart rate at the beginning of exercise. *Baroreceptors* (bar′-ō-re-SEP-tors) monitor changes in pressure and stretch in the walls of blood ves-

sels, and *chemoreceptors* (kē′-mō-rē-SEP-tors) monitor the concentration of various chemicals in the blood.

Output from the cardiovascular center flows along sympathetic and parasympathetic neurons of the ANS (Figure 21.12). Sympathetic impulses reach the heart via the **cardiac accelerator nerves.** An increase in sympathetic stimulation increases heart rate and contractility; a decrease in sympathetic stimulation decreases heart rate and contractility. Parasympathetic stimulation, conveyed along the **vagus (X) nerves,** decreases heart rate. Thus, opposing sympathetic (stimulatory) and parasympathetic (inhibitory) influences control the heart.

The cardiovascular center also continually sends impulses to smooth muscle in blood vessel walls via **vasomotor nerves** (vā-so-MŌ-tor). These sympathetic neurons exit the spinal cord through all thoracic and the first one or two lumbar spinal nerves and then pass into the sympathetic trunk ganglia (see Figure 15.2). From there, impulses propagate along sympathetic neurons that innervate blood vessels in viscera and peripheral areas. The vasomotor region of the cardiovascular center continually sends impulses over these routes to arterioles throughout the body, but especially to those in the skin and abdominal viscera. The result is a moderate state of tonic contraction or vasoconstriction, called **vasomotor tone,** that sets the resting level of systemic vascular resistance. Sympathetic stimulation of most veins causes constriction that moves blood out of venous blood reservoirs and increases blood pressure.

Figure 21.12 Location and function of the cardiovascular (CV) center in the medulla oblongata. The CV center receives input from higher brain centers, proprioceptors, baroreceptors, and chemoreceptors. Then, it provides output to the sympathetic and parasympathetic divisions of the autonomic nervous system (ANS).

The cardiovascular center is the main region for nervous system regulation of the heart and blood vessels.

? **What types of effector tissues are regulated by the cardiovascular center?**

Neural Regulation of Blood Pressure

The nervous system regulates blood pressure via negative feedback loops that occur as two types of reflexes: baroreceptor reflexes and chemoreceptor reflexes.

Baroreceptor Reflexes

Baroreceptors, pressure-sensitive sensory receptors, are located in the aorta, internal carotid arteries (arteries in the neck that supply blood to the brain), and other large arteries in the neck and chest. They send impulses to the cardiovascular center to help regulate blood pressure. The two most important **baroreceptor reflexes** are the carotid sinus reflex and the aortic reflex.

Baroreceptors in the wall of the carotid sinuses initiate the **carotid sinus reflex** (ka-ROT-id), which helps regulate blood pressure in the brain. The **carotid sinuses** are small widenings of the right and left internal carotid arteries just above the point where they branch from the common carotid arteries (Figure 21.13). Blood pressure stretches the wall of the carotid sinus, which stimulates the baroreceptors. Nerve impulses propagate from the carotid sinus baroreceptors over sensory axons in the **glossopharyngeal (IX) nerves** (glos´-ō-fa-RIN-jē-al) to the cardiovascular center in the medulla oblongata. Baroreceptors in the wall of the ascending aorta and arch of the aorta initiate the **aortic reflex,** which regulates systemic blood pressure. Nerve impulses from aortic baroreceptors reach the cardiovascular center via sensory axons of the **vagus (X) nerves.**

When blood pressure falls, the baroreceptors are stretched less, and they send nerve impulses at a slower rate to the cardiovascular center (Figure 21.14). In response, the CV center decreases parasympathetic stimulation of the heart by way of motor axons of the vagus nerves and increases sympathetic stimulation of the heart via cardiac accelerator nerves. Another consequence of increased sympathetic stimulation is increased secretion of epinephrine and norepinephrine by the adrenal medulla. As the heart beats faster and more forcefully, and as systemic vascular resistance increases, cardiac output and systemic vascular resistance rise, and blood pressure increases to the normal level.

Conversely, when an increase in pressure is detected, the baroreceptors send impulses at a faster rate. The CV center responds by increasing parasympathetic stimulation and decreasing sympathetic stimulation. The resulting decreases in heart rate and force of contraction reduce the cardiac output. The cardiovascular center also slows the rate at which it sends sympathetic impulses along vasomotor neurons that normally cause vasoconstriction. The resulting vasodilation lowers systemic vascular resistance. Decreased cardiac output and decreased systemic vascular resistance both lower systemic arterial blood pressure to the normal level.

Figure 21.13 ANS innervation of the heart and the baroreceptor reflexes that help regulate blood pressure.

Baroreceptors are pressure-sensitive neurons that monitor stretching.

Key:
Sensory (afferent) neurons ←
Motor (efferent) neurons →

Which cranial nerves conduct impulses to the cardiovascular center from baroreceptors in the carotid sinuses and the arch of the aorta?

Figure 21.14 Negative feedback regulation of blood pressure via baroreceptor reflexes.

When blood pressure decreases, heart rate increases.

Some stimulus disrupts homeostasis by

Decreasing

Blood pressure

Receptors

Baroreceptors in arch of aorta and carotid sinus are stretched less

Input — Decreased rate of nerve impulses

Control centers

CV center in medulla oblongata

and adrenal medulla

Return to homeostasis when increased cardiac output and increased vascular resistance bring blood pressure back to normal

Output

Increased sympathetic, decreased parasympathetic stimulation

Increased secretion of epinephrine and norepinephrine from adrenal medulla

Effectors

Increased stroke volume and heart rate lead to increased cardiac output (CO)

Constriction of blood vessels increases systemic vascular resistance (SVR)

Increased blood pressure

Does this negative feedback cycle represent the changes that occur when you lie down or when you stand up?

Moving from a prone (lying down) to an erect position decreases blood pressure and blood flow in the head and upper part of the body. The baroreceptor reflexes, however, quickly counteract the drop in pressure. Sometimes these reflexes operate more slowly than normal, especially in the elderly, in which case a person can faint due to reduced brain blood flow after standing up too quickly.

CLINICAL CONNECTION | *Carotid Sinus Massage and Carotid Sinus Syncope*

Because the carotid sinus is close to the anterior surface of the neck, it is possible to stimulate the baroreceptors there by putting pressure on the neck. Physicians sometimes use **carotid sinus massage,** which involves carefully massaging the neck over the carotid sinus, to slow heart rate in a person who has paroxysmal supraventricular tachycardia, a type of tachycardia that originates in the atria. Anything that stretches or puts pressure on the carotid sinus, such as hyperextension of the head, tight collars, or carrying heavy shoulder loads, may also slow heart rate and can cause **carotid sinus syncope,** fainting due to inappropriate stimulation of the carotid sinus baroreceptors. •

Chemoreceptor Reflexes

Chemoreceptors, sensory receptors that monitor the chemical composition of blood, are located close to the baroreceptors of the carotid sinus and arch of the aorta in small structures called **carotid bodies** and **aortic bodies,** respectively. These chemoreceptors detect changes in blood level of O_2, CO_2, and H^+. *Hypoxia* (lowered O_2 availability), *acidosis* (an increase in H^+ concentration), or *hypercapnia* (excess CO_2) stimulates the chemoreceptors to send impulses to the cardiovascular center. In response, the CV center increases sympathetic stimulation to arterioles and veins, producing vasoconstriction and an increase in blood pressure. These chemoreceptors also provide input to the respiratory center in the brain stem to adjust the rate of breathing.

Hormonal Regulation of Blood Pressure

As you learned in Chapter 18, several hormones help regulate blood pressure and blood flow by altering cardiac output, changing systemic vascular resistance, or adjusting the total blood volume:

1. ***Renin–angiotensin–aldosterone (RAA) system.*** When blood volume falls or blood flow to the kidneys decreases, juxtaglomerular cells in the kidneys secrete **renin** into the bloodstream. In sequence, renin and angiotensin-converting enzyme (ACE) act on their substrates to produce the active hormone **angiotensin II** (an′-jē-ō-TEN-sin), which raises blood pressure in two ways. First, angiotensin II is a potent vasoconstrictor; it raises blood pressure by increasing systemic vascular resistance. Second, it stimulates secretion of **aldosterone,** which increases reabsorption of sodium ions (Na^+) and water by the kidneys. The water reabsorption increases total blood volume, which increases blood pressure. (See Section 21.6.)

2. ***Epinephrine and norepinephrine.*** In response to sympathetic stimulation, the adrenal medulla releases epinephrine and

norepinephrine. These hormones increase cardiac output by increasing the rate and force of heart contractions. They also cause vasoconstriction of arterioles and veins in the skin and abdominal organs and vasodilation of arterioles in cardiac and skeletal muscle, which helps increase blood flow to muscle during exercise. (See Figure 18.20.)

3. *Antidiuretic hormone (ADH).* ADH is produced by the hypothalamus and released from the posterior pituitary in response to dehydration or decreased blood volume. Among other actions, ADH causes vasoconstriction, which increases blood pressure. For this reason ADH is also called **vasopressin.** (See Figure 18.9.) ADH also promotes movement of water from the lumen of kidney tubules into the bloodstream. This results in an increase in blood volume and a decrease in urine output.

4. *Atrial natriuretic peptide (ANP).* Released by cells in the atria of the heart, ANP lowers blood pressure by causing vasodilation and by promoting the loss of salt and water in the urine, which reduces blood volume.

Table 21.2 summarizes the regulation of blood pressure by hormones.

Autoregulation of Blood Pressure

In each capillary bed, local changes can regulate vasomotion. When vasodilators produce local dilation of arterioles and relaxation of precapillary sphincters, blood flow into capillary network is increased, which increases O_2 level. Vasoconstrictors have the opposite effect. The ability of a tissue to automatically adjust its blood flow to match its metabolic demands is called **autoregulation** (aw′-tō-reg′-u-LĀ-shun). In tissues such as the heart and skeletal muscle, where the demand for O_2 and nutrients and for the removal of wastes can increase as much as tenfold during physical activity, autoregulation is an important contributor to increased blood flow through the tissue. Autoregulation also controls regional blood flow in the brain; blood distribution to various parts of the brain changes dramatically for different mental and physical activities. During a conversation, for example, blood flow increases to your motor speech areas when you are talking and increases to the auditory areas when you are listening.

Two general types of stimuli cause autoregulatory changes in blood flow:

1. *Physical changes.* Warming promotes vasodilation, and cooling causes vasoconstriction. In addition, smooth muscle in arteriole walls exhibits a **myogenic response** (mī-ō-JEN-ik)—it contracts more forcefully when it is stretched and relaxes when stretching lessens. If, for example, blood flow through an arteriole decreases, stretching of the arteriole walls decreases. As a result, the smooth muscle relaxes and produces vasodilation, which increases blood flow.

2. *Vasodilating and vasoconstricting chemicals.* Several types of cells—including white blood cells, platelets, smooth muscle fibers, macrophages, and endothelial cells—release a wide variety of chemicals that alter blood-vessel diameter. Vasodilating chemicals released by metabolically active tissue cells include K^+, H^+, lactic acid (lactate), and adenosine (from ATP). Another important vasodilator released by endothelial cells is nitric oxide (NO). Tissue trauma or inflammation causes release of vasodilating kinins and histamine. Vasoconstrictors include thromboxane A2, superoxide radicals, serotonin (from platelets), and endothelins (from endothelial cells).

An important difference between the pulmonary and systemic circulations is their autoregulatory response to changes in O_2 level. The walls of blood vessels in the systemic circulation *dilate* in response to low O_2. With vasodilation, O_2 delivery increases, which restores the normal O_2 level. By contrast, the walls of blood vessels in the pulmonary circulation *constrict* in response to low levels of O_2. This response ensures that blood mostly bypasses those alveoli (air sacs) in the lungs that are poorly ventilated by fresh air. Thus, most blood flows to better-ventilated areas of the lung.

✔CHECKPOINT

13. What are the principal inputs to and outputs from the cardiovascular center?

14. Explain the operation of the carotid sinus reflex and the aortic reflex.

15. What is the role of chemoreceptors in the regulation of blood pressure?

16. How do hormones regulate blood pressure?

17. What is autoregulation, and how does it differ in the systemic and pulmonary circulations?

TABLE 21.2

Blood Pressure Regulation by Hormones

FACTOR INFLUENCING BLOOD PRESSURE	HORMONE	EFFECT ON BLOOD PRESSURE
CARDIAC OUTPUT		
Increased heart rate and contractility	Norepinephrine, epinephrine.	Increase.
SYSTEMIC VASCULAR RESISTANCE		
Vasoconstriction	Angiotensin II, antidiuretic hormone (ADH) or vasopressin, norepinephrine,* epinephrine.†	Increase.
Vasodilation	Atrial natriuretic peptide (ANP), epinephrine,† nitric oxide.	Decrease.
BLOOD VOLUME		
Blood volume increase	Aldosterone, antidiuretic hormone.	Increase.
Blood volume decrease	Atrial natriuretic peptide.	Decrease.

*Acts at α_1 receptors in arterioles of abdomen and skin.
†Acts at β_2 receptors in arterioles of cardiac and skeletal muscle; norepinephrine has a much smaller vasodilating effect.

21.5 CHECKING CIRCULATION

◉ **OBJECTIVE**
• Define pulse, and define systolic, diastolic, and pulse pressures.

Pulse

The alternate expansion and recoil of elastic arteries after each systole of the left ventricle creates a traveling pressure wave that is called the **pulse.** The pulse is strongest in the arteries closest to the heart, becomes weaker in the arterioles, and disappears altogether in the capillaries. The pulse may be felt in any artery that lies near the surface of the body that can be compressed against a bone or other firm structure. Table 21.3 depicts some common pulse points.

The pulse rate normally is the same as the heart rate, about 70 to 80 beats per minute at rest. **Tachycardia** (tak′-i-KAR-dē-a; *tachy-*

= fast) is a rapid resting heart or pulse rate over 100 beats/min. **Bradycardia** (brād′-i-KAR-dē-a; *brady-* = slow) is a slow resting heart or pulse rate under 50 beats/min. Endurance-trained athletes normally exhibit bradycardia.

Measuring Blood Pressure

In clinical use, the term **blood pressure** usually refers to the pressure in arteries generated by the left ventricle during systole and the pressure remaining in the arteries when the ventricle is in diastole. Blood pressure is usually measured in the brachial artery in the left arm (Table 21.3). The device used to measure blood pressure is a **sphygmomanometer** (sfig′-mō-ma-NOM-e-ter; *sphygmo-* = pulse; *-manometer* = instrument used to measure pressure). It consists of a rubber cuff connected to a rubber bulb that is used to inflate the cuff and a meter that registers the pressure in the cuff. With the arm resting on a table so that it is about the same level as the heart, the cuff of the sphygmomanometer is

TABLE 21.3

Pulse Points

STRUCTURE	LOCATION	STRUCTURE	LOCATION
Superficial temporal artery	Medial to ear.	**Femoral artery**	Inferior to inguinal ligament.
Facial artery	Mandible (lower jawbone) on line with corners of mouth.	**Popliteal artery**	Posterior to knee.
		Radial artery	Distal aspect of wrist.
Common carotid artery	Lateral to larynx (voice box).	**Dorsal artery of foot (dorsalis pedis artery)**	Superior to instep of foot.
Brachial artery	Medial side of biceps brachii muscle.		

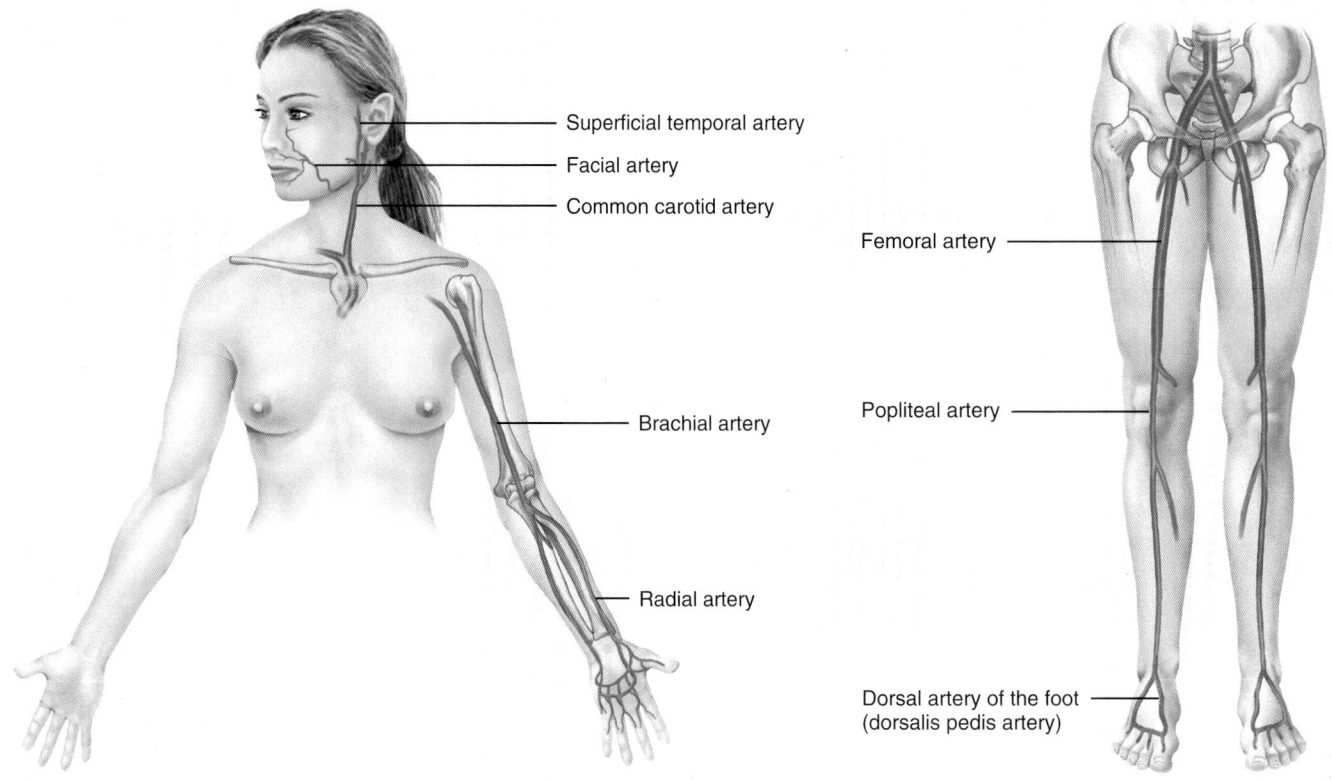

wrapped around a bared arm. The cuff is inflated by squeezing the bulb until the brachial artery is compressed and blood flow stops, about 30 mmHg higher than the person's usual systolic pressure. The technician places a stethoscope below the cuff on the brachial artery, and slowly deflates the cuff. When the cuff is deflated enough to allow the artery to open, a spurt of blood passes through, resulting in the first sound heard through the stethoscope. This sound corresponds to **systolic blood pressure (SBP),** the force of blood pressure on arterial walls just after ventricular contraction (Figure 21.15). As the cuff is deflated further, the sounds suddenly become too faint to be heard through the stethoscope. This level, called the **diastolic blood pressure (DBP),** represents the force exerted by the blood remaining in arteries during ventricular relaxation. At pressures below diastolic blood pressure, sounds disappear altogether. The various sounds that are heard while taking blood pressure are called **Korotkoff sounds** (kō-ROT-kof).

The normal blood pressure of an adult male is less than 120 mmHg systolic and less than 80 mmHg diastolic. For example, "110 over 70" (written as 110/70) is a normal blood pressure. In young adult females, the pressures are 8 to 10 mmHg less. People who exercise regularly and are in good physical condition may have even lower blood pressures. Thus, blood pressure slightly lower than 120/80 may be a sign of good health and fitness.

The difference between systolic and diastolic pressure is called **pulse pressure.** This pressure, normally about 40 mmHg, provides information about the condition of the cardiovascular system. For example, conditions such as atherosclerosis and patent (open) ductus arteriosus greatly increase pulse pressure. The normal ratio of systolic pressure to diastolic pressure to pulse pressure is about 3:2:1.

Figure 21.15 Relationship of blood pressure changes to cuff pressure.

As the cuff is deflated, sounds first occur at the systolic blood pressure; the sounds suddenly become faint at the diastolic blood pressure.

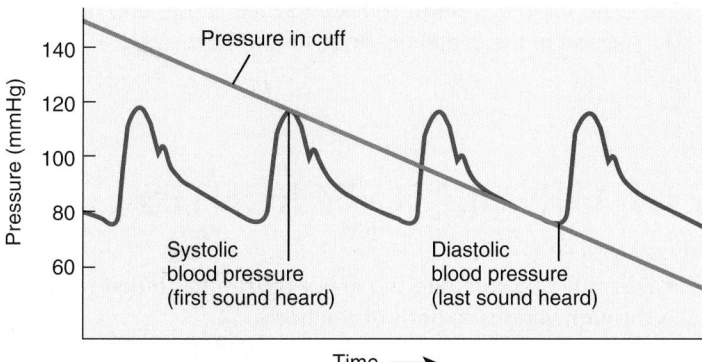

? If a blood pressure is reported as "142 over 95," what are the diastolic, systolic, and pulse pressures? Does this person have hypertension as defined in Disorders: Homeostatic Imbalances at the end of the chapter?

✔CHECKPOINT
18. Where may the pulse be felt?
19. What do tachycardia and bradycardia mean?
20. How are systolic and diastolic blood pressures measured with a sphygmomanometer?

21.6 SHOCK AND HOMEOSTASIS

◉ OBJECTIVES

- Define shock, and describe the four types of shock.
- Explain how the body's response to shock is regulated by negative feedback.

Shock is a failure of the cardiovascular system to deliver enough O_2 and nutrients to meet cellular metabolic needs. The causes of shock are many and varied, but all are characterized by inadequate blood flow to body tissues. With inadequate oxygen delivery, cells switch from aerobic to anaerobic production of ATP, and lactic acid accumulates in body fluids. If shock persists, cells and organs become damaged, and cells may die unless proper treatment begins quickly.

Types of Shock

Shock can be of four different types: (1) **hypovolemic shock** (hī-pō-vō-LĒ-mik; *hypo-* = low; *-volemic* = volume) due to decreased blood volume, (2) **cardiogenic shock** (kar′-dē-ō-JEN-ik) due to poor heart function, (3) **vascular shock** due to inappropriate vasodilation, and (4) **obstructive shock** due to obstruction of blood flow.

A common cause of hypovolemic shock is acute (sudden) hemorrhage. The blood loss may be external, as occurs in trauma, or internal, as in rupture of an aortic aneurysm. Loss of body fluids through excessive sweating, diarrhea, or vomiting also can cause hypovolemic shock. Other conditions—for instance, diabetes mellitus—may cause excessive loss of fluid in the urine. Sometimes, hypovolemic shock is due to inadequate intake of fluid. Whatever the cause, when the volume of body fluids falls, venous return to the heart declines, filling of the heart lessens, stroke volume decreases, and cardiac output decreases. Replacing fluid volume as quickly as possible is essential in managing hypovolemic shock.

In cardiogenic shock, the heart fails to pump adequately, most often because of a myocardial infarction (heart attack). Other causes of cardiogenic shock include poor perfusion of the heart (ischemia), heart valve problems, excessive preload or afterload, impaired contractility of heart muscle fibers, and arrhythmias.

Even with normal blood volume and cardiac output, shock may occur if blood pressure drops due to a decrease in systemic vascular resistance. A variety of conditions can cause inappropriate dilation of arterioles or venules. In *anaphylactic shock* (AN-a-fil-lak′-tik), a severe allergic reaction—for example, to a bee sting—releases histamine and other mediators that cause vasodilation. In *neurogenic shock,* vasodilation may occur following trauma to the head that causes malfunction of the cardiovascular center in the medulla. Shock stemming from certain bacterial toxins that produce vasodilation is termed *septic shock.* In the United States, septic shock causes more than 100,000 deaths per year and is the most common cause of death in hospital critical care units.

Obstructive shock occurs when blood flow through a portion of the circulation is blocked. The most common cause is *pulmonary embolism,* a blood clot lodged in a blood vessel of the lungs.

Homeostatic Responses to Shock

The major mechanisms of compensation in shock are *negative feedback systems* that work to return cardiac output and arterial blood pressure to normal. When shock is mild, compensation by homeostatic mechanisms prevents serious damage. In an otherwise healthy person, compensatory mechanisms can maintain adequate blood flow and blood pressure despite an acute blood loss of as much as 10% of total volume. Figure 21.16 shows several of the negative feedback systems that respond to hypovolemic shock.

1. ***Activation of the renin–angiotensin–aldosterone system.*** Decreased blood flow to the kidneys causes the kidneys to secrete renin and initiates the renin–angiotensin–aldosterone system (see Figure 18.16). Recall that angiotensin II causes vasoconstriction and stimulates the adrenal cortex to secrete aldosterone, a hormone that increases reabsorption of Na^+ and water by the kidneys. The increases in systemic vascular resistance and blood volume help raise blood pressure.

2. ***Secretion of antidiuretic hormone.*** In response to decreased blood pressure, the posterior pituitary releases more antidiuretic hormone (ADH). ADH enhances water reabsorption by the kidneys, which conserves remaining blood volume. It also causes vasoconstriction, which increases systemic vascular resistance. (See Figure 18.9.)

3. ***Activation of the sympathetic division of the ANS.*** As blood pressure decreases, the aortic and carotid baroreceptors initiate powerful sympathetic responses throughout the body. One result is marked vasoconstriction of arterioles and veins of the skin, kidneys, and other abdominal viscera. (Vasoconstriction does not occur in the brain or heart.) The constriction of arterioles increases systemic vascular resistance, and the constriction of veins increases venous return. Both effects help maintain an adequate blood pressure. Sympathetic stimulation also increases heart rate and contractility and increases secretion of epinephrine and norepinephrine by the adrenal medulla. These hormones intensify vasoconstriction and increase heart rate and contractility, all of which help raise blood pressure.

4. ***Release of local vasodilators.*** In response to *hypoxia,* cells liberate vasodilators—including K^+, H^+, lactic acid, adenosine, and nitric oxide—that dilate arterioles and relax precapillary sphincters. Such vasodilation increases local blood flow and may restore O_2 level to normal in part of the body.

However, vasodilation also has the potentially harmful effect of decreasing systemic vascular resistance and thus lowering the blood pressure.

If blood volume drops more than 10–20%, or if the heart cannot bring blood pressure up sufficiently, compensatory mechanisms may fail to maintain adequate blood flow to tissues. At this point, shock becomes life-threatening as damaged cells start to die.

Signs and Symptoms of Shock

Even though the signs and symptoms of shock vary with the severity of the condition, most can be predicted in light of the responses generated by the negative feedback systems that attempt to correct the problem. Among the signs and symptoms of shock are the following:

• Systolic blood pressure is lower than 90 mmHg.
• Resting heart rate is rapid due to sympathetic stimulation and increased blood levels of epinephrine and norepinephrine.
• Pulse is weak and rapid due to reduced cardiac output and fast heart rate.
• Skin is cool, pale, and clammy due to sympathetic constriction of skin blood vessels and sympathetic stimulation of sweating.
• Mental state is altered due to reduced oxygen supply to the brain.
• Urine formation is reduced due to increased levels of aldosterone and antidiuretic hormone (ADH).
• The person is thirsty due to loss of extracellular fluid.
• The pH of blood is low (acidosis) due to buildup of lactic acid.
• The person may have nausea because of impaired blood flow to the digestive organs from sympathetic vasoconstriction.

✔ CHECKPOINT

21. Which symptoms of hypovolemic shock relate to actual body fluid loss, and which relate to the negative feedback systems that attempt to maintain blood pressure and blood flow?
22. Describe the types of shock and their causes and how a person in hypovolemic shock should be treated.

21.7 CIRCULATORY ROUTES

◉ OBJECTIVE

• Describe and compare the major routes that blood takes through various regions of the body.

Arteries, arterioles, capillaries, venules, and veins are organized into **circulatory routes** that deliver blood throughout the body. Now that you understand the structures of each of these vessel types, we can look at the basic routes the blood takes as it is transported throughout the body.

Figure 21.16 Negative feedback systems that can restore normal blood pressure during hypovolemic shock.

Homeostatic mechanisms can compensate for an acute blood loss of as much as 10% of total blood volume.

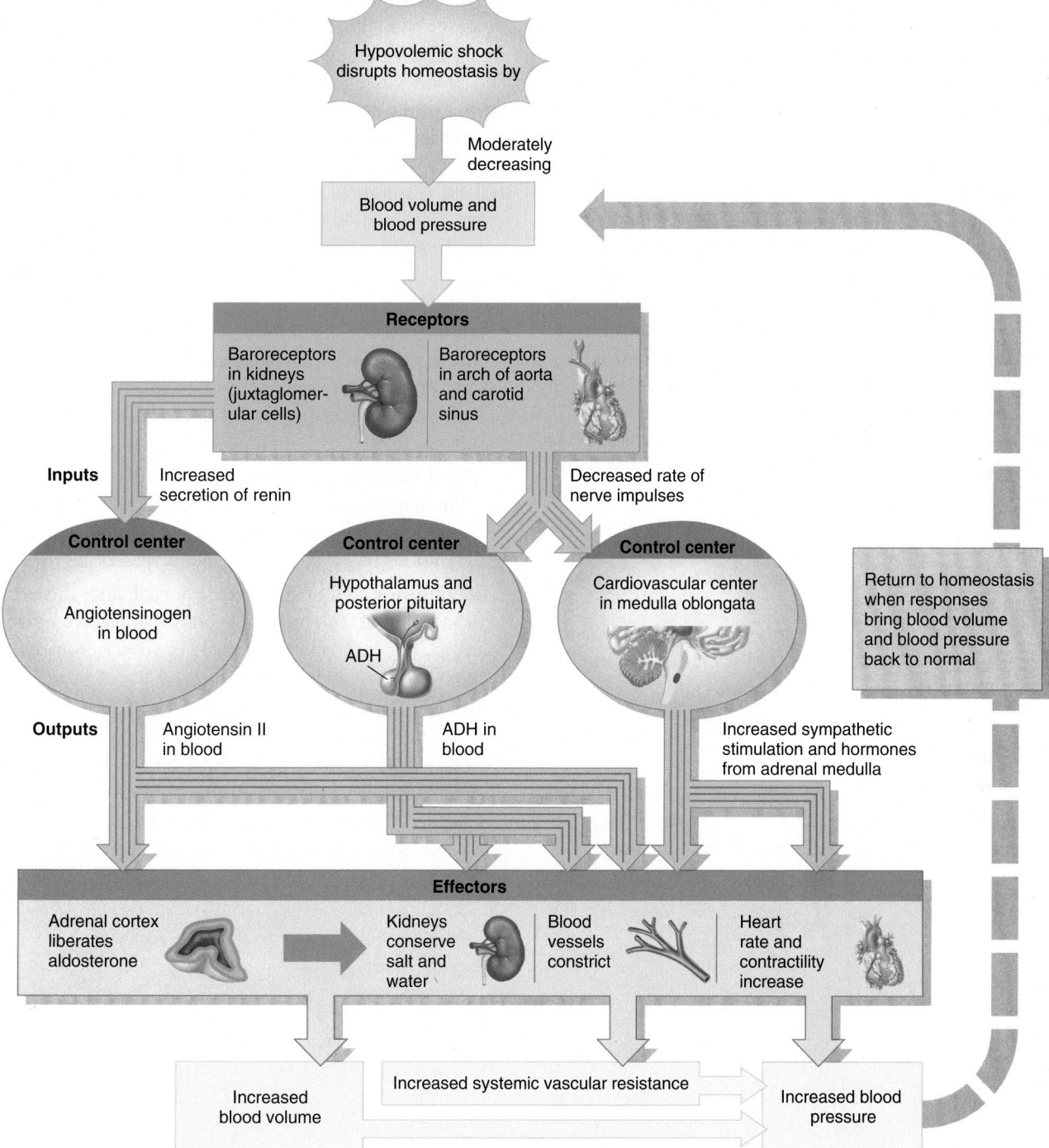

Does almost-normal blood pressure in a person who has lost blood indicate that the patient's tissues are receiving adequate perfusion (blood flow)?

Figure 21.17 shows the circulatory routes for blood flow. The routes are parallel; that is, in most cases a portion of the cardiac output flows separately to each tissue of the body. Thus, each organ receives its own supply of freshly oxygenated blood. The two basic postnatal (after birth) routes for blood flow are the systemic circulation and the pulmonary circulation. The **systemic circulation** includes all arteries and arterioles that carry oxygenated blood from the left ventricle to systemic capillaries, plus the veins and venules that carry deoxygenated blood returning to the right atrium after flowing through body organs. Blood leaving the aorta and flowing through the systemic arteries is a bright red color. As it moves through capillaries, it loses some of its oxygen and picks up carbon dioxide, so that blood in systemic veins is dark red.

Subdivisions of the systemic circulation include the **coronary (cardiac) circulation** (see Figure 20.8), which supplies the myocardium of the heart; **cerebral circulation,** which supplies the brain (see Figure 21.19c); and the **hepatic portal circulation,** which extends from the gastrointestinal tract to the liver (see Figure 21.28). The nutrient arteries to the lungs, such as the bronchial arteries, are also part of the systemic circulation.

When blood returns to the heart from the systemic route, it is pumped out of the right ventricle through the **pulmonary circulation** to the lungs (see Figure 21.29). In capillaries of the air sacs (alveoli) of the lungs, the blood loses some of its carbon dioxide and takes on oxygen. Bright red again, it returns to the left atrium of the heart and reenters the systemic circulation as it is pumped out by the left ventricle.

Another major route—the **fetal circulation**—exists only in the fetus and contains special structures that allow the developing fetus to exchange materials with its mother (see Figure 21.30).

The Systemic Circulation

The **systemic circulation** carries oxygen and nutrients to body tissues and removes carbon dioxide and other wastes and heat from the tissues. All systemic arteries branch from the **aorta.** De-oxygenated blood returns to the heart through the systemic veins. All the veins of the systemic circulation drain into the **superior vena cava** (KĀ-va), **inferior vena cava,** or **coronary sinus,** which in turn empty into the right atrium.

The principal arteries and veins of the systemic circulation are described and illustrated in Exhibits 21.A through 21.L and Figures 21.18 through 21.27 to assist you in learning their names. The blood vessels are organized in the exhibits according to regions of the body. Figure 21.18a shows an overview of the major arteries, and Figure 21.23 shows an overview of the major veins. As you study the various blood vessels in the exhibits, refer to these two figures to see the relationships of the blood vessels under consideration to other regions of the body.

Each of the exhibits contains the following information:

- **An overview.** This information provides a general orientation to the blood vessels under consideration, with emphasis on how the blood vessels are organized into various regions as well as distinguishing and/or interesting features of the blood vessels.
- **Blood vessel names.** Students often have difficulty with the pronunciations and meanings of blood vessels' names. To learn them more easily, study the phonetic pronunciations and word derivations that indicate how blood vessels get their names.
- **Region supplied or drained.** For each artery listed, there is a description of the parts of the body that receive blood from the vessel. For each vein listed, there is a description of the parts of the body that are drained by the vessel.
- **Illustrations and photographs.** The figures that accompany the exhibits contain several elements. Many include illustrations of the blood vessels under consideration and flow diagrams to indicate the patterns of blood distribution or drainage. Cadaver photographs are also included in selected exhibits to provide more realistic views of the blood vessels.

Figure 21.17 Circulatory routes. Long black arrows indicate the systemic circulation (detailed in Exhibits 21.C–21.L), short black arrows the pulmonary circulation (detailed in Figure 21.29), and red arrows the hepatic portal circulation (detailed in Figure 21.28). Refer to Figure 20.8 for details of the coronary circulation, and to Figure 21.30 for details of the fetal circulation.

Blood vessels are organized into various routes that deliver blood to tissues of the body.

- = Oxygenated blood
- = Deoxygenated blood

Systemic capillaries of head, neck, and upper limbs

Internal jugular vein

Common carotid artery

Aorta

Left pulmonary artery

Pulmonary trunk

Pulmonary capillaries of left lung

Superior vena cava

Left pulmonary veins

Right atrium

Left atrium

Right ventricle

Left ventricle

Inferior vena cava

Celiac trunk

Hepatic vein

Common hepatic artery

Splenic artery

Left gastric artery

Sinusoids of liver

Capillaries of spleen

Capillaries of stomach

Hepatic portal vein

Superior mesenteric artery

Systemic capillaries of gastrointestinal tract

Inferior mesenteric artery

Common iliac vein

Internal iliac vein

Common iliac artery

External iliac vein

Internal iliac artery

Systemic capillaries of pelvis

External iliac artery

Venules

Arterioles

Systemic capillaries of lower limbs

? **What are the two main circulatory routes?**

EXHIBIT 21.A The Aorta and Its Branches *(Figure 21.18)*

◉ **OBJECTIVE**

- Identify the four principal divisions of the aorta.
- Locate the major arterial branches arising from each division.

The **aorta** (ā-OR-ta = to lift up) is the largest artery of the body, with a diameter of 2–3 cm (about 1 in.). Its four principal divisions are the ascending aorta, arch of the aorta, thoracic aorta, and abdominal aorta (Figure 21.18). The portion of the aorta that emerges from the left ventricle posterior to the pulmonary trunk is the **ascending aorta.** The beginning of the aorta contains the aortic valve (see Figure 20.4a). The ascending aorta gives off two coronary arteries that supply the myocardium of the heart. Then the ascending aorta arches to the left, forming the **arch of the aorta,** which descends and ends at the level of the intervertebral disc between the fourth and fifth thoracic vertebrae. As

the aorta continues to descend, it lies close to the vertebral bodies and is called the **thoracic aorta.** When the thoracic aorta reaches the bottom of the thorax it passes through the aortic hiatus of the diaphragm to become the **abdominal aorta.** The abdominal aorta descends to the level of the fourth lumbar vertebra where it divides into two **common iliac arteries,** which carry blood to the pelvis and lower limbs. Each division of the aorta gives off arteries that branch into distributing arteries that lead to various organs. Within the organs, the arteries divide into arterioles and then into capillaries that service the systemic tissues (all tissues except the alveoli of the lungs).

✔ **CHECKPOINT**

What general regions do each of the four principal divisions of the aorta supply?

DIVISION AND BRANCHES	REGION SUPPLIED
ASCENDING AORTA	
Right and left coronary arteries	Heart.
ARCH OF THE AORTA	
Brachiocephalic trunk (brā′-kē-ō-se-FAL-ik)	
Right common carotid artery (ka-ROT-id)	Right side of head and neck.
Right subclavian artery (sub-KLĀ-vē-an)	Right upper limb.
Left common carotid artery	Left side of head and neck.
Left subclavian artery	Left upper limb.
THORACIC AORTA (*THORAC-* = CHEST)	
Pericardial arteries (per-i-KAR-dē-al)	Pericardium.
Bronchial arteries (BRONG-kē-al)	Bronchi of lungs.
Esophageal arteries (e-sof′-a-JĒ-al)	Esophagus.
Mediastinal arteries (mē′-dē-as-TĪ-nal)	Structures in mediastinum.
Posterior intercostal arteries (in′-ter-KOS-tal)	Intercostal and chest muscles.
Subcostal arteries (sub-KOS-tal)	Same as posterior intercostals.
Superior phrenic arteries (FREN-ik)	Superior and posterior surfaces of diaphragm.
ABDOMINAL AORTA	
Inferior phrenic arteries (FREN-ik)	Inferior surface of diaphragm.
Celiac trunk (SĒ-lē-ak)	
Common hepatic artery (he-PAT-ik)	Liver, stomach, duodenum, and pancreas.
Left gastric artery (GAS-trik)	Stomach and esophagus.
Splenic artery (SPLĒN-ik)	Spleen, pancreas, and stomach.
Superior mesenteric artery (MES-en-ter′-ik)	Small intestine, cecum, ascending and transverse colons, and pancreas.
Suprarenal arteries (soo-pra-RĒ-nal)	Adrenal (suprarenal) glands.
Renal arteries (RĒ-nal)	Kidneys.
Gonadal arteries (gō-NAD-al)	
Testicular arteries (tes-TIK-ū-lar)	Testes (male).
Ovarian arteries (ō-VAR-ē-an)	Ovaries (female).
Inferior mesenteric artery	Transverse, descending, and sigmoid colons; rectum.
Common iliac arteries (IL-ē-ak)	
External iliac arteries	Lower limbs.
Internal iliac arteries	Uterus (female), prostate (male), muscles of buttocks, and urinary bladder.

Figure 21.18 Aorta and its principal branches.

All systemic arteries branch from the aorta.

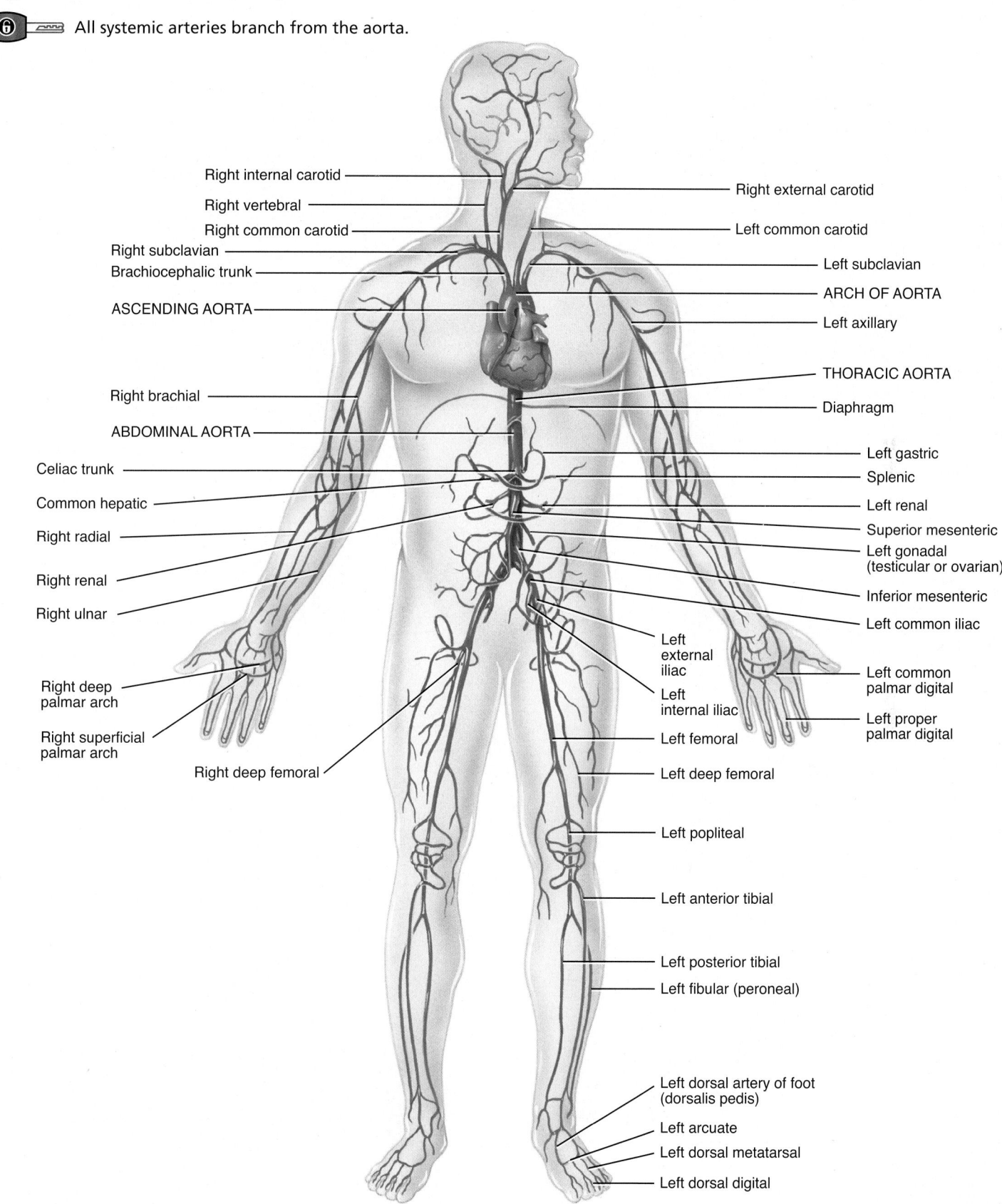

Right internal carotid

Right vertebral

Right common carotid

Right subclavian

Brachiocephalic trunk

ASCENDING AORTA

Right brachial

ABDOMINAL AORTA

Celiac trunk

Common hepatic

Right radial

Right renal

Right ulnar

Right deep palmar arch

Right superficial palmar arch

Right deep femoral

Right external carotid

Left common carotid

Left subclavian

ARCH OF AORTA

Left axillary

THORACIC AORTA

Diaphragm

Left gastric

Splenic

Left renal

Superior mesenteric

Left gonadal (testicular or ovarian)

Inferior mesenteric

Left common iliac

Left external iliac

Left internal iliac

Left common palmar digital

Left proper palmar digital

Left femoral

Left deep femoral

Left popliteal

Left anterior tibial

Left posterior tibial

Left fibular (peroneal)

Left dorsal artery of foot (dorsalis pedis)

Left arcuate

Left dorsal metatarsal

Left dorsal digital

(a) Overall anterior view of the principal branches of aorta

EXHIBIT 21.A CONTINUES

EXHIBIT 21.A **829**

FIGURE 21.18 CONTINUED

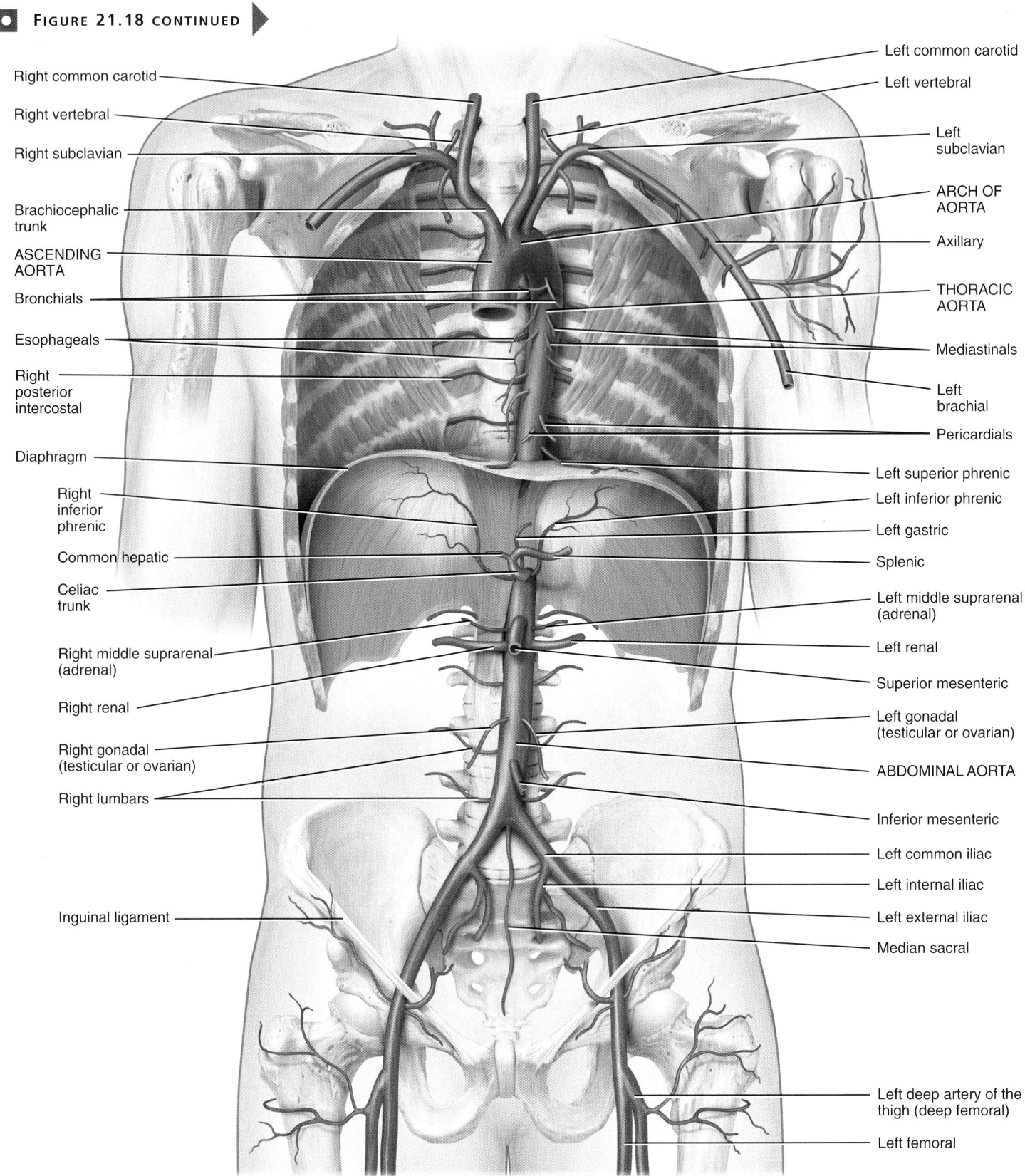

Right common carotid

Right vertebral

Right subclavian

Brachiocephalic trunk

ASCENDING AORTA

Bronchials

Esophageals

Right posterior intercostal

Diaphragm

Right inferior phrenic

Common hepatic

Celiac trunk

Right middle suprarenal (adrenal)

Right renal

Right gonadal (testicular or ovarian)

Right lumbars

Inguinal ligament

Left common carotid

Left vertebral

Left subclavian

ARCH OF AORTA

Axillary

THORACIC AORTA

Mediastinals

Left brachial

Pericardials

Left superior phrenic

Left inferior phrenic

Left gastric

Splenic

Left middle suprarenal (adrenal)

Left renal

Superior mesenteric

Left gonadal (testicular or ovarian)

ABDOMINAL AORTA

Inferior mesenteric

Left common iliac

Left internal iliac

Left external iliac

Median sacral

Left deep artery of the thigh (deep femoral)

Left femoral

(b) Detailed anterior view of the principal branches of the aorta

? **What are the four subdivisions of the aorta?**

EXHIBIT 21.B   Ascending Aorta

OBJECTIVE
- Identify the two primary arterial branches of the ascending aorta.

The **ascending aorta** is about 5 cm (2 in.) in length and begins at the aortic valve (see Figure 20.8). It is directed superiorly, slightly anteriorly, and to the right. It ends at the level of the sternal angle, where it becomes the arch of the aorta. The beginning of the ascending aorta is posterior to the pulmonary trunk and right auricle; the right pulmonary artery is posterior to it. At its origin, the ascending aorta contains three dilations called *aortic sinuses*. Two of these, the right and left sinuses, give rise to the right and left coronary arteries, respectively.

The right and left **coronary arteries** (*coron-* = crown) arise from the ascending aorta just superior to the aortic valve (see Figure 20.8). They

form a crownlike ring around the heart, giving off branches to the atrial and ventricular myocardium. The **posterior interventricular branch** (in-ter-ven-TRIK-ū-lar; *inter-* = between) of the right coronary artery supplies both ventricles, and the **marginal branch** supplies the right ventricle. The **anterior interventricular branch,** also known as the **left anterior descending (LAD) branch,** of the left coronary artery supplies both ventricles, and the **circumflex branch** (SER-kum-flex; *circum-* = around; *-flex* = to bend) supplies the left atrium and left ventricle.

✔CHECKPOINT
Which branches of the coronary arteries supply the left ventricle? Why does the left ventricle have such an extensive arterial blood supply?

SCHEME OF DISTRIBUTION

Ascending aorta

→ Right coronary artery
→ Left coronary artery

Right coronary artery → Posterior interventricular branch / Marginal branch

Left coronary artery → Anterior interventricular branch / Circumflex branch

Arch of aorta

ASCENDING AORTA

Pulmonary trunk

Right atrium

POSTERIOR INTER-VENTRICULAR BRANCH

Right ventricle

LEFT CORONARY

RIGHT CORONARY

Left auricle

Right auricle

CIRCUMFLEX BRANCH

ANTERIOR INTERVENTRICULAR BRANCH

ANTERIOR CARDIAC VEIN

MARGINAL BRANCH

Left ventricle

Arch of aorta

Left pulmonary artery

ASCENDING AORTA

Pulmonary trunk

Left auricle

GREAT CARDIAC VEIN

LEFT MARGINAL BRANCH

TRIBUTARY TO GREAT CARDIAC VEIN

Left ventricle

Right ventricle

Anterior view of coronary arteries and their major branches

EXHIBIT 21.B   **831**

EXHIBIT 21.C

The Arch of the Aorta *(Figure 21.19)*

OBJECTIVE

- Identify the three principal arteries that branch from the arch of the aorta.

The **arch of the aorta** is 4–5 cm (almost 2 in.) in length and is the continuation of the ascending aorta. It emerges from the pericardium posterior to the sternum at the level of the sternal angle (Figure 21.19). The arch of the aorta is directed superiorly and posteriorly to the left and then inferiorly; it ends at the intervertebral disc between the fourth and fifth thoracic vertebrae, where it becomes the thoracic aorta. Three major arteries branch from the superior aspect of the arch of the aorta: the brachiocephalic trunk, the left common carotid, and the left subclavian. The first and largest branch from the arch of the aorta is the **brachiocephalic trunk** (brā′-kē-ō-se-FAL-ik; *brachio-* = arm; *-cephalic* = head). It extends superiorly, bending slightly to the right, and divides at the right sternoclavicular joint to form the right subclavian artery and right common carotid artery. The second branch from the arch of the aorta is the **left common carotid artery** (ka-ROT-id), which divides into the same branches with the same names as the right common carotid artery. The third branch from the arch of the aorta is the **left subclavian artery** (sub-KLĀ-vē-an), which distributes blood to the left vertebral artery and vessels of the left upper limb. Arteries branching from the left subclavian artery are similar in distribution and name to those branching from the right subclavian artery.

✔ CHECKPOINT

What general regions do the arteries that arise from the arch of the aorta supply?

BRANCH	DESCRIPTION AND BRANCHES	REGIONS SUPPLIED
Brachiocephalic	First branch of arch of the aorta; divides to form right subclavian artery and right common carotid artery (Figure 21.19a).	Head, neck, upper limb, and thoracic wall.
Right subclavian artery* (sub-KLĀ-ve-an)	Extends from brachiocephalic artery to inferior border of first rib; gives rise to a number of branches at base of neck.	Brain, spinal cord, neck, shoulder, thoracic muscle wall, and scapular muscles.
Internal thoracic (mammary) artery (thor-AS-ik; *thorac-* = chest)	Arises from first part of subclavian artery and descends posterior to costal cartilages of superior six ribs just lateral to sternum; terminates at sixth intercostal space by bifurcating (branching into two arteries) and sends branches into intercostal spaces. **Clinical note:** In **coronary artery bypass grafting,** if only a single vessel is obstructed, the internal thoracic (usually the left) is used to create the bypass. The upper end of the artery is left attached to the subclavian artery and the cut end is connected to the coronary artery at a point distal to the blockage. The lower end of the internal thoracic is tied off. Artery grafts are preferred over vein grafts because arteries can withstand the greater pressure of blood flowing through coronary arteries and are less likely to become obstructed over time.	Anterior thoracic wall.
Vertebral artery (VER-te-bral)	Major branch to brain of right subclavian artery before it passes into axilla (Figure 21.19b); ascends through neck, passes through transverse foramina of cervical vertebrae, and enters skull via foramen magnum to reach inferior surface of brain. Unites with left vertebral artery to form **basilar** (BĀS-i-lar) **artery.** Basilar artery passes along midline of anterior aspect of brain stem and gives off several branches (**posterior cerebral** and **cerebellar arteries**).	Posterior portion of cerebrum, cerebellum, pons, and inner ear.
Axillary artery* (AK-sil-ar-ē = armpit)	Continuation of right subclavian artery into axilla; begins where subclavian artery passes inferior border of first rib and ends as it crosses distal margin of teres major muscle; gives rise to numerous branches in axilla.	Thoracic, shoulder, and scapular muscles and humerus.
Brachial artery* (BRĀ-kē-al = arm)	Continuation of axillary artery into arm; begins at distal border of teres major muscle and terminates by bifurcating into radial and ulnar arteries just distal to bend of elbow; superficial and palpable along medial side of arm. As it descends toward elbow it curves laterally and passes through cubital fossa, a triangular depression anterior to elbow, where you can easily detect pulse of brachial artery and listen to various sounds when taking a person's blood pressure. **Clinical note: Blood pressure** is usually measured in the brachial artery. In order to control hemorrhage, the best place to compress the brachial artery is near the middle of the arm where it is superficial and easily pressed against the humerus.	Muscles of arm, humerus, elbow joint.

*This is an example of the practice of giving the same vessel different names as it passes through different regions. See the axillary and brachial arteries.

BRANCH	DESCRIPTION AND BRANCHES	REGIONS SUPPLIED
Radial artery (RĀ-dē-al = radius)	Smaller branch of brachial bifurcation; a direct continuation of brachial artery. Passes along lateral (radial) aspect of forearm and enters wrist where it bifurcates into superficial and deep branches that anastomose with corresponding branches of ulnar artery to form palmar arches of hand. Makes contact with distal end of radius at wrist, where it is covered only by fascia and skin. **Clinical note:** Because of its superficial location at this point, it is a common site for measuring **radial pulse.**	Major blood source to muscles of posterior compartment of forearm.
Ulnar artery (UL-nar = ulna)	Larger branch of brachial artery passes along medial (ulnar) aspect of forearm and then into wrist, where it branches into superficial and deep branches that enter hand. These branches anastomose with corresponding branches of radial artery to form palmar arches of hand.	Major blood source to muscles of anterior compartment of forearm.
Superficial palmar arch (*palma* = palm)	Formed mainly by superficial branch of ulnar artery, with contribution from superficial branch of radial artery; superficial to long flexor tendons of fingers and extends across palm at bases of metacarpals; gives rise to **common palmar digital arteries,** each of which divides into **proper palmar digital arteries.**	Muscles, bones, joints, and skin of palm and fingers.
Deep palmar arch	Arises mainly from deep branch of radial artery, but receives contribution from deep branch of ulnar artery; deep to long flexor tendons of fingers and extends across palm just distal to bases of metacarpals; gives rise to **palmar metacarpal arteries,** which anastomose with common palmar digital arteries from superficial arch.	Muscles, bones, and joints of palm and fingers.
Right common carotid	Begins at bifurcation of brachiocephalic trunk, posterior to right sternoclavicular joint; passes superiorly into neck to supply structures in head (Figure 21.19c); divides into right external and right internal carotid arteries at superior border of larynx (voice box). **Clinical note: Pulse** may be detected in the common carotid artery, just lateral to the larynx. It is convenient to detect a carotid pulse when exercising or when administering cardiopulmonary resuscitation.	Head and neck.
External carotid artery	Begins at superior border of larynx and terminates near temporomandibular joint of parotid gland, where it divides into two branches: superficial temporal and maxillary arteries. **Clinical note:** The **carotid pulse** can be detected in the external carotid artery just anterior to the sternocleidomastoid muscle at the superior border of the larynx.	Major blood source to all structures of head except brain. Supplies skin, connective tissues, muscles, bones, joints, dura and arachnoid mater in head and supplies much of neck anatomy.
Internal carotid artery	Arises from common carotid artery; enters cranial cavity through carotid foramen in temporal bone and emerges in cranial cavity near base of sella turcica of sphenoid bone; gives rise to numerous branches inside cranial cavity and terminates as anterior cerebral arteries. The **anterior cerebral artery** passes forward toward frontal lobe of and cerebrum and **middle cerebral artery** passes laterally between temporal and parietal lobes of cerebrum. Inside cranium (Figure 21.19c), anastomoses of left and right internal carotid arteries via anterior communicating artery between two anterior cerebral arteries, along with internal carotid–basilar artery anastomoses, form an arrangement of blood vessels at base of brain called **cerebral arterial circle (circle of Willis)** (Figure 21.19c). Internal carotid–basilar anastomosis occurs where **posterior communicating arteries** arising from internal carotid artery anastomose with posterior cerebral arteries from basilar artery to link internal carotid blood supply with vertebral blood supply. Cerebral arterial circle equalizes blood pressure to brain and provides alternate routes for blood flow to brain, should arteries become damaged.	Eyeball and other orbital structures, ear, and parts of nose and nasal cavity. Frontal, temporal, parietal lobes of the cerebrum of brain, pituitary gland, and pia mater.
Left common carotid artery	Arises as second branch of arch of the aorta and ascends through mediastinum to enter neck deep to clavicle, then follows similar path to right common carotid artery.	Distribution similar to right common carotid artery.
Left subclavian artery	Arises as third and final branch of arch of the aorta; passes superior and lateral through mediastinum and deep to clavicle at base of neck as it courses toward upper limb; has similar course to right subclavian artery after leaving mediastinum.	Distribution similar to right subclavian artery.

EXHIBIT 21.C CONTINUES

EXHIBIT 21.C **833**

EXHIBIT 21.C   The Arch of the Aorta *(Figure 21.19)*   CONTINUED

SCHEME OF DISTRIBUTION

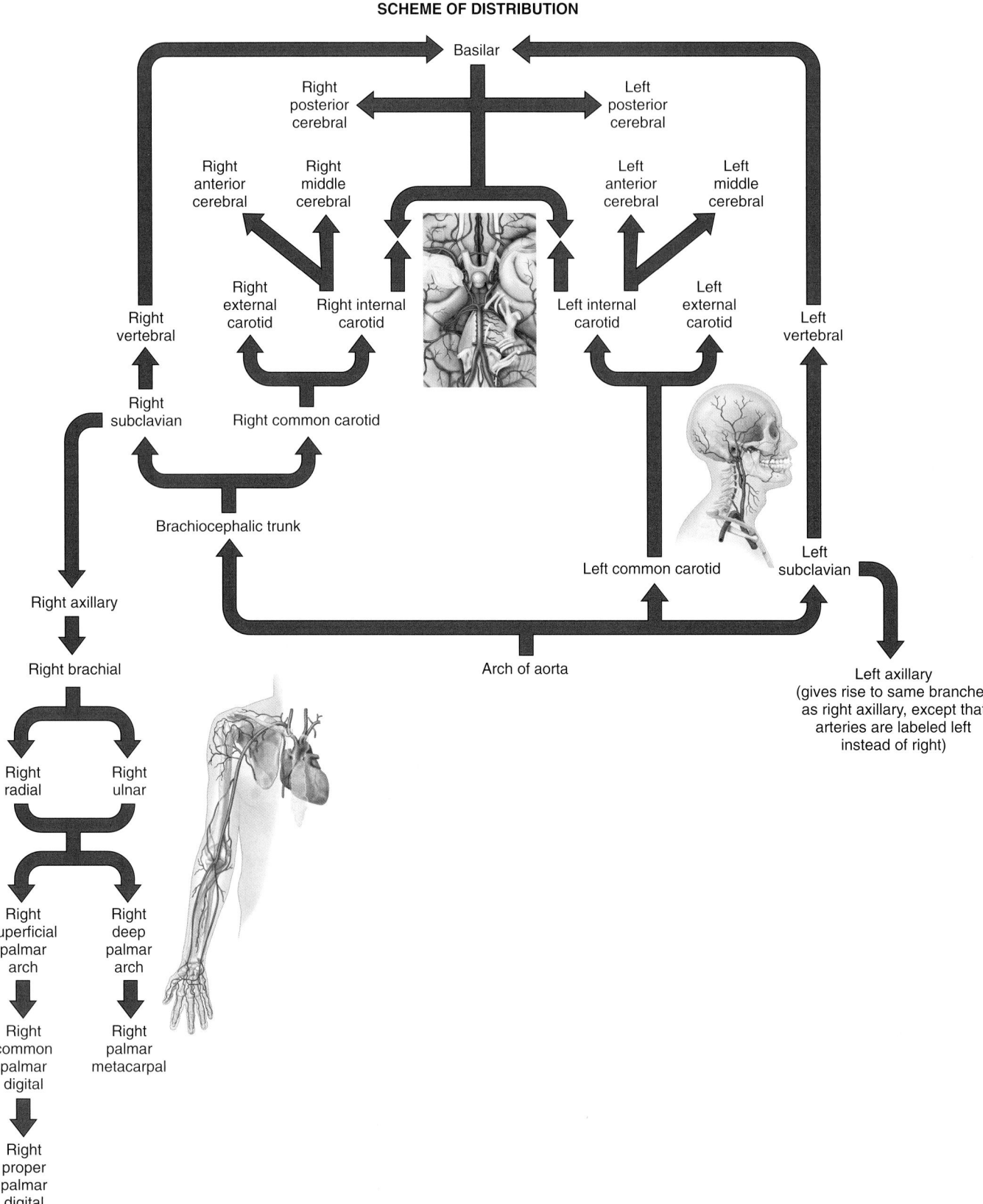

Figure 21.19 Arch of the aorta and its branches. Note in (c) the arteries that constitute the cerebral arterial circle (circle of Willis).

🔒 The arch of the aorta ends at the level of the intervertebral disc between the fourth and fifth thoracic vertebrae.

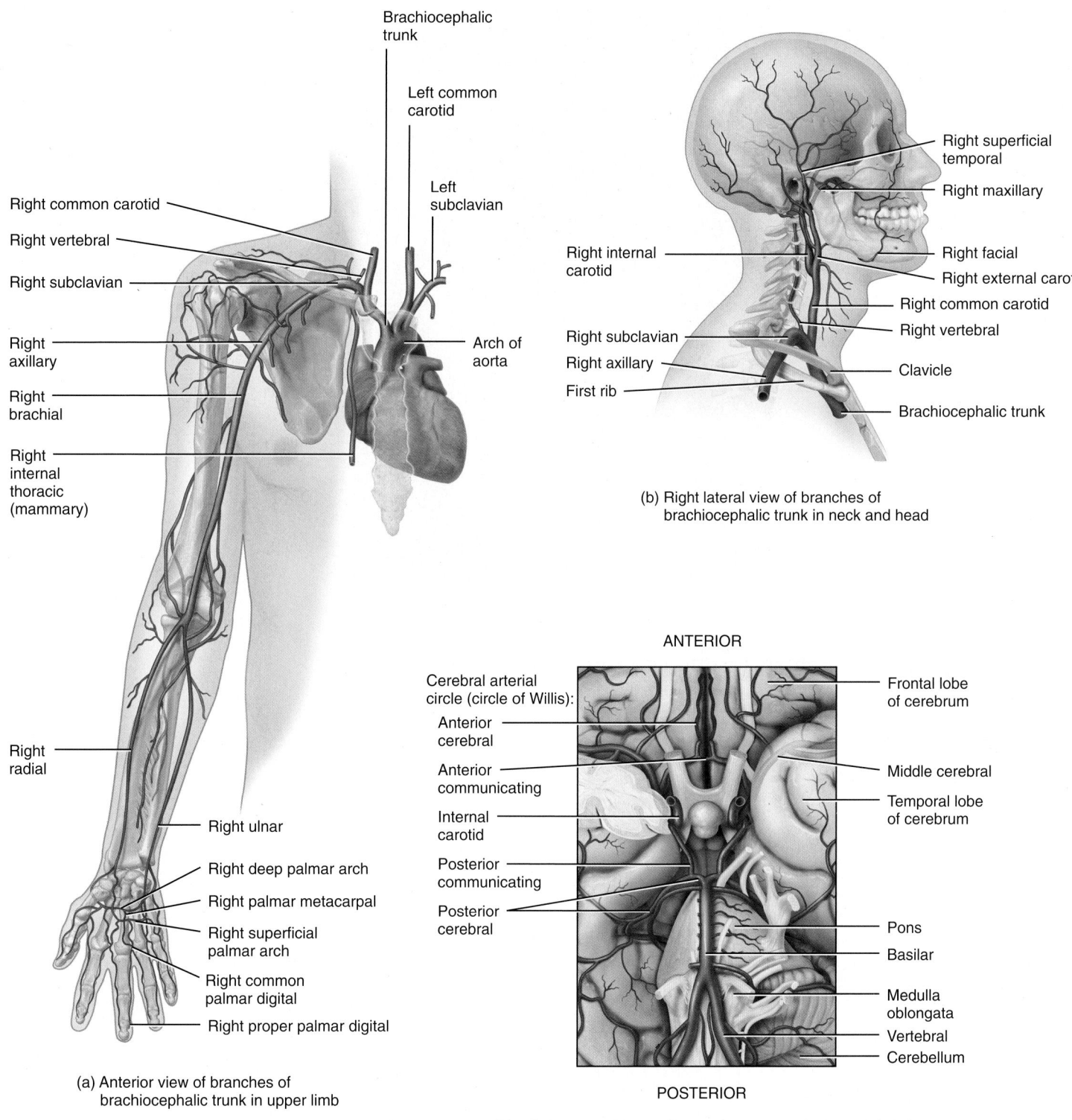

Brachiocephalic trunk

Left common carotid

Right common carotid

Left subclavian

Right vertebral

Right subclavian

Right axillary

Arch of aorta

Right brachial

Right internal thoracic (mammary)

Right radial

Right ulnar

Right deep palmar arch

Right palmar metacarpal

Right superficial palmar arch

Right common palmar digital

Right proper palmar digital

(a) Anterior view of branches of brachiocephalic trunk in upper limb

Right superficial temporal

Right maxillary

Right internal carotid

Right facial

Right external carotid

Right common carotid

Right vertebral

Right subclavian

Right axillary

Clavicle

First rib

Brachiocephalic trunk

(b) Right lateral view of branches of brachiocephalic trunk in neck and head

ANTERIOR

Cerebral arterial circle (circle of Willis):

Anterior cerebral

Anterior communicating

Internal carotid

Posterior communicating

Posterior cerebral

Frontal lobe of cerebrum

Middle cerebral

Temporal lobe of cerebrum

Pons

Basilar

Medulla oblongata

Vertebral

Cerebellum

POSTERIOR

(c) Inferior view of base of brain showing cerebral arterial circle

EXHIBIT 21.C CONTINUES ▶

EXHIBIT 21.C **835**

EXHIBIT 21.C   The Arch of the Aorta *(Figure 21.19)* CONTINUED

● FIGURE 21.19 CONTINUED ▶

Right common carotid

Right subclavian

Trachea

Brachiocephalic

Right internal thoracic
(mammary)

Arch of aorta

Left subclavian

Clavicle (cut)

First rib (cut)

Left axillary

Left common carotid

Vagus nerve (recurrent
laryngeal branch)

Phrenic nerve

Left lung

Pulmonary trunk

Heart

(d) Anterior view of branches of arch of aorta

? **What are the three major branches of the arch of the aorta, in order of their origination?**

EXHIBIT 21.D   Thoracic Aorta   *(Figure 21.20)*

◉ **OBJECTIVE**

• Identify the visceral and parietal branches of the thoracic aorta.

The **thoracic aorta** is about 20 cm (8 in.) long and is a continuation of the arch of the aorta (Figure 21.20). It begins at the level of the intervertebral disc between the fourth and fifth thoracic vertebrae, where it lies to the left of the vertebral column. As it descends, it moves closer to the midline and extends through an opening in the diaphragm (aortic hiatus), which is located anterior to the vertebral column at the level of the intervertebral disc between the twelfth thoracic and first lumbar vertebrae.

Along its course, the thoracic aorta sends off numerous small arteries, **visceral branches** (VIS-er-al) to viscera, and **parietal branches** (pa-RĪ-e-tal) to body wall structures.

✔ **CHECKPOINT**

What general regions do the visceral and parietal branches of the thoracic aorta supply?

BRANCH	DESCRIPTION AND BRANCHES	REGIONS SUPPLIED
VISCERAL BRANCHES		
Pericardial arteries (per'-i-KAR-dē-al; *peri-* = around; *-cardia* = heart)	Two to three small arteries that arise from variable levels of thoracic aorta and pass forward to pericardial sac surrounding heart.	Tissues of pericardial sac.
Bronchial arteries (BRONG-kē-al = windpipe)	Arise from thoracic aorta or one of its branches. Right bronchial artery typically arises from third posterior intercostal artery; two left bronchial arteries arise from upper end of thoracic aorta. All follow bronchial tree into lungs.	Supply tissues of bronchial tree and surrounding lung tissue down to level of alveolar ducts.
Esophageal arteries (e-sof'-a-JĒ-al; *eso-* = to carry; *phage* = food)	Four to five arteries that arise from anterior surface of thoracic aorta and pass forward to branch onto esophagus.	All tissues of esophagus.
Mediastinal arteries (mē-dē-as-TĪ-nal)	Arise from various points on thoracic aorta.	Assorted tissues within mediastinum, primarily connective tissue and lymph nodes.
PARIETAL BRANCHES		
Posterior intercostal arteries (in'ter-KOS-tal; *inter-* = between; *-costa* = rib)	Typically, nine pairs of arteries that arise from posterolateral aspect on each side of thoracic aorta. Each passes laterally and then anteriorly through intercostal space, where they will eventually anastomose with anterior branches from internal thoracic arteries.	Skin, muscles, and ribs of thoracic wall. Thoracic vertebrae, meninges, and spinal cord. Mammary glands.
Subcostal arteries (sub-KOS-tal; *sub-* = under)	The lowest segmental branches of thoracic aorta; one on each side passes into thoracic body wall inferior to twelfth rib and courses forward into upper abdominal region of body wall.	Skin, muscles, and ribs. Twelfth thoracic vertebra, meninges, and spinal cord.
Superior phrenic arteries (FREN-ik = pertaining to diaphragm)	Arise from lower end of thoracic aorta and pass onto superior surface of diaphragm.	Diaphragm muscle and pleura covering diaphragm.

SCHEME OF DISTRIBUTION

EXHIBIT 21.D CONTINUES

EXHIBIT 21.D **837**

EXHIBIT 21.D   Thoracic Aorta   *(Figure 21.20)*   CONTINUED

Figure 21.20 **Thoracic aorta and abdominal aorta and their principal branches.**

The thoracic aorta is the continuation of the ascending aorta.

Detailed anterior view of principal branches of aorta

Where does the thoracic aorta begin?

| EXHIBIT 21.E | Abdominal Aorta *(Figure 21.21)* |

○ OBJECTIVE

- Identify the visceral and parietal branches of the abdominal aorta.

The **abdominal aorta** is the continuation of the thoracic aorta after it passes through the diaphragm (Figure 21.21). It begins at the aortic hiatus in the diaphragm and ends at about the level of the fourth lumbar

BRANCH	DESCRIPTION AND BRANCHES	REGIONS SUPPLIED
UNPAIRED VISCERAL BRANCHES		
Celiac trunk (artery) (SĒ-lē-ak)	First visceral branch of aorta inferior to diaphragm; arises from abdominal aorta at level of twelfth thoracic vertebra as aorta passes through hiatus in diaphragm; divides into three branches: left gastric, splenic, and common hepatic arteries (Figure 21.21a).	Supplies all organs of gastrointestinal tract that arise from embryonic foregut, that is, from abdominal part of esophagus to duodenum, and also spleen.
	1. Left gastric artery (GAS-trik = stomach). Smallest of three celiac branches arises superiorly to left toward esophagus and then turns to follow lesser curvature of stomach. On lesser curvature of stomach it anastomoses with right gastric artery.	Abdominal part of esophagus and lesser curvature of stomach.
	2. Splenic artery (SPLEN-ik = spleen). Largest branch of celiac trunk arises from left side of celiac trunk distal to left gastric artery, and passes horizontally to left along pancreas. Before reaching spleen, it gives rise to three named arteries:	Spleen, pancreas, fundus and greater curvature of stomach, and greater omentum.
	• **Pancreatic arteries** (pan-krē-AT-ik), a series of small arteries that arise from splenic and descend into tissue of pancreas.	Pancreas.
	• **Left gastroepiploic (gastro-omental) artery** (gas′-trō-ep′-i-PLŌ-ik; *epiplo-* = omentum) arises from terminal end of splenic artery and passes from left to right along greater curvature of stomach.	Greater curvature of stomach and greater omentum.
	• **Short gastric arteries** arise from terminal end of splenic artery and pass onto fundus of stomach.	Fundus of stomach.
	3. Common hepatic artery (he-PAT-ik = liver). Intermediate in size between left gastric and splenic arteries; arises from right side of celiac trunk and gives rise to three arteries:	Liver, gallbladder, lesser omentum, stomach, pancreas, and duodenum.
	• **Proper hepatic artery** branches from common hepatic artery and ascends along bile ducts into liver and gallbladder.	Liver, gallbladder, and lesser omentum.
	• **Right gastric artery** arises from common hepatic artery and curves back to left along lesser curvature of stomach, where it anastomoses with left gastric artery.	Stomach and lesser omentum.
	• **Gastroduodenal artery** (gas′-trō-doo′-ō-DĒ-nal) passes inferiorly toward stomach and duodenum and sends branches along greater curvature of stomach.	Stomach, duodenum, and pancreas.
Superior mesenteric artery (MES-en-ter′-ik; *meso-* = middle; *-enteric* = pertaining to intestines)	Arises from anterior surface of abdominal aorta about 1 cm inferior to celiac trunk at level of first lumbar vertebra (Figure 21.21b); extends inferiorly and anteriorly between layers of mesentery (portion of peritoneum that attaches small intestine to posterior abdominal wall). It anastomoses extensively and has five branches:	Supplies all organs of gastrointestinal tract that arise from embryonic midgut, that is, from duodenum to transverse colon.
	1. Inferior pancreaticoduodenal artery (pan′-krē-at′-i-kō-doo′-ō-DĒ-nal) passes superiorly and to right toward head of pancreas and duodenum.	Pancreas and duodenum.
	2. Jejunal (je-JOO-nal) and **ileal arteries** (IL-ē-al) spread through mesentery to pass to loops of jejunum and ileum (small intestine).	Jejunum and ileum, which is majority of small intestine.
	3. Ileocolic artery (il′-ē-ō-KOL-ik) passes inferiorly and laterally toward right side toward terminal part of ileum, cecum, appendix, and first part of ascending colon.	Terminal part of ileum, cecum, appendix, and first part of ascending colon.
	4. Right colic artery (KOL-ik) passes laterally to right toward ascending colon.	Ascending colon and first part of transverse colon.
	5. Middle colic artery ascends slightly to right toward transverse colon.	Most of transverse colon.
Inferior mesenteric artery	Arises from anterior aspect of abdominal aorta at level of third lumbar vertebra and then passes inferiorly to left of aorta (Figure 21.21c). It anastomoses extensively and has three branches:	Supplies all organs of gastrointestinal tract that arise from embryonic hindgut from transverse colon to rectum.
	1. Left colic artery ascends laterally to left toward distal end of transverse colon and descending colon.	End of transverse colon and descending colon.
	2. Sigmoid arteries (SIG-moyd) descend laterally to left toward sigmoid colon.	Sigmoid colon.
	3. Superior rectal artery (REK-tal) passes inferiorly to superior part of rectum.	Upper part of rectum.

EXHIBIT 21.E CONTINUES ▶

EXHIBIT 21.E **839**

EXHIBIT 21.E Abdominal Aorta *(Figure 21.21)* CONTINUED

vertebra, where it divides into the right and left common iliac arteries. The abdominal aorta lies anterior to the vertebral column.

As with the thoracic aorta, the abdominal aorta gives off **visceral** and **parietal branches.** The unpaired visceral branches arise from the anterior surface of the aorta and include the **celiac trunk** and the **superior mesenteric** and **inferior mesenteric arteries** (see Figure 21.20).

The paired visceral branches arise from the lateral surfaces of the aorta and include the **suprarenal, renal,** and **gonadal arteries.** The lone unpaired parietal branch is the **median sacral artery.** The paired parietal branches arise from the posterolateral surfaces of the aorta and include the **inferior phrenic** and **lumbar arteries.**

 CHECKPOINT

Name the paired visceral and parietal branches and the unpaired visceral and parietal branches of the abdominal aorta, and indicate the general regions they supply.

BRANCH	DESCRIPTION AND BRANCHES	REGIONS SUPPLIED
PAIRED VISCERAL BRANCHES		
Suprarenal arteries (soo′-pra-RĒ-nal; *supra-* = above; *-ren-* = kidney)	There are typically three pairs (superior, middle, and inferior), but only middle pair originates directly from abdominal aorta (see Figure 21.20). Middle suprarenal arteries arise from abdominal aorta at level of first lumbar vertebra at or superior to renal arteries. Superior suprarenal arteries arise from inferior phrenic arteries, and inferior suprarenal arteries originate from renal arteries.	Suprarenal (adrenal) glands.
Renal arteries (RĒ-nal; *ren* = kidney)	Right and left **renal arteries** usually arise from lateral aspects of abdominal aorta at superior border of second lumbar vertebra, about 1 cm inferior to superior mesenteric artery (see Figure 21.20). Right renal artery, which is longer than left, arises slightly lower than left and passes posterior to right renal vein and inferior vena cava. Left renal artery is posterior to left renal vein and is crossed by inferior mesenteric vein.	All tissues of kidneys.
Gonadal (gō-NAD-al; *gon-* = seed) **arteries** [**testicular** (tes-TIK-ū-lar) or **ovarian** (ō -VAR-ē-an)]	Arise from anterior aspect of abdominal aorta at level of second lumbar vertebra just inferior to renal arteries (see Figure 21.20). In males, gonadal arteries are specifically referred to as **testicular arteries.** They descend along posterior abdominal wall to pass through inguinal canal and descend into scrotum. In females, gonadal arteries are called **ovarian arteries.** They are much shorter than testicular arteries and remain within abdominal cavity.	Males: testis, epididymis, ductus deferens, and ureters. Females: ovaries, uterine (fallopian) tubes, and ureters.
UNPAIRED PARIETAL BRANCH		
Median sacral artery (SĀ-kral = pertaining to sacrum)	Arises from posterior surface of abdominal aorta about 1 cm superior to bifurcation (division into two branches) of aorta into right and left common iliac arteries (see Figure 21.20).	Sacrum, coccyx, sacral spinal nerves, and piriformis muscle.
PAIRED PARIETAL BRANCH		
Inferior phrenic arteries (FREN-ik = pertaining to diaphragm)	First paired branches of abdominal aorta; arise immediately superior to origin of celiac trunk (see Figure 21.20). (They may also arise from renal arteries.)	Diaphragm and suprarenal (adrenal) glands.
Lumbar arteries (LUM-bar = pertaining to loin)	Four pairs arise from posterolateral surface of abdominal aorta in a way similar to posterior intercostal arteries of thorax (see Figure 21.20); pass laterally into abdominal muscle wall and curve toward anterior aspect of wall.	Lumbar vertebrae, spinal cord and meninges, skin and muscles of posterior and lateral part of abdominal wall.

SCHEME OF DISTRIBUTION

Abdominal aorta

Visceral branches
1. Celiac trunk

Parietal branches

1. Inferior phrenic
2. Lumbar
3. Median sacral

Left gastric — Splenic — Common hepatic

Pancreatic — Left gastroepiploic — Short gastric

Hepatic artery proper — Right gastric — Gastro-duodenal

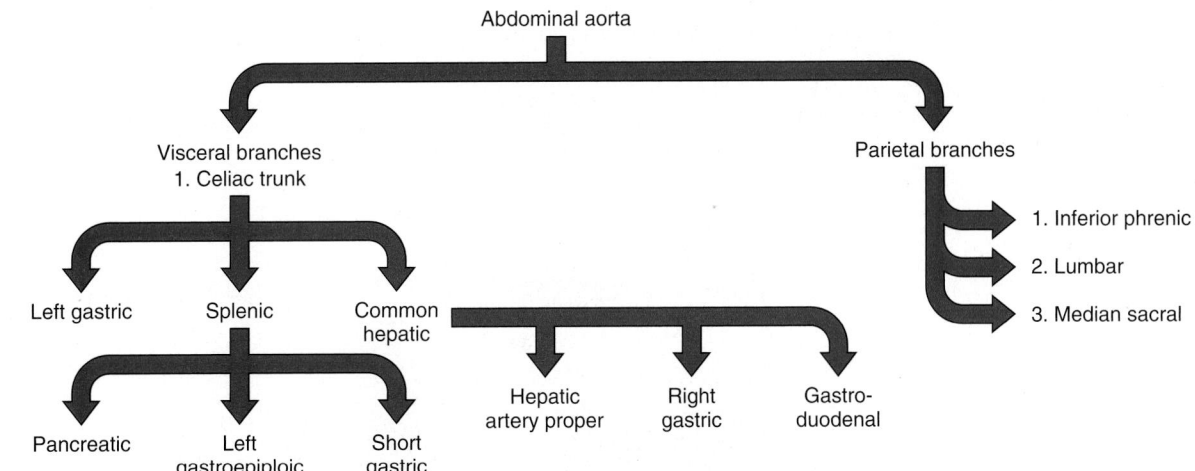

2. Superior mesenteric

Inferior pancreatico-duodenal — Jejunal — Ileocolic — Right colic — Middle colic

3. Suprarenal

4. Renal

5. Gonadal (testicular or ovarian)

6. Inferior mesenteric

Left colic — Sigmoid — Superior rectal

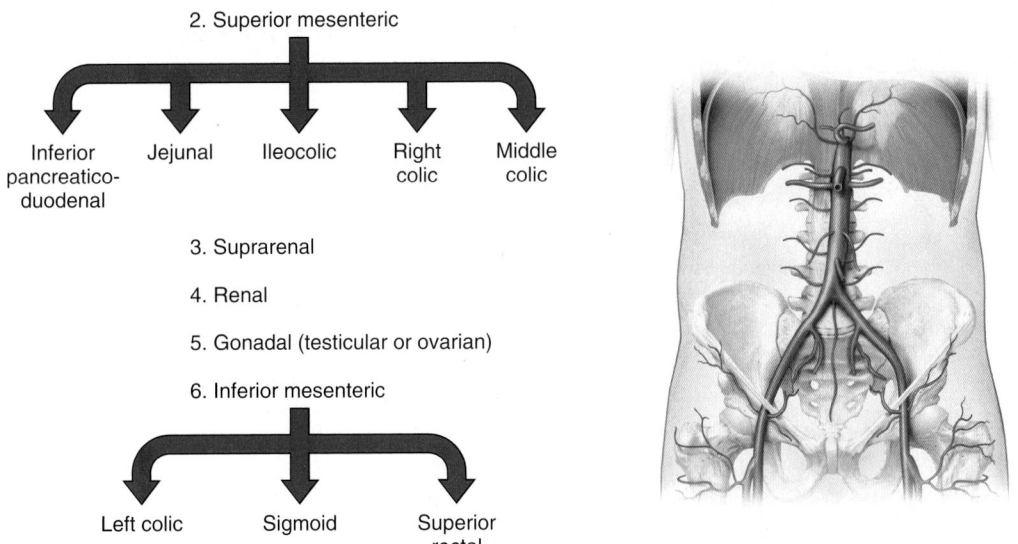

EXHIBIT 21.E CONTINUES ▶

EXHIBIT 21.E **841**

Figure 21.21   **Abdominal aorta and its principal branches.**

The abdominal aorta is the continuation of the thoracic aorta.

Stomach (turned upward)

Left gastroepiploic

Right gastroepiploic

Right gastric

Celiac trunk

Proper hepatic

Common hepatic

Gastroduodenal

Superior pancreaticoduodenal

Duodenum of small intestine

Short gastrics

Diaphragm

Left gastric

Spleen

Splenic

Pancreatic

Pancreas

Superior mesenteric

Jejunum of small intestine

Abdominal aorta

(a) Anterior view of celiac trunk and its branches

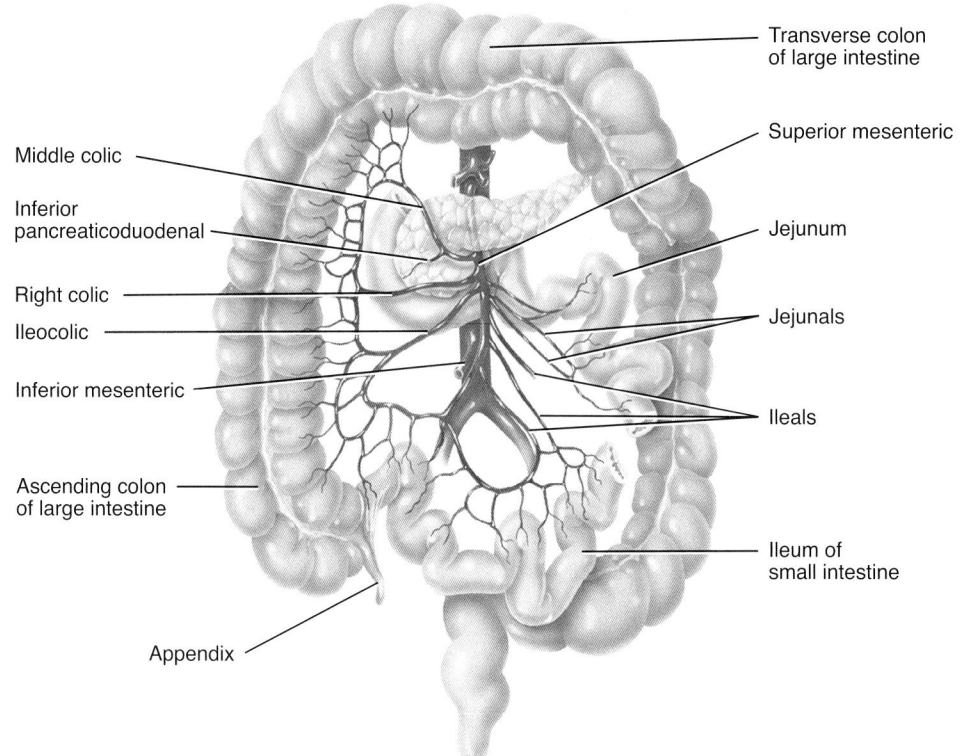

Middle colic

Inferior pancreaticoduodenal

Right colic

Ileocolic

Inferior mesenteric

Ascending colon of large intestine

Appendix

Transverse colon of large intestine

Superior mesenteric

Jejunum

Jejunals

Ileals

Ileum of small intestine

(b) Anterior view of superior mesenteric artery and its branches

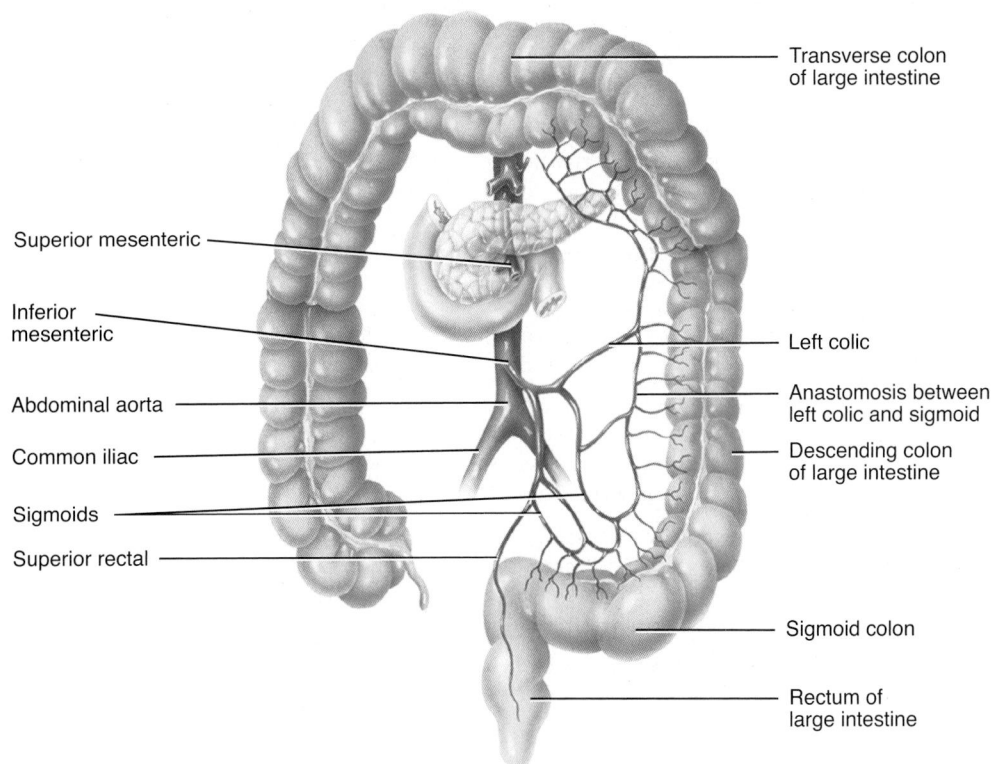

Transverse colon
of large intestine

Superior mesenteric

Inferior
mesenteric

Abdominal aorta

Common iliac

Sigmoids

Superior rectal

Left colic

Anastomosis between
left colic and sigmoid

Descending colon
of large intestine

Sigmoid colon

Rectum of
large intestine

(c) Anterior view of inferior mesenteric artery and its branches

Diaphragm

Inferior vena cava

Right adrenal (suprarenal) gland

Right kidney

Superior mesenteric

Right common iliac

Right internal iliac

Right external iliac

Hepatic veins

Esophageal hiatus

Common hepatic

Splenic

Celiac trunk

Right renal

Abdominal aorta

Inferior mesenteric

Ureter

? **Where does the abdominal aorta begin?**

(d) Anterior view of arteries of abdomen and pelvis

EXHIBIT 21.E **843**

EXHIBIT 21.F   Arteries of the Pelvis and Lower Limbs   *(Figure 21.22)*

● **OBJECTIVE**

• Identify the two major branches of the common iliac arteries.

The abdominal aorta ends by dividing into the right and left **common iliac arteries** (Figure 21.22). These, in turn, divide into the **internal** and **external iliac arteries.** In sequence, the external iliacs become the **femoral arteries** in the thighs, the **popliteal arteries** posterior to the knee, and the **anterior** and **posterior tibial arteries** in the legs.

✔**CHECKPOINT**

What general regions do the internal and external iliac arteries supply?

BRANCH	DESCRIPTION AND BRANCHES	REGIONS SUPPLIED
Common iliac arteries (IL-ē-ak = pertaining to ilium)	Arise from abdominal aorta at about level of fourth lumbar vertebra. Each common iliac artery passes inferiorly and slightly laterally for about 5 cm (2 in.) and gives rise to two branches: internal and external iliac arteries.	Pelvic muscle wall, pelvic organs, external genitals, and lower limbs.
Internal iliac arteries	Primary arteries of pelvis. Begin at bifurcation (division into two branches) of common iliac arteries anterior to sacroiliac joint at level of lumbosacral intervertebral disc. Pass posteriorly as they descend into pelvis and divide into anterior and posterior divisions.	Pelvic muscle wall, pelvic organs, buttocks, external genitals, and medial muscles of thigh.
External iliac arteries	Larger than internal iliac arteries and begin at bifurcation of common iliac arteries. Descend along medial border of psoas major muscles following pelvic brim, pass posterior to midportion of inguinal ligaments, and become femoral arteries as they pass beneath inguinal ligament and enter thigh.	Lower abdominal wall, cremaster muscle in males and round ligament of uterus in females, and lower limb.
Femoral arteries (FEM-o-ral=pertaining to thigh)	Continuations of external iliac arteries as they enter thigh. In *femoral triangle* of upper thighs they are superficial along with femoral vein and nerve and deep inguinal lymph nodes (see Figure 11.20a). Pass beneath sartorius muscle as they descend along anteromedial aspects of thighs and follow its course to distal end of thigh where they pass through opening in tendon of adductor magnus muscle to end at posterior aspect of knee, where they become popliteal arteries. **Clinical note:** In **cardiac catheterization,** a catheter is inserted through a blood vessel and advanced into the major vessels to access a heart chamber. A catheter often contains a measuring instrument or other device at its tip. To reach the left side of the heart, the catheter is inserted into the femoral artery and passed into the aorta to the coronary arteries or heart chamber.	Muscles of thigh (quadriceps, adductors, and hamstrings), femur, and ligaments and tendons around knee joint.
Popliteal arteries (pop'-li-TĒ-al = posterior surface of knee)	Continuation of femoral arteries through popliteal fossa (space behind knee). Descend to inferior border of popliteus muscles, where they divide into anterior and posterior tibial arteries.	Muscles of distal thigh, skin of knee region, muscles of proximal leg, knee joint, femur, patella, tibia, and fibula.
Anterior tibial arteries (TIB-ē-al = pertaining to shin)	Descend from bifurcation of popliteal arteries at distal border of popliteus muscles. Smaller than posterior tibial arteries; pass over interosseous membrane of tibia and fibula to descend through anterior muscle compartment of leg; become **dorsal arteries of foot (dorsalis pedis arteries)** at ankles. On dorsum of feet, dorsal arteries of foot give off transverse branch at first medial cuneiform bone called **arcuate arteries** (*arcuat-* = bowed) that run laterally over bases of metatarsals. From arcuate arteries branch **dorsal metatarsal arteries,** which course along metatarsal bones. Dorsal metatarsal arteries terminate by dividing into **dorsal digital arteries,** which pass into toes.	Tibia, fibula, anterior muscles of leg, dorsal muscles of foot, tarsal bones, metatarsal bones, and phalanges.
Posterior tibial arteries	Direct continuations of popliteal arteries, descend from bifurcation of popliteal arteries. Pass down posterior muscular compartment of legs deep to soleus muscles. Pass posterior to medial malleolus at distal end of leg and curve forward toward plantar surface of feet; pass deep to flexor retinaculum on medial side of feet and terminate by branching into medial and lateral plantar arteries. Give rise to **fibular (peroneal) arteries** in middle of leg, which course laterally as they descend into lateral compartment of leg. **Medial plantar arteries** (PLAN-tar = sole) pass along medial side of sole and **lateral plantar arteries** angle toward lateral side of sole and unite with branch of dorsal arteries of foot to form **plantar arch.** Arch begins at base of fifth metatarsal and extends medially across metatarsals. As arch crosses foot, it gives off **plantar metatarsal arteries,** which course along plantar surface of metatarsal bones. These arteries terminate by dividing into **plantar digital arteries** that pass into toes.	Posterior and lateral muscle compartments of leg, plantar muscles of foot, tibia, fibula, tarsal, metatarsal, and phalangeal bones.

SCHEME OF DISTRIBUTION

Abdominal aorta

Right common iliac

Left common iliac
(gives rise to same branches
as right common iliac, except
that arteries are labeled left
instead of right)

Right external iliac

Right internal iliac

Right femoral

Right deep artery of thigh (deep peroneal)

Right popliteal

Right anterior tibial

Right posterior tibial

Right dorsal artery
of foot (dorsalis pedis)

Right fibular (peroneal)

Right arcuate

Right dorsal
metatarsal

Right lateral
plantar

Right medial
plantar

Right dorsal
digital

Right plantar
arch

Right plantar
metatarsal

Right plantar
digital

Exhibit 21.F continues ▶

EXHIBIT 21.F **845**

Figure 21.22   **Arteries of the pelvis and right lower limb.**

The internal iliac arteries carry most of the blood supply to the pelvic viscera and wall.

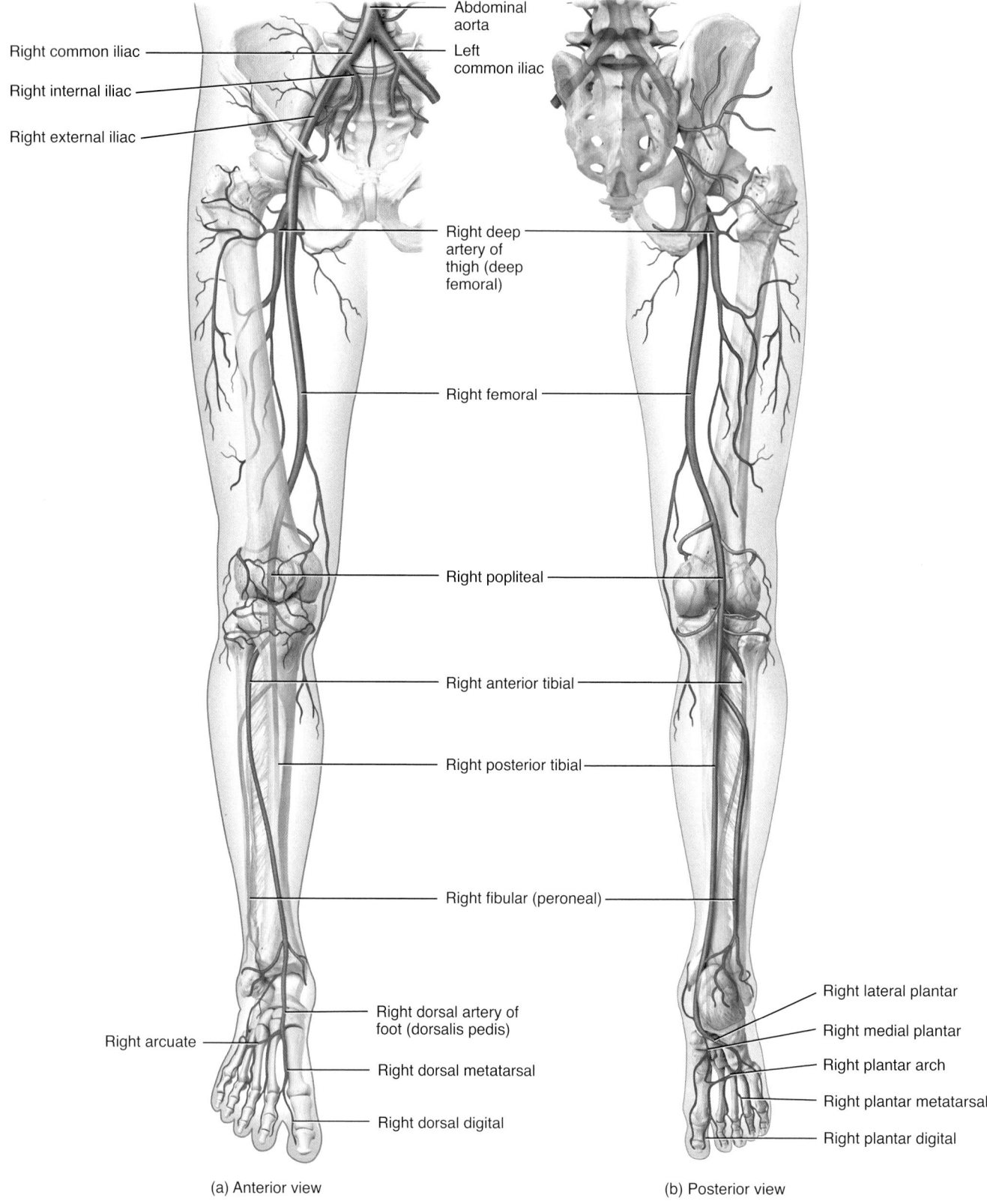

Abdominal aorta

Right common iliac

Right internal iliac

Right external iliac

Left common iliac

Right deep artery of thigh (deep femoral)

Right femoral

Right popliteal

Right anterior tibial

Right posterior tibial

Right fibular (peroneal)

Right dorsal artery of foot (dorsalis pedis)

Right arcuate

Right dorsal metatarsal

Right dorsal digital

Right lateral plantar

Right medial plantar

Right plantar arch

Right plantar metatarsal

Right plantar digital

(a) Anterior view

(b) Posterior view

At what point does the abdominal aorta divide into the common iliac arteries?

EXHIBIT 21.G   Veins of the Systemic Circulation   *(Figure 21.23)*

◉ OBJECTIVE

- Identify the three systemic veins that return deoxygenated blood to the heart.

As you have already learned, arteries distribute blood from the heart to various parts of the body, and veins drain blood away from the various parts and return the blood to the heart. In general, arteries are deep; veins may be superficial or deep. Superficial veins are located just beneath the skin and can be seen easily. Because there are no large superficial arteries, the names of superficial veins do not correspond to those of arteries. Superficial veins are clinically important as sites for withdrawing blood or giving injections. Deep veins generally travel alongside arteries and usually bear the same name. Arteries usually follow definite pathways; veins are more difficult to follow because they connect in irregular networks in which many tributaries merge to form a large vein. Although only one systemic artery, the aorta, takes oxygenated blood away from the heart (left ventricle), three systemic veins, the **coronary sinus, superior vena cava,** and **inferior vena cava,** return deoxygenated blood to the heart (right atrium) (Figure 21.23). The coronary sinus receives blood from the cardiac veins; the superior vena cava receives blood from other veins superior to the diaphragm, except the air sacs (alveoli) of the lungs; the inferior vena cava receives blood from veins inferior to the diaphragm.

✔ CHECKPOINT

What are the three tributaries of the coronary sinus?

VEINS	DESCRIPTION AND TRIBUTARIES	REGIONS DRAINED
Coronary sinus (KOR-ō-nar-ē; *corona* = crown)	Main vein of heart; receives almost all venous blood from myocardium; located in coronary sulcus (see Figure 20.3c) on posterior aspect of heart and opens into right atrium between orifice of inferior vena cava and tricuspid valve. Wide venous channel into which three veins drain. Receives **great cardiac vein** (from anterior interventricular sulcus) into its left end, and **middle cardiac vein** (from posterior interventricular sulcus) and **small cardiac vein** into its right end. Several **anterior cardiac veins** drain directly into right atrium.	All tissues of heart.
Superior vena cava (SVC) (VĒ-na KĀ-va; *vena* = vein; *cava* = cavelike)	About 7.5 cm (3 in.) long and 2 cm (1 in.) in diameter; empties its blood into superior part of right atrium. Begins posterior to right first costal cartilage by union of right and left brachiocephalic veins and ends at level of right third costal cartilage, where it enters right atrium.	Head, neck, upper limbs, and thorax.
Inferior vena cava (IVC)	Largest vein in body, about 3.5 cm (1.4 in.) in diameter. Begins anterior to fifth lumbar vertebra by union of common iliac veins, ascends behind peritoneum to right of midline, pierces caval opening of diaphragm at level of eighth thoracic vertebra, and enters inferior part of right atrium. 🜀 **Clinical note:** The inferior vena cava is commonly **compressed during the later stages of pregnancy** by the enlarging uterus, producing edema of the ankles and feet and temporary varicose veins.	Abdomen, pelvis, and lower limbs.

EXHIBIT 21.G CONTINUES ▶

EXHIBIT 21.G   **847**

EXHIBIT 21.G   Veins of the Systemic Circulation   *(Figure 21.23)*   CONTINUED

Figure 21.23 Principal veins.

Deoxygenated blood returns to the heart via the superior vena cava, inferior vena cava, and coronary sinus.

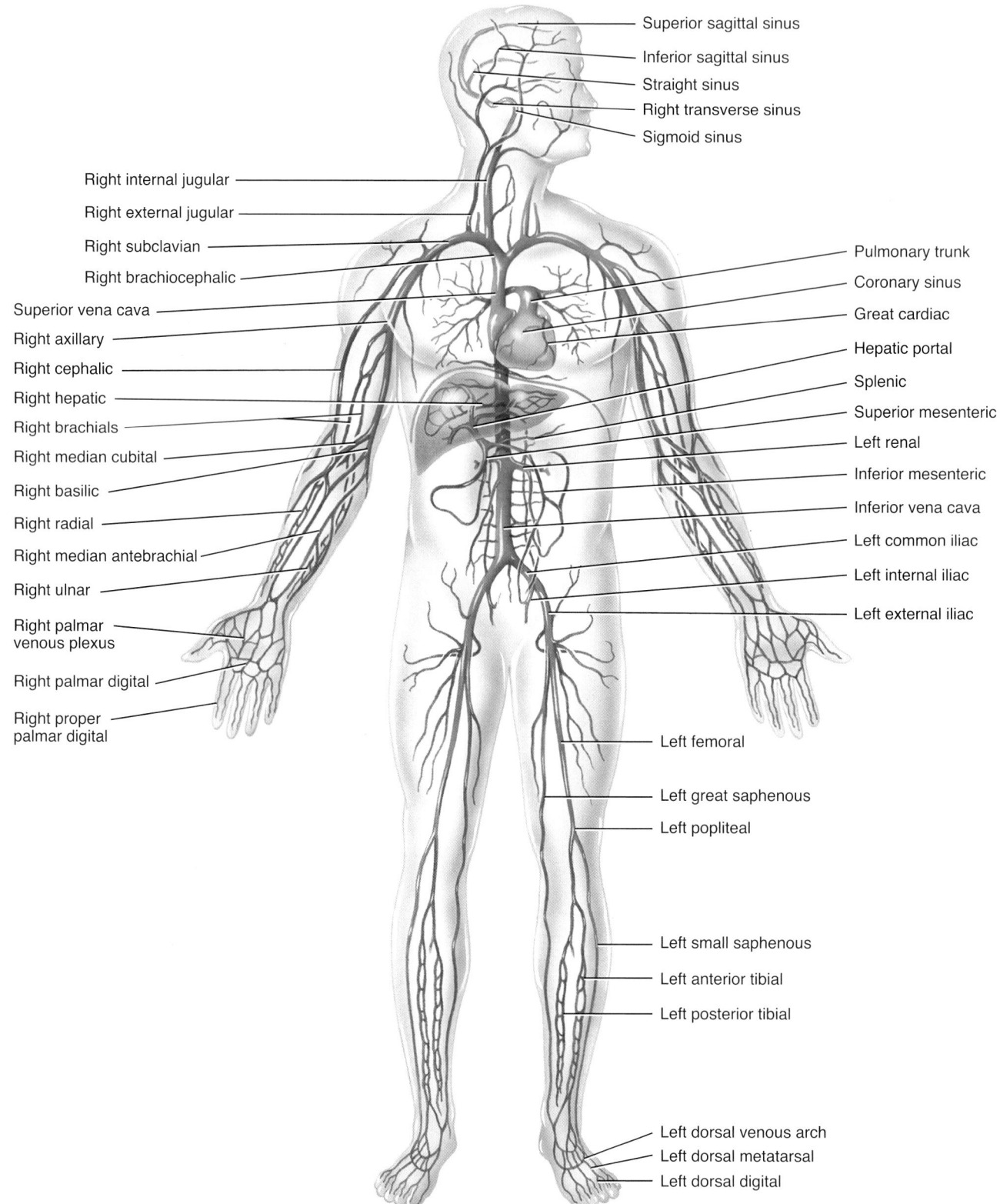

Superior sagittal sinus
Inferior sagittal sinus
Straight sinus
Right transverse sinus
Sigmoid sinus

Right internal jugular
Right external jugular
Right subclavian
Right brachiocephalic
Superior vena cava
Right axillary
Right cephalic
Right hepatic
Right brachials
Right median cubital
Right basilic
Right radial
Right median antebrachial
Right ulnar
Right palmar venous plexus
Right palmar digital
Right proper palmar digital

Pulmonary trunk
Coronary sinus
Great cardiac
Hepatic portal
Splenic
Superior mesenteric
Left renal
Inferior mesenteric
Inferior vena cava
Left common iliac
Left internal iliac
Left external iliac

Left femoral
Left great saphenous
Left popliteal
Left small saphenous
Left anterior tibial
Left posterior tibial
Left dorsal venous arch
Left dorsal metatarsal
Left dorsal digital

Overall anterior view of principal veins

Which general regions of the body are drained by the superior vena cava and the inferior vena cava?

EXHIBIT 21.H Veins of the Head and Neck *(Figure 21.24)*

● OBJECTIVE

• Identify the three major veins that drain blood from the head.

Most blood draining from the head passes into three pairs of veins: the **internal jugular** (JUG-ū-lar), **external jugular**, and **vertebral veins** (Figure 21.24). Within the cranial cavity, all veins drain into dural ve-nous sinuses and then into the internal jugular veins. **Dural venous sinuses** are endothelial-lined venous channels between layers of the cranial dura mater.

✔ CHECKPOINT

Which general areas are drained by the internal jugular, external jugular, and vertebral veins?

VEINS	DESCRIPTION AND TRIBUTARIES	REGIONS DRAINED
Brachiocephalic veins	(See Exhibit 21.C.)	
Internal jugular veins (JUG-ū-lar = throat)	Begin at base of cranium as sigmoid sinus and inferior petrosal sinus converge at opening of the jugular foramen. Descend within carotid sheath lateral to internal and common carotid arteries, deep to sternocleidomastoid muscles. Receive numerous tributaries from the face and neck. Internal jugular veins anastomose with subclavian veins to form brachiocephalic veins (brā′-kē-ō-se-FAL-ik; *brachi-* = arm; *-cephal-* = head) deep and slightly lateral to sternoclavicular joints. Major dural sinuses that contribute to internal jugular vein are as follows:	Brain, meninges, bones of cranium, muscles and tissues of face and neck.
	1. **Superior sagittal sinus** (SAJ-i-tal = arrow) begins at frontal bone, where it receives vein from nasal cavity, and passes posteriorly to occipital bone along midline of skull deep to sagittal sinus. It usually angles to right and drains into right transverse sinus.	Nasal cavity; superior, lateral, and medial aspects of cerebrum; skull bones; meninges.
	2. **Inferior sagittal sinus** is much smaller than superior sagittal sinus. It begins posterior to attachment of falx cerebri and receives great cerebral vein to become straight sinus.	Medial aspects of cerebrum and diencephalon.
	3. **Straight sinus** runs in tentorium cerebelli and is formed by union of inferior sagittal sinus and great cerebral vein. It typically drains into left transverse sinus.	Medial and inferior aspects of cerebrum and the cerebellum.
	4. **Sigmoid sinuses** (SIG-moyd = S-shaped) are located along posterior aspect of petrous temporal bone. They begin where transverse sinuses and superior petrosal sinuses anastomose and terminate in internal jugular vein at jugular foramen.	Lateral and posterior aspect of cerebrum and the cerebellum.
	5. **Cavernous sinuses** (KAV-er-nus = cavelike) are located on either side of sphenoid bone. Ophthalmic veins from orbits and cerebral veins from cerebral hemispheres, along with other small sinuses, empty into cavernous sinuses. They drain posteriorly to petrosal sinuses to eventually return to internal jugular veins. Cavernous sinuses are unique because they have major blood vessels and nerves passing through them on their way to orbit and face. Oculomotor (III) nerve, trochlear (IV) nerve, ophthalmic and maxillary branches of the trigeminal (V) nerve, abducens (VI) nerve, and internal carotid arteries pass through cavernous sinuses.	Orbits, nasal cavity, frontal regions of cerebrum, and superior aspect of brain stem.
Subclavian veins	(See Exhibit 21.I.)	
External jugular veins	Begin in parotid glands near angle of the mandible. Descend through neck across sternocleidomastoid muscles. Terminate at point opposite middle of clavicles, where they empty into subclavian veins. Become very prominent along side of neck when venous pressure rises, for example, during heavy coughing or straining or in cases of heart failure.	Scalp and skin of head and neck, muscles of face and neck, and oral cavity and pharynx.
Vertebral veins (VER-te-bral = of vertebrae)	Right and left **vertebral veins** originate inferior to occipital condyles. They descend through successive transverse foramina of first six cervical vertebrae and emerge from foramina of sixth cervical vertebra to enter brachiocephalic veins in root of neck.	Cervical vertebrae, cervical spinal cord and meninges, and some deep muscles in neck.

EXHIBIT 21.H CONTINUES ▶

EXHIBIT 21.H **849**

EXHIBIT 21.H **Veins of the Head and Neck** *(Figure 21.24)* **CONTINUED**

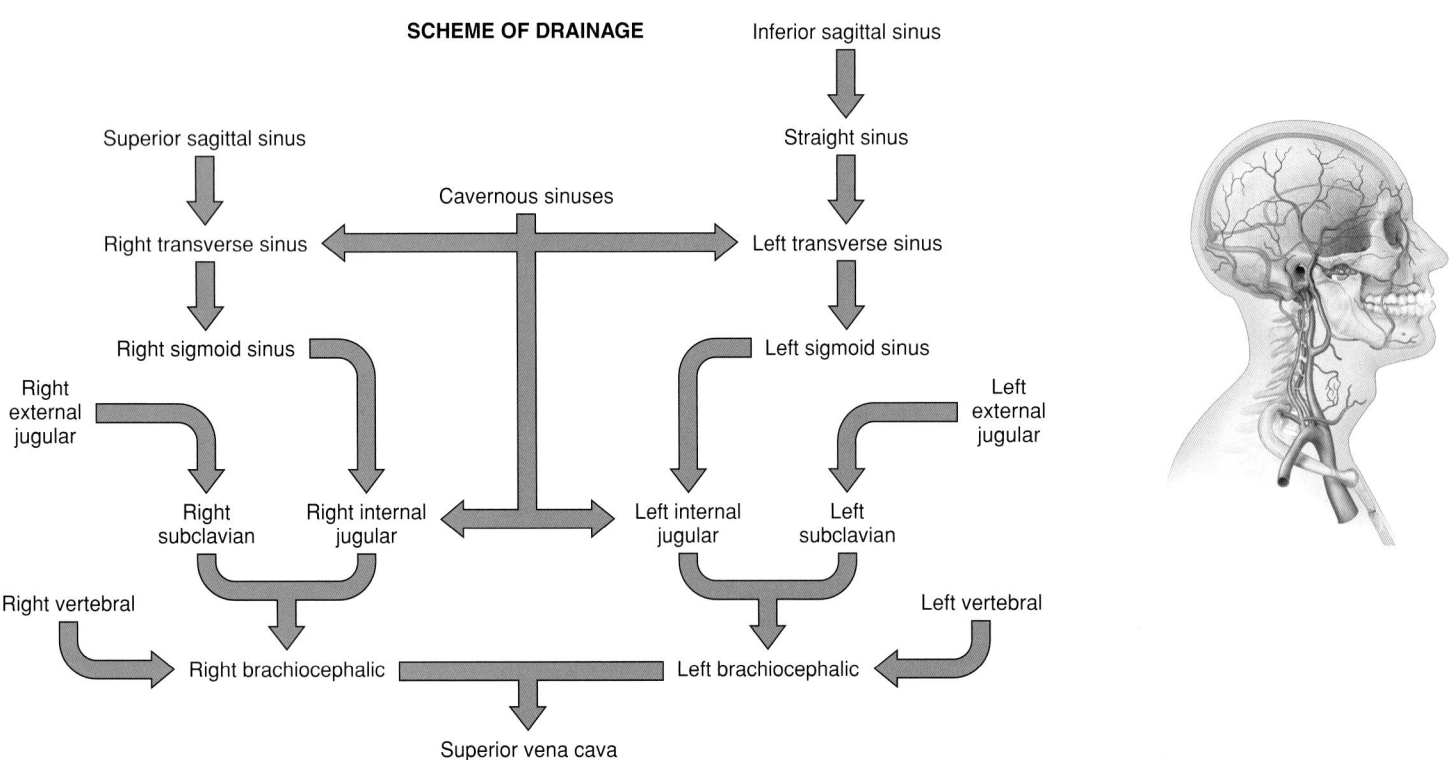

SCHEME OF DRAINAGE

Figure 21.24 Principal veins of the head and neck.

Blood draining from the head passes into the internal jugular, external jugular, and vertebral veins.

Right lateral view

Into which veins in the neck does all venous blood in the brain drain?

EXHIBIT 21.1 | Veins of the Upper Limbs *(Figure 21.25)*

OBJECTIVE

- Identify the principal veins that drain the upper limbs.

Both superficial and deep veins return blood from the upper limbs to the heart (Figure 21.25). **Superficial veins** are located just deep to the skin and are often visible. They anastomose extensively with one another and with deep veins, and they do not accompany arteries. Superficial veins are larger than deep veins and return most of the blood from the upper limbs. **Deep veins** are located deep in the body. They usually accompany arteries and have the same names as the corresponding arteries. Both superficial and deep veins have valves, but valves are more numerous in the deep veins.

CHECKPOINT

Where do the cephalic, basilic, median antebrachial, radial, and ulnar veins originate?

VEINS	DESCRIPTION AND TRIBUTARIES	REGIONS DRAINED
DEEP VEINS		
Brachiocephalic veins	(See Exhibit 21.J.)	
Subclavian veins (sub-KLĀ-vē-an; *sub-* = under; *-clavian* = pertaining to clavicle)	Continuations of axillary veins. Pass over first rib deep to clavicle to terminate at sternal end of clavicle, where they unite with internal jugular veins to form brachiocephalic veins. Thoracic duct of lymphatic system delivers lymph into junction between left subclavian and left internal jugular veins. Right lymphatic duct delivers lymph into junction between right subclavian and right internal jugular veins (see Figure 22.3). **Clinical note:** In a procedure called **central line placement,** the right subclavian vein is frequently used to administer nutrients and medication and measure venous pressure.	Skin, muscles, bones of arms, shoulders, neck, and superior thoracic wall.
Axillary veins (AK-sil-ār-ē; *axilla* = armpit)	Arise as brachial veins and basilic veins unite near base of axilla (armpit). Ascend to outer borders of first ribs, where they become subclavian veins. Receive numerous tributaries in axilla that correspond to branches of axillary arteries.	Skin, muscles, bones of arm, axilla, shoulder, and superolateral chest wall.
Brachial veins (BRĀ-kē-al; *brachi-* = arm)	Accompany brachial arteries. Begin in anterior aspect of elbow region where radial and ulnar veins join one another. As they ascend through arm, basilic veins join them to form axillary vein near distal border of teres major muscle.	Muscles and bones of elbow and brachial regions.
Ulnar veins (UL-nar = pertaining to ulna)	Begin at **superficial palmar venous arches,** which drain **common palmar digital veins** and **proper palmar digital veins** in fingers. Course along medial aspect of forearms, pass alongside ulnar arteries, and join with radial veins to form brachial veins.	Muscles, bones, and skin of hand, and muscles of medial aspect of forearm.
Radial veins (RĀ-dē-al = pertaining to radius)	Begin at **deep palmar venous arches** (Figure 21.25c), which drain **palmar metacarpal veins** in palms. Drain lateral aspects of forearms and pass alongside radial arteries. Unite with ulnar veins to form brachial veins just inferior to elbow joint.	Muscles and bones of lateral hand and forearm.
SUPERFICIAL VEINS		
Cephalic veins (se-FAL-ik = pertaining to head)	Begin on lateral aspect of **dorsal venous networks of hands (dorsal venous arches),** networks of veins on dorsum of hands formed by dorsal metacarpal veins (Figure 21.25a). These veins in turn drain **dorsal digital veins,** which pass along sides of fingers. Arch around radial side of forearms to anterior surface and ascend through entire limbs along anterolateral surface. End where they join axillary veins, just inferior to clavicles. **Accessory cephalic veins** originate either from venous plexus on dorsum of forearms or from medial aspects of dorsal venous networks of hands, and unite with cephalic veins just inferior to elbow. After receiving median cubital veins, basilic veins continue ascending until they reach middle of arm. There they penetrate tissues deeply and run alongside brachial arteries until they join with brachial veins to form axillary veins.	Integument and superficial muscles of lateral aspect of upper limb.
Basilic veins (ba-SIL-ik = royal, of prime importance)	Begin on medial aspects of dorsal venous networks of hands and ascend along posteromedial surface of forearm and anteromedial surface of arm (Figure 21.25b). Connected to cephalic veins anterior to elbow by **median cubital veins** (*cubital* = pertaining to elbow). **Clinical note:** If veins must be **punctured** for an injection, transfusion, or removal of a blood sample, the median cubital veins are preferred.	Integument and superficial muscles of medial aspect of upper limb.
Median antebrachial veins (median veins of forearm) (an'-tē-BRĀ-kē-al; *ante-* = before, in front of; *brachi-* = arm)	Begin in **palmar venous plexuses,** networks of veins on palms. Drain **palmar digital veins** in fingers. Ascend anteriorly in forearms to join basilic or median cubital veins, sometimes both.	Integument and superficial muscles of palm and anterior aspect of upper limb.

EXHIBIT 21.1 CONTINUES ▶

EXHIBIT 21.1 **851**

EXHIBIT 21.I Veins of the Upper Limbs *(Figure 21.25)* CONTINUED

SCHEME OF DRAINAGE

Superior vena cava

Right brachiocephalic

Right internal jugular   Right subclavian

Right axillary   Right brachial

Right cephalic   Right basilic   Right median cubital   Right radial   Right ulnar

Right accessory cephalic

Right median antebrachial

Right dorsal venous network of hand (dorsal venous arch)

Right palmar venous plexus

Right deep palmar venous arch

Right superficial palmar venous arch

Right dorsal metacarpal

Right palmar digital

Right palmar metacarpal

Right common palmar digital

Right dorsal digital

Right proper palmar digital

Figure 21.25 Principal veins of the right upper limb.

Deep veins usually accompany arteries that have similar names.

Right cephalic

Right dorsal venous network of hand (dorsal venous arch)

Right dorsal metacarpal

Right dorsal digital

(a) Posterior view of superficial veins of hand

Right external jugular

Right subclavian

Right brachiocephalic

Right axillary

Right basilic

Right cephalic

Right internal jugular

Superior vena cava

Sternum

Right accessory cephalic

Right cephalic

Right median cubital

Right basilic

Right median antebrachial

Right palmar venous plexus

Right palmar digital

(b) Anterior view of superficial veins

Right external jugular

Right subclavian

Right brachiocephalic

Right axillary

Right brachial

Right internal jugular

Superior vena cava

Right radials

Anastomotic veins

Right ulnars

Right deep palmar venous arch

Right superficial palmar venous arch

Right common palmar digital

Right proper palmar digital

(c) Anterior view of deep veins

Biceps brachii muscle

Brachial

Median cubital

Basilic

Cephalic

Median antebrachial

(d) Anteromedial view of superficial veins of arm and forearm

? **From which vein in the upper limb is a blood sample often taken?**

EXHIBIT 21.1   **853**

EXHIBIT 21.J | **Veins of the Thorax** *(Figure 21.26)*

⊙ OBJECTIVE

• Identify the components of the azygos system of veins.

Although the brachiocephalic veins drain some portions of the thorax, most thoracic structures are drained by a network of veins, called the **azygos system** (az-Ī-gus or ā-ZĪ-gus), that runs on either side of the vertebral column (Figure 21.26). The system consists of three veins—the **azygos, hemiazygos,** and **accessory hemiazygos veins**—that show considerable variation in origin, course, tributaries, anastomoses, and termination. Ultimately they empty into the superior vena cava.

The azygos system, besides collecting blood from the thorax and abdominal wall, may serve as a bypass for the inferior vena cava, which drains blood from the lower body. Several small veins directly link the azygos system with the inferior vena cava. Larger veins that drain the lower limbs and abdomen also connect into the azygos system. If the inferior vena cava or hepatic portal vein becomes obstructed, blood that typically passes through the inferior vena cava can detour into the azygos system to return blood from the lower body to the superior vena cava.

✔ CHECKPOINT

What is the importance of the azygos system relative to the inferior vena cava?

VEINS	DESCRIPTION AND TRIBUTARIES	REGIONS DRAINED
Brachiocephalic veins (brā′-kē-ō-se-FAL-ik; brachio- = arm; -cephalic = pertaining to head)	Form by union of subclavian and internal jugular veins. Two brachiocephalic veins unite to form superior vena cava. Because superior vena cava is to right of body's midline, left brachiocephalic vein is longer than right. Right brachiocephalic vein is anterior and to right of brachiocephalic trunk and follows more vertical course. Left brachiocephalic vein is anterior to brachiocephalic trunk, left common carotid and left subclavian arteries, trachea, left vagus (X) nerve, and phrenic nerve. It approaches more horizontal position as it passes from left to right.	Head, neck, upper limbs, mammary glands, and superior thorax.
Azygos vein (az-Ī-gus = unpaired)	An unpaired vein that is anterior to vertebral column, slightly to right of midline. Usually begins at junction of right ascending lumbar and right subcostal veins near diaphragm. Arches over root of right lung at level of fourth thoracic vertebra to end in superior vena cava. Receives the following tributaries: **right posterior intercostal, hemiazygos, accessory hemiazygos, esophageal, mediastinal, pericardial,** and **bronchial veins.**	Right side of thoracic wall, thoracic viscera, and posterior abdominal wall.
Hemiazygos vein (HEM-ē-ā-zī-gus; hemi- = half)	Anterior to vertebral column and slightly to left of midline. Often begins at junction of left ascending lumbar and left subcostal veins. Terminates by joining azygos vein at about level of ninth thoracic vertebra. Receives following tributaries: ninth through eleventh **left posterior intercostal, esophageal, mediastinal,** and sometimes **accessory hemiazygos veins.**	Left side of lower thoracic wall, thoracic viscera, and left posterior abdominal wall.
Accessory hemiazygos vein	Anterior to vertebral column and to left of midline. Begins at fourth or fifth intercostal space and descends from fifth to eighth thoracic vertebra or ends in hemiazygos vein. Terminates by joining azygos vein at about level of eighth thoracic vertebra. Receives the following tributaries: fourth through eighth **left posterior intercostal veins** (first through third posterior intercostal veins drain into left brachiocephalic vein), **left bronchial,** and **mediastinal veins.**	Left side of upper thoracic wall and thoracic viscera.

SCHEME OF DRAINAGE

Figure 21.26 Principal veins of the thorax, abdomen, and pelvis.

🔑 Most thoracic structures are drained by the azygos system of veins.

Right internal jugular

Right external jugular

Right brachiocephalic

Right superior intercostal

Superior vena cava

Right posterior intercostal

Azygos

Mediastinals

Bronchial

Pericardial

Diaphragm

Hepatics

Right suprarenal

Right subcostal

Right renal

Right ascending lumbar

Right gonadal (testicular or ovarian)

Right lumbar

Right common iliac

Right internal iliac

Right external iliac

Left internal jugular

Left external jugular

Left subclavian

Left brachiocephalic

Left superior intercostal

Left axillary

Left cephalic

Left posterior intercostal

Left brachial

Accessory hemiazygos

Left basilic

Esophageals

Hemiazygos

Left inferior phrenics

Left suprarenal

Left renal

Left ascending lumbar

Left gonadal (testicular or ovarian)

Inferior vena cava

Left lumbar

Left common iliac

Middle sacral

Left internal iliac

Inguinal ligament

Left external iliac

Left femoral

Anterior view

❓ **Which vein returns blood from the abdominopelvic viscera to the heart?**

EXHIBIT 21.J **855**

EXHIBIT 21.K

Veins of the Abdomen and Pelvis

EXHIBIT 21.K

OBJECTIVE

- Identify the principal veins that drain the abdomen and pelvis.

Blood from the abdominal and pelvic viscera and abdominal wall returns to the heart via the inferior vena cava. Many small veins enter the inferior vena cava. Most carry return flow from parietal branches of the abdominal aorta, and their names correspond to the names of the arteries (see also Figure 21.26).

The inferior vena cava does not receive veins directly from the gastrointestinal tract, spleen, pancreas, and gallbladder. These organs pass their blood into a common vein, the **hepatic portal vein,** which delivers the blood to the liver. The superior mesenteric and splenic veins unite to form the hepatic portal vein (see Figure 21.28). This special flow of venous blood, called the **hepatic portal circulation,** is described shortly. After passing through the liver for processing, blood drains into the hepatic veins, which empty into the inferior vena cava.

✔CHECKPOINT

What structures do the lumbar, gonadal, renal, suprarenal, inferior phrenic, and hepatic veins drain?

VEINS	DESCRIPTION AND TRIBUTARIES	REGIONS DRAINED
Inferior vena cava	(See Exhibit 21.G.)	
Inferior phrenic veins (FREN-ik = pertaining to diaphragm)	Arise on inferior surface of diaphragm. Left inferior phrenic vein usually sends one tributary to left suprarenal vein, which empties into left renal vein, and another tributary into inferior vena cava. Right inferior phrenic vein empties into inferior vena cava.	Inferior surface of diaphragm and adjoining peritoneal tissues.
Hepatic veins (he-PAT-ik = pertaining to liver)	Typically two or three in number. Drain sinusoidal capillaries of liver. Capillaries of liver receive venous blood from capillaries of gastrointestinal organs via hepatic portal vein. **Hepatic portal vein** receives the following tributaries from gastrointestinal organs:	
	1. **Left gastric vein** arises from left side of lesser curvature of stomach and joins left side of hepatic portal vein in lesser omentum.	Terminal esophagus, stomach, liver, gallbladder, spleen, pancreas, small intestine, and large intestine.
	2. **Right gastric vein** arises from right aspect of lesser curvature of stomach and joins hepatic portal vein on its anterior surface within lesser omentum.	Lesser curvature of stomach, abdominal portion of esophagus, stomach, and duodenum.
	3. **Splenic vein** arises in spleen and crosses abdomen transversely posterior to stomach to anastomose with superior mesenteric vein to form hepatic portal vein. It receives **inferior mesenteric vein** near its junction with hepatic portal vein.	Spleen, fundus and greater curvature of stomach, pancreas, greater omentum, descending colon, sigmoid colon, and rectum.
	4. **Superior mesenteric vein** arises from numerous tributaries from most of small intestine and first half of large intestine and ascends to join splenic vein to form hepatic portal vein.	Duodenum, jejunum, ileum, cecum, appendix, ascending colon, and transverse colon.
Lumbar veins (LUM-bar = pertaining to loin)	Usually four on each side; course horizontally through posterior abdominal wall with lumbar arteries. Connect at right angles with right and left **ascending lumbar veins,** which form origin of corresponding azygos or hemiazygos vein. Join ascending lumbar veins and then connect from ascending lumbar veins to inferior vena cava.	Posterior abdominal muscle wall, lumbar vertebrae, spinal cord and spinal nerves (cauda equina) within vertebral canal, and meninges.
Suprarenal veins (soo'-pra-RĒ-nal; supra- = above)	Pass medially from adrenal (suprarenal) glands (left suprarenal vein joins left renal vein, and right suprarenal vein joins inferior vena cava).	Adrenal (suprarenal) glands.
Renal veins (RĒ-nal; ren- = kidney)	Pass anterior to renal arteries. Left renal vein is longer than right renal vein and passes anterior to abdominal aorta. It receives left testicular (or ovarian), left inferior phrenic, and usually left suprarenal veins. Right renal vein empties into inferior vena cava posterior to duodenum.	Kidneys.

VEINS	DESCRIPTION AND TRIBUTARIES	REGIONS DRAINED
Gonadal veins (gō-NAD-al; *gon-* = seed) [**testicular** (tes-TIK-ū-lar) or **ovarian** (ō-VAR-ē-an)]	Ascend with gonadal arteries along posterior abdominal wall. Called testicular veins in male. **Testicular veins** drain testes (left testicular vein joins left renal vein, and right testicular vein joins inferior vena cava). Called ovarian veins in female. **Ovarian veins** drain ovaries. Left ovarian vein joins left renal vein, and right ovarian vein joins inferior vena cava.	Testes, epididymis, ductus deferens, ovaries, and ureters.
Common iliac veins (IL-ē-ak = pertaining to ilium)	Formed by union of internal and external iliac veins anterior to sacroiliac joint and anastomose anterior to fifth lumbar vertebra to form inferior vena cava. Right common iliac is much shorter than left and is also more vertical, as inferior vena cava sits to right of midline.	Pelvis, external genitals, and lower limbs.
Internal iliac veins	Begin near superior portion of greater sciatic notch and run medial to their corresponding arteries	Muscles of pelvic wall and gluteal region, pelvic viscera, and external genitals.
External iliac veins	Companions of internal iliac arteries. Begin at inguinal ligaments as continuations of femoral veins. End anterior to sacroiliac joints where they join with internal iliac veins to form common iliac veins.	Lower abdominal wall anteriorly, cremaster muscle in males, and external genitals and lower limb.

SCHEME OF DRAINAGE

EXHIBIT 21.K **857**

EXHIBIT 21.L

Veins of the Lower Limbs *(Figure 21.27)*

- Identify the principal superficial and deep veins that drain the lower limbs.

As with the upper limbs, blood from the lower limbs is drained by both **superficial** and **deep veins.** The superficial veins often anastomose with one another and with deep veins along their length. Deep veins, for the

VEINS	DESCRIPTION AND TRIBUTARIES	REGIONS DRAINED
DEEP VEINS		
Common iliac veins	(See Exhibit 21.K.)	
External iliac veins	(See Exhibit 21.K.)	
Femoral veins (FEM-o-ral)	Accompany femoral arteries and are continuations of popliteal veins just superior to knee where veins pass through opening in adductor magnus muscle. Ascend deep to sartorius muscle and emerge from beneath muscle in femoral triangle at proximal end of thigh. Receive **deep veins of thigh (deep femoral veins)** and great saphenous veins just before penetrating abdominal wall. Pass below inguinal ligament and enter abdominopelvic region to become external iliac veins.	Skin, lymph nodes, muscles, and bones of thigh, and external genitals.
	Clinical note: In order to take **blood samples** or **pressure recordings** from the right side of the heart, a catheter is inserted into the femoral vein as it passes through the femoral triangle. The catheter passes through the external and common iliac veins, then into the inferior vena cava, and finally into the right atrium.	
Popliteal veins (pop′-li-TĒ-al = pertaining to hollow behind knee)	Formed by union of anterior and posterior tibial veins at proximal end of leg; ascend through popliteal fossa with popliteal arteries and tibial nerve. Terminate where they pass through window in adductor magnus muscle and pass to front of knee to become femoral veins. Also receive blood from small saphenous veins and tributaries that correspond to branches of popliteal artery.	Knee joint and skin, muscles, and bones around knee joint.
Posterior tibial veins (TIB-ē-al)	Begin posterior to medial malleolus at union of **medial** and **lateral plantar veins** from plantar surface of foot. Ascend through leg with posterior tibial artery and tibial nerve deep to soleus muscle. Join posterior tibial veins about two-thirds of way up leg. Join anterior tibial veins near top of interosseous membrane to form popliteal veins. On plantar surface of foot **plantar digital veins** unite to form **plantar metatarsal veins,** which parallel metatarsals. They in turn unite to form **deep plantar venous arches.** Medial and lateral plantar veins emerge from deep plantar venous arches.	Skin, muscles, and bones on plantar surface of foot, and skin, muscles, and bones from posterior and lateral aspects of leg.
Anterior tibial veins	Arise in dorsal venous arch and accompany anterior tibial artery. Ascend deep to tibialis anterior muscle on anterior surface of interosseous membrane. Pass through opening at superior end of interosseous membrane to join posterior tibial veins to form popliteal veins.	Dorsal surface of foot, ankle joint, anterior aspect of leg, knee joint, and tibiofibular joint.
SUPERFICIAL VEINS		
Great (long) saphenous veins (sa-FĒ-nus = clearly visible)	Longest veins in body; ascend from foot to groin in subcutaneous layer. Begin at medial end of dorsal venous arches of foot. **Dorsal venous arches** (VĒ-nus) are networks of veins on dorsum of foot formed by **dorsal digital veins,** which collect blood from toes, and then unite in pairs to form **dorsal metatarsal veins,** which parallel metatarsals. As dorsal metatarsal veins approach foot, they combine to form dorsal venous arches. Pass anterior to medial malleolus of tibia and then superiorly along medial aspect of leg and thigh just deep to skin. Receive tributaries from superficial tissues and connect with deep veins as well. Empty into femoral veins at groin. Have from 10 to 20 valves along their length, with more located in leg than thigh.	Integumentary tissues and superficial muscles of lower limbs, groin, and lower abdominal wall.
	Clinical note: These veins are more likely to be subject to **varicosities** than other veins in the lower limbs because they must support a long column of blood and are not well supported by skeletal muscles. The great saphenous veins are often used for prolonged administration of intravenous fluids. This is particularly important in very young children and in patients of any age who are in shock and whose veins are collapsed. In **coronary artery bypass grafting,** if multiple blood vessels need to be grafted, sections of the great saphenous vein are used along with at least one artery as a graft (see first **Clinical Note** in Exhibit 21.C). After the great saphenous vein is removed and divided into sections, the sections are used to bypass the blockages. The vein grafts are reversed so that the valves do not obstruct the flow of blood.	
Small saphenous veins	Begin at lateral aspect of dorsal venous arches of foot. Pass posterior to lateral malleolus of fibula and ascend deep to skin along posterior aspect of leg. Empty into popliteal veins in popliteal fossa, posterior to knee. Have from 9 to 12 valves. May communicate with great saphenous veins in proximal thigh.	Integumentary tissues and superficial muscles of foot and posterior aspect of leg.

most part, have the same names as corresponding arteries (Figure 21.27). All veins of the lower limbs have valves, which are more numerous than in veins of the upper limbs.

✔CHECKPOINT

What is the clinical importance of the great saphenous veins?

SCHEME OF DRAINAGE

Inferior vena cava
↑
Right common iliac
↑
Right external iliac ← Right internal iliac
↑
Right femoral ← Right deep vein of the thigh (deep femoral)
↑
Right popliteal
↑
Right posterior tibial | Right anterior tibial

Right great saphenous | Right small saphenous | Right fibular (peroneal)

Right medial and lateral plantar | Right dorsal venous arch

Right dorsal venous arch

Right dorsal metatarsal

Right dorsal digital

Right deep plantar venous arch

Right plantar metatarsal

Right plantar digital

EXHIBIT 21.L CONTINUES ▶

EXHIBIT 21.L **859**

Figure 21.27 Principal veins of the pelvis and lower limbs.

Deep veins usually bear the names of their companion arteries.

Inferior vena cava

Right common iliac

Right internal iliac

Right external iliac

Left common iliac

Right deep vein of the thigh (deep femoral)

Right femoral

Right accessory saphenous

Right great saphenous

Right popliteal

Right anterior tibial

Right posterior tibial

Right small saphenous

Right fibular (peroneal)

Right anterior tibial

Right great saphenous

Right small saphenous

Right dorsal venous arch

Right dorsal metatarsal

Right dorsal digital

Right medial plantar

Right deep plantar venous arch

Right plantar digital

Right lateral plantar

Right plantar metatarsal

(a) Anterior view

(b) Posterior view

Which veins of the lower limb are superficial?

The Hepatic Portal Circulation

The **hepatic portal circulation** (he-PAT-ik) carries venous blood from the gastrointestinal organs and spleen to the liver. A vein that carries blood from one capillary network to another is called a **portal vein.** The **hepatic portal vein** (*hepat-* = liver) receives blood from capillaries of gastrointestinal organs and the spleen and delivers it to the sinusoids of the liver (Figure 21.28). After a meal, hepatic portal blood is rich in nutrients absorbed from the gastrointestinal tract. The liver stores some of them and modifies others before they pass into the general circulation. For example, the liver converts glucose into glycogen for storage, reducing blood glucose level shortly after a meal. The liver also detoxifies harmful substances, such as alcohol, that have been absorbed from the gastrointestinal tract and destroys bacteria by phagocytosis.

The superior mesenteric and splenic veins unite to form the hepatic portal vein. The **superior mesenteric vein** (mez-en-TER-ik) drains blood from the small intestine and portions of the large intestine, stomach, and pancreas through the *jejunal, ileal, ileocolic* (il′-ē-ō-KOL-ik), *right colic, middle colic, pancreaticoduodenal* (pan-krē-at′-i-kō-doo′-ō-DĒ-nal), and *right gastroepiploic veins* (gas′-trō-ep′-i-PLŌ-ik). The **splenic vein** drains blood from the stomach, pancreas, and portions of the large intestine through the *short gastric, left gastroepiploic, pancreatic,* and *inferior*

Figure 21.28 Hepatic portal circulation. A schematic diagram of blood flow through the liver, including arterial circulation, is shown in (b). As usual, deoxygenated blood is indicated in blue, and oxygenated blood in red.

The hepatic portal circulation delivers venous blood from the organs of the gastrointestinal tract and spleen to the liver.

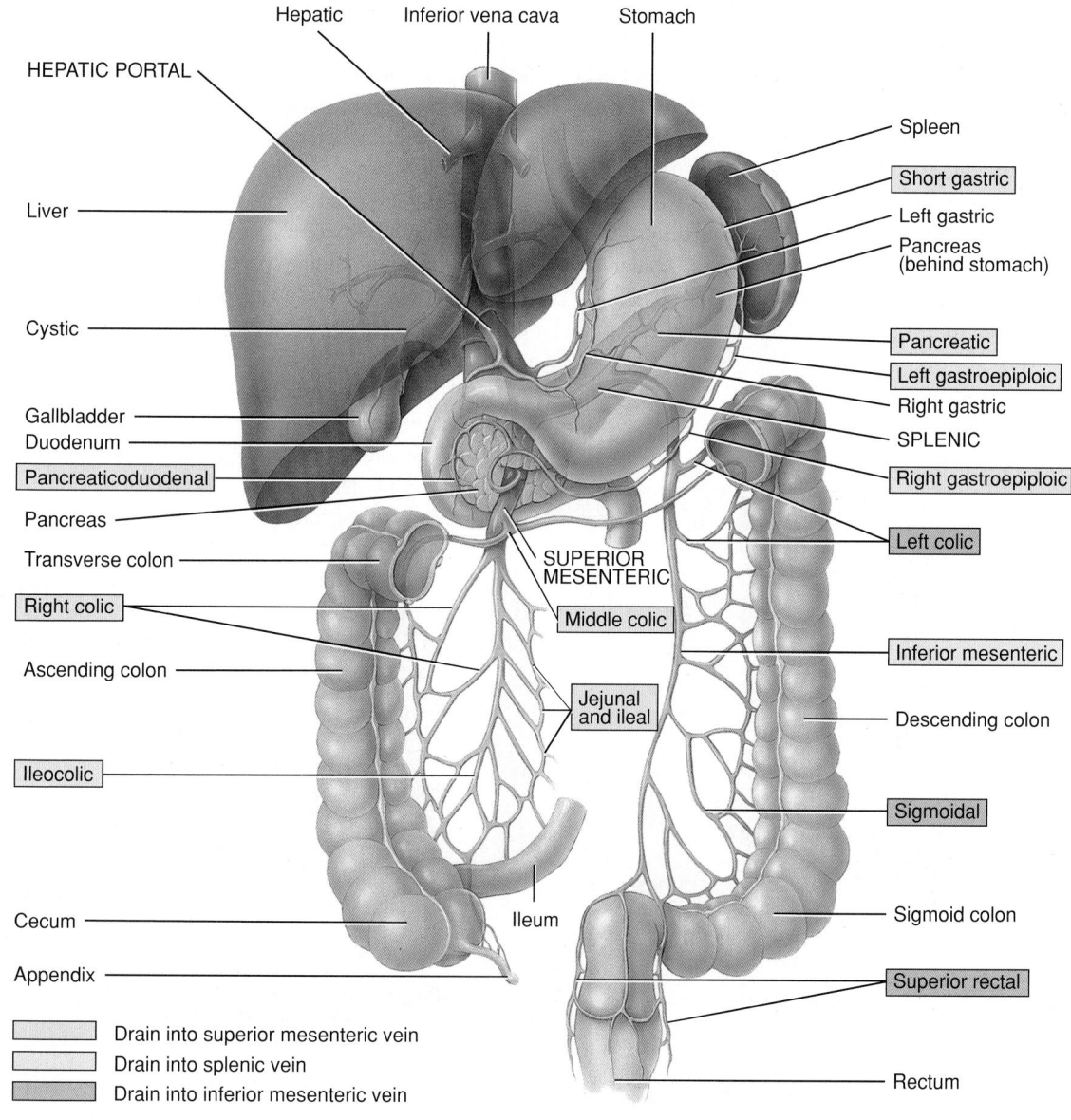

(a) Anterior view of veins draining into hepatic portal vein

FIGURE 21.28 CONTINUES ▶

mesenteric veins. The inferior mesenteric vein, which passes into the splenic vein, drains portions of the large intestine through the *superior rectal, sigmoidal,* and *left colic veins.* The *right* and *left gastric veins,* which open directly into the hepatic portal vein, drain the stomach. The *cystic vein,* which also opens into the hepatic portal vein, drains the gallbladder.

At the same time the liver is receiving nutrient-rich but deoxygenated blood via the hepatic portal vein, it is also receiving oxygenated blood via the hepatic artery, a branch of the celiac trunk. The oxygenated blood mixes with the deoxygenated blood in sinusoids. Eventually, blood leaves the sinusoids of the liver through the **hepatic veins,** which drain into the inferior vena cava.

The Pulmonary Circulation

The **pulmonary circulation** (PUL-mō-ner′-ē; *pulmo-* = lung) carries deoxygenated blood from the right ventricle to the air sacs (alveoli) within the lungs and returns oxygenated blood from the air sacs to the left atrium (Figure 21.29). The **pulmonary trunk** emerges from the right ventricle and passes superiorly, posteriorly, and to the left. It then divides into two branches: the **right pulmonary artery** to the right lung and the **left pulmonary artery** to the left lung. After birth, the pulmonary arteries are the only arteries that carry deoxygenated blood. On entering the lungs, the branches divide and subdivide until finally they form capillaries around the air sacs (alveoli) within the lungs. CO_2 passes from the blood into the air sacs and is exhaled. Inhaled O_2 passes from the air within the lungs into the blood. The pulmonary capillaries unite to form venules and eventually **pul-**

monary veins, which exit the lungs and carry the oxygenated blood to the left atrium. Two left and two right pulmonary veins enter the left atrium. After birth, the pulmonary veins are the only veins that carry oxygenated blood. Contractions of the left ventricle then eject the oxygenated blood into the systemic circulation.

The Fetal Circulation

The circulatory system of a fetus, called the **fetal circulation,** exists only in the fetus and contains special structures that allow the developing fetus to exchange materials with its mother (Figure 21.30). It differs from the postnatal (after birth) circulation because the lungs, kidneys, and gastrointestinal organs do not begin to function until birth. The fetus obtains O_2 and nutrients from the maternal blood and eliminates CO_2 and other wastes into it.

The exchange of materials between fetal and maternal circulations occurs through the **placenta** (pla-SEN-ta), which forms inside the mother's uterus and attaches to the umbilicus (navel) of the fetus by the **umbilical cord** (um-BIL-i-kal). The placenta communicates with the mother's cardiovascular system through many small blood vessels that emerge from the uterine wall. The umbilical cord contains blood vessels that branch into capillaries in the placenta. Wastes from the fetal blood diffuse out of the capillaries, into spaces containing maternal blood (intervillous spaces) in the placenta, and finally into the mother's uterine veins. Nutrients travel the opposite route—from the maternal blood vessels to the intervillous spaces to the fetal capillaries. Normally, there is no direct mixing of maternal and fetal blood because all exchanges occur by diffusion through capillary walls.

Blood passes from the fetus to the placenta via two **umbilical arteries** in the umbilical cord (Figure 21.30a, c). These branches of the internal iliac (hypogastric) arteries are within the umbilical cord. At the placenta, fetal blood picks up O_2 and nutrients and eliminates CO_2 and wastes. The oxygenated blood returns from the placenta via a single **umbilical vein** in the umbilical cord. This vein ascends to the liver of the fetus, where it divides into two branches. Some blood flows through the branch that joins the hepatic portal vein and enters the liver, but most of the blood flows into the second branch, the **ductus venosus** (DUK-tus ve-NŌ-sus), which drains into the inferior vena cava.

FIGURE 21.28 CONTINUED

(b) Scheme of principal blood vessels of hepatic portal circulation and arterial supply and venous drainage of liver

Which veins carry blood away from the liver?

Figure 21.29 Pulmonary circulation.

The pulmonary circulation brings deoxygenated blood from the right ventricle to the lungs and returns oxygenated blood from the lungs to the left atrium.

Superior vena cava

Right pulmonary artery

Pulmonary trunk

Right pulmonary veins

Right lung

Inferior vena cava

Arch of aorta

Ascending aorta

Left pulmonary artery

Left pulmonary veins

Left lung

Diaphragm

Abdominal aorta

(a) Anterior view

Right pulmonary artery

Left pulmonary artery

Pulmonary trunk

Air sacs (alveoli) of right lung

Air sacs (alveoli) of left lung

Right ventricle

Right pulmonary veins

Left pulmonary veins

Left atrium

(b) Scheme of pulmonary circulation

? After birth, which are the only arteries that carry deoxygenated blood?

Deoxygenated blood returning from lower body regions of the fetus mingles with oxygenated blood from the ductus venosus in the inferior vena cava. This mixed blood then enters the right atrium. Deoxygenated blood returning from upper body regions of the fetus enters the superior vena cava and also passes into the right atrium.

Most of the fetal blood does not pass from the right ventricle to the lungs, as it does in postnatal circulation, because an opening called the **foramen ovale** (fō-RĀ-men ō-VAL-ē) exists in the septum between the right and left atria. Most of the blood that enters the right atrium passes through the foramen ovale into the left atrium and joins the systemic circulation. The blood that does pass into the right ventricle is pumped into the pulmonary trunk, but little of this blood reaches the nonfunctioning fetal lungs. In-

stead, most is sent through the **ductus arteriosus** (ar-tē-rē-Ō-sus), a vessel that connects the pulmonary trunk with the aorta. The blood in the aorta is carried to all fetal tissues through the systemic circulation. When the common iliac arteries branch into the external and internal iliacs, part of the blood flows into the internal iliacs, into the umbilical arteries, and back to the placenta for another exchange of materials.

After birth, when pulmonary (lung), renal (kidney), and digestive functions begin, the following vascular changes occur (Figure 21.30b):

1. When the umbilical cord is tied off, blood no longer flows through the umbilical arteries, they fill with connective tissue,

Figure 21.30 Fetal circulation and changes at birth. The gold boxes between parts (a) and (b) describe the fate of certain fetal structures once postnatal circulation is established.

The lungs and gastrointestinal organs do not begin to function until birth.

Right atrium

Arch of aorta

Superior vena cava

Left atrium

DUCTUS ARTERIOSUS becomes Ligamentum arteriosum

Lung

Pulmonary artery

Pulmonary veins

Right ventricle

Heart

FORAMEN OVALE becomes Fossa ovalis

Left ventricle

Inferior vena cava

Liver

DUCTUS VENOSUS becomes Ligamentum venosum

Hepatic portal vein

UMBILICAL VEIN becomes Ligamentum teres

Umbilicus

Inferior vena cava

Abdominal aorta

Common iliac artery

UMBILICAL ARTERIES become Medial umbilical ligaments

Urinary bladder

Urethra

UMBILICAL CORD

Placenta

(a) Fetal circulation

(b) Circulation at birth

■ Oxygenated blood

■ Mixed oxygenated and deoxygenated blood

■ Deoxygenated blood

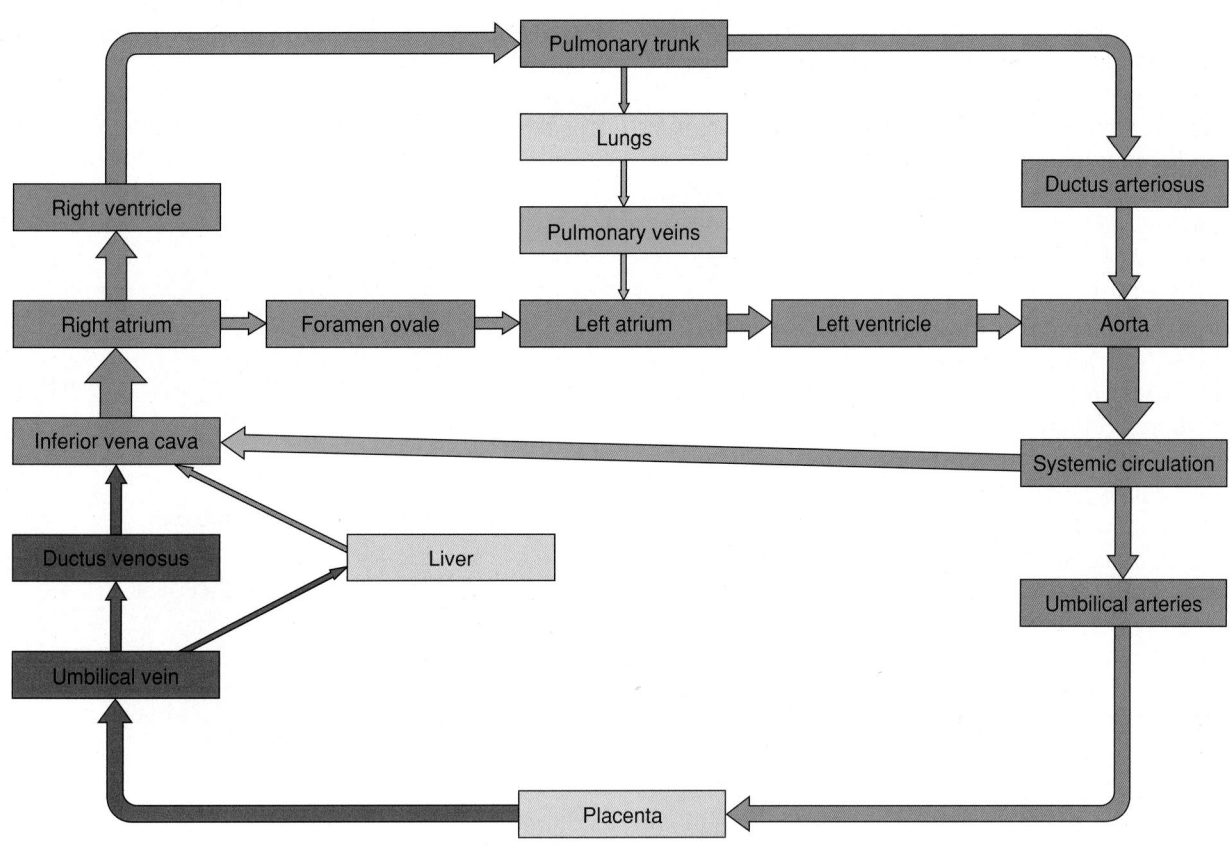

(c) Scheme of fetal circulation

? **Which structure provides for exchange of materials between mother and fetus?**

and the distal portions of the umbilical arteries become fibrous cords called the **medial umbilical ligaments.** Although the arteries are closed functionally only a few minutes after birth, complete obliteration of the lumens may take 2 to 3 months.

2. The umbilical vein collapses but remains as the **ligamentum teres** (TE-rez) **(round ligament),** a structure that attaches the umbilicus to the liver.

3. The ductus venosus collapses but remains as the **ligamentum venosum** (ve-NŌ-sum), a fibrous cord on the inferior surface of the liver.

4. The placenta is expelled as the **"afterbirth."**

5. The foramen ovale normally closes shortly after birth to become the **fossa ovalis,** a depression in the interatrial septum. When an infant takes its first breath, the lungs expand and blood flow to the lungs increases. Blood returning from the lungs to the heart increases pressure in the left atrium. This closes the foramen ovale by pushing the valve that guards it against the interatrial septum. Permanent closure occurs in about a year.

6. The ductus arteriosus closes by vasoconstriction almost immediately after birth and becomes the **ligamentum arteriosum** (ar-tēr′-ē-Ō-sum). Complete anatomical obliteration of the lumen takes 1 to 3 months.

✔ **CHECKPOINT**

23. Diagram the hepatic portal circulation. Why is this route important?
24. Diagram the route of the pulmonary circulation.
25. Discuss the anatomy and physiology of the fetal circulation. Indicate the function of the umbilical arteries, umbilical vein, ductus venosus, foramen ovale, and ductus arteriosus.

21.8 DEVELOPMENT OF BLOOD VESSELS AND BLOOD

⦿ **OBJECTIVE**

• Describe the development of blood vessels and blood.

The development of blood cells and the formation of blood vessels begins outside the embryo as early as 15 to 16 days in the **mesoderm** of the wall of the yolk sac, chorion, and connecting stalk. About 2 days later, blood vessels form within the embryo. The early formation of the cardiovascular system is linked to the small amount of yolk in the ovum and yolk sac. As the embryo develops rapidly during the third week, there is a greater need to develop a cardiovascular system to supply sufficient nutrients to the embryo and remove wastes from it.

Blood vessels and blood cells develop from the same precursor cell, called a **hemangioblast** (hē-MAN-jē-ō-blast; *hema-* = blood; *-blast* = immature stage). Once mesenchyme develops into hemangioblasts, they can give rise to cells that produce blood vessels (angioblasts) or cells that produce blood cells (pluripotent stem cells).

Blood vessels develop from **angioblasts** (AN-jē-ō-blasts), which are derived from hemangioblasts. Angioblasts aggregate to form isolated masses and cords throughout the embryonic discs called **blood islands** (Figure 21.31). Spaces soon appear in the islands and become the lumens of the blood vessels. Some of the angioblasts immediately around the spaces give rise to the *endothelial lining of the blood vessels.* Angioblasts around the endothelium form the *tunics* (interna, media, and externa) of the larger blood vessels. Growth and fusion of blood islands form an extensive network of blood vessels throughout the embryo. By continuous branching, blood vessels outside the embryo connect with those inside the embryo, linking the embryo with the placenta.

Blood cells develop from **pluripotent stem cells** (ploo-RIP-ō-tent) derived from hemangioblasts. This development occurs in the walls of blood vessels in the yolk sac, chorion, and allantois at about 3 weeks after fertilization. Blood formation in the embryo itself begins at about the fifth week in the liver and the twelfth week in the spleen, red bone marrow, and thymus.

✔ CHECKPOINT

26. What are the sites of blood cell production outside the embryo and within the embryo?

21.9 AGING AND THE CARDIOVASCULAR SYSTEM

◉ OBJECTIVE

• Explain the effects of aging on the cardiovascular system.

General changes in the cardiovascular system associated with aging include decreased compliance (distensibility) of the aorta, reduction in cardiac muscle fiber size, progressive loss of cardiac muscular strength, reduced cardiac output, a decline in maximum heart rate, and an increase in systolic blood pressure. Total blood cholesterol tends to increase with age, as does low-density lipoprotein (LDL); high-density lipoprotein (HDL) tends to decrease. There is an increase in the incidence of coronary artery disease (CAD), the major cause of heart disease and death in older Americans. Congestive heart failure (CHF), a set of symptoms associated with impaired pumping of the heart, is also prevalent in older individuals. Changes in blood vessels that serve brain tissue—for example, atherosclerosis—reduce nourishment to the brain and result in malfunction or death of brain cells. By

Figure 21.31 Development of blood vessels and blood cells from blood islands.

 Blood vessel development begins in the embryo on about the fifteenth or sixteenth day.

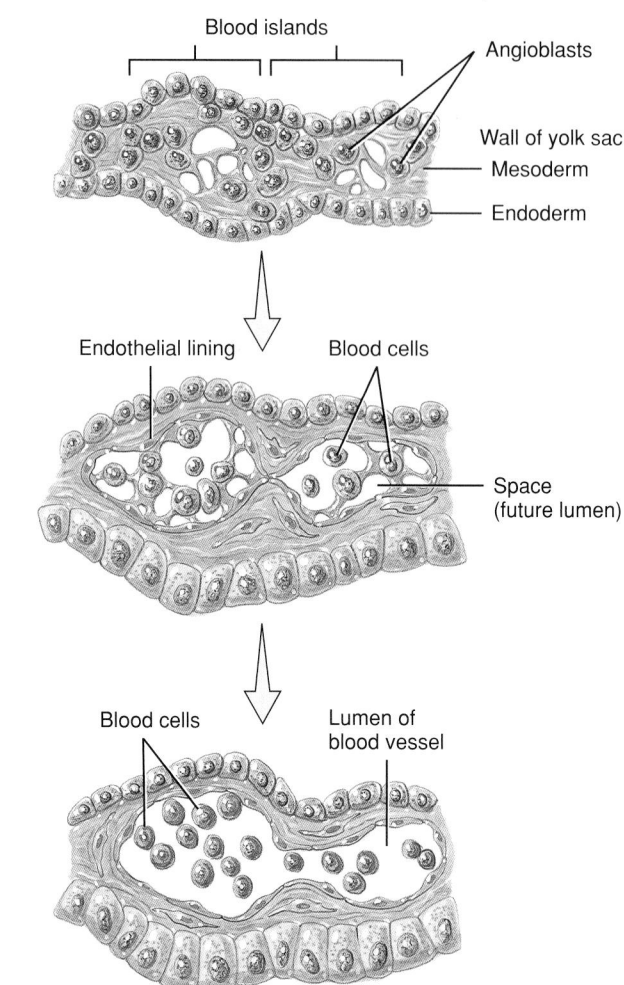

? From which germ cell layer are blood vessels and blood derived?

age 80, cerebral blood flow is 20% less and renal blood flow is 50% less than in the same person at age 30 because of the effects of aging on blood vessels.

✔ CHECKPOINT

27. How does aging affect the heart?

• • •

To appreciate the many ways the blood, heart, and blood vessels contribute to homeostasis of other body systems, examine *Focus on Homeostasis: The Cardiovascular System.*

BODY SYSTEM	CONTRIBUTION OF THE CARDIOVASCULAR SYSTEM
For all body systems	The heart pumps blood through blood vessels to body tissues, delivering oxygen and nutrients and removing wastes by means of capillary exchange. Circulating blood keeps body tissues at a proper temperature.
Integumentary system	Blood delivers clotting factors and white blood cells that aid in hemostasis when skin is damaged and contribute to repair of injured skin. Changes in skin blood flow contribute to body temperature regulation by adjusting the amount of heat loss via the skin. Blood flowing in skin may give skin a pink hue.
Skeletal system	Blood delivers calcium and phosphate ions that are needed for building bone extracellular matrix, hormones that govern building and breakdown of bone extracellular matrix, and erythropoietin that stimulates production of red blood cells by red bone marrow.
Muscular system	Blood circulating through exercising muscle removes heat and lactic acid.
Nervous system	Endothelial cells lining choroid plexuses in brain ventricles help produce cerebrospinal fluid (CSF) and contribute to the blood–brain barrier.
Endocrine system	Circulating blood delivers most hormones to their target tissues. Atrial cells secrete atrial natriuretic peptide.
Lymphatic system and immunity	Circulating blood distributes lymphocytes, antibodies, and macrophages that carry out immune functions. Lymph forms from excess interstitial fluid, which filters from blood plasma due to blood pressure generated by the heart.
Respiratory system	Circulating blood transports oxygen from the lungs to body tissues and carbon dioxide to the lungs for exhalation.
Digestive system	Blood carries newly absorbed nutrients and water to liver. Blood distributes hormones that aid digestion.
Urinary system	Heart and blood vessels deliver 20% of the resting cardiac output to the kidneys, where blood is filtered, needed substances are reabsorbed, and unneeded substances remain as part of urine, which is excreted.
Reproductive systems	Vasodilation of arterioles in penis and clitoris cause erection during sexual intercourse. Blood distributes hormones that regulate reproductive functions.

THE CARDIOVASCULAR SYSTEM

867

Hypertension

About 50 million Americans have **hypertension** (hī′-per-TEN-shun), or persistently high blood pressure. It is the most common disorder affecting the heart and blood vessels and is the major cause of heart failure, kidney disease, and stroke. In May 2003, the Joint National Committee on Prevention, Detection, Evaluation, and Treatment of High Blood Pressure published new guidelines for hypertension because clinical studies have linked what were once considered fairly low blood pressure readings to an increased risk of cardiovascular disease. The new guidelines are as follows:

Category	Systolic (mmHg)	Diastolic (mmHg)
Normal	Less than 120 *and*	Less than 80
Prehypertension	120–139 *or*	80–89
Stage 1 hypertension	140–159 *or*	90–99
Stage 2 hypertension	Greater than 160 *or*	Greater than 100

Using the new guidelines, the normal classification was previously considered optimal; prehypertension now includes many more individuals previously classified as normal or high-normal; stage 1 hypertension is the same as in previous guidelines; and stage 2 hypertension now combines the previous stage 2 and stage 3 categories since treatment options are the same for the former stages 2 and 3.

Types and Causes of Hypertension

Between 90 and 95% of all cases of hypertension are **primary hypertension,** a persistently elevated blood pressure that cannot be attributed to any identifiable cause. The remaining 5–10% of cases are **secondary hypertension,** which has an identifiable underlying cause. Several disorders cause secondary hypertension:

• *Obstruction of renal blood flow* or disorders that damage renal tissue may cause the kidneys to release excessive amounts of renin into the blood. The resulting high level of angiotensin II causes vasoconstriction, thus increasing systemic vascular resistance.

• *Hypersecretion of aldosterone*—resulting, for instance, from a tumor of the adrenal cortex—stimulates excess reabsorption of salt and water by the kidneys, which increases the volume of body fluids.

• *Hypersecretion of epinephrine and norepinephrine* may occur by a **pheochromocytoma** (fē-ō-krō′-mō-sī-TŌ-ma), a tumor of the adrenal medulla. Epinephrine and norepinephrine increase heart rate and contractility and increase systemic vascular resistance.

Damaging Effects of Untreated Hypertension

High blood pressure is known as the "silent killer" because it can cause considerable damage to the blood vessels, heart, brain, and kidneys before it causes pain or other noticeable symptoms. It is a major risk factor for the number-one (heart disease) and number-three (stroke) causes of death in the United States. In blood vessels, hypertension causes thickening of the tunica media, accelerates development of atherosclerosis and coronary artery disease, and increases systemic vascular resistance. In the heart, hypertension increases the afterload, which forces the ventricles to work harder to eject blood.

The normal response to an increased workload due to vigorous and regular exercise is hypertrophy of the myocardium, especially in the wall of the left ventricle. This is a positive effect that makes the heart a more efficient pump. An increased afterload, however, leads to myocardial hypertrophy that is accompanied by muscle damage and fibrosis (a buildup of collagen fibers between the muscle fibers). As a result, the left ventricle enlarges, weakens, and dilates. Because arteries in the brain are usually less protected by surrounding tissues than are the major arteries in other parts of the body, prolonged hypertension can eventually cause them to rupture, resulting in a stroke. Hypertension also damages kidney arterioles, causing them to thicken, which narrows the lumen; because the blood supply to the kidneys is thereby reduced, the kidneys secrete more renin, which elevates the blood pressure even more.

Lifestyle Changes to Reduce Hypertension

Although several categories of drugs (described next) can reduce elevated blood pressure, the following lifestyle changes are also effective in managing hypertension:

• *Lose weight.* This is the best treatment for high blood pressure short of using drugs. Loss of even a few pounds helps reduce blood pressure in overweight hypertensive individuals.

• *Limit alcohol intake.* Drinking in moderation may lower the risk of coronary heart disease, mainly among males over 45 and females over 55. Moderation is defined as no more than one 12-oz beer per day for females and no more than two 12-oz beers per day for males.

• *Exercise.* Becoming more physically fit by engaging in moderate activity (such as brisk walking) several times a week for 30 to 45 minutes can lower systolic blood pressure by about 10 mmHg.

• *Reduce intake of sodium (salt).* Roughly half the people with hypertension are "salt sensitive." For them, a high-salt diet appears to promote hypertension, and a low-salt diet can lower their blood pressure.

• *Maintain recommended dietary intake of potassium, calcium, and magnesium.* Higher levels of potassium, calcium, and magnesium in the diet are associated with a lower risk of hypertension.

• *Don't smoke or quit smoking.* Smoking has devastating effects on the heart and can augment the damaging effects of high blood pressure by promoting vasoconstriction.

• *Manage stress.* Various meditation and biofeedback techniques help some people reduce high blood pressure. These methods may work by decreasing the daily release of epinephrine and norepinephrine by the adrenal medulla.

Drug Treatment of Hypertension

Drugs having several different mechanisms of action are effective in lowering blood pressure. Many people are successfully treated with *diuretics* (dī-ū-RET-iks), agents that decrease blood pressure by decreasing blood volume, because they increase elimination of water and salt in the urine. *ACE (angiotensin-converting enzyme) inhibitors* block formation of angiotensin II and thereby promote vasodilation and decrease the secretion of aldosterone. *Beta blockers* (BĀ-ta) reduce blood pressure by inhibiting the secretion of renin and by decreasing heart rate and contractility. *Vasodilators* relax the smooth muscle in arterial walls, causing vasodilation and lowering blood pressure by lowering systemic vascular resistance. An important category of vasodilators are the *calcium channel blockers,* which slow the inflow of Ca^{2+} into vascular smooth muscle cells. They reduce the heart's workload by slowing Ca^{2+} entry into pacemaker cells and regular myocardial fibers, thereby decreasing heart rate and the force of myocardial contraction.

MEDICAL TERMINOLOGY

Aneurysm (AN-ū-rizm) A thin, weakened section of the wall of an artery or a vein that bulges outward, forming a balloonlike sac. Common causes are atherosclerosis, syphilis, congenital blood vessel defects, and trauma. If untreated, the aneurysm enlarges and the blood vessel wall becomes so thin that it bursts. The result is massive hemorrhage with shock, severe pain, stroke, or death. Treatment may involve surgery in which the weakened area of the blood vessel is removed and replaced with a graft of synthetic material.

Aortography (ā′-or-TOG-ra-fē) X-ray examination of the aorta and its main branches after injection of a radiopaque dye.

Carotid endarterectomy (ka-ROT-id end′-ar-ter-EK-tō-mē) The removal of atherosclerotic plaque from the carotid artery to restore greater blood flow to the brain.

Claudication (klaw′-di-KĀ-shun) Pain and lameness or limping caused by defective circulation of the blood in the vessels of the limbs.

Deep vein thrombosis (DVT) The presence of a thrombus (blood clot) in a deep vein of the lower limbs. It may lead to (1) pulmonary embolism, if the thrombus dislodges and then lodges within the pulmonary arterial blood flow, and (2) postphlebitic syndrome, which consists of edema, pain, and skin changes due to destruction of venous valves.

Doppler ultrasound scanning Imaging technique commonly used to measure blood flow. A transducer is placed on the skin and an image is displayed on a monitor that provides the exact position and severity of a blockage.

Femoral angiography An imaging technique in which a contrast medium is injected into the femoral artery and spreads to other arteries in the lower limb, and then a series of radiographs are taken of one or more sites. It is used to diagnose narrowing or blockage of arteries in the lower limbs.

Hypotension (hī-pō-TEN-shun) Low blood pressure; most commonly used to describe an acute drop in blood pressure, as occurs during excessive blood loss.

Normotensive (nor′-mō-TEN-siv) Characterized by normal blood pressure.

Occlusion (ō-KLOO-zhun) The closure or obstruction of the lumen of a structure such as a blood vessel. An example is an atherosclerotic plaque in an artery.

Orthostatic hypotension (or′-thō-STAT-ik; *ortho-* = straight; *-static* = causing to stand) An excessive lowering of systemic blood pressure when a person assumes an erect or semierect posture; it is usually a sign of a disease. May be caused by excessive fluid loss, certain drugs, and cardiovascular or neurogenic factors. Also called **postural hypotension.**

Phlebitis (fle-BĪ-tis; *phleb-* = vein) Inflammation of a vein, often in a leg.

Thrombectomy (throm-BEK-tō-mē; *thrombo-* = clot) An operation to remove a blood clot from a blood vessel.

Thrombophlebitis (throm′-bō-fle-BĪ-tis) Inflammation of a vein involving clot formation. Superficial thrombophlebitis occurs in veins under the skin, especially in the calf.

Venipuncture (VEN-i-punk-chur; *vena-* = vein) The puncture of a vein, usually to withdraw blood for analysis or to introduce a solution, for example, an antibiotic. The median cubital vein is frequently used.

White coat (office) hypertension A stress-induced syndrome found in patients who have elevated blood pressure when being examined by health-care personnel, but otherwise have normal blood pressure.

CHAPTER REVIEW AND RESOURCE SUMMARY

Review

Resource

21.1 Structure and Function of Blood Vessels

1. Arteries carry blood away from the heart. The wall of an artery consists of a tunica interna, a tunica media (which maintains elasticity and contractility), and a tunica externa. Large arteries are termed elastic (conducting) arteries, and medium-sized arteries are called muscular (distributing) arteries.

2. Many arteries anastomose: The distal ends of two or more vessels unite. An alternative blood route from an anastomosis is called collateral circulation. Arteries that do not anastomose are called end arteries.

3. Arterioles are small arteries that deliver blood to capillaries. Through constriction and dilation, arterioles assume a key role in regulating blood flow from arteries into capillaries and in altering arterial blood pressure.

4. Capillaries are microscopic blood vessels through which materials are exchanged between blood and tissue cells; some capillaries are continuous, and others are fenestrated. Capillaries branch to form an extensive network throughout a tissue. This network increases surface area, allowing a rapid exchange of large quantities of materials.

5. Precapillary sphincters regulate blood flow through capillaries.

6. Microscopic blood vessels in the liver are called sinusoids.

7. Venules are small vessels that continue from capillaries and merge to form veins.

8. Veins consist of the same three tunics as arteries but have a thinner tunica interna and a thinner tunica media. The lumen of a vein is also larger than that of a comparable artery. Veins contain valves to prevent backflow of blood. Weak valves can lead to varicose veins.

9. Vascular (venous) sinuses are veins with very thin walls.

10. Systemic veins are collectively called blood reservoirs because they hold a large volume of blood. If the need arises, this blood can be shifted into other blood vessels through vasoconstriction of veins. The principal blood reservoirs are the veins of the abdominal organs (liver and spleen) and skin.

Resource column:
Anatomy Overview - The Cardiovascular System
Anatomy Overview - Arteries and Arterioles
Anatomy Overview - Capillaries
Figure 21.1 - Comparative Structure of Blood Vessels
Exercise - Vein Archery
Exercise - Concentrate on Blood and Vessels

Review	Resource

21.2 Capillary Exchange

1. Substances enter and leave capillaries by diffusion, transcytosis, or bulk flow.
2. The movement of water and solutes (except proteins) through capillary walls depends on hydrostatic and osmotic pressures.
3. The near equilibrium between filtration and reabsorption in capillaries is called Starling's law of the capillaries.
4. Edema is an abnormal increase in interstitial fluid.

21.3 Hemodynamics: Factors Affecting Blood Flow

1. The velocity of blood flow is inversely related to the cross-sectional area of blood vessels; blood flows slowest where cross-sectional area is greatest. The velocity of blood flow decreases from the aorta to arteries to capillaries and increases in venules and veins.
2. Blood pressure and resistance determine blood flow.
3. Blood flows from regions of higher to lower pressure. The higher the resistance, however, the lower the blood flow.
4. Cardiac output equals the mean arterial pressure divided by total resistance (CO = MAP ÷ R).
5. Blood pressure is the pressure exerted on the walls of a blood vessel.
6. Factors that affect blood pressure are cardiac output, blood volume, viscosity, resistance, and the elasticity of arteries.
7. As blood leaves the aorta and flows through the systemic circulation, its pressure progressively falls to 0 mmHg by the time it reaches the right ventricle.
8. Resistance depends on blood vessel diameter, blood viscosity, and total blood vessel length.
9. Venous return depends on pressure differences between the venules and the right ventricle.
10. Blood return to the heart is maintained by several factors, including skeletal muscular contractions, valves in veins (especially in the limbs), and pressure changes associated with breathing.

21.4 Control of Blood Pressure and Blood Flow

1. The cardiovascular (CV) center is a group of neurons in the medulla oblongata that regulates heart rate, contractility, and blood vessel diameter.
2. The cardiovascular center receives input from higher brain regions and sensory receptors (baroreceptors and chemoreceptors).
3. Output from the cardiovascular center flows along sympathetic and parasympathetic axons. Sympathetic impulses propagated along cardioaccelerator nerves increase heart rate and contractility; parasympathetic impulses propagated along vagus nerves decrease heart rate.
4. Baroreceptors monitor blood pressure, and chemoreceptors monitor blood levels of O_2, CO_2, and hydrogen ions. The carotid sinus reflex helps regulate blood pressure in the brain. The aortic reflex regulates general systemic blood pressure.
5. Hormones that help regulate blood pressure are epinephrine, norepinephrine, ADH (antidiuretic hormone), angiotensin II, and ANP (atrial natriuretic peptide).
6. Autoregulation refers to local, automatic adjustments of blood flow in a given region to meet a particular tissue's need.
7. O_2 level is the principal stimulus for autoregulation.

21.5 Checking Circulation

1. Pulse is the alternate expansion and elastic recoil of an artery wall with each heartbeat. It may be felt in any artery that lies near the surface or over a hard tissue.
2. A normal resting pulse (heart) rate is 70–80 beats/min.
3. Blood pressure is the pressure exerted by blood on the wall of an artery when the left ventricle undergoes systole and then diastole. It is measured by the use of a sphygmomanometer.
4. Systolic blood pressure (SBP) is the arterial blood pressure during ventricular contraction. Diastolic blood pressure (DBP) is the arterial blood pressure during ventricular relaxation. Normal blood pressure is less than 120/80.
5. Pulse pressure is the difference between systolic and diastolic blood pressure. It normally is about 40 mmHg.

Review	Resource

21.6 Shock and Homeostasis

1. Shock is a failure of the cardiovascular system to deliver enough O_2 and nutrients to meet the metabolic needs of cells.
2. Types of shock include hypovolemic, cardiogenic, vascular, and obstructive.
3. Signs and symptoms of shock include systolic blood pressure less than 90 mmHg; rapid resting heart rate; weak, rapid pulse; clammy, cool, pale skin; sweating; hypotension; altered mental state; decreased urinary output; thirst; and acidosis.

21.7 Circulatory Routes

1. The two main circulatory routes are the systemic and pulmonary circulations.
2. Among the subdivisions of the systemic circulation are the coronary (cardiac) circulation and the hepatic portal circulation.
3. The systemic circulation carries oxygenated blood from the left ventricle through the aorta to all parts of the body, including some lung tissue, but *not* the air sacs (alveoli) of the lungs, and returns the deoxygenated blood to the right atrium.
4. The aorta is divided into the ascending aorta, the arch of the aorta, and the descending aorta. Each section gives off arteries that branch to supply the whole body.
5. Blood returns to the heart through the systemic veins. All veins of the systemic circulation drain into the superior or inferior venae cavae or the coronary sinus, which in turn empty into the right atrium.
6. The major blood vessels of the systemic circulation may be reviewed in Exhibits 21.A–21.L.
7. The hepatic portal circulation directs venous blood from the gastrointestinal organs and spleen into the hepatic portal vein of the liver before it returns to the heart. It enables the liver to utilize nutrients and detoxify harmful substances in the blood.
8. The pulmonary circulation takes deoxygenated blood from the right ventricle to the alveoli within the lungs and returns oxygenated blood from the alveoli to the left atrium.
9. Fetal circulation exists only in the fetus. It involves the exchange of materials between fetus and mother via the placenta.
10. The fetus derives O_2 and nutrients from and eliminates CO_2 and wastes into maternal blood. At birth, when pulmonary (lung), digestive, and liver functions begin to operate, the special structures of fetal circulation are no longer needed.

Anatomy Overview - Comparison of Circulatory Routes
Anatomy Overview - Pulmonary Circulation
Figure 21.28 - Hepatic Portal Circulation
Figure 21.30 - Fetal Circulation and Changes at Birth

21.8 Development of Blood Vessels and Blood

1. Blood vessels develop from mesenchyme (hemangioblasts → angioblasts → blood islands) in mesoderm called blood islands.
2. Blood cells also develop from mesenchyme (hemangioblasts → pluripotent stem cells).
3. The development of blood cells from pluripotent stem cells derived from angioblasts occurs in the walls of blood vessels in the yolk sac, chorion, and allantois at about 3 weeks after fertilization. Within the embryo, blood is produced by the liver at about the fifth week and in the spleen, red bone marrow, and thymus at about the twelfth week.

21.9 Aging and the Cardiovascular System

Animation - Regulating Blood Pressure

1. General changes associated with aging include reduced compliance (distensibility) of blood vessels, reduction in cardiac muscle size, reduced cardiac output, and increased systolic blood pressure.
2. The incidence of coronary artery disease (CAD), congestive heart failure (CHF), and atherosclerosis increases with age.

SELF-QUIZ QUESTIONS

Fill in the blanks in the following statements.

1. The _____ reflex helps maintain normal blood pressure in the brain; the _____ reflex governs general systemic blood pressure.

2. In addition to the pressure created by contraction of the left ventricle, venous return is aided by the _____ and the _____, both of which depend on the presence of valves in the veins.

Indicate whether the following statements are true or false.

3. Baroreceptors and chemoreceptors are located in the aorta and carotid arteries.

4. The most important method of capillary exchange is simple diffusion.

Choose the one best answer to the following questions.

5. Which of the following are *not* true? (1) Muscular arteries are also known as conducting arteries. (2) Capillaries play a key role in regulating resistance. (3) The flow of blood through true capillaries is controlled by precapillary sphincters. (4) The lumen of an artery is larger than in a comparable vein. (5) Elastic arteries help propel blood. (6) The tunica media of arteries is thicker than the tunica media of veins.
   (a) 2, 3, and 6   (b) 1, 2, and 4   (c) 1, 2, 4, and 6
   (d) 3, 4, and 5   (e) 1, 2, 3, and 4

6. Which of the following are *true* concerning capillary exchange? (1) Large, lipid-insoluble molecules cross capillary walls by transcytosis. (2) The blood hydrostatic pressure promotes reabsorption of fluid into the capillaries. (3) If the pressures that promote filtration are greater than the pressures that promote reabsorption, fluid will move out of a capillary and into interstitial spaces. (4) A negative net filtration pressure results in reabsorption of fluid from interstitial spaces into a capillary. (5) The difference in osmotic pressure across a capillary wall is due primarily to red blood cells.
   (a) 1, 3, and 4   (b) 1, 2, 3, 4, and 5   (c) 1, 2, 3, and 4
   (d) 3 and 4   (e) 2, 4, and 5

7. Which of the following would *not* increase vascular resistance? (1) vasodilation, (2) polycythemia, (3) obesity, (4) dehydration, (5) anemia.
   (a) 1 and 2   (b) 1, 3, and 4   (c) 1 and 5
   (d) 1, 4, and 5   (e) 1 only

8. Capillary exchange is enhanced by (1) the slow rate of flow through the capillaries, (2) a small cross-sectional area, (3) the thinness of capillary walls, (4) the respiratory pump, (5) extensive branching, which increases the surface area.
   (a) 1, 2, 3, 4, and 5   (b) 1, 2, 3, and 5   (c) 1 and 3
   (d) 3 and 5   (e) 1, 3, and 5

9. Systemic vascular resistance depends on which of the following factors? (1) blood viscosity, (2) total blood vessel length, (3) size of the lumen, (4) type of blood vessel, (5) oxygen concentration of the blood.
   (a) 1, 2, and 3   (b) 2, 3, and 4   (c) 3, 4, and 5
   (d) 1, 3, and 5   (e) 2, 4, and 5

10. Which of the following help regulate blood pressure and help control regional blood flow? (1) baroreceptor and chemoreceptor reflexes, (2) hormones, (3) autoregulation, (4) H^+ concentration of blood, (5) oxygen concentration of the blood.
    (a) 1, 2, and 4   (b) 2, 4, and 5   (c) 1, 4, and 5
    (d) 1, 2, 3, 4, and 5 (e) 3, 4, and 5

11. For each of the following, indicate if it causes vasoconstriction or vasodilation. Use D for vasodilation and C for vasoconstriction.
    (a) atrial natriuretic peptide   (b) ADH
    (c) decrease in body temperature   (d) lactic acid
    (e) histamine   (f) hypoxia
    (g) hypercapnia   (h) angiotensin II
    (i) nitric oxide   (j) decreased sympathetic
    (k) acidosis   impulses

12. Match the following:
    _2_ (a) pressure generated by the pumping of the heart; pushes fluids out of capillaries
    _5_ (b) pressure created by proteins present in the interstitial fluid; pulls fluid out of capillaries
    _1_ (c) balance of pressure; determines whether blood volume and interstitial fluid remain steady or change
    _4_ (d) force due to presence of plasma proteins; pulls fluid into capillaries from interstitial spaces
    _3_ (e) pressure due to fluid in interstitial spaces; pushes fluid back into capillaries

    (1) net filtration pressure
    (2) blood hydrostatic pressure
    (3) interstitial fluid hydrostatic pressure
    (4) blood colloid osmotic pressure
    (5) interstitial fluid osmotic pressure

13. Match the following:
    _11_ (a) supplies blood to the kidney
    _1_ (b) drains blood from the small intestine, portions of the large intestine, stomach, and pancreas
    _4_ (c) main blood supply to arm; commonly used to measure blood pressure
    _9_ (d) supply blood to the free lower limbs
    _3_ (e) drain oxygenated blood from the lungs and send it to the left atrium
    _8_ (f) supplies blood to the stomach, liver, and pancreas
    _6_ (g) supply blood to the brain
    _2_ (h) supplies blood to the large intestine
    _7_ (i) drain blood from the head
    _5_ (j) detours venous blood from the gastrointestinal organs and spleen through the liver before it returns to the heart
    _10_ (k) drains most of the thorax and abdominal wall; can serve as a bypass for the inferior vena cava
    _12_ (l) a part of the venous circulation of the leg; a vessel used in heart bypass surgery
    _13_ (m) carry deoxygenated blood from the right ventricle to the lungs

    (1) superior mesenteric vein
    (2) inferior mesenteric artery
    (3) pulmonary veins
    (4) brachial artery
    (5) hepatic portal circulation
    (6) carotid arteries
    (7) jugular veins
    (8) celiac trunk
    (9) common iliac arteries
    (10) azygos veins
    (11) renal artery
    (12) great saphenous vein
    (13) pulmonary arteries

14. Match the following:
    __2__ (a) a traveling pressure wave created by the alternate expansion and recoil of elastic arteries after each systole of the left ventricle
    __6__ (b) the lowest blood pressure in arteries during ventricular relaxation
    __4__ (c) a slow resting heart rate or pulse rate
    __1__ (d) an inadequate cardiac output that results in a failure of the cardiovascular system to deliver enough oxygen and nutrients to meet the metabolic needs of body cells
    __3__ (e) a rapid resting heart rate or pulse rate
    __5__ (f) the highest force with which blood pushes against arterial walls as a result of ventricular contraction

    (1) shock
    (2) pulse
    (3) tachycardia
    (4) bradycardia
    (5) systolic blood pressure
    (6) diastolic blood pressure

15. Match the following (some answers will be used more than once):
    __5__ (a) returns oxygenated blood from the placenta to the fetal liver
    __3__ (b) an opening in the septum between the right and left atria
    __1__ (c) becomes the ligamentum venosum after birth
    __4__ (d) pass blood from the fetus to the placenta
    __2__ (e) bypasses the nonfunctioning lungs; becomes the ligamentum arteriosum at birth
    __4__ (f) become the medial umbilical ligaments at birth
    __1__ (g) transports oxygenated blood into the inferior vena cava
    __5__ (h) becomes the ligamentum teres at birth
    __3__ (i) becomes the fossa ovalis after birth

    (1) ductus venosus
    (2) ductus arteriosus
    (3) foramen ovale
    (4) umbilical arteries
    (5) umbilical vein

CRITICAL THINKING QUESTIONS

1. Kim Sung was told that her baby was born with a hole in the upper chambers of his heart. Is this something Kim Sung should worry about?

2. Michael was brought into the emergency room suffering from a gunshot wound. He is bleeding profusely and exhibits the following: systolic blood pressure is 40 mmHg; weak pulse of 200 beats per minute; cool, pale, and clammy skin. Michael is not producing urine but is asking for water. He is confused and disoriented. What is his diagnosis and what, specifically, is causing these symptoms?

3. Maureen's job entails standing on a concrete floor for 10-hour days on an assembly line. Lately she has noticed swelling in her ankles at the end of the day and some tenderness in her calves. What do you suspect is Maureen's problem and how could she help counteract the problem?

❓ ANSWERS TO FIGURE QUESTIONS

21.1 The femoral artery has the thicker wall; the femoral vein has the wider lumen.

21.2 Due to atherosclerosis, less energy is stored in the less-compliant elastic arteries during systole; thus, the heart must pump harder to maintain the same rate of blood flow.

21.3 Metabolically active tissues use O_2 and produce wastes more rapidly than inactive tissues, so they require more extensive capillary networks.

21.4 Materials cross capillary walls through intercellular clefts and fenestrations, via transcytosis in pinocytic vesicles, and through the plasma membranes of endothelial cells.

21.5 Valves are more important in arm veins and leg veins than in neck veins because, when you are standing, gravity causes pooling of blood in the veins of the free limbs but aids the flow of blood in neck veins back toward the heart.

21.6 Blood volume in venules and veins is about 64% of 5 liters, or 3.2 liters; blood volume in capillaries is about 7% of 5 liters, or 350 mL.

21.7 Blood colloid osmotic pressure is lower than normal in a person with a low level of plasma proteins, and therefore capillary reabsorption is low. The result is edema.

21.8 Mean blood pressure in the aorta is closer to diastolic than to systolic pressure.

21.9 The skeletal muscle pump and respiratory pump aid venous return.

21.10 Vasodilation and vasoconstriction of arterioles are the main regulators of systemic vascular resistance.

21.11 Velocity of blood flow is fastest in the aorta and arteries.

21.12 The effector tissues regulated by the cardiovascular center are cardiac muscle in the heart and smooth muscle in blood vessel walls.

21.13 Impulses to the cardiovascular center pass from baroreceptors in the carotid sinuses via the glossopharyngeal (IX) nerves and from baroreceptors in the arch of the aorta via the vagus (X) nerves.

21.14 It represents a change that occurs when you stand up because gravity causes pooling of blood in leg veins once you are upright, decreasing the blood pressure in your upper body.

21.15 Diastolic blood pressure = 95 mmHg; systolic blood pressure = 142 mmHg; pulse pressure = 47 mmHg. This person has stage I hypertension because the systolic blood pressure is greater than 140 mmHg and the diastolic blood pressure is greater than 90 mmHg.

21.16 Almost-normal blood pressure in a person who has lost blood does not necessarily indicate that the patient's tissues are receiving adequate blood flow; if systemic vascular resistance has increased greatly, tissue perfusion may be inadequate.

21.17 The two main circulatory routes are the systemic circulation and the pulmonary circulation.

21.18 The subdivisions of the aorta are the ascending aorta, arch of the aorta, thoracic aorta, and abdominal aorta.

21.19 Branches of the arch of the aorta (in order of origination) are the brachiocephalic trunk, left common carotid artery, and left subclavian artery.

21.20 The thoracic aorta begins at the level of the intervertebral disc between T4 and T5.

21.21 The abdominal aorta begins at the aortic hiatus in the diaphragm.

21.22 The abdominal aorta divides into the common iliac arteries at about the level of L4.

21.23 The superior vena cava drains regions above the diaphragm, and the inferior vena cava drains regions below the diaphragm.

21.24 All venous blood in the brain drains into the internal jugular veins.

21.25 The median cubital vein of the upper limb is often used for withdrawing blood.

21.26 The inferior vena cava returns blood from abdominopelvic viscera to the heart.

21.27 Superficial veins of the lower limbs are the dorsal venous arches and the great saphenous and small saphenous veins.

21.28 The hepatic veins carry blood away from the liver.

21.29 After birth, the pulmonary arteries are the only arteries that carry deoxygenated blood.

21.30 Exchange of materials between mother and fetus occurs across the placenta.

21.31 Blood vessels and blood are derived from mesoderm.

22

THE LYMPHATIC SYSTEM AND IMMUNITY

THE LYMPHATIC SYSTEM, DISEASE RESISTANCE, AND HOMEOSTASIS *The lymphatic system contributes to homeostasis by draining interstitial fluid as well as providing the mechanisms for defense against disease.*

Maintaining homeostasis in the body requires continual combat against harmful agents in our internal and external environment. Despite constant exposure to a variety of **pathogens** (PATH-ō-jens)—disease-producing microbes such as bacteria and viruses—most people remain healthy. The body surface also endures cuts and bumps, exposure to ultraviolet rays, chemical toxins, and minor burns with an array of defensive ploys.

Immunity (i-MŪ-ni-tē) or **resistance** is the ability to ward off damage or disease through our defenses. Vulnerability or lack of resistance is termed **susceptibility.** The two general types of immunity are (1) innate and (2) adaptive. **Innate (nonspecific) immunity** refers to defenses that are present at birth. Innate immunity does not involve specific recognition of a microbe and acts against all microbes in the same way. Among the components of innate immunity are the first line of defense (the physical and chemical barriers of the skin and mucous membranes) and the second line of defense (antimicrobial substances, natural killer cells, phagocytes, inflammation, and fever). Innate immune responses represent immunity's early warning system and are designed to prevent microbes from gaining access into the body and to help eliminate those that do gain access.

Adaptive (specific) immunity refers to defenses that involve specific recognition of a microbe once it has breached the innate immunity defenses. Adaptive immunity is based on a specific response to a specific microbe; that is, it adapts or adjusts to handle a specific microbe. Adaptive immunity involves lymphocytes (a type of white blood cell) called T lymphocytes (T cells) and B lymphocytes (B cells).

The body system responsible for adaptive immunity (and some aspects of innate immunity) is the lymphatic system. This system is closely allied with the cardiovascular system, and it also functions with the digestive system in the absorption of fatty foods. In this chapter, we explore the mechanisms that provide defenses against intruders and promote the repair of damaged body tissues.

? *Did you ever wonder how cancer can spread from one part of the body to another?*

22.1 LYMPHATIC SYSTEM STRUCTURE AND FUNCTION

OBJECTIVES

- List the components and major functions of the lymphatic system.
- Describe the organization of lymphatic vessels.
- Explain the formation and flow of lymph.
- Compare the structure and functions of the primary and secondary lymphatic organs and tissues.

The **lymphatic system** (lim-FAT-ik) consists of a fluid called lymph, vessels called lymphatic vessels that transport the lymph, a number of structures and organs containing lymphatic tissue (lymphocytes within a filtering tissue), and red bone marrow (Figure 22.1). The lymphatic system assists in circulating body fluids and helps defend the body against disease-causing agents. As you will see shortly, most components of blood plasma filter through blood capillary walls to form interstitial fluid. After interstitial fluid passes into lymphatic vessels, it is called **lymph** (LIMF = clear fluid). The major difference between interstitial fluid and lymph is location: Interstitial fluid is found between cells, and lymph is located within lymphatic vessels and lymphatic tissue.

Lymphatic tissue is a specialized form of reticular connective tissue (see Table 4.4) that contains large numbers of lymphocytes. Recall from Chapter 19 that lymphocytes are agranular white blood cells (see Section 19.4). Two types of lymphocytes participate in adaptive immune responses: B cells and T cells.

Functions of the Lymphatic System

The lymphatic system has three primary functions:

1. *Drains excess interstitial fluid.* Lymphatic vessels drain excess interstitial fluid from tissue spaces and return it to the blood.
2. *Transports dietary lipids.* Lymphatic vessels transport lipids and lipid-soluble vitamins (A, D, E, and K) absorbed by the gastrointestinal tract.
3. *Carries out immune responses.* Lymphatic tissue initiates highly specific responses directed against particular microbes or abnormal cells.

Lymphatic Vessels and Lymph Circulation

Lymphatic vessels begin as **lymphatic capillaries.** These capillaries, which are located in the spaces between cells, are closed at one end (Figure 22.2). Just as blood capillaries converge to form venules and then veins, lymphatic capillaries unite to form larger **lymphatic vessels** (see Figure 22.1), which resemble small veins in structure but have thinner walls and more valves. At intervals along the lymphatic vessels, lymph flows through lymph nodes, encapsulated bean-shaped organs consisting of masses of B cells and T cells. In the skin, lymphatic vessels lie in the subcutaneous tissue and generally follow the same route as veins; lymphatic vessels of the viscera generally follow arteries, forming plexuses (networks) around them. Tissues that lack lymphatic capillaries include avascular tissues (such as cartilage, the epidermis, and the cornea of the eye), the central nervous system, portions of the spleen, and red bone marrow.

Lymphatic Capillaries

Lymphatic capillaries have greater permeability than blood capillaries and thus can absorb large molecules such as proteins and lipids. Lymphatic capillaries are also slightly larger in diameter than blood capillaries and have a unique one-way structure that permits interstitial fluid to flow into them but not out. The ends of endothelial cells that make up the wall of a lymphatic capillary overlap (Figure 22.2b). When pressure is greater in the interstitial fluid than in lymph, the cells separate slightly, like the opening of a one-way swinging door, and interstitial fluid enters the lymphatic capillary. When pressure is greater inside the lymphatic capillary, the cells adhere more closely, and lymph cannot escape back into interstitial fluid. The pressure is relieved as lymph moves further down the lymphatic capillary. Attached to the lymphatic capillaries are *anchoring filaments,* which contain elastic fibers. They extend out from the lymphatic capillary, attaching lymphatic endothelial cells to surrounding tissues. When excess interstitial fluid accumulates and causes tissue swelling, the anchoring filaments are pulled, making the openings between cells even larger so that more fluid can flow into the lymphatic capillary.

In the small intestine, specialized lymphatic capillaries called **lacteals** (LAK-tē-als; *lact-* = milky) carry dietary lipids into lymphatic vessels and ultimately into the blood (see Figure 24.20). The presence of these lipids causes the lymph draining from the small intestine to appear creamy white; such lymph is referred to as **chyle** (KĪL = juice). Elsewhere, lymph is a clear, pale-yellow fluid.

Lymph Trunks and Ducts

As you have already learned, lymph passes from lymphatic capillaries into lymphatic vessels and then through lymph nodes. As lymphatic vessels exit lymph nodes in a particular region of the body, they unite to form **lymph trunks.** The principal trunks are the lumbar, intestinal, bronchomediastinal, subclavian, and jugular trunks (Figure 22.3). The **lumbar trunks** drain lymph from the lower limbs, the wall and viscera of the pelvis, the kidneys, the adrenal glands, and the abdominal wall. The **intestinal trunk** drains lymph from the stomach, intestines, pancreas, spleen, and part of the liver. The **bronchomediastinal trunks** (brong-kō-mē′-dē-as-TĪ-nal) drain lymph from the thoracic wall, lung, and heart. The **subclavian trunks** drain the upper limbs. The **jugular trunks** drain the head and neck.

Lymph passes from lymph trunks into two main channels, the thoracic duct and the right lymphatic duct, and then drains into venous blood. The **thoracic (left lymphatic) duct** is about 38–45 cm (15–18 in.) long and begins as a dilation called the **cisterna chyli** (sis-TER-na KĪ-lē; *cisterna* = cavity or reservoir) anterior to the second lumbar vertebra. The thoracic duct is the main duct for the return of lymph to blood. The cisterna chyli receives

Figure 22.1 Components of the lymphatic system.

The lymphatic system consists of lymph, lymphatic vessels, lymphatic tissues, and red bone marrow.

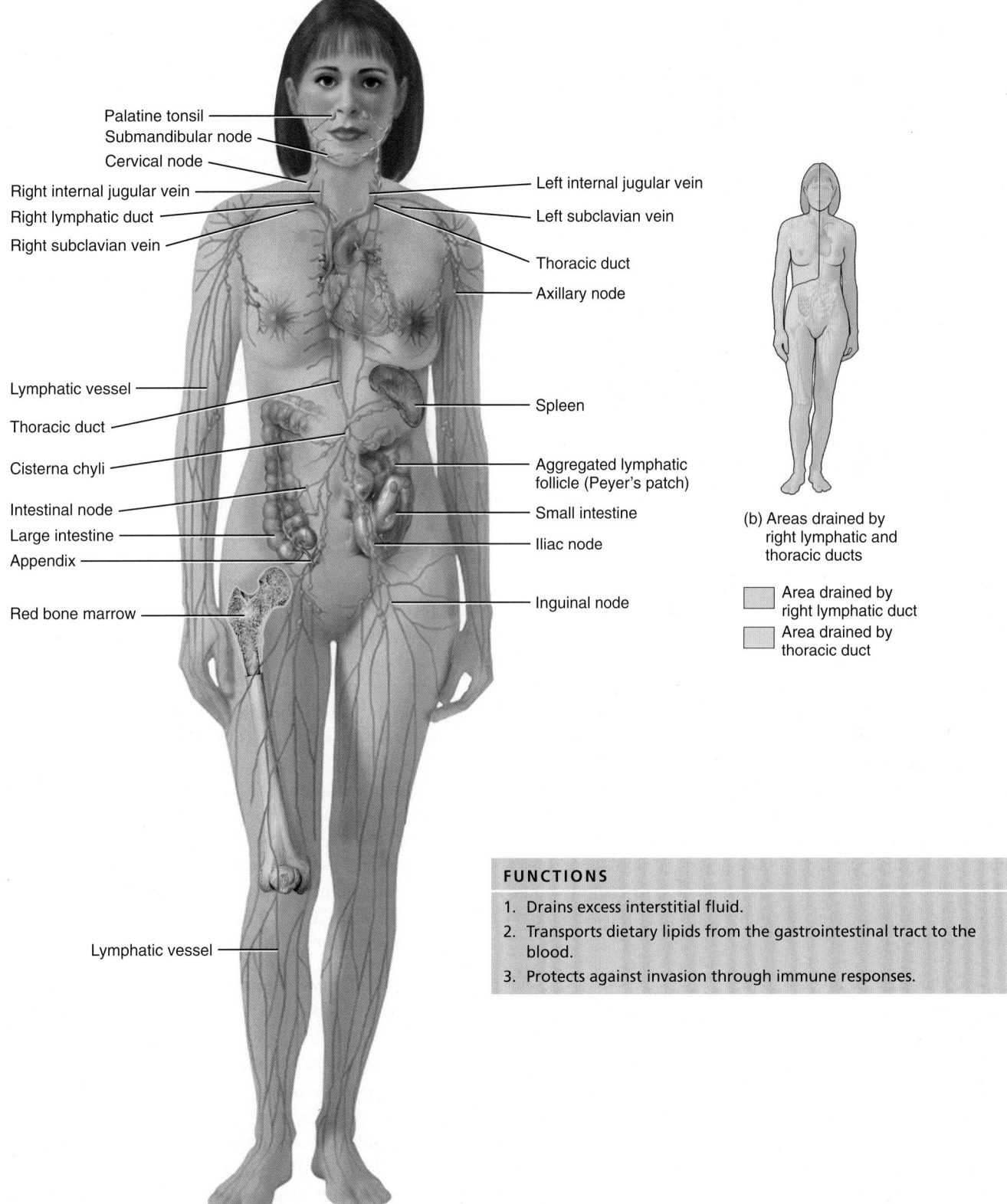

Palatine tonsil
Submandibular node
Cervical node
Right internal jugular vein
Right lymphatic duct
Right subclavian vein

Left internal jugular vein
Left subclavian vein
Thoracic duct
Axillary node

Lymphatic vessel
Thoracic duct
Cisterna chyli
Intestinal node
Large intestine
Appendix
Red bone marrow

Spleen
Aggregated lymphatic follicle (Peyer's patch)
Small intestine
Iliac node
Inguinal node

Lymphatic vessel

(b) Areas drained by right lymphatic and thoracic ducts

Area drained by right lymphatic duct

Area drained by thoracic duct

FUNCTIONS

1. Drains excess interstitial fluid.
2. Transports dietary lipids from the gastrointestinal tract to the blood.
3. Protects against invasion through immune responses.

(a) Anterior view of principal components of lymphatic system

? **What tissue contains stem cells that develop into lymphocytes?**

Figure 22.2 Lymphatic capillaries.

Lymphatic capillaries are found throughout the body except in avascular tissues, the central nervous system, portions of the spleen, and bone marrow.

(a) Relationship of lymphatic capillaries to tissue cells and blood capillaries

(b) Details of lymphatic capillary

Is lymph more similar to blood plasma or to interstitial fluid? Why?

lymph from the right and left lumbar trunks and from the intestinal trunk. In the neck, the thoracic duct also receives lymph from the left jugular, left subclavian, and left bronchomediastinal trunks. Therefore, the thoracic duct receives lymph from the left side of the head, neck, and chest, the left upper limb, and the entire body inferior to the ribs (see Figure 22.1b). The thoracic duct in turn drains lymph into venous blood at the junction of the left internal jugular and left subclavian veins.

The **right lymphatic duct** (Figure 22.3) is about 1.2 cm (0.5 in.) long and receives lymph from the right jugular, right subclavian, and right bronchomediastinal trunks. Thus, the right lymphatic duct receives lymph from the upper right side of the body (see Figure 22.1b). From the right lymphatic duct, lymph drains into venous blood at the junction of the right internal jugular and right subclavian veins.

Formation and Flow of Lymph

Most components of blood plasma, such as nutrients, gases, and hormones, filter freely through the capillary walls to form interstitial fluid, but more fluid filters out of blood capillaries than returns to them by reabsorption (see Figure 21.7). The excess filtered fluid—about 3 liters per day—drains into lymphatic vessels and becomes lymph. Because most plasma proteins are too large to leave blood vessels, interstitial fluid contains only a small amount of protein. Proteins that do leave blood plasma cannot return to the blood by diffusion because the concentration gradient (high level of proteins inside blood capillaries, low level outside) opposes such movement. The proteins can, however, move readily through the more permeable lymphatic capillaries into lymph.

Thus, an important function of lymphatic vessels is to return the lost plasma proteins and plasma to the bloodstream.

Like veins, lymphatic vessels contain valves, which ensure the one-way movement of lymph. As noted previously, lymph drains into venous blood through the right lymphatic duct and the thoracic duct at the junction of the internal jugular and subclavian veins (Figure 22.3). Thus, the sequence of fluid flow is blood capillaries (blood) → interstitial spaces (interstitial fluid) → lymphatic capillaries (lymph) → lymphatic vessels (lymph) → lymphatic ducts (lymph) → junction of the internal jugular and subclavian veins (blood). Figure 22.4 illustrates this sequence, along with the relationship of the lymphatic and cardiovascular systems. Both systems form a very efficient circulatory system.

The same two "pumps" that aid the return of venous blood to the heart maintain the flow of lymph.

1. **Skeletal muscle pump.** The "milking action" of skeletal muscle contractions (see Figure 21.9) compresses lymphatic vessels (as well as veins) and forces lymph toward the junction of the internal jugular and subclavian veins.

2. **Respiratory pump.** Lymph flow is also maintained by pressure changes that occur during inhalation (breathing in). Lymph flows from the abdominal region, where the pressure is higher, toward the thoracic region, where it is lower. When the pressures reverse during exhalation (breathing out), the valves in lymphatic vessels prevent backflow of lymph. In addition, when a lymphatic vessel distends, the smooth muscle in its wall contracts, which helps move lymph from one segment of the vessel to the next.

Figure 22.3 Routes for drainage of lymph from lymph trunks into the thoracic and right lymphatic ducts.

All lymph returns to the bloodstream through the thoracic (left) lymphatic duct and right lymphatic duct.

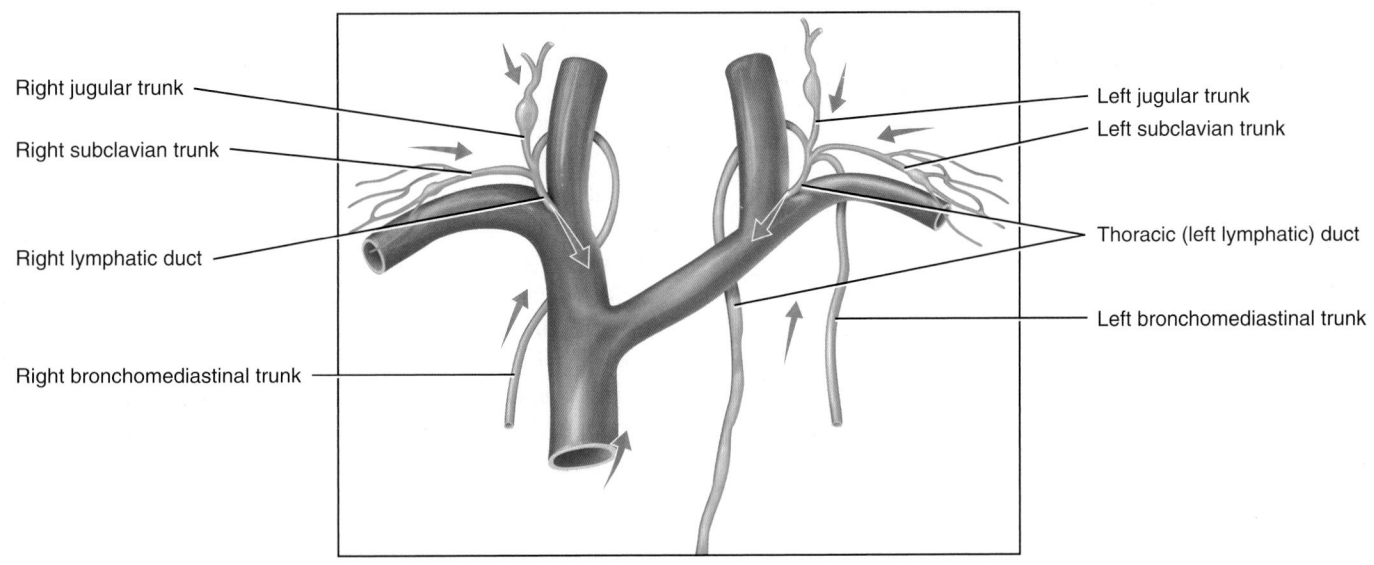

Right internal jugular vein
RIGHT JUGULAR TRUNK
RIGHT SUBCLAVIAN TRUNK
Right subclavian vein
RIGHT LYMPHATIC DUCT
Right brachiocephalic vein
RIGHT BRONCHO-MEDIASTINAL TRUNK
Superior vena cava
Rib
Intercostal muscle
Azygos vein
CISTERNA CHYLI
RIGHT LUMBAR TRUNK
Inferior vena cava

Left internal jugular vein
LEFT JUGULAR TRUNK
THORACIC (LEFT LYMPHATIC) DUCT
LEFT SUBCLAVIAN TRUNK
Left subclavian vein
First rib
Left brachiocephalic vein
Accessory hemiazygos vein
LEFT BRONCHO-MEDIASTINAL TRUNK
THORACIC (LEFT LYMPHATIC) DUCT
Hemiazygos vein
LEFT LUMBAR TRUNK
INTESTINAL TRUNK

(a) Overall anterior view

Right jugular trunk
Right subclavian trunk
Right lymphatic duct
Right bronchomediastinal trunk

Left jugular trunk
Left subclavian trunk
Thoracic (left lymphatic) duct
Left bronchomediastinal trunk

(b) Detailed anterior view

? Which lymphatic vessels empty into the cisterna chyli, and which duct receives lymph from the cisterna chyli?

Figure 22.4 **Schematic diagram showing the relationship of the lymphatic system to the cardiovascular system.** Arrows indicate the direction of flow of lymph and blood.

 The sequence of fluid flow is blood capillaries (blood) → interstitial spaces (interstitial fluid) → lymphatic capillaries (lymph) → lymphatic vessels (lymph) → lymphatic ducts (lymph) → junction of the internal jugular and subclavian veins (blood).

? **Does inhalation promote or hinder the flow of lymph?**

Lymphatic Organs and Tissues

The widely distributed lymphatic organs and tissues are classified into two groups based on their functions. **Primary lymphatic organs** are the sites where stem cells divide and become **immunocompetent** (im′-ū-nō-KOM-pe-tent), that is, capable of mounting an immune response. The primary lymphatic organs are the red bone marrow (in flat bones and the epiphyses of long bones of adults) and the thymus. Pluripotent stem cells in red bone marrow give rise to mature, immunocompetent B cells and to pre-T cells. The pre-T cells in turn migrate to the thymus, where they become immunocompetent T cells. The **secondary lymphatic organs** and **tissues** are the sites where most immune responses occur. They include lymph nodes, the spleen, and lymphatic nodules (follicles). The thymus, lymph nodes, and spleen are considered organs because each is surrounded by a connective tissue capsule; lymphatic nodules, in contrast, are not considered organs because they lack a capsule.

Thymus

The **thymus** is a bilobed organ located in the mediastinum between the sternum and the aorta (Figure 22.5a). An enveloping layer of connective tissue holds the two lobes closely together, but a connective tissue **capsule** separates the two. Extensions of the capsule, called **trabeculae** (tra-BEK-ū-lē = little beams), penetrate inward and divide each lobe into **lobules** (Figure 22.5b).

Each thymic lobule consists of a deeply staining outer cortex and a lighter-staining central medulla (Figure 22.5b). The **cortex** is composed of large numbers of T cells and scattered dendritic cells, epithelial cells, and macrophages. Immature T cells (pre-T cells) migrate from red bone marrow to the cortex of the thymus, where they proliferate and begin to mature. **Dendritic cells** (den-DRIT-ik; *dendr-* = a tree), which are derived from monocytes, and so named because they have long, branched projections that resemble the dendrites of a neuron, assist the maturation process. As you will see shortly, dendritic cells in other parts of the body, such as lymph nodes, play another key role in immune responses. Each of the specialized **epithelial cells** in the cortex has several long processes that surround and serve as a framework for as many as 50 T cells. These epithelial cells help "educate" the pre-T cells in a process known as positive selection (see Figure 22.22). Additionally, they produce thymic hormones that are thought to aid in the maturation of T cells. Only about 2% of developing T cells survive in the cortex. The remaining cells die via apoptosis (programmed cell death). Thymic **macrophages** help clear out the debris of dead and dying cells. The surviving T cells enter the medulla.

Figure 22.5 Thymus.

The bilobed thymus is largest at puberty and then the functional portion atrophies with age.

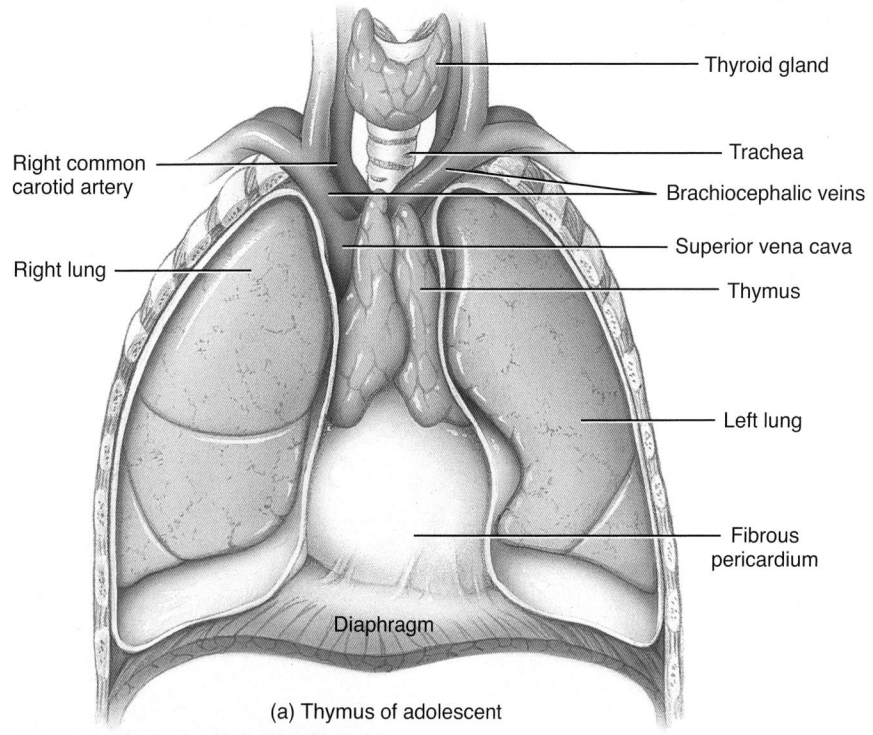

(a) Thymus of adolescent

Thyroid gland

Right common carotid artery

Trachea

Brachiocephalic veins

Superior vena cava

Right lung

Thymus

Left lung

Fibrous pericardium

Diaphragm

Capsule

Lobule:
  Cortex

Thymic (Hassall's) corpuscle

Medulla

Trabecula

LM 30x

(b) Thymic lobules

T cell

Thymic (Hassall's) corpuscle

Epithelial cell

LM 385x

(c) Details of thymic medulla

? **Which type of lymphocytes mature in the thymus?**

The **medulla** consists of widely scattered, more mature T cells, epithelial cells, dendritic cells, and macrophages (Figure 22.5c). Some of the epithelial cells become arranged into concentric layers of flat cells that degenerate and become filled with keratohyalin granules and keratin. These clusters are called **thymic (Hassall's) corpuscles.** Although their role is uncertain, they may serve as sites of T cell death in the medulla. T cells that leave the thymus via the blood migrate to lymph nodes, the spleen, and other lymphatic tissues, where they colonize parts of these organs and tissues.

Because of its high content of lymphoid tissue and a rich blood supply, the thymus has a reddish appearance in a living body. With age, however, fatty infiltrations replace the lymphoid tissue and the thymus takes on more of the yellowish color of the invading fat, giving the false impression of reduced size. However, the actual size of the thymus, defined by its connective tissue capsule, does not change. In infants, the thymus has a mass of about 70 g (2.3 oz). It is after puberty that adipose and areolar connective tissue begin to replace the thymic tissue. By the time a person reaches maturity, the functional portion of the gland is reduced

Figure 22.6 Structure of a lymph node. Arrows indicate the direction of lymph flow through a lymph node.

🔑 ⊂⊃ Lymph nodes are present throughout the body, usually clustered in groups.

Cells of inner cortex

T cells Dendritic cells

Cells around germinal center of outer cortex

B cells

Cells in germinal center of outer cortex

B cells Follicular dendritic cells Macrophages

Cells of medulla

B cells Plasma cells Macrophages

Valve

Afferent lymphatic vessels

Subcapsular sinus

Reticular fiber

Trabecula

Trabecular sinus

Outer cortex:

 Germinal center in secondary lymphatic nodule

 Cells around germinal center

Inner cortex

Medulla

Medullary sinus

Reticular fiber

Efferent lymphatic vessels

Valve

Hilum

Afferent lymphatic vessels

Capsule

Route of lymph flow through lymph node:

Afferent lymphatic vessel
↓
Subcapsular sinus
↓
Trabecular sinus
↓
Medullary sinus
↓
Efferent lymphatic vessel

(a) Partially sectioned lymph node

considerably, and in old age the functional portion may weigh only 3 g (0.1 oz). Before the thymus atrophies, it populates the secondary lymphatic organs and tissues with T cells. However, some T cells continue to proliferate in the thymus throughout an individual's lifetime, but this number decreases with age.

Lymph Nodes

Located along lymphatic vessels are about 600 bean-shaped **lymph nodes.** They are scattered throughout the body, both superficially and deep, and usually occur in groups (see Figure 22.1). Large groups of lymph nodes are present near the mammary glands and in the axillae and groin.

Lymph nodes are 1–25 mm (0.04–1 in.) long and, like the thymus, are covered by a **capsule** of dense connective tissue that extends into the node (Figure 22.6). The capsular extensions, called **trabeculae** (tra-BEK-ū-lē), divide the node into compartments, provide support, and provide a route for blood vessels into the interior of a node. Internal to the capsule is a supporting network of reticular fibers and fibroblasts. The capsule, trabeculae, reticular fibers, and fibroblasts constitute the *stroma* (supporting framework of connective tissue) of a lymph node.

The *parenchyma* (functioning part) of a lymph node is divided into a superficial cortex and a deep medulla. The cortex consists of an outer cortex and an inner cortex. Within the **outer cortex** are

egg-shaped aggregates of B cells called **lymphatic nodules (follicles).** A lymphatic nodule consisting chiefly of B cells is called a *primary lymphatic nodule.* Most lymphatic nodules in the outer cortex are *secondary lymphatic nodules* (Figure 22.6), which form in response to an antigen (a foreign substance) and are sites of plasma cell and memory B cell formation. After B cells in a primary lymphatic nodule recognize an antigen, the primary lymphatic nodule develops into a secondary lymphatic nodule. The center of a secondary lymphatic nodule contains a region of light-staining cells called a *germinal center.* In the germinal center are B cells, follicular dendritic cells (a special type of dendritic cell), and macrophages. When follicular dendritic cells "present" an antigen (described later in the chapter), B cells proliferate and develop into antibody-producing plasma cells or develop into memory B cells. Memory B cells persist after an initial immune response and "remember" having encountered a specific antigen. B cells that do not develop properly undergo apoptosis (programmed cell death) and are destroyed by macrophages. The region of a secondary lymphatic nodule surrounding the germinal center is composed of dense accumulations of B cells that have migrated away from their site of origin within the nodule.

The **inner cortex** does not contain lymphatic nodules. It consists mainly of T cells and dendritic cells that enter a lymph node from other tissues. The dendritic cells present antigens to T cells, causing their proliferation. The newly formed T cells then migrate from the lymph node to areas of the body where there is antigenic activity.

The **medulla** of a lymph node contains B cells, antibody-producing plasma cells that have migrated out of the cortex into the medulla, and macrophages. The various cells are embedded in a network of reticular fibers and reticular cells.

As you have already learned, lymph flows through a node in one direction only (Figure 22.6a). It enters through several

Capsule
Subcapsular sinus
Outer cortex
Trabecular sinus
Germinal center in secondary lymphatic nodule
Trabecula
Inner cortex
Medullary sinus
Medulla

LM 40x

(b) Portion of a lymph node

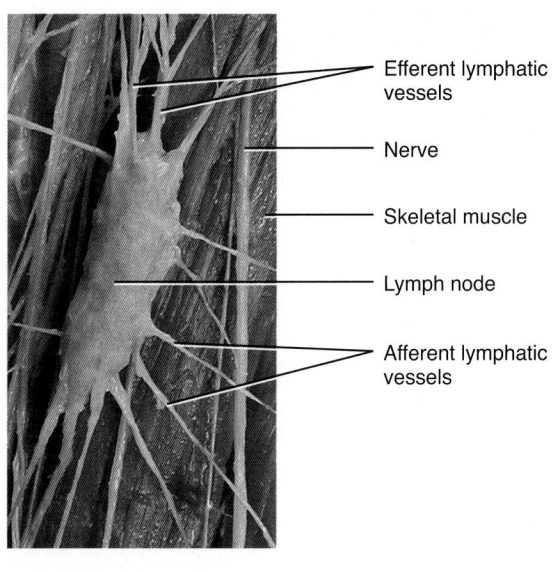

Efferent lymphatic vessels
Nerve
Skeletal muscle
Lymph node
Afferent lymphatic vessels

(c) Anterior view of inguinal lymph node

? **What happens to foreign substances in lymph that enter a lymph node?**

afferent lymphatic vessels (AF-er-ent; *afferent* = to carry toward), which penetrate the convex surface of the node at several points. The afferent vessels contain valves that open toward the center of the node, directing the lymph *inward*. Within the node, lymph enters **sinuses,** a series of irregular channels that contain branching reticular fibers, lymphocytes, and macrophages. From the afferent lymphatic vessels, lymph flows into the **subcapsular sinus** (sub-KAP-soo-lar), immediately beneath the capsule. From here the lymph flows through **trabecular sinuses** (tra-BEK-ū-lar), which extend through the cortex parallel to the trabeculae, and into **medullary sinuses,** which extend through the medulla. The medullary sinuses drain into one or two **efferent lymphatic vessels** (EF-er-ent; *efferent* = to carry away), which are wider and fewer in number than afferent vessels. They contain valves that open away from the center of the lymph node to convey lymph, antibodies secreted by plasma cells, and activated T cells *out* of the node. Efferent lymphatic vessels emerge from one side of the lymph node at a slight depression called a **hilum** (HĪ-lum) or **hilus.** Blood vessels also enter and leave the node at the hilum.

Lymph nodes function as a type of filter. As lymph enters one end of a lymph node, foreign substances are trapped by the reticular fibers within the sinuses of the node. Then macrophages destroy some foreign substances by phagocytosis, while lymphocytes destroy others by immune responses. The filtered lymph then leaves the other end of the lymph node. Since there are many afferent lymphatic vessels that bring lymph into a lymph node and only one or two efferent lymphatic vessels that transport lymph out of a lymph node, the slow flow of lymph within the lymph nodes allows additional time for lymph to be filtered. Additionally, all lymph flows through multiple lymph nodes on its path through the lymph vessels. This exposes the lymph to multiple filtering events before returning to the blood.

CLINICAL CONNECTION | Metastasis through Lymphatic Vessels

Metastasis (me-TAS-ta-sis; *meta-* = beyond; *-stasis* = to stand), the spread of a disease from one part of the body to another, can occur via lymphatic vessels. All malignant tumors eventually metastasize. Cancer cells may travel in the blood or lymph and establish new tumors where they lodge. When metastasis occurs via lymphatic vessels, secondary tumor sites can be predicted according to the direction of lymph flow from the primary tumor site. Cancerous lymph nodes feel enlarged, firm, nontender, and fixed to underlying structures. By contrast, most lymph nodes that are enlarged due to an infection are softer, tender, and movable. •

Spleen

The oval **spleen** is the largest single mass of lymphatic tissue in the body, measuring about 12 cm (5 in.) in length (Figure 22.7a). It is located in the left hypochondriac region between the stomach and diaphragm. The superior surface of the spleen is smooth and convex and conforms to the concave surface of the diaphragm. Neighboring organs make indentations in the visceral surface of the spleen—the *gastric impression* (stomach), the *renal impression* (left kidney), and the *colic impression* (left colic flexure of large intestine). Like lymph nodes, the spleen has a hilum. Through it pass the splenic artery, splenic vein, and efferent lymphatic vessels.

A capsule of dense connective tissue surrounds the spleen and is covered in turn by a serous membrane, the visceral peritoneum. Trabeculae extend inward from the capsule. The capsule plus trabeculae, reticular fibers, and fibroblasts constitute the stroma of the spleen; the parenchyma of the spleen consists of two different kinds of tissue called white pulp and red pulp (Figure 22.7b, c). **White pulp** is lymphatic tissue, consisting mostly of lymphocytes and macrophages arranged around branches of the splenic artery called **central arteries.** The **red pulp** consists of blood-filled **venous sinuses** and cords of splenic tissue called **splenic (Billroth's) cords.** Splenic cords consist of red blood cells, macrophages, lymphocytes, plasma cells, and granulocytes. Veins are closely associated with the red pulp.

Blood flowing into the spleen through the splenic artery enters the central arteries of the white pulp. Within the white pulp, B cells and T cells carry out immune functions, similar to lymph nodes, while spleen macrophages destroy blood-borne pathogens by phagocytosis. Within the red pulp, the spleen performs three functions related to blood cells: (1) removal by macrophages of ruptured, worn out, or defective blood cells and platelets; (2) storage of platelets, up to one-third of the body's supply; and (3) production of blood cells (hemopoiesis) during fetal life.

CLINICAL CONNECTION | Ruptured Spleen

The spleen is the organ most often damaged in cases of abdominal trauma. Severe blows over the inferior left chest or superior abdomen can fracture the protecting ribs. Such crushing injury may result in a **ruptured spleen,** which causes significant hemorrhage and shock. Prompt removal of the spleen, called a **splenectomy** (splē-NEK-tō-mē), is needed to prevent death due to bleeding. Other structures, particularly red bone marrow and the liver, can take over some functions normally carried out by the spleen. Immune functions, however, decrease in the absence of a spleen. The spleen's absence also places the patient at higher risk for **sepsis** (a blood infection) due to loss of the filtering and phagocytic functions of the spleen. To reduce the risk of sepsis, patients who have undergone a splenectomy take prophylactic (preventive) antibiotics before any invasive procedures. •

Lymphatic Nodules

Lymphatic nodules (follicles) are egg-shaped masses of lymphatic tissue that are not surrounded by a capsule. Because they are scattered throughout the lamina propria (connective tissue) of mucous membranes lining the gastrointestinal, urinary, and reproductive tracts and the respiratory airways, lymphatic nodules in these areas are also referred to as **mucosa-associated lymphatic tissue (MALT).**

Although many lymphatic nodules are small and solitary, some occur in multiple large aggregations in specific parts of the body. Among these are the tonsils in the pharyngeal region

Figure 22.7 Structure of the spleen.

The spleen is the largest single mass of lymphatic tissue in the body.

(a) Visceral surface

(b) Internal structure

(c) Portion of the spleen

LM 25x

After birth, what are the main functions of the spleen?

and the aggregated lymphatic follicles (Peyer's patches) in the ileum of the small intestine. Aggregations of lymphatic nodules also occur in the appendix. Usually there are five **tonsils,** which form a ring at the junction of the oral cavity and oropharynx and at the junction of the nasal cavity and nasopharynx (see Figure 23.2b). The tonsils are strategically positioned to participate in immune responses against inhaled or ingested foreign substances. The single **pharyngeal tonsil** (fa-RIN-jē-al) or **adenoid** is embedded in the posterior wall of the nasopharynx. The two **palatine tonsils** (PAL-a-tīn) lie at the posterior region of the oral cavity, one on either side; these are the tonsils commonly removed in a tonsillectomy. The paired **lingual tonsils** (LIN-gwal), located at the base of the tongue, may also require removal during a tonsillectomy.

CLINICAL CONNECTION | *Tonsillitis*

Tonsillitis is an infection or inflammation of the tonsils. Most often, it is caused by a virus, but it may also be caused by the same bacteria that cause strep throat. The principal symptom of tonsillitis is a sore throat. Additionally, fever, swollen lymph nodes, nasal congestion, difficulty in swallowing, and headache may also occur. Tonsillitis of viral origin usually resolves on its own. Bacterial tonsillitis is typically treated with antibiotics. **Tonsillectomy** (ton-si-LEK-tō-mē; *ectomy* = incision), the removal of a tonsil, may be indicated for individuals who do not respond to other treatments. Such individuals usually have tonsillitis lasting for more than 3 months (despite medication), obstructed air pathways, and difficulty in swallowing and talking. It appears that tonsillectomy does not interfere with a person's response to subsequent infections. •

✔ CHECKPOINT

1. How are interstitial fluid and lymph similar, and how do they differ?
2. How do lymphatic vessels differ in structure from veins?
3. Diagram the route of lymph circulation.
4. What is the role of the thymus in immunity?
5. What functions do lymph nodes, the spleen, and the tonsils serve?

22.2 DEVELOPMENT OF LYMPHATIC TISSUES

◉ OBJECTIVE

• Describe the development of lymphatic tissues.

Lymphatic tissues begin to develop by the end of the fifth week of embryonic life. *Lymphatic vessels* develop from **lymph sacs** that arise from developing veins, which are derived from **mesoderm.**

The first lymph sacs to appear are the paired **jugular lymph sacs** at the junction of the internal jugular and subclavian veins (Figure 22.8). From the jugular lymph sacs, lymphatic capillary plexuses spread to the thorax, upper limbs, neck, and head. Some of the plexuses enlarge and form lymphatic vessels in their respective regions. Each jugular lymph sac retains at least one connection with its jugular vein, the left one developing into the superior portion of the thoracic duct (left lymphatic duct).

The next lymph sac to appear is the unpaired **retroperitoneal lymph sac** (ret′-rō-per′-i-tō-NĒ-al) at the root of the mesentery of the intestine. It develops from the primitive vena cava and mesonephric (primitive kidney) veins. Capillary plexuses and lymphatic vessels spread from the retroperitoneal lymph sac to

Figure 22.8 Development of lymphatic tissues.

🔑 Lymphatic tissues are derived from mesoderm.

Jugular lymph sac
Thoracic duct
Cisterna chyli
Retroperitoneal lymph sac
Posterior lymph sac

Internal jugular vein
Subclavian vein
Inferior vena cava

❓ **When do lymphatic tissues begin to develop?**

the abdominal viscera and diaphragm. The sac establishes connections with the cisterna chyli but loses its connections with neighboring veins.

At about the time the retroperitoneal lymph sac is developing, another lymph sac, the **cisterna chyli** (sis-TER-na KĪ-lē), develops inferior to the diaphragm on the posterior abdominal wall. It gives rise to the inferior portion of the *thoracic duct* and the *cisterna chyli* of the thoracic duct. Like the retroperitoneal lymph sac, the cisterna chyli also loses its connections with surrounding veins.

The last of the lymph sacs, the paired **posterior lymph sacs,** develop from the iliac veins. The posterior lymph sacs produce capillary plexuses and lymphatic vessels of the abdominal wall, pelvic region, and lower limbs. The posterior lymph sacs join the cisterna chyli and lose their connections with adjacent veins.

With the exception of the anterior part of the sac from which the cisterna chyli develops, all lymph sacs become invaded by **mesenchymal cells** (me-SENG-kī-mal) and are converted into groups of *lymph nodes.*

The *spleen* develops from mesenchymal cells between layers of the dorsal mesentery of the stomach. The *thymus* arises as an outgrowth of the **third pharyngeal pouch** (see Figure 18.21a).

✔ CHECKPOINT

6. What are the names of the four lymph sacs from which lymphatic vessels develop?

22.3 INNATE IMMUNITY

◉ OBJECTIVE

• Describe the components of innate immunity.

Innate (nonspecific) immunity includes the external physical and chemical barriers provided by the skin and mucous membranes. It also includes various internal defenses, such as antimicrobial substances, natural killer cells, phagocytes, inflammation, and fever.

First Line of Defense: Skin and Mucous Membranes

The skin and mucous membranes of the body are the first line of defense against pathogens. These structures provide both physical and chemical barriers that discourage pathogens and foreign substances from penetrating the body and causing disease.

With its many layers of closely packed, keratinized cells, the outer epithelial layer of the skin—the **epidermis**—provides a formidable physical barrier to the entrance of microbes (see Figure 5.1). In addition, periodic shedding of epidermal cells helps remove microbes at the skin surface. Bacteria rarely penetrate the intact surface of healthy epidermis. If this surface is broken by cuts, burns, or punctures, however, pathogens can penetrate the epidermis and invade adjacent tissues or circulate in the blood to other parts of the body.

The epithelial layer of **mucous membranes,** which line body cavities, secretes a fluid called **mucus** that lubricates and moistens the cavity surface. Because mucus is slightly viscous, it traps many microbes and foreign substances. The mucous membrane of the nose has mucus-coated **hairs** that trap and filter microbes, dust, and pollutants from inhaled air. The mucous membrane of the upper respiratory tract contains **cilia,** microscopic hairlike projections on the surface of the epithelial cells. The waving action of cilia propels inhaled dust and microbes that have become trapped in mucus toward the throat. Coughing and sneezing accelerate movement of mucus and its entrapped pathogens out of the body. Swallowing mucus sends pathogens to the stomach, where gastric juice destroys them.

Other fluids produced by various organs also help protect epithelial surfaces of the skin and mucous membranes. The **lacrimal apparatus** (LAK-ri-mal) of the eyes (see Figure 17.6) manufactures and drains away tears in response to irritants. Blinking spreads tears over the surface of the eyeball, and the continual washing action of tears helps to dilute microbes and keep them from settling on the surface of the eye. Tears also contain **lysozyme** (LĪ-so-zīm), an enzyme capable of breaking down the cell walls of certain bacteria. Besides tears, lysozyme is present in saliva, perspiration, nasal secretions, and tissue fluids. **Saliva,** produced by the salivary glands, washes microbes from the surfaces of the teeth and from the mucous membrane of the mouth, much as tears wash the eyes. The flow of saliva reduces colonization of the mouth by microbes.

The cleansing of the urethra by the **flow of urine** retards microbial colonization of the urinary system. **Vaginal secretions** likewise move microbes out of the body in females. **Defecation** and **vomiting** also expel microbes. For example, in response to some microbial toxins, the smooth muscle of the lower gastrointestinal tract contracts vigorously; the resulting diarrhea rapidly expels many of the microbes.

Certain chemicals also contribute to the high degree of resistance of the skin and mucous membranes to microbial invasion. Sebaceous (oil) glands of the skin secrete an oily substance called **sebum** that forms a protective film over the surface of the skin. The unsaturated fatty acids in sebum inhibit the growth of certain pathogenic bacteria and fungi. The acidity of the skin (pH 3–5) is caused in part by the secretion of fatty acids and lactic acid. **Perspiration** helps flush microbes from the surface of the skin. **Gastric juice,** produced by the glands of the stomach, is a mixture of hydrochloric acid, enzymes, and mucus. The strong acidity of gastric juice (pH 1.2–3.0) destroys many bacteria and most bacterial toxins. Vaginal secretions also are slightly acidic, which discourages bacterial growth.

Second Line of Defense: Internal Defenses

When pathogens penetrate the physical and chemical barriers of the skin and mucous membranes, they encounter a second line of defense: internal antimicrobial substances, phagocytes, natural killer cells, inflammation, and fever.

Antimicrobial Substances

There are four main types of **antimicrobial substances** that discourage microbial growth: interferons, complement, iron-binding proteins, and antimicrobial proteins.

1. Lymphocytes, macrophages, and fibroblasts infected with viruses produce proteins called **interferons** (in′-ter-FĒR-ons), or **IFNs.** Once released by virus-infected cells, IFNs diffuse to uninfected neighboring cells, where they induce synthesis of antiviral proteins that interfere with viral replication. Although IFNs do not prevent viruses from attaching to and penetrating host cells, they do stop replication. Viruses can cause disease only if they can replicate within body cells. IFNs are an important defense against infection by many different viruses. The three types of interferons are alpha-, beta-, and gamma-IFN.

2. A group of normally inactive proteins in blood plasma and on the plasma membranes makes up the **complement system.** When activated, these proteins "complement" or enhance certain immune reactions (see Section 22.6). The complement system causes cytolysis (bursting) of microbes, promotes phagocytosis, and contributes to inflammation.

3. **Iron-binding proteins** inhibit the growth of certain bacteria by reducing the amount of available iron. Examples include *transferrin* (found in blood and tissue fluids), *lactoferrin* (found in milk, saliva, and mucus), *ferritin* (found in the liver, spleen, and red bone marrow), and *hemoglobin* (found in red blood cells).

4. **Antimicrobial proteins (AMPs)** are short peptides that have a broad spectrum of antimicrobial activity. Examples of AMPs are *dermicidin* (der-ma-SĪ-din) (produced by sweat glands), *defensins* and *cathelicidins* (cath-el-i-SĪ-dins) (produced by neutrophils, macrophages, and epithelia), and *thrombocidin* (throm′-bō-SĪ-din) (produced by platelets). Besides killing a wide range of microbes, AMPs can attract dendritic cells and mast cells, which participate in immune responses. Interestingly enough, microbes exposed to AMPs do not appear to develop resistance, as often happens with antibiotics.

Natural Killer Cells and Phagocytes

When microbes penetrate the skin and mucous membranes or bypass the antimicrobial substances in blood, the next nonspecific defense consists of natural killer cells and phagocytes. About 5–10% of lymphocytes in the blood are **natural killer (NK) cells.** They are also present in the spleen, lymph nodes, and red bone marrow. NK cells lack the membrane molecules that identify B and T cells, but they have the ability to kill a wide variety of infected body cells and certain tumor cells. NK cells attack any body cells that display abnormal or unusual plasma membrane proteins.

The binding of NK cells to a target cell, such as an infected human cell, causes the release of granules containing toxic substances from NK cells. Some granules contain a protein called **perforin** (PER-for-in) that inserts into the plasma membrane of the target cell and creates channels (perforations) in the membrane. As a result, extracellular fluid flows into the target cell and the cell bursts, a process called **cytolysis** (sī-TOL-i-sis; *cyto-* = cell; *-lysis* = loosening). Other granules of NK cells release **granzymes** (GRAN-zīms), which are protein-digesting enzymes that induce the target cell to undergo apoptosis, or self-destruction. This type of attack kills infected cells, but not the microbes inside the cells; the released microbes, which may or may not be intact, can be destroyed by phagocytes.

Phagocytes (FAG-ō-sīts; *phago-* = eat; *-cytes* = cells) are specialized cells that perform **phagocytosis** (fag-ō-sī-TŌ-sis; *-osis* = process), the ingestion of microbes or other particles such as cellular debris (see Figure 3.13). The two major types of phagocytes are **neutrophils** and **macrophages** (MAK-rō-fā-jez). When an infection occurs, neutrophils and monocytes migrate to the infected area. During this migration, the monocytes enlarge and develop into actively phagocytic macrophages called **wandering macrophages.** Other macrophages, called **fixed macrophages,** stand guard in specific tissues. Among the fixed macrophages are *histiocytes* (HIS-tē-ō-sīts) (connective tissue macrophages), *stellate reticuloendothelial cells* (STEL-āt re-tik′-ū-lō-en-dō-THĒ-lē-al) or *Kupffer cells* (KOOP-fer) in the liver, *alveolar macrophages* in the lungs, *microglia* in the nervous system, and *tissue macrophages* in the spleen, lymph nodes, and red bone marrow. In addition to being an innate defense mechanism, phagocytosis plays a vital role in adaptive immunity, as discussed later in the chapter.

CLINICAL CONNECTION | *Microbial Evasion of Phagocytosis*

Some microbes, such as the bacteria that cause pneumonia, have extracellular structures called capsules that prevent adherence. This makes it physically difficult for phagocytes to engulf the microbes. Other microbes, such as the toxin-producing bacteria that cause one kind of food poisoning, may be ingested but not killed; instead, the toxins they produce (leukocidins) may kill the phagocytes by causing the release of the phagocyte's own lysosomal enzymes into its cytoplasm. Still other microbes—such as the bacteria that cause tuberculosis—inhibit fusion of phagosomes and lysosomes and thus prevent exposure of the microbes to lysosomal enzymes. These bacteria apparently can also use chemicals in their cell walls to counter the effects of lethal oxidants produced by phagocytes. Subsequent multiplication of the microbes within phagosomes may eventually destroy the phagocyte. •

Phagocytosis occurs in five phases: chemotaxis, adherence, ingestion, digestion, and killing (Figure 22.9):

1 *Chemotaxis.* Phagocytosis begins with **chemotaxis** (kē-mō-TAK-sis), a chemically stimulated movement of phagocytes to a site of damage. Chemicals that attract phagocytes might come from invading microbes, white blood cells, damaged tissue cells, or activated complement proteins.

2 *Adherence* (ad-HER-ents). Attachment of the phagocyte to the microbe or other foreign material is termed **adherence.** The binding of complement proteins to the invading pathogen enhances adherence.

3 *Ingestion.* The plasma membrane of the phagocyte extends projections, called **pseudopods** (SOO-dō-pods), that engulf the microbe in a process called **ingestion.** When the pseudopods meet, they fuse, surrounding the microorganism with a sac called a **phagosome** (FAG-ō-sōm).

4 *Digestion.* The phagosome enters the cytoplasm and merges with lysosomes to form a single, larger structure called a **phagolysosome** (fag-ō-LĪ-sō-sōm). The lysosome contributes lysozyme, which breaks down microbial cell walls, and other digestive enzymes that degrade carbohydrates, proteins, lipids, and nucleic acids. The phagocyte also forms lethal oxidants, such as superoxide anion (O_2^-), hypochlorite anion (OCl^-), and hydrogen peroxide (H_2O_2), in a process called an **oxidative burst.**

5 *Killing.* The chemical onslaught provided by lysozyme, digestive enzymes, and oxidants within a phagolysosome quickly kills many types of microbes. Any materials that cannot be degraded further remain in structures called **residual bodies.**

Inflammation

Inflammation is a nonspecific, defensive response of the body to tissue damage. Among the conditions that may produce inflammation are pathogens, abrasions, chemical irritations, distortion or disturbances of cells, and extreme temperatures. The four characteristic signs and symptoms of inflammation are **redness, pain, heat,** and **swelling.** Inflammation can also cause a **loss of function** in the injured area (for example, the inability to detect sensations), depending on the site and extent of the injury. Inflammation is an attempt to dispose of microbes, toxins, or foreign material at the site of injury, to prevent their spread to other tissues, and to prepare the site for tissue repair in an attempt to restore tissue homeostasis.

Because inflammation is one of the body's nonspecific defense mechanisms, the response of a tissue to a cut is similar to the response to damage caused by burns, radiation, or bacterial or viral invasion. In each case, the inflammatory response has three basic stages: (1) vasodilation and increased permeability of blood vessels, (2) emigration (movement) of phagocytes from the blood into interstitial fluid, and, ultimately, (3) tissue repair.

VASODILATION AND INCREASED BLOOD VESSEL PERMEABILITY
Two immediate changes occur in the blood vessels in a region of tissue injury: **vasodilation** (increase in the diameter) of arterioles and **increased permeability** of capillaries (Figure 22.10). Increased permeability means that substances normally retained in blood are permitted to pass from the blood vessels. Vasodilation allows more blood to flow through the damaged area, and increased permeability permits defensive proteins such as antibodies and clotting factors to enter the injured area from the blood. The increased blood flow also helps remove microbial toxins and dead cells.

Among the substances that contribute to vasodilation, increased permeability, and other aspects of the inflammatory response are the following:

• *Histamine.* In response to injury, mast cells in connective tissue and basophils and platelets in blood release histamine. Neutrophils and macrophages attracted to the site of injury also stimulate the release of histamine, which causes vasodilation and increased permeability of blood vessels.

• *Kinins.* These polypeptides, formed in blood from inactive precursors called kininogens, induce vasodilation and increased

Figure 22.9 Phagocytosis of a microbe.

The major types of phagocytes are neutrophils and macrophages.

(a) Phases of phagocytosis

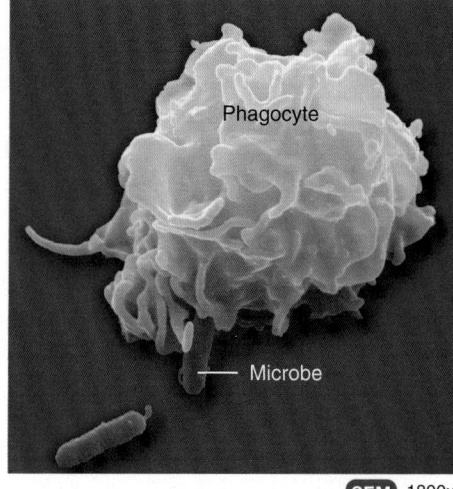

SEM 1800x

(b) Phagocyte (white blood cell) engulfing microbe.

? What chemicals are responsible for killing ingested microbes?

permeability and serve as chemotactic agents for phagocytes. An example of a kinin is bradykinin.

- **Prostaglandins (PGs)** (pros′-ta-GLAN-dins). These lipids, especially those of the E series, are released by damaged cells and intensify the effects of histamine and kinins. PGs also may stimulate the emigration of phagocytes through capillary walls.

- **Leukotrienes (LTs)** (loo′-kō-TRĪ-ēns). Produced by basophils and mast cells, LTs cause increased permeability; they also function in adherence of phagocytes to pathogens and as chemotactic agents that attract phagocytes.

- **Complement.** Different components of the complement system stimulate histamine release, attract neutrophils by chemotaxis, and promote phagocytosis; some components can also destroy bacteria.

Dilation of arterioles and increased permeability of capillaries produce three of the signs and symptoms of inflammation: heat, redness (erythema), and swelling (edema). Heat and redness result from the large amount of blood that accumulates in the damaged area. As the local temperature rises slightly, metabolic reactions proceed more rapidly and release additional heat. Edema results from increased permeability of blood vessels, which permits more fluid to move from blood plasma into tissue spaces.

Pain is a prime symptom of inflammation. It results from injury to neurons and from toxic chemicals released by microbes. Kinins affect some nerve endings, causing much of the pain associated with inflammation. Prostaglandins intensify and prolong the pain associated with inflammation. Pain may also be due to increased pressure from edema.

Figure 22.10 Inflammation.

The three stages of inflammation are as follows: (1) vasodilation and increased permeability of blood vessels, (2) phagocyte emigration, and (3) tissue repair.

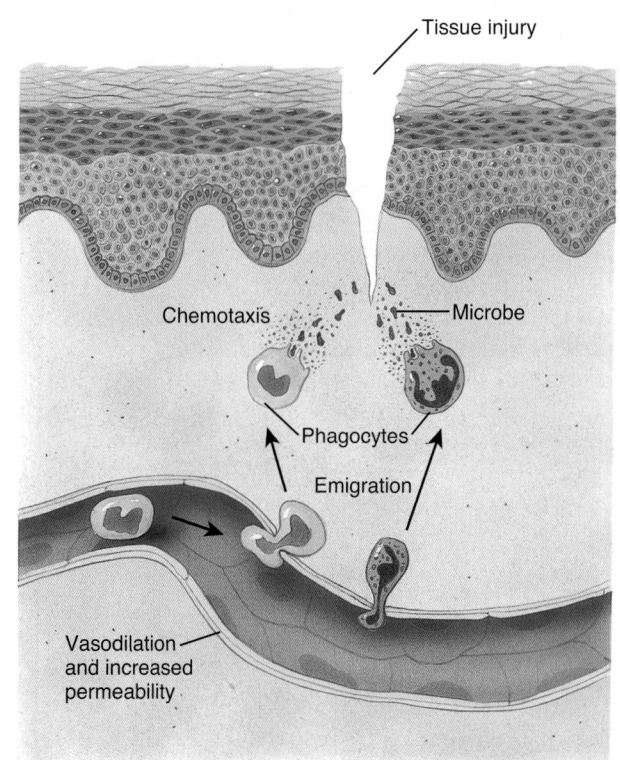

Phagocytes migrate from blood to site of tissue injury

? What causes each of the following signs and symptoms of inflammation: redness, pain, heat, and swelling?

The increased permeability of capillaries allows leakage of blood-clotting factors into tissues. The clotting sequence is set into motion, and fibrinogen is ultimately converted to an insoluble, thick mesh of fibrin threads that localizes and traps invading microbes and blocks their spread.

EMIGRATION OF PHAGOCYTES Within an hour after the inflammatory process starts, phagocytes appear on the scene. As large amounts of blood accumulate, neutrophils begin to stick to the inner surface of the endothelium (lining) of blood vessels (Figure 22.10). Then the neutrophils begin to squeeze through the wall of the blood vessel to reach the damaged area. This process, called **emigration** (em′-i-GRĀ-shun), depends on chemotaxis. Neutrophils attempt to destroy the invading microbes by phagocytosis. A steady stream of neutrophils is ensured by the production and release of additional cells from red bone marrow. Such an increase in white blood cells in the blood is termed **leukocytosis** (loo-kō-sī-TŌ-sis).

Although neutrophils predominate in the early stages of infection, they die off rapidly. As the inflammatory response continues, monocytes follow the neutrophils into the infected area. Once in the tissue, monocytes transform into wandering macrophages that add to the phagocytic activity of the fixed macrophages already present. True to their name, macrophages are much more potent phagocytes than neutrophils. They are large enough to engulf damaged tissue, worn-out neutrophils, and invading microbes.

Eventually, macrophages also die. Within a few days, a pocket of dead phagocytes and damaged tissue forms; this collection of dead cells and fluid is called **pus.** Pus formation occurs in most inflammatory responses and usually continues until the infection subsides. At times, pus reaches the surface of the body or drains into an internal cavity and is dispersed; on other occasions the pus remains even after the infection is terminated. In this case, the pus is gradually destroyed over a period of days and is absorbed.

CLINICAL CONNECTION | *Abscesses and Ulcers*

If pus cannot drain out of an inflamed region, the result is an **abscess**—an excessive accumulation of pus in a confined space. Common examples are pimples and boils. When superficial inflamed tissue sloughs off the surface of an organ or tissue, the resulting open sore is called an **ulcer.** People with poor circulation—for instance, diabetics with advanced atherosclerosis—are susceptible to ulcers in the tissues of their legs. These ulcers, which are called stasis ulcers, develop because of poor oxygen and nutrient supply to tissues that then become very susceptible to even a very mild injury or an infection. •

Fever

Fever is an abnormally high body temperature that occurs because the hypothalamic thermostat is reset. It commonly occurs during infection and inflammation. Many bacterial toxins elevate body temperature, sometimes by triggering release of fever-causing cytokines such as interleukin-1 from macrophages. Elevated body temperature intensifies the effects of interferons, inhibits the growth of some microbes, and speeds up body reactions that aid repair.

Table 22.1 summarizes the components of innate immunity.

✔ CHECKPOINT

7. What physical and chemical factors provide protection from disease in the skin and mucous membranes?
8. What internal defenses provide protection against microbes that penetrate the skin and mucous membranes?
9. How are the activities of natural killer cells and phagocytes similar and different?
10. What are the main signs, symptoms, and stages of inflammation?

22.4 ADAPTIVE IMMUNITY

◉ **OBJECTIVES**

• Define adaptive immunity, and describe how T cells and B cells arise.
• Explain the relationship between an antigen and an antibody.
• Compare the functions of cell-mediated immunity and antibody-mediated immunity.

The ability of the body to defend itself against specific invading agents such as bacteria, toxins, viruses, and foreign tissues is called **adaptive (specific) immunity.** Substances that are recognized as foreign and provoke immune responses are called **antigens (Ags)** (AN-ti-jens), meaning **anti**body **gen**erators. Two properties distinguish adaptive immunity from innate immunity: (1) *specificity* for particular foreign molecules (antigens), which also involves distinguishing self from nonself molecules, and (2) *memory* for most previously encountered antigens so that a second encounter prompts an even more rapid and vigorous response. The branch of science that deals with the responses of the body when challenged by antigens is called **immunology** (im′-ū-NOL-ō-jē; *immuno-* = free from service or exempt; *-logy* = study of). The **immune system** includes the cells and tissues that carry out immune responses.

Maturation of T Cells and B Cells

Adaptive immunity involves lymphocytes called **B cells** and **T cells.** Both develop in primary lymphatic organs (red bone marrow and the thymus) from pluripotent stem cells that originate in red bone marrow (see Figure 19.3). B cells complete their development in red bone marrow, a process that continues throughout life. T cells develop from pre-T cells that migrate from red bone marrow into the thymus, where they mature (Figure 22.11). Most T cells arise before puberty, but they continue to mature and leave the thymus throughout life. B cells and T cells are named based on where they mature. In birds, B cells mature in an organ called the *bursa of Fabricius*. Although this organ is not present in humans, the term *B cell* is still used, but the letter *B* stands for *bursa equivalent,* which is the red bone marrow since that is the location in humans where B cells mature. T cells are so named because they mature in the *thymus* gland.

Before T cells leave the thymus or B cells leave red bone marrow, they develop **immunocompetence** (im′-ū-nō-KOM-petens), the ability to carry out adaptive immune responses. This

TABLE 22.1

Summary of Innate Defenses

COMPONENT	FUNCTIONS
FIRST LINE OF DEFENSE: SKIN AND MUCOUS MEMBRANES	
Physical Factors	
Epidermis of skin	Forms physical barrier to entrance of microbes.
Mucous membranes	Inhibit entrance of many microbes, but not as effective as intact skin.
Mucus	Traps microbes in respiratory and gastrointestinal tracts.
Hairs	Filter out microbes and dust in nose.
Cilia	Together with mucus, trap and remove microbes and dust from upper respiratory tract.
Lacrimal apparatus	Tears dilute and wash away irritating substances and microbes.
Saliva	Washes microbes from surfaces of teeth and mucous membranes of mouth.
Urine	Washes microbes from urethra.
Defecation and vomiting	Expel microbes from body.
Chemical Factors	
Sebum	Forms protective acidic film over skin surface that inhibits growth of many microbes.
Lysozyme	Antimicrobial substance in perspiration, tears, saliva, nasal secretions, and tissue fluids.
Gastric juice	Destroys bacteria and most toxins in stomach.
Vaginal secretions	Slight acidity discourages bacterial growth; flush microbes out of vagina.
SECOND LINE OF DEFENSE: INTERNAL DEFENSES	
Antimicrobial Substances	
Interferons (IFNs)	Protect uninfected host cells from viral infection.
Complement system	Causes cytolysis of microbes; promotes phagocytosis; contributes to inflammation.
Iron-binding proteins	Inhibit growth of certain bacteria by reducing amount of available iron.
Antimicrobial proteins (AMPs)	Have broad-spectrum antimicrobial activities and attract dendritic cells and mast cells.
Natural killer (NK) cells	Kill infected target cells by releasing granules that contain perforin and granzymes; phagocytes then kill released microbes.
Phagocytes	Ingest foreign particulate matter.
Inflammation	Confines and destroys microbes; initiates tissue repair.
Fever	Intensifies effects of interferons; inhibits growth of some microbes; speeds up body reactions that aid repair.

means that B cells and T cells begin to make several distinctive proteins that are inserted into their plasma membranes. Some of these proteins function as **antigen receptors**—molecules capable of recognizing specific antigens (Figure 22.11).

There are two major types of mature T cells that exit the thymus: **helper T cells** and **cytotoxic T cells** (Figure 22.11). Helper T cells are also known as **CD4 T cells,** which means that, in addition to antigen receptors, their plasma membranes include a protein called CD4. Cytotoxic T cells are also referred to as **CD8 T cells** because their plasma membranes not only contain antigen receptors but also a protein known as CD8. As we will see later in this chapter, these two types of T cells have very different functions.

Types of Adaptive Immunity

There are two types of adaptive immunity: cell-mediated immunity and antibody-mediated immunity. Both types of adaptive immunity are triggered by antigens. In **cell-mediated immunity,** cytotoxic T cells directly attack invading antigens. In **antibody-mediated immunity,** B cells transform into plasma cells, which synthesize and secrete specific proteins called **antibodies (Abs)** or **immunoglobulins (Igs)** (im'-ū-nō-GLOB-ū-lins). A given antibody can bind to and inactivate a specific antigen. Helper T cells aid the immune responses of both cell-mediated and antibody-mediated immunity.

Cell-mediated immunity is particularly effective against (1) intracellular pathogens, which include any viruses, bacteria, or fungi that are inside cells; (2) some cancer cells; and (3) foreign tissue transplants. Thus, cell-mediated immunity always involves cells attacking cells. Antibody-mediated immunity works mainly against extracellular pathogens, which include any viruses, bacteria, or fungi that are in body fluids outside cells. Since antibody-mediated immunity involves antibodies that bind to antigens in body *humors* or fluids (such as blood and lymph), it is also referred to as *humoral immunity*.

In most cases, when a particular antigen initially enters the body, there is only a small group of lymphocytes with the correct antigen receptors to respond to that antigen; this small group of cells includes a few helper T cells, cytotoxic T cells, and B cells. Depending on its location, a given antigen can provoke both types of adaptive immune responses. This is due to the fact that when a specific antigen invades the body, there are usually many copies of that antigen spread throughout the body's tissues and fluids. Some copies of the antigen may be present inside body cells (which provokes a cell-mediated immune response by cytotoxic T cells), while other copies of the antigen may be present in extracellular fluid (which provokes an antibody-mediated immune response by B cells). Thus, cell-mediated and antibody-mediated immune responses often work together to get rid of the large number of copies of a particular antigen from the body.

Clonal Selection: The Principle

As you just learned, when a specific antigen is present in the body, there are usually many copies of that antigen located throughout the body's tissues and fluids. The numerous copies of the antigen

Figure 22.11 **B cells and pre-T cells arise from pluripotent stem cells in red bone marrow.** B cells and T cells develop in primary lymphatic tissues (red bone marrow and the thymus) and are activated in secondary lymphatic organs and tissues (lymph nodes, spleen, and lymphatic nodules). Once activated, each type of lymphocyte forms a clone of cells that can recognize a specific antigen. For simplicity, antigen receptors, CD4 proteins, and CD8 proteins are not shown in the plasma membranes of the cells of the lymphocyte clones.

The two types of adaptive immunity are cell-mediated immunity and antibody-mediated immunity.

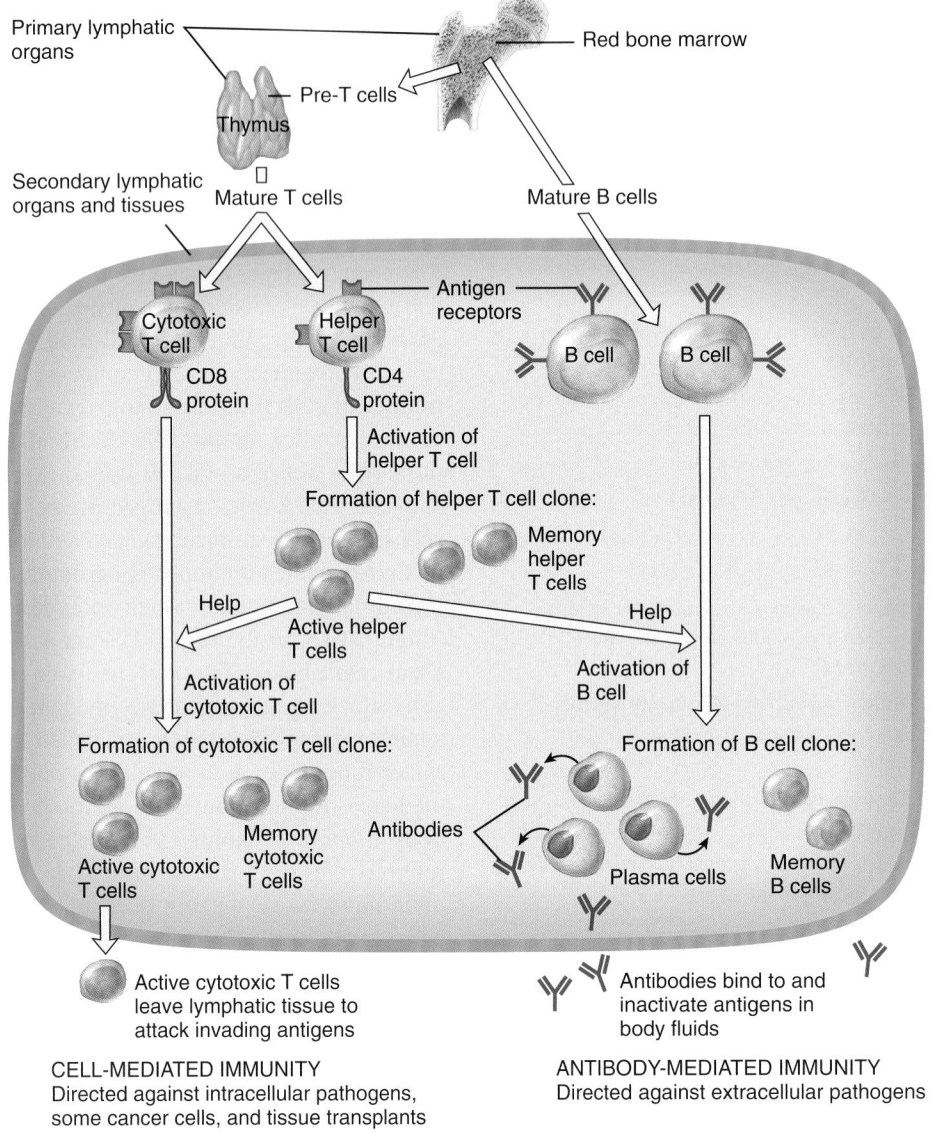

CELL-MEDIATED IMMUNITY
Directed against intracellular pathogens, some cancer cells, and tissue transplants

ANTIBODY-MEDIATED IMMUNITY
Directed against extracellular pathogens

? Which type of T cell participates in both cell-mediated and antibody-mediated immune responses?

initially outnumber the small group of helper T cells, cytotoxic T cells, and B cells with the correct antigen receptors to respond to that antigen. Therefore, once each of these lymphocytes encounters a copy of the antigen and receives stimulatory cues, it subsequently undergoes clonal selection. **Clonal selection** is the process by which a lymphocyte **proliferates** (divides) and **differentiates** (forms more highly specialized cells) in response to a specific antigen. The result of clonal selection is the formation of a population of identical cells, called a **clone,** that can recognize the same specific antigen as the original lymphocyte (Figure 22.11). Before the first exposure to a given antigen, only a few lymphocytes are able to recognize it, but once clonal selection

occurs, there are thousands of lymphocytes that can respond to that antigen. Clonal selection of lymphocytes occurs in the secondary lymphatic organs and tissues. The swollen tonsils or lymph nodes in your neck you experienced the last time you were sick were probably caused by clonal selection of lymphocytes participating in an immune response.

A lymphocyte that undergoes clonal selection gives rise to two major types of cells in the clone: effector cells and memory cells. The thousands of **effector cells** of a lymphocyte clone carry out immune responses that ultimately result in the destruction or inactivation of the antigen. Effector cells include **active helper T cells,** which are part of a helper T cell clone; **active cytotoxic T**

cells, which are part of a cytotoxic T cell clone; and **plasma cells,** which are part of a B cell clone. Most effector cells eventually die after the immune response has been completed.

Memory cells do not actively participate in the initial immune response to the antigen. However, if the same antigen enters the body again in the future, the thousands of memory cells of a lymphocyte clone are available to initiate a far swifter reaction than occurred during the first invasion. The memory cells respond to the antigen by proliferating and differentiating into more effector cells and more memory cells. Consequently, the second response to the antigen is usually so fast and so vigorous that the antigen is destroyed before any signs or symptoms of disease can occur. Memory cells include **memory helper T cells,** which are part of a helper T cell clone; **memory cytotoxic T cells,** which are part of a cytotoxic T cell clone; and **memory B cells,** which are part of a B cell clone. Most memory cells do not die at the end of an immune response. Instead, they have long life spans (often lasting for decades). The functions of effector cells and memory cells are described in more detail later in this chapter.

Antigens and Antigen Receptors

Antigens have two important characteristics: immunogenicity and reactivity. **Immunogenicity** (im-ū-nō-je-NIS-i-tē; *-genic* = producing) is the ability to provoke an immune response by stimulating the production of specific antibodies, the proliferation of specific T cells, or both. The term **antigen** derives from its function as an *anti*body *gen*erator. **Reactivity** is the ability of the antigen to react specifically with the antibodies or cells it provoked. Strictly speaking, immunologists define antigens as substances that have reactivity; substances with both immunogenicity and reactivity are considered **complete antigens.** Commonly, however, the term *antigen* implies both immunogenicity and reactivity, and we use the word in this way.

Entire microbes or parts of microbes may act as antigens. Chemical components of bacterial structures such as flagella, capsules, and cell walls are antigenic, as are bacterial toxins. Nonmicrobial examples of antigens include chemical components of pollen, egg white, incompatible blood cells, and transplanted tissues and organs. The huge variety of antigens in the environment provides myriad opportunities for provoking immune responses. Typically, just certain small parts of a large antigen molecule act as the triggers for immune responses. These small parts are called **epitopes** (EP-i-tōps), or *antigenic determinants* (Figure 22.12). Most antigens have many epitopes, each of which induces production of a specific antibody or activates a specific T cell.

Antigens that get past the innate defenses generally follow one of three routes into lymphatic tissue: (1) Most antigens that enter the bloodstream (for example, through an injured blood vessel) are trapped as they flow through the spleen. (2) Antigens that penetrate the skin enter lymphatic vessels and lodge in lymph nodes. (3) Antigens that penetrate mucous membranes are entrapped by mucosa-associated lymphatic tissue (MALT).

Chemical Nature of Antigens

Antigens are large, complex molecules. Most often, they are proteins. However, nucleic acids, lipoproteins, glycoproteins, and cer-

Figure 22.12 Epitopes (antigenic determinants).

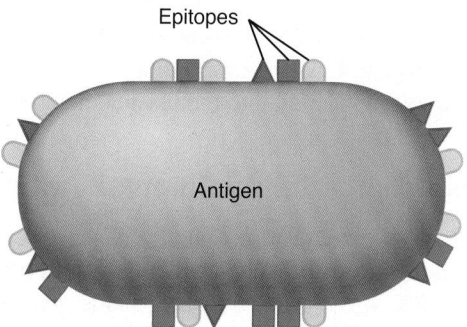

Most antigens have several epitopes that induce the production of different antibodies or activate different T cells.

? What is the difference between an epitope and a hapten?

tain large polysaccharides may also act as antigens. Complete antigens usually have large molecular weights of 10,000 daltons or more, but large molecules that have simple, repeating subunits—for example, cellulose and most plastics—are not usually antigenic. This is why plastic materials can be used in artificial heart valves or joints.

A smaller substance that has reactivity but lacks immunogenicity is called a **hapten** (HAP-ten = to grasp). A hapten can stimulate an immune response only if it is attached to a larger carrier molecule. An example is the small lipid toxin in poison ivy, which triggers an immune response after combining with a body protein. Likewise, some drugs, such as penicillin, may combine with proteins in the body to form immunogenic complexes. Such hapten-stimulated immune responses are responsible for some allergic reactions to drugs and other substances in the environment (see Disorders: Homeostatic Imbalances at the end of the chapter).

As a rule, antigens are foreign substances; they are not usually part of body tissues. However, sometimes the immune system fails to distinguish "friend" (self) from "foe" (nonself). The result is an autoimmune disorder (see Disorders: Homeostatic Imbalances at the end of the chapter) in which self-molecules or cells are attacked as though they were foreign.

Diversity of Antigen Receptors

An amazing feature of the human immune system is its ability to recognize and bind to at least a billion (10^9) different epitopes. Before a particular antigen ever enters the body, T cells and B cells that can recognize and respond to that intruder are ready and waiting. Cells of the immune system can even recognize artificially made molecules that do not exist in nature. The basis for the ability to recognize so many epitopes is an equally large diversity of antigen receptors. Given that human cells contain only about 35,000 genes, how could a billion or more different antigen receptors possibly be generated?

The answer to this puzzle turned out to be simple in concept. The diversity of antigen receptors in both B cells and T cells is the result of shuffling and rearranging a few hundred versions of several small gene segments. This process is called **genetic**

recombination. The gene segments are put together in different combinations as the lymphocytes are developing from stem cells in red bone marrow and the thymus. The situation is similar to shuffling a deck of 52 cards and then dealing out three cards. If you did this over and over, you could generate many more than 52 different sets of three cards. Because of genetic recombination, each B cell or T cell has a unique set of gene segments that codes for its unique antigen receptor. After transcription and translation, the receptor molecules are inserted into the plasma membrane.

Major Histocompatibility Complex Antigens

Located in the plasma membrane of body cells are "self-antigens," the **major histocompatibility complex (MHC) antigens** (his′-tō-kom-pat′-i-BIL-i-tē). These transmembrane glycoproteins are also called *human leukocyte antigens (HLA)* because they were first identified on white blood cells. Unless you have an identical twin, your MHC antigens are unique. Thousands to several hundred thousand MHC molecules mark the surface of each of your body cells except red blood cells. Although MHC antigens are the reason that tissues may be rejected when they are transplanted from one person to another, their normal function is to help T cells recognize that an antigen is foreign, not self. Such recognition is an important first step in any adaptive immune response.

The two types of major histocompatibility complex antigens are class I and class II. Class I MHC (MHC-I) molecules are built into the plasma membranes of all body cells except red blood cells. Class II MHC (MHC-II) molecules appear on the surface of antigen-presenting cells (described in the next section).

Pathways of Antigen Processing

For an immune response to occur, B cells and T cells must recognize that a foreign antigen is present. B cells can recognize and bind to antigens in lymph, interstitial fluid, or blood plasma. T cells only recognize fragments of antigenic proteins that are processed and presented in a certain way. In **antigen processing,** antigenic proteins are broken down into peptide fragments that then associate with MHC molecules. Next the antigen–MHC complex is inserted into the plasma membrane of a body cell. The insertion of the complex into the plasma membrane is called **antigen presentation.** When a peptide fragment comes from a *self-protein,* T cells ignore the antigen–MHC complex. However, if the peptide fragment comes from a *foreign protein,* T cells recognize the antigen–MHC complex as an intruder, and an immune response takes place. Antigen processing and presentation occurs in two ways, depending on whether the antigen is located outside or inside body cells.

Processing of Exogenous Antigens

Foreign antigens that are present in fluids *outside* body cells are termed **exogenous antigens** (ex-OG-e-nus). They include intruders such as bacteria and bacterial toxins, parasitic worms, inhaled pollen and dust, and viruses that have not yet infected a body cell. A special class of cells called **antigen-presenting cells (APCs)** process and present exogenous antigens. APCs include dendritic cells, macrophages, and B cells. They are strategically located in places where antigens are likely to penetrate the innate defenses and enter the body, such as the epidermis and dermis of the skin (Langerhans cells are a type of dendritic cell); mucous membranes that line the respiratory, gastrointestinal, urinary, and reproductive tracts; and lymph nodes. After processing and presenting an antigen, APCs migrate from tissues via lymphatic vessels to lymph nodes.

The steps in the processing and presenting of an exogenous antigen by an antigen-presenting cell occur as follows (Figure 22.13):

❶ *Ingestion of the antigen.* Antigen-presenting cells ingest exogenous antigens by phagocytosis or endocytosis. Ingestion could occur almost anywhere in the body that invaders, such as microbes, have penetrated the innate defenses.

❷ *Digestion of antigen into peptide fragments.* Within the phagosome or endosome, protein-digesting enzymes split large antigens into short peptide fragments.

❸ *Synthesis of MHC-II molecules.* At the same time, the APC synthesizes MHC-II molecules at the endoplasmic reticulum (ER).

❹ *Packaging of MHC-II molecules.* Once synthesized, the MHC-II molecules are packaged into vesicles.

❺ *Fusion of vesicles.* The vesicles containing antigen peptide fragments and MHC-II molecules merge and fuse.

❻ *Binding of peptide fragments to MHC-II molecules.* After fusion of the two types of vesicles, antigen peptide fragments bind to MHC-II molecules.

❼ *Insertion of antigen–MHC-II complexes into the plasma membrane.* The combined vesicle that contains antigen–MHC-II complexes undergoes exocytosis. As a result, the antigen–MHC-II complexes are inserted into the plasma membrane.

After processing an antigen, the antigen-presenting cell migrates to lymphatic tissue to present the antigen to T cells. Within lymphatic tissue, a small number of T cells that have compatibly shaped receptors recognize and bind to the antigen fragment–MHC-II complex, triggering an adaptive immune response. The presentation of exogenous antigen together with MHC-II molecules by antigen-presenting cells informs T cells that intruders are present in the body and that combative action should begin.

Processing of Endogenous Antigens

Foreign antigens that are present *inside* body cells are termed **endogenous antigens** (en-DOJ-e-nus). Such antigens may be viral proteins produced after a virus infects the cell and takes over the cell's metabolic machinery, toxins produced from intracellular bacteria, or abnormal proteins synthesized by a cancerous cell.

The steps in the processing and presenting of an endogenous antigen by an infected body cell occur as follows (Figure 22.14):

❶ *Digestion of antigen into peptide fragments.* Within the infected cell, protein-digesting enzymes split the endogenous antigen into short peptide fragments.

❷ *Synthesis of MHC-I molecules.* At the same time, the infected cell synthesizes MHC-I molecules at the endoplasmic reticulum (ER).

Figure 22.13 Processing and presenting of exogenous antigen by an antigen-presenting cell (APC).

🔑 Fragments of exogenous antigens are processed and then presented with MHC-II molecules on the surface of an antigen-presenting cell (APC).

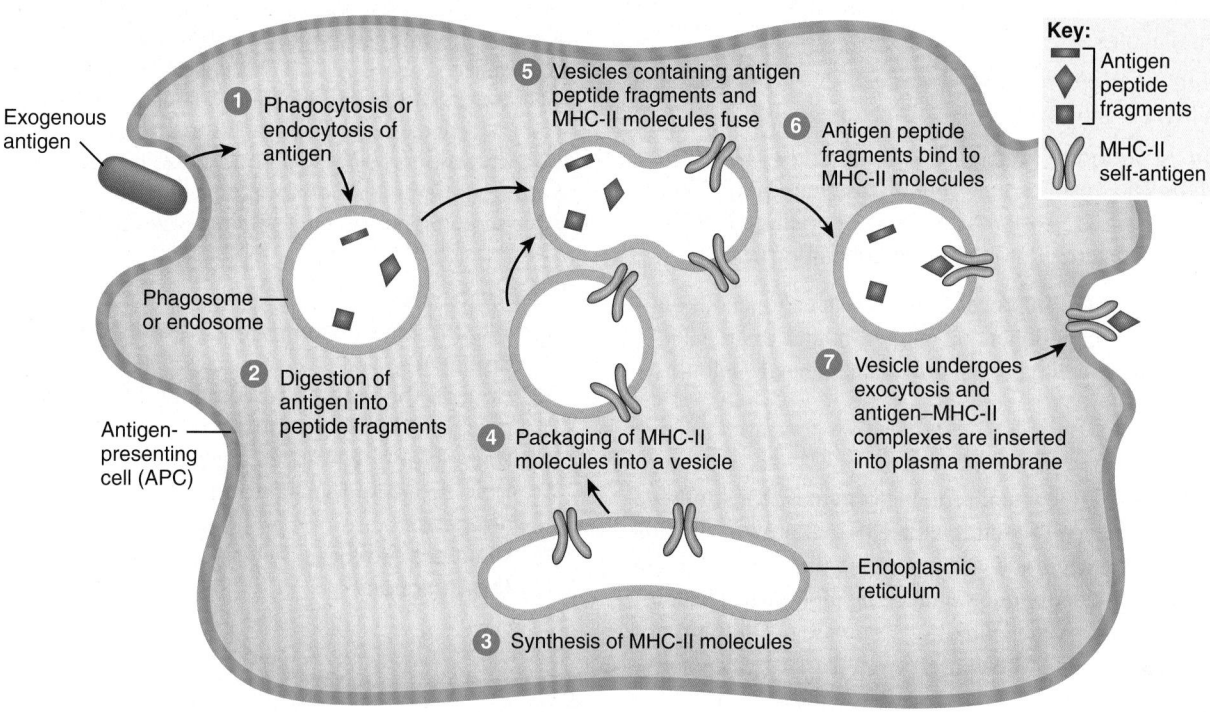

Key:
- ▬ ◆ ▪ Antigen peptide fragments
- ⋎ MHC-II self-antigen

① Phagocytosis or endocytosis of antigen

Exogenous antigen

Phagosome or endosome

② Digestion of antigen into peptide fragments

Antigen-presenting cell (APC)

⑤ Vesicles containing antigen peptide fragments and MHC-II molecules fuse

⑥ Antigen peptide fragments bind to MHC-II molecules

⑦ Vesicle undergoes exocytosis and antigen–MHC-II complexes are inserted into plasma membrane

④ Packaging of MHC-II molecules into a vesicle

Endoplasmic reticulum

③ Synthesis of MHC-II molecules

APCs present exogenous antigens in association with MHC-II molecules

❓ **What types of cells are APCs, and where in the body are they found?**

Figure 22.14 Processing and presenting of endogenous antigen by an infected body cell.

🔑 Fragments of endogenous antigens are processed and then presented with MHC-I proteins on the surface of an infected body cell.

⑤ Vesicle undergoes exocytosis and antigen–MHC-I complexes are inserted into plasma membrane

Endogenous antigen

① Digestion of antigen into peptide fragments

④ Packaging of antigen–MHC-I molecules into a vesicle

③ Antigen peptide fragments bind to MHC-I molecules

Key:
- ● ▶ ◗ Antigen peptide fragments
- ⋎ MHC-I self-antigen

Endoplasmic reticulum

② Synthesis of MHC-I molecules

Infected body cell

Infected body cells present endogenous antigens in association with MHC-I molecules

❓ **What are some examples of endogenous antigens?**

895

③ *Binding of peptide fragments to MHC-I molecules.* The antigen peptide fragments enter the ER and then bind to MHC-I molecules.

④ *Packaging of antigen–MHC-I molecules.* From the ER, antigen–MHC-I molecules are packaged into vesicles.

⑤ *Insertion of antigen–MHC-I complexes into the plasma membrane.* The vesicles that contain antigen–MHC-I complexes undergo exocytosis. As a result, the antigen–MHC-I complexes are inserted into the plasma membrane.

Most cells of the body can process and present endogenous antigens. The display of an endogenous antigen bound to an MHC-I molecule signals that a cell has been infected and needs help.

Cytokines

Cytokines (SĪ-tō-kīns) are small protein hormones that stimulate or inhibit many normal cell functions, such as cell growth and differentiation. Lymphocytes and antigen-presenting cells secrete cytokines, as do fibroblasts, endothelial cells, monocytes, hepatocytes, and kidney cells. Some cytokines stimulate proliferation of progenitor blood cells in red bone marrow. Others regulate activities of cells involved in innate defenses or adaptive immune responses, as described in Table 22.2.

CLINICAL CONNECTION | *Cytokine Therapy*

Cytokine therapy is the use of cytokines to treat medical conditions. Interferons were the first cytokines shown to have limited effects against some human cancers. Alpha-interferon (Intron A®) is approved in the United States for treating Kaposi (KAP-ō-sē) sarcoma, a cancer that often occurs in patients infected with HIV, the virus that causes AIDS. Other approved uses for alpha-interferon include treating genital herpes caused by the herpes virus; treating hepatitis B and C, caused by the hepatitis B and C viruses; and treating hairy cell leukemia. A form of beta-interferon (Betaseron®) slows the progression of multiple sclerosis (MS) and lessens the frequency and severity of MS attacks. Of the interleukins, the one most widely used to fight cancer is interleukin-2. Although this treatment is effective in causing tumor regression in some patients, it also can be very toxic. Among the adverse effects are high fever, severe weakness, difficulty breathing due to pulmonary edema, and hypotension leading to shock. •

✔ CHECKPOINT

11. What is immunocompetence, and which body cells display it?
12. How do the major histocompatibility complex class I and class II self-antigens function?
13. How do antigens arrive at lymphatic tissues?
14. How do antigen-presenting cells process exogenous antigens?
15. What are cytokines, where do they arise, and how do they function?

22.5 CELL-MEDIATED IMMUNITY

◉ OBJECTIVES

• Outline the steps in a cell-mediated immune response.
• Distinguish between the action of natural killer cells and cytotoxic T cells.
• Define immunological surveillance.

A cell-mediated immune response begins with *activation* of a small number of T cells by a specific antigen. Once a T cell has been activated, it undergoes clonal selection. Recall that **clonal selection** is the process by which a lymphocyte proliferates (divides

---

TABLE 22.2

Summary of Cytokines Participating in Immune Responses

CYTOKINE	ORIGINS AND FUNCTIONS
Interleukin-1 (IL-1) (in'-ter-LOO-kin)	Produced by macrophages; promotes proliferation of helper T cells; acts on hypothalamus to cause fever.
Interleukin-2 (IL-2)	Secreted by helper T cells; costimulates proliferation of helper T cells, cytotoxic T cells, and B cells; activates NK cells.
Interleukin-4 (IL-4) (B cell stimulating factor)	Produced by helper T cells; costimulator for B cells; causes plasma cells to secrete IgE antibodies (see Table 22.3); promotes growth of T cells.
Interleukin-5 (IL-5)	Produced by some helper T cells and mast cells; costimulator for B cells; causes plasma cells to secrete IgA antibodies.
Interleukin-6 (IL-6)	Produced by helper T cells; enhances B cell proliferation, B cell differentiation into plasma cells, and secretion of antibodies by plasma cells.
Tumor necrosis factor (TNF) (ne-KRŌ-sis)	Produced mainly by macrophages; stimulates accumulation of neutrophils and macrophages at sites of inflammation and stimulates their killing of microbes.
Interferons (IFNs) (in'-ter-FĒR-ons)	Produced by virus-infected cells to inhibit viral replication in uninfected cells; activate cytotoxic T cells and natural killer cells, inhibit cell division, and suppress the formation of tumors.
Macrophage migration inhibiting factor	Produced by cytotoxic T cells; prevents macrophages from leaving site of infection.

several times) and differentiates (forms more highly specialized cells) in response to a specific antigen. The result of clonal selection is the formation of a **clone** of cells that can recognize the same antigen as the original lymphocyte (see Figure 22.11). Some of the cells of a T cell clone become effector cells, while other cells of the clone become memory cells. The effector cells of a T cell clone carry out immune responses that ultimately result in *elimination* of the intruder.

Activation of T Cells

At any given time, most T cells are inactive. As you learned in the last section, antigen receptors on the surface of T cells, called **T-cell receptors (TCRs),** recognize and bind to specific foreign antigen fragments that are presented in antigen–MHC complexes. There are millions of different T cells; each has its own unique TCRs that can recognize a specific antigen–MHC complex. When an antigen enters the body, only a few T cells have TCRs that can recognize and bind to the antigen. Antigen recognition also involves other surface proteins on T cells, the CD4 or CD8 proteins. These proteins interact with the MHC antigens and help maintain the TCR–MHC coupling. For this reason, they are referred to as *coreceptors.* Antigen recognition by a TCR with CD4 or CD8 proteins is the *first signal* in activation of a T cell.

A T cell becomes activated only if it binds to the foreign antigen and at the same time receives a *second signal,* a process known as **costimulation.** Of the more than 20 known costimulators, some are cytokines, such as **interleukin-2.** Other costimulators include pairs of plasma membrane molecules, one on the surface of the T cell and a second on the surface of an antigen-presenting cell, that enable the two cells to adhere to one another for a period of time.

The need for two signals to activate a T cell is a little like starting and driving a car: When you insert the correct key (antigen) in the ignition (TCR) and turn it, the car starts (recognition of specific antigen), but it cannot move forward until you move the gear shift into drive (costimulation). The need for costimulation may prevent immune responses from occurring accidentally. Different costimulators affect the activated T cell in different ways, just as shifting a car into reverse has a different effect than shifting it into drive. Moreover, recognition (antigen binding to a receptor) without costimulation leads to a prolonged *state of inactivity* called **anergy** (AN-er-jē) in both T cells and B cells. Anergy is rather like leaving a car in neutral gear with its engine running until it's out of gas!

Once a T cell has received these two signals (antigen recognition and costimulation), it is **activated.** An activated T cell subsequently undergoes clonal selection.

Activation and Clonal Selection of Helper T Cells

Most T cells that display CD4 develop into **helper T cells,** also known as **CD4 T cells.** Inactive (resting) helper T cells recognize exogenous antigen fragments associated with major histocompatibility complex class II (MHC-II) molecules at the surface of an

APC (Figure 22.15). With the aid of the CD4 protein, the helper T cell and APC interact with each other (antigenic recognition), costimulation occurs, and the helper T cell becomes activated.

Once activated, the helper T cell undergoes clonal selection (Figure 22.15). The result is the formation of a clone of helper T cells that consists of active helper T cells and memory helper T cells. Within hours after costimulation, **active helper T cells** start secreting a variety of cytokines (see Table 22.2). One very important cytokine produced by helper T cells is interleukin-2 (IL-2), which is needed for virtually all immune responses and is the prime

Figure 22.15 Activation and clonal selection of a helper T cell.

 Once a helper T cell is activated, it forms a clone of active helper T cells and memory helper T cells.

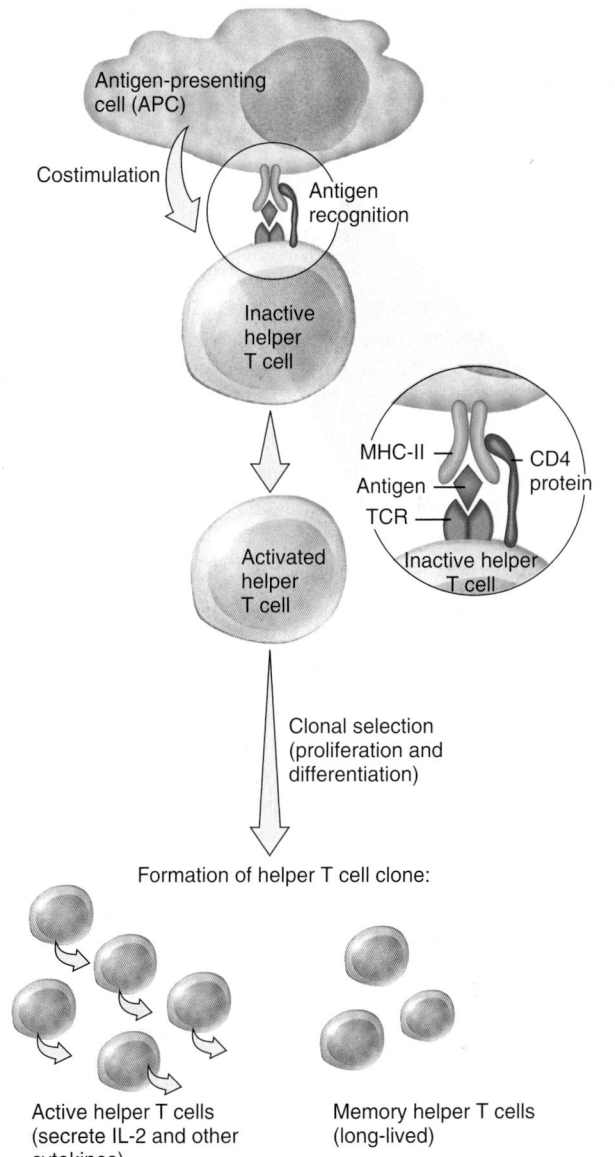

? **What are the first and second signals in activation of a T cell?**

trigger of T cell proliferation. IL-2 can act as a costimulator for resting helper T cells or cytotoxic T cells, and it enhances activation and proliferation of T cells, B cells, and natural killer cells. Some actions of interleukin-2 provide a good example of a beneficial positive feedback system. As noted earlier, activation of a helper T cell stimulates it to start secreting IL-2, which then acts in an autocrine manner by binding to IL-2 receptors on the plasma membrane of the cell that secreted it. One effect is stimulation of cell division. As the helper T cells proliferate, a positive feedback effect occurs because they secrete more IL-2, which causes further cell division. IL-2 may also act in a paracrine manner by binding to IL-2 receptors on neighboring helper T cells, cytotoxic T cells, or B cells. If any of these neighboring cells have already become bound to a copy of the same antigen, IL-2 serves as a costimulator.

The **memory helper T cells** of a helper T cell clone are not active cells. However, if the same antigen enters the body again in the future, memory helper T cells can quickly proliferate and differentiate into more active helper T cells and more memory helper T cells.

Activation and Clonal Selection of Cytotoxic T Cells

Most T cells that display CD8 develop into **cytotoxic T cells** (sī′-tō-TOK-sik), also termed **CD8 T cells.** Cytotoxic T cells recognize foreign antigens combined with major histocompatibility complex class I (MHC-I) molecules on the surface of (1) body cells infected by microbes, (2) some tumor cells, and (3) cells of a tissue transplant (Figure 22.16). Recognition requires the TCR and CD8 protein to maintain the coupling with MHC-I. Following antigenic recognition, costimulation occurs. In order to become activated, cytotoxic T cells require costimulation by interleukin-2 or other cytokines produced by active helper T cells that have already become bound to copies of the same antigen. (Recall that helper T cells are activated by antigen associated with MHC-II molecules.) Thus, *maximal activation* of cytotoxic T cells requires presentation of antigen associated with both MHC-I and MHC-II molecules.

Once activated, the cytotoxic T cell undergoes clonal selection. The result is the formation of a clone of cytotoxic T cells that consists of active cytotoxic T cells and memory cytotoxic T cells. **Active cytotoxic T cells** attack other body cells that have been infected with the antigen. **Memory cytotoxic T cells** do not attack infected body cells. Instead, they can quickly proliferate and differentiate into more active cytotoxic T cells and more memory cytotoxic T cells if the same antigen enters the body at a future time.

Elimination of Invaders

Cytotoxic T cells are the soldiers that march forth to do battle with foreign invaders in cell-mediated immune responses. They leave secondary lymphatic organs and tissues and migrate to seek out and destroy infected target cells, cancer cells, and transplanted cells (Figure 22.17). Cytotoxic T cells recognize and attach to target cells. Then, the cytotoxic T cells deliver a "lethal hit" that kills the target cells.

Figure 22.16 **Activation and clonal selection of a cytotoxic T cell.**

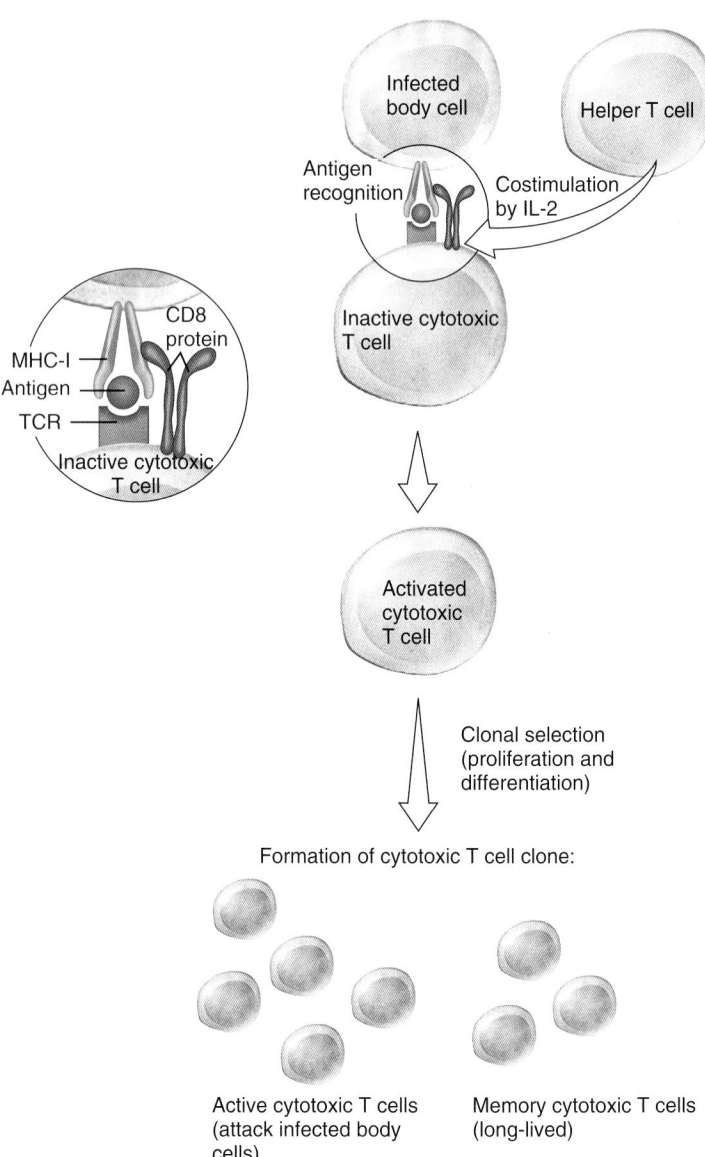

Once a cytotoxic T cell is activated, it forms a clone of active cytotoxic T cells and memory cytotoxic T cells.

? **What is the function of the CD8 protein of a cytotoxic T cell?**

Cytotoxic T cells kill infected target body cells much like natural killer cells do. The major difference is that cytotoxic T cells have receptors specific for a particular microbe and thus kill only target body cells infected with *one* particular type of microbe; natural killer cells can destroy a wide variety of microbe-infected body cells. Cytotoxic T cells have two principal mechanisms for killing infected target cells.

1. Cytotoxic T cells, using receptors on their surfaces, recognize and bind to infected target cells that have microbial antigens displayed on their surface. The cytotoxic T cell then releases

Figure 22.17 **Activity of cytotoxic T cells.** After delivering a "lethal hit," a cytotoxic T cell can detach and attack another infected target cell displaying the same antigen.

Cytotoxic T cells release granzymes that trigger apoptosis and perforin that triggers cytolysis of infected target cells.

(a) Cytotoxic T cell destruction of infected cell by release of granzymes that cause apoptosis; released microbes are destroyed by phagocyte

(b) Cytotoxic T cell destruction of infected cell by release of perforins that cause cytolysis; microbes are destroyed by granulysin

Key:
TCR   CD8 protein
Antigen–MHC-I complex

? In addition to cells infected by microbes, what other types of target cells are attacked by cytotoxic T cells?

granzymes, protein-digesting enzymes that trigger apoptosis (Figure 22.17a). Once the infected cell is destroyed, the released microbes are killed by phagocytes.

2. Alternatively, cytotoxic T cells bind to infected body cells and release two proteins from their granules: perforin and granulysin. **Perforin** inserts into the plasma membrane of the target cell and creates channels in the membrane (Figure 22.17b). As a result, extracellular fluid flows into the target cell and cytolysis (cell bursting) occurs. Other granules in cytotoxic T cells release **granulysin** (gran′-ū-LĪ-sin), which enters through the channels and destroys the microbes by creating holes in their plasma membranes. Cytotoxic T cells may also destroy target cells by releasing a toxic molecule called **lymphotoxin** (lim′-fō-TOK-sin), which activates enzymes in the target cell. These enzymes cause the target cell's DNA to fragment and the cell dies. In addition, cytotoxic T cells se-

crete gamma-interferon, which attracts and activates phagocytic cells, and macrophage migration inhibition factor, which prevents migration of phagocytes from the infection site. After detaching from a target cell, a cytotoxic T cell can seek out and destroy another target cell.

Immunological Surveillance

When a normal cell transforms into a cancerous cell, it often displays novel cell surface components called **tumor antigens.** These molecules are rarely, if ever, displayed on the surface of normal cells. If the immune system recognizes a tumor antigen as nonself, it can destroy any cancer cells carrying that antigen. Such immune responses, called **immunological surveillance** (im′-ū-nō-LOJ-i-kul sur-VĀ-lants), are carried out by cytotoxic T cells, macrophages, and natural killer cells. Immunological surveillance is most effective in eliminating tumor cells due to cancer-causing viruses. For this reason, transplant recipients who are taking immunosuppressive drugs to prevent transplant rejection have an increased incidence of virus-associated cancers. Their risk for other types of cancer is not increased.

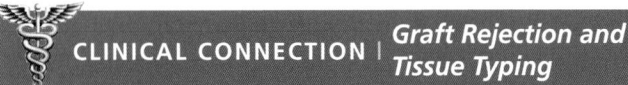

CLINICAL CONNECTION | *Graft Rejection and Tissue Typing*

Organ transplantation involves the replacement of an injured or diseased organ, such as the heart, liver, kidney, lungs, or pancreas, with an organ donated by another individual. Usually, the immune system recognizes the proteins in the transplanted organ as foreign and mounts both cell-mediated and antibody-mediated immune responses against them. This phenomenon is known as **graft rejection.**

The success of an organ or tissue transplant depends on **histocompatibility** (his′-tō-kom-pat-i-BIL-i-tē)—that is, the tissue compatibility between the donor and the recipient. The more similar the MHC antigens, the greater the histocompatibility, and thus the greater the probability that the transplant will not be rejected. **Tissue typing (histocompatibility testing)** is done before any organ transplant. In the United States, a nationwide computerized registry helps physicians select the most histocompatible and needy organ transplant recipients whenever donor organs become available. The closer the match between the major histocompatibility complex proteins of the donor and recipient, the weaker is the graft rejection response.

To reduce the risk of graft rejection, organ transplant recipients receive immunosuppressive drugs. One such drug is *cyclosporine*, derived from a fungus, which inhibits secretion of interleukin-2 by helper T cells but has only a minimal effect on B cells. Thus, the risk of rejection is diminished while resistance to some diseases is maintained. •

✔CHECKPOINT

16. What are the functions of helper, cytotoxic, and memory T cells?
17. How do cytotoxic T cells kill infected target cells?
18. How is immunological surveillance useful?

22.6 ANTIBODY-MEDIATED IMMUNITY

⦿ OBJECTIVES

• Describe the steps in an antibody-mediated immune response.
• List the chemical characteristics and actions of antibodies.
• Explain how the complement system operates.
• Distinguish between a primary response and a secondary response to infection.

The body contains not only millions of different T cells but also millions of different B cells, each capable of responding to a specific antigen. Cytotoxic T cells leave lymphatic tissues to seek out and destroy a foreign antigen, but B cells stay put. In the presence of a foreign antigen, a specific B cell in a lymph node, the spleen,

or mucosa-associated lymphatic tissue becomes activated. Then it undergoes clonal selection, forming a clone of plasma cells and memory cells. Plasma cells are the effector cells of a B cell clone; they secrete specific antibodies, which in turn circulate in the lymph and blood to reach the site of invasion.

Activation and Clonal Selection of B Cells

During activation of a B cell, an antigen binds to **B-cell receptors (BCRs)** (Figure 22.18). These integral transmembrane proteins are chemically similar to the antibodies that eventually are secreted by plasma cells. Although B cells can respond to an unprocessed antigen present in lymph or interstitial fluid, their response is much more intense when they process the antigen. Antigen processing in a B cell occurs in the following way: The antigen is taken into the B cell, broken down into peptide fragments and combined with MHC-II self-antigens, and moved to the B cell plasma membrane. Helper T cells recognize the antigen–MHC-II complex and deliver the costimulation needed for B cell proliferation and differentiation. The helper T cell produces interleukin-2 and other cytokines that function as costimulators to activate B cells.

Once activated, a B cell undergoes clonal selection (Figure 22.18). The result is the formation of a clone of B cells that consists of plasma cells and memory B cells. **Plasma cells** secrete antibodies. A few days after exposure to an antigen, a plasma cell secretes hundreds of millions of antibodies each day for about 4 or 5 days, until the plasma cell dies. Most antibodies travel in lymph and blood to the invasion site. Interleukin-4 and interleukin-6, also produced by helper T cells, enhance B cell proliferation, B cell differentiation into plasma cells, and secretion of antibodies by plasma cells. **Memory B cells** do not secrete antibodies. Instead, they can quickly proliferate and differentiate into more plasma cells and more memory B cells should the same antigen reappear at a future time.

Different antigens stimulate different B cells to develop into plasma cells and their accompanying memory B cells. All of the B cells of a particular clone are capable of secreting only one type of antibody, which is identical to the antigen receptor displayed by the B cell that first responded. Each specific antigen activates only those B cells that are predestined (by the combination of gene segments they carry) to secrete antibody specific to that antigen. Antibodies produced by a clone of plasma cells enter the circulation and form antigen–antibody complexes with the antigen that initiated their production.

Antibodies

An **antibody (Ab)** can combine specifically with the epitope on the antigen that triggered its production. The antibody's structure matches its antigen much as a lock accepts a specific key. In theory, plasma cells could secrete as many different antibodies as there are different B-cell receptors because the same recombined gene segments code for both the BCR and the antibodies eventually secreted by plasma cells.

Figure 22.18 **Activation and clonal selection of B cells.** Plasma cells are actually much larger than B cells.

6 Plasma cells secrete antibodies.

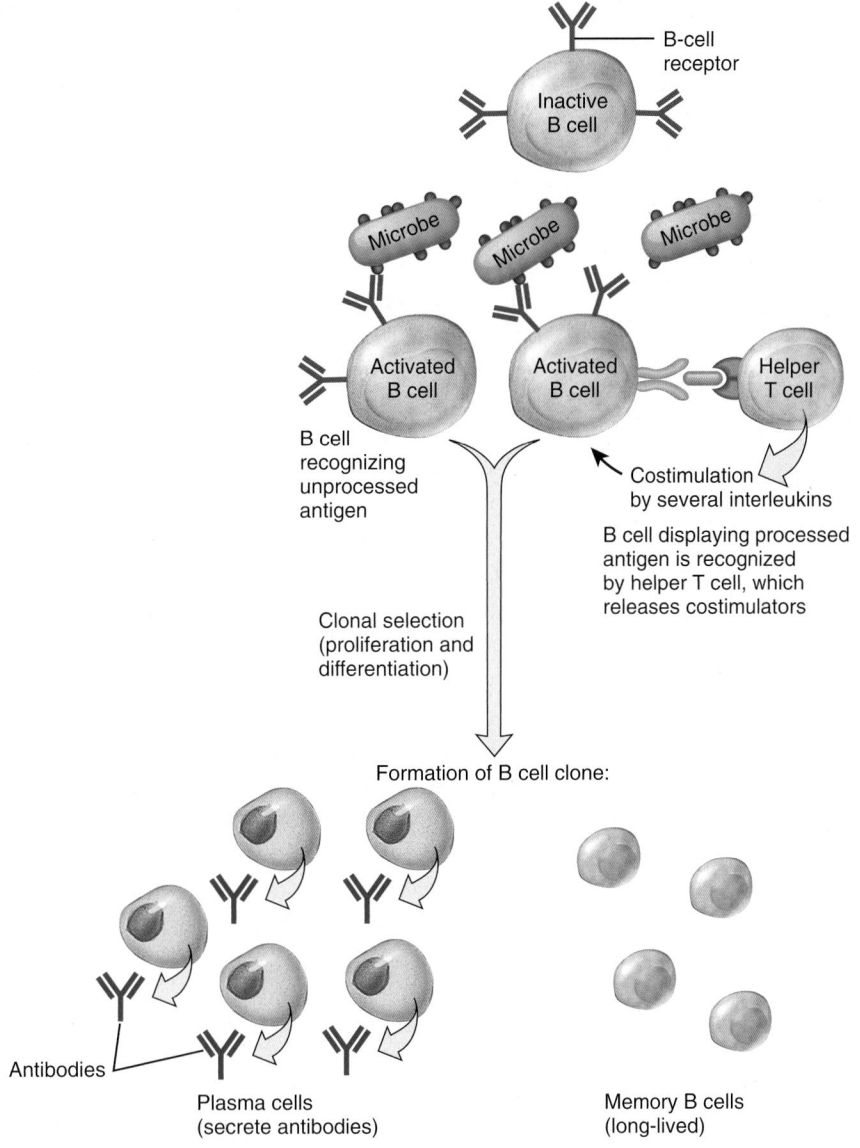

? How many different kinds of antibodies will be secreted by the plasma cells in the clone shown here?

Severe combined immunodeficiency disease (SCID) (im′-ū-nō-de-FISH-en-sē) is a rare inherited disorder in which both B cells and T cells are missing or inactive. Scientists have now identified mutations in several genes that are responsible for some types of SCID. In some cases, an infusion of red bone marrow cells from a sibling having very similar MHC (HLA) antigens can provide normal stem cells that give rise to normal B and T cells. The result can be a complete cure. Less than 30% of afflicted patients, however, have a compatible sibling who could serve as a donor. The disorder, which occurs more frequently in males, is also known as *bubble boy disease*, named for David Vetter, who was born with the condition and lived behind plastic barriers to protect him from microbes. He died at age 12 in 1984. The chances of a child born with SCID are about 1 in 500,000 and until recent years, it was always fatal. Children with SCID have virtually no defenses against microbes. Treatment consists of bringing any current infections under control, bolstering nutrition, bone marrow transplant (provides stem cells to make new B and T cells), enzymatic replacement therapy (injections of polyethylene glycol–linked adenosine deaminase, or PE-ADA), and gene therapy. In this technique, the most common approach is to insert a normal gene into a genome to replace a nonfunctional gene. The normal gene is usually delivered by a virus. Then, the normal gene would produce B and T cells to provide sufficient immunity. •

Antibody Structure

Antibodies belong to a group of glycoproteins called globulins, and for this reason they are also known as **immunoglobulins (Igs)** (im′-ū-nō-GLOB-ū-lins). Most antibodies contain four polypeptide chains (Figure 22.19). Two of the chains are identical to each other and are called **heavy (H) chains;** each consists of about 450 amino acids. Short carbohydrate chains are attached to each heavy polypeptide chain. The two other polypeptide chains, also identical to each other, are called **light (L) chains,** and each consists of about 220 amino acids. A disulfide bond (S—S) holds each light chain to a heavy chain. Two disulfide bonds also link the midregion of the two heavy chains; this part of the antibody displays considerable flexibility and is called the **hinge region.** Because the antibody "arms" can move somewhat as the hinge region bends, an antibody can assume either a T shape or a Y shape (Figure 22.19a, b). Beyond the hinge region, parts of the two heavy chains form the **stem region.**

Within each H and L chain are two distinct regions. The tips of the H and L chains, called the **variable (V) regions,** constitute the **antigen-binding site.** The variable region, which is different for each kind of antibody, is the part of the antibody that recognizes and attaches specifically to a particular antigen. Because most antibodies have two antigen-binding sites, they are said to be *bivalent.* Flexibility at the hinge allows the antibody to simultaneously bind to two epitopes that are some distance apart—for example, on the surface of a microbe.

The remainder of each H and L chain, called the **constant (C) region,** is nearly the same in all antibodies of the same class and is responsible for the type of antigen–antibody reaction that occurs. However, the constant region of the H chain differs from one class of antibody to another, and its structure serves as a basis for distinguishing five different classes, designated IgG, IgA, IgM, IgD, and IgE. Each class has a distinct chemical structure and a specific biological role. Because they appear first and are relatively short-lived, IgM antibodies indicate a recent invasion. In a sick patient, the responsible pathogen may be suggested by the presence of high levels of IgM specific to a particular organism. Resistance of the fetus and newborn baby to infection stems mainly from maternal IgG antibodies that cross the placenta before birth and IgA antibodies in breast milk after birth. Table 22.3 summarizes the structures and functions of the five classes of antibodies.

Antibody Actions

The actions of the five classes of immunoglobulins differ somewhat, but all of them act to disable antigens in some way. Actions of antibodies include the following:

- *Neutralizing antigen.* The reaction of antibody with antigen blocks or neutralizes some bacterial toxins and prevents attachment of some viruses to body cells.

- *Immobilizing bacteria.* If antibodies form against antigens on the cilia or flagella of motile bacteria, the antigen–antibody reaction may cause the bacteria to lose their motility, which limits their spread into nearby tissues.

- *Agglutinating and precipitating antigen.* Because antibodies have two or more sites for binding to antigen, the antigen–antibody reaction may cross-link pathogens to one another, causing agglutination (clumping together). Phagocytic cells ingest agglutinated microbes more readily. Likewise, soluble antigens may come out of solution and form a more easily phagocytized precipitate when cross-linked by antibodies.

Figure 22.19 Chemical structure of the immunoglobulin G (IgG) class of antibody. Each molecule is composed of four polypeptide chains (two heavy and two light) plus a short carbohydrate chain attached to each heavy chain. In (a), each circle represents one amino acid. In (b), V_L = variable regions of light chain, C_L = constant region of light chain, V_H = variable region of heavy chain, and C_H = constant region of heavy chain.

An antibody combines only with the epitope on the antigen that triggered its production.

(a) Model of IgG molecule

(b) Diagram of IgG heavy and light chains

? **What is the function of the variable regions in an antibody molecule?**

TABLE 22.3

Classes of Immunoglobulins (Igs)

NAME AND STRUCTURE	CHARACTERISTICS AND FUNCTIONS
IgG	Most abundant, about 80% of all antibodies in blood; found in blood, lymph, and intestines; monomer (one-unit) structure. Protects against bacteria and viruses by enhancing phagocytosis, neutralizing toxins, and triggering complement system. Is the only class of antibody to cross placenta from mother to fetus, conferring considerable immune protection in newborns.
IgA	Found mainly in sweat, tears, saliva, mucus, breast milk, and gastrointestinal secretions. Smaller quantities are present in blood and lymph. Makes up 10–15% of all antibodies in blood; occurs as monomers and dimers (two units). Levels decrease during stress, lowering resistance to infection. Provides localized protection of mucous membranes against bacteria and viruses.
IgM	About 5–10% of all antibodies in blood; also found in lymph. Occurs as pentamers (five units); first antibody class to be secreted by plasma cells after initial exposure to any antigen. Activates complement and causes agglutination and lysis of microbes. Also present as monomers on surfaces of B cells, where they serve as antigen receptors. In blood plasma, anti-A and anti-B antibodies of ABO blood group, which bind to A and B antigens during incompatible blood transfusions, are also IgM antibodies (see Figure 19.12).
IgD	Mainly found on surfaces of B cells as antigen receptors, where it occurs as monomers; involved in activation of B cells. About 0.2% of all antibodies in blood.
IgE	Less than 0.1% of all antibodies in blood; occurs as monomers; located on mast cells and basophils. Involved in allergic and hypersensitivity reactions; provides protection against parasitic worms.

- *Activating complement.* Antigen–antibody complexes initiate the classical pathway of the complement system (discussed shortly).

- *Enhancing phagocytosis.* The stem region of an antibody acts as a flag that attracts phagocytes once antigens have bound to the antibody's variable region. Antibodies enhance the activity of phagocytes by causing agglutination and precipitation, by activating complement, and by coating microbes so that they are more susceptible to phagocytosis.

CLINICAL CONNECTION | *Monoclonal Antibodies*

The antibodies produced against a given antigen by plasma cells can be harvested from an individual's blood. However, because an antigen typically has many epitopes, several different clones of plasma cells produce different antibodies against the antigen. If a single plasma cell could be isolated and induced to proliferate into a clone of identical plasma cells, then a large quantity of identical antibodies could be produced. Unfortunately, lymphocytes and plasma cells are difficult to grow in culture, so scientists sidestepped this difficulty by fusing B cells with tumor cells that grow easily and proliferate endlessly. The resulting hybrid cell is called a **hybridoma** (hī-bri-DŌ-ma). Hybridomas are long-term sources of large quantities of pure, identical antibodies, called

monoclonal antibodies (MAbs) (mon′-ō-KLŌ-nal) because they come from a single clone of identical cells. One clinical use of monoclonal antibodies is for measuring levels of a drug in a patient's blood. Other uses include the diagnosis of strep throat, pregnancy, allergies, and diseases such as hepatitis, rabies, and some sexually transmitted diseases. MAbs have also been used to detect cancer at an early stage and to ascertain the extent of metastasis. They may also be useful in preparing vaccines to counteract the rejection associated with transplants, to treat autoimmune diseases, and perhaps to treat AIDS. •

Role of the Complement System in Immunity

The **complement system** (KOM-ple-ment) is a defensive system made up of over 30 proteins produced by the liver and found circulating in blood plasma and within tissues throughout the body. Collectively, the complement proteins destroy microbes by causing phagocytosis, cytolysis, and inflammation; they also prevent excessive damage to body tissues.

Most complement proteins are designated by an uppercase letter C, numbered C1 through C9, named for the order in which they were discovered. The C1–C9 complement proteins are inactive and become activated only when split by enzymes into active fragments, which are indicated by lowercase letters *a* and *b*. For example, inactive complement protein C3 is split into the activated fragments, C3a and C3b. The active fragments carry out the destructive actions of the C1–C9 complement proteins. Other complement proteins are referred to as factors B, D, and P (properdin).

Complement proteins act in a *cascade*—one reaction triggers another reaction, which in turn triggers another reaction, and so on. With each succeeding reaction, more and more product is formed so that the net effect is amplified many times.

Complement activation may begin by three different pathways (described shortly), all of which activate C3. Once activated, C3 begins a cascade of reactions that brings about phagocytosis, cytolysis, and inflammation as follows (Figure 22.20):

① Inactivated C3 splits into activated C3a and C3b.

② C3b binds to the surface of a microbe and receptors on phagocytes attach to the C3b. Thus C3b enhances **phagocytosis** by coating a microbe, a process called **opsonization** (op-sō-ni-ZĀ-shun). Opsonization promotes attachment of a phagocyte to a microbe.

③ C3b also initiates a series of reactions that bring about cytolysis. First, C3b splits C5. The C5b fragment then binds to C6 and C7, which attach to the plasma membrane of an invading microbe. Then C8 and several C9 molecules join the other complement proteins and together form a cylinder-shaped **membrane attack complex,** which inserts into the plasma membrane.

④ The membrane attack complex creates channels in the plasma membrane that result in **cytolysis,** the bursting of the microbial cells due to the inflow of extracellular fluid through the channels.

⑤ C3a and C5a bind to mast cells and cause them to release histamine that increases blood vessel permeability during **inflammation.** C5a also attracts phagocytes to the site of inflammation (chemotaxis).

Figure 22.20 Complement activation and results of activation. (Adapted from Tortora, Funke, and Case, *Microbiology: An Introduction, Eighth Edition,* Figure 16.10, Pearson Benjamin-Cummings, 2004.)

🔑 When activated, complement proteins enhance phagocytosis, cytolysis, and inflammation.

? **Which pathway for activation of complement involves antibodies? Explain why.**

C3 can be activated in three ways: (1) The **classical pathway** starts when antibodies bind to antigens (microbes). The antigen–antibody complex binds and activates C1. Eventually, C3 is activated and the C3 fragments initiate phagocytosis, cytolysis, and inflammation. (2) The **alternative pathway** does not involve antibodies. It is initiated by an interaction between lipid–carbohydrate complexes on the surface of microbes and complement protein factors B, D, and P. This interaction activates C3. (3) In the **lectin pathway,** macrophages that digest microbes release chemicals that cause the liver to produce proteins called **lectins.** Lectins bind to the carbohydrates on the surface of microbes, ultimately causing the activation of C3.

Once complement is activated, proteins in blood and on body cells such as blood cells break down activated C3. In this way, its destructive capabilities cease very quickly so that damage to body cells is minimized.

Immunological Memory

A hallmark of immune responses is memory for specific antigens that have triggered immune responses in the past. Immunological memory is due to the presence of long-lasting antibodies and very long-lived lymphocytes that arise during clonal selection of antigen-stimulated B cells and T cells.

Immune responses, whether cell-mediated or antibody-mediated, are much quicker and more intense after a second or subsequent exposure to an antigen than after the first exposure. Initially, only a few cells have the correct specificity to respond, and the immune response may take several days to build to maximum intensity. Because thousands of memory cells exist after an initial encounter with an antigen, the next time the same antigen appears they can proliferate and differentiate into helper T cells, cytotoxic T cells, or plasma cells within hours.

One measure of immunological memory is *antibody titer* (TĪ-ter), the amount of antibody in serum. After an initial contact with an antigen, no antibodies are present for a period of several days. Then, a slow rise in the antibody titer occurs, first IgM and then IgG, followed by a gradual decline in antibody titer (Figure 22.21). This is the **primary response.**

Memory cells may remain for decades. Every new encounter with the same antigen results in a rapid proliferation of memory cells. After subsequent encounters, the antibody titer is far greater than during a primary response and consists mainly of IgG antibodies. This accelerated, more intense response is called the **secondary response.** Antibodies produced during a secondary response have an even higher affinity for the antigen than those produced during a primary response, and thus they are more successful in disposing of it.

Primary and secondary responses occur during microbial infection. When you recover from an infection without taking antimicrobial drugs, it is usually because of the primary response. If the same microbe infects you later, the secondary response could be so swift that the microbes are destroyed before you exhibit any signs or symptoms of infection.

Figure 22.21 Production of antibodies in the primary (after first exposure) and secondary (after second exposure) responses to a given antigen.

Immunological memory is the basis for successful immunization by vaccination.

According to this graph, how much more IgG is circulating in the blood in the secondary response than in the primary response? (Hint: Notice that each mark on the antibody titer axis represents a 10-fold increase.)

Immunological memory provides the basis for immunization by vaccination against certain diseases (for example, polio). When you receive the vaccine, which may contain **attenuated** (weakened) or killed whole microbes or portions of microbes, your B cells and T cells are activated. Should you subsequently encounter the living pathogen as an infecting microbe, your body initiates a secondary response.

Table 22.4 summarizes the various ways to acquire adaptive immunity.

TABLE 22.4

Ways to Acquire Adaptive Immunity

METHOD	DESCRIPTION
Naturally acquired active immunity	Following exposure to a microbe, antigen recognition by B cells and T cells and costimulation lead to formation of antibody-secreting plasma cells, cytotoxic T cells, and B and T memory cells.
Naturally acquired passive immunity	IgG antibodies are transferred from mother to fetus across placenta, or IgA antibodies are transferred from mother to baby in milk during breast-feeding.
Artificially acquired active immunity	Antigens introduced during vaccination stimulate cell-mediated and antibody-mediated immune responses, leading to production of memory cells. Antigens are pretreated to be immunogenic but not pathogenic (they will trigger an immune response but not cause significant illness).
Artificially acquired passive immunity	Intravenous injection of immunoglobulins (antibodies).

19. How do the five classes of antibodies differ in structure and function?
20. How are cell-mediated and antibody-mediated immune responses similar and different?
21. In what ways does the complement system augment antibody-mediated immune responses?
22. How is the secondary response to an antigen different from the primary response?

22.7 SELF-RECOGNITION AND SELF-TOLERANCE

⦿ **OBJECTIVE**

• Describe how self-recognition and self-tolerance develop.

To function properly, your T cells must have two traits: (1) They must be able to *recognize* your own major histocompatibility complex (MHC) proteins, a process known as **self-recognition,** and (2) they must *lack reactivity* to peptide fragments from your own proteins, a condition known as **self-tolerance** (Figure 22.22). B cells also display self-tolerance. Loss of self-tolerance leads to the development of autoimmune diseases (see Disorders: Homeostatic Imbalances at the end of the chapter).

Pre-T cells in the thymus develop the capability for self-recognition via **positive selection** (Figure 22.22a). In this process, some pre-T cells express T-cell receptors (TCRs) that

Figure 22.22 Development of self-recognition and self-tolerance. MHC = major histocompatibility complex, TCR = T-cell receptor.

🔑 Positive selection allows recognition of self-MHC proteins; negative selection provides self-tolerance of your own peptides and other self-antigens.

(a) Positive and negative selection of T cells in the thymus

(b) Selection of T cells after they emerge from the thymus

(c) Selection of B cells

Key:
⫷ Cell survival or activation
⫸ Cell death or anergy (inactivation)

❓ **How does deletion differ from anergy?**

interact with self-MHC proteins on epithelial cells in the thymic cortex. Because of this interaction, the T cells can recognize the MHC part of an antigen–MHC complex. These T cells survive. Other immature T cells that fail to interact with thymic epithelial cells are not able to recognize self-MHC proteins. These cells undergo apoptosis.

The development of self-tolerance occurs by a weeding-out process called **negative selection** in which the T cells interact with dendritic cells located at the junction of the cortex and medulla in the thymus. In this process, T cells with receptors that recognize self-peptide fragments or other self-antigens are eliminated or inactivated (Figure 22.22a). The T cells selected to survive do not respond to self-antigens, the fragments of molecules that are normally present in the body. Negative selection occurs via both deletion and anergy. In **deletion,** self-reactive T cells undergo apoptosis and die; in **anergy** they remain alive but are unresponsive to antigenic stimulation. Only 1–5% of the immature T cells in the thymus receive the proper signals to survive apoptosis during both positive and negative selection and emerge as mature, immunocompetent T cells.

Once T cells have emerged from the thymus, they may still encounter an unfamiliar self-protein; in such cases they may also become anergic if there is no costimulator (Figure 22.22b). Deletion of self-reactive T cells may also occur after they leave the thymus.

B cells also develop tolerance through deletion and anergy (Figure 22.22c). While B cells are developing in bone marrow, those cells exhibiting antigen receptors that recognize common self-antigens (such as MHC proteins or blood group antigens) are deleted. Once B cells are released into the blood, however, anergy appears to be the main mechanism for preventing responses to self-proteins. When B cells encounter an antigen not associated with an antigen-presenting cell, the necessary costimulation signal often is missing. In this case, the B cell is likely to become anergic (inactivated) rather than activated.

Table 22.5 summarizes the activities of cells involved in adaptive immune responses.

CLINICAL CONNECTION | *Cancer Immunology*

Although the immune system responds to cancerous cells, often immunity provides inadequate protection, as evidenced by the number of people dying each year from cancer. Over the past 25 years, considerable research has focused on *cancer immunology,* the study of ways to use immune responses for detecting, monitoring, and treating cancer. For example, some tumors of the colon release *carcinoembryonic antigen (CEA)* into the blood, and prostate cancer cells release *prostate-specific antigen (PSA)*. Detecting these antigens in blood does not provide definitive diagnosis of cancer, because both antigens are also released in certain noncancerous conditions. However, high levels of cancer-related antigens in the blood often do indicate the presence of a malignant tumor.

Finding ways to induce our immune system to mount vigorous attacks against cancerous cells has been an elusive goal. Many different techniques have been tried, with only modest success. In one method, inactive lymphocytes are removed in a blood sample and cultured with interleukin-2. The resulting *lymphokine-activated killer (LAK) cells* are then transfused back into the patient's blood. Although LAK cells have produced dramatic improvement in a few cases, severe complications affect most patients. In another method, lymphocytes procured from a small biopsy sample of a tumor are cultured with interleukin-2. After

---

TABLE 22.5

Summary of Functions of Cells Participating in Adaptive Immune Responses

CELL	FUNCTIONS
ANTIGEN-PRESENTING CELLS (APCs)	
Macrophage	Processing and presentation of foreign antigens to T cells; secretion of interleukin-1, which stimulates secretion of interleukin-2 by helper T cells and induces proliferation of B cells; secretion of interferons that stimulate T cell growth.
Dendritic cell	Processes and presents antigen to T cells and B cells; found in mucous membranes, skin, lymph nodes.
B cell	Processes and presents antigen to helper T cells.
LYMPHOCYTES	
Cytotoxic T cell	Kills host target cells by releasing granzymes that induce apoptosis, perforin that forms channels to cause cytolysis, granulysin that destroys microbes, lymphotoxin that destroys target cell DNA, gamma-interferon that attracts macrophages and increases their phagocytic activity, and macrophage migration inhibition factor that prevents macrophage migration from site of infection.
Helper T cell	Cooperates with B cells to amplify antibody production by plasma cells and secretes interleukin-2, which stimulates proliferation of T cells and B cells. May secrete gamma-IFN and tumor necrosis factor (TNF), which stimulate inflammatory response.
Memory T cell	Remains in lymphatic tissue and recognizes original invading antigens, even years after first encounter.
B cell	Differentiates into antibody-producing plasma cell.
Plasma cell	Descendant of B cell that produces and secretes antibodies.
Memory B cell	Descendant of B cell that remains after immune response and is ready to respond rapidly and forcefully should the same antigen enter body in future.

their proliferation in culture, such *tumor-infiltrating lymphocytes (TILs)* are reinjected. About a quarter of patients with malignant melanoma and renal-cell carcinoma who received TIL therapy showed significant improvement. The many studies currently under way provide reason to hope that immune-based methods will eventually lead to cures for cancer. •

✔**CHECKPOINT**

23. What do positive selection, negative selection, and anergy accomplish?

22.8 STRESS AND IMMUNITY

◉ **OBJECTIVE**

• Describe the effects of stress on immunity.

The field of **psychoneuroimmunology (PNI)** deals with communication pathways that link the nervous, endocrine, and immune systems. PNI research appears to justify what people have long observed: Your thoughts, feelings, moods, and beliefs influence your level of health and the course of disease. For example, cortisol, a hormone secreted by the adrenal cortex in association with the stress response, inhibits immune system activity.

If you want to observe the relationship between lifestyle and immune function, visit a college campus. As the semester progresses and the workload accumulates, an increasing number of students can be found in the waiting rooms of student health services. When work and stress pile up, health habits can change. Many people smoke or consume more alcohol when stressed, two habits detrimental to optimal immune function. Under stress, people are less likely to eat well or exercise regularly, two habits that enhance immunity.

People resistant to the negative health effects of stress are more likely to experience a sense of control over the future, a commitment to their work, expectations of generally positive outcomes for themselves, and feelings of social support. To increase your stress resistance, cultivate an optimistic outlook, get involved in your work, and build good relationships with others.

Adequate sleep and relaxation are especially important for a healthy immune system. But when there aren't enough hours in the day, you may be tempted to steal some from the night. While skipping sleep may give you a few more hours of productive time in the short run, in the long run you end up even farther behind, especially if getting sick keeps you out of commission for several days, blurs your concentration, and blocks your creativity.

Even if you make time to get 8 hours of sleep, stress can cause insomnia. If you find yourself tossing and turning at night, it's time to improve your stress management and relaxation skills! Be sure to unwind from the day before going to bed.

✔**CHECKPOINT**

24. Have you ever observed a connection between stress and illness in your own life?

22.9 AGING AND THE IMMUNE SYSTEM

◉ **OBJECTIVE**

• Describe the effects of aging on the immune system.

With advancing age, most people become more susceptible to all types of infections and malignancies. Their response to vaccines is decreased, and they tend to produce more autoantibodies (antibodies against their body's own molecules). In addition, the immune system exhibits lowered levels of function. For example, T cells become less responsive to antigens, and fewer T cells respond to infections. This may result from age-related atrophy of the thymus or decreased production of thymic hormones. Because the T cell population decreases with age, B cells are also less responsive. Consequently, antibody levels do not increase as rapidly in response to a challenge by an antigen, resulting in increased susceptibility to various infections. It is for this key reason that elderly individuals are encouraged to get influenza (flu) vaccinations each year.

✔**CHECKPOINT**

25. How are T cells affected by aging?

• • •

To appreciate the many ways that the lymphatic system contributes to homeostasis of other body systems, examine *Focus on Homeostasis: The Lymphatic System and Immunity*.

Next, in Chapter 23, we will explore the structure and function of the respiratory system and see how its operation is regulated by the nervous system. Most importantly, the respiratory system provides for gas exchange—taking in oxygen and blowing off carbon dioxide. The cardiovascular system aids gas exchange by transporting blood containing these gases between the lungs and tissue cells.

BODY SYSTEM	CONTRIBUTION OF THE LYMPHATIC SYSTEM AND IMMUNITY
For all body systems	B cells, T cells, and antibodies protect all body systems from attack by harmful foreign invaders (pathogens), foreign cells, and cancer cells.
Integumentary system	Lymphatic vessels drain excess interstitial fluid and leaked plasma proteins from dermis of skin. Immune system cells (Langerhans cells) in skin help protect skin. Lymphatic tissue provides IgA antibodies in sweat.
Skeletal system	Lymphatic vessels drain excess interstitial fluid and leaked plasma proteins from connective tissue around bones.
Muscular system	Lymphatic vessels drain excess interstitial fluid and leaked plasma proteins from muscles.
Endocrine system	Flow of lymph helps distribute some hormones and cytokines. Lymphatic vessels drain excess interstitial fluid and leaked plasma proteins from endocrine glands.
Cardiovascular system	Lymph returns excess fluid filtered from blood capillaries and leaked plasma proteins to venous blood. Macrophages in spleen destroy aged red blood cells and remove debris in blood.
Respiratory system	Tonsils, alveolar macrophages, and MALT (mucosa-associated lymphatic tissue) help protect lungs from pathogens. Lymphatic vessels drain excess interstitial fluid from lungs.
Digestive system	Tonsils and MALT help defend against toxins and pathogens that penetrate the body from the gastrointestinal tract. Digestive system provides IgA antibodies in saliva and gastrointestinal secretions. Lymphatic vessels pick up absorbed dietary lipids and fat-soluble vitamins from the small intestine and transport them to the blood. Lymphatic vessels drain excess interstitial fluid and leaked plasma proteins from organs of the digestive system.
Urinary system	Lymphatic vessels drain excess interstitial fluid and leaked plasma proteins from organs of the urinary system. MALT helps defend against toxins and pathogens that penetrate the body via the urethra.
Reproductive systems	Lymphatic vessels drain excess interstitial fluid and leaked plasma proteins from organs of the reproductive system. MALT helps defend against toxins and pathogens that penetrate the body via the vagina and penis. In females, sperm deposited in the vagina are not attacked as foreign invaders due to inhibition of immune responses. IgG antibodies can cross the placenta to provide protection to a developing fetus. Lymphatic tissue provides IgA antibodies in the milk of a nursing mother.

THE LYMPHATIC SYSTEM AND IMMUNITY

DISORDERS: HOMEOSTATIC IMBALANCES

AIDS: Acquired Immunodeficiency Syndrome

Acquired immunodeficiency syndrome (AIDS) is a condition in which a person experiences a telltale assortment of infections due to the progressive destruction of immune system cells by the **human immunodeficiency virus (HIV)**. AIDS represents the end stage of infection by HIV. A person who is infected with HIV may be symptom-free for many years, even while the virus is actively attacking the immune system. In the two decades after the first five cases were reported in 1981, 22 million people died of AIDS. Worldwide, 35 million to 40 million people are currently infected with HIV.

HIV Transmission

Because HIV is present in the blood and some body fluids, it is most effectively transmitted (spread from one person to another) by actions or practices that involve the exchange of blood or body fluids between people. HIV is transmitted in semen or vaginal fluid during unprotected (without a condom) anal, vaginal, or oral sex. HIV also is transmitted by direct blood-to-blood contact, such as occurs among intravenous drug users who share hypodermic needles or health-care professionals who may be accidentally stuck by HIV-contaminated hypodermic needles. In addition, HIV can be transmitted from an HIV-infected mother to her baby at birth or during breast-feeding.

The chance of transmitting or of being infected by HIV during vaginal or anal intercourse can be greatly reduced—although not entirely eliminated—by the use of latex condoms. Public health programs aimed at encouraging drug users not to share needles have proved effective at checking the increase in new HIV infections in this population. Also, giving certain drugs to pregnant HIV-infected women greatly reduces the risk of transmission of the virus to their babies.

HIV is a very fragile virus; it cannot survive for long outside the human body. The virus is not transmitted by insect bites. One cannot become infected by casual physical contact with an HIV-infected person, such as by hugging or sharing household items. The virus can be eliminated from personal care items and medical equipment by exposing them to heat (135°F for 10 minutes) or by cleaning them with common disinfectants such as hydrogen peroxide, rubbing alcohol, household bleach, or germicidal cleansers such as Betadine or Hibiclens. Standard dishwashing and clothes washing also kill HIV.

HIV: Structure and Infection

HIV consists of an inner core of ribonucleic acid (RNA) covered by a protein coat (capsid). HIV is classified as a **retrovirus** (RET-rō-vī-rus) since its genetic information is carried in RNA instead of DNA. Surrounding the HIV capsid is an envelope composed of a lipid bilayer that is penetrated by glycoproteins (Figure 22.23).

Outside a living host cell, a virus is unable to replicate. However, when a virus infects and enters a host cell, it uses the host cell's enzymes and ribosomes to make thousands of copies of the virus. New viruses eventually leave and then infect other cells. HIV infection of a host cell begins with the binding of HIV glycoproteins to receptors in the host cell's plasma membrane. This causes the cell to transport the virus into its cytoplasm via receptor-mediated endocytosis. Once inside the host cell, HIV sheds its protein coat, and a viral enzyme called **reverse transcriptase** (tran-SKRIP-tās′) reads the viral RNA strand and makes a DNA copy. The viral DNA copy then becomes integrated into the host cell's DNA. Thus, the viral DNA is duplicated along with the host cell's DNA during normal cell division. In addition, the viral DNA can cause the infected cell to begin producing millions of copies of viral RNA and to assemble new protein coats for each copy. The

Figure 22.23 Human immunodeficiency virus (HIV), the causative agent of AIDS.

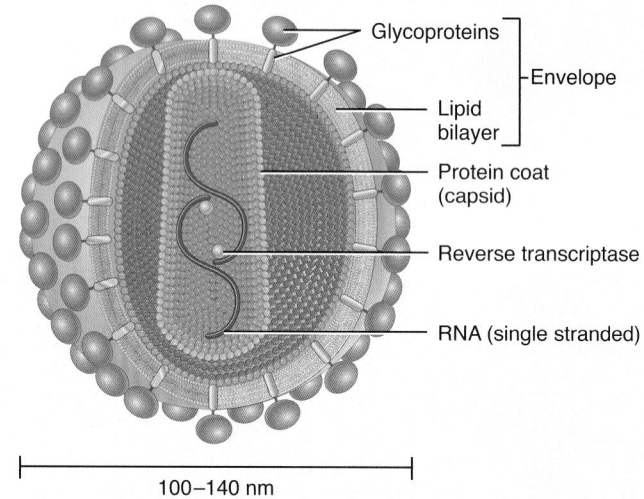

HIV is most effectively transmitted by practices that involve the exchange of body fluids.

Human immunodeficiency virus (HIV)

? Which cells of the immune system are attacked by HIV?

new HIV copies bud off from the cell's plasma membrane and circulate in the blood to infect other cells.

HIV mainly damages helper T cells, and it does so in various ways. Over 10 billion viral copies may be produced each day. The viruses bud so rapidly from an infected cell's plasma membrane that cell lysis eventually occurs. In addition, the body's defenses attack the infected cells, killing them but not all the viruses they harbor. In most HIV-infected individuals, helper T cells are initially replaced as fast as they are destroyed. After several years, however, the body's ability to replace helper T cells is slowly exhausted, and the number of helper T cells in circulation progressively declines.

Signs, Symptoms, and Diagnosis of HIV Infection

Soon after being infected with HIV, most people experience a brief flulike illness. Common signs and symptoms are fever, fatigue, rash, headache, joint pain, sore throat, and swollen lymph nodes. About 50% of infected people also experience night sweats. As early as 3 to 4 weeks after HIV infection, plasma cells begin secreting antibodies against HIV. These antibodies are detectable in blood plasma and form the basis for some of the screening tests for HIV. When people test "HIV-positive," it usually means they have antibodies to HIV antigens in their bloodstream.

Progression to AIDS

After a period of 2 to 10 years, HIV destroys enough helper T cells that most infected people begin to experience symptoms of immunodeficiency. HIV-infected people commonly have enlarged lymph nodes and experience persistent fatigue, involuntary weight loss, night sweats, skin rashes, diarrhea, and various lesions of the mouth and gums. In addition, the virus may begin to infect neurons in the brain, affecting the person's memory and producing visual disturbances.

As the immune system slowly collapses, an HIV-infected person becomes susceptible to a host of *opportunistic infections*. These are diseases caused by microorganisms that are normally held in check but now proliferate because of the defective immune system. AIDS is diagnosed when the helper T cell count drops below 200 cells per microliter (= cubic millimeter) of blood or when opportunistic infections arise, whichever occurs first. In time, opportunistic infections usually are the cause of death.

Treatment of HIV Infection

At present, infection with HIV cannot be cured. Vaccines designed to block new HIV infections and to reduce the viral load (the number of copies of HIV RNA in a microliter of blood plasma) in those who are already infected are in clinical trials. Meanwhile, two categories of drugs have proved successful in extending the life of many HIV-infected people:

1. **Reverse transcriptase inhibitors** interfere with the action of the reverse transcriptase enzyme, the enzyme that the virus uses to convert its RNA into a DNA copy. Among the drugs in this category are zidovudine (ZDV, previously called AZT), didanosine (ddI), and stavudine (trade name d4T®). Trizivir, approved in 2000 for treatment of HIV infection, combines three reverse transcriptase inhibitors in one pill.

2. **Protease inhibitors** interfere with the action of protease, a viral enzyme that cuts proteins into pieces to assemble the protein coat of newly produced HIV particles. Drugs in this category include nelfinavir, saquinavir, ritonavir, and indinavir.

In 1996, physicians treating HIV-infected patients widely adopted *highly active antiretroviral therapy (HAART)*—a combination of two differently acting reverse transcriptase inhibitors and one protease inhibitor. Most HIV-infected individuals receiving HAART experience a drastic reduction in viral load and an increase in the number of helper T cells in their blood. Not only does HAART delay the progression of HIV infection to AIDS, but many individuals with AIDS have seen the remission or disappearance of opportunistic infections and an apparent return to health. Unfortunately, HAART is very costly (exceeding $10,000 per year), the dosing schedule is grueling, and not all people can tolerate the toxic side effects of these drugs. Although HIV may virtually disappear from the blood with drug treatment (and thus a blood test may be "negative" for HIV), the virus typically still lurks in various lymphatic tissues. In such cases, the infected person can still transmit the virus to another person.

Allergic Reactions

A person who is overly reactive to a substance that is tolerated by most other people is said to be **allergic** or **hypersensitive.** Whenever an allergic reaction takes place, some tissue injury occurs. The antigens that induce an allergic reaction are called **allergens** (AL-er-jens). Common allergens include certain foods (milk, peanuts, shellfish, eggs), antibiotics (penicillin, tetracycline), vaccines (pertussis, typhoid), venoms (honeybee, wasp, snake), cosmetics, chemicals in plants such as poison ivy, pollens, dust, molds, iodine-containing dyes used in certain x-ray procedures, and even microbes.

There are four basic types of hypersensitivity reactions: type I (anaphylactic), type II (cytotoxic), type III (immune-complex), and type IV (cell-mediated). The first three are antibody-mediated immune responses; the last is a cell-mediated immune response.

Type I (anaphylactic) reactions (AN-a-fil-lak´-tik) are the most common and occur within a few minutes after a person sensitized to an allergen is re-exposed to it. In response to the first exposure to certain allergens, some people produce IgE antibodies that bind to the surface of mast cells and basophils. The next time the same allergen enters the body, it attaches to the IgE antibodies already present. In response, the mast cells and basophils release histamine, prostaglandins, leukotrienes, and kinins. Collectively, these mediators cause vasodilation, increased blood capillary permeability, increased smooth muscle contraction in the airways of the lungs, and increased mucus secretion. As a result, a person may experience inflammatory responses, difficulty in breathing through the constricted airways, and a runny nose from excess mucus secretion. In **anaphylactic shock,** which may occur in a susceptible individual who has just received a triggering drug or been stung by a wasp, wheezing and shortness of breath as airways constrict are usually accompanied by shock due to vasodilation and fluid loss from blood. This life-threatening emergency is usually treated by injecting epinephrine to dilate the airways and strengthen the heartbeat.

Type II (cytotoxic) reactions are caused by antibodies (IgG or IgM) directed against antigens on a person's blood cells (red blood cells, lymphocytes, or platelets) or tissue cells. The reaction of antibodies and antigens usually leads to activation of complement. Type II reactions, which may occur in incompatible blood transfusion reactions, damage cells by causing lysis.

Type III (immune-complex) reactions involve antigens, antibodies (IgA or IgM), and complement. When certain ratios of antigen to antibody occur, the immune complexes are small enough to escape phagocytosis, but they become trapped in the basement membrane under the endothelium of blood vessels, where they activate complement and cause inflammation. Glomerulonephritis and rheumatoid arthritis (RA) arise in this way.

Type IV (cell-mediated) reactions or **delayed hypersensitivity reactions** usually appear 12–72 hours after exposure to an allergen. Type IV reactions occur when allergens are taken up by antigen-presenting cells (such as Langerhans cells in the skin) that migrate to lymph nodes and present the allergen to T cells, which then proliferate. Some of the new T cells return to the site of allergen entry into the body, where they produce gamma-interferon, which activates macrophages, and tumor necrosis factor, which stimulates an inflammatory response. Intracellular bacteria such as *Mycobacterium tuberculosis* (mī-kō-bak-TE-rē-um too-ber´-ku-LŌ-sis) trigger this type of cell-mediated immune response, as do certain haptens, such as poison ivy toxin. The skin test for tuberculosis also is a delayed hypersensitivity reaction.

Autoimmune Diseases

In an **autoimmune disease** (aw-tō-i-MŪN) or **autoimmunity,** the immune system fails to display self-tolerance and attacks the person's own tissues. Autoimmune diseases usually arise in early adulthood and are common, afflicting an estimated 5% of adults in North America and Europe. Females suffer autoimmune diseases twice as often as males. Recall that self-reactive B cells and T cells normally are deleted or undergo anergy during negative selection (see Figure 22.22). Apparently, this process is not 100% effective. Under the influence of unknown environmental triggers and certain genes that make some people more susceptible, self-tolerance breaks down, leading to activation of self-reactive clones of T cells and B cells. These cells then generate cell-mediated or antibody-mediated immune responses against self-antigens.

A variety of mechanisms produce different autoimmune diseases. Some involve production of **autoantibodies,** antibodies that bind to and stimulate or block self-antigens. For example, autoantibodies that mimic TSH (thyroid-stimulating hormone) are present in Graves disease and stimulate secretion of thyroid hormones (thus producing hyperthyroidism); autoantibodies that bind to and block acetylcholine receptors cause the muscle weakness characteristic of myasthenia gravis. Other autoimmune diseases involve activation of cytotoxic T cells that destroy certain body cells. Examples include type 1 diabetes mellitus, in which T cells attack the insulin-producing pancreatic beta cells, and multiple sclerosis (MS), in which T cells attack myelin sheaths around axons of neurons. Inappropriate activation of helper T cells or excessive production of gamma-interferon also occur in certain autoimmune diseases. Other autoimmune disorders include rheumatoid arthritis (RA), systemic lupus erythematosus (SLE), rheumatic fever, hemolytic and pernicious anemias, Addison's disease, Hashimoto's thyroiditis, and ulcerative colitis.

Therapies for various autoimmune diseases include removal of the thymus gland (thymectomy), injections of beta-interferon, immunosuppressive drugs, and plasmapheresis, in which the person's blood plasma is filtered to remove antibodies and antigen–antibody complexes.

Infectious Mononucleosis

Infectious mononucleosis (mon'-ō-noo-klē-Ō-sis) or "mono" is a contagious disease caused by the *Epstein–Barr virus (EBV)*. It occurs mainly in children and young adults, and more often in females than in males. The virus most commonly enters the body through intimate oral contact such as kissing, which accounts for its common name, the "kissing disease." EBV then multiplies in lymphatic tissues and filters into the blood, where it infects and multiplies in B cells, the primary host cells. Because of this infection, the B cells become so enlarged and abnormal in appearance that they resemble monocytes, the primary reason for the term **mononucleosis.** In addition to an elevated white blood cell count with an abnormally high percentage of lymphocytes, signs and symptoms include fatigue, headache, dizziness, sore throat, enlarged and tender lymph nodes, and fever. There is no cure for infectious mononucleosis, but the disease usually runs its course in a few weeks.

Lymphomas

Lymphomas (lim-FŌ-mas; *lymph-* = clear water; *-oma* = tumor) are cancers of the lymphatic organs, especially the lymph nodes. Most have no known cause. The two main types of lymphomas are Hodgkin disease and non-Hodgkin lymphoma.

Hodgkin disease (HD) (HOJ-kin) is characterized by a painless, nontender enlargement of one or more lymph nodes, most commonly in the neck, chest, and axilla. If the disease has metastasized from these sites, fever, night sweats, weight loss, and bone pain also occur. HD primarily affects individuals between ages 15 and 35 and those over 60, and it is more common in males. If diagnosed early, HD has a 90–95% cure rate.

Non-Hodgkin lymphoma (NHL), which is more common than HD, occurs in all age groups, the incidence increasing with age to a maximum between ages 45 and 70. NHL may start the same way as HD but may also include an enlarged spleen, anemia, and general malaise. Up to half of all individuals with NHL are cured or survive for a lengthy period. Treatment options for both HD and NHL include radiation therapy, chemotherapy, and bone marrow transplantation.

Systemic Lupus Erythematosus

Systemic lupus erythematosus (er'-e-thē'-ma-TŌ-sus), **SLE,** or simply **lupus** (*lupus* = wolf) is a chronic autoimmune, inflammatory disease that affects multiple body systems. Lupus is characterized by periods of active disease and remission; symptoms range from mild to life-threatening. Lupus most often develops between ages 15 and 44 and is 10–15 times more common in females than in males. It is also 2–3 times more common in African Americans, Hispanics, Asian Americans, and Native Americans than in European Americans. Although the cause of SLE is not known, both a genetic predisposition to the disease and environmental factors (infections, antibiotics, ultraviolet light, stress, and hormones) may trigger it. Sex hormones appear to influence the development of SLE. The disorder often occurs in females who exhibit extremely low levels of androgens.

Signs and symptoms of SLE include joint pain, muscle pain, chest pain with deep breaths, headaches, pale or purple fingers or toes, kidney dysfunction, low blood cell count, nerve or brain dysfunction, slight fever, fatigue, oral ulcers, weight loss, swelling in the legs or around the eyes, enlarged lymph nodes and spleen, photosensitivity, rapid loss of large amounts of scalp hair, and sometimes an eruption across the bridge of the nose and cheeks called a "butterfly rash." The erosive nature of some of the SLE skin lesions was thought to resemble the damage inflicted by the bite of a wolf—thus, the term *lupus.*

Two immunological features of SLE are excessive activation of B cells and inappropriate production of autoantibodies against DNA (anti-DNA antibodies) and other components of cellular nuclei such as histone proteins. Triggers of B cell activation are thought to include various chemicals and drugs, viral and bacterial antigens, and exposure to sunlight. Circulating complexes of abnormal autoantibodies and their "antigens" cause damage in tissues throughout the body. Kidney damage occurs as the complexes become trapped in the basement membrane of kidney capillaries, obstructing blood filtering. Renal failure is the most common cause of death.

There is no cure for lupus, but drug therapy can minimize symptoms, reduce inflammation, and forestall flare-ups. The most commonly used lupus medications are pain relievers (nonsteroidal anti-inflammatory drugs such as aspirin and ibuprofen), antimalarial drugs (hydroxychloroquine), and corticosteroids (prednisone and hydrocortisone).

MEDICAL TERMINOLOGY

Adenitis (ad'-e-NĪ-tis; *aden-* = gland; *-itis* = inflammation of) Enlarged, tender, and inflamed lymph nodes resulting from an infection.

Allograft (AL-ō-graft; *allo-* = other) A transplant between genetically distinct individuals of the same species. Skin transplants from other people and blood transfusions are allografts.

Autograft (AW-tō-graft; *auto-* = self) A transplant in which one's own tissue is grafted to another part of the body (such as skin grafts for burn treatment or plastic surgery).

Chronic fatigue syndrome (CFS) A disorder, usually occurring in young adults and primarily in females, characterized by (1) extreme fatigue that impairs normal activities for at least 6 months and (2) the absence of other known diseases (cancer, infections, drug abuse, toxicity, or psychiatric disorders) that might produce similar symptoms.

Gamma globulin (GLOB-ū-lin) Suspension of immunoglobulins from blood consisting of antibodies that react with a specific pathogen. It is prepared by injecting the pathogen into animals, removing blood from the animals after antibodies have been produced, isolating the antibodies, and injecting them into a human to provide short-term immunity.

Hypersplenism (hī-per-SPLĒN-izm; *hyper-* = over) Abnormal splenic activity due to splenic enlargement and associated with an increased rate of destruction of normal blood cells.

Lymphadenopathy (lim-fad'-e-NOP-a-thē; *lymph-* = clear fluid; *-pathy* = disease) Enlarged, sometimes tender lymph glands as a response to infection; also called swollen glands.

Lymphangitis (lim-fan-JĪ-tis; *-itis* = inflammation of) Inflammation of lymphatic vessels.

Lymphedema (lim'-fe-DĒ-ma; *edema* = swelling) Accumulation of lymph in lymphatic vessels, causing painless swelling of a limb.

Splenomegaly (splē'-nō-MEG-a-lē; *mega-* = large) Enlarged spleen.

Xenograft (ZEN-ō-graft; *xeno-* = strange or foreign) A transplant between animals of different species. Xenografts from porcine (pig) or bovine (cow) tissue may be used in humans as a physiological dressing for severe burns. Other xenografts include pig heart valves and baboon hearts.

CHAPTER REVIEW AND RESOURCE SUMMARY

Review

Resource

Introduction

1. The ability to ward off disease is called immunity (resistance). Lack of resistance is called susceptibility.
2. The two general types of immunity are (a) innate and (b) adaptive.
3. Innate immunity refers to a wide variety of body responses to a wide range of pathogens.
4. Adaptive immunity involves activation of specific lymphocytes to combat a particular foreign substance.

Anatomy Overview - The Lymphatic System and Disease Resistance

22.1 Lymphatic System Structure and Function

1. The lymphatic system carries out immune responses and consists of lymph, lymphatic vessels, and structures and organs that contain lymphatic tissue (specialized reticular tissue containing many lymphocytes). The lymphatic system drains interstitial fluid, transports dietary lipids, and protects against invasion through immune responses.
2. Lymphatic vessels begin as closed-ended lymphatic capillaries in tissue spaces between cells. Interstitial fluid drains into lymphatic capillaries, thus forming lymph. Lymphatic capillaries merge to form larger vessels, called lymphatic vessels, which convey lymph into and out of lymph nodes.
3. The route of lymph flow is from lymphatic capillaries to lymphatic vessels to lymph trunks to the thoracic duct (left lymphatic duct) and right lymphatic duct to the subclavian veins.
4. Lymph flows because of skeletal muscle contractions and respiratory movements. Valves in lymphatic vessels also aid flow of lymph.
5. The primary lymphatic organs are red bone marrow and the thymus. Secondary lymphatic organs are lymph nodes, the spleen, and lymphatic nodules.
6. The thymus lies between the sternum and the large blood vessels above the heart. It is the site of T cell maturation.
7. Lymph nodes are encapsulated, egg-shaped structures located along lymphatic vessels. Lymph enters lymph nodes through afferent lymphatic vessels, is filtered, and exits through efferent lymphatic vessels. Lymph nodes are the site of proliferation of B cells and T cells.
8. The spleen is the largest single mass of lymphatic tissue in the body. Within the spleen, B cells and T cells carry out immune functions and macrophages destroy blood-borne pathogens and worn-out red blood cells by phagocytosis.
9. Lymphatic nodules are scattered throughout the mucosa of the gastrointestinal, respiratory, urinary, and reproductive tracts. This lymphatic tissue is termed mucosa-associated lymphatic tissue (MALT).

Anatomy Overview - Lymphatic Vessels
Anatomy Overview - The Thymus
Anatomy Overview - The Spleen and Lymph Nodes
Animation - Lymph Formation and Flow
Animation - Lymphatic System Functions
Figure 22.5 - Thymus
Figure 22.6 - Structure of a Lymph Node
Exercise - Lymphatic Highway
Concepts and Connections - Lymphatic System

22.2 Development of Lymphatic Tissues

1. Lymphatic vessels develop from lymph sacs, which arise from developing veins. Thus, they are derived from mesoderm.
2. Lymph nodes develop from lymph sacs that become invaded by mesenchymal cells.

22.3 Innate Immunity

1. Innate immunity includes physical factors, chemical factors, antimicrobial proteins, natural killer cells, phagocytes, inflammation, and fever.
2. The skin and mucous membranes are the first line of defense against entry of pathogens.
3. Antimicrobial substances include interferons, the complement system, iron-binding proteins, and antimicrobial proteins.
4. Natural killer cells and phagocytes attack and kill pathogens and defective cells in the body.
5. Inflammation aids disposal of microbes, toxins, or foreign material at the site of an injury, and prepares the site for tissue repair.

Anatomy Overview - The Integument and Disease Resistance
Animation - Introduction to Disease Resistance
Animation - Complement Proteins
Concepts and Connections: Complement Proteins
Exercise - Integument vs. Disease
Exercise - Microbe Massacre
Concepts and Connections - Nonspecific Disease Resistance

Review

6. Fever intensifies the antiviral effects of interferons, inhibits growth of some microbes, and speeds up body reactions that aid repair.

7. Table 22.1 summarizes the innate defenses.

22.4 Adaptive Immunity

1. Adaptive immunity involves lymphocytes called B cells and T cells. B cells and T cells arise from stem cells in red bone marrow. B cells mature in red bone marrow, whereas T cells mature in the thymus gland.

2. Before B cells leave the red bone marrow or T cells leave the thymus, they develop immunocompetence, the ability to carry out adaptive immune responses. This process involves the insertion of antigen receptors into their plasma membranes. Antigen receptors are molecules that are capable of recognizing specific antigens.

3. Two major types of mature T cells exit the thymus: helper T cells (also known as CD4 T cells) and cytotoxic T cells (also referred to as CD8 T cells).

4. There are two types of adaptive immunity: cell-mediated immunity and antibody-mediated immunity. In cell-mediated immune responses, cytotoxic T cells directly attack invading antigens; in antibody-mediated immune responses, B cells transform into plasma cells that secrete antibodies.

5. Clonal selection is the process by which a lymphocyte proliferates and differentiates in response to a specific antigen. The result of clonal selection is the formation of a clone of cells that can recognize the same specific antigen as the original lymphocyte.

6. A lymphocyte that undergoes clonal selection gives rise to two major types of cells in the clone: effector cells and memory cells. The effector cells of a lymphocyte clone carry out immune responses that ultimately result in the destruction or inactivation of the antigen. Effector cells include active helper T cells, which are part of a helper T cell clone; active cytotoxic T cells, which are part of a cytotoxic T cell clone; and plasma cells, which are part of a B cell clone. The memory cells of a lymphocyte clone do not actively participate in the initial immune response. However, if the antigen reappears in the body in the future, the memory cells can quickly respond to the antigen by proliferating and differentiating into more effector cells and more memory cells. Memory cells include memory helper T cells, which are part of a helper T cell clone; memory cytotoxic T cells, which are part of a cytotoxic T cell clone; and memory B cells, which are part of a B cell clone.

7. Antigens (Ags) are chemical substances that are recognized as foreign by the immune system. Antigen receptors exhibit great diversity due to genetic recombination.

8. "Self-antigens" called major histocompatibility complex (MHC) antigens are unique to each person's body cells. All cells except red blood cells display MHC-I molecules. Antigen-presenting cells (APCs) display MHC-II molecules. APCs include macrophages, B cells, and dendritic cells.

9. Exogenous antigens (formed outside body cells) are presented with MHC-II molecules; endogenous antigens (formed inside body cells) are presented with MHC-I molecules.

10. Cytokines are small protein hormones that may stimulate or inhibit many normal cell functions such as growth and differentiation. Other cytokines regulate immune responses (see Table 22.2).

22.5 Cell-Mediated Immunity

Animation - Introduction to Cell-Mediated Immunity
Figure 22.15 - Activation and Clonal Selection of a Helper T Cell
Figure 22.16 - Activation and Clonal Selection of a Cytotoxic T Cell
Interactive Exercise - Activate the Attack

1. A cell-mediated immune response begins with activation of a small number of T cells by a specific antigen.

2. During the activation process, T-cell receptors (TCRs) recognize antigen fragments associated with MHC molecules on the surface of a body cell.

3. Activation of T cells also requires costimulation, either by cytokines such as interleukin-2 or by pairs of plasma membrane molecules.

4. Once a T cell has been activated, it undergoes clonal selection. The result of clonal selection is the formation of a clone of effector cells and memory cells. The effector cells of a T cell clone carry out immune responses that ultimately result in elimination of the antigen.

5. Helper T cells display CD4 protein, recognize antigen fragments associated with MHC-II molecules, and secrete several cytokines, most importantly interleukin-2, which acts as a costimulator for other helper T cells, cytotoxic T cells, and B cells.

6. Cytotoxic T cells display CD8 protein and recognize antigen fragments associated with MHC-I molecules.

7. Active cytotoxic T cells eliminate invaders by (1) releasing granzymes that cause target cell apoptosis (phagocytes then kill the microbes) and (2) releasing perforin, which causes cytolysis, and granulysin that destroys the microbes.

8. Cytotoxic T cells, macrophages, and natural killer cells carry out immunological surveillance, recognizing and destroying cancerous cells that display tumor antigens.

Review

22.6 Antibody-Mediated Immunity

1. An antibody-mediated immune response begins with activation of a B cell by a specific antigen.
2. B cells can respond to unprocessed antigens, but their response is more intense when they process the antigen. Interleukin-2 and other cytokines secreted by helper T cells provide costimulation for activation of B cells.
3. Once activated, a B cell undergoes clonal selection, forming a clone of plasma cells and memory cells. Plasma cells are the effector cells of a B cell clone; they secrete antibodies.
4. An antibody (Ab) is a protein that combines specifically with the antigen that triggered its production.
5. Antibodies consist of heavy and light chains and variable and constant regions.
6. Based on chemistry and structure, antibodies are grouped into five principal classes (IgG, IgA, IgM, IgD, and IgE), each with specific biological roles.
7. Actions of antibodies include neutralization of antigen, immobilization of bacteria, agglutination and precipitation of antigen, activation of complement, and enhancement of phagocytosis.
8. Complement is a group of proteins that complement immune responses and help clear antigens from the body.
9. Immunization against certain microbes is possible because memory B cells and memory T cells remain after a primary response to an antigen. The secondary response provides protection should the same microbe enter the body again.

Animation - Antibody-Mediated Immunity
Animation - Primary and Secondary Infections
Animation - Complement Proteins
Figure 22.18 - Activation and Clonal Selection of B Cells
Exercise - Antibody Ambush
Exercise - Disease Resistance Fighters
Exercise - Concentrate on Disease Resistance

22.7 Self-Recognition and Self-Tolerance

1. T cells undergo positive selection to ensure that they can recognize self-MHC proteins (self-recognition), and negative selection to ensure that they do not react to other self-proteins (self-tolerance). Negative selection involves both deletion and anergy.
2. B cells develop tolerance through deletion and anergy.

Anatomy Overview - Antigens and Antibodies

22.8 Stress and Immunity

1. Psychoneuroimmunology (PNI) deals with communication pathways that link the nervous, endocrine, and immune systems. Thoughts, feelings, moods, and beliefs influence health and the course of disease.
2. Under stress, people are less likely to eat well or exercise regularly, two habits that enhance immunity.

22.9 Aging and the Immune System

1. With advancing age, individuals become more susceptible to infections and malignancies, respond less well to vaccines, and produce more autoantibodies.
2. Immune responses also diminish with age.

SELF-QUIZ QUESTIONS

Fill in the blanks in the following statements.

1. The first line of defense of innate immunity against pathogens consists of the _____ and _____; the second line of defense consists of _____, _____, and _____.

2. Substances that are recognized as foreign and provoke immune responses are known as _____.

Indicate whether the following statements are true or false.

3. Your body's ability to ward off damage or disease through your defenses is known as resistance; vulnerability to disease is susceptibility.

4. Your T cells must be able to recognize your own MHC molecules, a process known as self-recognition, and lack reactivity to peptide fragments from your own proteins, a condition known as self-tolerance.

Choose the one best answer to the following questions.

5. Trace the sequence of fluid from blood vessel to blood vessel by way of the lymphatic system. (1) lymphatic vessels, (2) blood capillaries, (3) subclavian veins, (4) lymphatic capillaries, (5) interstitial spaces, (6) arteries, (7) lymphatic ducts.
(a) 2, 5, 4, 1, 7, 6, 3   (b) 3, 6, 2, 4, 5, 1, 7   (c) 6, 2, 5, 4, 1, 7, 3
(d) 6, 2, 5, 4, 7, 1, 3   (e) 2, 5, 4, 7, 1, 3, 6

6. Which of the following describe lymph nodes? (1) Lymph enters the nodes through efferent lymphatic vessels and leaves through afferent lymphatic vessels. (2) The outer cortex consists of lymphatic nodules that contain B cells and are the sites of plasma cell and memory B cell formation. (3) The inner cortex contains lymphatic nodules with mature T cells. (4) The reticular fibers within the sinuses of the lymph nodes trap foreign substances in the lymph. (5) The sinuses of lymph nodes are known as red pulp.
(a) 1, 2, 3, and 4   (b) 2, 4, and 5   (c) 1, 2, 3, 4, and 5
(d) 2 and 4   (e) 1, 2, and 4

7. Which of the following statements are *correct*? (1) Lymphatic vessels are found throughout the body, except in avascular tissues, the CNS, portions of the spleen, and red bone marrow. (2) Lymphatic capillaries allow interstitial fluid to flow into them but not out of them. (3) Anchoring filaments attach lymphatic endothelial cells to surrounding tissues. (4) Lymphatic vessels freely receive all the components of blood, including the formed elements. (5) Lymph ducts directly connect to blood vessels by way of the subclavian veins.

(a) 1, 3, 4, and 5   (b) 2, 3, 4, and 5   (c) 1, 2, 3, and 4
(d) 1, 2, 4, and 5   (e) 1, 2, 3, and 5

8. Which of the following are physical factors that help fight pathogens and disease? (1) numerous layers of the epidermis, (2) mucus of mucous membranes, (3) saliva, (4) interferons, (5) complement.

(a) 1, 3, and 4   (b) 2, 4, and 5   (c) 1, 4, and 5
(d) 1, 2, and 3   (e) 1, 2, and 4

9. Which of the following are functions of antibodies? (1) neutralization of antigens, (2) immobilization of bacteria, (3) agglutination and precipitation of antigens, (4) activation of complement, (5) enhancement of phagocytosis.

(a) 1, 3, and 4   (b) 2, 4, and 5   (c) 1, 2, 3, and 4
(d) 1, 2, 3, and 5   (e) 1, 2, 3, 4, and 5

10. Which of the following are *true*? (1) Lymphatic vessels resemble arteries. (2) Lymph is very similar to interstitial fluid. (3) Lacteals are specialized lymphatic capillaries responsible for transporting dietary lipids. (4) Lymph is normally a cloudy, pale yellow fluid. (5) The thoracic duct drains lymph from the upper right side of the body. (6) Lymph flow is maintained by skeletal muscle contractions, one-way valves, and breathing movements.

(a) 1, 2, 5, and 6   (b) 2, 3, and 6   (c) 2, 3, 4, and 6
(d) 2, 4, and 6   (e) 3, 5, and 6

11. Place the stages of phagocytosis in the correct order of occurrence. (1) formation of phagolysosome, (2) adherence to microbe, (3) destruction of microbe, (4) ingestion to form a phagosome, (5) chemotactic attraction of phagocyte.

(a) 2, 5, 4, 1, 3   (b) 4, 5, 2, 1, 3   (c) 5, 2, 4, 1, 3
(d) 5, 4, 2, 3, 1   (e) 2, 5, 1, 4, 3

12. Place in order the steps involved in cell-mediated immune response to an exogenous antigen.
(a) costimulation and activation of helper T cells
(b) presentation of antigen to helper T cells
(c) elimination of invaders through the release of granzymes, perforin, granulysin, or lymphotoxin or by attraction and activation of phagocytes
(d) proliferation and differentiation of helper T cells to produce a helper T cell clone
(e) antigen processing by dendritic cells, macrophages, or B cells
(f) recognition of antigen fragments associated with MHC-II molecules by T-cell receptors
(g) secretion of cytokines such as interleukin-2 by activated helper T cells
(h) migration of antigen-presenting cells to lymphatic tissue
(i) activation of cytotoxic T cells

13. Match the following:
_____ (a) encapsulated bean-shaped structures located along the length of lymphatic vessels; contain T and B cells, macrophages, and follicular dendritic cells; filter lymph
_____ (b) produces pre-T cells and B cells; found in flat bones and epiphyses of long bones
_____ (c) clusters of lymphatic nodules involved in immune responses against inhaled or ingested foreign substances
_____ (d) the single largest mass of lymphatic tissue in the body; consists of red and white pulp
_____ (e) responsible for the maturation of T cells
_____ (f) lymphatic nodules associated with mucous membranes of the digestive, urinary, reproductive, and respiratory systems
_____ (g) nonencapsulated clusters of lymphocytes

(1) red bone marrow
(2) thymus
(3) lymph nodes
(4) spleen
(5) mucosa-associated lymphatic tissue
(6) lymphatic nodules
(7) tonsils

14. Match the following:
_____ (a) recognize foreign antigens combined with MHC-I molecules on the surface of body cells infected by microbes, some tumor cells, and cells of a tissue transplant; display CD8 proteins
_____ (b) are programmed to recognize the reappearance of a previously encountered antigen
_____ (c) differentiate into plasma cells that secrete specific antibodies
_____ (d) process and present exogenous antigens; include macrophages, B cells, and dendritic cells
_____ (e) secrete cytokines as costimulators; display CD4 proteins
_____ (f) ingest microbes or any foreign particulate matter; include neutrophils and macrophages
_____ (g) lymphocytes that have the ability to kill a wide variety of infectious microbes plus certain spontaneously arising tumor cells; lack antigen receptors

(1) active helper T cells
(2) cytotoxic T cells
(3) memory T cells
(4) B cells
(5) NK cells
(6) phagocytes
(7) antigen-presenting cells

CRITICAL THINKING QUESTIONS

1. Esperanza watched as her mother got her "flu shot." "Why do you need a shot if you're not sick?" she asked. "So I won't get sick," answered her mom. Explain how the influenza vaccination prevents illness.

2. Due to the presence of breast cancer, Mrs. Franco had a right radical mastectomy in which her right breast, underlying muscle, and right axillary lymph nodes and vessels were removed. Now she is experiencing severe swelling in her right arm. Why did the surgeon remove lymph tissue as well as the breast? Why is Mrs. Franco's right arm swollen?

3. Tariq's little sister has the mumps. Tariq can't remember if he has had mumps or not, but he is feeling slightly feverish. How could Tariq's doctor determine if he is getting sick with mumps or if he has previously had mumps?

? ANSWERS TO FIGURE QUESTIONS

22.1 Red bone marrow contains stem cells that develop into lymphocytes.

22.2 Lymph is more similar to interstitial fluid than to blood plasma because the protein content of lymph is low.

22.3 The left and right lumbar trunks and the intestinal trunk empty into the cisterna chyli, which then drains into the thoracic duct.

22.4 Inhalation promotes the movement of lymph from abdominal lymphatic vessels toward the thoracic region because the pressure in the vessels of the thoracic region is lower than the pressure in the abdominal region when a person inhales.

22.5 T cells mature in the thymus.

22.6 Foreign substances in lymph that enter a lymph node may be phagocytized by macrophages or attacked by lymphocytes that mount immune responses.

22.7 White pulp of the spleen functions in immunity; red pulp of the spleen performs functions related to blood cells.

22.8 Lymphatic tissues begin to develop by the end of the fifth week of gestation.

22.9 Lysozyme, digestive enzymes, and oxidants can kill microbes ingested during phagocytosis.

22.10 Redness results from increased blood flow due to vasodilation; pain, from injury of nerve fibers, irritation by microbial toxins, kinins, and prostaglandins, and pressure due to edema; heat, from increased blood flow and heat released by locally increased metabolic reactions; swelling, from leakage of fluid from capillaries due to increased permeability.

22.11 Helper T cells participate in both cell-mediated and antibody-mediated immune responses.

22.12 Epitopes are small immunogenic parts of a larger antigen; haptens are small molecules that become immunogenic only when they attach to a body protein.

22.13 APCs include macrophages in tissues throughout the body, B cells in blood and lymphatic tissue, and dendritic cells in mucous membranes and the skin.

22.14 Endogenous antigens include viral proteins, toxins from intracellular bacteria, and abnormal proteins synthesized by a cancerous cell.

22.15 The first signal in T cell activation is antigen binding to a TCR; the second signal is a costimulator, such as a cytokine or another pair of plasma membrane molecules.

22.16 The CD8 protein of a cytotoxic T cell binds to the MHC-I molecule of an infected body cell to help anchor the T cell receptor (TCR)–antigen interaction so that antigen recognition can occur.

22.17 Cytotoxic T cells attack some tumor cells and transplanted tissue cells, as well as cells infected by microbes.

22.18 Since all of the plasma cells in this figure are part of the same clone, they secrete just one kind of antibody.

22.19 The variable regions recognize and bind to a specific antigen.

22.20 The classical pathway for the activation of complement is linked to antibody-mediated immunity because Ag–Ab complexes activate C1.

22.21 At peak secretion, approximately 1000 times more IgG is produced in the secondary response than in the primary response.

22.22 In deletion, self-reactive T cells or B cells die; in anergy, T cells or B cells are alive but are unresponsive to antigenic stimulation.

22.23 HIV attacks helper T cells.

23

THE RESPIRATORY SYSTEM

THE RESPIRATORY SYSTEM AND HOMEOSTASIS *The respiratory system contributes to homeostasis by providing for the exchange of gases—oxygen and carbon dioxide—between the atmospheric air, blood, and tissue cells. It also helps adjust the pH of body fluids.*

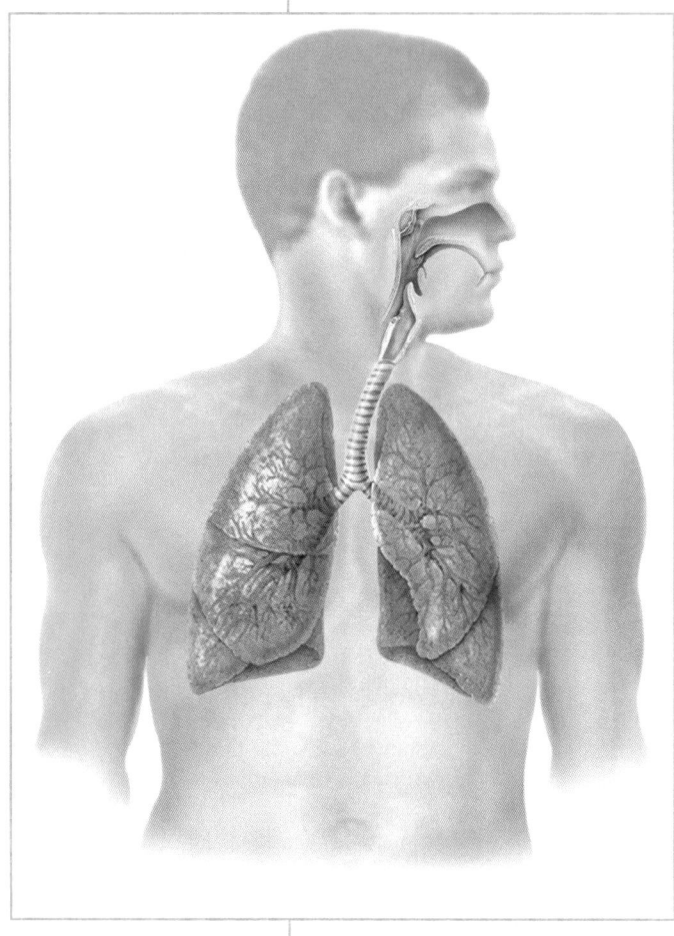

Your body's cells continually use oxygen (O_2) for the metabolic reactions that release energy from nutrient molecules and produce ATP. At the same time, these reactions release carbon dioxide (CO_2). Because an excessive amount of CO_2 produces acidity that can be toxic to cells, excess CO_2 must be eliminated quickly and efficiently. The cardiovascular and respiratory systems cooperate to supply O_2 and eliminate CO_2. The respiratory system provides for gas exchange—intake of O_2 and elimination of CO_2—and the cardiovascular system transports blood containing the gases between the lungs and body cells. Failure of either system disrupts homeostasis by causing rapid death of cells from oxygen starvation and buildup of waste products. In addition to functioning in gas exchange, the respiratory system also participates in regulating blood pH, contains receptors for the sense of smell, filters inspired air, produces sounds, and rids the body of some water and heat in exhaled air. As in the digestive and urinary systems, which will be covered in subsequent chapters, in the respiratory system there is an extensive area of contact between the external environment and capillary blood vessels.

? *Did you ever wonder how smoking affects the respiratory system?*

23.1 RESPIRATORY SYSTEM ANATOMY

◉ OBJECTIVES

- Describe the anatomy and histology of the nose, pharynx, larynx, trachea, bronchi, and lungs.
- Identify the functions of each respiratory system structure.

The **respiratory system** (RES-pi-ra-tōr-ē) consists of the nose, pharynx (throat), larynx (voice box), trachea (windpipe), bronchi, and lungs (Figure 23.1). Its parts can be classified according to either structure or function. *Structurally,* the respiratory system consists of two parts: (1) The **upper respiratory system** includes the nose, nasal cavity, pharynx, and associated structures. (2) The **lower respiratory system** includes the larynx, trachea, bronchi, and lungs. *Functionally,* the respiratory system also consists of two parts: (1) The **conducting zone** consists of a series of interconnecting cavities and tubes both outside and within the lungs. These include the nose, nasal cavity, pharynx, larynx, trachea, bronchi, bronchioles, and terminal bronchioles; their function is to filter, warm, and moisten air and conduct it into the lungs. (2) The **respiratory zone** consists of tubes and tissues within the lungs where gas exchange occurs. These include the respiratory bronchioles, alveolar ducts, alveolar sacs, and alveoli and are the main sites of gas exchange between air and blood.

The branch of medicine that deals with the diagnosis and treatment of diseases of the ears, nose, and throat (ENT) is called **otorhinolaryngology** (ō'-tō-rī'-nō-lar'-in-GOL-ō-jē; *oto-* = ear; *-rhino-* = nose; *-laryngo-* = voice box; *-logy* = study of). A **pulmonologist** (pul'-mo-NOL-ō-jist) is a specialist in the diagnosis and treatment of diseases of the lungs.

Nose

The **nose** is a specialized organ at the entrance of the respiratory system that is divided into an external portion and an internal portion called the nasal cavity. The **external nose** is the portion of the nose visible on the face and consists of a supporting framework of bone and hyaline cartilage covered with muscle and skin and lined by a mucous membrane. The frontal bone, nasal bones, and maxillae form the *bony framework* of the external nose (Figure 23.2a). The *cartilaginous framework* of the external nose consists of the **septal nasal cartilage,** which forms the anterior portion of the nasal septum; the **lateral nasal cartilages** inferior to the nasal bones; and the **alar cartilages** (Ā-lar), which form a portion of the walls of the nostrils. Because it consists of pliable hyaline cartilage, the cartilaginous framework of the external nose is somewhat flexible. On the undersurface of the external nose are two openings called the **external nares** (NĀ-rez; singular is **naris**) or **nostrils.** Figure 23.3 shows the surface anatomy of the nose.

The interior structures of the external nose have three functions: (1) warming, moistening, and filtering incoming air; (2) detecting olfactory stimuli; and (3) modifying speech vibrations as they pass through the large, hollow resonating chambers. *Resonance* refers to prolonging, amplifying, or modifying a sound by vibration.

The **nasal cavity** is a large space in the anterior aspect of the skull that lies inferior to the nasal bone and superior to the oral cavity; it is lined with muscle and mucous membrane. Anteriorly, the nasal cavity merges with the external nose, and posteriorly it communicates with the pharynx through two openings called the **internal nares** or **choanae** (kō-Ā-nē) (see Figure 23.2b). Ducts from the *paranasal sinuses* (which drain mucus) and the *nasolacrimal ducts* (which drain tears) also open into the nasal cavity. Recall from Chapter 7 that the paranasal sinuses are cavities in certain cranial and facial bones lined with mucous membranes that are continuous with the lining of the nasal cavity. Skull bones containing the paranasal sinuses are the frontal, sphenoid, ethmoid, and maxillae. Besides producing mucus, the paranasal sinuses serve as resonating chambers for sound as we speak or sing. The lateral walls of the internal nose are formed by the ethmoid, maxillae, lacrimal, palatine, and inferior nasal conchae bones (see Figure 7.9); the ethmoid bone also forms the roof. The palatine bones and palatine processes of the maxillae, which together constitute the hard palate, form the floor of the internal nose.

The bony and cartilaginous framework of the nose help to keep the vestibule and nasal cavity **patent,** that is, open or unobstructed. The nasal cavity is divided into a larger, inferior *respiratory region* and a smaller, superior *olfactory region.* The respiratory region is lined with pseudostratified ciliated columnar epithelium with numerous goblet cells, which is frequently called the **respiratory epithelium** (see Table 4.1). The anterior portion of the nasal cavity just inside the nostrils, called the **nasal vestibule,** is surrounded by cartilage; the superior part of the nasal cavity is surrounded by bone. A vertical partition, the **nasal septum,** divides the nasal cavity into right and left sides. The anterior portion of the nasal septum consists primarily of hyaline cartilage; the remainder is formed by the vomer, perpendicular plate of the ethmoid, maxillae, and palatine bones (see Figure 7.11).

When air enters the nostrils, it passes first through the vestibule, which is lined by skin containing coarse hairs that filter out large dust particles. Three shelves formed by projections of the superior, middle, and inferior **nasal conchae** extend out of each lateral wall of the nasal cavity. The conchae, almost reaching the nasal septum, subdivide each side of the nasal cavity into a series of groovelike passageways—the **superior, middle,** and **inferior meatuses** (mē-Ā-tus-ēz = openings or passages; singular is **meatus**). Mucous membrane lines the nasal cavity and its

Figure 23.1 Structures of the respiratory system.

The upper respiratory system includes the nose, nasal cavity, pharynx, and associated structures; the lower respiratory system includes the larynx, trachea, bronchi, and lungs.

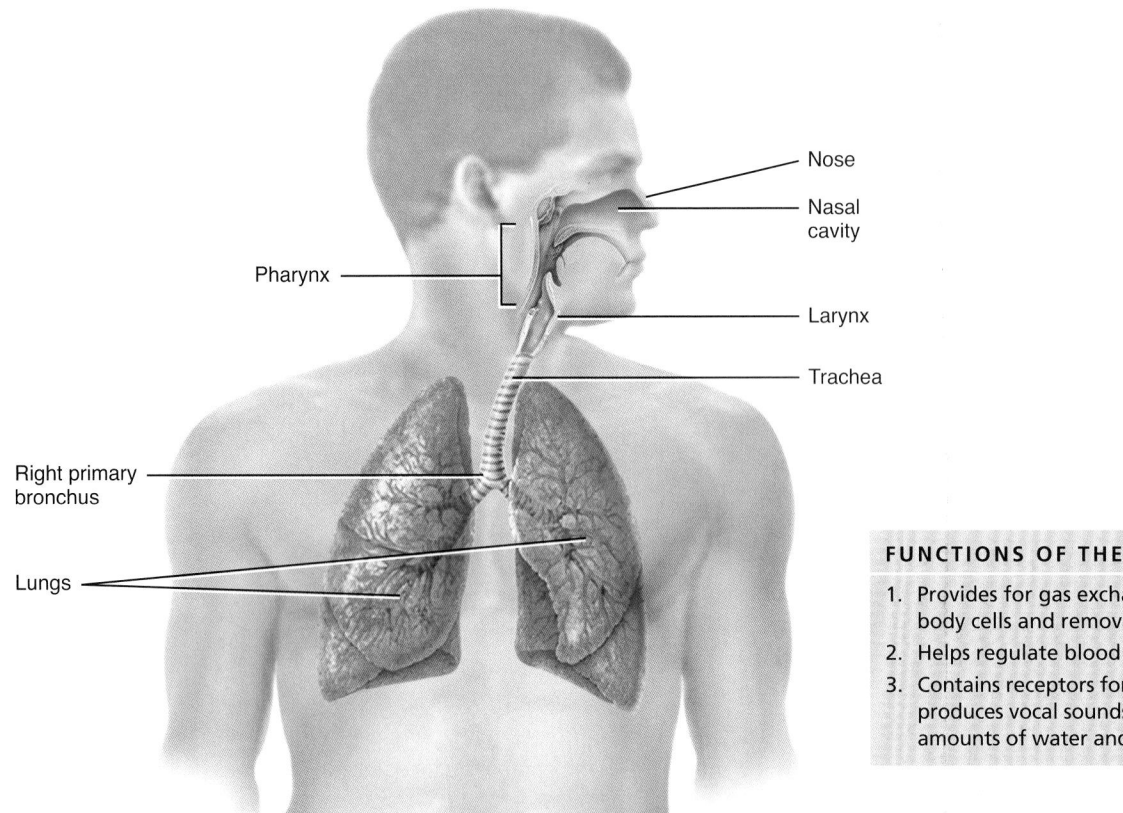

Nose

Nasal cavity

Pharynx

Larynx

Trachea

Right primary bronchus

Lungs

FUNCTIONS OF THE RESPIRATORY SYSTEM

1. Provides for gas exchange: intake of O_2 for delivery to body cells and removal of CO_2 produced by body cells.
2. Helps regulate blood pH.
3. Contains receptors for sense of smell, filters inspired air, produces vocal sounds (phonation), and excretes small amounts of water and heat.

(a) Anterior view showing organs of respiration

Larynx

Right common carotid artery

Thyroid gland

Trachea

Right subclavian artery

Subclavian artery

Brachiocephalic artery

Phrenic nerve

Superior vena cava

Left common carotid artery

Arch of aorta

Rib (cut)

Right lung

Left lung

Heart in pericardial sac

Liver

Diaphragm

(b) Anterior view of lungs and heart after removal of anterolateral thoracic wall and pleura

Which structures are part of the conducting zone of the respiratory system?

Figure 23.2 Respiratory structures in the head and neck.

As air passes through the nose; it is warmed, filtered, and moistened; and olfaction occurs.

Bony framework:

Frontal bone

Nasal bones

Maxilla

Cartilaginous framework:

Lateral nasal cartilages

Septal cartilage

Alar cartilage

Dense fibrous connective and adipose tissue

(a) Anterolateral view of external portion of nose showing cartilaginous and bony framework

Parasagittal plane

Nasal meatuses
- Superior
- Middle
- Inferior

Frontal sinus

Frontal bone

Olfactory epithelium

Sphenoid bone

Sphenoidal sinus

Internal naris

Pharyngeal tonsil

NASOPHARYNX

Opening of auditory tube

Uvula

Palatine tonsil

Fauces

OROPHARYNX

Epiglottis

LARYNGOPHARYNX

Esophagus

Trachea

Nasal conchae
- Superior
- Middle
- Inferior

Nasal vestibule

External naris

Maxilla

Oral cavity

Palatine bone

Tongue

Soft palate

Lingual tonsil

Mandible

Hyoid bone

Ventricular fold (false vocal cord)

Vocal fold (true vocal cord)

Larynx

Thyroid cartilage

Cricoid cartilage

Thyroid gland

(b) Parasagittal section of left side of head and neck showing location of respiratory structures

Regions of the pharynx

Nasopharynx

Oropharynx

Laryngopharynx

FIGURE 23.2 CONTINUES

FIGURE 23.2 CONTINUED

Frontal plane

View

Periorbital fat

Ethmoidal cell

Eyeball

Superior nasal concha

Middle nasal concha

Nasal septum:
Perpendicular plate of ethmoid

Vomer

Hard palate

Inferior nasal concha

Maxillary sinus

(c) Frontal section showing conchae

? What is the path taken by air molecules into and through the nose?

Figure 23.3 Surface anatomy of the nose.

The external nose has a cartilaginous framework and a bony framework.

Anterior view

1. **Root**: Superior attachment of the nose to the frontal bone
2. **Apex**: Tip of nose
3. **Bridge**: Bony framework of nose formed by nasal bones
4. **External naris**: Nostril; external opening into nasal cavity

? Which part of the nose is attached to the frontal bone?

shelves. The arrangement of conchae and meatuses increases surface area in the internal nose and prevents dehydration by trapping water droplets during exhalation.

As inhaled air whirls around the conchae and meatuses, it is warmed by blood in the capillaries. Mucus secreted by the goblet cells moistens the air and traps dust particles. Drainage from the nasolacrimal ducts also helps moisten the air, and is sometimes assisted by secretions from the paranasal sinuses. The cilia move the mucus and trapped dust particles toward the pharynx, at which point they can be swallowed or spit out, thus removing the particles from the respiratory tract.

The olfactory receptors, supporting cells, and basal cells lie in the respiratory region, which is near the superior nasal conchae and adjacent septum. These cells make up the **olfactory epithelium.** It contains cilia but no goblet cells.

✔CHECKPOINT

1. What functions do the respiratory and cardiovascular systems have in common?
2. What structural and functional features are different in the upper and lower respiratory systems? Which are the same?
3. Compare the structure and functions of the external nose and the internal nose.

Pharynx

The **pharynx** (FAR-inks), or throat, is a funnel-shaped tube about 13 cm (5 in.) long that starts at the internal nares and extends to the level of the cricoid cartilage, the most inferior cartilage of the larynx (voice box) (see Figure 23.2b). The pharynx lies just posterior to the nasal and oral cavities, superior to the larynx, and just anterior to the cervical vertebrae. Its wall is composed of skeletal muscles and is lined with a mucous membrane. Relaxed skeletal muscles help keep the pharynx patent. Contraction of the skeletal muscles assists in deglutition (swallowing). The pharynx functions as a passageway for air and food, provides a resonating chamber for speech sounds, and houses the tonsils, which participate in immunological reactions against foreign invaders.

The pharynx can be divided into three anatomical regions: (1) nasopharynx, (2) oropharynx, and (3) laryngopharynx. (See the lower orientation diagram in Figure 23.2b.) The muscles of the entire pharynx are arranged in two layers, an outer circular layer and an inner longitudinal layer.

The superior portion of the pharynx, called the **nasopharynx,** lies posterior to the nasal cavity and extends to the soft palate. The **soft palate,** which forms the posterior portion of the roof of the mouth, is an arch-shaped muscular partition between the nasopharynx and oropharynx that is lined by mucous membrane. There are five openings in its wall: two internal nares, two openings that lead into the *auditory* (*pharyngotympanic*) *tubes* (commonly known as the *eustachian tubes*), and the opening into the oropharynx. The posterior wall also contains the **pharyngeal tonsil** (fa-RIN-je-al), or **adenoid.** Through the internal nares, the nasopharynx receives air from the nasal cavity along with packages of dust-laden mucus. The nasopharynx is lined with pseudostratified ciliated columnar epithelium, and the cilia move the mucus down toward the most inferior part of the pharynx. The nasopharynx also exchanges small amounts of air with the auditory tubes to equalize air pressure between the pharynx and the middle ear.

The intermediate portion of the pharynx, the **oropharynx,** lies posterior to the oral cavity and extends from the soft palate inferiorly to the level of the hyoid bone. It has only one opening into it, the **fauces** (FAW-sēz = throat), the opening from the mouth. This portion of the pharynx has both respiratory and digestive functions, serving as a common passageway for air, food, and drink. Because the oropharynx is subject to abrasion by food particles, it is lined with nonkeratinized stratified squamous epithelium. Two pairs of tonsils, the **palatine** and **lingual tonsils,** are found in the oropharynx.

The inferior portion of the pharynx, the **laryngopharynx** (la-ring′-gō-FAR-ingks), or **hypopharynx,** begins at the level of the hyoid bone. At its inferior end it opens into the esophagus (food tube) posteriorly and the larynx (voice box) anteriorly. Like the oropharynx, the laryngopharynx is both a respiratory and a digestive pathway and is lined by nonkeratinized stratified squamous epithelium.

CLINICAL CONNECTION | *Tonsillectomy*

Tonsillectomy (ton-si-LEK-tō-mē-; -*ektome* = excision or to cut out) is surgical removal of the tonsils. The procedure is usually performed under general anesthesia on an outpatient basis. Tonsillectomies are performed in individuals who have frequent *tonsillitis* (ton′-si-Lῑ-tis), that is, inflammation of the tonsils; tonsils that develop an abscess or tumor; or tonsils that obstruct breathing during sleep. •

Larynx

The **larynx** (LAR-inks), or voice box, is a short passageway that connects the laryngopharynx with the trachea. It lies in the midline of the neck anterior to the esophagus and the fourth through sixth cervical vertebrae (C4–C6).

The wall of the larynx is composed of nine pieces of cartilage (Figure 23.4). Three occur singly (thyroid cartilage, epiglottis, and cricoid cartilage), and three occur in pairs (arytenoid, cuneiform, and corniculate cartilages). Of the paired cartilages, the arytenoid cartilages are the most important because they influence changes in position and tension of the vocal folds (true vocal cords for speech). The extrinsic muscles of the larynx connect the cartilages to other structures in the throat; the intrinsic muscles connect the cartilages to one another. The **cavity of the larynx** is the space that extends from the entrance into the larynx down to the inferior border of the cricoid cartilage (described shortly). The portion of the cavity of the larynx above the vocal folds (true vocal cords) is called the **vestibule of the larynx** (Figure 23.4d).

The **thyroid cartilage (Adam's apple)** consists of two fused plates of hyaline cartilage that form the anterior wall of the larynx and give it a triangular shape. It is present in both males and females but is usually larger in males due to the influence of male sex hormones on its growth during puberty. The ligament that connects the thyroid cartilage to the hyoid bone is called the **thyrohyoid membrane.**

The **epiglottis** (*epi-* = over; -*glottis* = tongue) is a large, leaf-shaped piece of elastic cartilage that is covered with epithelium (see also Figure 23.2b). The "stem" of the epiglottis is the tapered inferior portion that is attached to the anterior rim of the thyroid cartilage and hyoid bone. The broad superior "leaf" portion of the epiglottis is unattached and is free to move up and down like a trap door. During swallowing, the pharynx and larynx rise. Elevation of the pharynx widens it to receive food or drink; elevation of the larynx causes the epiglottis to move down and form a lid over the glottis, closing it off. The **glottis** consists of a pair of folds of mucous membrane, the vocal folds (true vocal cords) in the larynx, and the space between them called the **rima glottidis** (RῑΙ-ma GLOT-ti-dis). The closing of the larynx in this way during swallowing routes liquids and foods into the esophagus and keeps them out of the larynx and airways. When small particles of dust, smoke, food, or liquids pass into the larynx, a cough reflex occurs, usually expelling the material.

The **cricoid cartilage** (KRῑ-koyd = ringlike) is a ring of hyaline cartilage that forms the inferior wall of the larynx. It is attached to the first ring of cartilage of the trachea by the **cricotracheal ligament** (krῑ′-kō-TRĀ-kē-al). The thyroid cartilage is connected to the cricoid cartilage by the **cricothyroid ligament.** The cricoid cartilage is the landmark for making an emergency airway called a tracheotomy (see Clinical Connection: Tracheotomy and Intubation).

The paired **arytenoid cartilages** (ar′-i-TĒ-noyd = ladlelike) are triangular pieces of mostly hyaline cartilage located at the posterior, superior border of the cricoid cartilage. They form synovial joints with the cricoid cartilage and have a wide range of mobility.

The paired **corniculate cartilages** (kor-NIK-ū-lāt = shaped like a small horn), horn-shaped pieces of elastic cartilage, are located at the apex of each arytenoid cartilage. The paired **cuneiform cartilages** (KŪ-nē-i-form = wedge-shaped), club-shaped elastic cartilages anterior to the corniculate cartilages, support the vocal folds and lateral aspects of the epiglottis.

The lining of the larynx superior to the vocal folds is nonkeratinized stratified squamous epithelium. The lining of the larynx inferior to the vocal folds is pseudostratified ciliated columnar epithelium consisting of ciliated columnar cells, goblet cells, and basal cells. The mucus produced by the goblet cells helps trap dust not removed in the upper passages. The cilia in the upper respiratory tract

Figure 23.4 Larynx.

The larynx is composed of nine pieces of cartilage.

Larynx Thyroid gland

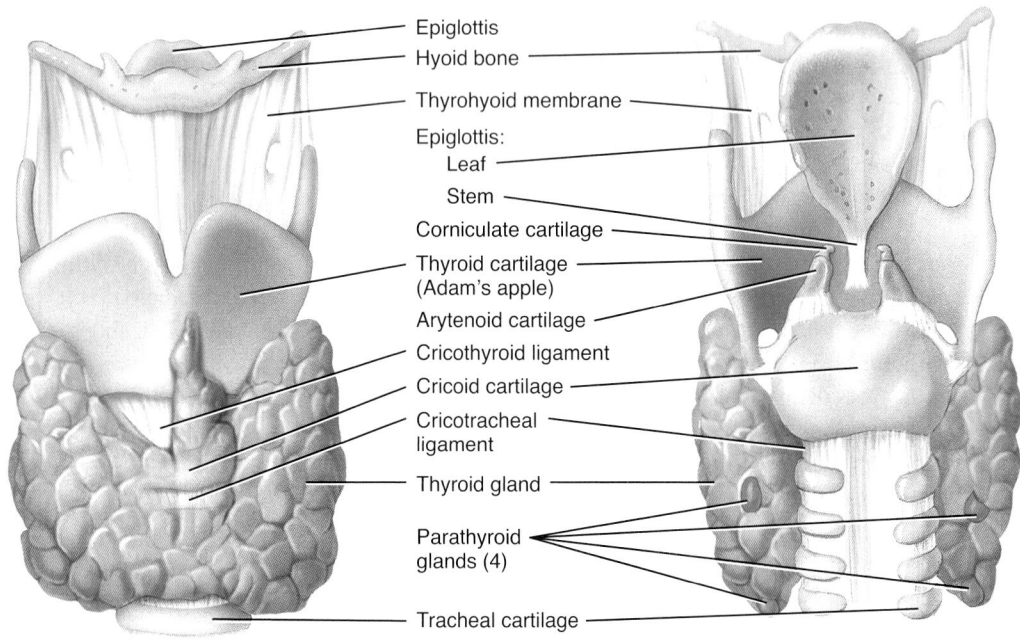

(a) Anterior view

(b) Posterior view

Epiglottis
Hyoid bone
Thyrohyoid membrane
Epiglottis:
    Leaf
    Stem
Corniculate cartilage
Thyroid cartilage (Adam's apple)
Arytenoid cartilage
Cricothyroid ligament
Cricoid cartilage
Cricotracheal ligament
Thyroid gland
Parathyroid glands (4)
Tracheal cartilage

Sagittal plane

Epiglottis
Thyrohyoid membrane
Cuneiform cartilage
Corniculate cartilage
Arytenoid cartilage
Laryngeal sinus
Cricoid cartilage
Tracheal cartilage

Hyoid bone
Thyrohyoid membrane
Fat body
Ventricular fold (false vocal cord)
Thyroid cartilage
Vocal fold (true vocal cord)
Cricothyroid ligament
Cricotracheal ligament

(c) Sagittal section

move mucus and trapped particles *down* toward the pharynx; the cilia in the lower respiratory tract move them *up* toward the pharynx.

The Structures of Voice Production

The mucous membrane of the larynx forms two pairs of folds (Figure 23.4c): a superior pair called the **ventricular folds (false vocal cords)** and an inferior pair called the **vocal folds (true vocal cords).** The space between the ventricular folds is known as the **rima vestibuli.** The **laryngeal sinus (ventricle)** is a lateral expansion of the middle portion of the laryngeal cavity inferior to the ventricular folds and superior to the vocal folds (see Figure 23.2b). While the ventricular folds do not function in voice production, they do have other important functional roles. When the ventricular folds are brought together, they function in holding the breath against pressure in the thoracic cavity, such as might occur when a person strains to lift a heavy object.

The vocal folds are the principal structures of voice production. Deep to the mucous membrane of the vocal folds, which is nonkeratinized stratified squamous epithelium, are bands of elastic ligaments stretched between the rigid cartilages of the larynx like the strings on a guitar. Intrinsic laryngeal muscles attach to both the rigid cartilages and the vocal folds. When the muscles contract they move the cartilages, which pulls the elastic ligaments tight, and this stretches the vocal folds out into the airways so that the rima glottidis is narrowed. Contracting and relaxing the muscles varies the tension in the vocal folds, much like loosening or tightening a guitar string. Air passing through the larynx vibrates the folds and produces sound (phonation) by setting up sound waves in the column of air in the pharynx, nose, and mouth. The variation in the pitch of the sound is related to the tension in the vocal folds. The greater the pressure of air, the louder the sound produced by the vibrating vocal folds.

When the intrinsic muscles of the larynx contract, they pull on the arytenoid cartilages, which causes the cartilages to pivot and slide. Contraction of the posterior cricoarytenoid muscles, for example, moves the vocal folds apart (abduction), thereby opening the rima glottidis (Figure 23.5a). By contrast, contraction of the lateral cricoarytenoid muscles moves the vocal folds together (adduction), thereby closing the rima glottidis (Figure 23.5b). Other intrinsic muscles can elongate (and place tension on) or shorten (and relax) the vocal folds.

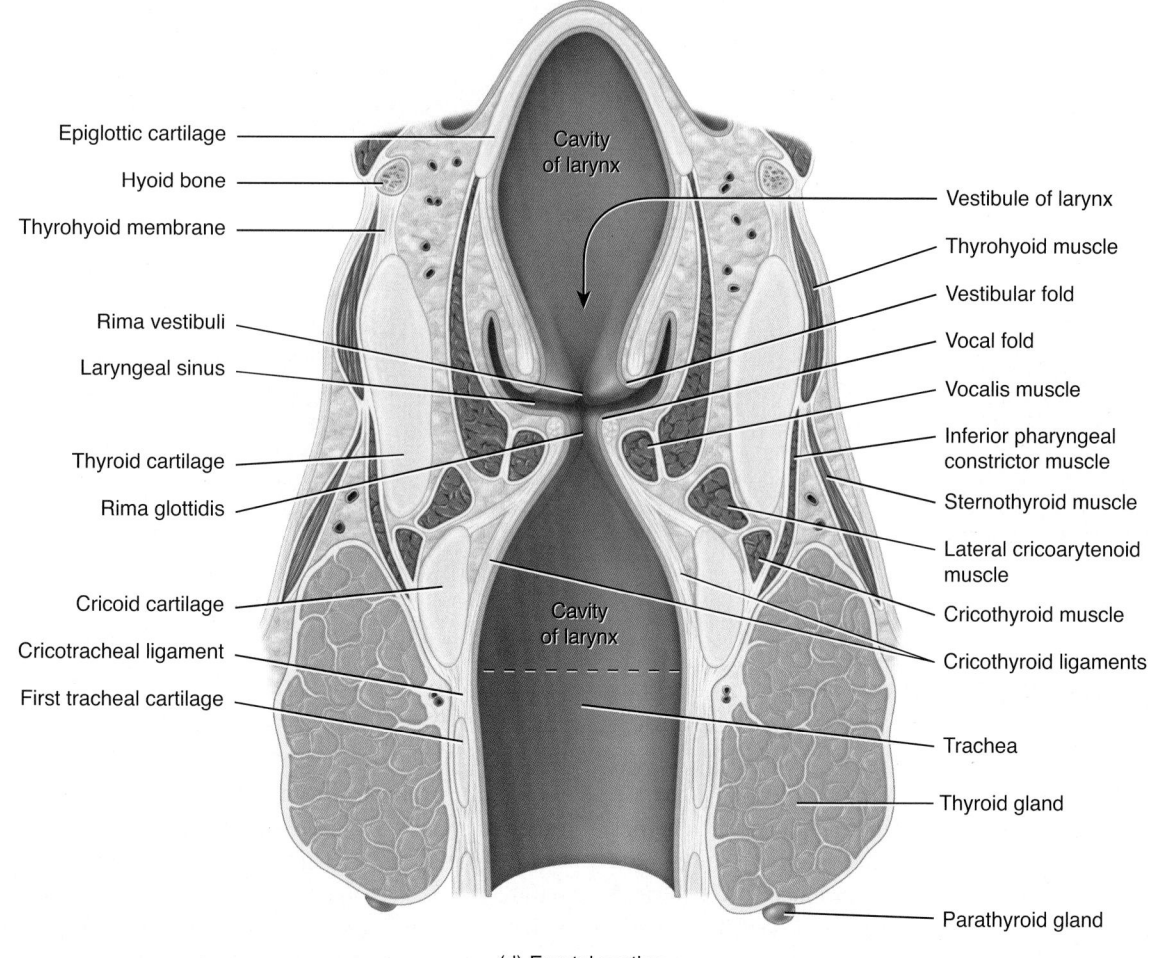

(d) Frontal section

? **How does the epiglottis prevent aspiration of foods and liquids?**

Figure 23.5 Movement of the vocal folds.

The glottis consists of a pair of folds of mucous membrane in the larynx (the vocal folds) and the space between them (the rima glottidis).

Thyroid cartilage

Cricoid cartilage

Vocal fold

Arytenoid cartilage

Posterior cricoarytenoid muscle

Superior view of cartilages and muscles

Tongue

Epiglottis

Glottis:
Vocal folds (true vocal cords)

Rima glottidis

Ventricular folds (false vocal cords)

Cuneiform cartilage

Corniculate cartilage

View through a laryngoscope

(a) Movement of vocal folds apart (abduction)

Lateral cricoarytenoid muscle

(b) Movement of vocal folds together (adduction)

View

Larynx

Epiglottis

Vocal folds (true vocal cords)

Rima glottidis

Cuneiform cartilage

Corniculate cartilage

Ventricular folds (false vocal cords)

(c) Superior view

❓ What is the main function of the vocal folds?

Pitch is controlled by the tension on the vocal folds. If they are pulled taut by the muscles, they vibrate more rapidly, and a higher pitch results. Decreasing the muscular tension on the vocal folds causes them to vibrate more slowly and produce lower-pitched sounds. Due to the influence of androgens (male sex hormones), vocal folds are usually thicker and longer in males than in females, and therefore they vibrate more slowly. This is why a man's voice generally has a lower range of pitch than that of a woman.

Sound originates from the vibration of the vocal folds, but other structures are necessary for converting the sound into recognizable speech. The pharynx, mouth, nasal cavity, and paranasal sinuses all act as resonating chambers that give the voice its human and individual quality. We produce the vowel sounds by constricting and relaxing the muscles in the wall of the pharynx. Muscles of the face, tongue, and lips help us enunciate words.

Whispering is accomplished by closing all but the posterior portion of the rima glottidis. Because the vocal folds do not vibrate during whispering, there is no pitch to this form of speech. However, we can still produce intelligible speech while whispering by changing the shape of the oral cavity as we enunciate. As the size of the oral cavity changes, its resonance qualities change, which imparts a vowel-like pitch to the air as it rushes toward the lips.

Trachea

The **trachea** (TRĀ-kē-a = sturdy), or **windpipe,** is a tubular passageway for air that is about 12 cm (5 in.) long and 2.5 cm (1 in.) in diameter. It is located anterior to the esophagus (Figure 23.6)

and extends from the larynx to the superior border of the fifth thoracic vertebra (T5), where it divides into right and left primary bronchi (see Figure 23.7).

The layers of the tracheal wall, from deep to superficial, are the (1) mucosa, (2) submucosa, (3) hyaline cartilage, and (4) adventitia (composed of areolar connective tissue). The mucosa of the trachea consists of an epithelial layer of pseudostratified ciliated columnar epithelium and an underlying layer of lamina propria that contains elastic and reticular fibers. It provides the same protection against dust as the membrane lining the nasal cavity and larynx. The submucosa consists of areolar connective tissue that contains seromucous glands and their ducts.

The 16–20 incomplete, horizontal rings of hyaline cartilage resemble the letter C, are stacked one above another, and are connected together by dense connective tissue. They may be felt through the skin inferior to the larynx. The open part of each C-shaped cartilage ring faces posteriorly toward the esophagus (Figure 23.6) and is spanned by a *fibromuscular membrane*. Within this membrane are transverse smooth muscle fibers, called the **trachealis muscle** (trā-kē-Ā-lis), and elastic connective tissue that allow the diameter of the trachea to change subtly during inhalation and exhalation, which is important in maintaining efficient airflow. The solid C-shaped cartilage rings provide a semirigid support to maintain patency so that the tracheal wall does not collapse inward (especially during inhalation) and obstruct the air passageway. The adventitia of the trachea consists of areolar connective tissue that joins the trachea to surrounding tissues.

Figure 23.6 Location of the trachea in relation to the esophagus.

The trachea is anterior to the esophagus and extends from the larynx to the superior border of the fifth thoracic vertebra.

Superior view of transverse section of thyroid gland, trachea, and esophagus

? **What is the benefit of not having complete rings of tracheal cartilage between the trachea and the esophagus?**

Several conditions may block airflow by obstructing the trachea. The rings of cartilage that support the trachea may be accidentally crushed, the mucous membrane may become inflamed and swell so much that it closes off the passageway, excess mucus secreted by inflamed membranes may clog the lower respiratory passages, a large object may be aspirated (breathed in), or a cancerous tumor may protrude into the airway. Two methods are used to reestablish airflow past a tracheal obstruction. If the obstruction is above the level of the larynx, a **tracheotomy** (trā-kē-O-tō-mē) may be performed. In this procedure, also called a *tracheostomy*, a skin incision is followed by a short longitudinal incision into the trachea below the cricoid cartilage. A tracheal tube is then inserted to create an emergency air passageway. The second method is **intubation** (in'-too-BĀ-shun), in which a tube is inserted into the mouth or nose and passed inferiorly through the larynx and trachea. The firm wall of the tube pushes aside any flexible obstruction, and the lumen of the tube provides a passageway for air; any mucus clogging the trachea can be suctioned out through the tube. •

Bronchi

At the superior border of the fifth thoracic vertebra, the trachea divides into a **right primary bronchus** (BRONG-kus = windpipe), which goes into the right lung, and a **left primary bronchus,** which goes into the left lung (Figure 23.7). The right primary bronchus is more vertical, shorter, and wider than the left. As a result, an aspirated object is more likely to enter and lodge in the right primary bronchus than the left. Like the trachea, the primary

Figure 23.7 Branching of airways from the trachea: the bronchial tree.

The bronchial tree begins at the trachea and ends at the terminal bronchioles.

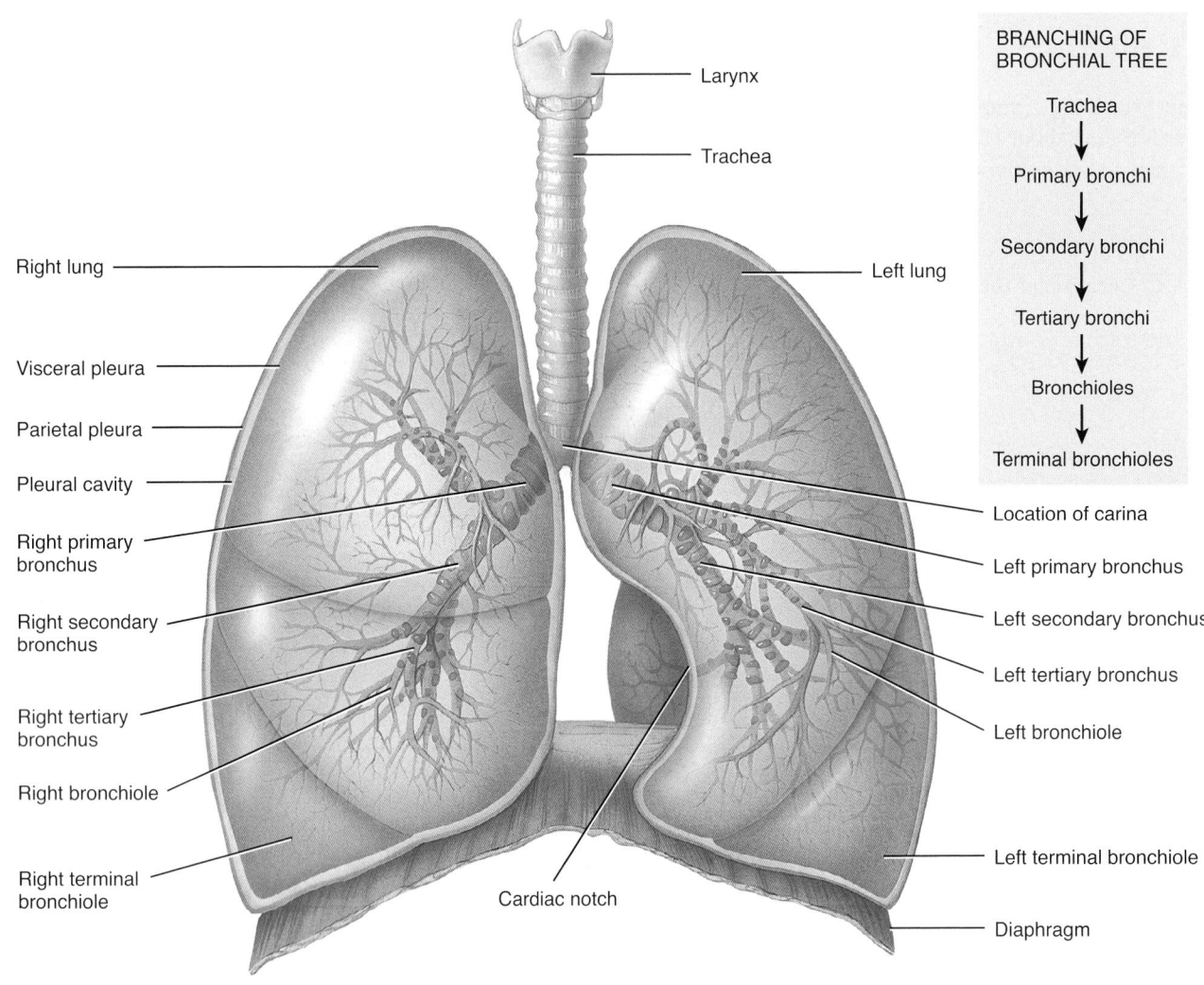

BRANCHING OF BRONCHIAL TREE

Trachea
↓
Primary bronchi
↓
Secondary bronchi
↓
Tertiary bronchi
↓
Bronchioles
↓
Terminal bronchioles

Larynx

Trachea

Right lung

Visceral pleura

Parietal pleura

Pleural cavity

Right primary bronchus

Right secondary bronchus

Right tertiary bronchus

Right bronchiole

Right terminal bronchiole

Left lung

Location of carina

Left primary bronchus

Left secondary bronchus

Left tertiary bronchus

Left bronchiole

Left terminal bronchiole

Cardiac notch

Diaphragm

Anterior view

? **How many lobes and secondary bronchi are present in each lung?**

bronchi (BRONG-kī) contain incomplete rings of cartilage and are lined by pseudostratified ciliated columnar epithelium.

At the point where the trachea divides into right and left primary bronchi an internal ridge called the **carina** (ka-RĪ-na = keel of a boat) is formed by a posterior and somewhat inferior projection of the last tracheal cartilage. The mucous membrane of the carina is one of the most sensitive areas of the entire larynx and trachea for triggering a cough reflex. Widening and distortion of the carina is a serious sign because it usually indicates a carcinoma of the lymph nodes around the region where the trachea divides.

On entering the lungs, the primary bronchi divide to form smaller bronchi—the **secondary (lobar) bronchi,** one for each lobe of the lung. (The right lung has three lobes; the left lung has two.) The secondary bronchi continue to branch, forming still smaller bronchi, called **tertiary (segmental) bronchi** (TER-shē-e-rē), that divide into **bronchioles.** Bronchioles in turn branch repeatedly, and the smallest ones branch into even smaller tubes called **terminal bronchioles.** These bronchioles contain *Clara cells,* columnar, nonciliated cells interspersed among the epithelial cells. Clara cells may protect against harmful effects of inhaled toxins and carcinogens, produce surfactant (discussed shortly), and function as stem cells (reserve cells), which give rise to various cells of the epithelium. The terminal bronchioles represent the end of the conducting zone of the respiratory system. This extensive branching from the trachea through the terminal bronchioles resembles an inverted tree and is commonly referred to as the **bronchial tree.**

As the branching becomes more extensive in the bronchial tree, several structural changes may be noted.

1. The mucous membrane in the bronchial tree changes from pseudostratified ciliated columnar epithelium in the primary bronchi, secondary bronchi, and tertiary bronchi to ciliated simple columnar epithelium with some goblet cells in larger bronchioles, to mostly ciliated simple cuboidal epithelium with no goblet cells in smaller bronchioles, to mostly nonciliated simple cuboidal epithelium in terminal bronchioles. Recall that ciliated epithelium of the respiratory membrane removes inhaled particles in two ways. Mucus produced by goblet cells traps the particles, and the cilia move the mucus and trapped particles toward the pharynx for removal. In regions where nonciliated simple cuboidal epithelium is present, inhaled particles are removed by macrophages.

2. Plates of cartilage gradually replace the incomplete rings of cartilage in primary bronchi and finally disappear in the distal bronchioles.

3. As the amount of cartilage decreases, the amount of smooth muscle increases. Smooth muscle encircles the lumen in spiral bands and helps maintain patency. However, because there is no supporting cartilage, muscle spasms can close off the airways. This is what happens during an asthma attack, which can be a life-threatening situation.

During exercise, activity in the sympathetic division of the autonomic nervous system (ANS) increases and the adrenal medulla releases the hormones epinephrine and norepinephrine; both of these events cause relaxation of smooth muscle in the bronchioles, which dilates the airways. Because air reaches the alveoli more quickly, lung ventilation improves. The parasympathetic division of the ANS and mediators of allergic reactions such as histamine have the opposite effect, causing contraction of bronchiolar smooth muscle, which results in constriction of distal bronchioles.

✔ CHECKPOINT

4. List the roles of each of the three anatomical regions of the pharynx in respiration.
5. How does the larynx function in respiration and voice production?
6. Describe the location, structure, and function of the trachea.
7. Describe the structure of the bronchial tree.

Lungs

The **lungs** (= lightweights, because they float) are paired cone-shaped organs in the thoracic cavity. They are separated from each other by the heart and other structures of the mediastinum, which divides the thoracic cavity into two anatomically distinct chambers. As a result, if trauma causes one lung to collapse, the other may remain expanded. Each lung is enclosed and protected by a double-layered serous membrane called the **pleural membrane** (PLOOR-al; *pleur-* = side). The superficial layer, called the **parietal pleura,** lines the wall of the thoracic cavity; the deep layer, the **visceral pleura,** covers the lungs themselves (Figure 23.8). Between the visceral and parietal pleurae is a small space, the **pleural cavity,** which contains a small amount of lubricating fluid secreted by the membranes. This pleural fluid reduces friction between the membranes, allowing them to slide easily over one another during breathing. Pleural fluid also causes the two membranes to adhere to one another just as a film of water causes two glass microscope slides to stick together, a phenomenon called surface tension. Separate pleural cavities surround the left and right lungs. Inflammation of the pleural membrane, called **pleurisy** or **pleuritis,** may in its early stages cause pain due to friction between the parietal and visceral layers of the pleura. If the inflammation persists, excess fluid accumulates in the pleural space, a condition known as **pleural effusion.**

CLINICAL CONNECTION | *Pneumothorax and Hemothorax*

In certain conditions, the pleural cavities may fill with air (**pneumothorax;** noo'-mō-THOR-aks; *pneumo-* = air or breath), blood (**hemothorax**), or pus. Air in the pleural cavities, most commonly introduced in a surgical opening of the chest or as a result of a stab or gunshot wound, may cause the lungs to collapse. This collapse of a part of a lung, or rarely an entire lung, is called **atelectasis** (at'-e-LEK-ta-sis; *ateles-* = incomplete; *-ectasis* = expansion). The goal of treatment is the evacuation of air (or blood) from the pleural space, which allows the lung to reinflate. A small pneumothorax may resolve on its own, but it is often necessary to insert a chest tube to assist in evacuation. •

Figure 23.8 Relationship of the pleural membranes to the lungs.

The parietal pleura lines the thoracic cavity, and the visceral pleura covers the lungs.

Inferior view of transverse section through thoracic cavity showing pleural cavity and pleural membranes

? **What type of membrane is the pleural membrane?**

The lungs extend from the diaphragm to just slightly superior to the clavicles and lie against the ribs anteriorly and posteriorly (Figure 23.9a). The broad inferior portion of the lung, the **base,** is concave and fits over the convex area of the diaphragm. The narrow superior portion of the lung is the **apex.** The surface of the lung lying against the ribs, the **costal surface,** matches the rounded curvature of the ribs. The **mediastinal (medial) surface** of each lung contains a region, the **hilum** (or **hilus**), through which bronchi, pulmonary blood vessels, lymphatic vessels, and nerves enter and exit (Figure 23.9e). These structures are held together by the pleura and connective tissue and constitute the **root** of the lung. Medially, the left lung also contains a concavity, the **cardiac notch,** in which the apex of the heart lies. Due to the space occupied by the heart, the left lung is about 10% smaller than the right lung. Although the right lung is thicker and broader, it is also somewhat shorter than the left lung because the diaphragm is higher on the right side, accommodating the liver that lies inferior to it.

The lungs almost fill the thorax (Figure 23.9a). The apex of the lungs lies superior to the medial third of the clavicles, and this is the only area that can be palpated. The anterior, lateral, and posterior surfaces of the lungs lie against the ribs. The base of the lungs extends from the sixth costal cartilage anteriorly to the spinous process of the tenth thoracic vertebra posteriorly. The pleura extends about 5 cm (2 in.) below the base from the sixth costal cartilage anteriorly to the twelfth rib posteriorly. Thus, the lungs do not completely fill the pleural cavity in this area. Removal of excessive fluid in the pleural cavity can be accomplished without injuring lung tissue by inserting a needle anteriorly through the seventh intercostal space, a procedure called **thoracentesis** (thor′-a-sen-TĒ-sis; -*centesis* = puncture). The needle is passed along the superior border of the lower rib to avoid damage to the intercostal nerves and blood vessels. Inferior to the seventh intercostal space there is danger of penetrating the diaphragm.

Lobes, Fissures, and Lobules

One or two fissures divide each lung into lobes (Figure 23.9b–e). Both lungs have an **oblique fissure,** which extends inferiorly and anteriorly; the right lung also has a **horizontal fissure.** The oblique fissure in the left lung separates the **superior lobe** from the **inferior lobe.** In the right lung, the superior part of the oblique fissure separates the superior lobe from the inferior lobe; the inferior part of the oblique fissure separates the inferior lobe from the **middle lobe,** which is bordered superiorly by the horizontal fissure.

Each lobe receives its own secondary (lobar) bronchus. Thus, the right primary bronchus gives rise to three secondary (lobar) bronchi called the **superior, middle,** and **inferior secondary (lobar) bronchi,** and the left primary bronchus gives rise to **superior** and **inferior secondary (lobar) bronchi.** Within the lung, the secondary bronchi give rise to the **tertiary (segmental) bronchi,** which are constant in both origin and distribution—there

Figure 23.9 Surface anatomy of the lungs.

The oblique fissure divides the left lung into two lobes. The oblique and horizontal fissures divide the right lung into three lobes.

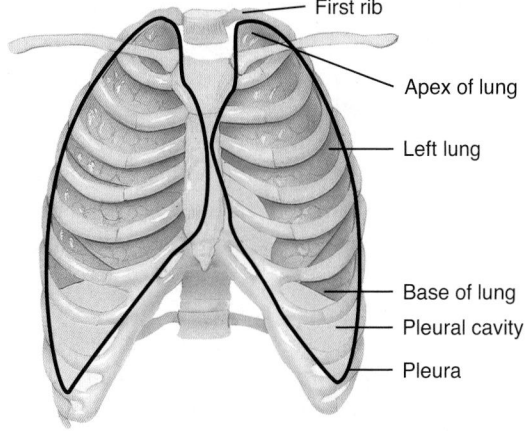

(a) Anterior view of lungs and pleurae in thorax

First rib
Apex of lung
Left lung
Base of lung
Pleural cavity
Pleura

View (b)   View (c)

Apex
Superior lobe

ANTERIOR

Horizontal fissure

Oblique fissure

Cardiac notch

Oblique fissure

Inferior lobe

Middle lobe

Inferior lobe

POSTERIOR

POSTERIOR

Base

(b) Lateral view of right lung

(c) Lateral view of left lung

View (d)

View (e)

Apex
Superior lobe
Oblique fissure

POSTERIOR

Hilum and its contents (root)

Horizontal fissure

Middle lobe

Inferior lobe

Oblique fissure

Cardiac notch

ANTERIOR

Base

ANTERIOR

(d) Medial view of right lung

(e) Medial view of left lung

? **Why are the right and left lungs slightly different in size and shape?**

are 10 tertiary bronchi in each lung. The segment of lung tissue that each tertiary bronchus supplies is called a **bronchopulmonary segment** (brong′-kō-PUL-mō-nar′-ē). Bronchial and pulmonary disorders (such as tumors or abscesses) that are localized in a bronchopulmonary segment may be surgically removed without seriously disrupting the surrounding lung tissue.

Each bronchopulmonary segment of the lungs has many small compartments called **lobules;** each lobule is wrapped in elastic connective tissue and contains a lymphatic vessel, an arteriole, a venule, and a branch from a terminal bronchiole (Figure 23.10a). Terminal bronchioles subdivide into microscopic branches called **respiratory bronchioles** (Figure 23.10b). They also have alveoli (described shortly) budding from their walls. Alveoli participate in gas exchange, and thus respiratory bronchioles begin the respiratory zone of the respiratory system. As the respiratory bronchioles penetrate more deeply into the lungs, the epithelial lining changes from simple cuboidal to simple squamous. Respiratory bronchioles in turn subdivide into several (2–11) **alveolar**

ducts (al-VĒ-ō-lar), which consist of simple squamous epithelium. The respiratory passages from the trachea to the alveolar ducts contain about 25 orders of branching; branching that from the trachea into primary bronchi is called first-order branching, that from primary bronchi into secondary bronchi is called second-order branching, and so on down to the alveolar ducts.

Alveoli

Around the circumference of the alveolar ducts are numerous alveoli and alveolar sacs. An **alveolus** (al-VĒ-ō-lus) is a cup-shaped outpouching lined by simple squamous epithelium and supported by a thin elastic basement membrane; an **alveolar sac** consists of two or more alveoli that share a common opening (Figure 23.10a, b). The walls of alveoli consist of two types of alveolar epithelial cells (Figure 23.11). The more numerous **type I alveolar cells** are simple squamous epithelial cells that form a nearly continuous lining of the alveolar wall. **Type II alveolar cells,** also called **septal cells,** are fewer in number and are found

Figure 23.10 Microscopic anatomy of a lobule of the lungs.

Alveolar sacs consist of two or more alveoli that share a common opening.

(a) Diagram of portion of lobule of lung

(b) Lung lobule

? What types of cells make up the wall of an alveolus?

Figure 23.11 Structural components of an alveolus. The respiratory membrane consists of a layer of type I and type II alveolar cells, an epithelial basement membrane, a capillary basement membrane, and the capillary endothelium.

🔑 The exchange of respiratory gases occurs by diffusion across the respiratory membrane.

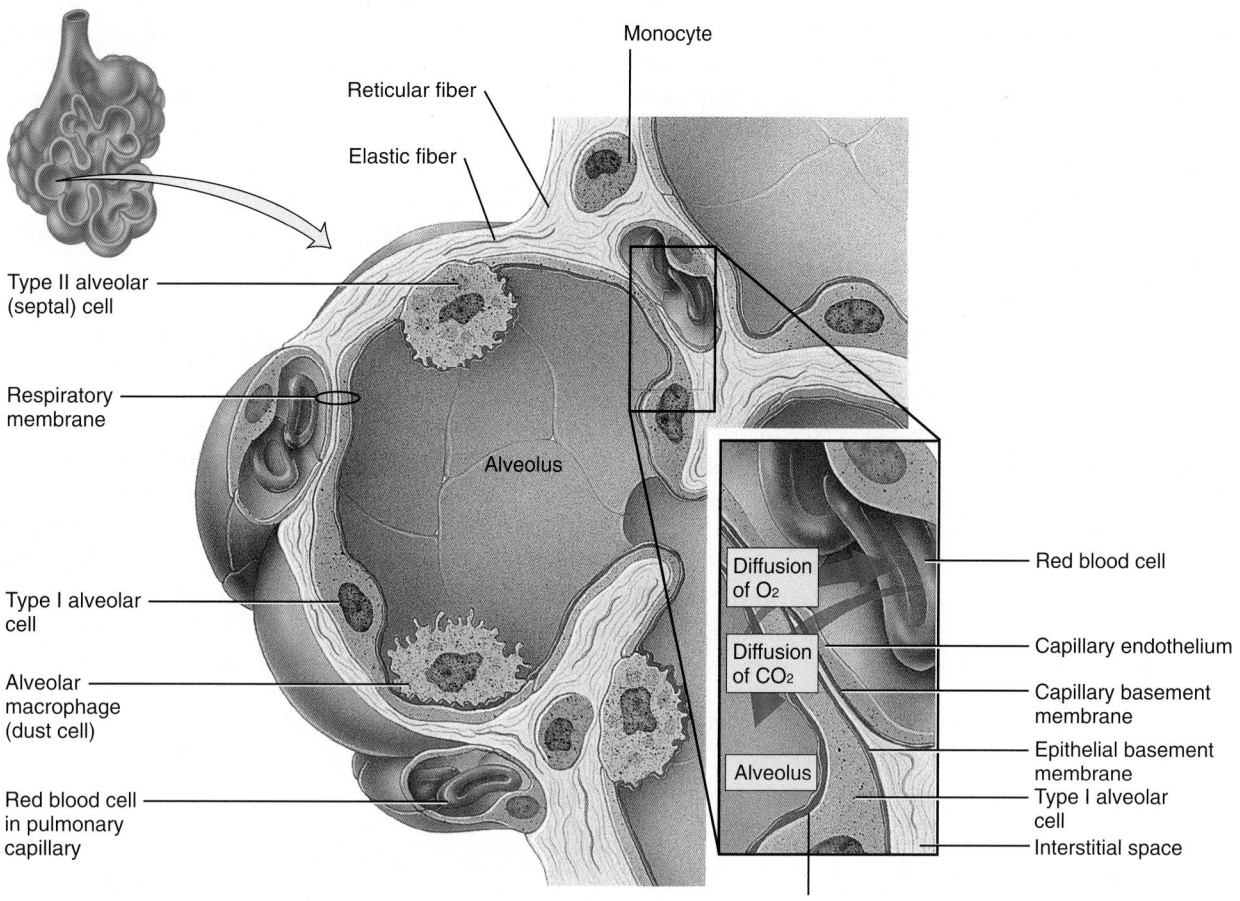

Monocyte

Reticular fiber

Elastic fiber

Type II alveolar (septal) cell

Respiratory membrane

Alveolus

Type I alveolar cell

Alveolar macrophage (dust cell)

Red blood cell in pulmonary capillary

Diffusion of O_2

Diffusion of CO_2

Alveolus

Red blood cell

Capillary endothelium

Capillary basement membrane

Epithelial basement membrane

Type I alveolar cell

Interstitial space

Alveolar fluid with surfactant

(a) Section through alveolus showing cellular components

(b) Details of respiratory membrane

Alveolus

Alveolar macrophage (dust cell)

Type II alveolar (septal) cell

Type I alveolar (squamous pulmonary epithelial) cell

Alveolus

LM 1000x

(c) Details of several alveoli

❓ **How thick is the respiratory membrane?**

933

between type I alveolar cells. The thin type I alveolar cells are the main sites of gas exchange. Type II alveolar cells, rounded or cuboidal epithelial cells with free surfaces containing microvilli, secrete alveolar fluid, which keeps the surface between the cells and the air moist. Included in the alveolar fluid is **surfactant** (sur-FAK-tant), a complex mixture of phospholipids and lipoproteins. Surfactant lowers the surface tension of alveolar fluid, which reduces the tendency of alveoli to collapse and thus maintains their patency (described later).

Associated with the alveolar wall are **alveolar macrophages (dust cells),** phagocytes that remove fine dust particles and other debris from the alveolar spaces. Also present are fibroblasts that produce reticular and elastic fibers. Underlying the layer of type I alveolar cells is an elastic basement membrane. On the outer surface of the alveoli, the lobule's arteriole and venule disperse into a network of blood capillaries (see Figure 23.10a) that consist of a single layer of endothelial cells and basement membrane.

The exchange of O_2 and CO_2 between the air spaces in the lungs and the blood takes place by diffusion across the alveolar and capillary walls, which together form the **respiratory membrane.** Extending from the alveolar air space to blood plasma, the respiratory membrane consists of four layers (Figure 23.11b):

1. A layer of type I and type II alveolar cells and associated alveolar macrophages that constitutes the **alveolar wall**

2. An **epithelial basement membrane** underlying the alveolar wall

3. A **capillary basement membrane** that is often fused to the epithelial basement membrane

4. The **capillary endothelium**

Despite having several layers, the respiratory membrane is very thin—only 0.5 μm thick, about one-sixteenth the diameter of a red blood cell—to allow rapid diffusion of gases. It has been estimated that the lungs contain 300 million alveoli, providing an immense surface area of 70 m^2 (750 ft^2)—about the size of a racquetball court—for gas exchange.

Blood Supply to the Lungs

The lungs receive blood via two sets of arteries: pulmonary arteries and bronchial arteries. Deoxygenated blood passes through the pulmonary trunk, which divides into a left pulmonary artery that enters the left lung and a right pulmonary artery that enters the right lung. (The pulmonary arteries are the only arteries in the body that carry deoxygenated blood.) Return of the oxygenated blood to the heart occurs by way of the four pulmonary veins, which drain into the left atrium (see Figure 21.29). A unique feature of pulmonary blood vessels is their constriction in response to localized hypoxia (low O_2 level). In all other body tissues, hypoxia causes dilation of blood vessels to increase blood flow. In the lungs, however, vasoconstriction in response to hypoxia diverts pulmonary blood from poorly ventilated areas of the lungs to well-ventilated regions for more efficient gas exchange. This phenomenon is known as **ventilation–perfusion coupling** (per-FYU-zhun) because the perfusion (blood flow) to

each area of the lungs matches the extent of ventilation (airflow) to alveoli in that area.

Bronchial arteries, which branch from the aorta, deliver oxygenated blood to the lungs. This blood mainly perfuses the muscular walls of the bronchi and bronchioles. Connections exist between branches of the bronchial arteries and branches of the pulmonary arteries, however; most blood returns to the heart via pulmonary veins. Some blood, however, drains into bronchial veins, branches of the azygos system, and returns to the heart via the superior vena cava.

Patency of the Respiratory System

Throughout the discussion of the respiratory organs, several examples were given of structures or secretions that help to maintain patency of the system so that air passageways are kept free of obstruction. These included the bony and cartilaginous frameworks of the nose, skeletal muscles of the pharynx, cartilages of the larynx, C-shaped rings of cartilage in the trachea and bronchi, smooth muscle in the bronchioles, and surfactant in the alveoli.

Unfortunately, there are also factors that can compromise patency. These include crushing injuries to bone and cartilage, a deviated nasal septum, nasal polyps, inflammation of mucous membranes, spasms of smooth muscle, and a deficiency of surfactant.

A summary of the epithelial linings and special features of the organs of the respiratory system is presented in Table 23.1.

CLINICAL CONNECTION | *Coryza, Seasonal Influenza, and H1N1 Influenza*

Hundreds of viruses can cause coryza (ko-RĪ-za), or the **common cold,** but a group of viruses called *rhinoviruses* (RĪ-nō-vī-rus-es) is responsible for about 40% of all colds in adults. Typical symptoms include sneezing, excessive nasal secretion, dry cough, and congestion. The uncomplicated common cold is not usually accompanied by a fever. Complications include sinusitis, asthma, bronchitis, ear infections, and laryngitis. Recent investigations suggest an association between emotional stress and the common cold. The higher the stress level, the greater the frequency and duration of colds. **Seasonal influenza (flu)** is also caused by a virus. Its symptoms include chills, fever (usually higher than 101°F = 39°C), headache, and muscular aches. Seasonal influenza can become life-threatening and may develop into pneumonia. It is important to recognize that influenza is a respiratory disease, not a gastrointestinal (GI) disease. Many people mistakenly report having seasonal flu when they are suffering from a GI illness.

H1N1 influenza (flu), also known as *swine flu*, is a type of influenza caused by a new virus called *influenza H1N1*. The term swine flu originated because early laboratory testing indicated that many of the genes in the new virus were similar to ones found in pigs (swine) in North America. However, subsequent testing revealed that the new virus is very different from the one that circulates in North American pigs.

H1N1 flu is a respiratory disorder first detected in the United States in April 2009. In June 2009, the World Health Organization declared H1N1 flu to be a *global pandemic disease* (a disease that affects large numbers of individuals in a short period of time and that occurs world-

TABLE 23.1

Summary of the Respiratory System

STRUCTURE	EPITHELIUM	CILIA	GOBLET CELLS	SPECIAL FEATURES
NOSE				
Vestibule	Nonkeratinized stratified squamous.	No.	No.	Contains numerous hairs.
Respiratory region	Pseudostratified ciliated columnar.	Yes.	Yes.	Contains conchae and meatuses.
Olfactory region	Olfactory epithelium (olfactory receptors).	Yes.	No.	Functions in olfaction.
PHARYNX				
Nasopharynx	Pseudostratified ciliated columnar.	Yes.	Yes.	Passageway for air; contains internal nares, openings for auditory tubes, and pharyngeal tonsil.
Oropharynx	Nonkeratinized stratified squamous.	No.	No.	Passageway for both air and food and drink; contains opening from mouth (fauces).
Laryngopharynx	Nonkeratinized stratified squamous.	No.	No.	Passageway for both air and food and drink.
LARYNX	Nonkeratinized stratified squamous above the vocal folds; pseudostratified ciliated columnar below the vocal folds.	No above folds; yes below folds.	No above folds; yes below folds.	Passageway for air; contains vocal folds for voice production.
TRACHEA	Pseudostratified ciliated columnar.	Yes.	Yes.	Passageway for air; contains C-shaped rings of cartilage to keep trachea open.
BRONCHI				
Primary bronchi	Pseudostratified ciliated columnar.	Yes.	Yes.	Passageway for air; contain C-shaped rings of cartilage to maintain patency.
Secondary bronchi	Pseudostratified ciliated columnar.	Yes.	Yes.	Passageway for air; contain plates of cartilage to maintain patency.
Tertiary bronchi	Pseudostratified ciliated columnar.	Yes.	Yes.	Passageway for air; contain plates of cartilage to maintain patency.
Larger bronchioles	Ciliated simple columnar.	Yes.	Yes.	Passageway for air; contain more smooth muscle than in the bronchi.
Smaller bronchioles	Ciliated simple columnar.	Yes.	No.	Passageway for air; contain more smooth muscle than in the larger bronchioles.
Terminal bronchioles	Nonciliated simple columnar.	No.	No.	Passageway for air; contain more smooth muscle than in the smaller bronchioles.
LUNGS				
Respiratory bronchioles	Simple cuboidal to simple squamous.	No.	No.	Passageway for air; gas exchange.
Alveolar ducts	Simple squamous.	No.	No.	Passageway for air; gas exchange; produce surfactant.
Alveoli	Simple squamous.	No.	No.	Passageway for air; gas exchange; produce surfactant to maintain patency.

☐ Conducting structures

☐ Respiratory structures

wide). The virus is spread in the same way that seasonal flu spreads: from person to person through coughing or sneezing or by touching infected objects and then touching one's mouth or nose. Most individuals infected with the virus have mild disease and recover without medical treatment, but some people have severe disease and have even died. The symptoms of H1N1 flu include fever, cough, runny or stuffy nose, headache, body aches, chills, and fatigue. Some people also have vomiting and diarrhea. Most people who have been hospitalized for H1N1 flu have had one or more preexisting medical conditions such as diabetes, heart disease, asthma, kidney disease, or pregnancy.

People infected with the virus can infect others from 1 day before symptoms occur to 5–7 days or more after they occur. Treatment of H1N1 flu involves taking antiviral drugs, such as Tamiflu® and Relenza®. A vaccine is also available, but the H1N1 flu vaccine is not a substitute for seasonal flu vaccines. In order to prevent infection, the Centers for Disease Control and Prevention (CDC) recommends washing your hands often with soap and water or with an alcohol-based hand cleaner; covering your mouth and nose with a tissue when coughing or sneezing and disposing of the tissue; avoiding touching your mouth, nose, or eyes; avoiding close contact (within 6 feet) with people who have flu-like symptoms; and staying home for 7 days after symptoms begin or for 24 hours after being symptom-free, whichever is longer. •

✔ CHECKPOINT

8. Where are the lungs located? Distinguish the parietal pleura from the visceral pleura.
9. Define each of the following parts of a lung: base, apex, costal surface, medial surface, hilum, root, cardiac notch, lobe, and lobule.
10. What is a bronchopulmonary segment?
11. Describe the histology and function of the respiratory membrane.

23.2 PULMONARY VENTILATION

⦿ OBJECTIVE

• Describe the events that cause inhalation and exhalation.

The process of gas exchange in the body, called **respiration,** has three basic steps:

1. **Pulmonary ventilation** (*pulmon-* = lung), or **breathing,** is the inhalation (inflow) and exhalation (outflow) of air and involves the exchange of air between the atmosphere and the alveoli of the lungs.

2. **External (pulmonary) respiration** is the exchange of gases between the alveoli of the lungs and the blood in pulmonary capillaries across the respiratory membrane. In this process, pulmonary capillary blood gains O_2 and loses CO_2.

3. **Internal (tissue) respiration** is the exchange of gases between blood in systemic capillaries and tissue cells. In this step the blood loses O_2 and gains CO_2. Within cells, the metabolic reactions that consume O_2 and give off CO_2 during the production of ATP are termed *cellular respiration* (discussed in Chapter 25).

In pulmonary ventilation, air flows between the atmosphere and the alveoli of the lungs because of alternating pressure differences created by contraction and relaxation of respiratory muscles. The rate of airflow and the amount of effort needed for breathing are also influenced by alveolar surface tension, compliance of the lungs, and airway resistance.

Pressure Changes during Pulmonary Ventilation

Air moves into the lungs when the air pressure inside the lungs is less than the air pressure in the atmosphere. Air moves out of the lungs when the air pressure inside the lungs is greater than the air pressure in the atmosphere.

Inhalation

Breathing in is called **inhalation (inspiration).** Just before each inhalation, the air pressure inside the lungs is equal to the air pressure of the atmosphere, which at sea level is about 760 millimeters of mercury (mmHg), or 1 atmosphere (atm). For air to flow into the lungs, the pressure inside the alveoli must become lower than the atmospheric pressure. This condition is achieved by increasing the size of the lungs.

The pressure of a gas in a closed container is inversely proportional to the volume of the container. This means that if the size of a closed container is increased, the pressure of the gas inside the container decreases, and that if the size of the container is decreased, then the pressure inside it increases. This inverse relationship between volume and pressure, called **Boyle's law,** may be demonstrated as follows (Figure 23.12): Suppose we place a gas in a cylinder that has a movable piston and a pressure gauge, and that the initial pressure created by the gas molecules striking the wall of the container is 1 atm. If the piston is pushed down, the gas is compressed into a smaller volume, so that the same number of gas molecules strike less wall area. The gauge shows that the pressure doubles as the gas is compressed to half its original volume. In other words, the same number of molecules in half the volume produces twice the pressure. Conversely, if the piston is raised to increase the volume, the pressure decreases. Thus, the pressure of a gas varies inversely with volume.

Differences in pressure caused by changes in lung volume force air into our lungs when we inhale and out when we exhale. For inhalation to occur, the lungs must expand, which increases lung volume and thus decreases the pressure in the lungs to below atmospheric pressure. The first step in expanding the lungs during normal quiet inhalation involves contraction of the main muscles of inhalation, the diaphragm and external intercostals (Figure 23.13).

Figure 23.12 Boyle's law.

🔑 ▭ The volume of a gas varies inversely with its pressure.

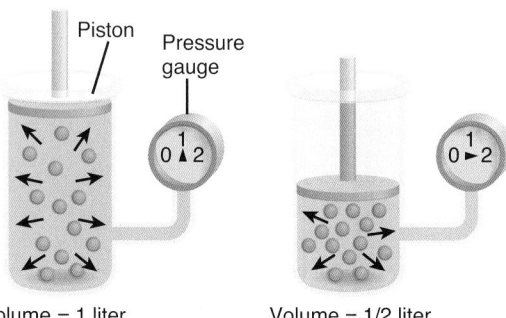

Volume = 1 liter
Pressure = 1 atm

Volume = 1/2 liter
Pressure = 2 atm

❓ **If the volume is decreased from 1 liter to ¼ liter, how would the pressure change?**

Figure 23.13 Muscles of inhalation and exhalation and their actions. The pectoralis minor muscle (not shown here) is illustrated in Figure 11.14a.

During deep, labored breathing, accessory muscles of inhalation (sternocleidomastoids, scalenes, and pectoralis minors) participate.

MUSCLES OF INHALATION
MUSCLES OF EXHALATION

Sternocleidomastoid
Scalenes
External intercostals
Diaphragm

Internal intercostals
External oblique
Internal oblique
Transversus abdominis
Rectus abdominis

(a) Muscles of inhalation and their actions (left); muscles of exhalation and their actions (right)

Sternum:
Exhalation
Inhalation

Diaphragm:
Exhalation
Inhalation

(b) Changes in size of thoracic cavity during inhalation and exhalation

(c) During inhalation, the ribs move upward and outward like the handle on a bucket

Right now, what is the main muscle that is powering your breathing?

The most important muscle of inhalation is the diaphragm, the dome-shaped skeletal muscle that forms the floor of the thoracic cavity. It is innervated by fibers of the phrenic nerves, which emerge from the spinal cord at cervical levels 3, 4, and 5. Contraction of the diaphragm causes it to flatten, lowering its dome. This increases the vertical diameter of the thoracic cavity. During normal quiet inhalation, the diaphragm descends about 1 cm (0.4 in.), producing a pressure difference of 1–3 mmHg and the inhalation of about 500 mL of air. In strenuous breathing, the diaphragm may descend 10 cm (4 in.), which produces a pressure difference of 100 mmHg and the inhalation of 2–3 liters of air. Contraction of the diaphragm is responsible for about 75% of the air that enters the lungs during quiet breathing. Advanced pregnancy, excessive obesity, or confining abdominal clothing can prevent complete descent of the diaphragm.

The next most important muscles of inhalation are the external intercostals. When these muscles contract, they elevate the ribs. As a result, there is an increase in the anteroposterior and lateral diameters of the chest cavity. Contraction of the external intercostals is responsible for about 25% of the air that enters the lungs during normal quiet breathing.

During quiet inhalations, the pressure between the two pleural layers in the pleural cavity, called **intrapleural (intrathoracic) pressure,** is always subatmospheric (lower than atmospheric pressure). Just before inhalation, it is about 4 mmHg less than the atmospheric pressure, or about 756 mmHg at an atmospheric pressure of 760 mmHg (Figure 23.14). As the diaphragm and external intercostals contract and the overall size of the thoracic cavity increases, the volume of the pleural cavity also increases, which causes intrapleural pressure to decrease to about 754 mmHg. During expansion of the thorax, the parietal and visceral pleurae normally adhere tightly because of the subatmospheric pressure between them and because of the surface tension created by their moist adjoining surfaces. As the thoracic cavity expands, the pari-

etal pleura lining the cavity is pulled outward in all directions, and the visceral pleura and lungs are pulled along with it.

As the volume of the lungs increases in this way, the pressure inside the lungs, called the **alveolar (intrapulmonic) pressure,** drops from 760 to 758 mmHg. A pressure difference is thus established between the atmosphere and the alveoli. Because air always flows from a region of higher pressure to a region of lower pressure, inhalation takes place. Air continues to flow into the lungs as long as a pressure difference exists. During deep, forceful inhalations, accessory muscles of inspiration also participate in increasing the size of the thoracic cavity (see Figure 23.13a). The muscles are so named because they make little, if any, contribution during normal quiet inhalation, but during exercise or forced ventilation they may contract vigorously. The accessory muscles of inhalation include the sternocleidomastoid muscles, which elevate the sternum; the scalene muscles, which elevate the

Figure 23.14 Pressure changes in pulmonary ventilation. During inhalation, the diaphragm contracts, the chest expands, the lungs are pulled outward, and alveolar pressure decreases. During exhalation, the diaphragm relaxes, the lungs recoil inward, and alveolar pressure increases, forcing air out of the lungs.

 Air moves into the lungs when alveolar pressure is less than atmospheric pressure, and out of the lungs when alveolar pressure is greater than atmospheric pressure.

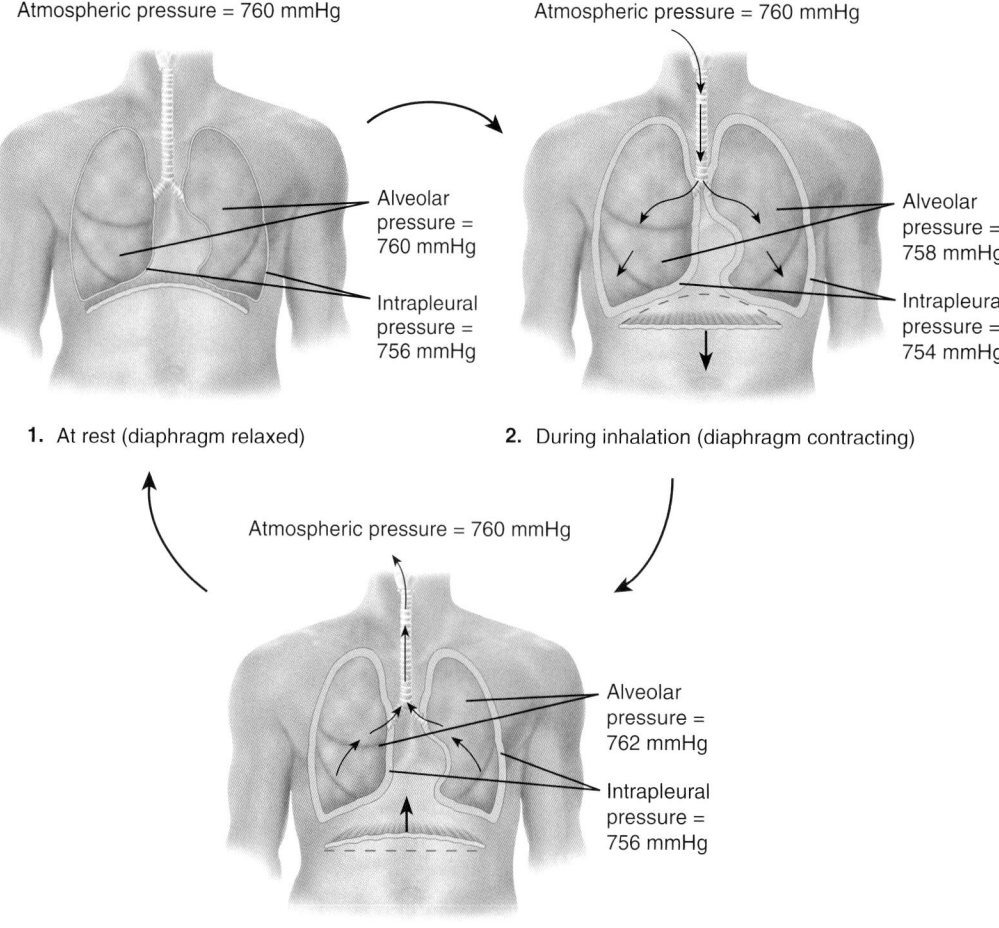

1. At rest (diaphragm relaxed)

Atmospheric pressure = 760 mmHg

Alveolar pressure = 760 mmHg

Intrapleural pressure = 756 mmHg

2. During inhalation (diaphragm contracting)

Atmospheric pressure = 760 mmHg

Alveolar pressure = 758 mmHg

Intrapleural pressure = 754 mmHg

3. During exhalation (diaphragm relaxing)

Atmospheric pressure = 760 mmHg

Alveolar pressure = 762 mmHg

Intrapleural pressure = 756 mmHg

? How does the intrapleural pressure change during a normal, quiet breath?

first two ribs; and the pectoralis minor muscles, which elevate the third through fifth ribs. Because both normal quiet inhalation and inhalation during exercise or forced ventilation involve muscular contraction, the process of inhalation is said to be *active*.

Figure 23.15a summarizes the events of inhalation.

Exhalation

Breathing out, called **exhalation (expiration),** is also due to a pressure gradient, but in this case the gradient is in the opposite direction: The pressure in the lungs is greater than the pressure of the atmosphere. Normal exhalation during quiet breathing, unlike inhalation, is a *passive process* because no muscular contractions are involved. Instead, exhalation results from **elastic recoil** of the chest wall and lungs, both of which have a natural tendency to spring back after they have been stretched. Two inwardly directed forces contribute to elastic recoil: (1) the recoil of elastic fibers that were stretched during inhalation and (2) the inward pull of surface tension due to the film of alveolar fluid.

Exhalation starts when the inspiratory muscles relax. As the diaphragm relaxes, its dome moves superiorly owing to its elasticity. As the external intercostals relax, the ribs are depressed. These movements decrease the vertical, lateral, and anteroposterior diameters of the thoracic cavity, which decreases lung volume. In turn, the alveolar pressure increases to about 762 mmHg. Air then flows from the area of higher pressure in the alveoli to the area of lower pressure in the atmosphere (see Figure 23.14).

Exhalation becomes active only during forceful breathing, as occurs while playing a wind instrument or during exercise. During these times, muscles of exhalation—the abdominals and internal intercostals (see Figure 23.13a)—contract, which increases pressure in the abdominal region and thorax. Contraction of the abdominal muscles moves the inferior ribs downward and compresses the abdominal viscera, thereby forcing the diaphragm superiorly. Contraction of the internal intercostals, which extend inferiorly and posteriorly between adjacent ribs, pulls the ribs inferiorly. Although intrapleural pressure is always less than alveolar pressure, it may briefly exceed atmospheric pressure during a forceful exhalation, such as during a cough.

Figure 23.15b summarizes the events of exhalation.

Other Factors Affecting Pulmonary Ventilation

As you have just learned, air pressure differences drive airflow during inhalation and exhalation. However, three other factors affect the rate of airflow and the ease of pulmonary ventilation: surface tension of the alveolar fluid, compliance of the lungs, and airway resistance.

Surface Tension of Alveolar Fluid

As noted earlier, a thin layer of alveolar fluid coats the luminal surface of alveoli and exerts a force known as **surface tension.** Surface tension arises at all air–water interfaces because the polar

Figure 23.15 Summary of events of inhalation and exhalation.

Inhalation and exhalation are caused by changes in alveolar pressure.

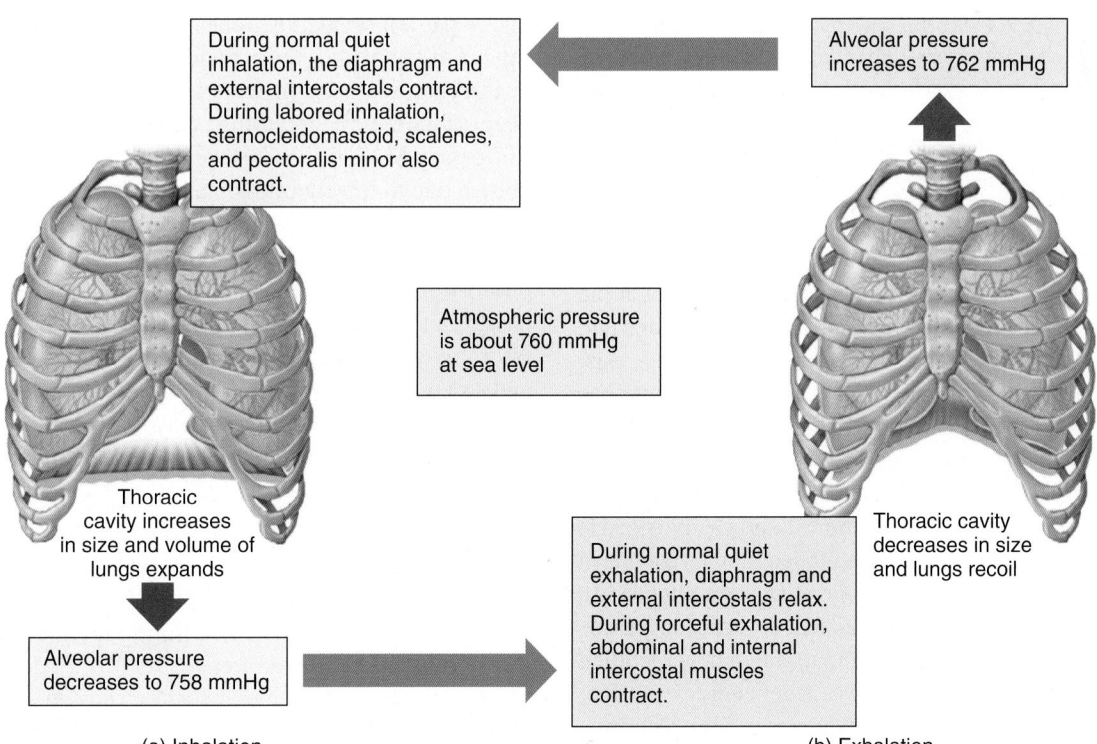

During normal quiet inhalation, the diaphragm and external intercostals contract. During labored inhalation, sternocleidomastoid, scalenes, and pectoralis minor also contract.

Alveolar pressure increases to 762 mmHg

Atmospheric pressure is about 760 mmHg at sea level

Thoracic cavity increases in size and volume of lungs expands

Alveolar pressure decreases to 758 mmHg

During normal quiet exhalation, diaphragm and external intercostals relax. During forceful exhalation, abdominal and internal intercostal muscles contract.

Thoracic cavity decreases in size and lungs recoil

(a) Inhalation

(b) Exhalation

? What is normal atmospheric pressure at sea level?

water molecules are more strongly attracted to each other than they are to gas molecules in the air. When liquid surrounds a sphere of air, as in an alveolus or a soap bubble, surface tension produces an inwardly directed force. Soap bubbles "burst" because they collapse inward due to surface tension. In the lungs, surface tension causes the alveoli to assume the smallest possible diameter. During breathing, surface tension must be overcome to expand the lungs during each inhalation. Surface tension also accounts for two-thirds of lung elastic recoil, which decreases the size of alveoli during exhalation.

The surfactant (a mixture of phospholipids and lipoproteins) present in alveolar fluid reduces its surface tension below the surface tension of pure water. A deficiency of surfactant in premature infants causes *respiratory distress syndrome*, in which the surface tension of alveolar fluid is greatly increased, so that many alveoli collapse at the end of each exhalation. Great effort is then needed at the next inhalation to reopen the collapsed alveoli.

CLINICAL CONNECTION | *Respiratory Distress Syndrome*

Respiratory distress syndrome (RDS) is a breathing disorder of premature newborns in which the alveoli do not remain open due to a lack of surfactant. Recall that surfactant reduces surface tension and is necessary to prevent the collapse of alveoli during exhalation. The more premature the newborn, the greater the chance that RDS will develop. The condition is also more common in infants whose mothers have diabetes and in males; it also occurs more often in European Americans than African Americans. Symptoms of RDS include labored and irregular breathing, flaring of the nostrils during inhalation, grunting during exhalation, and perhaps a blue skin color. Besides the symptoms, RDS is diagnosed on the basis of chest radiographs and a blood test. A newborn with mild RDS may require only supplemental oxygen administered through an oxygen hood or through a tube placed in the nose. In severe cases oxygen may be delivered by continuous positive airway pressure (CPAP) through tubes in the nostrils or a mask on the face. In such cases surfactant may be administered directly into the lungs. •

Compliance of the Lungs

Compliance refers to how much effort is required to stretch the lungs and chest wall. High compliance means that the lungs and chest wall expand easily; low compliance means that they resist expansion. By analogy, a thin balloon that is easy to inflate has high compliance, and a heavy and stiff balloon that takes a lot of effort to inflate has low compliance. In the lungs, compliance is related to two principal factors: elasticity and surface tension. The lungs normally have high compliance and expand easily because elastic fibers in lung tissue are easily stretched and surfactant in alveolar fluid reduces surface tension. Decreased compliance is a common feature in pulmonary conditions that (1) scar lung tissue (for example, tuberculosis), (2) cause lung tissue to become filled with fluid (pulmonary edema), (3) produce a deficiency in surfactant, or (4) impede lung expansion in any way (for example, paralysis of the intercostal muscles). Decreased lung compliance occurs in emphysema (see Disorders: Homeostatic Imbalances

at the end of the chapter) due to destruction of elastic fibers in alveolar walls.

Airway Resistance

Like the flow of blood through blood vessels, the rate of airflow through the airways depends on both the pressure difference and the resistance: Airflow equals the pressure difference between the alveoli and the atmosphere divided by the resistance. The walls of the airways, especially the bronchioles, offer some resistance to the normal flow of air into and out of the lungs. As the lungs expand during inhalation, the bronchioles enlarge because their walls are pulled outward in all directions. Larger-diameter airways have decreased resistance. Airway resistance then increases during exhalation as the diameter of bronchioles decreases. Airway diameter is also regulated by the degree of contraction or relaxation of smooth muscle in the walls of the airways. Signals from the sympathetic division of the autonomic nervous system cause relaxation of this smooth muscle, which results in bronchodilation and decreased resistance.

Any condition that narrows or obstructs the airways increases resistance, so that more pressure is required to maintain the same airflow. The hallmark of asthma or chronic obstructive pulmonary disease (COPD)—emphysema or chronic bronchitis—is increased airway resistance due to obstruction or collapse of airways.

Breathing Patterns and Modified Respiratory Movements

The term for the normal pattern of quiet breathing is **eupnea** (ūp-NĒ-a; *eu-* = good, easy, or normal; *-pnea* = breath). Eupnea can consist of shallow, deep, or combined shallow and deep breathing. A pattern of shallow (chest) breathing, called **costal breathing,** consists of an upward and outward movement of the chest due to contraction of the external intercostal muscles. A pattern of deep (abdominal) breathing, called **diaphragmatic breathing** (dī′-a-frag-MAT-ik), consists of the outward movement of the abdomen due to the contraction and descent of the diaphragm.

Respirations also provide humans with methods for expressing emotions such as laughing, sighing, and sobbing. Moreover, respiratory air can be used to expel foreign matter from the lower air passages through actions such as sneezing and coughing. Respiratory movements are also modified and controlled during talking and singing. Some of the modified respiratory movements that express emotion or clear the airways are listed in Table 23.2. All these movements are reflexes, but some of them also can be initiated voluntarily.

✔ CHECKPOINT

12. What are the basic differences among pulmonary ventilation, external respiration, and internal respiration?
13. Compare what happens during quiet versus forceful pulmonary ventilation.
14. Describe how alveolar surface tension, compliance, and airway resistance affect pulmonary ventilation.
15. Demonstrate the various types of modified respiratory movements.

TABLE 23.2

Modified Respiratory Movements

MOVEMENT	DESCRIPTION
Coughing	A long-drawn and deep inhalation followed by a complete closure of the rima glottidis, which results in a strong exhalation that suddenly pushes the rima glottidis open and sends a blast of air through the upper respiratory passages. Stimulus for this reflex act may be a foreign body lodged in the larynx, trachea, or epiglottis.
Sneezing	Spasmodic contraction of muscles of exhalation that forcefully expels air through the nose and mouth. Stimulus may be an irritation of the nasal mucosa.
Sighing	A long-drawn and deep inhalation immediately followed by a shorter but forceful exhalation.
Yawning	A deep inhalation through the widely opened mouth producing an exaggerated depression of the mandible. It may be stimulated by drowsiness, or someone else's yawning, but the precise cause is unknown.
Sobbing	A series of convulsive inhalations followed by a single prolonged exhalation. The rima glottidis closes earlier than normal after each inhalation so only a little air enters the lungs with each inhalation.
Crying	An inhalation followed by many short convulsive exhalations, during which the rima glottidis remains open and the vocal folds vibrate; accompanied by characteristic facial expressions and tears.
Laughing	The same basic movements as crying, but the rhythm of the movements and the facial expressions usually differ from those of crying. Laughing and crying are sometimes indistinguishable.
Hiccupping	Spasmodic contraction of the diaphragm followed by a spasmodic closure of the rima glottidis, which produces a sharp sound on inhalation. Stimulus is usually irritation of the sensory nerve endings of the gastrointestinal tract.
Valsalva (val-SAL-va) maneuver	Forced exhalation against a closed rima glottidis as may occur during periods of straining while defecating.
Pressurizing the middle ear	The nose and mouth are held closed and air from the lungs is forced through the pharyngotympanic tube into the middle ear. Employed by those snorkeling or scuba diving during descent to equalize the pressure of the middle ear with that of the external environment.

23.3 LUNG VOLUMES AND CAPACITIES

OBJECTIVES
• Explain the difference between tidal volume, inspiratory reserve volume, expiratory reserve volume, and residual volume.
• Differentiate between inspiratory capacity, functional residual capacity, vital capacity, and total lung capacity.

While at rest, a healthy adult averages 12 breaths a minute, with each inhalation and exhalation moving about 500 mL of air into and out of the lungs. The volume of one breath is called the **tidal volume (V_T).** The **minute ventilation (MV)**—the total volume of air inhaled and exhaled each minute—is respiratory rate multiplied by tidal volume:

$$MV = 12 \text{ breaths/min} \times 500 \text{ mL/breath}$$
$$= 6 \text{ liters/min}$$

A lower-than-normal minute ventilation usually is a sign of pulmonary malfunction. The apparatus commonly used to measure the volume of air exchanged during breathing and the respiratory rate is a **spirometer** (spī-ROM-e-ter; *spiro-* = breathe; *-meter* = measuring device) or **respirometer** (res'-pi-ROM-e-ter). The record is called a **spirogram.** Inhalation is recorded as an upward deflection, and exhalation is recorded as a downward deflection (Figure 23.16).

Tidal volume varies considerably from one person to another and in the same person at different times. In a typical adult, about 70% of the tidal volume (350 mL) actually reaches the respiratory zone of the respiratory system—the respiratory bronchioles, alveolar ducts, alveolar sacs, and alveoli—and participates in external respiration. The other 30% (150 mL) remains in the conducting airways of the nose, pharynx, larynx, trachea, bronchi, bronchioles, and terminal bronchioles. Collectively, the conducting airways with air that does not undergo respiratory exchange are known as the **anatomic (respiratory) dead space.** (An easy rule of thumb for determining the volume of your anatomic dead space is that it is about the same in milliliters as your ideal weight in pounds.) Not all of the minute ventilation can be used in gas exchange because some of it remains in the anatomic dead space. The **alveolar ventilation rate** is the volume of air per minute that actually reaches the respiratory zone. In the example just given, alveolar ventilation rate would be 350 mL/breath × 12 breaths/min = 4200 mL/min.

Several other lung volumes are defined relative to forceful breathing. In general, these volumes are larger in males, taller individuals, and younger adults, and smaller in females, shorter individuals, and the elderly. Various disorders also may be diagnosed by comparison of actual and predicted normal values for a

Figure 23.16 Spirogram of lung volumes and capacities. The average values for a healthy adult male and female are indicated, with the values for a female in parentheses. Note that the spirogram is read from right (start of record) to left (end of record).

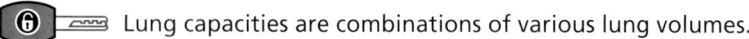 Lung capacities are combinations of various lung volumes.

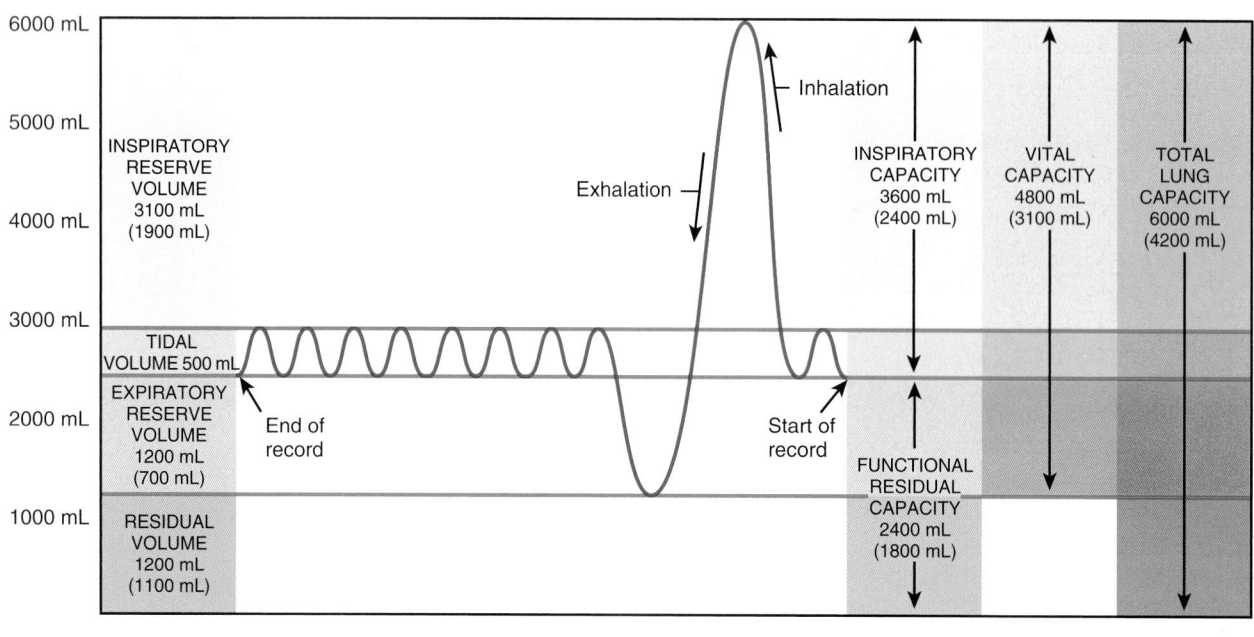

LUNG VOLUMES   LUNG CAPACITIES

❓ If you breathe in as deeply as possible and then exhale as much air as you can, which lung capacity have you demonstrated?

patient's gender, height, and age. The values given here are averages for young adults.

By taking a very deep breath, you can inhale a good deal more than 500 mL. This additional inhaled air, called the **inspiratory reserve volume,** is about 3100 mL in an average adult male and 1900 mL in an average adult female (Figure 23.16). Even more air can be inhaled if inhalation follows forced exhalation. If you inhale normally and then exhale as forcibly as possible, you should be able to push out considerably more air in addition to the 500 mL of tidal volume. The extra 1200 mL in males and 700 mL in females is called the **expiratory reserve volume.** The $FEV_{1.0}$ is the **forced expiratory volume in 1 second,** the volume of air that can be exhaled from the lungs in 1 second with maximal effort following a maximal inhalation. Typically, chronic obstructive pulmonary disease (COPD) greatly reduces $FEV_{1.0}$ because COPD increases airway resistance.

Even after the expiratory reserve volume is exhaled, considerable air remains in the lungs because the subatmospheric intrapleural pressure keeps the alveoli slightly inflated, and some air also remains in the noncollapsible airways. This volume, which cannot be measured by spirometry, is called the **residual volume** (re-ZID-u-al) and amounts to about 1200 mL in males and 1100 mL in females.

If the thoracic cavity is opened, the intrapleural pressure rises to equal the atmospheric pressure and forces out some of the residual volume. The air remaining is called the **minimal volume.** Minimal volume provides a medical and legal tool for determining whether a baby is born dead (stillborn) or died after birth. The presence of minimal volume can be demonstrated by placing a piece of lung in water and observing if it floats. Fetal lungs contain no air, so the lung of a stillborn baby will not float in water.

Lung capacities are combinations of specific lung volumes (Figure 23.16). **Inspiratory capacity** is the sum of tidal volume and inspiratory reserve volume (500 mL + 3100 mL = 3600 mL in males and 500 mL + 1900 mL = 2400 mL in females). **Functional residual capacity** is the sum of residual volume and expiratory reserve volume (1200 mL + 1200 mL = 2400 mL in males and 1100 mL + 700 mL = 1800 mL in females). **Vital capacity** is the sum of inspiratory reserve volume, tidal volume, and expiratory reserve volume (4800 mL in males and 3100 mL in females). Finally, **total lung capacity** is the sum of vital capacity and residual volume (4800 mL + 1200 mL = 6000 mL in males and 3100 mL + 1100 mL = 4200 mL in females).

✔ CHECKPOINT

16. What is a spirometer?
17. What is the difference between a lung volume and a lung capacity?
18. How is minute ventilation calculated?
19. Define alveolar ventilation rate and $FEV_{1.0}$.

23.4 EXCHANGE OF OXYGEN AND CARBON DIOXIDE

● **OBJECTIVES**

- Explain Dalton's law and Henry's law.
- Describe the exchange of oxygen and carbon dioxide in external and internal respiration.

The exchange of oxygen and carbon dioxide between alveolar air and pulmonary blood occurs via passive diffusion, which is governed by the behavior of gases as described by two gas laws, Dalton's law and Henry's law. Dalton's law is important for understanding how gases move down their pressure differences by diffusion, and Henry's law helps explain how the solubility of a gas relates to its diffusion.

Gas Laws: Dalton's Law and Henry's Law

According to **Dalton's law,** each gas in a mixture of gases exerts its own pressure as if no other gases were present. The pressure of a specific gas in a mixture is called its *partial pressure* (P_x); the subscript is the chemical formula of the gas. The total pressure of the mixture is calculated simply by adding all the partial pressures. Atmospheric air is a mixture of gases—nitrogen (N_2), oxygen (O_2), argon (Ar), carbon dioxide (CO_2), variable amounts of water vapor (H_2O), plus other gases present in small quantities. Atmospheric pressure is the sum of the pressures of all these gases:

$$\text{Atmospheric pressure (760 mmHg)}$$
$$= P_{N_2} + P_{O_2} + P_{Ar} + P_{H_2O} + P_{CO_2} + P_{\text{other gases}}$$

We can determine the partial pressure exerted by each component in the mixture by multiplying the percentage of the gas in the mixture by the total pressure of the mixture. Atmospheric air is 78.6% nitrogen, 20.9% oxygen, 0.93% argon, 0.04% carbon dioxide, and 0.06% other gases; and a variable amount of water vapor is also present. The amount of water varies from practically 0% over a desert to 4% over the ocean, about 0.4% on a cool, dry day. Thus, the partial pressures of the gases in inhaled air are as follows:

$$P_{N_2} = 0.786 \times 760 \text{ mmHg} = 597.4 \text{ mmHg}$$
$$P_{O_2} = 0.209 \times 760 \text{ mmHg} = 158.8 \text{ mmHg}$$
$$P_{Ar} = 0.0009 \times 760 \text{ mmHg} = 0.7 \text{ mmHg}$$
$$P_{H_2O} = 0.003 \times 760 \text{ mmHg} = 2.3 \text{ mmHg}$$
$$P_{CO_2} = 0.0004 \times 760 \text{ mmHg} = 0.3 \text{ mmHg}$$
$$P_{\text{other gases}} = 0.0006 \times 760 \text{ mmHg} = 0.5 \text{ mmHg}$$
$$\text{Total} = 760.0 \text{ mmHg}$$

These partial pressures determine the movement of O_2 and CO_2 between the atmosphere and lungs, between the lungs and blood, and between the blood and body cells. Each gas diffuses across a permeable membrane from the area where its partial pressure is greater to the area where its partial pressure is less. The greater the difference in partial pressure, the faster the rate of diffusion.

Compared with inhaled air, alveolar air has less O_2 (13.6% versus 20.9%) and more CO_2 (5.2% versus 0.04%) for two reasons. First, gas exchange in the alveoli increases the CO_2 content and decreases the O_2 content of alveolar air. Second, when air is inhaled it becomes humidified as it passes along the moist mucosal linings. As water vapor content of the air increases, the relative percentage that is O_2 decreases. In contrast, exhaled air contains more O_2 than alveolar air (16% versus 13.6%) and less CO_2 (4.5% versus 5.2%) because some of the exhaled air was in the anatomic dead space and did not participate in gas exchange. Exhaled air is a mixture of alveolar air and inhaled air that was in the anatomic dead space.

Henry's law states that the quantity of a gas that will dissolve in a liquid is proportional to the partial pressure of the gas and its solubility. In body fluids, the ability of a gas to stay in solution is greater when its partial pressure is higher and when it has a high solubility in water. The higher the partial pressure of a gas over a liquid and the higher the solubility, the more gas will stay in solution. In comparison to oxygen, much more CO_2 is dissolved in blood plasma because the solubility of CO_2 is 24 times greater than that of O_2. Even though the air we breathe contains mostly N_2, this gas has no known effect on bodily functions, and at sea level pressure very little of it dissolves in blood plasma because its solubility is very low.

An everyday experience gives a demonstration of Henry's law. You have probably noticed that a soft drink makes a hissing sound when the top of the container is removed, and bubbles rise to the surface for some time afterward. The gas dissolved in carbonated beverages is CO_2. Because the soft drink is bottled or canned under high pressure and capped, the CO_2 remains dissolved as long as the container is unopened. Once you remove the cap, the pressure decreases and the gas begins to bubble out of solution.

Henry's law explains two conditions resulting from changes in the solubility of nitrogen in body fluids. Even though the air we breathe contains about 79% nitrogen, this gas has no known effect on bodily functions, and very little of it dissolves in blood plasma because of its low solubility at sea level pressure. As the total air pressure increases, the partial pressures of all its gases increase. When a scuba diver breathes air under high pressure, the nitrogen in the mixture can have serious negative effects. Because the partial pressure of nitrogen is higher in a mixture of compressed air than in air at sea level pressure, a considerable amount of nitrogen dissolves in plasma and interstitial fluid. Excessive amounts of dissolved nitrogen may produce giddiness and other symptoms similar to alcohol intoxication. The condition is called **nitrogen narcosis** or "rapture of the deep."

If a diver comes to the surface slowly, the dissolved nitrogen can be eliminated by exhaling it. However, if the ascent is too rapid, nitrogen comes out of solution too quickly and forms gas bubbles in the tissues, resulting in **decompression sickness** (the **bends**). The effects of decompression sickness typically result from bubbles in nervous tissue and can be mild or severe, depending on the number of bubbles formed. Symptoms include joint pain, especially in the arms and legs, dizziness, shortness of breath, extreme fatigue, paralysis, and unconsciousness.

A major clinical application of Henry's law is **hyperbaric oxygenation** (*hyper-* = over; *-baros* = pressure), the use of pressure to cause more O_2 to dissolve in the blood. It is an effective technique in treating patients infected by anaerobic bacteria, such as those that cause tetanus and gangrene. (Anaerobic bacteria cannot live in the presence of free O_2.) A person undergoing hyperbaric oxygenation is placed in a hyperbaric chamber, which contains O_2 at a pressure greater than 1 atmosphere (760 mmHg). As body tissues pick up the O_2, the bacteria are killed. Hyperbaric chambers may also be used for treating certain heart disorders, carbon monoxide poisoning, gas embolisms, crush injuries, cerebral edema, certain hard-to-treat bone infections caused by anaerobic bacteria, smoke inhalation, near-drowning, asphyxia, vascular insufficiencies, and burns. •

External and Internal Respiration

External respiration or **pulmonary gas exchange** is the diffusion of O_2 from air in the alveoli of the lungs to blood in pulmonary capillaries and the diffusion of CO_2 in the opposite direction (Figure 23.17a). External respiration in the lungs converts **deoxygenated blood** (depleted of some O_2) coming from the right side of the heart into **oxygenated blood** (saturated with O_2) that returns to the left side of the heart (see Figure 21.29). As blood flows through the pulmonary capillaries, it picks up O_2 from alveolar air and unloads CO_2 into alveolar air. Although this process is commonly called an "exchange" of gases, each gas diffuses independently from the area where its partial pressure is higher to the area where its partial pressure is lower.

As Figure 23.17a shows, O_2 diffuses from alveolar air, where its partial pressure is 105 mmHg, into the blood in pulmonary capillaries, where P_{O_2} is only 40 mmHg in a resting person. If you have been exercising, the P_{O_2} will be even lower because contracting muscle fibers are using more O_2. Diffusion continues until the P_{O_2} of pulmonary capillary blood increases to match the P_{O_2} of alveolar air, 105 mmHg. Because blood leaving pulmonary capillaries near alveolar air spaces mixes with a small volume of blood that has flowed through conducting portions of the respiratory system, where gas exchange does not occur, the P_{O_2} of blood in the pulmonary veins is slightly less than the P_{O_2} in pulmonary capillaries, about 100 mmHg.

While O_2 is diffusing from alveolar air into deoxygenated blood, CO_2 is diffusing in the opposite direction. The P_{CO_2} of deoxygenated blood is 45 mmHg in a resting person, and the P_{CO_2} of alveolar air is 40 mmHg. Because of this difference in P_{CO_2}, carbon dioxide diffuses from deoxygenated blood into the alveoli until the P_{CO_2} of the blood decreases to 40 mmHg. Exhalation keeps alveolar P_{CO_2} at 40 mmHg. Oxygenated blood returning to the left side of the heart in the pulmonary veins thus has a P_{CO_2} of 40 mmHg.

The number of capillaries near alveoli in the lungs is very large, and blood flows slowly enough through these capillaries that it picks up a maximal amount of O_2. During vigorous exercise, when cardiac output is increased, blood flows more rapidly through both the systemic and pulmonary circulations. As a result, blood's transit time in the pulmonary capillaries is shorter. Still, the P_{O_2} of blood in the pulmonary veins normally reaches 100 mmHg. In diseases that decrease the rate of gas diffusion, however, the blood may not come into full equilibrium with alveolar air, especially during exercise. When this happens, the P_{O_2} declines and P_{CO_2} rises in systemic arterial blood.

The left ventricle pumps oxygenated blood into the aorta and through the systemic arteries to systemic capillaries. The exchange of O_2 and CO_2 between systemic capillaries and tissue cells is called **internal respiration** or **systemic gas exchange** (Figure 23.17b). As O_2 leaves the bloodstream, oxygenated blood is converted into deoxygenated blood. Unlike external respiration, which occurs only in the lungs, internal respiration occurs in tissues throughout the body.

The P_{O_2} of blood pumped into systemic capillaries is higher (100 mmHg) than the P_{O_2} in tissue cells (40 mmHg at rest) because the cells constantly use O_2 to produce ATP. Due to this pressure difference, oxygen diffuses out of the capillaries into tissue cells and blood P_{O_2} drops to 40 mmHg by the time the blood exits systemic capillaries.

While O_2 diffuses from the systemic capillaries into tissue cells, CO_2 diffuses in the opposite direction. Because tissue cells are constantly producing CO_2, the P_{CO_2} of cells (45 mmHg at rest) is higher than that of systemic capillary blood (40 mmHg). As a result, CO_2 diffuses from tissue cells through interstitial fluid into systemic capillaries until the P_{CO_2} in the blood increases to 45 mmHg. The deoxygenated blood then returns to the heart and is pumped to the lungs for another cycle of external respiration.

In a person at rest, tissue cells, on average, need only 25% of the available O_2 in oxygenated blood; despite its name, deoxygenated blood retains 75% of its O_2 content. During exercise, more O_2 diffuses from the blood into metabolically active cells, such as contracting skeletal muscle fibers. Active cells use more O_2 for ATP production, causing the O_2 content of deoxygenated blood to drop below 75%.

The *rate* of pulmonary and systemic gas exchange depends on several factors.

• ***Partial pressure difference of the gases.*** Alveolar P_{O_2} must be higher than blood P_{O_2} for oxygen to diffuse from alveolar air into the blood. The rate of diffusion is faster when the difference between P_{O_2} in alveolar air and pulmonary capillary blood is larger; diffusion is slower when the difference is smaller. The differences between P_{O_2} and P_{CO_2} in alveolar air versus pulmonary blood increase during exercise. The larger partial pressure differences accelerate the rates of gas diffusion. The partial pressures of O_2 and CO_2 in alveolar air also depend on the rate of airflow into and out of the lungs. Certain drugs (such as morphine) slow ventilation, thereby decreasing the amount of O_2 and CO_2 that can be exchanged between alveolar air and blood. With increasing altitude, the total atmospheric pressure decreases, as does the partial pressure of O_2—from 159 mmHg at sea level, to 110 mmHg at 10,000 ft, to 73 mmHg at 20,000 ft. Although O_2 still is 20.9% of the

Figure 23.17 Changes in partial pressures of oxygen and carbon dioxide (in mmHg) during external and internal respiration.

Gases diffuse from areas of higher partial pressure to areas of lower partial pressure.

Atmospheric air:
$P_{O_2} = 159$ mmHg
$P_{CO_2} = 0.3$ mmHg

CO_2 exhaled

O_2 inhaled

Alveoli

CO_2 O_2

Alveolar air:
$P_{O_2} = 105$ mmHg
$P_{CO_2} = 40$ mmHg

Pulmonary capillaries

(a) External respiration:
pulmonary gas
exchange

To lungs

To left atrium

Deoxygenated blood:
$P_{O_2} = 40$ mmHg
$P_{CO_2} = 45$ mmHg

Oxygenated blood:
$P_{O_2} = 100$ mmHg
$P_{CO_2} = 40$ mmHg

To right atrium

To tissue cells

(b) Internal respiration:
systemic gas
exchange

Systemic capillaries

CO_2

O_2

Systemic tissue cells:
$P_{O_2} = 40$ mmHg
$P_{CO_2} = 45$ mmHg

? What causes oxygen to enter pulmonary capillaries from alveoli and to enter tissue cells from systemic capillaries?

total, the P_{O_2} of inhaled air decreases with increasing altitude. Alveolar P_{O_2} decreases correspondingly, and O_2 diffuses into the blood more slowly. The common signs and symptoms of **high altitude sickness**—shortness of breath, headache, fatigue, insomnia, nausea, and dizziness—are due to a lower level of oxygen in the blood.

- *Surface area available for gas exchange.* As you learned earlier in the chapter, the surface area of the alveoli is huge (about 70 m² or 750 ft²). In addition, many capillaries surround each alveolus, so many that as much as 900 mL of blood is able to participate in gas exchange at any instant. Any pulmonary disorder that decreases the functional surface area of the respiratory membranes decreases the rate of external respiration. In emphysema (see Disorders: Homeostatic Imbalances at the end of the chapter), for example, alveolar walls disintegrate, so surface area is smaller than normal and pulmonary gas exchange is slowed.

- *Diffusion distance.* The respiratory membrane is very thin, so diffusion occurs quickly. Also, the capillaries are so narrow that the red blood cells must pass through them in single file, which minimizes the diffusion distance from an alveolar air space to hemoglobin inside red blood cells. Buildup of interstitial fluid between alveoli, as occurs in pulmonary edema (see Disorders: Homeostatic Imbalances at the end of the chapter), slows the rate of gas exchange because it increases diffusion distance.

- *Molecular weight and solubility of the gases.* Because O_2 has a lower molecular weight than CO_2, it could be expected to diffuse across the respiratory membrane about 1.2 times faster. However, the solubility of CO_2 in the fluid portions of the respiratory membrane is about 24 times greater than that of O_2. Taking both of these factors into account, net outward CO_2 diffusion occurs 20 times more rapidly than net inward O_2 diffusion. Consequently, when diffusion is slower than normal—for example, in emphysema or pulmonary edema—O_2 insufficiency (hypoxia) typically occurs before there is significant retention of CO_2 (hypercapnia).

✔ CHECKPOINT

20. Distinguish between Dalton's law and Henry's law and give a practical application of each.
21. How does the partial pressure of oxygen change as altitude changes?
22. What are the diffusion paths of oxygen and carbon dioxide during external and internal respiration?
23. What factors affect the rate of diffusion of oxygen and carbon dioxide?

23.5 TRANSPORT OF OXYGEN AND CARBON DIOXIDE

◉ OBJECTIVE

- Describe how the blood transports oxygen and carbon dioxide.

As you have already learned, the blood transports gases between the lungs and body tissues. When O_2 and CO_2 enter the blood, certain chemical reactions occur that aid in gas transport and gas exchange.

Oxygen Transport

Oxygen does not dissolve easily in water, so only about 1.5% of inhaled O_2 is dissolved in blood plasma, which is mostly water. About 98.5% of blood O_2 is bound to hemoglobin in red blood cells (Figure 23.18). Each 100 mL of oxygenated blood contains the equivalent of 20 mL of gaseous O_2. Using the percentages just given, the amount dissolved in the plasma is 0.3 mL and the amount bound to hemoglobin is 19.7 mL.

The heme portion of hemoglobin contains four atoms of iron, each capable of binding to a molecule of O_2 (see Figure 19.4b, c). Oxygen and hemoglobin bind in an easily reversible reaction to form **oxyhemoglobin:**

$$\text{Hb} \; + \; O_2 \; \underset{\substack{\text{Dissociation} \\ \text{of } O_2}}{\overset{\substack{\text{Binding of } O_2}}{\rightleftharpoons}} \; \text{Hb—}O_2$$

Reduced hemoglobin (deoxyhemoglobin)   Oxygen   Oxyhemoglobin

The 98.5% of the O_2 that is bound to hemoglobin is trapped inside RBCs, so only the dissolved O_2 (1.5%) can diffuse out of tissue capillaries into tissue cells. Thus, it is important to understand the factors that promote O_2 binding to and dissociation (separation) from hemoglobin.

The Relationship between Hemoglobin and Oxygen Partial Pressure

The most important factor that determines how much O_2 binds to hemoglobin is the P_{O_2}; the higher the P_{O_2}, the more O_2 combines with Hb. When reduced hemoglobin (Hb) is completely converted to oxyhemoglobin (Hb—O_2), the hemoglobin is said to be **fully saturated;** when hemoglobin consists of a mixture of Hb and Hb—O_2, it is **partially saturated.** The **percent saturation of hemoglobin** expresses the average saturation of hemoglobin with oxygen. For instance, if each hemoglobin molecule has bound two O_2 molecules, then the hemoglobin is 50% saturated because each Hb can bind a maximum of four O_2.

The relationship between the percent saturation of hemoglobin and P_{O_2} is illustrated in the oxygen–hemoglobin dissociation curve in Figure 23.19. Note that when the P_{O_2} is high, hemoglobin binds with large amounts of O_2 and is almost 100% saturated. When P_{O_2} is low, hemoglobin is only partially saturated. In other words, the greater the P_{O_2}, the more O_2 will bind to hemoglobin, until all the available hemoglobin molecules are saturated. Therefore, in pulmonary capillaries, where P_{O_2} is high, a lot of O_2 binds to hemoglobin. In tissue capillaries, where the P_{O_2} is lower, hemoglobin does not hold as much O_2, and the dissolved O_2 is unloaded via diffusion into tissue cells (see Figure 23.18b). Note that hemoglobin is still 75% saturated with O_2 at a P_{O_2} of 40 mmHg, the average P_{O_2} of tissue cells in a person at rest. This is the basis for the earlier statement that only 25% of the available O_2 unloads from hemoglobin and is used by tissue cells under resting conditions.

When the P_{O_2} is between 60 and 100 mmHg, hemoglobin is 90% or more saturated with O_2 (Figure 23.19). Thus, blood picks up a nearly full load of O_2 from the lungs even when the P_{O_2} of alveolar air is as low as 60 mmHg. The Hb–P_{O_2} curve explains why people can still perform well at high altitudes or when they

Figure 23.18 Transport of oxygen (O_2) and carbon dioxide (CO_2) in the blood.

Most O_2 is transported by hemoglobin as oxyhemoglobin (Hb—O_2) within red blood cells; most CO_2 is transported in blood plasma as bicarbonate ions (HCO_3^-).

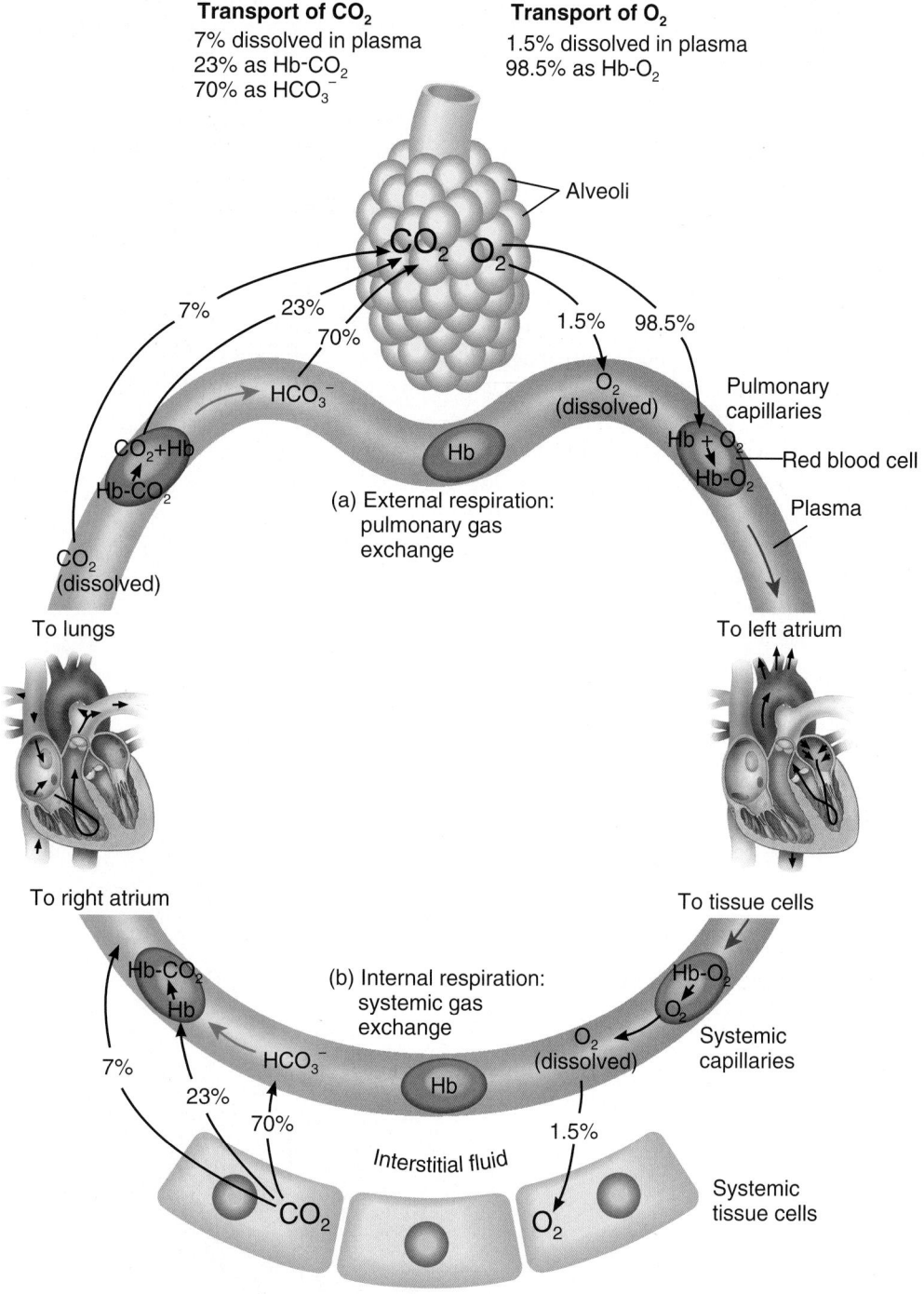

Transport of CO_2
7% dissolved in plasma
23% as Hb-CO_2
70% as HCO_3^-

Transport of O_2
1.5% dissolved in plasma
98.5% as Hb-O_2

? **What is the most important factor that determines how much O_2 binds to hemoglobin?**

have certain cardiac and pulmonary diseases, even though P_{O_2} may drop as low as 60 mmHg. Note also in the curve that at a considerably lower P_{O_2} of 40 mmHg, hemoglobin is still 75% saturated with O_2. However, oxygen saturation of Hb drops to 35% at 20 mmHg. Between 40 and 20 mmHg, large amounts of O_2 are released from hemoglobin in response to only small decreases in P_{O_2}. In active tissues such as contracting muscles, P_{O_2} may drop well below 40 mmHg. Then, a large percentage of the O_2 is released from hemoglobin, providing more O_2 to metabolically active tissues.

Other Factors Affecting the Affinity of Hemoglobin for Oxygen

Although P_{O_2} is the most important factor that determines the percent O_2 saturation of hemoglobin, several other factors influence the tightness or **affinity** with which hemoglobin binds O_2. In effect, these factors shift the entire curve either to the left (higher affinity) or to the right (lower affinity). The changing affinity of hemoglobin for O_2 is another example of how homeostatic mechanisms adjust body activities to cellular needs. Each one makes sense if you keep in mind that metabolically active tissue cells need O_2 and produce acids, CO_2, and heat as wastes. The following four factors affect the affinity of hemoglobin for O_2:

1. *Acidity (pH).* As acidity increases (pH decreases), the affinity of hemoglobin for O_2 decreases, and O_2 dissociates more readily from hemoglobin (Figure 23.20a). In other words, increasing acidity enhances the unloading of oxygen from hemoglobin. The main acids produced by metabolically active tissues are lactic acid and carbonic acid. When pH decreases, the entire oxygen–hemoglobin dissociation curve shifts to the right; at any given P_{O_2}, Hb is less saturated with O_2, a change termed the **Bohr effect** (BŌR). The Bohr effect works both ways: An increase in H^+ in blood causes O_2 to unload from hemoglobin, and the binding of O_2 to hemoglobin causes unloading of H^+ from hemoglobin. The explanation for the Bohr effect is that hemoglobin can act as a buffer for hydrogen ions

(H^+). But when H^+ ions bind to amino acids in hemoglobin, they alter its structure slightly, decreasing its oxygen-carrying capacity. Thus, lowered pH drives O_2 off hemoglobin, making more O_2 available for tissue cells. By contrast, elevated pH increases the affinity of hemoglobin for O_2 and shifts the oxygen–hemoglobin dissociation curve to the left.

Figure 23.20 Oxygen–hemoglobin dissociation curves showing the relationship of (a) pH and (b) P_{CO_2} to hemoglobin saturation at normal body temperature. As pH increases or P_{CO_2} decreases, O_2 combines more tightly with hemoglobin, so that less is available to tissues. The broken lines emphasize these relationships.

 As pH decreases or P_{CO_2} increases, the affinity of hemoglobin for O_2 declines, so less O_2 combines with hemoglobin and more is available to tissues.

(a) Effect of pH on affinity of hemoglobin for oxygen

(b) Effect of P_{CO_2} on affinity of hemoglobin for oxygen

Figure 23.19 Oxygen-hemoglobin dissociation curve showing the relationship between hemoglobin saturation and P_{O_2} at normal body temperature.

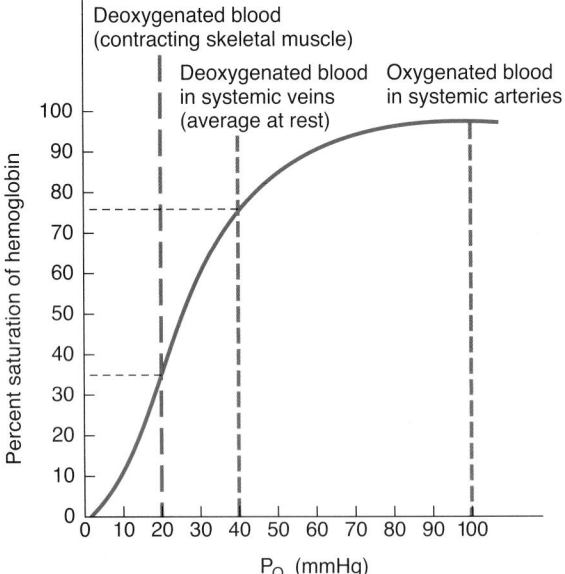 As P_{O_2} increases, more O_2 combines with hemoglobin.

? What point on the curve represents blood in your pulmonary veins right now? In your pulmonary veins if you were jogging?

? In comparison to the value when you are sitting, is the affinity of your hemoglobin for O_2 higher or lower when you are exercising? How does this benefit you?

2. *Partial pressure of carbon dioxide.* CO_2 also can bind to hemoglobin, and the effect is similar to that of H^+ (shifting the curve to the right). As P_{CO_2} rises, hemoglobin releases O_2 more readily (Figure 23.20b). P_{CO_2} and pH are related factors because low blood pH (acidity) results from high P_{CO_2}. As CO_2 enters the blood, much of it is temporarily converted to carbonic acid (H_2CO_3), a reaction catalyzed by an enzyme in red blood cells called *carbonic anhydrase (CA)*:

$$CO_2 + H_2O \underset{CA}{\rightleftharpoons} H_2CO_3 \rightleftharpoons H^+ + HCO_3^-$$

Carbon dioxide · Water · Carbonic acid · Hydrogen ion · Bicarbonate ion

The carbonic acid thus formed in red blood cells dissociates into hydrogen ions and bicarbonate ions. As the H^+ concentration increases, pH decreases. Thus, an increased P_{CO_2} produces a more acidic environment, which helps release O_2 from hemoglobin. During exercise, lactic acid—a by-product of anaerobic metabolism within muscles—also decreases blood pH. Decreased P_{CO_2} (and elevated pH) shifts the saturation curve to the left.

3. *Temperature.* Within limits, as temperature increases, so does the amount of O_2 released from hemoglobin (Figure 23.21). Heat is a by-product of the metabolic reactions of all cells, and the heat released by contracting muscle fibers tends to raise body temperature. Metabolically active cells require more O_2 and liberate more acids and heat. The acids and heat in turn promote release of O_2 from oxyhemoglobin. Fever produces a similar result. In contrast, during hypothermia (lowered body temperature) cellular metabolism slows, the need for O_2 is re-

duced, and more O_2 remains bound to hemoglobin (a shift to the left in the saturation curve).

4. *BPG.* A substance found in red blood cells called **2,3-bisphosphoglycerate (BPG)** (bis′-fos-fō-GLIS-e-rāt), previously called diphosphoglycerate (DPG), decreases the affinity of hemoglobin for O_2 and thus helps unload O_2 from hemoglobin. BPG is formed in red blood cells when they break down glucose to produce ATP in a process called glycolysis (described in Section 25.3). When BPG combines with hemoglobin by binding to the terminal amino groups of the two beta globin chains, the hemoglobin binds O_2 less tightly at the heme group sites. The greater the level of BPG, the more O_2 is unloaded from hemoglobin. Certain hormones, such as thyroxine, human growth hormone, epinephrine, norepinephrine, and testosterone, increase the formation of BPG. The level of BPG also is higher in people living at higher altitudes.

Oxygen Affinity of Fetal and Adult Hemoglobin

Fetal hemoglobin (Hb-F) differs from **adult hemoglobin (Hb-A)** in structure and in its affinity for O_2. Hb-F has a higher affinity for O_2 because it binds BPG less strongly. Thus, when P_{O_2} is low, Hb-F can carry up to 30% more O_2 than maternal Hb-A (Figure 23.22). As the maternal blood enters the placenta, O_2 is readily transferred to fetal blood. This is very important because the O_2 saturation in maternal blood in the placenta is quite low, and the fetus might suffer hypoxia were it not for the greater affinity of fetal hemoglobin for O_2.

Figure 23.21 Oxygen–hemoglobin dissociation curves showing the effect of temperature changes.

As temperature increases, the affinity of hemoglobin for O_2 decreases.

Is O_2 more available or less available to tissue cells when you have a fever? Why?

Figure 23.22 Oxygen–hemoglobin dissociation curves comparing fetal and maternal hemoglobin.

Fetal hemoglobin has a higher affinity for O_2 than does adult hemoglobin.

The P_{O_2} of placental blood is about 40 mmHg. What are the O_2 saturations of maternal and fetal hemoglobin at this P_{O_2}?

Carbon monoxide (CO) is a colorless and odorless gas found in exhaust fumes from automobiles, gas furnaces, and space heaters and in tobacco smoke. It is a by-product of the combustion of carbon-containing materials such as coal, gas, and wood. CO binds to the heme group of hemoglobin, just as O_2 does, except that the binding of carbon monoxide to hemoglobin is over 200 times as strong as the binding of O_2 to hemoglobin. Thus, at a concentration as small as 0.1% ($P_{CO} = 0.5$ mmHg), CO will combine with half the available hemoglobin molecules and reduce the oxygen-carrying capacity of the blood by 50%. Elevated blood levels of CO cause **carbon monoxide poisoning,** which can cause the lips and oral mucosa to appear bright, cherry-red (the color of hemoglobin with carbon monoxide bound to it). Without prompt treatment, carbon monoxide poisoning is fatal. It is possible to rescue a victim of CO poisoning by administering pure oxygen, which speeds up the separation of carbon monoxide from hemoglobin. •

Carbon Dioxide Transport

Under normal resting conditions, each 100 mL of deoxygenated blood contains the equivalent of 53 mL of gaseous CO_2, which is transported in the blood in three main forms (see Figure 23.18):

1. *Dissolved CO_2.* The smallest percentage—about 7%—is dissolved in blood plasma. On reaching the lungs, it diffuses into alveolar air and is exhaled.

2. *Carbamino compounds.* A somewhat higher percentage, about 23%, combines with the amino groups of amino acids and proteins in blood to form **carbamino compounds** (kar-BAM-i-nō). Because the most prevalent protein in blood is hemoglobin (inside red blood cells), most of the CO_2 transported in this manner is bound to hemoglobin. The main CO_2 binding sites are the terminal amino acids in the two alpha and two beta globin chains. Hemoglobin that has bound CO_2 is termed **carbaminohemoglobin (Hb—CO_2):**

$$\text{Hb} + \underset{\text{Carbon dioxide}}{CO_2} \rightleftharpoons \underset{\text{Carbaminohemoglobin}}{\text{Hb—}CO_2}$$
$$\underset{\text{Hemoglobin}}{}$$

The formation of carbaminohemoglobin is greatly influenced by P_{CO_2}. For example, in tissue capillaries P_{CO_2} is relatively high, which promotes formation of carbaminohemoglobin. But in pulmonary capillaries, P_{CO_2} is relatively low, and the CO_2 readily splits apart from globin and enters the alveoli by diffusion.

3. *Bicarbonate ions.* The greatest percentage of CO_2—about 70%—is transported in blood plasma as **bicarbonate ions** (HCO_3^-) (bī′-KAR-bo-nāt). As CO_2 diffuses into systemic capillaries and enters red blood cells, it reacts with water in the presence of the enzyme carbonic anhydrase (CA) to form carbonic acid, which dissociates into H^+ and HCO_3^-:

$$\underset{\substack{\text{Carbon}\\\text{dioxide}}}{CO_2} + \underset{\text{Water}}{H_2O} \overset{CA}{\rightleftharpoons} \underset{\substack{\text{Carbonic}\\\text{acid}}}{H_2CO_3} \rightleftharpoons \underset{\substack{\text{Hydrogen}\\\text{ion}}}{H^+} + \underset{\substack{\text{Bicarbonate}\\\text{ion}}}{HCO_3^-}$$

Thus, as blood picks up CO_2, HCO_3^- accumulates inside RBCs. Some HCO_3^- moves out into the blood plasma, down its concentration gradient. In exchange, chloride ions (Cl^-) move from plasma into the RBCs. This exchange of negative ions, which maintains the electrical balance between blood plasma and RBC cytosol, is known as the **chloride shift** (Figure 23.23b). The net effect of these reactions is that CO_2 is removed from tissue cells and transported in blood plasma as HCO_3^-. As blood passes through pulmonary capillaries in the lungs, all these reactions reverse and CO_2 is exhaled.

The amount of CO_2 that can be transported in the blood is influenced by the percent saturation of hemoglobin with oxygen. The lower the amount of oxyhemoglobin (Hb—O_2), the higher the CO_2-carrying capacity of the blood, a relationship known as the **Haldane effect.** Two characteristics of deoxyhemoglobin give rise to the Haldane effect: (1) Deoxyhemoglobin binds to and thus transports more CO_2 than does Hb—O_2. (2) Deoxyhemoglobin also buffers more H^+ than does Hb—O_2, thereby removing H^+ from solution and promoting conversion of CO_2 to HCO_3^- via the reaction catalyzed by carbonic anhydrase.

Summary of Gas Exchange and Transport in Lungs and Tissues

Deoxygenated blood returning to the pulmonary capillaries in the lungs (Figure 23.23a) contains CO_2 dissolved in blood plasma, CO_2 combined with globin as carbaminohemoglobin (Hb—CO_2), and CO_2 incorporated into HCO_3^- within RBCs. The RBCs have also picked up H^+, some of which binds to and therefore is buffered by hemoglobin (Hb—H). As blood passes through the pulmonary capillaries, molecules of CO_2 dissolved in blood plasma and CO_2 that dissociates from the globin portion of hemoglobin diffuse into alveolar air and are exhaled. At the same time, inhaled O_2 is diffusing from alveolar air into RBCs and is binding to hemoglobin to form oxyhemoglobin (Hb—O_2). Carbon dioxide also is released from HCO_3^- when H^+ combines with HCO_3^- inside RBCs. The H_2CO_3 formed from this reaction then splits into CO_2, which is exhaled, and H_2O. As the concentration of HCO_3^- declines inside RBCs in pulmonary capillaries, HCO_3^- diffuses in from the blood plasma, in exchange for Cl^-. In sum, oxygenated blood leaving the lungs has increased O_2 content and decreased amounts of CO_2 and H^+. In systemic capillaries, as cells use O_2 and produce CO_2, the chemical reactions reverse (Figure 23.23b).

✔ **CHECKPOINT**

24. In a resting person, how many O_2 molecules are attached to each hemoglobin molecule, on average, in blood in the pulmonary arteries? In blood in the pulmonary veins?
25. What is the relationship between hemoglobin and P_{O_2}? How do temperature, H^+, P_{CO_2}, and BPG influence the affinity of Hb for O_2?
26. Why can hemoglobin unload more oxygen as blood flows through capillaries of metabolically active tissues, such as skeletal muscle during exercise, than is unloaded at rest?

Figure 23.23 **Summary of chemical reactions that occur during gas exchange.** (a) As carbon dioxide (CO_2) is exhaled, hemoglobin (Hb) inside red blood cells in pulmonary capillaries unloads CO_2 and picks up O_2 from alveolar air. Binding of O_2 to Hb—H releases hydrogen ions (H^+). Bicarbonate ions (HCO_3^-) pass into the RBC and bind to released H^+, forming carbonic acid (H_2CO_3). The H_2CO_3 dissociates into water (H_2O) and CO_2, and the CO_2 diffuses from blood into alveolar air. To maintain electrical balance, a chloride ion (Cl^-) exits the RBC for each HCO_3^- that enters (reverse chloride shift). (b) CO_2 diffuses out of tissue cells that produce it and enters red blood cells, where some of it binds to hemoglobin, forming carbaminohemoglobin (Hb—CO_2). This reaction causes O_2 to dissociate from oxyhemoglobin (Hb—O_2). Other molecules of CO_2 combine with water to produce bicarbonate ions (HCO_3^-) and hydrogen ions (H^+). As Hb buffers H^+, the Hb releases O_2 (Bohr effect). To maintain electrical balance, a chloride ion (Cl^-) enters the RBC for each HCO_3^- that exits (chloride shift).

🔑 Hemoglobin inside red blood cells transports O_2, CO_2, and H^+.

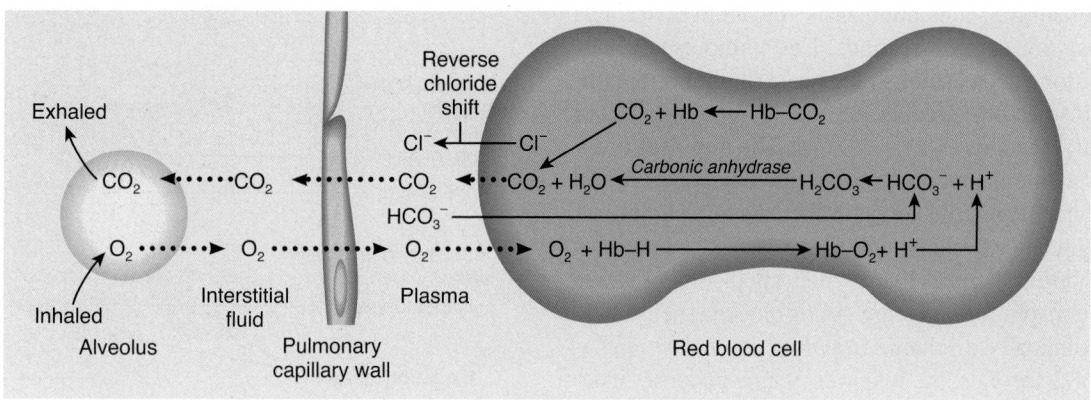

(a) Exchange of O_2 and CO_2 in pulmonary capillaries (external respiration)

(b) Exchange of O_2 and CO_2 in systemic capillaries (internal respiration)

❓ **Would you expect the concentration of HCO_3^- to be higher in blood plasma taken from a systemic artery or a systemic vein?**

23.6 CONTROL OF RESPIRATION

● **OBJECTIVES**

• Explain how the nervous system controls breathing.
• List the factors that can alter the rate and depth of breathing.

At rest, about 200 mL of O_2 is used each minute by body cells. During strenuous exercise, however, O_2 use typically increases 15- to 20-fold in normal healthy adults, and as much as 30-fold in elite endurance-trained athletes. Several mechanisms help match respiratory effort to metabolic demand.

Respiratory Center

The size of the thorax is altered by the action of the respiratory muscles, which contract as a result of nerve impulses transmitted to them from centers in the brain and relax in the absence of nerve impulses. These nerve impulses are sent from clusters of neurons located bilaterally in the medulla oblongata and pons of the brain stem. This widely dispersed group of neurons, collectively called the **respiratory center,** can be divided into three areas on the basis of their functions: (1) the medullary rhythmicity area in the medulla oblongata; (2) the pneumotaxic area in the pons; and (3) the apneustic area, also in the pons (Figure 23.24).

Medullary Rhythmicity Area

The function of the **medullary rhythmicity area** (rith-MIS-i-tē) is to control the basic rhythm of respiration. There are inspiratory and expiratory areas within the medullary rhythmicity area. Figure 23.25 shows the relationships of the inspiratory and expiratory areas during normal quiet breathing and forceful breathing.

During quiet breathing, inhalation lasts for about 2 seconds and exhalation lasts for about 3 seconds. Nerve impulses generated in the **inspiratory area** establish the basic rhythm of breathing. While the inspiratory area is active, it generates nerve impulses for about 2 seconds (Figure 23.25a). The impulses propagate to the external intercostal muscles via intercostal nerves and to the diaphragm via the phrenic nerves. When the nerve impulses reach the diaphragm and external intercostal muscles, the muscles contract and inhalation occurs. Even when all incoming nerve connections to the inspiratory area are cut or blocked, neurons in this area still rhythmically discharge impulses that cause inhalation. At the end of 2 seconds, the inspiratory area becomes inactive and nerve impulses cease. With no impulses arriving, the diaphragm and external intercostal muscles relax for about 3 seconds, allowing passive elastic recoil of the lungs and thoracic wall. Then, the cycle repeats.

The neurons of the **expiratory area** remain inactive during quiet breathing. However, during forceful breathing nerve impulses from the inspiratory area activate the expiratory area

Figure 23.24 Locations of areas of the respiratory center.

The respiratory center is composed of neurons in the medullary rhythmicity area in the medulla oblongata plus the pneumotaxic and apneustic areas in the pons.

Sagittal section of brain stem

? **Which area contains neurons that are active and then inactive in a repeating cycle?**

Figure 23.25 Roles of the medullary rhythmicity area in controlling (a) the basic rhythm of respiration and (b) forceful breathing.

 During normal quiet breathing, the expiratory area is inactive; during forceful breathing, the inspiratory area activates the expiratory area.

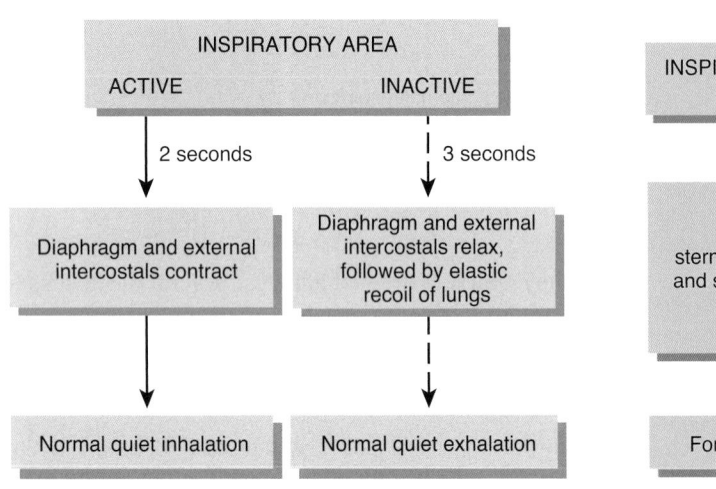

(a) During normal quiet breathing

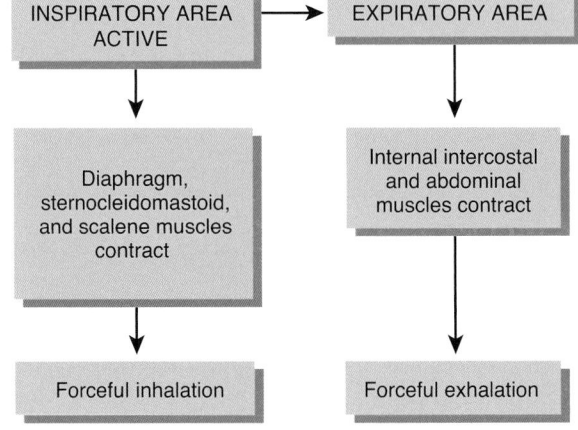

(b) During forceful breathing

? **Which nerves convey impulses from the respiratory center to the diaphragm?**

(Figure 23.25b). Impulses from the expiratory area cause contraction of the internal intercostal and abdominal muscles, which decreases the size of the thoracic cavity and causes forceful exhalation.

Pneumotaxic Area

Although the medullary rhythmicity area controls the basic rhythm of respiration, other sites in the brain stem help coordinate the transition between inhalation and exhalation. One of these sites is the **pneumotaxic area** (noo-mō-TAK-sik; *pneumo-* = air or breath; *-taxic* = arrangement) in the upper pons (see Figure 23.24), which transmits inhibitory impulses to the inspiratory area. The major effect of these nerve impulses is to help turn off the inspiratory area before the lungs become too full of air. In other words, the impulses shorten the duration of inhalation. When the pneumotaxic area is more active, breathing rate is more rapid.

Apneustic Area

Another part of the brain stem that coordinates the transition between inhalation and exhalation is the **apneustic area** (ap-NOO-stik) in the lower pons (see Figure 23.24). This area sends stimulatory impulses to the inspiratory area that activate it and prolong inhalation. The result is a long, deep inhalation. When the pneumotaxic area is active, it overrides signals from the apneustic area.

Regulation of the Respiratory Center

The basic rhythm of respiration set and coordinated by the inspiratory area can be modified in response to inputs from other brain regions, receptors in the peripheral nervous system, and other factors.

Cortical Influences on Respiration

Because the cerebral cortex has connections with the respiratory center, we can voluntarily alter our pattern of breathing. We can even refuse to breathe at all for a short time. Voluntary control is protective because it enables us to prevent water or irritating gases from entering the lungs. The ability to not breathe, however, is limited by the buildup of CO_2 and H^+ in the body. When P_{CO_2} and H^+ concentrations increase to a certain level, the inspiratory area is strongly stimulated, nerve impulses are sent along the phrenic and intercostal nerves to inspiratory muscles, and breathing resumes, whether the person wants it to or not. It is impossible for small children to kill themselves by voluntarily holding their breath, even though many have tried in order to get their way. If breath is held long enough to cause fainting, breathing resumes when consciousness is lost. Nerve impulses from the hypothalamus and limbic system also stimulate the respiratory center, allowing emotional stimuli to alter respirations as, for example, in laughing and crying.

Chemoreceptor Regulation of Respiration

Certain chemical stimuli modulate how quickly and how deeply we breathe. The respiratory system functions to maintain proper levels of CO_2 and O_2 and is very responsive to changes in the lev-

els of these gases in body fluids. We introduced sensory neurons that are responsive to chemicals, called **chemoreceptors** (kē'-mō-rē-SEP-tors), in Chapter 21. Chemoreceptors in two locations of the respiratory system monitor levels of CO_2, H^+, and O_2 and provide input to the respiratory center (Figure 23.26). **Central chemoreceptors** are located in or near the medulla oblongata in

Figure 23.26 Locations of peripheral chemoreceptors.

Chemoreceptors are sensory neurons that respond to changes in the levels of certain chemicals in the body.

Medulla oblongata

Sensory axons in glossopharyngeal nerve (cranial nerve IX)

Internal carotid artery

Carotid body

External carotid artery

Carotid sinus

Common carotid artery

Sensory axons in vagus nerve (cranial nerve X)

Arch of aorta

Aortic bodies

Heart

? **Which chemicals stimulate peripheral chemoreceptors?**

the *central* nervous system. They respond to changes in H^+ concentration or P_{CO_2}, or both, in cerebrospinal fluid. **Peripheral chemoreceptors** are located in the **aortic bodies,** clusters of chemoreceptors located in the wall of the arch of the aorta, and in the **carotid bodies,** which are oval nodules in the wall of the left and right common carotid arteries where they divide into the internal and external carotid arteries. (The chemoreceptors of the aortic bodies are located close to the aortic baroreceptors, and the carotid bodies are located close to the carotid sinus baroreceptors. Recall from Chapter 21 that baroreceptors are sensory receptors that monitor blood pressure.) These chemoreceptors are part of the *peripheral* nervous system and are sensitive to changes in P_{O_2}, H^+, and P_{CO_2} in the blood. Axons of sensory neurons from the aortic bodies are part of the vagus (X) nerves, and those from the carotid bodies are part of the right and left glossopharyngeal (IX) nerves. Recall from Chapter 17 that olfactory receptors for the sense of smell and gustatory receptor cells for the sense of taste are also chemoreceptors. Both respond to external stimuli.

Because CO_2 is lipid-soluble, it easily diffuses into cells where, in the presence of carbonic anhydrase, it combines with water (H_2O) to form carbonic acid (H_2CO_3). Carbonic acid quickly breaks down into H^+ and HCO_3^-. Thus, an increase in CO_2 in the blood causes an increase in H^+ inside cells, and a decrease in CO_2 causes a decrease in H^+.

Normally, the P_{CO_2} in arterial blood is 40 mmHg. If even a slight increase in P_{CO_2} occurs—a condition called **hypercapnia** (hī′-per-KAP-nē-a) or **hypercarbia**—the central chemoreceptors are stimulated and respond vigorously to the resulting increase in H^+ level. The peripheral chemoreceptors also are stimulated by both the high P_{CO_2} and the rise in H^+. In addition, the peripheral chemoreceptors (but not the central chemoreceptors) respond to a deficiency of O_2. When P_{O_2} in arterial blood falls from a normal level of 100 mmHg but is still above 50 mmHg, the peripheral chemoreceptors are stimulated. Severe deficiency of O_2 depresses activity of the central chemoreceptors and inspiratory area, which then do not respond well to any inputs and send fewer impulses to the muscles of inhalation. As the breathing rate decreases or breathing ceases altogether, P_{O_2} falls lower and lower, establishing a positive feedback cycle with a possibly fatal result.

The chemoreceptors participate in a negative feedback system that regulates the levels of CO_2, O_2, and H^+ in the blood (Figure 23.27). As a result of increased P_{CO_2}, decreased pH (increased H^+), or decreased P_{O_2}, input from the central and peripheral chemoreceptors causes the inspiratory area to become highly active, and the rate and depth of breathing increase. Rapid and deep breathing, called **hyperventilation,** allows the inhalation of more O_2 and exhalation of more CO_2 until P_{CO_2} and H^+ are lowered to normal.

If arterial P_{CO_2} is lower than 40 mmHg—a condition called **hypocapnia** or **hypocarbia**—the central and peripheral chemoreceptors are not stimulated, and stimulatory impulses are not sent to the inspiratory area. As a result, the area sets its own moderate pace until CO_2 accumulates and the P_{CO_2} rises to 40 mmHg. The inspiratory center is more strongly stimulated when P_{CO_2} is rising above normal than when P_{O_2} is falling below normal. As a result,

Figure 23.27 Regulation of breathing in response to changes in blood P_{CO_2}, P_{O_2}, and pH (H^+) via negative feedback control.

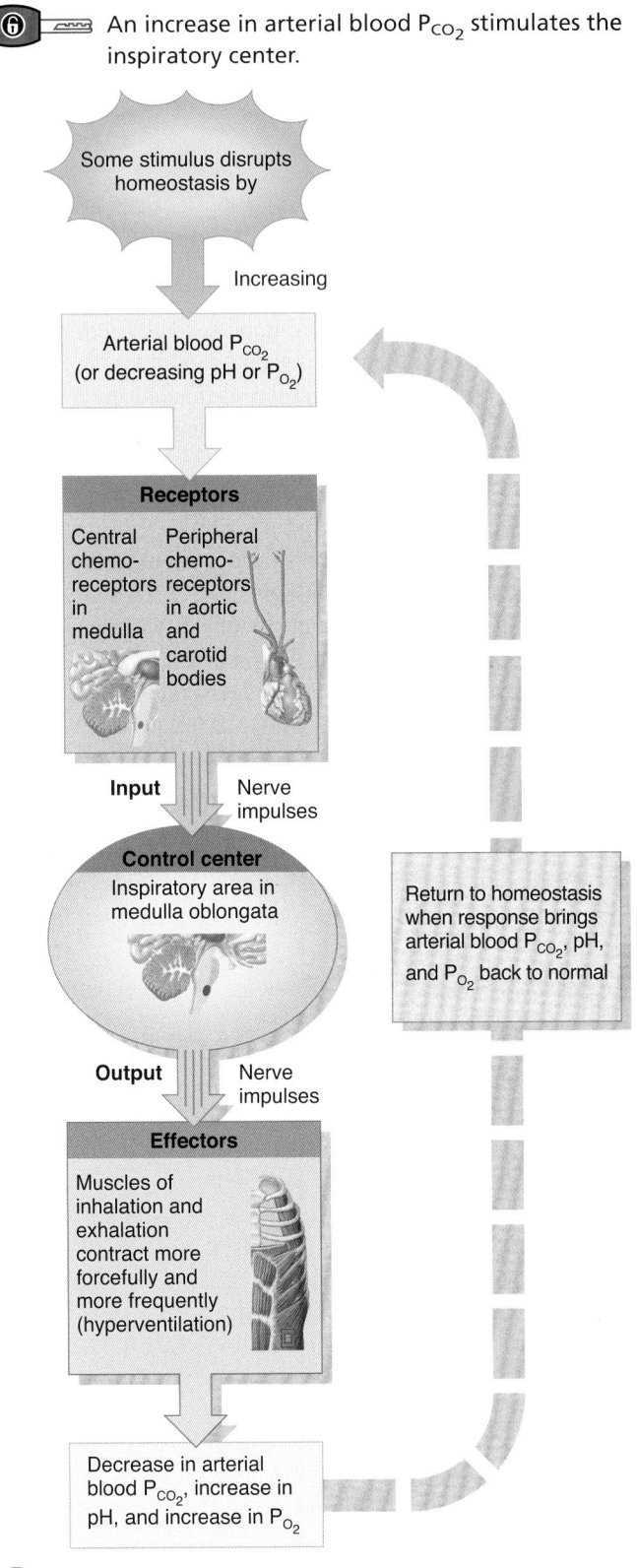

An increase in arterial blood P_{CO_2} stimulates the inspiratory center.

? What is the normal arterial blood P_{CO_2}?

people who hyperventilate voluntarily and cause hypocapnia can hold their breath for an unusually long period. Swimmers were once encouraged to hyperventilate just before diving in to compete. However, this practice is risky because the O_2 level may fall dangerously low and cause fainting before the P_{CO_2} rises high enough to stimulate inhalation. If you faint on land you may suffer bumps and bruises, but if you faint in the water you could drown.

CLINICAL CONNECTION | *Hypoxia*

Hypoxia (hī-POK-sē-a; *hypo-* = under) is a deficiency of O_2 at the tissue level. Based on the cause, we can classify hypoxia into four types, as follows:

1. **Hypoxic hypoxia** is caused by a low P_{O_2} in arterial blood as a result of high altitude, airway obstruction, or fluid in the lungs.

2. In **anemic hypoxia**, too little functioning hemoglobin is present in the blood, which reduces O_2 transport to tissue cells. Among the causes are hemorrhage, anemia, and failure of hemoglobin to carry its normal complement of O_2, as in carbon monoxide poisoning.

3. In **ischemic hypoxia** (is-KĒ-mik), blood flow to a tissue is so reduced that too little O_2 is delivered to it, even though P_{O_2} and oxyhemoglobin levels are normal.

4. In **histotoxic hypoxia** (his-tō-TOK-sik), the blood delivers adequate O_2 to tissues, but the tissues are unable to use it properly because of the action of some toxic agent. One cause is cyanide poisoning, in which cyanide blocks an enzyme required for the use of O_2 during ATP synthesis. •

Proprioceptor Stimulation of Respiration

As soon as you start exercising, your rate and depth of breathing increase, even before changes in P_{O_2}, P_{CO_2}, or H^+ level occur. The main stimulus for these quick changes in respiratory effort is input from proprioceptors, which monitor movement of joints and muscles. Nerve impulses from the proprioceptors stimulate the inspiratory area of the medulla oblongata. At the same time, axon collaterals (branches) of upper motor neurons that originate in the primary motor cortex (precentral gyrus) also feed excitatory impulses into the inspiratory area.

The Inflation Reflex

Similar to those in the blood vessels, stretch-sensitive receptors called **baroreceptors** (bar′-ō-re-SEP-tors) or **stretch receptors** are located in the walls of bronchi and bronchioles. When these receptors become stretched during overinflation of the lungs, nerve impulses are sent along the vagus (X) nerves to the inspiratory and apneustic areas. In response, the inspiratory area is inhibited directly, and the apneustic area is inhibited from activating the inspiratory area. As a result, exhalation begins. As air leaves the lungs during exhalation, the lungs deflate and the stretch receptors are no longer stimulated. Thus, the inspiratory and apneustic areas are no longer inhibited, and a new inhalation begins. This reflex, referred to as the **inflation (Hering–Breuer) reflex** (HER-ing BROY-er), is mainly a protective mechanism for preventing excessive inflation of the lungs rather than a key component in the normal regulation of respiration.

Other Influences on Respiration

Other factors that contribute to regulation of respiration include the following:

- *Limbic system stimulation.* Anticipation of activity or emotional anxiety may stimulate the limbic system, which then sends excitatory input to the inspiratory area, increasing the rate and depth of ventilation.

- *Temperature.* An increase in body temperature, as occurs during a fever or vigorous muscular exercise, increases the rate of respiration. A decrease in body temperature decreases respiratory rate. A sudden cold stimulus (such as plunging into cold water) causes temporary **apnea** (AP-nē-a; *a-* = without; *-pnea* = breath), an absence of breathing.

- *Pain.* A sudden, severe pain brings about brief apnea, but a prolonged somatic pain increases respiratory rate. Visceral pain may slow the rate of respiration.

- *Stretching the anal sphincter muscle.* This action increases the respiratory rate and is sometimes used to stimulate respiration in a newborn baby or a person who has stopped breathing.

- *Irritation of airways.* Physical or chemical irritation of the pharynx or larynx brings about an immediate cessation of breathing followed by coughing or sneezing.

- *Blood pressure.* The carotid and aortic baroreceptors that detect changes in blood pressure have a small effect on respiration. A sudden rise in blood pressure decreases the rate of respiration, and a drop in blood pressure increases the respiratory rate.

Table 23.3 summarizes the stimuli that affect the rate and depth of ventilation.

✔CHECKPOINT

27. How does the medullary rhythmicity area regulate respiration?

28. How are the apneustic and pneumotaxic areas related to the control of respiration?

29. How do the cerebral cortex, levels of CO_2 and O_2, proprioceptors, inflation reflex, temperature changes, pain, and irritation of the airways modify respiration?

23.7 EXERCISE AND THE RESPIRATORY SYSTEM

◉ OBJECTIVE

- Describe the effects of exercise on the respiratory system.

The respiratory and cardiovascular systems make adjustments in response to both the intensity and duration of exercise. The effects of exercise on the heart are discussed in Chapter 20. Here we focus on how exercise affects the respiratory system.

Recall that the heart pumps the same amount of blood to the lungs as to all the rest of the body. Thus, as cardiac output rises, the blood flow to the lungs, termed **pulmonary perfusion,** increases as well. In addition, the O_2 **diffusing capacity,** a

TABLE 23.3	
Summary of Stimuli That Affect Ventilation Rate and Depth	
STIMULI THAT INCREASE VENTILATION RATE AND DEPTH	**STIMULI THAT DECREASE VENTILATION RATE AND DEPTH**
Voluntary hyperventilation controlled by cerebral cortex and anticipation of activity by stimulation of limbic system.	Voluntary hypoventilation controlled by cerebral cortex.
Increase in arterial blood P_{CO_2} above 40 mmHg (causes an increase in H^+) detected by peripheral and central chemoreceptors.	Decrease in arterial blood P_{CO_2} below 40 mmHg (causes a decrease in H^+) detected by peripheral and central chemoreceptors.
Decrease in arterial blood P_{O_2} from 105 mmHg to 50 mmHg.	Decrease in arterial blood P_{O_2} below 50 mmHg.
Increased activity of proprioceptors.	Decreased activity of proprioceptors.
Increase in body temperature.	Decrease in body temperature (decreases respiration rate), sudden cold stimulus (causes apnea).
Prolonged pain.	Severe pain (causes apnea).
Decrease in blood pressure.	Increase in blood pressure.
Stretching of anal sphincter.	Irritation of pharynx or larynx by touch or chemicals (causes brief apnea followed by coughing or sneezing).

measure of the rate at which O_2 can diffuse from alveolar air into the blood, may increase threefold during maximal exercise because more pulmonary capillaries become maximally perfused. As a result, there is a greater surface area available for diffusion of O_2 into pulmonary blood capillaries.

When muscles contract during exercise, they consume large amounts of O_2 and produce large amounts of CO_2. During vigorous exercise, O_2 consumption and pulmonary ventilation both increase dramatically. At the onset of exercise, an abrupt increase in pulmonary ventilation is followed by a more gradual increase. With moderate exercise, the increase is due mostly to an increase in the depth of ventilation rather than to increased breathing rate. When exercise is more strenuous, the frequency of breathing also increases.

The abrupt increase in ventilation at the start of exercise is due to *neural* changes that send excitatory impulses to the inspiratory area in the medulla oblongata. These changes include (1) anticipation of the activity, which stimulates the limbic system; (2) sensory impulses from proprioceptors in muscles, tendons, and joints; and (3) motor impulses from the primary motor cortex (precentral gyrus). The more gradual increase in ventilation during moderate exercise is due to *chemical* and *physical* changes in the bloodstream, including (1) slightly decreased P_{O_2}, due to increased O_2 consumption; (2) slightly increased P_{CO_2}, due to increased CO_2 production by contracting muscle fibers; and (3) increased temperature, due to liberation of more heat as more O_2 is utilized. During strenuous exercise, HCO_3^- buffers H^+ released by lactic acid in a reaction that liberates CO_2, which further increases P_{CO_2}.

At the end of an exercise session, an abrupt decrease in pulmonary ventilation is followed by a more gradual decline to the resting level. The initial decrease is due mainly to changes in neural factors when movement stops or slows; the more gradual phase reflects the slower return of blood chemistry levels and temperature to the resting state.

CLINICAL CONNECTION | *Effects of Smoking on the Respiratory System*

Smoking may cause a person to become easily "winded" during even moderate exercise because several factors decrease respiratory efficiency in smokers: (1) Nicotine constricts terminal bronchioles, which decreases airflow into and out of the lungs. (2) Carbon monoxide in smoke binds to hemoglobin and reduces its oxygen-carrying capability. (3) Irritants in smoke cause increased mucus secretion by the mucosa of the bronchial tree and swelling of the mucosal lining, both of which impede airflow into and out of the lungs. (4) Irritants in smoke also inhibit the movement of cilia and destroy cilia in the lining of the respiratory system. Thus, excess mucus and foreign debris are not easily removed, which further adds to the difficulty in breathing. The irritants can also convert the normal respiratory epithelium into stratified squamous epithelium, which lacks cilia and goblet cells. (5) With time, smoking leads to destruction of elastic fibers in the lungs and is the prime cause of emphysema (described in Disorders: Homeostatic Imbalances at the end of the chapter). These changes cause collapse of small bronchioles and trapping of air in alveoli at the end of exhalation. The result is less efficient gas exchange. •

✔ **CHECKPOINT**

30. How does exercise affect the inspiratory area?

23.8 DEVELOPMENT OF THE RESPIRATORY SYSTEM

◉ **OBJECTIVE**

• Describe the development of the respiratory system.

The development of the mouth and pharynx are discussed in Chapter 24. Here we consider the development of the other structures of the respiratory system that you learned about in this chapter.

At about 4 weeks of development, the respiratory system begins as an outgrowth of the foregut (precursor of some digestive organs) just anterior to the pharynx. This outgrowth is called the **respiratory diverticulum** (dī-ver-TIK-ū-lum) or **lung bud** (Figure 23.28). The **endoderm** lining the respiratory diverticulum gives rise to the epithelium and glands of the trachea, bronchi, and alveoli. **Mesoderm** surrounding the respiratory diverticulum gives rise to the connective tissue, cartilage, and smooth muscle of these structures.

The epithelial lining of the *larynx* develops from the **endoderm** of the respiratory diverticulum; the cartilages and muscles originate from the **fourth** and **sixth pharyngeal arches,** swellings on the surface of the embryo.

As the respiratory diverticulum elongates, its distal end enlarges to form a globular **tracheal bud,** which gives rise to the *trachea.* Soon after, the tracheal bud divides into **bronchial buds,** which branch repeatedly and develop with the *bronchi.* By 24 weeks, 17 orders of branches have formed and *respiratory bronchioles* have developed.

During weeks 6 to 16, all major elements of the *lungs* have formed, except for those involved in gaseous exchange (respiratory bronchioles, alveolar ducts, and alveoli). Since respiration is not possible at this stage, fetuses born during this time cannot survive.

During weeks 16 to 26, lung tissue becomes highly vascular and respiratory bronchioles, alveolar ducts, and some primitive alveoli develop. Although it is possible for a fetus born near the end of this period to survive if given intensive care, death frequently occurs due to the immaturity of the respiratory and other systems.

From 26 weeks to birth, many more primitive alveoli develop; they consist of type I alveolar cells (main sites of gaseous exchange) and type II surfactant-producing cells. Blood capillaries also establish close contact with the primitive alveoli. Recall that surfactant is necessary to lower surface tension of alveolar fluid and thus reduce the tendency of alveoli to collapse on exhalation. Although surfactant production begins by 20 weeks, it is present in only small quantities. Amounts sufficient to permit survival of a premature (preterm) infant are not produced until 26 to 28 weeks' gestation. Infants born before 26 to 28 weeks are at high risk of respiratory distress syndrome (RDS), in which the alveoli collapse during exhalation and must be reinflated during inhalation (see Clinical Connection: Respiratory Distress Syndrome in Section 23.2).

At about 30 weeks, mature alveoli develop. However, it is estimated that only about one-sixth of the full complement of alveoli develop before birth; the remainder develop after birth during the first 8 years.

As the lungs develop, they acquire their *pleural sacs.* The *visceral pleura* and the *parietal pleura* develop from **mesoderm.** The space between the pleural layers is the *pleural cavity.*

During development, breathing movements of the fetus cause the aspiration of fluid into the lungs. This fluid is a mixture of amniotic fluid, mucus from bronchial glands, and surfactant. At birth, the lungs are about half-filled with fluid. When breathing begins at birth, most of the fluid is rapidly reabsorbed by blood and lymph capillaries and a small amount is expelled through the nose and mouth during delivery.

Figure 23.28 Development of the bronchial tubes and lungs.

🔑 The respiratory system develops from endoderm and mesoderm.

Pharynx
Respiratory diverticulum
Tracheal bud
Esophagus

Pharynx
Trachea
Bronchial buds
Esophagus

Fourth week

Left primary bronchus
Trachea
Left secondary bronchi
Right primary bronchus
Right secondary bronchi

Fifth week

Right tertiary bronchi
Left tertiary bronchi

Sixth week

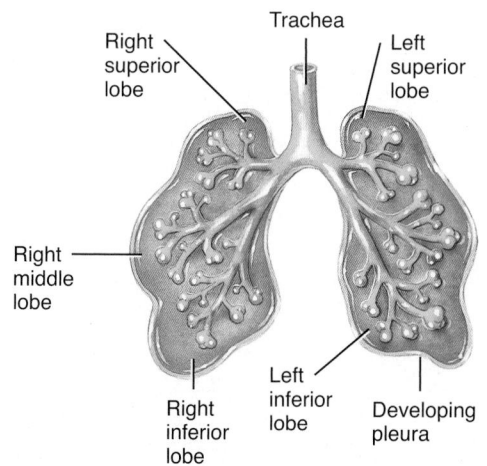

Trachea
Right superior lobe
Left superior lobe
Right middle lobe
Right inferior lobe
Left inferior lobe
Developing pleura

Eighth week

❓ **When does the respiratory system begin to develop in an embryo?**

✔**CHECKPOINT**

31. What structures develop from the laryngotracheal bud?

Focus on Homeostasis

BODY SYSTEM	CONTRIBUTION OF THE RESPIRATORY SYSTEM

For all body systems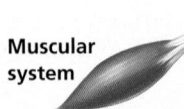
Provides oxygen and removes carbon dioxide. Helps adjust pH of body fluids through exhalation of carbon dioxide.

Muscular system
Increased rate and depth of breathing support increased activity of skeletal muscles during exercise.

Nervous system
Nose contains receptors for sense of smell (olfaction). Vibrations of air flowing across vocal folds produce sounds for speech.

Endocrine system
Angiotensin-converting enzyme (ACE) in lungs catalyzes formation of the hormone angiotensin II from angiotensin I.

Cardiovascular system
During inhalations, respiratory pump aids return of venous blood to the heart.

Lymphatic system and immunity
Hairs in nose, cilia and mucus in trachea, bronchi, and smaller airways, and alveolar macrophages contribute to nonspecific resistance to disease. Pharynx (throat) contains lymphatic tissue (tonsils). Respiratory pump (during inhalation) promotes flow of lymph.

Digestive system
Forceful contraction of respiratory muscles can assist in defecation.

Urinary system
Together, respiratory and urinary systems regulate pH of body fluids.

Reproductive systems
Increased rate and depth of breathing support activity during sexual intercourse. Internal respiration provides oxygen to developing fetus.

THE RESPIRATORY SYSTEM

23.9 AGING AND THE RESPIRATORY SYSTEM

◉ **OBJECTIVE**

- Describe the effects of aging on the respiratory system.

With advancing age, the airways and tissues of the respiratory tract, including the alveoli, become less elastic and more rigid; the chest wall becomes more rigid as well. The result is a decrease in lung capacity. In fact, vital capacity (the maximum amount of air that can be expired after maximal inhalation) can decrease as much as 35% by age 70. A decrease in blood level of O_2, decreased activity of alveolar macrophages, and diminished ciliary action of the epithelium lining the respiratory tract occur. Because of these age-related factors, elderly people are more susceptible to pneumonia, bronchitis, emphysema, and other pulmonary disorders. Age-related changes in the structure and functions of the lung can also contribute to an older person's reduced ability to perform vigorous exercises, such as running.

✔ **CHECKPOINT**

32. What accounts for the decrease in lung capacity with aging?

• • •

To appreciate the many ways that the respiratory system contributes to homeostasis of other body systems, examine *Focus on Homeostasis: The Respiratory System.* Next, in Chapter 24, we will see how the digestive system makes nutrients available to body cells so that oxygen provided by the respiratory system can be used for ATP production.

DISORDERS: HOMEOSTATIC IMBALANCES

Asthma

Asthma (AZ-ma = panting) is a disorder characterized by chronic airway inflammation, airway hypersensitivity to a variety of stimuli, and airway obstruction. It is at least partially reversible, either spontaneously or with treatment. Asthma affects 3–5% of the U.S. population and is more common in children than in adults. Airway obstruction may be due to smooth muscle spasms in the walls of smaller bronchi and bronchioles, edema of the mucosa of the airways, increased mucus secretion, and/or damage to the epithelium of the airway.

Individuals with asthma typically react to concentrations of agents too low to cause symptoms in people without asthma. Sometimes the trigger is an allergen such as pollen, house dust mites, molds, or a particular food. Other common triggers of asthma attacks are emotional upset, aspirin, sulfiting agents (used in wine and beer and to keep greens fresh in salad bars), exercise, and breathing cold air or cigarette smoke. In the early phase (acute) response, smooth muscle spasm is accompanied by excessive secretion of mucus that may clog the bronchi and bronchioles and worsen the attack. The late phase (chronic) response is characterized by inflammation, fibrosis, edema, and necrosis (death) of bronchial epithelial cells. A host of mediator chemicals, including leukotrienes, prostaglandins, thromboxane, platelet-activating factor, and histamine, take part.

Symptoms include difficult breathing, coughing, wheezing, chest tightness, tachycardia, fatigue, moist skin, and anxiety. An acute attack is treated by giving an inhaled beta$_2$-adrenergic agonist (albuterol) to help relax smooth muscle in the bronchioles and open up the airways. This drug mimics the effect of sympathetic stimulation, that is, it causes bronchodilation. However, long-term therapy of asthma strives to suppress the underlying inflammation. The anti-inflammatory drugs that are used most often are inhaled corticosteroids (glucocorticoids), cromolyn sodium (Intal®), and leukotriene blockers (Accolate®).

Chronic Obstructive Pulmonary Disease

Chronic obstructive pulmonary disease (COPD) is a type of respiratory disorder characterized by chronic and recurrent obstruction of airflow, which increases airway resistance. COPD affects about 30 million Americans and is the fourth leading cause of death behind heart disease, cancer, and cerebrovascular disease. The principal types of COPD are emphysema and chronic bronchitis. In most cases, COPD is preventable because its most common cause is cigarette smoking or breathing secondhand smoke. Other causes include air pollution, pulmonary infection, occupational exposure to dusts and gases, and genetic factors. Because men, on average, have more years of exposure to cigarette smoke than women, they are twice as likely as women to suffer from COPD; still, the incidence of COPD in women has risen sixfold in the past 50 years, a reflection of increased smoking among women.

Emphysema

Emphysema (em-fi-SĒ-ma = blown up or full of air) is a disorder characterized by destruction of the walls of the alveoli, producing abnormally large air spaces that remain filled with air during exhalation. With less surface area for gas exchange, O_2 diffusion across the damaged respiratory membrane is reduced. Blood O_2 level is somewhat lowered, and any mild exercise that raises the O_2 requirements of the cells leaves the patient breathless. As increasing numbers of alveolar walls are damaged, lung elastic recoil decreases due to loss of elastic fibers, and an increasing amount of air becomes trapped in the lungs at the end of exhalation. Over several years, added exertion during inhalation increases the size of the chest cage, resulting in a "barrel chest."

Emphysema is generally caused by a long-term irritation; cigarette smoke, air pollution, and occupational exposure to industrial dust are the most common irritants. Some destruction of alveolar sacs may be caused by an enzyme imbalance. Treatment consists of cessation of smoking, removal of other environmental irritants, exercise training under careful medical supervision, breathing exercises, use of bronchodilators, and oxygen therapy.

Chronic Bronchitis

Chronic bronchitis is a disorder characterized by excessive secretion of bronchial mucus accompanied by a productive cough (sputum is raised) that lasts for at least 3 months of the year for two successive years. Cigarette smoking is the leading cause of chronic bronchitis. Inhaled irritants lead to chronic inflammation with an increase in the size and number of mucous glands and goblet cells in the airway epithelium. The thickened and excessive mucus produced narrows the airway and impairs ciliary function. Thus, inhaled pathogens become embedded in airway secretions and multiply rapidly. Besides a productive cough, symptoms of chronic bronchitis are shortness of breath, wheezing, cyanosis, and pulmonary hypertension. Treatment for chronic bronchitis is similar to that for emphysema.

Lung Cancer

In the United States, **lung cancer** is the leading cause of cancer death in both males and females, accounting for 160,000 deaths annually. At the time of diagnosis, lung cancer is usually well advanced, with distant metastases present in about 55% of patients, and regional lymph node involvement in an additional 25%. Most people with lung cancer die within a year of the initial diagnosis; the overall survival rate is only 10–15%. Cigarette smoke is the most common cause of lung cancer. Roughly 85% of lung cancer cases are related to smoking, and the disease is 10 to 30 times more common in smokers than nonsmokers. Exposure to secondhand smoke is also associated with lung cancer and heart disease. In the United States, secondhand smoke causes an estimated 4000 deaths a year from lung cancer, and nearly 40,000 deaths a year from heart disease. Other causes of lung cancer are ionizing radiation and inhaled irritants, such as asbestos and radon gas. Emphysema is a common precursor to the development of lung cancer.

The most common type of lung cancer, **bronchogenic carcinoma** (brong'-kō-JEN-ik), starts in the epithelium of the bronchial tubes. Bronchogenic tumors are named based on where they arise. For example, *adenocarcinomas* (ad-en-ō-kar-si-NŌ-mas; *adeno-* = gland) develop in peripheral areas of the lungs from bronchial glands and alveolar cells, *squamous cell carcinomas* develop from the squamous cells in the epithelium of larger bronchial tubes, and *small (oat) cell carcinomas* develop from epithelial cells in primary bronchi near the hilus of the lungs that get their name due to their flat cell shape with little cytoplasm. They tend to involve the mediastinum early on. Depending on the type, bronchogenic tumors may be aggressive, locally invasive, and undergo widespread metastasis. The tumors begin as epithelial lesions that grow to form masses that obstruct the bronchial tubes or invade adjacent lung tissue. Bronchogenic carcinomas metastasize to lymph nodes, the brain, bones, liver, and other organs.

Symptoms of lung cancer are related to the location of the tumor. These may include a chronic cough, spitting blood from the respiratory tract, wheezing, shortness of breath, chest pain, hoarseness, difficulty swallowing, weight loss, anorexia, fatigue, bone pain, confusion, problems with balance, headache, anemia, thrombocytopenia, and jaundice.

Treatment consists of partial or complete surgical removal of a diseased lung (pulmonectomy), radiation therapy, and chemotherapy.

Pneumonia

Pneumonia (noo-MŌ-ne-a) is an acute infection or inflammation of the alveoli. It is the most common infectious cause of death in the United States, where an estimated 4 million cases occur annually. When certain microbes enter the lungs of susceptible individuals, they release damaging toxins, stimulating inflammation and immune responses that have damaging side effects. The toxins and immune response damage alveoli and bronchial mucous membranes; inflammation and edema cause the alveoli to fill with fluid, interfering with ventilation and gas exchange.

The most common cause of pneumonia is the pneumococcal bacterium *Streptococcus pneumoniae* (strep'-tō-KOK-us noo-MŌ-ne-i), but other microbes may also cause pneumonia. Those who are most susceptible to pneumonia are the elderly, infants, immunocompromised individuals (AIDS or cancer patients, or those taking immunosuppressive drugs), cigarette smokers, and individuals with an obstructive lung disease. Most cases of pneumonia are preceded by an upper respiratory infection that often is viral. Individuals then develop fever, chills, productive or dry cough, malaise, chest pain, and sometimes dyspnea (difficult breathing) and hemoptysis (spitting blood).

Treatment may involve antibiotics, bronchodilators, oxygen therapy, increased fluid intake, and chest physiotherapy (percussion, vibration, and postural drainage).

Tuberculosis

The bacterium *Mycobacterium tuberculosis* (mī'-kō-bak'-TI-rē-um) produces an infectious, communicable disease called **tuberculosis (TB)** that most often affects the lungs and the pleurae but may involve other parts of the body. Once the bacteria are inside the lungs, they multiply and cause inflammation, which stimulates neutrophils and macrophages to migrate to the area and engulf the bacteria to prevent their spread. If the immune system is not impaired, the bacteria remain dormant for life, but impaired immunity may enable the bacteria to escape into blood and lymph to infect other organs. In many people, symptoms—fatigue, weight loss, lethargy, anorexia, a low-grade fever, night sweats, cough, dyspnea, chest pain, and hemoptysis—do not develop until the disease is advanced.

During the past several years, the incidence of TB in the United States has risen dramatically. Perhaps the single most important factor related to this increase is the spread of the human immunodeficiency virus (HIV). People infected with HIV are much more likely to develop tuberculosis because their immune systems are impaired. Among the other factors that have contributed to the increased number of cases are homelessness, increased drug abuse, increased immigration from countries with a high prevalence of tuberculosis, increased crowding in housing among the poor, and airborne transmission of tuberculosis in prisons and shelters. In addition, recent outbreaks of tuberculosis involving multi-drug-resistant strains of *Mycobacterium tuberculosis* have occurred because patients fail to complete their antibiotic and other treatment regimens. TB is treated with the medication isoniazid.

Pulmonary Edema

Pulmonary edema is an abnormal accumulation of fluid in the interstitial spaces and alveoli of the lungs. The edema may arise from increased permeability of the pulmonary capillaries (pulmonary origin) or increased pressure in the pulmonary capillaries (cardiac origin); the latter cause may coincide with congestive heart failure. The most common symptom is dyspnea. Others include wheezing, tachypnea (rapid breathing rate), restlessness, a feeling of suffocation, cyanosis, pallor (paleness), diaphoresis (excessive perspiration), and pulmonary hypertension. Treatment consists of administering oxygen, drugs that dilate the bronchioles and lower blood pressure, diuretics to rid the body of excess fluid, and drugs that correct acid–base imbalance; suctioning of airways; and mechanical ventilation. One of the recent culprits in the development of pulmonary edema was found to be "phen-fen" diet pills.

Asbestos-Related Diseases

Asbestos-related diseases are serious lung disorders that develop as a result of inhaling asbestos particles decades earlier. When asbestos particles are inhaled, they penetrate lung tissue. In response, white blood cells attempt to destroy them by phagocytosis. However, the fibers usually destroy the white blood cells and scarring of lung tissue may follow. Asbestos-related diseases include **asbestosis** (as'-bes-TŌ-sis) (widespread scarring of lung tissue), **diffuse pleural thickening** (thickening of the pleurae), and **mesothelioma** (mez'-o-thē-lē-Ō-ma) (cancer of the pleurae or, less commonly, the peritoneum).

Sudden Infant Death Syndrome

Sudden infant death syndrome (SIDS) is the sudden, unexpected death of an apparently healthy infant during sleep. It rarely occurs before 2 weeks or after 6 months of age, with the peak incidence between the second and fourth months. SIDS is more common in premature infants, male babies, low-birth-weight babies, babies of drug users or smokers, babies who have stopped breathing and have had to be resuscitated, babies with upper respiratory tract infections, and babies who have had a sibling die of SIDS. African-American and Native American babies are at higher risk. The exact cause of SIDS is unknown. However, it may be due to an abnormality in the mechanisms that control respiration or low levels of oxygen in the blood. SIDS may also be linked to hypoxia while sleeping in a prone position (on the stomach) and the rebreathing of exhaled air trapped in a depression of a mattress. It is recommended that for the first 6 months infants be placed on their backs for sleeping ("back to sleep").

Severe Acute Respiratory Syndrome

Severe acute respiratory syndrome (SARS) is an example of an *emerging infectious disease,* that is, a disease that is new or changing. Other examples of emerging infectious diseases are West Nile encephalitis, mad cow disease, and AIDS. SARS first appeared in southern China in late 2002 and has subsequently spread worldwide. It is a respiratory illness caused by a new variety of coronavirus. Symptoms of SARS include fever, malaise, muscle aches, nonproductive (dry) cough, difficulty in breathing, chills, headache, and diarrhea. About 10–20% of patients require mechanical ventilation and in some cases death may result. The disease is primarily spread through person-to-person contact. There is no effective treatment for SARS and the death rate is 5–10%, usually among the elderly and in persons with other medical problems.

MEDICAL TERMINOLOGY

Abdominal thrust maneuver First-aid procedure designed to clear the airways of obstructing objects. It is performed by applying a quick upward thrust between the navel and costal margin that causes sudden elevation of the diaphragm and forceful, rapid expulsion of air in the lungs; this action forces air out the trachea to eject the obstructing object. The abdominal thrust maneuver is also used to expel water from the lungs of near-drowning victims before resuscitation is begun.

Asphyxia (as-FIK-sē-a; *sphyxia* = pulse) Oxygen starvation due to low atmospheric oxygen or interference with ventilation, external respiration, or internal respiration.

Aspiration (as'-pi-RĀ-shun) Inhalation of a foreign substance such as water, food, or a foreign body into the bronchial tree; also, the drawing of a substance in or out by suction.

Avian influenza A respiratory disorder that has resulted in the deaths of hundreds of millions of birds worldwide. It is usually transmitted from one bird to another bird through their droppings, saliva, and nasal secretions. Currently, avian influenza is difficult to transmit from birds to humans; the few humans who have died from avian influenza have had close contact with infected birds. Also called **bird flu.**

Black lung disease A condition in which the lungs appear black instead of pink due to inhalation of coal dust over a period of many years. Most often it affects people who work in the coal industry.

Bronchiectasis (bron'-kē-EK-ta-sis; *-ektasis* = stretching) A chronic dilation of the bronchi or bronchioles resulting from damage to the bronchial wall, for example, from respiratory infections.

Bronchography (bron-KOG-ra-fē) An imaging technique used to visualize the bronchial tree using x-rays. After an opaque contrast medium is inhaled through an intratracheal catheter, radiographs of the chest in various positions are taken, and the developed film, a **bronchogram** (BRON-kō-gram), provides a picture of the bronchial tree.

Bronchoscopy (bron-KOS-kō-pē) Visual examination of the bronchi through a **bronchoscope,** an illuminated, flexible tubular instrument that is passed through the mouth (or nose), larynx, and trachea into the bronchi. The examiner can view the interior of the trachea and bronchi to biopsy a tumor, clear an obstructing object or secretions from an airway, take cultures or smears for microscopic examination, stop bleeding, or deliver drugs.

Cheyne–Stokes respiration (CHĀN STŌKS res'-pi-RĀ-shun) A repeated cycle of irregular breathing that begins with shallow breaths that increase in depth and rapidity and then decrease and cease altogether for 15 to 20 seconds. Cheyne–Stokes is normal in infants; it is also often seen just before death from pulmonary, cerebral, cardiac, and kidney disease.

Dyspnea (DISP-nē-a; *dys-* = painful, difficult) Painful or labored breathing.

Epistaxis (ep'-i-STAK-sis) Loss of blood from the nose due to trauma, infection, allergy, malignant growths, or bleeding disorders. It can be arrested by cautery with silver nitrate, electrocautery, or firm packing. Also called **nosebleed.**

Hypoventilation (*hypo-* = below) Slow and shallow breathing.

Mechanical ventilation The use of an automatically cycling device (ventilator or respirator) to assist breathing. A plastic tube is inserted into the nose or mouth and the tube is attached to a device that forces air into the lungs. Exhalation occurs passively due to the elastic recoil of the lungs.

Rales (RĀLS) Sounds sometimes heard in the lungs that resemble bubbling or rattling. Rales are to the lungs what murmurs are to the heart. Different types are due to the presence of an abnormal type or amount of fluid or mucus within the bronchi or alveoli, or to bronchoconstriction that causes turbulent airflow.

Respirator (RES-pi-rā'-tor) An apparatus fitted to a mask over the nose and mouth, or hooked directly to an endotracheal or tracheotomy tube, that is used to assist or support ventilation or to provide nebulized medication to the air passages.

Respiratory failure A condition in which the respiratory system either cannot supply sufficient O_2 to maintain metabolism or cannot eliminate enough CO_2 to prevent respiratory acidosis (a lower-than-normal pH in interstitial fluid).

Rhinitis (rī-NĪ-tis; *rhin-* = nose) Chronic or acute inflammation of the mucous membrane of the nose due to viruses, bacteria, or irritants. Excessive mucus production produces a runny nose, nasal congestion, and postnasal drip.

Sleep apnea (AP-nē-a; *a-* = without; *-pnea* = breath) A disorder in which a person repeatedly stops breathing for 10 or more seconds while sleeping. Most often, it occurs because loss of muscle tone in pharyngeal muscles allows the airway to collapse.

Sputum (SPŪ-tum = to spit) Mucus and other fluids from the air passages that is expectorated (expelled by coughing).

Strep throat Inflammation of the pharynx caused by the bacterium *Streptococcus pyogenes.* It may also involve the tonsils and middle ear.

Tachypnea (tak′-ip-NĒ-a; *tachy-* = rapid; *-pnea* = breath) Rapid breathing rate.

Wheeze (HWĒZ) A whistling, squeaking, or musical high-pitched sound during breathing resulting from a partially obstructed airway.

CHAPTER REVIEW AND RESOURCE SUMMARY

Review

Resource

23.1 Respiratory System Anatomy

1. The respiratory system consists of the nose, pharynx, larynx, trachea, bronchi, and lungs. They act with the cardiovascular system to supply oxygen (O_2) and remove carbon dioxide (CO_2) from the blood.

2. The external portion of the nose is made of cartilage and skin and is lined with a mucous membrane. Openings to the exterior are the external nares. The internal portion of the nose communicates with the paranasal sinuses and nasopharynx through the internal nares. The nasal cavity is divided by a septum. The anterior portion of the cavity is called the vestibule. The nose warms, moistens, and filters air and functions in olfaction and speech.

3. The pharynx (throat) is a muscular tube lined by a mucous membrane. The anatomic regions are the nasopharynx, oropharynx, and laryngopharynx. The nasopharynx functions in respiration. The oropharynx and laryngopharynx function both in digestion and in respiration.

4. The larynx (voice box) is a passageway that connects the pharynx with the trachea. It contains the thyroid cartilage (Adam's apple); the epiglottis, which prevents food from entering the larynx; the cricoid cartilage, which connects the larynx and trachea; and the paired arytenoid, corniculate, and cuneiform cartilages. The larynx contains vocal folds, which produce sound as they vibrate. Taut folds produce high pitches, and relaxed ones produce low pitches.

5. The trachea (windpipe) extends from the larynx to the primary bronchi. It is composed of C-shaped rings of cartilage and smooth muscle and is lined with pseudostratified ciliated columnar epithelium.

6. The bronchial tree consists of the trachea, primary bronchi, secondary bronchi, tertiary bronchi, bronchioles, and terminal bronchioles. Walls of bronchi contain rings of cartilage; walls of bronchioles contain increasingly smaller plates of cartilage and increasing amounts of smooth muscle.

7. Lungs are paired organs in the thoracic cavity enclosed by the pleural membrane. The parietal pleura is the superficial layer that lines the thoracic cavity; the visceral pleura is the deep layer that covers the lungs. The right lung has three lobes separated by two fissures; the left lung has two lobes separated by one fissure and a depression, the cardiac notch.

8. Secondary bronchi give rise to branches called segmental bronchi, which supply segments of lung tissue called bronchopulmonary segments. Each bronchopulmonary segment consists of lobules, which contain lymphatics, arterioles, venules, terminal bronchioles, respiratory bronchioles, alveolar ducts, alveolar sacs, and alveoli.

9. Alveolar walls consist of type I alveolar cells, type II alveolar cells, and associated alveolar macrophages.

10. Gas exchange occurs across the respiratory membranes.

Anatomy Overview - Overview of Respiratory Organs
Anatomy Overview - Respiratory Tissues
Figure 23.4 - The Larynx
Figure 23.8 - Relationship of the Pleural Membranes to the Lungs
Figure 23.11 - Structural Components of an Alveolus
Exercise - Concentrate on Respiratory Structures
Exercise - Paint the Lung
Concepts and Connections - Functional Anatomy of the Respiratory System

23.2 Pulmonary Ventilation

1. Pulmonary ventilation, or breathing, consists of inhalation and exhalation.

2. The movement of air into and out of the lungs depends on pressure changes governed in part by Boyle's law, which states that the volume of a gas varies inversely with pressure, assuming that temperature remains constant.

3. Inhalation occurs when alveolar pressure falls below atmospheric pressure. Contraction of the diaphragm and external intercostals increases the size of the thorax, thereby decreasing the intrapleural pressure so that the lungs expand. Expansion of the lungs decreases alveolar pressure so that air moves down a pressure gradient from the atmosphere into the lungs.

4. During forceful inhalation, accessory muscles of inhalation (sternocleidomastoids, scalenes, and pectoralis minors) are also used.

5. Exhalation occurs when alveolar pressure is higher than atmospheric pressure. Relaxation of the diaphragm and external intercostals results in elastic recoil of the chest wall and lungs, which increases intrapleural pressure; lung volume decreases and alveolar pressure increases, so air moves from the lungs to the atmosphere.

Animation - Pulmonary Ventilation
Exercise - The Airway
Exercise - Build an Airway
Concepts and Connections - Ventilation

Review	Resource

6. Forceful exhalation involves contraction of the internal intercostal and abdominal muscles.

7. The surface tension exerted by alveolar fluid is decreased by the presence of surfactant.

8. Compliance is the ease with which the lungs and thoracic wall can expand.

9. The walls of the airways offer some resistance to breathing.

10. Normal quiet breathing is termed eupnea; other patterns are costal breathing and diaphragmatic breathing. Modified respiratory movements, such as coughing, sneezing, sighing, yawning, sobbing, crying, laughing, and hiccupping, are used to express emotions and to clear the airways. (See Table 23.2.)

23.3 Lung Volumes and Capacities

Animation - Regulation of Ventilation: Basic Rhythms

1. Lung volumes exchanged during breathing and the rate of respiration are measured with a spirometer.

2. Lung volumes measured by spirometry include tidal volume, minute ventilation, alveolar ventilation rate, inspiratory reserve volume, expiratory reserve volume, and $FEV_{1.0}$. Other lung volumes include anatomic dead space, residual volume, and minimal volume.

3. Lung capacities, the sum of two or more lung volumes, include inspiratory, functional, residual, vital, and total lung capacities.

23.4 Exchange of Oxygen and Carbon Dioxide

Animation - Gas Exchange Introduction
Animation - Gas Exchange: Internal and External Respiration
Figure 23.17 - Changes in Partial Pressures of Oxygen and Carbon Dioxide
Exercise - Gas Exchange Match-Up

1. The partial pressure of a gas is the pressure exerted by that gas in a mixture of gases. It is symbolized by P_x, where the subscript is the chemical formula of the gas.

2. According to Dalton's law, each gas in a mixture of gases exerts its own pressure as if all the other gases were not present.

3. Henry's law states that the quantity of a gas that will dissolve in a liquid is proportional to the partial pressure of the gas and its solubility (given that the temperature remains constant).

4. In internal and external respiration, O_2 and CO_2 diffuse from areas of higher partial pressures to areas of lower partial pressures.

5. External respiration or pulmonary gas exchange is the exchange of gases between alveoli and pulmonary blood capillaries. It depends on partial pressure differences, a large surface area for gas exchange, a small diffusion distance across the respiratory membrane, and the rate of airflow into and out of the lungs.

6. Internal respiration or systemic gas exchange is the exchange of gases between systemic blood capillaries and tissue cells.

23.5 Transport of Oxygen and Carbon Dioxide

Animation - Gas Transport
Figure 23.23 - Summary of Chemical Reactions That Occur during Gas Exchange
Exercise - Carbon Dioxide Transport Try-Out
Exercise - Oxygen Transport Try-Out
Exercise - Concentrate on Respiration
Concepts and Connections - Carbon Dioxide Transport
Concepts and Connections - Oxygen Transport

1. In each 100 mL of oxygenated blood, 1.5% of the O_2 is dissolved in blood plasma and 98.5% is bound to hemoglobin as oxyhemoglobin (Hb—O_2).

2. The binding of O_2 to hemoglobin is affected by P_{O_2}, acidity (pH), P_{CO_2}, temperature, and 2,3-bisphosphoglycerate (BPG).

3. Fetal hemoglobin differs from adult hemoglobin in structure and has a higher affinity for O_2.

4. In each 100 mL of deoxygenated blood, 7% of CO_2 is dissolved in blood plasma, 23% combines with hemoglobin as carbaminohemoglobin (Hb—CO_2), and 70% is converted to bicarbonate ions (HCO_3^-).

5. In an acidic environment, hemoglobin's affinity for O_2 is lower, and O_2 dissociates more readily from it (Bohr effect).

6. In the presence of O_2, less CO_2 binds to hemoglobin (Haldane effect).

23.6 Control of Respiration

Anatomy Overview - Respiratory Control Center
Animation - Regulation of Ventilation
Animation - Control of Ventilation Rate and Blood Chemistry
Animation - Role of Respiratory System in pH Regulation
Exercise - Respiration and pH Reflex
Concepts and Connections - Ventilation

1. The respiratory center consists of a medullary rhythmicity area in the medulla oblongata and a pneumotaxic area and an apneustic area in the pons.

2. The inspiratory area sets the basic rhythm of respiration.

3. The pneumotaxic and apneustic areas coordinate the transition between inhalation and exhalation.

4. Respirations may be modified by a number of factors, including cortical influences; the inflation reflex; chemical stimuli, such as O_2 and CO_2 and H^+ levels; proprioceptor input; blood pressure changes; limbic system stimulation; temperature; pain; and irritation to the airways. (See Table 23.3.)

23.7 Exercise and the Respiratory System

1. The rate and depth of ventilation change in response to both the intensity and duration of exercise.

2. An increase in pulmonary perfusion and O_2-diffusing capacity occurs during exercise.

3. The abrupt increase in ventilation at the start of exercise is due to neural changes that send excitatory

Review   **Resource**

impulses to the inspiratory area in the medulla oblongata. The more gradual increase in ventilation during moderate exercise is due to chemical and physical changes in the bloodstream.

23.8 Development of the Respiratory System

1. The respiratory system begins as an outgrowth of endoderm called the respiratory diverticulum.
2. Smooth muscle, cartilage, and connective tissue of the bronchial tubes and pleural sacs develop from mesoderm.

23.9 Aging and the Respiratory System

1. Aging results in decreased vital capacity, decreased blood level of O_2, and diminished alveolar macrophage activity.
2. Elderly people are more susceptible to pneumonia, emphysema, bronchitis, and other pulmonary disorders.

SELF-QUIZ QUESTIONS

Fill in the blanks in the following statements.

1. Oxygen in blood is carried primarily in the form of _____; carbon dioxide is carried as _____, _____, and _____.
   oxyhemoglobin
   dissolved CO₂, carbamino cmpds, bicarbonate ions

2. Write the equation for the chemical reaction that occurs for the transport of carbon dioxide as bicarbonate ions in blood: _____.
   CO₂ + H₂O → H₂CO₃ → H⁺ + HCO₃⁻

Indicate whether the following statements are true or false.

3. The three basic steps of respiration are pulmonary ventilation, external respiration, and cellular respiration. *F*

4. For inhalation to occur, air pressure in the alveoli must be less than atmospheric pressure; for exhalation to occur, air pressure in the alveoli must be greater than atmospheric pressure. *T*

Choose the one best answer to the following questions.

5. What structural changes occur from primary bronchi to terminal bronchioles? (1) The mucous membrane changes from pseudostratified ciliated columnar epithelium to nonciliated simple cuboidal epithelium. (2) The number of goblet cells increases. (3) The amount of smooth muscle increases. (4) Incomplete rings of cartilage disappear. (5) The amount of branching decreases.
   (a) 1, 2, 3, 4, and 5 (b) 2, 3, and 4 (c) 1, 3, and 4
   (d) 1, 3, 4, and 5 (e) 1, 2, 3, and 4

6. Which of the following would cause oxygen to dissociate more readily from hemoglobin? (1) low P_{O_2}, (2) an increase in H^+ in blood, (3) hypercapnia, (4) hypothermia, (5) low levels of BPG (2,3-bisphosphoglycerate).
   (a) 1 and 2 (b) 2, 3, and 4 (c) 1, 2, 3, and 5
   (d) 1, 3, and 5 (e) 1, 2, and 3

7. Which of the following statements are *correct?* (1) Normal exhalation during quiet breathing is an active process involving intensive muscle contraction. (2) Passive exhalation results from elastic recoil of the chest wall and lungs. (3) Air flow during breathing is due to a pressure gradient between the lungs and the atmospheric air. (4) During normal breathing, the pressure between the two pleural layers (intrapleural pressure) is always subatmospheric. (5) Surface tension of alveolar fluid facilitates inhalation.
   (a) 1, 2, and 3 (b) 2, 3, and 4 (c) 3, 4, and 5
   (d) 1, 3, and 5 (e) 2, 3, and 5

8. Which of the following factors affect the rate of external respiration? (1) partial pressure differences of the gases, (2) surface area for gas exchange, (3) diffusion distance, (4) solubility and molecular weight of the gases, (5) presence of bisphosphoglycerate (BPG).
   (a) 1, 2, and 3 (b) 2, 4, and 5 (c) 1, 2, 4, and 5
   (d) 1, 2, 3, and 4 (e) 2, 3, 4, and 5

9. The most important factor in determining the percent oxygen saturation of hemoglobin is
   (a) the partial pressure of oxygen.
   (b) acidity.
   (c) the partial pressure of carbon dioxide.
   (d) temperature.
   (e) BPG.

10. Which of the following statements are *true?* (1) Central chemoreceptors are stimulated by changes in P_{CO_2}, H^+, and P_{O_2}. (2) Respiratory rate increases during the initial onset of exercise due to input to the inspiratory area from proprioceptors. (3) When baroreceptors in the lungs are stimulated, the expiratory area is activated. (4) Stimulation of the limbic system can result in excitation of the inspiratory area. (5) Sudden severe pain causes brief apnea, while prolonged somatic pain causes an increase in respiratory rate. (6) The respiratory rate increases during fever.
    (a) 1, 2, 3, and 6 (b) 1, 4, and 5 (c) 1, 2, 4, 5, and 6
    (d) 2, 3, 4, 5, and 6 (e) 2, 4, 5, and 6

11. Place the steps for normal inhalation in order.
    5 (a) decrease in intrapleural pressure to 754 mmHg
    3 (b) increase in the size of the thoracic cavity
    8 (c) flow of air from higher to lower pressure
    6 (d) outward pull of pleurae, resulting in lung expansion
    1 (e) stimulation of primary breathing muscles by phrenic and intercostal nerves
    7 (f) decrease in alveolar pressure to 758 mmHg
    2 (g) contraction of the diaphragm and external intercostals
    4 (h) increase in the volume of the pleural cavity

12. Match the following:

_3_ (a) total volume of air inhaled and exhaled each minute

_8_ (b) tidal volume + inspiratory reserve volume + expiratory reserve volume

_5_ (c) additional amount of air inhaled beyond tidal volume when taking a very deep breath

_9_ (d) residual volume + expiratory reserve volume

_2_ (e) amount of air remaining in lungs after expiratory reserve volume is expelled

_7_ (f) tidal volume + inspiratory reserve volume

_10_ (g) vital capacity + residual volume

_1_ (h) volume of air in one breath

_4_ (i) amount of air exhaled in forced exhalation following a normal exhalation

_6_ (j) provides a medical and legal tool for determining if a baby was born dead or died after birth

(1) tidal volume
(2) residual volume
(3) minute ventilation
(4) expiratory reserve volume
(5) inspiratory reserve volume
(6) minimal volume
(7) inspiratory capacity
(8) vital capacity
(9) functional residual capacity
(10) total lung capacity

13. Match the following:

_2_ (a) functions as a passageway for air and food, provides a resonating chamber for speech sounds, and houses the tonsils

_11_ (b) site of external respiration

_3_ (c) connects the laryngopharynx with the trachea; houses the vocal cords

_9_ (d) serous membrane that surrounds the lungs

_1_ (e) functions in warming, moistening, and filtering air; receives olfactory stimuli; is a resonating chamber for sound

_12_ (f) simple squamous epithelial cells that form a continuous lining of the alveolar wall; sites of gas exchange

_10_ (g) forms anterior wall of the larynx

_5_ (h) a tubular passageway for air connecting the larynx to the bronchi

_13_ (i) secrete alveolar fluid and surfactant

_8_ (j) forms inferior wall of larynx; landmark for tracheotomy

_4_ (k) prevents food or fluid from entering the airways

_6_ (l) air passageways entering the lungs

_7_ (m) ridge covered by a sensitive mucous membrane; irritation triggers cough reflex

(1) nose
(2) pharynx
(3) larynx
(4) epiglottis
(5) trachea
(6) bronchi
(7) carina
(8) cricoid cartilage
(9) pleura
(10) thyroid cartilage
(11) alveoli
(12) type I alveolar cells
(13) type II alveolar cells

14. Match the following:

_7_ (a) a deficiency of oxygen at the tissue level

_8_ (b) above-normal partial pressure of carbon dioxide

_1_ (c) normal quiet breathing

_5_ (d) deep, abdominal breathing

_6_ (e) the ease with which the lungs and thoracic wall can be expanded

_9_ (f) hypoxia-induced vasoconstriction to divert pulmonary blood from poorly ventilated to well-ventilated regions of the lungs

_2_ (g) absence of breathing

_3_ (h) rapid and deep breathing

_4_ (i) shallow, chest breathing

(1) eupnea
(2) apnea
(3) hyperventilation
(4) costal breathing
(5) diaphragmatic breathing
(6) compliance
(7) hypoxia
(8) hypercapnia
(9) ventilation-- perfusion coupling

15. Match the following:

_9_ (a) prevents excessive inflation of the lungs

_11_ (b) the lower the amount of oxyhemoglobin, the higher the carbon dioxide–carrying capacity of the blood

_3_ (c) controls the basic rhythm of respiration

_4_ (d) active during normal inhalation; sends nerve impulses to external intercostals and diaphragm

_6_ (e) sends stimulatory impulses to the inspiratory area that activate it and prolong inhalation

_1_ (f) as acidity increases, the affinity of hemoglobin for oxygen decreases and oxygen dissociates more readily from hemoglobin; shifts oxygen-dissociation curve to the right

_5_ (g) active during forceful exhalation

_10_ (h) pressure of a gas in a closed container is inversely proportional to the volume of the container

_7_ (i) transmits inhibitory impulses to turn off the inspiratory area before the lungs become too full of air

_8_ (j) the quantity of a gas that dissolves in a liquid is proportional to the partial pressure of the gas and its solubility

_2_ (k) relates to the partial pressure of a gas in a mixture of gases whereby each gas in a mixture exerts its own pressure as if all the other gases were not present

(1) Bohr effect
(2) Dalton's law
(3) medullary rhythmicity area
(4) inspiratory area
(5) expiratory area
(6) apneustic area
(7) pneumotaxic area
(8) Henry's law
(9) inflation (Hering–Breuer) reflex
(10) Boyle's law
(11) Haldane effect

CRITICAL THINKING QUESTIONS

1. Aretha loves to sing. Right now she has a cold, a severely runny nose, and a "sore throat" that is affecting her ability to sing and talk. What structures are involved and how are they affected by her cold?

2. Ms. Brown has smoked cigarettes for years and is having breathing difficulties. She has been diagnosed with emphysema. Describe specific kinds of structural changes you would expect to observe in Mrs. Brown's respiratory system. How are air flow and gas exchange affected by these structural changes?

3. The Robinson family went to bed one frigid winter night and were found deceased the next day. A squirrel's nest was found in their chimney. What happened to the Robinsons?

? ANSWERS TO FIGURE QUESTIONS

23.1 The conducting zone of the respiratory system includes the nose, pharynx, larynx, trachea, bronchi, and bronchioles (except the respiratory bronchioles).

23.2 The path of air is external nares → vestibule → nasal cavity → internal nares.

23.3 The root of the nose attaches it to the frontal bone.

23.4 During swallowing, the epiglottis closes over the rima glottidis, the entrance to the trachea, to prevent aspiration of food and liquids into the lungs.

23.5 The main function of the vocal folds is voice production.

23.6 Because the tissues between the esophagus and trachea are soft, the esophagus can bulge and press against the trachea during swallowing.

23.7 The left lung has two lobes and two secondary bronchi; the right lung has three of each.

23.8 The pleural membrane is a serous membrane.

23.9 Because two-thirds of the heart lies to the left of the midline, the left lung contains a cardiac notch to accommodate the presence of the heart. The right lung is shorter than the left because the diaphragm is higher on the right side to accommodate the liver.

23.10 The wall of an alveolus is made up of type I alveolar cells, type II alveolar cells, and associated alveolar macrophages.

23.11 The respiratory membrane averages 0.5 μm in thickness.

23.12 The pressure would increase fourfold, to 4 atm.

23.13 If you are at rest while reading, your diaphragm is responsible for about 75% of each inhalation.

23.14 At the start of inhalation, intrapleural pressure is about 756 mmHg. With contraction of the diaphragm, it decreases to about 754 mmHg as the volume of the space between the two pleural layers expands. With relaxation of the diaphragm, it increases back to 756 mmHg.

23.15 Normal atmospheric pressure at sea level is 760 mmHg.

23.16 Breathing in and then exhaling as much air as possible demonstrates vital capacity.

23.17 A difference in P_{O_2} promotes oxygen diffusion into pulmonary capillaries from alveoli and into tissue cells from systemic capillaries.

23.18 The most important factor that determines how much O_2 binds to hemoglobin is the P_{O_2}.

23.19 Both during exercise and at rest, hemoglobin in your pulmonary veins would be fully saturated with O_2, a point that is at the upper right of the curve.

23.20 Because lactic acid (lactate) and CO_2 are produced by active skeletal muscles, blood pH decreases slightly and P_{CO_2} increases when you are actively exercising. The result is lowered affinity of hemoglobin for O_2, so more O_2 is available to the working muscles.

23.21 O_2 is more available to your tissue cells when you have a fever because the affinity of hemoglobin for O_2 decreases with increasing temperature.

23.22 At a P_{O_2} of 40 mmHg, fetal Hb is 80% saturated with O_2 and maternal Hb is about 75% saturated.

23.23 Blood in a systemic vein would have a higher concentration of HCO_3^-.

23.24 The medullary inspiratory area contains autorhythmic neurons that are active and then inactive in a repeating cycle.

23.25 The phrenic nerves innervate the diaphragm.

23.26 Peripheral chemoreceptors are responsive to changes in blood levels of oxygen, carbon dioxide, and H^+.

23.27 Normal arterial P_{CO_2} is 40 mmHg.

23.28 The respiratory system begins to develop about 4 weeks after fertilization.

24 | THE DIGESTIVE SYSTEM

THE DIGESTIVE SYSTEM AND HOMEOSTASIS *The digestive system contributes to homeostasis by breaking down food into forms that can be absorbed and used by body cells. It also absorbs water, vitamins, and minerals, and it eliminates wastes from the body.*

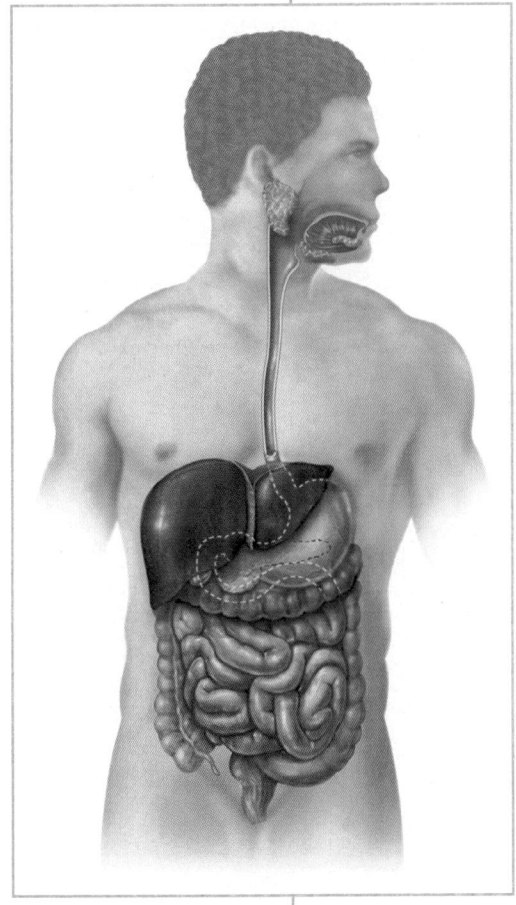

The food we eat contains a variety of nutrients, which are used for building new body tissues and repairing damaged tissues. Food is also vital to life because it is our only source of chemical energy. However, most of the food we eat consists of molecules that are too large to be used by body cells. Therefore, foods must be broken down into molecules that are small enough to enter body cells, a process known as **digestion** (*dis-* = apart; *-gerere* = to carry). The organs involved in the breakdown of food—collectively called the **digestive system**—are the focus of this chapter. Like the respiratory system, the digestive system is a tubular system. It extends from the mouth to the anus, forms an extensive surface area in contact with the external environment, and is closely associated with the cardiovascular system. The combination of extensive environmental exposure and close association with blood vessels is essential for processing the food that we eat.

The medical specialty that deals with the structure, function, diagnosis, and treatment of diseases of the stomach and intestines is called **gastroenterology** (gas′-trō-en′-ter-OL-ō-jē; *gastro-* = stomach; *-entero-* = intestines; *-logy* = study of). The medical specialty that deals with the diagnosis and treatment of disorders of the rectum and anus is called **proctology** (prok-TOL-ō-jē; *proct-* = rectum).

? *Did you ever wonder why some people are sensitive to dairy products?*

24.1 OVERVIEW OF THE DIGESTIVE SYSTEM

◉ OBJECTIVES

- Identify the organs of the digestive system.
- Describe the basic processes performed by the digestive system.

Two groups of organs compose the digestive system (Figure 24.1): the gastrointestinal (GI) tract and the accessory digestive organs. The **gastrointestinal (GI) tract,** or **alimentary canal** (*alimentary = nourishment*), is a continuous tube that extends from the mouth to the anus through the thoracic and abdominopelvic cavities. Organs of the gastrointestinal tract include the mouth, most of the pharynx, esophagus, stomach, small intestine, and large intestine. The length of the GI tract is about 5–7 meters (16.5–23 ft) in a living person when the muscles along the wall of the GI tract organs are in a state of *tonus* (sustained contraction). It is longer in a cadaver (about 7–9 meters or 23–29.5 ft) because of the loss of muscle tone after death. The **accessory digestive organs** include the teeth, tongue, salivary glands, liver, gallbladder, and pancreas. Teeth aid in the physical breakdown of food, and the tongue assists in chewing and swallowing. The other accessory digestive organs, however, never come into direct contact with food. They produce or store secretions that flow into the GI tract through ducts; the secretions aid in the chemical breakdown of food.

The GI tract contains food from the time it is eaten until it is digested and absorbed or eliminated. Muscular contractions in the wall of the GI tract physically break down the food by churning it and propel the food along the tract, from the esophagus to the anus. The contractions also help to dissolve foods by mixing them with fluids secreted into the tract. Enzymes secreted by accessory digestive organs and cells that line the tract break down the food chemically.

Overall, the digestive system performs six basic processes:

1. *Ingestion.* This process involves taking foods and liquids into the mouth (eating).
2. *Secretion.* Each day, cells within the walls of the GI tract and accessory digestive organs secrete a total of about 7 liters of water, acid, buffers, and enzymes into the lumen (interior space) of the tract.
3. *Mixing and propulsion.* Alternating contractions and relaxations of smooth muscle in the walls of the GI tract mix food and secretions and propel them toward the anus. This capability of the GI tract to mix and move material along its length is called **motility** (mō-TIL-i-tē).
4. *Digestion.* Mechanical and chemical processes break down ingested food into small molecules. In **mechanical digestion** the teeth cut and grind food before it is swallowed, and then smooth muscles of the stomach and small intestine churn the food. As a result, food molecules become dissolved and thoroughly mixed with digestive enzymes. In **chemical digestion** the large carbohydrate, lipid, protein, and nucleic acid molecules in food are split into smaller molecules by hydrolysis

(see Figure 2.15). Digestive enzymes produced by the salivary glands, tongue, stomach, pancreas, and small intestine catalyze these catabolic reactions. A few substances in food can be absorbed without chemical digestion. These include vitamins, ions, cholesterol, and water.

5. *Absorption.* The entrance of ingested and secreted fluids, ions, and the products of digestion into the epithelial cells lining the lumen of the GI tract is called **absorption** (ab-SŌRP-shun).

Figure 24.1 Organs of the digestive system.

 Organs of the gastrointestinal (GI) tract are the mouth, pharynx, esophagus, stomach, small intestine, and large intestine. Accessory digestive organs include the teeth, tongue, salivary glands, liver, gallbladder, and pancreas.

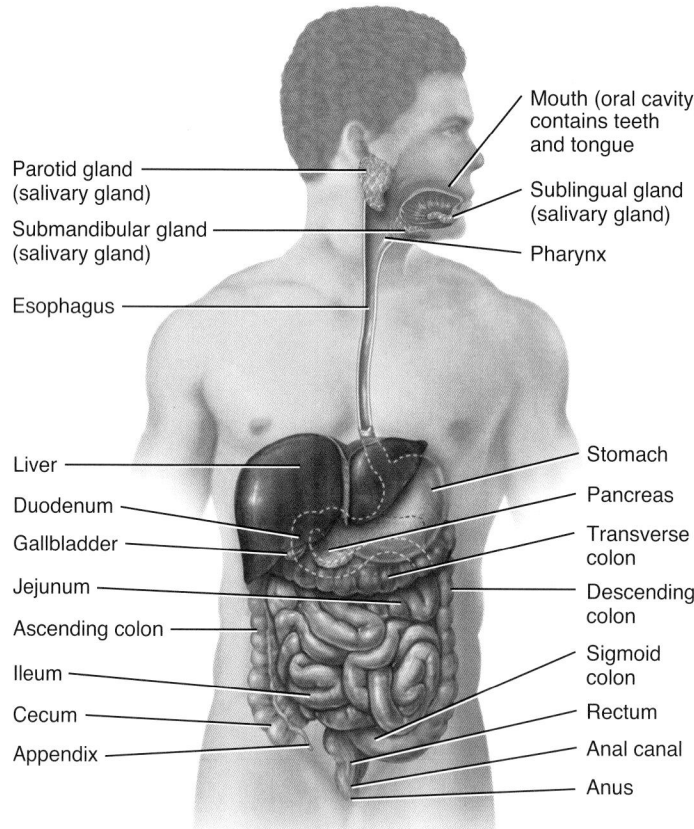

(a) Right lateral view of head and neck and anterior view of trunk

FUNCTIONS OF THE DIGESTIVE SYSTEM

1. Ingestion: taking food into mouth.
2. Secretion: release of water, acid, buffers, and enzymes into lumen of GI tract.
3. Mixing and propulsion: churning and propulsion of food through GI tract.
4. Digestion: mechanical and chemical breakdown of food.
5. Absorption: passage of digested products from GI tract into blood and lymph.
6. Defecation: elimination of feces from GI tract.

The absorbed substances pass into blood or lymph and circulate to cells throughout the body.

6. ***Defecation.*** Wastes, indigestible substances, bacteria, cells sloughed from the lining of the GI tract, and digested materials that were not absorbed in their journey through the digestive tract leave the body through the anus in a process called **defecation** (def′-e-KĀ-shun). The eliminated material is termed **feces** (FĒ-sēz) or **stool.**

✔ **CHECKPOINT**

1. Which components of the digestive system are GI tract organs, and which are accessory digestive organs?
2. Which organs of the digestive system come in contact with food, and what are some of their digestive functions?
3. Which kinds of food molecules undergo chemical digestion, and which do not?

24.2 LAYERS OF THE GI TRACT

◉ **OBJECTIVE**

- Describe the structure and function of the layers that form the wall of the gastrointestinal tract.

The wall of the GI tract from the lower esophagus to the anal canal has the same basic, four-layered arrangement of tissues.

The four layers of the tract, from deep to superficial, are the mucosa, submucosa, muscularis, and serosa/adventitia (Figure 24.2).

Mucosa

The **mucosa,** or inner lining of the GI tract, is a mucous membrane. It is composed of (1) a layer of epithelium in direct contact with the contents of the GI tract, (2) a layer of connective tissue called the lamina propria, and (3) a thin layer of smooth muscle (muscularis mucosae).

1. The **epithelium** in the mouth, pharynx, esophagus, and anal canal is mainly nonkeratinized stratified squamous epithelium that serves a protective function. Simple columnar epithelium, which functions in secretion and absorption, lines the stomach and intestines. The tight junctions that firmly seal neighboring simple columnar epithelial cells to one another restrict leakage between the cells. The rate of renewal of GI tract epithelial cells is rapid: Every 5 to 7 days they slough off and are replaced by new cells. Located among the epithelial cells are exocrine cells that secrete mucus and fluid into the lumen of the tract, and several types of endocrine cells, collectively called **enteroendocrine cells** (en′-ter-ō-EN-dō-krin), which secrete hormones.

2. The **lamina propria** (*lamina* = thin, flat plate; *propria* = one's own) is areolar connective tissue containing many

SUPERIOR

(b) Anterior view

❓ **Which structures of the digestive system secrete digestive enzymes?**

Figure 24.2 Layers of the gastrointestinal tract. Variations in this basic plan may be seen in the esophagus (Figure 24.9), stomach (Figure 24.12), small intestine (Figure 24.19), and large intestine (Figure 24.24).

The four layers of the GI tract, from deep to superficial, are the mucosa, submucosa, muscularis, and serosa.

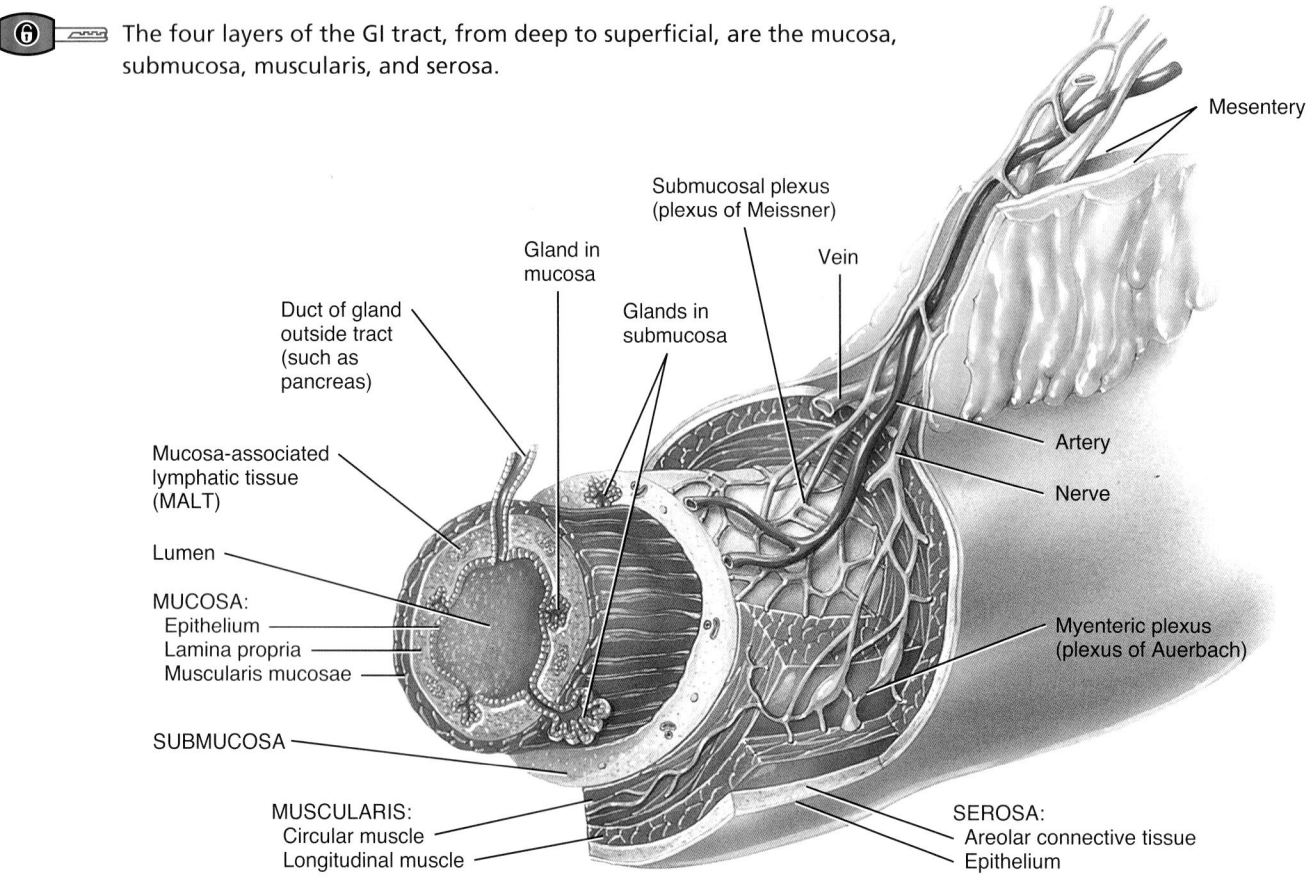

What are the functions of the lamina propria?

blood and lymphatic vessels, which are the routes by which nutrients absorbed into the GI tract reach the other tissues of the body. This layer supports the epithelium and binds it to the muscularis mucosae (discussed next). The lamina propria also contains the majority of the cells of the **mucosa-associated lymphatic tissue (MALT).** These prominent lymphatic nodules contain immune system cells that protect against disease (see Chapter 22). MALT is present all along the GI tract, especially in the tonsils, small intestine, appendix, and large intestine.

3. A thin layer of smooth muscle fibers called the **muscularis mucosae** (mū-KŌ-sē) throws the mucous membrane of the stomach and small intestine into many small folds, which increase the surface area for digestion and absorption. Movements of the muscularis mucosae ensure that all absorptive cells are fully exposed to the contents of the GI tract.

Submucosa

The **submucosa** consists of areolar connective tissue that binds the mucosa to the muscularis. It contains many blood and lymphatic vessels that receive absorbed food molecules. Also located in the submucosa is an extensive network of neurons known as the

submucosal plexus (to be described shortly). The submucosa may also contain glands and lymphatic tissue.

Muscularis

The **muscularis** of the mouth, pharynx, and superior and middle parts of the esophagus contains *skeletal muscle* that produces voluntary swallowing. Skeletal muscle also forms the external anal sphincter, which permits voluntary control of defecation. Throughout the rest of the tract, the muscularis consists of *smooth muscle* that is generally found in two sheets: an inner sheet of circular fibers and an outer sheet of longitudinal fibers. Involuntary contractions of the smooth muscle help break down food, mix it with digestive secretions, and propel it along the tract. Between the layers of the muscularis is a second plexus of neurons—the myenteric plexus (to be described shortly).

Serosa

Those portions of the GI tract that are suspended in the abdominopelvic cavity have a superficial layer called the **serosa.** As its name implies, the serosa is a serous membrane composed of areolar connective tissue and simple squamous epithelium (mesothelium). The serosa is also called the *visceral peritoneum* because it forms a portion of the peritoneum, which we examine in detail shortly. The esophagus lacks a serosa; instead only a single layer of areolar connective tissue called the *adventitia* forms the superficial layer of this organ.

✔CHECKPOINT

4. Where along the GI tract is the muscularis composed of skeletal muscle? Is control of this skeletal muscle voluntary or involuntary?
5. Name the four layers of the gastrointestinal tract, and describe their functions.

24.3 NEURAL INNERVATION OF THE GI TRACT

◉ OBJECTIVE

• Describe the nerve supply of the GI tract.

The gastrointestinal tract is regulated by an intrinsic set of nerves known as the enteric nervous system and by an extrinsic set of nerves that are part of the autonomic nervous system.

Enteric Nervous System

We first introduced you to the **enteric nervous system (ENS),** the "brain of the gut," in Chapter 12. It consists of about 100 million neurons that extend from the esophagus to the anus. The neurons of the ENS are arranged into two plexuses: the myenteric plexus and submucosal plexus (see Figure 24.2). The **myenteric plexus** (*myo-* = muscle), or *plexus of Auerbach* (OW-er-bak), is located between the longitudinal and circular smooth muscle layers of the muscularis. The **submucosal plexus,** or *plexus of Meissner* (MĪS-ner), is found within the submucosa. The plexuses of the ENS consist of motor neurons, interneurons, and sensory neurons (Figure 24.3). Because the motor neurons of the myenteric plexus supply the longitudinal and circular smooth muscle layers of the muscularis, this plexus mostly controls GI tract motility (movement), particularly the frequency and strength of contraction of the muscularis. The motor neurons of the submucosal plexus supply the secretory cells of the mucosal epithelium, controlling the secretions of the organs of the GI tract. The interneurons of the ENS interconnect the neurons of the myenteric and submucosal plexuses. The sensory neurons of the ENS supply the mucosal epithelium. Some of these sensory neurons function as *chemoreceptors,* receptors that are activated by the presence of certain chemicals in food located in the lumen of a GI organ. Other sensory neurons function as *stretch receptors,* receptors that are activated when food distends (stretches) the wall of a GI organ.

Autonomic Nervous System

Although the neurons of the ENS can function independently, they are subject to regulation by the neurons of the autonomic nervous system. The vagus (X) nerves supply parasympathetic fibers to

Figure 24.3 Organization of the enteric nervous system.

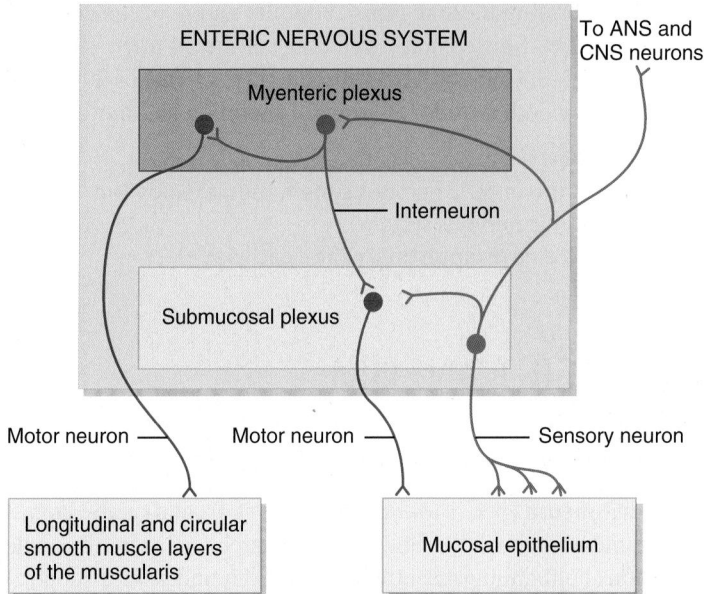

The enteric nervous system consists of neurons arranged into the myenteric and submucosal plexuses.

? What are the functions of the myenteric and submucosal plexuses of the enteric nervous system?

most parts of the GI tract, with the exception of the last half of the large intestine, which is supplied with parasympathetic fibers from the sacral spinal cord. The parasympathetic nerves that supply the GI tract form neural connections with the ENS. Parasympathetic preganglionic neurons of the vagus or pelvic splanchnic nerves synapse with parasympathetic postganglionic neurons located in the myenteric and submucosal plexuses. Some of the parasympathetic postganglionic neurons in turn synapse with neurons in the ENS; others directly innervate smooth muscle and glands within the wall of the GI tract. In general, stimulation of the parasympathetic nerves that innervate the GI tract causes an increase in GI secretion and motility by increasing the activity of ENS neurons.

Sympathetic nerves that supply the GI tract arise from the thoracic and upper lumbar regions of the spinal cord. Like the parasympathetic nerves, these sympathetic nerves form neural connections with the ENS. Sympathetic postganglionic neurons synapse with neurons located in the myenteric plexus and the submucosal plexus. In general, the sympathetic nerves that supply the GI tract cause a decrease in GI secretion and motility by inhibiting the neurons of the ENS. Emotions such as anger, fear, and anxiety may slow digestion because they stimulate the sympathetic nerves that supply the GI tract.

Gastrointestinal Reflex Pathways

Many neurons of the ENS are components of *GI (gastrointestinal) reflex pathways* that regulate GI secretion and motility in response to stimuli present in the lumen of the GI tract. The initial components of a typical GI reflex pathway are sensory receptors (such as chemoreceptors and stretch receptors) that are associated with the sensory neurons of the ENS. The axons of these sensory neurons can synapse with other neurons located in the ENS, CNS, or ANS, informing these regions about the nature of the contents and the degree of distension (stretching) of the GI tract. The neurons of the ENS, CNS, or ANS subsequently activate or inhibit GI glands and smooth muscle, altering GI secretion and motility.

✔ **CHECKPOINT**

6. How is the enteric nervous system regulated by the autonomic nervous system?
7. What is a gastrointestinal reflex pathway?

24.4 PERITONEUM

◉ **OBJECTIVE**

• Describe the peritoneum and its folds.

The **peritoneum** (per′-i-tō-NĒ-um; *peri-* = around) is the largest serous membrane of the body; it consists of a layer of simple squamous epithelium (mesothelium) with an underlying support-

ing layer of areolar connective tissue. The peritoneum is divided into the **parietal peritoneum,** which lines the wall of the abdominopelvic cavity, and the **visceral peritoneum,** which covers some of the organs in the cavity and is their serosa (Figure 24.4a). The slim space containing lubricating serous fluid that is between the parietal and visceral portions of the peritoneum is called the **peritoneal cavity.** In certain diseases, the peritoneal cavity may become distended by the accumulation of several liters of fluid, a condition called **ascites** (a-SĪ-tēz).

As you will see shortly, some organs lie on the posterior abdominal wall and are covered by peritoneum only on their anterior surfaces; they are not in the peritoneal cavity. Such organs, including the kidneys, ascending and descending colons of the large intestine, duodenum of the small intestine, and pancreas, are said to be **retroperitoneal** (*retro-* = behind).

Unlike the pericardium and pleurae, which smoothly cover the heart and lungs, the peritoneum contains large folds that weave between the viscera. The folds bind the organs to one another and to the walls of the abdominal cavity. They also contain blood vessels, lymphatic vessels, and nerves that supply the abdominal organs. There are five major peritoneal folds: the greater omentum, falciform ligament, lesser omentum, mesentery, and mesocolon:

Figure 24.4 Relationship of the peritoneal folds to one another and to organs of the digestive system. The size of the peritoneal cavity in (a) is exaggerated for emphasis.

The peritoneum is the largest serous membrane in the body.

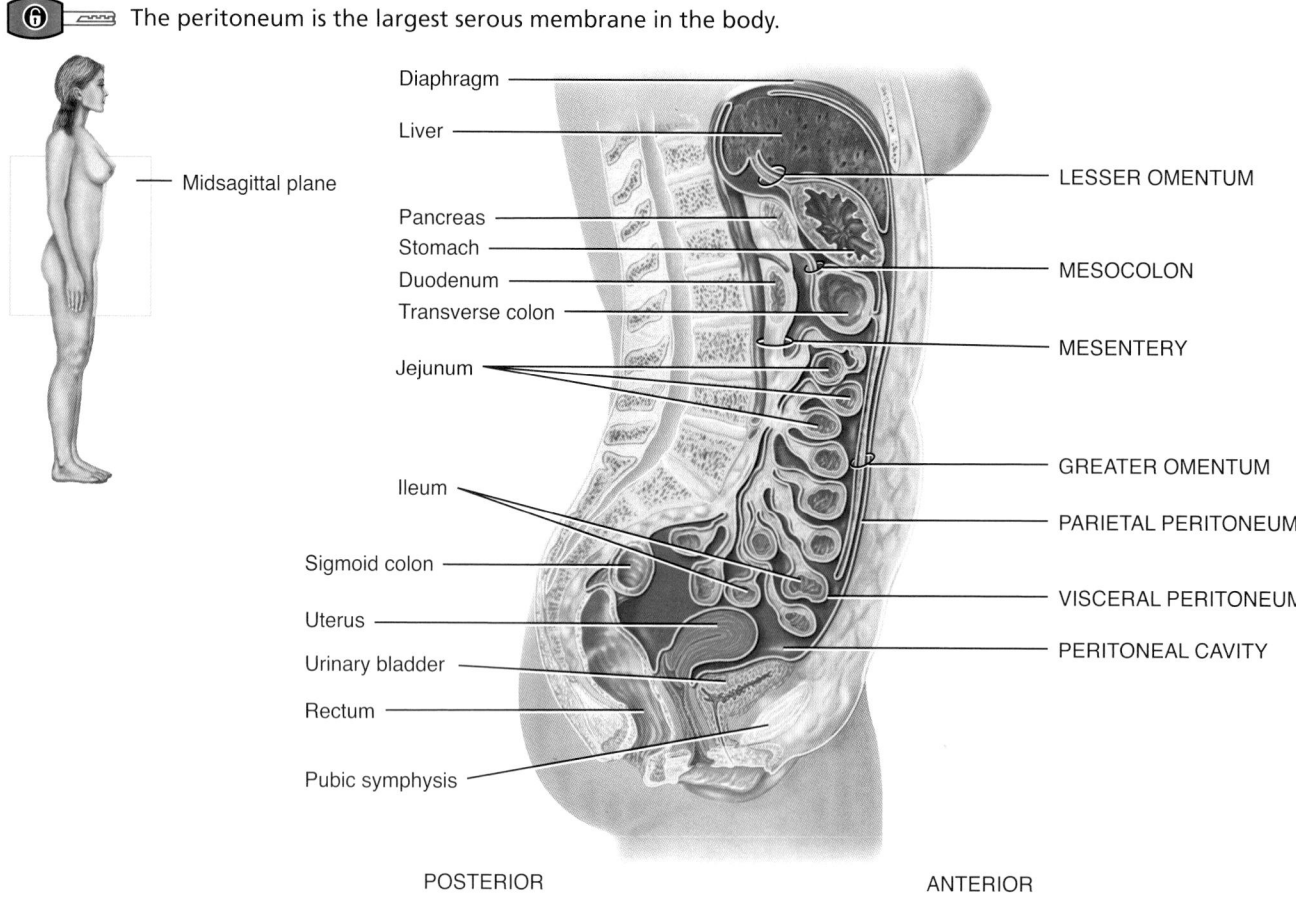

(a) Midsagittal section showing the peritoneal folds

1. The **greater omentum** (ō-MEN-tum = fat skin), the largest peritoneal fold, drapes over the transverse colon and coils of the small intestine like a "fatty apron" (Figure 24.4a, d). The greater omentum is a double sheet that folds back on itself, giving it a total of four layers. From attachments along the stomach and duodenum, the greater omentum extends downward anterior to the small intestine, then turns and extends upward and attaches to the transverse colon. The greater omentum normally contains a considerable amount of adipose tissue. Its adipose tissue content can greatly expand with weight gain, contributing to the characteristic "beer belly" seen in some overweight individuals.

The many lymph nodes of the greater omentum contribute macrophages and antibody-producing plasma cells that help combat and contain infections of the GI tract.

2. The **falciform ligament** (FAL-si-form; *falc-* = sickle-shaped) attaches the liver to the anterior abdominal wall and diaphragm (Figure 24.4b). The liver is the only digestive organ that is attached to the anterior abdominal wall.

3. The **lesser omentum** arises as an anterior fold in the serosa of the stomach and duodenum, and it suspends the stomach and duodenum from the liver (Figure 24.4a, c). It is the pathway

(b) Anterior view

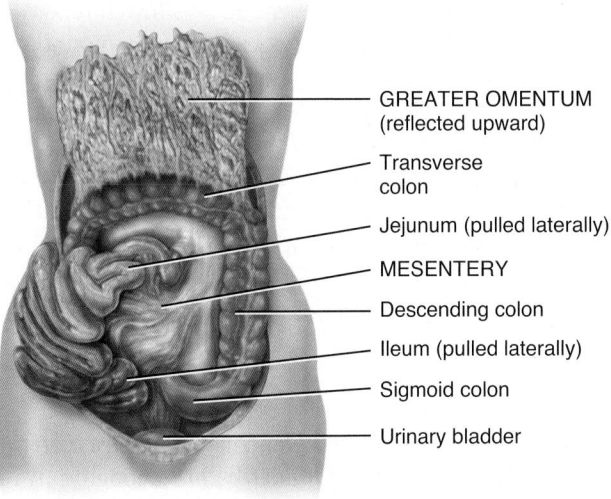

(d) Anterior view (greater omentum lifted and small intestine reflected to right side)

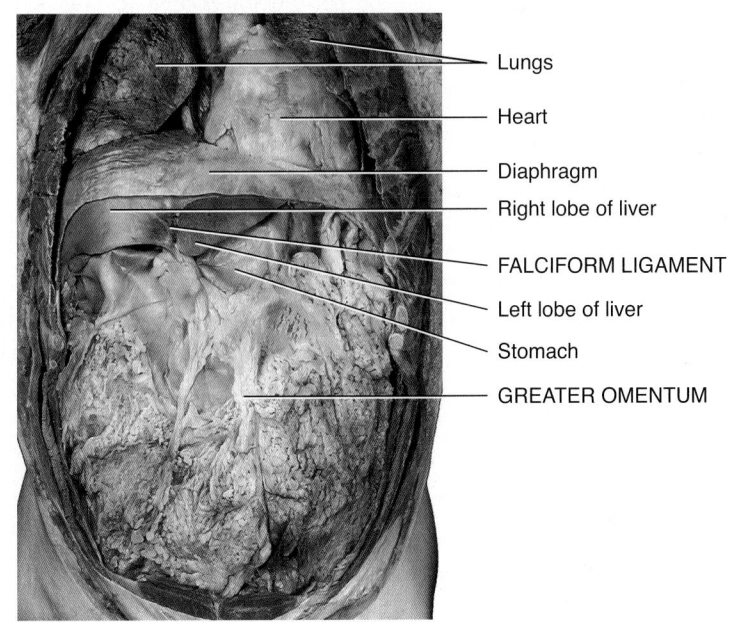

(c) Lesser omentum, anterior view (liver and gallbladder lifted)

(e) Anterior view

? **Which peritoneal fold binds the small intestine to the posterior abdominal wall?**

for blood vessels entering the liver and contains the hepatic portal vein, common hepatic artery, and common bile duct, along with some lymph nodes.

4. A fan-shaped fold of the peritoneum, called the **mesentery** (MEZ-en-ter′-ē; *mes-* = middle), binds the jejunum and ileum of the small intestine to the posterior abdominal wall (Figure 24.4a, d). This is the largest peritoneal fold and is typically laden with fat and contributes extensively to the large abdomen in obese individuals. It extends from the posterior abdominal wall to wrap around the small intestine and then returns to its origin, forming a double-layered structure. Between the two layers are blood and lymphatic vessels and lymph nodes.

5. Two separate folds of peritoneum, called the **mesocolon** (mez′-ō-KŌ-lon), bind the transverse colon (transverse mesocolon) and sigmoid colon (sigmoid mesocolon) of the large intestine to the posterior abdominal wall (Figure 24.4a). It also carries blood and lymphatic vessels to the intestines. Together, the mesentery and mesocolon hold the intestines loosely in place, allowing movement as muscular contractions mix and move the luminal contents along the GI tract.

CLINICAL CONNECTION | *Peritonitis*

A common cause of **peritonitis** (per′-i-tō-NĪ-tis), an acute inflammation of the peritoneum, is contamination of the peritoneum by infectious microbes, which can result from accidental or surgical wounds in the abdominal wall, or from perforation or rupture of abdominal organs. If, for example, bacteria gain access to the peritoneal cavity through an intestinal perforation or rupture of the appendix, they can produce an acute, life-threatening form of peritonitis. A less serious (but still painful) form of peritonitis can result from the rubbing together of inflamed peritoneal surfaces. The increased risk of peritonitis is of particular concern to those who rely on peritoneal dialysis, a procedure in which the peritoneum is used to filter the blood when the kidneys do not function properly (see Clinical Connection: Dialysis in Chapter 26). •

✔ CHECKPOINT

8. Where are the visceral peritoneum and parietal peritoneum located?
9. Describe the attachment sites and functions of the mesentery, mesocolon, falciform ligament, lesser omentum, and greater omentum.

24.5 MOUTH

◉ OBJECTIVES

• Identify the locations of the salivary glands, and describe the functions of their secretions.
• Describe the structure and functions of the tongue.
• Identify the parts of a typical tooth, and compare deciduous and permanent dentitions.

The **mouth,** also referred to as the **oral** or **buccal cavity** (BUK-al; *bucca* = cheeks), is formed by the cheeks, hard and soft

palates, and tongue (Figure 24.5). The **cheeks** form the lateral walls of the oral cavity. They are covered externally by skin and internally by a mucous membrane, which consists of nonkeratinized stratified squamous epithelium. Buccinator muscles and connective tissue lie between the skin and mucous membranes of the cheeks. The anterior portions of the cheeks end at the lips.

The **lips** or **labia** (= fleshy borders) are fleshy folds surrounding the opening of the mouth. They contain the orbicularis oris muscle and are covered externally by skin and internally by a mucous membrane. The inner surface of each lip is attached to its corresponding gum by a midline fold of mucous membrane called the **labial frenulum** (LĀ-bē-al FREN-ū-lum; *frenulum* = small bridle). During chewing, contraction of the buccinator muscles in the cheeks and orbicularis oris muscle in the lips helps keep food between the upper and lower teeth. These muscles also assist in speech.

The **oral vestibule** (= entrance to a canal) of the oral cavity is the space bounded externally by the cheeks and lips and internally by the gums and teeth. The **oral cavity proper** is the space that extends from the gums and teeth to the **fauces** (FAW-sēs = passages), the opening between the oral cavity and the oropharynx (throat).

The **palate** is a wall or septum that separates the oral cavity from the nasal cavity, and forms the roof of the mouth. This important structure makes it possible to chew and breathe at the same time. The **hard palate**—the anterior portion of the roof of the mouth—is formed by the maxillae and palatine bones and is covered by a mucous membrane; it forms a bony partition between the oral and nasal cavities. The **soft palate,** which forms the posterior portion of the roof of the mouth, is an arch-shaped muscular partition between the oropharynx and nasopharynx that is lined with mucous membrane.

Hanging from the free border of the soft palate is a conical muscular process called the **uvula** (Ū-vū-la = little grape). During swallowing, the soft palate and uvula are drawn superiorly, closing off the nasopharynx and preventing swallowed foods and liquids from entering the nasal cavity. Lateral to the base of the uvula are two muscular folds that run down the lateral sides of the soft palate: Anteriorly, the **palatoglossal arch** (pal-a-tō-GLOS-al) extends to the side of the base of the tongue; posteriorly, the **palatopharyngeal arch** (pal-a-tō-fa-RIN-jē-al) extends to the side of the pharynx. The palatine tonsils are situated between the arches, and the lingual tonsils are situated at the base of the tongue. At the posterior border of the soft palate, the mouth opens into the oropharynx through the fauces (Figure 24.5).

Salivary Glands

A **salivary gland** (SAL-i-vār-ē) is a gland that releases a secretion called saliva into the oral cavity. Ordinarily, just enough saliva is secreted to keep the mucous membranes of the mouth and pharynx moist and to cleanse the mouth and teeth. When food enters the mouth, however, secretion of saliva increases, and it lubricates, dissolves, and begins the chemical breakdown of the food.

The mucous membrane of the mouth and tongue contains many small salivary glands that open directly, or indirectly via short ducts, to the oral cavity. These glands include *labial, buccal,*

Figure 24.5 **Structures of the mouth (oral cavity).**

 The mouth is formed by the cheeks, hard and soft palates, and tongue.

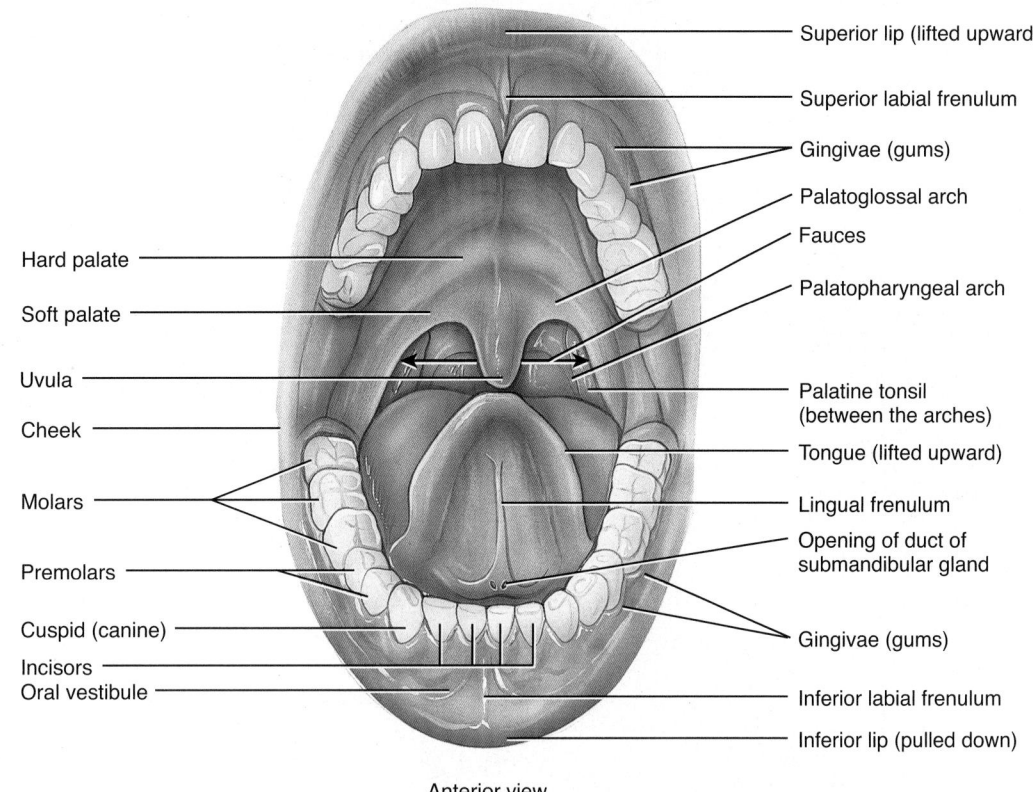

- Superior lip (lifted upward)
- Superior labial frenulum
- Gingivae (gums)
- Palatoglossal arch
- Fauces
- Palatopharyngeal arch
- Palatine tonsil (between the arches)
- Tongue (lifted upward)
- Lingual frenulum
- Opening of duct of submandibular gland
- Gingivae (gums)
- Inferior labial frenulum
- Inferior lip (pulled down)

- Hard palate
- Soft palate
- Uvula
- Cheek
- Molars
- Premolars
- Cuspid (canine)
- Incisors
- Oral vestibule

Anterior view

? **What is the function of the uvula?**

and *palatal glands* in the lips, cheeks, and palate, respectively, and *lingual glands* in the tongue, all of which make a small contribution to saliva.

However, most saliva is secreted by the **major salivary glands,** which lie beyond the oral mucosa, into ducts that lead to the oral cavity. There are three pairs of major salivary glands: the parotid, submandibular, and sublingual glands (Figure 24.6a). The **parotid glands** (pa-ROT-id; *par-* = near; *to-* = ear) are located inferior and anterior to the ears, between the skin and the masseter muscle. Each secretes saliva into the oral cavity via a **parotid duct** that pierces the buccinator muscle to open into the vestibule opposite the second maxillary (upper) molar tooth. The **submandibular glands** (sub'-man-DIB-ū-lar) are found in the floor of the mouth; they are medial and partly inferior to the body of the mandible. Their ducts, the **submandibular ducts,** run under the mucosa on either side of the midline of the floor of the mouth and enter the oral cavity proper lateral to the lingual frenulum. The **sublingual glands** (sub-LING-gwal) are beneath the tongue and superior to the submandibular glands. Their ducts, the **lesser sublingual ducts,** open into the floor of the mouth in the oral cavity proper.

Composition and Functions of Saliva

Chemically, **saliva** is 99.5% water and 0.5% solutes. Among the solutes are ions, including sodium, potassium, chloride, bicarbon-

ate, and phosphate. Also present are some dissolved gases and various organic substances, including urea and uric acid, mucus, immunoglobulin A, the bacteriolytic enzyme lysozyme, and salivary amylase, a digestive enzyme that acts on starch.

Not all salivary glands supply the same ingredients. The parotid glands secrete a watery (serous) liquid containing salivary amylase. Because the submandibular glands contain cells similar to those found in the parotid glands, plus some mucous cells, they secrete a fluid that contains amylase but is thickened with mucus. The sublingual glands contain mostly mucous cells, so they secrete a much thicker fluid that contributes only a small amount of salivary amylase.

The water in saliva provides a medium for dissolving foods so that they can be tasted by gustatory receptors and so that digestive reactions can begin. Chloride ions in the saliva activate **salivary amylase** (AM-i-lās), an enzyme that starts the breakdown of starch in the mouth into maltose, maltotriose, and α-dextrin. Bicarbonate and phosphate ions buffer acidic foods that enter the mouth, so saliva is only slightly acidic (pH 6.35–6.85). Salivary glands (like the sweat glands of the skin) help remove waste molecules from the body, which accounts for the presence of urea and uric acid in saliva. Mucus lubricates food so it can be moved around easily in the mouth, formed into a ball, and swallowed. Immunoglobulin A (IgA) prevents attachment of microbes so

Figure 24.6 **The three major salivary glands—parotid, sublingual, and submandibular.** The submandibular glands, shown in the light micrograph (b), consist mostly of serous acini (serous fluid–secreting portions of gland) and a few mucous acini (mucus-secreting portions of gland); the parotid glands consist of serous acini only; and the sublingual glands consist of mostly mucous acini and a few serous acini.

🔑 Saliva lubricates and dissolves foods and begins the chemical breakdown of carbohydrates and lipids.

(a) Location of salivary glands

(b) Portion of submandibular gland

LM 240x

❓ **What is the function of the chloride ions in saliva?**

they cannot penetrate the epithelium, and the enzyme lysozyme kills bacteria; however, these substances are not present in large enough quantities to eliminate all oral bacteria.

Salivation

The secretion of saliva, called **salivation** (sal-i-VĀ-shun), is controlled by the autonomic nervous system. Amounts of saliva secreted daily vary considerably but average 1000–1500 mL (1–1.6 qt). Normally, parasympathetic stimulation promotes continuous secretion of a moderate amount of saliva, which keeps the mucous membranes moist and lubricates the movements of the tongue and lips during speech. The saliva is then swallowed and helps moisten the esophagus. Eventually, most components of

saliva are reabsorbed, which prevents fluid loss. Sympathetic stimulation dominates during stress, resulting in dryness of the mouth. If the body becomes dehydrated, the salivary glands stop secreting saliva to conserve water; the resulting dryness in the mouth contributes to the sensation of thirst. Drinking not only restores the homeostasis of body water but also moistens the mouth.

The feel and taste of food also are potent stimulators of salivary gland secretions. Chemicals in the food stimulate receptors in taste buds on the tongue, and impulses are conveyed from the taste buds to two salivary nuclei in the brain stem (**superior** and **inferior salivatory nuclei**). Returning parasympathetic impulses in fibers of the facial (VII) and glossopharyngeal (IX) nerves stimulate the secretion of saliva. Saliva continues to be secreted heavily for some time after food is swallowed; this flow of saliva washes out the mouth and dilutes and buffers the remnants of irritating chemicals such as that tasty (but hot!) salsa. The smell, sight, sound, or thought of food may also stimulate secretion of saliva.

CLINICAL CONNECTION | Mumps

Although any of the salivary glands may be the target of a nasopharyngeal infection, the mumps virus (paramyxovirus) typically attacks the parotid glands. **Mumps** is an inflammation and enlargement of the parotid glands accompanied by moderate fever, malaise (general discomfort), and extreme pain in the throat, especially when swallowing sour foods or acidic juices. Swelling occurs on one or both sides of the face, just anterior to the ramus of the mandible. In about 30% of males past puberty, the testes may also become inflamed; sterility rarely occurs because testicular involvement is usually unilateral (one testis only). Since a vaccine became available for mumps in 1967, the incidence of the disease has declined dramatically. •

Tongue

The **tongue** is an accessory digestive organ composed of skeletal muscle covered with mucous membrane. Together with its associated muscles, it forms the floor of the oral cavity. The tongue is divided into symmetrical lateral halves by a median septum that extends its entire length, and it is attached inferiorly to the hyoid bone, styloid process of the temporal bone, and mandible. Each half of the tongue consists of an identical complement of extrinsic and intrinsic muscles.

The **extrinsic muscles** of the tongue, which originate outside the tongue (attach to bones in the area) and insert into connective tissues in the tongue, include the *hyoglossus, genioglossus,* and *styloglossus muscles* (see Figure 11.7). The extrinsic muscles move the tongue from side to side and in and out to maneuver food for chewing, shape the food into a rounded mass, and force the food to the back of the mouth for swallowing. They also form the floor of the mouth and hold the tongue in position. The **intrinsic muscles** originate in and insert into connective tissue within the tongue. They alter the shape and size of the tongue for speech and swallowing. The intrinsic muscles include the *longitudinalis superior, longitudinalis inferior, transversus linguae,* and *verticalis linguae muscles.* The **lingual frenulum** (*lingua* = the tongue), a fold of mucous membrane in the midline of the undersurface of the tongue, is attached to the floor of the mouth and aids in limiting the movement of the tongue posteriorly (see Figures 24.5 and 24.6). If a person's lingual frenulum is abnormally short or rigid—a condition called **ankyloglossia** (ang′-kē-lō-GLOSS-ē-a)—the person is said to be "tongue-tied" because of the resulting impairment to speech.

The dorsum (upper surface) and lateral surfaces of the tongue are covered with **papillae** (pa-PIL-ē = nipple-shaped projections), projections of the lamina propria covered with stratified squamous epithelium (see Figure 17.3). Many papillae contain taste buds, the receptors for gustation (taste). Some papillae lack taste buds, but they contain receptors for touch and increase friction between the tongue and food, making it easier for the tongue to move food in the oral cavity. The different types of taste buds are described in detail in Chapter 17. **Lingual glands** in the lamina propria of the tongue secrete both mucus and a watery serous fluid that contains the enzyme **lingual lipase,** which acts on as much as 30% of dietary triglycerides (fats and oils) and converts them to simpler fatty acids and diglycerides.

Teeth

The **teeth,** or **dentes** (Figure 24.7), are accessory digestive organs located in sockets of the alveolar processes of the mandible and maxillae. The alveolar processes are covered by the **gingivae** (JIN-ji-vē), or gums, which extend slightly into each socket. The sockets are lined by the **periodontal ligament** (per′-ē-ō-DON-tal; *odont-* = tooth) or **membrane,** which consists of dense fibrous connective tissue that anchors the teeth to the socket walls and acts as a shock absorber during chewing.

A typical tooth has three major external regions: the crown, root, and neck. The **crown** is the visible portion above the level of

---

Figure 24.7 A typical tooth and surrounding structures.

Teeth are anchored in sockets of the alveolar processes of the mandible and maxillae.

Sagittal plane

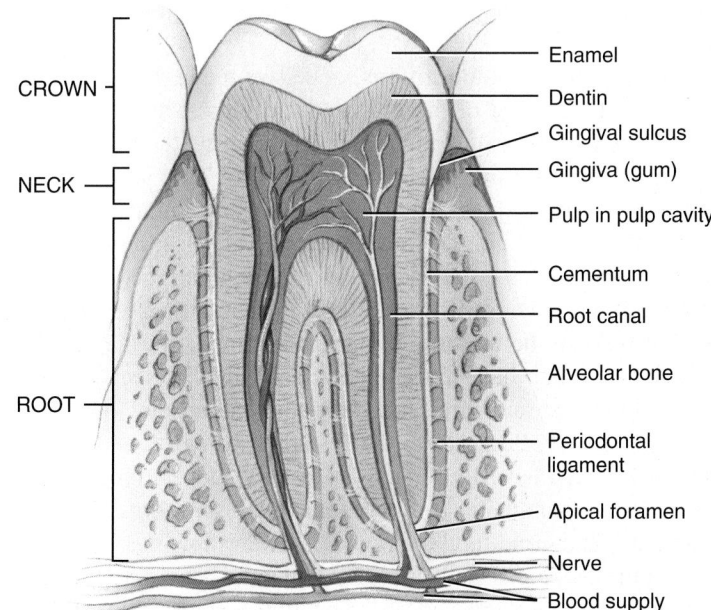

Sagittal section of a mandibular (lower) molar

CROWN
NECK
ROOT

Enamel
Dentin
Gingival sulcus
Gingiva (gum)
Pulp in pulp cavity
Cementum
Root canal
Alveolar bone
Periodontal ligament
Apical foramen
Nerve
Blood supply

? **What type of tissue is the main component of teeth?**

the gums. Embedded in the socket are one to three **roots.** The **neck** is the constricted junction of the crown and root near the gum line.

Internally, **dentin** forms the majority of the tooth. Dentin consists of a calcified connective tissue that gives the tooth its basic shape and rigidity. It is harder than bone because of its higher content of calcium salts (70% of dry weight).

The dentin of the crown is covered by **enamel,** which consists primarily of calcium phosphate and calcium carbonate. Enamel is also harder than bone because of its even higher content of calcium salts (about 95% of dry weight). In fact, enamel is the hardest substance in the body. It serves to protect the tooth from the wear and tear of chewing. It also protects against acids that can easily dissolve dentin. The dentin of the root is covered by **cementum,** another bonelike substance, which attaches the root to the periodontal ligament.

The dentin of a tooth encloses a space. The enlarged part of the space, the **pulp cavity,** lies within the crown and is filled with **pulp,** a connective tissue containing blood vessels, nerves, and lymphatic vessels. Narrow extensions of the pulp cavity, called **root canals,** run through the root of the tooth. Each root canal has an opening at its base, the **apical foramen,** through which blood vessels, lymphatic vessels, and nerves extend. The blood vessels bring nourishment, the lymphatic vessels offer protection, and the nerves provide sensation.

The branch of dentistry that is concerned with the prevention, diagnosis, and treatment of diseases that affect the pulp, root, periodontal ligament, and alveolar bone is known as **endodontics** (en′-dō-DON-tiks; *endo-* = within). **Orthodontics** (or′-thō-DON-tiks; *ortho-* = straight) is a branch of dentistry that is concerned with the prevention and correction of abnormally aligned teeth; **periodontics** (per′-ē-ō-DON-tiks) is a branch of dentistry concerned with the treatment of abnormal conditions of the tissues immediately surrounding the teeth, such as gingivitis (gum disease).

Humans have two **dentitions,** or sets of teeth: deciduous and permanent. The first of these—the **deciduous teeth** (*decidu-* = falling out), also called **primary teeth, milk teeth,** or **baby teeth**—begin to erupt at about 6 months of age, and approximately two teeth appear each month thereafter, until all 20 are present (Figure 24.8a). The **incisors,** which are closest to the midline, are chisel-shaped and adapted for cutting into food. They are referred to as either **central** or **lateral incisors** based on their position. Next to the incisors, moving posteriorly, are the **cuspids (canines),** which have a pointed surface called a *cusp.* Cuspids are used to tear and shred food. Incisors and cuspids have only one root apiece. Posterior to the cuspids lie the **first** and **second molars,** which have four cusps. Maxillary (upper) molars have three roots; mandibular (lower) molars have two roots. The molars crush and grind food to prepare it for swallowing.

All the deciduous teeth are lost—generally between ages 6 and 12 years—and are replaced by the **permanent (secondary) teeth** (Figure 24.8b). The permanent dentition contains 32 teeth that erupt between age 6 and adulthood. The pattern resembles the deciduous dentition, with the following exceptions. The deciduous

molars are replaced by the **first** and **second premolars (bicuspids),** which have two cusps and one root (upper first premolars have two roots) and are used for crushing and grinding. The permanent molars, which erupt into the mouth posterior to the premolars, do not replace any deciduous teeth and erupt as the jaw grows to accommodate them—the **first molars** at age 6 (six-year molars), the **second molars** at age 12 (twelve-year molars), and the **third molars (wisdom teeth)** after age 17 or not at all.

Often the human jaw does not have enough room posterior to the second molars to accommodate the eruption of the third molars. In this case, the third molars remain embedded in the alveolar bone and are said to be *impacted.* They often cause pressure and pain and must be removed surgically. In some people, third molars may be dwarfed in size or may not develop at all.

CLINICAL CONNECTION | Root Canal Therapy

Root canal therapy is a multistep procedure in which all traces of pulp tissue are removed from the pulp cavity and root canals of a badly diseased tooth. After a hole is made in the tooth, the root canals are filed out and irrigated to remove bacteria. Then, the canals are treated with medication and sealed tightly. The damaged crown is then repaired. •

Mechanical and Chemical Digestion in the Mouth

Mechanical digestion in the mouth results from chewing, or **mastication** (mas′-ti-KĀ-shun = to chew), in which food is manipulated by the tongue, ground by the teeth, and mixed with saliva. As a result, the food is reduced to a soft, flexible, easily swallowed mass called a **bolus** (= lump). Food molecules begin to dissolve in the water in saliva, an important activity because enzymes can react with food molecules in a liquid medium only.

Two enzymes, salivary amylase and lingual lipase, contribute to chemical digestion in the mouth. Salivary amylase, which is secreted by the salivary glands, initiates the breakdown of starch. Dietary carbohydrates are either monosaccharide and disaccharide sugars or complex polysaccharides such as starches. Most of the carbohydrates we eat are starches, but only monosaccharides can be absorbed into the bloodstream. Thus, ingested disaccharides and starches must be broken down into monosaccharides. The function of salivary amylase is to begin starch digestion by breaking down starch into smaller molecules such as the disaccharide maltose, the trisaccharide maltotriose, and short-chain glucose polymers called α-dextrins. Even though food is usually swallowed too quickly for all the starches to be broken down in the mouth, salivary amylase in the swallowed food continues to act on the starches for about another hour, at which time stomach acids inactivate it. Saliva also contains **lingual lipase** (LĪ-pās), which is secreted by lingual glands in the tongue. This enzyme becomes activated in the acidic environment of the stomach and thus starts to work after food is swallowed. It breaks down dietary triglycerides (fats and oils) into fatty acids and diglycerides. A diglyceride consists of a glycerol molecule that is attached to two fatty acids.

Table 24.1 summarizes the digestive activities in the mouth.

Figure 24.8 Dentitions and times of eruptions (indicated in parentheses). A designated letter (deciduous teeth) or number (permanent teeth) uniquely identifies each tooth. Deciduous teeth begin to erupt at 6 months of age, and approximately two teeth appear each month thereafter, until all 20 are present.

There are 20 teeth in a complete deciduous set and 32 teeth in a complete permanent set.

Central incisor (8–12 mo.)

Lateral incisor (12–24 mo.)

Cuspid or canine (16–24 mo.)

First molar (12–16 mo.)

Second molar (24–32 mo.)

Upper Teeth

Second molar (24–32 mo.)

First molar (12–16 mo.)

Cuspid or canine (16–24 mo.)

Lateral incisor (12–15 mo.)

Central incisor (6–8 mo.)

Lower Teeth

(a) Deciduous (primary) dentition; teeth are designated by letters (with times of eruption)

Central incisor (7–8 yr.)

Lateral incisor (8–9 yr.)

Cuspid or canine (11–12 yr.)

First premolar or bicuspid (9–10 yr.)

Second premolar or bicuspid (10–12 yr.)

First molar (6–7 yr.)

Second molar (12–13 yr.)

Third molar or wisdom tooth (17–21 yr.)

Upper Teeth

Third molar or wisdom tooth (17–21 yr.)

Second molar (11–13 yr.)

First molar (6–7 yr.)

Second premolar or bicuspid (11–12 yr.)

First premolar or bicuspid (9–10 yr.)

Cuspid or canine (9–10 yr.)

Lateral incisor (7–8 yr.)

Central incisor (7–8 yr.)

Lower Teeth

(b) Permanent (secondary) dentition; teeth are designated by numbers (with times of eruption)

? **Which permanent teeth do not replace any deciduous teeth?**

TABLE 24.1

Summary of Digestive Activities in the Mouth

STRUCTURE	ACTIVITY	RESULT
Cheeks and lips	Keep food between teeth.	Foods uniformly chewed during mastication.
Salivary glands	Secrete saliva.	Lining of mouth and pharynx moistened and lubricated. Saliva softens, moistens, and dissolves food and cleanses mouth and teeth. Salivary amylase splits starch into smaller fragments (maltose, maltotriose, and α-dextrins).
Tongue		
Extrinsic tongue muscles	Move tongue from side to side and in and out.	Food maneuvered for mastication, shaped into bolus, and maneuvered for swallowing.
Intrinsic tongue muscles	Alter shape of tongue.	Swallowing and speech.
Taste buds	Serve as receptors for gustation (taste) and presence of food in mouth.	Secretion of saliva stimulated by nerve impulses from taste buds to salivatory nuclei in brain stem to salivary glands.
Lingual glands	Secrete lingual lipase.	Triglycerides broken down into fatty acids and diglycerides.
Teeth	Cut, tear, and pulverize food.	Solid foods reduced to smaller particles for swallowing.

10. What structures form the mouth?
11. How are the major salivary glands distinguished on the basis of location?
12. How is the secretion of saliva regulated?
13. What functions do incisors, cuspids, premolars, and molars perform?

24.6 PHARYNX

◉ OBJECTIVE
- Describe the location and function of the pharynx.

When food is first swallowed, it passes from the mouth into the **pharynx** (= throat), a funnel-shaped tube that extends from the internal nares to the esophagus posteriorly and to the larynx anteriorly (see Figure 23.2). The pharynx is composed of skeletal muscle and lined by mucous membrane, and is divided into three parts: the nasopharynx, the oropharynx, and the laryngopharynx. The nasopharynx functions only in respiration, but both the oropharynx and laryngopharynx have digestive as well as respiratory functions. Swallowed food passes from the mouth into the oropharynx and laryngopharynx; the muscular contractions of these areas help propel food into the esophagus and then into the stomach.

✔CHECKPOINT

14. To which two organ systems does the pharynx belong?

24.7 ESOPHAGUS

◉ OBJECTIVE
- Describe the location, anatomy, histology, and functions of the esophagus.

The **esophagus** (e-SOF-a-gus = eating gullet) is a collapsible muscular tube, about 25 cm (10 in.) long, that lies posterior to the trachea. The esophagus begins at the inferior end of the laryngopharynx, passes through the inferior aspect of the neck, and enters the mediastinum anterior to the vertebral column. Then it pierces the diaphragm through an opening called the **esophageal hiatus** (e-sof-a-JĒ-al hī-Ā-tus), and ends in the superior portion of the stomach (see Figure 24.1). Sometimes, part of the stomach protrudes above the diaphragm through the esophageal hiatus. This condition, termed a **hiatus hernia** (HER-nē-a), is described in the Medical Terminology section at the end of the chapter.

Histology of the Esophagus

The **mucosa** of the esophagus consists of nonkeratinized stratified squamous epithelium, lamina propria (areolar connective tissue), and a muscularis mucosae (smooth muscle) (Figure 24.9). Near the stomach, the mucosa of the esophagus also contains mucous glands. The stratified squamous epithelium associated with the lips, mouth, tongue, oropharynx, laryngopharynx, and esophagus affords considerable protection against abrasion and wear and tear

Figure 24.9 Histology of the esophagus. A higher-magnification view of nonkeratinized stratified squamous epithelium is shown in Table 4.1.

🔑 The esophagus secretes mucus and transports food to the stomach.

Lumen of esophagus
Mucosa:
Nonkeratinized stratified squamous epithelium
Lamina propria
Muscularis mucosae
Submucosa
Muscularis (circular layer)
Muscularis (longitudinal layer)
Adventitia
Transverse plane

LM 20x

Wall of the esophagus

❓ In which layers of the esophagus are the glands that secrete lubricating mucus located?

from food particles that are chewed, mixed with secretions, and swallowed. The **submucosa** contains areolar connective tissue, blood vessels, and mucous glands. The **muscularis** of the superior third of the esophagus is skeletal muscle, the intermediate third is skeletal and smooth muscle, and the inferior third is smooth muscle. At each end of the esophagus, the muscularis becomes slightly more prominent and forms two sphincters—the **upper esophageal sphincter (UES)** (e-sof′-a-JĒ-al) or **valve**, which consists of skeletal muscle, and the **lower esophageal sphincter (LES)** or **valve**, which consists of smooth muscle. The upper esophageal sphincter regulates the movement of food from the pharynx into the esophagus; the lower esophageal sphincter regulates the movement of food from the esophagus into the stomach. The superficial layer of the esophagus is known as the **adventitia** (ad-ven-TISH-a), rather than the serosa as in the stomach and intestines, because the areolar connective tissue of this layer is not covered by mesothelium and because the connective tissue merges with the connective tissue of surrounding structures of the mediastinum through which it passes. The adventitia attaches the esophagus to surrounding structures.

Physiology of the Esophagus

The esophagus secretes mucus and transports food into the stomach. It does not produce digestive enzymes, and it does not carry on absorption.

✔ CHECKPOINT

15. Describe the location and histology of the esophagus. What is its role in digestion?
16. What are the functions of the upper and lower esophageal sphincters?

24.8 DEGLUTITION

◉ OBJECTIVE

• Describe the three phases of deglutition.

The movement of food from the mouth into the stomach is achieved by the act of swallowing, or **deglutition** (dē-gloo-TISH-un) (Figure 24.10). Deglutition is facilitated by the secretion of saliva and mucus and involves the mouth, pharynx, and esophagus. Swallowing occurs in three stages: (1) the voluntary stage, in which the bolus is passed into the oropharynx; (2) the pharyngeal stage, the involuntary passage of the bolus through the pharynx into the esophagus; and (3) the esophageal stage, the involuntary passage of the bolus through the esophagus into the stomach.

Swallowing starts when the bolus is forced to the back of the oral cavity and into the oropharynx by the movement of the tongue upward and backward against the palate; these actions constitute the **voluntary stage** of swallowing. With the passage of the bolus into the oropharynx, the involuntary **pharyngeal stage** of swallowing begins (Figure 24.10b). The bolus stimulates receptors in the oropharynx, which send impulses to the **deglutition center** in the medulla oblongata and lower pons of the brain stem. The returning impulses cause the soft palate and uvula to move upward to close off the nasopharynx, which prevents swallowed foods and liquids from entering the nasal cavity. In addition, the epiglottis closes off the opening to the larynx, which prevents the bolus from entering the rest of the respiratory tract. The bolus moves through the oropharynx and the laryngopharynx. Once the upper esophageal sphincter relaxes, the bolus moves into the esophagus.

The **esophageal stage** of swallowing begins once the bolus enters the esophagus. During this phase, **peristalsis** (per′-i-STAL-sis; *stalsis* = constriction), a progression of coordinated contractions and relaxations of the circular and longitudinal layers of the muscularis, pushes the bolus onward (Figure 24.10c). (Peristalsis occurs in other tubular structures, including other parts of the GI tract and the ureters, bile ducts, and uterine tubes; in the esophagus it is controlled by the medulla oblongata.) In the section of the esophagus just superior to the bolus, the circular muscle fibers contract, constricting the esophageal wall and squeezing the bolus

Figure 24.10 Deglutition (swallowing). During the pharyngeal stage (b) the tongue rises against the palate, the nasopharynx is closed off, the larynx rises, the epiglottis seals off the larynx, and the bolus is passed into the esophagus. During the esophageal stage (c), food moves through the esophagus into the stomach via peristalsis.

🔑 Deglutition is a mechanism that moves food from the mouth into the stomach.

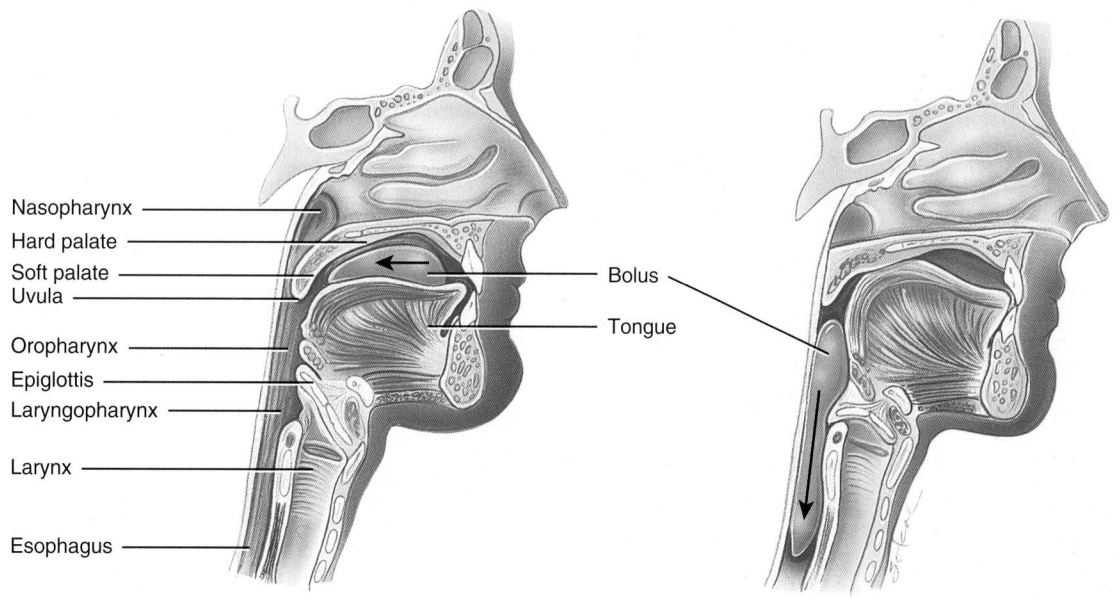

(a) Position of structures before swallowing

(b) During pharyngeal stage of swallowing

FIGURE 24.10 CONTINUES ▶

■ **FIGURE 24.10 CONTINUED** ▶

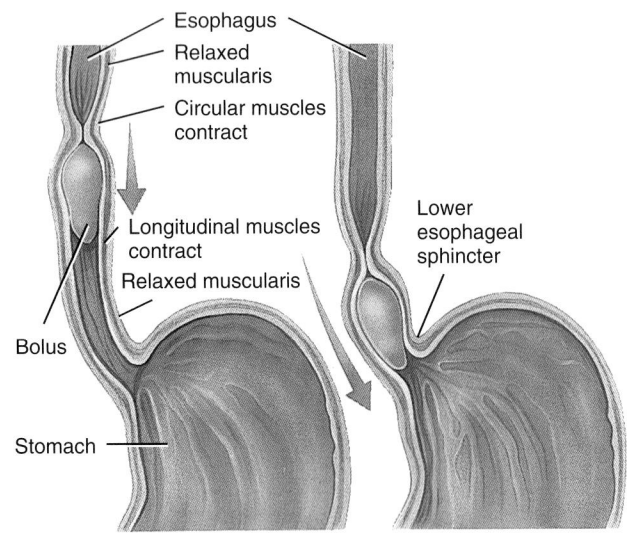

(c) Anterior view of frontal sections of peristalsis in esophagus

❓ **Is swallowing a voluntary action or an involuntary action?**

toward the stomach. Meanwhile, longitudinal fibers inferior to the bolus also contract, which shortens this inferior section and pushes its walls outward so it can receive the bolus. The contractions are repeated in waves that push the food toward the stomach. As the bolus approaches the end of the esophagus, the lower esophageal sphincter relaxes and the bolus moves into the stomach. Mucus secreted by esophageal glands lubricates the bolus and reduces friction. The passage of solid or semisolid food from the mouth to the stomach takes 4 to 8 seconds; very soft foods and liquids pass through in about 1 second.

Table 24.2 summarizes the digestive activities of the pharynx and esophagus.

TABLE 24.2

Summary of Digestive Activities in the Pharynx and Esophagus

STRUCTURE	ACTIVITY	RESULT
Pharynx	Pharyngeal stage of deglutition.	Moves bolus from oropharynx to laryngopharynx and into esophagus; closes air passageways.
Esophagus	Relaxation of upper esophageal sphincter.	Permits entry of bolus from laryngopharynx into esophagus.
	Esophageal stage of deglutition (peristalsis).	Pushes bolus down esophagus.
	Relaxation of lower esophageal sphincter.	Permits entry of bolus into stomach.
	Secretion of mucus.	Lubricates esophagus for smooth passage of bolus.

CLINICAL CONNECTION | *Gastroesophageal Reflux Disease*

If the lower esophageal sphincter fails to close adequately after food has entered the stomach, the stomach contents can reflux (back up) into the inferior portion of the esophagus. This condition is known as **gastroesophageal reflux disease (GERD)** (gas′-trō-e-sof-a-JĒ-al). Hydrochloric acid (HCl) from the stomach contents can irritate the esophageal wall, resulting in a burning sensation that is called **heartburn** because it is experienced in a region very near the heart; it is unrelated to any cardiac problem. Drinking alcohol and smoking can cause the sphincter to relax, worsening the problem. The symptoms of GERD often can be controlled by avoiding foods that strongly stimulate stomach acid secretion (coffee, chocolate, tomatoes, fatty foods, orange juice, peppermint, spearmint, and onions). Other acid-reducing strategies include taking over-the-counter histamine-2 (H_2) blockers such as Tagamet HB® or Pepcid AC® 30 to 60 minutes before eating to block acid secretion, and neutralizing acid that has already been secreted with antacids such as Tums® or Maalox®. Symptoms are less likely to occur if food is eaten in smaller amounts and if the person does not lie down immediately after a meal. GERD may be associated with cancer of the esophagus. •

✔ **CHECKPOINT**

17. What does deglutition mean?
18. What occurs during the voluntary and pharyngeal phases of swallowing?
19. Does peristalsis "push" or "pull" food along the gastrointestinal tract?

24.9 STOMACH

◉ **OBJECTIVE**

• Describe the location, anatomy, histology, and functions of the stomach.

The **stomach** is a J-shaped enlargement of the GI tract directly inferior to the diaphragm in the abdomen. The stomach connects the esophagus to the duodenum, the first part of the small intestine (Figure 24.11). Because a meal can be eaten much more quickly than the intestines can digest and absorb it, one of the functions of the stomach is to serve as a mixing chamber and holding reservoir. At appropriate intervals after food is ingested, the stomach forces a small quantity of material into the first portion of the small intestine. The position and size of the stomach vary continually; the diaphragm pushes it inferiorly with each inhalation and pulls it superiorly with each exhalation. Empty, it is about the size of a large sausage, but it is the most distensible part of the GI tract and can accommodate a large quantity of food. In the stomach, digestion of starch and triglycerides continues, digestion of proteins begins, the semisolid bolus is converted to a liquid, and certain substances are absorbed.

Anatomy of the Stomach

The stomach has four main regions: the cardia, fundus, body, and pyloric part (Figure 24.11). The **cardia** (KAR-dē-a) surrounds the

Figure 24.11 External and internal anatomy of the stomach.

The four regions of the stomach are the cardia, fundus, body, and pyloric part.

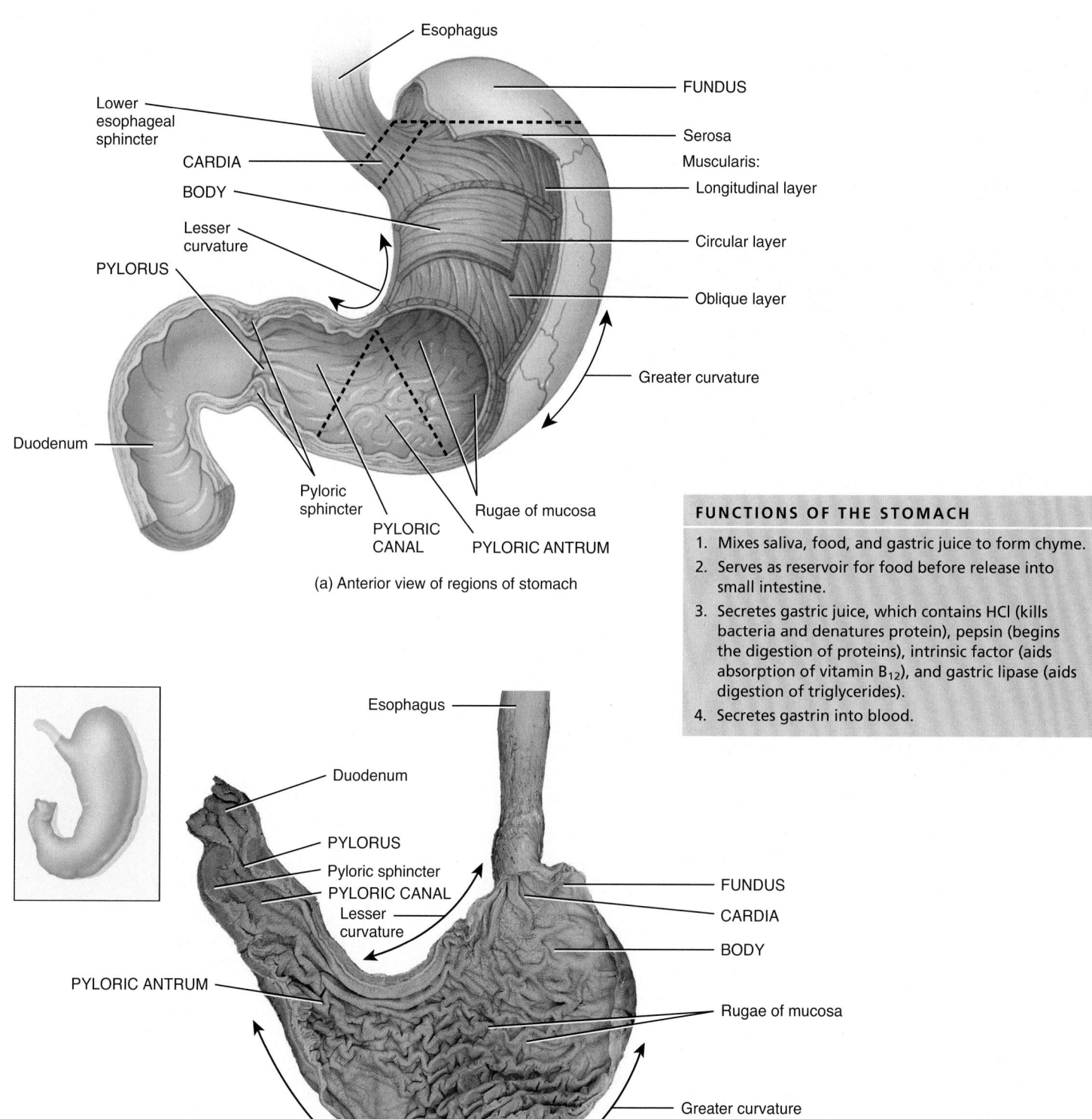

(a) Anterior view of regions of stomach

FUNCTIONS OF THE STOMACH

1. Mixes saliva, food, and gastric juice to form chyme.
2. Serves as reservoir for food before release into small intestine.
3. Secretes gastric juice, which contains HCl (kills bacteria and denatures protein), pepsin (begins the digestion of proteins), intrinsic factor (aids absorption of vitamin B_{12}), and gastric lipase (aids digestion of triglycerides).
4. Secretes gastrin into blood.

(b) Anterior view of internal anatomy

? After a very large meal, does your stomach still have rugae?

superior opening of the stomach. The rounded portion superior to and to the left of the cardia is the **fundus** (FUN-dus). Inferior to the fundus is the large central portion of the stomach, the **body.** The **pyloric part** is divisible into three regions. The first region, the **pyloric antrum,** connects to the body of the stomach. The second region, the **pyloric canal,** leads to the third region, the **pylorus** (pī-LOR-us; *pyl-* = gate; *-orus* = guard), which in turn connects to the duodenun. When the stomach is empty, the mucosa lies in large folds, or **rugae** (ROO-gē = wrinkles), that can be seen with the unaided eye. The pylorus communicates with the duodenum of the small intestine via a smooth muscle sphincter called the **pyloric sphincter.** The concave medial border of the stomach is called the **lesser curvature;** the convex lateral border is called the **greater curvature.**

CLINICAL CONNECTION | *Pylorospasm and Pyloric Stenosis*

Two abnormalities of the pyloric sphincter can occur in infants. In **pylorospasm** (pī-LOR-ō-spazm), the smooth muscle fibers of the sphincter fail to relax normally, so food does not pass easily from the stomach to the small intestine, the stomach becomes overly full, and the infant vomits often to relieve the pressure. Pylorospasm is treated by drugs that relax the muscle fibers of the pyloric sphincter. **Pyloric stenosis** (ste-NŌ-sis) is a narrowing of the pyloric sphincter that must be corrected surgically. The hallmark symptom is *projectile vomiting*—the spraying of liquid vomitus some distance from the infant. •

Histology of the Stomach

The stomach wall is composed of the same basic layers as the rest of the GI tract, with certain modifications. The surface of the **mucosa** is a layer of simple columnar epithelial cells called **surface mucous cells** (Figure 24.12). The mucosa contains a **lamina propria** (areolar connective tissue) and a **muscularis mucosae** (smooth muscle) (Figure 24.12). Epithelial cells extend down into the lamina propria, where they form columns of secretory cells called **gastric glands.** Several gastric glands open into the bottom of narrow channels called **gastric pits.** Secretions from several gastric glands flow into each gastric pit and then into the lumen of the stomach.

The gastric glands contain three types of *exocrine gland cells* that secrete their products into the stomach lumen: mucous neck

Figure 24.12 Histology of the stomach.

Gastric juice is the combined secretions of mucous cells, parietal cells, and chief cells.

Lumen of stomach

Gastric pits

Surface mucous cell

Lamina propria

Mucous neck cell

Parietal cell

Chief cell

Gastric gland

G cell

Lymphatic nodule

Muscularis mucosae

Lymphatic vessel

Venule

Arteriole

Oblique layer of muscle

Circular layer of muscle

Myenteric plexus

Longitudinal layer of muscle

MUCOSA

SUBMUCOSA

MUSCULARIS

SEROSA

(a) Three-dimensional view of layers of stomach

cells, chief cells, and parietal cells. Both surface mucous cells and **mucous neck cells** secrete mucus (Figure 24.12b). **Parietal cells** produce intrinsic factor (needed for absorption of vitamin B$_{12}$) and hydrochloric acid. The **chief cells** secrete pepsinogen and gastric lipase. The secretions of the mucous, parietal, and chief cells form

gastric juice, which totals 2000–3000 mL (roughly 2–3 qt) per day. In addition, gastric glands include a type of enteroendocrine cell, the **G cell,** which is located mainly in the pyloric antrum and secretes the hormone gastrin into the bloodstream. As we will see shortly, this hormone stimulates several aspects of gastric activity.

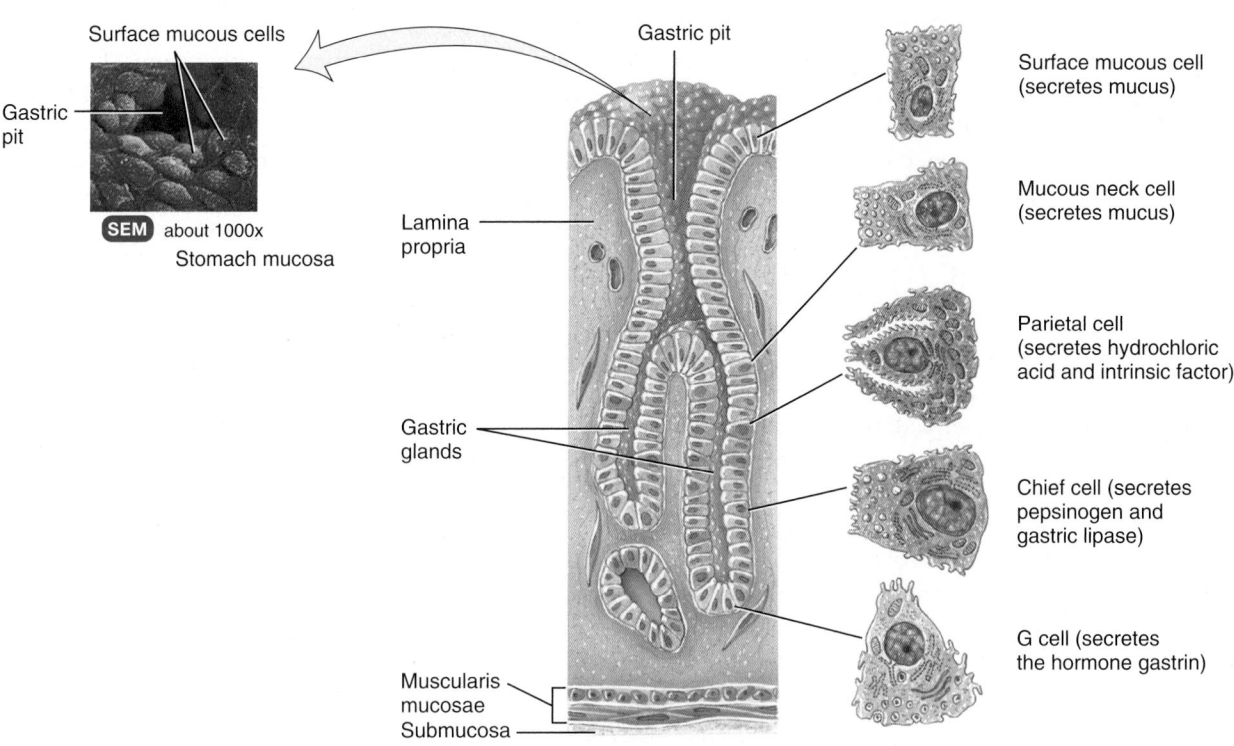

Surface mucous cells

Gastric pit

Gastric pit

SEM about 1000x
Stomach mucosa

Lamina propria

Gastric glands

Muscularis mucosae

Submucosa

Surface mucous cell (secretes mucus)

Mucous neck cell (secretes mucus)

Parietal cell (secretes hydrochloric acid and intrinsic factor)

Chief cell (secretes pepsinogen and gastric lipase)

G cell (secretes the hormone gastrin)

(b) Sectional view of stomach mucosa showing gastric glands and cell types

Gastric pit
Lamina propria
Surface mucous cell

Lumen of gastric gland

Mucous neck cell

Parietal cell

G cells

Lumen of gastric gland

Chief cells

LM 180x

(c) Fundic mucosa

? **Where is HCl secreted, and what are its functions?**

Three additional layers lie deep to the mucosa. The **submucosa** of the stomach is composed of areolar connective tissue. The **muscularis** has three layers of smooth muscle (rather than the two found in the esophagus and small and large intestines): an outer longitudinal layer, a middle circular layer, and an inner oblique layer. The oblique layer is limited primarily to the body of the stomach. The **serosa** is composed of simple squamous epithelium (mesothelium) and areolar connective tissue; the portion of the serosa covering the stomach is part of the visceral peritoneum. At the lesser curvature of the stomach, the visceral peritoneum extends upward to the liver as the lesser omentum. At the greater curvature of the stomach, the visceral peritoneum continues downward as the greater omentum and drapes over the intestines.

Mechanical and Chemical Digestion in the Stomach

Several minutes after food enters the stomach, gentle, rippling, peristaltic movements called **mixing waves** pass over the stomach every 15 to 25 seconds. These waves macerate food, mix it with secretions of the gastric glands, and reduce it to a soupy liquid called **chyme** (KĪM = juice). Few mixing waves are observed in the fundus, which primarily has a storage function. As digestion proceeds in the stomach, more vigorous mixing waves begin at the body of the stomach and intensify as they reach the pylorus. The pyloric sphincter normally remains almost, but not completely, closed. As food reaches the pylorus, each mixing wave periodically forces about 3 mL of chyme into the duodenum through the pyloric sphincter, a phenomenon known as **gastric emptying.** Most of the chyme is forced back into the body of the stomach, where mixing continues. The next wave pushes the chyme forward again and forces a little more into the duodenum. These forward and backward movements of the gastric contents are responsible for most mixing in the stomach.

Foods may remain in the fundus for about an hour without becoming mixed with gastric juice. During this time, digestion by salivary amylase continues. Soon, however, the churning action mixes chyme with acidic gastric juice, inactivating salivary amylase and activating lingual lipase, which starts to digest triglycerides into fatty acids and diglycerides.

Although parietal cells secrete hydrogen ions (H^+) and chloride ions (Cl^-) separately into the stomach lumen, the net effect is secretion of hydrochloric acid (HCl). **Proton pumps** powered by H^+/K^+ ATPases actively transport H^+ into the lumen while bringing potassium ions (K^+) into the cell (Figure 24.13). At the same time, Cl^- and K^+ diffuse out into the lumen through Cl^- and K^+ channels in the apical membrane. The enzyme *carbonic anhydrase*, which is especially plentiful in parietal cells, catalyzes the formation of carbonic acid (H_2CO_3) from water (H_2O) and carbon dioxide (CO_2). As carbonic acid dissociates, it provides a ready source of H^+ for the proton pumps but also generates bicarbonate ions (HCO_3^-). As HCO_3^- builds up in the cytosol, it exits the parietal cell in exchange for Cl^- via Cl^-/HCO_3^- antiporters in the basolateral membrane (next to the lamina propria). HCO_3^- diffuses into nearby blood capillaries.

This "alkaline tide" of bicarbonate ions entering the bloodstream after a meal may be large enough to elevate blood pH slightly and make urine more alkaline.

HCl secretion by parietal cells can be stimulated by several sources: acetylcholine (ACh) released by parasympathetic neurons, gastrin secreted by G cells, and histamine, which is a paracrine substance released by mast cells in the nearby lamina propria (Figure 24.14). Acetylcholine and gastrin stimulate parietal cells to secrete more HCl in the presence of histamine. In other words, histamine acts synergistically, enhancing the effects of acetylcholine and gastrin. Receptors for all three substances are present in the plasma membrane of parietal cells. The histamine receptors on parietal cells are called H_2 receptors; they mediate different responses than do the H_1 receptors involved in allergic responses.

Figure 24.13 Secretion of HCl (hydrochloric acid) by parietal cells in the stomach.

Proton pumps, powered by ATP, secrete the H^+; Cl^- diffuses into the stomach lumen through Cl^- channels.

Key:

Proton pump (H^+/K^+ ATPase)	CA	Carbonic anhydrase
K^+ (potassium ion) channel	• • • • • ▶	Diffusion
Cl^- (chloride ion) channel	HCO_3^- / Cl^-	HCO_3^-/Cl^- antiporter

? **What molecule is the source of the hydrogen ions that are secreted into gastric juice?**

The strongly acidic fluid of the stomach kills many microbes in food. HCl partially denatures (unfolds) proteins in food and stimulates the secretion of hormones that promote the flow of bile and pancreatic juice. Enzymatic digestion of proteins also begins in the stomach. The only proteolytic (protein-digesting) enzyme in the stomach is **pepsin,** which is secreted by chief cells. Pepsin severs certain peptide bonds between amino acids, breaking down a protein chain of many amino acids into smaller peptide fragments. Pepsin is most effective in the very acidic environment of the stomach (pH 2); it becomes inactive at a higher pH.

What keeps pepsin from digesting the protein in stomach cells along with the food? First, pepsin is secreted in an inactive form called *pepsinogen;* in this form, it cannot digest the proteins in the chief cells that produce it. Pepsinogen is not converted into active pepsin until it comes in contact with hydrochloric acid secreted by parietal cells or active pepsin molecules. Second, the stomach epithelial cells are protected from gastric juices by a layer 1–3 mm thick of alkaline mucus secreted by surface mucous cells and mucous neck cells.

Another enzyme of the stomach is **gastric lipase,** which splits the short-chain triglycerides (fats and oils) in fat molecules (such as those found in milk) into fatty acids and monoglycerides. A monoglyceride consists of a glycerol molecule that is attached to one fatty acid molecule. This enzyme, which has a limited role in the adult stomach, operates best at a pH of 5–6. More important than either lingual lipase or gastric lipase is pancreatic lipase, an enzyme secreted by the pancreas into the small intestine.

Only a small amount of nutrients are absorbed in the stomach because its epithelial cells are impermeable to most materials. However, mucous cells of the stomach absorb some water, ions, and short-chain fatty acids, as well as certain drugs (especially aspirin) and alcohol.

Within 2 to 4 hours after eating a meal, the stomach has emptied its contents into the duodenum. Foods rich in carbohydrate spend the least time in the stomach; high-protein foods remain somewhat longer, and emptying is slowest after a fat-laden meal containing large amounts of triglycerides.

Table 24.3 summarizes the digestive activities of the stomach.

Figure 24.14 Regulation of HCl secretion.

🔑 HCl secretion by parietal cells can be stimulated by several sources: acetylcholine (ACh), gastrin, and histamine.

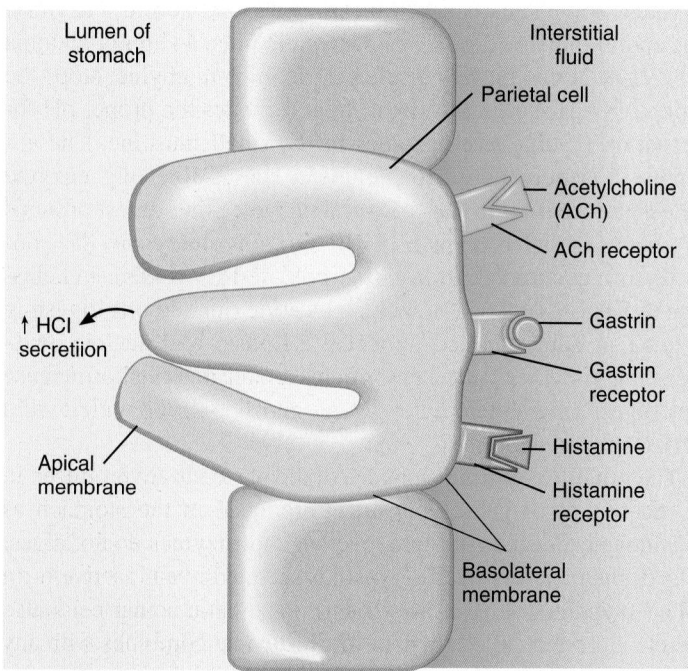

❓ **Among the sources that stimulate HCl secretion, which one is a paracrine agent that is released by mast cells in the lamina propria?**

TABLE 24.3		
Summary of Digestive Activities in the Stomach		
STRUCTURE	ACTIVITY	RESULT
Mucosa		
Surface mucous cells and mucous neck cells	Secrete mucus.	Forms protective barrier that prevents digestion of stomach wall.
	Absorption.	Small quantity of water, ions, short-chain fatty acids, and some drugs enter bloodstream.
Parietal cells	Secrete intrinsic factor.	Needed for absorption of vitamin B_{12} (used in red blood cell formation, or erythropoiesis).
	Secrete hydrochloric acid.	Kills microbes in food; denatures proteins; converts pepsinogen into pepsin.
Chief cells	Secrete pepsinogen.	Pepsin (activated form) breaks down proteins into peptides.
	Secrete gastric lipase.	Splits triglycerides into fatty acids and monoglycerides.
G cells	Secrete gastrin.	Stimulates parietal cells to secrete HCl and chief cells to secrete pepsinogen; contracts lower esophageal sphincter, increases motility of stomach, and relaxes pyloric sphincter.
Muscularis	Mixing waves (gentle peristaltic movements).	Churns and physically breaks down food and mixes it with gastric juice, forming chyme. Forces chyme through pyloric sphincter.
Pyloric sphincter	Opens to permit passage of chyme into duodenum.	Regulates passage of chyme from stomach to duodenum; prevents backflow of chyme from duodenum to stomach.

Vomiting or *emesis* is the forcible expulsion of the contents of the upper GI tract (stomach and sometimes duodenum) through the mouth. The strongest stimuli for vomiting are irritation and distension of the stomach; other stimuli include unpleasant sights, general anesthesia, dizziness, and certain drugs such as morphine and derivatives of digitalis. Nerve impulses are transmitted to the vomiting center in the medulla oblongata, and returning impulses propagate to the upper GI tract organs, diaphragm, and abdominal muscles. Vomiting involves squeezing the stomach between the diaphragm and abdominal muscles and expelling the contents through open esophageal sphincters. Prolonged vomiting, especially in infants and elderly people, can be serious because the loss of acidic gastric juice can lead to alkalosis (higher than normal blood pH), dehydration, and damage to the esophagus and teeth. •

✔ CHECKPOINT

20. Compare the epithelium of the esophagus with that of the stomach. How is each adapted to the function of the organ?
21. What is the importance of rugae, surface mucous cells, mucous neck cells, chief cells, parietal cells, and G cells in the stomach?
22. What is the role of pepsin? Why is it secreted in an inactive form?
23. What are the functions of gastric lipase and lingual lipase in the stomach?

24.10 PANCREAS

◉ OBJECTIVE

• Describe the location, anatomy, histology, and function of the pancreas.

From the stomach, chyme passes into the small intestine. Because chemical digestion in the small intestine depends on activities of the pancreas, liver, and gallbladder, we first consider the activities of these accessory digestive organs and their contributions to digestion in the small intestine.

Anatomy of the Pancreas

The **pancreas** (*pan-* = all; *-creas* = flesh), a retroperitoneal gland that is about 12–15 cm (5–6 in.) long and 2.5 cm (1 in.) thick, lies posterior to the greater curvature of the stomach. The pancreas consists of a head, a body, and a tail and is usually connected to the duodenum by two ducts (Figure 24.15a). The **head** is the expanded portion of the organ near the curve of the duodenum; superior to and to the left of the head are the central **body** and the tapering **tail.**

Pancreatic juices are secreted by exocrine cells into small ducts that ultimately unite to form two larger ducts, the pancreatic duct and the accessory duct. These in turn convey the secretions into the small intestine. The **pancreatic duct, or duct of Wirsung** (VĒR-sung), is the larger of the two ducts. In most people, the pancreatic duct joins the common bile duct from the liver and gallbladder and enters the duodenum as a dilated common duct called the **hepatopancreatic ampulla** (hep′-a-tō-pan-krē-A-tik), or **ampulla of Vater** (FAH-ter). The ampulla opens on an elevation of the duodenal mucosa known as the **major duodenal papilla,** which lies about 10 cm (4 in.) inferior to the pyloric sphincter of the stomach. The passage of pancreatic juice and bile through the hepatopancreatic ampulla into the small intestine is regulated by a mass of smooth muscle surrounding the ampulla known as the **sphincter of the hepatopancreatic ampulla,** or **sphincter of Oddi** (OD-ē). The other major duct of the pancreas, the **accessory duct (duct of Santorini),** leads from the pancreas and empties into the duodenum about 2.5 cm (1 in.) superior to the hepatopancreatic ampulla.

Histology of the Pancreas

The pancreas is made up of small clusters of glandular epithelial cells. About 99% of the clusters, called **acini** (AS-i-nī), constitute the *exocrine* portion of the organ (see Figure 18.18b, c). The cells within acini secrete a mixture of fluid and digestive enzymes called pancreatic juice. The remaining 1% of the clusters, called **pancreatic islets** (Ī-lets) **(islets of Langerhans),** form the *endocrine* portion of the pancreas. These cells secrete the hormones glucagon, insulin, somatostatin, and pancreatic polypeptide. The functions of these hormones are discussed in Chapter 18.

Composition and Functions of Pancreatic Juice

Each day the pancreas produces 1200–1500 mL (about 1.2–1.5 qt) of **pancreatic juice,** a clear, colorless liquid consisting mostly of water, some salts, sodium bicarbonate, and several enzymes. The sodium bicarbonate gives pancreatic juice a slightly alkaline pH (7.1–8.2) that buffers acidic gastric juice in chyme, stops the action of pepsin from the stomach, and creates the proper pH for the action of digestive enzymes in the small intestine. The enzymes in pancreatic juice include a starch-digesting enzyme called **pancreatic amylase;** several enzymes that digest proteins into peptides called **trypsin** (TRIP-sin), **chymotrypsin** (kī′-mō-TRIP-sin), **carboxypeptidase** (kar-bok′-sē-PEP-ti-dās), and **elastase** (ē-LAS-tās); the principal triglyceride (fat and oil)-digesting enzyme in adults, called **pancreatic lipase;** and nucleic acid–digesting enzymes called **ribonuclease** and **deoxyribonuclease** that digest ribonucleic acid (RNA) and deoxyribonucleic acid (DNA) into nucleotides.

The protein-digesting enzymes of the pancreas are produced in an inactive form just as pepsin is produced in the stomach as pepsinogen. Because they are inactive, the enzymes do not digest cells of the pancreas itself. Trypsin is secreted in an inactive form called **trypsinogen** (trip-SIN-ō-jen). Pancreatic acinar cells also secrete a protein called **trypsin inhibitor** that combines with any trypsin formed accidentally in the pancreas or in pancreatic juice and blocks its enzymatic activity. When trypsinogen reaches the lumen of the small intestine, it encounters an activating brush-border enzyme called **enterokinase** (en′-ter-ō-KĪ-nās), which

Figure 24.15 **Relation of the pancreas to the liver, gallbladder, and duodenum.** The inset (b) shows details of the common bile duct and pancreatic duct forming the hepatopancreatic ampulla (ampulla of Vater) and emptying into the duodenum.

🔑 Pancreatic enzymes digest starches (polysaccharides), proteins, triglycerides, and nucleic acids.

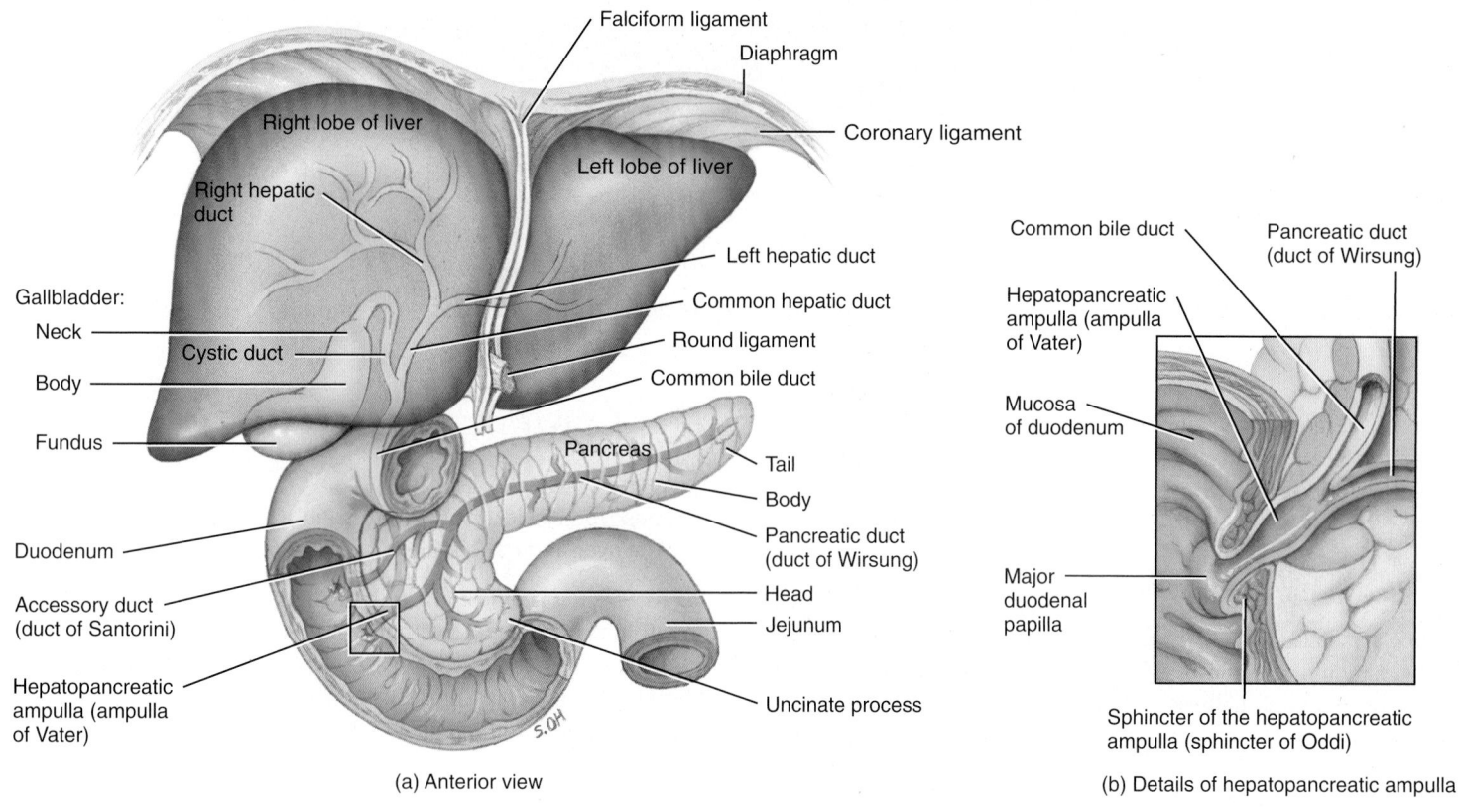

(a) Anterior view

(b) Details of hepatopancreatic ampulla

(c) Ducts carrying bile from liver and gallbladder and pancreatic juice from pancreas to duodenum

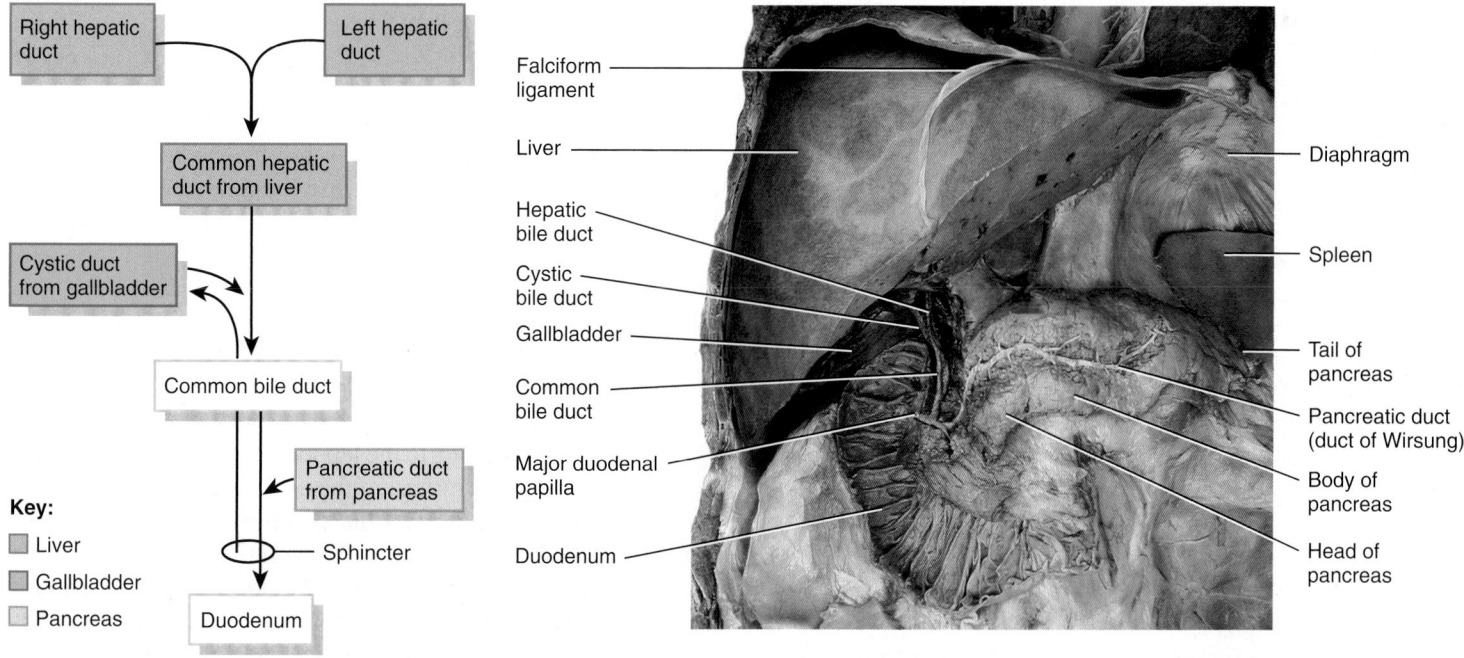

(d) Anterior view

FIGURE 24.15 CONTINUES ▶

FIGURE 24.15 CONTINUED

SUPERIOR

Duodenum

Common bile duct

Major duodenal papilla

Head of pancreas

MEDIAL

Uncinate process

Tail of pancreas

Body of pancreas

Pancreatic duct (duct of Wirsung)

LATERAL

(e) Anterior view

What type of fluid is found in the pancreatic duct? The common bile duct? The hepatopancreatic ampulla?

splits off part of the trypsinogen molecule to form trypsin. In turn, trypsin acts on the inactive precursors (called **chymotrypsino-gen, procarboxypeptidase,** and **proelastase**) to produce chymotrypsin, carboxypeptidase, and elastase, respectively.

CLINICAL CONNECTION | *Pancreatitis and Pancreatic Cancer*

Inflammation of the pancreas, as may occur in association with alcohol abuse or chronic gallstones, is called **pancreatitis** (pan'-krē-a-TĪ-tis). In a more severe condition known as **acute pancreatitis,** which is associated with heavy alcohol intake or biliary tract obstruction, the pancreatic cells may release either trypsin instead of trypsinogen or insufficient amounts of trypsin inhibitor, and the trypsin begins to digest the pancreatic cells. Patients with acute pancreatitis usually respond to treatment, but recurrent attacks are the rule. In some people pancreatitis is idiopathic, meaning that the cause is unknown. Other causes of pancreatitis include cystic fibrosis, high levels of calcium in the blood (hypercalcemia), high levels of blood fats (hyperlipidemia or hypertriglyceridemia), some drugs, and certain autoimmune conditions. However, in roughly 70% of adults with pancreatitis, the cause is alcoholism. Often the first episode happens between ages 30 and 40.

Pancreatic cancer usually affects people over 50 years of age and occurs more frequently in males. Typically, there are few symptoms until the disorder reaches an advanced stage and often not until it has metastasized to other parts of the body such as the lymph nodes, liver, or lungs. The disease is nearly always fatal and is the fourth most common cause of death from cancer in the United States. Pancreatic cancer has been linked to fatty foods, high alcohol consumption, genetic factors, smoking, and chronic pancreatitis. •

✔CHECKPOINT

24. Describe the duct system connecting the pancreas to the duodenum.

25. What are pancreatic acini? How do their functions differ from those of the pancreatic islets (islets of Langerhans)?

26. What are the digestive functions of the components of pancreatic juice?

24.11 LIVER AND GALLBLADDER

OBJECTIVE

• Describe the location, anatomy, histology, and functions of the liver and gallbladder.

The **liver** is the heaviest gland of the body, weighing about 1.4 kg (about 3 lb) in an average adult. Of all of the organs of the body, it is second only to the skin in size. The liver is inferior to the diaphragm and occupies most of the right hypochondriac and part of the epigastric regions of the abdominopelvic cavity (see Figure 1.12b).

The **gallbladder** (*gall-* = bile) is a pear-shaped sac that is located in a depression of the posterior surface of the liver. It is 7–10 cm (3–4 in.) long and typically hangs from the anterior inferior margin of the liver (Figure 24.15a).

Anatomy of the Liver and Gallbladder

The liver is almost completely covered by visceral peritoneum and is completely covered by a dense irregular connective tissue layer that lies deep to the peritoneum. The liver is divided into two principal lobes—a large **right lobe** and a smaller **left lobe**—by the **falciform ligament,** a fold of the mesentery (Figure 24.15a). Although the right lobe is considered by many anatomists to include an inferior **quadrate lobe** (kwa-DRĀT) and a posterior **caudate lobe** (KAW-dāt), based on internal morphology (primarily the distribution of blood vessels), the quadrate and caudate lobes more appropriately belong to the left lobe. The falciform ligament extends from the undersurface of the diaphragm between the two principal lobes of the liver to the superior surface of the liver, helping to suspend the liver in the abdominal cavity. In the free border of the falciform ligament is the **ligamentum teres (round ligament),** a remnant of the umbilical vein of the fetus (see Figure 21.30a, b); this fibrous cord extends from the liver to the umbilicus. The right and left **coronary ligaments** are narrow extensions of the parietal peritoneum that suspend the liver from the diaphragm.

The parts of the gallbladder include the broad **fundus,** which projects inferiorly beyond the inferior border of the liver; the **body,** the central portion; and the **neck,** the tapered portion. The body and neck project superiorly.

Histology of the Liver and Gallbladder

Histologically, the liver is composed of several components (Figure 24.16a–c):

1. **Hepatocytes** (*hepat-* = liver; *-cytes* = cells). Hepatocytes are the major functional cells of the liver and perform a wide array of metabolic, secretory, and endocrine functions. These are specialized epithelial cells with 5 to 12 sides that make up about 80% of the volume of the liver. Hepatocytes form complex three-dimensional arrangements called **hepatic laminae** (LAM-i-nē). The hepatic laminae are plates of hepatocytes one cell thick bordered on either side by the endothelial-lined vascular spaces called hepatic sinusoids. The hepatic laminae are highly branched, irregular structures. Grooves in the cell membranes between neighboring hepatocytes provide spaces for canaliculi (described next) into which the hepatocytes secrete bile. Bile, a yellow, brownish, or olive-green liquid secreted by hepatocytes, serves as both an excretory product and a digestive secretion.

2. **Bile canaliculi** (kan-a-LIK-ū-li = small canals). These are small ducts between hepatocytes that collect bile produced by the hepatocytes. From bile canaliculi, bile passes into **bile ductules** and then **bile ducts.** The bile ducts merge and eventually form the larger **right** and **left hepatic ducts,** which unite and exit the liver as the **common hepatic duct** (see Figure 24.15). The common hepatic duct joins the **cystic duct** (*cystic* = bladder) from the gallbladder to form the **common bile duct.** From here, bile enters the small intestine to participate in digestion.

3. **Hepatic sinusoids.** These are highly permeable blood capillaries between rows of hepatocytes that receive oxygenated blood from branches of the hepatic artery and nutrient-rich deoxygenated blood from branches of the hepatic portal vein. Recall that the hepatic portal vein brings venous blood from the gastrointestinal organs and spleen into the liver. Hepatic sinusoids converge and deliver blood into a **central vein.** From central veins the blood flows into the **hepatic veins,** which

Figure 24.16 Histology of the liver.

Histologically, the liver is composed of hepatocytes, bile canaliculi, and hepatic sinusoids.

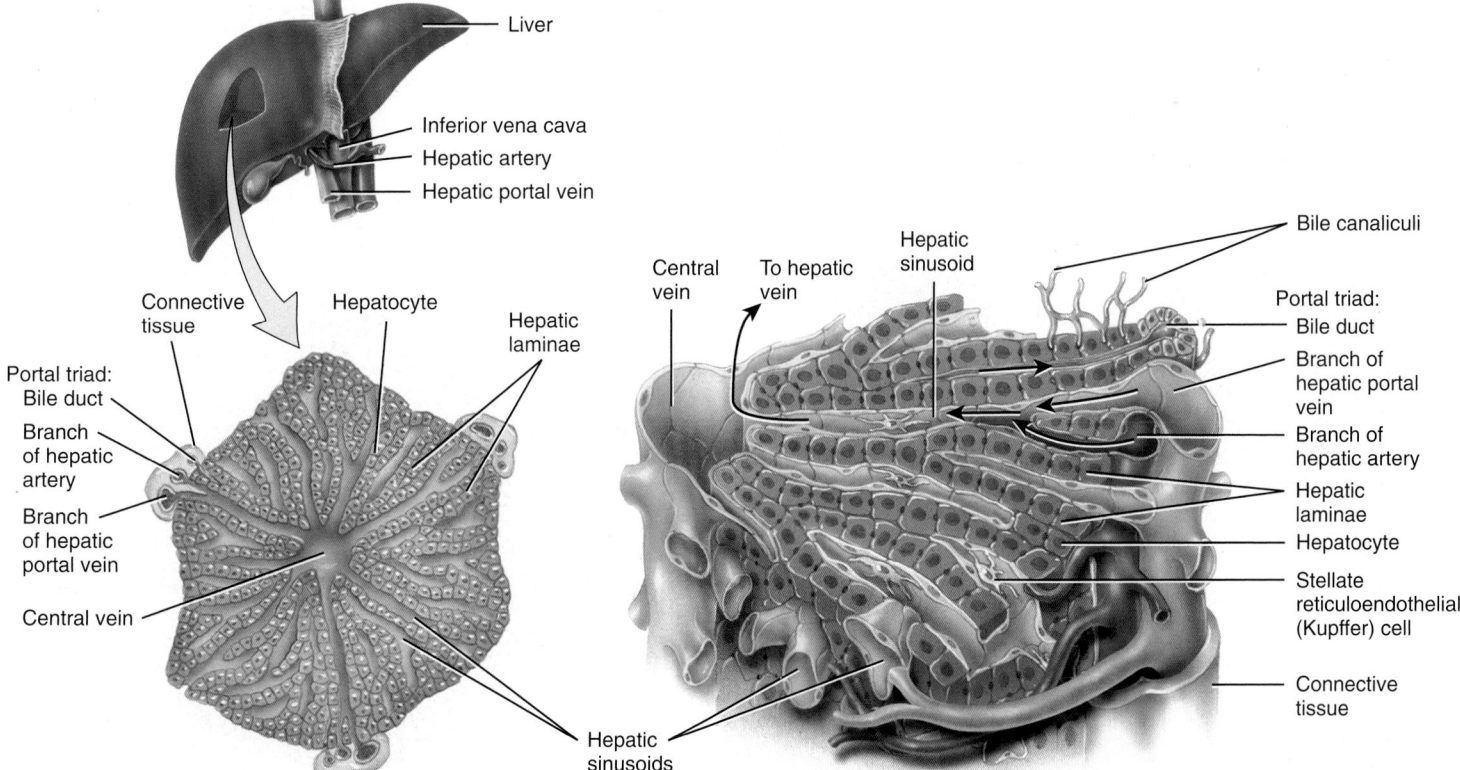

(a) Overview of histological components of liver

(b) Details of histological components of liver

FIGURE 24.16 CONTINUES ▶

FIGURE 24.16 CONTINUED ▶

Hepatocyte

Central vein

Sinusoid

LM 100x

LM 50x

Portal triad:
Branch of hepatic artery

Bile duct

Branch of hepatic portal vein

LM 150x

(c) Photomicrographs

Central vein

Portal triad

Hepatic lobule

Portal lobule

Hepatic acinus

(d) Comparison of three units of liver structure and function

Portal triad

Central vein

Central vein

Zone 3

Zone 2

Zone 1

(e) Details of hepatic acinus

? **Which type of cell in the liver is phagocytic?**

drain into the inferior vena cava (see Figure 21.28). In contrast to blood, which flows toward a central vein, bile flows in the opposite direction. Also present in the hepatic sinusoids are fixed phagocytes called **stellate reticuloendothelial (Kupffer) cells** (STEL-āt re-tik′-ū-lō-en′-dō-THĒ-lē-al [KŪP-fer]), which destroy worn-out white and red blood cells, bacteria, and other foreign matter in the venous blood draining from the gastrointestinal tract.

Together, a bile duct, branch of the hepatic artery, and branch of the hepatic vein are referred to as a **portal triad** (*tri* = three). The hepatocytes, bile duct system, and hepatic sinusoids can be organized into anatomical and functional units in three different ways:

1. **Hepatic lobule.** For years, anatomists described the hepatic lobule as the functional unit of the liver. According to this model, each hepatic lobule is shaped like a hexagon (six-sided structure) (Figure 24.16d, left). At its center is the central vein, and radiating out from it are rows of hepatocytes and hepatic sinusoids. Located at three corners of the hexagon is a portal triad. This model is based on a description of the liver of adult pigs. In the human liver it is difficult to find such well-defined hepatic lobules surrounded by thick layers of connective tissue.

2. **Portal lobule.** This model emphasized the exocrine function of the liver, that is, bile secretion. Accordingly, the bile duct of a portal triad is taken as the center of the portal lobule. The portal lobule is triangular in shape and is defined by three imaginary straight lines that connect three central veins that are closest to the portal triad (Figure 24.16d, center). This model has not gained widespread acceptance.

3. **Hepatic acinus.** In recent years, the preferred structural and functional unit of the liver is the hepatic **acinus** (AS-i-nus). Each hepatic acinus is an approximately oval mass that includes portions of two neighboring hepatic lobules. The short axis of the hepatic acinus is defined by branches of the portal triad—branches of the hepatic artery, vein, and bile ducts—that run along the border of the hepatic lobules. The long axis of the acinus is defined by two imaginary curved lines, which connect the two central veins closest to the short axis (Figure 24.16d, right). Hepatocytes in the hepatic acinus are arranged in three zones around the short axis, with no sharp boundaries between them (Figure 24.16e). Cells in zone 1 are closest to the branches of the portal triad and the first to receive incoming oxygen, nutrients, and toxins from incoming blood. These cells are the first ones to take up glucose and store it as glycogen after a meal and break down glycogen to glucose during fasting. They are also the first to show morphological changes following bile duct obstruction or exposure to toxic substances. Zone 1 cells are the last ones to die if circulation is impaired and the first ones to regenerate. Cells in zone 3 are farthest from branches of the portal triad and are the last to show the effects of bile obstruction or exposure to toxins, the first ones to show the effects of impaired circulation, and the

last ones to regenerate. Zone 3 cells also are the first to show evidence of fat accumulation. Cells in zone 2 have structural and functional characteristics intermediate between the cells in zones 1 and 3.

The hepatic acinus is the smallest structural and functional unit of the liver. Its popularity and appeal are based on the fact that it provides a logical description and interpretation of (1) patterns of glycogen storage and release and (2) toxic effects, degeneration, and regeneration relative to the proximity of the acinar zones to branches of the portal triad.

The mucosa of the gallbladder consists of simple columnar epithelium arranged in rugae resembling those of the stomach. The wall of the gallbladder lacks a submucosa. The middle, muscular coat of the wall consists of smooth muscle fibers. Contraction of the smooth muscle fibers ejects the contents of the gallbladder into the **cystic duct.** The gallbladder's outer coat is the visceral peritoneum. The functions of the gallbladder are to store and concentrate the bile produced by the liver (up to tenfold) until it is needed in the small intestine. In the concentration process, water and ions are absorbed by the gallbladder mucosa.

CLINICAL CONNECTION | *Jaundice*

Jaundice (JAWN-dis = yellowed) is a yellowish coloration of the sclerae (whites of the eyes), skin, and mucous membranes due to a buildup of a yellow compound called bilirubin. After bilirubin is formed from the breakdown of the heme pigment in aged red blood cells, it is transported to the liver, where it is processed and eventually excreted into bile. The three main categories of jaundice are (1) *prehepatic jaundice,* due to excess production of bilirubin; (2) *hepatic jaundice,* due to congenital liver disease, cirrhosis of the liver, or hepatitis; and (3) *extrahepatic jaundice,* due to blockage of bile drainage by gallstones or cancer of the bowel or the pancreas.

Because the liver of a newborn functions poorly for the first week or so, many babies experience a mild form of jaundice called *neonatal (physiological) jaundice* that disappears as the liver matures. Usually, it is treated by exposing the infant to blue light, which converts bilirubin into substances the kidneys can excrete. •

Blood Supply of the Liver

The liver receives blood from two sources (Figure 24.17). From the hepatic artery it obtains oxygenated blood, and from the hepatic portal vein it receives deoxygenated blood containing newly absorbed nutrients, drugs, and possibly microbes and toxins from the gastrointestinal tract (see Figure 21.28). Branches of both the hepatic artery and the hepatic portal vein carry blood into liver sinusoids, where oxygen, most of the nutrients, and certain toxic substances are taken up by the hepatocytes. Products manufactured by the hepatocytes and nutrients needed by other cells are secreted back into the blood, which then drains into the central vein and eventually passes into a hepatic vein. Because blood from the gastrointestinal tract passes through the liver as part of the hepatic portal circulation, the liver is often a site for metastasis of cancer that originates in the GI tract.

Figure 24.17 Hepatic blood flow: sources, path through the liver, and return to the heart.

The liver receives oxygenated blood via the hepatic artery and nutrient-rich deoxygenated blood via the hepatic portal vein.

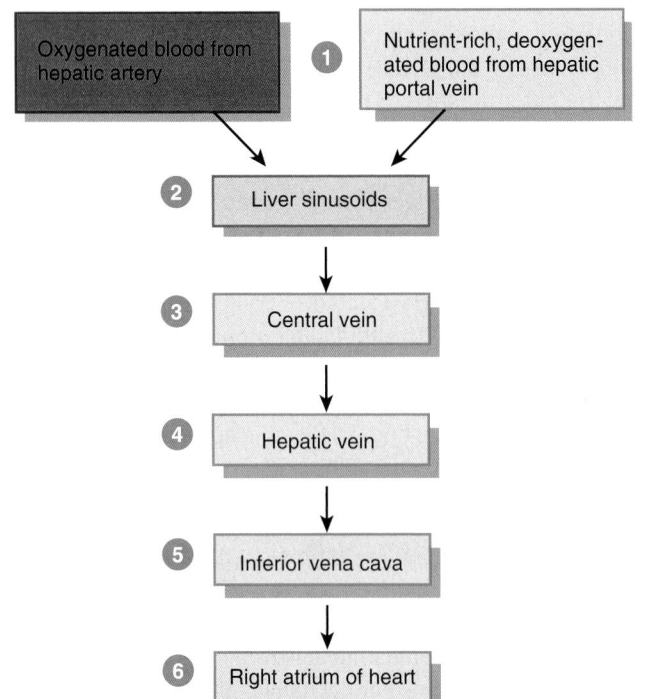

? During the first few hours after a meal, how does the chemical composition of blood change as it flows through the liver sinusoids?

CLINICAL CONNECTION | *Liver Function Tests*

Liver function tests are blood tests designed to determine the presence of certain chemicals released by liver cells. These include albumin globulinase, alanine aminotransferase (ALT), aspartate aminotransferase (AST), alkaline phosphatase (ALP), gamma-glutamyl-transpeptidase (GGT), and bilirubin. The tests are used to evaluate and monitor liver disease or damage. Common causes of elevated liver enzymes include nonsteroidal anti-inflammatory drugs, cholesterol-lowering medications, some antibiotics, alcohol, diabetes, infections (viral hepatitis and mononucleosis), gallstones, tumors of the liver, and excessive use of herbal supplements such as kava, comfrey, pennyroyal, dandelion root, skullcap, and ephedra. •

Functions of the Liver and Gallbladder

Each day, hepatocytes secrete 800–1000 mL (about 1 qt) of **bile,** a yellow, brownish, or olive-green liquid. It has a pH of 7.6–8.6 and consists mostly of water, bile salts, cholesterol, a phospholipid called lecithin, bile pigments, and several ions.

The principal bile pigment is **bilirubin** (bil-i-ROO-bin). The phagocytosis of aged red blood cells liberates iron, globin, and bilirubin (derived from heme) (see Figure 19.5). The iron and globin are recycled; the bilirubin is secreted into the bile and is eventually broken down in the intestine. One of its breakdown products—**stercobilin** (ster-kō-BĪ-lin)—gives feces their normal brown color.

Bile is partially an excretory product and partially a digestive secretion. Bile salts, which are sodium salts and potassium salts of bile acids (mostly chenodeoxycholic acid and cholic acid), play a role in **emulsification** (ē-mul'-si-fi-KĀ-shun), the breakdown of large lipid globules into a suspension of small lipid globules. The small lipid globules present a very large surface area that allows pancreatic lipase to more rapidly accomplish digestion of triglycerides. Bile salts also aid in the absorption of lipids following their digestion.

Although hepatocytes continually release bile, they increase production and secretion when the portal blood contains more bile acids; thus, as digestion and absorption continue in the small intestine, bile release increases. Between meals, after most absorption has occurred, bile flows into the gallbladder for storage because the sphincter of the hepatopancreatic ampulla (sphincter of Oddi; see Figure 24.15) closes off the entrance to the duodenum. The sphincter surrounds the hepatopancreatic ampulla.

CLINICAL CONNECTION | *Gallstones*

If bile contains either insufficient bile salts or lecithin or excessive cholesterol, the cholesterol may crystallize to form **gallstones.** As they grow in size and number, gallstones may cause minimal, intermittent, or complete obstruction to the flow of bile from the gallbladder into the duodenum. Treatment consists of using gallstone-dissolving drugs, lithotripsy (shock-wave therapy), or surgery. For people with recurrent gallstones or for whom drugs or lithotripsy is not indicated, **cholecystectomy** (kō'-lē-sis-TEK-tō-mē)—the removal of the gallbladder and its contents—is necessary. More than half a million cholecystectomies are performed each year in the United States. To prevent side effects resulting from a loss of the gallbladder, patients should make lifestyle and dietary changes, including the following: (1) limiting the intake of saturated fat; (2) avoiding the consumption of alcoholic beverages; (3) eating smaller amounts of food during a meal and eating five to six smaller meals per day instead of two to three larger meals; and (4) taking vitamin and mineral supplements. •

In addition to secreting bile, which is needed for absorption of dietary fats, the liver performs many other vital functions:

• ***Carbohydrate metabolism.*** The liver is especially important in maintaining a normal blood glucose level. When blood glucose is low, the liver can break down glycogen to glucose and release the glucose into the bloodstream. The liver can also convert certain amino acids and lactic acid to glucose, and it can convert other sugars, such as fructose and galactose, into glucose. When blood glucose is high, as occurs just after eating a meal, the liver converts glucose to glycogen and triglycerides for storage.

- **Lipid metabolism.** Hepatocytes store some triglycerides; break down fatty acids to generate ATP; synthesize lipoproteins, which transport fatty acids, triglycerides, and cholesterol to and from body cells; synthesize cholesterol; and use cholesterol to make bile salts.

- **Protein metabolism.** Hepatocytes *deaminate* (remove the amino group, NH_2, from) amino acids so that the amino acids can be used for ATP production or converted to carbohydrates or fats. The resulting toxic ammonia (NH_3) is then converted into the much less toxic urea, which is excreted in urine. Hepatocytes also synthesize most plasma proteins, such as alpha and beta globulins, albumin, prothrombin, and fibrinogen.

- **Processing of drugs and hormones.** The liver can detoxify substances such as alcohol and excrete drugs such as penicillin, erythromycin, and sulfonamides into bile. It can also chemically alter or excrete thyroid hormones and steroid hormones such as estrogens and aldosterone.

- **Excretion of bilirubin.** As previously noted, bilirubin, derived from the heme of aged red blood cells, is absorbed by the liver from the blood and secreted into bile. Most of the bilirubin in bile is metabolized in the small intestine by bacteria and eliminated in feces.

- **Synthesis of bile salts.** Bile salts are used in the small intestine for the emulsification and absorption of lipids.

- **Storage.** In addition to glycogen, the liver is a prime storage site for certain vitamins (A, B_{12}, D, E, and K) and minerals (iron and copper), which are released from the liver when needed elsewhere in the body.

- **Phagocytosis.** The stellate reticuloendothelial (Kupffer) cells of the liver phagocytize aged red blood cells, white blood cells, and some bacteria.

- **Activation of vitamin D.** The skin, liver, and kidneys participate in synthesizing the active form of vitamin D.

The liver functions related to metabolism are discussed more fully in Chapter 25.

✔**CHECKPOINT**

27. Draw and label a diagram of the cell zones of a hepatic acinus.
28. Describe the pathways of blood flow into, through, and out of the liver.
29. How are the liver and gallbladder connected to the duodenum?
30. Once bile has been formed by the liver, how is it collected and transported to the gallbladder for storage?
31. Describe the major functions of the liver and gallbladder.

24.12 SMALL INTESTINE

◉ **OBJECTIVES**

- Describe the location and structure of the small intestine.
- Identify the functions of the small intestine.

Most digestion and absorption of nutrients occur in a long tube called the **small intestine.** Because of this, its structure is specially adapted for these functions. Its length alone provides a large surface area for digestion and absorption, and that area is further increased by circular folds, villi, and microvilli. The small intestine begins at the pyloric sphincter of the stomach, coils through the central and inferior part of the abdominal cavity, and eventually opens into the large intestine. It averages 2.5 cm (1 in.) in diameter; its length is about 3 m (10 ft) in a living person and about 6.5 m (21 ft) in a cadaver due to the loss of smooth muscle tone after death.

Anatomy of the Small Intestine

The small intestine is divided into three regions (Figure 24.18). The **duodenum** (doo′-ō-DĒ-num or doo-OD-e-num), the shortest region, is retroperitoneal. It starts at the pyloric sphincter of the stomach and is in the form of a C-shaped tube that extends about 25 cm (10 in.) until it merges with the jejunum. *Duodenum* means "12"; it is so named because it is about as long as the width of 12 fingers. The **jejunum** (je-JOO-num) is about 1 m (3 ft) long and extends to the ileum. *Jejunum* means "empty," which is how it is found at death. The final and longest region of the small intestine, the **ileum** (IL-ē-um = twisted), measures about 2 m (6 ft) and joins the large intestine at a smooth muscle sphincter called the **ileocecal sphincter (valve)** (il′-ē-ō-SĒ-kal).

Histology of the Small Intestine

The wall of the small intestine is composed of the same four layers that make up most of the GI tract: mucosa, submucosa, muscularis, and serosa (Figure 24.19b). The **mucosa** is composed of a layer of epithelium, lamina propria, and muscularis mucosae. The epithelial layer of the small intestinal mucosa consists of simple columnar epithelium that contains many types of cells (Figure 24.19c). **Absorptive cells** of the epithelium digest and absorb nutrients in small intestinal chyme. Also present in the epithelium are **goblet cells,** which secrete mucus. The small intestinal mucosa contains many deep crevices lined with glandular epithelium. Cells lining the crevices form the **intestinal glands,** or **crypts of Lieberkühn** (LĒ-ber-kūn), and secrete intestinal juice (to be discussed shortly). Besides absorptive cells and goblet cells, the intestinal glands also contain paneth cells and enteroendocrine cells. **Paneth cells** secrete lysozyme, a bactericidal enzyme, and are capable of phagocytosis. Paneth cells may have a role in regulating the microbial population in the small intestine. Three types of enteroendocrine cells are found in the intestinal glands of the small intestine: **S cells, CCK cells,** and **K cells,** which secrete the hormones **secretin** (se-KRĒ-tin), **cholecystokinin** (kō-lē-sis′-tō-KĪN-in)

Figure 24.18 **Anatomy of the small intestine.** (a) Regions of the small intestine are the duodenum, jejunum, and ileum. (b) Circular folds increase the surface area for digestion and absorption in the small intestine.

🔑 Most digestion and absorption occur in the small intestine.

> **FUNCTIONS OF THE SMALL INTESTINE**
> 1. Segmentations mix chyme with digestive juices and bring food into contact with mucosa for absorption; peristalsis propels chyme through small intestine.
> 2. Completes digestion of carbohydrates, proteins, and lipids; begins and completes digestion of nucleic acids.
> 3. Absorbs about 90% of nutrients and water that pass through digestive system.

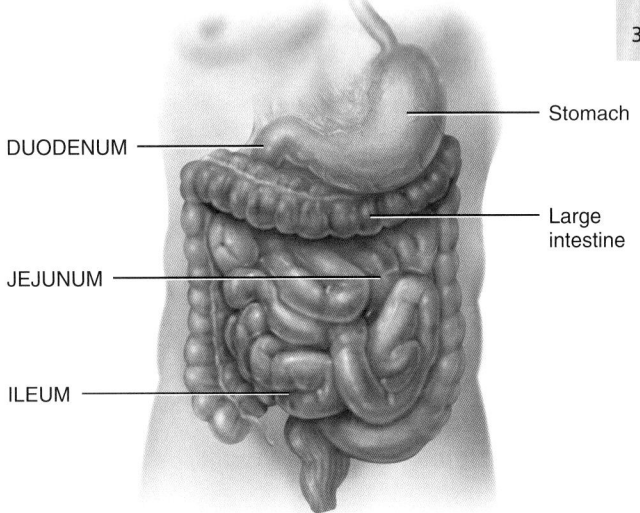

DUODENUM

Stomach

Large intestine

JEJUNUM

ILEUM

(a) Anterior view of external anatomy

Circular folds (plicae circulares)

(b) Internal anatomy of jejunum

❓ **Which portion of the small intestine is the longest?**

or **CCK,** and **glucose-dependent insulinotropic peptide** (in-soo-lin′-ō-TRŌ-pik) or **GIP,** respectively.

The lamina propria of the small intestinal mucosa contains areolar connective tissue and has an abundance of mucosa-associated lymphoid tissue (MALT). **Solitary lymphatic nodules** are most numerous in the distal part of the ileum (see Figure 24.20c). Groups of lymphatic nodules referred to as **aggregated lymphatic follicles,** or **Peyer's patches** (PĪ-erz), are also present in the ileum. The muscularis mucosae of the small intestinal mucosa consists of smooth muscle.

The **submucosa** of the duodenum contains **duodenal glands,** also called **Brunner's glands** (BRUN-erz) (Figure 24.20a), which secrete an alkaline mucus that helps neutralize gastric acid in the chyme. Sometimes the lymphatic tissue of the lamina propria extends through the muscularis mucosae into the submucosa. The **muscularis** of the small intestine consists of two layers of smooth muscle. The outer, thinner layer contains longitudinal fibers; the inner, thicker layer contains circular fibers. Except for a major portion of the duodenum, the **serosa** (or visceral peritoneum) completely surrounds the small intestine.

Even though the wall of the small intestine is composed of the same four basic layers as the rest of the GI tract, special structural features of the small intestine facilitate the process of digestion and absorption. These structural features include circular folds, villi, and microvilli. **Circular folds** or **plicae circulares** are folds of the mucosa and submucosa (see Figures 24.18b and 24.19a). These permanent ridges, which are about 10 mm (0.4 in.) long, begin near the proximal portion of the duodenum and end at about the midportion of the ileum. Some extend all the way around the circumference of the intestine; others extend only part of the way around. Circular folds enhance absorption by increasing surface area and causing the chyme to spiral, rather than move in a straight line, as it passes through the small intestine.

Also present in the small intestine are **villi** (= tufts of hair), which are fingerlike projections of the mucosa that are 0.5–1 mm long (see Figure 24.19b, c). The large number of villi (20–40 per square millimeter) vastly increases the surface area of the epithelium available for absorption and digestion and gives the intestinal mucosa a velvety appearance. Each villus (singular form) is covered by epithelium and has a core of lamina propria; embedded in the connective tissue of the lamina propria are an arteriole, a venule, a blood capillary network, and a **lacteal** (LAK-tē-al = milky), which is a lymphatic capillary (see Figure 24.19c). Nutrients absorbed by the epithelial cells covering the villus pass through the wall of a capillary or a lacteal to enter blood or lymph, respectively.

Figure 24.19 **Histology of the small intestine.**

Circular folds, villi, and microvilli increase the surface area of the small intestine for digestion and absorption.

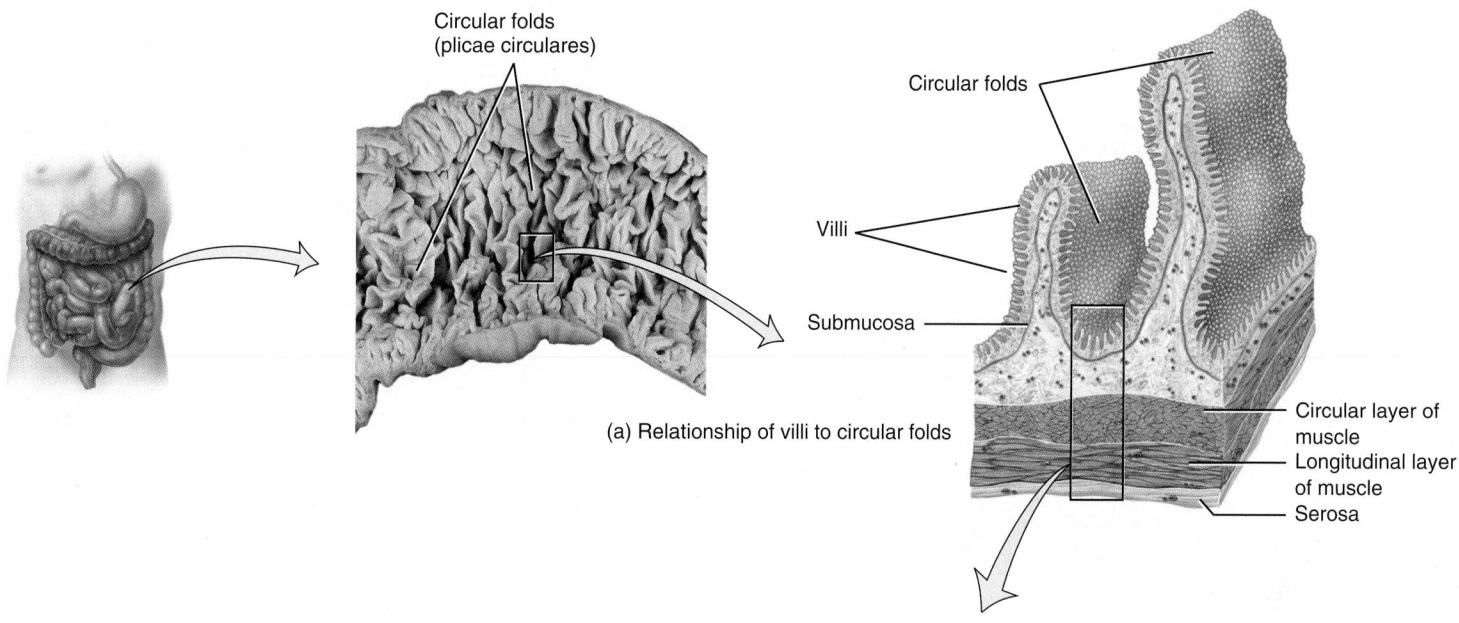

Circular folds
(plicae circulares)

Circular folds

Villi

Submucosa

Circular layer of muscle

Longitudinal layer of muscle

Serosa

(a) Relationship of villi to circular folds

Lumen of small intestine

Blood capillary

Lacteal

Villi

Opening of intestinal gland

Absorptive cell

Goblet cell

Lacteal

Lamina propria

Enteroendocrine cell

Paneth cell

Lymphatic nodule

Muscularis mucosae

Arteriole

Venule

Lymphatic vessel

Circular layer of muscle

Myenteric plexus

Longitudinal layer of muscle

MUCOSA

SUBMUCOSA

MUSCULARIS

SEROSA

(b) Three-dimensional view of layers of the small intestine showing villi

FIGURE 24.19 CONTINUES

■ FIGURE 24.19 CONTINUED ▶

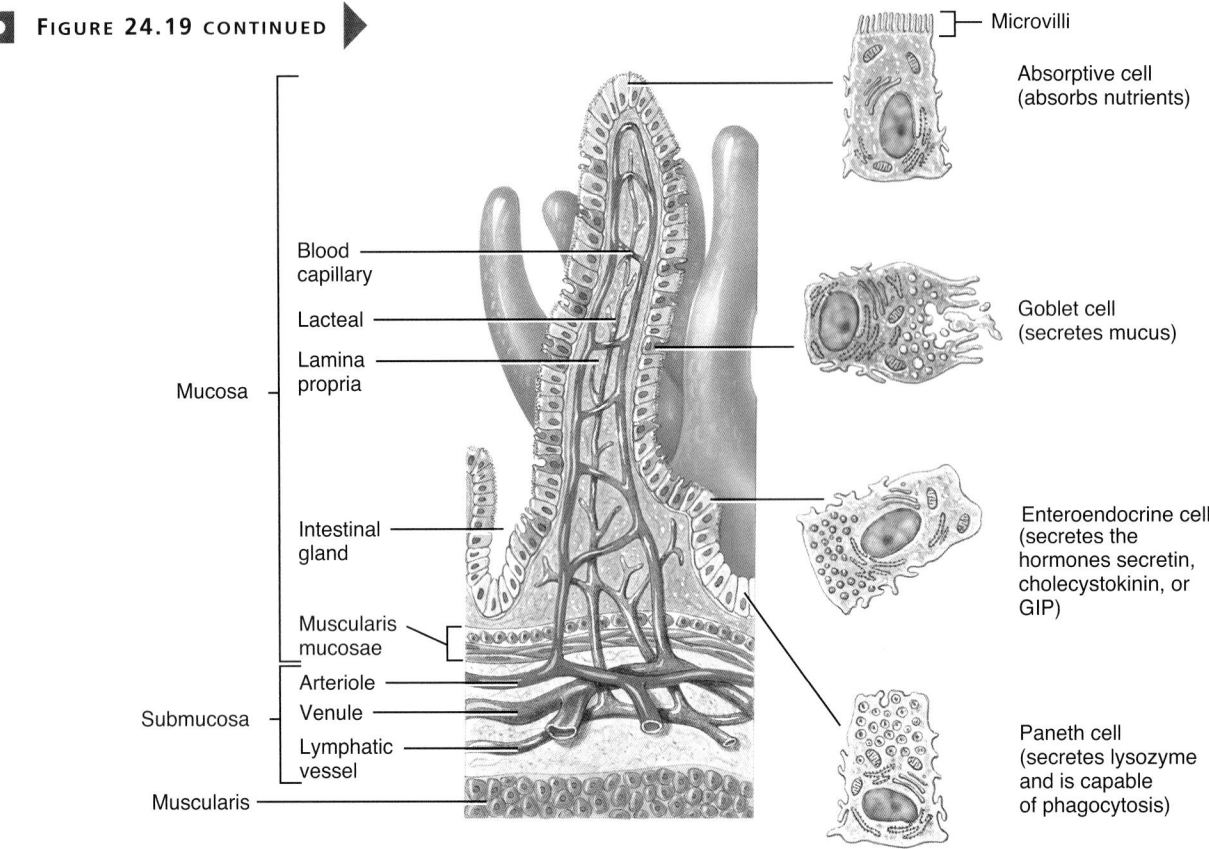

(c) Enlarged villus showing lacteal, capillaries, intestinal glands, and cell types

? **What is the functional significance of the blood capillary network and lacteal in the center of each villus?**

Besides circular folds and villi, the small intestine also has **microvilli** (mī-krō-VIL-ī; *micro-* = small), which are projections of the apical (free) membrane of the absorptive cells. Each microvillus is a 1-μm-long cylindrical, membrane-covered projection that contains a bundle of 20–30 actin filaments. When viewed through a light microscope, the microvilli are too small to be seen individually; instead they form a fuzzy line, called the **brush border,** extending into the lumen of the small intestine (Figure 24.20d). There are an estimated 200 million microvilli per square millimeter of small intestine. Because the microvilli greatly increase the surface area of the plasma membrane, larger amounts of digested nutrients can diffuse into absorptive cells in a given period. The brush border also contains several brush-border enzymes that have digestive functions (discussed shortly).

Role of Intestinal Juice and Brush-Border Enzymes

About 1–2 liters (1–2 qt) of **intestinal juice,** a clear yellow fluid, is secreted each day. Intestinal juice contains water and mucus and is slightly alkaline (pH 7.6). Together, pancreatic and intestinal juices provide a liquid medium that aids the absorption of substances from chyme in the small intestine. The absorptive cells of

the small intestine synthesize several digestive enzymes, called **brush-border enzymes,** and insert them in the plasma membrane of the microvilli. Thus, some enzymatic digestion occurs at the surface of the absorptive cells that line the villi, rather than in the lumen exclusively, as occurs in other parts of the GI tract. Among the brush-border enzymes are four carbohydrate-digesting enzymes called α-dextrinase, maltase, sucrase, and lactase; protein-digesting enzymes called peptidases (aminopeptidase and dipeptidase); and two types of nucleotide-digesting enzymes, nucleosidases and phosphatases. Also, as absorptive cells slough off into the lumen of the small intestine, they break apart and release enzymes that help digest nutrients in the chyme.

Mechanical Digestion in the Small Intestine

The two types of movements of the small intestine—segmentations and a type of peristalsis called migrating motility complexes—are governed mainly by the myenteric plexus. **Segmentations** are localized, mixing contractions that occur in portions of intestine distended by a large volume of chyme. Segmentations mix chyme with the digestive juices and bring the particles of food into contact with the mucosa for absorption; they do not push the intestinal contents along the tract. A segmentation starts with the contractions

Figure 24.20 **Histology of the duodenum and ileum.**

Microvilli in the small intestine contain several brush-border enzymes that help digest nutrients.

Lumen of duodenum
Villi
Mucosa
Intestinal glands
Muscularis mucosae
Duodenal gland
Submucosa
Muscularis

LM 45x

(a) Wall of the duodenum

Villi
Lumen of duodenum
Brush border
Simple columnar epithelium
Goblet cell
Absorptive cell
Lamina propria
Intestinal glands
Muscularis mucosae
Duodenum
Duodenal gland in submucosa

LM 160x

(b) Three villi from duodenum

FIGURE 24.20 CONTINUES ▶

FIGURE 24.20 CONTINUED

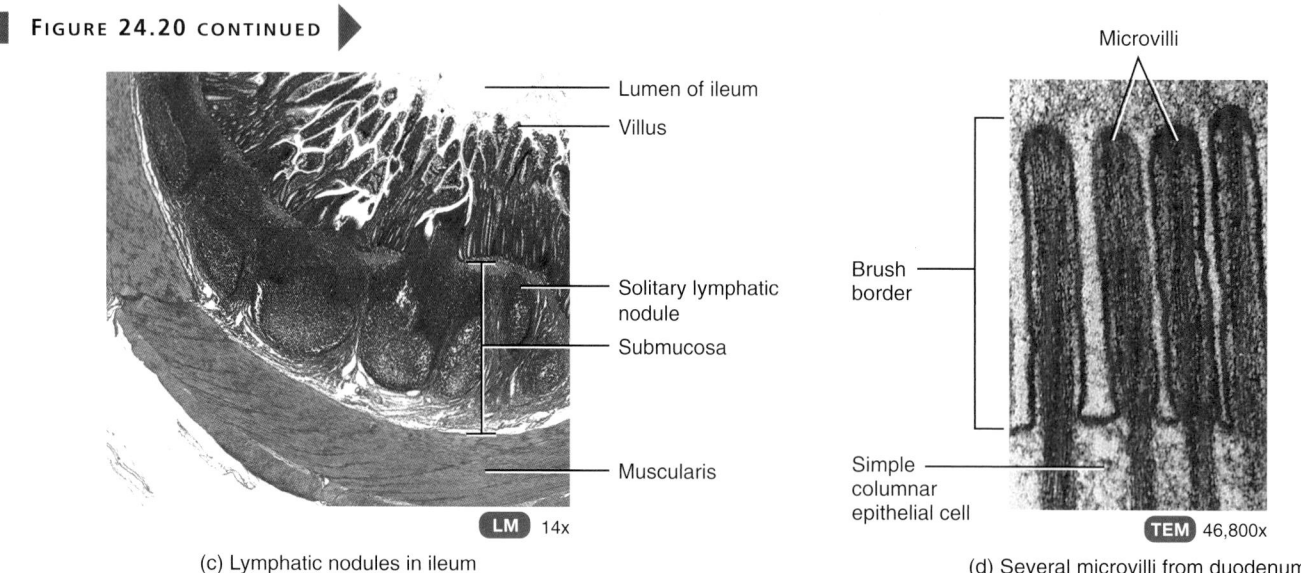

(c) Lymphatic nodules in ileum

(d) Several microvilli from duodenum

? What is the function of the fluid secreted by duodenal (Brunner's) glands?

of circular muscle fibers in a portion of the small intestine, an action that constricts the intestine into segments. Next, muscle fibers that encircle the middle of each segment also contract, dividing each segment again. Finally, the fibers that first contracted relax, and each small segment unites with an adjoining small segment so that large segments are formed again. As this sequence of events repeats, the chyme sloshes back and forth. Segmentations occur most rapidly in the duodenum, about 12 times per minute, and progressively slow to about 8 times per minute in the ileum. This movement is similar to alternately squeezing the middle and then the ends of a capped tube of toothpaste.

After most of a meal has been absorbed, which lessens distension of the wall of the small intestine, segmentation stops and peristalsis begins. The type of peristalsis that occurs in the small intestine, termed a **migrating motility complex (MMC),** begins in the lower portion of the stomach and pushes chyme forward along a short stretch of small intestine before dying out. The MMC slowly migrates down the small intestine, reaching the end of the ileum in 90–120 minutes. Then another MMC begins in the stomach. Altogether, chyme remains in the small intestine for 3–5 hours.

Chemical Digestion in the Small Intestine

In the mouth, salivary amylase converts starch (a polysaccharide) to maltose (a disaccharide), maltotriose (a trisaccharide), and α-dextrins (short-chain, branched fragments of starch with 5–10 glucose units). In the stomach, pepsin converts proteins to peptides (small fragments of proteins), and lingual and gastric lipases convert some triglycerides into fatty acids, diglycerides, and monoglycerides. Thus, chyme entering the small intestine contains partially digested carbohydrates, proteins, and lipids. The completion of the digestion of carbohydrates, proteins, and lipids is a collective effort of pancreatic juice, bile, and intestinal juice in the small intestine.

Digestion of Carbohydrates

Even though the action of **salivary amylase** may continue in the stomach for a while, the acidic pH of the stomach destroys salivary amylase and ends its activity. Thus, only a few starches are broken down by the time chyme leaves the stomach. Those starches not already broken down into maltose, maltotriose, and α-dextrins are cleaved by **pancreatic amylase,** an enzyme in pancreatic juice that acts in the small intestine. Although pancreatic amylase acts on both glycogen and starches, it has no effect on another polysaccharide called cellulose, an indigestible plant fiber that is commonly referred to as "roughage" as it moves through the digestive system. After amylase (either salivary or pancreatic) has split starch into smaller fragments, a brush-border enzyme called **α-dextrinase** acts on the resulting α-dextrins, clipping off one glucose unit at a time.

Ingested molecules of sucrose, lactose, and maltose—three disaccharides—are not acted on until they reach the small intestine. Three brush-border enzymes digest the disaccharides into monosaccharides. **Sucrase** breaks sucrose into a molecule of glucose and a molecule of fructose; **lactase** digests lactose into a molecule of glucose and a molecule of galactose; and **maltase** splits maltose and maltotriose into two or three molecules of glucose, respectively. Digestion of carbohydrates ends with the production of monosaccharides, which the digestive system is able to absorb.

CLINICAL CONNECTION | *Lactose Intolerance*

In some people the absorptive cells of the small intestine fail to produce enough lactase, which, as you just learned, is essential for the digestion of lactose. This results in a condition called **lactose intolerance,** in which undigested lactose in chyme causes fluid to be retained in the feces; bacterial fermentation of the undigested lactose results in the production of gases. Symptoms of lactose intolerance include diarrhea, gas, bloating, and abdominal cramps after consumption of milk and other dairy products. The symptoms can be relatively minor or serious enough to require medical attention. The *hydrogen breath test* is often used to aid in diagnosis of lactose intolerance. Very little hydrogen can be detected in the breath of a normal person, but hydrogen is among the gases produced when undigested lactose in the colon is fermented by bacteria. The hydrogen is absorbed from the intestines and carried through the bloodstream to the lungs, where it is exhaled. Persons with lactose intolerance should select a diet that restricts lactose (but not calcium) and take dietary supplements to aid in the digestion of lactose. •

Digestion of Proteins

Recall that protein digestion starts in the stomach, where proteins are fragmented into peptides by the action of **pepsin.** Enzymes in pancreatic juice—**trypsin, chymotrypsin, carboxypeptidase,** and **elastase**—continue to break down proteins into peptides. Although all these enzymes convert whole proteins into peptides, their actions differ somewhat because each splits peptide bonds between different amino acids. Trypsin, chymotrypsin, and elastase all cleave the peptide bond between a specific amino acid and its neighbor; carboxypeptidase splits off the amino acid at the carboxyl end of a peptide. Protein digestion is completed by two **peptidases** in the brush border: aminopeptidase and dipeptidase. **Aminopeptidase** cleaves off the amino acid at the amino end of a peptide. **Dipeptidase** splits dipeptides (two amino acids joined by a peptide bond) into single amino acids.

Digestion of Lipids

The most abundant lipids in the diet are triglycerides, which consist of a molecule of glycerol bonded to three fatty acid molecules (see Figure 2.17). Enzymes that split triglycerides and phospholipids are called **lipases.** Recall that there are three types of lipases that can participate in lipid digestion: **lingual lipase, gastric lipase,** and **pancreatic lipase.** Although some lipid digestion occurs in the stomach through the action of lingual and gastric lipases, most occurs in the small intestine through the action of pancreatic lipase. Triglycerides are broken down by pancreatic lipase into fatty acids and monoglycerides. The liberated fatty acids can be either short-chain fatty acids (with fewer than 10–12 carbons) or long-chain fatty acids.

Before a large lipid globule containing triglycerides can be digested in the small intestine, it must first undergo **emulsification**—a process in which the large lipid globule is broken down into several small lipid globules. Recall that bile contains bile salts, the sodium salts and potassium salts of bile acids (mainly chenodeoxycholic acid and cholic acid). Bile salts are **amphipathic** (am′-fē-PATH-ik), which means that each bile salt has a hydrophobic (nonpolar) region and a hydrophilic (polar) region. The amphipathic nature of bile salts allows them to emulsify a large lipid globule: The hydrophobic regions of bile salts interact with the large lipid globule, while the hydrophilic regions of bile salts interact with the watery intestinal chyme. Consequently, the large lipid globule is broken apart into several small lipid globules, each about 1 μm in diameter. The small lipid globules formed from emulsification provide a large surface area that allows pancreatic lipase to function more effectively.

Digestion of Nucleic Acids

Pancreatic juice contains two nucleases: **ribonuclease,** which digests RNA, and **deoxyribonuclease,** which digests DNA. The nucleotides that result from the action of the two nucleases are further digested by brush-border enzymes called **nucleosidases** (noo′-klē-ō-SĪ-dās-ez) and **phosphatases** (FOS-fa-tās′-ez) into pentoses, phosphates, and nitrogenous bases. These products are absorbed via active transport.

Absorption in the Small Intestine

All the chemical and mechanical phases of digestion from the mouth through the small intestine are directed toward changing food into forms that can pass through the absorptive epithelial cells lining the mucosa and into the underlying blood and lymphatic vessels. These forms are monosaccharides (glucose, fructose, and galactose) from carbohydrates; single amino acids, dipeptides, and tripeptides from proteins; and fatty acids, glycerol, and monoglycerides from triglycerides. Passage of these digested nutrients from the gastrointestinal tract into the blood or lymph is called **absorption.**

Absorption of materials occurs via diffusion, facilitated diffusion, osmosis, and active transport. About 90% of all absorption of nutrients occurs in the small intestine; the other 10% occurs in the stomach and large intestine. Any undigested or unabsorbed material left in the small intestine passes on to the large intestine.

Absorption of Monosaccharides

All carbohydrates are absorbed as monosaccharides. The capacity of the small intestine to absorb monosaccharides is huge—an estimated 120 grams per hour. As a result, all dietary carbohydrates that are digested normally are absorbed, leaving only indigestible cellulose and fibers in the feces. Monosaccharides pass from the lumen through the apical membrane via *facilitated diffusion* or *active transport.* Fructose, a monosaccharide found in fruits, is transported via *facilitated diffusion;* glucose and galactose are transported into absorptive cells of the villi via *secondary active transport* that is coupled to the active transport of Na$^+$ (Figure 24.21a). The transporter has binding sites for one glucose molecule and two sodium ions; unless all three sites are filled, neither substance is transported. Galactose competes with glucose to ride the same transporter. (Because both Na$^+$ and glucose or galactose move in the same direction, this is a *symporter.*) Monosaccharides then move out of the absorptive cells through their basolateral surfaces via *facilitated diffusion* and enter the capillaries of the villi (Figure 24.21b).

Absorption of Amino Acids, Dipeptides, and Tripeptides

Most proteins are absorbed as amino acids via *active transport* processes that occur mainly in the duodenum and jejunum. About half of the absorbed amino acids are present in food; the other half come from the body itself as proteins in digestive juices and dead cells that slough off the mucosal surface! Normally, 95–98% of the protein present in the small intestine is digested and absorbed. Different transporters carry different types of amino acids. Some

Figure 24.21 Absorption of digested nutrients in the small intestine. For simplicity, all digested foods are shown in the lumen of the small intestine, even though some nutrients are digested by brush-border enzymes.

🔑 Long-chain fatty acids and monoglycerides are absorbed into lacteals; other products of digestion enter blood capillaries.

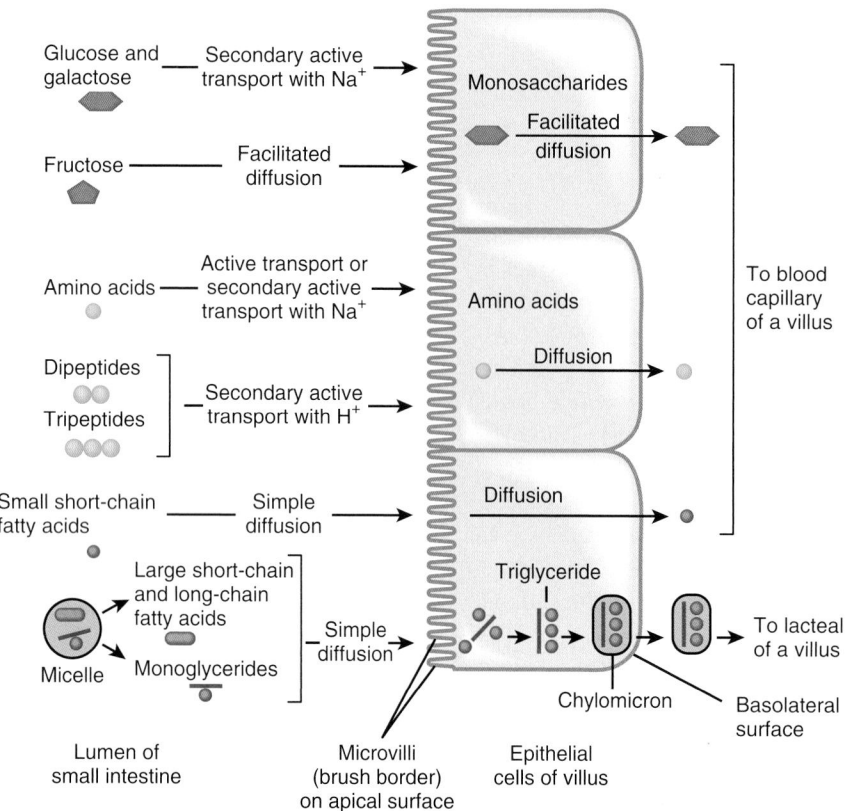

(a) Mechanisms for movement of nutrients through absorptive epithelial cells of villi

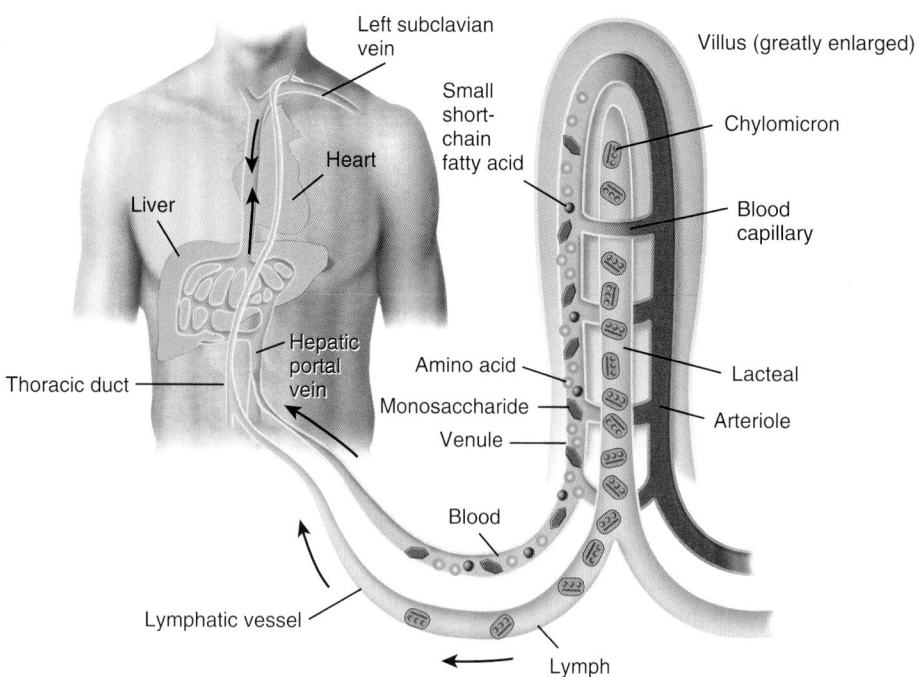

(b) Movement of absorbed nutrients into blood and lymph

? **A monoglyceride may be larger than an amino acid. Why can monoglycerides be absorbed by simple diffusion, but amino acids cannot?**

amino acids enter absorptive cells of the villi via Na^+-dependent secondary active transport processes that are similar to the glucose transporter; other amino acids are actively transported by themselves. At least one symporter brings in dipeptides and tripeptides together with H^+; the peptides then are hydrolyzed to single amino acids inside the absorptive cells. Amino acids move out of the absorptive cells via diffusion and enter capillaries of the villus (Figure 24.21). Both monosaccharides and amino acids are transported in the blood to the liver by way of the hepatic portal system. If not removed by hepatocytes, they enter the general circulation.

Absorption of Lipids

All dietary lipids are absorbed via *simple diffusion*. Adults absorb about 95% of the lipids present in the small intestine; due to their lower production of bile, newborn infants absorb only about 85% of lipids. As a result of their emulsification and digestion, triglycerides are mainly broken down into monoglycerides and fatty acids, which can be either short-chain fatty acids or long-chain fatty acids. Although short-chain fatty acids are hydrophobic, they are very small but do vary in size. Small short-chain fatty acids contain less than 10–12 carbon atoms and are more water-soluble. Thus, they can dissolve in the watery intestinal chyme, pass through the absorptive cells via simple diffusion, and follow the same route taken by monosaccharides and amino acids into a blood capillary of a villus (Figure 24.21a).

Large short-chain fatty acids (with more than 10–12 carbon atoms), long-chain fatty acids, and monoglycerides are larger and hydrophobic, and since they are not water-soluble, they have difficulty being suspended in the watery environment of the intestinal chyme. Besides their role in emulsification, bile salts also help to make these large short-chain fatty acids, long-chain fatty acids, and monoglycerides more soluble. The bile salts in intestinal chyme surround them, forming tiny spheres called **micelles** (mī-SELZ = small morsels), each of which is 2–10 nm in diameter and includes 20–50 bile salt molecules (Figure 24.21a). Micelles are formed due to the amphipathic nature of bile salts: The hydrophobic regions of bile salts interact with the large short-chain fatty acids, long-chain fatty acids, and monoglycerides, and the hydrophilic regions of bile salts interact with the watery intestinal chyme. Once formed, the micelles move from the interior of the small intestinal lumen to the brush border of the absorptive cells. At that point, the large short-chain fatty acids, long-chain fatty acids, and monoglycerides diffuse out of the micelles into the absorptive cells, leaving the micelles behind in the chyme. The micelles continually repeat this ferrying function as they move from the brush border back through the chyme to the interior of the small intestinal lumen to pick up more of the large short-chain fatty acids, long-chain fatty acids, and monoglycerides. Micelles also solubilize other large hydrophobic molecules such as fat-soluble vitamins (A, D, E, and K) and cholesterol that may be present in intestinal chyme, and aid in their absorption. These fat-soluble vitamins and cholesterol molecules are packed in the micelles along with the long-chain fatty acids and monoglycerides.

Once inside the absorptive cells, long-chain fatty acids and monoglycerides are recombined to form triglycerides, which ag-gregate into globules along with phospholipids and cholesterol and become coated with proteins. These large spherical masses, about 80 nm in diameter, are called **chylomicrons** (kī-lō-MĪ-krons). Chylomicrons leave the absorptive cell via exocytosis. Because they are so large and bulky, chylomicrons cannot enter blood capillaries—the pores in the walls of blood capillaries are too small. Instead, chylomicrons enter lacteals, which have much larger pores than blood capillaries. From lacteals, chylomicrons are transported by way of lymphatic vessels to the thoracic duct and enter the blood at the left subclavian vein (Figure 24.21b). The hydrophilic protein coat that surrounds each chylomicron keeps the chylomicrons suspended in blood and prevents them from sticking to each other.

Within 10 minutes after absorption, about half of the chylomicrons have already been removed from the blood as they pass through blood capillaries in the liver and adipose tissue. This removal is accomplished by an enzyme attached to the apical surface of capillary endothelial cells, called **lipoprotein lipase,** that breaks down triglycerides in chylomicrons and other lipoproteins into fatty acids and glycerol. The fatty acids diffuse into hepatocytes and adipose cells and combine with glycerol during resynthesis of triglycerides. Two or three hours after a meal, few chylomicrons remain in the blood.

After participating in the emulsification and absorption of lipids, 90–95% of the bile salts are reabsorbed by active transport in the final segment of the small intestine (ileum) and returned by the blood to the liver through the hepatic portal system for recycling. This cycle of bile salt secretion by hepatocytes into bile, reabsorption by the ileum, and resecretion into bile is called the **enterohepatic circulation** (en′-ter-ō-he-PAT-ik). Insufficient bile salts, due either to obstruction of the bile ducts or removal of the gallbladder, can result in the loss of up to 40% of dietary lipids in feces due to diminished lipid absorption. When lipids are not absorbed properly, the fat-soluble vitamins are not adequately absorbed.

Absorption of Electrolytes

Many of the electrolytes absorbed by the small intestine come from gastrointestinal secretions, and some are part of ingested foods and liquids. Recall that electrolytes are compounds that separate into ions in water and conduct electricity. Sodium ions are actively transported out of absorptive cells by basolateral sodium–potassium pumps (Na^+/K^+ ATPases) after they have moved into absorptive cells via diffusion and secondary active transport. Thus, most of the sodium ions (Na^+) in gastrointestinal secretions are reclaimed and not lost in the feces. Negatively charged bicarbonate, chloride, iodide, and nitrate ions can passively follow Na^+ or be actively transported. Calcium ions also are absorbed actively in a process stimulated by calcitriol. Other electrolytes such as iron, potassium, magnesium, and phosphate ions also are absorbed via active transport mechanisms.

Absorption of Vitamins

As you have just learned, the fat-soluble vitamins A, D, E, and K are included with ingested dietary lipids in micelles and are absorbed via simple diffusion. Most water-soluble vitamins, such as most B vitamins and vitamin C, also are absorbed via simple diffusion. Vitamin B₁₂, however, combines with intrinsic factor produced by the stomach, and the combination is absorbed in the ileum via an active transport mechanism.

Absorption of Water

The total volume of fluid that enters the small intestine each day—about 9.3 liters (9.8 qt)—comes from ingestion of liquids (about 2.3 liters) and from various gastrointestinal secretions (about 7.0 liters). Figure 24.22 depicts the amounts of fluid ingested, secreted, absorbed, and excreted by the GI tract. The small intestine absorbs about 8.3 liters of the fluid; the remainder passes into the large intestine, where most of the rest of it—about 0.9 liter—is also absorbed. Only 0.1 liter (100 mL) of water is excreted in the feces each day.

All water absorption in the GI tract occurs via *osmosis* from the lumen of the intestines through absorptive cells and into blood

capillaries. Because water can move across the intestinal mucosa in both directions, the absorption of water from the small intestine depends on the absorption of electrolytes and nutrients to maintain

Figure 24.22 **Daily volumes of fluid ingested, secreted, absorbed, and excreted from the GI tract.**

All water absorption in the GI tract occurs via osmosis.

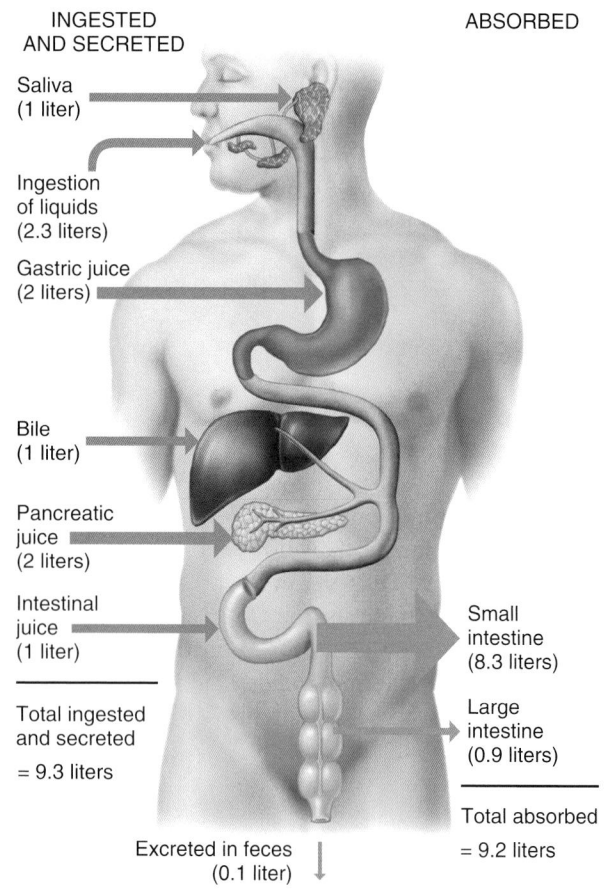

Fluid balance in GI tract

❓ **Which two organs of the digestive system secrete the most fluid?**

TABLE 24.4	
Summary of Digestive Activities in the Pancreas, Liver, Gallbladder, and Small Intestine	
STRUCTURE	**ACTIVITY**
Pancreas	Delivers pancreatic juice into duodenum via pancreatic duct to assist absorption (see Table 24.5 for pancreatic enzymes and their functions).
Liver	Produces bile (bile salts) necessary for emulsification and absorption of lipids.
Gallbladder	Stores, concentrates, and delivers bile into duodenum via common bile duct.
Small intestine	Major site of digestion and absorption of nutrients and water in gastrointestinal tract.
Mucosa/submucosa	
Intestinal glands	Secrete intestinal juice to assist absorption.
Absorptive cells	Digest and absorb nutrients.
Goblet cells	Secrete mucus.
Enteroendocrine cells (S, CCK, K)	Secrete secretin, cholecystokinin, and glucose-dependent insulinotropic peptide.
Paneth cells	Secrete lysozyme (bactericidal enzyme) and phagocytosis.
Duodenal (Brunner's) glands	Secrete alkaline fluid to buffer stomach acids, and mucus for protection and lubrication.
Circular folds	Folds of mucosa and submucosa that increase surface area for digestion and absorption.
Villi	Fingerlike projections of mucosa that are sites of absorption of digested food and increase surface area for digestion and absorption.
Microvilli	Microscopic, membrane-covered projections of absorptive epithelial cells that contain brush-border enzymes (listed in Table 24.5) and that increase surface area for digestion and absorption.
Muscularis	
Segmentation	Type of peristalsis: alternating contractions of circular smooth muscle fibers that produce segmentation and resegmentation of sections of small intestine; mixes chyme with digestive juices and brings food into contact with mucosa for absorption.
Migrating motility complex (MMC)	Type of peristalsis: waves of contraction and relaxation of circular and longitudinal smooth muscle fibers passing the length of the small intestine; moves chyme toward ileocecal sphincter.

an osmotic balance with the blood. The absorbed electrolytes, monosaccharides, and amino acids establish a concentration gradient for water that promotes water absorption via osmosis.

Table 24.4 summarizes the digestive activities of the pancreas, liver, gallbladder, and small intestine and Table 24.5 summarizes the digestive enzymes and their functions in the digestive system.

CLINICAL CONNECTION | *Absorption of Alcohol*

The intoxicating and incapacitating effects of alcohol depend on the blood alcohol level. Because it is lipid-soluble, alcohol begins to be absorbed in the stomach. However, the surface area available for absorption is much greater in the small intestine than in the

TABLE 24.5

Summary of Digestive Enzymes

ENZYME	SOURCE	SUBSTRATES	PRODUCTS
SALIVA			
Salivary amylase	Salivary glands.	Starches (polysaccharides).	Maltose (disaccharide), maltotriose (trisaccharide), and α-dextrins.
Lingual lipase	Lingual glands in tongue.	Triglycerides (fats and oils) and other lipids.	Fatty acids and diglycerides.
GASTRIC JUICE			
Pepsin (activated from pepsinogen by pepsin and hydrochloric acid)	Stomach chief cells.	Proteins.	Peptides.
Gastric lipase	Stomach chief cells.	Triglycerides (fats and oils).	Fatty acids and monoglycerides.
PANCREATIC JUICE			
Pancreatic amylase	Pancreatic acinar cells.	Starches (polysaccharides).	Maltose (disaccharide), maltotriose (trisaccharide), and α-dextrins.
Trypsin (activated from trypsinogen by enterokinase)	Pancreatic acinar cells.	Proteins.	Peptides.
Chymotrypsin (activated from chymotrypsinogen by trypsin)	Pancreatic acinar cells.	Proteins.	Peptides.
Elastase (activated from proelastase by trypsin)	Pancreatic acinar cells.	Proteins.	Peptides.
Carboxypeptidase (activated from procarboxypeptidase by trypsin)	Pancreatic acinar cells.	Amino acid at carboxyl end of peptides.	Amino acids and peptides.
Pancreatic lipase	Pancreatic acinar cells.	Triglycerides (fats and oils) that have been emulsified by bile salts.	Fatty acids and monoglycerides.
Nucleases			
Ribonuclease	Pancreatic acinar cells.	Ribonucleic acid.	Nucleotides.
Deoxyribonuclease	Pancreatic acinar cells.	Deoxyribonucleic acid.	Nucleotides.
BRUSH-BORDER ENZYMES IN MICROVILLI PLASMA MEMBRANE			
α-**Dextrinase**	Small intestine.	α-Dextrins.	Glucose.
Maltase	Small intestine.	Maltose.	Glucose.
Sucrase	Small intestine.	Sucrose.	Glucose and fructose.
Lactase	Small intestine.	Lactose.	Glucose and galactose.
Enterokinase	Small intestine.	Trypsinogen.	Trypsin.
Peptidases			
Aminopeptidase	Small intestine.	Amino acid at amino end of peptides.	Amino acids and peptides.
Dipeptidase	Small intestine.	Dipeptides.	Amino acids.
Nucleosidases and phosphatases	Small intestine.	Nucleotides.	Nitrogenous bases, pentoses, and phosphates.

stomach, so when alcohol passes into the duodenum, it is absorbed more rapidly. Thus, the longer the alcohol remains in the stomach, the more slowly blood alcohol level rises. Because fatty acids in chyme slow gastric emptying, blood alcohol level will rise more slowly when fat-rich foods, such as pizza, hamburgers, or nachos, are consumed with alcoholic beverages. Also, the enzyme alcohol dehydrogenase, which is present in gastric mucosa cells, breaks down some of the alcohol to acetaldehyde, which is not intoxicating. When the rate of gastric emptying is slower, proportionally more alcohol will be absorbed and converted to acetaldehyde in the stomach, and thus less alcohol will reach the bloodstream. Given identical consumption of alcohol, females often develop higher blood alcohol levels (and therefore experience greater intoxication) than males of comparable size because the activity of gastric alcohol dehydrogenase is up to 60% lower in females than in males. Asian males may also have lower levels of this gastric enzyme. •

✔CHECKPOINT

32. List the regions of the small intestine and describe their functions.
33. In what ways are the mucosa and submucosa of the small intestine adapted for digestion and absorption?
34. Describe the types of movement that occur in the small intestine.
35. Explain the functions of pancreatic amylase, aminopeptidase, gastric lipase, and deoxyribonuclease.
36. What is the difference between digestion and absorption? How are the end products of carbohydrate, protein, and lipid digestion absorbed?
37. By what routes do absorbed nutrients reach the liver?
38. Describe the absorption of electrolytes, vitamins, and water by the small intestine.

24.13 LARGE INTESTINE

⦿ OBJECTIVE

• Describe the anatomy, histology, and functions of the large intestine.

The large intestine is the terminal portion of the GI tract. The overall functions of the large intestine are the completion of absorption, the production of certain vitamins, the formation of feces, and the expulsion of feces from the body.

Anatomy of the Large Intestine

The **large intestine,** which is about 1.5 m (5 ft) long and 6.5 cm (2.5 in.) in diameter, extends from the ileum to the anus. It is attached to the posterior abdominal wall by its **mesocolon** (mez'-ō-KŌ-lon), which is a double layer of peritoneum (see Figure 24.4a). Structurally, the four major regions of the large intestine are the cecum, colon, rectum, and anal canal (Figure 24.23a).

The opening from the ileum into the large intestine is guarded by a fold of mucous membrane called the ileocecal sphincter (valve), which allows materials from the small intestine to pass into the large intestine. Hanging inferior to the ileocecal valve is

the **cecum,** a small pouch about 6 cm (2.4 in.) long. Attached to the cecum is a twisted, coiled tube, measuring about 8 cm (3 in.) in length, called the **appendix** or **vermiform appendix** (VER-mi-form; *vermiform* = worm-shaped; *appendix* = appendage). The mesentery of the appendix, called the **mesoappendix** (mez-ō-a-PEN-diks), attaches the appendix to the inferior part of the mesentery of the ileum.

CLINICAL CONNECTION | *Appendicitis*

Inflammation of the appendix, termed **appendicitis,** is preceded by obstruction of the lumen of the appendix by chyme, inflammation, a foreign body, a carcinoma of the cecum, stenosis, or kinking of the organ. It is characterized by high fever, elevated white blood cell count, and a neutrophil count higher than 75%. The infection that follows may result in edema and ischemia and may progress to gangrene and perforation within 24 hours. Typically, appendicitis begins with referred pain in the umbilical region of the abdomen, followed by anorexia (loss of appetite for food), nausea, and vomiting. After several hours the pain localizes in the right lower quadrant (RLQ) and is continuous, dull or severe, and intensified by coughing, sneezing, or body movements. Early appendectomy (removal of the appendix) is recommended because it is safer to operate than to risk rupture, peritonitis, and gangrene. Although it required major abdominal surgery in the past, today appendectomies are usually performed laparoscopically. •

The open end of the cecum merges with a long tube called the **colon** (= food passage), which is divided into ascending, transverse, descending, and sigmoid portions. Both the ascending and descending colon are retroperitoneal; the transverse and sigmoid colon are not. True to its name, the **ascending colon** ascends on the right side of the abdomen, reaches the inferior surface of the liver, and turns abruptly to the left to form the **right colic (hepatic) flexure.** The colon continues across the abdomen to the left side as the **transverse colon.** It curves beneath the inferior end of the spleen on the left side as the **left colic (splenic) flexure** and passes inferiorly to the level of the iliac crest as the **descending colon.** The **sigmoid colon** (*sigm-* = S-shaped) begins near the left iliac crest, projects medially to the midline, and terminates as the rectum at about the level of the third sacral vertebra.

The **rectum,** the last 20 cm (8 in.) of the GI tract, lies anterior to the sacrum and coccyx. The terminal 2–3 cm (1 in.) of the rectum is called the **anal canal** (Figure 24.23b). The mucous membrane of the anal canal is arranged in longitudinal folds called **anal columns** that contain a network of arteries and veins. The opening of the anal canal to the exterior, called the **anus,** is guarded by an **internal anal sphincter** of smooth muscle (involuntary) and an **external anal sphincter** of skeletal muscle (voluntary). Normally these sphincters keep the anus closed except during the elimination of feces.

Histology of the Large Intestine

The wall of the large intestine contains the typical four layers found in the rest of the GI tract: mucosa, submucosa, muscularis,

Figure 24.23 Anatomy of the large intestine.

The regions of the large intestine are the cecum, colon, rectum, and anal canal.

(a) Anterior view of large intestine showing major regions

(b) Frontal section of anal canal

FUNCTIONS OF THE LARGE INTESTINE

1. Haustral churning, peristalsis, and mass peristalsis drive contents of colon into rectum.
2. Bacteria in large intestine convert proteins to amino acids, break down amino acids, and produce some B vitamins and vitamin K.
3. Absorbing some water, ions, and vitamins.
4. Forming feces.
5. Defecating (emptying rectum).

? Which portions of the colon are retroperitoneal?

and serosa. The **mucosa** consists of simple columnar epithelium, lamina propria (areolar connective tissue), and muscularis mucosae (smooth muscle) (Figure 24.24a). The epithelium contains mostly absorptive and goblet cells (Figure 24.24b, c). The absorptive cells function primarily in water absorption; the goblet cells secrete mucus that lubricates the passage of the colonic contents. Both absorptive and goblet cells are located in long, straight, tubular intestinal glands (crypts of Lieberkühn) that extend the full thickness of the mucosa. Solitary lymphatic nodules are also found in the lamina propria of the mucosa and may extend through the muscularis mucosae into the submucosa. Compared to the small intestine, the mucosa of the large intestine does not have as many structural adaptations that increase surface area. There are no circular folds or villi; however, microvilli are present on the absorptive cells. Consequently, much more absorption occurs in the small intestine than in the large intestine.

The **submucosa** of the large intestine consists of areolar connective tissue. The **muscularis** consists of an external layer of longitudinal smooth muscle and an internal layer of circular smooth muscle. Unlike other parts of the GI tract, portions of the longitudinal muscles are thickened, forming three conspicuous bands called the **teniae coli** (TĒ-nē-ē KŌ-lī; *teniae* = flat bands) that run most of the length of the large intestine (see Figure 24.23a). The teniae coli are separated by portions of the wall with less or no longitudinal muscle. Tonic contractions of the bands gather the colon into a series of pouches called **haustra** (HAWS-tra = shaped like pouches; singular is *haustrum*), which give the colon a puckered appearance. A single layer of circular smooth muscle lies between teniae coli. The **serosa** of the large intestine is part of the visceral peritoneum. Small pouches of visceral peritoneum filled with fat are attached to teniae coli and are called **omental (fatty) appendices.**

Figure 24.24 Histology of the large intestine.

Intestinal glands formed by simple columnar epithelial cells and goblet cells extend the full thickness of the mucosa.

Lumen of large intestine

Openings of intestinal glands
Absorptive cell
Goblet cell
Intestinal gland
Lamina propria
Lymphatic nodule
Muscularis mucosae
Lymphatic vessel
Arteriole
Venule
Circular layer of muscle
Myenteric plexus
Longitudinal layer of muscle

MUCOSA
SUBMUCOSA
MUSCULARIS
SEROSA

(a) Three-dimensional view of layers of large intestine

Openings of intestinal glands
Lamina propria
Microvilli
Intestinal gland
Absorptive cell (absorbs water)
Goblet cell (secretes mucus)
Lymphatic nodule
Muscularis mucosae
Submucosa

(b) Sectional view of intestinal glands and cell types

point, the walls contract and squeeze the contents into the next haustrum. **Peristalsis** also occurs, although at a slower rate (3–12 contractions per minute) than in more proximal portions of the tract. A final type of movement is **mass peristalsis,** a strong peristaltic wave that begins at about the middle of the transverse colon and quickly drives the contents of the colon into the rectum. Because food in the stomach initiates this **gastrocolic reflex** in the colon, mass peristalsis usually takes place three or four times a day, during or immediately after a meal.

Mechanical Digestion in the Large Intestine

The passage of chyme from the ileum into the cecum is regulated by the action of the ileocecal sphincter. Normally, the valve remains partially closed so that the passage of chyme into the cecum usually occurs slowly. Immediately after a meal, a **gastroileal reflex** (gas'-trō-IL-ē-al) intensifies peristalsis in the ileum and forces any chyme into the cecum. The hormone gastrin also relaxes the sphincter. Whenever the cecum is distended, the degree of contraction of the ileocecal sphincter intensifies.

Movements of the colon begin when substances pass the ileocecal sphincter. Because chyme moves through the small intestine at a fairly constant rate, the time required for a meal to pass into the colon is determined by gastric emptying time. As food passes through the ileocecal sphincter, it fills the cecum and accumulates in the ascending colon.

One movement characteristic of the large intestine is **haustral churning.** In this process, the haustra remain relaxed and become distended while they fill up. When the distension reaches a certain

Chemical Digestion in the Large Intestine

The final stage of digestion occurs in the colon through the activity of bacteria that inhabit the lumen. Mucus is secreted by the glands of the large intestine, but no enzymes are secreted. Chyme is prepared for elimination by the action of bacteria, which ferment any remaining carbohydrates and release hydrogen, carbon dioxide, and methane gases. These gases contribute to flatus (gas) in the colon, termed *flatulence* when it is excessive. Bacteria also convert any remaining proteins to amino acids and break down the amino acids into simpler substances: indole, skatole, hydrogen sulfide, and fatty acids. Some of the indole and skatole is eliminated in the feces and contributes to their odor; the rest is absorbed and transported to the liver, where these compounds are converted to less toxic compounds and excreted in the urine. Bacteria also decompose bilirubin to simpler pigments, including stercobilin, which gives feces their brown color. Bacterial products that are absorbed in the colon include several vitamins needed for normal metabolism, among them some B vitamins and vitamin K.

Mucosa

Submucosa

Muscularis

Serosa

Lumen of large intestine

Lamina propria

Intestinal gland

Lymphatic nodule

Muscularis mucosae

LM 315x

(c) Portion of wall of large intestine

Opening of intestinal gland

Lumen of large intestine

Absorptive cell

Goblet cell

Lamina propria

Intestinal gland

LM 300x

(d) Details of mucosa of large intestine

? **What is the function of the goblet cells in the large intestine?**

Absorption and Feces Formation in the Large Intestine

By the time chyme has remained in the large intestine 3–10 hours, it has become solid or semisolid because of water absorption and is now called **feces.** Chemically, feces consist of water, inorganic salts, sloughed-off epithelial cells from the mucosa of the gastrointestinal tract, bacteria, products of bacterial decomposition, unabsorbed digested materials, and indigestible parts of food.

Although 90% of all water absorption occurs in the small intestine, the large intestine absorbs enough to make it an important organ in maintaining the body's water balance. Of the 0.5–1.0 liter of water that enters the large intestine, all but about 100–200 mL is normally absorbed via osmosis. The large intestine also absorbs ions, including sodium and chloride, and some vitamins.

CLINICAL CONNECTION | *Occult Blood*

The term **occult blood** refers to blood that is hidden; it is not detectable by the human eye. The main diagnostic value of occult blood testing is to screen for colorectal cancer. Two substances often examined for occult blood are feces and urine. Several types of products are available for at-home testing for hidden blood in feces. The tests are based on color changes when reagents are added to feces. The presence of occult blood in urine may be detected at home by using dip-and-read reagent strips. •

The Defecation Reflex

Mass peristaltic movements push fecal material from the sigmoid colon into the rectum. The resulting distension of the rectal wall stimulates stretch receptors, which initiates a **defecation reflex** that empties the rectum. The defecation reflex occurs as follows: In response to distension of the rectal wall, the receptors send sensory nerve impulses to the sacral spinal cord. Motor impulses from the cord travel along parasympathetic nerves back to the descending colon, sigmoid colon, rectum, and anus. The resulting contraction of the longitudinal rectal muscles shortens the rectum, thereby increasing the pressure within it. This pressure, along with voluntary contractions of the diaphragm and abdominal muscles, plus parasympathetic stimulation, opens the internal anal sphincter.

The external anal sphincter is voluntarily controlled. If it is voluntarily relaxed, defecation occurs and the feces are expelled through the anus; if it is voluntarily constricted, defecation can be postponed. Voluntary contractions of the diaphragm and abdominal muscles aid defecation by increasing the pressure within the abdomen, which pushes the walls of the sigmoid colon and rectum inward. If defecation does not occur, the feces back up into the sigmoid colon until the next wave of mass peristalsis stimulates the stretch receptors, again creating the urge to defecate. In infants, the defecation reflex causes automatic emptying of the rectum because voluntary control of the external anal sphincter has not yet developed.

The amount of bowel movements that a person has over a given period of time depends on various factors such as diet, health, and stress. The normal range of bowel activity varies from two or three bowel movements per day to three or four bowel movements per week.

Diarrhea (dī-a-RĒ-a; *dia-* = through; *-rrhea* = flow) is an increase in the frequency, volume, and fluid content of the feces caused by increased motility of and decreased absorption by the intestines. When chyme passes too quickly through the small intestine and feces pass too quickly through the large intestine, there is not enough time for absorption. Frequent diarrhea can result in dehydration and electrolyte imbalances. Excessive motility may be caused by lactose intolerance, stress, and microbes that irritate the gastrointestinal mucosa.

Constipation (kon-sti-PĀ-shun; *con-* = together; *-stip-* = to press) refers to infrequent or difficult defecation caused by decreased motility of the intestines. Because the feces remain in the colon for prolonged periods, excessive water absorption occurs, and the feces become dry and hard. Constipation may be caused by poor habits (delaying defecation), spasms of the colon, insufficient fiber in the diet, inadequate fluid intake, lack of exercise, emotional stress, and certain drugs. A common treatment is a mild laxative, such as milk of magnesia, which induces defecation. However, many physicians maintain that laxatives are habit-forming, and that adding fiber to the diet, increasing the amount of exercise, and increasing fluid intake are safer ways of controlling this common problem.

Table 24.6 summarizes the digestive activities in the large intestine, and Table 24.7 summarizes the functions of all digestive system organs.

TABLE 24.6

Summary of Digestive Activities in the Large Intestine

STRUCTURE	ACTIVITY	FUNCTION(S)
Lumen	Bacterial activity.	Breaks down undigested carbohydrates, proteins, and amino acids into products that can be expelled in feces or absorbed and detoxified by liver; synthesizes certain B vitamins and vitamin K.
Mucosa	Secretes mucus.	Lubricates colon; protects mucosa.
	Absorption.	Water absorption solidifies feces and contributes to body's water balance; solutes absorbed include ions and some vitamins.
Muscularis	Haustral churning.	Moves contents from haustrum to haustrum by muscular contractions.
	Peristalsis.	Moves contents along length of colon by contractions of circular and longitudinal muscles.
	Mass peristalsis.	Forces contents into sigmoid colon and rectum.
	Defecation reflex.	Eliminates feces by contractions in sigmoid colon and rectum.

TABLE 24.7

Summary of Organs of the Digestive System and Their Functions

ORGAN	FUNCTION(S)
Tongue	Maneuvers food for mastication, shapes food into a bolus, maneuvers food for deglutition, detects sensations for taste, and initiates digestion of triglycerides.
Salivary glands	Saliva produced by these glands softens, moistens, and dissolves foods; cleanses mouth and teeth; initiates the digestion of starch.
Teeth	Cut, tear, and pulverize food to reduce solids to smaller particles for swallowing.
Pancreas	Pancreatic juice buffers acidic gastric juice in chyme, stops the action of pepsin from the stomach, creates the proper pH for digestion in the small intestine, and participates in the digestion of carbohydrates, proteins, triglycerides, and nucleic acids.
Liver	Produces bile, which is required for the emulsification and absorption of lipids in the small intestine.
Gallbladder	Stores and concentrates bile and releases it into the small intestine.
Mouth	See the functions of the tongue, salivary glands, and teeth, all of which are in the mouth. Additionally, the lips and cheeks keep food between the teeth during mastication, and buccal glands lining the mouth produce saliva.
Pharynx	Receives a bolus from the oral cavity and passes it into the esophagus.
Esophagus	Receives a bolus from the pharynx and moves it into the stomach; this requires relaxation of the upper esophageal sphincter and secretion of mucus.
Stomach	Mixing waves combine saliva, food, and gastric juice, which activates pepsin, initiates protein digestion, kills microbes in food, helps absorb vitamin B_{12}, contracts the lower esophageal sphincter, increases stomach motility, relaxes the pyloric sphincter, and moves chyme into the small intestine.
Small intestine	Segmentation mixes chyme with digestive juices; peristalsis propels chyme toward the ileocecal sphincter; digestive secretions from the small intestine, pancreas, and liver complete the digestion of carbohydrates, proteins, lipids, and nucleic acids; circular folds, villi, and microvilli help absorb about 90 percent of digested nutrients.
Large intestine	Haustral churning, peristalsis, and mass peristalsis drive the colonic contents into the rectum; bacteria produce some B vitamins and vitamin K; absorption of some water, ions, and vitamins occurs; defecation.

CLINICAL CONNECTION | *Dietary Fiber*

Dietary fiber consists of indigestible plant carbohydrates—such as cellulose, lignin, and pectin—found in fruits, vegetables, grains, and beans. **Insoluble fiber,** which does not dissolve in water, includes the woody or structural parts of plants such as the skins of fruits and vegetables and the bran coating around wheat and corn kernels. Insoluble fiber passes through the GI tract largely unchanged but speeds up the passage of material through the tract. **Soluble fiber,** which does dissolve in water, forms a gel that slows the passage of material through the tract. It is found in abundance in beans, oats, barley, broccoli, prunes, apples, and citrus fruits.

People who choose a fiber-rich diet may reduce their risk of developing obesity, diabetes, atherosclerosis, gallstones, hemorrhoids, diverticulitis, appendicitis, and colorectal cancer. Soluble fiber also may help lower blood cholesterol. The liver normally converts cholesterol to bile salts, which are released into the small intestine to help fat digestion. Having accomplished their task, the bile salts are reabsorbed by the small intestine and recycled back to the liver. Since soluble fiber binds to bile salts to prevent their reabsorption, the liver makes more bile salts to replace those lost in feces. Thus, the liver uses more cholesterol to make more bile salts and blood cholesterol level is lowered. •

✔ CHECKPOINT

39. What are the major regions of the large intestine?
40. How does the muscularis of the large intestine differ from that of the rest of the gastrointestinal tract? What are haustra?
41. Describe the mechanical movements that occur in the large intestine.
42. What is defecation and how does it occur?
43. What activities occur in the large intestine to change its contents into feces?

24.14 PHASES OF DIGESTION

◉ **OBJECTIVES**

• Describe the three phases of digestion.
• Describe the major hormones regulating digestive activities.

Digestive activities occur in three overlapping phases: the cephalic phase, the gastric phase, and the intestinal phase.

Cephalic Phase

During the **cephalic phase** of digestion, the smell, sight, thought, or initial taste of food activates neural centers in the cerebral cortex, hypothalamus, and brain stem. The brain stem then activates the facial (VII), glossopharyngeal (IX), and vagus (X) nerves. The facial and glossopharyngeal nerves stimulate the salivary glands to secrete saliva, while the vagus nerves stimulate the gastric glands to secrete gastric juice. The purpose of the cephalic phase of digestion is to prepare the mouth and stomach for food that is about to be eaten.

Gastric Phase

Once food reaches the stomach, the **gastric phase** of digestion begins. Neural and hormonal mechanisms regulate the gastric phase of digestion to promote gastric secretion and gastric motility.

• *Neural regulation.* Food of any kind distends the stomach and stimulates stretch receptors in its walls. Chemoreceptors in the stomach monitor the pH of the stomach chyme. When the stomach walls are distended or pH increases because proteins have entered the stomach and buffered some of the stomach acid, the stretch receptors and chemoreceptors are activated, and a neural negative feedback loop is set in motion (Figure 24.25). From the stretch receptors and chemoreceptors, nerve impulses propagate to the submucosal plexus, where they activate parasympathetic and enteric neurons. The resulting nerve impulses cause waves of peristalsis and continue to stimulate the flow of gastric juice from gastric glands. The peristaltic waves mix the food with gastric juice; when the waves become strong enough, a small quantity of chyme undergoes gastric emptying into the duodenum. The pH of the stomach chyme decreases (becomes more acidic) and the distension of the stomach walls lessens because chyme has passed into the small intestine, suppressing secretion of gastric juice.

• *Hormonal regulation.* Gastric secretion during the gastric phase is also regulated by the hormone **gastrin.** Gastrin is released from the **G cells** of the gastric glands in response to several stimuli: distension of the stomach by chyme, partially digested proteins in chyme, the high pH of chyme due to the presence of food in the stomach, caffeine in gastric chyme, and acetylcholine released from parasympathetic neurons. Once it is released, gastrin enters the bloodstream, makes a round-trip through the body, and finally reaches its target organs in the digestive system. Gastrin stimulates gastric glands to secrete large amounts of gastric juice. It also strengthens the contraction of the lower esophageal sphincter to prevent reflux of acid chyme into the esophagus, increases motility of the stomach, and relaxes the pyloric sphincter, which promotes gastric emptying. Gastrin secretion is inhibited when the pH of gastric juice drops below 2.0 and is stimulated when the pH rises. This negative feedback mechanism helps provide an optimal low pH for the functioning of pepsin, the killing of microbes, and the denaturing of proteins in the stomach.

Intestinal Phase

The **intestinal phase** of digestion begins once food enters the small intestine. In contrast to reflexes initiated during the cephalic

Figure 24.25 Neural negative feedback regulation of the pH of gastric juice and gastric motility during the gastric phase of digestion.

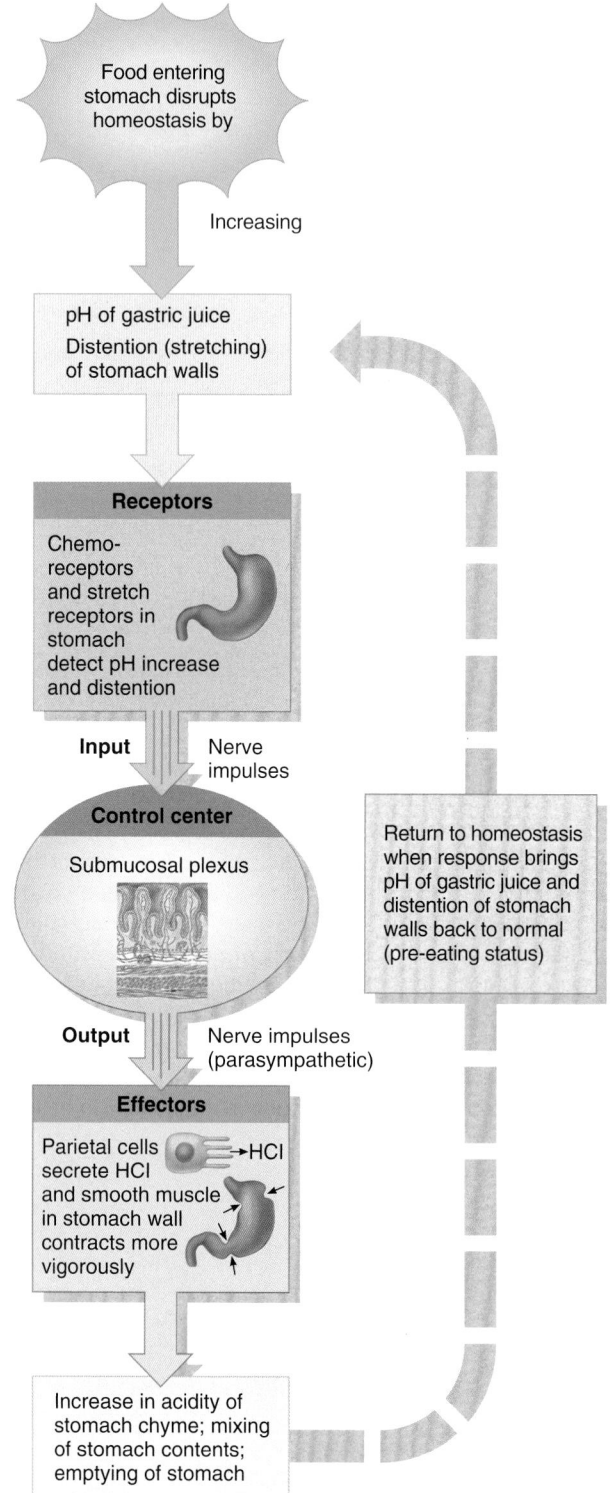

Food entering the stomach stimulates secretion of gastric juice and causes vigorous waves of peristalsis.

Why does food initially cause the pH of the gastric juice to rise?

and gastric phases, which stimulate stomach secretory activity and motility, those occurring during the intestinal phase have inhibitory effects that slow the exit of chyme from the stomach. This prevents the duodenum from being overloaded with more chyme than it can handle. In addition, responses occurring during the intestinal phase promote the continued digestion of foods that have reached the small intestine. These activities of the intestinal phase of digestion are regulated by neural and hormonal mechanisms.

- *Neural regulation.* Distension of the duodenum by the presence of chyme causes the **enterogastric reflex** (en′-ter-ō-GAS-trik). Stretch receptors in the duodenal wall send nerve impulses to the medulla oblongata, where they inhibit parasympathetic stimulation and stimulate the sympathetic nerves to the stomach. As a result, gastric motility is inhibited and there is an increase in the contraction of the pyloric sphincter, which decreases gastric emptying.

- *Hormonal regulation.* The intestinal phase of digestion is mediated by two major hormones secreted by the small intestine: cholecystokinin and secretin. **Cholecystokinin (CCK)** is secreted by the **CCK cells** of the small intestinal crypts of Lieberkühn in response to chyme containing amino acids from partially digested proteins and fatty acids from partially digested triglycerides. CCK stimulates secretion of pancreatic juice that is rich in digestive enzymes. It also causes contraction of the wall of the gallbladder, which squeezes stored bile out of the gallbladder into the cystic duct and through the common bile duct. In addition, CCK causes relaxation of the sphincter of the hepatopancreatic ampulla (sphincter of Oddi), which allows pancreatic juice and bile to flow into the duodenum. CCK also slows gastric emptying by promoting contraction of the pyloric sphincter, produces satiety (a feeling of fullness) by acting on the hypothalamus in the brain, promotes normal growth and maintenance of the pancreas, and enhances the effects of secretin. Acidic chyme entering the duodenum stimulates the release of **secretin** from the **S cells** of the small intestinal crypts of Lieberkühn. In turn, secretin stimulates the flow of pancreatic juice that is rich in bicarbonate (HCO_3^-) ions to buffer the acidic chyme that enters the duodenum from the small intestine. Besides this major effect, secretin inhibits secretion of gastric juice, promotes normal growth and maintenance of the pancreas, and enhances the effects of CCK. Overall, secretin causes buffering of acid in chyme that reaches the duodenum and slows production of acid in the stomach.

Other Hormones of the Digestive System

Besides gastrin, CCK, and secretin, at least 10 other so-called gut hormones are secreted by and have effects on the GI tract. They include *motilin, substance P,* and *bombesin,* which stimulate motility of the intestines; *vasoactive intestinal polypeptide (VIP),* which stimulates secretion of ions and water by the intestines and inhibits gastric acid secretion; *gastrin-releasing peptide,* which stimulates release of gastrin; and *somatostatin,* which inhibits gastrin release. Some of these hormones are thought to act as local hormones (paracrines), whereas others are secreted into the blood or even into the lumen of the GI tract. The physiological roles of these and other gut hormones are still under investigation.

Table 24.8 summarizes the major hormones that control digestion.

✔ CHECKPOINT

44. What is the purpose of the cephalic phase of digestion?
45. Describe the role of gastrin in the gastric phase of digestion.
46. Outline the steps of the enterogastric reflex.
47. Explain the roles of CCK and secretin in the intestinal phase of digestion.

TABLE 24.8

Major Hormones That Control Digestion

HORMONE	STIMULUS AND SITE OF SECRETION	ACTIONS
Gastrin	Distension of stomach, partially digested proteins and caffeine in stomach, and high pH of stomach chyme stimulate gastrin secretion by enteroendocrine G cells, located mainly in mucosa of pyloric antrum of stomach.	*Major effects:* Promotes secretion of gastric juice, increases gastric motility, promotes growth of gastric mucosa. *Minor effects:* Constricts lower esophageal sphincter, relaxes pyloric sphincter.
Secretin	Acidic (high H^+ level) chyme that enters small intestine stimulates secretion of secretin by enteroendocrine S cells in the mucosa of duodenum.	*Major effects:* Stimulates secretion of pancreatic juice and bile that are rich in HCO_3^- (bicarbonate ions). *Minor effects:* Inhibits secretion of gastric juice, promotes normal growth and maintenance of pancreas, enhances effects of CCK.
Cholecystokinin (CCK)	Partially digested proteins (amino acids), triglycerides, and fatty acids that enter small intestine stimulate secretion of CCK by enteroendocrine CCK cells in mucosa of small intestine; CCK is also released in brain.	*Major effects:* Stimulates secretion of pancreatic juice rich in digestive enzymes, causes ejection of bile from gallbladder and opening of sphincter of the hepatopancreatic ampulla (sphincter of Oddi), induces satiety (feeling full to satisfaction). *Minor effects:* Inhibits gastric emptying, promotes normal growth and maintenance of pancreas, enhances effects of secretin.

24.15 DEVELOPMENT OF THE DIGESTIVE SYSTEM

◉ OBJECTIVE

• Describe the development of the digestive system.

During the fourth week of development, the cells of the **endoderm** form a cavity called the **primitive gut,** the forerunner of the gastrointestinal tract (see Figure 29.12b). Soon afterward the mesoderm forms and splits into two layers (somatic and splanchnic), as shown in Figure 29.9d. The splanchnic mesoderm associates with the endoderm of the primitive gut; as a result, the primitive gut has a double-layered wall. The **endodermal layer** gives rise to the *epithelial lining* and *glands* of most of the gastrointestinal tract; the **mesodermal layer** produces the *smooth muscle* and *connective tissue* of the tract.

The primitive gut elongates and differentiates into an anterior **foregut,** an intermediate **midgut,** and a posterior **hindgut** (see Figure 29.12c). Until the fifth week of development, the midgut opens into the yolk sac; after that time, the yolk sac constricts and detaches from the midgut, and the midgut seals. In the region of the foregut, a depression consisting of ectoderm, the **stomodeum** (stō-mō-DĒ-um), appears (see Figure 29.12d), which develops into the *oral cavity*. The **oropharyngeal membrane** (or'-ō-fa-RIN-jē-al) is a depression of fused ectoderm and endoderm on the surface of the embryo that separates the foregut from the stomodeum. The membrane ruptures during the fourth week of development, so that the foregut is continuous with the outside of the embryo through the oral cavity. Another depression consisting of ectoderm, the **proctodeum** (prok-tō-DĒ-um), forms in the hindgut and goes on to develop into the *anus* (see Figure 29.12d). The **cloacal membrane** (klō-Ā-kul) is a fused membrane of ectoderm and endoderm that separates the hindgut from the proctodeum. After it ruptures during the seventh week, the hindgut is continuous with the outside of the embryo through the anus. Thus, the gastrointestinal tract forms a continuous tube from mouth to anus.

The foregut develops into the *pharynx, esophagus, stomach,* and *part of the duodenum*. The midgut is transformed into the *remainder of the duodenum,* the *jejunum,* the *ileum,* and *portions of the large intestine* (cecum, appendix, ascending colon, and most of the transverse colon). The hindgut develops into the *remainder of the large intestine,* except for a portion of the anal canal that is derived from the proctodeum.

As development progresses, the endoderm at various places along the foregut develops into hollow buds that grow into the mesoderm. These buds will develop into the *salivary glands, liver, gallbladder,* and *pancreas*. Each of these organs retains a connection with the gastrointestinal tract through ducts.

✔CHECKPOINT

48. What structures develop from the foregut, midgut, and hindgut?

24.16 AGING AND THE DIGESTIVE SYSTEM

◉ OBJECTIVE

• Describe the effects of aging on the digestive system.

Overall changes of the digestive system associated with aging include decreased secretory mechanisms, decreased motility of the digestive organs, loss of strength and tone of the muscular tissue and its supporting structures, changes in neurosensory feedback regarding enzyme and hormone release, and diminished response to pain and internal sensations. In the upper portion of the GI tract, common changes include reduced sensitivity to mouth irritations and sores, loss of taste, periodontal disease, difficulty in swallowing, hiatal hernia, gastritis, and peptic ulcer disease. Changes that may appear in the small intestine include duodenal ulcers, malabsorption, and maldigestion. Other pathologies that increase in incidence with age are appendicitis, gallbladder problems, jaundice, cirrhosis, and acute pancreatitis. Large intestinal changes such as constipation, hemorrhoids, and diverticular disease may also occur. Cancer of the colon or rectum is quite common, as are bowel obstructions and impactions.

✔CHECKPOINT

49. What are the general effects of aging on the digestive system?

• • •

Now that our exploration of the digestive system is complete, you can appreciate the many ways that this system contributes to homeostasis of other body systems by examining *Focus on Homeostasis: The Digestive System*. Next, in Chapter 25, you will discover how the nutrients absorbed by the GI tract enter into metabolic reactions in the body tissues.

Focus on Homeostasis

BODY SYSTEM	CONTRIBUTION OF THE DIGESTIVE SYSTEM
For all body systems	The digestive system breaks down dietary nutrients into forms that can be absorbed and used by body cells for producing ATP and building body tissues. Absorbs water, minerals, and vitamins needed for growth and function of body tissues; and eliminates wastes from body tissues in feces.
Integumentary system	Small intestine absorbs vitamin D, which skin and kidneys modify to produce the hormone calcitriol. Excess dietary calories are stored as triglycerides in adipose cells in dermis and subcutaneous layer.
Skeletal system	Small intestine absorbs dietary calcium and phosphorus salts needed to build bone extracellular matrix.
Muscular system	Liver can convert lactic acid (produced by muscles during exercise) to glucose.
Nervous system	Gluconeogenesis (synthesis of new glucose molecules) in liver plus digestion and absorption of dietary carbohydrates provide glucose, needed for ATP production by neurons.
Endocrine system	Liver inactivates some hormones, ending their activity. Pancreatic islets release insulin and glucagon. Cells in mucosa of stomach and small intestine release hormones that regulate digestive activities. Liver produces angiotensinogen.
Cardiovascular system	GI tract absorbs water that helps maintain blood volume and iron that is needed for synthesis of hemoglobin in red blood cells. Bilirubin from hemoglobin breakdown is partially excreted in feces. Liver synthesizes most plasma proteins.
Lymphatic system and immunity	Acidity of gastric juice destroys bacteria and most toxins in stomach. Lymphatic nodules in lamina propria of mucosa of gastrointestinal tract (MALT) destroy microbes.
Respiratory system	Pressure of abdominal organs against diaphragm helps expel air quickly during forced exhalation.
Urinary system	Absorption of water by GI tract provides water needed to excrete waste products in urine.
Reproductive systems	Digestion and absorption provide adequate nutrients, including fats, for normal development of reproductive structures, for production of gametes (oocytes and sperm), and for fetal growth and development during pregnancy.

THE DIGESTIVE SYSTEM

Dental Caries

Dental caries (KAR-ēz), or tooth decay, involves a gradual demineralization (softening) of the enamel and dentin. If untreated, microorganisms may invade the pulp, causing inflammation and infection, with subsequent death of the pulp and abscess of the alveolar bone surrounding the root's apex, requiring root canal therapy (see Section 24.5).

Dental caries begin when bacteria, acting on sugars, produce acids that demineralize the enamel. **Dextran,** a sticky polysaccharide produced from sucrose, causes the bacteria to stick to the teeth. Masses of bacterial cells, dextran, and other debris adhering to teeth constitute **dental plaque** (PLAK). Saliva cannot reach the tooth surface to buffer the acid because the plaque covers the teeth. Brushing the teeth after eating removes the plaque from flat surfaces before the bacteria can produce acids. Dentists also recommend that the plaque between the teeth be removed every 24 hours with dental floss.

Periodontal Disease

Periodontal disease is a collective term for a variety of conditions characterized by inflammation and degeneration of the gingivae, alveolar bone, periodontal ligament, and cementum. In one such condition, called **pyorrhea,** initial symptoms include enlargement and inflammation of the soft tissue and bleeding of the gums. Without treatment, the soft tissue may deteriorate and the alveolar bone may be resorbed, causing loosening of the teeth and recession of the gums. Periodontal diseases are often caused by poor oral hygiene; by local irritants, such as bacteria, impacted food, and cigarette smoke; or by a poor "bite."

Peptic Ulcer Disease

In the United States, 5–10% of the population develops **peptic ulcer disease (PUD).** An **ulcer** is a craterlike lesion in a membrane; ulcers that develop in areas of the GI tract exposed to acidic gastric juice are called **peptic ulcers.** The most common complication of peptic ulcers is bleeding, which can lead to anemia if enough blood is lost. In acute cases, peptic ulcers can lead to shock and death. Three distinct causes of PUD are recognized: (1) the bacterium *Helicobacter pylori* (hel-i-kō-BAK-ter pī-LŌ-rē); (2) nonsteroidal anti-inflammatory drugs (NSAIDs) such as aspirin; and (3) hypersecretion of HCl, as occurs in Zollinger–Ellison syndrome (ZOL-in-jer EL-i-son), a gastrin-producing tumor, usually of the pancreas.

Helicobacter pylori (previously named *Campylobacter pylori*) is the most frequent cause of PUD. The bacterium produces an enzyme called urease, which splits urea into ammonia and carbon dioxide. While shielding the bacterium from the acidity of the stomach, the ammonia also damages the protective mucous layer of the stomach and the underlying gastric cells. The microbe also produces catalase, an enzyme that may protect *H. pylori* from phagocytosis by neutrophils, plus several adhesion proteins that allow the bacterium to attach itself to gastric cells.

Several therapeutic approaches are helpful in the treatment of PUD. Cigarette smoke, alcohol, caffeine, and NSAIDs should be avoided because they can impair mucosal defensive mechanisms, which increases mucosal susceptibility to the damaging effects of HCl. In cases associated with *H. pylori,* treatment with an antibiotic drug often resolves the problem. Oral antacids such as Tums® or Maalox® can help temporarily by buffering gastric acid. When hypersecretion of HCl is the cause of PUD, H_2 blockers (such as Tagamet®) or proton pump inhibitors such as omeprazole (Prilosec®), which block secretion of H^+ from parietal cells, may be used.

Diverticular Disease

In **diverticular disease** (dī′-ver-TIK-ū-lar), saclike outpouchings of the wall of the colon, termed **diverticula,** occur in places where the muscularis has weakened and may become inflamed. Development of diverticula is known as **diverticulosis.** Many people who develop diverticulosis have no symptoms and experience no complications. Of those people known to have diverticulosis, 10–25% eventually develop an inflammation known as **diverticulitis** (dī′-ver-tik-ū-LĪ-tis). This condition may be characterized by pain, either constipation or increased frequency of defecation, nausea, vomiting, and low-grade fever. Because diets low in fiber contribute to development of diverticulitis, patients who change to high-fiber diets show marked relief of symptoms. In severe cases, affected portions of the colon may require surgical removal. If diverticula rupture, the release of bacteria into the abdominal cavity can cause peritonitis.

Colorectal Cancer

Colorectal cancer is among the deadliest of malignancies, ranking second to lung cancer in males and third after lung cancer and breast cancer in females. Genetics plays a very important role; an inherited predisposition contributes to more than half of all cases of colorectal cancer. Intake of alcohol and diets high in animal fat and protein are associated with increased risk of colorectal cancer; dietary fiber, retinoids, calcium, and selenium may be protective. Signs and symptoms of colorectal cancer include diarrhea, constipation, cramping, abdominal pain, and rectal bleeding, either visible or occult (hidden in feces). Precancerous growths on the mucosal surface, called **polyps,** also increase the risk of developing colorectal cancer. Screening for colorectal cancer includes testing for blood in the feces, digital rectal examination, sigmoidoscopy, colonoscopy, and barium enema. Tumors may be removed endoscopically or surgically.

Hepatitis

Hepatitis is an inflammation of the liver that can be caused by viruses, drugs, and chemicals, including alcohol. Clinically, several types of viral hepatitis are recognized.

Hepatitis A (infectious hepatitis) is caused by the hepatitis A virus and is spread via fecal contamination of objects such as food, clothing, toys, and eating utensils (fecal–oral route). It is generally a mild disease of children and young adults characterized by loss of appetite, malaise, nausea, diarrhea, fever, and chills. Eventually, jaundice appears. This type of hepatitis does not cause lasting liver damage. Most people recover in 4 to 6 weeks.

Hepatitis B is caused by the hepatitis B virus and is spread primarily by sexual contact and contaminated syringes and transfusion equipment. It can also be spread via saliva and tears. Hepatitis B virus can be present for years or even a lifetime, and it can produce cirrhosis and possibly cancer of the liver. Individuals who harbor the active hepatitis B virus also become carriers. Vaccines produced through recombinant DNA technology (for example, Recombivax HB®) are available to prevent hepatitis B infection.

Hepatitis C, caused by the hepatitis C virus, is clinically similar to hepatitis B. Hepatitis C can cause cirrhosis and possibly liver cancer. In developed nations, donated blood is screened for the presence of hepatitis B and C.

Hepatitis D is caused by the hepatitis D virus. It is transmitted like hepatitis B, and in fact a person must have been co-infected with hepatitis B before contracting hepatitis D. Hepatitis D results in severe liver damage and has a higher fatality rate than infection with hepatitis B virus alone.

Hepatitis E is caused by the hepatitis E virus and is spread like hepatitis A. Although it does not cause chronic liver disease, hepatitis E virus has a very high mortality rate among pregnant women.

MEDICAL TERMINOLOGY

Achalasia (ak'-a-LĀ-zē-a; *a-* = without; *-chalasis* = relaxation) A condition caused by malfunction of the myenteric plexus in which the lower esophageal sphincter fails to relax normally as food approaches. A whole meal may become lodged in the esophagus and enter the stomach very slowly. Distension of the esophagus results in chest pain that is often confused with pain originating from the heart.

Bariatric surgery (bar'-ē-AT-rik; *baros-* = weight; *-iatreia* = medical treatment) A surgical procedure that limits the amount of food that can be ingested and absorbed in order to bring about a significant weight loss in obese individuals. The most commonly performed type of bariatric surgery is called *gastric bypass surgery*. In one variation of this procedure, the stomach is reduced in size by making a small pouch at the top of the stomach about the size of a walnut. The pouch, which is only 5–10% of the stomach, is sealed off from the rest of the stomach using surgical staples or a plastic band. The pouch is connected to the jejunum of the small intestine, thus bypassing the rest of the stomach and the duodenum. The result is that smaller amounts of food are ingested and fewer nutrients are absorbed in the small intestine. This leads to weight loss.

Borborygmus (bor'-bō-RIG-mus) A rumbling noise caused by the propulsion of gas through the intestines.

Canker sore (KANG-ker) Painful ulcer on the mucous membrane of the mouth that affects females more often than males, usually between ages 10 and 40; may be an autoimmune reaction or a food allergy.

Cirrhosis Distorted or scarred liver as a result of chronic inflammation due to hepatitis, chemicals that destroy hepatocytes, parasites that infect the liver, or alcoholism; the hepatocytes are replaced by fibrous or adipose connective tissue. Symptoms include jaundice, edema in the legs, uncontrolled bleeding, and increased sensitivity to drugs.

Colitis (kō-LĪ-tis) Inflammation of the mucosa of the colon and rectum in which absorption of water and salts is reduced, producing watery, bloody feces and, in severe cases, dehydration and salt depletion. Spasms of the irritated muscularis produce cramps. It is thought to be an autoimmune condition.

Colonoscopy (kō-lon-OS-kō-pē; *-skopes* = to view) The visual examination of the lining of the colon using an elongated, flexible, fiber-optic endoscope called a *colonoscope*. It is used to detect disorders such as polyps, cancer, and diverticulosis; to take tissue samples; and to remove small polyps. Most tumors of the large intestine occur in the rectum.

Colostomy (kō-LOS-tō-mē; *-stomy* = provide an opening) The diversion of feces through an opening in the colon, creating a surgical "stoma" (artificial opening) that is made in the exterior of the abdominal wall. This opening serves as a substitute anus through which feces are eliminated into a bag worn on the abdomen.

Dysphagia (dis-FĀ-jē-a; *dys-* = abnormal; *-phagia* = to eat) Difficulty in swallowing that may be caused by inflammation, paralysis, obstruction, or trauma.

Flatus (FLĀ-tus) Air (gas) in the stomach or intestine, usually expelled through the anus. If the gas is expelled through the mouth, it is called **eructation** or **belching** (burping). Flatus may result from gas released during the breakdown of foods in the stomach or from swallowing air or gas-containing substances such as carbonated drinks.

Food poisoning A sudden illness caused by ingesting food or drink contaminated by an infectious microbe (bacterium, virus, or protozoan) or a toxin (poison). The most common cause of food poisoning is the toxin produced by the bacterium *Staphylococcus aureus.* Most types of food poisoning cause diarrhea and/or vomiting, often associated with abdominal pain.

Gastroenteritis (gas'-trō-en-ter-Ī-tis; *gastro-* = stomach; *-enteron-* = intestine; *-itis* = inflammation) Inflammation of the lining of the stomach and intestine (especially the small intestine). It is usually caused by a viral or bacterial infection that may be acquired by contaminated food or water or by people in close contact. Symptoms include diarrhea, vomiting, fever, loss of appetite, cramps, and abdominal discomfort.

Gastroscopy (gas-TROS-kō-pē; *-scopy* = to view with a lighted instrument) Endoscopic examination of the stomach in which the examiner can view the interior of the stomach directly to evaluate an ulcer, tumor, inflammation, or source of bleeding.

Halitosis (hal'-i-TŌ-sis; *halitus-* = breath; *-osis* = condition) A foul odor from the mouth; also called **bad breath.**

Heartburn A burning sensation in a region near the heart due to irritation of the mucosa of the esophagus from hydrochloric acid in stomach contents. It is caused by failure of the lower esophageal sphincter to close properly, so that the stomach contents enter the inferior esophagus. It is not related to any cardiac problem.

Hemorrhoids (HEM-ō-royds; *hemo-* = blood; *-rhoia* = flow) Varicosed (enlarged and inflamed) superior rectal veins. Hemorrhoids develop when the veins are put under pressure and become engorged with blood. If the pressure continues, the wall of the vein stretches. Such a distended vessel oozes blood; bleeding or itching is usually the first sign that a hemorrhoid has developed. Stretching of a vein also favors clot formation, further aggravating swelling and pain. Hemorrhoids may be caused by constipation, which may be brought on by low-fiber diets. Also, repeated straining during defecation forces blood down into the rectal veins, increasing pressure in those veins and possibly causing hemorrhoids. Also called **piles.**

Hernia (HER-nē-a) Protrusion of all or part of an organ through a membrane or cavity wall, usually the abdominal cavity. *Hiatus (diaphragmatic) hernia* is the protrusion of a part of the stomach into the thoracic cavity through the esophageal hiatus of the diaphragm. *Inguinal hernia* is the protrusion of the hernial sac into the inguinal opening; it may contain a portion of the bowel in an advanced stage and may extend into the scrotal compartment in males, causing strangulation of the herniated part.

Inflammatory bowel disease (in-FLAM-a-tō'-rē BOW-el) Inflammation of the gastrointestinal tract that exists in two forms. (1) *Crohn's disease* is an inflammation of any part of the gastrointestinal tract in which the inflammation extends from the mucosa through the submucosa, muscularis, and serosa. (2) *Ulcerative colitis* is an inflammation of the mucosa of the colon and rectum, usually accompanied by rectal bleeding. Curiously, cigarette smoking increases the risk of Crohn's disease but decreases the risk of ulcerative colitis.

Irritable bowel syndrome (IBS) Disease of the entire gastrointestinal tract in which a person reacts to stress by developing symptoms (such as cramping and abdominal pain) associated with alternating patterns of diarrhea and constipation. Excessive amounts of mucus may appear in feces; other symptoms include flatulence, nausea, and loss of appetite. The condition is also known as **irritable colon** or **spastic colitis.**

Malabsorption (mal-ab-SORP-shun; *mal-* = bad) A number of disorders in which nutrients from food are not absorbed properly. It may be due to disorders that result in the inadequate breakdown of food during digestion (due to inadequate digestive enzymes or juices), damage to the lining of the small intestine (from surgery, infections, and drugs like neomycin and alcohol), and impairment of motility. Symptoms may include diarrhea, weight loss, weakness, vitamin deficiencies, and bone demineralization.

Malocclusion (mal′-ō-KLOO-zhun; *mal-* = bad; *-occlusion* = to fit together) Condition in which the surfaces of the maxillary (upper) and mandibular (lower) teeth fit together poorly.

Nausea (NAW-sē-a; *nausia* = seasickness) Discomfort characterized by a loss of appetite and the sensation of impending vomiting. Its causes include local irritation of the gastrointestinal tract, a systemic disease, brain disease or injury, overexertion, or the effects of medication or drug overdosage.

Traveler's diarrhea Infectious disease of the gastrointestinal tract that results in loose, urgent bowel movements, cramping, abdominal pain, malaise, nausea, and occasionally fever and dehydration. It is acquired through ingestion of food or water contaminated with fecal material typically containing bacteria (especially *Escherichia coli*); viruses or protozoan parasites are less common causes.

CHAPTER REVIEW AND RESOURCE SUMMARY

Review

Resource

Introduction

1. The breaking down of larger food molecules into smaller molecules is called digestion.
2. The organs involved in the breakdown of food are collectively known as the digestive system.

24.1 Overview of the Digestive System

1. The digestive system is composed of two main groups of organs: the gastrointestinal (GI) tract and accessory digestive organs.
2. The GI tract is a continuous tube extending from the mouth to the anus.
3. The accessory digestive organs include the teeth, tongue, salivary glands, liver, gallbladder, and pancreas.
4. Digestion includes six basic processes: ingestion, secretion, mixing and propulsion, mechanical and chemical digestion, absorption, and defecation.
5. Mechanical digestion consists of mastication and movements of the gastrointestinal tract that aid chemical digestion.
6. Chemical digestion is a series of hydrolysis reactions that break down large carbohydrates, lipids, proteins, and nucleic acids in foods into smaller molecules that are usable by body cells.

Anatomy Overview - The Digestive System
Animation - Chemical Digestion: Enzymes
Exercise - Concentrate on Digestion

24.2 Layers of the GI Tract

1. The basic arrangement of layers in most of the gastrointestinal tract, from deep to superficial, is the mucosa, submucosa, muscularis, and serosa.
2. Associated with the lamina propria of the mucosa are extensive patches of lymphatic tissue called mucosa-associated lymphoid tissue (MALT).

Anatomy Overview - GI Tract Histology
Figure 24.2 - Layers of the Gastrointestinal Tract
Exercise - Paint the Gastrointestinal Tract

24.3 Neural Innervation of the GI Tract

1. The gastrointestinal tract is regulated by an intrinsic set of nerves known as the enteric nervous system (ENS) and by an extrinsic set of nerves that are part of the autonomic nervous system (ANS).
2. The ENS consists of neurons arranged into two plexuses: the myenteric plexus and the submucosal plexus.
3. The myenteric plexus, which is located between the longitudinal and circular smooth muscle layers of the muscularis, regulates GI tract motility.
4. The submucosal plexus, which is located in the submucosa, regulates GI secretion.
5. Although the neurons of the ENS can function independently, they are subject to regulation by the neurons of the ANS.
6. Parasympathetic fibers of the vagus (X) nerves and pelvic splanchnic nerves increase GI tract secretion and motility by increasing the activity of ENS neurons.
7. Sympathetic fibers from the thoracic and upper lumbar regions of the spinal cord decrease GI tract secretion and motility by inhibiting ENS neurons.

.Animation - Neural Regulation of Mechanical Digestion
Animation - Enterogastric Reflex

Review	Resource

24.4 Peritoneum

1. The peritoneum is the largest serous membrane of the body; it lines the wall of the abdominal cavity and covers some abdominal organs.
2. Folds of the peritoneum include the mesentery, mesocolon, falciform ligament, lesser omentum, and greater omentum.

24.5 Mouth

1. The mouth is formed by the cheeks, hard and soft palates, lips, and tongue.
2. The vestibule is the space bounded externally by the cheeks and lips and internally by the teeth and gums.
3. The oral cavity proper extends from the vestibule to the fauces.
4. The tongue, together with its associated muscles, forms the floor of the oral cavity. It is composed of skeletal muscle covered with mucous membrane. The upper surface and sides of the tongue are covered with papillae, some of which contain taste buds.
5. The major portion of saliva is secreted by the major salivary glands, which lie outside the mouth and pour their contents into ducts that empty into the oral cavity. There are three pairs of major salivary glands: parotid, submandibular, and sublingual glands.
6. Saliva lubricates food and starts the chemical digestion of carbohydrates. Salivation is controlled by the nervous system.
7. The teeth (dentes) project into the mouth and are adapted for mechanical digestion.
8. A typical tooth consists of three principal regions: crown, root, and neck. Teeth are composed primarily of dentin and are covered by enamel, the hardest substance in the body. There are two dentitions: deciduous and permanent.
9. Through mastication, food is mixed with saliva and shaped into a soft, flexible mass called a bolus. Salivary amylase then begins the digestion of starches, and lingual lipase acts on triglycerides.

Resource:
Anatomy Overview - Oral Cavity
Anatomy Overview - Salivary Glands
Animation - Mastication
Figure 24.8 - Dentitions and Times of Eruption

24.6 Pharynx

1. The pharynx is a funnel-shaped tube that extends from the internal nares to the esophagus posteriorly and to the larynx anteriorly.
2. The pharynx has both respiratory and digestive functions.

Resource:
Anatomy Overview - Pharynx and Esophagus

24.7 Esophagus

1. The esophagus is a collapsible, muscular tube that connects the pharynx to the stomach.
2. It contains an upper and a lower esophageal sphincter.

Resource:
Anatomy Overview - Esophagus Histology

24.8 Deglutition

1. Deglutition, or swallowing, moves a bolus from the mouth to the stomach.
2. Swallowing consists of a voluntary stage, a pharyngeal stage (involuntary), and an esophageal stage (involuntary).

Resource:
Animation - Deglutition
Animation - Neural Regulation of Mechanical Digestion

24.9 Stomach

1. The stomach connects the esophagus to the duodenum.
2. The principal anatomical regions of the stomach are the cardia, fundus, body, and pylorus.
3. Adaptations of the stomach for digestion include rugae; glands that produce mucus, hydrochloric acid, pepsin, gastric lipase, and intrinsic factor; and a three-layered muscularis.
4. Mechanical digestion consists of mixing waves.
5. Chemical digestion consists mostly of the conversion of proteins into peptides by pepsin.
6. The stomach wall is impermeable to most substances.
7. Among the substances the stomach can absorb are water, certain ions, drugs, and alcohol.

Resource:
Anatomy Overviews - The Stomach; Stomach Histology
Animations - Stomach Peristalsis; Protein Digestion in the Stomach; Lipid Digestion in the Stomach; Chemical Digestion: Gastric Acid
Figure 24.12 - Histology of the Stomach
Figure 24.13 - Secretion of HCl by Parietal Cells in the Stomach
Exercise - Role of Digestive Chemicals

24.10 Pancreas

1. The pancreas consists of a head, a body, and a tail and is connected to the duodenum via the pancreatic duct and accessory duct.
2. Endocrine pancreatic islets (islets of Langerhans) secrete hormones, and exocrine acini secrete pancreatic juice.
3. Pancreatic juice contains enzymes that digest starch (pancreatic amylase), proteins (trypsin, chymotrypsin, carboxypeptidase, and elastase), triglycerides (pancreatic lipase), and nucleic acids (ribonuclease and deoxyribonuclease).

Resource:
Anatomy Overviews - Pancreas; Pancreas Histology
Animations - Carbohydrate Digestion: Pancreas; Protein Digestion: Pancreatic Juice; Lipid Digestion: Bile Salts and Pancreatic Lipase
Figure 24.15 - Relationship of the Pancreas to the Liver, Gallbladder, and Duodenum

Review

24.11 Liver and Gallbladder

1. The liver has left and right lobes; the left lobe includes a quadrate lobe and a caudate lobe. The gallbladder is a sac located in a depression on the posterior surface of the liver that stores and concentrates bile.
2. The lobes of the liver are made up of lobules that contain hepatocytes (liver cells), sinusoids, stellate reticuloendothelial (Kupffer) cells, and a central vein.
3. Hepatocytes produce bile that is carried by a duct system to the gallbladder for concentration and temporary storage.
4. Bile's contribution to digestion is the emulsification of dietary lipids.
5. The liver also functions in carbohydrate, lipid, and protein metabolism; processing of drugs and hormones; excretion of bilirubin; synthesis of bile salts; storage of vitamins and minerals; phagocytosis; and activation of vitamin D.

24.12 Small Intestine

1. The small intestine extends from the pyloric sphincter to the ileocecal sphincter. It is divided into duodenum, jejunum, and ileum.
2. Its glands secrete fluid and mucus, and the circular folds, villi, and microvilli of its wall provide a large surface area for digestion and absorption.
3. Brush-border enzymes digest α-dextrins, maltose, sucrose, lactose, peptides, and nucleotides at the surface of mucosal epithelial cells.
4. Pancreatic and intestinal brush-border enzymes break down starches into maltose, maltotriose, and α-dextrins (pancreatic amylase), α-dextrins into glucose (α-dextrinase), maltose to glucose (maltase), sucrose to glucose and fructose (sucrase), lactose to glucose and galactose (lactase), and proteins into peptides (trypsin, chymotrypsin, and elastase). Also, enzymes break off amino acids at the carboxyl ends of peptides (carboxypeptidases) and break off amino acids at the amino ends of peptides (aminopeptidases). Finally, enzymes split dipeptides into amino acids (dipeptidases), triglycerides to fatty acids and monoglycerides (lipases), and nucleotides to pentoses and nitrogenous bases (nucleosidases and phosphatases).
5. Mechanical digestion in the small intestine involves segmentation and migrating motility complexes.
6. Absorption occurs via diffusion, facilitated diffusion, osmosis, and active transport; most absorption occurs in the small intestine.
7. Monosaccharides, amino acids, and short-chain fatty acids pass into the blood capillaries.
8. Long-chain fatty acids and monoglycerides are absorbed from micelles, resynthesized to triglycerides, and formed into chylomicrons.
9. Chylomicrons move into lymph in the lacteal of a villus.
10. The small intestine also absorbs electrolytes, vitamins, and water.

Anatomy Overviews - Small Intestine; Small Intestine Histology
Animations - Segmentation; Carbohydrate Digestion in the Small Intestine; Protein Digestion in the Small Intestine; Nucleic Acid Digestion in the Small Intestine; Carbohydrate, Protein, Nucleic Acid, and Lipid Absorption in the Small Intestine
Figure 24.21 - Absorption in the Small Intestine
Exercises - Digestive Enzyme Activity; Nutrient Absorption Routes

24.13 Large Intestine

1. The large intestine extends from the ileocecal sphincter to the anus.
2. Its regions include the cecum, colon, rectum, and anal canal.
3. The mucosa contains many goblet cells, and the muscularis consists of teniae coli and haustra.
4. Mechanical movements of the large intestine include haustral churning, peristalsis, and mass peristalsis.
5. The last stages of chemical digestion occur in the large intestine through bacterial action. Substances are further broken down, and some vitamins are synthesized.
6. The large intestine absorbs water, ions, and vitamins.
7. Feces consist of water, inorganic salts, epithelial cells, bacteria, and undigested foods.
8. The elimination of feces from the rectum is called defecation.
9. Defecation is a reflex action aided by voluntary contractions of the diaphragm and abdominal muscles and relaxation of the external anal sphincter.

24.14 Phases of Digestion

1. Digestive activities occur in three overlapping phases: cephalic phase, gastric phase, and intestinal phase.
2. During the cephalic phase of digestion, salivary glands secrete saliva and gastric glands secrete gastric juice in order to prepare the mouth and stomach for food that is about to be eaten.
3. The presence of food in the stomach causes the gastric phase of digestion, which promotes gastric juice secretion and gastric motility.

Review   **Resource**

4. During the intestinal phase of digestion, food is digested in the small intestine. In addition, gastric motility and gastric secretion decrease in order to slow the exit of chyme from the stomach, which prevents the small intestine from being overloaded with more chyme than it can handle.

5. The activities that occur during the various phases of digestion are coordinated by neural pathways and by hormones. Table 24.8 summarizes the major hormones that control digestion.

24.15 Development of the Digestive System

1. The endoderm of the primitive gut forms the epithelium and glands of most of the gastrointestinal tract.

2. The mesoderm of the primitive gut forms the smooth muscle and connective tissue of the gastrointestinal tract.

24.16 Aging and the Digestive System

1. General changes include decreased secretory mechanisms, decreased motility, and loss of tone.

2. Specific changes may include loss of taste, pyorrhea, hernias, peptic ulcer disease, constipation, hemorrhoids, and diverticular diseases.

SELF-QUIZ QUESTIONS

Fill in the blanks in the following statements.

1. The end products of chemical digestion of carbohydrates are _____, of proteins are _____, of lipids are _____ and _____, and of nucleic acids are _____, _____, and _____.

2. List the mechanisms of absorption of materials in the small intestine: _____, _____, _____, and _____.

Indicate whether the following statements are true or false.

3. The soft palate, uvula, and epiglottis prevent swallowed foods and liquids from entering the respiratory passages.

4. The coordinated contractions and relaxations of the muscularis, which propels materials through the GI tract, is known as peristalsis.

Choose the one best answer to the following questions.

5. Which of the following are mismatched?
   (a) chemical digestion: splitting food molecules into simple substances by hydrolysis with the assistance of digestive enzymes
   (b) motility: mechanical processes that break apart ingested food into small molecules
   (c) ingestion: taking foods and liquids into the mouth
   (d) propulsion: movement of food through GI tract due to smooth muscle contraction
   (e) absorption: passage into blood or lymph of ions, fluids and small molecules via the epithelial lining of the GI tract lumen

6. Which of the following are *true* concerning the peritoneum? (1) The kidneys and pancreas are retroperitoneal. (2) The greater omentum is the largest of the peritoneal folds. (3) The lesser omentum binds the large intestine to the posterior abdominal wall. (4) The falciform ligament attaches the liver to the anterior abdominal wall and diaphragm. (5) The mesentery is associated with the jejunum and ileum.
   (a) 1, 2, 3, and 5   (b) 1, 2, and 5   (c) 2 and 5
   (d) 1, 2, 4, and 5   (e) 3, 4, and 5

7. When a surgeon makes an incision in the small intestine, in what order would the physician encounter these structures? (1) epithelium, (2) submucosa, (3) serosa, (4) muscularis, (5) lamina propria, (6) muscularis mucosae.
   (a) 3, 4, 5, 6, 2, 1   (b) 1, 2, 3, 4, 6, 5   (c) 1, 5, 6, 2, 4, 3
   (d) 5, 1, 2, 6, 4, 3   (e) 3, 4, 2, 6, 5, 1

8. Which of the following are functions of the liver? (1) carbohydrate, lipid, and protein metabolism, (2) nucleic acid metabolism, (3) excretion of bilirubin, (4) synthesis of bile salts, (5) activation of vitamin D.
   (a) 1, 2, 3, and 5   (b) 1, 2, 3, and 4   (c) 1, 3, 4, and 5
   (d) 2, 3, 4, and 5   (e) 1, 2, 4, and 5

9. Which of the following statements regarding the regulation of gastric secretion and motility are *true*? (1) The sight, smell, taste, or thought of food can initiate the cephalic phase of gastric activity. (2) The gastric phase begins when food enters the small intestine. (3) Once activated, stretch receptors and chemoreceptors in the stomach trigger the flow of gastric juice and peristalsis. (4) The intestinal phase reflexes inhibit gastric activity. (5) The enterogastric reflex stimulates gastric emptying.
   (a) 1, 3, and 4   (b) 2, 4, and 5   (c) 1, 3, 4, and 5
   (d) 1, 2, and 5   (e) 1, 2, 3, and 4

10. Which of the following are *true*? (1) Segmentations in the small intestine help propel chyme through the intestinal tract. (2) The migrating motility complex is a type of peristalsis in the small intestine. (3) The large surface area for absorption in the small intestine is due to the presence of circular folds, villi, and microvilli. (4) The mucus-producing cells of the small intestine are paneth cells. (5) Most long-chain fatty acid and monoglyceride absorption in the small intestine requires the presence of bile salts.
   (a) 1, 2, and 3   (b) 2, 3, and 5   (c) 1, 2, 3, 4, and 5
   (d) 1, 3, and 5   (e) 1, 2, 3, and 5

11. The release of feces from the large intestine is dependent on (1) stretching of the rectal walls, (2) voluntary relaxation of the

external anal sphincter, (3) involuntary contraction of the diaphragm and abdominal muscles, (4) activity of the intestinal bacteria, (5) sympathetic stimulation of the internal sphincter.

(a) 2, 4, and 5   (b) 1, 2, and 5   (c) 1, 2, 3, and 5
(d) 1 and 2   (e) 3, 4, and 5

12. Which of the following is *not* true concerning the liver?
    (a) The left hepatic duct joins the cystic duct from the gallbladder.
    (b) As blood passes through the sinusoids, it is processed by hepatocytes and phagocytes.
    (c) Processed blood returns from the liver to systemic circulation through the hepatic veins.
    (d) The liver receives oxygenated blood through the hepatic artery.
    (e) The hepatic portal vein delivers deoxygenated blood from the GI tract to the liver.

13. Match the following:
    ____ (a) collapsed, muscular tube involved in deglutition and peristalsis
    ____ (b) coiled tube attached to the cecum
    ____ (c) contains duodenal glands in the submucosa
    ____ (d) produces and secretes bile
    ____ (e) contains aggregated lymphatic follicles in the mucosa
    ____ (f) responsible for ingestion, mastication, and deglutition
    ____ (g) responsible for churning, peristalsis, storage, and chemical digestion with the enzyme pepsin
    ____ (h) storage area for bile
    ____ (i) contain acini that release juices containing several digestive enzymes for protein, carbohydrate, lipid, and nucleic acid digestion and sodium bicarbonate to buffer stomach acid
    ____ (j) composed of enamel, dentin, and pulp cavity; used in mastication
    ____ (k) passageway for food, fluid, and air; involved in deglutition
    ____ (l) forms a semisolid waste material through haustral churning and peristalsis
    ____ (m) forces the food to the back of the mouth for swallowing; places food in contact with the teeth
    ____ (n) produce a fluid in the mouth that helps cleanse the mouth and teeth and that lubricates, dissolves, and begins the chemical breakdown of food

    (1) mouth
    (2) teeth
    (3) salivary glands
    (4) pharynx
    (5) esophagus
    (6) tongue
    (7) stomach
    (8) duodenum
    (9) ileum
    (10) colon
    (11) liver
    (12) gallbladder
    (13) appendix
    (14) pancreas

14. Match the following:
    ____ (a) an activating brush-border enzyme that splits off part of the trypsinogen molecule to form trypsin, a protease
    ____ (b) an enzyme that initiates carbohydrate digestion in the mouth
    ____ (c) the principal triglyceride-digesting enzyme in adults
    ____ (d) stimulates secretion of gastric juices and promotes gastric emptying
    ____ (e) secreted by chief cells in the stomach; a proteolytic enzyme
    ____ (f) stimulates the flow of pancreatic juice rich in bicarbonates; decreases gastric secretions
    ____ (g) a nonenzymatic fat-emulsifying agent
    ____ (h) causes contraction of the gallbladder and stimulates the production of pancreatic juice rich in digestive enzymes
    ____ (i) inhibits gastrin release
    ____ (j) stimulates secretion of ions and water by the intestines and inhibits gastric acid secretion
    ____ (k) secreted by glands in the tongue; begins breakdown of triglycerides in the stomach

    (1) gastrin
    (2) cholecystokinin
    (3) secretin
    (4) enterokinase
    (5) pepsin
    (6) salivary amylase
    (7) pancreatic lipase
    (8) lingual lipase
    (9) bile
    (10) vasoactive intestinal polypeptide
    (11) somatostatin

CRITICAL THINKING QUESTIONS

1. Why would you *not* want to completely suppress HCl secretion in the stomach?

2. Trey has cystic fibrosis, a genetic disorder that is characterized by the production of excessive mucus, affecting several body systems (e.g., respiratory, digestive, reproductive). In the digestive system, the excess mucus blocks bile ducts in the liver and pancreatic ducts. How would this affect Trey's digestive processes?

3. Antonio had dinner at his favorite Italian restaurant. His menu consisted of a salad, a large plate of spaghetti, garlic bread, and wine. For dessert, he consumed "death by chocolate" cake and a cup of coffee. He topped off his evening with a cigarette and brandy. He returned home and, while lying on his couch watching television, he experienced a pain in his chest. He called 911 because he was certain he was having a heart attack. Antonio was told his heart was fine, but he needed to watch his diet. What happened to Antonio?

? ANSWERS TO FIGURE QUESTIONS

24.1 Digestive enzymes are produced by the salivary glands, tongue, stomach, pancreas, and small intestine.

24.2 The lamina propria has the following functions: (1) It contains blood vessels and lymphatic vessels, which are the routes by which nutrients are absorbed from the GI tract; (2) it supports the mucosal epithelium and binds it to the muscularis mucosae; and (3) it contains mucosa-associated lymphatic tissue (MALT), which helps protect against disease.

24.3 The neurons of the myenteric plexus regulate GI tract motility, and the neurons of the submucosal plexus regulate GI secretion.

24.4 Mesentery binds the small intestine to the posterior abdominal wall.

24.5 The uvula helps prevent foods and liquids from entering the nasal cavity during swallowing.

24.6 Chloride ions in saliva activate salivary amylase.

24.7 The main component of teeth is connective tissue, specifically dentin.

24.8 The first, second, and third molars do not replace any deciduous teeth.

24.9 The esophageal mucosa and submucosa contain mucus-secreting glands.

24.10 Both. Initiation of swallowing is voluntary and the action is carried out by skeletal muscles. Completion of swallowing—moving a bolus along the esophagus and into the stomach—is involuntary and involves peristalsis by smooth muscle.

24.11 After a large meal, the rugae stretch and disappear as the stomach fills.

24.12 Parietal cells in gastric glands secrete HCl, which is a component of gastric juice. HCl kills microbes in food, denatures proteins, and converts pepsinogen into pepsin.

24.13 Hydrogen ions secreted into gastric juice are derived from carbonic acid (H_2CO_3).

24.14 Histamine is a paracrine agent released by mast cells in the lamina propria.

24.15 The pancreatic duct contains pancreatic juice (fluid and digestive enzymes); the common bile duct contains bile; the hepatopancreatic ampulla contains pancreatic juice and bile.

24.16 The phagocytic cell in the liver is the stellate reticuloendothelial (Kupffer) cell.

24.17 While a meal is being absorbed, nutrients, O_2, and certain toxic substances are removed by hepatocytes from blood flowing through liver sinusoids.

24.18 The ileum is the longest part of the small intestine.

24.19 Nutrients being absorbed in the small intestine enter the blood via capillaries or the lymph via lacteals.

24.20 The fluid secreted by duodenal (Brunner's) glands—alkaline mucus—neutralizes gastric acid and protects the mucosal lining of the duodenum.

24.21 Because monoglycerides are hydrophobic (nonpolar) molecules, they can dissolve in and diffuse through the lipid bilayer of the plasma membrane.

24.22 The stomach and pancreas are the two digestive system organs that secrete the largest volumes of fluid.

24.23 The ascending and descending portions of the colon are retroperitoneal.

24.24 Goblet cells in the large intestine secrete mucus to lubricate colonic contents.

24.25 The pH of gastric juice rises due to the buffering action of some amino acids in food proteins.

25

METABOLISM AND NUTRITION

METABOLISM, NUTRITION, AND HOMEOSTASIS *Metabolic reactions contribute to homeostasis by harvesting chemical energy from consumed nutrients for use in the body's growth, repair, and normal functioning.*

The food we eat is our only source of energy for running, walking, and even breathing. Many molecules needed to maintain cells and tissues can be made from simpler precursors by the body's metabolic reactions; others—the essential amino acids, essential fatty acids, vitamins, and minerals—must be obtained from our food. As you learned in Chapter 24, carbohydrates, lipids, and proteins in food are digested by enzymes and absorbed in the gastrointestinal tract. The products of digestion that reach body cells are monosaccharides, fatty acids, glycerol, monoglycerides, and amino acids. Some minerals and many vitamins are part of enzyme systems that catalyze the breakdown and synthesis of carbohydrates, lipids, and proteins. Food molecules absorbed by the gastrointestinal (GI) tract have three main fates:

1. Most food molecules are used to *supply energy* for sustaining life processes, such as active transport, DNA replication, protein synthesis, muscle contraction, maintenance of body temperature, and mitosis.

2. Some food molecules *serve as building blocks* for the synthesis of more complex structural or functional molecules, such as muscle proteins, hormones, and enzymes.

3. Other food molecules are *stored for future use.* For example, glycogen is stored in liver cells, and triglycerides are stored in adipose cells.

In this chapter we discuss how metabolic reactions harvest the chemical energy stored in foods; how each group of food molecules contributes to the body's growth, repair, and energy needs; and how heat and energy balance is maintained in the body. Finally, we explore some aspects of nutrition to discover why you should opt for fish instead of a burger the next time you eat out.

? *Did you ever wonder how fasting and starvation affect the body?*

25.1 METABOLIC REACTIONS

◉ OBJECTIVES

- Define metabolism.
- Explain the role of ATP in anabolism and catabolism.

Metabolism (me-TAB-ō-lizm; *metabol-* = change) refers to all of the chemical reactions that occur in the body. There are two types of metabolism: catabolism and anabolism. Those chemical reactions that break down complex organic molecules into simpler ones are collectively known as **catabolism** (ka-TAB-ō-lizm; *cata-* = downward). Overall, catabolic (decomposition) reactions are *exergonic;* they produce more energy than they consume, releasing the chemical energy stored in organic molecules. Important sets of catabolic reactions occur in glycolysis, the Krebs cycle, and the electron transport chain, each of which will be discussed later in the chapter.

Chemical reactions that combine simple molecules and monomers to form the body's complex structural and functional components are collectively known as **anabolism** (a-NAB-ō-lizm; *ana-* = upward). Examples of anabolic reactions are the formation of peptide bonds between amino acids during protein synthesis, the building of fatty acids into phospholipids that form the plasma membrane bilayer, and the linkage of glucose monomers to form glycogen. Anabolic reactions are *endergonic;* they consume more energy than they produce.

Metabolism is an energy-balancing act between catabolic (decomposition) reactions and anabolic (synthesis) reactions. The molecule that participates most often in energy exchanges in living cells is **ATP (adenosine triphosphate),** which couples energy-releasing catabolic reactions to energy-requiring anabolic reactions.

The metabolic reactions that occur depend on which enzymes are active in a particular cell at a particular time, or even in a particular part of the cell. Catabolic reactions can be occurring in the mitochondria of a cell at the same time as anabolic reactions are taking place in the endoplasmic reticulum.

A molecule synthesized in an anabolic reaction has a limited lifetime. With few exceptions, it will eventually be broken down and its component atoms recycled into other molecules or excreted from the body. Recycling of biological molecules occurs continuously in living tissues, more rapidly in some than in others. Individual cells may be refurbished molecule by molecule, or a whole tissue may be rebuilt cell by cell.

Coupling of Catabolism and Anabolism by ATP

The chemical reactions of living systems depend on the efficient transfer of manageable amounts of energy from one molecule to another. The molecule that most often performs this task is ATP, the "energy currency" of a living cell. Like money, it is readily available to "buy" cellular activities; it is spent and earned over and over. A typical cell has about a billion molecules of ATP, each of which typically lasts for less than a minute before being used. Thus, ATP is not a long-term storage form of currency, like gold in a vault, but rather convenient cash for moment-to-moment transactions.

Recall from Chapter 2 that a molecule of ATP consists of an adenine molecule, a ribose molecule, and three phosphate groups bonded to one another (see Figure 2.25). Figure 25.1 shows how ATP links anabolic and catabolic reactions. When the terminal phosphate group is split off ATP, adenosine diphosphate (ADP) and a phosphate group (symbolized as ⓟ) are formed. Some of the energy released is used to drive anabolic reactions such as the formation of glycogen from glucose. In addition, energy from complex molecules is used in catabolic reactions to combine ADP and a phosphate group to resynthesize ATP:

$$ADP + ⓟ + energy \longrightarrow ATP$$

About 40% of the energy released in catabolism is used for cellular functions; the rest is converted to heat, some of which helps maintain normal body temperature. Excess heat is lost to the environment. Compared with machines, which typically convert only 10–20% of energy into work, the 40% efficiency of the body's metabolism is impressive. Still, the body has a continuous need to take in and process external sources of energy so that cells can synthesize enough ATP to sustain life.

✔CHECKPOINT

1. What is metabolism? Distinguish between anabolism and catabolism, and give examples of each.
2. How does ATP link anabolism and catabolism?

Figure 25.1 Role of ATP in linking anabolic and catabolic reactions. When complex molecules and polymers are split apart (catabolism, at left), some of the energy is transferred to form ATP and the rest is given off as heat. When simple molecules and monomers are combined to form complex molecules (anabolism, at right), ATP provides the energy for synthesis, and again some energy is given off as heat.

🔑 The coupling of energy-releasing and energy-requiring reactions is achieved through ATP.

? **In a pancreatic cell that produces digestive enzymes, does anabolism or catabolism predominate?**

25.2 ENERGY TRANSFER

◉ OBJECTIVES

- Describe oxidation–reduction reactions.
- Explain the role of ATP in metabolism.

Various catabolic reactions transfer energy into the "high-energy" phosphate bonds of ATP. Although the amount of energy in these bonds is not exceptionally large, it can be released quickly and easily. Before discussing metabolic pathways, it is important to understand how this transfer of energy occurs. Two important aspects of energy transfer are oxidation–reduction reactions and mechanisms of ATP generation.

Oxidation–Reduction Reactions

Oxidation (ok′-si-DĀ-shun) is the *removal of electrons* from an atom or molecule; the result is a *decrease* in the potential energy of the atom or molecule. Because most biological oxidation reactions involve the loss of hydrogen atoms, they are called *dehydrogenation reactions.* An example of an oxidation reaction is the conversion of lactic acid into pyruvic acid:

$$\underset{\text{Lactic acid}}{\underset{\displaystyle CH_3}{\overset{\displaystyle COOH}{H-\overset{\displaystyle |}{\underset{\displaystyle |}{C}}-OH}}} \xrightarrow[\text{Remove 2 H (H}^+ + \text{H}^-)]{\text{Oxidation}} \underset{\text{Pyruvic acid}}{\underset{\displaystyle CH_3}{\overset{\displaystyle COOH}{\overset{\displaystyle |}{\underset{\displaystyle |}{C}}=O}}}$$

In the preceding reaction, 2 H (H^+ + H^-) means that two neutral hydrogen atoms (2 H) are removed as one hydrogen ion (H^+) plus one hydride ion (H^-).

Reduction (rē-DUK-shun) is the opposite of oxidation; it is the *addition of electrons* to a molecule. Reduction results in an *increase* in the potential energy of the molecule. An example of a reduction reaction is the conversion of pyruvic acid into lactic acid:

$$\underset{\text{Pyruvic acid}}{\underset{\displaystyle CH_3}{\overset{\displaystyle COOH}{\overset{\displaystyle |}{\underset{\displaystyle |}{C}}=O}}} \xrightarrow[\text{Add 2 H (H}^+ + \text{H}^-)]{\text{Reduction}} \underset{\text{Lactic acid}}{\underset{\displaystyle CH_3}{\overset{\displaystyle COOH}{H-\overset{\displaystyle |}{\underset{\displaystyle |}{C}}-OH}}}$$

When a substance is oxidized, the liberated hydrogen atoms do not remain free in the cell but are transferred immediately by coenzymes to another compound. Two coenzymes are commonly used by animal cells to carry hydrogen atoms: **nicotinamide adenine dinucleotide (NAD),** a derivative of the B vitamin niacin, and **flavin adenine dinucleotide (FAD),** a derivative of vitamin B_2 (riboflavin). The oxidation and reduction states of NAD^+ and FAD can be represented as follows:

$$\underset{\text{Oxidized}}{NAD^+} \underset{\xleftarrow{-2 \text{ H (H}^+ + \text{H}^-)}}{\xrightarrow{+2 \text{ H (H}^+ + \text{H}^-)}} \underset{\text{Reduced}}{NADH + H^+}$$

$$\underset{\text{Oxidized}}{FAD} \underset{\xleftarrow{-2 \text{ H (H}^+ + \text{H}^-)}}{\xrightarrow{+2 \text{ H (H}^+ + \text{H}^-)}} \underset{\text{Reduced}}{FADH_2}$$

When NAD^+ is reduced to NADH + H^+, the NAD^+ gains a hydride ion (H^-), neutralizing its charge, and the H^+ is released into the surrounding solution. When NADH is oxidized to NAD^+, the loss of the hydride ion results in one less hydrogen atom and an additional positive charge. FAD is reduced to $FADH_2$ when it gains a hydrogen ion and a hydride ion, and $FADH_2$ is oxidized to FAD when it loses the same two ions.

Oxidation and reduction reactions are always coupled; each time one substance is oxidized, another is simultaneously reduced. Such paired reactions are called **oxidation–reduction** or **redox reactions.** For example, when lactic acid is *oxidized* to form pyruvic acid, the two hydrogen atoms removed in the reaction are used to *reduce* NAD^+. This coupled redox reaction may be written as follows:

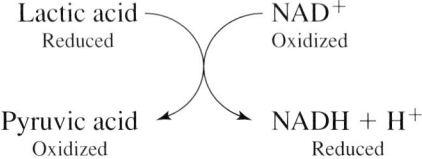

An important point to remember about oxidation–reduction reactions is that oxidation is usually an exergonic (energy-releasing) reaction. Cells use multistep biochemical reactions to release energy from energy-rich, highly reduced compounds (with many hydrogen atoms) to lower-energy, highly oxidized compounds (with many oxygen atoms or multiple bonds). For example, when a cell oxidizes a molecule of glucose ($C_6H_{12}O_6$), the energy in the glucose molecule is removed in a stepwise manner. Ultimately, some of the energy is captured by transferring it to ATP, which then serves as an energy source for energy-requiring reactions within the cell. Compounds with many hydrogen atoms such as glucose contain more chemical potential energy than oxidized compounds. For this reason, glucose is a valuable nutrient.

Mechanisms of ATP Generation

Some of the energy released during oxidation reactions is captured within a cell when ATP is formed. Briefly, a phosphate group (Ⓟ) is added to ADP, with an input of energy, to form ATP. The two high-energy phosphate bonds that can be used to transfer energy are indicated by "squiggles" (∼):

$$\underset{ADP}{Adenosine - Ⓟ \sim Ⓟ} + Ⓟ + energy \longrightarrow$$

$$\underset{ATP}{Adenosine - Ⓟ \sim Ⓟ \sim Ⓟ}$$

The high-energy phosphate bond that attaches the third phosphate group contains the energy stored in this reaction. The addition of a phosphate group to a molecule, called **phosphorylation** (fos′-for-i-LĀ-shun), increases its potential energy. Organisms use three mechanisms of phosphorylation to generate ATP:

1. **Substrate-level phosphorylation** generates ATP by transferring a high-energy phosphate group from an intermediate phosphorylated metabolic compound—a substrate—directly to ADP. In human cells, this process occurs in the cytosol.

2. **Oxidative phosphorylation** removes electrons from organic compounds and passes them through a series of electron acceptors, called the **electron transport chain,** to molecules of oxygen (O_2). This process occurs in the inner mitochondrial membrane of cells.

3. **Photophosphorylation** occurs only in chlorophyll-containing plant cells or in certain bacteria that contain other light-absorbing pigments.

✔ CHECKPOINT

3. How is a hydride ion different from a hydrogen ion? What is the involvement of both ions in redox reactions?
4. What are three ways that ATP can be generated?

25.3 CARBOHYDRATE METABOLISM

◉ OBJECTIVE

- Describe the fate, metabolism, and functions of carbohydrates.

As you learned in Chapter 24, both polysaccharides and disaccharides are hydrolyzed into the monosaccharides glucose (about 80%), fructose, and galactose during the digestion of carbohydrates. (Some fructose is converted into glucose as it is absorbed through the intestinal epithelial cells.) Hepatocytes (liver cells) convert most of the remaining fructose and practically all the galactose to glucose. So the story of carbohydrate metabolism is really the story of glucose metabolism. Because negative feedback systems maintain blood glucose at about 90 mg/100 mL of plasma (5 mmol/liter), a total of 2–3 g of glucose normally circulates in the blood.

The Fate of Glucose

Because glucose is the body's preferred source for synthesizing ATP, its use depends on the needs of body cells, which include the following:

- **ATP production.** In body cells that require immediate energy, glucose is oxidized to produce ATP. Glucose not needed for immediate ATP production can enter one of several other metabolic pathways.

- **Amino acid synthesis.** Cells throughout the body can use glucose to form several amino acids, which then can be incorporated into proteins.

- **Glycogen synthesis.** Hepatocytes and muscle fibers can perform **glycogenesis** (glī′-kō-JEN-e-sis; glyco- = sugar or sweet; -genesis = to generate), in which hundreds of glucose monomers are combined to form the polysaccharide glycogen. Total storage capacity of glycogen is about 125 g in the liver and 375 g in skeletal muscles.

- **Triglyceride synthesis.** When the glycogen storage areas are filled up, hepatocytes can transform the glucose to glycerol and fatty acids that can be used for **lipogenesis** (lip-ō-JEN-e-sis), the synthesis of triglycerides. Triglycerides then are deposited in adipose tissue, which has virtually unlimited storage capacity.

Glucose Movement into Cells

Before glucose can be used by body cells, it must first pass through the plasma membrane and enter the cytosol. Glucose absorption in the gastrointestinal tract (and kidney tubules) is accomplished via secondary active transport (Na^+–glucose symporters). Glucose entry into most other body cells occurs via GluT molecules, a family of transporters that bring glucose into cells via facilitated diffusion (see Section 3.3). A high level of insulin increases the insertion of one type of GluT, called GluT4, into the plasma membranes of most body cells, thereby increasing the rate of facilitated diffusion of glucose into cells. In neurons and hepatocytes, however, another type of GluT is always present in the plasma membrane, so glucose entry is always "turned on." On entering a cell, glucose becomes phosphorylated. Because GluT cannot transport phosphorylated glucose, this reaction traps glucose within the cell.

Glucose Catabolism

The oxidation of glucose to produce ATP is also known as **cellular respiration,** and it involves four sets of reactions: glycolysis, the formation of acetyl coenzyme A, the Krebs cycle, and the electron transport chain (Figure 25.2).

❶ **Glycolysis** is a set of reactions in which one glucose molecule is oxidized and two molecules of pyruvic acid are produced. The reactions also produce two molecules of ATP and two energy-containing NADH + H^+. Because glycolysis does not require oxygen, it is a way to produce ATP anaerobically (without oxygen) and is known as **anaerobic cellular respiration** (an-ar-Ō-bik; an- = not; -aer- = air; -bios = life).

❷ **Formation of acetyl coenzyme A** is a transition step that prepares pyruvic acid for entrance into the Krebs cycle. This step also produces energy-containing NADH + H^+ plus carbon dioxide (CO_2).

❸ **Krebs cycle reactions** oxidize acetyl coenzyme A and produce CO_2, ATP, NADH + H^+, and $FADH_2$.

❹ **Electron transport chain reactions** oxidize NADH + H^+ and $FADH_2$ and transfer their electrons through a series of electron carriers. The Krebs cycle and the electron transport chain both require oxygen to produce ATP and are collectively known as **aerobic cellular respiration.**

Glycolysis

During **glycolysis** (glī-KOL-i-sis; -lysis = breakdown), chemical reactions split a 6-carbon molecule of glucose into two 3-carbon molecules of pyruvic acid (Figure 25.3). Even though glycolysis consumes two ATP molecules, it produces four ATP molecules, for a net gain of two ATP molecules for each glucose molecule that is oxidized.

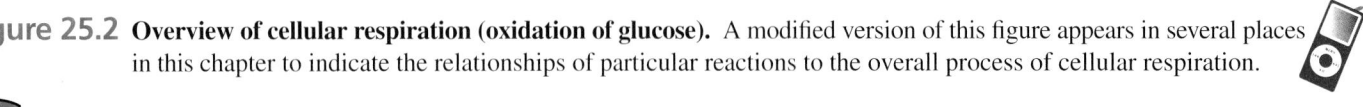

Figure 25.2 Overview of cellular respiration (oxidation of glucose). A modified version of this figure appears in several places in this chapter to indicate the relationships of particular reactions to the overall process of cellular respiration.

🔑 The oxidation of glucose involves glycolysis, the formation of acetyl coenzyme A, the Krebs cycle, and the electron transport chain.

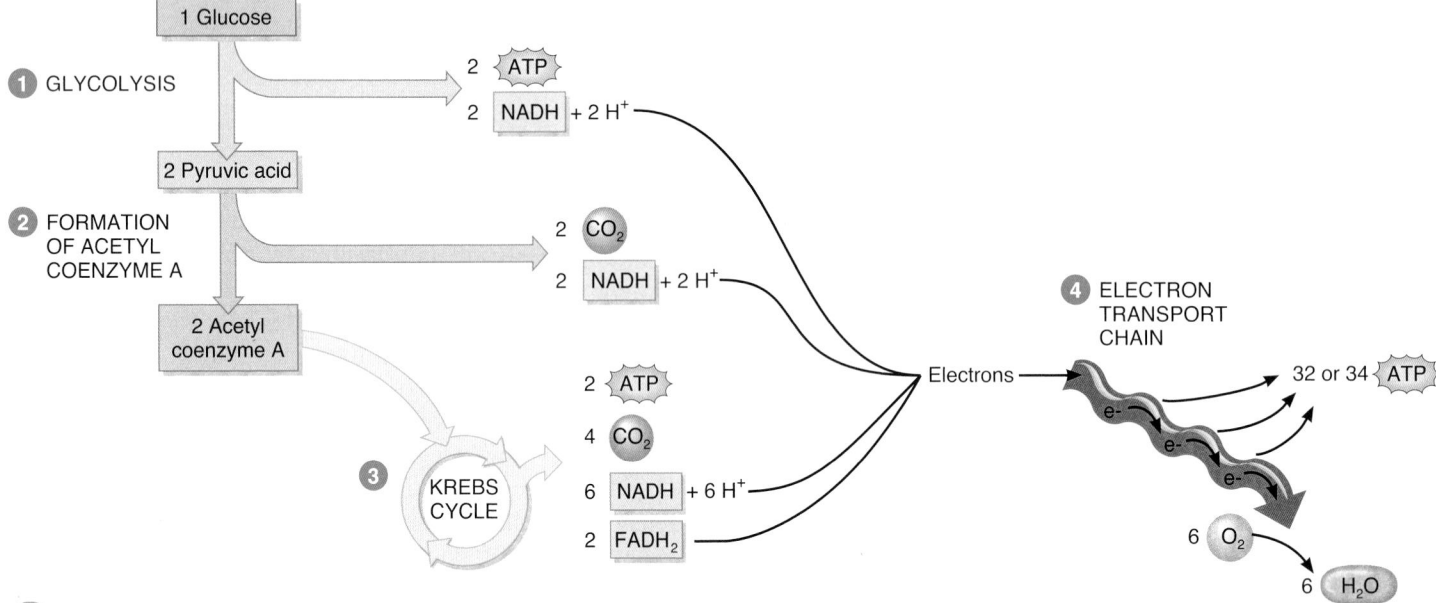

❓ Which of the four processes shown here is also called anaerobic cellular respiration?

Figure 25.3 Cellular respiration begins with glycolysis.

🔑 During glycolysis, each molecule of glucose is converted to two molecules of pyruvic acid.

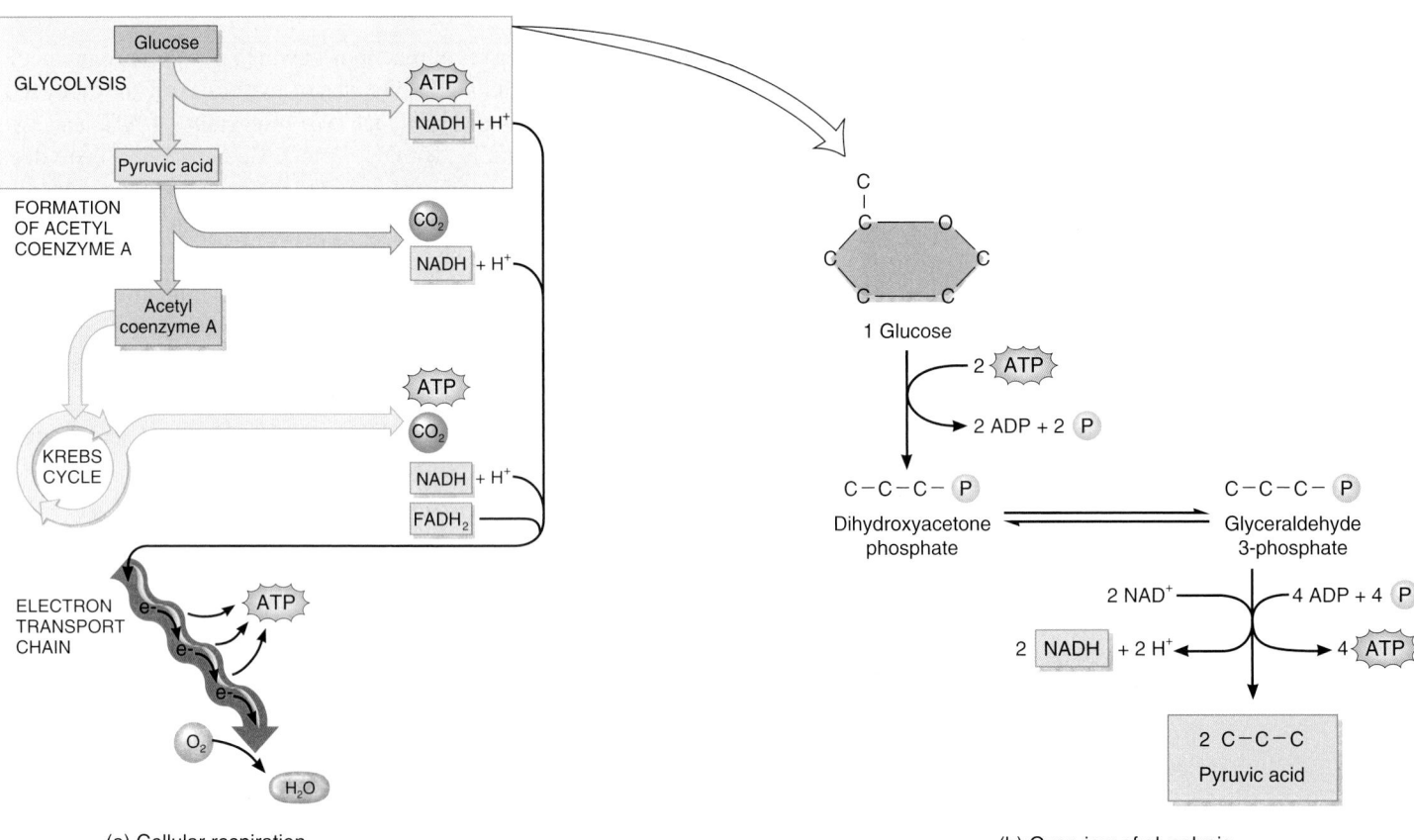

(a) Cellular respiration

(b) Overview of glycolysis

❓ For each glucose molecule that undergoes glycolysis, how many ATP molecules are generated?

Figure 25.4 shows the 10 reactions that glycolysis comprises. In the first half of the sequence (reactions ❶ through ❺), energy in the form of ATP is "invested" and the 6-carbon glucose is split into two 3-carbon molecules of glyceraldehyde 3-phosphate.

Phosphofructokinase (fos′-fō-fruk′-tō-KĪ-nās), the enzyme that catalyzes step ❸, is the key regulator of the rate of glycolysis. The activity of this enzyme is high when ADP concentration is high, in which case ATP is produced rapidly. When the activity of

Figure 25.4 The 10 reactions of glycolysis. ❶ Glucose is phosphorylated, using a phosphate group from an ATP molecule to form glucose 6-phosphate. **❷** Glucose 6-phosphate is converted to fructose 6-phosphate. **❸** A second ATP is used to add a second phosphate group to fructose 6-phosphate to form fructose 1,6-bisphosphate. **❹** and **❺** Fructose splits into two 3-carbon molecules, glyceraldehyde 3-phosphate (G 3-P) and dihydroxyacetone phosphate, each having one phosphate group. **❻** Oxidation occurs as two molecules of NAD$^+$ accept two pairs of electrons and hydrogen ions from two molecules of G 3-P to form two molecules of NADH. Many body cells use the two NADH produced in this step to generate four ATPs in the electron transport chain. Hepatocytes, kidney cells, and cardiac muscle fibers can generate six ATPs from the two NADH. A second phosphate group attaches to G 3-P, forming 1,3-bisphosphoglyceric acid (BPG). **❼** through **❿** These reactions generate four molecules of ATP and produce two molecules of pyruvic acid (pyruvate*).

🔑 Glycolysis results in a net gain of two ATP, two NADH, and two H$^+$.

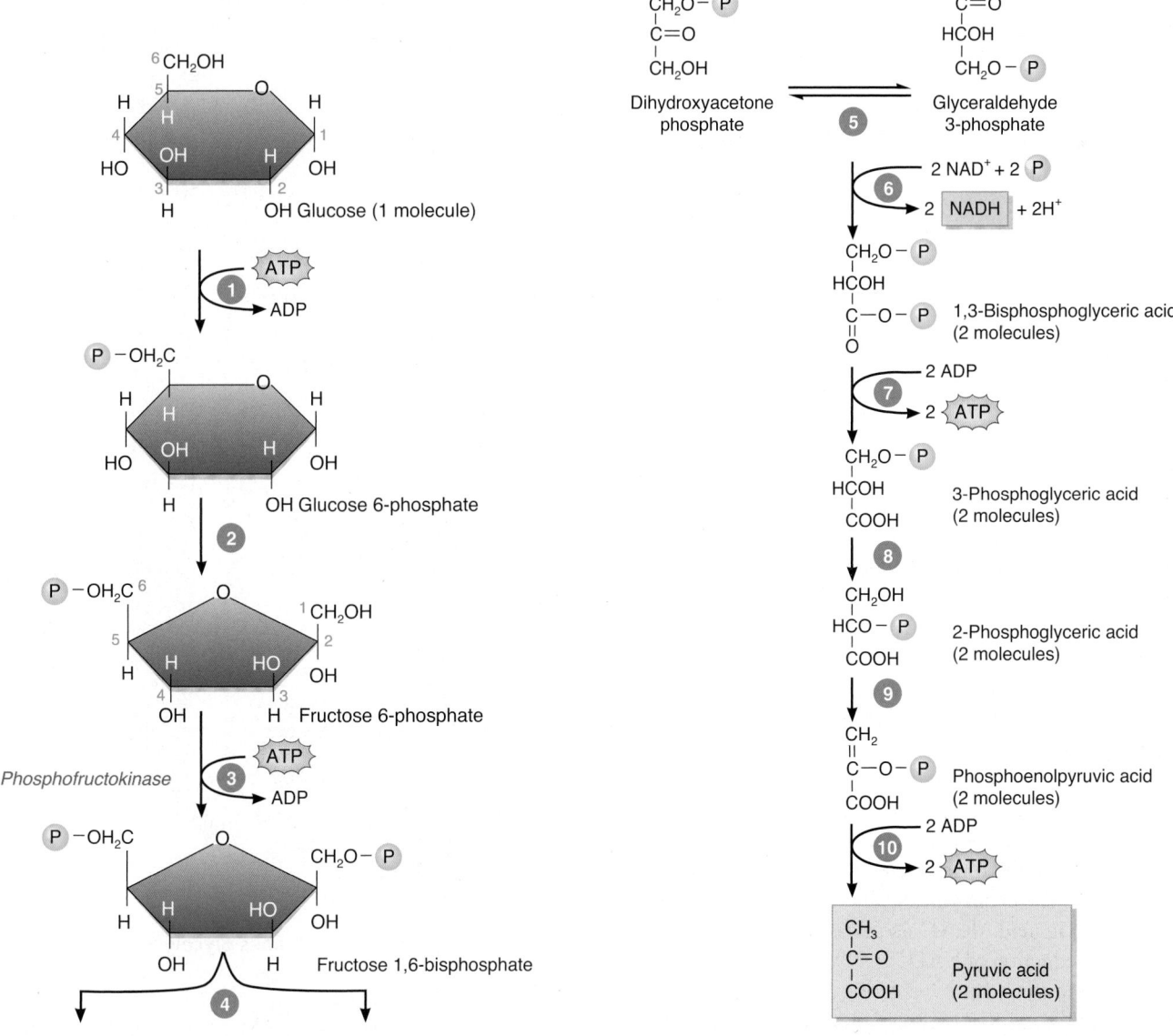

? **Why is the enzyme that catalyzes step ❸ called a kinase?**

*The carboxyl groups (—COOH) of intermediates in glycolysis and in the citric acid cycle are mostly ionized at the pH of body fluids to —COO$^-$. The suffix "-ic acid" indicates the non-ionized form, whereas the ending "-ate" indicates the ionized form. Although the "-ate" names are more correct, we will use the "acid" names because these terms are more familiar.

phosphofructokinase is low, most glucose does not enter the reactions of glycolysis but instead undergoes conversion to glycogen for storage. In the second half of the sequence (reactions ⑥ through ⑩), the two glyceraldehyde 3-phosphate molecules are converted to two pyruvic acid molecules and ATP is generated.

The Fate of Pyruvic Acid

The fate of pyruvic acid produced during glycolysis depends on the availability of oxygen (Figure 25.5). If oxygen is scarce (anaerobic conditions)—for example, in skeletal muscle fibers during strenuous exercise—then pyruvic acid is reduced via an anaerobic pathway by the addition of two hydrogen atoms to form lactic acid (lactate):

$$2\,\text{Pyruvic acid} + 2\,\text{NADH} + 2\,\text{H}^+ \longrightarrow 2\,\text{Lactic acid} + 2\,\text{NAD}^+$$

Oxidized   Reduced

This reaction regenerates the NAD$^+$ that was used in the oxidation of glyceraldehyde 3-phosphate (see step ⑥ in Figure 25.4) and thus allows glycolysis to continue. As lactic acid is produced, it rapidly diffuses out of the cell and enters the blood. Hepatocytes remove lactic acid from the blood and convert it back to pyruvic acid. Recall that a buildup of lactic acid is one factor that contributes to muscle fatigue.

When oxygen is plentiful (aerobic conditions), most cells convert pyruvic acid to acetyl coenzyme A. This molecule links glycolysis, which occurs in the cytosol, with the Krebs cycle, which occurs in the matrix of mitochondria. Pyruvic acid enters the mitochondrial matrix with the help of a special transporter protein. Because they lack mitochondria, red blood cells can only produce ATP through glycolysis.

Formation of Acetyl Coenzyme A

Each step in the oxidation of glucose requires a different enzyme, and often a coenzyme as well. The coenzyme used at this point in cellular respiration is **coenzyme A (CoA),** which is derived from pantothenic acid, a B vitamin. During the transitional step between glycolysis and the Krebs cycle, pyruvic acid is prepared for entrance into the cycle. The enzyme *pyruvate dehydrogenase* (pī-ROO-vāt dē-HĪ-drō-jen-ās), which is located exclusively in the mitochondrial matrix, converts pyruvic acid to a 2-carbon fragment called an **acetyl group** (AS-e-til) by removing a molecule of carbon dioxide (Figure 25.5). The loss of a molecule of CO$_2$ by a substance is called **decarboxylation** (dē-kar-bok′-si-LĀ-shun). This is the first reaction in cellular respiration that releases CO$_2$. During this reaction, pyruvic acid is also oxidized. Each pyruvic acid loses two hydrogen atoms in the form of one hydride ion (H$^-$) plus one hydrogen ion (H$^+$). The coenzyme NAD$^+$ is reduced as it picks up the H$^-$ from pyruvic acid; the H$^+$ is released into the mitochondrial matrix. The reduction of NAD$^+$ to NADH + H$^+$ is indicated in Figure 25.5 by the curved arrow entering and then leaving the reaction. Recall that the oxidation of one glucose molecule produces two molecules of pyruvic acid, so for each molecule of glucose, two molecules of carbon dioxide are lost and two NADH + H$^+$ are produced. The acetyl group attaches to coenzyme A, producing a molecule called **acetyl coenzyme A (acetyl CoA).**

Figure 25.5 Fate of pyruvic acid.

When oxygen is plentiful, pyruvic acid enters mitochondria, is converted to acetyl coenzyme A, and enters the Krebs cycle (aerobic pathway). When oxygen is scarce, most pyruvic acid is converted to lactic acid via an anaerobic pathway.

? In which part of the cell does glycolysis occur?

The Krebs Cycle

Once the pyruvic acid has undergone decarboxylation and the remaining acetyl group has attached to CoA, the resulting compound (acetyl CoA) is ready to enter the Krebs cycle (Figure 25.6). The

Figure 25.6 **After formation of acetyl coenzyme A, the next stage of cellular respiration is the Krebs cycle.**

Reactions of the Krebs cycle occur in the matrix of mitochondria.

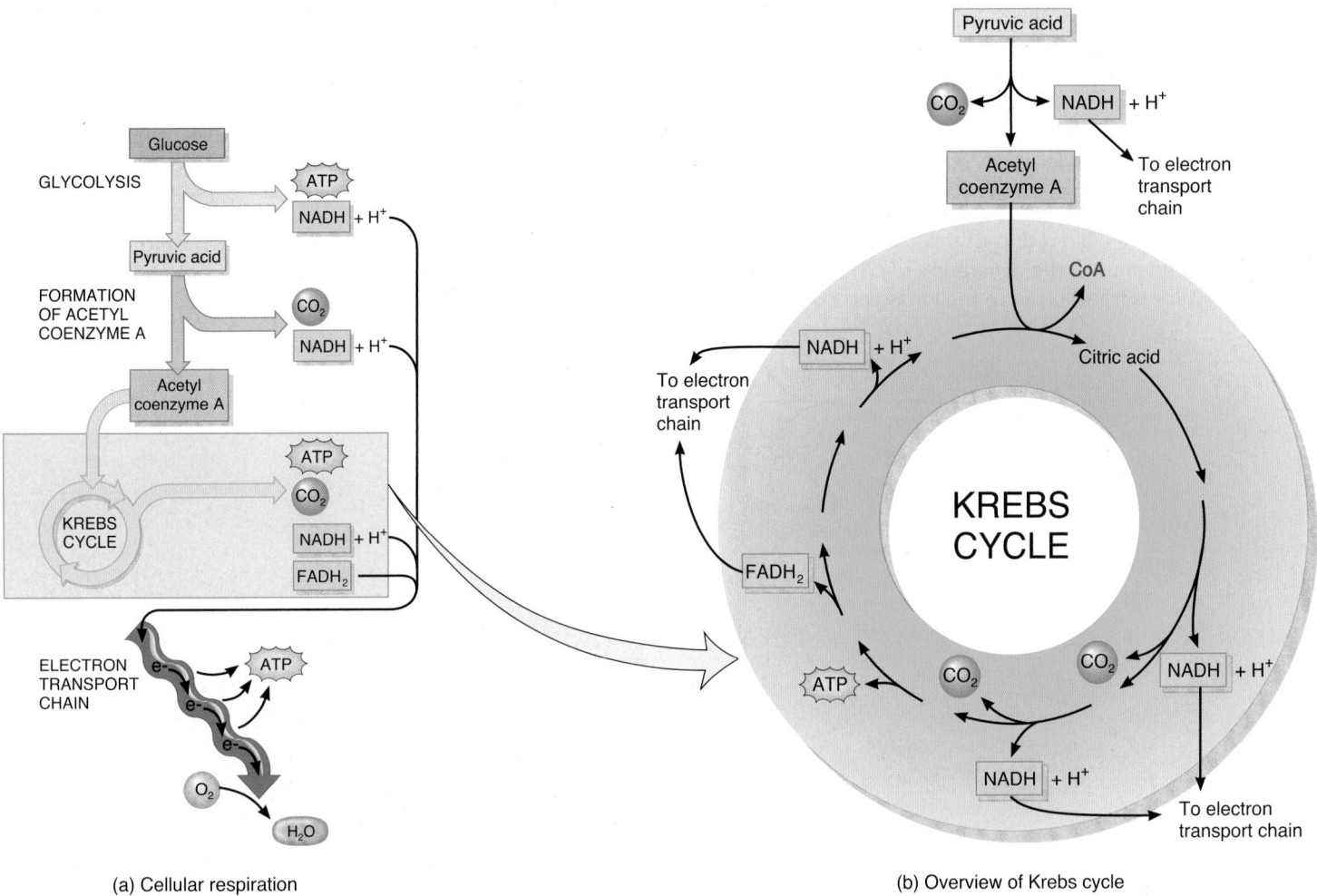

(a) Cellular respiration

(b) Overview of Krebs cycle

When in cellular respiration is carbon dioxide given off? What happens to this gas?

Krebs cycle—named for the biochemist Hans Krebs, who described these reactions in the 1930s—is also known as the **citric acid cycle,** for the first molecule formed when an acetyl group joins the cycle. The reactions occur in the matrix of mitochondria and consist of a series of oxidation–reduction reactions and decarboxylation reactions that release CO_2. In the Krebs cycle, the oxidation–reduction reactions transfer chemical energy, in the form of electrons, to two coenzymes—NAD^+ and FAD. The pyruvic acid derivatives are oxidized, and the coenzymes are reduced. In addition, one step generates ATP. Figure 25.7 shows the reactions of the Krebs cycle in more detail.

The reduced coenzymes (NADH and $FADH_2$) are the most important outcome of the Krebs cycle because they contain the energy originally stored in glucose and then in pyruvic acid. Overall, for every acetyl CoA that enters the Krebs cycle, three NADH, three H^+, and one $FADH_2$ are produced by oxidation–reduction reactions, and one molecule of ATP is generated by

substrate-level phosphorylation (see Figure 25.6). In the electron transport chain, the three NADH + three H^+ will later yield nine ATP molecules, and the $FADH_2$ will later yield two ATP molecules. Thus, each "turn" of the Krebs cycle eventually generates 12 molecules of ATP. Because each glucose molecule provides two acetyl CoA molecules, glucose catabolism via the Krebs cycle and the electron transport chain yields 24 molecules of ATP per glucose molecule.

Liberation of CO_2 occurs as pyruvic acid is converted to acetyl CoA and during the two decarboxylation reactions of the Krebs cycle (see Figure 25.6). Because each molecule of glucose generates two molecules of pyruvic acid, six molecules of CO_2 are liberated from each original glucose molecule catabolized along this pathway. The molecules of CO_2 diffuse out of the mitochondria, through the cytosol and plasma membrane, and then into the blood. Blood transports the CO_2 to the lungs, where it eventually is exhaled.

Figure 25.7 The eight reactions of the Krebs cycle. ❶ *Entry of the acetyl group.* The chemical bond that attaches the acetyl group to coenzyme A (CoA) breaks, and the 2-carbon acetyl group attaches to a 4-carbon molecule of oxaloacetic acid to form a 6-carbon molecule called citric acid. CoA is free to combine with another acetyl group from pyruvic acid and repeat the process. ❷ *Isomerization.* Citric acid undergoes isomerization to isocitric acid, which has the same molecular formula as citrate. Notice, however, that the hydroxyl group (—OH) is attached to a different carbon. ❸ *Oxidative decarboxylation.* Isocitric acid is oxidized and loses a molecule of CO_2, forming alpha-ketoglutaric acid. The H^+ from the oxidation is passed on to NAD^+, which is reduced to $NADH + H^+$. ❹ *Oxidative decarboxylation.* Alpha-ketoglutaric acid is oxidized, loses a molecule of CO_2, and picks up CoA to form succinyl-CoA. ❺ *Substrate-level phosphorylation.* CoA is displaced by a phosphate group, which is then transferred to guanosine diphosphate (GDP) to form guanosine triphosphate (GTP). GTP can donate a phosphate group to ADP to form ATP. ❻ *Dehydrogenation.* Succinic acid is oxidized to fumaric acid as two of its hydrogen atoms are transferred to the coenzyme flavin adenine dinucleotide (FAD), which is reduced to $FADH_2$. ❼ *Hydration.* Fumaric acid is converted to malic acid by the addition of a molecule of water. ❽ *Dehydrogenation.* In the final step in the cycle, malic acid is oxidized to re-form oxaloacetic acid. Two hydrogen atoms are removed and one is transferred to NAD^+, which is reduced to $NADH + H^+$. The regenerated oxaloacetic acid can combine with another molecule of acetyl CoA, beginning a new cycle.

🔑 The three main results of the Krebs cycle are the production of reduced coenzymes (NADH and $FADH_2$), which contain stored energy; the generation of GTP, a high-energy compound that is used to produce ATP; and the formation of CO_2, which is transported to the lungs and exhaled.

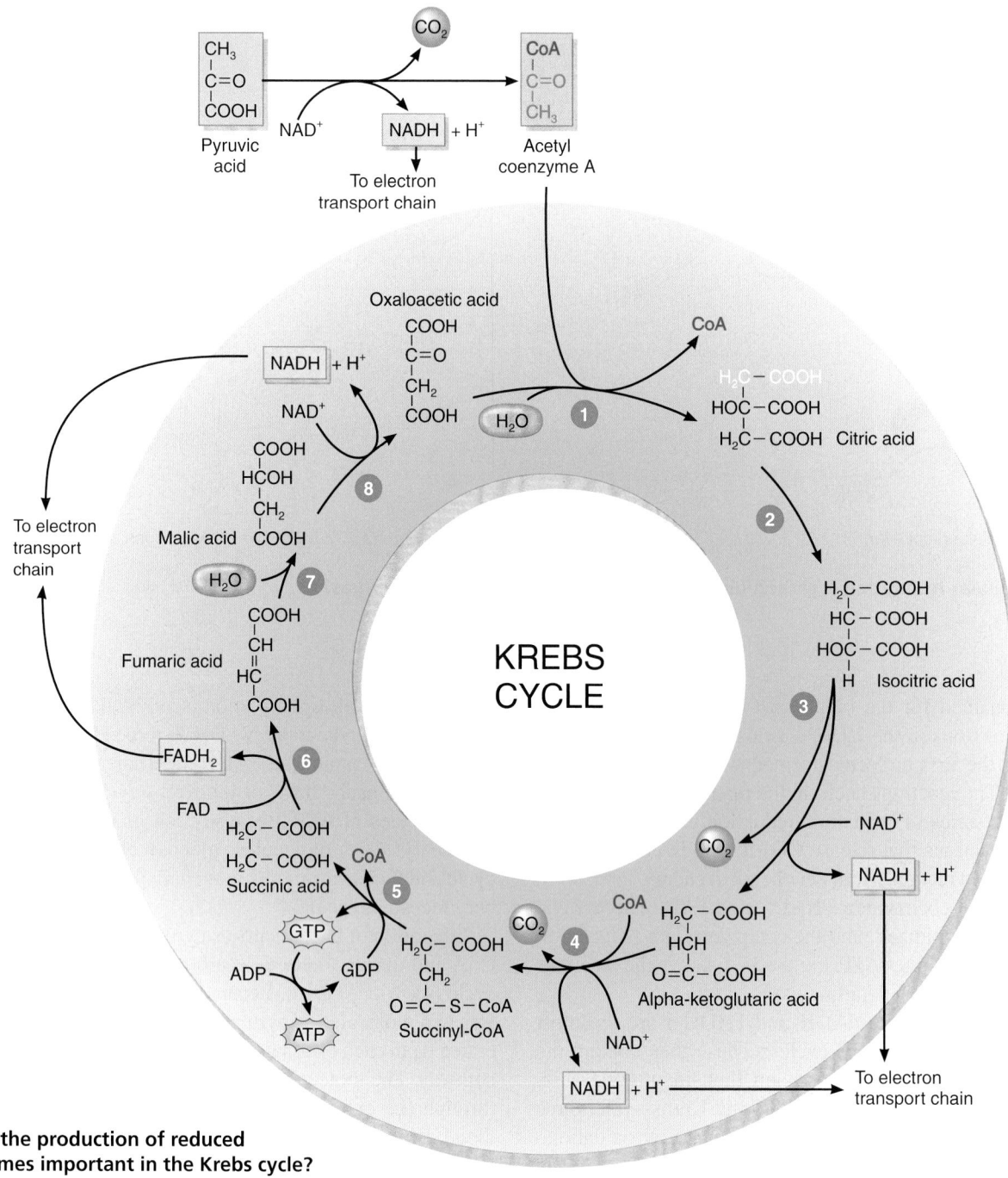

❓ Why is the production of reduced coenzymes important in the Krebs cycle?

The Electron Transport Chain

The **electron transport chain** is a series of **electron carriers,** integral membrane proteins in the inner mitochondrial membrane. This membrane is folded into cristae that increase its surface area, accommodating thousands of copies of the transport chain in each mitochondrion. Each carrier in the chain is reduced as it picks up electrons and oxidized as it gives up electrons. As electrons pass through the chain, a series of exergonic reactions release small amounts of energy; this energy is used to form ATP. In aerobic cellular respiration, the final electron acceptor of the chain is oxygen. Because this mechanism of ATP generation links chemical reactions (the passage of electrons along the transport chain) with the pumping of hydrogen ions, it is called **chemiosmosis** (kem′-ē-oz-MŌ-sis; *chemi-* = chemical; *-osmosis* = pushing). Briefly, chemiosmosis works as follows (Figure 25.8):

1. Energy from NADH + H⁺ passes along the electron transport chain and is used to pump H⁺ from the matrix of the mitochondrion into the space between the inner and outer mitochondrial membranes. This mechanism is called a **proton pump** because H⁺ ions consist of a single proton.

2. A high concentration of H⁺ accumulates between the inner and outer mitochondrial membranes.

3. ATP synthesis then occurs as hydrogen ions flow back into the mitochondrial matrix through a special type of H⁺ channel in the inner membrane.

ELECTRON CARRIERS Several types of molecules and atoms serve as electron carriers:

- **Flavin mononucleotide (FMN)** (FLĀ-vin mon′-ō-NOO-klē-ō-tīd) is a flavoprotein derived from riboflavin (vitamin B₂).

- **Cytochromes** (SĪ-tō-krōmz) are proteins with an iron-containing group (heme) capable of existing alternately in a reduced form (Fe^{2+}) and an oxidized form (Fe^{3+}). The cytochromes involved in the electron transport chain include cytochrome *b* (cyt *b*), cytochrome c_1 (cyt c_1), cytochrome *c* (cyt *c*), cytochrome *a* (cyt *a*), and cytochrome a_3 (cyt a_3).

- **Iron–sulfur (Fe-S) centers** contain either two or four iron atoms bound to sulfur atoms that form an electron transfer center within a protein.

- **Copper (Cu) atoms** bound to two proteins in the chain also participate in electron transfer.

- **Coenzyme Q,** symbolized **Q,** is a nonprotein, low-molecular-weight carrier that is mobile in the lipid bilayer of the inner membrane.

STEPS IN ELECTRON TRANSPORT AND CHEMIOSMOTIC ATP GENERATION Within the inner mitochondrial membrane, the carriers of the electron transport chain are clustered into three complexes, each of which acts as a proton pump that expels H⁺ from the mitochondrial matrix and helps create an electrochemical gradient of H⁺. Each of the three proton pumps transports electrons and pumps H⁺, as shown in Figure 25.9. Notice that oxygen is used to help form water in step 3. This is the only point in aerobic cellular respiration where O_2 is consumed. **Cyanide** is a deadly poison because it binds to the cytochrome oxidase complex and blocks this last step in electron transport.

The pumping of H⁺ produces both a concentration gradient of protons and an electrical gradient. The buildup of H⁺ makes one side of the inner mitochondrial membrane positively charged compared with the other side. The resulting electrochemical gradient has potential energy, called the *proton motive force.* Proton channels in the inner mitochondrial membrane allow H⁺ to flow back across the membrane, driven by the proton motive force. As H⁺ flow back, they generate ATP because the H⁺ channels also include an enzyme called **ATP synthase** (SIN-thās). The enzyme uses the proton motive force to synthesize ATP from ADP and Ⓟ. The process of chemiosmosis is responsible for most of the ATP produced during cellular respiration.

Summary of Cellular Respiration

The various electron transfers in the electron transport chain generate either 32 or 34 ATP molecules from each molecule of

Figure 25.8 Chemiosmosis.

In chemiosmosis, ATP is produced when hydrogen ions diffuse back into the mitochondrial matrix.

? **What is the energy source that powers the proton pumps?**

Figure 25.9 **The actions of the three proton pumps and ATP synthase in the inner membrane of mitochondria.** Each pump is a complex of three or more electron carriers. ❶ The first proton pump is the *NADH dehydrogenase complex,* which contains flavin mononucleotide (FMN) and five or more Fe-S centers. $NADH + H^+$ is oxidized to NAD^+, and FMN is reduced to $FMNH_2$, which in turn is oxidized as it passes electrons to the iron–sulfur centers. Q, which is mobile in the membrane, shuttles electrons to the second pump complex. ❷ The second proton pump is the *cytochrome b–c_1 complex,* which contains cytochromes and an iron–sulfur center. Electrons are passed successively from Q to cyt b, to Fe-S, to cyt c_1. The mobile shuttle that passes electrons from the second pump complex to the third is cytochrome c (cyt c). ❸ The third proton pump is the *cytochrome oxidase complex,* which contains cytochromes a and a_3 and two copper atoms. Electrons pass from cyt c, to Cu, to cyt a, and finally to cyt a_3. Cyt a_3 passes its electrons to one-half of a molecule of oxygen (O_2), which becomes negatively charged and then picks up two H^+ from the surrounding medium to form H_2O.

🔑 As the three proton pumps pass electrons from one carrier to the next, they also move protons (H^+) from the matrix into the space between the inner and outer mitochondrial membranes. As protons flow back into the mitochondrial matrix through the H^+ channel in ATP synthase, ATP is synthesized.

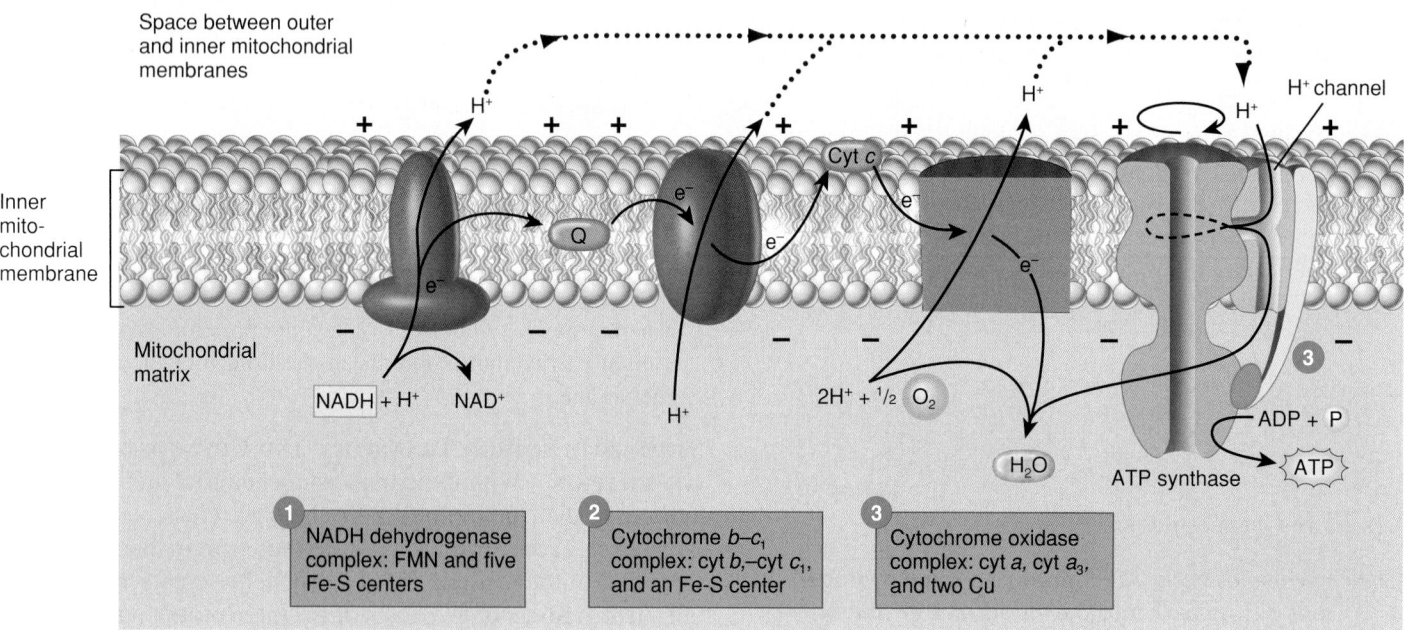

Where is the concentration of H^+ highest?

glucose that is oxidized: either 28 or 30* from the 10 molecules of $NADH + H^+$ and two from each of the two molecules of $FADH_2$ (four total). Thus, during cellular respiration, 36 or 38 ATPs can be generated from one molecule of glucose. Note that two of those ATPs come from substrate-level phosphorylation in glycolysis, and two come from substrate-level phosphorylation in the Krebs cycle. The overall reaction is

$$C_6H_{12}O_6 + 6\ O_2 + 36\ \text{or}\ 38\ ADPs + 36\ \text{or}\ 38\ \text{Ⓟ} \longrightarrow$$
Glucose   Oxygen

$$6\ CO_2 + 6\ H_2O + 36\ \text{or}\ 38\ ATPs$$
Carbon dioxide Water

*The two NADH produced in the cytosol during glycolysis cannot enter mitochondria. Instead, they donate their electrons to one of two transfer molecules, known as the malate shuttle and the glycerol phosphate shuttle. In cells of the liver, kidneys, and heart, use of the malate shuttle results in three ATPs synthesized for each NADH. In other body cells, such as skeletal muscle fibers and neurons, use of the glycerol phosphate shuttle results in two ATPs synthesized for each NADH.

Table 25.1 summarizes the ATP yield during cellular respiration. A schematic depiction of the principal reactions of cellular respiration is presented in Figure 25.10. The actual ATP yield may be lower than 36 or 38 ATPs per glucose. One uncertainty is the exact number of H^+ that must be pumped out to generate one ATP during chemiosmosis. In addition, the ATP generated in mitochondria must be transported out of these organelles into the cytosol for use elsewhere in a cell. Exporting ATP in exchange for the inward movement of ADP formed from metabolic reactions in the cytosol uses up part of the proton motive force.

Glycolysis, the Krebs cycle, and especially the electron transport chain provide all the ATP for cellular activities. Because the Krebs cycle and electron transport chain are aerobic processes, cells cannot carry on their activities for long if oxygen is lacking.

Glucose Anabolism

Even though most of the glucose in the body is catabolized to generate ATP, glucose may take part in or be formed via several

Figure 25.10 Summary of the principal reactions of cellular respiration. ETC = electron transport chain and chemiosmosis.

Except for glycolysis, which occurs in the cytosol, all other reactions of cellular respiration occur within mitochondria.

How many molecules of O_2 are used, and how many molecules of CO_2 are produced during the complete oxidation of one glucose molecule?

TABLE 25.1

Summary of ATP Produced in Cellular Respiration

SOURCE	ATP YIELD PER GLUCOSE MOLECULE (PROCESS)
GLYCOLYSIS	
Oxidation of one glucose molecule to two pyruvic acid molecules	2 ATPs (substrate-level phosphorylation).
Production of 2 NADH + H^+	4 or 6 ATPs (oxidative phosphorylation in electron transport chain).
FORMATION OF TWO MOLECULES OF ACETYL COENZYME A	
2 NADH + 2 H^+	6 ATPs (oxidative phosphorylation in electron transport chain).
KREBS CYCLE AND ELECTRON TRANSPORT CHAIN	
Oxidation of succinyl-CoA to succinic acid	2 GTPs that are converted to 2 ATPs (substrate-level phosphorylation).
Production of 6 NADH + 6 H^+	18 ATPs (oxidative phosphorylation in electron transport chain).
Production of 2 FADH$_2$	4 ATPs (oxidative phosphorylation in electron transport chain).
Total	36 or 38 ATPs per glucose molecule (theoretical maximum).

Glucose Storage: Glycogenesis

If glucose is not needed immediately for ATP production, it combines with many other molecules of glucose to form **glycogen,** a polysaccharide that is the only stored form of carbohydrate in our bodies. The hormone insulin, from pancreatic beta cells, stimulates hepatocytes and skeletal muscle cells to carry out **glycogenesis** (glī′-kō-JEN-e-sis), the synthesis of glycogen (Figure 25.11). The body can store about 500 g (about 1.1 lb) of glycogen, roughly 75% in skeletal muscle fibers and the rest in liver cells. During glycogenesis, glucose is first phosphorylated to glucose 6-phosphate by hexokinase. Glucose 6-phosphate is converted to glucose 1-phosphate, then to uridine diphosphate glucose, and finally to glycogen.

Glucose Release: Glycogenolysis

When body activities require ATP, glycogen stored in hepatocytes is broken down into glucose and released into the blood to be transported to cells, where it will be catabolized by the processes of cellular respiration already described. The process of splitting glycogen into its glucose subunits is called **glycogenolysis** (glī′-kō-je-NOL-e-sis). (Note: Do not confuse *glycogenolysis*, which is the breakdown of glycogen to glucose, with *glycolysis*, the 10 reactions that convert glucose to pyruvic acid.)

Glycogenolysis is not a simple reversal of the steps of glycogenesis (Figure 25.11). It begins by splitting glucose molecules off the branched glycogen molecule via phosphorylation to form glucose 1-phosphate. Phosphorylase, the enzyme that catalyzes this reaction, is activated by glucagon from pancreatic alpha cells and

anabolic reactions. One is the synthesis of glycogen; another is the synthesis of new glucose molecules from some of the products of protein and lipid breakdown.

Figure 25.11 Glycogenesis and glycogenolysis.

The glycogenesis pathway converts glucose into glycogen; the glycogenolysis pathway breaks down glycogen into glucose.

Key:

→ Glycogenesis (stimulated by insulin)

→ Glycogenolysis (stimulated by glucagon and epinephrine)

? **Besides hepatocytes, which body cells can synthesize glycogen? Why can't they release glucose into the blood?**

epinephrine from the adrenal medulla. Glucose 1-phosphate is then converted to glucose 6-phosphate and finally to glucose, which leaves hepatocytes via glucose transporters (GluT) in the plasma membrane. Phosphorylated glucose molecules cannot ride aboard the GluT transporters, however, and *phosphatase,* the enzyme that converts glucose 6-phosphate into glucose, is absent in skeletal muscle cells. Thus, hepatocytes, which have phosphatase, can release glucose derived from glycogen to the bloodstream, but skeletal muscle cells cannot. In skeletal muscle cells, glycogen is broken down into glucose 1-phosphate, which is then catabolized for ATP production via glycolysis and the Krebs cycle. However, the lactic acid produced by glycolysis in muscle cells can be converted to glucose in the liver. In this way, muscle glycogen can be an indirect source of blood glucose.

⚕ CLINICAL CONNECTION | *Carbohydrate Loading*

The amount of glycogen stored in the liver and skeletal muscles varies and can be completely exhausted during long-term athletic endeavors. Thus, many marathon runners and other endurance athletes follow a precise exercise and dietary regimen that includes eating large amounts of complex carbohydrates, such as pasta and potatoes, in the 3 days before an event. This practice, called **carbohydrate loading,** helps maximize the amount of glycogen available for ATP production in muscles. For athletic events lasting more than an hour, carbohydrate loading has been shown to increase an athlete's endurance. The increased endurance is due to increased glycogenolysis, which results in more glucose that can be catabolized for energy. •

Formation of Glucose from Proteins and Fats: Gluconeogenesis

When your liver runs low on glycogen, it is time to eat. If you don't, your body starts catabolizing triglycerides (fats) and proteins. Actually, the body normally catabolizes some of its triglycerides and proteins, but large-scale triglyceride and protein catabolism does not happen unless you are starving, eating very few carbohydrates, or suffering from an endocrine disorder.

The glycerol part of triglycerides, lactic acid, and certain amino acids can be converted in the liver to glucose (Figure 25.12). The process by which glucose is formed from these noncarbohydrate sources is called **gluconeogenesis** (gloo′-kō-nē′-ō-JEN-e-sis; *neo-* = new). An easy way to distinguish this term from glycogenesis or glycogenolysis is to remember that in this case glucose is not converted back from glycogen, but is instead *newly formed.* About 60% of the amino acids in the body can be used for gluconeogenesis. Lactic acid and amino acids such as alanine, cysteine, glycine, serine, and threonine are converted to pyruvic acid, which then may be synthesized into glucose or enter the Krebs cycle. Glycerol may be converted into glyceraldehyde 3-phosphate, which may form pyruvic acid or be used to synthesize glucose.

Gluconeogenesis is stimulated by cortisol, the main glucocorticoid hormone of the adrenal cortex, and by glucagon from the pancreas. In addition, cortisol stimulates the breakdown of proteins into amino acids, thus expanding the pool of amino acids available for gluconeogenesis. Thyroid hormones (thyroxine and triiodothyronine) also mobilize proteins and may mobilize triglycerides from adipose tissue, thereby making glycerol available for gluconeogenesis.

Figure 25.12 Gluconeogenesis, the conversion of noncarbohydrate molecules (amino acids, lactic acid, and glycerol) into glucose.

About 60% of the amino acids in the body can be used for gluconeogenesis.

Key:

→ Gluconeogenesis (stimulated by cortisol and glucagon)

? **What cells can carry out gluconeogenesis and glycogenesis?**

✔**CHECKPOINT**

5. How does glucose move into or out of body cells?
6. What happens during glycolysis?
7. How is acetyl coenzyme A formed?
8. Outline the principal events and outcomes of the Krebs cycle.
9. What happens in the electron transport chain and why is this process called chemiosmosis?
10. Which reactions produce ATP during the complete oxidation of a molecule of glucose?
11. Under what circumstances do glycogenesis and glycogenolysis occur?
12. What is gluconeogenesis, and why is it important?

25.4 LIPID METABOLISM

◉ **OBJECTIVES**

• Describe the lipoproteins that transport lipids in the blood.
• Describe the fate, metabolism, and functions of lipids.

Transport of Lipids by Lipoproteins

Most lipids, such as triglycerides, are nonpolar and therefore very hydrophobic molecules. They do not dissolve in water. To be transported in watery blood, such molecules first must be made more water-soluble by combining them with proteins produced by the liver and intestine. The lipid and protein combinations thus formed are **lipoproteins** (lip′-ō-PRŌ-tēns), spherical particles with an outer shell of proteins, phospholipids, and cholesterol molecules surrounding an inner core of triglycerides and other lipids (Figure 25.13). The proteins in the outer shell are called **apoproteins (apo)** (ap-ō-PRŌ-tēns) and are designated by the letters A, B, C, D, and E plus a number. In addition to helping solubilize the lipoprotein in body fluids, each apoprotein has specific functions.

Each of the several types of lipoproteins has different functions, but all essentially are transport vehicles. They provide delivery and pickup services so that lipids can be available when cells need them or removed from circulation when they are not needed. Lipoproteins are categorized and named mainly according to their density, which varies with the ratio of lipids (which have a low density) to proteins (which have a high density). From largest and lightest to smallest and heaviest, the four major classes of lipoproteins are chylomicrons, very-low-density lipoproteins (VLDLs), low-density lipoproteins (LDLs), and high-density lipoproteins (HDLs).

Chylomicrons (kī′-lō-MĪ-krons), which form in mucosal epithelial cells of the small intestine, transport *dietary* (ingested) lipids to adipose tissue for storage. They contain about 1–2% proteins, 85% triglycerides, 7% phospholipids, and 6–7% cholesterol, plus a small amount of fat-soluble vitamins. Chylomicrons enter lacteals of intestinal villi and are carried by lymph into venous blood and then into the systemic circulation. Their presence gives blood plasma a milky appearance, but they remain in the

blood for only a few minutes. As chylomicrons circulate through the capillaries of adipose tissue, one of their apoproteins, **apo C-2,** activates *endothelial lipoprotein lipase,* an enzyme that removes fatty acids from chylomicron triglycerides. The free fatty acids are then taken up by adipocytes for synthesis and storage as triglycerides and by muscle cells for ATP production. Hepatocytes remove chylomicron remnants from the blood via receptor-mediated endocytosis, in which another chylomicron apoprotein, **apo E,** is the docking protein.

Very-low-density lipoproteins (VLDLs), which form in hepatocytes, contain mainly *endogenous* (made in the body) lipids. VLDLs contain about 10% proteins, 50% triglycerides, 20% phospholipids, and 20% cholesterol. VLDLs transport triglycerides synthesized in hepatocytes to adipocytes for storage. Like chylomicrons, they lose triglycerides as their apo C-2 activates endothelial lipoprotein lipase, and the resulting fatty acids are taken up by adipocytes for storage and by muscle cells for ATP production. As they deposit some of their triglycerides in adipose cells, VLDLs are converted to LDLs.

Low-density lipoproteins (LDLs) contain 25% proteins, 5% triglycerides, 20% phospholipids, and 50% cholesterol. They carry about 75% of the total cholesterol in blood and deliver it to cells throughout the body for use in repair of cell membranes and synthesis of steroid hormones and bile salts. LDLs contain a single apoprotein, **apo B100,** which is the docking protein that binds to LDL receptors on the plasma membrane of body cells so that LDL can enter the cell via receptor-mediated endocytosis. Within

Figure 25.13 A lipoprotein. Shown here is a VLDL.

 A single layer of amphipathic phospholipids, cholesterol, and proteins surrounds a core of nonpolar lipids.

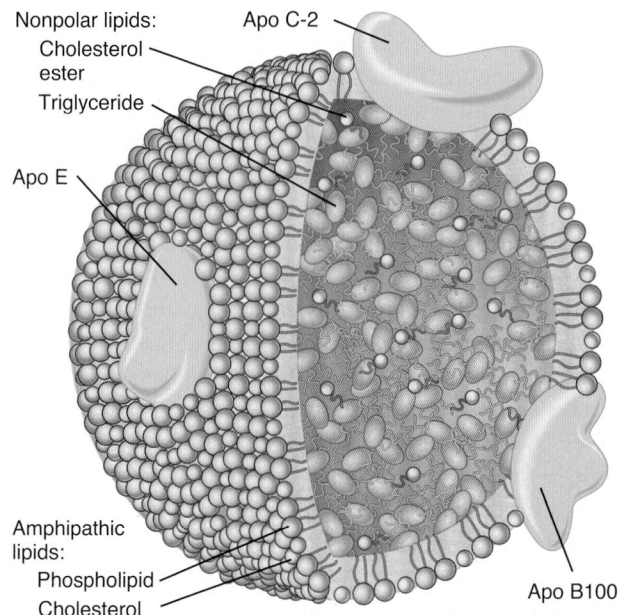

Nonpolar lipids:
 Cholesterol ester
 Triglyceride

Apo C-2

Apo E

Amphipathic lipids:
 Phospholipid
 Cholesterol

Apo B100

 ? **Which type of lipoprotein delivers cholesterol to body cells?**

the cell, the LDL is broken down, and the cholesterol is released to serve the cell's needs. Once a cell has sufficient cholesterol for its activities, a negative feedback system inhibits the cell's synthesis of new LDL receptors.

When present in excessive numbers, LDLs also deposit cholesterol in and around smooth muscle fibers in arteries, forming fatty plaques that increase the risk of coronary artery disease (see Disorders: Homeostatic Imbalances at the end of Chapter 20). For this reason, the cholesterol in LDLs, called LDL-cholesterol, is known as "bad" cholesterol. Because some people have too few LDL receptors, their body cells remove LDL from the blood less efficiently; as a result, their plasma LDL level is abnormally high, and they are more likely to develop fatty plaques. Eating a high-fat diet increases the production of VLDLs, which elevates the LDL level and increases the formation of fatty plaques.

High-density lipoproteins (HDLs), which contain 40–45% proteins, 5–10% triglycerides, 30% phospholipids, and 20% cholesterol, remove excess cholesterol from body cells and the blood and transport it to the liver for elimination. Because HDLs prevent accumulation of cholesterol in the blood, a high HDL level is associated with decreased risk of coronary artery disease. For this reason, HDL-cholesterol is known as "good" cholesterol.

Sources and Significance of Blood Cholesterol

There are two sources of cholesterol in the body. Some is present in foods (eggs, dairy products, organ meats, beef, pork, and processed luncheon meats), but most is synthesized by hepatocytes. Fatty foods that don't contain any cholesterol at all can still dramatically increase blood cholesterol level in two ways. First, a high intake of dietary fats stimulates reabsorption of cholesterol-containing bile back into the blood, so less cholesterol is lost in the feces. Second, when saturated fats are broken down in the body, hepatocytes use some of the breakdown products to make cholesterol.

A lipid profile test usually measures total cholesterol (TC), HDL-cholesterol, and triglycerides (VLDLs). LDL-cholesterol then is calculated by using the following formula: LDL-cholesterol = TC − HDL-cholesterol − (triglycerides/5). In the United States, blood cholesterol is usually measured in milligrams per deciliter (mg/dL); a deciliter is 0.1 liter or 100 mL. For adults, desirable levels of blood cholesterol are total cholesterol under 200 mg/dL, LDL-cholesterol under 130 mg/dL, and HDL-cholesterol over 40 mg/dL. Normally, triglycerides are in the range of 10–190 mg/dL.

As total cholesterol level increases, the risk of coronary artery disease begins to rise. When total cholesterol is above 200 mg/dL (5.2 mmol/liter), the risk of a heart attack doubles with every 50 mg/dL (1.3 mmol/liter) increase in total cholesterol. Total cholesterol of 200–239 mg/dL and LDL of 130–159 mg/dL are borderline-high; total cholesterol above 239 mg/dL and LDL above 159 mg/dL are classified as high blood cholesterol. The ratio of total cholesterol to HDL-cholesterol predicts the risk of developing coronary artery disease. For example, a person with a total cholesterol of 180 mg/dL and HDL of 60 mg/dL has a risk ratio of 3. Ratios above 4 are considered undesirable; the higher the ratio, the greater the risk of developing coronary artery disease.

Among the therapies used to reduce blood cholesterol level are exercise, diet, and drugs. Regular physical activity at aerobic and nearly aerobic levels raises HDL level. Dietary changes are aimed at reducing the intake of total fat, saturated fats, and cholesterol. Drugs used to treat high blood cholesterol levels include cholestyramine (Questran) and colestipol (Colestid), which promote excretion of bile in the feces; nicotinic acid (Liponicin); and the "statin" drugs—atorvastatin (Lipitor), lovastatin (Mevacor), and simvastatin (Zocor), which block the key enzyme (HMG-CoA reductase) needed for cholesterol synthesis.

The Fate of Lipids

Lipids, like carbohydrates, may be oxidized to produce ATP. If the body has no immediate need to use lipids in this way, they are stored in adipose tissue (fat depots) throughout the body and in the liver. A few lipids are used as structural molecules or to synthesize other essential substances. Some examples include phospholipids, which are constituents of plasma membranes; lipoproteins, which are used to transport cholesterol throughout the body; thromboplastin, which is needed for blood clotting; and myelin sheaths, which speed up nerve impulse conduction. Two **essential fatty acids** that the body cannot synthesize are linoleic acid and linolenic acid. Dietary sources include vegetable oils and leafy vegetables. The various functions of lipids in the body may be reviewed in Table 2.7.

Triglyceride Storage

A major function of adipose tissue is to remove triglycerides from chylomicrons and VLDLs and store them until they are needed for ATP production in other parts of the body. Triglycerides stored in adipose tissue constitute 98% of all body energy reserves. They are stored more readily than glycogen, in part because triglycerides are hydrophobic and do not exert osmotic pressure on cell membranes. Adipose tissue also insulates and protects various parts of the body. Adipocytes in the subcutaneous layer contain about 50% of the stored triglycerides. Other adipose tissues account for the other half: about 12% around the kidneys, 10–15% in the omenta, 15% in genital areas, 5–8% between muscles, and 5% behind the eyes, in the sulci of the heart, and attached to the outside of the large intestine. Triglycerides in adipose tissue are continually broken down and resynthesized. Thus, the triglycerides stored in adipose tissue today are not the same molecules that were present last month because they are continually released from storage, transported in the blood, and redeposited in other adipose tissue cells.

Lipid Catabolism: Lipolysis

In order for muscle, liver, and adipose tissue to oxidize the fatty acids derived from triglycerides to produce ATP, the triglycerides must first be split into glycerol and fatty acids, a process called **lipolysis** (li-POL-i-sis). Lipolysis is catalyzed by enzymes called **lipases.** Epinephrine and norepinephrine enhance triglyceride breakdown into fatty acids and glycerol. These hormones are released when sympathetic tone increases, as occurs, for example, during exercise. Other lipolytic hormones include cortisol, thyroid hormones, and insulinlike growth factors. By contrast, insulin inhibits lipolysis.

The glycerol and fatty acids that result from lipolysis are catabolized via different pathways (Figure 25.14). Glycerol is converted by many cells of the body to glyceraldehyde 3-phosphate, one of the compounds also formed during the catabolism of glucose. If ATP supply in a cell is high, glyceraldehyde 3-phosphate is converted into glucose, an example of gluconeogenesis. If ATP supply in a cell is low, glyceraldehyde 3-phosphate enters the catabolic pathway to pyruvic acid.

Fatty acids are catabolized differently than glycerol and yield more ATP. The first stage in fatty acid catabolism is a series of reactions, collectively called **beta oxidation** (BĀ-ta), that occurs in the matrix of mitochondria. Enzymes remove two carbon atoms at a time from the long chain of carbon atoms composing a fatty acid and attach the resulting two-carbon fragment to coenzyme A, forming acetyl CoA. Then, acetyl CoA enters the Krebs cycle (Figure 25.14). A 16-carbon fatty acid such as palmitic acid can yield as many as 129 ATPs on its complete oxidation via beta oxidation, the Krebs cycle, and the electron transport chain.

As part of normal fatty acid catabolism, hepatocytes can take two acetyl CoA molecules at a time and condense them to form **acetoacetic acid** (as'-ē-tō-a-SĒ-tik). This reaction liberates the

bulky CoA portion, which cannot diffuse out of cells. Some acetoacetic acid is converted into **beta-hydroxybutyric acid** (hī-drok-sē-bū-TIR-ik) and **acetone** (AS-e-tōn). The formation of these three substances, collectively known as **ketone bodies** (KĒ-tōn), is called **ketogenesis** (kē-tō-JEN-e-sis) (Figure 25.14). Because ketone bodies freely diffuse through plasma membranes, they leave hepatocytes and enter the bloodstream.

Other cells take up acetoacetic acid and attach its four carbons to two coenzyme A molecules to form two acetyl CoA molecules, which can then enter the Krebs cycle for oxidation. Heart muscle and the cortex (outer part) of the kidneys use acetoacetic acid in preference to glucose for generating ATP. Hepatocytes, which make acetoacetic acid, cannot use it for ATP production because they lack the enzyme that transfers acetoacetic acid back to coenzyme A.

Lipid Anabolism: Lipogenesis

Liver cells and adipose cells can synthesize lipids from glucose or amino acids through **lipogenesis** (lip-ō-JEN-e-sis) (Figure 25.14), which is stimulated by insulin. Lipogenesis occurs when individuals consume more calories than are needed to satisfy their ATP needs. Excess dietary carbohydrates, proteins, and fats all have the same fate—they are converted into triglycerides.

Figure 25.14 Pathways of lipid metabolism. Glycerol may be converted to glyceraldehyde 3-phosphate, which can then be converted to glucose or enter the Krebs cycle for oxidation. Fatty acids undergo beta oxidation and enter the Krebs cycle via acetyl coenzyme A. The synthesis of lipids from glucose or amino acids is called lipogenesis.

Glycerol and fatty acids are catabolized in separate pathways.

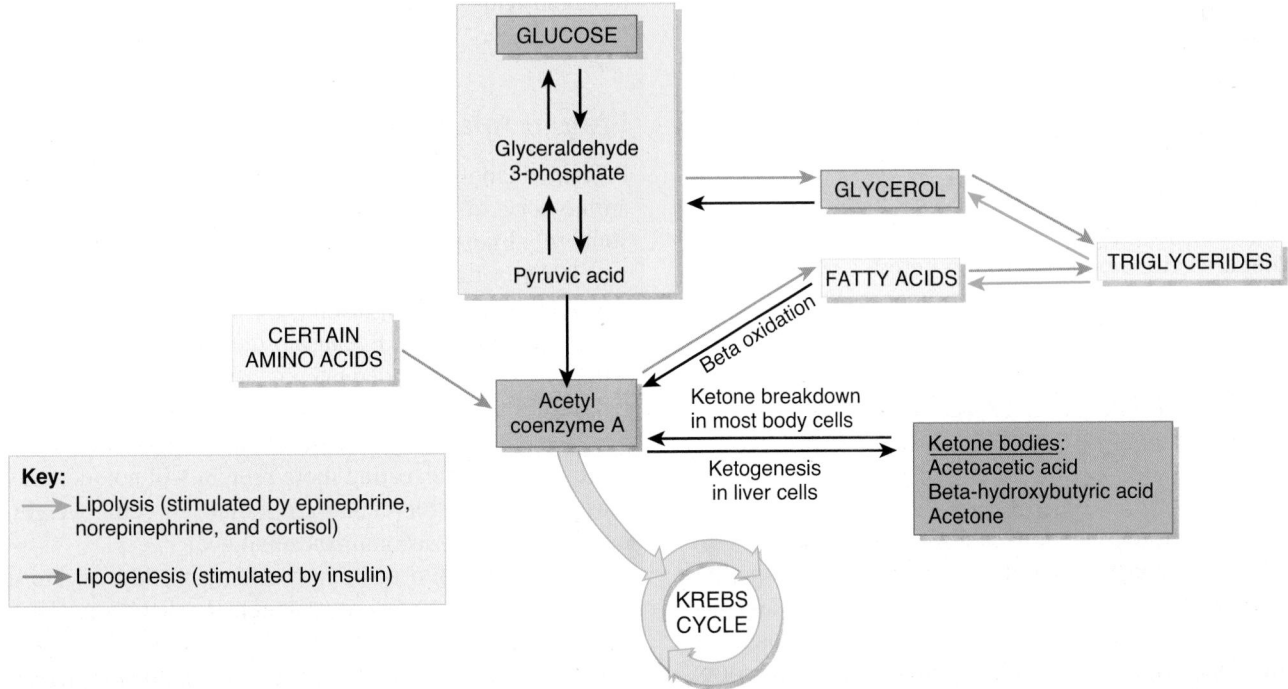

? What types of cells can carry out lipogenesis, beta oxidation, and lipolysis? What type of cell can carry out ketogenesis?

Certain amino acids can undergo the following reactions: amino acids → acetyl CoA → fatty acids → triglycerides. The use of glucose to form lipids takes place via two pathways: (1) glucose → glyceraldehyde 3-phosphate → glycerol and (2) glucose → glyceraldehyde 3-phosphate → acetyl CoA → fatty acids. The resulting glycerol and fatty acids can undergo anabolic reactions to become stored triglycerides, or they can go through a series of anabolic reactions to produce other lipids such as lipoproteins, phospholipids, and cholesterol.

CLINICAL CONNECTION | *Ketosis*

The level of ketone bodies in the blood normally is very low because other tissues use them for ATP production as fast as they are generated from the breakdown of fatty acids in the liver. During periods of excessive beta oxidation, however, the production of ketone bodies exceeds their uptake and use by body cells. This might occur after a meal rich in triglycerides, or during fasting or starvation, because few carbohydrates are available for catabolism. Excessive beta oxidation may also occur in poorly controlled or untreated diabetes mellitus for two reasons: (1) Because adequate glucose cannot get into cells, triglycerides are used for ATP production, and (2) because insulin normally inhibits lipolysis, a lack of insulin accelerates the pace of lipolysis. When the concentration of ketone bodies in the blood rises above normal—a condition called ketosis—the ketone bodies, most of which are acids, must be buffered. If too many accumulate, they decrease the concentration of buffers, such as bicarbonate ions, and blood pH falls. Extreme or prolonged ketosis can lead to **acidosis (ketoacidosis),** an abnormally low blood pH. The decreased blood pH in turn causes depression of the central nervous system, which can result in disorientation, coma, and even death if the condition is not treated. When a diabetic becomes seriously insulin-deficient, one of the telltale signs is the sweet smell on the breath from the ketone body acetone. •

✔ CHECKPOINT

13. What are the functions of the apoproteins in lipoproteins?
14. Which lipoprotein particles contain "good" and "bad" cholesterol, and why are these terms used?
15. Where are triglycerides stored in the body?
16. Explain the principal events of the catabolism of glycerol and fatty acids.
17. What are ketone bodies? What is ketosis?
18. Define lipogenesis and explain its importance.

25.5 PROTEIN METABOLISM

◉ OBJECTIVE

• Describe the fate, metabolism, and functions of proteins.

During digestion, proteins are broken down into amino acids. Unlike carbohydrates and triglycerides, which are stored, proteins are not warehoused for future use. Instead, amino acids are either oxidized to produce ATP or used to synthesize new proteins for body growth and repair. Excess dietary amino acids are not excreted in the urine or feces but instead are converted into glucose (gluconeogenesis) or triglycerides (lipogenesis).

The Fate of Proteins

The active transport of amino acids into body cells is stimulated by insulinlike growth factors (IGFs) and insulin. Almost immediately after digestion, amino acids are reassembled into proteins. Many proteins function as enzymes; others are involved in transportation (hemoglobin) or serve as antibodies, clotting chemicals (fibrinogen), hormones (insulin), or contractile elements in muscle fibers (actin and myosin). Several proteins serve as structural components of the body (collagen, elastin, and keratin). The various functions of proteins in the body may be reviewed in Table 2.8.

Protein Catabolism

A certain amount of protein catabolism occurs in the body each day, stimulated mainly by cortisol from the adrenal cortex. Proteins from worn-out cells (such as red blood cells) are broken down into amino acids. Some amino acids are converted into other amino acids, peptide bonds are re-formed, and new proteins are synthesized as part of the recycling process. Hepatocytes convert some amino acids to fatty acids, ketone bodies, or glucose. Cells throughout the body oxidize a small amount of amino acids to generate ATP via the Krebs cycle and the electron transport chain. However, before amino acids can be oxidized, they must first be converted to molecules that are part of the Krebs cycle or can enter the Krebs cycle, such as acetyl CoA (Figure 25.15). Before amino acids can enter the Krebs cycle, their amino group (NH_2) must first be removed—a process called **deamination** (dē-am′-i-NĀ-shun). Deamination occurs in hepatocytes and produces ammonia (NH_3). The liver cells then convert the highly toxic ammonia to urea, a relatively harmless substance that is excreted in the urine. The conversion of amino acids into glucose (gluconeogenesis) may be reviewed in Figure 25.12; the conversion of amino acids into fatty acids (lipogenesis) or ketone bodies (ketogenesis) is shown in Figure 25.14.

Protein Anabolism

Protein anabolism, the formation of peptide bonds between amino acids to produce new proteins, is carried out on the ribosomes of almost every cell in the body, directed by the cells' DNA and RNA (see Figure 3.29). Insulinlike growth factors, thyroid hormones (T_3 and T_4), insulin, estrogen, and testosterone stimulate protein synthesis. Because proteins are a main component of most cell structures, adequate dietary protein is especially essential during the growth years, during pregnancy, and when tissue has been damaged by disease or injury. Once dietary intake of protein is adequate, eating more protein will not increase bone or muscle mass; only a regular program of forceful, weight-bearing muscular activity accomplishes that goal.

Of the 20 amino acids in the human body, 10 are **essential amino acids:** They must be present in the diet because they cannot be synthesized in the body in adequate amounts. It is *essential* to include them in your diet. Humans are unable to synthesize eight amino acids (isoleucine, leucine, lysine, methionine, phenylalanine, threonine, tryptophan, and valine) and synthesize two others

Figure 25.15 **Points at which amino acids (yellow boxes) enter the Krebs cycle for oxidation.**

🔑 Before amino acids can be catabolized, they must first be converted to various substances that can enter the Krebs cycle.

Alanine
Cysteine
Glycine
Serine
Threonine

Phenylalanine

Pyruvic acid

Tyrosine
Leucine
Lysine
Tryptophan

Acetoacetyl CoA

Acetyl CoA

Isoleucine
Leucine
Tryptophan

Aspartic acid
Asparagine

Oxaloacetic acid

Citric acid

Malic acid

Phenylalanine

Isocitric acid

Tyrosine

Fumaric acid

Alpha-ketoglutaric acid

Glutamic acid

Succinic acid

Succinyl CoA

Arginine
Histidine
Glutamine
Proline

Isoleucine
Methionine
Valine

❓ **What group is removed from an amino acid before it can enter the Krebs cycle, and what is this process called?**

(arginine and histidine) in inadequate amounts, especially in childhood. A **complete protein** contains sufficient amounts of all essential amino acids. Beef, fish, poultry, eggs, and milk are examples of foods that contain complete proteins. An **incomplete protein** does not contain all essential amino acids. Examples of incomplete proteins are leafy green vegetables, legumes (beans and peas), and grains. **Nonessential amino acids** can be synthesized by body cells. They are formed by **transamination** (trans′-am-i-NĀ-shun), the transfer of an amino group from an amino acid to pyruvic acid or to an acid in the Krebs cycle. Once the ap-

propriate essential and nonessential amino acids are present in cells, protein synthesis occurs rapidly.

CLINICAL CONNECTION | *Phenylketonuria*

Phenylketonuria (fen′-il-kē′-tō-NOO-rē-a) or **PKU** is a genetic error of protein metabolism characterized by elevated blood levels of the amino acid phenylalanine. Most children with phenylketonuria have a mutation in the gene that codes for the enzyme phenylalanine hydroxylase, the enzyme needed to convert phenylalanine into the amino acid tyrosine, which can enter the Krebs cycle (Figure 25.15).

Because the enzyme is deficient, phenylalanine cannot be metabolized, and what is not used in protein synthesis builds up in the blood. If untreated, the disorder causes vomiting, rashes, seizures, growth deficiency, and severe mental retardation. Newborns are screened for PKU, and mental retardation can be prevented by restricting the affected child to a diet that supplies only the amount of phenylalanine needed for growth, although learning disabilities may still ensue. Because the artificial sweetener aspartame (NutraSweet) contains phenylalanine, its consumption must be restricted in children with PKU. •

✔ CHECKPOINT

19. What is deamination and why does it occur?
20. What are the possible fates of the amino acids from protein catabolism?
21. How are essential and nonessential amino acids different?

25.6 KEY MOLECULES AT METABOLIC CROSSROADS

⊙ OBJECTIVE

• Identify the key molecules in metabolism, and describe the reactions and the products they may form.

Although there are thousands of different chemicals in cells, three molecules—glucose 6-phosphate, pyruvic acid, and acetyl coenzyme A—play pivotal roles in metabolism (Figure 25.16). These molecules stand at "metabolic crossroads"; as you will learn shortly, the reactions that occur (or do not occur) depend on the nutritional or activity status of the individual. Reactions ❶ through ❼ in Figure 25.16 occur in the cytosol, reactions ❽ and ❾ occur inside mitochondria, and reactions indicated by ❿ occur on smooth endoplasmic reticulum.

Figure 25.16 Summary of the roles of the key molecules in metabolic pathways. Double-headed arrows indicate that reactions between two molecules may proceed in either direction, if the appropriate enzymes are present and the conditions are favorable; single-headed arrows signify the presence of an irreversible step.

🔑 Three molecules—glucose 6-phosphate, pyruvic acid, and acetyl coenzyme A—stand at "metabolic crossroads." They can undergo different reactions depending on your nutritional or activity status.

? **Which substance is the gateway into the Krebs cycle for molecules that are being oxidized to generate ATP?**

The Role of Glucose 6-Phosphate

Shortly after glucose enters a body cell, a kinase converts it to **glucose 6-phosphate.** Four possible fates await glucose 6-phosphate (see Figure 25.16):

1 *Synthesis of glycogen.* When glucose is abundant in the bloodstream, as it is just after a meal, a large amount of glucose 6-phosphate is used to synthesize glycogen, the storage form of carbohydrate in animals. Subsequent breakdown of glycogen into glucose 6-phosphate occurs through a slightly different series of reactions. Synthesis and breakdown of glycogen occur mainly in skeletal muscle fibers and hepatocytes.

2 *Release of glucose into the bloodstream.* If the enzyme glucose 6-phosphatase is present and active, glucose 6-phosphate can be dephosphorylated to glucose. Once glucose is released from the phosphate group, it can leave the cell and enter the bloodstream. Hepatocytes are the main cells that can provide glucose to the bloodstream in this way.

3 *Synthesis of nucleic acids.* Glucose 6-phosphate is the precursor used by cells throughout the body to make ribose 5-phosphate, a 5-carbon sugar that is needed for synthesis of RNA (ribonucleic acid) and DNA (deoxyribonucleic acid). The same sequence of reactions also produces NADPH. This molecule is a hydrogen and electron donor in certain reduction reactions, such as synthesis of fatty acids and steroid hormones.

4 *Glycolysis.* Some ATP is produced anaerobically via glycolysis, in which glucose 6-phosphate is converted to pyruvic acid, another key molecule in metabolism. Most body cells carry out glycolysis.

The Role of Pyruvic Acid

Each 6-carbon molecule of glucose that undergoes glycolysis yields two 3-carbon molecules of **pyruvic acid** (pī-ROO-vik). This molecule, like glucose 6-phosphate, stands at a metabolic crossroads: Given enough oxygen, the aerobic (oxygen-consuming) reactions of cellular respiration can proceed; if oxygen is in short supply, anaerobic reactions can occur (Figure 25.16):

5 *Production of lactic acid.* When oxygen is in short supply in a tissue, as in actively contracting skeletal or cardiac muscle, some pyruvic acid is changed to lactic acid. The lactic acid then diffuses into the bloodstream and is taken up by hepatocytes, which eventually convert it back to pyruvic acid.

6 *Production of alanine.* Carbohydrate and protein metabolism are linked by pyruvic acid. Through transamination, an amino group (—NH₂) can either be added to pyruvic acid (a carbohydrate) to produce the amino acid alanine, or be removed from alanine to generate pyruvic acid.

7 *Gluconeogenesis.* Pyruvic acid and certain amino acids also can be converted to oxaloacetic acid, one of the Krebs cycle intermediates, which in turn can be used to form glucose 6-phosphate. This sequence of gluconeogenesis reactions bypasses certain one-way reactions of glycolysis.

The Role of Acetyl Coenzyme A

8 When the ATP level in a cell is low but oxygen is plentiful, most pyruvic acid streams toward ATP-producing reactions—the Krebs cycle and electron transport chain—via conversion to **acetyl coenzyme A.**

9 *Entry into the Krebs cycle.* Acetyl CoA is the vehicle for 2-carbon acetyl groups to enter the Krebs cycle. Oxidative Krebs cycle reactions convert acetyl CoA to CO_2 and produce reduced coenzymes (NADH and $FADH_2$) that transfer electrons into the electron transport chain. Oxidative reactions in the electron transport chain in turn generate ATP. Most fuel molecules that will be oxidized to generate ATP—glucose, fatty acids, and ketone bodies—are first converted to acetyl CoA.

10 *Synthesis of lipids.* Acetyl CoA also can be used for synthesis of certain lipids, including fatty acids, ketone bodies, and cholesterol. Because pyruvic acid can be converted to acetyl CoA, carbohydrates can be turned into triglycerides; this metabolic pathway stores some excess carbohydrate calories as fat. Mammals, including humans, cannot reconvert acetyl CoA to pyruvic acid, however, so fatty acids cannot be used to generate glucose or other carbohydrate molecules.

Table 25.2 is a summary of carbohydrate, lipid, and protein metabolism.

✔ **CHECKPOINT**

22. What are the possible fates of glucose 6-phosphate, pyruvic acid, and acetyl coenzyme A in a cell?

25.7 METABOLIC ADAPTATIONS

◉ **OBJECTIVE**

• Compare metabolism during the absorptive and postabsorptive states.

Regulation of metabolic reactions depends both on the chemical environment within body cells, such as the levels of ATP and oxygen, and on signals from the nervous and endocrine systems. Some aspects of metabolism depend on how much time has passed since the last meal. During the **absorptive state,** ingested nutrients are entering the bloodstream, and glucose is readily available for ATP production. During the **postabsorptive state,** absorption of nutrients from the GI tract is complete, and energy needs must be met by fuels already in the body. A typical meal requires about 4 hours for complete absorption; given three meals a day, the absorptive state exists for about 12 hours each day. Assuming no between-meal snacks, the other 12 hours—typically late morning, late afternoon, and most of the night—are spent in the postabsorptive state.

Because the nervous system and red blood cells continue to depend on glucose for ATP production during the postabsorptive state, maintaining a steady blood glucose level is critical during this period. Hormones are the major regulators of metabolism in

TABLE 25.2

Summary of Metabolism

PROCESS	COMMENTS
CARBOHYDRATES	
Glucose catabolism	Complete oxidation of glucose (cellular respiration) is chief source of ATP in cells; consists of glycolysis, Krebs cycle, and electron transport chain. Complete oxidation of 1 molecule of glucose yields maximum of 36 or 38 molecules of ATP.
Glycolysis	Conversion of glucose into pyruvic acid results in production of some ATP. Reactions do not require oxygen (anaerobic cellular respiration).
Krebs cycle	Cycle includes series of oxidation–reduction reactions in which coenzymes (NAD^+ and FAD) pick up hydrogen ions and hydride ions from oxidized organic acids; some ATP produced. CO_2 and H_2O are by-products. Reactions are aerobic.
Electron transport chain	Third set of reactions in glucose catabolism: another series of oxidation–reduction reactions, in which electrons are passed from one carrier to next; most ATP produced. Reactions require oxygen (aerobic cellular respiration).
Glucose anabolism	Some glucose is converted into glycogen (glycogenesis) for storage if not needed immediately for ATP production. Glycogen can be reconverted to glucose (glycogenolysis). Conversion of amino acids, glycerol, and lactic acid into glucose is called gluconeogenesis.
LIPIDS	
Triglyceride catabolism	Triglycerides are broken down into glycerol and fatty acids. Glycerol may be converted into glucose (gluconeogenesis) or catabolized via glycolysis. Fatty acids are catabolized via beta oxidation into acetyl coenzyme A that can enter Krebs cycle for ATP production or be converted into ketone bodies (ketogenesis).
Triglyceride anabolism	Synthesis of triglycerides from glucose and fatty acids is called lipogenesis. Triglycerides are stored in adipose tissue.
PROTEINS	
Protein catabolism	Amino acids are oxidized via Krebs cycle after deamination. Ammonia resulting from deamination is converted into urea in liver, passed into blood, and excreted in urine. Amino acids may be converted into glucose (gluconeogenesis), fatty acids, or ketone bodies.
Protein anabolism	Protein synthesis is directed by DNA and utilizes cells' RNA and ribosomes.

each state. The effects of insulin dominate in the absorptive state; several other hormones regulate metabolism in the postabsorptive state. During fasting and starvation, many body cells turn to ketone bodies for ATP production, as noted in the Clinical Connection on Ketosis.

Metabolism during the Absorptive State

Soon after a meal, nutrients start to enter the blood. Recall that ingested food reaches the bloodstream mainly as glucose, amino acids, and triglycerides (in chylomicrons). Two metabolic hallmarks of the absorptive state are the oxidation of glucose for ATP production, which occurs in most body cells, and the storage of excess fuel molecules for future between-meal use, which occurs mainly in hepatocytes, adipocytes, and skeletal muscle fibers.

Absorptive State Reactions

The following reactions dominate during the absorptive state (Figure 25.17):

1. About 50% of the glucose absorbed from a typical meal is oxidized by cells throughout the body to produce ATP via glycolysis, the Krebs cycle, and the electron transport chain.

2. Most glucose that enters hepatocytes is converted to glycogen. Small amounts may be used for synthesis of fatty acids and glyceraldehyde 3-phosphate.

3. Some fatty acids and triglycerides synthesized in the liver remain there, but hepatocytes package most into VLDLs, which carry lipids to adipose tissue for storage.

4. Adipocytes also take up glucose not picked up by the liver and convert it into triglycerides for storage. Overall, about 40% of the glucose absorbed from a meal is converted to triglycerides, and about 10% is stored as glycogen in skeletal muscles and hepatocytes.

5. Most dietary lipids (mainly triglycerides and fatty acids) are stored in adipose tissue; only a small portion is used for synthesis reactions. Adipocytes obtain the lipids from chylomicrons, from VLDLs, and from their own synthesis reactions.

6. Many absorbed amino acids that enter hepatocytes are deaminated to keto acids, which can either enter the Krebs cycle for ATP production or be used to synthesize glucose or fatty acids.

7. Some amino acids that enter hepatocytes are used to synthesize proteins (for example, plasma proteins).

Figure 25.17 Principal metabolic pathways during the absorptive state.

During the absorptive state, most body cells produce ATP by oxidizing glucose to CO_2 and H_2O.

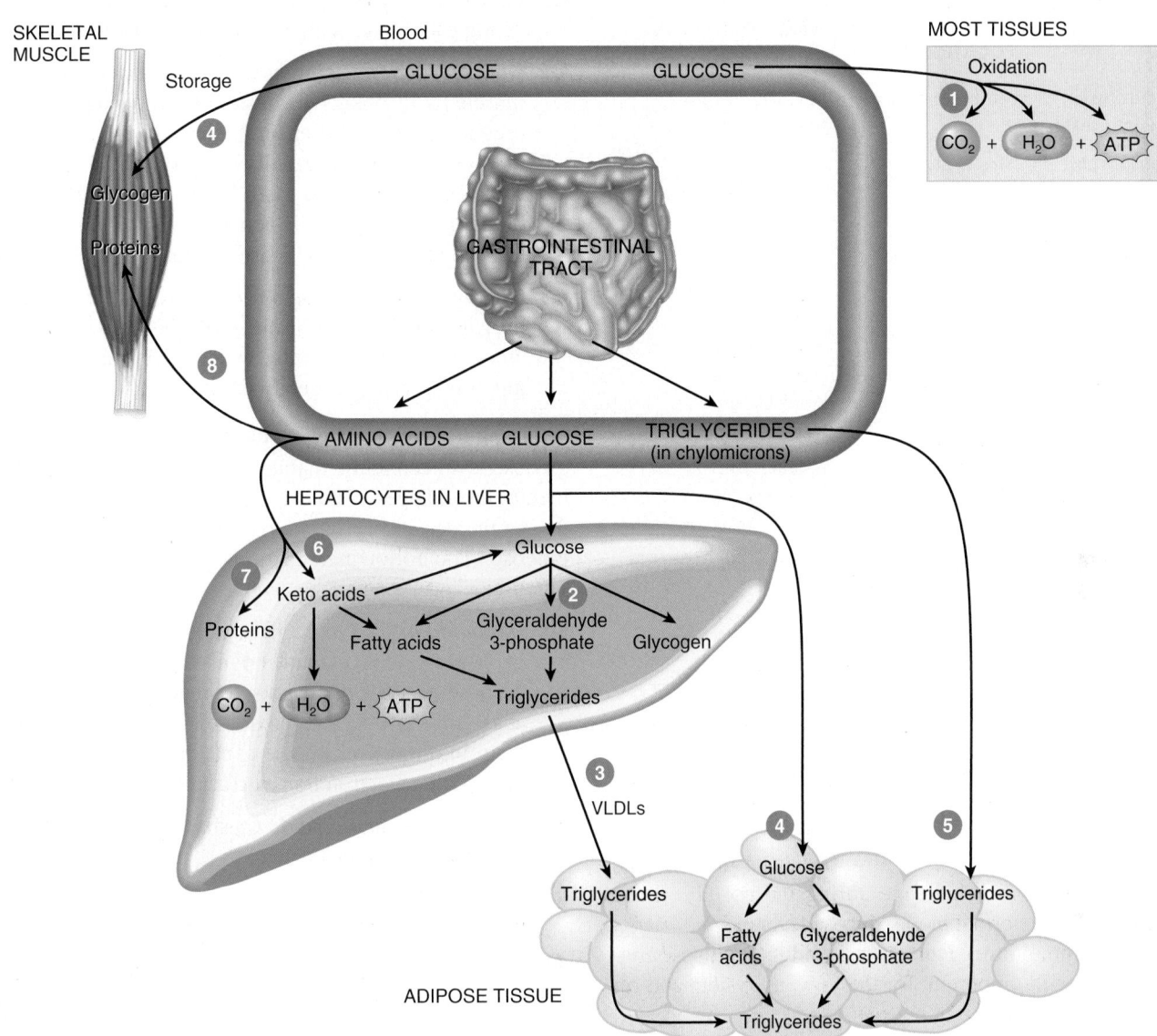

Are the reactions shown in this figure mainly anabolic or catabolic?

8️⃣ Amino acids not taken up by hepatocytes are used in other body cells (such as muscle cells) for synthesis of proteins or regulatory chemicals such as hormones or enzymes.

Regulation of Metabolism during the Absorptive State

Soon after a meal, glucose-dependent insulinotropic peptide (GIP), plus the rising blood levels of glucose and certain amino acids, stimulates pancreatic beta cells to release insulin. In general, insulin increases the activity of enzymes needed for anabolism and the synthesis of storage molecules; at the same time it decreases the activity of enzymes needed for catabolic or breakdown reactions. Insulin promotes the entry of glucose and amino acids into cells of many tissues, and it stimulates the phosphorylation of glucose in hepatocytes and the conversion of glucose 6-phosphate to glycogen in both liver and muscle cells. In liver and adipose tissue, insulin enhances the synthesis of triglycerides, and in cells throughout the body insulin stimulates protein synthesis. (See Section 18.10 to review the effects of insulin.) Insulinlike growth factors and the thyroid hormones (T_3 and T_4) also stimulate protein synthesis. Table 25.3 summarizes the hormonal regulation of metabolism in the absorptive state.

TABLE 25.3

Hormonal Regulation of Metabolism in the Absorptive State

PROCESS	LOCATION(S)	MAIN STIMULATING HORMONE(S)
Facilitated diffusion of glucose into cells	Most cells.	Insulin.*
Active transport of amino acids into cells	Most cells.	Insulin.
Glycogenesis (glycogen synthesis)	Hepatocytes and muscle fibers.	Insulin.
Protein synthesis	All body cells.	Insulin, thyroid hormones, and insulinlike growth factors.
Lipogenesis (triglyceride synthesis)	Adipose cells and hepatocytes.	Insulin.

*Facilitated diffusion of glucose into hepatocytes (liver cells) and neurons is always "turned on" and does not require insulin.

Metabolism during the Postabsorptive State

About 4 hours after the last meal, absorption of nutrients from the small intestine is nearly complete, and blood glucose level starts to fall because glucose continues to leave the bloodstream and enter body cells while none is being absorbed from the GI tract. Thus, the main metabolic challenge during the postabsorptive state is to maintain the normal blood glucose level of 70–110 mg/ 100 mL (3.9–6.1 mmol/liter). Homeostasis of blood glucose concentration is especially important for the nervous system and for red blood cells for the following reasons:

• The dominant fuel molecule for ATP production in the nervous system is glucose, because fatty acids are unable to pass the blood–brain barrier.

• Red blood cells derive all their ATP from glycolysis of glucose because they have no mitochondria, so the Krebs cycle and the electron transport chain are not available to them.

Postabsorptive State Reactions

During the postabsorptive state, both *glucose production* and *glucose conservation* help maintain blood glucose level: Hepatocytes produce glucose molecules and export them into the blood, and other body cells switch from glucose to alternative fuels for ATP production to conserve scarce glucose. The major reactions of the postabsorptive state that produce glucose are the following (Figure 25.18):

1 **Breakdown of liver glycogen.** During fasting, a major source of blood glucose is liver glycogen, which can provide about a

4-hour supply of glucose. Liver glycogen is continually being formed and broken down as needed.

2 **Lipolysis.** Glycerol, produced by breakdown of triglycerides in adipose tissue, is also used to form glucose.

3 **Gluconeogenesis using lactic acid.** During exercise, skeletal muscle tissue breaks down stored glycogen (see step 9) and produces some ATP anaerobically via glycolysis. Some of the pyruvic acid that results is converted to acetyl CoA, and some is converted to lactic acid, which diffuses into the blood. In the liver, lactic acid can be used for gluconeogenesis, and the resulting glucose is released into the blood.

4 **Gluconeogenesis using amino acids.** Modest breakdown of proteins in skeletal muscle and other tissues releases large amounts of amino acids, which then can be converted to glucose by gluconeogenesis in the liver.

Despite all of these ways the body produces glucose, blood glucose level cannot be maintained for very long without further metabolic changes. Thus, a major adjustment must be made during the postabsorptive state to produce ATP while conserving glucose. The following reactions produce ATP without using glucose:

5 **Oxidation of fatty acids.** The fatty acids released by lipolysis of triglycerides cannot be used for glucose production because acetyl CoA cannot be readily converted to pyruvic acid. But most cells can oxidize the fatty acids directly, feed them into the Krebs cycle as acetyl CoA, and produce ATP through the electron transport chain.

6 **Oxidation of lactic acid.** Cardiac muscle can produce ATP aerobically from lactic acid.

7 **Oxidation of amino acids.** In hepatocytes, amino acids may be oxidized directly to produce ATP.

8 **Oxidation of ketone bodies.** Hepatocytes also convert fatty acids to ketone bodies, which can be used by the heart, kidneys, and other tissues for ATP production.

9 **Breakdown of muscle glycogen.** Skeletal muscle cells break down glycogen to glucose 6-phosphate, which undergoes glycolysis and provides ATP for muscle contraction.

Regulation of Metabolism during the Postabsorptive State

Both hormones and the sympathetic division of the autonomic nervous system (ANS) regulate metabolism during the postabsorptive state. The hormones that regulate postabsorptive state metabolism sometimes are called anti-insulin hormones because they counter the effects of insulin during the absorptive state. As blood glucose level declines, the secretion of insulin falls and the release of anti-insulin hormones rises.

When blood glucose concentration starts to drop, the pancreatic alpha cells release glucagon at a faster rate, and the beta cells secrete insulin more slowly. The primary target tissue of glucagon is the liver; the major effect is increased release of glucose into the bloodstream due to gluconeogenesis and glycogenolysis.

Figure 25.18 Principal metabolic pathways during the postabsorptive state.

The principal function of postabsorptive state reactions is to maintain a normal blood glucose level.

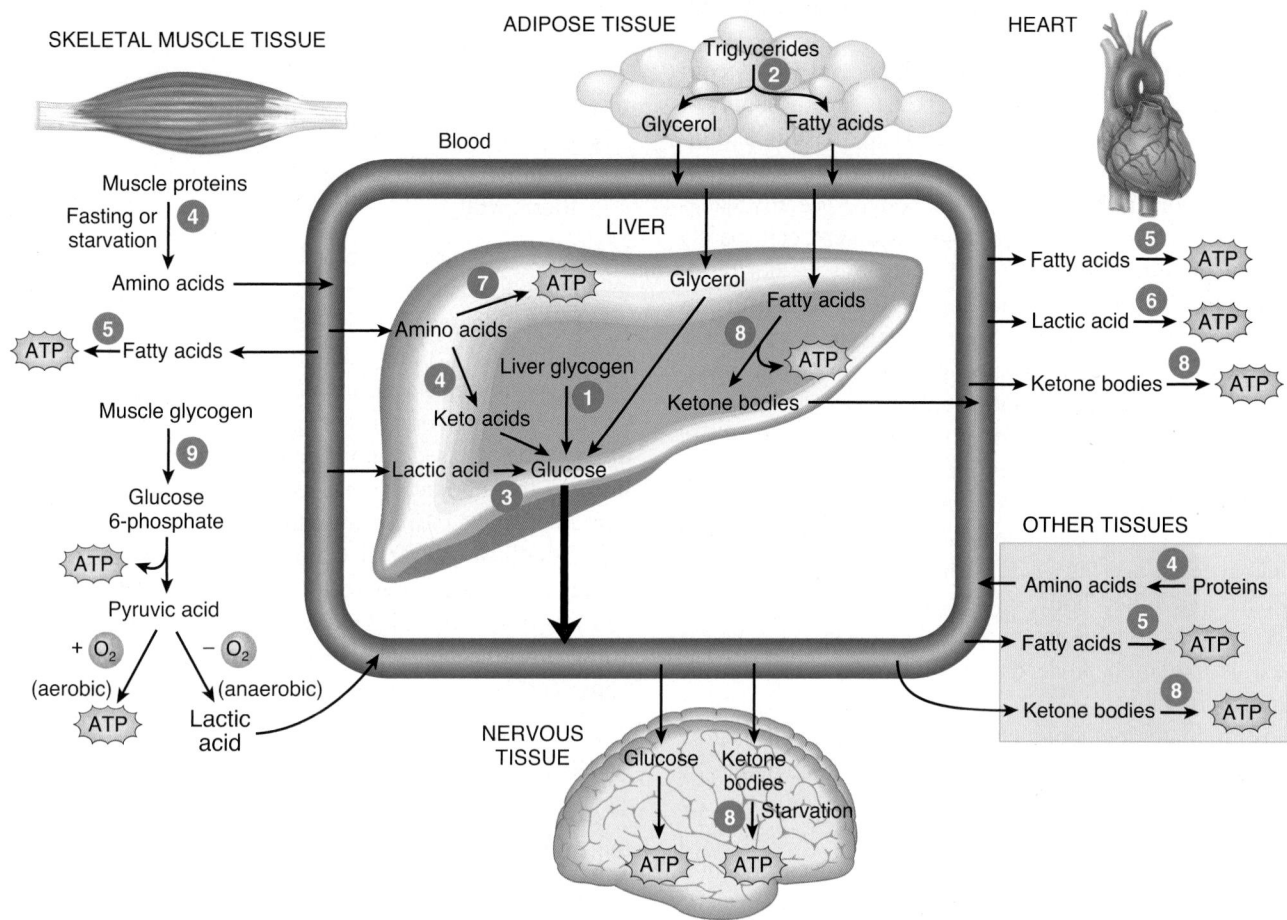

What processes directly elevate blood glucose during the postabsorptive state, and where does each occur?

Low blood glucose also activates the sympathetic branch of the ANS. Glucose-sensitive neurons in the hypothalamus detect low blood glucose and increase sympathetic output. As a result, sympathetic nerve endings release the neurotransmitter norepinephrine, and the adrenal medulla releases two catecholamine hormones—epinephrine and norepinephrine—into the bloodstream. Like glucagon, epinephrine stimulates glycogen breakdown. Epinephrine and norepinephrine are both potent stimulators of lipolysis. These actions of the catecholamines help to increase glucose and free fatty acid levels in the blood. As a result, muscle uses more fatty acids for ATP production, and more glucose is available to the nervous system. Table 25.4 summarizes the hormonal regulation of metabolism in the postabsorptive state.

Metabolism during Fasting and Starvation

The term **fasting** means going without food for many hours or a few days; **starvation** implies weeks or months of food deprivation or inadequate food intake. People can survive without food for 2 months or more if they drink enough water to prevent

TABLE 25.4

Hormonal Regulation of Metabolism in the Postabsorptive State

PROCESS	LOCATION(S)	MAIN STIMULATING HORMONE(S)
Glycogenolysis (glycogen breakdown)	Hepatocytes and skeletal muscle fibers.	Glucagon and epinephrine.
Lipolysis (triglyceride breakdown)	Adipocytes.	Epinephrine, norepinephrine, cortisol, insulinlike growth factors, thyroid hormones, and others.
Protein breakdown	Most body cells, but especially skeletal muscle fibers.	Cortisol.
Gluconeogenesis (synthesis of glucose from noncarbohydrates)	Hepatocytes and kidney cortex cells.	Glucagon and cortisol.

dehydration. Although glycogen stores are depleted within a few hours of beginning a fast, catabolism of stored triglycerides and structural proteins can provide energy for several weeks. The amount of adipose tissue the body contains determines the life-span possible without food.

During fasting and starvation, nervous tissue and RBCs continue to use glucose for ATP production. There is a ready supply of amino acids for gluconeogenesis because lowered insulin and increased cortisol levels slow the pace of protein synthesis and promote protein catabolism. Most cells in the body, especially skeletal muscle cells because of their high protein content, can spare a fair amount of protein before their performance is adversely affected. During the first few days of fasting, protein catabolism outpaces protein synthesis by about 75 grams daily as some of the "old" amino acids are being deaminated and used for gluconeogenesis and "new" (that is, dietary) amino acids are lacking.

By the second day of a fast, blood glucose level has stabilized at about 65 mg/100 mL (3.6 mmol/liter); at the same time the level of fatty acids in plasma has risen fourfold. Lipolysis of triglycerides in adipose tissue releases glycerol, which is used for gluconeogenesis, and fatty acids. The fatty acids diffuse into muscle fibers and other body cells, where they are used to produce acetyl CoA, which enters the Krebs cycle. ATP then is synthesized as oxidation proceeds via the Krebs cycle and the electron transport chain.

The most dramatic metabolic change that occurs with fasting and starvation is the increase in the formation of ketone bodies by hepatocytes. During fasting, only small amounts of glucose undergo glycolysis to pyruvic acid, which in turn can be converted to oxaloacetic acid. Acetyl CoA enters the Krebs cycle by combining with oxaloacetic acid (see Figure 25.16); when oxaloacetic acid is scarce due to fasting, only some of the available acetyl CoA can enter the Krebs cycle. Surplus acetyl CoA is used for ketogenesis, mainly in hepatocytes. Ketone body production thus increases as catabolism of fatty acids rises. Lipid-soluble ketone bodies can diffuse through plasma membranes and across the blood–brain barrier and be used as an alternative fuel for ATP production, especially by cardiac and skeletal muscle fibers and neurons. Normally, only a trace of ketone bodies (0.01 mmol/liter) are present in the blood, so they are a negligible fuel source. After 2 days of fasting, however, the level of ketones is 100–300 times higher and supplies roughly a third of the brain's fuel for ATP production. By 40 days of starvation, ketones provide up to two-thirds of the brain's energy needs. The presence of ketones actually reduces the use of glucose for ATP production, which in turn decreases the demand for gluconeogenesis and slows the catabolism of muscle proteins later in starvation to about 20 grams daily.

✔ CHECKPOINT

23. What are the roles of insulin, glucagon, epinephrine, insulinlike growth factors, thyroxine, cortisol, estrogen, and testosterone in regulation of metabolism?
24. Why is ketogenesis more significant during fasting or starvation than during normal absorptive and postabsorptive states?

25.8 HEAT AND ENERGY BALANCE

◉ OBJECTIVES

- Define basal metabolic rate (BMR), and explain several factors that affect it.
- Describe the factors that influence body heat production.
- Explain how normal body temperature is maintained by negative feedback loops involving the hypothalamic thermostat.

Your body produces more or less heat depending on the rates of its metabolic reactions. Because homeostasis of body temperature can be maintained only if the rate of heat loss from the body equals the rate of heat production by metabolism, it is important to understand the ways in which heat can be lost, gained, or conserved. **Heat** is a form of energy that can be measured as **temperature** and expressed in units called calories. A **calorie (cal)** is defined as the amount of heat required to raise the temperature of 1 gram of water 1°C. Because the calorie is a relatively small unit, the **kilocalorie (kcal)** or **Calorie (Cal)** (always spelled with an uppercase C) is often used to measure the body's metabolic rate and to express the energy content of foods. A kilocalorie equals 1000 calories. Thus, when we say that a particular food item contains 500 Calories, we are actually referring to kilocalories.

Metabolic Rate

The overall rate at which metabolic reactions use energy is termed the **metabolic rate.** As you have already learned, some of the energy is used to produce ATP, and some is released as heat. Because many factors affect metabolic rate, it is measured under standard conditions, with the body in a quiet, resting, and fasting condition called the **basal state.** The measurement obtained under these conditions is the **basal metabolic rate (BMR).** The most common way to determine BMR is by measuring the amount of oxygen used per kilocalorie of food metabolized. When the body uses 1 liter of oxygen to oxidize a typical dietary mixture of triglycerides, carbohydrates, and proteins, about 4.8 Cal of energy is released. BMR is 1200–1800 Cal/day in adults, or about 24 Cal/kg of body mass in adult males and 22 Cal/kg in adult females. The added calories needed to support daily activities, such as digestion and walking, range from 500 Cal for a small, relatively sedentary person to over 3000 Cal for a person in training for Olympic-level competitions or mountain climbing.

Body Temperature Homeostasis

Despite wide fluctuations in environmental temperature, homeostatic mechanisms can maintain a normal range for internal body temperature. If the rate of body heat production equals the rate of heat loss, the body maintains a constant core temperature near 37°C (98.6°F). **Core temperature** is the temperature in body structures deep to the skin and subcutaneous layer. **Shell temperature** is the temperature near the body surface—in the skin and subcutaneous layer. Depending on the environmental temperature, shell temperature is 1–6°C lower than core temperature. A core temperature that is too high kills by denaturing body proteins; a core temperature that is too low causes cardiac arrhythmias that result in death.

Heat Production

The production of body heat is proportional to metabolic rate. Several factors affect the metabolic rate and thus the rate of heat production:

- **Exercise.** During strenuous exercise, the metabolic rate may increase to as much as 15 times the basal rate. In well-trained athletes, the rate may increase up to 20 times.
- **Hormones.** Thyroid hormones (thyroxine and triiodothyronine) are the main regulators of BMR; BMR increases as the blood levels of thyroid hormones rise. The response to changing levels of thyroid hormones is slow, however, taking several days to appear. Thyroid hormones increase BMR in part by stimulating aerobic cellular respiration. As cells use more oxygen to produce ATP, more heat is given off, and body temperature rises. Other hormones have minor effects on BMR. Testosterone, insulin, and human growth hormone can increase the metabolic rate by 5–15%.
- **Nervous system.** During exercise or in a stressful situation, the sympathetic division of the autonomic nervous system is stimulated. Its postganglionic neurons release norepinephrine (NE), and it also stimulates release of the hormones epinephrine and norepinephrine by the adrenal medulla. Both epinephrine and norepinephrine increase the metabolic rate of body cells.
- **Body temperature.** The higher the body temperature, the higher the metabolic rate. Each 1°C rise in core temperature increases the rate of biochemical reactions by about 10%. As a result, metabolic rate may be increased substantially during a fever.
- **Ingestion of food.** The ingestion of food raises the metabolic rate 10–20% due to the energy "costs" of digesting, absorbing, and storing nutrients. This effect, *food-induced thermogenesis,* is greatest after eating a high-protein meal and is less after eating carbohydrates and lipids.
- **Age.** The metabolic rate of a child, in relation to its size, is about double that of an elderly person due to the high rates of reactions related to growth.
- **Other factors.** Other factors that affect metabolic rate are gender (lower in females, except during pregnancy and lactation), climate (lower in tropical regions), sleeping (lower), and malnutrition (lower).

Mechanisms of Heat Transfer

Maintaining normal body temperature depends on the ability to lose heat to the environment at the same rate as it is produced by metabolic reactions. Heat can be transferred from the body to its surroundings in four ways: via conduction, convection, radiation, and evaporation.

1. **Conduction** is the heat exchange that occurs between molecules of two materials that are in direct contact with each other. At rest, about 3% of body heat is lost via conduction to solid materials in contact with the body, such as a chair, clothing, and jewelry. Heat can also be gained via conduction—for example, while soaking in a hot tub. Because water conducts heat 20 times more effectively than air, heat loss or heat gain via conduction is much greater when the body is submerged in cold or hot water.

2. **Convection** (kon-VEK-shun) is the transfer of heat by the movement of a fluid (a gas or a liquid) between areas of different temperature. The contact of air or water with your body results in heat transfer by both conduction and convection. When cool air makes contact with the body, it becomes warmed and therefore less dense and is carried away by convection currents created as the less dense air rises. The faster the air moves—for example, by a breeze or a fan—the faster the rate of convection. At rest, about 15% of body heat is lost to the air via conduction and convection.

3. **Radiation** is the transfer of heat in the form of infrared rays between a warmer object and a cooler one without physical contact. Your body loses heat by radiating more infrared waves than it absorbs from cooler objects. If surrounding objects are warmer than you are, you absorb more heat than you lose by radiation. In a room at 21°C (70°F), about 60% of heat loss occurs via radiation in a resting person.

4. **Evaporation** is the conversion of a liquid to a vapor. Every milliliter of evaporating water takes with it a great deal of heat—about 0.58 Cal/mL. Under typical resting conditions, about 22% of heat loss occurs through evaporation of about 700 mL of water per day—300 mL in exhaled air and 400 mL from the skin surface. Because we are not normally aware of this water loss through the skin and mucous membranes of the mouth and respiratory system, it is termed **insensible water loss** (in-SEN-si-bel). The rate of evaporation is inversely related to relative humidity, the ratio of the actual amount of moisture in the air to the maximum amount it can hold at a given temperature. The higher the relative humidity, the lower the rate of evaporation. At 100% humidity, heat is gained via condensation of water on the skin surface as fast as heat is lost via evaporation. Evaporation provides the main defense against overheating during exercise. Under extreme conditions, a maximum of about 3 liters of sweat can be produced each hour, removing more than 1700 Calories of heat if all of it evaporates. (Note: Sweat that drips off the body rather than evaporating removes very little heat.)

Hypothalamic Thermostat

The control center that functions as the body's thermostat is a group of neurons in the anterior part of the hypothalamus, the **preoptic area.** This area receives impulses from thermoreceptors in the skin and mucous membranes and in the hypothalamus. Neurons of the preoptic area generate nerve impulses at a higher frequency when blood temperature increases, and at a lower frequency when blood temperature decreases.

Nerve impulses from the preoptic area propagate to two other parts of the hypothalamus known as the **heat-losing center** and the **heat-promoting center,** which, when stimulated by the preoptic area, set into operation a series of responses that lower body temperature and raise body temperature, respectively.

Thermoregulation

If core temperature declines, mechanisms that help conserve heat and increase heat production act via several negative feedback loops to raise the body temperature to normal (Figure 25.19). Thermoreceptors in the skin and hypothalamus send nerve impulses to the preoptic area and the heat-promoting center in the hypothalamus, and to hypothalamic neurosecretory cells that produce thyrotropin-releasing hormone (TRH). In response, the hypothalamus discharges nerve impulses and secretes TRH, which in turn stimulates thyrotrophs in the anterior pituitary gland to

Figure 25.19 **Negative feedback mechanisms that conserve heat and increase heat production.**

Core temperature is the temperature in body structures deep to the skin and subcutaneous layer; shell temperature is the temperature near the body surface.

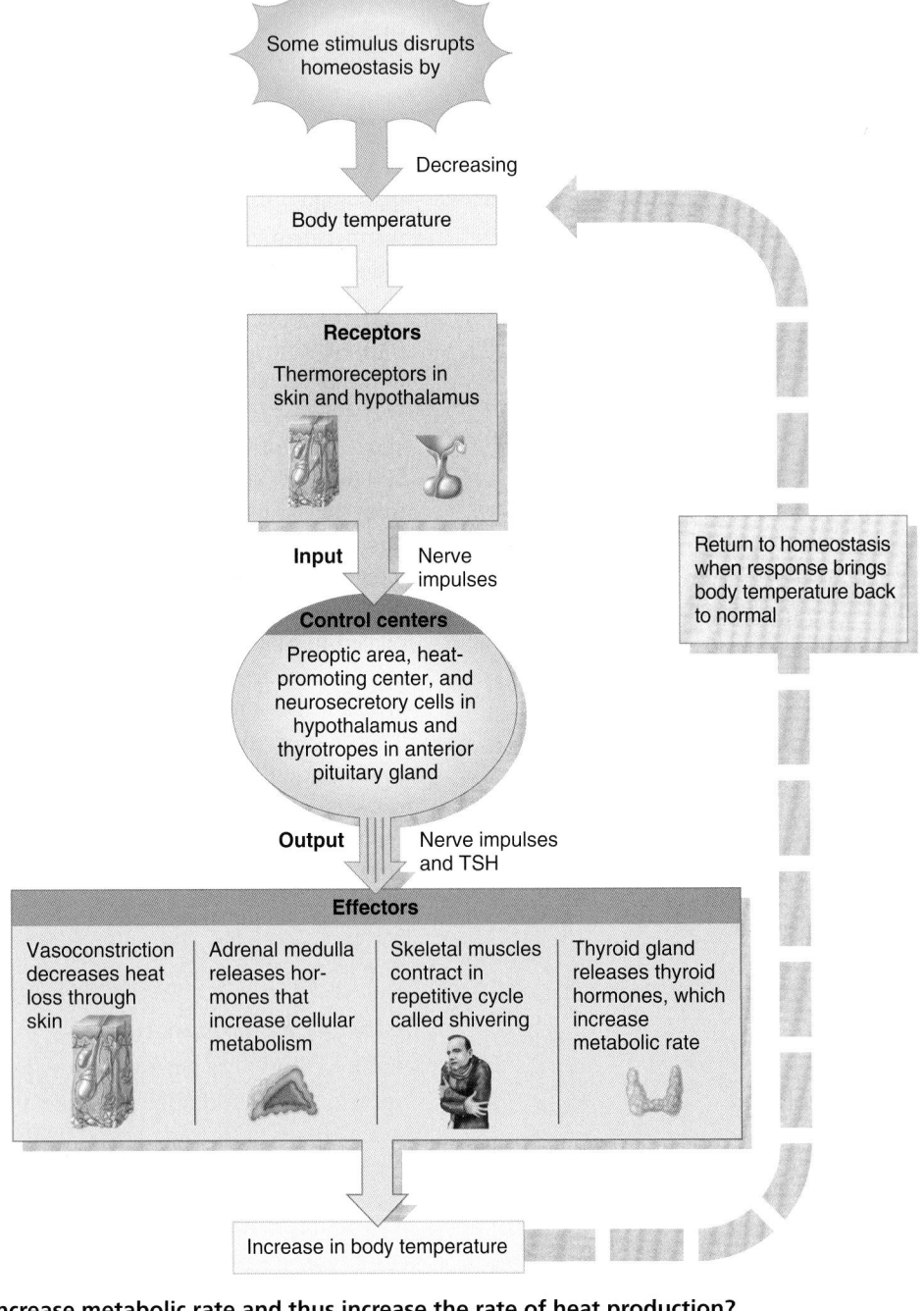

What factors can increase metabolic rate and thus increase the rate of heat production?

release thyroid-stimulating hormone (TSH). Nerve impulses from the hypothalamus and TSH then activate several effectors.

Each effector responds in a way that helps increase core temperature to the normal value:

- Nerve impulses from the heat-promoting center stimulate sympathetic nerves that cause blood vessels of the skin to constrict. Vasoconstriction decreases the flow of warm blood, and thus the transfer of heat, from the internal organs to the skin. Slowing the rate of heat loss allows the internal body temperature to increase as metabolic reactions continue to produce heat.

- Nerve impulses in sympathetic nerves leading to the adrenal medulla stimulate the release of epinephrine and norepinephrine into the blood. The hormones in turn bring about an increase in cellular metabolism, which increases heat production.

- The heat-promoting center stimulates parts of the brain that increase muscle tone and hence heat production. As muscle tone increases in one muscle (the agonist), the small contractions stretch muscle spindles in its antagonist, initiating a stretch reflex. The resulting contraction in the antagonist stretches muscle spindles in the agonist, and it too develops a stretch reflex. This repetitive cycle—called **shivering**—greatly increases the rate of heat production. During maximal shivering, body heat production can rise to about four times the basal rate in just a few minutes.

- The thyroid gland responds to TSH by releasing more thyroid hormones into the blood. As increased levels of thyroid hormones slowly increase the metabolic rate, body temperature rises.

If core body temperature rises above normal, a negative feedback loop opposite to the one depicted in Figure 25.19 goes into action. The higher temperature of the blood stimulates thermoreceptors that send nerve impulses to the preoptic area, which in turn stimulate the heat-losing center and inhibit the heat-promoting center. Nerve impulses from the heat-losing center cause dilation of blood vessels in the skin. The skin becomes warm, and the excess heat is lost to the environment via radiation and conduction as an increased volume of blood flows from the warmer core of the body into the cooler skin. At the same time, metabolic rate decreases, and shivering does not occur. The high temperature of the blood stimulates sweat glands of the skin via hypothalamic activation of sympathetic nerves. As the water in perspiration evaporates from the surface of the skin, the skin is cooled. All these responses counteract heat-promoting effects and help return body temperature to normal.

CLINICAL CONNECTION | Hypothermia

Hypothermia (hī′-pō-THER-mē-a) is a lowering of core body temperature to 35°C (95°F) or below. Causes of hypothermia include an overwhelming cold stress (immersion in icy water), metabolic diseases (hypoglycemia, adrenal insufficiency, or hypothyroidism), drugs (alcohol, antidepressants, sedatives, or tranquilizers), burns, and malnutrition. Hypothermia is characterized by the following as core body temperature falls: sensation of cold, shivering, confusion, vasoconstriction, muscle rigidity, bradycardia, acidosis, hypoventilation, hypotension, loss of spontaneous movement, coma, and death (usually caused by cardiac arrhythmias). Because the elderly have reduced metabolic protection against a cold environment coupled with a reduced perception of cold, they are at greater risk for developing hypothermia. •

Energy Homeostasis and Regulation of Food Intake

Most mature animals and many men and women maintain **energy homeostasis,** the precise matching of energy intake (in food) to energy expenditure over time. When the energy content of food balances the energy used by all the cells of the body, body weight remains constant (unless there is a gain or loss of water). In many people, weight stability persists despite large day-to-day variations in activity and food intake. In the more affluent nations, however, a large fraction of the population is overweight. Easy access to tasty, high-calorie foods and a "couch potato" lifestyle promote weight gain. Being overweight increases the risk of dying from a variety of cardiovascular and metabolic disorders, including hypertension, varicose veins, diabetes mellitus, arthritis, certain cancers, and gallbladder disease.

Energy intake depends only on the amount of food consumed (and absorbed), but three components contribute to total energy expenditure:

1. The basal metabolic rate accounts for about 60% of energy expenditure.

2. Physical activity typically adds 30–35% but can be lower in sedentary people. The energy expenditure is partly from voluntary exercise, such as walking, and partly from **nonexercise activity thermogenesis (NEAT)** (ther-mō-JEN-e-sis), the energy costs for maintaining muscle tone, posture while sitting or standing, and involuntary fidgeting movements.

3. **Food-induced thermogenesis,** the heat produced while food is being digested, absorbed, and stored, represents 5–10% of total energy expenditure.

The major site of stored chemical energy in the body is adipose tissue. When energy use exceeds energy input, triglycerides in adipose tissue are catabolized to provide the extra energy, and when energy input exceeds energy expenditure, triglycerides are stored. Over time, the amount of stored triglycerides indicates the excess of energy intake over energy expenditure. Even small differences add up over time. A gain of 20 lb (9 kg) between ages 25 and 55 represents only a tiny imbalance, about 0.3% more energy intake in food than energy expenditure.

Clearly, negative feedback mechanisms are regulating both our energy intake and our energy expenditure. But no sensory receptors exist to monitor our weight or size. How, then, is food intake regulated? The answer to this question is incomplete, but important advances in understanding regulation of food intake have occurred recently. It depends on many factors, including neural and endocrine signals, levels of certain nutrients in the blood, psychological elements such as stress or depression, signals from the GI tract and the special senses, and neural connections between the hypothalamus and other parts of the brain.

Within the hypothalamus are clusters of neurons that play key roles in regulating food intake. **Satiety** (sa-TĪ-i-tē) is a feeling of fullness accompanied by lack of desire to eat. Two hypothalamic areas involved in regulation of food intake are the *arcuate nucleus* and the *paraventricular nucleus* (see Figure 14.10). There is a mouse gene, named *obese*, that causes overeating and severe obesity in its mutated form. The product of this gene is the hormone **leptin.** In both mice and humans, leptin helps decrease **adiposity** (ad-i-POS-i-tē), total body-fat mass. Leptin is synthesized and secreted by adipocytes in proportion to adiposity; as more triglycerides are stored, more leptin is secreted into the bloodstream. Leptin acts on the hypothalamus to inhibit circuits that stimulate eating while also activating circuits that increase energy expenditure. The hormone insulin has a similar, but smaller, effect. Both leptin and insulin are able to pass through the blood–brain barrier.

When leptin and insulin levels are *low,* neurons that extend from the arcuate nucleus to the paraventricular nucleus release a neurotransmitter called **neuropeptide Y** that stimulates food intake. Other neurons that extend between the arcuate and paraventricular nuclei release a neurotransmitter called **melanocortin** (mel-an-ō-KOR-tin), which is similar to melanocyte-stimulating hormone (MSH). Leptin stimulates release of melanocortin, which acts to inhibit food intake. Although leptin, neuropeptide Y, and melanocortin are key signaling molecules for maintaining energy homeostasis, several other hormones and neurotransmitters also contribute. An understanding of the brain circuits involved is still far from complete. Other areas of the hypothalamus plus nuclei in the brain stem, limbic system, and cerebral cortex take part.

Achieving energy homeostasis requires regulation of energy intake. Most increases and decreases in food intake are due to changes in meal size rather than changes in number of meals. Many experiments have demonstrated the presence of satiety signals, chemical or neural changes that help terminate eating when "fullness" is attained. For example, an increase in blood glucose level, as occurs after a meal, decreases appetite. Several hormones, such as glucagon, cholecystokinin, estrogens, and epinephrine (acting via beta receptors), act to signal satiety and to increase energy expenditure. Distension of the GI tract, particularly the stomach and duodenum, also contributes to termination of food intake. Other hormones increase appetite and decrease energy expenditure. These include growth hormone–releasing hormone (GHRH), androgens, glucocorticoids, epinephrine (acting via alpha receptors), and progesterone.

CLINICAL CONNECTION | *Emotional Eating*

In addition to keeping us alive, eating serves countless psychological, social, and cultural purposes. We eat to celebrate, punish, comfort, defy, and deny. Eating in response to emotional drives, such as feeling stressed, bored, or tired, is called **emotional eating.**

Emotional eating is so common that, within limits, it is considered well within the range of normal behavior. Who hasn't at one time or another headed for the refrigerator after a bad day? Problems arise when emotional eating becomes so excessive that it interferes with health. Physical health problems include obesity and associated disorders such as hypertension and heart disease. Psychological health problems include poor self-esteem, an inability to cope effectively with feelings of stress, and in extreme cases, eating disorders such as anorexia nervosa, bulimia, and obesity.

Eating provides comfort and solace, numbing pain and "feeding the hungry heart." Eating may provide a biochemical "fix" as well. Emotional eaters typically overeat carbohydrate foods (sweets and starches), which may raise brain serotonin levels and lead to feelings of relaxation. Food becomes a way to self-medicate when negative emotions arise. •

✔ CHECKPOINT

25. Define a kilocalorie (kcal). How is the unit used? How does it relate to a calorie?
26. Distinguish between core temperature and shell temperature.
27. In what ways can a person lose heat to or gain heat from the surroundings? How is it possible for a person to lose heat on a sunny beach when the temperature is 40°C (104°F) and the humidity is 85%?
28. What does the term energy homeostasis mean?
29. How is food intake regulated?

25.9 NUTRITION

◉ OBJECTIVES

• Describe how to select foods to maintain a healthy diet.
• Compare the sources, functions, and importance of minerals and vitamins in metabolism.

Nutrients are chemical substances in food that body cells use for growth, maintenance, and repair. The six main types of nutrients are water, carbohydrates, lipids, proteins, minerals, and vitamins. Water is the nutrient needed in the largest amount—about 2–3 liters per day. As the most abundant compound in the body, water provides the medium in which most metabolic reactions occur, and it also participates in some reactions (for example, hydrolysis reactions). The important roles of water in the body can be reviewed in Section 2.4. Three organic nutrients—carbohydrates, lipids, and proteins—provide the energy needed for metabolic reactions and serve as building blocks to make body structures. Some minerals and many vitamins are components of the enzyme systems that catalyze metabolic reactions. *Essential nutrients* are specific nutrient molecules that the body cannot make in sufficient quantity to meet its needs and thus must be obtained from the diet. Some amino acids, fatty acids, vitamins, and minerals are essential nutrients.

Next, we discuss some guidelines for healthy eating and the roles of minerals and vitamins in metabolism.

Guidelines for Healthy Eating

Each gram of protein or carbohydrate in food provides about 4 Calories; 1 gram of fat (lipids) provides about 9 Calories. We do not know with certainty what levels and types of carbohydrate, fat, and protein are optimal in the diet. Different populations around the world eat radically different diets that are adapted to their particular lifestyles. However, many experts recommend the following distribution of calories: 50–60% from carbohydrates, with less than 15% from simple sugars; less than 30% from fats (triglycerides are the main type of dietary fat), with no more than 10% as saturated fats; and about 12–15% from proteins.

The guidelines for healthy eating are to:

- Eat a variety of foods.
- Maintain a healthy weight.
- Choose foods low in fat, saturated fat, and cholesterol.
- Eat plenty of vegetables, fruits, and grain products.
- Use sugars in moderation only.

In 2005, the United States Department of Agriculture (USDA) introduced a food pyramid called **MyPyramid,** which represents a *personalized* approach to making healthy food choices and maintaining regular physical activity. By consulting a chart, it is possible to determine your calorie level based on your gender, age, and activity level. Once this is determined, you can choose the type and amount of food to be consumed.

If you carefully examine the MyPyramid in Figure 25.20 you will note that the six color bands represent the five basic food groups plus oils. Foods from all bands are needed each day. Also note that the overall size of the bands suggests the proportion of food a person should choose on a daily basis. The wider base of each band represents foods with little or no solid fats or added sugars; these foods should be selected more often. The narrower top of each band represents foods with more added sugars and solid fats, which should be selected less frequently. The person climbing the steps is a reminder of the need for daily physical activity.

As an example of how MyPyramid works, let's assume based on consulting a chart that the calorie level of an 18-year-old moderately active female is 2000 Calories and that of an 18-year-old moderately active male is 2800 Calories. Accordingly, it is suggested that the following foods should be chosen in the following amounts:

Calorie Level	2000	2800
Fruits (includes all fresh, frozen, canned, and dried fruits and fruit juices)	2 cups	2.5 cups
Vegetables (includes all fresh, frozen, canned, and dried vegetables and vegetable juices)	2.5 cups	3.5 cups
Grains (includes all foods made from wheat, rice, oats, cornmeal, and barley such as bread, cereals, oatmeal, rice, pasta, crackers, tortillas, and grits)	6 oz	10 oz
Meats and beans (includes lean meat, poultry, fish, eggs, peanut butter, beans, nuts, and seeds)	5.5 oz	7 oz
Milk group (includes milk products and foods made from milk that retain their calcium content such as cheeses and yogurt)	3 cups	3 cups
Oils (choose mostly fats that contain monounsaturated and polyunsaturated fatty acids such as fish, nuts, seeds, and vegetable oils)	6 tsp	8 tsp

Figure 25.20 MyPyramid.

MyPyramid is a personalized approach to making healthy food choices and maintaining regular physical activity.

GRAINS   VEGETABLES   FRUITS   OILS   MILK   MEAT & BEANS

? **What does the wider base of each band mean?**

In addition, you should choose and prepare foods with little salt. In fact, sodium intake should be less than 2300 mg per day. If you choose to drink alcohol, it should be consumed in moderation (no more than 1 drink per day for women and 2 drinks per day for men). A drink is defined as 12 oz of regular beer, 5 oz of wine, or $1^{1}/_{2}$ oz of 80 proof distilled spirits.

Minerals

Minerals are inorganic elements that occur naturally in Earth's crust. In the body they appear in combination with one another, in combination with organic compounds, or as ions in solution. Minerals constitute about 4% of total body mass and are concentrated most heavily in the skeleton. Minerals with known functions in the body include calcium, phosphorus, potassium, sulfur, sodium, chloride, magnesium, iron, iodide, manganese, copper, cobalt, zinc, fluoride, selenium, and chromium. Table 25.5 describes the vital functions of these minerals. Note that the body generally uses the ions of the minerals rather than the non-ionized form. Some minerals, such as chlorine, are toxic or even fatal if ingested in the non-ionized form. Other minerals—aluminum, boron, silicon, and molybdenum—are present but their functions are unclear. Typical diets supply adequate amounts of potassium, sodium, chloride, and magnesium. Some attention must be paid to eating foods that provide enough calcium, phosphorus, iron, and iodide. Excess amounts of most minerals are excreted in the urine and feces.

Calcium and phosphorus form part of the matrix of bone. Because minerals do not form long-chain compounds, they are otherwise poor building materials. A major role of minerals is to help regulate enzymatic reactions. Calcium, iron, magnesium, and manganese are constituents of some coenzymes. Magnesium also serves as a catalyst for the conversion of ADP to ATP. Minerals such as sodium and phosphorus work in buffer systems, which help control the pH of body fluids. Sodium also helps regulate the osmosis of water and, along with other ions, is involved in the generation of nerve impulses.

Vitamins

Organic nutrients required in small amounts to maintain growth and normal metabolism are called **vitamins.** Unlike carbohydrates, lipids, or proteins, vitamins do not provide energy or serve as the body's building materials. Most vitamins with known functions are coenzymes.

Most vitamins cannot be synthesized by the body and must be ingested in food. Other vitamins, such as vitamin K, are produced by bacteria in the GI tract and then absorbed. The body can assemble some vitamins if the raw materials, called **provitamins,** are provided. For example, vitamin A is produced by the body from the provitamin beta-carotene, a chemical present in yellow vegetables such as carrots and in dark green vegetables such as spinach. No single food contains all the required vitamins—one of the best reasons to eat a varied diet.

Vitamins are divided into two main groups: fat-soluble and water-soluble. The **fat-soluble vitamins,** vitamins A, D, E, and K, are absorbed along with other dietary lipids in the small intestine and packaged into chylomicrons. They cannot be absorbed in adequate quantity unless they are ingested with other lipids. Fat-soluble vitamins may be stored in cells, particularly hepatocytes. The **water-soluble vitamins,** including several B vitamins and vitamin C, are dissolved in body fluids. Excess quantities of these vitamins are not stored but instead are excreted in the urine.

Besides their other functions, three vitamins—C, E, and beta-carotene (a provitamin)—are termed **antioxidant vitamins** because they inactivate oxygen free radicals. Recall that free radicals are highly reactive ions or molecules that carry an unpaired electron in their outermost electron shell (see Figure 2.3). Free radicals damage cell membranes, DNA, and other cellular structures and contribute to the formation of artery-narrowing atherosclerotic plaques. Some free radicals arise naturally in the body, and others come from environmental hazards such as tobacco smoke and radiation. Antioxidant vitamins are thought to play a role in protecting against some kinds of cancer, reducing the buildup of atherosclerotic plaque, delaying some effects of aging, and decreasing the chance of cataract formation in the lens of the eyes. Table 25.6 lists the major vitamins, their sources, their functions, and related deficiency disorders.

CLINICAL CONNECTION | Vitamin and Mineral Supplements

Most nutritionists recommend eating a balanced diet that includes a variety of foods rather than taking vitamin or mineral supplements, except in special circumstances. Common examples of necessary supplementations include iron for women who have excessive menstrual bleeding; iron and calcium for women who are pregnant or breast-feeding; folic acid (folate) for all women who may become pregnant, to reduce the risk of fetal neural tube defects; calcium for most adults, because they do not receive the recommended amount in their diets; and vitamin B_{12} for strict vegetarians, who eat no meat. Because most North Americans do not ingest in their food the high levels of antioxidant vitamins thought to have beneficial effects, some experts recommend supplementing vitamins C and E. More is not always better, however; larger doses of vitamins or minerals can be very harmful.

Hypervitaminosis (hī-per-vī-ta-mi-NŌ-sis; *hyper-* = too much or above) refers to dietary intake of a vitamin that exceeds the ability of the body to utilize, store, or excrete the vitamin. Since water-soluble vitamins are not stored in the body, few of them cause any problems related to hypervitaminosis. However, because lipid-soluble vitamins are stored in the body, excessive consumption may cause problems. For example, excess intake of vitamin A can cause drowsiness, general weakness, irritability, headache, vomiting, dry and peeling skin, partial hair loss, joint pain, liver and spleen enlargement, coma, and even death. Excessive intake of vitamin D may result in loss of appetite, nausea, vomiting, excessive thirst, general weakness, irritability, hypertension, and kidney damage and malfunction. **Hypovitaminosis** (*hypo-* = too little or below), or vitamin deficiency, is discussed in Table 25.6 for the various vitamins. •

TABLE 25.5

Minerals Vital to the Body

MINERAL	COMMENTS	IMPORTANCE
Calcium	Most abundant mineral in body. Appears in combination with phosphates. About 99% stored in bone and teeth. Blood Ca^{2+} level controlled by parathyroid hormone (PTH). Calcitriol promotes absorption of dietary calcium. Excess excreted in feces and urine. Sources: milk, egg yolk, shellfish, leafy green vegetables.	Formation of bones and teeth, blood clotting, normal muscle and nerve activity, endocytosis and exocytosis, cellular motility, chromosome movement during cell division, glycogen metabolism, release of neurotransmitters and hormones.
Phosphorus	About 80% found in bones and teeth as phosphate salts. Blood phosphate level controlled by parathyroid hormone (PTH). Excess excreted in urine; small amount eliminated in feces. Sources: dairy products, meat, fish, poultry, nuts.	Formation of bones and teeth. Phosphates ($H_2PO_4^-$, HPO_4^{2-}, and PO_4^{3-}) constitute a major buffer system of blood. Role in muscle contraction and nerve activity. Component of many enzymes. Involved in energy transfer (ATP). Component of DNA and RNA.
Potassium	Major cation (K^+) in intracellular fluid. Excess excreted in urine. Present in most foods (meats, fish, poultry, fruits, nuts).	Needed for generation and conduction of action potentials in neurons and muscle fibers.
Sulfur	Component of many proteins (such as insulin and chondroitin sulfate), electron carriers in electron transport chain, and some vitamins (thiamine and biotin). Excreted in urine. Sources: beef, liver, lamb, fish, poultry, eggs, cheese, beans.	As component of hormones and vitamins, regulates various body activities. Needed for ATP production by electron transport chain.
Sodium	Most abundant cation (Na^+) in extracellular fluids; some found in bones. Excreted in urine and perspiration. Normal intake of NaCl (table salt) supplies more than required amounts.	Strongly affects distribution of water through osmosis. Part of bicarbonate buffer system. Functions in nerve and muscle action potential conduction.
Chloride	Major anion (Cl^-) in extracellular fluid. Excess excreted in urine. Sources: table salt (NaCl), soy sauce, processed foods.	Role in acid–base balance of blood, water balance, and formation of HCl in stomach.
Magnesium	Important cation (Mg^{2+}) in intracellular fluid. Excreted in urine and feces. Widespread in various foods, such as green leafy vegetables, seafood, and whole-grain cereals.	Required for normal functioning of muscle and nervous tissue. Participates in bone formation. Constituent of many coenzymes.
Iron	About 66% found in hemoglobin of blood. Normal losses of iron occur by shedding of hair, epithelial cells, and mucosal cells, and in sweat, urine, feces, bile, and blood lost during menstruation. Sources: meat, liver, shellfish, egg yolk, beans, legumes, dried fruits, nuts, cereals.	As component of hemoglobin, reversibly binds O_2. Component of cytochromes involved in electron transport chain.
Iodide	Essential component of thyroid hormones. Excreted in urine. Sources: seafood, iodized salt, vegetables grown in iodine-rich soils.	Required by thyroid gland to synthesize thyroid hormones, which regulate metabolic rate.
Manganese	Some stored in liver and spleen. Most excreted in feces.	Activates several enzymes. Needed for hemoglobin synthesis, urea formation, growth, reproduction, lactation, bone formation, and possibly production and release of insulin, and inhibition of cell damage.
Copper	Some stored in liver and spleen. Most excreted in feces. Sources: eggs, whole-wheat flour, beans, beets, liver, fish, spinach, asparagus.	Required with iron for synthesis of hemoglobin. Component of coenzymes in electron transport chain and enzyme necessary for melanin formation.
Cobalt	Constituent of vitamin B_{12}.	As part of vitamin B_{12}, required for erythropoiesis.
Zinc	Important component of certain enzymes. Widespread in many foods, especially meats.	As component of carbonic anhydrase, important in carbon dioxide metabolism. Necessary for normal growth and wound healing, normal taste sensations and appetite, and normal sperm counts in males. As component of peptidases, involved in protein digestion.
Fluoride	Components of bones, teeth, other tissues.	Appears to improve tooth structure and inhibit tooth decay.
Selenium	Important component of certain enzymes. Sources: seafood, meat, chicken, tomatoes, egg yolk, milk, mushrooms, garlic, cereal grains grown in selenium-rich soil.	Needed for synthesis of thyroid hormones, sperm motility, and proper functioning of immune system. Also functions as antioxidant. Prevents chromosome breakage and may play role in preventing certain birth defects, miscarriage, prostate cancer, and coronary artery disease.
Chromium	Found in high concentrations in brewer's yeast. Also found in wine and some brands of beer.	Needed for normal activity of insulin in carbohydrate and lipid metabolism.

TABLE 25.6

The Principal Vitamins

VITAMIN	COMMENT AND SOURCE	FUNCTIONS	DEFICIENCY SYMPTOMS AND DISORDERS
Fat-soluble	All require bile salts and some dietary lipids for adequate absorption.		
A	Formed from provitamin beta-carotene (and other provitamins) in GI tract. Stored in liver. Sources of carotene and other provitamins: orange, yellow, and green vegetables. Sources of vitamin A: liver, milk.	Maintains general health and vigor of epithelial cells. Beta-carotene acts as antioxidant to inactivate free radicals. Essential for formation of light-sensitive pigments in photoreceptors of retina. Aids in growth of bones and teeth by helping to regulate activity of osteoblasts and osteoclasts.	Deficiency results in atrophy and keratinization of epithelium, leading to dry skin and hair; increased incidence of ear, sinus, respiratory, urinary, and digestive system infections; inability to gain weight; drying of cornea; and skin sores. **Night blindness** (decreased ability for dark adaptation). Slow and faulty development of bones and teeth.
D	Sunlight converts 7-dehydrocholesterol in skin to cholecalciferol (vitamin D_3). A liver enzyme then converts cholecalciferol to 25-hydroxycholecalciferol. A second enzyme in kidneys converts 25-hydroxycholecalciferol to calcitriol (1,25-dihydroxycalciferol), the active form of vitamin D. Most excreted in bile. Dietary sources: fish-liver oils, egg yolk, fortified milk.	Essential for absorption of calcium and phosphorus from GI tract. Works with parathyroid hormone (PTH) to maintain Ca^{2+} homeostasis.	Defective utilization of calcium by bones leads to **rickets** in children and **osteomalacia** in adults. Possible loss of muscle tone.
E (tocopherols)	Stored in liver, adipose tissue, and muscles. Sources: fresh nuts and wheat germ, seed oils, green leafy vegetables.	Inhibits catabolism of certain fatty acids that help form cell structures, especially membranes. Involved in formation of DNA, RNA, and red blood cells. May promote wound healing, contribute to normal structure and functioning of nervous system, and prevent scarring. May help protect liver from toxic chemicals such as carbon tetrachloride. Acts as antioxidant to inactivate free radicals.	May cause oxidation of monounsaturated fats, resulting in abnormal structure and function of mitochondria, lysosomes, and plasma membranes. Possible consequence is hemolytic anemia.
K	Produced by intestinal bacteria. Stored in liver and spleen. Dietary sources: spinach, cauliflower, cabbage, liver.	Coenzyme essential for synthesis of several clotting factors by liver, including prothrombin.	Delayed clotting time results in excessive bleeding.
Water-soluble	Dissolved in body fluids. Most not stored in body. Excess intake eliminated in urine.		
B_1 (thiamine)	Rapidly destroyed by heat. Sources: whole-grain products, eggs, pork, nuts, liver, yeast.	Acts as coenzyme for many different enzymes that break carbon-to-carbon bonds and are involved in carbohydrate metabolism of pyruvic acid to CO_2 and H_2O. Essential for synthesis of neurotransmitter acetylcholine.	Improper carbohydrate metabolism leads to buildup of pyruvic and lactic acids and insufficient production of ATP for muscle and nerve cells. Deficiency leads to (1) **beriberi,** partial paralysis of smooth muscle of GI tract, causing digestive disturbances; skeletal muscle paralysis; and atrophy of limbs; (2) **polyneuritis,** due to degeneration of myelin sheaths; impaired reflexes, impaired sense of touch, stunted growth in children, and poor appetite.
B_2 (riboflavin)	Small amounts supplied by bacteria of GI tract. Dietary sources: yeast, liver, beef, veal, lamb, eggs, whole-grain products, asparagus, peas, beets, peanuts.	Component of certain coenzymes (for example, FAD and FMN) in carbohydrate and protein metabolism, especially in cells of eye, integument, mucosa of intestine, and blood.	Deficiency may lead to improper utilization of oxygen, resulting in blurred vision, cataracts, and corneal ulcerations. Also dermatitis and cracking of skin, lesions of intestinal mucosa, and one type of anemia.

TABLE 25.6 CONTINUED

The Principal Vitamins

VITAMIN	COMMENT AND SOURCE	FUNCTIONS	DEFICIENCY SYMPTOMS AND DISORDERS
Water-soluble (continued)			
Niacin (nicotinamide)	Derived from amino acid tryptophan. Sources: yeast, meats, liver, fish, whole-grain products, peas, beans, nuts.	Essential component of NAD and NADP, coenzymes in oxidation–reduction reactions. In lipid metabolism, inhibits production of cholesterol and assists in triglyceride breakdown.	Principal deficiency is **pellagra,** characterized by dermatitis, diarrhea, and psychological disturbances.
B_6 (pyridoxine)	Synthesized by bacteria of GI tract. Stored in liver, muscle, and brain. Other sources: salmon, yeast, tomatoes, yellow corn, spinach, whole grain products, liver, yogurt.	Essential coenzyme for normal amino acid metabolism. Assists production of circulating antibodies. May function as coenzyme in triglyceride metabolism.	Most common deficiency symptom is dermatitis of eyes, nose, and mouth. Other symptoms are retarded growth and nausea.
B_{12} (cyanocobalamin)	Only B vitamin not found in vegetables; only vitamin containing cobalt. Absorption from GI tract depends on intrinsic factor secreted by gastric mucosa. Sources: liver, kidney, milk, eggs, cheese, meat.	Coenzyme necessary for red blood cell formation, formation of amino acid methionine, entrance of some amino acids into Krebs cycle, and manufacture of choline (used to synthesize acetylcholine).	Pernicious anemia, neuropsychiatric abnormalities (ataxia, memory loss, weakness, personality and mood changes, and abnormal sensations), and impaired activity of osteoblasts.
Pantothenic acid	Some produced by bacteria of GI tract. Stored primarily in liver and kidneys. Other sources: kidneys, liver, yeast, green vegetables, cereal.	Constituent of coenzyme A, which is essential for transfer of acetyl group from pyruvic acid into Krebs cycle, conversion of lipids and amino acids into glucose, and synthesis of cholesterol and steroid hormones.	Fatigue, muscle spasms, insufficient production of adrenal steroid hormones, vomiting, and insomnia.
Folic acid (folate, folacin)	Synthesized by bacteria of GI tract. Dietary sources: green leafy vegetables, broccoli, asparagus, breads, dried beans, citrus fruits.	Component of enzyme systems synthesizing nitrogenous bases of DNA and RNA. Essential for normal production of red and white blood cells.	Production of abnormally large red blood cells (macrocytic anemia). Higher risk of neural tube defects in babies born to folate-deficient mothers.
Biotin	Synthesized by bacteria of GI tract. Dietary sources include yeast, liver, egg yolk, kidneys.	Essential coenzyme for conversion of pyruvic acid to oxaloacetic acid and synthesis of fatty acids and purines.	Mental depression, muscular pain, dermatitis, fatigue, and nausea.
C (ascorbic acid)	Rapidly destroyed by heat. Some stored in glandular tissue and plasma. Sources: citrus fruits, tomatoes, green vegetables.	Promotes protein synthesis, including laying down of collagen in formation of connective tissue. As coenzyme, may combine with poisons, rendering them harmless until excreted. Works with antibodies, promotes wound healing, and functions as an antioxidant.	Scurvy; anemia; many symptoms related to poor collagen formation, including tender swollen gums, loosening of teeth (alveolar processes also deteriorate), poor wound healing, bleeding (vessel walls are fragile because of connective tissue degeneration), and retardation of growth.

✔ CHECKPOINT

30. What is a nutrient?
31. Describe the food guide pyramid and give examples of foods from each food group.
32. What is a mineral? Briefly describe the functions of the following minerals: calcium, phosphorus, potassium, sulfur, sodium, chloride, magnesium, iron, iodine, copper, zinc, fluoride, manganese, cobalt, chromium, and selenium.
33. Define vitamin. Explain how we obtain vitamins. Distinguish between a fat-soluble vitamin and a water-soluble vitamin.
34. For each of the following vitamins, indicate its principal function and the effect(s) of deficiency: A, D, E, K, B_1, B_2, niacin, B_6, B_{12}, pantothenic acid, folic acid, biotin, and C.

DISORDERS: HOMEOSTATIC IMBALANCES

Anorexia Nervosa

Anorexia nervosa is a chronic disorder characterized by self-induced weight loss, negative perception of body image, and physiological changes that result from nutritional depletion. Patients with anorexia nervosa have a fixation on weight control and often insist on having a bowel movement every day despite inadequate food intake. They often abuse laxatives, which worsens the fluid and electrolyte imbalances and nutrient deficiencies. The disorder is found predominantly in young, single females, and it may be inherited. Abnormal patterns of menstruation, amenorrhea (absence of menstruation), and a lowered basal metabolic rate reflect the depressant effects of starvation. Individuals may become emaciated and may ultimately die of starvation or one of its complications. Also associated with the disorder are osteoporosis, depression, and brain abnormalities coupled with impaired mental performance. Treatment consists of psychotherapy and dietary regulation.

Fever

A **fever** is an elevation of core temperature caused by a resetting of the hypothalamic thermostat. The most common causes of fever are viral or bacterial infections and bacterial toxins; other causes are ovulation, excessive secretion of thyroid hormones, tumors, and reactions to vaccines. When phagocytes ingest certain bacteria, they are stimulated to secrete a **pyrogen** (PĪ-rō-gen; *pyro-* = fire; *-gen* = produce), a fever-producing substance. One pyrogen is interleukin-1. It circulates to the hypothalamus and induces neurons of the preoptic area to secrete prostaglandins. Some prostaglandins can reset the hypothalamic thermostat at a higher temperature, and temperature-regulating reflex mechanisms then act to bring the core body temperature up to this new setting. *Antipyretics* are agents that relieve or reduce fever. Examples include aspirin, acetaminophen (Tylenol), and ibuprofen (Advil), all of which reduce fever by inhibiting synthesis of certain prostaglandins.

Suppose that due to production of pyrogens the thermostat is reset at 39°C (103°F). Now the heat-promoting mechanisms (vasoconstriction, increased metabolism, shivering) are operating at full force. Thus, even though core temperature is climbing higher than normal—say, 38°C (101°F)—the skin remains cold, and shivering occurs. This condition, called a **chill,** is a definite sign that core temperature is rising. After several hours, core temperature reaches the setting of the thermostat, and the chills disappear. But now the body will continue to regulate temperature at 39°C (103°F). When the pyrogens disappear, the thermostat is reset at normal—37.0°C (98.6°F). Because core temperature is high in the beginning, the heat-losing mechanisms (vasodilation and sweating) go into operation to decrease core temperature. The skin becomes warm, and the person begins to sweat. This phase of the fever is called the **crisis,** and it indicates that core temperature is falling.

Although death results if core temperature rises above 44–46°C (112–114°F), up to a point, fever is beneficial. For example, a higher temperature intensifies the effects of interferons and the phagocytic activities of macrophages while hindering replication of some pathogens. Because fever increases heart rate, infection-fighting white blood cells are delivered to sites of infection more rapidly. In addition, antibody production and T cell proliferation increase. Moreover, heat speeds up the rate of chemical reactions, which may help body cells repair themselves more quickly.

Obesity

Obesity is body weight more than 20% above a desirable standard due to an excessive accumulation of adipose tissue. About one-third of the adult population in the United States is obese. (An athlete may be *overweight* due to higher-than-normal amounts of muscle tissue without being obese.) Even moderate obesity is hazardous to health; it is a risk factor in cardiovascular disease, hypertension, pulmonary disease, non-insulin-dependent diabetes mellitus, arthritis, certain cancers (breast, uterus, and colon), varicose veins, and gallbladder disease.

In a few cases, obesity may result from trauma of or tumors in the food-regulating centers in the hypothalamus. In most cases of obesity, no specific cause can be identified. Contributing factors include genetic factors, eating habits taught early in life, overeating to relieve tension, and social customs. Studies indicate that some obese people burn fewer calories during digestion and absorption of a meal, a smaller food-induced thermogenesis effect. Additionally, obese people who lose weight require about 15% fewer calories to maintain normal body weight than do people who have never been obese. Interestingly, people who gain weight easily when deliberately fed excess calories exhibit less NEAT (nonexercise activity thermogenesis, such as occurs with fidgeting) than people who resist weight gains in the face of excess calories. Although leptin suppresses appetite and produces satiety in experimental animals, it is not deficient in most obese people.

Most surplus calories in the diet are converted to triglycerides and stored in adipose cells. Initially, the adipocytes increase in size, but at a maximal size, they divide. As a result, proliferation of adipocytes occurs in extreme obesity. The enzyme endothelial lipoprotein lipase regulates triglyceride storage. The enzyme is very active in abdominal fat but less active in hip fat. Accumulation of fat in the abdomen is associated with higher blood cholesterol level and other cardiac risk factors because adipose cells in this area appear to be more metabolically active.

Treatment of obesity is difficult because most people who are successful at losing weight gain it back within 2 years. Yet, even modest weight loss is associated with health benefits. Treatments for obesity include behavior modification programs, very-low-calorie diets, drugs, and surgery. Behavior modification programs, offered at many hospitals, strive to alter eating behaviors and increase exercise activity. The nutrition program includes a "heart-healthy" diet that includes abundant vegetables but is low in fats, especially saturated fats. A typical exercise program suggests walking for 30 minutes a day, five to seven times a week. Regular exercise enhances both weight loss and weight-loss maintenance. Very-low-calorie (VLC) diets include 400 to 800 kcal/day in a commercially made liquid mixture. The VLC diet is usually prescribed for 12 weeks, under close medical supervision. Two drugs are available to treat obesity. Sibutramine is an appetite suppressant that works by inhibiting reuptake of serotonin and norepinephrine in brain areas that govern eating behavior. Orlistat works by inhibiting the lipases released into the lumen of the GI tract. With less lipase activity, fewer dietary triglycerides are absorbed. For those with extreme obesity who have not responded to other treatments, a surgical procedure may be considered. The two operations most commonly performed—gastric bypass and gastroplasty—both greatly reduce the stomach size so that it can hold just a tiny quantity of food.

MEDICAL TERMINOLOGY

Bulimia (*bu-* = ox; *-limia* = hunger) or *binge–purge syndrome* A disorder that typically affects young, single, middle-class, white females, characterized by overeating at least twice a week followed by purging by self-induced vomiting, strict dieting or fasting, vigorous exercise, or use of laxatives or diuretics; it occurs in response to fears of being overweight or to stress, depression, and physiological disorders such as hypothalamic tumors.

Heat cramps Cramps that result from profuse sweating. The salt lost in sweat causes painful contractions of muscles; such cramps tend to occur in muscles used while working but do not appear until the person relaxes once the work is done. Drinking salted liquids usually leads to rapid improvement.

Heat exhaustion (heat prostration) A condition in which the core temperature is generally normal, or a little below, and the skin is cool and moist due to profuse perspiration. Heat exhaustion is usually characterized by loss of fluid and electrolytes, especially salt (NaCl). The salt loss results in muscle cramps, dizziness, vomiting, and fainting; fluid loss may cause low blood pressure. Complete rest, rehydration, and electrolyte replacement are recommended.

Heatstroke (sunstroke) A severe and often fatal disorder caused by exposure to high temperatures, especially when the relative humidity is high, which makes it difficult for the body to lose heat. Blood flow to the skin is decreased, perspiration is greatly reduced, and body temperature rises sharply because of failure of the hypothalamic thermostat. Body temperature may reach 43°C (110°F). Treatment, which must be undertaken immediately, consists of cooling the body by immersing the victim in cool water and by administering fluids and electrolytes.

Kwashiorkor (kwash-ē-OR-kor) A disorder in which protein intake is deficient despite normal or nearly normal caloric intake, characterized by edema of the abdomen, enlarged liver, decreased blood pressure, low pulse rate, lower-than-normal body temperature, and sometimes mental retardation. Because the main protein in corn (zein) lacks two essential amino acids, which are needed for growth and tissue repair, many African children whose diet consists largely of cornmeal develop kwashiorkor.

Malnutrition (*mal-* = bad) An imbalance of total caloric intake or intake of specific nutrients, which can be either inadequate or excessive.

Marasmus (mar-AZ-mus) A type of protein–calorie undernutrition that results from inadequate intake of both protein and calories. Its characteristics include retarded growth, low weight, muscle wasting, emaciation, dry skin, and thin, dry, dull hair.

CHAPTER REVIEW AND RESOURCE SUMMARY

Review	Resource
Introduction	Anatomy Overview - Role of Nutrients
1. Our only source of energy for performing biological work is the food we eat. Food also provides essential substances that we cannot synthesize.	
2. Most food molecules absorbed by the gastrointestinal tract are used to supply energy for life processes, serve as building blocks during synthesis of complex molecules, or are stored for future use.	
25.1 Metabolic Reactions	Animation - Introduction to Metabolism
1. Metabolism refers to all chemical reactions of the body and is of two types: catabolism and anabolism.	Animation - Regulation of Metabolism
2. Catabolism is the term for reactions that break down complex organic compounds into simple ones. Overall, catabolic reactions are exergonic; they produce more energy than they consume.	Animation - Carbohydrate Metabolism
3. Chemical reactions that combine simple molecules into more complex ones that form the body's structural and functional components are collectively known as anabolism. Overall, anabolic reactions are endergonic; they consume more energy than they produce.	Animation - Types of Reactions and Equilibrium
4. The coupling of anabolism and catabolism occurs via ATP.	
25.2 Energy Transfer	Animation - Types of Reactions and Equilibrium
1. Oxidation is the removal of electrons from a substance; reduction is the addition of electrons to a substance.	Animation - Enzyme Functions and ATP
2. Two coenzymes that carry hydrogen atoms during coupled oxidation–reduction reactions are nicotinamide adenine dinucleotide (NAD) and flavin adenine dinucleotide (FAD).	Exercise - Predict ATP Production
3. ATP can be generated via substrate-level phosphorylation, oxidative phosphorylation, and photophosphorylation.	
25.3 Carbohydrate Metabolism	Anatomy Overview - Role of Nutrients
1. During digestion, polysaccharides and disaccharides are hydrolyzed into the monosaccharides glucose (about 80%), fructose, and galactose; the latter two are then converted to glucose. Some glucose is oxidized by cells to provide ATP. Glucose also can be used to synthesize amino acids, glycogen, and triglycerides.	Animation - Cell Respiration Animation - Carbohydrate Metabolism Figure 25.2 - Overview of Cellular Respiration

Review

2. Glucose moves into most body cells via facilitated diffusion through glucose transporters (GluT) and becomes phosphorylated to glucose 6-phosphate. In muscle cells, this process is stimulated by insulin. Glucose entry into neurons and hepatocytes is always "turned on."

3. Cellular respiration, the complete oxidation of glucose to CO_2 and H_2O, involves glycolysis, the Krebs cycle, and the electron transport chain.

4. Glycolysis is the breakdown of glucose into two molecules of pyruvic acid; there is a net production of two molecules of ATP.

5. When oxygen is in short supply, pyruvic acid is reduced to lactic acid; under aerobic conditions, pyruvic acid enters the Krebs cycle. Pyruvic acid is prepared for entrance into the Krebs cycle by conversion to a 2-carbon acetyl group followed by the addition of coenzyme A to form acetyl coenzyme A. The Krebs cycle involves decarboxylations, oxidations, and reductions of various organic acids. Each molecule of pyruvic acid that is converted to acetyl coenzyme A and then enters the Krebs cycle produces three molecules of CO_2, four molecules of NADH and four H^+, one molecule of $FADH_2$, and one molecule of ATP. The energy originally stored in glucose and then in pyruvic acid is transferred primarily to the reduced coenzymes NADH and $FADH_2$.

6. The electron transport chain involves a series of oxidation–reduction reactions in which the energy in NADH and $FADH_2$ is liberated and transferred to ATP. The electron carriers include FMN, cytochromes, iron–sulfur centers, copper atoms, and coenzyme Q. The electron transport chain yields a maximum of 32 or 34 molecules of ATP and six molecules of H_2O.

7. Table 25.1 summarizes the ATP yield during cellular respiration. The complete oxidation of glucose can be represented as follows:

$$C_6H_{12}O_6 + 6\ O_2 + 36 \text{ or } 38 \text{ ADPs} + 36 \text{ or } 38 \text{ \textcircled{P}} \longrightarrow 6\ CO_2 + 6\ H_2O + 36 \text{ or } 38 \text{ ATPs}$$

8. The conversion of glucose to glycogen for storage in the liver and skeletal muscle is called glycogenesis. It is stimulated by insulin.

9. The conversion of glycogen to glucose is called glycogenolysis. It occurs between meals and is stimulated by glucagon and epinephrine.

10. Gluconeogenesis is the conversion of noncarbohydrate molecules into glucose. It is stimulated by cortisol and glucagon.

25.4 Lipid Metabolism

1. Lipoproteins transport lipids in the bloodstream. Types of lipoproteins include chylomicrons, which carry dietary lipids to adipose tissue; very-low-density lipoproteins (VLDLs), which carry triglycerides from the liver to adipose tissue; low-density lipoproteins (LDLs), which deliver cholesterol to body cells; and high-density lipoproteins (HDLs), which remove excess cholesterol from body cells and transport it to the liver for elimination.

2. Cholesterol in the blood comes from two sources: from food and from synthesis by the liver.

3. Lipids may be oxidized to produce ATP or stored as triglycerides in adipose tissue, mostly in the subcutaneous layer.

4. A few lipids are used as structural molecules or to synthesize essential molecules.

5. Adipose tissue contains lipases that catalyze the deposition of triglycerides from chylomicrons and hydrolyze triglycerides into fatty acids and glycerol.

6. In lipolysis, triglycerides are split into fatty acids and glycerol and released from adipose tissue under the influence of epinephrine, norepinephrine, cortisol, thyroid hormones, and insulinlike growth factors.

7. Glycerol can be converted into glucose by conversion into glyceraldehyde 3-phosphate.

8. In beta oxidation of fatty acids, carbon atoms are removed in pairs from fatty acid chains; the resulting molecules of acetyl coenzyme A enter the Krebs cycle.

9. The conversion of glucose or amino acids into lipids is called lipogenesis; it is stimulated by insulin.

25.5 Protein Metabolism

1. During digestion, proteins are hydrolyzed into amino acids, which enter the liver via the hepatic portal vein.

2. Amino acids, under the influence of insulinlike growth factors and insulin, enter body cells via active transport.

3. Inside cells, amino acids are synthesized into proteins that function as enzymes, hormones, structural elements, and so forth; stored as fat or glycogen; or used for energy.

Review	Resource

Review

4. Before amino acids can be catabolized, they must be deaminated and converted to substances that can enter the Krebs cycle.
5. Amino acids may also be converted into glucose, fatty acids, and ketone bodies.
6. Protein synthesis is stimulated by insulinlike growth factors, thyroid hormones, insulin, estrogen, and testosterone.
7. Table 25.2 summarizes carbohydrate, lipid, and protein metabolism.

25.6 Key Molecules at Metabolic Crossroads

1. Three molecules play a key role in metabolism: glucose 6-phosphate, pyruvic acid, and acetyl coenzyme A.
2. Glucose 6-phosphate may be converted to glucose, glycogen, ribose 5-phosphate, and pyruvic acid.
3. When ATP is low and oxygen is plentiful, pyruvic acid is converted to acetyl coenzyme A; when oxygen supply is low, pyruvic acid is converted to lactic acid. Carbohydrate and protein metabolism are linked by pyruvic acid.
4. Acetyl coenzyme A is the molecule that enters the Krebs cycle; it is also used to synthesize fatty acids, ketone bodies, and cholesterol.

Resource:
Animation - Cell Respiration
Figure 25.16 - Roles of Key Molecules in Metabolic Pathways

25.7 Metabolic Adaptations

1. During the absorptive state, ingested nutrients enter the blood and lymph from the GI tract.
2. During the absorptive state, blood glucose is oxidized to form ATP, and glucose transported to the liver is converted to glycogen or triglycerides. Most triglycerides are stored in adipose tissue. Amino acids in hepatocytes are converted to carbohydrates, fats, and proteins. Table 25.3 summarizes the hormonal regulation of metabolism during the absorptive state.
3. During the postabsorptive state, absorption is complete and the ATP needs of the body are satisfied by nutrients already present in the body. The major task is to maintain normal blood glucose level by converting glycogen in the liver and skeletal muscle into glucose, converting glycerol into glucose, and converting amino acids into glucose. Fatty acids, ketone bodies, and amino acids are oxidized to supply ATP. Table 25.4 summarizes the hormonal regulation of metabolism during the postabsorptive state.
4. Fasting is going without food for a few days; starvation implies weeks or months of inadequate food intake. During fasting and starvation, fatty acids and ketone bodies are increasingly utilized for ATP production.

Resource:
Animation - Maintaining Normal Blood Glucose Levels
Exercise - Alternative Metabolic Pathways

25.8 Heat and Energy Balance

1. Measurement of the metabolic rate under basal conditions is called the basal metabolic rate (BMR).
2. A calorie (cal) is the amount of energy required to raise the temperature of 1 g of water 1°C.
3. Because the calorie is a relatively small unit, the kilocalorie (kcal) or Calorie (Cal) is often used to measure the body's metabolic rate and to express the energy content of foods; a kilocalorie equals 1000 Calories.
4. Normal core temperature is maintained by a delicate balance between heat-producing and heat-losing mechanisms.
5. Exercise, hormones, the nervous system, body temperature, ingestion of food, age, gender, climate, sleep, and malnutrition affect metabolic rate.
6. Mechanisms of heat transfer include conduction, convection, radiation, and evaporation. Conduction is the transfer of heat between two substances or objects in contact with each other. Convection is the transfer of heat by a liquid or gas between areas of different temperatures. Radiation is the transfer of heat from a warmer object to a cooler object without physical contact. Evaporation is the conversion of a liquid to a vapor; in the process, heat is lost.
7. The hypothalamic thermostat is in the preoptic area.
8. Responses that produce, conserve, or retain heat when core temperature falls are vasoconstriction; release of epinephrine, norepinephrine, and thyroid hormones; and shivering.
9. Responses that increase heat loss when core temperature increases include vasodilation, decreased metabolic rate, and evaporation of perspiration.
10. Two nuclei in the hypothalamus that help regulate food intake are the arcuate and paraventricular nuclei. The hormone leptin, released by adipocytes, inhibits release of neuropeptide Y from the arcuate nucleus and thereby decreases food intake. Melanocortin also decreases food intake.

Resource:
Animation - Metabolic Rate, Heat, and Temperature Regulation
Animation - Regulation of Metabolism
Concepts and Connections - Hormonal Control of Metabolic Activities
Concepts and Connections - Nutrient Utilization Pathways

Review	Resource
25.9 Nutrition	BBC Video - Study on Vitamin Supplements
1. Nutrients include water, carbohydrates, lipids, proteins, minerals, and vitamins.	Animation - Enzyme Functions and ATP Part 1
2. Nutrition experts suggest dietary calories be 50–60% from carbohydrates, 30% or less from fats, and 12–15% from proteins.	
3. The MyPyramid guide represents a personalized approach to making healthy food choices and maintaining regular physical activity.	
4. Minerals known to perform essential functions include calcium, phosphorus, potassium, sulfur, sodium, chloride, magnesium, iron, iodide, manganese, copper, cobalt, zinc, fluoride, selenium, and chromium. Their functions are summarized in Table 25.5.	
5. Vitamins are organic nutrients that maintain growth and normal metabolism. Many function in enzyme systems.	
6. Fat-soluble vitamins are absorbed with fats and include vitamins A, D, E, and K; water-soluble vitamins include the B vitamins and vitamin C.	
7. The functions and deficiency disorders of the principal vitamins are summarized in Table 25.6.	

SELF-QUIZ QUESTIONS

Fill in the blanks in the following statements.

1. The thermostat and food intake regulating center of the body is in the _____ of the brain.

2. The three key molecules of metabolism are _____, _____, and _____.

Indicate whether the following statements are true or false.

3. Foods that we eat are used to supply energy for life processes, serve as building blocks for synthesis reactions, or are stored for future use.

4. Vitamins A, B, D, and K are fat-soluble vitamins.

Choose the one best answer to the following questions.

5. NAD^+ and FAD (1) are both derivatives of B vitamins, (2) are used to carry hydrogen atoms released during oxidation reactions, (3) become NADH and $FADH_2$ in their reduced forms, (4) act as coenzymes in the Krebs cycle, (5) are the final electron acceptors in the electron transport chain.
   (a) 1, 2, 3, 4, and 5   (b) 2, 3, and 4   (c) 2 and 4
   (d) 1, 2, and 3   (e) 1, 2, 3, and 4

6. During glycolysis, (1) a 6-carbon glucose is split into two 3-carbon pyruvic acids, (2) there is a net gain of two ATP molecules, (3) two NADH molecules are oxidized, (4) moderately high levels of oxygen are needed, (5) the activity of phosphofructokinase determines the rate of the chemical reactions.
   (a) 1, 2, and 3   (b) 1 and 2   (c) 1, 2, and 5
   (d) 2, 3, 4, and 5   (e) 1, 2, 3, 4, and 5

7. If glucose is not needed for immediate ATP production, it can be used for (1) vitamin synthesis, (2) amino acid synthesis, (3) gluconeogenesis, (4) glycogenesis, (5) lipogenesis.
   (a) 1, 3, and 5   (b) 2, 4, and 5   (c) 2, 3, 4, and 5
   (d) 1, 2, and 3   (e) 2 and 5

8. Which of the following is the *correct* sequence for the oxidation of glucose to produce ATP?
   (a) electron transport chain, Krebs cycle, glycolysis, formation of acetyl CoA

   (b) Krebs cycle, formation of acetyl CoA, electron transport chain, glycolysis
   (c) glycolysis, electron transport chain, Krebs cycle, formation of acetyl CoA
   (d) glycolysis, formation of acetyl CoA, Krebs cycle, electron transport chain
   (e) formation of acetyl CoA, Krebs cycle, glycolysis, electron transport chain.

9. Which of the following would you *not* expect to experience during fasting or starvation?
   (a) decrease in plasma fatty acid levels
   (b) increase in ketone body formation
   (c) lipolysis
   (d) increased use of ketones for ATP production in the brain
   (e) depletion of glycogen

10. If core body temperature rises above normal, which of the following would occur to cool the body? (1) dilation of vessels in the skin, (2) increased radiation and conduction of heat to the environment, (3) increased metabolic rate, (4) evaporation of perspiration, (5) increased secretion of thyroid hormones.
    (a) 3, 4, and 5   (b) 1, 2, and 4   (c) 1, 2, and 5
    (d) 1, 2, 3, 4 and 5   (e) 1, 2, 4, and 5

11. In which of the following situations would the metabolic rate increase? (1) sleep, (2) after ingesting food, (3) increased secretion of thyroid hormones, (4) parasympathetic nervous system stimulation, (5) fever.
    (a) 3 and 4   (b) 1, 3, and 5   (c) 2 and 3
    (d) 2, 3, and 4   (e) 2, 3, and 5

12. Which of the following are absorptive state reactions? (1) aerobic cellular respiration, (2) glycogenesis, (3) glycogenolysis, (4) gluconeogenesis using lactic acid, (5) lipolysis.
    (a) 1 and 2   (b) 2 and 3   (c) 3 and 4
    (d) 4 and 5   (e) 1 and 5

13. Match the hormones with the reactions they regulate (answers may be used more than once; some reactions have more than one answer):

_____ (a) gluconeogenesis
_____ (b) glycogenesis
_____ (c) glycogenolysis
_____ (d) lipolysis
_____ (e) lipogenesis
_____ (f) protein catabolism
_____ (g) protein anabolism

(1) insulin
(2) cortisol
(3) glucagon
(4) thyroid hormones
(5) epinephrine
(6) insulinlike growth factors

14. Match the following:

_____ (a) deliver cholesterol to body cells for use in repair of membranes and synthesis of steroid hormones and bile salts
_____ (b) remove excess cholesterol from body cells and transport it to the liver for elimination
_____ (c) organic nutrients required in small amounts for growth and normal metabolism
_____ (d) the energy-transferring molecule of the body
_____ (e) nutrient molecules that can be oxidized to produce ATP or stored in adipose tissue
_____ (f) transport endogenous lipids to adipocytes for storage
_____ (g) the body's preferred source for synthesizing ATP
_____ (h) composed of amino acids and are the primary regulatory molecules in the body
_____ (i) acetoacetic acid, beta-hydroxybutyric acid, and acetone
_____ (j) hormone secreted by adipocytes that acts to decrease total body-fat mass
_____ (k) neurotransmitter that stimulates food intake
_____ (l) inorganic substances that perform many vital functions in the body
_____ (m) carriers of electrons in the electron transport chain

(1) leptin
(2) minerals
(3) glucose
(4) lipids
(5) proteins
(6) neuropeptide Y
(7) cytochromes
(8) ketone bodies
(9) low-density lipoproteins
(10) ATP
(11) vitamins
(12) high-density lipoproteins
(13) very-low-density lipoproteins

15. Match the following:

_____ (a) the mechanism of ATP generation that links chemical reactions with pumping of hydrogen ions
_____ (b) the removal of electrons from an atom or molecule resulting in a decrease in potential energy
_____ (c) the transfer of an amino group from an amino acid to a substance such as pyruvic acid
_____ (d) the formation of glucose from noncarbohydrate sources
_____ (e) refers to all the chemical reactions in the body
_____ (f) the oxidation of glucose to produce ATP
_____ (g) the splitting of a triglyceride into glycerol and fatty acids
_____ (h) the synthesis of lipids
_____ (i) the addition of electrons to a molecule resulting in an increase in potential energy content of the molecule
_____ (j) the formation of ketone bodies
_____ (k) the breakdown of glycogen back to glucose
_____ (l) exergonic chemical reactions that break down complex organic molecules into simpler ones
_____ (m) overall rate at which metabolic reactions use energy
_____ (n) the breakdown of glucose into two molecules of pyruvic acid
_____ (o) removal of CO_2 from a molecule
_____ (p) endergonic chemical reactions that combine simple molecules and monomers to make more complex ones
_____ (q) the addition of a phosphate group to a molecule
_____ (r) the removal of the amino group from an amino acid
_____ (s) the cleavage of one pair of carbon atoms at a time from a fatty acid
_____ (t) the conversion of glucose into glycogen

(1) metabolism
(2) catabolism
(3) beta oxidation
(4) lipolysis
(5) phosphorylation
(6) glycolysis
(7) cellular respiration
(8) transamination
(9) anabolism
(10) lipogenesis
(11) glycogenolysis
(12) glycogenesis
(13) metabolic rate
(14) ketogenesis
(15) oxidation
(16) reduction
(17) chemiosmosis
(18) deamination
(19) gluconeogenesis
(20) decarboxylation

CRITICAL THINKING QUESTIONS

1. Jane Doe's deceased body was found at her dining room table. Her death was considered suspicious. Lab results from the medical investigation revealed cyanide in her blood. How did the cyanide cause her death?

2. During a recent physical, 55-year-old Glenn's blood serum lab results showed the following: total cholesterol = 300 mg/dL; LDL = 175 mg/dL; HDL = 20 mg/dL. Interpret these results for Glenn and indicate to him what changes, if any, he needs to make in his lifestyle. Why are these changes important?

3. Marissa has joined a weight loss program. As part of the program, she regularly submits a urine sample which is tested for ketones. She went to the clinic today, had her urine checked, and was confronted by the nurse who accused Marissa of "cheating" on her diet. How did the nurse know Marissa was not following her diet?

? ANSWERS TO FIGURE QUESTIONS

25.1 In pancreatic acinar cells, anabolism predominates because the primary activity is synthesis of complex molecules (digestive enzymes).

25.2 Glycolysis is also called anaerobic cellular respiration.

25.3 The reactions of glycolysis consume two molecules of ATP but generate four molecules of ATP, for a net gain of two.

25.4 Kinases are enzymes that phosphorylate (add phosphate to) their substrate.

25.5 Glycolysis occurs in the cytosol.

25.6 CO_2 is given off during the production of acetyl coenzyme A and during the Krebs cycle. It diffuses into the blood, is transported by the blood to the lungs, and is exhaled.

25.7 The production of reduced coenzymes is important in the Krebs cycle because they will subsequently yield ATP in the electron transport chain.

25.8 The energy source that powers the proton pumps is electrons provided by NADH + H^+.

25.9 The concentration of H^+ is highest in the space between the inner and outer mitochondrial membranes.

25.10 During the complete oxidation of one glucose molecule, six molecules of O_2 are used and six molecules of CO_2 are produced.

25.11 Skeletal muscle fibers can synthesize glycogen, but they cannot release glucose into the blood because they lack the enzyme phosphatase required to remove the phosphate group from glucose.

25.12 Hepatocytes can carry out gluconeogenesis and glycogenesis.

25.13 LDLs deliver cholesterol to body cells.

25.14 Hepatocytes and adipose cells carry out lipogenesis, beta oxidation, and lipolysis; hepatocytes carry out ketogenesis.

25.15 Before an amino acid can enter the Krebs cycle, an amino group must be removed via deamination.

25.16 Acetyl coenzyme A is the gateway into the Krebs cycle for molecules being oxidized to generate ATP.

25.17 Reactions of the absorptive state are mainly anabolic.

25.18 Processes that directly elevate blood glucose during the postabsorptive state include lipolysis (in adipocytes and hepatocytes), gluconeogenesis (in hepatocytes), and glycogenolysis (in hepatocytes).

25.19 Exercise, the sympathetic nervous system, hormones (epinephrine, norepinephrine, thyroxine, testosterone, human growth hormone), elevated body temperature, and ingestion of food increase metabolic rate, which results in an increase in body temperature.

25.20 The wider base of each band represents foods with little or no solid fats or added sugars.

26 THE URINARY SYSTEM

THE URINARY SYSTEM AND HOMEOSTASIS *The urinary system contributes to homeostasis by altering blood composition, pH, volume, and pressure; maintaining blood osmolarity; excreting wastes and foreign substances; and producing hormones.*

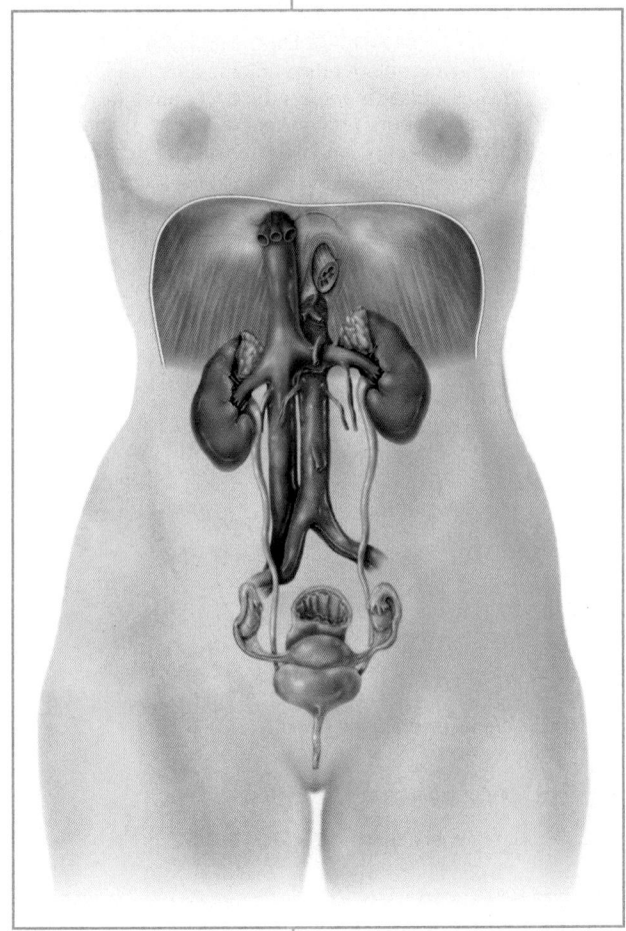

The **urinary system** consists of two kidneys, two ureters, one urinary bladder, and one urethra (Figure 26.1). After the kidneys filter blood plasma, they return most of the water and solutes to the bloodstream. The remaining water and solutes constitute **urine,** which passes through the ureters and is stored in the urinary bladder until it is excreted from the body through the urethra. **Nephrology** (nef-ROL-ō-jē; *nephr-* = kidney; *-ology* = study of) is the scientific study of the anatomy, physiology, and pathology of the kidneys. The branch of medicine that deals with the male and female urinary systems and the male reproductive system is called **urology** (ū-ROL-ō-jē; *uro-* = urine). A physician who specializes in this branch of medicine is called a **urologist** (ū-ROL-ō-jist).

? *Did you ever wonder how diuretics work and why they are used?*

Figure 26.1 Organs of the urinary system in a female.

Urine formed by the kidneys passes first into the ureters, then to the urinary bladder for storage, and finally through the urethra for elimination from the body.

RIGHT KIDNEY

Right renal artery

RIGHT URETER

URINARY
BLADDER

URETHRA

Diaphragm

Esophagus

Left adrenal
(suprarenal) gland

Left renal vein

LEFT KIDNEY

Abdominal aorta

Inferior vena cava

LEFT URETER

Rectum

Left ovary

Uterus

FUNCTIONS OF THE URINARY SYSTEM

1. Kidneys regulate blood volume and composition; help regulate blood pressure, pH, and glucose levels; produce two hormones (calcitriol and erythropoietin); and excrete wastes in urine.
2. Ureters transport urine from kidneys to urinary bladder.
3. Urinary bladder stores urine and expels it into urethra.
4. Urethra discharges urine from body.

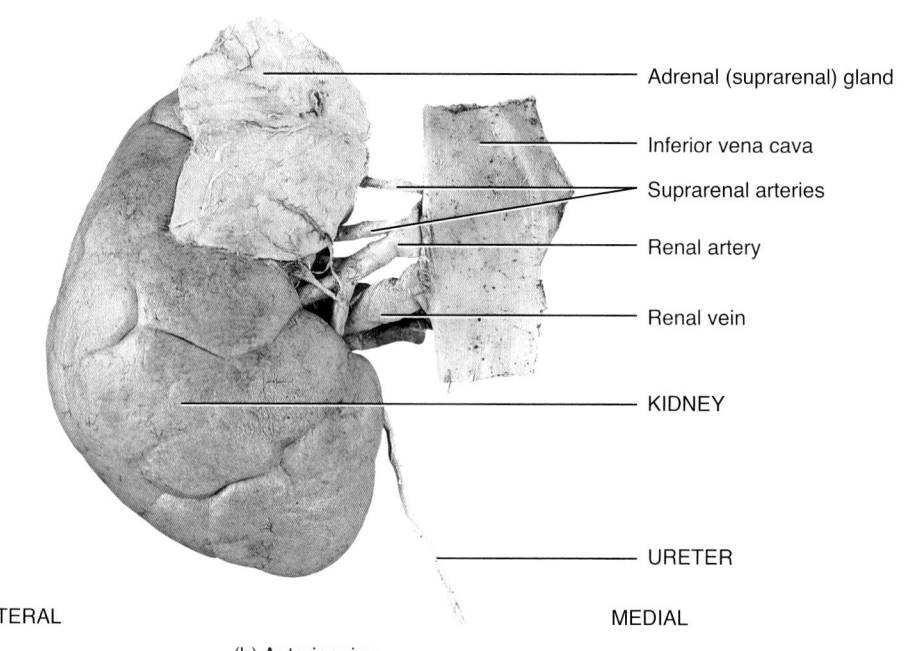

Adrenal (suprarenal) gland

Inferior vena cava

Suprarenal arteries

Renal artery

Renal vein

KIDNEY

URETER

LATERAL

MEDIAL

(b) Anterior view

? **Which organs constitute the urinary system?**

26.1 OVERVIEW OF KIDNEY FUNCTIONS

◉ **OBJECTIVE**

• List the functions of the kidneys.

The kidneys do the major work of the urinary system. The other parts of the system are mainly passageways and storage areas. Functions of the kidneys include the following:

• ***Regulation of blood ionic composition.*** The kidneys help regulate the blood levels of several ions, most importantly sodium ions (Na^+), potassium ions (K^+), calcium ions (Ca^{2+}), chloride ions (Cl^-), and phosphate ions (HPO_4^{2-}).

• ***Regulation of blood pH.*** The kidneys excrete a variable amount of hydrogen ions (H^+) into the urine and conserve bicarbonate ions (HCO_3^-), which are an important buffer of H^+ in the blood. Both of these activities help regulate blood pH.

• ***Regulation of blood volume.*** The kidneys adjust blood volume by conserving or eliminating water in the urine. An increase in blood volume increases blood pressure; a decrease in blood volume decreases blood pressure.

• ***Regulation of blood pressure.*** The kidneys also help regulate blood pressure by secreting the enzyme renin, which activates the renin–angiotensin–aldosterone pathway (see Figure 18.16). Increased renin causes an increase in blood pressure.

• ***Maintenance of blood osmolarity.*** By separately regulating loss of water and loss of solutes in the urine, the kidneys maintain a relatively constant blood osmolarity close to 300 milliosmoles per liter (mOsm/liter).*

• ***Production of hormones.*** The kidneys produce two hormones. *Calcitriol,* the active form of vitamin D, helps regulate calcium homeostasis (see Figure 18.14), and *erythropoietin* stimulates the production of red blood cells (see Figure 19.5).

• ***Regulation of blood glucose level.*** Like the liver, the kidneys can use the amino acid glutamine in *gluconeogenesis,* the synthesis of new glucose molecules. They can then release glucose into the blood to help maintain a normal blood glucose level.

• ***Excretion of wastes and foreign substances.*** By forming urine, the kidneys help excrete **wastes**—substances that have no useful function in the body. Some wastes excreted in urine result from metabolic reactions in the body. These include ammonia and urea from the deamination of amino acids; bilirubin

---

*The **osmolarity** of a solution is a measure of the total number of dissolved particles per liter of solution. The particles may be molecules, ions, or a mixture of both. To calculate osmolarity, multiply molarity (see Section 2.4) by the number of particles per molecule, once the molecule dissolves. A similar term, *osmolality,* is the number of particles of solute per *kilogram* of water. Because it is easier to measure volumes of solutions than to determine the mass of water they contain, osmolarity is used more commonly than osmolality. Most body fluids and solutions used clinically are dilute, in which case there is less than a 1% difference between the two measures.

from the catabolism of hemoglobin; creatinine from the breakdown of creatine phosphate in muscle fibers; and uric acid from the catabolism of nucleic acids. Other wastes excreted in urine are foreign substances from the diet, such as drugs and environmental toxins.

✔**CHECKPOINT**

1. What are wastes, and how do the kidneys participate in their removal from the body?

26.2 ANATOMY AND HISTOLOGY OF THE KIDNEYS

◉ **OBJECTIVES**

• Describe the external and internal gross anatomical features of the kidneys.
• Trace the path of blood flow through the kidneys.
• Describe the structure of renal corpuscles and renal tubules.

The paired **kidneys** are reddish, kidney bean–shaped organs located just above the waist between the peritoneum and the posterior wall of the abdomen. Because their position is posterior to the peritoneum of the abdominal cavity, they are said to be **retroperitoneal** (re′-trō-per-i-tō-NĒ-al; *retro-* = behind) organs (Figure 26.2). The kidneys are located between the levels of the last thoracic and third lumbar vertebrae, a position where they are partially protected by the eleventh and twelfth pairs of ribs. Unfortunately, if these lower ribs are fractured, they can puncture the kidneys and cause significant, and even life-threatening, damage. The right kidney is slightly lower than the left (see Figure 26.1) because the liver occupies considerable space on the right side superior to the kidney.

External Anatomy of the Kidneys

A typical adult kidney is 10–12 cm (4–5 in.) long, 5–7 cm (2–3 in.) wide, and 3 cm (1 in.) thick—about the size of a bar of bath soap—and has a mass of 135–150 g (4.5–5 oz). The concave medial border of each kidney faces the vertebral column (see Figure 26.1). Near the center of the concave border is an indentation called the **renal hilum** (RĒ-nal HĪ-lum; *renal* = kidney) or **hilus** (see Figure 26.3), through which the ureter emerges from the kidney along with blood vessels, lymphatic vessels, and nerves.

Three layers of tissue surround each kidney (Figure 26.2). The deep layer, the **renal capsule,** is a smooth, transparent sheet of dense irregular connective tissue that is continuous with the outer coat of the ureter. It serves as a barrier against trauma and helps maintain the shape of the kidney. The middle layer, the **adipose capsule,** is a mass of fatty tissue surrounding the renal capsule. It also protects the kidney from trauma and holds it firmly in place within the abdominal cavity. The superficial layer, the **renal fascia** (FASH-ē-a), is another thin layer of dense irregular connective tissue that anchors the kidney to the surrounding structures and to the abdominal wall. On the anterior surface of the kidneys, the renal fascia is deep to the peritoneum.

Figure 26.2 Position and coverings of the kidneys.

The kidneys are surrounded by a renal capsule, adipose capsule, and renal fascia.

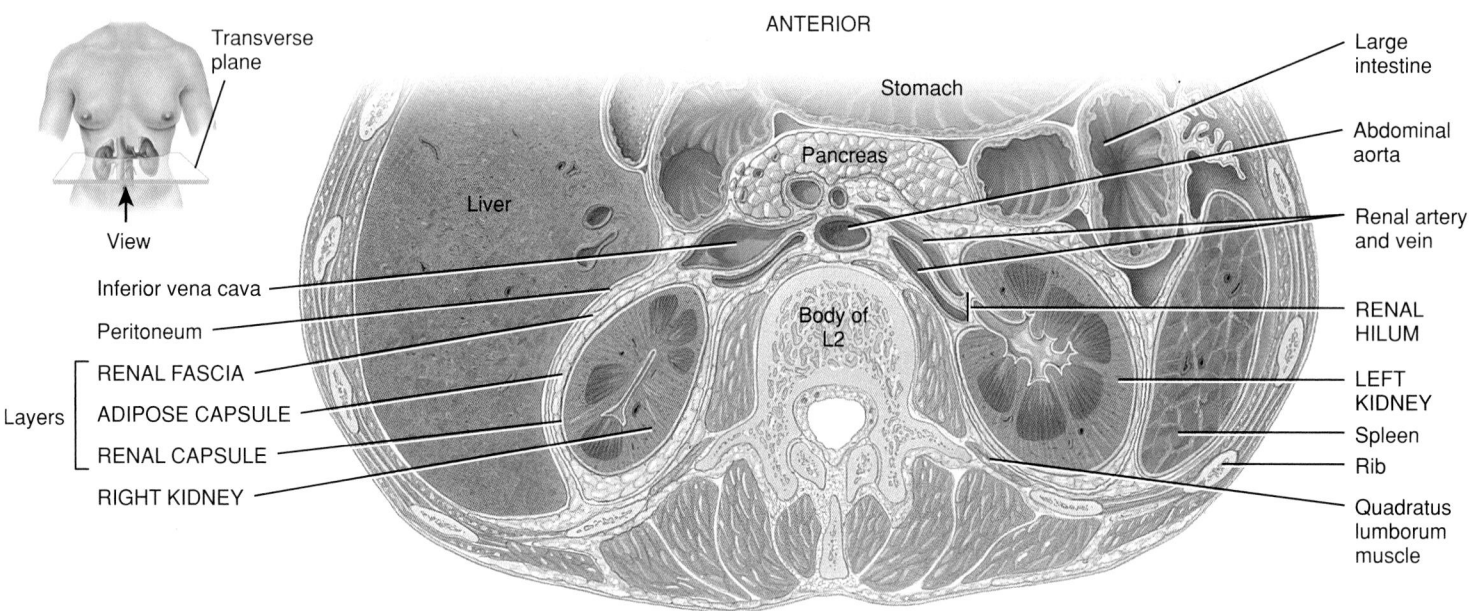

(a) Inferior view of transverse section of abdomen (L2)

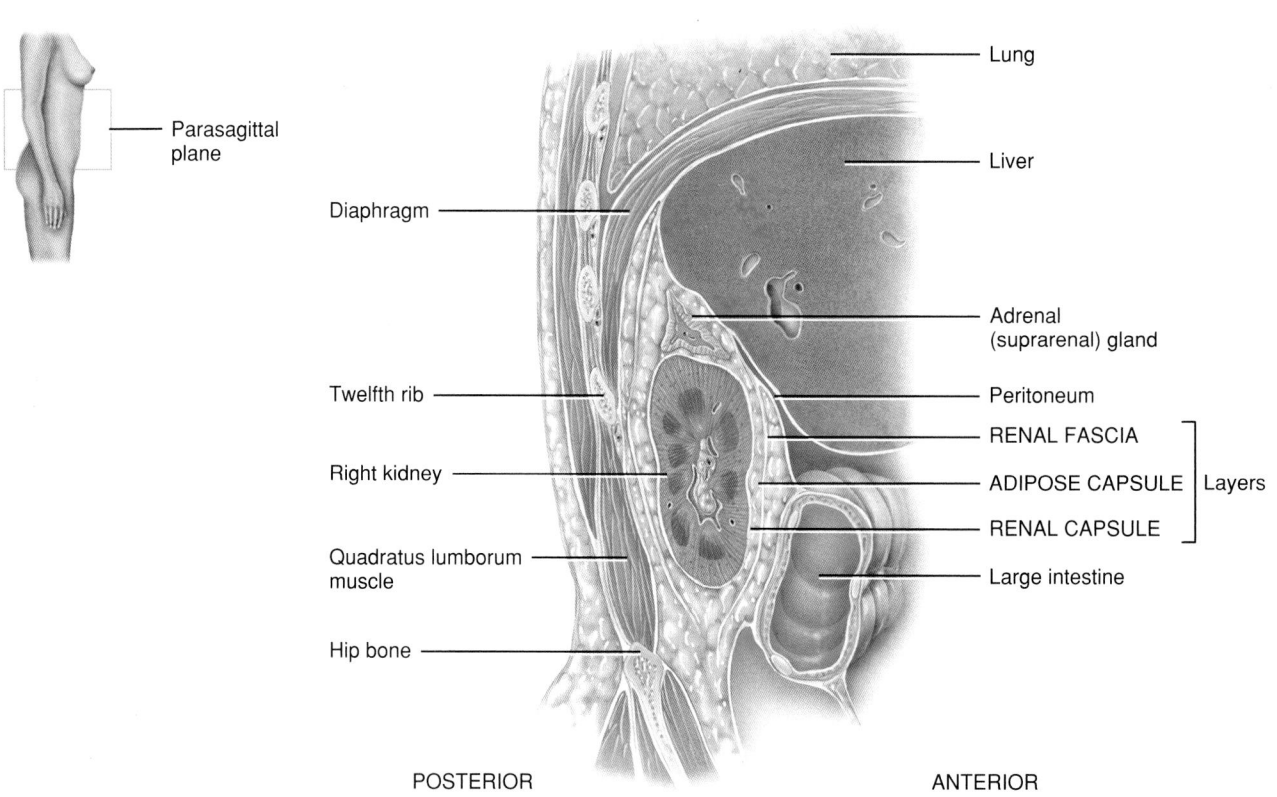

(b) Parasagittal section through right kidney

Why are the kidneys said to be retroperitoneal?

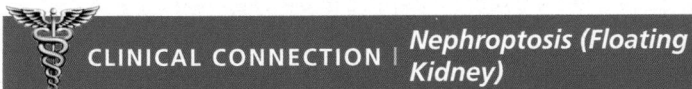

Nephroptosis (nef'-rōp-TŌ-sis; *-ptosis* = falling), or **floating kidney,** is an inferior displacement or dropping of the kidney. It occurs when the kidney slips from its normal position because it is not securely held in place by adjacent organs or its covering of fat. Nephroptosis develops most often in very thin people whose adipose capsule or renal fascia is deficient. It is dangerous because the ureter may kink and block urine flow. The resulting backup of urine puts pressure on the kidney, which damages the tissue. Twisting of the ureter also causes pain. Nephroptosis is very common; about one in four people has some degree of weakening of the fibrous bands that hold the kidney in place. It is 10 times more common in females than males. Because it happens during life it is very easy to distinguish from congenital anomalies. •

Internal Anatomy of the Kidneys

A frontal section through the kidney reveals two distinct regions: a superficial, light red area called the **renal cortex** (*cortex* = rind or bark) and a deep, darker reddish-brown inner region called the **renal medulla** (*medulla* = inner portion) (Figure 26.3). The renal medulla consists of several cone-shaped **renal pyramids.** The base (wider end) of each pyramid faces the renal cortex, and its apex (narrower end), called a **renal papilla,** points toward the renal hilum. The renal cortex is the smooth-textured area extending from the renal capsule to the bases of the renal pyramids and into the spaces between them. It is divided into an outer *cortical zone* and an inner *juxtamedullary zone* (juks'-ta-MED-ū-la-rē). Those portions of the renal cortex that extend between renal pyramids are called **renal columns.** A **renal lobe** consists of a renal pyramid, its overlying area of renal cortex, and one-half of each adjacent renal column.

Together, the renal cortex and renal pyramids of the renal medulla constitute the **parenchyma** (pa-RENG-ki-ma) or functional portion of the kidney. Within the parenchyma are the functional units of the kidney—about 1 million microscopic structures called **nephrons** (NEF-rons). Filtrate (filtered fluid) formed by the nephrons drains into large **papillary ducts,** which extend through the renal papillae of the pyramids. The papillary ducts drain into cuplike structures called **minor** and **major calyces** (KĀ-li-sēz = cups; singular is *calyx,* pronounced KĀ-liks). Each kidney has 8 to 18 minor calyces and 2 or 3 major calyces. A minor calyx receives urine from the papillary ducts of one renal papilla and delivers it to a major calyx. Once the filtrate enters the calyces it becomes urine because no further reabsorption can occur. The reason for this is that the simple epithelium of the nephron and ducts becomes transitional epithelium in the calyces. From the major calyces, urine drains into a single large cavity called the **renal pelvis** (*pelv-* = basin) and then out through the ureter to the urinary bladder.

The hilum expands into a cavity within the kidney called the **renal sinus,** which contains part of the renal pelvis, the calyces, and branches of the renal blood vessels and nerves. Adipose tissue helps stabilize the position of these structures in the renal sinus.

Blood and Nerve Supply of the Kidneys

Because the kidneys remove wastes from the blood and regulate its volume and ionic composition, it is not surprising that they are abundantly supplied with blood vessels. Although the kidneys constitute less than 0.5% of total body mass, they receive 20–25% of the resting cardiac output via the right and left **renal arteries** (Figure 26.4). In adults, **renal blood flow,** the blood flow through both kidneys, is about 1200 mL per minute.

Within the kidney, the renal artery divides into several **segmental arteries** (seg-MEN-tal), which supply different segments (areas) of the kidney. Each segmental artery gives off several branches that enter the parenchyma and pass through the renal columns between the renal pyramids as the **interlobar arteries** (in'-ter-LŌ-bar). At the bases of the renal pyramids, the interlobar arteries arch between the renal medulla and cortex; here they are known as the **arcuate arteries** (AR-kū-āt = shaped like a bow). Divisions of the arcuate arteries produce a series of **interlobular arteries** (in'-ter-LOB-ū-lar). These arteries are so named because they pass between renal lobules. Interlobular arteries enter the renal cortex and give off branches called **afferent arterioles** (AF-er-ent; *af-* = toward; *-ferrent* = to carry).

Each nephron receives one afferent arteriole, which divides into a tangled, ball-shaped capillary network called the **glomerulus** (glō-MER-ū-lus = little ball; plural is *glomeruli*). The glomerular capillaries then reunite to form an **efferent arteriole** (EF-er-ent; *ef-* = out) that carries blood out of the glomerulus. Glomerular capillaries are unique among capillaries in the body because they are positioned between two arterioles, rather than between an arteriole and a venule. Because they are capillary networks and they also play an important role in urine formation, the glomeruli are considered part of both the cardiovascular and the urinary systems.

The efferent arterioles divide to form the **peritubular capillaries** (per-i-TOOB-ū-lar; *peri-* = around), which surround tubular parts of the nephron in the renal cortex. Extending from some efferent arterioles are long loop-shaped capillaries called **vasa recta** (VĀ-sa REK-ta; *vasa* = vessels; *recta* = straight) that supply tubular portions of the nephron in the renal medulla (see Figure 26.5b).

The peritubular capillaries eventually reunite to form **peritubular venules** and then **interlobular veins,** which also receive blood from the vasa recta. Then the blood drains through the **arcuate veins** to the **interlobar veins** running between the renal pyramids. Blood leaves the kidney through a single **renal vein** that exits at the renal hilum and carries venous blood to the inferior vena cava.

Many renal nerves originate in the *renal ganglion* and pass through the *renal plexus* into the kidneys along with the renal arteries. Renal nerves are part of the sympathetic division of the autonomic nervous system. Most are vasomotor nerves that regulate the flow of blood through the kidney by causing vasodilation or vasoconstriction of renal arterioles.

Figure 26.3 Internal anatomy of the kidneys.

The two main regions of the kidney parenchyma are the renal cortex and the renal pyramids in the renal medulla.

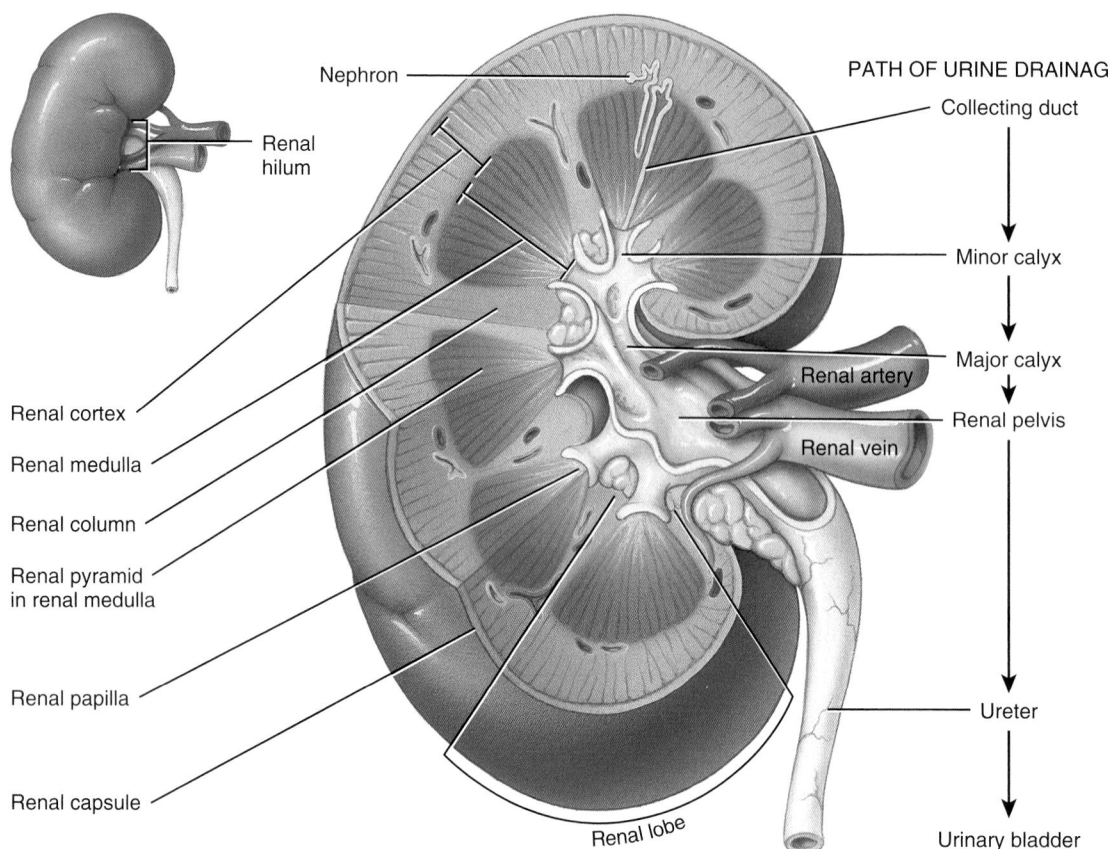

Nephron

Renal hilum

Renal cortex

Renal medulla

Renal column

Renal pyramid in renal medulla

Renal papilla

Renal capsule

Renal lobe

PATH OF URINE DRAINAGE:

Collecting duct

↓

Minor calyx

↓

Major calyx

Renal artery

Renal pelvis

Renal vein

↓

Ureter

↓

Urinary bladder

(a) Anterior view of dissection of right kidney

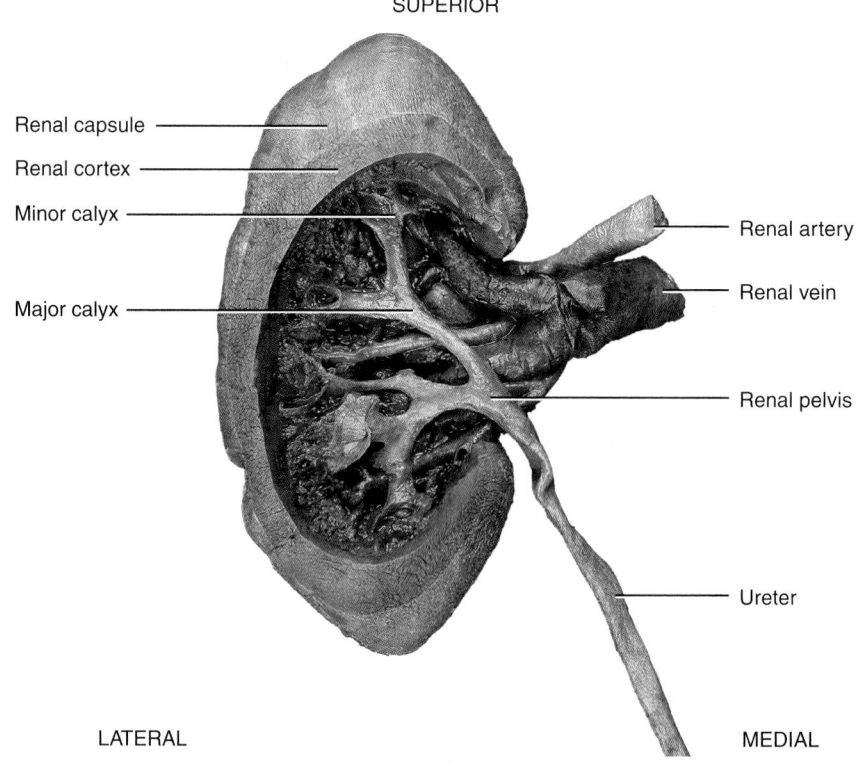

SUPERIOR

Renal capsule

Renal cortex

Minor calyx

Major calyx

Renal artery

Renal vein

Renal pelvis

Ureter

LATERAL

MEDIAL

(b) Posterior view of dissection of left kidney

? What structures pass through the renal hilum?

Figure 26.4 Blood supply of the kidneys.

The renal arteries deliver 20–25% of the resting cardiac output to the kidneys.

Blood supply of nephron

(b) Path of blood flow

(a) Frontal section of right kidney

? **What volume of blood enters the renal arteries per minute?**

The Nephron

Parts of a Nephron

Nephrons (NEF-rons) are the functional units of the kidneys. Each nephron (Figure 26.5) consists of two parts: a **renal corpuscle** (KOR-pus-sul = tiny body), where blood plasma is filtered, and a **renal tubule** into which the filtered fluid passes. The two components of a renal corpuscle are the **glomerulus** (glō-MER-ū-lus) (capillary network) and the **glomerular (Bowman's) capsule,** a double-walled epithelial cup that surrounds the glomerular capillaries. Blood plasma is filtered in the glomerular capsule, and then the filtered fluid passes into the renal tubule, which has three main sections. In the order that fluid passes through them, the renal tubule consists of a (1) **proximal convoluted**

Figure 26.5 The structure of nephrons and associated blood vessels. Note that the collecting duct and papillary duct are not part of a nephron.

🔑 Nephrons are the functional units of the kidneys.

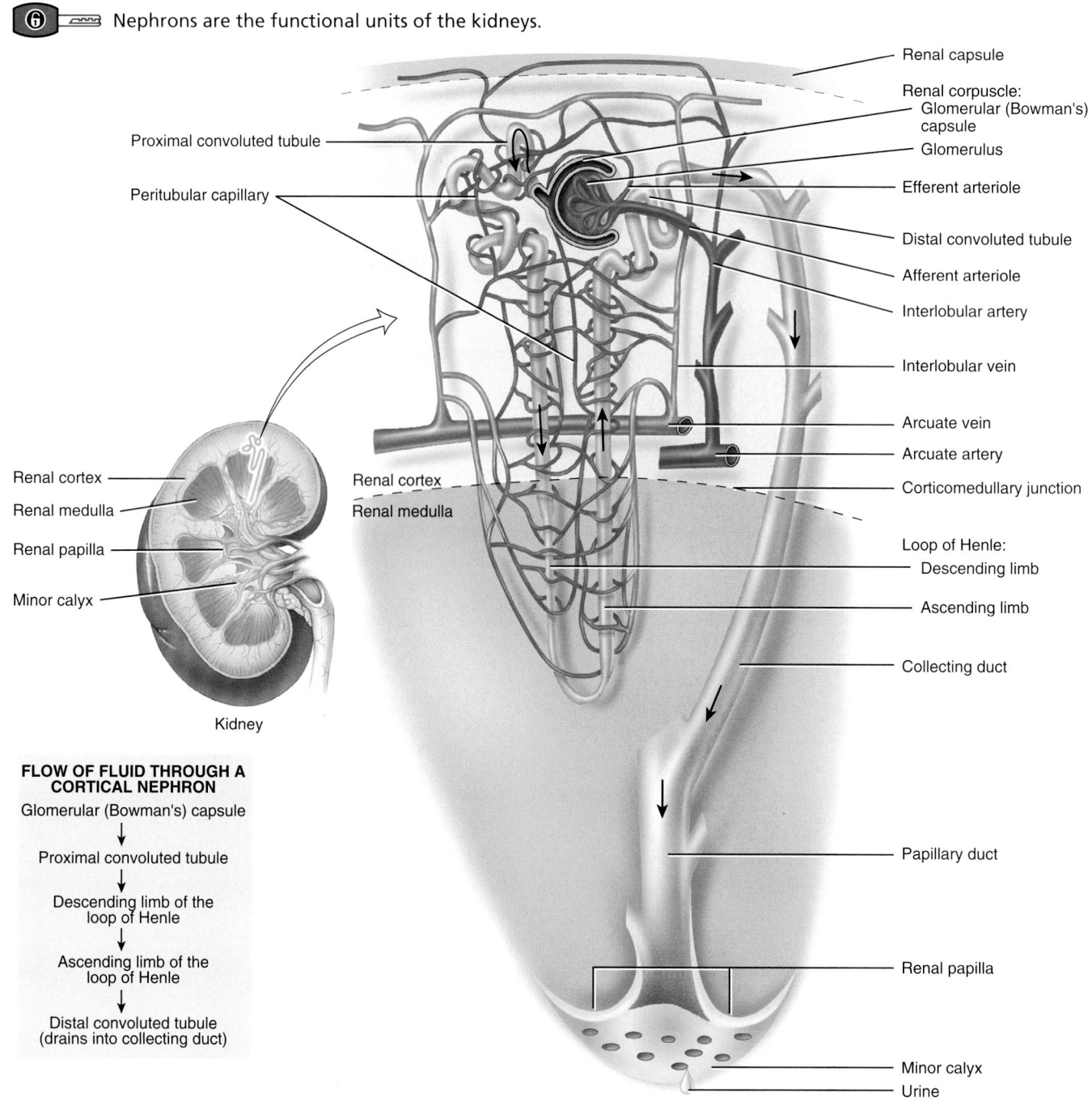

FLOW OF FLUID THROUGH A
CORTICAL NEPHRON

Glomerular (Bowman's) capsule
↓
Proximal convoluted tubule
↓
Descending limb of the
loop of Henle
↓
Ascending limb of the
loop of Henle
↓
Distal convoluted tubule
(drains into collecting duct)

(a) Cortical nephron and vascular supply

tubule or **PCT** (kon′-vō-LOOT-ed), (2) **loop of Henle (nephron loop),** and (3) **distal convoluted tubule** or **DCT.** *Proximal* denotes the part of the tubule attached to the glomerular capsule, and *distal* denotes the part that is further away. *Convoluted* means the tubule is tightly coiled rather than straight. The renal corpuscle and both convoluted tubules lie within the renal cortex; the loop of Henle extends into the renal medulla, makes a hairpin turn, and then returns to the renal cortex.

The distal convoluted tubules of several nephrons empty into a single **collecting duct.** Collecting ducts then unite and converge into several hundred large **papillary ducts** (PAP-i-lar′-ē), which drain into the minor calyces. The collecting ducts and papillary ducts extend from the renal cortex through the renal medulla to the renal pelvis. So one kidney has about 1 million nephrons, but a much smaller number of collecting ducts and even fewer papillary ducts.

Renal capsule

Distal convoluted tubule

Renal corpuscle:
Glomerular (Bowman's) capsule

Glomerulus

Afferent arteriole

Efferent arteriole

Interlobular artery

Interlobular vein

Arcuate vein

Arcuate artery

Corticomedullary junction

Collecting duct

Loop of Henle:
Descending limb

Thick ascending limb

Thin ascending limb

Papillary duct

Renal papilla

Minor calyx

Urine

Proximal convoluted tubule

Peritubular capillary

Renal cortex

Renal medulla

Peritubular capillary

Vasa recta

Renal cortex

Renal medulla

Renal papilla

Minor calyx

Kidney

FLOW OF FLUID THROUGH A JUXTAMEDULLARY NEPHRON

Glomerular (Bowman's) capsule
↓
Proximal convoluted tubule
↓
Descending limb of the loop of Henle
↓
Thin ascending limb of the loop of Henle
↓
Thick ascending limb of the loop of Henle
↓
Distal convoluted tubule (drains into collecting duct)

(b) Juxtamedullary nephron and vascular supply

? **What are the basic differences between cortical and juxtamedullary nephrons?**

In a nephron, the loop of Henle connects the proximal and distal convoluted tubules. The first part of the loop of Henle dips into the renal medulla, where it is called the **descending limb of the loop of Henle** (Figure 26.5). It then makes that hairpin turn and returns to the renal cortex as the **ascending limb of the loop of Henle.** About 80–85% of the nephrons are **cortical nephrons** (KOR-ti-kul). Their renal corpuscles lie in the outer portion of the renal cor-

tex, and they have *short* loops of Henle that lie mainly in the cortex and penetrate only into the outer region of the renal medulla (Figure 26.5a). The short loops of Henle receive their blood supply from peritubular capillaries that arise from efferent arterioles. The other 15–20% of the nephrons are **juxtamedullary nephrons** (juks′-ta-MED-ū-lar′-e; *juxta-* = near to). Their renal corpuscles lie deep in the cortex, close to the medulla, and they have a *long* loop of Henle that extends into the deepest region of the medulla (Figure 26.5b). Long loops of Henle receive their blood supply from per-itubular capillaries and from the vasa recta that arise from efferent arterioles. In addition, the ascending limb of the loop of Henle of juxtamedullary nephrons consists of two portions: a **thin ascending limb** followed by a **thick ascending limb** (Figure 26.5b). The lumen of the thin ascending limb is the same as in other areas of the renal tubule; it is only the epithelium that is thinner. Nephrons with long loops of Henle enable the kidneys to excrete very dilute or very concentrated urine (described in Section 26.6).

Histology of the Nephron and Collecting Duct

A single layer of epithelial cells forms the entire wall of the glomerular capsule, renal tubule, and ducts. However, each part has distinctive histological features that reflect its particular functions. We will discuss them in the order that fluid flows through them: glomerular capsule, renal tubule, and collecting duct.

GLOMERULAR CAPSULE The glomerular (Bowman's) capsule consists of visceral and parietal layers (Figure 26.6a). The vis-ceral layer consists of modified simple squamous epithelial cells called **podocytes** (PŌD-ō-sīts; *podo-* = foot; *-cytes* = cells). The many footlike projections of these cells (pedicels) wrap around the single layer of endothelial cells of the glomerular capillaries and form the inner wall of the capsule. The parietal layer of the glomerular capsule consists of simple squamous epithelium and forms the outer wall of the capsule. Fluid filtered from the glomerular capillaries enters the **capsular (Bowman's) space,** the space between the two layers of the glomerular capsule, which is the lumen of the urinary tube. Think of the relationship between the glomerulus and glomerular capsule in the following way. The glomerulus is a fist punched into a limp balloon (the glomerular capsule) until the fist is covered by two layers of balloon (the layer of the balloon touching the fist is the visceral layer and the layer not against the fist is the parietal layer) with a space in between (the inside of the balloon), the capsular space.

RENAL TUBULE AND COLLECTING DUCT Table 26.1 illustrates the histology of the cells that form the renal tubule and collecting duct. In the proximal convoluted tubule, the cells are

TABLE 26.1	
Histological Features of the Renal Tubule and Collecting Duct	
REGION AND HISTOLOGY	**DESCRIPTION**
Proximal convoluted tubule (PCT)	Simple cuboidal epithelial cells with prominent brush borders of microvilli.
Loop of Henle: descending limb and thin ascending limb	Simple squamous epithelial cells.
Loop of Henle: thick ascending limb	Simple cuboidal to low columnar epithelial cells.
Most of distal convoluted tubule (DCT)	Simple cuboidal epithelial cells.
Last part of DCT and all of collecting duct (CD)	Simple cuboidal epithelium consisting of principal cells and intercalated cells.

Figure 26.6 Histology of a renal corpuscle.

A renal corpuscle consists of a glomerular (Bowman's) capsule and a glomerulus.

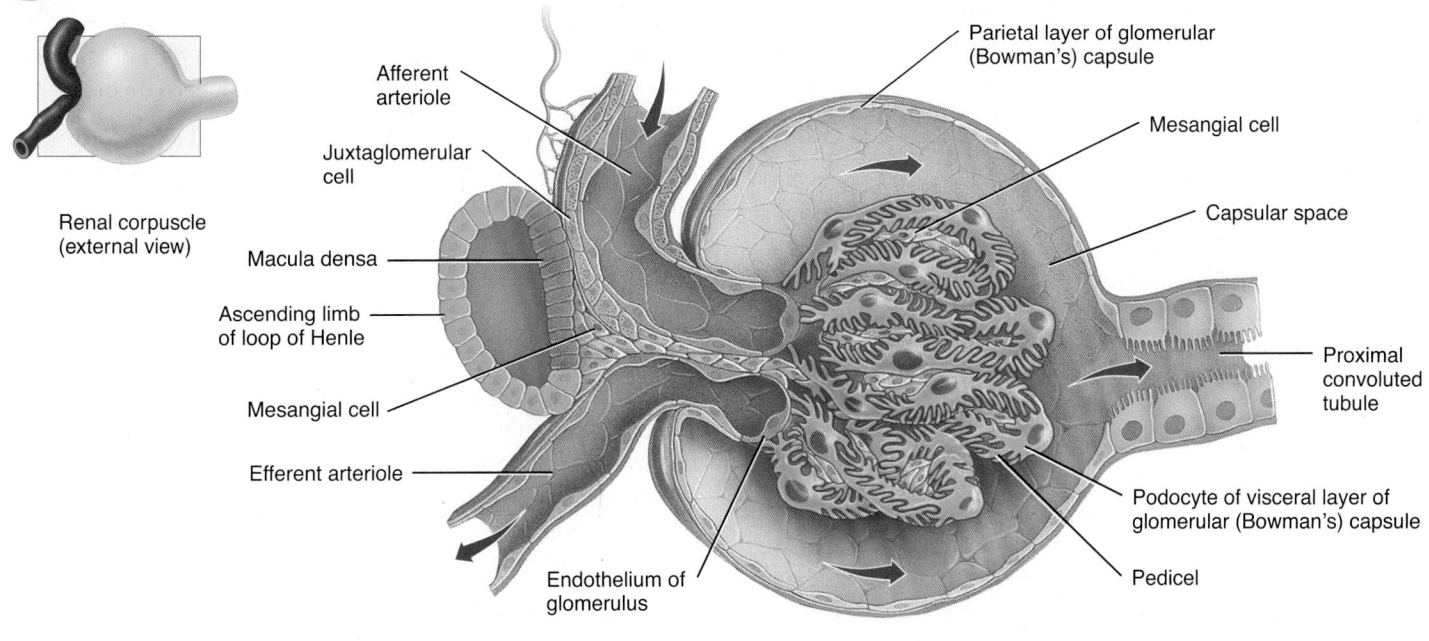

Renal corpuscle (external view)

Afferent arteriole

Juxtaglomerular cell

Macula densa

Ascending limb of loop of Henle

Mesangial cell

Efferent arteriole

Endothelium of glomerulus

Parietal layer of glomerular (Bowman's) capsule

Mesangial cell

Capsular space

Proximal convoluted tubule

Podocyte of visceral layer of glomerular (Bowman's) capsule

Pedicel

(a) Renal corpuscle (internal view)

Glomerular capsule:

Parietal layer

Visceral layer

Afferent arteriole

Juxtaglomerular cell

Ascending limb of loop of Henle

Macula densa cell

Efferent arteriole

Proximal convoluted tubule

Glomerulus

Podocytes of visceral layer of glomerular capsule

Capsular space

Simple squamous epithelial cells

LM 1380x

(b) Renal corpuscle

? Is the photomicrograph in (b) from a section through the renal cortex or renal medulla? How can you tell?

simple cuboidal epithelial cells with a prominent brush border of microvilli on their apical surface (surface facing the lumen). These microvilli, like those of the small intestine, increase the surface area for reabsorption and secretion. The descending limb of the loop of Henle and the first part of the ascending limb of the loop of Henle (the thin ascending limb) are composed of simple squamous epithelium. (Recall that cortical or short-loop nephrons

lack the thin ascending limb.) The thick ascending limb of the loop of Henle is composed of simple cuboidal to low columnar epithelium.

In each nephron, the final part of the ascending limb of the loop of Henle makes contact with the afferent arteriole serving that renal corpuscle (Figure 26.6a). Because the columnar tubule cells in this region are crowded together, they are known as the macula densa (MAK-ū-la DEN-sa; *macula* = spot; *densa* = dense). Alongside the macula densa, the wall of the afferent arteriole (and sometimes the efferent arteriole) contains modified smooth muscle fibers called **juxtaglomerular (JG) cells** (juks'-ta-glō-MER-ū-lar). Together with the macula densa, they constitute the **juxtaglomerular apparatus (JGA).** As you will see later, the JGA helps regulate blood pressure within the kidneys. The distal convoluted tubule (DCT) begins a short distance past the macula densa. In the last part of the DCT and continuing into the collecting ducts, two different types of cells are present. Most are **principal cells,** which have receptors for both antidiuretic hormone (ADH) and aldosterone, two hormones that regulate their functions. A smaller number are **intercalated cells** (in-TER-ka-lā-ted), which play a role in the homeostasis of blood pH. The collecting ducts drain into large papillary ducts, which are lined by simple columnar epithelium.

The number of nephrons is constant from birth. Any increase in kidney size is due solely to the growth of individual nephrons. If nephrons are injured or become diseased, new ones do not form. Signs of kidney dysfunction usually do not become apparent until function declines to less than 25% of normal because the remaining functional nephrons adapt to handle a larger-than-normal load. Surgical removal of one kidney, for example, stimulates hypertrophy (enlargement) of the remaining kidney, which eventually is able to filter blood at 80% of the rate of two normal kidneys.

✔**CHECKPOINT**

2. What is the renal capsule and why is it important?
3. What are the two main parts of a nephron?
4. What are the components of the renal tubule?
5. Where is the juxtaglomerular apparatus (JGA) located, and what is its structure?

26.3 OVERVIEW OF RENAL PHYSIOLOGY

◉ **OBJECTIVE**

• Identify the three basic functions performed by nephrons and collecting ducts, and indicate where each occurs.

To produce urine, nephrons and collecting ducts perform three basic processes—glomerular filtration, tubular reabsorption, and tubular secretion (Figure 26.7):

❶ *Glomerular filtration.* In the first step of urine production, water and most solutes in blood plasma move across the wall of glomerular capillaries, where they are filtered and move into the glomerular capsule and then into the renal tubule.

❷ *Tubular reabsorption.* As filtered fluid flows through the renal tubules and through the collecting ducts, tubule cells reabsorb about 99% of the filtered water and many useful solutes. The water and solutes return to the blood as it flows through the peritubular capillaries and vasa recta. Note that the term *reabsorption* refers to the return of substances to the bloodstream. The term *absorption,* by contrast, means entry of new substances into the body, as occurs in the gastrointestinal tract.

❸ *Tubular secretion.* As filtered fluid flows through the renal tubules and collecting ducts, the renal tubule and duct cells secrete other materials, such as wastes, drugs, and excess ions, into the fluid. Notice that tubular secretion *removes a substance from the blood.*

Solutes and the fluid that drain into the minor and major calyces and renal pelvis constitute urine and are excreted. The rate

Figure 26.7 **Relation of a nephron's structure to its three basic functions: glomerular filtration, tubular reabsorption, and tubular secretion.** Excreted substances remain in the urine and subsequently leave the body. For any substance S, excretion rate of S = filtration rate of S − reabsorption rate of S + secretion rate of S.

🔑 Glomerular filtration occurs in the renal corpuscle. Tubular reabsorption and tubular secretion occur all along the renal tubule and collecting duct.

? **When cells of the renal tubules secrete the drug penicillin, is the drug being added to or removed from the bloodstream?**

of urinary excretion of any solute is equal to its rate of glomerular filtration, plus its rate of secretion, minus its rate of reabsorption.

By filtering, reabsorbing, and secreting, nephrons help maintain homeostasis of the blood's volume and composition. The situation is somewhat analogous to a recycling center: Garbage trucks dump refuse into an input hopper, where the smaller refuse passes onto a conveyor belt (glomerular filtration of plasma). As the conveyor belt carries the garbage along, workers remove useful items, such as aluminum cans, plastics, and glass containers (reabsorption). Other workers place additional garbage left at the center and larger items onto the conveyor belt (secretion). At the end of the belt, all remaining garbage falls into a truck for transport to the landfill (excretion of wastes in urine).

✔CHECKPOINT

6. What is the difference between tubular reabsorption and tubular secretion?

26.4 GLOMERULAR FILTRATION

◉ OBJECTIVES

• Describe the filtration membrane.
• Discuss the pressures that promote and oppose glomerular filtration.

The fluid that enters the capsular space is called the **glomerular filtrate.** The fraction of blood plasma in the afferent arterioles

of the kidneys that becomes glomerular filtrate is the **filtration fraction.** Although a filtration fraction of 0.16–0.20 (16–20%) is typical, the value varies considerably in both health and disease. On average, the daily volume of glomerular filtrate in adults is 150 liters in females and 180 liters in males. More than 99% of the glomerular filtrate returns to the bloodstream via tubular reabsorption, so only 1–2 liters (about 1–2 qt) is excreted as urine.

The Filtration Membrane

Together, the glomerular capillaries and the podocytes, which completely encircle the capillaries, form a leaky barrier known as the **filtration membrane.** This sandwichlike assembly permits filtration of water and small solutes but prevents filtration of most plasma proteins, blood cells, and platelets. Substances filtered from the blood cross three filtration barriers—a glomerular endothelial cell, the basal lamina, and a filtration slit formed by a podocyte (Figure 26.8):

❶ Glomerular endothelial cells are quite leaky because they have large **fenestrations** (fen′-es-TRĀ-shuns) (pores) that measure 0.07–0.1 μm in diameter. This size permits all solutes in blood plasma to exit glomerular capillaries but prevents filtration of blood cells and platelets. Located among the glomerular capillaries and in the cleft between afferent and efferent arterioles are **mesangial cells** (mes-AN-jē-al; *mes-* =

Figure 26.8 The filtration membrane. The size of the endothelial fenestrations and filtration slits in (a) have been exaggerated for emphasis.

🔑 During glomerular filtration, water and solutes pass from blood plasma into the capsular space.

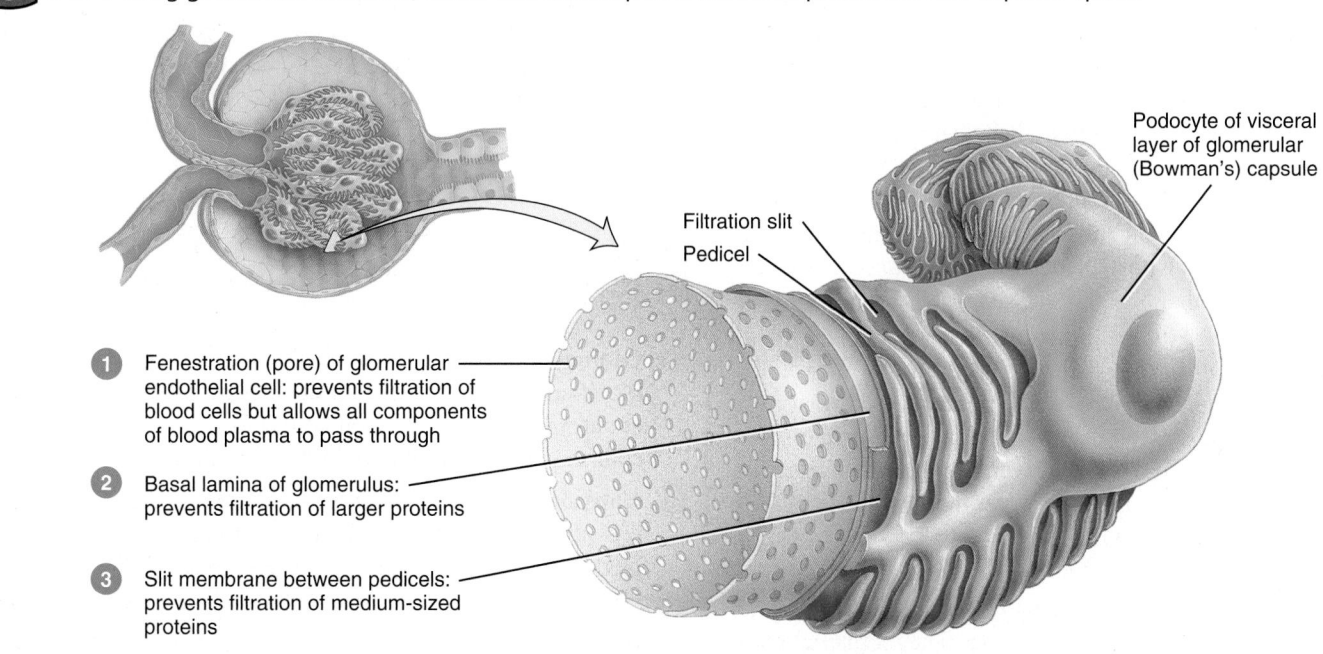

❶ Fenestration (pore) of glomerular endothelial cell: prevents filtration of blood cells but allows all components of blood plasma to pass through

❷ Basal lamina of glomerulus: prevents filtration of larger proteins

❸ Slit membrane between pedicels: prevents filtration of medium-sized proteins

Filtration slit
Pedicel
Podocyte of visceral layer of glomerular (Bowman's) capsule

(a) Details of filtration membrane

FIGURE 26.8 CONTINUES ▶

FIGURE 26.8 CONTINUED

Pedicel of podocyte

Filtration slit

Basal lamina

Lumen of glomerulus

TEM 78,000x

Fenestration (pore) of glomerular endothelial cell

(b) Filtration membrane

? **Which part of the filtration membrane prevents red blood cells from entering the capsular space?**

in the middle; *-angi-* = blood vessel) (see Figure 26.6a). These contractile cells help regulate glomerular filtration.

2 The **basal lamina,** a layer of acellular material between the endothelium and the podocytes, consists of minute collagen fibers and proteoglycans in a glycoprotein matrix; negative charges in the matrix prevent filtration of larger negatively charged plasma proteins.

3 Extending from each podocyte are thousands of footlike processes termed **pedicels** (PED-i-sels = little feet) that wrap around glomerular capillaries. The spaces between pedicels are the **filtration slits.** A thin membrane, the **slit membrane,** extends across each filtration slit; it permits the passage of molecules having a diameter smaller than 0.006–0.007 μm, including water, glucose, vitamins, amino acids, very small plasma proteins, ammonia, urea, and ions. Less than 1% of albumin, the most plentiful plasma protein, passes the slit membrane because, with a diameter of 0.007 μm, it is slightly too big to get through.

The principle of *filtration*—the use of pressure to force fluids and solutes through a membrane—is the same in glomerular capillaries as in capillaries elsewhere in the body (see Starling's law of the capillaries, Section 21.2). However, the volume of fluid filtered by the renal corpuscle is much larger than in other capillaries of the body for three reasons:

1. Glomerular capillaries present a large surface area for filtration because they are long and extensive. The mesangial cells regulate how much of this surface area is available for filtration. When mesangial cells are relaxed, surface area is maximal, and glomerular filtration is very high. Contraction of mesangial

cells reduces the available surface area, and glomerular filtration decreases.

2. The filtration membrane is thin and porous. Despite having several layers, the thickness of the filtration membrane is only 0.1 mm. Glomerular capillaries also are about 50 times leakier than capillaries in most other tissues, mainly because of their large fenestrations.

3. Glomerular capillary blood pressure is high. Because the efferent arteriole is smaller in diameter than the afferent arteriole, resistance to the outflow of blood from the glomerulus is high. As a result, blood pressure in glomerular capillaries is considerably higher than in capillaries elsewhere in the body.

Net Filtration Pressure

Glomerular filtration depends on three main pressures. One pressure *promotes* filtration and two pressures *oppose* filtration (Figure 26.9).

1 **Glomerular blood hydrostatic pressure (GBHP)** is the blood pressure in glomerular capillaries. Generally, GBHP is about 55 mmHg. It promotes filtration by forcing water and solutes in blood plasma through the filtration membrane.

2 **Capsular hydrostatic pressure (CHP)** is the hydrostatic pressure exerted against the filtration membrane by fluid already in the capsular space and renal tubule. CHP opposes filtration and represents a "back pressure" of about 15 mmHg.

3 **Blood colloid osmotic pressure (BCOP),** which is due to the presence of proteins such as albumin, globulins, and fibrinogen in blood plasma, also opposes filtration. The average BCOP in glomerular capillaries is 30 mmHg.

Figure 26.9 The pressures that drive glomerular filtration. Taken together, these pressures determine net filtration pressure (NFP).

Glomerular blood hydrostatic pressure promotes filtration, whereas capsular hydrostatic pressure and blood colloid osmotic pressure oppose filtration.

1 GLOMERULAR BLOOD HYDROSTATIC PRESSURE (GBHP) = 55 mmHg

2 CAPSULAR HYDROSTATIC PRESSURE (CHP) = 15 mmHg

3 BLOOD COLLOID OSMOTIC PRESSURE (BCOP) = 30 mmHg

Afferent arteriole

Proximal convoluted tubule

Efferent arteriole

NET FILTRATION PRESSURE (NFP)
= GBHP – CHP – BCOP
= 55 mmHg – 15 mmHg – 30 mmHg
= 10 mmHg

Glomerular (Bowman's) capsule

Capsular space

? Suppose a tumor is pressing on and obstructing the right ureter. What effect might this have on CHP and thus on NFP in the right kidney? Would the left kidney also be affected?

Net filtration pressure (NFP), the total pressure that promotes filtration, is determined as follows:

Net filtration pressure (NFP) = GBHP − CHP − BCOP

By substituting the values just given, normal NFP may be calculated:

$$NFP = 55 \text{ mmHg} - 15 \text{ mmHg} - 30 \text{ mmHg}$$
$$= 10 \text{ mmHg}$$

Thus, a pressure of only 10 mmHg causes a normal amount of blood plasma (minus plasma proteins) to filter from the glomerulus into the capsular space.

CLINICAL CONNECTION | Loss of Plasma Proteins in Urine Causes Edema

In some kidney diseases, glomerular capillaries are damaged and become so permeable that plasma proteins enter glomerular filtrate. As a result, the filtrate exerts a colloid osmotic pressure that draws water out of the blood. In this situation, the NFP increases, which means more fluid is filtered. At the same time, blood colloid osmotic pressure decreases because plasma proteins are being lost in the urine. Because more fluid filters out of blood capillaries into tissues throughout the body than returns via reabsorption, blood volume de-

creases and interstitial fluid volume increases. Thus, loss of plasma proteins in urine causes *edema*, an abnormally high volume of interstitial fluid. •

Glomerular Filtration Rate

The amount of filtrate formed in all the renal corpuscles of both kidneys each minute is the **glomerular filtration rate (GFR).** In adults, the GFR averages 125 mL/min in males and 105 mL/min in females. Homeostasis of body fluids requires that the kidneys maintain a relatively constant GFR. If the GFR is too high, needed substances may pass so quickly through the renal tubules that some are not reabsorbed and are lost in the urine. If the GFR is too low, nearly all the filtrate may be reabsorbed and certain waste products may not be adequately excreted.

GFR is directly related to the pressures that determine net filtration pressure; any change in net filtration pressure will affect GFR. Severe blood loss, for example, reduces mean arterial blood pressure and decreases the glomerular blood hydrostatic pressure. Filtration ceases if glomerular blood hydrostatic pressure drops to 45 mmHg because the opposing pressures add up to 45 mmHg. Amazingly, when systemic blood pressure rises above normal, net filtration pressure and GFR increase very little. GFR is nearly constant when the mean arterial blood pressure is anywhere between 80 and 180 mmHg.

The mechanisms that regulate glomerular filtration rate operate in two main ways: (1) by adjusting blood flow into and out of the glomerulus and (2) by altering the glomerular capillary surface area available for filtration. GFR increases when blood flow into the glomerular capillaries increases. Coordinated control of the diameter of both afferent and efferent arterioles regulates glomerular blood flow. Constriction of the afferent arteriole decreases blood flow into the glomerulus; dilation of the afferent arteriole increases it. Three mechanisms control GFR: renal autoregulation, neural regulation, and hormonal regulation.

Renal Autoregulation of GFR

The kidneys themselves help maintain a constant renal blood flow and GFR despite normal, everyday changes in blood pressure, like those that occur during exercise. This capability is called **renal autoregulation** (aw′-tō-reg′-ū-LĀ-shun) and consists of two mechanisms—the myogenic mechanism and tubuloglomerular feedback. Working together, they can maintain nearly constant GFR over a wide range of systemic blood pressures.

The **myogenic mechanism** (mī-ō-JEN-ik; *myo-* = muscle; *-genic* = producing) occurs when stretching triggers contraction of smooth muscle cells in the walls of afferent arterioles. As blood pressure rises, GFR also rises because renal blood flow increases. However, the elevated blood pressure stretches the walls of the afferent arterioles. In response, smooth muscle fibers in the wall of the afferent arteriole contract, which narrows the arteriole's lumen. As a result, renal blood flow decreases, thus reducing GFR to its previous level. Conversely, when arterial blood pressure drops, the smooth muscle cells are stretched less and thus relax. The afferent arterioles dilate, renal blood flow increases, and GFR increases. The myogenic mechanism normalizes renal blood flow and GFR within seconds after a change in blood pressure.

The second contributor to renal autoregulation, **tubuloglomerular feedback** (too′-bū-lō-glō-MER-ū-lar), is so named because part of the renal tubules—the macula densa—provides feedback to the glomerulus (Figure 26.10). When GFR is above normal due to elevated systemic blood pressure, filtered fluid flows more rapidly along the renal tubules. As a result, the proximal convoluted tubule and loop of Henle have less time to reabsorb Na$^+$, Cl$^-$, and water. Macula densa cells are thought to detect the increased delivery of Na$^+$, Cl$^-$, and water and to inhibit release of nitric oxide (NO) from cells in the juxtaglomerular apparatus (JGA). Because NO causes vasodilation, afferent arterioles constrict when the level of NO declines. As a result, less blood flows into the glomerular capillaries, and GFR decreases. When blood pressure falls, causing GFR to be lower than normal, the opposite sequence of events occurs, although to a lesser degree. Tubuloglomerular feedback operates more slowly than the myogenic mechanism.

Neural Regulation of GFR

Like most blood vessels of the body, those of the kidneys are supplied by sympathetic ANS fibers that release norepinephrine. Nor-

Figure 26.10 Tubuloglomerular feedback.

Macula densa cells of the juxtaglomerular apparatus (JGA) provide negative feedback regulation of the glomerular filtration rate.

Some stimulus disrupts homeostasis by

Increasing

Glomerular filtration rate (GFR)

Receptors

Macula densa cells of JGA detect increased delivery of Na$^+$, Cl$^-$, and water

Input

Control center

Juxtaglomerular apparatus

Return to homeostasis when response brings GFR back to normal

Output Decreased secretion of nitric oxide

Effectors

Afferent arteriole constricts, which decreases blood flow through glomerulus

Decrease in GFR

? **Why is this process termed autoregulation?**

epinephrine causes vasoconstriction through the activation of α_1 receptors, which are particularly plentiful in the smooth muscle fibers of afferent arterioles. At rest, sympathetic stimulation is moderately low, the afferent and efferent arterioles are dilated, and renal autoregulation of GFR prevails. With moderate sympathetic stimulation, both afferent and efferent arterioles constrict to the same degree. Blood flow into and out of the glomerulus is restricted to the same extent, which decreases GFR only slightly. With greater sympathetic stimulation, however, as occurs during exercise or hemorrhage, vasoconstriction of the afferent arterioles predominates. As a result, blood flow into glomerular capillaries is greatly decreased, and GFR drops. This lowering of renal blood flow has two consequences: (1) It reduces urine output, which helps conserve blood volume. (2) It permits greater blood flow to other body tissues.

Hormonal Regulation of GFR

Two hormones contribute to regulation of GFR. Angiotensin II reduces GFR; atrial natriuretic peptide (ANP) increases GFR. **Angiotensin II** (an'-jē-ō-TEN-sin) is a very potent vasoconstrictor that narrows both afferent and efferent arterioles and reduces renal blood flow, thereby decreasing GFR. Cells in the atria of the heart secrete **atrial natriuretic peptide (ANP)** (na'-trē-ū-RET-ik). Stretching of the atria, as occurs when blood volume increases, stimulates secretion of ANP. By causing relaxation of the glomerular mesangial cells, ANP increases the capillary surface area available for filtration. Glomerular filtration rate rises as the surface area increases.

Table 26.2 summarizes the regulation of glomerular filtration rate.

✔CHECKPOINT

7. If the urinary excretion rate of a drug such as penicillin is greater than the rate at which it is filtered at the glomerulus, how else is it getting into the urine?
8. What is the major chemical difference between blood plasma and glomerular filtrate?
9. Why is there much greater filtration through glomerular capillaries than through capillaries elsewhere in the body?
10. Write the equation for the calculation of net filtration pressure (NFP) and explain the meaning of each term.
11. How is glomerular filtration rate regulated?

26.5 TUBULAR REABSORPTION AND TUBULAR SECRETION

◉ OBJECTIVES

- Describe the routes and mechanisms of tubular reabsorption and secretion.
- Describe how specific segments of the renal tubule and collecting duct reabsorb water and solutes.
- Describe how specific segments of the renal tubule and collecting duct secrete solutes into the urine.

Principles of Tubular Reabsorption and Secretion

The volume of fluid entering the proximal convoluted tubules in just half an hour is greater than the total blood plasma volume because the normal rate of glomerular filtration is so high. Obviously some of this fluid must be returned somehow to the bloodstream. Reabsorption—the return of most of the filtered water and many of the filtered solutes to the bloodstream—is the second basic function of the nephron and collecting duct. Normally, about 99% of the filtered water is reabsorbed. Epithelial cells all along the renal tubule and duct carry out reabsorption, but proximal convoluted tubule cells make the largest contribution. Solutes that are reabsorbed by both active and passive processes include glucose, amino acids, urea, and ions such as Na^+ (sodium), K^+ (potassium), Ca^{2+} (calcium), Cl^- (chloride), HCO_3^- (bicarbonate), and HPO_4^{2-} (phosphate). Once fluid passes through the proximal convoluted tubule, cells located more distally fine-tune the reabsorption processes to maintain homeostatic balances of water and selected ions. Most small proteins and peptides that pass through the filter also are reabsorbed, usually via pinocytosis.

TABLE 26.2

Regulation of Glomerular Filtration Rate (GFR)

TYPE OF REGULATION	MAJOR STIMULUS	MECHANISM AND SITE OF ACTION	EFFECT ON GFR
Renal autoregulation			
Myogenic mechanism	Increased stretching of smooth muscle fibers in afferent arteriole walls due to increased blood pressure.	Stretched smooth muscle fibers contract, thereby narrowing lumen of afferent arterioles.	Decrease.
Tubuloglomerular feedback	Rapid delivery of Na^+ and Cl^- to the macula densa due to high systemic blood pressure.	Decreased release of nitric oxide (NO) by juxtaglomerular apparatus causes constriction of afferent arterioles.	Decrease.
Neural regulation	Increase in activity level of renal sympathetic nerves releases norepinephrine.	Constriction of afferent arterioles through activation of α_1 receptors and increased release of renin.	Decrease.
Hormone regulation			
Angiotensin II	Decreased blood volume or blood pressure stimulates production of angiotensin II.	Constriction of afferent and efferent arterioles.	Decrease.
Atrial natriuretic peptide (ANP)	Stretching of atria of heart stimulates secretion of ANP.	Relaxation of mesangial cells in glomerulus increases capillary surface area available for filtration.	Increase.

TABLE 26.3

Substances Filtered, Reabsorbed, and Excreted in Urine

SUBSTANCE	FILTERED* (ENTERS GLOMERULAR CAPSULE PER DAY)	REABSORBED (RETURNED TO BLOOD PER DAY)	URINE (EXCRETED PER DAY)
Water	180 liters	178–179 liters	1–2 liters
Proteins	2.0 g	1.9 g	0.1 g
Sodium ions (Na^+)	579 g	575 g	4 g
Chloride ions (Cl^-)	640 g	633.7 g	6.3 g
Bicarbonate ions (HCO_3^-)	275 g	274.97 g	0.03 g
Glucose	162 g	162 g	0 g
Urea	54 g	24 g	30 g[†]
Potassium ions (K^+)	29.6 g	29.6 g	2.0 g[‡]
Uric acid	8.5 g	7.7 g	0.8 g
Creatinine	1.6 g	0 g	1.6 g

*Assuming GFR is 180 liters per day.   [†]In addition to being filtered and reabsorbed, urea is secreted.
[‡]After virtually all filtered K^+ is reabsorbed in the convoluted tubules and loop of Henle, a variable amount of K^+ is secreted by principal cells in the collecting duct.

To appreciate the magnitude of tubular reabsorption, look at Table 26.3 and compare the amounts of substances that are filtered, reabsorbed, and excreted in urine.

The third function of nephrons and collecting ducts is tubular secretion, the transfer of materials from the blood and tubule cells into glomerular filtrate. Secreted substances include hydrogen ions (H^+), K^+, ammonium ions (NH_4^+), creatinine, and certain drugs such as penicillin. Tubular secretion has two important outcomes: (1) The secretion of H^+ helps control blood pH. (2) The secretion of other substances helps eliminate them from the body.

As a result of tubular secretion, certain substances pass from blood into urine and may be detected by a urinalysis (see Section 26.7). It is especially important to test athletes for the presence of performance-enhancing drugs such as anabolic steroids, plasma expanders, erythropoietin, hCG, hGH, and amphetamines. Urine tests can also be used to detect the presence of alcohol or illegal drugs such as marijuana, cocaine, and heroin.

Reabsorption Routes

A substance being reabsorbed from the fluid in the tubule lumen can take one of two routes before entering a peritubular capillary: It can move *between* adjacent tubule cells or *through* an individual tubule cell (Figure 26.11). Along the renal tubule, tight junctions surround and join neighboring cells to one another, much like the plastic rings that hold a six-pack of soda cans together. The **apical membrane** (the tops of the soda cans) contacts the tubular fluid, and the **basolateral membrane** (the bottoms and sides of the soda cans) contacts interstitial fluid at the base and sides of the cell.

Fluid can leak *between* the cells in a passive process known as **paracellular reabsorption** (par′-a-SEL-ū-lar; *para-* = beside). Even though the epithelial cells are connected by tight junctions, the tight junctions between cells in the proximal convoluted

Figure 26.11 Reabsorption routes: paracellular reabsorption and transcellular reabsorption.

 In paracellular reabsorption, water and solutes in tubular fluid return to the bloodstream by moving between tubule cells; in transcellular reabsorption, solutes and water in tubular fluid return to the bloodstream by passing through a tubule cell.

Key:

•••••▶ Diffusion

——▶ Active transport

⌀ Sodium–potassium pump (Na^+/K^+ ATPase)

❓ **What is the main function of the tight junctions between tubule cells?**

tubules are "leaky" and permit some reabsorbed substances to pass between cells into peritubular capillaries. In some parts of the renal tubule, the paracellular route is thought to account for up to 50% of the reabsorption of certain ions and the water that accompanies them via osmosis. In **transcellular reabsorption** (trans′-SEL-ū-lar; *trans-* = across), a substance passes from the fluid in the tubular lumen *through* the apical membrane of a tubule cell, across the cytosol, and out into interstitial fluid through the basolateral membrane.

Transport Mechanisms

When renal cells transport solutes out of or into tubular fluid, they move specific substances in one direction only. Not surprisingly, different types of transport proteins are present in the apical and basolateral membranes. The tight junctions form a barrier that prevents mixing of proteins in the apical and basolateral membrane compartments. Reabsorption of Na^+ by the renal tubules is especially important because of the large number of sodium ions that pass through the glomerular filters.

Cells lining the renal tubules, like other cells throughout the body, have a low concentration of Na^+ in their cytosol due to the activity of sodium–potassium pumps (Na^+/K^+ ATPases). These pumps are located in the basolateral membranes and eject Na^+ from the renal tubule cells (Figure 26.11). The absence of sodium–potassium pumps in the apical membrane ensures that reabsorption of Na^+ is a one-way process. Most sodium ions that cross the apical membrane will be pumped into interstitial fluid at the base and sides of the cell. The amount of ATP used by sodium–potassium pumps in the renal tubules is about 6% of the total ATP consumption of the body at rest. This may not sound like much, but it is about the same amount of energy used by the diaphragm as it contracts during quiet breathing.

As we noted in Chapter 3, transport of materials across membranes may be either active or passive. Recall that in **primary active transport** the energy derived from hydrolysis of ATP is used to "pump" a substance across a membrane; the sodium–potassium pump is one such pump. In **secondary active transport** the energy stored in an ion's electrochemical gradient, rather than hydrolysis of ATP, drives another substance across a membrane. Secondary active transport couples the movement of an ion down its electrochemical gradient to the "uphill" movement of a second substance against its electrochemical gradient. *Symporters* are membrane proteins that move two or more substances in the same direction across a membrane. *Antiporters* move two or more substances in opposite directions across a membrane. Each type of transporter has an upper limit on how fast it can work, just as an escalator has a limit on how many people it can carry from one level to another in a given period. This limit, called the **transport maximum (T_m),** is measured in mg/min.

Solute reabsorption drives water reabsorption because all water reabsorption occurs via osmosis. About 90% of the reabsorption of water filtered by the kidneys occurs along with the reabsorption of solutes such as Na^+, Cl^-, and glucose. Water reabsorbed with solutes in tubular fluid is termed **obligatory water reabsorption** (ob-LIG-a-tor′-ē) because the water is "obliged" to fol-

low the solutes when they are reabsorbed. This type of water reabsorption occurs in the proximal convoluted tubule and the descending limb of the loop of Henle because these segments of the nephron are always permeable to water. Reabsorption of the final 10% of the water, a total of 10–20 liters per day, is termed **facultative water reabsorption** (FAK-ul-tā′-tiv). The word *facultative* means "capable of adapting to a need." Facultative water reabsorption is regulated by antidiuretic hormone and occurs mainly in the collecting ducts.

CLINICAL CONNECTION | *Glucosuria*

When the blood concentration of glucose is above 200 mg/mL, the renal symporters cannot work fast enough to reabsorb all the glucose that enters the glomerular filtrate. As a result, some glucose remains in the urine, a condition called **glucosuria** (gloo′-kō-SOO-rē-a). The most common cause of glucosuria is diabetes mellitus, in which the blood glucose level may rise far above normal because insulin activity is deficient. Rare genetic mutations in the renal Na^+–glucose symporter greatly reduce its T_m and cause glucosuria. In these cases, glucose appears in the urine even though the blood glucose level is normal. Excessive glucose in the glomerular filtrate inhibits water reabsorption by kidney tubules. This leads to increased urinary output (polyuria), decreased blood volume, and dehydration. •

Now that we have discussed the principles of renal transport, we will follow the filtered fluid from the proximal convoluted tubule, into the loop of Henle, on to the distal convoluted tubule, and through the collecting ducts. In each segment, we will examine where and how specific substances are reabsorbed and secreted. The filtered fluid becomes *tubular fluid* once it enters the proximal convoluted tubule. The composition of tubular fluid changes as it flows along the nephron tubule and through the collecting duct due to reabsorption and secretion. The fluid that drains from papillary ducts into the renal pelvis is *urine*.

Reabsorption and Secretion in the Proximal Convoluted Tubule

The largest amount of solute and water reabsorption from filtered fluid occurs in the proximal convoluted tubules, which reabsorb 65% of the filtered water, Na^+, and K^+; 100% of most filtered organic solutes such as glucose and amino acids; 50% of the filtered Cl^-; 80–90% of the filtered HCO_3^-; 50% of the filtered urea; and a variable amount of the filtered Ca^{2+}, Mg^{2+}, and HPO_4^{2-} (phosphate). In addition, proximal convoluted tubules secrete a variable amount of H^+ ions, ammonium ions (NH_4^+), and urea.

Most solute reabsorption in the proximal convoluted tubule (PCT) involves Na^+. Na^+ transport occurs via symport and antiport mechanisms in the proximal convoluted tubule. Normally, filtered glucose, amino acids, lactic acid, water-soluble vitamins, and other nutrients are not lost in the urine. Rather, they are completely reabsorbed in the first half of the proximal convoluted tubule by several types of **Na^+ symporters** located in the apical membrane. Figure 26.12 depicts the operation of one such symporter, the **Na^+–glucose symporter** in the apical membrane of a

Figure 26.12 Reabsorption of glucose by Na⁺–glucose symporters in cells of the proximal convoluted tubule (PCT).

🔑 Normally, all filtered glucose is reabsorbed in the PCT.

Key:

▨ Na⁺–glucose symporter

⌣ Glucose facilitated diffusion transporter

•••▶ Diffusion

⊘ Sodium–potassium pump

❓ **How does filtered glucose enter and leave a PCT cell?**

cell in the PCT. Two Na^+ and a molecule of glucose attach to the symporter protein, which carries them from the tubular fluid into the tubule cell. The glucose molecules then exit the basolateral membrane via facilitated diffusion and they diffuse into peritubular capillaries. Other Na^+ symporters in the PCT reclaim filtered HPO_4^{2-} (phosphate) and SO_4^{2-} (sulfate) ions, all amino acids, and lactic acid in a similar way.

In another secondary active transport process, the **Na^+/H^+ antiporters** carry filtered Na^+ down its concentration gradient into a PCT cell as H^+ is moved from the cytosol into the lumen (Figure 26.13a), causing Na^+ to be reabsorbed into blood and H^+ to be secreted into tubular fluid. PCT cells produce the H^+ needed to keep the antiporters running in the following way. Carbon dioxide (CO_2) diffuses from peritubular blood or tubular fluid or is produced by metabolic reactions within the cells. As also occurs in red blood cells (see Figure 23.23), the enzyme *carbonic anhydrase (CA)* (an-HĪ-drās) catalyzes the reaction of CO_2 with

Figure 26.13 Actions of Na^+/H^+ antiporters in proximal convoluted tubule cells. (a) Reabsorption of sodium ions (Na^+) and secretion of hydrogen ions (H^+) via secondary active transport through the apical membrane; (b) reabsorption of bicarbonate ions (HCO_3^-) via facilitated diffusion through the basolateral membrane. CO_2 = carbon dioxide; H_2CO_3 = carbonic acid; CA = carbonic anhydrase.

🔑 Na^+/H^+ antiporters promote transcellular reabsorption of Na^+ and secretion of H^+.

(a) Na^+ reabsorption and H^+ secretion

(b) HCO_3^- reabsorption

Key:

▨ Na^+/H^+ antiporter

⌣ HCO_3^- facilitated diffusion transporter

•••▶ Diffusion

⊘ Sodium–potassium pump

❓ **Which step in Na^+ movement in part (a) is promoted by the electrochemical gradient?**

water (H_2O) to form carbonic acid (H_2CO_3), which then dissociates into H^+ and HCO_3^-:

$$CO_2 + H_2O \xrightarrow{\text{Carbonic anhydrase}} H_2CO_3 \longrightarrow H^+ + HCO_3^-$$

Most of the HCO_3^- in filtered fluid is reabsorbed in proximal convoluted tubules, thereby safeguarding the body's supply of an important buffer (Figure 26.13b). After H^+ is secreted into the fluid within the lumen of the proximal convoluted tubule, it reacts with filtered HCO_3^- to form H_2CO_3, which readily dissociates into CO_2 and H_2O. Carbon dioxide then diffuses into the tubule cells and joins with H_2O to form H_2CO_3, which dissociates into H^+ and HCO_3^-. As the level of HCO_3^- rises in the cytosol, it exits via facilitated diffusion transporters in the basolateral membrane and diffuses into the blood with Na^+. Thus, for every H^+ secreted into the tubular fluid of the proximal convoluted tubule, one HCO_3^- and one Na^+ are reabsorbed.

Solute reabsorption in proximal convoluted tubules promotes osmosis of water. Each reabsorbed solute increases the osmolarity, first inside the tubule cell, then in interstitial fluid, and finally in the blood. Water thus moves rapidly from the tubular fluid, via both the paracellular and transcellular routes, into the peritubular capillaries and restores osmotic balance (Figure 26.14). In other words, reabsorption of the solutes creates an osmotic gradient that promotes the reabsorption of water via osmosis. Cells lining the proximal convoluted tubule and the descending limb of the loop of Henle are especially permeable to water because they have many molecules of **aquaporin-1** (ak-kwa-PŌR-in). This integral protein in the plasma membrane is a water channel that greatly

increases the rate of water movement across the apical and basolateral membranes.

As water leaves the tubular fluid, the concentrations of the remaining filtered solutes increase. In the second half of the PCT, electrochemical gradients for Cl^-, K^+, Ca^{2+}, Mg^{2+}, and urea promote their passive diffusion into peritubular capillaries via both paracellular and transcellular routes. Among these ions, Cl^- is present in the highest concentration. Diffusion of negatively charged Cl^- into interstitial fluid via the paracellular route makes the interstitial fluid electrically more negative than the tubular fluid. This negativity promotes passive paracellular reabsorption of cations, such as K^+, Ca^{2+}, and Mg^{2+}.

Ammonia (NH_3) is a poisonous waste product derived from the deamination (removal of an amino group) of various amino acids, a reaction that occurs mainly in hepatocytes (liver cells). Hepatocytes convert most of this ammonia to urea, a less-toxic compound. Although tiny amounts of urea and ammonia are present in sweat, most excretion of these nitrogen-containing waste products occurs via the urine. Urea and ammonia in blood are both filtered at the glomerulus and secreted by proximal convoluted tubule cells into the tubular fluid.

Proximal convoluted tubule cells can produce additional NH_3 by deaminating the amino acid glutamine in a reaction that also generates HCO_3^-. The NH_3 quickly binds H^+ to become an ammonium ion (NH_4^+), which can substitute for H^+ aboard Na^+/H^+ antiporters in the apical membrane and be secreted into the tubular fluid. The HCO_3^- generated in this reaction moves through the basolateral membrane and then diffuses into the bloodstream, providing additional buffers in blood plasma.

Figure 26.14 Passive reabsorption of Cl^-, K^+, Ca^{2+}, Mg^{2+}, urea, and water in the second half of the proximal convoluted tubule.

⑥ 🔑 Electrochemical gradients promote passive reabsorption of solutes via both paracellular and transcellular routes.

❓ **By what mechanism is water reabsorbed from tubular fluid?**

Reabsorption in the Loop of Henle

Because all of the proximal convoluted tubules reabsorb about 65% of the filtered water (about 80 mL/min), fluid enters the next part of the nephron, the loop of Henle, at a rate of 40–45 mL/min. The chemical composition of the tubular fluid now is quite different from that of glomerular filtrate because glucose, amino acids, and other nutrients are no longer present. The osmolarity of the tubular fluid is still close to the osmolarity of blood, however, because reabsorption of water by osmosis keeps pace with reabsorption of solutes all along the proximal convoluted tubule.

The loop of Henle reabsorbs about 15% of the filtered water, 20–30% of the filtered Na^+ and K^+, 35% of the filtered Cl^-, 10–20% of the filtered HCO_3^-, and a variable amount of the filtered Ca^{2+} and Mg^{2+}. Here, for the first time, reabsorption of water via osmosis is *not* automatically coupled to reabsorption of filtered solutes because part of the loop of Henle is relatively impermeable to water. The loop of Henle thus sets the stage for *independent* regulation of both the *volume* and *osmolarity* of body fluids.

The apical membranes of cells in the thick ascending limb of the loop of Henle have **Na^+–K^+–$2Cl^-$ symporters** that simultaneously reclaim one Na^+, one K^+, and two Cl^- from the fluid in the tubular lumen (Figure 26.15). Na^+ that is actively transported into interstitial fluid at the base and sides of the cell diffuses into the vasa recta. Cl^- moves through leakage channels in the basolateral membrane into interstitial fluid and then into the vasa

Figure 26.15 Na^+–K^+–$2Cl^-$ symporter in the thick ascending limb of the loop of Henle.

 Cells in the thick ascending limb have symporters that simultaneously reabsorb one Na^+, one K^+, and two Cl^-.

Key:

 Na^+–K^+–$2Cl^-$ symporter

 Leakage channels

 Sodium–potassium pump

••••► Diffusion

? **Why is this process considered secondary active transport? Does water reabsorption accompany ion reabsorption in this region of the nephron?**

recta. Because many K^+ leakage channels are present in the apical membrane, most K^+ brought in by the symporters moves down its concentration gradient back into the tubular fluid. Thus, the main effect of the Na^+–K^+–$2Cl^-$ symporters is reabsorption of Na^+ and Cl^-.

The movement of positively charged K^+ into the tubular fluid through the apical membrane channels leaves the interstitial fluid and blood with more negative charges relative to fluid in the ascending limb of the loop of Henle. This relative negativity promotes reabsorption of cations—Na^+, K^+, Ca^{2+}, and Mg^{2+}—via the paracellular route.

Although about 15% of the filtered water is reabsorbed in the *descending* limb of the loop of Henle, little or no water is reabsorbed in the *ascending* limb. In this segment of the tubule, the apical membranes are virtually impermeable to water. Because ions but not water molecules are reabsorbed, the osmolarity of the tubular fluid decreases progressively as fluid flows toward the end of the ascending limb.

Reabsorption in the Early Distal Convoluted Tubule

Fluid enters the distal convoluted tubules at a rate of about 25 mL/min because 80% of the filtered water has now been reabsorbed. The early or initial part of the distal convoluted tubule (DCT) reabsorbs about 10–15% of the filtered water, 5% of the filtered Na^+, and 5% of the filtered Cl^-. Reabsorption of Na^+ and Cl^- occurs by means of Na^+–Cl^- **symporters** in the apical membranes. Sodium–potassium pumps and Cl^- leakage channels in the basolateral membranes then permit reabsorption of Na^+ and Cl^- into the peritubular capillaries. The early DCT also is a major site where parathyroid hormone (PTH) stimulates reabsorption of Ca^{2+}. The amount of Ca^{2+} reabsorption in the early DCT varies depending on the body's needs.

Reabsorption and Secretion in the Late Distal Convoluted Tubule and Collecting Duct

By the time fluid reaches the end of the distal convoluted tubule, 90–95% of the filtered solutes and water have returned to the bloodstream. Recall that two different types of cells—principal cells and intercalated cells—are present at the late or terminal part of the distal convoluted tubule and throughout the collecting duct. The principal cells reabsorb Na^+ and secrete K^+; the intercalated cells reabsorb K^+ and HCO_3^- and secrete H^+. In the late distal convoluted tubules and collecting ducts, the amount of water and solute reabsorption and the amount of solute secretion vary depending on the body's needs.

In contrast to earlier segments of the nephron, Na^+ passes through the apical membrane of principal cells via Na^+ leakage channels rather than by means of symporters or antiporters (Figure 26.16). The concentration of Na^+ in the cytosol remains low, as usual, because the sodium–potassium pumps actively transport Na^+ across the basolateral membranes. Then Na^+ passively diffuses into the peritubular capillaries from the interstitial spaces around the tubule cells.

Normally, transcellular and paracellular reabsorption in the proximal convoluted tubule and loop of Henle return most filtered K^+ to the bloodstream. To adjust for varying dietary intake of potassium and to maintain a stable level of K^+ in body fluids, principal cells secrete a variable amount of K^+ (Figure 26.16). Because the basolateral sodium–potassium pumps continually bring K^+ into principal cells, the intracellular concentration of K^+ remains high. K^+ leakage channels are present in both the apical and basolateral membranes. Thus, some K^+ diffuses down its concentration gradient into the tubular fluid, where the K^+

Figure 26.16 Reabsorption of Na⁺ and secretion of K⁺ by principal cells in the last part of the distal convoluted tubule and in the collecting duct.

 In the apical membrane of principal cells, Na⁺ leakage channels allow entry of Na⁺ while K⁺ leakage channels allow exit of K⁺ into the tubular fluid.

Key:

• • • • ▶ Diffusion

⊥⊥ ⊢⊢ Leakage channels

✗ Sodium–potassium pump

? **Which hormone stimulates reabsorption and secretion by principal cells, and how does this hormone exert its effect?**

concentration is very low. This secretion mechanism is the main source of K⁺ excreted in the urine.

Hormonal Regulation of Tubular Reabsorption and Tubular Secretion

Five hormones affect the extent of Na⁺, Cl⁻, Ca²⁺, and water reabsorption as well as K⁺ secretion by the renal tubules. These hormones include angiotensin II, aldosterone, antidiuretic hormone, atrial natriuretic peptide, and parathyroid hormone.

Renin–Angiotensin–Aldosterone System

When blood volume and blood pressure decrease, the walls of the afferent arterioles are stretched less, and the juxtaglomerular cells

secrete the enzyme **renin** (RĒ-nin) into the blood. Sympathetic stimulation also directly stimulates release of renin from juxtaglomerular cells. Renin clips off a 10–amino acid peptide called angiotensin I from angiotensinogen, which is synthesized by hepatocytes (see Figure 18.16). By clipping off two more amino acids, *angiotensin-converting enzyme (ACE)* (an′-jē-ō-TEN-sin) converts angiotensin I to **angiotensin II,** which is the active form of the hormone.

Angiotensin II affects renal physiology in three main ways:

1. It decreases the glomerular filtration rate by causing vasoconstriction of the afferent arterioles.

2. It enhances reabsorption of Na⁺, Cl⁻, and water in the proximal convoluted tubule by stimulating the activity of Na⁺/H⁺ antiporters.

3. It stimulates the adrenal cortex to release **aldosterone** (al-DOS-ter-ōn), a hormone that in turn stimulates the principal cells in the collecting ducts to reabsorb more Na⁺ and Cl⁻ and secrete more K⁺. The osmotic consequence of reabsorbing more Na⁺ and Cl⁻ is that more water is reabsorbed, which causes an increase in blood volume and blood pressure.

Antidiuretic Hormone

Antidiuretic hormone (ADH or **vasopressin)** is released by the posterior pituitary. It regulates facultative water reabsorption by increasing the water permeability of principal cells in the last part of the distal convoluted tubule and throughout the collecting duct. In the absence of ADH, the apical membranes of principal cells have a very low permeability to water. Within principal cells are tiny vesicles containing many copies of a water channel protein known as **aquaporin-2.*** ADH stimulates insertion of the aquaporin-2–containing vesicles into the apical membranes via exocytosis. As a result, the water permeability of the principal cell's apical membrane increases, and water molecules move more rapidly from the tubular fluid into the cells. Because the basolateral membranes are always relatively permeable to water, water molecules then move rapidly into the blood. The kidneys can produce as little as 400–500 mL of very concentrated urine each day when ADH concentration is maximal, for instance, during severe dehydration. When ADH level declines, the aquaporin-2 channels are removed from the apical membrane via endocytosis. The kidneys produce a large volume of dilute urine when ADH level is low.

A negative feedback system involving ADH regulates facultative water reabsorption (Figure 26.17). When the osmolarity or osmotic pressure of plasma and interstitial fluid increases—that is, when water concentration decreases—by as little as 1%, osmoreceptors in the hypothalamus detect the change. Their nerve impulses stimulate secretion of more ADH into the blood, and the principal cells become more permeable to water. As facultative water reabsorption increases, plasma osmolarity decreases to normal. A second powerful stimulus for ADH secretion is a decrease in blood volume, as occurs in hemorrhaging or severe

---

*ADH does not govern the previously mentioned water channel (aquaporin-1).

Figure 26.17 Negative feedback regulation of facultative water reabsorption by ADH.

Most water reabsorption (90%) is obligatory; 10% is facultative.

In addition to ADH, which other hormones contribute to the regulation of water reabsorption?

dehydration. In the pathological absence of ADH activity, a condition known as *diabetes insipidus*, a person may excrete up to 20 liters of very dilute urine daily.

Atrial Natriuretic Peptide

A large increase in blood volume promotes release of **atrial natriuretic peptide (ANP)** from the heart. Although the importance of ANP in normal regulation of tubular function is unclear, it can inhibit reabsorption of Na^+ and water in the proximal convoluted tubule and collecting duct. ANP also suppresses the secretion of aldosterone and ADH. These effects increase the excretion of Na^+ in urine (natriuresis) and increase urine output (diuresis), which decreases blood volume and blood pressure.

Parathyroid Hormone

A lower-than-normal level of Ca^{2+} in the blood stimulates the parathyroid glands to release **parathyroid hormone (PTH).** PTH in turn stimulates cells in the early distal convoluted tubules to reabsorb more Ca^{2+} into the blood. PTH also inhibits HPO_4^{2-} (phosphate) reabsorption in proximal convoluted tubules, thereby promoting phosphate excretion.

Table 26.4 summarizes hormonal regulation of tubular reabsorption and tubular secretion.

✔ CHECKPOINT

12. Diagram the reabsorption of substances via the transcellular and paracellular routes. Label the apical membrane and the basolateral membrane. Where are the sodium–potassium pumps located?
13. Describe two mechanisms in the PCT, one in the loop of Henle, one in the DCT, and one in the collecting duct for reabsorption of Na^+. What other solutes are reabsorbed or secreted with Na^+ in each mechanism?
14. How do intercalated cells secrete hydrogen ions?
15. Graph the percentages of filtered water and filtered Na^+ that are reabsorbed in the PCT, loop of Henle, DCT, and collecting duct. Indicate which hormones, if any, regulate reabsorption in each segment.

26.6 PRODUCTION OF DILUTE AND CONCENTRATED URINE

◉ OBJECTIVE

• Describe how the renal tubule and collecting ducts produce dilute and concentrated urine.

Even though your fluid intake can be highly variable, the total volume of fluid in your body normally remains stable. Homeostasis of body fluid volume depends in large part on the ability of the kidneys to regulate the rate of water loss in urine. Normally functioning kidneys produce a large volume of dilute urine when fluid intake is high, and a small volume of concentrated urine when fluid intake is low or fluid loss is large. ADH controls whether dilute urine or concentrated urine is formed. In the absence of ADH, urine is very dilute. However, a high level of ADH stimulates

TABLE 26.4

Hormonal Regulation of Tubular Reabsorption and Tubular Secretion

HORMONE	MAJOR STIMULI THAT TRIGGER RELEASE	MECHANISM AND SITE OF ACTION	EFFECTS
Angiotensin II	Low blood volume or low blood pressure stimulates renin-induced production of angiotensin II.	Stimulates activity of Na^+/H^+ antiporters in proximal tubule cells.	Increases reabsorption of Na^+, other solutes, and water, which increases blood volume and blood pressure.
Aldosterone	Increased angiotensin II level and increased level of plasma K^+ promote release of aldosterone by adrenal cortex.	Enhances activity of sodium–potassium pumps in basolateral membrane and Na^+ channels in apical membrane of principal cells in collecting duct.	Increases secretion of K^+ and reabsorption of Na^+, Cl^-; increases reabsorption of water, which increases blood volume and blood pressure.
Antidiuretic hormone (ADH) or vasopressin	Increased osmolarity of extracellular fluid or decreased blood volume promotes release of ADH from posterior pituitary gland.	Stimulates insertion of water channel proteins (aquaporin-2) into apical membranes of principal cells.	Increases facultative reabsorption of water, which decreases osmolarity of body fluids.
Atrial natriuretic peptide (ANP)	Stretching of atria of heart stimulates ANP secretion.	Suppresses reabsorption of Na^+ and water in proximal tubule and collecting duct; inhibits secretion of aldosterone and ADH.	Increases excretion of Na^+ in urine (natriuresis); increases urine output (diuresis) and thus decreases blood volume and blood pressure.
Parathyroid hormone (PTH)	Decreased level of plasma Ca^{2+} promotes release of PTH from parathyroid glands.	Stimulates opening of Ca^{2+} channels in apical membranes of early distal tubule cells.	Increases reabsorption of Ca^{2+}.

reabsorption of more water into blood, producing a concentrated urine.

Formation of Dilute Urine

Glomerular filtrate has the same ratio of water and solute particles as blood; its osmolarity is about 300 mOsm/liter. As previously noted, fluid leaving the proximal convoluted tubule is still isotonic to plasma. When *dilute* urine is being formed (Figure 26.18), the osmolarity of the fluid in the tubular lumen *increases* as it flows down the descending limb of the loop of Henle, *decreases* as it flows up the ascending limb, and *decreases* still more as it flows through the rest of the nephron and collecting duct. These changes in osmolarity result from the following conditions along the path of tubular fluid:

1. Because the osmolarity of the interstitial fluid of the renal medulla becomes progressively greater, more and more water is reabsorbed by osmosis as tubular fluid flows along the descending limb toward the tip of the loop. (The source of this medullary osmotic gradient is explained shortly.) As a result, the fluid remaining in the lumen becomes progressively more concentrated.

2. Cells lining the thick ascending limb of the loop have symporters that actively reabsorb Na^+, K^+, and Cl^- from the tubular fluid (see Figure 26.15). The ions pass from the tubular fluid into thick ascending limb cells, then into interstitial fluid, and finally some diffuse into the blood inside the vasa recta.

3. Although solutes are being reabsorbed in the thick ascending limb, the water permeability of this portion of the nephron is always quite low, so water cannot follow by osmosis. As solutes—but not water molecules—are leaving the tubular fluid, its osmolarity drops to about 150 mOsm/liter. The fluid entering the distal convoluted tubule is thus more dilute than plasma.

4. While the fluid continues flowing along the distal convoluted tubule, additional solutes but only a few water molecules are reabsorbed. The early distal convoluted tubule cells are not very permeable to water and are not regulated by ADH.

5. Finally, the principal cells of the late distal convoluted tubules and collecting ducts are impermeable to water when the ADH level is very low. Thus, tubular fluid becomes progressively more dilute as it flows onward. By the time the tubular fluid drains into the renal pelvis, its concentration can be as low as 65–70 mOsm/liter. This is four times more dilute than blood plasma or glomerular filtrate.

Formation of Concentrated Urine

When water intake is low or water loss is high (such as during heavy sweating), the kidneys must conserve water while still eliminating wastes and excess ions. Under the influence of ADH, the kidneys produce a small volume of highly concentrated urine. Urine can be four times more concentrated (up to 1200 mOsm/liter) than blood plasma or glomerular filtrate (300 mOsm/liter).

Figure 26.18 Formation of dilute urine. Numbers indicate osmolarity in milliosmoles per liter (mOsm/liter). Heavy brown lines in the ascending limb of the loop of Henle and in the distal convoluted tubule indicate impermeability to water; heavy blue lines indicate the last part of the distal convoluted tubule and the collecting duct, which are impermeable to water in the absence of ADH; light blue areas around the nephron represent interstitial fluid.

When the ADH level is low, urine is dilute and has an osmolarity less than the osmolarity of blood.

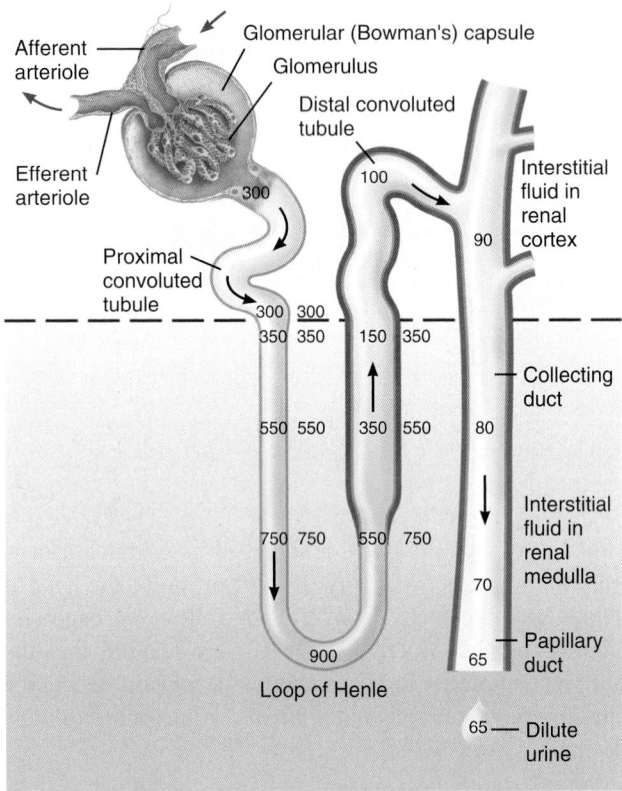

Which portions of the renal tubule and collecting duct reabsorb more solutes than water to produce dilute urine?

The ability of ADH to cause excretion of concentrated urine depends on the presence of an **osmotic gradient** of solutes in the interstitial fluid of the renal medulla. Notice in Figure 26.19 that the solute concentration of the interstitial fluid in the kidney increases from about 300 mOsm/liter in the renal cortex to about 1200 mOsm/liter deep in the renal medulla. The three major solutes that contribute to this high osmolarity are Na^+, Cl^-, and urea. Two main factors contribute to building and maintaining this osmotic gradient: (1) differences in solute and water permeability and reabsorption in different sections of the long loops of Henle and the collecting ducts, and (2) the countercurrent flow of fluid through tube-shaped structures in the renal medulla. *Countercurrent flow* refers to the flow of fluid in opposite directions. This occurs when fluid flowing in one tube runs counter (opposite) to fluid flowing in a nearby parallel tube. Examples of countercur-

rent flow include the flow of tubular fluid through the descending and ascending limbs of the loop of Henle and the flow of blood through the ascending and descending parts of the vasa recta. Two types of **countercurrent mechanisms** exist in the kidneys: countercurrent multiplication and countercurrent exchange.

Countercurrent Multiplication

Countercurrent multiplication is the process by which a progressively increasing osmotic gradient is formed in the interstitial fluid of the renal medulla as a result of countercurrent flow. Countercurrent multiplication involves the long loops of Henle of juxtamedullary nephrons. Note in Figure 26.19a that the descending limb of the loop of Henle carries tubular fluid from the renal cortex deep into the medulla, and the ascending limb carries it in the opposite direction. Since countercurrent flow through the descending and ascending limbs of the long loop of Henle establishes the osmotic gradient in the renal medulla, the long loop of Henle is said to function as a **countercurrent multiplier.** The kidneys use this osmotic gradient to excrete concentrated urine.

Production of concentrated urine by the kidneys occurs in the following way (Figure 26.19):

① *Symporters in thick ascending limb cells of the loop of Henle cause a buildup of Na^+ and Cl^- in the renal medulla.* In the thick ascending limb of the loop of Henle, the Na^+–K^+–$2Cl^-$ symporters reabsorb Na^+ and Cl^- from the tubular fluid (Figure 26.19a). Water is not reabsorbed in this segment, however, because the cells are impermeable to water. As a result, there is a buildup of Na^+ and Cl^- ions in the interstitial fluid of the medulla.

② *Countercurrent flow through the descending and ascending limbs of the loop of Henle establishes an osmotic gradient in the renal medulla.* Since tubular fluid constantly moves from the descending limb to the thick ascending limb of the loop of Henle, the thick ascending limb is constantly reabsorbing Na^+ and Cl^-. Consequently, the reabsorbed Na^+ and Cl^- become increasingly concentrated in the interstitial fluid of the medulla, which results in the formation of an osmotic gradient that ranges from 300 mOsm/liter in the outer medulla to 1200 mOsm/liter deep in the inner medulla. The descending limb of the loop of Henle is very permeable to water but impermeable to solutes except urea. Because the osmolarity of the interstitial fluid outside the descending limb is higher than the tubular fluid within it, water moves out of the descending limb via osmosis. This causes the osmolarity of the tubular fluid to increase. As the fluid continues along the descending limb, its osmolarity increases even more: At the hairpin turn of the loop, the osmolarity can be as high as 1200 mOsm/liter in juxtamedullary nephrons. As you have already learned, the ascending limb of the loop is impermeable to water, but its symporters reabsorb Na^+ and Cl^- from the tubular fluid into the interstitial fluid of the renal medulla, so the osmolarity of the tubular fluid progressively decreases as it flows through the ascending limb. At the junction of the medulla and cortex, the osmolarity of the tubular fluid has

Figure 26.19 **Mechanism of urine concentration in long-loop juxtamedullary nephrons.** The green line indicates the presence of Na+–K+–2Cl− symporters that simultaneously reabsorb these ions into the interstitial fluid of the renal medulla; this portion of the nephron is also relatively impermeable to water and urea. All concentrations are in milliosmoles per liter (mOsm/liter).

🔑 The formation of concentrated urine depends on high concentrations of solutes in interstitial fluid in the renal medulla.

Juxtamedullary nephron and its blood supply together

Vasa recta
Loop of Henle

Legend:
- - - ➤ H_2O
- ──➤ Na^+Cl^-
- ──➤ Blood flow
- ──➤ Presence of $Na^+–K^+–2Cl^-$ symporters
- ──➤ Flow of tubular fluid

Glomerular (Bowman's) capsule
Glomerulus
Afferent arteriole
Distal convoluted tubule
Efferent arteriole
Proximal convoluted tubule
Collecting duct

Interstitial fluid in renal cortex

Interstitial fluid in renal medulla

Osmotic gradient

Osmotic gradient scale: 300, 400, 600, 800, 1000, 1200

3 Principal cells in collecting duct reabsorb more water when ADH is present

1 Symporters in thick ascending limb cause buildup of Na+ and Cl− in renal medulla

Urea

4 Urea recycling causes buildup of urea in renal medulla

2 Countercurrent flow through loop of Henle establishes osmotic gradient

Loop of Henle

Papillary duct

Concentrated urine

(a) Reabsorption of Na+, Cl−, and water in long-loop juxtamedullary nephron

(b) Recycling of salts and urea in vasa recta

❓ **Which solutes are the main contributors to the high osmolarity of interstitial fluid in the renal medulla?**

fallen to about 100 mOsm/liter. Overall, tubular fluid becomes progressively more concentrated as it flows along the descending limb and progressively more dilute as it moves along the ascending limb.

❸ ***Cells in the collecting ducts reabsorb more water and urea.*** When ADH increases the water permeability of the principal cells, water quickly moves via osmosis out of the collecting duct tubular fluid, into the interstitial fluid of the inner medulla, and then into the vasa recta. With loss of water, the urea left behind in the tubular fluid of the collecting duct becomes increasingly concentrated. Because duct cells deep in the medulla are permeable to it, urea diffuses from the fluid in the duct into the interstitial fluid of the medulla.

❹ ***Urea recycling causes a buildup of urea in the renal medulla.*** As urea accumulates in the interstitial fluid, some of it diffuses into the tubular fluid in the descending and thin ascending limbs of the long loops of Henle, which also are permeable to urea (Figure 26.19a). However, while the fluid flows through the thick ascending limb, distal convoluted tubule, and cortical portion of the collecting duct, urea remains in the lumen because cells in these segments are impermeable to it. As fluid flows along the collecting ducts, water reabsorption continues via osmosis because ADH is present. This water reabsorption *further increases* the concentration of urea in the tubular fluid, more urea diffuses into the interstitial fluid of the inner renal medulla, and the cycle repeats. The constant transfer of urea between segments of the renal tubule and the interstitial fluid of the medulla is termed *urea recycling.* In this way, reabsorption of water from the tubular fluid of the ducts promotes the buildup of urea in the interstitial fluid of the renal medulla, which in turn promotes water reabsorption. The solutes left behind in the lumen thus become very concentrated, and a small volume of concentrated urine is excreted.

Countercurrent Exchange

Countercurrent exchange is the process by which solutes and water are passively exchanged between the blood of the vasa recta and interstitial fluid of the renal medulla as a result of countercurrent flow. Note in Figure 26.19b that the vasa recta also consists of descending and ascending limbs that are parallel to each other and to the loop of Henle. Just as tubular fluid flows in opposite directions in the loop of Henle, blood flows in opposite directions in the ascending and descending parts of the vasa recta. Since countercurrent flow between the descending and ascending limbs of the vasa recta allows for exchange of solutes and water between the blood and interstitial fluid of the renal medulla, the vasa recta is said to function as a **countercurrent exchanger.**

Blood entering the vasa recta has an osmolarity of about 300 mOsm/liter. As it flows along the descending part into the renal medulla, where the interstitial fluid becomes increasingly concentrated, Na^+, Cl^-, and urea diffuse from interstitial fluid into the blood and water diffuses from the blood into the interstitial fluid. But after its osmolarity increases, the blood flows into the ascending part of the vasa recta. Here blood flows through a region where the interstitial fluid becomes increasingly less concentrated. As a result Na^+, Cl^-, and urea diffuse from the blood back into interstitial fluid, and water diffuses from interstitial fluid back into the vasa recta. The osmolarity of blood leaving the vasa recta is only slightly higher than the osmolarity of blood entering the vasa recta. Thus, the vasa recta provides oxygen and nutrients to the renal medulla without washing out or diminishing the osmotic gradient. The long loop of Henle *establishes* the osmotic gradient in the renal medulla by countercurrent multiplication, but the vasa recta *maintains* the osmotic gradient in the renal medulla by countercurrent exchange.

Figure 26.20 summarizes the processes of filtration, reabsorption, and secretion in each segment of the nephron and collecting duct.

CLINICAL CONNECTION | *Diuretics*

Diuretics (dī-ū-RET-iks) are substances that slow renal reabsorption of water and thereby cause *diuresis,* an elevated urine flow rate, which in turn reduces blood volume. Diuretic drugs often are prescribed to treat *hypertension* (high blood pressure) because lowering blood volume usually reduces blood pressure. Naturally occurring diuretics include *caffeine* in coffee, tea, and sodas, which inhibits Na^+ reabsorption, and *alcohol* in beer, wine, and mixed drinks, which inhibits secretion of ADH. Most diuretic drugs act by interfering with a mechanism for reabsorption of filtered Na^+. For example, loop diuretics, such as furosemide (Lasix®), selectively inhibit the Na^+–K^+–$2Cl^-$ symporters in the thick ascending limb of the loop of Henle (see Figure 26.15). The thiazide diuretics, such as chlorothiazide (Diuril®), act in the distal convoluted tubule, where they promote loss of Na^+ and Cl^- in the urine by inhibiting Na^+–Cl^- symporters. •

✔CHECKPOINT

16. How do symporters in the ascending limb of the loop of Henle and principal cells in the collecting duct contribute to the formation of concentrated urine?
17. How does ADH regulate facultative water reabsorption?
18. What is the countercurrent mechanism? Why is it important?

26.7 EVALUATION OF KIDNEY FUNCTION

◉ OBJECTIVES

• Define urinalysis and describe its importance.
• Define renal plasma clearance and describe its importance.

Routine assessment of kidney function involves evaluating both the quantity and quality of urine and the levels of wastes in the blood.

Urinalysis

An analysis of the volume and physical, chemical, and microscopic properties of urine, called a **urinalysis** (ū-ri-NAL-i-sis), reveals much about the state of the body. Table 26.5 summarizes

Figure 26.20 Summary of filtration, reabsorption, and secretion in the nephron and collecting duct.

🔑 Filtration occurs in the renal corpuscle; reabsorption occurs all along the renal tubule and collecting ducts.

PROXIMAL CONVOLUTED TUBULE

Reabsorption (into blood) of filtered:

Water	65% (osmosis)
Na^+	65% (sodium–potassium pumps, symporters, antiporters)
K^+	65% (diffusion)
Glucose	100% (symporters and facilitated diffusion)
Amino acids	100% (symporters and facilitated diffusion)
Cl^-	50% (diffusion)
HCO_3^-	80–90% (facilitated diffusion)
Urea	50% (diffusion)
Ca^{2+}, Mg^{2+}	variable (diffusion)

Secretion (into urine) of:

H^+	variable (antiporters)
NH_4^+	variable, increases in acidosis (antiporters)
Urea	variable (diffusion)
Creatinine	small amount

At end of PCT, tubular fluid is still isotonic to blood (300 mOsm/liter).

LOOP OF HENLE

Reabsorption (into blood) of:

Water	15% (osmosis in descending limb)
Na^+	20–30% (symporters in ascending limb)
K^+	20–30% (symporters in ascending limb)
Cl^-	35% (symporters in ascending limb)
HCO_3^-	10–20% (facilitated diffusion)
Ca^{2+}, Mg^{2+}	variable (diffusion)

Secretion (into urine) of:

Urea	variable (recycling from collecting duct)

At end of loop of Henle, tubular fluid is hypotonic (100–150 mOsm/liter).

RENAL CORPUSCLE

Glomerular filtration rate:
105–125 mL/min of fluid that is isotonic to blood

Filtered substances: water and all solutes present in blood (except proteins) including ions, glucose, amino acids, creatinine, uric acid

EARLY DISTAL CONVOLUTED TUBULE

Reabsorption (into blood) of:

Water	10–15% (osmosis)
Na^+	5% (symporters)
Cl^-	5% (symporters)
Ca^{2+}	variable (stimulated by parathyroid hormone)

LATE DISTAL CONVOLUTED TUBULE AND COLLECTING DUCT

Reabsorption (into blood) of:

Water	5–9% (insertion of water channels stimulated by ADH)
Na^+	1–4% (sodium–potassium pumps and sodium channels stimulated by aldosterone)
HCO_3^-	variable amount, depends on H^+ secretion (antiporters)
Urea	variable (recycling to loop of Henle)

Secretion (into urine) of:

K^+	variable amount to adjust for dietary intake (leakage channels)
H^+	variable amounts to maintain acid–base homeostasis (H^+ pumps)

Tubular fluid leaving the collecting duct is dilute when ADH level is low and concentrated when ADH level is high.

Urine

❓ **In which segments of the nephron and collecting duct does secretion occur?**

TABLE 26.5

Characteristics of Normal Urine

CHARACTERISTIC	DESCRIPTION
Volume	One to two liters in 24 hours; varies considerably.
Color	Yellow or amber; varies with urine concentration and diet. Color due to urochrome (pigment produced from breakdown of bile) and urobilin (from breakdown of hemoglobin). Concentrated urine is darker in color. Color affected by diet (reddish from beets), medications, and certain diseases. Kidney stones may produce blood in urine.
Turbidity	Transparent when freshly voided; becomes turbid (cloudy) on standing.
Odor	Mildly aromatic; becomes ammonia-like on standing. Some people inherit ability to form methylmercaptan from digested asparagus, which gives characteristic odor. Urine of diabetics has fruity odor due to presence of ketone bodies.
pH	Ranges between 4.6 and 8.0; average 6.0; varies considerably with diet. High-protein diets increase acidity; vegetarian diets increase alkalinity.
Specific gravity (density)	Specific gravity (density) is ratio of weight of volume of substance to weight of equal volume of distilled water. In urine, 1.001–1.035. The higher the concentration of solutes, the higher the specific gravity.

the major characteristics of normal urine. The volume of urine eliminated per day in a normal adult is 1–2 liters (about 1–2 qt). Fluid intake, blood pressure, blood osmolarity, diet, body temperature, diuretics, mental state, and general health influence urine volume. For example, low blood pressure triggers the renin–angiotensin–aldosterone pathway. Aldosterone increases reabsorption of water and salts in the renal tubules and decreases urine volume. By contrast, when blood osmolarity decreases—for example, after drinking a large volume of water—secretion of ADH is inhibited and a larger volume of urine is excreted.

Water accounts for about 95% of the total volume of urine. The remaining 5% consists of electrolytes, solutes derived from cellular metabolism, and exogenous substances such as drugs. Normal urine is virtually protein-free. Typical solutes normally present in urine include filtered and secreted electrolytes that are not reabsorbed, urea (from breakdown of proteins), creatinine (from breakdown of creatine phosphate in muscle fibers), uric acid (from breakdown of nucleic acids), urobilinogen (from breakdown of hemoglobin), and small quantities of other substances, such as fatty acids, pigments, enzymes, and hormones.

If disease alters body metabolism or kidney function, traces of substances not normally present may appear in the urine, or normal constituents may appear in abnormal amounts. Table 26.6 lists several abnormal constituents in urine that may be detected as part of a urinalysis. Normal values of urine components and the

clinical implications of deviations from normal are listed in Appendix D.

Blood Tests

Two blood-screening tests can provide information about kidney function. One is the **blood urea nitrogen (BUN)** test, which measures the blood nitrogen that is part of the urea resulting from catabolism and deamination of amino acids. When glomerular filtration rate decreases severely, as may occur with renal disease or obstruction of the urinary tract, BUN rises steeply. One strategy in treating such patients is to minimize their protein intake, thereby reducing the rate of urea production.

Another test often used to evaluate kidney function is measurement of **plasma creatinine** (krē-AT-i-nin), which results from catabolism of creatine phosphate in skeletal muscle. Normally, the blood creatinine level remains steady because the rate of creatinine excretion in the urine equals its discharge from muscle. A creatinine level above 1.5 mg/dL (135 mmol/liter) usually is an indication of poor renal function. Normal values for selected blood tests are listed in Appendix C along with situations that may cause the values to increase or decrease.

Renal Plasma Clearance

Even more useful than BUN and blood creatinine values in the diagnosis of kidney problems is an evaluation of how effectively the kidneys are removing a given substance from blood plasma. **Renal plasma clearance** is the volume of blood that is "cleaned" or cleared of a substance per unit of time, usually expressed in units of *milliliters per minute*. High renal plasma clearance indicates efficient excretion of a substance in the urine; low clearance indicates inefficient excretion. For example, the clearance of glucose normally is zero because it is completely reabsorbed (see Table 26.3); therefore, glucose is not excreted at all. Knowing a drug's clearance is essential for determining the correct dosage. If clearance is high (one example is penicillin), then the dosage must also be high, and the drug must be given several times a day to maintain an adequate therapeutic level in the blood.

The following equation is used to calculate clearance:

$$\text{Renal plasma clearance of substance S} = \left(\frac{U \times V}{P}\right)$$

where U and P are the concentrations of the substance in urine and plasma, respectively (both expressed in the same units, such as mg/mL), and V is the urine flow rate in mL/min.

The clearance of a solute depends on the three basic processes of a nephron: glomerular filtration, tubular reabsorption, and tubular secretion. Consider a substance that is filtered but neither reabsorbed nor secreted. Its clearance equals the glomerular filtration rate because all the molecules that pass the filtration membrane appear in the urine. This is the situation for the plant polysaccharide **inulin** (IN-ū-lin); it easily passes the filter, it is not reabsorbed, and it is not secreted. (Do not confuse inulin with the hormone insulin, which is produced by the pancreas.) Typically,

TABLE 26.6

Summary of Abnormal Constituents in Urine

ABNORMAL CONSTITUENT	COMMENTS
Albumin	Normal constituent of plasma; usually appears in only very small amounts in urine because it is too large to pass through capillary fenestrations. Presence of excessive albumin in urine—**albuminuria** (al'-bū-mi-NOO-rē-a)—indicates increase in permeability of filtration membranes due to injury or disease, increased blood pressure, or irritation of kidney cells by substances such as bacterial toxins, ether, or heavy metals.
Glucose	Presence of glucose in urine—**glucosuria** (gloo-kō-SOO-rē-a)—usually indicates diabetes mellitus. Occasionally caused by stress, which can cause excessive epinephrine secretion. Epinephrine stimulates breakdown of glycogen and liberation of glucose from liver.
Red blood cells (erythrocytes)	Presence of red blood cells in urine—**hematuria** (hēm-a-TOO-rē-a)—generally indicates pathological condition. One cause is acute inflammation of urinary organs due to disease or irritation from kidney stones. Other causes: tumors, trauma, kidney disease, contamination of sample by menstrual blood.
Ketone bodies	High levels of ketone bodies in urine—**ketonuria** (kē-tō-NOO-rē-a)—may indicate diabetes mellitus, anorexia, starvation, or too little carbohydrate in diet.
Bilirubin	When red blood cells are destroyed by macrophages, the globin portion of hemoglobin is split off and heme is converted to biliverdin. Most biliverdin is converted to bilirubin, which gives bile its major pigmentation. Above-normal level of bilirubin in urine is called **bilirubinuria** (bil'-ē-roo-bi-NOO-rē-a).
Urobilinogen	Presence of urobilinogen (breakdown product of hemoglobin) in urine is called **urobilinogenuria** (ū'-rō-bi-lin'-ō-je-NOO-rē-a). Trace amounts are normal, but elevated urobilinogen may be due to hemolytic or pernicious anemia, infectious hepatitis, biliary obstruction, jaundice, cirrhosis, congestive heart failure, or infectious mononucleosis.
Casts	**Casts** are tiny masses of material that have hardened and assumed shape of lumen of tubule in which they formed, from which they are flushed when filtrate builds up behind them. Casts are named after cells or substances that compose them or based on appearance (for example, white blood cell casts, red blood cell casts, and epithelial cell casts that contain cells from walls of tubules).
Microbes	Number and type of bacteria vary with specific urinary tract infections. One of the most common is *E. coli*. Most common fungus is yeast *Candida albicans*, cause of vaginitis. Most frequent protozoan is *Trichomonas vaginalis*, cause of vaginitis in females and urethritis in males.

the clearance of inulin is about 125 mL/min, which equals the GFR. Clinically, the clearance of inulin can be used to determine the GFR. The clearance of inulin is obtained in the following way: Inulin is administered intravenously and then the concentrations of inulin in plasma and urine are measured along with the urine flow rate. Although using the clearance of inulin is an accurate method for determining the GFR, it has a few drawbacks: Inulin is not produced by the body and it must be infused continuously while clearance measurements are being determined. Measuring the creatinine clearance, however, is an easier way to assess the GFR because creatinine is a substance that is naturally produced by the body as an end product of muscle metabolism. Once creatinine is filtered, it is not reabsorbed, and is secreted only to a very small extent. Because there is a small amount of creatinine secretion, the creatinine clearance is only a close estimate of the GFR and is not as accurate as using the inulin clearance to determine the GFR. The creatinine clearance is normally about 120–140 mL/min.

The clearance of the organic anion *para*-**aminohippuric acid (PAH)** (par'-a-a-mē'-nō-hi-PYOOR-ik) is also of clinical importance. After PAH is administered intravenously, it is filtered and secreted in a single pass through the kidneys. Thus, the clearance of PAH is used to measure **renal plasma flow,** the amount of plasma that passes through the kidneys in one minute. Typically, the renal plasma flow is 650 mL per minute, which is about 55% of the renal blood flow (1200 mL per minute).

CLINICAL CONNECTION | *Dialysis*

If a person's kidneys are so impaired by disease or injury that he or she is unable to function adequately, then blood must be cleansed artificially by **dialysis** (dī-AL-i-sis; *dialyo* = to separate), the separation of large solutes from smaller ones by diffusion through a selectively permeable membrane. One method of dialysis is **hemodialysis** (hē-mō-dī-AL-i-sis; *hemo-* = blood), which directly filters the patient's blood by removing wastes and excess electrolytes and fluid and then returning the cleansed blood to the patient. Blood removed from the body is delivered to a *hemodialyzer* (artificial kidney). Inside the hemodialyzer, blood flows through a *dialysis membrane,* which contains pores large enough to permit the diffusion of small solutes. A special solution, called the *dialysate* (dī-AL-i-sāt), is pumped into the hemodialyzer so that it surrounds the dialysis membrane. The dialysate is specially formulated to maintain diffusion gradients that remove wastes from the blood (for example, urea, creatinine, uric acid, excess phosphate, potassium, and sulfate ions) and add needed substances (for example, glucose and bicarbonate ions) to it. The cleansed blood is passed through an air embolus detector to remove air and then returned to the body. An anticoagulant (heparin) is added to prevent blood from clotting in the hemodialyzer. As a rule, most people on hemodialysis require about 6–12 hours a week, typically divided into three sessions.

Another method of dialysis, called **peritoneal dialysis** (per'-i-tō-NĒ-al), uses the peritoneum of the abdominal cavity as the dialysis membrane to filter the blood. The peritoneum has a large surface area and numerous blood vessels, and is a very effective filter. A catheter is inserted into the peritoneal cavity and connected to a bag of dialysate. The fluid flows into the peritoneal cavity by gravity and is left there for sufficient time to permit wastes and excess electrolytes and fluids to diffuse into the dialysate. Then the dialysate is drained out into a bag, discarded, and replaced with fresh dialysate.

Each cycle is called an *exchange*. One variation of peritoneal dialysis, called **continuous ambulatory peritoneal dialysis (CAPD)**, can be performed at home. Usually, the dialysate is drained and replenished four times a day and once at night during sleep. Between exchanges the person can move about freely with the dialysate in the peritoneal cavity. •

✔CHECKPOINT

19. What are the characteristics of normal urine?
20. What chemical substances normally are present in urine?
21. How may kidney function be evaluated?
22. Why are the renal plasma clearances of glucose, urea, and creatinine different? How does each clearance compare to glomerular filtration rate?

26.8 URINE TRANSPORTATION, STORAGE, AND ELIMINATION

◉ OBJECTIVE

• Describe the anatomy, histology, and physiology of the ureters, urinary bladder, and urethra.

From collecting ducts, urine drains into the minor calyces, which join to become major calyces that unite to form the renal pelvis (see Figure 26.3). From the renal pelvis, urine first drains into the ureters and then into the urinary bladder. Urine is then discharged from the body through the single urethra (see Figure 26.1).

Ureters

Each of the two **ureters** (Ū-rē-ters) transports urine from the renal pelvis of one kidney to the urinary bladder. Peristaltic contractions of the muscular walls of the ureters push urine toward the urinary bladder, but hydrostatic pressure and gravity also contribute. Peristaltic waves that pass from the renal pelvis to the urinary bladder vary in frequency from one to five per minute, depending on how fast urine is being formed.

The ureters are 25–30 cm (10–12 in.) long and are thick-walled, narrow tubes that vary in diameter from 1 mm to 10 mm along their course between the renal pelvis and the urinary bladder. Like the kidneys, the ureters are retroperitoneal. At the base of the urinary bladder, the ureters curve medially and pass obliquely through the wall of the posterior aspect of the urinary bladder (Figure 26.21).

Figure 26.21 Ureters, urinary bladder, and urethra in a female.

🔒 Urine is stored in the urinary bladder before being expelled by micturition.

Anterior view of frontal section

❓ What is a lack of voluntary control over micturition called?

Even though there is no anatomical valve at the opening of each ureter into the urinary bladder, a physiological one is quite effective. As the urinary bladder fills with urine, pressure within it compresses the oblique openings into the ureters and prevents the backflow of urine. When this physiological valve is not operating properly, it is possible for microbes to travel up the ureters from the urinary bladder to infect one or both kidneys.

Three layers of tissue form the wall of the ureters. The deepest coat, the **mucosa,** is a mucous membrane with **transitional epithelium** (see Table 4.1) and an underlying **lamina propria** of areolar connective tissue with considerable collagen, elastic fibers, and lymphatic tissue. Transitional epithelium is able to stretch—a marked advantage for any organ that must accommodate a variable volume of fluid. Mucus secreted by the goblet cells of the mucosa prevents the cells from coming in contact with urine, the solute concentration and pH of which may differ drastically from the cytosol of cells that form the wall of the ureters.

Throughout most of the length of the ureters, the intermediate coat, the **muscularis,** is composed of inner longitudinal and outer circular layers of smooth muscle fibers. This arrangement is opposite to that of the gastrointestinal tract, which contains inner circular and outer longitudinal layers. The muscularis of the distal third of the ureters also contains an outer layer of longitudinal muscle fibers. Thus, the muscularis in the distal third of the ureter is inner longitudinal, middle circular, and outer longitudinal. Peristalsis is the major function of the muscularis.

The superficial coat of the ureters is the **adventitia,** a layer of areolar connective tissue containing blood vessels, lymphatic vessels, and nerves that serve the muscularis and mucosa. The adventitia blends in with surrounding connective tissue and anchors the ureters in place.

Urinary Bladder

The **urinary bladder** is a hollow, distensible muscular organ situated in the pelvic cavity posterior to the pubic symphysis. In males, it is directly anterior to the rectum; in females, it is anterior to the vagina and inferior to the uterus (see Figure 26.22). Folds of the peritoneum hold the urinary bladder in position. When slightly distended due to the accumulation of urine, the urinary bladder is spherical. When it is empty, it collapses. As urine volume increases, it becomes pear-shaped and rises into the abdominal cavity. Urinary bladder capacity averages 700–800 mL. It is smaller in females because the uterus occupies the space just superior to the urinary bladder.

Anatomy and Histology of the Urinary Bladder

In the floor of the urinary bladder is a small triangular area called the **trigone** (TRĪ-gōn = triangle). The two posterior corners of the trigone contain the two ureteral openings; the opening into the urethra, the **internal urethral orifice** (OR-i-fis), lies in the anterior corner (see Figure 26.21). Because its mucosa is firmly bound to the muscularis, the trigone has a smooth appearance.

Three coats make up the wall of the urinary bladder. The deepest is the **mucosa,** a mucous membrane composed of **transitional**

epithelium and an underlying **lamina propria** similar to that of the ureters. Rugae (the folds in the mucosa) are also present to permit expansion of the urinary bladder. Surrounding the mucosa is the intermediate **muscularis,** also called the **detrusor muscle** (de-TROO-ser = to push down), which consists of three layers of smooth muscle fibers: the inner longitudinal, middle circular, and outer longitudinal layers. Around the opening to the urethra the circular fibers form an **internal urethral sphincter;** inferior to it is the **external urethral sphincter,** which is composed of skeletal muscle and is a modification of the deep muscles of the perineum (see Figure 11.12). The most superficial coat of the urinary bladder on the posterior and inferior surfaces is the **adventitia,** a layer of areolar connective tissue that is continuous with that of the ureters. Over the superior surface of the urinary bladder is the **serosa,** a layer of visceral peritoneum.

The Micturition Reflex

Discharge of urine from the urinary bladder, called **micturition** (mik-choo-RISH-un; *mictur-* = urinate), is also known as *urination* or *voiding.* Micturition occurs via a combination of involuntary and voluntary muscle contractions. When the volume of urine in the urinary bladder exceeds 200–400 mL, pressure within the bladder increases considerably, and stretch receptors in its wall transmit nerve impulses into the spinal cord. These impulses propagate to the **micturition center** in sacral spinal cord segments S2 and S3 and trigger a spinal reflex called the **micturition reflex.** In this reflex arc, parasympathetic impulses from the micturition center propagate to the urinary bladder wall and internal urethral sphincter. The nerve impulses cause *contraction* of the detrusor muscle and *relaxation* of the internal urethral sphincter muscle. Simultaneously, the micturition center inhibits somatic motor neurons that innervate skeletal muscle in the external urethral sphincter. On contraction of the urinary bladder wall and relaxation of the sphincters, urination takes place. Urinary bladder filling causes a sensation of fullness that initiates a conscious desire to urinate before the micturition reflex actually occurs. Although emptying of the urinary bladder is a reflex, in early childhood we learn to initiate it and stop it voluntarily. Through learned control of the external urethral sphincter muscle and certain muscles of the pelvic floor, the cerebral cortex can initiate micturition or delay its occurrence for a limited period.

CLINICAL CONNECTION | Cystoscopy

Cystoscopy (sis-TOS-kō-pē; *cysto-* = bladder; *-skopy* = to examine) is a very important procedure for direct examination of the mucosa of the urethra and urinary bladder and prostate in males. In the procedure, a *cystoscope* (a flexible narrow tube with a light) is inserted into the urethra to examine the structures through which it passes. With special attachments, tissue samples can be removed for examination (biopsy) and small stones can be removed. Cystoscopy is useful for evaluating urinary bladder problems such as cancer and infections. It can also evaluate the degree of obstruction resulting from an enlarged prostate. •

Figure 26.22 **Comparison between female and male urethras.**

The female urethra is about 4 cm (1.5 in.) in length, while the male urethra is about 20 cm (8 in.) in length.

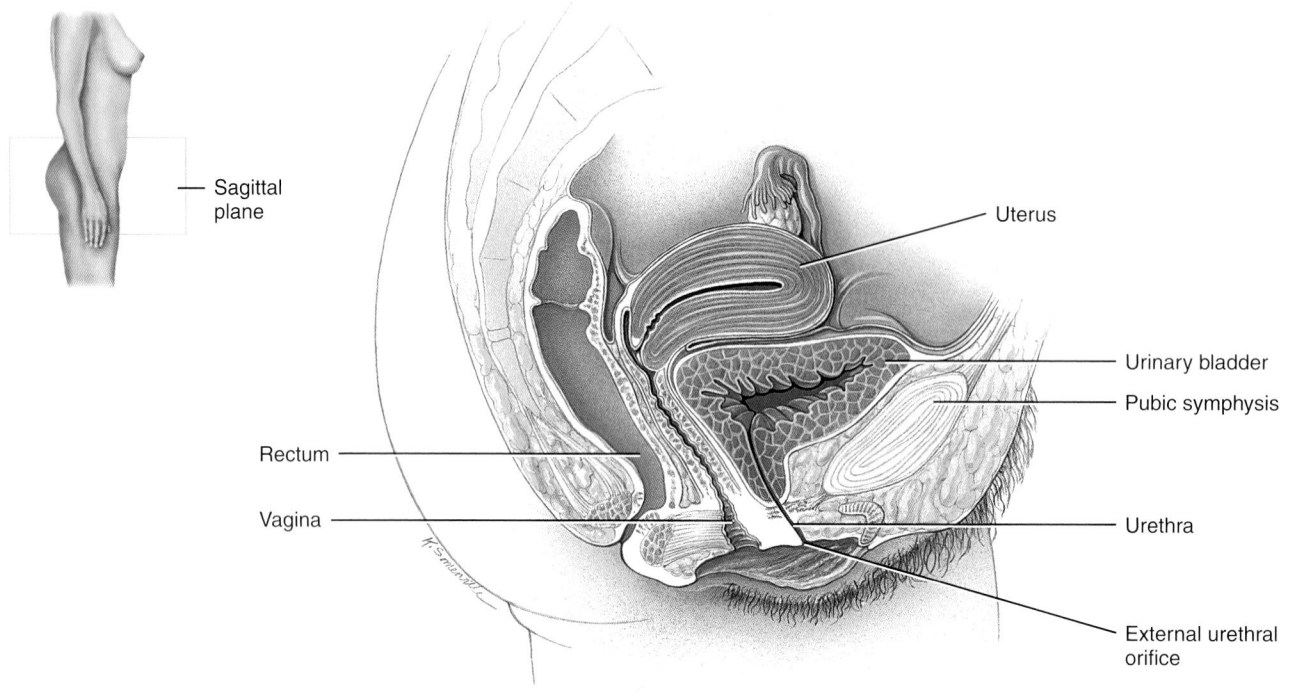

Sagittal plane

Uterus

Urinary bladder

Pubic symphysis

Rectum

Urethra

Vagina

External urethral orifice

(a) Sagittal section, female

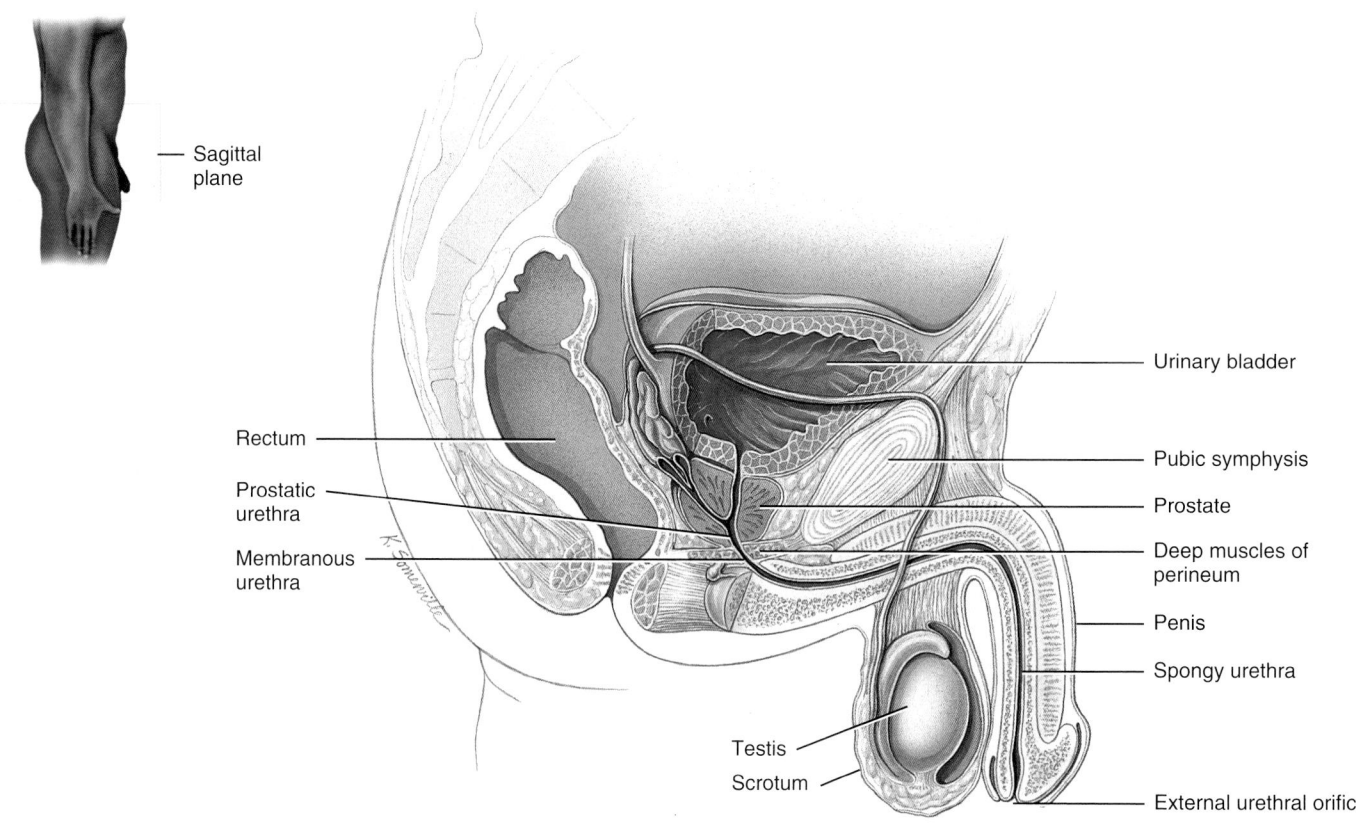

Sagittal plane

Urinary bladder

Rectum

Pubic symphysis

Prostatic urethra

Prostate

Membranous urethra

Deep muscles of perineum

Penis

Spongy urethra

Testis

Scrotum

External urethral orifice

(b) Sagittal section, male

What are the three subdivisions of the male urethra?

Urethra

The **urethra** (ū-RĒ-thra) is a small tube leading from the internal urethral orifice in the floor of the urinary bladder to the exterior of the body. In both males and females, the urethra is the terminal portion of the urinary system and the passageway for discharging urine from the body. In males, it discharges semen (fluid that contains sperm) as well.

In females, the urethra lies directly posterior to the pubic symphysis; is directed obliquely, inferiorly, and anteriorly; and has a length of 4 cm (1.5 in.) (Figure 26.22a). The opening of the urethra to the exterior, the **external urethral orifice,** is located between the clitoris and the vaginal opening (see Figure 28.11a). The wall of the female urethra consists of a deep **mucosa** and a superficial **muscularis.** The mucosa is a mucous membrane composed of **epithelium** and **lamina propria** (areolar connective tissue with elastic fibers and a plexus of veins). The muscularis consists of circularly arranged smooth muscle fibers and is continuous with that of the urinary bladder. Near the urinary bladder, the mucosa contains transitional epithelium that is continuous with that of the urinary bladder; near the external urethral orifice, the epithelium is nonkeratinized stratified squamous epithelium. Between these areas, the mucosa contains stratified columnar or pseudostratified columnar epithelium.

In males, the urethra also extends from the internal urethral orifice to the exterior, but its length and passage through the body are considerably different than in females (Figure 26.22b). The male urethra first passes through the prostate, then through the deep muscles of the perineum, and finally through the penis, a distance of about 20 cm (8 in.).

The male urethra, which also consists of a deep **mucosa** and a superficial **muscularis,** is subdivided into three anatomical regions: (1) The **prostatic urethra** passes through the prostate. (2) The **membranous (intermediate) urethra,** the shortest portion, passes through the deep muscles of the perineum. (3) The **spongy urethra,** the longest portion, passes through the penis. The epithelium of the prostatic urethra is continuous with that of the urinary bladder and consists of transitional epithelium that becomes stratified columnar or pseudostratified columnar epithelium more distally. The mucosa of the membranous urethra contains stratified columnar or pseudostratified columnar epithelium. The epithelium of the spongy urethra is stratified columnar or pseudostratified columnar epithelium, except near the external urethral orifice. There it is nonkeratinized stratified squamous epithelium. The **lamina propria** of the male urethra is areolar connective tissue with elastic fibers and a plexus of veins.

The muscularis of the prostatic urethra is composed of mostly circular smooth muscle fibers superficial to the lamina propria; these circular fibers help form the internal urethral sphincter of the urinary bladder. The muscularis of the membranous urethra consists of circularly arranged skeletal muscle fibers of the deep muscles of the perineum that help form the external urethral sphincter of the urinary bladder.

Several glands and other structures associated with reproduction deliver their contents into the male urethra (see Figure 28.9).

The prostatic urethra contains the openings of (1) ducts that transport secretions from the **prostate** and (2) the **seminal vesicles** and **ductus (vas) deferens,** which deliver sperm into the urethra and provide secretions that both neutralize the acidity of the female reproductive tract and contribute to sperm motility and viability. The openings of the ducts of the **bulbourethral (Cowper's) glands** (bul'-bō-ū-RĒ-thral) empty into the spongy urethra. They deliver an alkaline substance prior to ejaculation that neutralizes the acidity of the urethra. The glands also secrete mucus, which lubricates the end of the penis during sexual arousal. Throughout the urethra, but especially in the spongy urethra, the openings of the ducts of **urethral (Littré) glands** (LĒ-trē) discharge mucus during sexual arousal and ejaculation.

CLINICAL CONNECTION | *Urinary Incontinence*

A lack of voluntary control over micturition is called **urinary incontinence** (in-KON-ti-nens). In infants and children under 2–3 years old, incontinence is normal because neurons to the external urethral sphincter muscle are not completely developed; voiding occurs whenever the urinary bladder is sufficiently distended to stimulate the micturition reflex. Urinary incontinence also occurs in adults. There are four types of urinary incontinence—stress, urge, overflow, and functional. **Stress incontinence** is the most common type of incontinence in young and middle-aged females, and results from weakness of the deep muscles of the pelvic floor. As a result, any physical stress that increases abdominal pressure, such as coughing, sneezing, laughing, exercising, straining, lifting heavy objects, and pregnancy, causes leakage of urine from the urinary bladder. **Urge incontinence** is most common in older people and is characterized by an abrupt and intense urge to urinate followed by an involuntary loss of urine. It may be caused by irritation of the urinary bladder wall by infection or kidney stones, stroke, multiple sclerosis, spinal cord injury, or anxiety. **Overflow incontinence** refers to the involuntary leakage of small amounts of urine caused by some type of blockage or weak contractions of the musculature of the urinary bladder. When urine flow is blocked (for example, from an enlarged prostate or kidney stones) or the urinary bladder muscles can no longer contract, the urinary bladder becomes overfilled and the pressure inside increases until small amounts of urine dribble out. **Functional incontinence** is urine loss resulting from the inability to get to a toilet facility in time as a result of conditions such as stroke, severe arthritis, and Alzheimer disease. Choosing the right treatment option depends on correct diagnosis of the type of incontinence. Treatments include Kegel exercises (see Clinical Connection: Injury of Levator Ani and Urinary Stress Incontinence in Chapter 11), urinary bladder training, medication, and possibly even surgery. •

A summary of the organs of the urinary system is presented in Table 26.7.

✔CHECKPOINT

23. What forces help propel urine from the renal pelvis to the urinary bladder?
24. What is micturition? How does the micturition reflex occur?
25. How do the location, length, and histology of the urethra compare in males and females?

TABLE 26.7

Summary of Urinary System Organs

STRUCTURE	LOCATION	DESCRIPTION	FUNCTION
Kidneys	Posterior abdomen between last thoracic and third lumbar vertebrae posterior to peritoneum (retroperitoneal). Lie against eleventh and twelfth ribs.	Solid, reddish, bean-shaped organs. Internal structure: three tubular systems (arteries, veins, urinary tubes).	Regulate blood volume and composition, help regulate blood pressure, synthesize glucose, release erythropoietin, participate in vitamin D synthesis, excrete wastes in urine.
Ureters	Posterior to peritoneum (retroperitoneal); descend from kidney to urinary bladder along anterior surface of psoas major muscle and cross back of pelvis to reach inferoposterior surface of urinary bladder anterior to sacrum.	Thick, muscular walled tubes with three structural layers: mucosa of transitional epithelium, muscularis with circular and longitudinal layers of smooth muscle, adventitia of areolar connective tissue.	Transport tubes that move urine from kidneys to urinary bladder.
Urinary bladder	In pelvic cavity anterior to sacrum and rectum in males and sacrum, rectum, and vagina in females and posterior to pubis in both sexes. In males, superior surface covered with parietal peritoneum; in females, uterus covers superior aspect.	Hollow, distensible, muscular organ with variable shape depending on how much urine it contains. Three basic layers: inner mucosa of transitional epithelium, middle smooth muscle coat (detrusor muscle), outer adventitia or serosa over superior aspect in males.	Storage organ that temporarily stores urine until convenient to discharge from body.
Urethra	Exits urinary bladder in both sexes. In females, runs through perineal floor of pelvis to exit between labia minora. In males, passes through prostate, then perineal floor of pelvis, and then penis to exit at its tip.	Thin-walled tubes with three structural layers: inner mucosa that consists of transitional, stratified columnar, and stratified squamous epithelium; thin middle layer of circular smooth muscle; thin connective tissue exterior.	Drainage tube that transports stored urine from body.

26.9 WASTE MANAGEMENT IN OTHER BODY SYSTEMS

⦿ **OBJECTIVE**

• Describe the ways that body wastes are handled.

As we have seen, just one of the many functions of the urinary system is to help rid the body of some kinds of waste materials. Besides the kidneys, several other tissues, organs, and processes contribute to the temporary confinement of wastes, the transport of waste materials for disposal, the recycling of materials, and the excretion of excess or toxic substances in the body. These waste management systems include the following:

• **Body buffers.** Buffers in body fluids bind excess hydrogen ions (H^+), thereby preventing an increase in the acidity of body fluids. Buffers, like wastebaskets, have a limited capacity; eventually the H^+, like the paper in a wastebasket, must be eliminated from the body by excretion.

• **Blood.** The bloodstream provides pickup and delivery services for the transport of wastes, in much the same way that garbage trucks and sewer lines serve a community.

• **Liver.** The liver is the primary site for metabolic recycling, as occurs, for example, in the conversion of amino acids into glucose or of glucose into fatty acids. The liver also converts toxic substances into less toxic ones, such as ammonia into urea. These functions of the liver are described in Chapters 24 and 25.

• **Lungs.** With each exhalation, the lungs excrete CO_2, and expel heat and a little water vapor.

• **Sweat (sudoriferous) glands.** Especially during exercise, sweat glands in the skin help eliminate excess heat, water, and CO_2, plus small quantities of salts and urea as well.

• **Gastrointestinal tract.** Through defecation, the gastrointestinal tract excretes solid, undigested foods; wastes; some CO_2; water; salts; and heat.

✔ **CHECKPOINT**

26. What roles do the liver and lungs play in the elimination of wastes?

26.10 DEVELOPMENT OF THE URINARY SYSTEM

⦿ **OBJECTIVE**

• Describe the development of the urinary system.

Starting in the third week of fetal development, a portion of the mesoderm along the posterior aspect of the embryo, the **intermediate mesoderm,** differentiates into the kidneys. The intermedi-

Figure 26.23 Development of the urinary system.

Three pairs of kidneys form within intermediate mesoderm in succession: pronephros, mesonephros, and metanephros.

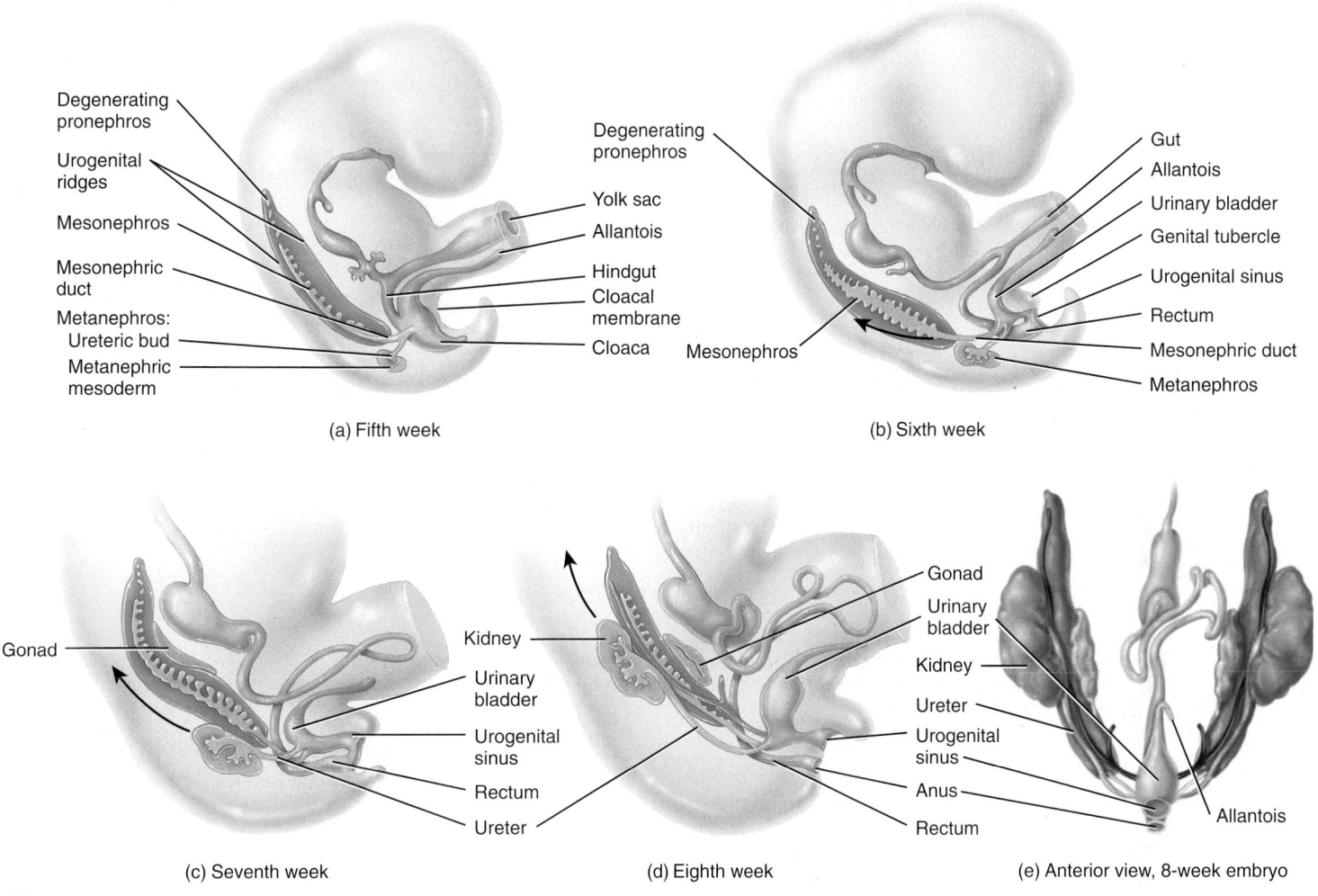

(a) Fifth week

(b) Sixth week

(c) Seventh week

(d) Eighth week

(e) Anterior view, 8-week embryo

When do the kidneys begin to develop?

ate mesoderm is located in paired elevations called **urogenital ridges** (ū-rō-JEN-i-tal). Three pairs of kidneys form within the intermediate mesoderm in succession: the pronephros, the mesonephros, and the metanephros (Figure 26.23). Only the last pair remains as the functional kidneys of the newborn.

The first kidney to form, the **pronephros** (prō-NEF-rōs; *pro- = before; -nephros = kidney*), is the most superior of the three and has an associated **pronephric duct.** This duct empties into the **cloaca** (klō-Ā-ka), the expanded terminal part of the hindgut, which functions as a common outlet for the urinary, digestive, and reproductive ducts. The pronephros begins to degenerate during the fourth week and is completely gone by the sixth week.

The second kidney, the **mesonephros** (mez′-ō-NEF-rōs; *meso- = middle*), replaces the pronephros. The retained portion of the pronephric duct, which connects to the mesonephros, develops into the **mesonephric duct.** The mesonephros begins to degenerate by the sixth week and is almost gone by the eighth week.

At about the fifth week, a mesodermal outgrowth, called a **ureteric bud** (ū-rē-TER-ik), develops from the distal portion of

the mesonephric duct near the cloaca. The **metanephros** (met-a-NEF-rōs; *meta- = after*), or ultimate kidney, develops from the ureteric bud and metanephric mesoderm. The ureteric bud forms the *collecting ducts, calyces, renal pelvis,* and *ureter.* The **metanephric mesoderm** (met′-a-NEF-rik) forms the *nephrons* of the kidneys. By the third month, the fetal kidneys begin excreting urine into the surrounding amniotic fluid; indeed, fetal urine makes up most of the amniotic fluid.

During development, the cloaca divides into a **urogenital sinus,** into which urinary and genital ducts empty, and a *rectum* that discharges into the anal canal. The *urinary bladder* develops from the urogenital sinus. In females, the *urethra* develops as a result of lengthening of the short duct that extends from the urinary bladder to the urogenital sinus. In males, the urethra is considerably longer and more complicated, but it is also derived from the urogenital sinus.

Although the metanephric kidneys form in the pelvis, they ascend to their ultimate destination in the abdomen. As they do so, they receive renal blood vessels. Although the inferior blood

BODY SYSTEM	CONTRIBUTION OF THE URINARY SYSTEM
For all body systems	Kidneys regulate the volume, composition, and pH of body fluids by removing wastes and excess substances from blood and excreting them in urine; the ureters transport urine from the kidneys to the urinary bladder, which stores urine until it is eliminated through the urethra.
Integumentary system	Kidneys and skin both contribute to synthesis of calcitriol, the active form of vitamin D.
Skeletal system	Kidneys help adjust levels of blood calcium and phosphates, needed for building extracellular bone matrix.
Muscular system	Kidneys help adjust level of blood calcium, needed for contraction of muscle.
Nervous system	Kidneys perform gluconeogenesis, which provides glucose for ATP production in neurons, especially during fasting or starvation.
Endocrine system	Kidneys participate in synthesis of calcitriol, the active form of vitamin D, and release erythropoietin, the hormone that stimulates production of red blood cells.
Cardiovascular system	By increasing or decreasing their reabsorption of water filtered from blood, the kidneys help adjust blood volume and blood pressure; renin released by juxtaglomerular cells in the kidneys raises blood pressure; some bilirubin from hemoglobin breakdown is converted to a yellow pigment (urobilin), which is excreted in urine.
Lymphatic system and immunity	By increasing or decreasing their reabsorption of water filtered from blood, the kidneys help adjust the volume of interstitial fluid and lymph; urine flushes microbes out of the urethra.
Respiratory system	Kidneys and lungs cooperate in adjusting pH of body fluids.
Digestive system	Kidneys help synthesize calcitriol, the active form of vitamin D, which is needed for absorption of dietary calcium.
Reproductive systems	In males, the portion of the urethra that extends through the prostate and penis is a passageway for semen as well as urine.

THE URINARY SYSTEM

vessels usually degenerate as superior ones appear, sometimes the inferior vessels do not degenerate. Consequently, some individuals (about 30%) develop multiple renal vessels.

In a condition called **unilateral renal agenesis** (a-JEN-e-sis; *a-* = without; *-genesis* = production; *unilateral* = one side) only one kidney develops (usually the right) due to the absence of a ureteric bud. The condition occurs once in every 1000 newborn infants and usually affects males more than females. Other kidney abnormalities that occur during development are **malrotated kidneys** (the hilum faces anteriorly, posteriorly, or laterally instead of medially); **ectopic kidney** (one or both kidneys may be in an abnormal position, usually inferior); and **horseshoe kidney** (the fusion of the two kidneys, usually inferiorly, into a single U-shaped kidney).

✔ CHECKPOINT

27. Which type of embryonic tissue develops into nephrons?
28. Which tissue gives rise to collecting ducts, calyces, renal pelves, and ureters?

26.11 AGING AND THE URINARY SYSTEM

◉ OBJECTIVE

• Describe the effects of aging on the urinary system.

With aging, the kidneys shrink in size, have a decreased blood flow, and filter less blood. These age-related changes in kidney size and function seem to be linked to a progressive reduction in blood supply to the kidneys as an individual gets older; for example, blood vessels such as the glomeruli become damaged or decrease in number. The mass of the two kidneys decreases from an average of nearly 300 g in 20-year-olds to less than 200 g by age 80, a decrease of about one-third. Likewise, renal blood flow and filtration rate decline by 50% between ages 40 and 70. By age 80, about 40% of glomeruli are not functioning and thus filtration, reabsorption, and secretion decrease. Kidney diseases that become more common with age include acute and chronic kidney inflammations and renal calculi (kidney stones). Because the sensation of thirst diminishes with age, older individuals also are susceptible to dehydration. Urinary bladder changes that occur with aging include a reduction in size and capacity and weakening of the muscles. Urinary tract infections are more common among the elderly, as are polyuria (excessive urine production), nocturia (excessive urination at night), increased frequency of urination, dysuria (painful urination), urinary retention or incontinence, and hematuria (blood in the urine).

✔ CHECKPOINT

29. To what extent do kidney mass and filtration rate decrease with age?

• • •

To appreciate the many ways that the urinary system contributes to homeostasis of other body systems, examine *Focus on Homeostasis: The Urinary System*. Next, in Chapter 27, we will see how the kidneys and lungs contribute to maintenance of homeostasis of body fluid volume, electrolyte levels in body fluids, and acid–base balance.

DISORDERS: HOMEOSTATIC IMBALANCES

Renal Calculi

The crystals of salts present in urine occasionally precipitate and solidify into insoluble stones called **renal calculi** (KAL-kū-lī = pebbles) or **kidney stones.** They commonly contain crystals of calcium oxalate, uric acid, or calcium phosphate. Conditions leading to calculus formation include the ingestion of excessive calcium, low water intake, abnormally alkaline or acidic urine, and overactivity of the parathyroid glands. When a stone lodges in a narrow passage, such as a ureter, the pain can be intense. **Shock-wave lithotripsy** (LITH-ō-trip′-sē; *litho-* = stone) is a procedure that uses high-energy shock waves to disintegrate kidney stones and offers an alternative to surgical removal. Once the kidney stone is located using x-rays, a device called a *lithotripter* delivers brief, high-intensity sound waves through a water- or gel-filled cushion placed under the back. Over a period of 30 to 60 minutes, 1000 or more shock waves pulverize the stone, creating fragments that are small enough to wash out in the urine.

Urinary Tract Infections

The term **urinary tract infection (UTI)** is used to describe either an infection of a part of the urinary system or the presence of large numbers of microbes in urine. UTIs are more common in females due to the shorter length of the urethra. Symptoms include painful or burning urination, urgent and frequent urination, low back pain, and bedwetting. UTIs include *urethritis* (ū-rē-THRĪ-tis), inflammation of the urethra; *cystitis* (sis-TĪ-tis), inflammation of the urinary bladder; and *pyelonephritis* (pī-e-lō-ne-FRĪ-tis), inflammation of the kidneys. If pyelonephritis becomes chronic, scar tissue can form in the kidneys and severely impair their function. Drinking cranberry juice can prevent the attachment of *E. coli* bacteria to the lining of the urinary bladder so that they are more readily flushed away during urination.

Glomerular Diseases

A variety of conditions may damage the kidney glomeruli, either directly or indirectly because of disease elsewhere in the body. Typically, the filtration membrane sustains damage, and its permeability increases.

Glomerulonephritis (glō-mer′-ū-lō-ne-FRĪ-tis) is an inflammation of the kidney that involves the glomeruli. One of the most common causes is an allergic reaction to the toxins produced by streptococcal bacteria that have recently infected another part of the body, especially the throat. The glomeruli become so inflamed, swollen, and engorged with blood that the filtration membranes allow blood cells and plasma proteins to enter the filtrate. As a result, the urine contains many erythrocytes (hematuria) and a lot of protein. The glomeruli may be permanently damaged, leading to chronic renal failure.

Nephrotic syndrome (nef-ROT-ik) is a condition characterized by *proteinuria* (prō-tēn-OO-rē-a), protein in the urine, and *hyperlipidemia* (hī′-per-lip-i-DĒ-mē-a), high blood levels of cholesterol, phospholipids, and triglycerides. The proteinuria is due to an increased permeability of the filtration membrane, which permits proteins, especially albumin, to escape from blood into urine. Loss of albumin results in *hypoalbuminemia* (hī'-pō-al-bū-mi-NĒ-mē-a), low blood albumin level, once liver production of albumin fails to meet increased urinary losses. Edema, usually seen around the eyes, ankles, feet, and abdomen, occurs in nephrotic syndrome because loss of albumin from the blood decreases blood colloid osmotic pressure. Nephrotic syndrome is associated with several glomerular diseases of unknown cause, as well as with systemic disorders such as diabetes mellitus, systemic lupus erythematosus (SLE), a variety of cancers, and AIDS.

Renal Failure

Renal failure is a decrease or cessation of glomerular filtration. In **acute renal failure (ARF),** the kidneys abruptly stop working entirely (or almost entirely). The main feature of ARF is the suppression of urine flow, usually characterized either by *oliguria* (ol-i-GŪ-rē-a), daily urine output between 50 mL and 250 mL, or by *anuria* (an-Ū-rē-a), daily urine output less than 50 mL. Causes include low blood volume (for example, due to hemorrhage), decreased cardiac output, damaged renal tubules, kidney stones, the dyes used to visualize blood vessels in angiograms, nonsteroidal anti-inflammatory drugs, and some antibiotic drugs. It is also common in people who suffer a devastating illness or overwhelming traumatic injury; in such cases it may be related to a more general organ failure known as *multiple organ dysfunction syndrome (MODS).*

Renal failure causes a multitude of problems. There is edema due to salt and water retention and metabolic acidosis due to an inability of the kidneys to excrete acidic substances. In the blood, urea builds up due to impaired renal excretion of metabolic waste products and potassium level rises, which can lead to cardiac arrest. Often, there is anemia because the kidneys no longer produce enough erythropoietin for adequate red blood cell production. Because the kidneys are no longer able to convert vitamin D to calcitriol, which is needed for adequate calcium absorption from the small intestine, osteomalacia also may occur.

Chronic renal failure (CRF) refers to a progressive and usually irreversible decline in glomerular filtration rate (GFR). CRF may result from chronic glomerulonephritis, pyelonephritis, polycystic kidney disease, or traumatic loss of kidney tissue. CRF develops in three stages. In the first stage, *diminished renal reserve,* nephrons are destroyed until about 75% of the functioning nephrons are lost. At this stage, a person may have no signs or symptoms because the remaining nephrons enlarge and take over the function of those that have been lost. Once 75% of the nephrons are lost, the person enters the second stage, called *renal insufficiency,* characterized by a decrease in

GFR and increased blood levels of nitrogen-containing wastes and creatinine. Also, the kidneys cannot effectively concentrate or dilute the urine. The final stage, called *end-stage renal failure,* occurs when about 90% of the nephrons have been lost. At this stage, GFR diminishes to 10–15% of normal, oliguria is present, and blood levels of nitrogen-containing wastes and creatinine increase further. People with end-stage renal failure need dialysis therapy and are possible candidates for a kidney transplant operation.

Polycystic Kidney Disease

Polycystic kidney disease (PKD) (pol'-ē-SIS-tik) is one of the most common inherited disorders. In PKD, the kidney tubules become riddled with hundreds or thousands of cysts (fluid-filled cavities). In addition, inappropriate apoptosis (programmed cell death) of cells in noncystic tubules leads to progressive impairment of renal function and eventually to end-stage renal failure.

People with PKD also may have cysts and apoptosis in the liver, pancreas, spleen, and gonads; increased risk of cerebral aneurysms; heart valve defects; and diverticula in the colon. Typically, symptoms are not noticed until adulthood, when patients may have back pain, urinary tract infections, blood in the urine, hypertension, and large abdominal masses. Using drugs to restore normal blood pressure, restricting protein and salt in the diet, and controlling urinary tract infections may slow progression to renal failure.

Urinary Bladder Cancer

Each year, nearly 12,000 Americans die from **urinary bladder cancer.** It generally strikes people over 50 years of age and is three times more likely to develop in males than females. The disease is typically painless as it develops, but in most cases blood in the urine is a primary sign of the disease. Less often, people experience painful and/or frequent urination.

As long as the disease is identified early and treated promptly, the prognosis is favorable. Fortunately, about 75% of the urinary bladder cancers are confined to the epithelium of the urinary bladder and are easily removed by surgery. The lesions tend to be low-grade, meaning that they have only a small potential for metastasis.

Urinary bladder cancer is frequently the result of a carcinogen. About half of all cases occur in people who smoke or have at some time smoked cigarettes. The cancer also tends to develop in people who are exposed to chemicals called aromatic amines. Workers in the leather, dye, rubber, and aluminum industries, as well as painters, are often exposed to these chemicals.

MEDICAL TERMINOLOGY

Azotemia (az-ō-TĒ-mē-a; *azot-* = nitrogen; *-emia* = condition of blood) Presence of urea or other nitrogen-containing substances in the blood.

Cystocele (SIS-tō-sēl; *cysto-* = bladder; *-cele* = hernia or rupture) Hernia of the urinary bladder.

Diabetic kidney disease A disorder caused by diabetes mellitus in which glomeruli are damaged. The result is the leakage of proteins into the urine and a reduction in the ability of the kidney to remove water and waste.

Dysuria (dis-Ū-rē-a; *dys-* = painful; *-uria* = urine) Painful urination.

Enuresis (en'-ū-RĒ-sis = to void urine) Involuntary voiding of urine after the age at which voluntary control has typically been attained.

Hydronephrosis (hī'-drō-ne-FRŌ-sis; *hydro-* = water; *-nephros-* = kidney; *-osis* = condition) Swelling of the kidney due to dilation of

the renal pelvis and calyces as a result of an obstruction to the flow of urine. It may be due to a congenital abnormality, a narrowing of the ureter, a kidney stone, or an enlarged prostate.

Intravenous pyelogram (in'-tra-VĒ-nus PĪ-e-lō-gram'; *intra-* = within; *-veno-* = vein; *pyelo-* = pelvis of kidney; *-gram* = record) (or **IVP**) Radiograph (x-ray) of the kidneys, ureters, and urinary bladder after venous injection of a radiopaque contrast medium.

Nephropathy (ne-FROP-a-thē; *nephro-* = kidney; *-pathos* = suffering) Any disease of the kidneys. Types include analgesic (from long-term and excessive use of drugs such as ibuprofen), lead (from ingestion of lead-based paint), and solvent (from carbon tetrachloride and other solvents).

Nocturnal enuresis (nok-TUR-nal en'-ū-RĒ-sis) Discharge of urine during sleep, resulting in bed-wetting; occurs in about 15% of

5-year-old children and generally resolves spontaneously, afflicting only about 1% of adults. It may have a genetic basis, as bedwetting occurs more often in identical twins than in fraternal twins and more often in children whose parents or siblings were bed-wetters. Possible causes include smaller-than-normal bladder capacity, failure to awaken in response to a full bladder, and above-normal production of urine at night. Also referred to as **nocturia.**

Polyuria (pol′-ē-Ū-rē-a; *poly-* = too much) Excessive urine formation. It may occur in conditions such as diabetes mellitus and glomerulonephritis.

Stricture (STRIK-chur) Narrowing of the lumen of a canal or hollow organ, as may occur in the ureter, urethra, or any other tubular structure in the body.

Uremia (ū-RĒ-mē-a; *-emia* = condition of blood) Toxic levels of urea in the blood resulting from severe malfunction of the kidneys.

Urinary retention A failure to completely or normally void urine; may be due to an obstruction in the urethra or neck of the urinary bladder, to nervous contraction of the urethra, or to lack of urge to urinate. In men, an enlarged prostate may constrict the urethra and cause urinary retention. If urinary retention is prolonged, a catheter (slender rubber drainage tube) must be placed into the urethra to drain the urine.

CHAPTER REVIEW AND RESOURCE SUMMARY

Review	Resource
Introduction	Anatomy Overview - The Urinary System Overview
1. The organs of the urinary system are the kidneys, ureters, urinary bladder, and urethra.	
2. After the kidneys filter blood and return most water and many solutes to the bloodstream, the remaining water and solutes constitute urine.	
26.1 Overview of Kidney Functions	Anatomy Overview - Kidney Overview
1. The kidneys regulate blood ionic composition, blood osmolarity, blood volume, blood pressure, and blood pH.	
2. The kidneys also perform gluconeogenesis, release calcitriol and erythropoietin, and excrete wastes and foreign substances.	

26.2 Anatomy and Histology of the Kidneys

Anatomy Overview - Blood Supply to the Kidney

Figure 26.3 - Internal Anatomy of the Kidneys

Figure 26.6 - Histology of a Renal Corpuscle

Exercise - Assemble the Urinary Tract

Exercise - Concentrate on Urinary System

Exercise - Paint the Nephron

1. The kidneys are retroperitoneal organs attached to the posterior abdominal wall.
2. Three layers of tissue surround the kidneys: renal capsule, adipose capsule, and renal fascia.
3. Internally, the kidneys consist of a renal cortex, a renal medulla, renal pyramids, renal papillae, renal columns, major and minor calyces, and a renal pelvis.
4. Blood flows into the kidney through the renal artery and successively into segmental, interlobar, arcuate, and interlobular arteries; afferent arterioles; glomerular capillaries; efferent arterioles; peritubular capillaries and vasa recta; and interlobular, arcuate, and interlobar veins before flowing out of the kidney through the renal vein.
5. Vasomotor nerves from the sympathetic division of the autonomic nervous system supply kidney blood vessels; they help regulate the flow of blood through the kidney.
6. The nephron is the functional unit of the kidneys. A nephron consists of a renal corpuscle (glomerulus and glomerular or Bowman's capsule) and a renal tubule.
7. A renal tubule consists of a proximal convoluted tubule, a loop of Henle, and a distal convoluted tubule, which drains into a collecting duct (shared by several nephrons). The loop of Henle consists of a descending limb and an ascending limb.
8. A cortical nephron has a short loop that dips only into the superficial region of the renal medulla; a juxtamedullary nephron has a long loop of Henle that stretches through the renal medulla almost to the renal papilla.
9. The wall of the entire glomerular capsule, renal tubule, and ducts consists of a single layer of epithelial cells. The epithelium has distinctive histological features in different parts of the tubule. Table 26.1 summarizes the histological features of the renal tubule and collecting duct.
10. The juxtaglomerular apparatus (JGA) consists of the juxtaglomerular cells of an afferent arteriole and the macula densa of the final portion of the ascending limb of the loop of Henle.

26.3 Overview of Renal Physiology

Animation - Renal Filtration

Filtration Finale

Exercise - Magical Renal Ride

Exercise - Renal Regulation

Exercise - The Renal Trip

1. Nephrons perform three basic tasks: glomerular filtration, tubular secretion, and tubular reabsorption.

Review

26.4 Glomerular Filtration

1. Fluid that is filtered by glomeruli enters the capsular space and is called glomerular filtrate.
2. The filtration membrane consists of the glomerular endothelium, basal lamina, and filtration slits between pedicels of podocytes.
3. Most substances in blood plasma easily pass through the glomerular filter. However, blood cells and most proteins normally are not filtered.
4. Glomerular filtrate amounts to up to 180 liters of fluid per day. This large amount of fluid is filtered because the filter is porous and thin, the glomerular capillaries are long, and the capillary blood pressure is high.
5. Glomerular blood hydrostatic pressure (GBHP) promotes filtration; capsular hydrostatic pressure (CHP) and blood colloid osmotic pressure (BCOP) oppose filtration. Net filtration pressure (NFP) = GBHP – CHP – BCOP. NFP is about 10 mmHg.
6. Glomerular filtration rate (GFR) is the amount of filtrate formed in both kidneys per minute; it is normally 105–125 mL/min.
7. Glomerular filtration rate depends on renal autoregulation, neural regulation, and hormonal regulation. Table 26.2 summarizes regulation of GFR.

Animation - Renal Filtration
Animation - Hormonal Control of Blood Volume and Pressure
Animation - Water and Fluid Flow
Figure 26.9 - The Pressures That Drive Glomerular Filtration
Homeostatic Imbalances - The Case of the Thirsty Woman

26.5 Tubular Reabsorption and Tubular Secretion

1. Tubular reabsorption is a selective process that reclaims materials from tubular fluid and returns them to the bloodstream. Reabsorbed substances include water, glucose, amino acids, urea, and ions, such as sodium, chloride, potassium, bicarbonate, and phosphate (Table 26.3).
2. Some substances not needed by the body are removed from the blood and discharged into the urine via tubular secretion. Included are ions (K^+, H^+, and NH_4^+), urea, creatinine, and certain drugs.
3. Reabsorption routes include both paracellular (between tubule cells) and transcellular (across tubule cells) routes. The maximum amount of a substance that can be reabsorbed per unit time is called the transport maximum (T_m).
4. About 90% of water reabsorption is obligatory; it occurs via osmosis, together with reabsorption of solutes, and is not hormonally regulated. The remaining 10% is facultative water reabsorption, which varies according to body needs and is regulated by ADH.
5. Sodium ions are reabsorbed throughout the basolateral membrane via primary active transport.
6. In the proximal convoluted tubule, Na^+ ions are reabsorbed through the apical membranes via Na^+–glucose symporters and Na^+/H^+ antiporters; water is reabsorbed via osmosis; Cl^-, K^+, Ca^{2+}, Mg^{2+}, and urea are reabsorbed via passive diffusion; and NH_3 and NH_4^+ are secreted.
7. The loop of Henle reabsorbs 20–30% of the filtered Na^+, K^+, Ca^{2+}, and HCO_3^-; 35% of the filtered Cl^-; and 15% of the filtered water.
8. The distal convoluted tubule reabsorbs sodium and chloride ions via Na^+–Cl^- symporters.
9. In the collecting duct, principal cells reabsorb Na^+ and secrete K^+; intercalated cells reabsorb K^+ and HCO_3^- and secrete H^+.
10. Angiotensin II, aldosterone, antidiuretic hormone, atrial natriuretic peptide, and parathyroid hormone regulate solute and water reabsorption, as summarized in Table 26.4.

Animation - Renal Reabsorption and Secretion
Animation - Water Homeostasis
Animation - Hormonal Control of Blood Volume and Pressure

26.6 Production of Dilute and Concentrated Urine

1. In the absence of ADH, the kidneys produce dilute urine; renal tubules absorb more solutes than water.
2. In the presence of ADH, the kidneys produce concentrated urine; large amounts of water are reabsorbed from the tubular fluid into interstitial fluid, increasing solute concentration of the urine.
3. The countercurrent multiplier establishes an osmotic gradient in the interstitial fluid of the renal medulla that enables production of concentrated urine when ADH is present.

Animation - Water Homeostasis
Figure 26.19 - Mechanism of Urine Concentration

26.7 Evaluation of Kidney Function

1. A urinalysis is an analysis of the volume and physical, chemical, and microscopic properties of a urine sample. Table 26.5 summarizes the principal physical characteristics of normal urine.
2. Chemically, normal urine contains about 95% water and 5% solutes. The solutes normally include urea, creatinine, uric acid, urobilinogen, and various ions.
3. Table 26.6 lists several abnormal components that can be detected in a urinalysis, including albumin, glucose, red and white blood cells, ketone bodies, bilirubin, excessive urobilinogen, casts, and microbes.
4. Renal clearance refers to the ability of the kidneys to clear (remove) a specific substance from blood.

Homeostatic Imbalances - The Case of the Man with the Swollen Kidneys

Review

26.8 Urine Transportation, Storage, and Elimination

1. The ureters are retroperitoneal and consist of a mucosa, muscularis, and adventitia. They transport urine from the renal pelvis to the urinary bladder, primarily via peristalsis.
2. The urinary bladder is located in the pelvic cavity posterior to the pubic symphysis; its function is to store urine before micturition.
3. The urinary bladder consists of a mucosa with rugae, a muscularis (detrusor muscle), and an adventitia (serosa over the superior surface).
4. The micturition reflex discharges urine from the urinary bladder via parasympathetic impulses that cause contraction of the detrusor muscle and relaxation of the internal urethral sphincter muscle and via inhibition of impulses in somatic motor neurons to the external urethral sphincter.
5. The urethra is a tube leading from the floor of the urinary bladder to the exterior. Its anatomy and histology differ in females and males. In both sexes, the urethra functions to discharge urine from the body; in males, it discharges semen as well.

26.9 Waste Management in Other Body Systems

1. Besides the kidneys, several other tissues, organs, and processes temporarily confine wastes, transport waste materials for disposal, recycle materials, and excrete excess or toxic substances.
2. Buffers bind excess H^+, the blood transports wastes, the liver converts toxic substances into less toxic ones, the lungs exhale CO_2, sweat glands help eliminate excess heat, and the gastrointestinal tract eliminates solid wastes.

26.10 Development of the Urinary System

1. The kidneys develop from intermediate mesoderm.
2. The kidneys develop in the following sequence: pronephros, mesonephros, and metanephros. Only the metanephros remains and develops into a functional kidney.

26.11 Aging and the Urinary System

1. With aging, the kidneys shrink in size, have a decreased blood flow, and filter less blood.
2. Common problems related to aging include urinary tract infections, increased frequency of urination, urinary retention or incontinence, and renal calculi.

Resource

Anatomy Overview - Ureters, Urinary Bladder, and Urethra
Figure 26.22 - Comparison between Male and Female Urethras
Exercise - Pick the Urinary Process

Animation - Regulation of pH

Homeostatic Imbalances - The Case of the Consistently Full Bladder

SELF-QUIZ QUESTIONS

Fill in the blanks in the following statements.

1. The renal corpuscle consists of the _____ and _____.
2. Discharge of urine from the urinary bladder is called _____.

Indicate whether the following statements are true or false.

3. The most superficial region of the internal kidney is the renal medulla.
4. When dilute urine is being formed, the osmolarity of the fluid in the tubular lumen increases as it flows down the descending limb of the loop of Henle, decreases as it flows up the ascending limb, and continues to decrease as it flows through the rest of the nephron and collecting duct.

Choose the one best answer to the following questions.

5. Which of the following statements are *correct*? (1) Glomerular filtration rate (GFR) is directly related to the pressures that determine net filtration pressure. (2) Angiotensin II and atrial natriuretic peptide help regulate GFR. (3) Mechanisms that regulate GFR work by adjusting blood flow into and out of the glomerulus and by altering the glomerular capillary surface area available for filtration. (4) GFR increases when blood flow into glomerular capillaries decreases. (5) Normally, GFR increases very little when systemic blood pressure rises.
(a) 1, 2, and 3   (b) 2, 3, and 4   (c) 3, 4, and 5
(d) 1, 2, 3, and 5   (e) 2, 3, 4, and 5

6. Which of the following hormones affect Na^+, Cl^-, Ca^{2+}, and/or water reabsorption and/or K^+ secretion by the renal tubules? (1) angiotensin II, (2) aldosterone, (3) ADH, (4) atrial natriuretic peptide, (5) thyroid hormone, (6) parathyroid hormone.
(a) 1, 3, and 5   (b) 2, 3, and 6   (c) 2, 4, and 5
(d) 1, 2, 4, and 5   (e) 1, 2, 3, 4, and 6

7. Which of the following are features of the renal corpuscle that enhance its filtering capacity? (1) large glomerular capillary surface area, (2) thick, selectively permeable filtration membrane, (3) high capsular hydrostatic pressure, (4) high glomerular capillary pressure, (5) mesangial cells regulating the filtering surface area.
(a) 1, 2, and 3   (b) 2, 4, and 5   (c) 1, 4, and 5
(d) 2, 3, and 4   (e) 2, 3, and 5

8. Given the following values, calculate the net filtration pressure: (1) glomerular blood hydrostatic pressure = 40 mmHg, (2) capsular

hydrostatic pressure = 10 mmHg, (3) blood colloid osmotic pressure = 30 mmHg.
(a) −20 mmHg   (b) 0 mmHg   (c) 20 mmHg
(d) 60 mmHg   (e) 80 mmHg

9. The micturition reflex (1) is initiated by stretch receptors in the ureters, (2) relies on parasympathetic impulses from the micturition center in S2 and S3, (3) results in contraction of the detrusor muscle, (4) results in contraction of the internal urethral sphincter muscle, (5) inhibits motor neurons in the external urethral sphincter.
(a) 1, 2, 3, 4, and 5   (b) 1, 3, and 4   (c) 2, 3, 4, and 5
(d) 2 and 5   (e) 2, 3, and 5

10. Which of the following are mechanisms that control GFR? (1) renal autoregulation, (2) neural regulation, (3) hormonal regulation, (4) chemical regulation of ions, (5) presence or absence of a transporter.
(a) 1, 2, and 3   (b) 2, 3, and 4   (c) 3, 4, and 5
(d) 1, 3, and 5   (e) 1, 3, and 4

11. Place the route of blood flow through the kidney in the correct order:
(a) segmental arteries   (b) vasa recta
(c) arcuate arteries   (d) peritubular venules
(e) interlobular veins   (f) renal vein
(g) renal artery   (h) interlobar arteries
(i) peritubular capillaries (j) efferent arterioles
(k) interlobar veins   (l) glomeruli
(m) arcuate veins   (n) afferent arterioles
(o) interlobular arteries

12. Place the route of filtrate flow in the correct order from its origin to the ureter:
(a) minor calyx
(b) ascending limb of loop of Henle
(c) papillary duct
(d) distal convoluted tubule
(e) major calyx
(f) descending limb of loop of Henle
(g) proximal convoluted tubule
(h) collecting duct
(i) renal pelvis

13. Match the following:
____ (a) cells in the last portion of the distal convoluted tubule and in the collecting ducts; regulated by ADH and aldosterone
____ (b) a capillary network lying in the glomerular capsule and functioning in filtration
____ (c) the functional unit of the kidney
____ (d) drains into a collecting duct
____ (e) combined glomerulus and glomerular capsule; where plasma is filtered
____ (f) the visceral layer of the glomerular capsule consisting of modified simple squamous epithelial cells
____ (g) cells of the final portion of the ascending limb of the loop of Henle that make contact with the afferent arteriole
____ (h) site of obligatory water reabsorption
____ (i) pores in the glomerular endothelial cells that allow filtration of blood solutes but not blood cells and platelets
____ (j) can secrete H^+ against a concentration gradient
____ (k) modified smooth muscle cells in the wall of the afferent arteriole

(1) podocytes
(2) glomerulus
(3) renal corpuscle
(4) proximal convoluted tubule
(5) distal convoluted tubule
(6) juxtaglomerular cells
(7) macula densa
(8) principal cells
(9) intercalated cells
(10) nephron
(11) fenestrations

14. Match the following:
____ (a) measure of blood nitrogen resulting from the catabolism and deamination of amino acids
____ (b) produced from the catabolism of creatine phosphate in skeletal muscle
____ (c) volume of blood that is cleared of a substance per unit of time
____ (d) can result from diabetes mellitus
____ (e) insoluble stones of crystallized salts
____ (f) usually indicates a pathological condition
____ (g) lack of voluntary control of micturition
____ (h) can be caused by damage to the filtration membranes

(1) incontinence
(2) renal calculi
(3) plasma creatinine
(4) BUN test
(5) albuminuria
(6) glucosuria
(7) renal plasma clearance
(8) hematuria

15. Match the following:
    ____ (a) membrane proteins that function as water channels
    ____ (b) a secondary active transport process that achieves Na^+ reabsorption, returns filtered HCO_3^- and water to the peritubular capillaries, and secretes H^+
    ____ (c) stimulates principal cells to secrete more K^+ into tubular fluid and reabsorb more Na^+ and Cl^- from tubular fluid
    ____ (d) enzyme secreted by juxtaglomerular cells
    ____ (e) reduces glomerular filtration rate; increases blood volume and pressure
    ____ (f) inhibits Na^+ and H_2O reabsorption in the proximal convoluted tubules and collecting ducts
    ____ (g) regulates facultative water reabsorption by increasing the water permeability of principal cells in the distal convoluted tubules and collecting ducts
    ____ (h) reabsorb Na^+ together with a variety of other solutes
    ____ (i) stimulates cells in the distal convoluted tubule to reabsorb more calcium into the blood

    (1) angiotensin II
    (2) atrial natriuretic peptide
    (3) Na^+ symporters
    (4) Na^+/H^+ antiporters
    (5) aquaporins
    (6) aldosterone
    (7) ADH
    (8) renin
    (9) parathyroid hormone

CRITICAL THINKING QUESTIONS

1. Imagine the discovery of a new toxin that blocks renal tubule reabsorption but does not affect filtration. Predict the short-term effects of this toxin.

2. For each of the following urinalysis results, indicate whether you should be concerned or not and why: (a) dark yellow urine that is turbid; (b) ammonia-like odor of the urine; (c) presence of excessive albumin; (d) presence of epithelial cell casts; (e) pH of 5.5; (f) hematuria.

3. Bruce is experiencing sudden, rhythmic waves of pain in his groin area. He has noticed that, although he is consuming fluids, his urine output has decreased. From what condition is Bruce suffering? How is it treated? How can he prevent future episodes?

? ANSWERS TO FIGURE QUESTIONS

26.1 The kidneys, ureters, urinary bladder, and urethra are the components of the urinary system.

26.2 The kidneys are retroperitoneal because they are posterior to the peritoneum.

26.3 Blood vessels, lymphatic vessels, nerves, and a ureter pass through the renal hilum.

26.4 About 1200 mL of blood enters the renal arteries each minute.

26.5 Cortical nephrons have glomeruli in the superficial renal cortex, and their short loops of Henle penetrate only into the superficial renal medulla. Juxtamedullary nephrons have glomeruli deep in the renal cortex, and their long loops of Henle extend through the renal medulla nearly to the renal papilla.

26.6 This section must pass through the renal cortex because there are no renal corpuscles in the renal medulla.

26.7 Secreted penicillin is being removed from the bloodstream.

26.8 Endothelial fenestrations (pores) in glomerular capillaries are too small for red blood cells to pass through.

26.9 Obstruction of the right ureter would increase CHP and thus decrease NFP in the right kidney; the obstruction would have no effect on the left kidney.

26.10 *Auto-* means self; tubuloglomerular feedback is an example of autoregulation because it takes place entirely within the kidneys.

26.11 The tight junctions between tubule cells form a barrier that prevents diffusion of transporter, channel, and pump proteins between the apical and basolateral membranes.

26.12 Glucose enters a PCT cell via a Na^+–glucose symporter in the apical membrane and leaves via facilitated diffusion through the basolateral membrane.

26.13 The electrochemical gradient promotes movement of Na^+ into the tubule cell through the apical membrane antiporters.

26.14 Reabsorption of the solutes creates an osmotic gradient that promotes the reabsorption of water via osmosis.

26.15 This is considered secondary active transport because the symporter uses the energy stored in the concentration gradient of Na^+ between extracellular fluid and the cytosol. No water is reabsorbed here because the thick ascending limb of the loop of Henle is virtually impermeable to water.

26.16 In principal cells, aldosterone stimulates secretion of K^+ and reabsorption of Na^+ by increasing the activity of sodium–potassium pumps and number of leakage channels for Na^+ and K^+.

26.17 Aldosterone and atrial natriuretic peptide influence renal water reabsorption along with ADH.

26.18 Dilute urine is produced when the thick ascending limb of the loop of Henle, the distal convoluted tubule, and the collecting duct reabsorb more solutes than water.

26.19 The high osmolarity of interstitial fluid in the renal medulla is due mainly to Na^+, Cl^-, and urea.

26.20 Secretion occurs in the proximal convoluted tubule, the loop of Henle, and the collecting duct.

26.21 Lack of voluntary control over micturition is termed urinary incontinence.

26.22 The three subdivisions of the male urethra are the prostatic urethra, membranous urethra, and spongy urethra.

26.23 The kidneys start to form during the third week of development.

27 FLUID, ELECTROLYTE, AND ACID–BASE HOMEOSTASIS

FLUID, ELECTROLYTE, AND ACID–BASE HOMEOSTASIS *Regulating the volume and composition of body fluids, controlling their distribution throughout the body, and balancing the pH of body fluids are crucial to maintaining overall homeostasis and health.*

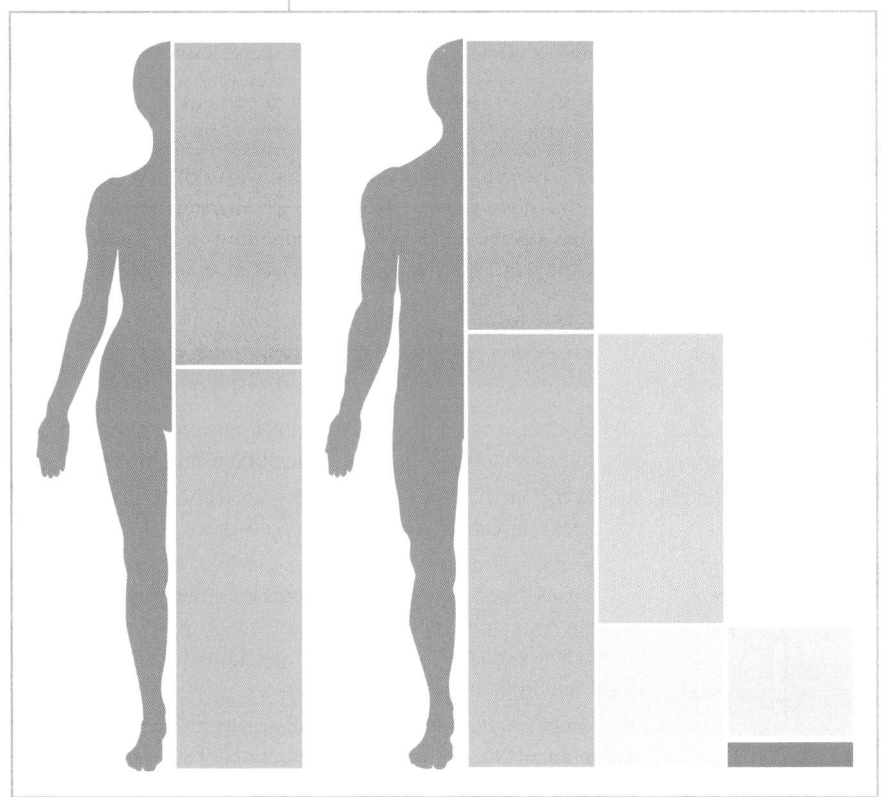

In Chapter 26 you learned how the kidneys form urine. One important function of the kidneys is to help maintain fluid balance in the body. The water and dissolved solutes throughout the body constitute the **body fluids.** Regulatory mechanisms involving the kidneys and other organs normally maintain homeostasis of the body fluids. Malfunction in any or all of them may seriously endanger the functioning of organs throughout the body. In this chapter, we will explore the mechanisms that regulate the volume and distribution of body fluids and examine the factors that determine the concentrations of solutes and the pH of body fluids.

Did you ever wonder how breathing can affect your body's pH?

27.1 FLUID COMPARTMENTS AND FLUID BALANCE

OBJECTIVES

- Compare the locations of intracellular fluid (ICF) and extracellular fluid (ECF), and describe the various fluid compartments of the body.
- Describe the sources of water and solute gain and loss, and explain how each is regulated.
- Explain how fluids move between compartments.

In lean adults, body fluids constitute between 55% and 60% of total body mass in females and males, respectively (Figure 27.1). Body fluids are present in two main "compartments"—inside cells and outside cells. About two-thirds of body fluid is **intracellular fluid (ICF)** (*intra-* = within) or **cytosol,** the fluid within cells. The other third, called **extracellular fluid (ECF)** (*extra-* = outside), is outside cells and includes all other body fluids. About 80% of the ECF is **interstitial fluid** (*inter-* = between), which occupies the microscopic spaces between tissue cells, and 20% of the ECF is **plasma,** the liquid portion of the blood. Other extracellular fluids that are grouped with interstitial fluid include lymph in lymphatic vessels; cerebrospinal fluid in the nervous system; synovial fluid in joints; aqueous humor and vitreous body in the eyes; endolymph and perilymph in the ears; and pleural, pericardial, and peritoneal fluids between serous membranes.

Two general "barriers" separate intracellular fluid, interstitial fluid, and blood plasma.

1. The *plasma membrane* of individual cells separates intracellular fluid from the surrounding interstitial fluid. You learned in Chapter 3 that the plasma membrane is a selectively permeable barrier: It allows some substances to cross but blocks the movement of other substances. In addition, active transport pumps work continuously to maintain different concentrations of certain ions in the cytosol and interstitial fluid.

2. *Blood vessel walls* divide the interstitial fluid from blood plasma. Only in capillaries, the smallest blood vessels, are the walls thin enough and leaky enough to permit the exchange of water and solutes between blood plasma and interstitial fluid.

The body is in **fluid balance** when the required amounts of water and solutes are present and are correctly proportioned among the various compartments. **Water** is by far the largest single component of the body, making up 45–75% of total body mass, depending on age and gender.

Figure 27.1 Body fluid compartments.

The term body fluid refers to body water and its dissolved substances.

(a) Distribution of body solids and fluids in average lean, adult female and male

(b) Exchange of water among body fluid compartments

? What is the approximate volume of blood plasma in a lean 60-kg male? In a lean 60-kg female? (Note: One liter of body fluid has a mass of 1 kilogram.)

The processes of filtration, reabsorption, diffusion, and osmosis allow continual exchange of water and solutes among body fluid compartments (Figure 27.1b). Yet the volume of fluid in each compartment remains remarkably stable. The pressures that promote filtration of fluid from blood capillaries and reabsorption of fluid back into capillaries can be reviewed in Figure 21.7. Because osmosis is the primary means of water movement between intracellular fluid and interstitial fluid, the concentration of solutes in these fluids determines the *direction* of water movement. Because most solutes in body fluids are **electrolytes,** inorganic compounds that dissociate into ions, fluid balance is closely related to electrolyte balance. Because intake of water and electrolytes rarely occurs in exactly the same proportions as their presence in body fluids, the ability of the kidneys to excrete excess water by producing dilute urine, or to excrete excess electrolytes by producing concentrated urine, is of utmost importance in the maintenance of homeostasis.

Sources of Body Water Gain and Loss

The body can gain water by ingestion and by metabolic synthesis (Figure 27.2). The main sources of body water are ingested liquids (about 1600 mL) and moist foods (about 700 mL) absorbed from the gastrointestinal (GI) tract, which total about 2300 mL/day. The other source of water is **metabolic water** that is produced in the body mainly when electrons are accepted by oxygen during

Figure 27.2 Sources of daily water gain and loss under normal conditions. Numbers are average volumes for adults.

Normally, daily water loss equals daily water gain.

How does each of the following affect fluid balance: Hyperventilation? Vomiting? Fever? Diuretics?

aerobic cellular respiration (see Figure 25.2) and to a smaller extent during dehydration synthesis reactions (see Figure 2.15). Metabolic water gain accounts for only 200 mL/day. Daily water gain from these two sources totals about 2500 mL.

Normally, body fluid volume remains constant because water loss equals water gain. Water loss occurs in four ways (Figure 27.2). Each day the kidneys excrete about 1500 mL in urine, the skin evaporates about 600 mL (400 mL through insensible perspiration—sweat that evaporates before it is perceived as moisture—and 200 mL as sweat), the lungs exhale about 300 mL as water vapor, and the gastrointestinal tract eliminates about 100 mL in feces. In women of reproductive age, additional water is lost in menstrual flow. On average, daily water loss totals about 2500 mL. The amount of water lost by a given route can vary considerably over time. For example, water may literally pour from the skin in the form of sweat during strenuous exertion. In other cases, water may be lost in diarrhea during a GI tract infection.

Regulation of Body Water Gain

The volume of metabolic water formed in the body depends entirely on the level of aerobic cellular respiration, which reflects the demand for ATP in body cells. When more ATP is produced, more water is formed. Body water gain is regulated mainly by the volume of water intake, or how much fluid you drink. An area in the hypothalamus known as the **thirst center** governs the urge to drink.

When water loss is greater than water gain, **dehydration**—a decrease in volume and an increase in osmolarity of body fluids—stimulates thirst (Figure 27.3). When body mass decreases by 2% due to fluid loss, mild dehydration exists. A decrease in blood volume causes blood pressure to fall. This change stimulates the kidneys to release renin, which promotes the formation of angiotensin II. Increased nerve impulses from osmoreceptors in the hypothalamus, triggered by increased blood osmolarity, and increased angiotensin II in the blood both stimulate the thirst center in the hypothalamus. Other signals that stimulate thirst come from (1) neurons in the mouth that detect dryness due to a decreased flow of saliva and (2) baroreceptors that detect lowered blood pressure in the heart and blood vessels. As a result, the sensation of thirst increases, which usually leads to increased fluid intake (if fluids are available) and restoration of normal fluid volume. Overall, fluid gain balances fluid loss. Sometimes, however, the sensation of thirst does not occur quickly enough or access to fluids is restricted, and significant dehydration ensues. This happens most often in elderly people, in infants, and in those who are in a confused mental state. When heavy sweating or fluid loss from diarrhea or vomiting occurs, it is wise to start replacing body fluids by drinking fluids even before the sensation of thirst occurs.

Regulation of Water and Solute Loss

Even though the loss of water and solutes through sweating and exhalation increases during exercise, elimination of *excess* body water or solutes occurs mainly by control of their loss in urine. The extent of *urinary salt (NaCl) loss* is the main factor that

Figure 27.3 Pathways through which dehydration stimulates thirst.

 Dehydration occurs when water loss is greater than water gain.

? Does regulation of these pathways occur via negative or positive feedback? Why?

Figure 27.4 Hormonal regulation of renal Na$^+$ and Cl$^-$ reabsorption.

 The three main hormones that regulate renal Na$^+$ and Cl$^-$ reabsorption (and thus the amount lost in the urine) are angiotensin II, aldosterone, and atrial natriuretic peptide.

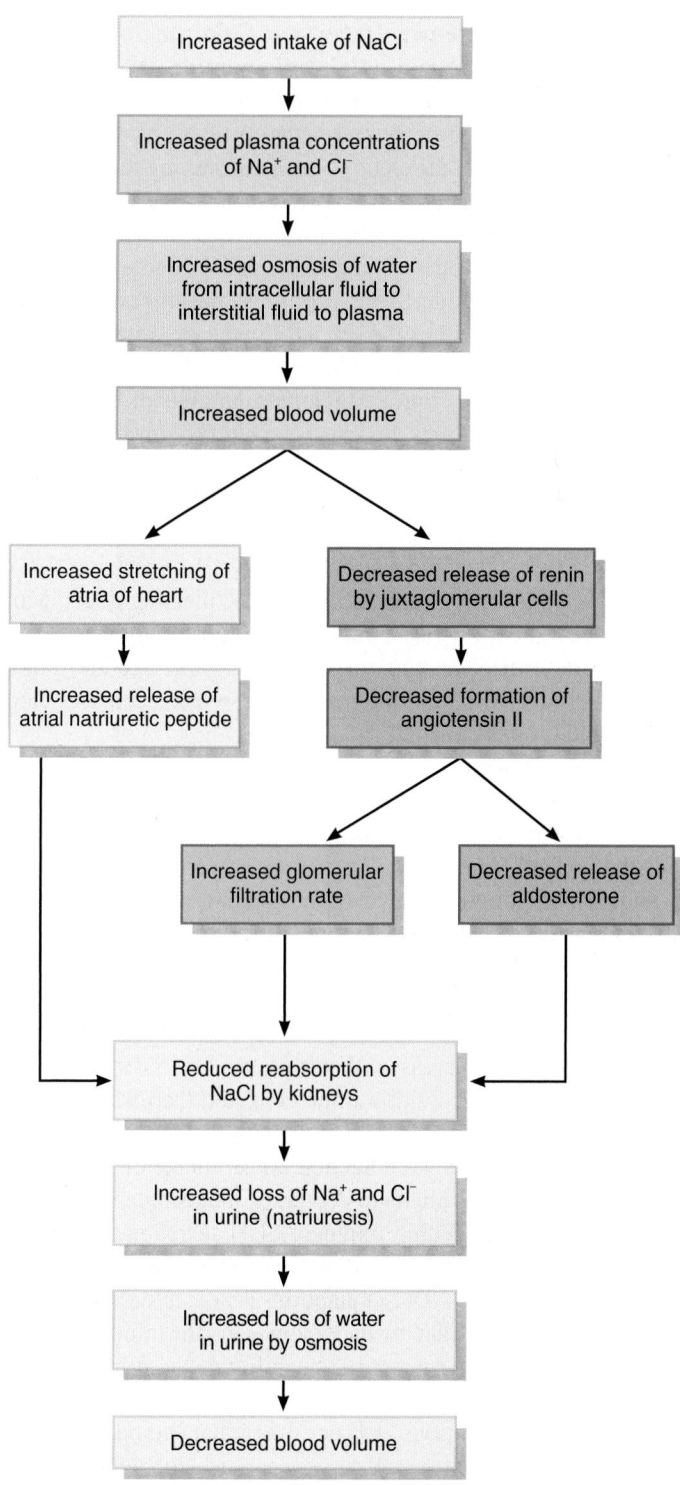

? How does hyperaldosteronism (excessive aldosterone secretion) cause edema?

determines body fluid *volume.* The reason for this is that "water follows solutes" in osmosis, and the two main solutes in extracellular fluid (and in urine) are sodium ions (Na$^+$) and chloride ions (Cl$^-$). In a similar way, the main factor that determines body fluid *osmolarity* is the extent of *urinary water loss.*

Because our daily diet contains a highly variable amount of NaCl, urinary excretion of Na$^+$ and Cl$^-$ must also vary to maintain homeostasis. Hormonal changes regulate the urinary loss of these ions, which in turn affects blood volume. Figure 27.4 depicts the sequence of changes that occur after a salty meal. The increased intake of NaCl produces an increase in plasma levels of Na^+ and Cl^- (the major contributors to osmolarity of extracellular fluid). As a result, the osmolarity of interstitial fluid increases, which causes movement of water from intracellular fluid into interstitial fluid and then into plasma. Such water movement increases blood volume.

The three most important hormones that regulate the extent of renal Na^+ and Cl^- reabsorption (and thus how much is lost in the urine) are **angiotensin II** (an′-jē-ō-TEN-sin), **aldosterone** (al-DOS-ter-ōn), and **atrial natriuretic peptide (ANP)** (nā′-trē-ū-RET-ik). When your body is dehydrated, angiotensin II and aldosterone promote urinary reabsorption of Na^+ and Cl^- (and water by osmosis with the electrolytes), conserving the volume of body fluids by reducing urinary loss. An increase in blood volume, as might occur after you finish one or more supersized drinks, stretches the atria of the heart and promotes release of atrial natriuretic peptide. ANP promotes **natriuresis,** elevated urinary excretion of Na^+ (and Cl^-) followed by water excretion, which decreases blood volume. An increase in blood volume also slows release of renin from juxtaglomerular cells of the kidneys. When renin level declines, less angiotensin II is formed. Decline in angiotensin II from a moderate level to a low level increases glomerular filtration rate and reduces Na^+, Cl^-, and water reabsorption in the kidney tubules. In addition, less angiotensin II leads to lower levels of aldosterone, which causes reabsorption of filtered Na^+ and Cl^- to slow in the renal collecting ducts. More filtered Na^+ and Cl^- thus remain in the tubular fluid to be excreted in the urine. The osmotic consequence of excreting more Na^+ and Cl^- is loss of more water in urine, which decreases blood volume and blood pressure.

The major hormone that regulates water loss is **antidiuretic hormone (ADH)** (an′-tē-dī-ū-RET-ik). This hormone, also known as **vasopressin** (vā-sō-PRES-in), is produced by neurosecretory cells that extend from the hypothalamus to the posterior pituitary. In addition to stimulating the thirst mechanism, an increase in the osmolarity of body fluids stimulates release of ADH (see Figure 26.17). ADH promotes the insertion of water-channel proteins (aquaporin-2) into the apical membranes of principal cells in the collecting ducts of the kidneys. As a result, the permeability of these cells to water increases. Water molecules move by osmosis from the renal tubular fluid into the cells and then from the cells into the bloodstream. The result is production of a small volume of very concentrated urine (see Section 26.6). Intake of water in response to the thirst mechanism decreases the osmolarity of blood and interstitial fluid. Within minutes, ADH secretion shuts down, and soon its blood level is close to zero. When the principal cells are not stimulated by ADH, aquaporin-2 molecules are removed from the apical membrane by endocytosis. As the number of water channels decreases, the water permeability of the principal cells' apical membrane falls, and more water is lost in the urine.

Under some conditions, factors other than blood osmolarity influence ADH secretion. A large decrease in blood volume, which is detected by baroreceptors (sensory neurons that respond to stretching) in the left atrium and in blood vessel walls, also stimulates ADH release. In severe dehydration, glomerular filtra-

tion rate decreases because blood pressure falls, so that less water is lost in the urine. Conversely, the intake of too much water increases blood pressure, causing the rate of glomerular filtration to rise and more water to be lost in the urine. Hyperventilation (abnormally fast and deep breathing) can increase fluid loss through the exhalation of more water vapor. Vomiting and diarrhea result in fluid loss from the GI tract. Finally, fever, heavy sweating, and destruction of extensive areas of the skin from burns can cause excessive water loss through the skin. In all of these conditions, an increase in ADH secretion will help conserve body fluids.

Table 27.1 summarizes the factors that maintain body water balance.

Movement of Water between Body Fluid Compartments

Normally, cells neither shrink nor swell because intracellular and interstitial fluids have the same osmolarity. Changes in the osmolarity of interstitial fluid, however, cause fluid imbalances. An increase in the osmolarity of interstitial fluid draws water out of cells, and they shrink slightly. A decrease in the osmolarity of interstitial fluid, by contrast, causes cells to swell. Changes in

TABLE 27.1

Summary of Factors That Maintain Body Water Balance

FACTOR	MECHANISM	EFFECT
Thirst center in hypothalamus	Stimulates desire to drink fluids.	Water gained if thirst is quenched.
Angiotensin II	Stimulates secretion of aldosterone.	Reduces loss of water in urine.
Aldosterone	By promoting urinary reabsorption of Na^+ and Cl^-, increases water reabsorption via osmosis.	Reduces loss of water in urine.
Atrial natriuretic peptide (ANP)	Promotes natriuresis, elevated urinary excretion of Na^+ (and Cl^-), accompanied by water.	Increases loss of water in urine.
Antidiuretic hormone (ADH), also known as vasopressin	Promotes insertion of water-channel proteins (aquaporin-2) into apical membranes of principal cells in collecting ducts of kidneys. As a result, water permeability of these cells increases and more water is reabsorbed.	Reduces loss of water in urine.

osmolarity most often result from changes in the concentration of Na$^+$.

A decrease in the osmolarity of interstitial fluid, as may occur after drinking a large volume of water, inhibits secretion of ADH. Normally, the kidneys then excrete a large volume of dilute urine, which restores the osmotic pressure of body fluids to normal. As a result, body cells swell only slightly, and only for a brief period. But when a person steadily consumes water faster than the kidneys can excrete it (the maximum urine flow rate is about 15 mL/min) or when renal function is poor, the result may be **water intoxication,** a state in which excessive body water causes cells to swell dangerously (Figure 27.5). If the body water and Na$^+$ lost during blood loss or excessive sweating, vomiting, or diarrhea is replaced by drinking plain water, then body fluids become more dilute. This dilution can cause the Na$^+$ concentration of plasma and then of interstitial fluid to fall below the normal range. When the Na$^+$ concentration of interstitial fluid decreases, its osmolarity also falls. The net result is osmosis of water from interstitial fluid into the cytosol. Water entering the cells causes them to swell, producing convulsions, coma, and possibly death. To prevent this dire sequence of events in cases of severe electrolyte and water loss, solutions given for intravenous or oral rehydration therapy (ORT) include a small amount of table salt (NaCl).

Figure 27.5 Series of events in water intoxication.

Water intoxication is a state in which excessive body water causes cells to swell.

? **Why do solutions used for oral rehydration therapy contain a small amount of table salt (NaCl)?**

CLINICAL CONNECTION | *Enemas and Fluid Balance*

An **enema** (EN-e-ma) is the introduction of a solution into the rectum to draw water (and electrolytes) into the colon osmotically. The increased volume increases peristalsis, which evacuates feces. Enemas are used to treat constipation. Repeated enemas, especially in young children, increase the risk of fluid and electrolyte imbalances. •

✔ **CHECKPOINT**

1. What is the approximate volume of each of your body fluid compartments?
2. How are the routes of water gain and loss from the body regulated?
3. By what mechanism does thirst help regulate water intake?
4. How do angiotensin II, aldosterone, atrial natriuretic peptide, and antidiuretic hormone regulate the volume and osmolarity of body fluids?
5. What factors control the movement of water between interstitial fluid and intracellular fluid?

27.2 ELECTROLYTES IN BODY FLUIDS

◉ **OBJECTIVES**

- Compare the electrolyte composition of the three major fluid compartments: plasma, interstitial fluid, and intracellular fluid.
- Discuss the functions of sodium, chloride, potassium, bicarbonate, calcium, phosphate, and magnesium ions, and explain how their concentrations are regulated.

The ions formed when electrolytes dissolve and dissociate serve four general functions in the body. (1) Because they are largely confined to particular fluid compartments and are more numerous than nonelectrolytes, certain ions *control the osmosis of water between fluid compartments.* (2) Ions *help maintain the acid–base balance* required for normal cellular activities. (3) Ions *carry electrical current,* which allows production of action potentials and graded potentials. (4) Several ions *serve as cofactors* needed for optimal activity of enzymes.

Concentrations of Electrolytes in Body Fluids

To compare the charge carried by ions in different solutions, the concentration of ions is typically expressed in units of **milliequivalents per liter (mEq/liter)** (mil′-i-ē-KWIV-a-lents). These units give the concentration of cations or anions in a given volume of solution. One equivalent is the positive or negative charge equal to the amount of charge in one mole of H$^+$; a milliequivalent is one-thousandth of an equivalent. Recall that a mole of a substance is its molecular weight expressed in grams. For ions such as sodium (Na$^+$), potassium (K$^+$), and bicarbonate (HCO$_3^-$), which have a single positive or negative charge, the number of mEq/liter is equal to the number of mmol/liter. For ions such as calcium (Ca^{2+}) or phosphate (HPO_4^{2-}), which have two positive or negative charges, the number of mEq/liter is twice the number of mmol/liter.

Figure 27.6 compares the concentrations of the main electrolytes and protein anions in blood plasma, interstitial fluid, and intracellular fluid. The chief difference between the two extracellular fluids—blood plasma and interstitial fluid—is that blood plasma contains many protein anions, in contrast to interstitial fluid, which has very few. Because normal capillary membranes are virtually impermeable to proteins, only a few plasma proteins leak out of blood vessels into the interstitial fluid. This difference in protein concentration is largely responsible for the blood colloid osmotic pressure exerted by blood plasma. In other respects, the two fluids are similar.

The electrolyte content of intracellular fluid differs considerably from that of extracellular fluid. In extracellular fluid, the most abundant cation is Na^+, and the most abundant anion is Cl^-. In intracellular fluid, the most abundant cation is K^+, and the most abundant anions are proteins and phosphates (HPO_4^{2-}). By actively transporting Na^+ out of cells and K^+ into cells, sodium–potassium pumps (Na^+/K^+ ATPases) play a major role in maintaining the high intracellular concentration of K^+ and high extracellular concentration of Na^+.

Sodium

Sodium ions (Na^+) are the most abundant ions in extracellular fluid, accounting for 90% of the extracellular cations. The normal blood plasma Na^+ concentration is 136–148 mEq/liter. As we have already seen, Na^+ plays a pivotal role in fluid and electrolyte balance because it accounts for almost half of the osmolarity of extracellular fluid (142 of about 300 mOsm/liter). The flow of Na^+ through voltage-gated channels in the plasma membrane also is necessary for the generation and conduction of action potentials in neurons and muscle fibers. The typical daily intake of Na^+ in North America often far exceeds the body's normal daily requirements, due largely to excess dietary salt. The kidneys excrete excess Na^+, but they also can conserve it during periods of shortage.

The Na^+ level in the blood is controlled by aldosterone, antidiuretic hormone (ADH), and atrial natriuretic peptide (ANP). Aldosterone increases renal reabsorption of Na^+. When the blood plasma concentration of Na^+ drops below 135 mEq/liter, a condition called *hyponatremia*, ADH release ceases. The lack of ADH in turn permits greater excretion of water in urine and restoration of the normal Na^+ level in ECF. Atrial natriuretic peptide increases Na^+ excretion by the kidneys when the Na^+ level is above normal, a condition called *hypernatremia*.

Figure 27.6 Electrolyte and protein anion concentrations in plasma, interstitial fluid, and intracellular fluid. The height of each column represents milliequivalents per liter (mEq/liter).

The electrolytes present in extracellular fluids are different from those present in intracellular fluid.

What cation and two anions are present in the highest concentrations in ECF and ICF?

CLINICAL CONNECTION | *Indicators of Na⁺ Imbalance*

If excess sodium ions remain in the body because the kidneys fail to excrete enough of them, water is also osmotically retained. The result is increased blood volume, increased blood pressure, and **edema,** an abnormal accumulation of interstitial fluid. Renal failure and hyperaldosteronism (excessive aldosterone secretion) are two causes of Na⁺ retention. Excessive urinary loss of Na⁺, by contrast, causes excessive water loss, which results in **hypovolemia** (hī′-pō-vō-LĒ-mē-a), an abnormally low blood volume. Hypovolemia related to Na⁺ loss is most frequently due to the inadequate secretion of aldosterone associated with adrenal insufficiency or overly vigorous therapy with diuretic drugs. •

Chloride

Chloride ions (Cl^-) are the most prevalent anions in extracellular fluid. The normal blood plasma Cl^- concentration is 95–105 mEq/liter. Cl^- moves relatively easily between the extracellular and intracellular compartments because most plasma membranes contain many Cl^- leakage channels and antiporters. For this reason, Cl^- can help balance the level of anions in different fluid compartments. One example is the chloride shift that occurs between red blood cells and blood plasma as the blood level of carbon dioxide either increases or decreases (see Figure 23.23). In this case, the antiporter exchange of Cl^- for HCO_3^- maintains the correct balance of anions between ECF and ICF. Chloride ions also are part of the hydrochloric acid secreted into gastric juice. ADH helps regulate Cl^- balance in body fluids because it governs the extent of water loss in urine. Processes that increase or decrease renal reabsorption of sodium ions also affect reabsorption of chloride ions. (Recall that reabsorption of Na^+ and Cl^- occurs by means of Na^+–Cl^- symporters.)

Potassium

Potassium ions (K^+) are the most abundant cations in intracellular fluid (140 mEq/liter). K^+ plays a key role in establishing the resting membrane potential and in the repolarization phase of action potentials in neurons and muscle fibers; K^+ also helps maintain normal intracellular fluid volume. When K^+ moves into or out of cells, it often is exchanged for H^+ and thereby helps regulate the pH of body fluids.

The normal blood plasma K^+ concentration is 3.5–5.0 mEq/liter and is controlled mainly by aldosterone. When blood plasma K^+ concentration is high, more aldosterone is secreted into the blood. Aldosterone then stimulates principal cells of the renal collecting ducts to secrete more K^+ so excess K^+ is lost in the urine. Conversely, when blood plasma K^+ concentration is low, aldosterone secretion decreases and less K^+ is excreted in urine. Because K^+ is needed during the repolarization phase of action potentials, abnormal K^+ levels can be lethal. For instance, *hyperkalemia* (above-normal concentration of K^+ in blood) can cause death due to ventricular fibrillation.

Bicarbonate

Bicarbonate ions (HCO_3^-) are the second most prevalent extracellular anions. Normal blood plasma HCO_3^- concentration is 22–26 mEq/liter in systemic arterial blood and 23–27 mEq/liter in systemic venous blood. HCO_3^- concentration increases as blood flows through systemic capillaries because the carbon dioxide released by metabolically active cells combines with water to form carbonic acid; the carbonic acid then dissociates into H^+ and HCO_3^-. As blood flows through pulmonary capillaries, however, the concentration of HCO_3^- decreases again as carbon dioxide is exhaled. (Figure 23.23 shows these reactions.) Intracellular fluid also contains a small amount of HCO_3^-. As previously noted, the exchange of Cl^- for HCO_3^- helps maintain the correct balance of anions in extracellular fluid and intracellular fluid.

The kidneys are the main regulators of blood HCO_3^- concentration. The intercalated cells of the renal tubule can either form HCO_3^- and release it into the blood when the blood level is low (see Figure 27.8) or excrete excess HCO_3^- in the urine when the level in blood is too high. Changes in the blood level of HCO_3^- are considered later in this chapter in the section on acid–base balance.

Calcium

Because such a large amount of calcium is stored in bone, it is the most abundant mineral in the body. About 98% of the calcium in adults is located in the skeleton and teeth, where it is combined with phosphates to form a crystal lattice of mineral salts. In body fluids, calcium is mainly an extracellular cation (Ca^{2+}). The normal concentration of free or unattached Ca^{2+} in blood plasma is 4.5–5.5 mEq/liter. About the same amount of Ca^{2+} is attached to various plasma proteins. Besides contributing to the hardness of bones and teeth, Ca^{2+} plays important roles in blood clotting, neurotransmitter release, maintenance of muscle tone, and excitability of nervous and muscle tissue.

The most important regulator of Ca^{2+} concentration in blood plasma is parathyroid hormone (PTH) (see Figure 18.14). A low level of Ca^{2+} in blood plasma promotes release of more PTH, which stimulates osteoclasts in bone tissue to release calcium (and phosphate) from bone extracellular matrix. Thus, PTH increases bone *resorption*. Parathyroid hormone also enhances *reabsorption* of Ca^{2+} from glomerular filtrate through renal tubule cells and back into blood, and increases production of calcitriol (the form of vitamin D that acts as a hormone), which in turn increases Ca^{2+} *absorption* from food in the gastrointestinal tract. Recall that calcitonin (CT) produced by the thyroid gland inhibits the activity of osteoclasts, accelerates Ca^{2+} deposition into bones, and thus lowers blood Ca^{2+} levels.

Phosphate

About 85% of the phosphate in adults is present as calcium phosphate salts, which are structural components of bone and teeth. The remaining 15% is ionized. Three phosphate ions ($H_2PO_4^-$, HPO_4^{2-}, and PO_4^{3-}) are important intracellular anions. At the normal pH of body fluids, HPO_4^{2-} is the most prevalent form. Phosphates contribute about 100 mEq/liter of anions to intracellular fluid. HPO_4^{2-} is an important buffer of H^+, both in body fluids and in the urine. Although some are "free," most phosphate ions are covalently bound to organic molecules such as lipids (phospholipids), proteins, carbohydrates, nucleic acids (DNA and RNA), and adenosine triphosphate (ATP).

The normal blood plasma concentration of ionized phosphate is only 1.7–2.6 mEq/liter. The same two hormones that govern calcium homeostasis—parathyroid hormone (PTH) and calcitriol—also regulate the level of HPO_4^{2-} in blood plasma. PTH stimulates resorption of bone extracellular matrix by osteoclasts, which releases both phosphate and calcium ions into the bloodstream. In the kidneys, however, PTH inhibits reabsorption of phosphate ions while stimulating reabsorption of calcium ions by renal tubular cells. Thus, PTH increases urinary excretion of phosphate and lowers blood phosphate level. Calcitriol promotes absorption of both phosphates and calcium from the gastrointestinal tract. Fibroblast growth factor 23 (FGF 23) is a polypeptide paracrine (local hormone) that also helps regulate blood plasma levels of HPO_4^{2-}. This hormone decreases HPO_4^{2-} blood levels by increasing HPO_4^{2-} excretion by the kidneys and decreasing absorption of HPO_4^{2-} by the gastrointestinal tract.

Magnesium

In adults, about 54% of the total body magnesium is part of bone matrix as magnesium salts. The remaining 46% occurs as magnesium ions (Mg^{2+}) in intracellular fluid (45%) and extracellular fluid (1%). Mg^{2+} is the second most common intracellular cation (35 mEq/liter). Functionally, Mg^{2+} is a cofactor for certain enzymes needed for the metabolism of carbohydrates and proteins and for the sodium–potassium pump. Mg^{2+} is essential for normal neuromuscular activity, synaptic transmission, and myocardial functioning. In addition, secretion of parathyroid hormone (PTH) depends on Mg^{2+}.

Normal blood plasma Mg^{2+} concentration is low, only 1.3–2.1 mEq/liter. Several factors regulate the blood plasma level of Mg^{2+} by varying the rate at which it is excreted in the urine. The kidneys increase urinary excretion of Mg^{2+} in response to hypercalcemia, hypermagnesemia, increases in extracellular fluid volume, decreases in parathyroid hormone, and acidosis. The opposite conditions decrease renal excretion of Mg^{2+}.

Table 27.2 describes the imbalances that result from the deficiency or excess of several electrolytes.

People at risk for fluid and electrolyte imbalances include those who depend on others for fluid and food, such as infants, the elderly, and the hospitalized; individuals undergoing medical treatment that involves intravenous infusions, drainages or suctions, and urinary catheters; and people who receive diuretics, experience excessive fluid losses and require increased fluid intake, or experience fluid retention and have fluid restrictions. Finally, athletes and military personnel in extremely hot environments, postoperative individuals, severe burn or trauma cases, individuals with chronic diseases (congestive heart failure, diabetes,

chronic obstructive lung disease, and cancer), people in confinement, and individuals with altered levels of consciousness who may be unable to communicate needs or respond to thirst are also subject to fluid and electrolyte imbalances.

✔ **CHECKPOINT**

6. What are the functions of electrolytes in the body?
7. Name three important extracellular electrolytes and three important intracellular electrolytes and indicate how each is regulated.

27.3 ACID–BASE BALANCE

◉ **OBJECTIVES**

• Compare the roles of buffers, exhalation of carbon dioxide, and kidney excretion of H^+ in maintaining pH of body fluids.
• Define acid–base imbalances, describe their effects on the body, and explain how they are treated.

From our discussion thus far, it should be clear that various ions play different roles that help maintain homeostasis. A major homeostatic challenge is keeping the H^+ concentration (pH) of body fluids at an appropriate level. This task—the maintenance of acid–base balance—is of critical importance to normal cellular function. For example, the three-dimensional shape of all body proteins, which enables them to perform specific functions, is very sensitive to pH changes. When the diet contains a large amount of protein, as is typical in North America, cellular metabolism produces more acids than bases, which tends to acidify the blood. Before proceeding with this section of the chapter, you may wish to review the discussion of acids, bases, and pH in Section 2.4.

In a healthy person, several mechanisms help maintain the pH of systemic arterial blood between 7.35 and 7.45. (A pH of 7.4 corresponds to a H^+ concentration of 0.00004 mEq/liter = 40 nEq /liter.) Because metabolic reactions often produce a huge excess of H^+, the lack of any mechanism for the disposal of H^+ would cause H^+ in body fluids to rise quickly to a lethal level. Homeostasis of H^+ concentration within a narrow range is thus essential to survival. The removal of H^+ from body fluids and its subsequent elimination from the body depend on the following three major mechanisms:

1. *Buffer systems.* Buffers act quickly to temporarily bind H^+, removing the highly reactive, excess H^+ from solution. Buffers thus raise pH of body fluids but do not remove H^+ from the body.

2. *Exhalation of carbon dioxide.* By increasing the rate and depth of breathing, more carbon dioxide can be exhaled. Within minutes this reduces the level of carbonic acid in blood, which raises the blood pH (reduces blood H^+ level).

3. *Kidney excretion of H^+.* The slowest mechanism, but the only way to eliminate acids other than carbonic acid, is through their excretion in urine.

TABLE 27.2

Blood Electrolyte Imbalances

ELECTROLYTE*	DEFICIENCY		EXCESS	
	NAME AND CAUSES	SIGNS AND SYMPTOMS	NAME AND CAUSES	SIGNS AND SYMPTOMS
Sodium (Na^+) 136–148 mEq/liter	**Hyponatremia** (hī-pō-na-TRĒ-mē-a) may be due to decreased sodium intake; increased sodium loss through vomiting, diarrhea, aldosterone deficiency, or taking certain diuretics; and excessive water intake.	Muscular weakness; dizziness, headache, and hypotension; tachycardia and shock; mental confusion, stupor, and coma.	**Hypernatremia** may occur with dehydration, water deprivation, or excessive sodium in diet or intravenous fluids; causes hypertonicity of ECF, which pulls water out of body cells into ECF, causing cellular dehydration.	Intense thirst, hypertension, edema, agitation, and convulsions.
Chloride (Cl^-) 95–105 mEq/liter	**Hypochloremia** (hī-pō-klō-RĒ-mē-a) may be due to excessive vomiting, overhydration, aldosterone deficiency, congestive heart failure, and therapy with certain diuretics such as furosemide (Lasix®).	Muscle spasms, metabolic alkalosis, shallow respirations, hypotension, and tetany.	**Hyperchloremia** may result from dehydration due to water loss or water deprivation; excessive chloride intake; or severe renal failure, hyperaldosteronism, certain types of acidosis, and some drugs.	Lethargy, weakness, metabolic acidosis, and rapid, deep breathing.
Potassium (K^+) 3.5–5.0 mEq/liter	**Hypokalemia** (hī-pō-ka-LĒ-mē-a) may result from excessive loss due to vomiting or diarrhea, decreased potassium intake, hyperaldosteronism, kidney disease, and therapy with some diuretics.	Muscle fatigue, flaccid paralysis, mental confusion, increased urine output, shallow respirations, and changes in electrocardiogram, including flattening of T wave.	**Hyperkalemia** may be due to excessive potassium intake, renal failure, aldosterone deficiency, crushing injuries to body tissues, or transfusion of hemolyzed blood.	Irritability, nausea, vomiting, diarrhea, muscular weakness; can cause death by inducing ventricular fibrillation.
Calcium (Ca^{2+}) Total = 9.0–10.5 mg/dL; ionized = 4.5–5.5 mEq/liter	**Hypocalcemia** (hī-pō-kal-SĒ-mē-a) may be due to increased calcium loss, reduced calcium intake, elevated phosphate levels, or hypoparathyroidism.	Numbness and tingling of fingers; hyperactive reflexes, muscle cramps, tetany, and convulsions; bone fractures; spasms of laryngeal muscles that can cause death by asphyxiation.	**Hypercalcemia** may result from hyperparathyroidism, some cancers, excessive intake of vitamin D, and Paget's disease of bone.	Lethargy, weakness, anorexia, nausea, vomiting, polyuria, itching, bone pain, depression, confusion, paresthesia, stupor, and coma.
Phosphate (HPO_4^{2-}) 1.7–2.6 mEq/liter	**Hypophosphatemia** (hī-pō-fos′-fa-TĒ-mē-a) may occur through increased urinary losses, decreased intestinal absorption, or increased utilization.	Confusion, seizures, coma, chest and muscle pain, numbness and tingling of fingers, decreased coordination, memory loss, and lethargy.	**Hyperphosphatemia** occurs when kidneys fail to excrete excess phosphate, as in renal failure; can also result from increased intake of phosphates or destruction of body cells, which releases phosphates into blood.	Anorexia, nausea, vomiting, muscular weakness, hyperactive reflexes, tetany, and tachycardia.
Magnesium (Mg^{2+}) 1.3–2.1 mEq/liter	**Hypomagnesemia** (hī′-pō-mag′-ne-SĒ-mē-a) may be due to inadequate intake or excessive loss in urine or feces; also occurs in alcoholism, malnutrition, diabetes mellitus, and diuretic therapy.	Weakness, irritability, tetany, delirium, convulsions, confusion, anorexia, nausea, vomiting, paresthesia, and cardiac arrhythmias.	**Hypermagnesemia** occurs in renal failure or due to increased intake of Mg^{2+}, such as Mg^{2+}-containing antacids; also occurs in aldosterone deficiency and hypothyroidism.	Hypotension, muscular weakness or paralysis, nausea, vomiting, and altered mental functioning.

*Values are normal ranges of blood plasma levels in adults.

We will examine each of these mechanisms in more detail in the following sections.

The Actions of Buffer Systems

Most buffer systems in the body consist of a weak acid and the salt of that acid, which functions as a weak base. Buffers prevent rapid, drastic changes in the pH of body fluids by converting strong acids and bases into weak acids and weak bases within fractions of a second. Strong acids lower pH more than weak acids because strong acids release H^+ more readily and thus contribute more free hydrogen ions. Similarly, strong bases raise pH more than weak ones. The principal buffer systems of the body fluids are the protein buffer system, the carbonic acid–bicarbonate buffer system, and the phosphate buffer system.

Protein Buffer System

The **protein buffer system** is the most abundant buffer in intracellular fluid and blood plasma. For example, the protein hemoglobin is an especially good buffer within red blood cells, and albumin is the main protein buffer in blood plasma. Proteins are composed of amino acids, organic molecules that contain at least one carboxyl group (—COOH) and at least one amino group (—NH₂); these groups are the functional components of the protein buffer system. The free carboxyl group at one end of a protein acts like an acid by releasing H^+ when pH rises; it dissociates as follows:

$$NH_2-\underset{\underset{H}{|}}{\overset{\overset{R}{|}}{C}}-COOH \longrightarrow NH_2-\underset{\underset{H}{|}}{\overset{\overset{R}{|}}{C}}-COO^- + H^+$$

The H^+ is then able to react with any excess OH^- in the solution to form water. The free amino group at the other end of a protein can act as a base by combining with H^+ when pH falls, as follows:

$$NH_2-\underset{\underset{H}{|}}{\overset{\overset{R}{|}}{C}}-COOH + H^+ \longrightarrow {}^+NH_3-\underset{\underset{H}{|}}{\overset{\overset{R}{|}}{C}}-COOH$$

So proteins can buffer both acids and bases. In addition to the terminal carboxyl and amino groups, side chains that can buffer H^+ are present on 7 of the 20 amino acids.

As we have already noted, the protein hemoglobin is an important buffer of H^+ in red blood cells (see Figure 23.23). As blood flows through the systemic capillaries, carbon dioxide (CO_2) passes from tissue cells into red blood cells, where it combines with water (H_2O) to form carbonic acid (H_2CO_3). Once formed, H_2CO_3 dissociates into H^+ and HCO_3^-. At the same time that CO_2 is entering red blood cells, oxyhemoglobin (Hb—O_2) is giving up its oxygen to tissue cells. Reduced hemoglobin (deoxyhemoglobin) picks up most of the H^+. For this reason, reduced he-

moglobin usually is written as Hb—H. The following reactions summarize these relations:

$$H_2O + CO_2 \longrightarrow H_2CO_3$$

Water Carbon dioxide Carbonic acid
(entering RBCs)

$$H_2CO_3 \longrightarrow H^+ + HCO_3^-$$

Carbonic acid Hydrogen ion Bicarbonate ion

$$Hb-O_2 + H^+ \longrightarrow Hb-H + O_2$$

Oxyhemoglobin Hydrogen ion Reduced Oxygen
(in RBCs) (from carbonic hemoglobin (released to
acid) tissue cells)

Carbonic Acid–Bicarbonate Buffer System

The **carbonic acid–bicarbonate buffer system** is based on the *bicarbonate ion* (HCO_3^-), which can act as a weak base, and *carbonic acid* (H_2CO_3), which can act as a weak acid. As you have already learned, HCO_3^- is a significant anion in both intracellular and extracellular fluids (see Figure 27.6). Because the kidneys also synthesize new HCO_3^- and reabsorb filtered HCO_3^-, this important buffer is not lost in the urine. If there is an excess of H^+, the HCO_3^- can function as a weak base and remove the excess H^+ as follows:

$$H^+ + HCO_3^- \longrightarrow H_2CO_3$$

Hydrogen ion Bicarbonate ion Carbonic acid
(weak base)

Then, H_2CO_3 dissociates into water and carbon dioxide, and the CO_2 is exhaled from the lungs.

Conversely, if there is a shortage of H^+, the H_2CO_3 can function as a weak acid and provide H^+ as follows:

$$H_2CO_3 \longrightarrow H^+ + HCO_3^-$$

Carbonic acid Hydrogen ion Bicarbonate ion
(weak acid)

At a pH of 7.4, HCO_3^- concentration is about 24 mEq/liter and H_2CO_3 concentration is about 1.2 mmol/liter, so bicarbonate ions outnumber carbonic acid molecules by 20 to 1. Because CO_2 and H_2O combine to form H_2CO_3, this buffer system cannot protect against pH changes due to respiratory problems in which there is an excess or shortage of CO_2.

Phosphate Buffer System

The **phosphate buffer system** acts via a mechanism similar to the one for the carbonic acid–bicarbonate buffer system. The components of the phosphate buffer system are the ions *dihydrogen phosphate* ($H_2PO_4^-$) and *monohydrogen phosphate* (HPO_4^{2-}). Recall that phosphates are major anions in intracellular fluid and minor ones in extracellular fluids (see Figure 27.6). The dihydrogen phosphate ion acts as a weak acid and is capable of buffering strong bases such as OH^-, as follows:

$$OH^- + H_2PO_4^- \longrightarrow H_2O + HPO_4^{2-}$$

Hydroxide ion Dihydrogen Water Monohydrogen
(strong base) phosphate phosphate
(weak acid) (weak base)

The monohydrogen phosphate ion is capable of buffering the H^+ released by a strong acid such as hydrochloric acid (HCl) by acting as a weak base:

$$H^+ \; + \; HPO_4{}^{2-} \longrightarrow H_2PO_4{}^-$$
<table>
<tr><td>Hydrogen ion</td><td>Monohydrogen</td><td>Dihydrogen</td></tr>
<tr><td>(strong acid)</td><td>phosphate</td><td>phosphate</td></tr>
<tr><td></td><td>(weak base)</td><td>(weak acid)</td></tr>
</table>

Because the concentration of phosphates is highest in intracellular fluid, the phosphate buffer system is an important regulator of pH in the cytosol. It also acts to a smaller degree in extracellular fluids and buffers acids in urine. $H_2PO_4{}^-$ is formed when excess H^+ in the kidney tubule fluid combines with $HPO_4{}^{2-}$ (see Figure 27.8). The H^+ that becomes part of the $H_2PO_4{}^-$ passes into the urine. This reaction is one way the kidneys help maintain blood pH by excreting H^+ in the urine.

Exhalation of Carbon Dioxide

The simple act of breathing also plays an important role in maintaining the pH of body fluids. An increase in the carbon dioxide (CO_2) concentration in body fluids increases H^+ concentration and thus lowers the pH (makes body fluids more acidic). Because H_2CO_3 can be eliminated by exhaling CO_2, it is called a **volatile acid.** Conversely, a decrease in the CO_2 concentration of body fluids raises the pH (makes body fluids more alkaline). This chemical interaction is illustrated by the following reversible reactions:

$$CO_2 + H_2O \rightleftharpoons H_2CO_3 \rightleftharpoons H^+ + HCO_3{}^-$$
<table>
<tr><td>Carbon</td><td>Water</td><td>Carbonic</td><td>Hydrogen</td><td>Bicarbonate</td></tr>
<tr><td>dioxide</td><td></td><td>acid</td><td>ion</td><td>ion</td></tr>
</table>

Changes in the rate and depth of breathing can alter the pH of body fluids within a couple of minutes. With increased ventilation, more CO_2 is exhaled. When CO_2 levels decrease, the reaction is driven to the left (blue arrows), H^+ concentration falls, and blood pH increases. Doubling the ventilation increases pH by about 0.23 units, from 7.4 to 7.63. If ventilation is slower than normal, less carbon dioxide is exhaled. When CO_2 levels increase, the reaction is driven to the right (red arrows), the H^+ concentration increases, and blood pH decreases. Reducing ventilation to one-quarter of normal lowers the pH by 0.4 units, from 7.4 to 7.0. These examples show the powerful effect of alterations in breathing on the pH of body fluids.

The pH of body fluids and the rate and depth of breathing interact via a negative feedback loop (Figure 27.7). When the blood acidity increases, the decrease in pH (increase in concentration of H^+) is detected by central chemoreceptors in the medulla oblongata and peripheral chemoreceptors in the aortic and carotid bodies, both of which stimulate the inspiratory area in the medulla oblongata. As a result, the diaphragm and other respiratory muscles contract more forcefully and frequently, so more CO_2 is exhaled. As less H_2CO_3 forms and fewer H^+ are present, blood pH increases. When the response brings blood pH (H^+ concentration) back to normal, there is a return to acid–base

Figure 27.7 Negative feedback regulation of blood pH by the respiratory system.

Exhalation of carbon dioxide lowers the H^+ concentration of blood.

? **If you hold your breath for 30 seconds, what is likely to happen to your blood pH?**

homeostasis. The same negative feedback loop operates if the blood level of CO_2 increases. Ventilation increases, which removes more CO_2, reducing the H^+ concentration and increasing the blood's pH.

By contrast, if the pH of the blood increases, the respiratory center is inhibited and the rate and depth of breathing decreases. A decrease in the CO_2 concentration of the blood has the same effect. When breathing decreases, CO_2 accumulates in the blood so its H^+ concentration increases.

Kidney Excretion of H^+

Metabolic reactions produce **nonvolatile acids** such as sulfuric acid at a rate of about 1 mEq of H^+ per day for every kilogram of body mass. The only way to eliminate this huge acid load is to excrete H^+ in the urine. Given the magnitude of these contributions to acid–base balance, it's not surprising that renal failure can quickly cause death.

As you learned in Chapter 26, cells in both the proximal convoluted tubules (PCT) and the collecting ducts of the kidneys secrete hydrogen ions into the tubular fluid. In the PCT, Na^+/H^+ antiporters secrete H^+ as they reabsorb Na^+ (see Figure 26.13). Even more important for regulation of pH of body fluids, however, are the intercalated cells of the collecting duct. The *apical* membranes of some intercalated cells include **proton pumps (H^+ ATPases)** that secrete H^+ into the tubular fluid (Figure 27.8). Intercalated cells can secrete H^+ against a concentration gradient so effectively that urine can be up to 1000 times (3 pH units) more acidic than blood. HCO_3^- produced by dissociation of H_2CO_3 inside intercalated cells crosses the basolateral membrane by means of Cl^-/HCO_3^- **antiporters** and then diffuses into peritubular capillaries (Figure 27.8a). The HCO_3^- that enters the blood in this way is *new* (not filtered). For this reason, blood leaving the kidney in the renal vein may have a higher HCO_3^- concentration than blood entering the kidney in the renal artery.

Interestingly, a second type of intercalated cell has proton pumps in its *basolateral* membrane and Cl^-/HCO_3^- antiporters in its apical membrane. These intercalated cells secrete HCO_3^- and reabsorb H^+. Thus, the two types of intercalated cells help maintain the pH of body fluids in two ways—by excreting excess H^+ when pH of body fluids is too low and by excreting excess HCO_3^- when pH is too high.

Some H^+ secreted into the tubular fluid of the collecting duct is buffered, but not by HCO_3^-, most of which has been filtered and reabsorbed. Two other buffers combine with H^+ in the collecting duct (Figure 27.8b). The most plentiful buffer in the tubular fluid of the collecting duct is HPO_4^{2-} (monohydrogen phosphate ion). In addition, a small amount of NH_3 (ammonia) also is present. H^+ combines with HPO_4^{2-} to form $H_2PO_4^-$ (dihydrogen phosphate ion) and with NH_3 to form NH_4^+ (ammonium ion). Because these ions cannot diffuse back into tubule cells, they are excreted in the urine.

Table 27.3 summarizes the mechanisms that maintain the pH of body fluids.

Figure 27.8 Secretion of H^+ by intercalated cells in the collecting duct. HCO_3^- = bicarbonate ion; CO_2 = carbon dioxide; H_2O = water; H_2CO_3 = carbonic acid; Cl^- = chloride ion; NH_3 = ammonia; NH_4^+ = ammonium ion; HPO_4^{2-} = monohydrogen phosphate ion; $H_2PO_4^-$ = dihydrogen phosphate ion.

Urine can be up to 1000 times more acidic than blood due to the operation of the proton pumps in the collecting ducts of the kidneys.

(a) Secretion of H^+

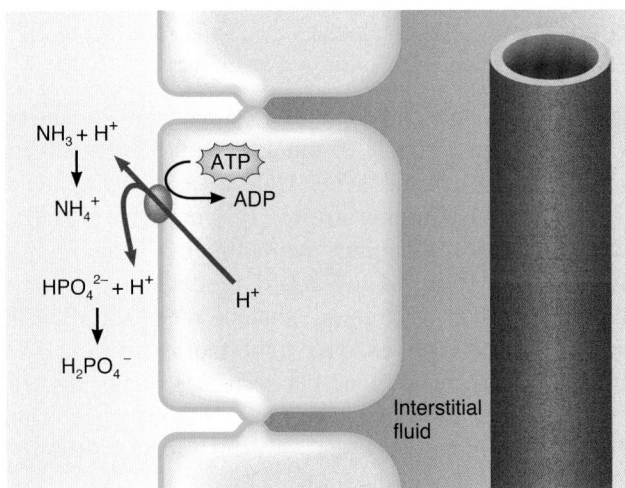

(b) Buffering of H^+ in urine

Key:

 Proton pump (H^+ ATPase) in apical membrane

 HCO_3^-/Cl^- antiporter in basolateral membrane

 Diffusion

 What would be the effects of a drug that blocks the activity of carbonic anhydrase?

TABLE 27.3

Mechanisms That Maintain pH of Body Fluids

MECHANISM	COMMENTS
Buffer systems	Most consist of a weak acid and its salt, which functions as a weak base. They prevent drastic changes in body fluid pH.
Proteins	The most abundant buffers in body cells and blood. Hemoglobin inside red blood cells is a good buffer.
Carbonic acid–bicarbonate	Important regulator of blood pH. The most abundant buffers in extracellular fluid (ECF).
Phosphates	Important buffers in intracellular fluid and urine.
Exhalation of CO_2	With increased exhalation of CO_2, pH rises (fewer H^+). With decreased exhalation of CO_2, pH falls (more H^+).
Kidneys	Renal tubules secrete H^+ into urine and reabsorb HCO_3^- so it is not lost in urine.

Acid–Base Imbalances

The normal pH range of systemic arterial blood is between 7.35 (= 45 nEq of H^+/liter) and 7.45 (= 35 nEq of H^+/liter). **Acidosis** (or **acidemia**) is a condition in which blood pH is below 7.35; **alkalosis** (or **alkalemia**) is a condition in which blood pH is higher than 7.45.

The major physiological effect of acidosis is depression of the central nervous system through depression of synaptic transmission. If the systemic arterial blood pH falls below 7, depression of the nervous system is so severe that the individual becomes disoriented, then comatose, and may die. Patients with severe acidosis usually die while in a coma. A major physiological effect of alkalosis, by contrast, is overexcitability in both the central nervous system and peripheral nerves. Neurons conduct impulses repetitively, even when not stimulated by normal stimuli; the results are nervousness, muscle spasms, and even convulsions and death.

A change in blood pH that leads to acidosis or alkalosis may be countered by **compensation,** the physiological response to an acid–base imbalance that acts to normalize arterial blood pH. Compensation may be either *complete,* if pH indeed is brought within the normal range, or *partial,* if systemic arterial blood pH is still lower than 7.35 or higher than 7.45. If a person has altered blood pH due to metabolic causes, hyperventilation or hypoventilation can help bring blood pH back toward the normal range; this form of compensation, termed **respiratory compensation,** occurs within minutes and reaches its maximum within hours. If, however, a person has altered blood pH due to respiratory causes, then **renal compensation**—changes in secretion of H^+ and reabsorption of HCO_3^- by the kidney tubules—can help reverse the change. Renal compensation may begin in minutes, but it takes days to reach maximum effectiveness.

In the discussion that follows, note that both respiratory acidosis and respiratory alkalosis are disorders resulting from changes in the partial pressure of CO_2 (P_{CO_2}) in systemic arterial blood (normal range is 35–45 mmHg). By contrast, both metabolic acidosis and metabolic alkalosis are disorders resulting from changes in HCO_3^- concentration (normal range is 22–26 mEq/liter in systemic arterial blood).

Respiratory Acidosis

The hallmark of **respiratory acidosis** is an abnormally high P_{CO_2} in systemic arterial blood—above 45 mmHg. Inadequate exhalation of CO_2 causes the blood pH to drop. Any condition that decreases the movement of CO_2 from the blood to the alveoli of the lungs to the atmosphere causes a buildup of CO_2, H_2CO_3, and H^+. Such conditions include emphysema, pulmonary edema, injury to the respiratory center of the medulla oblongata, airway obstruction, or disorders of the muscles involved in breathing. If the respiratory problem is not too severe, the kidneys can help raise the blood pH into the normal range by increasing excretion of H^+ and reabsorption of HCO_3^- (renal compensation). The goal in treatment of respiratory acidosis is to increase the exhalation of CO_2, as, for instance, by providing ventilation therapy. In addition, intravenous administration of HCO_3^- may be helpful.

Respiratory Alkalosis

In **respiratory alkalosis,** systemic arterial blood P_{CO_2} falls below 35 mmHg. The cause of the drop in P_{CO_2} and the resulting increase in pH is hyperventilation, which occurs in conditions that stimulate the inspiratory area in the brain stem. Such conditions include oxygen deficiency due to high altitude or pulmonary disease, cerebrovascular accident (stroke), or severe anxiety. Again, renal compensation may bring blood pH into the normal range if the kidneys are able to decrease excretion of H^+ and reabsorption of HCO_3^-. Treatment of respiratory alkalosis is aimed at increasing the level of CO_2 in the body. One simple treatment is to have the person inhale and exhale into a paper bag for a short period; as a result, the person inhales air containing a higher-than-normal concentration of CO_2.

Metabolic Acidosis

In **metabolic acidosis,** the systemic arterial blood HCO_3^- level drops below 22 mEq/liter. Such a decline in this important buffer causes the blood pH to decrease. Three situations may lower the blood level of HCO_3^-: (1) actual loss of HCO_3^-, such as may occur with severe diarrhea or renal dysfunction; (2) accumulation of an acid other than carbonic acid, as may occur in ketosis (described in Clinical Connection: Ketosis in Chapter 25); or (3) failure of the kidneys to excrete H^+ from metabolism of dietary proteins. If the problem is not too severe, hyperventilation can help bring blood pH into the normal range (respiratory compensation). Treatment of metabolic acidosis consists of administering intravenous solutions of sodium bicarbonate and correcting the cause of the acidosis.

Metabolic Alkalosis

In **metabolic alkalosis,** the systemic arterial blood HCO_3^- concentration is above 26 mEq/liter. A nonrespiratory loss of acid or excessive intake of alkaline drugs causes the blood pH to increase above 7.45. Excessive vomiting of gastric contents, which results in a substantial loss of hydrochloric acid, is probably the most frequent cause of metabolic alkalosis. Other causes include gastric suctioning, use of certain diuretics, endocrine disorders, excessive intake of alkaline drugs (antacids), and severe dehydration. Respiratory compensation through hypoventilation may bring blood pH into the normal range. Treatment of metabolic alkalosis consists of giving fluid solutions to correct Cl^-, K^+, and other electrolyte deficiencies plus correcting the cause of alkalosis.

Table 27.4 summarizes respiratory and metabolic acidosis and alkalosis.

CLINICAL CONNECTION | *Diagnosis of Acid–Base Imbalances*

One can often pinpoint the cause of an acid–base imbalance by careful evaluation of three factors in a sample of systemic arterial blood: pH, concentration of HCO_3^-, and P_{CO_2}. These three blood chemistry values are examined in the following four-step sequence:

1. Note whether the pH is high (alkalosis) or low (acidosis).
2. Then decide which value—P_{CO_2} or HCO_3^-—is out of the normal range and could be the *cause* of the pH change. For example, *elevated pH* could be caused by *low* P_{CO_2} or high HCO_3^-.
3. If the cause is a *change in P_{CO_2}*, the problem is *respiratory*; if the cause is a *change in HCO_3^-*, the problem is *metabolic*.
4. Now look at the value that doesn't correspond with the observed pH change. If it is within its normal range, there is no compensation. If it is outside the normal range, compensation is occurring and partially correcting the pH imbalance. •

✔CHECKPOINT

8. Explain how each of the following buffer systems helps to maintain the pH of body fluids: proteins, carbonic acid–bicarbonate buffers, and phosphates.
9. Define acidosis and alkalosis. Distinguish among respiratory and metabolic acidosis and alkalosis.
10. What are the principal physiological effects of acidosis and alkalosis?

27.4 AGING AND FLUID, ELECTROLYTE, AND ACID–BASE BALANCE

◉ OBJECTIVE
• Describe the changes in fluid, electrolyte, and acid–base balance that may occur with aging.

There are significant differences between adults and infants, especially premature infants, with respect to fluid distribution, regulation of fluid and electrolyte balance, and acid–base homeostasis. Accordingly, infants experience more problems than adults in these areas. The differences are related to the following conditions:

• **Proportion and distribution of water.** A newborn's total body mass is about 75% water (and can be as high as 90% in a premature infant); an adult's total body mass is about 55–60% water. (The "adult" percentage is achieved at about 2 years of age.) Adults have twice as much water in ICF as ECF, but the opposite is true in premature infants. Because ECF is subject to more changes than ICF, rapid losses or gains of body water are

TABLE 27.4

Summary of Acidosis and Alkalosis

CONDITION	DEFINITION	COMMON CAUSES	COMPENSATORY MECHANISM
Respiratory acidosis	Increased P_{CO_2} (above 45 mmHg) and decreased pH (below 7.35) if no compensation.	Hypoventilation due to emphysema, pulmonary edema, trauma to respiratory center, airway obstructions, or dysfunction of muscles of respiration.	Renal: increased excretion of H^+; increased reabsorption of HCO_3^-. If compensation is complete, pH will be within normal range but P_{CO_2} will be high.
Respiratory alkalosis	Decreased P_{CO_2} (below 35 mmHg) and increased pH (above 7.45) if no compensation.	Hyperventilation due to oxygen deficiency, pulmonary disease, cerebrovascular accident (CVA), or severe anxiety.	Renal: decreased excretion of H^+; decreased reabsorption of HCO_3^-. If compensation is complete, pH will be within normal range but P_{CO_2} will be low.
Metabolic acidosis	Decreased HCO_3^- (below 22 mEq/liter) and decreased pH (below 7.35) if no compensation.	Loss of bicarbonate ions due to diarrhea, accumulation of acid (ketosis), renal dysfunction.	Respiratory: hyperventilation, which increases loss of CO_2. If compensation is complete, pH will be within normal range but HCO_3^- will be low.
Metabolic alkalosis	Increased HCO_3^- (above 26 mEq/liter) and increased pH (above 7.45) if no compensation.	Loss of acid due to vomiting, gastric suctioning, or use of certain diuretics; excessive intake of alkaline drugs.	Respiratory: hypoventilation, which slows loss of CO_2. If compensation is complete, pH will be within normal range but HCO_3^- will be high.

much more critical in infants. Given that the rate of fluid intake and output is about seven times higher in infants than in adults, the slightest changes in fluid balance can result in severe abnormalities.

- *Metabolic rate.* The metabolic rate of infants is about double that of adults. This results in the production of more metabolic wastes and acids, which can lead to the development of acidosis in infants.

- *Functional development of the kidneys.* Infant kidneys are only about half as efficient in concentrating urine as those of adults. (Functional development is not complete until close to the end of the first month after birth.) As a result, the kidneys of newborns can neither concentrate urine nor rid the body of excess acids as effectively as those of adults.

- *Body surface area.* The ratio of body surface area to body volume of infants is about three times greater than that of adults. Water loss through the skin is significantly higher in infants than in adults.

- *Breathing rate.* The higher breathing rate of infants (about 30 to 80 times a minute) causes greater water loss from the lungs. Respiratory alkalosis may occur because greater ventilation eliminates more CO_2 and lowers the P_{CO_2}.

- *Ion concentrations.* Newborns have higher K^+ and Cl^- concentrations than adults. This creates a tendency toward metabolic acidosis.

By comparison with children and younger adults, older adults often have an impaired ability to maintain fluid, electrolyte, and acid–base balance. With increasing age, many people have a decreased volume of intracellular fluid and decreased total body K^+ due to declining skeletal muscle mass and increasing mass of adipose tissue (which contains very little water). Age-related decreases in respiratory and renal functioning may compromise acid–base balance by slowing the exhalation of CO_2 and the excretion of excess acids in urine. Other kidney changes, such as decreased blood flow, decreased glomerular filtration rate, and reduced sensitivity to antidiuretic hormone, have an adverse effect on the ability to maintain fluid and electrolyte balance. Due to a decrease in the number and efficiency of sweat glands, water loss from the skin declines with age. Because of these age-related changes, older adults are susceptible to several fluid and electrolyte disorders:

- *Dehydration* and *hypernatremia* often occur due to inadequate fluid intake or loss of more water than Na^+ in vomit, feces, or urine.

- *Hyponatremia* may occur due to inadequate intake of Na^+; elevated loss of Na^+ in urine, vomit, or diarrhea; or impaired ability of the kidneys to produce dilute urine.

- *Hypokalemia* often occurs in older adults who chronically use laxatives to relieve constipation or who take K^+-depleting diuretic drugs for treatment of hypertension or heart disease.

- *Acidosis* may occur due to impaired ability of the lungs and kidneys to compensate for acid–base imbalances. One cause of acidosis is decreased production of ammonia (NH_3) by renal tubule cells, which then is not available to combine with H^+ and be excreted in urine as NH_4^+; another cause is reduced exhalation of CO_2.

✔ **CHECKPOINT**

11. Why do infants experience greater problems with fluid, electrolyte, and acid–base balance than adults?

CHAPTER REVIEW AND RESOURCE SUMMARY

Review

Resource

27.1 Fluid Compartments and Fluid Balance

1. Body fluid includes water and dissolved solutes. About two-thirds of the body's fluid is located within cells and is called intracellular fluid (ICF). The other one-third, called extracellular fluid (ECF), includes interstitial fluid; blood plasma and lymph; cerebrospinal fluid; gastrointestinal tract fluids; synovial fluid; fluids of the eyes and ears; pleural, pericardial, and peritoneal fluids; and glomerular filtrate.
2. Fluid balance means that the required amounts of water and solutes are present and are correctly proportioned among the various compartments.
3. An inorganic substance that dissociates into ions in solution is called an electrolyte.
4. Water is the largest single constituent in the body. It makes up 45–75% of total body mass, depending on age, gender, and the amount of adipose tissue present.
5. Daily water gain and loss are each about 2500 mL. Sources of water gain are ingested liquids and foods, and water produced by cellular respiration and dehydration synthesis reactions (metabolic water). Water is lost from the body via urination, evaporation from the skin surface, exhalation of water vapor, and defecation. In women, menstrual flow is an additional route for loss of body water.
6. Body water gain is regulated by adjusting the volume of water intake, mainly by drinking more or less fluid. The thirst center in the hypothalamus governs the urge to drink. Although increased amounts of water and solutes are lost through sweating and exhalation during exercise, loss of excess body water or excess solutes depends mainly on regulating excretion in the urine. The extent of urinary NaCl loss is the main determinant of body fluid volume; the extent of urinary water loss is the main determinant

Overview of Fluids
Animation - Water Homeostasis
Animation - Water and Fluid Flow

Review

Resource

of body fluid osmolarity. Table 27.1 summarizes the factors that regulate water gain and water loss in the body.

7. Angiotensin II and aldosterone reduce urinary loss of Na^+ and Cl^- and thereby increase the volume of body fluids. ANP promotes natriuresis, elevated excretion of Na^+ (and Cl^-), which decreases blood volume.

8. The major hormone that regulates water loss and thus body fluid osmolarity is antidiuretic hormone (ADH).

9. An increase in the osmolarity of interstitial fluid draws water out of cells, and they shrink slightly. A decrease in the osmolarity of interstitial fluid causes cells to swell. Most often a change in osmolarity is due to a change in the concentration of Na^+, the dominant solute in interstitial fluid.

10. When a person consumes water faster than the kidneys can excrete it or when renal function is poor, the result may be water intoxication, in which cells swell dangerously.

27.2 Electrolytes in Body Fluids

Concepts and Connections - Functions of Ions

1. Ions formed when electrolytes dissolve in body fluids control the osmosis of water between fluid compartments, help maintain acid–base balance, and carry electrical current.

2. The concentrations of cations and anions is expressed in units of milliequivalents/liter (mEq/liter). Blood plasma, interstitial fluid, and intracellular fluid contain varying types and amounts of ions.

3. Sodium ions (Na^+) are the most abundant extracellular ions. They are involved in impulse transmission, muscle contraction, and fluid and electrolyte balance. Na^+ level is controlled by aldosterone, antidiuretic hormone, and atrial natriuretic peptide.

4. Chloride ions (Cl^-) are the major extracellular anions. They play a role in regulating osmotic pressure and forming HCl in gastric juice. Cl^- level is controlled indirectly by antidiuretic hormone and by processes that increase or decrease renal reabsorption of Na^+.

5. Potassium ions (K^+) are the most abundant cations in intracellular fluid. They play a key role in the resting membrane potential and action potential of neurons and muscle fibers; help maintain intracellular fluid volume; and contribute to regulation of pH. K^+ level is controlled by aldosterone.

6. Bicarbonate ions (HCO_3^-) are the second most abundant anions in extracellular fluid. They are the most important buffer in blood plasma.

7. Calcium is the most abundant mineral in the body. Calcium salts are structural components of bones and teeth. Ca^{2+}, which are principally extracellular cations, function in blood clotting, neurotransmitter release, and contraction of muscle. Ca^{2+} level is controlled mainly by parathyroid hormone and calcitriol.

8. Phosphate ions ($H_2PO_4^-$, HPO_4^{2-}, and PO_4^{3-}) are principally intracellular anions, and their salts are structural components of bones and teeth. They are also required for the synthesis of nucleic acids and ATP and participate in buffer reactions. Their level is controlled by parathyroid hormone and calcitriol.

9. Magnesium ions (Mg^{2+}) are primarily intracellular cations. They act as cofactors in several enzyme systems.

10. Table 27.2 describes the imbalances that result from deficiency or excess of important body electrolytes.

27.3 Acid–Base Balance

Animation - Role of the Respiratory System in pH Regulation
Figure 27.7 - Negative Feedback Regulation of Blood pH
Figure 27.8 - Secretion of H^+ by Intercalated Cells
Exercise - Phix the pH
Concepts and Connections - Mechanisms of pH Balance
Concepts and Connections - Urinary System Regulation of pH

1. The overall acid–base balance of the body is maintained by controlling the H^+ concentration of body fluids, especially extracellular fluid.

2. The normal pH of systemic arterial blood is 7.35–7.45.

3. Homeostasis of pH is maintained by buffer systems, via exhalation of carbon dioxide, and via kidney excretion of H^+ and reabsorption of HCO_3^-. The important buffer systems include proteins, carbonic acid–bicarbonate buffers, and phosphates.

4. An increase in exhalation of carbon dioxide increases blood pH; a decrease in exhalation of CO_2 decreases blood pH.

5. In the proximal convoluted tubules of the kidneys, Na^+/H^+ antiporters secrete H^+ as they reabsorb Na^+. In the collecting ducts of the kidneys, some intercalated cells reabsorb K^+ and HCO_3^- and secrete H^+; other intercalated cells secrete HCO_3^-. In these ways, the kidneys can increase or decrease the pH of body fluids.

6. Table 27.3 summarizes the mechanisms that maintain pH of body fluids.

7. Acidosis is a systemic arterial blood pH below 7.35; its principal effect is depression of the central nervous system (CNS). Alkalosis is a systemic arterial blood pH above 7.45; its principal effect is overexcitability of the CNS.

Review   **Resource**

8. Respiratory acidosis and alkalosis are disorders due to changes in blood P_{CO_2}; metabolic acidosis and alkalosis are disorders associated with changes in blood HCO_3^- concentration.

9. Metabolic acidosis or alkalosis can be compensated by respiratory mechanisms (respiratory compensation); respiratory acidosis or alkalosis can be compensated by renal mechanisms (renal compensation). Table 27.4 summarizes the effects of respiratory and metabolic acidosis and alkalosis.

10. By examining systemic arterial blood pH, HCO_3^-, and P_{CO_2} values, it is possible to pinpoint the cause of an acid–base imbalance.

27.4 Aging and Fluid, Electrolyte, and Acid–Base Balance

1. With increasing age, there is decreased intracellular fluid volume and decreased K^+ due to declining skeletal muscle mass.

2. Decreased kidney function with aging adversely affects fluid and electrolyte balance.

SELF-QUIZ QUESTIONS

Fill in the blanks in the following statements.

1. The source of water that is derived from aerobic cellular respiration and dehydration synthesis reactions is _____ water.

2. In the carbonic acid–bicarbonate buffer system, the _____ acts as a weak base, and _____ acts as a weak acid.

Indicate whether the following statements are true or false.

3. The phosphate buffer system is an important regulator of pH in the cytosol.

4. The two compartments in which water can be found are plasma and cytosol.

Choose the one best answer to the following questions.

5. Which of the following statements are *true*? (1) Buffers prevent rapid, drastic changes in pH of a body fluid. (2) Buffers work slowly. (3) Strong acids lower pH more than weak acids because strong acids contribute fewer H^+. (4) Most buffers consist of a weak acid and the salt of that acid, which acts as a weak base. (5) Hemoglobin is an important buffer.
   (a) 1, 2, 3, and 5   (b) 1, 3, 4, and 5   (c) 1, 3, and 5
   (d) 1, 4, and 5   (e) 2, 3, and 5

6. Which of the following are *true* concerning ions in the body? (1) They control osmosis of water between fluid compartments. (2) They help maintain acid–base balance. (3) They carry electrical current. (4) They serve as cofactors for enzyme activity. (5) They serve as neurotransmitters under special circumstances.
   (a) 1, 3, and 5   (b) 2, 4, and 5   (c) 1, 4, and 5
   (d) 1, 2, and 4   (e) 1, 2, 3, and 4

7. Which of the following statements are *true*? (1) An increase in the carbon dioxide concentration in body fluids increases H^+ concentration and thus lowers pH. (2) Breath holding results in a decline in blood pH. (3) The respiratory buffer mechanism can eliminate a single volatile acid: carbonic acid. (4) The only way to eliminate nonvolatile acids is to excrete H^+ in the urine. (5) When the diet contains a large amount of protein, normal metabolism produces more acids than bases.
   (a) 1, 2, 3, 4, and 5   (b) 1, 3, 4, and 5   (c) 1, 2, 3, and 4
   (d) 1, 2, 4, and 5   (e) 1, 3, and 4

8. Concerning acid–base imbalances: (1) Acidosis can cause depression of the central nervous system through depression of synaptic transmission. (2) Renal compensation can resolve respiratory alkalosis or acidosis. (3) A major physiological effect of alkalosis is lack of excitability in the central nervous system and peripheral nerves. (4) Resolution of metabolic acidosis and alkalosis occurs through renal compensation. (5) In adjusting blood pH, renal compensation occurs quickly, but respiratory compensation takes days.
   (a) 1, 2, and 5   (b) 1 and 2   (c) 2, 3, and 4
   (d) 2, 3, and 5   (e) 1, 2, 3, and 5

9. Match the following:
   _____ (a) the most abundant cation in intracellular fluid; plays a key role in establishing the resting membrane potential
   _____ (b) the most abundant mineral in the body; plays important roles in blood clotting, neurotransmitter release, maintenance of muscle tone, and excitability of nervous and muscle tissue
   _____ (c) second most common intracellular cation; is a cofactor for enzymes involved in carbohydrate, protein, and Na^+/K^+ ATPase metabolism
   _____ (d) the most abundant extracellular cation; essential in fluid and electrolyte balance
   _____ (e) ions that are mostly combined with lipids, proteins, carbohydrates, nucleic acids, and ATP inside cells
   _____ (f) most prevalent extracellular anion; can help balance the level of anions in different fluid compartments
   _____ (g) second most prevalent extracellular anion; mainly regulated by the kidneys; important for acid–base balance
   _____ (h) substances that act to prevent rapid, drastic changes in the pH of a body fluid
   _____ (i) inorganic substances that dissociate into ions when in solution

   (1) sodium
   (2) chloride
   (3) electrolytes
   (4) bicarbonate
   (5) buffers
   (6) phosphate
   (7) magnesium
   (8) potassium
   (9) calcium

10. Match the following:

_____ (a) an abnormal increase in the volume of interstitial fluid

_____ (b) can occur during renal failure or destruction of body cells, which releases phosphates into the blood

_____ (c) the swelling of cells due to water moving from plasma into interstitial fluid and then into cells

_____ (d) occurs when water loss is greater than water gain

_____ (e) can be caused by excessive sodium in diet or with dehydration

_____ (f) condition that can occur as water moves out of plasma into interstitial fluid and blood volume decreases

_____ (g) can be caused by decreased potassium intake or kidney disease; results in muscle fatigue, increased urine output, changes in electrocardiogram

_____ (h) can occur from hypoparathyroidism

_____ (i) can be caused by emphysema, pulmonary edema, injury to the respiratory center of the medulla oblongata, airway destruction, or disorders of the muscles involved in breathing

_____ (j) can be caused by excessive water intake, excessive vomiting, or aldosterone deficiency

_____ (k) can be caused by actual loss of bicarbonate ions, ketosis, or failure of kidneys to excrete H^+

_____ (l) can be caused by excessive vomiting of gastric contents, gastric suctioning, use of certain diuretics, severe dehydration, or excessive intake of alkaline drugs

_____ (m) can be caused by oxygen deficiency at high altitude, stroke, or severe anxiety

(1) respiratory acidosis
(2) respiratory alkalosis
(3) metabolic acidosis
(4) metabolic alkalosis
(5) dehydration
(6) hypovolemia
(7) water intoxication
(8) edema
(9) hypokalemia
(10) hypernatremia
(11) hyponatremia
(12) hyperphosphatemia
(13) hypocalcemia

CRITICAL THINKING QUESTIONS

1. Robin is in the early stages of pregnancy and has been vomiting excessively for several days. She became weak, was confused, and was taken to the emergency room. What do you suspect has happened to Robin's acid–base balance? How would her body attempt to compensate? What electrolytes would be affected by her vomiting, and how do her symptoms reflect those imbalances?

2. Henry is in the intensive care unit because he suffered a severe myocardial infarction three days ago. The lab reports the following values from an arterial blood sample: pH 7.30, $HCO_3^- = 20$ mEq/liter, $P_{CO_2} = 32$ mmHg. Diagnose Henry's acid–base status and decide whether compensation is occurring.

3. This summer, Sam is training for a marathon by running 10 miles a day. Describe changes in his fluid balance as he trains.

? ANSWERS TO FIGURE QUESTIONS

27.1 Plasma volume equals body mass × percent of body mass that is body fluid × proportion of body fluid that is ECF × proportion of ECF that is plasma × a conversion factor (1 liter/kg). For males, blood plasma volume = 60 kg × 0.60 × 1/3 × 0.20 × 1 liter/kg = 2.4 liters. Using similar calculations, female blood plasma volume is 2.2 liters.

27.2 Hyperventilation, vomiting, fever, and diuretics all increase fluid loss.

27.3 Negative feedback is in operation because the result (an increase in fluid intake) is opposite to the initiating stimulus (dehydration).

27.4 An elevated aldosterone level promotes abnormally high renal reabsorption of NaCl and water, which expands blood volume and increases blood pressure. Because of the increased blood pressure, more fluid filters out of capillaries and accumulates in the interstitial fluid, causing edema.

27.5 If a solution used for oral rehydration therapy contains a small amount of salt, both the salt and water are absorbed in the gastrointestinal tract, blood volume increases without a decrease in osmolarity, and water intoxication does not occur.

27.6 In ECF, the major cation is Na^+, and the major anions are Cl^- and HCO_3^-. In ICF, the major cation is K^+, and the major anions are proteins and organic phosphates (for example, ATP).

27.7 Holding your breath causes blood pH to decrease slightly as CO_2 and H^+ accumulate in the blood.

27.8 A carbonic anhydrase inhibitor reduces secretion of H^+ into the urine and reduces reabsorption of Na^+ and HCO_3^- into the blood. It has a diuretic effect and can cause acidosis (lowered pH of the blood) due to loss of HCO_3^- in the urine.

28

THE REPRODUCTIVE SYSTEMS

THE REPRODUCTIVE SYSTEMS AND HOMEOSTASIS *The male and female reproductive organs work together to produce offspring. In addition, the female reproductive organs contribute to sustaining the growth of embryos and fetuses.*

Sexual reproduction is the process by which organisms produce offspring by making germ cells called **gametes** (GAM-ēts = spouses). After the male gamete (sperm cell) unites with the female gamete (secondary oocyte)—an event called **fertilization** (fer′-til-i-ZĀ-shun)—the resulting cell contains one set of chromosomes from each parent. Males and females have anatomically distinct reproductive organs that are adapted for producing gametes, facilitating fertilization, and, in females, sustaining the growth of the embryo and fetus.

The male and female reproductive organs can be grouped by function. The **gonads**—testes in males and ovaries in females—produce gametes and secrete sex hormones. Various **ducts** then store and transport the gametes, and **accessory sex glands** produce substances that protect the gametes and facilitate their movement. Finally, **supporting structures,** such as the penis in males and the uterus in females, assist the delivery of gametes, and the uterus is also the site for the growth of the embryo and fetus during pregnancy.

Gynecology (gī-ne-KOL-ō-jē; *gyneco-* = woman; *-logy* = study of) is the specialized branch of medicine concerned with the diagnosis and treatment of diseases of the female reproductive system. As noted in Chapter 26, **urology** (ū-ROL-ō-jē) is the study of the urinary system. Urologists also diagnose and treat diseases and disorders of the male reproductive system. The branch of medicine that deals with male disorders, especially infertility and sexual dysfunction, is called **andrology** (an-DROL-ō-jē; *andro-* = masculine).

? *Did you ever wonder how breast augmentation and breast reduction are performed?*

28.1 MALE REPRODUCTIVE SYSTEM

⊙ **OBJECTIVES**

- Describe the location, structure, and functions of the organs of the male reproductive system.
- Discuss the process of spermatogenesis in the testes.

The organs of the **male reproductive system** include the testes, a system of ducts (including the epididymis, ductus deferens, ejaculatory ducts, and urethra), accessory sex glands (seminal vesicles, prostate, and bulbourethral glands), and several supporting structures, including the scrotum and the penis (Figure 28.1). The testes (male gonads) produce sperm and secrete hormones. The duct system transports and stores sperm, assists in their maturation, and conveys them to the exterior. Semen contains sperm plus the secretions provided by the accessory sex glands. The supporting structures have various functions. The penis delivers sperm into the female reproductive tract and the scrotum supports the testes.

Scrotum

The **scrotum** (SKRŌ-tum = bag), the supporting structure for the testes, consists of loose skin and underlying subcutaneous layer that hangs from the root (attached portion) of the penis (Figure 28.1a). Externally, the scrotum looks like a single pouch of skin separated into lateral portions by a median ridge called the **raphe** (RĀ-fē = seam). Internally, the **scrotal septum** divides the scrotum into two sacs, each containing a single testis (Figure 28.2). The septum is made up of a subcutaneous layer and muscle tissue called the **dartos muscle** (DAR-tōs = skinned), which is composed of bundles of smooth muscle fibers. The dartos muscle is also found in the subcutaneous layer of the scrotum. Associated with each testis in the scrotum is the **cremaster muscle** (krē-MAS-ter = suspender), a series of small bands of skeletal muscle that descend as an extension of the internal oblique muscle through the spermatic cord to surround the testes.

The location of the scrotum and the contraction of its muscle fibers regulate the temperature of the testes. Normal sperm production requires a temperature about 2–3°C below core body tempera-

Figure 28.1 Male organs of reproduction and surrounding structures.

🔑 ⊂▭⊃ Reproductive organs are adapted for producing new individuals and passing on genetic material from one generation to the next.

(a) Sagittal section

ture. This lowered temperature is maintained within the scrotum because it is outside the pelvic cavity. In response to cold temperatures, the cremaster and dartos muscles contract. Contraction of the cremaster muscles moves the testes closer to the body, where they can absorb body heat. Contraction of the dartos muscle causes the scrotum to become tight (wrinkled in appearance), which reduces heat loss. Exposure to warmth reverses these actions.

Testes

The **testes** (TES-tēz = witness), or **testicles,** are paired oval glands in the scrotum measuring about 5 cm (2 in.) long and 2.5 cm (1 in.) in diameter (Figure 28.3). Each testis (singular) has a mass of 10–15 grams. The testes develop near the kidneys, in the posterior portion of the abdomen, and they usually begin their descent into the scrotum through the inguinal canals (passageways in the anterior abdominal wall; see Figure 28.2) during the latter half of the seventh month of fetal development.

A serous membrane called the **tunica vaginalis** (TOO-nik-a vaj-i-NAL-is; *tunica* = sheath), which is derived from the peritoneum and forms during the descent of the testes, partially covers the testes. A collection of serous fluid in the tunica vaginalis is called a **hydrocele** (HĪ-drō-sēl; *hydro-* = water; *-kele* = hernia).

It may be caused by injury to the testes or inflammation of the epididymis. Usually, no treatment is required. Internal to the tunica vaginalis is a white fibrous capsule composed of dense irregular connective tissue, the **tunica albuginea** (al'-bū-JIN-ē-a; *albu-* = white); it extends inward, forming septa that divide the testis into a series of internal compartments called **lobules.** Each of the 200–300 lobules contains one to three tightly coiled tubules, the **seminiferous tubules** (sem'-i-NIF-er-us; *semin-* = seed; *-fer-* = to carry), where sperm are produced. The process by which the seminiferous tubules of the testes produce sperm is called spermatogenesis (sper'-ma-tō-JEN-e-sis; *genesis* = to be born).

The seminiferous tubules contain two types of cells: **spermatogenic cells** (sper'-ma-tō-JEN-ik), the sperm-forming cells, and **Sertoli cells** (ser-TŌ-lē), which have several functions in supporting spermatogenesis (Figure 28.4). Stem cells called **spermatogonia** (sper'-ma-tō-GŌ-nē-a; *-gonia* = offspring; singular is *spermatogonium*) develop from **primordial germ cells** (prī-MŌR-dē-al = primitive or early form) that arise from the yolk sac and enter the testes during the fifth week of development. In the embryonic testes, the primordial germ cells differentiate into spermatogonia, which remain dormant during childhood and actively begin producing sperm at puberty. Toward the lumen of the seminiferous tubule are layers of progressively more mature cells.

FUNCTIONS OF THE MALE REPRODUCTIVE SYSTEM

1. The testes produce sperm and the male sex hormone testosterone.
2. The ducts transport, store, and assist in maturation of sperm.
3. The accessory sex glands secrete most of the liquid portion of semen.
4. The penis contains the urethra, a passageway for ejaculation of semen and excretion of urine.

SUPERIOR

Seminal vesicle

Ductus (vas) deferens

Prostatic urethra

Ejaculatory duct

Membranous urethra

Rectum

Root of penis

Bulbospongiosus muscle

Spongy (penile) urethra

Urinary bladder (opened)

Prostate

Pubic symphysis

Corpus cavernosum penis

Corpus spongiosum penis

Body of penis

Testis

Corona

Glans penis

POSTERIOR

ANTERIOR

(b) Sagittal section

 What are the groups of reproductive organs in males, and what are the functions of each group?

Figure 28.2 The scrotum, the supporting structure for the testes.

The scrotum consists of loose skin and an underlying subcutaneous layer and supports the testes.

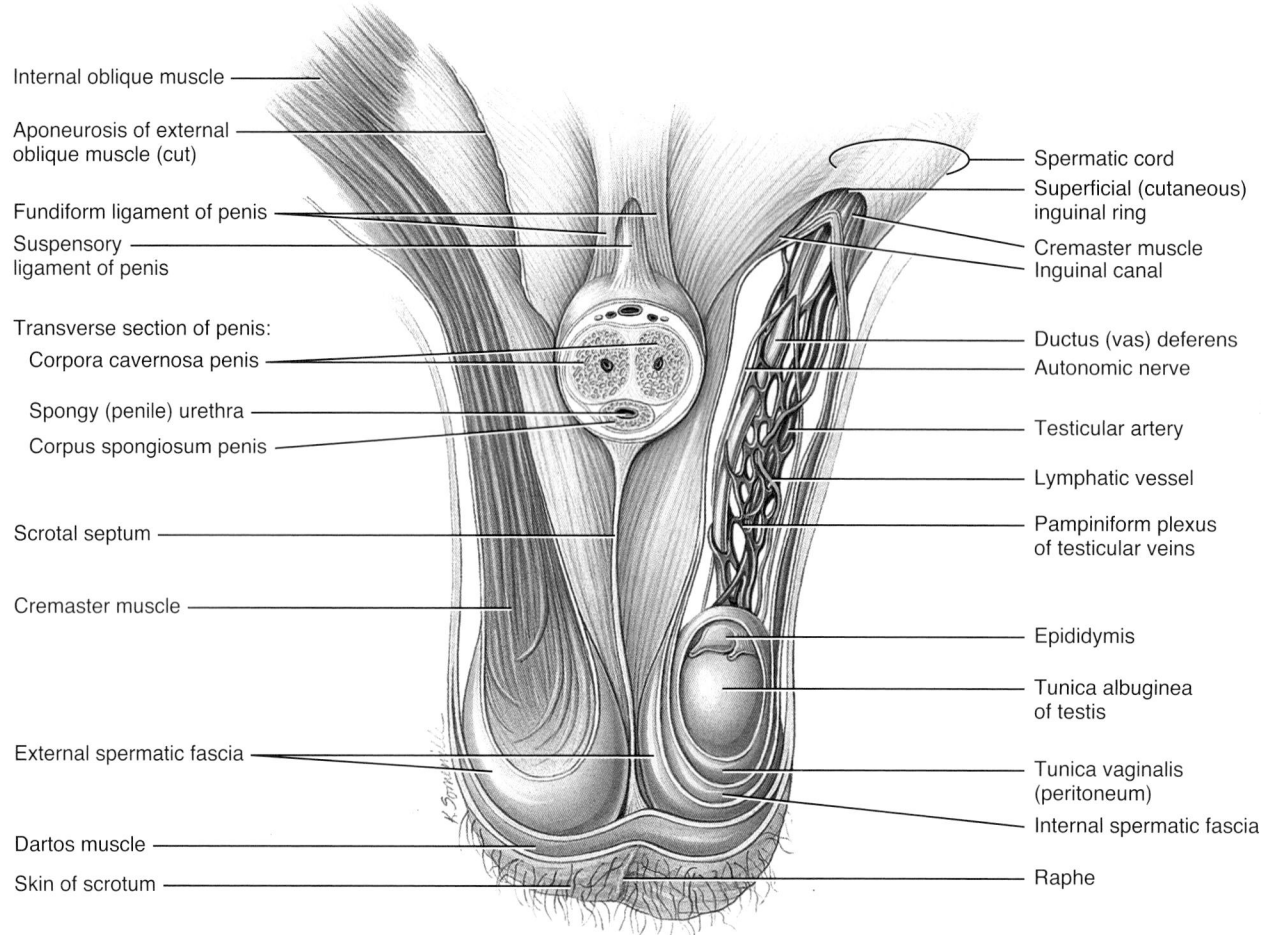

Internal oblique muscle

Aponeurosis of external oblique muscle (cut)

Fundiform ligament of penis

Suspensory ligament of penis

Transverse section of penis:
Corpora cavernosa penis

Spongy (penile) urethra
Corpus spongiosum penis

Scrotal septum

Cremaster muscle

External spermatic fascia

Dartos muscle
Skin of scrotum

Spermatic cord
Superficial (cutaneous) inguinal ring
Cremaster muscle
Inguinal canal

Ductus (vas) deferens
Autonomic nerve

Testicular artery

Lymphatic vessel

Pampiniform plexus of testicular veins

Epididymis

Tunica albuginea of testis

Tunica vaginalis (peritoneum)
Internal spermatic fascia

Raphe

Anterior view of scrotum and testes and transverse section of penis

Which muscles help regulate the temperature of the testes?

In order of advancing maturity, these are primary spermatocytes, secondary spermatocytes, spermatids, and sperm cells. After a **sperm cell,** or **spermatozoon** (sper′-ma-tō-ZŌ-on; *zoon* = life), has formed, it is released into the lumen of the seminiferous tubule. (The plural terms are *sperm* and *spermatozoa*.)

Embedded among the spermatogenic cells in the seminiferous tubules are large Sertoli cells or **sustentacular cells** (sus′-ten-TAK-ū-lar), which extend from the basement membrane to the lumen of the tubule. Internal to the basement membrane and spermatogonia, tight junctions join neighboring Sertoli cells to one another. These junctions form an obstruction known as the **blood–testis barrier** because substances must first pass through the Sertoli cells before they can reach the developing sperm. By isolating the developing gametes from the blood, the blood–testis barrier prevents an immune response against the spermatogenic cell's surface antigens, which are recognized as "foreign" by the immune system. The blood–testis barrier does not include spermatogonia.

Sertoli cells support and protect developing spermatogenic cells in several ways. They nourish spermatocytes, spermatids, and sperm; phagocytize excess spermatid cytoplasm as development proceeds; and control movements of spermatogenic cells and the release of sperm into the lumen of the seminiferous tubule. They also produce fluid for sperm transport, secrete the hormone inhibin, and regulate the effects of testosterone and FSH (follicle-stimulating hormone).

In the spaces between adjacent seminiferous tubules are clusters of cells called **Leydig (interstitial) cells** (LĪ-dig) (Figure 28.4). These cells secrete testosterone, the most prevalent androgen. An **androgen** is a hormone that promotes the development of masculine characteristics. Testosterone also promotes a man's *libido* (sexual drive).

Figure 28.3 Internal and external anatomy of a testis.

🔑 ⊂▭⊃ The testes are the male gonads, which produce haploid sperm.

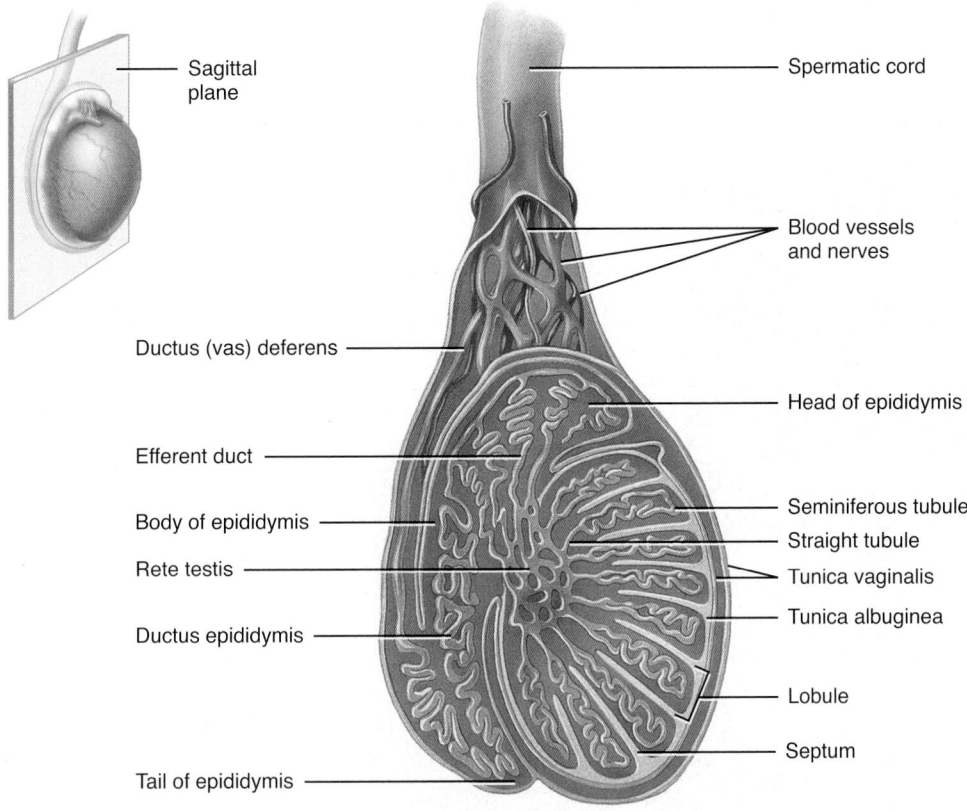

- Sagittal plane
- Spermatic cord
- Blood vessels and nerves
- Ductus (vas) deferens
- Head of epididymis
- Efferent duct
- Seminiferous tubule
- Body of epididymis
- Straight tubule
- Rete testis
- Tunica vaginalis
- Tunica albuginea
- Ductus epididymis
- Lobule
- Septum
- Tail of epididymis

(a) Sagittal section of testis showing seminiferous tubules

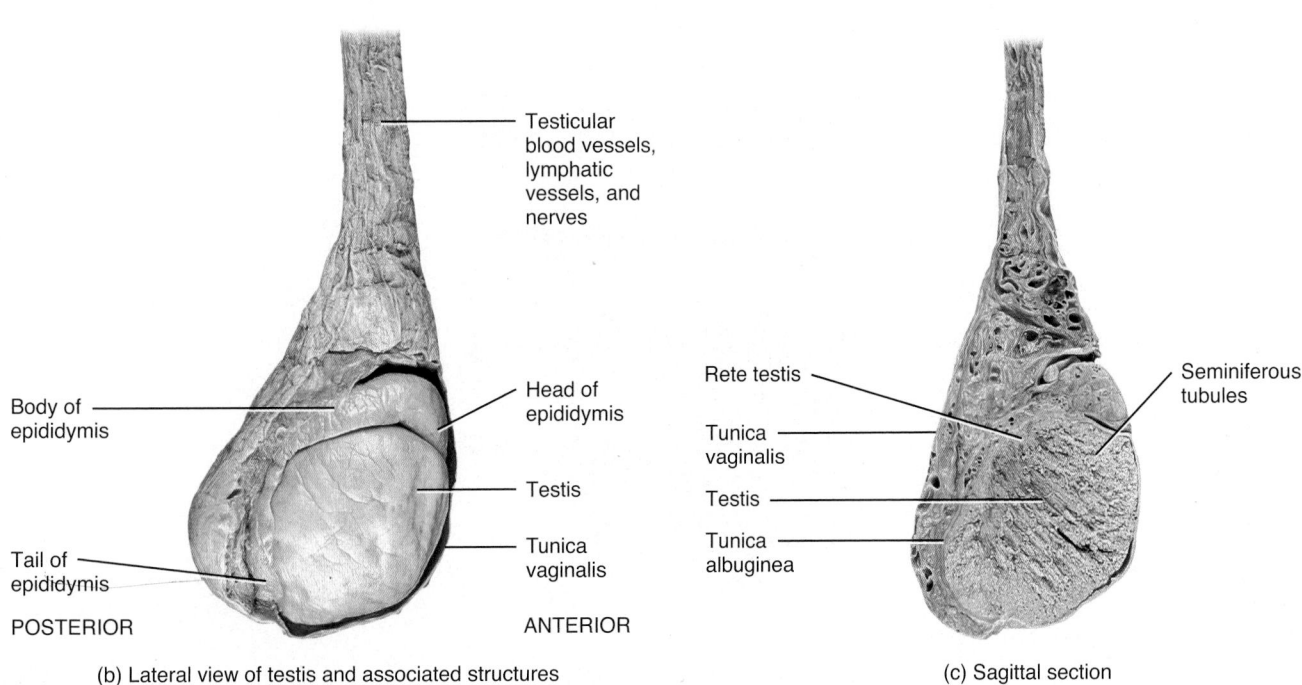

SUPERIOR

- Testicular blood vessels, lymphatic vessels, and nerves
- Body of epididymis
- Head of epididymis
- Testis
- Tail of epididymis
- Tunica vaginalis

POSTERIOR   ANTERIOR

(b) Lateral view of testis and associated structures

- Rete testis
- Seminiferous tubules
- Tunica vaginalis
- Testis
- Tunica albuginea

(c) Sagittal section

❓ **What tissue layers cover and protect the testes?**

Figure 28.4 Microscopic anatomy of the seminiferous tubules and stages of sperm production (spermatogenesis). Arrows in (b) indicate the progression of spermatogenic cells from least mature to most mature. The (*n*) and (*2n*) refer to haploid and diploid numbers of chromosomes, respectively.

Spermatogenesis occurs in the seminiferous tubules of the testes.

Transverse plane

Leydig cell

Basement membrane

Basement membrane

Sertoli cell

Spermatid (*n*)

Secondary spermatocyte (*n*)

Primary spermatocyte (*2n*)

Spermatogonium (*2n*) (stem cell)

LM 40x

LM 160x

(a) Transverse section of several seminiferous tubules

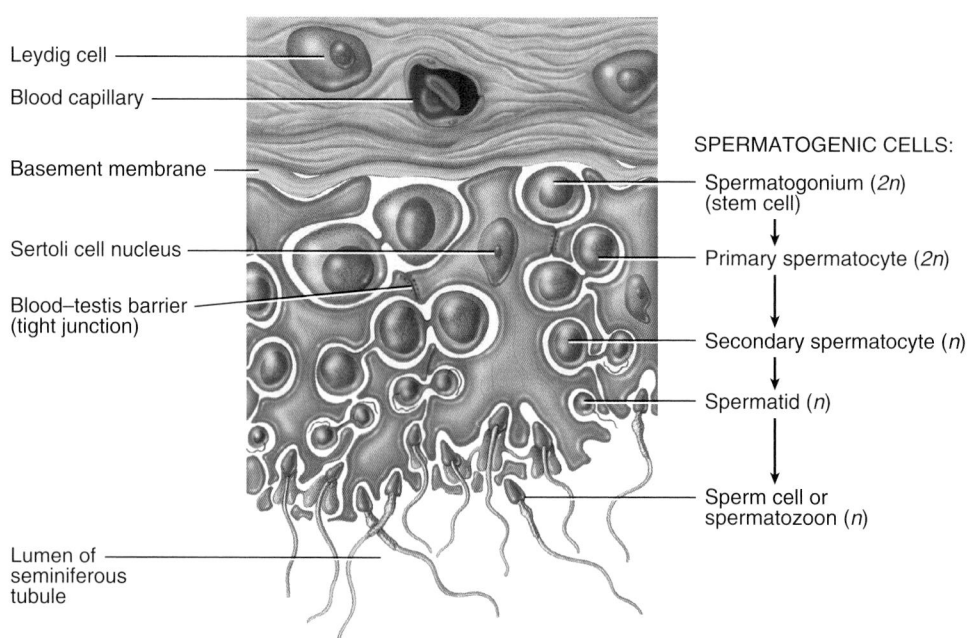

Leydig cell

Blood capillary

Basement membrane

Sertoli cell nucleus

Blood–testis barrier (tight junction)

Lumen of seminiferous tubule

SPERMATOGENIC CELLS:

Spermatogonium (*2n*) (stem cell)

↓

Primary spermatocyte (*2n*)

↓

Secondary spermatocyte (*n*)

↓

Spermatid (*n*)

↓

Sperm cell or spermatozoon (*n*)

? **Which cells secrete testosterone?**

(b) Transverse section of part of seminiferous tubule

The condition in which the testes do not descend into the scrotum is called **cryptorchidism** (krip-TOR-ki-dizm; *crypt-* = hidden; *-orchid* = testis); it occurs in about 3% of full-term infants and about 30% of premature infants. Untreated bilateral cryptorchidism results in sterility because the cells involved in the initial stages of spermatogenesis are destroyed by the higher temperature of the pelvic cavity. The chance of testicular cancer is 30–50 times greater in cryptorchid testes. The testes of about 80% of boys with cryptorchidism will descend spontaneously during the first year of life. When the testes remain undescended, the condition can be corrected surgically, ideally before 18 months of age. •

Spermatogenesis

Before you read this section, please review the topic of reproductive cell division in Chapter 3 in Section 3.7. Pay particular attention to Figures 3.33 and 3.34.

In humans, spermatogenesis takes 65–75 days. It begins with the spermatogonia, which contain the diploid ($2n$) number of chromosomes (Figure 28.5). Spermatogonia are types of *stem cells;* when they undergo mitosis, some spermatogonia remain near the basement membrane of the seminiferous tubule in an undifferentiated state to serve as a reservoir of cells for future cell division and subsequent sperm production. The rest of the spermatogonia lose contact with the basement membrane, squeeze through the tight junctions of the blood–testis barrier, undergo developmental changes, and differentiate into **primary spermatocytes** (SPER-ma-tō-sītz′). Primary spermatocytes, like spermatogonia, are diploid ($2n$); that is, they have 46 chromosomes.

Shortly after it forms, each primary spermatocyte replicates its DNA and then meiosis begins (Figure 28.5). In meiosis I, homologous pairs of chromosomes line up at the metaphase plate, and crossing-over occurs. Then, the meiotic spindle pulls one (duplicated) chromosome of each pair to an opposite pole of the dividing cell. The two cells formed by meiosis I are called **secondary spermatocytes.** Each secondary spermatocyte has 23 chromosomes, the haploid number (n). Each chromosome within a secondary spermatocyte, however, is made up of two chromatids (two copies of the DNA) still attached by a centromere. No replication of DNA occurs in the secondary spermatocytes.

In meiosis II, the chromosomes line up in single file along the metaphase plate, and the two chromatids of each chromosome separate. The four haploid cells resulting from meiosis II are called **spermatids** (SPER-ma-tids). A single primary spermatocyte therefore produces four spermatids via two rounds of cell division (meiosis I and meiosis II).

A unique process occurs during spermatogenesis. As spermatogenic cells proliferate, they fail to complete cytoplasmic separation (cytokinesis). The cells remain in contact via cytoplasmic bridges through their entire development (see Figures 28.4b and 28.5). This pattern of development most likely accounts for the synchronized production of sperm in any given area of the seminiferous tubule. It may also have survival value

in that half of the sperm contain an X chromosome and half contain a Y chromosome. The larger X chromosome may carry genes needed for spermatogenesis that are lacking on the smaller Y chromosome.

The final stage of spermatogenesis, **spermiogenesis** (sper′-mē-ō-JEN-e-sis), is the development of haploid spermatids into sperm. No cell division occurs in spermiogenesis; each spermatid becomes a single **sperm cell.** During this process, spherical spermatids transform into elongated, slender sperm. An acrosome (described shortly) forms atop the nucleus, which condenses and elongates, a flagellum develops, and mitochondria multiply. Sertoli cells dispose of the excess cytoplasm that sloughs off. Finally, sperm are released from their connections to Sertoli cells, an event known as **spermiation** (sper′-mē-Ā-shun). Sperm then enter the lumen of the seminiferous tubule. Fluid secreted by Sertoli cells pushes sperm along their way, toward the ducts of the testes. At this point, sperm are not yet able to swim.

Figure 28.5 Events in spermatogenesis. Diploid cells ($2n$) have 46 chromosomes; haploid cells (n) have 23 chromosomes.

Spermiogenesis involves the maturation of spermatids into sperm.

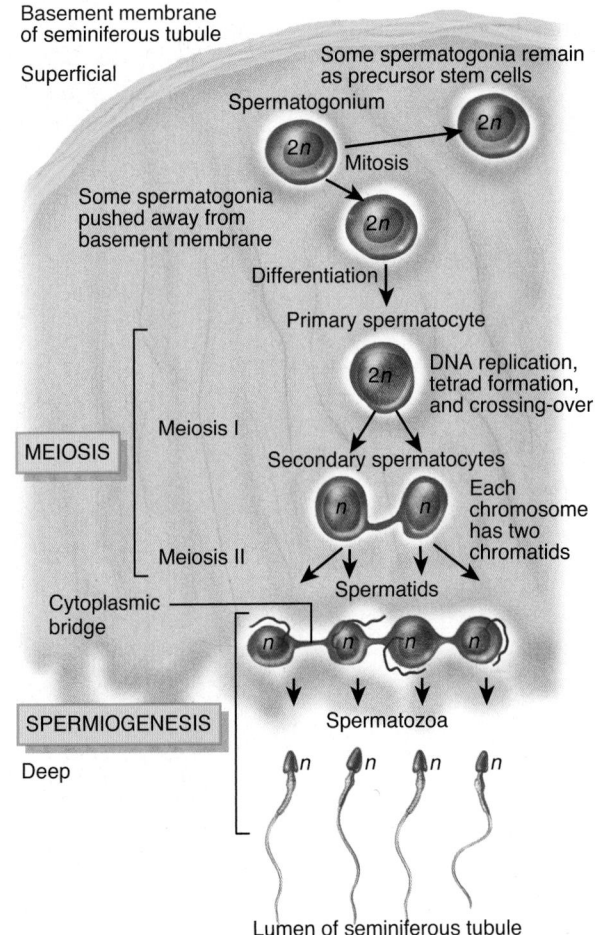

Basement membrane of seminiferous tubule

Superficial

Some spermatogonia remain as precursor stem cells

Spermatogonium

Mitosis

Some spermatogonia pushed away from basement membrane

Differentiation

Primary spermatocyte

DNA replication, tetrad formation, and crossing-over

MEIOSIS

Meiosis I

Meiosis II

Secondary spermatocytes

Each chromosome has two chromatids

Cytoplasmic bridge

Spermatids

SPERMIOGENESIS

Spermatozoa

Deep

Lumen of seminiferous tubule

 What is the outcome of meiosis I?

Sperm

Each day about 300 million sperm complete the process of spermatogenesis. A sperm is about 60 μm long and contains several structures that are highly adapted for reaching and penetrating a secondary oocyte (Figure 28.6). The major parts of a sperm are the head and the tail. The flattened, pointed **head** of the sperm is about 4–5 μm long. It contains a **nucleus** with 23 highly condensed chromosomes. Covering the anterior two-thirds of the nucleus is the **acrosome** (AK-rō-sōm; *acro-* = atop; *-some* = body), a caplike vesicle filled with enzymes that help a sperm to penetrate a secondary oocyte to bring about fertilization. Among the enzymes are hyaluronidase and proteases. The **tail** of a sperm is subdivided into four parts: neck, middle piece, principal piece, and end piece. The **neck** is the constricted region just behind the head that contains centrioles. The centrioles form the microtubules that comprise the remainder of the tail. The **middle piece** contains mitochondria arranged in a spiral, which provide the energy (ATP) for locomotion of sperm to the site of fertilization and for sperm metabolism. The **principal piece** is the longest portion of the tail, and the **end piece** is the terminal, tapering portion of the tail. Once ejaculated, most sperm do not survive more than 48 hours within the female reproductive tract.

Hormonal Control of the Testes

Although the initiating factors are unknown, at puberty certain hypothalamic neurosecretory cells increase their secretion of

Figure 28.6 Parts of a sperm cell.

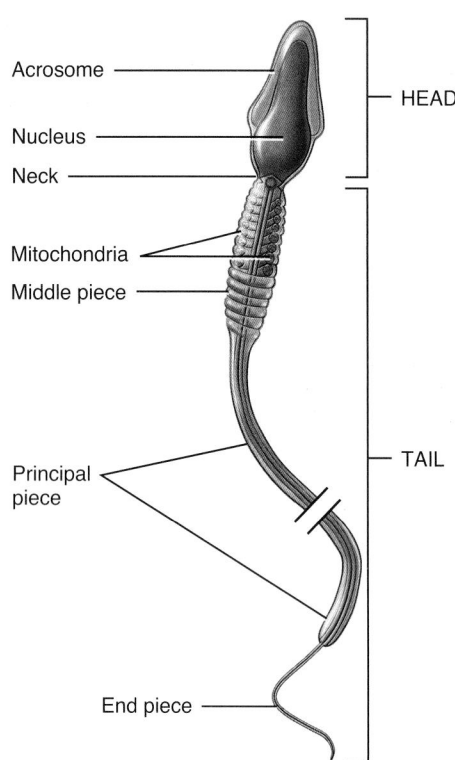

About 300 million sperm mature each day.

- Acrosome
- Nucleus
- Neck
- Mitochondria
- Middle piece
- Principal piece
- End piece
- HEAD
- TAIL

? **What are the functions of each part of a sperm cell?**

gonadotropin-releasing hormone (GnRH) (gō′-nad-ō-TRŌ-pin). This hormone in turn stimulates gonadotrophs in the anterior pituitary to increase their secretion of the two gonadotropins, **luteinizing hormone (LH)** (LOO-tē-in′-īz-ing) and **follicle-stimulating hormone (FSH).** Figure 28.7 shows the hormones and negative feedback loops that control secretion of testosterone and spermatogenesis.

LH stimulates Leydig cells, which are located between seminiferous tubules, to secrete the hormone **testosterone** (tes-TOS-te-rōn). This steroid hormone is synthesized from cholesterol in the testes and is the principal androgen. It is lipid-soluble and readily diffuses out of Leydig cells into the interstitial fluid and then into blood. Via negative feedback, testosterone suppresses secretion of LH by anterior pituitary gonadotrophs and suppresses secretion of GnRH by hypothalamic neurosecretory cells. In some target cells, such as those in the external genitals and prostate, the enzyme 5 alpha-reductase converts testosterone to another androgen called **dihydrotestosterone (DHT)** (dī-hī′-drō-tes-TOS-ter-ōn).

FSH acts indirectly to stimulate spermatogenesis (Figure 28.7). FSH and testosterone act synergistically on the Sertoli cells to stimulate secretion of **androgen-binding protein (ABP)** into the lumen of the seminiferous tubules and into the interstitial fluid around the spermatogenic cells. ABP binds to testosterone, keeping its concentration high. Testosterone stimulates the final steps of spermatogenesis in the seminiferous tubules. Once the degree of spermatogenesis required for male reproductive functions has been achieved, Sertoli cells release **inhibin,** a protein hormone named for its role in inhibiting FSH secretion by the anterior pituitary (Figure 28.7). If spermatogenesis is proceeding too slowly, less inhibin is released, which permits more FSH secretion and an increased rate of spermatogenesis.

Testosterone and dihydrotestosterone both bind to the same androgen receptors, which are found within the nuclei of target cells. The hormone–receptor complex regulates gene expression, turning some genes on and others off. Because of these changes, the androgens produce several effects:

- *Prenatal development.* Before birth, testosterone stimulates the male pattern of development of reproductive system ducts and the descent of the testes. Dihydrotestosterone stimulates development of the external genitals (described in Section 28.5). Testosterone also is converted in the brain to estrogens (feminizing hormones), which may play a role in the development of certain regions of the brain in males.

- *Development of male sexual characteristics.* At puberty, testosterone and dihydrotestosterone bring about development and enlargement of the male sex organs and the development of masculine secondary sexual characteristics. **Secondary sex characteristics** are traits that distinguish males and females but do not have a direct role in reproduction. These include muscular and skeletal growth that results in wide shoulders and narrow hips; facial and chest hair (within hereditary limits) and more hair on other parts of the body; thickening of the skin; increased sebaceous (oil) gland secretion; and enlargement of the larynx and consequent deepening of the voice.

- *Development of sexual function.* Androgens contribute to male sexual behavior and spermatogenesis and to sex drive (libido) in both males and females. Recall that the adrenal cortex is the main source of androgens in females.
- *Stimulation of anabolism.* Androgens are anabolic hormones; that is, they stimulate protein synthesis. This effect is

obvious in the heavier muscle and bone mass of most men as compared to women.

A negative feedback system regulates testosterone production (Figure 28.8). When testosterone concentration in the blood increases to a certain level, it inhibits the release of GnRH by cells in the hypothalamus. As a result, there is less GnRH in the portal blood that flows from the hypothalamus to the anterior pituitary. Gonadotrophs in the anterior pituitary then release less LH, so

Figure 28.7 **Hormonal control of spermatogenesis and actions of testosterone and dihydrotestosterone (DHT).** In response to stimulation by FSH and testosterone, Sertoli cells secrete androgen-binding protein (ABP). Dashed red lines indicate negative feedback inhibition.

Release of FSH is stimulated by GnRH and inhibited by inhibin; release of LH is stimulated by GnRH and inhibited by testosterone.

Figure 28.8 **Negative feedback control of blood level of testosterone.**

Gonadotrophs of the anterior pituitary produce luteinizing hormone (LH).

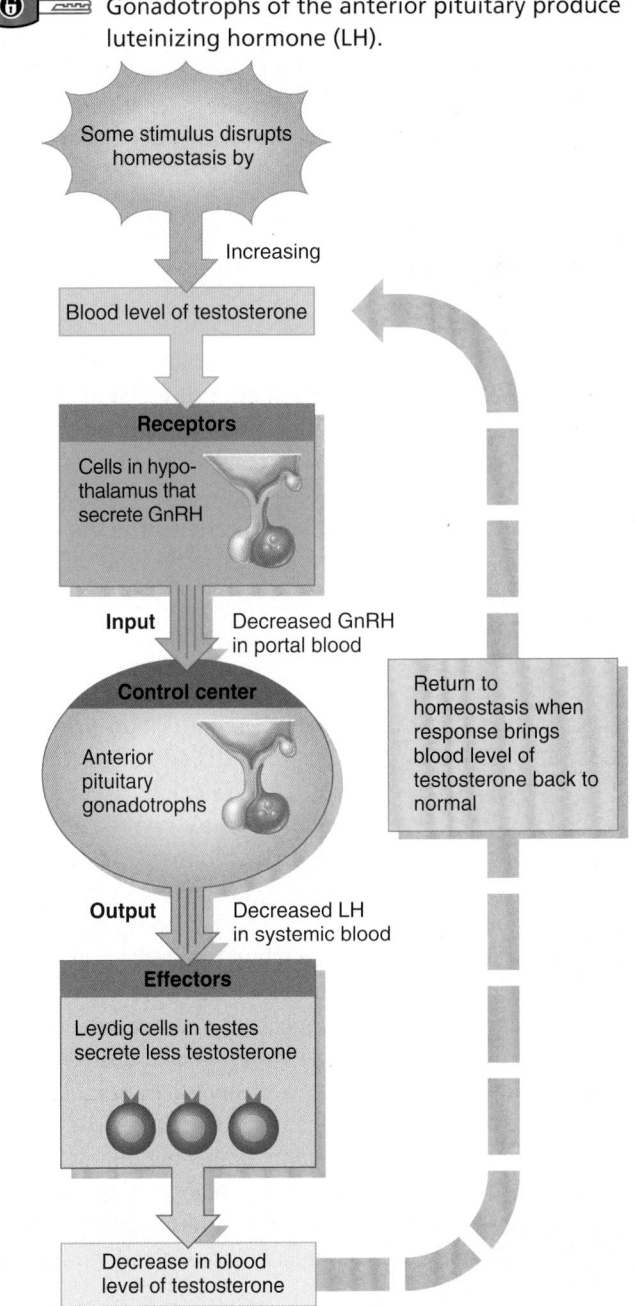

? Which cells secrete inhibin?

? Which hormones inhibit secretion of FSH and LH by the anterior pituitary?

the concentration of LH in systemic blood falls. With less stimulation by LH, the Leydig cells in the testes secrete less testosterone, and there is a return to homeostasis. If the testosterone concentration in the blood falls too low, however, GnRH is again released by the hypothalamus and stimulates secretion of LH by the anterior pituitary. LH in turn stimulates testosterone production by the testes.

✔CHECKPOINT

1. Describe the function of the scrotum in protecting the testes from temperature fluctuations.
2. Describe the internal structure of a testis. Where are sperm cells produced? What are the functions of Sertoli cells and Leydig cells?
3. Describe the principal events of spermatogenesis.
4. Which part of a sperm cell contains enzymes that help the sperm cell fertilize a secondary oocyte?
5. What are the roles of FSH, LH, testosterone, and inhibin in the male reproductive system? How is secretion of these hormones controlled?

Reproductive System Ducts in Males

Ducts of the Testis

Pressure generated by the fluid secreted by Sertoli cells pushes sperm and fluid along the lumen of seminiferous tubules and then into a series of very short ducts called **straight tubules** (see Figure 28.3a). The straight tubules lead to a network of ducts in the testis called the **rete testis** (RĒ-tē = network). From the rete testis, sperm move into a series of coiled **efferent ducts** (EF-erent) in the epididymis that empty into a single tube called the **ductus epididymis.**

Epididymis

The **epididymis** (ep′-i-DID-i-mis; *epi-* = above or over; *-didymis* = testis) is a comma-shaped organ about 4 cm (1.5 in.) long that lies along the posterior border of each testis (see Figure 28.3a). The plural is **epididymides** (ep′-i-di-DIM-i-dēz). Each epididymis consists mostly of the tightly coiled **ductus epididymis.** The efferent ducts from the testis join the ductus epididymis at the larger, superior portion of the epididymis called the **head.** The **body** is the narrow midportion of the epididymis, and the **tail** is the smaller, inferior portion. At its distal end, the tail of the epididymis continues as the ductus (vas) deferens (discussed shortly).

The ductus epididymis would measure about 6 m (20 ft) in length if it were uncoiled. It is lined with pseudostratified columnar epithelium and encircled by layers of smooth muscle. The free surfaces of the columnar cells contain **stereocilia** (ster′-ē-ō-SIL-ē-a), which despite their name are long, branching microvilli (not cilia) that increase the surface area for the reabsorption of degenerated sperm. Connective tissue around the muscle layer attaches the loops of the ductus epididymis and carries blood vessels and nerves.

Functionally, the epididymis is the site of **sperm maturation,** the process by which sperm acquire motility and the ability to fer-

tilize an ovum. This occurs over a period of about 14 days. The epididymis also helps propel sperm into the ductus (vas) deferens during sexual arousal by peristaltic contraction of its smooth muscle. In addition, the epididymis stores sperm, which remain viable here for up to several months. Any stored sperm that are not ejaculated by that time are eventually reabsorbed.

Ductus Deferens

Within the tail of the epididymis, the ductus epididymis becomes less convoluted, and its diameter increases. Beyond this point, the duct is known as the **ductus deferens** or **vas deferens** (DEF-er-enz) (see Figure 28.3a). The ductus deferens, which is about 45 cm (18 in.) long, ascends along the posterior border of the epididymis through the spermatic cord and then enters the pelvic cavity. There it loops over the ureter and passes over the side and down the posterior surface of the urinary bladder (see Figure 28.1a). The dilated terminal portion of the ductus deferens is the **ampulla** (am-PUL-la = little jar; Figure 28.9). The mucosa of the ductus deferens consists of pseudostratified columnar epithelium and lamina propria (areolar connective tissue). The muscularis is composed of three layers of smooth muscle; the inner and outer layers are longitudinal, and the middle layer is circular.

Functionally, the ductus deferens conveys sperm during sexual arousal from the epididymis toward the urethra by peristaltic contractions of its muscular coat. Like the epididymis, the ductus deferens also can store sperm for several months. Any stored sperm that are not ejaculated by that time are eventually reabsorbed.

Spermatic Cord

The **spermatic cord** is a supporting structure of the male reproductive system that ascends out of the scrotum (see Figure 28.2). It consists of the ductus (vas) deferens as it ascends through the scrotum, the testicular artery, veins that drain the testes and carry testosterone into circulation (the pampiniform plexus), autonomic nerves, lymphatic vessels, and the cremaster muscle. The spermatic cord and ilioinguinal nerve pass through the **inguinal canal** (ING-gwi-nal = groin), an oblique passageway in the anterior abdominal wall just superior and parallel to the medial half of the inguinal ligament. The canal, which is about 4–5 cm (about 2 in.) long, originates at the **deep (abdominal) inguinal ring,** a slitlike opening in the aponeurosis of the transversus abdominis muscle; the canal ends at the **superficial (subcutaneous) inguinal ring** (see Figure 28.2), a somewhat triangular opening in the aponeurosis of the external oblique muscle. In females, the round ligament of the uterus and ilioinguinal nerve pass through the inguinal canal.

The term **varicocele** (VAR-i-kō-sēl; *varico-* = varicose; *-kele* = hernia) refers to a swelling in the scrotum due to a dilation of the veins that drain the testes. It is usually more apparent when the person is standing and typically does not require treatment.

Ejaculatory Ducts

Each **ejaculatory duct** (ē-JAK-ū-la-tōr-ē; *ejacul-* = to expel) is about 2 cm (1 in.) long and is formed by the union of the duct from the seminal vesicle and the ampulla of the ductus (vas) deferens (Figure 28.9). The short ejaculatory ducts form just superior to the base (superior portion) of the prostate and pass inferiorly and anteriorly through the prostate. They terminate in the prostatic urethra, where they eject sperm and seminal vesicle secretions just before the release of semen from the urethra to the exterior.

Urethra

In males, the **urethra** (ū-RĒ-thra) is the shared terminal duct of the reproductive and urinary systems; it serves as a passageway for both semen and urine. About 20 cm (8 in.) long, it passes through the prostate, the deep muscles of the perineum, and the penis, and is subdivided into three parts (see Figures 28.1 and 26.22). The

Figure 28.9 Locations of several accessory reproductive organs in males. The prostate, urethra, and penis have been sectioned to show internal details.

The male urethra has three subdivisions: the prostatic, membranous, and spongy (penile) urethra.

FUNCTIONS OF ACCESSORY SEX GLAND SECRETIONS

1. The seminal vesicles secrete an alkaline, viscous fluid that helps neutralize acid in the female reproductive tract, provides fructose for ATP production by sperm, contributes to sperm motility and viability, and helps semen coagulate after ejaculation.
2. The prostate secretes a milky, slightly acidic fluid that helps semen coagulate after ejaculation and subsequently breaks down the clot.
3. The bulbourethral (Cowper's) glands secrete an alkaline fluid that neutralizes the acidic environment of the urethra and mucus that lubricates the lining of the urethra and the tip of the penis during sexual intercourse.

View

Left ureter

Hip bone (cut)

Prostate

Prostatic urethra

Membranous (intermediate) urethra

Crus of penis

Bulb of penis

Corpus spongiosum penis

Urinary bladder

Right ductus (vas) deferens

Ampulla of ductus (vas) deferens

Seminal vesicle

Seminal vesicle duct

Ejaculatory duct

Deep muscles of perineum

Bulbourethral (Cowper's) gland

Corpora cavernosa penis

Spongy (penile) urethra

(a) Posterior view of male accessory organs of reproduction

FIGURE 28.9 CONTINUES ▶

■ FIGURE 28.9 CONTINUED ▶

Left ureter

Pelvic diaphragm

Obturator internus
muscle

Deep muscles of perineum

Bulbospongiosus muscle

Urinary bladder

Hip bone

Right ductus
(vas) deferens

Seminal vesicle

Ampulla of ductus
(vas) deferens

Prostate

(b) Posterior view of male accessory organs of reproduction

? What accessory sex gland contributes the majority of the seminal fluid?

prostatic urethra (pros-TAT-ik) is 2–3 cm (1 in.) long and passes through the prostate. As this duct continues inferiorly, it passes through the deep muscles of the perineum, where it is known as the **membranous urethra** (MEM-bra-nus). The membranous urethra is about 1 cm (0.5 in.) in length. As this duct passes through the corpus spongiosum of the penis, it is known as the **spongy (penile) urethra,** which is about 15–20 cm (6–8 in.) long. The spongy urethra ends at the **external urethral orifice.** The histology of the male urethra may be reviewed in Section 26.8.

✔ CHECKPOINT

6. Which ducts transport sperm within the testes?
7. Describe the location, structure, and functions of the ductus epididymis, ductus (vas) deferens, and ejaculatory duct.
8. Give the locations of the three subdivisions of the male urethra.
9. Trace the course of sperm through the system of ducts from the seminiferous tubules to the urethra.
10. List the structures within the spermatic cord.

Accessory Sex Glands

The ducts of the male reproductive system store and transport sperm cells, but the **accessory sex glands** secrete most of the liquid portion of semen. The accessory sex glands include the seminal vesicles, the prostate, and the bulbourethral glands.

Seminal Vesicles

The paired **seminal vesicles** (VES-i-kuls) or **seminal glands** are convoluted pouchlike structures, about 5 cm (2 in.) in length, ly-

ing posterior to the base of the urinary bladder and anterior to the rectum (Figure 28.9). Through the seminal vesicle ducts they secrete an alkaline, viscous fluid that contains fructose (a monosaccharide sugar), prostaglandins, and clotting proteins that are different from those in blood. The alkaline nature of the seminal fluid helps to neutralize the acidic environment of the male urethra and female reproductive tract that otherwise would inactivate and kill sperm. The fructose is used for ATP production by sperm. Prostaglandins contribute to sperm motility and viability and may stimulate smooth muscle contractions within the female reproductive tract. The clotting proteins help semen coagulate after ejaculation. Fluid secreted by the seminal vesicles normally constitutes about 60% of the volume of semen.

Prostate

The **prostate** (PROS-tāt; *prostata* = one who stands before) is a single, doughnut-shaped gland about the size of a golf ball. It measures about 4 cm (1.6 in.) from side to side, about 3 cm (1.2 in.) from top to bottom, and about 2 cm (0.8 in.) from front to back. It is inferior to the urinary bladder and surrounds the prostatic urethra (Figure 28.9). The prostate slowly increases in size from birth to puberty. It then expands rapidly until about age 30, after which time its size typically remains stable until about age 45, when further enlargement may occur.

The prostate secretes a milky, slightly acidic fluid (pH about 6.5) that contains several substances. (1) *Citric acid* in prostatic fluid is used by sperm for ATP production via the Krebs cycle. (2) Several *proteolytic enzymes,* such as *prostate-specific antigen (PSA),* pepsinogen, lysozyme, amylase, and hyaluronidase, eventually break down the clotting proteins from the seminal vesicles. (3) The function of the *acid phosphatase* secreted by the prostate is unknown. (4) *Seminalplasmin* in prostatic fluid is an antibiotic that can destroy bacteria. Seminalplasmin may help decrease the number of naturally occurring bacteria in semen and in the lower female reproductive tract. Secretions of the prostate enter the prostatic urethra through many prostatic ducts. Prostatic secretions make up about 25% of the volume of semen and contribute to sperm motility and viability.

Bulbourethral Glands

The paired **bulbourethral glands** (bul′-bō-ū-RĒ-thral), or **Cowper's glands** (KOW-pers), are about the size of peas. They are located inferior to the prostate on either side of the membranous urethra within the deep muscles of the perineum, and their ducts open into the spongy urethra (Figure 28.9). During sexual arousal, the bulbourethral glands secrete an alkaline fluid into the urethra that protects the passing sperm by neutralizing acids from urine in the urethra. They also secrete mucus that lubricates the end of the penis and the lining of the urethra, decreasing the number of sperm damaged during ejaculation. Some males release a drop or two of this mucus upon sexual arousal and erection. The fluid does not contain sperm cells.

Semen

Semen (= seed) is a mixture of sperm and **seminal fluid,** a liquid that consists of the secretions of the seminiferous tubules, seminal vesicles, prostate, and bulbourethral glands. The volume of semen in a typical ejaculation is 2.5–5 milliliters (mL), with 50–150 million sperm per mL. When the number falls below 20 million/mL, the male is likely to be infertile. A very large number of sperm is required for successful fertilization because only a tiny fraction ever reaches the secondary oocyte.

Despite the slight acidity of prostatic fluid, semen has a slightly alkaline pH of 7.2–7.7 due to the higher pH and larger volume of fluid from the seminal vesicles. The prostatic secretion gives semen a milky appearance, and fluids from the seminal vesicles and bulbourethral glands give it a sticky consistency. Seminal fluid provides sperm with a transportation medium, nutrients, and protection from the hostile acidic environment of the male's urethra and the female's vagina.

Once ejaculated, liquid semen coagulates within 5 minutes due to the presence of clotting proteins from the seminal vesicles. The functional role of semen coagulation is not known, but the proteins involved are different from those that cause blood coagulation. After about 10 to 20 minutes, semen reliquefies because prostate-specific antigen (PSA) and other proteolytic enzymes produced by the prostate break down the clot. Abnormal or delayed liquefaction of clotted semen may cause complete or partial immobilization of sperm, thereby inhibiting their movement through the cervix of the uterus. After passing through the uterus and uterine tube, the sperm are affected by secretions of the uterine tube in a process called **capacitation** (see Section 28.2). The

presence of blood in semen is called **hemospermia** (hē-mō-SPER-mē-a; *hemo-* = blood; *-sperma* = seed). In most cases, it is caused by inflammation of the blood vessels lining the seminal vesicles; it is usually treated with antibiotics.

Penis

The **penis** (= tail) contains the urethra and is a passageway for the ejaculation of semen and the excretion of urine (Figure 28.10). It is cylindrical in shape and consists of a body, glans penis, and a root. The **body of the penis** is composed of three cylindrical masses of tissue, each surrounded by fibrous tissue called the **tunica albuginea** (al′-bū-JIN-ē-a) (Figure 28.10). The two dorsolateral masses are called the **corpora cavernosa penis** (*corpora* = main bodies; *cavernosa* = hollow). The smaller mid-ventral mass, the **corpus spongiosum penis,** contains the spongy urethra and keeps it open during ejaculation. Skin and a subcutaneous layer enclose all three masses, which consist of erectile tissue. *Erectile tissue* is composed of numerous blood sinuses (vascular spaces) lined by endothelial cells and surrounded by smooth muscle and elastic connective tissue.

The distal end of the corpus spongiosum penis is a slightly enlarged, acorn-shaped region called the **glans penis;** its margin is the **corona** (kō-RŌ-na). The distal urethra enlarges within the glans penis and forms a terminal slitlike opening, the **external urethral orifice.** Covering the glans in an uncircumcised penis is the loosely fitting **prepuce** (PRĒ-poos), or **foreskin.**

The **root of the penis** is the attached portion (proximal portion). It consists of the **bulb of the penis,** the expanded posterior continuation of the base of the corpus spongiosum penis, and the **crura of the penis** (KROO-ra; singular is *crus* = resembling a leg), the two separated and tapered portions of the corpora cavernosa penis. The bulb of the penis is attached to the inferior surface of the deep muscles of the perineum and is enclosed by the bulbospongiosus muscle, a muscle that aids ejaculation. Each crus of the penis bends laterally away from the bulb of the penis to attach to the ischial and inferior pubic rami and is surrounded by the ischiocavernosus muscle (see Figure 11.13). The weight of the penis is supported by two ligaments that are continuous with the fascia of the penis. (1) The **fundiform ligament** (FUN-di-form) arises from the inferior part of the linea alba. (2) The **suspensory ligament of the penis** arises from the pubic symphysis.

CLINICAL CONNECTION | *Circumcision*

Circumcision (= to cut around) is a surgical procedure in which part of or the entire prepuce is removed. It is usually performed several days after birth, and is done for social, cultural, religious, and (more rarely) medical reasons. Although most health-care professionals find no medical justification for circumcision, some feel that it has benefits, such as a lower risk of urinary tract infections, protection against penile cancer, and possibly a lower risk for sexually transmitted diseases. Indeed, studies in several African villages have found lower rates of HIV infection among circumcised men. •

Figure 28.10 Internal structure of the penis. The inset in (b) shows details of the skin and fasciae.

The penis contains the urethra, a common pathway for semen and urine.

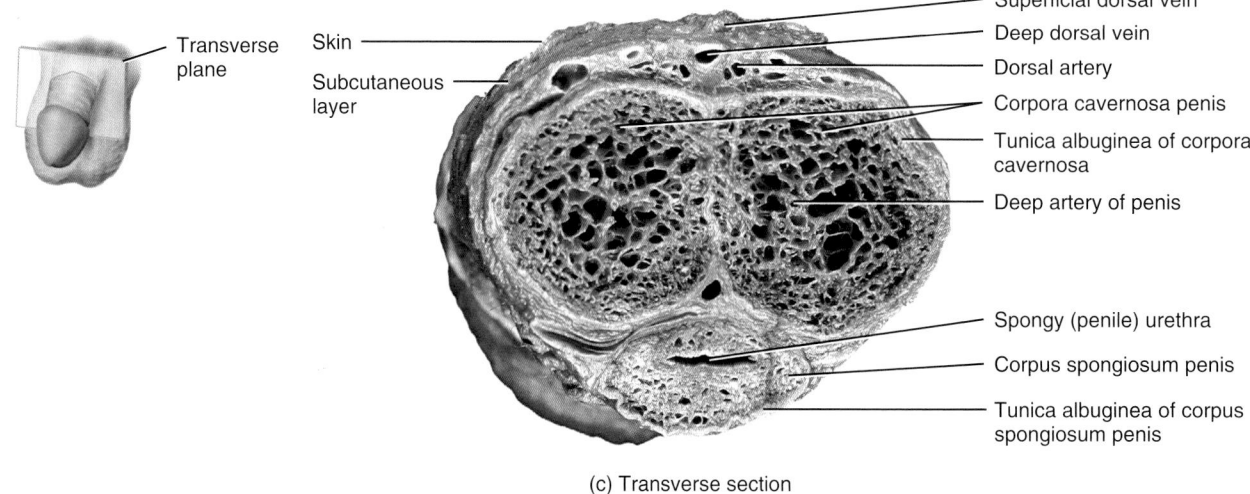

Internal urethral orifice
Prostatic urethra
Bulbourethral (Cowper's) gland
Deep muscles of perineum

Urinary bladder
Prostate
Orifice of ejaculatory duct
Membranous urethra

ROOT OF PENIS:
Bulb of penis
Crus of penis

BODY OF PENIS:
Corpora cavernosa penis
Corpus spongiosum penis
Spongy (penile) urethra

Frontal plane

Corona
GLANS PENIS
Prepuce (foreskin)
External urethral orifice

(a) Frontal section

Transverse plane

Deep dorsal vein
Dorsal artery

Skin
Superficial (subcutaneous) dorsal vein
Subcutaneous layer
Fascia

Dorsal

Corpora cavernosa penis
Tunica albuginea of corpora cavernosa
Deep artery of penis
Corpus spongiosum penis
Spongy (penile) urethra
Tunica albuginea of corpus spongiosum penis

Ventral

(b) Transverse section

Transverse plane

Skin
Subcutaneous layer

Superficial dorsal vein
Deep dorsal vein
Dorsal artery
Corpora cavernosa penis
Tunica albuginea of corpora cavernosa
Deep artery of penis
Spongy (penile) urethra
Corpus spongiosum penis
Tunica albuginea of corpus spongiosum penis

(c) Transverse section

Which tissue masses form the erectile tissue in the penis, and why do they become rigid during sexual arousal?

Upon sexual stimulation (visual, tactile, auditory, olfactory, or imagined), parasympathetic fibers from the sacral portion of the spinal cord initiate and maintain an **erection,** the enlargement and stiffening of the penis. The parasympathetic fibers produce and release nitric oxide (NO). The NO causes smooth muscle in the walls of arterioles supplying erectile tissue to relax, which allows these blood vessels to dilate. This in turn causes large amounts of blood to enter the erectile tissue of the penis. NO also causes the smooth muscle within the erectile tissue to relax, resulting in widening of the blood sinuses. The combination of increased blood flow and widening of the blood sinuses results in an erection. Expansion of the blood sinuses also compresses the veins that drain the penis; the slowing of blood outflow helps to maintain the erection.

The term **priapism** (PRĪ-a-pizm) refers to a persistent and usually painful erection of the penis that does not involve sexual desire or excitement. The condition may last up to several hours and is accompanied by pain and tenderness. It results from abnormalities of blood vessels and nerves, usually in response to medication used to produce erections in males who otherwise cannot attain them. Other causes include a spinal cord disorder, leukemia, sickle-cell disease, or a pelvic tumor.

Ejaculation (ē-jak-ū-LĀ-shun; *ejectus-* = to throw out), the powerful release of semen from the urethra to the exterior, is a sympathetic reflex coordinated by the lumbar portion of the spinal cord. As part of the reflex, the smooth muscle sphincter at the base of the urinary bladder closes, preventing urine from being expelled during ejaculation, and semen from entering the urinary bladder. Even before ejaculation occurs, peristaltic contractions in the epididymis, ductus (vas) deferens, seminal vesicles, ejaculatory ducts, and prostate propel semen into the penile portion of the urethra (spongy urethra). Typically, this leads to **emission** (ē-MISH-un), the discharge of a small volume of semen before ejaculation. Emission may also occur during sleep (nocturnal emission). The musculature of the penis (bulbospongiosus, ischiocavernosus, and superficial transverse perineus muscles), which is supplied by the pudendal nerve, also contracts at ejaculation (see Figure 11.13).

Once sexual stimulation of the penis has ended, the arterioles supplying the erectile tissue of the penis constrict and the smooth muscle within erectile tissue contracts, making the blood sinuses smaller. This relieves pressure on the veins supplying the penis and allows the blood to drain through them. Consequently, the penis returns to its flaccid (relaxed) state.

CLINICAL CONNECTION | *Premature Ejaculation*

A **premature ejaculation** is ejaculation that occurs too early, for example, during foreplay or on or shortly after penetration. It is usually caused by anxiety, other psychological causes, or an unusually sensitive foreskin or glans penis. For most males, premature ejaculation can be overcome by various techniques (such as squeezing the penis between the glans penis and shaft as ejaculation approaches), behavioral therapy, or medication. •

CHECKPOINT

11. Briefly explain the locations and functions of the seminal vesicles, the prostate, and the bulbourethral (Cowper's) glands.
12. What is semen? What is its function?
13. Explain the physiological processes involved in erection and ejaculation.

28.2 FEMALE REPRODUCTIVE SYSTEM

OBJECTIVES
• Describe the location, structure, and functions of the organs of the female reproductive system.
• Discuss the process of oogenesis in the ovaries.

The organs of the **female reproductive system** (Figure 28.11) include the ovaries (female gonads); the uterine (fallopian) tubes, or oviducts; the uterus; the vagina; and external organs, which are collectively called the vulva, or pudendum. The mammary glands are considered part of both the integumentary system and the female reproductive system.

Ovaries

The **ovaries** (= egg receptacles), which are the female gonads, are paired glands that resemble unshelled almonds in size and shape; they are homologous to the testes. (Here *homologous* means that two organs have the same embryonic origin.) The ovaries produce (1) gametes, secondary oocytes that develop into mature ova (eggs) after fertilization, and (2) hormones, including progesterone and estrogens (the female sex hormones), inhibin, and relaxin.

The ovaries, one on either side of the uterus, descend to the brim of the superior portion of the pelvic cavity during the third month of development. A series of ligaments holds them in position (Figure 28.12). The **broad ligament** of the uterus, which is a fold of the parietal peritoneum, attaches to the ovaries by a double-layered fold of peritoneum called the **mesovarium** (mez'-ō-VĀ-rē-um). The **ovarian ligament** anchors the ovaries to the uterus, and the **suspensory ligament** attaches them to the pelvic wall. Each ovary contains a **hilum** (HĪ-lum) or **hilus,** the point of entrance and exit for blood vessels and nerves along which the mesovarium is attached.

Histology of the Ovary

Each ovary consists of the following parts (Figure 28.13):

• The **germinal epithelium** (*germen* = sprout or bud) is a layer of simple epithelium (low cuboidal or squamous) that covers the surface of the ovary. We now know that the term germinal epithelium in humans is not accurate because it does not give rise to ova; the name came about because, at one time, people believed that it did. We have since learned that the cells that produce ova arise from the yolk sac and migrate to the ovaries during embryonic development.

Figure 28.11 **Female organs of reproduction and surrounding structures.**

The organs of reproduction in females include the ovaries, uterine (fallopian) tubes, uterus, vagina, vulva, and mammary glands.

FUNCTIONS OF THE FEMALE REPRODUCTIVE SYSTEM

1. The ovaries produce secondary oocytes and hormones, including progesterone and estrogens (female sex hormones), inhibin, and relaxin.
2. The uterine tubes transport a secondary oocyte to the uterus and normally are the sites where fertilization occurs.
3. The uterus is the site of implantation of a fertilized ovum, development of the fetus during pregnancy, and labor.
4. The vagina receives the penis during sexual intercourse and is a passageway for childbirth.
5. The mammary glands synthesize, secrete, and eject milk for nourishment of the newborn.

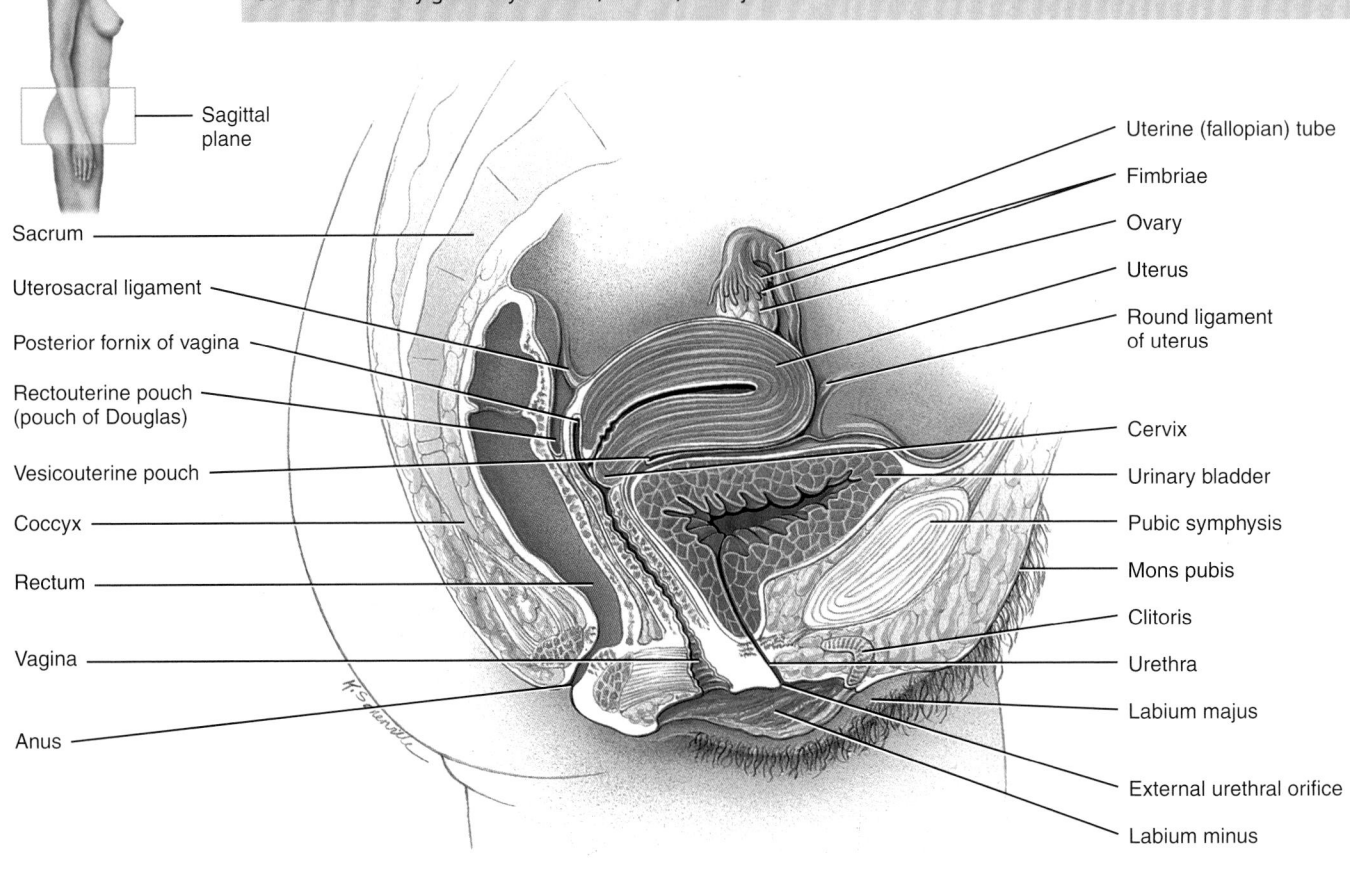

(a) Sagittal section

- The **tunica albuginea** is a whitish capsule of dense irregular connective tissue located immediately deep to the germinal epithelium.

- The **ovarian cortex** is a region just deep to the tunica albuginea. It consists of ovarian follicles (described shortly) surrounded by dense irregular connective tissue that contains collagen fibers and fibroblast-like cells called *stromal cells.*

- The **ovarian medulla** is deep to the ovarian cortex. The border between the cortex and medulla is indistinct, but the medulla consists of more loosely arranged connective tissue and contains blood vessels, lymphatic vessels, and nerves.

- **Ovarian follicles** (*folliculus* = little bag) are in the cortex and consist of **oocytes** (Ō-ō-sīts) in various stages of development,

plus the cells surrounding them. When the surrounding cells form a single layer, they are called **follicular cells** (fo-LIK-ū-lar); later in development, when they form several layers, they are referred to as **granulosa cells** (gran'-u-LŌ-sa). The surrounding cells nourish the developing oocyte and begin to secrete estrogens as the follicle grows larger.

- A **mature (graafian) follicle** (GRA-fē-an) is a large, fluid-filled follicle that is ready to rupture and expel its secondary oocyte, a process known as **ovulation.**

- A **corpus luteum** (= yellow body) contains the remnants of a mature follicle after ovulation. The corpus luteum produces progesterone, estrogens, relaxin, and inhibin until it degenerates into fibrous scar tissue called the **corpus albicans** (AL-bi-kanz = white body).

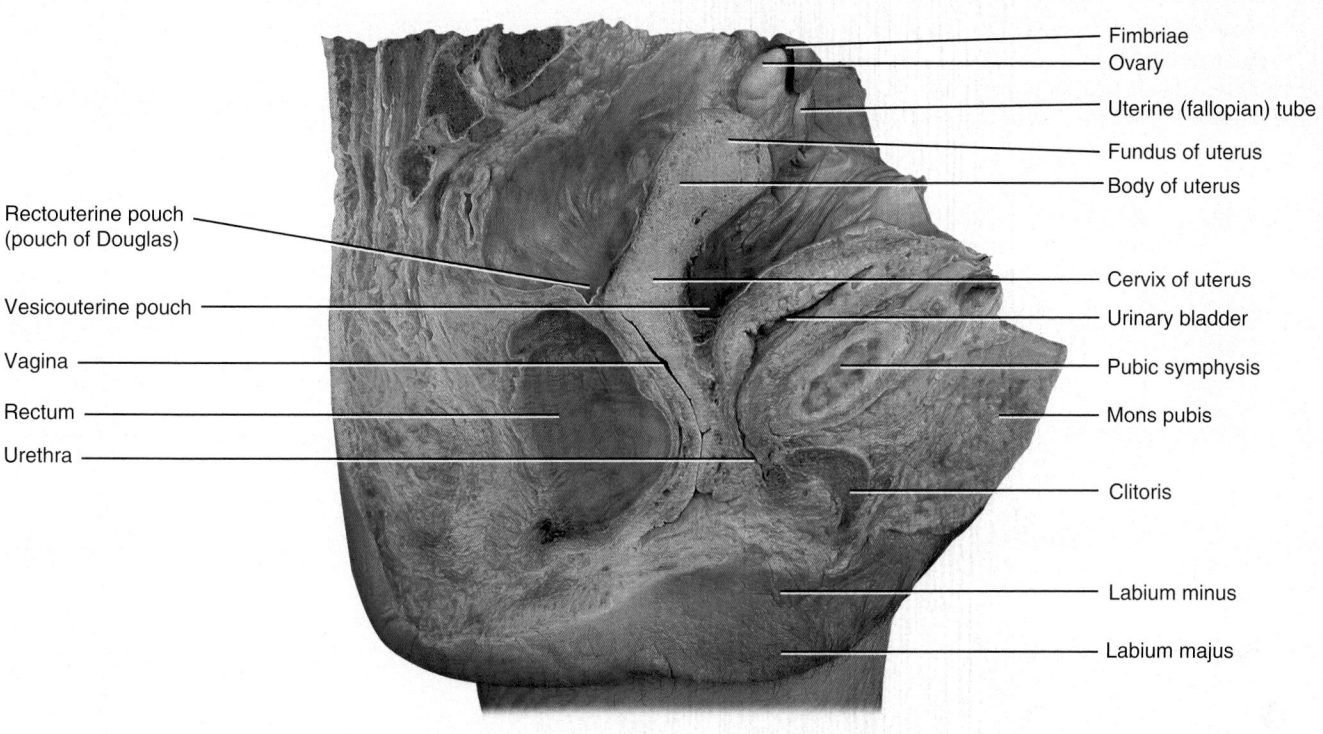

Fimbriae
Ovary
Uterine (fallopian) tube
Fundus of uterus
Body of uterus

Rectouterine pouch (pouch of Douglas)

Vesicouterine pouch

Vagina

Rectum

Urethra

Cervix of uterus
Urinary bladder
Pubic symphysis
Mons pubis
Clitoris

Labium minus

Labium majus

(b) Sagittal section

? Which structures in males are homologous to the ovaries, the clitoris, the paraurethral glands, and the greater vestibular glands?

Figure 28.12 **Relative positions of the ovaries, the uterus, and the ligaments that support them.**

🔑 Ligaments holding the ovaries in position are the mesovarium, the ovarian ligament, and the suspensory ligament.

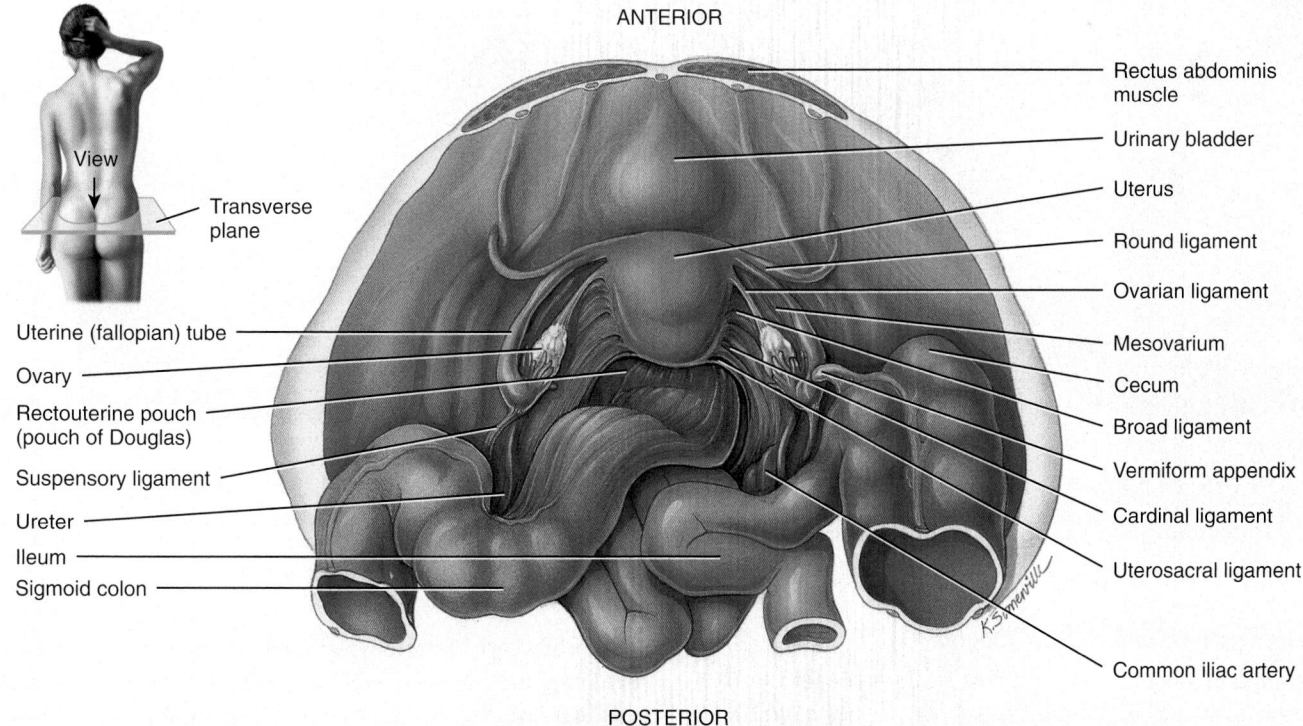

ANTERIOR

View

Transverse plane

Rectus abdominis muscle
Urinary bladder
Uterus
Round ligament
Ovarian ligament
Mesovarium
Cecum
Broad ligament
Vermiform appendix
Cardinal ligament
Uterosacral ligament
Common iliac artery

Uterine (fallopian) tube
Ovary
Rectouterine pouch (pouch of Douglas)
Suspensory ligament
Ureter
Ileum
Sigmoid colon

POSTERIOR

Superior view of transverse section

? To which structures do the mesovarium, ovarian ligament, and suspensory ligament anchor the ovary?

Figure 28.13 Histology of the ovary. The arrows in (a) indicate the sequence of developmental stages that occur as part of the maturation of an ovum during the ovarian cycle.

🔑 The ovaries are the female gonads; they produce haploid oocytes.

Primordial follicle

Primary follicle

Secondary follicle

Germinal epithelium

Frontal plane

Ovarian cortex

Follicular fluid

Mature (graafian) follicle

Ovarian medulla

Corpus hemorrhagicum (ruptured follicle)

Corona radiata

Ovulation discharges secondary oocyte

Blood vessels in hilum of ovary

Corpus albicans

Degenerating corpus luteum

Blood clot

Corpus luteum

(a) Frontal section

Germinal epithelium

Ovarian cortex

Ovarian medulla

Blood vessel in hilum of ovary

Degenerating mature (graafian) follicle

LM 20x

(b) Hemisection

Ovary

Secondary oocyte

LM 30x

(c) Ovulation of a secondary oocyte

❓ **What structures in the ovary contain endocrine tissue, and what hormones do they secrete?**

Oogenesis and Follicular Development

The formation of gametes in the ovaries is termed **oogenesis** (ō-ō-JEN-e-sis; *oo-* = egg). In contrast to spermatogenesis, which begins in males at puberty, oogenesis begins in females before they are even born. Oogenesis occurs in essentially the same manner as spermatogenesis; meiosis (see Chapter 3) takes place and the resulting germ cells undergo maturation.

During early fetal development, primordial (primitive) germ cells migrate from the yolk sac to the ovaries. There, germ cells differentiate within the ovaries into **oogonia** (ō-ō-GŌ-nē-a; singular is **oogonium**). Oogonia are diploid (2*n*) stem cells that divide mitotically to produce millions of germ cells. Even before birth, most of these germ cells degenerate in a process known as **atresia** (a-TRĒ-zē-a). A few, however, develop into larger cells called **primary oocytes** that enter prophase of meiosis I during fetal development but do not complete that phase until after puberty. During this arrested stage of development, each primary oocyte is surrounded by a single layer of flat follicular cells, and the entire structure is called a **primordial follicle** (Figure 28.14a).

The ovarian cortex surrounding the primordial follicles consists of collagen fibers and fibroblast-like **stromal cells.** At birth, approximately 200,000 to 2,000,000 primary oocytes remain in each ovary. Of these, about 40,000 are still present at puberty, and around 400 will mature and ovulate during a woman's

Figure 28.14 Ovarian follicles.

As an ovarian follicle enlarges, follicular fluid accumulates in a cavity called the antrum.

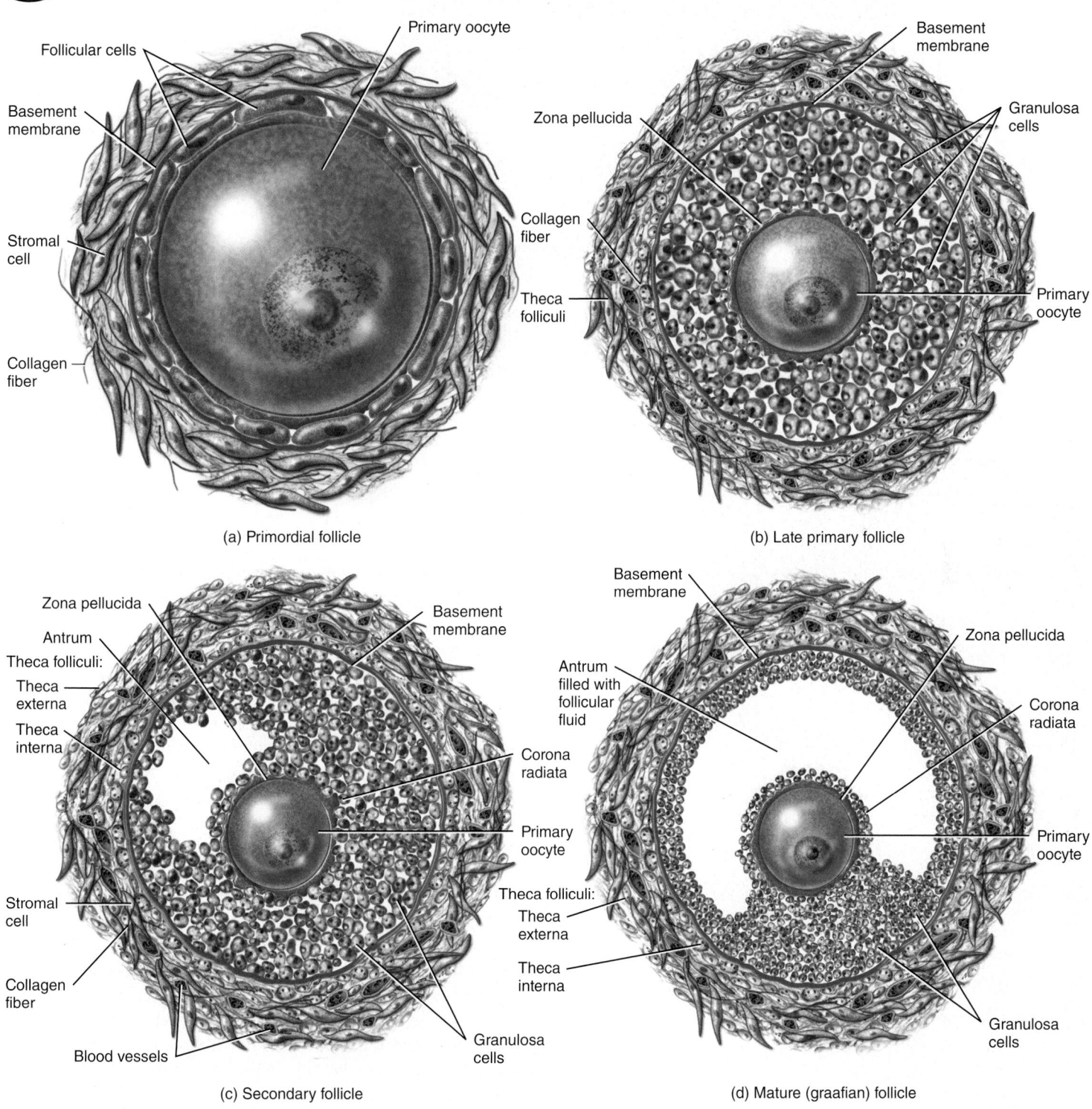

(a) Primordial follicle

(b) Late primary follicle

(c) Secondary follicle

(d) Mature (graafian) follicle

FIGURE **28.14** CONTINUES ▶

FIGURE 28.14 CONTINUED

- Germinal epithelium
- Tunica albuginea
- Primordial follicle
- Ovarian cortex
- Primary follicle granulosa cells
- Theca folliculi
- Zona pellucida
- Primary oocyte
- Secondary follicle granulosa cells
- Secondary follicle
- Corpus luteum

LM 30x

(e) Ovarian cortex

- Primary oocyte
- Zona pellucida
- Corona radiata
- Secondary follicle granulosa cells
- Antrum filled with follicular fluid
- Theca folliculi

LM 70x

(f) Secondary follicle

? **What happens to most ovarian follicles?**

reproductive lifetime. The remainder of the primary oocytes undergo atresia.

Each month after puberty until menopause, gonadotropins (FSH and LH) secreted by the anterior pituitary further stimulate the development of several primordial follicles, although only one will typically reach the maturity needed for ovulation. A few primordial follicles start to grow, developing into **primary follicles** (Figure 28.14b). Each primary follicle consists of a primary oocyte that is surrounded in a later stage of development by several layers of cuboidal and low-columnar cells called **granulosa cells.** The outermost granulosa cells rest on a basement membrane. As the primary follicle grows, it forms a clear glycoprotein layer called the **zona pellucida** (pe-LOO-si-da) between the primary oocyte and the granulosa cells. In addition, stromal cells surrounding the basement membrane begin to form an organized layer called the **theca folliculi** (THĒ-ka fo-LIK-ū-lī).

With continuing maturation, a primary follicle develops into a secondary follicle (Figure 28.14c). In a **secondary follicle,** the theca differentiates into two layers: (1) the **theca interna,** a highly vascularized internal layer of cuboidal secretory cells that secrete estrogens, and (2) the **theca externa,** an outer layer of stromal cells and collagen fibers. In addition, the granulosa cells begin to secrete follicular fluid, which builds up in a cavity called the **antrum** in the center of the secondary follicle. The innermost layer of granulosa cells becomes firmly attached to the zona pellucida and is now called the **corona radiata** (*corona* = crown; *radiata* = radiation) (Figure 28.14c).

The secondary follicle eventually becomes larger, turning into a **mature (graafian) follicle** (Figure 28.14d). While in this follicle, and just before ovulation, the diploid primary oocyte completes meiosis I, producing two haploid (*n*) cells of unequal size—each with 23 chromosomes (Figure 28.15). The smaller cell produced by meiosis I, called the **first polar body,** is essentially a packet of discarded nuclear material. The larger cell, known as the **secondary oocyte,** receives most of the cytoplasm. Once a secondary oocyte is formed, it begins meiosis II but then stops in metaphase. The mature (graafian) follicle soon ruptures and releases its secondary oocyte, a process known as **ovulation.**

At ovulation, the secondary oocyte is expelled into the pelvic cavity together with the first polar body and corona radiata. Normally these cells are swept into the uterine tube. If fertilization does not occur, the cells degenerate. If sperm are present in the uterine tube and one penetrates the secondary oocyte, however,

meiosis II resumes. The secondary oocyte splits into two haploid cells, again of unequal size. The larger cell is the **ovum,** or mature egg; the smaller one is the **second polar body.** The nuclei of the sperm cell and the ovum then unite, forming a diploid **zygote.** If the first polar body undergoes another division to produce two polar bodies, then the primary oocyte ultimately gives rise to three haploid polar bodies, which all degenerate, and a single haploid ovum. Thus, one primary oocyte gives rise to a single gamete (an ovum). By contrast, recall that in males one primary spermatocyte produces four gametes (sperm).

Table 28.1 summarizes the events of oogenesis and follicular development.

Figure 28.15 Oogenesis. Diploid cells (2*n*) have 46 chromosomes; haploid cells (*n*) have 23 chromosomes.

🔑 In a secondary oocyte, meiosis II is completed only if fertilization occurs.

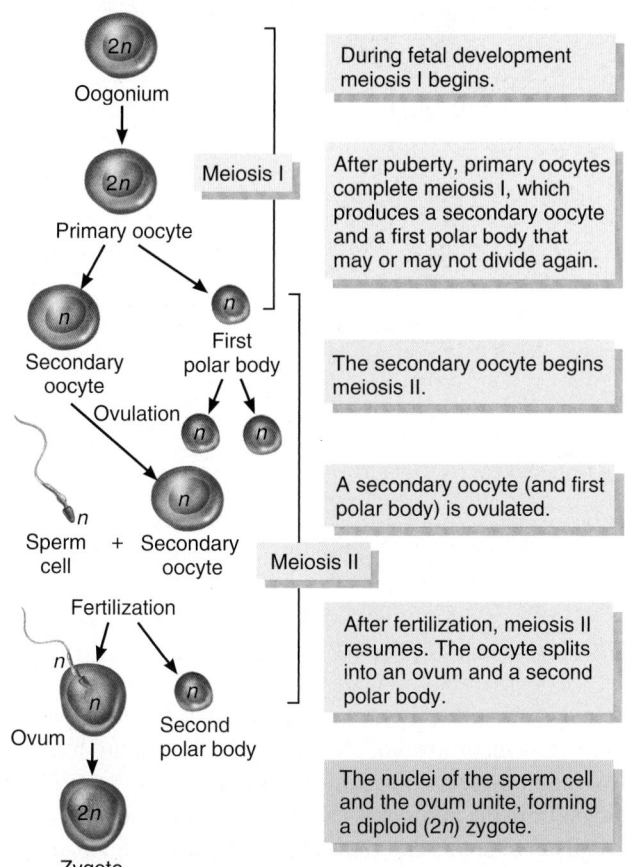

How does the age of a primary oocyte in a female compare with the age of a primary spermatocyte in a male?

⚕ **CLINICAL CONNECTION | *Ovarian Cysts***

An **ovarian cyst** is a fluid-filled sac in or on an ovary. Such cysts are relatively common, are usually noncancerous, and frequently disappear on their own. Cancerous cysts are more likely to occur in women over 40. Ovarian cysts may cause pain, pressure, a dull ache, or fullness in the abdomen; pain during sexual intercourse; delayed, painful, or irregular menstrual periods; abrupt onset of sharp pain in the lower abdomen; and/or vaginal bleeding. Most ovarian cysts require no treatment, but larger ones (more than 5 cm or 2 in.) may be removed surgically. •

✔ **CHECKPOINT**

14. How are the ovaries held in position in the pelvic cavity?
15. Describe the microscopic structure and functions of an ovary.
16. Describe the principal events of oogenesis.

Uterine Tubes

Females have two **uterine (fallopian) tubes,** or **oviducts,** that extend laterally from the uterus (Figure 28.16). The tubes, which measure about 10 cm (4 in.) long, lie within the folds of the broad ligaments of the uterus. They provide a route for sperm to reach an ovum and transport secondary oocytes and fertilized ova from the ovaries to the uterus. The funnel-shaped portion of each tube, called the **infundibulum** (in-fun-DIB-ū-lum), is close to the ovary but is open to the pelvic cavity. It ends in a fringe of fingerlike projections called **fimbriae** (FIM-brē-ē = fringe), one of which is attached to the lateral end of the ovary. From the infundibulum, the uterine tube extends medially and eventually inferiorly and attaches to the superior lateral angle of the uterus. The **ampulla** (am-PUL-la) of the uterine tube is the widest, longest portion, making up about the lateral two-thirds of its length. The **isthmus** (IS-mus) of the uterine tube is the more medial, short, narrow, thick-walled portion that joins the uterus.

Histologically, the uterine tubes are composed of three layers: mucosa, muscularis, and serosa. The mucosa consists of epithelium and lamina propria (areolar connective tissue). The epithelium contains ciliated simple columnar cells, which function as a "ciliary conveyor belt" to help move a fertilized ovum (or secondary oocyte) within the uterine tube toward the uterus, and nonciliated

TABLE 28.1

Summary of Oogenesis and Follicular Development

AGE	OOGENESIS	FOLLICULAR DEVELOPMENT

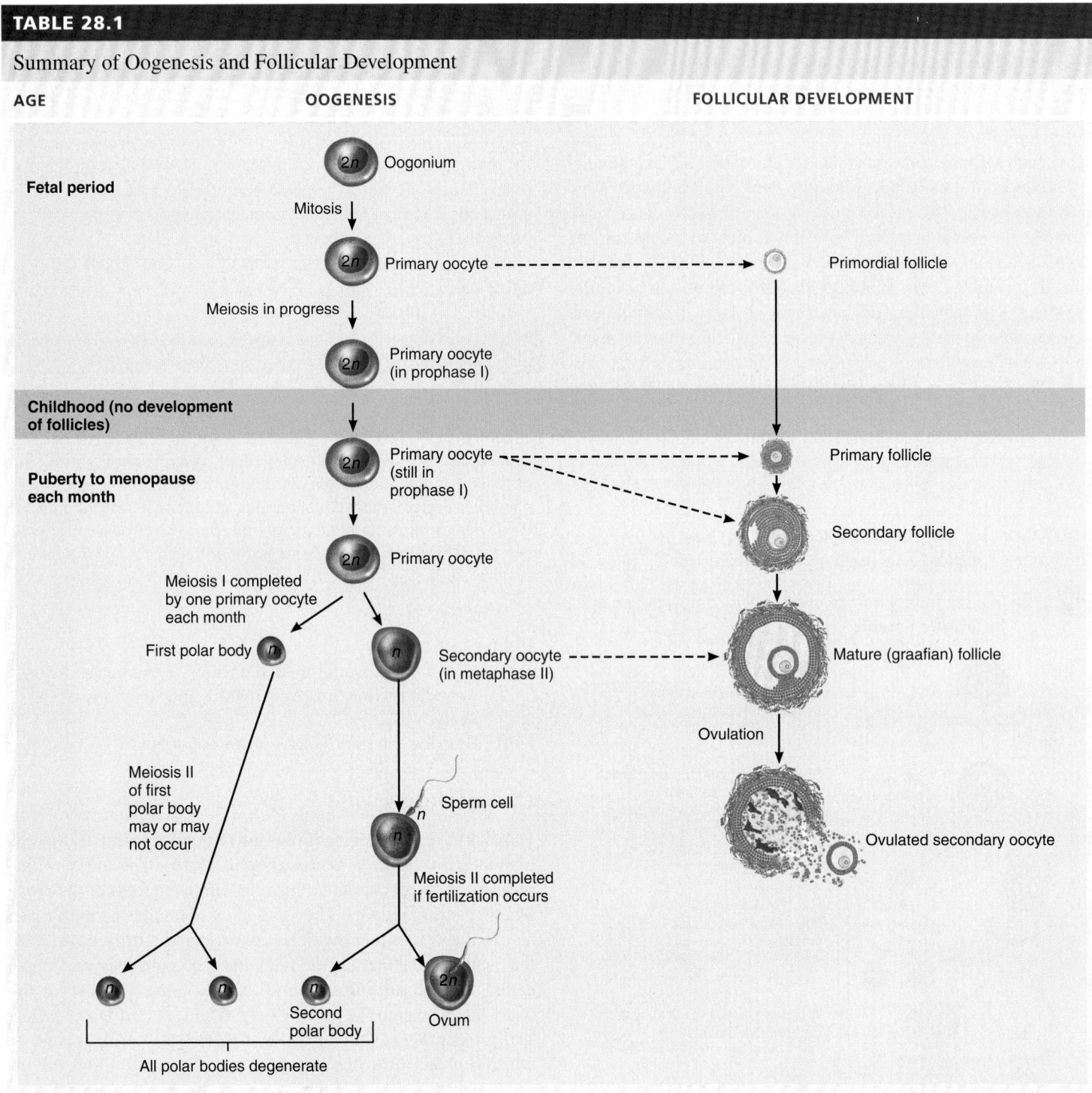

cells called **peg cells,** which have microvilli and secrete a fluid that provides nutrition for the ovum (Figure 28.17). The middle layer, the muscularis, is composed of an inner, thick, circular ring of smooth muscle and an outer, thin region of longitudinal smooth muscle. Peristaltic contractions of the muscularis and the ciliary action of the mucosa help move the oocyte or fertilized ovum toward the uterus. The outer layer of the uterine tubes is a serous membrane, the serosa.

After ovulation, local currents are produced by movements of the fimbriae, which surround the surface of the mature follicle just before ovulation occurs. These currents sweep the ovulated secondary oocyte from the peritoneal cavity into the uterine tube. A sperm cell usually encounters and fertilizes a secondary oocyte in the ampulla of the uterine tube, although fertilization in the peritoneal cavity is not uncommon. Fertilization can occur up to about 24 hours after ovulation. Some hours after fertilization, the nuclear materials

Figure 28.16 **Relationship of the uterine (fallopian) tubes to the ovaries, uterus, and associated structures.** In the left
side of the drawing the uterine tube and uterus have been sectioned to show internal structures.

 After ovulation, a secondary oocyte and its corona radiata move from the pelvic cavity into the infundibulum of the
uterine tube. The uterus is the site of menstruation, implantation of a fertilized ovum, development of the fetus, and
labor.

Ampulla of uterine tube

Isthmus of uterine tube

Fundus of uterus

Fimbriae of uterine tube

Infundibulum of uterine tube

Suspensory ligament

Infundibulum of uterine tube

Uterine cavity

Endometrium

Myometrium

Perimetrium

Internal os

Cervix of uterus

Cervical canal

Lateral fornix

Rugae

Uterine (fallopian) tube

Ovary

Ovarian ligament

Broad ligament

Body of uterus

Ureter

Isthmus

Uterosacral ligament

External os

Vagina

View

Posterior view of uterus and associated structures

? **Where does fertilization usually occur?**

of the haploid ovum and sperm unite. The diploid fertilized ovum is now called a **zygote** and begins to undergo cell divisions while moving toward the uterus. It arrives in the uterus 6 to 7 days after ovulation. Unfertilized secondary oocytes disintegrate.

Uterus

The **uterus** (womb) serves as part of the pathway for sperm deposited in the vagina to reach the uterine tubes. It is also the site of implantation of a fertilized ovum, development of the fetus during pregnancy, and labor. During reproductive cycles when implantation does not occur, the uterus is the source of menstrual flow.

Anatomy of the Uterus

Situated between the urinary bladder and the rectum, the uterus is the size and shape of an inverted pear (see Figure 28.16). In females

who have never been pregnant, it is about 7.5 cm (3 in.) long, 5 cm (2 in.) wide, and 2.5 cm (1 in.) thick. The uterus is larger in females who have recently been pregnant, and smaller (atrophied) when sex hormone levels are low, as occurs after menopause.

Anatomical subdivisions of the uterus include (1) a dome-shaped portion superior to the uterine tubes called the **fundus,** (2) a tapering central portion called the **body,** and (3) an inferior narrow portion called the **cervix** that opens into the vagina. Between the body of the uterus and the cervix is the **isthmus,** a constricted region about 1 cm (0.5 in.) long. The interior of the body of the uterus is called the **uterine cavity,** and the interior of the cervix is called the **cervical canal.** The cervical canal opens into the uterine cavity at the **internal os** (*os* = mouthlike opening) and into the vagina at the **external os.**

Normally, the body of the uterus projects anteriorly and superiorly over the urinary bladder in a position called **anteflexion**

Figure 28.17 Histology of the uterine (fallopian) tube.

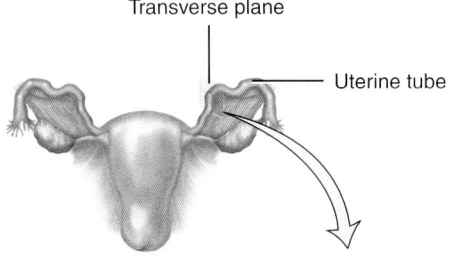

Peristaltic contractions of the muscularis and ciliary action of the mucosa of the uterine tube help move the oocyte or fertilized ovum toward the uterus.

Transverse plane

Uterine tube

Cilia

Peg cell (nonciliated) with microvilli

Ciliated simple columnar cell

Lamina propria (areolar connective tissue)

Lumen of uterine tube

LM 400x

(a) Details of epithelium

Cilia of ciliated simple columnar epithelial cell

Peg cell (nonciliated) with microvilli

SEM 4000x

(b) Details of epithelium in surface view

? **What types of cells line the uterine tubes?**

(an'-te-FLEK-shun; *ante-* = before). The cervix projects inferiorly and posteriorly and enters the anterior wall of the vagina at nearly a right angle (see Figure 28.11). Several ligaments that are either extensions of the parietal peritoneum or fibromuscular cords maintain the position of the uterus (see Figure 28.12). The paired **broad ligaments** are double folds of peritoneum attaching the uterus to either side of the pelvic cavity. The paired

uterosacral ligaments (ū'-ter-ō-SĀ-kral), also peritoneal extensions, lie on either side of the rectum and connect the uterus to the sacrum. The **cardinal (lateral cervical) ligaments** are located inferior to the bases of the broad ligaments and extend from the pelvic wall to the cervix and vagina. The **round ligaments** are bands of fibrous connective tissue between the layers of the broad ligament; they extend from a point on the uterus just inferior to the uterine tubes to a portion of the labia majora of the external genitalia. Although the ligaments normally maintain the anteflexed position of the uterus, they also allow the uterine body enough movement such that the uterus may become malpositioned. A posterior tilting of the uterus, called **retroflexion** (RET-rō-flek-shun; *retro-* = backward or behind), is a harmless variation of the normal position of the uterus. There is often no cause for the condition, but it may occur after childbirth.

CLINICAL CONNECTION | Uterine Prolapse

A condition called **uterine prolapse** (*prolapse* = falling down or downward displacement) may result from weakening of supporting ligaments and pelvic musculature associated with age or disease, traumatic vaginal delivery, chronic straining from coughing or difficult bowel movements, or pelvic tumors. The prolapse may be characterized as *first degree (mild),* in which the cervix remains within the vagina; *second degree (marked),* in which the cervix protrudes through the vagina to the exterior; and *third degree (complete),* in which the entire uterus is outside the vagina. Depending on the degree of prolapse, treatment may involve pelvic exercises, dieting if a patient is overweight, a stool softener to minimize straining during defecation, pessary therapy (placement of a rubber device around the uterine cervix that helps prop up the uterus), or surgery. •

Histology of the Uterus

Histologically, the uterus consists of three layers of tissue: perimetrium, myometrium, and endometrium (Figure 28.18). The outer layer—the **perimetrium** (per'-i-MĒ-trē-um; *peri-* = around; *-metrium* = uterus) or serosa—is part of the visceral peritoneum; it is composed of simple squamous epithelium and areolar connective tissue. Laterally, it becomes the broad ligament. Anteriorly, it covers the urinary bladder and forms a shallow pouch, the **vesicouterine pouch** (ves'-i-kō-Ū-ter-in; *vesico-* = bladder; see Figure 28.11). Posteriorly, it covers the rectum and forms a deep pouch between the uterus and urinary bladder, the **rectouterine pouch** (rek-tō-Ū-ter-in; *recto-* = rectum) or **pouch of Douglas**—the most inferior point in the pelvic cavity.

The middle layer of the uterus, the **myometrium** (*myo-* = muscle), consists of three layers of smooth muscle fibers that are thickest in the fundus and thinnest in the cervix. The thicker middle layer is circular; the inner and outer layers are longitudinal or oblique. During labor and childbirth, coordinated contractions of the myometrium in response to oxytocin from the posterior pituitary help expel the fetus from the uterus.

The inner layer of the uterus, the **endometrium** (*endo-* = within), is highly vascularized and has three components: (1) An innermost layer composed of simple columnar epithelium (ciliated

Figure 28.18 Histology of the uterus.

 The three layers of the uterus from superficial to deep are the perimetrium (serosa), the myometrium, and the endometrium.

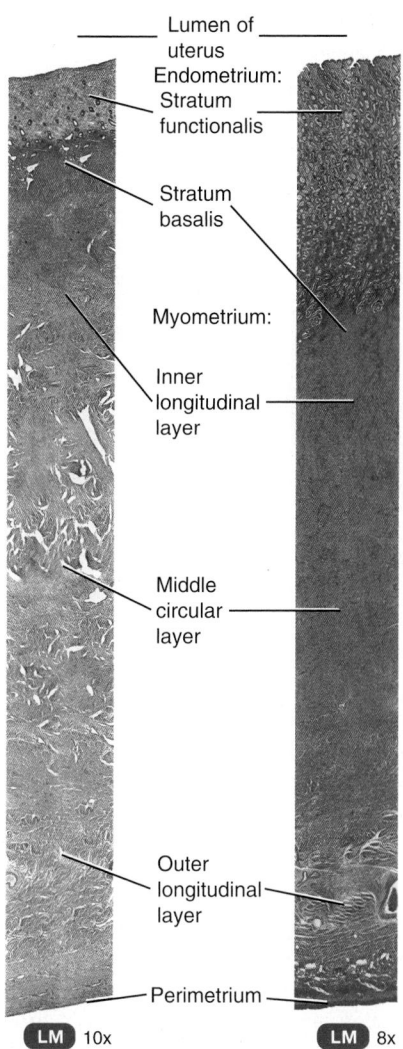

Lumen of uterus

Endometrium:
Stratum functionalis

Stratum basalis

Myometrium:

Inner longitudinal layer

Middle circular layer

Outer longitudinal layer

Perimetrium

LM 10x

LM 8x

(a) Transverse section through the uterine wall: second week of menstrual cycle (left) and third week of menstrual cycle (right)

Lumen of uterus

Simple columnar epithelium

Stratum functionalis

Endometrial gland

Stratum basalis

LM 115x

(b) Details of endometrium

 What structural features of the endometrium and myometrium contribute to their functions?

and secretory cells) lines the lumen. (2) An underlying endometrial stroma is a very thick region of lamina propria (areolar connective tissue). (3) Endometrial (uterine) glands develop as invaginations of the luminal epithelium and extend almost to the myometrium. The endometrium is divided into two layers. The **stratum functionalis** *(functional layer)* lines the uterine cavity and sloughs off during menstruation. The deeper layer, the **stratum basalis** *(basal layer),* is permanent and gives rise to a new stratum functionalis after each menstruation.

Branches of the internal iliac artery called **uterine arteries** (Figure 28.19) supply blood to the uterus. Uterine arteries give off branches called **arcuate arteries** (AR-kū-āt = shaped like a bow)

that are arranged in a circular fashion in the myometrium. These arteries branch into **radial arteries** that penetrate deeply into the myometrium. Just before the branches enter the endometrium, they divide into two kinds of arterioles: **Straight arterioles** supply the stratum basalis with the materials needed to regenerate the stratum functionalis; **spiral arterioles** supply the stratum functionalis and change markedly during the menstrual cycle. Blood leaving the uterus is drained by the **uterine veins** into the internal iliac veins. The extensive blood supply of the uterus is essential to support regrowth of a new stratum functionalis after menstruation, implantation of a fertilized ovum, and development of the placenta.

Figure 28.19 Blood supply of the uterus. The inset shows histological details of the blood vessels of the endometrium.

Straight arterioles supply the materials needed for regeneration of the stratum functionalis.

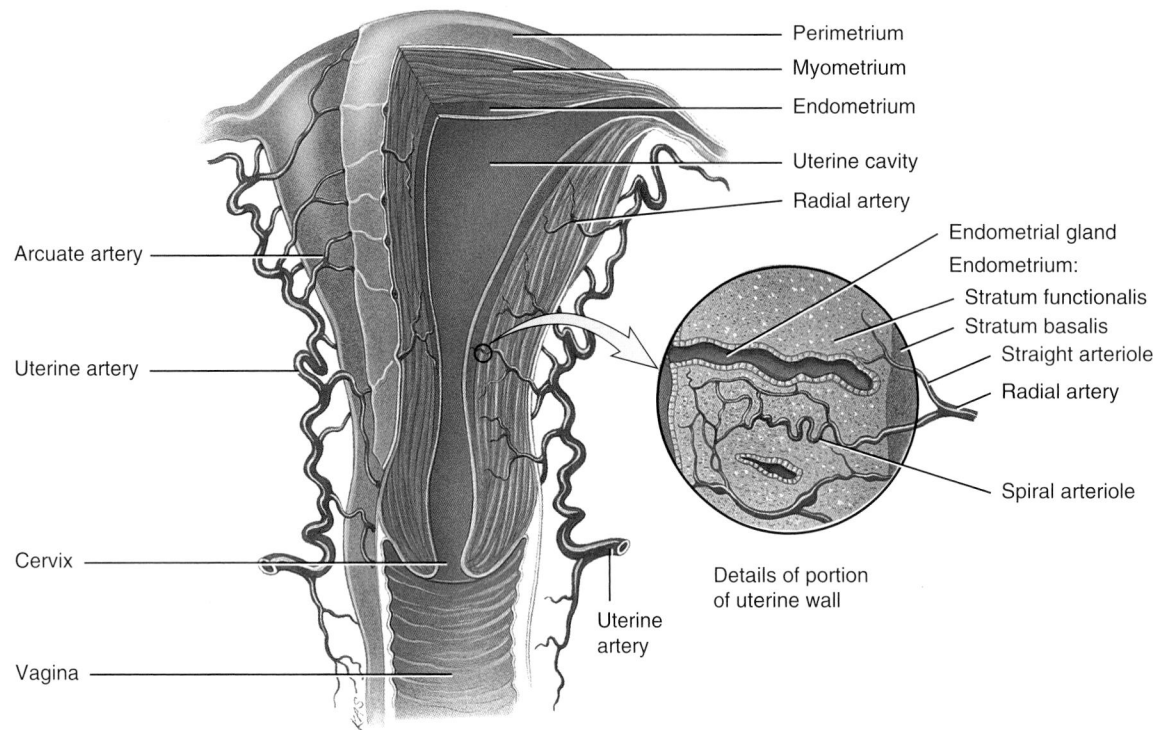

Anterior view with left side of uterus partially sectioned

What is the functional significance of the stratum basalis of the endometrium?

Cervical Mucus

The secretory cells of the mucosa of the cervix produce a secretion called **cervical mucus,** a mixture of water, glycoproteins, lipids, enzymes, and inorganic salts. During their reproductive years, females secrete 20–60 mL of cervical mucus per day. Cervical mucus is more hospitable to sperm at or near the time of ovulation because it is then less viscous and more alkaline (pH 8.5). At other times, a more viscous mucus forms a cervical plug that physically impedes sperm penetration. Cervical mucus supplements the energy needs of sperm, and both the cervix and cervical mucus protect sperm from phagocytes and the hostile environment of the vagina and uterus. Cervical mucus may also play a role in **capacitation** (ka-pas′-i-TĀ-shun)—a series of functional changes that sperm undergo in the female reproductive tract before they are able to fertilize a secondary oocyte. Capacitation causes a sperm cell's tail to beat even more vigorously, and it prepares the sperm cell's plasma membrane to fuse with the oocyte's plasma membrane.

CLINICAL CONNECTION | Hysterectomy

Hysterectomy (hiss-ter-EK-tō-mē; *hyster-* = uterus), the surgical removal of the uterus, is the most common gynecological operation. It may be indicated in conditions such as fibroids, which are noncancerous tumors composed of muscular and fibrous tissue; endometriosis; pelvic inflammatory disease; recurrent ovarian cysts; excessive uterine bleeding; and cancer of the cervix, uterus, or ovaries. In a *partial (subtotal) hysterectomy,* the body of the uterus is removed but the cervix is left in place. A *complete hysterectomy* is the removal of both the body and cervix of the uterus. A *radical hysterectomy* includes removal of the body and cervix of the uterus, uterine tubes, possibly the ovaries, the superior portion of the vagina, pelvic lymph nodes, and supporting structures, such as ligaments. A hysterectomy can be performed either through an incision in the abdominal wall or through the vagina. •

✔**CHECKPOINT**

17. Where are the uterine tubes located, and what is their function?
18. What are the principal parts of the uterus? Where are they located in relation to one another?
19. Describe the arrangement of ligaments that hold the uterus in its normal position.
20. Describe the histology of the uterus.
21. Why is an abundant blood supply important to the uterus?

Vagina

The **vagina** (= sheath) is a tubular, 10-cm (4-in.) long fibromuscular canal lined with mucous membrane that extends from the exterior of the body to the uterine cervix (see Figures 28.11 and 28.16). It is the receptacle for the penis during sexual intercourse, the outlet for menstrual flow, and the passageway for childbirth. Situated between the urinary bladder and the rectum, the vagina is directed superiorly and posteriorly, where it attaches to the uterus. A recess called the **fornix** (= arch or vault) surrounds the vaginal attachment to the cervix. When properly inserted, a contraceptive diaphragm rests in the fornix, where it is held in place as it covers the cervix.

The **mucosa** of the vagina is continuous with that of the uterus. Histologically, it consists of nonkeratinized stratified squamous epithelium and areolar connective tissue that lies in a series of transverse folds called **rugae** (ROO-gē). Dendritic cells in the mucosa are antigen-presenting cells (described in Section 22.4). Unfortunately, they also participate in the transmission of viruses—for example, HIV (the virus that causes AIDS)—to a female during intercourse with an infected male. The mucosa of the vagina contains large stores of glycogen, the decomposition of which produces organic acids. The resulting acidic environment retards microbial growth, but it also is harmful to sperm. Alkaline components of semen, mainly from the seminal vesicles, raise the pH of fluid in the vagina and increase viability of the sperm.

The **muscularis** is composed of an outer circular layer and an inner longitudinal layer of smooth muscle that can stretch considerably to accommodate the penis during sexual intercourse and a child during birth.

The **adventitia,** the superficial layer of the vagina, consists of areolar connective tissue. It anchors the vagina to adjacent organs such as the urethra and urinary bladder anteriorly and the rectum and anal canal posteriorly.

A thin fold of vascularized mucous membrane, called the **hymen** (= membrane), forms a border around and partially closes the inferior end of the vaginal opening to the exterior, the **vaginal orifice** (see Figure 28.20). After its rupture, usually following the first sexual intercourse, only remnants of the hymen remain. Sometimes the hymen completely covers the orifice, a condition called **imperforate hymen** (im-PER-fō-rāt). Surgery may be needed to open the orifice and permit the discharge of menstrual flow.

Figure 28.20 **Components of the vulva (pudendum).**

The vulva refers to the external genitals of the female.

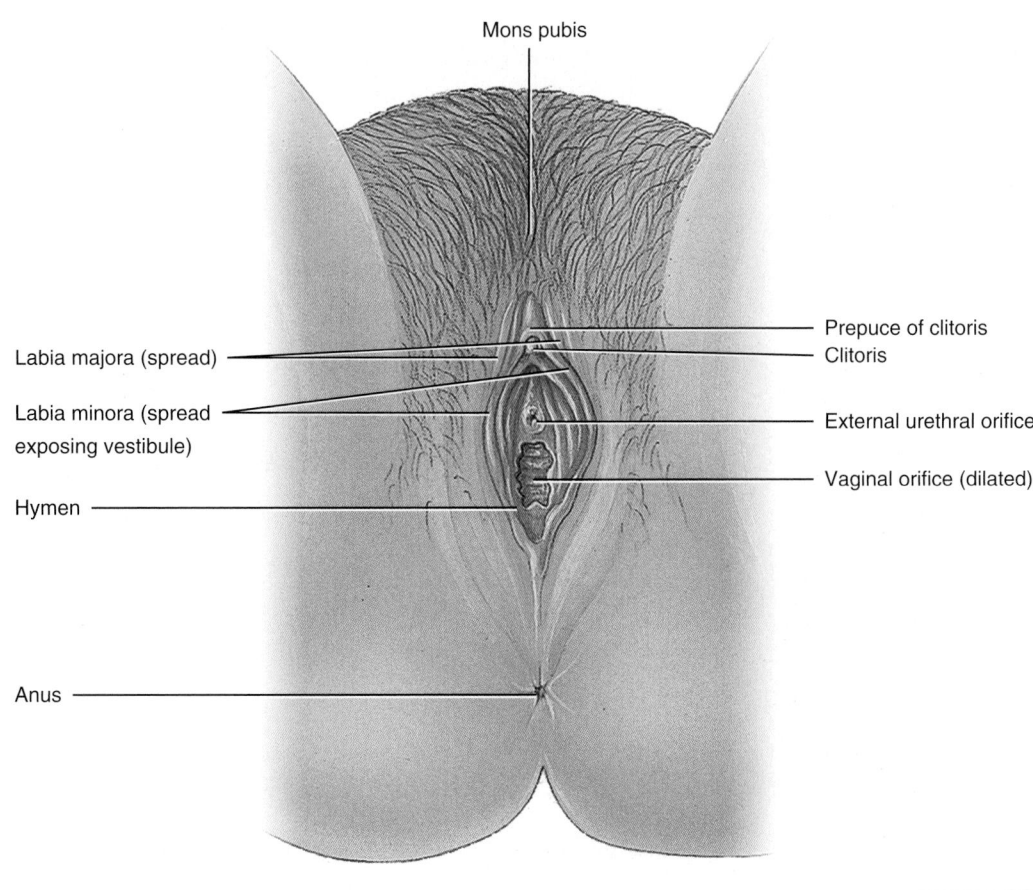

(a) Inferior view

FIGURE 28.20 CONTINUES ▶

○ FIGURE 28.20 CONTINUED ▶

Transverse plane

Vagina

Lumen of vagina

Mucosa:
Nonkeratinized stratified squamous epithelium

Lamina propria

Muscularis:

Inner circular layer

Outer longitudinal layer

Adventitia

LM 15x

(b) Transverse section through vaginal wall

Lumen of vagina

Mucosa:

Nonkeratinized stratified squamous epithelium

Lamina propria

LM 100x

(c) Details of mucosa

? **What surface structures are anterior to the vaginal opening? Lateral to it?**

Vulva

The term **vulva** (VUL-va = to wrap around), or **pudendum** (pū-DEN-dum), refers to the external genitals of the female (Figure 28.20). The following components make up the vulva:

- Anterior to the vaginal and urethral openings is the **mons pubis** (MONZ PŪ-bis; *mons* = mountain), an elevation of adipose tissue covered by skin and coarse pubic hair that cushions the pubic symphysis.

- From the mons pubis, two longitudinal folds of skin, the **labia majora** (LĀ-bē-a ma-JŌ-ra; *labia* = lips; *majora* = larger), extend inferiorly and posteriorly. The singular term is *labium majus.* The labia majora are covered by pubic hair and contain an abundance of adipose tissue, sebaceous (oil) glands, and apocrine sudoriferous (sweat) glands. They are homologous to the scrotum.

- Medial to the labia majora are two smaller folds of skin called the **labia minora** (min-OR-a = smaller). The singular term is *labium minus.* Unlike the labia majora, the labia minora are devoid of pubic hair and fat and have few sudoriferous glands, but they do contain many sebaceous glands. The labia minora are homologous to the spongy (penile) urethra.

- The **clitoris** (KLI-to-ris) is a small cylindrical mass composed of two small erectile bodies, the *corpora cavernosa,* and numerous nerves and blood vessels. The clitoris is located at the anterior junction of the labia minora. A layer of skin called the **prepuce of the clitoris** (PRĒ-poos) is formed at the point where the labia minora unite and covers the body of the clitoris. The exposed portion of the clitoris is the **glans clitoris.** The clitoris is homologous to the glans penis in males. Like the male structure, the clitoris is capable of enlargement on tactile stimulation and has a role in sexual excitement in the female.

- The region between the labia minora is the **vestibule.** Within the vestibule are the hymen (if still present), the vaginal orifice, the external urethral orifice, and the openings of the ducts of several glands. The vestibule is homologous to the membranous urethra of males. The **vaginal orifice,** the opening of the vagina to the exterior, occupies the greater portion of the vestibule and is bordered by the hymen. Anterior to the vaginal orifice and posterior to the clitoris is the **external urethral orifice,** the opening of the urethra to the exterior. On either side of the external urethral orifice are the openings of the ducts of the **paraurethral glands** (par'-a-ū-RĒ-thral) or **Skene's glands** (SKĒNZ). These mucus-secreting glands are embedded in the wall of the urethra. The paraurethral glands are homologous to the prostate. On either side of the vaginal orifice itself are the **greater vestibular (Bartholin's) glands** (BAR-to-linz) (see Figure 28.21), which open by ducts into a groove between the hymen and labia minora. They produce a small quantity of

Figure 28.21 Perineum of a female. (Figure 11.13 shows the perineum of a male.)

The perineum is a diamond-shaped area that includes the urogenital triangle and the anal triangle.

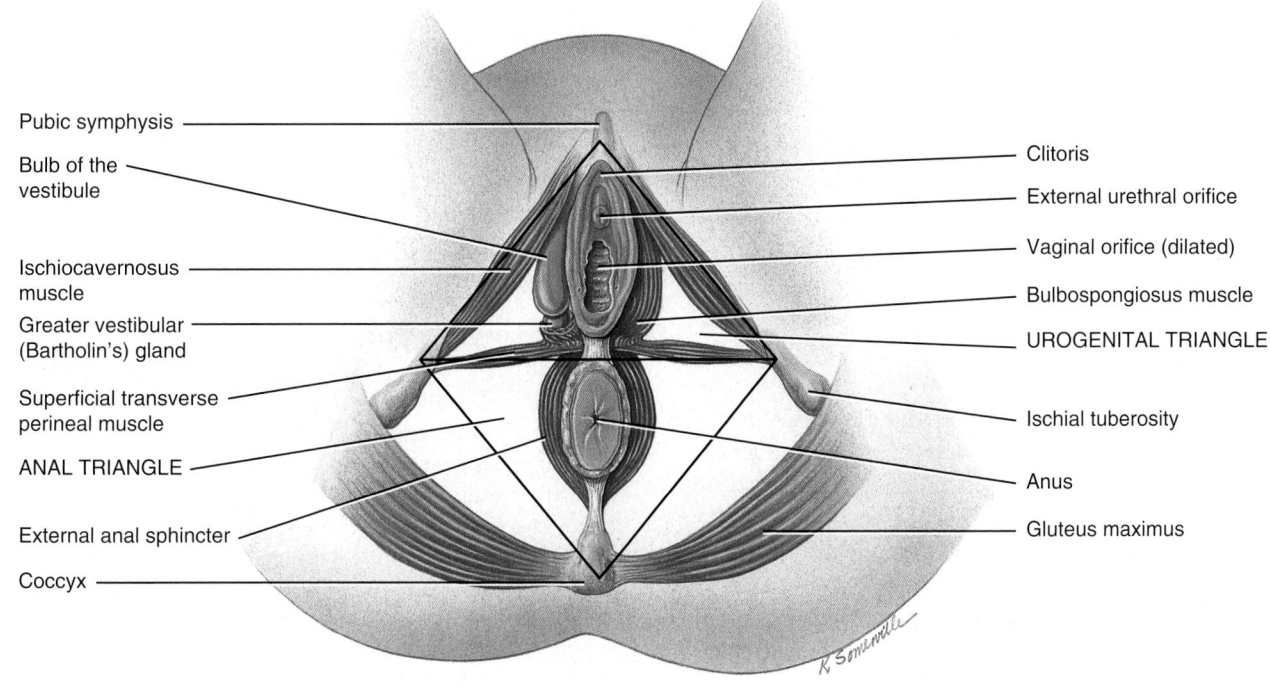

Pubic symphysis

Bulb of the vestibule

Ischiocavernosus muscle

Greater vestibular (Bartholin's) gland

Superficial transverse perineal muscle

ANAL TRIANGLE

External anal sphincter

Coccyx

Clitoris

External urethral orifice

Vaginal orifice (dilated)

Bulbospongiosus muscle

UROGENITAL TRIANGLE

Ischial tuberosity

Anus

Gluteus maximus

Inferior view

? Why is the anterior portion of the perineum called the urogenital triangle?

mucus during sexual arousal and intercourse that adds to cervical mucus and provides lubrication. The greater vestibular glands are homologous to the bulbourethral glands in males. Several **lesser vestibular glands** also open into the vestibule.

- The **bulb of the vestibule** (see Figure 28.21) consists of two elongated masses of erectile tissue just deep to the labia on either side of the vaginal orifice. The bulb of the vestibule becomes engorged with blood during sexual arousal, narrowing the vaginal orifice and placing pressure on the penis during intercourse. The bulb of the vestibule is homologous to the corpus spongiosum and bulb of the penis in males.

Table 28.2 summarizes the homologous structures of the female and male reproductive systems.

Perineum

The **perineum** (per′-i-NĒ-um) is the diamond-shaped area medial to the thighs and buttocks of both males and females (Figure 28.21). It contains the external genitals and anus. The perineum is bounded anteriorly by the pubic symphysis, laterally by the ischial tuberosities, and posteriorly by the coccyx. A transverse line drawn between the ischial tuberosities divides the perineum into an anterior **urogenital triangle** (ū′-rō-JEN-i-tal) that contains the external genitals and a posterior **anal triangle** that contains the anus.

> ### CLINICAL CONNECTION | *Episiotomy*
>
> In specific maternal-fetal situations that require a quick delivery, an **episiotomy** (e-piz-ē-OT-ō-mē; *episi-* = vulva or pubic region; *-otomy* = incision) may be performed. In the procedure a perineal cut made with surgical scissors. The cut may be made along the midline or at an angle of approximately 45 degrees to the midline. In effect, a straight, more easily sutured cut is substituted for the jagged tear that would otherwise be caused by passage of the fetus. The incision is closed in layers with sutures that are absorbed within a few weeks, so that the busy new mom does not have to worry about making time to have them removed. •

Mammary Glands

Each **breast** is a hemispheric projection of variable size anterior to the pectoralis major and serratus anterior muscles and attached to them by a layer of fascia composed of dense irregular connective tissue.

Each breast has one pigmented projection, the **nipple,** that has a series of closely spaced openings of ducts called **lactiferous ducts** (lak-TIF-e-rus), where milk emerges. The circular pigmented area of skin surrounding the nipple is called the **areola** (a-RĒ-ō-la = small space); it appears rough because it contains modified sebaceous (oil) glands. Strands of connective tissue called the **suspensory ligaments of the breast (Cooper's ligaments)** run between the skin and fascia and support the breast.

TABLE 28.2

Summary of Homologous Structures of the Female and Male Reproductive Systems

FEMALE STRUCTURES	MALE STRUCTURES
Ovaries	Testes
Ovum	Sperm cell
Labia majora	Scrotum
Labia minora	Spongy (penile) urethra
Vestibule	Membranous urethra
Bulb of vestibule	Corpus spongiosum penis and bulb of penis
Clitoris	Glans penis and corpora cavernosa
Paraurethral glands	Prostate
Greater vestibular glands	Bulbourethral (Cowper's) glands

These ligaments become looser with age or with the excessive strain that can occur in long-term jogging or high-impact aerobics. Wearing a supportive bra can slow this process and help maintain the strength of the suspensory ligaments.

Within each breast is a **mammary gland,** a modified sudoriferous (sweat) gland that produces milk (Figure 28.22). A mammary gland consists of 15 to 20 lobes, or compartments, separated by a variable amount of adipose tissue. In each lobe are several smaller compartments called **lobules,** composed of grapelike clusters of milk-secreting glands termed **alveoli** (al-VĒ-o-lī = small cavities) embedded in connective tissue. Contraction of **myoepithelial cells** (mī′-ō-ep′-i-THĒ-lē-al) surrounding the alveoli helps propel milk toward the nipples. When milk is being produced, it passes from the alveoli into a series of **secondary tubules** and then into the **mammary ducts.** Near the nipple, the mammary ducts expand to form sinuses called **lactiferous sinuses** (*lact-* = milk), where some milk may be stored before draining into a lactiferous duct. Each lactiferous duct typically carries milk from one of the lobes to the exterior.

> ### CLINICAL CONNECTION | *Breast Augmentation and Reduction*
>
> **Breast augmentation** (awg-men-TĀ-shun = enlargement), technically called **augmentation mammaplasty** (mam-a-PLAS-tē), is a surgical procedure to increase breast size and shape. It may be done to enhance breast size for females who feel that their breasts are too small, to restore breast volume due to weight loss or following pregnancy, to improve the shape of breasts that are sagging, and to improve breast appearance following surgery, trauma, or congenital abnormalities. The most commonly used implants are filled with either a saline solution or silicone gel. The incision for the implant is made under the breast, around the areola, in the armpit, or in the navel. Then a pocket is made to place the implant either directly behind the breast tissue or beneath the pectoralis major muscle.

Figure 28.22 Mammary glands within the breasts.

The mammary glands function in the synthesis, secretion, and ejection of milk (lactation).

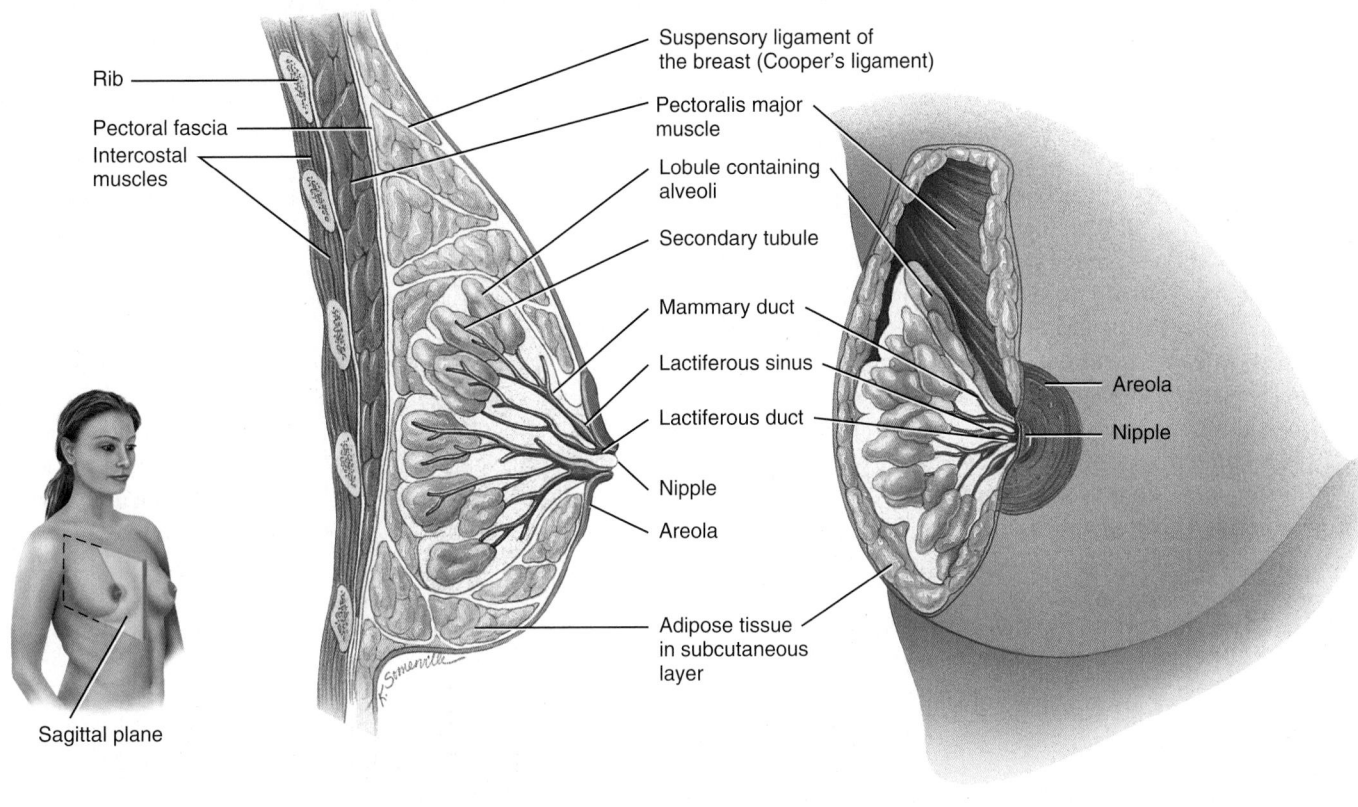

(a) Sagittal section

(b) Anterior view, partially sectioned

? **What hormones regulate the synthesis and ejection of milk?**

Breast reduction or **reduction mammaplasty** is a surgical procedure that involves decreasing breast size by removing fat, skin, and glandular tissue. This procedure is done because of chronic back, neck, and shoulder pain; poor posture; circulation or breathing problems; a skin rash under the breasts; restricted levels of activity; self-esteem problems; deep grooves in the shoulders from bra strap pressure; and difficulty wearing or fitting into certain bras and clothing. The most common procedure involves an incision around the areola, down the breast toward the crease between the breast and abdomen, and then along the crease. The surgeon removes excess tissue through the incision. In most cases, the nipple and areola remain attached to the breast. However, if the breasts are extremely large, the nipple and areola may have to be reattached at a higher position. •

The functions of the mammary glands are the synthesis, secretion, and ejection of milk; these functions, called **lactation** (lak-TĀ-shun), are associated with pregnancy and childbirth. Milk production is stimulated largely by the hormone prolactin from the anterior pituitary, with contributions from progesterone and estrogens. The ejection of milk is stimulated by oxytocin, which is released from the posterior pituitary in response to the sucking of an infant on the mother's nipple (suckling).

CLINICAL CONNECTION | *Fibrocystic Disease of the Breasts*

The breasts of females are highly susceptible to cysts and tumors. In **fibrocystic disease** (fī-brō-SIS-tik), the most common cause of breast lumps in females, one or more cysts (fluid-filled sacs) and thickenings of alveoli develop. The condition, which occurs mainly in females between the ages of 30 and 50, is probably due to a relative excess of estrogens or a deficiency of progesterone in the postovulatory (luteal) phase of the reproductive cycle (discussed shortly). Fibrocystic disease usually causes one or both breasts to become lumpy, swollen, and tender a week or so before menstruation begins. •

✔CHECKPOINT

22. How does the histology of the vagina contribute to its function?
23. What are the structures and functions of each part of the vulva?
24. Describe the components of the mammary glands and the structures that support them.
25. Outline the route milk takes from the alveoli of the mammary gland to the nipple.

28.3 THE FEMALE REPRODUCTIVE CYCLE

• Compare the major events of the ovarian and uterine cycles.

During their reproductive years, nonpregnant females normally exhibit cyclical changes in the ovaries and uterus. Each cycle takes about a month and involves both oogenesis and preparation of the uterus to receive a fertilized ovum. Hormones secreted by the hypothalamus, anterior pituitary, and ovaries control the main events. The **ovarian cycle** is a series of events in the ovaries that occur during and after the maturation of an oocyte. The **uterine (menstrual) cycle** is a concurrent series of changes in the endometrium of the uterus to prepare it for the arrival of a fertilized ovum that will develop there until birth. If fertilization does not occur, ovarian hormones wane, which causes the stratum functionalis of the endometrium to slough off. The general term **female reproductive cycle** encompasses the ovarian and uterine cycles, the hormonal changes that regulate them, and the related cyclical changes in the breasts and cervix.

Hormonal Regulation of the Female Reproductive Cycle

Gonadotropin-releasing hormone (GnRH) secreted by the hypothalamus controls the ovarian and uterine cycles (Figure 28.23). GnRH stimulates the release of **follicle-stimulating hormone (FSH)** and **luteinizing hormone (LH)** from the anterior pituitary. FSH initiates follicular growth, while LH stimulates further development of the ovarian follicles. In addition, both FSH and LH stimulate the ovarian follicles to secrete estrogens. LH stimulates the theca cells of a developing follicle to produce androgens. Under the influence of FSH, the androgens are taken up by the granulosa cells of the follicle and then converted into estrogens. At midcycle, LH triggers ovulation and then promotes formation of the corpus luteum, the reason for the name luteinizing hormone. Stimulated by LH, the corpus luteum produces and secretes estrogens, progesterone, relaxin, and inhibin.

Figure 28.23 Secretion and physiological effects of estrogens, progesterone, relaxin, and inhibin in the female reproductive cycle. Dashed red lines indicate negative feedback inhibition.

🔑 The uterine and ovarian cycles are controlled by gonadotropin-releasing hormone (GnRH) and ovarian hormones (estrogens and progesterone).

GnRH — Hypothalamus

GnRH stimulates release of FSH and LH

Anterior pituitary gland

FSH LH

FSH stimulates

LH stimulates

Growing follicles

Initial development of ovarian follicles

Ovaries

Ovulation

Further development of ovarian follicles and their secretion of estrogens and inhibin

Corpus luteum

Secretion of progesterone, estrogens, relaxin, and inhibin by corpus luteum

Estrogens
• Promote development and maintenance of female reproductive structures, feminine secondary sex characteristics, and breasts
• Increase protein anabolism
• Lower blood cholesterol
• Moderate levels inhibit release of GnRH, FSH, and LH

Progesterone
• Works with estrogens to prepare endometrium for implantation
• Prepares mammary glands to secrete milk
• Inhibits release of GnRH and LH

Relaxin
• Inhibits contractions of uterine smooth muscle
• During labor, increases flexibility of pubic symphysis and dilates uterine cervix

Inhibin
• Inhibits release of FSH and, to a lesser extent, LH

❓ **Of the several estrogens, which one exerts the major effect?**

At least six different estrogens have been isolated from the plasma of human females, but only three are present in significant quantities: *beta* (*β*)-*estradiol* (es-tra-DĪ-ol), *estrone,* and *estriol* (ES-trē-ol). In a nonpregnant woman, the most abundant estrogen is β-estradiol, which is synthesized from cholesterol in the ovaries.

Estrogens secreted by ovarian follicles have several important functions (Figure 28.23):

- Estrogens promote the development and maintenance of female reproductive structures, secondary sex characteristics, and the breasts. The secondary sex characteristics include distribution of adipose tissue in the breasts, abdomen, mons pubis, and hips; voice pitch; a broad pelvis; and pattern of hair growth on the head and body.

- Estrogens increase protein anabolism, including the building of strong bones. In this regard, estrogens are synergistic with human growth hormone (hGH).

- Estrogens lower blood cholesterol level, which is probably the reason that women under age 50 have a much lower risk of coronary artery disease than do men of comparable age.

- Moderate levels of estrogens in the blood inhibit both the release of GnRH by the hypothalamus and secretion of LH and FSH by the anterior pituitary.

Progesterone, secreted mainly by cells of the corpus luteum, cooperates with estrogens to prepare and maintain the endometrium for implantation of a fertilized ovum and to prepare the mammary glands for milk secretion. High levels of progesterone also inhibit secretion of GnRH and LH.

The small quantity of **relaxin** produced by the corpus luteum during each monthly cycle relaxes the uterus by inhibiting contractions of the myometrium. Presumably, implantation of a fertilized ovum occurs more readily in a "quiet" uterus. During pregnancy, the placenta produces much more relaxin, and it continues to relax uterine smooth muscle. At the end of pregnancy, relaxin also increases the flexibility of the pubic symphysis and may help dilate the uterine cervix, both of which ease delivery of the baby.

Inhibin is secreted by granulosa cells of growing follicles and by the corpus luteum after ovulation. It inhibits secretion of FSH and, to a lesser extent, LH.

Phases of the Female Reproductive Cycle

The duration of the female reproductive cycle typically ranges from 24 to 36 days. For this discussion, we assume a duration of 28 days and divide it into four phases: the menstrual phase, the preovulatory phase, ovulation, and the postovulatory phase (Figure 28.24).

Menstrual Phase

The **menstrual phase** (MEN-stroo-al), also called **menstruation** (men′-stroo-Ā-shun) or **menses** (MEN-sēz = month), lasts for roughly the first 5 days of the cycle. (By convention, the first day of menstruation is day 1 of a new cycle.)

EVENTS IN THE OVARIES Under the influence of FSH, several primordial follicles develop into primary follicles and then into secondary follicles. This developmental process may take several months to occur. Therefore, a follicle that begins to develop at the beginning of a particular menstrual cycle may not reach maturity and ovulate until several menstrual cycles later.

EVENTS IN THE UTERUS Menstrual flow from the uterus consists of 50–150 mL of blood, tissue fluid, mucus, and epithelial cells shed from the endometrium. This discharge occurs because the declining levels of progesterone and estrogens stimulate release of prostaglandins that cause the uterine spiral arterioles to constrict. As a result, the cells they supply become oxygen-deprived and start to die. Eventually, the entire stratum functionalis sloughs off. At this time the endometrium is very thin, about 2–5 mm, because only the stratum basalis remains. The menstrual flow passes from the uterine cavity through the cervix and vagina to the exterior.

Preovulatory Phase

The **preovulatory phase** (prē-OV-ū-la-tō-rē) is the time between the end of menstruation and ovulation. The preovulatory phase of the cycle is more variable in length than the other phases and accounts for most of the differences in length of the cycle. It lasts from days 6 to 13 in a 28-day cycle.

EVENTS IN THE OVARIES Some of the secondary follicles in the ovaries begin to secrete estrogens and inhibin. By about day 6, a single secondary follicle in one of the two ovaries has outgrown all the others to become the **dominant follicle.** Estrogens and inhibin secreted by the dominant follicle decrease the secretion of FSH, which causes other, less well-developed follicles to stop growing and undergo atresia. Fraternal (nonidentical) twins or triplets result when two or three secondary follicles become codominant and later are ovulated and fertilized at about the same time.

Normally, the one dominant secondary follicle becomes the **mature (graafian) follicle,** which continues to enlarge until it is more than 20 mm in diameter and ready for ovulation (see Figure 28.13). This follicle forms a blisterlike bulge due to the swelling antrum on the surface of the ovary. During the final maturation process, the mature follicle continues to increase its production of estrogens (Figure 28.24).

With reference to the ovarian cycle, the menstrual and preovulatory phases together are termed the **follicular phase** (fo-LIK-ū-lar) because ovarian follicles are growing and developing.

EVENTS IN THE UTERUS Estrogens liberated into the blood by growing ovarian follicles stimulate the repair of the endometrium; cells of the stratum basalis undergo mitosis and produce a new stratum functionalis. As the endometrium thickens, the short, straight endometrial glands develop, and the arterioles coil and lengthen as they penetrate the stratum functionalis. The thickness of the endometrium approximately doubles, to about 4–10 mm. With reference to the uterine cycle, the preovulatory phase is also termed the **proliferative phase** (prō-LIF-er-a-tiv) because the endometrium is proliferating.

Figure 28.24 The female reproductive cycle. The length of the female reproductive cycle typically is 24 to 36 days; the preovulatory phase is more variable in length than the other phases. (a) Events in the ovarian and uterine cycles and the release of anterior pituitary hormones are correlated with the sequence of the cycle's four phases. In the cycle shown, fertilization and implantation have not occurred. (b) Relative concentrations of anterior pituitary hormones (FSH and LH) and ovarian hormones (estrogens and progesterone) during the phases of a normal female reproductive cycle.

Estrogens are the primary ovarian hormones before ovulation; after ovulation, both progesterone and estrogens are secreted by the corpus luteum.

(a) Hormonal regulation of changes in the ovary and uterus

(b) Changes in concentration of anterior pituitary and ovarian hormones

? **Which hormones are responsible for the proliferative phase of endometrial growth, for ovulation, for growth of the corpus luteum, and for the surge of LH at midcycle?**

Ovulation

Ovulation (ov'-ū-LĀ-shun), the rupture of the mature (graafian) follicle and the release of the secondary oocyte into the pelvic cavity, usually occurs on day 14 in a 28-day cycle. During ovulation, the secondary oocyte remains surrounded by its zona pellucida and corona radiata.

The *high levels of estrogens* during the last part of the preovulatory phase exert a *positive* feedback effect on the cells that secrete LH and gonadotropin-releasing hormone (GnRH) and cause ovulation, as follows (Figure 28.25):

1 A high concentration of estrogens stimulates more frequent release of GnRH from the hypothalamus. It also directly stimulates gonadotrophs in the anterior pituitary to secrete LH.

2 GnRH promotes the release of FSH and additional LH by the anterior pituitary.

3 LH causes rupture of the mature (graafian) follicle and expulsion of a secondary oocyte about 9 hours after the peak of the LH surge. The ovulated oocyte and its corona radiata cells are usually swept into the uterine tube.

From time to time, an oocyte is lost into the pelvic cavity, where it later disintegrates. The small amount of blood that sometimes leaks into the pelvic cavity from the ruptured follicle can cause pain, known as **mittelschmerz** (MIT-el-shmārts = pain in the middle), at the time of ovulation.

An over-the-counter home test that detects a rising level of LH can be used to predict ovulation a day in advance.

Postovulatory Phase

The **postovulatory phase** of the female reproductive cycle is the time between ovulation and onset of the next menses. In duration, it is the most constant part of the female reproductive cycle. It lasts for 14 days in a 28-day cycle, from day 15 to day 28 (see Figure 28.24).

EVENTS IN ONE OVARY   After ovulation, the mature follicle collapses, and the basement membrane between the granulosa cells and theca interna breaks down. Once a blood clot forms from minor bleeding of the ruptured follicle, the follicle becomes the **corpus hemorrhagicum** (hem'-o-RAJ-i-kum; *hemo-* = blood; *rrhagic-* = bursting forth) (see Figure 28.13). Theca interna cells mix with the granulosa cells as they all become transformed into **corpus luteum cells** under the influence of LH. Stimulated by LH, the corpus luteum secretes progesterone, estrogen, relaxin, and inhibin. The luteal cells also absorb the blood clot. With reference to the ovarian cycle, this phase is also called the **luteal phase** (LOO-tē-al).

Later events in an ovary that has ovulated an oocyte depend on whether the oocyte is fertilized. If the oocyte *is not fertilized,* the corpus luteum has a lifespan of only 2 weeks. Then, its secretory activity declines, and it degenerates into a corpus albicans (see Figure 28.13). As the levels of progesterone, estrogens, and inhibin decrease, release of GnRH, FSH, and LH rises due to loss of negative feedback suppression by the ovarian hormones. Follicular growth resumes and a new ovarian cycle begins.

If the secondary oocyte *is fertilized* and begins to divide, the corpus luteum persists past its normal 2-week lifespan. It is "res-

Figure 28.25 High levels of estrogens exert a positive feedback effect (green arrows) on the hypothalamus and anterior pituitary, thereby increasing secretion of GnRH and LH.

6 At midcycle, a surge of LH triggers ovulation.

? What is the effect of rising but still moderate levels of estrogens on the secretion of GnRH, LH, and FSH?

cued" from degeneration by **human chorionic gonadotropin (hCG)** (kō-rē-ON-ik). This hormone is produced by the chorion of the embryo beginning about 8 days after fertilization. Like LH, hCG stimulates the secretory activity of the corpus luteum. The presence of hCG in maternal blood or urine is an indicator of pregnancy and is the hormone detected by home pregnancy tests.

EVENTS IN THE UTERUS   Progesterone and estrogens produced by the corpus luteum promote growth and coiling of the endometrial glands, vascularization of the superficial endometrium, and thickening of the endometrium to 12–18 mm (0.48–0.72 in.). Because of the secretory activity of the endometrial glands, which begin to secrete glycogen, this period is called the **secretory phase** of the uterine cycle. These preparatory changes peak about 1 week after ovulation, at the time a fertilized ovum might arrive in the uterus. If fertilization does not occur, the levels of progesterone and estrogens decline due to degeneration of the corpus luteum. Withdrawal of progesterone and estrogens causes menstruation.

Figure 28.26 summarizes the hormonal interactions and cyclical changes in the ovaries and uterus during the ovarian and uterine cycles.

CLINICAL CONNECTION | *Female Athlete Triad: Disordered Eating, Amenorrhea, and Premature Osteoporosis*

The female reproductive cycle can be disrupted by many factors, including weight loss, low body weight, disordered eating, and vigorous physical activity. The observation that three conditions—disordered eating, amenorrhea, and osteoporosis—often occur together in female athletes led researchers to coin the term **female athlete triad.**

Many athletes experience intense pressure from coaches, parents, peers, and themselves to lose weight to improve performance. Hence,

they may develop disordered eating behaviors and engage in other harmful weight-loss practices in a struggle to maintain a very low body weight. **Amenorrhea** (a-men-ō-RĒ-a; *a-* = without; *-men-* = month; *-rrhea* = a flow) is the absence of menstruation. The most common causes of amenorrhea are pregnancy and menopause. In female athletes, amenorrhea results from reduced secretion of gonadotropin-releasing hormone, which decreases the release of LH and FSH. As a result, ovarian follicles fail to develop, ovulation does not occur, synthesis of estrogens and progesterone wanes, and monthly menstrual bleeding ceases. Most cases of the female athlete triad occur in young women with very low amounts of body fat. Low levels of the hormone leptin, secreted by adipose cells, may be a contributing factor.

Figure 28.26 Summary of hormonal interactions in the ovarian and uterine cycles.

Hormones from the anterior pituitary regulate ovarian function, and hormones from the ovaries regulate the changes in the endometrial lining of the uterus.

When declining levels of estrogens and progesterone stimulate secretion of GnRH, is this a positive or a negative feedback effect? Why?

Because estrogens help bones retain calcium and other minerals, chronically low levels of estrogens are associated with loss of bone mineral density. The female athlete triad causes "old bones in young women." In one study, amenorrheic runners in their twenties had low bone mineral densities, similar to those of postmenopausal women 50 to 70 years old! Short periods of amenorrhea in young athletes may cause no lasting harm. However, long-term cessation of the reproductive cycle may be accompanied by a loss of bone mass, and adolescent athletes may fail to achieve an adequate bone mass; both of these situations can lead to premature osteoporosis and irreversible bone damage. •

✔ CHECKPOINT

26. Describe the function of each of the following hormones in the uterine and ovarian cycles: GnRH, FSH, LH, estrogens, progesterone, and inhibin.
27. Briefly outline the major events of each phase of the uterine cycle, and correlate them with the events of the ovarian cycle.
28. Prepare a labeled diagram of the major hormonal changes that occur during the uterine and ovarian cycles.

28.4 BIRTH CONTROL METHODS AND ABORTION

◉ OBJECTIVE

• Explain the differences among the various types of birth control methods and compare their effectiveness.

Birth control refers to restricting the number of children by various methods designed to control fertility and prevent conception. No single, ideal method of birth control exists. The only method of preventing pregnancy that is 100% reliable is **complete abstinence,** the avoidance of sexual intercourse. Several other methods are available; each has its advantages and disadvantages. These include surgical sterilization, hormonal methods, intrauterine devices, spermicides, barrier methods, and periodic abstinence. Table 28.3 provides the failure rates for various methods of birth control. Although it is not a form of birth control, in this section we will also discuss abortion, the premature expulsion of the products of conception from the uterus.

Birth Control Methods

Surgical Sterilization

Sterilization is a procedure that renders an individual incapable of further reproduction. The principal method for sterilization of males is a **vasectomy** (va-SEK-tō-mē; -*ectomy* = cut out), in which a portion of each ductus deferens is removed. In order to gain access to the ductus deferens, an incision is made with a scalpel (conventional procedure) or a puncture is made with special forceps (non-scalpel vasectomy). Next the ducts are located and cut, each is tied (ligated) in two places with stitches, and the portion between the ties is removed. Although sperm production continues in the testes, sperm can no longer reach the exterior. The

TABLE 28.3

Failure Rates for Several Birth Control Methods

METHOD	FAILURE RATES* (%)	
	PERFECT USE†	TYPICAL USE
Complete abstinence	0	0
Surgical sterilization		
Vasectomy	0.10	0.15
Tubal ligation	0.5	0.5
Non-incisional sterilization (Essure®)	0.2	0.2
Hormonal methods		
Oral contraceptives		
Combined pill	0.3	1–2
Seasonale®	0.3	1–2
Minipill	0.5	2
Non-oral contraceptives		
Contraceptive skin patch	0.1	1–2
Vaginal contraceptive ring	0.1	1–2
Emergency contraception	25	25
Hormone injections	0.3	1–2
Intrauterine devices (Copper T 380A®)	0.6	0.8
Spermicides (alone)	15	29
Barrier methods		
Male condom	2	15
Vaginal pouch	5	21
Diaphragm (with spermicide)	6	16
Cervical cap (with spermicide)	9	16
Periodic abstinence		
Rhythm	9	25
Sympto-thermal	2	20
No method	85	85

*Defined as percentage of women having an unintended pregnancy during the first year of use.
†Failure rate when the method is used correctly and consistently.

sperm degenerate and are destroyed by phagocytosis. Because the blood vessels are not cut, testosterone levels in the blood remain normal, so vasectomy has no effect on sexual desire or performance. If done correctly, it is close to 100% effective. The procedure can be reversed, but the chance of regaining fertility is only 30–40%. Sterilization in females most often is achieved by performing a **tubal ligation** (lī-GĀ-shun), in which both uterine tubes are tied closed and then cut. This can be achieved in a few different ways. "Clips" or "clamps" can be placed on the uterine tubes, the tubes can be tied and/or cut, and sometimes they are cauterized. In any case, the result is that the secondary oocyte cannot pass through the uterine tubes, and sperm cannot reach the oocyte.

Non-incisional Sterilization

Essure® is a non-incisional irreversible procedure that is an alternative to tubal ligation. In the Essure procedure, a soft micro-insert coil made of polyester fibers and metals (nickel–titanium and stainless steel) is inserted with a catheter into the vagina, through the uterus, and into each uterine tube. Over a three-month period, the insert stimulates tissue growth (scar tissue) in and around itself, blocking the uterine tubes. As with tubal ligation, the secondary oocyte cannot pass through the uterine tubes, and sperm cannot reach the oocyte. Unlike tubal ligation, Essure® does not require general anesthesia.

Hormonal Methods

Aside from complete abstinence or surgical sterilization, hormonal methods are the most effective means of birth control. **Oral contraceptives** (the pill) contain hormones designed to prevent pregnancy. Some, called *combined oral contraceptives (COCs),* contain both progestin (hormone with actions similar to progesterone) and estrogens. The primary action of COCs is to inhibit ovulation by suppressing the gonadotropins FSH and LH. The low levels of FSH and LH usually prevent the development of a dominant follicle in the ovary. As a result, levels of estrogens do not rise, the midcycle LH surge does not occur, and ovulation does not take place. Even if ovulation does occur, as it does in some cases, COCs may also block implantation in the uterus and inhibit the transport of ova and sperm in the uterine tubes.

Progestins thicken cervical mucus and make it more difficult for sperm to enter the uterus. *Progestin-only pills* thicken cervical mucus and may block implantation in the uterus, but they do not consistently inhibit ovulation.

Among the noncontraceptive benefits of oral contraceptives are regulation of the length of menstrual cycle and decreased menstrual flow (and therefore decreased risk of anemia). The pill also provides protection against endometrial and ovarian cancers and reduces the risk of endometriosis. However, oral contraceptives may not be advised for women with a history of blood clotting disorders, cerebral blood vessel damage, migraine headaches, hypertension, liver malfunction, or heart disease. Women who take the pill and smoke face far higher odds of having a heart attack or stroke than do nonsmoking pill users. Smokers should quit smoking or use an alternative method of birth control.

Following are several variations of *oral* hormonal methods of contraception:

- **Combined pill.** Contains both progestin and estrogens and is typically taken once a day for 3 weeks to prevent pregnancy and regulate the menstrual cycle. The pills taken during the fourth week are inactive (do not contain hormones) and permit menstruation to occur.
- **Seasonale®.** Contains both progestin and estrogens and is taken once a day in 3-month cycles of 12 weeks of hormone-containing pills followed by 1 week of inactive pills. Menstruation occurs during the thirteenth week.
- **Minipill.** Contains progestin only and is taken every day of the month.

Non-oral hormonal methods of contraception are also available. Among these are the following:

- **Contraceptive skin patch (Ortho Evra®).** Contains both progestin and estrogens delivered in a skin patch placed on the skin (upper outer arm, back, lower abdomen, or buttocks) once a week for 3 weeks. After 1 week, the patch is removed from one location and then a new one is placed elsewhere. During the fourth week no patch is used.
- **Vaginal contraceptive ring (NuvaRing®).** A flexible doughnut-shaped ring about 5 cm (2 in.) in diameter that contains estrogens and progesterone and is inserted by the female herself into the vagina. It is left in the vagina for 3 weeks to prevent conception and then removed for one week to permit menstruation.
- **Emergency contraception (EC) (morning after pill).** Consists of progestin and estrogens or progestin alone to prevent pregnancy following unprotected sexual intercourse. The relatively high levels of progestin and estrogens in EC pills provide inhibition of FSH and LH secretion. Loss of the stimulating effects of these gonadotropic hormones causes the ovaries to cease secretion of their own estrogens and progesterone. In turn, declining levels of estrogens and progesterone induce shedding of the uterine lining, thereby blocking implantation. One pill is taken as soon as possible but within 72 hours of unprotected sexual intercourse. The second pill must be taken 12 hours after the first. The pills work in the same way as regular birth control pills.
- **Hormone injections (Depo-provera®).** An injectable progestin given intramuscularly by a health-care practitioner once every 3 months.

Intrauterine Devices

An **intrauterine device (IUD)** is a small object made of plastic, copper, or stainless steel that is inserted by a health-care professional into the cavity of the uterus. IUDs prevent fertilization from taking place by blocking sperm from entering the uterine tubes. The IUD most commonly used in the United States today is the Copper T 380A®, which is approved for up to 10 years of use and has long-term effectiveness comparable to that of tubal ligation. Some women cannot use IUDs because of expulsion, bleeding, or discomfort.

Spermicides

Various foams, creams, jellies, suppositories, and douches that contain sperm-killing agents, or **spermicides** (SPER-mi-sīds), make the vagina and cervix unfavorable for sperm survival and are available without prescription. They are placed in the vagina before sexual intercourse. The most widely used spermicide is *nonoxynol-9,* which kills sperm by disrupting their plasma membranes. A spermicide is more effective when used with a barrier method such as a male condom, vaginal pouch, diaphragm, or cervical cap.

Barrier Methods

Barrier methods use a physical barrier and are designed to prevent sperm from gaining access to the uterine cavity and uterine tubes. In addition to preventing pregnancy, certain barrier methods (male condom and vaginal pouch) may also provide some protection against sexually transmitted diseases (STDs) such as

AIDS. In contrast, oral contraceptives and IUDs confer no such protection. Among the barrier methods are the male condom, vaginal pouch, diaphragm, and cervical cap.

A **male condom** is a nonporous, latex covering placed over the penis that prevents deposition of sperm in the female reproductive tract. A **vaginal pouch,** sometimes called a **female condom,** is designed to prevent sperm from entering the uterus. It is made of two flexible rings connected by a polyurethane sheath. One ring lies inside the sheath and is inserted to fit over the cervix; the other ring remains outside the vagina and covers the female external genitals. A **diaphragm** is a rubber, dome-shaped structure that fits over the cervix and is used in conjunction with a spermicide. It can be inserted by the female up to 6 hours before intercourse. The diaphragm stops most sperm from passing into the cervix and the spermicide kills most sperm that do get by. Although diaphragm use does decrease the risk of some STDs, it does not fully protect against HIV infection because the vagina is still exposed. A **cervical cap** resembles a diaphragm but is smaller and more rigid. It fits snugly over the cervix and must be fitted by a health-care professional. Spermicides should be used with the cervical cap.

Periodic Abstinence

A couple can use their knowledge of the physiological changes that occur during the female reproductive cycle to decide either to abstain from intercourse on those days when pregnancy is a likely result, or to plan intercourse on those days if they wish to conceive a child. In females with normal and regular menstrual cycles, these physiological events help to predict the day on which ovulation is likely to occur.

The first physiologically based method, developed in the 1930s, is known as the **rhythm method.** It involves abstaining from sexual activity on the days that ovulation is likely to occur in each reproductive cycle. During this time (3 days before ovulation, the day of ovulation, and 3 days after ovulation) the couple abstains from intercourse. The effectiveness of the rhythm method for birth control is poor in many women due to the irregularity of the female reproductive cycle.

Another system is the **sympto-thermal method,** in which couples are instructed to know and understand certain signs of fertility. The signs of ovulation include increased basal body temperature; the production of abundant clear, stretchy cervical mucus; and pain associated with ovulation (*mittelschmerz*). If a couple abstains from sexual intercourse when the signs of ovulation are present and for 3 days afterward, the chance of pregnancy is decreased. A big problem with this method is that fertilization is very likely if intercourse occurs 1 or 2 days *before* ovulation.

Abortion

Abortion refers to the premature expulsion of the products of conception from the uterus, usually before the twentieth week of pregnancy. An abortion may be **spontaneous** (naturally occurring; also called a *miscarriage*) or **induced** (intentionally performed).

There are several types of induced abortions. One involves **mifepristone** (MIF-pris-tōn), called **miniprex** in the United States and **RU 486** in Europe. It is a hormone approved only for pregnancies 9 weeks or less when taken with misoprostol (a prostaglandin). Mifepristone is an antiprogestin; it blocks the action of progesterone by binding to and blocking progesterone receptors. Progesterone prepares the uterine endometrium for implantation and then maintains the uterine lining after implantation. If the level of progesterone falls during pregnancy or if the action of the hormone is blocked, menstruation occurs, and the embryo sloughs off along with the uterine lining. Within 12 hours after taking mifepristone, the endometrium starts to degenerate, and within 72 hours it begins to slough off. Misoprostol stimulates uterine contractions and is given after mifepristone to aid in expulsion of the endometrium.

Another type of induced abortion is called **vacuum aspiration** (suction) and can be performed up to the sixteenth week of pregnancy. A small, flexible tube attached to a vacuum source is inserted into the uterus through the vagina. The embryo or fetus, placenta, and lining of the uterus are then removed by suction. For pregnancies between 13 and 16 weeks, a technique called **dilation and evacuation** is commonly used. After the cervix is dilated, suction and forceps are used to remove the fetus, placenta, and uterine lining. From the sixteenth to twenty-fourth week, a **late-stage abortion** may be employed using surgical methods similar to dilation and evacuation or through nonsurgical methods using a saline solution or medications to induce abortion. Labor may be induced by using vaginal suppositories, intravenous infusion, or injections into the amniotic fluid through the uterus.

✔**CHECKPOINT**

29. How do oral contraceptives reduce the likelihood of pregnancy?
30. How do some methods of birth control protect against sexually transmitted diseases?
31. What is the problem with developing an oral contraceptive pill for males?

28.5 DEVELOPMENT OF THE REPRODUCTIVE SYSTEMS

◉ **OBJECTIVE**

• Describe the development of the male and female reproductive systems.

The *gonads* develop from **gonadal ridges** that arise from growth of **intermediate mesoderm.** During the fifth week of development, the gonadal ridges appear as bulges just medial to the mesonephros (intermediate kidney) (Figure 28.27). Adjacent to the genital ridges are the **mesonephric ducts** (mez′-o-NEF-rik) or **Wolffian ducts** (WULF-ē-an), which eventually develop into structures of the reproductive system in males. A second pair of ducts, the **paramesonephric ducts** (par′-a-mes′-o-NEF-rik) or

Figure 28.27 Development of the internal reproductive systems.

🔒 The gonads develop from intermediate mesoderm.

Mesonephros

Gonadal ridges

Paramesonephric (Müllerian) duct

Mesonephric (Wolffian) duct

Urogenital sinus

Anterolateral view

Mesonephros

Gonadal ridges

Paramesonephric (Müllerian) duct

Mesonephric (Wolffian) duct

Urogenital sinus

Anterior view

Undifferentiated stage (fifth–sixth week)

♂ MALE DEVELOPMENT

♀ FEMALE DEVELOPMENT

Testis

Efferent duct

Epididymis

Paramesonephric (Müllerian) duct degenerating

Mesonephric (Wolffian) duct

Prostatic utricle

Seventh–eighth week

Ovary

Uterine (fallopian) tube

Mesonephric (Wolffian) duct degenerating

Fused paramesonephric (Müllerian) ducts (uterus)

Urogenital sinus

Eighth–ninth week

Seminal vesicle

Ductus (vas) deferens

Prostate

Urethra

Bulbourethral (Cowper's) gland

Epididymis

Efferent duct

Testis

At birth

Uterine (fallopian) tube

Remnant of mesonephric duct

Ovary

Uterus

Vagina

At birth

❓ Which gene is responsible for the development of the gonads into testes?

1168

Müllerian ducts (mil-E-rē-an), develop lateral to the mesonephric ducts and eventually form structures of the reproductive system in females. Both sets of ducts empty into the urogenital sinus. An early embryo has the potential to follow either the male or the female pattern of development because it contains both sets of ducts and genital ridges that can differentiate into either testes or ovaries.

Cells of a male embryo have one X chromosome and one Y chromosome. The male pattern of development is initiated by a "master switch" gene on the Y chromosome named **SRY,** which stands for *Sex-determining Region* of the *Y* chromosome. When the *SRY* gene is expressed during development, its protein product causes the primitive Sertoli cells to begin to differentiate in the *testes* during the seventh week. The developing Sertoli cells secrete a hormone called **Müllerian-inhibiting substance (MIS),** which causes apoptosis of cells within the paramesonephric (Müllerian) ducts. As a result, those cells do not contribute any functional structures to the male reproductive system. Stimulated by human chorionic gonadotropin (hCG), primitive Leydig cells in the testes begin to secrete the androgen **testosterone** during the eighth week. Testosterone then stimulates development of the mesonephric duct on each side into the *epididymis, ductus (vas) deferens, ejaculatory duct,* and *seminal vesicle.* The *testes* connect to the mesonephric duct through a series of tubules that eventually become the *seminiferous tubules.* The *prostate* and *bulbourethral glands* are **endodermal** outgrowths of the urethra.

Cells of a female embryo have two X chromosomes and no Y chromosome. Because *SRY* is absent, the genital ridges develop into *ovaries,* and because MIS is not produced, the paramesonephric ducts flourish. The distal ends of the paramesonephric ducts fuse to form the *uterus* and *vagina;* the unfused proximal portions of the ducts become the *uterine (fallopian) tubes.* The mesonephric ducts degenerate without contributing any functional structures to the female reproductive system because of the absence of testosterone. The *greater* and *lesser vestibular glands* develop from endodermal outgrowths of the vestibule.

The *external genitals* of both male and female embryos (penis and scrotum in males and clitoris, labia, and vaginal orifice in females) also remain undifferentiated until about the eighth week. Before differentiation, all embryos have the following external structures (Figure 28.28):

1. **Urethral (urogenital) folds.** These paired structures develop from mesoderm in the cloacal region (see Figure 26.23).

2. **Urethral groove.** An indentation between the urethral folds, which is the opening into the urogenital sinus.

Figure 28.28 Development of the external genitals.

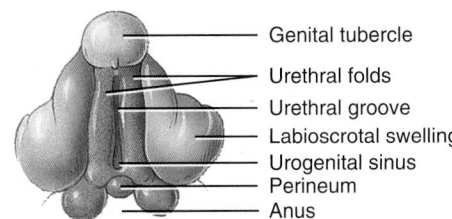
The external genitals of male and female embryos remain undifferentiated until about the eighth week.

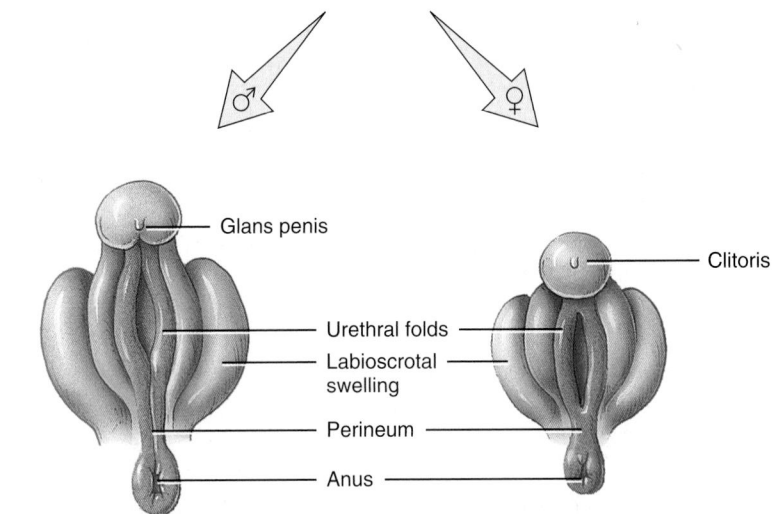

? Which hormone is responsible for the differentiation of the external genitals?

3. Genital tubercle. A rounded elevation just anterior to the urethral folds.

4. Labioscrotal swelling (lā-bē-ō-SKRŌ-tal). Paired, elevated structures lateral to the urethral folds.

In male embryos, some testosterone is converted to a second androgen called **dihydrotestosterone (DHT).** DHT stimulates development of the urethra, prostate, and external genitals (scrotum and penis). Part of the genital tubercle elongates and develops into a penis. Fusion of the urethral folds forms the *spongy (penile) urethra* and leaves an opening to the exterior only at the distal end of the penis, the *external urethral orifice.* The labioscrotal swellings develop into the *scrotum.* In the absence of DHT, the genital tubercle gives rise to the *clitoris* in female embryos. The urethral folds remain open as the *labia minora,* and the labioscrotal swellings become the *labia majora.* The urethral groove becomes the *vestibule.* After birth, androgen levels decline because hCG is no longer present to stimulate secretion of testosterone.

✔ CHECKPOINT

32. Describe the role of hormones in differentiation of the gonads, the mesonephric ducts, the paramesonephric ducts, and the external genitals.

28.6 AGING AND THE REPRODUCTIVE SYSTEMS

◉ OBJECTIVE

• Describe the effects of aging on the reproductive systems.

During the first decade of life, the reproductive system is in a juvenile state. At about age 10, hormone-directed changes start to occur in both sexes. **Puberty** (PŪ-ber-tē = a ripe age) is the period when secondary sexual characteristics begin to develop and the potential for sexual reproduction is reached. The onset of puberty is marked by pulses or bursts of LH and FSH secretion, each triggered by a pulse of GnRH. Most pulses occur during sleep. As puberty advances, the hormone pulses occur during the day as well as at night. The pulses increase in frequency during a 3- to 4-year period until the adult pattern is established. The stimuli that cause the GnRH pulses are still unclear, but a role for the hormone leptin is starting to unfold. Just before puberty, leptin levels rise in proportion to adipose tissue mass. Interestingly, leptin receptors are present in both the hypothalamus and anterior pituitary. Mice that lack a functional leptin gene from birth are sterile and remain in a prepubertal state. Giving leptin to such mice elicits secretion of gonadotropins, and they become fertile. Leptin may signal the hypothalamus that long-term energy stores (triglycerides in adipose tissue) are adequate for reproductive functions to begin.

In females, the reproductive cycle normally occurs once each month from **menarche** (me-NAR-kē), the first menses, to **menopause,** the permanent cessation of menses. Thus, the female reproductive system has a time-limited span of fertility between menarche and menopause. For the first 1 to 2 years after menarche, ovulation only occurs in about 10% of the cycles and the luteal phase is short. Gradually, the percentage of ovulatory cycles increases, and the luteal phase reaches its normal duration of 14 days. With age, fertility declines. Between the ages of 40 and 50 the pool of remaining ovarian follicles becomes exhausted. As a result, the ovaries become less responsive to hormonal stimulation. The production of estrogens declines, despite copious secretion of FSH and LH by the anterior pituitary. Many women experience hot flashes and heavy sweating, which coincide with bursts of GnRH release. Other symptoms of menopause are headache, hair loss, muscular pains, vaginal dryness, insomnia, depression, weight gain, and mood swings. Some atrophy of the ovaries, uterine tubes, uterus, vagina, external genitalia, and breasts occurs in postmenopausal women. Due to loss of estrogens, most women experience a decline in bone mineral density after menopause. Sexual desire (libido) does not show a parallel decline; it may be maintained by adrenal sex steroids. The risk of having uterine cancer peaks at about 65 years of age, but cervical cancer is more common in younger women.

In males, declining reproductive function is much more subtle than in females. Healthy men often retain reproductive capacity into their eighties or nineties. At about age 55 a decline in testosterone synthesis leads to reduced muscle strength, fewer viable sperm, and decreased sexual desire. Although sperm production decreases 50–70% between ages 60 and 80, abundant sperm may still be present even in old age.

Enlargement of the prostate to two to four times its normal size occurs in most males over age 60. This condition, called **benign prostatic hyperplasia (BPH)** (hī-per-PLĀ-zē-a), decreases the size of the prostatic urethra and is characterized by frequent urination, nocturia (bed-wetting), hesitancy in urination, decreased force of urinary stream, postvoiding dribbling, and a sensation of incomplete emptying.

✔ CHECKPOINT

33. What changes occur in males and females at puberty?
34. What do the terms menarche and menopause mean?

• • •

To appreciate the many ways that the reproductive systems contribute to homeostasis of other body systems, examine *Focus on Homeostasis: The Reproductive Systems.* Next, in Chapter 29, you will explore the major events that occur during pregnancy and you will discover how genetics (inheritance) plays a role in the development of a child.

BODY SYSTEM	CONTRIBUTION OF THE REPRODUCTIVE SYSTEMS

For all body systems
The male and female reproductive systems produce gametes (oocytes and sperm) that unite to form embryos and fetuses, which contain cells that divide and differentiate to form all of the organ systems of the body.

Integumentary systems
Androgens promote the growth of body hair. Estrogens stimulate the deposition of fat in the breasts, abdomen, and hips. Mammary glands produce milk. Skin stretches during pregnancy as the fetus enlarges.

Skeletal system
Androgens and estrogens stimulate the growth and maintenance of bones of the skeletal system.

Muscular system
Androgens stimulate the growth of skeletal muscles.

Nervous system
Androgens influence libido (sex drive). Estrogens may play a role in the development of certain regions of the brain in males.

Endocrine system
Testosterone and estrogens exert feedback effects on the hypothalamus and anterior pituitary gland.

Cardiovascular system
Estrogens lower blood cholesterol level and may reduce the risk of coronary artery disease in women under age 50.

Lymphatic systems and immunity
The presence of an antibiotic-like chemical in semen and the acidic pH of vaginal fluid provide innate immunity against microbes in the reproductive tract.

Respiratory system
Sexual arousal increases the rate and depth of breathing.

Digestive system
The presence of the fetus during pregnancy crowds the digestive organs, which leads to heartburn and constipation.

Urinary system
In males, the portion of the urethra that extends through the prostate and penis is a passageway for urine as well as semen.

THE REPRODUCTIVE SYSTEMS

1171

DISORDERS: HOMEOSTATIC IMBALANCES

Reproductive System Disorders in Males

Testicular Cancer

Testicular cancer is the most common cancer in males between the ages of 20 and 35. More than 95% of testicular cancers arise from spermatogenic cells within the seminiferous tubules. An early sign of testicular cancer is a mass in the testis, often associated with a sensation of testicular heaviness or a dull ache in the lower abdomen; pain usually does not occur. To increase the chance for early detection of a testicular cancer, all males should perform regular self-examinations of the testes. The examination should be done starting in the teen years and once each month thereafter. After a warm bath or shower (when the scrotal skin is loose and relaxed) each testis should be examined as follows. The testis is grasped and gently rolled between the index finger and thumb, feeling for lumps, swellings, hardness, or other changes. If a lump or other change is detected, a physician should be consulted as soon as possible.

Prostate Disorders

Because the prostate surrounds part of the urethra, any infection, enlargement, or tumor can obstruct the flow of urine. Acute and chronic infections of the prostate are common in postpubescent males, often in association with inflammation of the urethra. Symptoms may include fever, chills, urinary frequency, frequent urination at night, difficulty in urinating, burning or painful urination, low back pain, joint and muscle pain, blood in the urine, or painful ejaculation. However, often there are no symptoms. Antibiotics are used to treat most cases that result from a bacterial infection. In **acute prostatitis,** the prostate becomes swollen and tender. **Chronic prostatitis** is one of the most common chronic infections in men of the middle and later years. On examination, the prostate feels enlarged, soft, and very tender, and its surface outline is irregular.

Prostate cancer is the leading cause of death from cancer in men in the United States, having surpassed lung cancer in 1991. Each year it is diagnosed in almost 200,000 U.S. men and causes nearly 40,000 deaths. The amount of PSA (prostate-specific antigen), which is produced only by prostate epithelial cells, increases with enlargement of the prostate and may indicate infection, benign enlargement, or prostate cancer. A blood test can measure the level of PSA in the blood. Males over the age of 40 should have an annual examination of the prostate gland. In a **digital rectal exam,** a physician palpates the gland through the rectum with the fingers (digits). Many physicians also recommend an annual PSA test for males over age 50. Treatment for prostate cancer may involve surgery, cryotherapy, radiation, hormonal therapy, and chemotherapy. Because many prostate cancers grow very slowly, some urologists recommend "watchful waiting" before treating small tumors in men over age 70.

Erectile Dysfunction

Erectile dysfunction (ED), previously termed **impotence,** is the consistent inability of an adult male to ejaculate or to attain or hold an erection long enough for sexual intercourse. Many cases of impotence are caused by insufficient release of nitric oxide (NO), which relaxes the smooth muscle of the penile arterioles and erectile tissue. The drug *Viagra*® (sildenafil) enhances smooth muscle relaxation by nitric oxide in the penis. Other causes of erectile dysfunction include diabetes mellitus, physical abnormalities of the penis, systemic disorders such as syphilis, vascular disturbances (arterial or venous obstructions), neurological disorders, surgery, testosterone deficiency, and drugs (alcohol, antidepressants, antihistamines, antihypertensives, narcotics, nicotine, and tranquilizers). Psychological factors such as anxiety or depression, fear of causing pregnancy, fear of sexually transmitted diseases, religious inhibitions, and emotional immaturity may also cause ED.

Reproductive System Disorders in Females

Premenstrual Syndrome and Premenstrual Dysphoric Disorder

Premenstrual syndrome (PMS) is a cyclical disorder of severe physical and emotional distress. It appears during the postovulatory (luteal) phase of the female reproductive cycle and dramatically disappears when menstruation begins. The signs and symptoms are highly variable from one woman to another. They may include edema, weight gain, breast swelling and tenderness, abdominal distension, backache, joint pain, constipation, skin eruptions, fatigue and lethargy, greater need for sleep, depression or anxiety, irritability, mood swings, headache, poor coordination and clumsiness, and cravings for sweet or salty foods. The cause of PMS is unknown. For some women, getting regular exercise; avoiding caffeine, salt, and alcohol; and eating a diet that is high in complex carbohydrates and lean proteins can bring considerable relief.

Premenstrual dysphoric disorder (PMDD) is a more severe syndrome in which PMS-like signs and symptoms do not resolve after the onset of menstruation. Clinical research studies have found that suppression of the reproductive cycle by a drug that interferes with GnRH (leuprolide) decreases symptoms significantly. Because symptoms reappear when estradiol or progesterone is given together with leuprolide, researchers propose that PMDD is caused by abnormal responses to normal levels of these ovarian hormones. *SSRIs* (selective serotonin reuptake inhibitors) have shown promise in treating both PMS and PMDD.

Endometriosis

Endometriosis (en'-dō-MĒ-trē-o'-sis; *endo-* = within; *metri-* = uterus; *-osis* = condition) is characterized by the growth of endometrial tissue outside the uterus. The tissue enters the pelvic cavity via the open uterine tubes and may be found in any of several sites—on the ovaries, the rectouterine pouch, the outer surface of the uterus, the sigmoid colon, pelvic and abdominal lymph nodes, the cervix, the abdominal wall, the kidneys, and the urinary bladder. Endometrial tissue responds to hormonal fluctuations, whether it is inside or outside the uterus. With each reproductive cycle, the tissue proliferates and then breaks down and bleeds. When this occurs outside the uterus, it can cause inflammation, pain, scarring, and infertility. Symptoms include premenstrual pain or unusually severe menstrual pain.

Breast Cancer

One in eight women in the United States faces the prospect of **breast cancer.** After lung cancer, it is the second-leading cause of death from cancer in U.S. women. Breast cancer can occur in males but is rare. In females, breast cancer is seldom seen before age 30; its incidence rises rapidly after menopause. An estimated 5% of the 180,000 cases diagnosed each year in the United States, particularly those that arise in younger women, stem from inherited genetic mutations (changes in the DNA). Researchers have now identified two genes that increase susceptibility to breast cancer: *BRCA*1 (*breast cancer* 1) and *BRCA*2. Mutation of *BRCA*1 also confers a high risk for ovarian cancer. In addition, mutations of the *p*53 gene increase the risk of breast cancer in both males and females, and mutations of the androgen receptor gene are associated with the occurrence of breast cancer in some males. Because breast cancer generally is not painful until it becomes quite advanced, any lump, no matter how small, should be reported to a physician at once. Early detection—by breast self-examination and mammograms—is the best way to increase the chance of survival.

The most effective technique for detecting tumors less than 1 cm (0.4 in.) in diameter is **mammography** (mam-OG-ra-fē; -*graphy* = to record), a type of radiography using very sensitive x-ray film. The image of the breast, called a **mammogram** (see Table 1.3), is best obtained by compressing the breasts, one at a time, using flat plates. A supplementary procedure for evaluating breast abnormalities is **ultrasound.** Although ultrasound cannot detect tumors smaller than 1 cm in diameter (which mammography can detect), it can be used to determine whether a lump is a benign, fluid-filled cyst or a solid (and therefore possibly malignant) tumor.

Among the factors that increase the risk of developing breast cancer are (1) a family history of breast cancer, especially in a mother or sister; (2) nulliparity (never having borne a child) or having a first child after age 35; (3) previous cancer in one breast; (4) exposure to ionizing radiation, such as x-rays; (5) excessive alcohol intake; and (6) cigarette smoking.

The American Cancer Society recommends the following steps to help diagnose breast cancer as early as possible:

- All women over 20 should develop the habit of monthly breast self-examination.

- A physician should examine the breasts every 3 years when a woman is between the ages of 20 and 40, and every year after age 40.

- A mammogram should be taken in women between the ages of 35 and 39, to be used later for comparison (baseline mammogram).

- Women with no symptoms should have a mammogram every year after age 40.

- Women of any age with a history of breast cancer, a strong family history of the disease, or other risk factors should consult a physician to determine a schedule for mammography.

In November 2009, the United States Preventive Services Task Force (USPSTF) issued a series of recommendations relative to breast cancer screening for females at normal risk for breast cancer, that is, for females who have no signs or symptoms of breast cancer and who are not at a higher risk for breast cancer (for example, no family history). These recommendations are as follows:

- Women aged 50–74 should have a mammogram every 2 years.

- Women over 75 should not have mammograms.

- Breast self-examination is not required.

Treatment for breast cancer may involve hormone therapy, chemotherapy, radiation therapy, **lumpectomy** (lump-EK-tō-mē) (removal of the tumor and the immediate surrounding tissue), a modified or radical mastectomy, or a combination of these approaches. A **radical mastectomy** (mas-TEK-tō-mē; *mast-* = breast) involves removal of the affected breast along with the underlying pectoral muscles and the axillary lymph nodes. (Lymph nodes are removed because metastasis of cancerous cells usually occurs through lymphatic or blood vessels.) Radiation treatment and chemotherapy may follow the surgery to ensure the destruction of any stray cancer cells.

Several types of chemotherapeutic drugs are used to decrease the risk of relapse or disease progression. *Nolvadex® (tamoxifen)* is an antagonist to estrogens that binds to and blocks receptors for estrogens, thus decreasing the stimulating effect of estrogens on breast cancer cells. Tamoxifen has been used for 20 years and greatly reduces the risk of cancer recurrence. *Herceptin®*, a monoclonal antibody drug, targets an antigen on the surface of breast cancer cells. It is effective in causing regression of tumors and retarding progression of the disease. The early data from clinical trials of two new drugs, *Femara®* and *Amimidex®*, show relapse rates that are lower than those for tamoxifen. These drugs are inhibitors of aromatase, the enzyme needed for the final step in synthesis of estrogens. Finally, two drugs—tamoxifen and *Evista® (raloxifene)*—are being marketed for breast cancer *prevention*. Interestingly, raloxifene blocks estrogen receptors in the breasts and uterus but activates estrogen receptors in bone. Thus, it can be used to treat osteoporosis without increasing a woman's risk of breast or endometrial (uterine) cancer.

Ovarian Cancer

Even though **ovarian cancer** is the sixth most common form of cancer in females, it is the leading cause of death from all gynecological malignancies (excluding breast cancer) because it is difficult to detect before it metastasizes (spreads) beyond the ovaries. Risk factors associated with ovarian cancer include age (usually over age 50); race (whites are at highest risk); family history of ovarian cancer; more than 40 years of active ovulation; nulliparity or first pregnancy after age 30; a high-fat, low-fiber, vitamin A–deficient diet; and prolonged exposure to asbestos or talc. Early ovarian cancer has no symptoms or only mild ones associated with other common problems, such as abdominal discomfort, heartburn, nausea, loss of appetite, bloating, and flatulence. Later-stage signs and symptoms include an enlarged abdomen, abdominal and/or pelvic pain, persistent gastrointestinal disturbances, urinary complications, menstrual irregularities, and heavy menstrual bleeding.

Cervical Cancer

Cervical cancer, carcinoma of the cervix of the uterus, starts with **cervical dysplasia** (dis-PLĀ-sē-a), a change in the shape, growth, and number of cervical cells. The cells may either return to normal or progress to cancer. In most cases, cervical cancer may be detected in its earliest stages by a Pap test (see Clinical Connection: Papanicolaou Test in Chapter 4). Some evidence links cervical cancer to the virus that causes genital warts, human papillomavirus (HPV). Increased risk is associated with having a large number of sexual partners, having first intercourse at a young age, and smoking cigarettes.

Vulvovaginal Candidiasis

Candida albicans is a yeastlike fungus that commonly grows on mucous membranes of the gastrointestinal and genitourinary tracts. The organism is responsible for **vulvovaginal candidiasis** (vul-vō-VAJ-i-nal can-di-DĪ-a-sis), the most common form of **vaginitis** (vaj-i-NĪ-tis), inflammation of the vagina. Candidiasis is characterized by severe itching; a thick, yellow, cheesy discharge; a yeasty odor; and pain. The disorder, experienced at least once by about 75% of females, is usually a result of proliferation of the fungus following antibiotic therapy for another condition. Predisposing conditions include the use of oral contraceptives or cortisone-like medications, pregnancy, and diabetes.

Sexually Transmitted Diseases

A **sexually transmitted disease (STD)** is one that is spread by sexual contact. In most developed countries of the world, such as those of Western Europe, Japan, Australia, and New Zealand, the incidence of STDs has declined markedly during the past 25 years. In the United

States, by contrast, STDs have been rising to near-epidemic proportions; they currently affect more than 65 million people. AIDS and hepatitis B, which are sexually transmitted diseases that also may be contracted in other ways, are discussed in Chapters 22 and 24, respectively.

Chlamydia

Chlamydia (kla-MID-ē-a) is a sexually transmitted disease caused by the bacterium *Chlamydia trachomatis* (*chlamy-* = cloak). This unusual bacterium cannot reproduce outside body cells; it "cloaks" itself inside cells, where it divides. At present, chlamydia is the most prevalent sexually transmitted disease in the United States. In most cases, the initial infection is asymptomatic and thus difficult to recognize clinically. In males, urethritis is the principal result, causing a clear discharge, burning on urination, frequent urination, and painful urination. Without treatment, the epididymides may also become inflamed, leading to sterility. In 70% of females with chlamydia, symptoms are absent, but chlamydia is the leading cause of pelvic inflammatory disease. The uterine tubes may also become inflamed, which increases the risk of ectopic pregnancy (implantation of a fertilized ovum outside the uterus) and infertility due to the formation of scar tissue in the tubes.

Gonorrhea

Gonorrhea (gon-ō-RĒ-a) or **"the clap"** is caused by the bacterium *Neisseria gonorrhoeae*. In the United States, 1 million to 2 million new cases of gonorrhea appear each year, most among individuals aged 15–29 years. Discharges from infected mucous membranes are the source of transmission of the bacteria either during sexual contact or during the passage of a newborn through the birth canal. The infection site can be in the mouth and throat after oral–genital contact, in the vagina and penis after genital intercourse, or in the rectum after recto–genital contact.

Males usually experience urethritis with profuse pus drainage and painful urination. The prostate and epididymis may also become infected. In females, infection typically occurs in the vagina, often with a discharge of pus. Both infected males and females may harbor the disease without any symptoms, however, until it has progressed to a more advanced stage; about 5–10% of males and 50% of females are asymptomatic. In females, the infection and consequent inflammation can proceed from the vagina into the uterus, uterine tubes, and pelvic cavity. An estimated 50,000 to 80,000 women in the United States are made infertile by gonorrhea every year as a result of scar tissue formation that closes the uterine tubes. If bacteria in the birth canal are transmitted to the eyes of a newborn, blindness can result. Administration of a 1% silver nitrate solution in the infant's eyes prevents infection.

Syphilis

Syphilis, caused by the bacterium *Treponema pallidum* (trep-o-NĒ-ma PAL-i-dum), is transmitted through sexual contact or exchange of blood, or through the placenta to a fetus. The disease progresses through several stages. During the *primary stage,* the chief sign is a painless open sore, called a **chancre** (SHANG-ker), at the point of contact. The chancre heals within 1 to 5 weeks. From 6 to 24 weeks later, signs and symptoms such as a skin rash, fever, and aches in the joints and muscles usher in the *secondary stage,* which is systemic— the infection spreads to all major body systems. When signs of organ degeneration appear, the disease is said to be in the *tertiary stage.* If the nervous system is involved, the tertiary stage is called **neurosyphilis.** As motor areas become damaged extensively, victims may be unable to control urine and bowel movements. Eventually they may become bedridden and unable even to feed themselves. In addition, damage to the cerebral cortex produces memory loss and personality changes that range from irritability to hallucinations.

Genital Herpes

Genital herpes is an incurable STD. Type II herpes simplex virus (HSV-2) causes genital infections, producing painful blisters on the prepuce, glans penis, and penile shaft in males and on the vulva or sometimes high up in the vagina in females. The blisters disappear and reappear in most patients, but the virus itself remains in the body. A related virus, type I herpes simplex virus (HSV-1), causes cold sores on the mouth and lips. Infected individuals typically experience recurrences of symptoms several times a year.

Genital Warts

Warts are an infectious disease caused by viruses. *Human papillomavirus (HPV)* causes **genital warts,** which can be transmitted sexually. Nearly 1 million people a year develop genital warts in the United States. Patients with a history of genital warts may be at increased risk for cancers of the cervix, vagina, anus, vulva, and penis. There is no cure for genital warts. A vaccine (Gardasil®) against certain types of HPV that cause cervical cancer and genital warts is available and recommended for 11- and 12-year-old girls.

MEDICAL TERMINOLOGY

Castration (kas-TRĀ-shun = to prune) Removal, inactivation, or destruction of the gonads; commonly used in reference to removal of the testes only.

Colposcopy (kol-POS-kō-pē; *colpo-* = vagina; *-scopy* = to view) Visual inspection of the vagina and cervix of the uterus using a culposcope, an instrument that has a magnifying lens (between 5× and 50×) and a light. The procedure generally takes place after an unusual Pap smear.

Culdoscopy (kul-DOS-kō-pē; *-cul-* = cul-de-sac; *-scopy* = to examine) A procedure in which a culdoscope (endoscope) is inserted through the posterior wall of the vagina to view the rectouterine pouch in the pelvic cavity.

Dysmenorrhea (dis-men-ōr-Ē-a; *dys-* = difficult or painful) Pain associated with menstruation; the term is usually reserved to describe menstrual symptoms that are severe enough to prevent a woman from functioning normally for one or more days each month.

Some cases are caused by uterine tumors, ovarian cysts, pelvic inflammatory disease, or intrauterine devices.

Dyspareunia (dis-pa-ROO-nē-a; *dys-* = difficult; *-para-* = beside; *-enue* = bed) Pain during sexual intercourse. It may occur in the genital area or in the pelvic cavity, and may be due to inadequate lubrication, inflammation, infection, an improperly fitting diaphragm or cervical cap, endometriosis, pelvic inflammatory disease, pelvic tumors, or weakened uterine ligaments.

Endocervical curettage (kū-re-TAHZH; *curette* = scraper) A procedure in which the cervix is dilated and the endometrium of the uterus is scraped with a spoon-shaped instrument called a curette; commonly called a *D and C* (dilation and curettage).

Fibroids (FĪ-broyds; *fibro-* = fiber; *-eidos* = resemblance) Noncancerous tumors in the myometrium of the uterus composed of muscular and fibrous tissue. Their growth appears to be related to high levels of estrogens. They do not occur before puberty and usually stop growing after menopause. Symptoms include abnormal menstrual bleeding and pain or pressure in the pelvic area.

Hermaphroditism (her-MAF-rō-dĭt-izm) The presence of both ovarian and testicular tissue in one individual.

Hypospadias (hī′-pō-SPĀ-dē-as; *hypo-* = below) A common congenital abnormality in which the urethral opening is displaced. In males, the displaced opening may be on the underside of the penis, at the penoscrotal junction, between the scrotal folds, or in the perineum; in females, the urethra opens into the vagina. The problem can be corrected surgically.

Leukorrhea (loo′-kō-RĒ-a; *leuko-* = white) A whitish (nonbloody) vaginal discharge containing mucus and pus cells that may occur at any age and affects most women at some time.

Menorrhagia (men-ō-RA-jē-a; *meno-* = menstruation; *-rhage* = to burst forth) Excessively prolonged or profuse menstrual period. May be due to a disturbance in hormonal regulation of the menstrual cycle, pelvic infection, medications (anticoagulants), fibroids (noncancerous uterine tumors composed of muscle and fibrous tissue), endometriosis, or intrauterine devices.

Oophorectomy (ō′-of-ō-REK-tō-mē; *oophor-* = bearing eggs) Removal of the ovaries.

Orchitis (or-KĪ-tis; *orchi-* = testes; *-itis* = inflammation) Inflammation of the testes, for example, as a result of the mumps virus or a bacterial infection.

Ovarian cyst The most common form of ovarian tumor, in which a fluid-filled follicle or corpus luteum persists and continues growing.

Pelvic inflammatory disease (PID) A collective term for any extensive bacterial infection of the pelvic organs, especially the uterus, uterine tubes, or ovaries, which is characterized by pelvic soreness, lower back pain, abdominal pain, and urethritis. Often the early symptoms of PID occur just after menstruation. As infection spreads, fever may develop, along with painful abscesses of the reproductive organs.

Salpingectomy (sal′-pin-JEK-tō-mē; *salpingo* = tube) Removal of a uterine (fallopian) tube.

Smegma (SMEG-ma) The secretion, consisting principally of desquamated epithelial cells, found chiefly around the external genitals and especially under the foreskin of the male.

CHAPTER REVIEW AND RESOURCE SUMMARY

Review

Resource

Introduction

28.1 Male Reproductive System

1. Reproduction is the process by which new individuals of a species are produced and the genetic material is passed from generation to generation.

2. The organs of reproduction are grouped as gonads (produce gametes), ducts (transport and store gametes), accessory sex glands (produce materials that support gametes), and supporting structures (have various roles in reproduction). The male structures of reproduction include the testes (2), epididymidis (2), ducti (vasa) deferentia (2), ejaculatory ducts (2), seminal vesicles (2), urethra (1), prostate (1), bulbourethral (Cowper's) glands (2), and penis (1). The scrotum is a sac that hangs from the root of the penis and consists of loose skin and underlying subcutaneous layer; it supports the testes. The temperature of the testes is regulated by the cremaster muscles, which either contract to elevate the testes and move them closer to the pelvic cavity or relax and move them farther from the pelvic cavity. The dartos muscle causes the scrotum to become tight and wrinkled.

3. The testes are paired oval glands (gonads) in the scrotum containing seminiferous tubules, in which sperm cells are made; Sertoli cells (sustentacular cells), which nourish sperm cells and secrete inhibin; and Leydig (interstitial) cells, which produce the male sex hormone testosterone. The testes descend into the scrotum through the inguinal canals during the seventh month of fetal development. Failure of the testes to descend is called cryptorchidism.

4. Secondary oocytes and sperm, both of which are called gametes, are produced in the gonads. Spermatogenesis, which occurs in the testes, is the process whereby immature spermatogonia develop into sperm. The spermatogenesis sequence, which includes meiosis I, meiosis II, and spermiogenesis, results in the formation of four haploid sperm (spermatozoa) from each primary spermatocyte. Mature sperm consist of a head and a tail. Their function is to fertilize a secondary oocyte.

5. At puberty, gonadotropin-releasing hormone (GnRH) stimulates anterior pituitary secretion of FSH and LH. LH stimulates production of testosterone; FSH and testosterone stimulate spermatogenesis. Sertoli cells secrete androgen-binding protein (ABP), which binds to testosterone and keeps its concentration high in the seminiferous tubule. Testosterone controls the growth, development, and maintenance of sex organs; stimulates bone growth, protein anabolism, and sperm maturation; and stimulates development of masculine secondary sex characteristics. Inhibin is produced by Sertoli cells; its inhibition of FSH helps regulate the rate of spermatogenesis.

6. The duct system of the testes includes the seminiferous tubules, straight tubules, and rete testis. Sperm flow out of the testes through the efferent ducts. The ductus epididymis is the site of sperm maturation and storage. The ductus (vas) deferens stores sperm and propels them toward the urethra during ejaculation.

Review

7. Each ejaculatory duct, formed by the union of the duct from the seminal vesicle and ampulla of the ductus (vas) deferens, is the passageway for ejection of sperm and secretions of the seminal vesicles into the first portion of the urethra, the prostatic urethra.

8. The urethra in males is subdivided into three portions: the prostatic, membranous, and spongy (penile) urethra.

9. The seminal vesicles secrete an alkaline, viscous fluid that contains fructose (used by sperm for ATP production). Seminal fluid constitutes about 60% of the volume of semen and contributes to sperm viability. The prostate secretes a slightly acidic fluid that constitutes about 25% of the volume of semen and contributes to sperm motility. The bulbourethral (Cowper's) glands secrete mucus for lubrication and an alkaline substance that neutralizes acid. Semen is a mixture of sperm and seminal fluid; it provides the fluid in which sperm are transported, supplies nutrients, and neutralizes the acidity of the male urethra and the vagina.

10. The penis consists of a root, a body, and a glans penis. Engorgement of the penile blood sinuses under the influence of sexual excitation is called erection.

28.2 Female Reproductive System

1. The female organs of reproduction include the ovaries (gonads), uterine (fallopian) tubes or oviducts, uterus, vagina, and vulva. The mammary glands are part of the integumentary system and also are considered part of the reproductive system in females.

2. The ovaries, the female gonads, are located in the superior portion of the pelvic cavity, lateral to the uterus. Ovaries produce secondary oocytes, discharge secondary oocytes (the process of ovulation), and secrete estrogens, progesterone, relaxin, and inhibin.

3. Oogenesis (the production of haploid secondary oocytes) begins in the ovaries. The oogenesis sequence includes meiosis I and meiosis II, which goes to completion only after an ovulated secondary oocyte is fertilized by a sperm cell.

4. The uterine (fallopian) tubes transport secondary oocytes from the ovaries to the uterus and are the normal sites of fertilization. Ciliated cells and peristaltic contractions help move a secondary oocyte or fertilized ovum toward the uterus.

5. The uterus is an organ the size and shape of an inverted pear that functions in menstruation, implantation of a fertilized ovum, development of a fetus during pregnancy, and labor. It also is part of the pathway for sperm to reach the uterine tubes to fertilize a secondary oocyte. Normally, the uterus is held in position by a series of ligaments. Histologically, the layers of the uterus are an outer perimetrium (serosa), a middle myometrium, and an inner endometrium.

6. The vagina is a passageway for sperm and the menstrual flow, the receptacle of the penis during sexual intercourse, and the inferior portion of the birth canal. It is capable of considerable stretching.

7. The vulva, a collective term for the external genitals of the female, consists of the mons pubis, labia majora, labia minora, clitoris, vestibule, vaginal and urethral orifices, hymen, and bulb of the vestibule, as well as three sets of glands: the paraurethral (Skene's), greater vestibular (Bartholin's), and lesser vestibular glands.

8. The perineum is a diamond-shaped area at the inferior end of the trunk medial to the thighs and buttocks.

9. The mammary glands are modified sweat glands lying superficial to the pectoralis major muscles. Their function is to synthesize, secrete, and eject milk (lactation).

10. Mammary gland development depends on estrogens and progesterone. Milk production is stimulated by prolactin, estrogens, and progesterone; milk ejection is stimulated by oxytocin.

28.3 The Female Reproductive Cycle

1. The function of the ovarian cycle is to develop a secondary oocyte; the function of the uterine (menstrual) cycle is to prepare the endometrium each month to receive a fertilized egg. The female reproductive cycle includes both the ovarian and uterine cycles.

2. The uterine and ovarian cycles are controlled by GnRH from the hypothalamus, which stimulates the release of FSH and LH by the anterior pituitary. FSH and LH stimulate development of follicles and secretion of estrogens by the follicles. LH also stimulates ovulation, formation of the corpus luteum, and the secretion of progesterone and estrogens by the corpus luteum.

3. Estrogens stimulate the growth, development, and maintenance of female reproductive structures; stimulate the development of secondary sex characteristics; and stimulate protein synthesis. Progesterone works with estrogens to prepare the endometrium for implantation and the mammary glands for milk synthesis.

Review	Resource

4. Relaxin relaxes the myometrium at the time of possible implantation. At the end of a pregnancy, relaxin increases the flexibility of the pubic symphysis and helps dilate the uterine cervix to facilitate delivery.

5. During the menstrual phase, the stratum functionalis of the endometrium is shed, discharging blood, tissue fluid, mucus, and epithelial cells.

6. During the preovulatory phase, a group of follicles in the ovaries begins to undergo final maturation. One follicle outgrows the others and becomes dominant while the others degenerate. At the same time, endometrial repair occurs in the uterus. Estrogens are the dominant ovarian hormones during the preovulatory phase.

7. Ovulation is the rupture of the mature (graafian) follicle and the release of a secondary oocyte into the pelvic cavity. It is brought about by a surge of LH. Signs and symptoms of ovulation include increased basal body temperature; clear, stretchy cervical mucus; changes in the uterine cervix; and abdominal pain.

8. During the postovulatory phase, both progesterone and estrogens are secreted in large quantity by the corpus luteum of the ovary, and the uterine endometrium thickens in readiness for implantation.

9. If fertilization and implantation do not occur, the corpus luteum degenerates, and the resulting low levels of progesterone and estrogens allow discharge of the endometrium followed by the initiation of another reproductive cycle.

10. If fertilization and implantation do occur, the corpus luteum is maintained by hCG. The corpus luteum and later the placenta secrete progesterone and estrogens to support pregnancy and breast development for lactation.

Resource: *Concepts and Connections - Regulation of Female Reproduction*

28.4   Birth Control Methods and Abortion

1. Methods include complete abstinence, surgical sterilization (vasectomy, tubal ligation), non-incisional sterilization, hormonal methods (combined pill, minipill, contraceptive skin patch, vaginal contraceptive ring, emergency contraception, hormonal injections), intrauterine devices, spermicides, barrier methods (male condom, vaginal pouch, diaphragm, cervical cap), and periodic abstinence (rhythm and sympto-thermal methods).

2. Contraceptive pills of the combination type contain progestin and estrogens in concentrations that decrease the secretion of FSH and LH and thereby inhibit development of ovarian follicles and ovulation, inhibit transport of ova and sperm in the uterine tubes, and block implantation in the uterus.

3. An abortion is the premature expulsion from the uterus of the products of conception; it may be spontaneous or induced.

28.5   Development of the Reproductive Systems

1. The gonads develop from gonadal ridges that arise from growth of intermediate mesoderm. In the presence of the *SRY* gene, the gonads begin to differentiate into testes during the seventh week. The gonads differentiate into ovaries when the *SRY* gene is absent.

2. In males, testosterone stimulates development of each mesonephric duct into an epididymis, ductus (vas) deferens, ejaculatory duct, and seminal vesicle, and Müllerian-inhibiting substance (MIS) causes the paramesonephric duct cells to die. In females, testosterone and MIS are absent; the paramesonephric ducts develop into the uterine tubes, uterus, and vagina and the mesonephric ducts degenerate.

3. The external genitals develop from the genital tubercle and are stimulated to develop into typical male structures by the hormone dihydrotestosterone (DHT). The external genitals develop into female structures when DHT is not produced, the normal situation in female embryos.

Resource: *Anatomy Overview - Developmental Stages*

28.6   Aging and the Reproductive Systems

1. Puberty is the period when secondary sex characteristics begin to develop and the potential for sexual reproduction is reached.

2. The onset of puberty is marked by pulses or bursts of LH and FSH secretion, each triggered by a pulse of GnRH. The hormone leptin, released by adipose tissue, may signal the hypothalamus that long-term energy stores (triglycerides in adipose tissue) are adequate for reproductive functions to begin.

3. In females, the reproductive cycle normally occurs once each month from menarche, the first menses, to menopause, the permanent cessation of menses.

4. Between the ages of 40 and 50, the pool of remaining ovarian follicles becomes exhausted and levels of progesterone and estrogens decline. Most women experience a decline in bone mineral density after menopause, together with some atrophy of the ovaries, uterine tubes, uterus, vagina, external genitalia, and breasts. Uterine and breast cancer increase in incidence with age.

Review

5. In older males, decreased levels of testosterone are associated with decreased muscle strength, waning sexual desire, and fewer viable sperm; prostate disorders are common.

SELF-QUIZ QUESTIONS

Fill in the blanks in the following statement.

1. The period of time when secondary sexual characteristics begin to develop and the potential for sexual reproduction is reached is called _____. The first menses is called _____, and the permanent cessation of menses is called _____.

Indicate whether the following statements are true or false.

2. Spermatogenesis does not occur at normal core body temperature.

3. The route of sperm from the production in the testes to the exterior of the body is seminiferous tubules, straight tubules, rete testes, epididymis, ductus (vas) deferens, ejaculatory duct, prostatic urethra, membranous urethra, spongy urethra, external urethral orifice.

Choose the one best answer to the following questions.

4. Which of the following are functions of Sertoli cells? (1) protection of developing spermatogenic cells, (2) nourishment of spermatocytes, spermatids, and sperm, (3) phagocytosis of excess sperm cytoplasm as development proceeds, (4) mediation of the effects of testosterone and FSH, (5) control of movements of spermatogenic cells and release of sperm into the lumen of seminiferous tubules.
   (a) 1, 2, 4, and 5   (b) 1, 2, 3, and 5   (c) 2, 3, 4, and 5
   (d) 1, 2, 3, and 4   (e) 1, 2, 3, 4, and 5

5. Which of the following are *true*? (1) An erection is a sympathetic response initiated by sexual stimulation. (2) Dilation of blood vessels supplying erectile tissue results in erection. (3) Nitric oxide causes smooth muscle within erectile tissue to relax, which results in widening of blood sinuses. (4) Ejaculation is a sympathetic reflex coordinated by the sacral region of the spinal cord. (5) The purpose of the corpus cavernosa penis is to keep the spongy urethra open during ejaculation.
   (a) 1, 2, and 3   (b) 1, 2, 3, 4, and 5   (c) 2 and 3
   (d) 2, 4, and 5   (e) 1, 2, 3, and 4

6. Which of the following are *true* concerning estrogens? (1) They promote development and maintenance of female reproductive structures and secondary sex characteristics. (2) They help control fluid and electrolyte balance. (3) They increase protein catabolism. (4) They lower blood cholesterol. (5) In moderate levels, they inhibit the release of GnRH and the secretion of LH and FSH.
   (a) 1, 4, and 5   (b) 1, 3, 4, and 5   (c) 1, 2, 3, and 5
   (d) 1, 2, 3, and 4   (e) 1, 2, 3, 4, and 5

7. Which of the following statements are *correct?* (1) A sperm head contains DNA and an acrosome. (2) An acrosome is a specialized lysosome that contains enzymes that enable sperm to produce the ATP needed to propel themselves out of the male reproductive tract. (3) Mitochondria in the midpiece of a sperm produce ATP for sperm motility. (4) A sperm's tail, a flagellum, propels it along its way. (5) Once ejaculated, sperm are viable and normally are able to fertilize a secondary oocyte for 5 days.
   (a) 1, 2, 3, and 4   (b) 2, 3, 4, and 5   (c) 1, 3, and 4
   (d) 2, 4, and 5   (e) 2, 3, and 4

8. Which of the following statements are *correct?* (1) Spermatogonia are stem cells because when they undergo mitosis, some of the daughter cells remain to serve as a reservoir of cells for future mitosis. (2) Meiosis I is a division of pairs of chromosomes resulting in daughter cells with only one member of each chromosome pair. (3) Meiosis II separates the chromatids of each chromosome. (4) Spermiogenesis involves the maturation of spermatids into sperm. (5) The process by which the seminiferous tubules produce haploid sperm is called spermatogenesis.
   (a) 1, 2, 3, and 5   (b) 1, 2, 3, 4, and 5   (c) 1, 3, 4, and 5
   (d) 1, 2, 3, and 4   (e) 1, 3, and 5

9. Which of the following statements are *correct?* (1) Cells from the yolk sac give rise to oogonia. (2) Ova arise from the germinal epithelium of the ovary. (3) Primary oocytes enter prophase of meiosis I during fetal development but do not complete it until after puberty. (4) Once a secondary oocyte is formed, it proceeds to metaphase of meiosis II and stops at this stage. (5) The secondary oocyte resumes meiosis II and forms the ovum and a polar body only if fertilization occurs. (6) A primary oocyte gives rise to an ovum and four polar bodies.
   (a) 1, 3, 4, and 5   (b) 1, 3, 4, and 6   (c) 1, 2, 4, and 6
   (d) 1, 2, 4, and 5   (e) 1, 2, 5, and 6

10. Which of the following statements are *correct?* (1) The female reproductive cycle consists of a menstrual phase, a preovulatory phase, ovulation, and a postovulatory phase. (2) During the menstrual phase, small secondary follicles in the ovary begin to enlarge while the uterus is shedding its lining. (3) During the preovulatory phase, a dominant follicle continues to grow and begins to secrete estrogens and inhibin while the uterine lining begins to rebuild. (4) Ovulation results in the release of an ovum and the shedding of the uterine lining to nourish and support the released ovum. (5) After ovulation, a corpus luteum forms from the ruptured follicle and begins to secrete progesterone and estrogens, which it will continue to do throughout pregnancy if the egg is fertilized. (6) If pregnancy does not occur, then the corpus luteum degenerates into a scar called the corpus albicans, and the uterine lining is prepared to be shed again.
   (a) 1, 2, 4, and 5   (b) 2, 4, 5, and 6   (c) 1, 4, 5, and 6
   (d) 1, 3, 4, and 6   (e) 1, 2, 3, and 6

11. Oral contraceptives work by (1) causing a thickening of the cervical mucus, (2) blocking the uterine tubes, (3) inhibiting the release of FSH and LH, (4) preventing ovulation, (5) disrupting the plasma membranes of sperm, (6) irritating the endometrial lining so that it is inhospitable for fetal development.
   (a) 3 only   (b) 3 and 4   (c) 1, 2, and 5
   (d) 1, 3, and 4   (e) 1, 2, 3, 4, and 5

Au/Ed: Questions reordered to save # - OK? Pls. correct answer section.

SELF-QUIZ QUESTIONS **1179**

12. Match the following:

_____ (a) the process during meiosis when portions of homologous chromosomes may be exchanged with each other

_____ (b) refers to cells containing one-half the chromosome number

_____ (c) the cell produced by the union of an egg and a sperm

_____ (d) the degeneration of oogonia before and after birth

_____ (e) a packet of discarded nuclear material from the first or second meiotic division of the egg

_____ (f) refers to cells containing the full chromosome number

(1) zygote
(2) haploid
(3) diploid
(4) crossing-over
(5) polar body
(6) atresia

13. Match the following:

_____ (a) modified sudoriferous glands involved in lactation

_____ (b) a small, cylindrical mass of erectile tissue and nerves in the female; homologue of the male glans penis

_____ (c) produce mucus in the female during sexual arousal and intercourse; homologous to the male bulbourethral glands

_____ (d) the group of cells that nourish the developing oocyte and begin to secrete estrogens

_____ (e) a pathway for sperm to reach the uterine tubes; the site of menstruation; the site of implantation of a fertilized ovum; the womb

_____ (f) produces progesterone, estrogens, relaxin, and inhibin

_____ (g) draw the ovum into the uterine tube

_____ (h) the opening between the uterus and vagina

_____ (i) muscular layer of uterus; responsible for expulsion of fetus from uterus

_____ (j) mucus-secreting glands in the female that are homologous to the prostate gland

_____ (k) the female copulatory organ; the birth canal

_____ (l) passageway for the ovum to the uterus; usual site of fertilization; site of tubal ligation

_____ (m) refers to the external genitals of the female

_____ (n) the layer of the uterine lining that is partially shed during each monthly cycle

(1) follicle
(2) corpus luteum
(3) uterine tube
(4) fimbriae
(5) uterus
(6) cervix
(7) endometrium
(8) vagina
(9) vulva
(10) clitoris
(11) paraurethral glands
(12) greater vestibular glands
(13) mammary glands
(14) myometrium

14. Match the following:

_____ (a) site of sperm maturation

_____ (b) the male copulatory organ; a passageway for ejaculation of sperm and excretion of urine

_____ (c) sperm-forming cells

_____ (d) produce an alkaline substance that protects sperm by neutralizing acids in the urethra

_____ (e) ejects sperm into the urethra just before ejaculation

_____ (f) the supporting structure for the testes

_____ (g) carries the sperm from the scrotum into the abdominopelvic cavity for release by ejaculation; is cut and tied as a means of sterilization

_____ (h) the shared terminal duct of the reproductive and urinary systems in the male

_____ (i) surrounds the urethra at the base of the bladder; produces secretions that contribute to sperm motility and viability

_____ (j) produce testosterone

_____ (k) supporting structure that consists of the ductus deferens, testicular artery, autonomic nerves, veins that drain the testes, lymphatic vessels, and cremaster muscle

_____ (l) support and protect developing spermatogenic cells; secrete inhibin; form the blood–testis barrier

_____ (m) secrete an alkaline fluid to help neutralize acids in the female reproductive tract; secrete fructose for use in ATP production by sperm

_____ (n) contraction and relaxation move testes near to or away from pelvic cavity

_____ (o) site of spermatogenesis

(1) spermatogenic cells
(2) Sertoli cells
(3) Leydig cells
(4) penis
(5) scrotum
(6) epididymis
(7) ductus deferens
(8) ejaculatory duct
(9) seminiferous tubules
(10) seminal vesicles
(11) prostate gland
(12) bulbourethral glands
(13) urethra
(14) spermatic cord
(15) cremaster muscle

CRITICAL THINKING QUESTIONS

1. Twenty-three-year-old Monica and her husband Bill are ready to start a family. They are both avid bicyclists and weight-lifters who carefully watch what they eat and pride themselves on their "buff" bodies. However, Monica is having difficulty becoming pregnant. Monica hasn't had a menstrual period for some time but informs the doctor that is normal for her. After consulting with her physician, the doctor tells Monica that she needs to cut back on her exercise routine and "put on some weight" in order to get pregnant. Monica is outraged because she figures she will gain enough weight when she is pregnant! Explain to Monica what has happened to her and why weight gain could help her achieve her goal of pregnancy.

2. The term "progesterone" means "for gestation (or pregnancy)." Describe how progesterone helps prepare the female body for pregnancy and helps maintain pregnancy.

3. After having borne five children, Mark's wife, Isabella, insists that he have a vasectomy. Mark is afraid that he will "dry up" and won't be able to perform sexually. How can you reassure him that his reproductive organs will function fine?

? ANSWERS TO FIGURE QUESTIONS

28.1 The gonads (testes) produce gametes (sperm) and hormones; the ducts transport, store, and receive gametes; the accessory sex glands secrete materials that support gametes; and the penis assists in the delivery and joining of gametes.

28.2 The cremaster and dartos muscles help regulate the temperature of the testes.

28.3 The tunica vaginalis and tunica albuginea are tissue layers that cover and protect the testes.

28.4 The Leydig (interstitial) cells of the testes secrete testosterone.

28.5 As a result of meiosis I, the number of chromosomes in each cell is reduced by half.

28.6 The sperm head contains the nucleus with 23 highly condensed chromosomes and an acrosome that contains enzymes for penetration of a secondary oocyte; the neck contains centrioles that produce microtubules for the rest of the tail; the midpiece contains mitochondria for ATP production for locomotion and metabolism; the principal and end pieces of the tail provide motility.

28.7 Sertoli cells secrete inhibin.

28.8 Testosterone inhibits secretion of LH, and inhibin inhibits secretion of FSH.

28.9 The seminal vesicles are the accessory sex glands that contribute the largest volume to seminal fluid.

28.10 Two tissue masses called the corpora cavernosa penis and one corpus spongiosum penis contain blood sinuses that fill with blood that cannot flow out of the penis as quickly as it flows in. The trapped blood engorges and stiffens the tissue, producing an erection. The corpus spongiosum penis keeps the spongy urethra open so that ejaculation can occur.

28.11 The testes are homologous to the ovaries; the glans penis is homologous to the clitoris; the prostate is homologous to the paraurethral glands; and the bulbourethral glands are homologous to the greater vestibular glands.

28.12 The mesovarium anchors the ovary to the broad ligament of the uterus and the uterine tube; the ovarian ligament anchors it to the uterus; the suspensory ligament anchors it to the pelvic wall.

28.13 Ovarian follicles secrete estrogens; the corpus luteum secretes progesterone, estrogens, relaxin, and inhibin.

28.14 Most ovarian follicles undergo atresia (degeneration).

28.15 Primary oocytes are present in the ovary at birth, so they are as old as the woman. In males, primary spermatocytes are continu-

ally being formed from stem cells (spermatogonia) and thus are only a few days old.

28.16 Fertilization most often occurs in the ampulla of the uterine tube.

28.17 Ciliated columnar epithelial cells and nonciliated (peg) cells with microvilli line the uterine tubes.

28.18 The endometrium is a highly vascularized, secretory epithelium that provides the oxygen and nutrients needed to sustain a fertilized egg; the myometrium is a thick smooth muscle layer that supports the uterine wall during pregnancy and contracts to expel the fetus at birth.

28.19 The stratum basalis of the endometrium provides cells to replace those that are shed (the stratum functionalis) during each menstruation.

28.20 Anterior to the vaginal opening are the mons pubis, clitoris, prepuce, and external urethral orifice. Lateral to the vaginal opening are the labia minora and labia majora.

28.21 The anterior portion of the perineum is called the urogenital triangle because its borders form a triangle that encloses the urethral (uro-) and vaginal (-genital) orifices.

28.22 Prolactin, estrogens, and progesterone regulate the synthesis of milk. Oxytocin regulates the ejection of milk.

28.23 The principal estrogen is β-estradiol.

28.24 The hormones responsible for the proliferative phase of endometrial growth are estrogens; for ovulation, LH; for growth of the corpus luteum, LH; and for the midcycle surge of LH, estrogens.

28.25 The effect of rising but moderate levels of estrogens is negative feedback inhibition of the secretion of GnRH, LH, and FSH.

28.26 This is negative feedback, because the response is opposite to the stimulus. A reduced amount of negative feedback due to declining levels of estrogens and progesterone stimulates release of GnRH, which in turn increases the production and release of FSH and LH, ultimately stimulating the secretion of estrogens.

28.27 The *SRY* gene on the Y chromosome is responsible for the development of the gonads into testes.

28.28 The presence of dihydrotestosterone (DHT) stimulates differentiation of the external genitals in males; its absence allows differentiation of the external genitals in females.

29 DEVELOPMENT AND INHERITANCE

DEVELOPMENT, INHERITANCE, AND HOMEOSTASIS   Both the genetic material inherited from parents (heredity) and normal development in the uterus (environment) play important roles in determining the homeostasis of a developing embryo and fetus and the subsequent birth of a healthy child.

Developmental biology is the study of the sequence of events from the fertilization of a secondary oocyte by a sperm cell to the formation of an adult organism. **Pregnancy** is a sequence of events that begins with fertilization; proceeds to implantation, embryonic development, and fetal development; and ideally ends with birth about 38 weeks later, or 40 weeks after the last menstrual period.

Obstetrics (ob-STET-riks; *obstetrix* = midwife) deals with the management of pregnancy, labor, and the **neonatal period** (nē-ō-NĀ-tal), the first 28 days after birth. **Prenatal development** (prē-NĀ-tal; *pre-* = before; *-natal* = birth) is the time from fertilization to birth and is divided into three periods of three calendar months each, called **trimesters.**

1. The **first trimester** is the most critical stage of development, during which the rudiments of all the major organ systems appear, and also during which the developing organism is the most vulnerable to the effects of drugs, radiation, and microbes.
2. The **second trimester** is characterized by the nearly complete development of organ systems. By the end of this stage, the fetus assumes distinctively human features.
3. The **third trimester** represents a period of rapid fetal growth. During the early stages of this period, most of the organ systems are becoming fully functional.

In this chapter, we focus on the developmental sequence from fertilization through implantation, embryonic and fetal development, labor, birth, and the principles of inheritance (the passage of hereditary traits from one generation to another).

? *Did you ever wonder why the heart, blood vessels, and blood begin to form so early in the developmental process?*

29.1 EMBRYONIC PERIOD

◉ **OBJECTIVE**

- Explain the major developmental events that occur during the embryonic period.

First Week of Development

The **embryonic period** extends from fertilization through the eighth week. The first week of development is characterized by several significant events including fertilization, cleavage of the zygote, blastocyst formation, and implantation.

Fertilization

During **fertilization** (fer'-ti-li-ZĀ-shun; *fertil-* = fruitful), the genetic material from a haploid sperm cell (spermatozoon) and a haploid secondary oocyte merges into a single diploid nucleus. Of the about 200 million sperm introduced into the vagina, fewer than 2 million (1%) reach the cervix of the uterus and only about 200 reach the secondary oocyte. Fertilization normally occurs in the uterine (fallopian) tube within 12 to 24 hours after ovulation. Sperm can remain viable for about 48 hours after deposition in the vagina, although a secondary oocyte is viable for only about 24 hours after ovulation. Thus, pregnancy is *most likely* to occur if intercourse takes place during a 3-day window—from 2 days before ovulation to 1 day after ovulation.

Sperm swim from the vagina into the cervical canal by the whiplike movements of their tails (flagella). The passage of sperm through the rest of the uterus and then into the uterine tube results mainly from contractions of the walls of these organs. Prostaglandins in semen are believed to stimulate uterine motility at the time of intercourse and to aid in the movement of sperm through the uterus and into the uterine tube. Sperm that reach the vicinity of the oocyte within minutes after ejaculation *are not capable* of fertilizing it until about seven hours later. During this time in the female reproductive tract, mostly in the uterine tube, sperm undergo **capacitation** (ka-pas-i-TĀ-shun; *capacit-* = capable of), a series of functional changes that cause the sperm's tail to beat even more vigorously and prepare its plasma membrane to fuse with the oocyte's plasma membrane. During capacitation, sperm are acted on by secretions in the female reproductive tract that result in the removal of cholesterol, glycoproteins, and proteins from the plasma membrane around the head of the sperm cell. Only capacitated sperm are capable of being attracted by and responding to chemical factors produced by the surrounding cells of the ovulated oocyte.

For fertilization to occur, a sperm cell first must penetrate two layers: the **corona radiata** (kō-RŌ-na = crown; rā-dē-A-ta = to shine), the granulosa cells that surround the secondary oocyte, and the **zona pellucida** (ZŌ-na = zone; pe-LOO-si-da = allowing passage of light), the clear glycoprotein layer between the corona radiata and the oocyte's plasma membrane (Figure 29.1a). The **acrosome** (AK-rō-sōm), a helmetlike structure that covers the head of a sperm (see Figure 28.6), contains several enzymes. Acrosomal enzymes and strong tail movements by the sperm help it penetrate the cells of the corona radiata and come in contact

with the zona pellucida. One of the glycoproteins in the zona pellucida, called ZP3, acts as a sperm receptor. Its binding to specific membrane proteins in the sperm head triggers the **acrosomal reaction,** the release of the contents of the acrosome. The acrosomal enzymes digest a path through the zona pellucida as the lashing sperm tail pushes the sperm cell onward. Although many sperm bind to ZP3 molecules and undergo acrosomal reactions, only the first sperm cell to penetrate the entire zona pellucida and reach the oocyte's plasma membrane fuses with the oocyte.

The fusion of a sperm cell with a secondary oocyte sets in motion events that block **polyspermy** (POL-ē-sper'-mē), fertilization by more than one sperm cell. Within a few seconds, the cell membrane of the oocyte depolarizes, which acts as a *fast block to polyspermy*—

Figure 29.1 Selected structures and events in fertilization.

🔒 ▭ During fertilization, genetic material from a sperm cell and a secondary oocyte merge to form a single diploid nucleus.

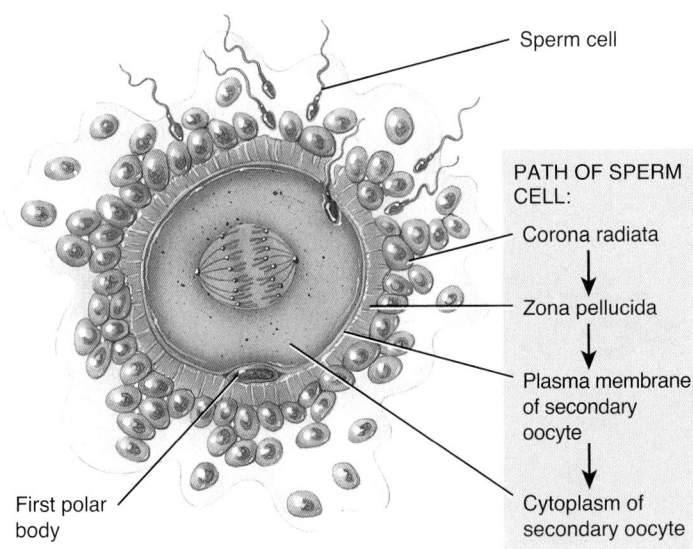

Sperm cell

PATH OF SPERM CELL:

Corona radiata
↓
Zona pellucida
↓
Plasma membrane of secondary oocyte
↓
Cytoplasm of secondary oocyte

First polar body

(a) Sperm cell penetrating secondary oocyte

Head of sperm cell   Secondary oocyte

Pronuclei

SEM 1100x

LM 250x

(b) Sperm cell in contact with secondary oocyte

(c) Male and female pronuclei

❓ **What is capacitation?**

a depolarized oocyte cannot fuse with another sperm. Depolarization also triggers the intracellular release of calcium ions, which stimulate exocytosis of secretory vesicles from the oocyte. The molecules released by exocytosis inactivate ZP3 and harden the entire zona pellucida, events called the *slow block to polyspermy.*

Once a sperm cell enters a secondary oocyte, the oocyte first must complete meiosis II. It divides into a larger ovum (mature egg) and a smaller second polar body that fragments and disintegrates (see Figure 28.15). The nucleus in the head of the sperm develops into the **male pronucleus,** and the nucleus of the fertilized ovum develops into the **female pronucleus** (Figure 29.1c). After the male and female pronuclei form, they fuse, producing a single diploid nucleus, a process known as **syngamy** (SIN-ga-mē). Thus, the fusion of the haploid (*n*) pronuclei restores the diploid number (2*n*) of 46 chromosomes. The fertilized ovum now is called a **zygote** (*zygon* = yolk).

Dizygotic (fraternal) twins are produced from the independent release of two secondary oocytes and the subsequent fertilization of each by different sperm. They are the same age and in the uterus at the same time, but genetically they are as dissimilar as any other siblings. Dizygotic twins may or may not be the same sex. Because **monozygotic (identical) twins** develop from a single fertilized ovum, they contain exactly the same genetic material and are always the same sex. Monozygotic twins arise from separation of the developing cells into two embryos, which in 99% of the cases occurs before 8 days have passed. Separations that occur later than 8 days are likely to produce **conjoined twins,** a situation in which the twins are joined together and share some body structures.

Cleavage of the Zygote

After fertilization, rapid mitotic cell divisions of the zygote called **cleavage** (KLĒV-ij) take place (Figure 29.2). The first division of the zygote begins about 24 hours after fertilization and is completed about 6 hours later. Each succeeding division takes slightly less time. By the second day after fertilization, the second cleavage is completed and there are four cells (Figure 29.2b). By the end of the third day, there are 16 cells. The progressively smaller cells produced by cleavage are called **blastomeres** (BLAS-tō-mērz; *blasto-* = germ or sprout; *-meres* = parts). Successive cleavages eventually produce a solid sphere of cells called the **morula** (MOR-ū-la = mulberry). The morula is still surrounded by the zona pellucida and is about the same size as the original zygote (Figure 29.2c).

Blastocyst Formation

By the end of the fourth day, the number of cells in the morula increases as it continues to move through the uterine tube toward the uterine cavity. When the morula enters the uterine cavity on day 4 or 5, a glycogen-rich secretion from the glands of the endometrium of the uterus passes into the uterine cavity and enters the morula through the zona pellucida. This fluid, called **uterine milk,** along with nutrients stored in the cytoplasm of the blastomeres of the morula, provides nourishment for the developing morula. At the 32-cell stage, the fluid enters the morula, collects between the blastomeres, and reorganizes them around a large fluid-filled cavity called the **blastocyst cavity** (BLAS-tō-sist; *blasto-* = germ or

sprout; *-cyst* = bag), also called the **blastocoel** (BLAS-tō-sēl) (Figure 29.2e). Once the blastocyst cavity is formed, the developing mass is called the **blastocyst.** Though it now has hundreds of cells, the blastocyst is still about the same size as the original zygote.

During the formation of the blastocyst two distinct cell populations arise: the embryoblast and trophoblast (Figure 29.2e). The **embryoblast** (EM-brē-ō-blast′), or **inner cell mass,** is located internally and eventually develops into the embryo. The **trophoblast** (TRŌF-ō-blast; *tropho-* = develop or nourish) is the outer superficial layer of cells that forms the spherelike wall of the blastocyst. It will ultimately develop into the outer chorionic sac that surrounds

Figure 29.2 Cleavage and the formation of the morula and blastocyst.

Cleavage refers to the early, rapid mitotic divisions of a zygote.

(a) Cleavage of zygote, two-cell stage (day 1) — Polar bodies, Blastomeres, Zona pellucida

(b) Cleavage, four-cell stage (day 2) — Nucleus, Cytoplasm

(c) Morula (day 4)

(d) Blastocyst, external view (day 5)

(e) Blastocyst, internal view (day 5) — Embryoblast (inner cell mass), Blastocyst cavity, Trophoblast

What is the histological difference between a morula and a blastocyst?

the fetus and the fetal portion of the placenta, the site of exchange of nutrients and wastes between the mother and fetus. On about the fifth day after fertilization, the blastocyst "hatches" from the zona pellucida by digesting a hole in it with an enzyme, and then squeezing through the hole. This shedding of the zona pellucida is necessary in order to permit the next step, implantation (attachment) into the vascular, glandular endometrial lining of the uterus.

Implantation

The blastocyst remains free within the uterine cavity for about 2 days before it attaches to the uterine wall. At this time the en-

dometrium is in its secretory phase. About 6 days after fertilization, the blastocyst loosely attaches to the endometrium in a process called **implantation** (im-plan-TĀ-shun) (Figure 29.3). As the blastocyst implants, usually in either the posterior portion

Figure 29.3 Relation of a blastocyst to the endometrium of the uterus at the time of implantation.

 Implantation, the attachment of a blastocyst to the endometrium, occurs about 6 days after fertilization.

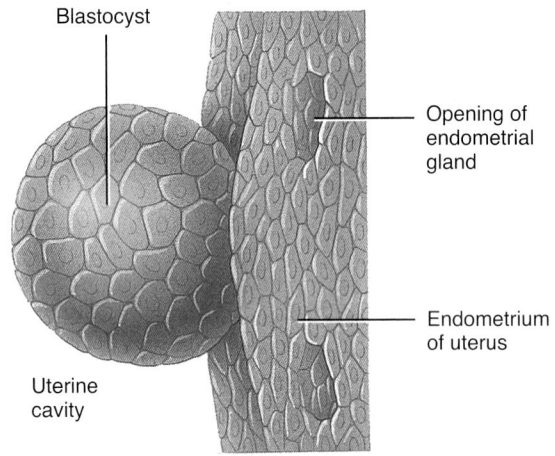

(a) External view of blastocyst, about 6 days after fertilization

Frontal section through uterus

(b) Frontal section through endometrium of uterus and blastocyst, about 6 days after fertilization

? How does the blastocyst merge with and burrow into the endometrium?

CLINICAL CONNECTION | Stem Cell Research and Therapeutic Cloning

Stem cells are unspecialized cells that have the ability to divide for indefinite periods and give rise to specialized cells. In the context of human development, a zygote (fertilized ovum) is a stem cell. Because it has the potential to form an entire organism, a zygote is known as a *totipotent stem cell* (tō-TIP-ō-tent; *totus-* = whole; *-potentia* = power). Inner cell mass cells, called *pluripotent stem cells* (ploo-RIP-ō-tent; *plur-* = several), can give rise to many (but not all) different types of cells. Later, pluripotent stem cells can undergo further specialization into *multipotent stem cells* (mul-TIP-ō-tent), stem cells with a specific function. Examples include keratinocytes that produce new skin cells, myeloid and lymphoid stem cells that develop into blood cells, and spermatogonia that give rise to sperm. Pluripotent stem cells currently used in research are derived from (1) the embryoblast of embryos in the blastocyst stage that were destined to be used for infertility treatments but were not needed and from (2) non-living fetuses terminated during the first 3 months of pregnancy.

On October 13, 2001, researchers reported cloning of the first human embryo to grow cells to treat human diseases. **Therapeutic cloning** is envisioned as a procedure in which the genetic material of a patient with a particular disease is used to create pluripotent stem cells to treat the disease. Using the principles of therapeutic cloning, scientists hope to make an embryo clone of a patient, remove the pluripotent stem cells from the embryo, and then use them to grow tissues to treat particular diseases and disorders, such as cancer, Parkinson disease, Alzheimer disease, spinal cord injury, diabetes, heart disease, stroke, burns, birth defects, osteoarthritis, and rheumatoid arthritis. Presumably, the tissues would not be rejected since they would contain the patient's own genetic material.

Scientists are also investigating the potential clinical applications of *adult stem cells*—stem cells that remain in the body throughout adulthood. Recent experiments suggest that the ovaries of adult mice contain stem cells that can develop into new ova (eggs). If these same types of stem cells are found in the ovaries of adult women, scientists could potentially harvest some of them from a woman about to undergo a sterilizing medical treatment (such as chemotherapy), store them, and then return the stem cells to her ovaries after the medical treatment is completed in order to restore fertility. Studies have also suggested that stem cells in human adult red bone marrow have the ability to differentiate into cells of the liver, kidney, heart, lung, skeletal muscle, skin, and organs of the gastrointestinal tract. In theory, adult stem cells from red bone marrow could be harvested from a patient and then used to repair other tissues and organs in that patient's body without having to use stem cells from embryos. •

of the fundus or the body of the uterus, it orients with the inner cell mass toward the endometrium (Figure 29.3b). About 7 days after fertilization, the blastocyst attaches to the endometrium more firmly, endometrial glands in the vicinity enlarge, and the endometrium becomes more vascularized (forms new blood vessels). The blastocyst eventually secretes enzymes and burrows into the endometrium and becomes surrounded by it.

Following implantation, the endometrium is known as the **decidua** (dē-SID-ū-a = falling off). The decidua separates from the endometrium after the fetus is delivered, much as it does in normal menstruation. Different regions of the decidua are named based on their positions relative to the site of the implanted blastocyst (Figure 29.4). The **decidua basalis** is the portion of the endometrium between the embryo and the stratum basalis of the uterus; it provides large amounts of glycogen and lipids for the developing embryo and fetus and later becomes the maternal part of the placenta. The **decidua capsularis** is the portion of the endometrium located between the embryo and the uterine cavity. The **decidua parietalis** (par-ri-e-TAL-is) is the remaining modified endometrium that lines the noninvolved areas of the rest of the uterus. As the embryo and later the fetus enlarges, the decidua capsularis bulges into the uterine cavity and fuses with the decidua parietalis, thereby obliterating the uterine cavity. By about 27 weeks, the decidua capsularis degenerates and disappears.

The major events associated with the first week of development are summarized in Figure 29.5.

Figure 29.4 Regions of the decidua.

 The decidua is a modified portion of the endometrium that develops after implantation.

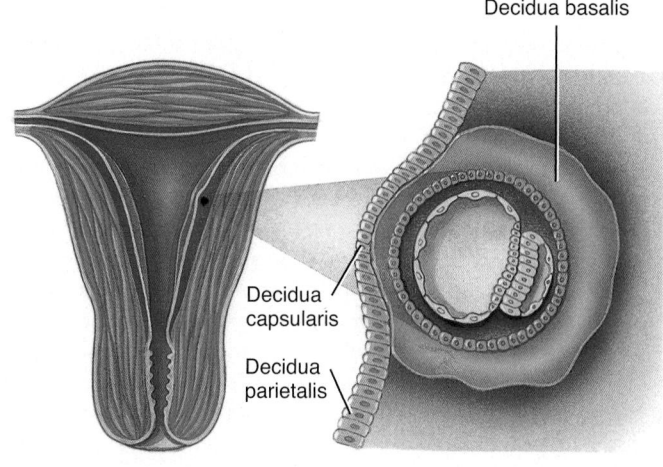

Frontal section of uterus   Details of decidua

Which part of the decidua helps form the maternal part of the placenta?

Figure 29.5 Summary of events associated with the first week of development.

Fertilization usually occurs in the uterine tube.

Frontal plane

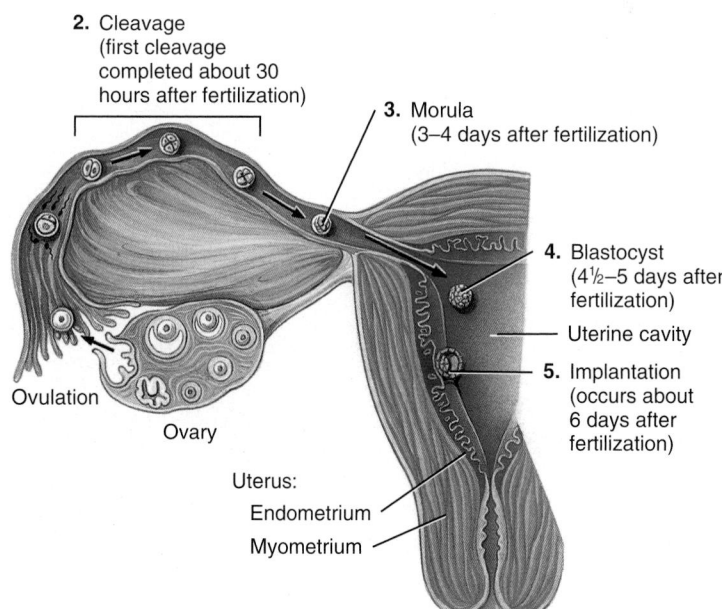

2. Cleavage (first cleavage completed about 30 hours after fertilization)

1. Fertilization (occurs within uterine tube 12–24 hours after ovulation)

3. Morula (3–4 days after fertilization)

Ovulation

Ovary

4. Blastocyst (4½–5 days after fertilization)

Uterine cavity

5. Implantation (occurs about 6 days after fertilization)

Uterus:
Endometrium
Myometrium

Frontal section through uterus, uterine tube, and ovary

 In which phase of the uterine cycle does implantation occur?

Ectopic pregnancy (ek-TOP-ik; *ec-* = out of; *-topic* = place) is the development of an embryo or fetus outside the uterine cavity. An ectopic pregnancy usually occurs when movement of the fertilized ovum through the uterine tube is impaired by scarring due to a prior tubal infection, decreased movement of the uterine tube smooth muscle, or abnormal tubal anatomy. Although the most common site of ectopic pregnancy is the uterine tube, ectopic pregnancies may also occur in the ovary, abdominal cavity, or uterine cervix. Women who smoke are twice as likely to have an ectopic pregnancy because nicotine in cigarette smoke paralyzes the cilia in the lining of the uterine tube (as it does those in the respiratory airways). Scars from pelvic inflammatory disease, previous uterine tube surgery, and previous ectopic pregnancy may also hinder movement of the fertilized ovum.

The signs and symptoms of ectopic pregnancy include one or two missed menstrual cycles followed by bleeding and acute abdominal and pelvic pain. Unless removed, the developing embryo can rupture the uterine tube, often resulting in death of the mother. Treatment options include surgery or the use of a cancer drug called methotrexate, which causes embryonic cells to stop dividing and eventually disappear. •

✔CHECKPOINT

1. Where does fertilization normally occur?
2. How is polyspermy prevented?
3. What is a morula, and how is it formed?
4. Describe the layers of a blastocyst and their eventual fates.
5. When, where, and how does implantation occur?

Second Week of Development

Development of the Trophoblast

About 8 days after fertilization, the trophoblast develops into two layers in the region of contact between the blastocyst and endometrium. These are a **syncytiotrophoblast** (sin-sīt′-ē-ō-TRŌF-ō-blast) that contains no distinct cell boundaries and a **cytotrophoblast** (sī-tō-TRŌF-ō-blast) between the embryoblast and syncytiotrophoblast that is composed of distinct cells (Figure 29.6a). The two layers of trophoblast become part of the chorion (one of the fetal membranes) as they undergo further growth (see Figure 29.11a inset). During implantation, the syncytiotrophoblast secretes enzymes that enable the blastocyst to penetrate the uterine lining by digesting and liquefying the endometrial cells. Eventually, the blastocyst becomes buried in the endometrium and inner one-third of the myometrium. Another secretion of the trophoblast is human chorionic gonadotropin (hCG), which has actions similar to LH. Human chorionic gonadotropin rescues the corpus luteum from degeneration and sustains its secretion of progesterone and estrogens. These hormones maintain the uterine lining in a secretory state, preventing menstruation. Peak secretion of hCG occurs about the ninth week of pregnancy, at which time the placenta is fully developed and produces the progesterone and estrogens that continue to sustain the pregnancy. The presence of hCG in maternal blood or urine is an indicator of pregnancy and is the hormone detected by home pregnancy tests.

Figure 29.6 Principal events of the second week of development.

About 8 days after fertilization, the trophoblast develops into a syncytiotrophoblast and a cytotrophoblast; the inner cell mass develops into a hypoblast and epiblast (bilaminar embryonic disc).

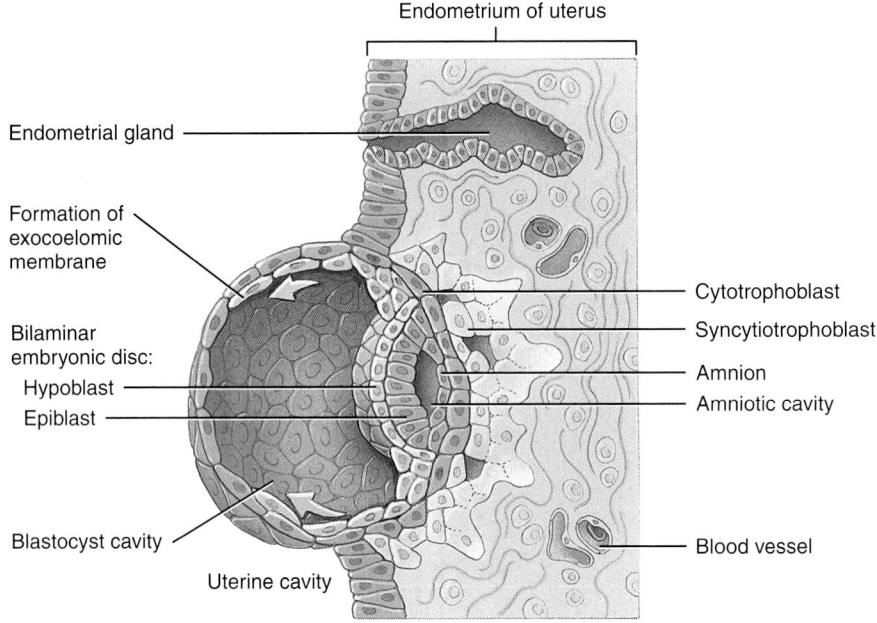

(a) Frontal section through endometrium of uterus showing blastocyst, about 8 days after fertilization

Development of the Bilaminar Embryonic Disc

Like those of the trophoblast, cells of the embryoblast also differentiate into two layers around 8 days after fertilization: a **hypoblast (primitive endoderm)** and **epiblast (primitive ectoderm)** (Figure 29.6a). Cells of the hypoblast and epiblast together form a flat disc referred to as the **bilaminar embryonic disc** (bī-LAM-in-ar = two-layered). Soon, a small cavity appears within the epiblast and eventually enlarges to form the **amniotic cavity** (am-nē-OT-ik; *amnio-* = lamb).

Development of the Amnion

As the amniotic cavity enlarges, a single layer of squamous cells forms a domelike roof above the epiblast cells called the **amnion** (AM-nē-on) (Figure 29.6a). Thus, the amnion forms the roof of the amniotic cavity, and the epiblast forms the floor. Initially, the amnion overlies only the bilaminar embryonic disc. However, as the embryonic disc increases in size and begins to fold, the amnion eventually surrounds the entire embryo (see Figure 29.11a inset), creating the amniotic cavity that becomes filled with **amniotic fluid.** Most amniotic fluid is initially derived from maternal blood. Later, the fetus contributes to the fluid by excreting urine into the amniotic cavity. Amniotic fluid serves as a shock absorber for the fetus, helps regulate fetal body temperature, helps prevent the fetus from drying out, and prevents adhesions between the skin of the fetus and surrounding tissues. The amnion usually ruptures just before birth; it and its fluid constitute the "bag of waters." Embryonic cells are normally sloughed off into amniotic fluid. They can be examined in a procedure called *amniocentesis,* which involves withdrawing some of the amniotic fluid that bathes the developing fetus and analyzing the fetal cells and dissolved substances (see Section 29.4).

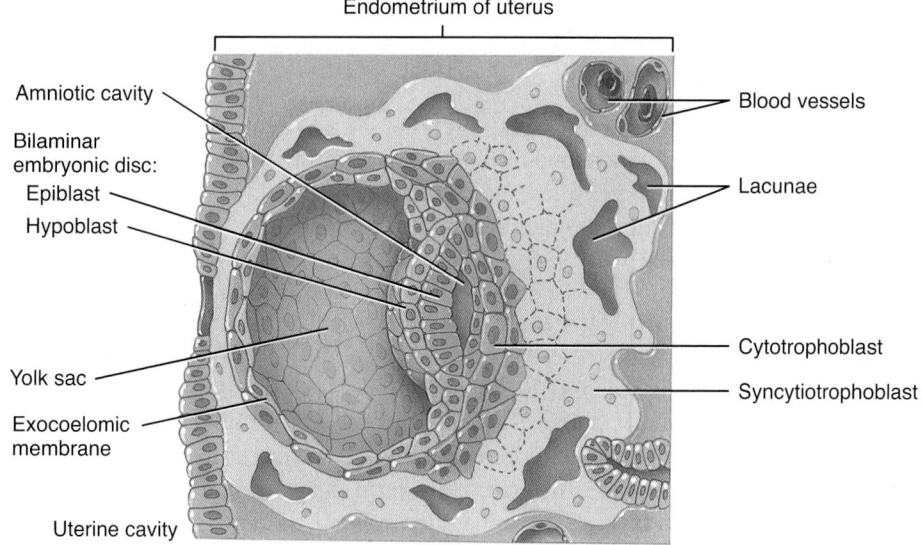

Endometrium of uterus

Amniotic cavity

Bilaminar embryonic disc:
Epiblast
Hypoblast

Yolk sac

Exocoelomic membrane

Uterine cavity

Blood vessels

Lacunae

Cytotrophoblast

Syncytiotrophoblast

(b) Frontal section through endometrium of uterus showing blastocyst, about 9 days after fertilization

Endometrium of uterus

Lacunae

Yolk sac

Lacunar network

Uterine cavity

Maternal sinusoids

Chorion:
Extraembryonic mesoderm

Syncytiotrophoblast
Cytotrophoblast

Amnion
Amniotic cavity

Bilaminar embryonic disc:
Epiblast
Hypoblast

Endometrial gland (right) and sinusoid (left) emptying into lacunar network

(c) Frontal section through endometrium of uterus showing blastocyst, about 12 days after fertilization

 How is the bilaminar embryonic disc connected to the trophoblast?

Development of the Yolk Sac

Also on the eighth day after fertilization, cells at the edge of the hypoblast migrate and cover the inner surface of the blastocyst wall (Figure 29.6a). The migrating columnar cells become squamous (flat) and then form a thin membrane referred to as the **exocoelomic membrane** (ek′-sō-sē-LŌ-mik; *exo-* = outside; *-koilos* = space). Together with the hypoblast, the exocoelomic membrane forms the wall of the **yolk sac,** the former blastocyst cavity during earlier development (Figure 29.6b). As a result, the bilaminar embryonic disc is now positioned between the amniotic cavity and yolk sac.

Since human embryos receive their nutrients from the endometrium, the yolk sac is relatively empty and small, and decreases in size as development progresses (see Figure 29.11a). Nevertheless, the yolk sac has several important functions in humans: supplies nutrients to the embryo during the second and third weeks of development; is the source of blood cells from the third through sixth weeks; contains the first cells (primordial germ cells) that will eventually migrate into the developing gonads, differentiate into the primitive germ cells, and form gametes; forms part of the gut (gastrointestinal tract); functions as a shock absorber; and helps prevent drying out of the embryo.

Development of Sinusoids

On the ninth day after fertilization, the blastocyst becomes completely embedded in the endometrium. As the syncytiotrophoblast expands, small spaces called **lacunae** (la-KOO-nē = little lakes) develop within it (Figure 29.6b).

By the twelfth day of development, the lacunae fuse to form larger, interconnecting spaces called **lacunar networks** (Figure 29.6c). Endometrial capillaries around the developing embryo become dilated and are referred to as **maternal sinusoids** (SĪ-nū-soyds). As the syncytiotrophoblast erodes some of the maternal sinusoids and endometrial glands, maternal blood and secretions from the glands enter the lacunar networks and flow through them. Maternal blood is both a rich source of materials for embryonic nutrition and a disposal site for the embryo's wastes.

Development of the Extraembryonic Coelom

About the twelfth day after fertilization, the **extraembryonic mesoderm** develops. These mesodermal cells are derived from the yolk sac and form a connective tissue layer (mesenchyme) around the amnion and yolk sac (Figure 29.6c). Soon a number of large cavities develop in the extraembryonic mesoderm, which then fuse to form a single, larger cavity called the **extraembryonic coelom** (SĒ-lom).

Development of the Chorion

The extraembryonic mesoderm, together with the two layers of the trophoblast (the cytotrophoblast and syncytiotrophoblast), forms the **chorion** (KŌ-rē-on = membrane) (Figure 29.6c). The chorion surrounds the embryo and, later, the fetus (see Figure 29.11a). Eventually it becomes the principal embryonic part of the placenta, the structure for exchange of materials between mother and fetus. The chorion also protects the embryo and fetus from the immune responses of the mother in two ways: (1) It secretes proteins that block antibody production by the mother. (2) It promotes the production of

T lymphocytes that suppress the normal immune response in the uterus. Finally, the chorion produces human chorionic gonadotropin (hCG), an important hormone of pregnancy (see Figure 29.16).

The inner layer of the chorion eventually fuses with the amnion. With the development of the chorion, the extraembryonic coelom is now referred to as the **chorionic cavity.** By the end of the second week of development, the bilaminar embryonic disc becomes connected to the trophoblast by a band of extraembryonic mesoderm called the **connecting (body) stalk** (see Figure 29.7). The connecting stalk is the future umbilical cord.

✔CHECKPOINT

6. What are the functions of the trophoblast?
7. How is the bilaminar embryonic disc formed?
8. Describe the formation of the amnion, yolk sac, and chorion and explain their functions.
9. Why are sinusoids important during embryonic development?

Third Week of Development

The third embryonic week begins a 6-week period of very rapid development and differentiation. During the third week, the three primary germ layers are established and lay the groundwork for organ development in weeks 4 through 8.

Gastrulation

The first major event of the third week of development, **gastrulation** (gas-troo-LĀ-shun), occurs about 15 days after fertilization. In this process, the bilaminar (two-layered) embryonic disc, consisting of epiblast and hypoblast, transforms into a **trilaminar** (three-layered) **embryonic disc** consisting of three primary germ layers: the ectoderm, mesoderm, and endoderm. The **primary germ layers** are the major embryonic tissues from which the various tissues and organs of the body develop.

Gastrulation involves the rearrangement and migration of cells from the epiblast. The first evidence of gastrulation is the formation of the **primitive streak,** a faint groove on the dorsal surface of the epiblast that elongates from the posterior to the anterior part of the embryo (Figure 29.7a). The primitive streak clearly establishes the head and tail ends of the embryo, as well as its right and left sides. At the head end of the primitive streak a small group of epiblastic cells forms a rounded structure called the **primitive node.**

Following formation of the primitive streak, cells of the epiblast move inward below the primitive streak and detach from the epiblast (Figure 29.7b) in a process called **invagination** (in-vaj′-i-NĀ-shun). Once the cells have invaginated, some of them displace the hypoblast, forming the **endoderm** (*endo-* = inside; *-derm* = skin). Other cells remain between the epiblast and newly formed endoderm to form the **mesoderm** (*meso-* = middle). Cells remaining in the epiblast then form the **ectoderm** (*ecto-* = outside). The ectoderm and endoderm are epithelia composed of tightly packed cells; the mesoderm is a loosely organized connective tissue (mesenchyme). As the embryo develops, the endoderm ultimately becomes the epithelial lining of the gastrointestinal tract, respiratory tract, and several other organs. The mesoderm gives rise to muscles, bones, and other connective tissues, and the

Figure 29.7 Gastrulation.

 Gastrulation involves the rearrangement and migration of cells from the epiblast.

Amnion

Amniotic cavity

Yolk sac

Cytotrophoblast

Connecting stalk

Bilaminar embryonic disc:
Epiblast
Hypoblast

Extraembryonic mesoderm

Uterine cavity

Dorsal surface of bilaminar embryonic disc

Oropharyngeal membrane (future site of mouth)

HEAD END

Yolk sac

Transverse plane

Primitive node

Amnion

Connecting stalk

TAIL END

Primitive streak

Bilaminar embryonic disc:
Epiblast
Hypoblast

(a) Dorsal and partial sectional views of embryonic disc, about 15 days after fertilization

Oropharyngeal membrane

Primitive streak

Trilaminar embryonic disc:
Ectoderm
Mesoderm
Endoderm

Yolk sac

(b) Transverse section of trilaminar embryonic disc, about 16 days after fertilization

? **What is the significance of gastrulation?**

peritoneum. The ectoderm develops into the epidermis of the skin and the nervous system. Table 29.1 provides more details about the fates of these primary germ layers.

About 16 days after fertilization, mesodermal cells from the primitive node migrate toward the head end of the embryo and form a hollow tube of cells in the midline called the **notochordal process** (nō-tō-KOR-dal) (Figure 29.8). By days 22–24, the notochordal process becomes a solid cylinder of cells called the **notochord** (nō-

tō-KORD; *noto-* = back; *-chord* = cord). This structure plays an extremely important role in **induction** (in-DUK-shun), the process by which one tissue *(inducing tissue)* stimulates the development of an adjacent unspecialized tissue *(responding tissue)* into a specialized one. An inducing tissue usually produces a chemical substance that influences the responding tissue. The notochord induces certain mesodermal cells to develop into the vertebral bodies. It also forms the nucleus pulposus of the intervertebral discs (see Figure 7.24).

TABLE 29.1

Structures Produced by the Three Primary Germ Layers

ENDODERM	MESODERM	ECTODERM
Epithelial lining of gastrointestinal tract (except oral cavity and anal canal) and epithelium of its glands.	All skeletal and cardiac muscle tissue and most smooth muscle tissue.	All nervous tissue.
Epithelial lining of urinary bladder, gallbladder, and liver.	Cartilage, bone, and other connective tissues.	Epidermis of skin.
Epithelial lining of pharynx, auditory (eustachian) tubes, tonsils, tympanic (middle ear) cavity, larynx, trachea, bronchi, and lungs.	Blood, red bone marrow, and lymphatic tissue.	Hair follicles, arrector pili muscles, nails, epithelium of skin glands (sebaceous and sudoriferous), and mammary glands.
	Blood vessels and lymphatic vessels.	Lens, cornea, and internal eye muscles.
	Dermis of skin.	Internal and external ear.
Epithelium of thyroid gland, parathyroid glands, pancreas, and thymus.	Fibrous tunic and vascular tunic of eye.	Neuroepithelium of sense organs.
	Mesothelium of thoracic, abdominal, and pelvic cavities.	Epithelium of oral cavity, nasal cavity, paranasal sinuses, salivary glands, and anal canal.
Epithelial lining of prostate and bulbourethral (Cowper's) glands, vagina, vestibule, urethra, and associated glands such as greater (Bartholin's) vestibular and lesser vestibular glands.	Kidneys and ureters.	Epithelium of pineal gland, pituitary gland, and adrenal medullae.
	Adrenal cortex.	
	Gonads and genital ducts (except germ cells).	Melanocytes (pigment cells).
Gametes (sperm and oocytes).	Dura mater.	Almost all skeletal and connective tissue components of head.
		Arachnoid mater and pia mater.

Also during the third week of development, two faint depressions appear on the dorsal surface of the embryo where the ectoderm and endoderm make contact but lack mesoderm between them. The structure closer to the head end is called the **oropharyngeal membrane** (or-ō-fa-RIN-jē-al; *oro-* = mouth;

-pharyngeal = pertaining to the pharynx) (Figure 29.8a, b). It breaks down during the fourth week to connect the mouth cavity to the pharynx and the remainder of the gastrointestinal tract. The structure closer to the tail end is called the **cloacal membrane** (klō-Ā-kul = sewer), which degenerates in the seventh

Figure 29.8   Development of the notochordal process.

The notochordal process develops from the primitive node and later becomes the notochord.

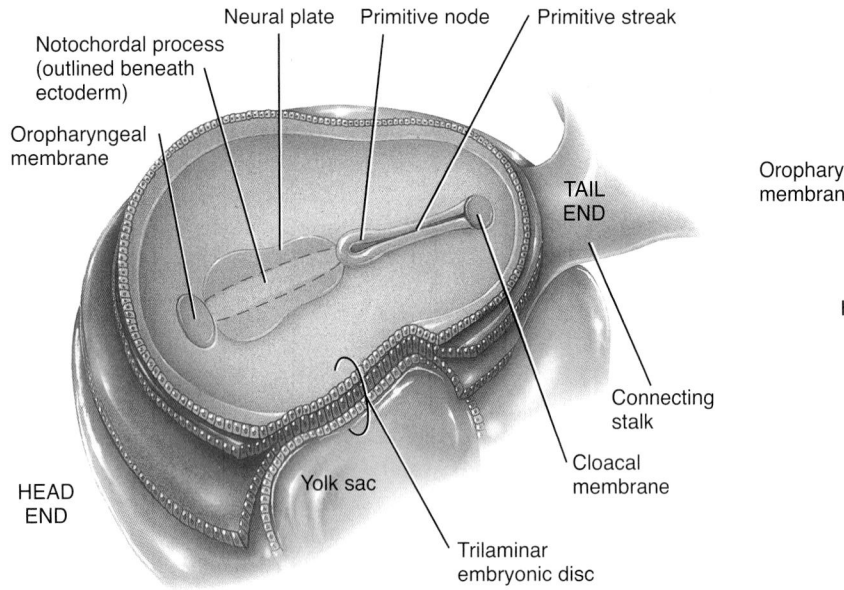

(a) Dorsal and partial sectional views of trilaminar embryonic disc, about 16 days after fertilization

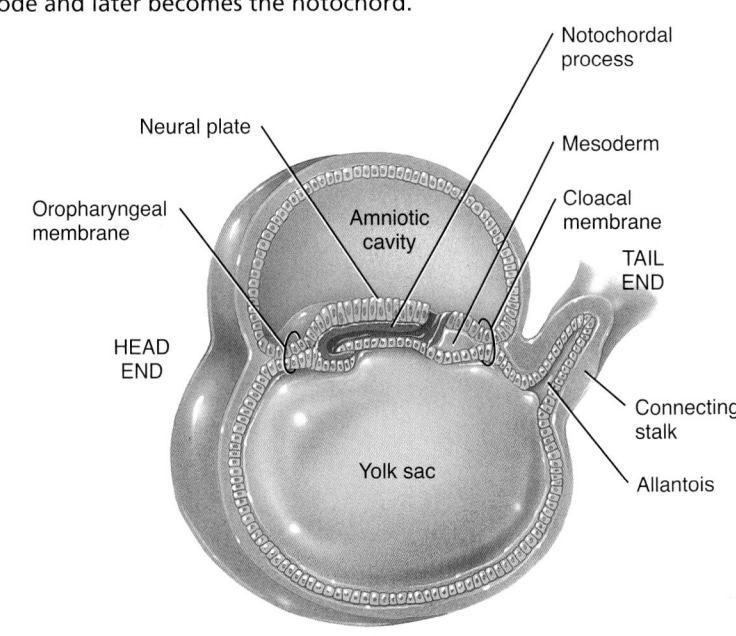

(b) Sagittal section of trilaminar embryonic disc, about 16 days after fertilization

? **What is the significance of the notochord?**

week to form the openings of the anus and urinary and reproductive tracts.

When the cloacal membrane appears, the wall of the yolk sac forms a small vascularized outpouching called the **allantois** (a-LAN-tō-is; *allant-* = sausage) that extends into the connecting stalk (Figure 29.8b). In nonmammalian organisms enclosed in an amnion, the allantois is used for gas exchange and waste removal. Because of the role of the human placenta in these activities, the allantois is not a prominent structure in humans (see Figure 29.11a, inset). Nevertheless, it does function in the early formation of blood and blood vessels, and it is associated with the development of the urinary bladder.

Neurulation

In addition to inducing mesodermal cells to develop into vertebral bodies, the notochord also induces ectodermal cells over it to form the **neural plate** (Figure 29.9a). (Also see Figure 14.27.) By the end of the third week, the lateral edges of the neural plate

Figure 29.9 Neurulation and the development of somites.

Neurulation is the process by which the neural plate, neural folds, and neural tube form.

(a) 17 days

(b) 19 days

(c) 20 days

Dorsal views (d) 22 days Transverse sections

? Which structures develop from the neural tube and somites?

become more elevated and form the **neural fold** (Figure 29.9b). The depressed midregion is called the **neural groove** (Figure 29.9c). Generally, the neural folds approach each other and fuse, thus converting the neural plate into a **neural tube** (Figure 29.9d). This occurs first near the middle of the embryo and then progresses toward the head and tail ends. Neural tube cells then develop into the brain and spinal cord. The process by which the neural plate, neural folds, and neural tube form is called **neurulation** (noor-oo-LĀ-shun).

As the neural tube forms, some of the ectodermal cells from the tube migrate to form several layers of cells called the **neural crest** (see Figure 14.27b). Neural crest cells give rise to all sensory neurons and postganglionic neurons of the peripheral nerves, the adrenal medullae, melanocytes (pigment cells) of the skin, arachnoid mater, and pia mater of the brain and spinal cord, and almost all of the skeletal and connective tissue components of the head.

At about 4 weeks after fertilization, the head end of the neural tube develops into three enlarged areas called **primary brain vesicles** (see Figure 14.28): the **prosencephalon** (PROS-en-sef'-a-lon) or **forebrain, mesencephalon** (mez-en-SEF-a-lon) or **midbrain,** and **rhombencephalon** (ROM-ben-sef'-a-lon) or **hindbrain.** At about 5 weeks, the prosencephalon develops into **secondary brain vesicles** called the **telencephalon** (TEL-en-sef'-a-lon) and **diencephalon** (dī-en-SEF-a-lon), and the rhombencephalon develops into secondary brain vesicles called the **metencephalon** (MET-en-sef'-a-lon) and **myelencephalon** (MĪ-el-en-sef'-a-lon). The areas of the neural tube adjacent to the myelencephalon develop into the spinal cord. The parts of the brain that develop from the various brain vesicles are described in Section 14.1.

CLINICAL CONNECTION | Anencephaly

Neural tube defects (NTDs) are caused by arrest of the normal development and closure of the neural tube. These include spina bifida (discussed in Disorders: Homeostatic Imbalances in Chapter 7) and **anencephaly** (an-en-SEF-a-lē; *an-* = without; *-encephal* = brain). In anencephaly, the cranial bones fail to develop and certain parts of the brain remain in contact with amniotic fluid and degenerate. Usually, a part of the brain that controls vital functions such as breathing and regulation of the heart is also affected. Infants with anencephaly are stillborn or die within a few days after birth. The condition occurs about once in every 1000 births and is 2 to 4 times more common in female infants than males. •

Development of Somites

By about the seventeenth day after fertilization, the mesoderm adjacent to the notochord and neural tube forms paired longitudinal columns of **paraxial mesoderm** (par-AK-sē-al; *para-* = near) (Figure 29.9b). The mesoderm lateral to the paraxial mesoderm forms paired cylindrical masses called **intermediate mesoderm.** The mesoderm lateral to the intermediate mesoderm consists of a pair of flattened sheets called **lateral plate mesoderm.** The paraxial mesoderm soon segments into a series of paired, cube-shaped structures called **somites** (SŌ-mīts = little bodies). By the

end of the fifth week, 42–44 pairs of somites are present. The number of somites that develop over a given period can be correlated to the approximate age of the embryo.

Each somite differentiates into three regions: a **myotome** (MĪ-ō-tōm), a **dermatome,** and a **sclerotome** (SKLĒR-ō-tōm) (see Figure 10.17b). The myotomes develop into the skeletal muscles of the neck, trunk, and limbs; the dermatomes form connective tissue, including the dermis of the skin; and the sclerotomes give rise to the vertebrae and ribs.

Development of the Intraembryonic Coelom

In the third week of development, small spaces appear in the lateral plate mesoderm. These spaces soon merge to form a larger cavity called the **intraembryonic coelom** (SĒ-lom = cavity). This cavity splits the lateral plate mesoderm into two parts called the splanchnic mesoderm and somatic mesoderm (Figure 29.9d). **Splanchnic mesoderm** (SPLANGK-nik = visceral) forms the heart and the visceral layer of the serous pericardium, blood vessels, the smooth muscle and connective tissues of the respiratory and digestive organs, and the visceral layer of the serous membrane of the pleurae and peritoneum. **Somatic mesoderm** (sō-MAT-ik; *soma-* = body) gives rise to the bones, ligaments, blood vessels, and connective tissue of the limbs and the parietal layer of the serous membrane of the pericardium, pleurae, and peritoneum.

Development of the Cardiovascular System

At the beginning of the third week, **angiogenesis** (an'-jē-ō-JEN-e-sis; *angio-* = vessel; *-genesis* = production), the formation of blood vessels, begins in the extraembryonic mesoderm in the yolk sac, connecting stalk, and chorion. This early development is necessary because there is insufficient yolk in the yolk sac and ovum to provide adequate nutrition for the rapidly developing embryo. Angiogenesis is initiated when mesodermal cells differentiate into **hemangioblasts** (hē-MAN-jē-ō-blasts). These then develop into cells called **angioblasts,** which aggregate to form isolated masses of cells referred to as **blood islands** (see Figure 21.31). Spaces soon develop in the blood islands and form the lumens of blood vessels. Some angioblasts arrange themselves around each space to form the endothelium and the tunics (layers) of the developing blood vessels. As the blood islands grow and fuse, they soon form an extensive system of blood vessels throughout the embryo.

About 3 weeks after fertilization, blood cells and blood plasma begin to develop *outside* the embryo from hemangioblasts in the blood vessels in the walls of the yolk sac, allantois, and chorion. These then develop into pluripotent stem cells that form blood cells. Blood formation begins *within* the embryo at about the fifth week in the liver and the twelfth week in the spleen, red bone marrow, and thymus.

The heart forms from splanchnic mesoderm in the head end of the embryo on days 18 and 19. This region of mesodermal cells is called the **cardiogenic area** (kar-dē-ō-JEN-ik; *cardio-* = heart; *-genic* = producing). In response to induction signals from the underlying endoderm, these mesodermal cells form a pair of **endocardial tubes** (see Figure 20.19). The tubes then fuse to form a single **primitive heart tube.** By the end of the third week, the primitive heart tube bends on itself, becomes S-shaped, and begins to beat. It then joins blood vessels in other parts of the embryo, connecting stalk, chorion, and yolk sac to form a primitive cardiovascular system.

Development of the Chorionic Villi and Placenta

As the embryonic tissue invades the uterine wall, maternal uterine vessels are eroded and maternal blood fills spaces, called **lacunae** (la-KOO-nē), within the invading tissue. By the end of the second week of development, **chorionic villi** (kō-rē-ON-ik VIL-lī) begin to develop. These fingerlike projections consist of chorion (syncytiotrophoblast surrounded by cytotrophoblast) that projects into the endometrial wall of the uterus (Figure 29.10a). By the end of the third week, blood capillaries develop in the chorionic villi (Figure 29.10b). Blood vessels in the chorionic villi connect to the embryonic heart by way of the umbilical arteries and umbilical vein through the connecting (body) stalk, which will eventually become the umbilical cord (Figure 29.10c). The fetal blood capillaries within the chorionic villi project into the lacunae, which unite to form the **intervillous spaces** (in′-ter-VIL-us) that bathe the chorionic villi with maternal blood. As a result, maternal blood bathes the chorion-covered fetal blood vessels. Note, however, that maternal and fetal blood vessels do not join, and the blood they carry does not normally mix. Instead, oxygen and nutrients in the blood of the mother's intervillous spaces, which are the spaces between chorionic villi, diffuse across the cell membranes into the capillaries of the villi. Waste products such as carbon dioxide diffuse in the opposite direction.

Figure 29.10 Development of chorionic villi.

Blood vessels in chorionic villi connect to the embryonic heart via the umbilical arteries and umbilical vein.

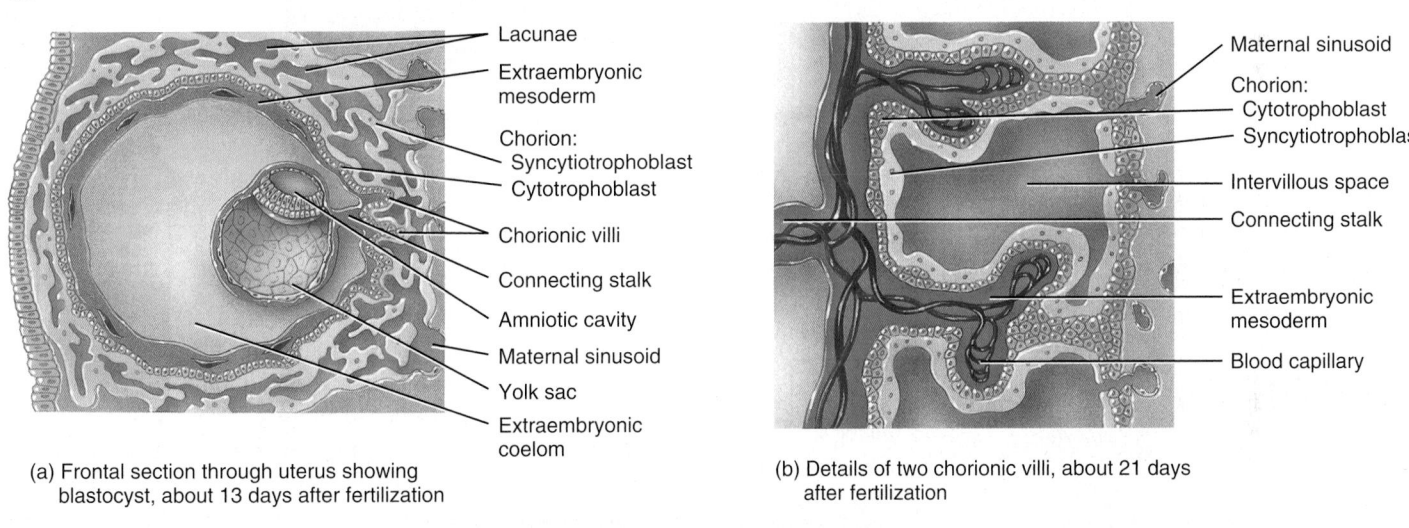

(a) Frontal section through uterus showing blastocyst, about 13 days after fertilization

(b) Details of two chorionic villi, about 21 days after fertilization

(c) Frontal section through uterus showing embryo and vascular supply, about 21 days after fertilization

Why is development of chorionic villi important?

Figure 29.11 Placenta and umbilical cord.

The placenta is formed by the chorionic villi of the embryo and the decidua basalis of the endometrium of the mother.

(a) Details of placenta and umbilical cord

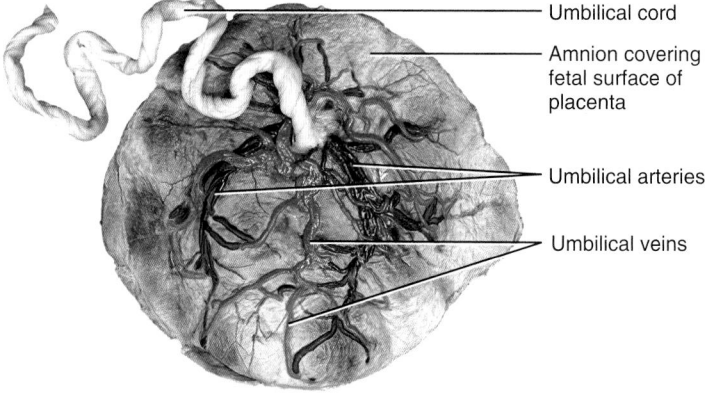

(b) Fetal surface of placenta

? **What is the function of the placenta?**

Placentation (plas-en-TĀ-shun) is the process of forming the **placenta** (pla-SEN-ta = flat cake), the site of exchange of nutrients and wastes between the mother and fetus. The placenta also produces hormones needed to sustain the pregnancy (see Figure 29.16). The placenta is unique because it develops from two separate individuals, the mother and the fetus.

By the beginning of the twelfth week, the placenta has two distinct parts: (1) the fetal portion formed by the chorionic villi of the chorion and (2) the maternal portion formed by the decidua basalis of the endometrium (Figure 29.11a). When fully developed, the placenta is shaped like a pancake (Figure 29.11b). Functionally, the placenta allows oxygen and nutrients to diffuse

from maternal blood into fetal blood while carbon dioxide and wastes diffuse from fetal blood into maternal blood. The placenta also is a protective barrier because most microorganisms cannot pass through it. However, certain viruses, such as those that cause AIDS, German measles, chickenpox, measles, encephalitis, and poliomyelitis, can cross the placenta. Many drugs, alcohol, and some substances that can cause birth defects also pass freely. The placenta stores nutrients such as carbohydrates, proteins, calcium, and iron, which are released into fetal circulation as required.

The actual connection between the placenta and embryo, and later the fetus, is through the **umbilical cord** (um-BIL-i-kul = navel), which develops from the connecting stalk and is usually about 2 cm (1 in.) wide and about 50–60 cm (20–24 in.) in length. The umbilical cord consists of two umbilical arteries that carry deoxygenated fetal blood to the placenta, one umbilical vein that carries oxygen and nutrients acquired from the mother's intervillous spaces into the fetus, and supporting mucous connective tissue called **Wharton's jelly** (WOR-tons) derived from the allantois. A layer of amnion surrounds the entire umbilical cord and gives it a shiny appearance (Figure 29.11). In some cases, the umbilical vein is used to transfuse blood into a fetus or to introduce drugs for various medical treatments.

In about 1 in 200 newborns, only one of the two umbilical arteries is present in the umbilical cord. It may be due to failure of the artery to develop or degeneration of the vessel early in development. Nearly 20% of infants with this condition develop cardiovascular defects.

After the birth of the baby, the placenta detaches from the uterus and is therefore termed the **afterbirth.** At this time, the umbilical cord is tied off and then severed. The small portion (about an inch) of the cord that remains attached to the infant begins to wither and falls off, usually within 12 to 15 days after birth. The area where the cord was attached becomes covered by a thin layer of skin, and scar tissue forms. The scar is the **umbilicus** (um-BIL-i-kus) or navel.

Pharmaceutical companies use human placentas as a source of hormones, drugs, and blood; portions of placentas are also used for burn coverage. The placental and umbilical cord veins can also be used in blood vessel grafts, and cord blood can be frozen to provide a future source of pluripotent stem cells, for example, to repopulate red bone marrow following radiotherapy for cancer.

CLINICAL CONNECTION | *Placenta Previa*

In some cases, the entire placenta or part of it may become implanted in the inferior portion of the uterus, near or covering the internal os of the cervix. This condition is called **placenta previa** (PRĒ-vē-a = before or in front of). Although placenta previa may lead to spontaneous abortion, it also occurs in approximately 1 in 250 live births. It is dangerous to the fetus because it may cause premature birth and intrauterine hypoxia due to maternal bleeding. Maternal mortality is increased due to hemorrhage and infection. The most important symptom is sudden, painless, bright-red vaginal bleeding in the third trimester. Cesarean section is the preferred method of delivery in placenta previa. •

✔CHECKPOINT

10. When does gastrulation occur?
11. How do the three primary germ layers form? Why are they important?
12. What is meant by the term *induction?*
13. Describe how neurulation occurs. Why is it important?
14. What are the functions of somites?
15. How does the cardiovascular system develop?
16. How does the placenta form?

Fourth Week of Development

The fourth through eighth weeks of development are very significant in embryonic development because all major organs appear during this time. The term **organogenesis** (or′-ga-nō-JEN-e-sis) refers to the formation of body organs and systems. By the end of the eighth week, all the major body systems have begun to develop, although their functions for the most part are minimal. Organogenesis requires the presence of blood vessels to supply developing organs with oxygen and other nutrients. However, recent studies suggest that blood vessels play a significant role in organogenesis even before blood begins to flow

within them. The endothelial cells of blood vessels apparently provide some type of developmental signal, either a secreted substance or a direct cell-to-cell interaction, that is necessary for organogenesis.

During the fourth week after fertilization, the embryo undergoes very dramatic changes in shape and size, nearly tripling its size. It is essentially converted from a flat, two-dimensional trilaminar embryonic disc to a three-dimensional cylinder, a process called **embryonic folding** (Figure 29.12a–d). The cylinder consists of endoderm in the center (gut), ectoderm on the outside (epidermis), and mesoderm in between. The main force responsible for embryonic folding is the different rates of growth of various parts of the embryo, especially the rapid longitudinal growth of the nervous system (neural tube). Folding in the median plane produces a **head fold** and a **tail fold;** folding in the horizontal plane results in the two **lateral folds.** Overall, due to the foldings, the embryo curves into a C-shape.

The head fold brings the developing heart and mouth into their eventual adult positions. The tail fold brings the developing anus into its eventual adult position. The lateral folds form as the lateral margins of the trilaminar embryonic disc bend ventrally. As they move toward the midline, the lateral folds incorporate the dorsal part of the yolk sac into the embryo as the **primitive gut,** the forerunner of the gastrointestinal tract (Figure 29.12b). The primitive gut differentiates into an anterior **foregut,** an intermediate **midgut,** and a posterior **hindgut** (Figure 29.12c). The fates of the foregut, midgut, and hindgut are described in Section 24.15. Recall that the oropharyngeal membrane is located in the head end of the embryo (see Figure 29.8). It separates the future pharyngeal (throat) region of the foregut from the **stomodeum** (stō-mō-DĒ-um; *stomo-* = mouth), the future oral cavity. Because of head folding, the oropharyngeal membrane moves downward and the foregut and stomodeum move closer to their final positions. When the oropharyngeal membrane ruptures during the fourth week, the pharyngeal region of the pharynx is brought into contact with the stomodeum.

In a developing embryo, the last part of the hindgut expands into a cavity called the **cloaca** (klo-Ā-ka) (see Figure 26.23). On the outside of the embryo is a small cavity in the tail region called the **proctodeum** (prok-tō-DĒ-um; *procto-* = anus) (Figure 29.12c). Separating the cloaca from the proctodeum is the **cloacal membrane** (see Figure 29.8). During embryonic development, the cloaca divides into a ventral urogenital sinus and a dorsal anorectal canal. As a result of tail folding, the cloacal membrane moves downward and the urogenital sinus, anorectal canal, and proctodeum move closer to their final positions. When the cloacal membrane ruptures during the seventh week of development, the urogenital and anal openings are created.

In addition to embryonic folding, development of somites, and development of the neural tube, five pairs of **pharyngeal** (fa-RIN-jē-al) or **branchial arches** (BRANG-kē-al; *branch* = gill) begin to develop on each side of the future head and neck regions (Figure 29.13) during the fourth week. These five paired structures begin their development on the twenty-second day after fertilization and form swellings on the surface of the embryo. Each

Figure 29.12 Embryonic folding.

Embryonic folding converts the two-dimensional trilaminar embryonic disc into a three-dimensional cylinder.

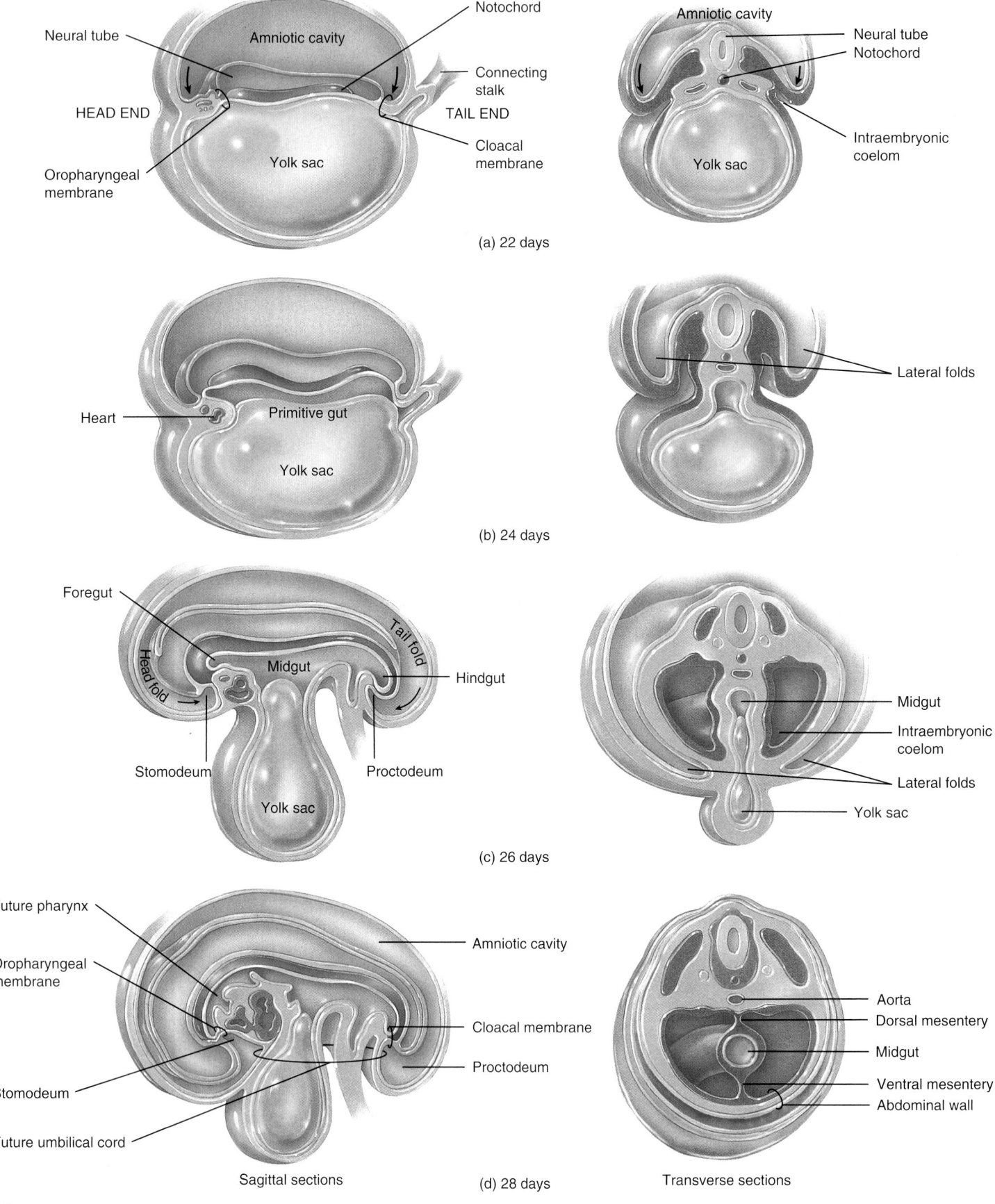

(a) 22 days

(b) 24 days

(c) 26 days

(d) 28 days

Sagittal sections

Transverse sections

? What are the results of embryonic folding?

pharyngeal arch consists of an outer covering of ectoderm and an inner covering of endoderm, with mesoderm in between. Within each pharyngeal arch there is an artery, a cranial nerve, skeletal cartilaginous rods that support the arch, and skeletal muscle tissue that attaches to and moves the cartilage rods. On the ectodermal surface of the pharyngeal region, each pharyngeal arch is separated by a groove called a **pharyngeal cleft** (Figure 29.13a). The pharyngeal clefts meet corresponding balloonlike outgrowths of the endodermal pharyngeal lining called **pharyngeal pouches.** Where the pharyngeal cleft and pouch meet to separate the arches, the outer ectoderm of the cleft contacts the inner endoderm of the pouch and there is no mesoderm between (Figure 29.13b).

Just as the somite gives rise to specified structures in the body wall, each pharyngeal arch, cleft, and pouch gives rise to specified structures in the head and neck. Each pharyngeal arch is a developmental unit and includes a skeletal component, muscle, nerve, and blood vessels. In the human embryo, there are four obvious pharyngeal arches and two less distinct arches. Each of these arches develops into a specific and unique component of the head and neck region. For example, the first pharyngeal arch is often called the *mandibular arch* because it forms the jaws (the *mandible* is the lower jawbone).

The first sign of a developing ear is a thickened area of ectoderm, the **otic placode** (PLAK-ōd), or future internal ear, which can be distinguished about 22 days after fertilization. A thickened area of ectoderm called the **lens placode,** which will become the eye, also appears at this time (see Figure 29.13a).

By the middle of the fourth week, the upper limbs begin their development as outgrowths of mesoderm covered by ectoderm called **upper limb buds** (see Figure 8.18b). By the end of the fourth week, the **lower limb buds** develop. The heart also forms a distinct projection on the ventral surface of the embryo called the **heart prominence** (see Figure 8.18b). At the end of the fourth week the embryo has a distinctive **tail** (see Figure 8.18b).

Figure 29.13 Development of pharyngeal arches, pharyngeal clefts, and pharyngeal pouches.

The five pairs of pharyngeal pouches consist of ectoderm, mesoderm, and endoderm and contain blood vessels, cranial nerves, cartilage, and muscle tissue.

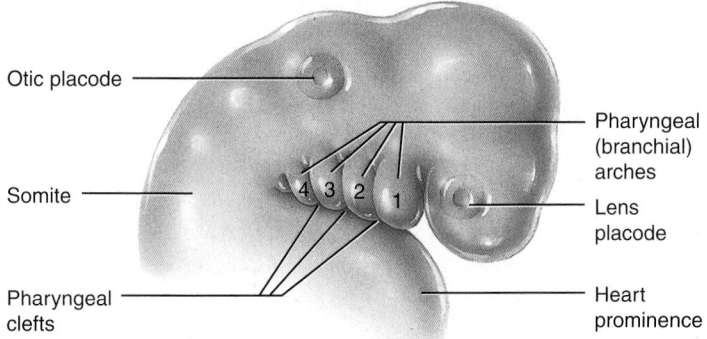

(a) External view, about 28-day embryo

(b) Sagittal section, about 28-day embryo

(c) Transverse section of the pharynx, about 28-day embryo

? **Why are pharyngeal arches, clefts, and pouches important?**

Fifth through Eighth Weeks of Development

During the fifth week of development, there is a very rapid development of the brain, so growth of the head is considerable. By the end of the sixth week, the head grows even larger relative to the trunk, and the limbs show substantial development (see Figure 8.18c). In addition, the neck and trunk begin to straighten, and the heart is now four-chambered. By the seventh week, the various regions of the limbs become distinct and the beginnings of digits appear (see Figure 8.18d). At the start of the eighth week (the final week of the embryonic period), the digits of the hands are short and webbed, the tail is shorter but still visible, the eyes are open, and the auricles of the ears are visible (see Figure 8.18c). By the end of the eighth week, all regions of limbs are apparent; the digits are distinct and no longer webbed due to removal of cells via apoptosis. Also, the eyelids come together and may fuse, the tail disappears, and the external genitals begin to differentiate. The embryo now has clearly human characteristics.

✔ CHECKPOINT

17. How does embryonic folding occur?
18. How does the primitive gut form, and what is its significance?
19. What is the origin of the structures of the head and neck?
20. What are limb buds?
21. What changes occur in the limbs during the second half of the embryonic period?

29.2 FETAL PERIOD

◉ **OBJECTIVE**

• Describe the major events of the fetal period.

During the **fetal period** (from the ninth week until birth), tissues and organs that developed during the embryonic period grow and differentiate. Very few new structures appear during the fetal period, but the rate of body growth is remarkable, especially during the second half of intrauterine life. For example, during the last 2.5 months of intrauterine life, half of the full-term weight is added. At the beginning of the fetal period, the head is half the length of the body. By the end of the fetal period, the head size is only one-quarter the length of the body. During the same period, the limbs also increase in size from one-eighth to one-half the fetal length. The fetus is also less vulnerable to the damaging effects of drugs, radiation, and microbes than it was as an embryo.

A summary of the major developmental events of the embryonic and fetal periods is presented in Table 29.2 and illustrated in Figure 29.14.

TABLE 29.2

Summary of Changes during Embryonic and Fetal Development

TIME	APPROXIMATE SIZE AND WEIGHT	REPRESENTATIVE CHANGES
EMBRYONIC PERIOD		
1–4 weeks	0.6 cm (3/16 in.)	Primary germ layers and notochord develop. Neurulation occurs. Primary brain vesicles, somites, and intraembryonic coelom develop. Blood vessel formation begins and blood forms in yolk sac, allantois, and chorion. Heart forms and begins to beat. Chorionic villi develop and placental formation begins. The embryo folds. The primitive gut, pharyngeal arches, and limb buds develop. Eyes and ears begin to develop, tail forms, and body systems begin to form.
5–8 weeks	3 cm (1.25 in.) 1 g (1/30 oz)	Limbs become distinct and digits appear. Heart becomes four-chambered. Eyes are far apart and eyelids are fused. Nose develops and is flat. Face is more humanlike. Bone formation begins. Blood cells start to form in liver. External genitals begin to differentiate. Tail disappears. Major blood vessels form. Many internal organs continue to develop.
FETAL PERIOD		
9–12 weeks	7.5 cm (3 in.) 30 g (1 oz)	Head constitutes about half the length of fetal body, and fetal length nearly doubles. Brain continues to enlarge. Face is broad, with eyes fully developed, closed, and widely separated. Nose develops a bridge. External ears develop and are low set. Bone formation continues. Upper limbs almost reach final relative length but lower limbs are not quite as well developed. Heartbeat can be detected. Gender is distinguishable from external genitals. Urine secreted by fetus is added to amniotic fluid. Red bone marrow, thymus, and spleen participate in blood cell formation. Fetus begins to move, but its movements cannot be felt yet by the mother. Body systems continue to develop.
13–16 weeks	18 cm (6.5–7 in.) 100 g (4 oz)	Head is relatively smaller than rest of body. Eyes move medially to final positions, and ears move to final positions on sides of head. Lower limbs lengthen. Fetus appears even more humanlike. Rapid development of body systems occurs.
17–20 weeks	25–30 cm (10–12 in.) 200–450 g (0.5–1 lb)	Head is more proportionate to rest of body. Eyebrows and head hair are visible. Growth slows but lower limbs continue to lengthen. Vernix caseosa (fatty secretions of oil glands and dead epithelial cells) and lanugo (delicate fetal hair) cover fetus. Brown fat forms and is the site of heat production. Fetal movements are commonly felt by mother (quickening).
21–25 weeks	27–35 cm (11–14 in.) 550–800 g (1.25–1.5 lb)	Head becomes even more proportionate to rest of body. Weight gain is substantial, and skin is pink and wrinkled. Fetuses 24 weeks and older usually survive if born prematurely.
26–29 weeks	32–42 cm (13–17 in.) 1100–1350 g (2.5–3 lb)	Head and body are more proportionate and eyes are open. Toenails are visible. Body fat is 3.5% of total body mass and additional subcutaneous fat smoothes out some wrinkles. Testes begin to descend toward scrotum at 28 to 32 weeks. Red bone marrow is major site of blood cell production. Many fetuses born prematurely during this period survive if given intensive care because lungs can provide adequate ventilation and central nervous system is developed enough to control breathing and body temperature.
30–34 weeks	41–45 cm (16.5–18 in.) 2000–2300 g (4.5–5 lb)	Skin is pink and smooth. Fetus assumes upside-down position. Body fat is 8% of total body mass.
35–38 weeks	50 cm (20 in.) 3200–3400 g (7–7.5 lb)	By 38 weeks, circumference of fetal abdomen is greater than that of head. Skin is usually bluish-pink, and growth slows as birth approaches. Body fat is 16% of total body mass. Testes are usually in scrotum in full-term male infants. Even after birth, an infant is not completely developed; an additional year is required, especially for complete development of nervous system.

Throughout the text we have discussed the developmental anatomy of the various body systems in their respective chapters. The following list of these sections is presented here for your review.

- Integumentary System (Section 5.6)
- Skeletal System (Section 8.7)
- Muscular System (Section 10.11)
- Nervous System (Section 14.9)
- Endocrine System (Section 18.16)

- Heart (Section 20.8)
- Blood and Blood Vessels (Section 21.8)
- Lymphatic System and Immunity (Section 22.2)
- Respiratory System (Section 23.8)
- Digestive System (Section 24.15)
- Urinary System (Section 26.10)
- Reproductive Systems (Section 28.5)

4   8   12   16   20   24   28   32   36   (weeks)

22. What are the general developmental trends during the fetal period?

23. Using Table 29.2 as a guide, select any one body structure in weeks 9 through 12 and trace its development through the remainder of the fetal period.

Figure 29.14 Summary of representative developmental events of the embryonic and fetal periods. The embryos and fetuses are not shown at their actual sizes.

🔑 Development during the fetal period is mostly concerned with the growth and differentiation of tissues and organs formed during the embryonic period.

(a) 20-day embryo

Neural plate
Neural groove
Cut edge of amnion
Somite
Yolk sac
Primitive streak

(b) 24-day embryo

Developing brain
Heart prominence
Developing spinal cord
Somite

(c) 32-day embryo

Pharyngeal arches
Lens placode
Heart prominence
Upper limb bud
Tail
Lower limb bud

(d) 44-day embryo

Otic placode
Developing nose
Upper limb
Lower limb
Umbilical cord

(e) 52-day embryo

Ear
Eye
Nose
Upper limb
Umbilical cord
Lower limb

(f) 10-week fetus

Ear
Eye
Nose
Upper limb
Yolk sac
Rib
Umbilical cord
Placenta
Lower limb

(g) 13-week fetus

Ear
Eye
Nose
Mouth
Upper limb
Umbilical cord
Lower limb

(h) 26-week fetus

Ear
Eye
Nose
Mouth
Upper limb
Lower limb

❓ **How does mid-fetal weight compare to end-fetal weight?**

29.3 TERATOGENS

• Define a teratogen and list several examples of teratogens.

Exposure of a developing embryo or fetus to certain environmental factors can damage the developing organism or even cause death. A **teratogen** (TER-a-tō-jen; *terato-* = monster; *-gen* = creating) is any agent or influence that causes developmental defects in the embryo. In the following sections we briefly discuss several examples.

Chemicals and Drugs

Because the placenta is not an absolute barrier between the maternal and fetal circulations, any drug or chemical that is dangerous to an infant should be considered potentially dangerous to the fetus when given to the mother. Alcohol is by far the number-one fetal teratogen. Intrauterine exposure to even a small amount of alcohol may result in **fetal alcohol syndrome (FAS),** one of the most common causes of mental retardation and the most common preventable cause of birth defects in the United States. The symptoms of FAS may include slow growth before and after birth, characteristic facial features (short palpebral fissures, a thin upper lip, and sunken nasal bridge), defective heart and other organs, malformed limbs, genital abnormalities, and central nervous system damage. Behavioral problems, such as hyperactivity, extreme nervousness, reduced ability to concentrate, and an inability to appreciate cause-and-effect relationships, are common.

Other teratogens include certain viruses (hepatitis B and C and certain papilloma viruses that cause sexually transmitted diseases); pesticides; defoliants (chemicals that cause plants to shed their leaves prematurely); industrial chemicals; some hormones; antibiotics; oral anticoagulants, anticonvulsants, antitumor agents, thyroid drugs, thalidomide, diethylstilbestrol (DES), and numerous other prescription drugs; LSD; and cocaine. A pregnant woman who uses cocaine, for example, subjects the fetus to higher risk of retarded growth, attention and orientation problems, hyperirritability, a tendency to stop breathing, malformed or missing organs, strokes, and seizures. The risks of spontaneous abortion, premature birth, and stillbirth also increase with fetal exposure to cocaine.

Cigarette Smoking

Strong evidence implicates cigarette smoking during pregnancy as a cause of low infant birth weight; there is also a strong association between smoking and a higher fetal and infant mortality rate. Women who smoke have a much higher risk of an ectopic pregnancy. Cigarette smoke may be teratogenic and may cause cardiac abnormalities as well as anencephaly (see Clinical Connection: Anencephaly). Maternal smoking also is a significant factor in the development of cleft lip and palate and has been linked with sudden infant death syndrome (SIDS). Infants nursing from smoking mothers have also been found to have an increased incidence of gastrointestinal disturbances. Even a mother's exposure to secondhand cigarette smoke (breathing air containing tobacco smoke) during pregnancy or while nursing predisposes her baby to increased incidence of respiratory problems, including bronchitis and pneumonia, during the first year of life.

Irradiation

Ionizing radiation of various kinds is a potent teratogen. Exposure of pregnant mothers to x-rays or radioactive isotopes during the embryo's susceptible period of development may cause microcephaly (small head size relative to the rest of the body), mental retardation, and skeletal malformations. Caution is advised, especially during the first trimester of pregnancy.

✔ **CHECKPOINT**

24. What are some of the symptoms of fetal alcohol syndrome?
25. How does cigarette smoking affect embryonic and fetal development?

29.4 PRENATAL DIAGNOSTIC TESTS

• Describe the procedures for fetal ultrasonography, amniocentesis, and chorionic villi sampling.

Several tests are available to detect genetic disorders and assess fetal well-being. Here we describe fetal ultrasonography, amniocentesis, and chorionic villi sampling (CVS).

Fetal Ultrasonography

If there is a question about the normal progress of a pregnancy, **fetal ultrasonography** (ul-tra-son-OG-ra-fē) may be performed. By far the most common use of diagnostic ultrasound is to determine a more accurate fetal age when the date of conception is unclear. It is also used to confirm pregnancy, evaluate fetal viability and growth, determine fetal position, identify multiple pregnancies, identify fetal–maternal abnormalities, and serve as an adjunct to special procedures such as amniocentesis. During fetal ultrasonography, a transducer, an instrument that emits high-frequency sound waves, is passed back and forth over the abdomen. The reflected sound waves from the developing fetus are picked up by the transducer and converted to an on-screen image called a **sonogram** (see Table 1.3). Because the urinary bladder serves as a landmark during the procedure, the patient needs to drink liquids before the procedure and not void urine to maintain a full bladder.

Amniocentesis

Amniocentesis (am′-nē-ō-sen-TĒ-sis; *amnio-* = amnion; *-centesis* = puncture to remove fluid) involves withdrawing some of the amniotic fluid that bathes the developing fetus and analyzing the fetal cells and dissolved substances. It is used to test for the presence of certain genetic disorders, such as Down syndrome (DS), hemophilia, Tay-Sachs disease, sickle-cell disease, and certain muscular dystrophies. It is also used to help determine survivability of the fetus. The test is usually done at 14–18 weeks of gestation. All gross chromosomal abnormalities and over 50 biochemical defects can be detected through amniocentesis. It can also reveal the baby's gender, which is important information for the diagnosis of sex-linked disorders, in which an abnormal gene carried by the mother affects her male offspring only (described in Section 29.10).

During amniocentesis, the position of the fetus and placenta is first identified using ultrasound and palpation. After the skin is prepared with an antiseptic and a local anesthetic is given, a hypodermic needle is inserted through the mother's abdominal wall and into the amniotic cavity within the uterus. Then, 10 to 30 mL of fluid and suspended cells are aspirated (Figure 29.15a) for microscopic examination and biochemical testing. Elevated levels of alpha-fetoprotein (AFP) and acetylcholinesterase may indicate failure of the nervous system to develop properly, as occurs in spina bifida or anencephaly (absence of the cerebrum), or may be due to other developmental or chromosomal problems. Chromosome studies, which require growing the cells for 2–4 weeks in a culture medium, may reveal rearranged, missing, or extra chromosomes. Amniocentesis is performed only when a risk for genetic defects is suspected, because there is about a 0.5% chance of spontaneous abortion after the procedure.

Chorionic Villi Sampling

In **chorionic villi sampling** or **CVS,** a catheter is guided through the vagina and cervix of the uterus and then advanced to the chorionic villi under ultrasound guidance (Figure 29.15b). About 30 milligrams of tissue is suctioned out and prepared for chromosomal analysis. Alternatively, the chorionic villi can be sampled by inserting a needle through the abdominal cavity, as performed in amniocentesis.

CVS can identify the same defects as amniocentesis because chorion cells and fetal cells contain the same genome. CVS offers several advantages over amniocentesis: It can be performed as early as 8 weeks of gestation, and test results are available in only a few days, permitting an earlier decision on whether to continue the pregnancy. However, CVS is slightly riskier than amniocentesis; after the procedure there is a 1–2% chance of spontaneous abortion.

Noninvasive Prenatal Tests

Currently, chorionic villi testing and amniocentesis are the only useful ways to obtain fetal tissue for prenatal testing of gene defects. While these invasive procedures pose relatively little risk when performed by experts, much work has been done to develop **noninvasive prenatal tests,** which do not require the penetration of any

Figure 29.15 Amniocentesis and chorionic villi sampling.

To detect genetic abnormalities, amniocentesis is performed at 14–16 weeks of gestation; chorionic villi sampling may be performed as early as 8 weeks of gestation.

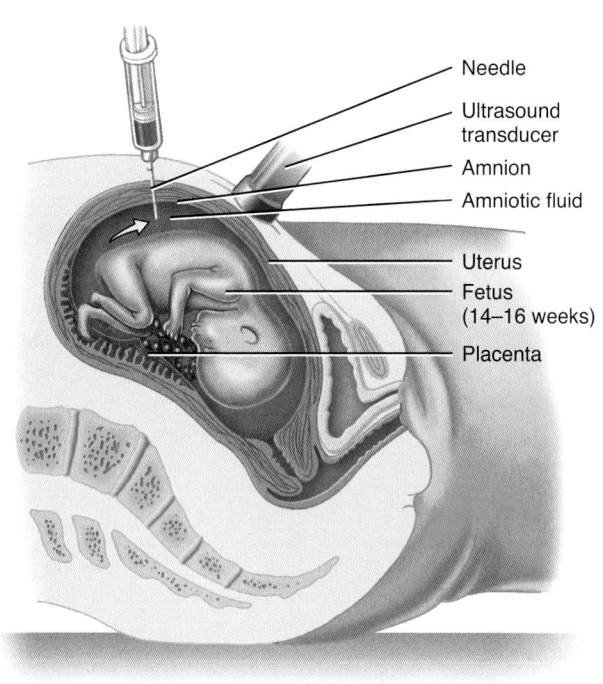

(a) Amniocentesis

(b) Chorionic villi sampling (CVS)

? **What information can be provided by amniocentesis?**

embryonic structure. The goal is to develop accurate, safe, more efficient, and less expensive tests for screening a large population.

The first such test developed was the **maternal alpha-fetoprotein (AFP) test** (AL-fa fē′-tō-PRŌ-tēn). In this test, the mother's blood is analyzed for the presence of AFP, a protein synthesized in the fetus that passes into the maternal circulation. The highest levels of AFP normally occur during weeks 12 through 15 of pregnancy. Later, AFP is not produced, and its concentration decreases to a very low level both in the fetus and in maternal blood. A high level of AFP after week 16 usually indicates that the fetus has a neural tube defect, such as spina bifida or anencephaly. Because the test is 95% accurate, it is now recommended that all pregnant women be tested for AFP. A newer test (Quad AFP Plus) probes maternal blood for AFP and three other molecules. The test permits prenatal screening for Down syndrome, trisomy 18, and neural tube defects; it also helps predict the delivery date and may reveal the presence of twins.

✔ CHECKPOINT

26. What conditions can be detected using fetal ultrasonography, amniocentesis, and chorionic villi sampling? What are the advantages of noninvasive prenatal tests?

29.5 MATERNAL CHANGES DURING PREGNANCY

◉ OBJECTIVES

• Describe the sources and functions of the hormones secreted during pregnancy.
• Describe the hormonal, anatomical, and physiological changes in the mother during pregnancy.

Hormones of Pregnancy

During the first 3 to 4 months of pregnancy, the corpus luteum in the ovary continues to secrete **progesterone** and **estrogens,** which maintain the lining of the uterus during pregnancy and prepare the mammary glands to secrete milk. The amounts secreted by the corpus luteum, however, are only slightly more than those produced after ovulation in a normal menstrual cycle. From the third month through the remainder of the pregnancy, the placenta itself provides the high levels of progesterone and estrogens required. As noted previously, the chorion secretes **human chorionic gonadotropin (hCG)** (kō-rē-ON-ik gō′-nad-ō-TRŌ-pin) into the blood. In turn, hCG stimulates the corpus luteum to continue production of progesterone and estrogens—an activity required to prevent menstruation and for the continued attachment of the embryo and fetus to the lining of the uterus (Figure 29.16a). By the eighth day after fertilization, hCG can be detected in the blood and urine of a pregnant woman. Peak secretion of hCG occurs at about the ninth week of pregnancy (Figure 29.16b). During the fourth and fifth months the hCG level decreases sharply and then levels off until childbirth.

The chorion begins to secrete estrogens after the first 3 or 4 weeks of pregnancy and progesterone by the sixth week. These hormones are secreted in increasing quantities until the time of birth (Figure 29.16b). By the fourth month, when the placenta is fully established, the secretion of hCG is greatly reduced, and the secretions of the corpus luteum are no longer essential. A high level of progesterone ensures that the uterine myometrium is relaxed and that the cervix is tightly closed. After delivery, estrogens and progesterone in the blood decrease to normal levels.

Relaxin, a hormone produced first by the corpus luteum of the ovary and later by the placenta, increases the flexibility of the pubic symphysis and ligaments of the sacroiliac and sacrococcygeal joints and helps dilate the uterine cervix during labor. Both of these actions ease delivery of the baby.

A third hormone produced by the chorion of the placenta is **human chorionic somatomammotropin (hCS)** (sō′-ma-tō-mam-ō-TRŌ-pin), also known as **human placental lactogen (hPL).** The rate of secretion of hCS increases in proportion to placental mass, reaching maximum levels after 32 weeks and remaining relatively constant after that. It is thought to help prepare the mammary glands for lactation, enhance maternal growth by increasing protein synthesis, and regulate certain aspects of metabolism in both mother and fetus. For example, hCS decreases the use of glucose by the mother and promotes the release of fatty acids from her adipose tissue, making more glucose available to the fetus.

The hormone most recently found to be produced by the placenta is **corticotropin-releasing hormone (CRH)** (kor′-ti-kō-TRŌ-pin), which in nonpregnant people is secreted only by neurosecretory cells in the hypothalamus. CRH is now thought to be part of the "clock" that establishes the timing of birth. Secretion of CRH by the placenta begins at about 12 weeks and increases enormously toward the end of pregnancy. Women who have higher levels of CRH earlier in pregnancy are more likely to deliver prematurely; those who have low levels are more likely to deliver after their due date. CRH from the placenta has a second important effect: It increases secretion of cortisol, which is needed for maturation of the fetal lungs and the production of surfactant (see "Alveoli" in Section 23.1).

⚕ **CLINICAL CONNECTION | *Early Pregnancy Tests***

Early pregnancy tests detect the tiny amounts of human chorionic gonadotropin (hCG) in the urine that begin to be excreted about 8 days after fertilization. The test kits can detect pregnancy as early as the first day of a missed menstrual period—that is, at about 14 days after fertilization. Chemicals in the kits produce a color change if a reaction occurs between hCG in the urine and hCG antibodies included in the kit.

Several of the test kits available at pharmacies are as sensitive and accurate as test methods used in many hospitals. Still, false-negative and false-positive results can occur. A false-negative result (the test is negative, but the woman is pregnant) may be due to testing too soon or to an ectopic pregnancy. A false-positive result (the test is positive, but the woman is not pregnant) may be due to excess protein or blood in the urine or to hCG production due to a rare type of uterine cancer. Thiazide diuretics, hormones, steroids, and thyroid drugs may also affect the outcome of an early pregnancy test. •

Figure 29.16 Hormones during pregnancy.

The corpus luteum produces progesterone and estrogens during the first 3–4 months of pregnancy, after which time the placenta assumes this function.

(a) Sources and functions of hormones

(b) Blood levels of hormones during pregnancy

? **Which hormone is detected by early pregnancy tests?**

Changes during Pregnancy

Near the end of the third month of pregnancy, the uterus occupies most of the pelvic cavity. As the fetus continues to grow, the uterus extends higher and higher into the abdominal cavity. Toward the end of a full-term pregnancy, the uterus fills nearly the entire abdominal cavity, reaching above the costal margin nearly to the xiphoid process of the sternum (Figure 29.17). It pushes the maternal intestines, liver, and stomach superiorly, elevates the diaphragm, and widens the thoracic cavity. Pressure on the stomach may force the stomach contents superiorly into the esophagus, resulting in heartburn. In the pelvic cavity, compression of the ureters and urinary bladder occurs.

Pregnancy-induced physiological changes also occur, including weight gain due to the fetus, amniotic fluid, the placenta, uterine enlargement, and increased total body water; increased storage of proteins, triglycerides, and minerals; marked breast enlargement in preparation for lactation; and lower back pain due to lordosis (hollow back).

Several changes occur in the maternal cardiovascular system. Stroke volume increases by about 30% and cardiac output rises by 20–30% due to increased maternal blood flow to the placenta and increased metabolism. Heart rate increases 10–15% and blood volume increases 30–50%, mostly during the second half of pregnancy. These increases are necessary to meet the additional demands of the fetus for nutrients and oxygen. When a pregnant woman is lying on her back, the enlarged uterus may compress the aorta, resulting in diminished blood flow to the uterus. Compression of the inferior vena cava also decreases venous return, which leads to edema in the lower limbs and may

Figure 29.17 **Normal fetal location and position at the end of a full-term pregnancy.**

 The gestation period is the time interval (about 38 weeks) from fertilization to birth.

Label (left)	Label (right)
Right lung	Left lung
	Heart
Right breast	Left breast
Gallbladder	
Liver	Stomach
Greater omentum	
Small intestine	
Uterine wall	Small intestine
Ascending colon	
Maternal umbilicus	Descending colon
Right uterine (fallopian) tube	
Right ovary	
Umbilical cord	
	Left ovary
Inguinal ligament	Left uterine (fallopian) tube
Round ligament of uterus	
	Head of fetus
Urinary bladder	
Pubic symphysis	

Anterior view of position of organs at end of full-term pregnancy

? **What hormone increases the flexibility of the pubic symphysis and helps dilate the cervix of the uterus to ease delivery of the baby?**

produce varicose veins. Compression of the renal artery can lead to renal hypertension.

Respiratory function is also altered during pregnancy to meet the added oxygen demands of the fetus. Tidal volume can increase by 30–40%, expiratory reserve volume can be reduced by up to 40%, functional residual capacity can decline by up to 25%, minute ventilation (the total volume of air inhaled and exhaled each minute) can increase by up to 40%, airway resistance in the bronchial tree can decline by 30–40%, and total body oxygen consumption can increase by about 10–20%. Dyspnea (difficult breathing) also occurs.

The digestive system also undergoes changes. Pregnant women experience an increase in appetite due to the added nutritional demands of the fetus. A general decrease in GI tract motility can cause constipation, delay gastric emptying time, and produce nausea, vomiting, and heartburn.

Pressure on the urinary bladder by the enlarging uterus can produce urinary symptoms, such as increased frequency and urgency of urination, and stress incontinence. An increase in renal plasma flow up to 35% and an increase in glomerular filtration rate up to 40% increase the renal filtering capacity, which allows faster elimination of the extra wastes produced by the fetus.

Changes in the skin during pregnancy are more apparent in some women than in others. Some women experience increased pigmentation around the eyes and cheekbones in a masklike pattern (chloasma), in the areolae of the breasts, and in the linea alba of the lower abdomen (linea nigra). Striae (stretch marks) over the abdomen can occur as the uterus enlarges, and hair loss increases.

Changes in the reproductive system include edema and increased vascularity of the vulva and increased pliability and vascularity of the vagina. The uterus increases from its nonpregnant mass of 60–80 g to 900–1200 g at term because of hyperplasia of muscle fibers in the myometrium in early pregnancy and hypertrophy of muscle fibers during the second and third trimesters.

CLINICAL CONNECTION | Pregnancy-Induced Hypertension

About 10–15% of all pregnant women in the United States experience **pregnancy-induced hypertension (PIH)**, an elevated blood pressure that is associated with pregnancy. The major cause is **preeclampsia** (prē-ē-KLAMP-sē-a), an abnormal condition of pregnancy characterized by sudden hypertension, large amounts of protein in the urine, and generalized edema that typically appears after the twentieth week of pregnancy. Other signs and symptoms are generalized edema, blurred vision, and headaches. Preeclampsia might be related to an autoimmune or allergic reaction resulting from the presence of a fetus. Treatment involves bed rest and various drugs. When the condition is also associated with convulsions and coma, it is termed **eclampsia.** •

✔**CHECKPOINT**

27. List the hormones involved in pregnancy, and describe the functions of each.
28. What structural and functional changes occur in the mother during pregnancy?

29.6 EXERCISE AND PREGNANCY

◉ **OBJECTIVE**

• Explain the effects of pregnancy on exercise and of exercise on pregnancy.

Only a few changes in early pregnancy affect the ability to exercise. A pregnant woman may tire more easily than usual, or morning sickness may interfere with regular exercise. As the pregnancy progresses, weight is gained and posture changes, so more energy is needed to perform activities, and certain maneuvers (sudden stopping, changes in direction, rapid movements) are more difficult to execute. In addition, certain joints, especially the pubic symphysis, become less stable in response to the increased level of the hormone relaxin. As compensation, many mothers-to-be walk with widely spread legs and a shuffling motion.

Although blood shifts from viscera (including the uterus) to the muscles and skin during exercise, there is no evidence of inadequate blood flow to the placenta. The heat generated during exercise may cause dehydration and further increase body temperature. Especially during early pregnancy, excessive exercise and heat buildup should be avoided because elevated body temperature has been implicated in neural tube defects. Exercise has no known effect on lactation, provided a woman remains hydrated and wears a bra that provides good support. Overall, moderate physical activity does not endanger the fetus of a healthy woman who has a normal pregnancy. However, any physical activity that might endanger the fetus should be avoided.

Among the benefits of exercise to the mother during pregnancy are a greater sense of well-being and fewer physical complaints.

✔**CHECKPOINT**

29. Which changes in pregnancy have an effect on the ability to exercise?

29.7 LABOR

◉ **OBJECTIVE**

• Explain the events associated with the three stages of labor.

Labor is the process by which the fetus is expelled from the uterus through the vagina, also referred to as giving birth. A synonym for labor is **parturition** (par-toor-ISH-un; *parturit-* = childbirth).

The onset of labor is determined by complex interactions of several placental and fetal hormones. Because progesterone inhibits uterine contractions, labor cannot take place until the effects of progesterone are diminished. Toward the end of gestation, the levels of estrogens in the mother's blood rise sharply, producing changes that overcome the inhibiting effects of progesterone. The rise in estrogens results from increasing secretion by the placenta of corticotropin-releasing hormone, which stimulates the anterior pituitary gland of the fetus to secrete ACTH (adrenocorticotropic hormone). In turn, ACTH stimulates the fetal adrenal gland to secrete cortisol and dehydroepiandrosterone (DHEA) (dē-hī′-drō-ep′-ē-an-DROS-ter-ōn), the major adrenal androgen. The placenta then converts DHEA into an estrogen. High levels of estrogens cause the number of receptors for oxytocin on uterine muscle fibers to increase, and cause uterine muscle fibers to form gap junctions with one another. Oxytocin released by the posterior pituitary stimulates uterine contractions, and relaxin from the placenta assists by increasing the flexibility of the pubic symphysis and helping dilate the uterine cervix. Estrogen also stimulates the placenta to release prostaglandins, which induce production of enzymes that digest collagen fibers in the cervix, causing it to soften.

Control of labor contractions during parturition occurs via a positive feedback cycle (see Figure 1.4). Contractions of the uterine myometrium force the baby's head or body into the cervix, distending (stretching) the cervix. Stretch receptors in the cervix send nerve impulses to neurosecretory cells in the hypothalamus, causing them to release oxytocin into blood capillaries of the posterior pituitary gland. Oxytocin then is carried by the blood to the uterus, where it stimulates the myometrium to contract more forcefully. As the contractions intensify, the baby's body stretches the cervix still more, and the resulting nerve impulses stimulate the secretion of yet more oxytocin. With birth of the infant, the positive feedback cycle is broken because cervical distension suddenly lessens.

Uterine contractions occur in waves (quite similar to the peristaltic waves of the gastrointestinal tract) that start at the top of the uterus and move downward, eventually expelling the fetus. **True labor** begins when uterine contractions occur at regular intervals, usually producing pain. As the interval between contractions shortens, the contractions intensify. Another symptom of true labor in some women is localization of pain in the back that is intensified by walking. The most reliable indicator of true labor is dilation of the cervix and the "show," a discharge of a blood-containing mucus into the cervical canal. In **false labor,** pain is felt in the abdomen at irregular intervals, but it does not intensify and walking does not alter it significantly. There is no "show" and no cervical dilation.

True labor can be divided into three stages (Figure 29.18):

1. *Stage of dilation.* The time from the onset of labor to the complete dilation of the cervix is the **stage of dilation.** This stage, which typically lasts 6–12 hours, features regular contractions of the uterus, usually a rupturing of the amniotic sac, and complete dilation (to 10 cm) of the cervix. If the amniotic sac does not rupture spontaneously, it is ruptured intentionally.

2. *Stage of expulsion.* The time (10 minutes to several hours) from complete cervical dilation to delivery of the baby is the **stage of expulsion.**

3. *Placental stage.* The time (5–30 minutes or more) after delivery until the placenta or "afterbirth" is expelled by powerful uterine contractions is the **placental stage.** These contractions also constrict blood vessels that were torn during delivery, reducing the likelihood of hemorrhage.

As a rule, labor lasts longer with first babies, typically about 14 hours. For women who have previously given birth, the average duration of labor is about 8 hours—although the time varies

Figure 29.18 Stages of true labor.

The term *parturition* refers to birth.

1 Stage of dilation

2 Stage of expulsion

3 Placental stage

What event marks the beginning of the stage of expulsion?

enormously among births. Because the fetus may be squeezed through the birth canal (cervix and vagina) for up to several hours, the fetus is stressed during childbirth: The fetal head is compressed, and the fetus undergoes some degree of intermittent hypoxia due to compression of the umbilical cord and the placenta during uterine contractions. In response to this stress, the fetal adrenal medullae secrete very high levels of epinephrine and norepinephrine, the "fight-or-flight" hormones. Much of the protection against the stresses of parturition, as well as preparation of the infant for surviving extrauterine life, is provided by these hormones.

Among other functions, epinephrine and norepinephrine clear the lungs and alter their physiology in readiness for breathing air, mobilize readily usable nutrients for cellular metabolism, and promote an increased flow of blood to the brain and heart.

About 7% of pregnant women do not deliver by 2 weeks after their due date. Such cases carry an increased risk of brain damage to the fetus, and even fetal death, due to inadequate supplies of oxygen and nutrients from an aging placenta. Post-term deliveries may be facilitated by inducing labor, initiated by administration of oxytocin (Pitocin®), or by surgical delivery (cesarean section).

Following the delivery of the baby and placenta is a 6-week period during which the maternal reproductive organs and physiology return to the prepregnancy state. This period is called the **puerperium** (pū-er-PER-ē-um). Through a process of tissue catabolism, the uterus undergoes a remarkable reduction in size, called **involution** (in-vō-LOO-shun), especially in lactating women. The cervix loses its elasticity and regains its prepregnancy firmness. For 2–4 weeks after delivery, women have a uterine discharge called **lochia** (LŌ-kē-a), which consists initially of blood and later of serous fluid derived from the former site of the placenta.

CLINICAL CONNECTION | Dystocia and Cesarean Section

Dystocia (dis-TŌ-sē-a; *dys-* = painful or difficult; *-toc-* = birth), or difficult labor, may result either from an abnormal position (presentation) of the fetus or a birth canal of inadequate size to permit vaginal delivery. In a **breech presentation,** for example, the fetal buttocks or lower limbs, rather than the head, enter the birth canal first; this occurs most often in premature births. If fetal or maternal distress prevents a vaginal birth, the baby may be delivered surgically through an abdominal incision. A low, horizontal cut is made through the abdominal wall and lower portion of the uterus, through which the baby and placenta are removed. Even though it is popularly associated with the birth of Julius Caesar, the true reason this procedure is termed a **cesarean section (C-section)** is because it was described in Roman law, *lex cesarea,* about 600 years before Julius Caesar was born. Even a history of multiple C-sections need not exclude a pregnant woman from attempting a vaginal delivery. •

✔ CHECKPOINT

30. What hormonal changes induce labor?
31. What is the difference between false labor and true labor?
32. What happens during the stage of dilation, the stage of expulsion, and the placental stage of true labor?

29.8 ADJUSTMENTS OF THE INFANT AT BIRTH

◉ OBJECTIVE

• Explain the respiratory and cardiovascular adjustments that occur in an infant at birth.

During pregnancy, the embryo (and later the fetus) is totally dependent on the mother for its existence. The mother supplies the fetus with oxygen and nutrients, eliminates its carbon dioxide and

other wastes, protects it against shocks and temperature changes, and provides antibodies that confer protection against certain harmful microbes. At birth, a physiologically mature baby becomes much more self-supporting, and the newborn's body systems must make various adjustments. The most dramatic changes occur in the respiratory and cardiovascular systems.

Respiratory Adjustments

The reason that the fetus depends entirely on the mother for obtaining oxygen and eliminating carbon dioxide is that the fetal lungs are either collapsed or partially filled with amniotic fluid. The production of surfactant begins by the end of the sixth month of development. Because the respiratory system is fairly well developed at least 2 months before birth, premature babies delivered at 7 months are able to breathe and cry. After delivery, the baby's supply of oxygen from the mother ceases, and any amniotic fluid in the fetal lungs is absorbed. Because carbon dioxide is no longer being removed, it builds up in the blood. A rising CO_2 level stimulates the respiratory center in the medulla oblongata, causing the respiratory muscles to contract, and the baby to draw his or her first breath. Because the first inspiration is unusually deep, as the lungs contain no air, the baby also exhales vigorously and naturally cries. A full-term baby may breathe 45 times a minute for the first 2 weeks after birth. Breathing rate gradually declines until it approaches a normal rate of 12 breaths per minute.

Cardiovascular Adjustments

After the baby's first inspiration, the cardiovascular system must make several adjustments (see Figure 21.30). Closure of the foramen ovale between the atria of the fetal heart, which occurs at the moment of birth, diverts deoxygenated blood to the lungs for the first time. The foramen ovale is closed by two flaps of septal heart tissue that fold together and permanently fuse. The remnant of the foramen ovale is the fossa ovalis.

Once the lungs begin to function, the ductus arteriosus shuts off due to contractions of smooth muscle in its wall, and it becomes the ligamentum arteriosum. The muscle contraction is probably mediated by the polypeptide bradykinin, released from the lungs during their initial inflation. The ductus arteriosus generally does not close completely until about 3 months after birth. Prolonged incomplete closure results in a condition called **patent ductus arteriosus** (see Figure 20.23b).

After the umbilical cord is tied off and severed and blood no longer flows through the umbilical arteries, they fill with connective tissue, and their distal portions become the medial umbilical ligaments. The umbilical vein then becomes the ligamentum teres (round ligament) of the liver.

In the fetus, the ductus venosus connects the umbilical vein directly with the inferior vena cava, allowing blood from the placenta to bypass the fetal liver. When the umbilical cord is severed, the ductus venosus collapses, and venous blood from the viscera of the fetus flows into the hepatic portal vein to the liver and then via the hepatic vein to the inferior vena cava. The remnant of the ductus venosus becomes the ligamentum venosum.

At birth, an infant's pulse may range from 120 to 160 beats per minute and may go as high as 180 on excitation. After birth, oxygen use increases, which stimulates an increase in the rate of red blood cell and hemoglobin production. The white blood cell count at birth is very high—sometimes as much as 45,000 cells per microliter—but the count decreases rapidly by the seventh day. Recall that the white blood cell count of an adult is 5000–10,000 cells per microliter.

CLINICAL CONNECTION | *Premature Infants*

Delivery of a physiologically immature baby carries certain risks. A **premature infant** or "preemie" is generally considered a baby who weighs less than 2500 g (5.5 lb) at birth. Poor prenatal care, drug abuse, history of a previous premature delivery, and mother's age below 16 or above 35 increase the chance of premature delivery. The body of a premature infant is not yet ready to sustain some critical functions, and thus its survival is uncertain without medical intervention. The major problem after delivery of an infant under 36 weeks of gestation is respiratory distress syndrome (RDS) of the newborn due to insufficient surfactant. RDS can be eased by use of artificial surfactant and a ventilator that delivers oxygen until the lungs can operate on their own. •

✔CHECKPOINT

33. Why are respiratory and cardiovascular adjustments so important at birth?

29.9 THE PHYSIOLOGY OF LACTATION

⦿ OBJECTIVE

• Discuss the physiology and hormonal control of lactation.

Lactation (lak-TĀ-shun) is the production and ejection of milk from the mammary glands. A principal hormone in promoting milk production is **prolactin (PRL),** which is secreted from the anterior pituitary gland. Even though prolactin levels increase as the pregnancy progresses, no milk production occurs because progesterone inhibits the effects of prolactin. After delivery, the levels of estrogens and progesterone in the mother's blood decrease, and the inhibition is removed. The principal stimulus in maintaining prolactin secretion during lactation is the sucking action of the infant. Suckling initiates nerve impulses from stretch receptors in the nipples to the hypothalamus; the impulses decrease hypothalamic release of prolactin-inhibiting hormone (PIH) and increase release of prolactin-releasing hormone (PRH), so more prolactin is released by the anterior pituitary.

Oxytocin causes release of milk into the mammary ducts via the **milk ejection reflex** (Figure 29.19). Milk formed by the glandular cells of the breasts is stored until the baby begins active suckling. Stimulation of touch receptors in the nipple initiates sensory nerve impulses that are relayed to the hypothalamus. In response, secretion of oxytocin from the posterior pituitary increases. Carried by the bloodstream to the mammary glands, oxytocin stimulates contraction of myoepithelial (smooth muscle–like)

Figure 29.19 The milk ejection reflex, a positive feedback cycle.

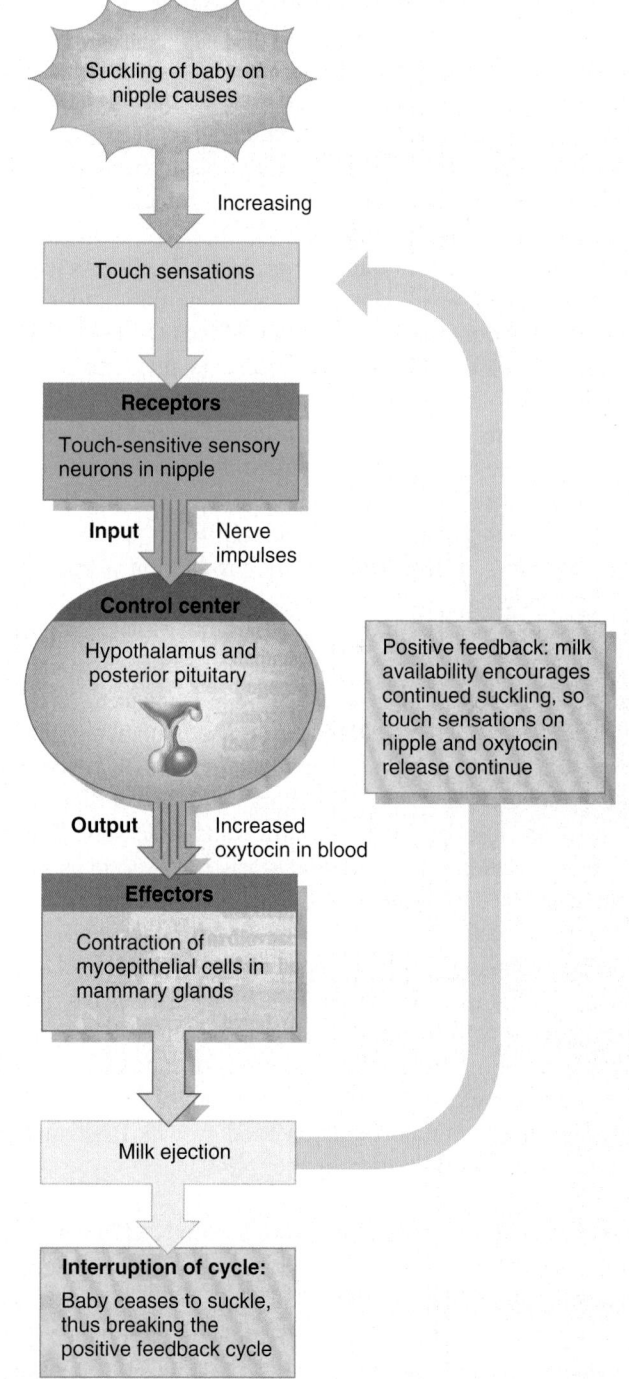

Oxytocin stimulates contraction of myoepithelial cells in the breasts, which squeezes the glandular and duct cells and causes milk ejection.

Suckling of baby on nipple causes

Increasing

Touch sensations

Receptors

Touch-sensitive sensory neurons in nipple

Input | Nerve impulses

Control center

Hypothalamus and posterior pituitary

Positive feedback: milk availability encourages continued suckling, so touch sensations on nipple and oxytocin release continue

Output | Increased oxytocin in blood

Effectors

Contraction of myoepithelial cells in mammary glands

Milk ejection

Interruption of cycle:
Baby ceases to suckle, thus breaking the positive feedback cycle

? **What is another function of oxytocin?**

cells surrounding the glandular cells and ducts. The resulting compression moves the milk from the alveoli of the mammary glands into the mammary ducts, where it can be suckled. This process is termed **milk ejection (let-down).** Even though the actual ejection of milk does not occur until 30–60 seconds after nursing begins (the latent period), some milk stored in lactiferous sinuses near the nipple is available during the latent period. Stimuli other than suckling, such as hearing a baby's cry or touching the mother's genitals, also can trigger oxytocin release and milk ejection. The suckling stimulation that produces the release of oxytocin also inhibits the release of PIH; this results in increased secretion of prolactin, which maintains lactation.

During late pregnancy and the first few days after birth, the mammary glands secrete a cloudy fluid called **colostrum.** Although it is not as nutritious as milk—it contains less lactose and virtually no fat—colostrum serves adequately until the appearance of true milk on about the fourth day. Colostrum and maternal milk contain important antibodies that protect the infant during the first few months of life.

Following birth of the infant, the prolactin level starts to return to the nonpregnant level. However, each time the mother nurses the infant, nerve impulses from the nipples to the hypothalamus increase the release of PRH (and decrease the release of PIH), resulting in a tenfold increase in prolactin secretion by the anterior pituitary that lasts about an hour. Prolactin acts on the mammary glands to provide milk for the next nursing period. If this surge of prolactin is blocked by injury or disease, or if nursing is discontinued, the mammary glands lose their ability to produce milk in only a few days. Even though milk production normally decreases considerably within 7–9 months after birth, it can continue for several years if nursing or **breast-feeding** continues.

Lactation often blocks ovarian cycles for the first few months following delivery, if the frequency of sucking is about 8–10 times a day. This effect is inconsistent, however, and ovulation commonly precedes the first menstrual period after delivery of a baby. As a result, the mother can never be certain she is not fertile. Breast-feeding is therefore an unreliable birth control measure. The suppression of ovulation during lactation is believed to occur as follows: During breast-feeding, neural input from the nipple reaches the hypothalamus and causes it to produce neurotransmitters that suppress the release of gonadotropin-releasing hormone (GnRH). As a result, production of LH and FSH decreases, and ovulation is inhibited.

A primary benefit of breast-feeding is nutritional: Human milk is a sterile solution that contains amounts of fatty acids, lactose, amino acids, minerals, vitamins, and water that are ideal for the baby's digestion, brain development, and growth. Breast-feeding also benefits infants by providing the following:

- *Beneficial cells.* Several types of white blood cells are present in breast milk. Neutrophils and macrophages serve as phagocytes, ingesting microbes in the baby's gastrointestinal tract. Macrophages also produce lysozyme and other immune system components. Plasma cells, which develop from B lymphocytes, produce antibodies against specific microbes, and T lymphocytes kill microbes directly or help mobilize other defenses.

- *Beneficial molecules.* Breast milk also contains an abundance of beneficial molecules. Maternal IgA antibodies in breast milk bind to microbes in the baby's gastrointestinal tract and prevent their migration into other body tissues. Because a mother produces antibodies to whatever disease-causing microbes are present in her environment, her breast milk affords protection against the specific infectious agents to which her baby is also exposed. Additionally, two milk proteins bind to nutrients that many bacteria need to grow and survive: B_{12}-binding protein ties up vitamin B_{12}, and lactoferrin ties up iron. Some fatty acids can kill certain viruses by disrupting their membranes, and lysozyme kills bacteria by disrupting their cell walls. Finally, interferons enhance the antimicrobial activity of immune cells.

- *Decreased incidence of diseases later in life.* Breast-feeding provides children with a slight reduction in risk of lymphoma, heart disease, allergies, respiratory and gastrointestinal infections, ear infections, diarrhea, diabetes mellitus, and meningitis.

- *Miscellaneous benefits.* Breast-feeding supports optimal infant growth, enhances intellectual and neurological development, and fosters mother–infant relations by establishing early and prolonged contact between them. Compared to cow's milk, the fats and iron in breast milk are more easily absorbed, the proteins in breast milk are more readily metabolized, and the lower sodium content of breast milk is more suited to an infant's needs. Premature infants benefit even more from breast-feeding because the milk produced by mothers of premature infants seems to be specially adapted to the infant's needs; it has a higher protein content than the milk of mothers of full-term infants. Finally, a baby is less likely to have an allergic reaction to its mother's milk than to milk from another source.

Years before oxytocin was discovered, it was common practice in midwifery to let a first-born twin nurse at the mother's breast to speed the birth of the second child. Now we know why this practice is helpful—it stimulates the release of oxytocin. Even after a single birth, nursing promotes expulsion of the placenta (afterbirth) and helps the uterus return to its normal size. Synthetic oxytocin (Pitocin) is often given to induce labor or to increase uterine tone and control hemorrhage just after parturition.

✔ CHECKPOINT

34. Which hormones contribute to lactation? What is the function of each?
35. What are the benefits of breast-feeding over bottle-feeding?

29.10 INHERITANCE

◉ OBJECTIVE

- Define inheritance, and explain the inheritance of dominant, recessive, complex, and sex-linked traits.

As previously indicated, the genetic material of a father and a mother unite when a sperm cell fuses with a secondary oocyte to form a zygote. Children resemble their parents because they inherit traits passed down from both parents. We now examine some of the principles involved in that process, called inheritance.

Inheritance is the passage of hereditary traits from one generation to the next. It is the process by which you acquired your characteristics from your parents and may transmit some of your traits to your children. The branch of biology that deals with inheritance is called **genetics** (je-NET-iks). The area of health care that offers advice on genetic problems (or potential problems) is called **genetic counseling.**

Genotype and Phenotype

As you have already learned, the nuclei of all human cells except gametes contain 23 pairs of chromosomes—the diploid number (2*n*). One chromosome in each pair came from the mother, and the other came from the father. Each of these two homologues contains genes that control the same traits. If one chromosome of the pair contains a gene for body hair, for example, its homologue will contain a gene for body hair in the same position. Alternative forms of a gene that code for the same trait and are at the same location on homologous chromosomes are called **alleles** (a-LĒLZ). One allele of the previously mentioned body hair gene might code for coarse hair, and another might code for fine hair. A **mutation** (mū-TĀ-shun; *muta-* = change) is a permanent heritable change in an allele that produces a different variant of the same trait.

The relationship of genes to heredity is illustrated by examining the alleles involved in a disorder called **phenylketonuria (PKU)** (fen′-il-kē′-ton-OO-rē-a). People with PKU (see Clinical Connection: Phenylketonuria in Chapter 25) are unable to manufacture the enzyme phenylalanine hydroxylase. The allele that codes for phenylalanine hydroxylase is symbolized as *P;* the mutated allele that fails to produce a functional enzyme is represented by *p*. The chart in Figure 29.20, which shows the possible combinations of gametes from two parents who each have one *P* and one *p* allele, is called a **Punnett square.** In constructing a Punnett square, the possible paternal alleles in sperm are written at the left side and the possible maternal alleles in ova (or secondary oocytes) are written at the top. The four spaces on the chart show how the alleles can combine in zygotes formed by the union of these sperm and ova to produce the three different combinations of genes, or **genotypes** (JĒ-nō-tīps): *PP, Pp,* or *pp*. Notice from the Punnett square that 25% of the offspring will have the *PP* genotype, 50% will have the *Pp* genotype, and 25% will have the *pp* genotype. (These percentages are probabilities only; parents who have four children won't necessarily end up with one with PKU.) People who inherit *PP* or *Pp* genotypes do not have PKU; those with a *pp* genotype suffer from the disorder. Although people with a *Pp* genotype have one PKU allele (*p*), the allele that codes for the normal trait (*P*) masks the presence of the PKU allele. An allele that dominates or masks the presence of another allele and is fully expressed (*P* in this example) is said to be a **dominant allele,** and the trait expressed is called a dominant trait. The allele whose presence is completely masked (*p* in this example) is

Figure 29.20 Inheritance of phenylketonuria (PKU).

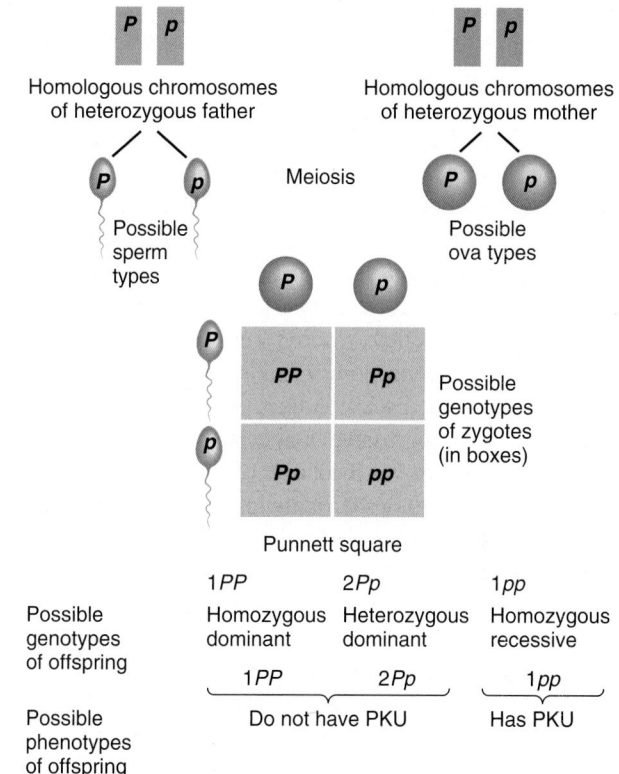

Genotype refers to genetic makeup; phenotype refers to the physical or outward expression of a gene.

 If parents have the genotypes shown here, what is the chance that their first child will have PKU? What is the chance of PKU occurring in their second child?

said to be a **recessive allele,** and the trait it controls is called a recessive trait.

By tradition, the symbols for genes are written in italics, with dominant alleles written in capital letters and recessive alleles in lowercase letters. A person with the same alleles on homologous chromosomes (for example, *PP* or *pp*) is said to be **homozygous** (hō-mō-ZĪ-gus) for the trait. *PP* is homozygous dominant, and *pp* is homozygous recessive. An individual with different alleles on homologous chromosomes (for example, *Pp*) is said to be **heterozygous** (het′-er-ō-ZĪ-gus) for the trait.

Phenotype (FĒ-nō-tīp; *pheno-* = showing) refers to how the genetic makeup is expressed in the body; it is the physical or outward expression of a gene. A person with *Pp* (a heterozygote) has a different *genotype* from a person with *PP* (a homozygote), but both have the same *phenotype*—normal production of phenylalanine hydroxylase. Heterozygous individuals who carry a recessive gene but do not express it (*Pp*) can pass the gene on to their offspring. Such individuals are called **carriers** of the recessive gene.

Most genes give rise to the same phenotype whether they are inherited from the mother or the father. In a few cases, however, the phenotype is dramatically different, depending on the parental origin. This surprising phenomenon, first appreciated in the 1980s, is called **genomic imprinting.** In humans, the abnormalities most clearly associated with mutation of an imprinted gene are *Angelman syndrome* (mental retardation, ataxia, seizures, and minimal speech), which results when the gene for a particular abnormal trait is inherited from the mother, and *Prader-Willi syndrome* (short stature, mental retardation, obesity, poor responsiveness to external stimuli, and sexual immaturity), which results when it is inherited from the father.

Alleles that code for normal traits do not always dominate over those that code for abnormal ones, but dominant alleles for severe disorders usually are lethal and cause death of the embryo or fetus. One exception is Huntington disease (HD) (see Clinical Connection: Disorders of the Basal Nuclei in Chapter 16), which is caused by a dominant allele whose effects do not become manifest until adulthood. Both homozygous dominant and heterozygous people exhibit the disease; homozygous recessive people are normal. HD causes progressive degeneration of the nervous system and eventual death, but because symptoms typically do not appear until after age 30 or 40, many afflicted individuals will already have passed on the allele for the condition to their children by the time they discover they have the disease.

Occasionally an error in cell division, called **nondisjunction** (non'-dis-JUNK-shun), results in an abnormal number of chromosomes. In this situation, homologous chromosomes (during meiosis I) or sister chromatids (during anaphase of mitosis or meiosis II) fail to separate properly. See Figure 3.34. A cell from which one or more chromosomes have been added or deleted is called an **aneuploid** (AN-ū-ployd). A monosomic cell $(2n - 1)$ is missing a chromosome; a trisomic cell $(2n + 1)$ has an extra chromosome. Most cases of Down syndrome (see Disorders: Homeostatic Imbalances at the end of this chapter) are aneuploid disorders in which there is trisomy of chromosome 21. Nondisjunction usually occurs during gametogenesis (meiosis), but about 2% of Down syndrome cases result from nondisjunction during mitotic divisions in early embryonic development.

Another error in meiosis is a **translocation.** In this case, two chromosomes that are *not* homologous break and interchange portions of their chromosomes. The individual who has a translocation may be perfectly normal if no loss of genetic material took place when the rearrangement occurred. However, some of the person's gametes may not contain the correct amount and type of genetic material. About 3% of Down syndrome cases result from a translocation of part of chromosome 21 to another chromosome, usually chromosome 14 or 15. The individual who has this translocation is normal and does not even know that he or she is a "carrier." When such a carrier produces gametes, however, some gametes end up with a whole chromosome 21 plus another chromosome with the translocated fragment of chromosome 21. On fertilization, the zygote then has three, rather than two, copies of that part of chromosome 21.

Table 29.3 lists some dominant and recessive inherited structural and functional traits in humans.

TABLE 29.3

Selected Hereditary Traits in Humans

DOMINANT	RECESSIVE
Normal skin pigmentation	Albinism
Near- or farsightedness	Normal vision
PTC taster*	PTC nontaster
Polydactyly (extra digits)	Normal digits
Brachydactyly (short digits)	Normal digits
Syndactylism (webbed digits)	Normal digits
Diabetes insipidus	Normal urine excretion
Huntington disease	Normal nervous system
Widow's peak	Straight hairline
Curved (hyperextended) thumb	Straight thumb
Normal Cl^- transport	Cystic fibrosis
Hypercholesterolemia (familial)	Normal cholesterol level

*Ability to taste a chemical compound called phenylthiocarbamide (PTC).

Variations on Dominant–Recessive Inheritance

Most patterns of inheritance do not conform to the simple **dominant–recessive inheritance** we have just described, in which only dominant and recessive alleles interact. The phenotypic expression of a particular gene may be influenced not only by which alleles are present, but also by other genes and by the environment. Most inherited traits are influenced by more than one gene, and, to complicate matters, most genes can influence more than one trait. Variations on dominant–recessive inheritance include incomplete dominance, multiple-allele inheritance, and complex inheritance.

Incomplete Dominance

In **incomplete dominance,** neither member of a pair of alleles is dominant over the other, and the heterozygote has a phenotype intermediate between the homozygous dominant and the homozygous recessive phenotypes. An example of incomplete dominance in humans is the inheritance of **sickle-cell disease (SCD)** (Figure 29.21). People with the homozygous dominant genotype $Hb^A Hb^A$ form normal hemoglobin; those with the homozygous recessive genotype $Hb^S Hb^S$ have sickle-cell disease and severe anemia. Although they are usually healthy, those with the heterozygous genotype $Hb^A Hb^S$ have minor problems with anemia because half their hemoglobin is normal and half is not. Heterozygotes are carriers, and they are said to have *sickle-cell trait.*

Multiple-Allele Inheritance

Although a single individual inherits only two alleles for each gene, some genes may have more than two alternative forms; this is the basis for **multiple-allele inheritance.** One example of multiple-allele inheritance is the inheritance of the ABO blood group. The four blood types (phenotypes) of the ABO group—A, B, AB, and O—result from the inheritance of six combinations of

Figure 29.21 Inheritance of sickle-cell disease.

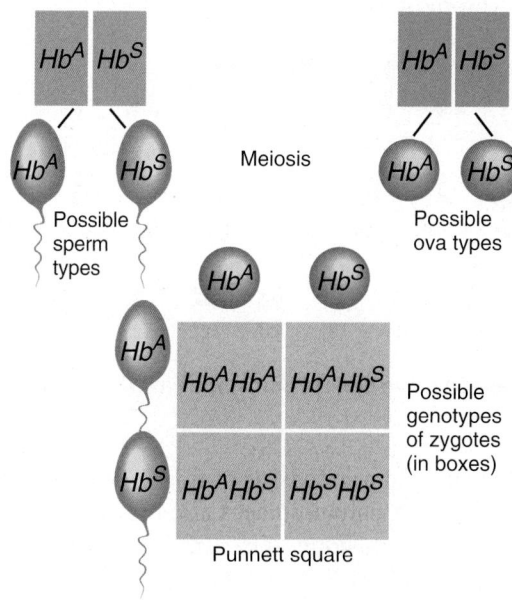

🔑 Sickle-cell disease is an example of incomplete dominance.

Meiosis

Possible sperm types

Possible ova types

Possible genotypes of zygotes (in boxes)

Punnett square

$Hb^A Hb^A$ = normal
$Hb^A Hb^S$ = carrier of sickle-cell disease
$Hb^S Hb^S$ = has sickle-cell disease

❓ **What are the distinguishing features of incomplete dominance?**

three different alleles of a single gene called the *I* gene: (1) allele I^A produces the A antigen, (2) allele I^B produces the B antigen, and (3) allele *i* produces neither A nor B antigen. Each person inherits two *I*-gene alleles, one from each parent, that give rise to the various phenotypes. The six possible genotypes produce four blood types, as follows:

Genotype	Blood type (phenotype)
$I^A I^A$ or $I^A i$	A
$I^B I^B$ or $I^B i$	B
$I^A I^B$	AB
ii	O

Notice that both I^A and I^B are inherited as dominant alleles, and *i* is inherited as a recessive allele. Because an individual with type AB blood has characteristics of both type A and type B red blood cells expressed in the phenotype, alleles I^A and I^B are said to be **codominant.** In other words, both genes are expressed equally in the heterozygote. Depending on the parental blood types, different offspring may have blood types different from each other. Figure 29.22 shows the blood types offspring could inherit, given the blood types of their parents.

Complex Inheritance

Most inherited traits are not controlled by one gene, but instead by the combined effects of two or more genes, a situation referred

Figure 29.22 The 10 possible combinations of parental ABO blood types and the blood types their offspring could inherit.

For each possible set of parents, the blue letters represent the blood types their offspring could inherit.

🔑 Inheritance of ABO blood types is an example of multiple-allele inheritance.

❓ **How is it possible for a baby to have type O blood if neither parent is type O?**

to as **polygenic inheritance** (pol-ē-JĒN-ik; *poly-* = many), or the combined effects of many genes and environmental factors, a situation referred to as **complex inheritance.** Examples of complex traits include skin color, hair color, eye color, height, metabolism rate, and body build. In complex inheritance, one genotype can have many possible phenotypes, depending on the environment, or one phenotype can include many possible genotypes. For example, even though a person inherits several genes for tallness, full height potential may not be reached due to environmental factors, such as disease or malnutrition during the growth years. You have already learned that the risk of having a child with a neural tube defect is greater in pregnant women who lack adequate folic acid in their diet; this is also considered an environmental factor. Because neural tube defects are more prevalent in some families than in others, however, one or more genes may also contribute.

Often, a complex trait shows a continuous gradation of small differences between extremes among individuals. It is relatively easy to predict the risk of passing on an undesirable trait that is due to a single dominant or recessive gene, but it is very difficult to make this prediction when the trait is complex. Such traits are difficult to follow in a family because the range of variation is large, the number of different genes involved usually is not known, and the impact of environmental factors may be incompletely understood.

Skin color is a good example of a complex trait. It depends on environmental factors such as sun exposure and nutrition, as well as on several genes. Suppose that skin color is controlled by three separate genes, each having two alleles: *A, a; B, b;* and *C, c*

Figure 29.23 Complex inheritance of skin color.

In complex inheritance, a trait is controlled by the combined effects of many genes and environmental factors.

? What other traits are transmitted by complex inheritance?

(Figure 29.23). A person with the genotype *AABBCC* is very dark skinned, an individual with the genotype *aabbcc* is very light skinned, and a person with the genotype *AaBbCc* has an intermediate skin color. Parents having an intermediate skin color may have children with very light, very dark, or intermediate skin color. Note that the **P generation** (parental generation) is the starting generation, the **F₁ generation** (first filial generation) is produced from the P generation, and the **F₂ generation** (second filial generation) is produced from the F₁ generation.

Autosomes, Sex Chromosomes, and Sex Determination

When viewed under a microscope, the 46 human chromosomes in a normal somatic cell can be identified by their size, shape, and staining pattern to be members of 23 different pairs. An entire set of chromosomes arranged in decreasing order of size and according to the position of the centromere is called a **karyotype** (KAR-ē-ō-tīp; *karyo-* = nucleus; *-typos* = model) (Figure 29.24). In 22 of the pairs, the homologous chromosomes look alike and have the same appearance in both males and females; these 22 pairs are called **autosomes** (AW-tō-sōms). The two members of the 23rd pair are termed the **sex chromosomes;** they look different in males and fe-

Figure 29.24 Human karyotype showing autosomes and sex chromosomes. The white circles are the centromeres.

Human somatic cells contain 23 different pairs of chromosomes.

? What are the two sex chromosomes in females and males?

males. In females, the pair consists of two chromosomes called X chromosomes. One X chromosome is also present in males, but its mate is a much smaller chromosome called a Y chromosome. The Y chromosome has only 231 genes, less than 10% of the 2968 genes present on chromosome 1, the largest autosome.

When a spermatocyte undergoes meiosis to reduce its chromosome number, it gives rise to two sperm that contain an X chromosome and two sperm that contain a Y chromosome. Oocytes have no Y chromosomes and produce only X-containing gametes. If the secondary oocyte is fertilized by an X-bearing sperm, the offspring normally is female (XX). Fertilization by a Y-bearing sperm produces a male (XY). Thus, an individual's sex is determined by the father's chromosomes (Figure 29.25).

Both female and male embryos develop identically until about 7 weeks after fertilization. At that point, one or more genes set into motion a cascade of events that leads to the development of a male; in the absence of normal expression of the gene or genes, the female pattern of development occurs. It has been known since 1959 that the Y chromosome is needed to initiate male development. Experiments published in 1991 established that the prime male-determining gene is one called **SRY (sex-determining region of the Y chromosome).** When a small DNA fragment containing this gene was inserted into 11 female mouse embryos, three of them developed as males. (The researchers suspected that the gene failed to be integrated into the genetic material in the other eight.) *SRY* acts as a molecular switch to turn on the male pattern of development. Only if the *SRY* gene is present and functional in a fertilized ovum will the fetus develop testes and differentiate into a male; in the absence of *SRY*, the fetus will develop ovaries and differentiate into a female.

Case studies have confirmed the key role of *SRY* in directing the male pattern of development in humans. In some cases, phenotypic females with an XY genotype were found to have mutated *SRY* genes. These individuals failed to develop normally as males because their *SRY* gene was defective. In other cases, phenotypic males with an XX genotype were found to have a small piece of the Y chromosome, including the *SRY* gene, inserted into one of their X chromosomes.

Figure 29.25 Sex determination.

 Sex is determined at the time of fertilization by the presence or absence of a Y chromosome in the sperm.

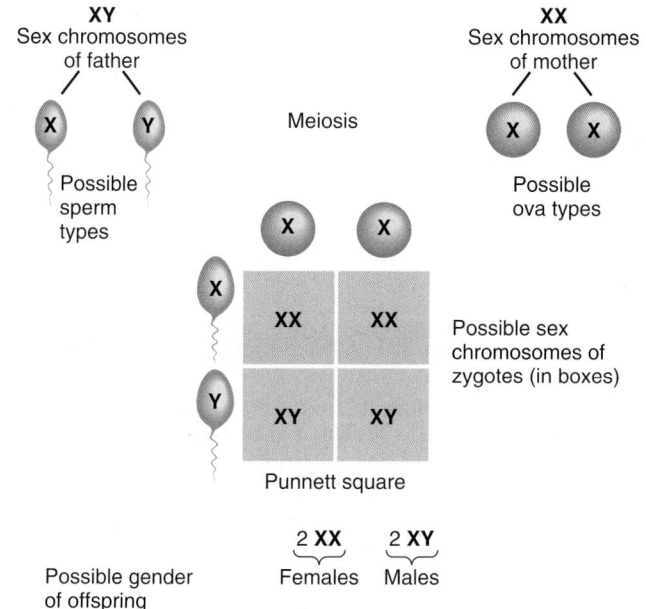

Punnett square

Possible gender of offspring: 2 **XX** Females, 2 **XY** Males

? **What are all chromosomes other than the sex chromosomes called?**

Sex-Linked Inheritance

In addition to determining the sex of the offspring, the sex chromosomes are responsible for the transmission of several nonsexual traits. Many of the genes for these traits are present on X chromosomes but are absent from Y chromosomes. This feature produces a pattern of heredity called **sex-linked inheritance** that is different from the patterns already described.

Red–Green Color Blindness

One example of sex-linked inheritance is **red–green color blindness,** the most common type of color blindness. This condition is characterized by a deficiency in either red- or green-sensitive cones, so red and green are seen as the same color (either red or green, depending on which cone is present). The gene for red–green color blindness is a recessive one designated c. Normal color vision, designated C, dominates. The C/c genes are located only on the X chromosome, so the ability to see colors depends entirely on the X chromosomes. The possible combinations are as follows:

Genotype	Phenotype
$X^C X^C$	Normal female
$X^C X^c$	Normal female (but carrier of recessive gene)
$X^c X^c$	Red–green color-blind female
$X^C Y$	Normal male
$X^c Y$	Red–green color-blind male

Only females who have two X^c genes are red–green color blind. This rare situation can result only from the mating of a color-blind male and a color-blind or carrier female. Because males do not have a second X chromosome that could mask the trait, all males with an X^c gene will be red–green color blind. Figure 29.26 illustrates the inheritance of red–green color blindness in the offspring of a normal male and a carrier female.

Traits inherited in the manner just described are called **sex-linked traits.** The most common type of **hemophilia** (hē-mō-FIL-ē-a)—a condition in which the blood fails to clot or clots very slowly after an injury—is also a sex-linked trait. Like the trait for red–green color blindness, hemophilia is caused by a recessive gene. Other sex-linked traits in humans are fragile X syndrome, nonfunctional sweat glands, certain forms of diabetes, some types of deafness, uncontrollable rolling of the eyeballs, absence of central incisors, night blindness, one form of cataract, juvenile glaucoma, and juvenile muscular dystrophy.

X-Chromosome Inactivation

Because they have two X chromosomes in every cell (except developing oocytes), females have a double set of all genes on the X chromosome. A mechanism termed **X-chromosome inactivation (lyonization)** in effect reduces the X-chromosome genes to a

Figure 29.26 An example of the inheritance of red–green color blindness.

 Red–green color blindness and hemophilia are examples of sex-linked traits.

Punnett square

Possible phenotypes of offspring	$X^C X^c$ Normal female	$X^C X^c$ Normal female (carrier)	$X^C Y$ Normal male	$X^c Y$ Color-blind male

? **What is the genotype of a red–green color-blind female?**

single set in females. In each cell of a female's body, one X chromosome is randomly and permanently inactivated early in development, and most of the genes of the inactivated X chromosome are not expressed (transcribed and translated). The nuclei of cells in female mammals contain a dark-staining body, called a **Barr body,** that is not present in the nuclei of cells in males. Geneticist Mary Lyon correctly predicted in 1961 that the Barr body is the inactivated X chromosome. During inactivation, chemical groups that prevent transcription into RNA are added to the X chromosome's DNA. As a result, an inactivated X chromosome reacts differently to histological stains and has a different appearance than the rest of the DNA. In nondividing (interphase) cells, it remains tightly coiled and can be seen as a dark-staining body

within the nucleus. In a blood smear, the Barr body of neutrophils is termed a "drumstick" because it looks like a tiny drumstick-shaped projection of the nucleus.

✔ **CHECKPOINT**

36. What do the terms genotype, phenotype, dominant, recessive, homozygous, and heterozygous mean?
37. What are genomic imprinting and nondisjunction?
38. Give an example of incomplete dominance.
39. What is multiple-allele inheritance? Give an example.
40. Define complex inheritance and give an example.
41. Why does X-chromosome inactivation occur?

 DISORDERS: HOMEOSTATIC IMBALANCES

Infertility

Female infertility, or the inability to conceive, occurs in about 10% of all women of reproductive age in the United States. Female infertility may be caused by ovarian disease, obstruction of the uterine tubes, or conditions in which the uterus is not adequately prepared to receive a fertilized ovum. **Male infertility (sterility)** is an inability to fertilize a secondary oocyte; it does not imply erectile dysfunction (impotence). Male fertility requires production of adequate quantities of viable, normal sperm by the testes, unobstructed transport of sperm though the ducts, and satisfactory deposition in the vagina. The seminiferous tubules of the testes are sensitive to many factors—x-rays, infections, toxins, malnutrition, and higher-than-normal scrotal temperatures—that may cause degenerative changes and produce male sterility.

One cause of infertility in females is inadequate body fat. To begin and maintain a normal reproductive cycle, a female must have a minimum amount of body fat. Even a moderate deficiency of fat—10% to 15% below normal weight for height—may delay the onset of menstruation (menarche), inhibit ovulation during the reproductive cycle, or cause amenorrhea (cessation of menstruation). Both dieting and intensive exercise may reduce body fat below the minimum amount and lead to infertility that is reversible, if weight gain or reduction of intensive exercise or both occur. Studies of very obese women indicate that they, like very lean ones, experience problems with amenorrhea and infertility. Males also experience reproductive problems in response to undernutrition and weight loss. For example, they produce less prostatic fluid and reduced numbers of sperm having decreased motility.

Many fertility-expanding techniques now exist for assisting infertile couples to have a baby. The birth of Louise Joy Brown on July 12, 1978, near Manchester, England, was the first recorded case of **in vitro fertilization (IVF)**—fertilization in a laboratory dish. In the IVF procedure, the mother-to-be is given follicle-stimulating hormone (FSH) soon after menstruation, so that several secondary oocytes, rather than the typical single oocyte, will be produced (superovulation). When several follicles have reached the appropriate size, a small incision is made near the umbilicus, and the secondary oocytes are aspirated from the stimulated follicles and transferred to a solution containing sperm, where the oocytes undergo fertilization. Alternatively, an oocyte may be fertilized in vitro by suctioning a sperm or even a spermatid obtained from the testis into a tiny pipette and then injecting it into the oocyte's cytoplasm. This procedure, termed **intracytoplasmic sperma injection (ICSI)** (in'-tra-sī-tō-PLAZ-mik), has been used when infertility is due to impairments in sperm motility or to the

failure of spermatids to develop into spermatozoa. When the zygote achieved by IVF reaches the 8-cell or 16-cell stage, it is introduced into the uterus for implantation and subsequent growth.

In **embryo transfer,** a man's semen is used to artificially inseminate a fertile secondary oocyte donor. After fertilization in the donor's uterine tube, the morula or blastocyst is transferred from the donor to the infertile woman, who then carries it (and subsequently the fetus) to term. Embryo transfer is indicated for women who are infertile or who do not want to pass on their own genes because they are carriers of a serious genetic disorder.

In **gamete intrafallopian transfer (GIFT)** the goal is to mimic the normal process of conception by uniting sperm and secondary oocyte in the prospective mother's uterine tubes. It is an attempt to bypass conditions in the female reproductive tract that might prevent fertilization, such as high acidity or inappropriate mucus. In this procedure, a woman is given FSH and LH to stimulate the production of several secondary oocytes, which are aspirated from the mature follicles, mixed outside the body with a solution containing sperm, and then immediately inserted into the uterine tubes.

Congenital Defects

An abnormality that is present at birth, and usually before, is called a **congenital defect.** Such defects occur during the formation of structures that develop during the period of organogenesis, the fourth through eighth weeks of development, when all major organs appear. During organogenesis stem cells are establishing the basic patterns of organ development, and it is during this time that developing structures are very susceptible to genetic and environmental influences.

Major structural defects occur in 2–3% of liveborn infants, and they are the leading cause of infant mortality, accounting for about 21% of infant deaths. Many congenital defects can be prevented by supplementation or avoidance of certain substances. For example, neural tube defects, such as spina bifida and anencephaly, can be prevented by having a pregnant female take folic acid. Iodine supplementation can prevent the mental retardation and bone deformation associated with cretinism. Avoidance of teratogens is also very important in preventing congenital defects.

Down Syndrome

Down syndrome (DS) is a disorder characterized by three, rather than two, copies of at least part of chromosome 21. Overall, one infant in 900 is born with Down syndrome. However, older women are more likely to have a DS baby. The chance of having a baby with this syndrome, which is less than 1 in 3000 for women under age 30, increases to 1 in 300 in the 35–39 age group and to 1 in 9 at age 48.

Down syndrome is characterized by mental retardation, retarded physical development (short stature and stubby fingers), distinctive facial structures (large tongue, flat profile, broad skull, slanting eyes, and round head), kidney defects, suppressed immune system, and malformations of the heart, ears, hands, and feet. Sexual maturity is rarely attained, and life expectancy is shorter.

MEDICAL TERMINOLOGY

Breech presentation A malpresentation in which the fetal buttocks or lower limbs present into the maternal pelvis; the most common cause is prematurity.

Conceptus (kon-SEP-tus) Includes all structures that develop from a zygote and includes an embryo plus the embryonic part of the placenta and associated membranes (chorion, amnion, yolk sac, and allantois).

Cryopreserved embryo (krī-ō-PRĒ-servd; *cryo-* = cold) An early embryo produced by in vitro fertilization (fertilization of a secondary oocyte in a laboratory dish) that is preserved for a long period by freezing it. After thawing, the early embryo is implanted into the uterine cavity. Also called a **frozen embryo.**

Deformation (dē-for-MĀ-shun; *de-* = without; *-forma* = form) A developmental abnormality due to mechanical forces that mold a part of the fetus over a prolonged period of time. Deformations usually involve the skeletal and/or muscular system and may be corrected after birth. An example is clubfeet.

Emesis gravidarum (EM-e-sis gra-VID-ar-um; *emeo* = to vomit; *gravida* = a pregnant woman) Episodes of nausea and possibly vomiting that are most likely to occur in the morning during the early weeks of pregnancy; also called **morning sickness.** Its cause is unknown, but the high levels of human chorionic gonadotropin (hCG) secreted by the placenta, and of progesterone secreted by the ovaries, have been implicated. If the severity of these symptoms requires hospitalization for intravenous feeding, the condition is known as **hyperemesis gravidarum.**

Epigenesis (ep-i-JEN-e-sis; *epi-* = upon; *-genesis* = creation) The development of an organism from an undifferentiated cell.

Fertilization age Two weeks less than the gestational age, since a secondary oocyte is not fertilized until about 2 weeks after the last normal menstrual period (LNMP).

Fetal alcohol syndrome (FAS) A specific pattern of fetal malformation due to intrauterine exposure to alcohol. FAS is one of the most common causes of mental retardation and the most common preventable cause of birth defects in the United States.

Fetal surgery A surgical procedure performed on a fetus; in some cases the uterus is opened and the fetus is operated on directly.

Fetal surgery has been used to repair diaphragmatic hernias and remove lesions in the lungs.

Gestational age (jes-TĀ-shun-al; *gestatus* = to bear) The age of an embryo or fetus calculated from the presumed first day of the last normal menstrual period (LNMP).

Karyotype (KAR-ē-ō-tīp; *karyo-* = nucleus) The chromosomal characteristics of an individual presented as a systematic arrangement of pairs of metaphase chromosomes arrayed in descending order of size and according to the position of the centromere (see Figure 29.24); useful in judging whether chromosomes are normal in number and structure.

Klinefelter's syndrome A sex chromosome aneuploidy, usually due to trisomy XXY, that occurs once in every 500 births. Such individuals are somewhat mentally disadvantaged, sterile males with undeveloped testes, scant body hair, and enlarged breasts.

Lethal gene (LĒ-thal JĒN; *lethum* = death) A gene that, when expressed, results in death either in the embryonic state or shortly after birth.

Metafemale syndrome A sex chromosome aneuploidy characterized by at least three X chromosomes (XXX) that occurs about once in every 700 births. These females have underdeveloped genital organs and limited fertility, and most are mentally retarded.

Primordium (prī-MOR-dē-um; *primus-* = first; *-ordior* = to begin) The beginning or first discernible indication of the development of an organ or structure.

Puerperal fever (pū-ER-per-al; *puer* = child) An infectious disease of childbirth, also called puerperal sepsis and childbed fever. The disease, which results from an infection originating in the birth canal, affects the mother's endometrium. It may spread to other pelvic structures and lead to septicemia.

Turner's syndrome A sex chromosome aneuploidy caused by the presence of a single X chromosome (designated XO); occurring about once in every 5000 births, it produces a sterile female with virtually no ovaries and limited development of secondary sex characteristics. Other features include short stature, webbed neck, underdeveloped breasts, and widely spaced nipples. Intelligence usually is normal.

CHAPTER REVIEW AND RESOURCE SUMMARY

Review

Resource

WILEY
PLUS

29.1 Embryonic Period

1. Pregnancy is a sequence of events that begins with fertilization and proceeds to implantation, embryonic development, and fetal development. It normally ends in birth.

2. During fertilization a sperm cell penetrates a secondary oocyte and their pronuclei unite. Penetration of the zona pellucida is facilitated by enzymes in the sperm's acrosome. The resulting cell is a zygote. Normally, only one sperm cell fertilizes a secondary oocyte because of the fast and slow blocks to polyspermy.

3. Early rapid cell division of a zygote is called cleavage, and the cells produced by cleavage are called blastomeres. The solid sphere of cells produced by cleavage is a morula. The morula develops into a blastocyst, a hollow ball of cells differentiated into a trophoblast and an inner cell mass. The attachment

Animation - Fertilization
Anatomy Overview - Developmental Stages
Animation - Embryonic and Fetal Development
Figure 29.2 - Cleavage and the Formation of the Morula and Blastocyst

Review

of a blastocyst to the endometrium is termed implantation; it occurs as a result of enzymatic degradation of the endometrium. After implantation, the endometrium becomes modified and is known as the decidua. The trophoblast develops into the syncytiotrophoblast and cytotrophoblast, both of which become part of the chorion. The inner cell mass differentiates into hypoblast and epiblast, the bilaminar (two-layered) embryonic disc. The amnion is a thin protective membrane that develops from the cytotrophoblast.

4. The exocoelomic membrane and hypoblast form the yolk sac, which transfers nutrients to the embryo, forms blood cells, produces primordial germ cells, and forms part of the gut. Erosion of sinusoids and endometrial glands provides blood and secretions, which enter lacunar networks to supply nutrition to and remove wastes from the embryo. The extraembryonic coelom forms within extraembryonic mesoderm. The extraembryonic mesoderm and trophoblast form the chorion, the principal embryonic part of the placenta.

5. The third week of development is characterized by gastrulation, the conversion of the bilaminar disc into a trilaminar (three-layered) embryo consisting of ectoderm, mesoderm, and endoderm. The first evidence of gastrulation is formation of the primitive streak, after which the primitive node, notochordal process, and notochord develop. The three primary germ layers form all tissues and organs of the developing organism. Table 29.1 summarizes the structures that develop from the primary germ layers. Also during the third week, the oropharyngeal and cloacal membranes form. The wall of the yolk sac forms a small vascularized outpouching called the allantois, which functions in blood formation and development of the urinary bladder.

6. The process by which the neural plate, neural folds, and neural tube form is called neurulation. The brain and spinal cord develop from the neural tube.

7. Paraxial mesoderm segments to form somites from which skeletal muscles of the neck, trunk, and limbs develop. Somites also form connective tissues and vertebrae.

8. Blood vessel formation, called angiogenesis, begins in mesodermal cells called angioblasts. The heart forms from mesodermal cells called the cardiogenic area. By the end of the third week, the primitive heart beats and circulates blood.

9. Chorionic villi, projections of the chorion, connect to the embryonic heart so that maternal and fetal blood vessels are brought into close proximity, allowing the exchange of nutrients and wastes between maternal and fetal blood. Placentation refers to formation of the placenta, the site of exchange of nutrients and wastes between the mother and fetus. The placenta also functions as a protective barrier, stores nutrients, and produces several hormones to maintain pregnancy. The actual connection between the placenta and embryo (and later the fetus) is the umbilical cord.

10. Organogenesis refers to the formation of body organs and systems and occurs during the fourth week of development. Conversion of the flat, two-dimensional trilaminar embryonic disc to a three-dimensional cylinder occurs by a process called embryonic folding. Embryonic folding brings various organs into their final adult positions and helps form the gastrointestinal tract. Pharyngeal arches, clefts, and pouches give rise to the structures of the head and neck. By the end of the fourth week, upper and lower limb buds develop, and by the end of the eighth week the embryo has clearly human features.

29.2 Fetal Period

1. The fetal period is primarily concerned with the growth and differentiation of tissues and organs that developed during the embryonic period.

2. The rate of body growth is remarkable, especially during the ninth and sixteenth weeks.

3. The principal changes associated with embryonic and fetal growth are summarized in Table 29.2.

29.3 Teratogens

1. Teratogens are agents that cause physical defects in developing embryos.

2. Among the more important teratogens are alcohol, pesticides, industrial chemicals, some prescription drugs, cocaine, LSD, nicotine, and ionizing radiation.

29.4 Prenatal Diagnostic Tests

1. Several prenatal diagnostic tests are used to detect genetic disorders and to assess fetal well-being. These include fetal ultrasonography, in which an image of a fetus is displayed on a screen; amniocentesis, the withdrawal and analysis of amniotic fluid and the fetal cells within it; and chorionic villi sampling (CVS), which involves withdrawal of chorionic villi tissue for chromosomal analysis.

2. CVS can be done earlier than amniocentesis, and the results are available more quickly, but it is slightly riskier than amniocentesis.

3. Noninvasive prenatal tests include the maternal alpha-fetoprotein (AFP) test to detect neural tube defects and the Quad AFP Plus test to detect Down syndrome, trisomy 18, and neural tube defects.

Review	Resource

29.5 Maternal Changes during Pregnancy

Animation - Hormonal Regulation of Pregnancy
Figure 29.16 - Hormones during Pregnancy

1. Pregnancy is maintained by human chorionic gonadotropin (hCG), estrogens, and progesterone.
2. Human chorionic somatomammotropin (hCS) contributes to breast development, protein anabolism, and catabolism of glucose and fatty acids.
3. Relaxin increases flexibility of the pubic symphysis and helps dilate the uterine cervix near the end of pregnancy.
4. Corticotropin-releasing hormone, produced by the placenta, is thought to establish the timing of birth, and stimulates the secretion of cortisol by the fetal adrenal gland.
5. During pregnancy, several anatomical and physiological changes occur in the mother.

29.6 Exercise and Pregnancy

1. During pregnancy, some joints become less stable, and certain physical activities are more difficult to execute.
2. Moderate physical activity does not endanger the fetus in a normal pregnancy.

29.7 Labor

Animation - Regulation of Labor and Birth
Animation - Positive Feedback Control of Labor
Figure 29.18 - Stages of True Labor
Exercise - Pregnancy, Birth, and Lactation

1. Labor is the process by which the fetus is expelled from the uterus through the vagina to the outside. True labor involves dilation of the cervix, expulsion of the fetus, and delivery of the placenta.
2. Oxytocin stimulates uterine contractions via a positive feedback cycle.

29.8 Adjustments of the Infant at Birth

1. The fetus depends on the mother for oxygen and nutrients, the removal of wastes, and protection.
2. Following birth, an infant's respiratory and cardiovascular systems undergo changes to enable them to become self-supporting during postnatal life.

29.9 The Physiology of Lactation

Animation - Regulation of Lactation

1. Lactation refers to the production and ejection of milk by the mammary glands.
2. Milk production is influenced by prolactin (PRL), estrogens, and progesterone.
3. Milk ejection is stimulated by oxytocin.
4. A few of the many benefits of breast-feeding include ideal nutrition for the infant, protection from disease, and decreased likelihood of developing allergies.

29.10 Inheritance

1. Inheritance is the passage of hereditary traits from one generation to the next.
2. The genetic makeup of an organism is called its genotype; the traits expressed are called its phenotype.
3. Dominant genes control a particular trait; expression of recessive genes is masked by dominant genes.
4. Many patterns of inheritance do not conform to the simple dominant–recessive patterns. In incomplete dominance, neither member of an allelic pair dominates; phenotypically, the heterozygote is intermediate between the homozygous dominant and the homozygous recessive. In multiple-allele inheritance, genes have more than two alternative forms. An example is the inheritance of ABO blood groups. In complex inheritance, a trait such as skin or eye color is controlled by the combined effects of two or more genes and may be influenced by environmental factors.
5. Each somatic cell has 46 chromosomes—22 pairs of autosomes and 1 pair of sex chromosomes.
6. In females, the sex chromosomes are two X chromosomes; in males, they are one X chromosome and a much smaller Y chromosome, which normally includes the prime male-determining gene, called *SRY*.
7. If the *SRY* gene is present and functional in a fertilized ovum, the fetus will develop testes and differentiate into a male. In the absence of *SRY*, the fetus will develop ovaries and differentiate into a female.
8. Red–green color blindness and hemophilia result from recessive genes located on the X chromosome. These sex-linked traits occur primarily in males because of the absence of any counterbalancing dominant genes on the Y chromosome.
9. A mechanism termed X-chromosome inactivation (lyonization) balances the difference in number of X chromosomes between males (one X) and females (two Xs). In each cell of a female's body, one X chromosome is randomly and permanently inactivated early in development and becomes a Barr body.
10. A given phenotype is the result of the interactions of genotype and the environment.

Fill in the blanks in the following statements.

1. The three stages of true labor, in order of occurrence, are _____, _____, and _____.

2. Hormones produced by the _____ are responsible for maintaining the pregnancy during the first 3–4 months. The hormone responsible for preventing degeneration of the corpus luteum is _____ produced by the trophoblast.

3. Indicate the germ layers responsible for development of the following structures: (a) muscle, bone, and peritoneum: _____; (b) nervous system and epidermis: _____; (c) epithelial linings of respiratory and gastrointestinal tracts: _____ .

Indicate whether the following statement is true or false.

4. Labor is an example of a negative feedback cycle that ends with the birth of the infant.

Choose the one best answer to the following questions.

5. Which of the following are *true*? (1) During implantation the outer cell mass of the blastocyst orients toward the endometrium. (2) The decidua basalis provides glycogen and lipids for the developing fetus. (3) The decidua parietalis becomes the maternal part of the placenta. (4) During implantation, the syncytiotrophoblast secretes enzymes that allow the blastocyst to penetrate the uterine lining. (5) After fetal delivery, the decidua separates from the endometrium and is released from the uterus.
(a) 2, 4, and 5 (b) 1, 2, and 3 (c) 2, 3, 4, and 5
(d) 1, 2, 3, 4, and 5 (e) 1, 3, and 5

6. Which of the following are maternal changes that occur during pregnancy? (1) altered pulmonary function; (2) increased stroke volume, cardiac output, and heart rate, and decreased blood volume; (3) weight gain; (4) increased gastric motility, causing a delay in gastric emptying time; (5) edema and possible varicose veins.
(a) 1, 2, 3, and 4 (b) 2, 3, 4, and 5 (c) 1, 3, 4, and 5
(d) 1, 3, and 5 (e) 2, 4, and 5

7. Which of the following statements is *correct*? (a) Normal traits always dominate over abnormal traits. (b) Occasionally an error in meiosis called nondisjunction results in an abnormal number of chromosomes. (c) The mother always determines the sex of the child because she has either an X or Y gene in her oocytes. (d) Most patterns of inheritance are simple dominant–recessive inheritances. (e) Genes are expressed normally regardless of any outside influence such as chemicals or radiation.

8. Which of the following are *true* concerning fertilization? (1) The sperm first penetrate the zona pellucida and then the corona radiata. (2) The binding of specific membrane proteins in the sperm head to ZP3 causes the release of acrosomal contents. (3) Sperm are able to fertilize the oocyte within minutes after ejaculation. (4) Depolarization of the cell membrane of the secondary oocyte inhibits fertilization by more than one sperm. (5) The oocyte completes meiosis II after fertilization.
(a) 1, 2, 4, and 5 (b) 1, 3, and 5 (c) 1, 2, 3, and 4
(d) 1, 4, and 5 (e) 2, 4, and 5

9. Amniotic fluid (1) is derived entirely from a filtrate of maternal blood, (2) acts as a fetal shock absorber, (3) provides nutrients to the fetus, (4) helps regulate fetal body temperature, (5) prevents adhesions between the skin of the fetus and surrounding tissues.
(a) 1, 2, 3, 4, and 5 (b) 2, 4, and 5 (c) 2, 3, 4, and 5
(d) 1, 4, and 5 (e) 1, 2, 4, and 5

10. Which of the following structures develop during the fourth week after fertilization? (1) embryonic folding, (2) the neural tube, (3) otic placode (beginning of the ear), (4) beginning of the eyes, (5) upper and lower limb buds.
(a) 1 and 2 (b) 1, 2, and 5 (c) 1, 2, 3, 4, and 5
(d) 2, 3, and 5 (e) 1, 3, 4, and 5

11. Match the following:
_____ (a) a fluid-filled sphere of cells that enters the uterine cavity
_____ (b) cells produced by cleavage
_____ (c) the developing individual from week nine of pregnancy until birth
_____ (d) the outer covering of cells of the blastocyst
_____ (e) membrane derived from trophoblast
_____ (f) early divisions of the zygote
_____ (g) a solid sphere of cells still surrounded by the zona pellucida
_____ (h) event in which differentiation into the three primary germ layers occurs
_____ (i) embryonic development of structures that will become the nervous system
_____ (j) the formation of blood vessels to support the developing embryo
_____ (k) result of the fusion of female and male pronuclei

(1) cleavage
(2) blastomeres
(3) morula
(4) angiogenesis
(5) trophoblast
(6) blastocyst
(7) zygote
(8) gastrulation
(9) neurulation
(10) chorion
(11) fetus

12. Match the following:

____ (a) stimulates the corpus luteum to continue production of progesterone and estrogens

____ (b) increases the flexibility of the pubic symphysis and helps dilate the uterine cervix during labor

____ (c) secreted by the placenta; helps establish the timing of birth and increases the secretion of cortisol for fetal lung maturation

____ (d) helps prepare mammary glands for lactation; regulates certain aspects of maternal and fetal metabolism

____ (e) stimulates uterine contractions; responsible for the milk ejection reflex

____ (f) promotes milk synthesis and secretion; inhibited by progesterone during pregnancy

(1) oxytocin
(2) human chorionic somatomammo- tropin
(3) human chorionic gonadotropin
(4) prolactin
(5) corticotropin- releasing hormone
(6) relaxin

13. Match the following:

____ (a) the penetration of a secondary oocyte by a single sperm cell

____ (b) fertilization of a secondary oocyte by more than one sperm

____ (c) the attachment of a blastocyst to the endometrium

____ (d) the fusion of the genetic mate- rial from a haploid sperm and a haploid secondary oocyte into a single diploid nucleus

____ (e) the induction by the female re- productive tract of functional changes in sperm that allow them to fertilize a secondary oocyte

____ (f) the examination of embryonic or fetal cells sloughed off into the amniotic fluid

____ (g) an abnormal condition of pregnancy characterized by sudden hypertension, large amounts of protein in urine, and generalized edema

____ (h) noninvasive test that can detect fetal neural tube defects

____ (i) the process of giving birth

____ (j) the period of time (about 6 weeks) during which the maternal reproductive organs and physiology return to the prepregnancy state

(1) fertilization
(2) capacitation
(3) syngamy
(4) polyspermy
(5) implantation
(6) amniocentesis
(7) preeclampsia
(8) parturition
(9) puerperium
(10) maternal AFP test

14. Match the following:

____ (a) the control of inherited traits by the combined effects of many genes

____ (b) the two alternative forms of a gene that code for the same trait and are at the same location on homologous chromosomes

____ (c) abnormal number of chromo- somes due to failure of homol- ogous chromosomes or chro- matids to separate

____ (d) inheritance based on genes that have more than two alter- native forms; an example is the inheritance of blood type

____ (e) a cell in which one or more chromosomes of a set is added or deleted

____ (f) refers to an individual with different alleles on homologous chromosomes

____ (g) traits linked to the X chromosome

____ (h) permanent inheritable change in an allele that produces a different variant of the same trait

____ (i) neither member of the allelic pair is dominant over the other, and the heterozygote has a phenotype intermediate between the homozygous dominant and the homozygous recessive

____ (j) refers to how the genetic makeup is expressed in the body; the physical or outward expression of a gene

____ (k) a homozygous dominant, homozygous recessive, or heterozygous genetic makeup; the actual gene arrangement

____ (l) refers to a person with the same alleles on homologous chromosomes

____ (m) inactivated X chromosome in females

____ (n) heterozygous individuals who possess a recessive gene (but do not express it) and can pass the gene on to their offspring

____ (o) interchange of portions of nonhomologous chromosomes

____ (p) an allele that masks the presence of another allele and is fully expressed

(1) genotype
(2) phenotype
(3) alleles
(4) aneuploid
(5) incomplete dominance
(6) multiple-allele inheritance
(7) polygenic inheritance
(8) sex-linked inheritance
(9) homozygous
(10) heterozygous
(11) carriers
(12) dominant trait
(13) mutation
(14) nondisjunction
(15) translocation
(16) Barr body

15. Match the following:

____ (a) the embryonic membrane that entirely surrounds the embryo

____ (b) functions as an early site of blood formation; contains cells that migrate into the gonads and differentiate into the primitive germ cells

____ (c) becomes the principal part of the embryonic placenta; produces human chorionic gonadotropin

____ (d) modified endometrium after implantation has occurred; separates from the endometrium after the fetus is delivered

____ (e) contains the vascular connections between mother and fetus

____ (f) the fetal portion is formed by the chorionic villi and the maternal portion is formed by the decidua basalis of the endometrium; allows oxygen and nutrients to diffuse from maternal blood into fetal blood

____ (g) serves as an early site of blood vessel formation

____ (h) fingerlike projections of the chorion that bring maternal and fetal blood vessels into close proximity

____ (i) plays an important role in induction whereby an inducing tissue stimulates the development of an unspecialized responding tissue into a specialized tissue

(1) decidua
(2) placenta
(3) amnion
(4) chorion
(5) allantois
(6) yolk sac
(7) notochord
(8) chorionic villi
(9) umbilical cord

CRITICAL THINKING QUESTIONS

1. Kathy is breast-feeding her infant and is experiencing what feels like early labor pains. What is causing these painful feelings? Is there a benefit to them?

2. Jack has hemophilia, which is a sex-linked blood clotting disorder. He blames his father for passing on the gene for hemophilia. Explain to Jack why his reasoning is wrong. How can Jack have hemophilia if his parents do not?

3. Alisa has asked her obstetrician to save and freeze her baby's cord blood after delivery in case the child needs a future bone marrow transplant. What is in the baby's cord blood that could be used to treat future disorders in the child?

? ANSWERS TO FIGURE QUESTIONS

29.1 Capacitation is the group of functional changes in sperm that enable them to fertilize a secondary oocyte, which occur after the sperm have been deposited in the female reproductive tract.

29.2 A morula is a solid ball of cells; a blastocyst consists of a rim of cells (trophoblast) surrounding a cavity (blastocyst cavity) and an inner cell mass.

29.3 The blastocyst secretes digestive enzymes that eat away the endometrial lining at the site of implantation.

29.4 The decidua basalis helps form the maternal part of the placenta.

29.5 Implantation occurs during the secretory phase of the uterine cycle.

29.6 The bilaminar embryonic disc is attached to the trophoblast by the connecting stalk.

29.7 Gastrulation converts a bilaminar embryonic disc into a trilaminar embryonic disc.

29.8 The notochord induces mesodermal cells to develop into vertebral bodies and forms the nucleus pulposus of intervertebral discs.

29.9 The neural tube forms the brain and spinal cord; somites develop into skeletal muscles, connective tissue, and the vertebrae.

29.10 Chorionic villi help to bring the fetal and maternal blood vessels into close proximity.

29.11 The placenta participates in the exchange of materials between fetus and mother, serves as a protective barrier against many microbes, and stores nutrients.

29.12 As a result of embryonic folding, the embryo curves into a C-shape, various organs are brought into their eventual adult positions, and the primitive gut is formed.

29.13 Pharyngeal arches, clefts, and pouches give rise to structures of the head and neck.

29.14 Fetal weight doubles between the midfetal period and birth.

29.15 Amniocentesis is used primarily to detect genetic disorders, but it also provides information concerning the maturity (and survivability) of the fetus.

29.16 Early pregnancy tests detect elevated levels of human chorionic gonadotropin (hCG).

29.17 Relaxin increases the flexibility of the pubic symphysis and helps dilate the cervix of the uterus to ease delivery.

29.18 Complete dilation of the cervix marks the onset of the stage of expulsion.

29.19 Oxytocin also stimulates contraction of the uterus during delivery of a baby.

29.20 The odds that a child will have PKU are the same for each child—25%.

29.21 In incomplete dominance, neither member of an allelic pair is dominant; the heterozygote has a phenotype intermediate between the homozygous dominant and the homozygous recessive phenotypes.

29.22 A baby can have blood type O if each parent is heterozygous and has one i allele.

29.23 Hair color, height, and body build, among others, are traits passed on by complex inheritance.

29.24 The female sex chromosomes are XX, and the male sex chromosomes are XY.

29.25 The chromosomes that are not sex chromosomes are called autosomes.

29.26 A red–green color-blind female has an $X^c X^c$ genotype.

MEASUREMENTS

U.S. Customary System

PARAMETER	UNIT	RELATION TO OTHER U.S. UNITS	SI (METRIC) EQUIVALENT
Length	inch	1/12 foot	2.54 centimeters
foot	12 inches	0.305 meter	
yard	36 inches	0.914 meter	
mile	5,280 feet	1.609 kilometers	
Mass	grain	1/1,000 pound	64.799 milligrams
dram	1/16 ounce	1.772 grams	
ounce	16 drams	28.350 grams	
pound	16 ounces	453.6 grams	
ton	2,000 pounds	907.18 kilograms	
Volume (Liquid)	ounce	1/16 pint	29.574 milliliters
pint	16 ounces	0.473 liter	
quart	2 pints	0.946 liter	
gallon	4 quarts	3.785 liters	
Volume (Dry)	pint	1/2 quart	0.551 liter
quart	2 pints	1.101 liters	
peck	8 quarts	8.810 liters	
bushel	4 pecks	35.239 liters	

International System (SI)

BASE UNITS			PREFIXES		
UNIT	QUANTITY	SYMBOL	PREFIX	MULTIPLIER	SYMBOL
meter	length	m	tera-	$10^{12} = 1,000,000,000,000$	T
kilogram	mass	kg	giga-	$10^{9} = 1,000,000,000$	G
second	time	s	mega-	$10^{6} = 1,000,000$	M
liter	volume	L	kilo-	$10^{3} = 1,000$	k
mole	amount of matter	mol	hecto-	$10^{2} = 100$	h
			deca-	$10^{1} = 10$	da
			deci-	$10^{-1} = 0.1$	d
			centi-	$10^{-2} = 0.01$	c
			milli-	$10^{-3} = 0.001$	m
			micro-	$10^{-6} = 0.000,001$	μ
			nano-	$10^{-9} = 0.000,000,001$	n
			pico-	$10^{-12} = 0.000,000,000,001$	p

Temperature Conversion

FAHRENHEIT (F) TO CELSIUS (C)

$$°C = (°F - 32) ÷ 1.8$$

CELSIUS (C) TO FAHRENHEIT (F)

$$°F = (°C × 1.8) + 32$$

U.S. to SI (Metric) Conversion

WHEN YOU KNOW	MULTIPLY BY	TO FIND
inches	2.54	centimeters
feet	30.48	centimeters
yards	0.91	meters
miles	1.61	kilometers
ounces	28.35	grams
pounds	0.45	kilograms
tons	0.91	metric tons
fluid ounces	29.57	milliliters
pints	0.47	liters
quarts	0.95	liters
gallons	3.79	liters

SI (Metric) to U.S. Conversion

WHEN YOU KNOW	MULTIPLY BY	TO FIND
millimeters	0.04	inches
centimeters	0.39	inches
meters	3.28	feet
kilometers	0.62	miles
liters	1.06	quarts
cubic meters	35.32	cubic feet
grams	0.035	ounces
kilograms	2.21	pounds

PERIODIC TABLE

The periodic table lists the known **chemical elements,** the basic units of matter. The elements in the table are arranged left-to-right in rows in order of their **atomic number,** the number of protons in the nucleus. Each horizontal row, numbered from 1 to 7, is a **period.** All elements in a given period have the same number of electron shells as their period number. For example, an atom of hydrogen or helium each has one electron shell, while an atom of potassium or calcium each has four electron shells. The elements in each column, or **group,** share chemical properties. For example, the elements in column IA are very chemically reactive, whereas the elements in column VIIIA have full electron shells and thus are chemically inert.

Scientists now recognize 117 different elements; 92 occur naturally on Earth, and the rest are produced from the natural elements using particle accelerators or nuclear reactors. Elements are designated by **chemical symbols,** which are the first one or two letters of the element's name in English, Latin, or another language.

Twenty-six of the 92 naturally occurring elements normally are present in your body. Of these, just four elements—oxygen (O), carbon (C), hydrogen (H), and nitrogen (N) (coded blue)—constitute about 96% of the body's mass. Eight others—calcium (Ca), phosphorus (P), potassium (K), sulfur (S), sodium (Na), chlorine (Cl), magnesium (Mg), and iron (Fe) (coded pink)—contribute 3.8% of the body's mass. An additional 14 elements, called **trace elements** because they are present in tiny amounts, account for the remaining 0.2% of the body's mass. The trace elements are aluminum, boron, chromium, cobalt, copper, fluorine, iodine, manganese, molybdenum, selenium, silicon, tin, vanadium, and zinc (coded yellow). Table 2.1 on page 30 provides information about the main chemical elements in the body.

NORMAL VALUES FOR SELECTED BLOOD TESTS

The system of international (SI) units (Système Internationale d'Unités) is used in most countries and in many medical and scientific journals. Clinical laboratories in the United States, by contrast, usually report values for blood and urine tests in conventional units. The laboratory values in this Appendix give conventional units first, followed by SI equivalents in parentheses. Values listed for various blood tests should be viewed as reference values rather than absolute "normal" values for all well people. Values may vary due to age, gender, diet, and environment of the subject or the equipment, methods, and standards of the lab performing the measurement.

KEY TO SYMBOLS

g = gram	mL = milliliter
mg = milligram = 10^{-3} gram	μL = microliter
μg = microgram = 10^{-6} gram	mEq/L = milliequivalents per liter
U = units	mmol/L = millimoles per liter
L = liter	μmol/L = micromoles per liter
dL = deciliter	> = greater than; < = less than

Blood Tests

TEST (SPECIMEN)	U.S. REFERENCE VALUES (SI UNITS)	VALUES INCREASE IN	VALUES DECREASE IN
Aminotransferases (serum)			
Alanine aminotransferase (ALT)	0–35 U/L (same)	Liver disease or liver damage due to toxic drugs.	
Aspartate aminotransferase (AST)	0–35 U/L (same)	Myocardial infarction, liver disease, trauma to skeletal muscles, severe burns.	Beriberi, uncontrolled diabetes mellitus with acidosis, pregnancy.
Ammonia (plasma)	20–120 μg/dL (12–55 μmol/L)	Liver disease, heart failure, emphysema, pneumonia, hemolytic disease of newborn.	Hypertension.
Bilirubin (serum)	Conjugated: <0.5 mg/dL (<5.0 μmol/L)	Conjugated bilirubin: liver dysfunction or gallstones.	
	Unconjugated: 0.2–1.0 mg/dL (18–20 μmol/L) Newborn: 1.0–12.0 mg/dL (<200 μmol/L)	Unconjugated bilirubin: excessive hemolysis of red blood cells.	
Blood urea nitrogen (BUN) (serum)	8–26 mg/dL (2.9–9.3 mmol/L)	Kidney disease, urinary tract obstruction, shock, diabetes, burns, dehydration, myocardial infarction.	Liver failure, malnutrition, overhydration, pregnancy.
Carbon dioxide content (bicarbonate + dissolved CO_2) (whole blood)	Arterial: 19–24 mEq/L (19–24 mmol/L) Venous: 22–26 mEq/L (22–26 mmol/L)	Severe diarrhea, severe vomiting, starvation, emphysema, aldosteronism.	Renal failure, diabetic ketoacidosis, shock.

Blood Tests   Continued

TEST (SPECIMEN)	U.S. REFERENCE VALUES (SI UNITS)	VALUES INCREASE IN	VALUES DECREASE IN
Cholesterol, total (plasma) **HDL cholesterol** (plasma) **LDL cholesterol** (plasma)	<200 mg/dL (<5.2 mmol/L) is desirable >40 mg/dL (>1.0 mmol/L) is desirable <130 mg/dL (<3.2 mmol/L) is desirable	Hypercholesterolemia, uncontrolled diabetes mellitus, hypothyroidism, hypertension, atherosclerosis, nephrosis.	Liver disease, hyperthyroidism, fat malabsorption, pernicious or hemolytic anemia, severe infections.
Creatine (serum)	Males: 0.15–0.5 mg/dL (10–40 μmol/L) Females: 0.35–0.9 mg/dL (30–70 μmol/L)	Muscular dystrophy, damage to muscle tissue, electric shock, chronic alcoholism.	
Creatine kinase (CK), also known as **creatine phosphokinase (CPK)** (serum)	0–130 U/L (same)	Myocardial infarction, progressive muscular dystrophy, hypothyroidism, pulmonary edema.	
Creatinine (serum)	0.5–1.2 mg/dL (45–105 μmol/L)	Impaired renal function, urinary tract obstruction, giantism, acromegaly.	Decreased muscle mass, as occurs in muscular dystrophy or myasthenia gravis.
Electrolytes (plasma)	See Table 27.2 on page 1119		
Gamma-glutamyl transferase (GGT) (serum)	0–30 U/L (same)	Bile duct obstruction, cirrhosis, alcoholism, metastatic liver cancer, congestive heart failure.	
Glucose (plasma)	70–110 mg/dL (3.9–6.1 mmol/L)	Diabetes mellitus, acute stress, hyperthyroidism, chronic liver disease, Cushing's disease.	Addison's disease, hypothyroidism, hyperinsulinism.
Hemoglobin (whole blood)	Males: 14–18 g/100 mL (140–180 g/L) Females: 12–16 g/100 mL (120–160 g/L) Newborns: 14–20 g/100 mL (140–200 g/L)	Polycythemia, congestive heart failure, chronic obstructive pulmonary disease, living at high altitude.	Anemia, severe hemorrhage, cancer, hemolysis, Hodgkin's disease, nutritional deficiency of vitamin B$_{12}$, systemic lupus erythematosus, kidney disease.
Iron, total (serum)	Males: 80–180 μg/dL (14–32 μmol/L) Females: 60–160 μg/dL (11–29 μmol/L)	Liver disease, hemolytic anemia, iron poisoning.	Iron-deficiency anemia, chronic blood loss, pregnancy (late), chronic heavy menstruation.
Lactic dehydrogenase (LDH) (serum)	71–207 U/L (same)	Myocardial infarction, liver disease, skeletal muscle necrosis, extensive cancer.	

Lipids (serum)
   Total
   Triglycerides
	
	400–850 mg/dL (4.0–8.5 g/L)
	10–190 mg/dL (0.1–1.9 g/L)
		Hyperlipidemia, diabetes mellitus.	Fat malabsorption, hypothyroidism.

Platelet (thrombocyte) count (whole blood)	150,000–400,000/μL	Cancer, trauma, leukemia, cirrhosis.	Anemias, allergic conditions, hemorrhage.

Protein (serum)
   Total
   Albumin
   Globulin
	
	6–8 g/dL (60–80 g/L)
	4–6 g/dL (40–60 g/L)
	2.3–3.5 g/dL (23–35 g/L)
		Dehydration, shock, chronic infections.	Liver disease, poor protein intake, hemorrhage, diarrhea, malabsorption, chronic renal failure, severe burns.

Red blood cell (erythrocyte) count (whole blood)	Males: 4.5–6.5 million/μL Females: 3.9–5.6 million/μL	Polycythemia, dehydration, living at high altitude.	Hemorrhage, hemolysis, anemias, cancer, overhydration.
Uric acid (urate) (serum)	2.0–7.0 mg/dL (120–420) μmol/L	Impaired renal function, gout, metastatic cancer, shock, starvation.	
White blood cell (leukocyte) count, total (whole blood)	5,000–10,000/μL (See Table 19.3 on page 742 for relative percentages of different types of WBCs.)	Acute infections, trauma, malignant diseases, cardiovascular diseases. (See also Table 19.2 on page 741.)	Diabetes mellitus, anemia. (See also Table 19.2 on page 741.)

NORMAL VALUES FOR SELECTED URINE TESTS

Urine Tests

TEST (SPECIMEN)	U.S. REFERENCE VALUES (SI UNITS)	CLINICAL IMPLICATIONS
Amylase (2 hour)	35–260 Somogyi units/hr (6.5–48.1 units/hr)	Values increase in inflammation of the pancreas (pancreatitis) or salivary glands, obstruction of the pancreatic duct, and perforated peptic ulcer.
Bilirubin* (random)	Negative	Values increase in liver disease and obstructive biliary disease.
Blood* (random)	Negative	Values increase in renal disease, extensive burns, transfusion reactions, and hemolytic anemia.
Calcium (Ca^{2+}) (random)	10 mg/dL (2.5 mmol/liter); up to 300 mg/24 hr (7.5 mmol/24 hr)	Amount depends on dietary intake; values increase in hyperparathyroidism, metastatic malignancies, and primary cancer of breasts and lungs; values decrease in hypoparathyroidism and vitamin D deficiency.
Casts (24 hour)		
Epithelial	Occasional	Values increase in nephrosis and heavy metal poisoning.
Granular	Occasional	Values increase in nephritis and pyelonephritis.
Hyaline	Occasional	Values increase in kidney infections.
Red blood cell	Occasional	Values increase in glomerular membrane damage and fever.
White blood cell	Occasional	Values increase in pyelonephritis, kidney stones, and cystitis.
Chloride (Cl$^-$) (24 hour)	140–250 mEq/24 hr (140–250 mmol/24 hr)	Amount depends on dietary salt intake; values increase in Addison's disease, dehydration, and starvation; values decrease in pyloric obstruction, diarrhea, and emphysema.
Color (random)	Yellow, straw, amber	Varies with many disease states, hydration, and diet.
Creatinine (24 hour)	Males: 1.0–2.0 g/24 hr (9–18 mmol/24 hr) Females: 0.8–1.8 g/24 hr (7–16 mmol/24 hr)	Values increase in infections; values decrease in muscular atrophy, anemia, and kidney diseases.
Glucose*	Negative	Values increase in diabetes mellitus, brain injury, and myocardial infarction.
Hydroxycorticosteroids (17-hydroxysteroids) (24 hour)	Males: 5–15 mg/24 hr (13–41 μmol/24 hr) Females: 2–13 mg/24 hr (5–36 μmol/24 hr)	Values increase in Cushing's syndrome, burns, and infections; values decrease in Addison's disease.
Ketone bodies* (random)	Negative	Values increase in diabetic acidosis, fever, anorexia, fasting, and starvation.
17-ketosteroids (24 hour)	Males: 8–25 mg/24 hr (28–87 μmol/24 hr) Females: 5–15 mg/24 hr (17–53 μmol/24 hr)	Values decrease in surgery, burns, infections, adrenogenital syndrome, and Cushing's syndrome.

TEST (SPECIMEN)	U.S. REFERENCE VALUES (SI UNITS)	CLINICAL IMPLICATIONS
Odor (random)	Aromatic	Becomes acetonelike in diabetic ketosis.
Osmolality (24 hour)	500–1400 mOsm/kg water (500–1400 mmol/kg water)	Values increase in cirrhosis, congestive heart failure (CHF), and high-protein diets; values decrease in aldosteronism, diabetes insipidus, and hypokalemia.
pH* (random)	4.6–8.0	Values increase in urinary tract infections and severe alkalosis; values decrease in acidosis, emphysema, starvation, and dehydration.
Phenylpyruvic acid (random)	Negative	Values increase in phenylketonuria (PKU).
Potassium (K$^+$) (24 hour)	40–80 mEq/24 hr (40–80 mmol/24 hr)	Values increase in chronic renal failure, dehydration, starvation, and Cushing's syndrome; values decrease in diarrhea, malabsorption syndrome, and adrenal cortical insufficiency.
Protein* (albumin) (random)	Negative	Values increase in nephritis, fever, severe anemias, trauma, and hyperthyroidism.
Sodium (Na$^+$) (24 hour)	75–200 mg/24 hr (75–200 mmol/24 hr)	Amount depends on dietary salt intake; values increase in dehydration, starvation, and diabetic acidosis; values decrease in diarrhea, acute renal failure, emphysema, and Cushing's syndrome.
Specific gravity* (random)	1.001–1.035 (same)	Values increase in diabetes mellitus and excessive water loss; values decrease in absence of antidiuretic hormone (ADH) and severe renal damage.
Urea (random)	25–35 g/24 hr (420–580 mmol/24 hr)	Values increase in response to increased protein intake; values decrease in impaired renal function.
Uric acid (24 hour)	0.4–1.0 g/24 hr (1.5–4.0 mmol/24 hr)	Values increase in gout, leukemia, and liver disease; values decrease in kidney disease.
Urobilinogen* (2 hour)	0.3–1.0 Ehrlich units (1.7–6.0 μmol/24 hr)	Values increase in anemias, hepatitis A (infectious), biliary disease, and cirrhosis; values decrease in cholelithiasis and renal insufficiency.
Volume, total (24 hour)	1000–2000 mL/24 hr (1.0–2.0 liters/24 hr)	Varies with many factors.

* Test often performed using a **dipstick,** a plastic strip impregnated with chemicals that is dipped into a urine specimen to detect particular substances. Certain colors indicate the presence or absence of a substance and sometimes give a rough estimate of the amount(s) present.

ANSWERS

Answers to Self-Quiz Questions

Chapter 1   1. tissue 2. metabolism, anabolism, catabolism 3. intracellular fluid (ICF), extracellular fluid (ECF) 4. true 5. false 6. false 7. e 8. d 9. a 10. c 11. c 12. (a) 4, (b) 6, (c) 8, (d) 1, (e) 9, (f) 5, (g) 2, (h) 7, (i) 3, (j) 10 13. (a) 1, (b) 12, (c) 1, 6, (d) 6, (e)   , (f) 8, (g) 7 (h) 3, (i) 2, (j) 10 14. (a) 4, (b) 1, (c) 3, (d) 6, (e) 5, (f) 7, (g) 2 15. (a) 6, (b) 1, (c) 11, (d) 5, (e) 10, (f) 8, (g) 7, (h) 9, (i) 4, (j) 3, (k) 2

Chapter 2   1. 8 2. solid, liquid, gas 3. monosaccharides, amino acids 4. true 5. false 6. true 7. c 8. a 9. d 10. b 11. e 12. a 13. e 14. (a) 1, (b) 2, (c) 1, (d) 4, (e) 3 15. (a) 11, (b) 1, (c) 8, (d) 3, (e) 7, (f) 4, (g) 5, (h) 9, (i) 10, (j) 12, (k) 6, (l) 2

Chapter 3   1. plasma membrane, cytoplasm, nucleus 2. apoptosis, necrosis 3. Telomeres 4. UAG 5. false 6. true 7. true 8. e 9. c 10. c, g, i, b, d, k, f, j, a, e, h 11. a 12. c 13. (a) 2, (b) 3, (c) 5, (d) 7, (e) 6, (f) 8, (g) 1, (h) 4 14. (a) 2, (b) 9, (c) 3, (d) 5, (e) 11, (f) 8, (g) 1, (h) 6, (i) 10, (j) 7, (k) 13, (l) 4, (m) 12 15. (a) 3, (b) 9, (c) 1, (d) 5, (e) 11, (f) 4, (g) 8, (h) 7, (i) 2, (j) 10, (k) 6

Chapter 4   1. epithelial, connective, muscle, nervous 2. arrangement of cells in layers, cell shape 3. true 4. true 5. e 6. b 7. a 8. c 9. e 10. b 11. d 12. c 13. (a) 4, (b) 8, (c) 5, (d) 2, (e) 6, (f) 3, (g) 1, (h) 7 14. (a) C, (b) M, (c) N, (d) E, (e) C, (f) E, (g) M, (h) E, (i) C, (j) M, (k) N, (l) E, (m) C, (n) E, (o) M and N 15. (a) 3, (b) 5, (c) 8, (d) 13, (e) 9, (f) 7, (g) 11, (h) 6, (i) 2, (j) 4, (k) 10, (l) 12, (m) 1

Chapter 5   1. stratum lucidum 2. eccrine, ceruminous, apocrine 3. false 4. true 5. c 6. e 7. a 8. c 9. b 10. e 11. a 12. c 13. (a) 3, (b) 5, (c) 4, (d) 1, (e) 6, (f) 11, (g) 2, (h) 8, (i) 9, (j) 10, (k) 7 14. (a) 3, (b) 4, (c) 1, (d) 2 15. (a) 4, (b) 3, (c) 2, (d) 1, inflammatory, migratory, proliferative, maturation

Chapter 6   1. interstitial, appositional 2. hardness, tensile strength 3. true 4. true 5. true 6. d 7. a 8. e 9. c 10. a 11. (a) 1, (b) 4, (c) 3, (d) 2 12. (a) 2, (b) 6, (c) 4, (d) 5, (e) 7, (f) 3, (g) 1 13. (a) 3, (b) 9, (c) 8, (d) 1, (e) 5, (f) 4, (g) 6, (h) 7, (i) 12, (j) 2, (k) 11 (l) 10 14. (a) 3, (b) 7, (c) 6, (d) 1, (e) 4, (f) 2, (g) 5, (h) 9, (i) 8, (j) 10 15. (a) 12, (b) 4, (c) 8, (d) 6, (e) 3, (f) 9, (g) 13, (h) 10, (i) 7, (j) 5, (k) 2, (l) 11, (m) 1

Chapter 7   1. fontanels 2. pituitary gland 3. sacrum, coccyx 4. false 5. false 6. c 7. c 8. a 9. e 10. d 11. e 12. (a) 4, (b) 9, (c) 7, (d) 5, (e) 3, (f) 1, (g) 2, (h) 8, (i) 6 13. (a) 7, (b) 5, (c) 1, (d) 6, (e) 2, (f) 4, (g) 8, (h) 9, (i) 3, (j) 10, (k) 11, (l) 13, (m) 12 14. (a) 2, (b) 3, (c) 5, (d) 6, (e) 4, (f) 1, (g) 5, (h) 4, (i) 2, (j) 4, (k) 3 15. (a) 3, (b) 1, (c) 6, (d) 9, (e) 13, (f) 12, (g) 2, (h) 4, (i) 5, (j) 7, (k) 10, (l) 15, (m) 8, (n) 11, (o) 14

Chapter 8   1. metacarpals 2. ilium, ischium, pubis 3. true (lesser), false (greater) 4. false 5. true 6. b 7. c 8. e 9. c 10. a 11. b and e 12. a 13. (a) 2, (b) 6, (c) 9, (d) 7, (e) 4, (f) 5, (g) 8, (h) 10, (i) 1, (j) 3 14. (a) 3, (b) 8, (c) 4, (d) 11, (e) 9, (f) 13, (g) 5, (h) 6, (i) 10, (j) 14, (k) 2, (l) 1, (m) 7, (n) 12 15. (a) 4, (b) 3, (c) 3, (d) 6, (e) 7, (f) 1, (g) 3, (h) 2, (i) 5, (j) 9, (k) 8, (l) 2, (m) 4, (n) 6, (o) 7, (p) 9, (q) 6, (r) 3, (s) 4, (t) 9, (u) 4 and 5

Chapter 9   1. joint, articulation or arthrosis 2. arthroplasty 3. false 4. false 5. false 6. e 7. d 8. b 9. c 10. a 11. c 12. e 13. (a) 5, (b) 3, (c) 7, (d) 2, (e) 6, (f) 4, (g) 1 14. (a) 6, (b) 4, (c) 5, (d) 1, (e) 3, (f) 2 15. (a) 8, (b) 11, (c) 10, (d) 13, (e) 15, (f) 9, (g) 6, (h) 12, (i) 3, (j) 4, (k) 16, (l) 2, (m) 18, (n) 1, (o) 7, (p) 14, (q) 17, (r) 5

Chapter 10   1. motor unit 2. muscular atrophy, fibrosis 3. acetylcholine 4. true 5. true 6. e 7. a 8. c 9. e 10. d 11. b 12. (a) 5, (b) 6, (c) 9, (d) 7, (e) 2, (f) 4, (g) 10, (h) 3, (i) 1, (j) 8 13. (a) 7, (b) 10, (c) 9, (d) 12, (e) 8, (f) 11, (g) 6, (h) 1,

(i) 2, (j) 3, (k) 4, (l) 13, (m) 5 14. (a) 2, (b) 3, (c) 1, (d) 1 and 2, (e) 3, (f) 2, (g) 1, (h) 3, (i) 1 and 2, (j) 3, (k) 2 and 3, (l) 3 15. (a) 10, (b) 2, (c) 4, (d) 3, (e) 6, (f) 5, (g) 1, (h) 12, (i) 7, (j) 9, (k) 11, (l) 8

Chapter 11   1. buccinator 2. gastrocnemius, soleus, plantaris, calcaneus 3. true 4. true 5. b 6. c 7. c 8. a 9. e 10. e 11. (a) 6, (b) 2, (c) 8, (d) 5, (e) 3, (f) 1, (g) 7, (h) 4 12. (a) 13, (b) 9, (c) 8, (d) 6, (e) 3, (f) 11, (g) 10, (h) 1, (i) 2, (j) 7, (k) 12, (l) 4, (m) 5 13. (a) 6, (b) 3, (c) 7, (d) 4, (e) 2, (f) 9, (g) 5, (h) 1, (i) 8 14. (a) 10, (b) 1, (c) 9, (d) 8, (e) 12, (f) 17, (g) 2, (h) 6, (i) 8, (j) 14, (k) 5, (l) 4, (m) 2, (n) 15, (o) 1, (p) 11, (q) 13, (r) 12, (s) 7, (t) 16, (u) 11, (v) 17, (w) 16, (x) 15, (y) 3, (z) 10 15. (a) 3, (b) 1, (c) 2, (d) 1, (e) 2, (f) 3, (g) 3

Chapter 12   1. somatic, autonomic, enteric 2. sympathetic, parasympathetic 3. false 4. false 5. c 6. d 7. c 8. e 9. e 10. b 11. e 12. b 13. (a) 6, (b) 12, (c) 1, (d) 2, (e) 9, (f) 14, (g) 4, (h) 8, (i) 7, (j) 13, (k) 5, (l) 3, (m) 10, (n) 15, (o) 11 14. (a) 2, (b) 1, (c) 10, (d) 9, (e) 6, (f) 3, (g) 4, (h) 5, (i) 12, (j) 8, (k) 7, (l) 13, (m) 11 15. (a) 4, (b) 5, (c) 16, (d) 8, (e) 7, (f) 1, (g) 2, (h) 10, (i) 15, (j) 6, (k) 3, (l) 13, (m) 9, (n) 11, (o) 14, (p) 12

Chapter 13   1. mixed 2. sensory receptor, sensory neuron, integrating center, motor neuron, effector 3. true 4. false 5. c 6. c 7. a 8. c 9. d 10. e 11. a 12. d 13. (a) 1, (b) 8, (c) 4, (d) 2, (e) 11, (f) 1, (g) 6, (h) 5, (i) 3, (j) 9, (k) 1, (l) 12, (m) 7, (n) 2, (o) 10 14. (a) 14, (b) 12, (c) 13, (d) 1, (e) 2, (f) 5, (g) 11, (h) 8, (i) 10, (j) 9, (k) 15, (l) 4, (m) 7, (n) 3, (o) 6 15. (a) 2, (b) 1, (c) 3, (d) 4, (e) 1, (f) 5, (g) 3, (h) 2, (i) 4, (j) 1, (k) 2, (l) 4, (m) 3, (n) 5, (o) 1

Chapter 14   1. corpus callosum 2. frontal, temporal, parietal, occipital, insula 3. longitudinal fissure 4. false 5. true 6. d 7. c 8. d 9. e 10. d 11. e 12. (a) 3, (b) 5, (c) 6, (d) 8, (e) 11, (f) 10, (g) 7, (h) 9, (i) 1, (j) 4, (k) 2, (l) 12, (m) 1, (n) 8, (o) 5, (p) 7, (q) 12, (r) 10, (s) 9, (t) 1 and 2, (u) 3, 4, and 6, (v) 11 13. (a) 9, (b) 2, (c) 6, (d) 10, (e) 4, (f) 11, (g) 1, (h) 2, (i) 5, (j) 8, (k) 12, (l) 7, (m) 3, (n) 6 and 8, (o) 13, (p) 7, (q) 1 14. (a) 5, (b) 9, (c) 11, (d) 6, (e) 3, (f) 1, (g) 10, (h) 8, (i) 2, (j) 4, (k) 7 15. (a) 10, (b) 2, (c) 6, (d) 8, (e) 7, (f) 5, (g) 3, (h) 11, (i) 14, (j) 13, (k) 4, (l) 1, (m) 12, (n) 9, (o) 15

Chapter 15   1. acetylcholine, epinephrine or norepinephrine 2. thoracolumbar, craniosacral 3. true 4. true 5. d 6. e 7. b 8. c 9. e 10. a 11. a 12. c 13. e, b, g, f, d, a, c 14. (a) 3, (b) 2, (c) 1, (d) 1, (e) 2, (f) 3, (g) 3, (h) 1, (i) 4, (j) 2, (k) 5 15. (a) 2, (b) 1, (c) 1, (d) 2, (e) 1, (f) 1, (g) 2, (h) 2

Chapter 16   1. sensation, perception 2. decussation 3. false 4. true 5. c 6. 1 7. d 8. b 9. d 10. e 11. e 12. d 13. (a) 3, (b) 2, (c) 5, (d) 7, (e) 1, (f) 9, (g) 3, (h) 11, (i) 8, (j) 4, (k) 6, (l) 10 14. (a) 10, (b) 8, (c) 7, (d) 1, (e) 4, (f) 3, (g) 5, (h) 6, (i) 9, (j) 2 15. (a) 9, (b) 8, (c) 4, (d) 7, (e) 10, (f) 2, (g) 3, (h) 1, (i) 5, (j) 6, (k) 11, (l) 12

Chapter 17   1. sweet, sour, salty, bitter, umami 2. static, dynamic 3. true 4. false 5. d 6. a 7. d 8. b 9. b 10. c, j, k, d, h, l, e, b, f, i, a, m, g 11. c 12. a 13. (a) 1, (b) 5, (c) 7, (d) 6, (e) 8, (f) 2, (g) 4, (h) 3 14. (a) 3, (b) 5, (c) 8, (d) 2, (e) 7, (f) 4, (g) 1, (h) 6, (i) 9 15. (a) 2, (b) 11, (c) 14, (d) 13, (e) 3, (f) 10, (g) 6, (h) 12, (i) 4, (j) 5, (k) 9, (l) 1, (m) 7, (n) 8

Chapter 18   1. fight-or-flight response, resistance reaction, exhaustion 2. hypothalamus 3. less, more 4. false 5. true 6. b 7. e 8. d 9. a 10. e 11. c 12. a 13. (a) 8, (b) 2 (and 13, 18, 22), (c) 7, (d) 1, (e) 12, (f) 20, (g) 5, (h) 18, (i) 22, (j) 15 (and 11), (k) 3, (l) 17, (m) 21, (n) 6, (o) 13, (p) 11 (and 15), (q) 4, (r) 10, (s) 14, (t) 9, (u) 16, (v) 19 14. (a) 10, (b) 8, (c) 2, (d) 12, (e) 15, (f) 4, (g) 1, (h) 16, (i) 6, (j) 9, (k) 13, (l) 7, (m) 5, (n) 14, (o) 3, (p) 11 15. (a) 12, (b) 1, (c) 11, (d) 7, (e) 3, (f) 10, (g) 2, (h) 9, (i) 4, (j) 8, (k) 5, (l) 6

Chapter 19   1. serum 2. clot retraction 3. true 4. true 5. e 6. a 7. b 8. c 9. d 10. a 11. d 12. e 13. (a) 4, (b) 7, (c) 6, (d) 1, (e) 3, (f) 5, (g) 2 14. (a) 4, (b) 6, (c) 2, (d) 7, (e) 1, (f) 5, (g) 3 15. (a) 4, (b) 6, (c) 1, (d) 3, (e) 5, (f) 2

Chapter 20 1. left ventricle 2. systole, diastole 3. false 4. true 5. a 6. c 7. d 8. b 9. b 10. e 11. c 12. (a) 3, (b) 6, (c) 1, (d) 5, (e) 2, (f) 4 13. (a) 8, (b) 4, (c) 11, (d) 5, (e) 1, (f) 9, (g) 7, (h) 2, (i) 10, (j) 6, (k) 3 14. (a) 3, (b) 2, (c) 9, (d) 14, (e) 8, (f) 7, (g) 11, (h) 12, (i) 15, (j) 4, (k) 5, (l) 1, (m) 6, (n) 21, (o) 22, (p) 19, (q) 17, (r) 18, (s) 20, (t) 16, (u) 13, (v) 10 15. (a) 3, (b) 7, (c) 2, (d) 5, (e) 1, (f) 6, (g) 4 and 7

Chapter 21 1. carotid sinus, aortic 2. skeletal muscle pump, respiratory pump 3. true 4. true 5. b 6. a 7. c 8. e 9. a 10. d 11. (a) D, (b) C, (c) C, (d) D, (e) D, (f) C, (g) C, (h) C, (i) D, (j) D, (k) C 12. (a) 2, (b) 5, (c) 1, (d) 4, (e) 3 13. (a) 11, (b) 1, (c) 4, (d) 9, (e) 3, (f) 8, (g) 6, (h) 2, (i) 7, (j) 5, (k) 10, (l) 12, (m) 13 14. (a) 2, (b) 6, (c) 4, (d) 1, (e) 3, (f) 5 15. (a) 5, (b) 3, (c) 1, (d) 4, (e) 2, (f) 4, (g) 1, (h) 5, (i) 3

Chapter 22 1. skin, mucous membranes, antimicrobial proteins, natural killer cells, phagocytes 2. antigens 3. true 4. true 5. c 6. d 7. e 8. d 9. e 10. b 11. c 12. e, h, b, f, a, d, g, i, c 13. (a) 3, (b) 1, (c) 7, (d) 4, (e) 2, (f) 5, (g) 6 14. (a) 2, (b) 3, (c) 4, (d) 7, (e) 1, (f) 6, (g) 5

Chapter 23 1. oxyhemoglobin; dissolved CO_2, carbamino compounds (primarily carbaminohemoglobin), and bicarbonate ion 2. $CO_2 + H_2O \rightarrow H_2CO_3 \rightarrow H^+ + HCO_3^-$ 3. false 4. true 5. c 6. e 7. b 8. d 9. a 10. e 11. e, g, b, h, a, d, f, c 12. (a) 3, (b) 8, (c) 5, (d) 9, (e) 2, (f) 7, (g) 10, (h) 1, (i) 4, (j) 6 13. (a) 2, (b) 11, (c) 3, (d) 9, (e) 1, (f) 12, (g) 10, (h) 5, (i) 13, (j) 8, (k) 4, (l) 6, (m) 7 14. (a) 7, (b) 8, (c) 1, (d) 5, (e) 6, (f) 9, (g) 2, (h) 3, (i) 4 15. (a) 9, (b) 11, (c) 3, (d) 4, (e) 6, (f) 1, (g) 5, (h) 10, (i) 7, (j) 8, (k) 2

Chapter 24 1. monosaccharides; amino acids; monoglycerides, fatty acids; pentoses, phosphates, nitrogenous bases 2. diffusion, facilitated diffusion, osmosis, active transport 3. true 4. true 5. b 6. d 7. e 8. c 9. a 10. b 11. d 12. a 13. (a) 5, (b) 13, (c) 8, (d) 11, (e) 9, (f) 1, (g) 7, (h) 12, (i) 14, (j) 2, (k) 4, (l) 10, (m) 6, (n) 3 14. (a) 4, (b) 6, (c) 7, (d) 1, (e) 5, (f) 3, (g) 9, (h) 2, (i) 11, (j) 10, (k) 8

Chapter 25 1. hypothalamus 2. glucose 6-phosphate, pyruvic acid, acetyl coenzyme A 3. true 4. false 5. e 6. c 7. b 8. d 9. a 10. b 11. e 12. a 13. (a) 2 and 3, (b) 1, (c) 3 and 5, (d) 2, 4, 5, and 6, (e) 1, (f) 2, (g) 1, 4, and 6 14. (a) 9, (b) 12, (c) 11, (d) 10, (e) 4, (f) 13, (g) 3, (h) 5, (i) 8, (j) 1, (k) 6, (l) 2, (m) 7 15. (a) 17, (b) 15, (c) 8, (d) 19, (e) 1, (f) 7, (g) 4, (h) 10, (i) 16, (j) 14, (k) 11, (l) 2, (m) 13, (n) 6, (o) 20, (p) 9, (q) 5, (r) 18, (s) 3, (t) 12

Chapter 26 1. glomerulus, glomerular (Bowman's) capsule 2. micturition 3. false 4. true 5. d 6. e 7. c 8. b 9. e 10. a 11. g, a, h, c, o, n, l, j, i, b, d, e, m, k, f 12. g, f, b, d, h, c, a, e, i 13. (a) 8, (b) 2, (c) 10, (d) 5, (e) 3, (f) 1, (g) 7, (h) 4, (i) 11, (j) 9, (k) 6 14. (a) 4, (b) 3, (c) 7, (d) 6, (e) 2, (f) 8, (g) 1, (h) 5 15. (a) 5, (b) 4, (c) 6, (d) 8, (e) 1, (f) 2, (g) 7, (h) 3, (i) 9

Chapter 27 1. metabolic 2. bicarbonate ion, carbonic acid 3. true 4. false 5. d 6. e 7. a 8. b 9. (a) 8, (b) 9, (c) 7, (d) 1, (e) 6, (f) 2, (g) 4, (h) 5, (i) 3 10. (a) 8, (b) 12, (c) 7, (d) 5, (e) 10, (f) 6, (g) 9, (h) 13, (i) 1, (j) 11, (k) 3, (l) 4, (m) 2

Chapter 28 1. puberty, menarche, menopause 2. true 3. true 4. e 5. c 6. a 7. c 8. b 9. a 10. e 11. d 12. (a) 4, (b) 2, (c) 1, (d) 6, (e) 5, (f) 3 13. (a) 13, (b) 10, (c) 12, (d) 1, (e) 5, (f) 2, (g) 4, (h) 6, (i) 14, (j) 11, (k) 8, (l) 3, (m) 9, (n) 7 14. (a) 6, (b) 4, (c) 1, (d) 12, (e) 8, (f) 5, (g) 7, (h) 13, (i) 11, (j) 3, (k) 14, (l) 2, (m) 10, (n) 15, (o) 9

Chapter 29 1. dilation, expulsion, placental 2. corpus luteum, human chorionic gonadotropin 3. mesoderm, ectoderm, endoderm 4. false 5. a 6. d 7. b 8. e 9. b 10. c 11. (a) 6, (b) 2, (c) 11, (d) 5, (e) 10, (f) 1, (g) 3, (h) 8, (i) 9, (j) 4, (k) 7 12. (a) 3, (b) 6, (c) 5, (d) 2, (e) 1, (f) 4 13. (a) 3, (b) 4, (c) 5, (d) 1, (e) 2, (f) 6, (g) 7, (h) 10, (i) 8, (j) 9 14. (a) 7, (b) 3, (c) 14, (d) 6, (e) 4, (f) 10, (g) 8, (h) 13, (i) 5, (j) 2, (k) 1, (l) 9, (m) 16, (n) 11, (o) 15, (p) 12 15. (a) 3, (b) 6, (c) 4, (d) 1, (e) 9, (f) 2, (g) 5, (h) 8, (i) 7

Answers to Critical Thinking Questions

Chapter 1

1. No. Computed tomography is used to look at differences in tissue density. To assess activity in an organ such as the brain, a positron emission tomography (PET) scan or a single-photo-emission computerized tomography (SPECT) scan would provide a colorized visual assessment of brain activity.

2. Stem cells are undifferentiated cells. Research using stem cells has shown that these undifferentiated cells may be prompted to differentiate into the specific cells needed to replace those which are damaged or malfunctioning.

3. Homeostasis is the relative constancy (or dynamic equilibrium) of the body's internal environment. Homeostasis is maintained as the body changes in response to shifting external and internal conditions, including those of temperature, pressure, fluid, electrolytes, and other chemicals.

Chapter 2

1. Neither butter nor margarine is a particularly good choice for frying eggs. Butter contains saturated fats that are associated with heart disease. However, many margarines contain hydrogenated or partially hydrogenated trans-fatty acids that also increase the risk of heart disease. An alternative would be frying the eggs in any of the mono- or polyunsaturated fats such as olive oil, peanut oil, or corn oil. Boiling or poaching eggs instead of frying them would reduce the fat content of his breakfast, as would eating only the egg whites (not the high-fat yolks).

2. High body temperatures can be life-threatening, especially in infants. The increased temperature can cause denaturing of structural proteins and vital enzymes. When this happens, the proteins become nonfunctional. If the denatured enzymes are required for reactions that are necessary for life, then the infant could die.

3. Simply adding water to the table sugar does not cause it to break apart into monosaccharides. The water acts as a solvent, dissolving the sucrose and forming a sugar-water solution. To complete the breakdown of table sugar to glucose and fructose would require the presence of the enzyme sucrase.

Chapter 3

1. Synthesis of mucin by ribosomes on rough endoplasmic reticulum, to transport vesicle, to entry face of Golgi complex, to transfer vesicle, to medial cisternae where protein is modified, to transfer vesicle, to exit face, to secretory vesicle, to plasma membrane where it undergoes exocytosis.

2. Since smooth ER inactivates or detoxifies drugs, and peroxisomes also destroy harmful substances such as alcohol, we would expect to see increased numbers of these organelles in Sebastian's liver cells.

3. In order to restore water balance to the cells, the runners need to consume hypotonic solutions. The water in the hypotonic solution will move from the blood, into the interstitial fluid, and then into the cells. Plain water works well; sports drinks contain water and some electrolytes (which may have been lost due to sweating) but will still be hypotonic in relation to the body cells.

Chapter 4

1. There are many possible adaptations, including: more adipose tissue for insulation; thicker bones for support; more red blood cells for oxygen transport; increased thickness of skin to prevent water loss; etc.

2. Infants tend to have a high proportion of brown fat, which contains many mitochondria and is highly vascularized. When broken down, brown fat produces heat that helps to maintain infants' body temperatures. This heat can also warm the blood, which then distributes the heat throughout the body.

3. Your bread-and-water diet is not providing you with the necessary nutrients to encourage tissue repair. You need proper amounts of many essential vitamins, especially vitamin C, which is required for repair of the matrix and blood vessels. Vitamin A is needed to help properly maintain epithelial tissue. Adequate protein is also needed in order to synthesize the structural proteins of the damaged tissue.

Chapter 5

1. The dust particles are primarily keratinocytes that are shed from the stratum corneum of the skin.

2. Tattoos are created by depositing ink into the dermis, which does not undergo shedding as the epidermis does. Although the tattoo will fade due to exposure to sunlight and the flushing away of ink particles by the lymphatic system, the tattoo is indeed permanent.

3. Chef Eduardo has damaged the nail matrix—the part of the nail that produces growth. Because the damaged area has not regrown properly, the nail matrix may be permanently damaged.

Chapter 6

1. Due to the strenuous, repetitive activity, Taryn has probably developed a stress fracture of her right tibia (lower leg bone). Stress fractures are due to repeated stress on a bone that causes microscopic breaks in the bone without any evidence of injury to other tissue. An x-ray would not reveal the stress fracture, but a bone scan would. Thus the bone scan would either confirm or negate the physician's diagnosis.

2. When Marcus broke his arm as a child, he injured his epiphyseal (growth) plate. Damage to the cartilage in the epiphyseal plate resulted in premature closure of the plate, which interfered with the lengthwise growth of the arm bone.

3. Exercise causes mechanical stress on bones, but because there is effectively zero gravity in space, the pull of gravity on bones is missing. The lack of stress from gravity results in bone demineralization and weakness.

Chapter 7

1. Inability to open mouth—damage to the mandible, probably at temporomandibular joint; black eye—trauma to the ridge over the supraorbital margin; broken nose—probably damage to the nasal septum (includes the vomer, septal cartilage, and perpendicular plate of the ethmoid) and possibly the nasal bones; broken cheek—fracture of zygomatic bone; broken upper jaw—fracture of maxilla; damaged eye socket—fracture of parts of the sphenoid, frontal, ethmoid, palatine, zygomatic, lacrimal and maxilla (all compose the eye socket); punctured lung—damage to the thoracic vertebrae, which have punctured the lung.

2. Due to the repeated and extensive tension on his bone surfaces, Bubba would experience deposition of new bone tissue. His arm bones would be thicker and with increased raised areas (projections) where the tendons attach his muscles to bone.

3. The "soft area" being referred to is the anterior fontanel, located between the parietal and frontal bones. This is one of several areas of fibrous connective tissue in the skull that has not ossified; it should complete its ossification at 18–24 months after birth. Fontanels allow flexibility of the skull for childbirth and for brain growth after birth. The connective tissue will not allow passage of water; thus no brain damage will occur through simply washing the baby's hair.

Chapter 8

1. There are several characteristics of the bony pelves that can be used to differentiate male from female: (1) The pelvis in the female is wider and more shallow than the male's; (2) the pelvic brim of the female is larger and more oval; (3) the pubic arch has an angle greater than 90º; (4) the pelvic outlet is wider than in a male; (5) the female's iliac crest is less curved and the ilium less vertical. Table 8.1 provides additional differences between female and male pelves. Age of the skeleton can be determined by the size of the bones, the presence or absence of epiphyseal plates, the degree of demineralization of the bones, and the general appearance of the "bumps" and ridges of bones.

2. Infants do have "flat feet" because their arches have not yet developed. As they begin to stand and walk, the arches should begin to develop in order to accommodate and support their body weight. The arches are usually fully developed by age 12 or 13, so Dad doesn't need to worry yet!

3. There are 14 phalanges in each hand: two bones in the thumb and three in each of the other fingers. Farmer White has lost five phalanges on his left hand (two in his thumb and three in his index finger), so he has nine remaining on his left and 14 remaining on his right for a total of 23.

Chapter 9

1. Katie's vertebral column, head, thighs, lower legs, lower arms, and fingers are flexed. Her forearms and shoulders are medially rotated. Her thighs and arms are adducted.

2. The knee joint is commonly injured, especially among athletes. The twisting of Jeremiah's leg could have resulted in a multitude of internal injuries to the knee joint but often football players suffer tearing of the anterior cruciate ligament and medial meniscus. The immediate swelling is due to blood from damaged blood vessels, damaged synovial membranes, and the torn meniscus. Continued swelling is a result of a buildup of synovial fluid, which can result in pain and decreased mobility. Jeremiah's doctor may aspirate some of the fluid ("draining the water off his knee") and might want to perform arthroscopy to check for the extent of the knee damage.

3. The condylar processes of the mandible passed anteriorly to the articular tubercles of the temporal bones, and this dislocated Antonio's mandible. It could be corrected by pressing the thumbs downward on the lower molar teeth and pushing the mandible backward.

Chapter 10

1. Muscle cells lose their ability to undergo cell division after birth. Therefore, the increase in size is not due to an increase in the number of muscle cells but rather is due to enlargement of the existing muscle fibers (hypertrophy). This enlargement can occur from forceful, repetitive muscular activity. It will cause the muscle fibers to increase their production of internal structures such as mitochondria and myofibrils and produce an increase in the muscle fiber diameter.

2. The "dark meat" of both chickens and ducks is composed primarily of slow oxidative (SO) muscle fibers. These fibers contain large amounts of myoglobin and capillaries, which accounts for their dark color. In addition, these fibers contain large numbers of mitochondria and generate ATP by aerobic respiration. SO fibers are resistant to fatigue and can produce sustained contractions for many hours. The legs of chickens and ducks are used for support, walking, and swimming (in ducks), all activities in which endurance is needed. In addition, migrating ducks require SO fibers in their breasts to enable them to have enough energy to fly for extremely long distances while migrating. There may be some fast oxidative–glycolytic (FOG) fibers in the dark meat. FOG fibers also contain large amounts of myoglobin and capillaries, contributing to the dark color. They can use aerobic or anaerobic cellular respiration to generate ATP and have high-to-moderate resistance to fatigue. These fibers would be good for the occasional "sprint" that ducks and chickens undergo to escape dangerous situations. In contrast, the white meat of a chicken breast is composed primarily of fast glycolytic (FG) fibers. FG fibers have lower amounts of myoglobin and capillaries that give the meat its white color. There are also few mitochondria in FG fibers, so these fibers generate ATP mainly by glycolysis. These fibers contract strongly and quickly and are adapted for intense anaerobic movements of short duration. Chickens occasionally use their breasts for flying extremely short distances, usually to escape prey or perceived danger, so FG fibers are appropriate for their breast muscle.

3. Destruction of the somatic motor neurons to skeletal muscle fibers will result in a loss of stimulation to the skeletal muscles. When not stimulated on a regular basis, a muscle begins to lose its muscle tone. Through lack of use, the muscle fibers will weaken, begin to decrease in size, and can be replaced by fibrous connective tissue, resulting in a type of denervation atrophy. A lack of stimulation of the breathing muscles (especially the diaphragm) from motor neurons can result in inability of the breathing muscles to contract, thus causing respiratory paralysis and possibly death of the individual from respiratory failure.

Chapter 11

1. All of the following could occur on the affected (right) side of the face: (1) drooping of eyelid—levator palpebrae superioris; (2) drooping of the mouth, drooling, keeping food in mouth—orbicularis oris, buccinator; (3) uneven smile—zygomaticus major, levator labii superioris, risorius; (4) inability to wrinkle forehead—occipitofrontalis; (5) trouble sucking through a straw—buccinator.

2. Bulbospongiosus, external urethral sphincter, and deep transverse perineal.

3. The rotator cuff is formed by a combination of the tendons of four deep muscles of the shoulder—subscapularis, supraspinatus, infraspinatus, and teres minor. These muscles add strength and stability to the shoulder joint. Although any of the muscles' tendons can be injured, the subscapularis is most often damaged. Dependent upon the injured muscle, Jose may have trouble medially rotating his arm (subscapularis), abducting his arm (supraspinatus), laterally rotating his arm (infraspinatus, teres minor), or extending his arm (teres minor).

Chapter 12

1. Smelling the coffee and hearing the alarm are somatic sensory, stretching and yawning are somatic motor, salivating is autonomic (parasympathetic) motor, stomach rumble is enteric motor.

2. Demyelinaton or destruction of the myelin sheath can lead to multiple problems, especially in infants and children whose myelin sheaths are still in the process of developing. The affected axons deteriorate, which will interfere with function in both the CNS and PNS. There will be lack of sensation and loss of motor control with less rapid and less coordinated body responses. Damage to the axons in the CNS can be permanent and Ming's brain development may be irreversibly affected.

3. Dr. Moro could develop a drug that: (1) is an agonist of substance P; (2) blocks the breakdown of substance P; (3) blocks the reuptake of substance P; (4) promotes the release of substance P; (5) suppresses the release of enkephalins.

Chapter 13

1. The needles will pierce the epidermis, dermis, and subcutaneous layer and then go between the vertebrae through the epidural space, the dura mater, the subdural space, the arachnoid mater, and into the CSF in the subarachnoid space. CSF is produced in the brain, and the spinal meninges are continuous with the cranial meninges.

2. The anterior gray horns contain cell bodies of somatic motor neurons and motor nuclei that are responsible for the nerve impulses for skeletal muscle contraction. Because the lower cervical region is affected (brachial plexus, C5–C8), you would expect that Sunil may have trouble with movement in his shoulder, arm, and hand on the affected side.

3. Allyson has damaged her posterior columns in the lower (lumbar) region of the spinal cord. The posterior columns are responsible for transmitting nerve impulses responsible for awareness of muscle position (proprioception) and touch—which are affected in Allyson—as well as other functions such as light pressure sensations and vibration sensations. Relating Allyson's symptoms to the distribution of dermatomes, it is likely that regions L4, L5, and S1 of her spinal cord were compressed.

Chapter 14

1. Movement of the right arm is controlled by the left hemisphere's primary motor area, located in the precentral gyrus. Speech is controlled by Broca's area in the left hemisphere's frontal lobe just superior to the lateral cerebral sulcus.

2. Nicky's right facial nerve has been affected; she is suffering from Bell's palsy due to the viral infection. The facial nerve controls contraction of skeletal muscles of the face, tear gland and salivary gland secretion, as well as conveying sensory impulses from many of the taste buds on the tongue.

3. You will need to design a drug that can get through the brain's blood–brain barrier (BBB). The drug should be lipid- or water-soluble. If the drug can open a gap between the tight junctions of the endothelial cells of the brain capillaries, it would be more likely to pass through the BBB. Targeting the drug to enter the brain in certain areas near the third ventricle (the circumventricular organs) might be an option as the BBB is entirely absent in those areas and the capillary endothelium is more permeable, allowing the blood-borne drug to more readily enter the brain tissue.

Chapter 15

1. Digestion and relaxation are controlled by increased stimulation of the parasympathetic division of the ANS. The salivary glands, pancreas, and liver will show increased secretion; the stomach and intestines will have increased activity; the gallbladder will have increased contractions; heart contractions will have decreased force and rate. Following is the nerve supply to each listed organ: salivary glands—facial nerves (cranial nerve VII) and glossopharyngeal nerves (cranial nerve IX); pancreas, liver, stomach, gallbladder, intestines and heart—vagus nerves (cranial nerve X).

2. Ciara experienced one of the "E situations" (emergency in her case), which has activated the fight-or-flight response. Some noticeable effects of increased sympathetic activity include an increase in heart rate, sweating on the palms, and contraction of the arrector pili muscles, which causes the goose flesh. Secretion of epinephrine and norepinphrine from the adrenal medullae will intensify and prolong the responses.

3. Mrs. Young needs to slow down the activity of her digestive system, which seems to be experiencing increased parasympathetic response. A parasympathetic blocking agent is needed. Because the stomach and intestines have muscarinic receptors, she needs to be provided with a muscarinic blocking agent (such as atropine), which will result in decreased motility in the stomach and intestines.

Chapter 16

1. Chemoreceptors in the nose detect odors. Proprioceptors detect body position and are involved in equilibrium. The chemoreceptors in the nose are rapidly adapting, whereas proprioceptors are slowly adapting. Thus the smell faded while the sensation of motion remained.

2. Thermal (heat) receptors in her left hand detect the stimulus. A nerve impulse is transmitted to the spinal cord through first-order neurons with cell bodies in posterior root ganglia. The impulses travel into the spinal cord, where the first-order neurons synapse with second-order neurons, whose cell bodies are located in the posterior gray horn of the spinal cord. The axons of the second-order neurons decussate to the right side in the spinal cord and then the impulses ascend through the lateral spinothalamic tract. The axons of the second-order neurons end in the ventral posterior nucleus of the right side of the thalamus, where they synapse with the third-order neurons. Axons of the third-order neurons transmit impulses to the specific primary somatosensory areas in the postcentral gyrus of the right parietal lobe.

3. When Marvin settled down for the night, he passed through Stages 1–3 of NREM sleep. Sleepwalking occurred when he was in Stage 4 (slow-wave sleep). Because this is the deepest stage of sleep, his mother was able to return him to his bed without awakening him. Marvin then cycled through REM and NREM sleep. His dreaming occurred during the REM phases of sleep. The noise of the alarm clock provided the sensory stimulus that stimulated the reticular activating system. Activation of this system sends numerous nerve impulses to widespread areas of the cerebral cortex, both directly and via the thalamus. The result is the state of wakefulness.

Chapter 17

1. Damage to the facial nerve would affect smell, taste, and hearing. Within the nasal epithelium and connective tissue, both the supporting cells and olfactory glands are innervated by branches of the facial nerve. Without input from the facial nerve, there will be a lack of mucus production required to dissolve odorants. The facial nerve also serves taste buds in the anterior two-thirds of the tongue, so damage can affect taste sensations. Hearing will be affected by a damaged facial nerve because the stapedius muscle, which is attached to the stapes, is innervated by the facial nerve. Contraction of the stapedius muscle helps to protect the inner ear from prolonged loud noises. Damage to the facial nerve will result in sounds that are excessively loud, resulting in more susceptibility to damage by prolonged loud noises.

2. With age, Gertrude has lost much of her sense of smell and taste due to a decline in olfactory and gustatory receptors. Since smell and taste are intimately linked, food no longer smells nor tastes as good to Gertrude. Gertrude has presbyopia, a loss of lens elasticity, which makes it difficult to read. She may also be experiencing age-related loss of sharpness of vision and depth perception. Gertrude's hearing difficulties could be a result of damage to hair cells in the organ of Corti or degeneration of the nerve pathway for hearing. The "buzzing" Gertrude hears may be tinnitus, which also occurs more frequently in the elderly.

3. Some of the eyedrops placed in the eye may pass through the nasolacrimal duct into the nasal cavity where olfactory receptors are stimulated. Because most "tastes" are actually smells, the child will "taste" the medicine from her eye.

Chapter 18

1. Yes, Amanda should visit the clinic, as these are serious signs and symptoms. She has an enlarged thyroid gland, or goiter. The goiter is probably due to hypothyroidism, which is causing the weight gain, fatigue, mental dullness, and other symptoms.

2. Amanda's problem is her pituitary gland, which is not secreting normal levels of TSH. Rising thyroxine (T_4) levels after the TSH injection indicates that her thyroid is functioning normally and able to respond to the increased TSH levels. If the thyroxine levels had not risen, then the problem would have been her thyroid gland.

3. Mr. Hernandez has diabetes insipidus caused by either insufficient production or release of ADH due to hypothalamus or posterior pituitary damage. He also could have defective ADH receptors in the kidneys. Diabetes insipidus is characterized by production of large volumes of urine, dehydration, and increased thirst, but with no glucose or ketones present in the urine (which would be indicative of diabetes mellitus rather than diabetes insipidus).

Chapter 19

1. The broad spectrum antibiotics may have destroyed the bacteria that caused Shilpa's bladder infection but also destroyed the naturally occurring large intestine bacteria that produce vitamin K. Vitamin K is required for the synthesis of four clotting factors (II, VII, IX, and X). Without these clotting factors present in normal amounts, Shilpa will experience clotting problems until the intestinal bacteria reach normal levels and produce additional vitamin K.

2. Mrs. Brown's kidney failure is interfering with her ability to produce erythropoietin (EPO). Her physician can prescribe Epoetin alfa, a recombinant EPO, which is very effective in treating the decline in RBC production with renal failure.

3. A primary problem Thomas may experience is with clotting. Clotting time becomes longer because the liver is responsible for producing many of the clotting factors and clotting proteins such as fibrinogen. Thrombopoietin, which stimulates the formation of platelets, is also produced in the liver. In addition, the liver is responsible for eliminating bilirubin, produced from the breakdown of RBCs. With a malfunctioning liver, the bilirubin will accumulate, resulting in jaundice. In addition, there can be decreased concentrations of the plasma protein albumin, which can affect blood pressure.

Chapter 20

1. The dental procedures introduced bacteria into Gerald's blood. The bacteria colonized his endocardium and heart valves, resulting in bacterial endocarditis. Gerald may have had a previously undetected heart murmur, or the heart murmur may have resulted from his endocarditis. His physician will want to monitor his heart to watch for further damage to the valve.

2. Extremely rapid heart rates can result in a decreased stroke volume due to insufficient ventricular filling. As a result, the cardiac output will decline to the point where there may not be enough blood reaching the central nervous system. She initially may experience light-headedness but could lose consciousness if the cardiac output declines dramatically.

3. Mr. Perkins is suffering from angina pectoris and has several risk factors for coronary artery disease such as smoking, obesity, sedentary lifestyle, and male gender. Cardiac angiography involves the use of a cardiac catheter to inject a radiopaque medium into the heart and its vessels. The angiogram may reveal blockages such as atherosclerotic plaques in his coronary arteries.

Chapter 21

1. The hole in the heart was the foramen ovale, which is an opening between the right and left atria. In fetal circulation it allows blood to bypass the right ventricle, enter the left atrium, and join systemic circulation. The "hole" should close shortly after birth to become the fossa ovalis. Closure of the foramen ovale after birth will allow the deoxygenated blood from the right atrium to enter pulmonary circulation so that the blood can become oxygenated prior to entering systemic circulation. If closure doesn't occur, surgery may be required.

2. Michael is suffering from hypovolemic shock due to the loss of blood. The low blood pressure is a result of low blood volume and a subsequent decrease in cardiac output. His rapid, weak pulse is an attempt of the heart to compensate for the decrease in cardiac output through sympathetic stimulation of the heart and increased blood levels of epinephrine and norepinephrine. His pale, cool, and clammy skin is a result of sympathetic constriction of the blood vessels of the skin and sympathetic stimulation of sweat glands. The lack of urine production is due to increased secretion of aldosterone and ADH, both of which are produced to increase blood volume in order to compensate for Michael's hypotension. The fluid loss from his bleeding results in activation of the thirst center in the hypothalamus. His confusion and disorientation are caused by a reduced oxygen supply to the brain from the decreased cardiac output.

3. Maureen has varicose veins, a condition in which the venous valves become leaky. The leaking valves allow the backflow of blood and an increased pressure that distends the veins and allows fluid to leak into the surrounding tissue. Standing on hard surfaces for long periods of time can cause varicosities to develop. Maureen needs to elevate her legs when possible to counteract the effects of gravity on the blood flow in the lower legs. She could also utilize support hose, which add external support for the superficial veins, much like

skeletal muscle does for deeper veins. If the varices become severe, Maureen may require more extensive treatment such as sclerotherapy, radiofrequency endovenous occlusion, laser occlusion, or stripping.

Chapter 22

1. The influenza vaccination introduces a weakened or killed virus (which will not cause the disease) to the body. The immune system recognizes the antigen and mounts a primary immune response. Upon exposure to the same flu virus that was in the vaccine, the body will produce a secondary response, which will usually prevent a case of the flu. This is artificially acquired active immunity.

2. Mrs. Franco's lymph nodes were removed because metastasis of cancerous cells can occur through the lymph nodes and lymphatic vessels. Mrs. Franco's swelling is a lymphedema that is occurring due to the buildup of interstitial fluid from interference with drainage in the lymph vessels.

3. Tariq's physician would need to perform an antibody titer, which is a measure of the amount of antibody in the serum. If Tariq has previously been exposed to mumps (or been vaccinated for mumps), his blood should have elevated levels of IgG antibodies after this exposure from his sister. His immune system would be experiencing a secondary response. If he has not previously had mumps and has contracted mumps from his sister, his immune system would initiate a primary response. In that case, his blood would show an elevated titer of IgM antibodies, which are secreted by plasma cells after an initial exposure to the mumps antigen.

Chapter 23

1. Aretha's excess mucus production is causing blockage of the paranasal sinuses, which are used as hollow resonating chambers for singing and speech. In addition, her sore throat could be due to inflammation of the pharynx and larynx, which will affect their normal functions. Normally, the pharynx also acts as a resonating chamber and the true vocal cords, located in the larynx, vibrate for speech and singing. Inflammation of the true vocal cords (laryngitis) interferes with their ability to freely vibrate, which will affect both singing and speech.

2. In emphysema, there is destruction of the alveolar walls, producing abnormally large air spaces that remain filled with air during exhalation. The destruction of alveoli decreases the surface area for gas exchange across the respiratory membrane, resulting in a decreased blood O_2 level. Damage to the alveolar walls also causes a loss of elasticity, making exhalation more difficult. This can result in a buildup of CO_2. Cigarette smoke contains nicotine, carbon monoxide, and a variety of irritants, all of which affect the lungs. Nicotine constricts terminal bronchioles, decreasing the air flow into and out of the lungs; carbon monoxide binds to hemoglobin, reducing its ability to carry oxygen; irritants such as tar and fine particulate matter destroy cilia and increase mucus secretion, interfering with the ability of the respiratory passages to cleanse themselves.

3. The squirrel's nest blocked the passage of exhaust gas from the furnace, causing an accumulation of carbon monoxide (CO), a colorless, odorless gas, in the home. As they were sleeping, their blood was saturated with CO, which has a stronger affinity for hemoglobin than oxygen. As a result, the Robinsons became oxygen deficient. Without adequate oxygenation of the brain, the Robinsons died during their sleep.

Chapter 24

1. HCl has several important roles in digestion. HCl stimulates the secretion of hormones that promote the flow of bile and pancreatic juice. The presence of HCl destroys certain microbes that may have been ingested with food. HCl begins denaturing proteins in food, and provides the proper chemical environment for activating pepsinogen into pepsin, which breaks apart certain peptide bonds in proteins. It also helps in the action of gastric lipase, which splits triglycerides in fat molecules found in milk into fatty acids and monoglycerides.

2. Blockage of the pancreatic and bile ducts prevents pancreatic digestive enzymes and bile from reaching the duodenum. As a consequence, there will be problems digesting carbohydrates, proteins, nucleic acids, and lipids. Of particular concern is lipid digestion since the pancreatic juices contain the primary lipid-digesting enzyme. Fats will not be adequately digested, and Trey's feces will contain larger than normal amounts of lipids. In addition, the lack of bile salts will affect the body's ability to emulsify lipids and to form micelles required for absorption of fatty acids and monoglycerides (from lipid breakdown).

When lipids are not absorbed properly, then there will be malabsorption of the lipid-soluble vitamins (A, D, E, and K).

3. Antonio experienced gastroesophageal reflux. The stomach contents backed up (refluxed) into Antonio's esophagus due to a failure of the lower esophageal sphincter to fully close. The HCl from the stomach irritated the esophageal wall, which resulted in the burning sensation he felt; this is commonly known as "heartburn," even though it is not related to the heart. Antonio's recent meal worsened the problem. Alcohol and smoking both can cause the sphincter to relax, while certain foods such tomatoes, chocolate, and coffee can stimulate stomach acid secretion. In addition, lying down immediately after a meal can exacerbate the problem.

Chapter 25

1. Ingestion of cyanide affects cellular respiration. The cyanide binds to the cytochrome oxidase complex in the inner membrane of mitochondria. Blocking this complex interferes with the last step in electron transport in aerobic ATP production. Jane Doe's body would quickly run out of energy to perform vital functions, resulting in her death.

2. Glenn's total cholesterol and LDL levels are very high, while his HDL levels are low. Total cholesterol above 239 mg/dL and LDL above 159 mg/dL are considered high. The ratio of total cholesterol (TC) to HDL-cholesterol is a predictor of the risk of developing coronary artery disease. Glenn's TC to HDL is 15; a ratio above 4 is undesirable. His ratio places him at high risk of developing coronary artery disease. In addition, for every 50 mg/dL TC over 200 mg/dL, the risk of a heart attack doubles. Glenn needs to reduce his TC and LDL-cholesterol while raising his HDL-cholesterol levels. LDLs contribute to fatty plaque formation on coronary artery walls. On the other hand, HDLs help remove excess cholesterol from the blood, which helps decrease the risk of coronary artery disease. Glenn will need to reduce his dietary intake of total fat, saturated fats, and cholesterol, all of which contribute to raising LDL levels. Exercise will raise HDL levels. If those changes are not successful, drug therapy may be required.

3. The goal of weight loss programs is to reduce caloric intake so that the body utilizes stored lipids as an energy source. As part of that desired lipid metabolism, ketone bodies are produced. Some of these ketone bodies will be excreted in the urine. If no ketones are present, then Marissa's body is not breaking down lipids. Only through using fewer calories than needed will her body break down the stored fat and release ketones. Thus, she must be eating more calories than needed to support her daily activities—she is "cheating."

Chapter 26

1. Without reabsorption, initially 105–125 mL of filtrate would be lost per minute, assuming normal glomerular filtration rate. Fluid loss from the blood would cause a decrease in blood pressure, and therefore a decrease in GBHP. When GBHP dropped below 45 mmHg, filtration would stop (assuming normal CHP and BCOP) because NFP would be zero.

2. a. Although normally pale yellow, urine color can vary based upon concentration, diet, drugs, and disease. A dark yellow color would not necessarily indicate a problem, but further investigation may be needed. Turbidity or cloudiness can be caused by urine that has been standing for a period of time, from certain foods, or from bacterial infections. Further investigation is needed. b. Ammonia-like odor occurs when the urine sample is allowed to stand. c. Albumin should not be present in urine (or be present only in very small amounts) because it is too large to pass through the filtration membranes. The presence of high levels of albumin is cause for concern as it indicates damage to the filtration membranes. d. Casts are hardened masses of material that are flushed out in the urine. The presence of casts is not normal and indicates a pathology. e. The pH of normal urine ranges from 4.8 to 8.0. A pH of 5.5 is in normal range. f. Hematuria is the presence of red blood cells in the urine. It can occur with certain pathological conditions or from kidney trauma. Hematuria may occur if the urine sample was contaminated with menstrual blood.

3. Bruce has developed renal calculi (kidney stones), which are blocking his ureters and interfering with the flow of urine from the kidneys to the bladder. The rhythmic pains are a result of the peristaltic contractions of the ureters as they attempt to move the stones toward the bladder. Bruce can wait for the stones to pass, can have them surgically removed, or can use shock-wave lithotripsy to break apart the stones into smaller fragments that can be elimi-

nated with urine. To prevent future episodes, Bruce needs to watch his diet (limit calcium) and drink fluids, and may need drug intervention.

Chapter 27

1. The loss of stomach acids can result in metabolic alkalosis. Robin's HCO_3^- levels would be higher than normal. She would be hypoventilating in order to decrease her pH by slowing the loss of CO_2. Excessive vomiting can result in hyponatremia, hypokalemia, and hypochloremia. Both hyponatremia and hypokalemia can cause mental confusion.

2. (Step 1) pH = 7.30 indicates slight acidosis, which could be caused by elevated PCO_2 or lowered HCO_3^-. (Step 2) The HCO_3^- is lower than normal (20 mEq/liter), so (Step 3) the cause is metabolic. (Step 4) The PCO_2 is lower than normal (32 mmHg), so hyperventilation is providing some compensation. Diagnosis: Henry has partially compensated metabolic acidosis. A possible cause is kidney damage that resulted from interruption of blood flow during the heart attack.

3. Sam will experience increased fluid loss through increased evaporation from the skin and water vapor from the respiratory system through his increased respiratory rate. His insensible water loss will also increase (loss of water from mucous membranes of the mouth and respiratory system). Sam will have a decrease in urine formation.

Chapter 28

1. Monica's excessive training has resulted in an abnormally low amount of body fat. A certain amount of body fat is needed in order to produce the hormones required for the ovarian cycle. Several hormones are involved. Her amenorrhea is due to a lack of gonadotropin-releasing hormone, which in turn reduces the release of LH and FSH. Her follicles with their enclosed ova fail to develop and ovulation will not occur. In addition, synthesis of estrogens and progesterone declines from the lack of hormonal feedback. Usually a gain of weight will allow normal hormonal feedback mechanisms to return.

2. Along with estrogens, progesterone helps to prepare the endometrium for possible implantation of a zygote by promoting growth of the endometrium. The endometrial glands secrete glycogen, which will help sustain an embryo if one should implant. If implantation occurs, progesterone helps maintain the endometrium for the developing fetus. In addition, it helps prepare mammary glands to secrete milk. It inhibits the release of GnRH and LH, which stops a new ovarian cycle from occurring.

3. The ductus deferens is cut and tied in a vasectomy. This stops the release of sperm into the ejaculatory duct and urethra. Mark will still produce the secretions from his accessory glands (prostate, seminal vesicles, bulbourethral glands) in his ejaculate. In addition, a vasectomy does not affect sexual performance; he will be able to achieve erection and ejaculation, as those events are nervous system responses.

Chapter 29

1. As part of the feedback mechanism for lactation, oxytocin is released from the posterior pituitary. It is carried to the mammary glands where it causes milk to be released into the mammary ducts (milk ejection). The oxytocin is also transported in the blood to the uterus, which contains oxytocin receptors on the myometrium. The oxytocin causes contraction of the myometrium, resulting in the painful sensations that Kathy is experiencing. The uterine contractions can help return the uterus back to its prepregnancy size.

2. Sex-linked genetic traits, such as hemophilia, are present on the X chromosomes but not on the Y chromosomes. In males, the X chromosome is always inherited from the mother, and the Y chromosome from the father. Thus, Jack's hemophilia gene was inherited from his mother on his X chromosome. The gene for hemophilia is a recessive gene. His mother would need two recessive genes, one on each of her X chromosomes, to be hemophiliac. His father must carry the dominant (nonhemophiliac) gene on his X chromosome, so he also would not have hemophilia.

3. The cord blood is a source of pluripotent stem cells, which are unspecialized cells that have the potential to specialize into cells with specific functions. The hope is that stem cells can be used to generate cells and tissues to treat a variety of disorders. It is assumed that the tissues would not be rejected since they would contain the same genetic material as the patient—in this case Alisa's baby.

GLOSSARY

Pronunciation Key

1. The most strongly accented syllable appears in capital letters, for example, bilateral (bī-LAT-er-al) and diagnosis (dī-ag-NŌ-sis).

2. If there is a secondary accent, it is noted by a prime ('), for example, constitution (kon'-sti-TOO-shun) and physiology (fiz'-ē-OL-ō-jē). Any additional secondary accents are also noted by a prime, for example, decarboxylation (dē'-kar-bok'-si-LĀ-shun).

3. Vowels marked by a line above the letter are pronounced with the long sound, as in the following common words:

ā as in māke   ō as in pōle
ē as in bē   ū as in cūte
ī as in īvy

4. Vowels not marked by a line above the letter are pronounced with the short sound, as in the following words:

a as in above or at   o as in not
e as in bet   u as in bud
i as in sip

5. Other vowel sounds are indicated as follows:

oy as in oil
oo as in root

6. Consonant sounds are pronounced as in the following words:

b as in bat   m as in mother
ch as in chair   n as in no
d as in dog   p as in pick
f as in father   r as in rib
g as in get   s as in so
h as in hat   t as in tea
j as in jump   v as in very
k as in can   w as in welcome
ks as in tax   z as in zero
kw as in quit   zh as in lesion
l as in let

A

Abdomen (ab-DŌ-men or AB-dō-men) The area between the diaphragm and pelvis.

Abdominal (ab-DOM-i-nal) **cavity** Superior portion of the abdominopelvic cavity that contains the stomach, spleen, liver, gallbladder, pancreas, kidneys, small intestine, and part of the large intestine.

Abdominal thrust maneuver A first-aid procedure for choking. Employs a quick, upward thrust against the diaphragm that forces air out of the lungs with sufficient force to eject any lodged material.

Abdominopelvic (ab-dom'-i-nō-PEL-vik) **cavity** Inferior to the diaphragm and subdivided into a superior abdominal cavity and an inferior pelvic cavity.

Abduction (ab-DUK-shun) Movement away from the midline of the body.

Abortion (a-BOR-shun) The premature loss (**spontaneous**) or removal (**induced**) of the embryo or nonviable fetus; miscarriage due to a failure in the normal process of developing or maturing.

Abscess (AB-ses) A localized collection of pus and liquefied tissue in a cavity.

Absorption (ab-SORP-shun) Intake of fluids or other substances by cells of the skin or mucous membranes; the passage of digested foods from the gastrointestinal tract into blood or lymph.

Accessory duct A duct of the pancreas that empties into the duodenum about 2.5 cm (1 in.) superior to the ampulla of Vater (hepatopancreatic ampulla). Also called the **duct of Santorini** (san'-tō-RĒ-nē).

Acetabulum (as'-e-TAB-ū-lum) The rounded cavity on the external surface of the hip bone that receives the head of the femur.

Acetylcholine (as'-ē-til-KŌ-lēn) **(ACh)** A neurotransmitter liberated by many peripheral nervous system neurons and some central nervous system neurons. It is excitatory at neuromuscular junctions but inhibitory at some other synapses (for example, it slows heart rate).

Achalasia (ak'-a-LĀ-zē-a) A condition, caused by malfunction of the myenteric plexus, in which the lower esophageal sphincter fails to relax normally as food approaches. A whole meal may become lodged in the esophagus and enter the stomach very slowly. Distension of the esophagus results in chest pain that is often confused with pain originating from the heart.

Achilles tendon See **Calcaneal tendon.**

Acini (AS-i-nē) Groups of cells in the pancreas that secrete digestive enzymes. *Singular* is **acinus** (AS-i-nus).

Acoustic (a-KOOS-tik) Pertaining to sound or the sense of hearing.

Acquired immunodeficiency syndrome (AIDS) A fatal disease caused by the human immunodeficiency virus (HIV). Characterized by a positive HIV-antibody test, low helper T cell count, and certain indicator diseases (for example, Kaposi's sarcoma, pneumocystis carinii pneumonia, tuberculosis, fungal diseases). Other symptoms include fever or night sweats, coughing, sore throat, fatigue, body aches, weight loss, and enlarged lymph nodes.

Acrosome (AK-rō-sōm) A lysosomelike organelle in the head of a sperm cell containing enzymes that facilitate the penetration of a sperm cell into a secondary oocyte.

Actin (AK-tin) A contractile protein that is part of thin filaments in muscle fibers.

Action potential (AP) An electrical signal that propagates along the membrane of a neuron or muscle fiber (cell); a rapid change in membrane potential that involves a depolarization followed by a repolarization. Also called a **nerve action potential** or **nerve impulse** as it relates to a neuron, and a **muscle action potential** as it relates to a muscle fiber.

Activation (ak'-ti-VĀ-shun) **energy** The minimum amount of energy required for a chemical reaction to occur.

Active transport The movement of substances across cell membranes against a concentration gradient, requiring the expenditure of cellular energy (ATP).

Acute (a-KŪT) Having rapid onset, severe symptoms, and a short course; not chronic.

Adaptation (ad'-ap-TĀ-shun) The adjustment of the pupil of the eye to changes in light intensity. The property by which a sensory neuron relays a decreased frequency of action potentials from a receptor, even though the strength of the stimulus remains constant; the decrease in perception of a sensation over time while the stimulus is still present.

Adduction (ad-DUK-shun) Movement toward the midline of the body.

Adenoids (AD-e-noyds) The pharyngeal tonsils.

Adenosine triphosphate (a-DEN-ō-sēn trī-FOS-fāt) **(ATP)** The main energy currency in living cells; used to transfer the chemical energy needed for metabolic reactions. ATP consists of the purine base *adenine* and the five-carbon sugar *ribose,* to which are added, in linear array, three *phosphate* groups.

Adhesion (ad-HĒ-zhun) Abnormal joining of parts to each other.

Adipocyte (AD-i-pō-sīt) Fat cell, derived from a fibroblast. Also called **fat cell** or **adipose cell.**

Adipose (AD-i-pōz) **tissue** Tissue composed of adipocytes specialized for triglyceride storage and present in the form of soft pads between various organs for support, protection, and insulation.

Adrenal cortex (a-DRĒ-nal KOR-teks) The outer portion of an adrenal gland, divided into three zones; the zona glomerulosa secretes mineralocorticoids, the zona fasciculata secretes glucocorticoids, and the zona reticularis secretes androgens.

Adrenal glands Two glands located superior to each kidney. Also called the **suprarenal** (soo′-pra-RĒ-nal) **glands.**

Adrenal medulla (me-DUL-a) The inner part of an adrenal gland, consisting of cells that secrete epinephrine, norepinephrine, and a small amount of dopamine in response to stimulation by sympathetic preganglionic neurons.

Adrenergic (ad′-ren-ER-jik) **neuron** A neuron that releases epinephrine (adrenaline) or norepinephrine (noradrenaline) as its neurotransmitter.

Adrenocorticotropic (ad-rē′-nō-kor-ti-kō-TRŌP-ik) **hormone (ACTH)** A hormone produced by the anterior pituitary that influences the production and secretion of certain hormones of the adrenal cortex. Also called **corticotropin** (kor′-ti-kō-TRŌ-pin).

Adventitia (ad-ven-TISH-a) The outermost covering of a structure or organ.

Aerobic (air-Ō-bik) Requiring molecular oxygen.

Aerobic (ār-Ō-bik) **cellular respiration** The production of ATP (36 molecules) from the complete oxidation of pyruvic acid in mitochondria. Carbon dioxide, water, and heat are also produced.

Afferent arteriole (AF-er-ent ar-TĒ-rē-ōl) A blood vessel of a kidney that divides into the capillary network called a glomerulus; there is one afferent arteriole for each glomerulus.

Agglutination (a-gloo-ti-NĀ-shun) Clumping of microorganisms or blood cells, typically due to an antigen–antibody reaction.

Aggregated lymphatic follicles Clusters of lymph nodules that are most numerous in the ileum. Also called **Peyer's** (PĪ-erz) **patches.**

Albinism (AL-bin-izm) Abnormal, nonpathological, partial, or total absence of pigment in skin, hair, and eyes.

Aldosterone (al-DOS-ter-ōn) A mineralocorticoid produced by the adrenal cortex that promotes sodium and water reabsorption by the kidneys and potassium excretion in urine.

All-or-none principle If a stimulus depolarizes a neuron to threshold, the neuron fires at its maximum voltage (all); if threshold is not reached, the neuron does not fire at all (none). Given above threshold, stronger stimuli do not produce stronger action potentials.

Allantois (a-LAN-tō-is) A small, vascularized outpouching of the yolk sac that serves as an early site for blood formation and development of the urinary bladder.

Alleles (a-LĒLZ) Alternate forms of a single gene that control the same inherited trait (such as type

A blood) and are located at the same position on homologous chromosomes.

Allergen (AL-er-jen) An antigen that evokes a hypersensitivity reaction.

Alopecia (al′-ō-PĒ-shē-a) The partial or complete lack of hair as a result of factors such as genetics, aging, endocrine disorders, chemotherapy, and skin diseases.

Alpha (AL-fa) **cell** A type of cell in the pancreatic islets (islets of Langerhans) that secretes the hormone glucagon. Also termed an **A cell.**

Alpha (α) receptor A type of receptor for norepinephrine and epinephrine; present on visceral effectors innervated by sympathetic postganglionic neurons.

Alveolar duct Branch of a respiratory bronchiole around which alveoli and alveolar sacs are arranged.

Alveolar macrophage (MAK-rō-fāj) Highly phagocytic cell found in the alveolar walls of the lungs. Also called a **dust cell.**

Alveolar sac A cluster of alveoli that share a common opening.

Alveolus (al-VĒ-ō-lus) A small hollow or cavity; an air sac in the lungs; milk-secreting portion of a mammary gland. *Plural* is **alveoli** (al-VĒ-ol-ī).

Alzheimer (ALTZ-hī-mer) **disease (AD)** Disabling neurological disorder characterized by dysfunction and death of specific cerebral neurons, resulting in widespread intellectual impairment, personality changes, and fluctuations in alertness.

Amenorrhea (ā-men-ō-RĒ-a) Absence of menstruation.

Amnesia (am-NĒ-zē-a) A lack or loss of memory.

Amnion (AM-nē-on) A thin, protective fetal membrane that develops from the epiblast; holds the fetus suspended in amniotic fluid. Also called the **"bag of waters."**

Amniotic (am′-nē-OT-ik) **fluid** Fluid in the amniotic cavity, the space between the developing embryo (or fetus) and amnion; the fluid is initially produced as a filtrate from maternal blood and later includes fetal urine. It functions as a shock absorber, helps regulate fetal body temperature, and helps prevent desiccation.

Amphiarthrosis (am′-fē-ar-THRŌ-sis) A slightly movable joint, in which the articulating bony surfaces are separated by fibrous connective tissue or fibrocartilage to which both are attached; types are syndesmosis and symphysis.

Ampulla (am-PUL-la) A saclike dilation of a canal or duct.

Ampulla of Vater *See* **Hepatopancreatic ampulla.**

Anabolism (a-NAB-ō-lizm) Synthetic, energy-requiring reactions whereby small molecules are built up into larger ones.

Anaerobic (an-ar-Ō-bik) Not requiring oxygen.

Anal (Ā-nal) **canal** The last 2 or 3 cm (1 in.) of the rectum; opens to the exterior through the anus.

Anal column A longitudinal fold in the mucous membrane of the anal canal that contains a network of arteries and veins.

Anal triangle The subdivision of the female or male perineum that contains the anus.

Analgesia (an-al-JĒ-zē-a) Pain relief; absence of the sensation of pain.

Anaphase (AN-a-fāz) The third stage of mitosis in which the chromatids that have separated at the centromeres move to opposite poles of the cell.

Anaphylaxis (an′-a-fi-LAK-sis) A hypersensitivity (allergic) reaction in which IgE antibodies attach to mast cells and basophils, causing them to produce mediators of anaphylaxis (histamine, leukotrienes, kinins, and prostaglandins) that bring about increased blood permeability, increased smooth muscle contraction, and increased mucus production. Examples are hay fever, hives, and anaphylactic shock.

Anastomosis (a-nas′-tō-MŌ-sis) An end-to-end union or joining of blood vessels, lymphatic vessels, or nerves.

Anatomic dead space Spaces of the nose, pharynx, larynx, trachea, bronchi, and bronchioles totaling about 150 mL of the 500 mL in a quiet breath (tidal volume); air in the anatomic dead space does not reach the alveoli to participate in gas exchange.

Anatomical (an′-a-TOM-i-kal) **position** A position of the body universally used in anatomical descriptions in which the body is erect, the head is level, the eyes face forward, the upper limbs are at the sides, the palms face forward, and the feet are flat on the floor.

Anatomy (a-NAT-ō-mē) The structure or study of the structure of the body and the relation of its parts to each other.

Androgens (AN-drō-jenz) Masculinizing sex hormones produced by the testes in males and the adrenal cortex in both sexes; responsible for libido (sexual desire); the two main androgens are testosterone and dihydrotestosterone.

Anemia (a-NĒ-mē-a) Condition of the blood in which the number of functional red blood cells or their hemoglobin content is below normal.

Anesthesia (an′-es-THĒ-zē-a) A total or partial loss of feeling or sensation; may be general or local.

Aneurysm (AN-ū-rizm) A saclike enlargement of a blood vessel caused by a weakening of its wall.

Angina pectoris (an-JĪ-na *or* AN-ji-na PEK-tō-ris) A pain in the chest related to reduced coronary circulation due to coronary artery disease (CAD) or spasms of vascular smooth muscle in coronary arteries.

Angiogenesis (an′-jē-ō-JEN-e-sis) The formation of blood vessels in the extraembryonic mesoderm of the yolk sac, connecting stalk, and chorion at the beginning of the third week of development.

Ankylosis (ang′-ki-LŌ-sis) Severe or complete loss of movement at a joint as the result of a disease process.

Antagonist (an-TAG-ō-nist) A muscle that has an action opposite that of the prime mover (agonist) and yields to the movement of the prime mover.

Antagonistic (an-tag-ō-NIST-ik) **effect** A hormonal interaction in which the effect of one hormone on a target cell is opposed by another hormone. For example, calcitonin (CT) lowers blood calcium level, whereas parathyroid hormone (PTH) raises it.

Anterior (an-TĒR-ē-or) Nearer to or at the front of the body. Equivalent to **ventral** in bipeds.

Anterior pituitary Anterior lobe of the pituitary gland. Also called the **adenohypophysis** (ad′-e-nō-hī-POF-i-sis).

Anterior root The structure composed of axons of motor (efferent) neurons that emerges from the anterior aspect of the spinal cord and extends laterally to join a posterior root, forming a spinal nerve. Also called a **ventral root.**

Anterolateral (an′-ter-ō-LAT-er-al) **pathway** Sensory pathway that conveys information related to pain, temperature, tickle, and itch. Also called **spinothalamic pathway.**

Antibody (AN-ti-bod′-ē) **(Ab)** A protein produced by plasma cells in response to a specific antigen; the antibody combines with that antigen to neutralize, inhibit, or destroy it. Also called an **immunoglobulin** (im-ū-nō-GLOB-ū-lin) or **Ig.**

Anticoagulant (an-tī-cō-AG-ū-lant) A substance that can delay, suppress, or prevent the clotting of blood.

Antidiuretic (an′-ti-dī-ū-RET-ik) Substance that inhibits urine formation.

Antidiuretic hormone (ADH) Hormone produced by neurosecretory cells in the paraventricular and supraoptic nuclei of the hypothalamus that stimulates water reabsorption from kidney tubule cells into the blood and vasoconstriction of arterioles. Also called **vasopressin** (vāz-ō-PRES-in).

Antigen (AN-ti-jen) **(Ag)** A substance that has immunogenicity (the ability to provoke an immune response) and reactivity (the ability to react with the antibodies or cells that result from the immune response); contraction of *anti*body *gen*erator. Also termed a **complete antigen.**

Antigen-presenting cell (APC) Special class of migratory cell that processes and presents antigens to T cells during an immune response; APCs include macrophages, B cells, and dendritic cells, which are present in the skin, mucous membranes, and lymph nodes.

Antioxidant A substance that inactivates oxygen-derived free radicals. Examples are selenium, zinc, beta carotene, and vitamins C and E.

Antrum (AN-trum) Any nearly closed cavity or chamber, especially one within a bone, such as a sinus.

Anuria (an-Ū-rē-a) Absence of urine formation or daily urine output of less than 50 mL.

Anus (Ā-nus) The distal end and outlet of the rectum.

Aorta (ā-OR-ta) The main systemic trunk of the arterial system of the body that emerges from the left ventricle.

Aortic (ā-OR-tik) **body** Cluster of chemoreceptors on or near the arch of the aorta that respond to changes in blood levels of oxygen, carbon dioxide, and hydrogen ions (H⁺).

Aortic reflex A reflex that helps maintain normal systemic blood pressure; initiated by baroreceptors in the wall of the ascending aorta and arch of the aorta. Nerve impulses from aortic baroreceptors reach the cardiovascular center via sensory axons of the vagus (X) nerves.

Apex (Ā-peks) The pointed end of a conical structure, such as the apex of the heart.

Aphasia (a-FĀ-zē-a) Loss of ability to express oneself properly through speech or loss of verbal comprehension.

Apnea (AP-nē-a) Temporary cessation of breathing.

Apneustic (ap-NOO-stik) **area** A part of the respiratory center in the pons that sends stimulatory nerve impulses to the inspiratory area that activate and prolong inhalation and inhibit exhalation.

Apocrine (AP-ō-krin) **gland** A type of gland in which the secretory products gather at the free end of the secreting cell and are pinched off, along with some of the cytoplasm, to become the secretion, as in mammary glands.

Aponeurosis (ap′-ō-noo-RŌ-sis) A sheetlike tendon joining one muscle with another or with bone.

Apoptosis (ap′-ōp-TŌ-sis *or* ap-ō-TŌ-sis) Programmed cell death; a normal type of cell death that removes unneeded cells during embryological development, regulates the number of cells in tissues, and eliminates many potentially dangerous cells such as cancer cells. During apoptosis, the DNA fragments, the nucleus condenses, mitochondria cease to function, and the cytoplasm shrinks, but the plasma membrane remains intact. Phagocytes engulf and digest the apoptotic cells, and an inflammatory response does not occur.

Appositional (a-pō-ZISH-o-nal) **growth** Growth due to surface deposition of material, as in the growth in diameter of cartilage and bone. Also called **exogenous** (eks-OJ-e-nus) **growth.**

Aqueous humor (A-kwē-us HŪ-mor) The watery fluid, similar in composition to cerebrospinal fluid, that fills the anterior cavity of the eye.

Arachnoid (a-RAK-noyd) **mater** The middle of the three meninges (coverings) of the brain and spinal cord. Also termed the **arachnoid.**

Arachnoid villus (VIL-us) Berrylike tuft of the arachnoid mater that protrudes into the superior sagittal sinus and through which cerebrospinal fluid is reabsorbed into the bloodstream.

Arbor vitae (AR-bor VĪ-tē) The white matter tracts of the cerebellum, which have a treelike appearance when seen in midsagittal section.

Arch of the aorta The most superior portion of the aorta, lying between the ascending and descending segments of the aorta.

Areola (a-RĒ-ō-la) Any tiny space in a tissue. The pigmented ring around the nipple of the breast.

Arm The part of the upper limb from the shoulder to the elbow.

Arousal (a-ROW-zal) Awakening from sleep, a response due to stimulation of the reticular activating system (RAS).

Arrector pili (a-REK-tor PĪ-lē) Smooth muscles attached to hairs; contraction pulls the hairs into a vertical position, resulting in "goose bumps."

Arrhythmia (a-RITH-mē-a) An irregular heart rhythm. Also called a **dysrhythmia.**

Arteriole (ar-TĒ-rē-ōl) A small, almost microscopic, artery that delivers blood to a capillary.

Arteriosclerosis (ar-tē-rē-ō-skle-RŌ-sis) Group of diseases characterized by thickening of the walls of arteries and loss of elasticity.

Artery (AR-ter-ē) A blood vessel that carries blood away from the heart.

Arthritis (ar-THRĪ-tis) Inflammation of a joint.

Arthrology (ar-THROL-ō-jē) The study or description of joints.

Arthroplasty (AR-thrō-plas′-tē) Surgical replacement of joints, for example, the hip and knee joints.

Arthroscopy (ar-THROS-kō-pē) A procedure for examining the interior of a joint, usually the knee, by inserting an arthroscope into a small incision; used to determine extent of damage, remove torn cartilage, repair cruciate ligaments, and obtain samples for analysis.

Arthrosis (ar-THRŌ-sis) A joint or articulation.

Articular (ar-TIK-ū-lar) **capsule** Sleevelike structure around a synovial joint composed of a fibrous capsule and a synovial membrane.

Articular cartilage (KAR-ti-lij) Hyaline cartilage attached to articular bone surfaces.

Articular disc Fibrocartilage pad between articular surfaces of bones of some synovial joints. Also called a **meniscus** (men-IS-kus).

Articulation (ar-tik-ū-LĀ-shun) A joint; a point of contact between bones, cartilage and bones, or teeth and bones.

Arytenoid (ar′-i-TĒ-noyd) **cartilages** A pair of small, pyramidal cartilages of the larynx that attach to the vocal folds and intrinsic pharyngeal muscles and can move the vocal folds.

Ascending colon (KŌ-lon) The part of the large intestine that passes superiorly from the cecum to the inferior border of the liver, where it bends at the right colic (hepatic) flexure to become the transverse colon.

Ascites (a-SĪ-tēz) Abnormal accumulation of serous fluid in the peritoneal cavity.

Association areas Large cortical regions on the lateral surfaces of the occipital, parietal, and temporal lobes and on the frontal lobes anterior to the motor areas connected by many motor and sensory axons to other parts of the cortex. The association areas are concerned with motor patterns, memory, concepts of word-hearing and word-seeing, reasoning, will, judgment, and personality traits.

Asthma (AZ-ma) Usually allergic reaction characterized by smooth muscle spasms in bronchi resulting in wheezing and difficult breathing. Also called **bronchial asthma.**

Astigmatism (a-STIG-ma-tizm) An irregularity of the lens or cornea of the eye causing the image to be out of focus and producing faulty vision.

Astrocyte (AS-trō-sīt) A neuroglial cell having a star shape that participates in brain development and the metabolism of neurotransmitters, helps form the blood–brain barrier, helps maintain the proper balance of K⁺ for generation of nerve impulses, and provides a link between neurons and blood vessels.

Ataxia (a-TAK-sē-a) A lack of muscular coordination, lack of precision.

Atherosclerosis (ath-er-ō-skle-RŌ-sis) A progressive disease characterized by the formation in the walls of large and medium-sized arteries of lesions called atherosclerotic plaques.

Atherosclerotic (ath′-er-ō-skle-RO-tik) **plaque** (PLAK) A lesion that results from accumulated cholesterol and smooth muscle fibers (cells) of the tunica media of an artery; may become obstructive.

Atom Unit of matter that makes up a chemical element; consists of a nucleus (containing positively charged protons and uncharged neutrons) and negatively charged electrons that orbit the nucleus.

Atresia (a-TRĒ-zē-a) Degeneration and reabsorption of an ovarian follicle before it fully matures and ruptures; abnormal closure of a passage, or absence of a normal body opening.

Atrial fibrillation (Ā-trē-al fib-ri-LĀ-shun) **(AF)** Asynchronous contraction of cardiac muscle fibers in the atria that results in the cessation of atrial pumping.

Atrial natriuretic (nā′-trē-ū-RET-ik) **peptide (ANP)** Peptide hormone, produced by the atria of the heart in response to stretching, that inhibits aldosterone production and thus lowers blood pressure; causes natriuresis, increased urinary excretion of sodium.

Atrioventricular (AV) (ā′-trē-ō-ven-TRIK-ū-lar) **bundle** The part of the conduction system of the heart that begins at the atrioventricular (AV) node, passes through the cardiac skeleton separating the atria and the ventricles, then extends a short distance down the interventricular septum before splitting into right and left bundle branches. Also called the **bundle of His** (HISS).

Atrioventricular (AV) node The part of the conduction system of the heart made up of a compact mass of conducting cells located in the septum between the two atria.

Atrioventricular (AV) valve A heart valve made up of membranous flaps or cusps that allows blood to flow in one direction only, from an atrium into a ventricle.

Atrium (Ā-trē-um) A superior chamber of the heart.

Atrophy (AT-rō-fē) Wasting away or decrease in size of a part, due to a failure, abnormality of nutrition, or lack of use.

Auditory ossicle (AW-di-tō-rē OS-si-kul) One of the three small bones of the middle ear called the malleus, incus, and stapes.

Auditory tube The tube that connects the middle ear with the nose and nasopharynx region of the throat. Also called the **eustachian** (ū-STĀ-kē-an *or* ū-STĀ-shun) **tube** or **pharyngotympanic tube.**

Auscultation (aws-kul-TĀ-shun) Examination by listening to sounds in the body.

Autoimmunity An immunological response against a person's own tissues.

Autolysis (aw-TOL-i-sis) Self-destruction of cells by their own lysosomal digestive enzymes after death or in a pathological process.

Autonomic ganglion (aw′-tō-NOM-ik GANG-lē-on) A cluster of cell bodies of sympathetic or parasympathetic neurons located outside the central nervous system.

Autonomic nervous system (ANS) Visceral sensory (afferent) and visceral motor (efferent) neurons. Autonomic motor neurons, both sympathetic and parasympathetic, conduct nerve impulses from the central nervous system to smooth muscle, cardiac muscle, and glands. So named because this part of the nervous system was thought to be self-governing or spontaneous.

Autonomic plexus (PLEK-sus) A network of sympathetic and parasympathetic axons; examples are the cardiac, celiac, and pelvic plexuses, which are located in the thorax, abdomen, and pelvis, respectively.

Autophagy (aw-TOF-a-jē) Process by which worn-out organelles are digested within lysosomes.

Autopsy (AW-top-sē) The examination of the body after death.

Autorhythmic (aw′-tō-RITH-mik) **cells** Cardiac or smooth muscle fibers that are self-excitable (generate impulses without an external stimulus); act as the heart's pacemaker and conduct the pacing impulse through the conduction system of the heart; self-excitable neurons in the central nervous system, as in the inspiratory area of the brain stem.

Autosome (AW-tō-sōm) Any chromosome other than the X and Y chromosomes (sex chromosomes).

Axilla (ak-SIL-a) The small hollow beneath the arm where it joins the body at the shoulders. Also called the **armpit.**

Axon (AK-son) The usually single, long process of a nerve cell that propagates a nerve impulse toward the axon terminals.

Axon terminal Terminal branch of an axon where synaptic vesicles undergo exocytosis to release neurotransmitter molecules. Also called **telodendria** (tel′-o-DEN-drea).

B

B cell A lymphocyte that can develop into a clone of antibody-producing plasma cells or memory cells when properly stimulated by a specific antigen.

Babinski (ba-BIN-skē) **sign** Extension of the great toe, with or without fanning of the other toes, in response to stimulation of the outer margin of the sole; normal up to 18 months of age and indicative of damage to descending motor pathways such as the corticospinal tracts after that.

Back The posterior part of the body; the dorsum.

Ball-and-socket joint A synovial joint in which the rounded surface of one bone moves within a cup-shaped depression or socket of another bone, as in the shoulder or hip joint. Also called a **spheroid** (SFĒ-royd) **joint.**

Baroreceptor (bar′-ō-re-SEP-tor) Neuron capable of responding to changes in blood or **stretch receptor** air or fluid pressure. Also called a **pressoreceptor.**

Basal nuclei Paired clusters of gray matter deep in each cerebral hemisphere including the globus pallidus, putamen, and caudate nucleus. Together, the caudate nucleus and putamen are known as the corpus striatum. Nearby structures that are functionally linked to the basal nuclei are the substantia nigra of the midbrain and the subthalamic nuclei of the diencephalon.

Basement membrane Thin, extracellular layer between epithelium and connective tissue consisting of a basal lamina and a reticular lamina.

Basilar (BĀS-i-lar) **membrane** A membrane in the cochlea of the internal ear that separates the cochlear duct from the scala tympani and on which the spiral organ (organ of Corti) rests.

Basophil (BĀ-sō-fil) A type of white blood cell characterized by a pale nucleus and large granules that stain blue-purple with basic dyes.

Belly The abdomen. The gaster or prominent, fleshy part of a skeletal muscle.

Beta (BĀ-ta) **cell** A type of cell in the pancreatic islets (islets of Langerhans) in the pancreas that secretes the hormone insulin. Also called a **B cell.**

Beta (β) receptor A type of adrenergic receptor for epinephrine and norepinephrine; found on visceral effectors innervated by sympathetic postganglionic neurons.

Bicuspid (bī-KUS-pid) **valve** Atrioventricular (AV) valve on the left side of the heart. Also called the **mitral valve.**

Bilateral (bī-LAT-er-al) Pertaining to two sides of the body.

Bile (BĪL) A secretion of the liver consisting of water, bile salts, bile pigments, cholesterol, lecithin, and several ions; it emulsifies lipids prior to their digestion.

Bilirubin (bil-ē-ROO-bin) An orange pigment that is one of the end products of hemoglobin breakdown in the hepatocytes and is excreted as a waste material in bile.

Biopsy (BĪ-op-sē) The removal of a sample of living tissue to help diagnose a disorder, for example, cancer.

Blastocyst (BLAS-tō-sist) In the development of an embryo, a hollow ball of cells that consists of a blastocele (the internal cavity), trophoblast (outer cells), and embryoblast (inner cell mass).

Blastocyst (BLAS-tō-sist) **cavity** The fluid-filled cavity within the blastocyst. Also called the **blastocele.**

Blastomere (BLAS-tō-mēr) One of the cells resulting from the cleavage of a fertilized ovum.

Blastula (BLAS-tyū-la) An early stage in the development of a zygote.

Blind spot Area in the retina at the end of the optic (II) nerve in which there are no photoreceptors. Also called **optic disc.**

Blood The fluid that circulates through the heart, arteries, capillaries, and veins and that constitutes the chief means of transport within the body.

Blood–brain barrier (BBB) A barrier consisting of specialized brain capillaries and astrocytes that prevents the passage of materials from the blood to the cerebrospinal fluid and brain.

Blood clot A gel that consists of the formed elements of blood trapped in a network of insoluble protein fibers.

Blood island Isolated mass of mesoderm derived from angioblasts and from which blood vessels develop.

Blood pressure (BP) Force exerted by blood against the walls of blood vessels due to contraction of the heart and influenced by the elasticity of the vessel walls; clinically, a measure of the pressure in arteries during ventricular systole and ventricular diastole.

Blood reservoir (REZ-er-vwar) Systemic veins and venules that contain large amounts of blood that can be moved quickly to parts of the body requiring the blood.

Blood–testis barrier (BTB) A barrier formed by Sertoli cells that prevents an immune response against antigens produced by spermatogenic cells by isolating the cells from the blood.

Body cavity A space within the body that contains various internal organs.

Bolus (BŌ-lus) A soft, rounded mass, usually food, that is swallowed.

Bone remodeling Replacement of old bone by new bone tissue.

Bony labyrinth (LAB-i-rinth) A series of cavities within the petrous portion of the temporal bone forming the vestibule, cochlea, and semicircular canals of the inner ear.

Bowman's capsule See **Glomerular capsule.**

Brachial plexus (BRĀ-kē-al PLEK-sus) A network of nerve axons of the ventral rami of spinal nerves C5, C6, C7, C8, and T1. The nerves that emerge from the brachial plexus supply the upper limb.

Bradycardia (brād′-i-KAR-dē-a) A slow resting heart or pulse rate (under 50 beats per minute).

Brain The part of the central nervous system contained within the cranial cavity.

Brain stem The portion of the brain immediately superior to the spinal cord, made up of the medulla oblongata, pons, and midbrain.

Brain waves Electrical signals that can be recorded from the skin of the head due to electrical activity of brain neurons.

Broad ligament A double fold of parietal peritoneum attaching the uterus to the side of the pelvic cavity.

Broca's (BRŌ-kaz) **speech area** Motor area of the brain in the frontal lobe that translates thoughts into speech. Also called the **motor speech area.**

Bronchi (BRON-kī) Branches of the respiratory passageway including primary bronchi (the two divisions of the trachea), secondary or lobar bronchi (divisions of the primary bronchi that are distributed to the lobes of the lung), and tertiary or segmental bronchi (divisions of the secondary bronchi that are distributed to bronchopulmonary segments of the lung). *Singular* is **bronchus.**

Bronchial (BRON-kē-al) **tree** The trachea, bronchi, and their branching structures up to and including the terminal bronchioles.

Bronchiole (BRONG-kē-ōl) Branch of a tertiary bronchus further dividing into terminal bronchioles (distributed to lobules of the lung), which divide into respiratory bronchioles (distributed to alveolar sacs).

Bronchitis (brong-KĪ-tis) Inflammation of the mucous membrane of the bronchial tree; characterized by hypertrophy and hyperplasia of seromucous glands and goblet cells that line the bronchi which results in a productive cough.

Bronchopulmonary (brong′-kō-PUL-mō-ner-ē) **segment** One of the smaller divisions of a lobe of a lung supplied by its own branches of a bronchus.

Brunner's gland See **Duodenal gland.**

Buccal (BUK-al) Pertaining to the cheek or mouth.

Buffer system A weak acid and the salt of that acid (that functions as a weak base). Buffers prevent drastic changes in pH by converting strong acids and bases to weak acids and bases.

Bulb of penis Expanded portion of the base of the corpus spongiosum penis.

Bulbourethral (bul′-bō-ū-RĒ-thral) **gland** One of a pair of glands located inferior to the prostate on either side of the urethra that secretes an alkaline fluid into the cavernous urethra. Also called a **Cowper's** (KOW-perz) **gland.**

Bulimia (boo-LIM-ē-a *or* boo-LĒ-mē-a) A disorder characterized by overeating at least twice a week followed by purging by self-induced vomiting, strict dieting or fasting, vigorous exercise, or use of laxatives or diuretics. Also called **binge–purge syndrome.**

Bulk-phase endocytosis A process by which most body cells can ingest membrane-surrounded droplets of interstitial fluid.

Bundle branch One of the two branches of the atrioventricular (AV) bundle made up of specialized muscle fibers (cells) that transmit electrical impulses to the ventricles.

Bundle of His See **Atrioventricular (AV) bundle.**

Burn Tissue damage caused by excessive heat, electricity, radioactivity, or corrosive chemicals that denature (break down) proteins in the skin.

Bursa (BUR-sa) A sac or pouch of synovial fluid located at friction points, especially about joints.

Bursitis (bur-SĪ-tis) Inflammation of a bursa.

Buttocks (BUT-oks) The two fleshy masses on the posterior aspect of the inferior trunk, formed by the gluteal muscles.

C

Calcaneal (kal-KĀ-nē-al) **tendon** The tendon of the soleus, gastrocnemius, and plantaris muscles at the back of the heel. Also called the **Achilles** (a-KIL-ēz) **tendon.**

Calcification (kal′-si-fi-KĀ-shun) Deposition of mineral salts, primarily hydroxyapatite, in a framework formed by collagen fibers in which the tissue hardens. Also called **mineralization** (min′-e-ral-i-ZĀ-shun).

Calcitonin (kal-si-TŌ-nin) **(CT)** A hormone produced by the parafollicular cells of the thyroid gland that can lower the amount of blood calcium and phosphates by inhibiting bone resorption (breakdown of bone extracellular matrix) and by accelerating uptake of calcium and phosphates into bone matrix.

Calculus (KAL-kū-lus) A stone, or insoluble mass of crystallized salts or other material, formed within the body, as in the gallbladder, kidney, or urinary bladder.

Callus (KAL-lus) A growth of new bone tissue in and around a fractured area, ultimately replaced by mature bone. An acquired, localized thickening.

Calyx (KĀL-iks) Any cuplike division of the kidney pelvis. *Plural* is **calyces** (KĀ-li-sēz).

Canal (ka-NAL) A narrow tube, channel, or passageway.

Canaliculus (kan′-a-LIK-ū-lus) A small channel or canal, as in bones, where they connect lacunae. *Plural* is **canaliculi** (kan′-a-LIK-ū-lī).

Canal of Schlemm See **Scleral venous sinus.**

Cancer A group of diseases characterized by uncontrolled or abnormal cell division.

Capacitation (ka-pas′-i-TĀ-shun) The functional changes that sperm undergo in the female reproductive tract that allow them to fertilize a secondary oocyte.

Capillary (KAP-i-lar′-ē) A microscopic blood vessel located between an arteriole and venule through which materials are exchanged between blood and interstitial fluid.

Carbohydrate Organic compound consisting of carbon, hydrogen, and oxygen; the ratio of hydrogen to oxygen atoms is usually 2:1. Examples include sugars, glycogen, starches, and glucose.

Carcinogen (kar-SIN-ō-jen) A chemical substance or radiation that causes cancer.

Cardiac (KAR-dē-ak) **arrest** Cessation of an effective heartbeat in which the heart is completely stopped or in ventricular fibrillation.

Cardiac cycle A complete heartbeat consisting of systole (contraction) and diastole (relaxation) of both atria plus systole and diastole of both ventricles.

Cardiac muscle Striated muscle fibers (cells) that form the wall of the heart; stimulated by an intrinsic conduction system and autonomic motor neurons.

Cardiac notch An angular notch in the anterior border of the left lung into which part of the heart fits.

Cardiac output (CO) The volume of blood ejected from the left ventricle (or the right ventricle) into the aorta (or pulmonary trunk) each minute.

Cardinal ligament A ligament of the uterus, extending laterally from the cervix and vagina as a continuation of the broad ligament.

Cardiogenic area (kar-dē-ō-JEN-ik) A group of mesodermal cells in the head end of an embryo that gives rise to the heart.

Cardiology (kar-dē-OL-ō-jē) The study of the heart and diseases associated with it.

Cardiovascular (kar-dē-ō-VAS-kū-lar) **(CV) center** Groups of neurons scattered within the medulla oblongata that regulate heart rate, force of contraction, and blood vessel diameter.

Cardiovascular physiology Study of the functions of the heart and blood vessels.

Cardiovascular system System that consists of blood, the heart, and blood vessels.

Carotene (KAR-ō-tēn) Antioxidant precursor of vitamin A, which is needed for synthesis of photopigments; yellow-orange pigment present in the stratum corneum of the epidermis. Accounts for the yellowish coloration of skin. Also termed **beta carotene.**

Carotid (ka-ROT-id) **body** Cluster of chemoreceptors on or near the carotid sinus that respond to changes in blood levels of oxygen, carbon dioxide, and hydrogen ions.

Carotid sinus A dilated region of the internal carotid artery just superior to where it branches from the common carotid artery; it contains baroreceptors that monitor blood pressure.

Carpal bones The eight bones of the wrist. Also called **carpals.**

Carpus (KAR-pus) A collective term for the eight bones of the wrist.

Cartilage (KAR-ti-lij) A type of connective tissue consisting of chondrocytes in lacunae embedded in a dense network of collagen and elastic fibers and an extracellular matrix of chondroitin sulfate.

Cartilaginous (kar′-ti-LAJ-i-nus) **joint** A joint without a synovial (joint) cavity where the articulating bones are held tightly together by cartilage, allowing little or no movement.

Catabolism (ka-TAB-ō-lizm) Chemical reactions that break down complex organic compounds into simple ones, with the net release of energy.

Cataract (KAT-a-rakt) Loss of transparency of the lens of the eye or its capsule or both.

Cauda equina (KAW-da ē-KWĪ-na) A tail-like array of roots of spinal nerves at the inferior end of the spinal cord.

Caudal (KAW-dal) Pertaining to any tail-like structure; inferior in position.

Cecum (SĒ-kum) A blind pouch at the proximal end of the large intestine that attaches to the ileum.

Celiac plexus (SĒ-lē-ak PLEK-sus) A large mass of autonomic ganglia and axons located at the level of the superior part of the first lumbar vertebra. Also called the **solar plexus.**

Cell The basic structural and functional unit of all organisms; the smallest structure capable of performing all the activities vital to life.

Cell biology The study of cellular structure and function. Also called **cytology.**

Cell cycle Growth and division of a single cell into two identical cells; consists of interphase and cell division.

Cell division Process by which a cell reproduces itself that consists of a nuclear division (mitosis) and a cytoplasmic division (cytokinesis); types include somatic and reproductive cell division.

Cell junction Point of contact between plasma membranes of tissue cells.

Cellular respiration The oxidation of glucose to produce ATP that involves glycolysis, acetyl coenzyme A formation, the Krebs cycle, and the electron transport chain.

Cementum (se-MEN-tum) Calcified tissue covering the root of a tooth.

Central canal A microscopic tube running the length of the spinal cord in the gray commissure. A circular channel running longitudinally in the center of an osteon (haversian system) of mature compact bone, containing blood and lymphatic vessels and nerves. Also called a **haversian** (ha-VER-shun) **canal.**

Central fovea (FŌ-vē-a) A depression in the center of the macula lutea of the retina, containing cones only and lacking blood vessels; the area of highest visual acuity (sharpness of vision).

Central nervous system (CNS) That portion of the nervous system that consists of the brain and spinal cord.

Centrioles (SEN-trē-ōlz) Paired, cylindrical structures of a centrosome, each consisting of a ring of microtubules and arranged at right angles to each other.

Centromere (SEN-trō-mēr) The constricted portion of a chromosome where the two chromatids are joined; serves as the point of attachment for the microtubules that pull chromatids during anaphase of cell division.

Centrosome (SEN-trō-sōm) A dense network of small protein fibers near the nucleus of a cell, containing a pair of centrioles and pericentriolar material.

Cephalic (se-FAL-ik) Pertaining to the head; superior in position.

Cerebellar peduncle (ser-e-BEL-ar pe-DUNG-kul) A bundle of nerve axons connecting the cerebellum with the brain stem.

Cerebellum (ser′-e-BEL-um) The part of the brain lying posterior to the medulla oblongata and pons; governs balance and coordinates skilled movements.

Cerebral aqueduct (SER-ē-bral AK-we-dukt) A channel through the midbrain connecting the third and fourth ventricles and containing cerebrospinal fluid. Also termed the **aqueduct of the midbrain.**

Cerebral arterial circle A ring of arteries forming an anastomosis at the base of the brain between the internal carotid and basilar arteries and arteries supplying the cerebral cortex. Also called the **circle of Willis.**

Cerebral cortex The surface of the cerebral hemispheres, 2–4 mm thick, consisting of gray matter; arranged in six layers of neuronal cell bodies in most areas.

Cerebral peduncle (pe-DUNG-kul or PĒ-dung-kul) One of a pair of nerve axon bundles located on the anterior surface of the midbrain, conducting nerve impulses between the pons and the cerebral hemispheres.

Cerebrospinal (se-rē′-brō-SPĪ-nal) **fluid (CSF)** A fluid produced by ependymal cells that cover choroid plexuses in the ventricles of the brain; the fluid circulates in the ventricles, the central canal, and the subarachnoid space around the brain and spinal cord.

Cerebrovascular (se- rē′-brō-VAS-kū-lar) **accident (CVA)** Destruction of brain tissue (infarction) resulting from obstruction or rupture of blood vessels that supply the brain. Also called a **stroke** or **brain attack.**

Cerebrum (se-RĒ-brum or SER-e-brum) The two hemispheres of the forebrain (derived from the telencephalon), making up the largest part of the brain.

Cerumen (se-ROO-men) Waxlike secretion produced by ceruminous glands in the external auditory meatus (ear canal). Also termed **earwax.**

Ceruminous (se-RŪ-mi-nus) **gland** A modified sudoriferous (sweat) gland in the external auditory meatus that secretes cerumen (ear wax).

Cervical ganglion (SER-vi-kul GANG-glē-on) A cluster of cell bodies of postganglionic sympathetic neurons located in the neck, near the vertebral column.

Cervical plexus (PLEK-sus) A network formed by nerve axons from the ventral rami of the first four cervical nerves and receiving gray rami communicantes from the superior cervical ganglion.

Cervix (SER-viks) Neck; any constricted portion of an organ, such as the inferior cylindrical part of the uterus.

Chemical reaction The formation of new chemical bonds or the breaking of old chemical bonds between atoms.

Chemistry (KEM-is-trē) The science of the structure and interactions of matter.

Chemoreceptor (kē′-mō-rē-SEP-tor) Sensory receptor that detects the presence of a specific chemical.

Chiasm (KĪ-azm) A crossing; especially the crossing of axons in the optic (II) nerve.

Chief cell The secreting cell of a gastric gland that produces pepsinogen, the precursor of the enzyme pepsin, and the enzyme gastric lipase. Also called a **zymogenic** (zī′-mō-JEN-ik) **cell.** Cell in the parathyroid glands that secretes parathyroid hormone (PTH). Also called a **principal cell.**

Cholecystectomy (kō′-lē-sis-TEK-tō-mē) Surgical removal of the gallbladder.

Cholecystitis (kō′-lē-sis-TĪ-tis) Inflammation of the gallbladder.

Cholesterol (kō-LES-te-rol) Classified as a lipid, the most abundant steroid in animal tissues; located in cell membranes and used for the synthesis of steroid hormones and bile salts.

Cholinergic (kō′-lin-ER-jik) **neuron** A neuron that liberates acetylcholine as its neurotransmitter.

Chondrocyte (KON-drō-sīt) Cell of mature cartilage.

Chondroitin (kon-DROY-tin) **sulfate** An amorphous extracellular matrix material found outside connective tissue cells.

Chordae tendineae (KOR-dē TEN-di-nē-ē) Tendonlike, fibrous cords that connect atrioventricular valves of the heart with papillary muscles.

Chorion (KŌ-rē-on) The most superficial fetal membrane that becomes the principal embryonic portion of the placenta; serves a protective and nutritive function.

Chorionic villi (kō-rē-ON-ik VIL-li) Fingerlike projections of the chorion that grow into the decidua basalis of the endometrium and contain fetal blood vessels.

Chorionic villi sampling (CVS) The removal of a sample of chorionic villus tissue by means of a catheter to analyze the tissue for prenatal genetic defects.

Choroid (KŌ-royd) One of the vascular coats of the eyeball.

Choroid plexus (PLEK-sus) A network of capillaries located in the roof of each of the four ventricles of the brain; ependymal cells around choroid plexuses produce cerebrospinal fluid.

Chromaffin (KRŌ-maf-in) **cell** Cell that has an affinity for chrome salts, due in part to the presence of the precursors of the neurotransmitter epinephrine; found, among other places, in the adrenal medulla.

Chromatid (KRŌ-ma-tid) One of a pair of identical connected nucleoprotein strands that are joined at the centromere and separate during cell division, each becoming a chromosome of one of the two daughter cells.

Chromatin (KRŌ-ma-tin) The threadlike mass of genetic material, consisting of DNA and histone proteins, that is present in the nucleus of a nondividing or interphase cell.

Chromatolysis (krō′-ma-TOL-i-sis) The breakdown of Nissl bodies into finely granular masses in the cell body of a neuron whose axon has been damaged.

Chromosome (KRŌ-mō-sōm) One of the small, threadlike structures in the nucleus of a cell, normally 46 in a human diploid cell, that bears the genetic material; composed of DNA and proteins (histones) that form a delicate chromatin thread during interphase; becomes packaged into compact rodlike structures that are visible under the light microscope during cell division.

Chronic (KRON-ik) Long term or frequently recurring; applied to a disease that is not acute.

Chronic obstructive pulmonary disease (COPD) A disease, such as bronchitis or emphysema, in which there is some degree of obstruction of airways and consequent increase in airway resistance.

Chyle (KĪL) The milky-appearing fluid found in the lacteals of the small intestine after absorption of lipids in food.

Chyme (KĪM) The semifluid mixture of partly digested food and digestive secretions found in the stomach and small intestine during digestion of a meal.

Ciliary (SIL-ē-ar′-ē) **body** One of the three parts of the vascular tunic of the eyeball, the others being the choroid and the iris; includes the ciliary muscle and the ciliary processes.

Ciliary ganglion (GANG-glē-on) A very small parasympathetic ganglion whose preganglionic axons come from the oculomotor (III) nerve and whose postganglionic axons carry nerve impulses to the ciliary muscle and the sphincter muscle of the iris.

Cilium (SIL-ē-um) A hair or hairlike process projecting from a cell that may be used to move the entire cell or to move substances along the surface of the cell. *Plural* is **cilia.**

Circadian (ser-KĀ-dē-an) **rhythm** The pattern of biological activity on a 24-hour cycle, such as the sleep–wake cycle.

Circle of Willis *See* **Cerebral arterial circle.**

Circular folds Permanent, deep, transverse folds in the mucosa and submucosa of the small intestine that increase the surface area for absorption. Also called **plicae circulares** (PLĪ-kē SER-kū-lar-ēs).

Circulation time The time required for a drop of blood to pass from the right atrium, through pulmonary circulation, back to the left atrium, through systemic circulation down to the foot, and back again to the right atrium.

Circumduction (ser-kum-DUK-shun) A movement at a synovial joint in which the distal end of a bone moves in a circle while the proximal end remains relatively stable.

Cirrhosis (si-RŌ-sis) A liver disorder in which the parenchymal cells are destroyed and replaced by connective tissue.

Cisterna chyli (sis-TER-na KĪ-lē) The origin of the thoracic duct.

Cleavage (KLĒV-ij) The rapid mitotic divisions following the fertilization of a secondary oocyte, resulting in an increased number of progressively smaller cells, called blastomeres.

Clitoris (KLI-to-ris) An erectile organ of the female, located at the anterior junction of the labia minora, that is homologous to the male penis.

Clone (KLŌN) A population of identical cells.

Coarctation (kō′-ark-TĀ-shun) **of the aorta** A congenital heart defect in which a segment of the aorta is too narrow. As a result, the flow of oxygenated blood to the body is reduced, the left ventricle is forced to pump harder, and high blood pressure develops.

Coccyx (KOK-siks) The fused bones at the inferior end of the vertebral column.

Cochlea (KOK-lē-a) A winding, cone-shaped tube forming a portion of the inner ear and containing the spiral organ (organ of Corti).

Cochlear duct The membranous cochlea consisting of a spirally arranged tube enclosed in the bony cochlea and lying along its outer wall. Also called the **scala media** (SCA-la MĒ-dē-a).

Collagen (KOL-a-jen) A protein that is the main organic constituent of connective tissue.

Collateral circuit The alternate route taken by blood through an anastomosis.

Colliculus (ko-LIK-ū-lus) A small elevation.

Colon The portion of the large intestine consisting of ascending, transverse, descending, and sigmoid portions.

Colony-stimulating factor (CSF) One of a group of molecules that stimulates development of white blood cells. Examples are macrophage CSF and granulocyte CSF.

Colostrum (kō-LOS-trum) A thin, cloudy fluid secreted by the mammary glands a few days prior to or after delivery before true milk is produced.

Column (KOL-um) Group of white matter tracts in the spinal cord.

Common bile duct A tube formed by the union of the common hepatic duct and the cystic duct that empties bile into the duodenum at the hepatopancreatic ampulla (ampulla of Vater).

Compact (dense) bone tissue Bone tissue that contains few spaces between osteons (haversian systems); forms the external portion of all bones and the bulk of the diaphysis (shaft) of long bones; is found immediately deep to the periosteum and external to spongy bone.

Concha (KON-ka) A scroll-like bone found in the skull. *Plural* is **conchae** (KON-kē).

Concussion (kon-KUSH-un) Traumatic injury to the brain that produces no visible bruising but may result in abrupt, temporary loss of consciousness.

Conduction system A group of autorhythmic cardiac muscle fibers that generates and distributes electrical impulses to stimulate coordinated contraction of the heart chambers; includes the sinoatrial (SA) node, the atrioventricular (AV) node, the atrioven-

tricular (AV) bundle, the right and left bundle branches, and the Purkinje fibers.

Condyloid (KON-di-loyd) **joint** A synovial joint structured so that an oval-shaped condyle of one bone fits into an elliptical cavity of another bone, permitting side-to-side and back-and-forth movements, such as the joint at the wrist between the radius and carpals. Also called an **ellipsoidal** (ē-lip-SOYD-al) **joint.**

Cone (KŌN) The type of photoreceptor in the retina that is specialized for highly acute color vision in bright light.

Congenital (kon-JEN-i-tal) Present at the time of birth.

Conjunctiva (kon′-junk-TĪ-va) The delicate membrane covering the eyeball and lining the eyes.

Connective tissue One of the most abundant of the four basic tissue types in the body, performing the functions of binding and supporting; consists of relatively few cells in a generous extracellular matrix (the ground substance and fibers between the cells).

Consciousness (KON-shus-nes) A state of wakefulness in which an individual is fully alert, aware, and oriented, partly as a result of feedback between the cerebral cortex and reticular activating system.

Continuous conduction (kon-DUK-shun) Propagation of an action potential (nerve impulse) in a step-by-step depolarization of each adjacent area of an axon membrane.

Contraception (kon′-tra-SEP-shun) The prevention of fertilization or impregnation without destroying fertility.

Contractility (kon′-trak-TIL-i-tē) The ability of cells or parts of cells to actively generate force to undergo shortening for movements. Muscle fibers (cells) exhibit a high degree of contractility.

Contralateral (KON-tra-lat-er-al) On the opposite side; affecting the opposite side of the body.

Conus medullaris (KŌ-nus med-ū-LAR-is) The tapered portion of the spinal cord inferior to the lumbar enlargement.

Convergence (con-VER-jens) A synaptic arrangement in which the synaptic end bulbs of several presynaptic neurons terminate on one postsynaptic neuron. The medial movement of the two eyeballs so that both are directed toward a near object being viewed in order to produce a single image.

Cornea (KOR-nē-a) The nonvascular, transparent fibrous coat through which the iris of the eye can be seen.

Corona (kō-RŌ-na) Margin of the glans penis.

Corona radiata The innermost layer of granulosa cells that is firmly attached to the zona pellucida around a secondary oocyte.

Coronary artery disease (CAD) A condition such as atherosclerosis that causes narrowing of coronary arteries so that blood flow to the heart is reduced. The result is **coronary heart disease (CHD),** in which the heart muscle receives inadequate blood flow due to an interruption of its blood supply.

Coronary circulation The pathway followed by the blood from the ascending aorta through the blood vessels supplying the heart and returning to the right atrium. Also called **cardiac circulation.**

Coronary sinus (SĪ-nus) A wide venous channel on the posterior surface of the heart that collects the blood from the coronary circulation and returns it to the right atrium.

Corpus albicans (KOR-pus AL-bi-kanz) A white fibrous patch in the ovary that forms after the corpus luteum regresses.

Corpus callosum (kal-LŌ-sum) The great commissure of the brain between the cerebral hemispheres.

Corpuscle of touch See **Meissner corpuscle.**

Corpus luteum (LOO-tē-um) A yellowish body in the ovary formed when a follicle has discharged its secondary oocyte; secretes estrogens, progesterone, relaxin, and inhibin.

Corpus striatum (strī-Ā-tum) An area in the interior of each cerebral hemisphere composed of the caudate and putamen of the basal ganglia and white matter of the internal capsule, arranged in a striated manner.

Cortex (KOR-teks) An outer layer of an organ. The convoluted layer of gray matter covering each cerebral hemisphere.

Costal (KOS-tal) Pertaining to a rib.

Cramp A spasmodic, usually painful contraction of a muscle.

Cranial (KRĀ-nē-al) **cavity** A body cavity formed by the cranial bones and containing the brain.

Cranial nerve One of 12 pairs of nerves that leave the brain; pass through foramina in the skull; and supply sensory and motor neurons to the head, neck, part of the trunk, and viscera of the thorax and abdomen. Each is designated by a Roman numeral and a name.

Craniosacral (krā-nē-ō-SĀK-ral) **outflow** The axons of parasympathetic preganglionic neurons, which have their cell bodies located in nuclei in the brain stem and in the lateral gray matter of the sacral portion of the spinal cord.

Cranium (KRĀ-nē-um) The skeleton of the skull that protects the brain and the organs of sight, hearing, and balance; includes the frontal, parietal, temporal, occipital, sphenoid, and ethmoid bones.

Crista (KRIS-ta) A crest or ridged structure. A small elevation in the ampulla of each semicircular duct that contains receptors for dynamic equilibrium. *Plural* is **cristae.**

Crossing-over The exchange of a portion of one chromatid with another during meiosis. It permits an exchange of genes among chromatids and is one factor that results in genetic variation of progeny.

Crus (KRUS) **of penis** Separated, tapered portion of the corpora cavernosa penis. *Plural* is **crura** (KROO-ra).

Crypt of Lieberkühn See **Intestinal gland.**

Cryptorchidism (krip-TOR-ki-dizm) The condition of undescended testes.

Cuneate (KŪ-nē-āt) **nucleus** A group of neurons in the inferior part of the medulla oblongata in which axons of the cuneate fasciculus terminate.

Cupula (KU-pū-la) A mass of gelatinous material covering the hair cells of a crista; a sensory receptor in the ampulla of a semicircular canal stimulated when the head moves.

Cushing's syndrome Condition caused by a hypersecretion of glucocorticoids characterized by spindly legs, "moon face," "buffalo hump," pendulous abdomen, flushed facial skin, poor wound healing, hyperglycemia, osteoporosis, hypertension, and increased susceptibility to disease.

Cutaneous (kū-TĀ-nē-us) Pertaining to the skin.

Cyanosis (sī-a-NŌ-sis) A blue or dark purple discoloration, most easily seen in nail beds and mucous membranes, that results from an increased concentration of deoxygenated (reduced) hemoglobin (more than 5 g/dL).

Cyst (SIST) A sac with a distinct connective tissue wall, containing a fluid or other material.

Cystic (SIS-tik) **duct** The duct that carries bile from the gallbladder to the common bile duct.

Cystitis (sis-TĪ-tis) Inflammation of the urinary bladder.

Cytokinesis (sī′-tō-ki-NĒ-sis) Distribution of the cytoplasm into two separate cells during cell division; coordinated with nuclear division (mitosis).

Cytolysis (sī-TOL-i-sis) The rupture of living cells in which the contents leak out.

Cytoplasm (SĪ-tō-plasm) Cytosol plus all organelles except the nucleus.

Cytoskeleton Complex internal structure of cytoplasm consisting of microfilaments, microtubules, and intermediate filaments.

Cytosol (SĪ-tō-sol) Semifluid portion of cytoplasm in which organelles and inclusions are suspended and solutes are dissolved. Also called **intracellular fluid.**

D

Dartos (DAR-tōs) The contractile smooth muscular tissue deep to the skin of the scrotum.

Decidua (dē-SID-ū-a) That portion of the endometrium of the uterus (all but the deepest layer) that is modified during pregnancy and shed after childbirth.

Deciduous (dē-SID-ū-us) Falling off or being shed seasonally or at a particular stage of development. In the body, referring to the first set of teeth.

Decussation (dē′-ku-SĀ-shun) A crossing-over to the opposite (contralateral) side; an example is the crossing of 90% of the axons in the large motor tracts to opposite sides in the medullary pyramids.

Decussation (dē′-ku-SĀ-shun) **of pyramids** The crossing of most axons (90%) in the left pyramid of the medulla to the right side and the crossing of most axons (90%) in the right pyramid to the left side.

Deep Away from the surface of the body or an organ.

Deep abdominal inguinal (IN-gwi-nal) **ring** A slit-like opening in the aponeurosis of the transversus abdominis muscle that represents the origin of the inguinal canal.

Deep-venous thrombosis (DVT) The presence of a thrombus in a vein, usually a deep vein of the lower limbs.

Defecation (def-e-KĀ-shun) The discharge of feces from the rectum.

Deglutition (dē-gloo-TISH-un) The act of swallowing.

Dehydration (dē-hī-DRĀ-shun) Excessive loss of water from the body or its parts.

Delta cell A cell in the pancreatic islets (islets of Langerhans) that secretes somatostatin. Also termed a **D cell.**

Demineralization (dē-min′-er-al-i-ZĀ-shun) Loss of calcium and phosphorus from bones.

Dendrite (DEN-drīt) A neuronal process that carries electrical signals, usually graded potentials, toward the cell body.

Dendritic (den-DRIT-ik) **cell** One type of antigen-presenting cell with long branchlike projections that commonly is present in mucosal linings such as the vagina, in the skin (Langerhans cells in the epidermis), and in lymph nodes (follicular dendritic cells).

Dental caries (KA-rēz) Gradual demineralization of the enamel and dentin of a tooth that may invade the pulp and alveolar bone. Also called **tooth decay.**

Denticulate (den-TIK-ū-lāt) Finely toothed or serrated; characterized by a series of small, pointed projections.

Dentin (DEN-tin) The bony tissues of a tooth enclosing the pulp cavity.

Dentition (den-TI-shun) The eruption of teeth. The number, shape, and arrangement of teeth.

Deoxyribonucleic (dē-ok′-sē-rī-bō-nū-KLĒ-ik) **acid (DNA)** A nucleic acid constructed of nucleotides consisting of one of four bases (adenine, cytosine, guanine, or thymine), deoxyribose, and a phosphate group; encoded in the nucleotides is genetic information.

Depression (de-PRESH-un) Movement in which a part of the body moves inferiorly.

Dermal papilla (pa-PIL-a) Fingerlike projection of the papillary region of the dermis that may contain blood capillaries or corpuscles of touch (Meissner corpuscles).

Dermatology (der′-ma-TOL-ō-jē) The medical specialty dealing with diseases of the skin.

Dermatome (DER-ma-tōm) The cutaneous area developed from one embryonic spinal cord segment and receiving most of its sensory innervation from one spinal nerve. An instrument for incising the skin or cutting thin transplants of skin.

Dermis (DER-mis) A layer of dense irregular connective tissue lying deep to the epidermis.

Descending colon (KŌ-lon) The part of the large intestine descending from the left colic (splenic) flexure to the level of the left iliac crest.

Detrusor (de-TROO-ser) **muscle** Smooth muscle that forms the wall of the urinary bladder.

Developmental biology The study of development from the fertilized egg to the adult form.

Deviated nasal septum A nasal septum that does not run along the midline of the nasal cavity. It deviates (bends) to one side.

Diabetes mellitus (dī-a-BĒ-tēz MEL-i-tus) An endocrine disorder caused by an inability to produce or use insulin. It is characterized by the

Diagnosis (dī′-ag-NŌ-sis) Distinguishing one disease from another or determining the nature of a disease from signs and symptoms by inspection, palpation, laboratory tests, and other means.

Dialysis (dī-AL-i-sis) The removal of waste products from blood by diffusion through a selectively permeable membrane.

Diaphragm (DĪ-a-fram) Any partition that separates one area from another, especially the dome-shaped skeletal muscle between the thoracic and abdominal cavities. Also a dome-shaped device that is placed over the cervix, usually with a spermicide, to prevent conception.

Diaphysis (dī-AF-i-sis) The shaft of a long bone.

Diarrhea (dī-a-RĒ-a) Frequent defecation of liquid feces caused by increased motility of the intestines.

Diarthrosis (dī-ar-THRŌ-sis) A freely movable joint; types are gliding, hinge, pivot, condyloid, saddle, and ball-and-socket.

Diastole (dī-AS-tō-lē) In the cardiac cycle, the phase of relaxation or dilation of the heart muscle, especially of the ventricles.

Diastolic (dī-as-TOL-ik) **blood pressure (DBP)** The force exerted by blood on arterial walls during ventricular relaxation; the lowest blood pressure measured in the large arteries, normally about 80 mmHg in a young adult.

Diencephalon (DĪ-en-sef′-a-lon) A part of the brain consisting of the thalamus, hypothalamus, and epithalamus.

Differentiation (dif′-er-en-shē-Ā-shun) Development of a cell from an unspecialized to a specialized one.

Diffusion (di-FŪ-zhun) A passive process in which there is a net or greater movement of molecules or ions from a region of high concentration to a region of low concentration until equilibrium is reached.

Digestion (dī-JES-chun) The mechanical and chemical breakdown of food to simple molecules that can be absorbed and used by body cells.

Digestive system A system that consists of the gastrointestinal tract (mouth, pharynx, esophagus, stomach, small intestine, and large intestine) and accessory digestive organs (teeth, tongue, salivary glands, liver, gallbladder, and pancreas). Its function is to break down foods into small molecules that can be used by body cells.

Dilate (DĪ-lāt) To expand or swell.

Diploid (DIP-loyd) **cell** Having the number of chromosomes characteristically found in the somatic cells of an organism; having two haploid sets of chromosomes, one each from the mother and father. Symbolized 2*n*.

Direct motor pathways Collections of upper motor neurons with cell bodies in the motor cortex that project axons into the spinal cord, where they synapse with lower motor neurons or interneurons in the anterior horns. Also called the **pyramidal pathways.**

Disease Any change from a state of health.

Disease Refers to an illness characterized by a recognizable set of signs and symptoms.

Dislocation (dis′-lō-KĀ-shun) Displacement of a bone from a joint with tearing of ligaments, tendons, and articular capsules. Also called **luxation** (luks-Ā-shun).

Dissect (di-SEKT) To separate tissues and parts of a cadaver or an organ for anatomical study.

Distal (DIS-tal) Farther from the attachment of a limb to the trunk; farther from the point of origin or attachment.

Diuretic (dī-ū-RET-ik) A chemical that increases urine volume by decreasing reabsorption of water, usually by inhibiting sodium reabsorption.

Divergence (dī-VER-jens) A synaptic arrangement in which the synaptic end bulbs of one presynaptic neuron terminate on several postsynaptic neurons.

Diverticulum (dī′-ver-TIK-ū-lum) A sac or pouch in the wall of a canal or organ, especially in the colon.

Dorsal ramus (RĀ-mus) A branch of a spinal nerve containing motor and sensory axons supplying the muscles, skin, and bones of the posterior part of the head, neck, and trunk.

Dorsiflexion (dor-si-FLEK-shun) Bending the foot in the direction of the dorsum (upper surface).

Down-regulation Phenomenon in which there is a decrease in the number of receptors in response to an excess of a hormone or neurotransmitter.

Dual innervation The concept by which most organs of the body receive impulses from sympathetic and parasympathetic neurons.

Duct of Santorini *See* **Accessory duct.**

Duct of Wirsung *See* **Pancreatic duct.**

Ductus arteriosus (DUK-tus ar-tē-rē-O-sus) A small vessel connecting the pulmonary trunk with the aorta; found only in the fetus.

Ductus (vas) deferens (DEF-er-ens) The duct that carries sperm from the epididymis to the ejaculatory duct. Also called the **seminal duct.**

Ductus epididymis (ep′-i-DID-i-mis) A tightly coiled tube inside the epididymis, distinguished into a head, body, and tail, in which sperm undergo maturation.

Ductus venosus (ve-NŌ-sus) A small vessel in the fetus that helps the circulation bypass the liver.

Duodenal (doo-ō-DĒ-nal) **gland** Gland in the submucosa of the duodenum that secretes an alkaline mucus to protect the lining of the small intestine from the action of enzymes and to help neutralize the acid in chyme. Also called a **Brunner's** (BRUN-erz) **gland.**

Duodenal papilla (pa-PIL-a) An elevation on the duodenal mucosa that receives the hepatopancreatic ampulla (ampulla of Vater).

Duodenum (doo′-ō-DĒ-num *or* doo-OD-e-num) The first 25 cm (10 in.) of the small intestine, which connects the stomach and the ileum.

Dura mater (DOO-ra MĀ-ter) The outermost of the three meninges (coverings) of the brain and spinal cord.

Dynamic equilibrium (ē-kwi-LIB-rē-um) The maintenance of body position, mainly the head, in response to sudden movements such as rotation.

Dysmenorrhea (dis′-men-ō-RĒ-a) Painful menstruation.

Dysplasia (dis-PLĀ-zē-a) Change in the size, shape, and organization of cells due to chronic irritation or inflammation; may either revert to normal if stress is removed or progress to neoplasia.

Dyspnea (DISP-nē-a) Shortness of breath; painful or labored breathing.

E

Ectoderm The primary germ layer that gives rise to the nervous system and the epidermis of skin and its derivatives.

Ectopic (ek-TOP-ik) Out of the normal location, as in ectopic pregnancy.

Edema (e-DĒ-ma) An abnormal accumulation of interstitial fluid.

Effector (e-FEK-tor) An organ of the body, either a muscle or a gland, that is innervated by somatic or autonomic motor neurons.

Efferent arteriole (EF-er-ent ar-TĒ-rē-ōl) A vessel of the renal vascular system that carries blood from a glomerulus to a peritubular capillary.

Efferent (EF-er-ent) **ducts** A series of coiled tubes that transport sperm from the rete testis to the epididymis.

Ejaculation (ē-jak-ū-LĀ-shun) The reflex ejection or expulsion of semen from the penis.

Ejaculatory (ē-JAK-ū-la-tō-rē) **duct** A tube that transports sperm from the ductus (vas) deferens to the prostatic urethra.

Elasticity (e-las-TIS-i-tē) The ability of tissue to return to its original shape after contraction or extension.

Electrocardiogram (e-lek′-trō-KAR-dē-ō-gram) **(ECG** or **EKG)** A recording of the electrical changes that accompany the cardiac cycle that can be detected at the surface of the body; may be resting, stress, or ambulatory.

Elevation (el-e-VĀ-shun) Movement in which a part of the body moves superiorly.

Embolus (EM-bō-lus) A blood clot, bubble of air or fat from broken bones, mass of bacteria, or other debris or foreign material transported by the blood.

Embryo (EM-brē-ō) The young of any organism in an early stage of development; in humans, the developing organism from fertilization to the end of the eighth week of development.

Embryoblast (EM-brē-ō-blast) A region of cells of a blastocyst that differentiates into the three primary germ layers—ectoderm, mesoderm, and endoderm—from which all tissues and organs develop; also called an **inner cell mass.**

Embryology (em′-brē-OL-ō-jē) The study of development from the fertilized egg to the end of the eighth week of development.

Emesis (EM-e-sis) Vomiting.

Emigration (em′-i-GRĀ-shun) Process whereby white blood cells (WBCs) leave the bloodstream by rolling along the endothelium, sticking to it, and squeezing between the endothelial cells. Adhesion molecules help WBCs stick to the endothelium. Also known as **migration** or **extravasation.**

Emission (ē-MISH-un) Propulsion of sperm into the urethra due to peristaltic contractions of the ducts of the testes, epididymides, and ductus (vas) deferens as a result of sympathetic stimulation.

Emphysema (em-fi-SĒ-ma) A lung disorder in which alveolar walls disintegrate, producing abnormally large air spaces and loss of elasticity in the lungs; typically caused by exposure to cigarette smoke.

Emulsification (e-mul-si-fi-KĀ-shun) The dispersion of large lipid globules into smaller, uniformly distributed particles in the presence of bile.

Enamel (e-NAM-el) The hard, white substance covering the crown of a tooth.

Endocardium (en-dō-KAR-dē-um) The layer of the heart wall, composed of endothelium and smooth muscle, that lines the inside of the heart and covers the valves and tendons that hold the valves open.

Endochondral (en′-dō-KON-dral) **ossification** Bone formation within hyaline cartilage that develops from mesenchyme.

Endocrine (EN-dō-krin) **gland** A gland that secretes hormones into interstitial fluid and then the blood; a ductless gland.

Endocrine system (EN-dō-krin) All endocrine glands and hormone-secreting cells.

Endocrinology (en′-dō-kri-NOL-ō-jē) The science concerned with the structure and functions of endocrine glands and the diagnosis and treatment of disorders of the endocrine system.

Endocytosis (en′-dō-sī-TŌ-sis) The uptake into a cell of large molecules and particles in which a segment of plasma membrane surrounds the substance, encloses it, and brings it in; includes phagocytosis, pinocytosis, and receptor-mediated endocytosis.

Endoderm (EN-dō-derm) A primary germ layer of the developing embryo; gives rise to the gastrointestinal tract, urinary bladder, urethra, and respiratory tract.

Endodontics (en′-dō-DON-tiks) The branch of dentistry concerned with the prevention, diagnosis, and treatment of diseases that affect the pulp, root, periodontal ligament, and alveolar bone.

Endolymph (EN-dō-limf′) The fluid within the membranous labyrinth of the internal ear.

Endometriosis (en′-dō-me′-trē-Ō-sis) The growth of endometrial tissue outside the uterus.

Endometrium (en′-dō-MĒ-trē-um) The mucous membrane lining the uterus.

Endomysium (en′-dō-MĪZ-ē-um) Invagination of the perimysium separating each individual muscle fiber (cell).

Endoneurium (en′-dō-NOO-rē-um) Connective tissue wrapping around individual nerve axons.

Endoplasmic reticulum (en′-dō-PLAS-mik re-TIK-ū-lum) **(ER)** A network of channels running through the cytoplasm of a cell that serves in intracellular transportation, support, storage, synthesis, and packaging of molecules. Portions of ER where ribosomes are attached to the outer surface are called **rough ER;** portions that have no ribosomes are called **smooth ER.**

End organ of Ruffini *See* **Type II cutaneous mechanoreceptor.**

Endosteum (end-OS-tē-um) The membrane that lines the medullary (marrow) cavity of bones, consisting of osteogenic cells and scattered osteoclasts.

Endothelium (en′-dō-THĒ-lē-um) The layer of simple squamous epithelium that lines the cavities of the heart, blood vessels, and lymphatic vessels.

Enteric (en-TER-ik) **nervous system** A portion of the autonomic nervous system within the wall of the gastrointestinal tract, pancreas, and gallbladder. Its sensory neurons monitor tension in the intestinal wall and assess the composition of intestinal contents; its motor neurons exert control over the motility and secretions of the gastrointestinal tract.

Enteroendocrine (en-ter-ō-EN-dō-krin) **cell** A cell of the mucosa of the gastrointestinal tract that secretes a hormone that governs function of the GI tract; hormones secreted include gastrin, cholecystokinin, glucose-dependent insulinotropic peptide (GIP), and secretin.

Enzyme (EN-zīm) A substance that accelerates chemical reactions; an organic catalyst, usually a protein.

Eosinophil (ē-ō-SIN-ō-fil) A type of white blood cell characterized by granules that stain red or pink with acid dyes.

Ependymal (ep-EN-de-mal) **cells** Neuroglial cells that cover choroid plexuses and produce cerebrospinal fluid (CSF); they also line the ventricles of the brain and probably assist in the circulation of CSF.

Epicardium (ep′-i-KAR-dē-um) The thin outer layer of the heart wall, composed of serous tissue and mesothelium. Also called the **visceral pericardium.**

Epidemiology (ep′-i-dē-mē-OL-ō-jē) Study of the occurrence and transmission of diseases and disorders in human populations.

Epidermis (ep′-i-DERM-is) The superficial, thinner layer of skin, composed of keratinized stratified squamous epithelium.

Epididymis (ep′-i-DID-i-mis) A comma-shaped organ that lies along the posterior border of the testis and contains the ductus epididymis, in which sperm undergo maturation. *Plural* is **epididymides** (ep′-i-di-DIM-i-dēz).

Epidural (ep′-i-DOO-ral) **space** A space between the spinal dura mater and the vertebral canal, containing areolar connective tissue and a plexus of veins.

Epiglottis (ep′-i-GLOT-is) A large, leaf-shaped piece of cartilage lying on top of the larynx, attached to the thyroid cartilage; its unattached portion is free to move up and down to cover the glottis (vocal folds and rima glottidis) during swallowing.

Epimysium (ep-i-MĪZ-ē-um) Fibrous connective tissue around muscles.

Epinephrine (ep-ē-NEF-rin) Hormone secreted by the adrenal medulla that produces actions similar to those that result from sympathetic stimulation. Also called **adrenaline** (a-DREN-a-lin).

Epineurium (ep′-i-NOO-rē-um) The superficial connective tissue covering around an entire nerve.

Epiphyseal (ep′-i-FIZ-ē-al) **line** The remnant of the epiphyseal plate in the metaphysis of a long bone.

Epiphyseal plate The hyaline cartilage plate in the metaphysis of a long bone; site of lengthwise growth of long bones.

Epiphysis (e-PIF-i-sis) The end of a long bone, usually larger in diameter than the shaft (diaphysis).

Epiphysis cerebri (se-RĒ-brē) Pineal gland.

Episiotomy (e-piz′-ē-OT-ō-mē) A cut made with surgical scissors to avoid tearing of the perineum at the end of the second stage of labor.

Epistaxis (ep′-i-STAK-sis) Loss of blood from the nose due to trauma, infection, allergy, neoplasm, and bleeding disorders. Also called **nosebleed.**

Epithalamus (ep′-i-THAL-a-mus) Part of the diencephalon superior and posterior to the thalamus, comprising the pineal gland and associated structures.

Epithelial (ep-i-THĒ-lē-al) **tissue** The tissue that forms the innermost and outermost surfaces of body structures and forms glands.

Eponychium (ep′-o-NIK-ē-um) Narrow band of stratum corneum at the proximal border of a nail that extends from the margin of the nail wall. Also called the **cuticle.**

Equilibrium (ē-kwi-LIB-rē-um) The state of being balanced.

Erectile dysfunction (ED) Failure to maintain an erection long enough for sexual intercourse. Previously known as **impotence** (IM-pō-tens).

Erection (ē-REK-shun) The enlarged and stiff state of the penis or clitoris resulting from the engorgement of the spongy erectile tissue with blood.

Eructation (e-ruk′-TĀ-shun) The forceful expulsion of gas from the stomach. Also called **belching.**

Erythema (er-e-THĒ-ma) Skin redness usually caused by dilation of the capillaries.

Erythrocyte (e-RITH-rō-sīt) A mature red blood cell.

Erythropoietin (e-rith′-rō-POY-e-tin) **(EPO)** A hormone released by the juxtaglomerular cells of the kidneys that stimulates red blood cell production.

Esophagus (e-SOF-a-gus) The hollow muscular tube that connects the pharynx and the stomach.

Estrogens (ES-tro-jenz) Feminizing sex hormones produced by the ovaries; govern development of oocytes, maintenance of female reproductive structures, and appearance of secondary sex characteristics; also affect fluid and electrolyte balance, and protein anabolism. Examples are β-estradiol, estrone, and estriol.

Eupnea (ŪP-nē-a) Normal quiet breathing.

Eustachian tube *See* **Auditory tube.**

Eversion (ē-VER-zhun) The movement of the sole laterally at the ankle joint or of an atrioventricular valve into an atrium during ventricular contraction.

Excitability (ek-sīt′-a-BIL-i-tē) The ability of muscle fibers to receive and respond to stimuli; the ability of neurons to respond to stimuli and generate nerve impulses.

Excretion (eks-KRĒ-shun) The process of eliminating waste products from the body; also the products excreted.

Exercise physiology Study of the changes in cell and organ function due to muscular activity.

Exhalation (eks-ha-LĀ-shun) Breathing out; expelling air from the lungs into the atmosphere. Also called **expiration.**

Exocrine (EK-sō-krin) **gland** A gland that secretes its products into ducts that carry the secretions into body cavities, into the lumen of an organ, or to the outer surface of the body.

Exocytosis (ek-sō-sī-TŌ-sis) A process in which membrane-enclosed secretory vesicles form inside the cell, fuse with the plasma membrane, and release their contents into the interstitial fluid; achieves secretion of materials from a cell.

Extensibility (ek-sten′-si-BIL-i-tē) The ability of muscle tissue to stretch when it is pulled.

Extension (eks-TEN-shun) An increase in the angle between two bones; restoring a body part to its anatomical position after flexion.

External Located on or near the surface.

External auditory (AW-di-tōr-ē) **canal** or **meatus** (mē-Ā-tus) A curved tube in the temporal bone that leads to the middle ear.

External ear The **outer ear,** consisting of the pinna, external auditory canal, and tympanic membrane (eardrum).

External nares (NĀ-rez) The openings into the nasal cavity on the exterior of the body. Also called the **nostrils.**

External respiration The exchange of respiratory gases between the lungs and blood. Also called **pulmonary respiration** or **pulmonary gas exchange.**

Exteroceptor (EKS-ter-ō-sep′-tor) A sensory receptor adapted for the reception of stimuli from outside the body.

Extracellular fluid (ECF) Fluid outside body cells, such as interstitial fluid and plasma.

Extracellular matrix (MĀ-triks) The ground substance and fibers between cells in a connective tissue.

Eyebrow The hairy ridge superior to the eye.

F

F cell A cell in the pancreatic islets (islets of Langerhans) that secretes pancreatic polypeptide.

Face The anterior aspect of the head.

Falciform ligament (FAL-si-form LIG-a-ment) A sheet of parietal peritoneum between the two principal lobes of the liver. The ligamentum teres, or remnant of the umbilical vein, lies within its fold.

Falx cerebelli (FALKS′ ser-e-BEL-lī) A small triangular process of the dura mater attached to the occipital bone in the posterior cranial fossa and projecting inward between the two cerebellar hemispheres.

Falx cerebri (FALKS SER-e-brē) A fold of the dura mater extending deep into the longitudinal fissure between the two cerebral hemispheres.

Fascia (FASH-ē-a) Large connective tissue sheet that wraps around groups of muscles.

Fascicle (FAS-i-kul) A small bundle or cluster, especially of nerve or muscle fibers (cells). Also called a **fasciculus** (fa-SIK-ū-lus). *Plural* is **fasciculi** (fa-SIK-yoo-lī).

Fasciculation (fa-sik-ū-LĀ-shun) Abnormal, spontaneous twitch of all skeletal muscle fibers in one motor unit that is visible at the skin surface; not associated with movement of the affected muscle; present in progressive diseases of motor neurons, for example, poliomyelitis.

Fat A triglyceride that is a solid at room temperature.

Fatty acid A simple lipid that consists of a carboxyl group and a hydrocarbon chain; used to synthesize triglyceride and phospholipids.

Fauces (FAW-sēs) The opening from the mouth into the pharynx.

Feces (FĒ-sēz) Material discharged from the rectum and made up of bacteria, excretions, and food residue. Also called **stool.**

Feedback system (loop) A cycle of events in which the status of a body condition is monitored, evaluated, changed, remonitored, reevaluated, and so on.

Female reproductive cycle General term for the ovarian and uterine cycles, the hormonal changes that accompany them, and cyclic changes in the breasts and cervix; includes changes in the endometrium of a nonpregnant female that prepares the lining of the uterus to receive a fertilized ovum. Less correctly termed the **menstrual cycle.**

Fertilization (fer-til-i-ZĀ-shun) Penetration of a secondary oocyte by a sperm cell, meiotic division of a secondary oocyte to form an ovum, and subsequent union of the nuclei of the gametes.

Fetal circulation The cardiovascular system of the fetus, including the placenta and special blood vessels involved in the exchange of materials between fetus and mother.

Fetus (FĒ-tus) In humans, the developing organism *in utero* from the beginning of the third month to birth.

Fever An elevation in body temperature above the normal temperature of 37°C (98.6°F) due to a resetting of the hypothalamic thermostat.

Fibroblast (FĪ-brō-blast) A large, flat cell that secretes most of the extracellular matrix of areolar and dense connective tissues.

Fibrosis The process by which fibroblasts synthesize collagen fibers and other extracellular matrix materials that aggregate to form scar tissue.

Fibrous (FĪ-brus) **joint** A joint that allows little or no movement, such as a suture or a syndesmosis.

Fibrous tunic (TOO-nik) The superficial coat of the eyeball, made up of the posterior sclera and the anterior cornea.

Fight-or-flight response The effects produced on stimulation of the sympathetic division of the autonomic nervous system.

Filiform papilla (FIL-i-form pa-PIL-a) One of the conical projections that are distributed in parallel rows over the anterior two-thirds of the tongue and lack taste buds.

Filtration (fil-TRĀ-shun) The flow of a liquid through a filter (or membrane that acts like a filter) due to a hydrostatic pressure; occurs in capillaries due to blood pressure.

Filum terminale (FĪ-lum ter-mi-NAL-ē) Non-nervous fibrous tissue of the spinal cord that extends inferiorly from the conus medullaris to the coccyx.

Fimbriae (FIM-brē-ē) Fingerlike structures, especially the lateral ends of the uterine (fallopian) tubes.

Fissure (FISH-ur) A groove, fold, or slit that may be normal or abnormal.

Fixator A muscle that stabilizes the origin of the prime mover so that the prime mover can act more efficiently.

Fixed macrophage (MAK-rō-fāj) Stationary phagocytic cell found in the liver, lungs, brain, spleen, lymph nodes, subcutaneous tissue, and red bone marrow. Also called a **tissue macrophage** or **histiocyte** (HIS-tē-ō-sīt).

Flaccid (FLAK-sid) Relaxed, flabby, or soft; lacking muscle tone.

Flagellum (fla-JEL-um) A hairlike, motile process on the extremity of a bacterium, protozoan, or sperm cell. *Plural* is **flagella** (fla-JEL-a).

Flatus (FLĀ-tus) Gas in the stomach or intestines; commonly used to denote expulsion of gas through the anus.

Flexion (FLEK-shun) Movement in which there is a decrease in the angle between two bones.

Follicle (FOL-i-kul) A small secretory sac or cavity; the group of cells that contains a developing oocyte in the ovaries.

Follicle-stimulating hormone (FSH) Hormone secreted by the anterior pituitary; it initiates development of ova and stimulates the ovaries to secrete estrogens in females, and initiates sperm production in males.

Fontanel (fon-ta-NEL) A mesenchyme-filled space where bone formation is not yet complete, especially between the cranial bones of an infant's skull.

Foot The terminal part of the lower limb, from the ankle to the toes.

Foramen (fō-RĀ-men) A passage or opening; a communication between two cavities of an organ, or a hole in a bone for passage of vessels or nerves. *Plural* is **foramina** (fō-RAM-i-na).

Foramen ovale (fō-RĀ-men ō-VAL-ē) An opening in the fetal heart in the septum between the right and left atria. A hole in the greater wing of the sphenoid bone that transmits the mandibular branch of the trigeminal (V) nerve.

Forearm (FOR-arm) The part of the upper limb between the elbow and the wrist.

Fornix (FOR-niks) An arch or fold; a tract in the brain made up of association fibers, connecting the hippocampus with the mammillary bodies; a recess around the cervix of the uterus where it protrudes into the vagina.

Fossa (FOS-a) A furrow or shallow depression.

Fourth ventricle (VEN-tri-kul) A cavity filled with cerebrospinal fluid within the brain lying between the cerebellum and the medulla oblongata and pons.

Fracture (FRAK-choor) Any break in a bone.

Free radical An atom or group of atoms with an unpaired electron in the outermost shell. It is unstable, highly reactive, and destroys nearby molecules.

Frontal plane A plane at a right angle to a midsagittal plane that divides the body or organs into anterior and posterior portions. Also called a **coronal** (kō-RŌ-nal) **plane.**

Fundus (FUN-dus) The part of a hollow organ farthest from the opening.

Fungiform papilla (FUN-ji-form pa-PIL-a) A mushroomlike elevation on the upper surface of the tongue appearing as a red dot; most contain taste buds.

Furuncle (FŪ-rung-kul) A boil; painful nodule caused by bacterial infection and inflammation of a hair follicle or sebaceous (oil) gland.

G

Gallbladder A small pouch, located inferior to the liver, that stores bile and empties by means of the cystic duct.

Gallstone A solid mass, usually containing cholesterol, in the gallbladder or a bile-containing duct; formed anywhere between bile canaliculi in the liver and the hepatopancreatic ampulla (ampulla of Vater), where bile enters the duodenum. Also called a **biliary calculus.**

Gamete (GAM-ēt) A male or female reproductive cell; a sperm cell or secondary oocyte.

Ganglion (GANG-glē-on) Usually, a group of neuronal cell bodies lying outside the central nervous system (CNS). *Plural* is **ganglia** (GANG-glē-a).

Gastric (GAS-trik) **glands** Glands in the mucosa of the stomach composed of cells that empty their secretions into narrow channels called gastric pits. Types of cells are chief cells (secrete pepsinogen), parietal cells (secrete hydrochloric acid and intrinsic factor), surface mucous and mucous neck cells (secrete mucus), and G cells (secrete gastrin).

Gastroenterology (gas′-trō-en′-ter-OL-ō-jē) The medical specialty that deals with the structure, function, diagnosis, and treatment of diseases of the stomach and intestines.

Gastrointestinal (gas-trō-in-TES-ti-nal) **(GI) tract** A continuous tube running through the ventral body cavity extending from the mouth to the anus. Also called the **alimentary** (al′-i-MEN-tar-ē) **canal.**

Gastrulation (gas-troo-LĀ-shun) The migration of groups of cells from the epiblast that transform a bilaminar embryonic disc into a trilaminar embryonic disc with three primary germ layers; transformation of the blastula into the gastrula.

Gene (JĒN) Biological unit of heredity; a segment of DNA located in a definite position on a particular chromosome; a sequence of DNA that codes for a particular mRNA, rRNA, or tRNA.

Genetic engineering The manufacture and manipulation of genetic material.

Genetics The study of genes and heredity.

Genome (JĒ-nōm) The complete set of genes of an organism.

Genotype (JĒ-nō-tīp) The genetic makeup of an individual; the combination of alleles present at one or more chromosomal locations, as distinguished from the appearance, or phenotype, that results from those alleles.

Geriatrics (jer′-ē-AT-riks) The branch of medicine devoted to the medical problems and care of elderly persons.

Gestation (jes-TĀ-shun) The period of development from fertilization to birth.

Gingivae (jin-JI-vē) Gums. They cover the alveolar processes of the mandible and maxilla and extend slightly into each socket.

Gland Specialized epithelial cell or cells that secrete substances; may be exocrine or endocrine.

Glans penis (glanz PĒ-nis) The slightly enlarged region at the distal end of the penis.

Glaucoma (glaw-KŌ-ma) An eye disorder in which there is increased intraocular pressure due to an excess of aqueous humor.

Glomerular (glō-MER-ū-lar) **capsule** A double-walled globe at the proximal end of a nephron that encloses the glomerular capillaries. Also called **Bowman's** (BŌ-manz) **capsule.**

Glomerular filtrate (glō-MER-ū-lar FIL-trāt) The fluid produced when blood is filtered by the filtration membrane in the glomeruli of the kidneys.

Glomerular filtration The first step in urine formation in which substances in blood pass through the filtration membrane and the filtrate enters the proximal convoluted tubule of a nephron.

Glomerular filtration rate The amount of filtrate formed in all renal corpuscles per minute. It averages 125 mL/min in males and 105 mL/min in females.

Glomerulus (glō-MER-ū-lus) A rounded mass of nerves or blood vessels, especially the microscopic tuft of capillaries that is surrounded by the glomerular (Bowman's) capsule of each kidney tubule. *Plural* is **glomeruli** (glō-MER-ū-li).

Glottis (GLOT-is) The vocal folds (true vocal cords) in the larynx plus the space between them (rima glottidis).

Glucagon (GLOO-ka-gon) A hormone produced by the alpha cells of the pancreatic islets (islets of Langerhans) that increases blood glucose level.

Glucocorticoids (gloo′-kō-KOR-ti-koyds) Hormones secreted by the cortex of the adrenal gland, especially cortisol, that influence glucose metabolism.

Glucose (GLOO-kōs) A hexose (six-carbon sugar), $C_6H_{12}O_6$, that is a major energy source for the production of ATP by body cells.

Glucosuria (gloo′-kō-SOO-rē-a) The presence of glucose in the urine; may be temporary or pathological. Also called **glycosuria.**

Glycogen (GLĪ-kō-jen) A highly branched polymer of glucose containing thousands of subunits; functions as a compact store of glucose molecules in liver and muscle fibers (cells).

Goblet cell A goblet-shaped unicellular gland that secretes mucus; present in epithelium of the airways and intestines.

Goiter (GOY-ter) An enlarged thyroid gland.

Golgi (GOL-jē) **complex** An organelle in the cytoplasm of cells consisting of four to six flattened sacs (cisternae), stacked on one another, with expanded areas at their ends; functions in processing, sorting, packaging, and delivering proteins and lipids to the plasma membrane, lysosomes, and secretory vesicles.

Golgi tendon organ *See* **Tendon organ.**

Gomphosis (gom-FŌ-sis) A fibrous joint in which a cone-shaped peg fits into a socket.

Gonad (GŌ-nad) A gland that produces gametes and hormones; the ovary in the female and the testis in the male.

Gonadotropic hormone Anterior pituitary hormone that affects the gonads.

Gout (GOWT) Hereditary condition associated with excessive uric acid in the blood; the acid crystallizes and deposits in joints, kidneys, and soft tissue.

Graafian follicle *See* **Mature follicle.**

Gracile (GRAS-īl) **nucleus** A group of nerve cells in the inferior part of the medulla oblongata in which axons of the gracile fasciculus terminate.

Gray commissure (KOM-mi-shur) A narrow strip of gray matter connecting the two lateral gray masses within the spinal cord.

Gray matter Areas in the central nervous system and ganglia containing neuronal cell bodies, dendrites, unmyelinated axons, axon terminals, and neuroglia; Nissl bodies impart a gray color and there is little or no myelin in gray matter.

Gray ramus communicans (RĀ-mus kō-MŪ-ni-kans) A short nerve containing axons of sympathetic postganglionic neurons; the cell bodies of the neurons are in a sympathetic chain ganglion, and the unmyelinated axons extend via the gray ramus to a spinal nerve and then to the periphery to supply smooth muscle in blood vessels, arrector pili muscles, and sweat glands. *Plural* is **rami communicantes** (RĀ-mē kō-mū-ni-KAN-tēz).

Greater omentum (ō-MEN-tum) A large fold in the serosa of the stomach that hangs down like an apron anterior to the intestines.

Greater vestibular (ves-TIB-ū-lar) **glands** A pair of glands on either side of the vaginal orifice that open by a duct into the space between the hymen and the labia minora. Also called **Bartholin's** (BAR-to-linz) **glands.**

Groin (GROYN) The depression between the thigh and the trunk; the inguinal region.

Gross anatomy The branch of anatomy that deals with structures that can be studied without using a microscope. Also called **macroscopic anatomy.**

Growth An increase in size due to an increase in (1) the number of cells, (2) the size of existing cells as internal components increase in size, or (3) the size of intercellular substances.

Gustation (gus-TĀ-shun). The sense of taste.

Gustatory (GUS-ta-tō′-rē) Pertaining to taste.

Gynecology (gī′-ne-KOL-ō-jē) The branch of medicine dealing with the study and treatment of disorders of the female reproductive system.

Gynecomastia (gīn′-e-kō-MAS-tē-a) Excessive growth (benign) of the male mammary glands due to secretion of estrogens by an adrenal gland tumor (feminizing adenoma).

Gyrus (JI-rus) One of the folds of the cerebral cortex of the brain. *Plural* is **gyri** (JĪ-rī). Also called a **convolution.**

H

Hair A threadlike structure produced by hair follicles that develops in the dermis. Also called a **pilus** (PĪ-lus).

Hair follicle (FOL-li-kul) Structure composed of epithelium and surrounding the root of a hair from which hair develops.

Hair root plexus (PLEK-sus) A network of dendrites arranged around the root of a hair as free or naked nerve endings that are stimulated when a hair shaft is moved.

Hand The terminal portion of an upper limb, including the carpals, metacarpals, and phalanges.

Haploid (HAP-loyd) **cell** Having half the number of chromosomes characteristically found in the somatic cells of an organism; characteristic of mature gametes. Symbolized *n*.

Hard palate (PAL-at) The anterior portion of the roof of the mouth, formed by the maxillae and palatine bones and lined by mucous membrane.

Haustra (HAWS-tra) A series of pouches that characterize the colon; caused by tonic contractions of the teniae coli. *Singular* is **haustrum.**

Haversian canal *See* **Central canal.**

Haversian system *See* **Osteon.**

Head The superior part of a human, cephalic to the neck. The superior or proximal part of a structure.

Hearing The ability to perceive sound.

Heart A hollow muscular organ lying slightly to the left of the midline of the chest that pumps the blood through the cardiovascular system.

Heart block An arrhythmia (dysrhythmia) of the heart in which the atria and ventricles contract independently because of a blocking of electrical impulses through the heart at some point in the conduction system.

Heart murmur (MER-mer) An abnormal sound that consists of a flow noise that is heard before, between, or after the normal heart sounds, or that may mask normal heart sounds.

Hemangioblast (hē-MAN-jē-ō-blast) A precursor mesodermal cell that develops into blood and blood vessels.

Hematocrit (he-MAT-ō-krit) **(Hct)** The percentage of blood made up of red blood cells. Usually measured by centrifuging a blood sample in a graduated tube and then reading the volume of red blood cells and dividing it by the total volume of blood in the sample.

Hematology (hēm-a-TOL-ō-jē) The study of blood.

Hematoma (hē′-ma-TŌ-ma) A tumor or swelling filled with blood.

Hemiplegia (hem-i-PLĒ-jē-a) Paralysis of the upper limb, trunk, and lower limb on one side of the body.

Hemodialysis (hē-mō-dī-AL-i-sis) Direct filtration of blood by removing wastes and excess electrolytes and fluid and then returning the cleansed blood.

Hemodynamics (hē-mō-dī-NAM-iks) The forces involved in circulating blood throughout the body.

Hemoglobin (hē′-mō-GLŌ-bin) **(Hb)** A substance in red blood cells consisting of the protein globin and the iron-containing red pigment heme that transports most of the oxygen and some carbon dioxide in blood.

Hemolysis (hē-MOL-i-sis) The escape of hemoglobin from the interior of a red blood cell into the surrounding medium; results from disruption of the cell membrane by toxins or drugs, freezing or thawing, or hypotonic solutions.

Hemolytic disease of the newborn (HDN) A hemolytic anemia of a newborn child that results from the destruction of the infant's erythrocytes (red blood cells) by antibodies produced by the mother; usually the antibodies are due to an Rh blood type incompatibility. Also called **erythroblastosis fetalis** (e-rith′-rō-blas-TŌ-sis fe-TAL-is).

Hemophilia (hē′-mō-FIL-ē-a) A hereditary blood disorder where there is a deficient production of certain factors involved in blood clotting, resulting in excessive bleeding into joints, deep tissues, and elsewhere.

Hemopoiesis (hēm-ō-poy-Ē-sis) Blood cell production, which occurs in red bone marrow after birth. Also called **hematopoiesis** (hem′-a-tō-poy-Ē-sis).

Hemorrhage (HEM-o-rij) Bleeding; the escape of blood from blood vessels, especially when the loss is profuse.

Hemorrhoids (HEM-ō-royds) Dilated or varicosed blood vessels (usually veins) in the anal region. Also called **piles.**

Hepatic (he-PAT-ik) Refers to the liver.

Hepatic duct A duct that receives bile from the bile capillaries. Small hepatic ducts merge to form the larger right and left hepatic ducts that unite to leave the liver as the common hepatic duct.

Hepatic portal circulation The flow of blood from the gastrointestinal organs to the liver before returning to the heart.

Hepatocyte (he-PAT-ō-cyte) A liver cell.

Hepatopancreatic (hep′-a-tō-pan′-krē-A-tik) **ampulla** A small, raised area in the duodenum where the combined common bile duct and main pancreatic duct empty into the duodenum. Also called the **ampulla of Vater** (FAH-ter).

Hernia (HER-nē-a) The protrusion or projection of an organ or part of an organ through a membrane or cavity wall, usually the abdominal cavity.

Herniated (HER-nē-ā′-ted) **disc** A rupture of an intervertebral disc so that the nucleus pulposus protrudes into the vertebral cavity. Also called a **slipped disc.**

Hiatus (hī-Ā-tus) An opening; a foramen.

Hilum (HĪ-lum) An area, depression, or pit where blood vessels and nerves enter or leave an organ. Also called a **hilus.**

Hinge joint A synovial joint in which a convex surface of one bone fits into a concave surface of another bone, such as the elbow, knee, ankle, and interphalangeal joints. Also called a **ginglymus** (JIN-gli-mus) **joint.**

Hirsutism (HER-soo-tizm) An excessive growth of hair in females and children, with a distribution similar to that in adult males, due to the conversion of vellus hairs into large terminal hairs in response to higher-than-normal levels of androgens.

Histamine (HISS-ta-mēn) Substance found in many cells, especially mast cells, basophils, and platelets, that is released when the cells are injured; results in vasodilation, increased permeability of blood vessels, and constriction of bronchioles.

Histology (his′-TOL-ō-jē) Microscopic study of the structure of tissues.

Holocrine (HŌ-lō-krin) **gland** A type of gland in which entire secretory cells, along with their accumulated secretions, make up the secretory product of the gland, as in the sebaceous (oil) glands.

Homeostasis (hō′-mē-ō-STĀ-sis) The condition in which the body's internal environment remains relatively constant within physiological limits.

Homologous (hō-MOL-ō-gus) **chromosomes** Two chromosomes that belong to a pair. Also called **homologs.**

Hormone (HOR-mōn) A secretion of endocrine cells that alters the physiological activity of target cells of the body.

Horn An area of gray matter (anterior, lateral, or posterior) in the spinal cord.

Human chorionic gonadotropin (kō-rē-ON-ik gō-nad-ō-TRŌ-pin) **(hCG)** A hormone produced by the developing placenta that maintains the corpus luteum.

Human chorionic somatomammotropin (sō-mat-ō-mam-ō-TRŌ-pin) **(hCS)** Hormone produced by the chorion of the placenta that stimulates breast tissue for lactation, enhances body growth, and regulates metabolism. Also called **human placental lactogen (hPL).**

Human growth hormone (hGH) Hormone secreted by the anterior pituitary that stimulates growth of body tissues, especially skeletal and muscular tissues. Also known as **somatotropin** (sō′-ma-tō-TRŌ-pin) and **somatotropic hormone (STH).**

Hyaluronic (hī′-a-loo-RON-ik) **acid** A viscous, amorphous extracellular material that binds cells together, lubricates joints, and maintains the shape of the eyeballs.

Hymen (HĪ-men) A thin fold of vascularized mucous membrane at the vaginal orifice.

Hyperextension (hī′-per-ek-STEN-shun) Continuation of extension beyond the anatomical position, as in bending the head backward.

Hyperplasia (hī-per-PLĀ-zē-a) An abnormal increase in the number of normal cells in a tissue or organ, increasing its size.

Hypersecretion (hī′-per-se-KRĒ-shun) Overactivity of glands resulting in excessive secretion.

Hypersensitivity (hī′-per-sen-si-TI-vi-tē) Overreaction to an allergen that results in pathological changes in tissues. Also called **allergy.**

Hypertension (hī′-per-TEN-shun) High blood pressure.

Hyperthermia (hī′-per-THERM-ē-a) An elevated body temperature.

Hypertonia (hī′-per-TŌ-nē-a) Increased muscle tone that is expressed as spasticity or rigidity.

Hypertonic (hī′-per-TON-ik) Solution that causes cells to shrink due to loss of water by osmosis.

Hypertrophy (hī-PER-trō-fē) An excessive enlargement or overgrowth of tissue without cell division.

Hyperventilation (hī′-per-ven-ti-LĀ-shun) A rate of inhalation and exhalation higher than that required to maintain a normal partial pressure of carbon dioxide in the blood.

Hyponychium (hī′-pō-NIK-ē-um) Free edge of the fingernail.

Hypophyseal fossa (hī′-pō-FIZ-ē-al FOS-a) A depression on the superior surface of the sphenoid bone that houses the pituitary gland.

Hypophyseal (hī′-pō-FIZ-ē-al) **pouch** An outgrowth of ectoderm from the roof of the mouth from which the anterior pituitary develops.

Hypophysis (hī-POF-i-sis) Pituitary gland.

Hyposecretion (hī′-pō-se-KRĒ-shun) Underactivity of glands resulting in diminished secretion.

Hypothalamohypophyseal (hī′-pō-thal′-a-mō-hī′-pō-FIZ-ē-al) **tract** A bundle of axons containing secretory vesicles filled with oxytocin or antidiuretic hormone that extend from the hypothalamus to the posterior pituitary.

Hypothalamus (hī′-pō-THAL-a-mus) A portion of the diencephalon, lying beneath the thalamus and forming the floor and part of the wall of the third ventricle.

Hypothermia (hī′-pō-THER-mē-a) Lowering of body temperature below 35°C (95°F); in surgical procedures, it refers to deliberate cooling of the body to slow down metabolism and reduce oxygen needs of tissues.

Hypotonia (hī′-pō-TŌ-nē-a) Decreased or lost muscle tone in which muscles appear flaccid.

Hypotonic (hī′-pō-TON-ik) Solution that causes cells to swell and perhaps rupture due to gain of water by osmosis.

Hypoventilation (hī-pō-ven-ti-LĀ-shun) A rate of inhalation and exhalation lower than that required to maintain a normal partial pressure of carbon dioxide in plasma.

Hypoxia (hī-POKS-ē-a) Lack of adequate oxygen at the tissue level.

Hysterectomy (hiss-te-REK-tō-mē) The surgical removal of the uterus.

I

Ileocecal (il-ē-ō-SĒ-kal) **sphincter** A fold of mucous membrane that guards the opening from the ileum into the large intestine. Also called the **ileocecal valve.**

Ileum (IL-ē-um) The terminal part of the small intestine.

Immunity (i-MŪ-ni-tē) The state of being resistant to injury, particularly by poisons, foreign proteins, and invading pathogens.

Immunoglobulin (im-ū-nō-GLOB-ū-lin) **(Ig)** An antibody synthesized by plasma cells derived from B lymphocytes in response to the introduction of an antigen. Immunoglobulins are divided into five kinds (IgG, IgM, IgA, IgD, IgE).

Immunology (im′-ū-NOL-ō-jē) The study of the responses of the body when challenged by antigens.

Imperforate (im′-PER-fō-rāt) Abnormally closed.

Implantation (im′-plan-TĀ-shun) The insertion of a tissue or a part into the body. The attachment of the blastocyst to the stratum basalis of the endometrium about 6 days after fertilization.

Incontinence (in-KON-ti-nens) Inability to retain urine, semen, or feces through loss of sphincter control.

Indirect motor pathways Motor tracts that convey information from the brain down the spinal cord for automatic movements, coordination of body movements with visual stimuli, skeletal muscle tone and posture, and balance. Also known as **extrapyramidal pathways.**

Induction (in-DUK-shun) The process by which one tissue (inducting tissue) stimulates the development of an adjacent unspecialized tissue (responding tissue) into a specialized one.

Infarction (in-FARK-shun) A localized area of necrotic tissue, produced by inadequate oxygenation of the tissue.

Infection (in-FEK-shun) Invasion and multiplication of microorganisms in body tissues, which may be inapparent or characterized by cellular injury.

Inferior (in-FĒR-ē-or) Away from the head or toward the lower part of a structure. Also called **caudal** (KAW-dal).

Inferior vena cava (VĒ-na KĀ-va) **(IVC)** Large vein that collects blood from parts of the body inferior to the heart and returns it to the right atrium.

Infertility Inability to conceive or to cause conception. Also called **sterility** in males.

Inflammation (in′-fla-MĀ-shun) Localized, protective response to tissue injury designed to destroy, dilute, or wall off the infecting agent or injured tissue; characterized by redness, pain, heat, swelling, and sometimes loss of function.

Infundibulum (in-fun-DIB-ū-lum) The stalklike structure that attaches the pituitary gland to the hypothalamus of the brain. The funnel-shaped, open, distal end of the uterine (fallopian) tube.

Ingestion (in-JES-chun) The taking in of food, liquids, or drugs, by mouth.

Inguinal (IN-gwin-al) Pertaining to the groin.

Inguinal canal An oblique passageway in the anterior abdominal wall just superior and parallel to the medial half of the inguinal ligament that transmits the spermatic cord and ilioinguinal nerve in the male and round ligament of the uterus and ilioinguinal nerve in the female.

Inhalation (in-ha-LĀ-shun) The act of drawing air into the lungs. Also termed **inspiration.**

Inheritance The acquisition of body traits by transmission of genetic information from parents to offspring.

Inhibin A hormone secreted by the gonads that inhibits release of follicle-stimulating hormone (FSH) by the anterior pituitary.

Inhibiting hormone Hormone secreted by the hypothalamus that can suppress secretion of hormones by the anterior pituitary.

Insertion (in-SER-shun) The attachment of a muscle tendon to a movable bone or the end opposite the origin.

Insula (IN-soo-la) A triangular area of the cerebral cortex that lies deep within the lateral cerebral fissure, under the parietal, frontal, and temporal lobes.

Insulin (IN-soo-lin) A hormone produced by the beta cells of a pancreatic islet (islet of Langerhans) that decreases the blood glucose level.

Integrins (IN-te-grinz) A family of transmembrane glycoproteins in plasma membranes that function in cell adhesion; they are present in hemidesmosomes, which anchor cells to a basement membrane, and they mediate adhesion of neutrophils to endothelial cells during emigration.

Integumentary (in-teg-ū-MEN-tar-ē) Relating to the skin.

Integumentary (in-teg-ū-MEN-tar-ē) **system** A system composed of organs such as the skin, hair, oil and sweat glands, nails, and sensory receptors.

Intercalated (in-TER-ka-lāt-ed) **disc** An irregular transverse thickening of sarcolemma that contains desmosomes, which hold cardiac muscle fibers (cells) together, and gap junctions, which aid in conduction of muscle action potentials from one fiber to the next.

Intercostal (in′-ter-KOS-tal) **nerve** A nerve supplying a muscle located between the ribs. Also called **thoracic nerve.**

Intermediate (in′-ter-MĒ-dē-at) Between two structures, one of which is medial and one of which is lateral.

Intermediate filament Protein filament, ranging from 8 to 12 nm in diameter, that may provide structural reinforcement, hold organelles in place, and give shape to a cell.

Internal Away from the surface of the body.

Internal capsule A large tract of projection fibers lateral to the thalamus that is the major connection between the cerebral cortex and the brain stem and spinal cord; contains axons of sensory neurons carrying auditory, visual, and somatic sensory signals to the cerebral cortex plus axons of motor neurons descending from the cerebral cortex to the thalamus, subthalamus, brain stem, and spinal cord.

Internal ear The inner ear or labyrinth, lying inside the temporal bone, containing the organs of hearing and balance.

Internal nares (NĀ-rez) The two openings posterior to the nasal cavities opening into the nasopharynx. Also called the **choanae** (kō-Ā-nē).

Internal respiration The exchange of respiratory gases between blood and body cells. Also called **tissue respiration** or **systemic gas exchange.**

Interneurons (in′-ter-NOO-ronz) Neurons whose axons extend only for a short distance and contact nearby neurons in the brain, spinal cord, or a ganglion; they comprise the vast majority of neurons in the body. Also called **association neurons.**

Interoceptor (IN-ter-ō-sep′-tor) Sensory receptor located in blood vessels and viscera that provides information about the body's internal environment. Also called **visceroceptor.**

Interphase (IN-ter-fāz) The period of the cell cycle between cell divisions, consisting of the G_0 phase; G_1 (gap or growth) phase, when the cell is engaged in growth, metabolism, and production of substances required for division; S (synthesis) phase, during which chromosomes are replicated; and G_2 phase.

Interstitial cell of Leydig *See* **Interstitial endocrinocyte.**

Interstitial (in′-ter-STISH-al) **endocrinocyte** A cell that is located in the connective tissue between seminiferous tubules in a mature testis that secretes testosterone. Also called an **interstitial cell of Leydig** (LĪ-dig).

Interstitial (in′-ter-STISH-al) **fluid** The portion of extracellular fluid that fills the microscopic spaces between the cells of tissues; the internal environment of the body. Also called **intercellular** or **tissue fluid.**

Interstitial growth Growth from within, as in the growth of cartilage. Also called **endogenous** (en-DOJ-e-nus) **growth.**

Interventricular (in′-ter-ven-TRIK-ū-lar) **foramen** A narrow, oval opening through which the lateral ventricles of the brain communicate with the third ventricle. Also called the **foramen of Monro.**

Intervertebral (in′-ter-VER-te-bral) **disc** A pad of fibrocartilage located between the bodies of two vertebrae.

Intestinal gland A gland that opens onto the surface of the intestinal mucosa and secretes digestive enzymes. Also called a **crypt of Lieberkühn** (LĒ-ber-kūn).

Intracellular (in′-tra-SEL-yū-lar) **fluid** (**ICF**) Fluid located within cells.

Intrafusal (in′-tra-FŪ-sal) **fibers** Three to ten specialized muscle fibers (cells), partially enclosed in a spindle-shaped connective tissue capsule, that make up a muscle spindle.

Intramembranous (in′-tra-MEM-bra-nus) **ossification** Bone formation within mesenchyme arranged in sheetlike layers that resemble membranes.

Intramuscular injection An injection that penetrates the skin and subcutaneous layer to enter a skeletal muscle. Common sites are the deltoid, gluteus medius, and vastus lateralis muscles.

Intraocular (in′-tra-OK-ū-lar) **pressure (IOP)** Pressure in the eyeball, produced mainly by aqueous humor.

Intrinsic factor (IF) A glycoprotein, synthesized and secreted by the parietal cells of the gastric mucosa, that facilitates vitamin B_{12} absorption in the small intestine.

Invagination (in-vaj′-i-NĀ-shun) The pushing of the wall of a cavity into the cavity itself.

Inversion (in-VER-zhun) The movement of the sole medially at the ankle joint.

In vitro (VĒ-trō) Literally, in glass; outside the living body and in an artificial environment such as a laboratory test tube.

Ipsilateral (ip-si-LAT-er-al) On the same side, affecting the same side of the body.

Iris The colored portion of the vascular tunic of the eyeball seen through the cornea that contains circular and radial smooth muscle; the hole in the center of the iris is the pupil.

Irritable bowel syndrome (IBS) Disease of the entire gastrointestinal tract in which a person reacts to stress by developing symptoms (such as cramping and abdominal pain) associated with alternating patterns of diarrhea and constipation.

Excessive amounts of mucus may appear in feces, and other symptoms include flatulence, nausea, and loss of appetite. Also known as **irritable colon** or **spastic colitis.**

Ischemia (is-KĒ-mē-a) A lack of sufficient blood to a body part due to obstruction or constriction of a blood vessel.

Islet of Langerhans *See* **Pancreatic islet.**

Isotonic (ī′-sō-TON-ik) Having equal tension or tone. A solution having the same concentration of impermeable solutes as cytosol.

Isthmus (IS-mus) A narrow strip of tissue or narrow passage connecting two larger parts.

J

Jaundice (JON-dis) A condition characterized by yellowness of the skin, the white of the eyes, mucous membranes, and body fluids because of a buildup of bilirubin.

Jejunum (je-JOO-num) The middle part of the small intestine.

Joint A point of contact between two bones, between bone and cartilage, or between bone and teeth. Also called an **articulation** or **arthrosis.**

Joint kinesthetic (kin′-es-THET-ik) **receptor** A proprioceptive receptor located in a joint, stimulated by joint movement.

Juxtaglomerular (juks-ta-glō-MER-ū-lar) **apparatus (JGA)** Consists of the macula densa (cells of the distal convoluted tubule adjacent to the afferent and efferent arteriole) and juxtaglomerular cells (modified cells of the afferent and sometimes efferent arteriole); secretes renin when blood pressure starts to fall.

K

Keratin (KER-a-tin) An insoluble protein found in the hair, nails, and other keratinized tissues of the epidermis.

Keratinocyte (ker-a-TIN-ō-sīt) The most numerous of the epidermal cells; produces keratin.

Kidney (KID-nē) One of the paired reddish organs located in the lumbar region that regulates the composition, volume, and pressure of blood and produces urine.

Kidney stone A solid mass, usually consisting of calcium oxalate, uric acid, or calcium phosphate crystals, that may form in any portion of the urinary tract. Also called a **renal calculus.**

Kinesiology (ki-nē-sē′-OL-ō-jē) The study of the movement of body parts.

Kinesthesia (kin′-es-THĒ-zē-a) The perception of the extent and direction of movement of body parts; this sense is possible due to nerve impulses generated by proprioceptors.

Kinetochore (ki-NET-ō-kor) Protein complex attached to the outside of a centromere to which kinetochore microtubules attach.

Kupffer's cell *See* **Stellate reticuloendothelial cell.**

Kyphosis (kī-FŌ-sis) An exaggeration of the thoracic curve of the vertebral column, resulting in a "round-shouldered" appearance. Also called **hunchback.**

L

Labial frenulum (LĀ-bē-al FREN-ū-lum) A medial fold of mucous membrane between the inner surface of the lip and the gums.

Labia majora (LĀ-bē-a ma-JŌ-ra) Two longitudinal folds of skin extending downward and backward from the mons pubis of the female.

Labia minora (min-OR-a) Two small folds of mucous membrane lying medial to the labia majora of the female.

Labium (LĀ-bē-um) A lip. A liplike structure. *Plural* is **labia** (LA-bē-a).

Labor The process of giving birth in which a fetus is expelled from the uterus through the vagina.

Labyrinth (LAB-i-rinth) Intricate communicating passageway, especially in the internal ear.

Lacrimal canal A duct, one on each eyelid, beginning at the punctum at the medial margin of an eyelid and conveying tears medially into the nasolacrimal sac.

Lacrimal gland Secretory cells, located at the superior anterolateral portion of each orbit, that secrete tears into excretory ducts that open onto the surface of the conjunctiva.

Lacrimal sac The superior expanded portion of the nasolacrimal duct that receives the tears from a lacrimal canal.

Lactation (lak-TĀ-shun) The secretion and ejection of milk by the mammary glands.

Lacteal (LAK-tē-al) One of many lymphatic vessels in villi of the intestines that absorb triglycerides and other lipids from digested food.

Lacuna (la-KOO-na) A small, hollow space, such as that found in bones in which the osteocytes lie. *Plural* is **lacunae** (la-KOO-nē).

Lambdoid (LAM-doyd) **suture** The joint in the skull between the parietal bones and the occipital bone; sometimes contains sutural (Wormian) bones.

Lamellae (la-MEL-ē) Concentric rings of hard, calcified extracellular matrix found in compact bone.

Lamellated corpuscle *See* **Pacinian corpuscle.**

Lamina (LAM-i-na) A thin, flat layer or membrane, as the flattened part of either side of the arch of a vertebra. *Plural* is **laminae** (LAM-i-nē).

Lamina propria (PRŌ-prē-a) The connective tissue layer of a mucosa.

Langerhans (LANG-er-hans) **cell** Epidermal dendritic cell that functions as an antigen-presenting cell (APC) during an immune response.

Lanugo (la-NOO-gō) Fine downy hairs that cover the fetus.

Large intestine The portion of the gastrointestinal tract extending from the ileum of the small intestine to the anus, divided structurally into the cecum, colon, rectum, and anal canal.

Laryngopharynx (la-rin′-gō-FAIR-inks) The inferior portion of the pharynx, extending downward from the level of the hyoid bone that divides posteriorly into the esophagus and anteriorly into the larynx. Also called the **hypopharynx.**

Larynx (LAIR-inks) The **voice box,** a short passageway that connects the pharynx with the trachea.

Lateral (LAT-er-al) Farther from the midline of the body or a structure.

Lateral ventricle (VEN-tri-kul) A cavity within a cerebral hemisphere that communicates with the lateral ventricle in the other cerebral hemisphere and with the third ventricle by way of the interventricular foramen.

Leg The part of the lower limb between the knee and the ankle.

Lens A transparent organ constructed of proteins (crystallins) lying posterior to the pupil and iris of the eyeball and anterior to the vitreous body.

Lesion (LĒ-zhun) Any localized, abnormal change in a body tissue.

Lesser omentum (ō-MEN-tum) A fold of the peritoneum that extends from the liver to the lesser curvature of the stomach and the first part of the duodenum.

Lesser vestibular (ves-TIB-ū-lar) **gland** One of the paired mucus-secreting glands with ducts that open on either side of the urethral orifice in the vestibule of the female.

Leukemia (loo-KĒ-mē-a) A malignant disease of the blood-forming tissues characterized by either uncontrolled production and accumulation of immature leukocytes in which many cells fail to reach maturity (acute) or an accumulation of mature leukocytes in the blood because they do not die at the end of their normal life span (chronic).

Leukocyte (LOO-kō-sīt) A white blood cell.

Leydig (LĪ-dig) **cell** A type of cell that secretes testosterone; located in the connective tissue between seminiferous tubules in a mature testis. Also known as **interstitial cell of Leydig** or **interstitial endocrinocyte.**

Ligament (LIG-a-ment) Dense regular connective tissue that attaches bone to bone.

Ligand (LĪ-gand) A chemical substance that binds to a specific receptor.

Limbic system A part of the forebrain, sometimes termed the visceral brain, concerned with various aspects of emotion and behavior; includes the limbic lobe, dentate gyrus, amygdala, septal nuclei, mammillary bodies, anterior thalamic nucleus, olfactory bulbs, and bundles of myelinated axons.

Lingual frenulum (LIN-gwal FREN-ū-lum) A fold of mucous membrane that connects the tongue to the floor of the mouth.

Lipase An enzyme that splits fatty acids from triglycerides and phospholipids.

Lipid (LIP-id) An organic compound composed of carbon, hydrogen, and oxygen that is usually insoluble in water, but soluble in alcohol, ether, and chloroform; examples include triglycerides (fats and oils), phospholipids, steroids, and eicosanoids.

Lipid bilayer Arrangement of phospholipid, glycolipid, and cholesterol molecules in two parallel sheets in which the hydrophilic "heads" face outward and the hydrophobic "tails" face inward; found in cellular membranes.

Lipoprotein (lip′-ō-PRŌ-tēn) One of several types of particles containing lipids (cholesterol and triglycerides) and proteins that make it water soluble for transport in the blood; high levels of

low-density lipoproteins (LDLs) are associated with increased risk of atherosclerosis, whereas high levels of **high-density lipoproteins (HDLs)** are associated with decreased risk of atherosclerosis.

Liver Large organ under the diaphragm that occupies most of the right hypochondriac region and part of the epigastric region. Functionally, it produces bile and synthesizes most plasma proteins; interconverts nutrients; detoxifies substances; stores glycogen, iron, and vitamins; carries on phagocytosis of worn-out blood cells and bacteria; and helps synthesize the active form of vitamin D.

Long-term potentiation (po-ten′-shē-Ā-shun) **(LTP)** Prolonged, enhanced synaptic transmission that occurs at certain synapses within the hippocampus of the brain; believed to underlie some aspects of memory.

Lordosis (lor-DŌ-sis) An exaggeration of the lumbar curve of the vertebral column. Also called **hollow back.**

Lower limb The appendage attached at the pelvic (hip) girdle, consisting of the thigh, knee, leg, ankle, foot, and toes. Also called the **lower extremity** or **lower appendage.**

Lumbar (LUM-bar) Region of the back and side between the ribs and pelvis; loin.

Lumbar plexus (PLEK-sus) A network formed by the anterior (ventral) branches of spinal nerves L1 through L4.

Lumen (LOO-men) The space within an artery, vein, intestine, renal tubule, or other tubular structure.

Lungs Main organs of respiration that lie on either side of the heart in the thoracic cavity.

Lunula (LOO-noo-la) The moon-shaped white area at the base of a nail.

Luteinizing (LOO-tē-in′-īz-ing) **hormone (LH)** A hormone secreted by the anterior pituitary that stimulates ovulation, stimulates progesterone secretion by the corpus luteum, and readies the mammary glands for milk secretion in females; stimulates testosterone secretion by the testes in males.

Lymph (LIMF) Fluid confined in lymphatic vessels and flowing through the lymphatic system until it is returned to the blood.

Lymph node An oval or bean-shaped structure located along lymphatic vessels.

Lymphatic (lim-FAT-ik) **capillary** Closed-ended microscopic lymphatic vessel that begins in spaces between cells and converges with other lymphatic capillaries to form lymphatic vessels.

Lymphatic system (lim-FAT-ik) A system consisting of a fluid called lymph, vessels called lymphatics that transport lymph, a number of organs containing lymphatic tissue (lymphocytes within a filtering tissue), and red bone marrow.

Lymphatic tissue A specialized form of reticular tissue that contains large numbers of lymphocytes.

Lymphatic vessel A large vessel that collects lymph from lymphatic capillaries and converges with other lymphatic vessels to form the thoracic and right lymphatic ducts.

Lymphocyte (LIM-fō-sīt) A type of white blood cell that helps carry out cell-mediated and antibody-mediated immune responses; found in blood and in lymphatic tissues.

Lysosome (LĪ-sō-sōm) An organelle in the cytoplasm of a cell, enclosed by a single membrane and containing powerful digestive enzymes.

Lysozyme (LĪ-sō-zīm) A bactericidal enzyme found in tears, saliva, and perspiration.

M

Macrophage (MAK-rō-fāj) Phagocytic cell derived from a monocyte; may be fixed or wandering.

Macula (MAK-ū-la) A discolored spot or a colored area. A small, thickened region on the wall of the utricle and saccule that contains receptors for static equilibrium.

Macula lutea (LOO-tē-a) The yellow spot in the center of the retina.

Major histocompatibility (MHC) antigens Surface proteins on white blood cells and other nucleated cells that are unique for each person (except for identical siblings); used to type tissues and help prevent rejection of transplanted tissues. Also known as **human leukocyte antigens (HLA).**

Malignant (ma-LIG-nant) Referring to diseases that tend to become worse and cause death, especially the invasion and spreading of cancer.

Mammary (MAM-ar-ē) **gland** Modified sudoriferous (sweat) gland of the female that produces milk for the nourishment of the young.

Mammillary (MAM-i-ler-ē) **bodies** Two small rounded bodies on the inferior aspect of the hypothalamus that are involved in reflexes related to the sense of smell.

Marrow (MAR-ō) Soft, spongelike material in the cavities of bone. *Red bone marrow* produces blood cells; *yellow bone marrow* contains adipose tissue that stores triglycerides.

Mast cell A cell found in areolar connective tissue that releases histamine, a dilator of small blood vessels, during inflammation.

Mastication (mas′-ti-KĀ-shun) Chewing.

Mature follicle A large, fluid-filled follicle containing a secondary oocyte and surrounding granulosa cells that secrete estrogens. Also called a **graafian** (GRAF-ē-an) **follicle.**

Meatus (mē-Ā-tus) A passage or opening, especially the external portion of a canal.

Mechanoreceptor (me-KAN-ō-rē-sep-tor) Sensory receptor that detects mechanical deformation of the receptor itself or adjacent cells; stimuli so detected include those related to touch, pressure, vibration, proprioception, hearing, equilibrium, and blood pressure.

Medial (MĒ-dē-al) Nearer the midline of the body or a structure.

Medial lemniscus (lem-NIS-kus) A white matter tract that originates in the gracile and cuneate nuclei of the medulla oblongata and extends to the thalamus on the same side; sensory axons in this tract conduct nerve impulses for the sensations of proprioception, fine touch, vibration, hearing, and equilibrium.

Median aperture (AP-er-choor) One of the three openings in the roof of the fourth ventricle through which cerebrospinal fluid enters the subarachnoid space of the brain and cord. Also called the **foramen of Magendie.**

Median plane A vertical plane dividing the body into right and left halves. Situated in the middle.

Mediastinum (mē′-dē-as-TĪ-num) The anatomical region on the thoracic cavity between the pleurae of the lungs that extends from the sternum to the vertebral column and from the first rib to the diaphragm.

Medulla (me-DOOL-la) An inner layer of an organ, such as the medulla of the kidneys.

Medulla oblongata (me-DOOL-la ob′-long-GA-ta) The most inferior part of the brain stem. Also termed the **medulla.**

Medullary (MED-ū-lar′-ē) **cavity** The space within the diaphysis of a bone that contains yellow bone marrow. Also called the **marrow cavity.**

Medullary rhythmicity (rith-MIS-i-tē) **area** The neurons of the respiratory center in the medulla oblongata that control the basic rhythm of respiration.

Meibomian gland *See* **Tarsal gland.**

Meiosis (mī-Ō-sis) A type of cell division that occurs during production of gametes, involving two successive nuclear divisions that result in cells with the haploid *(n)* number of chromosomes.

Meissner (MĪS-ner) **corpuscle** A sensory receptor for touch; found in dermal papillae, especially in the palms and soles. Also called a **corpuscle of touch.**

Melanin (MEL-a-nin) A dark black, brown, or yellow pigment found in some parts of the body such as the skin, hair, and pigmented layer of the retina.

Melanocyte (MEL-a-nō-sīt′) A pigmented cell, located between or beneath cells of the deepest layer of the epidermis, that synthesizes melanin.

Melanocyte-stimulating hormone (MSH) A hormone secreted by the anterior pituitary that stimulates the dispersion of melanin granules in melanocytes in amphibians; continued administration produces darkening of skin in humans.

Melatonin (mel-a-TŌN-in) A hormone secreted by the pineal gland that helps set the timing of the body's biological clock.

Membrane A thin, flexible sheet of tissue composed of an epithelial layer and an underlying connective tissue layer, as in an epithelial membrane, or of areolar connective tissue only, as in a synovial membrane.

Membranous labyrinth (mem-BRA-nus LAB-i-rinth) The part of the labyrinth of the internal ear that is located inside the bony labyrinth and separated from it by the perilymph; made up of the semicircular ducts, the saccule and utricle, and the cochlear duct.

Memory The ability to recall thoughts; commonly classified as short-term (activated) and long-term.

Menarche (me-NAR-kē) The first menses (menstrual flow) and beginning of ovarian and uterine cycles.

Meninges (me-NIN-jēz) Three membranes covering the brain and spinal cord, called the dura mater,

arachnoid mater, and pia mater. *Singular* is **meninx** (MEN-inks).

Menopause (MEN-ō-pawz) The termination of the menstrual cycles.

Menstruation (men′-stroo-Ā-shun) Periodic discharge of blood, tissue fluid, mucus, and epithelial cells that usually lasts for 5 days; caused by a sudden reduction in estrogens and progesterone. Also called the **menstrual phase** or **menses.**

Merkel (MER-kel) **cell** Type of cell in the epidermis of hairless skin that makes contact with a Merkel (tactile) disc, which functions in touch.

Merkel disc Saucer-shaped free nerve endings that make contact with Merkel cells in the epidermis and function as touch receptors. Also called a **tactile disc.**

Merocrine (MER-ō-krin) **gland** Gland made up of secretory cells that remain intact throughout the process of formation and discharge of the secretory product, as in the salivary and pancreatic glands.

Mesenchyme (MEZ-en-kīm) An embryonic connective tissue from which all other connective tissues arise.

Mesentery (MEZ-en-ter′-ē) A fold of peritoneum attaching the small intestine to the posterior abdominal wall.

Mesocolon (mez′-ō-KŌ-lon) A fold of peritoneum attaching the colon to the posterior abdominal wall.

Mesoderm The middle primary germ layer that gives rise to connective tissues, blood and blood vessels, and muscles.

Mesothelium (mez′-ō-THĒ-lē-um) The layer of simple squamous epithelium that lines serous membranes.

Mesovarium (mez′-ō-VAR-ē-um) A short fold of peritoneum that attaches an ovary to the broad ligament of the uterus.

Metabolism (me-TAB-ō-lizm) All the biochemical reactions that occur within an organism, including the synthetic (anabolic) reactions and decomposition (catabolic) reactions.

Metacarpus (met′-a-KAR-pus) A collective term for the five bones that make up the palm.

Metaphase (MET-a-fāz) The second stage of mitosis, in which chromatid pairs line up on the metaphase plate of the cell.

Metaphysis (me-TAF-i-sis) Region of a long bone between the diaphysis and epiphysis that contains the epiphyseal plate in a growing bone.

Metarteriole (met′-ar-TE-rē-ōl) A blood vessel that emerges from an arteriole, traverses a capillary network, and empties into a venule.

Metastasis (me-TAS-ta-sis) The spread of cancer to surrounding tissues (local) or to other body sites (distant).

Metatarsus (met′-a-TAR-sus) A collective term for the five bones located in the foot between the tarsals and the phalanges.

Microfilament (mī-krō-FIL-a-ment) Rodlike protein filament about 6 nm in diameter; constitutes contractile units in muscle fibers (cells) and provides support, shape, and movement in nonmuscle cells.

Microglia (mī-KROG-lē-a) Neuroglial cells that carry on phagocytosis.

Microtubule (mī-krō-TOO-būl) Cylindrical protein filament, from 18 to 30 nm in diameter, consisting of the protein tubulin; provides support, structure, and transportation.

Microvilli (mī-krō-VIL-ī) Microscopic, fingerlike projections of the plasma membranes of cells that increase surface area for absorption, especially in the small intestine and proximal convoluted tubules of the kidneys.

Micturition (mik′-choo-RISH-un) The act of expelling urine from the urinary bladder. Also called **urination** (ū-ri-NĀ-shun).

Midbrain The part of the brain between the pons and the diencephalon. Also called the **mesencephalon** (mes′-en-SEF-a-lon).

Middle ear A small, epithelial-lined cavity hollowed out of the temporal bone, separated from the external ear by the eardrum and from the internal ear by a thin bony partition containing the oval and round windows; extending across the middle ear are the three auditory ossicles. Also called the **tympanic** (tim-PAN-ik) **cavity.**

Midline An imaginary vertical line that divides the body into equal left and right sides.

Midsagittal plane A vertical plane through the midline of the body that divides the body or organs into *equal* right and left sides. Also called a **median plane.**

Mineralocorticoids (min′-er-al-ō-KOR-ti-koyds) A group of hormones of the adrenal cortex that help regulate sodium and potassium balance.

Mitochondrion (mī-tō-KON-drē-on) A double-membraned organelle that plays a central role in the production of ATP; known as the "powerhouse" of the cell. *Plural* is **mitochondria.**

Mitosis (mī-TŌ-sis) The orderly division of the nucleus of a cell that ensures that each new nucleus has the same number and kind of chromosomes as the original nucleus. The process includes the replication of chromosomes and the distribution of the two sets of chromosomes into two separate and equal nuclei.

Mitotic spindle Collective term for a football-shaped assembly of microtubules (nonkinetochore, kinetochore, and aster) that is responsible for the movement of chromosomes during cell division.

Modality (mō-DAL-i-tē) Any of the specific sensory entities, such as vision, smell, taste, or touch.

Modiolus (mō-DĪ-ō′-lus) The central pillar or column of the cochlea.

Molecule (mol′-e-KŪL) A substance composed of two or more atoms chemically combined.

Monocyte (MON-ō-sīt′) The largest type of white blood cell, characterized by agranular cytoplasm.

Monounsaturated fat A fatty acid that contains one double covalent bond between its carbon atoms; it is not completely saturated with hydrogen atoms. Plentiful in triglycerides of olive and peanut oils.

Mons pubis (MONZ PŪ-bis) The rounded, fatty prominence over the pubic symphysis, covered by coarse pubic hair.

Morula (MOR-ū-la) A solid sphere of cells produced by successive cleavages of a fertilized ovum about 4 days after fertilization.

Motor area The region of the cerebral cortex that governs muscular movement, particularly the precentral gyrus of the frontal lobe.

Motor end plate Region of the sarcolemma of a muscle fiber (cell) that includes acetylcholine (ACh) receptors, which bind ACh released by synaptic end bulbs of somatic motor neurons.

Motor neurons (NOO-ronz) Neurons that conduct impulses from the brain toward the spinal cord or out of the brain and spinal cord into cranial or spinal nerves to effectors that may be either muscles or glands. Also called **efferent neurons.**

Motor unit A motor neuron together with the muscle fibers (cells) it stimulates.

Mucosa-associated lymphatic tissue (MALT) Lymphatic nodules scattered throughout the lamina propria (connective tissue) of mucous membranes lining the gastrointestinal tract, respiratory airways, urinary tract, and reproductive tract.

Mucous (MŪ-kus) **cell** A unicellular gland that secretes mucus. Two types are mucous neck cells and surface mucous cells in the stomach.

Mucous membrane A membrane that lines a body cavity that opens to the exterior. Also called the **mucosa** (mū-KŌ-sa).

Mucus The thick fluid secretion of goblet cells, mucous cells, mucous glands, and mucous membranes.

Muscarinic (mus´-ka-RIN-ik) **receptor** Receptor for the neurotransmitter acetylcholine found on all effectors innervated by parasympathetic postganglionic axons and on sweat glands innervated by cholinergic sympathetic postganglionic axons; so named because muscarine activates these receptors but does not activate nicotinic receptors for acetylcholine.

Muscle An organ composed of one of three types of muscle tissue (skeletal, cardiac, or smooth), specialized for contraction to produce voluntary or involuntary movement of parts of the body.

Muscle action potential A stimulating impulse that propagates along the sarcolemma and transverse tubules; in skeletal muscle, it is generated by acetylcholine, which increases the permeability of the sarcolemma to cations, especially sodium ions (Na$^+$).

Muscle fatigue (fa-TĒG) Inability of a muscle to maintain its strength of contraction or tension; may be related to insufficient oxygen, depletion of glycogen, and/or lactic acid buildup.

Muscle spindle An encapsulated proprioceptor in a skeletal muscle, consisting of specialized intrafusal muscle fibers and nerve endings; stimulated by changes in length or tension of muscle fibers.

Muscle strain Tearing of skeletal muscle fibers or tendon. Also called a **muscle pull** or **muscle tear.**

Muscle tone A sustained, partial contraction of portions of a skeletal or smooth muscle in response to activation of stretch receptors or a baseline level of action potentials in the innervating motor neurons.

Muscular dystrophies (DIS-trō-fēz) Inherited muscle-destroying diseases, characterized by degeneration of muscle fibers (cells), which causes progressive atrophy of the skeletal muscle.

Muscularis (MUS-kū-la´-ris) A muscular layer (coat or tunic) of an organ.

Muscularis mucosae (mū-KŌ-sē) A thin layer of smooth muscle fibers that underlie the lamina propria of the mucosa of the gastrointestinal tract.

Muscular system Usually refers to the approximately 100 voluntary muscles of the body that are composed of skeletal muscle tissue.

Muscular system Usually refers to the voluntary muscles of the body that are composed of skeletal muscle tissue.

Muscular tissue A tissue specialized to produce motion in response to muscle action potentials by its qualities of contractility, extensibility, elasticity, and excitability; types include skeletal, cardiac, and smooth.

Musculoskeletal (mus´-kyū-lō-SKEL-e-tal) **system** An integrated body system consisting of bones, joints, and muscles.

Mutation (mū-TĀ-shun) Any change in the sequence of bases in a DNA molecule resulting in a permanent alteration in some inheritable trait.

Myasthenia (mī-as-THĒ-nē-a) **gravis** Weakness and fatigue of skeletal muscles caused by antibodies directed against acetylcholine receptors.

Myelin (MĪ-e-lin) **sheath** Multilayered lipid and protein covering, formed by Schwann cells and oligodendrocytes, around axons of many peripheral and central nervous system neurons.

Myenteric (mī-en-TER-ik) **plexus** A network of autonomic axons and postganglionic cell bodies located in the muscularis of the gastrointestinal tract. Also called the **plexus of Auerbach** (OW-er-bak).

Myocardial infarction (mī´-ō-KAR-dē-al in-FARK-shun) **(MI)** Gross necrosis of myocardial tissue due to interrupted blood supply. Also called a **heart attack.**

Myocardium (mī´-ō-KAR-dē-um) The middle layer of the heart wall, made up of cardiac muscle tissue, lying between the epicardium and the endocardium and constituting the bulk of the heart.

Myofibril (mī-ō-FĪ-bril) A threadlike structure, extending longitudinally through a muscle fiber (cell), consisting mainly of thick filaments (myosin) and thin filaments (actin, troponin, and tropomyosin).

Myoglobin (mī-ō-GLŌB-in) The oxygen-binding, iron-containing protein present in the sarcoplasm of muscle fibers (cells); contributes the red color to muscle.

Myogram (MĪ-ō-gram) The record or tracing produced by a myograph, an apparatus that measures and records the force of muscular contractions.

Myology (mī-OL-ō-jē) The study of muscles.

Myometrium (mī´-ō-MĒ-trē-um) The smooth muscle layer of the uterus.

Myopathy (mī-OP-a-thē) Any abnormal condition or disease of muscle tissue.

Myopia (mī-Ō-pē-a) Defect in vision in which objects can be seen distinctly only when very close to the eyes; nearsightedness.

Myosin (MĪ-ō-sin) The contractile protein that makes up the thick filaments of muscle fibers.

Myotome (MĪ-ō-tōm) A group of muscles innervated by the motor neurons of a single spinal segment. In an embryo, the portion of a somite that develops into some skeletal muscles.

N

Nail A hard plate, composed largely of keratin, that develops from the epidermis of the skin to form a protective covering on the dorsal surface of the distal phalanges of the fingers and toes.

Nail matrix (MĀ-triks) The part of the nail beneath the body and root from which the nail is produced.

Nasal (NĀ-zal) **cavity** A mucosa-lined cavity on either side of the nasal septum that opens onto the face at the external nares and into the nasopharynx at the internal nares.

Nasal septum (SEP-tum) A vertical partition composed of bone (perpendicular plate of ethmoid and vomer) and cartilage, covered with a mucous membrane, separating the nasal cavity into left and right sides.

Nasolacrimal (nā´-zō-LAK-ri-mal) **duct** A canal that transports the lacrimal secretion (tears) from the nasolacrimal sac into the nose.

Nasopharynx (nā´-zō-FAR-inks) The superior portion of the pharynx, lying posterior to the nose and extending inferiorly to the soft palate.

Neck The part of the body connecting the head and the trunk. A constricted portion of an organ, such as the neck of the femur or uterus.

Necrosis (ne-KRŌ-sis) A pathological type of cell death that results from disease, injury, or lack of blood supply in which many adjacent cells swell, burst, and spill their contents into the interstitial fluid, triggering an inflammatory response.

Negative feedback system A feedback cycle that reverses a change in a controlled condition.

Neonatal (nē-ō-NĀ-tal) Pertaining to the first 4 weeks after birth.

Neoplasm (NĒ-ō-plazm) A new growth that may be benign or malignant.

Nephron (NEF-ron) The functional unit of the kidney.

Nerve A cordlike bundle of neuronal axons and/or dendrites and associated connective tissue coursing together outside the central nervous system.

Nerve fiber General term for any process (axon or dendrite) projecting from the cell body of a neuron.

Nerve impulse A wave of depolarization and repolarization that self-propagates along the plasma membrane of a neuron; also called a **nerve action potential.**

Nervous system A network of billions of neurons and even more neuroglia that is organized into two main divisions. Central nervous system (brain and spinal cord) and peripheral nervous system (nerves, ganglia, enteric plexuses, and sensory receptors outside the central nervous system).

Nervous tissue Tissue containing neurons that initiate and conduct nerve impulses to coordinate homeostasis, and neuroglia that provide support and nourishment to neurons.

Neuralgia (noo-RAL-jē-a) Attacks of pain along the entire course or branch of a peripheral sensory nerve.

Neural plate A thickening of ectoderm, induced by the notochord, that forms early in the third week of development and represents the beginning of the development of the nervous system.

Neural tube defect (NTD) A developmental abnormality in which the neural tube does not close properly. Examples are spina bifida and anencephaly.

Neuritis (noo-RĪ-tis) Inflammation of one or more nerves.

Neurofibral node *See* **Node of Ranvier.**

Neuroglia (noo-RŌG-lē-a) Cells of the nervous system that perform various supportive functions. The neuroglia of the central nervous system are the astrocytes, oligodendrocytes, microglia, and ependymal cells; neuroglia of the peripheral nervous system include Schwann cells and satellite cells. Also called **glial** (GLĒ-al) **cells.**

Neurohypophyseal (noo´-rō-hī´-pō-FIZ-ē-al) **bud** An outgrowth of ectoderm located on the floor of the hypothalamus that gives rise to the posterior pituitary.

Neurolemma (noo-rō-LEM-ma) The peripheral, nucleated cytoplasmic layer of the Schwann cell. Also called **sheath of Schwann** (SCHWON).

Neurology (noo-ROL-ō-jē) The study of the normal functioning and disorders of the nervous system.

Neuromuscular (noo-rō-MUS-kū-lar) **junction (NMJ)** A synapse between the axon terminals of a motor neuron and the sarcolemma of a muscle fiber (cell).

Neuron (NOO-ron) A nerve cell, consisting of a cell body, dendrites, and an axon.

Neurophysiology (NOOR-ō-fiz-ē-ol´-ō-jē) Study of the functional properties of nerves.

Neurosecretory (noo-rō-SĒK-re-tō-rē) **cell** A neuron that secretes a hypothalamic releasing hormone or inhibiting hormone into blood capillaries of the hypothalmus; a neuron that secretes oxytocin or antidiuretic hormone into blood capillaries of the posterior pituitary.

Neurotransmitter (noo´-rō-trans´-MIT-er) One of a variety of molecules within axon terminals that are released into the synaptic cleft in response to a nerve impulse and that change the membrane potential of the postsynaptic neuron.

Neurulation (noor-oo-LĀ-shun) The process by which the neural plate, neural folds, and neural tube develop.

Neutrophil (NOO-trō-fil) A type of white blood cell characterized by granules that stain pale lilac with a combination of acidic and basic dyes.

Nicotinic (nik´-ō-TIN-ik) **receptor** Receptor for the neurotransmitter acetylcholine found on both sympathetic and parasympathetic postganglionic neurons and on skeletal muscle in the motor end plate; so named because nicotine activates these receptors but does not activate muscarinic receptors for acetylcholine.

Nipple A pigmented, wrinkled projection on the surface of the breast that is the location of the openings of the lactiferous ducts for milk release.

Nociceptor (nō´-sē-SEP-tor) A free (naked) nerve ending that detects painful stimuli.

Node of Ranvier (RON-vē-ā) A space along a myelinated axon between the individual Schwann cells that form the myelin sheath and the neurolemma. Also called a **neurofibral node.**

Norepinephrine (nor´-ep-ē-NEF-rin) **(NE)** A hormone secreted by the adrenal medulla that produces actions similar to those that result from sympathetic stimulation. Also called **noradrenaline** (nor-a-DREN-a-lin).

Notochord (NŌ-tō-cord) A flexible rod of mesodermal tissue that lies where the future vertebral column will develop and plays a role in induction.

Nucleic (noo-KLĒ-ik) **acid** An organic compound that is a long polymer of nucleotides, with each nucleotide containing a pentose sugar, a phosphate group, and one of four possible nitrogenous bases (adenine, cytosine, guanine, and thymine or uracil).

Nucleolus (noo´-KLĒ-ō-lus) Spherical body within a cell nucleus composed of protein, DNA, and RNA that is the site of the assembly of small and large ribosomal subunits. *Plural* is **nucleoli.**

Nucleosome (NOO-klē-ō-sōm) Structural subunit of a chromosome consisting of histones and DNA.

Nucleus (NOO-klē-us) A spherical or oval organelle of a cell that contains the hereditary factors of the cell, called genes. A cluster of unmyelinated nerve cell bodies in the central nervous system. The central part of an atom made up of protons and neutrons.

Nucleus pulposus (pul-PŌ-sus) A soft, pulpy, highly elastic substance in the center of an intervertebral disc; a remnant of the notochord.

Nutrient A chemical substance in food that provides energy, forms new body components, or assists in various body functions.

O

Obesity (ō-BĒS-i-tē) Body weight more than 20% above a desirable standard due to excessive accumulation of fat.

Oblique (ō-BLĒK) **plane** A plane that passes through the body or an organ at an angle between the transverse plane and either the midsagittal, parasagittal, or frontal plane.

Obstetrics (ob-STET-riks) The specialized branch of medicine that deals with pregnancy, labor, and the period of time immediately after delivery (about 6 weeks).

Olfaction (ōl-FAK-shun) The sense of smell.

Olfactory (ōl-FAK-tō-rē) Pertaining to smell.

Olfactory bulb A mass of gray matter containing cell bodies of neurons that form synapses with neurons of the olfactory (I) nerve, lying inferior to the frontal lobe of the cerebrum on either side of the crista galli of the ethmoid bone.

Olfactory receptor A bipolar neuron with its cell body lying between supporting cells located in the mucous membrane lining the superior portion of each nasal cavity; transduces odors into neural signals.

Olfactory tract A bundle of axons that extends from the olfactory bulb posteriorly to olfactory regions of the cerebral cortex.

Oligodendrocyte (OL-i-gō-den´-drō-sīt) A neuroglial cell that supports neurons and produces a myelin sheath around axons of neurons of the central nervous system.

Oliguria (ol´-i-GŪ-rē-a) Daily urinary output usually less than 250 mL.

Olive A prominent oval mass on each lateral surface of the superior part of the medulla oblongata.

Oncogene (ON-kō-jēn) Cancer-causing gene; it derives from a normal gene, termed a proto-oncogene, that encodes proteins involved in cell growth or cell regulation but has the ability to transform a normal cell into a cancerous cell when it is mutated or inappropriately activated. One example is *p53.*

Oncology (on-KOL-ō-jē) The study of tumors.

Oogenesis (ō-ō-JEN-e-sis) Formation and development of female gametes (oocytes).

Oophorectomy (ō´-of-ō-REK-tō-me) Surgical removal of the ovaries.

Ophthalmic (of-THAL-mik) Pertaining to the eye.

Ophthalmologist (of´-thal-MOL-ō-jist) A physician who specializes in the diagnosis and treatment of eye disorders using drugs, surgery, and corrective lenses.

Ophthalmology (of-thal-MOL-ō-jē) The study of the structure, function, and diseases of the eye.

Optic (OP-tik) Refers to the eye, vision, or properties of light.

Optic chiasm (kī-AZM) A crossing point of the two branches of the optic (II) nerve, anterior to the pituitary gland. Also called **optic chiasma.**

Optic disc A small area of the retina containing openings through which the axons of the ganglion cells emerge as the optic (II) nerve. Also called the **blind spot.**

Optic tract A bundle of axons that carry nerve impulses from the retina of the eye between the optic chiasm and the thalamus.

Ora serrata (Ō-ra ser-RĀ-ta) The irregular margin of the retina lying internal and slightly posterior to the junction of the choroid and ciliary body.

Orbit (OR-bit) The bony, pyramidal-shaped cavity of the skull that holds the eyeball.

Organ A structure composed of two or more different kinds of tissues with a specific function and usually a recognizable shape.

Organelle (or-ga-NEL) A permanent structure within a cell with characteristic morphology that is specialized to serve a specific function in cellular activities.

Organism (OR-ga-nizm) A total living form; one individual.

Organogenesis (or´-ga-nō-JEN-e-sis) The formation of body organs and systems. By the end of the eighth week of development, all major body systems have begun to develop.

Orifice (OR-i-fis) Any aperture or opening.

Origin (OR-i-jin) The attachment of a muscle tendon to a stationary bone or the end opposite the insertion.

Oropharynx (or´-ō-FAR-inks) The intermediate portion of the pharynx, lying posterior to the mouth and extending from the soft palate to the hyoid bone.

Orthopedics (or′-thō-PĒ-diks) The branch of medicine that deals with the preservation and restoration of the skeletal system, articulations, and associated structures.

Osmoreceptor (oz′-mō-rē-CEP-tor) Receptor in the hypothalamus that is sensitive to changes in blood osmolarity and, in response to high osmolarity (low water concentration), stimulates synthesis and release of antidiuretic hormone (ADH).

Osmosis (oz-MŌ-sis) The net movement of water molecules through a selectively permeable membrane from an area of higher water concentration to an area of lower water concentration until equilibrium is reached.

Osseous (OS-ē-us) Bony.

Ossicle (OS-si-kul) One of the small bones of the middle ear (malleus, incus, stapes).

Ossification (os′-i-fi-KĀ-shun) Formation of bone. Also called **osteogenesis.**

Ossification (os′-i-fi-KĀ-shun) **center** An area in the cartilage model of a future bone where the cartilage cells hypertrophy, secrete enzymes that calcify their extracellular matrix, and die, and the area they occupied is invaded by osteoblasts that then lay down bone.

Osteoblast (OS-tē-ō-blast′) Cell formed from an osteogenic cell that participates in bone formation by secreting some organic components and inorganic salts.

Osteoclast (OS-tē-ō-klast′) A large, multinuclear cell that resorbs (destroys) bone matrix.

Osteocyte (OS-tē-ō-sīt′) A mature bone cell that maintains the daily activities of bone tissue.

Osteogenic (os′-tē-ō-JEN-ik) **cell** Stem cell derived from mesenchyme that has mitotic potential and the ability to differentiate into an osteoblast.

Osteogenic layer The inner layer of the periosteum that contains cells responsible for forming new bone during growth and repair.

Osteology (os-tē-OL-ō-jē) The study of bones.

Osteon (OS-tē-on) The basic unit of structure in adult compact bone, consisting of a central (haversian) canal with its concentrically arranged lamellae, lacunae, osteocytes, and canaliculi. Also called a **haversian** (ha-VER-shun) **system.**

Osteoporosis (os′-tē-ō-pō-RŌ-sis) Age-related disorder characterized by decreased bone mass and increased susceptibility to fractures, often as a result of decreased levels of estrogens.

Otic (Ō-tik) Pertaining to the ear.

Otolith (Ō-tō-lith) A particle of calcium carbonate embedded in the otolithic membrane that functions in maintaining static equilibrium.

Otolithic (ō-tō-LITH-ik) **membrane** Thick, gelatinous, glycoprotein layer located directly over hair cells of the macula in the saccule and utricle of the internal ear.

Otorhinolaryngology (ō-tō-rī′-nō-lar-in-GOL-ō-jē) The branch of medicine that deals with the diagnosis and treatment of diseases of the ears, nose, and throat.

Oval window A small, membrane-covered opening between the middle ear and inner ear into which the footplate of the stapes fits.

Ovarian (ō-VAR-ē-an) **cycle** A monthly series of events in the ovary associated with the maturation of a secondary oocyte.

Ovarian follicle (FOL-i-kul) A general name for oocytes (immature ova) in any stage of development, along with their surrounding epithelial cells.

Ovarian ligament (LIG-a-ment) A rounded cord of connective tissue that attaches the ovary to the uterus.

Ovary (Ō-var-ē) Female gonad that produces oocytes and the estrogen, progesterone, inhibin, and relaxin hormones.

Ovulation (ov′-ū-LĀ-shun) The rupture of a mature ovarian (graafian) follicle with discharge of a secondary oocyte into the pelvic cavity.

Ovum (Ō-vum) The female reproductive or germ cell; an egg cell; arises through completion of meiosis in a secondary oocyte after penetration by a sperm.

Oxyhemoglobin (ok′-sē-HĒ-mō-glō-bin) **(Hb—O₂)** Hemoglobin combined with oxygen.

Oxytocin (ok′-sē-TŌ-sin) **(OT)** A hormone secreted by neurosecretory cells in the paraventricular and supraoptic nuclei of the hypothalamus that stimulates contraction of smooth muscle in the pregnant uterus and myoepithelial cells around the ducts of mammary glands.

P

P wave The deflection wave of an electrocardiogram that signifies atrial depolarization.

Pacinian corpuscle (pa-SIN-ē-an) Oval-shaped pressure receptor located in the dermis or subcutaneous tissue and consisting of concentric layers of a connective tissue wrapped around the dendrites of a sensory neuron. Also called a **lamellated corpuscle.**

Palate (PAL-at) The horizontal structure separating the oral and the nasal cavities; the roof of the mouth.

Palpate (PAL-pāt) To examine by touch; to feel.

Pancreas (PAN-krē-as) A soft, oblong organ lying along the greater curvature of the stomach and connected by a duct to the duodenum. It is both an exocrine gland (secreting pancreatic juice) and an endocrine gland (secreting insulin, glucagon, somatostatin, and pancreatic polypeptide).

Pancreatic (pan′-krē-AT-ik) **duct** A single large tube that unites with the common bile duct from the liver and gallbladder and drains pancreatic juice into the duodenum at the hepatopancreatic ampulla (ampulla of Vater). Also called the **duct of Wirsung** (VĒR-sung).

Pancreatic islet (Ī-let) Cluster of endocrine gland cells in the pancreas that secretes insulin, glucagon, somatostatin, and pancreatic polypeptide. Also called an **islet of Langerhans** (LAHNG-er-hanz).

Papanicolaou (pa-pa-NI-kō-lō) **test** A cytological staining test for the detection and diagnosis of premalignant and malignant conditions of the female genital tract. Cells scraped from the epithelium of the cervix of the uterus are examined microscopically. Also called a **Pap test** or **Pap smear.**

Papilla (pa-PIL-a) A small nipple-shaped projection or elevation.

Paralysis (pa-RAL-a-sis) Loss or impairment of motor function due to a lesion of nervous or muscular origin.

Paranasal sinus (par′-a-NĀ-zal SĪ-nus) A mucuslined air cavity in a skull bone that communicates with the nasal cavity. Paranasal sinuses are located in the frontal, maxillary, ethmoid, and sphenoid bones.

Paraplegia (par-a-PLĒ-jē-a) Paralysis of both lower limbs.

Parasagittal plane (par-a-SAJ-i-tal) A vertical plane that does not pass through the midline and that divides the body or organs into *unequal* left and right portions.

Parasympathetic (par′-a-sim-pa-THET-ik) **division** One of the two subdivisions of the autonomic nervous system, having cell bodies of preganglionic neurons in nuclei in the brain stem and in the lateral gray horn of the sacral portion of the spinal cord; primarily concerned with activities that conserve and restore body energy.

Parathyroid (par′-a-THĪ-royd) **gland** One of usually four small endocrine glands embedded in the posterior surfaces of the lateral lobes of the thyroid gland.

Parathyroid hormone (PTH) A hormone secreted by the chief (principal) cells of the parathyroid glands that increases blood calcium level and decreases blood phosphate level. Also called **parathormone.**

Paraurethral (par′-a-ū-RĒ-thral) **gland** Gland embedded in the wall of the urethra whose duct opens on either side of the urethral orifice and secretes mucus. Also called **Skene's** (SKĒNZ) **gland.**

Parenchyma (pa-RENG-ki-ma) The functional parts of any organ, as opposed to tissue that forms its stroma or framework.

Parietal (pa-RĪ-e-tal) Pertaining to or forming the outer wall of a body cavity.

Parietal cell A type of secretory cell in gastric glands that produces hydrochloric acid and intrinsic factor. Also called an **oxyntic cell.**

Parietal pleura (PLOO-ra) The outer layer of the serous pleural membrane that encloses and protects the lungs; the layer that is attached to the wall of the pleural cavity.

Parkinson disease (PD) Progressive degeneration of the basal nuclei and substantia nigra of the cerebrum resulting in decreased production of dopamine (DA) that leads to tremor, slowing of voluntary movements, and muscle weakness.

Parotid (pa-ROT-id) **gland** One of the paired salivary glands located inferior and anterior to the ears and connected to the oral cavity via a duct (Stensen's) that opens into the inside of the cheek opposite the maxillary (upper) second molar tooth.

Pars intermedia A small avascular zone between the anterior and posterior pituitary glands.

Parturition (par-toor-ISH-un) Act of giving birth to young; childbirth, delivery.

Patent (PĀ-tent) **ductus arteriosus (PDA)** A congenital heart defect in which the ductus arteriosus remains open. As a result, aortic blood flows into the lower-pressure pulmonary trunk,

increasing pulmonary trunk pressure and over-working both ventricles.

Pathogen (PATH-ō-jen) A disease-producing microbe.

Pathological (path′-ō-LOJ-i-kal) **anatomy** The study of structural changes caused by disease.

Pathophysiology (PATH-ō-fez-ē-ol-ō-jē) Study of functional changes associated with disease and aging.

Pectinate (PEK-ti-nāt) **muscles** Projecting muscle bundles of the anterior atrial walls and the lining of the auricles.

Pectoral (PEK-tō-ral) Pertaining to the chest or breast.

Pedicel (PED-i-sel) Footlike structure, as on podocytes of a glomerulus.

Pelvic (PEL-vik) **cavity** Inferior portion of the abdominopelvic cavity that contains the urinary bladder, sigmoid colon, rectum, and internal female and male reproductive structures.

Pelvic splanchnic (PEL-vik SPLANGK-nik) **nerves** Consist of preganglionic parasympathetic axons from the levels of S2, S3, and S4 that supply the urinary bladder, reproductive organs, and the descending and sigmoid colon and rectum.

Pelvis The basinlike structure formed by the two hip bones, the sacrum, and the coccyx. The expanded, proximal portion of the ureter, lying within the kidney and into which the major calyces open.

Penis (PĒ-nis) The organ of urination and copulation in males; used to deposit semen into the female vagina.

Pepsin Protein-digesting enzyme secreted by chief cells of the stomach in the inactive form pepsinogen, which is converted to active pepsin by hydrochloric acid.

Peptic ulcer An ulcer that develops in areas of the gastrointestinal tract exposed to hydrochloric acid; classified as a gastric ulcer if in the lesser curvature of the stomach and as a duodenal ulcer if in the first part of the duodenum.

Percussion (pur-KUSH-un) The act of striking (percussing) an underlying part of the body with short, sharp taps as an aid in diagnosing the part by the quality of the sound produced.

Perforating canal A minute passageway by means of which blood vessels and nerves from the periosteum penetrate into compact bone. Also called **Volkmann's** (FŌLK-mans) **canal.**

Pericardial (per′-i-KAR-dē-al) **cavity** Small potential space between the visceral and parietal layers of the serous pericardium that contains pericardial fluid.

Pericardium (per′-i-KAR-dē-um) A loose-fitting membrane that encloses the heart, consisting of a superficial fibrous layer and a deep serous layer.

Perichondrium (per′-i-KON-drē-um) A covering of dense irregular connective tissue that surrounds the surface of most cartilage.

Perilymph (PER-i-limf) The fluid contained between the bony and membranous labyrinths of the inner ear.

Perimetrium (per′-i-MĒ-trē-um) The serosa of the uterus.

Perimysium (per-i-MĪZ-ē-um) Invagination of the epimysium that divides muscles into bundles.

Perineum (per′-i-NĒ-um) The pelvic floor; the space between the anus and the scrotum in the male and between the anus and the vulva in the female.

Perineurium (per′-i-NOO-rē-um) Connective tissue wrapping around fascicles in a nerve.

Periodontal (per′-ē-ō-DON-tal) **disease** A collective term for conditions characterized by degeneration of gingivae, alveolar bone, periodontal ligament, and cementum.

Periodontal ligament The periosteum lining the alveoli (sockets) for the teeth in the alveolar processes of the mandible and maxillae.

Periosteum (per′-ē-OS-tē-um) The covering of a bone that consists of connective tissue, osteogenic cells, and osteoblasts; is essential for bone growth, repair, and nutrition.

Peripheral (pe-RIF-er-al) Located on the outer part or a surface of the body.

Peripheral nervous system (PNS) The part of the nervous system that lies outside the central nervous system, consisting of nerves and ganglia.

Peristalsis (per′-i-STAL-sis) Successive muscular contractions along the wall of a hollow muscular structure.

Peritoneum (per′-i-tō-NĒ-um) The largest serous membrane of the body that lines the abdominal cavity and covers the viscera within it.

Peritonitis (per′-i-tō-NĪ-tis) Inflammation of the peritoneum.

Peroxisome (pe-ROKS-i-sōm) Organelle similar in structure to a lysosome that contains enzymes that use molecular oxygen to oxidize various organic compounds; such reactions produce hydrogen peroxide; abundant in liver cells.

Perspiration Sweat; produced by sudoriferous (sweat) glands and containing water, salts, urea, uric acid, amino acids, ammonia, sugar, lactic acid, and ascorbic acid. Helps maintain body temperature and eliminate wastes.

Peyer's patches See **Aggregated lymphatic follicles.**

pH A measure of the concentration of hydrogen ions (H^+) in a solution. The **pH scale** extends from 0 to 14, with a value of 7 expressing neutrality, values lower than 7 expressing increasing acidity, and values higher than 7 expressing increasing alkalinity.

Phagocytosis (fag′-ō-sī-TŌ-sis) The process by which phagocytes ingest and destroy microbes, cell debris, and other foreign matter.

Phalanx (FĀ-lanks) The bone of a finger or toe. *Plural* is **phalanges** (fa-LAN-jēz).

Pharmacology (far′-ma-KOL-ō-jē) The science of the effects and uses of drugs in the treatment of disease.

Pharynx (FAR-inks) The throat; a tube that starts at the internal nares and runs partway down the neck, where it opens into the esophagus posteriorly and the larynx anteriorly.

Phenotype (FĒ-nō-tīp) The observable expression of genotype; physical characteristics of an organism determined by genetic makeup and influenced by interaction between genes and internal and external environmental factors.

Phlebitis (fle-BĪ-tis) Inflammation of a vein, usually in a lower limb.

Photopigment A substance that can absorb light and undergo structural changes that can lead to the development of a receptor potential. An example is rhodopsin. In the eye, also called **visual pigment.**

Photoreceptor Receptor that detects light shining on the retina of the eye.

Physiology (fiz′-ē-OL-ō-jē) Science that deals with the functions of an organism or its parts.

Pia mater (PĪ-a MĀ-ter *or* PĒ-a MA-ter) The innermost of the three meninges (coverings) of the brain and spinal cord.

Pineal (PĪN-ē-al) **gland** A cone-shaped gland located in the roof of the third ventricle that secretes melatonin. Also called the **epiphysis cerebri** (ē-PIF-i-sis se-RĒ-brē).

Pinealocyte (pin-ē-AL-ō-sīt) Secretory cell of the pineal gland that releases melatonin.

Pinna (PIN-na) The projecting part of the external ear composed of elastic cartilage and covered by skin and shaped like the flared end of a trumpet. Also called the **auricle** (OR-i-kul).

Pituicyte (pi-TOO-i-sīt) Supporting cell of the posterior pituitary.

Pituitary (pi-TOO-i-tār-ē) **gland** A small endocrine gland occupying the hypophyseal fossa of the sphenoid bone and attached to the hypothalamus by the infundibulum. Also called the **hypophysis** (hī-POF-i-sis).

Pivot joint A synovial joint in which a rounded, pointed, or conical surface of one bone articulates with a ring formed partly by another bone and partly by a ligament, as in the joint between the atlas and axis and between the proximal ends of the radius and ulna. Also called a **trochoid** (TRŌ-koyd) **joint.**

Placenta (pla-SEN-ta) The special structure through which the exchange of materials between fetal and maternal circulations occurs. Also called the **afterbirth.**

Plane joint A synovial joint having articulating surfaces that are usually flat, permitting only side-to-side and back-and-forth movements, as between carpal bones, tarsal bones, and the scapula and clavicle. Also called an **arthrodial** (ar-THRŌ-dē-al) **joint.**

Plantar flexion (PLAN-tar FLEK-shun) Bending the foot in the direction of the plantar surface (sole).

Plaque (PLAK) A layer of dense proteins on the inside of a plasma membrane in adherens junctions and desmosomes. A mass of bacterial cells, dextran (polysaccharide), and other debris that adheres to teeth (dental plaque). *See also* **Atherosclerotic plaque.**

Plasma (PLAZ-ma) The extracellular fluid found in blood vessels; blood minus the formed elements.

Plasma cell Cell that develops from a B cell (lymphocyte) and produces antibodies.

Plasma (cell) membrane Outer, limiting membrane that separates the cell's internal parts from extracellular fluid or the external environment.

Platelet (PLĀT-let) A fragment of cytoplasm enclosed in a cell membrane and lacking a nucleus; found in the circulating blood; plays a role in hemostasis. Also called a **thrombocyte** (THROM-bō-sīt).

Platelet plug Aggregation of platelets (thrombocytes) at a site where a blood vessel is damaged that helps stop or slow blood loss.

Pleura (PLOO-ra) The serous membrane that covers the lungs and lines the walls of the chest and the diaphragm.

Pleural cavity Small potential space between the visceral and parietal pleurae.

Plexus (PLEK-sus) A network of nerves, veins, or lymphatic vessels.

Plexus of Auerbach *See* **Myenteric plexus.**

Plexus of Meissner *See* **Submucosal plexus.**

Pluripotent (ploo-RI-pō-tent) **stem cell** Immature stem cell in red bone marrow that gives rise to precursors of all the different mature blood cells.

Pneumotaxic (noo-mō-TAK-sik) **area** A part of the respiratory center in the pons that continually sends inhibitory nerve impulses to the inspiratory area, limiting inhalation and facilitating exhalation.

Polycythemia (pol′-ē-sī-THĒ-mē-a) Disorder characterized by an above-normal hematocrit (above 55%) in which hypertension, thrombosis, and hemorrhage can occur.

Polyunsaturated fat A fatty acid that contains more than one double covalent bond between its carbon atoms; abundant in triglycerides of corn oil, safflower oil, and cottonseed oil.

Polyuria (pol′-ē-Ū-rē-a) An excessive production of urine.

Pons (PONZ) The part of the brain stem that forms a "bridge" between the medulla oblongata and the midbrain, anterior to the cerebellum.

Portal system The circulation of blood from one capillary network into another through a vein.

Positive feedback system A feedback cycle that strengthens or reinforces a change in a controlled condition.

Postcentral gyrus Gyrus of cerebral cortex located immediately posterior to the central sulcus; contains the primary somatosensory area.

Posterior (pos-TĒR-ē-or) Nearer to or at the back of the body. Equivalent to **dorsal** in bipeds.

Posterior column–medial lemniscus pathway Sensory pathway that carries information related to proprioception, fine touch, two-point discrimination, pressure, and vibration. First-order neurons project from the spinal cord to the ipsilateral medulla in the posterior columns (gracile fasciculus and cuneate fasciculus). Second-order neurons project from the medulla to the contralateral thalamus in the medial lemniscus. Third-order neurons project from the thalamus to the somatosensory cortex (postcentral gyrus) on the same side.

Posterior pituitary Posterior lobe of the pituitary gland. Also called the **neurohypophysis** (noo-rō-hī-POF-i-sis).

Posterior root The structure composed of sensory axons lying between a spinal nerve and the dorsolateral aspect of the spinal cord. Also called the **dorsal (sensory) root.**

Posterior root ganglion (GANG-glē-on) A group of cell bodies of sensory neurons and their supporting cells located along the posterior root of a

spinal nerve. Also called a **dorsal (sensory) root ganglion.**

Postganglionic neuron (pōst′-gang-lē-ON-ik NOO-ron) The second autonomic motor neuron in an autonomic pathway, having its cell body and dendrites located in an autonomic ganglion and its unmyelinated axon ending at cardiac muscle, smooth muscle, or a gland.

Postsynaptic (pōst-sin-AP-tik) **neuron** The nerve cell that is activated by the release of a neurotransmitter from another neuron and carries nerve impulses away from the synapse.

Pouch of Douglas *See* **Rectouterine pouch.**

Precapillary sphincter (SFINGK-ter) The distal-most muscle fiber (cell) at the metarteriole–capillary junction that regulates blood flow into capillaries.

Precentral gyrus Gyrus of cerebral cortex located immediately anterior to the central sulcus; contains the primary motor area.

Preganglionic (pre′-gang-lē-ON-ik) **neuron** The first autonomic motor neuron in an autonomic pathway, with its cell body and dendrites in the brain or spinal cord and its myelinated axon ending at an autonomic ganglion, where it synapses with a postganglionic neuron.

Pregnancy Sequence of events that normally includes fertilization, implantation, embryonic growth, and fetal growth and terminates in birth.

Premenstrual syndrome (PMS) Severe physical and emotional stress ocurring late in the postovulatory phase of the menstrual cycle and sometimes overlapping with menstruation.

Prepuce (PRĒ-poos) The loose-fitting skin covering the glans of the penis and clitoris. Also called the **foreskin.**

Presbyopia (prez-bē-Ō-pē-a) A loss of elasticity of the lens of the eye due to advancing age with resulting inability to focus clearly on near objects.

Presynaptic (prē-sin-AP-tik) **neuron** A neuron that propagates nerve impulses toward a synapse.

Prevertebral ganglion (prē-VER-te-bral GANG-glē-on) A cluster of cell bodies of postganglionic sympathetic neurons anterior to the spinal column and close to large abdominal arteries. Also called a **collateral ganglion.**

Primary germ layer One of three layers of embryonic tissue, called ectoderm, mesoderm, and endoderm, that give rise to all tissues and organs of the body.

Primary motor area A region of the cerebral cortex in the precentral gyrus of the frontal lobe of the cerebrum that controls specific muscles or groups of muscles.

Primary somatosensory area A region of the cerebral cortex posterior to the central sulcus in the postcentral gyrus of the parietal lobe of the cerebrum that localizes exactly the points of the body where somatic sensations originate.

Prime mover The muscle directly responsible for producing a desired motion. Also called an **agonist** (AG-ō-nist).

Primitive gut Embryonic structure formed from the dorsal part of the yolk sac that gives rise to most of the gastrointestinal tract.

Primordial (prī-MOR-dē-al) Existing first; especially primordial egg cells in the ovary.

Principal cell Cell type in the distal convoluted tubules and collecting ducts of the kidneys that is stimulated by aldosterone and antidiuretic hormone.

Proctology (prok-TOL-ō-jē) The branch of medicine concerned with the rectum and its disorders.

Progeny (PROJ-e-nē) Offspring or descendants.

Progesterone (prō-JES-te-rōn) A female sex hormone produced by the ovaries that helps prepare the endometrium of the uterus for implantation of a fertilized ovum and the mammary glands for milk secretion.

Prognosis (prog-NŌ-sis) A forecast of the probable results of a disorder; the outlook for recovery.

Prolactin (prō-LAK-tin) **(PRL)** A hormone secreted by the anterior pituitary that initiates and maintains milk secretion by the mammary glands.

Prolapse (PRŌ-laps) A dropping or falling down of an organ, especially the uterus or rectum.

Proliferation (prō-lif′-er-Ā-shun) Rapid and repeated reproduction of new parts, especially cells.

Pronation (prō-NĀ-shun) A movement of the forearm in which the palm is turned posteriorly.

Prophase (PRŌ-fāz) The first stage of mitosis during which chromatid pairs are formed and aggregate around the metaphase plate of the cell.

Proprioception (prō-prē-ō-SEP-shun) The perception of the position of body parts, especially the limbs, independent of vision; this sense is possible due to nerve impulses generated by proprioceptors.

Proprioceptor (PRŌ-prē-ō-sep′-tor) A receptor located in muscles, tendons, joints, or the internal ear (muscle spindles, tendon organs, joint kinesthetic receptors, and hair cells of the vestibular apparatus) that provides information about body position and movements.

Prostaglandin (pros′-ta-GLAN-din) **(PG)** A membrane-associated lipid; released in small quantities and acts as a local hormone.

Prostate (PROS-tāt) A doughnut-shaped gland inferior to the urinary bladder that surrounds the superior portion of the male urethra and secretes a slightly acidic solution that contributes to sperm motility and viability.

Proteasome (PRŌ-tē-a-sōm) Tiny cellular organelle in cytosol and nucleus containing proteases that destroy unneeded, damaged, or faulty proteins.

Protein An organic compound consisting of carbon, hydrogen, oxygen, nitrogen, and sometimes sulfur and phosphorus; synthesized on ribosomes and made up of amino acids linked by peptide bonds.

Prothrombin (prō-THROM-bin) An inactive blood-clotting factor synthesized by the liver, released into the blood, and converted to active thrombin in the process of blood clotting by the activated enzyme prothrombinase.

Proto-oncogene (prō′-tō-ON-kō-jēn) Gene responsible for some aspect of normal growth and development; it may transform into an oncogene, a gene capable of causing cancer.

Protraction (prō-TRAK-shun) The movement of the mandible or shoulder girdle forward on a plane parallel with the ground.

Proximal (PROK-si-mal) Nearer the attachment of a limb to the trunk; nearer to the point of origin or attachment.

Pseudopod (SOO-dō-pod) Temporary protrusion of the leading edge of a migrating cell; cellular projection that surrounds a particle undergoing phagocytosis.

Pterygopalatine ganglion (ter′-i-gō-PAL-a-tīn GANG-glē-on) A cluster of cell bodies of parasympathetic postganglionic neurons ending at the lacrimal and nasal glands.

Ptosis (TŌ-sis) Drooping, as of the eyelid or the kidney.

Puberty (PŪ-ber-tē) The time of life during which the secondary sex characteristics begin to appear and the capability for sexual reproduction is possible; usually occurs between the ages of 10 and 17.

Pubic symphysis A slightly movable cartilaginous joint between the anterior surfaces of the hip bones.

Puerperium (pū-er-PER-ē-um) The period immediately after childbirth, usually 4–6 weeks.

Pulmonary (PUL-mo-ner′-ē) Concerning or affected by the lungs.

Pulmonary circulation The flow of deoxygenated blood from the right ventricle to the lungs and the return of oxygenated blood from the lungs to the left atrium.

Pulmonary edema (e-DĒ-ma) An abnormal accumulation of interstitial fluid in the tissue spaces and alveoli of the lungs due to increased pulmonary capillary permeability or increased pulmonary capillary pressure.

Pulmonary embolism (EM-bō-lizm) **(PE)** The presence of a blood clot or a foreign substance in a pulmonary arterial blood vessel that obstructs circulation to lung tissue.

Pulmonary ventilation The inflow (inhalation) and outflow (exhalation) of air between the atmosphere and the lungs. Also called **breathing.**

Pulp cavity A cavity within the crown and neck of a tooth, which is filled with pulp, a connective tissue containing blood vessels, nerves, and lymphatic vessels.

Pulse (PULS) The rhythmic expansion and elastic recoil of a systemic artery after each contraction of the left ventricle.

Pupil The hole in the center of the iris, the area through which light enters the posterior cavity of the eyeball.

Purkinje (pur-KIN-jē) **fiber** Muscle fiber (cell) in the ventricular tissue of the heart specialized for conducting an action potential to the myocardium; part of the conduction system of the heart.

Pus The liquid product of inflammation containing leukocytes or their remains and debris of dead cells.

Pyloric (pī-LOR-ik) **sphincter** A thickened ring of smooth muscle through which the pylorus of the stomach communicates with the duodenum. Also called the **pyloric valve.**

Pyorrhea (pī-ō-RĒ-a) A discharge or flow of pus, especially in the alveoli (sockets) and the tissues of the gums.

Pyramid (PIR-a-mid) A pointed or cone-shaped structure. One of two roughly triangular structures on the anterior aspect of the medulla oblongata composed of the largest motor tracts that run from the cerebral cortex to the spinal cord. A triangular structure in the renal medulla.

Pyramidal (pi-RAM-i-dal) **tracts (pathways)** *See* **Direct motor pathways.**

Q

QRS complex The deflection waves of an electrocardiogram that represent onset of ventricular depolarization.

Quadrant (KWOD-rant) One of four parts.

Quadriplegia (kwod′-ri-PLĒ-jē-a) Paralysis of four limbs: two upper and two lower.

R

Radiographic (rā′-dē-ō-GRAF-ik) **anatomy** Diagnostic branch of anatomy that includes the use of x rays.

Rami communicantes (RĀ-mē kō-mū-ni-KAN-tēz) Branches of a spinal nerve. *Singular* is **ramus communicans** (RĀ-mus kō-MŪ-ni-kans).

Rathke's pouch *See* **Hypophyseal pouch.**

Receptor A specialized cell or a distal portion of a neuron that responds to a specific sensory modality, such as touch, pressure, cold, light, or sound, and converts it to an electrical signal (generator or receptor potential). A specific molecule or cluster of molecules that recognizes and binds a particular ligand.

Receptor-mediated endocytosis A highly selective process whereby cells take up specific ligands, which usually are large molecules or particles, by enveloping them within a sac of plasma membrane. Ligands are eventually broken down by enzymes in lysosomes.

Recombinant DNA Synthetic DNA, formed by joining a fragment of DNA from one source to a portion of DNA from another.

Rectouterine (rek-tō-Ū-ter-in) **pouch** A pocket formed by the parietal peritoneum as it moves posteriorly from the surface of the uterus and is reflected onto the rectum; the most inferior point in the pelvic cavity. Also called the **pouch of Douglas.**

Rectum (REK-tum) The last 20 cm (8 in.) of the gastrointestinal tract, from the sigmoid colon to the anus.

Recumbent (re-KUM-bent) Lying down.

Red bone marrow A highly vascularized connective tissue located in microscopic spaces between trabeculae of spongy bone tissue.

Red nucleus A cluster of cell bodies in the midbrain, occupying a large part of the tectum from which axons extend into the rubroreticular and rubrospinal tracts.

Red pulp That portion of the spleen that consists of venous sinuses filled with blood and thin plates of splenic tissue called splenic (Billroth's) cords.

Referred pain Pain that is felt at a site remote from the place of origin.

Reflex Fast response to a change (stimulus) in the internal or external environment that attempts to restore homeostasis.

Reflex arc The most basic conduction pathway through the nervous system, connecting a receptor and an effector and consisting of a receptor, a sensory neuron, an integrating center in the central nervous system, a motor neuron, and an effector. Also called **reflex circuit.**

Regional anatomy The division of anatomy dealing with a specific region of the body, such as the head, neck, chest, or abdomen.

Regurgitation (rē-gur′-ji-TĀ-shun) Return of solids or fluids to the mouth from the stomach; backward flow of blood through incompletely closed heart valves.

Relaxin (RLX) A female hormone produced by the ovaries and placenta that increases flexibility of the pubic symphysis and helps dilate the uterine cervix to ease delivery of a baby.

Releasing hormone Hormone secreted by the hypothalamus that can stimulate secretion of hormones of the anterior pituitary.

Renal (RĒ-nal) Pertaining to the kidneys.

Renal corpuscle (KOR-pus-l) A glomerular (Bowman's) capsule and its enclosed glomerulus.

Renal pelvis A cavity in the center of the kidney formed by the expanded, proximal portion of the ureter, lying within the kidney, and into which the major calyces open.

Renal physiology Study of the functions of the kidneys.

Renal pyramid A triangular structure in the renal medulla containing the straight segments of renal tubules and the vasa recta.

Reproduction (rē-prō-DUK-shun) The formation of new cells for growth, repair, or replacement; the production of a new individual.

Reproductive cell division Type of cell division in which gametes (sperm and oocytes) are produced; consists of meiosis and cytokinesis.

Respiration (res-pi-RĀ-shun) Overall exchange of gases between the atmosphere, blood, and body cells consisting of pulmonary ventilation, external respiration, and internal respiration.

Respiratory center Neurons in the pons and medulla oblongata of the brain stem that regulate the rate and depth of pulmonary ventilation.

Respiratory (RES-pir-a-tōr-ē) **physiology** Study of the functions of the air passageways and lungs.

Respiratory (RES-pi-ra-tōr-ē) **system** System composed of the nose, pharynx, larynx, trachea, bronchi, and lungs that obtains oxygen for body cells and eliminates carbon dioxide from them.

Retention (rē-TEN-shun) A failure to void urine due to obstruction, nervous contraction of the urethra, or absence of sensation of desire to urinate.

Rete (RĒ-tē) **testis** The network of ducts in the testes.

Reticular (re-TIK-ū-lar) **activating system (RAS)** A portion of the reticular formation that has many ascending connections with the cerebral cortex; when this area of the brain stem is active, nerve impulses pass to the thalamus and widespread areas of the cerebral cortex, resulting in generalized alertness or arousal from sleep.

Reticular formation A network of small groups of neuronal cell bodies scattered among bundles of axons (mixed gray and white matter) beginning in the medulla oblongata and extending superiorly through the central part of the brain stem.

Reticulocyte (re-TIK-ū-lō-sīt) An immature red blood cell.

Reticulum (re-TIK-ū-lum) A network.

Retina (RET-i-na) The deep coat of the posterior portion of the eyeball consisting of nervous tissue (where the process of vision begins) and a pigmented layer of epithelial cells that contact the choroid.

Retinaculum (ret-i-NAK-ū-lum) A thickening of fascia that holds structures in place, for example, the superior and inferior retinacula of the ankle.

Retraction (rē-TRAK-shun) The movement of a protracted part of the body posteriorly on a plane parallel to the ground, as in pulling the lower jaw back in line with the upper jaw.

Retroperitoneal (re′-trō-per-i-tō-NĒ-al) External to the peritoneal lining of the abdominal cavity.

Rh factor An inherited antigen on the surface of red blood cells in Rh⁺ individuals; not present in Rh⁻ individuals.

Rhinology (rī-NOL-ō-jē) The study of the nose and its disorders.

Ribonucleic (rī-bō-noo-KLĒ-ik) **acid (RNA)** A single-stranded nucleic acid made up of nucleotides, each consisting of a nitrogenous base (adenine, cytosine, guanine, or uracil), ribose, and a phosphate group; major types are messenger RNA (mRNA), transfer RNA (tRNA), and ribosomal RNA (rRNA), each of which has a specific role during protein synthesis.

Ribosome (RĪ-bō-sōm) A cellular structure in the cytoplasm of cells, composed of a small subunit and a large subunit that contain ribosomal RNA and ribosomal proteins; the site of protein synthesis.

Right lymphatic (lim-FAT-ik) **duct** A vessel of the lymphatic system that drains lymph from the upper right side of the body and empties it into the right subclavian vein.

Rigidity (ri-JID-i-tē) Hypertonia characterized by increased muscle tone, but reflexes are not affected.

Rigor mortis State of partial contraction of muscles after death due to lack of ATP; myosin heads (cross-bridges) remain attached to actin, thus preventing relaxation.

Rod One of two types of photoreceptors in the retina of the eye; specialized for vision in dim light.

Root canal A narrow extension of the pulp cavity lying within the root of a tooth.

Root of penis Attached portion of penis that consists of the bulb and crura.

Rotation (rō-TĀ-shun) Moving a bone around its own axis, with no other movement.

Rotator cuff Refers to the tendons of four deep shoulder muscles (subscapularis, supraspinatus, infraspinatus, and teres minor) that form a complete circle around the shoulder; they strengthen and stabilize the shoulder joint.

Round ligament (LIG-a-ment) A band of fibrous connective tissue enclosed between the folds of the broad ligament of the uterus, emerging from the uterus just inferior to the uterine tube, extending laterally along the pelvic wall and through the deep inguinal ring to end in the labia majora.

Round window A small opening between the middle and internal ear, directly inferior to the oval window, covered by the secondary tympanic membrane.

Ruffini corpuscle A sensory receptor embedded deeply in the dermis and deeper tissues that detects the stretching of the skin.

Rugae (ROO-gē) Large folds in the mucosa of an empty hollow organ, such as the stomach or vagina.

S

Saccule (SAK-ūl) The inferior and smaller of the two chambers in the membranous labyrinth inside the vestibule of the internal ear containing a receptor organ for static equilibrium.

Sacral plexus (SĀ-kral PLEK-sus) A network formed by the ventral branches of spinal nerves L4 through S3.

Sacral promontory (PROM-on-tor′-ē) The superior surface of the body of the first sacral vertebra that projects anteriorly into the pelvic cavity; a line from the sacral promontory to the superior border of the pubic symphysis divides the abdominal and pelvic cavities.

Saddle joint A synovial joint in which the articular surface of one bone is saddle-shaped and the articular surface of the other bone is shaped like the legs of the rider sitting in the saddle, as in the joint between the trapezium and the metacarpal of the thumb.

Sagittal (SAJ-i-tal) **plane** A plane that divides the body or organs into left and right portions. Such a plane may be **midsagittal (median),** in which the divisions are equal, or **parasagittal,** in which the divisions are unequal.

Saliva (sa-LĪ-va) A clear, alkaline, somewhat viscous secretion produced mostly by the three pairs of salivary glands; contains various salts, mucin, lysozyme, salivary amylase, and lingual lipase (produced by glands in the tongue).

Salivary amylase (SAL-i-ver-ē AM-i-lās) An enzyme in saliva that initiates the chemical breakdown of starch.

Salivary gland One of three pairs of glands that lie external to the mouth and pour their secretory product (saliva) into ducts that empty into the oral cavity; the parotid, submandibular, and sublingual glands.

Sarcolemma (sar′-kō-LEM-ma) The cell membrane of a muscle fiber (cell), especially of a skeletal muscle fiber.

Sarcomere (SAR-kō-mēr) A contractile unit in a striated muscle fiber (cell) extending from one Z disc to the next Z disc.

Sarcoplasm (SAR-kō-plazm) The cytoplasm of a muscle fiber (cell).

Sarcoplasmic reticulum (sar′-kō-PLAZ-mik re-TIK-ū-lum) **(SR)** A network of saccules and tubes surrounding myofibrils of a muscle fiber (cell), comparable to endoplasmic reticulum; functions to reabsorb calcium ions during relaxation and to release them to cause contraction.

Satellite (SAT-i-līt) **cell** Flat neuroglial cells that surround cell bodies of peripheral nervous system ganglia to provide structural support and regulate the exchange of material between a neuronal cell body and interstitial fluid.

Saturated fat A fatty acid that contains only single bonds (no double bonds) between its carbon atoms; all carbon atoms are bonded to the maximum number of hydrogen atoms; prevalent in triglycerides of animal products such as meat, milk, milk products, and eggs.

Scala tympani (SKA-la TIM-pan-ē) The inferior spiral-shaped channel of the bony cochlea, filled with perilymph.

Scala vestibuli (ves-TIB-ū-lē) The superior spiral-shaped channel of the bony cochlea, filled with perilymph.

Schwann (SCHVON or SCHWON) **cell** A neuroglial cell of the peripheral nervous system that forms the myelin sheath and neurolemma around a nerve axon by wrapping around the axon in a jellyroll fashion.

Sciatica (sī-AT-i-ka) Inflammation and pain along the sciatic nerve; felt along the posterior aspect of the thigh extending down the inside of the leg.

Sclera (SKLE-ra) The white coat of fibrous tissue that forms the superficial protective covering over the eyeball except in the most anterior portion; the posterior portion of the fibrous tunic.

Scleral venous sinus A circular venous sinus located at the junction of the sclera and the cornea through which aqueous humor drains from the anterior chamber of the eyeball into the blood. Also called the **canal of Schlemm** (SHLEM).

Sclerosis (skle-RŌ-sis) A hardening with loss of elasticity of tissues.

Scoliosis (skō-lē-Ō-sis) An abnormal lateral curvature from the normal vertical line of the backbone.

Scrotum (SKRŌ-tum) A skin-covered pouch that contains the testes and their accessory structures.

Sebaceous (se-BĀ-shus) **gland** An exocrine gland in the dermis of the skin, almost always associated with a hair follicle, that secretes sebum. Also called an **oil gland.**

Sebum (SĒ-bum) Secretion of sebaceous (oil) glands.

Secondary sex characteristic A characteristic of the male or female body that develops at puberty under the influence of sex hormones but is not directly involved in sexual reproduction; examples are distribution of body hair, voice pitch, body shape, and muscle development.

Secretion (se-KRĒ-shun) Production and release from a cell or a gland of a physiologically active substance.

Selective permeability (per′-mē-a-BIL-i-tē) The property of a membrane by which it permits the passage of certain substances but restricts the passage of others.

Semen (SĒ-men) A fluid discharged at ejaculation by a male that consists of a mixture of sperm and the secretions of the seminiferous tubules, seminal vesicles, prostate, and bulbourethral (Cowper's) glands.

Semicircular canals Three bony channels (anterior, posterior, lateral), filled with perilymph, in which lie the membranous semicircular canals filled with endolymph. They contain receptors for equilibrium.

Semicircular ducts The membranous semicircular canals filled with endolymph and floating in the perilymph of the bony semicircular canals; they contain cristae that are concerned with dynamic equilibrium.

Semilunar (sem′-ē-LOO-nar) **(SL) valve** A valve between the aorta or the pulmonary trunk and a ventricle of the heart.

Seminal vesicle (SEM-i-nal VES-i-kul) One of a pair of convoluted, pouchlike structures, lying posterior and inferior to the urinary bladder and anterior to the rectum, that secrete a component of semen into the ejaculatory ducts. Also termed **seminal gland.**

Seminiferous tubule (sem′-i-NI-fer-us TOO-būl) A tightly coiled duct, located in the testis, where sperm are produced.

Sensation A state of awareness of external or internal conditions of the body.

Sensory area A region of the cerebral cortex concerned with the interpretation of sensory impulses.

Sensory neurons (NOO-ronz) Neurons that carry sensory information from cranial and spinal nerves into the brain and spinal cord or from a lower to a higher level in the spinal cord and brain. Also called **afferent neurons.**

Septal defect An opening in the atrial septum (atrial septal defect) because the foramen ovale fails to close, or the ventricular septum (ventricular septal defect) due to incomplete development of the ventricular septum.

Septum (SEP-tum) A wall dividing two cavities.

Serous (SĒR-us) **membrane** A membrane that lines a body cavity that does not open to the exterior. The external layer of an organ formed by a serous membrane. The membrane that lines the pleural, pericardial, and peritoneal cavities. Also called a **serosa** (se-RŌ-sa).

Sertoli (ser-TŌ-lē) **cell** A supporting cell in the seminiferous tubules that secretes fluid for supplying nutrients to sperm and the hormone inhibin, removes excess cytoplasm from spermatogenic cells, and mediates the effects of FSH and testosterone on spermatogenesis. Also called a **sustentacular** (sus′-ten-TAK-ū-lar) **cell.**

Serum Blood plasma minus its clotting proteins.

Sesamoid (SES-a-moyd) **bones** Small bones usually found in tendons.

Sex chromosomes The twenty-third pair of chromosomes, designated X and Y, which determines the genetic sex of an individual; in males, the pair is XY; in females, XX.

Sexual intercourse The insertion of the erect penis of a male into the vagina of a female. Also called **coitus** (KŌ-i-tus).

Sheath of Schwann *See* **Neurolemma.**

Shock Failure of the cardiovascular system to deliver adequate amounts of oxygen and nutrients to meet the metabolic needs of the body due to inadequate cardiac output. It is characterized by hypotension; clammy, cool, and pale skin; sweating; reduced urine formation; altered mental state; acidosis; tachycardia; weak, rapid pulse; and thirst. Types include hypovolemic, cardiogenic, vascular, and obstructive.

Shoulder joint A synovial joint where the humerus articulates with the scapula.

Sigmoid colon (SIG-moyd KŌ-lon) The S-shaped part of the large intestine that begins at the level of the left iliac crest, projects medially, and terminates at the rectum at about the level of the third sacral vertebra.

Sign Any objective evidence of disease that can be observed or measured, such as a lesion, swelling, or fever.

Sinoatrial (si-nō-Ā-trē-al) **(SA) node** A small mass of cardiac muscle fibers (cells) located in the right atrium inferior to the opening of the superior vena cava that spontaneously depolarize and generate a cardiac action potential about 100 times per minute. Also called the natural **pacemaker.**

Sinus (SĪ-nus) A hollow in a bone (paranasal sinus) or other tissue; a channel for blood (vascular sinus); any cavity having a narrow opening.

Sinusoid (SĪ-nū-soyd) A large, thin-walled, and leaky type of capillary, having large intercellular clefts that may allow proteins and blood cells to pass from a tissue into the bloodstream; present in the liver, spleen, anterior pituitary, parathyroid glands, and red bone marrow.

Skeletal muscle An organ specialized for contraction, composed of striated muscle fibers (cells), supported by connective tissue, attached to a bone by a tendon or an aponeurosis, and stimulated by somatic motor neurons.

Skeletal system Framework of bones and their associated cartilages, ligaments, and tendons.

Skene's gland *See* **Paraurethral gland.**

Skin The external covering of the body that consists of a superficial, thinner epidermis (epithelial tissue) and a deep, thicker dermis (connective tissue) that is anchored to the subcutaneous layer. Also called **cutaneous membrane.**

Skin graft The transfer of a patch of healthy skin taken from a donor site to cover a wound.

Skull The skeleton of the head consisting of the cranial and facial bones.

Sleep A state of partial unconsciousness from which a person can be aroused; associated with a low level of activity in the reticular activating system.

Sliding filament mechanism A model that describes muscle contraction in which thin filaments slide past thick ones so that the filaments overlap, causing shortening of a sarcomere, and thus shortening of muscle fibers and alternately shortening of the entire muscle.

Small intestine A long tube of the gastrointestinal tract that begins at the pyloric sphincter of the stomach, coils through the central and inferior part of the abdominal cavity, and ends at the large intestine; divided into three segments: duodenum, jejunum, and ileum.

Smooth muscle A tissue specialized for contraction, composed of smooth muscle fibers (cells), located in the walls of hollow internal organs, and innervated by autonomic motor neurons.

Sodium–potassium ATPase An active transport pump located in the plasma membrane that transports sodium ions out of the cell and potassium ions into the cell at the expense of cellular ATP. It functions to keep the ionic concentrations of these ions at physiological levels. Also called the **sodium–potassium pump.**

Soft palate (PAL-at) The posterior portion of the roof of the mouth, extending from the palatine bones to the uvula. It is a muscular partition lined with mucous membrane.

Somatic (sō-MAT-ik) **cell division** Type of cell division in which a single starting cell duplicates itself to produce two identical cells; consists of mitosis and cytokinesis.

Somatic motor pathway Pathway that carries information from the cerebral cortex, basal nuclei, and cerebellum that stimulates contraction of skeletal muscles.

Somatic nervous system (SNS) The portion of the peripheral nervous system consisting of somatic sensory (afferent) neurons and somatic motor (efferent) neurons.

Somatic sensory pathway Pathway that carries information from somatic sensory receptor to the primary somatosensory area in the cerebral cortex and cerebellum.

Somite (SŌ-mīt) Block of mesodermal cells in a developing embryo that is distinguished into a myotome (which forms most of the skeletal muscles), dermatome (which forms connective tissues), and sclerotome (which forms the vertebrae).

Spasm (SPAZM) A sudden, involuntary contraction of large groups of muscles.

Spasticity (spas-TIS-i-tē) Hypertonia characterized by increased muscle tone, increased tendon reflexes, and pathological reflexes (Babinski sign).

Spermatic (sper-MAT-ik) **cord** A supporting structure of the male reproductive system, extending from a testis to the deep inguinal ring, that includes the ductus (vas) deferens, arteries, veins, lymphatic vessels, nerves, cremaster muscle, and connective tissue.

Spermatogenesis (sper′-ma-tō-JEN-e-sis) The formation and development of sperm in the seminiferous tubules of the testes.

Sperm cell A mature male gamete. Also termed **spermatozoon** (sper′-ma-tō-ZŌ-on).

Spermiogenesis (sper′-mē-ō-JEN-e-sis) The maturation of spermatids into sperm.

Sphincter (SFINGK-ter) A circular muscle that constricts an opening.

Sphincter of Oddi *See* **Sphincter of the hepatopancreatic ampulla.**

Sphincter of the hepatopancreatic ampulla A circular muscle at the opening of the common bile and main pancreatic ducts in the duodenum. Also called the **sphincter of Oddi** (OD-ē).

Spinal (SPĪ-nal) **cord** A mass of nerve tissue located in the vertebral canal from which 31 pairs of spinal nerves originate.

Spinal nerve One of the 31 pairs of nerves that originate on the spinal cord from posterior and anterior roots.

Spinal shock A period from several days to several weeks following transection of the spinal cord that is characterized by the abolition of all reflex activity.

Spinothalamic (spī-nō-tha-LAM-ik) **tract** Sensory (ascending) tract that conveys information up the spinal cord to the thalamus for sensations of pain, temperature, itch, and tickle.

Spinous (SPĪ-nus) **process** A sharp or thornlike process or projection. Also called a **spine.** A sharp ridge running diagonally across the posterior surface of the scapula.

Spiral organ The organ of hearing, consisting of supporting cells and hair cells that rest on the basilar membrane and extend into the endolymph of the cochlear duct. Also called the **organ of Corti** (KOR-tē).

Splanchnic (SPLANK-nik) Pertaining to the viscera.

Spleen (SPLĒN) Large mass of lymphatic tissue between the fundus of the stomach and the diaphragm that functions in formation of blood cells during early fetal development, phagocytosis of ruptured blood cells, and proliferation of B cells during immune responses.

Spongy (cancellous) bone tissue Bone tissue that consists of an irregular latticework of thin plates of bone called trabeculae; spaces between trabeculae of some bones are filled with red bone marrow; found inside short, flat, and irregular bones and in the epiphyses (ends) of long bones.

Sprain Forcible wrenching or twisting of a joint with partial rupture or other injury to its attachments without dislocation.

Squamous (SKWĀ-mus) Flat or scalelike.

Starvation (star-VĀ-shun) The loss of energy stores in the form of glycogen, triglycerides, and proteins due to inadequate intake of nutrients or inability to digest, absorb, or metabolize ingested nutrients.

Static equilibrium (ē-kwi-LIB-rē-um) The maintenance of posture in response to changes in the orientation of the body, mainly the head, relative to the ground.

Stellate reticuloendothelial (STEL-āt re-tik′-ū-lō-en′-dō-THĒ-lē-al) **cell** Phagocytic cell bordering a sinusoid of the liver. Also called a **Kupffer** (KOOP-fer) **cell.**

Stem cell An unspecialized cell that has the ability to divide for indefinite periods and give rise to a specialized cell.

Stenosis (sten-Ō-sis) An abnormal narrowing or constriction of a duct or opening.

Stereocilia (ste′-rē-ō-SIL-ē-a) Groups of extremely long, slender, nonmotile microvilli projecting from epithelial cells lining the epididymis.

Sterile (STE-ril) Free from any living microorganisms. Unable to conceive or produce offspring.

Sterilization (ster′-i-li-ZĀ-shun) Elimination of all living microorganisms. Any procedure that renders an individual incapable of reproduction (for example, castration, vasectomy, hysterectomy, or oophorectomy).

Stimulus Any stress that changes a controlled condition; any change in the internal or external environment that excites a sensory receptor, a neuron, or a muscle fiber.

Stomach The J-shaped enlargement of the gastrointestinal tract directly inferior to the diaphragm in the epigastric, umbilical, and left hypochondriac regions of the abdomen, between the esophagus and small intestine.

Straight tubule (TOO-būl) A duct in a testis leading from a convoluted seminiferous tubule to the rete testis.

Stratum (STRĀ-tum) A layer.

Stratum basalis (ba-SAL-is) The layer of the endometrium next to the myometrium that is maintained during menstruation and gestation and produces a new stratum functionalis following menstruation or parturition.

Stratum functionalis (funk′-shun-AL-is) The layer of the endometrium next to the uterine cavity that is shed during menstruation and that forms the maternal portion of the placenta during gestation.

Stretch receptor Receptor in the walls of blood vessels, airways, or organs that monitors the amount of stretching. Also termed **baroreceptor.**

Striae (STRĪ-ē) Internal scarring due to overstretching of the skin in which collagen fibers and blood vessels in the dermis are damaged. Also called **stretch marks.**

Stroma (STRŌ-ma) The tissue that forms the ground substance, foundation, or framework of an organ, as opposed to its functional parts (parenchyma).

Subarachnoid (sub′-a-RAK-noyd) **space** A space between the arachnoid mater and the pia mater that surrounds the brain and spinal cord and through which cerebrospinal fluid circulates.

Subcutaneous (sub′-kū-TĀ-nē-us) Beneath the skin. Also called **hypodermic** (hī-pō-DER-mik).

Subcutaneous (subQ) layer A continuous sheet of areolar connective tissue and adipose tissue between the dermis of the skin and the deep fascia of the muscles. Also called the **hypodermis.**

Subdural (sub-DOO-ral) **space** A space between the dura mater and the arachnoid mater of the brain and spinal cord that contains a small amount of fluid.

Sublingual (sub-LING-gwal) **gland** One of a pair of salivary glands situated in the floor of the mouth deep to the mucous membrane and to the side of the lingual frenulum, with a duct (Rivinus') that opens into the floor of the mouth.

Submandibular (sub′-man-DIB-ū-lar) **gland** One of a pair of salivary glands found inferior to the base of the tongue deep to the mucous membrane in the posterior part of the floor of the mouth, posterior to the sublingual glands, with a duct (Wharton's) situated to the side of the lingual frenulum. Also called the **submaxillary** (sub′-MAK-si-ler-ē) **gland.**

Submucosa (sub-mū-KŌ-sa) A layer of connective tissue located deep to a mucous membrane, as in the gastrointestinal tract or the urinary bladder; the submucosa connects the mucosa to the muscularis layer.

Submucosal plexus A network of autonomic nerve fibers located in the superficial part of the submucous layer of the small intestine. Also called the **plexus of Meissner** (MĪZ-ner).

Substrate A molecule on which an enzyme acts.

Subthalamus (sub-THAL-a-mus) Part of the diencephalon inferior to the thalamus; the substantia nigra and red nucleus extend from the midbrain into the subthalamus.

Sudoriferous (soo′-dor-IF-er-us) **gland** An apocrine or eccrine exocrine gland in the dermis or subcutaneous layer that produces perspiration. Also called a **sweat gland.**

Sulcus (SUL-kus) A groove or depression between parts, especially between the convolutions of the brain. *Plural* is **sulci** (SUL-sī).

Superficial (soo′-per-FISH-al) Located on or near the surface of the body or an organ. Also called **external.**

Superficial subcutaneous inguinal (IN-gwi-nal) **ring** A triangular opening in the aponeurosis of the external oblique muscle that represents the termination of the inguinal canal.

Superior (soo-PĒR-ē-or) Toward the head or upper part of a structure. Also called **cephalic** or **cranial.**

Superior vena cava (VĒ-na KĀ-va) **(SVC)** Large vein that collects blood from parts of the body superior to the heart and returns it to the right atrium.

Supination (soo-pi-NĀ-shun) A movement of the forearm in which the palm is turned anteriorly.

Surface anatomy The study of the structures that can be identified from the outside of the body.

Surfactant (sur-FAK-tant) Complex mixture of phospholipids and lipoproteins, produced by type II alveolar (septal) cells in the lungs, that decreases surface tension.

Suspensory ligament (sus-PEN-so-rē LIG-a-ment) A fold of peritoneum extending laterally from the surface of the ovary to the pelvic wall.

Sutural (SOO-chur-al) **bone** A small bone located within a suture between certain cranial bones. Also called **Wormian** (WER-mē-an) **bone.**

Suture (SOO-chur) An immovable fibrous joint that joins skull bones.

Sympathetic (sim′-pa-THET-ik) **division** One of the two subdivisions of the autonomic nervous system, having cell bodies of preganglionic neurons in the lateral gray columns of the thoracic segment and the first two or three lumbar segments of the spinal cord; primarily concerned with processes involving the expenditure of energy.

Sympathetic trunk ganglion (GANG-glē-on) A cluster of cell bodies of sympathetic postganglionic neurons lateral to the vertebral column, close to the body of a vertebra. These ganglia extend inferiorly through the neck, thorax, and abdomen to the coccyx on both sides of the vertebral column and are connected to one another to form a chain on each side of the vertebral column. Also called **sympathetic chain, vertebral chain ganglia,** or **paravertebral ganglia.**

Symphysis (SIM-fi-sis) A line of union. A slightly movable cartilaginous joint such as the pubic symphysis.

Symptom (SIMP-tum) A subjective change in body function not apparent to an observer, such as pain or nausea, that indicates the presence of a disease or disorder of the body.

Synapse (SIN-aps) The functional junction between two neurons or between a neuron and an effector, such as a muscle or gland; may be electrical or chemical.

Synapsis (sin-AP-sis) The pairing of homologous chromosomes during prophase I of meiosis.

Synaptic (sin-AP-tik) **cleft** The narrow gap at a chemical synapse that separates the axon terminal of one neuron from another neuron or muscle fiber (cell) and across which a neurotransmitter diffuses to affect the postsynaptic cell.

Synaptic end bulb Expanded distal end of an axon terminal that contains synaptic vesicles. Also called a **synaptic knob.**

Synaptic vesicle Membrane-enclosed sac in a synaptic end bulb that stores neurotransmitters.

Synarthrosis (sin′-ar-THRŌ-sis) An immovable joint such as a suture, gomphosis, or synchondrosis.

Synchondrosis (sin′-kon-DRŌ-sis) A cartilaginous joint in which the connecting material is hyaline cartilage.

Syndesmosis (sin′-dez-MŌ-sis) A slightly movable joint in which articulating bones are united by fibrous connective tissue.

Synergist (SIN-er-jist) A muscle that assists the prime mover by reducing undesired action or unnecessary movement.

Synergistic (syn-er-JIS-tik) **effect** A hormonal interaction in which the effects of two or more hormones acting together is greater or more extensive than the sum of each hormone acting alone.

Synostosis (sin′-os-TŌ-sis) A joint in which the dense fibrous connective tissue that unites bones at a suture has been replaced by bone, resulting in a complete fusion across the suture line.

Synovial (si-NŌ-vē-al) **cavity** The space between the articulating bones of a synovial joint, filled with synovial fluid. Also called a **joint cavity.**

Synovial fluid Secretion of synovial membranes that lubricates joints and nourishes articular cartilage.

Synovial joint A fully movable or diarthrotic joint in which a synovial (joint) cavity is present between the two articulating bones.

Synovial membrane The deeper of the two layers of the articular capsule of a synovial joint, composed of areolar connective tissue that secretes synovial fluid into the synovial (joint) cavity.

System An association of organs that have a common function.

Systemic (sis-TEM-ik) Affecting the whole body; generalized.

Systemic anatomy The anatomical study of particular systems of the body, such as the skeletal, muscular, nervous, cardiovascular, or urinary systems.

Systemic circulation The routes through which oxygenated blood flows from the left ventricle through the aorta to all the organs of the body and deoxygenated blood returns to the right atrium.

Systole (SIS-tō-lē) In the cardiac cycle, the phase of contraction of the heart muscle, especially of the ventricles.

Systolic (sis-TOL-ik) **blood pressure (SBP)** The force exerted by blood on arterial walls during ventricular contraction; the highest pressure measured in the large arteries, about 120 mmHg under normal conditions for a young adult.

T

T cell A lymphocyte that becomes immunocompetent in the thymus and can differentiate into a helper T cell or a cytotoxic T cell, both of which function in cell-mediated immunity.

T wave The deflection wave of an electrocardiogram that represents ventricular repolarization.

Tachycardia (tak′-i-KAR-dē-a) An abnormally rapid resting heartbeat or pulse rate (over 100 beats per minute).

Tactile (TAK-tīl) Pertaining to the sense of touch.

Tactile disc *See* **Merkel disc.**

Target cell A cell whose activity is affected by a particular hormone.

Tarsal bones The seven bones of the ankle. Also called **tarsals.**

Tarsal gland Sebaceous (oil) gland that opens on the edge of each eyelid. Also called a **Meibomian** (mī-BŌ-mē-an) **gland.**

Tarsal plate A thin, elongated sheet of connective tissue, one in each eyelid, giving the eyelid form and support. The aponeurosis of the levator palpebrae superioris is attached to the tarsal plate of the superior eyelid.

Tarsus (TAR-sus) A collective term for the seven bones of the ankle.

Tectorial (tek-TŌ-rē-al) **membrane** A gelatinous membrane projecting over and in contact with the hair cells of the spiral organ (organ of Corti) in the cochlear duct.

Teeth (TĒTH) Accessory structures of digestion, composed of calcified connective tissue and embedded in bony sockets of the mandible and maxilla, that cut, shred, crush, and grind food. Also called **dentes** (DEN-tēz).

Telophase (TEL-ō-fāz) The final stage of mitosis.

Tendon (TEN-don) A white fibrous cord of dense regular connective tissue that attaches muscle to bone.

Tendon organ A proprioceptive receptor, sensitive to changes in muscle tension and force of contraction, found chiefly near the junctions of tendons and muscles. Also called a **Golgi** (GOL-jē) **tendon organ.**

Tendon reflex A polysynaptic, ipsilateral reflex that protects tendons and their associated muscles from damage that might be brought about by excessive tension. The receptors involved are called tendon organs (Golgi tendon organs).

Teniae coli (TĒ-nē-ē KŌ-lī) The three flat bands of thickened, longitudinal smooth muscle running the length of the large intestine, except in the rectum. *Singular* is **tenia coli.**

Tentorium cerebelli (ten-TŌ-rē-um ser′-e-BEL-ī) A transverse shelf of dura mater that forms a partition between the occipital lobe of the cerebral hemispheres and the cerebellum and that covers the cerebellum.

Teratogen (TER-a-tō-jen) Any agent or factor that causes physical defects in a developing embryo.

Terminal ganglion (TER-min-al GANG-glē-on) A cluster of cell bodies of parasympathetic postganglionic neurons either lying very close to the visceral effectors or located within the walls of the visceral effectors supplied by the postganglionic neurons. Also called **intramural ganglion.**

Testis (TES-tis) Male gonad that produces sperm and the hormones testosterone and inhibin. Also called a **testicle.**

Testosterone (tes-TOS-te-rōn) A male sex hormone (androgen) secreted by interstitial (Leydig) cells of a mature testis; needed for development of sperm; together with a second androgen termed **dihydrotestosterone** (dī-hī-drō-tes-TOS-ter-ōn) **(DHT),** controls the growth and development of male reproductive organs, secondary sex characteristics, and body growth.

Tetralogy of Fallot (tet-RAL-ō-jē of fal-Ō) A combination of four congenital heart defects: (1) constricted pulmonary semilunar valve, (2) interventricular septal opening, (3) emergence of the aorta from both ventricles instead of from the left only, and (4) enlarged right ventricle.

Thalamus (THAL-a-mus) A large, oval structure located bilaterally on either side of the third ventricle, consisting of two masses of gray matter organized into nuclei; main relay center for sensory impulses ascending to the cerebral cortex.

Thermoreceptor (THER-mō-rē-sep-tor) Sensory receptor that detects changes in temperature.

Thermoregulation Homeostatic regulation of body temperature through sweating and adjustment of blood flow in the dermis.

Thigh The portion of the lower limb between the hip and the knee.

Third ventricle (VEN-tri-kul) A slitlike cavity between the right and left halves of the thalamus and between the lateral ventricles of the brain.

Thoracic (thor-AS-ik) **cavity** Cavity superior to the diaphragm that contains two pleural cavities, the mediastinum, and the pericardial cavity.

Thoracic duct A lymphatic vessel that begins as a dilation called the cisterna chyli, receives lymph from the left side of the head, neck, and chest, left arm, and the entire body below the ribs, and empties into the junction between the internal jugular and left subclavian veins. Also called the **left lymphatic** (lim-FAT-ik) **duct.**

Thoracolumbar (thōr′-a-kō-LUM-bar) **outflow** The axons of sympathetic preganglionic neurons, which have their cell bodies in the lateral gray columns of the thoracic segments and first two or three lumbar segments of the spinal cord.

Thorax (THŌ-raks) The chest.

Thrombosis (THROM-BŌ-sis) The formation of a clot in an unbroken blood vessel, usually a vein.

Thrombus (THROM-bus) A stationary clot formed in an unbroken blood vessel, usually a vein.

Thymus (THĪ-mus) A bilobed organ, located in the superior mediastinum posterior to the sternum and between the lungs, in which T cells develop immunocompetence.

Thyroid cartilage (THĪ-royd KAR-ti-lij) The largest single cartilage of the larynx, consisting of two fused plates that form the anterior wall of the larynx. Also called **Adam's apple.**

Thyroid follicle (FOL-i-kul) Spherical sac that forms the parenchyma of the thyroid gland and consists of follicular cells that produce thyroxine (T_4) and triiodothyronine (T_3).

Thyroid gland An endocrine gland with right and left lateral lobes on either side of the trachea connected by an isthmus; located anterior to the trachea just inferior to the cricoid cartilage; secretes thyroxine (T_4), triiodothyronine (T_3), and calcitonin.

Thyroid-stimulating hormone (TSH) A hormone secreted by the anterior pituitary that stimulates the synthesis and secretion of thyroxine (T_4) and triiodothyronine (T_3). Also called **thyrotropin** (THĪ-rō-TRŌ-pin).

Thyroxine (thī-ROK-sēn) (T_4) A hormone secreted by the thyroid gland that regulates metabolism, growth and development, and the activity of the nervous system. Also called **tetraiodothyronine** (tet-ra-ī-ō-dō-THĪ-rō-nēn).

Tic Spasmodic, involuntary twitching of muscles that are normally under voluntary control.

Tissue A group of similar cells and their intercellular substance joined together to perform a specific function.

Tissue rejection Phenomenon by which the body recognizes the protein (HLA antigens) in transplanted tissues or organs as foreign and produces antibodies against them.

Tongue A large skeletal muscle covered by a mucous membrane located on the floor of the oral cavity.

Tonsil (TON-sil) An aggregation of large lymphatic nodules embedded in the mucous membrane of the throat.

Topical (TOP-i-kal) Applied to the surface rather than ingested or injected.

Torn cartilage A tearing of an articular disc (meniscus) in the knee.

Trabecula (tra-BEK-ū-la) Irregular latticework of thin plates of spongy bone tissue. Fibrous cord of connective tissue serving as supporting fiber by forming a septum extending into an organ from its wall or capsule. *Plural* is **trabeculae** (tra-BEK-ū-lē).

Trabeculae carneae (KAR-nē-ē) Ridges and folds of the myocardium in the ventricles.

Trachea (TRĀ-kē-a) Tubular air passageway extending from the larynx to the fifth thoracic vertebra. Also called the **windpipe.**

Tract A bundle of nerve axons in the central nervous system.

Transplantation (tranz-plan-TĀ-shun) The transfer of living cells, tissues, or organs from a donor to a recipient or from one part of the body to another in order to restore a lost function.

Transverse colon (trans-VERS KŌ-lon) The portion of the large intestine extending across the abdomen from the right colic (hepatic) flexure to the left colic (splenic) flexure.

Transverse fissure (FISH-er) The deep cleft that separates the cerebrum from the cerebellum.

Transverse plane A plane that divides the body or organs into superior and inferior portions. Also called a **cross-sectional** or **horizontal plane.**

Transverse tubules (TOO-būls) **(T tubules)** Small, cylindrical invaginations of the sarcolemma of striated muscle fibers (cells) that conduct muscle action potentials toward the center of the muscle fiber.

Tremor (TREM-or) Rhythmic, involuntary, purposeless contraction of opposing muscle groups.

Triad (TRĪ-ad) A complex of three units in a muscle fiber composed of a transverse tubule and the sarcoplasmic reticulum terminal cisterns on both sides of it.

Tricuspid (trī-KUS-pid) **valve** Atrioventricular (AV) valve on the right side of the heart.

Triglyceride A lipid that consists of one glycerol molecule and three fatty acid molecules.

Triglyceride (trī-GLI-ser-īd) A lipid formed from one molecule of glycerol and three molecules of fatty acids that may be either solid (fats) or liquid (oils) at room temperature; the body's most highly concentrated source of chemical potential energy. Found mainly within adipocytes. Also called a **neutral fat** or a **triacylglycerol.**

Trigone (TRĪ-gōn) A triangular region at the base of the urinary bladder.

Triiodothyronine (trī-ī-ō-dō-THĪ-rō-nēn) **(T_3)** A hormone produced by the thyroid gland that regulates metabolism, growth and development, and the activity of the nervous system.

Trophoblast (TRŌF-ō-blast) The superficial covering of cells of the blastocyst.

Tropic (TRŌ-pik) **hormone** A hormone whose target is another endocrine gland.

Trunk The part of the body to which the upper and lower limbs are attached.

Tubal ligation (lī-GĀ-shun) A sterilization procedure in which the uterine (fallopian) tubes are tied and cut.

Tubular reabsorption The movement of filtrate from renal tubules back into blood in response to the body's specific needs.

Tubular secretion The movement of substances in blood into renal tubular fluid in response to the body's specific needs.

Tumor-suppressor gene A gene coding for a protein that normally inhibits cell division; loss or alteration of a tumor suppressor gene called *p53* is the most common genetic change in a wide variety of cancer cells.

Tunica albuginea (TOO-ni-ka al'-bū-JIN-ē-a) A dense white fibrous capsule covering a testis or deep to the surface of an ovary.

Tunica externa (eks-TER-na) The superficial coat of an artery or vein, composed mostly of elastic and collagen fibers. Also called the **adventitia.**

Tunica interna (in-TER-na) The deep coat of an artery or vein, consisting of a lining of endothelium, basement membrane, and internal elastic lamina. Also called the **tunica intima** (IN-ti-ma).

Tunica media (MĒ-dē-a) The intermediate coat of an artery or vein, composed of smooth muscle and elastic fibers.

Tympanic antrum (tim-PAN-ik AN-trum) An air space in the middle ear that leads into the mastoid air cells or sinus.

Tympanic (tim-PAN-ik) **membrane** A thin, semitransparent partition of fibrous connective tissue between the external auditory meatus and the middle ear. Also called the **eardrum.**

U

Umbilical (um-BIL-i-kul) **cord** The long, ropelike structure containing the umbilical arteries and vein that connect the fetus to the placenta.

Umbilicus (um-BIL-i-kus *or* um-bi-LĪ-kus) A small scar on the abdomen that marks the former attachment of the umbilical cord to the fetus. Also called the **navel.**

Upper limb The appendage attached at the shoulder girdle, consisting of the arm, forearm, wrist, hand, and fingers. Also called **upper extremity** or **upper appendage.**

Uremia (ū-RĒ-mē-a) Accumulation of toxic levels of urea and other nitrogenous waste products in the blood, usually resulting from severe kidney malfunction.

Ureter (Ū-rē-ter) One of two tubes that connect the kidney with the urinary bladder.

Urethra (ū-RĒ-thra) The duct from the urinary bladder to the exterior of the body that conveys urine in females and urine and semen in males.

Urinalysis (ū-ri-NAL-i-sis) An analysis of the volume and physical, chemical, and microscopic properties of urine.

Urinary (Ū-ri-ner-ē) **bladder** A hollow, muscular organ situated in the pelvic cavity posterior to the pubic symphysis; receives urine via two ureters and stores urine until it is excreted through the urethra.

Urinary system A system that consists of the kidneys, ureters, urinary bladder, and urethra. The system regulates the ionic composition, pH, volume, pressure, and osmolarity of blood.

Urine The fluid produced by the kidneys that contains wastes and excess materials; excreted from the body through the urethra.

Urogenital (ū'-rō-JEN-i-tal) **triangle** The region of the pelvic floor inferior to the pubic symphysis, bounded by the pubic symphysis and the ischial tuberosities, and containing the external genitalia.

Urology (ū-ROL-ō-jē) The specialized branch of medicine that deals with the structure, function, and diseases of the male and female urinary systems and the male reproductive system.

Uterine cycle A series of changes in the endometrium of a nonpregnant female that prepares the lining of the uterus to receive a fertilized ovum. Also called **menstrual cycle.**

Uterine (Ū-ter-in) **tube** Duct that transports ova from the ovary to the uterus. Also called the **fallopian** (fal-LŌ-pē-an) **tube** or **oviduct.**

Uterosacral ligament (ū-ter-ō-SĀ-kral LIG-a-ment) A fibrous band of tissue extending from the cervix of the uterus laterally to the sacrum.

Uterus (Ū-te-rus) The hollow, muscular organ in females that is the site of menstruation, implantation, development of the fetus, and labor. Also called the **womb.**

Utricle (Ū-tri-kul) The larger of the two divisions of the membranous labyrinth located inside the vestibule of the inner ear, containing a receptor organ for static equilibrium.

Uvea (Ū-vē-a) The three structures that together make up the vascular tunic of the eye.

Uvula (Ū-vū-la) A soft, fleshy mass, especially the U-shaped pendant part, descending from the soft palate.

V

Vagina (va-JĪ-na) A muscular, tubular organ that leads from the uterus to the vestibule, situated between the urinary bladder and the rectum of the female.

Vallate papilla (VAL-āt pa-PIL-a) One of the circular projections that is arranged in an inverted V-shaped row at the back of the tongue; the largest of the elevations on the upper surface of the tongue containing taste buds. Also called **circumvallate papilla.**

Varicocele (VAR-i-kō-sēl) A twisted vein; especially, the accumulation of blood in the veins of the spermatic cord.

Varicose (VAR-i-kōs) Pertaining to an unnatural swelling, as in the case of a varicose vein.

Vas A vessel or duct.

Vasa recta (VĀ-sa REK-ta) Extensions of the efferent arteriole of a juxtamedullary nephron that run alongside the loop of the nephron (loop of Henle) in the medullary region of the kidney.

Vasa vasorum (va-SŌ-rum) Blood vessels that supply nutrients to the larger arteries and veins.

Vascular (VAS-kū-lar) Pertaining to or containing many blood vessels.

Vascular (venous) sinus A vein with a thin endothelial wall that lacks a tunica media and externa and is supported by surrounding tissue.

Vascular spasm Contraction of the smooth muscle in the wall of a damaged blood vessel to prevent blood loss.

Vascular tunic (TOO-nik) The middle layer of the eyeball, composed of the choroid, ciliary body, and iris. Also called the **uvea** (Ū-vē-a).

Vasectomy (va-SEK-tō-mē) A means of sterilization of males in which a portion of each ductus (vas) deferens is removed.

Vasoconstriction (vāz-ō-kon-STRIK-shun) A decrease in the size of the lumen of a blood vessel caused by contraction of the smooth muscle in the wall of the vessel.

Vasodilation (vāz'-ō-dī-LĀ-shun) An increase in the size of the lumen of a blood vessel caused by relaxation of the smooth muscle in the wall of the vessel.

Vein A blood vessel that conveys blood from tissues back to the heart.

Vena cava (VĒ-na KĀ-va) One of two large veins that open into the right atrium, returning to the heart all of the deoxygenated blood from the systemic circulation except from the coronary circulation.

Ventral (VEN-tral) Pertaining to the anterior or front side of the body; opposite of dorsal.

Ventral ramus (RĀ-mus) The anterior branch of a spinal nerve, containing sensory and motor fibers to the muscles and skin of the anterior surface of the head, neck, trunk, and the limbs.

Ventricle (VEN-tri-kul) A cavity in the brain filled with cerebrospinal fluid. An inferior chamber of the heart.

Ventricular fibrillation (ven-TRIK-ū-lar fib-ri-LĀ-shun) (**VF** or **V-fib**) Asynchronous ventricular contractions; unless reversed by defibrillation, results in heart failure.

Venule (VEN-ūl) A small vein that collects blood from capillaries and delivers it to a vein.

Vermiform appendix (VER-mi-form a-PEN-diks) A twisted, coiled tube attached to the cecum.

Vermis (VER-mis) The central constricted area of the cerebellum that separates the two cerebellar hemispheres.

Vertebrae (VER-te-brē) Bones that make up the vertebral column.

Vertebral (VER-te-bral) **canal** A cavity within the vertebral column formed by the vertebral foramina of all the vertebrae and containing the spinal cord. Also called the **spinal canal.**

Vertebral column The 26 vertebrae of an adult and 33 vertebrae of a child; encloses and protects the spinal cord and serves as a point of attachment for the ribs and back muscles. Also called the **backbone, spine,** or **spinal column.**

Vesicle (VES-i-kul) A small bladder or sac containing liquid.

Vesicouterine (ves'-ik-ō-Ū-ter-in) **pouch** A shallow pouch formed by the reflection of the peritoneum from the anterior surface of the uterus, at the junction of the cervix and the body, to the posterior surface of the urinary bladder.

Vestibular (ves-TIB-ū-lar) **apparatus** Collective term for the organs of equilibrium, which include the saccule, utricle, and semicircular ducts.

Vestibular membrane The membrane that separates the cochlear duct from the scala vestibuli.

Vestibule (VES-ti-būl) A small space or cavity at the beginning of a canal, especially the inner ear, larynx, mouth, nose, and vagina.

Villus (VIL-lus) A projection of the intestinal mucosal cells containing connective tissue, blood vessels, and a lymphatic vessel; functions in the absorption of the end products of digestion. *Plural* is **villi** (VIL-ī).

Viscera (VIS-er-a) The organs inside the ventral body cavity. *Singular* is **viscus** (VIS-kus).

Visceral (VIS-er-al) Pertaining to the organs or to the covering of an organ.

Visceral effectors (e-FEK-torz) Organs of the ventral body cavity that respond to neural stimulation, including cardiac muscle, smooth muscle, and glands.

Vision The act of seeing.

Vitamin An organic molecule necessary in trace amounts that acts as a catalyst in normal metabolic processes in the body.

Vitreous (VIT-rē-us) **body** A soft, jellylike substance that fills the vitreous chamber of the eyeball, lying between the lens and the retina.

Vocal folds Pair of mucous membrane folds below the ventricular folds that function in voice production. Also called **true vocal cords.**

Volkmann's canal *See* **Perforating canal.**

Vulva (VUL-va) Collective designation for the external genitalia of the female. Also called the **pudendum** (poo-DEN-dum).

W

Wallerian (wal-LE-rē-an) **degeneration** Degeneration of the portion of the axon and myelin sheath of a neuron distal to the site of injury.

Wandering macrophage (MAK-rō-fāj) Phagocytic cell that develops from a monocyte, leaves the blood, and migrates to infected tissues.

White matter Aggregations or bundles of myelinated and unmyelinated axons located in the brain and spinal cord.

White pulp The regions of the spleen composed of lymphatic tissue, mostly B lymphocytes.

White ramus communicans (RĀ-mus kō-MŪ-ni-kans) The portion of a preganglionic sympathetic axon that branches from the anterior ramus of a spinal nerve to enter the nearest sympathetic trunk ganglion.

X

Xiphoid (ZĪ-foyd) Sword-shaped.

Xiphoid (ZĪ-foyd) **process** The inferior portion of the sternum (**breastbone).**

Y

Yolk sac An extraembryonic membrane composed of the exocoelomic membrane and hypoblast. It transfers nutrients to the embryo, is a source of blood cells, contains primordial germ cells that migrate into the gonads to form primitive germ cells, forms part of the gut, and helps prevent desiccation of the embryo.

Z

Zona fasciculata (ZŌ-na fa-sik'-ū-LA-ta) The middle zone of the adrenal cortex consisting of cells arranged in long, straight cords that secrete glucocorticoid hormones, mainly cortisol.

Zona glomerulosa (glo-mer'-ū-LŌ-sa) The outer zone of the adrenal cortex, directly under the connective tissue covering, consisting of cells arranged in arched loops or round balls that secrete mineralocorticoid hormones, mainly aldosterone.

Zona pellucida (pe-LOO-si-da) Clear glycoprotein layer between a secondary oocyte and the surrounding granulosa cells of the corona radiata.

Zona reticularis (ret-ik'-ū-LAR-is) The inner zone of the adrenal cortex, consisting of cords of branching cells that secrete sex hormones, chiefly androgens.

Zygote (ZĪ-gōt) The single cell resulting from the union of male and female gametes; the fertilized ovum.

CREDITS

Illustration Credits

Chapter 1 1.1, 1.11, 1.12: Kevin Somerville. 1.2–1.4, 1.8, 1.9: Imagineering. 1.5: Molly Borman. 1.6: Kevin Somerville/Imagineering. 1.7, Table 1.2: DNA Illustrations.

Chapter 2 2.1–2.25: Imagineering.

Chapter 3 3.1, 3.2, 3.16–3.20, 3.22–3.24, Table 3.2: Tomo Narashima. 3.3, 3.5–3.15, 3.21, 3.25–3.35: Imagineering.

Chapter 4 4.1, 4.9, Table 4.1–4.6: Kevin Somerville/Imagineering. 4.2, 4.4, 4.7, 4.8: Imagineering. 4.5, 4.6: Kevin Somerville.

Chapter 5 5.1a, 5.2–5.7: Kevin Somerville. 5.9: Kevin Somerville /Imagineering. 5.10: Imagineering.

Chapter 6 6.1, 6.4, 6.6a, Table 6.1: John Gibb. 6.2, 6.3, 6.5, 6.9: Kevin Somerville. 6.8, 6.10: Imagineering.

Chapter 7 7.1–7.24, Table 7.1, Table 7.4: John Gibb. 7.25: Imagineering.

Chapter 8 8.1–8.17, Table 8.1: John Gibb. 8.18: Kevin Somerville.

Chapter 9 9.1, 9.2, 9.3a, 9.10–9.16: John Gibb.

Chapter 10 10.1, 10.2, 10.17, 10.18: Kevin Somerville. 10.2d, 10.3, 10.4–10.8, 10.10–10.14: Imagineering. 10.9a-c: Kevin Somerville/Imagineering. 10.16: Imagineering.

Chapter 11 11.1, 11.2: John Gibb/Imagineering. 11.3–11.23: John Gibb.

Chapter 12 12.1a, 12.2a, 12.6–12.8: Kevin Somerville. 12.3–12.5, 12.9–12.28, Table 12.3: Imagineering.

Chapter 13 13.1–13.3, 13.5, 13.8b, 13.13: Kevin Somerville. 13.6, 13.10b: John Gibb. 13.7, 13.8a, 13.10a, 13.11: Steve Oh. 13.12 Imagineering. 13.14–13.18: Leonard Dank/Imagineering.

Chapter 14 14.1–14.3, 14.4b-c, 14.5–14.15, 14.27, 14.28, Table 14.3: Kevin Somerville. 14.4a, 14.4d, 14.16 Table 14.1: Imagineering. 14.17–14.26: Kevin Somerville/Richard Combs/Imagineering.

Chapter 15 15.1, 15.4, 15.6: Kevin Somerville/Imagineering. 15.2, 15.3, 15.7: Imagineering. 15.5: Kevin Somerville

Chapter 16 16.1, 16.3, 16.4, 16.8, 16.9, 16.12: Imagineering. 16.2, 16.13: Kevin Somerville. 16.5–16.7, 16.10–16.12: Kevin Somerville/Imagineering.

Chapter 17 17.1, 17.7, 17.14, 17.21b-d, 17.22: Tomo Narashima. 17.2, 17.10, 17.17, 17.23, 17.26, Table 17.1, Table 17.2: Imagineering. 17.3: Molly Borman. 17.6: Sharon Ellis. 17.12, 17.13, 17.15, 17.16: Imagineering. 17.18–17.21a, 17.25a: Tomo Narashima/Imagineering. 17.24, 17.25b: Tomo Narashima/Sharon Ellis/Imagineering. 17.27, 17.28: Kevin Somerville/Imagineering.

Chapter 18 18.1, 18.21: Kevin Somerville. 18.5, 18.8, 18.10, 18.13a, 18.15, 18.18: Lynn O'Kelley/Imagineering. 18.2–18.13c, 18.14, 18.16, 18.17, 18.19, 18.20: Imagineering.

Chapter 19 19.1, 19.3–19.6, 19.8, 19.9, 19.11–19.13, Table 19.2: Imagineering.

Chapter 20 20.1–20.5, 20.18–20.20, Table 20.1: Kevin Somerville. 20.7, 20.9, 20.14–20.16, 20.22, 20.23: Kevin Somerville/Imagineering. 20.10–20.13, 20.24: Imagineering

Chapter 21 21.1, 21.9, 21.12, 21.13, 21.17, 21.18a, 21.21, 21.23, 21.28, 21.29a, 21.30, 21.31: Kevin Somerville. 21.2, 20.5–20.8, 21.10, 21.11, 21.14–21.16 21.29b: Imagineering. 21.3, 21.4: Kevin Somerville/Imagineering. 21.18b, 21.19–21.22, 21.24–21.27: John Gibb. Exhibits: Imagineering/John Gibb.

Chapter 22 22.1: Kevin Somerville/Imagineering/Richard Coombs. 22.2, 22.4, 22.9, 22.11–22.18, 22.20, 22.23: Imagineering. 22.3: John Gibb. 22.5, 22.7: Steve Oh. 22.6, 22.10: Kevin Somerville/Imagineering. 22.8: Kevin Somerville. 22.19: Jared Schneidman Design.

Chapter 23 23.1, 23.2a, 23.10, 23.11: Kevin Somerville. 23.2b, 23.4a-c, 23.5, 23.6, 23.9: Molly Borman. 23.4d, 23.14–23.18, 23.23, 23.24: Imagineering. 23.13: John Gibb/Imagineering. 23.26: Kevin Somerville/Imagineering. 23.27: Imagineering.

Chapter 24 24.1, 24.2, 24.12, 24.16a-b, 24.18, 24.19, 24.24: Kevin Somerville. 24.4, 24.13, 24.16d-e, 24.21, 24.22, 24.25: Imagineering. 24.5, 24.7, 24.8, 24.10: Nadine Sokol. 24.6: DNA Illustrations. 24.11, 24.15: Steve Oh. 24.23: Molly Borman

Chapter 25 25.1–25.20: Imagineering.

Chapter 26 26.1, 26.2, 26.6, 26.8, 26.9, 26.23: Kevin Somerville. 26.3, 26.21: Steve Oh/Imagineering. 26.5, 26.10–26.20, Table 26.1: Imagineering. 26.22: Kevin Somerville/Imagineering.

Chapter 27 27.1–27.8: Imagineering

Chapter 28 28.1–28.4, 28.6, 28.9–28.14, 28.16, 28.19–28.22, 28.27, 28.28: Kevin Somerville. 28.5, 28.7, 28.8, 28.15, 28.23–28.26: Imagineering.

Chapter 29 29.1–29.13, 29.17, Table 29.2: Kevin Somerville. 29.15, 29.18: Kevin Somerville/Imagineering. 29.16, 29.19–29.26: Imagineering.

Photo Credits

Chapter 1 Fig. 1.1 Rubberball Productions/Getty Images. Fig. 1.8a Dissection Shawn Miller; Photograph Mark Nielsen. Fig. 1.8b Dissection Shawn Miller; Photograph Mark Nielsen. Fig. 1.8c Dissection Shawn Miller; Photograph Mark Nielsen. Fig. 1.12a Andy Washnik. Page 21 (left) Biophoto Associates/Photo Researchers. Page 21 (center) Breast Cancer Unit, Kings College Hospital, London/Photo Researchers, Inc. Page 21 (right) Zephyr/Photo Researchers, Inc. Page 22 (top left) Cardio-Thoracic Centre, Freeman Hospital, Newcastle-Upon-Tyne/Photo Researchers, Inc. Page 22 (top center) CNRI/Science Photo Library/Photo Researchers, Inc. Page 22 (top right) Science Photo Library/Photo Researchers, Inc. Page 22 (bottom left) Scott Camazine/Photo Researchers, Inc. Page 22 (bottom right) ISM/Phototake. Page 23 (center left) Simon Fraser/Photo Researchers, Inc. Page 23 (center right) ISM/Phototake. Page 23 (bottom) Courtesy Andrew Joseph Tortora and Damaris Soler. Page 24 (center left) Department of Nuclear Medicine, Charing Cross Hospital /Photo Researchers, Inc. Page 24 (center) SIU/Visuals Unlimited. Page 24 (center right) Dept. of Nuclear Medicine, Charing Cross Hospital/Photo Researchers, Inc. Page 24 (bottom right) ©Camal/Phototake.

Chapter 2 Fig. 2.23 Courtesy T.A. Steitz, Yale University.

Chapter 3 Fig. 3.4 Andy Washnik. Fig. 3.9 David Phillips/Photo Researchers, Inc. Fig. 3.13 ©Omikron/Photo Researchers. Fig. 3.15a Albert Tousson/Phototake. Fig. 3.15b Albert Tousson/Phototake. Fig. 3.15c Alexey Khodjakov/Photo Researchers, Inc. Fig. 3.16c David M. Phillips/Visuals Unlimited. Fig. 3.17b P. Motta/Photo Researchers, Inc. Fig. 3.17c David M. Phillips/Visuals Unlimited. Fig. 3.19b D. W. Fawcett/Photo Researchers, Inc. Fig. 3.20b Biophoto Associates/Photo Researchers. Fig. 3.22b David M. Phillips/Visuals Unlimited. Fig. 3.23b Don W. Fawcett/Visuals Unlimited. Fig. 3.24 D.W. Fawcett/Photo Researchers, Inc. Fig. 3.25b Andrew Syred/Photo Researchers, Inc. Fig. 3.32 Courtesy Michael Ross, University of Florida. Page 104 Steve Gschmeissner/Science Photo Library/Photo Researchers, Inc.

Chapter 4 All photos by Mark Nielsen except Fig 4.9a, which was taken by Michael Ross, University of Florida.

Chapter 5 Fig. 5.1b Courtesy Michael Ross, University of Florida. Fig. 5.1c David Becker/Photo Researchers, Inc. Fig. 5.1d Courtesy Andrew J. Kuntzman. Fig. 5.3b Mark Nielsen. Fig. 5.4b VVG/Science Photo Library/Photo Researchers, Inc. Fig. 5.8a Alain Dex/Photo Researchers, Inc. Fig. 5.8b

Biophoto Associates/Photo Researchers, Inc. Fig. 5.9a Sheila Terry/Science Photo Library/Photo Researchers, Inc. Fig. 5.9b St. Stephen's Hospital/ Science Photo Library/Photo Researchers, Inc. Fig. 5.9c St. Stephen's Hospital/Science Photo Library/Photo Researchers, Inc. Fig. 5.11 Dr. P. Marazzi/Science Photo Library/Photo Researchers, Inc.

Chapter 6 Fig. 6.1b Mark Nielsen. Fig. 6.2 (left) CNRI/Photo Researchers, Inc. Fig. 6.2 (center) Dr. Richard Kessel and Randy Kardon/Tissues & Organs/Visuals Unlimited. Fig. 6.2 (right) Dr. Richard Kessel and Randy Kardon/Tissues & Organs/Visuals Unlimited. Fig. 6.6 Ralph Hutchings/Visuals Unlimited. Fig. 6.7a The Bergman Collection. Fig. 6.7b Mark Nielsen. Fig. 6.11a P. Motta/Photo Researchers, Inc. Fig. 6.11b P. Motta/Photo Researchers, Inc. Table 6.1a Courtesy Dr. Brent Layton. Table 6.1b Courtesy Per Amundson, M.D. Table 6.1c Courtesy Dr. Brent Layton. Table 6.1d Courtesy Dr. Brent Layton. Table 6.1e Courtesy Dr. Brent Layton. Table 6.1f Watney Collection/Phototake.

Chapter 7 Fig. 7.3b Mark Nielsen. Fig. 7.4c Mark Nielsen. Fig. 7.18b Mark Nielsen. Fig. 7.18c Mark Nielsen. Fig. 7.18d Mark Nielsen. Fig. 7.18e Mark Nielsen. Fig. 7.20b Mark Nielsen. Fig. 7.20c Mark Nielsen. Fig. 7.19b Mark Nielsen. Fig. 7.19c Mark Nielsen. Fig. 7.22b Mark Nielsen. Fig. 7.25a Princess Margaret Rose Orthopaedic Hospital/Photo Researchers, Inc. Fig. 7.25b Dr. P. Marazzi/Photo Researchers, Inc. Fig. 7.25c Custom Medical Stock Photo, Inc. Fig. 7.26 Center for Disease Control.

Chapter 9 Fig. 9.3b Mark Nielsen. Fig. 9.4 Mark Nielsen. Fig. 9.5 John Wilson White. Fig. 9.6 John Wilson White. Fig. 9.7 John Wilson White. Fig. 9.8 John Wilson White. Fig. 9.9 John Wilson White. Fig. 9.12d Dissection Shawn Miller, Photograph Mark Nielsen. Fig. 9.15d Dissection Shawn Miller, Photograph Mark Nielsen. Fig. 9.15g Dissection Shawn Miller, Photograph Mark Nielsen. Fig. 9.15h Dissection Shawn Miller, Photograph Mark Nielsen. Fig. 9.15i Dissection Shawn Miller, Photograph Mark Nielsen. Fig. 9.15j Dissection Shawn Miller, Photograph Mark Nielsen. Fig. 9.16b SIU/Visuals Unlimited. Fig. 9.16c ISM/Phototake. Fig. 9.16f Charles McRae/Visuals Unlimited.

Chapter 10 Fig. 10.5 Courtesy Hiroyouki Sasaki, Yale E. Goldman and Clara Franzini-Armstrong. Fig. 10.9d Don Fawcett/Photo Researchers, Inc. Fig. 10.15 John Wiley & Sons. Table 10.1 Courtesy Denah Appelt and Clara Franzini-Armstrong. Table 10.4 Biophoto Associates/Photo Researchers, Inc.

Chapter 11 Fig. 11.5c Dissection Shawn Miller, Photograph Mark Nielsen. Fig. 11.10d Dissection Nathan Mortensen and Shawn Miller; Photograph Mark Nielsen. Fig. 11.18g Andy Washnik. Fig. 11.21b Dissection Shawn Miller, Photograph Mark Nielsen.

Chapter 12 Fig. 12.2b Mark Nielsen. Fig. 12.8c Dennis Kunkel/Phototake. Fig. 12.8d Martin Rotker/ Phototake.

Chapter 13 Fig. 13.1b Dissection Chris Roach; Photograph Mark Nielsen. Fig. 13.1c Dissection

Shawn Miller, Photograph Mark Nielsen. Fig. 13.3b Courtesy Michael Ross, University of Florida. Fig. 13.5b Dr. Richard Kessel and Randy Kardon/ Tissues & Organs/Visuals Unlimited. Fig. 13.6b Dissection Shawn Miller, Photograph Mark Nielsen. Table 13.1 Dissection Shawn Miller, Photograph Mark Nielsen. Table 13.1 Dissection Shawn Miller, Photograph Mark Nielsen. Table 13.1 Mark Nielsen and Shawn Miller. Table 13.1 Dissection Shawn Miller, Photograph Mark Nielsen. Page 529 (top) Bold Stock/ Age Fotostock America, Inc. Page 529 (just below top) Masterfile. Page 529 (center) Andersen Ross/ Getty Images, Inc. Page 529 (bottom) Masterfile.

Chapter 14 Fig. 14.1b Dissection Shawn Miller, Photograph Mark Nielsen. Fig. 14.4a Dissection Shawn Miller, Photograph Mark Nielsen. Fig. 14.8d Dissection Shawn Miller, Photograph Mark Nielsen. Fig. 14.11c Dissection Shawn Miller, Photograph Mark Nielsen. Fig. 14.12 N. Gluhbegovic and T.H. Williams, The Human Brain: A Photographic Guide, Harper and Row, Publishers, Inc. Hagerstown, MD, 1980. Table 14.2 Dissection Shawn Miller, Photograph Mark Nielsen. Table 14.2 Dissection Shawn Miller, Photograph Mark Nielsen. Table 14.2 Dissection Shawn Miller, Photograph Mark Nielsen. Table 14.2 Dissection Shawn Miller, Photograph Mark Nielsen. Table 14.2 Dissection Shawn Miller, Photograph Mark Nielsen. Table 14.2 Dissection Shawn Miller, Photograph Mark Nielsen. Fig. 14.17 Dissection Shawn Miller, Photograph Mark Nielsen. Fig. 14.18 Dissection Shawn Miller, Photograph Mark Nielsen. Fig. 14.19 Dissection Shawn Miller, Photograph Mark Nielsen. Fig. 14.19 Dissection Shawn Miller, Photograph Mark Nielsen. Fig. 14.19 Dissection Shawn Miller, Photograph Mark Nielsen. Fig. 14.20 Dissection Shawn Miller, Photograph Mark Nielsen. Fig. 14.21 Dissection Shawn Miller, Photograph Mark Nielsen. Fig. 14.22 Dissection Shawn Miller, Photograph Mark Nielsen. Fig. 14.23 Dissection Shawn Miller, Photograph Mark Nielsen. Fig. 14.24 Dissection Shawn Miller, Photograph Mark Nielsen. Fig. 14.25 Dissection Shawn Miller, Photograph Mark Nielsen. Fig. 14.26 Dissection Shawn Miller, Photograph Mark Nielsen.

Chapter 17 Fig. 17.1c Courtesy Michael Ross, University of Florida. Fig. 17.3d Mark Nielsen. Fig. 17.5 Geirge Diebold/Getty Images, Inc. Fig. 17.9 Paul Parker/Photo Researchers, Inc. Fig. 17.10c Mark Nielsen. Fig. 17.17 N. Gluhbegovic and T. H. Williams, The Human Brain: A Photographic Guide, Harper and Row, Publishers, Inc., Hagerstown, MD, 1980.

Chapter 18 Fig. 18.5 (top) Dissection Shawn Miller, Photograph Mark Nielsen. Fig. 18.5d Courtesy James Lowe, University of Nottingham, Nottingham, United Kingdom. Fig. 18.10b Mark Nielsen. Fig. 18.10c Dissection Shawn Miller, Photograph Mark Nielsen. Fig. 18.10d Dissection Shawn Miller, Photograph Mark Nielsen. Fig. 18.13b Mark Nielsen . Fig. 18.13d Dissection Shawn Miller, Photograph Mark Nielsen. Fig. 18.15c Dissection Shawn Miller, Photograph Mark Nielsen. Fig. 18.15d Mark Nielsen. Fig. 18.18 Mark Nielsen. Fig. 18.22a From New England Journal of Medicine, February 18, 1999,

vol. 340, No. 7, page 524. Photo provided courtesy of Robert Gagel, Department of Internal Medicine, University of Texas M.D. Anderson Cancer Center, Houston Texas Fig. 18.22b Lester Bergman/The Bergman Collection. Fig. 18.22c Lester Bergman/ The Bergman Collection. Fig. 18.22d Lester Bergman/ The Bergman Collection. Fig. 18.22e Biophoto Associates/Photo Researchers, Inc.

Chapter 19 Opener Juergen Berger/Photo Researchers, Inc. Fig. 19.2a Juergen Berger/Photo Researchers, Inc. Fig. 19.2b Mark Nielsen. Fig. 19.7a Courtesy Michael Ross, University of Florida. Fig. 19.7b Courtesy Michael Ross, University of Florida. Fig. 19.7c Courtesy Michael Ross, University of Florida. Fig. 19.7d Courtesy Michael Ross, University of Florida. Fig. 19.7e Courtesy Michael Ross, University of Florida. Fig. 19.10a Dennis Kunkel/ Visuals Unlimited. Fig. 19.10b Steve Gschmeissner/ Photo Researchers, Inc. Fig. 19.10c David Phillips/ Visuals Unlimited. Fig. 19.10d Dennis Kunkel/ Phototake. Fig. 19.14 Jean Claude Revy/Phototake. Fig. 19.15 Stanley Fleger/Visuals Unlimited.

Chapter 20 Fig. 20.1c Dissection Shawn Miller, Photograph Mark Nielsen. Fig. 20.3b Dissection Shawn Miller, Photograph Mark Nielsen. Fig. 20.4b Mark Nielsen. Fig. 20.4c Mark Nielsen. Fig. 20.6c Dissection Shawn Miller, Photograph Mark Nielsen. Fig. 20.6f Dissection Shawn Miller, Photograph Mark Nielsen. Fig. 20.6g Dissection Shawn Miller, Photograph Mark Nielsen. Fig. 20.8c Dissection Shawn Miller, Photograph Mark Nielsen. Fig. 20.17 Jim cummins/Taxi/Getty Images, Inc. Fig. 20.21 ©Vu/Cabisco/Visuals Unlimited. Fig. 20.21 W. Ober/Visuals Unlimited. Fig. 20.22d ©ISM/ Phototake.

Chapter 21 Fig. 21.1d Dennis Strete. Fig. 21.1e Courtesy Michael Ross, University of Florida. Fig. 21.5(top) Dissection Shawn Miller, Photograph Mark Nielsen. Fig. 21.5(bottom) Dissection Shawn Miller, Photograph Mark Nielsen. Exhibit 21.B Dissection Shawn Miller, Photograph Mark Nielsen. Fig. 21.19d Dissection Shawn Miller, Photograph Mark Nielsen. Fig. 21.21d Dissection Shawn Miller, Photograph Mark Nielsen. Fig. 21.25d Dissection Shawn Miller, Photograph Mark Nielsen.

Chapter 22 Fig. 22.5c Courtesy Michael Ross, University of Florida. Fig. 22.5b Courtesy Michael Ross, University of Florida. Fig. 22.6b Mark Nielsen. Fig. 22.6c Dissection Shawn Miller, Photograph Mark Nielsen. Fig. 22.7c Mark Nielsen. Fig. 22.9b Science Photo Library/Photo Researchers, Inc.

Chapter 23 Fig. 23.1b Dissection Shawn Miller, Photograph Mark Nielsen. Fig. 23.2c Dissection Shawn Miller, Photograph Mark Nielsen. Fig. 23.3 Courtesy Lynne Marie Barghesi. Fig. 23.5c CNRI/Photo Researchers, Inc. Fig. 23.6 Dissection Shawn Miller, Photograph Mark Nielsen. Fig. 23.8 Dissection Shawn Miller, Photograph Mark Nielsen. Fig. 23.10b Biophoto Associates/Photo Researchers, Inc. Fig. 23.11c Biophoto Associates/Photo Researchers, Inc.

Chapter 24 Fig. 24.1b Dissection Shawn Miller, Photograph Mark Nielsen. Fig. 24.4e Dissection

Shawn Miller, Photograph Mark Nielsen. Fig. 24.6b Mark Nielsen. Fig. 24.9 Mark Nielsen. Fig. 24.11b Dissection Shawn Miller, Photograph Mark Nielsen. Fig. 24.12 (inset) Hessler/Vu/Visuals Unlimited. Fig. 24.12c Mark Nielsen. Fig. 24.15d Dissection Shawn Miller, Photograph Mark Nielsen. Fig. 24.15e Dissection Shawn Miller, Photograph Mark Nielsen. Fig. 24.16c (top inset) Mark Nielsen. Fig. 24.16c Mark Nielsen. Fig. 24.18b Dissection Shawn Miller, Photograph Mark Nielsen. Fig. 24.19 Dissection Shawn Miller, Photograph Mark Nielsen. Fig. 24.20a Fred E. Hossler/Visuals Unlimited. Fig. 24.20b G. W. Willis/ Visuals Unlimited. Fig. 24.20c Courtesy Michael Ross, University of Florida. Fig. 24.20d Courtesy Michael Ross, University of Florida. Fig. 24.24c Courtesy Michael Ross, University of Florida. Fig. 24.24d Courtesy Michael Ross, University of Florida.

Chapter 26 Fig. 26.1b Dissection Shawn Miller, Photograph Mark Nielsen. Fig. 26.3b Mark Nielsen and Shawn Miller. Fig. 26.6b Dennis Strete. Fig. 26.8b Courtesy Michael Ross, University of Florida.

Chapter 28 Fig. 28.1b Dissection Shawn Miller, Photograph Mark Nielsen. Fig. 28.3b Dissection Shawn Miller, Photograph Mark Nielsen. Fig. 28.3c Dissection Shawn Miller, Photograph Mark Nielsen. Fig. 28.4a Mark Nielsen. Fig. 28.9b Dissection Shawn Miller, Photograph Mark Nielsen. Fig. 28.10c Dissection Shawn Miller, Photograph Mark Nielsen. Fig. 28.11b Dissection Shawn Miller, Photograph Mark Nielsen. Fig. 28.13b Courtesy Michael Ross, University of Florida. Fig. 28.13c Claude Edelmann/Photo Researchers, Inc. Fig. 28.14c Mark Nielsen. Fig. 28.14f Mark Nielsen. Fig. 28.17a Mark Nielsen. Fig. 28.17b P. Motta/Photo Researchers, Inc. Fig. 28.18a Mark Nielsen. Fig. 28.18b Courtesy Michael Ross, University of Florida. Fig. 28.20b Courtesy Michael Ross, University of Florida. Fig. 28.20c Mark Nielsen.

Chapter 29 Fig. 29.1b Don W. Fawcett/Photo Researchers, Inc. Fig. 29.1c Myriam Wharman/ Phototake. Fig. 29.11b Preparation and photography Mark Nielsen. Fig. 29.14a Photo provided courtesy of Kohei Shiota, Congenital Anomaly Research Center, Kyoto University, Graduate School of Medicine. Fig. 29.14b Courtesy National Museum of Health and Medicine, Armed Forces Institute of Pathology. Fig. 29.14c Courtesy National Museum of Health and Medicine, Armed Forces Institute of Pathology. Fig. 29.14d Courtesy National Museum of Health and Medicine, Armed Forces Institute of Pathology. Fig. 29.14e Courtesy National Museum of Health and Medicine, Armed Forces Institute of Pathology. Fig. 29.14f Photo by Lennart Nilsson/ Albert Bonniers Fîrlag AB, A Child is Born, Dell Publishing Company. Reproduced with permission. Fig. 29.14g Photo provided courtesy of Kohei Shiota, Congenital Anomaly Research Center, Kyoto University, Graduate School of Medicine. Fig. 29.14h Photo provided courtesy of Kohei Shiota, Congenital Anomaly Research Center, Kyoto University, Graduate School of Medicine.

INDEX